BIOLOGY

ROBERT J. BROOKER
University of Minnesota–Minneapolis

ERIC P. WIDMAIER
Boston University

LINDA E. GRAHAM
University of Wisconsin–Madison

PETER D. STILING
University of South Florida

**McGraw-Hill
Higher Education**

Boston Ridge, IL Dubuque, IA New York San Francisco St. Louis
Bangkok Bo Caracas Kuala Lumpur Lisbon London Madrid Mexico City
Milan Montre ew Delhi Santiago Seoul Singapore Sydney Taipei Toronto

McGraw-Hill
Higher Education

BIOLOGY

Published by McGraw-Hill, a business unit of The McGraw-Hill Companies, Inc., 1221 Avenue of the
Americas, New York, NY 10020. Copyright © 2008 by The McGraw-Hill Companies, Inc. All rights
reserved. No part of this publication may be reproduced or distributed in any form or by any means, or stored
in a database or retrieval system, without the prior written consent of The McGraw-Hill Companies, Inc.,
including, but not limited to, in any network or other electronic storage or transmission, or broadcast for
distance learning.

Some ancillaries, including electronic and print components, may not be available to customers outside the
United States.

This book is printed on acid-free paper.

2 3 4 5 6 7 8 9 0 DOW/DOW 0 9 8 7

ISBN 978–0–07–295620–7
MHID 0–07–295620–8

Publisher: *Janice Roerig-Blong*
Executive Editor: *Patrick E. Reidy*
Director of Development: *Kristine Tibbetts*
Senior Developmental Editor: *Lisa A. Bruflodt*
Director of Marketing: *Chad E. Grall*
Lead Project Manager: *Joyce M. Berendes*
Senior Production Supervisor: *Laura Fuller*
Senior Media Project Manager: *Tammy Juran*
Senior Media Producer: *Eric A. Weber*
Design Manager: *K. Wayne Harms*
Cover/Interior Designer: *John Joran*
Art Development and Rendering: *Imagineering Media Services, Inc.*
Senior Photo Research Coordinator: *John C. Leland*
Photo Research: *Pronk&Associates*
Supplement Producer: *Melissa M. Leick*
Compositor: *Lachina Publishing Services, Inc.*
Typeface: *9.5/12 Slimbach*
Printer: *R. R. Donnelley Willard, OH*

The credits section for this book begins on page C-1 and is considered an extension of the copyright page.

Library of Congress Cataloging-in-Publication Data

Biology / Robert J. Brooker . . . [et al.].—1st ed.
   p. cm.
   ISBN 978-0-07-295620-7—-ISBN 0-07-295620-8
   1.   Biology—Textbooks. I. Brooker, Robert J.

QH308.2.B564445 2008
570—dc22

                                            2006051688

www.mhhe.com

ABOUT THE AUTHORS

Robert J. Brooker

Robert J. Brooker (Ph.D., Yale University) received his B.A. in biology at Wittenberg University in 1978. At Harvard, he studied the lactose permease, the product of the *lacY* gene of the *lac* operon. He continues working on transporters at the University of Minnesota, where he is a Professor in the Department of Genetics, Cell Biology, and Development and has an active research laboratory. At the University of Minnesota, Dr. Brooker teaches undergraduate courses in biology, genetics, and cell biology. In addition to many other publications, he has written two editions of the undergraduate genetics text *Genetics: Analysis & Principles*, McGraw-Hill, copyright 2005.

Eric P. Widmaier

Eric P. Widmaier received his Ph.D. in 1984 in endocrinology from the University of California at San Francisco. His research is focused on the control of body mass and metabolism in mammals, the mechanisms of hormone action, and the postnatal development of adrenal gland function. Dr. Widmaier is currently Professor of Biology at Boston University. Among other publications, he is a co-author of *Vander's Human Physiology: The Mechanisms of Body Function*, 10th edition, published by McGraw-Hill, copyright 2006.

Linda E. Graham

Linda E. Graham received her Ph.D. in botany from the University of Michigan, Ann Arbor. Her research explores the evolutionary origin of land-adapted plants, focusing on their cell and molecular biology as well as ecological interactions. Dr. Graham is now Professor of Botany at the University of Wisconsin-Madison. She teaches undergraduate courses in biology and plant biology. She is the co-author of, among other publications, *Algae*, copyright 2000, a major's textbook on algal biology, and *Plant Biology*, copyright 2006, both published by Prentice Hall/Pearson.

Peter D. Stiling

Peter Stiling obtained his Ph.D. from University College, Cardiff, Wales in 1979. Subsequently, he became a Postdoc at Florida State University and later spent two years as a lecturer at the University of the West Indies, Trinidad. During this time, he began photographing and writing about butterflies and other insects, which led to publication of several books on local insects. Dr. Stiling is currently a Professor of Biology at the University of South Florida at Tampa. He teaches graduate and undergraduate courses in ecology and environmental science. He has published many scientific papers and is the author of *Ecology: Theories and Applications*, 4th edition by Prentice Hall/Pearson, copyright 2002. Dr. Stiling's research interests include plant-insect relationships, parasite-host relationships, biological control, restoration ecology, and the effects of elevated carbon dioxide levels on plant herbivore interactions.

The authors are grateful for the help, support, and patience of their families, friends, and students,

Deb, Dan, Nate, and Sarah Brooker,

Maria, Carrie, and Rick Widmaier,

Jim, Michael, and Melissa Graham, and

Jacqui, Zoe, Leah, and Jenna Stiling.

Left to right: Eric Widmaier, Linda Graham, Peter Stiling, and Rob Brooker

BRIEF CONTENTS

A New Biology Book with a Modern Perspective

In addition to being active researchers and experienced writers, our author team has taught majors biology for years. We have taught with the same books that you have. Our goal in creating something new is to offer something better—a comprehensive, modern textbook featuring an evolutionary focus with an emphasis on scientific inquiry.

Through our classroom experiences and research work, we became inspired by the prospect that a new Biology text could move biology education forward. In listening to educators and students, it became clear that we needed to concentrate our efforts on seven crucial areas. These are described briefly below. We will return to each in more detail in the pages that follow.

1. **Experimentation** During the 1970s and 1980s, biological information began to expand at an exponential rate. Biology textbooks grew and to some extent the content suffered as the scientific process was squeezed out by the avalanche of new details. We are committed to striking a better balance between general concepts and experimentation by showing the connection between scientific inquiry and the principles that are learned from such experimentation, especially through the Feature Investigation sections in every chapter.

2. **Modern content** Science is a moving target. Although the content and organization of our Biology textbook is not a dramatic departure from other books, we have added modern content that will better prepare students for future careers in biology. Toward this end, we have received content reviews from over three hundred faculty members from around the world. We are convinced they have helped us produce a book with the most up-to-date content possible.

Striking examples where we feel the content demonstrates a modern approach with an emphasis on recent experimentation include the following:

- Chapter 6 of our Cell Biology Unit explores cell biology at the level of "systems biology" in which the cell is viewed as a group of interacting parts. This allows students to understand how the parts of a cell work together.

- Our Genetics Unit takes a "molecular first" approach so that students will first understand what a gene is, and then consider how genes affect the traits of organisms.

- The Evolution Unit often takes a molecular perspective, and highlights cladistic methods to generate evolutionary relationships. This approach connects evolution at the molecular level and at the level of organisms in their native environments.

- The Diversity Unit has incorporated the newest information regarding evolutionary relationships among modern species. The connection between evolutionary innovation and reproductive success allows students to appreciate why organisms have certain types of traits.

- In our Plant Biology Unit, a much more modern connection has been made between plant structure, function and genetics. Recent information from *Arabidopsis* is often discussed.

- Each chapter in our Animal Biology Unit ends with a section on the modern Impact on Public Health, including the molecular basis of many diseases. In addition, Neuroscience is covered as a mini-Unit of its own, with three complete chapters incorporating some of the most recent information in this exciting area of biology.

- The Ecology Unit also incorporates an evolutionary theme and has an expanded discussion of species interactions. This approach provides students with a deeper understanding of evolutionary adaptations that organisms have.

3. **Evolutionary Perspective** A study of the processes and outcomes of evolution serves to unify the field of biology and the units of our text. Whether describing evolutionary mechanisms at the molecular level or surveying the diversity of life through a view of modern systematics, an understanding of evolution serves to connect and integrate the disciplines of biology.

4. **A Visual Outline** We were determined to create a new art program using both graphics and photography to serve as a "visual outline." We have worked with a large team of scientific illustrators, photographers, educators and students to build an accurate, up-to-date and visually appealing new illustration program that is easy to follow, realistic and instructive.

5. **Ensuring Accuracy** We chose to work as a team of authors to create this new Biology text because the information in our discipline is increasing so rapidly. Each member of our team has experience researching and writing in our respective areas, allowing us to combine efforts to stay abreast of the field. Likewise, we have worked with a much larger team of reviewers, advisors, editors, and accuracy checkers to ensure that this text is as current and accurate as humanly possible.

6. **A Learning System** Starting with a simple outline at the beginning of each chapter, we have focused on the crucial topics in a clear and easy to follow manner. We emphasize

critical thinking and active learning by constantly returning to how science is done, and through several pedagogical devices including our Biological Inquiry Questions which appear often in figure legends throughout the text. We end each chapter with a thorough review section which returns to our outline and emphasizes higher level learning through multiple question types.

7. **Media—Teaching and Learning with Technology**
Our new book is accompanied by a vast array of electronic teaching and learning tools. We have focused on creating new student content that is built upon learning outcomes and assessing student performance. We have also created an unprecedented array of presentation and course management tools to enable instructors to enhance their lectures and manage their classrooms more effectively. Finally, we are committed to offering several electronic book and customized print options to best fit your needs.

Experimentation in Biology Reveals General Principles

Biology is the study of life. The primary way that biologists study life is through experimentation. In this textbook, we have maintained a parallel focus on the general principles of biology and experimentation.

Each chapter is divided into a few sections focusing on general principles in biology. These sections begin with an overview of why the topic is important. We then describe the features of the topic that we and our reviewers have felt are the most important and sometimes the most difficult to grasp.

In describing the principles of biology, we have woven experimentation into each chapter. To be prepared for a career in biology, students need to understand the techniques that are used in biology and bolster their critical thinking skills. Each chapter has a Feature Investigation that shows the steps in the scientific process (for example, see Figure 10.13, p. 200). These investigations include a description of the methods and end with an analysis of the data. This deeper approach allows students to appreciate how the general principles of biology were derived from experiments.

In addition, many scientists are mentioned throughout each chapter with a brief description of how their work contributed to the general principles of biology, reinforcing a sense that biology is an enterprise carried out by scientists around the world. This will prepare students for their next step into the scientific literature in future courses.

An Emphasis on Evolution Provides a Modern Perspective

Evolution is the unifying theme that connects the various areas of biology. We have chosen to explain this theme from a modern perspective by relating the information in each chapter to the genetic material, namely the genomes of organisms. Like-wise, because most genes encode proteins, a logical extension is to also relate information to proteomes—the collection of proteins that a cell or organism can make.

We use our Genomes and Proteomes subsections as one way to integrate the various disciplines of biology. For example, let's consider our Cell Biology Unit. In this unit, we emphasize how gene regulation is responsible for the differences between a nerve and muscle cell and how descent with modification occurs at the protein level to produce families of proteins with related cellular functions. Likewise, let's consider our Ecology Unit. In this unit we explain how the modification of particular genes has enabled organisms to compete effectively in their environment. The Ecology Unit also considers how the genomes of organisms have evolved in response to environmental changes over many generations.

As a team, our authors are committed to the idea of integrating the various disciplines in Biology, not separating them. We feel it important, fun, engaging, and actually easy to high-light the evolutionary theme of biology by relating information in each chapter to Genomes and Proteomes. They are intended to provide "perspective". We keep returning to the idea that everything in biology stems from the evolution of genomes, and that genomes primarily encode proteins that ultimately provide organisms with their traits. We feel strongly that this approach provides a modern perspective that will serve our readers well in their future careers. The Genomes and Proteomes subsections are just one way that we integrate the various fields of biology. Genes, proteins, and the molecular mechanisms of life are discussed throughout the entire book.

Textbook Illustrations Are a Key to Learning

In discussions with many of our students, we have come to realize that many, probably most students, are visual learners. They read the textbook for information, but when it comes time to study, their main emphasis is on the figures. Likewise, instructors often rely heavily on good illustrations for their lectures. Therefore, a top priority in the development of our Biology textbook has been the conceptualization and rendering of the illustrations.

As you will see when you scan through this book, the illustrations are very easy to follow, particularly those that have multiple steps. We have taken the attitude that students should be able to look at the figures and understand what is going on, without having to glance back and forth between the text and art. Many figures contain text boxes that explain what the illustration is showing. In those figures with multiple steps, the boxes are numbered so that students will understand that the steps occur in a particular order. In some cases, the numbering was critical when the illustration involved features that could not be presented in a linear manner. For example, a description of hearing in Figure 45.7 is much easier to follow because we have guided the student through the process by using numbered text boxes.

Likewise, technology can help us to engage, to educate and even to inspire our students. As you will see when you skim through the pages, the drawings in this textbook are technologically advanced. They are primarily intended to educate the student and we have maintained a commitment to simplicity. Even so, the illustrations in this textbook are also aimed at being interesting and inspiring. Art elements are drawn with a strong sense of realism and three-dimensionality. The illustrations come to life, which, after all, is important to students who are interested in the study of life. We expect students to occasionally look at a figure and think, "Wow, that's cool!" We invite you to skim through the pages and see for yourself.

Accuracy Is a Top Priority

Inaccuracies in a science textbook come from two primary sources. The first is human error. Authors, editors, and illustrators occasionally make mistakes. Fortunately, such mistakes are rare. Each chapter of our Biology book has been read by dozens of people, including several accuracy checkers whose sole job was to find mistakes. Using our 360-degree developmental process (p. viii), we have worked to ensure an unprecedented level of content accuracy in our textbook.

A second and more common source of error is out-of-date material. Biology is continually changing. New information arrives on a daily basis. Some of this information makes us realize that past information was incorrect. Therefore, textbooks that fail to maintain a current perspective become progressively more inaccurate and misleading. Having an experienced author team with extensive research credentials helps to build a textbook with the most current and accurate information. In addition, over three-hundred faculty members have reviewed our Biology textbook for its content.

We are confident that our book has the most modern content that the industry has to offer. This modern perspective pushes the accuracy of our book to the highest standard possible. In future editions, we will continue to strive for cutting-edge content that maintains a high level of accuracy. In future editions, we will continue to employ the help of many reviewers and accuracy checkers to maintain our commitment to modern content and accuracy.

Our Review Process Ensures a Textbook with the Right Content

If the best writer in the world wrote a textbook single-handedly without input from others, the book would not turn out well. Extensive and open-minded reviews are essential to producing a book that is superior. As we developed our book, we took the attitude that we must always return to a previous draft, analyze it critically, and then revise it accordingly.

From an author's perspective, the review process can be pretty daunting. We turn in chapters that we think are letter perfect, and then receive back reviews that make us painfully aware that writing a textbook is harder than it looks. At the start, we created a process that would make it easier for the editors and reviewers to critically evaluate our work. The first-draft chapters were sent to many outside reviewers who are faculty that either teach General Biology, are experts in the topics found in the chapter, or both. Each first-draft chapter was reviewed by up to 15 different people.

The reviews were collected and provided to the author of a given chapter and the editorial staff. Our editors read the chapters and the reviews, and then gave each author advice on how to make the next draft better. A second important type of input also occurred at the first-draft stage. For each unit, we conducted Focus Groups in which faculty members who had reviewed the chapters of an entire unit came together for a two-day meeting to discuss the chapters with the authors and editors. While written feedback is great, face-to-face discussion often brings out bigger-picture issues that may not be found in written reviews.

At the second-draft stage, we decided upon an innovation that profoundly enhanced the quality of our Biology book. Although the illustrations in a Biology textbook are very expensive to make, we realized that a strong connection between the text and illustrations is critical to produce a superior textbook. Instead of making the illustrations at the final-draft stage, which is typical of textbook publishing, we made them very early in the process. This helped us in two ways. First, reviewers could see the art as the book developed, and make critical changes to it. Second, it allowed the authors and editors to develop a keen sense of consistency between the text and art.

Also at the second-draft stage, we assembled the text and illustrations into a format that looked like a chapter from an actual textbook. We had to keep reminding the reviewers that "These are not finalized chapters. These are early drafts that we want you to critically evaluate." This format allowed the reviewers, authors, and editors to understand how the pieces of the book would fit together. Although it was an exhaustive and rigorous process, adopting this step early in the writing process allowed us to produce a book with a sharp consistency between the text and figures.

We Are Committed to Serving Teachers and Learners

Writing a new textbook is a daunting task. To accurately and thoroughly cover a course as wide ranging as biology, we felt it was essential that our team reflect the diversity of the field. We saw an opportunity to reach students at an early stage in their education and provide their biology training with a solid and up-to-date foundation. We have worked to balance coverage of classic research with recent discoveries that extend biological concepts in surprising new directions or that forge new concepts. Some new discoveries were selected because they highlight scientific controversies, showing students that we don't have all the answers yet. There is still a lot of work for new generations of biologists. With this in mind, we've also spotlighted discoveries made by diverse people doing research in different countries to illustrate the global nature of modern biological science.

As active teachers and writers, one of the great joys of this process for us is that we have been able to meet many more educators and students during the creation of this textbook. It is humbling to see the level of dedication our peers bring to their teaching. Likewise, it is encouraging to see the energy and enthusiasm so many students bring to their studies. We hope this book and its media package will serve to aid both faculty and students in meeting the challenges of this dynamic and exciting course. For us, this remains a work in progress and we encourage you to let us know what you think of our efforts and what we can do to serve you better.

Rob Brooker   brook005@umn.edu
Eric Widmaier   widmaier@bu.edu
Linda Graham   lkgraham@wisc.edu
Peter Stiling   pstiling@cas.usf.edu

The Next Step in Textbook Development

- 10 developmental editors
- 7 developmental focus groups
- 7 art focus groups
- 25 + accuracy checkers
- more than 1,200 reviews by over 350 reviewers across the world
- 11 multiple-day symposia with over 215 majors biology educators participating
- an art development team who worked closely with the authors
- media board of consultants
- 3 photo consultants

The following groups of individuals have been instrumental in ensuring the highest standard of content and accuracy in this textbook. We are deeply indebted to them for their tireless efforts.

Developmental Focus Groups

Cell Unit

Russell Borski,
  North Carolina State University
Peter Fajer,
  Florida State University
Brad Mehrtens,
  University of Illinois–Urbana/Champaign
Randall Walikonis,
  University of Connecticut
Sue Simon Westendorf,
  Ohio University
Mark Staves,
  Grand Valley State University

Genetics Unit

Karl Aufderheide,
  Texas A&M University
John Doctor,
  Duquesne University
Arlene Larson,
  University of Colorado–Denver
Subhash Minocha,
  University of New Hampshire
John Osterman,
  University of Nebraska–Lincoln
Jill Reid,
  Virginia Commonwealth University

Evolution Unit

Mark Decker,
  University of Minnesota–Minneapolis
Robert Dill,
  Bergen Community College
Jennifer Regan,
  University of Southern Mississippi
Michelle Shuster,
  New Mexico State University
Fred Wasserman,
  Boston University

Diversity Unit

Ernest DuBrul,
  University of Toledo
Roland Dute,
  Auburn University
Florence Gleason,
  University of Minnesota–St. Paul
Ann Rushing,
  Baylor University
Randall Yoder,
  Lamar University

Plants Unit

Fred Essig,
  University of South Florida
Steve Herbert,
  University of Wyoming
Mike Muller,
  University of Illinois–Chicago
Stuart Reichler,
  University of Texas–Austin
Scott Russell,
  University of Oklahoma
Rani Vajravelu,
  University of Central Florida

Animals Unit

Linda Collins,
  University of Tennessee–Chattanooga
William Collins,
  Stony Brook University
David Kurjiaka,
  Ohio University
Phil Stephens,
  Villanova University
David Tam,
  University of North Texas
Charles Walcott,
  Cornell University

Ecology Unit

James Adams,
  Dalton State College
Stanley Faeth,
  Arizona State University
Barbara Frase,
  Bradley University
Daniel Moon,
  University of North Florida
Dan Tinker,
  University of Wyoming

Media Focus Group

Russell Borski,
  North Carolina State University
Mark Decker,
  University of Minnesota
Jon Glase,
  Cornell University
John Merrill,
  Michigan State University
Melissa Michael,
  University of Illinois–Urbana/Champaign
Randall Phillis,
  University of Massachusetts–Amherst
Mitch Price,
  Pennsylvania State University

Accuracy Checkers

David Asch,
Youngstown State University
Karl Aufderheide,
Texas A&M University
Deborah Brooker
Linda Collins,
University of Tennessee-Chattanooga
Mark Decker,
University of Minnesota
Laura DiCaprio,
Ohio University
Marjorie Doyle,
University of Wisconsin-Madison
Peter Fajer,
Florida State University
Pete Franco,
University of Minnesota
Barbara Frase,
Bradley University

John Graham,
Bowling Green State University
Eunsoo Kim,
University of Wisconsin-Madison
Arlene Larson,
University of Colorado-Denver
David Pennock,
Miami University
Anthony M. Rossi,
University of North Florida
Martin Silberberg,
McGraw-Hill chemistry author
Kevin Strang,
University of Wisconsin-Madison
Fred Wasserman,
Boston University
Jane E. Wissinger,
University of Minnesota

Class Testers

We would like to thank the students and faculty at Ohio University, UCLA, and Calvin College for class testing our book.

End-of-Chapter Questions

Robert Dill,
Bergen Community College
Arlene Larson,
University of Colorado-Denver

Jennifer Regan,
University of Southern Mississippi

Photo Consultants

John Osterman,
University of Nebraska-Lincoln
Sue Simon Westendorf,
Ohio University

Kevin Strang,
University of Wisconsin-Madison

General Biology Symposia

Every year McGraw-Hill conducts several General Biology Symposia, which are attended by instructors from across the country. These events are an opportunity for editors from McGraw-Hill to gather information about the needs and challenges of instructors teaching the major's biology course. It also offers a forum for the attendees to exchange ideas and experiences with colleagues they might not have otherwise met. The feedback we have received has been invaluable, and has contributed to the development of Biology and its supplements.

2006

Michael Bell, *Richland College*
Scott Bowling, *Auburn University*
Peter Busher, *Boston University*
Allison Cleveland,
University of South Florida-Tampa
Sehoya Cotner,
University of Minnesota
Kathyrn Dickson,
California State College-Fullerton
Cathy Donald-Whitney,
Collin County Community College
Stanley Faeth,
Arizona State University
Karen Gerhart,
University of California-Davis
William Glider,
University of Nebraska- Lincoln
Stan Guffey,
The University of Tennessee
Bernard Hauser,
University of Florida-Gainesville
Mark Hens,
University of North Carolina-Greensboro
James Hickey,
Miami University of Ohio-Oxford
Sherry Krayesky,
University of Louisiana-Lafayette
Brenda Leady,
University of Toledo
Michael Meighan,
University of California-Berkeley
Comer Patterson,
Texas A&M University
Debra Pires,
University of California-Los Angeles
Robert Simons,
University of California-Los Angeles
Steven D. Skopik,
University of Delaware
Ashok Upadhyaya,
University of South Florida-Tampa
Anthony Uzwiak,
Rutgers University
Dave Williams,
Valencia Community College-East Campus
Jay Zimmerman,
St. John's University

2005

Donald Buckley,
Quinnipiac University
Arthur Buikema,
Virginia Polytechnic Institute
Anne Bullerjahn,
Owens Community College
Garry Davies,
University of Alaska-Anchorage
Marilyn Hart,
Minnesota State University
Daniel Flisser,
Camden County College
Elizabeth Godrick,
Boston University
Miriam Golbert,
College of the Canyons
Sherry Harrel,
Eastern Kentucky University
William Hoese,
California State University-Fullerton
Margaret Horton,
University of North Carolina at Greensboro
Carol Hurney,
James Madison University
James Luken,
Coastal Carolina University
Mark Lyford,
University of Wyoming
Gail McKenzie,
Jefferson State Junior College
Melissa Michael,
University of Illinois at Urbana-Champaign
Subhash C. Minocha,
University of New Hampshire
Leonore Neary,
Joliet Junior College
K. Sata Sathasivan,
University of Texas at Austin
David Senseman,
University of Texas-San Antonio
Sukanya Subramanian,
Collin County Community College
Randall Terry,
Lamar University
Sharon Thoma,
University of Wisconsin-Madison
William Tyler,
Indian River Community College

2004

Jonathan Akin,
Northwestern State University of Louisiana

David Asch,
Youngstown State University

Diane Bassham,
Iowa State University

Donald Buckley,
Quinnipiac University

Ruth Buskirk,
University of Texas, Austin

Charles Creutz,
University of Toledo

Lydia Daniels,
University of Pittsburgh

Laura DiCaprio,
Ohio University

Michael Dini,
Texas Tech University

John Doctor,
Duquesne University

Ernest DuBrul,
University of Toledo

John Elam,
Florida State University

Samuel Hammer,
Boston University

Marilyn Hart,
Minnesota State University

Marc Hirrel,
University of Central Arkansas

Carol Johnson,
Texas A&M University

Dan Krane,
Wright State University

Karin Krieger,
University of Wisconsin–Green Bay

Josephine Kurdziel,
University of Michigan

Martha Lundell,
University of Texas, San Antonio

Roberta Maxwell,
University of North Carolina–Greensboro

John Merrill,
Michigan State University

Melissa Michael,
University of Illinois at Urbana-Champaign

Peter Niewarowski,
University of Akron

Ronald Patterson,
Michigan State University

Peggy Pollak,
Northern Arizona University

Uwe Pott,
University of Wisconsin, Green Bay

Mitch Price,
Pennsylvania State University

Steven Runge,
University of Central Arkansas

Thomas Shafer,
University of North Carolina, Wilmington

Richard Showman,
University of South Carolina

Michèle Shuster,
New Mexico State University

Dessie Underwood,
California State University–Long Beach

Mike Wade,
Indiana University

Elizabeth Willott,
University of Arizona

Carl Wolfe,
University of North Carolina, Charlotte

Reviewers

James K. Adams,
Dalton State College

Sylvester Allred,
Northern Arizona University

Jonathan W. Armbruster,
Auburn University

Joseph E. Armstrong,
Illinois State University

David K. Asch,
Youngstown State University

Amir M. Assadi-Rad,
Delta College

Karl J. Aufderheide,
Texas A&M University

Anita Davelos Baines,
University of Texas–Pan American

Lisa M. Baird,
University of San Diego

Diane Bassham,
Iowa State University

Donald Baud,
University of Memphis

Vernon W. Bauer,
Francis Marion University

Ruth E. Beattie,
University of Kentucky

Michael C. Bell,
Richland College

Steve Berg,
Winona State University

Arlene G. Billock,
University of Louisiana at Lafayette

Kristopher A. Blee,
California State University, Chico

Heidi B. Borgeas,
University of Tampa

Russell Borski,
North Carolina State University

Scott A. Bowling,
Auburn University

Robert Boyd,
Auburn University

Eldon J. Braun,
University of Arizona

Michael Breed,
University of Colorado, Boulder

Randy Brewton,
University of Tennessee, Knoxville

Peggy Brickman,
University of Georgia

Cheryl Briggs,
University of California, Berkeley

Peter S. Brown,
Mesa Community College

Mark Browning,
Purdue University

Cedric O. Buckley,
Jackson State University

Don Buckley,
Quinnipiac University

Arthur L. Buikema, Jr.,
Virginia Tech University

Anne Bullerjahn,
Owens Community College

Ray D. Burkett,
Southeast Tennessee Community College

Stephen P. Bush,
Coastal Carolina University

Peter E. Busher,
Boston University

Jeff Carmichael,
University of North Dakota

Clint E. Carter,
Vanderbilt University

Patrick A. Carter,
Washington State University

Merri Lynn Casem,
California State University, Fullerton

Domenic Castignetti,
Loyola University of Chicago

Maria V. Cattell

David T. Champlin,
University of Southern Maine

Jung H. Choi,
Georgia Institute of Technology

Curtis Clark,
Cal Poly Pomona

Allison Cleveland,
University of South Florida

Janice J. Clymer,
San Diego Mesa College

Linda T. Collins,
University of Tennessee at Chattanooga

Jay L. Comeaux,
Louisiana State University

Bob Connor II,
Owens Community College

Daniel Costa,
University of California at Santa Cruz

Sehoya Cotner,
University of Minnesota

Mack E. Crayton III,
Xavier University of Louisiana

Louis Crescitelli,
Bergen Community College

Charles Creutz,
University of Toledo

Karen A. Curto,
University of Pittsburgh

Mark A. Davis,
Macalester College

Mark D. Decker,
University of Minnesota

Jeffery P. Demuth,
Indiana University

Phil Denette,
Delgado Community College

Donald W. Deters,
Bowling Green State University

Hudson R. DeYoe,
University of Texas–Pan American

Laura DiCaprio,
Ohio University

Randy DiDomenico,
University of Colorado, Boulder

Robert S. Dill,
Bergen Community College

Kevin Dixon,
University of Illinois–Urbana/Champaign

John S. Doctor,
Duquesne University

Michael Meighan,
University of California, Berkeley

Douglas Meikle,
Miami University

Allen F. Mensinger,
University of Minnesota, Duluth

John Merrill,
Michigan State University

Richard Merritt,
Houston Community College

Brian T. Miller,
Middle Tennessee State University

Hugh A. Miller III,
East Tennessee State University

Thomas E. Miller,
Florida State University

Sarah L. Milton,
Florida Atlantic University

Dennis J. Minchella,
Purdue University

Subhash C. Minocha,
University of New Hampshire

Patricia Mire,
University of Louisiana at Lafayette

Daniela S. Monk,
Washington State University

Daniel C. Moon,
University of North Florida

Janice Moore,
Colorado State University

Mathew D. Moran,
Hendrix College

Jorge A. Moreno,
University of Colorado, Boulder

Roderick M. Morgan,
Grand Valley State University

James V. Moroney,
Louisiana State University

Molly R. Morris,
Ohio University

Michael Muller,
University of Illinois at Chicago

Michelle Mynlieff,
Marquette University

Allan D. Nelson,
Tarleton State University

Raymond L. Neubauer,
University of Texas at Austin

Jacalyn S. Newman,
University of Pittsburgh

Colleen J. Nolan,
St. Mary's University

Shawn E. Nordell,
St. Louis University

Margaret Nsofor,
Southern Illinois University, Carbondale

Dennis W. Nyberg,
University of Illinois at Chicago

Nicole S. Obert,
University of Illinois, Urbana-Champaign

David G. Oppenheimer,
University of Florida

John C. Osterman,
University of Nebraska–Lincoln

Brian Palestis,
Wagner College

Julie M. Palmer,
University of Texas at Austin

C. O. Patterson,
Texas A&M University

Ronald J. Patterson,
Michigan State University

Linda M. Peck,
University of Findlay

David Pennock,
Miami University

Shelley W. Penrod,
North Harris College

Beverly J. Perry,
Houston Community College System

Chris Petersen,
College of the Atlantic

Jay Phelan,
UCLA

Eric R. Pianka,
The University of Texas at Austin

Thomas Pitzer,
Florida International University

Peggy E. Pollak,
Northern Arizona University

Richard B. Primack,
Boston University

Lynda Randa,
College of Dupage

Marceau Ratard,
Delgado Community College

Robert S. Rawding,
Gannon University

Jennifer Regan,
University of Southern Mississippi

Stuart Reichler,
University of Texas at Austin

Jill D. Reid,
Virginia Commonwealth University

Anne E. Reilly,
Florida Atlantic University

Linda R. Richardson,
Blinn College

Laurel Roberts,
University of Pittsburgh

Kenneth R. Robinson,
Purdue University

Chris Ross,
Kansas State University

Anthony M. Rossi,
University of North Florida

Kenneth H. Roux,
Florida State University

Ann E. Rushing,
Baylor University

Scott Russell,
University of Oklahoma

Christina T. Russin,
Northwestern University

Charles L. Rutherford,
Virginia Tech University

Margaret Saha,
College of William and Mary

Kanagasabapathi Sathasivan,
The University of Texas at Austin

Stephen G. Saupe,
College of St. Benedict

Jon B. Scales,
Midwestern State University

Daniel C. Scheirer,
Northeastern University

H. Jochen Schenk,
California State University, Fullerton

John Schiefelbein,
University of Michigan

Deemah N. Schirf,
University of Texas at San Antonio

Mark Schlueter,
College of Saint Mary

Scott Schuette,
Southern Illinois University, Carbondale

Dean D. Schwartz,
Auburn University

Timothy E. Shannon,
Francis Marion University

Richard M. Showman,
University of South Carolina

Michele Shuster,
New Mexico State University

Robert Simons,
UCLA

J. Henry Slone,
Francis Marion University

Phillip Snider, Jr.,
Gadsden State Community College

Nancy G. Solomon,
Miami University

Lekha Sreedhar,
University of Missouri–Kansas City

Bruce Stallsmith,
University of Alabama, Huntsville

Susan J. Stamler,
College of Dupage

Mark P. Staves,
Grand Valley State University

William Stein,
Binghamton University

Philip J. Stephens,
Villanova University

Antony Stretton,
University of Wisconsin–Madison

Gregory W. Stunz,
Texas A&M University–Corpus Christi

Julie Sutherland,
College of Dupage

David Tam,
University of North Texas

Roy A. Tassava,
Ohio State University

Sharon Thoma,
University of Wisconsin–Madison

Shawn A. Thomas,
College of St. Benedict/ St. John's University

Daniel B. Tinker,
University of Wyoming

Marty Tracey,
Florida International University

Marsha Turell,
Houston Community College

J. M. Turbeville,
Virginia Commonwealth University

Rani Vajravelu,
University of Central Florida

Neal J. Voelz,
St. Cloud State University

Samuel E. Wages,
South Plains College

Jyoti R. Wagle,
Houston Community College System–Central

Charles Walcott,
Cornell University

Randall Walikonis,
University of Connecticut

Jeffrey A. Walker,
University of Southern Maine

Delon E. Washo-Krupps,
Arizona State University
Frederick Wasserman,
Boston University
Steven A. Wasserman,
*University of California,
San Diego*
R. Douglas Watson,
*University of Alabama
at Birmingham*
Cindy Martinez Wedig,
*University of Texas–
Pan American*

Arthur E. Weis,
University of California–Irvine
Sue Simon Westendorf,
Ohio University
Howard Whiteman,
Murray State University
Susan Whittemore,
Keene State College
David L. Wilson,
University of Miami
Robert Winning,
*Eastern Michigan
University*

Michelle D. Withers,
*Louisiana State
University*
Clarence C. Wolfe,
*Northern Virginia Community
College*
Gene K. Wong,
Quinnipiac University
Richard P. Wunderlin,
University of South Florida

Joanna Wysocka-Diller,
Auburn University
H. Randall Yoder,
Lamar University
Marilyn Yoder,
*University of Missouri–
Kansas City*
Scott D. Zimmerman,
*Southwest Missouri State
University*

International Reviewers

Heather Addy,
University of Calgary
Mari L. Acevedo,
*University of Puerto Rico
at Arecibo*
Heather E. Allison,
University of Liverpool, UK
David Backhouse,
University of New England
Andrew Bendall,
University of Guelph
Marinda Bloom,
*Stellenbosch University,
South Africa*
Tony Bradshaw,
Oxford-Brookes University, UK
Alison Campbell,
University of Waikato
Bruce Campbell,
Okanagan College

Clara E. Carrasco, Ph.D.,
*University of Puerto Rico–
Ponce Campus*
Keith Charnley,
University of Bath, UK
Ian Cock,
Griffith University
Margaret Cooley,
University of NSW
R. S. Currah,
University of Alberta
Logan Donaldson,
York University
Theo Elzenga,
*Rijks Universiteit Groningen,
Netherlands*
Neil C. Haave,
University of Alberta

Tom Haffie,
University of Western Ontario
Louise M. Hafner,
*Queensland University
of Technology*
Annika F. M. Haywood,
*Memorial University
of Newfoundland*
William Huddleston,
University of Calgary
Shin-Sung Kang,
KyungBuk University
Wendy J. Keenleyside,
University of Guelph
Christopher J. Kennedy,
Simon Fraser University
Bob Lauder,
Lancaster University

Richard C. Leegood,
Sheffield University, UK
Thomas H. MacRae,
Dalhousie University
R. Ian Menz,
Flinders University
Kirsten Poling,
University of Windsor
Jim Provan,
Queens University, Belfast, UK
Richard Roy,
McGill University
Han A.B. Wösten,
*Utrecht University,
Netherlands*

ACKNOWLEDGMENTS

The lives of most science-textbook authors do not revolve around an analysis of writing techniques. Instead, we are people who understand science and are inspired by it, and we want to communicate that information to other people. Simply put, we need a lot of help to get it right.

Editors are a key component that help the authors modify the content of their book so it is logical, easy to read, and inspiring. The editorial team for this Biology textbook has been a catalyst that kept this project rolling. The members played various roles in the editorial process. Lisa Bruflodt (Senior Developmental Editor) has been the master organizer. Frankly, this is a ridiculously hard job. Coordinating the efforts of dozens of people and keeping them on schedule is not always fun. Lisa's success at keeping us on schedule has been truly amazing. We are also grateful to Kris Tibbetts (Director of Development) who was involved in the early steps of the book, and kept the focus groups on track.

Our Biology book also has had 6 additional developmental editors who scrutinized each draft of their respective chapters with an emphasis on improving content, clarity, and readability. These developmental editors analyzed educational materials and reviewers' comments, and gave the authors advice on how to improve succeeding drafts. They also provided a list of the general principles that most instructors want in their Biology textbook. These general principles have been a cornerstone for the organization of our chapters.

Suzanne Olivier (Lead Freelance Developmental Editor) did an outstanding job of coordinating the staff of developmental editors. She also played an important role in editing chapters in the Genetics and Plant Biology units. Her early editing of the Genetics Unit, in particular, set the tone for many of the pedagogical features that became established throughout the entire textbook. Other developmental editors focused on particular units. Alice Fugate was involved with the Chemistry, Cell Biology, and Animal Biology Units. Her knack for getting the level of the writing appropriate for majors biology was invaluable, as was her attention to detail. Joni Fraser focused on the Cell Biology, Diversity, and Ecology Units. Somehow she successfully juggled the tasks of addressing all the reviewers' concerns while maintaining the necessary chapter length. Patricia Longoria played an important role in the early editing of the Diversity and Plant Biology Units. Patricia contributed many useful ideas for content and expression, and her unfailing enthusiasm smoothed the way over rocky parts of the process. Robin Fox edited three key chapters in the Genetics Unit. Her attention to detail and the explanation of mathematical principles were invaluable. And finally, Alan Titche was also involved with Animal Biology Unit, and played a major role in developing some of the most challenging chapters in that Unit. We would also like to thank Dr. Jim Deshler and Dr. Mary Erskine for their valuable contributions to the writing of several chapters in the Animal Biology Unit.

Deborah Brooker (Art/Text Coordinating Editor) analyzed all of the chapters in the textbook with one primary question in mind. Do the written text and figures tell a parallel story? With excruciating care, she made sure that the text and figures are consistent, and that the figures, by themselves, are accurate and easy to follow.

Imagineering Media Services Inc., of Ontario, Canada, did a fantastic job of illustrating our Biology book. They were involved early in the process by first making rough sketches based on the material in the first drafts, and then later progressed to drawings with finer detail. Their ability to make realistic, three-dimensional drawings is second to none. We're particularly grateful to Kierstan Hong, who provided a critical line of communication between the publisher, authors, and illustrators throughout most of this process, and also to Mark Mykytiuk, who also played a lead role in overseeing the art development. We would also like to gratefully acknowledge our photography researchers at Pronk & Associates of Ontario, Canada, and particularly to Fiona D'souza for keeping us on schedule. Likewise, we are grateful to John Leland, Photo Research Coordinator, at McGraw-Hill for his coordination of the photo selection process.

We would also like to thank our advisors and contributors:

Media Board of Advisors

Mark Decker,
University of Minnesota–Minneapolis

Naomi Friedman, *Developmental Editor*

Jon Glase,
Cornell University

John Merrill,
Michigan State University

Melissa Michael,
University of Illinois–Urbana/Champaign

Randall Phyllis,
University of Massachusetts

Mitch Price,
Pennsylvania State University

Tutorial Questions

Scott Bowling,
Auburn University

Don Buckley,
Quinnipiac University

Ernest DuBrul,
University of Toledo

Frederick B. Essig,
University of Florida

Jon Glase,
Cornell University

Norman A. Johnson
University of Massachusetts–Amherst

Kari Beth Krieger,
University of Wisconsin–Green Bay

Patricia Mire,
University of Louisiana–Lafayette

Allan Smits,
Quinnipiac University

Test Questions

Russell Borski,
*North Carolina State
University*

Robert Dunn,
*North Carolina State
University*

John Godwin,
*North Carolina State
University*

Mary Beth Hawkins,
*North Carolina State
University*

Harold Heatwole,
*North Carolina State
University*

James Mickle,
*North Carolina State
University*

Gerald Van Dyke,
*North Carolina State
University ARIS*

ARIS

Brad Mehrtens,
*University of Illinois–
Urbana/Champaign*

Instructor's Manual

Daniel Moon,
University of North Florida

Student Study Guide

Michelle Shuster,
New Mexico State University

Amy Marion,
New Mexico State University

Lecture Outlines

Brenda Leady,
University of Toledo

Animations

Kevin Dixon,
*University of Illinois–
Urbana/Champaign*

Another important aspect of the editorial process is the actual design, presentation, and layout of materials. It's confusing if the text and art aren't on the same page, or if a figure is too large or two small. We are indebted to the tireless efforts of Joyce Berendes (Lead Project Manager) and Wayne Harms (Design Manager) of McGraw-Hill. Their artistic talents, ability to size and arrange figures, and attention to the consistency of the figures have been remarkable. We also wish to thank John Joran (Designer) who cleverly crafted both the interior and exterior designs.

We would like to acknowledge the ongoing efforts of the superb marketing staff at McGraw-Hill. Kent Peterson (Vice President, Director of Marketing) oversees a talented staff of people who work tirelessly to promote our book. Special thanks to Chad Grall (Marketing Director), Wayne Vincent (Internet Marketing Manager), Debra Hash (Senior Marketing Manager) and Heather Wagner (Systems and Promotions Marketing Manager) for their ideas and enthusiasm for this book. The proposal of making a video website for the book *www.brookerbiology.com* was scary for the authors but actually turned out to be fun.

Finally, other staff members at McGraw-Hill Higher Education have ensured that the authors and editors were provided with adequate resources to achieve the goal of producing a superior textbook. These include Kurt Strand (President, Science, Engineering, and Math), Marty Lange (Vice President, Editor-in-Chief), Michael Lange (Vice President, New Product Launches), Janice Roerig-Blong (Publisher) and Patrick Reidy (Executive Editor). In particular, Michael and Patrick communicated with the authors on a regular basis regarding the progress of this project. They attended most of the focus groups and author meetings, and even provided occasional input regarding the content of the book. The bottom line is that the author team is grateful that you have believed in this project, and have provided us with the resources to make it happen.

Student Supplements

Designed to help students maximize their learning experience in biology—we offer the following options to students:

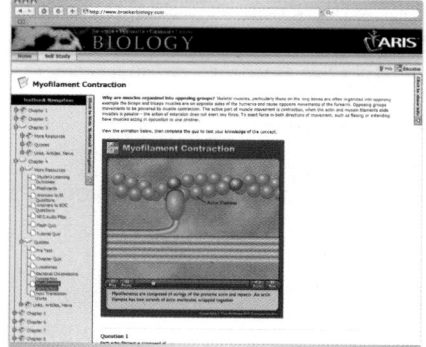

ARIS (Assessment, Review, and Instruction System) is an electronic study system that offers students a digital portal of knowledge. Students can readily access a variety of **digital learning objects** which include:

- chapter level quizzing
- pretests
- animations
- videos
- flashcards
- answers to Biological Inquiry Questions
- answers to all end-of-chapter questions
- MP3 and MP4 downloads of selected content
- learning outcomes and assessment capability woven around key content

Student Study Guide
ISBN: 0-07-299588-2

Helping students focus their time and energy on important concepts, the study guide offers students a variety of tools:

1. Practice Questions—approximately 10–12 multiple choice questions
2. Active Learning Questions—approximately 5–8 open-ended questions that ask the student to explore

something and delve into content a little deeper, reinforcing content through experiential learning.

3. Outline/Summary of Fundamental Concepts—efficient listing of key concepts.
4. Key Terms
5. Strategies for Difficult Concepts

Content Delivery Flexibility

Brooker et al., *Biology* is available in many formats in addition to the traditional textbook to give instructors and students more choices when deciding on the format of their biology text. Choices include:

Volumes

The complete text has been split into three natural segments to allow instructors more flexibility and students more purchasing options.

Volume 1—Units 1 (Chemistry), 2 (Cell), and 3 (Genetics)
ISBN 0-07-335332-9
Volume 2—Units 6 (Plants) and 7 (Animals)
ISBN 0-07-335331-0
Volume 3—Units 4 (Evolution), 5 (Diversity), and 8 (Ecology)
ISBN 0-07-335333-7

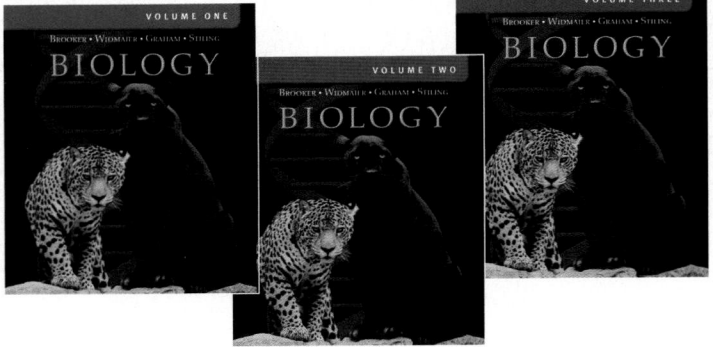

Color Custom by Chapter

For even more flexibility, we offer the Brooker: *Biology* text in a full-color, custom version that allows instructors to pick the chapters they want included. Students pay for only what the instructor chooses.

eBook

The entire text is available electronically through the ARIS website. This electronic text offers not only the text in a digital format but includes embedded links to figures, tables, animations, and videos to make full use of the digital tools available and further enhance student understanding.

Instructor Supplements

Dedicated to providing high quality and effective supplements for instructors, the following Instructor supplements were developed for *Biology:*

ARIS with Presentation Center

Assessment, Review, and Instruction System, also known as ARIS, is an electronic homework and course management system designed for greater flexibility, power, and ease of use than any other system. Whether you are looking for a preplanned course or one you can customize to fit your course needs, ARIS is your solution.

In addition to having access to all student digital learning objects, ARIS allows instructors to:

Build Assignments
- Choose from pre-built assignments or create your own custom content by importing your own content or editing an existing assignment from the pre-built assignment.

- Assignments can include quiz questions, animations, and videos . . . anything found on the website.
- Create announcements and utilize full course or individual student communication tools
- Assign **unique multi-level tutorial questions** developed by content experts that provide intelligent feedback through a series of questions to help students truly understand a concept; not just repeat an answer.

Track Student Progress
- Assignments are automatically graded
- Gradebook functionality allows full course management including:
  - Dropping the lowest grades
  - Weighting grades / manually adjusting grades
  - Exporting your gradebook to Excel, WebCT or BlackBoard

- Manipulating data allowing you to track student progress through multiple reports

Offer More Flexibility

- **Sharing Course Materials with Colleagues** — Instructors can create and share course materials and assignments with colleagues with a few clicks of the mouse allowing for multiple section courses with many instructors (and TAs) to continually be in synch if desired.
- **Integration with BlackBoard or WebCT**—once a student is registered in the course, all student activity within McGraw-Hill's ARIS is automatically recorded and available to the instructor through a fully integrated grade book that can be downloaded to Excel, WebCT, or Blackboard.

Presentation Center

Build instructional materials wherever, whenever, and however you want!

ARIS Presentation Center is an online digital library containing assets such as photos, artwork, animations, PowerPoints, and other media types that can be used to create customized lectures, visually enhanced tests and quizzes, compelling course websites, or attractive printed support materials.

Access to your book, access to all books!
The Presentation Center library includes thousands of assets from many McGraw-Hill titles. This ever-growing resource gives instructors the power to utilize assets specific to an adopted textbook as well as content from all other books in the library.

Nothing could be easier!
Accessed from the instructor side of your textbook's ARIS website, Presentation Center's dynamic search engine allows you to explore by discipline, course, textbook chapter, asset type, or keyword. Simply browse, select, and download the files you need to build engaging course materials. All assets are copyright McGraw-Hill Higher Education but can be used by instructors for classroom purposes.

Instructor's Testing and Resource CD-ROM
ISBN: 0-07-295658-5
This cross-platform CD-ROM provides these resources for instructors:

- **Instructor's Manual**—This manual contains instructional strategies and activities, student misconceptions,

etymology of key terms, "Beyond the Book" interesting facts, and sources for additional web resources.
- **Test Bank**—The test bank offers multiple-choice and true/false questions that can be used for homework assignments or the preparation of exams.
- **Computerized Test Bank**—This software can be utilized to quickly create customized exams. The user-friendly program allows instructors to sort questions by format or level of difficulty; edit existing questions or add new ones; and scramble questions and answer keys for multiple versions of the same test.

Student Response System
Wireless technology brings interactivity into the classroom or lecture hall. Instructors and students receive immediate feedback through wireless response pads that are easy to use and engage students. This system can be used by instructors to:

- Take attendance
- Administer quizzes and tests
- Create a lecture with intermittent questions
- Manage lectures and student comprehension through the use of the gradebook
- Integrate interactivity into their PowerPoint presentations

Transparencies
ISBN: 0-07-295657-7
This boxed set of overhead transparencies includes every piece of line art in the textbook. The images have been modified to ensure maximum readability in both small and large classroom settings.

BIOLOGY LABORATORY MANUAL
Darrell S. Vodopich, *Baylor University*
Randy Moore, *University of Minnesota*
ISBN: 0-07-332398-5

This laboratory manual is designed to accompany Brooker et al: *Biology*. The experiments and procedures are simple, safe, easy to perform, and especially appropriate for large classes. Few experiments require a second class-meeting to complete the procedure. Each exercise includes many photographs, traditional topics, and experiments that help students learn about life. Procedures within each exercise are numerous and discrete so that an exercise can be tailored to the needs of the students, the style of the instructor, and the facilities available.

BIOLOGICAL INVESTIGATIONS LAB MANUAL
Warren D. Dolphin, *Iowa State University*
ISBN: 0-07-332399-3

Developed to accompany Brooker et al: *Biology*, this lab manual focuses on labs that are investigative and ask students to use more critical thinking and hands-on learning. The author emphasizes investigative, quantitative, and comparative approaches to studying the life sciences.

A VISUAL JOURNEY

Our art program was painstakingly designed in conjunction with the text development to ensure 1) each concept is accurately portrayed, 2) consistency is maintained between the text and art, and 3) it's appropriately placed on the page. The art serves as a visual outline for students, often offering textboxes that explain difficult concepts. For multistep processes, these textboxes are numbered so that the student can easily follow the process from beginning to end.

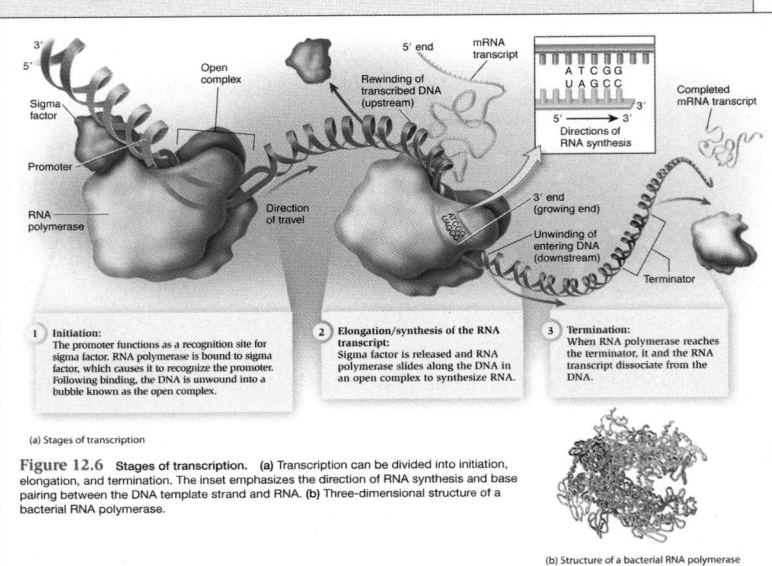

Figure 12.6 **Stages of transcription.** **(a)** Transcription can be divided into initiation, elongation, and termination. The inset emphasizes the direction of RNA synthesis and base pairing between the DNA template strand and RNA. **(b)** Three-dimensional structure of a bacterial RNA polymerase.

and birds maintain a relatively constant body temperature in spite of changing environmental temperatures (Figure 1.2d), while reptiles and amphibians do not. By comparison, all organisms continually regulate their cellular metabolism so that nutrient molecules are used at an appropriate rate, and new cellular components are synthesized when they are needed.

Growth and Development All living things grow and develop; **growth** produces more or larger cells, while **development** produces organisms with a defined set of characteristics. Among unicellular organisms such as bacteria, new cells are relatively small, and they increase in volume by the synthesis of additional cellular components. Multicellular organisms, such as

---

living organisms can be analyzed in a hierarchical manner, starting with the tiniest level of organization, and progressing to levels that are physically much larger and more complex. Figure 1.3 depicts a scientist's view of biological organization at different levels.

1. **Atoms:** An **atom** is the smallest component of an element that has the chemical properties of the element. All matter is composed of atoms.
2. **Molecules and macromolecules:** As discussed in Unit I, atoms bond with each other to form **molecules**. When many molecules bond together to form a polymer, this is called a **macromolecule**. Carbohydrates, proteins, and

The members of the same species are closely related genetically. In Units VI and VII, we will examine plants and animals at the level of cells, tissues, organs, and complete organisms.
7. **Population:** A group of organisms of the same species that occupy the same environment is called a **population.**
8. **Community:** A biological **community** is an assemblage of populations of different species. The types of species that are found in a community are determined by the environment and by the interactions of species with each other.
9. **Ecosystem:** Researchers may extend their work beyond living organisms and also study the environment. Ecologists analyze **ecosystems**, which are formed by

interactions of a community of organisms with their physical environment. Unit VIII considers biology from populations to ecosystems.
10. **Biosphere:** The **biosphere** includes all of the places on the Earth where living organisms exist, encompassing the air, water, and land.

Modern Forms of Life Are Connected by an Evolutionary History

Life began on Earth as primitive cells about 3.5 to 4 billion years ago. Since that time, those primitive cells underwent evolutionary changes that ultimately gave rise to the species we see today.

Figure 1.3   The levels of biological organization.

BIOLOGICAL INQUIRY QUESTIONS

These questions are designed to help students delve more deeply into a concept or experimental approach described in the art. These questions challenge a student to analyze the content of the figure they are looking at.

Figure 1.8 **The web of life.** This diagram of evolution includes both vertical evolution and ho[...] mechanisms in the evolution of life on Earth. Note: Archaea are unicellular species that are similar [...]

Biological inquiry: How does the concept of a tree of life differ from a web of life?

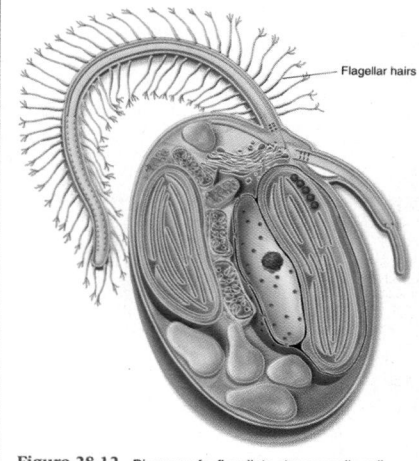

Figure 28.12 Diagram of a flagellate stramenopile cell, showing characteristic flagellar hairs.

Biological inquiry: How do the flagellar hairs aid cell motion?

Figure 17.7 Maternal inheritance in the four-o'clock plant. The genes for green pigment synthesis in plants are part of the chloroplast genome. The white phenotype in four o'clocks is due to chloroplasts with a mutant allele that greatly reduces green pigment production. The variegated phenotype is due to a mixture of normal and mutant chloroplasts. In four o'clocks, the egg contains all of the plastids that are inherited by the offspring, so the phenotype of the offspring is determined by the female parent.

Biological inquiry: In this example, where is the gene located that causes the green color of four-o'clock leaves? How is this gene transmitted from parent to offspring?

The unifying concept in biology is evolution. We can understand the unity of modern organisms by realizing that all living species evolved from an interrelated group of ancestors. However, from an experimental perspective, this realization presents a dilemma—we cannot take a time machine back over the course of 4 billion years to carefully study the characteristics of extinct organisms and fully appreciate the series of changes

Figure 5.4 **A method to measure the lateral movement of membrane proteins.**

Biological inquiry: Explain why the H-2 proteins are found only on one side of the cell when the cells were incubated at 0°C.

Figure 1.11 **Genomes and proteomes.** **(a)** The genome, which is composed of DNA, is the entire genetic complement of an organism. Most of the genetic material in eukaryotic cells is found in the cell nucleus. Its primary function is to encode the proteome. **(b)** The proteome is the entire protein complement of a cell or entire organism. Proteins are largely responsible for the structure and function of cells and complete organisms.

Biological inquiry: Biologists sometimes say that the genome is a storage unit, while the proteome is largely the functional unit of life. Explain this statement.

EXPERIMENTAL & MODERN CONTENT

FEATURE INVESTIGATION

Nirenberg and Leder Found That RNA Triplets Can Promote the Binding of tRNA to Ribosomes

In 1964, Nirenberg and Leder discovered that RNA molecules containing any three nucleotides (that is, any triplet) can stim-

later in the chapter, but for now just keep in mind that tRNAs interact with mRNA on a ribosome during the synthesis of polypeptides. This sample was divided into 20 tubes. To each tube, they next added a mixture of cellular tRNAs that already had amino acids attached to them. However, each mixture of

Figure 12.14   Nirenberg and Leder's use of triplet binding assays to decipher the genetic code.

Overview

HYPOTHESIS A triplet RNA can bind to a ribosome and promote the binding of the tRNA that carries the amino acid that the triplet RNA specifies.

STARTING MATERIALS Components of an in vitro translation system, including ribosomes and tRNAs. Preparations containing all of the different tRNA molecules were given 1 radiolabeled amino acid; the other 19 amino acids were nonlabeled. For example, in 1 sample, radiolabeled glycine was added and the other 19 amino acids were nonlabeled. In a different sample, radiolabeled proline was added and the other 19 amino acids were nonlabeled. The tRNA preparation also contained the enzymes that attach amino acids to tRNAs.

Steps

1 Mix together triplet RNAs of a specific sequence and ribosomes. In the example shown here, the triplet is 5'–CCC–3'. Add a tRNA sample to this mixture that contains 1 radiolabeled amino acid. (Note: Only 3 tubes are shown here. Because there are 20 different amino acids, this would be done in 20 different tubes.)

tRNAs with 1 radiolabeled amino acid (for example, proline)

2 Allow time for triplet RNA to bind to the ribosome, and for the appropriate tRNA to bind to the triplet RNA.

Radiolabeled proline
Proline tRNA
Triplet RNA that specifies proline

3 Pour mixture through a filter that allows the passage of unbound tRNA but does not allow the passage of ribosomes.

Ribosomes trapped on filter

Filter

Data & Analysis

5 THE DATA

Triplet	Radiolabeled amino acid trapped on the filter	Triplet
5' – AAA – 3'	Lysine	5' – GAC – 3'
5' – ACA – 3', 5' – ACC – 3'	Threonine	5' – GCC – 3'
5' – AGA – 3'	Arginine	5' – GGU – 3', 5' – GGC – 3'
5' – AUA – 3', 5' – AUU – 3'	Isoleucine	5' – GUU – 3'
5' – CCC – 3'	Proline	5' – UAU – 3'
5' – CGC – 3'	Arginine	5' – UGU – 3'
5' – GAA – 3'	Glutamic acid	5' – UUG – 3'

FEATURE INVESTIGATION

Focusing on hypothesis testing and discovery-based science, the Feature Investigations describe a key experiment, including 1) an overview of the hypothesis or goal of the experiment, 2) the steps of the experiment, and 3) ending with an analysis of data. This encourages an appreciation of the scientific process.

Later, we will examine the roles of these sites in the synthesis of a polypeptide.

GENOMES & PROTEOMES

Comparisons of Small Subunit rRNAs Among Different Species Provide a Basis for Establishing Evolutionary Relationships

Translation is a fundamental process that is vital for the existence of all living species. The components that are needed for translation arose very early in the evolution of life on our planet. In fact, they arose in an ancestor that gave rise to all known living species. For this reason, all organisms have translational components that are evolutionarily related to each other. For example, the rRNA found in the small subunit of ribosomes is similar in all forms of life, though it is slightly larger in eukaryotic species (18S) than in bacterial species (16S). In other words, the gene for the small subunit rRNA (SSU rRNA) is found in the genomes of all organisms.

One way that geneticists explore evolutionary relationships is to compare the sequences of evolutionarily related genes. At the molecular level, gene evolution involves changes in DNA sequences. After two different species have diverged from each other during evolution, the genes of each species have an opportunity to accumulate changes, or mutations, that alter the sequences of those genes. After many generations, evolutionarily related species contain genes that are similar but not identical to each other, because each species will accumulate different

in the primordial gene that gave rise to modern species and, because these sequences may have some critical function, have not been able to change over evolutionary time. Those sequences shaded in green are identical in all three mammals, but differ compared to one or more bacterial species. Actually, if you scan the mammalian species, you may notice that all three sequences are identical to each other in this region. The sequences shaded in red are identical in two or three bacterial species, but differ compared to the mammalian small subunit rRNA genes. The sequences from E. coli and Serratia marcescens are more similar to each other than the sequence from Bacillus subtilis is to either of them. This is consistent with the idea that E. coli and S. marcescens are more closely related evolutionarily than either of them is to B. subtilis.

12.6 The Stages of Translation

Like transcription, the process of translation occurs in three stages called initiation, elongation, and termination. Figure 12.20 provides an overview of the process. During initiation, mRNA, the first tRNA, and ribosomal subunits assemble into a complex. Next, in the elongation stage, the ribosome moves from the start codon in the mRNA toward the stop codon, synthesizing a polypeptide according to the sequence of codons in the mRNA. Finally, the process is terminated when the ribosome reaches a stop codon and the complex disassembles, releasing the completed polypeptide. In this section, we will examine the steps in this process as they occur in living cells.

```
GATTAAGAGGGACGGCCGGGGGCATTCGTATTGCGCCGCTAGAGGTGAAATTC   Human
GATTAAGAGGGACGGCCGGGGGCATTCGTATTGCGCCGCTAGAGGTGAAATTC   Mouse
GATTAAGAGGGACGGCCGGGGGCATTCGTATTGCGCCGCTAGAGGTGAAATTC   Rat
CAAGCTTGAGTCTCGTAGAGGGGGGTAGAATTCCAGGTGTAGCGGTGAAATGC   E. coli
CAAGCTAGAGTCTCGTAGAGGGGGGTAGAATTCCAGGTGTAGCGGTGAAATGC   S. marcescens
GAGAGACTTGAGTACAGAAGAGGAGAGTGAAATTCCACGTGTAGCGGTGAAATGC  B. subtilis
```

Figure 12.19 Comparison of small subunit rRNA gene sequences from three eukaryotes and three bacterial species. Note the many similarities (yellow) and differences (green and red) among the sequences.

GENOMES AND PROTEOMES

Providing an evolutionary foundation for our understanding of biology, each Genomes and Proteomes subsection describes modern information regarding the genomic composition of organisms and how this relates to proteomes (their protein composition) and evolution.

END-OF-CHAPTER MATERIALS

The end-of-chapter materials offer students many different opportunities to focus in on key concepts and help them work at improving their knowledge:

Chapter Summary

The Chapter Summary provides the student with an overview of the biological principles and experimental approaches that have been described in the chapter. The summary is organized according to the sections of each chapter, and presents a bulleted list of key concepts.

Test Yourself

These multiple-choice questions are designed to provide students with the sense of how well they understand the material in the chapter. Answers are provided on the ARIS website.

Conceptual Questions

The aim of conceptual questions is to test a student's knowledge of biological principles, such as how a biological mechanism works or the features of a biological process or structure.

Experimental Questions

These questions challenge the student to consider the experiments found in a chapter and to critically evaluate technical procedures and analyze biological data.

Collaborative Questions

Broad in nature, students may benefit by discussing these questions with their peers.

CHAPTER SUMMARY

3.1   The Carbon Atom and the Study of Organic Molecules

- Organic chemistry is the science of studying carbon-containing molecules, which are found in living organisms. Wöhler's work with urea marked the birth of organic chemistry. (Figure 3.1)
- One property of the carbon atom that makes life possible is its ability to form four covalent bonds with other atoms. Carbon can form both polar and nonpolar bonds. The combination of different elements and different types of bonds allows a vast number of organic compounds to be formed from only a few chemical elements. (Figures 3.2, 3.3)
- Organic molecules may occur in various shapes. The structures of molecules determine their functions.

3.2   Classes of Organic Molecules and Macromolecules

- The four major classes of organic molecules are carbohydrates, lipids, proteins, and nucleic acids. Macromolecules are large organic molecules that are composed of many thousands of atoms. Some macromolecules are polymers because they are formed by linking together many smaller molecules called monomers.
- Carbohydrates are composed of carbon, hydrogen, and oxygen atoms. Most cells can break down carbohydrates, releasing energy and storing it in newly created bonds in ATP.
- Carbohydrates include monosaccharides (the simplest sugars), disaccharides, and polysaccharides. The polysaccharides starch (in plant cells) and glycogen (in animal cells) provide an efficient means of storing energy. The plant polysaccharide cellulose serves a support or structural function. (Figures 3.6, 3.7, 3.8)

TEST YOURSELF

1. Molecules that contain the element _____ are considered organic molecules.
   a. hydrogen   d. nitrogen
   b. carbon   e. calcium
   c. oxygen

2. _____ was the first scientist to synthesize an organic molecule. The organic molecule synthesized was _____.
   a. Kolbe, urea   d. Kolbe, acetic acid
   b. Wöhler, urea   e. Wöhler, glucose
   c. Wöhler, acetic acid

3. The versatility of carbon to serve as the backbone for a variety of different molecules is due to
   a. the ability of carbon atoms to form four covalent bonds.
   b. the fact that carbon usually forms ionic bonds with many different atoms.
   c. the abundance of carbon in the environment.
   d. the ability of carbon to form covalent bonds with many different types of atoms.
   e. both a and d.

CONCEPTUAL QUESTIONS

1. Define isomers.
2. List the four classes of organic molecules and give a function of each.
3. Explain the difference between saturated and unsaturated fatty acids.
4. List the seven characteristics of life and explain a little about each.
5. Give the levels of organization from the simplest to most complex.
6. Discuss the difference between discovery-based science and hypothesis testing.
7. What are the steps in the scientific method, also called hypothesis testing?
8. When conducting an experiment, explain how a control sample and an experimental sample differ from each other.

EXPERIMENTAL QUESTIONS

1. Before the experiments conducted by Anfinsen, what were the common beliefs among scientists about protein folding?
2. Explain the hypothesis tested by Anfinsen.
3. Why did Anfinsen use urea and β-mercaptoethanol in his experiments? Explain the result that was crucial to the discovery that the tertiary structure of a protein is dependent on the primary structure.
4. List the seven characteristics of life and explain a little about each.
5. Give the levels of organization from the simplest to most complex.
6. List the taxonomic groups from most inclusive to least inclusive.
7. Explain how actin filaments are involved in movement.
8. Explain the function of the Golgi apparatus.

COLLABORATIVE QUESTIONS

1. Discuss several types of carbohydrates.
2. Discuss some of the roles that proteins play in organisms. Discuss several differences between plant and animal cells.
3. Discuss the relationship between the nucleus, the rough endoplasmic reticulum, and the Golgi apparatus.
4. Discuss the two categories of transport proteins found in plasma membranes.

---

www.brookerbiology.com
This website includes answers to the Biological Inquiry questions found in the figure legends and all end-of-chapter questions.

Chapter Summary

Test Yourself

Conceptual Questions

Experimental Questions

Collaborative Questions

CONTENTS

UNIT I   Chemistry

UNIT II   Cell

UNIT III   Genetics

UNIT IV   Evolution

UNIT V   Diversity

UNIT VI Plants

UNIT VII   Animals

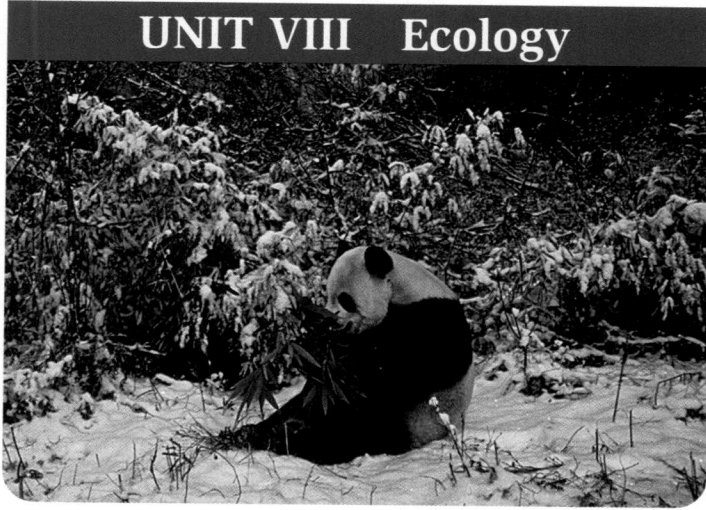

UNIT VIII Ecology

Chapter 54

An Introduction to Ecology and Biomes 1147

Chapter 55

Behavioral Ecology 1177

Chapter 56

Population Ecology 1199

Chapter 57

Species Interactions 1219

Chapter 58

Community Ecology 1241

1

AN INTRODUCTION TO BIOLOGY

Spotted and black jaguars.

Biology is the study of life. The diverse forms of life found on Earth provide biologists with an amazing array of organisms to study. In many cases, the investigation of living things leads to unforeseen discoveries that no one would have imagined. For example, researchers determined that the venom from certain poisonous snakes contains a chemical that lowers blood pressure in humans (**Figure 1.1a**). By analyzing that chemical, drugs were later developed to treat high blood pressure. Biologists also found that nine-banded armadillos usually give birth to identical quadruplets (**Figure 1.1b**). Because of this unique way of producing young, nine-banded armadillos are studied to learn more about multiple births and other reproductive issues. The ancient Romans discovered that the bark of the white willow tree can be used to fight fever (**Figure 1.1c**). Modern chemists determined that willow bark contains a substance called salicylic acid, which led to the development of the related compound acetylsalicylic acid, more commonly known as aspirin. In the last century, biologists studied soil bacteria that naturally produce "chemical weapons" to kill competing bacteria in their native environment (**Figure 1.1d**). These chemicals have been characterized and used to develop antibiotics such as streptomycin to treat bacterial infections. As you may have seen, jellyfish naturally produce a greenish glow (**Figure 1.1e**), which is due to a molecule they make called green fluorescent protein (GFP). Scientists have been able to transfer GFP to other organisms and use it as a research tool to study the functions of cells. GFP transferred to mice makes them glow in the dark! Finally, for many decades, biologists have known that the Pacific yew tree produces a toxin in its bark and needles that kills insects (**Figure 1.1f**). Since the 1990s, this toxin, known by the drug name Taxol®, has been used to treat patients with ovarian and breast cancer. These are but a few of the many discoveries that make biology an intriguing discipline. The study of life not only reveals the fascinating characteristics of living species but also leads to the development of drugs and research tools that benefit the lives of people.

To make new discoveries, biologists view life from many different perspectives. What is life made of? How is it organized? How do organisms reproduce? Sometimes, the questions posed by biologists are fundamental and even philosophical in nature. Where did we come from? Can we live forever? What is the physical basis for memory? Can we save endangered species? Biologists are scientific explorers looking for answers to some of the world's most enduring mysteries. Unraveling these mysteries presents an exciting challenge to the best and brightest minds. Our society has been substantially impacted by discoveries in biology, and future biologists will continue to make important advances. The rewards of a career in biology include the excitement of forging into uncharted territory, the thrill of making discoveries that affect the health and lives of people, and the impact of biology on the preservation of the environment and endangered species. For these and many other compelling reasons, students seeking challenging and rewarding careers may wish to choose biology as a lifelong pursuit.

In this chapter, we will begin our survey of biology by examining the basic features that are common to all living organisms. We will consider how evolution has led to the development of modern genomes—the entire genetic compositions of organisms—which can explain the unity and diversity that we observe among modern species. In the second section, we will explore the general approaches that scientists follow when making new discoveries.

(a) A chemical in the venom of the Brazilian arrowhead viper lowers blood pressure. Derivatives of this chemical, called angiotensin-converting enzyme (ACE) inhibitors, are now commonly used to treat high blood pressure in people.

(b) The nine-banded armadillo usually gives birth to identical quadruplets. Armadillos are studied by researchers to learn more about the mechanisms that cause multiple births.

(c) The bark of the white willow contains a chemical that is closely related to aspirin. Modern aspirin, acetylsalicylic acid, was developed after analyzing this chemical in willow trees.

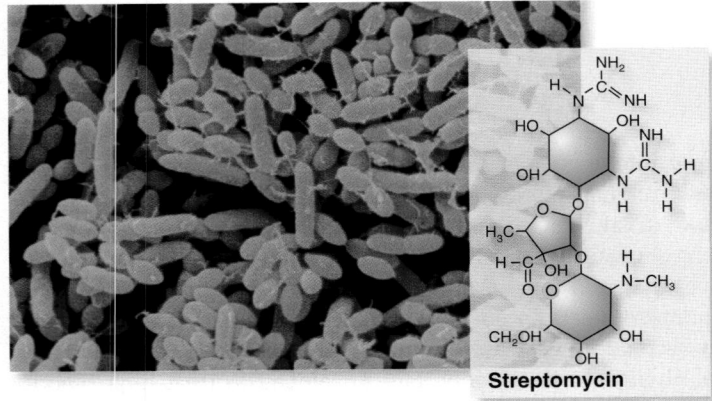

(d) This soil bacterium (*Streptomycin griseus*) naturally produces a molecule called streptomycin, which it uses to kill competing bacteria in the soil. Doctors administer streptomycin to people as an antibiotic to treat bacterial infections.

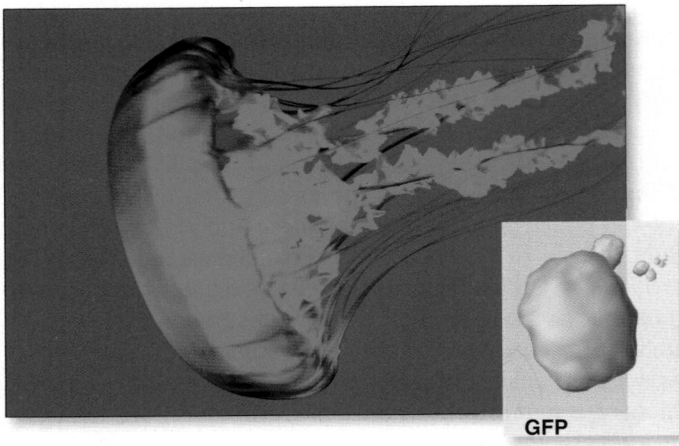

(e) Jellyfish naturally produce a green glow caused by green fluorescent protein (GFP). GFP can be transferred to other organisms, such as mice, and used as a research tool to study the functions of cells.

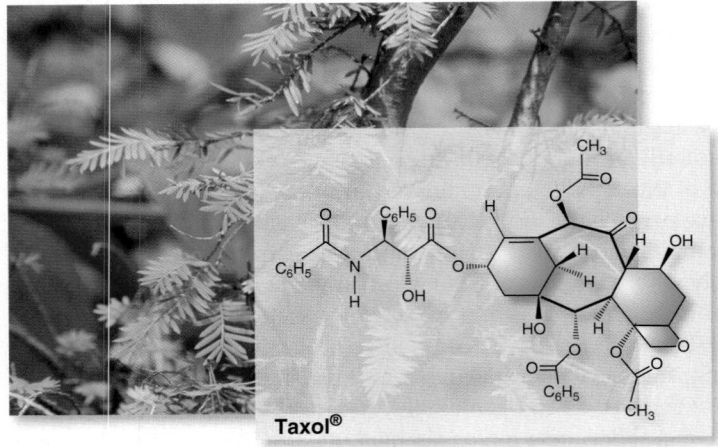

(f) The bark and needles of the Pacific yew tree produce a toxin that kills insects. This toxin, called Taxol, is effective in the treatment of ovarian and breast cancer.

Figure 1.1   Amazing discoveries in biology.

1.1 The Properties of Life: Past and Present

Unity and *diversity* are two words that often are used to describe the living world. Unity can be viewed from both modern and prehistorical perspectives. In this section, we first examine how all modern forms of life display a common set of characteristics that distinguish them from nonliving objects. This unity of common traits is rooted in the phenomenon of biological evolution. Life on Earth is united by an evolutionary past in which modern organisms have evolved from pre-existing organisms.

This evolutionary unity does not mean that organisms are exactly alike. Rather, the Earth has many different types of environments, ranging from tropical rain forests to salty oceans, hot and dry deserts, and cold mountaintops. Diverse forms of life have evolved to prosper in the myriad of environments that the Earth has to offer. In this section, we will also begin to examine the diversity that exists within the biological world.

A Set of Characteristics Is Common to All Forms of Modern Life

A fitting way to begin a biology textbook is to distinguish living organisms from nonliving objects. At first, the distinction might seem intuitively obvious. A person is alive, but a rock is not. However, the distinction between living and nonliving may seem less obvious when we consider microscopic entities. Is a bacterium alive? Is a virus alive? Is a chromosome alive? Biologists have wrestled with such questions and have determined that all living organisms display seven characteristics that set them apart from nonliving things.

Cells and Organization The concept of organization is so fundamental to biology that the term **organism** (which comes from the same Latin root, *organum*) can be applied to all living things. Organisms maintain an internal order that is separated from the environment (**Figure 1.2a**). The simplest unit of such organization is the **cell**, which we will examine in Unit II. The **cell theory** states that all organisms are made of cells. Unicellular organisms are composed of one cell, while multicellular organisms such as plants and animals contain many cells. In plants and animals, each cell has internal order, and the cells within the body have specific arrangements and functions.

Energy Use and Metabolism The maintenance of organization requires energy. Therefore, all living organisms acquire energy from the environment and use that energy to maintain their internal order. Cells use energy by catalyzing a variety of chemical reactions that are responsible for the breakdown of nutrients and the synthesis of the components that make up individual cells and living organisms. These chemical reactions are collectively known as **metabolism**. Plants, algae, and certain bacteria can directly harness light energy to produce their own nutrients in the process of **photosynthesis** (**Figure 1.2b**).

(a) Cells and organization: Organisms maintain an internal order. The simplest unit of organization is the cell.

(b) Energy use and metabolism: To maintain their internal order, energy is needed by organisms. Energy is utilized in chemical reactions collectively known as metabolism.

(c) Response to environmental changes: Organisms react to environmental changes to promote their survival.

(d) Regulation and homeostasis: Organisms regulate their cells and bodies to maintain relatively stable internal conditions, a process called homeostasis.

(e) Growth and development: Growth produces more or larger cells, while development produces organisms with a defined set of characteristics.

(f) Reproduction: To sustain life over many generations, organisms must reproduce. Due to genetic material, offspring tend to have traits like their parents.

(g) Biological evolution: Populations of organisms change over the course of many generations. Evolution results in traits that promote survival and reproductive success.

Figure 1.2 Seven characteristics that are common to life.

They are primary producers of food on Earth. In contrast, some organisms, such as animals and fungi, are consumers—they must eat other organisms as food to obtain energy.

Response to Environmental Changes To survive, living organisms must be able to respond to environmental changes. For example, bacterial cells have mechanisms to sense that certain nutrients in the environment are in short supply while others are readily available. Also, plants can respond to changes in the angle of the sun. If you place a plant in a window, it will grow toward the light (**Figure 1.2c**). In the winter, many species of mammals develop a thicker coat of fur to protect them from the cold temperatures. Responses to environmental changes are examples of **adaptations**—processes and structures by which organisms adjust to short-term or long-term changes in their environment.

Regulation and Homeostasis As we have just seen, one way that organisms can respond to environmental variation is to change themselves. The growth of thick fur in the wintertime is an example. A common reason for certain adaptations, including this example, is to maintain homeostasis (from the Greek meaning "to stay the same"). Although life is a dynamic process, living cells and organisms regulate their cells and bodies to maintain relatively stable internal conditions, a process called **homeostasis**. The degree to which homeostasis is achieved varies among different organisms. For example, most mammals and birds maintain a relatively constant body temperature in spite of changing environmental temperatures (**Figure 1.2d**), while reptiles and amphibians do not. By comparison, all organisms continually regulate their cellular metabolism so that nutrient molecules are used at an appropriate rate, and new cellular components are synthesized when they are needed.

Growth and Development All living things grow and develop; **growth** produces more or larger cells, while **development** produces organisms with a defined set of characteristics. Among unicellular organisms such as bacteria, new cells are relatively small, and they increase in volume by the synthesis of additional cellular components. Multicellular organisms, such as plants and animals, begin life at a single-cell stage (for example, a fertilized egg) and then undergo multiple cell divisions to develop into a complete organism with many cells (**Figure 1.2e**).

Reproduction All living organisms have a finite life span and will eventually die. To sustain life over many generations, organisms must **reproduce** (**Figure 1.2f**). A key feature of reproduction is that offspring tend to have characteristics that greatly resemble those of their parent(s). The reason for this is that all living organisms contain genetic material composed of **DNA (deoxyribonucleic acid)**, which provides a blueprint for the organization, development, and function of living things. As discussed in Unit III, DNA harbors **genes**, which contribute to the characteristics or traits of organisms. During reproduction, a copy of this blueprint is transmitted from parents to offspring. The central dogma of genetics is that most genes are transcribed into a type of RNA (ribonucleic acid) molecule called messenger RNA (mRNA) that is then translated into a polypeptide with a specific amino acid sequence. This process is called molecular gene expression. Polypeptides are the structural units of functional proteins. The functioning of proteins is largely responsible for the traits of living organisms.

Biological Evolution The first six characteristics of life, which we have just considered, apply to individual organisms over the short run. Over the long run, another universal characteristic of life is **biological evolution**, which refers to the phenomenon that populations of organisms change over the course of many generations. As a result of evolution, some organisms become more successful at survival and reproduction. Populations become better adapted to the environment in which they live. For example, the long snout of an anteater is an adaptation that enhances its ability to obtain food, namely ants (**Figure 1.2g**). The long snout occurred via biological evolution in which modern anteaters evolved from organisms that did not have such long snouts. Unit IV is devoted to the topic of evolution, while Unit V surveys the evolutionary diversity among different forms of life.

Living Organisms Can Be Viewed at Different Levels of Organization

As we have just learned, life exhibits a set of characteristics, beginning with the concept of organization. The organization of living organisms can be analyzed in a hierarchical manner, starting with the tiniest level of organization, and progressing to levels that are physically much larger and more complex. **Figure 1.3** depicts a scientist's view of biological organization at different levels.

1. **Atoms:** An **atom** is the smallest component of an element that has the chemical properties of the element. All matter is composed of atoms.
2. **Molecules and macromolecules:** As discussed in Unit I, atoms bond with each other to form **molecules**. When many molecules bond together to form a polymer, this is called a **macromolecule**. Carbohydrates, proteins, and nucleic acids (for example, DNA and RNA) are important macromolecules found in living organisms.
3. **Cells:** Molecules and macromolecules associate with each other to form larger structures such as membranes. A cell is formed from the association of these larger structures.
4. **Tissues:** In the case of multicellular organisms such as plants and animals, many cells of the same type associate with each other to form **tissues**. An example is muscle tissue.
5. **Organs:** An **organ** is composed of two or more types of tissue. For example, the heart of a parrot is composed of several types of tissues, including muscle, nervous, and connective tissue.
6. **Organism:** All living things can be called organisms. A single organism possesses the set of characteristics that define life. Biologists classify organisms as belonging to a particular **species**, which is a related group of organisms that share a distinctive form and set of attributes in nature.

The members of the same species are closely related genetically. In Units VI and VII, we will examine plants and animals at the level of cells, tissues, organs, and complete organisms.

7. **Population:** A group of organisms of the same species that occupy the same environment is called a **population**.

8. **Community:** A biological **community** is an assemblage of populations of different species. The types of species that are found in a community are determined by the environment and by the interactions of species with each other.

9. **Ecosystem:** Researchers may extend their work beyond living organisms and also study the environment. Ecologists analyze **ecosystems**, which are formed by interactions of a community of organisms with their physical environment. Unit VIII considers biology from populations to ecosystems.

10. **Biosphere:** The **biosphere** includes all of the places on the Earth where living organisms exist, encompassing the air, water, and land.

Modern Forms of Life Are Connected by an Evolutionary History

Life began on Earth as primitive cells about 3.5 to 4 billion years ago. Since that time, those primitive cells underwent evolutionary changes that ultimately gave rise to the species we see today.

Figure 1.3 The levels of biological organization.

Figure 1.4 **An example of modification of a structure for a new function.** The bird shown in the photograph has used a modified milk carton in which to build its nest. By analogy, evolution also involves the modification of pre-existing structures for a new function.

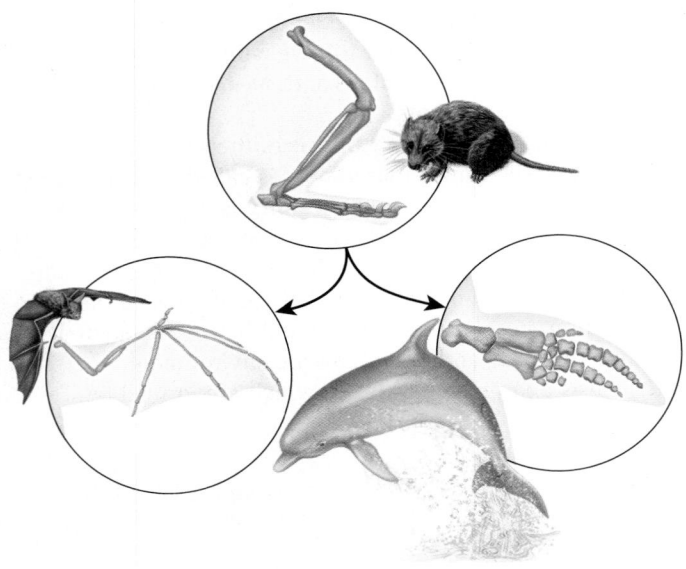

Figure 1.5 **An example showing a modification that has occurred as a result of biological evolution.** The wing of a bat and the flipper of a dolphin were modified from a limb that was used for walking in a pre-existing ancestor.

Biological inquiry: Among mammals, give two examples of how the tail has been modified for different purposes.

Understanding the evolutionary history of species often provides key insights regarding the structure and function of an organism's body. As a metaphor to help you appreciate this idea, **Figure 1.4** shows a photograph of a bird that is using a milk carton in which to build a nest. If we did not know that the milk carton had served an earlier purpose, namely to contain milk, we might wonder why the bird had made a nesting site that resembled a milk carton. Obviously, we do not worry about this because we immediately grasp that the milk carton had a previous history, and that it has been modified by a person to serve a new purpose—a nesting site for a bird. Understanding history allows us to make sense out of this nest.

Likewise, evolutionary change involves modifications of characteristics in pre-existing populations. Over long periods of time, populations may change such that structures with a particular function may become modified to serve a new function. For example, the wing of a bat is used for flying, while the flipper of a dolphin is used for swimming (**Figure 1.5**). Both structures were modified from a limb that was used for walking in a pre-existing ancestor.

Evolutionary change occurs by two mechanisms, vertical descent with mutation and horizontal gene transfer. Let's take a brief look at each of these mechanisms.

Vertical Descent with Mutation The traditional way to view evolution involves a progression of changes in a series of ancestors. Such a series is called a **lineage**. **Figure 1.6** shows a portion of the lineage that gave rise to modern horses. This type of evolution is called **vertical evolution** because biologists have traditionally depicted such evolutionary change in a vertical diagram like the one shown in Figure 1.6. In this mechanism of evolution, new species evolve from pre-existing species by the accumulation of **mutations**, which are changes in the genetic material of organisms. But why would some mutations accumulate in a population and eventually change the characteristics of an entire species? One reason is that a mutation may alter the traits of organisms in a way that increases their chances of survival or reproduction. When a mutation causes such a beneficial change, the mutation is more likely to increase in a population over the course of many generations, a process called **natural selection**, which is discussed in Units IV and V. Evolution also involves the accumulation of neutral changes that do not benefit a species, and even rare changes that may be harmful.

Horizontal Gene Transfer In addition to vertical evolution, which produces a lineage, species also evolve by another process that involves genetic exchanges between different species. Sexually reproducing species usually mate with members of their own species. Similarly, asexual species such as bacteria can occasionally transfer genetic material between cells, but again, that tends to occur most readily between members of the same bacterial species. However, on relatively rare occasions, genetic exchanges occur between different species. For example, you may have heard in the news media that resistance to antibiotics among bacteria is a growing medical problem. Genes that confer antibiotic resistance are sometimes transferred between different bacterial species (**Figure 1.7**).

When genes are transferred from one species to another, this event is called **horizontal gene transfer**. In a lineage in

Figure 1.6 **An example of vertical evolution: the horse.** The modern horse evolved from ancestors that were much smaller. The vertical evolution shown here occurred due to the accumulation of mutations that altered the traits of the species.

(Figure 1.6 — evolutionary tree of the horse, axis "Millions of years ago (mya)", CENOZOIC, with labels: Equus, Hippidium and other genera, Nannippus, Styohipparion, Neohipparion, Pliohippus, Hipparion, Sinohippus, Megahippus, Calippus, Archaeohippus, Merychippus, Anchitherium, Hypohippus, Parahippus, Miohippus, Mesohippus, Paleotherium, Propalaeotherium, Pachynolophus, Orohippus, Epihippus, Hyracotherium)

which the timescale is depicted on a vertical axis, horizontal gene transfer is shown as a horizontal line between two different species (**Figure 1.8**). Genes that are transferred horizontally may be acted upon by natural selection to eventually promote changes in an entire species. This has been an important mechanism of evolutionary change, particularly among bacterial species. In addition, during the early stages of evolution, which occurred a few billion years ago, horizontal gene transfer was an important part of the process that gave rise to all modern species.

Traditionally, biologists have described evolution using diagrams that depict the vertical evolution of species on a long timescale. This is the type of evolutionary tree that was shown in Figure 1.6. For many decades, the simplistic view held that all living organisms evolved from a common ancestor, resulting in a "tree of life," which could describe the vertical evolution that gave rise to all modern species. Now that we understand the great importance of horizontal gene transfer in the evolution of life on Earth, biologists have needed to re-evaluate the concept of evolution as it occurs over time. Rather than a tree of

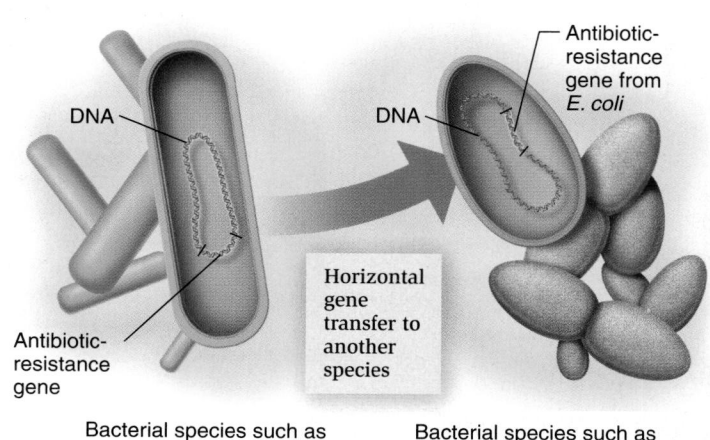

Figure 1.7 **An example of horizontal gene transfer: antibiotic resistance.** One bacterial species may transfer a gene to a different bacterial species, such as a gene that confers resistance to an antibiotic.

life, a more appropriate way to view the unity of living organisms is to describe it as a "web of life," which accounts for both vertical evolution and horizontal gene transfer. Figure 1.8 illustrates such a diagram.

The Classification of Living Organisms Allows Biologists to Appreciate the Unity and Diversity of Life

As biologists discover new species, they try to place them in groups based on their evolutionary history. This is an arduous task because researchers estimate that the Earth has between 10 and 100 million different species! The rationale for categorization is usually based on vertical descent. Species with a recent common ancestor are grouped together, while species whose common ancestor is in the very distant past are placed into different groups. The grouping of species is termed **taxonomy**.

Let's first consider taxonomy on a broad scale. You may have noticed that Figure 1.8 showed three main groups of organisms.

All forms of life can be placed into three large categories or domains called **Bacteria**, **Archaea**, and **Eukarya** (**Figure 1.9**). Bacteria and Archaea are microorganisms that are also termed **prokaryotic** because their cell structure is relatively simple. At the molecular level, bacterial and archaeal cells show significant differences in their lipid composition, metabolic pathways, and mechanisms of gene expression. By comparison, organisms in domain Eukarya are **eukaryotic** and have larger cells with internal compartments that serve various functions. A defining distinction between prokaryotic and eukaryotic cells is that eukaryotic cells have a **cell nucleus** in which the genetic material is surrounded by a membrane. The organisms in domain Eukarya have been further subdivided into four major categories or kingdoms called Animalia (animals), Plantae (plants), Protista (protists), and Fungi. However, as discussed in Chapter 26 and Unit V, the traditional view of four eukaryotic kingdoms is now under revision as biologists have gathered new information regarding the evolutionary relationships of these organisms.

Figure 1.8 **The web of life.** This diagram of evolution includes both vertical evolution and horizontal gene transfer as important mechanisms in the evolution of life on Earth. Note: Archaea are unicellular species that are similar in structure to bacteria.

Biological inquiry: How does the concept of a tree of life differ from a web of life?

(a) Domain Bacteria: Unicellular prokaryotes that inhabit many diverse environments on Earth.

(b) Domain Archaea: Unicellular prokaryotes that are less common than bacteria. Some live in extreme environments such as hot springs.

Kingdom Animalia: Multicellular organisms that usually have a nervous system and are capable of locomotion. They must eat other organisms to live.

Kingdom Plantae: Multicellular organisms that can carry out photosynthesis.

Kingdom Protista: Unicellular and small multicellular organisms that are now subdivided into several different kingdoms based on their evolutionary relationships.

Kingdom Fungi: Unicellular and multicellular organisms that have a cell wall but cannot carry out photosynthesis. Fungi usually survive on decaying organic material.

(c) Domain Eukarya

Figure 1.9   The three domains of life.   Two of these domains, **(a)** Bacteria and **(b)** Archaea, are prokaryotes, while the third domain, **(c)** Eukarya, comprises species that are eukaryotes.

Taxonomy involves multiple levels in which particular species are placed into progressively smaller and smaller groups of organisms that are more closely related to each other evolutionarily (**Figure 1.10**). Such an approach emphasizes the unity and diversity of different species. As an example, let's consider the jaguars, shown on the cover of your textbook. The broadest grouping for the jaguar is the domain, namely Eukarya, followed by progressively smaller divisions, from kingdom (Animalia) to species. In the animal kingdom, jaguars are part of a phylum, Chordata, which is subdivided into classes. Jaguars are in a class called Mammalia, which includes all mammals. The common ancestor that gave rise to mammals arose over 200 million years ago. Mammalia is subdivided into several smaller orders. The jaguar is in the order Carnivora. The order is in turn divided into families; the jaguar and all other cats belong to the family Felidae. The genus *Panthera* is the smallest group of different species that contains the jaguar. As you can see in Figure 1.10, the genus contains only four modern species, the jaguar and other types of large cats. Therefore, the genus has species that are very similar to each other in form, and have evolved from a common (extinct) ancestor that lived relatively recently on an evolutionary timescale, approximately 5 million years ago.

Biologists use a two-part description, called a **binomial**, to provide each species with a unique scientific name. The scientific name of the jaguar is *Panthera onca*. The first part is the genus and the second part is the specific epithet or species descriptor. By convention, the genus name is capitalized, while the specific epithet is not. Both names are italicized. All scientific names are Latinized.

Taxonomic group	Jaguar is found in:	Approximate time when the common ancestor for this group arose	Approximate number of modern species in this group	
Domain	Eukarya	2,000 mya	> 5,000,000	
Kingdom	Animalia	600 mya	> 1,000,000	
Phylum	Chordata	525 mya	50,000	
Class	Mammalia	200 mya	5,000	
Order	Carnivora	60 mya	270	
Family	Felidae	40 mya	38	
Genus	*Panthera*	5 mya	4	
Species	*onca*	1.5 mya	1	

Figure 1.10   Taxonomic and evolutionary groupings leading to the jaguar.

GENOMES & PROTEOMES

The Study of Genomes and Proteomes Provides an Evolutionary Foundation for Our Understanding of Biology

The unifying concept in biology is evolution. We can understand the unity of modern organisms by realizing that all living species evolved from an interrelated group of ancestors. However, from an experimental perspective, this realization presents a dilemma—we cannot take a time machine back over the course of 4 billion years to carefully study the characteristics of extinct organisms and fully appreciate the series of changes that have led to modern species. Fortunately though, evolution has given experimental biologists a wonderful puzzle to study, namely the genomes of modern species. The term **genome** refers to *the complete genetic makeup of an organism* (**Figure 1.11a**). The genome is critical to life because it performs these functions:

- *Acts as a stable informational unit:* The genome of every organism stores information that provides a blueprint to create their characteristics.
- *Provides continuity from generation to generation:* The genome is copied and transmitted from generation to generation.

Figure 1.11 **Genomes and proteomes.** (a) The genome, which is composed of DNA, is the entire genetic complement of an organism. Most of the genetic material in eukaryotic cells is found in the cell nucleus. Its primary function is to encode the proteome. (b) The proteome is the entire protein complement of a cell or entire organism. Proteins are largely responsible for the structure and function of cells and complete organisms.

Biological inquiry: Biologists sometimes say that the genome is a storage unit, while the proteome is largely the functional unit of life. Explain this statement.

• *Acts as an instrument of evolutionary change:* Every now and then, the genome undergoes a mutation that may alter the characteristics of an organism. In addition, a genome may acquire new genes by horizontal gene transfer. The accumulation of such changes over the course of many generations produces the evolutionary changes that alter species and produce new species.

The evolutionary history and relatedness of all living organisms can be illuminated by genome analysis. The genome of every organism carries the results and the evidence of millions of years of evolution. The genomes of prokaryotes usually contain a few thousand genes, while those of eukaryotes may contain tens of thousands. An exciting advance in biology over the past couple of decades has been the ability to analyze the DNA sequence of genomes, a technology called **genomics**. For instance, we can compare the genomes of a frog, a giraffe, and a petunia and discover intriguing similarities and differences. These comparisons help us to understand how new traits evolved. For example, all three types of organisms have the same kinds of genes that are needed for the breakdown of nutrients such as sugars. In contrast, only the petunia carries genes that allow it to carry out photosynthesis.

An extension of genome analysis is the study of **proteomes**, which refers to *the complete complement of proteins that a cell or organism can make.* The function of most genes is to encode polypeptides that become units in proteins. As shown in **Figure 1.11b**, these include transport proteins; extracellular proteins; proteins that function in cell organization, in cell signaling, and as enzymes; and proteins that form a cytoskeleton. Proteins are the key participants in maintaining cell structure and carrying out most cell functions. Therefore, the genome of each species carries the information to make its proteome, the hundreds or thousands of proteins that each cell of that species makes. Proteins are largely responsible for the structures and functions of cells and organisms. The technical approach called **proteomics** involves the analysis of the proteome of a single species and the comparison of the proteomes of different species. Proteomics helps us to understand how the various levels of biology are related to one another, from the molecular level—at the level of protein molecules—to the higher levels, such as how the functioning of proteins produces the characteristics of cells and organisms, and the ability of populations of organisms to survive in their natural environments.

As a concrete way to understand the unifying theme of evolution in biology, a recurring theme in the chapters that follow is a brief topic called "Genomes & Proteomes" that will allow you to appreciate how evolution produced the characteristics of modern species. These topics explore how the genomes of different species are similar to each other, and how they are different. You will learn how genome changes affect the proteome and thereby control the traits of modern species. Ultimately, these concepts provide you with a way to relate information at the molecular level to the traits of organisms and their survival within ecosystems.

Along these lines, the cover of your textbook provides food for thought. The cats on the cover are jaguars. A black jaguar is sometimes called a panther, but it is still the same species, *Panthera onca*, as a spotted jaguar. How are the genomes of the spotted and black jaguars different? How are their proteomes different? Can this information tell us anything about the ecosystem in which these animals live? Does this have anything to do with evolutionary change? If we analyzed the genomes of spotted and black jaguars, we would discover that they are overwhelmingly similar to each other (**Figure 1.12a**). Of the 20,000 or so genes, the majority would be identical to each other or nearly so. However, based on their differences in appearance, we would expect the DNA sequence within one particular gene to be different, namely a gene that plays a role in producing pigment in the fur. At the level of the proteome, this slight genome difference causes the spotted jaguar to make proteins (enzymes) that synthesize a background coat pigment that is tan, while a black jaguar makes a background coat pigment that is black. Under bright light, you can see that a black jaguar actually has spots, but the dark background pigment greatly masks their appearance (**Figure 1.12b**).

Do spotted versus black coats have any impact on the life of these animals? The answer is yes. The spotted animals are thought to be better hunters in lighter grassland environments. The black animals are more likely to survive in darker forests, where they are less likely to be seen by their prey. In nature, mixed populations of spotted and black jaguars are often observed on the edges of forests, where both light and dark environments exist. Therefore, biologists have speculated that this is an example of evolutionary change. Genetic mutations have occurred that promote the survival and reproductive success of these animals, which vary in light and dark environments.

1.2 Biology as a Scientific Discipline

What is science? Surprisingly, the definition of science is not easy to state. Most people have an idea of what science is, but actually articulating that idea proves difficult. In biology, we might define **science** as *the observation, identification, experimental investigation, and theoretical explanation of natural phenomena.*

Science is conducted in different ways and at different levels. Some biologists study the molecules that compose life, while others try to understand how organisms survive in their natural environments. In some cases, experiments are designed to test the validity of ideas that are suggested by researchers. In this section, we will examine how biologists follow a standard approach, called the **scientific method**, to test their ideas. We will learn that scientific insight is not based on intuition. Instead, scientific knowledge makes predictions that can be experimentally tested.

Even so, not all discoveries are the result of researchers following the scientific method. Some discoveries are simply made

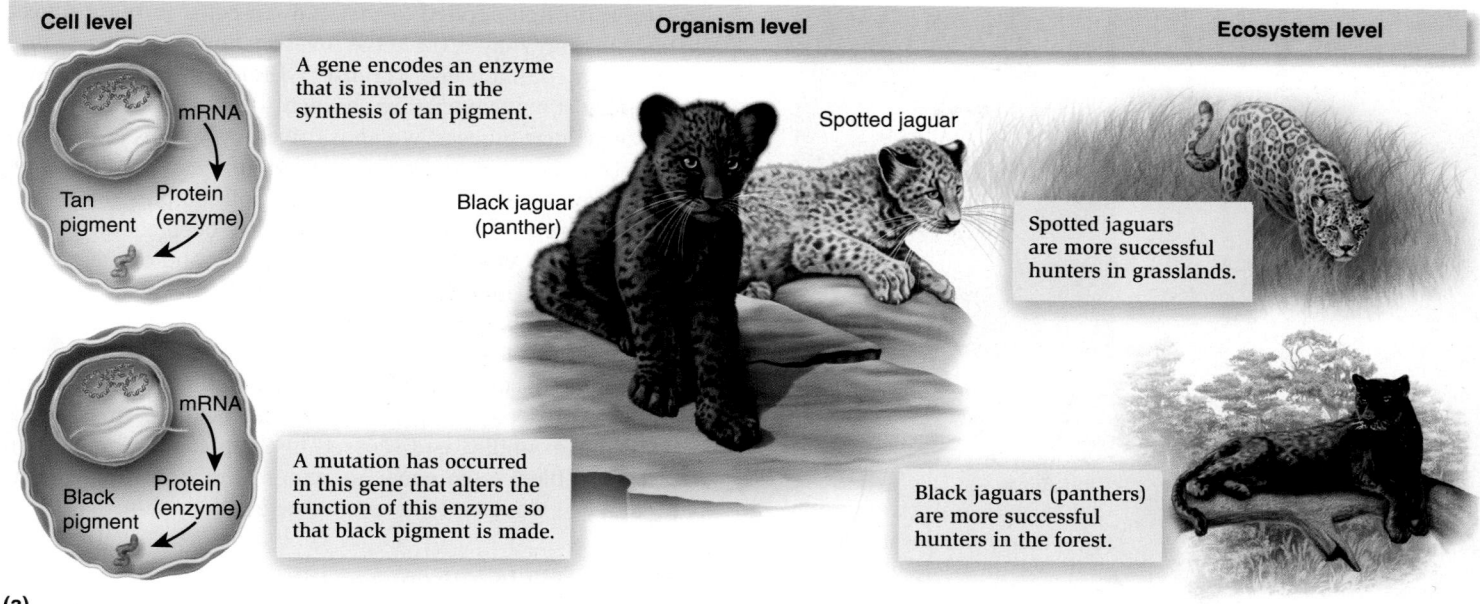

Cell level

A gene encodes an enzyme that is involved in the synthesis of tan pigment.

mRNA

Tan pigment   Protein (enzyme)

mRNA

Black pigment   Protein (enzyme)

A mutation has occurred in this gene that alters the function of this enzyme so that black pigment is made.

Organism level

Black jaguar (panther)

Spotted jaguar

Spotted jaguars are more successful hunters in grasslands.

Black jaguars (panthers) are more successful hunters in the forest.

Ecosystem level

(a)

(b)

Figure 1.12 **How the study of genomes and proteomes can provide us with connections to different biological levels.**   (a) Spotted jaguars differ from panthers because they make an enzyme that produces a tan pigment, while a mutation in the same gene in panthers results in black pigment. With regard to hunting success, the spotted jaguars are better in grasslands, while the panthers are better in the forest. (b) A close-up view of a panther, showing its spots.

by gathering new information. As described earlier in Figure 1.1, the characterization of many plants and animals has led to the development of many important medicines and research tools. In this section, we will also consider how researchers often set out on "fact-finding missions" that are aimed at uncovering new information that may eventually lead to new discoveries in biology.

Biologists Investigate Life at Different Levels of Organization

Earlier in Figure 1.3, we examined the various levels of biological organization. The study of these different levels depends not only on the scientific interests of biologists but also on the tools that are available to them. Prior to the development of high-quality microscopes, biologists primarily focused their attention on characteristics they could observe with their unaided eyes. They studied the activities of organisms in their natural environments, a branch of biology called **ecology** (**Figure 1.13a**). In addition, researchers have examined the structures and functions of plants and animals, which are disciplines called **anatomy** and **physiology** (**Figure 1.13b**). As microscope technology improved, many researchers shifted their attention to the study of

cells. **Cell biology**, which is the study of cells, became an important branch of biology in the early 1900s and remains so today (**Figure 1.13c**). In the 1970s, genetic tools became available to study single genes and the proteins they encode. This genetic technology enabled researchers to study individual molecules, such as proteins, in living cells. Genetic technology spawned the field of **molecular biology**. Together with the efforts of chemists and biochemists, molecular biologists focus their efforts on the structure and function of the molecules of life (**Figure 1.13d**). Such researchers want to understand how biology works at the molecular and even atomic levels. Overall, the 20th century saw a progressive increase in the number of biologists who used a reductionist approach to understand biology. **Reductionism** involves reducing complex systems to simpler components as a way to understand how the system works. In biology, reductionists study the parts of a cell or organism as individual units.

In the 1980s, the pendulum began to swing in the other direction. Scientists have invented new tools that allow us to study groups of genes (genomic techniques) and groups of proteins (proteomic techniques). Biologists now use the term **systems biology** to describe research that is aimed at understanding how the properties of life arise by complex interactions.

Ecologists study species in their native environments.

(a) Ecology—population/ community/ecosystem levels

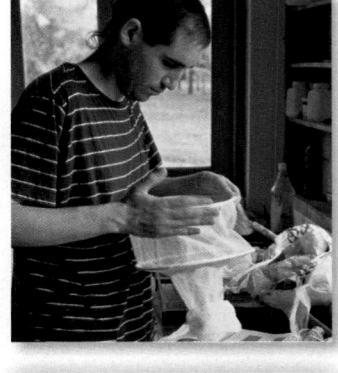

Anatomists and physiologists study how the structure of organisms are related to their functions.

(b) Anatomy and physiology— tissue/organ/organism levels

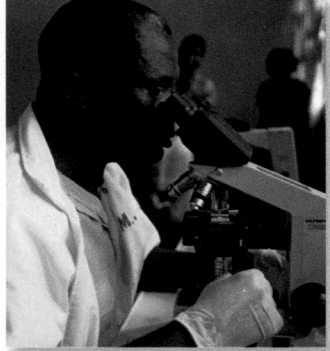

Cell biologists often use the microscope to learn how cells function.

(c) Cell biology—cellular levels

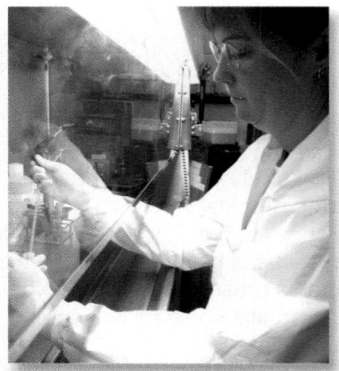

Molecular biologists and biochemists study the molecules and macromolecules that make up cells.

(d) Molecular biology— molecular/atomic levels

Systems biologists may study groups of molecules. The microarray shown in the inset determines the expression of many genes simultaneously.

(e) Systems biology—all levels, shown here at the molecular level

Figure 1.13   Biological investigation at different levels.

This term is often applied to the study of cells. In this context, systems biology may involve the investigation of groups of proteins with a common goal (**Figure 1.13e**). For example, a systems biologist may conduct experiments that try to characterize an entire cellular process, which is driven by dozens of different proteins. Systems biology is not new. Animal and plant physiologists have been studying the functions of complex organ systems for centuries. Likewise, ecologists have been characterizing ecosystems for a very long time. The novelty and excitement of systems biology in recent years has been the result of new experimental tools that allow us to study complex interactions at the molecular level. As described throughout this textbook, the investigation of genomes and proteomes has provided important insights regarding many interesting topics in systems biology. For example, as discussed in Chapter 6, systems biology has enabled researchers to understand how the various parts of a cell work together as an integrated system.

A Hypothesis Is a Proposed Idea, While a Theory Is a Broad Explanation Backed by Extensive Evidence

Let's now consider the process of science. In biology, a **hypothesis** is a proposed explanation for a natural phenomenon. It is a proposition based on previous observations or experimental studies. For example, with knowledge of seasonal changes, you might hypothesize that maple trees drop their leaves in the autumn because of the shortened amount of daylight. An alternative hypothesis might be that the trees drop their leaves because of colder temperatures. In biology, a hypothesis requires more work by researchers to either accept or reject it.

A useful hypothesis must make predictions that can be shown to be correct or incorrect. The validity of the predictions is usually determined by additional observations or experimentation. If the predictions do not agree with new data, the hypothesis is rejected. Alternatively, a hypothesis may be correct so that further work will not disprove it. Even so, a hypothesis is never really proven but rather always remains provisional. Researchers accept the possibility that perhaps they have not yet conceived of the correct hypothesis. However, after many experiments, biologists may say that they accept a hypothesis, but they should never say that the hypothesis is proven.

By comparison, the term **theory**, as it is used in biology, is a broad explanation of some aspect of the natural world that is substantiated by a large body of evidence. Biological theories incorporate observations, hypothesis testing, and the laws of other disciplines such as chemistry and physics. The power of theories is that they allow us to make many predictions regarding the properties of living organisms. As an example, let's consider the theory that DNA is the genetic material, and that it is organized into units called genes. An overwhelming body of evidence has substantiated this theory. Thousands of living species have been analyzed, and all of them have been found to use DNA as their genetic material, and to express genes that produce the proteins that lead to their characteristics. This theory makes many valid predictions. For example, certain types of mutations in genes are expected to affect the traits of organisms. This prediction has been confirmed experimentally. Similarly, genetic material is copied and transmitted from parents to offspring. By comparing the DNA of parents and offspring, this prediction has also been confirmed. Furthermore, the theory explains the observation that offspring resemble their parents. Overall, two key attributes of a scientific theory are consistency with a vast amount of known data, and the ability to make many correct predictions. Two other important biological theories that we have touched on in this chapter are the cell theory and the theory of evolution by natural selection.

The meaning of the term *theory* is sometimes muddled because it is used in different situations. In everyday language, a "theory" is often viewed as little more than a guess or a hypothesis. For example, a person might say, "My theory is that Professor Simpson did not come to class today because he went to the beach." However, in biology, a theory is much more than a guess. A theory is an established set of ideas that explains a vast amount of data and offers valid predictions that can be tested. Like a hypothesis, a theory can never be proven to be true. Scientists acknowledge that they do not know everything. Even so, biologists would say that theories are extremely likely to be true, based on all known information. In this regard, theories are viewed as **knowledge**, which is the awareness and understanding of information.

Discovery-Based Science and Hypothesis Testing Are Scientific Approaches That Help Us Understand Biology

The path that leads to an important discovery is rarely a straight line. Rather, scientists ask questions, make observations, ask modified questions, and may eventually conduct experiments to test their hypotheses. The first attempts at experimentation may fail, and new experimental approaches may be needed. To suggest that scientists follow a rigid scientific method is an oversimplification of the process of science. Scientific advances often occur as scientists dig deeper and deeper into a topic that interests them. Curiosity is the key phenomenon that sparks scientific inquiry. As discussed next, researchers typically follow two general types of approaches—discovery-based science and hypothesis testing.

Discovery-Based Science The collection and analysis of data without the need for a preconceived hypothesis is called **discovery-based science** or simply **discovery science**. The information gained from discovery-based science may have practical applications that benefit people. Drug companies, for example, may test hundreds or even thousands of drugs to determine if any of them are useful in the treatment of disease (**Figure 1.14a**). Once a drug has been discovered that is effective in disease treatment, researchers may dig deeper and try to understand how the drug exerts its effects. In this way, discovery-based science may help us learn about basic concepts in medicine

Drug companies may screen hundreds or thousands of different compounds trying to discover ones that may prove effective in the treatment of a particular disease.

(a) Drug discovery

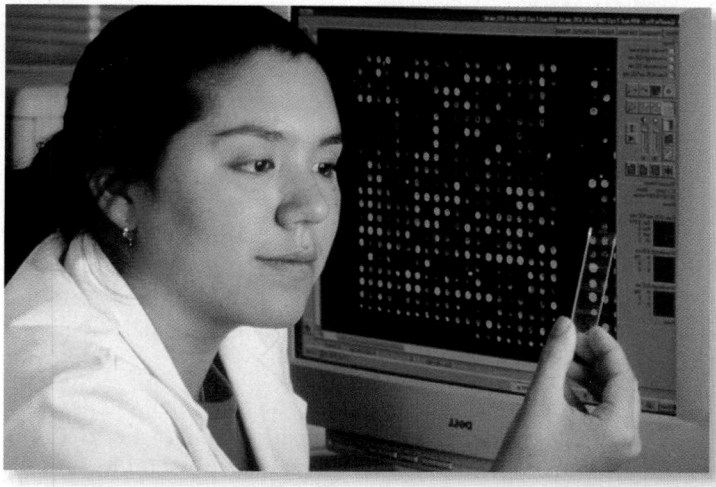

Genetic researchers search through the genomes of humans and other species trying to discover new genes. Such discoveries may help us understand molecular biology and provide insight into the causes of inherited diseases in people.

(b) Discovery of genes

Figure 1.14   Discovery-based science.

and biology. Another example involves the study of genomes (**Figure 1.14b**). Over the past few decades, researchers have identified and begun to investigate newly discovered genes within the human genome without already knowing the function of the gene they are studying. The goal is to gather additional clues that may eventually allow them to propose a hypothesis that explains the gene's function. Discovery-based science often leads to hypothesis testing.

1   **OBSERVATIONS** The leaves on maple trees fall in autumn when the days get colder and shorter.

2   **HYPOTHESIS** The shorter amount of daylight causes the leaves to fall.

3   **EXPERIMENTATION**
Small maple trees are grown in 2 greenhouses where the only variable is the length of light.

Control group:
Amount of daily light remains constant for 200 days.

Experimental group:
Amount of daily light becomes progressively shorter for 200 days.

4   **THE DATA**

A statistical analysis can determine if the control and the experimental data are significantly different. In this case, they are.

5   **CONCLUSION** The hypothesis is accepted.

Figure 1.15   **The steps of the scientific method, also known as hypothesis testing.**   In this example, the goal is to test the hypothesis that maple trees drop their leaves in the autumn due to shortening length of daylight.

Hypothesis Testing In biological science, the scientific method, also known as **hypothesis testing**, is often followed to test the validity of a hypothesis. This strategy may be described as a five-stage process:

1. Observations are made regarding natural phenomena.
2. These observations lead to a hypothesis that tries to explain the phenomena. As mentioned, a useful hypothesis is one that is testable because it makes specific predictions.
3. Experimentation is conducted to determine if the predictions are correct.
4. The data from the experiment are analyzed.
5. The hypothesis is accepted or rejected.

The scientific method is intended to be an objective way to gather knowledge.

As an example, let's return to our scenario of maple trees dropping their leaves in autumn. By observing the length of daylight throughout the year, and comparing that data with the time of the year when leaves fall, one hypothesis might be that shorter daylight causes the leaves to fall (**Figure 1.15**). This hypothesis makes a prediction—exposure of maple trees to shorter daylight will cause their leaves to fall. To test this prediction, researchers would design and conduct an experiment.

Although hypothesis testing may follow many paths, certain experimental features are common to this approach. First, data are often collected in two parallel manners. One set of experiments is done on the **control sample**, while another set is conducted on the **experimental sample**. In an ideal experiment, the control and experimental samples differ by only one factor. For example, an experiment could be conducted in which two groups of trees would be observed and the only difference between their environments would be the length of light each day. To conduct such an experiment, researchers would grow small trees in a greenhouse where they could keep factors such as temperature and water the same between the control and experimental samples, while providing them with different amounts of daylight. In the control group, the number of hours of light provided by light bulbs would be kept constant each day, while in the experimental group, the amount of light each day would become progressively shorter to mimic seasonal light changes. The researchers would then record the amount of leaves that were dropped by the two groups of trees over a certain period of time.

Another key feature of hypothesis testing is data analysis. The result of experimentation is a set of data from which a biologist tries to draw conclusions. Biology is a quantitative science. As such, data often come in the form of numbers that may or may not have important meaning. When experimentation involves a control and experimental sample, a common form of analysis is to determine if the data collected from the two samples are significantly different from each other. In this regard, the word *significant* means statistically significant. Biologists apply statistical analyses to their data to determine if the control and experimental samples are likely to be different from each other because of the single variable that is different between the two samples. When they are statistically significant, this means that the differences between the control and experimental data are not likely to have occurred as a matter of random chance. In our tree example shown in Figure 1.15, the trees in the control sample dropped far fewer leaves than did those in the experimental sample. A statistical analysis could determine if the data collected from the two greenhouses are significantly different from each other. If the two sets of data are found not to be significantly different, we would reject our hypothesis. Alternatively, if the differences between the two sets of data are significant, as shown in Figure 1.15, we would accept our hypothesis, though it is not proven.

As described next, discovery-based science and hypothesis testing are often used together to learn more about a particular scientific topic. As an example, let's look at how both approaches have led to successes in the study of the disease called cystic fibrosis.

The Study of Cystic Fibrosis Provides Examples of Both Discovery-Based Science and Hypothesis Testing

Let's consider how biologists made discoveries related to cystic fibrosis (CF), which affects about 1 in every 3,500 Americans. Persons with CF produce abnormally thick and sticky mucus that obstructs the lungs and causes life-threatening lung infections. The thick mucus also blocks the pancreas, which prevents the digestive enzymes that this organ produces from reaching the intestine. For this reason, CF patients tend to have excessive appetites but poor weight gain. Persons with this disease may also experience liver damage because the thick mucus can obstruct the liver. The average life span for people with CF is currently in their mid- to late 30s. Fortunately, as more advances have been made in treatment, this number has steadily increased.

Because of its medical significance, many scientists are interested in this disorder and have conducted studies aimed at gaining greater information regarding its underlying cause. The hope is that a better understanding of the disorder may lead to improved treatment options, and perhaps even a cure. As described next, discovery-based science and hypothesis testing have been critical to gaining a better understanding of this disorder.

The CF Gene and Discovery-Based Science In 1945, Dorothy Anderson determined that cystic fibrosis is a genetic disorder. Persons with CF have inherited two faulty *CF* genes, one from each parent. Over 40 years later, researchers used discovery-based science to identify the *CF* gene. Their search for the *CF* gene did not require any preconceived hypothesis regarding the function of the gene. Rather, they used genetic strategies that are similar to those described in Chapter 20. In 1989, research groups headed by Lap-Chi Tsui, Francis Collins, and John Riordan identified the *CF* gene.

The discovery of the gene made it possible to devise diagnostic testing methods to determine if a person carries a faulty *CF* gene. In addition, the identification of the *CF* gene may provide a potential treatment option for people with this disorder.

Figure 1.16   **A hypothesis that suggests an explanation of the function of the gene that is defective in patients with cystic fibrosis.**   The normal *CF* gene, which does not carry a mutation, encodes a transporter protein that transports chloride ions (Cl^-) across the plasma membrane to the outside of the cell. In persons with CF, this transporter is defective due to a mutation in the *CF* gene.

Biological inquiry: Explain how discovery-based science helped researchers to hypothesize that the CF gene encodes a transporter protein.

As discussed in Chapter 20, gene therapy is a technology in which persons with a faulty gene are given treatments that involve the introduction of the normal gene into their bodies. Clinical trials are underway to determine if the *CF* gene from normal individuals can be used to overcome the symptoms of the faulty gene in CF patients.

The characterization of the *CF* gene provided important clues regarding its function. Researchers observed striking similarities between the *CF* gene and other genes that were already known to encode proteins called transporters that function in the transport of substances across membranes. Based on this observation, as well as other kinds of data, the researchers hypothesized that the function of the normal *CF* gene is to encode a transporter. In this way, the identification of the *CF* gene led researchers to conduct experiments that were aimed at testing a hypothesis of its function.

The CF Gene and Hypothesis Testing Based on the characterization of the *CF* gene and on other studies showing that patients with the disorder have an abnormal regulation of salt balance across their plasma membranes, researchers hypothesized that the normal *CF* gene encodes a protein that transports chloride ions (Cl^-), a component of common table salt (NaCl), across the membrane of cells (**Figure 1.16**). This hypothesis led to experimentation in which researchers tested normal cells and cells from CF patients for their ability to transport Cl^-. The CF cells were found to be defective in chloride transport. In 1990, scientists successfully transferred the normal gene to CF cells in the laboratory.

The introduction of the normal *CF* gene into the cells from CF patients corrected the defect in chloride transport. Overall, the results showed that the *CF* gene encodes a protein that transports Cl⁻ across the plasma membrane. A mutation in this gene causes it to encode a defective transporter protein, leading to a salt imbalance that affects water levels outside the cell, which explains the thick and sticky mucus in CF patients. In this example, hypothesis testing has provided a way to accept or reject an idea regarding how a disease is caused by a genetic change.

FEATURE INVESTIGATION

Observation and Experimentation Form the Core of Biology

Because biology is the study of life, a biology textbook that focuses only on a description of living organisms would miss the main point. Biology is largely about the process of discovery. Therefore, a recurring theme of this textbook is discovery-based science and hypothesis testing. While each chapter contains many examples of data collection and experiments, a consistent element is a "Feature Investigation"—an actual study by current or past researchers. Some of these involve discovery-based science in which biologists collect and analyze data in an attempt to make discoveries that are not hypothesis driven. Alternatively, most Feature Investigations involve hypothesis testing in which a hypothesis is stated, and the experiment and resulting data are presented.

The Feature Investigations allow you to appreciate the connection between science and scientific theories. We hope you will find this a more interesting and rewarding way to learn about biology. As you read a Feature Investigation, you may find yourself thinking about different approaches and alternative hypotheses. Different people can view the same data and arrive at very different conclusions. As you progress through the experiments in this textbook, you will enjoy biology far more if you try to develop your own skills at formulating hypotheses, designing experiments, and interpreting data.

Finally, it is worthwhile to point out that science is a social discipline. After performing observations and experiments, scientists report their conclusions to the scientific community (**Figure 1.17**). They comment on each other's ideas and work, eventually shaping together the information that builds into scientific theories over many years. As you develop your skills at scrutinizing experiments, it is satisfying to discuss your ideas with other people, including fellow students and faculty members. Importantly, you do not need to "know all the answers" before you enter into a scientific discussion. Instead, a more rewarding way to view science is as an ongoing and never-ending series of questions.

Figure 1.17 **The social aspects of science.** At scientific meetings, researchers gather together to discuss new data and discoveries. Research that is conducted by professors, students, lab technicians, and industrial participants is sometimes hotly debated.

CHAPTER SUMMARY

- Biology is the study of life. Discoveries in biology help us understand how life exists, and they also have many practical applications such as the development of drugs to treat human diseases. (Figure 1.1)

1.1 The Properties of Life: Past and Present

- Seven characteristics are common to all forms of life. All living things (1) are composed of cells; (2) use energy; (3) respond to environmental changes; (4) regulate their internal conditions (homeostasis); (5) grow and develop; (6) reproduce; and (7) evolve over the course of many generations. (Figure 1.2)

- Living organisms can be viewed at different levels of complexity: atoms, molecules and macromolecules, cells, tissues, organs, organisms, populations, communities, ecosystems, and the biosphere. (Figure 1.3)

- Changes in species often occur as a result of modification of pre-existing structures. (Figures 1.4, 1.5)

- Vertical evolution involves mutations in a lineage that alter the characteristics of species over many generations. During this process, natural selection results in the survival of individuals with greater reproductive success. Over the long

AN INTRODUCTION TO BIOLOGY 19

run, this process alters species and may produce new species. (Figure 1.6)

- Horizontal gene transfer is the transfer of genes between different species. Along with vertical evolution, it is also an important force in biological evolution. (Figures 1.7, 1.8)

- Taxonomy involves the grouping of species according to their evolutionary relatedness to other species. Going from broad to narrow, each species is placed into a domain, kingdom, phylum, class, order, family, and genus. (Figures 1.9, 1.10)

- The genome is the genetic composition of a species. It provides a blueprint for the traits of an organism, is transmitted from parents to offspring, and acts as an instrument for evolutionary change. The proteome is the collection of proteins that a cell or organism can make. Each chapter in this textbook has a brief discussion of "Genomes & Proteomes" for you to understand this fundamental concept in biology. (Figure 1.11)

- An understanding of genomes and proteomes helps us to understand the characteristics of individuals and how they survive in their native environments. (Figure 1.12, and book cover)

1.2 Biology as a Scientific Discipline

- Biological science involves the observation, identification, experimental investigation, and theoretical explanation of natural phenomena.

- Biologists study life at different levels, ranging from ecosystems to molecular components in cells. (Figure 1.13)

- A hypothesis is a proposal to explain a natural phenomenon. A biological theory is a broad explanation that makes many valid predictions. A theory is based on vast amounts of data.

- Discovery-based science is an approach in which researchers conduct experiments without a preconceived hypothesis. It is a fact-finding mission. (Figure 1.14)

- The scientific method, also called hypothesis testing, is a series of steps to test the validity of a hypothesis. The experimentation often involves a comparison between control and experimental samples. (Figure 1.15)

- The study of cystic fibrosis is an interesting example in which both discovery-based science and hypothesis testing have provided key insights regarding the nature of the disease. (Figure 1.16)

- Each chapter in this textbook has a "Feature Investigation" to help you appreciate how science has led to key discoveries in biology.

- Advances in science often occur when scientists gather together and discuss their data. (Figure 1.17)

TEST YOURSELF

1. The process where living organisms maintain a relatively stable internal condition is
   a. adaptation.
   b. evolution.
   c. metabolism.
   d. homeostasis.
   e. development.

2. Populations of organisms change over the course of many generations. Many of these changes result in increased survival and reproduction. This phenomenon is
   a. evolution.
   b. homeostasis.
   c. development.
   d. genetics.
   e. metabolism.

3. All of the places on Earth where living organisms are found is
   a. the ecosystem.
   b. a community.
   c. the biosphere.
   d. a viable land mass.
   e. a population.

4. Horizontal evolution is the result of
   a. accumulation of mutations over many generations.
   b. transfer of genetic material between individuals of different species.
   c. transfer of genetic material from parent to offspring.
   d. all of the above.
   e. a and c only.

5. The scientific name for humans is *Homo sapiens*. The name *Homo* is the _____ to which humans are classified.
   a. kingdom
   b. phylum
   c. order
   d. genus
   e. species

6. The complete genetic makeup of an organism is called
   a. the genus.
   b. the genome.
   c. the proteome.
   d. the genotype.
   e. the phenotype.

7. A proposed explanation for a natural phenomenon is
   a. a theory.
   b. a law.
   c. a prediction.
   d. a hypothesis.
   e. an assay.

8. In science, a theory should
   a. be equated with knowledge.
   b. be supported by a substantial body of evidence.
   c. provide the ability to make many correct predictions.
   d. all of the above.
   e. b and c only.

9. Conducting research without a preconceived hypothesis is called
   a. discovery-based science.
   b. the scientific method.
   c. hypothesis testing.
   d. a control experiment.
   e. none of the above.

10. What is the purpose of using a control in scientific experiments?
   a. A control allows the researcher to practice the experiment first before actually conducting it.
   b. A researcher can compare the results in the experimental group and control group to determine if a single variable is causing a particular outcome in the experimental group.

c. A control provides the framework for the entire experiment so the researcher can recall the procedures that should be conducted.

d. A control allows the researcher to conduct other experimental changes without disturbing the original experiment.

e. All of the above.

CONCEPTUAL QUESTIONS

1. List the seven characteristics of life and explain a little about each.

2. Give the levels of organization from the simplest to most complex.

3. List the taxonomic groups from most inclusive to least inclusive.

EXPERIMENTAL QUESTIONS

1. Discuss the difference between discovery-based science and hypothesis testing.

2. What are the steps in the scientific method, also called hypothesis testing?

3. When conducting an experiment, explain how a control sample and an experimental sample differ from each other.

COLLABORATIVE QUESTIONS

1. Discuss the terms genomes and proteomes.

2. Discuss the levels of organization of life.

www.brookerbiology.com

This website includes answers to the Biological Inquiry questions found in the figure legends and all end-of-chapter questions.

2

THE CHEMICAL BASIS OF LIFE I: ATOMS, MOLECULES, AND WATER

CHAPTER OUTLINE

Crystals of sodium chloride (NaCl).

Biology—the study of life—is founded on the principles of chemistry and physics. All living organisms are a collection of atoms and molecules bound together and interacting with each other through the forces of nature. Throughout this textbook, we will see how chemistry can be applied to living organisms as we discuss the components of cells, the functions of proteins, the flow of nutrients in plants and animals, and the evolution of new genes. This chapter lays the groundwork for understanding these interactions. We begin with an overview of **inorganic chemistry**—that is, the nature of atoms and molecules with the exception of those that contain rings or chains of carbon. Such carbon-containing molecules form the basis of **organic chemistry** and are covered in Chapter 3.

2.1 Atoms

All life is composed of atoms, which in turn are composed of smaller, subatomic particles. A major role of the physicist is to uncover the properties of subatomic particles. Chemists, by contrast, are interested in the properties of atoms and molecules. Chemistry and physics merge when one attempts to understand the mechanisms by which atoms and molecules interact. When atoms and molecules are studied in the context of a living organism, the science of biochemistry emerges. No living creature is immortal, but atoms never "die." Instead, they exist *ad infinitum* as solitary atoms, or as components of a single molecule, or they shuttle between countless molecules over vast eons of time. In this section, we explore the physical properties of atoms so we can understand how atoms combine to form molecules of biological importance.

Atoms Are the Smallest Functional Units in Living Organisms

Atoms are the smallest functional units of matter that form all chemical substances and that cannot be further broken down into other substances by ordinary chemical or physical means. Many types of atoms are known. The simplest atom, hydrogen, is approximately 1 angstrom (10^{-10} meters) in diameter, roughly one-millionth the diameter of a human hair. Each specific type of atom—nitrogen, hydrogen, oxygen, and so on—occurs as a **chemical element**.

Three subatomic particles—**protons**, **neutrons**, and **electrons**—are found within atoms. The protons and neutrons are confined to a very small volume at the center of an atom, the **atomic nucleus**, whereas the electrons are found in regions at various distances from the nucleus. The numbers of protons and electrons in a given type of atom are identical, but the number of neutrons may vary. Each of the subatomic particles has a different electric charge. Protons have one unit of positive charge, electrons have one unit of negative charge, and neutrons are electrically neutral (**Table 2.1**).

Table 2.1	Characteristics of Major Subatomic Particles		
Particle		**Location**	**Charge**	**Mass relative to electron**
Electron	•⁻	Around the nucleus	−1	1
Proton	⊕	Nucleus	+1	1,836
Neutron	○	Nucleus	0	1,839

Because the protons are located in the atomic nucleus, the nucleus has a net positive charge equal to the number of protons it contains. The entire atom has no net electric charge, however, because the number of negatively charged electrons around the nucleus is equal to the number of positively charged protons in the nucleus. As shown in Table 2.1, the masses of protons and neutrons are similar to each other and much greater than the mass of electrons.

FEATURE INVESTIGATION

Rutherford Determined the Modern Model of the Atom

Nobel laureate Ernest Rutherford was born in 1871 in New Zealand, but he did his greatest work at McGill University in Montreal, Canada, and later at the University of Manchester in England. At that time, scientists knew that atoms contained charged particles but had no idea how those particles were arranged. Neutrons had not yet been discovered, and many scientists believed that the positive charge and the mass of an atom were evenly dispersed throughout the atom.

In a now-classic experiment, Rutherford aimed a fine beam of positively charged alpha particles (helium nuclei) at an extremely thin sheet of gold foil only 400 atoms thick (**Figure 2.1**). Surrounding the gold foil was a zinc sulfide screen that registered any alpha particles passing through or bouncing off the foil, much like film in a camera detects light. Rutherford hypothesized that if the positive charges of the gold atoms were uniformly distributed, most of the alpha particles would be slightly deflected as they passed through the foil, because one of the most important features of electric charge is that like charges repel each other. Due to their much smaller mass, he did not expect electrons to have any impact on the ability of an alpha particle to move through the metal foil.

Although some of the alpha particles were indeed deflected as they passed through the foil, more than 98% of them passed right through as if the foil was not there, and a few bounced nearly straight back! To explain the 98% that passed right through, Rutherford concluded that most of the volume of an atom is empty space. To explain the few alpha particles that bounced back, he postulated that most of the atom's positive charge was localized in a highly compact area. The existence of this small, dense region of highly concentrated positive charge—which today we call the atomic nucleus—explains how some alpha particles could be so strongly deflected by the gold foil. Alpha particles would bounce back if they directly collided with

Figure 2.1   Rutherford's gold foil experiment demonstrating that most of the volume of an atom is empty space.

HYPOTHESIS Atoms in gold foil are composed of diffuse, evenly distributed positive charges that should usually cause α-particles to be slightly deflected as they pass through.

STARTING MATERIALS Thin sheet of gold foil, α-particle emitter, zinc sulfide detection screen.

Experimental level   Conceptual level

1   Emit beam of α-particles.

α-particle emitter

α-particle

2   Pass beam through gold foil.

Gold foil

Gold atom   Gold foil   Positive charges of the gold atom

α-particle

Undeflected α-particles

Slightly deflected α-particle

α-particle that bounced back

3 Detect α-particles on zinc sulfide screens after they pass through foil or bounce back. Record number of α-particles detected on zinc sulfide screens and their locations.

α-particles that bounce back

α-particles that are undeflected

Zinc sulfide screens

α-particles that are slightly deflected

Detection of α-particles

4 THE DATA

% of α-particles detected on zinc sulfide screens	Location
98%	Undeflected
<2%	Slightly deflected
0.01%	Bounced back

an atomic nucleus. Therefore, based on these results, Rutherford rejected his original hypothesis that atoms are composed of diffuse, evenly distributed positive charges.

From a single experiment Rutherford proposed our modern model of an atom, with its small, positively charged nucleus surrounded at relatively great distances by negatively charged electrons. Today we know that more than 99.99% of an atom's volume is outside the nucleus. Indeed, the nucleus accounts for only about 1/10,000 of an atom's diameter—most of an atom is empty space!

Electrons Occupy Orbitals Around an Atom's Nucleus

After Rutherford's experiments, scientists initially visualized an atom as a mini–solar system, with the nucleus being the sun and the electrons traveling in clearly defined orbits around it. Electrons move at terrific speeds. Some estimates suggest that the electron in a typical hydrogen atom could circle the Earth in less than 20 seconds! **Figure 2.2** shows a diagram of the two simplest atoms, hydrogen and helium, which have the smallest numbers of protons. This model of the atom is now considered an oversimplification, because electrons do not actually orbit the nucleus in a defined path. However, this depiction of an atom remains a convenient way to diagram atoms in two dimensions.

At any given moment, it is impossible to precisely predict where a given electron will be located. Electrons travel within regions surrounding the nucleus in which the probability is high of finding that electron. These areas are called **orbitals**. Thus, another way of depicting atoms is a central nucleus surrounded by cloudlike orbitals. The cloud represents the region in which a given electron is most likely to be found. Some orbitals are spherical, called *s* orbitals, while others assume a shape that is often described as similar to a propeller or dumbbell and are called *p* orbitals (**Figure 2.3**). An orbital can contain a maximum of two electrons. Consequently, any atom with more than two electrons must contain additional orbitals.

Orbitals occupy so-called energy shells or energy levels. Atoms with progressively more electrons have orbitals within energy shells that are at greater and greater distances from the nucleus. These shells are numbered so that shell number 1 is closest to the nucleus. Different energy shells may contain one or more orbitals, each orbital with up to two electrons. The innermost shell of all atoms has room for only two electrons, which spin in opposite directions within a spherical orbital (1*s*). The second shell is composed of one spherical orbital (2*s*) and three dumbbell-shaped orbitals (2*p*). Thus, the second shell can hold up to four pairs of electrons (Figure 2.3).

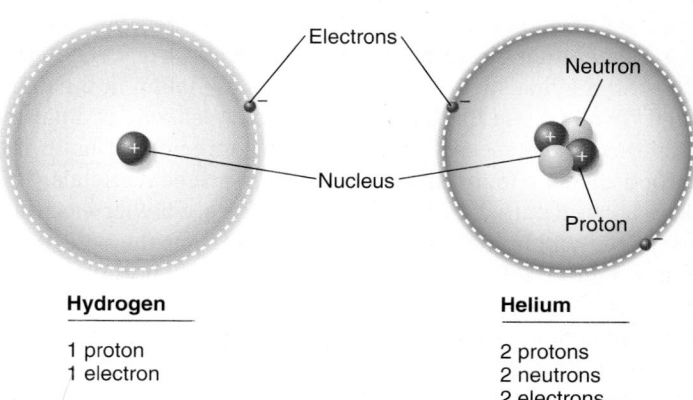

Electrons

Neutron

Nucleus

Proton

Hydrogen

1 proton
1 electron

Helium

2 protons
2 neutrons
2 electrons

Figure 2.2 **The nuclei of two simple atoms and their electrons.**   Early depictions of the atom envisioned a nucleus surrounded by electrons in discrete, measurable orbits, much like planets around the sun. This is a model of the two simplest atoms, hydrogen and helium. Note: In all figures of atoms, the sizes and distances are not to scale.

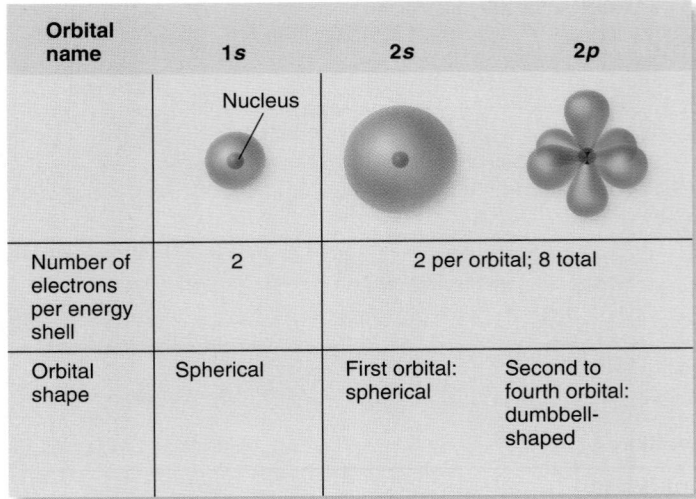

Orbital name	1s	2s	2p
	Nucleus		
Number of electrons per energy shell	2	2 per orbital; 8 total	
Orbital shape	Spherical	First orbital: spherical	Second to fourth orbital: dumbbell-shaped

Figure 2.3   **Diagrams of individual electron orbitals.** Electrons are found outside the nucleus in orbitals that may resemble spherical or dumbbell-shaped clouds. The orbital cloud represents a region of high probability of locating a particular electron.

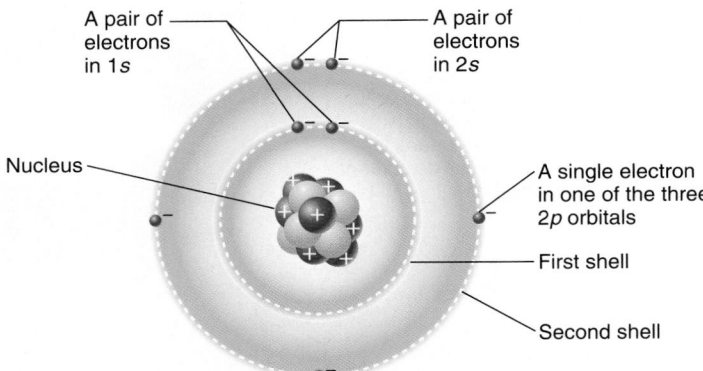

(a) Simplified depiction of a nitrogen atom (7 electrons; 2 electrons in first shell, 5 in second shell)

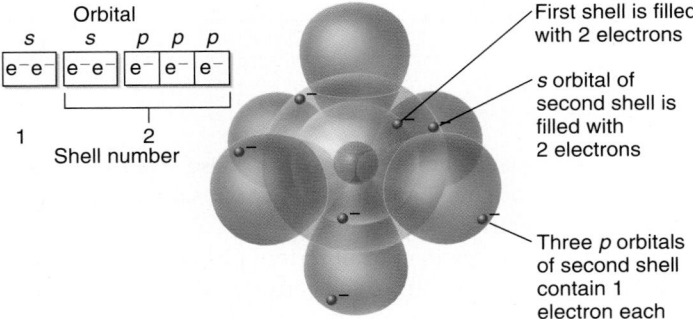

(b) Nitrogen atom showing electrons in orbitals

Figure 2.4   **Diagrams showing the multiple shells and orbitals of a nitrogen atom.**   The nitrogen atom shown **(a)** simplified and **(b)** with all of its orbitals and shells. An atom's shells fill up one by one. In shells containing more than one orbital, the orbital with lowest energy fills first. Subsequent orbitals gain one electron at a time, shown schematically in boxes, where *e* represents an electron.

Biological inquiry: What is the difference between an energy shell and an orbital?

Electrons vary in the amount of energy they have. The shell closest to the nucleus fills up with the lowest energy electrons first, and then each subsequent shell fills with higher and higher energy electrons, one shell at a time. Within a given shell, the energy of electrons can also vary among different orbitals. In the second shell, for example, the spherical orbital has lower energy, while the three dumbbell-shaped orbitals have slightly higher and roughly equal energies. In that case, therefore, two electrons fill the spherical orbital first. Any additional electrons fill the dumbbell-shaped orbitals one electron at a time.

Although electrons are actually found in orbitals of varying shapes, as shown in Figure 2.3, chemists often use more simplified diagrams when depicting the energy shells of electrons. **Figure 2.4a** illustrates an example involving nitrogen. An atom of this element has seven protons and seven electrons. Two electrons fill the first shell, and five electrons are found in the outer shell. Two of these fill the 2s orbital and are shown as a pair of electrons in the second shell. The other three electrons in the second shell are found singly in each of the three *p* orbitals. The diagram in Figure 2.4a makes it easy to see whether electrons are paired within the same orbital, and whether the outer shell is full. **Figure 2.4b** shows a more realistic depiction of a nitrogen atom, showing how the electrons actually occupy orbitals with different shapes.

Most atoms have outer shells that are not completely filled with electrons. Nitrogen, as we just saw, has a first shell filled with two electrons and a second shell with five electrons (Figure 2.4a). Because the second shell can actually hold eight electrons, the outer shell of a nitrogen atom is not full. As discussed later in this chapter, atoms that have unfilled energy shells tend to share, release, or obtain electrons to fill their outer shell. Those electrons in the outer shell that are available to combine with other atoms are called the **valence electrons**. Such electrons allow atoms to form chemical bonds with each other.

Each Element Has a Unique Number of Protons

Each chemical element has a specific and unique number of protons that distinguishes one element from another. The number of protons in an atom is its **atomic number**. For example, hydrogen, the simplest atom, has an atomic number of 1, corresponding to its single proton. Magnesium has an atomic number of 12, corresponding to its 12 protons. The term atom refers to a particle that is electrically neutral. Therefore, the atomic number is also equal to the number of electrons in the atom, resulting in a net charge of zero.

Atomic number and electron shells are a useful means of organizing the chemical elements. **Figure 2.5** shows the first three rows of the periodic table of the elements. A one- or two-letter symbol is used as an abbreviation for each element. (Three-letter

Figure 2.5   **A portion of the periodic table of the elements.**   The atoms are shown in models that depict the energy shells in different colors. The occupancy of orbitals is that of the elements in their pure state.

symbols are temporary until a two-letter symbol is chosen officially.) The elements are listed in the order of their atomic numbers. The rows indicate the number of energy shells. Hydrogen (H) has one shell, lithium (Li) has two shells, and sodium (Na) has three shells. The columns, from left to right, indicate the numbers of electrons in the outer shell. The outer shell of lithium (Li) has one electron, beryllium (Be) has two, boron (B) has three, and so forth. This organization of the periodic table tends to arrange elements based on similar chemical properties. For example, helium (He), neon (Ne), and argon (Ar) are inert gases and don't participate in chemical reactions. The inert gases have completely full outer shells. By comparison, beryllium (Be) and magnesium (Mg) are metals that have two electrons in their outer shell. The similarities of elements within a column occur because they have the same number of electrons in their outer shells, and therefore they have similar chemical bonding properties. These properties will be discussed later in this chapter.

Atoms Have a Small but Measurable Mass

Atoms are extremely minute and therefore have very little mass. A single hydrogen atom, for example, has a mass of only 1.67×10^{-24} g (grams). Protons and neutrons are nearly equal in mass, and both are more than 1,800 times the mass of an electron. Despite the great difference in mass of protons and electrons,

however, they nonetheless possess equal but opposite amounts of electric charge (see Table 2.1).

The **atomic mass** scale indicates an atom's mass relative to the mass of other atoms. By convention, the most common type of carbon atom, which has six protons and six neutrons, is assigned an atomic mass of exactly 12. On this scale, a hydrogen atom has an atomic mass of 1, indicating that it has 1/12 the mass of a carbon atom. A magnesium atom, with an atomic mass of 24, has twice the mass of a carbon atom. The term mass is sometimes confused with weight, but these two terms refer to different features of matter. Weight is derived from the gravitational pull on a given mass. If a man who weighs 154 pounds on Earth were standing on the moon, for example, he would only weigh about 25 pounds, but he would weigh 21 trillion pounds if he could stand on a neutron star. However, his mass is the same in all locations. Because we are discussing mass on Earth only, we can assume that the gravitational tug on all matter is roughly equivalent and thus the terms become essentially interchangeable for our purpose.

Atomic mass is measured in units called daltons, after the English chemist John Dalton, who postulated that matter is composed of minute, indivisible units he called atoms. One **dalton (Da)** equals 1/12 the mass of a carbon atom, or about the mass of a proton or a hydrogen atom. Thus, the most common type of carbon atom has an atomic mass of 12 daltons.

Because atoms such as hydrogen have a small mass, while atoms such as carbon have a larger mass, 1 g of hydrogen would have more atoms than 1 g of carbon. A **mole** of any substance contains the same number of particles as there are atoms in exactly 12 g of carbon. Twelve grams of carbon equals 1 mole, while 1 g of hydrogen equals 1 mole. As first described by Italian physicist Amedeo Avogadro, 1 mole of any element contains the same number of atoms—6.022×10^{23}. For example, 12 g of carbon contain 6.022×10^{23} atoms, and 1 g of hydrogen, whose atoms have 1/12 the mass of a carbon atom, also has 6.022×10^{23} atoms. This number, which is known today as **Avogadro's number**, is large enough to be somewhat mind-boggling, and thus gives us an idea of just how small atoms really are. To visualize the enormity of this number, imagine that people could move through a turnstile at a rate of 1 million people per second. It would require almost 20 billion years for 6.022×10^{23} people to move through that turnstile!

Isotopes Vary in Their Number of Neutrons

Although the number of neutrons in an atom is often equal to the number of protons, many elements can exist in multiple forms, called **isotopes**, that differ in the number of neutrons they contain. For example, the most abundant form of the carbon atom, ^{12}C, contains six protons and six neutrons, and thus has an atomic number of 6 and an atomic mass of 12 daltons, as described previously. The superscript placed to the left of ^{12}C is the sum of the protons and neutrons. The rare carbon isotope ^{14}C, however, contains six protons and eight neutrons, giving it an atomic number of 6 but an atomic mass of 14 Da. Nearly 99% of the carbon in living organisms is ^{12}C. Thus, the average atomic mass of carbon is very close to, but actually slightly greater than, 12 Da because of the existence of a small amount of heavier isotopes. This explains why the atomic masses given in the periodic table do not add up exactly to the predicted masses based on the atomic number and the number of neutrons of a given atom (Figure 2.5).

Many isotopes found in nature are inherently unstable and do not exist for long periods of time. Such isotopes are called **radioisotopes**, and they lose energy by emitting subatomic particles and/or radiation. At the very low amounts found in nature, radioisotopes usually pose no serious threat to life, but exposure of living organisms to high amounts of radioactivity can result in the disruption of cellular function and even death.

Modern medicine makes use of the high energy level of radioactive compounds in many ways. For example, solutions containing radioactive isotopes of iodine can be given to a person with an overactive thyroid gland. The thyroid is a gland in the neck that controls many important body functions. It is the only structure in the human body that uses iodine in large quantities, and so the isotope becomes concentrated in the gland. This localizes the radiation of the isotope to the thyroid, killing the hyperactive regions of the gland without harming other parts of the body. Another application makes use of the fact that radiation is easily detectable using various imaging techniques.

Figure 2.6 Diagnostic image of the human body using radioisotopes. A procedure called positron-emission tomography (PET) scanning highlights a region of the neck that contains the thyroid gland. A radioactive isotope of iodine was administered to the patient to reveal the size and activity of the gland. Radioactivity in this image shows up as a color.

These techniques, such as the PET scan shown in **Figure 2.6**, can detect the activities of body cells following injection of a compound that contains a radioactive isotope such as ^{131}I.

Researchers also use isotopes to study biological processes, such as tracking the movement of cellular compounds as they are shuttled between different cellular structures. For example, to determine which structures within a cell utilize cholesterol as part of their structure, researchers can incubate cells with cholesterol containing a radioactive isotope of hydrogen called tritium, ^3H, in place of one of its ordinary hydrogens. Cells are then examined at various times to determine where the tritium is located. For instance, much of the radioactivity will appear in the cell's plasma membrane, indicating that cholesterol is a component of the outer covering of cells.

Four Elements Constitute the Vast Majority of Living Organisms

Just four elements—oxygen, carbon, hydrogen, and nitrogen—account for the vast majority of atoms in living organisms (**Table 2.2**). These elements typically make up about 95% of the mass

Table 2.2	Chemical Elements Essential for Life in Most Organisms*	
Element	**Symbol**	**% Human body mass**
Most abundant in living organisms (approximately 95% of total mass)		
Oxygen	O	65
Carbon	C	18
Hydrogen	H	9
Nitrogen	N	3
Mineral elements (less than 1% of total mass)		
Calcium	Ca	
Chlorine	Cl	
Magnesium	Mg	
Phosphorus	P	
Potassium	K	
Sodium	Na	
Sulfur	S	
Trace elements (less than 0.01% of total mass)		
Chromium	Cr	
Cobalt	Co	
Copper	Cu	
Fluorine	F	
Iodine	I	
Iron	Fe	
Manganese	Mn	
Molybdenum	Mo	
Selenium	Se	
Silicon	Si	
Tin	Sn	
Vanadium	V	
Zinc	Zn	

* While these are the most common elements in living organisms, many other trace and mineral elements have reported functions. For example, aluminum is believed to be a cofactor for certain chemical reactions in animals, but it is generally toxic to plants.

of living organisms. Much of the hydrogen and oxygen occur in the form of water, which accounts for approximately 60% of the mass of most animals and up to 95% or more in some plants. Nitrogen is a vital element in all proteins, and carbon is a major building block of all living matter. Other vital elements in living organisms include the mineral elements, such as calcium and phosphorus, which are important constituents of the skeletons and shells of animals. Minerals like sodium and potassium are key regulators of water movement and electrical currents that occur across the surfaces of many cells.

In addition, all living organisms require trace elements. These atoms are present in extremely small quantities but still are essential for normal growth and function (Table 2.2). For example, iron plays an important role in how vertebrates store oxygen in their blood, and copper serves a similar role in some invertebrates.

2.2 Chemical Bonds and Molecules

The linkage of atoms with other atoms serves as the basis for life, and also gives life its great diversity. Two or more atoms bonded together make up a **molecule**. Atoms can combine with each other in several ways. For example, two oxygen atoms can combine to form one oxygen molecule, represented as O_2. This representation is called a **molecular formula**, and it consists of the chemical symbols for all of the atoms present (here, O for oxygen) and a subscript that tells you how many of those atoms are present in the molecule (in this case, two). The term **compound** refers to a molecule composed of two or more different elements. Examples include water (H_2O), with two hydrogen atoms and one oxygen atom, and the sugar glucose ($C_6H_{12}O_6$), which has 6 carbon atoms, 12 hydrogen atoms, and 6 oxygen atoms.

One of the most important features of compounds is their emerging physical properties. This means that the properties of a compound differ greatly from the elements that combined to form it. Let's consider sodium as an example. Pure sodium (Na), also called elemental sodium, is a soft, silvery white metal that you can cut with a knife. When sodium forms a compound with chlorine, table salt (NaCl) is made. NaCl is a white, relatively hard crystal that dissolves in water. Thus the properties of sodium in a compound can be dramatically different from its properties as an element.

The atoms in molecules are held together by chemical bonds. Important types of chemical bonds include covalent and ionic bonds. In this section, we will examine how these types of bonds form, and how they determine the structures of molecules.

Covalent Bonds Join Atoms Through the Sharing of Electrons

Covalent bonds, in which atoms share a pair of electrons, can occur between atoms whose outer shells are not full. A fundamental principle of chemistry is that *atoms tend to be most stable when their outer shells are full of electrons*. **Figure 2.7** shows this principle as it applies to the formation of hydrogen fluoride. The outer shell of a hydrogen atom is full when it contains two electrons, though a hydrogen atom has only one electron. The outer shell of a fluorine atom has seven electrons, though its outer shell would be full if it contained eight electrons. When hydrogen fluoride (HF) is made, the two atoms share a pair of electrons. This allows both of their outer shells to be full. Covalent bonds are strong chemical bonds, because the shared electrons behave as if they belong to each atom.

When the structure of a molecule is diagrammed, each covalent bond is represented by a line indicating a pair of shared electrons. For example, hydrogen fluoride is diagrammed as

H—F

A molecule of water can be diagrammed as

H—O—H

The structural formula of water indicates that the oxygen atom is covalently bound to two hydrogen atoms. Alternatively, as mentioned previously, water can be written by its shorthand molecular formula, H_2O.

Each atom forms a characteristic number of covalent bonds, which depends on the number of electrons that is needed to fill the outer shell. The atoms of some elements important for life, notably carbon, form more than one covalent bond and become linked simultaneously to two or more other atoms. **Figure 2.8** shows the number of covalent bonds that are formed by several atoms that are commonly found in the molecules of living cells.

Learning the **octet rule** may help you remember that atoms are stable when their outer shell is full. For many, but not all, types of atoms, their outer shell is full when they contain eight electrons, an octet. This rule applies to many types of atoms that are found in living organisms including carbon, nitrogen, oxygen, phosphorus, and sulfur. These atoms form a character-istic number of covalent bonds to make an octet in their outer shell (Figure 2.8). However, the octet rule does not always apply. For example, hydrogen has an outer shell that can contain only two electrons, not eight.

In some molecules, a **double bond** occurs when atoms share two pairs of electrons rather than one pair. As shown in **Figure 2.9**, this is the case for an oxygen molecule (O_2), which can be diagrammed as

$$O = O$$

Another common example occurs when two carbon atoms form bonds in compounds. They may share one pair of electrons (single bond) or two pairs (double bond), depending on how many other covalent bonds each carbon forms with other atoms. In rare cases, carbon can even form triple bonds, where three pairs of electrons are shared between two atoms.

Electrons Are Not Always Evenly Shared Between Atoms

Some atoms attract shared electrons more readily than do other atoms. The **electronegativity** of an atom is a measure of its ability to attract electrons in a bond from another atom. When two atoms with different electronegativities form a covalent bond, the shared electrons are more likely to be closer to the atom of higher electronegativity rather than the atom of lower electronegativity. Such bonds are called **polar covalent bonds**, because the distribution of electrons around the atoms creates a polarity, or difference in electric charge, across the molecule. Water is the classic example of a molecule containing polar covalent bonds. The shared electrons at any moment tend to be closer to the oxygen atom rather than either of the hydrogens. This unequal sharing of electrons gives the molecule a region of partial negative charge and two regions of partial positive charge (**Figure 2.10**).

Atoms with high electronegativity, such as oxygen and nitrogen, have a relatively strong attraction for electrons. These atoms

Figure 2.7 **The formation of covalent bonds.** In covalent bonds, electrons from the outer shell of two atoms are shared with each other, in order to complete the outer shells of both atoms. This simplified illustration shows hydrogen forming a covalent bond with fluorine.

Atom name	Hydrogen	Oxygen	Nitrogen	Carbon
Electron number needed to complete outer shell (typical number of covalent bonds)	1	2	3	4

Figure 2.8 **The most abundant elements found in living organisms.** These elements form different numbers of covalent bonds due to the electron configurations in their outer shells.

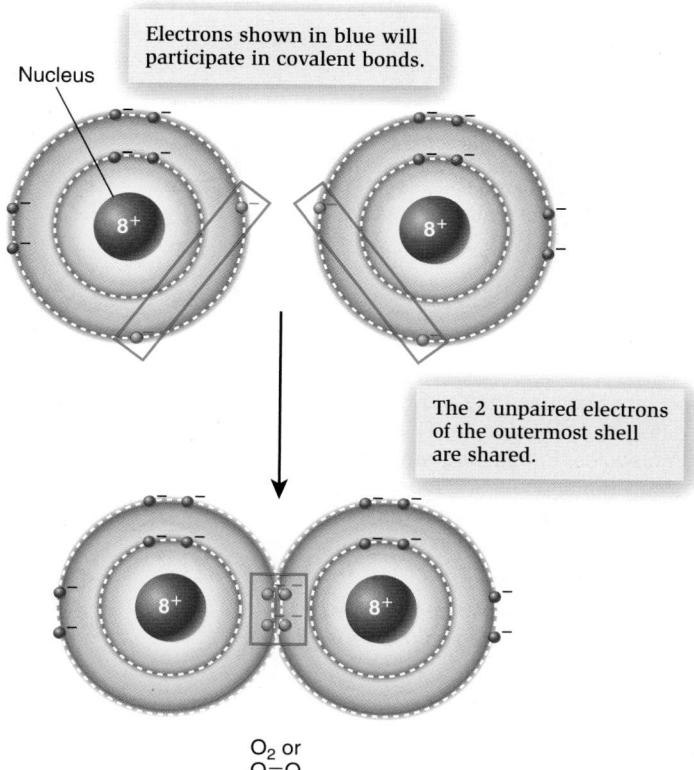

Figure 2.9 **A double bond between two oxygen atoms.**

Biological inquiry: Explain how an oxygen molecule obeys the octet rule.

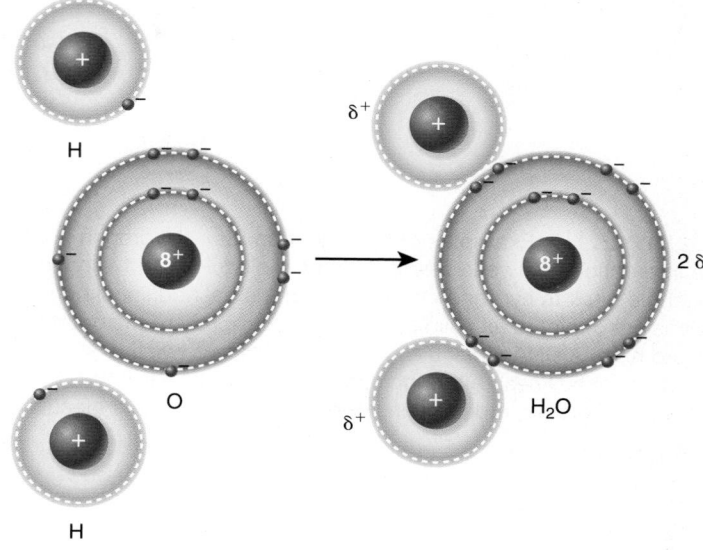

Figure 2.10 **Polar covalent bonds in water molecules.**
In a water molecule, two hydrogen atoms share electrons with an oxygen atom. Because oxygen has a higher electronegativity, the shared electrons spend more time closer to oxygen. This gives oxygen a partial negative charge, designated $2\delta^-$, and each hydrogen a partial positive charge, designated δ^+.

form polar covalent bonds with hydrogen atoms, which have low electronegativity. Examples of polar bonds include O—H and N—H. In contrast, bonds between carbon atoms (C—C) and between carbon and hydrogen atoms (C—H) are electrically neutral or nonpolar. Molecules containing significant numbers of polar bonds are known as **polar molecules**, whereas molecules composed predominantly of nonpolar bonds are called **nonpolar molecules**. A single molecule may have different regions with nonpolar bonds and polar bonds. For example, the detergent molecules found in soap have polar and nonpolar ends. The nonpolar ends dissolve in the oil of your skin, and the polar ends help the detergent rinse off in water. As you will see later, the physical characteristics of polar and nonpolar molecules, especially their solubility in water, are quite different.

Hydrogen Bonds Allow Interactions Within and Between Molecules

An important result of certain polar covalent bonds is the ability of one molecule to loosely associate with another molecule through a weak interaction called a **hydrogen bond**. A hydrogen bond forms when a hydrogen atom from one polar molecule becomes electrically attracted to an electronegative atom, such as an oxygen or nitrogen atom, in another polar molecule. Hydrogen bonds, like those between water molecules, are represented in diagrams by dashed or dotted lines to distinguish them from covalent bonds (**Figure 2.11a**). A single hydrogen bond is very weak. The strength of a hydrogen bond is only a few percent of the polar covalent bonds linking the hydrogen and oxygen within a water molecule.

Hydrogen bonds can also occur within a single large molecule. Many large molecules may have dozens, hundreds, or more hydrogen bonds within their structure. Collectively, many hydrogen bonds may add up to a strong force that helps maintain the three-dimensional structure of a molecule. This is particularly true in DNA—the molecule that makes up the genetic material of living organisms. DNA exists as two long, twisting strands of many thousands of atoms and molecules. The two strands are held together all along their length by hydrogen bonds (**Figure 2.11b**). Due to the large number of hydrogen bonds, considerable energy is needed to separate the two strands of DNA.

In contrast to the cumulative strength of many hydrogen bonds, the weakness of individual bonds is also important. When an interaction between two molecules involves relatively few hydrogen bonds, such interactions tend to be short-lived.

(a)

The hydrogen bond (H bond) is a weak attraction between a partially positive hydrogen and a partially negative atom such as oxygen.

H bonds

(b)

A DNA molecule consists of 2 twisted strands held together along its entire length by millions of hydrogen bonds.

H bonds

Figure 2.11 **Examples of hydrogen bonds.** Hydrogen bonds are important because they allow for interactions between atoms within a molecule or between atoms from different molecules. **(a)** This example depicts hydrogen bonds (shown as dashed lines) between water molecules. In this diagram, the atoms are depicted as solid spheres, which represent the outer shell. This is called a space-filling model for an atom. **(b)** A DNA molecule is composed of two twisting strands connected to each other by hydrogen bonds (dashed lines). Although each individual bond is weak, the sum of all the hydrogen bonds in a large molecule like DNA imparts considerable stability to the molecule.

Biological inquiry: As discussed in Chapter 11, DNA strands must separate for DNA to be replicated. Do you think the process of strand separation requires energy, or do you think the strands can separate spontaneously?

The reversible nature of hydrogen bonds allows molecules to interact and then to become separated again. For example, as discussed in Chapter 7, small molecules may bind to proteins called enzymes via hydrogen bonds. The small molecules are later released after the enzymes have changed their structure.

Ionic Bonds Involve an Attraction Between Positive and Negative Ions

Atoms are electrically neutral because they contain equal numbers of negative electrons and positive protons. If an atom or molecule gains or loses one or more electrons, it acquires a net electric charge and becomes an **ion**. For example, when a sodium atom (Na), which has 11 electrons, loses one electron, it becomes a sodium ion (Na^+) with a net positive charge. A sodium ion still has 11 protons, but only 10 electrons. Ions such as Na^+ are depicted with a superscript that indicates the net charge of the ion. Some atoms can gain or lose more than one electron. For instance, a calcium atom loses two electrons to become a calcium ion, depicted as Ca^{2+}. On the other hand, a chlorine atom (Cl), which has 17 electrons, can gain an electron and become a chloride ion (Cl^-) with a net negative charge—it has 18 electrons but only 17 protons. Hydrogen atoms and most mineral and trace element atoms readily form ions. **Table 2.3** lists the ionic forms of several elements. The ions listed in this table are relatively stable because the outer electron shells of the ions are full. For example, a sodium atom has one electron in its third (outer) shell. If it loses this electron to become Na^+, it no longer has a third shell, and the second shell, which is full, becomes its outer shell. Alternatively, a Cl atom has seven electrons in its third (outer) shell. If it gains an electron to become a chloride ion (Cl^-), its outer shell becomes full with eight electrons.

Ions that have a net positive charge are called **cations**, while those that have a net negative charge are **anions**. Ionization, the process of ion formation, can occur in single atoms or in atoms that are covalently linked in molecules.

An **ionic bond** occurs when a cation binds to an anion. **Figure 2.12a** shows an ionic bond between Na^+ and Cl^- to form NaCl. The general name, salt, is given to compounds that are formed from an attraction between a positively charged ion (a cation) and negatively charged ion (an anion). Examples of salts include NaCl, KCl, and $CaCl_2$. Salts may form crystals in which the cations and anions form a regular array. **Figure 2.12b** shows a NaCl crystal in which the sodium and chloride ions are held together by ionic bonds.

Table 2.3	Ionic Forms of Some Common Elements			
Atom	**Chemical symbol**	**Ion**	**Ion symbol**	**Electrons gained or lost**
Calcium	Ca	Calcium ion	Ca^{2+}	2 lost
Chlorine	Cl	Chloride ion	Cl^-	1 gained
Hydrogen	H	Hydrogen ion	H^+	1 lost
Magnesium	Mg	Magnesium ion	Mg^{2+}	2 lost
Potassium	K	Potassium ion	K^+	1 lost
Sodium	Na	Sodium ion	Na^+	1 lost

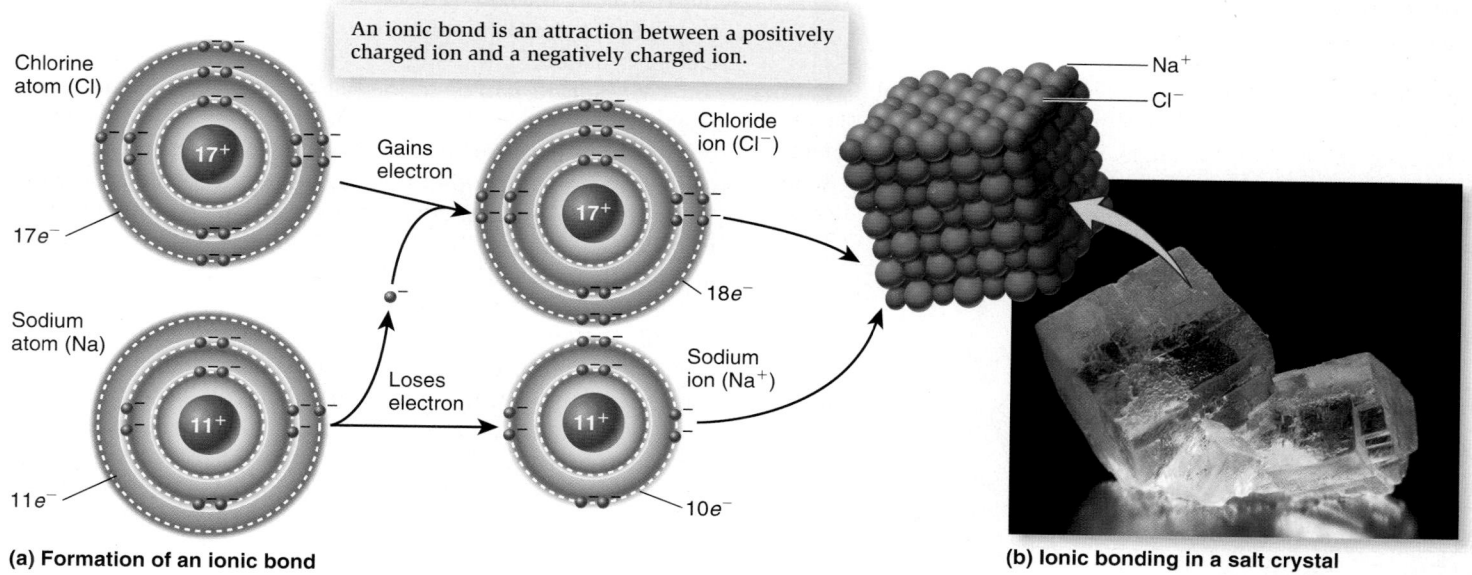

(a) Formation of an ionic bond

(b) Ionic bonding in a salt crystal

Figure 2.12 Ionic bonding in table salt (NaCl). (a) When an electron is transferred from a sodium atom to a chlorine atom, the resulting ions are attracted to each other via an ionic bond. (b) In a salt crystal, a lattice is formed in which the positively charged sodium ions (Na^+) are attracted to negatively charged chloride ions (Cl^-).

Molecules May Change Their Shapes

When atoms combine, they can form molecules with various three-dimensional shapes, depending on the arrangements and numbers of bonds between their atoms. As an example, let's consider the arrangements of covalent bonds in a few simple molecules (**Figure 2.13**). These molecules form new orbitals that cause the atoms to have defined angles relative to each other. This gives groups of atoms very specific shapes as shown in the three examples of Figure 2.13.

Molecules containing covalent bonds are not rigid, inflexible structures. Think of a covalent bond, for example, as an axle around which the joined atoms can rotate. Within certain limits, the shape of a molecule can change without breaking its covalent bonds. As illustrated in **Figure 2.14a**, a molecule of six carbon atoms bonded together can assume a number of shapes as a result of rotations around various covalent bonds. The three-dimensional, flexible shape of molecules contributes to their biological properties. As shown in **Figure 2.14b**, the binding of one molecule to another may affect the shape of one of the molecules. An animal can taste food, for instance, because food molecules interact with receptors on its tongue. When a food molecule encounters a receptor, the two molecules recognize each other by their unique shapes, much like a key fitting into a lock. As atoms in the receptor are attracted by hydrogen bonds to atoms in the food, the shape of the receptor changes. As we will see when we look at how the brain receives information from other parts of the body, the altered shape of the receptor initiates a signal that communicates to the animal's brain that the food tastes good (see Chapter 44).

Figure 2.13 Shapes of molecules. Molecules may assume different shapes depending on the types of bonds between their atoms. The angles between groups of atoms are well defined. For example, in liquid water at room temperature, the angle formed by the two bonds of each hydrogen to oxygen is approximately 104.5°. This bond angle can vary slightly depending on the temperature and degree of hydrogen bonding between adjacent water molecules.

Free Radicals Are a Special Class of Highly Reactive Molecules

Recall that an atom or ion is most stable when each of its orbitals is occupied by its full complement of electrons. A molecule containing an atom with a single, unpaired electron in its outer shell is known as a **free radical**. Free radicals can react with other molecules to "steal" an electron from one of their atoms,

Rotating this bond creates a new shape.

Subsequent bond rotations may create several additional shapes.

Molecule 2

Molecule 1

Shape changes in molecule 2

(a) Bond rotation in a small molecule   **(b) Noncovalent interactions that may alter the shape of molecules**

Figure 2.14 **Shape changes in molecules.**   A single molecule may assume different three-dimensional shapes without breaking any of the covalent bonds between its atoms, as shown in **(a)** for a six-carbon molecule. Hydrogen atoms above the blue plane are shown in white; those below the blue plane are blue. **(b)** Two molecules are shown schematically as having complementary shapes that permit them to interact. Upon interacting, the flexible nature of the molecules causes molecule 2 to twist sufficiently to assume a new shape. This change in shape is often an important mechanism by which one molecule influences the activity of another.

thereby filling the orbital in the free radical. In the process, this may create a new free radical in the donor molecule, setting off a chain reaction.

Free radicals can be formed in several ways, including exposure of cells to radiation and toxins. Free radicals can do considerable harm to living cells—for example, by causing a cell's membrane to rupture or damaging the genetic material. Surprisingly, the lethal effect of free radicals is sometimes put to good use. Some cells in animals' bodies create free radicals and use them to kill invading cells such as bacteria. Likewise, people use hydrogen peroxide to kill bacteria, as in a dirty skin wound. Hydrogen peroxide can break down to create free radicals, which can then attack bacteria in the wound.

Despite the exceptional case of fighting off bacteria, though, most free radicals that arise in an organism need to be inactivated so that they do not kill healthy cells. Protection from free radicals is afforded by molecules that can donate electrons to the free radicals without becoming highly reactive themselves. Examples of such protective compounds are certain vitamins found in fruits and vegetables, and the numerous plant compounds known as flavonoids. This is one reason why a diet rich in fruits and vegetables is beneficial to our health.

Free radicals are diagrammed with a dot next to the atomic symbol. Examples of biologically important free radicals are superoxide anion, $O_2 \bullet^-$; hydroxyl radical, $OH\bullet$; and nitric oxide, $NO\bullet$. Note that free radicals can be either charged or neutral.

Chemical Reactions Change Elements or Compounds into Different Compounds

A **chemical reaction** occurs when one or more substances are changed into other substances. This can happen when two or more elements or compounds combine with each other to form a compound, when one compound breaks down into two or more molecules, or when electrons are added to or taken away from an atom. Chemical reactions share many similar properties. First, they all require a source of energy for molecules to encounter each other. Such energy is provided partly by heat. In the complete absence of any heat (a temperature called absolute zero), atoms and molecules would be totally stationary and unable to interact. Heat energy causes atoms and molecules to vibrate and move, a phenomenon known as Brownian motion. Second, chemical reactions that occur in living organisms often require more than just Brownian motion to proceed at a reasonable rate. Such reactions need to be catalyzed. As discussed in Chapter 7, a catalyst is a substance that speeds up a chemical reaction. All cells contain many kinds of catalysts called enzymes. Third, chemical reactions tend to proceed in a particular direction, but will eventually reach a state of equilibrium. As an example, let's consider a chemical reaction between methane, a component found in natural gas, and oxygen. These molecules react with each other to produce carbon dioxide and water.

$$CH_4 + 2\,O_2 \rightleftharpoons CO_2 + 2\,H_2O$$
(methane) (oxygen)   (carbon dioxide) (water)

As it is written here, methane and oxygen are the **reactants**, while carbon dioxide and water are the **products**. Whether a chemical reaction is likely to proceed in a forward or reverse direction depends on changes in free energy, as described in Chapter 7. If we began with only methane and oxygen, the forward reaction is very favorable. The reaction would produce a large amount of carbon dioxide and water, as well as heat. This is why natural gas is used as a fuel to heat homes. However, all chemical reactions will eventually reach **equilibrium** in which the rate of the forward reaction is balanced by the rate of the reverse reaction. In the case of the reaction described above, this equilibrium would occur when nearly all of the reactants had been converted to products.

A final feature common to chemical reactions in living organisms is that they occur in water environments. Just as a crystal of sodium chloride will not dissolve in air but will dissolve in water, so too do the chemical reactions in organisms require water. Next, we will examine the properties of this amazing liquid.

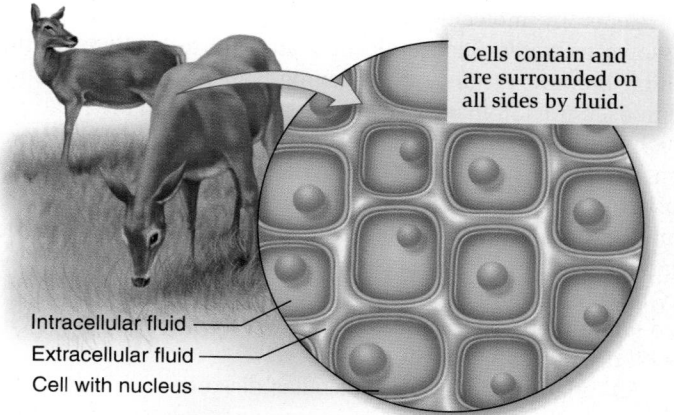

Figure 2.15 **Fluids inside and outside of cells.** Aqueous solutions exist in the intracellular fluid and in extracellular fluid. Chemical reactions are always ongoing in both fluids.

2.3 Properties of Water

It would be difficult to imagine life without water. People can survive for a month or more without food but usually die in less than a week without water. The bodies of organisms are composed largely of water. Up to 95% of the weight of certain plants comes from water. In people, typically 60–70% of body weight is from water. The brain is roughly 70% water, blood is about 80% water, and the lungs are nearly 90% water. Even our bones are about 20% water! In addition, water is an important liquid in the environments of living organisms. For example, many species are aquatic organisms that survive in a watery environment.

Thus far in this chapter we have considered the features of atoms and molecules, and the nature of chemical reactions between atoms and molecules. In this section, we will turn our attention to issues related to the liquid properties of living organisms, and the environment in which they live. Most of the chemical reactions that occur in nature involve molecules that are dissolved in water, including those reactions that happen inside cells and in the spaces that surround cells of living organisms (**Figure 2.15**). However, not all molecules dissolve in water. In this section, we will examine the properties of chemicals that influence whether they dissolve in water, and consider how biologists measure the amounts of dissolved substances. In addition, we examine some of the other special properties of water that make it a vital component of living organisms and their environments.

Polar Molecules and Ions Readily Dissolve in Water

Substances dissolved in a liquid are known as **solutes**, and the liquid in which they are dissolved is the **solvent**. Solutes dissolve in a solvent to form a **solution**. In all living organisms, the solvent for chemical reactions is water, which is the most abundant solvent in nature. Solutions made with water are called **aqueous solutions**. To understand why a substance dissolves in

Figure 2.16 **NaCl crystals dissolving in water.** The ability of water to dissolve sodium chloride crystals depends on the electrical attraction between the polar water molecules and the charged sodium and chloride ions. Water molecules surround each ion as it becomes dissolved.

water, we need to consider the chemical bonds in the solute molecule and those in water. As discussed earlier, the covalent bonds linking the two hydrogen atoms to the oxygen atom in a water molecule are polar. Therefore, the oxygen in water has a slight negative charge, and each hydrogen has a slight positive charge. To dissolve in water, a substance must be polar and electrically attracted to water molecules. For example, table salt (NaCl) is a solid crystalline substance because of the strong ionic bonds between positive sodium ions (Na^+) and negative chloride ions (Cl^-). When a crystal of sodium chloride is placed in water, the polar water molecules are attracted to the charged Na^+ and Cl^- (**Figure 2.16**). The ions become surrounded by clusters of water molecules, allowing the Na^+ and Cl^- to separate from each other and enter the water—that is, to dissolve.

Generally, molecules that contain polar covalent bonds will dissolve in water. Such molecules are said to be **hydrophilic**, which literally means "water-loving." In contrast, molecules composed predominantly of carbon and hydrogen are relatively insoluble in water, because carbon-carbon and carbon-hydrogen bonds are nonpolar. These molecules do not have partial positive and negative charges, and therefore are not attracted to water molecules. Such molecules are **hydrophobic**, or "water-fearing." Oils are a familiar example of hydrophobic molecules. Try mixing vegetable oil with water and observe the result. The two liquids separate into an oil phase and water phase. Very little oil dissolves in the water.

Although hydrophobic molecules dissolve poorly in water, they normally dissolve readily in nonpolar solvents. For example, cholesterol is a compound found in the blood and cells of animals. It is a hydrophobic molecule that is barely soluble in water but that easily dissolves in nonpolar solvents used in chemical laboratories, such as ether. Biological membranes like those that encase cells contain about 50% nonpolar compounds. Because of this, cholesterol also inserts into biological membranes, where it helps to maintain the membrane structure.

Molecules that have both polar or ionized regions at one or more sites and nonpolar regions at other sites are called **amphipathic**—consisting of two parts. When mixed with water, long amphipathic molecules may form spheres called **micelles**, with their polar (hydrophilic) regions at the surface of the micelle, where they are attracted to the surrounding water molecules. The nonpolar (hydrophobic) ends are oriented toward the interior of the micelle (**Figure 2.17**). Such an arrangement minimizes the interaction between water molecules and the nonpolar ends of the amphipathic molecules. Nonpolar molecules can dissolve in the central nonpolar regions of these clusters and thus exist in an aqueous environment in far higher amounts than would otherwise be possible based on their low solubility in water. We already considered one familiar example of amphipathic molecules, soap, which can form micelles that help to dissolve oils and nonpolar molecules found in dirt.

Instead of micelles, other amphipathic molecules form structures called bilayers. As you will learn in Chapter 5, lipid bilayers play a key role in membrane structure.

The Amount of a Dissolved Solute Is Its Concentration

Solute **concentration** is defined as the amount of a solute dissolved in a unit volume of solution. For example, if 1 gram (g) of NaCl were dissolved in enough water to make 1 liter of solution, we would say that its solute concentration is 1 g/L.

A comparison of the concentrations of two different substances on the basis of the number of grams per liter of solution does not directly indicate how many molecules of each substance are present. For example, let's compare 10 g each of glucose ($C_6H_{12}O_6$) and sodium chloride (NaCl). Because the individual molecules of glucose have more mass than those of NaCl,

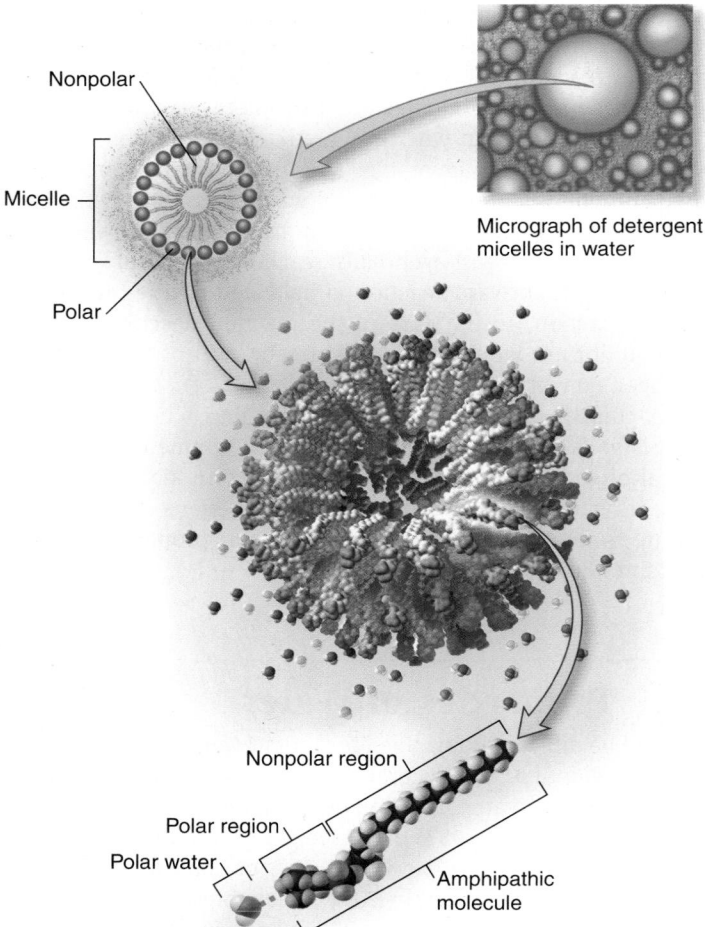

Micrograph of detergent micelles in water

Nonpolar

Micelle

Polar

Nonpolar region

Polar region

Polar water

Amphipathic molecule

Figure 2.17   **The formation of micelles by amphipathic molecules.**   In water, amphipathic molecules tend to arrange themselves so their nonpolar regions are directed away from water molecules, and the polar regions are directed toward the water and can form hydrogen bonds with it.

10 g of glucose will contain fewer molecules than 10 g of NaCl. Therefore, another way to describe solute concentration is according to the moles of dissolved solute per volume of solution. To make this calculation, we must know three things: the amount of dissolved solute, the molecular mass of the dissolved solute, and the volume of the solution.

The **molecular mass** of a molecule is equal to the sum of the atomic masses of all the atoms in the molecule. For example, glucose ($C_6H_{12}O_6$) has a molecular mass of 180 ([6 × 12] + [12 × 1] + [6 × 16] = 180). As mentioned earlier, 1 mole (abbreviated mol) of a substance is the amount of the substance in grams equal to its atomic or molecular mass. The **molarity** of a solution is defined as the number of moles of a solute dissolved in 1 L of solution. A solution containing 180 g of glucose (1 mol) dissolved in enough water to make 1 L is a 1 **molar** solution of glucose (1 mol/L). By convention, a 1 mol/L solution is usually written as 1 M, where the capital M stands for molar and is defined as mol/L. If 90 g of glucose (half its molecular mass) were dissolved in enough water to make 1 L, the solution would have a concentration of 0.5 mol/L, or 0.5 M.

The concentrations of solutes dissolved in the fluids of living organisms are usually much less than 1 mol/L. Many have concentrations in the range of millimoles per liter (1 mM = 0.001 M = 10^{-3} M), while others are present in even smaller concentrations—micromoles per liter (1 μM = 0.000001 M = 10^{-6} M) or nanomoles per liter (1 nM = 0.000000001 M = 10^{-9} M).

H_2O Exists as Ice, Water, and Water Vapor

Let's now consider some general features of water and how dissolved solutes affect its properties. H_2O is an abundant compound on Earth that exists in all three states of matter—solid (ice), liquid (water), and gas (water vapor). At the temperatures found over most regions of the planet, H_2O is found primarily as a liquid in which the weak hydrogen bonds between water molecules are continuously being formed, broken, and formed again. If the temperature rises, the rate at which hydrogen bonds break increases, and molecules of water escape into the gaseous state, becoming water vapor. If the temperature falls, hydrogen bonds are broken less frequently so that larger and larger clusters of water molecules are formed, until at 0°C water freezes into a crystalline matrix—ice. The H_2O molecules in ice tend to lie in a more "open" arrangement, which makes ice less dense than water. This is why ice floats on water (**Figure 2.18**). Compared to water, ice is also less likely to participate in most types of chemical reactions.

Changes in state, such as changes between the solid, liquid, and gaseous states of H_2O, involve an input or release of energy. For example, when energy is supplied to make water boil, it changes from the liquid to the gaseous state. This is called vaporization. The heat required to vaporize 1 mole of any substance at its boiling point under standard pressure is called the substance's **heat of vaporization**. For water, this value is very high. It takes more than five times as much heat to vaporize water than it does to raise the temperature of water from 0°C to 100°C. In contrast, energy is released when water freezes to form ice. Water has a high **heat of fusion**, which is the amount of heat energy that must be withdrawn or released from a substance to cause it to change from the liquid to the solid state. These two features, the high heats of vaporization and fusion, mean that water is extremely stable as a liquid. Not surprisingly, therefore, living organisms have evolved to function best within a range of temperatures consistent with the liquid phase of H_2O.

The temperature at which a solution freezes or vaporizes is influenced by the amounts of dissolved solutes. These are examples of **colligative properties**, which depend strictly on the concentration of dissolved solute particles, not on the specific type of particle. Pure water freezes at 0°C and vaporizes at 100°C. Addition of solutes to water lowers its freezing point below 0°C

Ice

Liquid water

Figure 2.18 **Structure of water and ice.** In its liquid form, the hydrogen bonds between water molecules continually form, break, and re-form, resulting in a changing arrangement of molecules from instant to instant. At temperatures at or below its freezing point, water forms a crystalline matrix called ice. In this solid form, hydrogen bonds are more stable. Ice has a hexagonally shaped crystal structure. The greater space between H_2O molecules in this crystal structure causes ice to have a lower density compared to water. For this reason, ice floats on water.

and raises its boiling point above 100°C. Adding a small amount of the compound ethylene glycol—antifreeze—to the water in your car's radiator, for instance, prevents the water from freezing in cold weather. Similarly, the presence of large amounts of solutes partly explains why the oceans do not freeze when the temperature falls below 0°C. Likewise, the colligative properties of water also account for the remarkable ability of certain ectothermic animals, which are unable to maintain warm body temperatures in cold environments, to nonetheless escape becoming frozen solid. Such "cold-blooded" animals produce antifreeze molecules that dissolve in their body fluids, thereby lowering the freezing point of the fluids and preventing their blood and cells from freezing in the extreme cold. The emerald rockcod (*Trematomus bernacchii*), found in the waters of Antarctica, for example, manages to live in ocean waters that are at or below 0°C (**Figure 2.19a**). Similarly, many insects such as the larvae of the parasitic wasp (*Brachon cephi*) also make use of natural antifreeze to stay alive in extreme conditions (**Figure 2.19b**).

Figure 2.19 **Antifreeze in living organisms.** Many animals, such as **(a)** the emerald rockcod (*Trematomus bernacchii*) and **(b)** the larvae of the parasitic wasp (*Brachon cephi*), can withstand extremely cold temperatures thanks to natural antifreeze molecules in their body fluids.

(a) Emerald rockcod in the waters of Antarctica

(b) Wasp larvae, which can withstand freezing temperatures

Water Performs Many Other Important Tasks in Living Organisms

As discussed earlier, water is the primary solvent in the fluids of all living organisms, from unicellular bacteria to the largest Sequoia tree. Water permits atoms and molecules to interact in ways that would be impossible in their nondissolved states. In Unit II we will consider a myriad of ions and molecules that are solutes in living cells. Even so, it is important to recognize that in addition to acting as a solvent, water serves many other remarkable functions that are critical for the survival of living organisms. For example, water molecules themselves take part in many chemical reactions of this general type:

$$R1{-}R2 + H{-}O{-}H \quad \rightarrow \quad R1{-}OH + H{-}R2$$

R is a general symbol to represent a group of atoms. In this equation, *R1* and *R2* are distinct groups of atoms. On the left side, *R1—R2* is a compound in which the groups of atoms are connected by a covalent bond. To be converted to products, a covalent bond is broken in each reactant, *R1—R2* and H—O—H, and OH and H (from water) form covalent bonds with R1 and R2, respectively. Reactions of this type are known as hydrolytic reactions (*hydro*, water; *lysis*, break apart), because water is used to break apart another molecule. This process is also called **hydrolysis** (**Figure 2.20a**). As discussed in later chapters, many large molecules are broken down into smaller units by hydrolysis. Alternatively, other chemical reactions in living organisms involve the removal of a water molecule so that a covalent bond can be formed between two separate molecules. For example, let's consider a chemical reaction that is the reverse of our previous hydrolytic reaction:

$$R1{-}OH + H{-}R2 \quad \rightarrow \quad R1{-}R2 + H{-}O{-}H$$

Such a reaction involves the removal of a water molecule, and the formation of a covalent bond between two separate molecules. This is termed a **dehydration** or **condensation reaction**. As discussed in later chapters, this is a common reaction that is used to build larger molecules in living organisms.

Another feature of water is that it is incompressible—its volume does not significantly decrease when subjected to high pressure. This has biological importance for many organisms that use water to provide force or support (**Figure 2.20b**). For example, water forms the so-called hydrostatic skeleton of worms and some other invertebrates, and it provides turgidity (stiffness) and support for plants.

Water is also the means by which unneeded and potentially toxic waste compounds are eliminated from an animal's body (**Figure 2.20c**). In mammals, for example, the kidneys filter out soluble waste products derived from the breakdown of proteins and other compounds. The filtered products remain in solution in the watery fluid, which eventually becomes urine and is excreted.

Recall from our discussion of water's properties that it takes considerable energy in the form of heat to convert water from a liquid to a gas. This feature has great biological significance. Although everyone is familiar with the fact that boiling water is converted to water vapor, water can vaporize into the gaseous state even at ordinary temperatures. This process is known as **evaporation**. The simplest way to understand this is to imagine that in any volume of water at any temperature, some water molecules will have higher energy than others. Not every molecule is vibrating identically with the same energy. Those with highest energy escape into the gaseous state. The important point, however, is that even at ordinary temperatures it still requires the same energy to change water from liquid to gas. Thus, the evaporation of sweat from an animal's skin requires considerable energy in the form of body heat, which is then lost to the environment. Evaporation is an important mechanism by which many animals cool themselves on hot days (**Figure 2.20d**).

Another important feature for living organisms is that water has a very high heat capacity, which means that it takes a lot of heat to raise its temperature. This accounts in part for the relatively stable temperatures of large bodies of water compared to inland temperatures. Large bodies of water tend to have a moderating effect on the temperature of nearby land masses.

The hydrogen-bonding properties of water affect its ability to form droplets and to adhere to surfaces. When the molecules within a substance tend to noncovalently attract each other, this phenomenon is called cohesion. Water exhibits strong cohesion due to hydrogen bonding. Cohesion aids in the movement of water through the vessels of plants (**Figure 2.20e**). A property that is similar to cohesion is adhesion, which refers to the ability of water to adhere to another surface. Water tends to cling to surfaces to which it can hydrogen-bond. For this reason, water can coat the surfaces of the digestive tract of animals and act as a lubricant for the passage of food (**Figure 2.20f**). Surface tension is a measure of how difficult it is to break the interface between a liquid and air. In the case of water, the attractive force between hydrogen-bonded water molecules at the interface between water and air is what causes water to form droplets. The surface water molecules attract each other into a configuration (a sphere) that reduces the number of water molecules in contact with air. Likewise, surface tension allows certain insects, such as water striders, to walk on the surface of a pond without sinking (**Figure 2.20g**).

Hydrogen Ion Concentrations Are Changed by Acids and Bases

Pure water has the ability to ionize to a very small extent into hydrogen ions (H^+) and **hydroxide ions** (OH^-). In pure water, the concentrations of H^+ and OH^- are both 10^{-7} mol/L, or 10^{-7} M. An inherent property of water is that the product of the concentrations of H^+ and OH^- is always 10^{-14} M at 25°C. Therefore, in pure water, $[H^+][OH^-] = [10^{-7}$ M$][10^{-7}$ M$] = 10^{-14}$ M. (The brackets around the symbols for the hydrogen and hydroxide ions indicate concentration.)

When certain substances are dissolved in water, they may release or absorb H^+ or OH^-, thereby altering the relative concentrations of these ions. Molecules that release hydrogen ions in solution are called **acids**. Two examples are:

$$HCl \rightarrow H^+ + Cl^-$$
(hydrochloric acid)   (chloride)

$$H_2CO_3 \rightleftharpoons H^+ + HCO_3^-$$
(carbonic acid)   (bicarbonate)

Hydrochloric acid is called a **strong acid** because it completely dissociates into H^+ and Cl^- when added to water. By comparison, carbonic acid is a **weak acid** because some of it will remain in the H_2CO_3 state when dissolved in water.

Compared to an acid, a **base** has the opposite effect when dissolved in water—it lowers the H^+ concentration. This can occur in different ways. Some bases, such as sodium hydroxide (NaOH), release OH^- when dissolved in water.

$$NaOH \rightarrow Na^+ + OH^-$$

Recall that the product of $[H^+]$ and $[OH^-]$ is always 10^{-14} M. When a base such as NaOH raises the OH^- concentration, some of the hydrogen ions bind to these hydroxide ions to form water. Therefore, increasing the OH^- concentration lowers the H^+ concentration. Alternatively, other bases, such as ammonia, react with water.

$$NH_3 + H_2O \rightleftharpoons NH_4^+ + OH^-$$
(ammonia)

H_2O

Hydrolysis

(a) Water participates in chemical reactions.

(b) Water provides support.

Blood enters and is purified by kidney cells.

Waste products are carried away in the urine.

(c) Water is used to eliminate soluble wastes.

(d) Evaporation helps animals dissipate body heat.

(e) The cohesive force of water molecules aids in the movement of fluid through vessels in plants.

(f) Water serves as a lubricant during feeding.

(g) The surface tension of water explains why this water strider doesn't sink.

Figure 2.20   Some amazing roles of water in biology. In addition to acting as a solvent, water serves many crucial functions in nature.

Both NaOH and ammonia have the same effect—they lower the concentration of H^+. NaOH achieves this by directly increasing the OH^- concentration, while NH_3 reacts with water to produce OH^-.

The addition of acids and bases to water can greatly change the H^+ and OH^- concentrations over a very broad range. Therefore, chemists and biologists use a log scale to describe the concentrations of these ions. The H^+ concentration is expressed as the solution's **pH**, which is defined as the negative logarithm to the base 10 of the H^+ concentration.

$$pH = -\log_{10} [H^+]$$

To understand what this equation means, let's consider a few examples. A solution with a H^+ concentration of 10^{-7} M has a pH of 7. A concentration of 10^{-7} M is the same as 0.1 μM. A solution in which $[H^+] = 10^{-6}$ M has a pH of 6. 10^{-6} M is the same as 1.0 μM. A solution at pH 6 is said to be more **acidic**, because the H^+ concentration is 10-fold higher than a solution at pH 7. Note that as the acidity increases, the pH decreases. A solution where the pH is 7 is said to be neutral because $[H^+]$ and $[OH^-]$ are equal. An acidic solution has a pH that is below 7, while an **alkaline** solution has a pH above 7. **Figure 2.21** considers the pH values of some familiar fluids.

Why is pH of importance to biologists? The answer lies in the observation that H^+ and OH^- can readily bind to many kinds of ions and molecules. For this reason, the pH of a solution can affect

- the shapes and functions of molecules;
- the rates of many chemical reactions;
- the ability of two molecules to bind to each other;
- the ability of ions or molecules to dissolve in water.

Due to the various effects of pH, many biological processes function best within very narrow ranges of pH, and even small shifts can have a negative effect. In living cells, the pH ranges from 6.5 to 7.8 and is carefully regulated to avoid major shifts in pH. The blood of the human body has a normal range of about pH 7.35 to 7.45 and is thus slightly alkaline. Certain diseases can reduce or increase blood pH by a few tenths of a unit. When this happens, the enzymes in the body needed for normal metabolism are rendered less functional, leading to illness and even death. As described next, living organisms have molecules called buffers to prevent such changes in pH.

Buffers Minimize Fluctuations in the pH of Fluids

What factors might alter the pH of an organism's fluids? External factors such as acid rain and other forms of pollution can reduce the pH of water entering the roots of plants. In animals, exercise generates lactic acid, and certain diseases can raise or lower the pH of blood.

Organisms have several ways to cope with changes in pH. Complex animals such as mammals, for example, can use struc-

Figure 2.21 **The pH scale and the relative acidities of common substances.**

Biological inquiry: What is the OH^- concentration at pH 8?

tures like the kidney to secrete acidic or alkaline compounds into the bloodstream when the blood pH becomes imbalanced. Similarly, the kidneys can transfer hydrogen ions from the body into the urine and adjust the body's pH in that way. Another mechanism by which pH balance is regulated in diverse organisms involves the actions of acid-base buffers. An acid-base **buffer** is composed of a weak acid and its related base. One such buffer is the bicarbonate pathway.

$$CO_2 + H_2O \rightleftharpoons H_2CO_3 \rightleftharpoons H^+ + HCO_3^-$$
$$\text{(carbonic acid)} \qquad \text{(bicarbonate)}$$

This buffer system can work in both directions. If the pH of an animal's blood were to increase (that is, the H^+ concentration decreased), the bicarbonate pathway would proceed from left to right. CO_2 would combine with water to make carbonic acid, and then the carbonic acid would dissociate into H^+ and bicarbonate. This would raise the H^+ concentration and thereby lower the pH. Alternatively, when the pH of an animal's blood decreases, this pathway runs in reverse. Bicarbonate combines with H^+ to make carbonic acid, which then dissociates to carbon

dioxide and water. This process removes H^+ from the blood, restoring it to its normal pH, and the CO_2 is exhaled from the lungs. Many buffers, including this example, exist in nature. Buffers found in living organisms are adapted to function most efficiently at the normal range of pH values seen in that organism.

CHAPTER SUMMARY

2.1 Atoms

- Atoms are the smallest functional units of matter that form all chemical elements and that cannot be further broken down into other substances by ordinary chemical or physical means. Atoms are composed of protons (positive charge), electrons (negative charge), and neutrons (electrically neutral). Electrons are found in orbitals around the nucleus. (Table 2.1, Figures 2.1, 2.2, 2.3, 2.4)

- Each element contains a unique number of protons, its atomic number. The periodic table organizes all known elements by atomic number and energy shells. (Figure 2.5)

- Each atom has a small but measurable mass, measured in daltons. The atomic mass scale indicates an atom's mass relative to the mass of other atoms.

- Many atoms exist as isotopes, which differ in the number of neutrons they contain. Some isotopes are unstable radioisotopes and emit radiation. (Figure 2.6)

- Four elements—oxygen, carbon, hydrogen, and nitrogen—account for the vast majority of atoms in living organisms. In addition, living organisms require mineral and trace elements that are essential for growth and function. (Table 2.2)

2.2 Chemical Bonds and Molecules

- A molecule consists of two or more atoms bonded together. The properties of a molecule are different from the properties of the atoms that combined to form it. A compound is composed of two or more different elements.

- Atoms tend to form bonds that fill their outer shell with electrons.

- Covalent bonds, in which atoms share electrons, are strong chemical bonds. Atoms form two covalent bonds—a double bond—when they share two pairs of electrons. (Figures 2.7, 2.8, 2.9)

- The electronegativity of an atom is a measure of its ability to attract bonded electrons. When two atoms with different electronegativities combine, the atoms form a polar covalent bond because the distribution of electrons around the atoms creates polarity, or difference in electric charge, across the molecule. Polar molecules, such as water, are largely composed of polar bonds, while most nonpolar molecules are composed predominantly of nonpolar bonds. (Figure 2.10)

- An important result of polar covalent bonds is the ability of one molecule to loosely associate with another molecule through weak interactions called hydrogen bonds. (Figure 2.11)

- If an atom or molecule gains or loses one or more electrons, it acquires a net electric charge and becomes an ion. The strong attraction between two oppositely charged ions forms an ionic bond. (Table 2.3, Figure 2.12)

- The three-dimensional, flexible shape of molecules allows them to interact and contributes to their biological properties. (Figures 2.13, 2.14)

- A free radical is an unstable molecule that interacts with other molecules by "stealing" electrons from their atoms.

2.3 Properties of Water

- Chemical reactions change compounds or elements into different compounds. All chemical reactions require energy. In living organisms, chemical reactions take place in a liquid environment, and many are readily reversible. (Figure 2.15)

- Solutes dissolve in a solvent to form a solution. Solute concentration refers to the amount of a solute dissolved in a unit volume of solution. The molarity of a solution is defined as the number of moles of a solute dissolved in 1 L of solution. (Figure 2.16)

- Polar molecules are hydrophilic, while nonpolar molecules composed predominantly of carbon and hydrogen are hydrophobic. Amphipathic molecules, such as detergents, have polar and nonpolar regions. (Figure 2.17)

- Water is the solvent for chemical reactions in all living organisms, which allows atoms and molecules to interact in ways that would be impossible in their nondissolved states.

- H_2O exists as ice, water, and water vapor. (Figure 2.18)

- The colligative properties of water allow it to function as an antifreeze in certain organisms. (Figure 2.19)

- Water's high heat of vaporization and high heat of fusion make it very stable in liquid form.

- Water molecules participate in many chemical reactions in living organisms. Hydrolysis breaks down large molecules into smaller units, and dehydration reactions combine two smaller molecules into one larger one. In living organisms, water provides support, is used to eliminate wastes, dissipates body heat, aids in the movement of liquid through vessels, and serves as a lubricant. Surface tension allows insects to walk on water. (Figure 2.20)

- The pH of a solution refers to its hydrogen ion concentration. The pH of pure water is 7 (a neutral solution). Alkaline solutions have a pH higher than 7, and acidic solutions have a pH lower than 7. (Figure 2.21)
- Buffers are compounds that act to minimize pH fluctuations in the fluids of living organisms. Buffer systems can raise or lower pH as needed.

TEST YOURSELF

1. _____ make up the nucleus of an atom.
   a. Protons and electrons   d. Neutrons and electrons
   b. Protons and neutrons   e. DNA only
   c. DNA and RNA

2. Living organisms are composed mainly of
   a. calcium, hydrogen, nitrogen, and oxygen.
   b. carbon, hydrogen, nitrogen, and oxygen.
   c. hydrogen, nitrogen, oxygen, and helium.
   d. carbon, helium, nitrogen, and oxygen.
   e. carbon, calcium, hydrogen, and oxygen.

3. The ability of an atom to attract bonded electrons is
   a. polarity.   d. valence.
   b. electronegativity.   e. both a and b.
   c. solubility.

4. Hydrogen bonds differ from covalent bonds in that
   a. covalent bonds can form between any type of atom and hydrogen bonds only form between H and O.
   b. covalent bonds involve sharing of electrons and hydrogen bonds involve the complete transfer of electrons.
   c. covalent bonds result from equal sharing of electrons but hydrogen bonds involve unequal sharing of electrons.
   d. covalent bonds involve sharing of electrons between atoms but hydrogen bonds are the result of weak attractions between a hydrogen atom of a polar molecule and an electronegative atom of another polar molecule.
   e. covalent bonds are weak bonds that break easily but hydrogen bonds are strong links between atoms that are not easily broken.

5. A free radical
   a. is a positively charged ion.
   b. is an atom with one unpaired electron in its outer shell.
   c. is a stable atom that is not bonded to another atom.
   d. can cause considerable cellular damage.
   e. both b and d.

6. Chemical reactions in living organisms
   a. require energy to begin.
   b. usually require a catalyst to initiate the process.
   c. are usually reversible.
   d. occur in liquid environments, such as water.
   e. all of the above.

7. Solutes that easily dissolve in water are said to be
   a. hydrophobic.   d. all of the above.
   b. hydrophilic.   e. b and c only.
   c. polar molecules.

8. The sum of the atomic masses of all the atoms of a molecule is its
   a. atomic weight.   d. concentration.
   b. molarity.   e. polarity.
   c. molecular mass.

9. Reactions that involve water in the breaking apart of other molecules are known as _____ reactions.
   a. hydrophilic   d. anabolic
   b. hydrophobic   e. hydrolytic
   c. dehydration

10. A difference between a strong acid and a weak acid is
    a. strong acids have a higher molecular mass than weak acids.
    b. strong acids completely ionize in solution, but weak acids do not completely ionize in solution.
    c. strong acids give off two hydrogen ions per molecule, but weak acids only give off one hydrogen ion per molecule.
    d. strong acids are water-soluble, but weak acids are not.
    e. strong acids give off hydrogen ions and weak acids give off hydroxyl groups.

CONCEPTUAL QUESTIONS

1. What are the types of bonds commonly found in biological molecules?

2. Distinguish between the terms hydrophobic and hydrophilic.

3. List the special properties of water that are ideally suited to life.

EXPERIMENTAL QUESTIONS

1. Before the experiment conducted by Ernest Rutherford, how did many scientists envision the structure of an atom?

2. What was the hypothesis tested by Rutherford?

3. What were the results of the experiment? How did Rutherford interpret the results?

COLLABORATIVE QUESTIONS

1. Discuss the three basic subatomic particles.

2. Discuss several properties of water that make it possible for life to exist.

---

www.brookerbiology.com

This website includes answers to the Biological Inquiry questions found in the figure legends and all end-of-chapter questions.

3

THE CHEMICAL BASIS OF LIFE II: ORGANIC MOLECULES

CHAPTER OUTLINE

A model showing the structure of a protein—a type of organic macromolecule.

I n Chapter 2, we learned that all life is composed of sub-atomic particles that form atoms, which in turn combine to form molecules. Molecules may be simple in atomic composition, as in water (H_2O) or hydrogen gas (H_2), or may bind with other molecules to form larger molecules. Of the countless possible molecules that can be produced from the known elements in nature, certain types contain carbon and are found in all forms of life. These carbon-containing molecules are collectively referred to as **organic molecules**, so named because they were first discovered in living organisms. Among these are lipids and large, complex compounds called **macromolecules**, which include carbohydrates, proteins, and nucleic acids. In this chapter, we will survey the structures of these molecules and examine their chief functions. We begin by examining the element whose chemical properties are fundamental to the formation of biologically important molecules: carbon. This element provides the atomic scaffold upon which life is built.

3.1 The Carbon Atom and the Study of Organic Molecules

The science of studying carbon-containing molecules is known as organic chemistry. This is a relatively young area of chemical exploration, considering that inorganic chemistry has been studied for hundreds of years, albeit in a rather primitive fashion in the beginning. In this section, we will examine the bonding properties of carbon that create groups of atoms with distinct functions and shapes.

Interestingly, the study of organic molecules was long considered a fruitless endeavor because of a concept called vitalism that persisted into the 19th century. Vitalism held that organic molecules were created by, and therefore imparted with, a vital life force contained within a plant or an animal's body. Supporters of vitalism argued there was no point in trying to synthesize an organic compound, because such molecules could arise only through the intervention of mysterious qualities associated with life. As described next, this would all change due to the pioneering experiments of Friedrich Wöhler in 1828.

Wöhler's Synthesis of an Organic Compound Transformed Misconceptions About Life's Molecules

Friedrich Wöhler (**Figure 3.1a**) was a German physician and chemist who was interested in the properties of inorganic and

(a)   (b)

Figure 3.1   Wöhler and his crystals of urea.   (a) Friedrich Wöhler. (b) Crystals of urea.

Biological inquiry: How did prior knowledge of urea allow Wöhler to realize he had made urea outside of the body?

organic compounds. He spent some time studying urea, which is found in urine. Urea is a natural organic product formed from the breakdown of proteins in an animal's body. In mammals, urea accumulates in the urine, which is formed by the kidneys, and then is excreted from the body. During the course of his studies, Wöhler purified urea from the urine of mammals. He noted the color, size, shape, and other characteristics of the crystals that formed when urea was isolated. This experience would serve him well in later years when he quite accidentally helped to put the concept of vitalism to rest.

In 1828, while exploring the reactive properties of ammonia and cyanic acid, Wöhler attempted to synthesize an inorganic molecule, ammonium cyanate (NH_4OCN). Instead, Wöhler discovered, to his surprise, that ammonia and cyanic acid reacted to produce a third compound, which, when heated formed familiar-looking crystals (**Figure 3.1b**). After careful analysis, he concluded that these crystals were in fact urea. He announced to the scientific community that he had synthesized urea, an organic compound, "without the use of kidneys, either man or dog." In other words, no mysterious life force was required to create this organic molecule. Subsequently, other scientists, such as Adolph Kolbe in 1845, would demonstrate that organic compounds such as acetic acid could be synthesized directly from their respective elements. These studies were a major breakthrough in the way in which scientists viewed life, and so began the field of science now called organic chemistry. From that time to the present, the fields of chemistry and biology have been understood to be intricately related.

Central to Wöhler's and Kolbe's reactions was the carbon atom. Urea and acetic acid, like all organic compounds, contain carbon atoms bound to other atoms. Let's now consider the chemical features that make carbon such an important element in living organisms.

Carbon Forms Four Covalent Bonds with Other Atoms

One of the properties of the carbon atom that makes life possible is its ability to form four covalent bonds with other atoms, including other carbon atoms. This occurs because carbon has four electrons in its outer shell, and it needs four additional electrons for its outer shell to be full (**Figure 3.2**). In living organisms, carbon atoms most commonly form covalent bonds with other carbons and with hydrogen, oxygen, nitrogen, and sulfur atoms. Bonds between two carbon atoms, between carbon and oxygen, or between carbon and nitrogen can be single or double. The combination of carbon with itself and with different elements and different types of bonds allows a vast number of organic compounds to be formed from only a few chemical elements. This is made all the more impressive because carbon bonds may occur in configurations that are linear, ringlike, or highly branched. Such molecular shapes can produce molecules with a variety of functions.

Because carbon and hydrogen have similar electronegativities, carbon-carbon and carbon-hydrogen bonds are nonpolar. As a consequence, molecules with predominantly hydrogen-

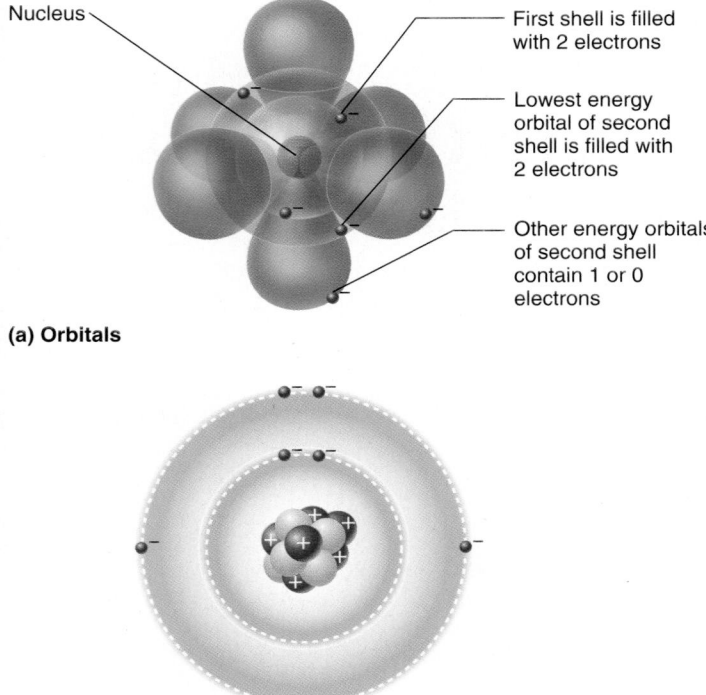

(a) Orbitals

Nucleus

First shell is filled with 2 electrons

Lowest energy orbital of second shell is filled with 2 electrons

Other energy orbitals of second shell contain 1 or 0 electrons

(b) Energy shells

Figure 3.2   Models for the electron orbitals and energy shells of carbon.   Carbon atoms have only four electrons in their outer (second) shell, which allows carbon to form four covalent bonds. When carbon forms four covalent bonds, the result is four hybrid orbitals of equal energy called sp^3 orbitals.

C—H and C—C bonds are electrically neutral and nonpolar.

Oxygen is more electronegative than carbon; thus, these are polar bonds.

Figure 3.3   Nonpolar and polar bonds in an organic molecule.   Carbon can form both nonpolar and polar bonds, and single and double bonds.

carbon bonds, called **hydrocarbons**, tend to be poorly soluble in water. In contrast, when carbon forms polar covalent bonds with oxygen or nitrogen, for example, the molecule is much more soluble in water due to the electrical attraction of polar water molecules. The ability of carbon to form both polar and nonpolar bonds contributes to its ability to serve as the backbone for an astonishing variety of molecules (**Figure 3.3**).

(a)

(b)

Figure 3.4 **The stability of carbon bonds makes life possible even in extreme environments.** **(a)** In the Russian Arctic, moss grows beneath the ice. **(b)** Deep-sea vents may reach temperatures greater than 650°F. Organisms can live near these vents, in very warm water.

Table 3.1	Some Biologically Important Functional Groups That Bond to Carbon	
Functional group	**Formula**	**Examples of where found***
Amino	$R-N\begin{smallmatrix}H\\H\end{smallmatrix}$	Amino acids (proteins)
Carbonyl**		
Ketone	$R-\overset{\overset{O}{\|\|}}{C}-R'$	Steroids, eicosanoids, waxes, and proteins
Aldehyde	$R-\overset{\overset{O}{\|\|}}{C}-H$	
Carboxyl	$R-C\overset{O}{\underset{OH}{}}$	Amino acids, fatty acids
Hydroxyl	$R-OH$	Steroids, alcohol, carbohydrates, some amino acids
Methyl	$R-\overset{\overset{H}{\|}}{\underset{\|}{C}}-H$	May be attached to DNA, proteins, and carbohydrates
Phosphate	$R-O-\overset{\overset{O}{\|\|}}{\underset{\|}{P}}-O^-$	Nucleic acids, ATP, attached to amino acids
Sulfate	$R-O-\overset{\overset{O}{\|\|}}{\underset{\|\|}{\underset{O}{S}}}-O^-$	May be attached to carbohydrates, proteins, and lipids
Sulfhydryl	$R-SH$	Proteins that contain the amino acid cysteine

* This list contains many of the functional groups that are important in biology. However, many more functional groups have been identified by biochemists.
** A carbonyl group is C=O. When the carbon is linked to a free hydrogen atom, this is an aldehyde. In a ketone, the carbon forms covalent bonds with two other carbon atoms.

One last feature of carbon that is important to biology is that carbon bonds are stable at the different temperatures associated with life. This property arises in part because the carbon atom is very small compared to most other atoms, and therefore the distance between carbon atoms forming a carbon-carbon bond is quite short. Shorter bonds tend to be stronger and more stable than longer bonds between two large atoms. Thus, carbon bonds are compatible with what we observe about life today, namely that living organisms can inhabit environments ranging from the Earth's icy poles to deep-sea vents (**Figure 3.4**).

Carbon Atoms Can Bond to Several Biologically Important Functional Groups

Aside from the simplest hydrocarbons, most organic molecules and macromolecules contain **functional groups**—groups of atoms with special chemical features that are functionally important. Each type of functional group exhibits the same properties in all molecules in which it occurs. For example, the amino group (NH_2) acts like a base. At the pH found in living organisms, amino groups readily bind H^+ to become NH_3^+, thereby removing H^+ from an aqueous solution and raising the pH. As discussed later in this chapter, amino groups are widely found in proteins and also in other types of organic molecules. **Table 3.1** describes examples of functional groups that are found in many different types of organic molecules. We will discuss each of these groups at numerous points throughout this textbook.

Carbon-Containing Molecules May Exist in Multiple Forms Called Isomers

When Wöhler did his famous experiment, he was surprised to discover that urea and ammonium cyanate apparently contained the exact same ratio of carbons, nitrogens, hydrogens, and oxygens, yet they were different molecules with distinct chemical and biological properties. Two structures with an identical molecular formula but different structures and characteristics are called **isomers**.

Figure 3.5 depicts three ways in which isomers may occur. **Structural isomers** contain the same atoms but in different

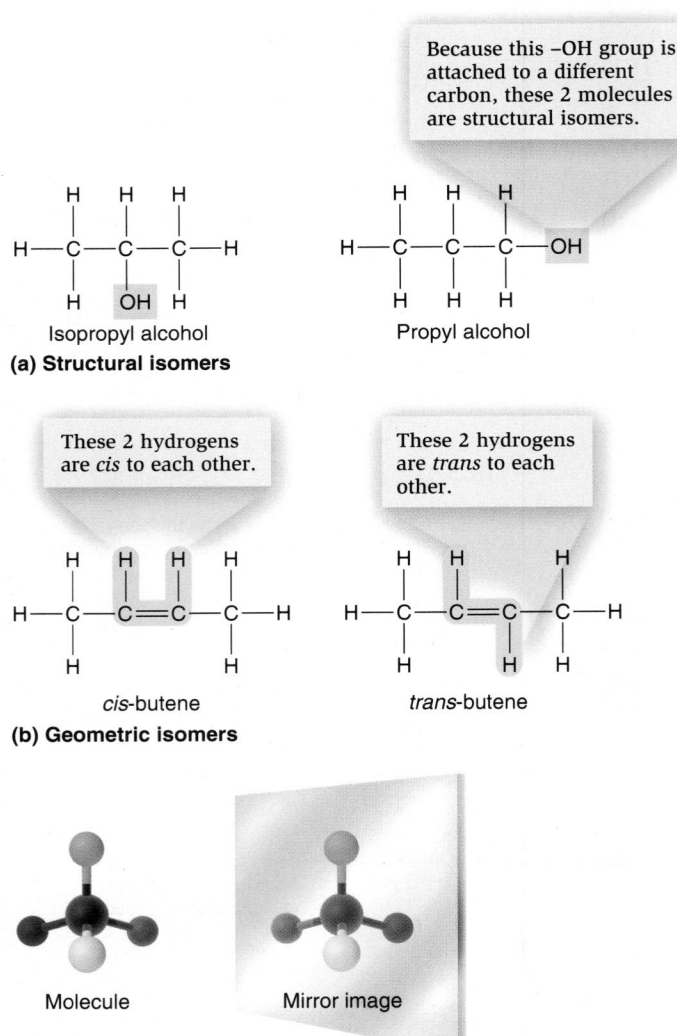

(a) Structural isomers

(b) Geometric isomers

(c) Enantiomers

Figure 3.5 **Types of isomers.** Isomers are compounds with the same molecular formula but different structures. The differences in structure, though small, are sufficient to result in very different biological properties.

bonding relationships. Wöhler's compounds fall into this category. **Stereoisomers** have identical bonding relationships, but the spatial positioning of the atoms differs in the two isomers. Two types of stereoisomers are geometric isomers and enantiomers. In **geometric isomers** like those shown in Figure 3.5b, the two hydrogen atoms linked to the two carbons of a C=C double bond may be on the same side of the carbons, in which case the C=C bond is called a *cis* double bond. If the hydrogens are on opposite sides, it is a *trans* double bond. *Cis-trans* stereoisomers may have very different chemical properties from each other, most notably their stability and sensitivity to heat and light. For instance, the light-sensitive region of your eye contains a molecule called retinal, which may exist in either a *cis* or *trans* form because of a pair of double-bonded carbons in its string of carbon atoms. In darkness, the *cis* form predominates. The energy of sunlight, however, causes retinal to isomerize to the *trans* form. The *trans*-retinal activates the light-capturing cells in the eye.

A second type of stereoisomer, called an **enantiomer**, exists as a pair of molecules that are mirror images. Four different atoms can bind to a single carbon atom in two possible ways, designated a left-handed and a right-handed structure. If the resulting structures are not identical, but instead are mirror images of each other, the molecules are enantiomers (Figure 3.5c). A given pair of enantiomers share identical chemical properties, such as solubility and melting point. However, due to the different orientation of atoms in space, their ability to noncovalently bind to other molecules can be strikingly different. For example, the right-handed form of glucose, called D-glucose, binds very well to certain enzymes in living cells, while the left-handed form, L-glucose, binds poorly. Most enzymes recognize only one type of enantiomer but not both.

3.2 Classes of Organic Molecules and Macromolecules

As we have seen, organic molecules have various shapes due to the bonding properties of carbon. During the past two centuries, biochemists have studied many organic molecules found in living organisms and determined their structures at the molecular level. Many of these compounds are relatively small molecules, containing a few or a few dozen atoms. However, some organic molecules are extremely large macromolecules, being composed of thousands or even millions of atoms. Such large molecules are formed by linking together many smaller molecules called monomers and are thus also known as **polymers** (meaning many small parts). The structure of macromolecules depends on the structure of their monomers, the number of monomers linked together, and the three-dimensional way in which the monomers are linked.

By analyzing the cells of many different species, researchers have determined that all forms of life have organic molecules that fall into four broad categories, based on their chemical and biological properties. The four major types of organic molecules and macromolecules found in all living organisms are carbohydrates, lipids, proteins, and nucleic acids. In this section, we will survey the structures of these organic compounds and begin to examine their biological functions.

Carbohydrates Exist as Sugars and Longer Polymers of Sugars

Carbohydrates are composed of carbon, hydrogen, and oxygen atoms in the proportions represented by the general formula $C_n(H_2O)_n$, where n is a whole number. This formula gives carbohydrates their name—carbon-containing compounds that are hydrated (that is, contain water). Most of the carbon atoms in a carbohydrate are linked to a hydrogen atom and a hydroxyl group. However, other functional groups, such as amino and carboxyl groups, are also found in certain carbohydrates. As discussed next, sugars are relatively small carbohydrates, while polysaccharides are large macromolecules.

Sugars Sugars are small carbohydrates that taste sweet. The simplest sugars are the **monosaccharides** (from the Greek, meaning single sugars). The most common types are molecules with five carbons, called pentoses, and six carbons, called hexoses. Important pentoses are ribose ($C_5H_{10}O_5$) and the closely related deoxyribose ($C_5H_{10}O_4$), which are part of RNA and DNA molecules, respectively. The most common hexose is glucose ($C_6H_{12}O_6$). Like other monosaccharides, glucose is very water-soluble and thus circulates in the blood of animals and the fluids of plants, where it can be transported across plasma membranes. Once inside a cell, glucose is broken down by enzymes. The energy released in this process is used to make many molecules of ATP (adenosine triphosphate), which powers a variety of cellular processes. In this way, sugar is often used as a source of energy by living organisms.

Figure 3.6a illustrates two traditional ways of depicting the bonds between atoms in a monosaccharide. The ring structure is a better approximation of the true shape of the molecule as it mostly exists in solution, with the carbon atoms numbered by convention as shown. The ring is made from the linear structure by an oxygen atom, which forms a bond that bridges two carbons. The hydrogen atoms and the hydroxyl groups may lie above or below the plane of the carbon ring structure.

Figure 3.6b compares different types of isomers of glucose. As mentioned earlier, glucose can exist as D- and L-glucose, which are mirror images of each other. These are enantiomers. Alternatively, other types of isomers are formed by changing the relative positions of the hydrogens and hydroxyl groups along the sugar ring. For example, glucose exists in two interconvertible forms, with the hydroxyl group attached to the number 1 carbon atom lying either above (the β form of glucose, Figure 3.6b) or below (the α form, Figure 3.6a) the plane of the ring. As discussed later, these different isoforms of glucose

have different biological properties. In another example, if the hydroxyl group on carbon atom number 4 of glucose is switched from below to above the plane of the ring, the sugar called galactose is created (Figure 3.6b).

Monosaccharides can join together to form larger carbohydrates. **Disaccharides** (meaning two sugars) are carbohydrates composed of two monosaccharides. A familiar disaccharide is sucrose, or table sugar, which is composed of glucose and fructose (**Figure 3.7**). The linking together of most monosaccharides involves the removal of a hydroxyl group from one monosaccharide and a hydrogen atom from the other, giving rise to a molecule of water and bonding the two sugars together through an oxygen atom. This occurs by a dehydration reaction, also known as a condensation reaction. The bond formed between two sugar molecules is a glycosidic bond. Conversely, hydrolysis of a glycosidic bond in a disaccharide breaks this linkage by adding back the water and thus uncouples the two monosaccharides. Other disaccharides frequently found in nature are maltose, formed in animals during the digestion of large carbohydrates in the intestinal tract, and lactose, present in the milk of mammals. Maltose is α-D-glucose linked to α-D-glucose, while lactose is β-D-galactose linked to β-D-glucose.

(a) Linear and ring structures of α-D-glucose

(b) Isomers

Figure 3.6 Monosaccharide structure. (a) A comparison of the linear and ring structures of glucose. In solution, such as the fluids of organisms, nearly all glucose is in the ring form. (b) Isomers of glucose. Glucose exists as stereoisomers designated α- and β-glucose, which differ in the position of the —OH group attached to carbon atom number 1. Glucose and galactose differ in the position of the —OH group attached to carbon atom number 4. Enantiomers of glucose, called D-glucose and L-glucose, are mirror images of each other. D-glucose is the form that is used by living cells. Note: The letters *D* and *L* are derived from dextrorotatory (rotating to the right) and levorotatory (rotating to the left).

Biological inquiry: With regard to their binding to enzymes, why do enantiomers such as D- and L-glucose have different biological properties?

Figure 3.7 **Formation of a disaccharide.** Two monosaccharides can bond to each other to form a disaccharide, such as sucrose, by a dehydration reaction.

Glucose   +   Fructose   ⟶   Sucrose   +   Water

Reactions resulting in the removal of 1 net molecule of water are called dehydration or condensation reactions.

Polysaccharides When many monosaccharides are linked together to form long polymers, **polysaccharides** (meaning many sugars) are made. **Starch**, found in plant cells, and **glycogen**, present in animal cells and sometimes called animal starch, are examples of polysaccharides (**Figure 3.8**). Both of these polysaccharides are composed of thousands of α-D-glucose molecules linked together in long, branched chains, differing only in the extent of branching along the chain. As you can see from the numbering system of the carbon atoms, the bonds that form in polysaccharides are not random but instead form between specific carbon atoms of each molecule. The higher degree of branching in glycogen contributes to its solubility in animal tissues, such as muscle. Starch, because it is less branched, is less soluble and contributes to the properties of plant structures (think of a tough, insoluble kernel of corn).

Certain polysaccharides, such as starch and glycogen, are used to store energy in cells. Like disaccharides, polysaccharides can be hydrolyzed in the presence of water to yield monosaccharides, which are broken down to make ATP. Starch and glycogen, the polymers of α-glucose, provide efficient means of storing energy for those times when a plant or animal cannot obtain sufficient energy for its needs from its environment or diet.

Other polysaccharides provide a structural role, rather than storing energy. The plant polysaccharide **cellulose** is a polymer of β-glucose, with a linear arrangement of carbon-carbon bonds (Figure 3.8). The bond orientations in β-glucose prevent cellulose from being hydrolyzed for ATP production in most organisms. Instead, cellulose forms part of the rigid cell-wall structure characteristic of plants. Cellulose accounts for up to half of all the carbon contained within a typical plant, making it the most common organic compound on Earth. A simple change in bond orientation of glucose molecules, namely α versus β, dramatically alters the biological properties of the resultant polymers, in one case yielding a form of stored fuel and in the other case providing a rigid, protective feature of plant cells.

Some bacteria present in the gastrointestinal tracts of grass and wood eaters, such as cows and termites, can digest cellulose into usable monosaccharides, because they contain an enzyme that can hydrolyze the β-glucose bonds. Humans lack this enzyme, and therefore we eliminate most of the cellulose ingested in our diet as fiber.

Other polysaccharides play structural roles. **Chitin**, a tough, structural polysaccharide, forms the external skeleton of many insects and the cell walls of fungi. The sugar monomers within chitin have nitrogen-containing groups attached to them. **Glycosaminoglycans** are large polysaccharides that play a structural role in animals. For example, glycosaminoglycans are abundantly found in cartilage. These polysaccharides tend to have sugar monomers with carboxyl and sulfate groups.

Lipids Store Energy and Form Membranes and Hormones

Lipids are molecules composed predominantly of hydrogen and carbon atoms. The defining feature of lipids is that they are nonpolar and therefore very insoluble in water. Lipids account for about 40% of the organic matter in the average human body and include fats, phospholipids, and steroids.

Fats Fats are a mixture of **triglycerides**, also known as triacylglycerols. Fats are formed by bonding glycerol to three fatty acids (**Figure 3.9**). A fatty acid is a chain of carbon and hydrogen atoms with a carboxyl group at the end. Because the carboxyl group ($-COOH$) releases an H^+ in water to become $-COO^-$, these molecules are called fatty acids. Each of the three hydroxyl groups ($-OH$) in glycerol is linked to the carboxyl group of a fatty acid by the removal of a molecule of water by a dehydration reaction. The resulting bond is called an ester bond.

The fatty acids found in fats and other lipids may differ with regard to their lengths and the presence of double bonds. Fatty acids are synthesized by the linking together of two-carbon fragments. Therefore, most fatty acids in nature have an even number of carbon atoms, with 16- and 18-carbon fatty acids being the most common in the cells of plants and animals. Fatty acids also differ with regard to the presence of double bonds.

Starch

α-1,4- glycosidic bonds form chains

Glycosidic bond

α-1,6-
glycosidic bonds
form branches

Branching patterns

Moderately branched

Glycogen

Highly branched

β-1,4-
glycosidic bonds
form chains

Cellulose

Unbranched

Figure 3.8   **Polysaccharides that are polymers of glucose.**   These polysaccharides differ in their arrangement, extent of branching, and type of glucose isomer.

The hydrogens from each hydroxyl group in glycerol are removed.

The hydroxyl groups from each carboxyl group of the 3 fatty acids are removed.

The new bond created is called an ester bond.

Dehydration

Glycerol

3 Fatty acids

Triglyceride (fat)

Figure 3.9   **The formation of a fat.**   The formation of a triglyceride requires three dehydration reactions in which fatty acids are bonded to glycerol.

When all the carbons in a fatty acid are linked by single covalent bonds, the fatty acid is said to be a **saturated fatty acid**, because all the carbons are saturated with covalently bound hydrogen. Alternatively, some fatty acids contain one or more C=C double bonds and are known as **unsaturated fatty acids**. A fatty acid with one C=C bond is a monounsaturated fatty acid, while two or more C=C bonds constitutes a polyunsaturated fatty acid (**Figure 3.10**). In organisms such as mammals, some fatty acids are necessary for good health but cannot be synthesized by the body. Such fatty acids are called essential fatty acids, because they must be obtained in the diet.

Figure 3.10 starts here visually.

Figure 3.10 **Examples of fatty acids.** Fatty acids are hydrocarbon chains with a carboxyl functional group at one end and either no double-bonded carbons (saturated) or one or more double bonds (unsaturated). Stearic acid, for example, is an abundant saturated fatty acid in animals, while linoleic acid is an unsaturated fatty acid found in plants.

Fats that contain high amounts of saturated fatty acids have high melting points, and therefore they tend to be solid at room temperature. Fats high in unsaturated fatty acids, on the other hand, usually have low melting points and thus are liquids at lower temperatures. Such fats are called oils. Animal fats generally contain a high proportion of saturated fatty acids, whereas vegetable oils contain more unsaturated fatty acids. When you heat a hamburger on the stove, the saturated animal fats melt, and liquid grease appears in the frying pan (**Figure 3.11**). When allowed to cool, however, the oily grease returns to its solid form. By comparison, oils derived from a plant, like olive oil, are liquid even at room temperature.

Like starch and glycogen, fats are important for storing energy. The hydrolysis of triglycerides releases the fatty acids from glycerol, and these products can then be metabolized to provide energy to make ATP. Certain organisms, most notably mammals, have the ability to store large amounts of energy by accumulating fats. One gram of fat stores twice as much energy as 1 g of glycogen or starch. For mobile animals, carrying around less weight is an advantage. In contrast, nonmobile organisms, such as plants, tend to store their energy in the form of polysaccharides. In animals, fats can also play a structural role by forming cushions that support organs. In addition, fats provide insulation under the skin that helps protect terrestrial animals during cold weather and marine mammals in cold water.

Phospholipids Another class of lipids, **phospholipids**, are similar in structure to triglycerides but with one important difference. The third hydroxyl group of glycerol is linked to a phosphate group instead of a fatty acid. In most phospholipids, a small polar or charged nitrogen-containing molecule is attached to this

| High temperature converts solid, saturated fats to oil. | After cooling, saturated fats return to their solid form. | Unsaturated fats are oils at room temperature and below. |

(a) Animal fats at high and low temperatures

(b) Vegetable fats at low temperature

Figure 3.11 **Fats at different temperatures.** Saturated fats found in animals tend to have high melting points compared to unsaturated fats found in plants.

Biological inquiry: Certain types of fats that are used in baking are called shortenings. They are solid at room temperature. Shortenings are often made from vegetable oils by a process called hydrogenation. What do you think happens to the structure of an oil when it is hydrogenated?

found in the center of the protein to avoid contact with water. As mentioned, some proteins have stretches of nonpolar amino acids that anchor them in the hydrophobic portion of membranes.

4. *Van der Waals forces*—Atoms within molecules have weak attractions for each other if they are an optimal distance apart. This optimal distance is called the van der Waals radius, and the weak attraction is the van der Waals force. If two atoms are too close together, their electron clouds will repel each other. If they are too far apart, the van der Waals force will diminish. Similar to hydrogen bonds, many van der Waals forces can contribute to protein folding and stability.

5. *Disulfide bridges*—The side chain of the amino acid cysteine (—CH$_2$—SH) contains a sulfhydryl group, which can react with a sulfhydryl group in another cysteine side chain to produce a disulfide bridge or bond, which links the two amino acid side chains together (—CH$_2$—S—S—CH$_2$—). Disulfide bonds are covalent bonds that can occur within a single polypeptide or between different polypeptides. Though other forces are usually more important in protein folding, the covalent nature of disulfide bonds can help to stabilize the structure of a protein.

The first four factors just described are also important in the ability of different proteins to interact with each other. As discussed throughout Unit II and other parts of this textbook, many cellular processes involve steps in which two or more different proteins interact with each other. For this to occur, the surface of one protein must bind to the surface of the other. Such binding is usually very specific. The surface of one protein precisely fits into the surface of another (**Figure 3.19**). Such **protein-protein interactions** are critically important so that cellular processes can occur in a series of defined steps. In addition, protein-protein interactions are also important in building large cellular structures that provide shape and organization to cells.

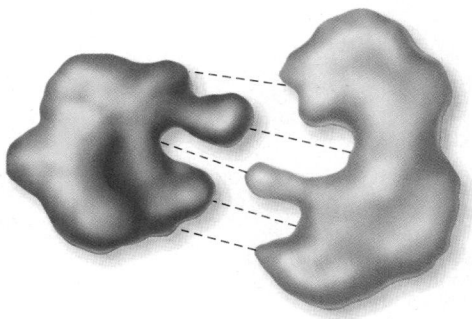

Protein 1   Protein 2

Figure 3.19   Protein-protein interaction.   Two different proteins may interact with each other due to hydrogen bonding, ionic bonding, the hydrophobic effect, and van der Waals forces.

FEATURE INVESTIGATION

Anfinsen Showed That the Primary Structure of Ribonuclease Determines Its Three-Dimensional Structure

Prior to the 1960s, the mechanisms by which proteins assume their three-dimensional structures were not understood. Scientists believed that correct folding required unknown cellular factors, or that ribosomes, the site where polypeptides are synthesized, somehow shaped proteins as they were being made. Christian Anfinsen, however, postulated that proteins contain all the information necessary to fold into their proper conformation without the need for organelles or cellular factors. He hypothesized that proteins spontaneously assume their most stable conformation based on the laws of chemistry and physics (**Figure 3.20**).

To test this hypothesis, Anfinsen studied ribonuclease, which we discussed earlier (see Figure 3.17). Biochemists had already determined that ribonuclease has four disulfide bonds between eight cysteine amino acids. Anfinsen began with purified ribonuclease—this is called an *in vitro* experiment, meaning under glass, as in a test tube. The key point is that other cellular components were not present, only the purified protein. He exposed ribonuclease to a chemical called urea that disrupted the hydrogen and ionic bonds, and to another chemical called β-mercaptoethanol that broke the S—S bonds. Following this treatment, he measured the ability of the treated enzyme to degrade RNA. The enzyme had lost nearly all of its ability to degrade RNA. Therefore, Anfinsen concluded that when ribonuclease was unfolded or denatured, it was no longer functional.

The key step in this experiment came when Anfinsen removed the urea and β-mercaptoethanol from the solution. Because these molecules are much smaller than ribonuclease, removing them from the solution was accomplished with a technique called dialysis. In dialysis, solutions are placed in a synthetic bag with microscopic pores that permit small molecules to pass through and leave the bag but retain large macromolecules such as ribonuclease. Anfinsen placed the entire bag in a large beaker of water, into which the urea and β-mercaptoethanol diffused. Then he retested the ribonuclease. The result revolutionized our understanding of proteins. The activity of the ribonuclease was almost completely restored! This meant that even in the complete absence of any cellular factors or organelles, an unfolded protein can refold into its functional structure. This was later confirmed by chemical analyses that demonstrated that the disulfide bonds had re-formed at the proper locations.

Since Anfinsen's time, we have also learned that his experiments with ribonuclease are not representative of all proteins. Some proteins do require certain enzymes and other proteins to assist them in their proper folding. Nonetheless, Anfinsen's experiments provided compelling evidence that the primary structure of a polypeptide is the key determinant of a protein's tertiary structure, a correct observation that earned him a Nobel Prize in 1972.

Figure 3.20   Anfinsen's experiments with ribonuclease demonstrating that the primary structure of a polypeptide plays a key role in protein folding.

HYPOTHESIS   Within their amino acid sequence, proteins contain all the information needed to fold into their correct, 3-dimensional shapes.

STARTING MATERIALS   Purified ribonuclease and RNA.

Experimental level	Conceptual level

1 Incubate purified ribonuclease in test tube with RNA, and measure its ability to degrade RNA.

Numerous H bonds and 4 S—S bonds. Protein is properly folded.

2 Denature protein shape by adding urea (breaks H bonds and ionic bonds) and β-mercaptoethanol (breaks S—S bonds). Measure its ability to degrade RNA.

β-mercaptoethanol + Urea

No more H bonds, ionic bonds, or S—S bonds. Protein is unfolded.

3 Place into dialysis bag. Pores allow urea and β-mercaptoethanol to escape, but ribonuclease is trapped inside. After urea and β-mercaptoethanol have exited the bag, measure the ability of ribonuclease to degrade RNA.

Add mixture to dialysis bag.

Denatured ribonuclease + β-mercapto-ethanol + Urea

Seal end when filled.

Sealed end

Pores

Ribonuclease   Urea   β-Mercaptoethanol

4   **THE DATA**

Activity restored

Biological activity (%) — Purified ribonuclease (step 1), Denatured ribonuclease (step 2), Ribonuclease after dialysis (step 3)

Figure 3.21   **The domain structure of a STAT protein.**   The boxes represent different domains, connected by chains of amino acids (the connecting lines in this figure).

GENOMES & PROTEOMES

Proteins Contain Functional Domains Within Their Structures

Modern research into the functions of proteins has revealed that many proteins have a modular design. This means that portions within proteins, called modules or **domains**, have distinct structures and functions. These units of amino acid sequences have been duplicated during evolution so that the same kind of domain may be found in several different proteins. When the same domain is found in different proteins, the domain has the same three-dimensional shape and performs a function that is characteristic of that domain.

As an example, **Figure 3.21** shows a protein that is known to play a critical role in regulating how certain genes are turned on and off in living cells. This protein bears the cumbersome name of *s*ignal *t*ransducer and *a*ctivator of *t*ranscription (STAT) protein. In the primary structure, each domain of the STAT protein is shown as a separate box, linked to the next domain by a straight line. These boxes and lines represent strings of amino acids, which have secondary and tertiary structures.

Each domain of this protein is involved in a distinct biological function, a common occurrence in proteins with multiple domains. For example, one of the domains is labeled the SH2 domain (Figure 3.21). In many different proteins, this domain allows proteins to recognize other proteins in a very specific way. The function of SH2 domains is to bind to tyrosines that are phosphorylated by cellular enzymes. Many proteins contain SH2 domains, and as might be predicted, they all bind to phosphorylated tyrosines in the proteins they recognize.

As a second example, a STAT protein has another domain called a DNA-binding domain. This portion of the protein has a structure that specifically binds to DNA. Overall, the domain structure of proteins enables them to have multiple regions, each with its own structure and purpose in the functioning of the protein.

Nucleic Acids Are the Source of Genetic Information

Nucleic acids account for only about 2% of the weight of animals like ourselves, yet these molecules are extremely important because they are responsible for the storage, expression, and transmission of genetic information. The expression of genetic information in the form of specific proteins determines whether one is a human, a mouse, an onion, or a bacterium. Likewise, genetic information determines whether a cell is part of a muscle or a leaf.

The two classes of nucleic acids are **deoxyribonucleic acid (DNA)** and **ribonucleic acid (RNA)**. DNA molecules store genetic information coded in the sequence of their monomer building blocks. RNA molecules are involved in decoding this information into instructions for linking together a specific sequence of amino acids to form a polypeptide chain.

Like other macromolecules, both types of nucleic acids are polymers and consist of linear sequences of repeating monomers.

Phosphate

Base (uracil)

Sugar (ribose)

OH OH

Example of a ribonucleotide

Phosphate

Base (cytosine)

Sugar (deoxyribose)

OH H

Example of a deoxyribonucleotide

Figure 3.22 Examples of two nucleotides. A nucleotide has a phosphate group, a five-carbon sugar, and a nitrogenous base.

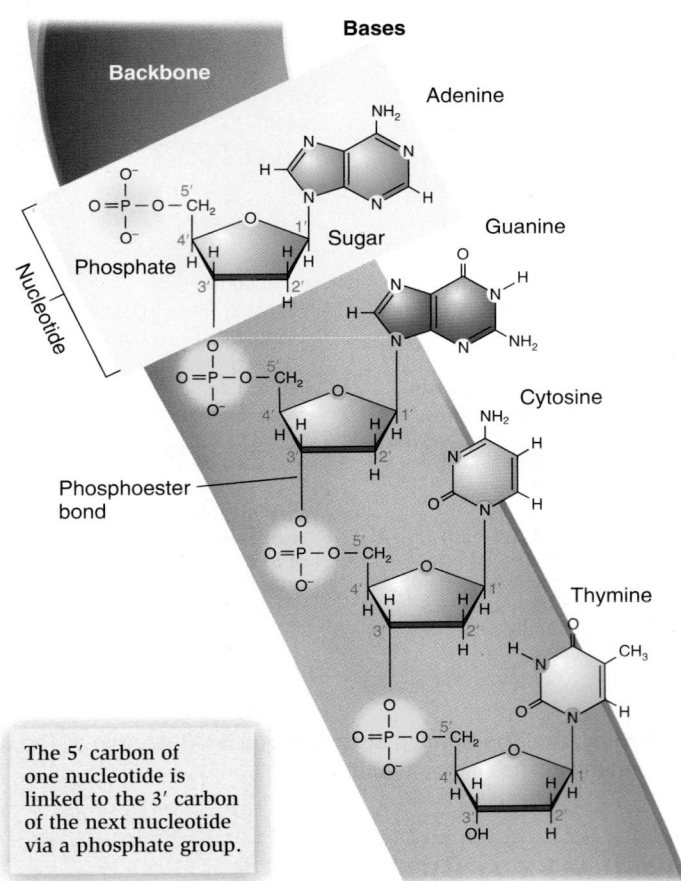

The 5' carbon of one nucleotide is linked to the 3' carbon of the next nucleotide via a phosphate group.

Figure 3.23 Structure of a DNA strand. Nucleotides are linked to each other to form a strand of DNA. The four bases found in DNA are shown. A strand of RNA would be similar except the sugar would be ribose and uracil would be substituted for thymine.

Each monomer, known as a **nucleotide**, has three components: a phosphate group, a five-carbon sugar (either ribose or deoxyribose), and a single or double ring of carbon and nitrogen atoms known as a **base** (**Figure 3.22**). The phosphate group of one nucleotide is linked to the sugar of the adjacent nucleotide to form a polynucleotide strand with the bases protruding from the side of the phosphate-sugar backbone (**Figure 3.23**). This is a phosphoester bond.

DNA The nucleotides in DNA contain the five-carbon sugar **deoxyribose**. Four different nucleotides are present in DNA, corresponding to the four different bases that can be linked to deoxyribose. The **purine** bases, **adenine (A)** and **guanine (G)**, have double (fused) rings of nitrogen and carbon atoms, and the **pyrimidine** bases, **cytosine (C)** and **thymine (T)**, have only a single ring (see Figure 3.23).

A DNA molecule consists of two strands of nucleotides coiled around each other to form a double helix (**Figure 3.24**). The two strands are held together by hydrogen bonds between a purine base in one strand and a pyrimidine base in the opposite strand. The ring structure of each base lies in a flat plane perpendicular to the sugar-phosphate backbone, somewhat like steps on a spiral staircase. This base pairing maintains a constant distance between the sugar-phosphate backbones of the two strands as they coil around each other.

As we will see in Chapter 11, only certain bases can pair with others, due to the location of the hydrogen-bonding groups in the four bases (see Figure 3.24). Two hydrogen bonds can be formed between adenine and thymine (A-T pairing), while three hydrogen bonds are formed between guanine and cytosine (G-C pairing). In a DNA molecule, A is always paired with T,

and G with C. If we know the amount of one type of base in a DNA molecule, we can predict the relative amounts of each of the other three bases. For example, if a DNA molecule were composed of 20% A bases, then there must also be 20% T bases. That leaves 60% of the bases that must be C and G combined. Because the amounts of C and G must be equal, this particular DNA molecule must be composed of 30% each of C and G. This specificity provides the mechanism for duplicating and transferring genetic information.

RNA RNA molecules differ in only a few respects from DNA. RNA consists of a single rather than double strand of nucleotides. In RNA, the sugar in each nucleotide is **ribose** rather than deoxyribose. Also, the pyrimidine base thymine in DNA is replaced in RNA with the pyrimidine base **uracil (U)** (see Figure 3.22). The other three bases—adenine, guanine, and cytosine—are found in both DNA and RNA. Certain forms of RNA called messenger RNA (mRNA) and transfer RNA (tRNA) are responsible for converting the information contained in DNA into the formation of a new polypeptide. This topic will be discussed in Chapter 12.

Figure 3.24   **The double-stranded structure of DNA.**   DNA consists of two strands coiled together into a double helix. The bases form hydrogen bonds in which A pairs with T, and G pairs with C.

CHAPTER SUMMARY

3.1   The Carbon Atom and the Study of Organic Molecules

- Organic chemistry is the science of studying carbon-containing molecules, which are found in living organisms. Wöhler's work with urea marked the birth of organic chemistry. (Figure 3.1)

- One property of the carbon atom that makes life possible is its ability to form four covalent bonds with other atoms. Carbon can form both polar and nonpolar bonds. The combination of different elements and different types of bonds allows a vast number of organic compounds to be formed from only a few chemical elements. (Figures 3.2, 3.3)

- Organic molecules may occur in various shapes. The structures of molecules determine their functions.

- Carbon bonds are stable at the different temperatures associated with life. (Figure 3.4)

- Organic compounds may contain functional groups. (Table 3.1)

- Carbon-containing molecules can exist as isomers, which have identical molecular composition but different structures and characteristics. Structural isomers contain the same atoms but in different bonding relationships. Stereoisomers have identical bonding relationships but different spatial positioning of their atoms. Enantiomers exist as mirror images of each other. (Figure 3.5)

3.2   Classes of Organic Molecules and Macromolecules

- The four major classes of organic molecules are carbohydrates, lipids, proteins, and nucleic acids. Macromolecules are large organic molecules that are composed of many thousands of atoms. Some macromolecules are polymers because they are formed by linking together many smaller molecules called monomers.

- Carbohydrates are composed of carbon, hydrogen, and oxygen atoms. Most cells can break down carbohydrates, releasing energy and storing it in newly created bonds in ATP.

- Carbohydrates include monosaccharides (the simplest sugars), disaccharides, and polysaccharides. The polysaccharides starch (in plant cells) and glycogen (in animal cells) provide an efficient means of storing energy. The plant polysaccharide cellulose serves a support or structural function. (Figures 3.6, 3.7, 3.8)

- Lipids, composed predominantly of hydrogen and carbon atoms, are nonpolar and very insoluble in water. Major classes of lipids include fats, phospholipids, steroids, and waxes.

- Fats, a mixture of triglycerides, are formed by bonding glycerol with three fatty acids. In a saturated fatty acid, all the carbons are linked by single covalent bonds. Unsaturated fatty acids contain one or more C=C double bonds. Animal fats generally contain a high proportion of saturated fatty acids, and vegetable fats contain more unsaturated fatty acids. (Figures 3.9, 3.10, 3.11)

- Phospholipids are similar in structure to triglycerides, except they are amphipathic because one fatty acid is replaced with a charged polar group. (Figure 3.12)

- Steroids are constructed of four fused rings of carbon atoms. Small differences in steroid structure can lead to profoundly different biological properties, such as the differences between estrogen and testosterone. (Figure 3.13)

- Waxes, another class of lipids, are nonpolar and repel water, and they are often found as protective coatings on the leaves of plants and the outer surfaces of animals' bodies.

- Proteins are composed of carbon, hydrogen, oxygen, nitrogen, and small amounts of other elements, such as sulfur. Proteins are macromolecules that play critical roles in almost all life processes. The proteins of living organisms are composed of the same set of 20 amino acids, corresponding to 20 different side chains. (Figure 3.14, Table 3.2)

- Amino acids are joined together by linking the carboxyl group of one amino acid to the amino group of another, forming a peptide bond. A polypeptide is a structural unit composed of amino acids, while a protein is a functional unit composed of one or more polypeptides that have been folded and twisted into precise three-dimensional shapes. (Figure 3.15)

- The four levels of protein structure are primary (its amino acid sequence), secondary (bending or twisting into helices or β sheets), tertiary (folding and refolding to assume a three-dimensional shape), and quaternary (multimeric proteins that consist of more than one polypeptide chain). If the primary structure of a protein changes, the other levels would change as a consequence. The three-dimensional structure of a protein determines its function, for example, by creating binding sites for other molecules. (Figures 3.16, 3.17, 3.18, 3.19, 3.20, 3.21)

- Nucleic acids are responsible for the storage, expression, and transmission of genetic information. The two types of nucleic acids are deoxyribonucleic acid (DNA) and ribonucleic acid (RNA). (Figures 3.22, 3.23)
- DNA molecules store genetic information coded in the sequence of their monomers. A DNA molecule consists of two strands of nucleotides coiled around each other to form a double helix, held together by hydrogen bonds between a purine base on one strand and a pyrimidine base on the opposite strand. (Figure 3.24)
- RNA molecules are involved in decoding this information into instructions for linking a specific sequence of amino acids to form a specific polypeptide chain. RNA consists of a single strand of nucleotides. The sugar in each nucleotide is ribose rather than deoxyribose, and the base uracil replaces thymine.

TEST YOURSELF

1. Molecules that contain the element _____ are considered organic molecules.
   a. hydrogen
   b. carbon
   c. oxygen
   d. nitrogen
   e. calcium

2. _____ was the first scientist to synthesize an organic molecule. The organic molecule synthesized was _____.
   a. Kolbe, urea
   b. Wöhler, urea
   c. Wöhler, acetic acid
   d. Kolbe, acetic acid
   e. Wöhler, glucose

3. The versatility of carbon to serve as the backbone for a variety of different molecules is due to
   a. the ability of carbon atoms to form four covalent bonds.
   b. the fact that carbon usually forms ionic bonds with many different atoms.
   c. the abundance of carbon in the environment.
   d. the ability of carbon to form covalent bonds with many different types of atoms.
   e. both a and d.

4. _____ are molecules that have the same molecular composition but differ in structure and/or bonding association.
   a. Isotopes
   b. Isomers
   c. Free radicals
   d. Analogues
   e. Ions

5. _____ is a storage polysaccharide commonly found in the cells of animals.
   a. Glucose
   b. Sucrose
   c. Glycogen
   d. Starch
   e. Cellulose

6. In contrast to other fatty acids, essential fatty acids
   a. are always saturated fats.
   b. cannot be synthesized by the organism and are necessary for survival.
   c. can act as building blocks for large, more complex macromolecules.
   d. are the simplest form of lipids found in plant cells.
   e. are structural components of plasma membranes.

7. Phospholipids are said to be amphipathic, which means these molecules
   a. are partially hydrolyzed during cellular metabolism.
   b. are composed of a hydrophilic portion and hydrophobic portion.
   c. may be poisonous to organisms if in combination with certain other molecules.
   d. are molecules composed of lipids and proteins.
   e. all of the above.

8. The monomers of proteins are _____ and these are linked by polar covalent bonds commonly referred to as _____ bonds.
   a. nucleotides, peptide
   b. amino acids, ester
   c. hydroxyl groups, phosphodiester
   d. amino acids, peptide
   e. monosaccharides, glycosidic

9. The _____ of a nucleotide determines whether it is a component of DNA or a component of RNA.
   a. phosphate group
   b. five-carbon sugar
   c. side chain
   d. fatty acid
   e. both b and d

10. A _____ is a portion of protein with a particular structure and function.
   a. peptide bond
   b. domain
   c. phospholipid
   d. wax
   e. monosaccharide

CONCEPTUAL QUESTIONS

1. Define isomers.
2. List the four classes of organic molecules and give a function of each.
3. Explain the difference between saturated and unsaturated fatty acids.

EXPERIMENTAL QUESTIONS

1. Before the experiments conducted by Anfinsen, what were the common beliefs among scientists about protein folding?
2. Explain the hypothesis tested by Anfinsen.
3. Why did Anfinsen use urea and β-mercaptoethanol in his experiments? Explain the result that was crucial to the discovery that the tertiary structure of a protein is dependent on the primary structure.

COLLABORATIVE QUESTIONS

1. Discuss several types of carbohydrates.
2. Discuss some of the roles that proteins play in organisms.

www.brookerbiology.com
This website includes answers to the Biological Inquiry questions found in the figure legends and all end-of-chapter questions.

4

GENERAL FEATURES OF CELLS

A cell from the pituitary gland. The cell in this micrograph was viewed by a technique called transmission electron microscopy, which is described in this chapter. The micrograph is artificially colored using a computer to enhance the visualization of certain cell structures.

When German botanist Matthias Schleiden studied plant material under the microscope, he was struck by certain consistent features. In particular, he noted the presence of many similar-looking compartments, each of which contained a dark area. (Today we call those compartments cells and the dark area is the nucleus.) In 1838, Schleiden speculated that cells are living entities and plants are aggregates of cells arranged according to definite laws.

Schleiden was a good friend of the German physiologist Theodor Schwann. Over dinner one evening, their conversation turned to the nuclei of plant cells, and Schwann remembered having seen similar structures in animal tissue. Schwann conducted additional studies that showed large numbers of nuclei in animal tissue, at regular intervals, and also located in cell-like compartments. In 1839, Schwann extended Schleiden's hypothesis to animals. In 1855, German biologist Rudolf Virchow proposed that *omnis cellula e cellula* ("every cell originates from another cell"). This idea arose from his research, which showed that diseased cells divide to produce more diseased cells.

According to the **cell theory** or **cell doctrine**, which is credited to both Schleiden and Schwann with contributions from Virchow:

1. all living things are composed of one or more cells;
2. cells are the smallest units of living organisms;
3. new cells come only from pre-existing cells by cell division.

Cell biology is the study of individual cells and their interactions with each other.

Most cells are so small that they cannot be seen with the naked eye. However, as cell biologists have begun to unravel

cell structure and function at the molecular level, the cell has emerged as a unit of wonderful complexity and adaptability. In this chapter, we will begin our examination of cells with an overview of their structures and functions. Later chapters in this unit will explore certain aspects of cell biology in greater detail. But first, let's look at the tools and techniques that allow us to observe these tiny entities.

4.1 Microscopy

The **microscope** is a magnification tool that enables researchers to study the structure and function of cells. The first compound microscope—a microscope with more than one lens—was invented in 1595 by Zacharias Jansen of Holland. In 1663 an English biologist, Robert Hooke, studied cork under a primitive compound microscope he had made. (He actually observed cell walls because cork cells are dead and have lost their internal components.) Hooke coined the word *cell* (from the Latin word *cellula*, meaning small compartment) to describe the structures he observed. Ten years later, the Dutch merchant Anton van Leeuwenhoek refined techniques of making lenses and was able to observe single-celled microorganisms such as bacteria. Among his many accomplishments, he discovered blood cells and was the first to see living sperm cells of animals.

Three important parameters in microscopy are magnification, resolution, and contrast. **Magnification** is the ratio between the size of an image produced by a microscope and its actual size. For example, if the image size is 100 times larger than its actual size, the magnification is designated 100×.

Figure 4.1 A comparison of the sizes of various chemical and biological structures, and the resolving power of the naked eye, light microscope, and electron microscope. The scale shown at the bottom is a logarithmic scale to accommodate the wide range of sizes shown in this drawing.

Depending on the quality of the lens and illumination source, every microscope has an optimal range of magnification before objects appear too blurry to be readily observed. **Resolution**, a measure of the clarity of an image, is the ability to observe two adjacent objects as distinct from one another. For example, a microscope with good resolution enables a researcher to distinguish two adjacent chromosomes as separate objects, which would appear as a single object under a microscope with poor resolution. The third important parameter in microscopy is **contrast**. The ability to visualize a particular cell structure may depend on how different it looks from an adjacent structure. If the object of interest, such as a particular protein, can be specifically stained with a colored dye, this makes viewing much easier. The application of stains, which selectively label individual components of the cell, greatly improves contrast. However, staining should not be confused with colorization. Many of the micrographs shown in this textbook are colorized to emphasize certain cellular structures (see the chapter opener, for example). In colorization, particular colors are added to micrographs with the aid of a computer. This is done for educational purposes. For example, colorization can help to emphasize different parts of a cell.

Microscopes are categorized into two groups based on the source of illumination. A **light microscope** utilizes light for illumination, while an **electron microscope** uses an electron beam. Very good light microscopes can resolve structures that are 0.2 μm (micron, or micrometer) apart or greater. The resolving power of a microscope depends on several factors including the wavelength of the source of illumination. Resolution is improved when the illumination source has a shorter wavelength.

A major advance in microscopy occurred in 1931 when Max Knoll and Ernst Ruska invented the first electron microscope. Because the wavelength of an electron beam is much shorter than visible light, the resolution of the electron microscope is far better than any light microscope. The limit is typically around 2 nm (nanometers), which is about 100 times better than the light microscope. **Figure 4.1** shows the range of resolving powers of the light and electron microscopes, and compares them to various cells and cell structures.

Over the past several decades, enormous technological advances have made light microscopy a powerful research tool. Improvements in lens technology, microscope organization, sample preparation, sample illumination, and computerized image processing have enabled researchers to create different types of light microscopes, each with its own advantages and disadvantages (**Figure 4.2**).

Similarly, improvements in electron microscopy occurred during the 1930s and 1940s, and by the 1950s the electron microscope was playing a major role in advancing our understanding of cell biology. During **transmission electron microscopy (TEM)**, a beam of electrons is transmitted through a biological sample. In preparation for TEM, a sample is treated with a chemical that binds to cellular molecules and fixes them in place. The sample is placed in a liquid resin and then the resin polymerizes to form a hardened block. The sample embedded within the block is sliced into very thin sections, typically 0.1–1.0 μm in thickness. To provide contrast, the sample is stained with a heavy metal. During staining, the metal binds to certain cellular structures such as membranes. The thin sections of the sample that have been stained with heavy metal are then adhered to a

Micrographs of nerve cells	Type of microscope	Micrographs of nerve cells	Type of microscope
(a) 30 μm	**Standard light microscope (bright field, unstained sample).** Light is passed directly through a sample, and the light is focused using glass lenses. Simple, inexpensive, and easy to use but offers little contrast.	**(b)** 46 μm	**Standard light microscope (bright field, stained sample).** Same type of microscope as in part (a) but the cell has been treated with a stain that selectively binds to certain regions of the cell. This provides better contrast.
(c) 30 μm	**Phase contrast microscope.** As an alternative to staining, this microscope controls the path of light and amplifies differences in the phase of light transmitted or reflected by a sample. The dense structures appear darker than the background, thereby improving the contrast in different parts of the specimen. Can be used to view living, unstained cells.	**(d)** 30 μm	**Differential-interference-contrast (Nomarski) microscope.** Similar to a phase contrast microscope in that it uses optical modifications to improve contrast in unstained specimens. Usually used to view whole cells or larger cell structures such as nuclei.
(e) 30 μm	**Fluorescence microscope.** Fluorescent molecules specifically label a particular type of cellular protein or organelle. A fluorescent molecule absorbs light at a particular wavelength and emits light at a longer wavelength. This microscope has filters that illuminate the sample with the wavelength of light that a fluorescent molecule absorbs, and then only the light that is emitted by the fluorescent molecules is allowed to reach the observer. This technique is often used to label specific cellular proteins, using fluorescent antibodies that bind specifically to a particular protein. The green filaments shown in this micrograph are due to the binding of fluorescently labeled antibody to a protein called actin that forms filaments in the cell.	**(f)** 23 μm	**Confocal microscope.** Uses lasers that illuminate various points in the sample. These points are processed by a computer to give a very clear three-dimensional image. This microscope technique can also be used in conjunction with fluorescence microscopy to view fluorescent molecules within a cell.

Figure 4.2   Micrographs of nerve cells using different types of light microscopes.

copper grid and placed in a transmission electron microscope. When the beam of electrons strikes the sample, some of them hit the heavy metal and are scattered, while those that pass through without being scattered are focused to form an image on a photographic plate or screen (**Figure 4.3a**). Because the scattered electrons are lost from the beam, the metal-stained regions of the sample that scatter electrons appear as areas of reduced electron penetration. These areas are darker than those that allow electrons to pass through them. Compared with other forms of microscopy, TEM has the best resolution of any microscope. TEM provides the greatest resolution of organelles and other cellular structures. However, such microscopes are expensive and are not commonly used to view living cells.

Scanning electron microscopy (SEM) is another type of electron microscopy that utilizes an electron beam to produce an image of the three-dimensional surface of biological samples (**Figure 4.3b**). A biological sample is coated with a thin layer of heavy metal such as gold or palladium and then exposed to an electron beam that scans the surface of the specimen.

(a) Transmission electron micrograph

(b) Scanning electron micrograph

Figure 4.3   **A comparison of transmission and scanning electron microscopy.**   **(a)** A developing human egg cell, observed by TEM, shortly before it was released from an ovary. **(b)** An egg cell, with a few attached sperm, was coated with heavy metal and observed via SEM. This SEM is colorized.

Biological inquiry: What is the primary advantage of SEM?

The electrons that are scattered from the surface of the sample are detected and create an image on a computer screen. SEM provides a three-dimensional image of the surface of a sample.

And now, let's turn our attention to what microscopes show us about cells.

4.2 Overview of Cell Structure

Cell structure relies on four critical phenomena: (1) matter, (2) energy, (3) organization, and (4) information. In Chapters 2 and 3, we considered the first factor. The matter that is found in living organisms is composed of atoms, molecules, and macromolecules. We will discuss the second factor, energy, throughout this unit, particularly in Chapters 7 and 8. Energy is needed to create molecules and macromolecules, and to carry out many cellular functions. Cells also need energy to become organized, the third phenomenon that underlies cell structure. As discussed throughout this chapter and Unit II, a cell is not a haphazard bag of components. The molecules and macromolecules that constitute cells have specific sites where they are found. For instance, if we compare the structure of a nerve cell in two different humans, or two nerve cells within the same individual, we would see striking similarities in their overall structures. Thus, a key attribute of all living cells is the ability to maintain a particular type of internal structure and organization.

Finally, a fourth critical issue is information. Cell organization requires instructions. These instructions are found in the blueprint of life, namely the genetic material, which is discussed in Unit III. Each living organism has a **genome**, which is defined as the entire complement of its genetic material. Every living cell has a copy of the genome; the **genes** within each species' genome contain the information to create cells with particular structures and functions. This information is passed from cell to cell and from parent to offspring to yield new generations of cells and new generations of life. In this section, we will explore the general structure of cells and examine how the genome contributes to cell structure and function.

Prokaryotic Cells Have a Simple Structure

All forms of life can be placed into two categories based on cell structure—prokaryotes and eukaryotes. We will first consider the **prokaryotes**, which have a relatively simple structure. The term comes from the Greek *pro* and *karyon,* which means "before a kernel"—a reference to the kernel-like appearance of what would later be named the cell nucleus. Prokaryotic cells lack a membrane-enclosed nucleus.

From an evolutionary perspective, the two categories of prokaryotes are **bacteria** and **archaea**. Both types are microorganisms that are relatively small. Bacteria are abundant throughout the world, being found in soil, water, and even our digestive tracts. Most bacterial species are not harmful to humans, but some species are pathogenic—they cause disease. Examples of pathogenic bacteria include *Vibrio cholerae*, the source of cholera, and *Bacillus anthracis*, which causes anthrax. Archaea are less common than bacteria and often occupy extreme environments such as hot springs and deep-sea vents. In this chapter, we will discuss the structure of bacterial cells. We will examine the genetics of bacteria in Chapter 18 and the evolutionary origins of bacteria and archaea in Chapter 22.

Figure 4.4 shows a typical bacterial cell. The **plasma membrane**—a double layer of phospholipids and embedded proteins—forms an important barrier between the cell and its external environment. The **cytoplasm** is the region of the cell that is contained within the plasma membrane. Certain structures in the bacterial cytoplasm are visible via microscopy. These include the **nucleoid**, which is the region of the cell where its genetic material (DNA) is located, and **ribosomes**, which are involved in polypeptide synthesis. We will examine the functions of ribosomes later in this chapter.

Many bacterial structures are located outside the plasma membrane. Nearly all species of prokaryotes have a relatively rigid **cell wall** that supports and protects the plasma membrane and cytoplasm. The cell wall is porous, so it does not prevent most nutrients in the environment from reaching the plasma membrane. Many bacteria also secrete a **glycocalyx**, an outer viscous covering surrounding the bacterium. The glycocalyx traps water and helps protect bacteria from drying out. Certain strains of bacteria that invade animals' bodies produce a very thick, gelatinous glycocalyx called a **capsule** that may help them avoid being destroyed by the animal's immune (defense) system. Finally, many prokaryotes have appendages such as **pili** and **flagella**. Pili allow prokaryotes to attach to surfaces and to each other. Flagella provide a way for prokaryotes to swim.

For many decades, the cytoplasm of bacterial cells was thought to be a fluid-filled space with relatively little defined organization, other than the nucleoid. The cell wall was believed to be solely involved with forming cell shape. During the past decade, however, these notions have been challenged by a molecular analysis of bacterial cells. Researchers have discovered that bacteria possess an architecture inside their cytoplasm that bears striking similarities to the cytoskeleton found in eukaryotic cells (described later in this chapter). **Figure 4.5** illustrates the functions of three proteins in the organization of bacterial cells. FtsZ, which is important in cell division, is found at the site where a cell divides into two cells. MreB plays a role in cell polarity. In many bacteria such as *Caulobacter crescentus*—an aquatic bacterium that survives in nutrient-poor water—cellular components are asymmetrically distributed throughout the cell. As suggested in Figure 4.5, the concentrations of components at one pole are different compared to the other pole. MreB forms a spiral structure in the cell, and evidence suggests it is necessary for this polarity. Finally, the localization of CreS in certain regions of the cell is critical for cell shape. Interestingly, the bacterial proteins described in Figure 4.5 are evolutionarily related to eukaryotic proteins described later in this chapter. In particular, FtsZ, MreB, and CreS are related to tubulin, actin, and intermediate filament proteins, respectively.

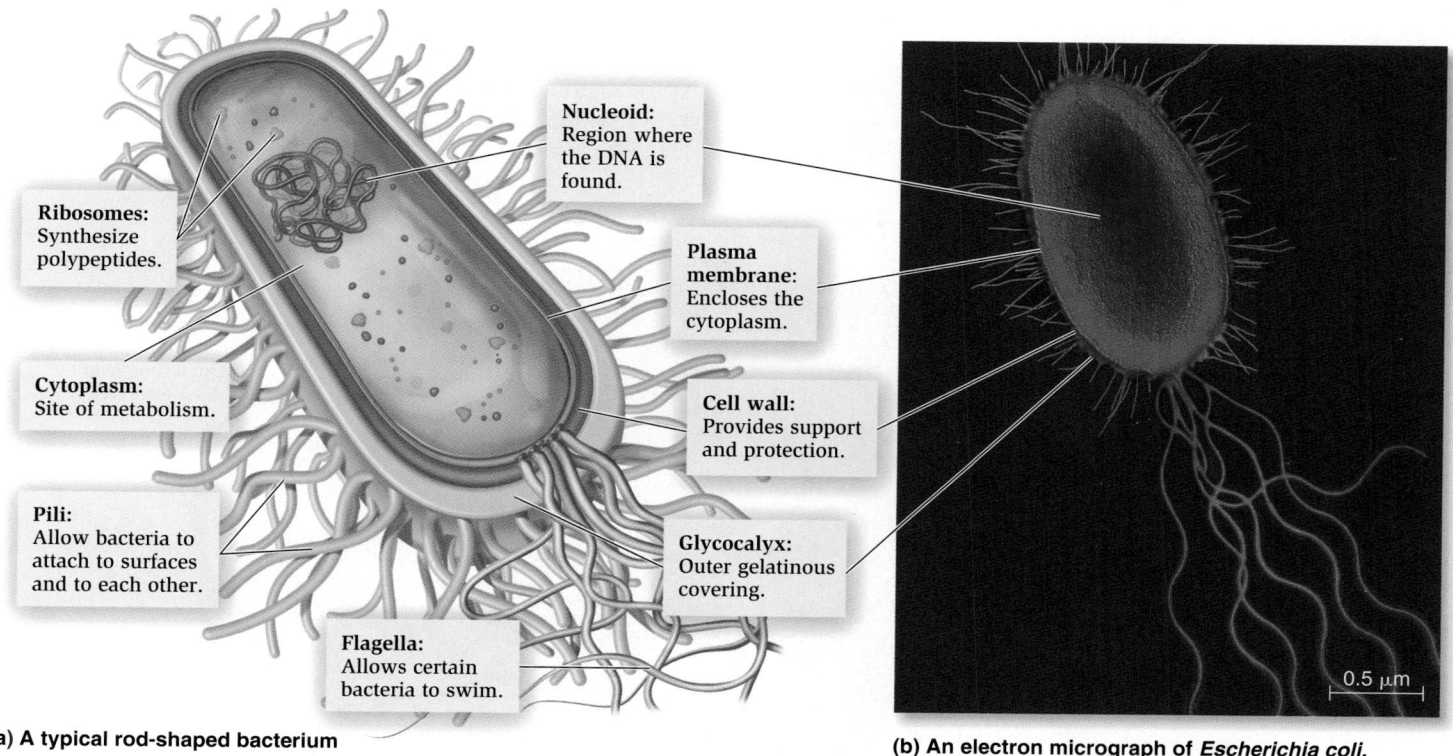

(a) A typical rod-shaped bacterium

Ribosomes: Synthesize polypeptides.

Nucleoid: Region where the DNA is found.

Cytoplasm: Site of metabolism.

Plasma membrane: Encloses the cytoplasm.

Pili: Allow bacteria to attach to surfaces and to each other.

Cell wall: Provides support and protection.

Flagella: Allows certain bacteria to swim.

Glycocalyx: Outer gelatinous covering.

(b) An electron micrograph of *Escherichia coli*.

0.5 μm

Figure 4.4 **Structure of a typical prokaryotic cell.** Prokaryotic cells, which include bacteria and archaea, lack internal compartmentalization.

Protein	Protein function	Protein localization in *C. cresentus*
FtsZ	Division	FtsZ
MreB	Polarity	MreB / Pole / Pole
CreS	Shape	CreS

Figure 4.5 **Proteins that play a role in the architecture of bacterial cells.** FtsZ plays a role in cell division, MreB is involved in cell polarity, and CreS influences cell shape.

Eukaryotic Cells Are Compartmentalized into Organelles

Aside from prokaryotes, all other species are **eukaryotes** (meaning "true nucleus"), which include protists, fungi, plants, and animals. Paramecia and algae are types of protists, while yeasts and molds are types of fungi. The DNA of all eukaryotic cells is housed in a distinct compartment called a **nucleus**, which is an example of an **organelle**—a subcellular structure or membrane-bounded compartment with its own unique structure and function. In contrast to prokaryotes, eukaryotic cells exhibit **compartmentalization**, which means that they have many organelles that separate the cell into different regions. Cellular compartmentalization allows a cell to carry out specialized chemical reactions in different places. For example, protein synthesis and protein breakdown occur in different compartments in the cell.

Figures 4.6 and **4.7** describe the morphology (form and structure) of typical animal and plant cells. Some general features of cell organization, such as a nucleus, are found in nearly all eukaryotic cells. Even so, be aware that the shape, size, and organization of cells vary considerably among different species and even among different cell types of the same species. Micrographs of a human skin cell and a human nerve cell show that, although these cells contain the same types of organelles, their overall morphologies are quite different (see Figure 4.6b).

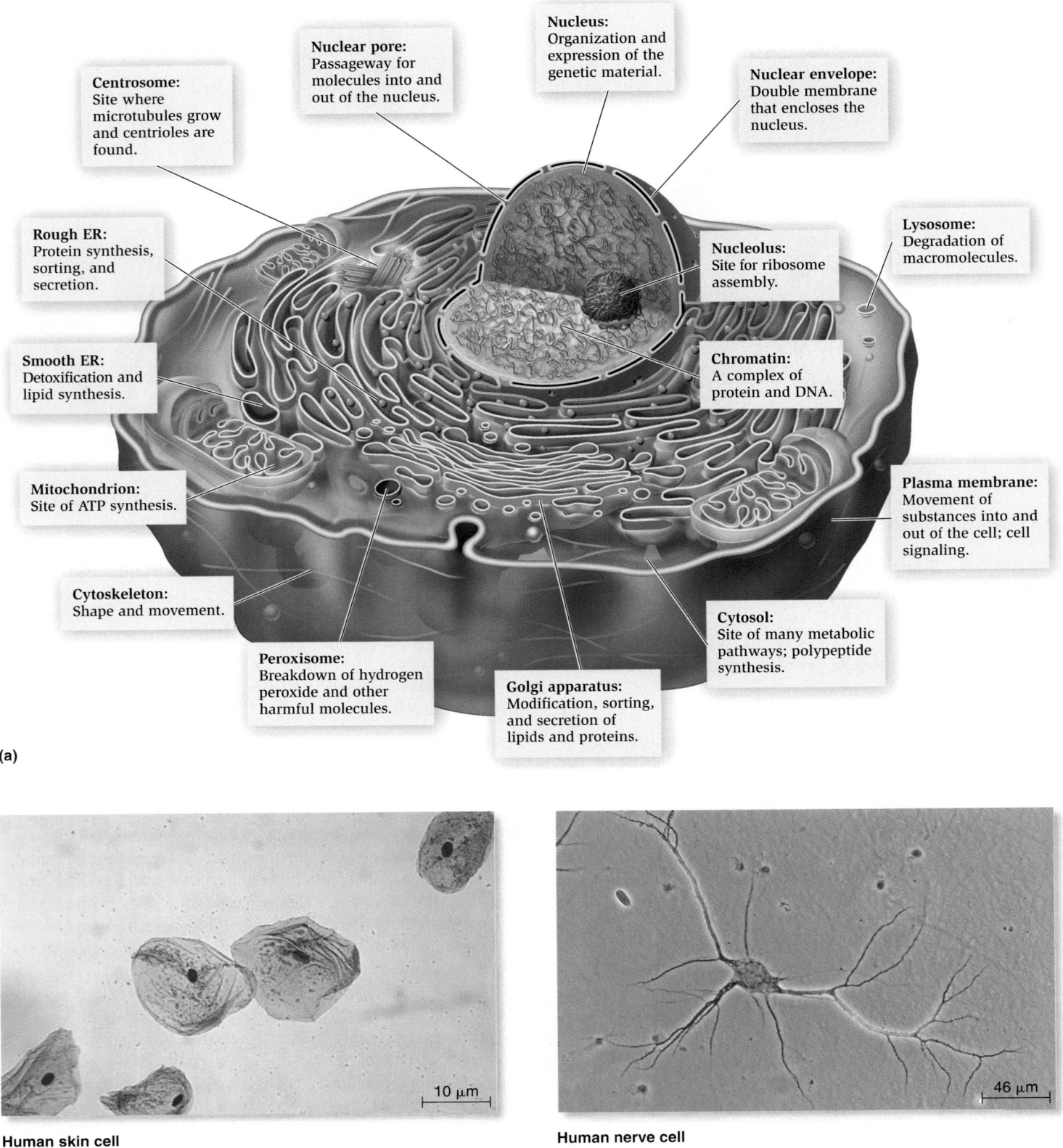

Centrosome:
Site where microtubules grow and centrioles are found.

Nuclear pore:
Passageway for molecules into and out of the nucleus.

Nucleus:
Organization and expression of the genetic material.

Nuclear envelope:
Double membrane that encloses the nucleus.

Rough ER:
Protein synthesis, sorting, and secretion.

Nucleolus:
Site for ribosome assembly.

Lysosome:
Degradation of macromolecules.

Smooth ER:
Detoxification and lipid synthesis.

Chromatin:
A complex of protein and DNA.

Mitochondrion:
Site of ATP synthesis.

Plasma membrane:
Movement of substances into and out of the cell; cell signaling.

Cytoskeleton:
Shape and movement.

Cytosol:
Site of many metabolic pathways; polypeptide synthesis.

Peroxisome:
Breakdown of hydrogen peroxide and other harmful molecules.

Golgi apparatus:
Modification, sorting, and secretion of lipids and proteins.

(a)

10 µm

Human skin cell

46 µm

Human nerve cell

(b)

Figure 4.6   **General structure of an animal cell.**   (a) A schematic drawing of a typical animal cell. (b) Micrographs of a human skin cell (*left*) and a human nerve cell (*right*). Although these cells have the same types of organelles, note that their general morphologies are quite different.

Nucleus:
Organization and
expression of the
genetic material.

Nuclear envelope:
Double membrane
that encloses the
nucleus.

Nuclear pore:
Passageway for
molecules into and
out of the nucleus.

Central vacuole:
Storage and
regulation of cell
volume.

Smooth ER:
Detoxification and
lipid synthesis.

Nucleolus:
Site for ribosome
assembly.

Rough ER:
Protein synthesis,
sorting, and
secretion.

Chromatin:
A complex of
protein and DNA.

Cytosol:
Site of many metabolic
pathways; polypeptide
synthesis.

Plasma membrane:
Movement of
substances into and
out of the cell; cell
signaling.

Mitochondrion:
Site of ATP synthesis.

Cell wall:
Cell support.

Chloroplast:
Site of photosynthesis.

Peroxisome:
Breakdown of hydrogen
peroxide and other
harmful molecules.

Cytoskeleton:
Shape and movement.

Golgi apparatus:
Modification, sorting,
and secretion of lipids
and proteins.

Figure 4.7 **General structure of a plant cell.** Plant cells lack lysosomes and centrioles. Unlike animal cells, plant cells have an outer cell wall; a large central vacuole that functions in storage, digestion, and cell volume; and chloroplasts, which carry out photosynthesis.

GENOMES & PROTEOMES

The Proteome Determines the Characteristics of a Cell

Many organisms, such as plants and animals, are multicellular, meaning that the body of a single organism is composed of many cells. However, the cells of a multicellular organism are not all identical. For example, your body contains nerve cells, muscle cells, skin cells, etcetera. An intriguing question, therefore, is, How does a single organism produce different types of cells?

To answer this question, we need to consider the distinction between genomes and proteomes. Recall that the genome constitutes all types of genetic material that an organism has. Most genes encode the production of polypeptides, which assemble into functional proteins. An emerging theme discussed in this unit is that the structures and functions of proteins are primarily responsible for the structures and functions of cells. A typical eukaryotic cell produces thousands of different types of proteins. For example, researchers estimate that one skeletal muscle cell in your arm produces about 15,000 different types of proteins; the total number of proteins in one cell is far more than 15,000 because many copies of some proteins are made. The **proteome** is defined as all of the types and relative amounts of proteins that are made in a particular cell at a particular time and under specific conditions.

As an example, let's consider muscle cells and nerve cells—two cell types that have dramatically different organization and structure. Actually, the genes in a human muscle cell of a particular individual are identical to those in a human nerve cell. However, their proteomes are different. An important principle in cell biology is that *the proteome of a cell determines its structure and function*. Several phenomena underlie the differences that are observed in the proteomes of different cell types. If we compare muscle and nerve cells, these phenomena include:

1. *Certain proteins that are found in muscle cells are not found in nerve cells, and vice versa.* For a protein to be expressed in a particular cell, the gene that corresponds to that protein must be "turned on," that is, actively expressing that protein. Due to gene regulation, which is described in Chapter 13, certain genes that are turned on in muscle cells are not turned on in nerve cells and vice versa.

2. *The relative amounts of certain proteins are different in muscle and nerve cells.* The amount of a given protein depends on many factors including how strongly the corresponding gene is turned on, how efficiently the protein is synthesized, and how long the protein lasts within a cell. Some proteins are found in both nerve and muscle cells, but in greatly different amounts. For example, a protein called actin, which is discussed later in this chapter, occurs abundantly in muscle cells—where it plays a key role in muscle contraction—but to a much lesser extent in nerve cells.

3. *The amino acid sequences of particular proteins can vary in different cell types.* As discussed in Chapter 13, the mRNA from a single gene can produce two or more polypeptides with slightly different amino acid sequences via a process called alternative splicing. For example, tropomyosin is a protein that regulates cell movement. The form of tropomyosin in muscle cells has a slightly different amino acid sequence from the type in nerve cells.

4. *Nerve and muscle cells may alter their proteins in different ways.* After a protein is made, its structure may be changed in a variety of ways. These include the covalent attachment of molecules such as phosphate and carbohydrate, and the cleavage of a protein to a smaller size.

These four phenomena enable nerve and muscle cells to produce different proteomes, and thus different structures and functions. Likewise, the proteomes of muscle and nerve cells differ from those of other cell types—skin, liver, etcetera.

During the last decade or so, researchers have also discovered an association between proteome changes and disease. For example, the proteomes of healthy cells are different from the proteomes of cancer cells. Furthermore, the proteomes of cancer cells change as the disease progresses. One reason for studying cancer-cell proteomes is to improve the early detection of cancer by identifying proteins that are made in the early stages, when the disease is most treatable. In addition, information about the ways that the proteomes of cancer cells change may help researchers uncover new treatment options. A key challenge for biologists is to understand the synthesis and function of proteomes in different cell types, and how proteome changes may lead to disease conditions.

4.3 The Cytosol

Thus far, we have focused on the general features of prokaryotic and eukaryotic cells. In the rest of this chapter we will survey the various compartments of eukaryotic cells with a greater emphasis on structure and function. **Figure 4.8** highlights a plant and animal cell according to four different regions. We will start with the **cytosol** (shown in yellow), the region of a eukaryotic cell that is outside the cell organelles but inside the plasma membrane. The other regions of the cell, which we will examine later in this chapter, include the interior of the nucleus (blue), the endomembrane system (purple/pink), and semiautonomous organelles (green). As in prokaryotes, the term cytoplasm refers to the region inside the plasma membrane. This includes the cytosol, the endomembrane system inside the plasma membrane, and the semiautonomous organelles.

(a) Animal cell

(b) Plant cell

Figure 4.8 Compartments within (a) animal and (b) plant cells. The cytosol, which is outside the organelles but inside the plasma membrane, is shown in yellow. The membranes of the endomembrane system are shown in purple, while the fluid filled interiors are pink. The interior of the nucleus is blue. Semiautonomous organelles, which include mitochondria, chloroplasts, and peroxisomes, are green.

Though the amount varies among different types of cells, the cytosol typically occupies about 50% of the total cell volume. In this section, we will consider the primary functions of the cytosol. First, it is the site of many chemical reactions that produce the materials and energy necessary for life, such as breaking down food molecules into smaller components. Such reactions release energy, and the components can be used as building blocks to create new cellular molecules and macromolecules. For example, we will explore a particularly important activity of the cytosol, which is the synthesis of cellular proteins. In addition, we will examine the structure and function of large protein filaments, found in the cytosol, that provide organization to the cell and allow cells to move.

Synthesis and Breakdown of Molecules Occur in the Cytosol

Metabolism is defined as the sum of the chemical reactions by which cells produce the materials and energy that are necessary to sustain life. Although specific steps of metabolism also occur in cell organelles, the cytosol is a central coordinating region for many metabolic activities of eukaryotic cells. Metabolism often involves a series of steps called a metabolic pathway. A specific **enzyme** is responsible for speeding up each step in a metabolic pathway. In Chapter 7, we will examine the functional properties of enzymes and consider a few metabolic pathways that occur in the cytosol and cellular organelles.

Some pathways involve the breakdown of a molecule into smaller components, a process termed **catabolism**. Such pathways are needed to capture energy for use by the cell, and also to generate molecules that provide the building blocks to construct cellular macromolecules. Conversely, other metabolic pathways

are involved in **anabolism**, the synthesis of cellular molecules and macromolecules. For example, polysaccharides are made by linking together sugar molecules. To create proteins, amino acids are covalently connected to form a polypeptide. This process is described next.

Translation Is the Process of Polypeptide Synthesis

A **polypeptide** is composed of a linear sequence of amino acids. The term polypeptide describes a unit of structure. By comparison, the term **protein** is a unit of function. One or more polypeptides assemble into a three-dimensional protein that performs a particular function. A critical activity of all cells is the synthesis of polypeptides. This process is called **translation** because the information within a gene is ultimately translated into the sequence of amino acids in a polypeptide. Chapter 12 describes the details of the process. In this chapter, we will briefly consider the general features of translation to help us appreciate how cells make proteins, which are vital to cell structure and function.

Translation requires many cellular components including a ribosome and two types of RNA molecules (**Figure 4.9**). The ribosome is the site where polypeptide synthesis occurs. Messenger RNA (mRNA) is produced from a gene and provides the information to make a polypeptide. Transfer RNA (tRNA) molecules, which carry amino acids, bind to the mRNA so that a polypeptide can be made, one amino acid at a time. The role of the ribosome is to facilitate the binding between the mRNA and tRNA molecules, and to catalyze the formation of covalent bonds between adjacent amino acids. Once the entire polypeptide is made, it is released from the ribosome.

Polypeptide—a chain of amino acids

tRNAs carrying amino acids

tRNA

5′

Ribosome

Direction of ribosome

mRNA

3′

A ribosome moves relative to an mRNA molecule, allowing tRNAs with specific amino acids to bind. This results in the synthesis of a polypeptide with a specific amino acid sequence.

Figure 4.9 The process of translation.

The Cytoskeleton Provides Cell Organization, Shape, and Movement

The **cytoskeleton** is a network of three different types of protein filaments: **microtubules, intermediate filaments**, and **actin filaments** (Table 4.1). Let's first consider the structure of these protein filaments and their roles in the construction and organization of cells. Later, we will examine how they are involved in cell movement.

Microtubules Microtubules are long, hollow, cylindrical structures about 25 nm in diameter composed of the protein tubulin. The assembly of tubulin to form a microtubule results in a polar structure with a plus end and a minus end (Table 4.1). Growth of microtubules occurs at the plus end, while shortening of microtubules can occur at either the plus or minus end. A single microtubule can oscillate between growing and shortening phases, a phenomenon termed **dynamic instability**. Dynamic instability is important in many cellular activities including the sorting of chromosomes during cell division.

The sites where microtubules form within a cell can vary among different types of organisms. Animal cells that are not preparing to divide contain a single structure near their nucleus called the **centrosome** or **microtubule-organizing center** (Table 4.1). Within the centrosome are the **centrioles**, a conspicuous pair of structures arranged perpendicular to each other. In animal cells, microtubule growth starts at the centrosome such that the minus end is anchored there. In contrast, most plant cells and many protists lack centrosomes and centrioles. Microtubules are created at many sites that are scattered throughout a plant cell.

Microtubules are important for cell shape and organization. Organelles such as the Golgi apparatus often are attached to microtubules. In addition, microtubules are involved in the organization of chromosomes during mitosis and in the orientation of cells during cell division. We will examine these events in Chapter 15.

Intermediate Filaments Intermediate filaments are another class of cytoskeletal filament found in the cells of many but not all animal species. Their name is derived from the observation

Table 4.1	Properties of Cytoskeletal Filaments		
Characteristic	**Microtubules**	**Intermediate filaments**	**Actin filaments**
Diameter	25 nm	10 nm	7 nm
Structure	Hollow tubule	Twisted filament	Spiral filament
	Plus end / Tubulin protein / Lumen / 25 nm / Minus end (may be anchored in the centrosome of animal cells)	10 nm / Staggered alignment of intermediate filament proteins	Plus end / Actin protein / Minus end / 7 nm
Protein composition	Tubule wall composed of the protein tubulin	Can be composed of different proteins including desmin, keratin, lamin, and others	Two intertwined strands composed of the protein actin
Cellular location	Centrosome		
Common functions	Cell shape / Organization of cell organelles / Chromosome sorting / Intracellular movement of cargo / Cell motility (cilia and flagella)	Cell shape / Provide cells with mechanical strength / Anchorage of cell and nuclear membranes	Cell shape / Muscle contraction / Cell movement (amoeboid movement) / Animal cell division / Intracellular movement of cargo

that they are intermediate in diameter between actin filaments and myosin filaments. (Myosin filaments are described in Chapter 46.) Intermediate filament proteins bind to each other in a staggered array to form a twisted, ropelike structure with a diameter of approximately 10 nm (Table 4.1). Intermediate filaments tend to be more stable than microtubules and actin filaments, which readily polymerize and depolymerize. They function as tension-bearing fibers that help maintain cell shape and rigidity.

Several types of related proteins can assemble into intermediate filaments. Desmins form intermediate filaments in muscle cells and provide mechanical strength. Keratins form intermediate filaments in skin, intestinal, and kidney cells, where they are important for mechanical strength and cell shape. They are also a major constituent of hair and nails. In addition, intermediate filaments are found inside the cell nucleus. As discussed later in this chapter, nuclear lamins form a network of intermediate filaments that line the inner nuclear membrane and provide anchorage points for the nuclear pores.

Actin Filaments Actin filaments—also known as **microfilaments** because they are the thinnest cytoskeletal filaments—are long, thin fibers approximately 7 nm in diameter (Table 4.1).

Each fiber is composed of two strands of actin monomers that spiral around each other. Like microtubules, actin filaments have plus and minus ends, and they are very dynamic structures that grow at the plus end.

Despite their thinness, actin filaments play a key role in cell strength and shape. Although actin filaments are dispersed throughout the cytosol, they tend to be highly concentrated near the plasma membrane. In many types of cells, actin filaments support the plasma membrane and provide strength and shape to the cell. The minus ends of actin filaments are usually anchored at the plasma membrane, which explains why actin filaments are typically found there.

Motor Proteins Interact with Microtubules or Actin Filaments to Promote Cellular Movements

Motor proteins are a category of cellular proteins that use ATP as a source of energy to promote movement. As shown in **Figure 4.10a**, a motor protein consists of three domains called the head, hinge, and tail. The head is the site where ATP binds and is hydrolyzed to ADP and P_i. ATP binding and hydrolysis cause a bend in the hinge, which results in movement. The tail region is attached to other proteins, or to other kinds of cellular molecules.

(a) Three-domain structure of a motor protein

(b) Movement of a motor protein along a cytoskeletal filament

(c) Three types of movements facilitated by motor proteins and cytoskeletal filaments

Figure 4.10 **Motor proteins and their interactions with cytoskeletal filaments.** (a) Three-domain structure of a motor protein. Note: The protein subunits of motor proteins often associate with each other along their tails, such that the motor has two tails, two hinges, and two heads. (b) Conformational changes in a motor protein that allow it to walk along a cytoskeletal filament. (c) Three ways that motor proteins and cytoskeletal filaments can cause movement.

To promote cellular movement, the head region of a motor protein interacts with a cytoskeletal filament (**Figure 4.10b**). When ATP binds and is hydrolyzed, the motor protein attempts to "walk" along the filament. The head of the motor protein is initially attached to a filament. To move forward, the head detaches from the filament, cocks forward, binds to the filament, and cocks backward. To imagine how this works, consider the act of walking in which the ground is a cytoskeletal filament, your leg is the head of the motor protein, and your hip is the hinge. To walk, you lift your leg up, you move it forward, you place it on the ground, and then you cock it backward (which propels you forward). This series of events is analogous to how a motor protein walks along a cytoskeletal filament.

Interestingly, cells have utilized the actions of motor proteins to promote three different kinds of movements. In the first example shown in **Figure 4.10c** (top), the tail region is attached to a cargo, so that the motor protein moves the cargo from one location to another. Alternatively, a motor protein can remain in place and cause the filament to move (Figure 4.10c, middle). As discussed in Chapter 46, this occurs during muscle contraction. A third possibility is that both the motor protein and filament are restricted in their movement. In this case, when the motor protein attempts to walk, it exerts a force that causes the filament to bend (Figure 4.10c, bottom). As described next, this occurs during the bending of cilia and flagella.

Let's now consider some specific examples where motor proteins and cytoskeletal filaments interact and result in movement. In certain kinds of cells, microtubules and motor proteins facilitate movement involving cell appendages called **cilia** and **flagella** (singular, *cilium* and *flagellum*). Flagella are usually longer than cilia and present singly or in pairs. A single flagellum may propel a cell such as a sperm cell with a whiplike motion (**Figure 4.11a**). Alternatively, a pair of flagella may move in a synchronized manner to pull a microorganism through the water (think of a human swimmer doing the breaststroke). Certain unicellular algae swim in this manner (**Figure 4.11b**). By comparison, cilia are often shorter than flagella and tend to cover all or part of the surface of a cell. Protists such as paramecia may have hundreds of adjacent cilia that beat in a coordinated fashion to propel the organism through water (**Figure 4.11c**).

Despite their differences in length, cilia and flagella share the same internal structure (**Figure 4.12**). This arrangement containing microtubules, the motor protein dynein, and linking proteins is called an **axoneme**. In the cilia and flagella of most eukaryotic organisms, the microtubules form an arrangement called a 9 + 2 array. Each of the two central microtubules consists of a single microtubule, while the outer nine are doublet microtubules, which are two merged microtubules. The microtubules in cilia and flagella emanate from **basal bodies**, which are anchored to the cytoplasmic side of the plasma membrane. Much like the centrosome of animal cells, the basal bodies provide a site for microtubules to grow. In addition to microtubules, the core structure of a cilium and flagellum also has motor proteins, namely dynein, and linking proteins such as nexin, that hold the axoneme together.

The movement of both cilia and flagella involves the propagation of a bend, which begins at the base of the structure and proceeds toward the tip (see Figure 4.11a). The bending occurs because dynein is activated to walk toward the basal body of the microtubules. ATP hydrolysis is required for this process.

10 μm

(a) Sperm with a long flagellum

5.6 μm

(b) *Chlamydomonas* with 2 flagella

168 μm

(c) Paramecium with many cilia

Figure 4.11   Cellular movements due to the actions of flagella and cilia.   Both flagella and cilia cause movement by a bending motion. In flagella, movement occurs by a whiplike motion that is due to the propagation of a bend from the base to the tip. In addition, the nature of swimming depends on the length of the appendage and whether it involves coordination among multiple flagella or cilia. **(a)** Sperm swim by means of a single, long flagellum that moves in a whiplike motion. **(b)** The swimming of *Chlamydomonas reinhardtii* also involves a whiplike motion at the base, but the motion is precisely coordinated between two flagella. This results in swimming behavior that resembles a breaststroke. **(c)** Ciliated protozoa swim via many shorter cilia. The bending motion is coordinated among multiple adjacent cilia.

Biological inquiry: During the movement of a cilium or flagellum, describe the type of movements that are occurring in dynein and microtubules.

(a) Human sperm cell

30 μm

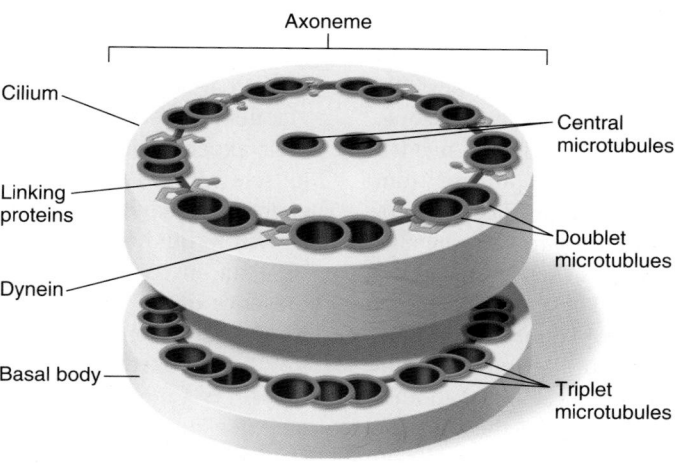

Axoneme

Cilium

Central microtubules

Linking proteins

Dynein

Basal body

Doublet microtublues

Triplet microtubules

(b) Molecular structure of a flagellum

Figure 4.12   Structure of a eukaryotic flagellum.   (a) SEM of a human sperm cell. **(b)** Drawing of the axoneme. The core structure consists of a 9 + 2 arrangement of nine outer doublet microtubules and two central microtubules. This structure is anchored to the basal body, which has nine triplet microtubules, in which three microtubules are fused together. Note: The structure of the basal body is very similar to centrioles in animal cells.

However, the microtubules and dynein are not free to move relative to each other because of linking proteins. Therefore, instead of dyneins freely walking along the microtubules, they exert a force that bends the microtubules (see Figure 4.10c, bottom). The dyneins at the base of the structure are activated first, followed by dyneins that are progressively closer to the tip of the appendage. The resulting movement propels the organism.

As a second example, let's consider how motor proteins interact with actin. As discussed in Chapter 46, actin and a motor protein called myosin are responsible for the movement observed in muscle cells. In protists such as the amoeba, movement occurs via the dynamic rearrangement of the actin cytoskeleton (**Figure 4.13**). Actin filaments are formed near the leading edge to create a projection called a lamellipodium. The cell is pulled toward the leading edge using motor proteins such as myosin, which tugs on actin filaments and promotes cellular movement. Actin filaments and myosin motors are also involved in other types of movement, as we will see next in our Feature Investigation.

Short actin filaments

Lamellipodium

Actin polymerization

1 The formation of long actin filaments at the leading edge extends the lamellipodium.

2 Rearrangement of the actin cytoskeleton in other regions of the cell causes the cell to be pulled toward the leading edge.

Figure 4.13   Amoeboid movement promoted by changes in the locations of actin filaments.

FEATURE INVESTIGATION

Sheetz and Spudich Showed That Myosin Walks Along Actin Filaments

Cell biologists often gain clues to protein function from studying living cells. In the early 1950s, Hugh Huxley, Andrew F. Huxley, and their colleagues studied muscle contraction via microscopy and other techniques. They proposed the sliding-filament model for muscle contraction, in which actin filaments slide past thicker filaments composed of myosin. This model was based on observations involving intact muscle cells. Research at that time and during subsequent decades indicated that actin and myosin were necessary for muscle movement.

Biologists want to understand how a process, such as muscle contraction, works at the molecular level. However, studying living cells has its drawbacks. Recall that the proteome of every living cell is composed of thousands of different proteins, making it difficult to establish the function of any one protein. As an alternative to studying a process in living cells, another approach is to isolate and purify cellular components and study their functions outside the cell. This is an ***in vitro*** approach (literally "in glass," such as in a glass test tube). By comparison, studying living cells is termed ***in vivo***, meaning "in life."

In 1983, approximately 30 years after the sliding-filament model was proposed, Michael Sheetz and James Spudich devised a clever approach to study myosin function *in vitro*.

Prior to their work, researchers had learned how to purify myosin protein from muscle cells. A fragment of myosin could be attached to a fluorescently labeled bead, making it possible to follow myosin movement using a fluorescence microscope. However, to study movement, Sheetz and Spudich also needed actin filaments. Although actin filaments can be purified from cells, they become a tangled mess during the purification process. As an alternative, the researchers were aware that the alga *Nitella axillaris* has arrays of actin filaments that lie inside the cell and parallel to the plasma membrane. These parallel arrays function in the phenomenon known as **cytoplasmic streaming**, in which the cytoplasm circulates throughout the cell to distribute resources efficiently in large cells. Along these lines, another advantage of *Nitella* for this experiment was that the cells are relatively large—several hundred microns in diameter and several centimeters in length.

Figure 4.14 illustrates the *in vitro* approach of Sheetz and Spudich. Their procedure involved cutting open *Nitella* cells and pinning down the plasma membrane to expose the actin filaments. Except for chloroplasts, which are found between the plasma membrane and the actin filaments, the rest of the cellular contents were washed away. Next, a solution containing purified myosin attached to fluorescent beads was added and observed via fluorescence microscopy. Sheetz and Spudich conducted their experiments with and without ATP, which is needed for muscle cell movement. In addition, the researchers tested the effects of N-ethylmaleimide (NEM), a chemical that was already known to bind to myosin and inhibit its function. As shown in the data, myosin was able to move along actin filaments in the presence of ATP. Furthermore, when myosin was treated with NEM, the movement was inhibited.

Taken together, these *in vitro* experiments confirmed that myosin is a motor protein that uses ATP to walk along actin filaments. Furthermore, this purified system provided evidence that all that is needed for movement are actin, myosin, and ATP. In addition, the results are consistent with the idea that different types of movement, such as cytoplasmic streaming in algae and muscle contraction in animals, use the same underlying molecular mechanism—in this case, an interaction between the motor protein myosin and actin filaments.

Figure 4.14   Movement of myosin-coated beads along actin filaments.

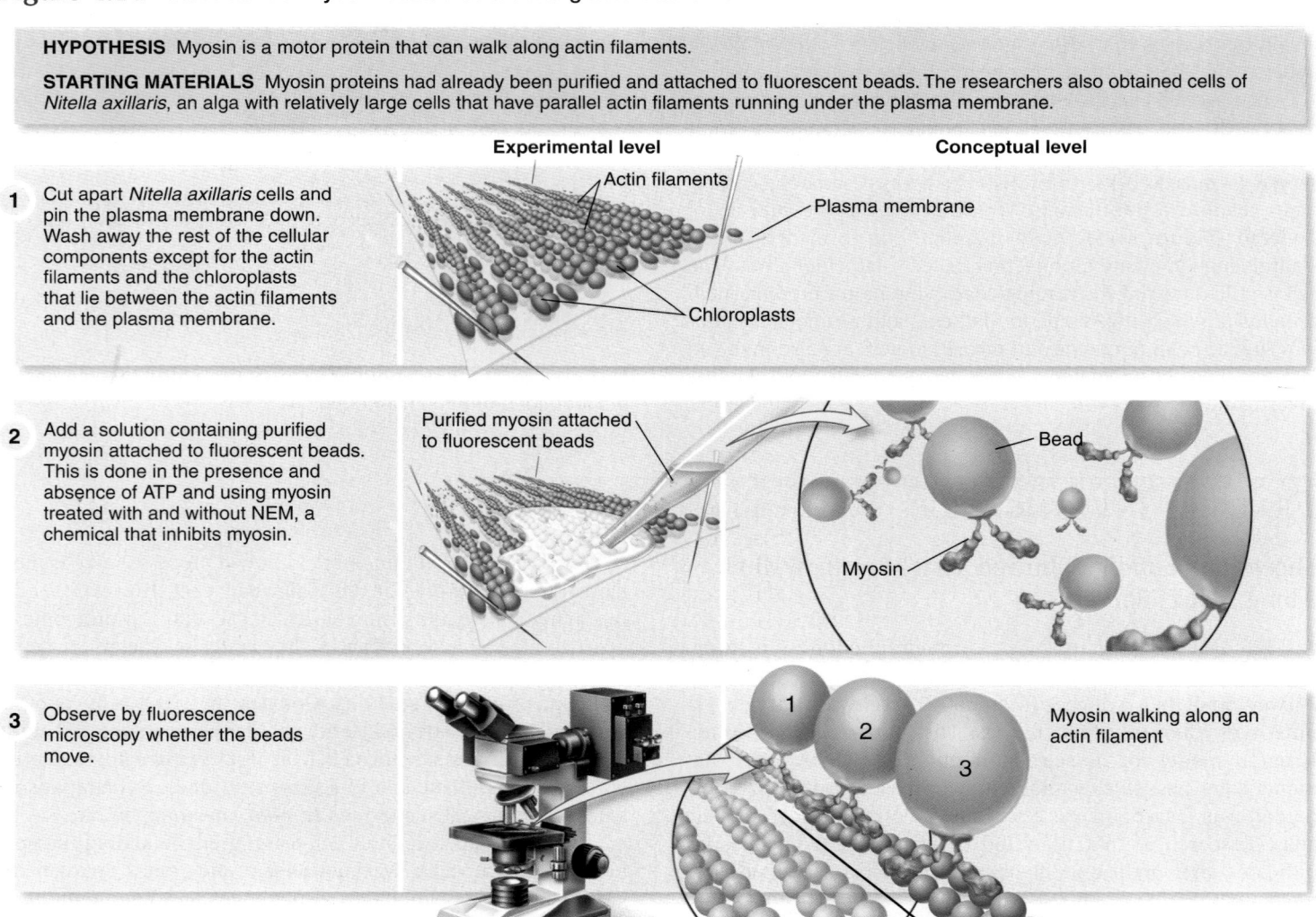

HYPOTHESIS Myosin is a motor protein that can walk along actin filaments.

STARTING MATERIALS Myosin proteins had already been purified and attached to fluorescent beads. The researchers also obtained cells of *Nitella axillaris*, an alga with relatively large cells that have parallel actin filaments running under the plasma membrane.

Experimental level   **Conceptual level**

1 Cut apart *Nitella axillaris* cells and pin the plasma membrane down. Wash away the rest of the cellular components except for the actin filaments and the chloroplasts that lie between the actin filaments and the plasma membrane.

Actin filaments
Plasma membrane
Chloroplasts

2 Add a solution containing purified myosin attached to fluorescent beads. This is done in the presence and absence of ATP and using myosin treated with and without NEM, a chemical that inhibits myosin.

Purified myosin attached to fluorescent beads
Bead
Myosin

3 Observe by fluorescence microscopy whether the beads move.

Myosin walking along an actin filament

4 THE DATA

Results from step 3:

Conditions	Movement observed?
Plus ATP, plus NEM	No
No ATP, no NEM	No
Plus ATP, no NEM	Yes

This is a time-lapse micrograph in which photos were taken every few seconds. It shows a single bead (yellow) moving from left to right. The red objects are stained chloroplasts, which are below the actin filaments. The actin filaments are not visible in this micrograph because they are not fluorescently labeled.

4.4 The Nucleus and Endomembrane System

In Chapter 2, we learned that the nucleus of an atom contains protons and neutrons. In cell biology, the term **nucleus** has a different meaning. It is an organelle found in eukaryotic cells that contains most of the cell's genetic material. A small amount of genetic material is also found in mitochondria and chloroplasts.

The membranes that enclose the nucleus are part of a larger network of membranes called the **endomembrane system**. These include not only the nuclear envelope, which encloses the nucleus, but also the endoplasmic reticulum, Golgi apparatus, lysosomes, and vacuoles. The prefix *endo*—meaning "inside"—originally referred only to these organelles and internal membranes. However, we now know that the plasma membrane is also part of this integrated membrane system (**Figure 4.15**). Some of these membranes, such as the nuclear envelope and the membrane of the endoplasmic reticulum, have direct connections to one another. Other organelles of the endomembrane system pass materials to each other via **vesicles**—small membrane-enclosed spheres. The movement of vesicles occurs in both directions. For example, vesicles that are formed from the endoplasmic reticulum can fuse with the Golgi apparatus and vesicles from the Golgi apparatus can fuse with the endoplasmic reticulum. Therefore, the endomembrane system forms a dynamic, integrated network of membranes that requires constant sorting to maintain the functional properties of each organelle.

In this section, we will survey the structures and functions of the organelles of the endomembrane system. We will also examine the plasma membrane, which is formed from the endomembrane system.

The Eukaryotic Nucleus Contains Chromosomes

The nucleus is the internal compartment that is enclosed by a double-membrane structure termed the **nuclear envelope** (**Figure 4.16**). In most cells, the nucleus is a relatively large organelle

Figure 4.15 The nucleus and endomembrane system. This figure highlights the internal compartment of the nucleus (blue), the membranes of the endomembrane system (purple), and the fluid-filled interiors of the endomembrane system (pink). The nuclear envelope is considered part of the endomembrane system, but the interior of the nucleus is not.

that typically occupies 10–20% of the total cell volume. The outer membrane of the nuclear envelope is continuous with the endoplasmic reticulum membrane. **Nuclear pores** are formed where the inner and outer nuclear membranes make contact with each other. The pores provide a passageway for the movement of molecules and macromolecules into and out of the nucleus. Although cell biologists view the nuclear envelope as part of the endomembrane system, the materials within the nucleus are not (see Figure 4.15).

Pore   Nucleus

Chromatin

Nucleolus

Nucleolus

Chromatin

Nuclear lamina

Nuclear envelope

Pore in nuclear envelope

5.4 μm

Pore complexes

0.4 μm

Two membranes of nuclear envelope

Chromatin in nucleus

Internal nuclear matrix

Inner membrane

Outer membrane

Nuclear envelope

Nucleus

Nuclear pore complex

Nuclear lamin proteins

Cytosol

Figure 4.16   The nucleus and nuclear envelope.   The nuclear envelope is composed of an inner and outer membrane that meet at the nuclear pores. The inner nuclear membrane is lined with lamin proteins to form the nuclear lamina. The interior of the nucleus contains chromatin, which is attached to the nuclear matrix, and a nucleolus, where ribosomes are assembled.

Inside the nucleus are the **chromosomes** and a filamentous network of proteins called the **nuclear matrix**. Each chromosome is composed of genetic material, namely DNA, and many types of proteins that help to compact the chromosome to fit inside the nucleus. The complex formed by DNA and such proteins is termed **chromatin**. The nuclear matrix consists of two parts: the nuclear lamina, which is composed of intermediate filaments that line the inner nuclear membrane, and an internal nuclear matrix, which is connected to the lamina and fills the interior of the nucleus. The nuclear matrix serves to organize the chromosomes within the nucleus. Each chromosome is located in a distinct, nonoverlapping **chromosome territory,** which is visible when cells are exposed to dyes that label each type of chromosome (**Figure 4.17**).

The primary function of the nucleus involves the protection, organization, and expression of the genetic material. Another important function is the assembly of ribosomes. Ribosome assembly occurs in the **nucleolus** (plural, *nucleoli*), a prominent region in the nucleus of nondividing cells. A ribosome is composed of two subunits, one small and one large; each subunit contains one or more RNA molecules and several types of proteins. The RNA molecules that are components of ribosomes are made at the nucleolus because the genes that encode these RNA molecules are located there. The ribosomal

proteins are produced in the cytosol, imported into the nucleus through the nuclear pores, and assembled with the RNA molecules to form the ribosomal subunits. The subunits then exit through the nuclear pores into the cytosol, where they are needed for protein synthesis.

3 μm

Figure 4.17   Chromosome territories in the cell nucleus. Chromosomes from a chicken were labeled with chromosome-specific probes. Each of the seven types of chicken chromosomes is colored with a different dye. Each chromosome occupies its own distinct, nonoverlapping territory within the cell nucleus.

The Endoplasmic Reticulum Initiates Protein Sorting and Carries Out Certain Metabolic Functions

The **endoplasmic reticulum (ER)** is a convoluted network of membranes that form flattened, fluid-filled tubules or **cisternae** (**Figure 4.18**). The terms endoplasmic (which means "in the cytoplasm") and reticulum ("little net") refer to the location and shape of this organelle when viewed under a microscope. The term **lumen** describes the internal space of an organelle. The ER membrane encloses a single compartment called the **ER lumen**. In some cells, the ER membrane makes up more than half of the total membrane in the cell. The rough ER has its outer surface studded with ribosomes, giving it a bumpy appearance. Once bound to the ER membrane, the ribosomes actively synthesize proteins through the ER membrane. The smooth ER lacks ribosomes.

Rough ER The **rough endoplasmic reticulum (rough ER)** plays a key role in the initial synthesis and sorting of proteins that are destined for the ER, Golgi apparatus, lysosomes, vacuoles, plasma membrane, or outside of the cell. To reach any of these locations, a protein must first be directed to the ER membrane. In conjunction with protein sorting, a second function of the rough ER is the insertion of certain newly made proteins into the membrane. A third important function of the rough ER is the attachment of carbohydrate to proteins and lipids. This process is called **glycosylation**. The topics of protein sorting, membrane protein insertion, and protein glycosylation will be discussed in Chapter 6, because they are important in the maintenance of cell organization.

Smooth ER The **smooth endoplasmic reticulum (smooth ER)**, which is continuous with the rough ER, functions in diverse metabolic processes. The extensive network of smooth ER membranes allows increased surface area for key enzymes that play important metabolic roles. In liver cells, enzymes in the smooth ER detoxify many potentially harmful organic molecules including barbiturate drugs and ethanol. These enzymes convert hydrophobic toxic molecules into more hydrophilic molecules, which are easily excreted from the body. Chronic alcohol consumption, as in alcoholics, leads to a greater amount of smooth ER in liver cells, which increases the rate of alcohol breakdown. This explains why people who consume alcohol regularly must ingest more alcohol to experience its effects.

The smooth ER of liver cells also plays a role in carbohydrate metabolism. The liver cells of animals store energy in the form of glycogen, which is a polymer of glucose. Glycogen granules, which are in the cytosol, sit very close to the smooth ER membrane. When chemical energy is needed, enzymes are activated that break down the glycogen to glucose-6-phosphate. Then, an enzyme in the smooth ER called glucose-6-phosphatase removes the phosphate group, and glucose is released into the bloodstream.

Another important function of the smooth ER in all eukaryotes is the accumulation of calcium ions. The smooth ER contains calcium pumps that transport Ca^{2+} into the ER lumen. The regulated release of Ca^{2+} into the cytosol is involved in many vital cellular processes, including muscle contraction in animals.

Finally, enzymes in the smooth ER are critical in the synthesis and modification of lipids. For example, steroid hormones such as estrogen and testosterone are derived from the lipid

Figure 4.18   Structure of the endoplasmic reticulum.   The ER is composed of a network of flattened tubules called cisternae that enclose a continuous ER lumen. The rough ER is studded with ribosomes, while the smooth ER lacks ribosomes. The rough ER is continuous with the outer nuclear membrane.

cholesterol. Enzymes in the smooth ER are necessary for certain modifications that are needed to produce these hormones. In addition, the smooth ER is the primary site for the synthesis of phospholipids, which are the main lipid component of eukaryotic cell membranes.

The Golgi Apparatus Directs the Secretion, Processing, and Sorting of Cellular Molecules

The **Golgi apparatus** (also called the Golgi body, Golgi complex, or simply Golgi) was discovered by the Italian microscopist Camillo Golgi in 1898. It consists of a stack of flattened, membranes; each flattened membrane encloses a single compartment (**Figure 4.19**). The Golgi stacks are named according to their orientation in the cell. The *cis* Golgi is close to the ER membrane, the *trans* Golgi is near the plasma membrane, and the *medial* Golgi is found in the middle. Materials are transported between the Golgi stacks via membrane vesicles that bud from one compartment in the Golgi (for example, the *cis* Golgi) and fuse with another compartment (for example, the *medial* Golgi).

The Golgi apparatus performs three overlapping functions: (1) secretion, (2) processing, and (3) protein sorting. The Golgi apparatus packages different types of materials into **secretory vesicles** that later fuse with the plasma membrane, thereby releasing their contents outside the plasma membrane (Figure 4.19). Proteins that are destined for secretion are synthesized into the ER, travel to the Golgi, and then to the plasma membrane for secretion. This route, called the secretory pathway, is described in Chapter 6.

Enzymes in the Golgi apparatus process, or modify, certain proteins and lipids. As mentioned earlier, carbohydrates can be attached to proteins and lipids in the endoplasmic reticulum. Glycosylation continues in the Golgi. For this to occur, a protein or lipid is transported via vesicles from the ER to the *cis* Golgi. Most of the glycosylation occurs in the *medial* Golgi. A second type of processing event is **proteolysis**—enzymes called **proteases** cut proteins into smaller polypeptides. For example, the hormone insulin is first made as a large precursor protein termed proinsulin. In the Golgi apparatus, proinsulin is packaged with proteases into vesicles. Prior to secretion, the proteases cut out a portion of the proinsulin to create a smaller insulin molecule that is a functional hormone.

The third function of the Golgi is protein sorting. After a protein enters the Golgi from the ER, it will be directed to one of six locations. Either it will stay in the Golgi, or it will be transported via vesicles to the ER, a lysosome (in animal cells), a vacuole (in plant cells), the plasma membrane, or the exterior of the cell.

Lysosomes Are Involved in Degrading Macromolecules

Lysosomes are small organelles found in animal cells that are able to lyse macromolecules, hence their name. Lysosomes contain many **acid hydrolases**, which are hydrolytic enzymes that use a molecule of water to break a covalent bond. This type of chemical reaction is called hydrolysis. The hydrolases found in a lysosome function optimally at an acidic pH. The fluid-filled

Figure 4.19 **The Golgi apparatus and secretory pathway.** The Golgi is composed of stacks of membranes that enclose separate compartments. Transport to and from the Golgi compartments occurs via membrane vesicles. Vesicles can bud from the ER and go to the Golgi, and vesicles from the Golgi can fuse with the plasma membrane to release cargo to the outside. The pathway from the ER to the Golgi to the plasma membrane is termed the secretory pathway.

Biological inquiry: If we consider the Golgi apparatus as three compartments (cis, medial, and trans), describe the compartments that a protein will travel through to be secreted.

interior of a lysosome has a pH of approximately 4.8. If a lysosomal membrane breaks, releasing acid hydrolases into the cytosol, the enzymes are not very active because the cytosolic pH is neutral (approximately pH 7.0). This prevents significant damage to the cell from accidental leakage.

Lysosomes contain many different types of acid hydrolases that can break down proteins, carbohydrates, nucleic acids, and lipids. This enzymatic function enables lysosomes to break down complex materials. One function of lysosomes involves the digestion of substances that are taken up from outside the cell. This process, called endocytosis, is described in Chapter 5. In addition, lysosomes help digest intracellular materials. In a process known as **autophagy** (meaning the eating of one's self), cellular material, such as a worn-out organelle, becomes enclosed in a double membrane (**Figure 4.20**). This **autophagosome** then fuses with a lysosome, and the material inside the autophagosome is digested. The small molecules that are released from this digestion are recycled back into the cytosol.

Vacuoles Are Specialized Compartments That Function in Storage, the Regulation of Cell Volume, and Degradation

The term **vacuole** (literally, "empty space") came from early microscopic observations of these compartments. We now know that vacuoles are not empty but instead contain fluid and sometimes even solid substances. Most vacuoles are made from the fusion of many smaller membrane vesicles. Vacuoles are prominent organelles in plant cells, fungal cells, and certain protists. In animal cells, vacuoles tend to be smaller and are more commonly used to temporarily store materials or transport substances. In animals, such vacuoles are sometimes called storage vesicles.

The functions of vacuoles are extremely varied, and they differ among cell types and even environmental conditions. The best way to appreciate vacuole function is to consider a few examples. Mature plant cells often have a large **central vacuole** that occupies 80% or more of the cell volume (**Figure 4.21a**).

1 Membrane begins to enclose an organelle.

2 Double membrane completely encloses an organelle to form an autophagosome.

3 Autophagosome fuses with a lysosome.

4 Contents are degraded and recycled back to cytosol.

Figure 4.20   Autophagy. In the example shown here, a double membrane surrounds a mitochondrion to form an autophagosome, which fuses with a lysosome. The contents of the autophagosome are degraded and released back into the cytosol.

Biological inquiry: Why do you think autophagy is useful to a cell?

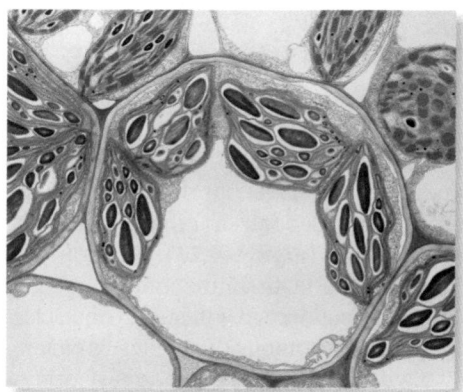

(a) Central vacuole in a plant cell

(b) Contractile vacuoles in an algal cell

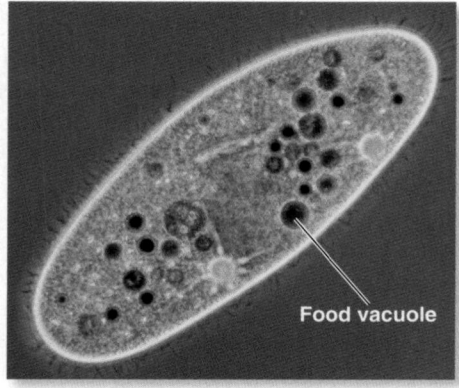

(c) Food vacuoles in a paramecium

Figure 4.21   Examples of vacuoles.

The membrane of this vacuole is called the **tonoplast**. The central vacuole serves two important purposes. First, it stores a large amount of water, enzymes, and inorganic ions such as calcium; it also stores other materials including proteins and pigments. Second, it performs a space-filling function. The large size of the vacuole exerts a pressure on the cell wall called turgor pressure. If a plant becomes dehydrated and this pressure is lost, a plant will wilt. Turgor pressure is important in maintaining the structure of plant cells and the plant itself, and it helps to drive the expansion of the cell wall, which is necessary for growth.

Certain species of protists also use vacuoles to maintain cell volume. Freshwater organisms such as the alga *Chlamydomonas reinhardtii* have small, water-filled **contractile vacuoles** that expand as water enters the cell (**Figure 4.21b**). Once they reach a certain size, the vacuoles suddenly contract, expelling their contents to the exterior of the cell. This mechanism is necessary to remove the excess water that continually enters the cell by diffusing across the plasma membrane.

Another function of vacuoles is degradation. Some protists engulf their food into large **phagocytic vacuoles** or **food vacuoles** (**Figure 4.21c**). As in the lysosomes of animal cells, food vacuoles contain digestive enzymes to break down the macromolecules within the food. Macrophages, a type of cell found in animals' immune systems, engulf bacterial cells into phagocytic vacuoles, where the bacteria are destroyed.

The Plasma Membrane Is the Interface Between a Cell and Its Environment

The cytoplasm of all cells is surrounded by a plasma membrane, which is part of the endomembrane system and provides a boundary between a cell and the extracellular environment. Proteins in the plasma membrane play many important roles that affect the activities inside the cell. First, many plasma membrane proteins are involved in **membrane transport** (**Figure 4.22**). Some of these proteins function to transport essential nutrients or ions into the cell, while others are involved in the export of substances. Due to the functioning of these transporters, the plasma membrane is selectively permeable; it allows only certain substances in and out. We will examine the structures and functions of a variety of transporters in Chapter 5.

A second vital function of the plasma membrane is **cell signaling**. To survive and adapt to changing conditions, cells must be able to sense changes in their environment. In addition, the cells of a multicellular organism need to communicate with each other to coordinate their activities. The plasma membrane of all cells contains receptors that recognize signaling molecules—either environmental agents or molecules secreted by other cells. Once signaling molecules bind to a receptor, this elicits a series of steps known as a signal cascade that causes the cell to respond to the signal (Figure 4.22). For example, when you eat a meal, the hormone insulin is secreted into your bloodstream. This hormone binds to receptors in the plasma membrane of your cells, which results in a cellular response that allows your cells to take up the glucose from the food into the cytosol.

Figure 4.22   Major functions of the plasma membrane. These include membrane transport, cell signaling, and cell adhesion.

We will explore the details of receptors and signal cascades in Chapter 9.

A third important role of the plasma membrane in animal cells is **cell adhesion**. Proteins in the plasma membranes of adjacent cells bind to each other and promote cell-to-cell adhesion (Figure 4.22). This phenomenon is critical for animal cells to properly interact to form a multicellular organism and allows cells to recognize each other. The structures and functions of proteins that are involved in cell adhesion will be examined in Chapter 10.

4.5 Semiautonomous Organelles

In the rest of this chapter, we will examine those organelles in eukaryotic cells that are considered semiautonomous: mitochondria, chloroplasts, and peroxisomes. These organelles can grow and divide to reproduce themselves, but they are not completely autonomous because they depend on other parts of the cell for their internal components (**Figure 4.23**). For example, most of the proteins that are found in mitochondria are imported from the cytosol. In this section, we will survey the structures and functions of the semiautonomous organelles in eukaryotic cells. In Chapter 6 we will consider the evolutionary origins of these organelles, and in Chapters 7 and 8 we will explore the functions of mitochondria and chloroplasts in greater depth.

Animal cell

Mitochondrion

Peroxisome

Chloroplast

Plant cell

Figure 4.23 **Semiautonomous organelles.** These are the mitochondria, chloroplasts, and peroxisomes.

Outer membrane

Inner membrane

Mitochondrial matrix

Cristae

Intermembrane space

Cytosol

0.3 μm

Figure 4.24 **Structure of a mitochondrion.** This organelle is enclosed in two membranes. Numerous cristae in the inner membrane increase the membrane's surface area. The mitochondrial matrix lies inside the inner membrane. The micrograph is a colorized TEM.

Mitochondria Supply Cells with Most of Their ATP

Mitochondrion (plural, *mitochondria*) literally means "thread granule," which is what mitochondria look like under a light microscope. They are similar in size to bacteria. Depending on its function, a cell may contain a few hundred to a few thousand mitochondria—cells with particularly heavy energy demands, such as muscle cells, have more mitochondria than other cells. Indeed, research has shown that regular exercise increases the number and size of mitochondria in human muscle cells to meet the expanded demand for energy.

A mitochondrion has an outer membrane and an inner membrane separated by a region called the intermembrane space (**Figure 4.24**). The inner membrane is highly invaginated (folded) to form projections called **cristae**. These invaginations greatly increase the surface area of the inner membrane, which is the site where ATP is made. The compartment inside the inner membrane is the **mitochondrial matrix**.

The primary role of mitochondria is to make ATP. Even though mitochondria produce most of a cell's ATP, it is incorrect to think that mitochondria create energy. Rather, their primary function is to convert chemical energy that is stored within the covalent bonds in organic molecules into a form that can be readily used by cells. Covalent bonds in sugars, fats, and amino acids store a large amount of energy. The breakdown of these molecules into simpler molecules releases energy that is used to make ATP. Many proteins in living cells utilize ATP to carry out their functions, such as muscle contraction, uptake of nutrients, cell division, and many other cellular processes.

Mitochondria perform other functions as well. They are involved in the synthesis, modification, and breakdown of several types of cellular molecules. For example, the synthesis of certain hormones requires enzymes that are found in mitochondria. Therefore, if mitochondria do not function properly, this affects not only ATP synthesis but the synthesis of other products as well. Another interesting role of mitochondria is to generate heat in specialized fat cells known as brown fat cells. Groups of brown fat cells serve as "heating pads" that help to revive hibernating animals and protect sensitive areas of young animals from the cold.

Chloroplasts Carry Out Photosynthesis

Chloroplasts are organelles that can capture light energy and use some of that energy to synthesize organic molecules such as glucose. This process, called **photosynthesis**, is described in Chapter 8. Chloroplasts are found in nearly all species of plants and algae. **Figure 4.25** shows the structure of a typical chloroplast. Like the mitochondrion, a chloroplast contains an outer and inner membrane. An intermembrane space lies between these two membranes. A third membrane, the **thylakoid membrane**, forms many flattened, fluid-filled tubules that enclose a single, convoluted compartment. These tubules tend to stack on top of each other to form a structure called a **granum** (plural, *grana*). The **stroma** is the compartment of the chloroplast that is inside the inner membrane but outside the thylakoid membrane.

The **thylakoid lumen** is enclosed by the thylakoid membrane. This organization facilitates the process of photosynthesis.

Chloroplasts are a specialized version of plant organelles that are more generally known as **plastids**. All plastids are derived from unspecialized **proplastids**. The various types of plastids are distinguished by their synthetic abilities and the types of pigments they contain. Chloroplasts, which carry out photosynthesis, contain the green pigment chlorophyll. The abundant number of chloroplasts in the leaves of plants gives them their green color (**Figure 4.26a**). Chromoplasts, a second type of plastid, function in synthesizing and storing yellow, orange, and red pigments. Chromoplasts give many fruits and flowers their colors (**Figure 4.26b**). In autumn, the chromoplasts also give many leaves their yellow, orange, and red colors. A third type of

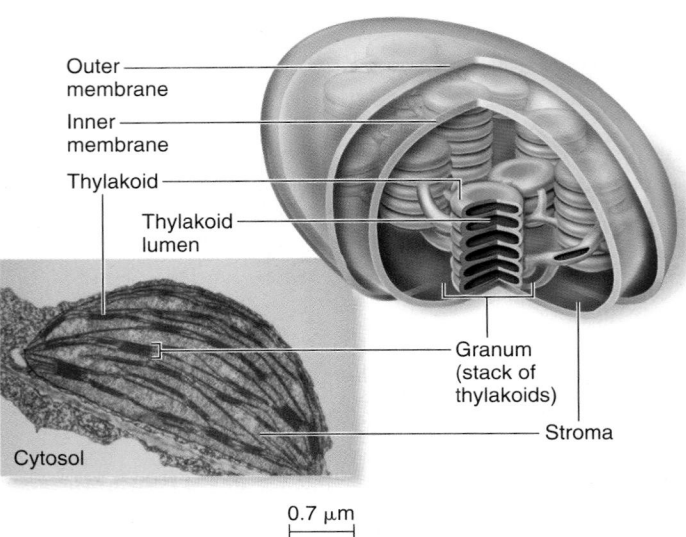

Figure 4.25   **Structure of a chloroplast.**   Like a mitochondrion, a chloroplast is enclosed in a double membrane. In addition, it has an internal thylakoid membrane system that forms flattened compartments. These compartments stack on each other to form grana. The stroma is located inside the inner membrane but outside the thylakoid membrane. The micrograph is a colorized TEM.

plastid, leucoplasts, typically lacks pigment molecules. An amyloplast is a leucoplast that synthesizes and stores starch. Amyloplasts are common in underground structures such as roots and tubers (**Figure 4.26c**).

Peroxisomes Catalyze Detoxifying Reactions

Peroxisomes, which were discovered by Christian de Duve in 1965, are relatively small organelles found in all eukaryotic cells. Peroxisomes consist of a single membrane that encloses a fluid-filled lumen (**Figure 4.27**). A typical eukaryotic cell contains several hundred of them. Peroxisomes are viewed as semi-autonomous because peroxisomal proteins are imported into the peroxisome in a manner that is very similar to the targeting of proteins to the mitochondria and chloroplasts. Another similarity is that new peroxisomes are produced by the division of pre-existing peroxisomes. However, the origin of peroxisomes remains controversial. As discussed in Chapter 6, mitochondria and chloroplasts have DNA because they are evolutionarily derived from bacteria that took up residence in a primordial eukaryotic cell. Likewise, some researchers speculate that peroxisomes could also have originated in this manner and lost their DNA during evolution. Alternatively, other scientists postulate that peroxisomes are derived from an invagination of cellular membranes around cytosolic enzymes. Further research is needed to resolve this controversy.

The general function of peroxisomes is to catalyze certain chemical reactions, typically those that break down molecules by removing hydrogen or adding oxygen. In mammals, for example, large numbers of peroxisomes can be found in the cells of the liver, where toxic molecules accumulate and are broken down. A by-product of this type of chemical reaction is hydrogen peroxide, H_2O_2, which is also broken down in this organelle (hence the name peroxisome). Hydrogen peroxide has the potential to be highly toxic. In the presence of metals such as iron (Fe^{2+}) that are found naturally in living cells, one way that hydrogen peroxide can be broken down causes the formation of a hydroxide ion (OH^-) and a molecule called a hydroxide free radical ($OH\cdot$).

$$Fe^{2+} + H_2O_2 \rightarrow Fe^{3+} + OH^- + OH\cdot \text{ (hydroxide free radical)}$$

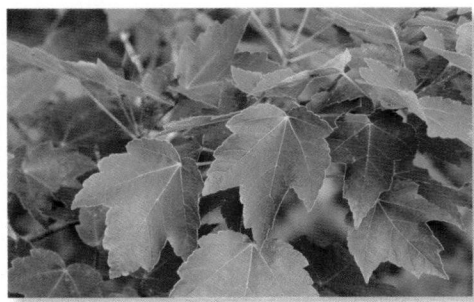

(a) Leaves, which contain chloroplasts

(b) Fruit, which contain chromoplasts

(c) Roots, which contain amyloplasts

Figure 4.26   **Types of plastids.**   **(a)** Chloroplasts are involved in photosynthesis and give plants their green color. **(b)** Chromoplasts store yellow, orange, and red pigments, typically found in fruit and flowers. **(c)** Amyloplasts are colorless plastids that store starch in roots.

The hydroxide free radical is highly reactive and can damage proteins, lipids, and DNA. Therefore, it is beneficial for cells to break down hydrogen peroxide in an alternative manner that does not form a hydroxide free radical. Peroxisomes contain an enzyme called **catalase** that breaks down hydrogen peroxide to make water and oxygen gas.

$$2\ H_2O_2 \xrightarrow{\text{Catalase}} 2\ H_2O + O_2$$

Depending on the cell type, peroxisomes also contain enzymes that detoxify other molecules. These reactions may generate hydrogen peroxide, which is broken down via catalase.

Aside from detoxification, peroxisomes usually contain enzymes involved in the metabolism of fats and amino acids. For instance, plant seeds contain specialized organelles called **glyoxysomes** that are similar to peroxisomes. Glyoxysomes contain enzymes that are needed to convert fats to sugars. These enzymes become active when a seed germinates and the seedling begins to grow.

0.2 μm

Figure 4.27   **Structure of a peroxisome.**   This micrograph is a TEM.

CHAPTER SUMMARY

4.1 Microscopy

- Three important parameters in microscopy are magnification, resolution, and contrast. A light microscope utilizes light for illumination, while an electron microscope uses an electron beam.

- Transmission electron microscopy (TEM) provides the best resolution of any form of microscopy, and scanning electron microscopy (SEM) produces an image of a three-dimensional surface. (Figures 4.1, 4.2, 4.3)

4.2 Overview of Cell Structure

- Cell structure relies on four phenomena: matter, energy, organization, and information. Every living organism has a genome. The genes within the genome contain the information to create cells with particular structures and functions.

- We can classify all forms of life into two categories based on cell structure: prokaryotes and eukaryotes.

- The prokaryotes have a relatively simple structure and lack a membrane-enclosed nucleus. The two categories of prokaryotes are bacteria and archaea. Structures in prokaryotic cells include the plasma membrane, cytoplasm, nucleoid, and ribosomes. Many prokaryotes also have a cell wall and a glycocalyx. (Figures 4.4, 4.5)

- Eukaryotic cells are compartmentalized into organelles and contain a nucleus that houses most of their DNA. (Figures 4.6, 4.7, 4.8)

- The proteome of a cell determines its structure and function.

4.3 The Cytosol

- The cytosol is a central coordinating region for many metabolic activities of eukaryotic cells. A critical activity of all cells is translation, the synthesis of polypeptides. (Figure 4.9)

- The cytoskeleton is a network of three different types of protein filaments: microtubules, intermediate filaments, and actin filaments. Microtubules are important for cell organization, shape, and movement.

Intermediate filaments help maintain cell shape and rigidity. Actin filaments support the plasma membrane and play a key role in cell strength, shape, and movement. (Table 4.1, Figures 4.10, 4.11, 4.12, 4.13, 4.14)

4.4 The Nucleus and Endomembrane System

- The primary function of the nucleus involves the organization and expression of the cell's genetic material. A second important function is the assembly of ribosomes. (Figures 4.15, 4.16, 4.17)

- The endomembrane system includes the nuclear envelope, endoplasmic reticulum, Golgi apparatus, lysosomes, vacuoles, and plasma membrane. The rough endoplasmic reticulum (rough ER) plays a key role in the initial sorting of proteins. The smooth endoplasmic reticulum (smooth ER) functions in metabolic processes such as detoxification, carbohydrate metabolism, accumulation of calcium ions, and synthesis and modification of lipids. The Golgi apparatus performs three overlapping functions: secretion, processing, and protein sorting. Lysosomes degrade macromolecules and help digest substances taken up from outside the cell (endocytosis) and inside the cell (autophagy). (Figures 4.18, 4.19, 4.20)

- Types and functions of vacuoles include central vacuoles, contractile vacuoles, and phagocytic or food vacuoles. (Figure 4.21)

- Proteins in the plasma membrane perform many important roles that affect activities inside the cell, including membrane transport, cell signaling, and cell adhesion. (Figure 4.22)

4.5 Semiautonomous Organelles

- Mitochondria, chloroplasts, and peroxisomes are considered semiautonomous because they can grow and divide, but they still depend on other parts of the cell for their internal components. (Figure 4.23)

- Mitochondria produce most of a cell's ATP, which is utilized by many proteins to carry out their functions. Other mitochondrial functions include the synthesis, modification, and breakdown of cellular molecules and the generation of heat in specialized fat cells. (Figure 4.24)

- Chloroplasts, which are found in nearly all species of plants and algae, carry out photosynthesis. (Figure 4.25)
- Plastids, such as chloroplasts, chromoplasts, and amyloplasts differ in their function and the pigments they store. (Figure 4.26)
- Peroxisomes catalyze certain chemical reactions, typically those that break down molecules by removing hydrogen or adding oxygen. Peroxisomes usually contain enzymes involved in the metabolism of fats and amino acids. (Figure 4.27)

TEST YOURSELF

1. The cell doctrine states that
   a. any living organism is composed of one or more cells.
   b. new cells are derived from organic molecules from the environment.
   c. the smallest units of living organisms are atoms.
   d. the function of cells depends on the shape of the cell.
   e. all of the above.

2. When using microscopes, the resolution refers to
   a. the ratio between the size of the image produced by the microscope and the actual size of the object.
   b. the degree to which a particular structure looks different from other structures around it.
   c. how well a structure takes up certain dyes.
   d. the clarity of an image.
   e. the degree to which the image is magnified.

3. A bacterial cell may possess a _____, which may protect it from the immune system of other multicellular organisms that it may infect.
   a. cell wall
   b. flagellum
   c. pili
   d. nucleoid
   e. capsule

4. Different cells of the same multicellular individual have different proteomes due to all of the following except
   a. differences in the types of proteins made in different cell types.
   b. differences in the genomes of the different cell types.
   c. the abundance of certain proteins may not be the same in different cell types.
   d. the amino acid sequences of proteins may be different in different cell types.
   e. different cell types may alter the proteins in different ways.

5. The process of polypeptide synthesis is called
   a. metabolism.
   b. transcription.
   c. translation.
   d. hydrolysis.
   e. both c and d.

6. Each of the following is part of the endomembrane system except
   a. the nuclear envelope.
   b. the endoplasmic reticulum.
   c. the Golgi apparatus.
   d. lysosomes.
   e. peroxisomes.

7. Molecules move into and out of the nucleus by
   a. diffusing through the nuclear membrane.
   b. transport proteins.
   c. moving through nuclear pores.
   d. attaching to the nucleolus.
   e. all of the above.

8. Functions of the smooth endoplasmic reticulum include
   a. detoxification of harmful organic molecules.
   b. metabolism of carbohydrates.
   c. protein sorting.
   d. all of the above.
   e. a and b only.

9. The central vacuole in many plant cells is important for
   a. storage.
   b. photosynthesis.
   c. structural support.
   d. all of the above.
   e. a and c only.

10. Peroxisomes
   a. are vesicles similar to lysosomes that break down different classes of macromolecules.
   b. play an important role in the synthesis of ATP.
   c. are vesicles that contain enzymes necessary for manufacturing complex sugars.
   d. are the organelles primarily involved in protein synthesis.
   e. contain the enzyme catalase, which breaks down hydrogen peroxide to water and oxygen.

CONCEPTUAL QUESTIONS

1. Define organelle.
2. Explain how actin filaments are involved in movement.
3. Explain the function of the Golgi apparatus.

EXPERIMENTAL QUESTIONS

1. What hypothesis was tested in the experiment of Figure 4.14? What observations led to the proposal of this hypothesis?
2. What is the benefit of purifying cellular components and studying them *in vitro* instead of in intact cells? What was the benefit of using *Nitella axillaris* to determine the function of myosin? What was the purpose of using the fluorescent beads in the experiment? What was the purpose of NEM in the experiment?
3. Explain the results of the experiment of Figure 4.14.

COLLABORATIVE QUESTIONS

1. Discuss several differences between plant and animal cells.
2. Discuss the relationship between the nucleus, the rough endoplasmic reticulum, and the Golgi apparatus.

---

www.brookerbiology.com

This website includes answers to the Biological Inquiry questions found in the figure legends and all end-of-chapter questions.

5

MEMBRANE STRUCTURE AND TRANSPORT

CHAPTER OUTLINE

A model for the structure of aquaporin. This protein, found in the plasma membrane of many cell types, such as red blood cells and plant cells, allows the rapid movement of water molecules across the membrane.

Cellular membranes, also known as biological membranes or biomembranes, are an essential characteristic of all living cells. The **plasma membrane** is the biomembrane that separates the internal contents of a cell from its external environment. With such a role, you might imagine that the plasma membrane would be thick and rigid. Remarkably, the opposite is true. Cellular membranes are thin, typically 5–10 nm thick, and somewhat fluid. It would take 5,000 to 10,000 membranes stacked on top of each other to equal the thickness of the page you are reading. Despite their thinness, cellular membranes are impressively dynamic structures that effectively maintain the separation between a cell and its surroundings, and they provide an interface to carry out many vital cellular activities. **Table 5.1** lists some of their key roles.

In this chapter, we will begin by considering the components that provide the structure of membranes. Then, we examine one of a membrane's primary functions, **membrane transport**.

Biomembranes regulate the traffic of substances into and out of the cell and its organelles. In the second part of this chapter, we will focus on the various ways to transport ions, small molecules, and large macromolecules across membranes. Chapters 6 through 10 will examine other functions of membranes.

5.1 Membrane Structure

As we progress through this textbook, a theme that will emerge is "structure determines function." This paradigm is particularly interesting when we consider how the structure of cellular membranes enables them to compartmentalize the cell while selectively importing and exporting vital substances. The two primary components are lipids, which form the basic matrix of a membrane, and proteins, which are embedded in the membrane or loosely attached to its surface. A third component is carbohydrate, which may be attached to membrane lipids and proteins. In this section, we are mainly concerned with the organization of these components to form a biological membrane, and how they are important in the overall function of membranes. We also consider several interesting experiments that provided insight into the dynamic properties of membranes.

Biological Membranes Are a Mosaic of Lipids, Proteins, and Carbohydrates

Figure 5.1 shows the biochemical organization of cellular membranes, which are similar in composition among all living organisms. The framework of the membrane is the **phospholipid bilayer**, which consists of two layers of lipids. The most abundant lipids found in membranes are the phospholipids. Recall

Table 5.1	Important Functions of Cellular Membranes
Function	
Selective uptake and export of ions and molecules	
Cell compartmentalization	
Protein sorting	
Anchoring of the cytoskeleton	
Production of energy intermediates such as ATP and NADPH	
Cell signaling	
Cell and nuclear division	
Adhesion of cells to each other and to the extracellular matrix	

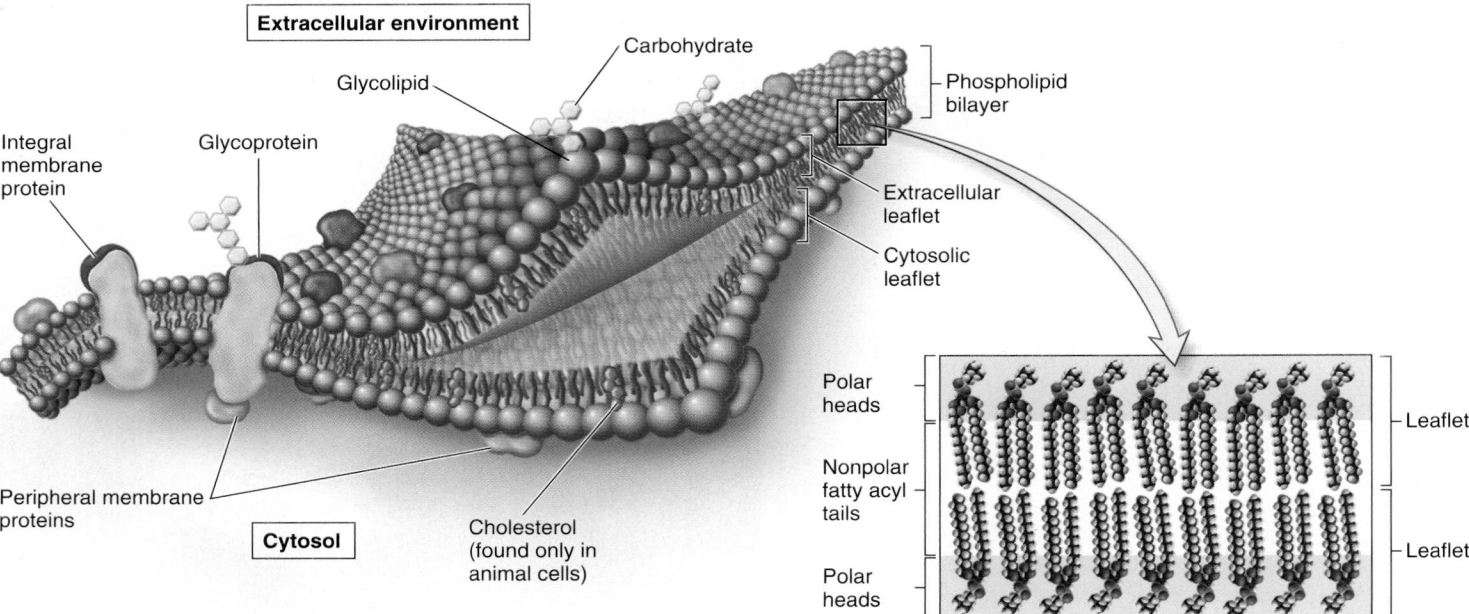

Figure 5.1   Fluid-mosaic model of membrane structure.
The basic framework of a plasma membrane is a phospholipid
bilayer. Proteins are inserted into the membrane and may be
bound on the surface to other proteins or to lipids. Proteins and
lipids, which have covalently bound carbohydrate, are called
glycoproteins and glycolipids, respectively.

from Chapter 3 that phospholipids are **amphipathic** molecules,
meaning they have a hydrophobic (water-fearing) region and a
hydrophilic (water-loving) region. The hydrophobic tails of the
lipids, referred to as fatty acyl tails, form the interior of the
membrane and the hydrophilic head groups are on the surface.
Experimentally, artificial membranes can be made from lipid
alone.

Cellular membranes contain proteins, and most membranes
also have carbohydrates attached to lipids and proteins. The relative amounts of lipids, proteins, and carbohydrates vary among
different membranes. Some membranes, such as the inner mitochondrial membrane, have no carbohydrate while the plasma
membrane of eukaryotic cells can have a large amount. A typical membrane found in cell organelles contains 50% protein by
mass; the remainder is mostly lipids. However, the smaller
lipid molecules outnumber the proteins by about 50 to 1
because the mass of one lipid molecule is much less than the
mass of a protein.

Overall, the membrane is considered a mosaic of lipid, protein, and carbohydrate molecules. The membrane structure illustrated in Figure 5.1 is referred to as the **fluid-mosaic model**,
originally proposed by Jonathan Singer and Garth Nicolson in
1972. As discussed later, the membrane exhibits properties that
resemble a fluid because lipids and proteins can move relative
to each other within the membrane. **Table 5.2** summarizes
some of the historical experiments that led to the formulation of
the fluid-mosaic model.

Table 5.2	Historical Developments That Led to the Formulation of the Fluid-Mosaic Model
Date	**Description**
1917	Irving Langmuir made artificial membranes experimentally by creating a monolayer of lipids on the surface of water. The polar heads interacted with water, and nonpolar tails projected into the air.
1925	Evert Gorter and F. Grendel proposed that lipids form bilayers around cells. This was based on careful measurements of lipid content enclosing red blood cells that showed there was just enough lipid to surround the cell with two layers.
1935	Because proteins were also found in membranes, Hugh Davson and James Danielli proposed incorrectly that a phospholipid bilayer was sandwiched between two layers of protein.
1950s	Electron microscopy revealed that membranes look like a train track—two dark lines separated by a light space. Initially, these results were misinterpreted. Researchers thought the two dark lines were layers of proteins and the light area was the phospholipids bilayer. Later, it was correctly determined that the dark lines in these experiments are the phospholipid heads, which were heavily stained, while the light region between them is their phospholipid tails.
1966	Using freeze fraction electron microscopy (described later in this chapter), Daniel Branton concluded that membranes are bilayers, because the freeze fracture procedure splits membranes in half, thus revealing proteins in the two membrane leaflets.
1972	Jonathan Singer and Garth Nicolson proposed the fluid-mosaic model described in Figure 5.1. Their model was consistent with the observation that membrane proteins are globular, and some are known to span the phospholipid bilayer and project from both sides.

Half of a phospholipid bilayer is termed a **leaflet**. Each leaflet faces a different region. For example, the plasma membrane contains a cytosolic leaflet and an extracellular leaflet (Figure 5.1). With regard to lipid composition, the two leaflets of cellular membranes are highly asymmetrical. The most striking asymmetry occurs with glycolipids—lipids with carbohydrate attached. These are found primarily in the extracellular leaflet so that the carbohydrate portion of a glycolipid protrudes into the extracellular medium.

Membrane Proteins Are Attached to or Embedded in the Phospholipid Bilayer

Although the phospholipid bilayer forms the basic foundation of cellular membranes, the protein component carries out most other functions (see Table 5.1). Proteins can bind to membranes in three different ways (**Figure 5.2**). **Transmembrane proteins** have one or more regions that are physically embedded in the hydrophobic region of the phospholipid bilayer. These regions, the **transmembrane segments**, are stretches of hydrophobic amino acids that span or traverse the membrane from one leaflet to the other. In most transmembrane proteins, the transmembrane segment is folded into an α-helix structure that is stabilized by hydrogen bonds. Such a segment is stable in a membrane because the nonpolar amino acids can interact favorably with the hydrophobic fatty acyl tails of the lipid molecules.

A second way for proteins to associate with the membrane is via **lipid anchors**. A lipid anchor involves the covalent attachment of a lipid to an amino acid side chain within a protein. The fatty acyl tails keep the protein firmly bound to the membrane. Both transmembrane proteins and lipid-anchored proteins are classified as **integral membrane proteins**, also called intrinsic membrane proteins, because they cannot be released from the membrane unless the membranes are dissolved with an organic solvent or detergent—in other words, you would have to disrupt the integrity of the membrane to remove them.

Peripheral membrane proteins, also called extrinsic proteins, are a third class of membrane proteins. These proteins do not interact with the hydrophobic interior of the phospholipid bilayer. Instead, they are noncovalently bound to regions of integral membrane proteins that project out from the membrane, or they are bound to the polar head groups of phospholipids. Peripheral membrane proteins are typically bound to the membrane by hydrogen and/or ionic bonds. For this reason, they usually can be removed from the membrane experimentally by exposing the membrane to high salt concentrations. Researchers can use this treatment to distinguish between peripheral and integral membrane proteins.

GENOMES & PROTEOMES

Approximately 25% of All Genes Encode Membrane Proteins

Membrane proteins participate in some of the most important and interesting cellular processes. These include transport, energy transduction, cell signaling, secretion, cell recognition, and cell-to-cell contact. Research studies have revealed that cells devote a sizeable fraction of their energy and metabolic machinery to the synthesis of membrane proteins. These proteins are particularly important in human medicine—approximately 70% of all medications exert their effects by binding to membrane proteins. Examples include the drugs aspirin, ibuprofen, and acetaminophen, which are widely used to relieve pain and inflammatory conditions such as arthritis. These drugs bind to cyclooxygenase, a protein in the ER membrane that is necessary for the synthesis of chemicals that play a role in inflammation and pain sensation.

Because membrane proteins are so important biologically and medically, researchers have analyzed the genomes of many species and asked the question, What percentage of genes encodes membrane proteins? They have developed tools to predict the likelihood that a gene encodes a protein that is a membrane protein. For example, the occurrence of transmembrane α helices can be predicted from the amino acid sequence of a protein. All 20 amino acids can be ranked according to their preference for a hydrophobic or hydrophilic environment. With these values, the amino acid sequence of a protein can be analyzed using computer software to determine the average hydrophobicity of short amino acid sequences within the protein. A stretch of 18 to 20 amino acids in an α helix is long enough to span the membrane.

Figure 5.2 **Three types of membrane proteins.** Integral membrane proteins may have transmembrane segments that traverse the phospholipid bilayer or they may contain lipid anchors. Peripheral membrane proteins are noncovalently bound to the hydrophilic regions of integral membrane proteins or to the polar head groups of lipids. Inset: The protein shown here contains seven transmembrane segments in an α-helix structure. The transmembrane α helices are depicted as cylinders. This particular protein, bacteriorhodopsin, functions as an ion pump in halophilic (salt-loving) archaea.

If such a stretch contains a high percentage of hydrophobic amino acids, it is predicted to be a transmembrane α helix. However, such computer predictions must eventually be verified by experimentation.

Using a computer approach, many research groups have attempted to calculate the percentage of genes that encode membrane proteins in various species. **Table 5.3** shows the results of one such study. Note that the estimated percentage

Table 5.3	Estimated Percentage of Genes That Encode Membrane Proteins*
Organism	**Percentage of genes that encode membrane proteins**
Archaea	
Archaeoglobus fulgidus	24.2
Methanococcus jannaschii	20.4
Pyrococcus horikoshii	29.9
Bacteria	
Escherichia coli	29.9
Bacillus subtilis	29.2
Haemophilus influenzae	25.3
Eukaryotes	
Homo sapiens	29.7
Drosophila melanogaster	24.9
Arabidopsis thaliana	30.5
Saccharomyces cerevisiae	28.2

* Data from Stevens and Arkin (2000) *Proteins: Structure, Function, and Genetics* 39: 417–420. While the numbers may vary due to different computer programs and estimation techniques, the same general trends have been observed in other similar studies.

of membrane proteins is substantial: 20–30% of all genes may encode membrane proteins. This trend is found throughout all domains of life including archaea, bacteria, and eukaryotes. For example, in *E. coli*, approximately 30% of its genes encode membrane proteins. The genome of this bacterium contains about 4,300 genes, so roughly 1,290 genes may encode different membrane proteins. The human genome also has about 30% of genes that encode membrane proteins. However, the human genome is larger, containing 20,000 to 25,000 different genes, so the total number of genes that encode different membrane proteins is estimated at 6,000 to 7,500. The functions of many of them have yet to be determined. Identifying their functions will help researchers gain a better understanding of human biology. Likewise, medical researchers and pharmaceutical companies are interested in the identification of new membrane proteins that could be targets for effective new medications.

Membranes Are Semifluid

Let's now turn our attention to the dynamic properties of membranes. Although a membrane provides a critical interface between a cell and its environment, it is not a solid, rigid structure. Rather, biomembranes exhibit properties of **fluidity**, which means that individual molecules remain in close association yet have the ability to readily move within the membrane. Though membranes are often described as fluid, it is more appropriate to say they are **semifluid**. In a fluid substance, molecules can move in three dimensions. By comparison, most lipids can rotate freely around their long axes and move laterally within the membrane leaflet (**Figure 5.3**). This type of motion is considered two-dimensional, which means that it occurs within the

(a)

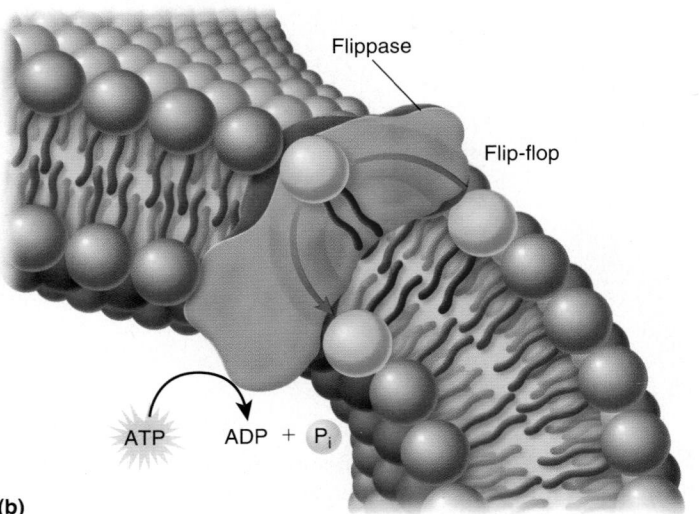

(b)

Figure 5.3   Semifluidity of the lipid bilayer.   **(a)** Spontaneous movements in the bilayer. Lipids can rotate (that is, move 360°) and move laterally (for example, from left to right in the plane of the bilayer). **(b)** Flip-flop does not happen spontaneously because the polar head group would pass through the hydrophobic region of the bilayer. Instead, the enzyme flippase uses ATP to flip phospholipids from one leaflet to the other.

plane of the membrane. Because rotational and lateral movements keep the fatty acyl tails within the hydrophobic interior, such movements are energetically favorable. In fact, the movements of lipids within cellular membranes are quite pronounced. At 37°C, a typical lipid molecule exchanges places with its neighbors about 10^7 times per second, and it can move several micrometers per second. At this rate, a lipid could traverse the length of a bacterial cell (approximately 1 μm) in only 1 second and the length of a typical animal cell in 10–20 seconds.

In contrast to rotational and lateral movements, the "flip-flop" of lipids from one leaflet to the opposite leaflet does not occur spontaneously (Figure 5.3). Energetically, such movements are extremely unfavorable because the polar head of a phospholipid would have to be transported through the hydrophobic interior of the membrane. For this reason, the transport of lipids from one leaflet to another requires the action of the enzyme flippase, which uses energy from the hydrolysis of ATP to flip a lipid from one leaflet to the other.

The biochemical properties of phospholipids have a profound effect on the fluidity of the phospholipid bilayer. One key factor is the length of their fatty acyl tails, which range from 14 to 24 carbon atoms with 18 to 20 carbons being the most common. Shorter acyl tails are less likely to interact, which makes the membrane more fluid. A second important factor is the presence of double bonds in the acyl tails. When a double bond is found, the lipid is said to be **unsaturated** with respect to the number of hydrogens that can be bound to the carbon atoms. A double bond creates a kink in the fatty acyl tail, making it more difficult for neighboring tails to interact and making the bilayer more fluid.

A third factor is the presence of cholesterol, which is a short and fairly rigid molecule that is produced by animal cells (refer back to Figure 3.13a). Cholesterol tends to stabilize membranes; its effects depend on temperature. At higher temperatures, such as those observed in mammals that maintain a constant body temperature, cholesterol makes the membrane less fluid. At lower temperatures, such as in icy water, cholesterol has the opposite effect. It makes the membrane more fluid and prevents it from freezing.

An optimal level of bilayer fluidity is essential for normal cell function, growth, and division. If a membrane is too fluid, which may occur at higher temperatures, it can become leaky. On the other hand, if a membrane becomes too solid, which may occur at lower temperatures, the functioning of membrane proteins will be inhibited. Cells adapt to changes in temperature by altering the lipid composition of their membranes. For example, when the water temperature drops, certain fish incorporate more cholesterol in their membranes. If a plant cell is exposed to high temperatures for many hours or days, it alters its lipid composition to have longer fatty acyl tails and fewer double bonds.

Like lipids, many integral membrane proteins may rotate and laterally move throughout the plane of a membrane. Because membrane proteins are larger than lipids, they move through the membrane at a much slower rate. Flip-flop of integral membrane proteins does not occur because the proteins also contain hydrophilic regions that project out from the phospholipid bilayer. It would be energetically unfavorable for the hydrophilic regions of membrane proteins to pass through the hydrophobic portion of the phospholipid bilayer.

Researchers can examine the lateral movements of lipids and integral membrane proteins by a variety of methods. In 1970, Larry Frye and Michael Edidin conducted an experiment that verified the lateral movement of membrane proteins (**Figure 5.4**). Mouse and human cells were mixed together and exposed to agents that caused them to fuse with each other. Some cells were cooled to 0°C while others were incubated at 37°C before being cooled. Both sets of cells were then exposed to fluorescently labeled antibodies that became specifically bound to a mouse membrane protein called H-2. The fluorescent label was observed with a fluorescence microscope. If the cells were maintained at 0°C, a temperature that greatly inhibits lateral movement, the fluorescence was seen on only one side of the fused cell. However, if the cells were incubated for several hours at 37°C

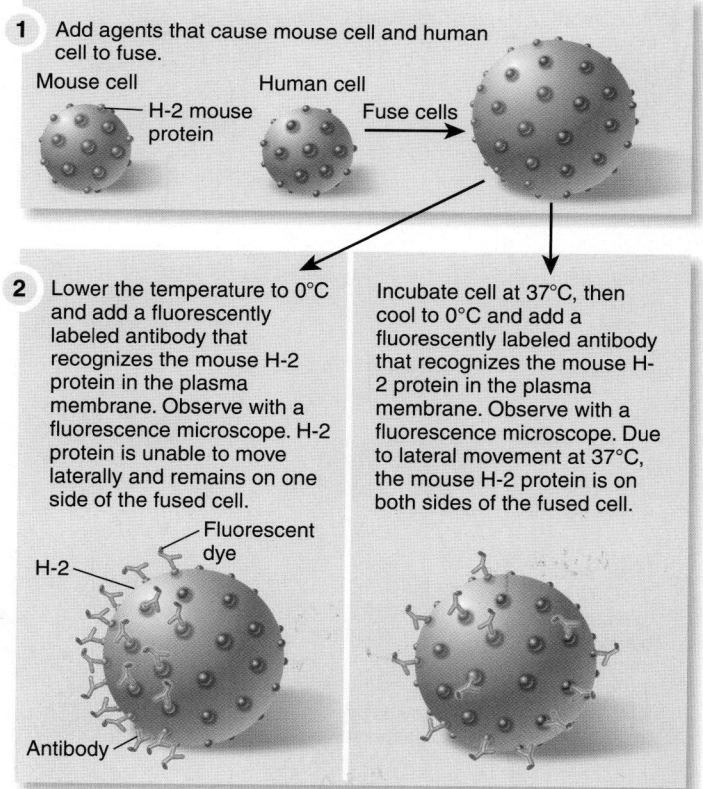

Figure 5.4 A method to measure the lateral movement of membrane proteins.

Biological inquiry: Explain why the H-2 proteins are found only on one side of the cell when the cells were incubated at 0°C.

and then cooled to 0°C, the fluorescence was distributed throughout the fused cell. This occurred because the higher temperature allowed the lateral movement of the H-2 protein throughout the plasma membrane.

A second approach to studying lateral movement, fluorescence recovery after photobleaching (FRAP), was developed in 1976 by Watt Webb and colleagues (**Figure 5.5**). In the experiment shown here, proteins on the surface of a cell were covalently labeled with a fluorescent molecule. In this example, the molecule emitted a red color so the entire surface of the cell appeared red. A laser beam was then focused on a small region of the cell surface. The energy of the laser beam altered the structure of the fluorescent molecules and eliminated the red color, a phenomenon called photobleaching. Immediately after photobleaching, a small region of the cell surface appeared white. Over time, bleached molecules within the white spot spread outward, and the white region filled in with red fluorescent molecules. These results indicate that proteins can laterally move in the membrane.

Unlike the examples shown in Figures 5.4 and 5.5, not all integral membrane proteins are capable of rotational and lateral movement. Depending on the cell type, 10–70% of membrane proteins may be restricted in their movement. Integral membrane proteins may be bound to components of the cytoskeleton, which restricts the proteins from moving laterally (**Figure 5.6**). Also, membrane proteins may be attached to molecules that are outside the cell, such as the interconnected network of proteins that forms the extracellular matrix of animal cells.

Glycosylation of Lipids and Proteins Serves a Variety of Cellular Functions

As mentioned earlier, the third constituent of biomembranes is carbohydrate. **Glycosylation** refers to the process of covalently attaching a carbohydrate to a protein or lipid. When a carbohydrate is attached to a lipid, this creates a **glycolipid**, while attachment to a protein produces a **glycoprotein**. We will discuss the mechanism of protein glycosylation in Chapter 6.

Though the roles of carbohydrate in cell structure and function are not entirely understood, some functional consequences of glycosylation are beginning to emerge. The carbohydrates that are attached to proteins and lipids have well-defined structures that serve in some cases as recognition signals for other cellular proteins. For example, proteins that are destined for the lysosome are glycosylated and have a sugar—mannose-6-phosphate—that is recognized by other proteins in the cell that target the glycosylated protein to the lysosome. Similarly, membrane glycolipids and glycoproteins often play a role in cell surface recognition.

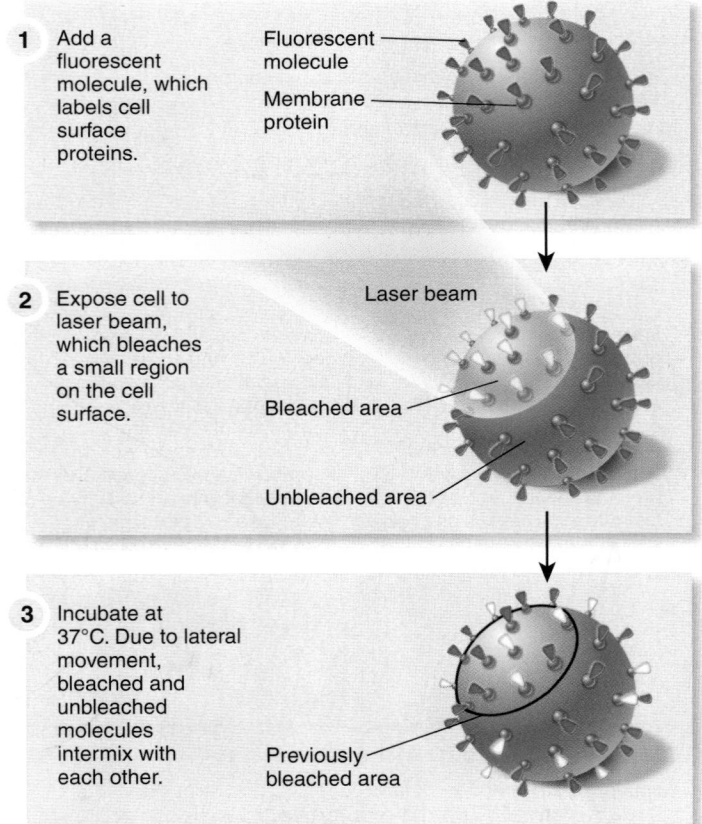

1   Add a fluorescent molecule, which labels cell surface proteins.

Fluorescent molecule

Membrane protein

2   Expose cell to laser beam, which bleaches a small region on the cell surface.

Laser beam

Bleached area

Unbleached area

3   Incubate at 37°C. Due to lateral movement, bleached and unbleached molecules intermix with each other.

Previously bleached area

Figure 5.5   **Fluorescence recovery after photobleaching (FRAP).**   In this method, the surface of cells was coated with fluorescent molecules. A laser beam bleached a small section shown in white. Over time, bleached molecules left the area and unbleached molecules invaded the bleached area by lateral movement.

Fiber in the extracellular matrix

Extracellular matrix

Plasma membrane

Cytosol

Linker protein

Cytoskeletal filament

Figure 5.6   **Attachment of transmembrane proteins to the cytoskeleton and extracellular matrix.**   Some membrane proteins have regions that project into the cytosol and are anchored to large cytoskeletal filaments (via linker proteins). The binding to these large filaments restricts lateral movement. Similarly, some integral membrane proteins may bind to large, immobile components in the extracellular matrix that also restrict the movement of the proteins.

During embryonic development in animals, significant cell movement occurs. Layers of cells slide over each other to create body structures such as the spinal cord and internal organs. The proper migration of individual cells and cell layers relies on the recognition of cell types via the carbohydrates on their cell surfaces.

In addition to its role as a recognition marker, carbohydrate can have a protective effect. The term **cell coat** or **glycocalyx** is used to describe the carbohydrate-rich zone on the cell surface that shields the cell from mechanical and physical damage (**Figure 5.7**). The carbohydrate portion of glycosylated proteins protects them from the harsh conditions of the extracellular environment and degradation by extracellular proteases, which are enzymes that digest proteins.

Membrane Structure Can Be Viewed with an Electron Microscope

Electron microscopy, discussed in Chapter 4, is a valuable tool to probe membrane structure and function. In transmission electron microscopy (TEM), a biological sample is thin sectioned and stained with heavy-metal dyes such as osmium tetroxide. This compound binds tightly to the polar head groups of phospholipids, but it does not bind well to the fatty acyl tails.

Figure 5.7 **A micrograph of the cell coat or glycocalyx.** This figure shows a lymphocyte—a type of white blood cell—stained with ethidium red, which emphasizes the thick carbohydrate layer that surrounds the cell.

As shown in **Figure 5.8a**, membranes stained with osmium tetroxide resemble a railroad track. Two thin dark lines, which are the stained polar head groups, are separated by a uniform light space about 2 nm thick. This railroad track morphology is seen consistently when cell membranes are subjected to electron microscopy.

(a) Transmission electron microscopy (TEM)

(b) Freeze fracture electron microscopy (FFEM)

Figure 5.8 **Electron micrographs of a cellular membrane.** **(a)** In the standard form of TEM, the membrane appears as two dark parallel lines. These lines are the lipid head groups, which stain darkly with osmium tetroxide. The fatty acyl tails do not stain well and appear as a light region sandwiched between the dark lines. **(b)** In the technique of freeze fracture electron microscopy, a sample is frozen in liquid nitrogen, split, and fractured. The sample is then coated with metal and viewed under the electron microscope. Note: The knife does not actually cut the two leaflets of the membrane apart. It simply separates the sample into two parts, and membranes that happen to be located at the fractured region will split along their bilayers.

Biological inquiry: If a heavy metal labeled the hydrophobic tails rather than the polar head groups (as osmium tetroxide does), do you think you would see a bilayer (that is, a railroad track) under TEM?

Due to the incredibly small size of biological membranes, scientists have not been able to invent instruments small enough to dissect them. However, a specialized form of TEM, freeze fracture electron microscopy (FFEM), can be used to analyze the interiors of phospholipid bilayers. Russell Steere invented this method in 1957. In FFEM, a sample is frozen in liquid nitrogen and split with a knife (**Figure 5.8b**). The knife does not actually cut through the bilayer, but it fractures the frozen sample. Due to the weakness of the central membrane region, the leaflets separate into a P face (the protoplasmic face that was next to the cytosol) and the E face (the extracellular face). Most transmembrane proteins do not break in half. They remain embedded within one of the leaflets, usually in the P face. The samples, which are under a vacuum, are then sprayed with a heavy metal such as platinum that coats the sample and reveals architectural features within each leaflet. When viewed with an electron microscope, membrane proteins are visible as bumps that provide significant three-dimensional detail about their form and shape.

5.2 Membrane Transport

If plasma membranes consisted of only a phospholipid bilayer, they would not permit the uptake of most nutrients and the export of waste products. However, the plasma membrane is a **selectively permeable** barrier between the cell and its external environment. As a protective envelope, its structure ensures that essential molecules such as glucose and amino acids enter the cell, metabolic intermediates remain in the cell, and waste products exit. The selective permeability of the plasma membrane allows the cell to maintain a favorable internal environment.

In this section, we begin with a discussion of the phospholipid bilayer, how it presents a barrier to the movement of ions and molecules across membranes, and the concept of gradients across membranes. We will then focus on **transport proteins**, which are embedded within the phospholipid bilayer. Transport proteins allow membranes to be selectively permeable by providing a passageway for the movement of some but not all substances across the membrane. Because different cells require various mixtures of ions and low-molecular-weight molecules, the plasma membrane of each cell type contains a specific set of transport proteins that allow only certain ions and molecules to cross. We will examine how transport proteins play a key role in the selective uptake and export of materials, and consider a mechanism found in eukaryotic cells for the transport of substances across membranes via membrane vesicles.

The Phospholipid Bilayer Is a Barrier to the Diffusion of Hydrophilic Substances

Because of their hydrophobic interiors, phospholipid bilayers present a formidable barrier to the movement of ions and hydrophilic molecules. **Diffusion** occurs when a solute (that is,

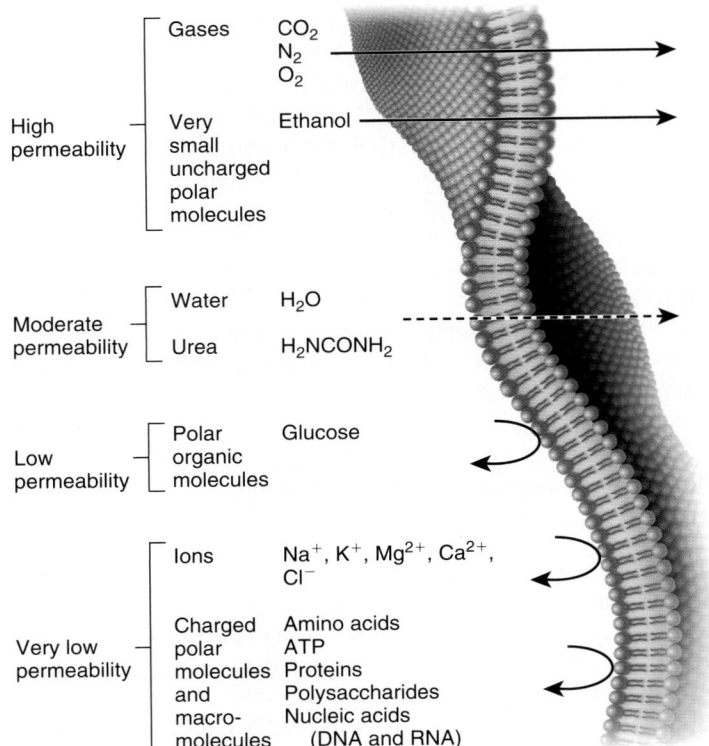

Figure 5.9 **Relative permeability of an artificial phospholipid bilayer to a variety of solutes.** Solutes that easily penetrate are shown with a *straight arrow* that passes through the bilayer. The *dotted line* indicates solutes that have moderate permeability. The remaining solutes shown at the bottom are relatively impermeable. Note: Permeability is also related to solute concentration. At high concentrations, solutes are more likely to pass through a membrane.

a dissolved substance) moves from a region of high concentration to a region of lower concentration. When diffusion occurs through a membrane without the aid of a transport protein, it is called **passive diffusion**. The rate of passive diffusion depends on the chemistry of the solute and its concentration. **Figure 5.9** compares the relative permeabilities of various solutes through an artificial phospholipid bilayer that does not contain any proteins or carbohydrates. Gases and a few small, uncharged polar molecules can passively diffuse across the bilayer. However, the rate of diffusion of ions and larger polar molecules such as sugars and amino acids is relatively slow. Similarly, macromolecules, such as proteins and large carbohydrates, do not readily cross a lipid bilayer.

When we consider the steps of passive diffusion among different solutes, the greatest variation occurs in the ability of solutes to enter the hydrophobic interior of the bilayer. As an example, let's compare urea and diethylurea. Diethylurea is much more hydrophobic because it contains two nonpolar ethyl groups ($-CH_2CH_3$) (**Figure 5.10**). For this reason, it can more readily pass through the hydrophobic region of the phospholipid

$$NH_2 \!-\! \overset{\overset{\displaystyle O}{\parallel}}{C} \!-\! NH_2$$

Urea

$$CH_3 \!-\! CH_2 \!-\! NH \!-\! \overset{\overset{\displaystyle O}{\parallel}}{C} \!-\! NH \!-\! CH_2 \!-\! CH_3$$

Diethylurea

Figure 5.10 Structures of urea and diethylurea.

Biological inquiry: Which molecule would you expect to pass through a phospholipid bilayer more quickly, methanol (CH_3OH) or methane (CH_4)?

bilayer. The rate of passive diffusion of diethylurea through a phospholipid bilayer is about 50 times faster than urea.

Cells Maintain Gradients Across Their Membranes

As we have just seen, phospholipid bilayers are quite impermeable to ions and most hydrophilic molecules. A hallmark of living cells is their ability to maintain a relatively constant internal environment that is distinctively different from their external environment. This involves establishing gradients of solutes across the plasma membrane and organellar membranes. When we speak of a **transmembrane gradient**, we mean that the concentration of a solute is higher on one side of a membrane than the other. For example, immediately after you eat a meal, a higher concentration of glucose is found outside your cells compared to inside. Gradients involving ions have two components—electrical and chemical. An **ion electrochemical gradient** is a dual gradient that has both an electrical gradient and chemical gradient. For example, let's consider a gradient involving Na^+. An electrical gradient could exist in which the amount of net positive charge outside a cell is greater than inside. At the same time, a chemical gradient could exist in which the concentration of Na^+ outside is greater than inside. These two gradients together constitute a Na^+ electrochemical gradient. Transmembrane gradients of ions and other solutes are a universal feature of all living cells.

One way to view the transport of solutes across membranes is to consider how the transport process affects the pre-existing gradients across membranes. **Passive transport** refers to the diffusion of a solute across a membrane in a process that is energetically favorable and does not require an input of energy. Passive transport tends to dissipate a pre-existing gradient. (Note: The adjective *passive* can have different meanings in biology. It can mean without an input of energy, as in passive transport, or without the aid of a transport protein, as in passive diffusion.)

Passive transport can occur in two ways, via passive diffusion or facilitated diffusion. As mentioned earlier, passive diffusion is the diffusion of a solute directly through the phospholipid bilayer to move across the membrane (**Figure 5.11a**). This is not a common way for most solutes to move. The second pathway for passive transport is **facilitated diffusion**, which involves the aid of transport proteins (**Figure 5.11b**). Transport proteins facilitate the movement of various nutrients and—in some types of cells—water across the membrane.

Passive diffusion is the movement of a solute down a gradient. A transport protein is not needed to cross a membrane.

(a) Passive diffusion

Facilitated diffusion is movement down a gradient with the aid of a transport protein.

(b) Facilitated diffusion

Figure 5.11 Types of passive transport. (a) Passive diffusion. (b) Facilitated diffusion.

Osmosis Is the Movement of Water Across Membranes to Balance Solute Concentrations

Lipid bilayers are relatively impermeable to many solutes, yet somewhat permeable to water. When the solute concentrations on both sides of the plasma membrane are equal, the two solutions are said to be **isotonic**. However, we have also seen that transmembrane gradients commonly exist across membranes. When the solute concentration inside the cell is higher, it is said to be **hypertonic** relative to the outside of the cell, which is **hypotonic**. If the solutes cannot move across the membrane, water will move and tend to balance the solute concentrations. In this process, called **osmosis**, water diffuses through a membrane from the hypotonic compartment into the hypertonic compartment.

Cells generally have a high internal concentration of a variety of solutes including ions, sugars, amino acids, etc. Animal cells,

(a) Osmosis in animal cells

(b) Osmosis in plant cells

Figure 5.12   **The phenomenon of osmosis.**   (a) In animal cells that lack a cell wall, osmosis may promote cell shrinkage (crenation) or swelling. (b) In plant cells that have a rigid cell wall, a hypertonic medium causes the plasma membrane to pull away from the cell wall, while a hypotonic medium causes only a minor amount of expansion.

Biological inquiry: The inside of a cell has 100 mM KCl, while the outside has 10 mM KCl. If the membrane is impermeable to KCl, which direction will water move?

which are not surrounded by a rigid cell wall, must maintain a balance between the extracellular and intracellular solute concentrations. As discussed later, animal cells contain a variety of transport proteins that can sense changes in cell volume and allow the necessary movements of solutes across the membrane to prevent osmotic changes and maintain normal cell shape. However, if animal cells are placed in a hypertonic medium, water will exit the cells via osmosis and equalize solute concentrations on both sides of the membrane, causing them to shrink in a process called **crenation** (**Figure 5.12a**). Alternatively, if cells are placed in a hypotonic medium, water will diffuse into them to equalize solute concentrations on both sides of the membrane. In extreme cases, a cell may take up so much water that it bursts, a phenomenon called osmotic lysis.

Osmosis can also affect cells with a rigid cell wall, such as bacteria, fungi, algae, and plant cells. If the extracellular fluid surrounding a plant cell is hypertonic, water will exit the cell and the plasma membrane will pull away from the cell wall (**Figure 5.12b**). Alternatively, if the extracellular fluid is hypotonic, a plant cell will take up a small amount of water, but the cell wall prevents major changes in cell size. The tendency

of water to move into any cell creates an **osmotic pressure**, which is defined as the hydrostatic pressure required to stop the net flow of water across a membrane due to osmosis. In plant cells, osmotic pressure is also called **turgor pressure** or simply cell turgor. The turgor pressure pushes the plasma membrane against the rigid cell wall. An appropriate level of turgor is needed for cells to maintain their proper structure (**Figure 5.13**). If a plant has insufficient water, the extracellular fluid surrounding plant cells becomes hypertonic, the plasma membrane pulls away from the cell wall, and the osmotic pressure drops. The result is plasmolysis, which is associated with wilting.

Some freshwater organisms, such as amoebae and paramecia, can exist in extremely hypotonic environments where the external solute concentration is always much lower than the concentration of solutes in their cytosol. Because of the great tendency for water to move into the cell by osmosis, some organisms contain a contractile vacuole to prevent osmotic lysis. A contractile vacuole takes up water from the cytosol and periodically discharges it by fusing the vacuole with the plasma membrane (**Figure 5.14**).

(a) Sufficient water

(b) Wilting

Figure 5.13   Wilting in plants.   (a) When a plant has plenty of water, the slightly hypotonic surroundings cause the vacuole to store water. The increased size of the vacuole influences the volume of the cytosol, thereby exerting a turgor pressure against the cell wall. **(b)** Under dry conditions, water is released from the cytosol into the extracellular medium. The vacuole also shrinks, because it loses water to the cytosol. Turgor pressure is lost. This causes the plant to wilt.

(a) Filled vacuole

(b) After releasing water

Figure 5.14   The contractile vacuole in *Paramecium caudatum*.   (a) A contractile vacuole is filled with water from radiating canals that collect fluid from the cytosol. **(b)** The contractile vacuole has recently fused with the plasma membrane (which would be above the plane of this page) and released the water from the cell.

FEATURE INVESTIGATION

Agre Discovered That Osmosis Occurs More Quickly in Cells with Transport Proteins That Allow the Facilitated Diffusion of Water

In living cells, the flow of water may occur by passive diffusion through the phospholipid bilayer. However, in the 1980s, researchers also discovered that certain cell types allow water to move across the plasma membrane at a much faster rate than would be predicted by passive diffusion. For example, water moves very quickly across the membrane of red blood cells, which causes them to shrink and swell in response to changes in extracellular solute concentrations. Likewise, bladder and kidney cells, which play a key role in regulating water balance in the bodies of vertebrates, allow the rapid movement of water across their membranes. Based on these observations, researchers speculated that certain cell types might have proteins in their plasma membranes that permit the rapid movement of water.

One approach that is used to characterize a new protein is to first identify a protein based on its relative abundance in a particular cell type, and then attempt to determine the protein's function. This rationale was applied to the discovery of proteins that allow the rapid movement of water across membranes.

Peter Agre and his colleagues first identified a protein that was abundant in red blood cells and kidney cells, but not found in many other cell types. Though they initially did not know the function of the protein, its physical structure was similar to other proteins that were already known to function as transport proteins. They named this protein CHIP28, which stands for channel-forming integral membrane protein with a molecular mass of <u>28</u>,000 Da. During the course of their studies, they also identified and isolated the gene that encodes CHIP28.

In 1992, Agre and his colleagues conducted experiments to determine if CHIP28 functions in the transport of water across membranes (**Figure 5.15**). Because they already had isolated the gene that encodes CHIP28, they could make many copies of this gene in a test tube (*in vitro*) using gene cloning techniques

Figure 5.15   The discovery of water channels.

HYPOTHESIS CHIP28 may function as a water channel.

STARTING MATERIALS Prior to this work, a protein called CHIP28 was identified that is abundant in red blood cells and kidney cells. The gene that encodes this protein was cloned, which means that many copies of the gene were made in a test tube.

Experimental level   Conceptual level

1 Add an enzyme (RNA polymerase) and nucleotides to a test tube that contains many copies of the CHIP28 gene. This results in the synthesis of many copies of CHIP28 mRNA.

Enzymes and nucleotides

CHIP28 mRNA   RNA polymerase

CHIP28 DNA

2 Inject the CHIP28 mRNA into frog oocytes. Wait several hours to allow time for the mRNA to be translated into CHIP28 protein, which is inserted into the plasma membrane.

Frog oocyte
Nucleus

CHIP28 mRNA

Cytosol

CHIP28 protein is inserted into the plasma membrane.

CHIP28 protein

Ribosome

3 Place oocytes into a hypotonic medium and observe under a light microscope. As a control, also place oocytes that have not been injected with CHIP28 mRNA into a hypotonic medium and observe by microscopy.

Control

CHIP28 protein

4 **THE DATA**

Oocyte

Oocyte rupturing

Control   CHIP28   Control   CHIP28

(see Chapter 20). Starting with many copies of the gene *in vitro*, they added an enzyme to transcribe the gene into mRNA that encodes the CHIP28 protein. This mRNA was then injected into frog oocytes, chosen because frog oocytes are large, easy to inject, and lack pre-existing proteins in their plasma membranes that allow the rapid movement of water. After injection, the mRNA was expected to be translated into CHIP28 proteins that should be inserted into the plasma membrane of the oocytes. After allowing sufficient time for this to occur, the oocytes were placed in a hypotonic medium. As a control, oocytes that had not been injected with CHIP28 mRNA were also exposed to a hypotonic medium.

As you can see in the data, a striking difference was observed between oocytes that expressed CHIP28 versus the control. Within minutes, oocytes that contained the CHIP28 protein were seen to swell due to the rapid uptake of water. Three to 5 minutes after being placed in a hypotonic medium, they actually burst! By comparison, the control oocytes did not swell as rapidly, and they did not rupture even after 1 hour. Taken together, these results are consistent with the hypothesis that CHIP28 functions as a transport protein that allows the facilitated diffusion of water across the membrane. Many subsequent studies confirmed this observation. Later, CHIP28 was renamed **aquaporin** to indicate its newly identified function of allowing water to diffuse through a pore in the membrane (**Figure 5.16**). More recently, the three-dimensional structure of aquaporin was determined (see inset to Figure 5.16). Agre was awarded the Nobel Prize in 2003 for this work.

Aquaporin is an example of a transport protein called a channel. Next, we will discuss the characteristics of channels and other types of transport proteins.

Figure 5.16 Function and structure of aquaporin. Aquaporin is found in the membrane of certain cell types and allows the rapid diffusion of water across the membrane. The inset shows the structure of aquaporin, which was determined by X-ray crystallography.

Transport Proteins Cause Biological Membranes to Be Selectively Permeable

Because the phospholipid bilayer is a physical barrier to the passive diffusion of ions and most hydrophilic molecules, cells are able to separate their internal contents from their external environment. However, this barrier also poses a severe problem because cells must take up nutrients from the environment and export waste products. To overcome this dilemma, species have evolved a multitude of transport proteins—transmembrane proteins that provide a passageway for the movement of ions and hydrophilic molecules across membranes. Transport proteins enable biological membranes to be selectively permeable. We can categorize transport proteins into two classes, channels and transporters, based on the manner in which they move solutes across the membrane.

Channels **Channels** are transmembrane proteins that form an open passageway for the facilitated diffusion of ions or molecules across the membrane (**Figure 5.17**). In other words, solutes move directly through a channel to get to the other side. Aquaporin, discussed in the Feature Investigation, is a channel that allows the movement of water across the membrane. When a

channel is open, the transmembrane movement of solutes can be extremely rapid, up to 100 million ions or molecules per second!

Most channels are **gated**, which means they can open to allow the diffusion of solutes and close to prohibit diffusion. The phenomenon of gating allows cells to regulate the movement of solutes. Researchers have discovered a variety of gating mechanisms (**Figure 15.18**). Gating sometimes involves the direct binding of a molecule to the channel protein itself. **Ligand-gated channels** are controlled by the noncovalent binding of small molecules—called ligands—such as hormones or neurotransmitters. These ligands are often important in the transmission of signals between nerve and muscle cells or between two nerve cells. Alternatively, intracellular proteins may bind noncovalently to channels and control their ability to open and close. For example, certain types of calcium channels that are found in nerve cells are controlled by regulatory proteins that bind to the channel and cause it to open. Another gating mechanism involves the covalent binding of a small molecule, such as a phosphate group, to the channel protein. Regulatory proteins involved in cell-signaling pathways may covalently attach phosphate to proteins as a way to regulate their function. In some cases, the targets of these regulatory proteins are channels.

When a channel is open, a solute directly diffuses through the channel to reach the other side of the membrane.

Gate opened

Gate closed

Figure 5.17 Mechanism of transport by a channel protein.

As an example, chloride channels in the human lung are opened by phosphorylation.

Channel gating can also occur by mechanisms that do not involve the direct binding of a molecule to the channel protein. For example, some channels are **voltage-gated**, meaning that the channel opens and closes in response to changes in the amount of electric charge across the membrane. Sodium and potassium channels in nerve cells are voltage-gated. Another interesting gating mechanism involves **mechanosensitive channels**, which are sensitive to changes in membrane tension. For example, our ability to hear depends, in part, on the functioning of mechanosensitive channels in the cells of the inner ear that are sensitive to different frequencies of sound. The opening of these ion channels in response to sound transmits signals to the brain that a particular frequency has been detected.

Transporters **Transporters**, also known as **carriers**, bind their solutes in a hydrophilic pocket and undergo a conformational change that switches the exposure of the pocket to the other side of the membrane (**Figure 5.19**). Transporters tend to be much slower than channels. Their rate of transport is typically 100 to 1,000 ions or molecules per second. Transporters provide the principal pathway for the uptake of organic molecules, such as sugars, amino acids, and nucleotides. In animals, they also allow cells to take up certain hormones and neurotransmitters. In addition, many transporters play a key role in export. Waste-products of cellular metabolism must be released from cells before they reach toxic levels. For example, a transporter removes lactic acid, a by-product of muscle cells during exercise. Excessive lactic acid buildup is partly responsible for the burning sensation you feel during a strenuous workout. Other transporters, which are involved with ion transport, play an important role in cellular processes such as regulating internal pH and controlling cell volume.

Channel closed ⟶ Channel open

Ligand

Regulatory protein

Phosphorylation

+++
+++
−
−−−

Change in voltage (less negative inside)

Sound waves

Figure 5.18 Different ways that channels are gated.

Transporters are subdivided according to the number of solutes they bind and the direction of transport (**Figure 5.20**). **Uniporters** bind a single molecule or ion and transport it across the membrane. **Symporters** or **cotransporters** bind two or more ions or molecules and transport them in the same direction. **Antiporters** bind two or more ions or molecules and transport them in opposite directions.

A **pump** is a transporter that directly couples its conformational changes to an energy source, such as ATP hydrolysis (**Figure 5.21**). The conformational changes of pumps are energetically driven. A common category of pumps found in all living cells are **ATP-driven pumps**, which have a binding site for ATP.

For transport to occur, a solute binds in a hydrophilic pocket exposed on one side of the membrane. The transporter then undergoes a conformational change that switches the exposure of the pocket to the other side of the membrane, where the solute is then released.

Figure 5.19 Mechanism of transport by a transporter, also called a carrier.

A Ca^{2+} pump uses ATP to transport Ca^{2+} from the cytosol to the ER lumen, against a gradient.

Figure 5.21 Transporters that function as pumps.

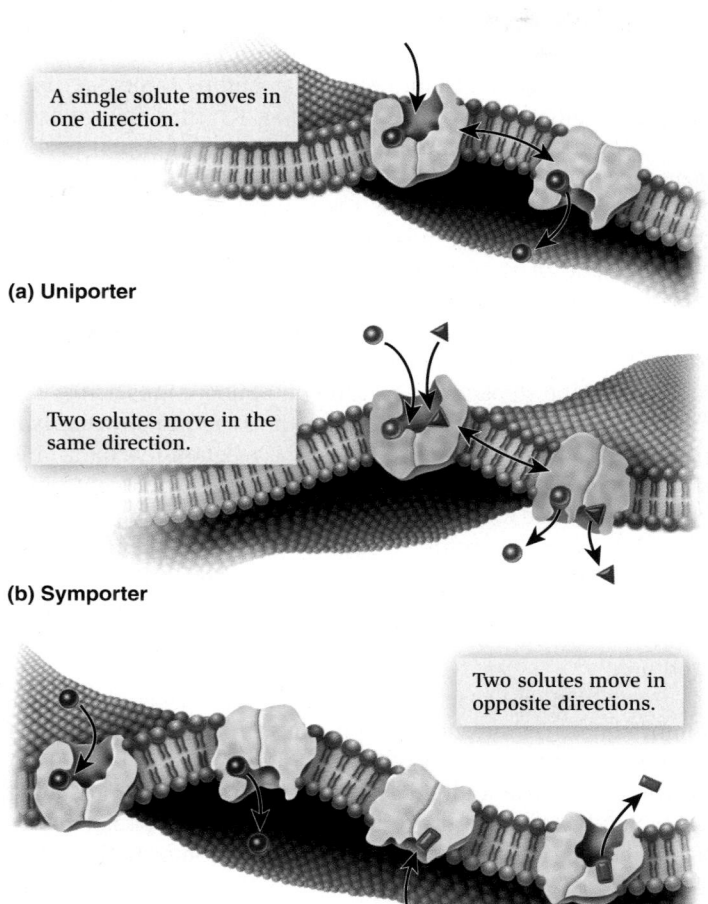

A single solute moves in one direction.

(a) Uniporter

Two solutes move in the same direction.

(b) Symporter

Two solutes move in opposite directions.

(c) Antiporter

Figure 5.20 Types of transporters based on the direction of transport. (a) Uniporter. (b) Symporter. (c) Antiporter.

The hydrolysis of ATP provides energy that controls the sequence of conformational changes. The energy obtained from ATP hydrolysis can be used to pump solutes against a gradient. Pumps can be uniporters, symporters, or antiporters. As discussed next, pumps use energy to achieve active transport.

Active Transport Is the Movement of Substances Against a Gradient

Active transport is the movement of a solute across a membrane against its gradient—that is, from a region of low concentration to higher concentration. Active transport is energetically unfavorable and requires the input of energy. **Primary active transport** involves the functioning of pumps that directly use energy to transport a solute against a gradient. **Figure 5.22** shows a pump that uses ATP to transport H^+ against a gradient. Such a pump can establish a large H^+ electrochemical gradient across a membrane.

Secondary active transport involves the utilization of a pre-existing gradient to drive the active transport of a solute (Figure 5.22). For example, an H^+/sucrose symporter can utilize an H^+ electrochemical gradient, established by an ion pump, to move sucrose against its concentration gradient. In this regard, only sucrose is actively transported. Hydrogen ions move down (with) their electrochemical gradient. H^+/solute symporters are more common in bacteria, fungi, algae, and plant cells, because H^+ pumps are found in their plasma membranes. In animal cells, a pump that exports Na^+ maintains the Na^+ gradient across the plasma membrane. Na^+/solute symporters are prevalent in animal cells.

Symport enables cells to actively import nutrients against a gradient. Symport proteins use the energy stored in the electrochemical gradient of Na^+ or H^+ to power the uphill movement of organic solutes such as sugars, amino acids, and other needed

A pump actively exports H⁺ against a gradient.

Extracellular environment

A H⁺/sucrose symporter can utilize the H⁺ gradient as a way to transport sucrose against a concentration gradient into the cell.

ATP   ADP + P$_i$

H⁺

Cytosol

Sucrose

H⁺

Primary active transport

Secondary active transport

Figure 5.22   **Types of active transport.**   During primary active transport, a pump directly uses energy, in this case from ATP, and generates a gradient. The pump shown here uses ATP to establish an H⁺ electrochemical gradient. Secondary active transport via symport involves the utilization of this gradient to drive the active transport of a solute, such as sucrose.

solutes. Therefore, with symporters in their plasma membrane, cells can scavenge nutrients from the extracellular environment and accumulate them to high levels within the cytoplasm.

ATP-Driven Ion Pumps Generate Ion Electrochemical Gradients

The concept of active transport was discovered in the 1940s based on the study of ion movements using radioisotopes of Na⁺ and K⁺. After analyzing the movement of these ions across the plasma membrane of muscle cells, nerve cells, and red blood cells, researchers determined that the export of sodium ions (Na⁺) is coupled to the import of potassium ions (K⁺) (**Figure 5.23a**). In the late 1950s, Danish biochemist Jens Skou proposed that a single transporter is responsible for this phenomenon. By studying the membranes of nerve cells from crabs, he was the first person to describe an ATP-driven ion pump, which was later named the Na⁺/K⁺-ATPase. This pump can actively transport Na⁺ and K⁺ against their gradients by using the energy from ATP hydrolysis.

Interestingly, Skou initially had trouble characterizing this pump. He focused his work on the large nerve cells found in the shore crab (*Carcinus maenas*). In membranes from these cells, he was able to identify a transporter that could hydrolyze ATP, but the rate of hydrolysis was too low to account for the known levels of ATP hydrolysis associated with Na⁺ and K⁺ pumping. When he added Na⁺ to his membranes, the ATP hydrolysis rate was not greatly affected. Then he tried adding K⁺, but the ATP hydrolysis rate still did not increase. This was a frustrating period for Skou. Eventually, he did the critical experiment in which he added both Na⁺ and K⁺ to his membranes. With both ions present, ATP hydrolysis soared dramatically. This observation led to the identification and purification of the Na⁺/K⁺-ATPase. Jens Skou was awarded the Nobel Prize in 1997, over 40 years after his original work.

Let's take a closer look at the Na⁺/K⁺-ATPase that Skou discovered. Every time one ATP is hydrolyzed, the Na⁺/K⁺-ATPase functions as an antiporter that pumps three Na⁺ into the extracellular environment and two K⁺ into the cytosol. Because one cycle of pumping results in the net export of one positive charge, the Na⁺/K⁺-ATPase also produces an electrical gradient across the membrane. For this reason, it is considered an **electrogenic pump**, meaning that it generates an electrical gradient. The plasma membrane of a typical animal cell contains thousands of Na⁺/K⁺-ATPase pumps.

By studying the interactions of Na⁺, K⁺, and ATP with the Na⁺/K⁺-ATPase, researchers have pieced together a molecular roadmap of the steps that direct the pumping of ions across the membrane. These steps are termed the **reaction mechanism** (**Figure 5.23b**). A central precept of the reaction mechanism is that the Na⁺/K⁺-ATPase can alternate between two conformations, designated E1 and E2. In E1, the ion-binding sites are accessible from the cytosol—sodium ions bind tightly to this conformation while potassium ions have a low affinity. In E2, the ion-binding sites are accessible from the extracellular environment—sodium ions have a low affinity and potassium ions bind tightly.

To examine the reaction mechanism, let's begin with the E1 conformation. Three sodium ions bind to the Na⁺/K⁺-ATPase from the cytosol (Figure 5.23b). When this occurs, ATP is hydrolyzed to ADP and phosphate. Temporarily, the phosphate is covalently bound to the pump, an event called phosphorylation. The pump then switches to the E2 conformation. The sodium ions are released into the extracellular environment because they have a lower affinity for the E2 conformation, and then two potassium ions bind from the outside. The binding of K⁺ causes the release of phosphate, which in turn causes a switch to E1. Because the E1 conformation has a low affinity for K⁺, the potassium ions are released into the cytosol. The Na⁺/K⁺-ATPase is now ready for another round of pumping.

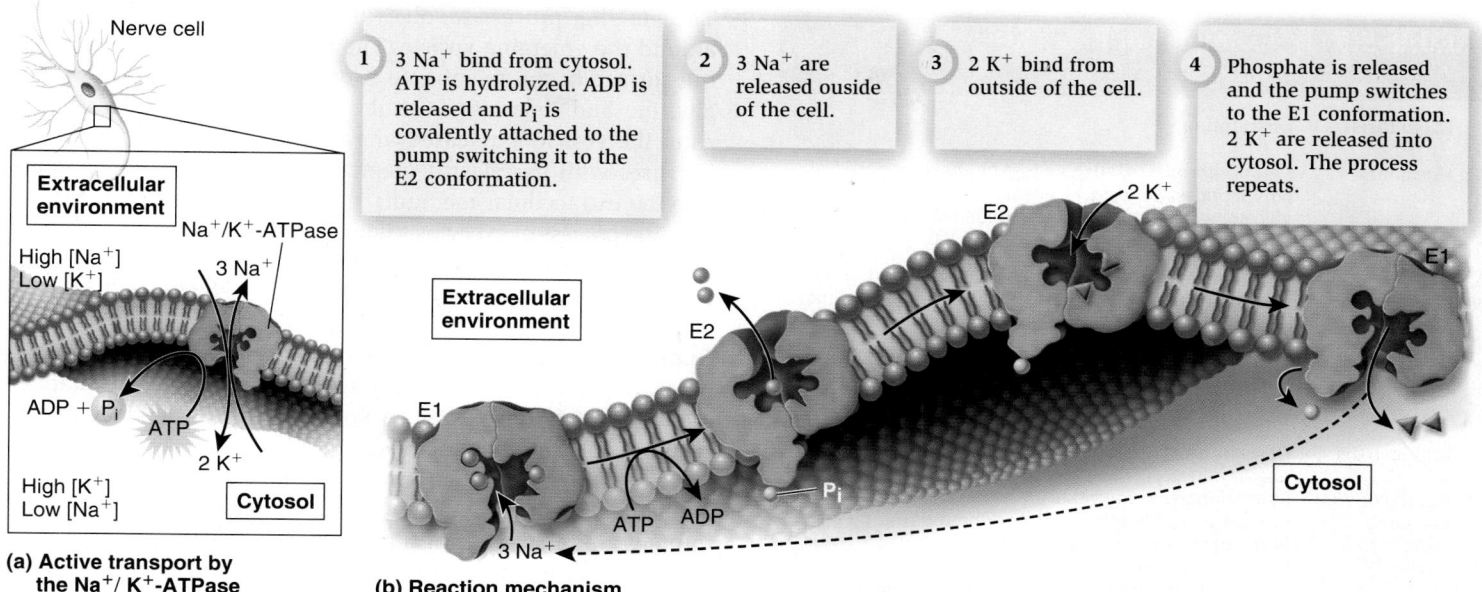

1 3 Na$^+$ bind from cytosol. ATP is hydrolyzed. ADP is released and P$_i$ is covalently attached to the pump switching it to the E2 conformation.

2 3 Na$^+$ are released ouside of the cell.

3 2 K$^+$ bind from outside of the cell.

4 Phosphate is released and the pump switches to the E1 conformation. 2 K$^+$ are released into cytosol. The process repeats.

(a) Active transport by the Na$^+$/ K$^+$-ATPase

(b) Reaction mechanism

Figure 5.23 **Structure and function of the Na$^+$/K$^+$-ATPase.** **(a)** Active transport by the Na$^+$/K$^+$-ATPase. Each time this protein hydrolyzes one ATP molecule, it pumps out three Na$^+$ and pumps in two K$^+$. **(b)** Reaction mechanism. The figure illustrates the protein conformational changes between E1 and E2. As this occurs, ATP is hydrolyzed to ADP and phosphate. During the reaction mechanism, phosphate is covalently attached to the protein but is released after potassium ions bind.

Biological inquiry: If a cell had ATP and sodium ions, but potassium ions were missing, how far through the reaction mechanism could the Na$^+$/K$^+$-ATPase proceed?

The Na$^+$/K$^+$-ATPase is a key cellular enzyme in animal cells because it functions as an ion pump that maintains Na$^+$ and K$^+$ gradients across the plasma membrane. Many other types of ion pumps are also found in the plasma membrane and in organellar membranes. Ion pumps play the primary role in the formation and maintenance of ion gradients that drive many important cellular processes (**Table 5.4**). Biologists have come to understand that the transport of ions against their gradients is a never-ending activity of all living cells. ATP is commonly the source of energy to drive ion pumps, and cells typically use a substantial portion of their cellular ATP to keep them working. For example, nerve cells use up to 70% of their ATP just to operate ion pumps!

Macromolecules and Large Particles Are Transported via Endocytosis and Exocytosis

We have seen that most small substances are transported via membrane proteins such as pumps, transporters, and channels, which provide a passageway for the movement of substances across the membrane. Eukaryotic cells have two other mechanisms, exocytosis and endocytosis, to transport larger molecules such as proteins and polysaccharides, and even very large particles. Both mechanisms involve the packaging of the transported substance, sometimes called the cargo, into a membrane vesicle or vacuole. **Table 5.5** describes important examples of exocytosis and endocytosis.

Table 5.4	Important Functions of Ion Electrochemical Gradients
Function	**Description**
Transport of ions and molecules	Symporters and antiporters use H$^+$ and Na$^+$ gradients to take up nutrients and export waste products.
Production of energy intermediates	In the mitochondrion and chloroplast, H$^+$ gradients are used to synthesize ATP.
Regulation of cystolic pH	Transporters sense pH changes and regulate the internal pH.
Osmotic regulation	Animal cells control their internal volume by regulating ion gradients between the cytosol and extracellular fluid.
Nerve signaling	Na$^+$ and K$^+$ gradients are involved in conducting action potentials, the signals transmitted by nerve cells.
Muscle contraction	Ca^{2+} gradients regulate the ability of muscle fibers to contract.
Bacterial swimming	H$^+$ gradients drive the rotation of bacterial flagella.

Exocytosis **Exocytosis** is a process in which material inside the cell, which is packaged into vesicles, is excreted into the extracellular environment (**Figure 5.24**). These vesicles are usually derived from the Golgi apparatus. As the vesicles form, a specific cargo is loaded into their interior. For example, in animal cells, large polysaccharides are made within the lumen of the Golgi and packaged within vesicles that bud from the Golgi.

Table 5.5	Examples of Exocytosis and Endocytosis
Exocytosis	**Description**
Hormones	Certain hormones, such as insulin, are composed of polypeptides. To exert its effect, insulin is secreted via exocytosis into the bloodstream from B cells of the pancreas.
Digestive enzymes	Digestive enzymes that function in the lumen of the small intestine are secreted via exocytosis from cells of the pancreas.
Extracellular matrix	Most of the components of the extracellular matrix that surrounds animal cells are secreted via exocytosis.
Endocytosis	**Description**
Uptake of vital nutrients	Many important nutrients are very insoluble in the bloodstream. Therefore, they are bound to proteins in the blood and then taken into cells via endocytosis. Examples include the uptake of lipids (bound to low-density lipoprotein) and iron (bound to transferrin protein).
Root nodules	Nitrogen-fixing root nodules found in certain species of plants such as legumes are formed by the endocytosis of bacteria. After endocytosis, the bacterial cells are contained within a membrane-enclosed compartment in the nitrogen-fixing tissue of functional nodules.
Immune system	Cells of the immune system, known as macrophages, engulf and destroy bacteria via endocytosis.

The budding process involves the formation of a protein coat around the emerging vesicle. The assembly of coat proteins on the surface of the membrane causes the bud to form. Eventually, the bud separates from the membrane to form a vesicle. After the vesicle is released, the coat is shed. Finally, the vesicle fuses with the plasma membrane and releases its contents into the extracellular medium.

The process of exocytosis illustrates the dynamic and fluid nature of membranes. They readily bud and fuse with other membranes. The same is true for endocytosis, which is described next.

Endocytosis During **endocytosis**, the plasma membrane invaginates, or folds inward, to form a vesicle that brings substances into the cell (**Figure 5.25**). A common form of endocytosis is **receptor-mediated endocytosis**, in which a receptor is specific for a given cargo. When a receptor binds to that cargo, this stimulates the binding of coat proteins to the membrane, which initiates the formation of a vesicle. Many receptors aggregate together as a protein coat forms around the vesicle. Once inside the cell, the vesicle sheds its coat. In most cases, the vesicle fuses with an internal membrane organelle, such as a lysosome, and the receptor releases its cargo. Depending on the cargo, the lysosome may release it directly into the cytosol or digest it into simpler building blocks before releasing it.

Other specialized forms of endocytosis occur in certain types of cells. **Pinocytosis** (meaning "cell drinking") involves the formation of membrane vesicles from the plasma membrane as a way for cells to internalize the extracellular fluid. This allows cells to sample the extracellular solutes. Pinocytosis is particularly important in cells that are actively involved in nutrient absorption, such as cells that line the intestine in animals.

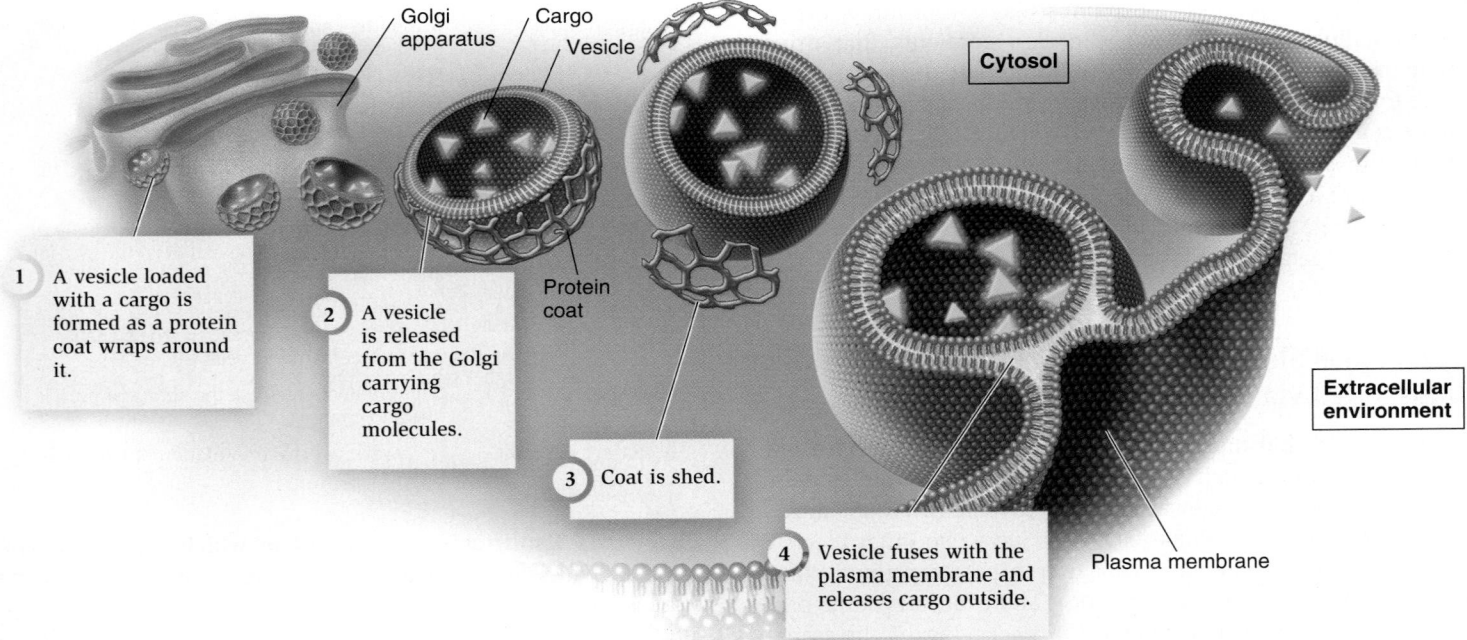

1 A vesicle loaded with a cargo is formed as a protein coat wraps around it.

2 A vesicle is released from the Golgi carrying cargo molecules.

3 Coat is shed.

4 Vesicle fuses with the plasma membrane and releases cargo outside.

Golgi apparatus

Cargo

Vesicle

Cytosol

Protein coat

Extracellular environment

Plasma membrane

Figure 5.24 Exocytosis. Membrane vesicles bud from an organelle inside the cell, such as the Golgi apparatus, and are loaded with a specific cargo. The budding process involves the formation of a protein coat around the emerging vesicle, which is later shed. When the vesicle fuses with the plasma membrane, it releases its contents into the extracellular environment.

Phagocytosis ("cell eating") is an extreme form of endocytosis. It involves the formation of an enormous membrane vesicle called a phagosome, or phagocytic vacuole, that engulfs a large particle such as a bacterium. Only certain kinds of cells can carry out phagocytosis. For example, macrophages, which are cells of the immune system in mammals, kill bacteria via phagocytosis. Once inside the cell, the phagosome fuses with lysosomes and the digestive enzymes within the lysosomes destroy the bacterium.

Figure 5.25   Receptor-mediated endocytosis.   A cargo binds to a receptor, which stimulates the aggregation of many receptors and the formation of a vesicle with a protein coat. Once inside the cell, the vesicle sheds its coat. In most cases, the vesicle fuses with an internal membrane organelle and releases its cargo into the cytosol.

CHAPTER SUMMARY

5.1 Membrane Structure

- Cellular membranes are dynamic structures that separate a cell from its surroundings yet provide an interface to carry out vital cellular activities. One of their crucial functions is membrane transport. (Table 5.1)

- The accepted model of the plasma membrane is the fluid-mosaic model, and its basic framework is the phospholipid bilayer. Cellular membranes also contain proteins, and most membranes have attached carbohydrates. (Figure 5.1, Table 5.2)

- Proteins can bind to membranes in three different ways: as transmembrane proteins, lipid-anchored proteins, or peripheral membrane proteins. Transmembrane proteins and lipid-anchored proteins are classified as integral membrane proteins. Researchers are working to identify new membrane proteins and their functions because these proteins are so important biologically and medically. (Figure 5.2, Table 5.3)

- Bilayer semifluidity is essential for normal cell function, growth, and division. The chemical properties of phospholipids, such as tail length, the presence of double bonds, and the presence of cholesterol, have a profound effect on the fluidity of the phospholipid bilayer. (Figures 5.3, 5.4, 5.5, 5.6)

- Glycosylation, which produces glycolipids or glycoproteins, has a variety of cellular functions. Carbohydrate can serve as a recognition marker or a protective cell coat. (Figure 5.7)

- Electron microscopy is a valuable tool for studying membrane structure and function. Freeze fracture electron microscopy (FFEM) can be used to analyze the interiors of phospholipid bilayers. (Figure 5.8)

5.2 Membrane Transport

- Living cells maintain a constant internal environment that is separated from their external environment. This involves establishing transmembrane gradients across the plasma membrane and organellar membranes. (Figure 5.9, Table 5.4)

- Diffusion occurs when a solute moves from a region of high concentration to a region of lower concentration. Passive diffusion occurs through a membrane without the aid of a transport protein. (Figure 5.10)

- Passive transport refers to the diffusion of a solute across a membrane in a process that does not require an input of energy. Passive transport can occur via passive diffusion or facilitated diffusion. (Figure 5.11)

- In the process of osmosis, water diffuses through a membrane from a solution that is hypotonic (lower solute concentration) into a solution that is hypertonic (higher solute concentration). Solutions with identical solute concentrations are isotonic. The tendency of water to move into any cell creates an osmotic (turgor) pressure. (Figures 5.12, 5.13, 5.14)

- Transport proteins enable biological membranes to be selectively permeable. The two classes of transport proteins are channels and transporters. Channels form an open passageway for the direct diffusion of solutes across the membrane; one example is aquaporin, which allows the movement of water. Most channels are gated, which allows cells to regulate the movement of solutes. Transporters (carriers), which tend to be slower than channels, bind their solutes in a hydrophilic pocket and undergo a conformational change that switches the exposure of the pocket to the other side of the membrane. A pump is a transporter that directly couples its conformational changes to an energy source. All living cells contain ATP-driven pumps. (Figures 5.15, 5.16, 5.17, 5.18, 5.19, 5.20, 5.21)

- Active transport is the movement of a solute across a membrane against its gradient. Primary active transport involves pumps that directly use energy and generate a solute gradient. Secondary active transport involves the utilization of a pre-existing gradient. (Figure 5.22)

- The Na^+/K^+-ATPase is an electrogenic ATP-driven pump. The reaction mechanism refers to the steps that direct the pumping of ions across the membrane. (Figure 5.23)

- Eukaryotic cells have two other mechanisms, exocytosis and endocytosis, to transport large molecules and particles. Exocytosis is a process in which material inside the cell is packaged into vesicles and excreted into the extracellular environment. During endocytosis, the plasma membrane folds inward to form a vesicle that brings substances into the cell. Forms of endocytosis include pinocytosis, phagocytosis, and receptor-mediated endocytosis. (Figures 5.24, 5.25, Table 5.5)

TEST YOURSELF

1. Which of the following statements best describes the chemical composition of biomembranes?
   a. Biomembranes are bilayers of proteins with associated lipids and carbohydrates.
   b. Biomembranes are composed of two layers, one layer of phospholipids and one layer of proteins.
   c. Biomembranes are bilayers of phospholipids with associated proteins and carbohydrates.
   d. Biomembranes are composed of equal numbers of phospholipids, proteins, and carbohydrates.
   e. Biomembranes are composed of lipids with proteins attached to the outer surface.

2. _____ is a lipid that helps stabilize membranes of animal cells by regulating fluidity as temperature changes.
   a. Cholesterol   d. ATP
   b. Prostaglandin   e. Acetone
   c. Glycerol

3. The presence of double bonds in the fatty acyl tail will
   a. decrease fluidity because of the attraction between the unsaturated tails.
   b. increase fluidity due to the difficulty of the kinked acyl tail to interact.
   c. decrease fluidity by increasing the space between the phospholipids.
   d. increase fluidity by allowing more room for cholesterol to move into the membrane.
   e. decrease fluidity by decreasing the amount of proteins in the membrane.

4. Carbohydrates of the plasma membrane
   a. are associated with a protein or lipid.
   b. are located on the outer surface of the membrane.
   c. can function as cell markers for recognition by other cells.
   d. all of the above.
   e. a and c only.

5. Which of the following can easily diffuse through a lipid bilayer?
   a. sodium ions   d. oxygen
   b. amino acids   e. DNA
   c. glucose

6. The tendency for Na^+ to move into the cell is due to
   a. the higher numbers of Na^+ outside the cell resulting in a chemical concentration gradient.
   b. the net negative charge inside the cell attracting the positively charged Na^+.
   c. the attractive force of K^+ inside the cell pulling Na^+ into the cell.
   d. all of the above.
   e. a and b only.

7. The hydrostatic pressure required to stop osmosis is
   a. an electrochemical gradient.   d. osmotic pressure.
   b. filtration.   e. partial pressure.
   c. tonicity.

8. The selectively permeable characteristic of plasma membranes is mainly due to the presence of
   a. phospholipids.
   b. transport proteins.
   c. glycolipids on the outer surface of the membrane.
   d. concentration gradients across the membrane.
   e. cholesterol.

9. During _____, materials are moved across the plasma membrane against their concentration gradient.
   a. facilitated diffusion   d. filtration
   b. osmosis   e. simple diffusion
   c. active transport

10. Large particles or large volumes of fluid can be brought into the cell by
    a. facilitated diffusion.   d. exocytosis.
    b. active transport.   e. all of the above.
    c. endocytosis.

CONCEPTUAL QUESTIONS

1. Explain the fluid-mosaic model of membrane structure.
2. What are three types of membrane proteins?
3. What is the difference between passive diffusion and passive transport?

EXPERIMENTAL QUESTIONS

1. What observations led to the experiment of Figure 5.15 to identify proteins that may increase water movement into the cell?
2. How did Agre and associates choose a candidate protein that may function as a water channel in plasma membranes? What was the hypothesis tested by the researchers? Briefly explain how they were able to test their hypothesis.
3. What were the results of the experiment of Figure 5.15? Do they support the proposed hypothesis?

COLLABORATIVE QUESTIONS

1. Discuss the concept of solute concentrations in cells.
2. Discuss the two categories of transport proteins found in plasma membranes.

---

www.brookerbiology.com

This website includes answers to the Biological Inquiry questions found in the figure legends and all end-of-chapter questions.

6

SYSTEMS BIOLOGY OF CELL ORGANIZATION

CHAPTER OUTLINE

Electron micrograph of a lymphocyte, a type of white blood cell. This cell has a large nucleus and several mitochondria (colored blue). A cell can be viewed as a system of four interacting parts: the nucleus, endomembrane system, cytosol, and semiautonomous organelles.

The first few chapters of this textbook laid the foundation for understanding cell structure and function. We learned that life depends on organic molecules, which form the building blocks for macromolecules such as proteins, nucleic acids, and carbohydrates. In addition, we considered cell organization at a higher level. Cells contain complex structures such as membranes, chromosomes, ribosomes, and a cytoskeleton. Eukaryotic cells have organelles that provide specialized compartments to carry out various cellular functions.

Thus far, we have surveyed the structures and functions of cells with an emphasis on describing the various parts of cells. We have taken a reductionist approach, which involves reducing a complex system to simpler components to understand how the system works. In cell biology, reductionists study the parts of a cell as individual units. In this chapter, however, we will view the cell from a broader perspective. In **systems biology**, researchers study living organisms in terms of their underlying network structure—groups of structural and functional connections—rather than their individual molecular components. A "system" can be anything from a metabolic pathway to a cell, an organ, or even an entire organism. In this chapter, we focus on the cell as a system. Our goal is to understand how the organization of a cell arises by complex interactions between its various components and compartments. In the first section, we will examine the general principles of cell organization. In the remaining two sections we focus on dynamic interactions among the different compartments of eukaryotic cells.

6.1 Principles of Cell Organization

Now that we are familiar with the general properties of cells and membranes, we can ask, How are the various parts of a cell formed and maintained? We usually think of a cell as having form and structure, and at any given instant, it does. However, a cell is also profoundly dynamic, which we can better appreciate when we look at its organization from the perspective of systems biology. Let's consider a football analogy. A living cell is less like a football field—with goalposts, lines, and turf—and more like the players who follow rules and have various formations and movements. Like the players in a football game, the components of a cell are dynamic entities. Like a successful team, these components work together as a system to maintain cell organization and carry out vital functions.

We begin this section by considering the phenomena that underlie the existence of cells. These include information, namely, a genome, functional molecules, and pre-existing organization. We will then examine how the genome allows cells to synthesize proteins at the correct time and in the correct cell type. Likewise, the genome provides proteins with sorting signals that are recognized by a sorting system, which directs them to the proper cellular location. This system, which is critical for cell organization, will be described in later sections.

Another factor that underlies cell organization is the ability of certain proteins to interact with each other to form larger complexes, sometimes called molecular machines. We will examine how molecular machines assemble from their individual components, and how they carry out complex functions and provide cell organization. Finally, the organization of living cells is a dynamic process that involves the turnover of cellular molecules and macromolecules. We will also explore how macromolecules such as proteins are broken down to their building blocks so that a cell can continually replenish its macromolecules.

Information, Functional Molecules, and Organization Must Already Exist for Cells to Maintain Their Structure

As discussed in Chapter 22, the first primitive cells arose from nonliving materials between 3.5 and 4.0 billion years ago. According to the cell doctrine, *all modern cells come from pre-existing cells by division*. Modern cells, which are much more complex than the first primitive cells, can grow, adapt, and modify their organization, but they do not arise spontaneously from nonliving materials. At least three key factors are responsible for this principle. First, all cells must possess a genome, which provides the information to make RNA and proteins. The genomes of modern species are the products of 3.5–4.0 billion years of evolution and contain the information to make the thousands of proteins necessary to maintain cell structure and function.

Even so, if a researcher were able to synthesize the genome of a living organism and put it in a test tube full of water, nothing would happen. Living cells require a second ingredient—pre-existing molecules such as enzymes and organic molecules—to make things happen. For instance, RNA polymerase, a protein that functions as an enzyme, is needed to make new RNA molecules, using nucleotides (organic molecules) as building blocks and DNA as a template. As another example, RNA molecules and proteins are needed to construct a ribosome, which is a critical component for synthesizing polypeptides. Cell vitality requires a genome to provide information and also active molecules such as RNA, proteins, and small molecules to access that information.

So let's suppose that a researcher could somehow synthesize the genome of a species and produce all of its RNA molecules and proteins. That would be an amazing feat! If the genome, RNAs, and proteins were combined in a test tube with other small molecules that are needed for energy and as building blocks, do you think a cell would arise? The answer is no. Some enzymatic activity would probably occur, but a living cell would not result. In addition to a genome and functional molecules, cells require a third key factor: pre-existing organization. For example, cells rely on a plasma membrane, which contains multiple transport proteins that allow the selective uptake and excretion of ions and molecules. In addition, organelles compartmentalize cells, and most organelles must be made from pre-existing organelles.

Thus, the formation of cells relies on a genome, functional molecules, and organization, all of which come from pre-existing cells. In the rest of this chapter, we will examine the interplay that occurs among these three phenomena to maintain cellular organization and promote the formation of cells.

Cell Organization Relies on Genetic Information, Which Produces a Proteome

The genome of every organism contains the information necessary to produce a system of RNA and protein molecules that provides the foundation for cell structure, function, and organization. An important paradigm in biology is that "structure determines function." The information in most genes is used to make mRNA molecules that encode the amino acid sequences of proteins. The amino acid sequence of proteins, together with chemical principles, governs their three-dimensional structure. The three-dimensional structure of proteins, in turn, determines their function.

Although the genome contains the information to make proteins with specific structures and functions, the study of individual proteins does not allow us to completely understand how the parts of cells are made, and how they maintain cellular organization. To appreciate the dynamic form and function of living cells, we need to take a broader, more integrative look at their molecular components.

1. *The proteome—the entire collection of proteins that a cell makes—is largely responsible for the structures and functions of living cells* (**Figure 6.1a**). Some proteins play a role in cell shape and organization. Other proteins and a few RNA molecules function as enzymes to synthesize or break down other cellular components. For example, as discussed later in this chapter, phospholipids are made by enzymes in the cell. The synthesis of most components that form the foundation for cell structure relies on enzymes. Some exceptions are water and inorganic ions, which are taken up from the environment. Likewise, transport proteins are important in the uptake and export of ions and molecules across cell membranes. Finally, the activities of cells are controlled by many proteins that are generally known as regulatory proteins. As described next, some of these proteins regulate genes in the cell nucleus, while others regulate the functions of other proteins.

2. *Gene and protein regulation cause the proteome to be dynamic.* In multicellular organisms, the regulation of genes causes the amounts of proteins to vary in different cell types. This allows a multicellular organism to have cells with specialized structures and functions (**Figure 6.1b**). Nerve and muscle cells have a different organization because their proteomes are different. In addition, the regulation of genes and proteins allows a cell to respond in a dynamic way to changes in the environment (**Figure 6.1c**). For example, when a cell is exposed to a higher temperature, it will produce proteins that help it cope with the increased heat.

Enzymes are involved in the synthesis and breakdown of cellular molecules and macromolecules.

Regulatory proteins control the expression of genes and the functions of other proteins.

Proteins play a structural role that influences cell shape, organization, and movement.

Transport proteins allow the movement of substances across membranes.

(a) The proteome is a diverse collection of proteins that carry out cell functions and promote cell organization.

Muscle fiber cell

Nerve cell

(b) Differences in cell morphology among specialized cells can be attributed to differences in the proteome.

Abrupt increase in temperature

Heat shock protein

Heat shock proteins are produced to help a cell cope with high temperature by keeping other cellular proteins in a properly folded condition.

(c) Cells adapt to environmental changes by altering the composition of their proteomes.

Nuclear proteins

ER proteins

Golgi proteins

Lysosomal proteins

Mitochondrial proteins

Cytosolic proteins

Secreted proteins

(d) Proteins have sorting signals within their amino acid sequences that direct them to the correct cellular compartment.

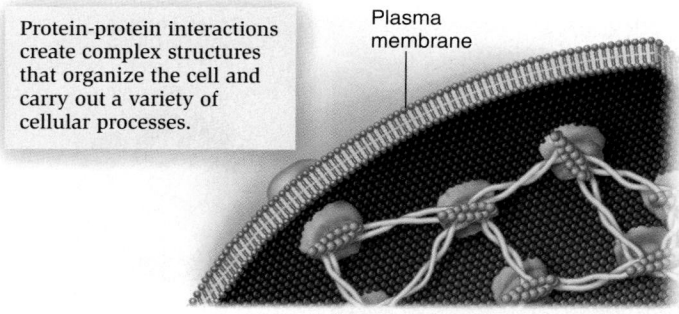

Protein-protein interactions create complex structures that organize the cell and carry out a variety of cellular processes.

Plasma membrane

(e) Proteins may have surfaces that cause them to interlock via protein-protein interactions to form larger cellular structures.

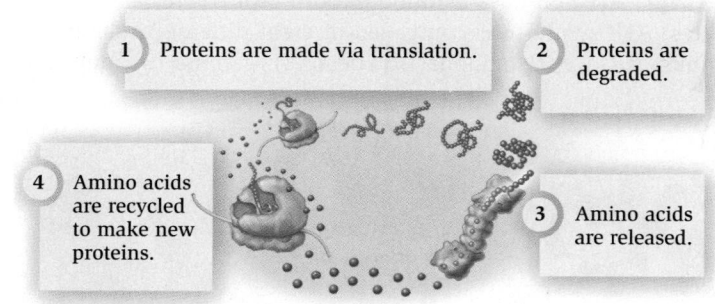

1 Proteins are made via translation.

2 Proteins are degraded.

4 Amino acids are recycled to make new proteins.

3 Amino acids are released.

(f) Cellular macromolecules are made and later broken down to recycle their building blocks.

Figure 6.1   Genomes and proteomes and their relationship to cell structure, function, and organization.

3. *Proteins have **sorting signals**—short amino acid sequences in their structure—that direct them to their correct location* (**Figure 6.1d**). For example, one sorting signal directs certain proteins to the mitochondria while a different sorting signal sends other proteins to the endoplasmic reticulum. The last two sections of this chapter will explore the molecular mechanisms that facilitate protein sorting, a key event that maintains cell organization.

4. *Because their structures may bind to each other and interlock, proteins often undergo **protein-protein interactions**.* These interactions are very specific and can build larger structures that provide organization to the cell (**Figure 6.1e**). An example is the cytoskeleton that forms filaments in the cell. In addition, protein-protein interactions can produce molecular machines that carry out complex cellular functions.

5. *Because most molecules and macromolecules in cells are short-lived, cells must continually synthesize new components and break down unwanted components.* Proteins, for example, are made during the process of translation and typically exist for just several minutes or a few hours (**Figure 6.1f**). As discussed later, cells have different mechanisms to degrade proteins to individual amino acids so that the amino acids can be recycled— that is, used to make new proteins.

Interactions Among Proteins and RNA Molecules Make Complex Molecular Machines

A machine is an object that has moving parts and does useful work. If the size of the machine is measured in nanometers, it is appropriately called a **molecular machine**. Even a single protein could be considered a molecular machine if it undergoes conformational changes as part of its function. The association of proteins with each other and also with RNA molecules may form larger molecular machines. These machines provide structure and organization to cells and enable them to carry out complicated processes.

Large molecular machines are formed by an assembly process. In some cases, interactions between the protein components of a machine may occur spontaneously, without requiring an input of energy. Sometimes additional proteins are needed for the machine to form but are not retained in its final structure. Some machines may require an input of energy, such as from ATP hydrolysis, to promote the assembly process.

Figure 6.2 describes the assembly of ATP synthase, a machine that makes ATP. Because all living organisms use ATP for energy, ATP synthase is found in both prokaryotes and eukaryotes. ATP synthase is composed of eight different protein subunits called a, b, c, alpha (α), beta (β), gamma (γ), delta (δ),

and epsilon (ϵ). The a, b, and c subunits are transmembrane proteins, found either in the bacterial plasma membrane or the mitochondrial and thylakoid membranes of eukaryotes. Nine to twelve of the c subunits assemble together to form a ring in the membrane, and one a subunit and two b subunits bind to this ring. The α, β, γ, δ, and ϵ subunits associate with each other to form a complex of three α subunits, three β subunits, and one subunit each of γ, δ, and ϵ. This complex then binds to the membrane components to complete the assembly process.

Why does ATP synthase assemble in this manner? The reason is **molecular recognition**—surfaces on the various protein subunits recognize each other in a very specific way, causing them to bind to each other and promote the assembly process. Said another way, the amino acid sequences of these proteins produce surfaces that fold in a way that causes the proteins to interlock. Thus, the amino acid sequences of proteins, which are stored in the genome, contain the information for protein-protein interactions.

Molecular machines carry out complex cellular functions. The ribosome (**Figure 6.3a**), a molecular machine composed of

(a) Ribosomes in the act of making polypeptides

(b) The flagella of sperm contain a molecular machine for locomotion

Figure 6.3 Other examples of molecular machines. (a) The ribosome, which plays a role in protein synthesis, is composed of many types of proteins and several large RNA molecules. In the electron micrograph shown here, many ribosomes are gliding along an mRNA molecule, synthesizing polypeptides as they go. (b) The flagellum of a sperm contains a molecular machine composed of microtubules and accessory proteins. The cross section shows the 9 + 2 arrangement of microtubules in which nine double microtubules form a ring with two central microtubules that are single microtubules.

Figure 6.2 Assembly of a molecular machine, the ATP synthase found in bacteria, mitochondria, and chloroplasts. Nine to twelve of the c subunits form a ring in the membrane, and one a subunit and two b subunits bind to the ring. The α, β, γ, δ, and ε subunits form a complex of three α subunits, three β subunits, and one subunit each of γ, δ, and ε. This complex binds to the membrane components.

Biological inquiry: Explain why the ATP synthase assembles the way it does.

many types of proteins and several large RNA molecules, functions as a molecular arena for the synthesis of new polypeptides. The flagellum (**Figure 6.3b**) is a molecular machine composed of microtubules, motor proteins, and many other proteins that enables eukaryotic cells, such as sperm cells, to move.

Molecular machines are also vital in promoting cell organization. Perhaps the most important example is the cytoskeleton, which we consider next.

The Cytoskeleton Is a Molecular Machine That Organizes the Cell and Directs Cellular Movements

Recall from Chapter 4 that the cytoskeleton plays a key role in cell organization and many of the dynamic processes that maintain it. For example, in many cell types, actin filaments form a band just inside the plasma membrane that provides mechanical strength and plays a role in cell shape. Protein-protein interactions occur between actin and many other cellular proteins. Actin filaments are sometimes linked to proteins that are embedded in the plasma membrane, an important factor in the shape of many cells such as the biconcave-disk appearance of human red blood cells (**Figure 6.4**).

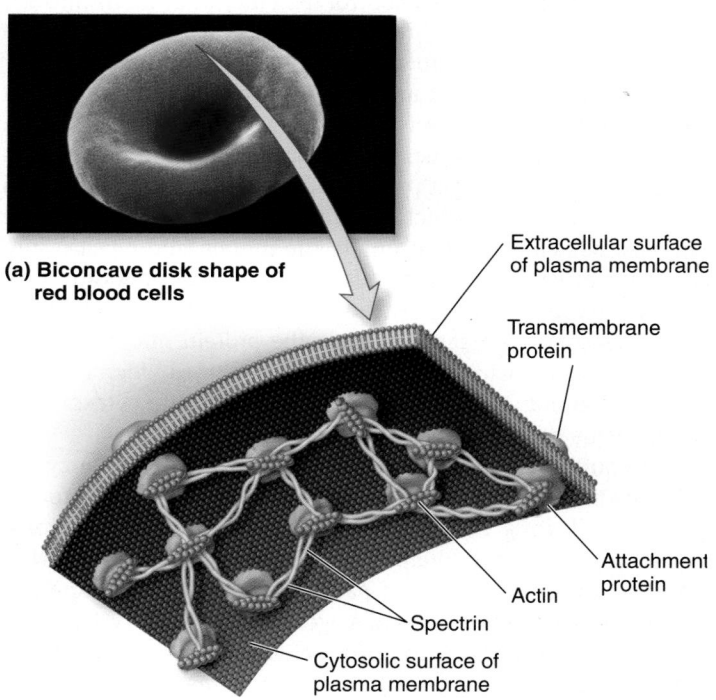

(a) Biconcave disk shape of red blood cells

(b) Cytoskeletal connections to the red blood cell membrane

Figure 6.4 **Role of the cytoskeleton in promoting cell shape.** Many different proteins are involved in forming the shape of a cell. **(a)** A micrograph of a red blood cell, which looks like a biconcave disk. **(b)** To create this shape, proteins within the red blood cell membrane are anchored to an intricate group of cytoskeletal proteins including actin and spectrin.

Biological inquiry: Describe the types of protein-protein interactions that produce the biconcave-disk shape of red blood cells.

The cytoskeleton also organizes and directs intracellular and cellular movements. For example, various types of intracellular cargo, including chromosomes and even organelles such as mitochondria and chloroplasts, are transported within the cell by moving along microtubules and actin filaments. For instance, chloroplasts are sometimes moved to the side of a plant cell that receives more light. **Figure 6.5** illustrates how a **vesicle**—a small, membrane-enclosed sac—moves along microtubules in a nerve cell. In this example, accessory proteins connect a motor protein to the vesicle. The motor protein uses the energy from ATP hydrolysis to "walk" along a microtubule.

Molecules Are Broken Down and Their Building Blocks Are Recycled

Thus far, we have considered how the interactions of cellular components provide organization and complex functions to living cells. The maintenance of cell organization is a dynamic process. Except for DNA, which is stably maintained and inherited from cell to cell, other large molecules such as RNA, proteins, lipids, and polysaccharides have finite lifetimes. Biologists often speak of the **half-life** of molecules, which is the time it takes for 50% of the molecules to be broken down and recycled. For example, a population of mRNA molecules in prokaryotes has an average half-life of about 5 minutes, while mRNAs in eukaryotes tend to exist for longer periods of time, on the order of 30 minutes to 24 hours or even several days.

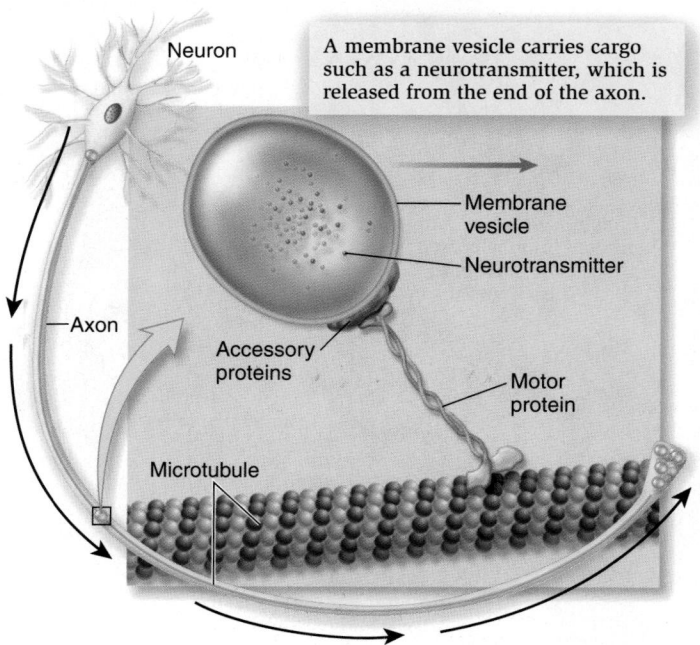

Figure 6.5 **Movement of a membrane vesicle along a microtubule.** In this example, a motor protein walks along a microtubule in the axon of a nerve cell. The vesicle carries a neurotransmitter, which is released at the end of the axon. Accessory proteins link the motor protein to the vesicle and also control the activity of the motor protein.

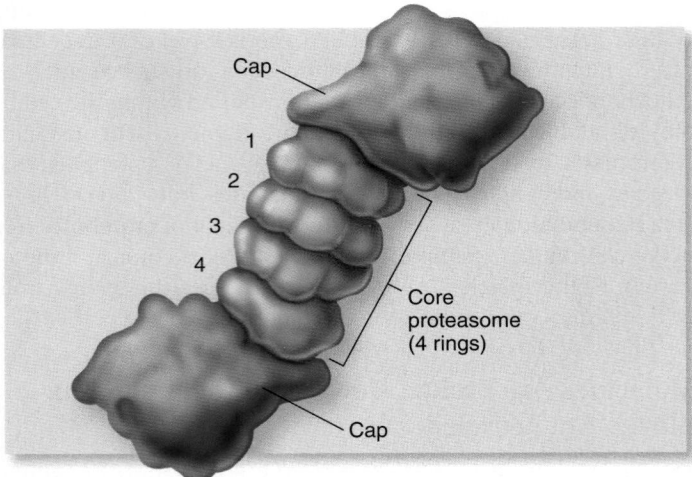

(a) Structure of the eukaryotic proteasome

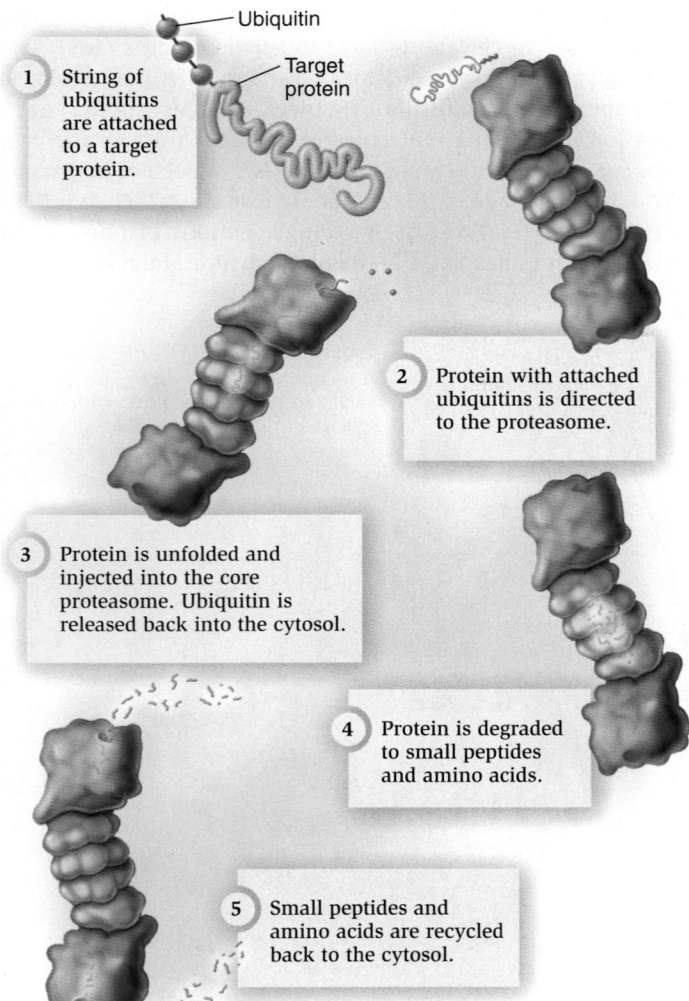

1. String of ubiquitins are attached to a target protein.

2. Protein with attached ubiquitins is directed to the proteasome.

3. Protein is unfolded and injected into the core proteasome. Ubiquitin is released back into the cytosol.

4. Protein is degraded to small peptides and amino acids.

5. Small peptides and amino acids are recycled back to the cytosol.

(b) Steps of protein degradation in eukaryotic cells

Figure 6.6 Protein degradation via the proteasome.

The breakdown of large, cellular molecules and the recycling of their building blocks occur by a variety of mechanisms. As discussed in Chapter 4, lysosomes can degrade materials in the cytosol by a process called autophagy (refer back to Figure 4.20). The components of a mitochondrion, for example, can be degraded in this manner, and building blocks are returned to the cytosol, where they can be used to make new macromolecules. Lysosomes also degrade proteins that are imported into the cell via endocytosis. In addition, cytosolic enzymes break down RNA, polysaccharides, and lipids into smaller building blocks. For example, ribonucleases are enzymes that degrade mRNA molecules.

To survive and respond to environmental changes, cells must continually degrade proteins that are faulty or nonfunctional and synthesize new ones. To be degraded, proteins are recognized by **proteases**—enzymes that cleave the bonds between adjacent amino acids. Although lysosomes in eukaryotic cells are involved in protein breakdown via autophagy and endocytosis, the primary pathway for protein degradation in archaea and eukaryotic cells is via a molecular machine called a **proteasome**. The core of the proteasome is formed from four stacked rings, each composed of seven protein subunits (**Figure 6.6a**). The proteasomes of eukaryotic cells also contain cap structures at each end that control the entry of proteins into the proteasome.

In eukaryotic cells, unwanted proteins are directed to a proteasome by the covalent attachment of a small protein called **ubiquitin**. **Figure 6.6b** describes the steps of protein degradation via eukaryotic proteasomes. First, a string of ubiquitin proteins are attached to the target protein. This event directs the protein to a proteasome cap, which has binding sites for ubiquitin. The cap also has enzymes that unfold the protein and inject it into the internal cavity of the proteasome core. The ubiquitin proteins are removed during entry and return to the cytosol. Inside the proteasome, proteases degrade the protein into small peptides and amino acids. The process is completed when the peptides and amino acids are recycled back into the cytosol. The amino acids can be used to make new proteins.

Ubiquitin-targeting has two advantages. First, the enzymes that attach ubiquitin to its target recognize improperly folded proteins, allowing cells to identify and degrade nonfunctional components. Second, changes in cellular conditions may warrant the rapid breakdown of particular proteins. For example, cell division requires a series of stages called the cell cycle, which depends on the degradation of specific proteins. Ubiquitin targeting directs these proteins to the proteasome for degradation.

Eukaryotic Cells Are a System of Four Interacting Parts That Work Together

We can view a eukaryotic cell as a system of four interacting parts: the interior of the nucleus, the cytosol, the endomembrane system, and the semiautonomous organelles (**Figure 6.7**). These four regions play a role in their own structure and organization, as well as the structure and organization of the entire cell.

Outer nuclear membrane

Inner nuclear membrane

ER

Nuclear pore

Lamins

Internal nuclear matrix proteins

Chromosome

Nucleus:
Lamins are fibrous proteins that organize the inner nuclear membrane and nuclear pores. Internal nuclear matrix proteins form a meshwork inside the nucleus. Each chromosome is organized into its own chromosome territory.

Endomembrane system:
The nuclear envelope, which helps protect and organize the chromosomes, is formed from the endomembrane system.

Endomembrane system:
Lipids are made in the endoplasmic reticulum membrane. The endomembrane system also attaches carbohydrates to lipids and proteins.

Nucleus:
Most of the genome is found in the nucleus. Gene regulation controls which genes in the genome will be expressed to make proteins, which are found throughout the entire cell.

Endomembrane system:
The endomembrane system plays a key role in the movement of larger substances into and out of the cell.

Cytosol:
The cytoskeleton, located in the cytosol, provides organization and facilitates cellular movements.

Cytosol:
The cytosol coordinates responses to the environment via signaling pathways composed of groups of regulatory proteins. Such regulatory proteins may control genes or cellular proteins.

Cytosol:
The cytosol controls the levels of small molecules as well as some macromolecules, such as proteins, via pathways for the synthesis and breakdown of molecules.

Semiautonomous organelles:
Semiautonomous organelles receive materials from the cytosol. Mitochondria provide the cell with most of its ATP, while chloroplasts provide organic molecules via photosynthesis. Peroxisomes help to synthesize certain cellular molecules and also detoxify the cell.

Endomembrane system:
The endomembrane system is involved in the storage and recycling of organic molecules. Lysosomes in animal cells and vacuoles in other organisms also function in the breakdown of the macromolecules whose building blocks are recycled back to the cytosol.

Figure 6.7 **The four interacting parts of eukaryotic cells.** These include the nucleus, cytosol, endomembrane system, and semiautonomous organelles.

Nucleus The nucleus houses the genome. Earlier in this chapter, we learned how the genome plays a key role in producing the proteome. The collection of proteins that a cell makes is largely responsible for the structure and function of the entire cell. Gene regulation, which occurs in the cell nucleus, is very important in creating specific cell types and enabling cells to respond to environmental changes. Chapters 12 and 13 will examine the topics of gene expression and regulation in greater detail.

The nucleus itself is organized by a collection of filamentous proteins called the nuclear matrix (see inset to Figure 6.7). These proteins form extensive protein-protein interactions that perform two roles. First, proteins known as lamins lie along the nuclear envelope and help to organize the inner nuclear membrane and nuclear pores. Second, internal nuclear matrix proteins form a meshwork inside the nucleus. Each chromosome is organized into its own chromosome territory (refer back to Figure 4.17).

Cytosol The cytosol is the region that is outside of the organelles and inside the plasma membrane. It is an important coordination center for cell function and organization. Along with the plasma membrane, the cytosol coordinates responses to the environment. Factors in the environment may stimulate signaling pathways in the cytosol that affect the functions of cellular proteins and change the expression of genes in the cell nucleus.

The cytosol also has a large impact on cell structure because it is the compartment where many small molecules are metabolized in the cell. The region receives molecules that are taken up from the environment. In addition, many pathways for the synthesis and breakdown of cellular molecules are found in the cytosol, and pathways in organelles are often regulated by events there.

A particularly important component of cell organization is the cytoskeleton, which is found in the cytosol. The formation and function of the cytoskeleton is caused by an amazing series of protein-protein interactions. As discussed earlier, the cytoskeleton provides organization to the cell and facilitates cellular movements. In most cells, the cytoskeleton is a dynamic structure, enabling its composition to respond to environmental and developmental changes.

Endomembrane System The endomembrane system can be viewed as a smaller system within the confines of a cell, which is a larger system. The endomembrane system includes the nuclear envelope, endoplasmic reticulum (ER), Golgi apparatus, lysosomes, vacuoles, secretory vesicles, and plasma membrane. This system forms a **secretory pathway** that is crucial in the movement of larger substances, such as proteins and carbohydrates out of the cell. The export of proteins and carbohydrates plays a key role in the organization of materials that surround cells. The structure of this extracellular matrix is discussed in Chapter 10. This pathway can also run in reverse, the **endocytic pathway**, to take substances into the cell.

A key feature of the endomembrane system is that the membranes are very dynamic and their structures change over time. In some cases, the change can be dramatic. For example, during cell division, the nuclear envelope breaks up into membrane vesicles, thereby allowing the release of chromosomes into the cytosol so they can be sorted to daughter cells. Later, vesicles coalesce to re-form a nuclear envelope in each daughter cell. The ability of the nuclear envelope to break up into vesicles and later re-form a nuclear envelope is a vital feature of cell division in eukaryotic cells.

The endomembrane system also contributes to the overall structure and organization of eukaryotic cells in other ways. Most of a cell's lipids are made in the endoplasmic reticulum membrane and distributed to other parts of the cell. In addition, the endomembrane system attaches carbohydrates to lipids and proteins. This process plays a role in protein sorting, and also helps to organize the materials that are found outside of cells.

Finally, another important function of the endomembrane system that serves the needs of the entire cell is the storage and recycling of organic molecules. Vacuoles often play a role in the storage of organic molecules such as proteins, carbohydrates, and fats. When needed, lysosomes in animal cells and vacuoles in the cells of other organisms also assist in breaking down these macromolecules. The building blocks are then recycled back to the cytosol and used to construct new macromolecules.

Semiautonomous Organelles The semiautonomous organelles include the mitochondria, chloroplasts, and peroxisomes. Regarding organization, these organelles tend to be rather independent. They exist in the cytosol much like a bacterium would grow in a laboratory medium. While a bacterium would take up essential nutrients from the growth medium, the semiautonomous organelles take up molecules from the cytosol. The organelles use these molecules to carry out their functions and maintain their organization. Like bacteria, the semiautonomous organelles divide by binary fission to produce more of themselves.

Although the semiautonomous organelles rely on the rest of the cell for many of their key components, they also give back to the cell in ways that are vital to maintaining cell organization. Mitochondria take up organic molecules from the cytosol and give back ATP, which is used throughout the cell to drive processes that are energetically unfavorable. This energy is crucial for cell organization. The chloroplasts capture light energy and synthesize organic molecules. These organic molecules also store energy and can be broken down when energy is needed. In addition, organic molecules, such as sugars and amino acids, are used as building blocks to synthesize many different types of cellular molecules, such as carbohydrate polymers and proteins.

Later in this chapter, we will focus on how mitochondria and chloroplasts are made and how they maintain their distinct organization. The functions of these organelles involve many interesting pathways that are described in greater detail in Chapters 7 and 8, respectively.

6.2 The Endomembrane System

Thus far, we have examined some of the general features that account for the maintenance of cell organization. In this section, we will take a closer look at the organization of the endomembrane system. As mentioned earlier, this system is a collection of membranes and cell compartments in eukaryotic cells that includes the nuclear envelope, endoplasmic reticulum (ER), Golgi apparatus, lysosomes, vacuoles, secretory vesicles, and plasma membrane (**Figure 6.8**). The intracellular organelles of the endomembrane system reside in the cytosol. The system is impressively dynamic. Much of its activity is related to the transport of membrane vesicles between the various compartments. Transport of vesicles occurs in both directions. For example, membrane vesicles bud from the ER and fuse with the Golgi, and vesicles from the Golgi can return to the ER. As we learned in Chapter 5, exocytosis involves the fusion of intracellular vesicles with the plasma membrane, while endocytosis

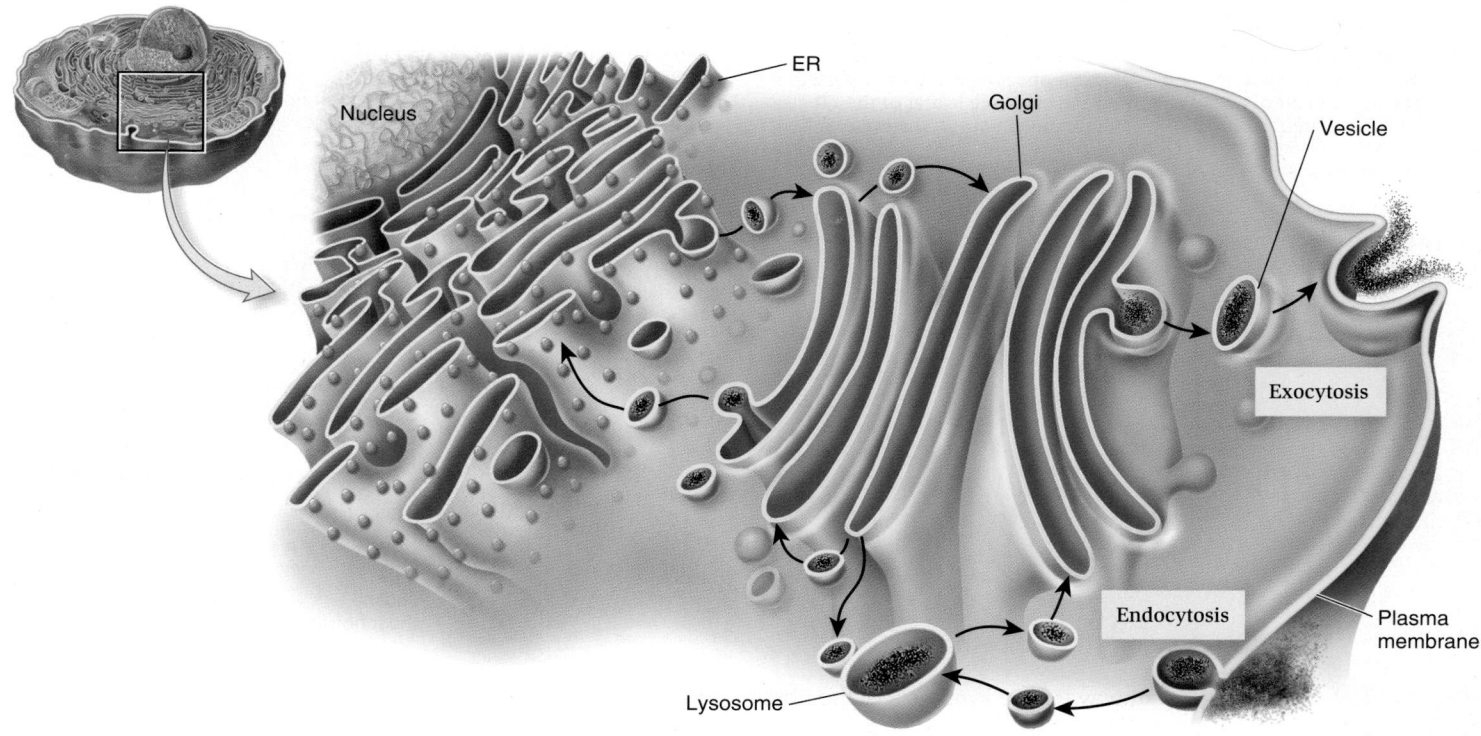

Figure 6.8   **The dynamic nature of the endomembrane system.**   Vesicle transport occurs in both directions, allowing materials to be shared among the various membranes and compartments of the endomembrane system.

involves the inward folding of the plasma membrane to form intracellular vesicles.

The endomembrane system is critical for lipid synthesis, protein synthesis and sorting, and the attachment of carbohydrates to lipids and proteins. Its dynamic nature serves to distribute these materials throughout the cell, modify them for use inside the cell, or process them for secretion to be used outside the cell. The past few decades have seen exciting advances in our understanding of how these events occur at the molecular level. In this section, we will examine the roles of the endomembrane system in cell structure, function, and organization.

Lipid Synthesis Occurs at the ER Membrane

In eukaryotic cells, the cytosol and endomembrane system work together to synthesize most lipids. This process occurs at the cytosolic leaflet of the ER membrane. **Figure 6.9** shows a simplified pathway for the synthesis of phospholipids, the main components of cell and organelle membranes. The building blocks for a phospholipid are two fatty acids each with an acyl tail, one glycerol molecule, one phosphate, and a polar group. These building blocks are made via enzymes in the cytosol or they are taken into cells from the diet. To begin the process of phospholipid synthesis, the fatty acids are activated by attachment to an organic molecule called coenzyme A (CoA). This acti-

vation promotes the bonding of the two fatty acids to a glycerol-phosphate molecule, and the resulting molecule is inserted into the cytosolic leaflet of the ER membrane. The phosphate is removed from glycerol, and then a polar molecule that is linked to phosphate is attached to glycerol. In this example, the polar head group contains choline, but many other types are possible. Phospholipids are initially made in the cytosolic leaflet, but flippases in the ER membrane transfer some to the other leaflet.

The lipids that are made in the ER membrane can be transferred to other membranes in the cell by a variety of mechanisms. Phospholipids in the ER can diffuse laterally to the nuclear envelope (**Figure 6.10a**). In addition, lipids can be transported through the cytosol via vesicles to the Golgi, lysosomes, vacuoles, and plasma membrane (**Figure 6.10b**). A third mode of lipid movement involves **lipid exchange proteins**, which extract a lipid from one membrane, diffuse through the cell, and insert the lipid into another membrane (**Figure 6.10c**). Such transfer can occur between any two membranes, even between the endomembrane system and semiautonomous organelles. For example, lipid exchange proteins can transfer lipids between the ER and mitochondria. In addition, chloroplasts and mitochondria can synthesize certain types of lipids that can be transferred from these organelles to other cellular membranes via lipid exchange proteins.

1. In the cytosol, fatty acids are activated by the attachment of a CoA molecule.

2. The activated fatty acids bind to glycerol-phosphate and are inserted into the cytosolic leaflet of the ER membrane via acyl transferase.

3. The phosphate is removed by a phosphatase enzyme.

4. A choline linked to phosphate is attached via choline phosphotransferase.

5. Flippases transfer some of the phospholipids to the other leaflet.

Figure 6.9   A simplified pathway for the synthesis of membrane phospholipids at the ER membrane.

Lipids in ER membrane can laterally diffuse to nuclear membrane.

(a) Lateral diffusion

A vesicle with a lipid bilayer can bud from one membrane of the endomembrane system and fuse with a different membrane of the endomembrane system.

(b) Vesicle transport

A lipid exchange protein can remove a lipid from one membrane and insert it into a different membrane.

(c) Lipid exchange proteins

Figure 6.10   **Mechanisms of lipid transfer throughout a eukaryotic cell.**   As shown in the insets, lipids can be distributed throughout the cell in various ways. **(a)** Lateral diffusion occurs between the ER and nuclear envelope. **(b)** Vesicle transport occurs among the membranes of the endomembrane system. **(c)** Lipid exchange proteins can transfer lipids between any two membranes, such as between the ER and a mitochondrion.

FEATURE INVESTIGATION

Palade Demonstrated That Secreted Proteins Move Sequentially Through Organelles of the Endomembrane System

Eukaryotic cells make thousands of different proteins. In most cases, a protein functions in only one compartment within a cell, or it functions only after it is secreted from the cell. Therefore, proteins must be sorted to the correct locations. One of the first indications that proteins are sorted intracellularly came from studies by George Palade and his collaborators in the 1960s.

Palade's team conducted **pulse-chase experiments**, in which the researchers administered a pulse of radioactive amino acids to cells so that they made radioactive proteins. A few minutes later, the cells were given a large amount of nonradioactive amino acids. This is called a "chase" because it chases away the ability of the cells to make any more radioactive proteins. In this way, radioactive proteins were produced only briefly. Because they were labeled with radioactivity, the fate of these proteins could be monitored over time. The goal of a pulse-chase experiment is to determine where the radioactive proteins are produced and the pathways they take as they travel through a cell.

Palade chose to study the cells of the pancreas. This organ secretes enzymes and protein hormones that play a role in digestion. Therefore, these cells were chosen because their primary activity is protein secretion. To study the pathway for protein secretion, Palade and colleagues injected a radioactive version of the amino acid leucine into the bloodstream of guinea pigs, followed 3 minutes later by an injection of nonradioactive leucine (**Figure 6.11**). At various times after the second injection, samples of pancreatic cells were removed from the animals. The cells were then prepared for transmission electron microscopy (see Chapter 4). The sample was stained with a heavy metal that became bound to membranes and showed the locations of the cell organelles. In addition, the sample was coated with a radiation-sensitive emulsion. When radiation was emitted from radioactive proteins, it interacted with the emulsion in a way that caused the precipitation of silver atoms, which became tightly bound to the sample. In this way, the precipitated silver atoms marked the location of the radiolabeled proteins. Unprecipitated silver chloride in the emulsion was later washed away. Because silver atoms are electron dense, they produce dark spots in a transmission electron micrograph. Therefore, dark spots revealed the locations of radioactive proteins.

Figure 6.11   Palade's use of the pulse-chase method to study protein secretion.

HYPOTHESIS Proteins that are to be secreted follow a particular intracellular pathway.

STARTING MATERIALS Male guinea pigs.

Experimental level   Conceptual level

1   Inject guinea pigs with a radioactive amino acid, [3H]-leucine. After 3 minutes, inject them with nonlabeled leucine, which is called a chase.

[3H]-leucine

Nonlabeled leucine

Pancreas

2   At various times after the second injection, remove samples of pancreatic cells.

Pancreatic cell

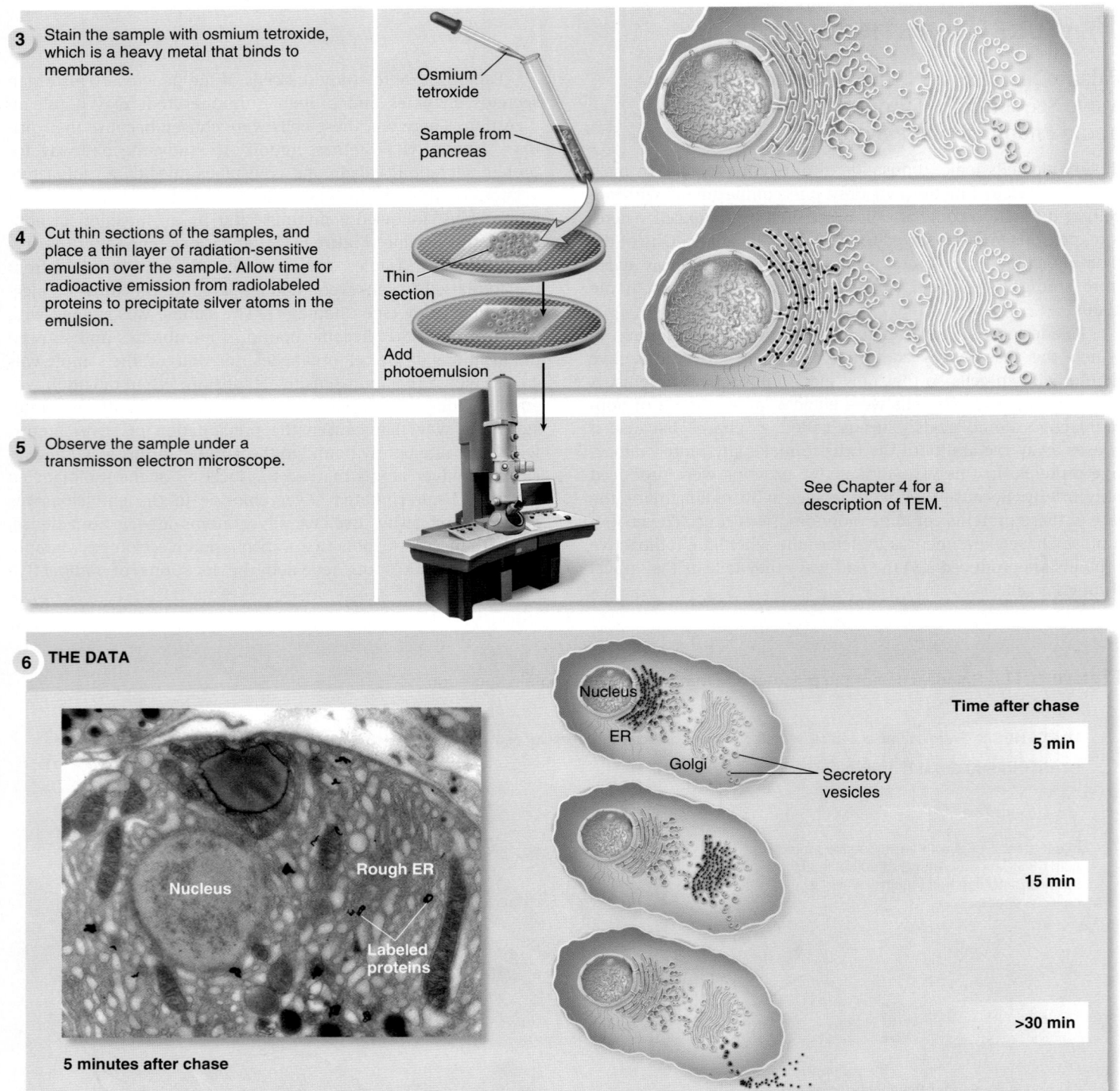

3 Stain the sample with osmium tetroxide, which is a heavy metal that binds to membranes.

Osmium tetroxide

Sample from pancreas

4 Cut thin sections of the samples, and place a thin layer of radiation-sensitive emulsion over the sample. Allow time for radioactive emission from radiolabeled proteins to precipitate silver atoms in the emulsion.

Thin section

Add photoemulsion

5 Observe the sample under a transmisson electron microscope.

See Chapter 4 for a description of TEM.

6 THE DATA

Nucleus

ER

Golgi

Secretory vesicles

Time after chase

5 min

15 min

>30 min

Nucleus

Rough ER

Labeled proteins

5 minutes after chase

The micrograph in the data of Figure 6.11 illustrates the results that were observed 5 minutes after the completion of the pulse-chase injections. Very dark objects, namely radioactive proteins, were observed in the rough ER. As shown schematically to the right of the actual data, later time points indicated that the radioactive proteins moved from the ER to the Golgi, and then to secretory vesicles near the plasma membrane. In this way, Palade followed the intracellular pathway of protein movement. His experiments provided the first evidence that secreted proteins were synthesized into the rough ER and moved through a series of cellular compartments before they were secreted. These findings caused researchers to wonder how proteins are targeted to particular organelles and how they move from one compartment to another. These topics are described next.

Protein Localization Involves Sorting Signals and Vesicle Transport

Since Palade's pioneering studies, scientists have learned a great deal about the localization of proteins. The process of evolution has produced a proteome in every organism in which each type of protein functions in a particular cellular compartment or performs its role after being secreted from the cell. For example, acid hydrolases function in the lysosome to degrade macromolecules. In eukaryotes, most proteins contain short stretches of amino acid sequences that direct them to their correct cellular location. These sequences are called **sorting signals** or **traffic signals**. Each sorting signal is recognized by specific cellular components that facilitate the proper routing of that protein. **Table 6.1** describes the sorting signals found in proteins that are destined for the nucleus, endomembrane system, or secretion. We will consider the signals involved in semiautonomous organelles later in this chapter.

Most eukaryotic proteins begin their synthesis on ribosomes in the cytosol. The cytosol provides amino acids, which are used as building blocks to make these proteins. The synthesis of a protein, a process called translation, is completed in the cytosol for those proteins destined for the cytosol, nucleus, mitochondria, chloroplasts, or peroxisomes. Cytosolic proteins lack any sorting signal, so they stay there. By comparison, the uptake of proteins into the nucleus, mitochondria, chloroplasts, and peroxisomes occurs after the protein is completely made (that is, completely translated). This is called **post-translational sorting** because sorting does not happen until translation is finished. We will examine the uptake of proteins into the mitochondrion later in the chapter. By comparison, the synthesis of other eukaryotic proteins begins in the cytosol and then halts temporarily until the ribosome has become bound to the ER membrane. After this occurs, translation resumes and the polypeptide is synthesized into the ER lumen or ER membrane. Proteins that are destined for the ER, Golgi, lysosome, vacuole, plasma membrane, or secretion are first directed to the ER. This is **cotranslational sorting** because the first step in the sorting process begins while translation is occurring.

The concept of sorting signals in proteins was first proposed by Günter Blobel in the 1970s. Blobel and colleagues discovered a sorting signal in proteins that sends them to the ER membrane, which is the first step in cotranslational sorting (**Figure 6.12**). To be directed to the rough ER membrane, a polypeptide must contain a sorting signal called an **ER signal sequence**, which is usually located near the amino terminus. As the ribosome is making the polypeptide in the cytosol, the ER signal sequence emerges from the ribosome and is recognized by a protein/RNA complex called a **signal recognition particle (SRP)**. The SRP has two functions. First, it recognizes the ER signal sequence and pauses translation. Second, SRP binds to a receptor in the ER membrane, which docks the ribosome over a channel. At this stage, SRP is released and translation resumes. The growing polypeptide is threaded through the channel to cross the ER membrane. In most cases, the ER signal sequence is removed by signal peptidase. If the protein is not a membrane protein, it will be released into the lumen of the ER. In 1999, Blobel won the Nobel Prize for his discovery of sorting signals in proteins. It is worth noting that the process shown in Figure 6.12 illustrates another important role of protein-protein interactions—a series of interactions causes the steps of a process to occur in a specific order.

Some proteins are meant to function in the ER. Such proteins contain ER retention signals in addition to the ER signal sequence. Alternatively, other proteins that are destined for the Golgi, lysosomes, vacuoles, plasma membrane, or secretion must be sorted to these other locations. Such proteins leave the ER and are transported to their correct location. This transport process occurs via membrane vesicles that are formed from one compartment and then move through the cytosol and fuse with another compartment. Vesicles from the ER may go to the Golgi, and then vesicles from the Golgi may go to the lysosomes, vacuoles, or plasma membrane. Sorting signals within their amino acid sequences are responsible for directing them to their correct location.

Table 6.1	Sorting Signals of Proteins That Are Destined for the Nucleus, Endomembrane System, or Secretion
Type of signal	**Description**
Nuclear-localization signal	Can be located almost anywhere in the polypeptide sequence. The signal sequence is four to eight amino acids in length and contains several positively charged residues and usually one or more prolines.
ER signal	A sequence of about 6 to 12 amino acids near the amino terminus that is composed of mostly nonpolar amino acids.
ER retention signal	A sequence of four amino acids, lysine-aspartic acid-glutamic acid-leucine, located at the carboxyl terminus of the protein.
Golgi retention signal	A sequence of 20 hydrophobic amino acids that form a transmembrane segment flanked by positively charged residues.
Lysosome-sorting signal	A grouping of amino acids in the tertiary structure of of a polypeptide. Positively charged residues within this grouping are thought to play an important role. This grouping causes lysosomal proteins to be covalently modified to contain a mannose-6-phosphate residue (a carbohydrate), which directs the protein to the lysosome.
Destination of a cellular protein	**Type of signal the protein contains within its amino acid sequence**
Nucleus	Nuclear-localization signal.
ER	ER signal and an ER retention signal.
Golgi	ER signal and a Golgi retention signal.
Lysosome	ER signal and a lysosomal sorting signal.
Plasma membrane	ER signal and a hydrophobic transmembrane domain that anchors it in the membrane.
Secretion	ER signal. No additional signal required.

1 SRP binds to ER signal sequence and pauses translation.

2 SRP binds to receptor in ER membrane.

3 SRP is released and translation resumes. The growing polypeptide is threaded into the channel.

5'

NH₃⁺

ER signal sequence

SRP

Cytosol

ER membrane

Channel protein

SRP receptor

ER lumen

Signal peptidase

Cleaved signal sequence

COO⁻

NH₃⁺

4 ER signal is cleaved by signal peptidase.

5 Protein is completely made and released into ER lumen.

Figure 6.12   First step in cotranslational protein localization: cotranslational sorting.

1 In this example, a cargo of proteins binds to receptors in the ER membrane. The binding of coat proteins helps a vesicle bud from the membrane, and v-snares are incorporated into vesicle.

t-snare

Golgi membrane

Coat proteins

Protein cargo

Cargo receptor

Dynamin

v-snare

2 When the vesicle is nearly complete, dynamin pinches the membrane and releases the vesicle.

ER membrane

3 After the vesicle is released, the coat is shed.

4 Vesicle binds to target membrane by a v-snare/ t-snare interaction.

5 Vesicle fuses with target membrane to deliver protein cargo to its target destination.

Figure 6.13   Second step in cotranslational protein localization: vesicle transport from the endoplasmic reticulum.

Figure 6.13 describes vesicle transport from the ER to the Golgi. A cargo, such as soluble protein molecules, is loaded into a developing vesicle by binding to cargo receptors in the ER membrane. Vesicle formation is facilitated by **coat proteins**, which help a vesicle to bud from a given membrane. As a vesicle forms, other proteins called **v-snares** are incorporated into the vesicle membrane, hence the name v-snare. Many types of v-snares are known to exist; the particular v-snare that is found in a vesicle membrane depends on the type of cargo that it carries. Finally, a protein called dynamin pinches the membrane and releases the vesicle. After a vesicle is released from one compartment such as the ER, the coat is shed. The vesicle then travels through the cytosol. But how does the vesicle know where to go? The answer is that the v-snares in the vesicle membrane are recognized by **t-snares** in a target membrane. After the v-snare recognizes a t-snare, the vesicle fuses with the membrane containing the t-snare. The recognition between v-snares and t-snares ensures that a vesicle carrying a specific cargo moves to the correct target membrane in the cell. Like the sorting of proteins to the ER membrane, the formation and sorting of vesicles also involves a series of protein-protein interactions that cause the steps to occur in a defined manner.

Most Transmembrane Proteins Are First Inserted into the ER Membrane

Our previous discussion of protein targeting involved a soluble protein (that is, a nonmembrane protein) that had an ER signal sequence. After the signal was removed, the protein was released into the ER lumen. With the exception of proteins destined for semiautonomous organelles, most membrane proteins are recognized by SRPs and synthesized into the ER membrane. From there, they may be transported via vesicles to other membranes of the endomembrane system.

As discussed in Chapter 5, if a sequence within a polypeptide contains a stretch of 20 amino acids that are mostly hydrophobic, this region will become a transmembrane segment. In the example shown in **Figure 6.14**, the polypeptide contains one such sequence. After the ER signal sequence is cleaved or removed, this will create a membrane protein with a single transmembrane segment. Other polypeptides may contain two or more transmembrane segments. Each time a polypeptide sequence contains a stretch of 20 hydrophobic amino acids, an additional transmembrane segment is synthesized into the membrane. For some membrane proteins, the ER signal sequence may not be removed. When it remains, it will usually function as a transmembrane segment.

Glycosylation of Proteins Occurs in the ER and Golgi Apparatus

Glycosylation is the attachment of carbohydrate to a protein, producing a glycoprotein. Carbohydrates may also be attached to lipids by glycosylation, but here we will focus on the glycosylation of proteins. In proteins, glycosylation may aid in protein folding, and it may protect a protein from extracellular factors that could harm its structure. In addition, glycosylation may play a role in protein sorting. For example, proteins destined for the lysosome have attached carbohydrate that serves as a sorting signal (see Table 6.1).

Two forms of protein glycosylation are known to occur in eukaryotes: N-linked and O-linked. N-linked glycosylation, which also occurs in archaea, involves the attachment of a carbohydrate to the amino acid asparagine in a polypeptide chain. It is called N-linked because the carbohydrate attaches to a nitrogen atom of the asparagine side chain. For this to occur, a group of 14 sugar molecules are built onto a lipid called dolichol. This carbohydrate tree is then transferred to an asparagine as a polypeptide is synthesized into the ER lumen (**Figure 6.15**). The carbohydrate tree is attached only to asparagines occurring in the sequence asparagine-X-serine or asparagine-X-threonine, where X could be any amino acid except proline. An enzyme in the ER, oligosaccharide transferase, recognizes this sequence and transfers the carbohydrate tree from dolichol to the asparagine.

Following this initial glycosylation step, the carbohydrate tree is further modified as other enzymes in the ER attach additional sugars or remove sugars. After a glycosylated protein is transferred to the Golgi by vesicle transport, enzymes in the Golgi usually modify the carbohydrate tree as well. N-linked glycosylation commonly occurs on membrane proteins

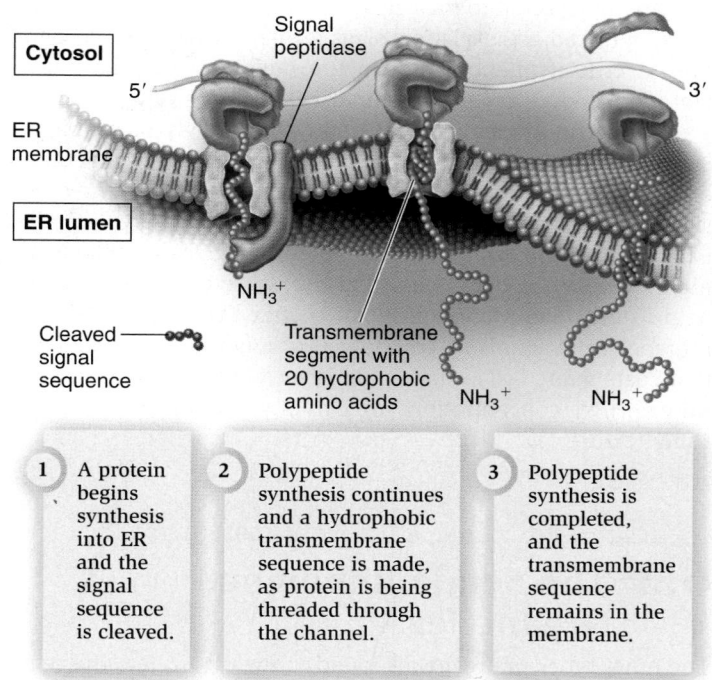

Figure 6.14   Insertion of membrane proteins into the ER membrane.

Biological inquiry: What structural feature of a protein causes a region to form a transmembrane segment?

Figure 6.15   N-linked glycosylation in the endoplasmic reticulum.

that are transported to the cell surface. In some cell surface proteins, N-linked glycosylation plays a role in cell-to-cell recognition, a crucial phenomenon in the migration of cells during embryonic development in animals.

The second form of glycosylation, O-linked glycosylation, occurs only in the Golgi apparatus. This form involves the addition of a string of sugars to the oxygen atom of serine or threonine side chains in polypeptides. In animals, O-linked glycosylation is important for the production of proteoglycans, which are highly glycosylated proteins that are secreted from cells and help to organize the extracellular matrix that surrounds cells. Proteoglycans are also a component of mucus, a slimy material that coats many cell surfaces and is secreted into fluids such as saliva. High concentrations of carbohydrates give mucus its slimy texture.

6.3 The Semiautonomous Organelles

The other types of organelles found in the cytosol are the semiautonomous organelles—mitochondria, chloroplasts, and peroxisomes—so named because they divide by fission to produce more of themselves. In Chapter 4, we examined their general structures and functions. Although they are compartments within the cytosol, the mitochondria, chloroplasts, and peroxisomes are somewhat like independent systems. For example, mitochondria and chloroplasts have their own genetic material, synthesize some proteins, and divide independently of when the cell divides. However, they are not entirely autonomous because they rely on the rest of the cell for raw materials and even import most of their proteins from the cytosol. As we will examine in this section, the seemingly mysterious behavior of these organelles can be traced to their evolutionary origin about 2 billion years ago.

Mitochondria and Chloroplasts Contain Their Own Genetic Material and Divide by Binary Fission

To appreciate the structure and organization of mitochondria and chloroplasts, we need to briefly examine their genetic properties. In 1951, Y. Chiba exposed plant cells to Feulgen, a DNA-specific dye, and discovered that the chloroplasts became stained. Based on this observation, he was the first to suggest that chloroplasts contain their own DNA. Researchers in the 1970s and 1980s isolated DNA from both chloroplasts and mitochondria. These studies revealed that the DNA of these organelles resembled smaller versions of bacterial chromosomes.

The chromosomes found in mitochondria and chloroplasts are referred to as the **mitochondrial genome** and **chloroplast genome**, while the chromosomes found in the nucleus of the cell constitute the **nuclear genome**. The genomes of most mitochondria and chloroplasts are composed of a single circular, double-stranded chromosome. Compared to the nuclear genome, they are very small. For example, the amount of DNA in the human nuclear genome (about 3 billion base pairs) is about 200,000 times greater than in the mitochondrial genome. In terms of genes, the human genome has approximately 20,000 to 25,000 different genes while the human mitochondrial genome has only several dozen. Chloroplast genomes tend to be larger than mitochondrial genomes, and they have a correspondingly greater number of genes. Depending on the particular species of plant or algae, a chloroplast genome is about 10 times larger than the mitochondrial genome of human cells.

Just as the genomes of mitochondria and chloroplasts resemble bacterial genomes, the production of new mitochondria and chloroplasts bears a striking resemblance to the division of bacterial cells. Like their bacterial counterparts, mitochondria and chloroplasts increase in number via **binary fission**, or splitting in two. **Figure 6.16** illustrates the process for mitochondria. The mitochondrial genome in the nucleoid is duplicated and the organelle divides into two separate organelles. Mitochondrial and chloroplast division are needed to maintain a full

Nucleoid

1 Mitochondrial genome replicates.

2 Mitochondrion begins to divide by binary fission.

3 Binary fission is completed.

(b) Micrographs of the process

(a) Binary fission of mitochondria

Figure 6.16   Division of mitochondria by binary fission. **(a)** The mitochondrial genome is duplicated and the organelle divides into two separate organelles. **(b)** Micrographs of the process.

complement of these organelles when cell growth occurs following cell division. In addition, environmental conditions may influence the sizes and numbers of these organelles. For example, when plants are exposed to more sunlight, the number of chloroplasts in leaf cells increases.

GENOMES & PROTEOMES

Mitochondria and Chloroplasts Are Derived from Ancient Symbiotic Relationships

The observation that mitochondria and chloroplasts contain their own genetic material may seem puzzling. Perhaps you might think that it would be simpler for a eukaryotic cell to have all of its genetic material in the nucleus. The distinct genomes of mitochondria and chloroplasts can be traced to their evolutionary origin, which involved an ancient symbiotic association.

A symbiotic relationship occurs when two different species live in direct contact with each other. **Endosymbiosis** describes a symbiotic relationship in which the smaller species—the symbiont—actually lives inside (*endo-*, inside) the larger species. In 1883, Andreas Schimper proposed that chloroplasts were descended from an endosymbiotic relationship between cyanobacteria (a bacterium capable of photosynthesis) and eukaryotic cells. In 1922, Ivan Wallin also hypothesized an endosymbiotic origin for mitochondria.

In spite of these interesting ideas, the question of endosymbiosis was largely ignored until the discovery that mitochondria and chloroplasts contain their own genetic material. In 1970, the issue of endosymbiosis as the origin of mitochondria and chloroplasts was revived by Lynn Margulis in her book *Origin of Eukaryotic Cells*. During the 1970s and 1980s, the advent of molecular genetic techniques allowed researchers to analyze genes from mitochondria, chloroplasts, bacteria, and eukaryotic nuclear genomes. Researchers discovered that genes in mitochondria and chloroplasts are very similar to bacterial genes. Likewise, mitochondria and chloroplasts are strikingly similar in size and shape to certain bacterial species. These observations provided strong support for the **endosymbiosis theory**, which proposes that mitochondria and chloroplasts originated from bacteria that took up residence within a primordial eukaryotic cell (**Figure 6.17**). Over the next 2 billion years, the characteristics of the intracellular bacterial cell gradually changed to those of a mitochondrion or chloroplast. We will return to this topic in Chapter 22.

Symbiosis occurs because the relationship is beneficial to one or both species. According to the endosymbiosis theory, this relationship provided eukaryotic cells with useful cellular characteristics. Chloroplasts, which were derived from cyanobacteria, have the ability to carry out photosynthesis. This benefits plant cells by giving them the ability to use the energy from sunlight. It is less clear how the relationship would have been beneficial to a cyanobacterium. By comparison, mitochondria are

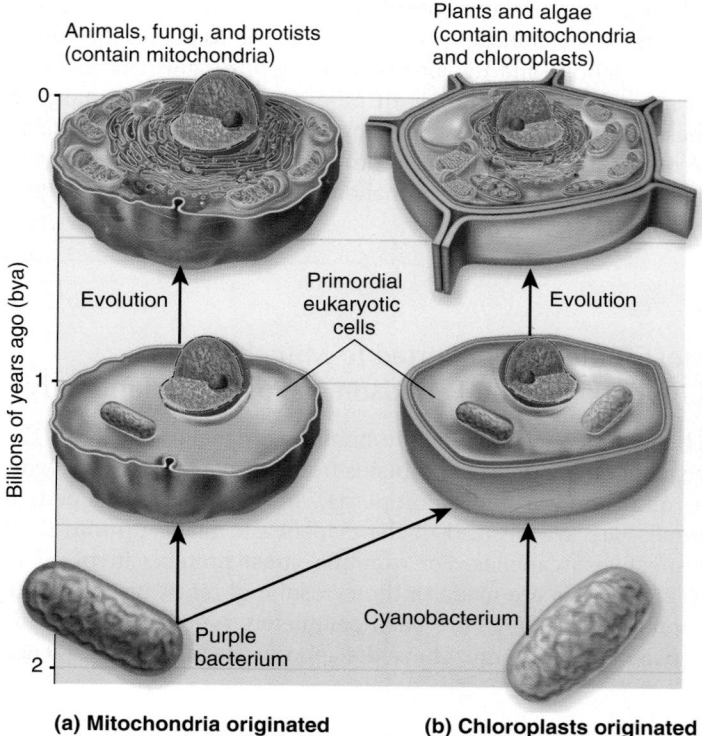

(a) Mitochondria originated from endosymbiotic purple bacteria.

(b) Chloroplasts originated from endosymbiotic cyanobacteria.

Figure 6.17 **The endosymbiosis theory.** **(a)** According to this concept, modern mitochondria were derived from purple bacteria, also called α-proteobacteria. Over the course of evolution, their characteristics changed into those found in mitochondria today. **(b)** A similar phenomenon occurred for chloroplasts, which were derived from cyanobacteria, a bacterium that is capable of photosynthesis.

Biological inquiry: Discuss the similarities and differences between modern bacteria and mitochondria.

thought to have been derived from a different type of bacteria known as purple bacteria or α-proteobacteria. In this case, the endosymbiotic relationship enabled eukaryotic cells to synthesize greater amounts of ATP.

During the evolution of eukaryotic species, genes that were originally found in the genome of the primordial cyanobacteria and purple bacteria have been transferred from the organelles to the nucleus. This has occurred many times throughout evolution, so that modern mitochondria and chloroplasts have lost most of the genes that still exist in present-day purple bacteria and cyanobacteria. Some biologists have proposed that peroxisomes may also have arisen by an endosymbiotic relationship but have lost all of their genetic material. Alternatively, others suggest that peroxisomes may have their origins in the endomembrane system. The evolutionary origin of peroxisomes remains unclear.

Some researchers speculate that the movement of genes into the nucleus makes it easier for the cell to control the structure, function, and division of mitochondria and chloroplasts. In modern cells, hundreds of different proteins that make up these organelles are encoded by genes that have been transferred to the nucleus. These proteins are made in the cytosol, and then taken up into mitochondria or chloroplasts. We will discuss this topic next.

Most Proteins Are Sorted to Mitochondria, Chloroplasts, and Peroxisomes Post-Translationally

The organization of semiautonomous organelles is largely dependent on the uptake of proteins from the cytosol. Most proteins in mitochondria and chloroplasts, and all proteins in peroxisomes, are synthesized in the cytosol and taken up into their respective organelles. For example, most proteins involved in ATP synthesis are made in the cytosol and taken up into mitochondria after they have been completely synthesized. For this to occur, a protein must have the appropriate targeting sequence as part of its amino acid sequence. **Table 6.2** summarizes the targeting sequences that direct proteins to mitochondria, chloroplasts, and peroxisomes.

As one example of post-translational sorting, let's consider how a protein is directed to the mitochondrial matrix. As described in **Figure 6.18**, this process involves a series of intricate protein-protein interactions. A protein that is destined for the mitochondrial matrix is first made in the cytosol, where proteins called chaperones keep it in an unfolded state. A receptor protein in the outer mitochondrial membrane recognizes the matrix targeting sequence. The protein is released from the chaperone as it is transferred to a channel in the outer mitochondrial membrane. Because it is in an unfolded state, the mitochondrial protein can be threaded through this channel, and then through another channel in the inner mitochondrial membrane. These channels lie close to each other at contact sites between the outer and inner membranes. As the protein emerges in the matrix, other chaperone proteins that were already in the matrix continue to keep it unfolded. Eventually, the matrix targeting sequence is cleaved and the entire protein is threaded into the matrix. At this stage, the chaperone proteins are released and the protein can adopt its three-dimensional structure.

Table 6.2	Targeting Sequences of Proteins That Are Destined for Semiautonomous Organelles*
Destination	**Mitochondria**
Matrix	Contains a matrix targeting sequence that is usually a short sequence at the amino terminus of a protein with several positively charged residues (that is, lysine and arginine). This sequence folds into an α-helix in which the positive charges are on one face of the helix.
Outer-membrane	Has a matrix targeting sequence followed by a stop transfer sequence that prevents passage through the outer membrane channel.
Intermembrane space	Contains a matrix targeting sequence followed by a cleavage site that causes the release of the protein into the intermembrane space.
Inner-membrane	Contains a matrix targeting sequence and one or more transmembrane regions.
	Chloroplasts
Stroma	Contains a stroma targeting sequence that is an uncharged α-helix with polar amino acids along one side.
Outer-membrane	Not well understood, but may have a stroma targeting sequence and a stop transfer sequence.
Intermembrane space	Not well understood, but may have a stroma targeting sequence and a cleavage site that causes the release of the protein into the intermembrane space.
Inner-membrane	Contains a stroma targeting sequence and one or more transmembrane regions.
Thylakoid membrane	Also has a stroma targeting sequence followed by a thylakoid targeting sequence that is a stretch of hydrophobic amino acids.
	Peroxisomes
Peroxisome membrane and lumen	Most peroxisomal proteins contain a specific sequence of three amino acids, serine-lysine-leucine, that is located near the carboxyl terminus of the protein. A few peroxisomal proteins have a 26 amino acid sorting sequence at the amino terminus.

*These are the most common types of targeting sequences, also called sorting signals. Variations are known to occur in the amino acid sequences that function as the sorting signals.

Figure 6.18 Post-translational sorting of a protein to the mitochondrial matrix.

Biological inquiry: What do you think would happen if chaperone proteins did not bind to a mitochondrial matrix protein before it was imported into the mitochondrion?

CHAPTER SUMMARY

6.1 Principles of Cell Organization

- In systems biology, researchers study living organisms in terms of their underlying network structure—groups of structural and functional connections—rather than their individual molecular components.

- The maintenance of cell structure relies on the genome, functional molecules, and pre-existing organization. Other factors include the ability of certain proteins to interact with each other to form molecular machines, and the dynamic turnover of cellular molecules and macromolecules.

- Modern cells can grow, adapt, and modify their organization, but they do not arise spontaneously from nonliving materials.

- Cell organization depends on genetic information in the cell nucleus, which produces a proteome. (Figure 6.1)

- Interactions among proteins and RNA molecules form molecular machines that carry out complex cellular functions. (Figures 6.2, 6.3)

- The cytoskeleton is a molecular machine that provides cell organization and directs cellular movements. (Figures 6.4, 6.5)

- The maintenance of cell organization is a dynamic process that involves breaking down molecules and recycling their building blocks. (Figure 6.6)

- In eukaryotic cells, four regions work together to produce dynamic organization. The nucleus houses the genome, which plays a key role in producing the proteome. The cytosol is an important coordination center for cell metabolism and organization. The organelles of the endomembrane system perform several important functions in eukaryotic cells. The semiautonomous organelles (mitochondria, chloroplasts, and peroxisomes) perform a variety of crucial functions. (Figure 6.7)

6.2 The Endomembrane System

- In eukaryotic cells, the cytosol and endomembrane system work together to synthesize most lipids. The endomembrane system includes the nuclear envelope, endoplasmic reticulum, Golgi apparatus, lysosomes, vacuoles, and plasma membrane. (Figures 6.8, 6.9, 6.10)

- Palade's pulse-chase experiments demonstrated that secreted proteins move sequentially through the organelles of the endomembrane system. (Figure 6.11)

- Protein localization involves sorting signals and vesicle transport. (Table 6.1, Figures 6.12, 6.13)

- Most transmembrane proteins are first inserted into the ER membrane and then transported via vesicles to other membranes of the endomembrane system. (Figure 6.14)

- Glycosylation of proteins occurs in the ER and Golgi apparatus. Glycosylation may help proteins fold properly, protect them from extracellular factors, and assist in protein sorting. (Figure 6.15)

6.3 The Semiautonomous Organelles

- Mitochondria and chloroplasts contain their own genetic material and divide by binary fission. (Figure 6.16)

- According to the endosymbiosis theory, mitochondria and chloroplasts have evolved from bacteria that took up residence in early eukaryotic cells. (Figure 6.17)

- Most proteins are sorted to mitochondria, chloroplasts, and peroxisomes post-translationally. (Table 6.2, Figure 6.18)

TEST YOURSELF

1. The main structural elements of cells that produce cellular organization are
   a. proteins.
   b. organelles.
   c. membranes.
   d. all of the above.
   e. a and c only.

2. Which of the following statements best supports the requirements to make new cells?
   a. The formation of new cells relies solely on the presence of the genome.
   b. New cell formation requires the correct genetic information and the building blocks necessary to produce the cellular components.
   c. New cell production requires the genome, functional molecules, and pre-existing organization.
   d. The formation of new cells requires the appropriate genetic information and functional molecules for cellular activity only.
   e. All of the above.

3. Large molecular complexes that perform different cellular activities that involve changes in molecular conformation are called molecular
   a. clocks.
   b. motors.
   c. machines.
   d. proteins.
   e. proteomes.

4. Protein conformation is important to protein function. The assembly of many complex proteins relies on _____ , where surfaces of subunits recognize each other and bind together.
   a. protein sorting
   b. traffic signaling
   c. post-transcriptional changes
   d. molecular recognition
   e. proteasome activity

5. In the nucleus, proteins help maintain organization by
   a. forming a meshwork in the nucleus that organizes each chromosome into its own chromosome territory.
   b. organizing the outer nuclear membrane.
   c. organizing the inner nuclear membrane.
   d. all of the above.
   e. a and c only.

6. The cytoskeleton is an important feature of the cytosol that provides organization by
   a. determining cell shape.
   b. determining the structure of the endomembrane system.
   c. transporting proteins to the endoplasmic reticulum for protein sorting.
   d. recycling membrane-bounded proteins.
   e. all of the above.

7. Proteins that function to move lipids from one membrane to another are called
   a. lipases.
   b. membrane-bounded lipoproteins.
   c. proteases.
   d. lipid exchange proteins.
   e. phospholipids.

8. Protein sorting in the cell is possible due to
   a. sorting signals in the amino acid sequences of proteins that determine protein destinations in the cell.
   b. chaperone proteins that function to direct all proteins to the proper location inside or outside the cell.

   c. formation of protein sorting vesicles that carry proteins from the Golgi to lysosomes.
   d. DNA sequences that remain part of the proteins that determine cellular destination.
   e. all of the above.

9. Proteins that remain in the cytosol as opposed to passing through the endomembrane system are sorted
   a. prior to translation by pretranslational sorting.
   b. during translation by cotranslational sorting.
   c. after translation by post-translational sorting.
   d. both b and c.
   e. none of the above.

10. Vesicles move to the appropriate membrane in the cell by the recognition of _____ , proteins in the target membrane that act as binding sites for the vesicle.
    a. v-snares
    b. traffic signals
    c. coat proteins
    d. chaperones
    e. t-snares

CONCEPTUAL QUESTIONS

1. List the components of the endomembrane system and briefly describe the functions of each.

2. Briefly explain how sorting signals function in protein localization.

3. Define glycosylation.

EXPERIMENTAL QUESTIONS

1. Explain the procedure of a pulse-chase experiment as described in Figure 6.11. What was the purpose of the approach?

2. Why were pancreatic cells used for the investigation of Figure 6.11?

3. What were the results of the experiment of Figure 6.11? What did the researchers conclude?

COLLABORATIVE QUESTIONS

1. What roles do the genome and proteome play in the organization of cells?

2. Discuss the theory of how the mitochondria and chloroplast were initially formed.

www.brookerbiology.com

This website includes answers to the Biological Inquiry questions found in the figure legends and all end-of-chapter questions.

7
ENZYMES, METABOLISM, AND CELLULAR RESPIRATION

CHAPTER OUTLINE

Metabolism of food. The food we eat contains a variety of organic molecules that we use for energy.

As discussed in Chapter 2, a **chemical reaction** is a process in which one or more substances are changed into other substances. Such reactions can result in molecules attaching to each other to form larger molecules, molecules breaking apart to form two or more smaller molecules, rearrangements of atoms within molecules, or the transfer of electrons from one atom to another. Every living cell continuously performs thousands of such chemical reactions to sustain life. The term **metabolism** is used to describe the sum total of all chemical reactions that occur within an organism. The term also refers to a specific set of chemical reactions occurring at the cellular level. For example, biologists may speak of sugar metabolism or fat metabolism. Most metabolism involves the breakdown or synthesis of organic molecules. In Chapter 6, we learned that cells maintain their structure by using organic molecules. Such molecules provide the building blocks to construct cells, and the chemical bonds within organic molecules store energy that can be used to drive cellular processes. A key emphasis of this chapter is how, through chemical reactions, cells utilize energy that is stored within the chemical bonds of organic molecules such as the sugar glucose.

When we eat food, we are using much of that food for energy. People often speak of "burning calories." While it is true that metabolism does generate some heat, the chemical reactions that take place in the cells of living organisms are uniquely different from those that occur, say, in a fire. When wood is burned, the reaction produces enormous amounts of heat in a short period of time; the reaction lacks control. In contrast, the metabolism that occurs in living cells is extremely controlled. The food molecules from which we harvest energy give up that energy in a very restrained manner rather than all at once, as in a fire. An underlying theme in metabolism involves the remarkable control that cells possess when coordinating chemical reactions that utilize energy.

In this chapter, we will begin with a general discussion of chemical reactions. We will examine what factors control the direction of a chemical reaction and what determines its rate, paying particular attention to the role of enzymes, a type of biological catalyst. We then consider metabolism at the cellular level. First, we will examine some of the general features of chemical reactions that are vital for the energy needs of living cells. We will also explore the variety of ways in which metabolic processes are regulated and survey a group of chemical reactions that involves the breakdown of carbohydrates, namely, the sugar glucose. As you will learn, cells carry out an intricate series of reactions so that glucose can be "burned" in a very controlled fashion. Finally, the last section explores the synthesis of natural products that play specialized roles in many species.

7.1 Energy, Chemical Reactions, and Enzymes

Two general factors govern the fate of a given chemical reaction in a living cell—its direction and rate. To illustrate this point, let's consider a generalized chemical reaction such as

$$aA + bB \rightleftharpoons cC + dD$$

where A and B are the reactants, C and D are the products, and a, b, c, and d are the number of moles of reactants and products. This reaction is reversible, which means that A + B could be converted to C + D, or C + D could be converted to A + B. The direction of the reaction, whether C + D are made (the forward direction) or A + B are made (the reverse direction), depends on energy and on the concentrations of A, B, C, and D.

In this section, we will begin by examining the interplay of energy and the concentration of reactants as they govern the direction of a chemical reaction. You will learn that cells use energy intermediate molecules, such as ATP, to drive chemical reactions in a desired direction. Many types of chemical reactions, particularly those that involve organic molecules, proceed at a very slow rate unless facilitated by a biological catalyst. In the second part of this section, we will examine how catalysts called enzymes are critical cellular components that speed up the rates of chemical reactions in living organisms.

Energy Exists in Many Forms

To understand why a chemical reaction occurs, we first need to consider **energy**, which we will define as the ability to promote change. Physicists often consider energy in two forms: kinetic energy and potential energy (**Figure 7.1**). **Kinetic energy** is energy associated with movement, such as the movement of a baseball bat from one location to another. By comparison, **potential energy** is the energy that a substance possesses due to its structure or location. The energy contained within covalent bonds in molecules is a type of potential energy called **chemical energy**. The breakage of those bonds is one way that living cells can harness this energy to perform cellular functions. **Table 7.1** summarizes chemical and other forms of energy that are important in biological systems.

An important issue in biology is the ability of energy to be converted from one form to another. The study of energy interconversions is called **thermodynamics**. Physicists have determined that two laws govern energy interconversions:

1. **The first law of thermodynamics**—The first law states that energy cannot be created or destroyed; it is also called the *law of conservation of energy*. However, energy can be transferred from one place to another and can be

Covalent bonds in glucose store energy.

(a) Kinetic energy   **(b) Potential energy**

Figure 7.1   Examples of energy.   (a) Kinetic energy, such as swinging a bat, is energy associated with motion. **(b)** Potential energy is stored energy. Chemical bonds in molecules such as glucose store large amounts of chemical energy.

transformed from one type to another (as when, for example, chemical energy is transformed into heat).

2. **The second law of thermodynamics**—The second law states that the transfer of energy or the transformation of energy from one form to another increases the **entropy**, or degree of disorder of a system (**Figure 7.2**). When energy is converted from one form to another, the increase in entropy causes some energy to become unusable by living organisms.

Next, we will see how these two laws place limits on the ways that living cells can use energy for their own needs.

The Change in Free Energy Determines the Direction of a Chemical Reaction or Any Other Cellular Process

From the perspective of living organisms, energy is a critical component that is necessary for life to exist. Energy powers many cellular processes including chemical reactions, cellular

Table 7.1	Types of Energy That Are Important in Biology	
Energy type	**Description**	**Biological example**
Light	Light is a form of electromagnetic radiation that is visible to the eye. The energy of light is packaged in photons.	During photosynthesis, light energy is captured by pigments in chloroplasts (Chapter 8). Ultimately, this energy is used to reduce carbon, thus producing organic molecules.
Heat	Heat is the transfer of kinetic energy from one object to another, or from an energy source to an object. In biology, heat is often viewed as energy that can be transferred due to a difference in temperature between two objects or locations.	Many organisms, such as humans, maintain their bodies at a constant temperature. This is achieved by chemical reactions that generate heat.
Mechanical	Mechanical energy is the energy that is possessed by an object due to its motion or its position relative to other objects.	In animals, mechanical energy is associated with movements due to muscle contraction such as walking.
Chemical	Chemical energy is energy stored in the chemical bonds of molecules. When the bonds are broken and rearranged, this can release large amounts of energy.	The covalent bonds in organic molecules, such as glucose and ATP, store large amounts of energy. When these bonds are broken, the chemical energy released can be used to drive cellular processes.
Electrical/Ion gradient	The movement of charge or the separation of charge can provide energy. Also, a difference in ion concentration across a membrane constitutes an electrochemical gradient, which is a source of potential energy.	High-energy electrons can release energy (that is, drop down to lower energy levels). The energy that is released can be used to drive cellular processes such as the pumping of H^+ across membranes (as discussed later in this chapter).

Figure 7.2 **Entropy.** Entropy is a measure of the disorder of a system. An increase in entropy means an increase in disorder.

Biological inquiry: Which do you think has more entropy, a NaCl crystal at the bottom of a beaker of water or the same beaker of water after the Na⁺ and Cl⁻ in the crystal have dissolved in the water?

Adenosine triphosphate (ATP)

Adenosine diphosphate (ADP)   **Phosphate (P_i)**

$$\Delta G = -7.3 \text{ kcal/mol}$$

Figure 7.3 **The hydrolysis of ATP to ADP and P_i.** As shown in this figure, ATP has a net charge of -4, while ADP and P_i are shown with net charges of -2 each. When these compounds are shown in chemical reactions with other molecules, the net charges will also be shown. Otherwise, these compounds will simply be designated ATP, ADP, and P_i. It should also be noted that at neutral pH, ADP^{2-} will dissociate to ADP^{3-} and H^+. Likewise, P_i^{2-} also dissociates to $P_i^{3-} + H^+$ at neutral pH.

movements such as those occurring in muscle contraction, and the maintenance of cell organization. To understand how living organisms use energy, we need to distinguish between the energy that can be used to do work (usable energy) and the energy that cannot do work (unusable energy).

$$\text{Total energy} = \text{Usable energy} + \text{Unusable energy}$$

Why is some energy unusable? The main culprit is entropy. As stated by the second law of thermodynamics, energy transformations involve an increase in entropy, a measure of the disorder that cannot be harnessed to do work. For living organisms, the total energy is termed **enthalpy (H)** and the usable energy—the amount of available energy that can be used to do work—is called the **free energy (G)**. The letter G is in recognition of J. Willard Gibbs, who proposed the concept of free energy in 1878. The unusable energy is the system's entropy (S). Gibbs proposed that these three factors are related to each other in the following way:

$$H = G + TS$$

where T is the absolute temperature in Kelvin (K). Because our focus is on free energy, we can rearrange this equation as

$$G = H - TS$$

A critical issue in biology is whether a process will or will not occur spontaneously. For example, will glucose be broken down into carbon dioxide and water? Another way of framing this question is to ask, Is the breakdown of glucose a spontaneous reaction? A spontaneous reaction or process is one that will occur without an additional input of energy. However, a spontaneous reaction does not necessarily proceed quickly. In some cases, the rate of a spontaneous reaction can be quite slow.

The key way to evaluate if a reaction is spontaneous is to determine the free energy change that occurs as a result of the chemical reaction.

$$\Delta G = \Delta H - T\Delta S$$

where the Δ sign indicates a change, such as before and after a chemical reaction. If the reaction has a negative free energy change ($\Delta G < 0$), this means that free energy is released. Such a reaction is said to be **exergonic.** Exergonic reactions are spontaneous. Alternatively, if a reaction has a positive free energy change ($\Delta G > 0$), requiring the addition of free energy from the environment, it is termed **endergonic.** An endergonic reaction is not a spontaneous reaction.

If ΔG for a chemical reaction is negative, the reaction favors the formation of products, while a reaction with a positive ΔG favors the formation of reactants. Chemists have determined free energy changes for a variety of chemical reactions. **Adenosine triphosphate (ATP)** is a molecule that is a common energy source for all cells. Let's look at the breakdown of ATP to adenosine diphosphate (ADP) and inorganic phosphate (P_i). Because water is used to remove a phosphate group, chemists refer to this as the hydrolysis of ATP (**Figure 7.3**). In the reaction of converting 1 mole of ATP to 1 mole of ADP and P_i, ΔG equals -7.3 kcal/mole. Because this is a negative value, the reaction strongly favors the formation of products. As discussed later in this chapter, the energy that is liberated by the hydrolysis of ATP is used to drive a variety of cellular processes.

Even when a chemical reaction is associated with a negative free energy change, not all of the reactants are converted to products. The reaction reaches a state of **chemical equilibrium** in which the rate of formation of products equals the rate of formation of reactants. According to the generalized equation:

$$aA + bB \rightleftharpoons cC + dD$$

An equilibrium occurs, such that:

$$K_{eq} = \frac{[C]^c[D]^d}{[A]^a[B]^b}$$

where K_{eq} is the equilibrium constant.

Biologists make two simplifying assumptions when determining values for equilibrium constants. First, the concentration of water does not change during the reaction and the pH remains constant at pH 7. The equilibrium constant under these conditions is designated K_{eq}'. If water is one of the reactants, as in a hydrolysis reaction, it is not included in the chemical equilibrium equation. As an example, let's consider the chemical equilibrium for the hydrolysis of ATP.

$$ATP^{4-} + H_2O \rightleftharpoons ADP^{2-} + P_i^{2-}$$

$$K_{eq}' = \frac{[ADP][P_i]}{[ATP]}$$

Experimentally, the value for K_{eq}' for this reaction has been determined and found to be approximately 1,650,000 M. Such a large value indicates that the equilibrium greatly favors the formation of products—ADP and P_i.

Cells Use ATP to Drive Endergonic Reactions

An important issue that faces living organisms is that many vital processes require the addition of free energy, that is, they are endergonic and will not occur spontaneously. Fortunately, organisms have a strategy to overcome this problem. If an endergonic reaction is coupled to an exergonic reaction, the endergonic reaction will proceed spontaneously if the net free energy change for both processes combined is negative. For example, consider the following reactions:

Glucose + phosphate^{2-} → Glucose-phosphate^{2-} + H$_2$O
                                    $\Delta G = +3.3$ kcal/mole

ATP^{4-} + H$_2$O → ATP^{2-} + P$_i^{2-}$ $\Delta G = -7.3$ kcal/mole

Coupled reaction:

Glucose + ATP^{4-} → Glucose-phosphate^{2-} + ADP^{2-}
                                    $\Delta G = -4.0$ kcal/mole

The first reaction, in which phosphate is covalently attached to glucose, is endergonic, while the second, the hydrolysis of ATP, is exergonic. By itself, the first reaction would not be spontaneous. If the two reactions are coupled, however, the net free energy change for both reactions combined is exergonic. In the coupled reaction, a phosphate is directly transferred from ATP to glucose; this coupled reaction proceeds spontaneously because the net free energy change is negative. As discussed later, the transfer of phosphate from ATP to glucose is a first step in the breakdown of glucose to smaller molecules. Exergonic reactions, such as the breakdown of ATP, are commonly coupled to cellular processes that would otherwise be endergonic.

Enzymes Increase the Rates of Chemical Reactions

Thus far we have examined aspects of energy and considered how the laws of physics are related to the direction of chemical reactions. If a chemical reaction has a negative free energy change, the reaction will be spontaneous; it will tend to proceed in the direction of reactants to products. While thermodynamics governs the direction of an energy transformation, it does not control the rate of a chemical reaction. For example, the breakdown of the molecules in gasoline to smaller molecules is a highly exergonic reaction. Even so, we could place gasoline and oxygen in a container and nothing much would happen (provided it wasn't near a flame). If we came back several days later, we would expect to see the gasoline still sitting there. Perhaps if we came back in a few million years, the gasoline would have been broken down. On a timescale of months or a few years, however, the chemical reaction would proceed very slowly.

For most chemical reactions in cells to proceed at a rapid pace, a catalyst is needed. A **catalyst** is an agent that speeds up the rate of a chemical reaction without being consumed during the reaction. In living cells, the most common catalysts are **enzymes**, protein molecules that accelerate chemical reactions. The term was coined in 1876 by a German physiologist, Wilhelm Kühne, who discovered the enzyme trypsin in pancreatic juice. Interestingly, a few biological catalysts are RNA molecules called **ribozymes**. For example, RNA molecules within ribosomes catalyze the formation of bonds that link amino acids to each other.

Why are catalysts necessary to speed up a chemical reaction? When a covalent bond is broken or formed, this process initially involves the straining or contortion of one or more bonds in the starting molecule(s) and it may involve the positioning of two molecules so that they interact with each other properly. Let's consider the reaction in which ATP is used to attach a phosphate to glucose.

Glucose + ATP^{4-} → Glucose-phosphate^{2-} + ADP^{2-}

For a reaction to occur between glucose and ATP, they must collide in the correct orientation and possess enough energy so that chemical bonds can be changed. As glucose and ATP approach each other, their electron clouds cause repulsion. To overcome this repulsion, an initial input of energy, called the **activation energy**, is required (**Figure 7.4**). Activation energy allows the molecules to get close enough to cause a rearrangement of bonds. With the input of activation energy, glucose and ATP can achieve a **transition state** in which the original bonds have stretched to their limit. Once the reactants have reached

Figure 7.4 **Activation energy of a chemical reaction.**
This figure depicts an exergonic reaction. The activation energy is needed for molecules to achieve a transition state. One way that enzymes lower the activation energy is by straining the reactants so that less energy is required to attain the transition state. A second way is by binding two reactants so they are close to each other and in a favorable orientation.

Biological inquiry: How does lowering the activation energy affect the rate of a chemical reaction?

the transition state, the chemical reaction can readily proceed to the formation of products.

The activation energy required to achieve the transition state is a barrier to the formation of products. This barrier is the reason why the rate of many chemical reactions is very slow. There are two common ways to overcome this barrier and thereby accelerate a chemical reaction. First, the reactants could be exposed to a large amount of heat. For example, as we noted previously, if gasoline is sitting at room temperature, nothing much happens. However, if the gasoline is exposed to a flame or spark, it breaks down rapidly, perhaps at an explosive rate! Alternatively, a second strategy is to lower the activation energy barrier. Enzymes lower the activation energy to a point where a small amount of available heat can push the reactants to a transition state.

How do enzymes work to lower the activation energy barrier of chemical reactions? Enzymes are large proteins that bind relatively small reactants (Figure 7.4). When bound to an enzyme, the bonds in the reactants can be strained (that is, stretched), thereby making it easier for them to achieve the transition state. This is one way that enzymes lower the activation energy. In addition, when a chemical reaction involves two or more reactants, the enzyme provides a site where the reactants are positioned very close to each other and in an orientation that facilitates the formation of new covalent bonds. This also lowers the necessary activation energy for a chemical reaction.

Straining the reactants and bringing them close together are two common ways that enzymes lower the activation energy barrier. In addition, enzymes may facilitate a chemical reaction by

changing the local environment of the reactants. For example, amino acids in an enzyme may have charges that affect the chemistry of the reactants. In some cases, enzymes lower the activation energy by directly participating in the chemical reaction. For example, certain enzymes that hydrolyze ATP form a covalent bond between phosphate and an amino acid in the enzyme. However, this is a very temporary condition. The covalent bond between phosphate and the amino acid is quickly broken, releasing phosphate and returning the amino acid back to its original condition. An important example of such an enzyme is Na^+/K^+-ATPase, described in Chapter 5 (refer back to Figure 5.23).

Enzymes Recognize Their Substrates with High Specificity and Undergo Conformational Changes

Thus far, we have considered how enzymes lower the activation energy of a chemical reaction and thereby increase its rate. Let's consider some other features of enzymes that enable them to serve as effective catalysts in chemical reactions. The **active site** is the location in an enzyme where the chemical reaction takes place. The **substrates** for an enzyme are the reactant molecules and/or ions that bind to an enzyme at the active site and participate in the chemical reaction. For example, hexokinase is an enzyme whose substrates are glucose and ATP (**Figure 7.5**). The binding between an enzyme and substrate produces an **enzyme-substrate complex**.

A key feature of nearly all enzymes is that they bind their substrates with a **high affinity** or high degree of specificity. For example, because hexokinase binds glucose very well, we say it has a high affinity for glucose. By comparison, hexokinase has a low affinity for other sugars, such as fructose and galactose, which have similar structures to glucose. In 1894, the German scientist Emil Fischer proposed that the recognition of a substrate by an enzyme resembles the interaction between a lock and key: only the right-sized key (the substrate) will fit into the keyhole (active site) of the lock (the enzyme). Further research revealed that the interaction between an enzyme and its substrates also involves movements or conformational changes in the enzyme itself. As shown in Figure 7.5, these conformational changes cause the substrates to bind more tightly to the enzyme, a phenomenon called **induced fit**. Only after this conformational change takes place does the enzyme catalyze the conversion of reactants to products.

Some enzymes require additional nonprotein molecules or ions to carry out their functions. **Prosthetic groups** are small molecules that are permanently attached to the surface of an enzyme and aid in catalysis. **Cofactors** are usually inorganic ions, such as Fe^{3+} or Zn^{2+}, that temporarily bind to the surface of an enzyme and promote a chemical reaction. Finally, some enzymes use **coenzymes**, organic molecules that participate in the chemical reaction but are left unchanged after the reaction is completed.

The ability of enzymes to increase the rate of a chemical reaction is also affected by the surrounding conditions. In particular, the temperature and pH play an important role in the proper functioning of enzymes. Most enzymes function maximally in a narrow range of temperature and pH. For example, many human

| 1 | ATP and glucose bind to enzyme (hexokinase). | 2 | Enzyme undergoes conformational change that strains the substrates and brings them close together (induced fit). | 3 | Substrates are converted to products. | 4 | Products are released. |

Figure 7.5 **The steps of an enzyme-catalyzed reaction.** The example shown here involves the enzyme hexokinase, which binds glucose and ATP. The products are glucose-phosphate and ADP, which are released from the enzyme.

enzymes work best at 37°C (98.6°F), which is the body's normal temperature. If the temperature was several degrees above or below this value, the function of many cytosolic enzymes would be greatly inhibited. Similarly, enzyme function is sensitive to pH. Many cytosolic enzymes function optimally at pH 7.2, the pH normally found in the cytosol of humans cells. If the pH was significantly above or below this value, enzyme function would be decreased.

7.2 Overview of Metabolism

In the previous section, we examined the underlying factors that govern individual chemical reactions. In living cells, chemical reactions are often coordinated with each other and occur in sequences called **metabolic pathways**, each step of which is catalyzed by a specific enzyme (**Figure 7.6**). These pathways are categorized according to whether the reactions lead to the breakdown or synthesis of substances. **Catabolic reactions** result in the breakdown of molecules into smaller molecules. Such reactions are often exergonic. By comparison, **anabolic reactions** promote the synthesis of larger molecules from smaller precursor molecules. This process usually is endergonic and, in living cells, must be coupled to an exergonic reaction. In this section, we will survey the general features of catabolic and anabolic reactions, and explore the ways in which these metabolic pathways are controlled.

Catabolic Reactions Recycle Organic Building Blocks and Produce Energy Intermediates Such as ATP and NADH

Catabolic reactions involve the breakdown of macromolecules or smaller organic molecules. One reason for the breakdown of these molecules is to recycle the building blocks needed to construct new macromolecules. For example, proteins are composed of a linear sequence of amino acids. When a protein is improp-

Figure 7.6 **A metabolic pathway.** In a metabolic pathway, a series of different enzymes catalyze the changes in the structure of a molecule, beginning with a starting substrate and ending with a final product.

erly folded or is no longer needed by a cell, the peptide bonds between amino acids in such a protein are broken by enzymes called proteases. This generates amino acids that can be used in the construction of new proteins.

$$\text{Protein} \xrightarrow[]{\text{Proteases}} \text{Many individual amino acids}$$

Similarly, the breakdown of RNA by enzymes called nucleases produces nucleotides that can be used in the synthesis of new RNA molecules.

$$\text{RNA} \xrightarrow[]{\text{Nucleases}} \text{Many individual nucleotides}$$

The breakdown of macromolecules, such as unneeded proteins and RNA molecules, allows a cell to recycle the building blocks that comprise those macromolecules and use them to make new macromolecules.

A second reason for the breakdown of macromolecules and smaller organic molecules is to obtain energy that can be used to drive endergonic processes in the cell. Covalent bonds store a large amount of energy. However, when cells break covalent bonds in organic molecules such as carbohydrates and proteins, they do not directly use the energy that is released in this process.

Instead, the released energy is stored in **energy intermediates**, molecules such as ATP and NADH, that are directly used to drive endergonic reactions in cells.

As an example, let's consider the breakdown of glucose into two molecules of pyruvate. As we'll discuss later in this chapter, the breakdown of glucose to pyruvate involves a catabolic pathway called glycolysis. Some of the energy that is released during the breakage of covalent bonds in glucose is harnessed to synthesize ATP. However, this does not occur in a single step. Rather, glycolysis involves a series of steps in which covalent bonds are broken and rearranged. This process creates molecules that can readily donate a phosphate group to ADP, thereby creating ATP. For example, phosphoenolpyruvate has a phosphate group attached to pyruvate. Due to the arrangement of bonds in phosphoenolpyruvate, this phosphate bond is easily broken. Therefore, the phosphate can be readily transferred to ADP

$$Phosphoenolpyruvate + ADP \rightarrow Pyruvate + ATP$$
$$\Delta G = -7.5 \text{ kcal/mole}$$

This is an exergonic reaction and therefore favors the formation of products. In this step of glycolysis, the breakdown of an organic molecule, namely, phosphoenolpyruvate, results in the synthesis of an energy intermediate molecule, ATP, which can then be used by a cell to drive endergonic reactions. This way of synthesizing ATP, termed **substrate-level phosphorylation**, occurs when an enzyme directly transfers a phosphate from one molecule to a different molecule. In this case, a phosphate is transferred from phosphoenolpyruvate to ADP. Another way to make ATP is via **chemiosmosis**. In this process, energy stored in an ion electrochemical gradient is used to make ATP from ADP and P_i. We will consider this mechanism later in the chapter.

An important event that may occur during the breakdown of small organic molecules is **oxidation**, a process that involves the removal of electrons. This event is called oxidation because oxygen is frequently involved in chemical reactions that remove electrons from other molecules. By comparison, **reduction** is the addition of electrons to an atom or molecule. Reduction is so named because the addition of a negatively charged electron reduces the net charge of a molecule.

Electrons do not exist freely in solution. When an atom or molecule is oxidized, the electron that is removed must be transferred to another atom or molecule, which becomes reduced. This type of reaction is termed a **redox reaction**, which is short for a reduction-oxidation reaction. As a generalized equation, an electron may be transferred from molecule A to molecule B as follows:

$$Ae^- + B \rightarrow A + Be^-$$

As shown in the right side of this reaction, A has been oxidized (that is, had an electron removed) and B has been reduced (that is, had an electron added). In general, a substance that has been oxidized has less energy, while a substance that has been reduced has more energy.

During the oxidation of organic molecules such as glucose, the electrons are used to create energy intermediates such as

Figure 7.7 **The reduction of NAD⁺ to create NADH.** NAD⁺ is composed of two nucleotides, one with an adenine base and one with a nicotinamide base. The oxidation of organic molecules releases electrons that can bind to NAD⁺ (and along with a hydrogen ion) result in the formation of NADH. The two electrons and H⁺ are incorporated into the nicotinamide ring. Note: the actual net charges of NAD⁺ and NADH are minus one and minus two, respectively. They are designated NAD⁺ and NADH to emphasize the net charge of the nicotinamide ring, which is involved in oxidation-reduction reactions.

NADH (**Figure 7.7**). In this process, an organic molecule has been oxidized and **NAD⁺** (**nicotinamide adenine dinucleotide**) has been reduced to NADH. Cells use NADH in two common ways. First, as we will see, the oxidation of NADH is a highly exergonic reaction that can be used to make ATP. Second, NADH can donate electrons to other organic molecules and thereby energize them. Such energized molecules can more readily form covalent bonds. Therefore, as described next, NADH is often needed in anabolic reactions that involve the synthesis of larger molecules through the formation of covalent bonds between smaller molecules.

Anabolic Reactions Require an Input of Energy to Make Larger Molecules

Anabolic reactions are also called **biosynthetic reactions**, because they are necessary to make larger molecules and macromolecules. We will examine the synthesis of macromolecules in several chapters of this textbook. For example, RNA and protein biosynthesis are described in Chapter 12. Cells also need to synthesize small organic molecules, such as amino acids and fats, if they are not readily available from food sources. Such molecules are made by the formation of covalent linkages between precursor molecules. For example, glutamate (an amino acid) is made by the covalent linkage between α-ketoglutarate (a product of sugar metabolism) and ammonium.

$$\begin{matrix} COO^- \\ | \\ CH_2 \\ | \\ CH_2 \\ | \\ C\!=\!O \\ | \\ COO^- \end{matrix} + NH_4^+ + NADH \longrightarrow \begin{matrix} \\ \\ \\ \\ H_3N^+\!-\!C\!-\!COO^- \\ | \\ H \end{matrix} + NAD^+ + H_2O$$

α-ketoglutarate   Ammonium   Glutamate

Another amino acid, glutamine, is made from glutamate and ammonium.

$$\begin{matrix} COO^- \\ | \\ CH_2 \\ | \\ CH_2 \\ | \\ H_3N^+\!-\!C\!-\!COO^- \\ | \\ H \end{matrix} + NH_4^+ + ATP^{4-} + H_2O \longrightarrow \begin{matrix} O\!=\!C\!-\!NH_2 \\ | \\ CH_2 \\ | \\ CH_2 \\ | \\ H_3N^+\!-\!C\!-\!COO^- \\ | \\ H \end{matrix} + ADP^{2-} + P_i^{2-}$$

Glutamate   Ammonium   Glutamine

In both reactions, an energy intermediate molecule such as NADH or ATP is needed to drive the reaction forward.

GENOMES & PROTEOMES

Many Proteins Use ATP as a Source of Energy

Over the past several decades, researchers have studied the functions of many types of proteins and discovered numerous examples in which a protein uses ATP to drive a cellular process (**Table 7.2**). In humans, a typical cell uses millions of ATP molecules per second. At the same time, the breakdown of food molecules releases energy that allows us to make more ATP from ADP and P_i. The turnover of ATP occurs at a remarkable pace. An average person hydrolyzes about 100 pounds of ATP per day, yet we do not have 100 pounds of ATP in our bodies. For this to happen, each ATP undergoes about 10,000 cycles of hydrolysis and resynthesis (from ADP and P_i) during an ordinary day (**Figure 7.8**).

By studying the structures of many proteins that use ATP, biochemists have discovered that particular amino acid sequences within proteins function as ATP-binding sites. This information has allowed researchers to predict whether a newly discovered protein uses ATP or not. When an entire genome sequence of a species is experimentally determined, the genes that encode proteins can be analyzed to find out if the encoded proteins have ATP-binding sites in their amino acid sequences. Using this approach, researchers have been able to analyze proteomes—all of the proteins that a given cell can make—and estimate the percentage of proteins that are able to bind ATP. (Most of these proteins are expected to use ATP as a source of energy, though some of them may simply bind ATP without hydrolyzing it to ADP and P_i.) This approach has been applied to the proteomes of bacteria, archaea, and eukaryotes. On average, over 20% of all proteins bind ATP. However, this number

Table 7.2	Examples of Proteins That Use ATP for Energy
Type	**Description**
Metabolic enzymes	Many enzymes use ATP to catalyze endergonic reactions. For example, hexokinase uses ATP to attach phosphate to glucose.
Transporters	Ion pumps, such as the Na$^+$/K$^+$-ATPase, use ATP to pump ions against a gradient (see Chapter 5).
Motor proteins	Motor proteins such as myosin use ATP to facilitate cellular movement, as in muscle contraction (see Chapter 46).
Chaperones	Chaperones are proteins that use ATP to aid in the folding and unfolding of cellular proteins (see Chapter 6).
DNA-modifying enzymes	Many proteins such as helicases and topoisomerases use ATP to modify the conformation of DNA (see Chapter 11).
Aminoacyl-tRNA synthetases	These enzymes use ATP to attach amino acids to tRNAs (see Chapter 12).
Protein kinases	Protein kinases are regulatory proteins that use ATP to attach a phosphate to proteins, thereby affecting the function of the phosphorylated protein (see Chapter 9).

The energy to make ATP comes from catabolic reactions that are exergonic.

Energy input

ADP + P_i

ATP

Energy release

ATP hydrolysis provides the energy for cellular processes that are endergonic.

Figure 7.8 **The ATP cycle.** Living cells continuously recycle ATP. The breakdown of food molecules is used to make ATP from ADP and P_i. The hydrolysis of ATP to ADP and P_i is used to drive many different endergonic reactions and processes that occur in cells.

Biological inquiry: If a large amount of ADP was broken down in the cell, how would this affect the ATP cycle?

is likely to be an underestimate of the total percentage of ATP-utilizing proteins because we may not have identified all of the types of ATP-binding sites in proteins. In humans, who have an estimated genome size of 20,000 to 25,000 different genes, a minimum of 4,000 to 5,000 of those genes encode proteins that use ATP. From these numbers, we can see the enormous importance of ATP as a source of energy for living cells.

Metabolic Pathways Are Regulated in Three General Ways

Before we end our general discussion of metabolism, let's consider the various ways in which chemical reactions are regulated in living cells. The regulation of catabolic pathways is important, so that a cell breaks down organic molecules when energy is needed but conserves them when an adequate supply of energy intermediates is available. The control of anabolic pathways is essential so that a cell does not waste energy making too much of the products of such pathways. The regulation of catabolic and anabolic pathways occurs at the genetic, cellular, and biochemical levels.

Gene Regulation Because enzymes in every metabolic pathway are encoded by genes, one way that cells control chemical reactions is via gene regulation. For example, if a bacterial cell is not exposed to a particular sugar in its environment, it will turn off the genes that encode the enzymes that are needed to break down that sugar. Alternatively, if the sugar becomes available, the genes are switched on. Chapter 13 examines the steps of gene regulation in detail.

Cellular Regulation Metabolism is also coordinated at the cellular level. Cells integrate signals from their environment and adjust their chemical reactions to adapt to those signals. As discussed in Chapter 9, cell-signaling pathways often lead to the activation of protein kinases that covalently attach a phosphate group to target proteins. For example, when people are frightened, they secrete a hormone called epinephrine into their bloodstream. This hormone binds to the surface of muscle cells and stimulates an intracellular pathway that leads to the phosphorylation of several intracellular proteins, including enzymes involved in carbohydrate metabolism. These activated enzymes promote the breakdown of carbohydrates, an event that supplies the frightened individual with more energy. Epinephrine is sometimes called the "fight-or-flight" hormone because the added energy prepares an individual to either stay and fight or run away. After a person is no longer frightened, hormone levels drop and other enzymes called phosphatases remove the phosphate groups from enzymes, thereby restoring the original level of carbohydrate metabolism.

Biochemical Regulation A third and very prominent way that metabolic pathways are controlled is at the biochemical level. In this case, the binding of a molecule to an enzyme directly regulates its function. Biochemical regulation is typically categorized according to the site where the regulatory molecule binds. Let's consider two types of regulation that involve regulatory molecules that inhibit enzyme function.

Competitive inhibitors are molecules that bind to the active site of an enzyme and inhibit the ability of the substrate to bind. In other words, such inhibitors "compete" with the substrate for the ability to bind to the enzyme. Competitive inhibitors usually have a structure that mimics the structure of the enzyme's substrate. Competitive inhibition can be overcome by increasing the concentration of the substrate and decreasing the concentration of the inhibitor.

Noncompetitive inhibitors bind to an enzyme at a location that is outside the active site and inhibit the enzyme's function. An example is a form of regulation called **feedback inhibition**, in which the product of a metabolic pathway inhibits an enzyme that acts early in the pathway, thus preventing the overaccumulation of the product (**Figure 7.9**). Many metabolic pathways use feedback inhibition as a form of biochemical regulation. In such cases, the inhibited enzyme has two binding sites. One site is the active site, where the reactants are converted to products. In addition, enzymes that are controlled by feedback inhibition also have an **allosteric site**, where a molecule can bind noncovalently and affect the function of the active site. The binding of a molecule to an allosteric site causes a conformational change in the enzyme that inhibits its catalytic function. Allosteric sites are often found in the enzymes that catalyze the early steps in a metabolic pathway. Such allosteric sites typically bind molecules that are the products of the metabolic pathway. When the products bind to these sites, they inhibit the function of these enzymes and thereby prevent the formation of too much product.

Figure 7.9 **Feedback inhibition.** In this process, the product of a metabolic pathway inhibits an enzyme that functions in the pathway, thereby preventing the overaccumulation of the product.

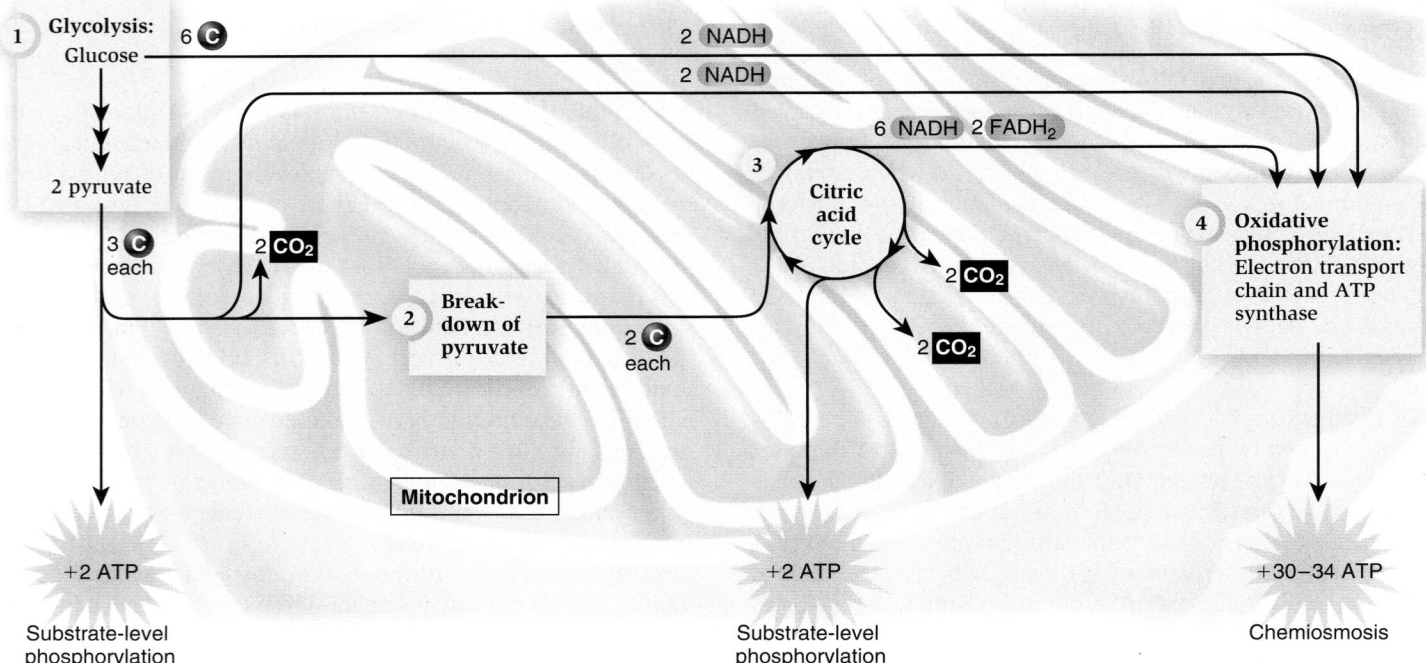

Figure 7.10   An overview of glucose metabolism.   Glycolysis occurs in the cytosol and results in the breakdown of glucose into two pyruvates, producing two ATP and two NADH molecules. The two pyruvates enter the mitochondrion, where they are broken down into two acetyl groups (each attached to CoA) and two CO_2 molecules. Two molecules of NADH are made in the process. The two acetyl groups then enter the citric acid cycle, where they are broken down to four CO_2 molecules. Two ATP, six NADH, and two $FADH_2$ are synthesized. The NADH and $FADH_2$ molecules that are made during these various steps are then used during oxidative phosphorylation to synthesize more ATP molecules. The maximum yield of ATP is 34 to 38 molecules for every glucose that is completely broken down.

Cellular and biochemical regulation are important and rapid ways to control chemical reactions in a cell. But when considering a metabolic pathway composed of several enzymes, which enzyme in a pathway should be controlled? In many cases, a metabolic pathway has a **rate-limiting step**, which is the slowest step in a pathway. If the rate-limiting step is inhibited or occurs at a faster rate, such changes will have the greatest impact on the formation of the product of the metabolic pathway. Rather than affecting all of the enzymes in a metabolic pathway, cellular and biochemical regulation are often directed at the enzyme that catalyzes the rate-limiting step. This is an efficient and rapid way to control the amount of product of a pathway.

7.3 Cellular Respiration

Cellular respiration is a process by which living cells obtain energy from organic molecules. A primary aim of cellular respiration is to make the energy intermediates ATP and NADH. When oxygen (O_2) is used, this process is termed **aerobic respiration**. During aerobic respiration, O_2 is consumed and CO_2 is released. When we breathe, we inhale the oxygen that is needed for aerobic respiration and exhale the CO_2 that is a by-product of the process. For this reason, the term respiration has a second meaning, which is the act of breathing.

Different types of organic molecules, such as carbohydrates, proteins, and fats, can be used as energy sources to drive aerobic respiration.

$$\text{Organic molecules} + O_2 \rightarrow CO_2 + H_2O + \text{Energy}$$

In this section, we will largely focus on the use of glucose as an energy source for cellular respiration.

$$\underset{\text{Glucose}}{C_6H_{12}O_6} + 6\,O_2 \rightarrow 6\,CO_2 + 6\,H_2O + \text{Energy intermediates} + \text{Heat}$$

We will examine the metabolic pathways in which glucose is broken down into carbon dioxide and water, thereby releasing a large amount of energy that is used to make many ATP molecules. In so doing, we will focus on four pathways: (1) glycolysis, (2) the breakdown of pyruvate, (3) the citric acid cycle, and (4) oxidative phosphorylation. We will conclude our discussion of cellular respiration with a consideration of the metabolism of other organic molecules, such as proteins and fats, and an examination of anaerobic respiration, a second form of respiration in which cells can oxidize fuel and generate ATP without using oxygen.

Several Metabolic Pathways Are Involved in the Breakdown of Glucose to CO_2

Before we examine the details of cellular respiration, let's take a look at the entire process. We will focus on the breakdown of glucose in a eukaryotic cell in the presence of oxygen. The covalent bonds within glucose contain a large amount of chemical bond energy. When glucose is broken down, ultimately to CO_2 and water, the energy within those bonds is released and used to make three types of energy intermediates: ATP, NADH, and $FADH_2$. The following is an overview of the stages that occur during the breakdown of glucose (**Figure 7.10**):

1. **Glycolysis:** In glycolysis, glucose (6 C, meaning a compound with six carbon atoms) is broken down in the cytosol to two pyruvate molecules (3 C each), producing a net gain of two ATP molecules, via substrate-level phosphorylation, and two NADH molecules.

2. **Breakdown of pyruvate to an acetyl group:** The two pyruvate molecules enter the mitochondrion, where each one is broken down to an acetyl group (2 C) and one CO_2 molecule. For each pyruvate broken down, one NADH molecule is made.

3. **Citric acid cycle:** Each acetyl group (2 C) is broken down to two CO_2 molecules. One ATP, three NADH, and one $FADH_2$ are made in this process. Because there are two acetyl groups, the total yield is four CO_2, two ATP via substrate-level phosphorylation, six NADH, and two $FADH_2$.

4. **Oxidative phosphorylation:** The NADH and $FADH_2$ made in the three previous stages contain high-energy electrons that can be readily transferred in a redox reaction to other molecules. Once removed, these high-energy electrons release some energy and that energy is harnessed to make approximately 30 to 34 ATP molecules via chemiosmosis. As discussed later, oxidative phosphorylation consists of two components, the electron transport chain and the ATP synthase.

Now, let's examine in detail the chemical changes that take place in each of these four stages.

Stage 1: Glycolysis Is a Metabolic Pathway That Breaks Down Glucose to Pyruvate

Glycolysis (from the Greek *glykos*, sweet, and *lysis*, splitting) involves the breakdown of glucose, a simple sugar. This process can occur in the presence or absence of oxygen, that is, under aerobic or anaerobic conditions. Our understanding of glycolysis has a rich history. In 1897, Hans Buchner and Eduard Buchner made an accidental discovery. They were interested in manufacturing cell-free extracts of yeast for possible clinical use. This cell-free extract contained only the internal contents of yeast cells, not the intact cells themselves. To preserve these extracts, they added sucrose, a commonly used preservative in 19th-century chemistry. To their great surprise, they discovered that the cell-free extract converted the sucrose to ethanol. The significance of this finding was extraordinary. The Buchners showed for the first time that metabolism could take place outside of living cells. This observation is considered by many as the birth of **biochemistry**, the study of the chemistry of living organisms.

The Buchners' findings paved the way for the in-depth investigation of the breakdown of glucose. During the 1930s, the efforts of several German biochemists, including Gustav Embden, Otto Meyerhof, and Jacob Parnas, determined that the process involved 10 steps, each one catalyzed by a different enzyme. The elucidation of these steps was a major achievement in the field of biochemistry. Researchers have since discovered that glycolysis is the common pathway for glucose breakdown in bacteria, archaea, and eukaryotes. Remarkably, the steps of glycolysis are virtually identical in nearly all living species, suggesting that glycolysis arose very early in the evolution of life on our planet.

The 10 steps of glycolysis can be grouped into three phases (**Figure 7.11**). The first phase (steps 1–3) involves an energy

Figure 7.11   Overview of glycolysis.

Figure 7.12 An inside look at the steps of glycolysis.

investment. Two ATP molecules are hydrolyzed to create fructose-1,6-bisphosphate. The cleavage phase (steps 4–5) breaks a six-carbon molecule into two molecules of glyceraldehyde-3-phosphate. The third phase (steps 6–10) liberates energy. The two glyceraldehyde-3-phosphate molecules are broken down to two pyruvate molecules. This produces two molecules of NADH and four molecules of ATP. Because two molecules of ATP are used in the energy investment phase, the net yield of ATP is two molecules. **Figure 7.12** describes the details of the 10 reactions of glycolysis. The net reaction of glycolysis is:

$$C_6H_{12}O_6 + 2\ NAD^+ + 2\ ADP^{2-} + 2\ P_i^{2-} \rightarrow$$
Glucose
$$2\ CH_3(C{=}O)COO^- + 2\ H^+ + 2\ NADH + 2\ ATP^{4-} + 2\ H_2O$$
Pyruvate

When a cell has a sufficient amount of ATP, feedback inhibition occurs in glycolysis. At high concentration, ATP binds to an allosteric site in phosphofructokinase, which catalyzes the third step in glycolysis and is thought to be the rate-limiting step. When ATP binds to this allosteric site, a conformational change occurs that renders the enzyme functionally inactive. This prevents the further breakdown of glucose and thereby inhibits the overproduction of ATP.

Stage 2: Pyruvate Enters the Mitochondrion and Is Broken Down to an Acetyl Group and CO_2

In eukaryotes, pyruvate is made in the cytosol and then transported into the mitochondrion. Once in the mitochondrial matrix, pyruvate molecules are broken down by an enzyme complex called pyruvate dehydrogenase (**Figure 7.13**). A molecule of CO_2 is removed from each pyruvate and the remaining acetyl group is attached to an organic molecule called coenzyme A (CoA) to create acetyl CoA. During this process, two high-energy electrons are removed from pyruvate and transferred to NAD^+ and together with H^+ create a molecule of NADH. For each pyruvate, the net reaction is:

$$\underset{\text{Pyruvate}}{{}^-O-\overset{\overset{\displaystyle O}{\|}}{C}-\overset{\overset{\displaystyle O}{\|}}{C}-CH_3} + \underset{\text{CoA}}{CoA\text{-}SH} + NAD^+ \rightarrow$$

$$\underset{\text{Acetyl CoA}}{CoA-S-\overset{\overset{\displaystyle O}{\|}}{C}-CH_3} + CO_2 + NADH$$

The acetyl group is attached to CoA via a covalent bond to a sulfur atom. The hydrolysis of this bond releases a large amount of

2 NAD$^+$ → 2 NADH + 2 H$^+$

Triose phosphate dehydrogenase 2 P$_i$

Unstable phosphate bond

$$P-O-C=O$$
$$|$$
$$CHOH$$
$$|$$
$$CH_2O-P$$

1,3-bisphosphoglycerate (× 2)

Phosphoglycero– kinase   2 ADP → 2 ATP

$$C=O$$
$$|$$
$$CHOH$$
$$|$$
$$CH_2O-P$$

(O$^-$)

3-phosphoglycerate (× 2)

Phosphoglycero– mutase

$$C=O$$
$$|$$
$$HCO-P$$
$$|$$
$$CH_2OH$$

(O$^-$)

2-phosphoglycerate (× 2)

Enolase   2 H$_2$O

Unstable phosphate bond

$$C=O$$
$$|$$
$$CO-P$$
$$||$$
$$CH_2$$

(O$^-$)

Phosphoenolpyruvate (× 2)

Pyruvate kinase   2 ADP → 2 ATP

$$C=O$$
$$|$$
$$C=O$$
$$|$$
$$CH_3$$

(O$^-$)

Pyruvate (× 2)

6 Glyceraldehyde-3-phosphate is oxidized to 1,3-bisphosphoglycerate. NADH is produced. In 1,3-bisphosphoglycerate, the phosphate group in the upper left is destabilized, meaning that the bond will break in a highly exergonic reaction.

7 A phosphate is removed from 1,3-bisphosphoglycerate to form 3-phosphoglycerate. The removed phosphate is transferred to ADP to make ATP.

8 The phosphate group in 3-phosphoglycerate is moved to a new location, creating 2-phosphoglycerate.

9 A water molecule is removed from 2-phosphoglycerate to form phosphoenol-pyruvate. In phosphoenol-pyruvate, the phosphate group is destabilized, meaning that the bond will break in a highly exergonic reaction.

10 A phosphate is removed from phosphoenolpyruvate to form pyruvate. The removed phosphate is transferred to ADP to make ATP.

free energy, making it possible for the acetyl group to be transferred to other organic molecules. As described next, the acetyl group attached to CoA enters the citric acid cycle.

Stage 3: During the Citric Acid Cycle, an Acetyl Group Is Oxidized to Two CO$_2$ Molecules

The third stage of sugar metabolism introduces a new concept, that of the **metabolic cycle**. During a metabolic cycle, particular molecules enter the cycle while others leave; the process is cyclical because it involves a series of organic molecules that are regenerated with each turn of the cycle. The idea of a metabolic cycle was first proposed in the early 1930s by German biochemist Hans Krebs. While studying carbohydrate metabolism, he analyzed cell extracts from pigeon muscle and determined that citric acid and other organic molecules participated in a cycle that resulted in the breakdown of carbohydrates to carbon dioxide. This cycle is called the **citric acid cycle** or the Krebs cycle, in honor of Krebs, who was awarded the Nobel Prize in 1953.

An overview of the citric acid cycle is shown in **Figure 7.14**. At the beginning of the cycle, the acetyl group is removed from acetyl CoA and attached to oxaloacetate (4 C) to form citrate (6 C), also called citric acid. In a series of several steps, two CO$_2$ molecules are released. As this occurs, three molecules of NADH,

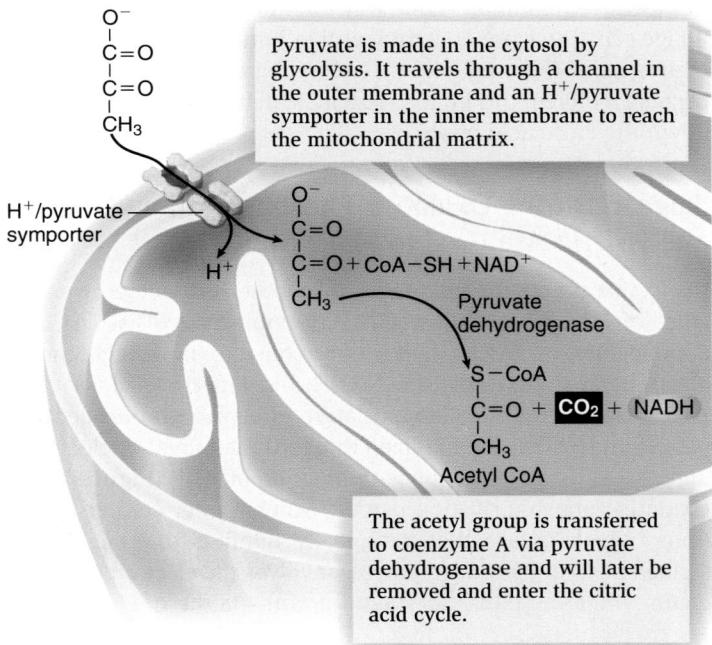

Pyruvate is made in the cytosol by glycolysis. It travels through a channel in the outer membrane and an H$^+$/pyruvate symporter in the inner membrane to reach the mitochondrial matrix.

H$^+$/pyruvate symporter

$$C=O$$
$$|$$
$$C=O$$
$$|$$
$$CH_3$$
(O$^-$)

$$C=O$$
$$|$$
$$C=O + CoA-SH + NAD^+$$
$$|$$
$$CH_3$$
(O$^-$)

Pyruvate dehydrogenase

$$S-CoA$$
$$|$$
$$C=O + \boxed{CO_2} + NADH$$
$$|$$
$$CH_3$$

Acetyl CoA

The acetyl group is transferred to coenzyme A via pyruvate dehydrogenase and will later be removed and enter the citric acid cycle.

Figure 7.13 **Breakdown of pyruvate and the attachment of an acetyl group to CoA.** Pyruvate enters the mitochondrion by traveling through a channel in the outer membrane and then through an H$^+$/pyruvate symporter in the inner membrane.

Figure 7.14 Overview of the citric acid cycle.

one molecule of FADH$_2$, and one molecule of GTP are made. The GTP is used to make ATP. After eight steps, oxaloacetate is regenerated so that the cycle can begin again, provided acetyl CoA is available. **Figure 7.15** shows a more detailed view of the citric acid cycle. For each acetyl group attached to CoA, the net reaction of the citric acid cycle is:

$$\text{Acetyl-CoA} + 3\,H_2O + 3\,NAD^+ + FAD + GDP^{2-} + P_i^{2-} \rightarrow$$
$$\text{CoA-SH} + 2\,CO_2 + 3\,NADH + FADH_2 + GTP^{4-}$$

Competitive inhibition is one way that the citric acid cycle is regulated. Oxaloacetate is a competitive inhibitor of succinate dehydrogenase (Figure 7.15). Therefore, when oxaloacetate levels become too high, this inhibits succinate dehydrogenase and slows down the citric acid cycle.

Stage 4: During Oxidative Phosphorylation, NADH and FADH2 Are Used to Make More ATP

Up to this point, the oxidation of glucose has yielded 6 molecules of CO$_2$, 4 molecules of ATP, 10 molecules of NADH, and 2 molecules of FADH$_2$. Let's now consider how high-energy electrons are removed from NADH and FADH$_2$ to make more ATP. This process is called **oxidative phosphorylation** because NADH and FADH$_2$ are oxidized due to the removal of electrons, and ATP is made by the phosphorylation of ADP (**Figure 7.16**). Oxidative phosphorylation usually occurs under aerobic conditions, which means that it typically requires oxygen. As described next, the oxidative process involves the electron transport chain, while the phosphorylation occurs via the ATP synthase.

Electron Transport Chain The **electron transport chain** consists of a group of protein complexes and small organic molecules embedded in the inner mitochondrial membrane. These components are referred to as an electron transport chain because the components can accept and donate electrons to each

other in a linear manner (Figure 7.16). Most of the members of the chain are protein complexes (designated I to IV) that have prosthetic groups. For example, cytochrome oxidase contains two prosthetic groups, each with an iron atom. The iron in each prosthetic group can readily accept and release an electron. One of the members of the electron transport chain, ubiquinone (Q), is not a protein. Rather, ubiquinone is a small organic molecule that can accept and release an electron and can diffuse through the lipid bilayer.

The red line in Figure 7.16 shows the path of electrons as they flow along the electron transport chain. This path is a series of redox reactions in which electrons are transferred to components with increasingly higher electronegativity. At the beginning of the chain, a pair of high-energy electrons from NADH are transferred one at a time to NADH dehydrogenase (complex I), and then to ubiquinone (Q), cytochrome b-c_1 (complex III), cytochrome c, cytochrome oxidase (complex IV), and finally O$_2$. At the end of the chain, 2 electrons, 2 H$^+$, and 1/2 O$_2$ combine to form a molecule of water. Similarly, FADH$_2$ transfers electrons to succinate reductase (complex II), then to ubiquinone, and the rest of the chain.

The electron transport chain is also called the **respiratory chain** because the oxygen we breathe is used in this process. One component of the electron transport chain, cytochrome oxidase, is inhibited by carbon monoxide. The deadly effects of carbon monoxide occur because the electron transport chain is shut down, preventing cells from making enough ATP for survival.

As shown in Figure 7.16, the movement of electrons results in the pumping of H$^+$ across the inner mitochondrial membrane to establish a large **H$^+$ electrochemical gradient** in which the concentration of H$^+$ is higher outside of the matrix, and an excess of positive charge exists outside the matrix. Because hydrogen ions consist of protons, the H$^+$ electrochemical gradient is also called the **proton-motive force**. NADH dehydrogenase, cytochrome b-c_1, and cytochrome oxidase are H$^+$ pumps.

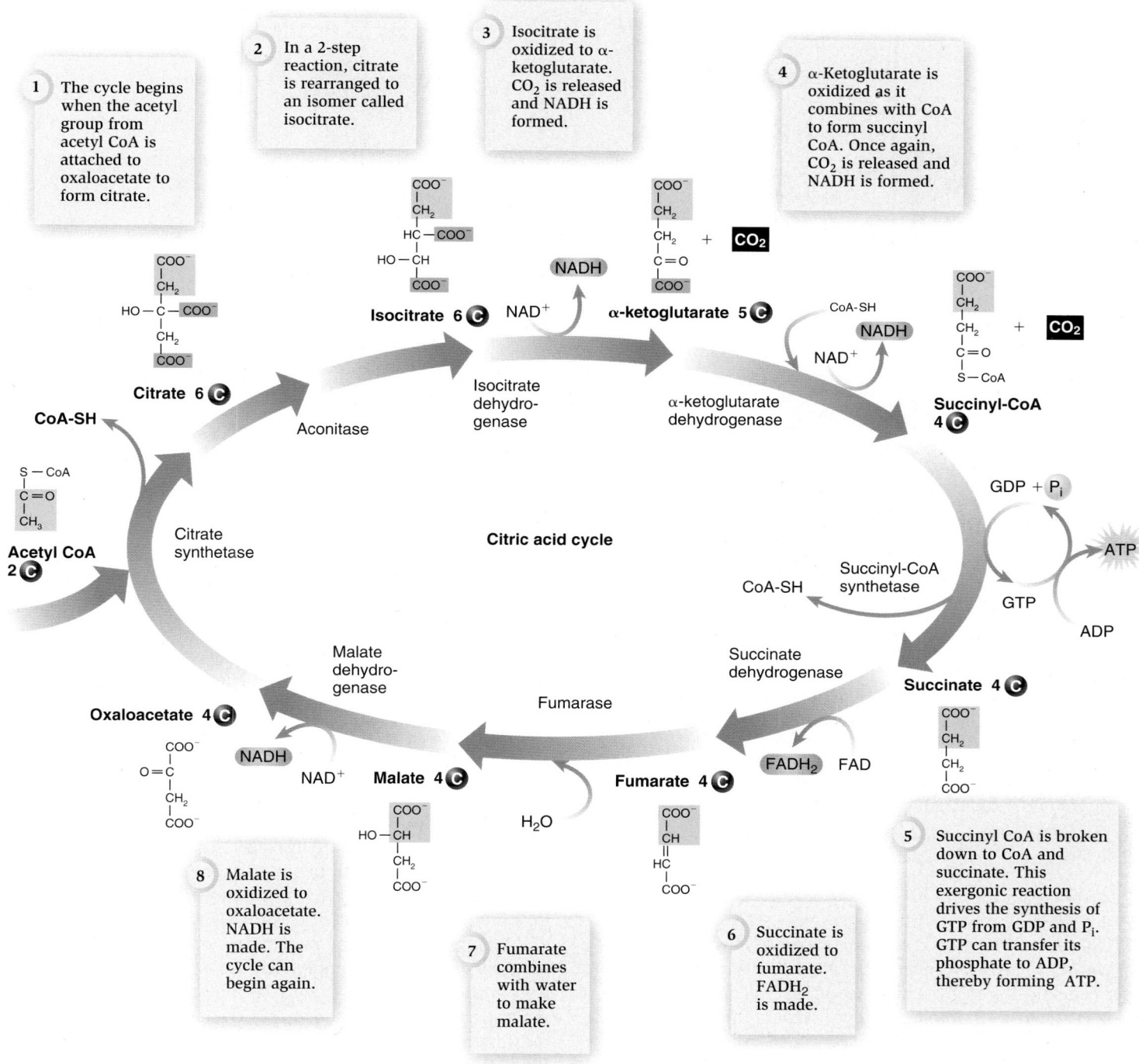

Figure 7.15   A detailed look at the steps of the citric acid cycle.

While traveling along the electron transport chain, electrons release free energy, and some of this energy is captured by these proteins to pump H^+ out of the matrix into the intermembrane space. Because the electrons from $FADH_2$ enter the chain at an intermediate step, fewer hydrogen ions are pumped out of the matrix compared to the release of electrons from NADH.

Why do electrons travel from NADH or $FADH_2$ to O_2? As you might expect, the answer lies in free energy changes. The electrons found on the energy intermediates have a high amount of potential energy. As they travel along the electron transport chain, free energy is released (**Figure 7.17**). The movement of one electron from NADH to O_2 results in a very negative free energy change of approximately -25 kcal/mole. That is why the process is spontaneous and proceeds in the forward direction. Because it is a highly exergonic reaction, some of the free energy can be harnessed to do cellular work. In this case, some energy is used to pump H^+ across the inner mitochondrial membrane and establish an H^+ electrochemical gradient.

Figure 7.16 **Oxidative phosphorylation.** This process consists of two distinct events involving the electron transport chain and the ATP synthase. The electron transport chain removes electrons from NADH or FADH₂ and pumps H⁺ across the inner mitochondrial membrane. The ATP synthase uses the energy in the H⁺ electrochemical gradient to synthesize ATP.

Biological inquiry: Can you explain the name of cytochrome oxidase? Can you think of another appropriate name?

Figure 7.17 **The relationship between free energy and electron movement along the electron transport chain.** As electrons hop from one site to another along the electron transport chain, they release energy. Some of this energy is harnessed to pump H⁺ across the inner mitochondrial membrane. The total energy released by a single electron is approximately −25 kcal/mole.

For each molecule of NADH that is oxidized and each molecule of ATP that is made, the two chemical reactions of oxidative phosphorylation can be represented as follows:

$$NADH + H^+ + 1/2\,O_2 \rightarrow NAD^+ + H_2O$$
$$ADP^{2-} + P_i^{2-} \rightarrow ATP^{4-} + H_2O$$

The oxidation of NADH to NAD⁺ results in an H⁺ electrochemical gradient in which more hydrogen ions are in the intermembrane space compared to the matrix. The synthesis of one ATP molecule is thought to require the movement of three to four ions into the matrix, down their H⁺ electrochemical gradient.

When we add up the maximal amount of ATP that can be made by oxidative phosphorylation, most researchers agree that it is in the range of 30 to 34 ATP molecules for each glucose molecule that is broken down to CO₂ and water. However, the maximum amount of ATP is rarely achieved for two reasons. First, although 10 NADH and 2 FADH₂ are available to create the H⁺ electrochemical gradient across the inner mitochondrial membrane, a cell may use some of these molecules for anabolic pathways. For example, NADH is used in the synthesis of organic molecules such as glycerol (a component of phospholipids) and lactic acid (which is secreted from muscle cells during strenuous exercise). Second, the mitochondrion may use some of the H⁺ electrochemical gradient for other purposes. For example, the gradient is used for the uptake of pyruvate into the matrix via an H⁺/pyruvate symporter. Therefore, the actual amount of ATP synthesis is usually a little less than the maximum number of 30 to 34. Even so, when we compare the amount of

ATP Synthase The second event of oxidative phosphorylation is the synthesis of ATP by an enzyme called **ATP synthase**. The H⁺ electrochemical gradient across the inner mitochondrial membrane is a source of potential energy. How is this energy used? The flow of H⁺ back into the matrix is an exergonic process. The lipid bilayer is relatively impermeable to H⁺. However, H⁺ can pass through the membrane-embedded portion of the ATP synthase. This enzyme harnesses some of the free energy that is released as the ions flow through its membrane-embedded region to synthesize ATP from ADP and Pᵢ (see Figure 7.16). This is an example of an energy conversion: energy in the form of an H⁺ electrochemical gradient, or proton-motive force, is converted to chemical bond energy in ATP. The chemical synthesis of ATP by pushing H⁺ across a membrane is called chemiosmosis (from the Greek *osmos*, to push), and the theory behind it was proposed by Peter Mitchell, a British chemist who was awarded the Nobel Prize in 1978.

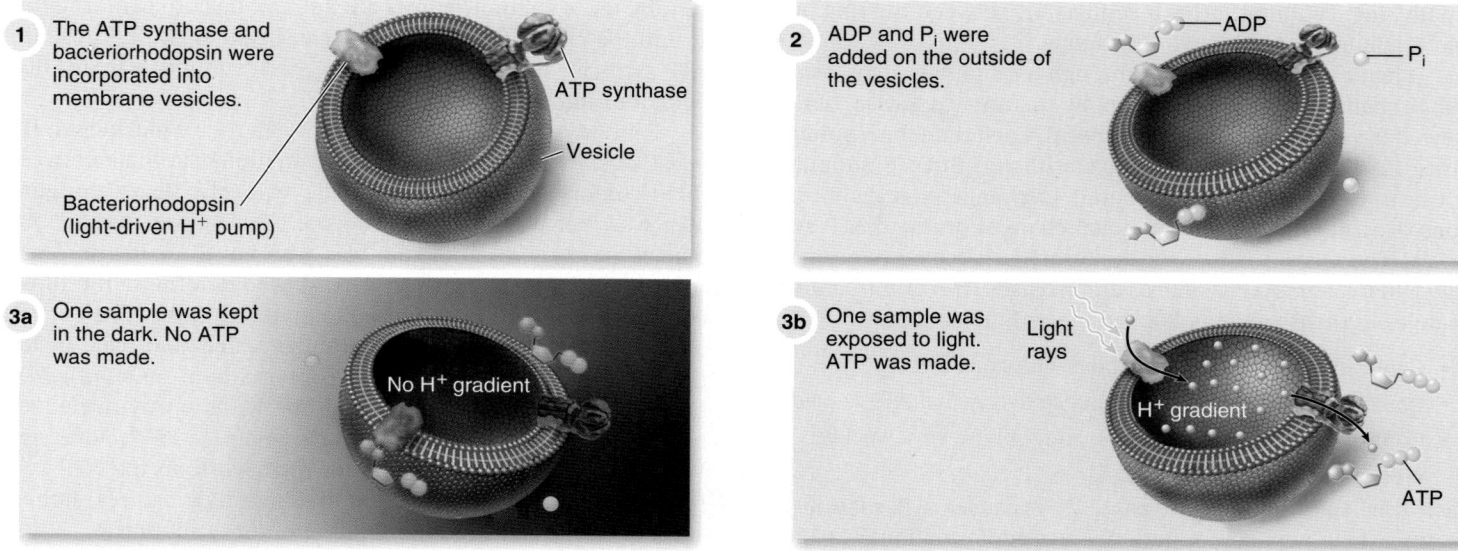

Figure 7.18   The Racker and Stoeckenius experiment showing that an H^+ electrochemical gradient drives ATP synthesis via the ATP synthase.

Biological inquiry: Is the functioning of the electron transport chain always needed to make ATP via the ATP synthase?

ATP that can be made by glycolysis (2), the citric acid cycle (2), and oxidative phosphorylation (30–34), we see that oxidative phosphorylation provides a cell with a much greater capacity to make ATP.

Experiments with Purified Proteins in Membrane Vesicles Verified Chemiosmosis

To show experimentally that the ATP synthase actually uses an H^+ electrochemical gradient to make ATP, researchers needed to purify the enzyme and study its function *in vitro*. In 1974, Ephraim Racker and Walther Stoeckenius purified the ATP synthase and another protein called bacteriorhodopsin, which is found in certain species of archaea. Previous research had shown that bacteriorhodopsin is a light-driven H^+ pump. Racker and Stoeckenius took both purified proteins and inserted them into membrane vesicles (**Figure 7.18**). The ATP synthase was oriented so that its ATP synthesizing region was on the outside of the vesicles. Bacteriorhodopsin was oriented so that it would pump H^+ into the vesicles. They added ADP and P_i on the outside of the vesicles. In the dark, no ATP was made. However, when they shone light on the vesicles, a substantial amount of ATP was made. Because bacteriorhodopsin was already known to be a light-driven H^+ pump, these results convinced researchers that the ATP synthase uses an H^+ electrochemical gradient as an energy source to make ATP.

The ATP Synthase Is a Rotary Machine That Makes ATP as It Spins

The structure and function of the ATP synthase are particularly intriguing and have received much attention over the past few decades (**Figure 7.19**). The ATP synthase is a rotary machine.

The nonmembrane-embedded portion consists of 1 ε, 1 γ, 1 δ, 3 α, and 3 β subunits. Movement of H^+ through the *c* subunits causes the γ subunit to rotate. The rotation, in 120° increments, causes the β subunits to progress through a series of 3 conformational changes that lead to the synthesis of ATP from ADP and P_i.

The membrane-embedded portion consists of a ring of 9–12 *c* subunits, 1 *a* subunit, and 2 *b* subunits. H^+ move through the *c* subunits.

Figure 7.19   The subunit structure and function of the ATP synthase.

The membrane-embedded region is composed of three types of subunits called *a*, *b*, and *c*. Approximately 9 to 12 *c* subunits form a ring in the membrane. Each *c* subunit is an H^+ channel. One *a* subunit is bound to this ring, and two *b* subunits are attached to the *a* subunit and protrude out of the membrane. The nonmembrane-embedded subunits are designated with Greek letters. One ε and one γ subunit bind to the ring of *c* subunits. The γ subunit forms a long stalk that pokes into the center of another ring of three α and three β subunits. The β subunits are the catalytic site where ATP is made.

Finally, the δ subunit forms a connection between the ring of α and β subunits and the two *b* subunits.

When hydrogen ions pass through a *c* subunit, a conformational change causes the γ subunit to turn clockwise (when viewed from the intermembrane space). Each time the γ subunit turns 120° it changes its contacts with the three β subunits, which in turn causes the β subunits to change their conformations. How do these conformational changes promote ATP synthesis? The answer is that the conformational changes occur in a way that favors ATP synthesis and release. The conformational changes in the β subunits happen in the following order:

- Conformation 1: ADP and P_i bind with good affinity.
- Conformation 2: ADP and P_i bind so tightly that ATP is made.
- Conformation 3: ATP (and ADP and P_i) bind very weakly, and ATP is released.

Each time the γ subunit turns 120°, it causes a β subunit to change to the next conformation. After conformation 3, a 120° turn by the γ subunit returns a β subunit back to conformation 1, and the cycle of ATP synthesis can begin again. Because the ATP synthase has three β subunits, each subunit is in a different conformation at any given time.

Paul Boyer proposed the concept of a rotary machine in the late 1970s. In his model, the three β subunits alternate between three conformations, as described previously. Boyer's original idea was met with great skepticism, because the concept that part of an enzyme could spin was very novel, to say the least. John Walker and colleagues were able to determine the three-dimensional structure of the nonmembrane-embedded portion of the ATP synthase. The structure revealed that each of the three β subunits had a different conformation—one with ADP bound, one with ATP bound, and one without any nucleotide bound. This result supported Boyer's model. In 1997, Boyer and Walker shared the Nobel Prize for their work on the ATP synthase. As described in the Feature Investigation, researchers subsequently visualized the rotation of the γ subunit.

FEATURE INVESTIGATION

Yoshida and Kinosita Demonstrated That the γ Subunit of the ATP Synthase Spins

In 1997, Masasuke Yoshida, Kazuhiko Kinosita, and colleagues set out to experimentally visualize the rotary nature of the ATP synthase (**Figure 7.20**). The membrane-embedded region of the ATP synthase can be separated from the rest of the protein by treatment of mitochondrial membranes with a high concentration of salt, releasing the portion of the protein containing the one γ, three α, and three β subunits. The researchers adhered the γα₃β₃ complex to a glass slide so that the γ subunit was protruding upwards. Because the γ subunit is too small to be seen with a light microscope, it is not possible to visualize the rotation of the γ subunit directly. To circumvent this problem, the researchers attached a large, fluorescently labeled actin filament to the γ subunit via a linker protein. The fluorescently labeled actin filament is very long compared to the γ subunit and can be readily seen with a fluorescence microscope.

Figure 7.20 Evidence that the ATP synthase is a rotary machine.

HYPOTHESIS The ATP synthase is a rotary machine.

STARTING MATERIALS Purified complex containing 1 γ, 3 α, and 3 β subunits.

Experimental level   **Conceptual level**

1 Adhere the purified γα₃β₃ complex to a glass slide so that the base of the γ subunit is protruding upwards.

Add purified complex.

γ
γα₃β₃ complex
α   β   α
Slide

2 Add linker proteins and fluorescently labeled actin filaments. The linker protein recognizes sites on both the γ subunit and the actin filament.

Add linker proteins and fluorescent actin filaments.

Fluorescent actin filament
Linker proteins
γ
α   β   α

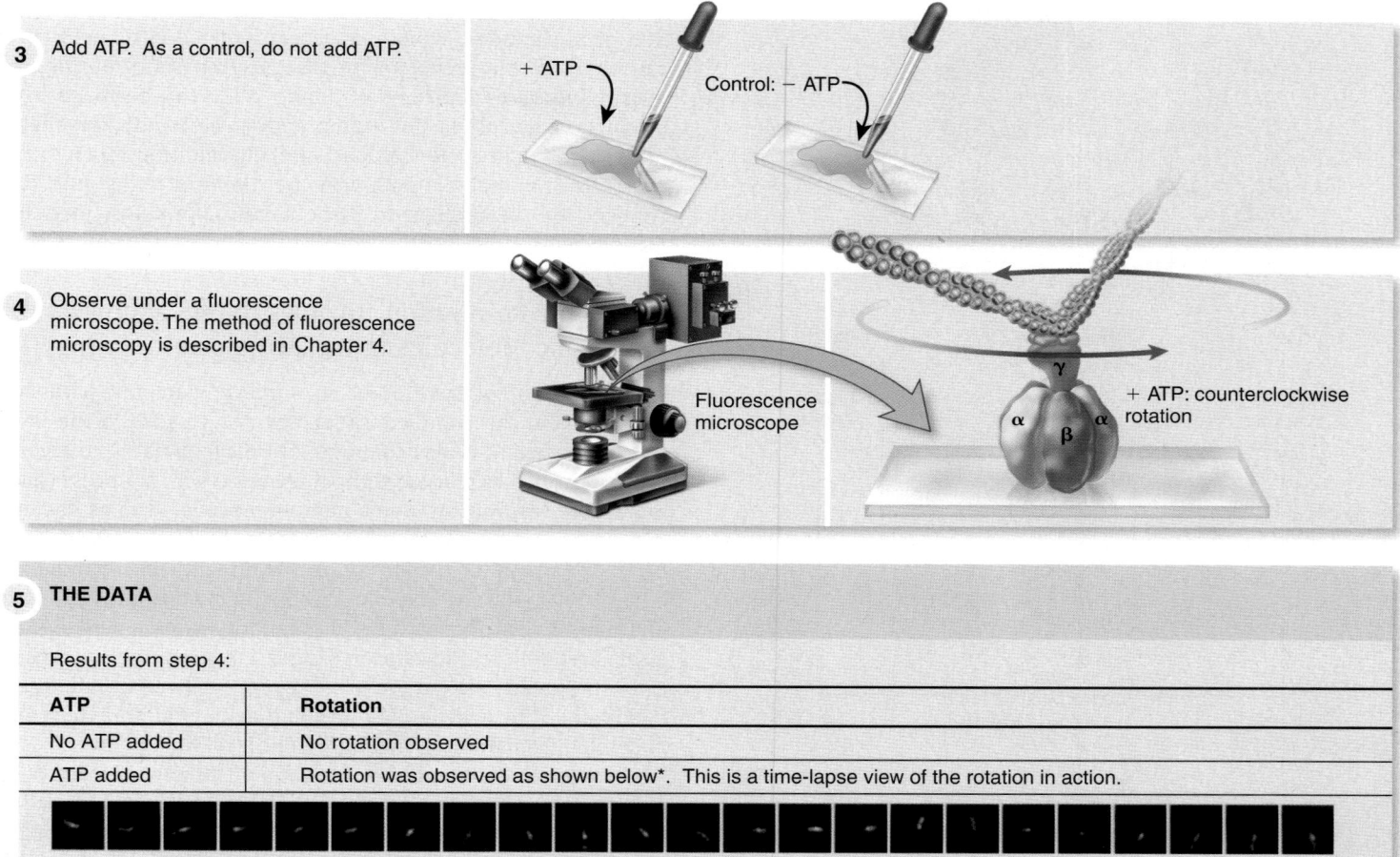

3 Add ATP. As a control, do not add ATP.

\+ ATP

Control: − ATP

4 Observe under a fluorescence microscope. The method of fluorescence microscopy is described in Chapter 4.

Fluorescence microscope

γ

α β α

\+ ATP: counterclockwise rotation

5 **THE DATA**

Results from step 4:

ATP	Rotation
No ATP added	No rotation observed
ATP added	Rotation was observed as shown below*. This is a time-lapse view of the rotation in action.

*This figure shows a series of micrographs from Noji et al. (1997) *Nature* 386, 299–303.

Because the membrane-embedded portion of the protein is missing, you may be wondering how the researchers could get the γ subunit to rotate. The answer is that they added ATP. Although the normal function of the ATP synthase is to make ATP, it can run backwards. In other words, the ATP synthase can hydrolyze ATP. As shown in the data for Figure 7.20, when the researchers added ATP, the fluorescently labeled actin fila- ment was observed to rotate in a counterclockwise direction, which is opposite to the direction that the γ subunit rotates when ATP is synthesized. In fact, actin filaments were observed to rotate for more than 100 revolutions in the presence of ATP. These results convinced the scientific community that the ATP synthase is indeed a rotary machine.

Metabolic Pathways for Carbohydrate Metabolism Are Interconnected to Pathways for Amino Acid and Fat Metabolism

Thus far, we have focused our attention on the stages of glucose breakdown that result in the release of CO_2 and the production of NADH, $FADH_2$ and ATP. In addition, cells use other organic molecules as a source of energy. When you eat a meal, it is likely to contain not only carbohydrates (including glucose) but also proteins and fats. Proteins and fats are also broken down by the enzymes involved with glucose metabolism.

As shown in **Figure 7.21**, proteins and fats can enter into glycolysis, or the citric acid cycle, at different points. Proteins are first acted upon by enzymes, either in digestive juices or within cells, that cleave the bonds connecting individual amino acids. Because the 20 amino acids differ in their side chains, amino acids and their breakdown products can enter at different points in the pathway. Breakdown products of amino acids can enter glycolysis, or an acetyl group can be removed from certain amino acids and become attached to CoA. Other amino acids can be modified and enter the citric acid cycle. Similarly, fats can be broken down to glycerol and fatty acids.

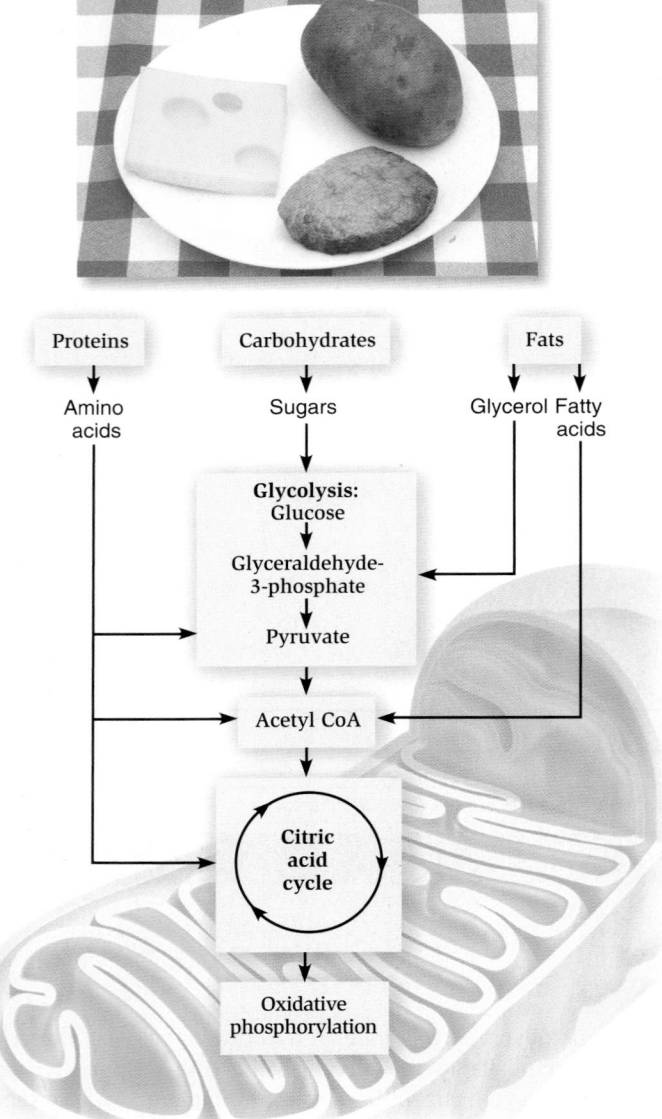

Figure 7.21 **Integration of protein, carbohydrate, and fat metabolism.** Breakdown products of amino acids and fats can enter the same pathway that is used to break down carbohydrates.

Glycerol can be modified to glyceraldehyde-3-phosphate and enter glycolysis at step 5 (see Figure 7.12). Fatty acyl tails can have two carbon acetyl units removed, which bind to CoA and then enter the citric acid cycle. By utilizing the same pathways for the breakdown of sugars, amino acids, and fats, cells are more efficient because they can use the same enzymes for the breakdown of different starting molecules.

Likewise, carbohydrate metabolism is connected to the metabolism of other cellular components at the anabolic level. Cells may use carbohydrates to manufacture parts of amino acids, fats, and nucleotides. For example, glucose-6-phosphate

of glycolysis is used to construct the sugar and phosphate portion of nucleotides, while oxaloacetate of the citric acid cycle can be used as a precursor for the biosynthesis of purine and pyrimidine bases. Portions of amino acids can be made from products of glycolysis (for example, pyruvate) and components of the citric acid cycle (oxaloacetate). In addition, many other catabolic and anabolic pathways are found in living cells that connect the metabolism of carbohydrates, proteins, fats, and nucleic acids.

Anaerobic Respiration Involves the Breakdown of Organic Molecules in the Absence of Oxygen

In this section, we have primarily surveyed catabolic pathways that result in the complete breakdown of glucose in the presence of oxygen. It is also common for cells to metabolize organic molecules such as glucose without using oxygen, a process called **anaerobic respiration**. Some microorganisms, such as bacteria and fungi, exist in environments that lack oxygen and have to carry out anaerobic respiration to obtain sufficient amounts of energy. Examples include microbes living in your intestinal tract and those living deep in the soil. Similarly, when a person exercises strenuously, the rate of oxygen consumption by muscle cells may greatly exceed the rate of oxygen delivery. Under these conditions, the muscle cells become anaerobic and must use anaerobic respiration to maintain their level of activity.

Organisms have evolved two different strategies to carry out anaerobic respiration. One strategy is to use a substance other than O_2 as the final electron acceptor of the electron transport chain. In the electron transport chain shown in Figure 7.16, cytochrome oxidase recognizes O_2 and catalyzes its reduction to H_2O. The final electron acceptor of the chain is O_2. Many species of bacteria that live under anaerobic conditions have evolved enzymes that function similarly to cytochrome oxidase, but recognize molecules other than O_2 and use them as the final electron acceptor. For example, *Escherichia coli*, which is found in your intestinal tract, has an enzyme called nitrate reductase that can use nitrate (NO_3^-) as the final electron acceptor of the electron transport chain under anaerobic conditions, and thus oxidative phosphorylation can occur.

Other organisms, including yeast and mammals, can use only O_2 as the final electron acceptor of their electron transport chains. When confronted with anaerobic conditions, these organisms must have a different strategy for anaerobic respiration. Such organisms can still carry out glycolysis, which can occur under anaerobic or aerobic conditions. However, a key issue is that glycolysis uses NAD^+ and generates NADH. As discussed earlier in this chapter, NADH is a good electron donor, because it readily gives up its electrons. Under aerobic conditions, oxygen acts as a final electron acceptor, and the high-energy electrons from NADH can be used to make more ATP. However, this cannot occur under anaerobic conditions in these organisms and, as a result, NADH builds up. At high concentrations, NADH will haphazardly donate its electrons to other molecules and

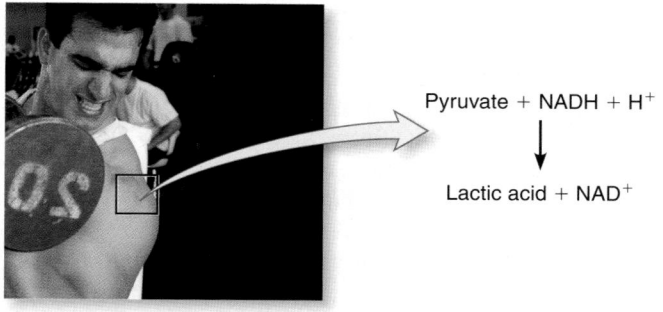

Pyruvate + NADH + H$^+$

↓

Lactic acid + NAD$^+$

(a) Production of lactic acid

Pyruvate

↓

Acetaldehyde + NADH + H$^+$ + CO$_2$

↓

Ethanol + NAD$^+$

(b) Production of ethanol

Figure 7.22   **Examples of anaerobic respiration.**   In these examples, NADH is produced by the oxidation of an organic molecule, and then the NADH is used up by donating electrons to a different organic molecule such as pyruvate **(a)** or acetaldehyde **(b)**. Note: at neutral pH, lactic acid releases H$^+$ to form lactate.

promote the formation of free radicals, highly reactive chemicals that can damage DNA and cellular proteins. For this reason, cells of mammals and yeast that are exposed to anaerobic conditions must have a strategy to remove the excess NADH normally generated from the breakdown of organic molecules such as glucose. In addition, cells need to regenerate NAD$^+$ to keep glycolysis running.

In the case of muscle cells working under anaerobic conditions, the pyruvate from glycolysis is converted by NADH to lactate, also known as lactic acid (**Figure 7.22a**). When a person exercises strenuously, the buildup of lactic acid causes a burning sensation or even cramps that often prompt the individual to slow down. Products of anaerobic respiration, such as lactic acid, are eventually secreted from cells. Similarly, during wine making, a yeast cell metabolizes sugar under anaerobic conditions. The pyruvate is broken down to CO$_2$ and a two-carbon molecule called acetaldehyde. The acetaldehyde is then reduced by NADH to make ethanol (**Figure 7.22b**). In these examples, NADH is used up by donating its electrons to pyruvate and acetaldehyde, thereby creating lactic acid and ethanol, respectively. These reactions also regenerate NAD$^+$, which allows cells to metabolize more glucose and keep making ATP.

The term **fermentation** is used to describe the breakdown of organic molecules to produce energy without any net oxida-

tion (that is, without any removal of electrons). The breakdown of glucose to lactic acid or ethanol are examples of fermentation. Although electrons are removed from an organic molecule such as glucose to make NADH and pyruvate, the electrons are donated back to an organic molecule in the creation of lactic acid or ethanol. Therefore, there is no net removal of electrons from an organic molecule. Compared with oxidative phosphorylation, fermentation produces far less ATP for two reasons. First, glucose is not oxidized completely to CO$_2$ and water. Second, the NADH that is made during glycolysis cannot be used to make more ATP. Overall, the complete breakdown of glucose in the presence of oxygen yields 34 to 38 ATP molecules. By comparison, the anaerobic breakdown of glucose to lactic acid or ethanol yields only two ATP molecules.

7.4 Secondary Metabolism

Primary metabolism is the synthesis and breakdown of molecules and macromolecules that are found in all forms of life and are essential for cell structure and function. These include compounds such as lipids, sugars, nucleotides, and amino acids and the macromolecules that are derived from them. Cellular respiration, which we considered earlier in this chapter, is an example of primary metabolism. By comparison, **secondary metabolism** involves the synthesis of molecules—**secondary metabolites**—that are not essential for cell structure and growth. Any given secondary metabolite is unique to one species or group of species and is not usually required for survival. Secondary metabolites, also called secondary compounds, are commonly made in plants, bacteria, and fungi, where they play a variety of roles. Many secondary metabolites taste bad. When produced in a plant, for example, such a molecule may prevent an animal from eating the plant. In some cases, secondary metabolites are toxic. Such molecules may act as a chemical weapon that inhibits the growth of nearby organisms. In addition, many secondary metabolites produce a strong smell or bright color that attracts or repels other organisms. For example, the scent from a rose is due to secondary metabolites. The scent attracts insects that aid in pollination. Alternatively, other secondary metabolites may smell bad and prevent an animal from ingesting a plant.

Biologists have discovered thousands of different secondary metabolites, though any given species tends to produce only one or a few types. As you will learn, many of these have been put to practical use by humans, including spices that are used in cooking and antibiotics that inhibit the growth of pathogenic microorganisms and are used to treat diseases. Plants are particularly diverse in the types of secondary metabolites they produce, perhaps because they need clever ways to defend themselves. Bacteria and fungi also produce a large array of these compounds, while animals tend to produce relatively few. In this section, we will survey four categories of secondary metabolites: phenolics, alkaloids, terpenoids, and polyketides.

Vanillin

Pelargonidin — an anthocyanin

(a) Flavonoids in vanilla provide flavor

(b) Anthocyanins such as pelargonidin give red color

Figure 7.23 Phenolic compounds as secondary metabolites. The two examples shown here are flavonoids, which are a type of phenolic compound. **(a)** The flavor of vanilla is largely produced by flavonoids, an example of which is vanillin. Commercially, vanillin is extracted from the seed capsule of *Vanilla planifolia*. **(b)** Another group of flavonoids that causes red, blue, or purple color are anthocyanins. The red color of strawberries is caused by pelargonidin, an anthocyanin.

Phenolic Compounds Are Antioxidants That Have Intense Flavors and Bright Colors

The **phenolic** compounds all contain a cyclic ring of carbon with three double bonds, known as a benzene ring, within their structure. When a benzene ring is covalently linked to a single hydroxyl group, the compound is known as phenol.

OH   OH

or

Phenol is the simplest of the phenolic compounds, though free phenol is not significantly accumulated in living organisms. However, more complex molecules that are derived from phenol are made in cells. Such phenolic compounds are synthesized using the side groups of the amino acids phenylalanine (which has a benzene ring) or tyrosine (which has a phenol ring). Common categories of phenolics are the flavonoids, tannins, and lignins. The tannins and lignins are large polymeric molecules composed of many phenolic units.

Flavonoids are produced by many plant species and create a variety of flavors and smells. These can play a role as deterrents to eating a plant or as attractants that promote pollination. The flavors of chocolate and vanilla largely come from a mixture of flavonoid molecules. Vanilla is produced by several spe-

cies of perennial vines of the genus *Vanilla* native to Mexico and tropical America (**Figure 7.23a**). The primary source of commercial vanilla comes from *V. planifolia*. Vanilla extract is obtained from the seed capsules. Another role of flavonoids is pigmentation. Anthocyanins (from the Greek *anthos*, flower, and *kyanos*, blue) produce the red, purple, and blue colors of many flowers, fruits, and vegetables (**Figure 7.23b**).

Biochemists have discovered that flavonoids have remarkable antioxidant properties that prevent the formation of damaging free radicals. In plants, flavonoids are thought to act as powerful antioxidants helping to protect plants from UV damage. In recent times, nutritionists have advocated the consumption of fruits and vegetables that have high amounts of flavonoids, such as broccoli and spinach. Dark chocolate is also rich in these antioxidants!

Tannins are large phenolic polymers, so named because they combine with the protein of animal skins to form leather. This process, known as tanning, also imparts a tan color to animal skins. Tannins are found in many plant species and typically act as a deterrent to animals, either because of a bitter taste or due to toxic effects. They also can inhibit the enzymes found in the digestive tracts of animals, if consumed in large amounts. Tannins are found abundantly in grape skins and play a key role in the flavor of red wine. Aging breaks down tannins, making the wine less bitter.

Lignins are also large phenolic polymers synthesized by plants. Lignin is found in plant cell walls and makes up about one-quarter to one-third of the weight of dry wood. The lignins form polymers that bond with other plant wall components such as cellulose. This strengthens plant cells and enables a plant to better withstand the rigors of environmental stress. To make paper, which is much more malleable than wood, the lignin is removed.

Alkaloids Form a Large Group of Bitter-Tasting Molecules That Also Provide Defense Mechanisms

Alkaloids are a group of structurally related molecules that all contain nitrogen and usually have a cyclic, ringlike structure. More than 12,000 different alkaloids have been discovered. Their name is derived from the observation that they are basic or alkaline molecules. Alkaloids are usually synthesized from amino acids precursors. Alkaloids are commonly made in plant species and occasionally in fungi (mushrooms) and animals (shellfish). Familiar examples include caffeine, nicotine, atropine, morphine, ergot, and quinine.

Like phenolics, many alkaloids serve a defense function in plants. Alkaloids are bitter-tasting molecules and often have an unpleasant odor. These features may prevent an animal from eating a plant or its fruit. For example, an alkaloid in chile peppers called capsaicin elicits a burning sensation. This molecule is so potent that one-millionth of a drop can be detected by the human tongue. Capsaicin may serve to discourage mammals from eating the peppers. Interestingly, however, birds do not experience the burning sensation of capsaicin and serve to disperse the seeds.

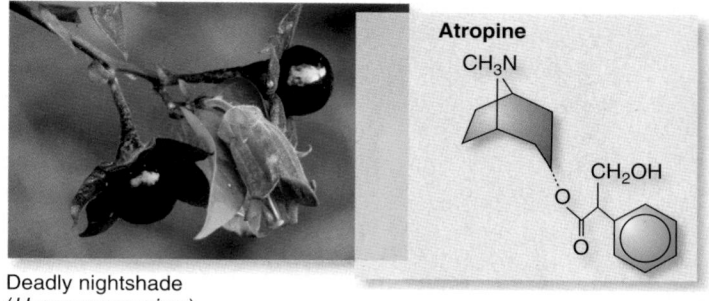

Deadly nightshade
(*Hyoscyamus niger*)

Figure 7.24 **Alkaloids as secondary metabolites.** Atropine is an alkaloid produced by the plant called deadly nightshade (*Hyoscyamus niger*). Atropine is toxic because it interferes with nerve transmission. In humans, atropine causes the heart to speed up to dangerous and possibly fatal rates.

Other alkaloids are poisonous, like the alkaloid **atropine**, a potent toxin derived from the deadly nightshade plant (**Figure 7.24**). Animals that eat this plant and consequently ingest atropine become very sick and may die. It is unlikely that an animal that eats deadly nightshade and survives would choose to eat it a second time. Atropine acts by interfering with nerve transmission. In humans, for example, atropine causes the heart to speed up to dangerous rates, because the nerve inputs that normally keep a check on heart rate are blocked by atropine. Other alkaloids are not necessarily toxic but can cause an animal that eats them to become overstimulated (caffeine), understimulated (any of the opium alkaloids like morphine), or simply nauseated because the compound interferes with nerves required for proper functioning of the gastrointestinal system.

Terpenoids Are Molecules with Intense Smells and Color

A third major class of secondary metabolites are the **terpenoids**, of which over 25,000 have been identified, more than any other family of naturally occurring products. Terpenoids are synthesized from five-carbon isoprene units (shown below) and are also called isoprenoids.

Isoprene units are linked to each other to form larger compounds with multiples of five-carbon atoms. In many cases, the isoprene units form cyclic structures.

Terpenoids have a wide array of functions in plants. Notably, because many terpenoids are volatile (they become gases), they are responsible for the odors emitted by many types of plants, such as menthol produced by mint. The odors of terpenoids may attract pollinators or repel animals that eat plants. In addition, terpenoids often impart an intense flavor to plant tissues. Many of the spices used in cooking are rich in different types of terpenoids. Examples include cinnamon, fennel, cloves, cumin, cara-

Flamingo (*Phoenicopterus ruber*)

β-carotene

Figure 7.25 **Terpenoids as secondary metabolites.** Carotenoids are a type of terpenoid with bright color. The example here is β-carotene, which gives many organisms an orange color. Flamingos (*Phoenicopterus ruber*) receive β-carotene in their diet, primarily from eating shellfish.

way, and tarragon. Terpenoids are found in many traditional herbal remedies and are under medical investigation for potential pharmaceutical effects.

Other terpenoids, like the carotenoids, are responsible for the coloration of many species. An example is β-carotene, which gives carrots their orange color. Carotenoids are also found in leaves, but their color is masked by chlorophyll, which is green. In the autumn, when chlorophyll breaks down, the color of the carotenoids becomes evident. Carotenoids give color to animals such as salmon, goldfish, and flamingos (**Figure 7.25**). Another role of terpenoids is cell signaling. All steroid hormones, which function as signaling molecules in animals and plants (as we will discuss in Chapter 9), are derived from terpenoids.

Polyketides Are Often Used as Chemical Weapons to Kill Competing Organisms

Polyketides are a group of secondary metabolites that are produced by bacteria, fungi, plants, insects, dinoflagellates, mollusks, and sponges. They are synthesized by the polymerization of acetyl (CH_3COOH) and propionyl (CH_3CH_2COOH) groups to create a diverse collection of molecules, often with many ringed structures. Polyketides are usually secreted by the organism that makes them and are often highly toxic to other organisms. For example, the polyketide known as streptomycin is made by the soil bacterium *Streptomyces griseus* (**Figure 7.26**). It is secreted by this bacterium and taken up by other species, where it disrupts protein synthesis and thereby inhibits their growth. In this way, *S. griseus* is able to kill or inhibit the growth of other species in its vicinity. This is an advantage for *S. griseus* because other species may be using limited resources that the bacterium could use for its own growth.

During the past several decades, over 10,000 polyketides have been identified and analyzed. Familiar examples include streptomycin, erythromycin, and tetracycline. The toxic effects of polyketides are often very selective, making them valuable medical tools. For example, streptomycin disrupts protein synthesis in many bacterial species, but it does not adversely affect protein synthesis in mammalian cells. Therefore, it has been used as an antibiotic to treat or prevent bacterial infections in humans and other mammals. Similarly, other polyketides inhibit the growth of fungi, parasites, and insects. More recently, researchers have even discovered that certain polyketides inhibit the growth of cancer cells. The production and sale of polyketides to treat and prevent diseases and as pesticides constitute an enormous industry, with annual sales in the U.S. at over $20 billion.

Streptomyces griseus, a soil bacterium

Streptomycin

Figure 7.26 **Polyketides as secondary metabolites.** Streptomycin, whose structure is shown here, is an antibiotic produced by *Streptomyces griseus*, a soil bacterium. The scanning electron micrograph shows *S. grisieus*.

CHAPTER SUMMARY

7.1 Energy, Chemical Reactions, and Enzymes

- The fate of a chemical reaction is determined by its direction and rate.

- Energy, the ability to promote change, exists in many forms. According to the first law of thermodynamics, energy cannot be created or destroyed but it can be converted from one form to another. The second law of thermodynamics states that energy interconversions involve an increase in entropy. (Figures 7.1, 7.2, Table 7.1)

- Free energy is the amount of available energy that can be used to do work. Spontaneous reactions release free energy, which means they have a negative free energy change. (Figure 7.3)

- An exergonic reaction has a negative free energy change, while an endergonic reaction has a positive change. Chemical reactions proceed until they reach a state of chemical equilibrium, where the rate of formation of products equals the rate of formation of reactants.

- Cells use energy intermediates such as ATP to drive endergonic reactions.

- Proteins that speed up the rate of a chemical reaction are called enzymes. They lower the activation energy that is needed to achieve a transition state. (Figure 7.4)

- Enzymes recognize the reactants, also called substrates, with a high specificity. Conformational changes are involved in lowering the activation energy for a chemical reaction. (Figure 7.5)

7.2 Overview of Metabolism

- Metabolism is the sum of the chemical reactions in a living organism. Enzymes often function in pathways that lead to the formation of a particular product. (Figure 7.6)

- Catabolic reactions involve the breakdown of larger molecules into smaller ones. These reactions regenerate small molecules that are used as building blocks to make new molecules. The small molecules are also broken down to make energy intermediates such as ATP and NADH. Such reactions are often redox reactions in which electrons are transferred from one molecule to another. (Figure 7.7)

- Anabolic reactions involve the synthesis of larger molecules and macromolecules.

- Estimates from genome analysis indicate that over 20% of a cell's proteins use ATP. (Table 7.2, Figure 7.8)

- Metabolic pathways are controlled by gene regulation, cell signaling, and feedback inhibition. (Figure 7.9)

7.3 Cellular Respiration

- Cells obtain energy via cellular respiration, which involves the breakdown of molecules.

- The breakdown of glucose occurs in four stages: glycolysis, pyruvate breakdown, citric acid cycle, and oxidative phosphorylation. (Figure 7.10)

- Glycolysis is the breakdown of glucose to two pyruvates, producing two ATP and two NADH. ATP is made by substrate-level phosphorylation. (Figures 7.11, 7.12)

- Pyruvate is broken down to CO_2 and an acetyl group that becomes attached to CoA. NADH is made during this process. (Figure 7.13)

- During the citric acid cycle, the acetyl group attached to CoA is broken down to two CO_2 molecules. Three NADH, one $FADH_2$, and one ATP are made during this process. (Figures 7.14, 7.15)

- Oxidative phosphorylation involves two events. The electron transport chain oxidizes NADH or $FADH_2$ and generates an H^+ electrochemical gradient. This gradient is utilized by the ATP synthase to make ATP via chemiosmosis. (Figures 7.16, 7.17)

- Racker and Stoeckenius showed that the ATP synthase uses an H^+ gradient by reconstituting the ATP synthase with a light-driven H^+ pump. (Figure 7.18)

- The ATP synthase is a rotary machine. The rotation is caused by the movement of H^+ through the c subunits that cause the γ subunit to spin, resulting in conformational changes in the β subunits that promote ATP synthesis. (Figure 7.19)

- Yoshida and Kinosita experimentally demonstrated rotation of the γ subunit by attaching a fluorescently labeled actin filament and watching it spin in the presence of ATP. (Figure 7.20)

- Proteins and fats can enter into glycolysis or the citric acid cycle at different points. (Figure 7.21)

- Anaerobic respiration occurs in the absence of oxygen. Either the final electron acceptor of the electron transport chain is not oxygen, or NAD^+ is regenerated by donating electrons to an organic molecule such as pyruvate (to make lactic acid). (Figure 7.22)

7.4 Secondary Metabolism

- Secondary metabolites are not usually necessary for cell structure and function, but they provide an advantage to an organism that may involve taste, smell, color, or poison. Four categories are phenolic compounds, alkaloids, terpenoids, and polyketides. (Figures 7.23, 7.24, 7.25, 7.26)

TEST YOURSELF

1. According to the second law of thermodynamics
   a. energy cannot be created or destroyed.
   b. each energy transfer decreases the disorder of a system.
   c. energy is constant in the universe.
   d. each energy transfer increases the level of disorder in a system.
   e. chemical energy is a form of potential energy.

2. _____ reactions release free energy.
   a. Exergonic   d. All of the above
   b. Spontaneous   e. Both a and b
   c. Endergonic

3. Enzymes speed up reactions by
   a. providing chemical energy to fuel a reaction.
   b. lowering the activation energy necessary to initiate the reaction.
   c. causing an endergonic reaction to become an exergonic reaction.
   d. substituting for one of the reactants necessary for the reaction.
   e. none of the above.

4. Which of the following factors will alter the function of enzymes?
   a. pH   d. all of the above
   b. temperature   e. b and c only
   c. cofactors

5. In biological systems, ATP functions by
   a. providing the energy necessary for an endergonic reaction by coupling it with an exergonic reaction.
   b. acting as an enzyme and lowering the activation energy of certain reactions.
   c. adjusting the pH of solutions to maintain optimal conditions for enzyme activity.
   d. regulating the speed at which endergonic reactions proceed.
   e. interacting with enzymes as a cofactor to stimulate chemical reactions.

6. During redox reactions, the molecule that donates an electron is said to be
   a. reduced.   d. catabolized.
   b. phosphorylated.   e. methylated.
   c. oxidized.

7. Currently scientists are identifying proteins that use ATP as an energy source by
   a. determining whether those proteins function in anabolic or catabolic reactions.
   b. determining if the protein has a known ATP-binding site.
   c. predicting the free energy necessary for the protein to function.
   d. determining if the protein has an ATP synthase subunit.
   e. all of the above.

8. During glycolysis, ATP is produced by
   a. oxidative phosphorylation.
   b. substrate-level phosphorylation.
   c. redox reactions.
   d. all of the above.
   e. both a and b.

9. The energy necessary to produce ATP during oxidative phosphorylation is provided by
   a. the H^+ concentration gradient produced by the electron transport chain.
   b. GTP produced during the citric acid cycle.
   c. lactic acid metabolism.
   d. the release of CO_2 from the mitochondria.
   e. all of the above.

10. Secondary metabolites
    a. help deter predation of certain organisms by causing the organism to taste bad.
    b. help attract pollinators by producing a pleasant smell.
    c. help organisms compete for resources by acting as a poison to competitors.
    d. provide protection from DNA damage.
    e. all of the above.

CONCEPTUAL QUESTIONS

1. Distinguish between endergonic and exergonic reactions.

2. Define feedback inhibition.

3. The electron transport chain is so named because electrons are transported from one component to another. Describe the purpose of the electron transport chain.

EXPERIMENTAL QUESTIONS

1. The components of the ATP synthase are too small to be visualized by light microscopy. For the experiment of Figure 7.20, how did the researchers observe the movement of the ATP synthase?

2. In the experiment of Figure 7.20, what observation did the researchers make that indicated that the ATP synthase is a rotary machine? What was the control of this experiment? What did it indicate?

3. Were the rotations seen by the researchers in the data of Figure 7.20 in the same direction as expected in the mitochondria during ATP synthesis? Why or why not?

COLLABORATIVE QUESTIONS

1. Discuss several ways in which metabolic pathways are controlled or regulated.

2. Discuss the concept of secondary metabolism and give an example.

www.brookerbiology.com

This website includes answers to the Biological Inquiry questions found in the figure legends and all end-of-chapter questions.

8

PHOTOSYNTHESIS

A tropical rain forest in Akaka Falls State Park, Hawaii. Plant life in tropical rain forests carries out a large amount of the world's photosynthesis and supplies the atmosphere with a sizeable fraction of its oxygen.

Across the Earth, the most visible color on land is green. We often associate this color with emerging life, as in the growth of plants in the spring. The green color of plants is due to a pigment called chlorophyll. This pigment provides the starting point for the process of **photosynthesis**, in which the energy within light is captured and used to synthesize carbohydrates. Nearly all living organisms ultimately rely on photosynthesis for their nourishment. Photosynthesis is also responsible for producing the oxygen that makes up a large portion of the Earth's atmosphere. Therefore, all aerobic organisms ultimately rely on photosynthesis for respiration.

Organisms can be categorized as heterotrophs and autotrophs. **Heterotrophs** must eat food, organic molecules from their environment, to sustain life. Examples of heterotrophs include most species of bacteria and protists, as well as all species of fungi and animals. In Chapter 7, we learned how cells use food molecules such as glucose for their energy needs and as building blocks to make new cellular molecules and macromolecules. In this chapter, we turn to **autotrophs**, organisms that make organic molecules from inorganic sources, and focus on **photoautotrophs**, those organisms that use light as a source of energy. These include green plants, algae, and some prokaryotic species such as cyanobacteria.

We begin this chapter with an overview of photosynthesis, with an emphasis on how it occurs in green plants. We will then explore the two stages of photosynthesis: the light reactions, in which light energy is captured by the chlorophyll pigments within plants and converted to chemical energy in the form of two compounds, ATP and NADPH; and the Calvin cycle, a series of steps in which these compounds drive the incorporation of CO_2 into carbohydrates. We conclude with a consideration of the variations in photosynthesis that occur in plants existing in hot and dry conditions.

8.1 Overview of Photosynthesis

In the mid-1600s, a Flemish physician, Jan Baptista Van Helmont, conducted an experiment in which he transplanted the shoot of a young willow tree into a bucket of soil of known weight. He watered the tree and allowed it to grow for five years. After this time, the willow tree had added 164 pounds to its original weight, but the soil had lost only 2 ounces. Van Helmont correctly concluded that the willow tree did not get most of its nutrients from the soil. However, he incorrectly concluded that the material that made up the bark, wood, roots, and leaves came from the water he had added over the five years. Although water does contribute to the growth and mass of plants, we now know that CO_2 from the air is also critically important.

In 1771, Joseph Priestley, an English chemist, carried out an experiment in which he placed a burning candle in a closed chamber. The candle burned out very quickly. He conducted another experiment in which he placed a burning candle and a sprig of mint in a second chamber. Similarly, the candle quickly went out. After several days, Priestley was able to relight each candle without opening the two chambers by focusing a beam of sunlight onto the wicks with a mirror. However, only in the chamber with the sprig of mint could the candle burn again. Priestley hypothesized that plants restore to the air whatever burning candles remove. His results occurred because plants release oxygen as a result of photosynthesis.

Shortly thereafter, Jan Ingenhousz, a Dutch physician, immersed green plants underwater and discovered that they released bubbles of oxygen. Moreover, Ingenhousz determined that sunlight was necessary for oxygen production. During this same period, Jean Senebier, a Swiss botanist, found that CO_2 is required for plant growth and Nicolas-Théodore de Saussure, a Swiss chemist, showed that water is also required. With this accumulating information, Julius von Mayer, a German physicist, proposed in 1845 that plants convert light energy from the sun into chemical energy.

For the next several decades, plant biologists studied photosynthesis in prokaryotes, algae, and green plants. Researchers discovered that some photosynthetic bacteria could use hydrogen sulfide (H_2S) instead of water (H_2O) for photosynthesis and that these organisms released sulfur instead of oxygen. In the 1930s, based on this information, Dutch American microbiologist Cornelis van Niel proposed a general equation for photosynthesis that applies to plants, algae, and photosynthetic bacteria alike.

$$CO_2 + 2\ H_2A + \text{Light energy} \rightarrow CH_2O + 2A + H_2O$$

where A is oxygen (O) or sulfur (S) and CH_2O is the general formula for a carbohydrate. This is a redox reaction in which H_2A is oxidized and CO_2 is reduced.

In green plants, A is oxygen and 2A is a molecule of oxygen that is designated O_2. Therefore this equation becomes:

$$CO_2 + 2\ H_2O + \text{Light energy} \rightarrow CH_2O + O_2 + H_2O$$

When the carbohydrate that is produced is glucose, we multiply each side of the equation by six to obtain:

$$6\ CO_2 + 12\ H_2O + \text{Light energy} \rightarrow C_6H_{12}O_6 + 6\ O_2 + 6\ H_2O$$

Glucose

$$\Delta G = +685\ \text{kcal/mole}$$

In this redox reaction, CO_2 is reduced during the formation of glucose, and H_2O is oxidized during the formation of O_2. The free energy change required for the production of 1 mole of glucose from carbon dioxide and water is a whopping +685 kcal/mole! As we learned in Chapter 7, endergonic reactions are driven forward by coupling the reaction with an exergonic process, a process that releases free energy. In this case, the energy from light ultimately drives the synthesis of glucose.

In this section, we will survey the general features of photosynthesis as it occurs in green plants. Later sections will examine the various steps in this process.

Photosynthesis Powers the Biosphere

The term **biosphere** describes the regions on the surface of the Earth and in the atmosphere where living organisms exist. Life in the biosphere is largely driven by the photosynthetic power of green plants. For most species to exist, a key energy cycle involves the interplay between organic molecules (such as glucose) and inorganic molecules, namely, CO_2 and O_2. Life on Earth involves a cycle in which cells use organic molecules for energy and plants replenish those molecules via photosynthesis. As we examined in Chapter 7, cellular respiration involves the breakdown of organic molecules to produce energy intermediates such as ATP. When glucose is broken down to CO_2, the net reaction of cellular respiration in the presence of oxygen can be summarized as:

$$C_6H_{12}O_6 + 6\ O_2 \rightarrow 6\ CO_2 + 6\ H_2O + \textbf{Energy}$$

The net reaction of photosynthesis (in which only six net molecules of H_2O are consumed) can be viewed as the opposite of respiration.

$$6\ CO_2 + 6\ H_2O + \textbf{Energy} \rightarrow C_6H_{12}O_6 + 6\ O_2$$

The breakdown of glucose during cell respiration is an energy-releasing process that drives the synthesis of ATP. By comparison, the energy that is needed to synthesize glucose during photosynthesis ultimately comes from sunlight.

Plants make a large proportion of the Earth's organic molecules via photosynthesis. At the same time, they also produce O_2. These organic molecules are metabolized by the plants themselves as well as by heterotrophs such as animals and fungi. This metabolism generates CO_2, which is released into the atmosphere and can be used by plants to make more organic molecules like glucose. In this way, a cycle exists between photosynthesis and cellular respiration that sustains life on our planet.

In Plants, Photosynthesis Occurs in the Chloroplast

Chloroplasts are organelles found in plant cells and algae that carry out photosynthesis. These organelles contain large quantities of **chlorophyll**, a pigment that gives plants their green color. All green parts of a plant contain chloroplasts and are capable of photosynthesis, although the majority of photosynthesis occurs in the leaves (**Figure 8.1**). The central part of the leaf, called the **mesophyll**, contains cells that carry out the bulk of photosynthesis in plants. For photosynthesis to occur, the mesophyll cells must obtain water and carbon dioxide. The water is taken up by the roots of the plant and is transported to the leaves by small veins. Carbon dioxide gas enters the leaf, and oxygen exits, via pores called **stomata** (singular, *stoma*), from the Greek, meaning mouth. The anatomy of leaves will be examined further in Chapter 35.

Like the mitochondrion, a chloroplast contains an outer and inner membrane, with an intermembrane space lying between the two. A third membrane, called the **thylakoid membrane**, contains pigment molecules, including chlorophyll. The thylakoid membrane forms many flattened, fluid-filled tubules called the **thylakoids**, which enclose a single, convoluted compartment known as the **thylakoid lumen**. Thylakoids stack on top of each other to form a structure called a **granum** (plural, *grana*). The **stroma** is the fluid-filled region of the chloroplast between the thylakoid membrane and the inner membrane.

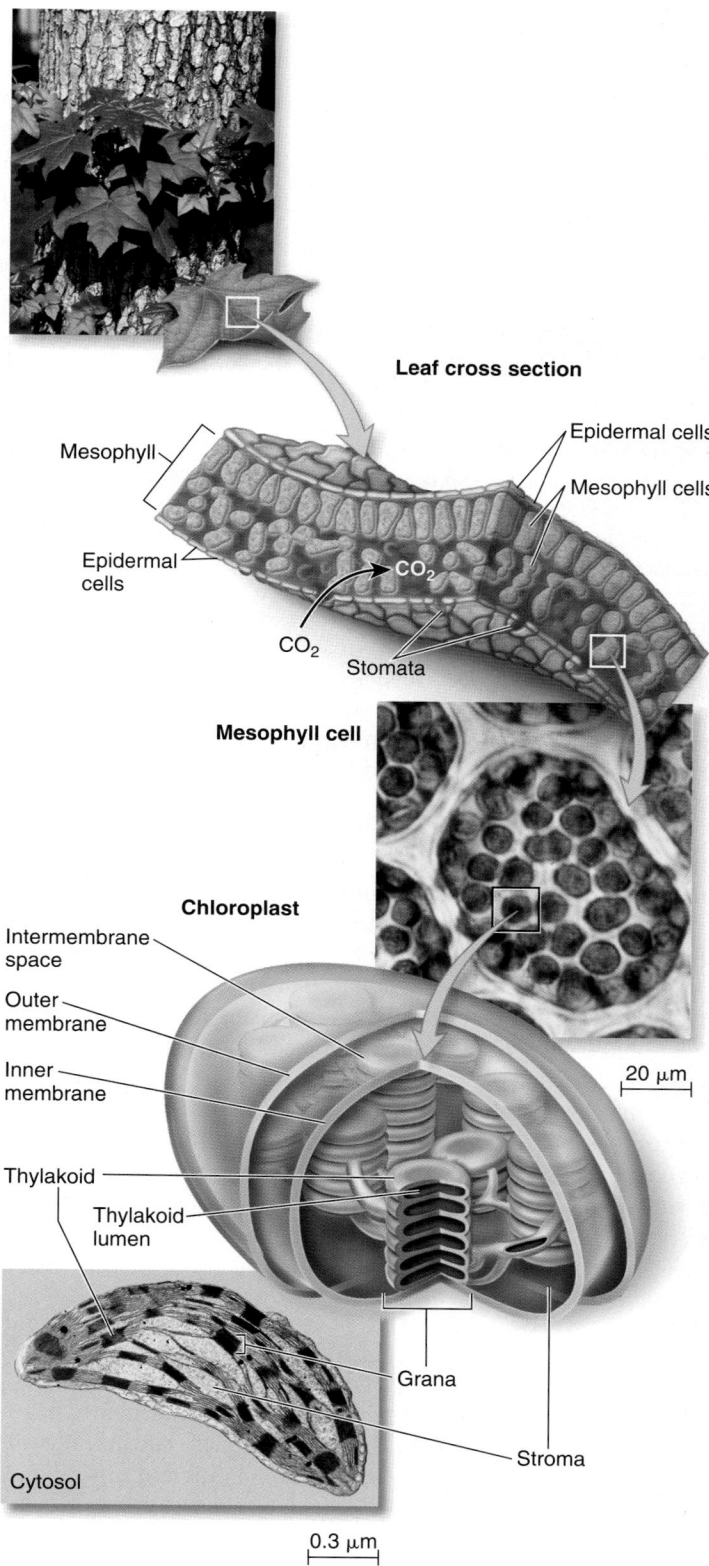

Figure 8.1 **Leaf organization.** Leaves are composed of layers of cells. The epidermal cells are on the outer surface, both top and bottom, with mesophyll cells sandwiched in the middle. The mesophyll cells in most plants are the primary sites of photosynthesis.

Photosynthesis Occurs in Two Stages

The process of photosynthesis can be divided into two stages called the **light reactions** and the **Calvin cycle**. The term photosynthesis is derived from the association between these two stages: The prefix photo refers to the light reactions that capture the energy needed for the synthesis of carbohydrates that occurs in the Calvin cycle. Each stage occurs at specific sites in the chloroplast: The light reactions take place at the thylakoid membrane, and the Calvin cycle occurs in the stroma (**Figure 8.2**).

The light reactions involve an amazing series of energy conversions, starting with light energy and ending with chemical energy in the form of covalent bonds. The light reactions produce three chemical products: ATP, NADPH, and O_2. ATP and NADPH are energy intermediates that provide the needed energy and electrons to drive the Calvin cycle. Like NADH, **NADPH (nicotinamide adenine dinucleotide phosphate)** is an electron carrier; its structure differs from NADH by the presence of an additional phosphate group. In the Calvin cycle, atmospheric carbon dioxide is incorporated into organic molecules, some of which are converted to carbohydrates.

O_2 is another important product of the light reactions. As described in Chapter 7, this molecule is vital to the process of aerobic respiration. Nearly all of the O_2 in the atmosphere is produced by photosynthesis from green plants and aquatic microorganisms. A large percentage of atmospheric oxygen is made by regions of the Earth that are rich in plant life, such as tropical rain forests. More than 20% of the world's oxygen is produced in the Amazon rain forest in South America alone.

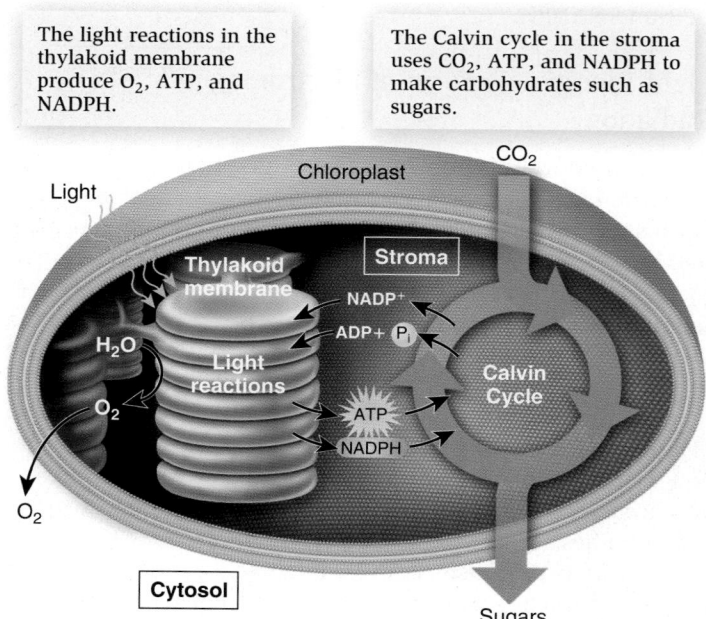

Figure 8.2 **An overview of the two stages of photosynthesis.** The light reactions, through which O_2, ATP, and NADPH are made, occur at the thylakoid membrane. The Calvin cycle, in which enzymes use ATP and NADPH to incorporate CO_2 into carbohydrate, occurs in the stroma.

Ecologists are alarmed about the rate at which such forests are being destroyed by human activities. Rain forests once covered 14% of the Earth's land surface but now occupy less than 6%. At their current rate of destruction, rain forests may be nearly eliminated in less than 40 years. Such an event may lower the level of oxygen in the atmosphere and thereby have a harmful impact on living organisms on a global scale.

8.2 Reactions That Harness Light Energy

Photosynthesis relies on the first law of thermodynamics. Recall from Chapter 7 that this law states that while energy cannot be created or destroyed, it can be transferred from one place to another and transformed from one type to another. During photosynthesis, energy in the form of light is transferred from the sun, some 92 million miles away, to a pigment molecule in a photosynthetic organism such as a plant. The next thing that happens is an interesting and complex series of energy transformations in which light energy is transformed into electrochemical energy and then into energy stored within chemical bonds.

In this section, we will explore this series of transformations, collectively called the light reactions of photosynthesis. We begin by examining the unique properties of light and then consider the features of chloroplasts that allow them to capture light energy. The remaining sections focus on how the light reactions of photosynthesis create three important products: O_2, ATP, and NADPH.

Light Energy Is a Form of Electromagnetic Radiation

Light is a critical phenomenon that is essential to support life on Earth. Light is a type of electromagnetic radiation, so named because it consists of energy in the form of electric and magnetic fields. Electromagnetic radiation travels as waves caused by the oscillation of the electric and magnetic fields. The **wavelength** is the distance between the peaks in a wave pattern. The **electromagnetic spectrum** encompasses all possible wavelengths of electromagnetic radiation, from relatively short wavelengths (gamma rays) to much longer wavelengths (radio waves) (**Figure 8.3**). Visible light is the range of wavelengths that are detected by the human eye, commonly in the range of 380–740 nm. As discussed later, visible light provides the energy to drive photosynthesis.

Physicists have also discovered that light exhibits behaviors that are characteristic of particles. Albert Einstein formulated the photon theory of light in which he proposed that light is composed of discrete particles called **photons**—massless particles each traveling in a wavelike pattern and moving at the speed of light. Each photon contains a specific amount of energy. An important difference between the various types of electro-

Figure 8.3 **The electromagnetic spectrum.** The bottom portion of this figure emphasizes visible light, the wavelengths of electromagnetic radiation that are visible to the human eye. Light in the visible portion of the electromagnetic spectrum drives photosynthesis.

magnetic radiation described in Figure 8.3 is the amount of energy found in the photons. Shorter wavelength radiation carries more energy per unit of time than longer wavelength radiation. The photons of gamma rays carry more energy than those of radio waves.

The sun radiates the entire spectrum of electromagnetic radiation. The atmosphere prevents much of this radiation from reaching the Earth's surface. For example, the ozone layer forms a thin shield in the upper atmosphere, protecting life on Earth from much of the sun's ultraviolet rays. Even so, a substantial amount of electromagnetic radiation does reach the Earth's surface. The effect of light on living organisms is critically dependent on the energy of the photons. The photons found in gamma rays, X-rays, and UV rays have very high energy. When molecules in cells absorb such energy, the effects can be devastating. Such types of radiation can cause mutations in DNA and even lead to cancer. By comparison, the energy of photons found in visible light is much milder. Molecules can absorb this energy in a way that does not cause permanent harm. Let's now consider how molecules in living cells absorb the energy within visible light.

Photosynthetic Pigments Absorb Light Energy

When light strikes an object, one of three things can happen. First, light may simply pass through an object. Second, the object may change the path of light toward a different direction. A third possibility is that the object may absorb the light. The term **pigment** is used to describe a molecule that can absorb light energy. When light strikes a pigment, some of the wavelengths of light energy are absorbed, while others are reflected. For example, we perceive that leaves are green because they are reflecting radiant energy of the green wavelength. Various pigments in the leaves absorb the other light energy wavelengths.

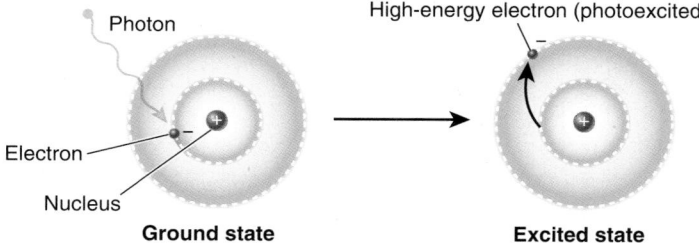

Photon

High-energy electron (photoexcited)

Electron

Nucleus

Ground state   **Excited state**

Figure 8.4   **Absorption of light energy by an electron.**   When a photon of light of the correct amount of energy strikes an electron, the electron is boosted from the ground (unexcited) state to a higher energy level. When this occurs, the electron occupies an orbital that is farther away from the nucleus of the atom. At this farther distance, the electron is held less firmly and is considered unstable.

Biological inquiry: For an electron to drop down to a lower orbital, describe the three things that could happen.

At the extremes of color reflection are white and black. A white object reflects nearly all of the light energy falling on it, whereas a black object absorbs nearly all of the light energy.

What do we mean when we say that light energy is absorbed? In the visible spectrum, light energy is usually absorbed by boosting electrons to higher energy levels (**Figure 8.4**). Recall from Chapter 2 that electrons are located around the nucleus of an atom. The location that an electron is likely to be found is called its orbital. Electrons in different orbitals possess different amounts of energy. For an electron to absorb light energy and be boosted to an orbital with a higher energy, it must overcome the difference in energy between the orbital it is in and the orbital to which it is going. For this to happen, an electron must absorb a photon that contains precisely that amount of energy. Among different pigment molecules, there exist a variety of electrons that can be shifted to different energy levels. Therefore, the wavelength of light that a pigment absorbs depends on the amount of energy that is needed to boost an electron to a higher orbital.

After an electron absorbs energy, it is said to be in an excited state. Usually, this is an unstable condition. The electron may release the energy in one of two forms. When an excited electron drops back down to a lower energy level, it may release heat. For example, on a sunny day, the sidewalk heats up because it absorbs light energy that is released as heat. A second way that an electron can release energy is in the form of light. Certain organisms, such as jellyfish, possess molecules that make them glow. This glow is due to the release of light when electrons drop down to lower energy levels, a phenomenon called fluorescence.

In the case of photosynthetic pigments, however, a different event happens that is critical for the process of photosynthesis. At a particular step, an excited electron in a photosynthetic pigment is removed from that molecule and transferred to another molecule where the electron is more stable. When this occurs, the energy in that electron is said to be "captured" because the

electron does not readily drop down to a lower energy level and release heat or light.

In plants, several different pigment molecules absorb the light energy used to drive photosynthesis. Two types of chlorophyll pigments, termed **chlorophyll *a*** and **chlorophyll *b***, are found in green algae and plants. Their structure was determined in the 1930s by German chemist Hans Fischer (**Figure 8.5a**). In the chloroplast, both chlorophylls *a* and *b* are bound to integral membrane proteins in the thylakoid membrane. The chlorophylls contain a structure called a porphyrin ring that has a bound magnesium ion (Mg^{2+}). In this ring, an electron can follow a path in which it spends some of its time around several different atoms. These electrons, called delocalized electrons because they aren't restricted to a single atom, absorb light energy. Chlorophyll also contains a long hydrocarbon structure called a phytol tail. This tail is hydrophobic and anchors the pigment to the surface of proteins within the thylakoid membrane.

Carotenoids are another type of pigment found in chloroplasts (**Figure 8.5b**). These pigments impart a color that ranges from yellow to orange to red. Carotenoids are often the major pigments in flowers and fruits. In leaves, the more abundant chlorophylls usually mask the colors of carotenoids. In temperate climates where the leaves change colors, the quantity of chlorophyll in the leaf declines during autumn. The carotenoids become readily visible and produce the yellows and oranges of autumn foliage.

An **absorption spectrum** is a diagram that depicts the wavelengths of electromagnetic radiation that are absorbed by a pigment. Each of the pigments shown in **Figure 8.5c** absorbs light in different regions of the visible spectrum. The absorption spectra of chlorophylls *a* and *b* are slightly different, though both chlorophylls absorb light most strongly in the red and violet parts of the visible spectrum and absorb green light poorly. Carotenoids absorb light in the blue and blue-green regions of the visible spectrum.

Having different pigments allows plants to absorb light at many different wavelengths. In this way, plants are more efficient at capturing the energy in sunlight. This phenomenon is highlighted in an **action spectrum**, which describes the rate of photosynthesis plotted as a function of different wavelengths of light (**Figure 8.5d**). The highest rates of photosynthesis correlate with the wavelengths that are strongly absorbed by the chlorophylls and carotenoids. Photosynthesis is poor in the green region of the spectrum, because these pigments do not readily absorb this wavelength of light.

Photosystem II Captures Light Energy and Produces O_2

Photosynthetic organisms have the unique ability not only to absorb light energy but also to capture that energy in a stable way. Many organic molecules can absorb light energy. For example, on a sunny day, molecules in your skin absorb light energy and release the energy as heat. The heat that is released,

(a) Chlorophylls *a* and *b*

(b) β-carotene (a carotenoid)

(c) Absorption spectra

(d) Action spectrum

Figure 8.5 Structures and properties of pigment molecules. (a) The structure of chlorophylls *a* and *b*. As indicated, chlorophylls *a* and *b* differ only at a single site, at which chlorophyll *a* has a —CH₃ group and chlorophyll *b* has a —CHO group. (b) The structure of β-carotene, an example of a carotenoid. The shaded areas are the regions where a delocalized electron can hop from one atom to another. (c) Absorption spectra that show the absorption of light by chlorophyll *a*, chlorophyll *b*, and β-carotene. (d) An action spectrum of photosynthesis depicting the relative rate of photosynthesis in green plants at different wavelengths of light.

however, cannot be harnessed to do useful work. A key feature of photosynthesis is the ability of pigments to capture light energy and transfer it to other molecules that can hold on to the energy in a stable fashion and ultimately produce energy intermediate molecules that can do cellular work.

Let's now consider how chloroplasts capture light energy. The thylakoid membrane contains two distinct complexes of proteins and pigment molecules called **photosystem I (PSI)** and **photosystem II (PSII)**. Photosystem I was discovered before photosystem II, but because photosystem II is the initial step in photosynthesis, we will examine its function first.

Photosystem II has two main components, a light-harvesting complex and a reaction center (**Figure 8.6**). In 1932, Robert Emerson and an undergraduate student, William Arnold, originally discovered the **light-harvesting complex** in the thylakoid membrane. It is composed of several dozen pigment molecules that are anchored to proteins. The role of the complex is to directly absorb photons of light. When a pigment molecule absorbs a photon, this boosts an electron to a higher energy level. As shown in Figure 8.6, the energy (not the electron itself) can be transferred to adjacent pigment molecules by a process called **resonance energy transfer**. Eventually, the energy may be transferred to a special pigment molecule designated P680, so called

because it is best at absorbing light at a wavelength of 680 nm. When an electron in P680 is excited, it is designated P680*. The P680 pigment molecule is located in the **reaction center** of PSII. The light-harvesting complex is also called the **antenna complex** because it acts like an antenna that absorbs energy from light and funnels that energy to P680 in the reaction center.

A high-energy (photoexcited) electron in a pigment molecule is relatively unstable. It may abruptly release its energy by giving off heat or light. Unlike the pigments in the antenna complex that undergo resonance energy transfer, P680* can actually release its high-energy electron. The role of the reaction center is to quickly remove the high-energy electron from P680* and transfer it to another molecule, where the electron will be more stable. This molecule is called the **primary electron acceptor** (Figure 8.6). The transfer of the electron from P680* to the primary electron acceptor is remarkably fast. It occurs in less than a few picoseconds! (One picosecond equals one-trillionth of a second, also noted as 10^{-12} s.) Because this occurs so quickly, the excited electron does not have much time to release its energy in the form of heat or light.

After the primary electron acceptor has received this high-energy electron, the light energy has been captured and can be used to perform cellular work. As we will discuss shortly, the

Photosystem II

Primary electron acceptor

Light

1 Light energy is absorbed by a pigment molecule. This boosts an electron in the pigment to a higher energy level.

Light-harvesting complex

Pigment molecule (chlorophyll)

P680

Reaction center

2 Energy is transferred among pigment molecules via resonance energy transfer until it reaches P680, converting it to P680*.

P680*

3 The high-energy electron on P680* is transferred to the primary electron acceptor, where it is very stable. P680* becomes P680⁺.

e⁻

P680⁺

Reduced primary acceptor (very stable)

4 A low-energy electron from water is transferred to P680⁺ to convert it to P680. O₂ is produced.

e^- P680

e^-

H_2O

$2 H^+ + \frac{1}{2} O_2$

Figure 8.6 The absorption of light energy by pigment molecules in the light-harvesting complex, and the path that leads to the capture of energy by the primary electron acceptor.

work that it first performs is to synthesize the energy intermediates ATP and NADPH. Later, these energy intermediates are used to make carbohydrates.

Before we examine the fate of the high-energy electron that was transferred to the primary electron acceptor, let's consider what happens to the P680 molecule after it has released its high-energy electron. Another function of the reaction center is to

replace the electrons that are removed from pigment molecules. They are replaced with electrons from water molecules (Figure 8.6). The reaction center of photosystem II removes electrons from water and transfers those electrons to oxidized pigment molecules (P680⁺).

$$H_2O \rightarrow 1/2\ O_2 + 2\ H^+ + 2\ e^-$$

$$2\ P680^+ + 2\ e^- \rightarrow 2\ P680$$

(from water)

The oxidation of water results in the formation of oxygen gas (O₂). Photosystem II is the only known protein complex that can oxidize water, resulting in the release of O₂ into the atmosphere.

Photosystem II Is an Amazing Redox Machine

Redox reactions are fundamentally important for cells to store and utilize energy and to form covalent bonds in organic molecules. Photosystem II is a particularly remarkable example of a redox machine. As we have learned, this complex of proteins removes electrons from a pigment molecule and transfers them to a primary electron acceptor. Perhaps even more remarkable is that photosystem II can remove electrons from water, a very stable molecule that holds onto its electrons tightly. The removal of electrons is how O₂ is made.

Many approaches have been used to study how photosystem II works. In recent years, much effort has been aimed at determining the biochemical composition of the protein complex and the roles of its individual components. The number of protein subunits varies somewhat from species to species and may vary due to environmental changes. Typically, photosystem II contains around 19 different protein subunits. Two subunits, designated D1 and D2, contain the reaction center that carries out the redox reactions (**Figure 8.7a**). Two other subunits, called CP43 and CP47, bind the pigment molecules that form the light-harvesting complex. Many additional subunits regulate the function of photosystem II and provide structural support.

The oxidation of water occurs in a region called the **manganese cluster**. This site is located on the side of D1 that faces the thylakoid lumen. The manganese cluster has four Mn²⁺, one Ca²⁺, and one Cl⁻. Two water molecules bind to this site. D1 catalyzes the removal of four electrons from the two water molecules to create four H⁺ and O₂. The electrons are transferred, one at a time, to a tyrosine (Tyr) in D1 and then to an oxidized pigment molecule (P680⁺) to produce P680. When the electron on P680 becomes excited, usually by resonance energy transfer, it then moves to the primary electron acceptor, which is an organic molecule called pheophytin (Pp) that is permanently bound to photosystem II. Pheophytin transfers its electron to a plastoquinone molecule, designated QA, which is also permanently bound to photosystem II. Next, the electron is transferred to another plastoquinone molecule designated QB, which can accept two high-energy electrons and bind two H⁺. QB can diffuse away from the reaction center.

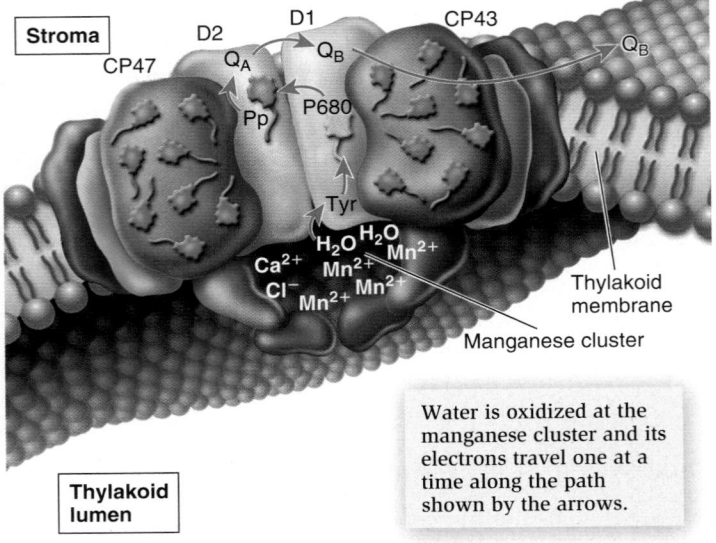

(a) The path of electron flow through photosystem II

This box encloses one photosystem II complex that contains 19 protein subunits.

(b) Three-dimensional structure of photosystem II determined by X-ray crystallography

Figure 8.7   **A closer look at the structure and function of photosystem II.**   (a) Schematic drawing showing the path of electron flow from water to Q_B. The CP47 and CP43 protein subunits wrap around D1 and D2 so that pigments in CP47 and CP43 can transfer energy to P680 by resonance energy transfer. **(b)** The three-dimensional structure of photosystem II as determined by X-ray crystallography. In the crystal structure, the colors are CP47 (red), D2 (orange), D1 (yellow), and CP43 (green).

Biological inquiry: According to this figure, how many redox reactions does photosystem II catalyze?

In 2004, So Iwata, James Barber, and colleagues determined the three-dimensional structure of photosystem II using a technique called **X-ray crystallography**. In this method, researchers must purify a protein or protein complex and expose it to conditions that cause the proteins to associate with each other in an ordered array. In other words, the proteins form a crystal. When a crystal is exposed to X-rays, the resulting pattern can be analyzed mathematically to determine the three-dimensional structure of the crystal's components. Major advances in this technique over the last couple of decades have enabled researchers to determine the structures of relatively large macromolecular machines such as photosystem II and ribosomes. **Figure 8.7b** shows the three-dimensional structure of photosystem II. The structure shown here is a dimer; it has two PSII complexes each with 19 protein subunits. As seen in this figure, the intricacy of the structure of photosystem II rivals the complexity of its function.

Photosystems II and I Work Together to Produce ATP and NADPH

Let's now consider what happens to the high-energy electrons that are transferred to the primary electron acceptor in photosystem II (**Figure 8.8**). After electrons reach Q_B, they enter an electron transport chain located in the thylakoid membrane. The electron transport chain functions similarly to the one found in mitochondria. From Q_B, electrons go to a cytochrome complex, then to plastocyanin (Pc), a small protein, and then

to a pigment molecule in the reaction center of photosystem I. Along the journey from photosystem II to photosystem I, the electron releases some of its energy at particular steps and is transferred to the next component that has a higher electronegativity. The energy that is released is harnessed to move H^+ into the thylakoid lumen. One result of the electron movement is to establish an H^+ electrochemical gradient. Additionally, the splitting of water also adds H^+ into the thylakoid lumen. The synthesis of ATP in chloroplasts is achieved by a chemiosmotic mechanism similar to that used to make ATP in mitochondria.

The key role of photosystem I is to make NADPH. When light strikes the light-harvesting complex of photosystem I, this energy is also transferred to a reaction center, where a high-energy electron is removed from a pigment molecule, designated P700, and transferred to a primary electron acceptor. A protein called ferredoxin (Fd) can accept two high-energy electrons, one at a time, from the primary electron acceptor. Fd then transfers the two electrons to the enzyme NADP$^+$ reductase. This enzyme transfers the two electrons to NADP$^+$ and together with an H^+ creates NADPH. The formation of NADPH consumes one H^+ in the stroma and thereby contributes to the formation of an H^+ electrochemical gradient because it results in fewer H^+ in the stroma.

A key difference between photosystem II and photosystem I is how the oxidized forms of P680 and P700 receive electrons. As discussed earlier, P680$^+$ receives an electron from water. By comparison, P700$^+$—the oxidized form of P700—receives an electron from Pc. Therefore, photosystem I does not need to split water to reduce P700$^+$ and thus does not generate oxygen.

1 Light excites electrons, and the excited electrons follow the path shown by the red arrow. This produces an H^+ electrochemical gradient. Also, electrons are removed from water and transferred to $P680^+$. This process creates O_2 and places additional H^+ in the lumen.

2 Electrons from PSII eventually reach PSI, where a second input of light boosts them to a very high level. They follow the path shown by the arrow.

3 Two high-energy electrons and one H^+ are transferred to $NADP^+$ to make NADPH. This removes H^+ from the stroma.

4 The production of O_2, the pumping of H^+ across the thylakoid membrane, and the synthesis of NADPH all contribute to the formation of an H^+ electrochemical gradient. This gradient is used to make ATP via an ATP synthase in the thylakoid membrane.

Figure 8.8 The synthesis of O_2, ATP, and NADPH by the concerted actions of photosystems II and I.

In summary, the steps of the light reactions of photosynthesis produce three chemical products:

1. O_2 is produced in the thylakoid lumen by the oxidation of water by photosystem II. Two electrons are removed from water, which creates two H^+ and $1/2\ O_2$. The two electrons are transferred to $P680^+$ molecules.
2. ATP is produced in the stroma by an H^+ electrochemical gradient. This gradient results from three events: (1) the splitting of water, which places H^+ in the thylakoid lumen, (2) the movement of high-energy electrons from photosystem II to photosystem I, which pumps H^+ into the lumen, and (3) the formation of NADPH, which consumes H^+ in the stroma.
3. NADPH is produced in the stroma from high-energy electrons that start in photosystem II and are boosted a second time in photosystem I. Two high-energy electrons and one H^+ are transferred to $NADP^+$ to create NADPH.

The combined action of photosystem II and photosystem I is termed **noncyclic electron flow** because the electrons move linearly from PSII to PSI and ultimately to $NADP^+$.

The Use of Light Flashes of Specific Wavelengths Provided Experimental Evidence for the Existence of PSII and PSI

The use of light flashes at particular wavelengths has been an important experimental technique for helping researchers understand the light reactions of photosynthesis. In this method, pioneered by Robert Emerson, a photosynthetic organism is exposed to a particular wavelength of light, after which the rate of photosynthesis is measured by the amount of CO_2 consumed or the amount of O_2 produced. In the 1950s, Emerson performed a particularly intriguing experiment that greatly stimulated photosynthesis research (**Figure 8.9**). He subjected algae to light flashes of different wavelengths and obtained a mysterious result. When he exposed algae to a wavelength of 680 nm, he observed a low rate of photosynthesis. A similarly low rate of photosynthesis occurred when he exposed algae to a wavelength of 700 nm. However, when he exposed the algae to both wavelengths of light simultaneously, the rate of photosynthesis was more than double the rates observed at only one wavelength. This phenomenon was termed the **enhancement effect**. We know now that it occurs because 680-nm light can readily activate the pigment (P680) in the reaction center in photosystem II, but is not very efficient at activating pigments in photosystem I. In contrast, light of 700-nm wavelength is optimal at activating photosystem I but not very good at activating photosystem II. When algae are exposed to both wavelengths, however, maximal activation of the pigments in both photosystems is achieved.

Figure 8.9   **The enhancement effect observed by Emerson.** When photosynthetic organisms such as green plants and algae are exposed to 680-nm and 700-nm light simultaneously, the rate of photosynthesis is much more than double the rate produced by each wavelength individually.

Biological inquiry: Would the enhancement effect be observed if two consecutive flashes of light occurred at 680 nm?

When researchers began to understand that photosynthesis results in the production of both ATP and NADPH, Robin Hill and Fay Bendall also proposed that photosynthesis involves two photoactivation events. According to their model, known as the **Z scheme**, an electron proceeds through a series of energy changes during photosynthesis. The Z refers to the zigzag shape of this energy curve. Based on our modern understanding of photosynthesis, we now know that these events involve increases and decreases in the energy of an electron as it moves from photosystem II to photosystem I (**Figure 8.10**). An electron on a nonexcited pigment molecule in photosystem II has the lowest energy. In photosystem II, light boosts an electron to a much higher energy level. As the electron travels from photosystem II to photosystem I, some of the energy is released. The input of light in photosystem I boosts the electron to an even higher energy than it attained in photosystem II. The electron releases a little energy before it is eventually transferred to $NADP^+$.

Cyclic Electron Flow Produces Only ATP

The mechanism of harvesting light energy, described in Figure 8.8, is called noncyclic electron flow because electrons begin at photosystem II and eventually are transferred to $NADP^+$. The electron flow is a linear process that produces ATP and NADPH in roughly equal amounts. However, as we will see, the Calvin cycle uses more ATP than NADPH. Therefore, in some photosynthetic organisms, excited electrons take an alternative path as a way to make more ATP. In 1959, Daniel Arnon discovered

Figure 8.10   **The Z scheme, which depicts the energy of an electron as it moves from photosystem II to $NADP^+$.**   During its journey from photosystem II to $NADP^+$, an electron varies in the amount of energy it contains. As seen here, the input of light boosts the energy of the electron two times. At the end of the pathway, 2 electrons are used to make NADPH.

When light strikes photosystem I, electrons are excited and sent to ferredoxin (Fd). From Fd, the electrons are then transferred to Q_B, to the cytochrome complex, to plastocyanin (Pc), and back to photosystem I. This produces an H^+ electrochemical gradient, which is used to make ATP via the ATP synthase.

Figure 8.11 **Cyclic photophosphorylation.** As shown in this figure, an electron follows a cyclic path that is powered by photosystem I. This contributes to the formation of an H^+ electrochemical gradient, which is used to make ATP.

a pattern of electron flow that is cyclic and generates ATP alone (**Figure 8.11**). Arnon termed the process **cyclic photophosphorylation** because (1) the path of electrons is cyclic, (2) light energizes the electrons, and (3) ATP is made via the phosphorylation of ADP. Due to the path of electrons, the mechanism is also called **cyclic electron flow**.

When light strikes photosystem I, high-energy electrons are sent to the primary electron acceptor and then to ferredoxin (Fd). The key difference in cyclic photophosphorylation is that the high-energy electrons are transferred from ferredoxin to Q_B. From Q_B, the electrons then go the cytochrome complex, to plastocyanin (Pc), and back to photosystem I. As the electrons travel along this cyclic route, they release energy and some of this energy is used to transport H^+ into the thylakoid lumen. The resulting H^+ gradient drives the synthesis of ATP via the ATP synthase.

GENOMES & PROTEOMES

The Cytochrome Complexes of Mitochondria and Chloroplasts Have Evolutionarily Related Proteins in Common

A recurring theme in cell biology is that evolution has resulted in groups of genes that encode proteins that play similar but specialized roles in cells. When two or more genes are similar because they are derived from the same ancestral gene, they are called **homologous genes**. As discussed in Chapter 23, homologous genes encode proteins that have similar amino acid sequences and perform similar functions.

A comparison of the electron transport chains of mitochondria and chloroplasts reveals homologous genes. In particular, let's consider the cytochrome complex found in the thylakoid membrane of plants and algae, called cytochrome b_6-f (**Figure 8.12a**) and cytochrome b-c_1, which is found in the electron transport chain of mitochondria (**Figure 8.12b**; refer back to Figure 7.16). Both cytochrome b_6-f and cytochrome b-c_1 are composed of several protein subunits. One of those proteins is called cytochrome b_6 in cytochrome b_6-f and cytochrome b in cytochrome b-c_1. By analyzing the gene sequences, researchers discovered that cytochrome b_6 and cytochrome b are homologous. These proteins carry out similar functions: both of them accept electrons from a quinone (plastoquinone or ubiquinone) and both donate an electron to another protein within their respective complexes (cytochrome f or cytochrome $c1$). Likewise, both of these proteins function as H^+ pumps that capture some of the energy that is released from electrons to transport H^+ across the membrane. Thus, evolution has produced a family of cytochrome b-type proteins that play similar but specialized roles. Cytochrome b functions as a redox protein and H^+ pump in the electron transport chain of mitochondria, while cytochrome b_6 plays the same role in chloroplasts.

Cytochrome b_6 is a protein subunit of cytochrome b_6-f.

Cytochrome b_6-f

Stroma

2 H$^+$

Thylakoid lumen

(a) Cytochrome b_6 in a chloroplast

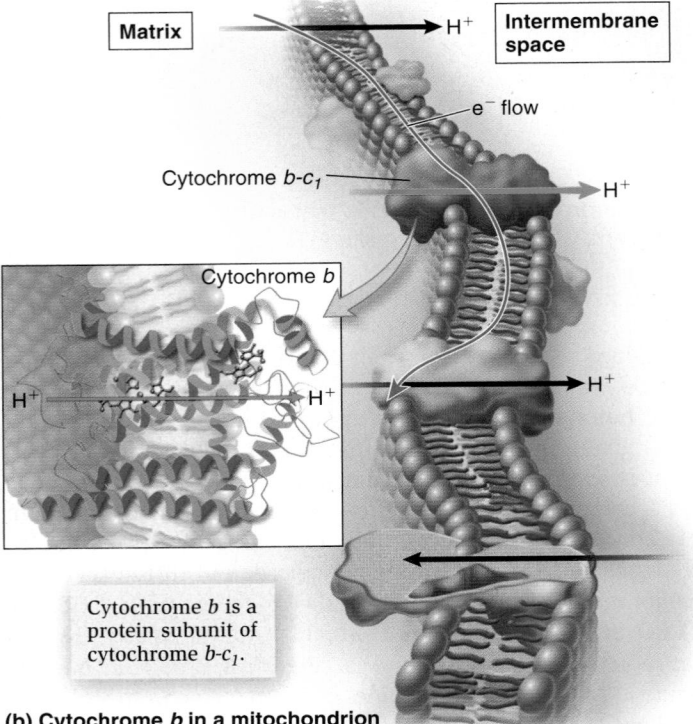

Matrix

H$^+$ **Intermembrane space**

e$^-$ flow

Cytochrome b-c$_1$

H$^+$

Cytochrome b

H$^+$ H$^+$ H$^+$

Cytochrome b is a protein subunit of cytochrome b-c$_1$.

(b) Cytochrome b in a mitochondrion

Figure 8.12 Homologous proteins in the electron transport chains of mitochondria and chloroplasts. **(a)** Cytochrome b_6, a subunit of the cytochrome b_6-f complex found in chloroplasts, and **(b)** cytochrome b, a subunit of the cytochrome b-c$_1$ complex found in mitochondria, are homologous proteins that play similar roles in their respective electron transport chains. The inset shows the three-dimensional structure of cytochrome b, which was determined by X-ray crystallography. It is an integral membrane protein with several transmembrane α helices and two heme groups, which are prosthetic groups involved in electron transfer. The structure of cytochrome b_6 has also been determined and found to be very similar. (Note: The orientation of cytochrome b in the inset is such that it would pump H$^+$ out of the matrix. The orientation of cytochrome b_6 is oriented so it would pump H$^+$ out of the stroma.)

Biological inquiry: Explain why the three-dimensional structures of cytochrome b and cytochrome b$_6$ are very similar.

8.3 Calvin Cycle

In the previous section, we learned how the light reactions of photosynthesis produce O$_2$, ATP, and NADPH. We will now turn our attention to the second phase of photosynthesis, the Calvin cycle, in which ATP and NADPH are used to make carbohydrates. The Calvin cycle consists of a series of steps that occur in a metabolic cycle somewhat similar to the citric acid cycle described in Chapter 7. However, while the citric acid cycle is catabolic, the Calvin cycle is an anabolic pathway leading to the biosynthesis of carbohydrates.

The Calvin cycle takes CO$_2$ from the atmosphere and incorporates the carbon into organic molecules, namely, carbohydrates. As mentioned earlier, carbohydrates are critical for two reasons. First, these organic molecules provide the precursors to make the organic molecules and macromolecules of nearly all living cells. The organic molecules in your body are ultimately derived from the operation of the Calvin cycle in algae and plants. The second key reason why the Calvin cycle is important involves the storage of energy. Recall that molecules such as glucose contain large amounts of chemical energy in their covalent bonds. The Calvin cycle produces carbohydrates, which store energy. These carbohydrates are accumulated inside plant cells. When a plant is in the dark and not carrying out photosynthesis, the stored carbohydrates can be used as a source of energy. Similarly, when an animal consumes a plant, it can use the carbohydrates as an energy source.

In this section, we will examine the three phases of the Calvin cycle and their components. We will also explore the experimental approach of Melvin Calvin and his colleagues that enabled them to elucidate these steps.

The Calvin Cycle Incorporates CO$_2$ into Carbohydrate

The Calvin cycle, also called the Calvin-Benson cycle, was determined by chemists Melvin Calvin and Andrew Adam Benson and their colleagues in the 1940s and 1950s. This cycle requires a massive input of energy. For every 6 carbon dioxide molecules that are incorporated into carbohydrate such as glucose, 18 ATP molecules are hydrolyzed and 12 NADPH molecules are oxidized.

$$6\ CO_2 + 12\ H_2O \rightarrow C_6H_{12}O_6 + 6\ O_2 + 6\ H_2O$$

Glucose

$$18\ ATP + 18\ H_2O \rightarrow 18\ ADP + 18\ P_i$$

$$12\ NADPH \rightarrow 12\ NADP^+ + 12\ H^+ + 24\ e^-$$

Although biologists commonly describe glucose as a product of photosynthesis, glucose is not directly made by the Calvin cycle. Instead, products of the Calvin cycle are used as starting materials for the synthesis of glucose and other molecules, including sucrose. After glucose molecules are made, they may be linked together to form a polymer of glucose called starch, which is

Figure 8.13   **The Calvin cycle.**   This cycle has three phases: (1) carbon fixation, (2) reduction and carbohydrate production, and (3) regeneration of RuBP.

stored in the chloroplast for later use. Alternatively, the disaccharide sucrose may be made and transported out of the leaf to other parts of the plant.

The Calvin cycle can be divided into three phases: carbon fixation, reduction and carbohydrate production, and regeneration of RuBP (**Figure 8.13**).

Carbon Fixation (Phase 1)   In **carbon fixation**, CO_2 becomes incorporated into ribulose bisphosphate (RuBP), a five-carbon sugar. The enzyme that catalyzes this step is named RuBP carboxylase/oxygenase, or **rubisco**. This enzyme, which constitutes the most abundant protein in chloroplasts, is perhaps the most abundant protein on Earth. The product of the reaction is a six-carbon intermediate that immediately splits in half to form two molecules of 3-phosphoglycerate (3PG).

Reduction and Carbohydrate Production (Phase 2)   In the second phase, ATP is used to convert 3PG to 1,3-bisphosphoglycerate. Next, electrons from NADPH reduce 1,3-bisphosphoglycerate to glyceraldehyde-3-phosphate (G3P). G3P is a carbohydrate with three carbon atoms. The key difference between 3PG and G3P is that G3P has a C—H bond, while the analogous carbon in 3PG forms a C—O bond (Figure 8.13). The C—H bond can occur because the G3P molecule has been reduced by the addition of two electrons from NADPH. Compared to 3PG, the bonds in G3P store more energy and enable G3P to readily form larger organic molecules such as glucose.

Only some of the G3P molecules are used to make glucose or other carbohydrates. Phase 1 began with 6 RuBP molecules and 6 CO_2 molecules. Twelve G3P molecules are made at the end of phase 2. Two of these G3P molecules are used in carbohydrate production. As described next, the other 10 G3P molecules are needed to keep the Calvin cycle turning.

Regeneration of RuBP (Phase 3) In the last phase of the Calvin cycle, a series of enzymatic steps converts the 10 G3P molecules into 6 RuBP molecules, using 6 molecules of ATP. After the RuBP molecules are regenerated, they can serve as acceptors for CO_2, thereby allowing the cycle to continue.

As we have just seen, the Calvin cycle begins by using carbon from an inorganic source, that is, CO_2, and ends with organic molecules that will be used by the plant to make other compounds. You may be wondering why it is not possible to directly link the CO_2 molecules together to form these larger molecules. The answer lies in the number of electrons that orbit carbon atoms. In CO_2, the carbon atom is considered electron poor.

Oxygen is a very electronegative atom that monopolizes the electrons it shares with other atoms. In a covalent bond between carbon and oxygen, the shared electrons are closer to the oxygen atom.

By comparison, in an organic molecule, the carbon atom is electron rich. During the Calvin cycle, ATP provides energy and NADPH donates high-energy electrons so that the carbon originally in CO_2 has been reduced. Put simply, the Calvin cycle places additional electrons onto carbon atoms. Compared to the carbon in CO_2, the carbon in an organic molecule can readily form C—H and C—C bonds, which allows the eventual synthesis of larger essential molecules including glucose, amino acids, and so on. In addition, the covalent bonds within these molecules are capable of storing large amounts of energy.

FEATURE INVESTIGATION

The Calvin Cycle Was Determined by Isotope Labeling Methods

The steps in the Calvin cycle involve the conversion of one type of molecule to another, eventually regenerating the starting material, RuBP. In the 1940s and 1950s, Calvin and his colleagues used ^{14}C, a radioisotope of carbon, as a way to label and trace molecules produced during the cycle (**Figure 8.14**). They injected ^{14}C-labeled CO_2 into cultures of the green algae *Chlorella pyrenoidosa* grown in an apparatus called a "lollipop" (because of its shape). The *Chlorella* cells were given different lengths of time to incorporate the ^{14}C-labeled carbon, ranging from fractions of a second to many minutes. After this incubation period, the cells were abruptly placed into a solution of alcohol to inhibit enzymatic reactions and thereby stop the cycle.

The researchers separated the newly made radiolabeled molecules by a variety of methods. The most commonly used method was two-dimensional paper chromatography. In this approach, a sample containing radiolabeled molecules was spotted onto a corner of the paper at a location called the origin. The edge of the paper was placed in a solvent, such as phenol-water, and the solvent would ascend to the top of the paper. As the solvent rose through the paper, so did the radiolabeled molecules. The rate at which they rose depended on their structures, which determined how strongly they interacted with the paper. This step separated the mixture of molecules spotted onto the paper at the origin. The paper was then dried, turned $90°$, and then the edge was placed in a different solvent, such as butanol-propionic acid-water. Again, the solvent would rise through the paper, thereby separating molecules that may not have been adequately separated during the first separation step. After this second separation step, the paper was dried and exposed to X-ray film, a procedure called autoradiography. Radioactive emission from the ^{14}C-labeled molecules caused dark spots to appear on the film.

Figure 8.14   The determination of the Calvin cycle using labeling of CO_2 with ^{14}C.

GOAL The incorporation of CO_2 into carbohydrate involves a biosynthetic pathway. The aim of this experiment was to identify the steps.

STARTING MATERIALS The green alga *Chlorella pyrenoidosa.*

Experimental level   **Conceptual level**

1 Grow *Chlorella* in an apparatus called a "lollipop." Add ^{14}C-labeled CO_2 and incubate for various lengths of time (from fractions of a second to many minutes). Stop the Calvin cycle by placing a sample of cells into a solution of alcohol.

Addition of $^{14}CO_2$

Chlorella

$^{14}CO_2$

Alcohol

Calvin cycle

Cycle stopped

Lollipop
Lamp
Alcohol solution

2 Take a sample of the internal cell contents and spot on the corner of chromatography paper. This spot is called the origin.

3 Place edge of paper in a solvent, such as phenol-water, and allow time for solvent to rise and separate the mixture of molecules that were spotted at the origin.

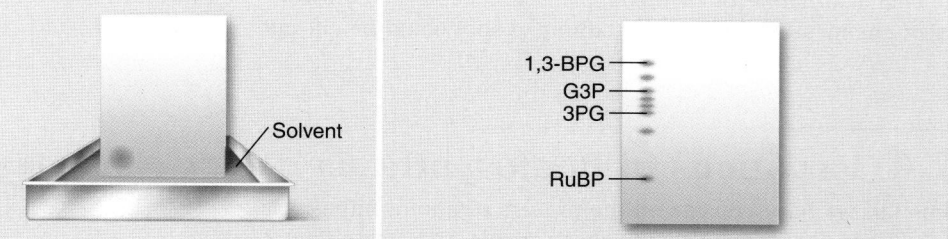

4 Dry paper, turn 90°, and then place the edge in a different solvent such as butanol-propionic acid-water. Allow time for solvent to rise.

5 Dry paper and place next to X-ray film. The developed film reveals dark spots where ^{14}C-labeled molecules were located. This procedure is called autoradiography.

6 **THE DATA**

The pattern of spots changed, depending on the length of time that the cells were incubated with ^{14}C-labeled CO_2. If the incubation period was short, only molecules that were made in the first steps of the Calvin cycle would be seen, while longer incubations revealed molecules synthesized in later steps. For example, after short incubations, 3-phosphoglycerate (3PG) and 1,3-bisphosphoglycerate (1,3-BPG) would be observed, while longer incubations would show glyceraldehyde-3-phosphate (G3P) and ribulose bisphosphate (RuBP).

A challenge for Calvin and his colleagues was to identify the chemical nature of each spot. This was achieved by a variety of chemical methods. For example, a spot could be cut out of the paper, the molecule within the paper could be washed out, or eluted, and then the eluted molecule could be subjected to the same procedure that included a radiolabeled molecule whose structure was already known. If the unknown molecule and known molecule migrated to the same spot in the paper, this indicated that they were the same molecule. During the late 1940s and 1950s, Calvin and his coworkers identified all of the ^{14}C-labeled spots and the order in which they appeared. In this way, they were able to determine the series of reactions of what we now know as the Calvin cycle. For this work, Calvin was awarded the Nobel Prize in 1961.

8.4 Variations in Photosynthesis

Thus far, we have considered the process of photosynthesis as it occurs in the chloroplasts of green plants and algae. Photosynthesis is a two-stage process in which the light reactions produce O_2, ATP, and NADPH, and the Calvin cycle uses the ATP and NADPH in the synthesis of carbohydrates. This two-stage process is a universal feature of photosynthesis in all green plants, algae, and cyanobacteria. However, certain environmental conditions such as light intensity, temperature, and water availability may influence both the efficiency of photosynthesis and the way in which the Calvin cycle operates. In this section, we begin by examining how hot and dry conditions may reduce the output of photosynthesis. We then explore two adaptations that certain plant species have evolved that conserve water and help to maximize photosynthetic efficiency in such environments.

Photorespiration Decreases the Efficiency of Photosynthesis

In the previous section, we learned that rubisco is a key enzyme of the Calvin cycle. Rubisco functions as a carboxylase because it adds a CO_2 molecule to RuBP, an organic molecule, to create two molecules of 3-phosphoglycerate (3PG).

$$RuBP + CO_2 \rightarrow 2\ 3PG$$

For most species of plants, the incorporation of CO_2 into RuBP is the only way for carbon fixation to occur. Because 3PG is a three-carbon molecule, these plants are called **C_3 plants**. Examples of C_3 plants include wheat and oak trees (**Figure 8.15**).

Researchers have discovered that the active site of rubisco can also function as an oxygenase. As mentioned earlier, that is why it is called rubisco, which stands for RuBP carboxylase/oxygenase. When rubisco adds an O_2 molecule to RuBP, this creates only one molecule of 3-phosphoglycerate and a two-carbon molecule called phosphoglycolate. The phosphoglycolate is then dephosphorylated to glycolate and released from the chloroplast. In a series of several steps, glycolate is eventually oxidized in other organelles to produce an organic molecule plus a molecule of CO_2.

$$RuBP + O_2 \rightarrow \text{3-phosphoglycerate} + \text{Phosphoglycolate}$$

$$\text{Phosphoglycolate} \rightarrow \text{Glycolate} \rightarrow\rightarrow\rightarrow CO_2$$

This process, called **photorespiration**, uses O_2 and liberates CO_2.

(a)

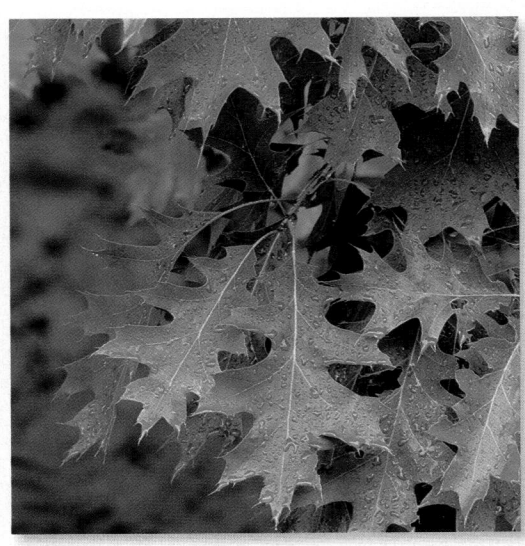

(b)

Figure 8.15 Examples of C_3 plants. The structures of (a) wheat and (b) white oak leaves are similar to that shown in Figure 8.1.

Photorespiration is considered wasteful because it reverses the effects of photosynthesis, thereby reducing the ability of a plant to make carbohydrates, which limits plant growth.

Photorespiration is more likely to occur when plants are exposed to a hot and dry environment. Under these conditions, the stomata of the leaves close, inhibiting the uptake of CO_2 from the air and trapping the O_2 that is produced by photosynthesis. When the level of CO_2 is low and O_2 is high, photorespiration is favored. If C_3 plants are subjected to hot and dry environmental conditions, as much as 25–50% of their photosynthetic work is reversed by the process of photorespiration.

Why do plants carry out photorespiration? The answer is not entirely clear. Photorespiration undoubtedly results in the disadvantage of lowering the efficiency of photosynthesis. One common view is that photorespiration does not offer any advantage and is an evolutionary relic. When rubisco first evolved some 3 billion years ago, the atmospheric oxygen level was low, so that phosphorespiration would not have been a problem. Another view is that photorespiration may have a protective advantage. On hot and dry days, CO_2 levels within a plant may fall and O_2 levels will rise. Under these conditions, highly toxic oxygen-containing molecules such as free radicals may be produced that could cause damage to the plant. Therefore, plant biologists have speculated that the role of photorespiration may be to protect the plant against the harmful effects of such toxic molecules by consuming O_2 and releasing CO_2. In addition, photorespiration may affect the metabolism of other compounds in plants. Recent research suggests that photorespiration may also help plants to assimilate nitrogen into organic molecules.

C_4 Plants Have Evolved a Mechanism to Minimize Photorespiration

Certain species of plants have developed an interesting way to prevent photorespiration. In the early 1960s, Hugo Kortschak discovered that the first product of photosynthesis in sugarcane is not 3-phosphoglycerate but instead is a compound with four carbon atoms. Species such as sugarcane are thus called **C_4 plants** because the first step in carbon fixation produces a four-carbon compound. Later, Marshall Hatch and Roger Slack confirmed this result and identified the compound as oxaloacetate. For this reason, the pathway is sometimes called the Hatch-Slack pathway.

C_4 plants employ an interesting cellular organization to avoid photorespiration (**Figure 8.16**). Unlike C_3 plants, an interior layer in the leaves of many C_4 plants has a two-cell organization composed of mesophyll cells and bundle-sheath cells. CO_2 from the atmosphere enters the mesophyll cells via stomata. Once inside, the enzyme **PEP carboxylase** adds CO_2 to phosphoenolpyruvate (PEP) to produce the four-carbon compound oxaloacetate. PEP carboxylase does not recognize O_2. Therefore, unlike rubisco, PEP carboxylase does not promote photorespiration when CO_2 is low and O_2 is high. Instead, PEP carboxylase continues to fix CO_2.

A key feature of these types of C_4 plants is that a four-carbon compound is transferred between cells. As shown in Figure 8.16, the compound oxaloacetate is converted to the four-carbon compound malate, which is transported into the bundle-sheath cell. Malate is then broken down into pyruvate and CO_2.

Figure 8.16 **Leaf structure and its relationship to the C_4 cycle.** C_4 plants have mesophyll cells that initially take up CO_2, and bundle-sheath cells, where much of the carbohydrate synthesis occurs. Compare this leaf structure with the structure of C_3 leaves shown in Figure 8.1.

The pyruvate returns to the mesophyll cell, where it is converted to PEP via ATP, and the cycle in the mesophyll cell can begin again. The main outcome of this C_4 cycle is that the mesophyll cell pumps CO_2 into the bundle-sheath cell. The Calvin cycle occurs in the chloroplasts of the bundle-sheath cell. Because the mesophyll cell supplies the bundle-sheath cell with a steady supply of CO_2, the concentration of CO_2 remains high in the bundle-sheath cell. This strategy minimizes photorespiration, which requires low CO_2 and high O_2 levels to proceed.

Which is better—being a C_3 or a C_4 plant? The answer is that it depends on the environment. In warm and dry climates, C_4 plants have an advantage because during the day they can keep their stomata partially closed to reduce water vaporization from the leaf and thereby conserve water. Furthermore, they can avoid photorespiration. C_4 plants are well adapted to habitats with high daytime temperatures and intense sunlight. Some examples of C_4 plants include crabgrass, corn, and sugarcane. In cooler climates, C_3 plants have the edge because it takes less energy for them to fix carbon dioxide. The process of carbon fixation that occurs in C_4 plants uses ATP to regenerate PEP from pyruvate (Figure 8.16), which C_3 plants do not have to expend. Biologists estimate that about 90% of the plant species on Earth are C_3 plants.

CAM Plants Are C_4 Plants That Take Up Carbon Dioxide at Night

We have just learned that certain C_4 plants prevent photorespiration by pumping CO_2 into the bundle-sheath cells, where the Calvin cycle occurs. Another strategy followed by other C_4 plants, called **CAM plants**, is to separate these processes in time. CAM stands for crassulacean acid metabolism, because the process was first studied in members of the plant family Crassulaceae. CAM plants are water-storing succulents such as cacti, bromeliads (including pineapple), and sedums. To avoid water loss, CAM plants keep their stomata closed during the day and open them during the night.

How, then, do CAM plants carry out photosynthesis? **Figure 8.17** compares CAM plants with the other type of C_4 plants we considered in Figure 8.16. During the night, the stomata of CAM plants open, thereby allowing the entry of CO_2 into mesophyll cells, which joins with PEP to form the four-carbon compound oxaloacetate. This is converted to malate, which accumulates during the night in the central vacuoles of the cells. In the morning, the stomata close to conserve moisture. The accumulated malate in the mesophyll cells leaves the vacuole and is broken down to release CO_2, which then drives the Calvin cycle during the daytime.

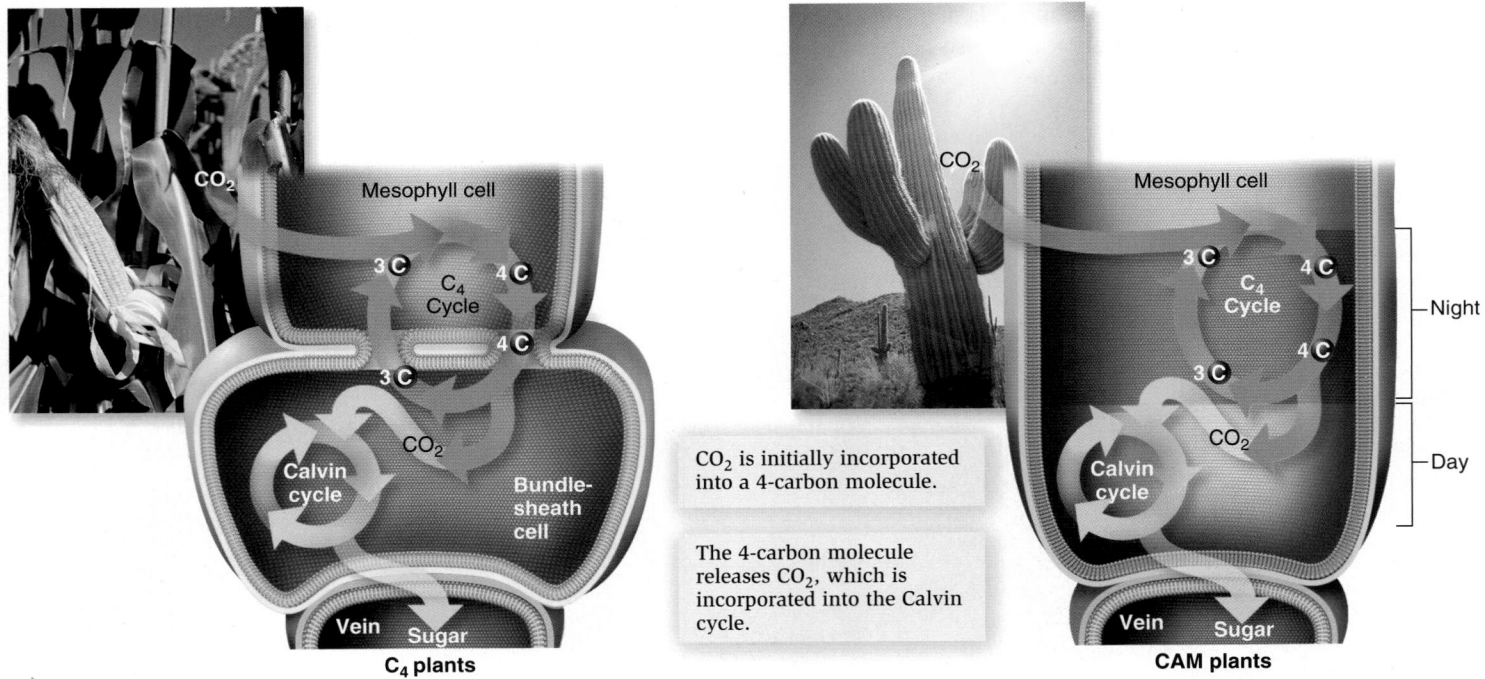

Figure 8.17 **A comparison of C_4 and CAM plants.** The name C_4 plant describes those plants in which the first organic product of carbon fixation is a four-carbon compound. Using this definition, CAM plants are a type of C_4 plant because they produce a four-carbon molecule when CO_2 is initially taken up. CAM plants, however, do not separate the functions of making a four-carbon molecule and the Calvin cycle into different types of cells. Instead, they make a four-carbon molecule at night and break down that molecule during the day, so that the CO_2 can be incorporated via the Calvin cycle.

CHAPTER SUMMARY

- Photosynthesis is the process by which plants, algae, and cyanobacteria capture light energy to synthesize carbohydrates.

- Heterotrophs must obtain organic molecules in their food, while autotrophs can make organic molecules from inorganic sources. Photoautotrophs use the energy from light to make organic molecules.

8.1 Overview of Photosynthesis

- The general formula for photosynthesis is that carbon dioxide, water, and energy are needed to make carbohydrates and oxygen.

- In plants and algae, photosynthesis occurs within chloroplasts, which have an outer membrane, inner membrane, and thylakoid membrane. The stroma is found between the thylakoid membrane and inner membrane. In plants, the leaves are the major site of photosynthesis. (Figure 8.1)

- The light reactions of photosynthesis capture light energy to make ATP, NADPH, and O_2. These reactions occur at the thylakoid membrane. Carbohydrate synthesis happens in the stroma and uses ATP and NADPH from the light reactions. (Figure 8.2)

8.2 Reactions That Harness Light Energy

- Light is a form of electromagnetic radiation that travels in waves and is composed of photons with discrete amounts of energy. (Figure 8.3)

- Electrons can absorb light energy and be boosted to a higher energy level. (Figure 8.4)

- Photosynthetic pigments include chlorophylls a and b and carotenoids. These pigments absorb light energy in the visible spectrum. (Figure 8.5)

- Pigment molecules in photosystem II capture light energy, and that energy is transferred to the reaction center via resonance energy transfer. A high-energy electron from P680* is transferred to a primary electron acceptor. An electron from water is then used to replenish the electron that is lost from P680*. (Figures 8.6, 8.7)

- During noncyclic electron flow, electrons from photosystem II follow a pathway along an electron transport chain in the thylakoid membrane. This pathway generates an H^+ gradient that is used to make ATP. In addition, light energy striking photosystem I boosts an electron to a very-high-energy level that allows the synthesis of NADPH. (Figure 8.8)

- Emerson showed that, compared to single light flashes at 680 nm and 700 nm, light flashes at both wavelengths more than doubled the amount of photosynthesis. This occurred because these wavelengths activate pigments in PSII and PSI, respectively. (Figure 8.9)

- Hill and Bendall proposed the Z scheme for electron activation during photosynthesis. According to this scheme, an electron absorbs light energy twice, and it loses some of that energy as it flows along the electron transport chain in the thylakoid membrane. (Figure 8.10)

- During cyclic photophosphorylation, electrons are activated in PSI and flow from Fd to Q_B to the cytochrome complex to Pc and back to PSI. This cyclic electron route produces an H^+ gradient that is used to make ATP. (Figure 8.11)

- Cytochrome b in mitochondria and cytochrome b_6 in chloroplasts are homologous proteins, both of which are involved in electron transport and H^+ pumping. (Figure 8.12)

8.3 Calvin Cycle

- The Calvin cycle can be divided into three phases: carbon fixation, reduction and carbohydrate production, and regeneration of ribulose bisphosphate (RuBP). During this process, ATP is used as a source of energy and NADPH is used as a source of high-energy electrons so that CO_2 can be incorporated into carbohydrate. (Figure 8.13)

- Calvin and Benson determined the steps in the Calvin cycle by isotope labeling methods in which products of the Calvin cycle were separated by chromatography. (Figure 8.14)

8.4 Variations in Photosynthesis

- C_3 plants can only incorporate CO_2 into organic molecules via RuBP to make 3PG, a three-carbon molecule. (Figure 8.15)

- Photorespiration can occur under conditions of high O_2 and low CO_2, which occur under hot and dry conditions. During this process, some CO_2 is liberated and O_2 is used.

- Plants generally known as C_4 plants avoid photorespiration because the CO_2 is first incorporated, via PEP carboxylase, into a four-carbon molecule, which is pumped from mesophyll cells into bundle-sheath cells. This maintains a high concentration of CO_2 in the bundle-sheath cells, where the Calvin cycle occurs. The high CO_2 concentration minimizes photorespiration. (Figure 8.16)

- CAM plants prevent photorespiration by fixing CO_2 into a four-carbon molecule at night and then running the Calvin cycle during the day with their stomata closed. (Figure 8.17)

TEST YOURSELF

1. The water necessary for photosynthesis
   a. is split into H_2 and O_2.
   b. is directly involved in the synthesis of carbohydrate.
   c. provides the electrons to replace lost electrons in photosystem II.
   d. provides H^+ needed to synthesize G3P.
   e. none of the above.

2. The reaction center pigment differs from the other pigment molecules of the light-harvesting complex in that
   a. the reaction center pigment is a carotenoid.
   b. the reaction center pigment absorbs light energy and transfers that energy to other molecules without the transfer of electrons.
   c. the reaction center pigment transfers excited electrons to other molecules.
   d. the other pigments transfer high-energy electrons to other pigment molecules.
   e. the reaction center acts as an ATP synthase to produce ATP.

3. The cyclic electron flow that occurs in photosystem I produces
   a. NADPH.
   b. oxygen.
   c. ATP.
   d. all of the above.
   e. a and c only.

4. During the light reactions, the high-energy electron from P680
   a. eventually moves to $NADP^+$.
   b. becomes incorporated in water molecules.
   c. is pumped into the thylakoid space to drive ATP production.
   d. provides the energy necessary to split water molecules.
   e. falls back to the low-energy state in photosystem II.

5. During the first phase of the Calvin cycle, carbon dioxide is incorporated into ribulose bisphosphate by
   a. oxaloacetate.
   b. rubisco.
   c. RuBP.
   d. quinone.
   e. G3P.

6. The NADPH produced during the light reactions is necessary for
   a. the carbon fixation phase, which incorporates carbon dioxide into an organic molecule of the Calvin cycle.
   b. the reduction phase, which produces carbohydrates in the Calvin cycle.
   c. the regeneration of RuBP of the Calvin cycle.
   d. all of the above.
   e. a and b only.

7. The majority of the G3P produced during the reduction and carbohydrate production phase is used to produce
   a. glucose.
   b. ATP.
   c. RuBP to continue the cycle.
   d. rubisco.
   e. all of the above.

8. Photorespiration
   a. is the process where plants use sunlight to make ATP.
   b. is an inefficient way plants can produce organic molecules and in the process use O_2 and release CO_2.
   c. is a process that plants use to convert light energy to NADPH.
   d. occurs in the thylakoid space.
   e. is the normal process of carbohydrate production in cool, moist environments.

9. Photorespiration is avoided in C_4 plants because
   a. these plants separate the formation of a four-carbon molecule from the rest of the Calvin cycle in different cells.
   b. these plants only carry out anaerobic respiration.
   c. the enzyme PEP functions to maintain high CO_2 concentrations in the bundle-sheath cells.
   d. all of the above.
   e. a and c only.

10. Plants that are commonly found in hot and dry environments that carry out carbon fixation at night are
    a. oak trees.
    b. C_3 plants.
    c. CAM plants.
    d. all of the above.
    e. a and b only.

CONCEPTUAL QUESTIONS

1. Define photosynthesis. Explain the formula for photosynthesis.
2. Explain the function of NADPH.
3. Describe the parts of the chloroplast.

EXPERIMENTAL QUESTIONS

1. What was the purpose of the study conducted by Calvin and his colleagues?
2. In the experiment of Figure 8.14, why did the researchers use ^{14}C? Why did they examine samples at several different time periods? How were the different molecules in the samples identified?
3. What were the results of Calvin's study?

COLLABORATIVE QUESTIONS

1. Discuss the terms heterotroph and photoautotroph.
2. Discuss some of the differences between C_3 and C_4 plants.

---

www.brookerbiology.com

This website includes answers to the Biological Inquiry questions found in the figure legends and all end-of-chapter questions.

---

9 CELL COMMUNICATION AND REGULATION OF THE CELL CYCLE

CHAPTER OUTLINE

A micrograph of human skin cells dividing. The proper division of skin cells depends on signaling molecules, such as epidermal growth factor, that promote cell division. In addition, the division process is highly regulated to avoid mutations that could harm the resulting daughter cells.

In Chapter 5, we learned that all living cells are surrounded by a plasma membrane, which separates the internal cell contents from the extracellular environment. The plasma membrane enables cells to maintain an internal environment that is well ordered and carries out a variety of cell functions, as described throughout this unit. However, the plasma membrane also tends to isolate a cell from its surroundings. A cell cannot survive if it cannot sense changes in the extracellular environment and respond to them. For example, if the nutrient conditions change, such as a change from carbohydrates to fats, a cell needs to respond in a way that facilitates the uptake and metabolism of fats. If it cannot adapt to environmental changes, a cell is destined to die.

Cell communication is the process through which cells can detect and respond to signals in their extracellular environment. In multicellular organisms, cell communication is also needed to coordinate cellular activities within the whole organism. One of the most important reasons why cells must sense environmental conditions and communicate with each other is to promote cell division. For example, cells should divide only when they have sufficient nutrients to support the division process. The uptake of nutrients acts as a signal to stimulate cell division. In addition, the cells of multicellular organisms must communicate with each other so that cell division in one part of the body is coordinated with other parts. In animals, for example, signaling molecules are released at specific stages of development and promote the division of particular cells.

In this chapter, we will examine how cells respond to environmental signals and also produce signals so they can communicate with other cells. As you will learn, cell communication involves an amazing diversity of cellular proteins, lipids, and steroids that are devoted to this process. We will also consider the cell cycle, which in eukaryotic organisms is a series of events that result in cell division. This chapter focuses on the regulation of the cell cycle, which links cell communication and cell division. Chapter 15 will examine the steps of the cell division process as they pertain to the sorting of chromosomes.

9.1 General Features of Cell Communication

All living cells, including bacteria, fungi, protists, plant cells, and animal cells, are capable of cell communication, also known as cell signaling, a phenomenon that involves both incoming and outgoing signals. A **signal** is an agent that can influence the properties of cells. For example, on a sunny day, cells can sense their exposure to ultraviolet (UV) light and respond accordingly. In humans, UV light acts as an incoming signal to promote the synthesis of melanin, a protective pigment that helps to prevent the harmful effects of UV radiation. In addition, organisms can produce outgoing signals that influence the behavior of neighboring cells. Plant cells, for example, produce hormones that influence the pattern of cell elongation so that the plant grows toward light. Cells of all living organisms both respond to incoming signals and elicit outgoing signals. Cell communication is thus a two-way street.

Communication at the cellular level involves not only sending and receiving signals but also their interpretation. For this to occur, a signal must affect the conformation, or shape, of a cellular protein called a **receptor**. When a signal and receptor interact, a conformational change occurs in the receptor, eventually leading to some type of response in the cell. In this section, we begin by considering why cells need to respond to signals. We will then examine various forms of signaling, which differ in part based on the distance between the cells. Finally, we will examine the main steps that occur when a cell is exposed to a signal and elicits a response to it.

Cells Respond to Signals from Their Environment and from Other Cells

Before getting into the details of cell communication, let's take a general look at why cells need to respond to signals. The first reason is that cells need to respond to a changing environment. Changes in the environment are a persistent feature of life, and living cells are continually faced with alterations in temperature, nutrient availability, and availability of water. A cell may even be exposed to a toxic chemical in its environment. Being able to respond to change, a phenomenon called **adaptation**, is critical for the survival of all living organisms. Adaptation at the cellular level is also referred to as a **cellular response**. As an example, let's consider the response of a yeast cell to glucose in its environment (**Figure 9.1**). Some of the glucose acts as a signaling molecule that causes the cell to respond accordingly. In this case, the cell increases the number of glucose transporters that are needed to take glucose into the cell and also increases the number of enzymes that are required to metabolize glucose once it is inside. The cellular response has therefore allowed the cell to utilize glucose efficiently. We can say that the cell has become adapted to the presence of glucose in its environment.

Figure 9.1 **Response of a yeast cell to glucose.** When glucose is absent from the extracellular environment, the cell is not well prepared to take up and metabolize this sugar. However, when glucose is present, some of that glucose binds to a receptor in the membrane, which changes the amounts and properties of intracellular and membrane proteins so that the cell can readily use glucose.

Figure 9.2 **Phototropism in plants.** This process involves a bend that occurs just beneath the actively growing tip. This bend is caused by cell elongation along the nonilluminated side. The amount of auxin is higher on the nonilluminated side.

Biological inquiry: Does auxin cause cells to elongate or shorten?

A second reason for cell signaling is the need for cells to communicate with each other—a process called cell-to-cell communication. In one of the earliest experiments demonstrating cell-to-cell communication, Charles Darwin and his son Francis Darwin studied phototropism, the phenomenon in which plants grow toward light (**Figure 9.2**). The Darwins observed that the actual bending occurs in a zone below the growing tip. They concluded that the transmission of a signal from the growing tip to cells below the tip had to take place for this to occur. Later research revealed that the signal is a molecule called auxin, which is transmitted from cell to cell. The higher amount of auxin present on the nonilluminated side promotes cell elongation on that side only, thereby causing the plant to bend toward the light source.

Cell-to-Cell Communication Can Occur Between Adjacent Cells and Between Cells That Are Long Distances Apart

Researchers have determined that organisms have evolved a variety of different mechanisms to achieve cell-to-cell communication. The mode of communication depends, in part, on the distance between the cells that need to communicate with each other. Let's first examine the various ways in which signals are transferred between cells. Later in this chapter, we will learn how such signals elicit a cellular response.

One way to categorize cell signaling is by the manner in which the signal is transmitted from one cell to another. Signals are relayed between cells in five common ways, all of which involve a cell that produces a signal and a target cell that receives the signal (**Figure 9.3**).

Direct Intercellular Signaling In a multicellular organism, cells that are adjacent to each other may have contacts, called

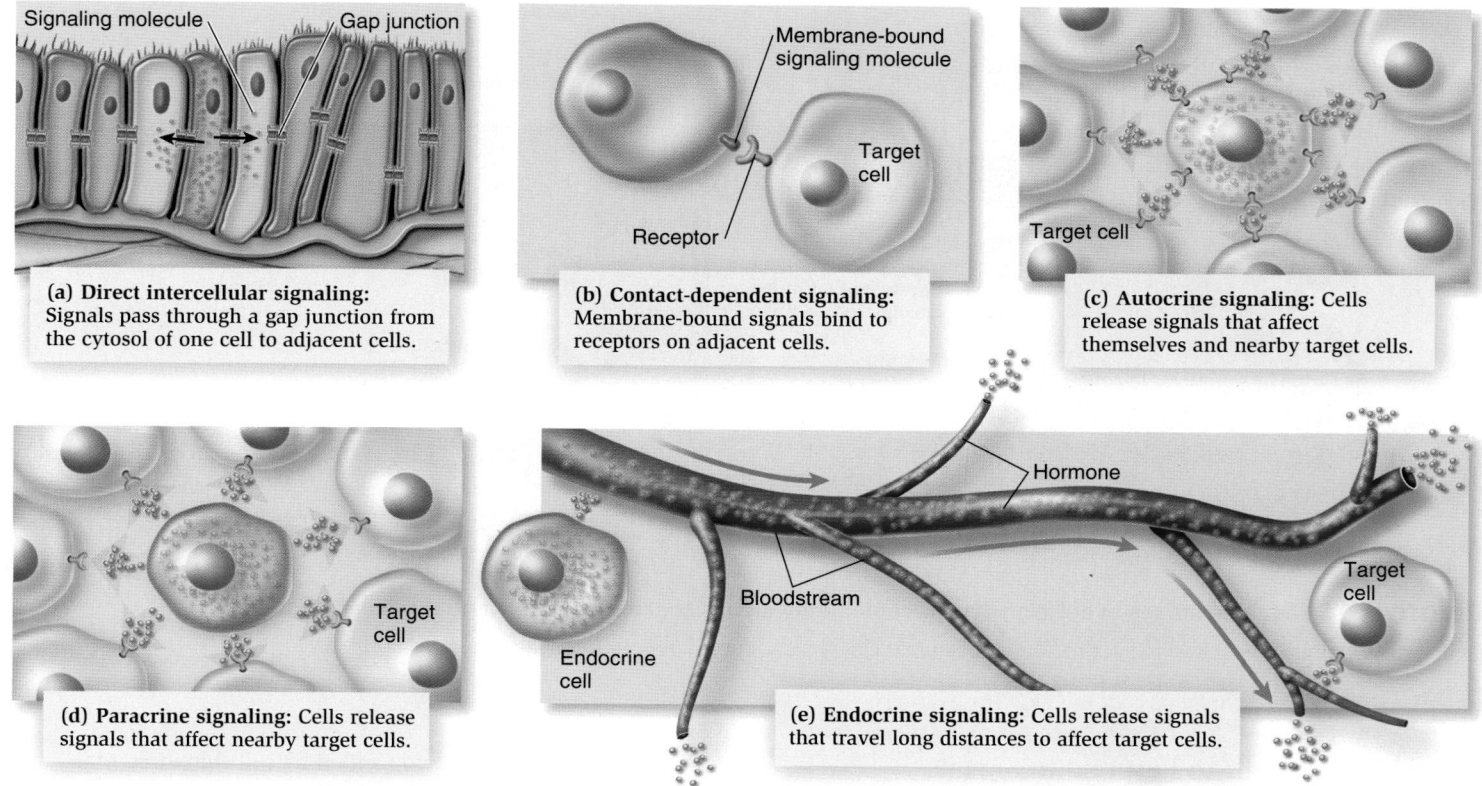

Figure 9.3 Examples of cell-to-cell communication based on the distance between cells.

cell junctions, that enable them to pass signaling molecules and other materials between the cytosol of one cell and the cytosol of another (Figure 9.3a). For example, certain cells that line your lungs have intercellular connections called gap junctions, which we will examine further in Chapter 10. One cell can transmit a signaling molecule to the cytosol of an adjacent cell and thereby affect that cell's behavior. This type of signaling influences the rhythmic movements of cilia on adjacent cells that sweep dust and other unwanted particles out of the lungs.

Contact-Dependent Signaling Not all signaling molecules can readily diffuse from one cell to another. Some molecules are bound to the surface of cells and provide a signal to other cells that make contact with the surface of that cell (Figure 9.3b). In this case, one cell has a membrane-bound signaling molecule that is recognized by a receptor on the surface of another cell. This occurs when portions of nerve cells grow and make contact with other nerve or muscle cells.

Autocrine Signaling In autocrine signaling, a cell secretes signaling molecules that bind to receptors on its own cell surface, stimulating a response (Figure 9.3c). In addition, the signaling molecule can affect neighboring cells of the same cell type. Autocrine signaling is often important for groups of cells to sense cell density. When cell density is high, the concentration of autocrine signals is also high.

Paracrine Signaling In paracrine signaling, a specific cell secretes a signaling molecule that does not affect the cell secreting the signal, but it does influence the behavior of target cells in close proximity (Figure 9.3d). Usually, the signal is broken down too quickly to be carried to other parts of the body and affect distant cells. A specialized form of paracrine signaling called **synaptic signaling** occurs in the nervous system of animals. Neurotransmitters—molecules made in nerve cells that transmit a signal to an adjacent cell—are released at the end of the nerve cell and traverse a narrow space called the synapse. The neurotransmitter then binds to a receptor in a target cell. We will discuss this mechanism of signaling in detail in Chapter 43.

Endocrine Signaling In contrast to the previous mechanisms of cell signaling, endocrine signaling occurs over long distances (Figure 9.3e). In both animals and plants, molecules involved in long-distance signaling are called **hormones**. They are usually longer lasting than signaling molecules involved in autocrine and paracrine signaling. In animals, endocrine signaling involves the secretion of hormones into the bloodstream that may affect virtually all cells of the body, including those that are far from the cells that secrete the signaling molecules. In plants, hormones move through vessels and can also move through adjacent cells. Some signaling molecules are even gases that are secreted into the air. Ethylene is a gas given off by plants that plays a variety of roles, such as the acceleration of fruit ripening.

1 **Receptor activation:** The binding of a signaling molecule causes a conformational change in a receptor that activates its function.

3 **Cellular response:** The signal transduction pathway affects the functions and/or amounts of cellular proteins, thereby producing a cellular response.

Signaling molecule

Receptor activation

2 **Signal transduction:** The activated receptor stimulates a series of proteins that forms a signal transduction pathway.

Unactivated receptor protein

Signal transduction pathway

Intracellular targets	Cellular response
Enzyme	Altered metabolism or other cell functions
Structural protein	Altered cell shape or movement
Gene regulatory protein	Altered gene expression, which changes the amounts of proteins in the cell

Nucleus

Figure 9.4   The three stages of cell signaling: receptor activation, signal transduction, and a cellular response.

Biological inquiry: For most signaling molecules, explain why a signal transduction pathway is necessary.

The Ability of Cells to Respond to Signals Is Usually a Three-Stage Process

Thus far, we have learned that signals influence the behavior of cells in close proximity or at long distances, usually interacting with receptors to elicit a cellular response. What events occur when a cell encounters a signal? In most cases, the binding of a signal to a receptor causes the receptor to activate a signal transduction pathway, which converts the signal to a cellular response. **Figure 9.4** diagrams the three stages of cell signaling: receptor activation, signal transduction, and a cellular response.

Stage 1: Receptor Activation   In the initial stage, a signaling molecule binds to a receptor, causing receptor activation. The binding of the signal causes a conformational change in the receptor that activates the receptor's function. In most cases, the activated receptor initiates a response by stimulating a sequence of changes in a series of proteins that collectively forms a signal transduction pathway, as described next.

Stage 2: Signal Transduction   During signal transduction, the initial signal is converted—or transduced—to a different signal inside the cell. This process is carried out by a group of proteins that form a **signal transduction pathway**. These proteins undergo a series of changes that may result in the production of another signaling molecule. For example, after a hormone binds to a receptor in the plasma membrane, the activated receptor may interact with intracellular proteins that produce new signals inside the cell.

Stage 3: Cellular Response   Cells can respond to signals in several different ways. Figure 9.4 shows three common cate-

gories of proteins that are controlled by cell signaling. Many signaling molecules exert their effects by altering the activity of one or more enzymes. For example, certain hormones provide a signal that the body needs energy. These hormones activate enzymes that are required for the breakdown of molecules such as carbohydrates.

A second way that cells respond to signals is by altering the functions of structural proteins in the cell. For example, when cells move during embryonic development or when an amoeba moves toward food, signals play a role in the rearrangement of actin filaments, which are components of the cytoskeleton. The coordination of signaling and changes in the cytoskeleton enable cells to move in the correct direction. In addition, changes in structural proteins may cause storage vesicles within a cell to fuse with the plasma membrane. Such vesicles may contain proteins or small molecules that are then secreted by the cell.

Cells may also respond to signals by changing the expression of genes. When exposed to sex hormones, cells can activate genes that change the properties of cells and even the characteristics of entire organisms. As discussed in Chapter 51, estrogens and androgens are responsible for the development of secondary sex characteristics in humans, including breast development in females and beard growth in males.

9.2 Cellular Receptors

As we have learned, signals are needed for cells to respond to environmental changes and to communicate with each other. Communication can occur between adjacent cells or over long distances. In most cases, the binding of a signaling molecule to its receptor activates a signal transduction pathway that ultimately leads to a cellular response. This response may involve effects on enzyme activities, the functions of structural proteins, and/or the transcription of genes.

In this section, we will take a closer look at receptors and how they interact with signaling molecules. We will focus on how these interactions affect the function of the receptor. As you will learn, cells contain many different types of receptors, and their molecular functions are quite diverse. In particular, we will look at differences among receptors based on whether they are located on the cell surface or inside the cell. In this chapter, our focus will be on receptors that respond to chemical signaling molecules. Other receptors discussed in Units VI and VII respond to mechanical motion (mechanoreceptors), temperature changes (thermoreceptors), and light (photoreceptors).

Signals Bind Specifically to Receptors and Alter Their Conformations

The ability of cells to respond to a signal usually requires precise recognition between a receptor and its signal. In many cases, the signal is a molecule—a steroid, peptide, or protein—that binds to the receptor. A signaling molecule binds to a receptor in much the same way that a substrate binds to the active site of an enzyme, as described in Chapter 7. The signaling molecule, which is called a **ligand**, binds noncovalently to the receptor molecule with a high degree of specificity. The binding occurs when the ligand and receptor collide in the correct orientation and with enough energy.

$$[\text{Ligand}] + [\text{Receptor}] \underset{k_{off}}{\overset{k_{on}}{\rightleftarrows}} [\text{Ligand} \cdot \text{Receptor complex}]$$

The value k_{on} is the rate at which binding occurs. After a complex forms between the ligand and its receptor, the noncovalent interaction between a ligand and receptor remains stable for a finite period of time. The term k_{off} is the rate at which the ligand • receptor complex falls apart or dissociates.

In general, the binding and release between a ligand and its receptor are relatively rapid, and therefore an equilibrium is reached when the rate of formation of new ligand • receptor complexes equals the rate at which existing ligand • receptor complexes dissociate:

$$k_{on} [\text{Ligand}][\text{Receptor}] = k_{off} [\text{Ligand} \cdot \text{Receptor complex}]$$

Rearranging,

$$\frac{[\text{Ligand}] [\text{Receptor}]}{[\text{Ligand} \cdot \text{Receptor complex}]} = \frac{k_{off}}{k_{on}} = K_d$$

K_d is called the **dissociation constant** between a ligand and its receptor. Let's look carefully at the left side of this equation and consider what it means. At a ligand concentration where half of the receptors are bound to a ligand, the concentration of the ligand • receptor complex equals the concentration of receptor that doesn't have ligand bound. At this ligand concentration, [Receptor] and [Ligand • Receptor complex] cancel out of the

equation because they are equal. Therefore, at a ligand concentration where half of the receptors have bound ligand:

$$K_d = [\text{Ligand}]$$

When the ligand concentration is above the K_d value, most of the receptors are likely to have ligand bound to them. In contrast, if the ligand concentration is substantially below the K_d value, most receptors will not be bound by their ligand. The K_d values for many different ligands and their receptors have been experimentally determined. This information allows researchers to predict when a signaling molecule is likely to cause a cellular response. If the concentration of a signaling molecule is far below the K_d value, a cellular response is not likely because relatively few receptors will form a complex with the signaling molecule.

Unlike enzymes, which convert their substrates into products, receptors do not usually alter the structure of their ligands. Instead, the ligands alter the structure of their receptors, causing a conformational change. This concept is shown in **Figure 9.5**, in which the ligand is a hormone. Once the hormone binds, the receptor undergoes a conformational change. After binding, the hormone • receptor complex is called an activated receptor, because the binding of the hormone to its receptor has changed the receptor in a way that will activate its ability to initiate a cellular response.

Because cell communication is critical for cell survival, the evolution of organisms over the past 3.5 to 4.0 billion years has produced a variety of ways in which cells can respond to signals. Previously, we learned that one way to categorize cell communication is based on the distance that the signal travels. Another way to categorize cell signaling is based on the location and activity of the receptor, as we will explore next.

Cells Contain a Variety of Cell Surface Receptors That Respond to Signals in Different Ways

Most signals, either environmental agents or signals that are secreted from cells, are small hydrophilic molecules or large molecules that do not readily pass through the plasma membrane of cells.

Hormone

Activated receptor

Cytosol

The binding of a hormone to a receptor causes a conformational change in the receptor, resulting in receptor activation.

Figure 9.5 A conformational change induced by a ligand binding to a receptor.

To respond to such signals, cells possess several different types of **cell surface receptors**—receptors that are found in the plasma membrane. A typical cell is expected to contain dozens or even hundreds of different cell surface receptors that enable the cell to respond to different kinds of signaling molecules.

By analyzing the functions of cell surface receptors from many different organisms, researchers have determined that most fall into one of three categories: enzyme-linked receptors, G-protein-coupled receptors, and ligand-gated ion channels.

Enzyme-Linked Receptors   Receptors known as **enzyme-linked receptors** are found in all living species. They typically have two important domains: an extracellular domain, which binds a signaling molecule, and an intracellular domain, which has a catalytic function (**Figure 9.6a**). When a signaling molecule binds to the extracellular domain, a conformational change is transmitted through the membrane-embedded portion of the protein that affects the conformation of the catalytic domain. In most cases, this conformational change causes the catalytic domain to become functionally active.

Most types of enzyme-linked receptors function as **protein kinases**, enzymes that transfer phosphate groups from ATP to a

(a) Structure of enzyme-linked receptors

(b) A receptor that functions as a protein kinase

Figure 9.6 **Enzyme-linked receptors.** (a) The enzyme-linked receptor protein has an extracellular domain, which binds a signaling molecule, and an intracellular domain, which has a catalytic function. (b) An enzyme-linked receptor can function as a protein kinase, which catalyzes the transfer of a phosphate group from ATP to an intracellular protein.

protein (**Figure 9.6b**). In the absence of a signaling molecule, the catalytic domain remains inactive. However, when a signal binds to the extracellular domain, this transmits a conformational change to the catalytic domain, making it active. Under these conditions, the cell surface receptor may phosphorylate itself or intracellular proteins. Later in this chapter, we will explore in more detail how this event leads to a cellular response.

G-Protein-Coupled Receptors   Receptors called **G-protein-coupled receptors (GPCRs)** are commonly found in the cells of eukaryotic species. GPCRs exhibit a typical structure in which the protein contains seven transmembrane segments that wind back and forth through the plasma membrane. The receptors interact with intracellular proteins called **G proteins**, which are so named because of their ability to bind guanosine triphosphate (GTP) and guanosine diphosphate (GDP). GTP is similar in structure to ATP except it has guanine as a base instead of adenine. In the 1970s, the existence of G proteins was first proposed by Martin Rodbell and colleagues, who found that GTP was needed for certain hormone receptors to cause an intracellular response. Later, Alfred Gilman and coworkers used genetic and biochemical techniques to identify and purify a G protein. In 1994, Rodbell and Gilman won the Nobel Prize for their pioneering work.

Figure 9.7 shows how a GPCR and G protein interact. At the cell surface, the binding of a signaling molecule to a GPCR activates the receptor by causing a conformational change. The activated receptor then binds to a membrane-bound G protein, causing the G protein to release GDP and bind GTP instead. GTP binding changes the conformation of the G protein, causing it to dissociate into its α subunit and a β/γ dimer. Eventually, the extracellular signaling molecule will be degraded, thereby lowering its concentration and resulting in the dissociation between the signaling molecule and the GPCR. When this occurs, the GPCR is no longer activated and the cellular response will be reversed, because the α subunit will eventually hydrolyze its bound GTP to GDP and P_i. When this occurs, the α and β/γ subunits reassociate with each other to form an inactive complex. Later in this chapter, we will examine how the α subunit and β/γ dimer interact with other proteins in a signal transduction pathway to elicit a cellular response.

Ligand-Gated Ion Channels   Ion channels are proteins that allow the diffusion of ions across cellular membranes. **Ligand-gated ion channels** are a third type of cell surface receptor found in the plasma membrane of animal and plant cells. When a signaling molecule (ligand) binds to this type of receptor, the channel opens and allows the flow of ions through the membrane (**Figure 9.8**).

In animals, ligand-gated ion channels are important in the transmission of signals between nerve and muscle cells and between two nerve cells. In addition, ligand-gated ion channels in the plasma membrane allow the uptake of Ca^{2+} into the cytosol. As discussed later in this chapter, changes in the cytosolic concentration of Ca^{2+} play a role in signal transduction.

1 Signal binds to GPCR, causing it to bind to a G protein.

2 G protein exchanges GDP for GTP. It dissociates from the receptor and separates into an active α subunit and a β/γ dimer.

Receptor protein (GPCR)

Signaling molecule

GDP

α

β

γ

Inactive G protein

GTP

GDP released

Activated G protein α subunit

Activated G protein β/γ dimer

P_i

Cytosol

3 The signaling molecule is eventually degraded and the α subunit hydrolyzes GTP into GDP + P_i. The α subunit and the β/γ dimer reassociate.

Figure 9.7   The activation of G-protein-coupled receptors.

Signaling molecule

Ions

Cytosol

The binding of two extracellular signaling molecules (a ligand) opens the ion channel, permitting ions to pass through the membrane.

Figure 9.8   The function of a ligand-gated ion channel.

Some Signaling Molecules Pass Through the Plasma Membrane and Activate Intracellular Receptors

Although most receptors for signaling molecules are located in the plasma membrane, some are found inside the cell. In these cases, the signaling molecule must pass through the plasma membrane to gain access to its receptor. For example, two types of signaling molecules—steroids and auxins—diffuse into the cell, where they bind to internal receptors. As described in Chapter 51, steroid hormones, such as estrogens and androgens, are secreted into the bloodstream from cells of endocrine glands.

The behavior of estrogen is typical of many steroid hormones (**Figure 9.9a**). Because of its hydrophobic nature, estrogen is capable of passing through the plasma membrane of a target cell and binding to a receptor in the nucleus. Some steroids bind to receptors in the cytosol, which then travel into the nucleus, while other steroid hormones bind to receptors that are in the nucleus; estrogen is one of the latter. After binding, the estrogen • receptor complex undergoes a conformational change that enables it to form a dimer with another estrogen • receptor complex. The dimer then binds to the DNA and activates the transcription of specific genes. The estrogen receptor is an example of a transcription factor, a protein that regulates the transcription of genes.

A second example of an intracellular receptor is the receptor for a group of related plant hormones called auxins, which are important in many plant signaling pathways, including the growth of shoots toward light, the growth of roots into the soil, and the flowering process. In 2005, two research groups led by Ottoline Leyser and Mark Estelle determined that auxin binds to an intracellular receptor called TIR1 (TIR stands for transport inhibitor response, so named because mutations in this receptor alter the response of drugs that inhibit auxin transport in plants). The binding of auxin in the cytosol activates TIR1, and it travels into the nucleus. Like steroid receptors, TIR1 also causes the activation of specific genes. However, TIR1 does not do this directly. After auxin binds, the receptor causes the breakdown of a protein that inhibits several genes (**Figure 9.9b**). When this inhibitory protein is broken down, the gene inhibition is relieved and the genes are transcribed into mRNA.

Estrogen

2 Estrogen receptors form a dimer, bind next to specific genes, and activate their transcription. The mRNAs are then translated into proteins that affect the structure and function of the cell.

Active estrogen receptor dimer

Protein that affects cell structure and function

1 Estrogen penetrates the plasma membrane, enters the nucleus, and binds to the estrogen receptors. The receptors undergo a conformational change.

Inactive estrogen receptor

Chromosomal DNA

mRNA

Nucleus

(a) Estrogen receptor

Auxin

Proteins that inhibit certain genes that are involved with cell division

2 Activated TIR1 enters the nucleus and causes the breakdown of a protein that inhibits several genes that are needed for cell division. These genes are then transcribed into mRNA, and the mRNA is translated into proteins that promote cell division.

TIR1

1 Auxin enters the cell and binds to TIR1 in the cytosol, thereby activating it.

mRNA

Protein that promotes cell division

DNA

Nucleus

(b) Auxin receptor

Figure 9.9 Examples of intracellular receptors.

9.3 Signal Transduction and the Cellular Response

In this section, we turn our attention to the intracellular events that enable a cell to respond to a signaling molecule that binds to a cell surface receptor. In most cases, the binding of a signaling molecule to its receptor stimulates a signal transduction pathway. We begin by examining the components of different signal transduction pathways, looking in detail at the two-component regulatory systems common to bacteria and the signaling transduction pathways involving molecules called second messengers. We will use the activities of the hormone epinephrine as a way to explore how signal transduction pathways affect cellular proteins in a way that leads to a cellular response. Lastly, we'll explore the reasons why different types of cells can respond in distinct ways to the same hormone.

Two-Component Regulatory Systems in Bacteria Sense Environmental Changes

The survival of bacteria is largely dependent on their ability to adapt to environmental changes. By studying many different bacterial species, researchers have discovered that a common way for bacteria to adapt to a changing environment is via **two-component regulatory systems**. In such systems, one component, an enzyme-linked receptor called a **sensor kinase**, recognizes a signal found in its environment. The sensor kinase also has the ability to hydrolyze ATP and phosphorylate itself. The phosphate group is then transferred to a second component, a protein called the **response regulator**, which is usually a protein that regulates the expression of many genes.

Figure 9.10 considers a two-component regulatory system, found in many species of bacteria, that senses the presence of

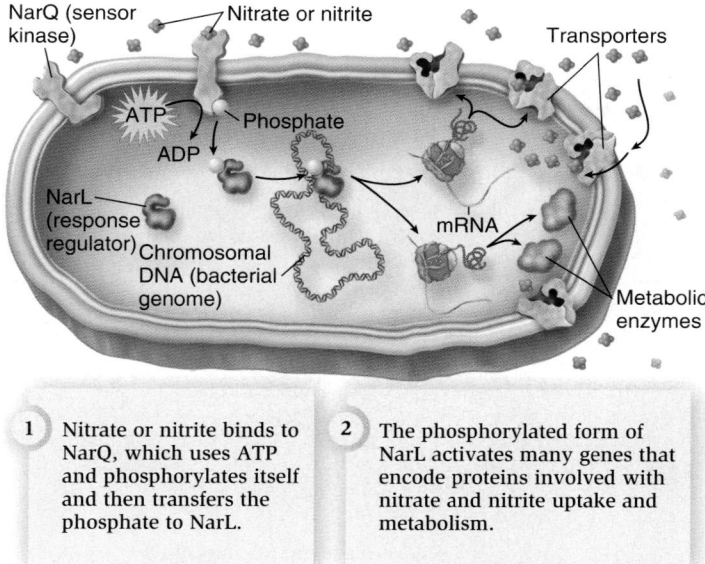

1. Nitrate or nitrite binds to NarQ, which uses ATP and phosphorylates itself and then transfers the phosphate to NarL.

2. The phosphorylated form of NarL activates many genes that encode proteins involved with nitrate and nitrite uptake and metabolism.

Figure 9.10 **An example of a two-component regulatory system in bacteria.** The system shown here regulates nitrate and nitrite metabolism. NarQ is a receptor, and NarL is a single component of a signal transduction pathway. When phosphorylated by NarQ, NarL functions as a transcriptional activator.

nitrate (NO_3^-) and nitrite (NO_2^-) in the environment. NarQ is a sensor kinase that recognizes either nitrate or nitrite. When either of these binds to NarQ, NarQ phosphorylates itself via ATP and then transfers a phosphate to the response regulator, NarL. The phosphorylated form of NarL activates the transcription of many genes involved in nitrate and nitrite metabolism and transport. Once these genes are transcribed into mRNA and then translated into specific proteins, the bacterium has become better adapted to the presence of nitrate or nitrite in its environment.

Two-component regulatory systems are also found in fungi and plants, but they do not appear to exist in animals. Why animals lack this form of signaling is not understood.

Second Messengers Are Key Components of Many Signal Transduction Pathways

Cell biologists call signals that bind to a cell surface receptor the first messengers. After first messengers bind to receptors, many signal transduction pathways lead to the production of **second messengers**, small molecules or ions that relay signals inside the cell. The following examples illustrate the roles of second messengers in signal transduction pathways.

Signal Transduction via cAMP As we saw earlier, the binding of a signaling molecule to a G-protein-coupled receptor (GPCR) activates an intracellular G protein by causing it to bind GTP and dissociate into an α subunit and a β/γ dimer (see Figure 9.7). Let's now follow the role of these subunits in signal

transduction pathways. The α subunit and the β/γ dimer can activate several different kinds of proteins that are components of signal transduction pathways.

Mammalian and plant cells make several different types of G protein α subunits. One type binds to **adenylyl cyclase**, an enzyme in the plasma membrane (**Figure 9.11a**). This stimulates adenylyl cyclase to synthesize **cyclic adenosine monophosphate (cAMP)** from ATP. The molecule, cAMP, is an example of a second messenger.

cAMP

One effect of cAMP is to activate protein kinase A (PKA), which is composed of four subunits: two catalytic subunits that phosphorylate specific cellular proteins, and two regulatory subunits that inhibit the catalytic subunits when they are bound to each other. Cyclic AMP binds to the regulatory subunits of PKA. The binding of cAMP separates the regulatory and catalytic subunits, which allows each catalytic subunit to be active.

The catalytic subunit of PKA then phosphorylates specific cellular proteins such as enzymes, structural proteins, and transcription factors. When a protein kinase attaches a phosphate group onto a protein, the shape and function of that protein is altered. Often, phosphorylation activates protein function. In 1955, Edmond Fischer and Edwin Krebs discovered that the phosphorylation of enzymes is important in regulating glycogen breakdown. Since that time, researchers have determined that phosphorylation is a widespread mechanism that regulates protein function in all species. For their key discovery, Fischer and Krebs were awarded the Nobel Prize in 1992.

Let's now consider how the effects of a signaling molecule are reversed. When the signaling molecule is no longer produced and its extracellular levels fall, the molecule dissociates from the receptor, because the binding to its receptor is a reversible process (**Figure 9.11b**). Once the signaling molecule is released, the receptor is no longer activated. Intracellularly, the α subunit hydrolyzes its GTP to GDP, and the α subunit and β/γ dimer reassociate to an inactive G protein. The levels of cAMP decrease due to the action of an enzyme called **phosphodiesterase**, which converts cAMP to AMP. As cAMP levels fall, the regulatory subunits of PKA release cAMP, and the regulatory and catalytic subunits reassociate, thereby inhibiting PKA. Finally, enzymes called **protein phosphatases** are responsible for removing phosphate groups from proteins, thereby reversing the effects of PKA.

In the 1950s, Earl Sutherland determined that many different hormones cause the formation of cAMP in a variety of cell types.

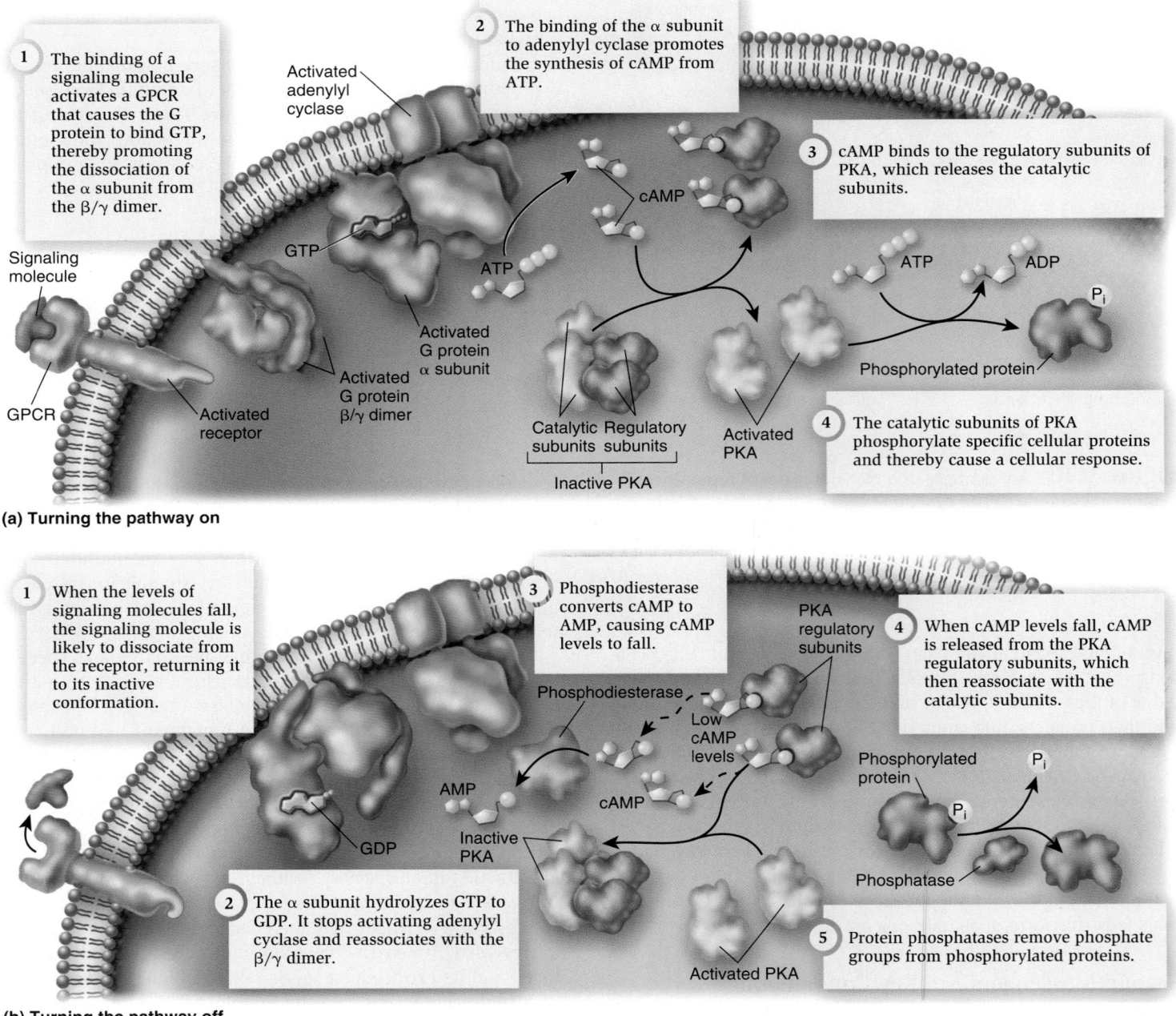

1 The binding of a signaling molecule activates a GPCR that causes the G protein to bind GTP, thereby promoting the dissociation of the α subunit from the β/γ dimer.

2 The binding of the α subunit to adenylyl cyclase promotes the synthesis of cAMP from ATP.

3 cAMP binds to the regulatory subunits of PKA, which releases the catalytic subunits.

4 The catalytic subunits of PKA phosphorylate specific cellular proteins and thereby cause a cellular response.

Activated adenylyl cyclase

Signaling molecule

GTP

ATP

cAMP

GPCR

Activated receptor

Activated G protein α subunit

Activated G protein β/γ dimer

Catalytic subunits Regulatory subunits

Inactive PKA

Activated PKA

ATP ADP

Pᵢ

Phosphorylated protein

(a) Turning the pathway on

1 When the levels of signaling molecules fall, the signaling molecule is likely to dissociate from the receptor, returning it to its inactive conformation.

3 Phosphodiesterase converts cAMP to AMP, causing cAMP levels to fall.

4 When cAMP levels fall, cAMP is released from the PKA regulatory subunits, which then reassociate with the catalytic subunits.

Phosphodiesterase

Low cAMP levels

PKA regulatory subunits

AMP

cAMP

GDP

Inactive PKA

Phosphorylated protein

Pᵢ

Pᵢ

Phosphatase

2 The α subunit hydrolyzes GTP to GDP. It stops activating adenylyl cyclase and reassociates with the β/γ dimer.

Activated PKA

5 Protein phosphatases remove phosphate groups from phosphorylated proteins.

(b) Turning the pathway off

Figure 9.11 **The signal transduction pathway involving cAMP.** (a) The pathway leading to the formation of cAMP and subsequent activation of PKA, which is mediated by a G-protein-coupled receptor (GPCR). (b) The steps that shut the pathway off when the signaling molecule is no longer present in the environment.

This observation, for which he won the Nobel prize, stimulated great interest in the study of signal transduction pathways. Since Sutherland's discovery, the production of second messengers such as cAMP has been found to have two important advantages. One benefit is the amplification of the signal; the binding of a signal to a single receptor can cause the synthesis of many cAMP molecules that activate PKA (**Figure 9.12**). Likewise, each molecule of PKA can phosphorylate many proteins in the cell to promote a cellular response. In this way, a signal can have a dramatic effect on the functioning of many proteins in a target cell.

A second advantage of second messengers such as cAMP is speed. Brian Bacskai and colleagues studied the response of nerve cells to a signaling molecule called serotonin, which is a neurotransmitter that binds to a GPCR. In humans, serotonin is believed to play a role in depression, anxiety, and sexual drive. To monitor cAMP levels, nerve cells grown in a laboratory were injected with a fluorescent protein that changes its fluorescence

Figure 9.12 **Signal amplification.** An advantage of a signal transduction pathway is the amplification of a signal. In this case, a single signaling molecule can lead to the phosphorylation of many target proteins.

Figure 9.13 **The rapid speed of cAMP production.** The micrograph on the *left* shows a nerve cell prior to its exposure to serotonin, while the micrograph on the *right* shows the same cell 20 seconds after exposure. Blue indicates a low level of cAMP, yellow is an intermediate level, and red/purple is a high level.

when cAMP is made. As shown in the micrograph on the right in **Figure 9.13**, a substantial amount of cAMP was made within 20 seconds after the addition of serotonin.

Signal Transduction via Ca^{2+} Cells use many different types of second messengers, and more than one type may be used at the same time. Calcium ions are another common second messenger. Cells maintain very large Ca^{2+} gradients (**Figure 9.14**). Two types of calcium pumps are found in the membranes of cells. Ca^{2+}-ATPases use the energy from ATP hydrolysis to pump Ca^{2+} against a gradient, while Na^{+}/Ca^{2+} and H^{+}/Ca^{2+} exchangers or antiporters use the energy within Na^{+} or H^{+} gradients to pump Ca^{2+} across the membrane. These pumps produce a cytosolic Ca^{2+} concentration that is very low, typically 0.1–1.0 μM. By comparison, the Ca^{2+} concentration found outside of cells and inside the endoplasmic reticulum (ER) lumen and mitochondrial matrix is far higher, as much as 1,000- to 10,000-fold higher!

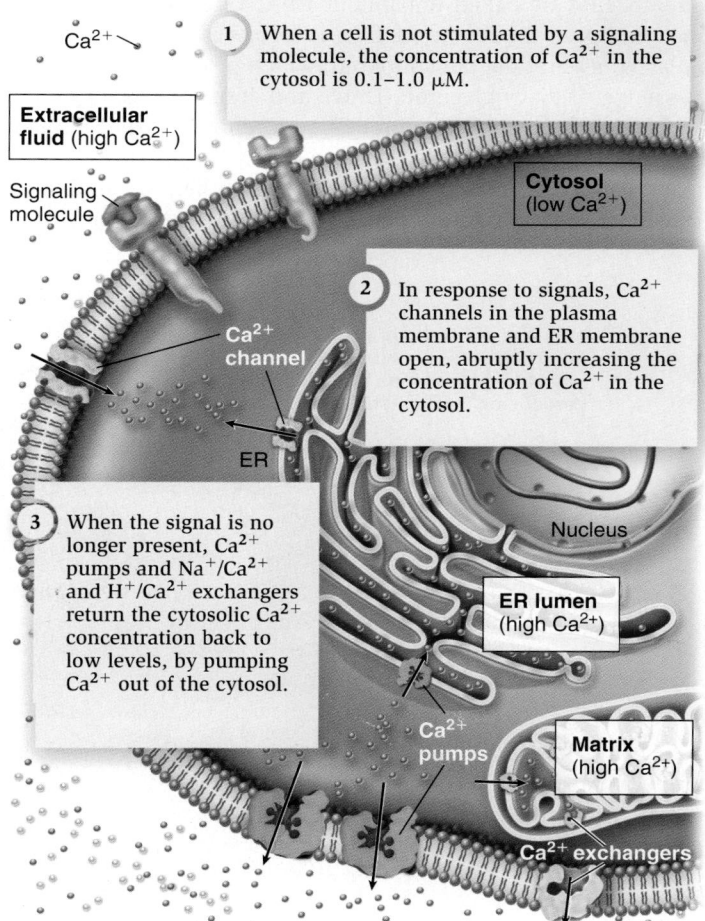

Figure 9.14 **Calcium levels inside and outside of eukaryotic cells.**

Calcium channels that are found in the plasma membrane and ER membrane play a key role in the ability of calcium to act as a second messenger. When such channels open in response to signaling, the cytosolic concentration of Ca^{2+} can increase dramatically and quickly, in a matter of seconds. The increase in Ca^{2+} concentration in the cytosol acts as a second messenger that elicits a cellular response. In plants, calcium signaling is involved in many processes, including cell elongation in response to light, the opening and closing of stomata, and the ability to sense gravity. In animals, calcium-signaling pathways are also important in many cellular events such as nerve transmission, muscle contraction, and the secretion of digestive enzymes by the pancreas. As we will explore later, such responses are mediated by calcium-binding proteins that are activated when the cytosolic concentration of Ca^{2+} is increased.

The calcium channels that are found in the ER membrane are opened via a second messenger called inositol trisphosphate, a pathway that is described next.

Signal Transduction via Diacylglycerol and Inositol Trisphosphate   We have already examined how a G protein can activate adenylyl cyclase and thereby promote the synthesis of cAMP. Let's now consider a second way that an activated G protein can influence a signal transduction pathway. The α subunit can activate an enzyme called phospholipase C (**Figure 9.15**). When phospholipase C becomes active, it breaks a covalent bond in a plasma membrane phospholipid, which produces two second messengers, diacylglycerol (DAG) and inositol trisphosphate (IP_3). DAG binds to a membrane-bound enzyme called protein kinase C (PKC). IP_3 is released into the cytosol and binds to a ligand-gated Ca^{2+} channel in the ER membrane. The binding of IP_3 causes the channel to open, releasing Ca^{2+} into the cytosol.

Therefore, this pathway involves another messenger, the calcium ions that we considered previously.

Calcium ions can affect the behavior of cells in a variety of ways, two of which are shown in Figure 9.15. Ca^{2+} can bind to PKC, which, in combination with DAG, causes this enzyme to become activated. Once activated, PKC can phosphorylate specific cellular proteins. In addition, Ca^{2+} can bind to a protein called calmodulin, which is a <u>ca</u>lcium-<u>modul</u>ated <u>prote</u>in. The Ca^{2+}-calmodulin complex can then interact with specific cellular proteins and alter their functions. For example, calmodulin regulates proteins involved in carbohydrate breakdown in liver cells.

A Cellular Response Depends on Which Proteins Are Controlled by a Signal Transduction Pathway

As we have seen, signaling molecules exert their effects on eukaryotic cells via signal transduction pathways that control the activities and/or synthesis of specific proteins. In multicellular organisms, one of the amazing effects of signaling molecules such as hormones is their ability to coordinate cellular activities. One example is the hormone epinephrine (also called adrenaline), which is secreted from endocrine cells. Epinephrine is also called the fight-or-flight hormone because it quickly prepares the body for strenuous physical activity. In humans, for example, epinephrine is secreted into the bloodstream when someone is confronted with a stressful event or is exercising vigorously.

Epinephrine has different effects throughout the body (**Figure 9.16**). In the heart, it stimulates heart muscle cells so that the heart beats faster. In the liver, epinephrine promotes the release of glucose into the bloodstream so that an individual

Figure 9.15   Signal transduction pathway involving diacylglycerol (DAG) and inositol trisphosphate (IP_3).

Growth factors such as EGF cause a rapid increase in the expression of many genes in mammals, perhaps as many as 100. As we will discuss in Chapter 14, growth factor signaling pathways are often involved in cancer. Mutations that cause proteins in these pathways to become hyperactive result in cells that divide uncontrollably. These genes encode proteins that are necessary for a cell to divide, including proteins called cyclins, whose function is described next.

The Cell Cycle Is Controlled by Checkpoint Proteins

The progression through the cell cycle is a highly regulated process that ensures that the nuclear genome is intact and that the conditions are appropriate for a cell to divide. Proteins called **cyclins** and **cyclin-dependent kinases (cdks)** are responsible for advancing a cell through the phases of the cell cycle. Cyclins are so named because their amount is cyclical, fluctuating with the cell cycle. To be active, the kinases controlling the cell cycle must bind to (are dependent on) a cyclin. The number of cyclins and cdks varies from species to species.

Figure 9.21 gives a simplified description of how cyclins and cdks work together to advance a cell through G_1 and mitosis. During G_1, the amount of a particular cyclin termed G_1 cyclin increases. The G_1 cyclin binds to cdk to form an activated G_1 cyclin/cdk complex. This complex phosphorylates proteins that are needed to advance the cell to the next stage in the cell cycle. For example, certain proteins involved with DNA synthesis are phosphorylated and activated, thereby allowing the cell to carry on events in S phase. When the cell passes into the S phase, G_1 cyclin is degraded. Similar events advance the cell through other stages of the cell cycle. A different cyclin, called mitotic cyclin, accumulates late in G_2. It binds to cdk to form an activated mitotic cyclin/cdk complex. This complex phosphorylates proteins that are needed to advance into M phase.

Three critical regulatory points called **checkpoints** are found in the cell cycle of eukaryotic cells (Figure 9.21). At these checkpoints, a variety of proteins, referred to as checkpoint proteins, act as sensors to determine if a cell is in the proper condition to divide. The G_1 checkpoint, also called the restriction point, determines if conditions are favorable for cell division. In addition, G_1-checkpoint proteins can sense if the DNA has incurred damage. If so, these checkpoint proteins will prevent the formation of active cyclin/cdk complexes, and thereby stop the progression of the cell cycle. A second checkpoint exists in G_2. This checkpoint also checks the DNA for damage and ensures that all of the DNA has been replicated. In addition, the G_2 checkpoint monitors the levels of proteins that are needed to progress through M phase. A third checkpoint, called the metaphase checkpoint, senses the integrity of the spindle apparatus. As we will see in Chapter 15, the spindle apparatus is involved in chromosome sorting. Metaphase is a step in mitosis during which all of the chromosomes should be attached to the spindle apparatus. If a chromosome is not correctly attached, the metaphase checkpoint will stop the cell cycle. This checkpoint prevents cells from incorrectly sorting their chromosomes during division.

Checkpoint proteins delay the cell cycle until problems are fixed or even prevent cell division when problems cannot be fixed. A primary aim of checkpoint proteins is to prevent the division of a cell that may have incurred DNA damage or harbors abnormalities in chromosome number. When the functions of checkpoint genes are lost due to mutation, cell division may not be directly accelerated. However, as discussed in Chapter 14, the loss of checkpoint protein function increases the likelihood that undesirable genetic changes will occur that can cause mutation and cancerous growth.

Figure 9.21 **Checkpoints in the cell cycle.** This is a general diagram of the eukaryotic cell cycle. Progression through the cell cycle requires the formation of activated cyclin/cdk complexes. There are different types of cyclin proteins, which are typically degraded after the cell has progressed to the next phase. The formation of activated cyclin/cdk complexes is regulated by checkpoint proteins.

FEATURE INVESTIGATION

Masui and Markert's Study of Oocyte Maturation Led to the Identification of Cyclin and Cyclin-Dependent Kinase

During the 1960s, researchers were intensely searching for the factors that promote cell division. In 1971, Yoshio Masui and Clement Markert developed a way to test whether a substance causes a cell to progress from one phase of the cell cycle to the next. They chose to study frog oocytes—cells produced by female frogs that develop or mature into egg cells. At the time of their work, researchers had already determined that frog oocytes naturally become dormant in the G_2 stage of the cell cycle for up to eight months (**Figure 9.22**). During mating season, female frogs produce a hormone called progesterone. After progesterone binds to receptors in dormant egg cells, they progress from G_2 to the beginning of M phase, where the chromosomes condense and become visible under the microscope. This phenomenon is called maturation. When a sperm fertilizes the egg, M phase is completed, and the zygote continues to undergo cellular divisions.

Because progesterone is a signaling molecule, Masui and Markert speculated that this hormone affects the functions and/or amounts of proteins that trigger the oocyte to undergo maturation. To test this hypothesis, they developed the procedure described in **Figure 9.23**, using the oocytes of the leopard frog (*Rana pipiens*). They began by exposing oocytes to progesterone *in vitro* and then incubating these oocytes for 2 hours or 12 hours. As a control, they also used oocytes that had not been exposed to progesterone. These three types of cells were called the donor oocytes.

Oocyte dormant in the G_2 phase.

Oocyte advances to the beginning of M phase where chromosomes condense (maturation).

Figure 9.22 Oocyte maturation in certain species of frogs.

Next, they used a micropipette to transfer a small amount of cytosol from the three types of donor oocytes to recipient oocytes that had not been exposed to progesterone. As seen in the data, the recipient oocytes that had been injected with cytosol from the control donor oocytes or from oocytes that had been incubated with progesterone for only 2 hours did not progress to M phase. However, cytosol from donor oocytes that had been incubated with progesterone for 12 hours caused the recipient oocytes to advance to M phase. Masui and Markert concluded that a cytosolic factor, which required more than 2 hours to be synthesized after progesterone treatment, had been

Figure 9.23 The experimental approach used by Masui and Markert to identify cyclin and cyclin-dependent kinase (cdk).

HYPOTHESIS Progesterone induces the synthesis of a factor(s) that advances frog oocytes through the cell cycle from G_2 to M phase.

STARTING MATERIALS Oocytes from *Rana pipiens*.

Experimental level | Conceptual level

1 Expose oocytes to progesterone. Incubate for 2 or 12 hours after progesterone exposure. As a control, also use oocytes that have not been exposed to progesterone. All 3 are called donor oocytes.

Progesterone Progesterone No progesterone (control)
02:00 12:00
Donor oocytes

Progesterone — Activates intracellular receptor

Donor oocyte

Factors are made that advance oocyte to M phase. One such factor is called maturation promoting factor (MPF).

2 Using a micropipette, transfer some cytosol from the three types of donor oocytes to recipient oocytes that have not been exposed to progesterone.

Micropipette tip
Donor cytosol
Donor oocytes Recipient oocyte

Recipient oocyte
MPF

Recipient oocyte received MPF from donor oocyte if donor oocyte was incubated for 12 hours with progesterone.

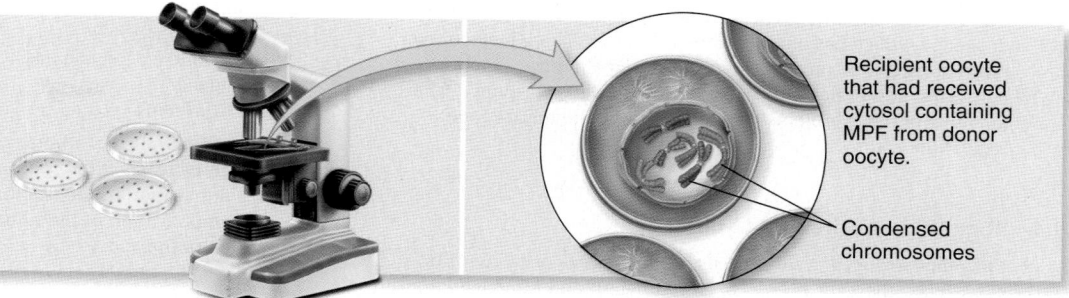

3 Incubate for several hours and observe the recipient oocytes under the microscope to determine if the recipient oocytes advance to M phase. Advancement to M phase can be determined by the condensation of the chromosomes.

Recipient oocyte that had received cytosol containing MPF from donor oocyte.

Condensed chromosomes

4 **THE DATA**

Donor oocytes	Recipient oocytes proceeded to M phase?
Control, no progesterone exposure	No
Progesterone exposure, incubation for 2 hours	No
Progesterone exposure, incubation for 12 hours	Yes

transferred to the recipient oocytes and induced maturation. The factor that caused the oocytes to progress (or mature) from G_2 to M phase was originally called the **maturation promoting factor (MPF)**.

After MPF was discovered in frogs, it was found in all eukaryotic species that researchers studied. MPF is important in the division of all types of cells, not just oocytes. It took another 17 years before Manfred Lohka, Marianne Hayes, and James Maller were able to purify the components that make up MPF. This was a difficult undertaking because these components are found in very small amounts in the cytosol, and they are easily degraded during purification procedures. We now know that MPF is a complex made of mitotic cyclin and cyclin-dependent kinase (cdk), as described in Figure 9.21.

Chapter Summary

9.1 General Features of Cell Communication

- A signal is an agent that can influence the properties of cells. Cell signaling is needed so that cells can sense environmental changes and communicate with each other.

- When a cell responds to an environmental signal, it has become adapted to its environment. (Figure 9.1)

- Cell-to-cell communication also allows cells to adapt, as when plants grow toward light. (Figure 9.2)

- Cell-to-cell communication can vary in the mechanism and distance that a signal travels. Signals are relayed between cells in five common ways: direct intercellular, contact-dependent, autocrine, paracrine, and endocrine signaling. (Figure 9.3)

- Cell signaling is usually a three-stage process involving receptor activation, signal transduction, and a cellular response. (Figure 9.4)

9.2 Cellular Receptors

- A signaling molecule, also called a ligand, binds to a receptor with an affinity that is measured as a K_d value. The binding of a ligand to a receptor is usually very specific and alters the conformation of the receptor. (Figure 9.5)

- Most receptors involved in cell signaling are found on the cell surface.

- Enzyme-linked receptors have some type of catalytic function. Many of them are protein kinases that can phosphorylate proteins. (Figure 9.6)

- G-protein-coupled receptors (GPCRs) interact with G proteins to initiate a cellular response. (Figure 9.7)

- Some receptors are ligand-gated ion channels that allow the flow of ions across cellular membranes (Figure 9.8)

- Some receptors, such as the estrogen receptor and auxin receptor, are intracellular receptors. (Figure 9.9)

9.3 Signal Transduction and the Cellular Response

- Two-component regulatory systems involve a sensor kinase and a response regulator. An example is NarQ and NarL, which regulate nitrate and nitrite utilization in bacteria. (Figure 9.10)

- Second messengers, such as cAMP, play a key role in signal transduction pathways. These pathways are reversible once the signal is degraded. (Figure 9.11)

- Second messenger pathways amplify the signal and occur with great speed. (Figures 9.12, 9.13)

- Changes in Ca^{2+} concentration play a key role in many signal transduction pathways. (Figure 9.14)

- Inositol trisphosphate (IP_3) and diacylglycerol (DAG) are other examples of second messengers involved in signal transduction. (Figure 9.15)

- Hormones such as epinephrine exert different effects throughout the body. (Figure 9.16)

- In heart muscle cells, epinephrine increases the rate of contraction due to signaling pathways that phosphorylate troponin and phospholamban. (Figure 9.17)

- The way in which any particular cell responds to a signaling molecule depends on the types of proteins that it makes. These include the types of receptors, proteins involved in signaling transduction pathways, and proteins that carry out the cellular response. The amounts of these proteins are controlled by differential gene regulation.

9.4   Regulation of the Cell Cycle

- The eukaryotic cell cycle consists of four phases called G_1 (first gap), S (synthesis of DNA), G_2 (second gap), and M phase (mitosis and cytokinesis). The G_1, S, and G_2 phases are collectively known as interphase. (Figure 9.18)
- Once a cell passes a restriction point in G_1, it is destined to duplicate its DNA and to divide. During S phase, chromosomes are replicated and form pairs of sister chromatids. (Figure 9.19)
- Signaling pathways influence whether or not a cell will divide. An example is the pathway that is stimulated by epidermal growth factor. (Figure 9.20)
- An interaction between cyclin and cyclin-dependent kinase is necessary for cells to progress through the cell cycle. Checkpoint proteins sense the environmental conditions and the integrity of the genome, and they control whether or not the cell progresses through the cell cycle. (Figure 9.21)
- Masui and Markert studied the maturation of frog oocytes to identify a substance that was necessary for oocytes to progress through the cell cycle. This substance was later identified as a complex of mitotic cyclin and cyclin-dependent kinase and called maturation promoting factor (MPF). (Figures 9.22, 9.23)

TEST YOURSELF

1. The ability of a cell to respond to changes in its environment is termed
   a. signaling.
   d. cell communication.
   b. adaptation.
   e. stimulation.
   c. irritability.

2. When a cell secretes a signaling molecule that binds to receptors on neighboring cells as well as the same cell, this is called _____ signaling.
   a. direct intercellular
   d. paracrine
   b. contact-dependent
   e. endocrine
   c. autocrine

3. Which of the following does not describe a typical cellular response to signaling molecules?
   a. activation of enzymes within the cell
   b. change in the function of structural proteins, which determine cell shape
   c. alteration of levels of certain proteins in the cell by changing the level of gene expression
   d. change in a gene sequence that encodes a particular protein
   e. all of the above are examples of cellular responses

4. A cell's ability to respond to a particular signal depends on
   a. whether or not the cell possesses the appropriate receptor for the signal.
   b. the chemical nature of the signal molecule.
   c. whether or not the signal molecule is water soluble.
   d. the concentration of ATP in the cell.
   e. none of the above.

5. _____ bind to receptors inside cells.
   a. Steroid hormones
   d. All of the above
   b. Ions
   e. Both a and c
   c. Auxins

6. Small molecules, such as cAMP, that relay signals within the cell are called
   a. secondary metabolites.
   d. second messengers.
   b. ligands.
   e. transcription factors.
   c. G proteins.

7. The benefit of second messengers in signal transduction pathways is
   a. an increase in the speed of a cellular response.
   b. duplication of the ligands in the system.
   c. amplification of the signal.
   d. all of the above.
   e. a and c only.

8. All cells of a multicellular organism may not respond in the same way to a particular signal that binds to a cell surface receptor. The difference in response may be due to
   a. the concentration of the signal molecule in the cytoplasm.
   b. the functional differences of the receptors.
   c. the structural differences that may occur in ligands.
   d. mutations that occur during development.
   e. none of the above.

9. Whether or not a cell divides depends on
   a. nutrient availability.
   b. environmental conditions.
   c. the presence of signal molecules that regulate cell division.
   d. cellular proteins that regulate cell division.
   e. all of the above.

10. Checkpoints during the cell cycle are important because they
    a. allow the organelle activity to catch up to cellular demands.
    b. ensure the integrity of the cell's DNA.
    c. allow the cell to generate sufficient ATP for cellular division.
    d. are the only time DNA replication can occur.
    e. all of the above.

CONCEPTUAL QUESTIONS

1. What are the two general reasons that cells need to communicate?
2. Explain the three stages of cell signaling.
3. What are protein kinases?

EXPERIMENTAL QUESTIONS

1. At the time of Masui and Markert's study shown in Figure 9.23, what was known about the effects of progesterone on oocytes?
2. What hypothesis did Masui and Markert propose to explain the function of progesterone? Explain the procedure used to test the hypothesis.
3. How did the researchers explain the difference between the results using 2-hour-exposed donor oocytes versus 12-hour-exposed donor oocytes?

COLLABORATIVE QUESTIONS

1. Discuss several different types of cell-to-cell communication.
2. Discuss how differential gene regulation enables various cell types to respond differently to the same signaling molecule.

www.brookerbiology.com
This website includes answers to the Biological Inquiry questions found in the figure legends and all end-of-chapter questions.

10

MULTICELLULARITY

An oak tree. The wood of an oak tree is largely composed of material that is outside of cells. This material connects cells to each other and provides structural support for the tree.

In unicellular species, an entire organism consists of a single cell. Prokaryotes, which include bacteria and archaea, are usually unicellular. Likewise, most protists are unicellular. As discussed in Chapter 22, multicellular organisms came into being approximately 1 billion years ago. In a species that is **multicellular**, a single organism is composed of more than one cell. Some species of protists are multicellular, as are most species of fungi. In this chapter we will focus on plants and animals, which are always multicellular organisms.

The main benefit of multicellularity arises from the division of labor between different cells of the body. For example, the intestinal cells of animals and the root cells of plants have become specialized for nutrient uptake. Other cells in a multicellular organism perform different roles, such as reproduction. In animals, most of the cells of the body are devoted to the growth, development, and survival of the organism, while specialized cells called gametes function in reproduction.

Multicellular species usually have much larger genomes than unicellular species. The increase in genome size is associated with an increase in proteome size; multicellular organisms produce a larger array of proteins than do unicellular species. The additional proteins play a role in three general phenomena. First, in a multicellular organism, cell communication is vital for the proper organization and functioning of cells. Many more proteins involved in cell signaling are made in multicellular species. Second, both the arrangement of cells within the body and the attachment of cells to each other require a greater variety of proteins in multicellular species than in unicellular species. Finally, additional proteins play a role in cell specialization because proteins that are needed for the structure and

function of one cell type may not be needed in a different cell type, and vice versa.

In this chapter, we consider characteristics specific to the cell biology of multicellular organisms. We will begin by exploring the material that is produced by animal and plant cells and then secreted into the surrounding medium to form an extracellular matrix. This matrix plays many important roles in the organization and functioning of cells within multicellular organisms. We will then turn our attention to cell junctions, specialized structures that enable cells to make physical contact with one another. Cells within multicellular organisms form junctions that help to create a cohesive and well-organized body. Finally, we examine the organization and function of tissues, groups of cells that have a similar structure and function. In this chapter, we will survey the general features of tissues from a cellular perspective. Units VI and VII will explore the characteristics of plant and animal tissues in greater detail.

10.1 Extracellular Matrix

Organisms are not composed solely of cells. A large portion of an animal or plant consists of a network of material that is secreted from cells and forms a complex meshwork outside of cells called the **extracellular matrix (ECM)**. The ECM is a major component of certain parts of animals and plants. For example, bones and cartilage in animals and the woody portions of plants are composed largely of ECM. In fact, while the cells within wood eventually die, the ECM that they have produced provides a rigid structure that can support the plant for years or even centuries.

Over the past few decades, cell biologists have examined the synthesis, composition, and function of the ECM in animals and plants. In this section, we will begin by examining the structure and role of the ECM in animal cells, focusing on the function of the major ECM protein and carbohydrate macromolecules. We will then explore the cell wall, the extracellular component of plant cells, and consider how it differs in structure and function from the ECM of animal cells.

The Extracellular Matrix in Animals Supports and Organizes Cells and Plays a Role in Cell Signaling

Unlike the cells of bacteria, fungi, and plants, the cells of animals are not surrounded by a rigid cell wall that provides structure and support. However, animal cells secrete materials that form an extracellular matrix that serves a similar purpose. Certain animal cells are completely embedded within an extensive ECM, while other cells may only adhere to the ECM on one side. **Figure 10.1** illustrates the general features of the ECM and its relationship to cells. The major macromolecules of the ECM are proteins and polysaccharides. The most abundant proteins are those that form large fibers; the polysaccharides give the ECM a gel-like character.

As we will see, the ECM found in animals performs many important roles, including strength, structural support, organization, and cell signaling.

- **Strength:** The ECM is the "tough stuff" of animals' bodies. In the skin of mammals, the strength of the ECM prevents tearing. The ECM found in cartilage resists compression and provides protection to the joints. Similarly, the ECM protects the soft parts of the body, such as the internal organs.

- **Structural support:** The skeletons of many animals are composed primarily of ECM. Skeletons not only provide structural support but also facilitate movement via the functioning of the attached muscles.

- **Organization:** The attachment of cells to the ECM plays a key role in the proper arrangement of cells throughout the body. In addition, the ECM binds many body parts together, such as tendons to bones.

- **Cell signaling:** A newly discovered role of the ECM is cell signaling. One way that cells in multicellular organisms sense their environment is via changes in the ECM.

Let's now consider the synthesis and structure of ECM components found in animals.

Adhesive and Structural Proteins Are Major Components of the ECM of Animals

The idea that fibers are important components of living organisms has a long history. In the Middle Ages, living organisms were thought to be composed of fibers rather than cells. This belief was eventually debunked by the cell theory, which is described in Chapter 4. In the 1850s, Rudolf Virchow, who proposed that all cells come from pre-existing cells, also suggested that all extracellular materials are made and secreted by cells. Around the same time, biologists realized that gelatin and glue, which are produced by the boiling of animal tissues, must contain a common fibrous substance. This substance was named **collagen**, from the Greek for "glue producing." Since that time, the advent of experimental techniques in the areas of chemistry, microscopy, and biophysics has enabled scientists to probe the structure of the ECM. We now understand that the ECM contains a mixture of several different components, including proteins such as collagen, that form fibers.

The proteins that are found in the ECM can be grouped into adhesive proteins such as fibronectin and laminin, and structural proteins, such as collagen and elastin (**Table 10.1**). Fibronectin and laminin have multiple binding sites that bind to other components in the ECM such as protein fibers and carbohydrates. These same proteins also have binding sites for receptors on the surfaces of cells. Therefore, adhesive proteins are so named because they adhere cells to the ECM.

Structural proteins such as collagen and elastin form large fibers that give the ECM its strength and elasticity. As described in Chapter 6, proteins that are secreted from eukaryotic cells

Some cells are attached to the ECM on one side.

Some cells are embedded within the ECM.

ECM

Protein fiber

Carbohydrate polymer

Protein fibers give strength and elasticity to the ECM.

Carbohydrates help the ECM resist compression.

Figure 10.1   Animal cells and the extracellular matrix (ECM).

Table 10.1	Proteins in the ECM of Animals	
General type	**Example**	**Function**
Adhesive	Fibronectin	Connects cells to the ECM and helps to organize components in the ECM.
	Laminin	Connects cells to the ECM and helps to organize components in the basal lamina, a specialized ECM found next to epithelial cells (described in Section 10.3).
Structural	Collagen	Forms large fibers and interconnected fibrous networks in the ECM. Provides tensile strength.
	Elastin	Forms elastic fibers in the ECM that can stretch and recoil.

Figure 10.2   Formation of collagen fibers.

1   Procollagen polypeptides are synthesized into the ER lumen, where they assemble into a triple helix.

Procollagen polypeptide (α chain)

ER lumen

Procollagen triple helix

Extension sequences

Collagen molecule

2   Procollagen is secreted from the cell, and the extension sequences are removed. The protein is now called collagen.

3   The removal of extension sequences allows collagen to assemble into fibrils.

Collagen fibril

4   Collagen fibrils assemble into larger collagen fibers.

Collagen fiber

Elastic fiber

In the absence of a stretching force, the elastin proteins are in a compact conformation.

Single elastin protein

Force

Force

Crosslink

When subjected to a force, the elastin proteins elongate but remain attached to each other via crosslinks.

Figure 10.3   Structure and function of elastic fibers.

Biological inquiry: Let's suppose you started with an unstretched elastic fiber and treated it with a chemical that breaks the cross-links between adjacent elastin proteins. What would happen when the fiber is stretched?

are first directed to the endoplasmic reticulum (ER), then to the Golgi apparatus, and subsequently are secreted from the cell via vesicles that fuse with the plasma membrane. **Figure 10.2** depicts the synthesis and assembly of collagen. Individual procollagen polypeptides are first synthesized into the lumen (inside) of the ER. Three procollagen polypeptides then associate with each other to form a procollagen triple helix. The amino acid sequences at both ends of the polypeptides, termed extension sequences, serve to promote the formation of procollagen and prevent the formation of a much larger fiber. After procollagen is secreted out of the cell, extracellular enzymes remove the extension sequences. Once this occurs, the protein, now called collagen, can form larger structures. Collagen proteins assemble in a staggered way to form relatively thin collagen fibrils, which then align and create large collagen fibers.

A key function of collagen is to provide tensile strength, which is a measure of how much stretching force a material can bear without tearing apart. Collagen provides high tensile strength to many parts of the animal body. Collagen is the main protein found in bones, cartilage, tendons, and skin and is also found lining blood vessels and internal organs. In mammals, collagen consists of more than 25% of the total protein mass, much more than any other protein. Approximately 75% of the protein in

mammalian skin is composed of collagen. Leather is largely a pickled and tanned form of collagen.

In addition to tensile strength, elasticity is needed in regions of the body such as the lungs and blood vessels, which regularly expand and return to their original shape. In these places, the ECM contains an abundance of elastic fibers composed primarily of a protein called **elastin** (**Figure 10.3**). Elastin proteins form many covalent cross links to create a fiber. In the absence of a stretching force, each protein tends to adopt a compact conformation. When subjected to a stretching force, the compact proteins become more linear, with the covalent cross links holding the fiber together. When the stretching force has ended, the proteins naturally return to their compact conformation. In this way, elastic fibers behave much like a rubber band, stretching under tension and snapping back when the tension is released.

GENOMES & PROTEOMES

Collagens Are a Family of Proteins That Give Animal Cells a Variety of ECM Properties

As we have seen, proteins are important constituents of the ECM of animals. By analyzing genomes and the biochemical composition of cells, researchers have determined that many different types of collagen fibers are made. These are designated type I, type II, and so on. At least 27 different types of collagens have been identified in humans. Therefore, the human genome, as well as the genomes of other animals, has many different genes that encode collagen polypeptides.

Collagens have a common structure, in which three collagen polypeptides wind around each other to form a triple helix (see Figure 10.2). Each polypeptide is called an α chain. In some collagens, all three α chains are identical, while in others the α chains may be encoded by different collagen genes. Nevertheless, the triple helix structure remains common to all collagen proteins.

Each of the many different types of collagen polypeptides has a similar yet distinctive amino acid sequence that affects the structure and function of collagen fibers. For example, within the triple helix, the amino acid sequence of an α chain may cause collagen proteins to bind to each other very tightly, thereby creating a relatively stiff fiber. Such collagen fibers are found in bone and cartilage. In addition, amino acid side chains in the α chains influence the interactions between collagen proteins within a fiber. For example, the amino acid sequences of certain α chains may promote a looser interaction that produces a more bendable or thin fiber. More flexible collagen fibers support the lining of your lungs and intestines. In addition, domains within the collagen polypeptide may affect the spatial arrangement of collagen proteins. The collagen shown earlier in Figure 10.2 forms fibers in which the collagen proteins align themselves in parallel arrays. Not all collagen proteins form long fibers. For example, type IV collagen proteins interact with each other in a meshwork pattern. This meshwork acts as a filtration unit around capillaries.

Differential gene regulation controls which types of collagens are made throughout the body and in what amounts. Of the 27 types of collagens, Table 10.2 considers types I to IV, each of which varies with regard to where it is primarily synthesized and its structure and function. Collagen genes are regulated so that collagen is made in the correct sites of the body. In skin cells, for example, the genes that encode the collagen polypeptides that make up collagen types I, III, and IV are turned on, while the synthesis of type II collagen is minimal.

The regulation of collagen synthesis has received a great deal of attention due to the phenomenon of wrinkling. As we age, the amount of collagen that is synthesized in our skin significantly decreases. The underlying network of collagen fibers, which provides scaffolding for the surface of our skin, loosens and unravels. This is one of the factors that cause the skin of older people to sink, sag, and form wrinkles. Various therapeu-

tic and cosmetic agents have been developed to prevent or reverse the appearance of wrinkles, most with limited benefits. One approach is collagen injections, in which small amounts of collagen (from cows) are injected into areas where the body's collagen has weakened, filling the depressions to the level of the surrounding skin. Because collagen is naturally broken down in the skin, the injections are not permanent and last only about three to six months.

Animal Cells Also Secrete Polysaccharides into the ECM

Polysaccharides are the second major component of the extracellular matrix of animals. Among vertebrates, the most abundant types of polysaccharides in the ECM are **glycosaminoglycans (GAGs)**. These molecules are long, unbranched polysaccharides containing a repeating disaccharide unit (**Figure 10.4a**). GAGs are highly negatively charged molecules that tend to attract positively charged ions and water. The majority of GAGs in the ECM are linked to core proteins, forming **proteoglycans** (**Figure 10.4b**).

Providing resistance to compression is the primary function of GAGs and proteoglycans. Once secreted from cells, these macromolecules form a gel-like component in the ECM. Due to its high water content, the ECM is difficult to compress and thereby serves to protect cells. GAGs and proteoglycans are abundantly found in regions of the body that are subjected to harsh mechanical forces, such as the joints of the human body.

Table 10.2	Examples of Collagen Types	
Type	**Sites of synthesis***	**Structure and function**
I	Tendons, ligaments, bones, and skin	Forms a relatively rigid and thick fiber. Very abundant, provides most of the tensile strength to the ECM.
II	Cartilage, disks between vertebrae	Forms a fairly thick and rigid fiber but is more flexible than type I. Permits smooth movements of joints.
III	Arteries, skin, internal organs, and around muscles	Forms thin fibers, often arranged in a netlike pattern. Allows for greater elasticity in tissues.
IV	Skin, intestine, and kidneys; also found around capillaries	Does not form long fibers. Instead, the proteins are arranged in a meshwork pattern that provides organization and support to cell layers. Functions as a filter around capillaries.

*The sites of synthesis denote where a large amount of the collagen type is made.

(a) Structure of chondroitin sulfate, a glycosaminoglycan

(b) General structure of a proteoglycan

Figure 10.4 Structures of glycosaminoglycans and proteoglycans. (a) Glycosaminoglycans (GAGs) are composed of repeating disaccharide units. The length can range from several dozen to 25,000 disaccharide units. The GAG shown here is chondroitin sulfate, which is commonly found in cartilage. (b) Proteoglycans are composed of a long, linear core protein with many GAGs attached. Each GAG is typically 80 disaccharide units long.

Two examples of GAGs are chondroitin sulfate, which is a major component of cartilage, and hyaluronic acid, which is found in the skin, eyes, and joint fluid.

Among many invertebrates, an important ECM component is **chitin**, a nitrogen-containing polysaccharide. Chitin forms the hard protective outer covering (called an exoskeleton) of insects, such as crickets and grasshoppers, and shellfish, such as lobsters and shrimp. In fact, the chitin exoskeleton is so rigid that as these animals grow, they must periodically shed this outer layer and secrete a new, larger one—a process called molting (look ahead to Figure 32.11).

The Cell Wall of Plants Provides Strength and Resistance to Compression

Let's now turn our attention to the extracellular matrix of plants. Plants cells are surrounded by an ECM called the **cell wall**, a protective layer that forms outside of the plasma membrane of the plant cell. Like animal cells, the cells of plants are surrounded by material that provides tensile strength and resistance to compression. The cell walls of plants, however, are usually thicker, stronger, and more rigid than the ECM found in animals. Plant cell walls provide rigidity for mechanical support, the maintenance of cell shape, and the direction of cell growth. As we learned in Chapter 5, the cell wall also prevents expansion when water enters the cell. During the evolution of plants, these structural features of the plant cell wall may have been key factors that caused plants to thrive in a sedentary lifestyle.

The cell walls of plants are composed of a primary cell wall and a secondary cell wall, so named according to the timing of their synthesis (**Figure 10.5**). The **primary cell wall** is made before the secondary cell wall. During cell division, the primary cell wall develops between two newly made daughter cells. It is usually very flexible and allows new cells to increase in size. The main macromolecule of the primary cell wall is **cellulose**, a polymer made of repeating molecules of glucose attached end to end. These glucose polymers associate with each other via hydrogen bonding to form microfibrils that provide great tensile strength (**Figure 10.6**).

Cellulose was discovered in 1838 by the French chemist Anselme Payen, who was the first scientist to try to separate wood into its component parts. After treating different types of wood with nitric acid, Payen obtained a fibrous substance that was also found in cotton and other plants. His chemical analysis revealed that the fibers were made of the carbohydrate glucose. Payen called this substance cellulose, which means consisting of cells. Cellulose is probably the single most abundant organic molecule on Earth. Wood consists mostly of cellulose, and cotton and paper (including the page you are reading now) are almost pure cellulose.

In addition to cellulose, other components found in the primary cell wall include hemicellulose, glycans, and pectins (see Figure 10.5). Hemicellulose is another linear carbohydrate with a structure similar to that of cellulose, but it contains sugars other than glucose in its structure and usually forms thinner microfibrils. Glycans, carbohydrates with branching structures, are also important in cell wall structure. The cross-linking glycans bind to cellulose and provide organization to the cellulose microfibrils. Pectins, which are highly negatively charged, attract water and have a gel-like character that provides the cell wall with the ability to resist compression. Besides carbohydrates, the primary cell wall also contains small amounts of protein. Some of these proteins may increase the strength of the cell wall, while others are enzymes involved in the synthesis and organization of the carbohydrate polymers. The mechanism of cellulose synthesis via extracellular enzymes is described in Chapter 30.

The **secondary cell wall** is synthesized and deposited between the plasma membrane and the primary cell wall after a plant cell matures and has stopped increasing in size. It is made in layers by the successive deposition of cellulose microfibrils and other components. While the primary wall structure is relatively similar in nearly all cell types and species, the structure of the secondary cell wall is more variable. The secondary cell wall

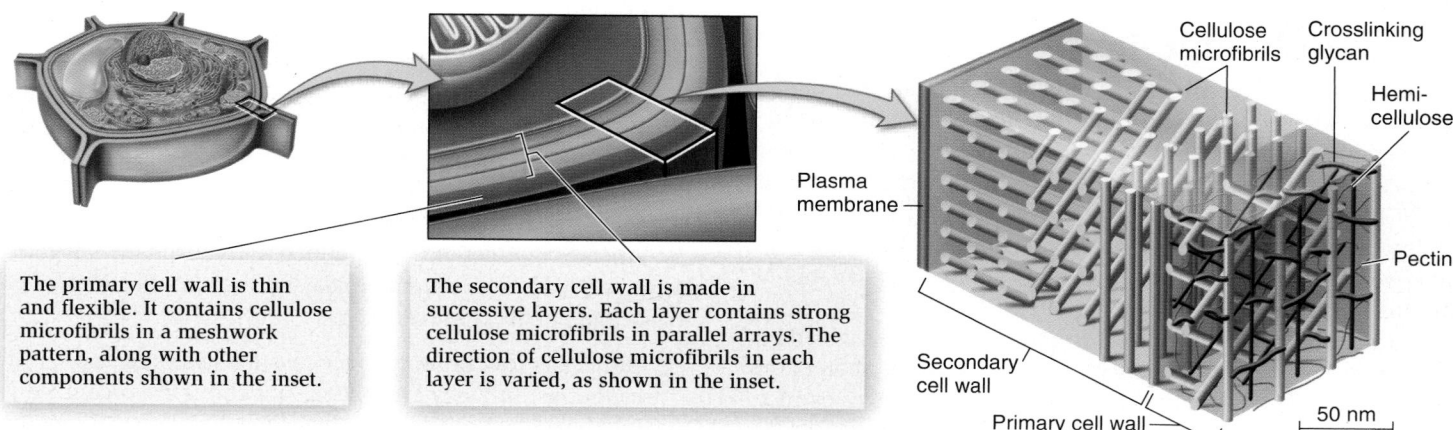

The primary cell wall is thin and flexible. It contains cellulose microfibrils in a meshwork pattern, along with other components shown in the inset.

The secondary cell wall is made in successive layers. Each layer contains strong cellulose microfibrils in parallel arrays. The direction of cellulose microfibrils in each layer is varied, as shown in the inset.

Figure 10.5 **Structure of the plant cell wall.** The primary cell wall is relatively thin and flexible. The secondary cell wall, which is produced only by certain plant cells, is made after the primary cell wall and is synthesized in successive layers.

Biological inquiry: With regard to cell growth, what would happen if the secondary cell wall was made too soon?

Many polymers associate with each other to form a microfibril.

Microfibril

20 nm

Figure 10.6 **Structure of cellulose.** Cellulose is made of repeating glucose units linked end to end that hydrogen-bond to each other to form microfibrils (SEM).

often contains components in addition to those found in the primary cell wall. For example, phenolic compounds called lignins are very hard and impart considerable strength to the secondary wall structure. Lignin, a type of secondary metabolite described in Chapter 7, is found in the woody parts of plants.

10.2 Cell Junctions

Thus far, we have learned that the cells of animals and plants create an extracellular matrix that provides strength, support, and organization. For an organism to become a multicellular unit, cells within the body must be linked to each other and to the ECM. In animals and plants, this is accomplished by specialized structures called **cell junctions** (**Table 10.3**). In this section, we will examine different types of cell junctions in animal and plant cells.

Animal cells, which lack the structural support provided by the cell wall, have a more varied group of junctions than plant cells. In animal cells, junctions called anchoring junctions play a role in anchoring cells to each other or to the extracellular matrix. In other words, they hold cells in their proper place in the body. Other junctions, termed tight junctions, seal cells together to prevent small molecules from leaking through one

Table 10.3	Common Types of Cell Junctions
Type	**Description**
Animals	
Anchoring junctions	Cell junctions that hold adjacent cells together or bond cells to the ECM. Anchoring junctions are mechanically strong.
Tight junctions	Junctions between adjacent cells that prevent the leakage of material between cell layers.
Gap junctions	Channels that permit the direct exchange of ions and small molecules between the cytosol of adjacent cells.
Plants	
Middle lamella	A carbohydrate layer that cements together the cell walls of adjacent cells.
Plasmodesmata	Open passageways between the cell walls of adjacent cells that permit the direct diffusion of ions and molecules between cells.

cell layer into another. Still another type of junction known as a gap junction allows cells to communicate directly with each other. In this section, we will examine all three of these types of junctions.

In plants, cellular organization is somewhat different because plants cells are surrounded by a rigid cell wall. As you will learn, plants cells are connected to each other by a component called the middle lamella, which cements their cell walls together. They also have junctions termed plasmodesmata that allow adjacent cells to communicate with each other.

Anchoring Junctions Link Animal Cells to Each Other and to the ECM

The advent of electron microscopy allowed researchers to explore the types of junctions that occur between cells and within the extracellular matrix. In the 1960s, Marilyn Farquhar, George Palade, and colleagues conducted several studies showing that various types of cellular junctions connect cells to each other. Over the past few decades, researchers have begun to unravel the functions and molecular structures of these junctions, called **anchoring junctions**, which attach cells to each other and to the extracellular matrix. Anchoring junctions rely on the functioning of membrane proteins called **cell adhesion molecules (CAMs)**. Two types of CAMs are cadherin and integrin.

Anchoring junctions are grouped into four main categories, according to their functional roles and their connections to cellular components (**Figure 10.7**)

1. **Adherens junctions** connect cells to each other via cadherins. In many cases, these junctions are organized into bands around cells. In the cytosol, adherens junctions bind to cytoskeletal filaments called actin filaments.
2. **Desmosomes** also connect cells to each other via cadherins. They are spotlike points of intercellular contact that rivet

In adherens junctions, cadherins connect cells to each other and to the actin filaments.

In desmosomes, cadherins connect cells to each other and to intermediate filaments.

Ring of actin filaments

Intermediate filaments

Cadherins

Linker proteins

Integrin

Actin filament

ECM

In hemidesmosomes, integrins connect the ECM to intermediate filaments.

In focal adhesions, integrins connect the ECM to actin filaments.

Figure 10.7   Types of anchoring junctions.

Cadherin dimer

Cadherins link cells to each other.

Actin

Ca^{2+}

Plasma membrane

Linker protein

(a) Cadherins

ECM

Cytosol

Linker protein

Actin

Integrins link cells to the extracellular matrix.

Integrin

Collagen fiber

Fibronectin

Plasma membrane

ECM

(b) Integrins

Figure 10.8   Types of cell adhesion molecules (CAMs). **(a)** A cadherin in one cell binds to a cadherin of an identical type in an adjacent cell. The binding requires Ca^{2+}. In the cytosol, cadherins bind to the actin or intermediate filaments of the cytoskeleton. **(b)** Integrins link cells to the extracellular matrix and form intracellular connections to actin or intermediate filaments.

cells together. Desmosomes are connected to cytoskeletal filaments called intermediate filaments.

3. **Focal adhesions** connect cells to the extracellular matrix via integrins. In the cytosol, focal adhesions bind to actin filaments.

4. **Hemidesmosomes** also connect cells to the extracellular matrix via integrins. Like desmosomes, they interact with intermediate filaments.

Let's now consider the molecular components of anchoring junctions. **Cadherins** are CAMs that create cell-to-cell junctions (**Figure 10.8a**). Each cadherin is a dimer of identical subunits. The extracellular domains of two cadherin dimers, each in adjacent cells, bind to each other to promote cell-to-cell adhesion. This binding requires the presence of calcium ions, which change the conformation of cadherin, so that cadherins in adjacent cells can bind to each other. On the inside of the cell, linker proteins connect cadherins to actin or intermediate filaments

of the cytoskeleton. This promotes a more stable interaction between two cells because their strong cytoskeletons are connected to each other.

Cadherins are the major CAMs in vertebrate species. The genomes of vertebrates contain several different cadherin genes, which encode slightly different cadherin proteins. Having different types of cadherins allows different types of cells to recognize each other. Dimer formation follows a homophilic, or like-to-like, binding mechanism. To understand the concept of homophilic binding, let's consider an example. One cadherin is called E-cadherin and another is N-cadherin. E-cadherin in one cell will bind to E-cadherin but not to N-cadherin in an adjacent cell. Similarly, N-cadherin will bind to N-cadherin but not to E-cadherin in an adjacent cell. By expressing only certain types of cadherins, each cell will bind only to other cells that express the same cadherin types. This phenomenon is important in the proper arrangement of cells throughout the body.

Integrins, a group of cell-surface receptor proteins, are a second type of CAM, one that creates connections between cells and the extracellular matrix. In the example shown in **Figure 10.8b**, an integrin is bound to fibronectin, an ECM protein that binds to other ECM components such as collagen fibers. Like cadherins, integrins also bind to actin or intermediate filaments in the cytosol of the cell, via linker proteins, to promote a strong association between the cytoskeleton of a cell and the extracellular matrix. Thus, integrins have an extracellular domain for the binding of ECM components and an intracellular domain for the binding of cytosolic proteins.

When these CAMs were first discovered, researchers imagined that cadherins and integrins played only a mechanical role in cell biology. In other words, their functions were described as holding cells together or to the ECM. More recently, however, experiments have shown that cadherins and integrins are important in cell communication. When cell-to-cell and cell-to-ECM junctions are formed or broken, this affects signaling pathways within the cell. Similarly, intracellular signaling pathways can affect cadherins and integrins in ways that alter intercellular junctions and the binding of cells to ECM components.

With regard to cell signaling, integrins are particularly interesting because they are capable of both outside-in and inside-out signaling (**Figure 10.9**). Integrins are so named because they integrate changes in the ECM to changes in the cytoskeleton. When the extracellular domain of an integrin binds to components of the ECM, this causes a conformational change

(a) Outside-in signaling

Cytosol — Actin filament
— Linker protein
— Integrin
ECM — Fibronectin
— Collagen

1. The binding of ECM components to integrins transmits a conformational change from outside to inside the cell.

2. A change in the intracellular domain causes it to bind to the actin filament. Note the change in the shape of the actin filament.

(b) Inside-out signaling

1. Intracellular proteins, such as actin, change the conformation of the extracellular domain.

2. The change in the extracellular domain alters its affinity for ECM components. In this case, the affinity is lower, and the ECM components are released.

Figure 10.9 **Cell signaling via integrins.** (a) Outside-in signaling occurs when an integrin binds to a component in the ECM, which transmits a signal to the cytosol, thereby affecting activities inside the cell. (b) Inside-out signaling occurs when the cytosol affects the structure of an integrin and thereby changes the integrin's ability to bind to components in the ECM. In the example shown here, the effect is to lower the affinity for an ECM component, causing it to be released from the integrin.

Biological inquiry: When an animal receives a wound, this causes outside-in signaling. Explain how.

that affects the structure of the intracellular domain, altering its interaction with cytoskeletal proteins. This outside-in signaling regulates cell adhesion (the ability of cells to adhere to the ECM), cell growth, and cell migration. In addition, signals generated inside the cell can alter integrins and affect their ability to bind to components in the ECM, either lowering or increasing their affinity. This phenomenon, termed inside-out signaling, is also important for cell adhesion and migration, and it provides a way for cells to contribute to the organization of the ECM.

Abnormalities in CAMs such as integrins are often associated with the ability of cancer cells to metastasize, that is, to move to other parts of the body. Cell adhesion molecules are critical for keeping cells in their correct locations. When their function becomes abnormal due to cancer-causing mutations, cells lose their proper connections with the ECM and adjacent cells and may spread to other parts of the body. This topic is considered in more detail in Chapter 14.

Tight Junctions Prevent the Leakage of Materials Across Animal Cell Layers

In animals, **tight junctions**, or occluding junctions, are a second type of junction, one that forms a tight seal between adjacent cells and thereby prevents extracellular material from leaking between cells. As an example, let's consider the intestine. The cells that line the intestine form a sheet that is one cell thick; one side faces the intestinal lumen, while the other faces the blood (**Figure 10.10**). Tight junctions between these cells ensure that nutrients pass through the plasma membranes of the intestinal cells before entering the blood and prevent the transport of materials from the blood into the intestine.

Tight junctions are made by membrane proteins, called occludin and claudin, that form interlaced strands in the plasma membrane (see inset to Figure 10.10). These strands of proteins,

each in adjacent cells, bind to each other and thereby form a tight seal between cells. Occluding junctions are not mechanically strong like anchoring junctions because they do not have strong connections with the cytoskeleton. Therefore, adjacent cells that have occluding junctions also have anchoring junctions to hold them in place.

The amazing ability of tight junctions to prevent the leakage of material across cell layers has been demonstrated by dye-injection studies. In 1972, Daniel Friend and Norton Gilula injected lanthanum, which is electron dense and can be visualized under the electron microscope, into the bloodstream of a rat. A few minutes later, a sample of a cell layer in the digestive tract was removed and visualized by electron microscopy. As seen in the micrograph in **Figure 10.11**, lanthanum diffused into the region between the cells that faces the blood, but it could not move past the tight junction to the side of the cell layer facing the lumen of the intestine.

Gap Junctions in Animal Cells Provide a Passageway for Intercellular Transport

A third type of junction found in animals is called a **gap junction** because a small gap occurs between the plasma membranes of cells connected by these junctions (**Figure 10.12**). In vertebrates, gap junctions are composed of a membrane protein called connexin. Invertebrates have a structurally similar protein called innexin. Six connexin proteins in one cell align with six connexin proteins in an adjacent cell to form a channel called a **connexon** (see inset to Figure 10.12).

The connexons allow the passage of ions and small molecules with a molecular mass that is less than 1,000 Daltons,

Figure 10.10 **Tight junctions between adjacent intestinal cells.** In this example, tight junctions form a seal that prevents the movement of material between cells, from the intestinal lumen into the blood, and vice versa. The inset shows the interconnected network of occludin and claudin that forms the tight junction.

Figure 10.11 **An experiment demonstrating the function of a tight junction.** When lanthanum was injected into the bloodstream of a rat, it diffused between the cells in the region up to a tight junction, but could not diffuse past the junction to the other side of the cell layer.

Biological inquiry: What results would you expect if a rat was fed lanthanum and then a sample of intestinal cells was observed under the EM?

Figure 10.12 **Gap junctions between adjacent cells.** Gap junctions form intercellular channels that allow the passage of small solutes with masses less than 1,000 Daltons. A transmembrane channel called a connexon consists of 12 proteins called connexins, 6 in each cell. The micrograph shows a gap junction between intestinal cells.

including amino acids, sugars, and signaling molecules like cAMP and IP$_3$. In this way, gap junctions allow adjacent cells to share metabolites and directly signal each other. At the same time, gap-junction channels are too small to allow the passage of RNA, proteins, or large carbohydrates. Therefore, cells that communicate via gap junctions still maintain their own distinctive set of macromolecules.

Because gap junctions allow the passage of ions, electrical changes in one cell are easily transmitted to an adjacent cell that is connected via gap junctions. In 1959, Edwin Furshpan and David Potter first postulated the existence of gap junctions. Their results, which showed that certain cells in the crayfish are electrically coupled, indicated that ions can directly move from the cytosol of one cell to the cytosol of an adjacent cell.

FEATURE INVESTIGATION

Loewenstein and Colleagues Followed the Transfer of Fluorescent Dyes to Determine the Size of Gap-Junction Channels

As we have seen, gap junctions allow the passage of small molecules, those with a mass up to about 1,000 Daltons. This property of gap junctions was determined in experiments involving the transfer of fluorescent dyes. During the 1960s, several research groups began using fluorescent dyes to study cell mor-

phology and function. As discussed in Chapter 4, the location of fluorescent dyes within cells can be seen via fluorescence microscopy. In 1964, Werner Loewenstein and colleagues observed that a fluorescent dye could move from one cell to an adjacent cell, which prompted them to investigate this phenomenon further.

In the experiment shown in **Figure 10.13**, Loewenstein and colleagues grew rat liver cells in the laboratory, where they formed a single layer (a monolayer). The adjacent cells formed

Figure 10.13 Use of fluorescent molecules to determine the size of gap-junction channels.

HYPOTHESIS Gap-junction channels allow the passage of ions and molecules, but there is a limit to how large the molecules can be.

STARTING MATERIALS Rat liver cells grown in laboratory.

Experimental level **Conceptual level**

1 Grow rat liver cells in a laboratory on solid growth media until they become a monolayer. At this point, adjacent cells have formed gap junctions.

Tissue culture bottle

Gap junction

Rat liver cells

2 Inject 1 or 2 cells in the monolayer with fluorescently labeled amino acids or peptides. Note: Several dyes with different molecular masses were tested.

Rat liver cell

3 Incubate for various lengths of time (for example, 40–45 minutes). Observe cell layer under the fluorescence microscope to determine if the dye has moved to adjacent cells.

Note: In this case, the dye was transferred to adjacent cells.

4 THE DATA

Mass of Dye	Transfer to adjacent cells *	Mass of Dye	Transfer to adjacent cells *
376	++++	851**	−
464	++++	901	+++
536	+++	946	−
559	++++	1004	−
665	+	1158	−
688	++++	1678	−
817	+++	1830	−

* The number of pluses indicates the relative speed of transfer. Four pluses denote fast transfer, while one plus is slow transfer. A minus indicates that transfer between cells did not occur. ** In some cases, molecules with less mass did not pass between cells compared to molecules with a higher mass. This may be due to differences in their structures (e.g., charges) that influence whether or not they can easily penetrate the channel.

gap junctions. The cells were injected with various dyes composed of fluorescently labeled amino acids or peptide molecules with different masses and then observed via fluorescence microscopy. As shown in the data, dyes with a molecular mass up to 901 Daltons were observed to pass from cell to cell. Larger dyes, however, did not move intercellularly. Loewenstein and other researchers subsequently investigated dye transfer in other cell types and species. Though some variation is found when comparing different cell types and species, the researchers generally observed that molecules with a mass greater than 1,000 Daltons do not pass through gap junctions.

The Middle Lamella Cements Adjacent Plant Cell Walls Together

In animal cells, we have seen that cell-to-cell contact, via anchoring junctions, tight junctions, and gap junctions, involves interactions between membrane proteins in adjacent cells. In plants, cell junctions are quite different. Rather than using membrane proteins to form cell-to-cell connections, an additional component in the ECM is made called the **middle lamella**. When plant cells are dividing, the middle lamella is the first layer that is formed. Next, the primary cell wall is made inside the middle lamella (**Figure 10.14**). The middle lamella is rich in pectins, negatively charged carbohydrate polymers that are also found in the primary cell wall (see Figure 10.5). These polymers attract water and make a hydrated gel. Ca^{2+} and Mg^{2+} interact with the negative charges in the carbohydrates and cement the cell walls of adjacent cells together.

The process of fruit ripening illustrates the importance of pectins in holding plant cells together. An unripened fruit, such as a green tomato, is very firm because the rigid cell walls of adjacent cells are firmly attached to each other. During ripening, the cells secrete a group of enzymes called pectinases, which digest pectins in the middle lamella as well as pectins in the primary cell wall. As this process continues, the attachments between cells are broken, and the cell walls become less rigid. For this reason, a red ripe tomato is much less firm than an unripe tomato.

The middle lamella is a layer made outside of the primary cell wall and composed largely of negatively charged carbohydrates such as pectins. Ca^{2+} and Mg^{2+} bind to these carbohydrates and fuse the cell walls of adjacent cells.

Primary cell wall

Middle lamella

1 μm

Plant cell walls

Figure 10.14 Cell-to-cell junctions known as middle lamella.

Plasmodesmata Are Channels Connecting the Cytoplasm of Adjacent Plant Cells

In 1879, Eduard Tangl, a Russian botanist, observed intercellular cytoplasmic connections in the seeds of the strychnine tree and hypothesized that the cytoplasm of adjacent cells is connected by ducts in the cell walls. He first proposed the concept that direct cell-to-cell communication integrates the functioning of plant cells. The ducts or intercellular channels that Tangl observed are now known as **plasmodesmata** (singular, *plasmodesma*).

Plasmodesmata are functionally similar to gap junctions in animal cells, in that they allow the passage of ions, water, sugars, amino acids, and signaling molecules between cells. However, the structure of plasmodesmata is quite different from that of gap junctions. As shown in **Figure 10.15**, plasmodesmata are open channels in the cell walls of adjacent cells. At these sites, the plasma membrane of one cell is continuous with the plasma membrane of the other cell, which permits the diffusion of molecules from the cytosol of one cell to the cytosol of the other. In addition to a cytosolic connection, plasmodesmata also have a central tubule, called a desmotubule, connecting the ER membranes of adjacent cells.

10.3 Tissues

A **tissue** is a part of an animal or plant consisting of a group of cells having a similar structure and function. Because tissues are fundamental units within multicellular organisms, understanding the characteristics of tissues is essential to many areas of biology, particularly plant and animal development. For example,

Plasmodesmata

0.6 μm

Plasma membrane

Cell walls of adjacent plant cells

Smooth endoplasmic reticulum

Desmotubule passing through a plasmodesma

Cytosol
Cell 1

Middle Lamella

Cytosol
Cell 2

Figure 10.15 Structure of plasmodesmata. Plasmodesmata are cell junctions connecting the cytosol of adjacent plant cells, allowing water and small molecules to pass from cell to cell. At these sites, the plasma membrane of one cell is continuous with the plasma membrane of an adjacent cell. In addition, the ER from one cell is connected to that of the adjacent cell via a desmotubule.

geneticists want to know how genes play a role in the formation and arrangement of tissues to create an entire organism. Likewise, plant and animal biologists are interested in understanding the genetic and morphological changes that occur as the tissues of young plants and animals develop from fertilized eggs into an adult organism. These topics will be examined in Chapters 19, 39, and 52.

In this section, we will view tissues from the perspective of cell biology. Animals and plants contain many different types of cells. Humans, for example, contain over 200 different cell types, each with a specific structure and function. Even so, these cells can be grouped into a few general categories. For example, muscle cells found in your heart (cardiac muscle cells), in your biceps (skeletal muscle cells), and around your arteries (smooth muscle cells) look somewhat different under the microscope and have unique roles in the body. Yet due to structural and functional similarities, all three types can be categorized as muscle tissue cells. In this section, we will begin by surveying the basic processes that cells undergo in order to create tissues. Next, we will examine the main categories of both animal and plant tissues. We will conclude by taking a more in-depth look at some similarities and differences between selected animal and plant tissues, focusing in particular on the functions of the ECM and cell junctions.

Six Basic Cell Processes Create Tissues and Organs

A multicellular organism such as a plant or animal contains many cells. For example, an adult human has somewhere between 10 trillion and 100 trillion cells in her or his body. Cells are organized into tissues, and tissues are organized into organs. An **organ** is a collection of two or more tissues that performs a specific function or set of functions. The heart is an organ found in the bodies of complex animals, while a leaf is an organ found in plants. We will examine the structures and functions of organs in Units VI and VII.

To create tissues and organs, cells undergo six basic processes that influence their morphology, arrangement, and number:

1. *Cell division*: As discussed in Chapters 9 and 15, eukaryotic cells progress through a cell cycle that leads to cell division.
2. *Cell growth*: Following cell division, cells take up nutrients and usually expand in volume.
3. *Differentiation*: Due to gene regulation, cells differentiate into specialized types of cells. Cell differentiation is described in Chapter 19.
4. *Migration*: During embryonic development in animals, cells migrate to their appropriate positions within the body. Cell migration does not occur during plant development.
5. *Apoptosis*: Cell death, also known as apoptosis, is necessary to produce certain morphological features of the body. For example, during development in mammals, the formation of individual fingers and toes requires the removal, by apoptosis, of the skin cells between them.
6. *Cell connections*: In the first two sections of this chapter, we learned that cells secrete an extracellular matrix that provides strength and support. In animals, the ECM serves to organize cells within tissues and organs. In plants, the cell wall forms the ECM that shapes plant tissues. Different types of cell junctions in both animal and plant cells enable cells to make physical contact and communicate with one another.

Animals Are Composed of Epithelial, Connective, Muscle, and Nervous Tissues

The body of an animal contains four general types of tissue—epithelial, connective, muscle, and nervous tissue—that serve very different purposes (**Figure 10.16**).

Epithelial Tissue Epithelial tissue is composed of cells that are joined together via tight junctions and form continuous

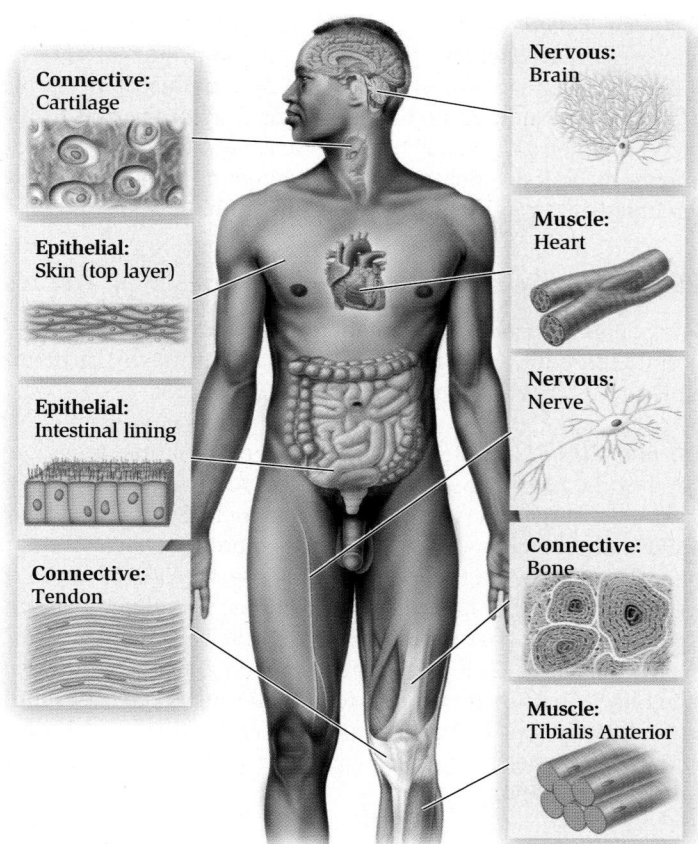

Figure 10.16 Examples of the four general types of tissues—epithelial, connective, muscle, and nervous—found in animals.

sheets. Epithelial tissue covers or forms the lining of all internal and external body surfaces. For example, epithelial tissue lines organs such as the lungs and digestive tract. In addition, epithelial tissue forms skin, a protective surface that shields the body from the outside environment.

Connective Tissue Most connective tissue provides support to the body and/or helps to connect different tissues to each other. Connective tissue is rich in extracellular matrix. In some cases, the tissue contains only a sparse population of cells that are embedded in the ECM. Examples of connective tissue include cartilage, tendons, bone, fat tissue, and the inner layers of the skin. Blood is also considered a form of connective tissue because it provides liquid connections to various regions of the body.

Muscle Tissue Muscle tissue can generate a force that facilitates movement. Muscle contraction is needed for bodily movements such as walking and running, and also plays a role in the movement of materials throughout the body. For example, contraction of heart muscle propels blood through your body. The properties of muscle tissue in animals are examined in Chapter 46.

Nervous Tissue Nervous tissue receives, generates, and conducts electrical signals throughout the body. In vertebrates, these electrical signals are integrated by nervous tissue in the brain and transmitted down the spinal cord to the rest of the body. Chapter 43 considers the cellular basis for nerve signals, known as action potentials, and Chapters 44 and 45 examine the organization of nervous systems in animals.

Plants Contain Dermal, Ground, and Vascular Tissues

The bodies of most plants contain three general types of tissue—dermal, ground, and vascular—each with a different structure suited to its functions, as shown in **Figure 10.17**.

Dermal Tissue The **dermal tissue** forms a covering on various parts of the plant. The **epidermis** refers to the newly made tissue on the surfaces of leaves, stems, and roots. Surfaces of leaves are usually coated with a waxy cuticle to prevent water loss and often have hairs, or trichomes, which are specialized types of epidermal cells that have diverse functions, including the secretion of oils and leaf protection. Epidermal cells called guard cells form pores in leaves known as stomata that permit gas exchange. The function of the root epidermis is the absorption of water and nutrients. The root epidermis does not have a waxy cuticle, because such a cuticle would inhibit water and nutrient absorption.

Leaf

Stem

Root

| Dermal | Vascular | Ground |

Figure 10.17   Locations of the three general types of tissues—dermal, ground, and vascular—found in plants.

Ground Tissue Most of a plant's body is made of **ground tissue**, which has a variety of functions, including photosynthesis, storage of carbohydrates, and support. Ground tissue can be subdivided into three types: parenchyma, collenchyma, and sclerenchyma. Let's look briefly at each of these types of ground tissue.

1. Parenchyma tissue is very active metabolically. The mesophyll, the central part of the leaf, which contains the cells that carry out the bulk of photosynthesis, is parenchyma tissue. Parenchyma tissue also functions in the storage of carbohydrates. The cells of parenchyma tissue usually lack a secondary cell wall or, if present, it is very thin.
2. Collenchyma tissue provides structural support to the plant body, particularly to growing regions such as the periphery of the stems and leaves. Collenchyma cells tend to have thick, secondary cell walls but do not contain much lignin. Therefore, they provide support but are also able to stretch.
3. Sclerenchyma tissue also provides structural support to the plant body, particularly to those parts that are no longer growing. The secondary cell walls of sclerenchyma cells tend to have large amounts of lignin and thereby provide rigid support. In many cases, sclerenchyma cells are dead at maturity but their cell walls continue to provide structural support during the life of the plant.

Vascular Tissue Most plants are vascular plants. In these species, which include ferns and flowering plants, the vascular tissue is composed of cells that are interconnected and form conducting vessels for water and nutrients. Two vascular tissues are found in flowering plants: xylem and phloem. The xylem transports water and mineral ions from the root to the rest of the plant, while the phloem distributes the products of photosynthesis and a variety of other nutrients throughout the plant. Some types of plants, such as mosses, are nonvascular plants that lack conducting vessels. These plants tend to be small and live in damp, shady places.

Plant and Animal Tissues Have Striking Differences and Similarities

Because plants and animals appear strikingly different, it is not too surprising that their cells and tissues show conspicuous differences. For example, the vascular tissue of plants (which transports water and nutrients) does not resemble any one tissue in animals. The blood vessels of animals (which transport blood carrying oxygen and nutrients throughout the body) are hollow tubes that contain both connective and muscle tissue. In addition, animals have two tissue types that are not found in plants: muscle and nervous tissue. Even so, plants are capable of movement and the transmission of signals via action potentials. For example, the Venus flytrap (*Dionaea muscipula*) has two modified leaves that resemble a clamshell. When an insect touches the surface of these leaves, an action potential is triggered across the leaf cells that causes the leaves to move closer to each other, thereby trapping the unsuspecting insect.

Although cellular differences are prominent between plants and animals, certain tissues show intriguing similarities. The epithelial tissue of animals and the dermal tissue of plants both form a protective covering over the organism. Also, the connective tissue of animals and the ground tissue of plants both play a role in structural support. Let's take a closer look at the similarities between these tissues in animals and plants.

A Comparison of Epithelial and Dermal Tissues Both the epithelial tissue in animals (also called an **epithelium**) and dermal tissue in plants form layers of cells. An epithelium can be classified according to its number of layers. Simple epithelium is one cell layer thick, while stratified epithelium has several layers (**Figure 10.18**). In both cases, the epithelium has a polarity, which is due to an asymmetry to its organization. The outer or apical side of an epithelium is exposed to air or to a watery fluid such as the lumen of the intestine. The inner or basal side rests on some type of support, such as another type of tissue or on a form of ECM called the basal lamina.

A hallmark of epithelial cells is that they form many connections with each other. For example, in the simple epithelium lining the intestine (**Figure 10.19**), adjacent cells form anchoring junctions with each other and with the basal lamina. These anchoring junctions hold the cells firmly in place. Tight junctions, found near the apical surface, prevent the leakage of materials from the lumen of the intestine into the blood. Instead, as described in Chapter 41, nutrients are selectively transported from the intestinal lumen into the cytosol of the epithelial cell and then are exported across the basal side of the cell into the blood. This phenomenon, called transepithelial transport, allows the body to take up the nutrients it needs, while preventing unwanted materials from getting into the bloodstream. Epithelial cells are also connected via gap junctions, which allow the exchange of nutrients and signaling molecules throughout the epithelium.

In flowering plants, the epidermis covers all of the newly made parts of a plant. For example, the upper and lower sides of leaves are covered by epidermis, which is usually a single layer of closely packed cells (**Figure 10.20a**). Epidermal cells have a thick primary cell wall and are tightly interlocked by their middle lamella.

Figure 10.19   **A closer look at a simple epithelium that lines the intestine.**   This figure emphasizes that the three major types of cell junctions are common in epithelial tissue.

(a) Epidermis in a young leaf

(b) Periderm in an older stem

Figure 10.20   Dermal tissues in plants.   (a) Epidermis in a young leaf. (b) Periderm in an older stem.

Biological inquiry: Do you notice any parallels between simple epithelium and epidermis, and between stratified epithelium and periderm?

Figure 10.18   Simple and stratified epithelia.

Figure 10.21   **An example of connective tissue in animals that is rich in extracellular matrix.**   This micrograph of cartilage shows chondrocytes in the ECM. The chondrocytes, which are responsible for making the components of cartilage, are found in cavities called lacunae.

Figure 10.22   **An example of ground tissue in plants, which is rich in extracellular matrix.**   This scanning electron micrograph shows cells from *Arabidopsis thaliana*. The cells themselves are dead; only their thick cell walls remain. These cell walls of the ECM provide support to the plant.

Thus, plant epidermal cells are tightly woven together, much like epithelial cell layers in animals. In leaves, the cells are often polygonal in shape, which means they are flattened cells that have several angles. The epidermal cells of new stems tend to have a more rectangular shape.

As a plant ages, the epidermis may be replaced by another type of dermal tissue called the periderm (**Figure 10.20b**). In woody plants, an example of periderm is the bark on trees. The tissue protects the plant from pathogens, prevents excessive water loss, and provides insulation. The periderm consists of interconnected cork cells, which may be several layers thick. The cork cells have extremely thick cell walls. When cork cells reach maturity, they die, but the cell walls continue to provide support.

A Comparison of Connective and Ground Tissue   In contrast to epithelial tissue, which is almost entirely composed of cells, the connective tissue of animals is largely composed of extracellular matrix and has relatively few cells. In animal connective tissue, cell-to-cell contact is somewhat infrequent. Instead, cells are usually adhered to the ECM via integrins, as shown earlier in Figure 10.8b. In some cases, the primary function of cells within connective tissue is to synthesize the components of the ECM. For example, let's consider cartilage, a connective tissue found in joints such as your knees. The cells that synthesize cartilage, known as chondrocytes, actually represent a small proportion of the total volume of cartilage. As shown in **Figure 10.21**, the chondrocytes are found in small cavities within the cartilage called lacunae (singular, *lacuna*). In some types of cartilage, the chondrocytes represent only 1–2% of the total volume of the tissue! Chondrocytes are the only cells found in cartilage; they are solely responsible for the synthesis of protein fibers, such as collagen, as well as glycosaminoglycans and proteoglycans.

Similar to connective tissue in animals, ground tissue in plants provides structural support and is also rich in ECM. **Figure 10.22** shows a scanning electron micrograph of sclerenchyma cells found in *Arabidopsis thaliana*, a model plant studied by plant biologists. At maturity, the cells are dead, but the thick secondary cell walls continue to provide rigid support for the stem. However, not all ground tissue in plants is rich in ECM. For example, mesophyll cells are a type of parenchymal cell in the leaf that carry out photosynthesis. Because a thick cell wall would inhibit the transmission of light, mesophyll cells have thin cell walls with only a small amount of ECM.

CHAPTER SUMMARY

10.1 Extracellular Matrix

- The extracellular matrix (ECM) is a network of material that is secreted from plant and animal cells and forms a complex meshwork outside of cells.
- In the ECM of animals, proteins and polysaccharides are the major constituents. These materials are involved in strength, structural support, organization, and cell signaling. (Figure 10.1)

- Adhesive proteins, such as fibronectin and laminin, help to adhere cells to the ECM. Structural proteins form fibers. Collagen fibers provide tensile strength, while elastic fibers allow regions of the body to stretch. (Table 10.1, Figures 10.2, 10.3)
- Differential gene regulation controls where in the body different types of collagen fibers are made. (Table 10.2)

- Glycosaminoglycans (GAGs) are polysaccharides of repeating disaccharide units that give a gel-like character to the ECM of animals. Proteoglycans consist of a core protein with attached GAGs. (Figure 10.4)
- Plant cells are surrounded by an ECM called the cell wall. The primary cell wall is made first. It is composed largely of cellulose. The secondary cell wall is made after the primary cell wall and is often quite thick and rigid. (Figures 10.5, 10.6)

10.2 Cell Junctions

- The three common types of cell junctions found in animals are anchoring, tight, and gap junctions. Plant junctions include middle lamella and plasmodesmata. (Table 10.3)
- Anchoring junctions involve cell adhesion molecules (CAMs), which bind cells to each other or to the ECM. The four types are adherens junctions, desmosomes, focal adhesions, and hemidesmosomes. (Figure 10.7)
- Two types of CAMs are cadherin and integrin. Cadherins link cells to each other, while integrins link cells to the ECM. In the cytosol, CAMs bind to actin or intermediate filaments. (Figure 10.8)
- Integrins can perform outside-in and inside-out signaling. This enables cells to communicate with the ECM. (Figure 10.9)
- Tight junctions, composed of occludin and claudin, prevent the leakage of materials between cells. (Figures 10.10, 10.11)
- Gap junctions form channels called connexons that permit the direct passage of materials between adjacent cells. (Figure 10.12)
- Experiments of Loewenstein and colleagues involving the transfer of fluorescent dyes showed that gap junctions permit the passage of substances with a molecular mass of less than 1,000 Daltons. (Figure 10.13)
- The cell walls of adjacent plant cells are cemented together via middle lamella. (Figure 10.14)
- Adjacent plant cells usually have direct connections called plasmodesmata, which are open channels in the cell walls. The ER of adjacent cells is also connected via plasmodesmata. (Figure 10.15)

10.3 Tissues

- A tissue is a group of cells that have a similar structure and function. An organ is composed of two or more tissues and carries out a particular function or functions.
- Six processes—cell division, cell growth, differentiation, migration, apoptosis, and cell connections—create tissues and organs.
- The four general kinds of tissues found in animals are epithelial, connective, muscle, and nervous tissues. (Figure 10.16)
- The three general kinds of tissues found in plants are dermal, ground, and vascular tissues. (Figure 10.17)
- Epithelial and dermal tissues form layers of cells that are highly interconnected. These layers can be one cell thick or several cells thick, and they serve as protective coverings for various parts of animal and plant bodies. (Figures 10.18, 10.19, 10.20)
- Connective and ground tissues are often rich in ECM and play a structural role in animals and plants. (Figures 10.21, 10.22)

TEST YOURSELF

1. The function of the extracellular matrix (ECM) in most multicellular organisms is
   a. to provide strength.
   b. to provide structural support.
   c. to organize cells and other body parts.
   d. cell signaling.
   e. all of the above.

2. The protein found in the ECM of animals that provides strength and resistance to tearing when stretched is
   a. elastin.
   b. cellulose.
   c. collagen.
   d. laminin.
   e. fibronectin.

3. The polysaccharide that forms the hard outer covering of many invertebrates is
   a. collagen.
   b. chitin.
   c. chondroitin sulfate.
   d. pectin.
   e. cellulose.

4. The single most abundant organic molecule of Earth is _____, and it is the main macromolecule of the _____.
   a. collagen, connective tissue of animals
   b. chitin, muscle tissue of animals
   c. cellulose, primary cell wall of plants
   d. integrins, cell junctions in plants
   e. pectin, secondary cell wall of plants

5. _____ are proteins that attach animal cells to the ECM.
   a. Cadherins
   b. Integrins
   c. Occludins
   d. Tight junctions
   e. Desmosomes

6. The gap junctions of animal cells differ from the plasmodesmata of plant cells in that
   a. gap junctions serve as communicating junctions and plasmodesmata serve as adhesion junctions.
   b. gap junctions prevent extracellular material from moving between adjacent cells, but the plasmodesmata do not.
   c. gap junctions allow for direct exchange of cellular material between cells, but plasmodesmata cannot allow the same type of exchange.
   d. gap junctions are formed by specialized proteins that form channels through the membranes of adjacent cells, but plasmodesmata are not formed by specialized proteins.
   e. all of the above.

7. Which of the following is involved in the process of tissue and organ formation in multicellular organisms?
   a. cell division
   b. cell differentiation
   c. cell connections
   d. cell growth
   e. all of the above

8. The tissue type common to animals that functions in the conduction of electrical signals is
   a. epithelial.
   b. dermal.
   c. muscle.
   d. nervous.
   e. ground.

9. Photosynthesis occurs mainly in the _____ tissue of plants.
   a. vascular
   b. dermal
   c. parenchyma
   d. collenchyma
   e. sclerenchyma

10. Which of the following is not a correct statement in the comparison of plant tissues to animal tissues?
   a. Nervous tissue of animals plays the same role as vascular tissue in plants.
   b. The dermal tissue of plants is similar to epithelial tissue of animals in that both provide a covering for the organism.
   c. The epithelial tissue of animals and the dermal tissue of plants have special characteristics that limit the movement of extracellular material between cell layers.
   d. The ground tissue of plants and the connective tissue of animals provide structural support for the organism.
   e. All of the above are correct comparisons between plant and animal tissues.

CONCEPTUAL QUESTIONS

1. List and explain the four characteristics of the extracellular matrix.
2. Distinguish between the primary cell wall and the secondary cell wall.
3. List and explain the four types of animal tissues.

EXPERIMENTAL QUESTIONS

1. What was the purpose of the study conducted by Loewenstein and colleagues?
2. Explain the experimental procedure used by Loewenstein to determine the size of gap-junction channels.
3. What did the results of Figure 10.13 indicate about the size of gap-junction channels?

COLLABORATIVE QUESTIONS

1. What role does the extracellular matrix play in animals?
2. What are the six basic cell processes required to create tissues and organs?

www.brookerbiology.com

This website includes answers to the Biological Inquiry questions found in the figure legends and all end-of-chapter questions.

11

NUCLEIC ACID STRUCTURE AND DNA REPLICATION

CHAPTER OUTLINE

A molecular model for the structure of a DNA double helix.

We begin our survey of genetics by examining DNA at the molecular level. Once we understand how DNA works at this level, it becomes easier to see how DNA function affects the properties of cells, and ultimately the characteristics or traits of both unicellular and multicellular organisms. The past several decades have seen exciting advances in techniques and approaches to investigate and even to alter the genetic material. Not only have these advances greatly expanded our understanding of molecular genetics, such technologies are also widely used in related disciplines such as biochemistry, cell biology, and microbiology.

To a large extent, our understanding of genetics comes from our knowledge of the molecular structure of DNA (deoxyribonucleic acid) and RNA (ribonucleic acid). In this chapter, we begin by considering some classic experiments that were consistent with the theory that DNA is the genetic material. We will then survey the molecular features of DNA, which will allow us to appreciate how DNA can store information and be accurately copied. Though this chapter is largely concerned with DNA, we will also consider the components of RNA, which bear some striking similarities to DNA.

11.1 Biochemical Identification of the Genetic Material

The genetic material functions as a blueprint for the construction of living organisms. In the case of multicellular organisms such as plants and animals, the information stored in the genetic material enables a fertilized egg to develop into an embryo, and eventually into an adult organism. In addition, the genetic material allows organisms to survive in their native environments. For example, an individual's DNA provides the blueprint to produce enzymes that are needed to metabolize food. To fulfill its role, the genetic material must meet several criteria:

1. **Information:** The genetic material must contain the information necessary to construct an entire organism.
2. **Transmission:** The genetic material must be passed from parent to offspring. It also must be passed from cell to cell during the process of cell division.
3. **Replication:** For transmission to occur, the genetic material must be accurately copied.
4. **Variation:** Differences in the genetic material must account for the known variation within each species and among different species.

The quest to identify the genetic material really began in the late 1800s, when a few scientists postulated that living organisms possess a blueprint that has a biochemical basis. In 1883, August Weismann and Karl Nägeli championed the idea that a chemical substance exists within living cells that is responsible for the transmission of traits from parents to offspring. During the next 30 years, experimentation along these lines centered on the behavior of **chromosomes**, which are structures found in cells that we now know contain the genetic material. The name *chromosome* literally means colored (*chromo*) body (*soma*), referring to the observation of early microscopists that the chromosomes are easily stained by colored dyes. By studying the transmission patterns of chromosomes from cell to cell and from parent to offspring, researchers were convinced that chromosomes carry the determinants that control the outcome of traits.

Ironically, the study of chromosomes initially misled researchers regarding the biochemical identity of the genetic material. Chromosomes contain two classes of macromolecules, namely, proteins and DNA. Scientists of this era viewed proteins as being more biochemically complex because they are made from 20 different amino acids. Furthermore, biochemists already knew that proteins perform an amazingly wide range of functions. Complexity seemed an important prerequisite for the blueprint of an organism. By comparison, DNA seemed less complex, because it contains only four types of repeating units called **nucleotides**, which will be described later in this chapter. In addition, the functional role of DNA in the nucleus had not been extensively investigated prior to the 1920s. Therefore, from the 1920s to the 1940s, most scientists were expecting that research studies would reveal that proteins are the genetic material. Contrary to this expectation, however, the experiments described in this section were pivotal in showing that DNA carries out this critical role.

Griffith's Bacterial Transformation Experiments Indicated the Existence of a Biochemical Genetic Material

Studies in microbiology were important in developing an experimental strategy to identify the genetic material. In the late 1920s, Frederick Griffith studied a type of bacterium known then as pneumococci and now classified as *Streptococcus pneumoniae*. Some strains of *S. pneumoniae* secrete a polysaccharide capsule, while other strains do not. When streaked on petri plates containing solid growth media, capsule-secreting strains have a smooth colony morphology, or form, and so look smooth to the naked eye, whereas those strains unable to secrete a capsule have a colony morphology that looks rough. In mammals, smooth strains of *S. pneumoniae* may cause pneumonia and other symptoms. In mice, such infections are often fatal.

As shown in **Figure 11.1**, Griffith injected live and/or heat-killed bacteria into mice and then observed whether or not the bacteria caused them to die. He investigated the effects of two strains of *S. pneumoniae*, type S for smooth and type R for rough. When injected into a live mouse, the smooth strain killed the mouse (Figure 11.1, step 1). A capsule is necessary for this to occur because it prevents the mouse's immune system from killing the bacterial cells. Following the death of the mouse, many type S bacteria were found in the mouse's blood. By comparison, when type R bacteria were injected into a mouse, they did not kill the mouse and living bacteria were not found in the live mouse's blood (Figure 11.1, step 2). In a follow-up to these results, Griffith also heat-killed the smooth bacteria and then injected them into the mouse. As expected, the mouse survived (Figure 11.1, step 3).

A surprising result occurred when Griffith mixed live type R bacteria with heat-killed type S bacteria, and then injected them into a mouse—the mouse died (Figure 11.1, step 4). The blood from the dead mouse contained living type S bacteria! These results indicated that a substance from dead type S bacteria was transforming the type R bacteria into type S. Griffith called this process **transformation**, and he termed the uniden-

Figure 11.1 Griffith's experiments that showed the transformation of bacteria by a "transformation principle."

Biological inquiry: Let's suppose that the type R strain used by Griffith was resistant to killing by an antibiotic, while the type S strain lacked this trait. For the experiment described in step 4, would you expect the living type S bacteria found in the dead mouse's blood to be resistant to the antibiotic?

tified material that was responsible for this phenomenon the *transformation principle*.

Now that we have examined Griffith's experiments, it's helpful if we consider what these observations mean in genetic terms. Among different streptococcal strains, researchers knew that *variation* exists in the presence or absence of a capsule. According to Griffith's results, the transformed bacteria had acquired the *information* to make a capsule from the heat-killed cells. For the transformed bacteria to proliferate and thereby kill the mouse, the substance conferring the ability to make a capsule must be *replicated* so that it can be *transmitted* from mother to daughter cells during cell division. Taken together, these observations are consistent with the idea that the formation of a capsule is governed by genetic material, because it meets the four criteria described at the beginning of this section. Thus, the experiment of Figure 11.1, step 4 was consistent with the idea that some genetic material from the dead bacteria had been transferred to the living rough bacteria and provided those bacteria with a new trait. At the time of his studies, however, Griffith could not determine the biochemical composition of the transforming substance.

FEATURE INVESTIGATION

Avery, MacLeod, and McCarty Used Purification Methods to Reveal That DNA Is the Genetic Material

Exciting discoveries sometimes occur when researchers recognize that another scientist's experimental approach may be modified and then used to dig deeper into a scientific question. In the 1940s, Oswald Avery, Colin MacLeod, and Maclyn McCarty were also interested in the process of bacterial transformation. During the course of their studies, they realized that Griffith's observations could be used as part of an experimental strategy to biochemically identify the genetic material. They asked the question, What substance is being transferred from the dead type S bacteria to the live type R?

Avery, MacLeod, and McCarty used established biochemical procedures to purify classes of macromolecules, such as proteins, DNA, and RNA, from a type S streptococcal strain. Initially, they discovered that only the purified DNA could convert type R bacteria into type S. To further verify that DNA is

the genetic material, they performed the investigation outlined in **Figure 11.2**. They purified DNA from a type S strain and mixed it with type R bacteria. After allowing time for DNA uptake, they added an antibody that aggregated any nontransformed type R bacteria, which were removed by centrifugation. The remaining bacteria were incubated overnight on petri plates.

When they mixed their DNA extract with type R bacteria, some of the bacteria were converted to type S (see plate B in step 5 of Figure 11.2). As a control, if no DNA extract was added, no type S bacterial colonies were observed on the petri plates (see plate A in step 5). Though this result was consistent with the idea that DNA is the genetic material, a careful biochemist might argue that the DNA extract may not have been 100% pure. Realistically, any purified extract is likely to contain small traces of other substances. For this reason, the researchers realized that a small amount of contaminating material in the DNA extract could actually be the genetic material. The most likely contaminating substances in this case would be protein or RNA. To address this possibility, Avery, MacLeod,

Figure 11.2   The Avery, MacLeod, and McCarty experiments that identified DNA as Griffith's "transformation principle"—the genetic material.

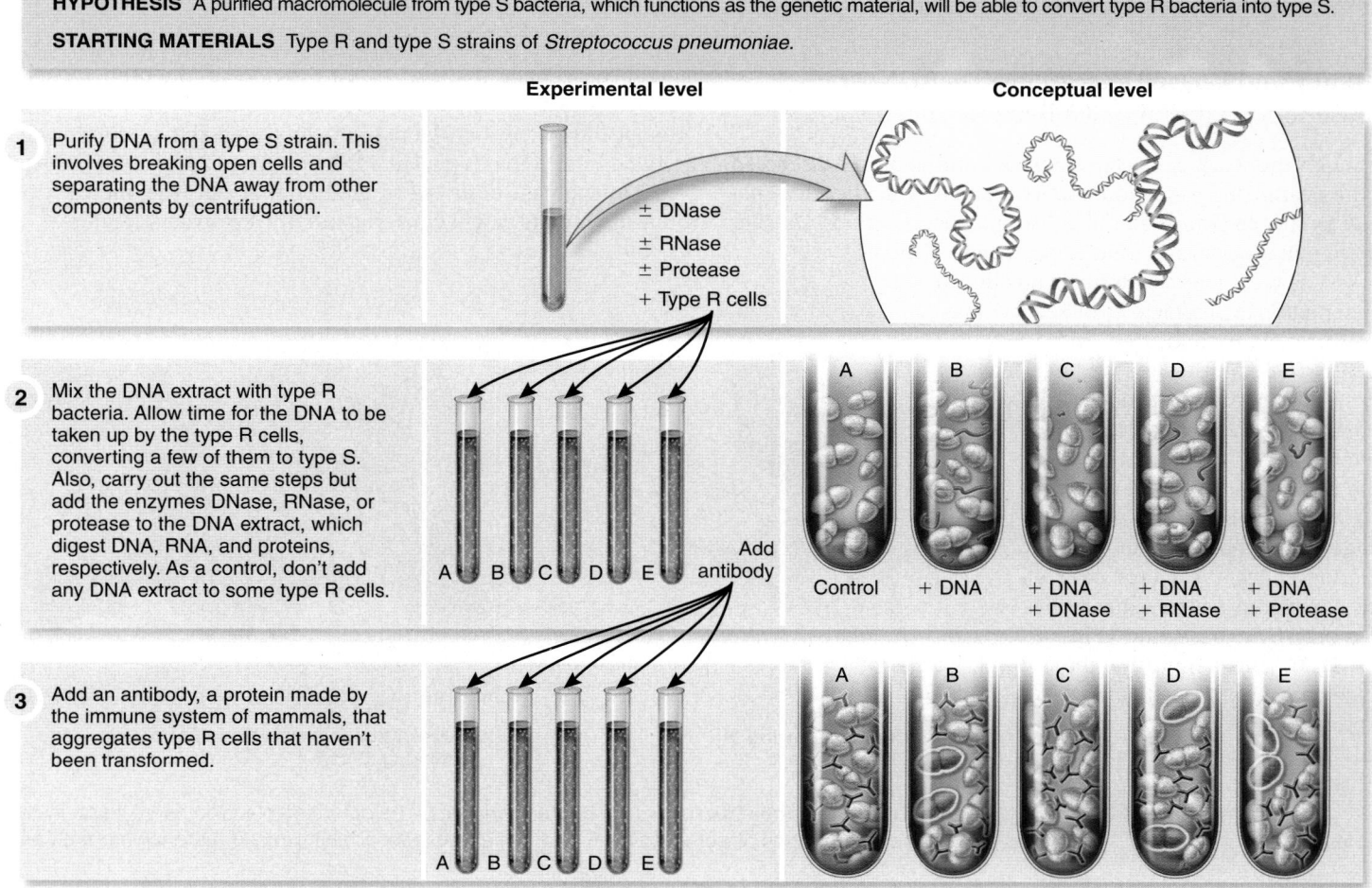

HYPOTHESIS A purified macromolecule from type S bacteria, which functions as the genetic material, will be able to convert type R bacteria into type S.

STARTING MATERIALS Type R and type S strains of *Streptococcus pneumoniae*.

Experimental level　　**Conceptual level**

1 Purify DNA from a type S strain. This involves breaking open cells and separating the DNA away from other components by centrifugation.

± DNase
± RNase
± Protease
+ Type R cells

2 Mix the DNA extract with type R bacteria. Allow time for the DNA to be taken up by the type R cells, converting a few of them to type S. Also, carry out the same steps but add the enzymes DNase, RNase, or protease to the DNA extract, which digest DNA, RNA, and proteins, respectively. As a control, don't add any DNA extract to some type R cells.

A B C D E   Add antibody

Control   + DNA   + DNA + DNase   + DNA + RNase   + DNA + Protease

3 Add an antibody, a protein made by the immune system of mammals, that aggregates type R cells that haven't been transformed.

A B C D E

4 Remove type R cells by centrifugation. Plate the remaining bacteria (if any) that are in the supernatant onto petri plates. Incubate overnight.

Centrifuge

Type S cells in supernatant

Type R cells in pellet

5 **THE DATA**

A

Control

B

DNA extract

C

DNA extract + DNase

D

DNA extract + RNase

E

DNA extract + protease

and McCarty treated the DNA extract with enzymes that digest DNA (called **DNase**), RNA (**RNase**), or protein (**protease**) (see step 2, Figure 11.2). When the DNA extracts were treated with RNase or protease, the type R bacteria were still converted into type S, suggesting that contaminating RNA or protein in the

extract was not acting as the genetic material (see plates D and E). Moreover, when the extract was treated with DNase, its ability to convert type R bacteria into type S was lost (see plate C). Taken together, these results indicated that DNA is the genetic material.

Hershey and Chase Determined That DNA Is the Genetic Material of T2 Bacteriophage

Although the work of Avery, MacLeod, and McCarty discussed in the Feature Investigation was consistent with the idea that DNA is the genetic material of streptococcal strains, further experimentation was needed to determine if the use of DNA as the genetic material is a widespread phenomenon in biology. In a second avenue of research conducted in 1952, the efforts of Alfred Hershey and Martha Chase centered on the study of a virus named T2. This virus infects bacterial cells, in this case *Escherichia coli*, and is therefore known as a **bacteriophage** or simply a **phage**. A T2 phage has an outer covering called the phage coat that contains a capsid, sheath, tail fibers, and baseplate (**Figure 11.3**). Biochemically, the phage coat is composed entirely of protein. DNA is found inside the T2 capsid. From a biochemical perspective, T2 is very simple because it is composed of only DNA and proteins.

The genetic material of T2 provides a blueprint to make new phages. However, as discussed in Chapter 18, all viruses must introduce their genetic material into the cytoplasm of a living host cell to replicate. In the case of T2, this first involves the attachment of its tail to the bacterial cell wall and then the injection of its genetic material into the cytoplasm (Figure 11.3).

Phage capsid

DNA

Sheath

Tail fiber

Base plate

E. coli cell

T2 genetic material being injected into *E. coli*.

50 nm

(a) Schematic drawing of T2 bacteriophage

(b) An electron micrograph of T2 bacteriophage infecting *E. coli*

Figure 11.3 **The structure of T2 bacteriophage.** The colorized electron micrograph in part **(b)** shows T2 phages attached to an *E. coli* cell and injecting their genetic material into the cell.

To determine if DNA is the genetic material of T2, Hershey and Chase devised a method to separate the phage coat, which is attached to the outside of the bacterium, from the genetic material, which is injected into the cytoplasm. They reasoned that the attachment of T2 on the surface of the bacterium could be disrupted if the cells were subjected to high shear forces such as those produced by a blender. In addition, they needed a way to distinguish T2 DNA from T2 proteins. Hershey and Chase used radioisotopes, which are described in Chapter 2, as a way to label these molecules. Sulfur atoms are found in proteins but not in DNA, whereas phosphorus atoms are found in DNA but not in phage proteins. They exposed T2-infected bacterial cells to ^{35}S (a radioisotope of sulfur) or to ^{32}P (a radioisotope of phosphorus). These infected cells produced phages that had incorporated ^{35}S into their proteins or ^{32}P into their DNA, a process called labeling. The labeled phages were then used in the experiment shown in **Figure 11.4**.

Let's now consider the steps in this experiment. In separate tubes, they took samples of T2 phage, one labeled with ^{35}S and the other with ^{32}P, and mixed them with *E. coli* cells for a short period of time. This allowed the phages enough time to inject their genetic material into the bacterial cells. The samples were then subjected to a shearing force using a blender for up to 8 minutes. This treatment removed the phage coat from the surface of the bacterial cell without causing cell lysis. Each sample

was then subjected to centrifugation at a speed that caused the heavier bacterial cells to form a pellet at the bottom of the tube, while the lighter phage coats remained in the supernatant, which is the solution above the pellet. The amount of radioactivity in the supernatant (emitted from either ^{35}S or ^{32}P) was determined using an instrument called a scintillation counter.

As you can see in the data of Figure 11.4, most of the ^{35}S isotope (80%) was found in the supernatant. Because the shearing force removed only the phage coat, this result indicates that the empty phages contain primarily protein. In contrast, only about 35% of the ^{32}P was found in the supernatant following shearing. This indicates that most of the phage DNA was located within the bacterial cells in the pellet. Taken together, these results suggest that the phage DNA is injected into the bacterial cytoplasm during infection. This is the expected outcome if DNA is the genetic material of T2.

11.2 Nucleic Acid Structure

The two experiments we've examined by Avery, MacLeod, and McCarty and by Hershey and Chase provided compelling evidence to support the role of DNA as the genetic material. Since these pioneering studies, thousands of researchers have studied the role of the genetic material as the blueprint of life.

Figure 11.4 Hershey and Chase experiment showing that the genetic material of T2 phage is DNA.

Biological inquiry: In these experiments, which isotope do you expect to find in the pellet, ^{32}P or ^{35}S?

As discussed in other chapters, scientists have uncovered an important principle in biology: "structure determines function." When biologists want to understand the function of a material at the molecular and cellular level, they focus some of their efforts on the investigation of its biochemical structure. In this regard, an understanding of DNA structure has proven to be particularly informative because the structure makes it easy for us to understand how DNA can store information, how variation in structure can occur, and also how DNA can be replicated and transmitted to future generations.

DNA and its molecular cousin, RNA, are known as **nucleic acids**. This term is derived from the discovery of DNA by Friedrich Miescher in 1869. He isolated a novel phosphorus-containing substance from the nuclei of white blood cells found in waste surgical bandages. He named this substance nuclein. As the structure of DNA and RNA became better understood, it was found that they are acidic molecules, which means that they release hydrogen ions (H^+) in solution, and have a net negative charge at neutral pH. Thus, the name nucleic acid was coined.

DNA is a very large macromolecule composed of smaller building blocks. We can consider the structural features of DNA at different levels of complexity (**Figure 11.5**):

1. **Nucleotides** are the building blocks of DNA (and RNA).
2. A **strand** of DNA (or RNA) is formed by the covalent linkage of nucleotides in a linear manner.
3. Two strands of DNA can hydrogen-bond with each other to form a **double helix**. In a DNA double helix, two DNA strands are twisted together to form a structure that resembles a spiral staircase.
4. In living cells, DNA is associated with an array of different proteins to form **chromosomes**. The association of proteins with DNA organizes the long strands into a compact structure.
5. A **genome** is the complete complement of an organism's genetic material. For example, the genome of most bacteria is a single circular chromosome, while the genomes of eukaryotic species are sets of linear chromosomes.

The first three levels of complexity will be the focus of this section. Levels 4 and 5 will be discussed in Chapters 15 and 21.

Nucleotides Contain a Phosphate, a Sugar, and a Base

The structures of nucleotides were determined by P. A. Levene in the 1920s. A nucleotide has three components: a phosphate group, a pentose sugar, and a nitrogenous base. The base and phosphate group are attached to the sugar at different sites to form a nucleotide (**Figure 11.6**). The nucleotides found in DNA and RNA contain different sugars. Deoxyribose is found in DNA, while ribose is found in RNA. Compared to ribose, deoxyribose is missing a single oxygen atom; the prefix *deoxy-* (meaning without oxygen) refers to this missing atom.

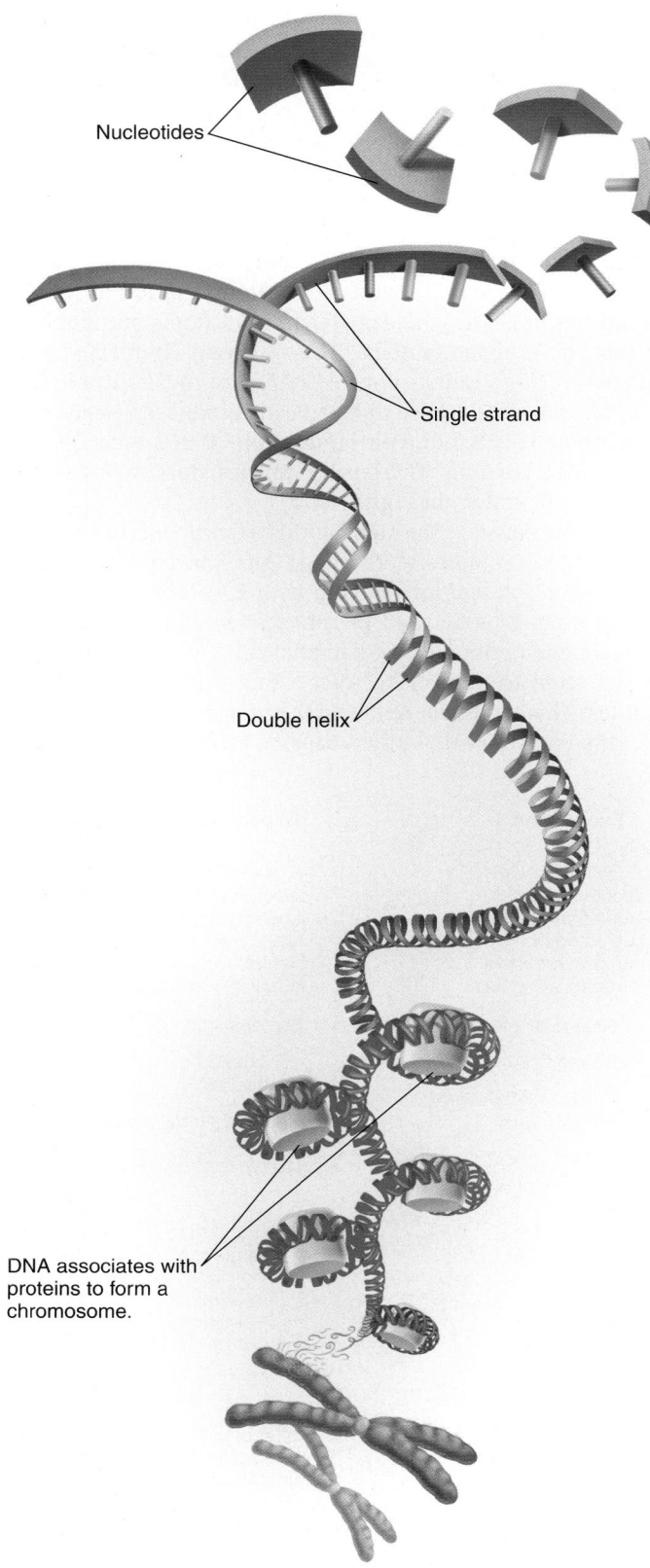

Nucleotides

Single strand

Double helix

DNA associates with proteins to form a chromosome.

Figure 11.5 Levels of DNA structure to create a chromosome.

(a) DNA nucleotide

(b) RNA nucleotide

Figure 11.6 Nucleotides and their components. For simplicity, the carbon atoms in the ring structures are not shown.

Figure 11.7 Conventional numbering in a DNA nucleotide. The carbons in the sugar are given a prime designation, while those in the base are not.

A Strand Is a Linear Linkage of Nucleotides with Directionality

The next level of nucleotide structure is the formation of a strand of DNA or RNA in which nucleotides are covalently attached to each other in a linear fashion. **Figure 11.8** depicts a short strand of DNA with four nucleotides. The linkage is a phosphoester bond (a covalent bond between phosphorus and oxygen) involving a sugar molecule in one nucleotide and a phosphate group in the next nucleotide. Another way of viewing this linkage is to notice that a phosphate group connects two sugar molecules together. From this perspective, the linkage in DNA and RNA strands is called a **phosphodiester linkage**, which has two phosphoester bonds. The phosphates and sugar molecules form the **backbone** of a DNA or RNA strand, while the bases project from the backbone. The backbone is negatively charged due to the negative charges of the phosphate groups.

An important structural feature of a nucleic acid strand is the orientation of the nucleotides. Each phosphate in a phosphodiester linkage is covalently bound to the 5′ carbon in one nucleotide and to the 3′ carbon in the other. In a strand, all sugar molecules are oriented in the same direction. For example, in the strand shown in Figure 11.8, all of the 5′ carbons in every sugar molecule are above the 3′ carbons. A strand has a **directionality** based on the orientation of the sugar molecules within that strand. In Figure 11.8, the direction of the strand is said to be 5′ to 3′ when going from top to bottom. The 5′ end of a DNA strand has a phosphate group while the 3′ end has an —OH group.

From the perspective of function, a key feature of DNA and RNA structure is that a strand contains a specific sequence of bases. In Figure 11.8, the sequence of bases is thymine–adenine–cytosine–guanine. This sequence is abbreviated TACG. To indicate its directionality, the strand should be abbreviated 5′–TACG–3′. Because the nucleotides within a strand are attached to each other by stable covalent bonds, the sequence of bases cannot shuffle around and become rearranged. Therefore, the sequence of bases in a DNA strand will remain the same over time, except in rare cases when mutations occur. The base sequence of DNA and RNA is the critical feature that allows them to store and transmit information.

Five different bases are found in nucleotides, although any given nucleotide contains only one base. The five bases are subdivided into two categories, the **purines** and the **pyrimidines**, due to differences in their structures (Figure 11.6). The purine bases, **adenine** (**A**) and **guanine** (**G**), have a double-ring structure; the pyrimidine bases, **cytosine** (**C**), **thymine** (**T**), and **uracil** (**U**), have a single-ring structure. Thymine is found only in DNA, while uracil is found only in RNA. Adenine, guanine, and cytosine are found in both DNA and RNA.

A conventional numbering system has been created to describe the attachment sites of the base and phosphate to the sugar molecule (**Figure 11.7**). In the sugar ring, carbon atoms are numbered in a clockwise direction starting with the carbon atom to the right of the ring oxygen atom. The fifth carbon is outside the ring. The prime symbol (′) is used to distinguish the numbering of carbons in the sugar, while the atoms in the ring structures of the bases are not given the prime designation. The sugar carbons are designated 1′ (that is, "one prime"), 2′, 3′, 4′, and 5′. A base is attached to the 1′ carbon atom and a phosphate group is attached at the 5′ position.

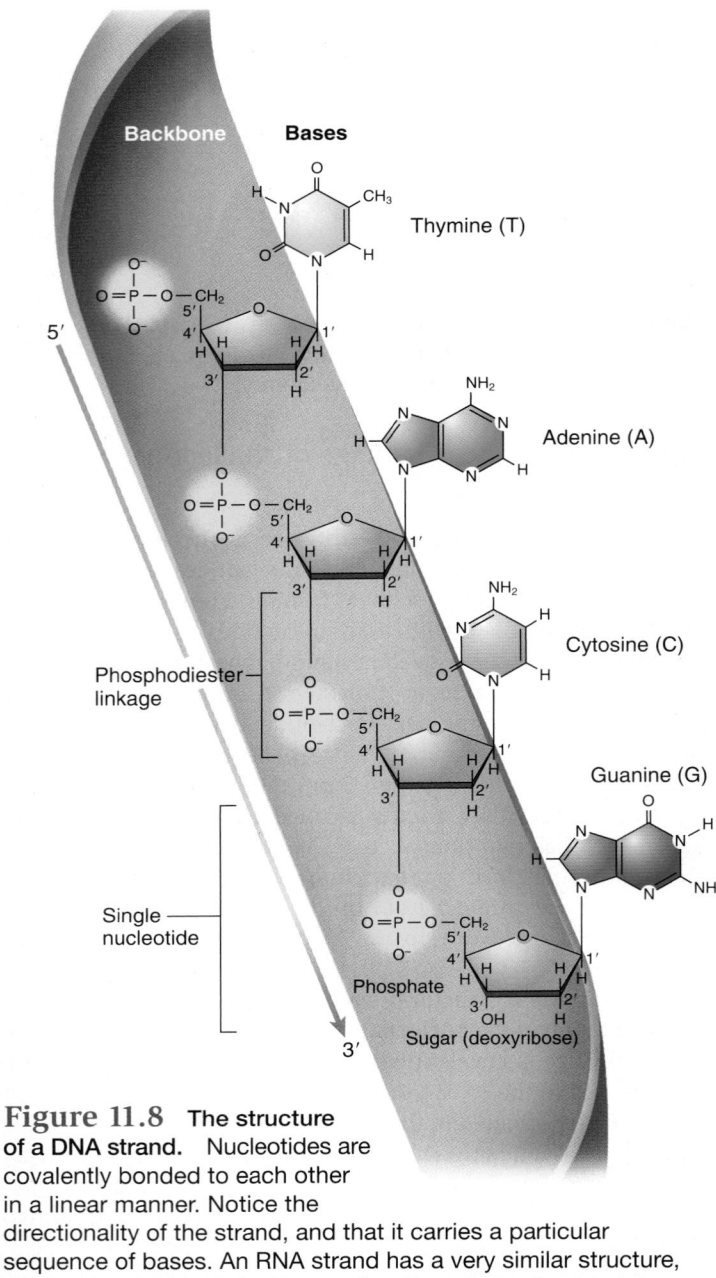

Figure 11.8   The structure of a DNA strand.   Nucleotides are covalently bonded to each other in a linear manner. Notice the directionality of the strand, and that it carries a particular sequence of bases. An RNA strand has a very similar structure, except that the sugar is ribose rather than deoxyribose, and uracil is substituted for thymine.

A Few Key Experiments Paved the Way to Solving the Structure of DNA

In 1953, James Watson and Francis Crick, along with their collaborator Maurice Wilkins, proposed the structure of the DNA double helix. Prior to their work, DNA was already known to be composed of nucleotides. However, scientists did not know how the nucleotides are bonded together to form the structure of DNA. Watson and Crick wanted to determine the structure of

DNA because they thought this knowledge would provide insights regarding the function of genes. Before we examine the characteristics of the double helix, let's consider the events that led to the discovery of the double helix structure.

In the early 1950s, more information was known about the structure of proteins than that of nucleic acids. Linus Pauling correctly proposed that regions of proteins can fold into a structure known as an α helix (alpha helix). To determine the structure of the α helix, Pauling built large models by linking together simple ball-and-stick units. In this way, he could see if atoms fit together properly in a complicated three-dimensional structure. This approach is still widely used today, except that now researchers construct three-dimensional models on computers. Watson and Crick used a ball-and-stick approach to solve the structure of the DNA double helix.

X-ray diffraction was also a key experimental tool that led to the discovery of the DNA double helix. When a substance is exposed to radiation, such as X-rays, the atoms in the substance will cause the X-rays to be scattered (**Figure 11.9a**). If the substance has a repeating structure, the pattern of scattering, known as the diffraction pattern, is mathematically related to the structural arrangement of the atoms causing the scattering. The diffraction pattern is analyzed using mathematical theory to provide information regarding the three-dimensional structure of the molecule. Rosalind Franklin (**Figure 11.9b**), working in the same laboratory as Maurice Wilkins, was a gifted experimentalist who made marked advances in X-ray diffraction techniques involving DNA. The diffraction pattern of DNA fibers produced by Franklin is shown in **Figure 11.9c**. This pattern suggested a helical structure with a diameter that is relatively uniform and too wide to be a single-stranded helix. In addition, the pattern provided information regarding the number of nucleotides per turn, and was consistent with the 2-nm (nanometers) spacing between the strands in which a purine (A or G) bonds with a pyrimidine (T or C). These observations were instrumental in solving the structure of DNA.

Another piece of information that proved to be critical for the determination of the double helix structure came from the studies of Erwin Chargaff. He pioneered many of the biochemical techniques for the isolation, purification, and measurement of nucleic acids from living cells. Chargaff analyzed the base composition of DNA that was isolated from many different species. As shown in **Table 11.1**, he consistently observed from his experiments that the amount of adenine in each sample was similar to thymine and the amount of cytosine was similar to guanine. As we will see, this observation became crucial evidence that helped Watson and Crick develop the double helix model of DNA.

Watson and Crick Deduced the Double Helix Structure of DNA

Thus far, we have considered the experimental studies that led to the determination of the DNA double helix. These included the biochemical modeling approach of Pauling, the X-ray diffraction work of Franklin, and the base composition studies of Chargaff.

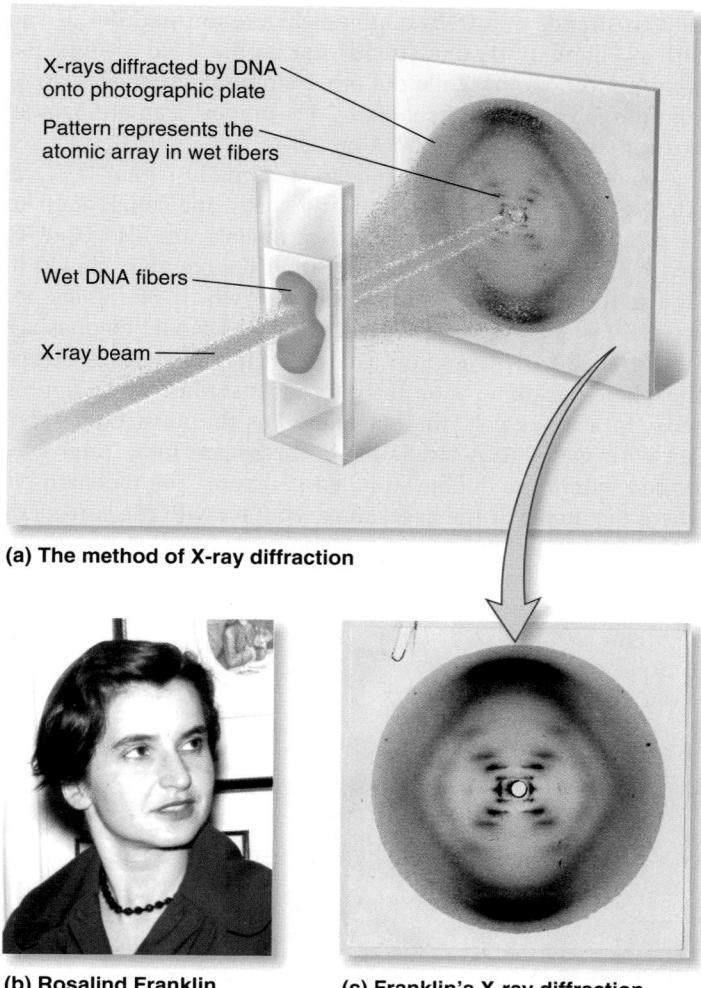

(a) The method of X-ray diffraction

X-rays diffracted by DNA onto photographic plate

Pattern represents the atomic array in wet fibers

Wet DNA fibers

X-ray beam

(b) Rosalind Franklin

(c) Franklin's X-ray diffraction pattern of wet DNA fibers

Figure 11.9   Rosalind Franklin and X-ray diffraction applied to DNA wet fibers.   **(a)** The exposure of X-rays to DNA causes the X-rays to be scattered. **(b)** Rosalind Franklin. **(c)** Because the DNA forms a repeating structure, the X-rays are scattered in a way that produces an X-ray diffraction pattern.

Table 11.1	Base Content in the DNA from a Variety of Organisms			
	% of bases			
Organism	**Adenine**	**Thymine**	**Guanine**	**Cytosine**
Escherichia coli (bacterium)	26.0	23.9	24.9	25.2
Streptococcus pneumoniae (bacterium)	29.8	31.6	20.5	18.0
Saccharomyces cerevisiae (yeast)	31.7	32.6	18.3	17.4
Turtle	28.7	27.9	22.0	21.3
Salmon	29.7	29.1	20.8	20.4
Chicken	28.0	28.4	22.0	21.6
Human	30.3	30.3	19.5	19.9

Watson and Crick assumed that nucleotides are linked together in a linear fashion and that the chemical linkage between two nucleotides is always the same. Then, they set out to build ball-and-stick models that incorporated all of the known experimental observations.

Modeling of chemical structures often involves trial and error. Watson and Crick initially considered several incorrect models. One model was a double helix in which the bases were on the outside of the helix. Another model showed the bases forming hydrogen bonds with the identical base in the opposite strand (that is, A to A, T to T, G to G, and C to C). However, the model-building revealed that the bases could not fit together this way. Finally, they realized that the hydrogen-bonding of adenine to thymine was structurally similar to that of cytosine to guanine. In both cases, a purine (A or G) bonds with a pyrimidine (T or C). With an interaction between A and T and between G and C, the ball-and-stick models showed that the two strands would form a double helix structure in which all atoms would fit together properly. This ball-and-stick model was consistent with the known data (**Figure 11.10**).

Watson, Crick, and Wilkins proposed the structure of DNA, which was published in the journal *Nature* in 1953. In 1962, they were awarded the Nobel Prize. Unfortunately, Rosalind Franklin had died before this time, and the Nobel Prize is not awarded posthumously.

DNA Has a Repeating, Antiparallel Helical Structure Formed by the Complementary Base Pairing of Nucleotides

The structure that Watson and Crick proposed is a double-stranded, repeating helical structure with the sugar-phosphate backbone on the outside and the bases on the inside (**Figure 11.11a**). This structure is stabilized by hydrogen bonding between

Figure 11.10   Watson and Crick and their model of the DNA double helix.

the bases in opposite strands to form **base pairs**. A distinguishing feature of base pairing is its specificity. An adenine base in one strand forms two hydrogen bonds with a thymine base in the opposite strand, or a guanine base forms three hydrogen bonds with a cytosine (**Figure 11.11b**). This **AT/GC rule** (also known as Chargaff's rule) explains the previous data of Chargaff in which there were equal amounts of A and T, and equal amounts of G and C, in DNA. According to the AT/GC rule, purines (A and G) always bond with pyrimidines (T and C) (recall that purines both have a double-ring structure, whereas pyrimidines have single rings). This keeps the width of the double helix relatively constant. One complete turn of the double helix is composed of 10 base pairs.

Due to the AT/GC rule, the base sequences of two DNA strands are **complementary** to each other. That is, you can predict the sequence in one DNA strand if you know the sequence in the opposite strand. For example, if one strand has the sequence of 5′–GCGGATTT–3′, the opposite strand must be 3′–CGCCTAAA–5′. With regard to their 5′ and 3′ directionality,

the two strands of a DNA double helix are **antiparallel**. If you look at Figure 11.11, one strand runs in the 5′ to 3′ direction from top to bottom, while the other strand is oriented 3′ to 5′ from top to bottom. Watson and Crick proposed an antiparallel structure in their original DNA model.

The DNA model we have been studying in Figure 11.11a is called a ribbon model, which clearly shows the components of the DNA molecule. However, other models are also used to visualize DNA. The model for the DNA double helix shown in **Figure 11.12** is a space-filling model in which the atoms are depicted as spheres. This type of structural model emphasizes the surface of DNA. As you can see in this model, the sugar-phosphate backbone is on the outermost surface of the double helix. In a living cell, the backbone has the most direct contact with water. The atoms of the bases are more internally located within the double-stranded structure. The indentations where the atoms of the bases make contact with the surrounding water are termed **grooves**. Two grooves, called the **major groove** and the **minor groove**, spiral around the double helix.

(a) Double helix

(b) Base pairing

Key Features
• Two strands of DNA form a double helix.
• The bases in opposite strands hydrogen-bond according to the AT/GC rule.
• The 2 strands are antiparallel.
• There are ~10 nucleotides in each strand per complete turn of the helix.

Figure 11.11 **Structure of the DNA double helix.** As seen in part (a), DNA is a helix composed of two antiparallel strands. Part (b) shows the AT/GC base pairing that holds the strands together via hydrogen bonds.

Biological inquiry: If one DNA strand is 5′–GATTCGTTC–3′, what is the complementary strand?

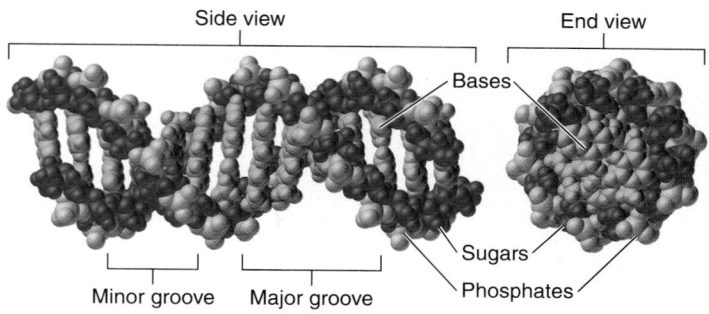

Side view   End view

Bases

Sugars

Phosphates

Minor groove   Major groove

Figure 11.12   A space-filling model of the DNA double helix.
In the sugar-phosphate backbone, sugar molecules are shown in
blue and phosphate groups are shown in yellow. This backbone
is on the outermost surface of the double helix. The atoms of the
bases, shown in green, are more internally located within the
double-stranded structure. Notice the major and minor grooves
that are formed by this arrangement.

As discussed in later chapters, the major groove provides a location where a protein can bind to a particular sequence of bases and affect the expression of a gene.

11.3 DNA Replication

In the previous section, we considered the structure of the genetic material. DNA is a double helix that obeys the AT/GC rule. From this structure, Watson and Crick immediately were able to suggest a mechanism by which DNA can be copied. During this process, known as **DNA replication**, they proposed that the original DNA strands are used as templates for the synthesis of new DNA strands. In this section, we will begin by looking at an early experiment that helped to determine the method of DNA replication. Next, we will examine the structural characteristics that enable a double helix to be faithfully copied. Finally, we will consider the features of DNA replication that allow it to occur in living cells. As you will learn, several cellular proteins are needed to initiate DNA replication and allow it to proceed accurately and at a rapid pace.

Meselson and Stahl Used Density Measurements to Investigate Three Proposed Mechanisms of DNA Replication

Researchers in the late 1950s considered three different models for the mechanism of DNA replication (**Figure 11.13**). In all of these models, the two newly made strands are called the **daughter strands**, while the original strands are the **parental strands**. The first model is a **semiconservative mechanism** (Figure 11.13a). In this model, the double-stranded DNA is half conserved following the replication process such that the new double-stranded DNA contains one parental strand and one daughter strand. This mechanism is consistent with the ideas of Watson and Crick. Even so, other models were possible and had to be ruled out. According to a second model, called a **conservative mechanism**, both parental strands of DNA remain

together following DNA replication (Figure 11.13b). The original arrangement of parental strands is completely conserved, while the two newly made daughter strands are also together following replication. Finally, a third possibility, called a **dispersive mechanism**, proposed that segments of parental DNA and newly made DNA are interspersed in both strands following the replication process (Figure 11.13c).

In 1958, Matthew Meselson and Franklin Stahl devised an experimental approach to distinguish among these three mechanisms. An important feature of their research was the use of isotope labeling. Nitrogen, which is found in DNA, occurs in a common light (^{14}N) form and a rare heavy (^{15}N) form. Meselson and Stahl studied DNA replication in the bacterium *E. coli*. They grew *E. coli* cells for many generations in a medium that

Original
double helix

First round
of replication

Second round
of replication

Parental strand
Daughter strand

(a) Semiconservative mechanism. DNA replication produces DNA molecules with 1 parental strand and 1 newly made strand.

(b) Conservative mechanism. DNA replication produces 1 double helix with both parental strands, and the other with 2 new daughter strands.

(c) Dispersive mechanism. DNA replication produces DNA strands in which segments of new DNA are interspersed with the parental DNA.

Figure 11.13   Three mechanisms for DNA replication.   The strands of the original double helix are shown in red. New strands are produced for two rounds of replication; these new strands are shown in blue.

contained only the ^{15}N form of nitrogen (**Figure 11.14**). This produced a population of bacterial cells in which all of the DNA was heavy labeled. Then, they switched the bacteria to a medium that contained only ^{14}N as its nitrogen source. The cells were allowed to divide, and samples were collected after one generation (that is, one round of DNA replication), two generations, and so on. Because the bacteria were doubling in a medium that contained only ^{14}N, all of the newly made DNA strands would be labeled with light nitrogen, while the original strands would remain labeled with the heavy form.

Meselson and Stahl then used centrifugation to separate DNA molecules based on differences in density. Samples were placed on the top of a solution that contained a salt gradient, in this case cesium chloride. A double helix containing all heavy nitrogen has a higher density and will travel closer to the bottom of the gradient. By comparison, if both DNA strands contained ^{14}N, the DNA would have a light density and remain closer to the top of the gradient. If one strand contained ^{14}N and the other strand contained ^{15}N, the DNA would be half-heavy and have an intermediate density, ending up near the middle of the gradient.

After one cell doubling (that is, one round of DNA replication), all of the DNA exhibited a density that was half-heavy (Figure 11.14, step 5). These results are consistent with both the semiconservative and dispersive models. In contrast, the conservative mechanism predicts two different DNA types: a light type and a heavy type. Because the DNA was found in a single (half-heavy) band after one doubling, the conservative model was disproved. After two cell doublings, both light DNA and half-heavy DNA were observed. This result was also predicted by a semiconservative mechanism of DNA replication, because some DNA molecules should contain all light DNA while other molecules should be half-heavy (see Figure 11.13a). However, in a dispersive mechanism, all of the DNA strands after two generations would have been 1/4 heavy. This mechanism predicts that the heavy nitrogen would be evenly dispersed among four double helices, each strand containing 1/4 heavy nitrogen and 3/4 light nitrogen (see Figure 11.13c). This result was not obtained. Taken together, the results of the Meselson and Stahl experiment are consistent only with a semiconservative mechanism for DNA replication.

Semiconservative DNA Replication Proceeds According to the AT/GC Rule

As originally proposed by Watson and Crick, semiconservative DNA replication relies on the complementarity of DNA strands according to the AT/GC rule. During the replication process, the two complementary strands of DNA separate and serve as **template strands** for the synthesis of new strands of DNA (**Figure 11.15**). After the double helix has separated, individual nucleotides have access to the template strands. Hydrogen bonding between individual nucleotides and the template strands must obey the AT/GC rule. A covalent bond is formed between the phosphate of one nucleotide and the sugar of the previous nucleotide. The end result is that two double helices are made that have the same base sequence as the original DNA molecule. This is a critical feature of DNA replication, because it enables

1 Grow bacteria in ^{15}N media.

^{15}N medium (heavy)

2 Transfer to ^{14}N media and continue growth for <1, 1.0, 2.0, or 3 generations.

^{14}N medium (light)

3 Isolate DNA after each generation. Transfer DNA to CsCl gradient, and centrifuge.

DNA

CsCl gradient

Centrifuge

4 Observe stained DNA under UV light.

5 THE DATA

Approximate generations after transfer to ^{14}N medium.

| < 1.0 | 1.0 | 2.0 | 3.0 |

Light
Half-heavy
Heavy

Figure 11.14 **The Meselson and Stahl experiment showing that DNA replication is semiconservative.**

Biological inquiry: If this experiment were conducted for four rounds of DNA replication (that is, four generations), what would be the expected fractions of light DNA and half-heavy DNA, according to the semiconservative model?

the replicated DNA molecules to retain the same information (that is, the same base sequence) as the original molecule. In this way, DNA has the striking ability to direct its own duplication.

DNA Replication Begins at an Origin of Replication, Where DNA Replication Forks Are Formed

An **origin of replication** is a site within a chromosome that serves as a starting point for DNA replication. At the origin, the two DNA strands unwind, and DNA replication proceeds outward from the origin in opposite directions, a process termed **bidirectional replication**. The number of origins of replication varies among different organisms. In bacteria, which have a small circular chromosome, a single origin of replication is found. Bidirectional replication starts at the origin of replication and proceeds until the new strands meet each other on the opposite side of the chromosome (**Figure 11.16a**). Eukaryotes have larger chromosomes that are linear. They require multiple origins of replication so that the DNA can be replicated in a reasonable length of time. The newly made strands from each origin eventually make contact with each other to complete the replication process (**Figure 11.16b**).

An origin of replication provides an opening called a replication bubble that forms two DNA **replication forks**. DNA replication occurs near the opening that forms each replication fork (**Figure 11.17**, step 1). As discussed later, the synthesis of a strand always begins with a primer and the new DNA is made in the 5′ to 3′ direction. The manner in which the two daughter strands are synthesized is strikingly different (Figure 11.17, step 2). One strand, called the **leading strand**, is made in the

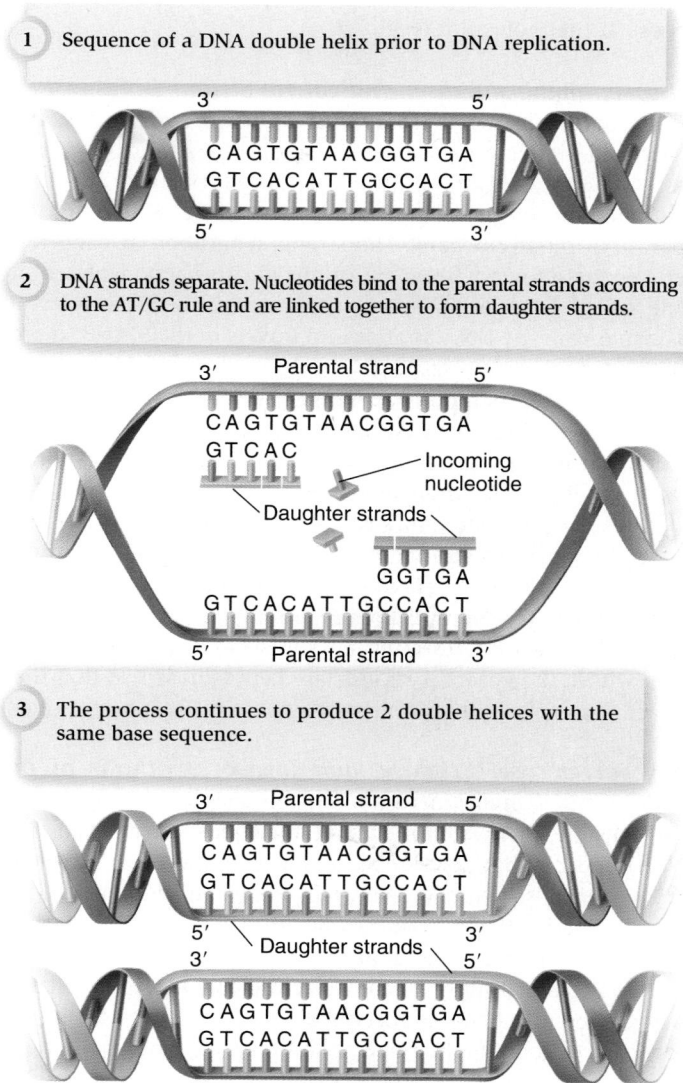

Figure 11.15 DNA replication according to the AT/GC rule.

(a) Bacterial chromosome replication

(b) Eukaryotic chromosome replication

Figure 11.16 Origins of replication in the chromosomes of different organisms. (a) A single origin of replication is found in bacteria. (b) Multiple origins are found in a eukaryotic chromosome. A description of centromeres, kinetochores (proteins bound to centromeres) and sister chromatids can be found in Chapter 15.

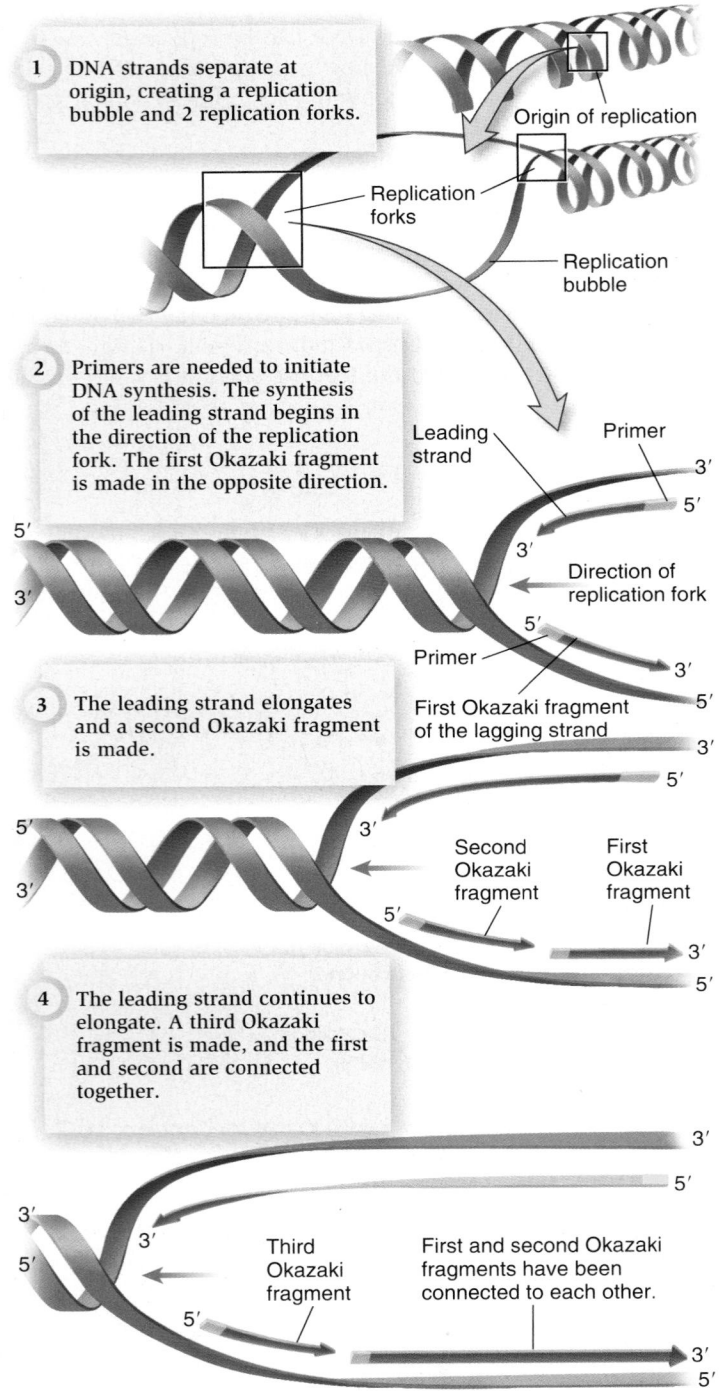

1 DNA strands separate at origin, creating a replication bubble and 2 replication forks.

Origin of replication

Replication forks

Replication bubble

2 Primers are needed to initiate DNA synthesis. The synthesis of the leading strand begins in the direction of the replication fork. The first Okazaki fragment is made in the opposite direction.

Leading strand

Primer

3′
5′

3′

Direction of replication fork

5′

Primer

3′
5′

First Okazaki fragment of the lagging strand

3 The leading strand elongates and a second Okazaki fragment is made.

3′
5′

Second Okazaki fragment

First Okazaki fragment

3′
5′

4 The leading strand continues to elongate. A third Okazaki fragment is made, and the first and second are connected together.

3′
5′

Third Okazaki fragment

First and second Okazaki fragments have been connected to each other.

3′
5′

Figure 11.17 **The bidirectional synthesis of DNA from an origin of replication.** The separation of DNA at the origin of replication produces two replication forks that move in opposite directions. New DNA strands are made near the opening of each fork. The leading strand is made continuously in the same direction that the fork is moving. The lagging strand is made as small pieces in the opposite direction. Eventually, these small pieces are connected to each other to form a continuous lagging strand.

same direction that the fork is moving. The leading strand is synthesized as one long continuous molecule. By comparison, the other daughter strand, termed the **lagging strand**, is made as a series of small fragments that are eventually connected to each other to form a continuous strand. The synthesis of these fragments occurs in the direction away from the fork. For example, the single fragment seen in Figure 11.17, step 2, is synthesized from left to right. Notice that this fragment connects to a second fragment in step 4 of Figure 11.17. These DNA fragments are known as **Okazaki fragments**, after Reiji and Tuneko Okazaki, who initially discovered them in the late 1960s.

DNA Replication Requires the Action of Many Different Proteins

Thus far, we have considered how DNA replication occurs outward from an origin of replication in a region called a DNA replication fork. In all living species, a set of many different proteins is involved in this process. An understanding of the functions of these proteins is critical to explaining the replication process at the molecular level.

Helicase, Topoisomerase, and Single-Strand Binding Proteins: Formation and Movement of the Replication Fork

To act as a template for DNA replication, the strands of a double helix must separate and the resulting fork must move. As mentioned, an origin of replication serves as a site where this separation initially occurs. The strand separation at each fork then moves outward from the origin via the action of an enzyme called **DNA helicase**. At each fork, a helicase enzyme binds to one of the DNA strands and travels in the 5′ to 3′ direction toward the fork (**Figure 11.18**). It uses energy from ATP to separate the DNA strands and keeps the fork moving forward. The action of DNA helicase generates additional coiling just ahead of the replication fork that is alleviated by another enzyme called **DNA topoisomerase**.

After the two parental DNA strands have separated, they must remain that way until the complementary daughter strands have been made. The function of **single-strand binding proteins** is to bind to both of the single strands of parental DNA and prevent them from re-forming a double helix. In this way, the bases within the parental strands are kept exposed so that they can act as templates for the synthesis of complementary strands.

DNA Polymerase, Primase, and Ligase: Synthesis of the Leading and Lagging Strands
The enzyme **DNA polymerase** is responsible for covalently linking nucleotides together to form DNA strands. Arthur Kornberg originally identified this enzyme in the 1950s. The structure of DNA polymerase in action resembles a human hand with the DNA threaded through it (**Figure 11.19a**). As DNA polymerase slides along the DNA, free nucleotides with three phosphate groups, called **deoxynucleoside triphosphates**, hydrogen-bond to the exposed bases in the template strand according to the AT/GC rule. At the catalytic site, DNA polymerase breaks a bond at the middle phosphate and then attaches

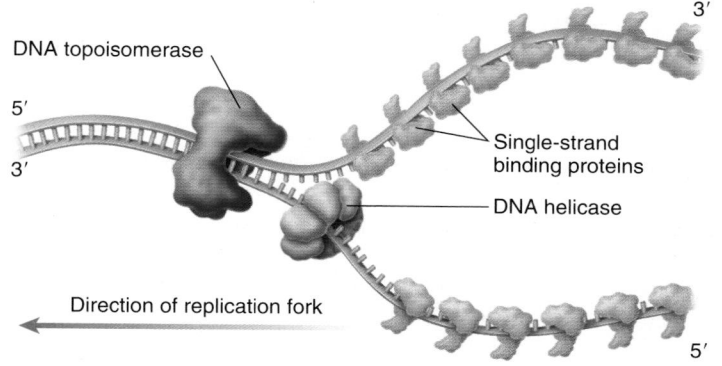

Figure 11.18 **Proteins that facilitate the formation and movement of a replication fork.** DNA helicase travels along one DNA strand in the 5′ to 3′ direction and separates the DNA strands. After they are separated, single-strand binding proteins coat the DNA strands to prevent them from re-forming a double helix. Topoisomerase travels slightly ahead of the fork and alleviates coiling that is caused by the action of helicase.

(a) Action of DNA polymerase

Figure 11.19 **Enzymatic synthesis of DNA.** **(a)** DNA polymerase recognizes free nucleotides called deoxynucleoside triphosphates and attaches a deoxynucleoside monophosphate to the 3′ end of a growing strand. **(b)** DNA polymerase breaks the bond at the middle phosphate in a deoxynucleoside triphosphate, causing the release of pyrophosphate. This provides the energy to form a covalent phosphoester bond between the resulting deoxynucleoside monophosphate and the previous nucleotide in the growing strand.

(b) Chemistry of DNA replication

the resulting nucleotide with one phosphate group (a deoxynucleoside monophosphate) to the 3′ end of a growing strand via a phosphoester bond. The breakage of the covalent bond that releases pyrophosphate (two phosphate groups) is an exergonic reaction that provides the energy to covalently connect adjacent nucleotides (**Figure 11.19b**). The rate of synthesis is truly remarkable. In bacteria, DNA polymerase can synthesize DNA at a rate of 500 nucleotides per second, while eukaryotic species can make DNA at a rate of about 50 nucleotides per second.

DNA polymerase has two specific enzymatic features that explain why the leading and lagging strands are made differently. First, DNA polymerase is unable to begin DNA synthesis on a bare template strand. However, if a DNA or RNA strand is already attached to a template strand, DNA polymerase can elongate such a pre-existing strand by making DNA. A different enzyme called **DNA primase** is required if the template strand is bare. DNA primase makes a complementary **primer** that is actually a short segment of RNA, typically 10 to 12 nucleotides in length. These short RNA strands start, or prime, the process of DNA replication (**Figure 11.20a**). At a later stage in DNA replication, the RNA primers are removed and replaced with DNA. A second feature of DNA polymerase is that once synthesis has begun, it can synthesize new DNA only in a 5′ to 3′ direction (**Figure 11.20b**).

In the leading strand, DNA primase makes one RNA primer at the origin and then DNA polymerase attaches nucleotides in a 5′ to 3′ direction as it slides toward the opening of the replication fork (**Figure 11.21**). In the lagging strand, DNA is also synthesized in a 5′ to 3′ direction, but this synthesis occurs in the direction away from the replication fork. In the lagging strand, short segments of DNA, the Okazaki fragments, are made. Each fragment contains a short RNA primer made by DNA primase at the 5′ end, and then the remainder of the fragment is a strand of DNA made by DNA polymerase.

To complete the synthesis of Okazaki fragments within the lagging strand, three additional events must occur: removal of the RNA primers, synthesis of DNA in the area where the primers have been removed, and the covalent joining of adjacent fragments of DNA (Figure 11.21, steps 3 and 4). The RNA primers are

removed by DNA polymerase, which can also digest the linkages between nucleotides in a 5′ to 3′ direction. After the RNA primer is removed, DNA polymerase then fills in the vacant region with DNA. However, once the DNA has been completely filled in, a covalent bond is missing between the last nucleotide added by DNA polymerase and the first nucleotide in the next DNA fragment. An enzyme known as **DNA ligase** catalyzes the formation of a covalent bond between these adjacent DNA fragments to complete the replication process in the lagging strand (Figure 11.21, step 4). Similarly, the single RNA primer of the leading strand is removed, DNA polymerase fills in the vacant region, and ligase completes the process. **Table 11.2** provides a summary of the actions of the proteins we've discussed in this section.

DNA Replication Is Very Accurate

Although errors can happen during DNA replication, permanent mistakes are extraordinarily rare. For example, during bacterial DNA replication, only one mistake per 100 million nucleotides is made. This high level of **fidelity** occurs for several reasons. First, hydrogen bonding between A and T or between G and C is more stable than between mismatched pairs. Second, the active site of DNA polymerase is unlikely to catalyze bond formation between adjacent nucleotides if a mismatched base pair is formed. A third way that DNA polymerase minimizes mistakes is by the enzymatic removal of mismatched nucleotides.

DNA polymerase can identify a mismatched nucleotide and remove it from the daughter strand. This event, called **proofreading**, occurs when DNA polymerase detects a mismatch, and then reverses its direction and digests the linkages between nucleotides at the end of a newly made strand in the 3′ to 5′ direction. Once it passes the mismatched base and removes it, DNA polymerase then changes direction again and continues to synthesize DNA in the 5′ to 3′ direction.

In addition to the proofreading ability of DNA polymerase, living cells also possess several other enzymes that function solely to repair abnormalities in DNA. If such lesions are not fixed, a permanent change in the genetic material, namely a mutation, can occur. These topics are considered in Chapter 14.

Table 11.2	Proteins Involved in DNA Replication
Common name	**Function**
DNA helicase	Separates double-stranded DNA into single strands
Single-strand binding protein	Binds to single-stranded DNA, and prevents it from re-forming a double helix
Topoisomerase	Removes tightened coils ahead of the replication fork
DNA primase	Synthesizes short RNA primers
DNA polymerase	Synthesizes DNA in the leading and lagging strands, removes RNA primers, and fills in gaps
DNA ligase	Covalently attaches adjacent Okazaki fragments

(a) Need for a primer   **(b) Directional synthesis**

Figure 11.20 Enzymatic feature of DNA polymerase. **(a)** DNA polymerase needs a primer to begin DNA synthesis and **(b)** once begun, it can synthesize DNA only in the 5′ to 3′ direction.

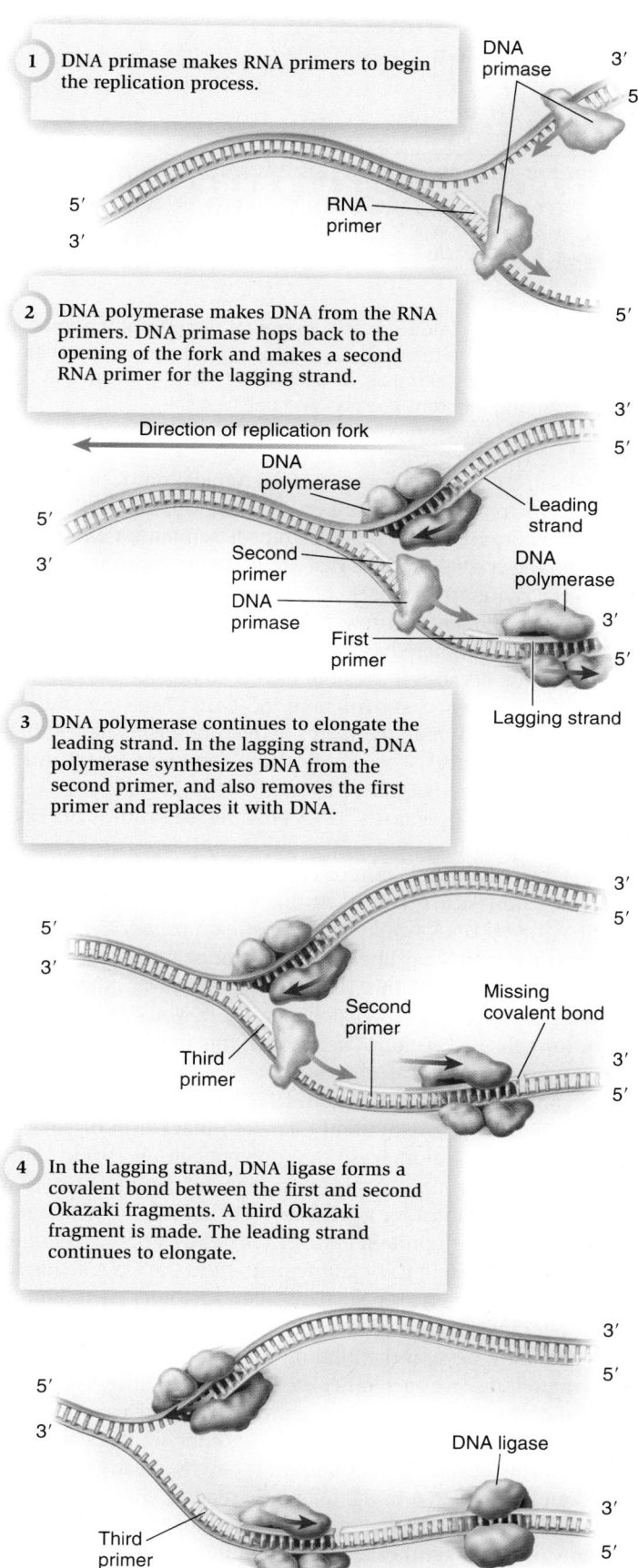

1. DNA primase makes RNA primers to begin the replication process.

2. DNA polymerase makes DNA from the RNA primers. DNA primase hops back to the opening of the fork and makes a second RNA primer for the lagging strand.

3. DNA polymerase continues to elongate the leading strand. In the lagging strand, DNA polymerase synthesizes DNA from the second primer, and also removes the first primer and replaces it with DNA.

4. In the lagging strand, DNA ligase forms a covalent bond between the first and second Okazaki fragments. A third Okazaki fragment is made. The leading strand continues to elongate.

GENOMES & PROTEOMES

DNA Polymerases Are a Family of Enzymes with Specialized Functions

Thus far, we have examined the general properties of DNA replication. Three important issues are speed, fidelity, and completeness. DNA replication must proceed quickly, with great accuracy, and gaps should not be left in the newly made strands. To ensure that these three requirements are met, nearly all living species produce more than one type of DNA polymerase. Each type may differ with regard to the rate of DNA synthesis, the accuracy of replication, and/or the ability to prevent the formation of DNA gaps. Let's first consider how evolution produced these different forms of DNA polymerase, and then examine how their functions are finely tuned to the process of DNA replication.

The genomes of most species have several DNA polymerase genes due to the phenomenon of gene duplication, which is described in Chapter 21. During evolution, independent genetic changes have altered the properties of each gene to produce a family of DNA polymerase enzymes with more specialized functions. These changes are suited to the organism in which they are found. For comparison, let's consider the bacterium *E. coli* and humans. *E. coli* has five different DNA polymerases, designated I, II, III, IV, and V. In humans, over a dozen different DNA polymerases have been identified (**Table 11.3**). Why does *E. coli* need 5 DNA polymerases while humans need 12 or more? The answer lies in specialization and the unique functional needs of each species.

In *E. coli*, DNA polymerase III is responsible for the majority of DNA replication. This enzyme is composed of multiple subunits, each with its own functional role. In addition to the catalytic subunit that actually synthesizes DNA, DNA polymerase III has other subunits that allow it to clamp onto the template DNA and synthesize new DNA very rapidly and with high fidelity. By comparison, DNA polymerase I is composed of a single subunit. Its role during DNA replication is to rapidly remove the RNA primers and fill in the short vacant regions with DNA. DNA polymerases II, IV, and V are involved in repairing DNA, and also in replicating DNA that has been damaged. DNA polymerases I and III become stalled when they encounter DNA damage and may be unable to make a complementary strand at such a site.

Figure 11.21 **Synthesis of the leading and lagging strands.** In the leading strand, DNA primase synthesizes an RNA primer at the origin, and then the leading strand is made in the same direction as the movement of the fork. Later, the RNA primer is removed, and the resulting gap is filled in by DNA polymerase, and connected by DNA ligase. In the lagging strand, primase periodically synthesizes an RNA primer near the opening of the replication fork, and then DNA polymerase synthesizes a short DNA fragment in a direction away from the fork. The RNA primers are removed by DNA polymerase, which also fills in the vacant region with DNA. Finally, DNA ligase connects adjacent Okazaki fragments.

Biological inquiry: Briefly describe the movement of primase in the lagging strand in this figure. When does it move from left to right, and when does it move from right to left?

Table 11.3	DNA Polymerases in *E. coli* and Humans
Polymerase types*	**Function**
E. coli:	
III	Replicates most of the DNA during cell division
I	Removes RNA primers and fills in the gaps
II, IV, and V	Repairs damaged DNA and replicates over DNA damage
Humans:	
α (alpha)	Primes DNA and synthesizes short DNA strands
δ (delta), ε (epsilon)	Displaces DNA polymerase α and then replicates DNA at a rapid rate
γ (gamma)	Replicates the mitochondrial DNA
η (eta), κ (kappa), ι (iota), ζ (zeta)	Replicates over damaged DNA
α, β (beta), δ, ε, σ (sigma), λ (lambda), μ (mu), φ (phi), θ (theta)	Repairs DNA or has other functions

*Certain DNA polymerases may have more than one function.

Figure 11.22   **Telomere sequences at the end of a human chromosome.**   The telomere sequence shown here is found in humans and other mammals. The length of the 3′ overhang is variable among different species and cell types.

By comparison, DNA polymerases II, IV, and V don't stall. Therefore, though their rate of synthesis is not as rapid as DNA polymerases I and III, they ensure that DNA replication is complete.

In human cells, the DNA polymerases are designated with Greek letters (Table 11.3). DNA polymerase α has its own "built-in" primase subunit. It synthesizes short RNA regions followed by short DNA regions. Two other DNA polymerases, δ (delta) and ε (epsilon), then extend the DNA at a faster rate. DNA polymerase γ (gamma) functions in the mitochondria to replicate mitochondrial DNA.

Certain DNA polymerases function as lesion-replicating enzymes. Like DNA polymerases I and III in *E. coli*, when the general DNA polymerases (α, δ, or ε) encounter abnormalities in DNA structure (that is, a lesion), they may be unable to replicate over the damage. If this happens, lesion-replicating polymerases are attracted to the damaged DNA. These polymerases have special properties that enable them to synthesize a complementary strand over the abnormal region. Each type of lesion-replicating polymerase may be able to replicate over different kinds of DNA damage. Similarly, other human DNA polymerases play an important role in DNA repair. The need for multiple repair enzymes is rooted in the various ways that DNA can be damaged, as described in Chapter 14. Multicellular organisms must be particularly vigilant about repairing DNA or cancer may occur.

Telomerase Attaches DNA Sequences at the Ends of Eukaryotic Chromosomes

We will end our discussion of DNA replication by considering a specialized form of DNA replication that happens only in eukaryotic cells. This unusual form of DNA replication occurs at the ends of eukaryotic chromosomes. This region, called the **telomere**, is composed of a series of repeat sequences within the DNA and the special proteins that are bound to those sequences. Telomeric sequences consist of a short sequence that is repeated a few dozen to a few hundred times in a row (**Figure 11.22**). Also, a telomere has a region at the 3′ end that is termed a 3′ overhang, because it does not have a complementary strand. The repeat sequence shown here, 5′–GGGTTA–3′, is the sequence found in human telomeres. Other organisms have different repeat sequences. For example, the sequence found in the telomeres of maize is 5′–GGGTTTA–3′.

As discussed previously, DNA polymerase synthesizes DNA only in a 5′ to 3′ direction, and requires a primer. For these reasons, DNA polymerase cannot copy the tip of the DNA strand with a 3′ end (**Figure 11.23**). At this location, there is no place on the parental DNA strand for an upstream primer to be made. Conceivably, an RNA primer could be made that is complementary to the very tip of this DNA strand. However, after such a primer is removed by DNA polymerase, it would still leave a short, unreplicated region. Therefore, if this replication problem was not solved, a linear chromosome would become progressively shorter with each round of DNA replication.

In 1984, Carol Greider and Elizabeth Blackburn discovered an enzyme called **telomerase** that prevents chromosome shortening. It attaches many copies of a repeated DNA sequence to the ends of chromosomes (**Figure 11.24**). The telomerase enzyme contains both protein and RNA. The RNA part of telomerase has a sequence that is complementary to the DNA sequence found in the telomeric repeat. This allows telomerase to bind to the 3′ overhang region of the telomere. Following binding, the RNA sequence beyond the binding site functions as a template, allowing telomerase to synthesize a six-nucleotide sequence at the end of the DNA strand. The enzyme then moves to the new end of this DNA strand and attaches another six nucleotides to the end. This occurs many times and thereby greatly lengthens the 3′ end of the DNA in the telomeric region. This lengthening provides an upstream site for an RNA primer to be made. DNA polymerase then synthesizes the complementary DNA strand.

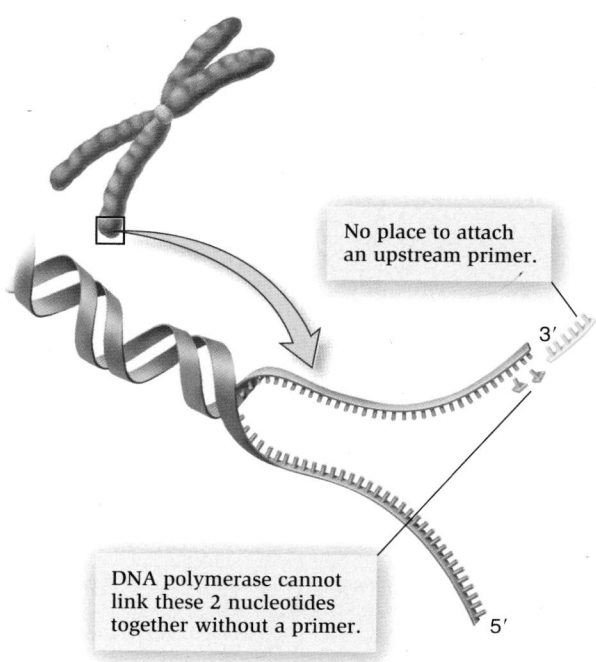

No place to attach an upstream primer.

DNA polymerase cannot link these 2 nucleotides together without a primer.

Figure 11.23 Enzymatic features of DNA polymerase that account for its inability to copy one of the DNA strands at the end of a linear chromosome.

In this way, the progressive shortening of eukaryotic chromosomes is prevented.

Researchers have discovered an interesting connection between telomeres and cellular aging. In humans and other mammals, the cells of the body have a predetermined life span. For example, if a small sample of skin is removed from a person's body and grown in the laboratory, the cells will double a finite number of times. Furthermore, the number of doublings depends on the age of the person from which the sample is taken. If a sample is from an infant, the cells will typically double about 80 times, while if a sample is from an older person, the cells will double only 10 to 20 times before division ceases. Cells that have doubled many times and have reached a point where they have lost the capacity to divide any further are termed **senescent**.

The progressive shortening of telomeres is correlated with cellular senescence, though the relationship between the two phenomena is not well understood. The telomerase enzyme is normally present in germ-line cells, which give rise to gametes, and also in many rapidly dividing somatic cells. However, telomerase function is typically reduced as an organism ages. In 1998, Andrea Bodnar and her colleagues inserted a gene that encodes a highly active telomerase into human cells grown in the laboratory, using techniques described in Chapter 20. The results were amazing. The expression of telomerase prevented telomere shortening and cellular senescence. The cells expressing telomerase continued to divide, just like younger, healthy cells!

Figure 11.24 Mechanism of DNA replication by telomerase.

Telomerase function is also associated with cancer. When cells become cancerous, they continue to divide uncontrollably. In 90% of all types of human cancers, telomerase has been found to be present at high levels in the cancerous cells. This prevents telomere shortening and may play a role in the continued growth of cancer cells. The mechanism whereby cancer cells are able to increase the function of telomerase is not well understood and is a topic of active research.

CHAPTER SUMMARY

11.1 Biochemical Identification of the Genetic Material

- The genetic material carries information to produce the traits of organisms. It is replicated and transmitted from cell to cell and generation to generation, and it has differences that explain the variation among different organisms.

- Griffith's work with type R and type S bacteria was consistent with the transfer of genetic material, which he called the transformation principle. (Figure 11.1)

- Avery, MacLeod, and McCarty used biochemical methods to show that DNA is the transformation principle. (Figure 11.2)

- Hershey and Chase labeled T2 phage with ^{35}S and ^{32}P and determined that the ^{32}P-labeled DNA is the genetic material of this phage. (Figures 11.3, 11.4)

11.2 Nucleic Acid Structure

- DNA is composed of nucleotides, which covalently link together to form DNA strands. Two DNA strands are held together by hydrogen bonds between the bases. Chromosomes are made of DNA and proteins. (Figure 11.5)

- Nucleotides are composed of a phosphate, sugar, and nitrogenous base. The sugar can be deoxyribose (DNA) or ribose (RNA). The purine bases are adenine and guanine, while the pyrimidine bases are thymine (DNA only), cytosine, and uracil (RNA only). (Figure 11.6)

- The atoms in a nucleotide are numbered in a conventional way. (Figure 11.7)

- In a strand of DNA (or RNA), the sugars are connected by covalent bonds in a 5′ to 3′ direction. (Figure 11.8)

- Watson, Crick, and Wilkins used the X-ray diffraction data of Franklin and the biochemical data of Chargaff (that is, A = T, G = C), and constructed ball-and-stick models to reveal the double helix structure of DNA. (Figures 11.9, 11.10, Table 11.1)

- DNA is a double helix in which the DNA strands are antiparallel and obey the AT/GC rule. (Figures 11.11, 11.12)

11.3 DNA Replication

- Meselson and Stahl used ^{15}N- and ^{14}N-isotope labeling methods to show that DNA is replicated by a semiconservative mechanism in which the product of DNA replication is one original strand and one new strand. (Figures 11.13, 11.14)

- New DNA strands are made according to the AT/GC rule in which parental strands serve as templates for the synthesis of new daughter strands. (Figure 11.15)

- Within chromosomes, DNA replication begins at sites known as origins of replication, where the DNA unwinds to make replication forks. Bacterial chromosomes have a single origin, while eukaryotic chromosomes have many. (Figure 11.16)

- DNA synthesis occurs bidirectionally from an origin of replication. The synthesis of new DNA strands happens near each replication fork. The leading strand is made continuously, in the same direction that the fork is moving. The lagging strand is made in the opposite direction as short Okazaki fragments that are connected together. (Figure 11.17)

- DNA helicase separates DNA strands, single-strand binding proteins keep them separated, and DNA topoisomerase alleviates coiling ahead of the fork. (Figure 11.18, Table 11.2)

- Deoxynucleoside triphosphates bind to the template strands, and DNA polymerase catalyzes the formation of a phosphoester bond between the 3′ end of the strand and a deoxynucleoside monophosphate. Pyrophosphate is released. (Figure 11.19)

- DNA polymerase requires a primer and can make new DNA strands only in the 5′ to 3′ direction. (Figure 11.20)

- Living organisms have several different types of DNA polymerases with specialized functions. (Table 11.3)

- DNA primase makes one primer in the leading strand and multiple primers in the lagging strand. DNA polymerase extends these primers with DNA, and it removes the primers when they are no longer needed. DNA ligase connects adjacent Okazaki fragments. (Figure 11.21, Table 11.2)

- The ends of linear, eukaryotic chromosomes have telomeres composed of repeat sequences. DNA polymerase cannot synthesize the tip of one of these DNA strands (Figures 11.22, 11.23)

- Telomerase binds to the telomere repeat sequence and synthesizes a six-nucleotide repeat. This happens many times in a row to lengthen one DNA strand of the telomere. DNA primase, DNA polymerase, and DNA ligase are needed to synthesize the complementary DNA strand. (Figure 11.24)

TEST YOURSELF

1. Why did researchers initially believe the genetic material was protein?
   a. Proteins are more biochemically complex than DNA.
   b. Proteins are found only in the nucleus, but DNA is found in many areas of the cell.
   c. Proteins are much larger molecules and can store more information than DNA.
   d. All of the above.
   e. Both a and c.

2. Considering the components of a nucleotide, what component always determines whether the nucleotide would be incorporated into a DNA strand or an RNA strand?
   a. phosphate group
   b. pentose sugar
   c. nitrogenous base
   d. Both b and c

3. Which of the following equations would be appropriate when considering DNA base composition?
   a. $\%A + \%T = \%G + \%C$
   b. $\%A = \%G$
   c. $\%A = \%G = \%T = \%C$
   d. $\%A + \%G = \%T + \%C$

4. If the sequence of a section of DNA is 5′–CGCAACTAC–3′, what is the appropriate sequence for the opposite strand?
   a. 5′–GCGTTGATG–3′
   b. 3′–ATACCAGCA–5′
   c. 5′–ATACCAGCA–3′
   d. 3′–GCGTTGATG–5′

5. Of the following statements, which is correct when considering the process of DNA replication?
   a. New DNA molecules are composed of two completely new strands.
   b. New DNA molecules are composed of one strand from the old molecule and one new strand.
   c. New DNA molecules are composed of strands that are a mixture of sections from the old molecule and sections that are new.
   d. None of the above.

6. Meselson and Stahl were able to demonstrate semiconservative replication in *E. coli* by
   a. using radioactive isotopes of phosphorus to label the old strand and visually determining the relationship of old and new DNA strands.
   b. using different enzymes to eliminate old strands from DNA.
   c. using isotopes of nitrogen to label the DNA and determining the relationship of old and new DNA strands by density differences of the new molecules.
   d. labeling viral DNA before it was incorporated into a bacterial cell and visually determining the location of the DNA after centrifugation.

7. During replication of a DNA molecule, the daughter strands are not produced in exactly the same manner. One strand, the leading strand, is made toward the replication fork, while the lagging strand is made in fragments in the opposite direction. This difference in the synthesis of the two strands is the result of which of the following?
   a. DNA polymerase is not efficient enough to make two "good" strands of DNA.
   b. The two parental strands are antiparallel, and DNA polymerase makes DNA only in the 5′ to 3′ direction.
   c. The lagging strand is the result of DNA breakage due to UV light.
   d. The cell does not contain enough nucleotides to make two complete strands.

8. Considering the different proteins involved in DNA replication and their functions, which of the following *does not* represent an accurate relationship?
   a. DNA helicase unwinds the DNA molecule.
   b. DNA primase produces the Okazaki fragments of the lagging strand.
   c. DNA topoisomerase reduces the coiling ahead of the replication fork.
   d. Single-strand binding proteins prevent the double helix from re-forming.

9. Which of the following *does not* promote the fidelity of DNA replication?
   a. DNA polymerase will not form bonds between nucleotides if there is a mismatched base.
   b. DNA polymerase has the ability to recognize mismatched bases and remove them.
   c. Hydrogen bonds that hold the two strands together are more stable in correctly matched bases than between mismatched bases.
   d. All of the above are correct.

10. Most organisms have more than one form of DNA polymerase. Some forms play specific roles in replication, while others play roles in DNA repair. What is the proposed mechanism for the evolution of the multiple forms of DNA polymerase?
    a. Increased copies of polymerase genes due to chromosome number change over several generations.
    b. Species hybridization may introduce new forms of polymerases.
    c. Gene duplication followed by mutations alters the activity of the polymerase.
    d. Mutations change other genes into DNA polymerases.

CONCEPTUAL QUESTIONS

1. List and explain the four characteristics of the genetic material.

2. Explain or describe the essential features of the Watson and Crick model of the structure of DNA.

3. Explain the function of telomerase.

EXPERIMENTAL QUESTIONS

1. Avery, MacLeod, and McCarty worked with two strains of *Streptococcus pneumoniae* to determine the biochemical identity of the genetic material. Explain the characteristics of the *Streptococcus pneumoniae* strains that made them particularly well suited for such an experiment.

2. In the experiment of Avery, MacLeod, and McCarty, what was the purpose of using the protease, RNase, and DNase if only the DNA extract caused transformation?

3. The Hershey and Chase experiment used radioactive isotopes to track the DNA and protein of phages as they infected bacterial cells. Explain how this procedure allowed them to detect that DNA was the genetic material of this particular virus.

COLLABORATIVE QUESTIONS

1. Discuss the reasons why there are very few errors made during DNA replication.

2. What was Frederick Griffith's contribution to the study of DNA and why was it so important?

---

www.brookerbiology.com

This website includes answers to the Biological Inquiry questions found in the figure legends and all end-of-chapter questions.

12

GENE EXPRESSION AT THE MOLECULAR LEVEL

CHAPTER OUTLINE

A micrograph of many ribosomes in the act of translating two mRNA molecules into polypeptides. The short polypeptides are seen emerging from the ribosomes.

A gene is defined as a unit of heredity. We can consider gene function at different biological levels. In Chapter 16, we will examine how genes affect the traits or characteristics of individuals. For example, we will consider how the transmission of genes from parents to offspring affects the color of the offspring's eyes and their likelihood of becoming bald. By comparison, in this chapter we will explore how genes work at the molecular level. You will learn how DNA sequences are organized to form genes, and how those genes are used as a template to make RNA copies, ultimately leading to the synthesis of a functional protein. The term **gene expression** can refer to gene function both at the level of traits and at the molecular level. In reality, the two phenomena are intricately woven together. The expression of genes at the molecular level affects the structure and function of cells, which, in turn, determine the traits that an organism expresses.

In this chapter, we will first consider how researchers came to realize that most genes store the information to make proteins. Then, we will explore the steps of gene expression as they occur at the molecular level. These steps include the use of a gene as a template to make an RNA molecule, the processing of the RNA into a functional molecule, and the use of RNA molecules to make proteins.

12.1 Overview of Gene Expression

Even before DNA was known to be the genetic material, scientists had asked the question, How do genes function to create the traits of living organisms? At the molecular level, a similar question can be asked, namely, How do genes affect the composition and/or function of molecules that are found within living

cells? The approach that was successful at answering this question involved the study of **mutations**, which are changes in the genetic material. Mutations can affect the genetic blueprint by altering gene function. For this reason, research that was focused on the effects of mutations proved instrumental in determining the molecular function of genes.

In this section, we will begin by considering two early experiments in which researchers studied the effects of mutations in humans and in a bread mold. Even though they involved different organisms, both studies led to the conclusion that the role of certain genes is to encode enzymes. Then we will examine the general features of gene expression at the molecular level.

The Study of Inborn Errors of Metabolism Suggested That Genes Encode Enzymes

In 1908, Archibald Garrod, a British physician, proposed a relationship between genes and the production of enzymes. Prior to his work, biochemists had studied many metabolic pathways that consist of a series of conversions of one molecule to another, each step catalyzed by an enzyme. **Figure 12.1** illustrates part of the metabolic pathway for the breakdown of phenylalanine, an amino acid commonly found in human diets. The enzyme phenylalanine hydroxylase catalyzes the conversion of phenylalanine to tyrosine, another amino acid. A different enzyme, tyrosine aminotransferase, converts tyrosine into the next molecule, called p-hydroxyphenylpyruvic acid. In each case, a specific enzyme catalyzes a single chemical reaction.

Garrod studied patients with defects in the ability to metabolize certain compounds. Much of his work centered on the inherited disease alkaptonuria, in which the patient's body accumulates abnormal levels of homogentisic acid (also called alkapton).

Figure 12.1 **The metabolic pathway that breaks down phenylalanine and its relationship to certain genetic diseases.** Each step in the pathway is catalyzed by a different enzyme, shown in the *right* boxes. If one of the enzymes is not functioning, the previous compound builds up, causing the conditions named in the *left* boxes. In this pathway, defects in phenylalanine hydroxylase, hydroxyphenylpyruvate oxidase, and homogentisic acid oxidase cause phenylketonuria, tyrosinosis, and alkaptonuria, respectively.

This compound, which is bluish black, results in discoloration of the skin and cartilage and causes the urine to appear black. Garrod hypothesized that the accumulation of homogentisic acid in these patients is due to a missing enzyme, namely homogentisic acid oxidase (Figure 12.1). Furthermore, he already knew that alkaptonuria is an inherited condition that follows what is called a recessive pattern of inheritance. As discussed in Chapter 16, if a disorder is inherited in a recessive manner, an individual with the disease has inherited the mutant (defective) gene that causes the disorder from each parent. Based on these observations, Garrod proposed a relationship between the inheritance of a mutant gene and a defect in metabolism. In the case of alkaptonuria, if an individual inherited the mutant gene from both parents, he or she would not produce any normal enzyme and would be unable to metabolize homogentisic acid. Garrod described alkaptonuria as an **inborn error of metabolism**. At the turn of the last century, this was a particularly insightful idea, because the structure and function of the genetic material were completely unknown.

Beadle and Tatum Proposed the One Gene–One Enzyme Hypothesis

Although Garrod published a book and several papers on inherited metabolic disorders, his work was largely ignored until the early 1940s, when it was rediscovered by George Beadle and Edward Tatum. Beadle and Tatum were also interested in the relationships among genes, enzymes, and traits. They decided to study the genetics of simple nutritional requirements in *Neurospora crassa*, a common bread mold. *Neurospora* is easily grown in the laboratory and has few nutritional requirements. The minimum requirements for growth are a carbon source (namely, sugar), inorganic salts, and one vitamin known as biotin. Otherwise, *Neurospora* has many different cellular enzymes that synthesize the small molecules, such as amino acids and many vitamins, that are essential for growth.

Hypothesizing that genes encode enzymes, Beadle and Tatum reasoned that a mutation or change in a gene might cause a defect in an enzyme that is needed for the synthesis of an essential molecule, such as a vitamin or amino acid. A mutant strain (one that carries such a mutation) would be unable to grow unless the vitamin or amino acid was supplemented in the growth medium. Strains without a mutation are called wild-type strains. In their original study of 1941, Beadle and Tatum isolated mutant strains that required vitamins for growth. In each case, a single mutation resulted in the requirement for a single type of vitamin in the growth medium. This early study by Beadle and Tatum led to additional research to study enzymes involved with the synthesis of other substances, such as the amino acid arginine. At that time, the pathway leading to arginine synthesis was thought to involve certain precursor molecules, including ornithine and citrulline. A simplified pathway for arginine synthesis is shown in **Figure 12.2a**. Each step is catalyzed by a different enzyme.

Researchers first isolated several different mutants that required arginine for growth. They hypothesized that each mutant strain might be blocked at only a single step in the consecutive series of reactions that lead to arginine synthesis. To test this hypothesis, the mutant strains were examined for their ability to grow in the presence of ornithine, citrulline, and/or arginine. A simplified depiction of the results is shown in **Figure 12.2b**. Based on their growth properties, the mutant strains that had been originally identified as requiring arginine for growth could be placed into three groups, designated 1, 2, and 3. Group 1 mutants were missing enzyme 1, needed for the conversion of a precursor molecule into ornithine. They could grow only if ornithine, citrulline, or arginine was added to the growth medium. Group 2 mutants were missing the second enzyme in this pathway that is needed for the conversion of ornithine into citrulline. The group 2 mutants would not grow if only ornithine was added but could grow if citrulline or arginine was added. Finally, the group 3 mutants were missing the enzyme needed for the conversion of citrulline into arginine. These mutants could grow only if arginine was added. Based on these results, the researchers were able to order the functions of the genes involved in arginine synthesis in the following way:

From these results and earlier studies, Beadle and Tatum concluded that a single gene controlled the synthesis of a single enzyme. This was referred to as the *one gene–one enzyme hypothesis*.

In later decades, this idea had to be modified in two ways. First, enzymes are only one category of cellular proteins. All proteins are encoded by genes, and many do not function as enzymes. Second, some proteins are composed of two or more different polypeptides. Therefore, a more accurate statement is that a gene encodes a polypeptide. The term **polypeptide** denotes structure; it is a linear sequence of amino acids. By comparison, the term **protein** denotes function. Some proteins are composed of one polypeptide. In such cases, a single gene does encode a single protein. In other cases, however, a functional protein is composed of two or more different polypeptides. An example is hemoglobin, the protein that carries oxygen in red blood cells, which is composed of two α-globin and two β-globin polypeptides. In this case, the expression of two genes (that is, the α-globin and β-globin genes) is needed to create a functional protein. Based on this more recent information, the original concept proposed by Beadle and Tatum has been modified to the **one gene–one polypeptide theory**.

Molecular Gene Expression Involves the Processes of Transcription and Translation

Thus far, we have considered two classic studies that led researchers to conclude that genes encode enzymes, or more broadly that genes encode polypeptides. Let's now consider the

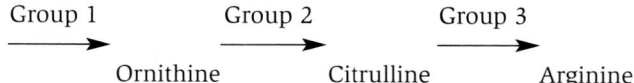

Group 1   Group 2   Group 3

Ornithine   Citrulline   Arginine

(a) Simplified pathway for arginine synthesis

(b) Growth of strains on minimal and supplemented growth media

Figure 12.2   **An example of an experiment that supported Beadle and Tatum's one gene–one enzyme hypothesis.**   (a) Simplified pathway for arginine synthesis. (b) Growth of wild-type (WT) and mutant (groups 1, 2, and 3) *Neurospora* strains in the presence of ornithine, citrulline, or arginine. As shown on the *left*, the mutant strains did not grow on minimal media lacking arginine. However, all mutants could grow if arginine was added, and certain mutants could grow if either ornithine or citrulline was added.

Biological inquiry: What type of enzyme function is missing in group 2 mutants?

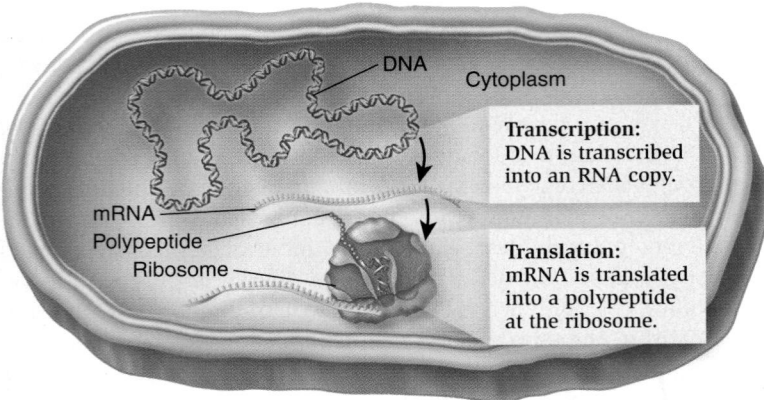

(a) Molecular gene expression in prokaryotes

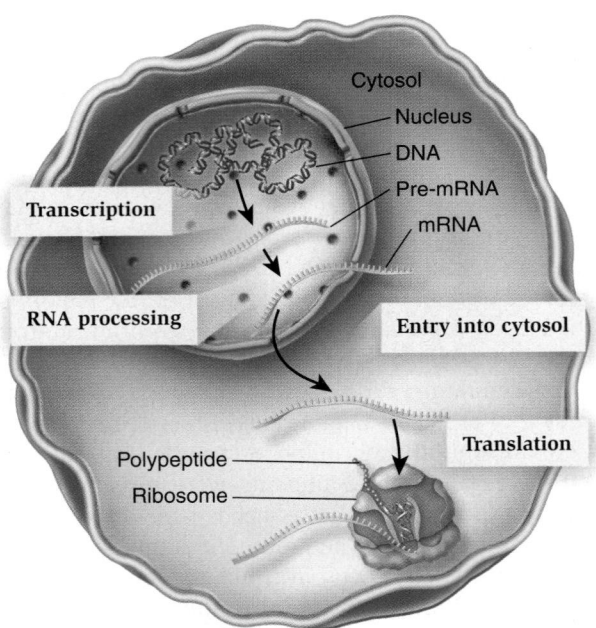

(b) Molecular gene expression in eukaryotes

Figure 12.3 **The central dogma of gene expression at the molecular level.** **(a)** In bacteria that lack a cell nucleus, transcription and translation occur in the cytoplasm. **(b)** In eukaryotes, transcription and RNA processing occur in the nucleus, while translation takes place in the cytosol.

general steps of gene expression at the molecular level. The first step, known as **transcription**, produces an RNA copy of a gene, also called an RNA transcript (**Figure 12.3a**). The term transcription literally means the act of making a copy. Most genes, which are termed **structural genes**, produce an RNA molecule that contains the information to specify a polypeptide with a specific amino acid sequence. This type of RNA is termed **messenger RNA** (abbreviated **mRNA**), because its job is to carry information from the DNA to the ribosome. As discussed later, **ribosomes** play a key role in the synthesis of polypeptides. The second step, termed **translation**, is the process of synthesizing a specific polypeptide on a ribosome. The term translation is used because a nucleotide sequence in mRNA is "translated" into an amino acid sequence of a polypeptide.

The transcription of DNA into mRNA and the translation of mRNA into a polypeptide have been called the **central dogma** of gene expression at the molecular level (Figure 12.3a). This central dogma applies equally to prokaryotes and eukaryotes. However, in eukaryotes an additional step occurs between transcription and translation. This step, called **RNA processing**, occurs when the RNA transcript, termed **pre-mRNA**, is modified in ways that make it a functionally active mRNA (**Figure 12.3b**). The processing events will be described later in this chapter.

The Protein Products of Genes Determine an Organism's Characteristics

The genes that constitute the genetic material provide a blueprint for the characteristics of every organism. They contain the information necessary to create an organism and allow it to favorably interact with its environment. Each structural gene stores the information for the production of a polypeptide, which then

becomes a unit within a functional protein. The activities of proteins determine the structure and function of cells. Furthermore, the traits or characteristics of all organisms are rooted in the activities of cellular proteins. As an example, **Figure 12.4** shows two fruit flies, one with red eyes and one with white eyes. A normal gene necessary for red eye color encodes a protein that transports pigment precursor molecules into the cell. The white-eye trait is due to a mutant gene. Depending on where the mutation is found, it could prevent the transcription of a functional mRNA or the translation of a functional polypeptide. In either case, a functional transporter would not be produced, and so the precursor molecule would not be transported across the membrane and into the cell. Therefore, the red pigment cannot be synthesized and the eyes are white. In this example, the striking difference between red and white eyes is due to the variation in the function of a single protein!

The main purpose of the genetic material is to encode the production of cellular proteins in the correct cell, at the proper time, and in suitable amounts. This is an intricate task, because living cells make thousands of different kinds of proteins. Genetic analyses have shown that a typical bacterium can make a few thousand different proteins, and estimates for eukaryotes range from several thousand in simpler eukaryotes to tens of thousands in more complex eukaryotes.

12.2 Transcription

Thus far we have considered the central dogma of gene expression. According to this theory, DNA is used as a template to make mRNA, which contains the information to make a polypeptide. This section will focus on the first step of this process,

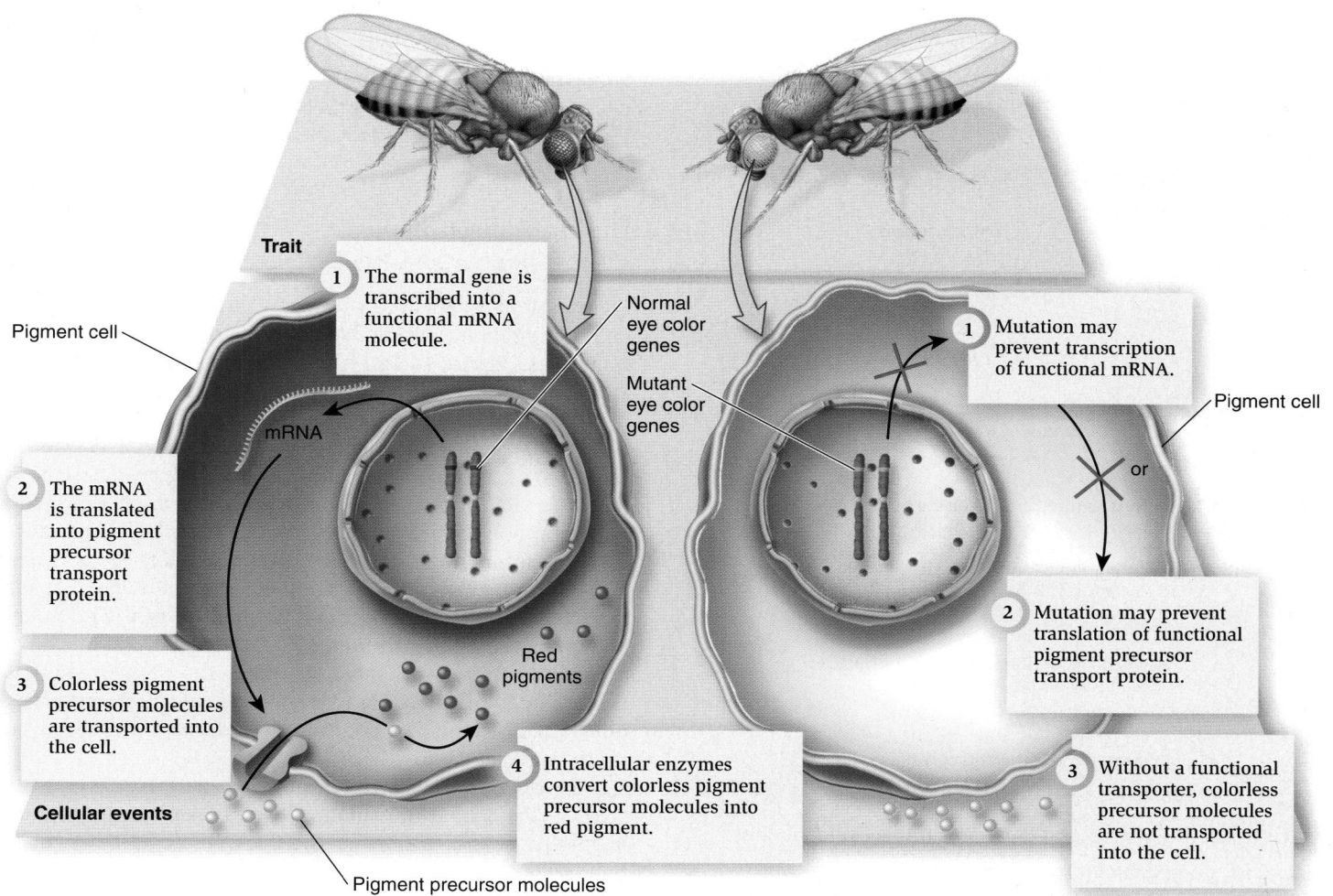

Trait

Pigment cell

1 The normal gene is transcribed into a functional mRNA molecule.

mRNA

2 The mRNA is translated into pigment precursor transport protein.

3 Colorless pigment precursor molecules are transported into the cell.

Cellular events

Normal eye color genes

Mutant eye color genes

Red pigments

4 Intracellular enzymes convert colorless pigment precursor molecules into red pigment.

Pigment precursor molecules

1 Mutation may prevent transcription of functional mRNA.

Pigment cell

or

2 Mutation may prevent translation of functional pigment precursor transport protein.

3 Without a functional transporter, colorless precursor molecules are not transported into the cell.

Figure 12.4 **An example of how a mutation in a single gene can affect the traits of an organism.**   On the *left*, the red-eye trait occurs when the normal eye color gene is expressed and produces a functional protein. However, as shown on the *right*, a mutation that renders the eye color gene inactive causes a fly to have white eyes.

namely transcription. Our understanding of transcription initially came from studies involving bacteria and bacteriophages (viruses that infect bacterial cells). In 1956, Eliot Volkin and Lazarus Astrachan first proposed that RNA is derived from the transcription of DNA. When they infected *Escherichia coli* cells with a bacteriophage (or phage for short), they discovered that the RNA made immediately after infection had a base composition that was very similar to the base composition of the phage DNA. They suggested that the bacteriophage DNA was used as a template to make bacteriophage RNA.

We now know that DNA is indeed an information storage unit. For genes to be expressed, the information in them must be accessed at the molecular level. Rather than accessing the information directly, however, a working copy of the DNA, composed of RNA, is made. This occurs by the process of transcription, in which a DNA sequence is copied into an RNA sequence. Importantly, transcription does not permanently alter the struc-

ture of DNA. Therefore, the same DNA can continue to store information even after an RNA copy has been made. In this section, we will examine the steps that are necessary for genes to act as transcriptional units. We will also consider some differences in these steps between prokaryotes and eukaryotes.

At the Molecular Level, a Gene Can Be Transcribed and Produces a Functional Product

A molecular definition of a **gene** is *an organized unit of DNA sequences that enables a segment of DNA to be transcribed into RNA and ultimately results in the formation of a functional product.* When a structural gene is transcribed, an mRNA is made that specifies the amino acid sequence of a polypeptide. In this case, the polypeptide is considered to be the functional product, while the mRNA is an intermediary in polypeptide synthesis. Among all species, over 90% of all genes are structural genes.

DNA

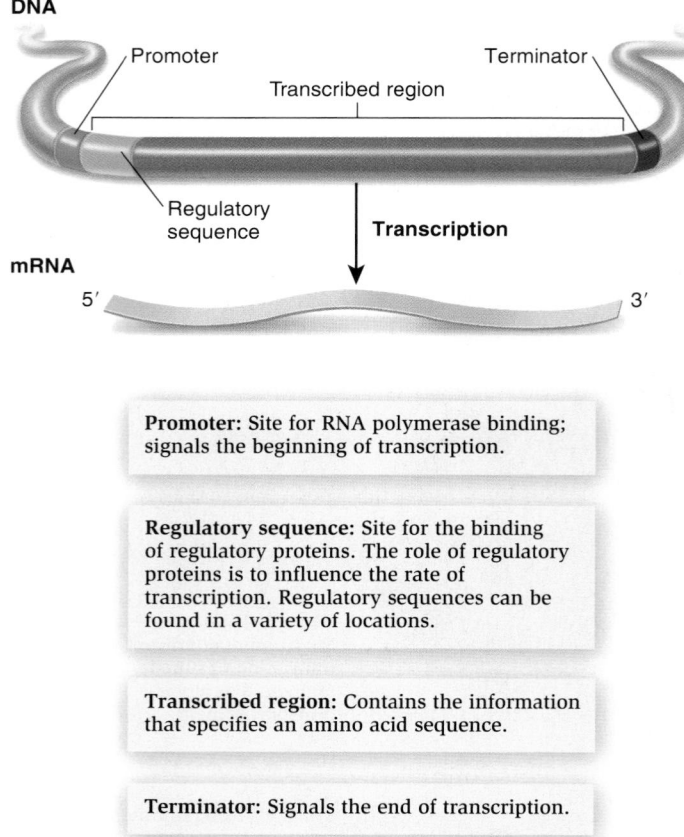

mRNA

5′   3′

Promoter: Site for RNA polymerase binding; signals the beginning of transcription.

Regulatory sequence: Site for the binding of regulatory proteins. The role of regulatory proteins is to influence the rate of transcription. Regulatory sequences can be found in a variety of locations.

Transcribed region: Contains the information that specifies an amino acid sequence.

Terminator: Signals the end of transcription.

Figure 12.5   A structural gene as a transcriptional unit.

Biological inquiry: If a terminator was removed from a gene, how would this affect transcription? Where would transcription end?

However, for other genes, the functional product is the RNA itself. The RNA from a nonstructural gene is never translated. Two important products of nonstructural genes are transfer RNA and ribosomal RNA. **Transfer RNA (tRNA)** translates the language of mRNA into that of amino acids, while **ribosomal RNA (rRNA)** forms part of ribosomes, which provide the arena in which translation occurs. We'll learn more about these two types of RNA later in this chapter.

A gene is composed of specific base sequences organized in a way that allows the DNA to be transcribed into RNA. **Figure 12.5** shows the general organization of sequences that are needed to form a structural gene. Transcription begins near a site in the DNA called the **promoter**, while the **terminator** specifies the end of transcription. Therefore, these two sequences provide the boundaries for RNA to be synthesized within a defined location. Within this transcribed region is the information that will specify the amino acid sequence of the polypeptide when the mRNA is translated. As shown in Figure 12.5, the DNA is transcribed into mRNA from the end of the promoter through the coding sequence to the terminator.

Other DNA sequences are involved in the regulation of transcription. These **regulatory sequences** function as sites for

genetic regulatory proteins. When a regulatory protein binds to a regulatory sequence, this affects the rate of transcription. Some regulatory proteins enhance the rate of transcription, while others inhibit it. This topic will be considered in Chapter 13. Now let's look at the process of transcription in prokaryotes.

During Transcription, RNA Polymerase Uses a DNA Template to Make RNA

Transcription occurs in three stages called initiation, elongation, and termination, during which proteins interact with DNA sequences (**Figure 12.6a**). The **initiation stage** is a recognition step. In bacteria such as *E. coli*, a protein called **sigma factor** binds to **RNA polymerase**, which is the enzyme that synthesizes strands of RNA. Sigma factor also recognizes the base sequence of a promoter and binds there. In this way, sigma factor causes RNA polymerase to recognize a promoter sequence.

RNA polymerase synthesizes the RNA transcript during the **elongation stage**. For this to occur, sigma factor is released and RNA polymerase slides along the DNA in a way that separates the DNA strands, forming a small bubble-like structure known as the **open complex** (or transcription bubble) that is approximately 10 to 15 base pairs long. The DNA strand that is used as a template for RNA synthesis is called the **template** or **noncoding strand**. The opposite DNA strand is called the **coding strand**.

As shown in Figure 12.6, a new molecule of RNA is synthesized in the 5′ to 3′ direction. The complementarity rule used in this process is similar to the AT/GC rule of DNA replication, except that uracil substitutes for thymine in RNA. For example, an RNA with a sequence of nucleotides reading 5′–AUGUUACAUCGG–3′ will be transcribed from a DNA template with a sequence of 3′–TACAATGTAGCC–5′. In bacteria, the rate of RNA synthesis is about 40 nucleotides per second! Behind the open complex, the DNA rewinds back into a double helix. Eventually, RNA polymerase reaches a termination sequence, which causes it and the newly made RNA transcript to dissociate from the DNA. This event constitutes the **termination stage** of transcription.

The catalytic portion of RNA polymerase that is responsible for the synthesis of RNA has a similar structure in all species, including prokaryotes and eukaryotes. The structure of a bacterial RNA polymerase is shown in **Figure 12.6b**. RNA polymerase contains a cavity that allows it to slide along the DNA. The DNA strands enter at the side of the protein and are separated at a region where the cavity makes a right-angle turn. Nucleotides can enter this region through a small pore, and RNA is made in a 5′ to 3′ direction. Both the DNA and the newly made strand of RNA then exit from the top of the protein.

When considering the transcription of multiple genes within a chromosome, the direction of transcription and the DNA strand that is used as a template vary among different genes. **Figure 12.7** shows three genes that are adjacent to each other within a chromosome. Genes A and B are transcribed from left to right, using the bottom DNA strand as a template.

3′
5′

Sigma
factor

Promoter

RNA
polymerase

Open
complex

Direction
of travel

Rewinding of
transcribed DNA
(upstream)

5′ end

mRNA
transcript

A T C G G
U A G C C

5′ ——→ 3′
Directions of
RNA synthesis

3′

Completed
mRNA transcript

ATCGG
UAGCC

3′ end
(growing end)

Unwinding of
entering DNA
(downstream)

Terminator

1 | **Initiation:**
The promoter functions as a recognition site for sigma factor. RNA polymerase is bound to sigma factor, which causes it to recognize the promoter. Following binding, the DNA is unwound into a bubble known as the open complex.

2 | **Elongation/synthesis of the RNA transcript:**
Sigma factor is released and RNA polymerase slides along the DNA in an open complex to synthesize RNA.

3 | **Termination:**
When RNA polymerase reaches the terminator, it and the RNA transcript dissociate from the DNA.

(a) Stages of transcription

Figure 12.6 **Stages of transcription.** **(a)** Transcription can be divided into initiation, elongation, and termination. The inset emphasizes the direction of RNA synthesis and base pairing between the DNA template strand and RNA. **(b)** Three-dimensional structure of a bacterial RNA polymerase.

(b) Structure of a bacterial RNA polymerase

Figure 12.7 **The transcription of three different genes that are found in the same chromosome.** RNA polymerase synthesizes each RNA transcript in a 5′ to 3′ direction, sliding along a DNA template strand in a 3′ to 5′ direction. However, the use of the template strand can vary from gene to gene. For example, genes A and B use the bottom strand, while gene C uses the top strand.

By comparison, gene C is transcribed from right to left, using the top DNA strand as a template. Notice that in all three cases the synthesis of the RNA transcript occurs in a 5′ to 3′ direction, and the template strand is read in the 3′ to 5′ direction.

Eukaryotic Transcription Is Fundamentally Similar to Prokaryotic Except That More Proteins Are Involved in the Process

The basic features of transcription are identical between prokaryotic and eukaryotic organisms. Eukaryotic genes have promoters and the transcription process involves initiation, elongation, and termination. However, in eukaryotes, each step tends to involve a greater complexity of protein components. For example, three forms of RNA polymerase are found in eukaryotes, designated I, II, and III. RNA polymerase II is responsible for transcribing the mRNA from eukaryotic structural genes, while RNA polymerases I and III transcribe nonstructural genes such as the genes that encode tRNAs and rRNAs. By comparison, bacteria have a single type of RNA polymerase that transcribes all genes.

With regard to the initiation of transcription, the eukaryotic process is also more complex. Recall that bacteria such as *E. coli* use a single protein, sigma factor, to recognize the promoter of genes. By comparison, RNA polymerase II of eukaryotes always requires five general transcription factors to initiate transcription. **Transcription factors** are proteins that influence the ability of RNA polymerase to transcribe genes. In addition, the regulation of gene transcription in eukaryotes typically involves the function of several different proteins. This topic will be considered in Chapter 13.

12.3 RNA Processing in Eukaryotes

During the 1960s and 1970s, the physical structure of the gene became well established based largely on studies of bacterial genes. Most bacterial mRNAs can be translated into polypeptides as soon as they are made. By comparison, eukaryotic mRNA transcripts undergo RNA processing or modification that is needed for their proper translation. In eukaryotes, transcription produces a longer RNA, pre-mRNA, which undergoes certain processing events before it exits the nucleus. The final product is called a **mature mRNA**.

In the late 1970s, when the experimental tools became available to study eukaryotic genes at the molecular level, the scientific community was astonished by the discovery that the coding sequences within many eukaryotic structural genes are separated by DNA sequences that are transcribed but not translated into protein. These intervening sequences that are not translated are called **introns**, while coding sequences are found within **exons** contained in the mature mRNA. To create a functional mRNA, the RNA undergoes a process known as **splicing** in which the introns are removed and the remaining exons are

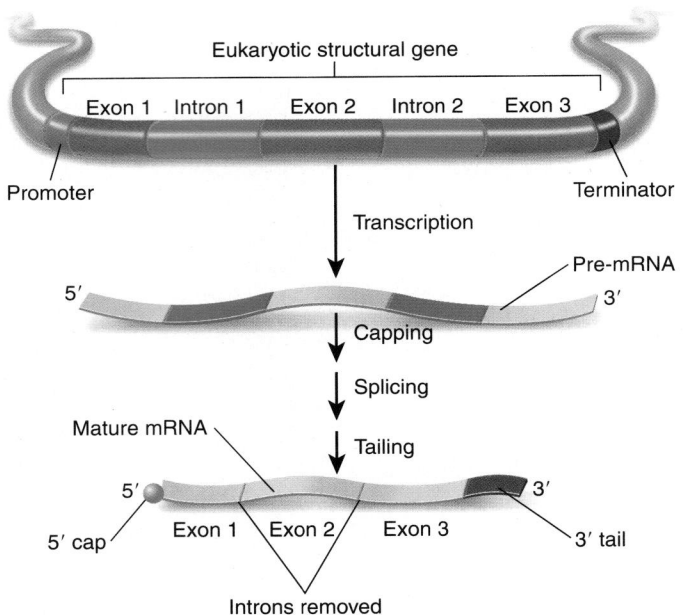

Figure 12.8 Modifications to eukaryotic pre-mRNA that are needed to create a functional (mature) mRNA molecule.

connected to each other (**Figure 12.8**). In addition to splicing, research has shown that pre-mRNA transcripts are modified in other ways, including the addition of caps and tails to the ends of the mRNA. After these modifications have been completed, the mRNA leaves the nucleus and enters the cytosol, where translation occurs. In this section, we will examine the molecular mechanisms that account for these RNA processing events and consider why they are functionally important.

Splicing Involves the Removal of Introns and the Linkage of Exons

In the late 1970s, several research groups, including those of Pierre Chambon, Bert O'Malley, and Phillip Leder, investigated the presence of introns in eukaryotic structural genes. Leder's experiments used electron microscopy to identify introns in the β-globin gene. β-globin is a polypeptide that is a subunit of hemoglobin. To identify introns, Leder used a strategy that involved the binding of mRNA to DNA. In this approach, the double-stranded DNA of the β-globin gene was first separated into single strands, and then mixed with mature mRNA from the β-globin gene that had been isolated from red blood cells. The mature mRNA, which already had its introns removed, was allowed to bind to the template strand of DNA, causing the intron in the DNA to loop out. Next, the coding strand was allowed to bind to the template strand, although it could not bind in places where the RNA was already bound. Leder and his colleagues realized that if mRNA binds to a region of a gene containing one intron, two single-stranded DNA loops

1 Mature mRNA binds to template strand, which causes the intron DNA to loop out.

Coding strand   Intron
5′
3′
Template strand
Discontinuous exons that are complementary to mRNA

Transcription

Pre-mRNA

Splicing

+   5′   Mature mRNA   3′

3′
5′

Intron DNA of template strand

3′   5′
Template strand   mRNA bound to template strand

2 The coding strand then binds to the template strand, but it loops out where the RNA is already bound.

Coding strand

Intron DNA
Single-stranded loop

5′   3′
3′   5′

(a) Schematic drawing of mature mRNA binding to DNA containing an intron

Intron DNA
Bound mRNA
Single-stranded loop

(b) Results after electron microscopy

Figure 12.9   Leder's experiment showing the binding of mRNA to DNA that contains an intron.   **(a)** The steps in the experimental procedure. **(b)** A micrograph of the results.

Biological inquiry: Explain why the double-stranded intron loops out of the structure shown in part (b).

will form that are separated by a double-stranded DNA region (**Figure 12.9a**, step 2). The intervening double-stranded region occurs because an intron has been spliced out of the mature mRNA, so that the mRNA cannot bind to this segment of the gene. In the actual experiment, samples were observed under the electron microscope. As seen in **Figure 12.9b**, the mRNA was bound to the DNA to form two single-stranded DNA loops separated by a double-stranded DNA region that is the intron. Chambon and O'Malley obtained similar results for other structural genes.

Since these initial discoveries, introns have been found in many eukaryotic genes. Splicing is less frequent among unicellular eukaryotic species, such as yeast, but is a widespread phenomenon among more complex eukaryotes. In mammals and flowering plants, most structural genes have one or more introns. As an extreme example, the human dystrophin gene, which harbors a mutation in people with Duchenne muscular dystrophy, has 79 exons punctuated by 78 introns. Introns are also found in a few bacterial genes, but they are very rare among all prokaryotic species.

Introns are removed from eukaryotic pre-mRNA by a large complex called a **spliceosome** that is composed of several subunits known as snRNPs (pronounced "snurps"). Each snRNP contains small nuclear RNA and a set of proteins. This small nuclear RNA is the product of a nonstructural gene. Intron RNA is defined by particular sequences within the intron and at the intron-exon boundaries (**Figure 12.10**). These include a 5′ splice site, a branch site, and a 3′ splice site. Spliceosome subunits bind to specific sequences at these three locations. This binding causes the intron to loop outward, and the two exons are brought closer together. The 5′ splice site is then cut, and the 5′ end of the intron becomes covalently attached to the branch site. In the final step, the 3′ splice site is cut, and then the exons are covalently attached to each other. The intron is released and will be eventually degraded.

In some cases, the function of the spliceosome can be regulated so that the splicing of exons for a given mRNA can occur in two or more ways. This phenomenon, called **alternative splicing**, enables a single gene to encode two or more polypeptides with differences in their amino acid sequences. This topic will be considered in Chapter 13.

Although primarily in mRNAs, introns can be found in rRNA and tRNA molecules of certain species. These introns, however, are not removed by the action of a spliceosome. Instead, rRNA and tRNAs are **self-splicing**, which means that the RNA itself can catalyze the removal of its own intron. Portions of the RNA act like an enzyme to cleave the covalent bonds at the intron-exon boundaries and connect the exons together. An RNA molecule that catalyzes a chemical reaction is termed a **ribozyme**.

RNA Processing Also Involves Adding a 5′ Cap and a 3′ Poly A Tail to the Ends of Eukaryotic mRNAs

Mature mRNAs of eukaryotes have a 7-methylguanosine covalently attached at the 5′ end, an event known as **capping** (**Figure 12.11a**). Capping occurs while a pre-mRNA is being made by RNA polymerase, usually when the transcript is only 20 to 25 nucleotides in length. The 7-methylguanosine cap structure, called a **5′ cap**, is recognized by cap-binding proteins, which are needed for the proper exit of mRNA from the nucleus. Also, after an mRNA is in the cytosol, the cap structure is recognized by other cap-binding proteins that enable the mRNA to bind to a ribosome for translation.

At the 3′ end, most mature eukaryotic mRNAs have a string of adenine nucleotides, typically 100 to 200 nucleotides in length, referred to as a **poly A tail** (**Figure 12.11b**). A long poly A tail causes a eukaryotic mRNA to be more stable and thereby exist for a longer period of time in the cytosol. The poly A tail is not encoded in the gene sequence. Instead, the tail is added enzymatically after a pre-mRNA has been completely transcribed. Interestingly, new research has shown that some bacterial mRNAs also have poly A tails attached to them. However, the poly A tail has an opposite effect in bacteria; it causes the mRNA to be rapidly degraded. The importance of a poly A tail in bacterial mRNAs is not well understood.

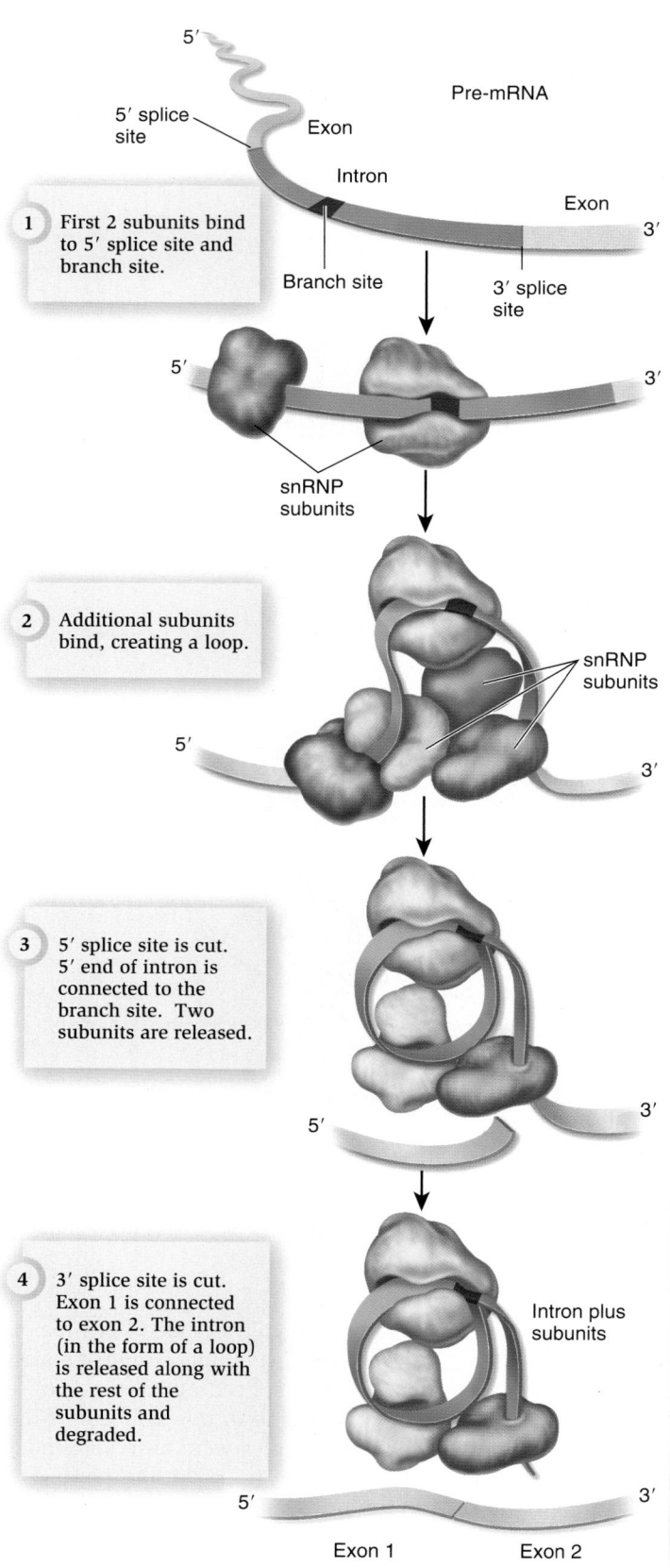

Figure 12.10 The splicing of a eukaryotic pre-mRNA by a spliceosome.

(a) Cap structure at the 5′ end of eukaryotic mRNA

7-methylguanosine cap

Figure 12.11
Modifications that occur at the ends of mRNA in eukaryotic cells. (a) A 7-methylguanosine cap is attached to the 5′ end. (b) A poly A tail is added to the 3′ end.

A poly A tail consisting of 100–200 adenine nucleotides is added after transcription.

Poly A tail

(b) Addition of a poly A tail at the 3′ end of eukaryotic mRNA

12.4 Translation and the Genetic Code

In the two previous sections, we considered how an RNA transcript is made, and how eukaryotes process that transcript. Now we will begin to examine the next process, that of translation, at the molecular level. In 1960, Matthew Meselson and Francois Jacob found that proteins are synthesized on cellular structures known as ribosomes. One year later, Francois Jacob and Jacques Monod made an insightful hypothesis. They proposed that RNA, which is transcribed from DNA, provides the information for protein synthesis via ribosomes. This type of RNA, which they named messenger RNA (mRNA), carries information from the DNA to the ribosome, where proteins are made during the process of translation. Since these early studies, much has been learned about the details of translation.

Translation involves an interpretation of one language into another. In this case, the language of mRNA, which is a nucleotide sequence, is translated into the language of a polypeptide, which is an amino acid sequence. To understand the process of translation, we will first examine the **genetic code**, which specifies the relationship between the sequence of nucleotides in the mRNA and the sequence of amino acids in a polypeptide.

During Translation, the Genetic Code Is Used to Make a Polypeptide with a Specific Amino Acid Sequence

The ability of mRNA to be translated into a polypeptide relies on the genetic code. This code is contained in the sequence of bases in an mRNA molecule. The code is read in groups of three nucleotide bases known as **codons** (**Table 12.1**). The sequence of three bases in most codons specifies a particular amino acid. For example, the codon CCC specifies the amino acid proline, while the codon GGC encodes the amino acid glycine. From the analysis of many different species, including bacteria, protists, fungi, plants, and animals, researchers have found that the genetic code is nearly universal. Only a few rare exceptions to the genetic code have been noted (see note to Table 12.1).

Because there are 20 types of amino acids, at least 20 codons are needed to specify each type. With four types of bases in mRNA (A, U, G, and C), a genetic code containing two bases in a codon would not be sufficient, because only 4^2, or 16, different codons would be possible. By comparison, a three-base codon system can specify 4^3, or 64, different codons. Because the number of possible codons exceeds 20, the genetic code is said to be **degenerate**. This means that more than one codon can specify the same amino acid (Table 12.1). For example, the codons GGU, GGC, GGA, and GGG all specify the amino acid glycine. In most instances, the third base in the codon is the degenerate or variable base.

Let's look at the organization of a bacterial mRNA to see how translation occurs (**Figure 12.12**). The 5′ end of the mRNA contains a ribosomal-binding site. Beyond the 5′ end, a large portion of an mRNA functions as a coding sequence that specifies the amino acid sequence of a polypeptide. This coding sequence consists of a series of codons. The **start codon**, which is only a few nucleotides from the ribosomal-binding site, is usually AUG. This codon specifies methionine. The many codons that follow the start codon dictate the linear sequence of amino acids within a given polypeptide. A typical polypeptide is a few hundred amino acids in length. Finally, one of three **stop codons** signals the end of translation. These codons, also known as **termination** or **nonsense codons**, are UAA, UAG, and UGA.

Table 12.1 — The Genetic Code*

		U	C	A	G	
				Second position		
		U	**C**	**A**	**G**	
First position	U	UUU } Phe UUC UUA } Leu UUG	UCU } UCC } Ser UCA UCG	UAU } Tyr UAC UAA Stop UAG Stop	UGU } Cys UGC UGA Stop UGG Trp	U C A G
	C	CUU } CUC } Leu CUA CUG	CCU } CCC } Pro CCA CCG	CAU } His CAC CAA } Gln CAG	CGU } CGC } Arg CGA CGG	U C A G
	A	AUU } AUC } Ile AUA AUG Met/ start	ACU } ACC } Thr ACA ACG	AAU } Asn AAC AAA } Lys AAG	AGU } Ser AGC AGA } Arg AGG	U C A G
	G	GUU } GUC } Val GUA GUG	GCU } GCC } Ala GCA GCG	GAU } Asp GAC GAA } Glu GAG	GGU } GGC } Gly GGA GGG	U C A G

*Exceptions to the genetic code are sporadically found among various species. A few examples are as follows: AUA encodes methionine in yeast and mammalian mitochondria; UGA encodes tryptophan in mammalian mitochondria; CUU, CUA, CUC, and CUG encode threonine in yeast mitochondria; AGA and AGG act as stop codons in ciliated protozoa and in yeast and mammalian mitochondria; and UAA and UAG encode glutamine in ciliated protozoa.

The start codon also defines the **reading frame** of an mRNA. Beginning at the start codon, each adjacent codon is read as a group of three bases, that is, a **triplet**, in the 5′ to 3′ direction. For example, look at the following two mRNA sequences and their corresponding amino acid sequences. The first sequence shows how the mRNA codons would be correctly translated into amino acids. In the second sequence, an additional U has been added to the same sequence after the start codon. This shifts the reading frame and thereby changes the codons as they occur in the 5′ to 3′ direction. This change results in the translation of the sequence into different amino acids.

	Ribosomal- binding site	Start codon

mRNA 5′–<u>AUAAGGAGG</u>UUACG (<u>AUG</u>)(CAG)(CAG)(GGC)(UUU)(ACC)–3′

Polypeptide   Met - Gln - Gln - Gly - Phe - Thr

	Ribosomal- binding site	Start codon

mRNA 5′–<u>AUAAGGAGG</u>UUACG(<u>AUG</u>)(UCA)(GCA)(GGG)(CUU)(UAC)C–3′

Polypeptide   Met - Ser - Ala - Gly - Leu - Tyr

From this comparison, we can also see that the genetic code is nonoverlapping, which means that each base functions within a single codon.

The relationships among the coding sequence of a gene, the codons in mRNA, and a polypeptide sequence are shown

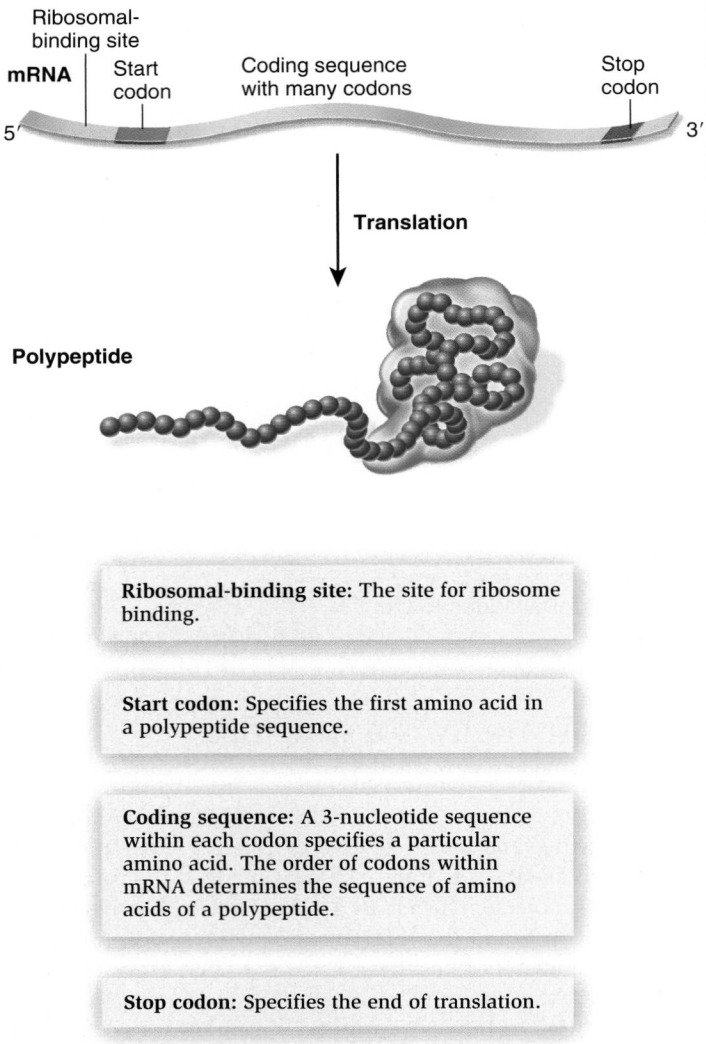

Ribosomal-binding site: The site for ribosome binding.

Start codon: Specifies the first amino acid in a polypeptide sequence.

Coding sequence: A 3-nucleotide sequence within each codon specifies a particular amino acid. The order of codons within mRNA determines the sequence of amino acids of a polypeptide.

Stop codon: Specifies the end of translation.

Figure 12.12 The organization of mRNA as a translational unit.

schematically in **Figure 12.13**. The coding strand of DNA corresponds to the mRNA strand, except that U in the mRNA is substituted for T in the DNA. The template strand is used to make mRNA. To translate a nucleotide sequence of mRNA into an amino acid sequence, recognition occurs between mRNA and transfer RNA (tRNA) molecules. Transfer RNA functions as the "translator" or intermediary between an mRNA codon and an amino acid. The **anticodon** allows a tRNA molecule to bind to a complementary codon in mRNA. Furthermore, the anticodon in a tRNA corresponds to the amino acid that it carries. For example, if the anticodon in a tRNA is 3′–AAG–5′, it is complementary to a 5′–UUC–3′ codon. According to the genetic code, a UUC codon specifies phenylalanine. Therefore, a tRNA with a 3′–AAG–5′ anticodon must carry phenylalanine. As another example, a tRNA with a 3′–GGG–5′ anticodon is complementary to a 5′– CCC–3′ codon, which specifies proline. This tRNA must carry proline.

Figure 12.13   Relationships among the coding sequence of a gene, the codon sequence of an mRNA, the anticodons of tRNA, and the amino acid sequence of a polypeptide.

Biological inquiry: If an anticodon in a tRNA molecule has the sequence 3′–ACC–5′, which amino acid does it carry?

Synthetic RNA Helped to Decipher the Genetic Code

Now let's look at some early experiments that allowed scientists to decipher the genetic code. During the early 1960s, the genetic code was determined by the collective efforts of several research groups headed by Marshall Nirenberg, Severo Ochoa, H. Gobind Khorana, and Phillip Leder. Prior to their studies, other scientists had discovered that bacterial cells can be broken open and components from the cytoplasm can synthesize polypeptides. This is termed an *in vitro* or cell-free translation system. Nirenberg and Ochoa made synthetic RNA molecules using an enzyme that covalently connects nucleotides together. Then, using this synthetic mRNA, they determined which amino acids were incorporated into polypeptides. For example, if an RNA molecule had only adenine-containing nucleotides (for example, 5′–AAAAAAAAAAAAAAAAAAAA–3′), a polypeptide was produced that contained only lysine. This result indicated that the AAA codon specifies lysine.

Khorana and his collaborators developed a novel method to synthesize RNA. They first created short RNA molecules that were two to four nucleotides in length and had a defined sequence. For example, RNA molecules with the sequence 5′–AUC–3′ were synthesized chemically. These short RNAs were then linked together enzymatically, in a 5′ to 3′ manner, to create a longer RNA molecule:

5′–AUCAUCAUCAUCAUCAUCAUCAUCAUCAUC–3′

Depending on whether the reading frame begins with the first, second, or third base in this sequence, this RNA contains three different codons: AUC (isoleucine), UCA (serine), and CAU (histidine). Using a cell-free translation system, this RNA encoded a polypeptide containing only isoleucine, serine, and histidine.

Another method used to decipher the genetic code involved the chemical synthesis of short RNA molecules. This method is described in the Feature Investigation.

FEATURE INVESTIGATION

Nirenberg and Leder Found That RNA Triplets Can Promote the Binding of tRNA to Ribosomes

In 1964, Nirenberg and Leder discovered that RNA molecules containing any three nucleotides (that is, any triplet) can stimulate ribosomes to bind a tRNA molecule. In other words, an RNA triplet can act like a codon within an mRNA molecule. Ribosomes bind RNA triplets, and then a tRNA with the appropriate anticodon subsequently binds to the ribosome.

To establish the relationship between triplet sequences and specific amino acids, Nirenberg and Leder began with samples containing ribosomes and a particular triplet (**Figure 12.14**). For example, in one experiment they began with a sample of ribosomes that were mixed with 5′–CCC–3′ triplets. The triplets became bound to the ribosomes just like the binding of mRNA to a ribosome; we'll go through the whole process of translation

later in the chapter, but for now just keep in mind that tRNAs interact with mRNA on a ribosome during the synthesis of polypeptides. This sample was divided into 20 tubes. To each tube, they next added a mixture of cellular tRNAs that already had amino acids attached to them. However, each mixture of tRNAs had only one type of radiolabeled amino acid. For example, one mixture of tRNAs had only proline that was radiolabeled, a second mixture of tRNAs had only serine that was radiolabeled, and so on. The tRNA with the anticodon that was complementary to the added triplet would bind to the triplet, which was already bound to the ribosome. For example, if the triplet was 5′–CCC–3′, a tRNA with a 3′–GGG–5′ anticodon would be bound to the triplet/ribosome complex. This tRNA carries proline.

To determine which tRNA had bound, the samples from each tube were poured through a filter that trapped the large

Figure 12.14 Nirenberg and Leder's use of triplet binding assays to decipher the genetic code.

> **HYPOTHESIS** A triplet RNA can bind to a ribosome and promote the binding of the tRNA that carries the amino acid that the triplet RNA specifies.
>
> **STARTING MATERIALS** Components of an *in vitro* translation system, including ribosomes and tRNAs. Preparations containing all of the different tRNA molecules were given 1 radiolabeled amino acid; the other 19 amino acids were nonlabeled. For example, in 1 sample, radiolabeled glycine was added and the other 19 amino acids were nonlabeled. In a different sample, radiolabeled proline was added and the other 19 amino acids were nonlabeled. The tRNA preparation also contained the enzymes that attach amino acids to tRNAs.

Experimental level — **Conceptual level**

1. Mix together triplet RNAs of a specific sequence and ribosomes. In the example shown here, the triplet is 5′–CCC–3′. Add a tRNA sample to this mixture that contains 1 radiolabeled amino acid. (Note: Only 3 tubes are shown here. Because there are 20 different amino acids, this would be done in 20 different tubes.)

   tRNAs with 1 radiolabeled amino acid (for example, proline)

   Proline / Ribosome

2. Allow time for triplet RNA to bind to the ribosome, and for the appropriate tRNA to bind to the triplet RNA.

   Radiolabeled proline / Proline tRNA / Triplet RNA that specifies proline

3. Pour mixture through a filter that allows the passage of unbound tRNA but does not allow the passage of ribosomes.

   Ribosomes trapped on filter / Filter / Filter

4   Count radioactivity on the filter.

Scintillation counter

5   **THE DATA**

Triplet	Radiolabeled amino acid trapped on the filter	Triplet	Radiolabeled amino acid trapped on the filter
5′ – AAA – 3′	Lysine	5′ – GAC – 3′	Aspartic acid
5′ – ACA – 3′, 5′ – ACC – 3′	Threonine	5′ – GCC – 3′	Alanine
5′ – AGA – 3′	Arginine	5′ – GGU – 3′, 5′ – GGC – 3′	Glycine
5′ – AUA – 3′, 5′ – AUU – 3′	Isoleucine	5′ – GUU – 3′	Valine
5′ – CCC – 3′	Proline	5′ – UAU – 3′	Tyrosine
5′ – CGC – 3′	Arginine	5′ – UGU – 3′	Cysteine
5′ – GAA – 3′	Glutamic acid	5′ – UUG – 3′	Leucine

ribosomes but did not trap tRNAs that were not bound to ribosomes. The researchers then determined the amount of radioactivity on each filter. If the tRNA carrying the radiolabeled amino acid was bound to the triplet/ribosome complex, radioactivity would be trapped on the filter. Because only one amino acid was radiolabeled in each tRNA sample, they could determine which triplet corresponded to which amino acid. In the example shown here, CCC corresponds to proline. Therefore, the tRNA sample containing radiolabeled proline showed a large amount of radioactivity on the filter. By studying triplets with different sequences, Nirenberg and Leder could identify many codons of the genetic code.

12.5 The Machinery of Translation

Let's now turn our attention to the components found in living cells that are needed to use the genetic code and translate mRNA into polypeptides. Earlier in this chapter, we considered the first step in gene expression, namely, transcription. To transcribe an RNA molecule, a pre-existing DNA template strand is used to make a complementary RNA strand. A single enzyme, RNA polymerase, can catalyze this reaction. By comparison, translation requires more components because the sequence of codons in an mRNA molecule must be translated into a sequence of amino acids according to the genetic code. A single protein cannot accomplish such a task. Instead, many different proteins and RNA molecules must interact in an intricate series of steps to achieve the synthesis of a polypeptide.

Due to its complexity, translation is a costly process from an energy point of view. A cell must make many different components, including mRNAs, tRNAs, ribosomes, and translation factors, so that polypeptides can be made (**Table 12.2**). The synthesis of these components uses a great deal of cellular energy. In addition, the mechanism of making a polypeptide, which we will examine later in this chapter, also uses a large amount of energy. Though the estimates vary from cell to cell

Table 12.2	Components of the Translation Machinery
Component	**Function**
mRNA	Contains the information for a polypeptide sequence according to the genetic code.
tRNA	A molecule with two functional sites. One site, termed the anticodon, recognizes a codon in mRNA. A second site has the appropriate amino acid attached to it.
Ribosomes	Composed of many proteins and rRNA molecules. The ribosome provides a location where mRNA and tRNA molecules can properly interact with each other. The ribosome also catalyzes the formation of covalent bonds between adjacent amino acids so that a polypeptide can be made.
Translation factors	Three categories of translation factors are needed for the three stages of translation that are described in the last section of this chapter. Initiation factors are required for the assembly of mRNA, the first tRNA, and ribosomal subunits. Elongation factors are needed to synthesize the polypeptide. And release factors are needed to recognize the stop codon and disassemble the translation machinery. Several translation factors use GTP as an energy source to carry out their functions.

and from species to species, most cells use a substantial amount of their energy to translate mRNA into polypeptides. In *E. coli*, for example, approximately 90% of the cellular energy is used for this process. This value underscores the complexity and importance of translation in living organisms. In this section, we will focus on the components of the translation machinery. The last section of the chapter will describe the steps of translation as they occur in living cells.

Transfer RNAs Share Common Structural Features

To understand how tRNAs act as carriers of the correct amino acids during translation, researchers have examined their structural characteristics. The cells of every organism make many different tRNA molecules, each encoded by a different gene. A tRNA is named according to the amino acid it carries. For example, tRNA^ser carries a serine. Because the genetic code contains six different serine codons, a cell produces more than one type of tRNA^ser.

The tRNAs of both prokaryotes and eukaryotes share common features. As originally proposed by Robert Holley in 1965, the two-dimensional structure of tRNAs exhibits a cloverleaf pattern. The structure has three stem-loops and a 3′ single-stranded region (**Figure 12.15a**). The stems are regions where the RNA is double stranded due to complementary base pairing, while the loops are regions without base pairing. The anticodon is located in the loop of the second stem-loop region. The 3′ single-stranded region is called the acceptor stem because it accepts the attachment of an amino acid. The three-dimensional structure of tRNA molecules involves additional folding of the secondary structure (**Figure 12.15b**).

Aminoacyl-tRNA Synthetases Charge tRNAs by Attaching an Appropriate Amino Acid

To perform its role during translation, a tRNA must have the appropriate amino acid attached to its 3′ end. The enzymes that catalyze the attachment of amino acids to tRNA molecules are known as **aminoacyl-tRNA synthetases**. Cells make 20 distinct aminoacyl-tRNA synthetase enzymes, one for each of the 20 different amino acids. Each aminoacyl-tRNA synthetase is named for the specific amino acid it attaches to tRNA. For example, alanyl-tRNA synthetase recognizes a tRNA with an alanine anticodon (that is, tRNA^ala) and attaches an alanine to it.

Aminoacyl-tRNA synthetases catalyze two chemical reactions involving an amino acid, a tRNA molecule, and an ATP (**Figure 12.16**). After an amino acid and ATP have bound to the enzyme, the amino acid is activated by the covalent attachment of an AMP molecule. Pyrophosphate is released. Next, the activated amino acid is covalently attached to the 3′ end of a tRNA molecule and AMP is released. Finally, the tRNA with its attached amino acid, called a **charged tRNA** or an **aminoacyl tRNA**, is released from the enzyme.

The ability of each aminoacyl-tRNA synthetase to recognize an appropriate tRNA has been called the second genetic code. A precise recognition process is necessary to maintain

(a) Secondary structure of tRNA

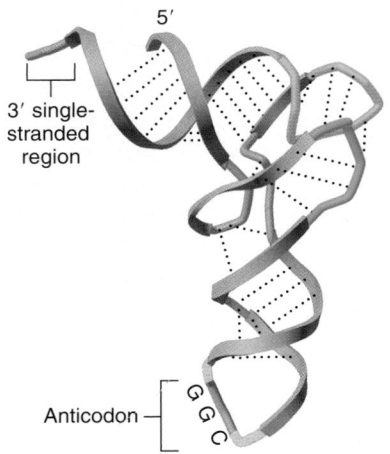

(b) Three-dimensional structure of tRNA

Figure 12.15 **Structure of tRNA.** (a) The two-dimensional or secondary structure is that of a cloverleaf. Notice the anticodon on the second of the three stem-loop structures. The single-stranded region (acceptor stem) is where an amino acid can attach. (b) The actual three-dimensional structure folds in on itself.

the fidelity of genetic information. If the wrong amino acid was attached to a tRNA, the amino acid sequence of the translated polypeptide would be incorrect. To prevent this from happening, aminoacyl-tRNA synthetases are amazingly accurate enzymes. The wrong amino acid is attached to a tRNA less than once in 100,000 times! The anticodon region of the tRNA is usually important for recognition by the correct aminoacyl-tRNA synthetase. In addition, the acceptor stem and base sequences in other stem-loop regions may facilitate binding to an aminoacyl-tRNA synthetase.

Ribosomes Are Assembled from rRNA and Proteins

As described earlier in Figure 12.13, the act of translation involves an interaction between mRNA and tRNA molecules. The ribosome can be thought of as a macromolecular arena where

Figure 12.16   Aminoacyl-tRNA synthetase charging a tRNA.

1 A specific amino acid and ATP bind to the aminoacyl-tRNA synthetase.

2 AMP is covalently bound to the amino acid thereby activating it, and pyrophosphate is released.

3 The correct tRNA binds to the synthetase. The amino acid is covalently attached to the tRNA. AMP is released.

4 The "charged" tRNA is released.

translation takes place. Bacterial cells have one type of ribosome, which translates all mRNAs in the cytoplasm. However, because eukaryotic cells are compartmentalized into cellular organelles that are bounded by membranes, their translation machinery is more complex. Biochemically distinct ribosomes are found in different cellular compartments. The most abundant type of eukaryotic ribosome functions in the cytosol. In addition, all eukaryotic cells have ribosomes in their mitochondria. Plant and algal cells also have ribosomes in their chloroplasts. The compositions of mitochondrial and chloroplast ribosomes are more similar to bacterial ribosomes than they are to eukaryotic cytosolic ribosomes. Unless otherwise noted, the term eukaryotic ribosome refers to ribosomes in the cytosol, not to those found in organelles.

A ribosome is a large complex composed of structures called the large and small subunits. The term subunit is perhaps misleading, because each ribosomal subunit is itself formed from the assembly of many different proteins and RNA molecules. In ribosomes found in the bacterium *E. coli*, the small subunit is called 30S, and the large subunit is 50S (**Figure 12.17a**). The designations 30S and 50S refer to the rate at which these subunits sediment when subjected to a centrifugal force. This rate is described as a sedimentation coefficient in Svedberg units (S) in honor of Theodor Svedberg, who invented the ultracentrifuge. The 30S subunit is formed from the assembly of 21 different ribosomal proteins and one 16S rRNA molecule. The 50S subunit contains 34 different proteins and two different rRNA

molecules, called 5S and 23S (Figure 12.17a). Together, the 30S and 50S subunits form a 70S ribosome (Svedberg units don't add up linearly, because they are a function of both size and shape). In bacteria, the ribosomal proteins and rRNA molecules are synthesized in the cytoplasm, and the ribosomal subunits are assembled there as well.

Eukaryotic ribosomes consist of subunits that are slightly larger than their bacterial counterparts (**Figure 12.17b**). In eukaryotes, 40S and 60S subunits are combined to form an 80S ribosome. The 40S subunit is composed of 33 proteins and an 18S rRNA. The 60S subunit has 49 proteins and 5S, 5.8S, and 28S rRNAs. The synthesis of eukaryotic rRNA occurs in the nucleolus, a region of the nucleus that is specialized for that purpose. The ribosomal proteins are made in the cytosol and imported into the nucleus. The rRNAs and ribosomal proteins are then assembled to make the 40S and 60S subunits within the nucleolus. The 40S and 60S subunits are exported into the cytosol, where they associate to form an 80S ribosome during translation.

Due to structural differences between bacterial and eukaryotic ribosomes, certain chemicals may bind to bacterial ribosomes but not to eukaryotic ribosomes, and vice versa. Some **antibiotics**, which are chemicals that inhibit the growth of certain microorganisms, bind only to bacterial ribosomes and inhibit translation. Examples include erythromycin and chloramphenicol. Because these chemicals do not inhibit eukaryotic ribosomes, they have been effective drugs for the treatment of bacterial infections in humans and domesticated animals.

(a) Bacterial ribosomes

16S rRNA

23S rRNA 5S rRNA

21 proteins

34 proteins

30S

50S

70S

18S rRNA

28S rRNA
5S rRNA | 5.8S rRNA

33 proteins

49 proteins

40S

60S

80S

(b) Eukaryotic ribosomes

Figure 12.17 **Composition of ribosomes.** **(a)** Bacterial ribosomes. **(b)** Eukaryotic ribosomes found in the cytosol.

Components of Ribosomal Subunits Form Functional Sites for Translation

To understand the structure and function of the ribosome at the molecular level, researchers have determined the locations and functional roles of the individual ribosomal proteins and rRNAs. In recent years, a few research groups have succeeded in purifying ribosomes and causing them to crystallize in a test tube. Using the technique of X-ray diffraction, the crystallized ribosomes provide detailed information about ribosome structure. **Figure 12.18a** shows a model of a bacterial ribosome. The overall shape of each subunit is largely determined by the structure of the rRNAs, which constitute most of the mass of the ribosome.

50S subunit

E site

P site

A site

30S subunit

(a) Bacterial ribosome model based on X-ray diffraction studies

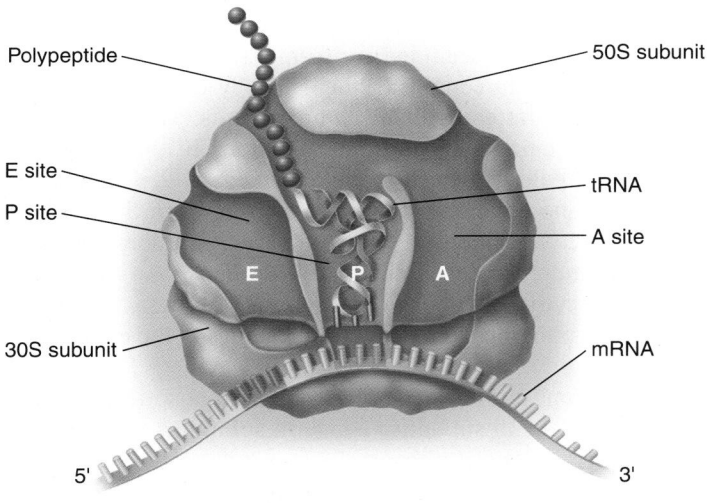

Polypeptide

50S subunit

E site

P site

tRNA

A site

30S subunit

mRNA

5'

3'

(b) Schematic model for ribosome structure

Figure 12.18 **Ribosome structure.**   **(a)** A model for the structure of the ribosome based on X-ray diffraction studies, showing the large and small subunits and the major binding sites. The rRNA is shown in gray (large subunit) and light blue (small subunit), while the ribosomal proteins are violet (large subunit) and dark blue (small subunit). **(b)** A schematic model emphasizing functional sites in the ribosome, also showing bound mRNA and tRNA with an attached polypeptide.

During bacterial translation, the mRNA lies on the surface of the 30S subunit, within a space between the 30S and 50S subunits (**Figure 12.18b**). As a polypeptide is synthesized, it exits through a hole within the 50S subunit. Ribosomes contain discrete sites where tRNAs bind and the polypeptide is synthesized. In 1964, James Watson proposed a two-site model for tRNA binding to the ribosome. These sites are known as the **peptidyl site** (**P site**) and **aminoacyl site** (**A site**). In 1984, Knud Nierhaus and Hans-Jorg Rheinberger expanded this to a three-site model (Figure 12.18b). The third site is known as the **exit site** (**E site**). Later, we will examine the roles of these sites in the synthesis of a polypeptide.

GENOMES & PROTEOMES

Comparisons of Small Subunit rRNAs Among Different Species Provide a Basis for Establishing Evolutionary Relationships

Translation is a fundamental process that is vital for the existence of all living species. The components that are needed for translation arose very early in the evolution of life on our planet. In fact, they arose in an ancestor that gave rise to all known living species. For this reason, all organisms have translational components that are evolutionarily related to each other. For example, the rRNA found in the small subunit of ribosomes is similar in all forms of life, though it is slightly larger in eukaryotic species (18S) than in bacterial species (16S). In other words, the gene for the small subunit rRNA (SSU rRNA) is found in the genomes of all organisms.

One way that geneticists explore evolutionary relationships is to compare the sequences of evolutionarily related genes. At the molecular level, gene evolution involves changes in DNA sequences. After two different species have diverged from each other during evolution, the genes of each species have an opportunity to accumulate changes, or mutations, that alter the sequences of those genes. After many generations, evolutionarily related species contain genes that are similar but not identical to each other, because each species will accumulate different

mutations. In general, if a very long time has elapsed since two species diverged evolutionarily, their genes tend to be quite different. In contrast, if two species diverged relatively recently on an evolutionary time scale, their genes tend to be more similar.

Figure 12.19 compares a portion of the sequence of the small subunit rRNA gene from three mammalian and three bacterial species. The colors highlight different types of comparisons. The sequences shaded in yellow are identical in five or six species. Sequences that are identical in different species are said to be **evolutionarily conserved**. Presumably, these sequences were found in the primordial gene that gave rise to modern species and, because these sequences may have some critical function, have not been able to change over evolutionary time. Those sequences shaded in green are identical in all three mammals, but differ compared to one or more bacterial species. Actually, if you scan the mammalian species, you may notice that all three sequences are identical to each other in this region. The sequences shaded in red are identical in two or three bacterial species, but differ compared to the mammalian small subunit rRNA genes. The sequences from *E. coli* and *Serratia marcescens* are more similar to each other than the sequence from *Bacillus subtilis* is to either of them. This is consistent with the idea that *E. coli* and *S. marcescens* are more closely related evolutionarily than either of them is to *B. subtilis*.

12.6 The Stages of Translation

Like transcription, the process of translation occurs in three stages called initiation, elongation, and termination. **Figure 12.20** provides an overview of the process. During initiation, mRNA, the first tRNA, and ribosomal subunits assemble into a complex. Next, in the elongation stage, the ribosome moves from the start codon in the mRNA toward the stop codon, synthesizing a polypeptide according to the sequence of codons in the mRNA. Finally, the process is terminated when the ribosome reaches a stop codon and the complex disassembles, releasing the completed polypeptide. In this section, we will examine the steps in this process as they occur in living cells.

GATTAAGAGGGACGGCCGGGGGGCATTCGTATTGCGCCGCTAGAGGTGAAATTC — Human

GATTAAGAGGGACGGCCGGGGGGCATTCGTATTGCGCCGCTAGAGGTGAAATTC — Mouse

GATTAAGAGGGACGGCCGGGGGGCATTCGTATTGCGCCGCTAGAGGTGAAATTC — Rat

CAAGCTTGAGTCTCGTAGAGGGGGGTAGAATTCCAGGTGTAGCGGTGAAATGC — E. coli

CAAGCTAGAGTCTCGTAGAGGGGGGTAGAATTCCAGGTGTAGCGGTGAAATGC — S. marcescens

GAGACTTGAGTACAGAAGAAGAGAGTGGAATTCCACGTGTAGCGGTGAAATGC — B. subtilis

Figure 12.19 Comparison of small subunit rRNA gene sequences from three eukaryotes and three bacterial species. Note the many similarities (yellow) and differences (green and red) among the sequences.

Translation Is Initiated with the Assembly of mRNA, tRNA, and the Ribosomal Subunits

During the **initiation stage**, mRNA, the first tRNA, and the ribosomal subunits assemble into a functional complex. In all species, this assembly requires the help of proteins called **ribosomal initiation factors** that facilitate the interactions between these components (see Table 12.2). The assembly also requires an input of energy. Guanosine triphosphate (GTP) is hydrolyzed by the initiation factors to provide the necessary energy.

In the absence of translation, the small and large ribosomal subunits exist separately. To begin assembly in bacteria, mRNA binds to the small ribosomal subunit (**Figure 12.21**). The binding of mRNA to this subunit is facilitated by a short ribosomal-binding sequence near the 5′ end of the mRNA. This sequence is complementary to a portion of the 16S rRNA within the small ribosomal subunit. For this reason, the mRNA and rRNA hydrogen-bond to each other by base pairing. The start codon is usually just a few nucleotides downstream (that is, toward the 3′ end) from the ribosomal-binding sequence. A specific tRNA, which functions as the **initiator tRNA**, recognizes the start codon AUG in mRNA and binds to it. To complete the initiation stage, the large ribosomal subunit associates with the small subunit.

At the end of this stage, the initiator tRNA is located in the P site of the ribosome.

In eukaryotic species, the initiation phase has two differences compared to their bacterial counterparts. First, instead of a ribosomal-binding sequence, eukaryotic mRNAs have a 7-methylguanosine cap at their 5′ end. This cap is recognized by cap-binding proteins that promote the binding of the mRNA to the small ribosomal subunit. Also, unlike bacteria, in which the start codon is very close to a ribosomal-binding sequence, the location of start codons in eukaryotes is more variable. To identify a start codon, Marilyn Kozak proposed in 1978 that the small ribosomal subunit begins at the 5′ end and then scans along the mRNA in the 3′ direction in search of an AUG sequence. In many, but not all, cases the first AUG codon is used as a start codon. By analyzing the sequences of many eukaryotic mRNAs, Kozak and her colleagues discovered that the sequence around an AUG codon is important for it to be used as a start codon. The sequence for optimal start codon recognition is shown here:

Upstream of start codon	Start codon	Downstream coding region

. . . G C C (A or G) C C (**A U G**) G . . . .

Translation: mRNA is translated at the ribosome into a polypeptide.

Completed polypeptide

Large subunit

Initiator tRNA

Polypeptide

Small subunit

5′

mRNA

Start codon

3′

Stop codon

1 **Initiation:** mRNA, tRNA, and the ribosomal subunits form a complex.

2 **Elongation:** The ribosome travels in the 5′ to 3′ direction and synthesizes a polypeptide.

3 **Termination:** The ribosome reaches a stop codon and all of the components disassemble, releasing a completed polypeptide.

Figure 12.20 An overview of the stages of translation.

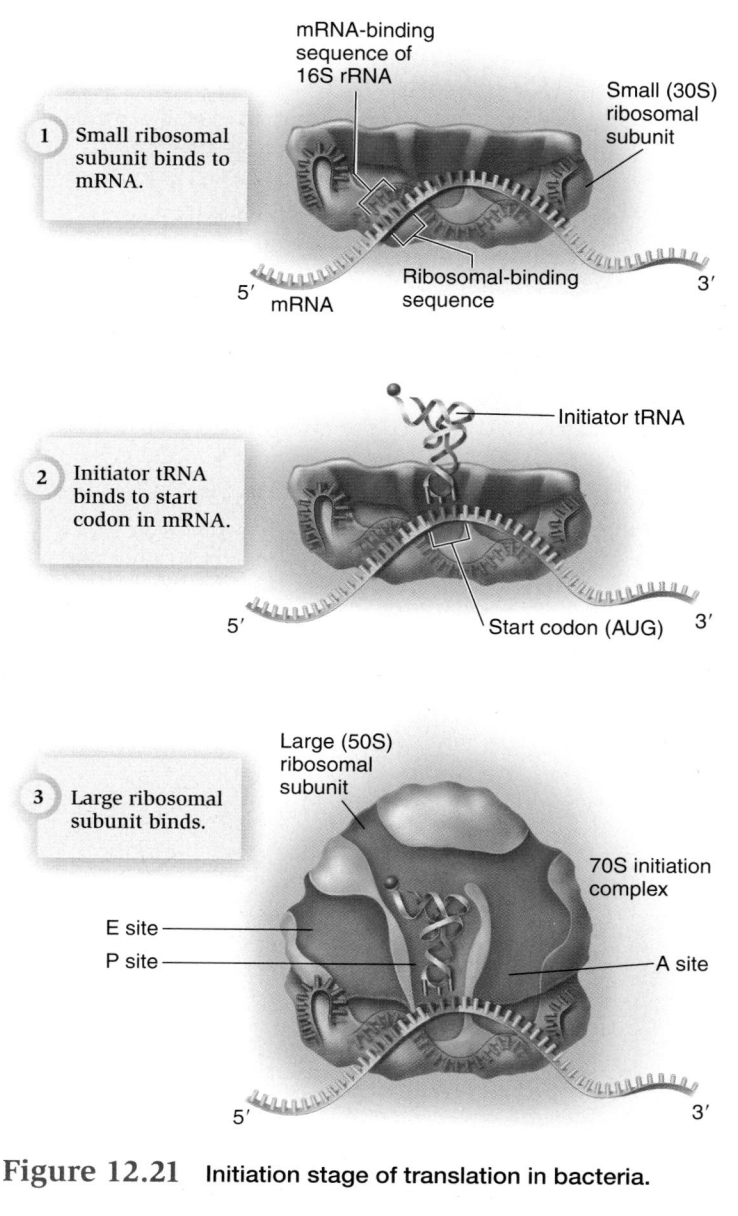

Figure 12.21 Initiation stage of translation in bacteria.

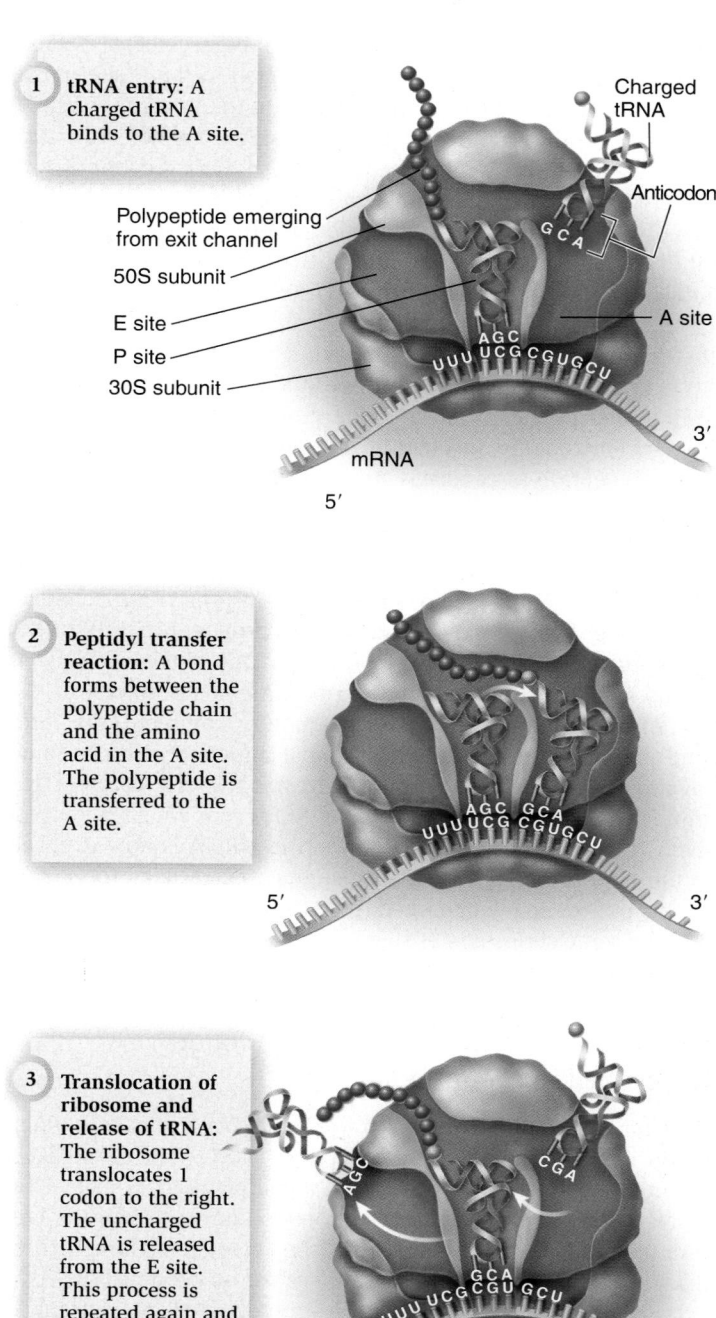

Figure 12.22 Elongation stage of translation in bacteria.

Aside from an AUG codon itself, a guanosine just past the start codon and the sequence of six bases directly upstream from the start codon are important for start codon selection. If an AUG codon is within a site that deviates markedly from this optimal sequence, the small subunit may skip this codon and instead use another AUG codon farther downstream. Once the small subunit selects a start codon, an initiator tRNA binds to the start codon, and then the large ribosomal subunit associates with the small subunit to complete the assembly process.

Polypeptide Synthesis Occurs During the Elongation Stage

As its name suggests, the **elongation stage** involves the covalent bonding of amino acids to each other, one at a time, to create a polypeptide. Even though this process involves several different

components, translation occurs at a remarkable rate. Under normal cellular conditions, the translation machinery can elongate a polypeptide chain at a rate of 15 to 18 amino acids per second in bacteria and 6 amino acids per second in eukaryotes!

To elongate a polypeptide by one amino acid, a tRNA brings a new amino acid to the ribosome, where it is attached to the end of a growing polypeptide chain. In step 1 of **Figure 12.22**, translation has already proceeded to a point where a short polypeptide is attached to the tRNA located in the P site of the ribosome. This is called peptidyl tRNA. In the first step of elongation, an aminoacyl tRNA carrying a single amino acid binds to the A site. This binding occurs because the anticodon in the tRNA is complementary to the codon in the mRNA. The hydrolysis of GTP by proteins that function as **elongation factors** provides the energy for the binding of the tRNA to the A site (see Table 12.2). At this stage of translation, a p̲eptidyl tRNA is at the P site and an a̲minoacyl tRNA is at the A site. This is how the P and A sites came to be named.

In the second step, a **peptide bond** is formed between the amino acid at the A site and the growing polypeptide chain, thereby lengthening the chain by one amino acid. As this occurs, the polyp̲eptide is removed from the tRNA in the P site and transf̲erred to the amino acid at the A site, an event termed a **peptidyl transfer reaction**. This reaction is catalyzed by a region of the 50S subunit known as the peptidyltransferase complex, which is composed of several proteins and rRNA. Thomas Steitz, Peter Moore, and their colleagues proposed that the rRNA is responsible for catalyzing bond formation between adjacent amino acids. In other words, the ribosome is a ribozyme!

After the peptidyl transfer reaction is complete, the third step involves the movement or translocation of the ribosome toward the 3′ end of the mRNA by exactly one codon. This shifts the tRNAs at the P and A sites to the E and P sites, respectively. Notice that the next codon in the mRNA is now exposed in the unoccupied A site. The uncharged tRNA exits the E site. At this point, the next charged tRNA can enter the empty A site, and the same series of steps will add the next amino acid to the polypeptide chain.

Termination Occurs When a Stop Codon Is Reached in the mRNA

When a stop codon is found in the A site of a ribosome, translation is ended. The three stop codons, UAA, UAG, and UGA, are not recognized by a tRNA with a complementary sequence. Instead, they are recognized by proteins known as **release factors** (see Table 12.2). Interestingly, the three-dimensional structure of a release factor protein mimics the structure of tRNAs.

Figure 12.23 illustrates the termination stage of translation. In step 1 of this figure, the completed polypeptide chain is attached to a tRNA in the P site. A stop codon is located at the A site. In the first step, a release factor binds to the stop codon at the A site. In the second step, the bond between the polypeptide and the tRNA is hydrolyzed, causing the polypeptide and tRNA to be released from the ribosome. Finally, in the third step the ribosomal subunits, mRNA and release factor dissociate.

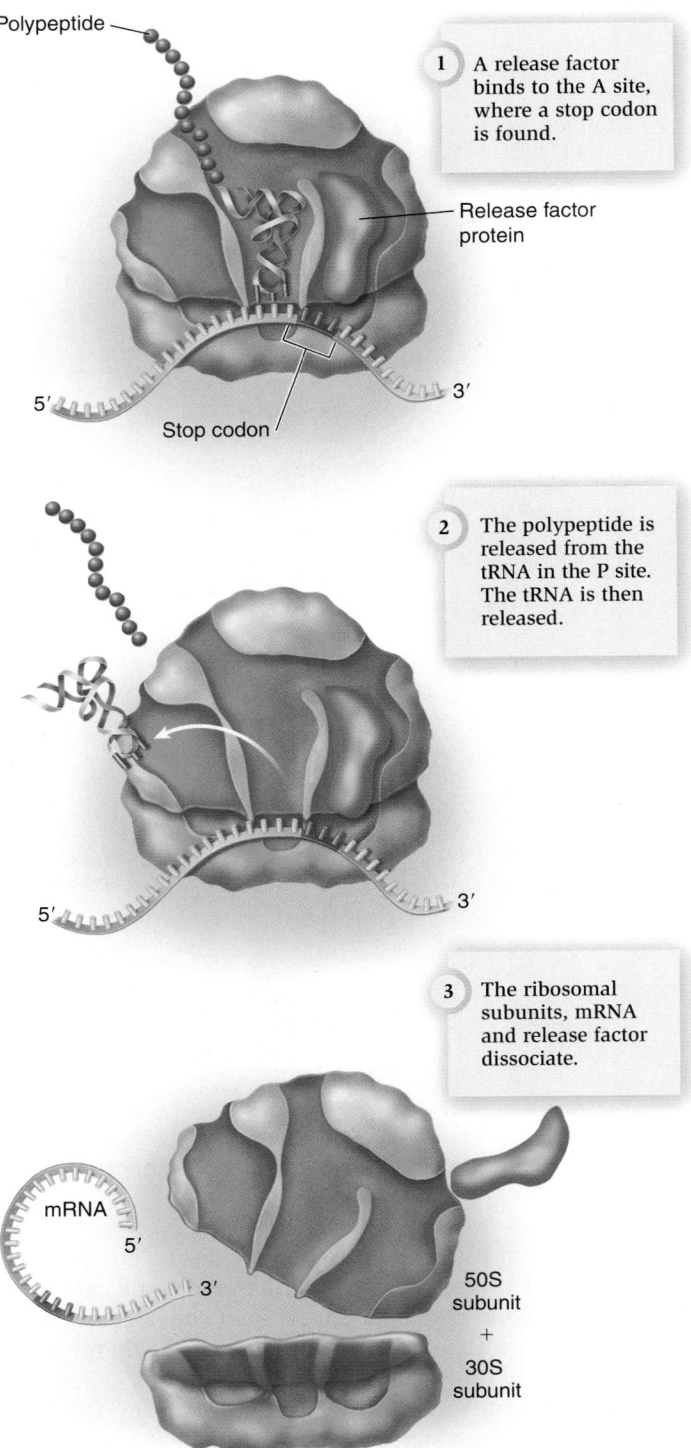

Figure 12.23   Termination stage of translation in bacteria.

A Polypeptide Chain Has Directionality from Its Amino End to Its Carboxyl End

Polypeptide synthesis has a directionality that parallels the 5′ to 3′ orientation of mRNA. **Figure 12.24** compares the sequence of a very short polypeptide with the mRNA that encodes it. The first amino acid is said to be at the **N-terminus** or **amino terminus** of the polypeptide. The term N-terminus refers to the presence of a nitrogen atom (N) at this end, while amino terminus indicates the presence of a free amino group. Peptide bonds connect the amino acids together. These bonds form between the carboxyl group of the previous amino acid and the amino group of the next amino acid. The last amino acid in a completed polypeptide does not have another amino acid attached to its carboxyl group. This last amino acid is said to be located at the **C-terminus** or **carboxyl terminus**. A carboxyl group (COOH) is always found at this end of the polypeptide chain. Note: At neutral pH, the amino group is positively charged (NH_3^+), while the carboxyl group is negatively charged (COO^-).

Figure 12.24   **The directionality of a polypeptide compared to an mRNA.**

Biological inquiry: Is the codon for the last amino acid in a polypeptide closer to the 5′ or 3′ end of the mRNA?

CHAPTER SUMMARY

12.1 Overview of Gene Expression

- Based on his studies of inborn errors of metabolism, Garrod hypothesized that genes encode enzymes. (Figure 12.1)
- By studying the nutritional requirements of bread mold, Beadle and Tatum proposed the one gene–one enzyme hypothesis. Later, this idea was expanded to the one gene–one polypeptide theory. (Figure 12.2)
- A polypeptide is a unit of structure. A protein, composed of one or more polypeptides, is a unit of function.
- At the molecular level, the central dogma of genetics states that most genes are transcribed into mRNA, and then the mRNA is translated into polypeptides. Eukaryotes modify their RNA transcripts to make them functional. (Figure 12.3)
- The molecular expression of genes is fundamental to the characteristics of an organism's traits. (Figure 12.4)

12.2 Transcription

- The promoter of a gene signals the beginning of transcription while the terminator specifies where transcription will end for a given gene. Regulatory sequences control whether a gene is turned on or off. (Figure 12.5)
- In bacteria, sigma factor binds to RNA polymerase and to a promoter, thereby promoting the initiation of transcription. The RNA transcript is made during the elongation stage due to base pairing of nucleotides to the template strand of DNA. RNA polymerase is released from the DNA at the termination site. (Figure 12.6)
- The genes along a chromosome are transcribed in different directions. (Figure 12.7)

12.3 RNA Processing in Eukaryotes

- Eukaryotic mRNA is first made as a pre-mRNA that is capped, spliced, and given a poly A tail. (Figures 12.8, 12.11)
- Eukaryotic genes often have introns. These are intervening sequences that are found in between the coding sequences, which are called exons. Microscopy studies with mature mRNA revealed the presence of introns in eukaryotic genes. (Figure 12.9)
- Introns are removed from eukaryotic pre-mRNA by a spliceosome. The components of a spliceosome first recognize the intron boundaries and the branch site, and then remove the intron and connect the adjacent exons. (Figure 12.10)

12.4 Translation and the Genetic Code

- The genetic code is found in mRNA molecules and determines the amino acid sequences of polypeptides. Each of the 64 codons specifies a start codon (methionine), other amino acids, or a stop codon. (Table 12.1, Figure 12.12)
- The template strand of DNA is used to make mRNA with a series of codons. Recognition between mRNA and many tRNA molecules determines the amino acid sequence of a polypeptide. (Figure 12.13)
- Nirenberg and Leder used the ability of RNA triplets to promote the binding of tRNA to ribosomes as a way to determine many of the codons of the genetic code. (Figure 12.14)

12.5 The Machinery of Translation

- Translation requires mRNA, tRNAs, ribosomes, and many translation factors. (Table 12.2)
- tRNA molecules have a cloverleaf structure. Two important sites are the 3′ end, which covalently binds an amino acid, and the anticodon, which base-pairs with a codon in mRNA. (Figure 12.15)

- The enzyme aminoacyl-tRNA synthetase attaches the correct amino acid to a tRNA molecule. (Figure 12.16)
- Ribosomes are composed of rRNA molecules and many proteins to produce a small and large subunit. Bacterial and eukaryotic ribosomes differ in their composition. (Figure 12.17)
- Ribosomes have three sites called the A, P, and E sites, which are locations for the binding and release of tRNA molecules. (Figure 12.18)
- The gene that encodes the small subunit rRNA has been extensively used in the evolutionary comparisons of different species. (Figure 12.19)

12.6 The Stages of Translation

- Translation occurs in three stages called initiation, elongation, and termination. (Figure 12.20)
- During initiation of translation, the mRNA assembles with the ribosomal subunits and the first tRNA molecule. (Figure 12.21)
- Polypeptide synthesis occurs during the elongation stage, one amino acid at a time. (Figure 12.22)
- During the termination of translation, the binding of a release factor to the stop codon causes the release of the completed polypeptide and the disassembly of the mRNA and ribosomal subunits. (Figure 12.23)
- A polypeptide has a directionality in which the first amino acid is at the N- or amino-terminus, while the last amino acid is at the C- or carboxyl-terminus. (Figure 12.24)

TEST YOURSELF

1. Which of the following best represents the central dogma of gene expression?
   a. During transcription, DNA codes for polypeptides.
   b. During transcription, DNA codes for RNA, which codes for polypeptides during translation.
   c. During translation, DNA codes for RNA, which codes for polypeptides during transcription.
   d. None of the above.

2. Transcription of a gene begins at a site on DNA called _____ and ends at a site on DNA known as _____.
   a. an initiation codon, the termination codon
   b. a promoter, the termination codon
   c. an initiation codon, the terminator
   d. a promoter, the terminator
   e. an initiator, the terminator

3. The product of transcription of a structural gene is
   a. tRNA.
   b. mRNA.
   c. rRNA.
   d. polypeptide.
   e. a, b, and c.

4. During eukaryotic RNA processing, the nontranslated sequences that are removed are called
   a. exons.
   b. introns.
   c. promoters.
   d. codons.
   e. ribozymes.

5. Ribozymes are
   a. the organelles where translation takes place.
   b. the RNA molecules that are components of ribosomes.
   c. the proteins that are components of ribosomes.
   d. the portions of the pre-mRNA that are removed.
   e. RNA molecules that catalyze chemical reactions.

6. The _____ is the organelle where the translation process takes place.
   a. mitochondria
   b. nucleus
   c. ribosome
   d. lysosome
   e. ribozyme

7. The region of the tRNA that is complementary to the triplet on the mRNA is
   a. the acceptor stem.
   b. the codon.
   c. the peptidyl site.
   d. the anticodon.
   e. the adaptor loop.

8. During the initiation step of translation, the first codon, ____, will enter the _____ and associate with the initiator tRNA.
   a. UAG, A site
   b. AUG, A site
   c. UAG, P site
   d. AUG, P site
   e. AUG, E site

9. The movement of the polypeptide from the tRNA in the P site to the tRNA in the A site is referred to as
   a. peptide bonding.
   b. aminoacyl binding.
   c. translation.
   d. peptidyl transfer reaction.
   e. elongation.

10. The synthesis of a polypeptide occurs during which stage of translation?
    a. initiation
    b. elongation
    c. termination
    d. splicing

CONCEPTUAL QUESTIONS

1. Define mutation.
2. Explain the one gene–one enzyme hypothesis and the more modern modifications of this hypothesis.
3. What is the function of an aminoacyl-tRNA synthetase?

EXPERIMENTAL QUESTIONS

1. Briefly explain how studying the pathway that leads to arginine synthesis allowed Beadle and Tatum to conclude that one gene encodes one enzyme.
2. What was the benefit of using radiolabeled amino acids in the Nirenberg and Leder experiment?
3. Predict the results that Nirenberg and Leder would have found for the following triplets: AUG, UAA, UAG, UGA.

COLLABORATIVE QUESTIONS

1. Discuss RNA processing in eukaryotes.
2. How can we analyze rRNA to establish evolutionary relationships between different species of organisms?

www.brookerbiology.com

This website includes answers to the Biological Inquiry questions found in the figure legends and all end-of-chapter questions.

13

GENE REGULATION

A model for a protein that binds to DNA and regulates genes.
The protein shown in red and pink is binding to the DNA double helix, shown in orange and white. This protein, described later in Figure 13.10, activates gene transcription.

The term **gene regulation** refers to the ability of cells to control their level of gene expression. The majority of structural genes in all species are regulated so that the proteins they encode are produced at certain times and in specific amounts. By comparison, unregulated genes have essentially constant levels of expression in all conditions over time. These are called **constitutive genes**. Frequently, constitutive genes, also referred to as housekeeping genes, encode proteins that are always necessary for the survival of an organism.

The importance of gene regulation is underscored by the number of genes devoted to this process in an organism. A sizeable portion of the genomes of complex organisms such as plants and animals is devoted to the process of gene regulation. For example, in *Arabidopsis thaliana*, a plant that has been studied as a model organism by plant geneticists, over 5% of its genome is involved with regulating gene transcription. This species has more than 1,500 different genes that encode proteins that regulate the transcription of other genes.

In this chapter, we will begin with an overview that emphasizes the benefits of gene regulation and the general mechanisms that achieve such regulation. Later sections will describe the ways that prokaryotes and eukaryotes regulate their genes at the molecular level.

13.1 Overview of Gene Regulation

Living organisms derive many benefits from regulating genes. This process ensures that proteins will be produced only when they are required. For example, certain proteins help an organism to survive environmental stress such as drastic changes in temperature or exposure to ultraviolet (UV) light. Because these proteins are required only when the organism is confronted with the stress, a cell conserves energy and resources if such proteins are made just when the stressor is present. In addition, some proteins function in the metabolism of small molecules that may or may not be present in the environment. For example, certain proteins are needed for a bacterium to take up and metabolize particular sugars. These proteins are required only when the bacterium is exposed to such sugars in its environment.

Like bacteria, eukaryotic organisms such as protists, fungi, plants, and animals need to adapt to changes in their environment. For example, all eukaryotic species can respond to environmental stresses such as UV radiation by turning on genes that provide protection against this harmful environmental agent. In humans, exposure to UV light turns on genes that result in a tanning response.

In addition to responding to environmental conditions, eukaryotes also regulate genes when cells are dividing and specializing, and when they are going through developmental stages. For example, some proteins are needed just for cell division. These are necessary only when a cell is preparing to divide and during the division process. In multicellular organisms, certain proteins are made only in particular cell types, or their amounts may vary from cell to cell. In humans, for example, some proteins are needed only in muscle cells but not in nerve cells, and vice versa. Similarly, in multicellular organisms that progress through different developmental stages (fertilized egg, embryo, and adult), certain proteins are needed only at particular stages of development. In this section, we will examine a few examples that illustrate the important consequences of gene regulation.

We will also survey the points in the gene expression process at which genes are regulated in prokaryotic and eukaryotic cells.

Bacteria Regulate Genes to Respond to Nutrients in Their Environment

As mentioned, gene regulation in prokaryotes is often used to respond to changes in the environment. To fully appreciate how genetic regulation helps bacteria survive, let's look at an example. The bacterium *Escherichia coli* can use many types of sugars as food sources, thereby increasing its chances of survival. Let's consider the process of how it uses lactose, which is the sugar found in milk. *E. coli* can use lactose because it carries genes that code for proteins that enable it to take up lactose and metabolize it.

Figure 13.1 illustrates the effects of lactose on the regulation of those genes. Lactose use requires a transporter, called the lactose permease, that facilitates the uptake of lactose into the cell, and an enzyme, called β-galactosidase, that catalyzes the breakdown of lactose. When lactose is not present in the environment, an *E. coli* cell makes very little of these proteins. However, when lactose becomes available, gene regulation causes the synthesis of many more of these proteins, enabling the bacterium to efficiently use lactose from its environment. Eventually, all of the lactose in the environment will be used up. At this point, the genes encoding these proteins will be shut off, and most of the proteins will be degraded. In the case of lactose use,

the main benefit of gene regulation is that the proteins that are needed for this process are made only when lactose is present in the environment. Therefore, *E. coli* does not waste valuable energy making these proteins when they are not needed.

Eukaryotic Gene Regulation Produces Different Cell Types in a Single Organism

One of the most amazing examples of gene regulation is the phenomenon of **cell differentiation**. In multicellular organisms such as plants and animals, cells become specialized, or differentiated, into particular types. In humans, for example, cells may become specialized into muscle cells, nerve cells, skin cells, or other types. **Figure 13.2** shows micrographs of three types of cells found in humans. As seen here, their morphologies are strikingly different. Likewise, their functions within the body are also quite different. Muscle cells are important in body movements, nerve cells function in cell signaling, and skin cells form a protective outer surface to the body.

Gene regulation is responsible for creating different types of cells within a multicellular organism. The three cell types shown in Figure 13.2 contain the same genome, meaning they carry the same set of genes. However, their proteomes are quite different due to gene regulation. Certain proteins are found in particular cell types, but not in others. Alternatively, a protein may be present in all three cell types, but the relative amounts of the protein may be different. The amount of a given protein depends

Figure 13.1   Gene regulation of lactose utilization in *E. coli.*

Biological inquiry: What is the advantage to E. coli *for regulating the genes involved with lactose utilization?*

(a) Skeletal muscle cell

(b) Nerve cell

(c) Skin cell

Figure 13.2 Examples of different cell types in humans.

on many factors, including how strongly the corresponding gene is turned on and how much protein is synthesized from mRNA. Gene regulation at the level of transcription and translation plays a major role in determining the proteome of each cell type. These forms of gene regulation will be discussed later in this chapter. In addition, factors described in Unit II such as protein degradation, post-translational modifications, and feedback inhibition are important in controlling the functions and relative amounts of proteins in a cell.

Eukaryotic Gene Regulation Enables Multicellular Organisms to Progress Through Developmental Stages

In multicellular organisms that progress through developmental stages, certain genes are expressed at one stage of development but not another. This topic is discussed in Chapter 19. Let's consider an example of such developmental gene regulation in mammals. Early stages of mammalian development occur in the uterus of female mammals. Following fertilization, an embryo develops inside the uterus. In humans, the embryonic stage is from fertilization to 8 weeks. During this stage, major developmental changes produce the various body parts. The fetal stage occurs from 8 weeks to birth (41 weeks). This stage is characterized by continued refinement of body parts and a large increase in size.

Because of this internal development, a system has evolved to provide both the embryonic and fetal cells with the oxygen they require for cellular respiration. The oxygen demands of a rapidly growing embryo and fetus are quite different from the needs of the mother. Gene regulation plays a vital role in ensuring that an embryo and fetus get the proper amount of oxygen. As described in Chapter 48, hemoglobin is the main protein that delivers oxygen to the cells of a mammal's body. The genomes of mammals carry several genes (designated with Greek letters) that encode slightly different globin polypeptides. A hemoglobin protein is composed of four globin polypeptides, two encoded by one globin gene and two encoded by another globin gene (**Figure 13.3**). During the embryonic stage of development, the epsilon (ϵ)-globin and zeta (ζ)-globin genes are turned on. At the fetal stage, these genes are turned off, and the alpha (α)-globin and gamma (γ)-globin genes are turned on. Finally, at birth, the γ-globin gene is turned off, and the beta (β)-globin gene is turned

	Embryo	Fetus	Adult
Hemoglobin protein	2 ζ-globins 2 ϵ-globins	2 α-globins 2 γ-globins	2 α-globins 2 β-globins
Oxygen affinity	highest	high	moderate
Gene expression α-globin gene β-globin gene γ-globin gene ζ-globin gene ϵ-globin gene	off off off on on	on off on off off	on on off off off

Figure 13.3 Developmental regulation of human globin genes.

on. The hemoglobin that is produced during the embryonic and fetal stages has a much higher binding affinity for oxygen than does the hemoglobin that is produced after birth. Therefore, the embryo and fetus can remove oxygen from the mother's bloodstream and use that oxygen for their own needs. This occurs across the placenta, where the mother's bloodstream is adjacent to the bloodstream of the embryo or fetus. In this way, gene regulation enables mammals to develop internally, even though the embryo and fetus are not breathing on their own.

Gene Regulation Can Occur at Different Points in the Process from DNA to Protein

Thus far we have learned that gene regulation can have a dramatic impact on the ability of organisms to respond to environmental changes, differentiate cells, and progress through developmental stages. Because structural genes encode proteins, the regulation of gene expression can occur at any of the steps that are needed to produce a functional protein.

(a) Prokaryotic gene regulation

(b) Eukaryotic gene regulation

Figure 13.4 Overview of points of control for gene regulation in (a) prokaryotes and (b) eukaryotes. The relative sizes of the red arrows indicate the prominence with which regulation at that point is used to control the production of functional proteins.

In bacteria, gene regulation most commonly occurs at the level of transcription, which means that bacteria regulate how much mRNA is made from most genes (**Figure 13.4a**). When geneticists say a gene is "turned off," they mean that no mRNA is made from that gene, while a gene that is "turned on" is transcribed into mRNA. A second way for bacteria to regulate gene expression is to control the rate at which mRNA is translated into protein. This form of gene regulation is relatively uncommon in bacteria. Finally, gene expression can be regulated at the protein or post-translational level. Various types of protein regulation are best understood within the context of cell biology, so they are discussed in the unit on cells. For example, in Chapter 7, we learned that enzymes are regulated by feedback inhibition.

In eukaryotes, gene regulation occurs at many levels, including transcription, RNA processing, translation, and post-translation (**Figure 13.4b**). As in their bacterial counterparts, transcriptional regulation is a prominent form of eukaryotic gene regulation. As discussed later in this chapter, eukaryotic genes are transcriptionally regulated in several different ways, some of which are not found in bacteria. Also, as discussed in Chapter 12, eukaryotes process their mRNA transcripts in ways that do not commonly occur in bacteria. For example, RNA splicing is a widespread phenomenon in eukaryotes. Later in this chapter, we will examine how this process is regulated to create two or more different types of mRNA from a single gene. Eukaryotes also regulate mRNA after its modification. The amount of mRNA may be regulated by controlling its degra-

dation. In addition, the translation of an mRNA may be regulated by small, inhibitory RNA molecules or by RNA-binding proteins that prevent translation from occurring. As in prokaryotes, eukaryotic proteins can be regulated in a variety of ways including feedback inhibition, post-translational modification, and degradation via the proteasome. These mechanisms are discussed in Unit II. Now let's take a look at transcriptional regulation in bacteria.

13.2 Regulation of Transcription in Bacteria

Due to gene regulation, bacteria can respond specifically to changes in their cellular and environmental conditions. As we have seen, when a bacterium is exposed to a particular nutrient in its environment such as a sugar, the genes are expressed that encode proteins that are needed for the uptake and metabolism of the sugar. Alternatively, bacteria have enzymes that synthesize cellular molecules such as particular amino acids. It is to their advantage for bacteria to turn off the genes that encode those enzymes when a sufficient amount of the amino acid is present in the cytoplasm. In this section, we will examine the underlying molecular mechanisms that bring about transcriptional regulation in bacteria.

Transcriptional Regulation Usually Involves Regulatory Transcription Factors and Small Effector Molecules

In most cases, transcriptional regulation involves the actions of **regulatory transcription factors**, which are proteins that bind to DNA in the vicinity of a promoter and affect the rate of transcription of one or more nearby genes. These transcription factors can either decrease or increase the rate of transcription of a gene. **Repressors** are transcription factors that bind to DNA and inhibit transcription, whereas **activators** increase the rate of transcription. The term **negative control** refers to transcriptional regulation by repressor proteins; **positive control** refers to regulation by activator proteins (**Figure 13.5a**).

In conjunction with regulatory transcription factors, molecules called **small effector molecules** often play a critical role in transcriptional regulation. A small effector molecule exerts its effects by binding to a regulatory transcription factor and causing a conformational change in the protein. In some cases, the effect of the conformational change determines whether or not the protein can bind to the DNA. **Figure 13.5b** illustrates an example. When the small effector molecule is not present in the cytoplasm, the repressor binds to the DNA and inhibits transcription. However, at a later time when the small effector molecule is found in the cytoplasm, it will bind to the repressor and cause a conformational change that inhibits the ability of the protein to bind the DNA. The gene is turned on because the repressor is released. Regulatory transcription factors that respond

Negative control: A repressor inhibits transcription.

Positive control: An activator promotes transcription.

(a) Actions of regulatory transcription factors

The repressor protein is bound to the promoter region when the small effector molecule is not present.

The binding of the small effector molecule causes a conformational change in the repressor protein that prevents it from binding to the DNA.

(b) Action of a small effector molecule on a repressor protein

Figure 13.5   Actions of regulatory transcription factors and small effector molecules.   (a) Regulatory transcription factors may exert positive or negative control. **(b)** One way that a small effector molecule may exert its effects is by preventing a repressor protein from binding to the DNA.

to small effector molecules have two functional regions called **domains**. One domain is a site where the protein binds to the DNA. The other domain is the binding site for the small effector molecule.

The *lac* Operon Contains Genes That Encode Proteins Involved in Lactose Metabolism

In bacteria, genes are sometimes clustered together in units that are under the transcriptional control of a single promoter. This arrangement, known as an **operon**, is transcribed into an mRNA, called a **polycistronic mRNA**, that contains the coding sequences for two or more structural genes. The transcription of the structural genes occurs as a single unit. An operon organization allows a bacterium to coordinately regulate a group of genes that encode proteins with a common function.

The genome of *E. coli* carries an operon called the ***lac* operon** that contains the genes for the enzymes we discussed earlier that allow it to metabolize lactose (see Figure 13.1). **Figure 13.6a** shows the organization of this operon as it is found in the *E. coli* chromosome, as well as the polycistronic mRNA that is

transcribed from it. The *lac* operon contains a promoter that is used to transcribe three structural genes, *lacZ*, *lacY*, and *lacA*. *LacZ* encodes β-galactosidase, which as you may remember is an enzyme that breaks down lactose (**Figure 13.6b**). As a side reaction, β-galactosidase also converts a small percentage of lactose into allolactose, a structurally similar sugar or lactose analogue. As described later, allolactose is important in the regulation of the *lac* operon. The *lacY* gene encodes lactose permease, which is a membrane protein required for the transport of lactose into the cytoplasm of the bacterium. The *lacA* gene encodes galactoside transacetylase, which covalently modifies lactose and lactose analogues by attaching an acetyl group (—COCH$_3$). Although the functional necessity of this enzyme remains unclear, the attachment of acetyl groups to nonmetabolizable lactose analogues may prevent their toxic buildup in the cytoplasm.

Near the *lac* promoter are two regulatory sites designated the operator and the CAP site (see Figure 13.6a). The **operator**, or *lacO* site, is a sequence of nucleotides that provides a binding site for a repressor protein. The **CAP site** is a DNA sequence recognized by an activator protein.

Transcription

(a) Organization of DNA sequences in the *lac* region of the *E. coli* chromosome

(b) Functions of lactose permease and β-galactosidase

Figure 13.6 **The *lac* operon.** **(a)** This diagram depicts a region of the *E. coli* chromosome that contains the *lacI* regulatory gene and the adjacent *lac* operon, as well as the polycistronic mRNA transcribed from the operon. **(b)** Function of lactose permease and β-galactosidase. Lactose permease cotransports H⁺ with lactose. Bacteria maintain an H⁺ gradient across their cytoplasmic membrane that drives the active transport of lactose into the cytoplasm. β-galactosidase cleaves lactose into galactose and glucose. As a side reaction, it can also convert lactose into allolactose.

Adjacent to the *lac* operon is the *lacI* gene, which encodes the **lac repressor**. This repressor protein is important for the regulation of the *lac* operon. The *lacI* gene, which is constitutively expressed at fairly low levels, has its own promoter called the *i* promoter. It is considered to be a **regulatory gene** because the sole function of the encoded protein is to regulate the expression of other genes. The *lacI* gene has its own promoter and is not considered a part of the *lac* operon. Let's now take a look at how the *lac* operon is regulated by the *lac* repressor.

The *lac* Operon Is Under Negative Control by a Repressor Protein

In the late 1950s, the first researchers to investigate genetic regulation were Francois Jacob and Jacques Monod at the Pasteur Institute in Paris, France. Their focus on gene regulation stemmed from an interest in the phenomenon known as enzyme adaptation, which had been identified early in the 20th century. Enzyme adaptation refers to the observation that a particular enzyme appears within a living cell only after the cell has been exposed to the substrate for that enzyme. Jacob and Monod studied lactose metabolism in *E. coli* to investigate this phenomenon. When they exposed bacteria to lactose, the levels of lactose-using enzymes in the cells increased by 1,000- to 10,000-fold. After lactose was removed, the synthesis of the enzymes abruptly stopped.

The first mechanism of regulation that Jacob and Monod discovered involved the *lac* repressor protein, which binds to the sequence of nucleotides found at the *lac* operator site. Once bound, the *lac* repressor prevents RNA polymerase from transcribing the *lacZ*, *lacY*, and *lacA* genes (**Figure 13.7a**). RNA polymerase can bind to the promoter when the *lac* repressor is bound to the operator site, but RNA polymerase cannot move past the operator to transcribe the *lacZ*, *lacY*, and *lacA* genes.

Whether or not the *lac* repressor binds to the operator site depends on allolactose, which is the previously mentioned side product of the β-galactosidase enzyme (see Figure 13.6b).

When lactose is absent from the environment, allolactose is not made and the *lac* repressor is free to bind to the *lac* operator, inhibiting the transcription of the operon.

lac regulatory gene

lac operon

RNA polymerase

lac promoter

Operator

lacI   *lacP*   *lacO*   *lacZ*   *lacY*   *lacA*

mRNA

lac repressor binds to the operator and inhibits transcription.

lac repressor (active)

(a) Lactose absent from the environment

When lactose is present, the binding of allolactose to the *lac* repressor prevents it from binding to the *lac* operator site. This permits the transcription of the *lac* operon.

RNA polymerase

lacI

lacP   *lacO*   *lacZ*   *lacY*   *lacA*

Transcription

mRNA

Polycistronic mRNA

Translation

β-galactosidase   Lactose permease   Galactoside transacetylase

Allolactose

Conformational change

The binding of allolactose causes a conformational change that prevents the *lac* repressor from binding to the operator site.

lac repressor (inactive)

(b) Lactose present

Figure 13.7   Negative control of an inducible set of genes: function of the *lac* repressor in regulating the *lac* operon.

Allolactose is an example of a small effector molecule. The *lac* repressor protein contains four identical subunits, each one recognizing a single allolactose molecule. When four allolactose molecules bind to the *lac* repressor, this causes a conformational change that prevents the repressor from binding to the operator site. Under these conditions, RNA polymerase is free to transcribe the operon (**Figure 13.7b**). Because transcription has been turned on by the presence of a small effector molecule, this process is called **induction**. Similarly, the *lac* operon is said to be **inducible**.

The regulation of the *lac* operon enables *E. coli* to efficiently respond to changes in the environment. When the bacterium is not exposed to lactose, no allolactose **inducer** is available to bind to the *lac* repressor. Therefore, the *lac* repressor binds to the operator site and inhibits transcription. In reality, the repressor does not completely inhibit transcription, so that very small amounts of β-galactosidase, lactose permease, and galactoside transacetylase are made. Even so, the levels are far too low for the bacterium to readily use lactose. When the bacterium is exposed to lactose, a small amount can be transported into the cytoplasm via lactose permease, and β-galactosidase will convert some of it to allolactose. The cytoplasmic level of allolactose will gradually rise until allolactose binds to the *lac* repressor, which induces the *lac* operon and promotes the transcription of the *lacZ*, *lacY*, and *lacA* genes. Translation of the encoded polypeptides will produce the proteins needed for lactose uptake and metabolism as described previously in Figure 13.1.

FEATURE INVESTIGATION

Jacob, Monod, and Pardee Studied a Constitutive Bacterial Mutant to Determine the Function of the *lac* Repressor

Thus far, we have learned that the *lac* repressor binds to the *lac* operator site to exert its effects. Let's now take a look back at experiments that helped researchers to determine the function of the *lac* repressor. Our understanding of *lac* operon regulation came from studies involving *E. coli* strains that showed abnormalities in the process. In the 1950s, Jacob and Monod, and their colleague Arthur Pardee, had identified a few rare mutant bacteria that had abnormal lactose use. The mutants expressed the genes of the *lac* operon constitutively, meaning that the *lacZ*, *lacY*, and *lacA* genes were expressed even in the absence of lactose in the environment. The researchers discovered that some mutations that caused this abnormality had occurred in the *lacI* region. Such strains were termed *lacI⁻* to indicate that the *lacI* region was not functioning properly. Normal or wild-type *lacI* strains of *E. coli* are called *lacI⁺* in comparison.

The researchers hypothesized that the *lacI⁻* mutation resulted in the synthesis of an internal inducer, making it unnecessary for cells to be exposed to lactose for induction (**Figure 13.8a**). By comparison, **Figure 13.8b** shows the correct explanation. A loss-of-function mutation in the *lacI* gene prevented the *lac* repressor protein from inhibiting transcription. At the time of their work, however, the function of the *lac* repressor was not yet known.

To further understand the nature of this mutation, Jacob, Monod, and Pardee applied a genetic approach. Although bacterial conjugation is described in Chapter 18, let's briefly examine this process to understand this experiment. The earliest studies of Jacob, Monod, and Pardee in 1959 involved matings

(a)

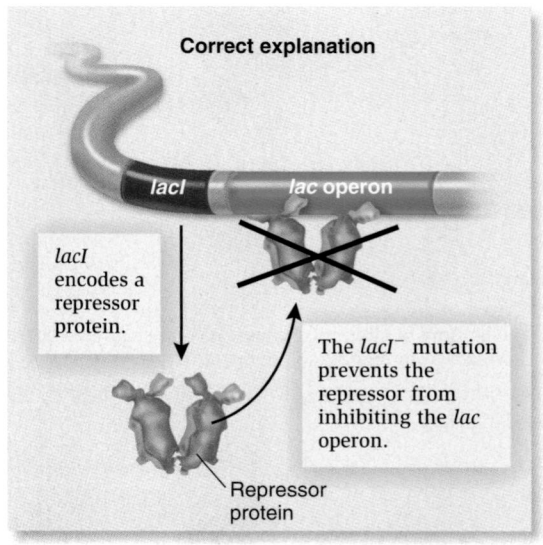

(b)

Figure 13.8 Jacob, Monod, and Pardee's hypothesis for the function of the *lacI* region.

between recipient cells, termed F⁻, and donor cells, which were Hfr strains that transferred a portion of the bacterial chromosome. Later experiments in 1961 involved the transfer of circular segments of DNA known as F factors. We will consider this later type of experiment here. Sometimes an F factor also carries genes that were originally found within the bacterial chromosome. These types of F factors are called F′ factors (F prime factors). A strain of bacteria containing F′ factor genes is called a **merozygote**, or partial diploid. The production of merozygotes was instrumental in allowing Jacob, Monod, and Pardee to elucidate the function of the *lacI* gene.

As shown in **Figure 13.9**, these researchers studied the *lac* operon in a bacterial strain carrying a *lacI⁻* mutation that caused constitutive expression of the *lac* operon. In addition, the mutant strain was subjected to mating to create a merozygote that also carried a normal *lac* operon and normal *lacI⁺* gene on an F′ factor. The constitutive mutant and corresponding merozygote were allowed to grow separately in liquid media and then divided into two tubes each. In half of the tubes, the cells were incubated with lactose to determine if lactose was needed to induce the expression of the operon. In the other tubes, lactose was omitted.

To monitor the expression of the *lac* operon, the cells were broken open and then tested for the amount of β-galactosidase they released by measuring its ability to convert a colorless compound into a yellow product.

The data table of Figure 13.9 summarizes the effects of this constitutive mutation and its analysis in a merozygote. As Jacob, Monod, and Pardee already knew, the *lacI⁻* mutant strain expressed the *lac* operon constitutively, in the presence and absence of lactose. However, when a normal *lac* operon and *lacI⁺* gene were introduced on an F′ factor into a cell harboring the mutant *lacI⁻* gene on the chromosome, the normal *lacI⁺* gene could regulate both operons. In the absence of lactose, both operons were shut off. This occurred because a single *lacI⁺* gene on the F′ factor can produce enough repressor protein to bind to both operator sites. Furthermore, this protein is diffusible (can spread through the cytoplasm) and can bind to *lac* operons that are on the F′ factor and on the bacterial chromosome. The hypothesis that the *lacI⁻* mutation resulted in the synthesis of an internal inducer was rejected. If this hypothesis had been correct, the inducer would have been made in the absence of lactose, and the *lac* operon would have been expressed.

Figure 13.9   The experiment performed by Jacob, Monod, and Pardee to study a *lacI* constitutive mutant.

4 Burst the cells with a sonicater which makes strong sound waves. This releases β-galactosidase.

Sound waves

β-galactosidase

Broken cell

β-galactosidase present
A

β-galactosidase present
B

β-galactosidase present
C

β-galactosidase absent
D

A   B   C   D

5 Measure the function of β-galactosidase. This is done by adding a colorless lactose analogue that β-galactosidase converts to a yellow product.

Add colorless lactose analogue to all 4 tubes.

Lactose analogue

NO₂   NO₂   NO₂   NO₂

No change

Galactose   +   NO₂

Yellow compound

+ NO₂   + NO₂

A   B   C   D

A   B   C   D

6 The amount of yellow color is measured with a spectrophotometer; the deeper the yellow color, the more β-galactosidase was produced.

More β-galactosidase produced

7 THE DATA

Results from step 6:

	Expression of the *lac* operon	
	With lactose	**Without lactose**
Mutant strain	100%	100%
Merozygote strain	220%	<1%

This result was not obtained. Taken together, the data indicated that the normal *lacI* gene encodes a diffusible protein that represses the *lac* operon.

The interactions between regulatory proteins and DNA sequences illustrated in this experiment have led to the definition of two genetic terms. In both prokaryotes and eukaryotes, a **trans-effect** is a form of genetic regulation that can occur even though two DNA segments are not physically adjacent. The action of the *lac* repressor on the *lac* operon is a *trans*-effect. In contrast, a **cis-effect** or a **cis-acting element** is a DNA segment that must be adjacent to the gene(s) that it regulates. The *lac* operator site is an example of a *cis*-acting element. A *trans*-effect is mediated by genes that encode diffusible regulatory proteins, whereas a *cis*-effect is mediated by DNA sequences that bind regulatory proteins.

The *lac* Operon Is Also Under Positive Control by an Activator Protein

In addition to negative control by a repressor protein, the *lac* operon is also positively regulated by an activator protein called the **catabolite activator protein (CAP)**, also known as the **c**AMP **r**eceptor **p**rotein (CRP). In a process known as **catabolite repression**, this form of transcriptional regulation is influenced by the presence of glucose, which is a catabolite (it is broken down—catabolized—inside the cell). The ability of glucose to repress the *lac* operon depends on a small effector molecule, **cyclic AMP (cAMP)**, that is produced from ATP via an enzyme known as adenylyl cyclase. When cAMP is made, it binds to CAP, which causes CAP to bind to the DNA. Because CAP is needed for activation of the *lac* operon, the operon is turned off when CAP is not bound to the DNA. Glucose inhibits the production of cAMP and thereby prevents the binding of CAP to the DNA.

The genetic regulation involving CAP and cAMP is an example of positive control (**Figure 13.10**). When cAMP binds to CAP, the cAMP-CAP complex binds to the CAP site near the *lac* promoter. This causes a bend in the DNA that enhances the ability of RNA polymerase to bind to the promoter. In this way, the rate of transcription is increased. Though it may seem puzzling, the term catabolite repression was coined before the action of the

cAMP-CAP complex was understood at the molecular level. Historically, the primary observation of researchers was that glucose (a catabolite) inhibited (repressed) lactose metabolism. Further experimentation revealed that CAP is an activator protein.

Let's now consider how the concerted actions of the *lac* repressor and CAP allow bacteria to use the sugars in their environment most efficiently. When a bacterium is exposed to glucose, the uptake of glucose into the cell inhibits adenylyl cyclase, which causes cAMP levels to drop. Without cAMP, CAP cannot bind to the CAP site. When a bacterium is exposed to lactose, allolactose levels in the cytoplasm rise. The binding of allolactose prevents the lac repressor from binding to the operator site. With these ideas in mind, **Figure 13.11** considers the four possible environmental conditions that an *E. coli* bacterium might experience with regard to these sugars. When both lactose and glucose levels are high, the *lac* operon is shut off, because CAP does not activate transcription. Under these conditions, the bacterium uses glucose rather than lactose. Greater efficiency is achieved if the bacterium uses one type of sugar at a time. At another time, if lactose levels are high and glucose is low, the *lac* operon is turned on because CAP is bound to the CAP site and the *lac* repressor is not bound to the operator site. Under these conditions, the bacterium metabolizes lactose. When lactose levels are low, the *lac* repressor prevents transcription of the *lac* operon, whether glucose levels are high or low.

The *trp* Operon Is Also Under Negative Control by a Repressor Protein

So far in this section, we have examined the regulation of the *lac* operon. Let's now consider an example of an operon that encodes enzymes involved in biosynthesis rather than degradation. Our example is the ***trp* operon** of *E. coli*, which encodes enzymes that are required to make the amino acid tryptophan, a building block of cellular proteins. More specifically, the *trpE*, *trpD*, *trpC*, *trpB*, and *trpA* genes encode enzymes that are involved in a pathway that leads to tryptophan synthesis.

The *trp* operon is regulated by a repressor protein that is encoded by the *trpR* gene. The binding of the repressor to the *trp* operator site inhibits transcription. The ability of the *trp* repressor to bind to the *trp* operator is controlled by tryptophan, which is the product of the enzymes that are encoded by the operon. When tryptophan levels within the cell are very low, the *trp* repressor cannot bind to the operator site. Under these conditions, RNA polymerase readily transcribes the operon (**Figure 13.12a**). In this way, the cell expresses the genes that encode enzymes that result in the synthesis of tryptophan, which is in short supply. Alternatively, when the tryptophan levels within the cell are high, tryptophan turns off the *trp* operon. Tryptophan acts as a small effector molecule or **corepressor** by binding to the *trp* repressor protein. This causes a conformational change in the repressor that allows it to bind to the *trp* operator site, inhibiting the ability of RNA polymerase to transcribe the operon (**Figure 13.12b**). Therefore, the bacterium does not waste energy making tryptophan when it is abundant.

Figure 13.10 **Positive regulation of the *lac* operon by the catabolite activator protein (CAP).** When cAMP is bound to CAP, CAP binds to the DNA and causes it to bend. This bend facilitates the binding of RNA polymerase.

(a) Lactose high, glucose high

(b) Lactose high, glucose low

(c) Lactose low, glucose high

(d) Lactose low, glucose low

Figure 13.11   Effects of lactose and glucose on the expression of the *lac* operon.

(a) Low tryptophan

(b) High tryptophan

Figure 13.12   Negative control of a repressible set of genes: function of the *trp* repressor and corepressor in regulating the *trp* operon.

Biological inquiry: How are the functions of the lac *repressor and* trp *repressor similar to each other, and how are they different?*

When comparing the *lac* and *trp* operons, the actions of their small effector molecules are quite different. The *lac* repressor binds to its operator in the absence of its small effector molecule, while the *trp* repressor binds to its operator only in the presence of its small effector molecule. The *lac* operon is categorized as inducible because its small effector molecule, namely allolactose, induces transcription. By comparison, the *trp* operon is considered to be **repressible** because its small effector molecule, namely tryptophan, represses transcription.

Repressible Operons Usually Encode Anabolic Enzymes, and Inducible Operons Encode Catabolic Enzymes

Thus far, we have seen that bacterial operons can be transcriptionally regulated in a positive or negative way—or sometimes both. By comparing the mechanisms of regulation among many bacterial operons, geneticists have noticed a general trend. When the genes in an operon encode proteins that function in the breakdown, or **catabolism**, of a substance, the substance to be broken down (or a related compound) often acts as the inducer. This keeps the genes turned off unless the appropriate substance is available. For example, allolactose, which is a product of lactose metabolism, acts as an inducer of the *lac* operon. An inducible form of regulation allows the bacterium to express the appropriate genes only when they are needed to metabolize lactose.

Other cellular enzymes are important for synthesizing organic molecules, a process termed **anabolism**. Because these molecules are generally needed for the functioning of the cell, the genes that encode these anabolic enzymes tend to be regulated by a repressible mechanism, allowing the genes to be transcribed unless they are turned off. The small effector molecule is commonly a product of the enzymes' biosynthetic activities. For example, as we learned, tryptophan is produced by the enzymes that are encoded by the *trp* operon. When enough of this amino acid has been made, tryptophan itself acts as a corepressor, turning off the genes required for tryptophan biosynthesis. Therefore, a repressible form of regulation provides the bacterium with a way to prevent the overproduction of the product of a biosynthetic pathway. Now let's take a look at how transcription is regulated in eukaryotes.

13.3 Regulation of Transcription in Eukaryotes

Transcriptional regulation in eukaryotes follows some of the same principles as those that are found in prokaryotes. For example, activator and repressor proteins are involved in regulating genes by influencing the ability of RNA polymerase to initiate transcription. In addition, many eukaryotic genes are regulated by small effector molecules. However, some important differences

also occur. In eukaryotic species, genes are almost always organized individually, not in operons. In addition, eukaryotic gene regulation tends to be more intricate, because eukaryotes are faced with complexities that differ from their prokaryotic counterparts. For example, eukaryotes have more complicated cell structures that include many more proteins and a variety of cell organelles. Many eukaryotes such as animals and plants are multicellular and contain different cell types. As discussed earlier in this chapter, animal cells may differentiate into nerve cells, muscle cells, and skin cells, among others. Furthermore, animals and plants progress through developmental stages that require changes in gene expression. For these reasons, gene regulation in eukaryotes requires much more coordination and integration.

By studying transcriptional regulation, researchers have discovered that most eukaryotic genes, particularly those found in multicellular species, are regulated by many factors. This phenomenon is called **combinatorial control** because the combination of many factors determines the expression of any given gene. At the level of transcription, common factors that contribute to combinatorial control include the following:

1. One or more activator proteins may stimulate the ability of RNA polymerase to initiate transcription.
2. One or more repressor proteins may inhibit the ability of RNA polymerase to initiate transcription.
3. The function of activators and repressors may be modulated in a variety of ways. These include the binding of small effector molecules, protein–protein interactions, and covalent modifications.
4. Activator proteins may promote the loosening up of the region in the chromosome where a gene is located, thereby making it easier for the gene to be recognized and transcribed by RNA polymerase.
5. DNA methylation (usually) inhibits transcription, either by preventing the binding of an activator protein or by recruiting proteins that cause the DNA to become more compact.

All five of these factors may contribute to the regulation of a single gene, or possibly only three or four will play a role. In most cases, transcriptional regulation is aimed at controlling the initiation of transcription at the promoter. In this section, we will survey these basic kinds of gene regulation in eukaryotic species.

Eukaryotic Structural Genes Have a Core Promoter and Response Elements

To understand gene regulation in eukaryotes, we first need to consider the DNA sequences that are needed to initiate transcription. For eukaryotic structural genes that encode proteins, three features are found in most promoters: a **transcriptional start site**, a **TATA box**, and **response elements** (**Figure 13.13**).

Figure 13.13 **A common organization of sequences for the promoter of a eukaryotic structural gene.** The core promoter is made up of a TATA box and a transcriptional start site. The TATA box sequence is 5′–TATAAAA–3′. The sequence of the transcriptional start site in the 5′ to 3′ direction is two pyrimidines (C or T), then CA, then five pyrimidines. The A marks the site of the first A in the RNA transcript. Note: The sequences shown for the TATA box and transcriptional start site are those found in the coding strand. *Py* refers to either pyrimidine, cytosine or thymine. Upstream from the core promoter, response elements such as enhancers and silencers are usually found.

The TATA box and transcriptional start site form the **core promoter**. The transcriptional start site is the place in the DNA where transcription actually begins. The TATA box, which is a 5′–TATAAAA–3′ sequence, is usually about 25 base pairs upstream from a transcriptional start site. The TATA box is important in determining the precise starting point for transcription. If it is missing from the core promoter, the transcriptional start site does not function properly, and transcription may start at a variety of different locations, producing unusable transcripts. The core promoter, by itself, results in a low level of transcription that is termed **basal transcription**.

Response elements (or regulatory elements) are DNA segments that regulate eukaryotic genes. As described later, response elements are recognized by regulatory proteins that control the ability of RNA polymerase to initiate transcription at the core promoter. Some response elements, known as **enhancers**, play a role in the ability of RNA polymerase to begin transcription and thereby enhance the rate of transcription. When enhancers are not functioning, most eukaryotic genes have very low levels of basal transcription. Other response elements, known as **silencers**, prevent transcription of a given gene when its expression is not needed. When these sequences function, the rate of transcription is decreased.

A common location for response elements is the region that is 50 to 100 base pairs upstream from the transcriptional start site (Figure 13.13). However, the locations of response elements are quite variable among different eukaryotic genes. Response elements can be quite distant from the promoter, even 100,000 base pairs away, yet exert strong effects on the ability of RNA polymerase to initiate transcription at the core promoter! Response elements were first discovered by Susumu Tonegawa and coworkers in the 1980s. While studying genes that play a

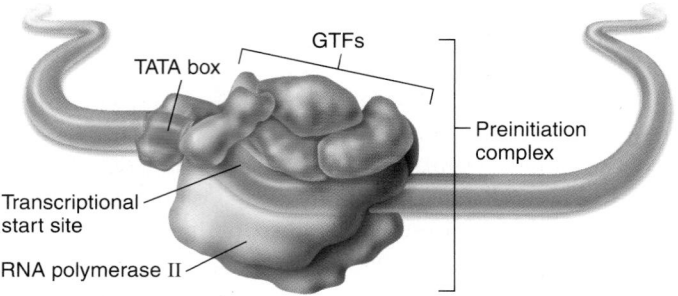

Figure 13.14 **The preinitiation complex.** General transcription factors (GTFs) and RNA polymerase II assemble into the preinitiation complex at the core promoter in eukaryotic structural genes.

role in immunity, they identified a region that was far away from the core promoter but was needed for high levels of transcription to take place.

RNA Polymerase II, General Transcription Factors, and Mediator Are Needed to Transcribe Eukaryotic Structural Genes

By studying transcription in a variety of eukaryotic species, researchers have identified three types of proteins that play a role in initiating transcription at the core promoter of structural genes. These are RNA polymerase II, five different proteins called **general transcription factors (GTFs)**, and a large protein complex called **mediator**. GTFs are needed for DNA binding at the core promoter and for initiation of transcription. Mediator is also needed for initiation.

The GTFs and RNA polymerase II must come together at the core promoter before transcription can be initiated. A series of interactions must occur between these proteins so that RNA polymerase II can bind to the DNA. **Figure 13.14** shows the structure of the completed assembly of GTFs and RNA polymerase II at the TATA box, which is known as the **preinitiation complex**. *In vitro*, when researchers mix together GTFs, RNA polymerase II, and a DNA sequence containing a TATA box and a transcriptional start site, the DNA is transcribed into RNA. Therefore, these components are referred to as the **basal transcription apparatus**. In a living cell, however, additional components such as regulatory transcription factors control the assembly of GTFs and RNA polymerase II at the core promoter, and are responsible for causing transcription to begin at a fast or slow rate.

A third component needed for transcription in eukaryotes is the mediator protein complex. Mediator is composed of several proteins that bind to each other to form an elliptical-shaped complex that partially wraps around RNA polymerase II and the GTFs. Mediator derives its name from the observation that it mediates interactions between RNA polymerase II and regulatory transcription factors such as activators or repressors that bind to enhancers or silencers. The function of mediator is to control the rate at which RNA polymerase can begin to transcribe RNA at the transcriptional start site.

Activators and Repressors May Influence the Function of GTFs or Mediator

In eukaryotes, regulatory transcription factor proteins called activators and repressors bind to enhancers or silencers, respectively, and regulate the rate of transcription of a nearby gene. Most activators and repressors do not bind directly to RNA polymerase II. Instead, some of them regulate the function of RNA polymerase II by interacting with other proteins that bind to RNA polymerase II. Certain activators and repressors exert their effects by binding to GTFs or mediator.

As shown in **Figure 13.15a**, some activators bind to an enhancer and then influence the function of GTFs. For example, an activator may improve the ability of a GTF called TFIID to initiate transcription. The function of TFIID is to recognize the TATA box and begin the assembly process. An activator may recruit TFIID to the TATA box, thereby promoting the assembly of GTFs and RNA polymerase II into the preinitiation complex. In contrast, repressors may inhibit the function of TFIID. Certain repressors exert their effects by preventing the binding of TFIID to the TATA box, or by inhibiting the ability of TFIID to assemble other GTFs and RNA polymerase II at the core promoter.

A second way that regulatory transcription factors control RNA polymerase II is via mediator. Activators stimulate the function of mediator and thereby cause RNA polymerase II to proceed to the elongation phase of transcription more quickly (**Figure 13.15b**). Alternatively, repressors have the opposite effect to those seen in this part of the figure. When a repressor inhibits mediator, RNA polymerase II cannot progress to the elongation stage.

A third way that regulatory transcription factors influence transcription is by recruiting proteins that affect the compaction or packing of DNA to the promoter region. As described in Chapter 15, the DNA found in eukaryotic chromosomes is highly compacted. To be accessible to RNA polymerase II and GTFs, the DNA must be loosened up. This topic is described next.

Gene Accessibility Is Controlled by Changes in Chromatin Structure

The regulation of gene transcription is not simply dependent on the activity of activators and repressors that interact with GTFs and mediator. In eukaryotes, DNA is associated with proteins to form a compact structure called **chromatin**. The

1 An activator protein binds to enhancer.

Activator

Enhancer

2 The activator protein enhances the ability of a GTF called TFIID to bind to the TATA box.

TFIID

TATA box

3 TFIID promotes the assembly of the preinitiation complex.

TFIID

TATA box

Preinitiation complex

(a) Effect of an activator via TFIID, a general transcription factor

1 Mediator binds to the preinitiation complex, but transcriptional initiation does not occur.

Enhancer

Mediator

Preinitiation complex

2 An activator binds to a distant enhancer. A bend in the DNA allows the bound activator to interact with mediator. This interaction causes RNA polymerase to proceed to the elongation stage of transcription.

3 RNA polymerase begins to transcribe an RNA molecule.

Activator

(b) Effect of an activator via mediator

Figure 13.15   Two mechanisms of eukaryotic activators.

Figure 13.16 Effects of an activator on chromatin compaction.

three-dimensional packing of chromatin is an important parameter affecting gene expression. If the chromatin is very tightly packed, in what is called a **closed conformation**, transcription may be difficult or impossible. Transcription requires changes in chromatin structure that allow transcription factors to gain access to and bind to the DNA in the promoter region. Such loosely packed chromatin, said to be in an **open conformation**, is accessible to transcription factors and RNA polymerase so that transcription can take place.

An important role of some activators is to diminish the level of chromatin compaction where a gene is located. To do this, an activator first binds to an accessible enhancer site (**Figure 13.16**). Next, the binding of the activator recruits proteins to the region that loosen the level of compaction. In some cases, an activator protein attracts **histone acetyltransferase** to the region. This enzyme attaches acetyl groups (—COCH₃) to histone proteins. As described in Chapter 15, histone proteins are critical in the compaction of eukaryotic DNA. When acetylated, histone proteins do not bind as tightly to the DNA. A second effect of an activator protein is to recruit **ATP-dependent chromatin remodeling enzymes** to the site. The overall effect of histone acetyltransferase and ATP-dependent chromatin remodeling enzymes is to loosen the compaction of chromatin, sometimes over a fairly long distance such as several hundred or several thousand base pairs of DNA. This loosening facilitates the ability of RNA polymerase II to recognize and transcribe a gene.

Steroid Hormones Exert Their Effects by Binding to a Regulatory Transcription Factor and Controlling the Transcription of Nearby Genes

Thus far, we have considered the general ways that regulatory transcription factors control transcription. Let's now turn to a specific example that illustrates how a regulatory transcription factor functions within living cells. Our example involves a transcriptional activator that responds to steroid hormones. This factor is known as a **steroid receptor**, because it binds

directly to the steroid hormone. The hormone is an example of a small effector molecule.

As discussed in Chapter 50, steroid hormones are a category of hormones that are synthesized by specialized cells of many organisms, including the endocrine glands of mammals, and then secreted into the bloodstream. The hormones are then taken up by cells that respond to the hormones in different ways. For example, glucocorticoid hormones influence nutrient metabolism in most body cells by promoting the metabolism of glucose, proteins, and fats.

The effect of glucocorticoid hormones is to activate the transcription of specific genes. Under certain conditions, such as when an animal is fasting and needs to regulate its blood levels of glucose, fats, and amino acids, glucocorticoids are released from endocrine cells and secreted into the bloodstream. The hormone diffuses across the plasma membrane of target cells and binds to glucocorticoid receptors (**Figure 13.17**). This binding releases proteins called chaperones and thereby exposes an amino acid sequence within the receptor called a nuclear localization signal (NLS). This signal directs the receptor to travel into the nucleus through a nuclear pore. Two glucocorticoid receptors bind to each other noncovalently to form a dimer and then travel through the nuclear pore into the nucleus. The glucocorticoid receptor dimer binds to two adjacent glucocorticoid response elements (GREs) that are next to particular genes. The GREs function as enhancer sequences. The binding of the glucocorticoid receptor dimer to GREs activates the transcription of the adjacent gene, eventually leading to the synthesis of the encoded protein.

Mammalian cells usually have a large number of glucocorticoid receptors within their cytosol. Because GREs are located near several different genes, the uptake of hormone molecules can activate many glucocorticoid receptors and thereby enhance the transcription of several different genes that encode proteins involved with the metabolism of glucose, proteins, and fats. For this reason, glucocorticoid hormones facilitate the coordinated expression of genes that play a role in nutrient metabolism.

Figure 13.17 Action of the glucocorticoid receptor as a transcriptional activator.

Biological inquiry: If a GRE next to a gene was deleted, how would that affect the regulation of the gene?

Structural Features of Regulatory Transcription Factors Allow Them to Bind to DNA

Thus far, we have considered various ways that transcription factors bind to DNA to influence transcription. Let's now take a closer look at the structural features that are common to many types of transcription factor proteins. General and regulatory transcription factors have been identified from many organisms, including prokaryotes, yeast, plants, and animals. Transcription factor proteins contain domains that have specific functions. When a domain or portion of a domain has a very similar structure in many different proteins, such a structure is called a **motif**.

Figure 13.18 depicts several different motif structures commonly found in transcription factor proteins. An α helix is frequently important in the recognition of the DNA double helix. In helix-turn-helix and helix-loop-helix motifs, an α helix called the recognition helix makes contact with and recognizes a base sequence along the major groove of the DNA (Figure 13.18a,b). As discussed in Chapter 11, the major groove is a region of the DNA double helix where the bases contact the surrounding water in the cell. Hydrogen bonding between an α helix and nucleotide bases enables the protein to recognize a specific DNA sequence. Similarly, a zinc finger motif is composed of one α helix and a β pleated sheet structure that are held together by a zinc metal ion (Zn^{2+}) (Figure 13.18c). The zinc finger also can recognize DNA sequences within the major groove. The glucocorticoid receptor described in Figure 13.17 binds to GREs via zinc fingers.

(a) Helix-turn-helix motif

(b) Helix-loop-helix motif

(c) Zinc finger motif

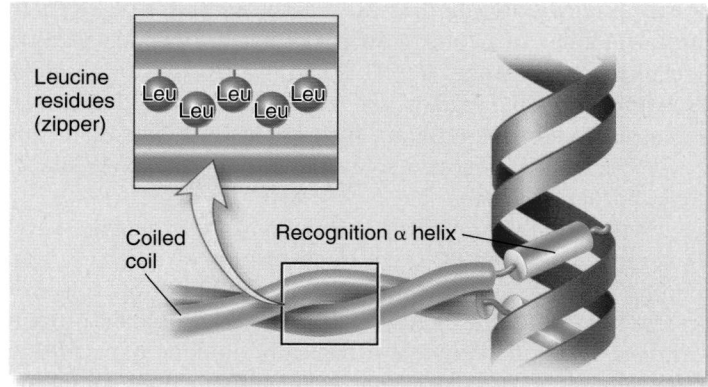

(d) Leucine zipper motif

Figure 13.18 Motif structures of transcription factors. In this figure, α helices are represented as cylinders, while β pleated sheets are represented as flattened arrows.

An interesting feature of certain motifs is that they promote protein dimerization, two proteins binding to each other as did the glucocorticoid receptors we discussed earlier. The leucine zipper and helix-loop-helix motifs mediate protein dimerization. For example, Figure 13.18d depicts the dimerization and DNA binding of two proteins that have leucine zippers. Alternating leucine residues in both proteins interact ("zip up"), resulting in protein dimerization. In some cases, two identical transcription factor proteins will come together to form a **homodimer**, or two different transcription factors will form a **heterodimer**. Dimerization of transcription factors can be an important way to modulate their function.

DNA Methylation Inhibits Gene Transcription

Let's now turn our attention to a mechanism that silences gene expression. DNA structure can be modified by the covalent attachment of methyl groups ($-CH_3$) by an enzyme called **DNA methylase**. Such **DNA methylation** is common in some eukaryotic species but not all. For example, yeast and *Drosophila* have little or no detectable methylation of their DNA, whereas DNA methylation in vertebrates and plants is relatively abundant. In mammals, approximately 5% of the DNA is methylated. Eukaryotic DNA methylation occurs on the cytosine base. The sequence that is methylated is shown here:

$$CH_3$$
$$|$$
$$5'- CG - 3'$$
$$3'- GC - 5'$$
$$|$$
$$CH_3$$

Note that this sequence contains cytosines in both strands. When only one strand is methylated, this is called hemimethylation; the methylation of the cytosines in both strands is termed full methylation.

DNA methylation usually inhibits the transcription of eukaryotic genes, particularly when it occurs in the vicinity of the promoter. In vertebrates and plants, many genes contain sequences called **CpG islands** near their promoters. CpG refers to the nucleotides of <u>C</u> and <u>G</u> in DNA that are connected by a phosphodiester linkage. A CpG island is a cluster of CpG sites. Unmethylated CpG islands are correlated with active genes, while repressed genes contain methylated CpG islands. In this way, DNA methylation may play an important role in the silencing of particular genes.

Methylation can inhibit transcription in two general ways. First, methylation of CpG islands may prevent an activator from binding to an enhancer element. The inability of an activator protein to bind to the DNA would inhibit the initiation of transcription. A second way that methylation inhibits transcription is by converting chromatin from an open to a closed conformation. Proteins known as **methyl-CpG-binding proteins** bind methylated sequences. Once bound to the DNA, the methyl-CpG-binding protein recruits other proteins to the region that cause the chromatin to become very compact.

13.4 Regulation of RNA Processing and Translation in Eukaryotes

In the first three sections of this chapter, we have mostly focused on the regulation of gene transcription in prokaryotes and eukaryotes. Unlike bacteria, eukaryotic gene expression is commonly regulated at the levels of RNA processing and translation. These added levels of regulation provide benefits that are important to eukaryotic species. First, by regulating RNA processing, eukaryotes can produce more than one mRNA transcript from a single gene. This allows a gene to encode two or more polypeptides, and thereby increases the complexity of eukaryotic proteomes. A second issue is timing. Transcriptional regulation in eukaryotes takes a fair amount of time before its effects are observed at the cellular level. During transcriptional regulation, (1) the chromatin must be converted to an open conformation, (2) the gene must be transcribed, (3) the RNA must be processed and exported from the nucleus, and (4) the protein must be made via translation. All of these steps take time, on the order of several minutes. One way to achieve faster regulation is to control steps that occur after an RNA transcript is made. In eukaryotes, translational regulation provides a faster way to regulate the levels of gene products, namely proteins. Translation can be regulated by controlling the stability of an mRNA transcript, causing it to remain in the cytosol for a long time or causing it to be rapidly degraded. Alternatively, small RNA molecules or RNA-binding proteins can bind to mRNAs and control whether or not a ribosome can translate the mRNA into a polypeptide.

During the past few decades, many critical advances have been made regarding our knowledge of the regulation of RNA processing and translation. Even so, molecular geneticists are still finding new forms of regulation, making this an exciting area of modern research. In this section, we will survey a few of the known mechanisms of RNA processing and translational regulation.

Alternative Splicing of pre-mRNAs Creates Protein Diversity

In eukaryotes, a pre-mRNA transcript is processed before it becomes a mature mRNA. When a pre-mRNA has multiple introns and exons, splicing may occur in more than one way. Such **alternative splicing** causes mRNAs to contain different patterns of exons. Alternative splicing is a form of gene regulation that allows an organism to use the same gene to make different proteins at different stages of development, in different cells types, and/or in response to a change in the environmental conditions. Alternative splicing is an important form of gene regulation in complex eukaryotes such as animals and plants.

As an example of how alternative splicing occurs, let's suppose a human pre-mRNA contains seven exons (**Figure 13.19**). In nerve cells, it is spliced to contain the following pattern of exons: 1-2-3-5-6-7. In muscle cells, it is alternatively spliced to have a different pattern: 1-2-4-5-6-7. The linear sequence of the exons is maintained in both patterns. In this example, the mRNA

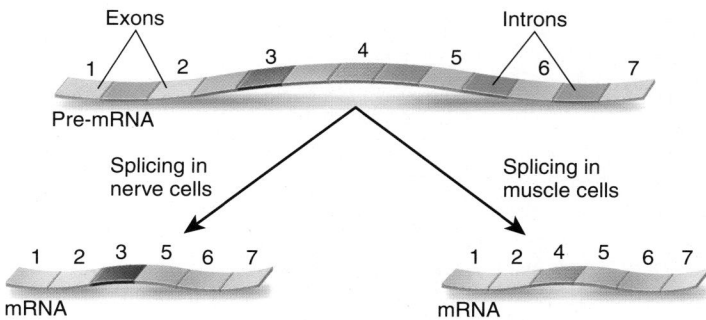

Figure 13.19 Alternative splicing as a regulatory method for RNA processing. In this example, the pre-mRNA transcript can be spliced to contain exon 3 (in nerve cells) or exon 4 (in muscle cells), but not both.

from nerve cells contains exon 3, while the mRNA from muscle cells contains exon 4. When alternative splicing occurs, proteins with significant differences in their amino acid sequences are produced.

In most cases, the alternative versions of a protein will have similar functions, because much of their amino acid sequences will be identical to each other. Nevertheless, alternative splicing produces differences in amino acid sequences that will provide each protein with its own unique characteristics. The alternatively spliced versions tend to be expressed in different cell types (for example, nerve versus muscle cells) or at different stages of development (for example, embryonic versus adult). This provides a way for multicellular organisms to "fine-tune" a given protein to function optimally in a given cell type or stage of development. The advantage of alternative splicing is that two (or more) different polypeptides can be derived from a single gene, thereby increasing the size of the proteome while minimizing the size of the genome. A small genome size is beneficial because less energy is spent replicating the DNA and the DNA more easily fits within the nucleus of the cell.

Table 13.1	Genome Size and Biological Complexity			
Species	**Level of complexity**	**Genome size (million bp)**	**Approximate number of genes**	**Percentage of genes alternatively spliced**
Escherichia coli	A unicellular prokaryote	4.2	4,000	0
Saccharomyces cerevisiae	A unicellular eukaryote	12	6,000	<1
Caenorhabditis elegans	A tiny worm (about 1,000 cells)	97	19,000	2
Drosophila melanogaster	An insect	137	14,000	7
Arabidopsis thaliana	A flowering plant	142	26,000	11
Homo sapiens	A complex mammal	3,000	25,000	70

fruit fly (*Drosophila melanogaster*), a small flowering plant (*Arabidopsis thaliana*), and humans (*Homo sapiens*). A general trend is that less complex organisms tend to have fewer genes. For example, unicellular organisms have only a few thousand genes, whereas multicellular species have tens of thousands. Even so, the trend is by no means a linear one. If we compare *C. elegans* and *D. melanogaster*, the fly actually has fewer genes even though it is biologically more complex. A second trend you can see in Table 13.1 concerns alternative splicing. This phenomenon does not occur in bacteria and is rare in *S. cerevisiae*. The frequency of alternative splicing increases from worms to flies to humans. For example, the level of alternative splicing is ten-fold higher in humans compared to *Drosophila*. This trend can partially explain the increase in biological complexity among these species.

GENOMES & PROTEOMES

Increases in Biological Complexity Are Correlated with Greater Sizes of Genomes and Proteomes

As we have just seen, alternative splicing can increase the proteome size without increasing the total number of genes. For organisms to become more complex, as in higher plants and animals, evolution has produced more complex proteomes. In the past few decades, many technical advances have improved our ability to analyze the genomes and proteomes of many different species. Researchers have been able to determine the amount of DNA from several species and estimate the total number of genes. In addition, scientists can also estimate the number of polypeptides if information is available concerning the degree of alternative splicing in a given species.

Table 13.1 compares six species, a bacterium (*Escherichia coli*), a eukaryotic single-celled organism (yeast—*Saccharomyces cerevisiae*), a small nematode worm (*Caenorhabditis elegans*), a

MicroRNAs Inhibit mRNA by Translational Repression or mRNA Degradation

Let's now turn our attention to regulatory mechanisms that affect translation. One method involves the silencing of pre-existing mRNA. **MicroRNAs** (abbreviated **miRNAs**) are small RNA molecules, typically 22 nucleotides in length, that silence the expression of specific mRNAs. In 1993, Victor Ambros and his colleagues, who were interested in the developmental stages that occur in the worm *C. elegans*, determined that the transcription of a particular gene produced a small RNA, now called a microRNA, that does not encode a protein. Instead, this miRNA was found to be complementary to an mRNA and inhibit its translation.

Insight into the mechanism of miRNA inhibition came from the surprising results of Andrew Fire and Craig Mello who showed that double-stranded RNA is very potent at silencing the expression of specific genes. In 2006, Fire and Mello were awarded the Nobel Prize for their studies aimed at understanding the effects

Pre-miRNA

Figure 13.20   Mechanism of action of microRNA (miRNA).

(a) Low iron levels

(b) High iron levels

Figure 13.21   Translational regulation of ferritin mRNA by the iron regulatory protein (IRP).

Biological inquiry: Poisoning may occur when a young child finds a bottle of vitamins, such as those that taste like candy, and eats a large number of them. One of the toxic effects involves the ingestion of too much iron. How does the IRP protect people from the toxic effects of too much iron?

of miRNA (**Figure 13.20**). MiRNAs are first synthesized as a pre-miRNA, which forms a hairpin structure. The double-stranded region is cut by an enzyme called dicer. This releases a miRNA that associates with cellular proteins to become part of a complex called the **RNA-induced silencing complex (RISC)**. The miRNA, which is complementary to specific cellular mRNAs, will direct the RISC to those mRNAs. Upon binding, two different things may happen. In some cases, the binding inhibits translation. This is the common effect of miRNAs in animals. In plants, the RISC directs the degradation of the mRNA. In either case, the expression of the mRNA is silenced. Fire and Mello called this **RNA interference (RNAi)**, because the miRNA interferes with the proper expression of an mRNA.

Since this study, researchers have discovered that genes encoding miRNAs are widely found in animals and plants. In humans, for example, approximately 200 different genes encode miRNAs. MiRNAs represent an important mechanism of mRNA silencing.

The Prevention of Iron Toxicity in Mammals Involves the Regulation of Translation

Another way to regulate translation involves RNA-binding proteins that directly affect translational initiation. The regulation of iron absorption provides a well-studied example. Iron is a

vital cofactor for many cellular enzymes. However, it is toxic at high levels. To prevent toxicity, mammalian cells synthesize a protein called ferritin, which forms a hollow, spherical complex that can store excess iron.

The mRNA that encodes ferritin is controlled by an RNA-binding protein known as the **iron regulatory protein (IRP)**. When iron levels in the cytosol are low and more ferritin is not needed, IRP binds to a response element within the ferritin mRNA known as the **iron regulatory element (IRE)**. The IRE is located upstream from the start codon in a region called the 5'-untranslated region (5'-UTR). Due to base pairing, it forms a stem-loop structure. The binding of IRP to the IRE inhibits translation of the ferritin mRNA (**Figure 13.21a**). However, when iron is abundant in the cytosol, the iron binds directly to IRP and prevents it from binding to the IRE. Under these conditions, the ferritin mRNA is translated to make more ferritin protein (**Figure 13.21b**). This mechanism of translational control allows cells to rapidly respond to changes in their environment. When cells are confronted with high levels of iron, they can quickly make more ferritin protein to prevent the toxic buildup of iron. This mechanism is faster than transcriptional regulation, which would require the activation of the ferritin gene and the transcription of ferritin mRNA prior to the synthesis of more ferritin protein.

Chapter Summary

13.1 Overview of Gene Regulation

- Constitutive genes are expressed at constant levels while most genes are regulated so that the levels of gene expression can vary under different conditions.

- Organisms regulate genes so that gene products are made only when they are needed. An example is the synthesis of the gene products needed for lactose utilization in bacteria. (Figure 13.1)

- Multicellular eukaryotes regulate genes to produce different cell types, such as muscle, nerve, and skin cells. (Figure 13.2)

- Eukaryotes also regulate genes so the gene products are produced at different developmental stages. An example is the group of globin genes in mammals. (Figure 13.3)

- All organisms regulate gene expression at a variety of levels, including transcription, translation, and post-translation. Eukaryotes also regulate RNA processing. (Figure 13.4)

13.2 Regulation of Transcription in Bacteria

- Activators and repressors are regulatory proteins that bind to the DNA and regulate the expression of genes. Small effector molecules control the ability of regulatory proteins to bind to DNA. (Figure 13.5)

- The *lac* operon found in *E. coli* is an arrangement of three structural genes controlled by a single promoter. The operon is transcribed into a polycistronic mRNA. The operator and CAP site are involved with gene regulation via the *lac* repressor and CAP, respectively. (Figure 13.6)

- The *lac* repressor binds to the operator site and prevents RNA polymerase from transcribing the operon. When allolactose binds to the repressor, this causes a conformational change that prevents the repressor from binding to the operator site. (Figure 13.7)

- Mutations in *lacI* can result in the constitutive expression of the *lac* operon. Jacob, Monod, and Pardee proposed an alternate hypothesis to explain this observation. (Figure 13.8)

- By constructing a merozygote, Jacob, Monod, and Pardee determined that *lacI* encodes a diffusible protein that represses the *lac* operon. (Figure 13.9)

- The catabolite activator protein (CAP) binds to the CAP site in the presence of cAMP. This causes a bend in the DNA, which allows RNA polymerase to bind to the promoter. (Figure 13.10)

- Glucose inhibits cAMP production. This inhibits the expression of the *lac* operon because CAP cannot bind to the CAP site. This form of regulation provides bacteria with a more efficient utilization of their resources because the bacteria use one sugar at a time. (Figure 13.11)

- The *trp* operon is repressible. The presence of tryptophan causes the *trp* repressor to bind to the *trp* operator and stop transcription. This prevents the excessive buildup of tryptophan in the cell, which would be a waste of energy. (Figure 13.12)

13.3 Regulation of Transcription in Eukaryotes

- Eukaryotic genes exhibit combinatorial control, meaning that many factors control the expression of a single gene. (See list on p. 267.)

- Eukaryotic promoters consist of a core promoter and response elements, such as enhancers or silencers, that regulate the rate of transcription. (Figure 13.13)

- Eukaryotes have multiple forms of RNA polymerase. The form named RNA polymerase II transcribes structural genes.

- General transcription factors (GTFs) are needed for RNA polymerase II to bind to the core promoter. (Figure 13.14)

- Activators and repressors may regulate RNA polymerase II by interacting with GTFs, such as TFIID, or via mediator, a protein complex that wraps around RNA polymerase II. (Figure 13.15)

- A loosening of chromatin is needed for eukaryotic genes to be transcribed. (Figure 13.16)

- Steroid hormones bind to receptors that function as activator proteins. (Figure 13.17)

- General and regulatory transcription factors have motifs, which are common structural features that enable them to bind to the DNA and to each other. (Figure 13.18)

- DNA methylation, which occurs at CpG islands near promoters, usually inhibits transcription by preventing the binding of activator proteins or by promoting the compaction of chromatin.

13.4 Regulation of RNA Processing and Translation in Eukaryotes

- Alternative splicing occurs when a single type of pre-mRNA can be spliced in more than one way, producing polypeptides with somewhat different sequences. This is a common way for complex eukaryotes to increase the size of their proteomes. (Figure 13.19, Table 13.1)

- MicroRNAs (miRNAs) inhibit mRNAs, either by inhibiting translation or by promoting the degradation of mRNAs. (Figure 13.20)

- RNA-binding proteins can regulate the translation of specific mRNAs. An example is iron regulatory protein (IRP), which regulates the translation of ferritin mRNA. (Figure 13.21)

Test Yourself

1. Genes that are expressed at all times at relatively constant levels are known as ____ genes.
   a. induced
   b. repressed
   c. positive
   d. constitutive
   e. structural

2. Which of the following is not considered a level of gene regulation in prokaryotes?
   a. transcriptional
   b. RNA processing
   c. translational
   d. post-translational
   e. All of the above are levels at which prokaryotes are able to regulate gene expression.

3. Transcription factors that bind to DNA and stimulate transcription are
   a. repressors.
   b. small effector molecules.
   c. activators.
   d. promoters.
   e. operators.

4. In prokaryotes the unit of DNA that contains multiple structural genes under the control of a single promoter is called _____. The mRNA produced from this unit is referred to as _____ mRNA.
   a. an operator, a polycistronic
   b. a template, a structural
   c. an operon, a polycistronic
   d. an operon, a monocistronic
   e. a template, a monocistronic

5. In the *lac* operon, the repressor protein binds to _____ to inhibit transcription.
   a. the promoter
   b. the operator
   c. the CAP site
   d. the enhancer
   e. lactose

6. The presence of _____ in the medium prevents the CAP from binding to the DNA, resulting in _____ in transcription of the *lac* operon.
   a. lactose, an increase
   b. glucose, an increase
   c. cAMP, a decrease
   d. glucose, a decrease
   e. lactose, a decrease

7. The *trp* operon is considered _____ operon because the structural genes necessary for tryptophan synthesis are not expressed when the levels of tryptophan in the cell are high.
   a. an inducible
   b. a positive
   c. a repressible
   d. a negative
   e. both c and d

8. Response elements that function to increase transcription levels in eukaryotes are called
   a. promoters.
   b. silencers.
   c. enhancers.
   d. transcriptional start sites.
   e. activators.

9. DNA methylation in many eukaryotic organisms seems to relate to
   a. increased translation levels.
   b. introns that will be removed.
   c. regions of DNA that do not contain structural genes.
   d. decreased transcription levels.
   e. response elements that are not necessary for transcription.

10. _____ refers to the phenomenon where a single type of pre-mRNA may give rise to multiple mRNAs due to different patterns of intron and exon removal.
   a. Spliceosomes
   b. Variable expression
   c. Alternative splicing
   d. Polycistronic mRNA
   e. Induced silencing

CONCEPTUAL QUESTIONS

1. List the components of the *lac* operon and explain the function of each.

2. Explain the difference between inducible and repressible operons.

3. Explain how DNA methylation inhibits transcription.

EXPERIMENTAL QUESTIONS

1. What were the key observations made by Jacob, Monod, and Pardee that led to the development of their hypothesis regarding the *lacI* gene and the regulation of the *lac* operon?

2. What was the hypothesis proposed by the researchers to explain the function of the *lacI* and the regulation of the *lac* operon?

3. How did Jacob, Monod, and Pardee test the hypothesis? What were the results of the experiment? How do these results support the idea that the *lacI* gene produces a repressor protein and does not function as an operator?

COLLABORATIVE QUESTIONS

1. Discuss the different points at which gene regulation can take place to produce a functional protein.

2. Discuss several factors that control the regulation of transcription in eukaryotes.

www.brookerbiology.com

This website includes answers to the Biological Inquiry questions found in the figure legends and all end-of-chapter questions.

14

MUTATION, DNA REPAIR, AND CANCER

CHAPTER OUTLINE

Tanning and cigarette smoking are two activities that expose people to agents that cause mutations. Such mutations may lead to cancer.

As we have learned in Chapters 11 through 13, DNA stores the information for the synthesis of proteins. When a gene is expressed, the DNA itself does not normally change. However, on rare occasions, a mutation may occur. A **mutation** is a heritable change in the genetic material. The structure of DNA, its base sequence, has been changed permanently and this alteration can be passed from mother to daughter cells during cell division. Likewise, if a mutation occurs in a reproductive cell, it may also be passed from parent to offspring.

On the positive side, mutations are essential to the continuity of life. They supply the variation that enables species to evolve and become better adapted to their environments. Mutations provide the foundation for evolutionary change. On the negative side, however, new mutations are more likely to be harmful than beneficial to the individual. The genes within modern species are the products of billions of years of evolution and have evolved to work properly. Random mutations are more likely to disrupt genes than to enhance their function. Many forms of inherited diseases, such as cystic fibrosis and sickle-cell anemia, are caused by gene mutations. For this and many other reasons, understanding the molecular nature of mutations is a compelling area of research.

Because mutations can be quite harmful, all species have evolved several ways to repair damaged DNA. Such **DNA repair systems** reverse DNA damage before a permanent mutation can occur. DNA repair systems are vital to the survival of all organisms. If these systems did not exist, mutations would be so prevalent that few species, if any, would survive. In this chapter, we will examine a few ways that these systems operate.

The final topic we discuss in this chapter is **cancer**, a disease that is caused by gene mutations that lead to uncontrolled cell growth. DNA repair systems may prevent the occurrence of cancer by minimizing the number of mutations in any given cell. As we will learn later, multicellular organisms also possess other safeguards to help prevent cancer. Unfortunately, however, many individuals develop cancer due to naturally occurring mutations that are unavoidable or due to exposure to environmental agents such as cigarette smoke. At the end of this chapter, we will explore the various types of genetic changes that lead to the progression of cancer.

14.1 Mutation

To appreciate why mutations are beneficial or detrimental, we must understand how changes in DNA structure can ultimately affect DNA function. Most of our understanding of mutation has come from the study of experimental organisms, such as bacteria and *Drosophila*. Researchers can expose these organisms to environmental agents that cause mutations and then study the consequences of the mutations that arise. In addition, because these organisms have a short generation time, researchers can investigate the effects of mutations when they are passed from parent to offspring.

The structure and amount of genetic material can be altered in a variety of ways. For example, chromosome structure and number can change. These types of genetic changes will be considered in Chapter 15. In this section, we will focus our attention on **gene mutations**, which are relatively small changes in DNA structure that alter a particular gene. We will be primarily concerned with the ways that mutations may affect the expression of single genes. We will also consider how the timing of mutations during an organism's development has important consequences.

In addition, we will explore how environmental agents may bring about mutations, and examine a testing method that can determine if an agent causes mutations.

Gene Mutations Alter the DNA Sequence of a Gene

Mutations can cause two basic types of changes to a gene: (1) the base sequence within a gene can be changed; and (2) one or more nucleotides can be added to or removed from the gene. A **point mutation** affects only a single base pair within the DNA. For example, the DNA sequence shown here has been altered by a **base substitution**:

5′—CCCGCTAGATA—3′   5′—CCCGCGAGATA—3′
3′—GGGCGATCTAT—5′ →   3′—GGGCGCTCTAT—5′

A point mutation could also involve the addition or deletion of a single base pair to a DNA sequence. For example, in the following sequence, a single base pair has been added to the DNA:

5′—GGCGCTAGATC—3′   5′—GGCAGCTAGATC—3′
3′—CCGCGATCTAG—5′ →   3′—CCGTCGATCTAG—5′

Though point mutations may seem like small changes to a DNA sequence, they can have important consequences when genes are expressed. This topic is discussed next.

Gene Mutations May Affect the Amino Acid Sequence of a Polypeptide

If a mutation occurs within the region of a structural gene that specifies the amino acid sequence, such a mutation may alter that sequence in a variety of ways. **Table 14.1** considers the potential effects of point mutations. **Silent mutations** do not alter the amino acid sequence of the polypeptide, even though the nucleotide sequence has changed. Because the genetic code is degenerate, silent mutations can occur in the third base within most codons without changing the type of amino acid it encodes. Silent mutations are also considered **neutral mutations**, because they do not affect the function of the encoded protein.

A **missense mutation** is a base substitution that changes a single amino acid in a polypeptide sequence. Missense mutations may not alter protein function because they change only a single amino acid within polypeptides that typically are hundreds of amino acids in length. When a missense mutation has no detectable effect on protein function, it is also referred to as a neutral mutation. A missense mutation that substitutes an amino acid with a chemistry similar to the original amino acid is likely to be neutral. For example, a missense mutation that substitutes a glutamic acid for an aspartic acid may not alter protein function because both amino acids are negatively charged and have similar side chain structures.

Alternatively, some missense mutations have a dramatic effect on protein function. A striking example of such a missense mutation occurs in the human disease known as **sickle-cell**

Table 14.1 Consequences of Point Mutations Within the Coding Sequence of a Structural Gene

*DNA sequence in the coding strand. This sequence is the same as the mRNA sequence, except that RNA contains uracil (U) instead of thymine (T).

anemia. This disease involves a mutation in the β-globin gene, which encodes one of the polypeptide subunits that make up hemoglobin, the oxygen-carrying protein in red blood cells. In the most common form of this disease, a missense mutation alters the polypeptide sequence so that the sixth amino acid is changed from a glutamic acid to a valine. Because glutamic acid is hydrophilic but valine is hydrophobic, this single amino acid substitution alters the structure and function of the hemoglobin protein. The mutant hemoglobin subunits tend to stick to one another when the oxygen concentration is low. The aggregated proteins form fiber-like structures within red blood cells, which causes the cells to lose their normal morphology and become sickle-shaped (**Figure 14.1**). It's rather amazing that a single amino acid substitution could have such a profound effect on the structure of cells.

Two other types of point mutations cause more dramatic changes to a polypeptide sequence. A **nonsense mutation** involves a change from a normal codon to a stop or termination codon. This causes translation to be terminated earlier than expected, producing a truncated polypeptide (see Table 14.1). A shorter polypeptide is much less likely than a normal polypeptide to function properly. Finally, a **frameshift mutation** involves the addition or deletion of nucleotides that are not in multiples of three nucleotides. For example, a frameshift mutation could involve the addition or deletion of one, two, four, or five nucleotides. Because the codons are read in multiples of three, this shifts the reading frame so that a completely different

Figure 14.1 A missense mutation that causes red blood cells to sickle in sickle-cell anemia.   Scanning electron micrographs of **(a)** normal red blood cells and **(b)** sickled red blood cells. As shown above the micrographs, a missense mutation in the β-globin gene (which codes for a subunit of hemoglobin) changes the sixth amino acid in the β-globin polypeptide from a glutamic acid to a valine. **(c)** This micrograph shows how this alteration to the structure of β-globin causes the formation of abnormal fiber-like structures.

Biological inquiry: Based on the fiber-like structures seen in part (c), what aspect of hemoglobin structure does a glutamic acid at the sixth position in β-globin prevent? Speculate on how the charge of this amino acid may play a role.

amino acid sequence occurs downstream from the mutation (see Table 14.1). Such a large change in polypeptide structure is also likely to inhibit protein function.

Except for silent mutations, new mutations are more likely to produce polypeptides that have reduced rather than enhanced function. However, mutations can occasionally produce a polypeptide that has a better ability to function. While these favorable mutations are relatively rare, they may result in an organism with a greater likelihood to survive and to reproduce. The favorable effect of a mutation may cause it to increase in frequency in a population over the course of many generations. This topic is discussed in Chapter 24.

Gene Mutations Can Occur Outside of Coding Sequences and Influence Gene Expression

Thus far, we have focused our attention on mutations in the coding regions of structural genes. In Chapters 12 and 13, we learned how other DNA sequences play important roles during gene expression. A mutation can occur within noncoding sequences and affect gene expression (**Table 14.2**). For example, a mutation may alter the sequence within the promoter of a gene. A mutation that causes an increase in the rate of transcription is called an up promoter mutation. Such a mutation may improve the ability of transcription factors and RNA polymerase to recognize the promoter. In contrast, a down promoter mutation causes a decrease in transcription.

Table 14.2	Effects of Mutations Outside of the Coding Sequence of a Gene
Sequence	**Effect of mutation**
Promoter	May increase or decrease the rate of transcription
Transcriptional response element/ operator site	May alter the regulation of transcription
Splice junctions	May alter the ability of pre-mRNA to be properly spliced
Translational response element	May alter the ability of mRNA to be translationally regulated

Mutations in transcriptional response elements or operator sites can alter the regulation of gene transcription. For example, in Chapter 13, we considered the roles of regulatory sequences such as the *lac* operator site in *E. coli*, which is recognized by the *lac* repressor protein (refer back to Figure 13.7). Mutations in the *lac* operator site can disrupt the proper regulation of the *lac* operon. An operator mutation may change the DNA sequence so that the *lac* repressor protein does not bind to it. This mutation would cause the operon to be constitutively expressed even in the absence of lactose.

Mutations can also occur in other noncoding regions of a gene and alter gene expression in a way that may affect the encoded polypeptide sequence or its expression. For example, in Chapter 12 we discussed the splicing of RNA during RNA processing in eukaryotes (refer back to Figure 12.10). Mutations can occur at the boundaries between introns and exons and prevent proper splicing. When this occurs, the mRNA may not contain the correct order and number of exons. In Chapter 13, we also discussed response elements that are recognized by proteins that control translation. For example, the iron regulatory protein recognizes response elements within the mRNA that encodes ferritin (refer back to Figure 13.21). A mutation that alters this response element may prevent the proper translational regulation of ferritin.

FEATURE INVESTIGATION

The Lederbergs Used Replica Plating to Show That Mutations Are Random Events

Thus far, we have considered how mutations can affect the expression of genes. Let's now ask the question, How do mutations occur? For many centuries, biologists have wondered whether mutations that affect the traits of an individual occur as a result of pre-existing circumstances or whether they are unplanned events that happen randomly in any gene of any individual. This question has an interesting history. In the 19th century, French naturalist Jean Baptiste Lamarck proposed that physiological events (for example, use and disuse) determine whether traits are passed along to offspring. For example, his hypothesis suggested that an individual who practiced and became adept at a physical activity, such as the long jump, would pass that quality on to his or her offspring. Alternatively, geneticists in the early 1900s suggested that genetic variation occurs as a matter of chance and that natural selection results in the differential reproductive success of organisms that are better adapted to their environments. According to this view, those individuals whose genes happen to contain beneficial mutations are more likely to survive and to pass those genes to their offspring.

These opposing views were tested in bacterial studies in the 1940s and 1950s. One such study, by Joshua and Ester Lederberg, focused on the occurrence of mutations in bacteria (**Figure 14.2**). First, they placed a large number of *E. coli* bacteria onto a master plate that was incubated overnight, so that each bacterial cell divided many times to form a bacterial colony composed of millions of cells. Using a technique known as **replica plating**, a sterile piece of velvet cloth was lightly touched to this plate to pick up a few bacterial cells from each colony. They then transferred this replica to two secondary plates that contained an agent that selected for the growth of bacterial cells with a particular mutation.

Figure 14.2   The experiment performed by the Lederbergs that used replica plating to show that mutations are random events.

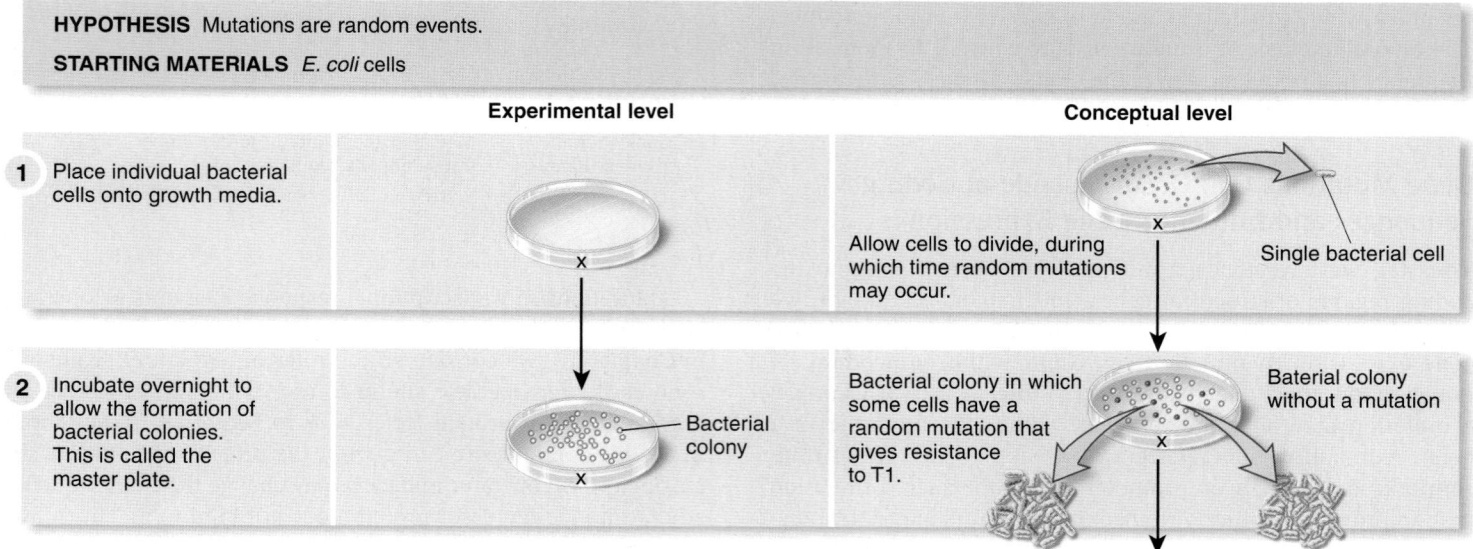

HYPOTHESIS Mutations are random events.

STARTING MATERIALS *E. coli* cells

Experimental level   Conceptual level

1 Place individual bacterial cells onto growth media.

Allow cells to divide, during which time random mutations may occur.

Single bacterial cell

2 Incubate overnight to allow the formation of bacterial colonies. This is called the master plate.

Bacterial colony

Bacterial colony in which some cells have a random mutation that gives resistance to T1.

Baterial colony without a mutation

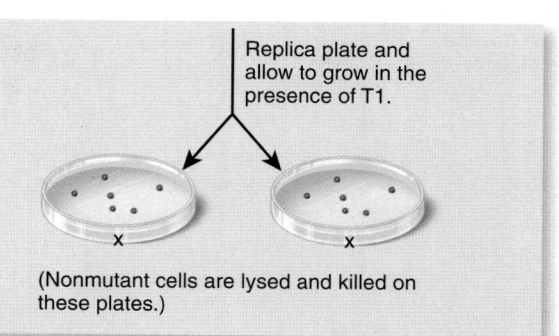

3 Press a velvet cloth (wrapped over a cylinder) onto the master plate, and then lift gently to obtain a replica of each bacterial colony. Press the replica onto 2 secondary plates that contain T1 phage. Incubate overnight to allow bacterial growth.

Master plate

Secondary plates containing T1 phage

Replica plate and allow to grow in the presence of T1.

(Nonmutant cells are lysed and killed on these plates.)

4 THE DATA

Colonies on each plate are in the same locations.

In the example shown in Figure 14.2, the secondary plates contained T1 bacteriophages, or phages, which are viruses that infect bacteria and cause them to burst or lyse. On these plates, only those rare cells that had acquired a mutation conferring resistance to T1, termed *ton^r*, could grow. All other cells were lysed by the proliferation of phages in the bacterial cytoplasm. Therefore, only a few colonies were observed on the secondary plates. Strikingly, these colonies occupied the same locations on each plate. These results indicated that the *ton^r* mutations occurred randomly while the bacterial cells were forming colonies on the nonselective master plate. The presence of T1 bacteriophages in the secondary plates simply selected for the growth of *ton^r* mutants that had occurred previously. These results supported the idea that mutations are random events.

Mutations Can Occur in Germ-Line or Somatic Cells

Let's now consider how the timing of a mutation may have an important impact on its potential effects. Multicellular organisms begin their lives as a single fertilized egg cell and by cell division produce the many cells of the adult organism. For example, humans begin as a single fertilized egg and become an adult with approximately 10 trillion cells. These cells carry copies of the genome. When a mutation occurs in a single gene, however, the mutation may or may not be found in all the cells in a multicellular organism. A mutation can occur in any cell of the body, either very early in life, such as in a gamete (eggs or sperm) or a fertilized egg, or later in life, such as in the embryonic or adult stages. The exact time and location of a mutation is critical both to the severity of the genetic effect and to whether the mutation can be passed on to offspring.

Geneticists classify the cells of animals into two types: germ-line and somatic cells. The term **germ line** refers to cells that give rise to gametes such as egg and sperm cells. A germ-line mutation can occur directly in a sperm or egg cell, or it can occur in a precursor cell that produces the gametes. If a mutant gamete participates in fertilization, all the cells of the resulting offspring will contain the mutation (**Figure 14.3a**). Likewise, when such an individual produces gametes, the mutation may be transmitted to future generations of offspring. Because an individual carries two copies of most genes, a new mutation in a single gene has a 50% chance of being transmitted from parent to offspring.

The **somatic cells** constitute all cells of the body excluding the germ-line cells. Examples include skin cells, muscle cells, and so forth. Mutations can also occur within somatic cells at early or late stages of development. **Figure 14.3b** illustrates the consequences of a mutation that happened during the embryonic stage. In this example, a mutation occurred within a single embryonic cell. This single cell was the precursor for many cells of the adult organism. Therefore, in the adult, a patch of tissue was made up of cells that carried the mutation. The size of any patch would depend on the timing of a mutation. In general, the earlier a mutation occurs during development, the larger the patch. An individual with somatic regions that are genetically different from each other is called a **genetic mosaic.**

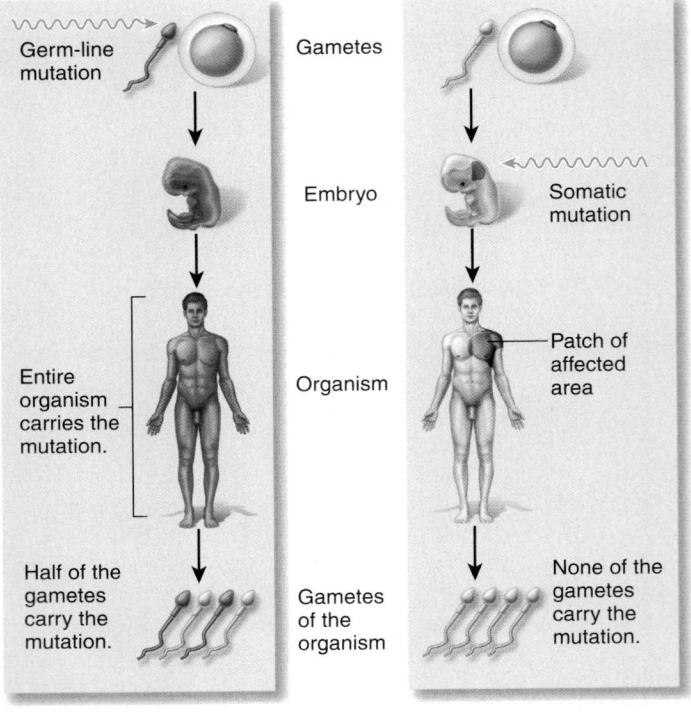

(a) Germe-line mutation   **(b) Somatic cell mutation**

Figure 14.3   The effects of germ-line versus somatic cell mutations.   Because humans have two copies of each gene, a germ-line mutation in one of those two copies will be transmitted to only half of the gametes.

Figure 14.4   Example of a somatic mutation.   This person has a patch of gray hair because a somatic mutation occurred in a single cell during embryonic development that prevented pigmentation of the hair. This cell continued to divide to produce a patch of gray hair.

Figure 14.4 illustrates an individual who had a somatic mutation during an early stage of development. In this case, the person has a patch of gray hair while the rest of the hair is pigmented. Presumably, this individual initially had a single mutation happen in an embryonic cell that ultimately gave rise to a patch of scalp that produced the gray hair. Although a patch of gray hair is not a harmful consequence, mutations during early stages of life can be quite harmful, especially if they disrupt essential developmental processes. For example, a somatic mutation during embryonic development is the cause of a human disease called McCune-Albright syndrome, which is characterized by bone defects, hormonal imbalances, and skin abnormalities. Therefore, even though it is sensible to avoid environmental agents that cause mutations during all stages of life, the possibility of somatic mutations is a rather compelling reason to avoid such agents during the early stages of life such as embryonic and fetal development, infancy, and early childhood. Let's now consider mutations caused by agents in our environment as well as those that occur naturally.

Mutations May Be Spontaneous or Induced

Biologists categorize the causes of mutation as being either spontaneous or induced. **Spontaneous mutations** result from abnormalities in biological processes (**Table 14.3**). Spontaneous mutations reflect the fact that biology isn't perfect. Enzymes, for example, can function abnormally. In Chapter 11, we learned that DNA polymerase can make a mistake during DNA replication by putting the wrong base in a newly synthesized daughter strand. Though such errors are rare, they do occur at a low rate. In addition, normal metabolic processes within the cell may produce toxic chemicals such as free radicals that can react directly with the DNA and alter its structure. Finally, the structure of nucleotides is not absolutely stable. On occasion, the structure may spontaneously isomerize, and such a change may cause a mutation if it occurs immediately prior to DNA replication. (Structural isomers are discussed in Chapter 3.)

The rates of mutations vary from species to species, and from gene to gene. Larger genes are usually more likely to incur a mutation than are smaller genes. Each time the DNA replicates, a common rate of spontaneous mutation is approximately 1 mutation for every 1 million genes, which equals 1 in 10^6, or simply 10^{-6}. This is the expected rate of "background" mutation, which creates the variation that is the raw material of evolution.

Induced mutations are brought about by environmental agents that enter the cell and then alter the structure of DNA. They cause the mutation rate to be higher than the spontaneous mutation rate. Such agents that are known to cause mutation are called **mutagens**. These can be chemical substances or physical agents that ultimately lead to changes in DNA structure. We will consider their effects next.

Table 14.3	Some Common Causes of Gene Mutations
Common causes of mutations	**Description**
Spontaneous:	
Errors in DNA replication	A mistake by DNA polymerase may cause a point mutation.
Toxic metabolic products	The products of normal metabolic processes may be reactive chemicals such as free radicals that can alter the structure of DNA.
Spontaneous changes in nucleotide structure	On rare occasions, the linkage between purines and deoxyribose can spontaneously break. Also, changes in base structure (isomerization) may cause mispairing during DNA replication.
Transposons	As discussed in Chapter 21, transposons are small segments of DNA that can insert at various sites in the genome. If they insert into a gene, they may inactivate the gene.
Induced:	
Chemical agents	Chemical substances, such as benzo(a)-pyrene, a chemical found in cigarette smoke, may cause changes in the structure of DNA.
Physical agents	Physical agents such as UV (ultraviolet) light and X-rays can damage the DNA.

Table 14.4	Examples of Mutagens
Mutagen	**Effect(s) on DNA structure**
Chemical:	
Nitrous acid	Deaminates bases
5-Bromouracil	Acts as a base analogue
2-Aminopurine	Acts as a base analogue
Nitrogen mustard	Alkylates bases
Ethyl methanesulfonate	Alkylates bases
Proflavin	Inserts next to bases in the DNA double helix and causes additions or deletions
Physical:	
X-rays	Causes base deletions, single nicks in DNA strands, cross-linking, and chromosomal breaks
UV light	Promotes pyrimidine dimer formation, which involves covalent bonds between adjacent pyrimidines (C or T)

Mutagens Alter DNA Structure in Different Ways

Researchers have discovered that an enormous array of agents can act as mutagens to permanently alter the structure of DNA. We often hear in the news media that we should avoid these agents in our foods and living environments. We even use products such as sunscreens that help us avoid the mutagenic effects of ultraviolet (UV) light from the sun. The public is concerned about mutagens for two important reasons. First, mutagenic agents are often involved in the development of human cancers. Second, because new mutations may be deleterious, people want to avoid mutagens to prevent gene mutations that may have harmful effects in their future offspring. Mutagenic agents are categorized as **chemical** or **physical mutagens** (**Table 14.4**).

Mutagens can alter the structure of DNA in various ways. Some chemical mutagens act by covalently modifying the structure of nucleotides. For example, nitrous acid (HNO_2) deaminates bases by replacing amino groups with keto groups ($-NH_2$ to $=O$). This can change cytosine to uracil, and adenine to a base called hypoxanthine. When this altered DNA replicates, the modified bases do not pair with the appropriate nucleotides in the newly made strand. Instead, uracil pairs with adenine, and hypoxanthine pairs with cytosine (**Figure 14.5**). Similarly, 5-bromouracil and 2-aminopurine, which are called base analogues, have structures that are similar to particular bases. When incorporated into DNA, they also cause errors in DNA replication. Other chemical mutagens can also disrupt the appropriate

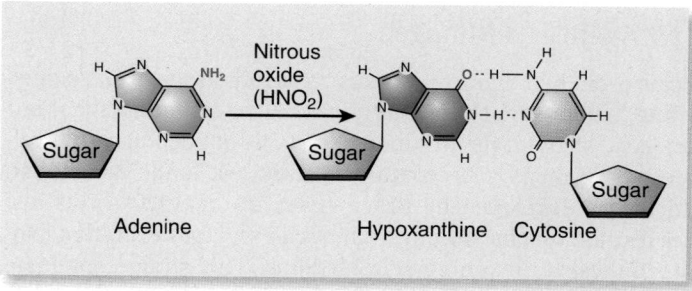

Figure 14.5 **Deamination and mispairing of modified bases by a chemical mutagen.** Nitrous acid changes cytosine to uracil and adenine to hypoxanthine by replacing NH_2 with an oxygen. During DNA replication, uracil will pair with adenine, and hypoxanthine will pair with cytosine. These incorrect bases will create mutations in the newly replicated strand.

pairing between nucleotides by alkylating bases within the DNA. During alkylation, methyl or ethyl groups are covalently attached to the bases. Examples of alkylating agents include nitrogen mustards (used as a chemical weapon during World War I) and ethyl methanesulfonate (EMS).

Some chemical mutagens exert their effects by interfering with DNA replication. For example, acridine dyes, such as proflavin, contain flat planar structures that insert in between the bases of the double helix, thereby distorting the helical structure. When DNA containing these mutagens is replicated, single nucleotide additions and/or deletions can be incorporated into the newly made strands.

DNA molecules are also sensitive to physical agents such as radiation. In particular, radiation of short wavelength and high energy, known as ionizing radiation, is known to alter DNA structure. Ionizing radiation includes X-rays and gamma rays. This type of radiation can penetrate deeply into biological materials, where it creates free radicals. These molecules can alter the structure of DNA in a variety of ways. Exposure to high doses of ionizing radiation can cause base deletions, breaks in one DNA strand, or even a break in both DNA strands.

Nonionizing radiation, such as UV light, contains less energy, and so it only penetrates the surface of biological materials such as the skin. Nevertheless, UV light is known to cause DNA mutations. For example, UV light can cause the formation of a **thymine dimer**, one type of pyrimidine dimer, which is a site where two adjacent thymine bases become covalently cross-linked to each other (**Figure 14.6**). A thymine dimer within a DNA strand may cause a mutation when that DNA strand is replicated. When DNA polymerase attempts to replicate over a thymine dimer, proper base pairing does not occur between the template strand and the incoming nucleotides. This mispairing can cause gaps in the newly made strand or the incorporation of incorrect bases. Plants, in particular, must have effective ways to prevent UV damage because they are exposed to sunlight throughout the day.

Testing Methods Can Determine If an Agent Is a Mutagen

Because of their harmful effects, researchers have developed testing methods that can ascertain whether or not an agent is a mutagen. Many different kinds of tests have been used to evaluate mutagenicity. One commonly used test is the **Ames test**, which was developed by Bruce Ames in the 1970s. This test uses a strain of a bacterium, *Salmonella typhimurium*, that cannot synthesize the amino acid histidine. This strain contains a point mutation within a gene that encodes an enzyme required for histidine biosynthesis. The mutation renders the enzyme inactive. The bacteria cannot grow unless histidine has been added to the growth medium. However, a second mutation in a bacterial cell may correct the first mutation and thereby restore the ability to synthesize histidine. The Ames test monitors the rate at which this second mutation occurs and thereby indicates whether an agent increases the mutation rate above the spontaneous rate.

Figure 14.7 outlines the steps in the Ames test. The suspected mutagen is mixed with a rat liver extract and the bacterial strain of *S. typhimurium*, which cannot synthesize histidine. Because some potential mutagens may require activation by

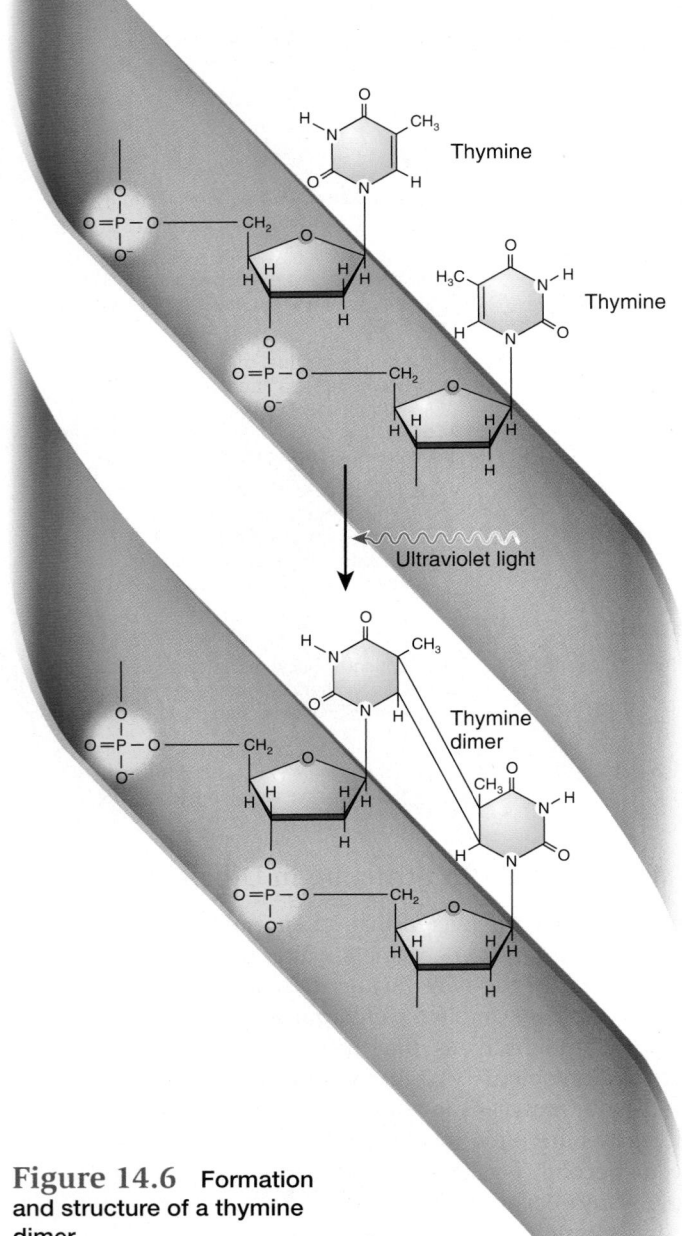

Figure 14.6 Formation and structure of a thymine dimer.

cellular enzymes, the rat liver extract provides a mixture of enzymes that may cause such activation. This step improves the ability to identify agents that may cause mutation in mammals. As a control, bacteria that have not been exposed to the mutagen are also tested. After an incubation period in which mutations may occur, a large number of bacteria are plated on a growth medium that does not contain histidine. The *Salmonella* strain is not expected to grow on these plates. However, if a mutation has occurred that allows a cell to synthesize histidine, the bacterium harboring this second mutation will proliferate during an overnight incubation period to form a visible bacterial colony.

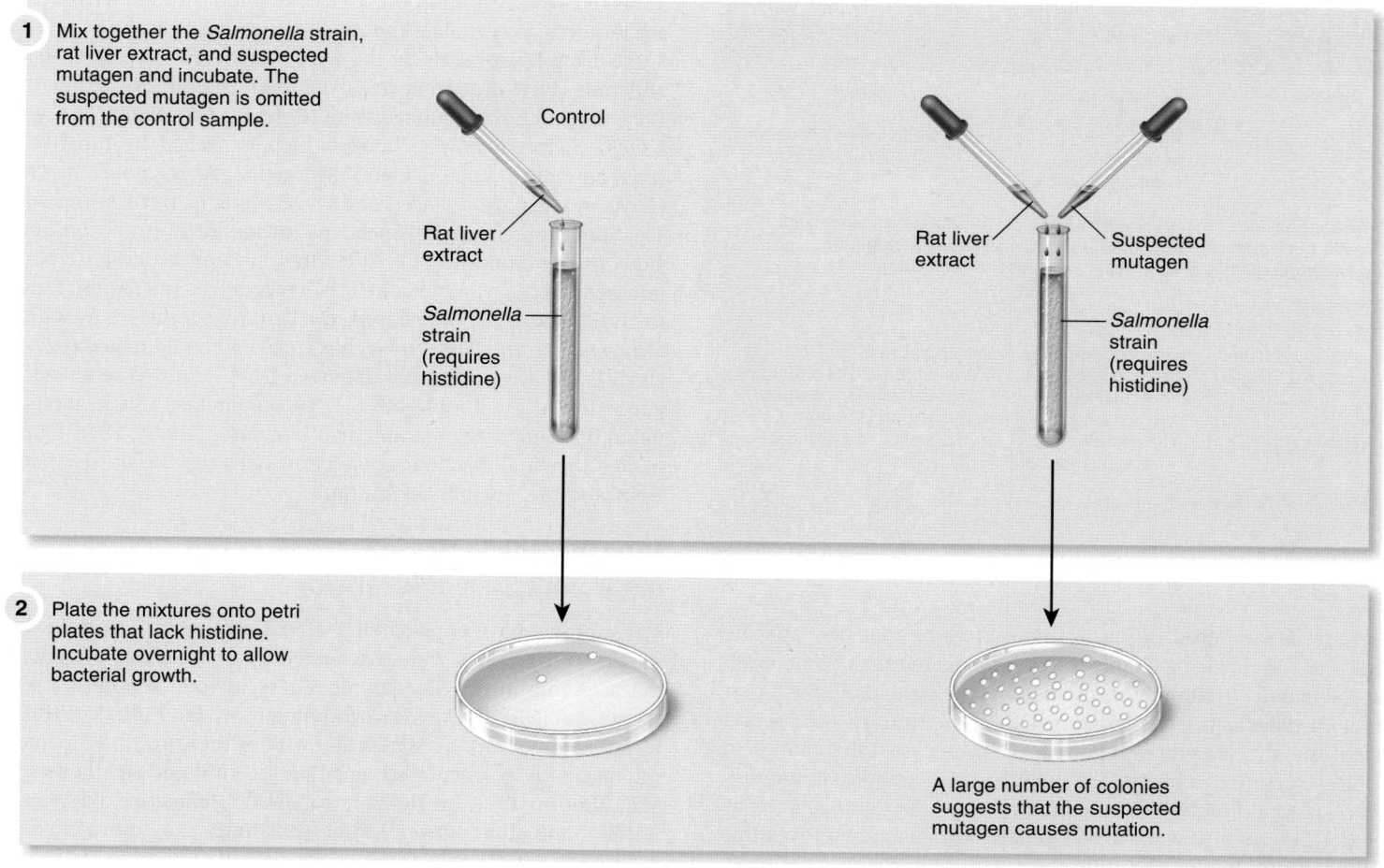

1   Mix together the *Salmonella* strain, rat liver extract, and suspected mutagen and incubate. The suspected mutagen is omitted from the control sample.

Control

Rat liver extract

Salmonella strain (requires histidine)

Rat liver extract

Suspected mutagen

Salmonella strain (requires histidine)

2   Plate the mixtures onto petri plates that lack histidine. Incubate overnight to allow bacterial growth.

A large number of colonies suggests that the suspected mutagen causes mutation.

Figure 14.7   **The Ames test for mutagenicity.**   In this example, two million bacterial cells were placed on each plate. Two colonies were observed from the control sample while 44 were observed from the sample exposed to a suspected mutagen.

Biological Inquiry: Based on the results seen in this figure, what is the rate of mutation that is caused by the suspected mutagen?

To estimate the mutation rate, the colonies that grow in the absence of histidine are counted and compared with the total number of bacterial cells that were originally placed on the plate for both the test and the control. The control condition is a measure of the spontaneous mutation rate, while the test condition monitors the rate of mutation in the presence of the suspected mutagen. As an example, let's suppose that 2 million bacteria were placed in both the test and control tubes. In the control experiment, 2 bacterial colonies were observed. The spontaneous mutation rate is calculated by dividing 2 (the number of mutants) by 2 million (the number of original cells). This equals 1 in 1 million, or 1×10^{-6}. By comparison, 44 colonies were observed in the test condition (Figure 14.7). In this case, the mutation rate would be 44 divided by 2 million, which equals 2.2×10^{-5}. The mutation rate in the presence of the mutagen is over 20 times higher than the spontaneous mutation rate. These results suggest that the suspected mutagen is causing mutations to occur. Now let's look at how organisms can repair changes in DNA structure.

14.2 DNA Repair

In the previous section, we considered the causes and consequences of mutation. As we have seen, mutations are random events that often have negative consequences. For this reason, all living organisms must have the ability to repair changes that occur in the structure of DNA in order to minimize mutation. Such DNA repair systems have been studied extensively in many organisms, particularly *E. coli*, yeast, and mammals. The diverse ways of repairing DNA underscore the extreme necessity for the structure of DNA to be maintained properly. The importance of these systems becomes evident when they are missing. For example, bacteria contain several different DNA repair systems. When even a single system is absent, the bacteria have a much higher rate of mutation. In fact, the rate of mutation can be so high that these bacterial strains are sometimes called "mutator strains"!

Living cells contain several DNA repair systems that can fix different types of DNA alterations (**Table 14.5**). Each repair

Table 14.5	Common Types of DNA Repair Systems*
System	**Description**
Direct repair	A repair enzyme recognizes an incorrect structure in the DNA and directly converts it back to a correct structure.
Base excision and nucleotide excision repair	An abnormal base or nucleotide is recognized and a portion of the strand containing the abnormality is removed. The complementary DNA strand is then used as a template to synthesize a normal DNA strand.
Methyl-directed mismatch repair	Similar to excision repair except that the DNA defect is a base pair mismatch in the DNA, not an abnormal nucleotide. The mismatch is recognized, and a strand of DNA in this region is removed. The complementary strand is used to synthesize a normal strand of DNA.

*Other types of repair systems exist; these are common examples.

system is composed of one or more proteins that play specific roles in the repair mechanism. DNA repair requires two coordinated events. In the first step, one or more proteins in the repair system detect an irregularity in DNA structure. In the second step, the abnormality is repaired. In some cases, the change in DNA structure can be directly repaired. For example, sometimes DNA is modified by the attachment of an alkyl group, such as —CH$_2$CH$_3$, to a base. In **direct repair**, an enzyme removes this alkyl group, thereby restoring the structure of the original base. More commonly, however, the altered DNA is removed and a new segment of DNA is synthesized. In this section, we will examine two of these repair systems, namely nucleotide excision repair and methyl-directed mismatch repair, as examples of how such systems operate. These two systems, which are found in all species, represent important mechanisms of DNA repair.

Nucleotide Excision Repair Systems Remove Segments of Damaged DNA

One of the most common types of DNA repair systems is **nucleotide excision repair (NER)**, which can fix many different types of DNA damage, including UV-induced damage, chemically modified bases, missing bases, and various types of cross-links (such as pyrimidine dimers). In NER, a region encompassing several nucleotides in the damaged strand is removed from the DNA, and the intact undamaged strand is used as a template for resynthesis of a normal complementary strand. NER is found in all eukaryotes and prokaryotes, although its molecular mechanism is better understood in prokaryotic species.

In *E. coli*, the NER system is composed of four key proteins: UvrA, UvrB, UvrC, and UvrD. They are named Uvr because they are involved in ultraviolet light repair of pyrimidine dimers, although these proteins are also important in repairing chemically

damaged DNA. In addition, DNA polymerase and DNA ligase are required to complete the repair process.

During DNA repair by the NER system, two UvrA proteins and one UvrB protein form a complex that tracks along the DNA in search of a damaged site (**Figure 14.8**). Such DNA will have a distorted double helix, which is sensed by the UvrA-UvrB complex. When it identifies a damaged segment, the two UvrA proteins are released and UvrC binds to UvrB at the site. The UvrC protein makes incisions in one DNA strand on both sides of the damaged site. After this incision process, UvrC is released. UvrD, which is a helicase, recognizes the region, binds to UvrB, and UvrB is released. UvrD unravels the DNA, which removes a short DNA strand that contains the damaged region. UvrD is released. After the damaged DNA strand is removed, a gap is left in the double helix. DNA polymerase fills in the gap using the undamaged strand as a template. Finally, DNA ligase makes the final covalent connection between the newly made DNA and the original DNA strand.

Human Genetic Diseases Occur When a Component of the NER System Is Missing

Thus far, we have considered the NER system in *E. coli*. In humans, NER systems were discovered by the analysis of genetic diseases that affect DNA repair. These include xeroderma pigmentosum (XP), Cockayne's syndrome (CS), and PIBIDS. (PIBIDS is an acronym for a syndrome with symptoms that include photosensitivity [increased sensitivity to sunlight], ichthyosis [a skin abnormality], brittle hair, impaired intelligence, decreased fertility, and short stature.) Photosensitivity is a common characteristic in all three syndromes because of an inability to repair UV-induced lesions. Therefore, people with these syndromes must avoid prolonged exposure to sunlight. **Figure 14.9** shows a photograph of a person with XP. Such individuals may have pigmentation abnormalities, many precancerous lesions, and a high predisposition to developing skin cancer.

Researchers have found that XP, CS, and PIBIDS result from defects in different genes that encode NER proteins. For example, XP can be caused by defects in seven different NER genes. In all cases, individuals have a defective NER system. In recent years, several human NER genes have been successfully identified. Although more research is needed to completely understand the mechanisms of DNA repair, the identification of NER genes has helped to unravel the complexities of NER systems in human cells.

Mismatch Repair Systems Recognize and Correct a Base Pair Mismatch

Another type of abnormality that should not occur in DNA is a base mismatch, in which the structure of the DNA double helix does not obey the AT/GC rule of base pairing. During the normal course of DNA replication, DNA polymerase may add an incorrect nucleotide to the growing strand by mistake. This creates a mismatch between a nucleotide in the parental

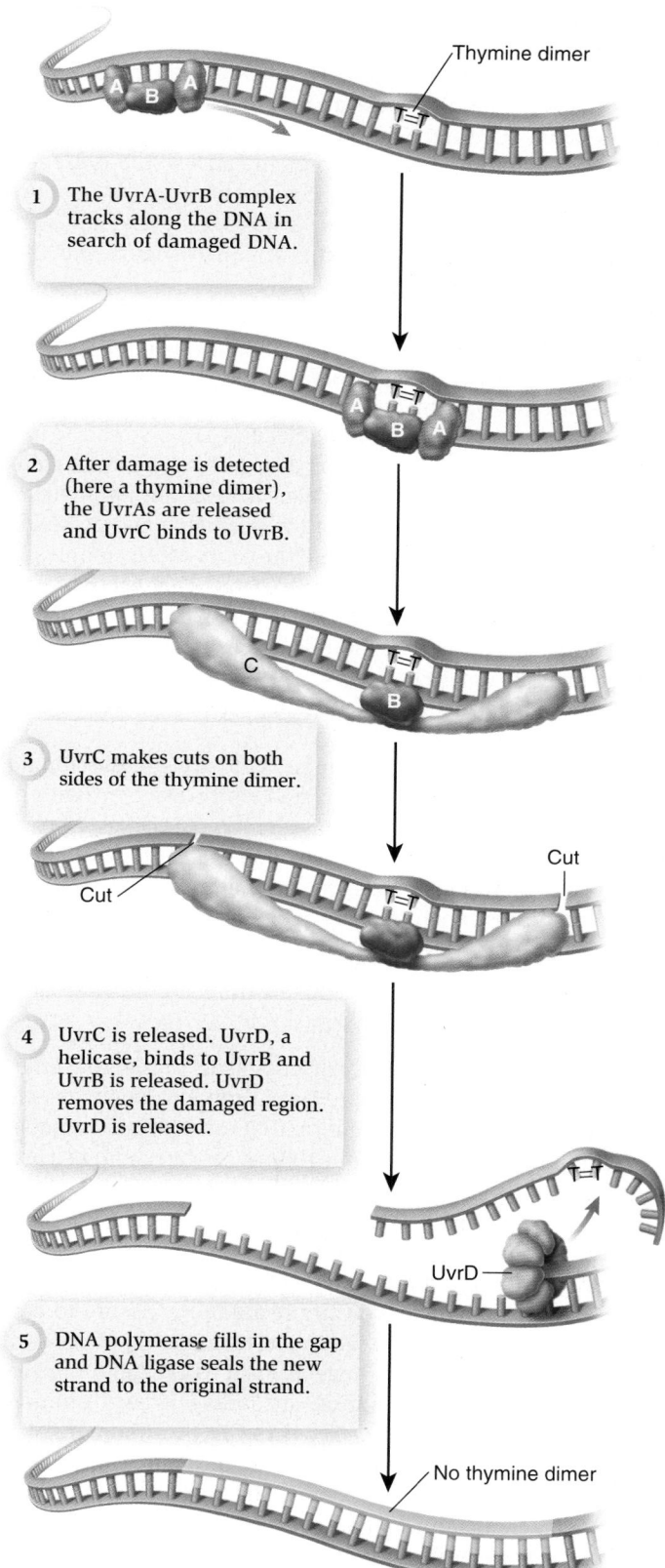

Thymine dimer

1 The UvrA-UvrB complex tracks along the DNA in search of damaged DNA.

2 After damage is detected (here a thymine dimer), the UvrAs are released and UvrC binds to UvrB.

3 UvrC makes cuts on both sides of the thymine dimer.

Cut

Cut

4 UvrC is released. UvrD, a helicase, binds to UvrB and UvrB is released. UvrD removes the damaged region. UvrD is released.

UvrD

5 DNA polymerase fills in the gap and DNA ligase seals the new strand to the original strand.

No thymine dimer

Figure 14.8   Nucleotide excision repair in *E. coli*.

Figure 14.9   An individual affected by xeroderma pigmentosum.

Biological inquiry: Why is this person so sensitive to the sun?

strand and one in the newly made strand. Various DNA repair mechanisms may recognize and remove this mismatch. For example, as we learned in Chapter 11, DNA polymerase itself has a proofreading ability that can detect mismatches and remove them. However, if this proofreading ability fails, cells contain additional DNA repair systems that can detect base mismatches and fix them.

One of these DNA repair systems is the **methyl-directed mismatch repair** system, which exists in all species. The molecular mechanism of mismatch repair has been studied extensively in *E. coli*. This system involves the participation of several proteins that detect the mismatch and specifically remove a segment from the newly made strand. Keep in mind that if DNA polymerase makes a mistake, the new strand contains the incorrect base, while the parental strand is normal. Therefore, a critical aspect of methyl-directed mismatch repair is that it specifically repairs the new strand rather than the parental template strand.

Three proteins in *E. coli*, designated MutH, MutL, and MutS, detect the mismatch and direct the removal of a portion of the new strand carrying the mismatched base (**Figure 14.10**). These proteins are named Mut because their absence leads to a much higher <u>mut</u>ation rate than that of normal strains of *E. coli*.

1   The MutS protein finds a mismatch. The MutS-MutL complex binds to MutH, which is already bound to a hemimethylated sequence.

MutH

Methylated bases in parental strand

MutL

Newly made strand

MutS

Incorrect base

2   MutH makes a cut in the nonmethylated strand. An exonuclease begins at this cleavage site and then digests the nonmethylated strand just beyond the base mismatch.

MutH cleavage site

3   DNA polymerase fills in the vacant region. DNA ligase seals the end.

The mismatch has been repaired correctly.

Figure 14.10   Methyl-directed mismatch repair in *E. coli.*

The role of the MutH protein is to distinguish between the parental and new DNA strands. In *E. coli*, adenine and cytosine bases are commonly methylated. Therefore, prior to DNA replication, the parental DNA has already been methylated. Immediately after DNA replication, the new strand is not yet methylated. Therefore, the MutH protein can use the presence or absence of methylation to distinguish between the parental strand and the new strand. MutH binds to a hemimethylated sequence in which the parental strand is methylated and the new strand is not methylated.

The role of MutS is to find mismatches. When MutS has located a DNA mismatch, it forms a complex with MutL. This MutS-MutL complex interacts with the previously bound MutH; the DNA must loop for this to occur. This interaction stimulates MutH to make a cut in the nonmethylated DNA strand. After the strand is cut, an exonuclease is attracted to the site where MutH is bound. The exonuclease then digests the nonmethylated DNA strand in the direction of the mismatch and proceeds just beyond the mismatch site where the MutS-MutL complex is bound. The MutS-MutL complex is released. This leaves a gap in the new strand that is repaired by DNA polymerase and DNA ligase. The net result is that the mismatch has been corrected by removing the incorrect region in the new strand and resynthesizing the correct sequence using the parental DNA as a template.

Similar to defects in NER systems, mutations in the human mismatch repair systems are associated with particular types of cancer. Mutations in two human mismatch repair genes, *hMSH2* and *hMLH1*, play a role in the development of a type of colon cancer. These mutations cause the mismatch repair system to function improperly, thereby increasing the rate of mutation.

14.3 Cancer

Cancer is a disease of multicellular organisms that is characterized by uncontrolled cell division. Over 1 million Americans are diagnosed with cancer each year, and about half that number will die from the disease. In about 10% of cancers, a higher predisposition to develop the disease is an inherited trait. Most cancers, though, perhaps 90%, do not involve genetic changes that are passed from parent to offspring. Rather, cancer is usually an acquired condition that typically occurs later in life. At least 80% of all human cancers are related to exposure to **carcinogens**, which are agents that increase the likelihood of developing cancer. Most carcinogens, such as UV light and certain chemicals in cigarette smoke, are mutagens that promote genetic changes in somatic cells. These DNA alterations can lead to effects on gene expression that ultimately affect cell division, and thereby lead to cancer. In this section, we will explore such genetic abnormalities.

In most cases, the development of cancer is a multistep process (**Figure 14.11**). Cancers originate from a single cell. This single cell and its line of daughter cells undergo a series of mutations that causes the cells to grow abnormally. At an early

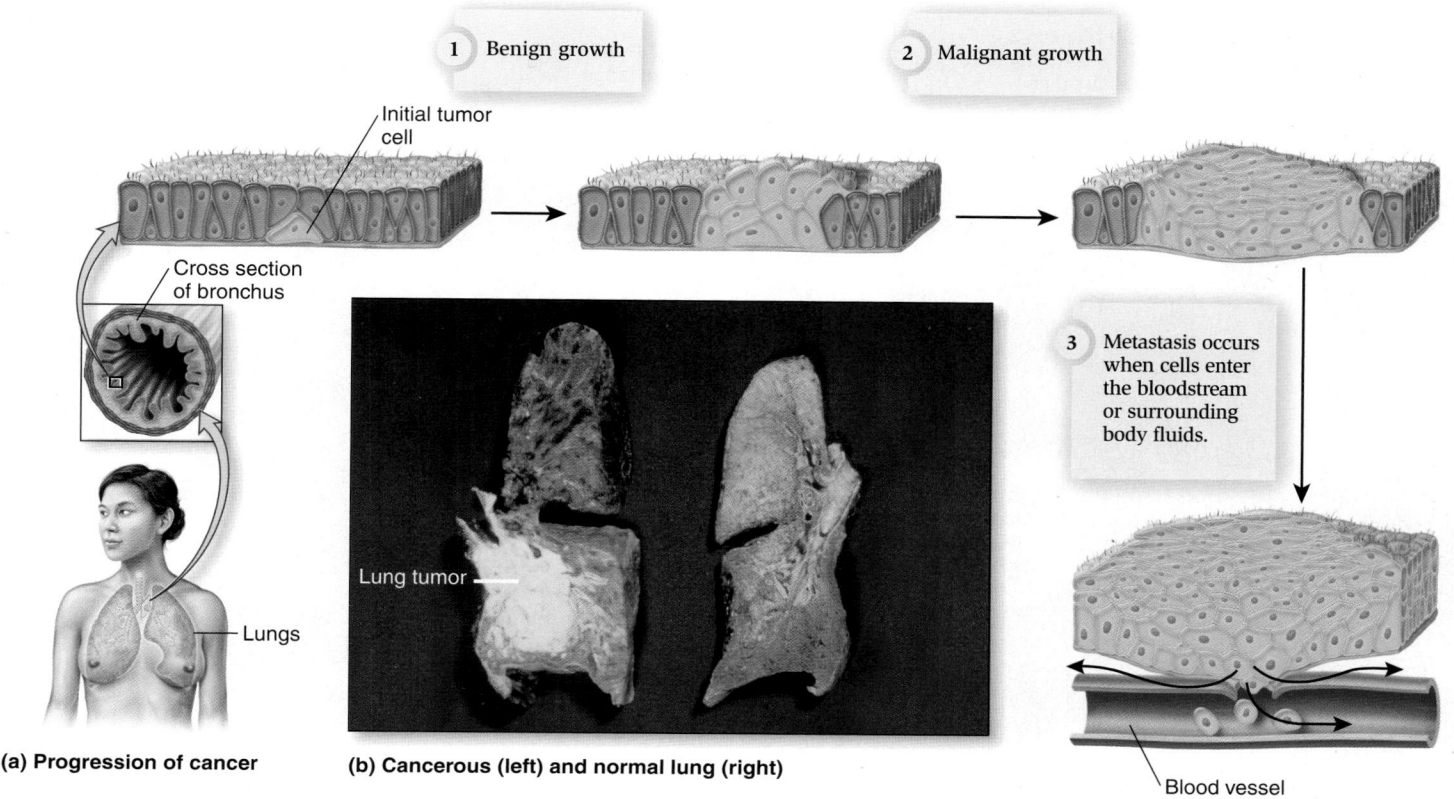

Figure 14.11   Cancer: Its progression and effects.   (a) In a healthy individual, an initial gene mutation converts a normal cell into a tumor cell. This tumor cell divides to produce a benign tumor. Additional genetic changes in the tumor cells may occur, leading to a malignant tumor. At a later stage in malignancy, the tumor cells will invade surrounding tissues, and some malignant cells may metastasize by traveling through the bloodstream to other parts of the body. **(b)** On the *right* is a photo of a human lung that was obtained from a healthy nonsmoker. The lung shown on the *left* has been ravaged by lung cancer. This lung was taken from a person who was a heavy smoker.

stage, the cells form a **tumor**, which is an overgrowth of cells that serves no useful purpose. For most types of cancer, a tumor begins as a precancerous or **benign** growth. This may be followed by additional mutations that cause cells in the tumor to progress to the cancerous stage. The tumor is now **malignant** because the cancer cells have lost their normal growth regulation and are **invasive** (that is, they can invade healthy tissues) and **metastatic** (that is, they can migrate to other parts of the body). The final stage of cancer involves these cancerous cells spreading through the bloodstream or surrounding body fluids and establishing more tumors elsewhere. If left untreated, malignant cells will cause the death of the organism.

Over the past few decades, researchers have identified many genes that promote cancer when they are mutant. By comparing the function of each mutant, cancer-causing gene with the corresponding nonmutant gene found in healthy cells, these genes have been placed into two categories. In some cases, a mutation causes a gene to be overactive. This overactivity contributes to the uncontrolled cell growth that is observed in cancer cells. This type of mutant gene is called an **oncogene**.

Alternatively, when a **tumor-suppressor gene** is normal (that is, not mutant), it encodes a protein that prevents cancer. However, when a mutation eliminates its function, cancer may occur. Thus, the two categories of cancer-causing genes are based on the effects of mutations. Oncogenes are the result of mutations that cause overactivity, while loss-of-function mutations in tumor-suppressor genes promote cancer. In this section, we will begin with a discussion of oncogenes and then consider tumor-suppressor genes.

Oncogenes Cause the Overactivity of Proteins That Promote Cell Division

Over the past three decades, researchers have identified many oncogenes. In parallel with our increased understanding of cell division, it has become apparent that most oncogenes encode proteins that function in cell growth signaling pathways. Cell division is regulated, in part, by hormones known as growth factors that bind to cell surface receptors and initiate a cascade of cellular events that lead eventually to cell division (**Figure 14.12**).

Figure 14.12   **The epidermal growth factor signaling pathway.**   (*Inset*) When GTP is bound, the activated Ras protein promotes cell division. When GTP is hydrolyzed to GDP and P_i, Ras is inactivated and cell division is inhibited.

In this example, a growth hormone known as epidermal growth factor (EGF) binds to its receptor, ErbB, thereby leading to the activation of an intracellular signaling pathway. This involves the sequential activation of six proteins called GRB2, Sos, Ras, Raf-1, MEK, and MAPK. This pathway, also known as a signal cascade, ultimately transmits a change in gene transcription by activating the transcription factors Myc, Fos, and Jun. The transcription of specific genes is activated in response to the growth hormone. The activated gene products promote the cell's progression through the cell cycle. Figure 14.12 is just one example of a pathway between a growth factor and gene activation. Eukaryotic species produce many different growth factors, and the signaling pathways are often more complex than the one shown here.

Mutations in the genes that produce these cell growth signaling proteins can change them into oncogenes. For example, **Table 14.6** lists oncogenes that can affect the EGF signaling pathway, which was shown in Figure 14.12. Oncogenes result in an abnormally high level of activity in these proteins, which can include growth factor receptors, intracellular signaling proteins, and transcription factors.

An oncogene may promote cancer by keeping the cell division signaling pathway in a permanent "on" position. This can occur in two ways. First, in some cancers the amount of gene product is abnormally high; in other words, the affected cell makes too much of the protein encoded by the oncogene.

Table 14.6	Examples of Genes Encoding Proteins of the Epidermal Growth Factor Signaling Pathway That Can Mutate to Become Oncogenes
Gene*	**Cellular function**
erbB	Growth factor receptor for EGF (epidermal growth factor)
ras	GTP/GDP-binding protein
raf	Serine kinase
myc	Transcription factor
fos	Transcription factor
jun	Transcription factor

* The genes described in this table are found in humans as well as other vertebrate species. Most of the genes have been given three-letter names that are abbreviations for the type of cancer the oncogene causes or the type of virus in which the gene was first identified.

In 1982, research groups headed by Robert Gallo and Mark Groudine showed that a *myc* gene, which codes for a transcription factor in the EGF signaling pathway, was present in 10 times as many copies as normal in a form of leukemia (a cancer of white blood cells) called promyelocytic leukemia (hence the name *myc*). Since that time, researchers have found that *myc* genes are overexpressed in many forms of cancer, including those of the breast, lung, and colon. The overexpression of this transcription factor leads to the transcriptional activation of genes that promote cell division.

A second way oncogenes can keep cell division turned on is by producing a functionally hyperactive protein. For example, mutations that alter the amino acid sequence of the Ras protein (an intracellular signaling protein in the EGF signaling pathway) have been shown to cause functional abnormalities. The Ras protein is a GTPase that hydrolyzes GTP to GDP + P_i (see Figure 14.12, inset). When GTP is bound, the activated Ras protein promotes cell division. After it has been activated, the Ras protein returns to its inactive state by hydrolyzing its bound GTP, and cell division is inhibited. Mutations that convert the normal *ras* gene into an oncogenic *ras* either decrease the GTPase activity of the Ras protein or increase the rate of exchange of bound GDP for GTP. Both of these functional changes result in a greater amount of the active GTP-bound form of the Ras protein. In this way, these mutations keep the signaling pathway turned on.

Mutations in Proto-Oncogenes Convert Them to Oncogenes

Thus far, we have examined the functions of proteins that cause cancer when they become overactive. In some cases the amount of the protein increases, while in other cases the activity of the protein increases. Such overactivity promotes uncontrolled cell division. Let's now consider the common types of genetic changes that create such oncogenes. A **proto-oncogene** is a normal gene that, if mutated, can become an oncogene. Four common genetic changes may convert a proto-oncogene into an oncogene (**Figure 14.13**). These changes are categorized as missense mutations, gene amplifications, chromosomal translocations, and retroviral insertions.

A missense mutation (Figure 14.13a), which we considered earlier in this chapter, is responsible for the conversion of the *ras* gene into an oncogene. For example, a change of a glycine to a valine in the Ras protein converts the *ras* gene into an oncogene (**Figure 14.14**). This mutation decreases the ability of the Ras protein to hydrolyze its GTP. Experimentally, chemical mutagens have been shown to cause this missense mutation and thereby lead to cancer.

Another genetic event that occurs in cancer cells is **gene amplification**, in this case an increase in the copy number of a proto-oncogene (see Figure 14.13b). An abnormal increase in copy number results in too much of the encoded protein. As mentioned previously, Gallo and Groudine discovered that *myc* was amplified in a human leukemia. Many human cancers are associated with the amplification of particular proto-oncogenes. In some cases, the extent of gene amplification is correlated with the progression of tumors to increasing malignancy. A malignancy may become more difficult to treat as the copy number of a proto-oncogene increases. In other types of malignancies, gene amplification is more random and may be a secondary event that increases the expression of oncogenes previously activated by other genetic changes.

Proto-oncogene

Promoter   Coding sequence

A change in the amino acid sequence of a proto-oncogene protein may cause it to function in an abnormal way. For example, missense mutations can convert *ras* genes into oncogenes.

Missense mutation

(a) Missense mutation

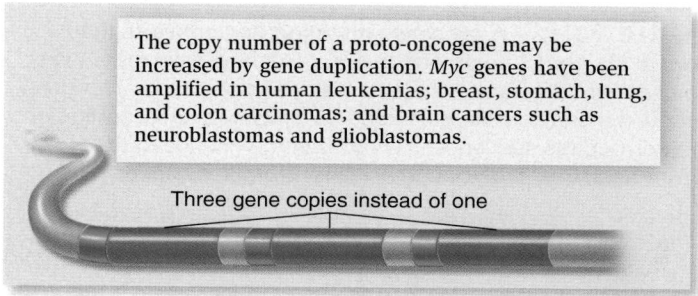

The copy number of a proto-oncogene may be increased by gene duplication. *Myc* genes have been amplified in human leukemias; breast, stomach, lung, and colon carcinomas; and brain cancers such as neuroblastomas and glioblastomas.

Three gene copies instead of one

(b) Gene amplification

A piece of chromosome may be translocated to another chromosome and affect the expression of genes at the breakpoint site. In one form of leukemia, for example, a translocation causes parts of the *bcr* and *abl* genes to fuse, thereby creating a chimeric oncogene.

Fused, chimeric gene

(c) Chromosomal translocation

When a virus integrates into the chromosome, viral response elements may enhance the expression of a nearby proto-oncogene, converting it to an oncogene.

Viral regulatory sequences

Inserted retroviral genome

(d) Retroviral insertion

Figure 14.13   Genetic changes that convert proto-oncogenes to oncogenes.

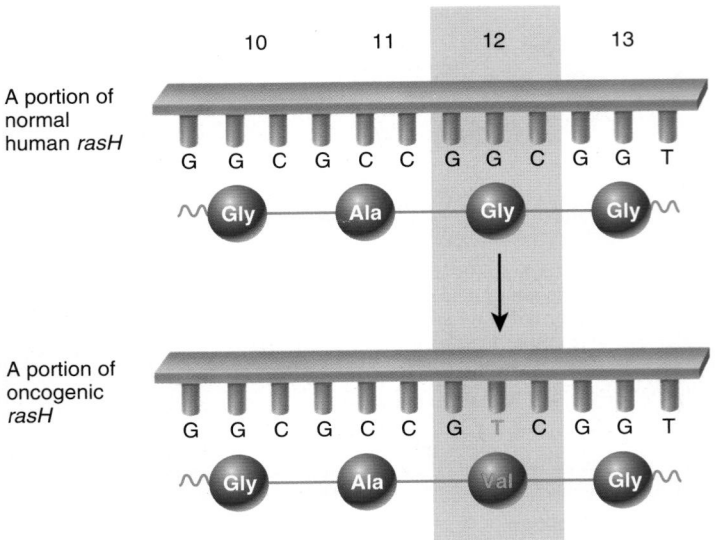

Figure 14.14   **A missense mutation that converts the normal *ras* gene into an oncogene.**   Humans have more than one type of *ras* gene. One of them is designated *rasH*. This gene encodes a protein with 189 amino acids. Only a short portion of the polypeptide sequence is shown here. A change of the twelfth amino acid, from a glycine to a valine, promotes cancer.

Biological inquiry: How does this mutation affect Ras protein function? How does this alteration in function contribute to cancer?

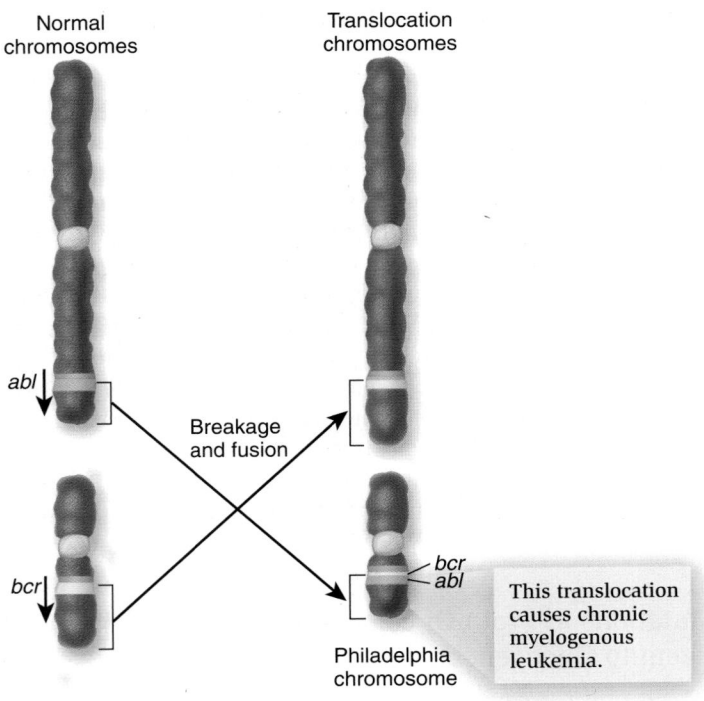

Figure 14.15   **The formation of a chimeric gene that is found in people with certain forms of leukemia.**   The fusion of the *abl* and *bcr* genes creates a chimeric gene that encodes an abnormal fusion protein, leading to leukemia. The blue regions are the promoters for the *abl* and *bcr* genes.

A third type of genetic alteration that can lead to cancer is a chromosomal translocation (see Figure 14.13c). This occurs when two different chromosomes break, and the ends of the broken chromosomes fuse with each other incorrectly. Very specific types of chromosomal translocations have been identified in certain types of tumors. In 1960, Peter Nowell discovered that a certain form of leukemia was correlated with the presence of a shortened version of a human chromosome, which he called the Philadelphia chromosome after the city where it was discovered. Rather than a deletion, this shortened chromosome is the result of a chromosome translocation in which pieces of two different chromosomes fused with each other. Later studies revealed that this activates a proto-oncogene, *abl*, in an unusual way (**Figure 14.15**). In healthy individuals, the *bcr* gene and the *abl* gene are located on different chromosomes. In certain forms of leukemia, these chromosomes break and fuse together, an abnormal event that causes the promoter and the first part of *bcr* to fuse with part of *abl*. This fusion creates a **chimeric gene** composed of two gene fragments. This chimeric gene acts as an oncogene that encodes an abnormal fusion protein, whose functional hyperactivity leads to leukemia.

Finally, certain types of viruses can convert proto-oncogenes into oncogenes during the viral replication cycle (see Figure 14.13d). Retroviruses insert their DNA into the chromosomal DNA of their host cell. The viral genome contains promoter and response elements that cause a high level of expression of viral genes. On occasion, the viral DNA may insert into a host chromosome in such a way that a viral promoter and response elements are next to a proto-oncogene. This may result in the overexpression of the proto-oncogene, thereby promoting cancer. This is one way for a virus to cause cancer. Alternatively, a virus may cause cancer because it carries an oncogene in the viral genome. This phenomenon is described next.

Some Types of Cancer Are Caused by Viruses

The great majority of cancers are caused by mutagens that alter the structure and expression of genes that are found in cells. A few viruses, however, are known to cause cancer in plants, animals, and humans (**Table 14.7**).

In 1911, the first cancer-causing virus to be discovered was isolated from chicken sarcomas by Peyton Rous. A **sarcoma** is a tumor of connective tissue such as bone or cartilage. During the 1970s, Rous sarcoma virus (RSV) research led to the identification of a viral gene that acts as an oncogene. Researchers investigated RSV by using it to infect chicken cells grown in the laboratory. This causes the chicken cells to grow like cancer cells, continuously and in an uncontrolled manner. Researchers identified mutant RSV strains that infected and proliferated within chicken cells without transforming them into malignant cells. These RSV strains were missing a gene that is found in the form

Table 14.7	Examples of Viruses That Cause Cancer
Virus	**Description**
Rous sarcoma virus	Causes sarcomas in chickens
Simian sarcoma virus	Causes sarcomas in monkeys
Abelson leukemia virus	Causes leukemia in mice
Hardy-Zuckerman-4 feline sarcoma virus	Causes sarcomas in cats
Hepatitis B	Causes liver cancer in several species, including humans
Papillomavirus	Causes benign tumors and malignant carcinomas in several species including humans; causes cervical cancer in humans
Epstein-Barr virus	Causes Burkitt's lymphoma, which primarily occurs in immunosuppressed individuals such as AIDS patients

Table 14.8	Functions of Selected Tumor-Suppressor Genes
Gene	**Function of encoded protein**
	Maintain genome integrity:
p53	p53 is a transcription factor that positively regulates a few specific target genes and negatively regulates others in a general manner. It acts as a sensor of DNA damage. It can prevent the progression through the cell cycle and also can promote apoptosis.
BRCA-1 BRCA-2	Brca-1 and Brca-2 proteins are both involved in the cellular defense against DNA damage. They may play a role in sensing DNA damage, or they may act to facilitate DNA repair. These genes are sometimes mutant in persons with inherited forms of breast cancer.
XPD	This represents several different genes whose products function in DNA repair. These genes are defective in patients with xeroderma pigmentosum.
	Inhibit cell division (negative regulators):
Rb	The Rb protein is a negative regulator that represses the transcription of genes required for DNA replication and cell division.
NF1	The NF1 protein stimulates Ras to hydrolyze its GTP to GDP. Loss of NF1 function causes the Ras protein to be overactive, which promotes cell division.
p16	A negative regulator of cyclin-dependent protein kinase.

of the virus that does cause cancer. This gene was called the *src* gene because it causes sarcoma.

Harold Varmus and Michael Bishop, in collaboration with Peter Vogt, later discovered that normal (non-viral-infected) cells contain a copy of the *src* gene in their chromosomes. It is a proto-oncogene. Researchers have speculated that RSV may have acquired the *src* gene during the viral replication cycle. The viral DNA integrates into the host cell DNA and later excises itself. An RSV may have integrated next to the *src* gene in a host cell, and later brought the *src* gene along with the viral DNA during the excision process. Once incorporated into a viral genome, the *src* gene is overexpressed because it is transcribed from a very active viral promoter. This ultimately produces too much of the Src protein and promotes uncontrolled cell division.

Tumor-Suppressor Genes Prevent Mutation or Cell Proliferation

Thus far, we have examined the first category of genes that can promote cancer, namely oncogenes. We now turn our attention to the second category of genes, those called tumor-suppressor genes. The role of a normal (nonmutant) tumor-suppressor gene is to prevent cancerous growth. The proteins encoded by tumor-suppressor genes usually have one of two functions (**Table 14.8**). Some tumor-suppressor genes encode proteins that maintain the integrity of the genome. Such proteins monitor and/or repair alterations in the genome. The proteins encoded by these genes are vital for the prevention of abnormalities such as DNA breaks and improperly segregated chromosomes. Therefore, when these proteins are functioning properly, they minimize the chance that a cancer-causing mutation will occur. In some cases, the proteins will prevent a cell from progressing through the cell cycle. These are termed **checkpoint proteins** because their role is to check the integrity of the genome and prevent a

cell from progressing past a certain point in the cell cycle if a genetic abnormality is detected. Checkpoint proteins are not usually required to regulate normal, healthy cell division, but they can stop cell division if an abnormality is detected. A second type of tumor-suppressor gene encodes proteins that are negative regulators or inhibitors of cell division. Their function is necessary to properly halt cell division. If their function is lost, cell division is abnormally accelerated. Let's now consider one example of a checkpoint protein and one example of a negative regulator. The function of checkpoint proteins is also discussed in Chapter 9 and shown in Figure 9.23.

Proteins called cyclins and cyclin-dependent protein kinases (cdks) are responsible for advancing a cell through the four phases of the cell cycle (**Figure 14.16**). The formation of activated cyclin/cdk complexes can be stopped by checkpoint proteins. A specific example of a tumor-suppressor gene that encodes a checkpoint protein is *p53*. Its name refers to the molecular mass of the p53 protein, which is 53 kDa (kilodaltons). About 50% of all human cancers are associated with defects in this gene, including malignant tumors of the lung, breast, esophagus, liver, bladder, and brain, as well as leukemias and lymphomas (cancer of the lymphatic system).

As shown in Figure 14.16, p53 is a G_1 checkpoint protein. The expression of the *p53* gene is induced when DNA is damaged. The p53 protein is a regulatory transcription factor that activates several different genes, leading to the synthesis of proteins that stop the cell cycle and other proteins that repair the DNA.

Figure 14.16   The cell cycle and checkpoints.   As discussed in Chapter 9, eukaryotic cells progress through a cell cycle composed of G_1, S, G_2, and M phases. The red bars indicate common checkpoints that will stop the cell cycle if genetic abnormalities are detected. The p53 protein will stop a cell at the G_1 checkpoint if it senses DNA damage.

When p53 is activated, a cell cannot progress from G_1 to the S, or synthesis, phase of the cell cycle. If the DNA is eventually repaired, a cell may later proceed through the cell cycle.

Alternatively, if the DNA damage is too severe, the p53 protein will also activate other genes that promote programmed cell death. This process, called **apoptosis**, involves cell shrinkage and DNA degradation. Enzymes known as **caspases** are activated during apoptosis. They function as proteases that are sometimes called the "executioners" of the cell. Caspases digest selected cellular proteins such as microfilaments, which are components of the cytoskeleton. This causes the cell to break down into small vesicles that are eventually phagocytized by cells of the immune system. It is beneficial for a multicellular organism to kill an occasional cell with cancer-causing potential.

When checkpoint genes are rendered inactive by mutation, the division of normal healthy cells may not be adversely affected. For example, mice that are missing the *p53* gene are born healthy. This indicates that cell division leading to normal growth is regulated properly. Therefore, checkpoint proteins such as p53 are not necessary for normal cell growth and division. However, these mice are very sensitive to mutagens such as UV light and easily develop cancer. This result suggests that checkpoint proteins are needed to stop cell division only when there is a strong likelihood that a mutation may have occurred. The loss of checkpoint protein function makes it more likely that undesirable genetic changes will occur that could cause cancerous growth.

Our second example of a tumor-suppressor gene is the *Rb* gene, which is a negative regulator of cell division. It was the first tumor-suppressor gene to be identified in humans by studying patients with a disease called retinoblastoma, a cancerous tumor that occurs in the retina of the eye. Some people have an inherited predisposition to develop this disease within the first few years of life. By comparison, the noninherited form of retinoblastoma, which is caused by environmental agents, is more likely to occur later in life. Based on these differences, in 1971 Alfred Knudson proposed a "two-hit" model for retinoblastoma. According to this idea, retinoblastoma requires two mutations to occur. People have two copies of the *Rb* gene, one from each parent. Individuals with the inherited form of the disease already have received one mutant gene from one of their parents. They need only one additional mutation to develop the disease. Because the retina has more than 1 million cells, it is somewhat likely that a mutation may occur in one of these cells at an early age, leading to the disease. However, people with the noninherited form of the disease must have two mutations in the same retinal cell to cause the disease. Because two mutations are less likely than a single mutation, the noninherited form of this disease is expected to occur much later in life, and only rarely. Since Knudsen's original hypothesis, molecular studies have confirmed the two-hit hypothesis.

The Rb protein negatively controls a regulatory transcription factor called E2F that activates genes required for cell cycle progression from G_1 to S phase. The binding of the Rb protein to E2F inhibits its activity and prevents cell division (**Figure 14.17**). When a normal cell is supposed to divide, cyclins bind to cyclin-dependent protein kinases. This binding activates the kinases, which then leads to the phosphorylation of the Rb protein. The phosphorylated form of the Rb protein is released from E2F, thereby allowing E2F to bind to enhancers and activate genes needed to progress through the cell cycle. By comparison, we can imagine how the cell cycle becomes unregulated without functional Rb protein. When both copies of Rb are defective, the E2F protein is always active. This explains why uncontrolled cell division occurs in retinoblastoma.

Gene Mutations, Chromosome Loss, and DNA Methylation Can Inhibit the Expression of Tumor-Suppressor Genes

Cancer biologists would also like to understand how tumor-suppressor genes are inactivated, because this knowledge may ultimately help them to prevent cancer. The function of tumor-suppressor genes is lost in three common ways. First, a mutation can occur specifically within a tumor-suppressor gene to inactivate its function. For example, a mutation could abolish the function of the promoter for a tumor-suppressor gene or introduce an early stop codon in its coding sequence. Either of these would prevent the expression of a functional protein. Chromosome loss is a second way that the function of a tumor-suppressor gene is lost. Chromosome loss may contribute to the progression of cancer if the missing chromosome carries one or more tumor-suppressor genes.

Recently, researchers have discovered a third way that these genes may be inactivated. Tumor-suppressor genes found in cancer cells are sometimes abnormally methylated. As discussed in Chapter 13, transcription is inhibited when CpG islands near

1 When E2F is bound to Rb, E2F is inhibited and cell division is prevented.

2 Phosphorylation of Rb via cyclin-dependent protein kinase causes it to dissociate from E2F.

Activated

Target gene

Gene product promotes cell division.

3 Unbound E2F becomes activated and can then bind to DNA causing target gene transcription.

Figure 14.17 **Function of the Rb protein.** The Rb protein inhibits the function of a regulatory transcription factor called E2F that turns on genes that cause a cell to divide. When cells are supposed to divide, Rb is phosphorylated by cyclin-dependent protein kinases, which allows E2F to function. If Rb protein is not made or is not functional due to a mutation, E2F will always be active, and the cell will be stimulated to divide uncontrollably.

a promoter region are methylated. Such methylation near the promoters of tumor-suppressor genes has been found in many types of tumors, suggesting that this form of gene inactivation plays an important role in the formation and/or progression of malignancy.

Most Forms of Cancer Are Caused by a Series of Genetic Changes That Progressively Alter the Growth Properties of Cells

The discovery of oncogenes and tumor-suppressor genes has allowed researchers to study the progression of certain forms of cancer at the molecular level. Cancer usually requires multiple genetic changes to the same cell, perhaps in the range of 10 or more. Many cancers begin with a benign genetic alteration that, over time and with additional mutations, leads to malignancy. Furthermore, a malignancy can continue to accumulate genetic changes that make it even more difficult to treat because the cells divide faster or invade surrounding tissues more readily.

Lung cancer is a type of cancer that progresses through different stages of abnormal cell proliferation. Lung cancer is diagnosed in approximately 170,000 men and women each year in the U.S. Worldwide, more than 1.2 million cases are diagnosed. Nearly 90% of these cases are caused by smoking and are thus preventable. Unlike other cancers for which early diagnosis is possible, lung cancer is usually detected only after it has become advanced and is difficult if not impossible to cure. The five-year survival rate for lung cancer patients is less than 15%.

Most cancers in the lung are **carcinomas**—cancers of epithelial cells (**Figure 14.18a**). Epithelial cells are described in Chapter 10. The top images in this figure show the normal epithelium found in a healthy lung. The rest of the figure shows the progression of cancer that is due to mutations in basal cells—

a type of epithelial cell. As mutations accumulate in basal cells, their numbers increase dramatically. This causes a thickening of the epithelium, a condition called hyperplasia. The proliferation of basal cells causes the loss of the ciliated, columnar epithelial cells that normally line the airways. As additional mutations accumulate, the basal cells develop more abnormal morphologies, a condition known as dysplasia. In the early stages of dysplasia, the abnormal basal cells are precancerous. If the source of chronic irritation (usually cigarette smoke) is eliminated, the abnormal cells are likely to disappear. Alternatively, if smoking continues, these abnormal cells may accumulate additional genetic changes and lose the ability to stop dividing. Such cells have become cancerous—the person has basal cell carcinoma.

The basement membrane is a sheetlike layer of extracellular matrix components that provides a barrier between the lung cells and the bloodstream. If the cancer cells have not yet metastasized by penetrating the basement membrane, they will not have spread into the blood and to other parts of the body. If the tumor is removed at this stage, the patient should be cured. The lower images in Figure 14.18a show a tumor that has broken through the basement membrane. The metastasis of these cells to other parts of the body will likely kill the patient, usually within a year of being diagnosed.

The cellular changes that lead to lung cancer are correlated with genetic changes (**Figure 14.18b**). These include the occurrence of mutations that create oncogenes and inhibit tumor-suppressor genes. The order of mutations is not absolute. Rather, it is the total number of genetic changes, not their exact order, that is important. It takes time for such multiple changes to accumulate, so cancer is usually a disease of older people. Reducing your exposure to mutagens throughout your lifetime will minimize the risk of mutations to your genes that could promote cancer.

Normal lung epithelium

Hyperplasia

Loss of ciliated cells

Dysplasia (initially precancerous, then cancerous)

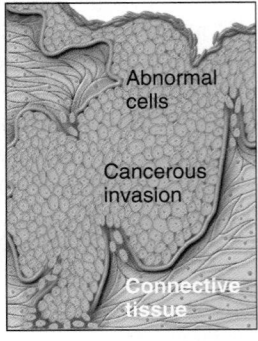

Invasive cancerous cells that can metastasize

(a) Cellular changes

Genes that are commonly mutated in lung cancer include:

Oncogenes	**Tumor-suppressor genes**
erbB – epidermal growth factor receptor	*p53* – checkpoint
ras – cell signaling	*Rb* – negative regulator
myc – transcription factor	*p16* – negative regulator
Cyclin D1 – promotes the cell cycle	*XPD* – DNA repair

(b) Genetic changes

Figure 14.18   Progression of changes leading to lung cancer.   In addition to connective tissue, lung tissue is largely composed of different types of epithelial cells, including columnar and basal cells. The general properties of connective and epithelial tissues are described in Chapter 10. **(a)** A progression of cellular changes in basal cells, caused by the accumulation of mutations, leads to basal cell carcinoma, a common type of lung cancer. **(b)** Mutations in several different genes contribute to lung cancer.

GENOMES & PROTEOMES

Mutations in Approximately 300 Human Genes May Promote Cancer

Researchers have identified a large number of genes that are mutated in cancer cells. Though not all of these mutant genes have been directly shown to affect the growth rate of cells, such mutations are likely to be found in tumors because they provide some type of growth advantage for the cell population from which the cancer developed. For example, certain mutations may enable cells to metastasize to neighboring locations. These mutations may not affect growth rate, but they provide the growth advantage that cancer cells are not limited to growing in a particular location. They can migrate to new locations.

Researchers have estimated that about 300 different genes may play a role in the development of human cancer. With an approximate genome size of 20,000 to 25,000 genes, this observation indicates that over 1% of our genes have the potential to promote cancer if their function is altered by a mutation.

In addition to mutations within specific genes, another common genetic change associated with cancer is abnormalities in chromosome structure and number. **Figure 14.19** compares the chromosome composition of a normal male cell and a tumor cell taken from the same person. The normal composition for this person is 22 pairs of chromosomes plus two sex chromosomes (X and Y). By comparison, the chromosome composition of the tumor cell is quite bizarre, including the fact that the

Figure 14.19 A comparison between chromosomes found in a normal human cell and a cancer cell from the same person. The set found in the cancer cell on the right is highly abnormal, with extra copies of some chromosomes and lost copies of others. Chromosomes made of fused pieces of chromosomes (designated *mar* in this figure) are also common in cancer cells.

tumor cell has two X chromosomes, which is characteristic of females. The tumor cells are missing several chromosomes. If tumor-suppressor genes were located on these missing chromosomes, their function is lost as well. Figure 14.19 also shows a few cases of extra chromosomes. If these chromosomes contain proto-oncogenes, the expression of those genes may be over-active. Finally, tumor cells often contain chromosomes that have translocations. This is the most common class of mutation that creates cancer-causing genes. Such translocations may create chimeric genes (as in the case of the Philadelphia chromosome discussed earlier in this chapter) or they may place a gene next to the regulatory sequences of another gene.

CHAPTER SUMMARY

14.1 Mutation

- A mutation is a heritable change in the genetic material.
- Point mutations, which affect a single nucleotide, can alter the coding sequence of genes in several ways. These include silent, missense, nonsense, and frameshift mutations. (Table 14.1)
- Sickle-cell anemia is caused by a missense mutation that changes a single amino acid in hemoglobin. (Figure 14.1)
- Gene mutations can also alter gene function by changing DNA sequences that are not within the coding region. (Table 14.2)
- The Lederbergs used replica plating and showed that mutations conferring resistance to T1 bacteriophage occurred randomly while the bacteria were not being exposed to the phage. (Figure 14.2)
- Germ-line mutations affect gametes while somatic mutations affect only a part of the body. (Figures 14.3. 14.4)
- Spontaneous mutations are the result of errors in natural biological processes, while induced mutations are due to agents in the environment that cause changes in DNA structure. (Table 14.3)
- Mutagens are agents that lead to mutations in the DNA. These can be chemical or physical agents. (Table 14.4, Figures 14.5, 14.6)
- Testing methods, such as the Ames test, can determine whether an agent is a mutagen. (Figure 14.7)

14.2 DNA Repair

- DNA repair systems involve proteins that sense DNA damage and repair it before a mutation occurs. (Table 14.5)

- Nucleotide excision repair systems recognize various types of DNA damage, such as thymine dimers. This type of system excises the damaged strand, and then a new strand is made. (Figure 14.8)
- Certain inherited diseases in humans are due to defects in nucleotide excision repair. (Figure 14.9)
- Mismatch repair systems recognize a base mismatch that is due to an error in DNA replication. A portion of the daughter strand is removed, and then a new region without the error is made. (Figure 14.10)

14.3 Cancer

- Cancer is due to the accumulation of mutations in a line of cells that leads to uncontrolled cell growth. (Figure 14.11)
- Mutations in proto-oncogenes that result in overactivity produce cancer-causing genes called oncogenes.
- Oncogenes often encode proteins involved in cell-signaling pathways that promote cell division. (Figure 14.12, Table 14.6)
- Four common types of genetic changes, namely missense mutations, gene amplifications, chromosomal translocations, and retroviral insertions, can change proto-oncogenes into oncogenes. (Figure 14.13)
- A mutation that inhibits the ability of the Ras protein to hydrolyze its GTP converts the *ras* gene into an oncogene. (Figure 14.14)
- A chromosome translocation that fuses parts of the *bcr* gene and the *abl* gene also creates an oncogene that causes leukemia. (Figure 14.15)
- Some types of cancer are caused by viruses. (Table 14.7)

- The normal function of tumor-suppressor genes is to prevent cancer. Loss-of-function mutations in such genes can promote cancer. Tumor-suppressor genes often encode proteins that are checkpoint proteins or negative regulators of cell division. (Table 14.8)
- Checkpoint proteins monitor the integrity of the genome and prevent the cell from progressing through the cell cycle if abnormalities are detected. (Figure 14.16)
- The Rb protein is a negative regulator of cell division because it inhibits E2F, a transcription factor that promotes cell division. (Figure 14.17)
- Gene mutations, chromosome loss, and DNA methylation are common ways that tumor-suppressor genes are inactivated.
- Most forms of cancer, such as lung cancer, involve multiple genetic changes that lead to malignancy. (Figure 14.18)
- Over 300 human genes are known to be associated with cancer when they become mutant. In addition, changes in chromosome number and structure are commonly found in cancer cells. (Figure 14.19)

TEST YOURSELF

1. Point mutations that do not alter the amino acid sequence of the resulting gene product are called _____ mutations.
   - a. frameshift
   - b. natural
   - c. silent
   - d. nonsense
   - e. missense

2. Some point mutations will lead to an mRNA that produces a much shorter polypeptide. This type of mutation is known as a _____ mutation.
   - a. neutral
   - b. silent
   - c. missense
   - d. nonsense
   - e. chromosomal

3. The type of mutation that alters the entire amino acid sequence from the site of the mutation is known as a _____ mutation.
   - a. neutral
   - b. silent
   - c. missense
   - d. nonsense
   - e. frameshift

4. Mutagens can cause mutations by
   - a. chemically altering DNA nucleotides.
   - b. disrupting DNA replication.
   - c. altering the genetic code of an organism.
   - d. all of the above.
   - e. a and b only.

5. The mutagenic effect of UV light is
   - a. the alteration of cytosine bases to adenine bases.
   - b. the formation of purine dimers that interfere with genetic expression.
   - c. the breaking of the sugar-phosphate backbone of the DNA molecule.
   - d. the formation of pyrimidine dimers that disrupt DNA replication.
   - e. the deletion of thymine bases along the DNA molecule.

6. The Ames test
   - a. provides a way to determine if any type of cell has experienced a mutation.
   - b. provides a way to determine the mutagenic effect of certain types of agents.
   - c. allows researchers to experimentally disrupt gene activity by causing a mutation in a specific gene.
   - d. provides a way to repair mutations in bacterial cells.
   - e. all of the above.

7. Xeroderma pigmentosum
   - a. is a genetic disorder that results in uncontrolled cell growth.
   - b. is a genetic disorder where normal NER systems are not fully functional.
   - c. is a genetic disorder that results in the loss of pigment in certain patches of skin.
   - d. results from the lack of DNA polymerase proofreading.
   - e. both b and d.

8. During mismatch repair, the parental strand is distinguishable from the new strand by
   - a. the lack of mutations in the parental strand.
   - b. the presence of methyl groups on the new strand.
   - c. the presence of methyl groups on the parental strand.
   - d. the 3′ to 5′ orientation of the strand.
   - e. the AUG codon on the new strand.

9. Cancer cells are said to be metastatic when they
   - a. begin to divide uncontrollably.
   - b. invade healthy tissue.
   - c. migrate to other parts of the body.
   - d. cause mutations in other healthy cells.
   - e. all of the above.

10. Oncogenes are
    - a. mutations in genes that normally inhibit the progression of a cell through the cell cycle.
    - b. mutations that cause the overexpression of genes that normally stimulate cell division.
    - c. viruses that cause cancer.
    - d. mutations in genes that cause metastasis.
    - e. all of the above.

CONCEPTUAL QUESTIONS

1. Explain the difference between a missense mutation and a nonsense mutation.
2. Explain how a frameshift mutation can lead to premature termination of a polypeptide.
3. Define oncogene, tumor-suppressor gene, and proto-oncogene.

EXPERIMENTAL QUESTIONS

1. Explain the difference between the opposing views of mutation prior to the Lederbergs' study.
2. What hypothesis was being tested by the Lederbergs? What were the results of the experiment?
3. How did the results of the Lederbergs support the idea that mutations are random events?

COLLABORATIVE QUESTIONS

1. Discuss the pros and cons of mutation.
2. Discuss three ways that alterations in DNA structure can be repaired.

---

www.brookerbiology.com
This website includes answers to the Biological Inquiry questions found in the figure legends and all end-of-chapter questions.

15

EUKARYOTIC CHROMOSOMES, MITOSIS, AND MEIOSIS

CHAPTER OUTLINE

A scanning electron micrograph of highly compacted chromosomes found in a dividing cell.

The **chromosomes** are structures in living cells that contain the genetic material. Genes are physically located within the chromosomes. Biochemically, chromosomes are composed of a very long molecule of DNA, which is the genetic material, and proteins, which are bound to the DNA and provide it with an organized structure.

The primary function of the genetic material in the chromosomes is to store the information needed to produce the characteristics of an organism. To fulfill their role at the molecular level, DNA sequences facilitate four important processes. These are (1) the synthesis of RNA and cellular proteins, (2) the replication of chromosomes, (3) the compaction of chromosomes so they can fit within living cells, and (4) the proper segregation of chromosomes between dividing cells. In this chapter, we will examine the last two of these topics.

This chapter begins with a discussion of the structure of eukaryotic chromosomes at the molecular level. Then we will turn to the process of reproduction in eukaryotic species at the cellular level. In these discussions we will be concerned with two phenomena. First, we will consider how cells divide to produce new daughter cells. Second, we will examine sexual reproduction from a cellular and genetic perspective. We will pay close attention to the sorting of chromosomes during cell division. Lastly, we will examine variation in the structure and number of chromosomes. As you will learn, a variety of mechanisms that alter chromosome structure and number can have important consequences for the organisms that carry them.

15.1 Molecular Structure of Eukaryotic Chromosomes

We now turn our attention to the ways that eukaryotic chromosomes are folded to fit in a living cell. A typical eukaryotic chromosome contains a single, linear, double-stranded DNA molecule that may be hundreds of millions of base pairs in length. If the DNA from a single set of human chromosomes were stretched from end to end, the length would be over 1 meter! By comparison, most eukaryotic cells are only 10–100 μm (micrometers) in diameter, and the cell nucleus is only about 2–4 μm in diameter. Therefore, to fit inside the nucleus, the DNA in a eukaryotic cell must be folded and packaged by a staggering amount.

Before biologists understood chromosome structure, they described the genetic material according to its appearance under the microscope. When a cell is dividing, the nuclear membrane is no longer present, and the chromosomes become very compact. Such chromosomes are readily stained with colored dyes. The term chromosome literally means colored body. This is the form of the genetic material that we are accustomed to seeing in photomicrographs. The term chromatin was first used to describe the genetic material that is found in the nucleus of nondividing cells. While in the nucleus, the genetic material is much less compact and appears to be in a twisted, spaghetti-like configuration.

Over the past couple of decades, as researchers have gained a more complete understanding of genetic material, the meaning of these two terms, namely chromosome and chromatin, have changed. The term chromosome is now used to describe a discrete unit of genetic material. For example, a human somatic cell contains 46 chromosomes. It would also be correct to say there are 46 chromosomes in the nucleus of a nondividing cell.

By comparison, the term chromatin has taken on a biochemical meaning. **Chromatin** is now used to describe the DNA-protein complex that makes up eukaryotic chromosomes. The chromosomes found in the nucleus are composed of chromatin, as are the highly condensed chromosomes found in dividing cells. Chromosomes are very dynamic structures that alternate between tight and loose compaction states in response to changes in protein composition. In this section, we will focus our attention on two issues of chromosome structure. First, we will consider how chromosomes are compacted and organized within the cell nucleus. Then, we will examine the additional compaction that is necessary to produce the highly condensed chromosomes that occur during cell division.

DNA Wraps Around Histone Proteins to Form Nucleosomes

The first way that DNA is compacted is by wrapping itself around a group of proteins called **histones**. As shown in **Figure 15.1**, a repeating structural unit of eukaryotic chromatin is the **nucleosome**, which is 11 nm (nanometers) in diameter and composed of double-stranded DNA wrapped around an octamer of histone proteins. Each octamer contains eight histone subunits. Different kinds of histone proteins form the histone octamer, two each of four kinds called H2A, H2B, H3, and H4. H2A and H2B are so named because they are similar in structure. Histone proteins are very basic proteins because they contain a large number of positively charged lysine and arginine amino acids. The negative charges that are found in the phosphate of DNA are attracted to the positive charges on histone proteins. The DNA lies on the surface of the histone octamer and makes 1.65 turns around it. The amount of DNA that is required to wrap around the histone octamer is 146 or 147 bp (base pairs). The amino terminal tail of each histone protein protrudes from the histone octamer. As discussed later, covalent modifications of these tails is one way to control the degree of chromatin compaction.

Figure 15.1 **Structure of a nucleosome.** A nucleosome is composed of double-stranded DNA wrapped around an octamer of histone proteins. A linker region connects two adjacent nucleosomes. Histone H1 is bound to the linker region, as are other proteins not shown in this figure.

The nucleosomes are connected by linker regions of DNA that vary in length from 20 to 100 bp, depending on the species and cell type. A particular histone named histone H1 is bound to the linker region, as are other types of proteins. The overall structure of connected nucleosomes resembles beads on a string. This structure shortens the length of the DNA molecule about sevenfold. Evidence for the beads-on-a-string structure is described next.

FEATURE INVESTIGATION

Noll Confirmed Kornberg's Beads-on-a-String Model by Digestion of the Linker Region

The beads-on-a-string model of nucleosome structure was originally proposed by Roger Kornberg in 1974. Markus Noll decided to test Kornberg's model by digesting chromatin with DNase-I, an enzyme that cuts the DNA backbone. He reasoned that if the model was correct, the linker region of DNA would be more accessible to DNase-I than would the 146-bp region that is tightly bound to the histones. Therefore, he expected incubation with DNase-I to make cuts in the linker region and produce DNA pieces that would be approximately 200 bp in length, given that the DNA used had linker regions of about 50 bp.

(Note: The size of the DNA fragments was expected to vary somewhat, since the linker region is not of constant length and the cut within the linker region may occur at different sites.)

Figure 15.2 describes Noll's experimental protocol. He began with nuclei from rat liver cells and incubated them with low, medium, or high concentrations of DNase-I. At high concentrations, the enzyme should make a cut in every linker region, while at medium and low concentrations, DNase-I may occasionally miss cutting a linker region, which would produce larger DNA fragments in multiples of 200 bp. Following digestion, the DNA was extracted from the cell nuclei, and then analyzed by gel electrophoresis to determine the sizes of the DNA fragments. (Gel electrophoresis is described in Chapter 20.)

Figure 15.2 Noll's DNase-I digestion experiment, which verified the beads-on-a-string model of DNA compaction.

HYPOTHESIS DNA wraps around histone proteins in a regular, repeating pattern.

STARTING MATERIAL Nuclei from rat liver cells.

Experimental level

Conceptual level

1 Incubate nuclei with low, medium, and high concentrations of DNase-I. The conceptual level shows a low DNase-I concentration.

DNase-I

37°C   37°C   37°C

Low   Medium   High

Treat with detergent and add phenol.

Before digestion (beads on a string)

After digestion (DNA is cut in linker region)

2 Extract the DNA from the cell nucleus by dissolving the membranes with detergent and adding phenol. DNA goes into aqueous phase.

Aqueous phase (contains DNA)

Phenol phase (contains membranes and proteins)

Marker   Low   Medium   High

DNA in aqueous solution (low DNase-I)

3 Load the DNA into a well of a gel and separate the DNA according to size. The marker lane contains a mixture of DNA fragments of known sizes.

Gel

Low

600 bp
400 bp
200 bp

4 Visualize the DNA fragments by staining the DNA with ethidium bromide, which binds to DNA and is fluorescent when excited by UV light.

Stain gel

Solution with ethidium bromide

UV light

Photograph gel

Results from the experiment:

Low   Medium   High

600 bp →
400 bp →
200 bp →

DNase-I
concentration: 30 units ml^{-1} 150 units ml^{-1} 600 units ml^{-1}

As shown in the data, at high DNase-I concentrations, the chromosomal DNA was digested into fragments of approximately 200 bp in length. This result is predicted by the beads-on-a-string model. At lower DNase-I concentrations, longer pieces were observed that were in multiples of 200 bp (400, 600, and so on). These longer pieces are explained by occasional uncut linker regions. For example, a DNA piece might contain two nucleosomes and be 400 bp in length. Taken together, these results strongly supported the nucleosome model for chromatin structure.

Nucleosomes Compact to Form a 30-nm Fiber

Nucleosome units are organized into a more compact structure that is 30 nm in diameter, known as the **30-nm fiber** (**Figure 15.3a**). Histone H1 and other proteins are important in the formation of the 30-nm fiber, which shortens the nucleosome structure another sevenfold. The structure of the 30-nm fiber has proven difficult to determine because the conformation of the DNA may be substantially altered when extracted from living cells. A current model for the 30-nm fiber was proposed by Rachel Horowitz and Christopher Woodcock in the 1990s (**Figure 15.3b**). According to their model, linker regions in the 30-nm structure are variably bent and twisted, and little direct contact is observed between nucleosomes. The 30-nm fiber forms an asymmetric, three-dimensional zigzag of nucleosomes. At this level of compaction, the overall picture of chromatin that emerges is an irregular, fluctuating structure with stable nucleosome units connected by bendable linker regions.

Chromatin Loops Are Anchored to the Nuclear Matrix

Thus far, we have examined two mechanisms that compact eukaryotic DNA, the formation of nucleosomes and their arrangement into a 30-nm fiber. Taken together, these two events shorten the folded DNA about 49-fold. A third level of compaction involves interactions between the 30-nm fibers and a filamentous network of proteins in the nucleus called the **nuclear matrix**. This matrix consists of the **nuclear lamina**, which is a collection of protein fibers that line the inner nuclear membrane, and an internal nuclear matrix that is connected to the lamina (**Figure 15.4a**). The internal matrix is an intricate network of irregular protein fibers plus many other proteins that bind to these fibers.

The nuclear matrix is involved in the compaction of the 30-nm fiber by participating in the formation of **radial loop domains**. These loops, often 25,000 to 200,000 base pairs in size,

(a) Micrograph of a 30-nm fiber

30 nm

(b) Three-dimensional zigzag model

Figure 15.3 The 30-nm fiber. **(a)** A photomicrograph of the 30-nm fiber. **(b)** In this three-dimensional zigzag model, the linker DNA forms a bendable structure with little contact between adjacent nucleosomes.

are anchored to the nuclear matrix (**Figure 15.4b**). In this way, the nuclear matrix organizes the chromosomes within the nucleus.

Each chromosome in the cell nucleus is located in a discrete and nonoverlapping **chromosome territory**, which can be experimentally viewed in nondividing cells (refer back to Chapter 4, Figure 4.17). **Figure 15.5** shows a model comparing human chromosomes in their fully compacted state, when a cell is preparing to divide, with chromosomes in nondividing cells. Each chromosome in nondividing cells occupies its own discrete region in the cell nucleus that usually does not overlap with the territory of adjacent chromosomes. In other words, different chromosomes are not substantially intertwined with each other, even when they are in a noncompacted condition.

(a) Proteins that form the nuclear matrix

(b) Radial loop domain bound to a protein fiber

Figure 15.4   **Structure of the nuclear matrix and its attachment to the 30-nm fiber.**   **(a)** This schematic drawing shows the arrangement of the matrix within a cell nucleus. The nuclear lamina is a collection of fibrous proteins that line the inner nuclear membrane. The internal nuclear matrix is also composed of protein fibers and many other proteins associated with them. **(b)** The radial loops are attached to the protein fibers of the nuclear matrix.

(a)

(b)

Figure 15.5   **A molecular model showing chromosome territories in the cell nucleus of humans.**   Each of the 23 pairs of human chromosomes is labeled with a different color. **(a)** Compacted chromosomes in a cell that is preparing to divide. **(b)** Chromosomes in the cell nucleus of a nondividing cell. Each of these chromosomes occupies its own distinct, nonoverlapping territory within the nucleus.

The compaction level of chromosomes in the cell nucleus is not completely uniform. This variability can be seen with a light microscope and was first observed by the German cytologist E. Heitz in 1928. He used the term **heterochromatin** to describe the highly compacted regions of chromosomes. In general, these regions are transcriptionally inactive because of their tight conformation, which prevents transcription factors and RNA polymerase from gaining access to genes. By compar-

ison, the less condensed regions, known as **euchromatin**, reflect areas that are capable of gene transcription. Euchromatin is the form of chromatin in which the 30-nm fiber forms radial loop domains. In heterochromatin, these radial loop domains are compacted even further. In nondividing cells, most chromosomal regions are euchromatic and some localized regions are heterochromatic.

During Cell Division, Chromosomes Undergo Maximum Compaction

When cells prepare to divide, the chromosomes become even more compacted or condensed. This aids in their proper sorting and movement during metaphase, which is a stage of cell division described in the next section. **Figure 15.6** illustrates the levels of compaction that contribute to the formation of a metaphase chromosome. DNA in the nucleus is always compacted by forming nucleosomes and condensing into a 30-nm fiber

(Figure 15.6a,b,c). In euchromatin, the 30-nm fibers are arranged in radial loop domains that are relatively loose, meaning that a fair amount of space is between the 30-nm fibers (Figure 15.6d). The average width of such loops is about 300 nm.

By comparison, heterochromatin involves a much tighter packing of the loops, so little space is between the 30-nm fibers (Figure 15.6e). Heterochromatic regions tend to be wider, in the range of 700 nm. When cells prepare to divide, all of the euchromatin is converted to heterochromatin. Because the 30-nm fibers are much closer together in heterochromatin, the conversion of

(a) DNA double helix

(b) Nucleosomes ("beads on a string")

(c) 30-nm fiber

(d) Radial loop domains

(e) Heterochromatin

(f) Metaphase chromosome

DNA double helix

2 nm

11 nm

Histones

Nucleosome

Histone H1

30 nm

300 nm

700 nm

1,400 nm

1. Wrapping of DNA around histone proteins.

2. Formation of a three-dimensional zigzag structure via histone H1 and other DNA-binding proteins.

3. Anchoring of radial loop domains to the nuclear matrix.

4. Further compaction of radial loops to form heterochromatin.

5. Formation of a scaffold from the nuclear matrix. All chromatin is heterochromatin.

Figure 15.6 The steps in eukaryotic chromosomal compaction leading to the metaphase chromosome.

(a) Metaphase chromosome

Scaffold

DNA strand

280 nm

(b) Metaphase chromosome treated with high salt to remove proteins

Figure 15.7 The importance of histones and scaffolding proteins in the compaction of eukaryotic chromosomes. **(a)** Transmission electron micrograph of a metaphase chromosome. **(b)** This photomicrograph shows a metaphase chromosome following treatment with a high salt concentration to remove the histone proteins. The label on the *left* points to the scaffold that anchors the bases of the radial loops. The *right* label points to an elongated strand of DNA.

Biological inquiry: After they have replicated and become compacted in preparation for cell division, chromosomes are often shaped like an X, as in part (a) of this figure. Which proteins are primarily responsible for this X shape?

euchromatin to heterochromatin greatly shortens the chromosomes. In a metaphase chromosome, which contains two copies of the DNA (Figure 15.6f), the width averages about 1,400 nm, but the length of a metaphase chromosome is much shorter than the same chromosome in the cell nucleus during interphase. These highly condensed chromosomes undergo little gene transcription because it is difficult for transcription proteins to gain access to the compacted DNA. Therefore, most transcriptional activity ceases during cell division, which usually lasts for a relatively short time.

In metaphase chromosomes, the highly compacted radial loops remain anchored to a **scaffold**, which is formed from proteins in the nuclear matrix (**Figure 15.7**). Experimentally, the scaffold proteins that hold the loops in place can be separated into an observable form. If a metaphase chromosome is treated with a high concentration of salt to remove histone proteins, the highly compact configuration is lost, but the bottoms of the elongated DNA loops remain attached to the scaffold. In the photomicrograph shown in Figure 15.7b, a label points to an elongated DNA strand emanating from the darkly stained scaffold. The scaffold retains the shape of the original metaphase chromosome even though the DNA strands have become greatly elongated. These remarkable results illustrate the importance of both nuclear matrix proteins (which form the scaffold) and histones (which are needed to compact the DNA) for the structure of chromosomes.

GENOMES & PROTEOMES

The Histone Code Controls Chromatin Compaction

In this section, we have learned that the genomes of eukaryotic species are greatly compacted to fit inside the cell nucleus. Even euchromatin, which is looser than heterochromatin, is still compacted to such a degree that it is difficult for transcription fac-

tors and RNA polymerase to access and transcribe genes. As described in Chapters 12 and 13, chromatin must be loosened up so that genes can be transcribed into RNA.

As discussed earlier, each of the histone proteins consists of a globular domain and a flexible, charged amino terminus called an amino terminal tail. The DNA wraps around the globular domains, as depicted in Figure 15.1, and the amino terminal tails protrude from the chromatin. In recent years, researchers have discovered that particular amino acids in the amino terminal tails are subject to several types of covalent modifications, including acetylation, methylation, and phosphorylation. Over 50 different enzymes have been identified in mammals that selectively modify amino terminal tails. **Figure 15.8** shows examples of sites in the tails of H2A, H2B, H3, and H4 that can be modified.

These tail modifications can have two effects. First, they may directly influence interactions between nucleosomes. Second, histone modifications provide binding sites that are recognized by proteins. According to the **histone code hypothesis**, proposed by Brian Strahl and David Allis in 2000, the pattern of histone modification is recognized by particular proteins, much like a language or code. For example, one pattern might involve phosphorylation of the serine at the first amino acid in H2A and acetylation of the fifth and eighth amino acids in H4, which are lysines. A different pattern could involve acetylation of the fifth amino acid, a lysine, in H2B and methylation of the third amino acid in H4, which is an arginine.

The pattern of covalent modifications of amino terminal tails provides binding sites for proteins that subsequently affect the degree of chromatin compaction. One pattern of histone modification may attract proteins that cause the chromatin to become even more compact. This would silence the transcription of genes in the region. Alternatively, a different combination of histone modifications may attract proteins, such as chromatin remodeling enzymes discussed in Chapter 13, that serve to loosen the chromatin and thereby promote gene transcription.

Figure 15.8 Examples of covalent modifications that occur to the amino terminal tails of histone proteins. The amino acids are numbered from the amino terminus. The modifications shown here are *m* for methylation, *p* for phosphorylation, and *ac* for acetylation. Many more modifications can occur to the amino terminal tails; the ones shown here represent common examples.

In this way, the histone code plays a key role in accessing the information within the genomes of eukaryotic species. Researchers are trying to unravel which patterns of histone modifications promote compaction, and which promote a loosening of chromatin structure. In other words, they are trying to decipher the histone code.

15.2 Mitotic Cell Division

We now turn our attention to the mechanism of cell division and its relationship to chromosome structure and replication. During the process of **mitotic cell division**, a cell divides to produce two new cells that are genetically identical to the original cell. By convention, the original cell is usually called the mother cell, and the new cells are the two daughter cells. Mitotic cell division involves **mitosis**, which is the division of the nucleus into two nuclei, and **cytokinesis**, which is the division of one mother cell into two daughter cells. One purpose of mitotic cell division is **asexual reproduction**. Certain unicellular eukaryotic organisms, such as baker's yeast (*Saccharomyces cerevisiae*) and the amoeba, increase their numbers in this manner.

A second important reason for mitotic cell division is the production and maintenance of multicellularity. Organisms such as plants, animals, and most fungi are derived from a single cell

that subsequently undergoes repeated cellular divisions to become a multicellular organism. Humans, for example, begin as a single fertilized egg and repeated cellular divisions produce an adult with several trillion cells. As you might imagine, the precise transmission of chromosomes is critical during every cell division so that all cells of the body receive the correct amount of genetic material.

In this section, we will explore how the process of mitotic cell division requires the duplication, organization, and sorting of chromosomes. We will also examine how a single cell is separated into two distinct cells by cytokinesis. But first, we need to consider some general features of chromosomes in eukaryotic species.

Eukaryotic Chromosomes Are Inherited in Sets

To understand the chromosomal composition of cells and the behavior of chromosomes during cell division, scientists observe chromosomes with the use of microscopes. **Cytogenetics** is the field of genetics that involves the microscopic examination of chromosomes. As discussed earlier in this chapter, when a cell prepares to divide the chromosomes become more tightly compacted, which shortens them and thereby increases their diameter. A consequence of this shortening is that distinctive shapes and numbers of chromosomes become visible with a light microscope.

Figure 15.9 shows the general procedure for preparing and viewing chromosomes from a eukaryotic cell. In this example, the cells are obtained from a sample of human blood. In particular, the chromosomes within lymphocytes (a type of white blood cell) are examined. A sample of the blood cells is obtained and treated with drugs that stimulate the cells to divide. The actively dividing cells are centrifuged to concentrate them and then mixed with a hypotonic solution that makes the cells swell. The expansion in cell structure causes the chromosomes to spread out from each other, making it easier to see each individual chromosome. Next, the cells are concentrated by a second centrifugation and treated with a fixative, which chemically freezes them so that the chromosomes will no longer move around. The cells are then exposed to a chemical dye that binds to the chromosomes and stains them. As we will learn later, this gives chromosomes a distinctive banding pattern that greatly enhances their contrast and ability to be uniquely identified. The cells are then placed on a slide and viewed with a light microscope. In a cytogenetics laboratory, the microscopes are equipped with an electronic camera to photograph the chromosomes. On a computer screen, the chromosomes can be organized in a standard way, usually from largest to smallest. A photographic representation of the chromosomes, as in the photo in step 5 of Figure 15.9, is called a **karyotype**. A karyotype reveals how many chromosomes are found within an actively dividing cell.

By studying the karyotypes of many species, scientists have discovered that eukaryotic chromosomes occur in sets; each set is composed of several different types of chromosomes. For example, one set of human chromosomes contains 23 different types of chromosomes. By convention, the chromosomes are

1. A sample of blood is collected and treated with drugs that stimulate cell division. The sample is then subjected to centrifugation.

Supernatant

Blood cells — Pellet

2. The supernatant is discarded, and the cell pellet is suspended in a hypotonic solution. This causes the cells to swell.

Hypotonic solution

3. The sample is subjected to centrifugation a second time to concentrate the cells. The cells are suspended in a fixative, stained, and placed on a slide.

Fix   Stain

Blood cells

4. The slide is viewed by a light microscope equipped with a camera; the sample is seen on a computer screen. The chromosomes can be photographed and arranged electronically on the screen.

5.

For a diploid human cell, 2 complete sets of chromosomes from a single cell constitute a karyotype of that cell.

Figure 15.9   The procedure for making a karyotype.

Biological inquiry: Researchers usually treat cells with drugs that stimulate them to divide prior to the procedure for making a karyotype. Why would this be useful?

numbered according to size, with the largest chromosomes having the smallest numbers. For example, human chromosomes 1, 2, and 3 are relatively large, whereas 21 and 22 are the two smallest (Figure 15.9). This numbering system does not apply to the sex chromosomes, which determine the sex of the individual. Sex chromosomes are designated with the letters X and Y in humans.

A second feature of many eukaryotic species is that an individual has two sets of chromosomes. Again, if we consider humans as an example, the karyotype shown in Figure 15.9 contains two sets of chromosomes, with 23 different chromosomes in each set. Therefore, this human cell contains a total of 46 chromosomes. A person's cells have 46 chromosomes each because the individual inherited one set from the father and one set from the mother. When the cells of an organism carry two sets of chromosomes, that organism is said to be **diploid**. Geneticists use the letter n to represent a set of chromosomes, so diploid organisms are referred to as $2n$. For example, humans are $2n$,

where $n = 23$. Most human cells are diploid. An exception involves **gametes**, namely sperm and egg cells. Gametes are **haploid** or $1n$, which means they contain one set of chromosomes.

When a species is diploid, the members of a pair of chromosomes are called **homologues**. As you can see in Figure 15.9, a cell has two copies of chromosome 1, two copies of chromosome 2, and so forth. Within each pair, the chromosome on the left is a homologue to the one on the right and vice versa. In the case of animals, one of each of these pairs comes from an organism's mother, and one comes from the father; these are referred to as maternal and paternal chromosomes, respectively.

Homologous chromosomes are very similar to each other. Each of the two chromosomes in a homologous pair is nearly identical in size and contains a similar composition of genetic material. A particular gene found on one copy of a chromosome is also found on the homologue. Because one homologue is received from each parent, the two homologues may vary with regard to the way that a gene affects an organism's traits.

As an example, let's consider an eye color gene in humans. One chromosome might carry the form of an eye color gene that confers brown eyes, while the gene on the homologue could confer blue eyes. This topic will be considered in Chapter 16.

The DNA sequences on homologous chromosomes are very similar. In most cases, the sequence of bases on one homologue would differ by less than 1% from the sequence on the other homologue. For example, the DNA sequence of chromosome 1 that you inherited from your mother would be greater than 99% identical to the DNA sequence of chromosome 1 that you inherited from your father. Nevertheless, keep in mind that the sequences are not identical. The slight differences in DNA sequence provide important variation in gene function. Again, if we use an eye color gene as an example, a minor difference in DNA sequence distinguishes two forms of the gene, brown versus blue.

The striking similarity between homologous chromosomes does not apply to pairs of sex chromosomes (for example, X and Y). These chromosomes differ in size and genetic composition. Certain genes that are found on the X chromosome are not found on the Y chromosome, and vice versa. The X and Y chromosomes are not considered homologous chromosomes, although they do have short regions of homology.

In Preparation for Cell Division, Eukaryotic Chromosomes Are Replicated and Compacted to Produce Pairs Called Sister Chromatids

Now that we understand that chromosomes are found in sets, and that many eukaryotic species are diploid, we will now turn our attention to how those chromosomes are replicated and sorted during cell division. Let's begin with the process of chromosome replication. In Chapter 11, we examined the molecular process of DNA replication. **Figure 15.10** describes the process at the chromosomal level. Prior to DNA replication, the DNA of each eukaryotic chromosome consists of a linear DNA double helix that is found in the nucleus and is not highly compacted. When the DNA is replicated, two identical copies of the original double helix are created. These copies, along with associated proteins, lie side by side and are termed **sister chromatids**. When a cell prepares to divide, the sister chromatids become highly compacted and readily visible under the microscope. As shown in the inset to Figure 15.10, the two sister chromatids are tightly associated at a region called the **centromere**. The centromere serves as an attachment site for a group of proteins that form the **kinetochore**, which is necessary for sorting each chromosome.

With regard to the cell cycle, which is described in Chapter 9, **Figure 15.11** provides an overview that relates chromosome replication and cell division. In the G_1 phase, the original cell had three pairs of chromosomes, for a total of six individual chromosomes. Such a cell is diploid ($2n$) and contains three chromosomes per set ($n = 3$). The paternal set is shown in blue, and the homologous maternal set is shown in red. In G_1, the chromosomes are not highly compacted. During the S, or synthesis, phase, these chromosomes replicate to yield 12 chromatids (that is, 6 pairs of sister chromatids). At the start of mito-

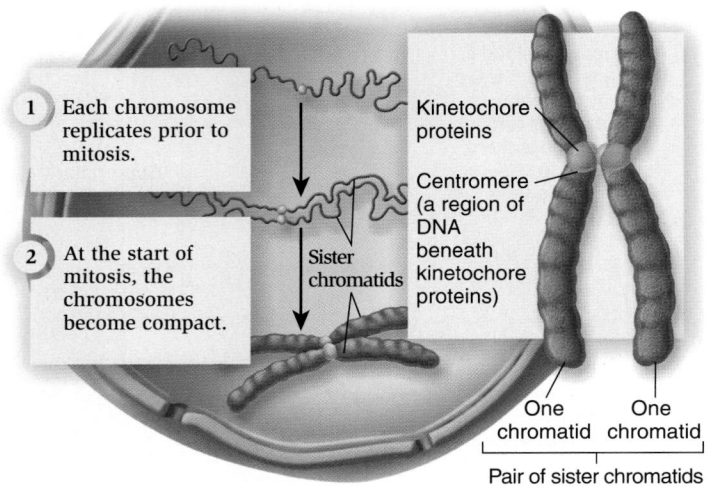

(a) **Chromosome replication and compaction**   (b) **Schematic drawing of a metaphase chromosome**

Figure 15.10 Replication and compaction of chromosomes into pairs of sister chromatids.   (a) Chromosomal replication producing a pair of sister chromatids. While the chromosomes are elongated, they are replicated to produce two copies that are connected and lie parallel to each other. This is a pair of sister chromatids. Later, when the cell is preparing to divide, the sister chromatids condense into more compact structures that are easily seen with a light microscope. (b) A schematic drawing of a metaphase chromosome. This structure has two chromatids that lie side by side. The two chromatids are held together by kinetochore proteins that bind to each other and to the centromeres on each chromatid.

sis, the chromatids become highly compacted, and during the process of cell division they are divided equally into two daughter cells. The term **M phase** refers to the sequential events of mitosis and cytokinesis. After cell division is completed, these two daughter cells each contain six chromosomes (three pairs of homologues).

The Transmission of Chromosomes Requires a Sorting Process Known as Mitosis

Mitosis is the sorting process that ensures that each daughter cell will obtain the correct number and types of chromosomes. Mitosis was first observed microscopically in the 1870s by a German biologist named Walter Flemming, who coined the term mitosis (from the Greek *mitos*, meaning thread). He studied the large, transparent skin cells of salamander larvae, as they were dividing, and noticed that chromosomes are constructed of "threads" that are doubled in appearance along their length. These double threads divided and moved apart, one going to each of the two daughter nuclei. By this mechanism, Flemming pointed out, the two daughter cells receive an identical group of threads, the same as the number of threads in the mother cell.

We now know that the **mitotic spindle apparatus** (also known simply as the **mitotic spindle**) is responsible for orga-

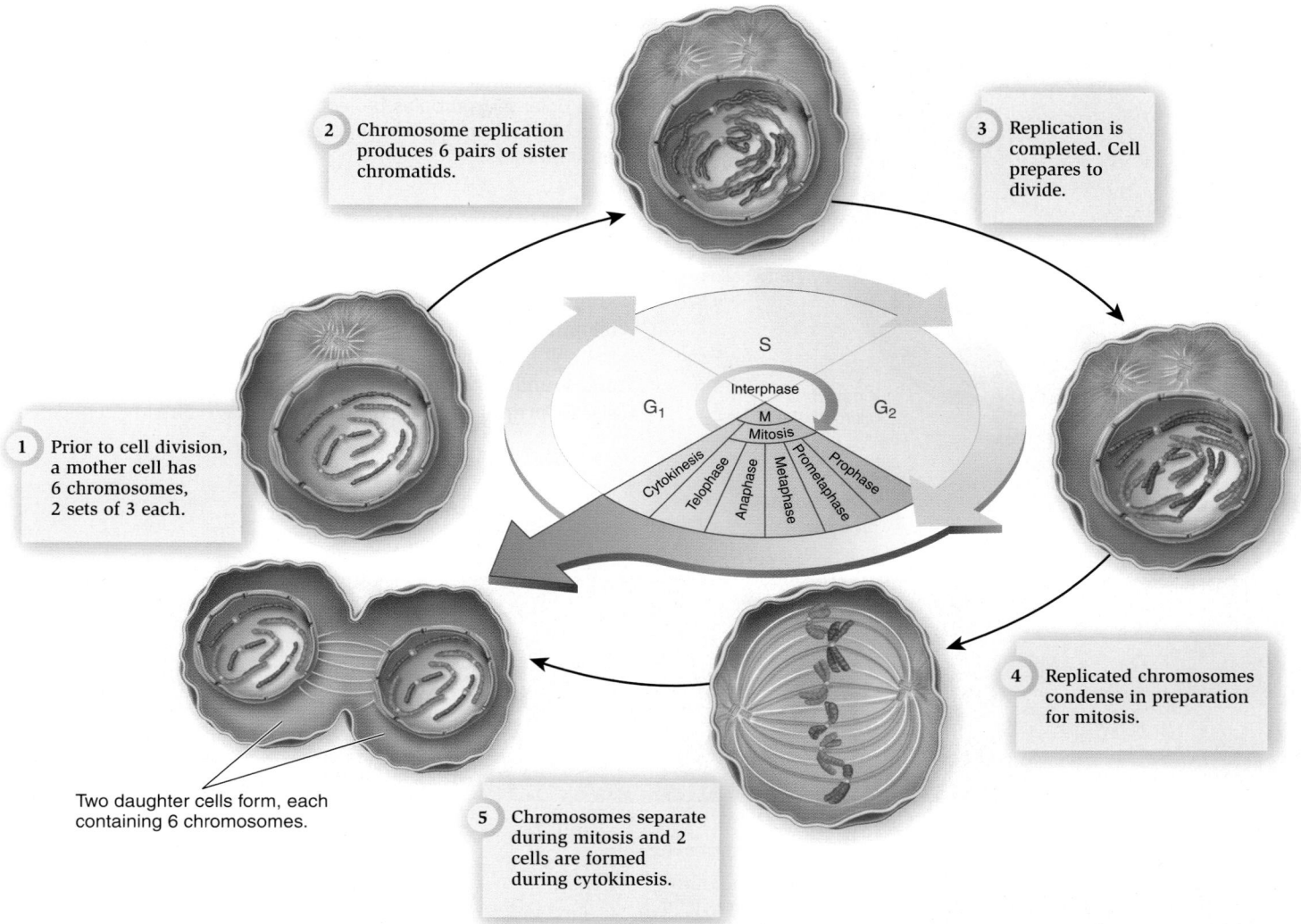

2 Chromosome replication produces 6 pairs of sister chromatids.

3 Replication is completed. Cell prepares to divide.

1 Prior to cell division, a mother cell has 6 chromosomes, 2 sets of 3 each.

S

Interphase

G_1

M

Mitosis

G_2

Cytokinesis

Telophase

Anaphase

Metaphase

Prometaphase

Prophase

4 Replicated chromosomes condense in preparation for mitosis.

Two daughter cells form, each containing 6 chromosomes.

5 Chromosomes separate during mitosis and 2 cells are formed during cytokinesis.

Figure 15.11   **The eukaryotic cell cycle and cell division.**   Dividing cells progress through a series of stages. G_1, S, and G_2 are collectively known as interphase, and M phase includes mitosis and cytokinesis. This diagram shows the progression of a cell through the cell cycle to produce two daughter cells. Note: The width of the phases shown in this figure is not meant to reflect their actual length. As discussed in Chapter 9, G_1 is typically the longest phase of the cell cycle, while M phase is relatively short.

nizing and sorting the chromosomes during mitosis. The structure of the mitotic spindle in animal cells is shown in **Figure 15.12**. The mitotic spindle is formed from two structures called the **centrosomes**. Because they are a site for organizing microtubules, centrosomes are also referred to as microtubule organizing centers (MTOCs). A single centrosome duplicates during interphase. After they separate from each other during mitosis, each centrosome defines a **pole** of the spindle apparatus, one within each of the future daughter cells. In animal cells, a conspicuous structure found in the centrosome is a pair of centrioles. However, centrioles are not found in many other eukaryotic species, such as plants, and are not required for spindle formation.

The spindle is formed from protein fibers called microtubules that are rooted in the centrosomes. Each centrosome organizes the construction of the microtubules by rapidly poly-merizing tubulin proteins. The three types of spindle microtubules are termed astral, polar, and kinetochore microtubules. The astral microtubules, which extend away from the chromosomes, are important for positioning the spindle apparatus within the cell. The polar microtubules project into the region between the two poles. Polar microtubules that overlap with each other play a role in the separation of the two poles. Finally, the kinetochore microtubules are attached to kinetochores, which are bound to the centromere of each chromosome.

Now that we understand the structure of the mitotic spindle, we can examine the sequence of events that occurs during mitosis. **Figure 15.13** depicts the process of mitosis in an animal cell, though the process is quite similar in a plant cell. In the simplified diagrams shown along the bottom of this figure, the original mother cell contains six chromosomes, as in Figure 15.11.

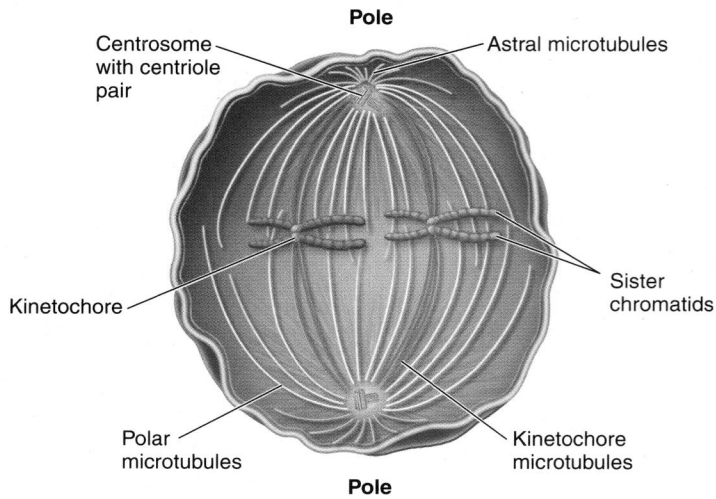

Pole

Centrosome with centriole pair

Astral microtubules

Kinetochore

Sister chromatids

Polar microtubules

Kinetochore microtubules

Pole

Figure 15.12   **The structure of the mitotic spindle.**   The mitotic spindle is formed by the centrosomes from three types of microtubules. The astral microtubules emanate away from the region between the poles. The polar microtubules project into the region between the two poles. The kinetochore microtubules are attached to the kinetochores of sister chromatids.

One set of chromosomes is again depicted in red, while the homologous set is blue; remember that these represent maternal and paternal chromosomes. Mitosis is subdivided into phases known as prophase, prometaphase, metaphase, anaphase, and telophase. Prior to mitosis, the cells are in **interphase**, a phase of the cell cycle during which the chromosomes are decondensed and found in the nucleus (Figure 15.13a). At the start of mitosis, in **prophase**, the chromosomes have already replicated to

(a) Interphase	**(b) Prophase**	**(c) Prometaphase**

Chromosomes

Nuclear membrane

Sister chromatids

Mitotic spindle

Spindle pole

Two centrosomes, each with centriole pairs

Kinetochore microtubule

1 Chromosomes have already replicated during interphase.

2 Sister chromatids condense and spindle starts to form. Nuclear membrane begins to dissociate into vesicles.

3 Nuclear membrane has completely dissociated into vesicles and the spindle is fully formed. Sister chromatids attach to spindle via kinetochore microtubules.

Figure 15.13   **The process of mitosis in an animal cell.**   The *top* panels illustrate the cells of a newt progressing through mitosis. The *bottom* panels are schematic drawings that emphasize the sorting and separation of the chromosomes in which the original diploid cell had 6 chromosomes (3 in each set). At the start of mitosis, these have already replicated into 12 chromatids. The final result is 2 daughter cells each containing 6 chromosomes.

produce 12 chromatids, joined as six pairs of sister chromatids (Figure 15.13b). As prophase proceeds, the nuclear membrane begins to dissociate into small vesicles. At the same time, the chromatids condense into highly compacted structures that are readily visible by light microscopy.

The mitotic spindle is completely formed during **prometaphase** (Figure 15.13c). As mitosis progresses, the centrosomes move apart and demarcate the two poles. Once the nuclear membrane has dissociated, the spindle fibers can interact with the sister chromatids. Initially, kinetochore microtubules are rapidly formed and can be seen under a microscope growing out from the two poles. As it grows, if a kinetochore microtubule happens to make contact with a kinetochore, it is said to be "captured" and remains firmly attached to the kinetochore. This seemingly random process is how sister chromatids become attached to kinetochore microtubules. Alternatively, if a kinetochore microtubule does not collide with a kinetochore, the microtubule will eventually depolymerize and retract to the centrosome. As the end of prometaphase nears, the two kinetochores on each pair of sister chromatids are attached to kinetochore microtubules from opposite poles. As these events are occurring, the sister chromatids are seen under the microscope to undergo jerky movements as they are tugged, back and forth, between the two poles by the kinetochore microtubules.

Eventually, the pairs of sister chromatids are aligned along a plane halfway between the poles called the **metaphase plate**. The pairs of sister chromatids have become organized into a single row within this plane. When this alignment is complete, the cell is in **metaphase** of mitosis (Figure 15.13d). The chromatids can then be equally distributed into two daughter cells.

(d) Metaphase

Metaphase plate

4 Sister chromatids align along the metaphase plate.

(e) Anaphase

Individual chromosomes

Polar microtubule

5 Sister chromotids separate and individual chromosomes move toward poles as kinetochore microtubles shorten. Polar microtubles lengthen and push poles apart.

(f) Telophase and cytokinesis

Cleavage furrow

6 Chromosomes decondense and nuclear membranes re-form. Cleavage furrow separates the 2 cells.

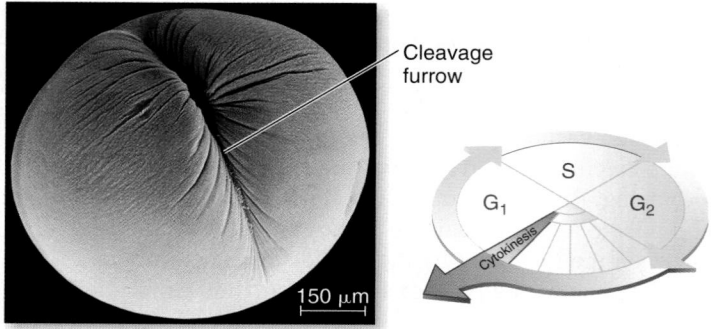

(a) Cleavage of an animal cell

(b) Formation of a cell plate in a plant cell

Figure 15.14 Micrographs showing cytokinesis in animal and plant cells.

The next step in the sorting process occurs during **anaphase** (Figure 15.13e). At this stage, the connections between the pairs of sister chromatids are broken. Each chromatid, now an individual chromosome, is linked to only one of the two poles by one or more kinetochore microtubules. As anaphase proceeds, the kinetochore microtubules shorten, pulling the chromosomes toward the pole to which they are attached. In addition, the two poles move farther away from each other. This occurs because the overlapping polar microtubules lengthen and push against each other, thereby pushing the poles farther apart. During **telophase**, the chromosomes have reached their respective poles and decondense. The nuclear membranes now re-form to produce two separate nuclei. In Figure 15.13f, two nuclei are being produced that contain six chromosomes each.

In most cases, mitosis is quickly followed by cytokinesis, in which the two nuclei are segregated into separate daughter cells. While the stages of mitosis are similar between plant and animal cells, the process of cytokinesis is quite different. In animal cells, cytokinesis involves the formation of a **cleavage furrow**, which constricts like a drawstring to separate the cells (**Figure 15.14a**). In plants, the two daughter cells are separated by the formation of a **cell plate** (**Figure 15.14b**), which forms a cell wall between the two daughter cells.

Mitosis and cytokinesis ultimately produce two daughter cells having the same number of chromosomes as the mother cell. Barring rare mutations, the two daughter cells are genetically identical to each other and to the mother cell from which they were derived. Thus, the critical consequence of this sorting process is to ensure genetic consistency from one cell to the next. The development of multicellularity relies on the repeated process of mitosis and cytokinesis. For diploid organisms that are multicellular, most of the somatic cells are diploid and genetically identical to each other. Next, we will consider how diploid cells can divide to produce haploid cells.

15.3 Meiosis and Sexual Reproduction

We now turn our attention to sexual reproduction. As discussed earlier, a diploid cell contains two homologous sets of chromosomes, while a haploid cell contains a single set. For example, a diploid human cell contains 46 chromosomes, but a human gamete (sperm or egg cell) is a haploid cell that contains only 23 chromosomes, one from each of the 23 pairs. **Sexual reproduction** requires a **fertilization** event in which two haploid gametes unite to create a diploid cell called a **zygote**. In the case of many multicellular species, the zygote then grows and divides by mitosis into a multicellular organism with many diploid cells.

Meiosis is the process by which haploid cells are produced from a cell that was originally diploid. The term *meiosis*, which means "to make smaller," refers to the fewer chromosomes found in cells following this process. For this to occur, the chromosomes must be correctly sorted and distributed in a way that reduces the chromosome number to half its original diploid value. In the case of human gametes, for example, each gamete must receive half the total number of chromosomes, but not just any 23 chromosomes will do. A gamete must receive one chromosome from each of the 23 pairs. For this to happen, two rounds of divisions are necessary, termed meiosis I and meiosis II (**Figure 15.15**). When a cell begins meiosis, it contains chromosomes that are found in homologous pairs. When meiosis is completed, a single diploid cell with homologous pairs of chromosomes has produced four haploid cells. In this section, we will examine the cellular events of meiosis that reduce the chromosome number from diploid to haploid. In addition, we will briefly consider how this process plays a role in the life cycles of fungi, plants, and animals.

The First Meiotic Division, Meiosis I, Separates Homologous Chromosomes

Like mitosis, meiosis begins after a cell has progressed through the G_1, S, and G_2 phases of the cell cycle. However, two key events occur at the beginning of meiosis that do not occur in mitosis. First, homologous pairs of sister chromatids associate

Figure 15.15 **How the process of meiosis reduces chromosome number.** This simplified diagram emphasizes the reduction in chromosome number as a diploid cell divides by meiosis to produce 4 haploid cells.

with each other, lying side by side to form a **bivalent**, also called a **tetrad** (**Figure 15.16**). The process of forming a bivalent is termed **synapsis**. In most eukaryotic species, a synaptonemal complex is formed between homologous chromosomes. However, the synaptonemal complex is not required for the pairing of homologous chromosomes because some species of fungi completely lack such a complex, yet their chromosomes associate with each other correctly. At present, the precise role of the synaptonemal complex is not clearly understood.

Figure 15.16 **Formation of a bivalent and crossing over during meiosis I.** At the beginning of meiosis, homologous chromosomes pair with each other to form a bivalent, usually with a synaptonemal complex between them. Crossing over then occurs between homologous chromatids within the bivalent. During this process, homologues exchange segments of chromosomes.

The second event that occurs at the beginning of meiosis, but not usually during mitosis, is **crossing over**, which involves a physical exchange between chromosome pieces of the bivalent (Figure 15.16). As discussed in Chapter 17, crossing over may increase the genetic variation of a species. After crossing over occurs, the arms of the chromosomes tend to separate but remain adhered at a crossover site. This connection is called a **chiasma** (plural, chiasmata), because it physically re-

sembles the Greek letter chi, χ. The number of crossovers is carefully controlled by cells and depends on the size of the chromosome and the species. The range of crossovers for eukaryotic chromosomes is typically one or two to a couple dozen. During the formation of sperm in humans, for example, an average chromosome undergoes slightly more than two crossovers, while chromosomes in certain plant species may undergo 20 or more crossovers.

Figure 15.17   The stages of meiosis in an animal cell.

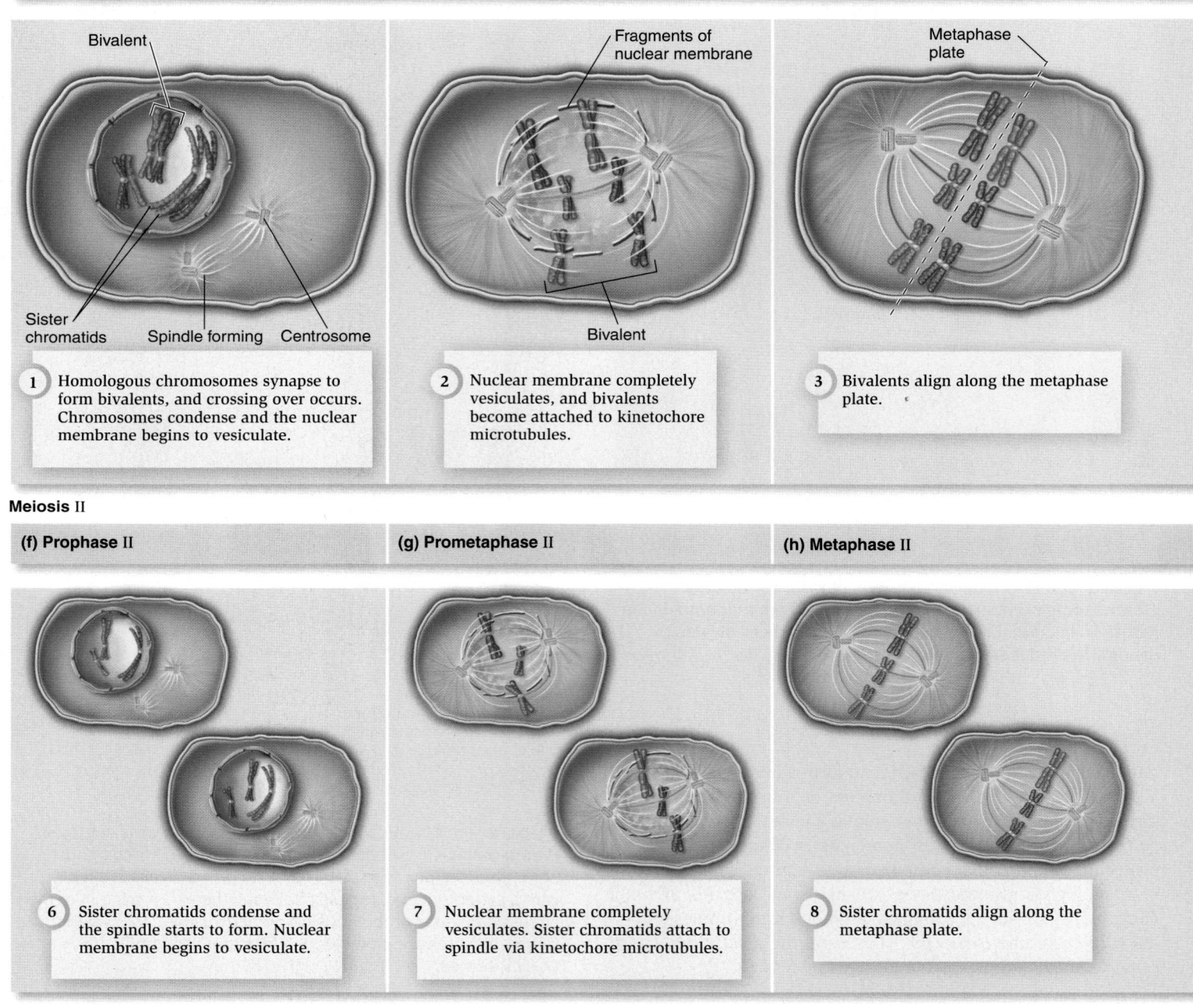

Meiosis I

(a) Prophase I

Bivalent

Sister chromatids   Spindle forming   Centrosome

1 Homologous chromosomes synapse to form bivalents, and crossing over occurs. Chromosomes condense and the nuclear membrane begins to vesiculate.

(b) Prometaphase I

Fragments of nuclear membrane

Bivalent

2 Nuclear membrane completely vesiculates, and bivalents become attached to kinetochore microtubules.

(c) Metaphase I

Metaphase plate

3 Bivalents align along the metaphase plate.

Meiosis II

(f) Prophase II

6 Sister chromatids condense and the spindle starts to form. Nuclear membrane begins to vesiculate.

(g) Prometaphase II

7 Nuclear membrane completely vesiculates. Sister chromatids attach to spindle via kinetochore microtubules.

(h) Metaphase II

8 Sister chromatids align along the metaphase plate.

Now that we have an understanding of bivalent formation and crossing over, we are ready to consider the phases of meiosis (**Figure 15.17**). These simplified diagrams depict a diploid cell ($2n$) that contains a total of six chromosomes (as in our look at mitosis in Figure 15.13). Prior to meiosis, the chromosomes are replicated in S phase to produce pairs of sister chromatids. This single replication event is then followed by the two sequential cell divisions called meiosis I and II. Like mitosis, each of these is subdivided into prophase, prometaphase, metaphase, anaphase, and telophase.

The sorting that occurs during **meiosis I** separates homologues from each other (Figure 15.17). In prophase I, the replicated chromosomes condense and the homologous chromosomes form bivalents as the nuclear membrane starts to vesiculate (fragment into small vesicles). In prometaphase I, the spindle apparatus is complete, and the chromatids are attached to kinetochore microtubules. At metaphase I, the bivalents are organized along the metaphase plate. However, notice how this pattern of alignment is strikingly different from that observed during mitosis (see Figure 15.13d). In particular, the sister chromatids are aligned in a double row rather than a single row (as in mitosis). Furthermore, the arrangement of sister chromatids within this double row is random with regard to the (blue and red) homologues. In Figure 15.17c, one of the red homologues is to the left of the metaphase plate and the other two are to the right, while two of the blue homologues are to the left of the metaphase plate and other one is to the right.

In other cells, homologues could be arranged differently along the metaphase plate (for example, three blues to the left and none to the right, or none to the left and three to the right). Because eukaryotic species typically have many chromosomes per set, homologues can be randomly aligned along the metaphase plate in a variety of ways. For example, consider that humans have 23 chromosomes per set. The possible number of different, random alignments equals 2^n, where n equals the number of chromosomes per set. Thus, in humans, this equals 2^{23}, or over 8 million possibilities. Because the homologues are genetically similar but not identical, we see from this calculation that the random alignment of homologous chromosomes provides a mechanism to promote a vast amount of genetic diversity among the resulting haploid cells. When meiosis is complete, it is very unlikely that any two human gametes will have the same combination of homologous chromosomes.

The segregation of homologues occurs during anaphase I (Figure 15.17d). The connections between bivalents break, but not the connections that hold sister chromatids together. Each joined pair of chromatids migrates to one pole, and the homologous pair of chromatids moves to the opposite pole, both pulled by kinetochore microtubules. Finally, at telophase I, the sister chromatids have reached their

(d) Anaphase I

4 Homologous chromosomes separate and move toward opposite poles.

(e) Telophase I and cytokinesis

Cleavage furrow

5 Nuclear membranes re-form and the chromosomes decondense. The 2 cells are separated by a cleavage furrow.

(i) Anaphase II

9 Sister chromatids separate and individual chromosomes move toward poles as kinetochore microtubules shorten. Polar microtubules lengthen and push poles apart.

(j) Telophase II and cytokinesis

Four haploid cells

10 Chromosomes decondense and nuclear membranes re-form. Cleavage furrow separates the 2 cells into 4 cells.

Table 15.1	A Comparison of Mitosis, Meiosis I, and Meiosis II		
Event	**Mitosis**	**Meiosis I**	**Meiosis II**
Synapsis during prophase:	No	Yes, bivalents are formed	No
Crossing over during prophase:	Rarely	Commonly	Rarely
Attachment to poles at prometaphase:	A pair of sister chromatids is attached to both poles.	A pair of sister chromatids is attached to just one pole.	A pair of sister chromatids is attached to both poles.
Alignment along the metaphase plate:	Sister chromatids	Bivalents	Sister chromatids
Type of separation at anaphase:	Sister chromatids separate. A single chromatid, now called a chromosome, moves to each pole.	Bivalents separate. A pair of sister chromatids moves to each pole.	Sister chromatids separate. A single chromatid, now called a chromosome, moves to each pole.

respective poles, and they then decondense. The nuclear membranes now re-form to produce two separate nuclei. If we consider the end result of meiosis I, we see that two cells are produced, each with three pairs of sister chromatids; it is a reduction division. The original diploid cell had its chromosomes in homologous pairs, while the two cells produced at the end of meiosis I are considered haploid—they do not have pairs of homologous chromosomes.

The Second Meiotic Division, Meiosis II, Separates Sister Chromatids

Meiosis I is followed by cytokinesis and then **meiosis II** (see Figure 15.17). An S phase does not occur between meiosis I and meiosis II. The sorting events of meiosis II are similar to those of mitosis, but the starting point is different. For a diploid cell with six chromosomes, mitosis begins with 12 chromatids that are joined as six pairs of sister chromatids (see Figure 15.13). By comparison, the two cells that begin meiosis II each have six chromatids that are joined as three pairs of sister chromatids. Otherwise, the steps that occur during prophase, prometaphase, metaphase, anaphase, and telophase of meiosis II are analogous to a mitotic division. Sister chromatids are separated during anaphase II, unlike anaphase I in which bivalents are separated.

Changes in a Few Key Steps in Meiosis and Mitosis Account for the Different Outcomes of These Two Processes

If we compare the outcome of meiosis with that of mitosis, the results are quite different. Mitosis produces two diploid daughter cells that are genetically identical. In our previous example shown in Figure 15.13, the starting cell had six chromosomes (three homologous pairs of chromosomes) and both daughter cells had copies of the same six chromosomes. By comparison, meiosis reduces the number of sets of chromosomes. In the

example shown in Figure 15.17, the starting cell also had six chromosomes while the four daughter cells had only three chromosomes. But the daughter cells did not contain a random mix of three chromosomes. Each haploid daughter cell contained one complete set of chromosomes, while the original diploid mother cell had two complete sets.

Table 15.1 emphasizes the differences between certain key steps in meiosis and mitosis that account for the different outcomes of these two processes. During prophase of meiosis I, the homologues synapse to form bivalents. This explains why crossing over occurs commonly during meiosis, but rarely during mitosis. During prometaphase of mitosis and meiosis II, pairs of sister chromatids are attached to both poles. In contrast, during meiosis I, each pair of sister chromatids (within a bivalent) is attached to a single pole. This affects their alignment during metaphase. Bivalents align along the metaphase plate during metaphase of meiosis I, whereas sister chromatids align along the metaphase plate during metaphase of mitosis and meiosis II. At anaphase of meiosis I, the homologues separate, while the sister chromatids remain together. In contrast, sister chromatid separation occurs during anaphase of mitosis and meiosis II. Taken together, the steps of meiosis result in a process in which two sequential cell divisions create four haploid cells, while the steps of mitosis create two diploid cells.

Mitosis, Meiosis, and Fertilization Allow Sexually Reproducing Species to Produce Haploid and Diploid Cells at Different Times in Their Life Cycles

Let's now turn our attention to the relationship between mitosis, meiosis, and sexual reproduction in animals, plants, fungi, and protists. For any given species, the sequence of events that produces another generation of organisms is known as a **life cycle**. For sexually reproducing organisms, this involves an alternation between haploid cells or organisms and diploid cells or organisms (**Figure 15.18**).

1 Meiosis occurs in cells within testes or ovaries to produce gametes (1*n*).

2 During fertilization, sperm and egg unite to create a zygote.

Sperm (1*n*)
Egg (1*n*)

Diploid adult (2*n*)

Somatic cells are diploid (2*n*).

Zygote (2*n*)

3 Repeated mitotic cell divisions produce a diploid multicellular organism.

(a) Animal life cycle (mouse)—diploid dominant

Key
Diploid (2*n*)
Haploid (1*n*)

1 Certain haploid cells act as reproductive cells.

2 Haploid reproductive cells unite to form a dipoid zygote (2*n*).

Multicellular haploid (1*n*) organism

Reproductive cells (1*n*)

Somatic cells are haploid (1*n*).

Spore (1*n*)

Diploid zygote (2*n*)

4 Repeated mitotic cell divisions produce a haploid multicellular organism.

3 Meiosis of the zygote produces 4 haploid spores.

(b) Fungal life cycle (bread mold)—haploid dominant

1 Certain cells in sporophyte undergo meiosis to produce haploid spores.

2 Repeated mitotic cell divisions produce a haploid multicellular organism.

Diploid plant— sporophyte (2*n*)

Young leaf (fiddlehead)

Sori

Haploid spores (1*n*)

Somatic cells are diploid (2*n*).

Haploid plant— gametophyte (1*n*)

Somatic cells are haploid (1*n*).

Diploid zygote (2*n*)

Sperm (1*n*)

3 Certain cells within the gametophyte differentiate into gametes.

5 Repeated mitotic cell divisions produce a diploid multicellular organism.

Egg (1*n*)

4 Two gametes unite during fertilization to create a diploid zygote.

Figure 15.18 **A comparison of three types of sexual life cycles.**

Biological inquiry: What is the main reason for meiosis in animals? What is the reason for mitosis in animals?

(c) Plant life cycle (fern)—alternation of generations

Most species of animals are diploid, and their haploid gametes are considered to be a specialized type of cell. For this reason, animals are viewed as **diploid-dominant species** (Figure 15.18a). By comparison, many fungi and some protists are just the opposite, they are **haploid-dominant species** (Figure 15.18b). In fungi, such as bread mold, the multicellular organism is haploid (1*n*). Haploid fungal cells are most commonly produced by mitosis. During sexual reproduction, haploid cells unite to create a diploid zygote, which then immediately proceeds through meiosis to create four haploid cells called spores. Each spore goes through mitotic cellular divisions to create a multicellular haploid organism.

Plants and some algae have life cycles that are intermediate between the extreme cases of diploid or haploid dominance. Such species exhibit an **alternation of generations** (Figure 15.18c). The species alternate between multicellular diploid organisms,

called **sporophytes**, and multicellular haploid organisms, called **gametophytes**. Meiosis in certain cells within the sporophyte produces haploid spores, which divide by mitosis to produce the gametophyte. Particular cells within the gametophyte differentiate into haploid gametes. Fertilization occurs between two gametes, producing a diploid zygote that then undergoes repeated mitotic cell divisions to create a sporophyte.

Among different plant species, the relative sizes of the haploid and diploid organisms vary greatly. In bryophytes, such as mosses, the haploid gametophyte is a visible multicellular organism, while the diploid sporophyte is tiny and survives within the haploid organism. In other plants, such as ferns, as shown in Figure 15.18c, both the diploid sporophyte and haploid gametophyte can grow independently. The sporophyte is considerably larger and is the organism that we commonly think of as a fern.

In seed-bearing plants, such as roses and oak trees, the diploid sporophyte is the large multicellular plant, while the gametophyte is composed of only a few cells and is formed within the sporophyte.

When comparing animals, plants, and fungi, it's interesting to consider how gametes are made. Animals produce gametes by meiosis. In contrast, plants and fungi produce reproductive cells by mitosis. The gametophyte of plants is a haploid multicellular organism that is created by mitotic cellular divisions of a haploid spore. Within the multicellular gametophyte, certain cells become specialized as gametes.

15.4 Variation in Chromosome Structure and Number

In the three previous sections of this chapter, we have examined two important features of chromosomes. First, we considered their structure, and second, we explored two sorting processes that determine the chromosome number following cell division. In this section, we will examine how the structures and numbers of chromosomes can vary between different species and within the same species.

The study of chromosomal variation is important for several reasons. Geneticists have discovered that variations in chromosome structure and number can have major effects on the characteristics of an organism. For example, we now know that several human genetic diseases are caused by such changes. In addition, changes in chromosome structure and number have been an important force in the evolution of new species. This topic is considered in Chapter 25.

Chromosome variation can be viewed in two ways. On relatively rare occasions, the structure or number of chromosomes can be changed so that an individual is different from most other members of the same species. This is generally viewed as an abnormality. Alternatively, some types of variation are normal. For example, the structure and number of chromosomes among different species tend to show wide variation. In this section, we will examine both abnormal and normal types of variation. Let's begin with natural (normal) variation.

Natural Variation Exists in Chromosome Structure and Number

Before we begin to examine chromosome variation, we need to have a reference point for a normal set of chromosomes. To determine what the normal chromosomes of a species look like, a cytogeneticist microscopically examines the chromosomes from several members of the species. Chromosome composition within a given species tends to remain relatively constant. In most cases, normal individuals of the same species will have the same number and types of chromosomes.

For example, as mentioned previously, the normal chromosome composition of human cells is two sets of 23 chromosomes for a total of 46. Other diploid species may have different numbers of chromosomes. The dog has 78 chromosomes (39 per set), the fruit fly has 8 chromosomes (4 per set), and the tomato has 24 chromosomes (12 per set). When comparing distantly related species, such as humans and fruit flies, major differences in chromosomal composition are observed.

The chromosomes of a given species can also vary considerably in size and shape. Cytogeneticists have various ways to classify and identify chromosomes in their metaphase form. The three most commonly used features are size, location of the centromere, and banding patterns that are revealed when the chromosomes are treated with stains. Based on centromere location, each chromosome is classified as **metacentric** (near the middle), **submetacentric** (off center), **acrocentric** (near one end), or **telocentric** (at the end) (**Figure 15.19a**). Because the centromere is not exactly in the center of a chromosome, each chromosome has a short arm and a long arm. The short arm is designated with the letter *p* (for the French *petite*), while the long arm is designated with the letter *q*. In the case of telocentric chromosomes, the short arm may be nearly nonexistent. When preparing a karyotype, the chromosomes are aligned with the short arms on top and the long arms on the bottom.

Because different chromosomes often have similar sizes and centromeric locations, cytogeneticists must use additional methods to accurately identify each type of chromosome within a karyotype. For detailed identification, chromosomes are treated with stains to produce characteristic banding patterns. Cytogeneticists use several different staining procedures to identify specific chromosomes. An example is Giemsa stain, which produces **G banding** (**Figure 15.19b**). The alternating pattern of G bands is unique for each type of chromosome.

The banding pattern of eukaryotic chromosomes is useful in two ways. First, individual chromosomes can be distinguished from each other, even if they have similar sizes and centromeric locations. Also, banding patterns are used to detect changes in chromosome structure that occur as a result of mutation.

Mutations Can Alter Chromosome Structure

Let's now consider how the structures of normal chromosomes can be modified by mutations. These chromosomal mutations are categorized as deficiencies, duplications, inversions, and translocations (**Figure 15.20**).

Deficiencies and duplications are changes in the total amount of genetic material in a single chromosome. When a **deficiency** occurs, a segment of chromosomal material is missing. In other words, the affected chromosome is deficient in a significant amount of genetic material. The term **deletion** is also used to describe a missing region of a chromosome. In contrast, a **duplication** occurs when a section of a chromosome occurs two or more times in a row.

The consequences of a deficiency or duplication depend on their size and whether they include genes or portions of genes that are vital to the development of the organism. When deficiencies or duplications have an effect, they are usually detrimental. Larger changes in the amount of genetic material

(a) A comparison of centromeric locations

Metacentric   Submetacentric   Acrocentric   Telocentric

A pair of sister chromatids

Centromere

G-band

Homologous chromosomes each existing as a pair of sister chromatids

(b) Giemsa staining of human chromosomes

Figure 15.19 Features of normal metaphase chromosomes. **(a)** A comparison of centromeric locations. **(b)** Illustrations of human chromosomes that have been stained with Giemsa to show G bands.

tend to be more harmful because more genes are missing or duplicated.

Inversions and translocations are chromosomal rearrangements. An **inversion** is a change in the direction of the genetic material along a single chromosome. When a segment of one chromosome has been inverted, the order of G bands is opposite to that of a normal chromosome (see Figure 15.20). A **translocation** occurs when one segment of a chromosome becomes attached to a different chromosome. In a **simple translocation**, a single piece of chromosome is attached to another chromosome. In a **reciprocal translocation**, two different types of chromosomes exchange pieces, thereby producing two abnormal chromosomes carrying translocations.

Deficiency

Effect: Removes a segment of chromosome

(a) Deficiency (deletion)

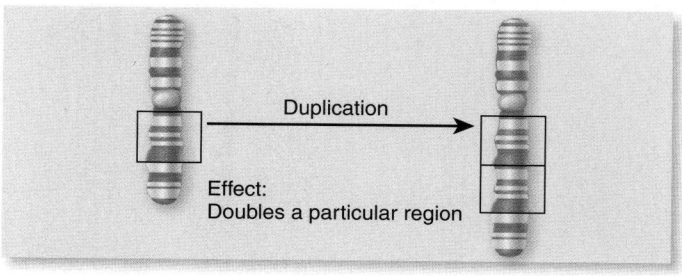

Duplication

Effect: Doubles a particular region

(b) Duplication

Inversion

Effect: Flips a region to the opposite orientation

(c) Inversion

Simple translocation

Effect: Moves a segment of 1 chromosome to a different chromosome

(d) Simple translocation

Reciprocal translocation

Effect: Exchange of pieces between 2 different chromosomes

(e) Reciprocal translocation

Figure 15.20 Types of changes in chromosome structure.

(a) Normal fruit fly chromosome composition

(b) Polyploidy

(c) Aneuploidy

Figure 15.21 **Types of variation in chromosome number.**
(a) The normal diploid number of chromosomes in *Drosophila*.
The X chromosome is also called chromosome 1. Examples of
chromosomes of **(b)** polyploid flies and **(c)** aneuploid flies.

Variation Occurs in the Number of Individual Chromosomes and in the Number of Chromosome Sets

Let's now turn our attention to changes in chromosome number. Variations in chromosome number can be categorized in two ways: variation in the number of sets of chromosomes, and variation in the number of particular chromosomes within a set. The suffix "ploid" or "ploidy" refers to a complete set of chromosomes. Organisms that are **euploid** (the prefix "eu-" means true) have a chromosome number that is viewed as the normal number. For example, in a species that is diploid, a euploid organism would have two sets of chromosomes in its somatic cells. In *Drosophila melanogaster*, for example, a normal indi-

vidual has eight chromosomes. The species is diploid, having two sets of four chromosomes each (**Figure 15.21a**). Organisms can vary with regard to the number of sets of chromosomes they have. For example, on rare occasions, an abnormal fruit fly can be produced with 12 chromosomes, containing three sets of 4 chromosomes each (**Figure 15.21b**). Organisms with three or more sets of chromosomes are called **polyploid**. A diploid organism is referred to as $2n$, a **triploid** organism as $3n$, a **tetraploid** organism as $4n$, and so forth.

The second way that chromosome number can vary is a phenomenon called **aneuploidy**. This refers to an alteration in the number of particular chromosomes, so that the total number of chromosomes is not an exact multiple of a set. For example, an abnormal fruit fly could contain nine chromosomes instead of eight because it had three copies of chromosome 2 instead of the normal two copies (**Figure 15.21c**). Such an animal is said to have trisomy 2 or to be **trisomic**. Instead of being perfectly diploid, a trisomic animal is $2n + 1$. By comparison, a fruit fly could be lacking a single chromosome, such as chromosome 1, and contain a total of seven chromosomes ($2n - 1$). This animal is **monosomic** and would be described as having monosomy 1.

Nondisjunction and Interspecies Breeding Are Common Mechanisms That Change Chromosome Number

Variations in chromosome number are fairly widespread and have a significant impact on the characteristics of plants and animals. For these reasons, researchers have wanted to understand the mechanisms that cause these variations. In some cases, a change in chromosome number is the result of the abnormal sorting of chromosomes during cell division. The term **nondisjunction** refers to an event in which the chromosomes do not sort properly during cell division. **Meiotic nondisjunction** can occur during meiosis I or meiosis II and produce haploid cells that have too many or too few chromosomes (**Figure 15.22**). If such a gamete fuses with a normal gamete during fertilization, the resulting organism will be aneuploid and will have an abnormal chromosomal composition in all of its cells.

Another mechanism that leads to variation in chromosome number is interspecies breeding. An **alloploid** organism contains a least one set of chromosomes from two or more different species (**Figure 15.23a**). This term refers to the occurrence of chromosome sets (*ploidy*) from the genomes of different (*allo-*) species. An alloploid that has only one set of chromosomes from two different species is an **allodiploid**. Species that are close evolutionary relatives are most likely to breed and produce allodiploid offspring. For example, closely related species of grasses may interbreed to produce allodiploids. When an organism contains two or more complete sets of chromosomes from two or more different species, such an organism is called an **allopolyploid**. An **allotetraploid** is a type of allopolyploid that contains two complete sets of chromosomes from two species for a total of four sets (**Figure 15.23b**).

Figure 15.22 **Nondisjunction during the formation of haploid cells.**   The cell on the *left* is diploid, containing three pairs of chromosomes. During the formation of haploid cells, one copy of each chromosome should be sorted into each cell. Nondisjunction, however, can produce a cell with too many or too few chromosomes.

Biological inquiry: Could nondisjunction produce a polyploid individual?

(a) Alloploidy (allodiploid)

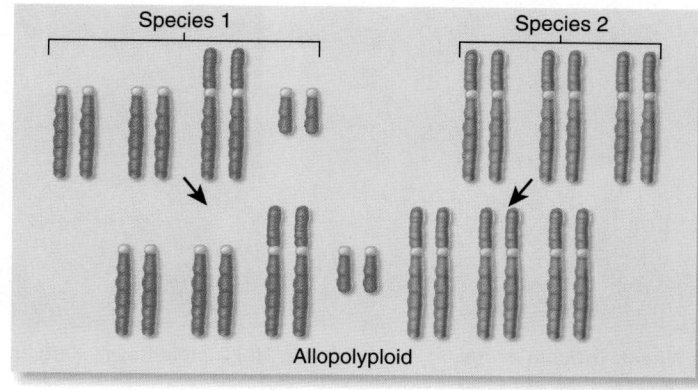

(b) Allopolyploidy (allotetraploid)

Figure 15.23   **A comparison of alloploidy and allopolyploidy.**

Changes in Chromosome Number Have Important Consequences

Let's now consider how changes in chromosome number may impact the characteristics of animals and plants. In many cases, animals do not tolerate deviations from diploidy well. For example, polyploidy in mammals is generally a lethal condition. However, a few cases of naturally occurring variations from diploidy do occur in animals. Male bees, which are called drones, contain a single set of chromosomes. They are produced from unfertilized eggs. By comparison, female bees are diploid. A few examples of vertebrate polyploid animals have been discovered. Interestingly, on rare occasions, animals that are morphologically very similar to each other can be found as a diploid species as well as a separate polyploid species. This situation occurs among certain amphibians and reptiles. **Figure 15.24** shows photographs of a diploid and a tetraploid frog. As you can see, they look very similar to each other. Their differences in chromosome number can be revealed only by a microscopic examination of the chromosomes in the somatic cells of the animals.

In contrast to animals, plants commonly exhibit polyploidy. Among ferns and flowering plants, about 30–35% of species are polyploid. Polyploidy is also important in agriculture. In many instances, polyploid strains of plants display characteristics that are helpful to humans. They are often larger in size and more robust. These traits are clearly advantageous in the production of food. Many of the fruits and grains we eat are produced from polyploid plants. For example, the species of wheat that we use to make bread, *Triticum aestivum*, is an allohexaploid (containing six sets of chromosomes) that arose from the union of diploid genomes from three closely related species (**Figure 15.25a**). During the course of its cultivation, two diploid species must have interbred to create an allotetraploid, and then a third species interbred with the allotetraploid to create an allohexaploid. Plant polyploids tend to exhibit a greater adaptability, which allows them to withstand harsher environmental conditions. Polyploid ornamental plants commonly produce larger flowers than their diploid counterparts (**Figure 15.25b**).

Although polyploidy is often beneficial in plants, aneuploidy in all eukaryotic species usually has detrimental consequences on the characteristics of an organism. To understand why, consider the relationship between gene expression and chromosome number. For many, but not all, genes, the level of gene expression is correlated with the number of genes per cell.

(a) *Hyla chrysorelis*

(b) *Hyla versicolor*

Figure 15.24 Differences in chromosome number in two closely related frog species.
The frog in **(a)** is diploid, whereas the frog in **(b)** is tetraploid. These frogs are in the act of performing their mating calls, which is why the skin under their mouths is protruding as a large bubble.

(a) Wheat, *Triticum aestivum*

(b) Diploid daylily (left) and tetraploid (right)

Figure 15.25 Examples of polyploid plants.
(a) Cultivated wheat, *Triticum aestivum*, is an allohexaploid. It was derived from three different diploid species of grasses that originally were found in the Middle East and were cultivated by ancient farmers in this region. Modern varieties of wheat have been produced from this allohexaploid species. **(b)** Differences in euploidy may exist in two closely related daylily species. The flower stems on the *left* are diploid, whereas the stems with the larger flowers on the *right* are tetraploid.

Compared to a diploid cell, if a gene is carried on a chromosome that is present in three copies instead of two, approximately 150% of the normal amount of gene product will be made. Alternatively, if only one copy of that gene is present due to a missing chromosome, only 50% of the gene product is usually made. Therefore, in trisomic and monosomic individuals, an imbalance in the level of gene expression occurs. This imbalance interferes with the proper functioning of cells.

One important reason that geneticists are so interested in aneuploidy is its relationship to certain inherited disorders in humans. Even though most people are born with a normal number of chromosomes, alterations in chromosome number occur at a surprising frequency during gamete formation. About 5–10% of all fertilized human eggs result in an embryo with an abnormality in chromosome number. In most cases, these abnormal embryos do not develop properly and result in a spontaneous abortion very early in pregnancy. Approximately 50% of all spontaneous abortions are due to alterations in chromosome number.

In some cases, an abnormality in chromosome number produces an offspring that can survive. Several human disorders are the result of abnormalities in chromosome number. The most common are trisomies of chromosomes 21, 18, or 13, or abnormalities in the number of the sex chromosomes (**Table 15.2**). Most of the known trisomies involve chromosomes that are relatively small, so they carry fewer genes. Trisomies of the other human chromosomes and most monosomies are presumed to be lethal and have been found in spontaneously aborted embryos and fetuses.

Table 15.2	Aneuploid Conditions in Humans		
Condition	**Frequency (# of live births)**	**Syndrome**	**Characteristics**
Autosomal:			
Trisomy 21	1/800	Down	Mental retardation, abnormal pattern of palm creases, slanted eyes, flattened face, short stature
Trisomy 18	1/6,000	Edward	Mental and physical retardation, facial abnormalities, extreme muscle tone, early death
Trisomy 13	1/15,000	Patau	Mental and physical retardation, wide variety of defects in organs, large triangular nose, early death
Sex chromosomal:			
XXY	1/1,000 (males)	Klinefelter	Sexual immaturity (no sperm), breast swelling (males)
XYY	1/1,000 (males)	Jacobs	Tall
XXX	1/1,500 (females)	Triple X	Tall and thin, menstrual irregularity
XO	1/5,000 (females)	Turner	Short stature, webbed neck, sexually undeveloped

Human abnormalities in chromosome number are influenced by the age of the parents. Older parents are more likely to produce children with abnormalities in chromosome number, because meiotic nondisjunction is more likely to occur in older cells. **Down syndrome**, which was first described by the English physician John Langdon Down in 1866, provides an example. This disorder is caused by the inheritance of three copies of chromo-

some 21 (Table 15.2). The incidence of Down syndrome rises with the age of either parent. In males, however, the rise occurs relatively late in life, usually past the age when most men have children. By comparison, the likelihood of having a child with Down syndrome rises dramatically during the later reproductive ages of women.

CHAPTER SUMMARY

15.1 Molecular Structure of Eukaryotic Chromosomes

- Chromosomes are structures in living cells that carry the genetic material.
- Chromatin is the name given to the DNA-protein complex that makes up chromosomes.
- In eukaryotic chromosomes, the DNA is wrapped around histone proteins to form nucleosomes. (Figure 15.1)
- Noll tested the nucleosome model for chromatin structure by digesting chromosomes with DNase-I and showing that the DNA was obtained in multiples of 200 bp. (Figure 15.2)
- Nucleosomes are further compacted into 30-nm fibers because the linker regions are variably twisted and bent into a zigzag pattern. (Figure 15.3)
- A third level of compaction of eukaryotic chromosomes involves the formation of radial loop domains in which the bases of 30-nm fibers are anchored to a network of proteins called the nuclear matrix. This level of compaction is called euchromatin. In heterochromatin, the loops are even more closely packed together. (Figure 15.4)
- Within the nucleus of a nondividing cell, each chromosome occupies its own nonoverlapping chromosome territory. (Figure 15.5)
- Chromosome compaction to produce a metaphase chromosome also involves the conversion of all euchromatin into heterochromatin. (Figure 15.6)
- In a metaphase chromosome, the DNA loops are anchored to a protein scaffold that is formed from the nuclear matrix. (Figure 15.7)
- Gene expression is controlled by the level of compaction of the chromatin. The pattern of covalent modification of the amino terminal tails of histone proteins, also called the histone code, is one way to control the level of compaction. (Figure 15.8).

15.2 Mitotic Cell Division

- The process of mitosis involves the sorting of chromosomes to produce two nuclei with the same number and types of chromosomes.
- Cytogeneticists examine cells microscopically to determine their chromosome composition. A micrograph that shows the alignment of chromosomes from a given cell is called a karyotype. (Figure 15.9)
- Diploid cells have two sets of chromosomes while haploid cells have only one set.
- During S phase, eukaryotic chromosomes are replicated to produce a pair of identical sister chromatids that remain attached to each other. (Figure 15.10)

- The cell cycle is a series of stages needed for cell division. G_1, S, and G_2 are known as interphase. Mitosis and cytokinesis occur during M phase. (Figure 15.11)
- The mitotic spindle is composed of astral, kinetochore, and polar microtubules. The spindle organizes the process of cell division and plays a central role in chromosome sorting. (Figure 15.12)
- Mitosis occurs in five phases called prophase, prometaphase, metaphase, anaphase, and telophase. During prophase, the chromosomes condense and the nuclear membrane fragments. The spindle apparatus is completely formed by the end of prometaphase. During metaphase, the chromosomes are aligned in a single row along the metaphase plate. At anaphase, the sister chromatids separate from each other and move to opposite poles; the poles themselves also move farther apart. During telophase, the chromosomes decondense and the nuclear membranes re-form. (Figure 15.13)
- Cytokinesis in animal cells involves the formation of a cleavage furrow. In plant cells, two separated cells are produced by the formation of a cell plate. (Figure 15.14)

15.3 Meiosis and Sexual Reproduction

- The process of meiosis begins with a diploid cell and produces four haploid cells with one set of chromosomes each. (Figure 15.15)
- During prophase of meiosis, homologous pairs of sister chromosomes synapse, and crossing over occurs. After crossing over, chiasmata are observed. (Figure 15.16)
- Meiosis is divided into meiosis I and II, each composed of prophase, prometaphase, metaphase, anaphase, and telophase. During meiosis I, the homologues are separated to different cells, and during meiosis II the sister chromatids are separated to different cells. (Figure 15.17, Table 15.1)
- The life cycle of animals is diploid dominant, while many fungi and protists show a haploid dominant life cycle. Plants alternate between diploid and haploid forms. (Figure 15.18)

15.4 Variation in Chromosome Structure and Number

- Chromosomes are named metacentric, submetacentric, acrocentric, and telocentric according to their centromere location. Each type of chromosome can be uniquely identified by its banding pattern after staining. (Figure 15.19)
- Deficiencies, duplications, inversions, and translocations are different ways to alter chromosome structure. (Figure 15.20)
- A euploid organism has the correct number of chromosomes. A polyploid organism has three or more sets of chromosomes. An aneuploid organism has one too many (trisomy) or one too few (monosomy) chromosomes. (Figure 15.21)
- Nondisjunction and interspecies matings can alter chromosome number. An alloploid has chromosome sets from two or more different species. (Figures 15.22, 15.23)

- Polyploid animals are relatively rare, but polyploid plants are common and tend to be more robust than their diploid counterparts. (Figures 15.24, 15.25)
- Aneuploidy in humans is responsible for several types of human genetic diseases including Down syndrome. (Table 15.2)

TEST YOURSELF

1. In eukaryotic cells, chromosomes consist of
   a. DNA and RNA.
   b. DNA only.
   c. RNA and proteins.
   d. DNA and proteins.
   e. RNA only.

2. A nucleosome is
   a. a dark-staining body composed of RNA and proteins found in the nucleus.
   b. a protein that helps organize the structure of chromosomes.
   c. another word for a chromosome.
   d. a structure composed of eight histones wrapped with DNA.
   e. the short arm of a chromosome.

3. Compaction of chromosomes is not uniform. Some areas are more compact than others. Which of the following statements is true about the differing levels of organization in the eukaryotic chromosome?
   a. More compact areas of the chromosome are called euchromatin and contain genes that are not expressed.
   b. More compact areas of the chromosome are called heterochromatin and contain genes that are not expressed.
   c. More compact areas of the chromosome are called heterochromatin and contain genes that are expressed at high levels.
   d. More compact areas of the chromosome are called euchromatin and contain genes that are expressed at high levels.
   e. None of the above is correct.

4. Which of the following is a reason for mitotic cell division?
   a. reproduction
   b. gamete formation in animals
   c. multicellularity
   d. all of the above
   e. both a and c

5. A replicated chromosome is composed of
   a. two homologous chromosomes held together at the centromere.
   b. four sister chromatids held together at the centromere.
   c. two sister chromatids held together at the centromere.
   d. four homologous chromosomes held together at the centromere.
   e. one chromosome with a centromere.

6. Which of the following is not an event of anaphase of mitosis?
   a. The nuclear envelope breaks down.
   b. Sister chromatids separate.
   c. Kinetochore microtubules shorten, pulling the chromosomes to the pole.
   d. Polar microtubules push against each other, moving the poles farther apart.
   e. All of the above occur during anaphase.

7. A student is looking at cells under the microscope. The cells are from an organism that has a diploid number of 14. For one particular slide, the cell has seven replicated chromosomes aligned at the metaphase plate of the cell. Which of the following statements accurately describes this particular cell?
   a. The cell is in metaphase of mitosis.
   b. The cell is in metaphase I of meiosis.
   c. The cell is in metaphase II of meiosis.
   d. All of the above are correct.
   e. Both b and c are correct.

8. Which of the following statements accurately describes a difference between meiosis and mitosis?
   a. Mitosis produces diploid cells and meiosis produces haploid cells.
   b. Homologous chromosomes synapse during meiosis but do not synapse during mitosis.
   c. Crossing over commonly occurs during meiosis, but it does not commonly occur during mitosis.
   d. All of the above are correct.
   e. Both a and c are correct.

9. During crossing over in meiosis I,
   a. homologous chromosomes are not altered.
   b. homologous chromosomes exchange genetic material.
   c. chromosomal damage occurs.
   d. genetic information is lost.
   e. cytokinesis occurs.

10. Aneuploidy is the result of
    a. duplication of a region of a chromosome.
    b. inversion of a region of a chromosome.
    c. nondisjunction during meiosis.
    d. interspecies breeding.
    e. all of the above.

CONCEPTUAL QUESTIONS

1. Distinguish between heterochromatin and euchromatin.
2. Explain the histone code hypothesis.
3. Distinguish between homologous chromosomes and sister chromatids.

EXPERIMENTAL QUESTIONS

1. For the Feature Investigation by Markus Noll, what is DNase-I and why was it used in this experiment?
2. Noll hypothesized that DNA fragments of 200 bp would support the beads-on-a-string model of nucleosome structure for this particular species. Explain the rationale for this particular fragment length. The experiment consisted of three assays with differing concentrations of DNase-I. How did the different concentrations of DNase-I affect the results?
3. Explain the results of the experiment shown in Figure 15.7b.

COLLABORATIVE QUESTIONS

1. Discuss several variations in chromosome structure and number.
2. What is crossing over, when does it occur, and what is its significance?

---

www.brookerbiology.com

This website includes answers to the Biological Inquiry questions found in the figure legends and all end-of-chapter questions.

16

SIMPLE PATTERNS OF INHERITANCE

CHAPTER OUTLINE

The pea plant, *Pisum sativum*, studied by Mendel.

Long before people knew anything about cells or chromosomes, they observed patterns of heredity and speculated about them. The ancient Greek physician Hippocrates, famous for his authorship of the physician's oath, provided the first known explanation for the transmission of hereditary traits (ca. 400 BCE). He suggested that "seeds" produced by all parts of the body are collected and transmitted to offspring at the time of conception, and that these seeds cause offspring to resemble their parents. This idea, known as **pangenesis**, influenced the thinking of scientists for many centuries.

The first systematic studies of genetic crosses were carried out by the plant breeder Joseph Kolreuter between 1761 and 1766. In crosses between different strains of tobacco plants, Kolreuter found that the offspring were usually intermediate in appearance between the two parents. He concluded that parents make equal genetic contributions to their offspring and that their genetic material blends together as it is passed to the next generation. This interpretation was consistent with the concept known as **blending inheritance**, which was widely accepted at that time. In the late 1700s, Jean-Baptiste Lamarck, a French naturalist, hypothesized that species change over the course of many generations by adapting to new environments. According to Lamarck, behavioral changes modify traits, and such modified traits were inherited by offspring. For example, an individual who became adept at archery would pass that skill to his or her offspring. Overall, the prevailing view prior to the 1800s was that hereditary traits were rather malleable and could change and blend over the course of one or two generations.

In the last chapter, we considered the process of cell division and how chromosomes are transmitted during mitosis and meiosis. Observations of these processes in the second half of the 19th century provided compelling evidence for **particulate inheritance**—the idea that the determinants of hereditary traits are transmitted intact from one generation to the next. Remarkably, this idea was first proposed in the 1860s by a researcher who knew nothing about chromosomes (**Figure 16.1**). Gregor Mendel, remembered today as the "father of genetics," used statistical analysis of carefully designed breeding experiments to arrive at the concept of a gene. Forty years later, through the convergence of Mendel's work and that of cell biologists, this concept became the foundation of the modern science of genetics.

In this chapter we will consider inheritance patterns, and how the transmission of genes is related to the transmission of chromosomes. In the first section, we consider the fundamental genetic patterns known as Mendelian inheritance and the relationship of these patterns to the behavior of chromosomes during meiosis. In the second section, we examine the distinctive inheritance patterns of genes located on the X chromosome, paying special attention to the work of Thomas Hunt Morgan, whose investigation of these patterns confirmed that genes are on chromosomes. Finally, drawing on what you have already learned about genes and their expression, we will discuss the molecular basis of Mendelian inheritance and its variations.

16.1 Mendel's Laws and the Chromosome Theory of Inheritance

Gregor Johann Mendel (1822–1884) grew up on a small farm in northern Moravia, then a part of the Austrian Empire and now in the Czech Republic. At the age of 21 he entered the Augustinian monastery of St. Thomas in Brno, and he was ordained as a priest in 1847. Mendel then worked for a short time as a substitute teacher, but to continue teaching he needed a license.

Figure 16.1 Gregor Johann Mendel, the father of genetics.

Surprisingly, he failed the licensing exam due to poor answers in physics and natural history, so he enrolled at the University of Vienna to expand his knowledge in these two areas. Mendel's training in physics and mathematics taught him to perceive the world as an orderly place, governed by natural laws that could be stated as simple mathematical relationships.

In 1856, Mendel began his historic studies on pea plants. For eight years, he analyzed thousands of pea plants that he grew on a small plot in his monastery garden. He kept meticulously accurate records that included quantitative data concerning the outcome of his studies. He published his work, entitled "Experiments on Plant Hybrids," in 1866. This paper was largely ignored by scientists at that time, partly because of its title and because it was published in a rather obscure journal (*The Proceedings of the Brünn Society of Natural History*). Also, Mendel was clearly ahead of his time. During this period, biology had not yet become a quantitative, experimental science. In addition, the behavior of chromosomes during mitosis and meiosis, which provides a framework for understanding inheritance patterns, had yet to be studied. Prior to his death in 1884, Mendel reflected, "My scientific work has brought me a great deal of satisfaction and I am convinced that it will be appreciated before long by the whole world." Sixteen years later, in 1900, Mendel's work was independently rediscovered by three biologists with an interest in plant genetics: Hugo de Vries of Holland, Carl Correns of Germany, and Erich von Tschermak of Austria. Within a few years, the impact of Mendel's studies was felt around the world.

In this section, we will examine Mendel's experiments and how they led to the formulation of the basic genetic principles we

call Mendel's laws. We will see that these principles apply not only to the pea plants Mendel studied but also to a wide variety of sexually reproducing organisms, including humans. Next, we will consider how the study of chromosomes in the late 19th and early 20th centuries provided a physical explanation for Mendel's laws. At the end of the section, we will briefly examine the laws of probability on which Mendel based his analysis and how they are used to predict simple patterns of inheritance.

Analysis of Inheritance Patterns in Pea Plants Led Mendel to Formulate Two Basic Laws of Genetics

When two individuals with different characteristics are mated or crossed to each other, this is called a **hybridization** experiment, and the offspring are referred to as hybrids. For example, a hybridization experiment could involve a cross between a purple-flowered plant and a white-flowered plant. Mendel was particularly intrigued by the consistency with which offspring of such crosses showed characteristics of one or the other parent in successive generations. His intellectual foundation in physics and the natural sciences led him to consider that this regularity might be rooted in natural laws that could be expressed mathematically. To uncover these laws, he carried out quantitative hybridization experiments in which he carefully analyzed the numbers of offspring carrying specific traits. This analysis led him to formulate two fundamental genetic principles, known today as the law of segregation and the law of independent assortment.

Mendel chose the garden pea, *Pisum sativum*, to investigate the natural laws that govern plant hybrids. Several properties of this species were particularly advantageous for studying plant hybridization. First, it had many readily available varieties that differed in visible characteristics such as the appearance of seeds, pods, flowers, and stems. Such features of an organism are called **characters** or **traits**. **Figure 16.2** illustrates the seven characteristics that Mendel eventually chose to follow in his breeding experiments. Each of these traits was found in two variants. For example, one trait he followed was height, which had the variants known as tall and dwarf. Another was seed color, which had the variants yellow and green.

A second important feature of garden peas is that they are normally self-fertilizing. In **self-fertilization**, a female gamete is fertilized by a male gamete from the same plant. Like many flowering plants, peas have male and female sex organs in the same flower (**Figure 16.3**). Male gametes (sperm cells) are produced within pollen grains, which are formed in the male structures called stamens. Female gametes (egg cells) are produced in structures called ovules that form within an organ called an ovary. For fertilization to occur, a pollen grain must land on the receptacle called a stigma, enabling a sperm to migrate to an ovule and fuse with an egg cell. In peas, the stamens and the ovaries are enclosed by a modified petal, an arrangement that greatly favors self-fertilization. Self-fertilization makes it easy to produce plants that breed true for a given trait, meaning that the trait does not vary from generation to generation.

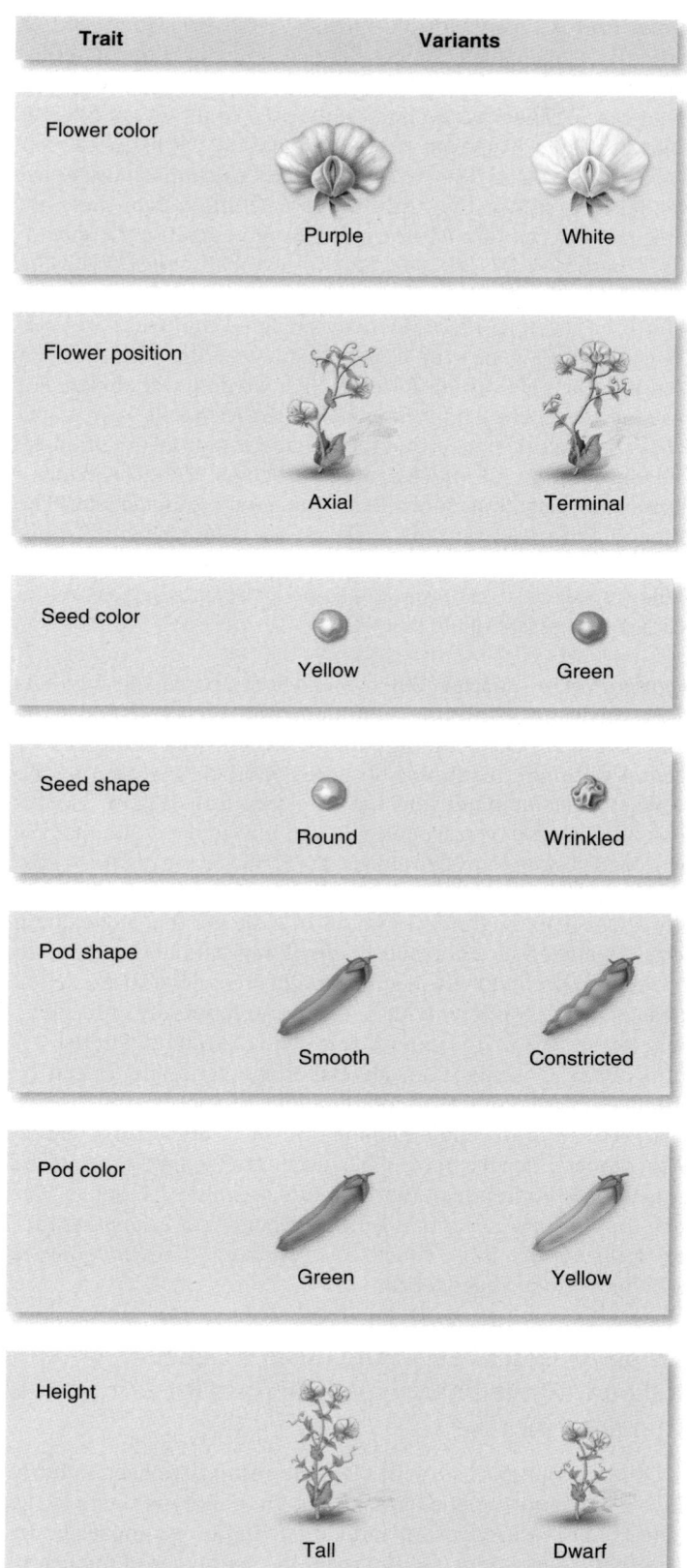

Trait	Variants	
Flower color	Purple	White
Flower position	Axial	Terminal
Seed color	Yellow	Green
Seed shape	Round	Wrinkled
Pod shape	Smooth	Constricted
Pod color	Green	Yellow
Height	Tall	Dwarf

Figure 16.2   The seven traits that Mendel studied.

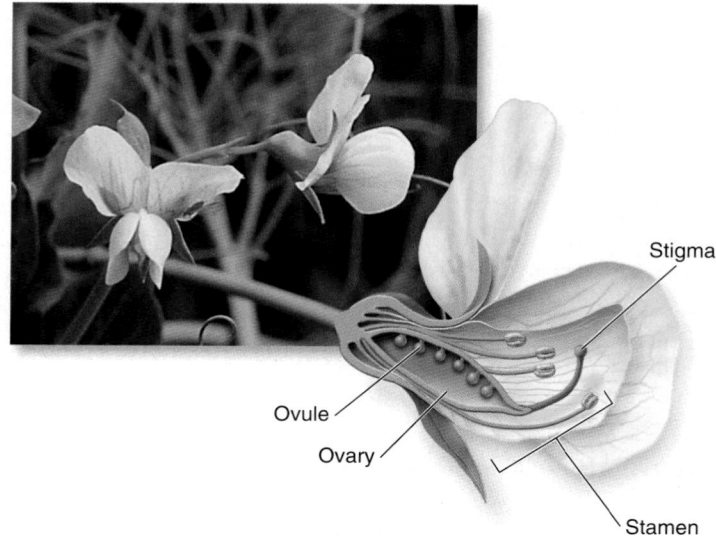

Figure 16.3   **Flower structure in pea plants.** The pea flower produces both male and female gametes. Sperm form in the pollen produced within the stamens; egg cells form in ovules within the ovary. A modified petal encloses the stamens and stigma, encouraging self-fertilization.

For example, if a pea plant with yellow seeds breeds true for seed color, all the plants that grow from these seeds will also produce yellow seeds. A variety that continues to exhibit the same trait after several generations of self-fertilization is called a **true-breeding line**. Prior to conducting the studies described in this chapter, Mendel had already established that the seven traits that he chose to study were true-breeding in the strains of pea plants he had obtained.

A third reason for using garden peas in hybridization experiments is the ease of making crosses: the flowers are quite large and easy to manipulate. In some cases Mendel wanted his pea plants to self-fertilize, but in others he wanted to cross plants that differed with respect to some trait, a process called **cross-fertilization** or hybridization. In garden peas, cross-fertilization requires placing pollen from one plant on the stigma of another plant's flower. Mendel's cross-fertilization procedure is shown in **Figure 16.4**. He would pry open an immature flower and remove the stamens before they produced pollen, so that the flower could not self-fertilize. He then used a paintbrush to transfer pollen from another plant to the stigma of the flower that had its stamens removed. In this way, Mendel was able to cross-fertilize any two of his true-breeding pea plants and obtain any type of hybrid he wanted.

By Following the Inheritance Pattern of Single Traits, Mendel's Work Revealed the Law of Segregation

Mendel began his investigations by studying the inheritance patterns of pea plants that differed with regard to a single trait.

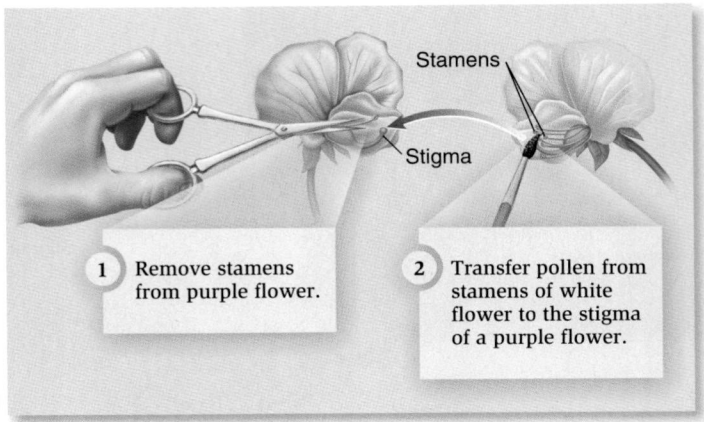

Figure 16.4 A procedure for cross-fertilizing pea plants.

A cross in which an experimenter follows the variants of only one trait is called a **single-factor cross**. As an example, we will consider a single-factor cross in which Mendel followed the tall and dwarf variants for height (**Figure 16.5**). The left-hand side of Figure 16.5a shows his experimental approach. The true-breeding parents are termed the **P generation** (parental generation), and a cross of these plants is called a P cross. The first-generation offspring of a P cross constitute the **F₁ generation** (first filial generation, from the Latin *filius*, son). When the true-breeding parents differ with regard to a single trait, their F₁ offspring are called single-trait hybrids, or **monohybrids**. When Mendel crossed true-breeding tall and dwarf plants, he observed that all plants of the F₁ generation were tall.

Next, Mendel followed the transmission of this trait for a second generation. To do so, he allowed the F₁ monohybrids to self-fertilize, producing a generation called the **F₂ generation** (second filial generation). The dwarf trait reappeared in the F₂ offspring: three-fourths of the plants were tall and one-fourth were dwarf. Mendel obtained similar results for each of the traits that he studied, as shown in the data of Figure 16.5b. A quantitative analysis of his data allowed Mendel to postulate three important ideas regarding the properties and transmission of these traits from parents to offspring:

Dominant and Recessive Traits Perhaps the most surprising outcome of Mendel's work was that the data argued strongly against the prevailing notion of a blending mechanism of heredity. In all seven cases, the F₁ generation displayed traits distinctly like one of the two parents rather than intermediate traits. Using genetic terms that Mendel originated, we describe the alternative traits as dominant and recessive. The term **dominant** describes the displayed trait, while the term **recessive** describes a trait that is masked by the presence of a dominant trait. Tall stems and green pods are examples of dominant traits; dwarf stems and yellow pods are examples of recessive traits. We say that tall is dominant over dwarf and green is dominant over yellow.

Genes and Alleles Mendel's results were consistent with a particulate mechanism of inheritance, in which the determinants of traits are inherited as unchanging, discrete units. In all seven cases, the recessive trait reappeared in the F₂ generation: some F₂ plants displayed the dominant trait, while a smaller proportion showed the recessive trait. This observation led Mendel to conclude that the genetic determinants of traits are "unit factors" that are passed intact from generation to generation. These unit factors are what we now call **genes** (from the Greek *genos*, birth), a term coined by the Danish botanist Wilhelm Johannsen in 1911. Mendel postulated that every individual carries two genes for a given trait, and that the gene for each trait has two variant forms, which we now call **alleles**. For example, the gene controlling height in Mendel's pea plants occurs in two variants, called the tall allele and the dwarf allele. The right-hand side of Figure 16.5a shows Mendel's conclusions, using genetic symbols (letters) that were adopted later. The letters T and t represent the alleles of the gene for plant height. By convention, the uppercase letter represents the dominant allele (in this case, tall) and the same letter in lowercase represents the recessive allele (dwarf).

Segregation of Alleles When Mendel compared the numbers of F₂ offspring exhibiting dominant and recessive traits, he noticed a recurring pattern. Although there was some experimental variation, he always observed approximately a 3:1 ratio between the dominant and the recessive trait (Figure 16.5b). This quantitative observation allowed him to conclude that the two copies of a gene carried by an F₁ plant **segregate** (separate) from each other, so that each sperm or egg carries only one allele. The diagram in **Figure 16.6** shows that segregation of the F₁ alleles should result in equal numbers of gametes carrying the dominant allele (T) and the recessive allele (t). If these gametes combine with one another randomly at fertilization, as shown in the figure, this would account for the 3:1 ratio of the F₂ generation. Note that the genotype Tt can be produced by two different combinations of alleles—the T allele can come from the male gamete and the t allele from the female gamete, or vice versa. This accounts for the fact that the Tt genotype is produced twice as often as either TT or tt. The idea that *the two copies of a gene segregate from each other during transmission from parent to offspring* is known today as Mendel's **law of segregation**.

Genotype Describes an Organism's Genetic Makeup, While Phenotype Describes Its Characteristics

To continue our discussion of Mendel's results, we need to introduce a few more genetic terms. The term **genotype** refers to the genetic composition of an individual. In the example shown earlier in Figure 16.5a, TT and tt are the genotypes of the P generation and Tt is the genotype of the F₁ generation. In a P cross, both parents are true-breeding plants, which means that each has identical copies of the gene for height. An individual with

(a) Mendel's protocol for making single-factor crosses

THE DATA			
P cross	**F₁ generation**	**F₂ generation**	**Ratio**
Purple × white flowers	All purple	705 purple, 224 white	3.15:1
Axial × terminal flowers	All axial	651 axial, 207 terminal	3.14:1
Yellow × green seeds	All yellow	6,022 yellow, 2,001 green	3.01:1
Round × wrinkled seeds	All round	5,474 round, 1,850 wrinkled	2.96:1
Smooth × constricted pods	All smooth	882 smooth, 299 constricted	2.95:1
Green × yellow pods	All green	428 green, 152 yellow	2.82:1
Tall × dwarf stem	All tall	787 tall, 277 dwarf	2.84:1
Total	**All dominant**	**14,949 dominant, 5,010 recessive**	**2.98:1**

(b) Mendel's observed data for all 7 traits

Figure 16.5   Mendel's analyses of single-factor crosses.

two identical copies of a gene is said to be **homozygous** with respect to that gene. In the specific P cross we are considering, the tall plant is homozygous for T and the dwarf plant is homozygous for t. In contrast, a **heterozygous** individual carries two different alleles of the same gene. Plants of the F₁ generation are heterozygous, with the genotype Tt, because every individual carries one copy of the tall allele and one copy of the dwarf allele.

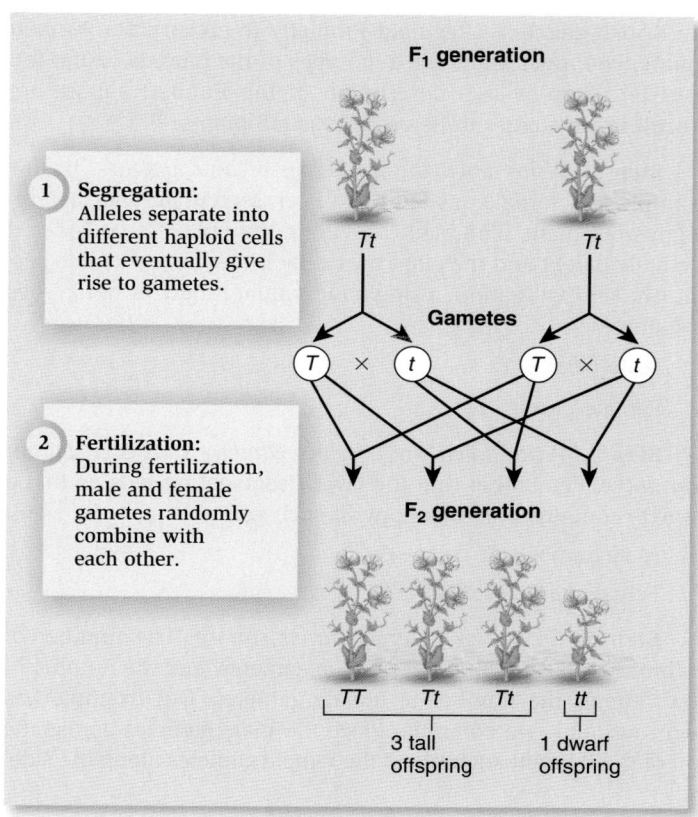

Figure 16.6   **How the law of segregation explains Mendel's observed ratios.**   The segregation of alleles in the F₁ generation gives rise to gametes that carry just one of the two alleles. These gametes combine randomly during fertilization, producing the allele combinations TT, Tt, and tt in the F₂ offspring. The combination Tt occurs twice as often as either of the other two combinations because it can be produced in two different ways. The TT and Tt offspring are tall, while the tt offspring are dwarf.

The F₂ generation includes both homozygous individuals (homozygotes) and heterozygous individuals (heterozygotes).

The term **phenotype** refers to the characteristics of an organism that are the result of the expression of its genes. In the example in Figure 16.5a, one of the parent plants is phenotypically tall and the other is phenotypically dwarf. Although the F₁ offspring are heterozygous (Tt), their phenotypes are tall because each of them has a copy of the dominant tall allele. In contrast, the F₂ plants display both phenotypes in a ratio of 3:1. Later in the chapter we will examine the underlying molecular mechanisms that produce phenotypes, but in our discussion of Mendel's results the term simply refers to a visible trait such as flower color or height.

A Punnett Square Can Be Used to Predict the Outcome of Crosses

A common way to predict the outcome of simple genetic crosses is to make a **Punnett square**, a method originally proposed by the British geneticist Reginald Punnett. To construct a Punnett square, you must know the genotypes of the parents. What follows is a step-by-step description of the Punnett square approach using a cross of heterozygous tall plants.

Step 1. *Write down the genotypes of both parents.* In this example, a heterozygous tall plant is crossed to another heterozygous tall plant. The plant providing the pollen is considered the male parent and the plant providing the eggs, the female parent. (In self-pollination, a single individual produces both types of gametes.)

Male parent: *Tt*

Female parent: *Tt*

Step 2. *Write down the possible gametes that each parent can make.* Remember that the law of segregation tells us that a gamete contains only one copy of each gene.

Male gametes: *T* or *t*

Female gametes: *T* or *t*

Step 3. *Create an empty Punnett square.* The number of columns equals the number of male gametes, and the number of rows equals the number of female gametes. Our example has two rows and two columns. Place the male gametes across the top of the Punnett square and the female gametes along the side.

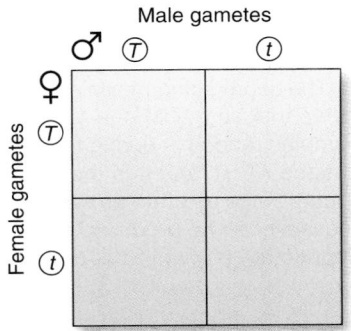

Step 4. *Fill in the possible genotypes of the offspring by combining the alleles of the gametes in the empty boxes.*

Male gametes

	♂ ⓣ ⒯	ⓣ
♀ ⓣ	**TT**	**Tt**
ⓣ	**Tt**	**tt**

Female gametes

Step 5. *Determine the relative proportions of genotypes and phenotypes of the offspring.* The genotypes are obtained directly from the Punnett square. In this example, the geno-

types are *TT*, *Tt*, and *tt* in a 1:2:1 ratio. To determine the phenotypes, you must know which allele is dominant. For plant height, *T* (tall) is dominant to *t* (dwarf). The genotypes *TT* and *Tt* are tall, whereas the genotype *tt* is dwarf. Therefore, our Punnett square shows us that the ratio of phenotypes is expected to be 3:1, or 3 tall plants to 1 dwarf plant. Keep in mind, however, that these are predicted ratios for large numbers of offspring. If only a few offspring are produced, the observed ratios could deviate significantly from the predicted ratios. We will examine the question of sample size and genetic prediction later in this chapter.

A Testcross Can Be Used to Determine an Individual's Genotype

When a trait has two variants, one of which is dominant over the other, we know that an individual with a recessive phenotype is homozygous for the recessive allele. A dwarf pea plant, for example, must have the genotype *tt*. But an individual with a dominant phenotype may be either homozygous or heterozygous—a tall pea plant may have the genotype *TT* or *Tt*. To distinguish between these two possibilities, Mendel devised a method called a **testcross** that is still used today. In a testcross, the researcher crosses the individual of interest to a homozygous recessive individual and observes the phenotypes of the offspring. **Figure 16.7** shows how this procedure can be used to determine the genotype of a tall pea plant. If the testcross produces some dwarf offspring, as shown on the right side of the figure, these offspring must have two copies of the recessive allele, one inherited from each parent. Therefore, the tall parent must be a heterozygote, with the genotype *Tt*. Alternatively, if all of the offspring are tall, as shown on the left, the tall parent is likely to be a homozygote, with the genotype *TT*.

Analyzing the Inheritance Pattern of Two Traits Simultaneously Demonstrated the Law of Independent Assortment

Mendel's analysis of single-factor crosses suggested that traits are inherited as discrete units and that the alleles for a given gene segregate during the formation of haploid cells. To obtain additional insights into how genes are transmitted from parents to offspring, Mendel conducted crosses in which he simultaneously followed the inheritance of two different traits. A cross of this type is called a **two-factor cross**. We will examine a two-factor cross in which Mendel simultaneously followed the inheritance of seed color and seed shape (**Figure 16.8**). He began by crossing pea plants from strains that bred true for both traits. The plants of one strain had yellow, round seeds and plants of the other strain had green, wrinkled seeds. He then allowed the F₁ offspring to self-fertilize and observed the phenotypes of the F₂ generation.

Before we discuss Mendel's results, let's consider possible patterns of inheritance for two traits. One possibility is that the two genes are linked in some way, so that variants that occur together in the parents are always inherited as a unit. In our

Figure 16.7 A testcross. The purpose of this experiment is to determine if the organism with the dominant phenotype, in this case a tall pea plant, is a homozygote (*TT*) or a heterozygote (*Tt*).

Biological inquiry: Let's suppose we had a plant with purple flowers and unknown genotype and conducted a testcross to determine its genotype. We obtained 41 plants, 20 of which had white flowers and 21 with purple flowers. What was the genotype of the original purple-flowered plant?

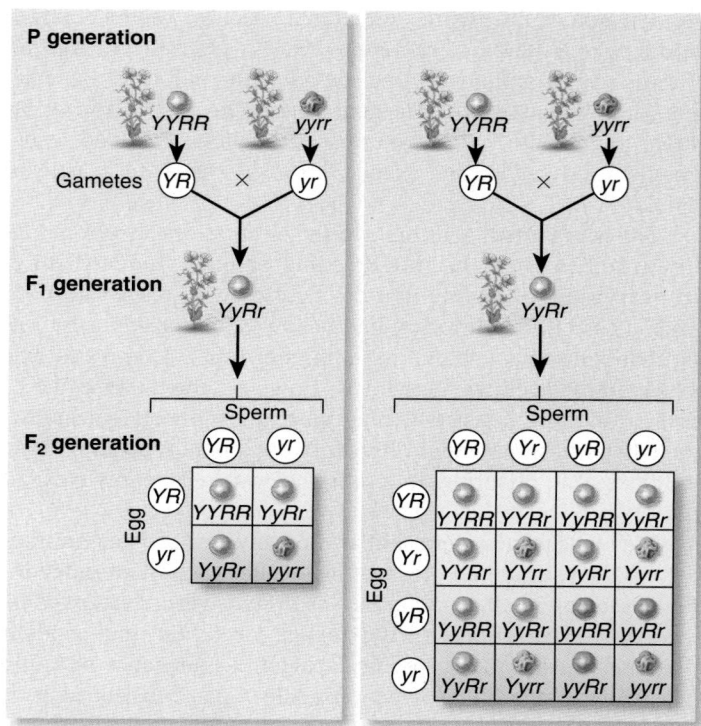

(a) Hypothesis:
Linked assortment

(b) Hypothesis:
Independent assortment

P Cross	F_1 generation	F_2 generation
Yellow, round seeds × Green, wrinkled seeds	Yellow, round seeds	315 yellow, round seeds 101 yellow, wrinkled seeds 108 green, round seeds 32 green, wrinkled seeds

(c) The data observed by Mendel

Figure 16.8 Two hypotheses for the assortment of two different genes. This figure shows a cross between two true-breeding pea plants, one with yellow, round seeds and one with green, wrinkled seeds. All of the F_1 offspring have yellow, round seeds. When the F_1 offspring self-fertilize, the two hypotheses predict different phenotypes in the F_2 generation. (a) The linkage hypothesis proposes that the two parental alleles always stay associated with each other. In this case, all of the F_2 offspring will have either yellow, round seeds or green, wrinkled seeds. (b) The independent assortment hypothesis proposes that each allele assorts independently. In this case, the F_2 generation will display four different phenotypes. (c) Mendel's observations supported the independent assortment hypothesis.

example, the allele for yellow seeds (*Y*) would always be inherited with the allele for round seeds (*R*), as shown in Figure 16.8a. The recessive alleles are green seeds (*y*) and wrinkled seeds (*r*). A second possibility is that the two genes are independent of one another, so that their alleles are randomly distributed into gametes (Figure 16.8b). By following the transmission pattern of two traits simultaneously, Mendel could determine whether the genes that determine seed shape and seed color **assort** (are distributed) together as a unit or independently of each other.

What experimental results could Mendel predict for each of these two models? The homozygous P generation can produce only two kinds of gametes, *YR* and *yr*, so in either case the F_1 offspring would be heterozygous for both traits; that is, they would have the genotypes *YyRr*. Because Mendel knew from his earlier experiments that yellow was dominant over green and round over wrinkled, he could predict that all the F_1 plants would have yellow, round seeds. In contrast, as shown in Figure 16.8, the ratios he obtained in the F_2 generation would depend on whether the traits assort together or independently.

If the parental traits are linked, as in Figure 16.8a, the F_1 plants could only produce gametes that are *YR* or *yr*. These gametes would combine to create offspring with the genotypes

YYRR (yellow, round), *YyRr* (yellow, round), or *yyrr* (green, wrinkled). The ratio of phenotypes would be 3 yellow, round to 1 green, wrinkled. Every F_2 plant would be phenotypically like one P-generation parent or the other; none would display a new combination of the parental traits. However, if the alleles assort independently, the F_2 generation would show a wider range of genotypes and phenotypes, as shown by the large

Punnett square in Figure 16.8b. In this case, each F₁ parent produces four kinds of gametes—*YR, Yr, yR,* and *yr*—instead of two, so the square is constructed with four rows on each side and shows 16 possible genotypes. The F₂ generation includes plants with yellow, round seeds; yellow, wrinkled seeds; green, round seeds; and green, wrinkled seeds, in a ratio of 9:3:3:1.

The actual results of this two-factor cross are shown in Figure 16.8c. Crossing the true-breeding parents produced **dihybrid** offspring—offspring that are hybrids with respect to both traits. These F₁ dihybrids all had yellow, round seeds, confirming that yellow and round are dominant traits. This result was consistent with either hypothesis. However, the data for the F₂ generation were consistent only with the independent assortment hypothesis. Mendel observed four phenotypically different types of F₂ offspring, in a ratio that was reasonably close to 9:3:3:1.

Mendel's results were similar for every pair of traits he studied. His data supported the idea, now called the **law of independent assortment**, that *the alleles of different genes assort independently of each other during gamete formation.* Independent assortment means that a specific allele for one gene may be found in a gamete regardless of which allele for a different gene is found in the same gamete. In our example, the yellow and green alleles assort independently of the round and wrinkled alleles. The union of gametes from F₁ plants carrying these alleles produces the F₂ genotypic and phenotypic ratios shown in Figure 16.8b.

The Chromosome Theory of Inheritance Relates Mendel's Observations to the Behavior of Chromosomes

Mendel's studies with pea plants led to the concept of a gene, which is the foundation for our understanding of inheritance. However, at the time of Mendel's work, the physical nature and location of genes was a complete mystery. In fact, the idea that inheritance has a physical basis was not even addressed until 1883, when the German biologist August Weismann and the Swiss botanist Carl Nägeli championed the idea that a substance in living cells is responsible for the transmission of hereditary traits. Nägeli also suggested that parents contribute equal amounts of this substance to their offspring. This idea challenged other researchers to identify the genetic material. Several scientists, including the German biologists Eduard Strasburger and Walter Flemming, observed dividing cells under the microscope and suggested that the chromosomes are the carriers of the genetic material. As we now know, the genetic material is the DNA within chromosomes.

In the early 1900s, the idea that chromosomes carry the genetic material dramatically unfolded as researchers continued to study the processes of fertilization, mitosis, and meiosis. It became increasingly clear that the characteristics of organisms are rooted in the continuity of cells during the life of an organism and from one generation to the next. Several scientists noted striking parallels between the segregation and assortment

of traits noted by Mendel and the behavior of chromosomes during meiosis. Among these scientists were the German biologist Theodor Boveri and the American biologist Walter Sutton, who independently proposed the chromosome theory of inheritance. According to this theory, the inheritance patterns of traits can be explained by the transmission of chromosomes during meiosis and fertilization. Thomas Hunt Morgan's studies of fruit flies, which we will examine later in this chapter, were instrumental in supporting this theory.

The **chromosome theory of inheritance** consists of a few fundamental principles:

1. Chromosomes contain the genetic material, which is transmitted from parent to offspring and from cell to cell. Genes are found in the chromosomes.
2. Chromosomes are replicated and passed from parent to offspring. They are also passed from cell to cell during the multicellular development of an organism. Each type of chromosome retains its individuality during cell division and gamete formation.
3. The nucleus of a diploid cell contains two sets of chromosomes, which are found in homologous pairs. One member of each pair is inherited from the mother and the other from the father. The maternal and paternal sets of homologous chromosomes are functionally equivalent; each set carries a full complement of genes.
4. At meiosis, one member of each chromosome pair segregates into one daughter nucleus and its homologue segregates into the other daughter nucleus. Each of the resulting haploid cells contains only one set of chromosomes. During the formation of haploid cells, the members of different chromosome pairs segregate independently of each other.
5. Gametes are haploid cells that combine to form a diploid cell during fertilization, with each gamete transmitting one set of chromosomes to the offspring. In animals, one set comes from the mother and the other set comes from the father.

Now that you have an understanding of the basic tenets of the chromosome theory, let's relate these ideas to Mendel's laws of inheritance.

Chromosomes and Segregation Mendel's law of segregation can be explained by the pairing and segregation of homologous chromosomes during meiosis. Before we examine this idea, it will be helpful to introduce another genetic term. The physical location of a gene on a chromosome is called the gene's **locus** (plural, loci). As shown in **Figure 16.9**, each member of a homologous chromosome pair carries an allele of the same gene at the same locus. The individual in this example is heterozygous (*Tt*), so each homologue has a different allele.

Figure 16.10 follows a homologous chromosome pair through the events of meiosis. This example involves a pea plant, heterozygous for height, *Tt.* The top of Figure 16.10 shows the two homologues prior to DNA replication. When a cell prepares

Figure 16.9 **A gene locus.** The locus (location) of a gene is the same for each member of a homologous pair, whether the individual is homozygous or heterozygous for that gene. This individual is heterozygous (*Tt*) for a gene for plant height in peas.

to divide, the homologues replicate to produce pairs of sister chromatids. Each chromatid carries a copy of the allele found on the original homologue, either *T* or *t*. The homologues, each consisting of two sister chromatids, pair during metaphase I and then segregate into two daughter cells during later phases of meiosis I. One of these cells has two copies of the *T* allele, and the other has two copies of the *t* allele. The sister chromatids separate during meiosis II, which produces four haploid cells. The end result of meiosis is that each haploid cell has a copy of just one of the two original homologues. Two of the cells have a chromosome carrying the *T* allele, while the other two have a chromosome carrying the *t* allele at the same locus. If the haploid cells shown at the bottom of Figure 16.10 combine randomly during fertilization, they produce diploid offspring with the genotypic and phenotypic ratios shown earlier in Figure 16.6.

Chromosomes and Independent Assortment The law of independent assortment can also be explained by the behavior of chromosomes during meiosis. **Figure 16.11** shows the segregation of two pairs of homologous chromosomes in a pea plant. One pair carries the gene for seed color: the yellow allele (*Y*) is on one chromosome, and the green allele (*y*) is on its homologue. The other pair of chromosomes carries the gene for seed shape: one member of the pair has the round allele (*R*), while its homologue carries the wrinkled allele (*r*). Thus, this individual is heterozygous for both genes, with the genotype *YyRr*.

When meiosis begins, each of the chromosomes has already replicated and consists of two sister chromatids. At metaphase I of meiosis, the two pairs of chromosomes randomly align themselves along the metaphase plate. This alignment can occur in two equally probable ways, shown on the two sides of the figure. On the left, the chromosome carrying the *y* allele is aligned on the same side of the metaphase plate as the chromosome carrying the *R* allele; *Y* is aligned with *r*. On the right, the opposite has occurred: *Y* is aligned with *R* and *y* is with *r*. In each case, the chromosomes that aligned on the same side of the metaphase plate segregate into the same daughter cell. In this way, the random alignment of chromosome pairs during meiosis I leads to the independent assortment of alleles found

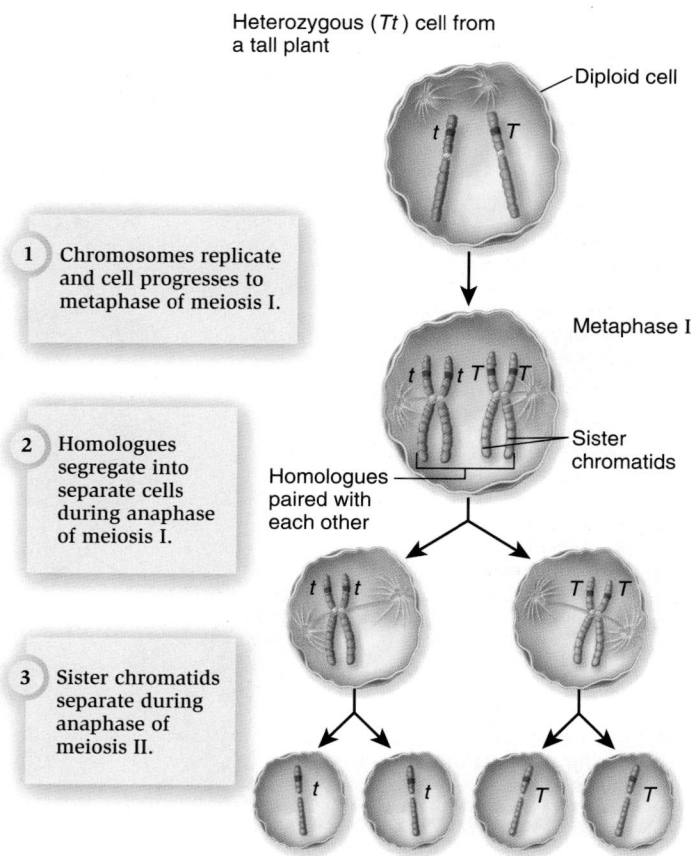

Heterozygous (*Tt*) cell from a tall plant

1 Chromosomes replicate and cell progresses to metaphase of meiosis I.

2 Homologues segregate into separate cells during anaphase of meiosis I.

3 Sister chromatids separate during anaphase of meiosis II.

Diploid cell

Metaphase I

Sister chromatids

Homologues paired with each other

Four haploid cells

Figure 16.10 **The chromosomal basis of allele segregation.** This example shows a pair of homologous chromosomes in a cell of a pea plant. The blue chromosome was inherited from the male parent and the red chromosome was inherited from the female parent. This individual is heterozygous for a height gene (*Tt*). The two homologues segregate from each other during meiosis, leading to segregation of the tall allele (*T*) and the dwarf allele (*t*) into different haploid cells.

Biological inquiry: When we say that alleles segregate, what does the word segregate *mean? How is this related to meiosis, described in Chapter 15?*

on different chromosomes. For two loci found on different chromosomes, each with two variant alleles, meiosis produces four allele combinations in equal numbers, as seen at the bottom of the figure.

If a *YyRr* (dihybrid) plant undergoes self-fertilization, any two gametes can combine randomly during fertilization. Because four kinds of gametes are made, this allows for 16 possible allele combinations in the offspring. These genotypes, in turn, produce four phenotypes in a 9:3:3:1 ratio, as seen earlier in Figure 16.8. This ratio is the expected outcome when a heterozygote for two genes on different chromosomes undergoes self-fertilization.

But what if two genes are located on the same chromosome? In this case, the transmission pattern may not conform to the law of independent assortment. We will discuss this phenomenon, known as linkage, in Chapter 17.

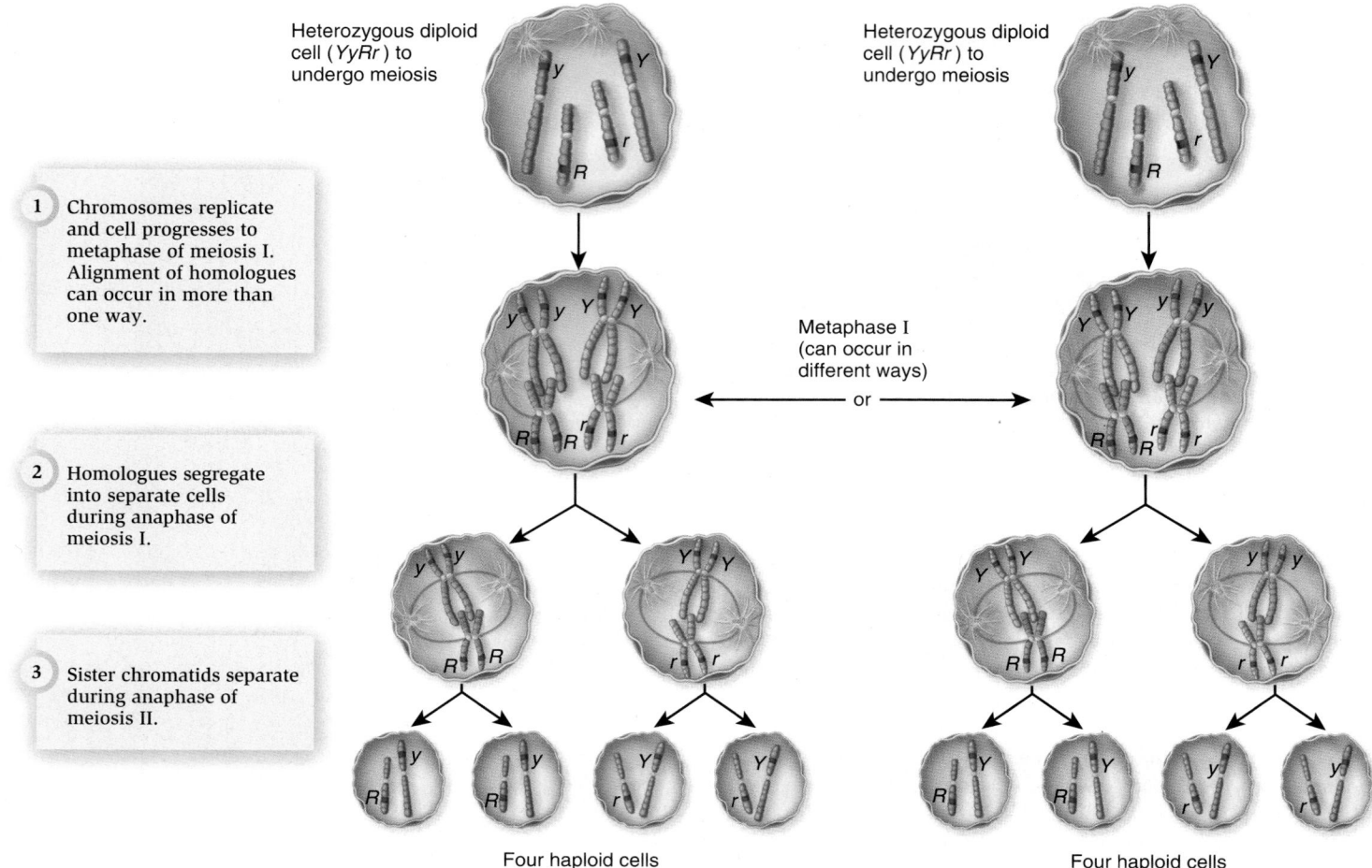

1 Chromosomes replicate and cell progresses to metaphase of meiosis I. Alignment of homologues can occur in more than one way.

2 Homologues segregate into separate cells during anaphase of meiosis I.

3 Sister chromatids separate during anaphase of meiosis II.

Heterozygous diploid cell (*YyRr*) to undergo meiosis

Metaphase I (can occur in different ways) — or —

Heterozygous diploid cell (*YyRr*) to undergo meiosis

Four haploid cells

Four haploid cells

Figure 16.11   **The chromosomal basis of independent assortment.**   The genes for seed color (*Y* or *y*) and seed shape (*R* or *r*) in peas are on different chromosomes. During metaphase of meiosis I, different arrangements of the two chromosome pairs can lead to different combinations of the alleles in the resulting haploid cells. On the *left,* the chromosome carrying the dominant *R* allele has segregated with the chromosome carrying the recessive *y* allele; on the *right,* the two chromosomes carrying the dominant alleles (*R* and *Y*) have segregated together.

Genetic Predictions Are Based on Probability

As you have seen, Mendel's laws of inheritance can be used to predict the outcome of genetic crosses. This is useful in many ways. In agriculture, for example, plant and animal breeders use predictions about the types of offspring their crosses will produce to develop commercially important crops and livestock. In addition, people are often interested in the potential characteristics of their future children. This information is particularly important to individuals who may carry alleles that cause inherited diseases. Of course, no one can see into the future and definitively predict what will happen. Nevertheless, genetic counselors can often help couples predict the likelihood of having an affected child. This probability is one factor that may influence a couple's decision about whether to have children.

Genetic predictions are based on the mathematical rules of probability. The chance that an event will have a particular outcome is called the **probability** of that outcome. The probability of a given outcome depends on the number of possible outcomes. For example, if you draw a card at random from a 52-card deck, the probability that you will get the jack of diamonds is 1 in 52, because there are 52 possible outcomes for the draw. In contrast, only two outcomes are possible when you flip a coin, so the probability is one in two (1/2, or 0.5, or 50%) that the heads side will be showing when the coin lands. The general formula for the probability (*P*) that a random event will have a specific outcome is

$$P = \frac{\text{Number of times an event occurs}}{\text{Total number of possible outcomes}}$$

Thus, for a single coin toss, the chance of getting heads is

$$P_{heads} = \frac{1 \text{ heads}}{(1 \text{ heads} + 1 \text{ tails})} = \frac{1}{2}$$

Earlier in this chapter, we considered the use of Punnett squares to predict the fractions of offspring with a given genotype or phenotype. Our example was self-fertilization of a pea plant that was heterozygous for the height gene (*Tt*), and our Punnett square predicted that one-fourth of the offspring would be dwarf. We can make the same prediction by using a probability calculation.

$$P_{dwarf} = \frac{1 \ tt}{(1 \ TT + 2 \ Tt + 1 \ tt)} = \frac{1}{4}$$

Probability and Sample Size A probability calculation allows us to predict the likelihood that a future event will have a specific outcome. However, the accuracy of this prediction depends to a great extent on the number of events we observe, or in other words, on the size of our sample. For example, if we toss a coin six times, the calculation we just presented for P_{heads} suggests that we should get heads three times and tails three times. However, each coin toss is an independent event, meaning that every time we toss the coin there is a random chance that it will come up heads or tails, regardless of the outcome of the previous toss. With only six tosses, we would not be too surprised if we got four heads and two tails. The deviation between the observed and expected outcomes is called the **random sampling error**. With a small sample, the random sampling error may cause the observed data to be quite different from the expected outcome. By comparison, if we flipped a coin 1,000 times, the percentage of heads would be fairly close to the predicted 50%. With a larger sample, we expect the sampling error to be smaller.

Earlier in this chapter, we examined Mendel's data for pea plants and learned that his observations were very close to the outcome that we would predict from a Punnett square. Mendel counted a large number of pea plants, which made his sampling error quite small. However, when we apply probability to humans, the small size of human families may cause the observed data to be quite different from the expected outcome. Consider, for example, a couple who are both heterozygous for an allele that affects eye color (*Bb*). The dominant allele is brown, so both of these parents have brown eyes. We can use a Punnett square to predict that one-fourth of their offspring will have blue eyes, which is the recessive phenotype for a *bb* homozygote. However, the genotype or phenotype of each child is independent of each other. Each child has one chance in four of having the recessive phenotype; the birth of a blue-eyed child does not make it more or less likely that the next child will have blue eyes. Thus, in a family with four children, we would not be too surprised if two of them had blue eyes—twice the predicted number. In this case, a large deviation occurs between the observed outcome (50% blue-eyed children) and the expected outcome (25% blue-eyed children). This large random sampling error can be attributed to the small sample size.

The Product Rule and the Sum Rule Punnett squares allow us to predict the likelihood that a genetic cross will produce an offspring with a particular genotype or phenotype. To predict the likelihood of producing multiple offspring with particular genotypes or phenotypes, we can use the **product rule**, which states that *the probability that two or more independent events will occur is equal to the product of their individual probabilities.* As we have already discussed, events are independent if the outcome of one event does not affect the outcome of another. In our previous coin-toss example, each toss is an independent event— if one toss comes up heads, another toss still has an equal chance of coming up either heads or tails. If we toss a coin twice, what is the probability that we will get heads both times? The product rule says that it is equal to the probability of getting heads on the first toss (1/2) times the probability of getting heads on the second toss (1/2), or one in four (1/2 × 1/2 = 1/4).

To see how the product rule can be applied to a genetics problem, let's consider a rare, recessive human trait known as congenital analgesia. (Congenital refers to a condition present at birth; analgesia means insensitivity to pain.) People with this trait can distinguish between sensations such as sharp and dull, or hot and cold, but they do not perceive extremes of sensation as painful. The first known case of congenital analgesia, described in 1932, was a man who made his living entertaining the public as a "human pincushion." For a phenotypically normal couple, each heterozygous for the recessive allele causing congenital analgesia, we can ask, What is the probability that their first three offspring will have the disorder? To answer this question, we must first determine the probability of a single offspring having the abnormal phenotype. By using a Punnett square, we would find that the probability of an individual offspring being homozygous recessive is 1/4. Thus, each of this couple's children has one chance in four of having the disorder.

We can now use the product rule to calculate the probability of this couple having three affected offspring in a row. The phenotypes of the first, second, and third offspring are independent events; that is, the phenotype of the first offspring does not affect the phenotype of the second or third offspring. The product rule tells us that the probability of all three children having the abnormal phenotype is

$$\frac{1}{4} \times \frac{1}{4} \times \frac{1}{4} = \frac{1}{64} = 0.016$$

The probability of the first three offspring having the disorder is 0.016, or 1.6%. In other words, we can say that this couple's chance of having three children in a row with congenital analgesia is very small—only 1.6 out of 100. The phenotypes of the first, second, and third child are independent of each other.

Let's now consider a second way to predict the outcome of particular crosses. In a cross between two heterozygous (*Tt*) pea plants, we may want to know the probability of a particular offspring being a homozygote. In this case we are asking, What is the chance that this individual will be either homozygous *TT* or homozygous *tt*? To answer an "either/or" question we use the sum rule, which applies to events with mutually exclusive outcomes. When we say that outcomes are mutually exclusive, we mean that they cannot occur at the same time. A pea plant can be tall or dwarf, but not both at the same time: the tall and dwarf phenotypes are mutually exclusive. Similarly, a plant with the genotype *TT* cannot be *Tt* or *tt*. Each of these genotypes is mutually exclusive with the other two. According to the **sum rule**, *the probability that one of two or more mutually exclusive outcomes will occur is the sum of the probabilities of the possible outcomes.* This means that to find the probability that an offspring will be either homozygous *TT* or homozygous *tt*, we add together the probability that it will be *TT* and the probability that it will be *tt*. Using a Punnett square, we find that the probability for each of these genotypes is one in four. We can now use the sum rule to determine the probability of an individual having one of these genotypes.

$$\frac{1}{4} \quad + \quad \frac{1}{4} \quad = \quad \frac{1}{2}$$

(probability of *TT*)   (probability of *tt*)   (probability of either *TT* or *tt*)

This calculation predicts that in crosses of two *Tt* parents, half of the offspring will be homozygotes—either *TT* or *tt*.

Pedigree Analysis Examines the Inheritance of Human Traits

As we have seen, Mendel conducted experiments by making selective crosses of pea plants and analyzing large numbers of offspring. Later geneticists also relied on crosses of experimental organisms, especially fruit flies. Obviously, geneticists studying human traits cannot use this approach, for ethical and practical reasons. Instead, human geneticists must rely on information from family trees, or pedigrees. In this approach, called **pedigree analysis**, an inherited trait is analyzed over the course of a few generations in one family. The results of this method may be less definitive than the results of breeding experiments because the small size of human families may lead to large sampling errors. Nevertheless, a pedigree analysis can often provide important clues concerning human inheritance.

Pedigree analysis has been used to understand the inheritance of human genetic diseases that follow simple Mendelian patterns. Many genes that play a role in disease exist in two forms—the normal allele and an abnormal allele that has arisen by mutation. The disease symptoms are associated with the mutant allele. Pedigree analysis allows us to determine whether the mutant allele is dominant or recessive and to predict the likelihood of an individual being affected.

We have already considered one human abnormality caused by a recessive allele, congenital analgesia. We will use another recessive condition to illustrate pedigree analysis. The pedigree in **Figure 16.12** concerns a human genetic disease known as cystic fibrosis (CF). Approximately 3% of Americans of European descent are heterozygous carriers of the recessive CF allele. These carriers are phenotypically normal. Individuals who are homozygous for the CF allele exhibit the disease symptoms, which include abnormalities of the pancreas, intestine, sweat glands, and lungs. (You will see how a single gene can have such far-reaching effects when we discuss the molecular basis

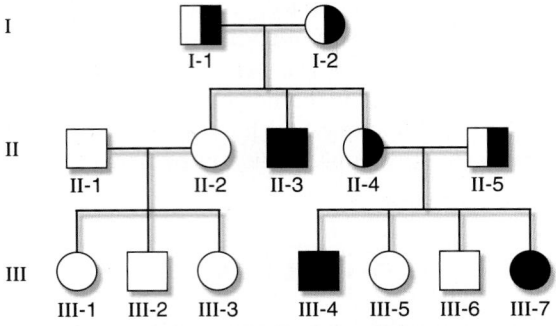

(a) **Human pedigree showing cystic fibrosis**

(b) **Symbols used in a human pedigree**

Figure 16.12   A family pedigree of a recessive trait. Some members of the family in this pedigree are affected with cystic fibrosis. Phenotypically normal individuals I-1, I-2, II-4, and II-5 are presumed to be heterozygotes because they have produced affected offspring.

Biological inquiry: Let's suppose a genetic disease is caused by a mutant allele. If two affected parents produce an unaffected offspring, can the mutant allele be recessive?

of Mendelian inheritance later in the chapter.) A human pedigree like the one in Figure 16.12 shows the oldest generation (designated by the Roman numeral I) at the top, with later generations (II and III) below it. A man (represented by a square) and a woman (represented by a circle) who produce offspring are connected by a horizontal line; a vertical line connects parents with their offspring. Siblings (brothers and sisters) are denoted by downward projections from a single horizontal line, from left to right in the order of their birth. For example, individuals I-1 and I-2 are the parents of individuals II-2, II-3, and II-4, who are all siblings. Individuals affected by the disease, such as individual II-3, are depicted by filled symbols.

The pattern of affected and unaffected individuals in this pedigree is consistent with a recessive mode of inheritance for CF: two unaffected individuals can produce an affected offspring. Such individuals are presumed to be heterozygotes (designated by a half-filled symbol). However, the same unaffected parents can also produce unaffected offspring, because an individual must inherit two copies of the mutant allele to exhibit the disease. A recessive mode of inheritance is also characterized

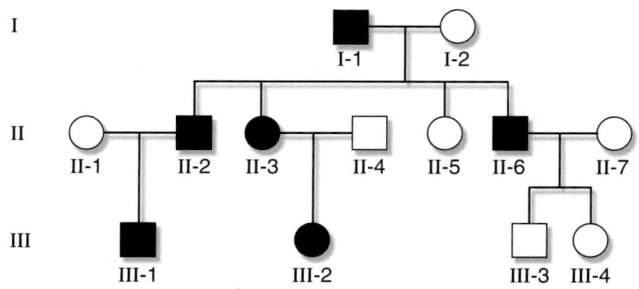

Figure 16.13   A family pedigree of a dominant trait. Huntington disease is caused by a dominant allele. Note that each affected offspring in this pedigree has an affected parent.

by the observation that all of the offspring of two affected individuals will be affected. However, for genetic diseases like CF that limit survival or fertility, there may rarely or never be cases where two affected individuals produce offspring.

Although many of the alleles causing human genetic diseases are recessive, some are known to be dominant. **Figure 16.13** shows a family pedigree involving Huntington disease, a condition that causes the degeneration of brain cells involved in emotions, intellect, and movement. If you examine this pedigree, you will see that every affected individual has one affected parent. This pattern is characteristic of most dominant disorders. The symptoms of Huntington disease, which usually begin to appear when people are 30 to 50 years old, include uncontrollable jerking movements of the limbs, trunk, and face; progressive loss of mental abilities; and the development of psychiatric problems. In 1993, researchers identified the gene involved in this disorder. The normal allele encodes a protein called huntingtin, which functions in nerve cells. The mutant allele encodes an abnormal form of the protein, which aggregates within nerve cells in the brain. Further research is needed to determine how this aggregation contributes to the disease.

Most human genes are found on the paired chromosomes known as **autosomes**, which are the same in both sexes. Mendelian inheritance patterns involving these autosomal genes are described as autosomal inheritance patterns. Huntington disease is an example of a trait with an autosomal dominant inheritance pattern, while cystic fibrosis illustrates the pattern called autosomal recessive. However, some human genes are located on sex chromosomes, which are different in males and females. These genes have their own characteristic inheritance patterns, which we will consider in the next section.

16.2 Sex Chromosomes and X-Linked Inheritance Patterns

In the first part of this chapter we discussed Mendel's experiments that established the basis for understanding how traits are transmitted from parents to offspring. We also examined the chromosome theory of inheritance, which provided a framework for explaining Mendel's observations. Mendelian patterns of gene transmission are observed for many genes located on autosomes in a wide variety of eukaryotic species.

We will now turn our attention to genes located on **sex chromosomes**. As you learned in Chapter 15, this term refers to a distinctive pair of chromosomes that are different in males and females. Sex chromosomes are found in many but not all species with two sexes. The study of sex chromosomes proved pivotal in confirming the chromosome theory. The distinctive transmission patterns of genes on sex chromosomes helped early geneticists show that particular genes are located on particular chromosomes. Later, other researchers became interested in these genes because some of them were found to cause inherited diseases.

In this section, we will consider several mechanisms by which the presence or composition of sex chromosomes determines an individual's sex in different species. We will then examine some of the early research involving sex chromosomes that provided convincing evidence for the chromosome theory of inheritance. Finally, we will consider the inheritance patterns of genes on sex chromosomes and why recessive alleles are expressed more frequently in males than in females.

In Many Species, Sex Differences Are Due to the Presence of Sex Chromosomes

According to the chromosome theory of inheritance, chromosomes carry the genes that determine an organism's traits. Some early evidence supporting this theory involved a consideration of sex determination. In 1901, the American biologist C. E. McClung, who studied fruit flies, suggested that male and female sexes in insects are due to the inheritance of particular chromosomes. Following McClung's initial observations, several mechanisms of sex determination were found in different species of animals. Some examples are described in **Figure 16.14**. All of these mechanisms involve chromosomal differences between the sexes, and most involve a difference in a single pair of sex chromosomes.

In the X-Y system of sex determination, which operates in mammals, the somatic cells of males have one X and one Y chromosome, while female somatic cells contain two X chromosomes (Figure 16.14a). For example, the 46 chromosomes carried by human cells consist of one pair of sex chromosomes (either XY or XX) and 22 pairs of autosomes. The presence of the Y chromosome causes maleness in mammals. This is known from the analysis of rare individuals who carry chromosomal abnormalities. For example, mistakes that occasionally occur during meiosis may produce an individual who carries two X chromosomes and one Y chromosome. Such an individual develops into a male. A gene called the *SRY* gene located on the Y chromosome of mammals plays a key role in the developmental pathway that leads to maleness.

The X-O system operates in many insects (Figure 16.14b). Females in this system have a pair of sex chromosomes and are designated XX. In some insect species that follow the X-O system, the male has only one sex chromosome, the X, and is designated XO. In other X-O insect species, such as *Drosophila melanogaster*, the male has both an X chromosome and a Y chromosome and is designated XY. Unlike the Y chromosome of mammals, the Y chromosome in the X-O system does not determine maleness. In both types of insect species (XO or XY males), the individual's sex is determined by the ratio between its X chro-

(a) The X-Y system in mammals

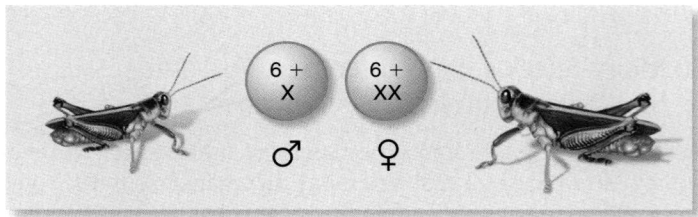

(b) The X-O system in certain insects

(c) The Z-W system in birds

(d) The haplo-diploid system in bees

Figure 16.14   Different mechanisms of sex determination in animals.   The numbers shown in the circles indicate the numbers of autosomes.

Biological inquiry: If a person is born with only one X chromosome and no Y chromosome, would you expect that person to be a male or a female? Explain your answer.

mosomes and its sets of autosomes. If a fly has one X chromosome and is diploid for the autosomes (2*n*), this ratio is 1/2, or 0.5. This fly will become a male whether or not it receives a Y chromosome. On the other hand, if a diploid fly receives two X chromosomes, the ratio is 2/2, or 1.0, and the fly becomes a female.

Thus far, we have considered examples where females have two similar copies of a sex chromosome, the X. However, in some animal species, such as birds and some fish, the male carries two similar chromosomes (Figure 16.14c). This is called the Z-W system to distinguish it from the X-Y system found in mammals. The male is ZZ and the female is ZW.

Not all chromosomal mechanisms of sex determination involve a special pair of sex chromosomes. An interesting mechanism known as the haplo-diploid system is found in bees (Figure 16.14d). The male bee, or drone, is produced from an unfertilized haploid egg. Thus, male bees are haploid individuals. Females, both worker bees and queen bees, are produced from fertilized eggs and therefore are diploid.

Although sex in many species of animals is determined by chromosomes, other mechanisms are also known. In certain reptiles and fish, sex is controlled by environmental factors such as temperature. For example, in the American alligator (*Alligator mississippiensis*), temperature controls sex development. When eggs of this alligator are incubated at 33°C, 100% of them produce male individuals. When the eggs are incubated at a temperature below 33°C, they produce 100% females, while at a temperature above 33°C, they produce 95% females.

Most species of flowering plants, including pea plants, have a single type of diploid plant, or sporophyte, that makes both male and female gametophytes. However, the sporophytes of some species have two sexually distinct types of individuals, one with flowers that produce male gametophytes, and the other with flowers that produce female gametophytes. Examples include hollies, willows, poplars, and date palms. Sex chromosomes, designated X and Y, are responsible for sex determination in many such species. The male plant is XY, while the female plant is XX. However, in some species with separate sexes, microscopic examination of the chromosomes does not reveal distinct types of sex chromosomes.

In Humans, Recessive X-Linked Traits Are More Likely to Occur in Males

In humans, the X chromosome is rather large and carries many genes, while the Y chromosome is quite small and has relatively few genes. Therefore, many genes are found on the X chromosome but not on the Y; these are known as **X-linked genes**. By comparison, only a few genes are known to be Y linked, meaning that they are found on the Y chromosome but not on the X. The term **sex linked** refers to genes that are found on one sex chromosome but not on the other. Because few genes are found on the Y chromosome, the term usually refers to X-linked genes.

In mammals, a male cannot be described as being homozygous or heterozygous for an X-linked gene, because these terms describe genes that are present in two copies. Instead, the term **hemizygous** is used to describe the single copy of an X-linked gene in a male. Many recessive X-linked alleles cause diseases in humans, and these diseases occur more frequently in males than in females.

As an example, let's consider the X-linked recessive disorder called classical hemophilia (hemophilia A). In individuals with hemophilia, blood does not clot normally and a minor cut may bleed for a long time. Small bumps can lead to large bruises because broken capillaries may leak blood profusely into surrounding tissues before the capillaries are repaired. Common accidental injuries pose a threat of severe internal or external bleeding for hemophiliacs. Hemophilia A is caused by a recessive X-linked gene that encodes a defective form of a clotting protein. If a mother is a heterozygous carrier of hemophilia A, each of her children has a 50% chance of inheriting the recessive allele. A Punnett square shows a cross between a normal father and a heterozygous mother. X^{h-A} is the chromosome that carries the recessive allele for hemophilia A.

Although each child has a 50% chance of inheriting the hemophilia allele from the mother, only sons will exhibit the disorder. Because sons do not inherit an X chromosome from their fathers, a son who inherits the abnormal allele from his mother will have hemophilia. However, a daughter who inherits the hemophilia allele from her mother will also inherit a normal allele from her father. This daughter will have a normal phenotype, but if she passes the abnormal allele to her sons they will have hemophilia.

FEATURE INVESTIGATION

Morgan's Experiments Showed a Correlation Between a Genetic Trait and the Inheritance of a Sex Chromosome in *Drosophila*

The distinctive inheritance pattern of X-linked recessive alleles provides a way of demonstrating that a specific gene is on an X chromosome. In fact, an X-linked gene was the first gene to be located on a specific chromosome. In 1910, the American geneticist Thomas Hunt Morgan began work on a project in which he reared large populations of fruit flies, *Drosophila melanogaster*, in the dark to determine if their eyes would atrophy from disuse and disappear in future generations. Even after many consecutive generations, the flies showed no noticeable changes. After two years, however, Morgan finally obtained an interesting result: a true-breeding line of *Drosophila* produced a male fly with white eyes rather than the normal red eyes. The white-eye trait must have arisen from a new mutation that converted a red-eye allele into a white-eye allele.

Using an approach similar to Mendel's, Morgan studied the inheritance of the white-eye trait by making crosses and quantitatively analyzing their outcome. In the experiment described in **Figure 16.15**, Morgan crossed his white-eyed male to a red-eyed female. All of the F_1 offspring had red eyes, indicating that red is dominant to white. The F_1 offspring were then mated to each other to obtain an F_2 generation. As seen in the data table, this cross produced 2,459 red-eyed females, 1,011 red-eyed males, and 782 white-eyed males. Surprisingly, no white-eyed females were observed in the F_2 generation.

Figure 16.15 Morgan's crosses of red-eyed and white-eyed *Drosophila*.

GOAL A quantitative analysis of genetic crosses may reveal the pattern of inheritance of a particular gene.

STARTING MATERIALS A true-breeding line of red-eyed fruit flies plus one white-eyed male fly that was discovered in the population.

2 Record the results of the F_1 generation. This involves noting the eye color and sexes of several thousand flies.

F_1 generation

♂   ♀

$X^{w^+}Y$: $X^{w^+}X^w$

3 Cross F_1 offspring with each other to obtain F_2 generation of offspring.

♂ × ♀

♂   ♀

$X^{w^+}Y$ × $X^{w^+}X^w$

4 Record the eye color and sex of F_2 offspring.

F_2 generation

♂   ♂   ♀   ♀

♂   ♂   ♀   ♀

$X^{w^+}Y$: X^wY : $X^{w^+}X^{w^+}$: $X^{w^+}X^w$

---

5   THE DATA

Cross	Results
Original white-eyed male to red-eyed females	F_1 generation All red-eyed flies
F_1 males to F_1 females	F_2 generation 2,459 red-eyed females 1,011 red-eyed males 0 white-eyed females 782 white-eyed males

Morgan's results suggested a connection between the alleles for eye color and the sex of the offspring. As shown in the conceptual column of Figure 16.15 and in the Punnett square, the data are consistent with the idea that the eye-color alleles in *Drosophila* are located on the X chromosome. X^{w^+} is the chromosome carrying the normal allele for red eyes and X^w is the chromosome carrying the mutant allele for white eyes.

The Punnett square predicts that the F_2 generation will not have any white-eyed females, a prediction that was confirmed by Morgan's experimental data. However, it should also be pointed out that the experimental ratio of red eyes to white eyes in the F_2 generation is (2,459 + 1,011):782, which equals 4.4:1.

This ratio deviates significantly from the ratio of 3:1 predicted in the Punnett square. The lower than expected number of white-eyed flies is explained by a decreased survival of white-eyed flies.

F_1 male is $X^{w^+}Y$
F_1 female is $X^{w^+}X^w$

Male gametes

♂ X^{w^+} Y

Female gametes

♀ X^{w^+}

$X^{w^+}X^{w^+}$ Red, female	$X^{w^+}Y$ Red, male
$X^{w^+}X^w$ Red, female	X^wY White, male

♀ X^w

Following this initial discovery, Morgan carried out many experimental crosses that located specific genes on the *Drosophila* X chromosome. This research provided some of the most persuasive evidence for the Mendelian gene concept and the chromosome theory of inheritance, which are the foundations of modern genetics. In 1933, Morgan became the first geneticist to receive a Nobel Prize.

16.3 Variations in Inheritance Patterns and Their Molecular Basis

The term **Mendelian inheritance** describes the inheritance patterns of genes that segregate and assort independently. In the first section of this chapter, we considered the inheritance pattern of traits affected by a single gene that is found in two variants, one of which is completely dominant over the other. This pattern is called **simple Mendelian inheritance** because the phenotypic ratios in the offspring clearly demonstrate Mendel's laws. In the second section, we examined **X-linked inheritance**, the pattern displayed by pairs of dominant and recessive alleles located on X chromosomes. Early geneticists observed these Mendelian inheritance patterns without knowing why one trait was dominant over another.

In this section we will discuss the molecular basis of dominant and recessive traits, and we will see how the molecular expression of a gene can have widespread effects on an organism's phenotype. In addition, we will examine the inheritance patterns of genes that segregate and assort independently but do not display a simple dominant/recessive relationship. The transmission of these genes from parents to offspring does not usually produce the ratios of phenotypes we would expect on the basis of Mendel's observations. This does not mean that Mendel was wrong. Rather, the inheritance patterns of many traits are more intricate and interesting than the simple patterns he chose to study. As described in **Table 16.1**, our understanding of gene function at the molecular level explains both simple Mendelian inheritance and other, more complex inheritance patterns that conform to Mendel's laws. This modern knowledge also sheds light on the role of the environment in producing an organism's phenotype, which we will discuss at the end of the section.

Protein Function Explains the Phenomenon of Dominance

As we learned at the beginning of this chapter, Mendel studied seven traits that were found in two variants each. The dominant variants are the common alleles for these traits in pea plants. For any given gene, geneticists refer to a prevalent allele in a population as a **wild-type allele** (see Figure 16.2). In most cases, a wild-type allele encodes a protein that is made in the proper amount and functions normally. By comparison, alleles that have been altered by mutation are called **mutant alleles**; these tend to be rare in natural populations. In the case of Mendel's seven traits, the recessive alleles are due to rare mutations.

By studying genes and their gene products at the molecular level, researchers have discovered that a mutant allele is often defective in its ability to express a functional protein. In other words, mutations that produce mutant alleles are likely to decrease or eliminate the synthesis or functional activity of a protein. Such mutations are often inherited in a recessive fashion. To understand why many defective alleles are recessive, we need to take a quantitative look at protein function.

Table 16.1	Different Types of Mendelian Inheritance Patterns and Their Molecular Basis
Type	**Description**
Simple Mendelian inheritance	**Inheritance pattern:** Pattern of traits determined by a pair of alleles that display a dominant/recessive relationship and are located on an autosome. The presence of the dominant allele masks the presence of the recessive allele.
	Molecular basis: In many cases, the amount of protein produced by a heterozygote, which may be 50% of that produced by a dominant homozygote, is sufficient to produce the dominant trait.
X-linked inheritance	**Inheritance pattern:** Pattern of traits determined by genes that display a dominant/recessive relationship and are located on the X chromosome. In mammals and fruit flies, males are hemizygous for X-linked genes. In these species, X-linked recessive traits occur more frequently in males than in females.
	Molecular basis: In a female with one recessive X-linked allele (a heterozygote), the protein encoded by the dominant allele is sufficient to produce the dominant trait. A male with a recessive X-linked allele (a hemizygote) does not have a dominant allele and does not make any of the functional protein.
Incomplete dominance	**Inheritance pattern:** Pattern that occurs when the heterozygote has a phenotype intermediate to the phenotypes of the dominant and recessive homozygotes, as when a cross between red-flowered and white-flowered plants produces pink-flowered offspring.
	Molecular basis: 50% of the protein encoded by the normal (wild-type) allele is not sufficient to produce the normal trait.
Codominance	**Inheritance pattern:** Pattern that occurs when the heterozygote expresses both alleles simultaneously. For example, a human carrying the A and B alleles for the ABO antigens of red blood cells produces both the A and the B antigens (has an AB blood type).
	Molecular basis: The codominant alleles encode proteins that function slightly differently from each other. In a heterozygote, the function of each protein affects the phenotype uniquely.
Sex-influenced inheritance	**Inheritance pattern:** Pattern that occurs when an allele is recessive in one sex and dominant in the other. An example is pattern baldness in humans.
	Molecular basis: Sex hormones affect the molecular expression of genes, which can have an impact on the phenotype.

In a simple dominant/recessive relationship, the recessive allele does not affect the phenotype of the heterozygote. In this type of relationship, a single copy of the dominant (wild-type) allele is sufficient to mask the effects of the recessive allele. But if the recessive allele cannot produce a functional protein, how do we explain the dominant phenotype of the heterozygote? **Figure 16.16** considers the example of flower color in a pea plant. As shown at the top of this figure, the gene encodes an enzyme that is needed to make a purple pigment. The *P* allele is dominant because one *P* allele encodes enough of the functional protein—50% of the amount found in a normal homozygote—to provide a normal phenotype. Thus, the *PP* homozygote and the *Pp* heterozygote both make enough of the purple pigment to yield purple flowers. The *pp* heterozygote cannot make any of the functional enzyme required for pigment synthesis, so its flowers are white.

The explanation "50% of the normal protein is enough" is true for many dominant alleles. In such cases, the normal homozygote is making much more of the wild-type protein than necessary, so if the amount is reduced to 50%, as it is in the heterozygote, the individual still has plenty of this protein to accomplish whatever cellular function it performs. In other cases, however, an allele may be dominant because the heterozygote actually produces more than 50% of the normal amount of functional protein. This increased production is due to the phenomenon of gene regulation, which is discussed in Chapter 13. The normal gene is "up-regulated" in the heterozygote to compensate for the lack of function of the defective allele.

GENOMES & PROTEOMES

Single-Gene Mutations Cause Many Inherited Diseases and Have Pleiotropic Effects

The idea that recessive alleles usually cause a substantial decrease in the expression of a functional protein is supported by analyses of many human genetic diseases. Keep in mind that a genetic disease is caused by a rare mutant allele. **Table 16.2** lists several examples of human genetic diseases in which a recessive allele fails to produce a specific cellular protein in its active form.

Over 7,000 human disorders are caused by mutations in single genes. With a human genome size of 20,000 to 25,000 genes, this means that roughly one-third of our genes are known to cause some kind of abnormality when mutations alter their expression. Any particular single-gene disorder is relatively rare.

Protein P functions as an enzyme. The amount of functional protein P is the molecular connection between the genotype and the phenotype.

Genotype	*PP*	*Pp*	*pp*
Amount of functional protein P produced	100%	50%	0%
Phenotype	Purple	Purple	White

The relationship of the normal and mutant alleles is simple dominant/recessive

Colorless precursor molecule → Protein P → Purple pigment

Figure 16.16 **How genes give rise to traits in a simple dominant/recessive relationship.** In many cases, the amount of protein encoded by a single dominant allele is sufficient to produce the normal phenotype. In this example, the normal phenotype is purple flower color in a pea plant. The normal allele (*P*) encodes protein P, an enzyme needed for the synthesis of purple pigment. A plant with one or two copies of the normal allele produces enough pigment to produce purple flowers. In a *pp* homozygote, the complete lack of the normal protein results in white flowers.

Table 16.2	Examples of Recessive Human Genetic Diseases	
Disease	**Protein produced by the normal gene***	**Description**
Phenylketonuria	Phenylalanine hydroxylase	Inability to metabolize phenylalanine. Can lead to severe mental retardation and physical degeneration. The disease can be prevented by following a phenylalanine-free diet beginning early in life.
Cystic fibrosis	A chloride-ion transporter	Inability to regulate ion balance in epithelial cells. Leads to a variety of abnormalities, including production of thick lung mucus and chronic lung infections.
Tay-Sachs disease	Hexosaminidase A	Defect in lipid metabolism. Leads to paralysis, blindness, and early death.
Alpha-1 antitrypsin deficiency	Alpha-1 antitrypsin	Inability to prevent the activity of protease enzymes. Causes liver damage and emphysema.
Hemophilia A	Coagulation Factor VIII	A defect in blood clotting due to a missing clotting factor. An accident may cause excessive bleeding or internal hemorrhaging.

* Individuals who exhibit the disease are homozygous (or hemizygous) for a recessive allele that results in a defect in the amount or function of the normal protein.

But taken together, about one individual in 100 has a disorder that is due to a single-gene mutation. Such diseases generally have simple inheritance patterns in family pedigrees. Although the majority of these diseases follow a recessive inheritance pattern, some are known to be dominant. We have already discussed Huntington disease as an example of a dominant human disorder (see Figure 16.13). Other examples of diseases caused by dominant alleles include achondroplasia (a form of dwarfism) and osteogenesis imperfecta (brittle bone disease).

Single-gene disorders illustrate the phenomenon of **pleiotropy**, which means that a mutation in a single gene can have multiple effects on an individual's phenotype. Pleiotropy occurs for several reasons, including:

1. The expression of a single gene can affect cell function in more than one way. For example, a defect in a microtubule protein may affect cell division and cell movement.
2. A gene may be expressed in different cell types in a multicellular organism.
3. A gene may be expressed at different stages of development.

In this genetics unit, we tend to discuss genes as they affect a single trait. This educational approach allows us to appreciate how genes function, and how they are transmitted from parents to offspring. However, this focus may also obscure how amazing genes really are. In all or nearly all cases, the expression of a gene is pleiotropic with regard to the characteristics of an organism. The expression of any given gene influences the expression of many other genes in the genome, and vice versa. Pleiotropy is revealed when researchers study the effects of gene mutations.

As an example of a pleiotropic mutation, let's consider cystic fibrosis (CF), which we considered earlier as an example of a recessive human disorder (see Figure 16.12). In the late 1980s, the gene for CF was identified. The normal allele encodes a protein called the cystic fibrosis transmembrane conductance regulator (CFTR) that regulates ionic balance by allowing the transport of chloride ions (Cl⁻) across epithelial-cell membranes. The mutation that causes CF diminishes the function of this Cl⁻ transporter, affecting several parts of the body in different ways. Because the movement of Cl⁻ affects water transport across membranes, the most severe symptom of CF is thick mucus in the lungs, which occurs because of a water imbalance. In sweat glands, the normal Cl⁻ transporter has the function of recycling salt out of the glands and back into the skin before it can be lost to the outside world. Persons with CF have excessively salty sweat due to their inability to recycle salt back into their skin cells. A common test for CF is measurement of salt on the skin. Another effect is seen in the reproductive systems of males who are homozygous for the CF allele. Most males with CF are infertile because the vas deferens, the tubules that transport sperm from the testes, are absent or undeveloped. Presumably, a normally functioning Cl⁻ transporter is needed for the proper development of the vas deferens in the embryo. Taken together, we can see that a defect in CFTR has multiple effects throughout the body.

Incomplete Dominance Results in an Intermediate Phenotype

For certain traits, a heterozygote that carries two different alleles exhibits a phenotype that is intermediate between the corresponding homozygous individuals. This phenomenon is known as **incomplete dominance**. In 1905, Carl Correns discovered this pattern of inheritance for alleles affecting flower color in the four-o'clock plant (*Mirabilis jalapa*). **Figure 16.17** shows a cross between two four-o'clock plants, a red-flowered homozygote and a white-flowered homozygote. The allele for red flower color is designated C^R and the white allele is C^W. These alleles are designated with superscripts rather than upper- and lowercase letters because neither allele is dominant.

Figure 16.17 Incomplete dominance in the four-o'clock plant. When red-flowered and white-flowered homozygotes ($C^R C^R$ and $C^W C^W$) are crossed, the resulting heterozygote ($C^R C^W$) has an intermediate phenotype of pink flowers.

The offspring of this cross have pink flowers—that is, they are $C^R C^W$ heterozygotes with an intermediate phenotype. If these F_1 offspring are allowed to self-fertilize, the F_2 generation shows a ratio of 1/4 red-flowered plants, 1/2 pink-flowered plants, and 1/4 white-flowered plants. As noted in the Punnett square, this is a 1:2:1 phenotypic ratio rather than the 3:1 ratio observed for simple Mendelian inheritance. In this case, 50% of the protein encoded by the C^R gene is not sufficient to produce the red-flower phenotype.

The degree to which we judge an allele to exhibit incomplete dominance may depend on how closely we examine an individual's phenotype. An example is an inherited human disease called phenylketonuria (PKU). This disorder is caused by a rare mutation in a gene that encodes an enzyme called phenylalanine hydroxylase. This enzyme is needed to metabolize the amino acid phenylalanine. If left untreated, homozygotes carrying the mutant allele suffer severe symptoms, including mental retardation, seizures, microcephaly (small head), poor development of tooth enamel, and decreased body growth. By comparison, heterozygotes appear phenotypically normal. For this reason, geneticists consider PKU to be a recessive disorder. However, biochemical analysis of the blood of heterozygotes shows that they typically have a phenylalanine blood level double that of an individual carrying two normal copies of the gene. Therefore, at this closer level of examination, heterozygotes exhibit incomplete dominance.

ABO Blood Type Is an Example of Multiple Alleles and Codominance

Although diploid individuals have only two copies of most genes, many genes have three or more variants in one population. We describe such a gene as occurring in **multiple alleles**. The phenotype depends on which two alleles an individual inherits. ABO blood types in humans are an example of phenotypes produced by multiple alleles.

As shown in **Table 16.3**, human red blood cells have structures on their plasma membrane known as surface antigens,

which are constructed from several sugar molecules that are connected to form a carbohydrate tree. Antigens are substances (in this case, carbohydrates) that are recognized as foreign by antibodies produced by the immune system. Three types of surface antigens, known as A, B, and O, are found on red blood cells. The synthesis of these antigens is determined by enzymes that are encoded by a gene that exists in three alleles designated I^A, I^B, and i, respectively. The i allele is recessive to both I^A and I^B. A person who is ii homozygous will have red blood cells with the surface antigen O (blood type O). The red blood cells of an $I^A I^A$ homozygous or $I^A i$ heterozygous individual will have surface antigen A (blood type A). Similarly, a homozygous $I^B I^B$ or heterozygous $I^B i$ individual will produce surface antigen B (blood type B). A person who is $I^A I^B$ heterozygous makes both antigens, A and B, on every red blood cell (blood type AB). The phenomenon in which a single individual expresses both alleles is called **codominance**.

What is the molecular explanation for codominance? Biochemists have analyzed the carbohydrate tree produced in people of differing blood types. The differences are shown schematically in Table 16.3. In type O, the carbohydrate tree is smaller than in type A or type B because a sugar has not been attached to a specific site on the tree. People with blood type O have a loss-of-function mutation in the gene that encodes the enzyme that attaches a sugar at this site. This enzyme, called a glycosyl transferase, is inactive in type O individuals. In contrast, the type A and type B antigens have sugars attached to this site, but each of them has a different sugar. This difference occurs because the enzymes encoded by the I^A allele and the I^B allele have slightly different active sites. As a result, the enzyme encoded by the I^A allele attaches a sugar called N-acetylgalactosamine to the carbohydrate tree, while the enzyme encoded by the I^B allele attaches galactose. N-acetylgalactosamine is represented by an orange hexagon in Table 16.3, and galactose by a green triangle.

The attachment of two different sugars gives surface antigens A and B significantly different molecular structures. Such differences in shape allow antibodies to recognize and bind very

Table 16.3	The ABO Blood Group			
	Antigen O	Antigen A	Antigen B	Antigen A Antigen B
Blood Type:	O	A	B	AB
Genotype:	ii	$I^A I^A$ or $I^A i$	$I^B I^B$ or $I^B i$	$I^A I^B$
Surface antigen:	O	A	B	AB
Serum antibodies:	against A and B	against B	against A	none

specifically to certain antigens. The blood of type A individuals has antibodies, called serum antibodies, that bind to the B antigen. Similarly, type B individuals produce antibodies against the A antigen. Type O individuals produce both kinds of antibodies and type AB individuals produce neither. (No antibodies are produced against antigen O.) When a person receives a blood transfusion, the donor's blood must be an appropriate match with the recipient's blood to avoid a dangerous antigen-antibody reaction. For example, if a person with type O blood is given type A blood, the recipient's anti-A antibodies will react with the donated blood cells and cause them to agglutinate (clump together). This situation is life-threatening because it will cause the blood vessels to clog. Identification of the donor and recipient blood types, called blood typing, is essential for safe transfusions.

The Expression of Certain Traits Is Influenced by the Sex of the Individual

Certain autosomal genes are expressed differently in heterozygous males and females. The term **sex-influenced inheritance** refers to the phenomenon in which an allele is dominant in one sex but recessive in the other. Pattern baldness is an example of a sex-influenced trait in humans. This trait is characterized by a balding pattern in which hair loss occurs on the front and top but not on the sides (**Figure 16.18**). A male who is heterozygous for the pattern-baldness allele (designated *B*) will become bald, but a heterozygous female will not. In other words, the baldness allele is dominant in males but recessive in females:

Genotype	Phenotype	
	Females	Males
BB	bald	bald
Bb	nonbald	bald
bb	nonbald	nonbald

A woman who is homozygous for the baldness allele will develop the trait (although in women it is usually characterized by a significant thinning of the hair that occurs relatively late in life).

As you can see from the pedigree in Figure 16.18, a bald male may have inherited the baldness allele from either parent. Thus, a striking observation is that fathers with pattern baldness can pass this trait to their sons. This could not occur if the trait was X linked, because fathers transmit only Y chromosomes to their sons.

The sex-influenced nature of pattern baldness is related to the production of the male sex hormone testosterone. The gene that affects pattern baldness encodes an enzyme called 5 α-reductase, which converts testosterone to 5-α-dihydrotestosterone (DHT). DHT binds to cellular receptors and affects the expression of many genes, including those in the cells of the scalp. The allele that causes pattern baldness results in an overexpres-

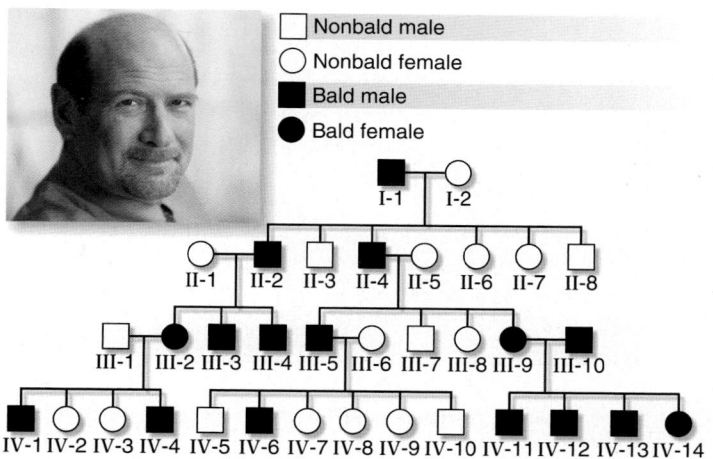

Figure 16.18   Pattern baldness. Pattern baldness, shown in an adult male in the photograph, is an example of the sex-influenced expression of an autosomal gene. Bald individuals are represented by filled symbols in the pedigree.

Biological inquiry: With regard to baldness, which phenotypes have a single genotype?

sion of this enzyme. Because mature males normally make more testosterone than females, this allele has a greater phenotypic impact in males. However, a rare tumor of the adrenal gland can cause the secretion of abnormally large amounts of testosterone in females. If this occurs in a woman who is heterozygous *Bb*, she will become bald. If the tumor is removed surgically, her hair will return to its normal condition.

The Environment Plays a Vital Role in the Making of a Phenotype

In this chapter, we have been concerned mainly with the effects of genes on phenotypes. However, phenotypes are shaped by an organism's environment as well as by its genes. An organism cannot exist without its genes or without an environment in which to live. Both are indispensable for life. An organism's genotype provides the information for environmental conditions to create a phenotype.

The term **norm of reaction** refers to the effects of environmental variation on a phenotype. Specifically, it is the phenotypic range seen in individuals with a particular genotype. To evaluate the norm of reaction, researchers study members of true-breeding strains that have the same genotypes, and subject them to different environmental conditions. For example, **Figure 16.19** shows the norm of reaction for genetically identical plants raised at different temperatures. As shown in the figure, these plants attain a maximal height when raised at 75°F. At 55°F and 85°F, the plants are substantially shorter. Growth cannot occur below 40°F or above 95°F.

The norm of reaction can be quite dramatic when we consider environmental influences on certain inherited diseases.

Figure 16.19 **The norm of reaction.** The norm of reaction is the range of phenotypes that an organism with a particular genotype exhibits under different environmental conditions. In this example, genetically identical plants were grown at different temperatures in a greenhouse and then measured for height.

A striking example is the human genetic disease phenylketonuria (PKU). As we discussed earlier in the chapter, this disorder is caused by a rare mutation in the gene for phenylalanine hydroxylase, an enzyme needed to metabolize the amino acid phenylalanine. People with one or two functional copies of the gene can eat foods containing the amino acid phenylalanine and metabolize it correctly. However, homozygotes for the defective allele cannot metabolize phenylalanine. When these individuals eat a standard diet containing phenylalanine, this amino acid accumulates within their bodies and becomes highly toxic. Under these conditions, PKU homozygotes manifest a variety of detrimental symptoms, including mental impairment, underdeveloped teeth, and foul-smelling urine.

Figure 16.20 **Environmental influences on the expression of PKU within a single family.** All three children in this photo have inherited the alleles that cause PKU. The child in the middle was raised on a phenylalanine-free diet and developed normally. The other two children, born before the benefits of such a diet were known, were raised on diets containing phenylalanine. These two children have symptoms of PKU, including mental impairment.

Photo from the March of Dimes Birth Defects Foundation.

In contrast, when these individuals are identified at birth and given a restricted diet that is free of phenylalanine, they develop normally (**Figure 16.20**). This is a dramatic example of how genes and the environment can interact to determine an individual's phenotype. In the U.S., most newborns are tested for PKU, which occurs in about 1 in 10,000 babies. A newborn who is found to have this disorder can be raised on a phenylalanine-free diet and develop normally.

CHAPTER SUMMARY

16.1 Mendel's Laws and the Chromosome Theory of Inheritance

- Mendel focused his attention on seven traits found in garden peas that existed in two variants each. (Figures 16.1, 16.2)

- Mendel could allow his peas to self-fertilize or he could carry out cross-fertilization, also known as hybridization. (Figures 16.3, 16.4)

- By following the inheritance pattern of a single trait (a single-factor cross) for two generations, Mendel determined the law of segregation. This law tells us that two alleles of a gene segregate from each other when passed from parents to offspring. (Figures 16.5, 16.6)

- The genotype is the genetic makeup of an organism. Alleles are alternative versions of the same gene. Phenotype is a description of the traits that an organism displays.

- A Punnett square can be constructed to predict the outcome of crosses.

- A testcross can be conducted to determine if an individual displaying a dominant trait is a homozygote or heterozygote. (Figure 16.7)

- By conducting a dihybrid cross, Mendel determined the law of independent assortment, which says that the alleles for two different genes assort independently of each other. In a dihybrid cross, this yields a 9:3:3:1 ratio in the F_2 generation. (Figure 16.8)

- The chromosome theory of inheritance explains how the steps of meiosis account for the inheritance patterns observed by Mendel. Each gene is located at a particular locus on a chromosome. (Figures 16.9, 16.10, 16.11)

- The product rule and sum rule allow us to predict the outcome of crosses based on probability. Random sampling error is the deviation between observed and predicted values.

- Instead of conducting crosses, the inheritance patterns in humans are determined from a pedigree analysis. (Figures 16.12, 16.13)

16.2 Sex Chromosomes and X-Linked Inheritance Patterns

- Many species of animals and a few species of plants have separate male and female sexes. In many cases, sex is determined by differences in sex chromosomes. (Figure 16.14)

- In mammals, recessive X-linked traits are more likely to occur in males. An example is hemophilia.

- Morgan carried out crosses that showed that an eye-color gene in *Drosophila* is located on the X chromosome. (Figure 16.15)

16.3 Variations in Inheritance Patterns and Their Molecular Basis

- Several inheritance patterns have been discovered that obey Mendel's laws but yield differing ratios of offspring compared to Mendel's crosses. (Table 16.1)

- Recessive inheritance is often due to a loss-of-function allele. The heterozygote has a dominant phenotype because 50% of the normal protein is sufficient to produce that phenotype. (Figure 16.16)

- Mutant genes are responsible for many inherited diseases in humans. In many cases, the effects of a mutant gene are pleiotropic, meaning that the gene affects several different aspects of bodily structure and function. (Table 16.2)

- Incomplete dominance occurs when a heterozygote has a phenotype that is intermediate between either homozygote. This occurs because 50% of the normal protein is not enough to give the same phenotype as a normal homozygote. (Figure 16.17)

- ABO blood type is an example of multiple alleles in which a gene exists in three alleles in a population. The A and B alleles are codominant, which means that both are expressed in the same individual. These alleles encode enzymes with different specificities for attaching sugar molecules to make antigens. (Table 16.3)

- Pattern baldness in people is a sex-influenced trait that is dominant in males and recessive in females. This pattern occurs because sex hormones influence the expression of certain genes. (Figure 16.18)

- All traits are influenced by the environment. The norm of reaction is a description of how a trait may change depending on the environmental conditions. (Figures 16.19, 16.20)

TEST YOURSELF

1. Based on Mendel's experimental crosses, what is the expected F_2 phenotypic ratio of a monohybrid cross?
   a. 1:2:1
   b. 2:1
   c. 3:1
   d. 9:3:3:1
   e. 4:1

2. During which phase of cellular division does Mendel's law of segregation physically occur?
   a. mitosis
   b. meiosis I
   c. meiosis II
   d. all of the above
   e. b and c only

3. An individual that has two different alleles of a particular gene is said to be
   a. dihybrid.
   b. recessive.
   c. homozygous.
   d. heterozygous.
   e. hemizygous.

4. Which of Mendel's laws cannot be observed in a monohybrid cross?
   a. segregation
   b. dominance/recessiveness
   c. independent assortment
   d. codominance
   e. All of the above can be observed in a monohybrid cross.

5. During a ____ cross, an individual with the dominant phenotype and unknown genotype is crossed with a _____ individual to determine the unknown genotype.
   a. monohybrid, homozygous recessive
   b. dihybrid, heterozygous
   c. test, homozygous dominant
   d. monohybrid, homozygous dominant
   e. test, homozygous recessive

6. In humans, males are said to be _____ at X-linked loci.
   a. dominant
   b. homozygous
   c. recessive
   d. heterozygous
   e. hemizygous

7. A gene that affects more than one phenotypic trait is said to be
   a. dominant.
   b. wild type.
   c. dihybrid.
   d. pleiotropic.
   e. heterozygous.

8. A hypothetical flowering plant species produces red, pink, and white flowers. To determine the inheritance pattern, the following crosses were conducted with the results indicated:
   red × red → all red
   white × white → all white
   red × white → all pink

   What type of inheritance pattern does this represent?
   a. dominance/recessiveness
   b. X-linked
   c. codominance
   d. incomplete dominance
   e. pleiotropy

9. Genes located on a sex chromosome are said to be
   a. X-linked.
   b. dominant.
   c. hemizygous.
   d. sex-linked.
   e. sex-influenced.

10. Genes that are expressed differently depending on whether the individual is male or female are
    a. sex-linked.
    b. pleiotropic.
    c. sex-influenced.
    d. incomplete dominant.
    e. hemizygous.

CONCEPTUAL QUESTIONS

1. Define genotype and phenotype.

2. Define autosome.

3. Explain why recessive X-linked traits in humans are more likely to occur in males.

EXPERIMENTAL QUESTIONS

1. Prior to the Feature Investigation, what was the original purpose of Morgan's experiments with *Drosophila*?

2. How was Morgan able to demonstrate that red-eye color is dominant to white-eye color?

3. What results led Morgan to conclude that eye color was associated with the sex of the individual?

COLLABORATIVE QUESTIONS

1. Discuss Mendel's two laws and why they are important.

2. What are the fundamental principles of the chromosome theory of inheritance?

www.brookerbiology.com
This website includes answers to the Biological Inquiry questions found in the figure legends and all end-of-chapter questions.

17

COMPLEX PATTERNS OF INHERITANCE

Snail shells (*Lymnaea peregra*) that coil to the right or left. The direction of coiling of a snail's shell is an example of a complex inheritance pattern.

I n Chapter 16, we examined inheritance patterns in which the outcome of a single trait was governed by a single gene. In the cases we considered, the alleles segregated and assorted independently, allowing us to predict the phenotypes of offspring from the genotypes of their parents. These phenotypes occurred in definite ratios and they did not overlap—a pea plant was either tall or dwarf; a blood type was either A, B, or O. The inheritance patterns of most traits are more complex, however, and in this chapter we will examine some of the factors that complicate the prediction of phenotypes.

In the first section of the chapter, we will consider how two or more different genes may affect the outcome of a single trait. For example, we examine continuously varying traits like human skin color, and you will see how the interaction of multiple genes and environmental influences can produce such a continuum. In the rest of the chapter we will consider inheritance patterns that defy Mendel's laws of inheritance. First we discuss genes that are linked on the same chromosome and therefore do not assort independently. Next we consider the genes found in chloroplasts and mitochondria, which defy the law of segregation. Don't worry, Mendel's laws do describe most inheritance patterns, and they accurately reflect the behavior of chromosomes during meiosis. However, as you will learn in this chapter, they simply don't apply to all of the genes that eukaryotic organisms possess. We will end the chapter with a discussion of three inheritance patterns, X inactivation, genomic imprinting, and maternal effect, that were not easily explained until researchers began to unravel genetic events that occur at the cellular and molecular levels. As you will learn, males and females don't always regulate their genes in the same way, and this can lead to seemingly bizarre patterns of inheritance that are distinct from X-linked and sex-influenced inheritance patterns,

which we considered in the previous chapter. An exciting advance over the past few decades has been a better understanding of such unusual patterns of inheritance.

Studies of complex inheritance patterns such as those described in this chapter have helped us appreciate more fully how genes influence phenotypes. These studies have revealed an astounding variety in the ways that inheritance occurs. The picture that emerges is of a wonderful web of diverse mechanisms by which genes give rise to phenotypes. **Table 17.1** provides a summary of the most common patterns of inheritance.

17.1 Gene Interactions

The study of single genes was pivotal in establishing the science of genetics. This focus allowed Mendel to formulate the basic laws of inheritance for traits with a simple dominant/recessive inheritance pattern. Likewise, this approach helped later researchers understand inheritance patterns involving incomplete dominance and codominance, as well as traits that are influenced by an individual's sex. In reality, however, all or nearly all traits are influenced by many genes. For example, in both plants and animals, height is affected by genes that encode proteins involved in the production of growth hormones, cell division, the uptake of nutrients, metabolism, and many other functions. A defect in any of these genes is likely to have a negative impact on an individual's height.

If height is controlled by many genes, you may be wondering how Mendel was able to study the effects of a single gene that produced tall or dwarf pea plants. The answer lies in the genotypes of his strains. Although many genes affect the height of pea plants, Mendel chose true-breeding strains that differed

Table 17.1	Different Types of Inheritance Patterns
Type	**Description**
Mendelian	Inheritance patterns in which a single gene affects a single trait, and the alleles segregate and assort independently. These patterns include simple dominant/recessive traits, X-linked traits controlled by a single gene, incomplete dominance, codominance, and sex-influenced traits (refer back to Table 16.1).
Epistasis	A type of gene interaction in which the alleles of one gene mask the effects of a dominant allele of another gene.
Continuous variation	Inheritance pattern in which the offspring display a continuous range of phenotypes. This pattern is produced by the additive interactions of several genes, together with environmental influences.
Linkage	Inheritance patterns involving two or more genes that are close together on the same chromosome. These genes do not assort independently.
Extranuclear inheritance	Transmission pattern of genes found in the DNA of mitochondria or chloroplasts, which are inherited independently of genes in the nucleus and do not segregate during meiosis. Usually these genes are inherited from the mother.
X inactivation	Phenomenon of female mammals in which one X chromosome is inactivated in every somatic cell, producing a mosaic phenotype. Most genes on the inactivated X chromosome are not expressed.
Genomic imprinting	Inheritance pattern in which an allele from one parent is inactivated in the somatic cells of the offspring, while the allele from the other parent is expressed.
Maternal effect	Inheritance pattern in which the genotype of the mother determines the phenotype of the offspring. This occurs because maternal effect genes of the mother provide gene products to developing egg cells.

with regard to only one of these genes. As a hypothetical example, let's suppose that pea plants have 10 genes affecting height, which we will call K, L, M, N, O, P, Q, R, S, and T. The genotypes of two hypothetical strains of pea plants may be:

Tall strain: *KK LL MM NN OO PP QQ RR SS TT*

Dwarf strain: *KK LL MM NN OO PP QQ RR SS tt*

In this example, the tall and dwarf strains differ at only a single locus. One strain is *TT* and the other is *tt*, and this accounts for the difference in their height. If we make crosses of tall and dwarf plants, the genotypes of the F_2 offspring will differ with regard to only one gene; the other nine genes will be identical in all of them. This approach allows a researcher to study the effects of a single gene even though many genes may affect a single trait.

In this section, we will examine situations in which a single trait is controlled by two or more genes, each of which has two

or more alleles. This phenomenon is called a **gene interaction**. As you will see, allelic variation at two or more loci may affect the outcome of traits in different ways. First we will look at interactions in which an allele of one gene prevents the expression of an allele of a different gene. Then we will discuss interactions in which multiple genes have additive effects on a single trait. These additive effects, together with environmental influences, account for the continuous phenotypic variation that we see for most traits.

An Epistatic Gene Interaction Occurs When the Allele of One Gene Masks the Phenotypic Effects of a Different Gene

In some gene interactions, the alleles of one gene mask the expression of the alleles of another gene. This phenomenon is called **epistasis** (Greek *epistanai*, stopping). An example is the unexpected gene interaction discovered by William Bateson and Reginald Punnett in the early 1900s, when they were studying crosses involving the sweet pea, *Lathyrus odoratus*. A cross between a true-breeding purple-flowered plant and a true-breeding white-flowered plant produced an F_1 generation with all purple-flowered plants and an F_2 generation with a 3:1 ratio of purple- to white-flowered plants. Of course, Mendel's laws predicted this result. The surprise came when the researchers crossed two different varieties of white-flowered sweet peas (**Figure 17.1**). All of the F_1 generation plants had purple flowers! When these plants were allowed to self-fertilize, the F_2 generation had purple-flowered and white-flowered plants in a 9:7 ratio. From these results, Bateson and Punnett deduced that two different genes were involved. To have purple flowers, a plant must have one or two dominant alleles for each of these genes. The relationships among the alleles are as follows:

C (one allele for purple) is dominant to *c* (white)

P (an allele for purple of a different gene) is dominant to *p* (white)

cc masks *P*, or *pp* masks *C*, in either case producing white flowers

A plant that was homozygous for either *c* or *p* would have white flowers even if it had a purple-producing allele at the other locus.

Epistatic interactions often arise because two or more different proteins are involved in a single cellular function. For example, two or more proteins may be part of an enzymatic pathway leading to the formation of a single product. This is the case for the formation of a purple pigment in the sweet pea strains we have been discussing:

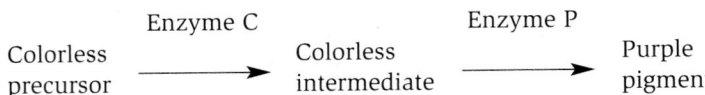

In this example, a colorless precursor molecule must be acted on by two different enzymes to produce the purple pig-

Figure 17.1 Epistasis in the sweet pea.   The color of the sweet pea flower is controlled by two genes, each with a dominant and a recessive allele. Each of the dominant alleles (*C* and *P*) encodes an enzyme required for the synthesis of purple pigment. A plant that is homozygous recessive for either gene (*cc* or *pp*) cannot synthesize the pigment and will have white flowers.

Biological inquiry: In a Ccpp individual, which functional enzyme is missing? Is it the enzyme encoded by the C or P gene?

or enzyme P. When either of these enzymes is missing, the plant cannot make the purple pigment and has white flowers.

Polygenic Inheritance and Environmental Influences Produce Continuous Phenotypic Variation

As we have just seen, an epistatic interaction causes the alleles of one gene to mask the effects of a different gene. Let's now turn to another way that the alleles of different genes may affect the phenotype of a single trait. In many cases, the effects of alleles may be additive. This has been observed for many traits, particularly those that are quantitative in nature.

Until now we have discussed the inheritance of traits with clearly defined phenotypic variants, such as red or white eyes in fruit flies. These are known as discrete traits, or **discontinuous traits**, because the phenotypes do not overlap. For most traits, however, the phenotypes cannot be sorted into discrete categories. The majority of traits in all organisms are **continuous traits**, also called **quantitative traits**, which show continuous variation over a range of phenotypes. In humans, quantitative traits include height, weight, skin color, metabolic rate, and heart size, to mention a few. In the case of domestic animals and plant crops, many of the traits that people consider desirable are quantitative in nature, such as the number of eggs a chicken lays, the amount of milk a cow produces, and the number of apples on an apple tree. Consequently, much of our modern understanding of quantitative traits comes from agricultural research.

Quantitative traits are **polygenic**, which means that several or many genes contribute to the outcome of the trait. For many polygenic traits, genes contribute to the phenotype in an additive way. Another important factor is the environment. As we saw in Chapter 16, the environment plays a vital role in the phenotypic expression of genes. Environmental factors often have a major impact on quantitative traits. For example, an animal's diet affects its weight, and the amount of rain and sunlight that fall on an apple tree affect how many apples it produces.

Because quantitative traits are polygenic and greatly influenced by environmental conditions, the phenotypes among different individuals may vary substantially in any given population. As an example, let's consider skin pigmentation in people. This trait is influenced by several genes that tend to interact in an additive way. As a simplified example, let's consider a population in which this trait is controlled by three genes, which we will designate *A*, *B*, and *C*. Each gene has a dark allele, designated A^D, B^D, or C^D, and a light allele, designated A^L, B^L, or C^L, respectively. All of the alleles encode enzymes that cause the synthesis of skin pigment, but the enzymes encoded by dark alleles cause more pigment synthesis than the enzymes encoded by light alleles. **Figure 17.2** considers a hypothetical case in which people who were heterozygous for all three genes produced a large population of offspring. The bar graph shows the genotypes of the offspring, grouped according to the total number of dark alleles. As shown by the shading of the figure, skin pigmentation increases as the number of dark alleles increases.

ment. Gene *C* encodes a functional protein, enzyme C, that converts the colorless precursor into a colorless intermediate. The recessive *c* allele results in a lack of production of enzyme C in the homozygote. Gene *P* encodes the functional enzyme P, which converts the colorless intermediate into the purple pigment. Like the *c* allele, the *p* allele results in an inability to produce a functional protein. A plant that is homozygous for either of the recessive alleles will not make any functional enzyme C

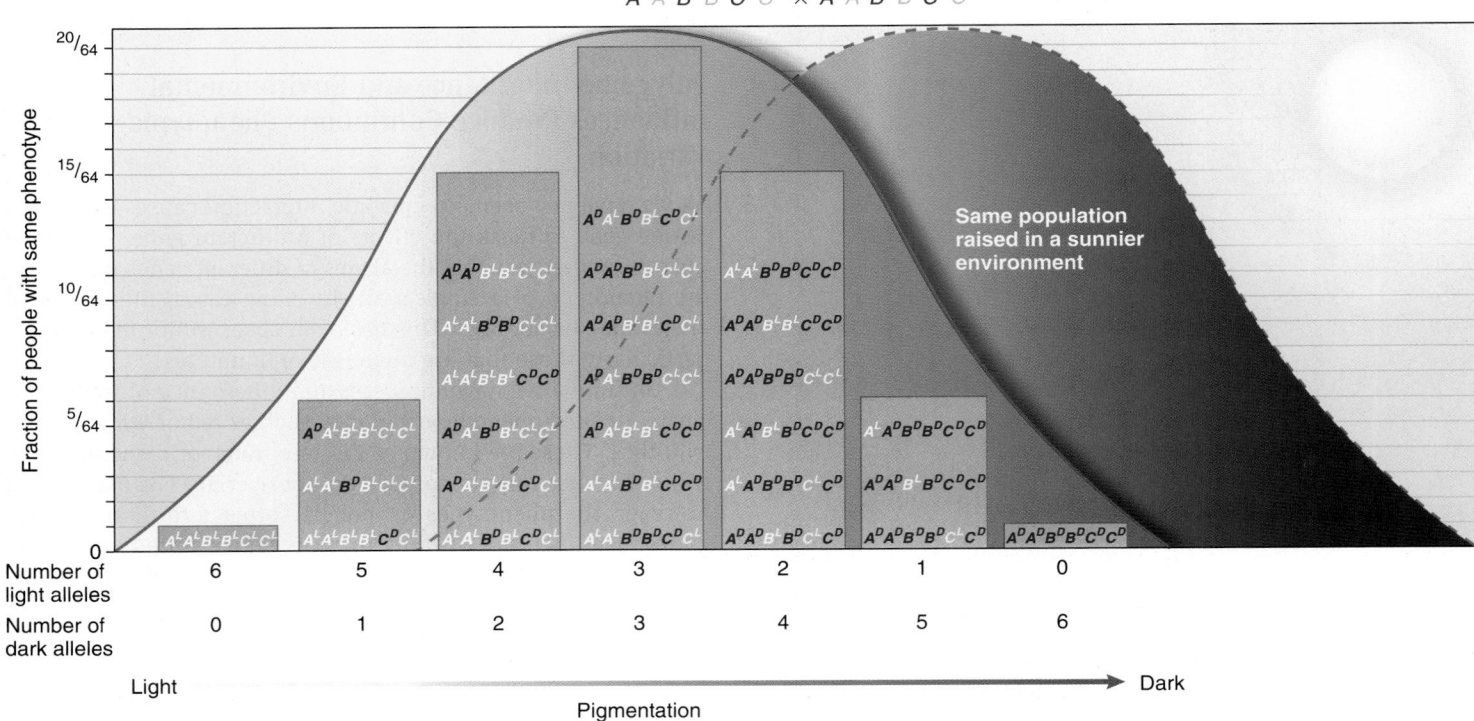

Figure 17.2 Continuous variation in a polygenic trait.   Skin color is a polygenic trait that can display a continuum of phenotypes. The bell curve on the *left* (solid line) shows the range of skin pigmentation in a hypothetical human population. The bar graphs below the curve show the additive effects of three genes that affect pigment production in this population; each bar shows the fraction of people with a particular number of dark alleles (A^D, B^D, and C^D) and light alleles (A^L, B^L, and C^L). The bell curve on the *right* (dashed line) represents the expected range of phenotypes if the same population was raised in a sunnier environment.

Offspring who have no dark alleles or no light alleles—that is, who are homozygous for all three genes—are fewer in number than those with some combination of dark and light alleles. As seen in the bell-shaped curve above the bar graph, the phenotypes of the offspring fall along a continuum. This continuous phenotypic variation, which is typical of quantitative traits, is produced by genotypic differences together with environmental effects. A second bell-shaped curve (the dashed line) depicts the expected phenotypic range if the same population of offspring had been raised in a sunnier environment, which increases pigment production. These two curves illustrate how the environment can also have a significant influence on the range of phenotypes.

In our discussion of genetics, we tend to focus on discrete traits because this makes it easier to relate a specific genotype with a phenotype. This is usually not possible for continuous traits. For example, as depicted in the middle bar of Figure 17.2, seven different genotypes can produce individuals with a medium amount of pigmentation. Nevertheless, it is important to emphasize that the majority of traits in all organisms are continuous, not discrete. Most traits are influenced by multiple genes, and the environment has an important impact on the phenotypic outcome.

17.2 Genes on the Same Chromosome: Linkage, Recombination, and Mapping

In all of the inheritance patterns we have studied so far, the alleles segregate and assort independently as predicted by Mendel's laws. As we have seen, phenotypes can be influenced by a variety of factors, including gene interactions and environmental effects, that make it difficult to relate genotype to phenotype. Even so, if we understand all of these factors and take them into account, we can see that each of the genes is transmitted according to Mendel's laws.

In the rest of this chapter, we will consider inheritance patterns in which the outcome of a cross violates one of Mendel's laws. In this section, we focus on transmission patterns that do not conform to the law of independent assortment. We will begin by examining the first experimental cross that demonstrated this pattern. You will learn that this pattern was explained by Thomas Hunt Morgan, who proposed that genes located close to each other on the same chromosome tend to be inherited as a group. Finally, we will see how crossing over between such genes provided the first method of mapping genes on chromosomes.

FEATURE INVESTIGATION

Bateson and Punnett's Crosses of Sweet Peas Showed That Genes Do Not Always Assort Independently

In Chapter 16, we learned that the independent assortment of alleles is due to the random alignment of homologous chromosomes during meiosis (refer back to Figure 16.11). But what happens when the alleles of different genes are on the same chromosome? A typical chromosome contains many hundreds or even a few thousand different genes. When two genes are close together on the same chromosome, they tend to be transmitted as a unit, a phenomenon known as **linkage**. A group of genes that usu-

ally stay together during meiosis is called a **linkage group**, and the genes in the group are said to be linked. In a two-factor cross, linked genes do not follow the law of independent assortment.

The first study showing linkage between two different genes was a cross of sweet peas carried out by William Bateson and Reginald Punnett in 1905. A surprising result occurred when they conducted a cross involving two different traits, flower color and pollen shape (**Figure 17.3**). One of the parent plants had purple flowers (*PP*) and long pollen (*LL*); the other had red flowers (*pp*) and round pollen (*ll*). As Bateson and Punnett expected, the F_1 plants all had purple flowers and long pollen (*PpLl*). The unexpected result came in the F_2 generation.

Figure 17.3 A cross of sweet peas showing that independent assortment does not always occur.

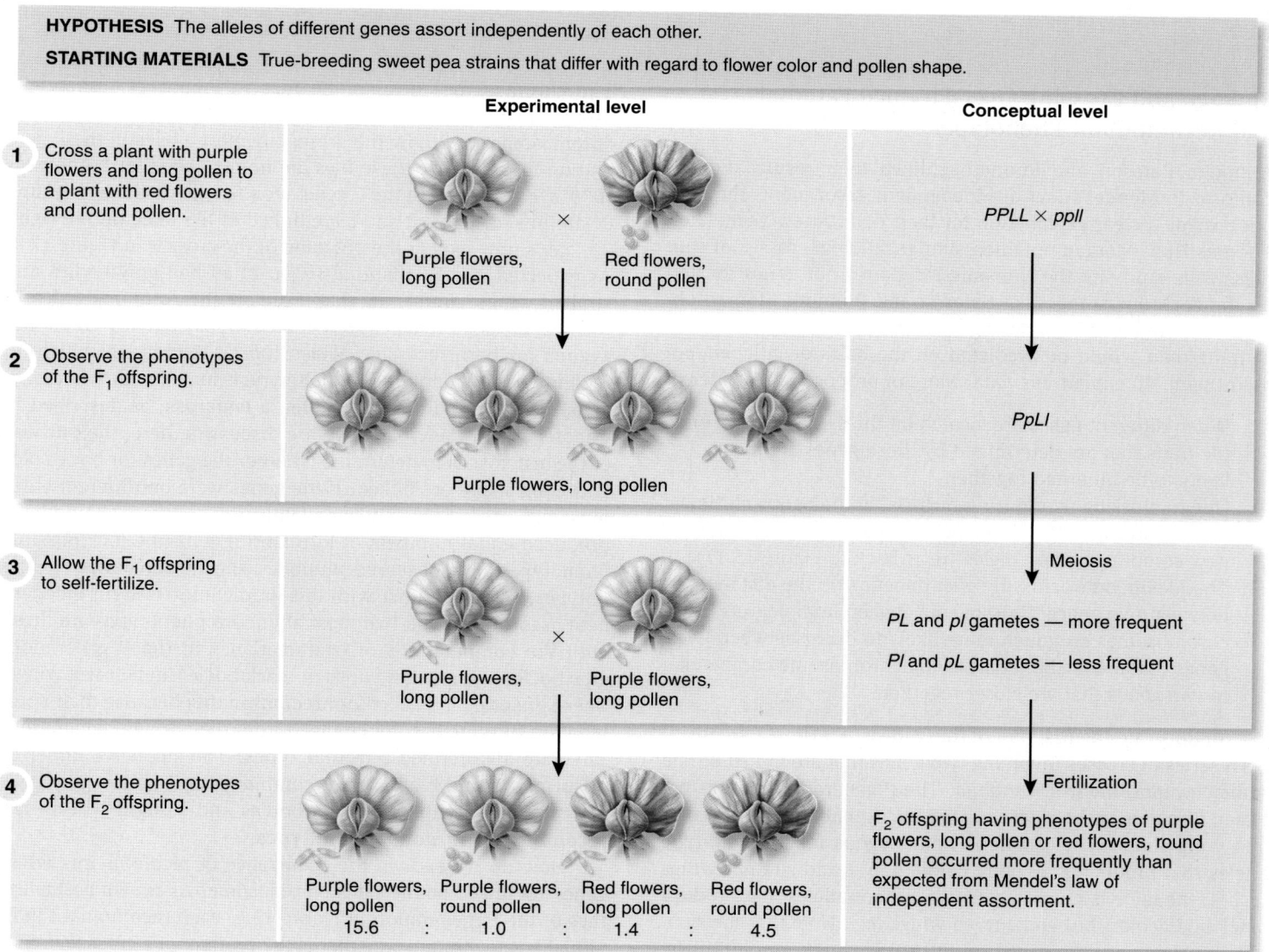

HYPOTHESIS The alleles of different genes assort independently of each other.

STARTING MATERIALS True-breeding sweet pea strains that differ with regard to flower color and pollen shape.

Experimental level | Conceptual level

1 Cross a plant with purple flowers and long pollen to a plant with red flowers and round pollen.

Purple flowers, long pollen × Red flowers, round pollen

$PPLL \times ppll$

2 Observe the phenotypes of the F_1 offspring.

Purple flowers, long pollen

$PpLl$

3 Allow the F_1 offspring to self-fertilize.

Purple flowers, long pollen × Purple flowers, long pollen

Meiosis

PL and *pl* gametes — more frequent

Pl and *pL* gametes — less frequent

4 Observe the phenotypes of the F_2 offspring.

Purple flowers, long pollen 15.6 : Purple flowers, round pollen 1.0 : Red flowers, long pollen 1.4 : Red flowers, round pollen 4.5

Fertilization

F_2 offspring having phenotypes of purple flowers, long pollen or red flowers, round pollen occurred more frequently than expected from Mendel's law of independent assortment.

THE DATA

Phenotypes of F₂ offspring	Observed number	Observed ratio	Expected number	Expected ratio
Purple flowers, long pollen	296	15.6	240	9
Purple flowers, round pollen	19	1.0	80	3
Red flowers, long pollen	27	1.4	80	3
Red flowers, round pollen	85	4.5	27	1

Although the offspring displayed the four phenotypes predicted by Mendel's laws, the observed numbers of offspring did not conform to the predicted 9:3:3:1 ratio. Rather, as seen in the data in Figure 17.3, the F_2 generation had a much higher proportion of the two phenotypes found in the parental generation: purple flowers with long pollen, and red flowers with round pollen.

These results did not support the law of independent assortment. Bateson and Punnett suggested that the transmission of flower color and pollen shape was somehow coupled, so that these traits did not always assort independently. Although the law of independent assortment applies to many other genes, in this example, the hypothesis of independent assortment was rejected.

Linkage and Crossing Over Produce Parental and Recombinant Phenotypes

Although Bateson and Punnett realized their results did not conform to Mendel's law of independent assortment, they did not provide a clear explanation for their data. A few years later, Thomas Hunt Morgan obtained similar ratios in crosses of fruit flies while studying the transmission pattern of genes located on the X chromosome. Like Bateson and Punnett, Morgan observed many more F_2 offspring with the parental combination of traits than would be predicted on the basis of independent assortment. To explain his data, Morgan proposed these ideas:

1. When different genes are located on the same chromosome, the traits that are determined by those genes are most likely to be inherited together.
2. Due to crossing over during meiosis, homologous chromosomes can exchange pieces of chromosomes and create new combinations of alleles (refer back to Figure 15.17).
3. The likelihood of crossing over depends on the distance between two genes. Crossovers between homologous chromosomes are much more likely to occur between two genes that are farther apart in the chromosome compared to two genes that are closer together.

To illustrate the first two of these ideas, **Figure 17.4** considers a series of crosses involving two genes that are linked on the same chromosome in *Drosophila*. The P generation cross is between flies that are homozygous for alleles that affect body color and wing shape. The female is homozygous for the wild-type alleles that produce gray body color (b^+b^+) and straight wings (c^+c^+); the male is homozygous for mutant alleles that produce black body color (bb) and curved wings (cc). Note that the symbols for the genes are based on the name of the mutant allele; the wild-type allele is indicated by a superscript plus sign ($+$).

The chromosomes next to the flies in Figure 17.4 show the arrangement of these alleles. If the two genes are on the same chromosome, we know the arrangement of alleles in the P generation flies because these flies are homozygous for both genes ($b^+b^+c^+c^+$ or $bbcc$). In the P generation female on the left, b^+ and c^+ are linked, while b and c are linked in the male on the right.

Let's now look at the outcome of the crosses in Figure 17.4. As expected, the F_1 offspring (b^+bc^+c) all had gray bodies and straight wings, confirming that these are the dominant traits. In the next cross, F_1 females were mated to males that were homozygous for both recessive alleles ($bbcc$). A cross in which an individual with a dominant phenotype is mated with a homozygous recessive individual is called a **testcross**, as described in Chapter 16. In the crosses we are discussing here, the purpose of the testcross is to determine whether the genes for body color and wing shape are linked. If the genes were on different chromosomes and assorted independently, this testcross should have produced equal numbers of F_2 offspring with the four possible phenotypes. The observed numbers, shown above the F_2 phenotypes, clearly conflict with this prediction based on independent assortment. The two most abundant phenotypes are those with the combinations of characteristics in the P generation: gray bodies and straight wings or black bodies and curved wings. These offspring are called **nonrecombinants** because their combination of traits has not changed from the parental generation. They are also termed **parental types**. The smaller number of offspring that have a different combination of traits—gray bodies and curved wings or black bodies and straight wings—are **recombinants** or **nonparental types**.

How do we explain the occurrence of recombinants when genes are linked on the same chromosome? As shown beside the flies of the F_2 generation in Figure 17.4, each recombinant individual has a chromosome that is the product of a crossover. The crossover occurred while the F_1 female fly was making egg cells.

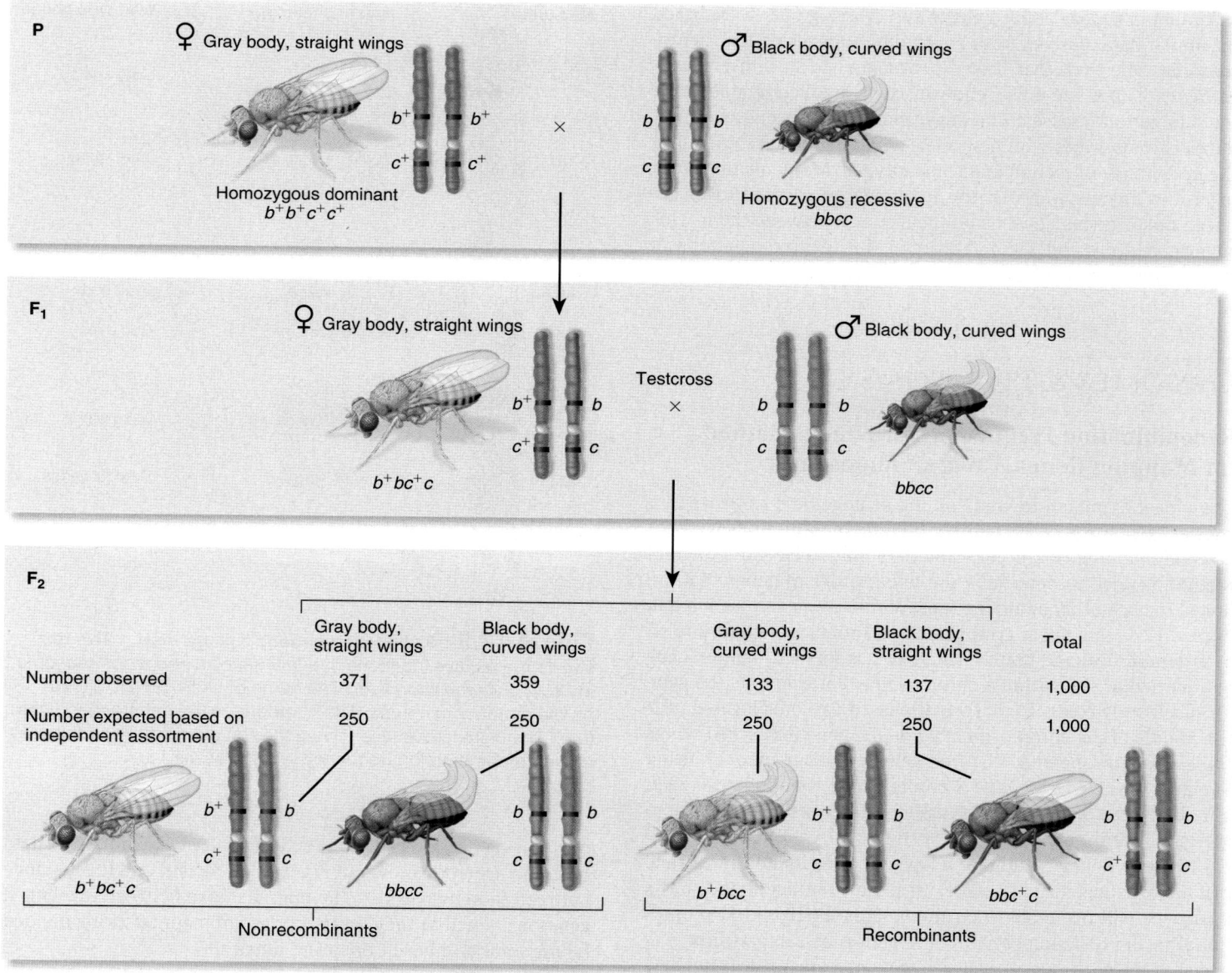

Figure 17.4   Linkage and recombination of alleles.   An experimenter crossed $b^+b^+c^+c^+$ and $bbcc$ flies to produce F_1 heterozygotes. F_1 females were then testcrossed to $bbcc$ males. The large number of parental phenotypes in the F_2 generation suggests that the two genes are linked on the same chromosome. F_2 recombinant phenotypes occur because the alleles can be rearranged by crossing over.

As shown below, four different egg cells are possible:

Due to crossing over, two of the four egg cells produced by meiosis have recombinant chromosomes. What happens when eggs containing such chromosomes are fertilized in the testcross? Each of the male fly's sperm cells carries a chromosome with the two recessive alleles. If the egg contains the recombinant chromosome carrying the b^+ and c alleles, the testcross will produce an F_2 offspring with a gray body and curved wings. If the egg contains the recombinant chromosome carrying the b and c^+ alleles, F_2 offspring will have a black body and straight wings. Therefore, crossing over in the F_1 female can explain the occurrence of both types of F_2 recombinant offspring.

Morgan's ideas about linkage and crossing over were based on similar data, derived from his studies of genes on the X chromosome. The idea that linked genes tend to be inherited together explained the high frequency of parental combinations of traits in certain crosses. The suggestion that crossing over produces chromosomes with new allele combinations accounted for the occurrence of recombinant phenotypes. Morgan's third idea regarding linkage was that the frequency of crossing over between linked genes depends on the distance between them. This suggested a method for determining the relative positions of genes on a chromosome, as we will see next.

GENOMES & PROTEOMES

Recombination Frequencies Provide a Method for Mapping Genes Along Chromosomes

The oldest approach to studying the arrangement of genes in a species' genome is called **genetic linkage mapping** (also known as gene mapping or chromosome mapping). This experimental method is used to determine the linear order of genes that are linked to each other along the same chromosome. As depicted in **Figure 17.5**, this linear arrangement is shown in a chart known as a **genetic linkage map**. Each gene has its own unique locus at a particular site within a chromosome. For example, the gene for black body color (*b*) that we discussed earlier is located near the middle of the chromosome, while the gene for curved wings (*c*) is closer to one end. The first genetic linkage map, showing five genes on the *Drosophila* X chromosome, was constructed in 1911 by Alfred Sturtevant, an undergraduate who spent time in Morgan's laboratory.

Genetic linkage mapping allows us to estimate the relative distances between linked genes based on the likelihood that a crossover will occur between them. This likelihood is proportional to the distance between the genes, as Morgan first proposed. If the genes are very close together, a crossover is unlikely to begin in the region between them. However, if the genes are very far apart, a crossover is more likely to be initiated between them and thereby recombine their alleles. Therefore, in a cross involving two genes on the same chromosome, the percentage of recombinant offspring is correlated with the distance between the genes. This correlation provides the experimental basis for gene mapping. If a two-factor testcross produces many recombinant offspring, the experimenter concludes that the genes are far apart. If very few recombinant offspring are observed, the genes must be close together.

To find the distance between two genes, the experimenter must determine the frequency of crossing over between them, called their **recombination frequency**. This is accomplished by conducting a testcross. As an example, let's refer back to the *Drosophila* testcross described in Figure 17.4. As we discussed, the genes for body color and wing shape are on the same chromosome; the recombinant offspring are the result of crossing

Map units	Mutant phenotype	Wild-type phenotype
0.0	Aristaless, *al*	Long aristae
13.0	Dumpy wings, *dp*	Long wings
48.5	Black body, *b*	Gray body
54.5	Purple eyes, *pr*	Red eyes
67.0	Vestigial wings, *vg*	Long wings
75.5	Curved wings, *c*	Straight wings
104.5	Brown eyes, *bw*	Red eyes

Figure 17.5 **A simplified genetic linkage map.** This map shows the relative locations of a few genes along a chromosome in *Drosophila melanogaster*. The name of each gene is based on the mutant phenotype. The numbers on the left are map units (mu). The distance between two genes, in map units, corresponds to their recombination frequency in testcrosses.

over during egg formation in the F$_1$ female. We can use the data from the testcross shown in Figure 17.4 to estimate the distance between these two genes. The **map distance** between two linked genes is defined as the number of recombinant offspring divided by the total number of offspring times 100.

$$\text{Map distance} = \frac{\text{Number of recombinant offspring}}{\text{Total number of offspring}} \times 100$$

$$= \frac{133 + 137}{371 + 359 + 133 + 137} \times 100$$

$$= 27.0 \text{ map units}$$

The units of distance are called **map units (mu)**, or sometimes **centiMorgans (cM)** in honor of Thomas Hunt Morgan. One map unit is equivalent to a 1% recombination frequency. In this example, 270 out of 1,000 offspring are recombinants, so the recombination frequency is 27% and the two genes are 27.0 mu apart.

Genetic linkage mapping has been useful for analyzing the genes of organisms that are easily crossed and produce many offspring in a short time. It has been used to map the genes of several plant species and of certain species of animals, such as

Drosophila. However, for most organisms, including humans, linkage mapping is impractical due to long generation times or the inability to carry out experimental crosses. Fortunately, many alternative methods of gene mapping have been developed in the past few decades that are faster and do not depend on crosses. These newer cytological and molecular approaches, which we will discuss in Chapter 20, are now used to map genes in a wide variety of organisms.

17.3 Extranuclear Inheritance: Organelle Genomes

In the previous section, we examined the inheritance patterns of linked genes that violate the law of independent assortment. In this section, we will explore inheritance patterns that violate the law of segregation. Gene transmission may defy this law because some genes are not found on the chromosomes in the cell nucleus. The segregation of genes is explained by the pairing and segregation of homologous chromosomes during meiosis; genes found elsewhere in the cell do not segregate in the same way. The transmission of genes that are located outside the cell nucleus is called **extranuclear inheritance**.

Two important types of extranuclear inheritance patterns involve genes that are found in mitochondria and chloroplasts (**Figure 17.6**). Extranuclear inheritance is also called **cytoplasmic inheritance** because these organelles are in the cytoplasm of the cell. As we discussed in Chapter 6, mitochondria and chloroplasts are found in eukaryotic cells because of an ancient endosymbiotic relationship. They contain their own genetic material, or genomes. Although these **organelle genomes** are much smaller than nuclear genomes, researchers have discovered that they are critically important in the phenotypes of organisms. In plants, for example, the chloroplast genome carries many genes that are vital for photosynthesis. Mitochondrial genes are critical for respiration. In humans, mutations in the mitochondrial genome may cause inherited diseases. In this section, we will examine the transmission patterns observed for genes found in the chloroplast and mitochondrial genomes and consider how mutations in these genes may affect an individual's traits.

Chloroplast Genomes Are Often Maternally Inherited

One of the first experiments showing an extranuclear inheritance pattern was carried out by Carl Correns in 1909. Correns discovered that leaf pigmentation in the four-o'clock plant (*Mirabilis jalapa*) follows a pattern of inheritance that does not obey Mendel's law of segregation. Four-o'clock leaves may be green, white, or variegated, as shown in **Figure 17.7**. Correns observed that the pigmentation of the offspring depended solely on the pigmentation of the maternal parent, a phenomenon called **maternal inheritance**. If the female parent had white leaves,

(a) An animal cell

(b) A plant cell

Figure 17.6   **The locations of genetic material in animal and plant cells.**   The chromosomes in the cell nucleus are collectively known as the nuclear genome. Mitochondria and chloroplasts have small circular chromosomes, which are called the mitochondrial and chloroplast genomes.

all of the offspring had white leaves. Similarly, if the female was green, so were all of the offspring. The offspring of a variegated female parent could be green, white, or variegated.

At the time, Correns did not understand that chloroplasts contain some genes. We now know that the pigmentation of four-o'clock leaves can be explained by the occurrence of genetically different types of chloroplasts in the leaf cells. As discussed in Chapter 8, chloroplasts are the site of photosynthesis, and their green color is due to the presence of the pigment called chlorophyll. Certain genes required for chlorophyll synthesis are found within the chloroplast DNA. The green phenotype is due to the presence of chloroplasts that have normal genes and synthesize the usual quantity of chlorophyll. The white phenotype is caused by a mutation in a gene within the chloroplast DNA that prevents the synthesis of most of the chlorophyll.

Correns' crosses

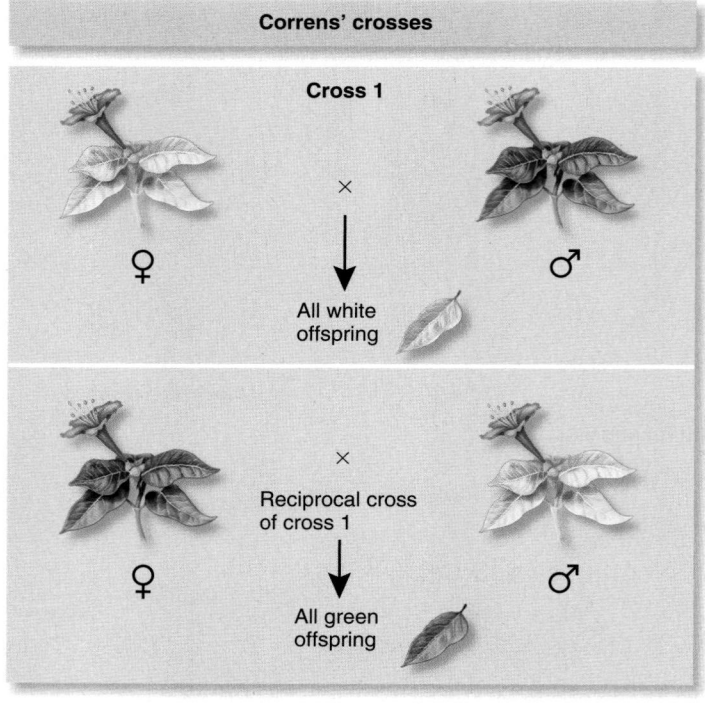

Cross 1

♀ × ♂

All white
offspring

♀ × ♂

Reciprocal cross
of cross 1

All green
offspring

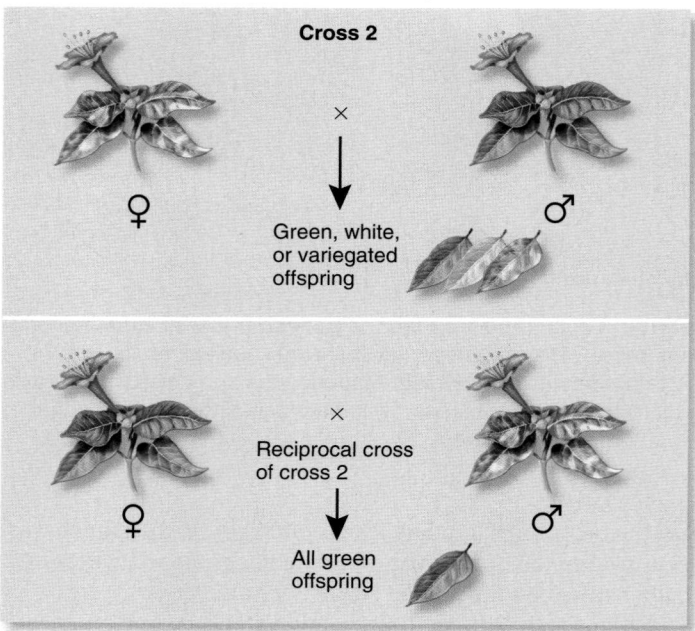

Cross 2

♀ × ♂

Green, white,
or variegated
offspring

♀ × ♂

Reciprocal cross
of cross 2

All green
offspring

Figure 17.7 **Maternal inheritance in the four-o'clock plant.** The genes for green pigment synthesis in plants are part of the chloroplast genome. The white phenotype in four o'clocks is due to chloroplasts with a mutant allele that greatly reduces green pigment production. The variegated phenotype is due to a mixture of normal and mutant chloroplasts. In four o'clocks, the egg contains all of the plastids that are inherited by the offspring, so the phenotype of the offspring is determined by the female parent.

Biological inquiry: In this example, where is the gene located that causes the green color of four-o'clock leaves? How is this gene transmitted from parent to offspring?

(Enough chlorophyll is made for the plant to survive.) The variegated phenotype occurs in leaves that have a mixture of the two types of chloroplasts.

Leaf pigmentation follows a maternal inheritance pattern because the chloroplasts in four o'clocks are inherited only through the cytoplasm of the egg (**Figure 17.8**). During plant fertilization, a sperm cell from a pollen grain fertilizes an egg cell to create a zygote, which eventually develops into a plant. In four o'clocks, the egg cell contains several proplastids that are inherited by the offspring, while the sperm cell does not contribute any proplastids. As discussed in Chapter 4, proplastids develop into various types of plastids, including chloroplasts. Thus, the phenotype of a four-o'clock plant depends on the types of proplastids it inherits from the maternal parent. If the maternal parent transmits only normal proplastids, all offspring will have green leaves (Figure 17.8a). Alternatively, if the maternal parent transmits only mutant proplastids, all offspring will have white leaves (Figure 17.8b). The genetic composition of the paternal parent does not affect the outcome. Because an egg cell contains several proplastids, an offspring from a variegated maternal parent may inherit only normal proplastids, only

Normal proplastid will produce chloroplasts with a normal amount of green pigment.

Mutant proplastid will produce chloroplasts with very little pigment.

Egg cell

(a) Egg cell from a maternal parent with green leaves

(b) Egg cell from a maternal parent with white leaves

(c) Possible egg cells from a maternal parent with variegated leaves

Figure 17.8 **Plastid composition of egg cells from green, white, and variegated four-o'clock plants.** In this drawing of four-o'clock egg cells, normal proplastids are represented as green and mutant proplastids as white. Proplastids do not differentiate into chloroplasts in egg cells, and they are not actually green. **(a)** A green plant produces eggs carrying normal proplastids. **(b)** A white plant produces eggs carrying mutant proplastids. **(c)** A variegated plant produces eggs that may contain either or both types of proplastids.

mutant proplastids, or a mixture of normal and mutant proplastids. Consequently, the offspring of a variegated maternal parent can be green, white, or variegated individuals (Figure 17.8c).

The variegated phenotype is due to segregation events that occur after fertilization. As a zygote containing both types of chloroplasts divides to produce a multicellular plant, some cells may receive mostly normal chloroplasts. Further division of these cells gives rise to a patch of green tissue. Alternatively, as a matter of chance, other cells may receive mostly mutant chloroplasts that are defective in chlorophyll synthesis. This results in a patch of tissue that is white.

In most species of plants, the egg cell provides most of the zygote's cytoplasm, while the much smaller male gamete often provides little more than a nucleus. Therefore, chloroplasts are most often inherited via the egg. In seed-bearing plants, maternal inheritance of chloroplasts is the most common transmission pattern. However, certain species exhibit a pattern called **biparental inheritance**, in which both the pollen and the egg contribute chloroplasts to the offspring. Others exhibit **paternal inheritance**, in which only the pollen contributes these organelles. For example, most types of pine trees show paternal inheritance of chloroplasts.

Mitochondrial Genomes Are Maternally Inherited in Humans and Most Other Species

Mitochondria are found in nearly all eukaryotic species. Similar to the transmission of chloroplasts in plants, maternal inheritance is the most common pattern of mitochondrial transmission in eukaryotic species, although some species do exhibit biparental or paternal inheritance. The mitochondrial genome of many mammalian species has been analyzed and usually contains a total of 37 genes. Twenty-four genes encode tRNAs and rRNAs, which are needed for translation inside the mitochondrion. Thirteen genes encode proteins that are involved in oxidative phosphorylation. As discussed in Chapter 7, the primary function of the mitochondrion is the synthesis of ATP via oxidative phosphorylation.

In humans, as in most species, mitochondria are maternally inherited. Researchers have discovered that mutations in human mitochondrial genes can cause a variety of rare diseases (**Table 17.2**). These are usually chronic degenerative disorders that affect the brain, eyes, heart, muscle, kidney, and endocrine glands. For example, Leber's hereditary optic neuropathy (LHON) affects the optic nerve. It may lead to the progressive loss of vision in one or both eyes. LHON can be caused by a mutation in one of several different mitochondrial genes.

17.4 X Inactivation, Genomic Imprinting, and Maternal Effect

We will end our discussion of complex inheritance patterns by considering examples in which the timing and control of gene expression create inheritance patterns that are determined by

Table 17.2	Examples of Human Mitochondrial Diseases
Disease	**Description**
Leber's hereditary optic neuropathy	Caused by a mutation in one of several mitochondrial genes that encode electron-transport proteins. The main symptom is loss of vision.
Neurogenic muscle weakness	Caused by a mutation in a mitochondrial gene that encodes a subunit of mitochondrial ATP synthase, which is required for ATP synthesis. Symptoms involve abnormalities in the nervous system that affect the muscles and eyes.
Mitochondrial encephalomyopathy, lactic acidosis, and strokelike episodes	Mutations in mitochondrial genes that encode tRNAs for leucine and lysine. Symptoms include strokelike episodes, secretion of lactic acid into the bloodstream, seizures, migraine headaches, and lack of coordination.
Maternal myopathy and cardiomyopathy	A mutation in a mitochondrial gene that encodes a tRNA for leucine. The primary symptoms involve muscle abnormalities, most notably in the heart.
Myoclonic epilepsy and ragged-red muscle fibers	A mutation in a mitochondrial gene that encodes a tRNA for lysine. Symptoms include epilepsy, dementia, blindness, deafness, and heart and kidney malfunctions.

the sex of the individual or by the sex of the parents. The first two patterns, called X inactivation and genomic imprinting, are types of **epigenetic inheritance**. In epigenetic inheritance, modification of a gene or chromosome during egg formation, sperm formation, or early stages of embryo growth alters gene expression in a way that is fixed during an individual's lifetime. Epigenetic changes permanently affect the phenotype of the individual, but they are not permanent over the course of many generations and they do not change the actual DNA sequence. For example, a gene may undergo an epigenetic change that inactivates it for an individual's entire life, so it is never expressed in that individual. However, when the same individual makes gametes, the gene may become activated and remain active during the lifetime of an offspring that inherits the gene.

At the end of this section, we will also consider genes that exhibit a bizarre inheritance pattern called the **maternal effect**, in which the genotype of the mother directly determines the phenotype of her offspring. Surprisingly, for maternal effect genes, the genotypes of the father and of the offspring themselves do not affect the offspring's phenotype. As you will learn, this phenomenon is explained by the accumulation of gene products that the mother provides to her developing eggs.

In Female Mammals, One X Chromosome Is Inactivated in Each Somatic Cell

In 1961, the British geneticist Mary Lyon proposed the phenomenon of **X inactivation**, in which one X chromosome in the somatic cells of female mammals is inactivated, meaning that its genes

Figure 17.9   **X-chromosome inactivation in female mammals.** This light micrograph shows the nucleus of a human female cell. The label shows the Barr body, a condensed, inactivated X chromosome found just inside the nuclear envelope in the somatic cells of female mammals.

are not expressed. The **Lyon hypothesis**, as X inactivation also came to be known, was based on two lines of evidence. The first evidence came from microscopic studies of mammalian cells. In 1949, Murray Barr and Ewart Bertram identified a highly condensed structure in the cells of female cats that was not found in the cells of male cats. This structure was named a **Barr body** after one of its discoverers (**Figure 17.9**). In 1960, Susumu Ohno correctly proposed that a Barr body is a highly condensed X chromosome. Lyon's second line of evidence was the inheritance pattern of variegated coat colors in certain mammals. A classic case is the calico cat, which has randomly distributed patches of black and orange fur (**Figure 17.10a**).

According to the Lyon hypothesis, the calico pattern is explained by the permanent inactivation of one X chromosome in each cell that forms a patch of the cat's skin, as shown in **Figure 17.10b**. The gene involved is an X-linked gene that occurs as an orange allele, X^O, and a black allele, X^B. A female cat that is heterozygous for this gene will be calico. (The white underside is due to a dominant allele of a different autosomal gene.) At an early stage of embryonic development, one of the two X chromosomes is randomly inactivated in each of the cat's somatic cells, including those that will give rise to the hair-producing skin cells. As the embryo grows and matures, the pattern of X inactivation is maintained during subsequent cell divisions. For example, skin cells derived from a single embryonic cell in which the X^B-carrying chromosome has been inactivated will produce a patch of orange fur, because they express only the X^O allele that is carried on the active chromosome. Alternatively, a group of skin cells in which the chromosome carrying X^O has been inactivated will express only the X^B allele, producing a patch of black fur. Because the primary event of X inactivation is a random process that occurs at an early stage of development, the result is an animal with randomly distributed patches of black and orange fur.

(a) Calico cat

Orange fur allele   Black fur allele

1   In the early embryo, all X chromosomes are initially active.

2   In each embryonic cell, random inactivation occurs for one of the X chromosomes, which becomes a Barr body.

Barr bodies

3   As development proceeds, the pattern of X inactivation is maintained during cell division.

(b) Process of X inactivation

Figure 17.10   **Random X-chromosome inactivation in a calico cat.**   (a) A calico cat. (b) X inactivation during embryonic development. The calico pattern is due to random X-chromosome inactivation in a female that is heterozygous for the X-linked gene with black and orange alleles. The cells at the *top* of this figure represent a small mass of cells making up the very early embryo. In these cells, both X chromosomes are active. At an early stage of embryonic development, one X chromosome is randomly inactivated in each cell. The initial inactivation pattern is maintained in the descendents of each cell as the embryo matures into an adult. The pattern of orange and black fur in the adult cat reflects the pattern of X inactivation in the embryo.

Biological inquiry: If a female cat is homozygous for the orange allele, would it show a calico phenotype?

In female mammals that are heterozygous for X-linked genes, approximately half of their somatic cells will express one allele, while the rest of their somatic cells will express the other allele. These heterozygotes are called **mosaics** because they are composed of two types of cells, analogous to the different-colored pieces in the pictures called mosaics. The phenomenon of mosaicism is readily apparent in calico cats, in which the alleles affect fur color. Likewise, human females who are heterozygous for X-linked genes are mosaics, with one allele expressed in some cells and the alternative allele in other cells. Women who are heterozygous for recessive X-linked alleles usually show the dominant trait because the expression of the dominant allele in 50% of their cells is sufficient to produce the dominant phenotype.

On rare occasions, a female who is heterozygous for a recessive X-linked disease-causing allele may show mild or even severe disease symptoms. Because the pattern of X-chromosome inactivation is random, there will be a small percentage of heterozygous women who happen to inactivate the X chromosome carrying the normal allele in a large percentage of their cells, as a matter of bad luck. As an example, let's consider the recessive X-linked form of hemophilia that we discussed in Chapter 16. This type of hemophilia is caused by a defect in a gene that encodes a blood-clotting factor, called factor VIII, that is made by cells in the liver and secreted into the bloodstream. X inactivation in humans occurs when an embryo is 10 days old. At this stage, the liver contains only about a dozen cells. In most females who are heterozygous for the normal and hemophilia alleles, roughly half of their liver cells will express the normal allele. However, on rare occasions, all or most of the dozen embryonic liver cells might happen to inactivate the X chromosome carrying the dominant normal allele. Following growth and development, such a female will have a very low level of factor VIII and as a result will show symptoms of hemophilia.

At this point, you may be wondering why X inactivation occurs. Researchers have proposed that X inactivation achieves **dosage compensation** between male and female mammals. The X chromosome carries many genes, while the Y chromosome has only a few. The inactivation of one X chromosome in the female reduces the number of expressed copies (doses) of X-linked genes from two to one. As a result, the expression of X-linked genes in females and males is roughly equal.

The X Chromosome Has an X Inactivation Center That Controls Compaction into a Barr Body

After the Lyon hypothesis was confirmed, researchers became interested in the genetic control of X inactivation. The cells of humans and other mammals have the ability to count their X chromosomes and allow only one of them to remain active. Additional X chromosomes are converted to Barr bodies. In normal females, two X chromosomes are counted and one is inactivated. In normal males, one X chromosome is counted and none inactivated. On occasion, however, people are born with abnormalities in the number of their sex chromosomes. In these disorders, known as Turner syndrome, Triple X syndrome, and Klinefelter syndrome, the cells inactivate the number of X chromosomes necessary to leave a single active chromosome.

Phenotype	Chromosome Composition	Number of Barr Bodies
Normal female	XX	1
Normal male	XY	0
Turner syndrome (female)	XO	0
Triple X syndrome (female)	XXX	2
Klinefelter syndrome (male)	XXY	1

Although the genetic control of inactivation is not entirely understood at the molecular level, a short region on the X chromosome called the **X inactivation center (Xic)** is known to play a critical role. Eeva Therman and Klaus Patau identified Xic from its key role in X inactivation. The counting of human X chromosomes is accomplished by counting the number of Xics. The Xic on each X chromosome is necessary for inactivation to occur. Therman and Patau found that in cells with two X chromosomes, if one of them is missing its Xic due to a chromosome mutation, neither X chromosome will be inactivated. This is a lethal condition for a human female embryo.

The expression of a specific gene within the X inactivation center is required for compaction of the X chromosome into a Barr body. This gene, discovered in 1991, is named *Xist* (for X inactive specific transcript). The *Xist* gene product is a long RNA molecule that does not encode a protein. Instead, the role of *Xist* RNA is to coat one of the two X chromosomes during the process of X inactivation. The *Xist* gene on the inactivated X chromosome continues to be expressed after other genes on this chromosome have been silenced.

The process of X inactivation can be divided into three phases: initiation, spreading, and maintenance (**Figure 17.11**). During initiation, one of the X chromosomes is targeted for inactivation. This chromosome is inactivated during the spreading phase, so called because inactivation begins near the X inactivation center and spreads in both directions along the chromosome. Spreading requires the transcription of the *Xist* gene and coating of the X chromosome with *Xist* RNA. After coating, proteins associate with the *Xist* RNA and promote compaction of the chromosome into a Barr body. Maintenance refers to replication of the compacted chromosome during subsequent cell divisions. While initiation and spreading occur only during embryonic development, maintenance occurs throughout the individual's life. Continued activity of the *Xist* gene on an inactivated X chromosome maintains this chromosome as a Barr body during cell division. Whenever a somatic cell divides in a female mammal, the Barr body is replicated to produce two Barr bodies.

The Transcription of an Imprinted Gene Depends on the Sex of the Parent

As we have seen, X inactivation is a type of epigenetic inheritance in which a chromosome is modified in the early embryo, permanently altering gene expression in that individual. Other types of epigenetic inheritance occur in which genes or chromosomes

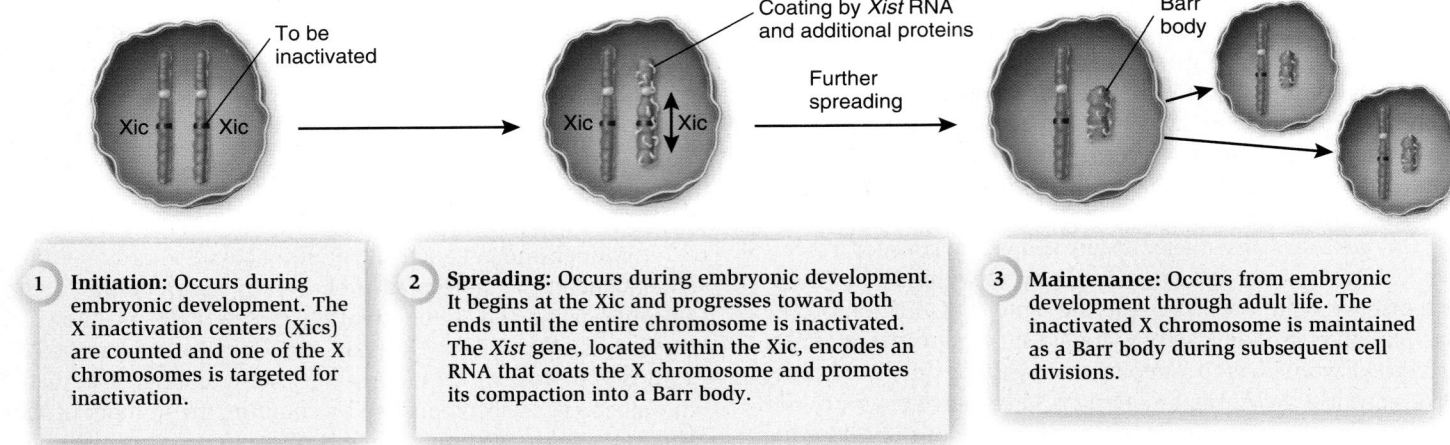

Figure 17.11   The process of X inactivation.

are modified in the gametes of a parent, permanently altering gene expression in the offspring. **Genomic imprinting** refers to a phenomenon in which a segment of DNA is **imprinted**, or marked, in a way that affects gene expression throughout the life of the individual who inherits that DNA.

Genomic imprinting occurs in numerous species, including insects, plants, and mammals. Imprinting may involve a single gene, a part of a chromosome, an entire chromosome, or even all of the chromosomes inherited from one parent. It is permanent in the somatic cells of a given individual, but the marking of the DNA is altered from generation to generation. Imprinted genes do not follow a Mendelian pattern of inheritance because imprinting causes the offspring to distinguish between maternally and paternally inherited alleles. Depending on how a par-

ticular gene is marked by each parent, the offspring will express either the maternal or the paternal allele, but not both.

Let's consider a specific example of imprinting that involves a gene called *Igf-2* that is found in mice and other mammals. This gene encodes a growth hormone called insulin-like growth factor 2 that is needed for proper growth. If a normal copy of this gene is not expressed, a mouse will be dwarf. The *Igf-2* gene is known to be located on an autosome, not on a sex chromosome. Because mice are diploid, they have two copies of this gene, one from each parent.

Researchers have discovered that mutations can occur in the *Igf-2* gene that block the function of the Igf-2 hormone. When mice carrying normal or mutant alleles are crossed to each other, a bizarre result is obtained (**Figure 17.12**). If the male parent

Figure 17.12   An example of genomic imprinting in the mouse.   In the cross on the *left*, a homozygous male with the normal *Igf-2* allele is crossed to a homozygous female carrying a defective allele, *Igf-2m*. An offspring is phenotypically normal because the paternal allele is expressed. In the cross on the *right*, a homozygous male carrying the defective allele is crossed to a homozygous normal female. In this case, an offspring is dwarf because the paternal allele is defective and the maternal allele is not expressed.

is homozygous for the normal allele and the female is homozygous for the mutant allele, all the offspring grow to a normal size. In contrast, if the male is homozygous for the mutant allele and the female is homozygous for the normal allele, all the offspring are dwarf. The reason this result is so surprising is that the normal and dwarf offspring have the same genotype but different phenotypes! These phenotypes are not the result of any external influence on the offspring's development. Rather, the allele that is expressed in their somatic cells depends on which parent contributed which allele. In mice, the *Igf-2* gene inherited from the mother is imprinted in such a way that it cannot be transcribed into mRNA. Therefore, only the paternal gene is expressed. The mouse on the left side of Figure 17.12 is normal because it expresses a functional paternal gene. In contrast, the mouse on the right is dwarf because the paternal gene is a mutant allele that results in a nonfunctional hormone. In both cases, the maternal gene is inactive due to imprinting.

Why is the maternal gene not transcribed into mRNA? To answer this question we need to consider the molecular function of genes. As discussed in Chapter 13, the attachment of methyl (—CH_3) groups to the bases of DNA can alter gene transcription. For most genes, methylation silences gene expression by causing the DNA to become more compact. For a few genes, methylation may enhance gene expression by attracting activator proteins to the promoter. Researchers have discovered that DNA methylation is the marking process that occurs during the imprinting of certain genes, including the *Igf-2* gene.

Figure 17.13 shows the imprinting process in which a maternal gene is methylated. The left side of the figure follows the marking process during the life of a female individual; the right side follows the same process in a male. Both individuals received a methylated gene from their mother and a nonmethylated copy of the same gene from their father. Via cell division, the zygote develops into a multicellular organism. Each time a somatic cell divides, enzymes in the cell maintain the methylation of the maternal gene, while the paternal gene remains unmethylated. If the methylation inhibits transcription of this gene, only the paternal copy will be expressed in the somatic cells of both the male and female offspring.

The methylation state of an imprinted gene may be altered when individuals make gametes. First, as shown in Figure 17.13, the methylation is erased. Next, the gene may be methylated again, but that depends on whether the individual is a female or male. In females making eggs, both copies of the gene are methylated; in males making sperm, neither copy is methylated. When we consider the effects of methylation over the course of two or more generations, we can see how this phenomenon creates an epigenetic transmission pattern. The male in Figure 17.13 has inherited a methylated gene from his mother that is transcriptionally silenced in his somatic cells. Although he does not express this gene during his lifetime, he can pass on an active, nonmethylated copy of this exact same gene to his offspring.

Genomic imprinting is a recently discovered phenomenon that has been shown to occur for a few genes in mammals. For some genes, such as *Igf-2*, the maternal allele is silenced, while for other genes the paternal allele is silenced. Biologists are still trying to understand the reason for this curious marking process.

Figure 17.13 **Genomic imprinting via DNA methylation.** The cells at the *top* of this figure have a methylated gene inherited from the mother and a nonmethylated version of the same gene inherited from the father. This pattern of methylation is the same in male and female offspring and is maintained in their somatic cells. The methylation is erased during gamete formation, but in females the gene is methylated again at a later stage in the formation of eggs. Therefore, females always transmit a methylated, transcriptionally silent copy of this gene, while males transmit a nonmethylated, active copy.

For Maternal Effect Genes, the Genotype of the Mother Determines the Phenotype of the Offspring

In epigenetic inheritance, genes are altered in ways that affect their expression in an individual or the individual's offspring. As we have seen, some of these alterations produce strange inheritance patterns, in which organisms with the same genotype have different phenotypes. Another strange inheritance pattern, with a very different explanation, involves a category of genes called **maternal effect genes**.

Inheritance patterns due to maternal effect genes were first identified in the 1920s by A. E. Boycott, in his studies of the freshwater snail *Lymnaea peregra*. In this species, the shell and internal organs can be arranged in either a right-handed (dextral) or a left-handed (sinistral) direction. The dextral orientation is more common and is dominant to the sinistral orientation. Whether a snail's body curves in a dextral or a sinistral direction depends on the pattern of cell division immediately following fertilization. **Figure 17.14** shows the results of Boycott's crosses of true-breeding strains of snails with either a dextral or a sinistral orientation. When a dextral female (*DD*) was crossed to a sinistral male (*dd*), all of the offspring were dextral. However, crossing a sinistral female (*dd*) to a dextral male (*DD*) produced the opposite result: all of the offspring were sinistral. These seemingly contradictory outcomes could not be explained in terms of Mendelian inheritance.

Alfred Sturtevant later suggested that snail coiling is due to a maternal effect gene that exists as a dextral (*D*) and a sinistral (*d*) allele. In the cross shown on the left, the P generation female is dextral (*DD*) and the male is sinistral (*dd*). In the cross on the right, the female is sinistral (*dd*) and the male is dextral (*DD*). In either case, the F_1 offspring are *Dd*. When the F_1 individuals from these two crosses are mated to each other, a genotypic ratio of 1 *DD* : 2 *Dd* : 1 *dd* is predicted for the F_2 generation. Because the *D* allele is dominant to the *d* allele, a Mendelian inheritance pattern would produce a 3:1 phenotypic ratio of dextral to sinistral snails. Instead, the snails of the F_2 generation were all dextral. To explain this observed result, Sturtevant proposed that the phenotype of the F_2 offspring depended solely on the genotype of the F_1 mother. Because the F_1 mothers were *Dd*, and the *D* allele is dominant, the F_2 offspring were dextral even if their genotype was *dd*!

Sturtevant's hypothesis is supported by the ratio of phenotypes seen in the F_3 generation. When members of the F_2 generation were crossed, the F_3 generation exhibited a 3:1 ratio of dextral to sinistral snails. These F_3 phenotypes reflect the genotypes of the F_2 mothers. The ratio of genotypes for the F_2 females was 1 *DD* : 2 *Dd* : 1 *dd*. The *DD* and *Dd* females produced dextral offspring, while the *dd* females produced sinistral offspring. This is consistent with the 3:1 phenotypic ratio in the F_3 generation.

The peculiar inheritance pattern of maternal effect genes can be explained by the process of egg maturation in female animals (**Figure 17.15**). Maternal cells called nurse cells surround a de-

Figure 17.14 **The inheritance of snail coiling direction as an example of a maternal effect gene.** In the snails shown in this experiment, the direction of body coiling is controlled by a single pair of genes. *D* (dextral, or right-handed) is dominant to *d* (sinistral, or left-handed). The genotype of the mother determines the phenotype of the offspring. A *DD* or *Dd* mother will produce dextral offspring and a *dd* mother will produce sinistral offspring, regardless of the genotypes of the father and of the offspring themselves.

Biological inquiry: An offspring has a genotype of Dd and coils to the left. What is the genotype of its mother?

veloping egg cell and provide it with nutrients. Within these diploid nurse cells, both copies of a maternal effect gene are activated to produce their gene products. The gene products are transported into the egg, where they persist for a significant time during embryonic development. The *D* and *d* gene products influence the pattern of cell division during the early stages of the snail's embryonic development. If an egg receives only the *D* gene product, the snail will develop a dextral orientation, while an egg that receives only the *d* gene product will produce a

Mother is *DD*.

Nurse cells

D gene products

Egg

All offspring are dextral because the egg received the gene products of the *D* allele.

Mother is *Dd*.

D and *d* gene products

All offspring are dextral because the egg received the gene products of the *D* and *d* alleles, but the *D* gene products are dominant.

Mother is *dd*.

d gene products

All offspring are sinistral because the egg received the gene products of the *d* allele.

Figure 17.15   **The mechanism of maternal effect in snail coiling.**   In this simplified diagram, the mother's diploid nurse cells transfer gene products to the egg as it matures. These gene products persist after fertilization, affecting development of the early embryo. If the nurse cells are *DD* or *Dd*, they will transfer the dominant *D* gene product to the egg, causing the offspring to be dextral. If the nurse cells are *dd*, only the *d* gene product will be transferred to the egg and the offspring will be sinistral.

snail with a sinistral orientation. If an egg receives both *D* and *d* gene products, the snail will be dextral because the *D* gene product is dominant over *d*. In this way, the gene products of nurse cells, which are determined by the mother's genotype, influence the development of the offspring.

Several dozen maternal effect genes have been identified in experimental organisms, such as *Drosophila*. Recently, they have also been found in mice and humans. As we will discuss in Chapter 19, the products of maternal effect genes are critically important in the early stages of animal development.

CHAPTER SUMMARY

- A variety of inheritance patterns are more complex than Mendel had realized. Many of these do not obey one or both of his laws of inheritance. (Table 17.1)

17.1 Gene Interactions

- When the alleles of one gene mask the effects of the alleles of a different gene, this type of gene interaction is called epistasis. (Figure 17.1)
- Quantitative traits such as height and weight are polygenic, which means that several genes govern the trait. Often, the alleles of such genes contribute in an additive way to the phenotype. This produces continuous variation in the trait, which is graphed as a bell curve. (Figure 17.2)

17.2 Genes on the Same Chromosome: Linkage, Recombination, and Mapping

- When two different genes are on the same chromosome, they are said to be linked. Linked genes tend to be inherited as a unit, unless crossing over separates them. (Figures 17.3, 17.4)
- The percentage of offspring produced in a two-factor testcross can be used to map the relative locations of genes along a chromosome. (Figure 17.5)

17.3 Extranuclear Inheritance: Organelle Genomes

- Mitochondria and chloroplasts carry a small number of genes. The inheritance of such genes is called extranuclear inheritance. (Figure 17.6)
- Chloroplasts in the four-o'clock plant are transmitted via the egg, a pattern called maternal inheritance. (Figures 17.7, 17.8)
- Several human diseases are known to be caused by mutations in mitochondrial genes, which follow a maternal inheritance pattern. (Table 17.2)

17.4 X Inactivation, Genomic Imprinting, and Maternal Effect

- Epigenetic inheritance refers to patterns in which a gene is inactivated during the life of an organism, but not over the course of many generations.
- X inactivation in mammals occurs when one X chromosome is randomly inactivated in females. If the female is heterozygous for an X-linked gene, this can lead to a variegated phenotype. (Figures 17.9, 17.10)
- X inactivation occurs in three phases: initiation, spreading, and maintenance. (Figure 17.11)
- Imprinted genes are inactivated by one parent but not both. The offspring expresses only one of the two alleles. (Figure 17.12)
- During gamete formation, methylation of a gene from one parent is a mechanism to achieve imprinting. (Figure 17.13)

• For maternal effect genes, the genotype of the mother determines the phenotype of the offspring. This is explained by the phenomenon that the mother's nurse cells contribute gene products to egg cells that are needed for early stages of development. (Figures 17.14, 17.15)

TEST YOURSELF

1. Quantitative traits such as height and weight are governed by several genes that usually contribute in an additive way to the trait. This is called
   a. independent assortment.
   b. discontinuous inheritance.
   c. maternal inheritance.
   d. linkage.
   e. polygenic inheritance.

2. When two genes are located on the same chromosome they are said to be
   a. homologous.   d. linked.
   b. allelic.   e. polygenic.
   c. epistatic.

3. Based on the ideas proposed by Morgan, which of the following statements concerning linkage is *not* true?
   a. Traits determined by genes located on the same chromosome are likely to be inherited together.
   b. Crossing over between homologous chromosomes can create new gene combinations.
   c. Crossing over is more likely to occur between genes that are closer together.
   d. The probability of crossing over depends on the distance between the genes.
   e. All but one of the above statements are correct.

4. In genetic linkage mapping, 1 map unit is equivalent to
   a. 100 base pairs.
   b. 1 base pair.
   c. 10% recombination frequency.
   d. 1% recombination frequency.
   e. 1% the length of the chromosome.

5. Organelle heredity is possible because
   a. gene products may be stored in organelles.
   b. mRNA may be stored in organelles.
   c. some organelles contain genetic information.
   d. conjugation of nuclei occurs before cellular division.
   e. both a and c.

6. In many organisms, organelles such as the mitochondria are contributed by only the egg. This phenomenon is known as
   a. biparental inheritance.
   b. paternal inheritance.
   c. maternal effect.
   d. maternal inheritance.
   e. both c and d.

7. Modification of a gene during gamete formation or early development that alters the way the gene is expressed during the individual's lifetime is called
   a. maternal inheritance.
   b. epigenetic inheritance.
   c. epistasis.
   d. multiple allelism.
   e. alternative splicing.

8. When a gene is inactivated during gamete formation and that gene is maintained in an inactivated state in the somatic cells of offspring, such an inheritance pattern is called
   a. linkage.
   b. X inactivation.
   c. maternal effect.
   d. genomic imprinting.
   e. polygenic inheritance.

9. Calico coat pattern in cats is the result of
   a. X inactivation.   d. genomic imprinting.
   b. epistasis.   e. maternal inheritance.
   c. organelle heredity.

10. Maternal effect inheritance can be explained by
    a. gene products that are given to an egg by the nurse cells.
    b. the methylation of genes during gamete formation.
    c. the spreading of X inactivation from the Xic locus.
    d. the inheritance of alleles that contribute additively to a trait.
    e. none of the above.

CONCEPTUAL QUESTIONS

1. Define linkage and linkage group.
2. Explain extranuclear inheritance and give two examples.
3. Define genomic imprinting.

EXPERIMENTAL QUESTIONS

1. What hypothesis were Bateson and Punnett testing when conducting the crosses in the sweet pea?
2. What were the expected results of Bateson and Punnett's cross?
3. How did the observed results differ from the predicted results? How did Bateson and Punnett explain the results of this particular cross?

COLLABORATIVE QUESTIONS

1. Discuss two types of gene interactions.
2. Discuss the concept of linkage.

---

www.brookerbiology.com

This website includes answers to the Biological Inquiry questions found in the figure legends and all end-of-chapter questions.

18

GENETICS OF BACTERIA AND VIRUSES

An electron micrograph showing the budding of human immunodeficiency viruses from the surface of a human cell.

In Chapters 15 through 17, we focused on the genetics of eukaryotic species. We considered the structure of chromosomes and their transmission during meiosis and mitosis, and examined the inheritance patterns that occur when genes are passed from parents to offspring. Plants and animals are amenable to inheritance studies for two reasons. First, their allelic differences give rise to readily discernable phenotypic differences between individuals, such as white versus red eyes in *Drosophila* or tall versus dwarf pea plants. Second, because eukaryotic species reproduce sexually, researchers can make crosses and analyze the transmission patterns of traits that are passed from parents to offspring. The ability to follow allelic differences in a genetic cross is a basic tool in the genetic examination of eukaryotic species.

In this chapter, we turn our attention to the genetic analyses of bacteria and viruses. From a human perspective, this is a compelling topic due to its impact on health. Infectious diseases caused by bacteria and viruses are a leading cause of human death, accounting for a quarter to a third of deaths worldwide. The spread of infectious diseases results from human behavior, and in recent times it has been accelerated by changes in land use patterns, increased trade and travel, and the inappropriate use of antibiotic drugs. Although the incidence of fatal infectious diseases in the U.S. is low compared to the worldwide average, an alarming increase in more deadly strains of bacteria and viruses has occurred over the past few decades. Since 1980, the number of deaths in the U.S. due to infectious diseases has approximately doubled.

In the first part of this chapter, we will examine the bacterial genome and the methods used in its investigation. Like their eukaryotic counterparts, bacteria have allelic differences that affect their cellular traits, and the techniques of modern microbiology make many of these differences easy to detect. Readily discernable differences among bacteria include sensitivity to antibiotics and differences in nutritional requirements for growth. Although bacteria reproduce asexually by cell division, their genetic variety is enhanced by the phenomenon called genetic transfer, in which genes are passed from one bacterial cell to another. Like sexual reproduction in eukaryotes, genetic transfer enhances the genetic diversity observed among different species and within the same species. A transferred gene may be beneficial to the recipient. For example, a bacterial cell carrying a gene that provides antibiotic resistance may transfer this gene to another bacterium, allowing the recipient cell to survive exposure to the antibiotic. In this chapter, we will explore three interesting ways that bacteria can transfer genetic material.

In the second part of this chapter, we examine viruses and other nonliving particles that infect living cells. All organisms are susceptible to infection by one or more types of viruses, which use cellular machinery to replicate their own genomes. Once a cell is infected, the genetic material of a virus orchestrates a series of events that ultimately leads to the production of new virus particles. In this chapter, we will consider the biological complexity of viruses and explore viral reproductive cycles.

18.1 Bacterial Genetics

Prokaryotes, which include bacteria and archaea, are usually unicellular organisms that lack cell nuclei. Individual prokaryotic cells may exist as single units or remain associated with each other after cell division, forming pairs, chains, or clumps.

Bacteria are widespread on Earth, and several species are known to cause many types of infectious diseases. Many species of archaea are also known, though they are less common than bacteria. In this chapter, we focus on bacteria. We will compare the genomes of bacteria, archaea, and eukaryotes in Chapter 21.

Much of our knowledge of bacterial genetics comes from studies of *Escherichia coli*, which is a natural inhabitant of the colon of humans and other species. As you learned in previous chapters, this bacterium has been an important model organism for studying molecular genetics. Indeed, we know more about the molecular biology of this species than any other species on our planet.

We begin this section by exploring the structure and replication of the bacterial genome. We will examine the organization of DNA sequences along a bacterial chromosome, how the chromosome is compacted to fit inside a bacterium, and how it is transmitted by asexual reproduction. We will then look at three types of genetic transfer called conjugation, transformation, and transduction in which DNA from one bacterium is transferred to another. Finally, we will consider gene transfer between different species, a process called horizontal gene transfer.

Bacteria Typically Have One or Several Copies of a Single Circular Chromosome

The genes of bacteria are found within structures known as bacterial chromosomes. Although a bacterial cell usually has a single type of chromosome, it may have more than one copy of that chromosome. The number of copies depends on the bacterial species and on growth conditions, but a bacterium typically has one to four identical chromosomes. Each bacterial chromosome is tightly packed within a distinct **nucleoid** region of the cell (**Figure 18.1**). Unlike the eukaryotic nucleus, the bacterial nucleoid is not a separate cellular compartment bounded by a membrane. The DNA in the nucleoid is in direct contact with the cytoplasm of the cell.

Like eukaryotic chromosomes, bacterial chromosomes are molecules of double-stranded DNA. Unlike eukaryotic chromosomes, however, bacterial chromosomes are usually circular and tend to be much shorter, typically only a few million nucleotides long. For example, the chromosome of *E. coli* has approximately 4.6 million base pairs, and the *Hemophilus influenzae* chromosome has roughly 1.8 million. By comparison, an average eukaryotic chromosome may be 100 million bp in length.

A typical bacterial chromosome contains a few thousand unique genes that are found throughout the chromosome (**Figure 18.2**). Structural gene sequences (nucleotide sequences that encode proteins) account for the largest part of bacterial DNA. Other nucleotide sequences in the chromosome influence DNA replication, gene expression, and chromosome structure. One of these sequences is the origin of replication, which is a few hundred nucleotides long. Bacterial chromosomes have a single origin of replication that functions as an initiation site for the assembly of several proteins that are required for DNA replication (refer back to Figure 11.16a).

A variety of short, repetitive sequences (less than 10 nucleotides long) have been identified in many bacterial species. These sequences are found in multiple copies and are usually interspersed within the intergenic regions throughout the bacterial chromosome, as shown in Figure 18.2. Some of these short, repetitive sequences appear to play no useful role, and they are not transcribed into RNA. Others may play a role in a variety of genetic processes, including DNA folding, DNA replication, and gene expression.

Origin of replication

■ Genes
⌇ Repetitive sequences

Key features

- Most, but not all, bacterial species contain circular chromosomal DNA.

- A typical chromosome is a few million base pairs in length.

- Most bacterial species contain a single type of chromosome, but it may be present in multiple copies.

- Several thousand different genes are interspersed throughout the chromosome.

- One origin of replication is required to initiate DNA replication.

- Short, repetitive sequences may be interspersed throughout the chromosome.

Figure 18.2 The organization of nucleotide sequences in bacterial chromosomal DNA.

Figure 18.1 Nucleoids within the bacterium *Bacillus subtilis.* In the light micrograph shown here, the nucleoids are fluorescently labeled and seen as bright, oval-shaped regions within the bacterial cytoplasm. Two or more nucleoids are usually found within each cell.

The Formation of Chromosomal Loops and DNA Supercoiling Makes the Bacterial Chromosome Compact

Bacterial cells are much smaller than most eukaryotic cells (refer back to Figure 4.1). *E. coli* cells, for example, are approximately 1 μm wide and 2 μm long. To fit within a bacterial cell, the DNA of a typical bacterial chromosome must be compacted about 1,000-fold. The compaction process, shown in **Figure 18.3**, involves the formation of loops and the supercoiling of the looped DNA.

Unlike eukaryotic DNA, bacterial DNA is not wound around histone proteins to form nucleosomes. However, the binding of proteins to bacterial DNA is important in the formation of **loop domains**, which are chromosomal segments that are folded into loops. As seen in Figure 18.3, DNA-binding proteins anchor the bases of the loops in place. The number of loops varies according to the size of a bacterial chromosome and the species. The *E. coli* chromosome has 50 to 100 loop domains, each with about 40,000 to 80,000 bp. This looping compacts the circular chromosome about 10-fold. A similar process of loop-domain formation occurs in eukaryotic chromatin compaction, which is described in Chapter 15.

DNA **supercoiling** is a second important way to compact the bacterial chromosome. Because DNA is a long, thin molecule, twisting can dramatically change its conformation. This compaction is similar to what happens to a rubber band if you twist it in one direction. Because the two strands of DNA already coil around each other, the formation of additional coils due to twisting is referred to as supercoiling. Bacterial enzymes called topoisomerases twist the DNA and control the degree of DNA supercoiling.

Plasmids Are Small Pieces of Extrachromosomal DNA

In addition to chromosomal DNA, bacterial cells commonly contain **plasmids**, which are small, circular pieces of DNA that exist independently of the bacterial chromosome (**Figure 18.4**). Plasmids occur naturally in many strains of bacteria and in a few types of eukaryotic cells, such as yeast. The smallest plasmids consist of just a few thousand base pairs and carry only a gene or two; the largest are in the range of 100,000 to 500,000 bp and carry several dozen or even hundreds of genes. A plasmid has its own origin of replication that allows it to be replicated independently of the bacterial chromosome. The DNA sequence of the origin of replication influences how many copies of the plasmid are found within a cell. Some origins are said to be very strong because they result in many copies of the plasmid, perhaps as many as 100 per cell. Other origins of replication have sequences that are much weaker, so that the number of copies is relatively low, such as one or two per cell.

Why do bacteria have plasmids? Plasmids are not usually necessary for survival. However, in many cases, certain genes within a plasmid provide some type of growth advantage to the cell. By studying plasmids in many different species, researchers have discovered that most plasmids fall into five different categories:

1. Resistance plasmids, also known as R factors, contain genes that confer resistance against antibiotics and other types of toxins.
2. Degradative plasmids carry genes that enable the bacterium to digest and utilize an unusual substance. For example, a degradative plasmid may carry genes that allow a bacterium to digest an organic solvent such as toluene.
3. Col-plasmids contain genes that encode colicines, which are proteins that kill other bacteria.
4. Virulence plasmids carry genes that turn a bacterium into a pathogenic strain.
5. Fertility plasmids, also known as F factors, allow bacteria to mate with each other, a topic described later in this chapter.

On occasion, a plasmid may integrate into the bacterial chromosome. Plasmids that can integrate are termed **episomes**.

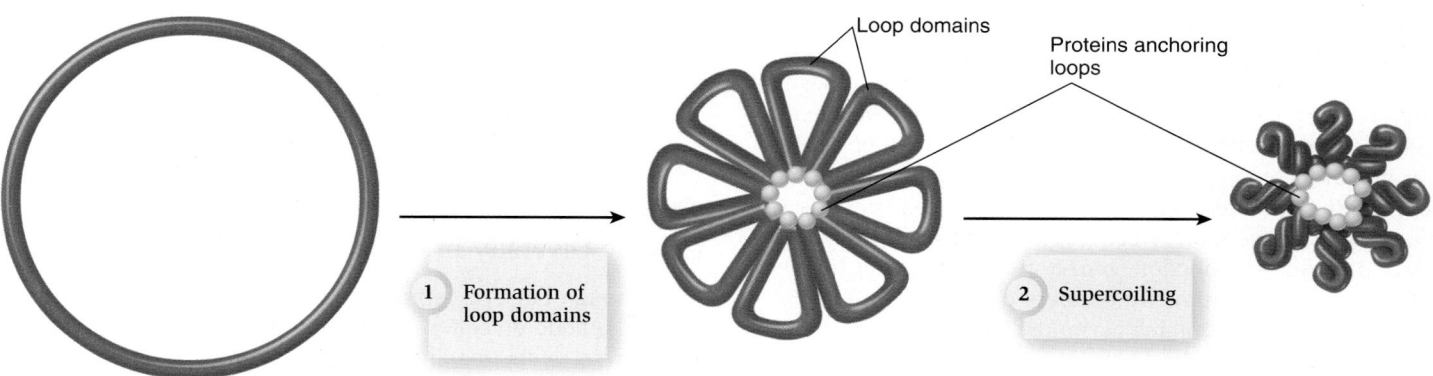

Circular chromosomal DNA · Looped chromosomal DNA with associated proteins · Supercoiled and looped DNA

Loop domains
Proteins anchoring loops
1 Formation of loop domains
2 Supercoiling

Figure 18.3 **The compaction of a bacterial chromosome.** As a way to compact the large, circular chromosome, segments are organized into smaller loop domains by binding to proteins at the bases of the loops. These loops are made more compact by DNA supercoiling.

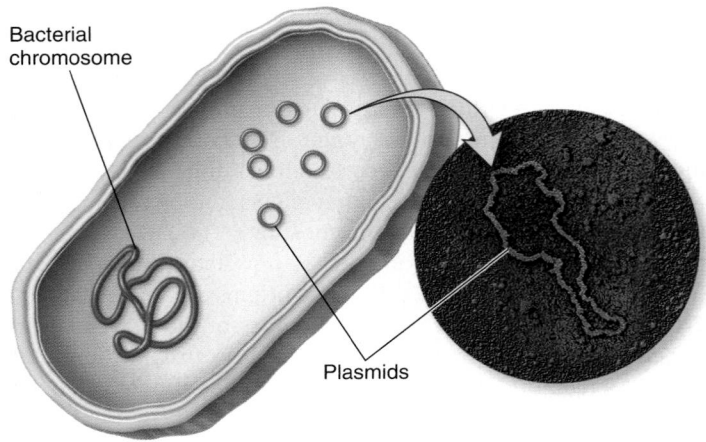

Figure 18.4   **Plasmids in a bacterial cell.**   Plasmids are small, circular DNA molecules that exist independently of the bacterial chromosome.

Bacteria Reproduce Asexually by Binary Fission

Thus far, we have considered the genetic material of bacteria, and how the bacterial chromosome is compacted to fit inside the cell. Let's now turn our attention to the process of cell division. The capacity of bacteria to divide is really quite astounding. The cells of some species, such as *E. coli*, can divide every 20–30 minutes. When placed on a solid growth medium in a petri dish, an *E. coli* cell and its daughter cells will undergo repeated cellular divisions and form a clone of genetically identical cells called a **bacterial colony** (**Figure 18.5**). Starting with a single cell that is invisible to the naked eye, a visible bacterial colony containing 10 to 100 million cells will form in less than a day!

As described in Chapter 15, the division of eukaryotic cells requires a sorting process called mitosis, because eukaryotic chromosomes occur in sets, and each daughter cell must receive the correct number and types of chromosomes. By comparison, a bacterial cell has only a single type of chromosome and a much simpler division process called **binary fission**. **Figure 18.6** shows this process for a cell with a single chromosome. Before it divides, the cell replicates its DNA. This produces two identical copies of the chromosome. Next, the cell's plasma membrane is drawn inward and deposits new cell-wall material, separating the two daughter cells. Each daughter cell receives one of the copies of the original chromosome. Therefore, except when a mutation occurs, each daughter cell contains an identical copy of the mother cell's genetic material. Like other types of asexual reproduction, binary fission does not involve genetic contributions from two different parents.

If a bacterial cell contains plasmids, these will replicate independently of the bacterial chromosome. When a plasmid is found in multiple copies in a cell, binary fission usually results in each daughter cell containing one or more copies of the plasmid.

Figure 18.5   **Growth of a bacterial colony.**   Through successive cell divisions, a single bacterial cell of *E. coli* forms a genetically identical group of cells called a bacterial colony.

Biological inquiry: Let's suppose a bacterial strain divides every 30 minutes. If a single cell is placed on a plate, how many cells will be in the colony after 16 hours?

Three Modes of Genetic Transfer Enhance Genetic Diversity in Bacteria

Even though bacteria reproduce asexually, they exhibit a great deal of genetic diversity. Within a given bacterial species, the term **strain** refers to a lineage that has genetic differences compared to another strain. For example, one strain of *E. coli* may be resistant to an antibiotic while another strain may be sensitive to the same antibiotic. Genetic diversity in bacteria comes

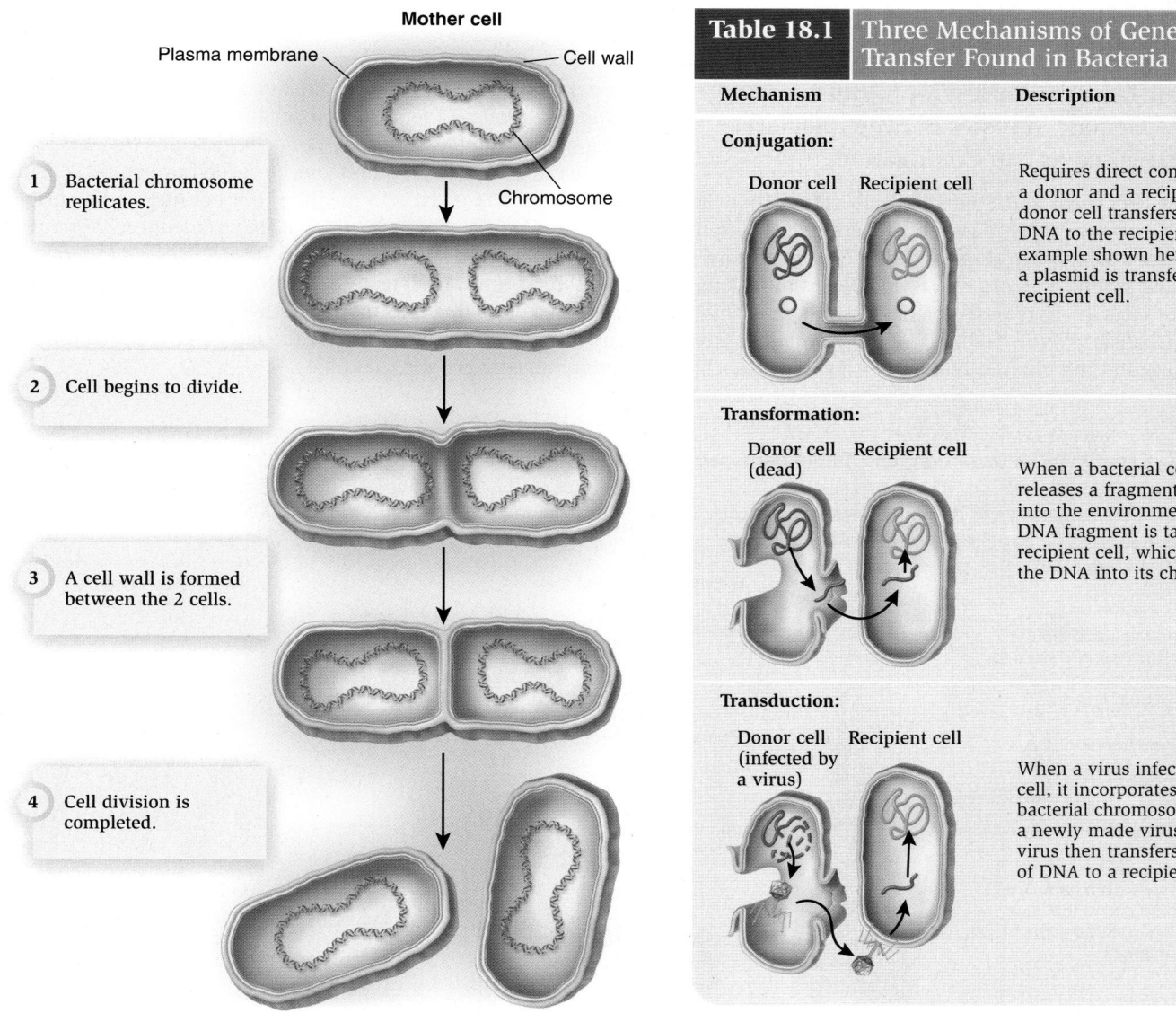

Figure 18.6 **Bacterial cell division.**   Bacteria reproduce by a type of cell division called binary fission. Before a bacterium divides, the bacterial chromosome is replicated to produce two identical copies. These two copies segregate from each other during cell division, with one copy going to each daughter cell.

Table 18.1	Three Mechanisms of Genetic Transfer Found in Bacteria	
Mechanism		**Description**
Conjugation:		Requires direct contact between a donor and a recipient cell. The donor cell transfers a strand of DNA to the recipient. In the example shown here, DNA from a plasmid is transferred to the recipient cell.
Transformation:		When a bacterial cell dies, it releases a fragment of its DNA into the environment. This DNA fragment is taken up by a recipient cell, which incorporates the DNA into its chromosome.
Transduction:		When a virus infects a donor cell, it incorporates a fragment of bacterial chromosomal DNA into a newly made virus particle. The virus then transfers this fragment of DNA to a recipient cell.

primarily from two sources. First, mutations can occur that alter the bacterial genome and affect the traits of bacterial cells. For example, a mutation may give rise to a bacterial strain that requires the amino acid histidine from an outside source for growth, although other strains of the same species can make this amino acid.

The second way that diversity can arise is by **genetic transfer**, in which genetic material is transferred from one bacterial cell to another. For example, a plasmid carrying an antibiotic-resistance gene may be transferred from a resistant strain to a sensitive strain. Genetic transfer occurs in three very different ways (**Table 18.1**). The process known as **conjugation** involves a direct physical interaction between two bacterial cells. During conjugation, one bacterium acts as a donor and transfers DNA to a recipient cell. In the process of **transformation**, DNA that is released into the environment when a bacterium dies is taken up by another bacterial cell. **Transduction** occurs when a virus infects a bacterial cell and then transfers some of that cell's DNA to another bacterium. These three types of genetic transfer have been extensively investigated in research laboratories, and their molecular pathways continue to be studied with great interest. Let's now examine these mechanisms in greater detail and consider the experiments that led to their discovery.

FEATURE INVESTIGATION

Lederberg and Tatum's Work with *E. coli* Demonstrated Genetic Transfer Between Bacteria and Led to the Discovery of Conjugation

In 1946, Joshua Lederberg and Edward Tatum carried out the first experiment that clearly demonstrated genetic transfer from one bacterial strain to another (**Figure 18.7**). The researchers had been studying strains of *E. coli* that had different nutritional requirements for growth. They designated one strain *met⁻ bio⁻ thr⁺ leu⁺ thi⁺* because its growth required that the amino acid methionine (met) and the vitamin biotin (bio) be added to the growth medium. This strain did not require the amino acids threonine (thr) and leucine (leu) or the vitamin thiamine (thi) for growth. Another strain, designated *met⁺ bio⁺ thr⁻ leu⁻ thi⁻*, had just the opposite requirement. It needed threonine, leucine, and thiamine in its growth medium, but not methionine or biotin. These differences in nutritional requirements correspond to allelic differences between the two strains. The *met⁻ bio⁻ thr⁺ leu⁺ thi⁺* strain had defective genes encoding enzymes necessary for methionine and biotin synthesis, while *met⁺ bio⁺ thr⁻ leu⁻ thi⁻* had defective genes for the enzymes required to make threonine, leucine, and thiamine.

Figure 18.7 Experiment of Lederberg and Tatum demonstrating genetic transfer in *E. coli*.

HYPOTHESIS Genetic material can be transferred from one bacterial strain to another.
STARTING MATERIALS Two bacterial strains, one that was *met⁻ bio⁻ thr⁺ leu⁺ thi⁺* and the other that was *met⁺ bio⁺ thr⁻ leu⁻ thi⁻*.

Experimental level   Conceptual level

1 In 3 separate tubes, add either the *met⁻ bio⁻ thr⁺ leu⁺ thi⁺* strain, the *met⁺ bio⁺ thr⁻ leu⁻ thi⁻* strain, or a mixture of both strains.

   Incubate several hours.

   met⁻ bio⁻ thr⁺ leu⁺ thi⁺   *met⁺ bio⁺ thr⁻ leu⁻ thi⁻*

2 Remove 10⁸ cells from each tube and spread onto plates that lack methionine, biotin, threonine, leucine, and thiamine.

   10⁸   10⁸   10⁸

   Nutrient agar plates lacking amino acids, biotin, and thiamine.

3 Incubate overnight to allow growth of bacterial colonies.

   No colonies   Bacterial colonies (*met⁺ bio⁺ thr⁺ leu⁺ thi⁺*)   No colonies

4 **THE DATA**

Strain	Number of colonies after overnight growth
met⁻ bio⁻ thr⁺ leu⁺ thi⁺	0
met⁺ bio⁺ thr⁻ leu⁻ thi⁻	0
Both strains together	~10

Figure 18.7 compares the results of mixing the two *E. coli* strains with the results when they were not mixed. The tube shown on the left contains only *met⁻ bio⁻ thr⁺ leu⁺ thi⁺* cells, and the tube on the right has only *met⁺ bio⁺ thr⁻ leu⁻ thi⁻* cells. The middle tube contains a mixture of the two kinds of cells. In each case, the researchers applied about 100 million (10⁸) cells to plates containing a growth medium lacking amino acids, biotin, and thiamine. When the unmixed strains were applied to

these plates, no colonies were observed to grow. This result is expected because the plates did not contain the methionine and biotin that the met^- bio^- thr^+ leu^+ thi^+ cells need for growth, or the threonine, leucine, and thiamine that the met^+ bio^+ thr^- leu^- thi^- cells require. The striking result occurred when the researchers plated 10^8 cells from the tube containing the mixture of the two strains. In this case, approximately 10 cells multiplied and formed visible bacterial colonies on the plates. Because these cells were able to reproduce without supplemental amino acids or vitamins, their genotype must have been met^+ bio^+ thr^+ leu^+ thi^+. Mutation cannot account for the occurrence of this new genotype because colonies were not observed on the other two plates, which had the same number of cells and also could have incurred mutations.

To explain the results of their experiment, Lederberg and Tatum hypothesized that some genetic material was transferred between the two strains when they were mixed. Note that this transfer could have occurred in two ways. One possibility is that the genes providing the ability to synthesize threonine, leucine, and thiamine (thr^+ leu^+ thi^+) were transferred to the met^+ bio^+ thr^- leu^- thi^- strain. Alternatively, the genes providing the ability to synthesize methionine and biotin (met^+ bio^+) may have been transferred to the met^- bio^- thr^+ leu^+ thi^+ cells. The experimental results cannot distinguish between these two possibilities, but they provide compelling evidence that at least one of them occurred.

How did the bacteria in Lederberg and Tatum's experiment transfer genes between strains? Two mechanisms seemed plausible. Either genetic material was released from one strain and taken up by the other, or cells of the two different strains made contact with each other and directly transferred genetic material. To distinguish these two scenarios, Bernard Davis conducted experiments using the same two strains of *E. coli*. The apparatus he used, known as a U-tube, is shown in **Figure 18.8**. The tube had a filter with pores big enough for pieces of DNA to pass through, but too small to permit the passage of bacteria. After filling the tube with a liquid medium, Davis added met^- bio^- thr^+ leu^+ thi^+ bacteria on one side of the filter and met^+ bio^+ thr^- leu^- thi^- bacteria on the other. The application of pressure or suction promoted the movement of liquid through the

Figure 18.8   A U-tube apparatus like the one used by Bernard Davis.   Bacteria of two different strains are suspended in the liquid in the tube and separated by a filter. The liquid is forced through the filter by alternating suction and pressure. The pores in the filter are too small for the passage of bacteria, but they allow the passage of DNA.

Biological inquiry: Would the results have been different if the pore size was larger and allowed the passage of bacterial cells?

pores. Although the two kinds of bacteria could not mix, any genetic material released by one of them would be available to the other.

After allowing the bacteria to incubate in the U-tube, Davis placed cells from each side of the tube on growth plates lacking methionine, biotin, threonine, leucine, and thiamine. No bacterial colonies grew on these plates. This result showed that without physical contact, the two *E. coli* strains do not transfer genetic material from one to the other. The conceptual level of Figure 18.7 shows the physical connection that explains Lederberg and Tatum's results. Genetic transfer that requires direct cell-to-cell contact is now called conjugation, and it was subsequently observed in other species of bacteria. Many, but not all, species of bacteria can conjugate.

During Conjugation, DNA Is Transferred from a Donor Cell to a Recipient Cell

In the early 1950s, Joshua and Esther Lederberg, William Hayes, and Luca Cavalli-Sforza independently discovered that only certain bacterial strains can donate genetic material during conjugation. For example, only about 5% of *E. coli* strains found in nature can act as donor strains. Further research showed that a strain that is incapable of acting as a donor can acquire this ability after being mixed with a donor strain. Hayes correctly proposed that donor strains contain a fertility factor, or **F factor**, that can be transferred to recipient strains. Also, other donor *E. coli* strains were later identified that can transfer portions of

the bacterial chromosome at high frequencies. After a segment of the chromosome is transferred, it then inserts, or recombines, into the chromosome of the recipient cell. Such donor strains were named Hfr (for High frequency of recombination).

The micrograph in **Figure 18.9a** shows two conjugating *E. coli* cells. The cell on the left is designated F^+, meaning that it has an F factor. This donor cell is transferring genetic material to the recipient cell on the right, which lacks an F factor and is designated F^-. As discussed earlier, the F factor, or fertility factor, is a type of plasmid found in many bacterial species. F factors carry several genes that are required for conjugation and also may carry genes that confer a growth advantage for the bacterium.

(a) Micrograph of conjugating cells

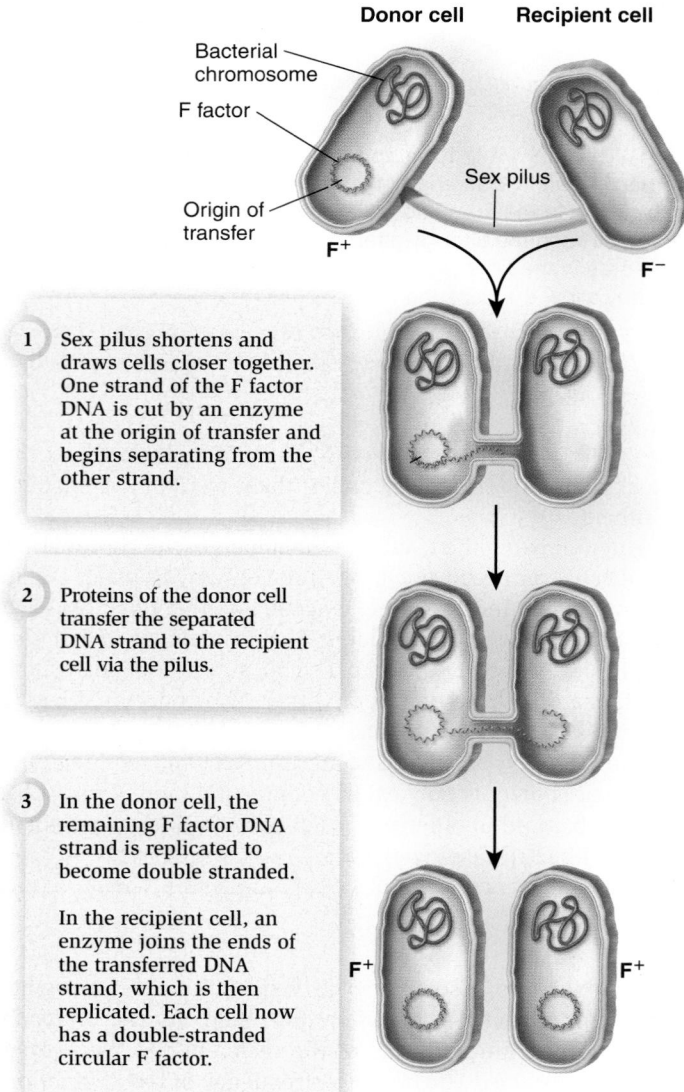

(b) Transfer of an F factor

Figure 18.9   **Bacterial conjugation.**   (a) Two *E. coli* cells conjugating. The cell on the *left*, designated F^+, is the donor; the cell on the *right*, designated F^-, is the recipient. The two cells make contact via sex pili made by the F^+ cell. (b) The transfer of an F factor during conjugation. At the end of conjugation, both the donor cell and the recipient cell are F^+.

Figure 18.9b describes the events that occur during conjugation in *E. coli*. The process is similar in other bacteria that are capable of conjugating, although the details vary somewhat from one species to another. Contact between donor and recipient cells is a key step that initiates the conjugation process. Some bacteria have hairlike appendages called pili (singular, pilus), which are hollow tubes that promote the attachment of bacteria to surfaces and to each other.

Sex pili are pili made by F^+ cells that bind specifically to F^- cells. They are so named because conjugation has sometimes been called bacterial mating or bacterial sex. However, these terms are a bit misleading because the process does not involve equal genetic contributions from two gametes and it does not produce offspring. Instead, bacterial mating is a form of genetic transfer that alters the genetic composition of the recipient cell. The genes responsible for the formation of sex pili are located on the F factor. In *E. coli* and some other species, F^+ cells make very long pili that attempt to make contact with nearby F^- cells. Once contact is made, the pili shorten, drawing the donor and recipient cells closer together.

Successful contact between a donor and a recipient cell stimulates the donor cell to begin the transfer process. Genes within the F factor encode proteins that promote the transfer of one strand of F factor DNA. This DNA strand is cut at the origin of transfer, and then the strand travels through the pilus into the recipient cell. The other strand remains in the donor cell and is replicated, restoring the F factor DNA to its original double-stranded condition. In the recipient cell, the two ends of the newly acquired F factor DNA strand are joined to form a circular molecule, which is then replicated to become double stranded. The end result of conjugation is that the recipient cell has acquired an F factor, converting it from an F^- to an F^+ cell. The genetic composition of the donor strain has not been changed.

In Transformation, Bacteria Take Up DNA from the Environment

In contrast to conjugation, the process of genetic transfer known as bacterial transformation does not require direct contact between bacterial cells. Frederick Griffith first discovered this process in 1928 while working with strains of *Streptococcus pneumoniae*. He called the unknown substance that transferred properties of one strain to another the transforming principle. Experimental work in the 1940s and 1950s showed that the transforming principle is DNA and that transformation involves the transfer of genes. We discussed some of these important experiments in Chapter 11.

During transformation, a living bacterial cell imports a strand of DNA that another bacterium released into the environment when it died. This DNA strand may then insert or recombine into the bacterial chromosome, so that the live bacterium is now carrying genes from the dead bacterium. Not all bacterial strains have the ability to take up DNA. Those that do have this ability are described as **naturally competent**, and they have genes that encode proteins called competence factors.

Competence factors facilitate the binding of DNA fragments to the bacterial cell surface, the uptake of DNA into the cytoplasm, and the incorporation of the imported DNA into the bacterial chromosome. Temperature, ionic conditions, and the availability of nutrients also affect whether or not a bacterium will be competent to take up genetic material. Thus, environmental conditions influence the expression of competence genes.

In recent years, biologists have unraveled some of the steps that occur when competent bacterial cells are transformed by taking up genetic material released from a dead bacterium. In the example shown in **Figure 18.10**, the foreign DNA carries a gene, tet^R, that confers resistance to the antibiotic tetracycline. First, a large fragment of the foreign DNA binds to a surface receptor on the outside of a bacterial cell. Enzymes secreted by the bacterium cut this large fragment into fragments small enough to enter the cell. The next step is for a small DNA fragment to begin its entry into the bacterial cytoplasm. One of the two DNA strands of this fragment is degraded. The other strand enters the bacterial cytoplasm via a DNA uptake system that transports the DNA across the plasma membrane. Finally, the imported DNA strand is incorporated into the bacterial chromosome and the complementary strand is synthesized. Following transformation, the recipient cell has been transformed from a tetracycline-sensitive to a tetracycline-resistant cell.

In Transduction, Viruses Transfer Genetic Material from One Bacterium to Another

Perhaps the most curious method of genetic transfer is transduction, in which viruses that infect bacteria transfer bacterial genes from one bacterium to another. As you will learn in the second part of this chapter, the viruses that we will consider are DNA-containing particles that use the bacterial cellular machinery for their own replication. The new viral particles made in this way usually contain only viral genes. In some cases, however, a virus may pick up a piece of DNA from the bacterial chromosome. When a virus carrying a segment of bacterial DNA infects another bacterium, it transfers this segment into the chromosome of its new bacterial host. This is the genetic transfer process called transduction. We will consider transduction in more detail when we look at viral reproductive cycles in the second part of this chapter.

GENOMES & PROTEOMES

Horizontal Gene Transfer Is the Transfer of Genes Between Different Species

So far we have considered gene transfer from one bacterial strain to another strain of the same species. In addition, conjugation, transformation, and transduction occasionally occur between cells of different bacterial species. Gene transfer between different species is termed **horizontal gene transfer**, in contrast to the

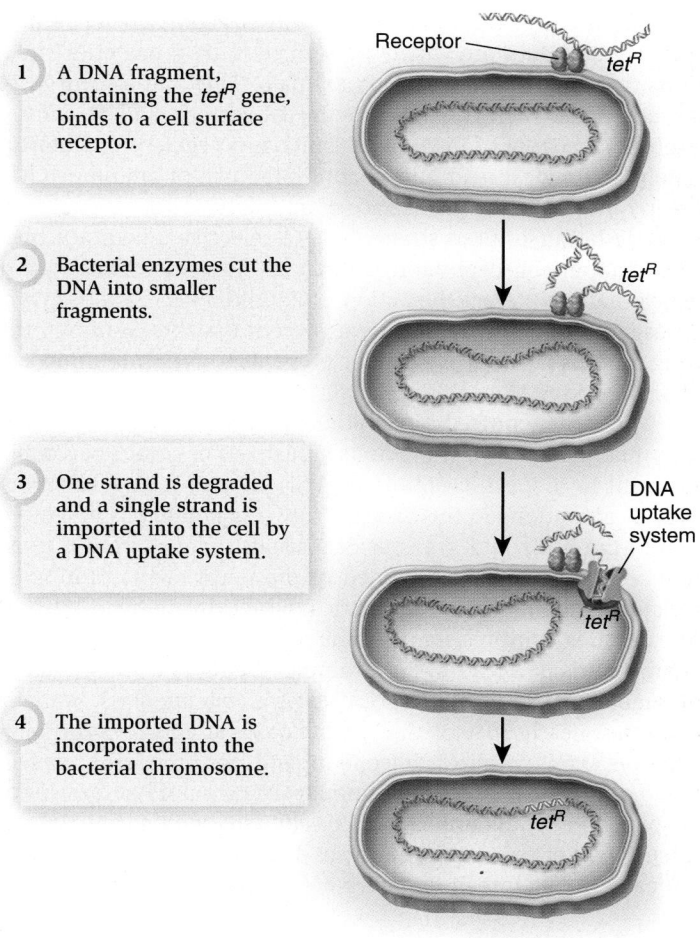

1. A DNA fragment, containing the tet^R gene, binds to a cell surface receptor.

2. Bacterial enzymes cut the DNA into smaller fragments.

3. One strand is degraded and a single strand is imported into the cell by a DNA uptake system.

4. The imported DNA is incorporated into the bacterial chromosome.

Transformed cell that is resistant to the antibiotic tetracycline

Figure 18.10 **The steps of bacterial transformation.** This process has transformed a bacterium that was sensitive to the antibiotic tetracycline into one that can grow in the presence of this antibiotic.

vertical gene transfer that occurs when genes are passed from one generation to the next among individuals of the same species.

In recent years, analyses of bacterial genomes have shown that a sizeable fraction of their genes are derived from horizontal gene transfer. For example, roughly 17% of the genes of *E. coli* and of *Salmonella typhimurium* have been acquired by horizontal transfer during the past 100 million years. Many of these acquired genes are for traits that give cells a selective advantage, including genes that confer antibiotic resistance, the ability to degrade toxic compounds, and the ability to withstand extreme environments. Some horizontally transferred genes confer pathogenicity, turning a harmless bacterial strain into one that can cause disease. Geneticists have suggested that horizontal gene transfer has played a major role in the evolution of different bacterial species. In many cases, the acquisition of new genes allows a bacterium to survive in a new type of environment and can eventually lead to the formation of a new species.

The medical relevance of horizontal gene transfer is profound. Antibiotics are widely prescribed to treat bacterial infections in humans and domestic animals. They are also used in agriculture to increase the growth of livestock and to control bacterial diseases of high-value fruits and vegetables. Unfortunately, the widespread and uncontrolled use of antibiotics has greatly increased the prevalence of antibiotic-resistant strains of bacteria because these strains have a selective advantage over those that are susceptible to antibiotics. Resistant strains carry genes that counteract the action of antibiotics in various ways. A resistance gene may encode a protein that breaks down the drug, pumps it out of the cell, or prevents it from inhibiting cellular processes.

The term **acquired antibiotic resistance** refers to the common phenomenon of a previously susceptible strain becoming resistant to a specific antibiotic. This change may result from genetic alterations in the bacteria's own genome, but it is often due to the horizontal transfer of resistance genes from a resistant strain. As often mentioned in the news media, antibiotic resistance has increased dramatically worldwide over the past few decades, with resistant strains reported in almost all pathogenic strains of bacteria. For example, the most common cause of pneumonia is infection by *Streptococcus pneumoniae*. In many countries, nearly 50% of all *S. pneumoniae* strains are now penicillin resistant, with resistance to other drugs increasing as well. Some of the most severe antibiotic resistance problems occur in hospitals. Resistant strains of *Klebsiella pneumoniae* and *Enterococcus faecium* are significant causes of infection and death among critically ill patients in intensive care units.

18.2 Viral Genetics, Viroids, and Prions

In earlier chapters and the first section of this chapter, we examined the replication and expression of eukaryotic and prokaryotic genes. Because all living organisms are either eukaryotes or prokaryotes, you may be thinking that we have considered every type of genome. Strangely, however, nonliving things also have genomes. Viruses and viroids are nonliving particles with nucleic acid genomes that require the assistance of living cells to reproduce.

The first virus to be discovered was tobacco mosaic virus (TMV). This virus infects several species of plants and causes mosaic-like patterns in which normal-colored patches are interspersed with light green or yellowish patches on the leaves (**Figure 18.11**). TMV damages leaves, flowers, and fruit but almost never kills the plant. In 1883, the German scientist Adolf Mayer determined that this disease could be spread by spraying the sap from one plant onto another. By subjecting this sap to filtration, the Russian scientist Dmitri Ivanovski demonstrated that the disease-causing agent was not a bacterium. Sap that had been passed through filters with pores small enough to prevent the passage of bacterial cells was still able to spread the disease. At first, some researchers suggested that the agent was a chemical

Figure 18.11   A plant infected with tobacco mosaic virus.

toxin. However, the Dutch botanist Martinus Beijerinck ruled out this possibility by showing that sap could continue to transmit the disease after many plant generations. A toxin would have been diluted after many generations, but Beijerinck's results indicated that the disease agent was multiplying in the plant. Around the same time, animal viruses were discovered in connection with a disease of cattle called foot-and-mouth disease. In 1900, the first human virus, the virus that causes yellow fever, was identified. Since these early studies, microbiologists, geneticists, and molecular biologists have taken a great interest in the study of viruses.

In this section, we will consider the structure, genetic composition, and replication of viruses. We will examine viral reproductive cycles, paying particular attention to HIV (human immunodeficiency virus), the virus that causes AIDS (acquired immunodeficiency syndrome) in humans. We will also look briefly at simpler nucleic acid particles called viroids that cause diseases in plants. Finally, we will discuss the infectious proteins known as prions, which cause devastating neurological diseases in humans and other mammals. Unlike other agents of infection, prions have no genes and cannot be copied by the replication machinery of a cell. Instead, they increase their numbers by inducing changes in other protein molecules within living cells.

Viruses Are Remarkably Varied, Despite Their Simple Structure

A **virus** is a small infectious particle that consists of nucleic acid enclosed in a protein coat. Researchers have identified and studied over 4,000 different types of viruses. Although all viruses share some similarities, such as small size and the reliance on a living cell for replication, they vary greatly in their characteristics, including their host range, structure, and genome composition. Some of the major differences are described next and in **Table 18.2**.

Differences in Host Range   A cell that is infected by a virus is called a **host cell**, and a species that can be infected by a specific

Table 18.2 Hosts and Characteristics of Selected Viruses

Virus or group of viruses	Host	Effect on host	Nucleic acid*	Genome size (kb)†	Number of genes†
Phage fd	E. coli	Slows growth	ssDNA	6.4	10
Phage λ	E. coli	Causes lysis	dsDNA	48.5	36
Phage T4	E. coli	Causes lysis	dsDNA	169	>190
Phage Qβ	E. coli	Slows growth	ssRNA	4.2	4
Tobacco mosaic virus	Many plants	Causes mottling and necrosis of leaves and other plant parts	ssRNA	6.4	6
Cauliflower mosaic virus	A few plants	Similar to tobacco mosaic virus	dsDNA	8.0	1
Baculoviruses	Insects	Most baculoviruses are species specific; they usually kill the insect	dsDNA	133.9	154
Parvovirus	Mammals	Causes respiratory, flulike symptoms	ssDNA	5.0	5
Influenza virus	Mammals	Causes classical "flu," with fever, cough, sore throat, and headache	ssRNA	13.5	12
Epstein-Barr virus	Humans	Causes mononucleosis, with fever, sore throat, and fatigue	dsDNA	172	80
Adenoviruses	Humans	Cause respiratory symptoms and diarrhea	dsDNA	34	35
Herpes simplex type II	Humans	Causes blistering sores around the genital region	dsDNA	155	74
HIV (type I)	Humans	Causes AIDS, an immunodeficiency syndrome eventually leading to death	ssRNA	9.2	9

* The abbreviations *ss* and *ds* refer to single stranded and double stranded, respectively.

† Several of the viruses listed in this table are found in different strains that show some variation with regard to genome size and number of genes. The numbers reported in this table are typical values. The abbreviation *kb* refers to kilobase, which equals 1,000 bases.

virus is called a host species for that virus. Viruses differ greatly in their **host range**, meaning the number of species and cell types they can infect. Table 18.2 lists a few examples of viruses with widely different ranges of host species. Tobacco mosaic virus, which we discussed earlier, has a broad host range. TMV is known to infect over 150 different species of plants. By comparison, other viruses have a limited host range. Some viruses infect only a single species, such as humans. Furthermore, a virus may infect only a specific cell type in that species. **Figure 18.12** shows some viruses that infect particular human cells and cause disease.

Structural Differences Although the existence of viruses was postulated in the 1890s, viruses were not observed until the 1930s, when the electron microscope was invented. Viruses cannot be resolved by even the best light microscope. In fact, most of them are smaller than the wavelength of visible light. Viruses range in size from about 20–400 nm in diameter (1 nanometer = 10^{-9} meters). For comparison, a typical bacterium is 1,000 nm in diameter, and the diameter of most eukaryotic cells is 10 to 1,000 times that of a bacterium. Over 50 million adenoviruses could fit into an average-sized human cell.

Figure 18.13 shows basic structural variations of viruses. All viruses have a protein coat called a **capsid** enclosing a genome that consists of one or more molecules of nucleic acid. Capsids are composed of one or several different proteins and may be a variety of shapes, including helical and polyhedral.

Figure 18.13a shows the structure of tobacco mosaic virus, which has a helical capsid made of identical protein subunits. Figure 18.13b shows an adenovirus, which has a polyhedral capsid. Many viruses that infect animal cells, such as the influenza virus shown in Figure 18.13c, have a **viral envelope** enclosing the capsid. The envelope consists of a membrane that is derived from the plasma membrane of the host cell and is embedded with virally encoded spike glycoproteins. You will see how such viruses acquire their envelopes when we examine viral reproductive cycles later in the chapter.

In addition to encasing and protecting the genetic material, the capsid enables viruses to infect their hosts. The capsids of many viruses have specialized proteins that help them bind to the surface of a host cell. The adenovirus in Figure 18.13b has glycoproteins called spike glycoproteins that serve this purpose. Similarly, the envelope and spike glycoproteins of the influenza virus are important in the process of viral attachment to the host cell. Viruses that infect bacteria, called **bacteriophages**, or **phages**, may have much more complex capsids, with accessory structures used for anchoring the virus to a host cell and injecting the viral nucleic acid (Figure 18.13d). As we will discuss later, the tail fibers of such bacteriophages are needed to attach the virus to the bacterial cell wall.

Genome Differences The genetic material in a virus is called a **viral genome**. The composition of viral genomes varies markedly among different types of viruses, as suggested by the examples

Viruses Reproduce by Mobilizing Their Host Cells to Produce New Viruses

Brain and CNS:
Flavivirus—yellow fever
Rhabdovirus—rabies

Respiratory tract:
Influenza virus—flu
Rhinovirus—common cold

Immune system:
Rubella virus—measles
Human immunodeficiency virus—AIDS
Epstein-Barr virus—mononucleosis

Digestive system:
Hepatitis B virus—viral hepatitis
Rotavirus—viral gastroenteritis
Norwalk virus—viral gastroenteritis

Skin:
Herpes simplex I—cold sores
Variola virus—smallpox

Reproductive system:
Herpes simplex II—genital herpes
Papillomavirus—warts, cervical cancer

Blood:
Ebola virus—hemorrhagic fever
Hantavirus—hemorrhagic fever
with renal syndrome

Figure 18.12   **Some viruses that cause human diseases.**   Most viruses that cause disease in humans infect cells of specific tissues, as illustrated by the examples in this figure. Note: herpes simplex I and II infect nerve cells of the peripheral nervous system that are found in the skin and genital region, respectively.

Viruses do not exhibit the properties that are found in living organisms, as described in Chapter 1. They are not cells or composed of cells, and they cannot carry out metabolism on their own. When a virus infects a cell, the expression of viral genes leads to a series of steps, called a **viral reproductive cycle**, that result in the production of new viruses. The details of the steps may be quite different among various types of viruses, and even the same virus may have the capacity to follow alternative cycles. However, by studying the reproductive cycles of hundreds of different viruses, researchers have determined that most viruses follow five or six basic steps.

To illustrate general features of viral reproductive cycles, **Figure 18.14** considers these steps for two types of viruses. Figure 18.14a shows the cycle of phage λ (lambda), a bacteriophage with double-stranded DNA as its genome. Figure 18.14b shows the cycle of HIV, an enveloped animal virus containing single-stranded RNA. The descriptions that follow compare the reproductive cycles of these two very different viruses. Later in the chapter, we will further examine HIV, which is the causative agent of AIDS.

Step 1: Attachment   In the first step of a viral reproductive cycle, the virus must attach to the surface of a host cell. This attachment is usually specific for one kind of cell because proteins in the virus recognize and bind to specific molecules on the cell surface. In the case of phage λ, the phage tail fibers bind to proteins in the outer bacterial cell membrane of *E. coli*. In the case of HIV, spike glycoproteins in the viral envelope bind to protein receptors in the plasma membrane of human blood cells called T cells.

Step 2: Entry   After attachment, the viral genome must enter the host cell. Attachment of phage λ stimulates a conformational change in the phage coat proteins, so that the shaft contracts and the phage injects its DNA into the bacterial cytoplasm. In contrast, the envelope of HIV fuses with the plasma membrane of the host cell, so that both the capsid and its contents are released into the cytosol. Some of the HIV capsid proteins are then removed by host cell enzymes, allowing the viral RNA and reverse transcriptase to escape from the capsid.

Once a viral genome has entered the cell, one or several viral genes are expressed immediately due to the action of host cell enzymes and ribosomes. Expression of these key genes leads quickly to either step 3 or step 4 of the reproductive cycle, depending on the specific virus. The genome of some viruses, including both phage λ and HIV, can integrate into a chromosome of the host cell.

in Table 18.2. The nucleic acid of some viruses is DNA, while in others it is RNA. These are referred to as DNA viruses and RNA viruses, respectively. It is striking that some viruses use RNA for their genome, whereas all living organisms use DNA. In some viruses the nucleic acid is single stranded, whereas in others it is double stranded. The genome can be linear or circular, depending on the type of virus. Some kinds of viruses have more than one copy of the genome.

Viral genomes also vary considerably in size, ranging from a few thousand to more than a hundred thousand nucleotides in length (see Table 18.2). For example, the genomes of some simple viruses, such as phage Qβ, are only a few thousand nucleotides in length and contain only a few genes. Other viruses, particularly those with a complex structure, contain many more genes. Phage T4 is an example of a complex virus having many genes. These extra genes are largely involved in the formation of the elaborate protein coat shown in Figure 18.13d.

(a) Tobacco mosaic virus, a nonenveloped virus with a helical capsid

(b) Adenovirus, a nonenveloped virus with a polyhedral capsid

(c) Influenza virus, an enveloped virus with spikes

(d) T4, a bacteriophage

Figure 18.13 **Variations in the structure of viruses.**   All viruses contain nucleic acid surrounded by a protein capsid. They may or may not have an outer envelope surrounding the capsid. **(a)** Tobacco mosaic virus (TMV) has a capsid made of 2,130 identical protein subunits, helically arranged around a strand of nucleic acid (RNA). **(b)** Adenoviruses have polyhedral capsids with glycoprotein spikes. **(c)** Many animal viruses, including the influenza virus, have an envelope composed of a lipid bilayer and spike proteins. The lipid bilayer is obtained from the host cell when the virus buds through the plasma membrane. **(d)** Some bacteriophages, such as T4, have capsids with accessory structures that facilitate invasion of a bacterial cell.

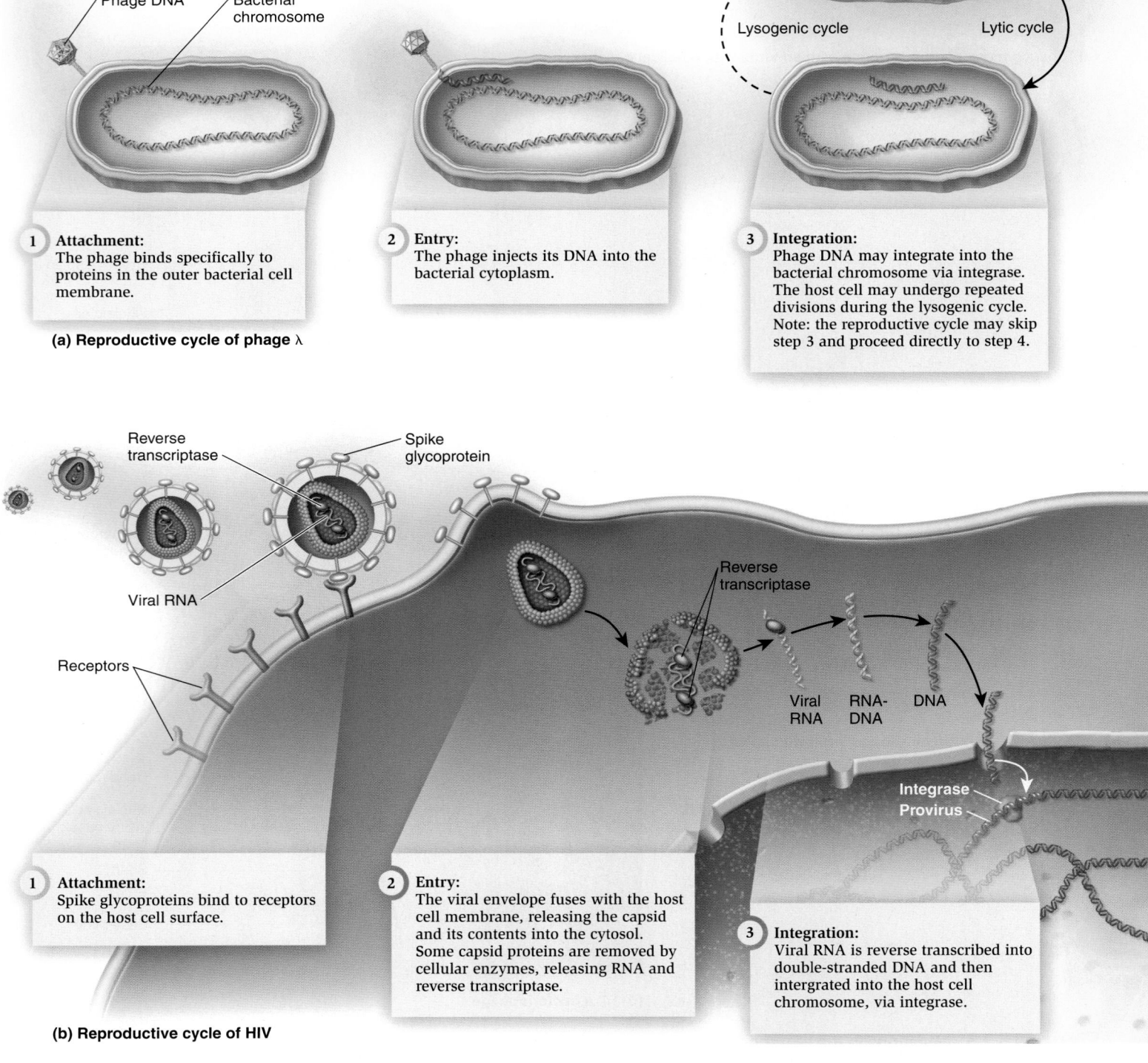

(a) Reproductive cycle of phage λ

1. **Attachment:** The phage binds specifically to proteins in the outer bacterial cell membrane.

2. **Entry:** The phage injects its DNA into the bacterial cytoplasm.

3. **Integration:** Phage DNA may integrate into the bacterial chromosome via integrase. The host cell may undergo repeated divisions during the lysogenic cycle. Note: the reproductive cycle may skip step 3 and proceed directly to step 4.

(b) Reproductive cycle of HIV

1. **Attachment:** Spike glycoproteins bind to receptors on the host cell surface.

2. **Entry:** The viral envelope fuses with the host cell membrane, releasing the capsid and its contents into the cytosol. Some capsid proteins are removed by cellular enzymes, releasing RNA and reverse transcriptase.

3. **Integration:** Viral RNA is reverse transcribed into double-stranded DNA and then intergrated into the host cell chromosome, via integrase.

Figure 18.14 **Comparison of the steps of two viral reproductive cycles.** (a) The reproductive cycle of phage λ, a bacteriophage with a double-stranded DNA genome. (b) The reproductive cycle of HIV, an enveloped animal virus with a single-stranded RNA genome.

4 **Synthesis of viral components:**
In the lytic cycle, phage DNA directs the synthesis of viral components. During this process, the host chromosomal DNA is degraded.

5 **Viral assembly:**
Phage components are assembled with the help of noncapsid proteins to make many new phages.

6 **Release:**
The viral enzyme called lysozyme causes cell lysis, and new phages are released from the broken cell.

Capsid proteins

Spike glycoproteins

Viral RNA

Translation

Reverse transcriptase

Viral RNA

Transcription

4 **Synthesis of viral components:**
Proviral DNA is transcribed into viral RNA, which is translated into viral proteins.

5 **Viral assembly:**
Capsid proteins enclose 2 RNA molecules and 2 molecules of reverse transcriptase. Capsid assembles with spike glycoproteins during budding.

6 **Release:**
Virus buds from the plasma membrane of the host cell and is released.

For such viruses, the cycle may proceed from step 2 to step 3 as described next, delaying the production of new viruses. Alternatively, the cycle may proceed directly from step 2 to step 4 and quickly lead to the production of new viruses.

Step 3: Integration Viruses that are capable of integration carry a gene that encodes an enzyme called **integrase**. For integration to occur, this gene is expressed soon after entry so that integrase protein is made. Integrase cuts the host's chromosomal DNA and inserts the viral genome into the chromosome. In the case of the bacteriophage, the double-stranded DNA that entered the cell can be directly integrated into the double-stranded DNA of the chromosome. Once integrated, the phage DNA in a bacterium is called a **prophage**. At a later time, the prophage may be excised from the bacterial chromosome and the cycle then proceeds to step 4.

In the case of an RNA virus, the viral genome must be copied into DNA for integration to occur. HIV accomplishes this by means of a viral enzyme called **reverse transcriptase**, which is carried within the capsid and released into the host cell along with the viral RNA. Reverse transcriptase uses the viral RNA strand to make a complementary copy of DNA, and it then uses the DNA strand as a template to make double-stranded viral DNA. This process is called reverse transcription because it is the reverse of the usual transcription process, in which a DNA strand is used to make a complementary strand of RNA. The viral double-stranded DNA enters the host cell nucleus and is inserted into a host chromosome via integrase. Once integrated, the viral DNA in a eukaryotic cell is called a **provirus**.

Step 4: Synthesis of Viral Components The production of new viruses by a host cell always involves replication of the viral genome and synthesis of the capsid proteins that make up the viral coat. For some viruses, including HIV, the host cell must synthesize additional viral proteins, such as integrase, that facilitate the viral reproductive cycle.

In the case of a bacteriophage that has been integrated into the host chromosome, the prophage must be excised as described in step 3 before synthesis of new viral components can occur. Following excision, host cell enzymes such as DNA polymerase make many copies of the phage DNA and transcribe the genes within these copies into mRNA. Host cell ribosomes translate this viral mRNA into viral proteins. For many bacteriophages, the expression of phage genes leads to the degradation of the host chromosomal DNA.

In the case of HIV, the DNA provirus is not excised from the host chromosome. Instead, it is transcribed in the nucleus to produce many copies of viral RNA. These RNA molecules enter the cytosol, where they serve two purposes. First, they are translated to make viral proteins using host cell ribosomes. These proteins include capsid proteins, reverse transcriptase, and HIV spike proteins that are inserted into the cell's plasma membrane. The spike glycoproteins are destined to be part of the HIV viral envelope. Second, the RNA molecules serve as the genome for new viral particles.

Step 5: Viral Assembly After all of the necessary components have been synthesized, they must be assembled into new viruses. Some viruses with a simple structure self-assemble, meaning that viral components made by the host cell spontaneously bind to each other to form a complete virus particle. An example of a self-assembling virus is the tobacco mosaic virus we examined earlier (see Figure 18.13a). TMV capsid proteins assemble around a TMV RNA molecule, which becomes trapped inside the hollow capsid.

Other viruses, including the two shown in Figure 18.14, have more complicated structures that do not self-assemble. The correct assembly of phage λ requires the help of noncapsid proteins that are not found in the completed phage particle. Some of these noncapsid proteins function as enzymes that modify capsid proteins, while others serve as scaffolding for the assembly of the capsid.

The assembly of an HIV virus occurs in two stages. First, capsid proteins assemble around two molecules of HIV RNA and two molecules of reverse transcriptase. Next, the newly formed capsid acquires its outer envelope in a budding process. This second phase of assembly occurs during step 6, as the virus is released from the cell.

Step 6: Release The final stage of a viral reproductive cycle is the release of new viruses from the host cell. The release of bacteriophages is a dramatic event. Because bacteria are surrounded by a rigid cell wall, the phages must lyse their host cell in order to escape. After the phages have been assembled, a phage-encoded enzyme called lysozyme digests the bacterial cell wall, causing the cell to burst. Lysis releases many new phages into the environment, where they can infect other bacteria and begin the cycle again.

The release of enveloped viruses from an animal cell is far less dramatic. This type of virus escapes by a mechanism called budding that does not lyse the cell. In the case of HIV, a newly assembled virus particle associates with a portion of the plasma membrane containing HIV spike glycoproteins. The membrane enfolds the viral capsid and eventually buds from the surface of the cell. This is how the virus acquires its envelope, which is a piece of host cell membrane studded with viral glycoproteins. The photograph at the beginning of this chapter shows the budding of new HIV particles from a T cell.

Latency As we saw in step 3, some viruses can integrate their genomes into a host chromosome. In some cases the prophage or provirus may remain inactive, or **latent**, for a long time. Most of the viral genes are silent during latency, and the viral reproductive cycle does not progress to step 4. Examples of latent viruses include phage λ as well as human viruses such as HIV and different types of herpes viruses, which cause genital herpes and cold sores.

Latency in bacteriophages is usually called **lysogeny**, and both the prophage and its host cell are said to be lysogenic. When a lysogenic bacterium prepares to divide, it copies the prophage DNA along with its own, so that each daughter cell inherits a copy of the prophage. A prophage can be replicated repeatedly in this way without killing the host cell or producing new phage particles. However, a lysogenic prophage may eventually become activated, and it is then excised from the host chromosome. This sequence of events—integration, prophage replication, and excision—is called a bacteriophage **lysogenic cycle** (see Figure 18.14a, step 3). The excised phage proceeds to steps 4, 5, and 6—synthesis, assembly, and release—of the reproductive cycle. For bacteriophages, these steps are called a **lytic cycle** because the production and release of new viruses lyses the host cell.

Many bacteriophages can alternate between lysogenic and lytic cycles (**Figure 18.15**). A bacteriophage that may spend some of its time in the lysogenic cycle is called a **temperate phage**. Phage λ is an example of a temperate phage. Upon infection, it can either enter the lysogenic cycle or proceed directly to the lytic cycle. Other phages, called **virulent phages**, have only lytic cycles. The genome of a virulent phage is not capable of integration into a host chromosome. Phage T2, which we examined in Chapter 11, is a virulent phage that infects *E. coli*. Unlike phage λ, which may coexist harmlessly with *E. coli*, T2 always lyses the infected cell.

For phages such as λ that can follow either cycle, environmental conditions influence whether or not viral DNA is integrated into a host chromosome and how long the virus remains latent or lysogenic. If nutrients are readily available, phage λ usually proceeds directly to the lytic cycle after its DNA enters

the cell. Alternatively, if nutrients are in short supply, latency is often favored because there may not be sufficient material to make new viruses. If more nutrients become available later, this may cause the prophage to become activated. At this point, the viral reproductive cycle will switch to the lytic cycle, and new viruses will be made, assembled, and released.

Bacterial Transduction Results from an Error in Phage Replication

Earlier in this chapter we considered transduction, in which gene transfer occurs between two bacterial cells with the aid of a bacteriophage. Now that we have examined the steps in the viral reproductive cycle, let's take a closer look at this type of gene transfer.

The mechanism of transduction is actually an error in a phage lytic cycle, as shown in **Figure 18.16**. During the synthesis of phage DNA and proteins, the bacterial chromosome is degraded into small pieces. When the new viruses are assembled, coat proteins occasionally surround a piece of bacterial DNA instead of phage genetic material. This mistake creates an abnormal phage carrying bacterial chromosomal DNA. In the example shown here, a phage called P1 infects an *E. coli* cell that has a gene (*his*+) for histidine synthesis. Phage P1 causes the host cell chromosome to break up, and one of the new phages encloses a piece of host DNA that carries this gene.

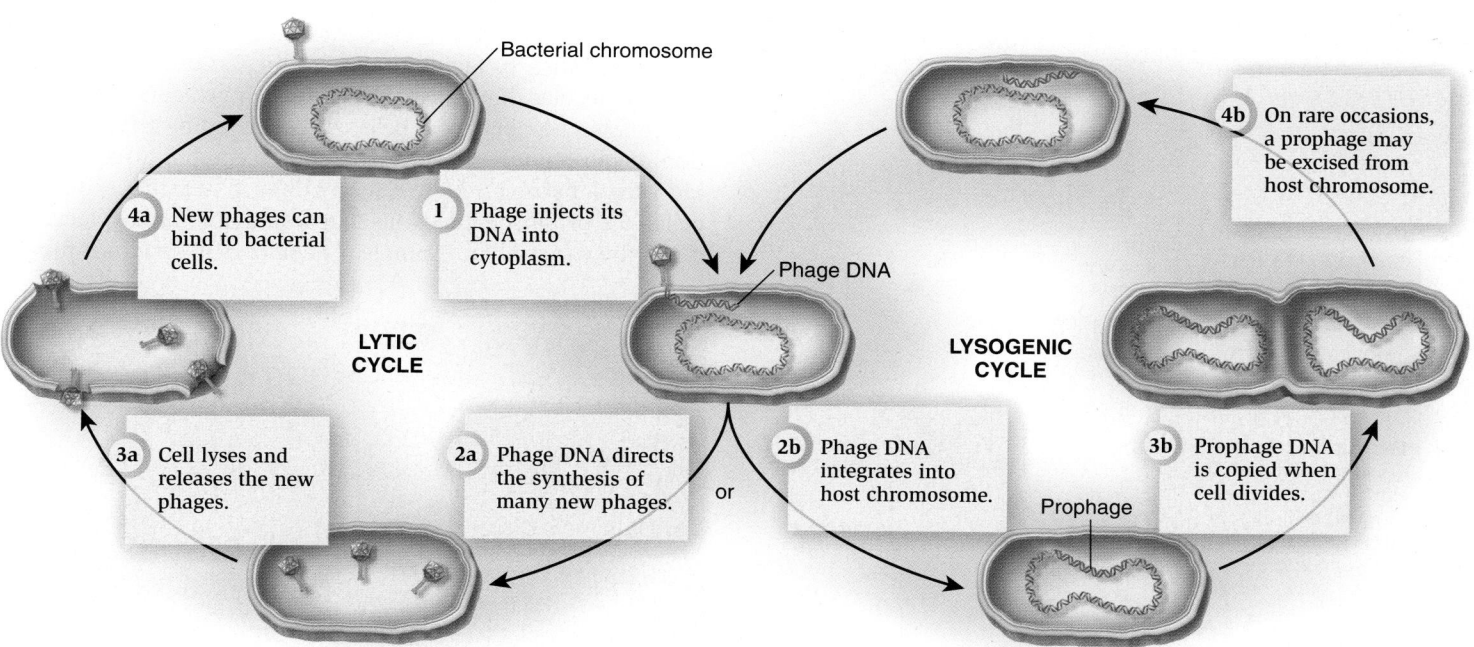

Figure 18.15 **Lytic and lysogenic cycles of bacteriophages.** Some phages, such as phage λ, may follow either a lytic or a lysogenic reproductive cycle. During the lysogenic cycle, the integrated phage DNA, or prophage, is replicated along with the DNA of the host cell. Other phages, such as T2, follow only lytic cycles.

Biological inquiry: From the perspective of the virus, what are the primary advantages of the lytic and lysogenic cycles?

When this abnormal phage is released, it binds to an *E. coli* cell that lacks the *his⁺* gene. It inserts the bacterial DNA fragment into the recipient cell, which then incorporates this fragment into its own chromosome by crossing over. In this example, genetic transfer by transduction converts a *his⁻* strain of *E. coli* to a *his⁺* strain.

AIDS Is a Fatal Immunodeficiency Syndrome Caused by HIV

A primary reason researchers have been interested in viral reproductive cycles is the ability of many viruses to cause diseases in

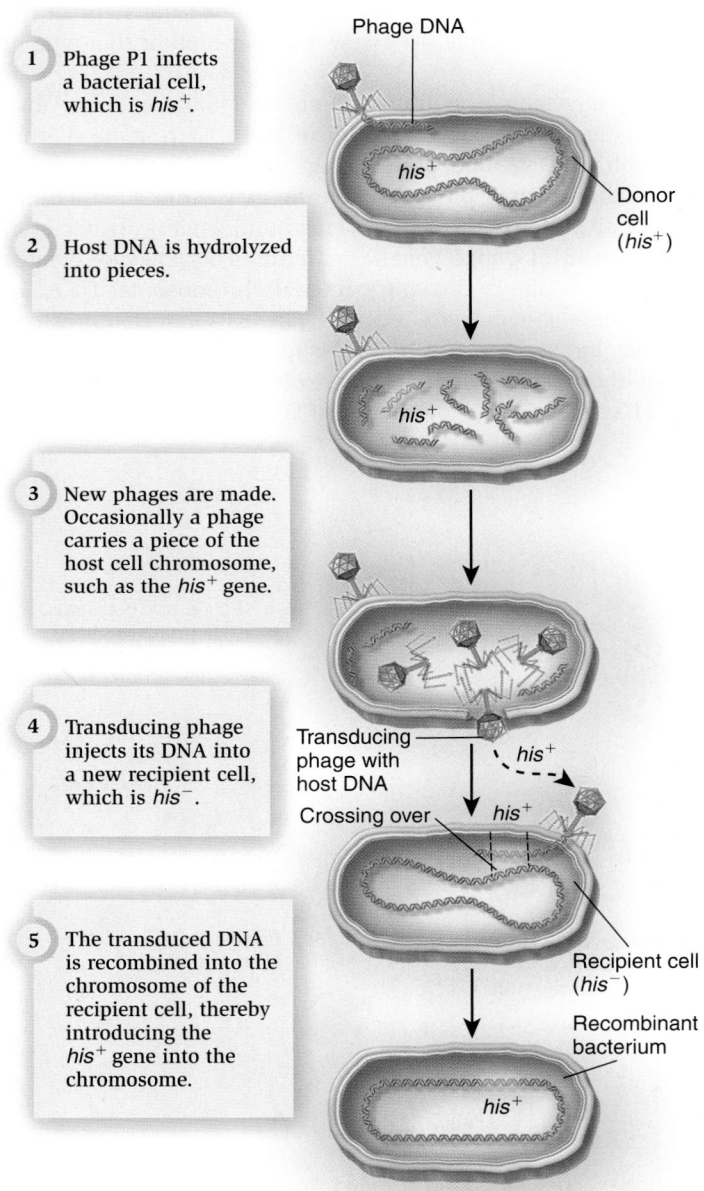

1 Phage P1 infects a bacterial cell, which is *his⁺*.

Phage DNA

his⁺

Donor cell (*his⁺*)

2 Host DNA is hydrolyzed into pieces.

his⁺

3 New phages are made. Occasionally a phage carries a piece of the host cell chromosome, such as the *his⁺* gene.

4 Transducing phage injects its DNA into a new recipient cell, which is *his⁻*.

Transducing phage with host DNA

his⁺

Crossing over   *his⁺*

Recipient cell (*his⁻*)

5 The transduced DNA is recombined into the chromosome of the recipient cell, thereby introducing the *his⁺* gene into the chromosome.

Recombinant bacterium

his⁺

The recombinant bacterium has a genotype (*his⁺*) that is different from the original recipient bacterial cell (*his⁻*).

Figure 18.16   Bacterial transduction by P1 phage.

humans and other mammalian hosts. Some examples of human disease-causing viruses were presented in Figure 18.12. Newly arising viruses, called **emerging viruses**, often cause public alarm and may lead to a significant loss of human life. New strains of influenza virus arise fairly regularly due to new mutations. In the U.S., despite attempts to minimize influenza deaths by vaccination, over 30,000 people die annually from this disease. Another emerging virus causes a severe acute respiratory syndrome (SARS). This enveloped RNA virus was first identified by researchers in Hong Kong, the U.S., and Germany in 2003. The origin of SARS is not understood, but the virus is likely to have arisen from viruses that infect humans and livestock.

During the past few decades, the most devastating example of an emerging virus has been **human immunodeficiency virus (HIV)**, the causative agent of **acquired immune deficiency syndrome (AIDS)**. AIDS is primarily spread by sexual contact between infected and uninfected individuals, but it can also be spread by the transfusion of HIV-infected blood, by the sharing of needles among drug users, and from mother to unborn child. The total number of AIDS deaths between 1981 and the end of 2003 was 20 million; approximately 0.5 million of these deaths occurred in the U.S. During 2004, around 5 million adults and children became infected with HIV. Worldwide, nearly 1 in every 100 adults between 15 and 49 is infected. In the U.S. about 40,000 new HIV infections occur each year. Seventy percent of these new infections are in men and 30% in women.

The devastating effects of AIDS result from viral destruction of a type of white blood cell termed a helper T cell, which plays an essential role in the immune system of mammals. **Figure 18.17** shows HIV virus particles invading a T cell. As described in Chapter 53, T cells interact with other cells of the immune system to facilitate the production of antibodies and other molecules that target and kill foreign invaders of the body. When large numbers of T cells are destroyed by HIV, the function of the immune system is seriously compromised and the individual becomes highly susceptible to infectious diseases called opportunistic infections that would not normally occur in a healthy person. For example, *Pneumocystis carinii*, a fungus

HIV

T cell

Figure 18.17   Micrograph of HIV invading a human T cell. This is a colorized scanning electron micrograph. The surface of the T cell is purple and HIV particles are red.

that causes pneumonia, is easily destroyed by a normal immune system. However, in people with AIDS, infection by this fungus can be fatal.

The mechanism by which HIV kills T cells is not well understood. However, a great deal is known about how the virus invades a T cell and uses it to make more HIV particles. **Figure 18.18** takes a closer look at the HIV reproductive cycle we considered earlier. The structure of HIV consists of two copies of single-stranded RNA and two molecules of reverse transcriptase, which are packaged into a capsid surrounded by a viral envelope. This structure is characteristic of the RNA viruses called **retroviruses**, which utilize reverse transcription to produce viral DNA that can be integrated into the host cell genome.

Let's take a detailed look at the events of the HIV reproductive cycle, as shown in Figure 18.18. The cycle begins when a viral particle binds to the surface of a T cell. Spike glycoproteins

in the viral envelope specifically recognize receptors that are abundant on human T cells, making these cells targets for HIV infection. The viral envelope fuses with the plasma membrane of the T cell, and the capsid is released into the cytosol. The capsid proteins are partially removed by host cell enzymes, releasing the viral RNA and reverse transcriptase. Reverse transcription then occurs, producing HIV DNA that enters the nucleus and is integrated into the host chromosome. Later, the proviral DNA is transcribed into viral RNA in the nucleus. The HIV RNA molecules enter the cytosol, where they serve as the genetic material for new virus particles and are also translated into HIV proteins. When the viral components have been synthesized, capsid proteins surround viral RNA and reverse transcriptase molecules. The newly assembled particles bud from the T cell, wrapped in pieces of T-cell membrane studded with viral spike glycoproteins.

Figure 18.18   A closer look at the reproductive cycle of human immunodeficiency virus (HIV).

An insidious feature of reverse transcriptase is that it lacks a proofreading function. In Chapter 11, we learned that DNA polymerase has a proofreading function that causes it to replicate DNA with few errors. Because reverse transcriptase lacks this function, it makes more errors and thereby tends to create mutant strains of HIV. This undermines the ability of the body to combat HIV because mutant strains may be resistant to the body's defenses, and also makes it difficult to create a vaccine that would provide immunity to all strains of the virus. In addition, mutant strains of HIV may be resistant to antiviral drugs.

Drugs Have Been Developed to Combat the Proliferation of HIV

A compelling reason to understand the reproductive cycle of HIV and other disease-causing viruses is that such knowledge may be used to invent drugs that stop viral proliferation. In the U.S., the estimated annual number of AIDS-related deaths fell approximately 14% from 1998 to 2002, owing in part to the development of new antiviral drugs. These drugs inhibit viral proliferation, though they don't completely eliminate the virus from the body.

One approach to the design of antiviral treatments is to create drugs that specifically bind to proteins encoded by the viral genome. For example, azidothymidine (AZT) mimics the structure of a normal nucleotide and can bind to reverse transcriptase. In this way, AZT inhibits reverse transcription, thereby retarding viral replication. Other drugs used to combat HIV inhibit other proteins that are critical in the HIV reproductive cycle. Proteases, which are enzymes that cut other proteins, are needed during the assembly of the HIV capsid. Certain proteases cleave capsid proteins, making them smaller, active HIV proteins. If the larger capsid proteins are not broken apart, they cannot assemble themselves into new functional HIV particles. Several drugs known as protease inhibitors have been developed that bind to the HIV proteases and inhibit their function.

A major challenge in AIDS research is to discover drugs that inhibit viral proteins without also binding to host cell proteins and inhibiting normal cellular functions. Researchers are optimistic, however, that as we learn more about the molecular structures and functions of viral proteins we will be able to design antiviral drugs with a minimum of harmful side effects. A second challenge is to develop drugs to which mutant strains will not become resistant. As we discussed earlier, HIV readily accumulates mutations during viral replication. A current strategy is to treat HIV patients with a cocktail of three or more drugs, making it less likely that any mutant strain will overcome all of the inhibitory effects.

Another approach to fighting HIV and other infectious diseases is vaccination. A vaccine is a substance or group of substances that causes the immune system to respond to and destroy infectious agents such as bacteria or viruses. The ideal HIV vaccine should be both inexpensive and easy to store and administer, and it must confer long-lasting protection against HIV infection by sexual contact or by exposure to infected blood. Importantly, due to the high mutation rate of HIV, the vaccine must protect against exposure to many different strains of the virus. Despite great advances in our understanding of HIV and the human immune system, such a vaccine has yet to be developed. However, many clinical trials are underway, and researchers hope eventually to create a vaccine that will prevent the spread of AIDS.

Viroids Are RNA Particles That Infect Plant Cells

Thus far in this section, we have examined the reproductive cycles of viruses. Some nonliving infectious agents are even simpler than viruses. In 1971, Theodor Diener discovered that the agent of potato spindle tuber disease is a small RNA molecule devoid of any protein. He coined the term **viroid** for this newly discovered infectious particle.

Viroids are composed solely of a single-stranded circular RNA molecule a few hundred nucleotides in length. They infect plant cells, where they depend entirely on host enzymes for their replication. Some viroids are replicated in the host cell nucleus, others in the chloroplast. In contrast to viral genomes, the RNA genomes of viroids do not code for any proteins. The RNA of some viroids is known to possess ribozyme activity, and some researchers think that this activity may damage plants by interfering with the function of host cell molecules. However, the mechanism by which viroids induce disease is not well understood.

Since Diener's initial discovery, many more viroids have been characterized as the agents of diseases affecting many economically important plants, including potato, tomato, cucumber, orange, coconut, grape, avocado, peach, apple, pear, and plum (**Figure 18.19**). Some viroids have devastating effects, as

Figure 18.19 Effects of the tomato apical stunt viroid. The plant on the left is infected, while the one on the right is healthy.

illustrated by the case of the coconut cadang-cadang viroid, which has killed more than 20 million coconut trees in Southeast Asia. Other viroids cause less severe damage, causing necrosis on leaves, shortening of stems, bark cracking, and delays in foliation, flowering, and fruit ripening. A few viroids induce mild symptoms or no symptoms at all.

Several Hypotheses Have Been Proposed to Explain the Origin of Viruses

Because viruses and viroids are such small particles, there is no fossil record of their evolution, and researchers must rely on analyses of modern viruses to propose hypotheses about their origin. Viral genomes follow the same rules of gene expression as the genomes of their host cells. Viral genes have promoter sequences that are similar to those of their host cells, and the translation of viral proteins relies on the genetic code. Viruses depend entirely on host cells for their proliferation. No known virus makes its own ribosomes or generates the energy it requires to make new viruses. Therefore, many biologists have argued that cells must have evolved before viruses.

A common hypothesis for the origin of viruses is that they evolved from macromolecules inside living cells. The precursors of the first viruses may have been plasmids or similar small nucleic acid fragments. Biologists have hypothesized that such fragments may have acquired genes that code for proteins that facilitate their own replication. Though many biologists favor the idea that viruses originated from primitive plasmids or other chromosomal elements, some have suggested that they are an example of regressive evolution. This hypothesis proposes that viruses are degenerate cells that have retained the minimal genetic information essential for reproduction.

A new and interesting hypothesis is that viruses did not evolve from living cells but instead evolved in parallel with cellular organisms. As discussed in Chapter 22, the precursors of cellular DNA genomes may have been RNA molecules that could replicate independently of cells. This stage of evolution, termed the RNA world, could have involved the parallel evolution of both viruses and cellular organisms.

Prions Are Infectious Proteins That Cause Neurodegenerative Diseases

Before we end our discussion of infectious particles, let's consider an unusual mechanism in which agents known as **prions** cause a group of rare, fatal brain diseases affecting humans and other mammals. Until the 1980s, biologists thought that any infectious agent, whether living or nonliving, must have genetic material. It would seem that genetic material is needed to store the information to create new infectious particles.

In the 1960s, British researchers Tikva Alper and J. S. Griffith discovered that preparations from animals with certain neurodegenerative diseases remained infectious even after exposure to radiation that would destroy any DNA or RNA. They suggested that the infectious agent was a protein. Furthermore, Alper and Griffith speculated that the protein usually preferred one folding pattern, but it could sometimes misfold and then catalyze other proteins to misfold, creating a chain reaction of misfolding. In the early 1970s, Stanley Prusiner, moved by the death of a patient from a neurodegenerative disease, began to search for the causative agent. In 1982, Prusiner isolated a disease-causing particle composed entirely of protein, which he called a prion. The term was based on his characterization of the particle as a proteinaceous infectious agent. Further experiments carried out by Prusiner and others conclusively demonstrated the particle's infectious character. In 1997, Prusiner was awarded the Nobel Prize in Physiology or Medicine for his work on prions.

Prion diseases arise from the ability of the prion protein to induce abnormal folding in normal protein molecules (**Figure 18.20**). The prion protein exists in a disease-causing conformation designated PrP^{Sc}. The superscript Sc refers to scrapie, an example of a prion disease. A normal conformation of this same protein, which does not cause disease, is termed PrP^{C}. The superscript C stands for cellular. The normal protein is encoded by an individual's genome, and the protein is expressed at low levels in certain types of nerve cells. A healthy person may become "infected" with the abnormal protein by eating meat of an animal with the disease. Unlike most other proteins in the diet, the prion protein escapes digestion in the stomach and small intestine and is absorbed into the bloodstream. The prion gradually converts the cell's normal proteins to the abnormal conformation. As a prion disease progresses, the PrP^{Sc} protein is deposited as dense aggregates in the cells of the brain and peripheral nervous tissues, causing the disease symptoms. Some of the abnormal prion protein is excreted from infected cells, where it travels through the bloodstream. In this way, a prion disease can spread through the body like many viral diseases.

Prions are now known to cause several types of neurodegenerative diseases affecting humans and livestock (**Table 18.3**).

Table 18.3	Examples of Neurodegenerative Diseases Caused by Infectious Prions
Disease	**Description**
Kuru	A human disease that was once common in certain native tribes of New Guinea. It begins with a loss of coordination, usually followed by dementia. Infection was spread by ritual cannibalism, a practice that ended in 1958.
Scrapie	A disease of sheep and pigs characterized by intense itching in which the animals tend to scrape themselves against trees or other objects, followed by neurodegeneration.
Mad cow disease	Begins with changes in posture and temperament, followed by loss of coordination and neurodegeneration.

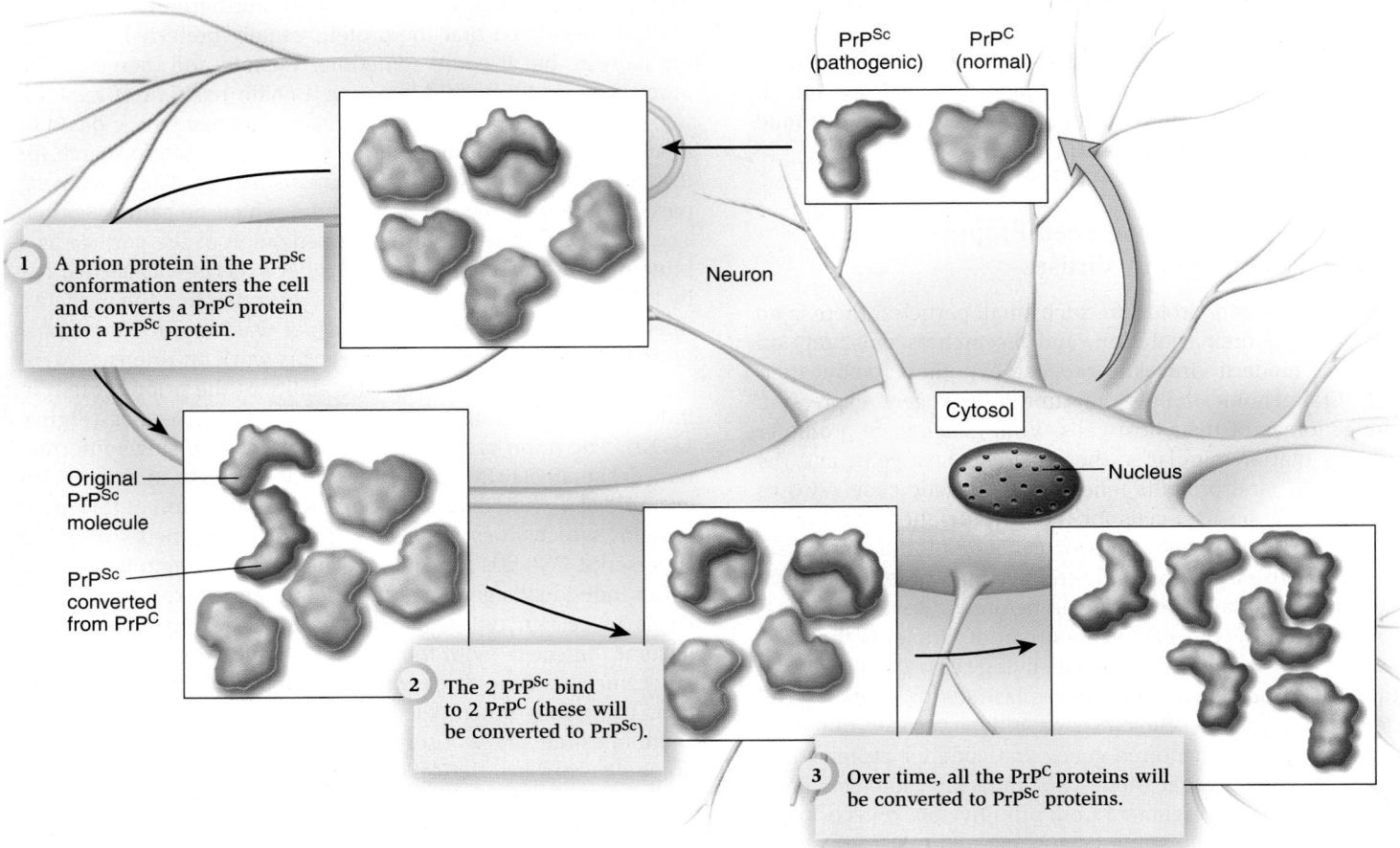

Figure 18.20 **A proposed molecular mechanism of prion diseases.** A healthy neuron normally contains only the PrP^C conformation of the prion protein. The abnormal PrP^{Sc} conformation catalyzes the conversion of PrP^C into PrP^{Sc}. Over time, the PrP^{Sc} protein accumulates at high levels, leading to symptoms of prion disease.

Biological inquiry: Researchers are trying to discover drugs that prevent prion diseases. What are possible effects of a drug that would prevent the spread of the disease?

As a group, prion diseases are termed transmissible spongiform encephalopathies (TSE). The postmortem examination of the brains of affected individuals reveals a substantial destruction of brain tissue; the brain has a spongy appearance. Most prion diseases progress fairly slowly. Over the course of several months to a few years, symptoms proceed from a loss of motor control to dementia, paralysis, wasting, and eventually death.

These symptoms are correlated with an increase in the level of prion protein in the nerve cells of infected individuals. No current treatment can halt progression of any of the TSEs. For this reason, great public alarm occurs when an outbreak of a TSE is reported. For example, in the U.S., a single report of mad cow disease in 2003 prompted several countries to restrict the import of American beef.

CHAPTER SUMMARY

18.1 Bacterial Genetics

- Bacteria typically have a single type of circular chromosome found in the nucleoid region of the cell. The chromosome contains many genes, an origin of replication, and repetitive sequences. (Figures 18.1, 18.2)
- The bacterial chromosome is made more compact by the formation of loops and DNA supercoiling. (Figure 18.3)

- Plasmids are small DNA molecules that exist independently of the bacterial chromosome. Examples are resistant, degradative, col-, virulence, and fertility plasmids. (Figure 18.4)
- When placed on solid growth media, a single bacterial cell will divide many times to produce a colony composed of many cells. (Figure 18.5)
- Bacterial cells reproduce by a process called binary fission. (Figure 18.6)

- Three common modes of genetic transfer among bacteria are conjugation, transformation, and transduction. (Table 18.1)
- Beadle and Tatum's work demonstrated the transfer of bacterial genes between different strains of *E. coli*. (Figure 18.7)
- Using a U-tube, Davis showed that direct contact was needed for the type of genetic transfer observed by Beadle and Tatum. (Figure 18.8)
- During the mechanism of conjugation, a strand of DNA from an F factor is transferred from a donor to a recipient cell through a sex pilus. (Figure 18.9)
- Transformation is the process in which a segment of DNA from the environment is taken up by a competent cell and incorporated into the bacterial chromosome. (Figure 18.10)
- Horizontal gene transfer is the transfer of genes between different species.

18.2 Viral Genetics, Viroids, and Prions

- Tobacco mosaic virus was the first virus to be discovered. It infects many species of plants. (Figure 18.11)
- Viruses vary with regard to their host range, structure, and genome composition. (Table 18.2, Figures 18.12, 18.13)
- Viruses follow a series of steps called a reproductive cycle to make new viruses. These steps typically include attachment, entry, integration, synthesis, assembly, and release. (Figure 18.14)
- Some bacteriophages can alternate between the lytic and lysogenic cycles. (Figure 18.15)
- Bacterial transduction is a form of genetic transfer in which a bacteriophage transfers a segment of bacterial chromosomal DNA to another cell. (Figure 18.16)
- The disease AIDS is caused by a virus called HIV. The virus is a retrovirus whose reproductive cycle involves the integration of the viral genome into a chromosome in the host cell. (Figures 18.17, 18.18)
- Drugs to combat viral diseases are often directed against viral proteins.
- Viroids are RNA molecules that infect plant cells. (Figure 18.19)
- Prions are proteins that exist in an abnormal conformation that can cause disease. (Figure 18.20, Table 3.3)

TEST YOURSELF

1. Genetic diversity is maintained in bacterial populations by all of the following *except*
   a. binary fission.
   b. mutation.
   c. transformation.
   d. transduction.
   e. conjugation.

2. Bacterial cells divide by a process known as
   a. mitosis.
   b. cytokinesis.
   c. meiosis.
   d. binary fission.
   e. glycolysis.

3. Genetic transfer, whereby a bacterial cell takes up bacterial DNA from the environment, is
   a. conjugation.
   b. binary fission.
   c. recombination.
   d. transformation.
   e. transduction.

4. A bacterial cell can donate DNA during conjugation when it
   a. produces competence factors.
   b. contains an F factor.
   c. is virulent.
   d. has been infected by a bacteriophage.
   e. all of the above.

5. A bacterial species that becomes resistant to certain antibiotics may have acquired the resistance genes from another bacterial species. This phenomenon of acquiring genes from another species is known as
   a. hybridization.
   b. conjugation.
   c. horizontal gene transfer.
   d. vertical gene transfer.
   e. competence.

6. The _____ is the protein coat of a virus.
   a. host
   b. prion
   c. capsid
   d. viroid
   e. capsule

7. Among the viruses identified, the characteristics of their genomes show many variations. Which of the following does not describe a typical characteristic of viral genomes?
   a. The genetic material may be DNA or RNA.
   b. The nucleic acid may be single stranded or double stranded.
   c. The genome may carry just a few genes or several dozen.
   d. The number of copies of the genome may vary.
   e. All of the above describe typical variation in viral genomes.

8. During viral infection, attachment is usually specific to a particular cell type because
   a. the virus is attracted to the appropriate host cells by proteins secreted into the extracellular fluid.
   b. the virus recognizes and binds to specific molecules in the cytoplasm of the host cell.
   c. the virus recognizes and binds to specific molecules on the surface of the host cell.
   d. the host cell produces channel proteins that provide passageways for viruses to enter the cytoplasm.
   e. the virus releases specific proteins that make holes in the membrane large enough for the virus to enter.

9. HIV, a retrovirus, has a high mutation rate because
   a. the DNA of the viral genome is less stable than other viral genomes.
   b. the viral enzyme reverse transcriptase has a high likelihood of making replication errors.
   c. the viral genome is altered every time it is incorporated into the host genome.
   d. antibodies produced by the host cell mutate the viral genome when infection occurs.
   e. all of the above.

10. A _____ is an infectious agent solely composed of RNA, whereas a _____ is an infectious agent solely composed of protein.
    a. retrovirus, bacteriophage
    b. viroid, virus
    c. prion, virus
    d. retrovirus, prion
    e. viroid, prion

CONCEPTUAL QUESTIONS

1. Define plasmid.
2. Explain acquired antibiotic resistance and its significance for disease.
3. Explain the steps in a viral reproductive cycle.

EXPERIMENTAL QUESTIONS

1. What was the hypothesis tested by Lederberg and Tatum?
2. During the Lederberg and Tatum experiment, the researchers compared the growth of mutant strains under two scenarios: mixed strains or unmixed strains. When the unmixed strains were plated on the experimental growth medium, why were no colonies observed to grow? When the mixed strains were plated on the experimental growth medium, a number of colonies were seen to grow. What was the significance of the growth of these colonies?
3. The genetic transfer seen in the Lederberg and Tatum experiment could have occurred in one of two ways: taking up DNA released into the environment or contact between two bacterial cells allowing for direct transfer. Bernard Davis conducted an experiment to determine the correct process. Explain how his results indicated the correct gene transfer process.

COLLABORATIVE QUESTIONS

1. Discuss the three mechanisms of genetic transfer in bacteria. What is the significance of horizontal genetic transfer?
2. Discuss the steps of viral reproduction.

www.brookerbiology.com
This website includes answers to the Biological Inquiry questions found in the figure legends and all end-of-chapter questions.

19

DEVELOPMENTAL GENETICS

CHAPTER OUTLINE

Experimental organisms for studying development. *Drosophila melanogaster*, the fruit fly, has served as a model organism to study animal development. This is a scanning electron micrograph.

In biology, the term **development** refers to a series of changes in the state of a cell, tissue, organ, or organism. Development is the underlying process that gives rise to the structure and function of living organisms. The structure or form of an organism is called its morphology. Researchers interested in morphology are fascinated by the changes that give rise to the shapes of cells, tissues, organs, and entire organisms. As we have learned throughout this textbook, an important paradigm in biology is that structure (morphology) determines function. Scientists interested in biological functions are called physiologists. Development is also of great importance to physiologists because the functions of a specialized cell, like a nerve cell, and a complex organ, such as a heart, are the result of development.

Biologists came to realize that plants and animals undergo amazing changes in development to create the structure and function that is found in adult organisms. Such observations motivated researchers to uncover the underlying processes that promote these developmental changes. Since the 1940s, genetics has emerged as the fundamental force behind development. It is currently one of the most intensely researched fields of biology.

Developmental genetics is aimed at understanding how gene expression controls the process of development. In this chapter, we will learn how the sequential actions of genes provide a program for the development of an organism from a fertilized egg to an adult. The last couple of decades have seen exciting advances in our understanding of developmental genetics at the molecular level. Scientists have chosen a few experimental organisms, such as the fruit fly, a nematode worm, the mouse, and the plant *Arabidopsis*, and worked toward the identification and characterization of the genes required for running their developmental programs. In certain organisms, notably the fruit fly, most of the genes that play a critical role in the embryonic stages of development have been identified. Researchers are now exploring how the proteins encoded by these genes control the course of development. In this chapter, we will begin with an overview that emphasizes the general principles of development. We will then examine the details of development in plants and animals, with an emphasis on genetics. Chapters 39 and 51 also consider plant and animal development, with an emphasis on structure and function.

19.1 General Themes in Development

Plants and animals usually begin to develop when sperm and egg unite to produce a **zygote**, a diploid cell that divides and develops into an **embryo**, and eventually into an adult organism. During the early stages of development, cells divide and begin to arrange themselves into ordered units. As this occurs, each cell also becomes **determined**, which means it is destined to become a particular cell type, such as a muscle or intestinal cell. This commitment occurs long before a cell becomes **differentiated**, meaning that the cell's morphology and function have changed, usually permanently, into a highly specialized cell type. In an adult, each cell type plays its own particular role for the good of the entire individual. In animals, for example, muscle cells allow an organism to move, while intestinal cells facilitate the absorption of nutrients. This division of labor among various cells of an organism works collectively to promote its survival.

The genomes of living organisms contain a set of genes that constitute a program of development. In unicellular species, the program controls the structure and function of the cell.

In multicellular species such as animals and plants, the program not only controls cellular features but also determines the arrangement of cells in the organisms. In this section, we will examine some of the general issues associated with the development of multicellular species. Later sections will focus on specific examples of development in animals and plants.

Developmental Biologists Have Chosen a Few Model Organisms to Study Development

The development of even a simple multicellular organism involves many types of changes in form and function. For this reason, the research community has focused its efforts on only a few model organisms with the expectation that knowledge of the developmental programs in these model organisms will also apply to many other species. With regard to animal development, the two organisms that have been the most extensively investigated are two invertebrate species: the fruit fly *Drosophila melanogaster* and the nematode worm *Caenorhabditis elegans* (**Figure 19.1a,b**). *Drosophila* has been studied for a variety of reasons. First, researchers have exposed this organism to mutagens and identified many mutant organisms with altered developmental pathways. The techniques for generating and analyzing mutants in this organism are more advanced than in any other animal. Second, in all of its life stages, *Drosophila* is large enough and has distinct morphological features to easily identify the effects of mutations, yet it is small enough to determine where particular genes are expressed at critical stages of development. *C. elegans* is used by developmental geneticists for its simplicity. The adult organism is a small transparent worm composed of only about a thousand somatic cells. Starting with a fertilized egg, the pattern of cell division and the fate of each cell within the embryo are completely known.

Embryologists have also studied the morphological features of development in many vertebrate species. Historically, amphibians and birds have been studied extensively, because their eggs are rather large and easy to manipulate. From a morphological point of view, the developmental stages of the chicken (*Gallus gallus*) and the African clawed frog (*Xenopus laevis*) have been described in great detail. More recently, several vertebrate species have been the subject of genetic studies of development. These include the house mouse (*Mus musculus*) and the small aquarium zebrafish (*Brachydanio rerio*) (**Figure 19.1c,d**).

In the study of plant development, the model organism for genetic analysis is thale cress (*Arabidopsis thaliana*), more commonly called *Arabidopsis* by researchers (**Figure 19.1e**). *Arabidopsis* is a small annual weed that belongs to the wild mustard family. It occurs naturally throughout temperate regions of the world. *Arabidopsis* has a short generation time of about two months and a small genome size of 14×10^7 bp, which is similar to *Drosophila* and *C. elegans*. A flowering *Arabidopsis* plant is small enough to be grown in the laboratory and produces a large number of seeds.

Both Animals and Plants Develop by Pattern Formation

Development in plants and animals produces a body plan or pattern. At the cellular level, the body pattern is due to the arrangement of cells and their differentiation. The coordination of these events leads to the formation of a body with a particular pattern, a process called **pattern formation**. The end result of animal development is usually the formation of an adult body pattern organized along three axes: the **dorsoventral axis**, the **anteroposterior axis**, and the **right-left axis** (**Figure 19.2a**). In addition, many animal bodies are then **segmented** into separate sections containing specific body parts such as wings or legs.

(a) *Drosophila melanogaster*

(b) *Caenorhabditis elegans*

(c) *Mus musculus*

(d) *Brachydanio rerio*

(e) *Arabidopsis thaliana*

Figure 19.1 Model organisms used to study the genetics of development.

By comparison, the body pattern of plants is quite different, being formed along a **root-shoot axis** and in a **radial pattern** (**Figure 19.2b**). The root and shoot axes are determined at the first division of the fertilized egg, and growth occurs in a radial pattern around those axes. As we'll see later, the identification of mutant alleles that disrupt development has permitted great insight into the genes controlling pattern formation.

Pattern Formation Depends on Positional Information

Before we examine how genes affect pattern formation, let's consider a central concept in developmental biology—**positional information**. For an organism to develop the correct morphological features or pattern, each cell of the body must become the appropriate cell type based on its position relative to other cells. Each cell receives positional information that provides it with instructions on where to go and what type of cell to become. Later in this chapter, we will examine how the expression of genes and the function of gene products provide this information.

A cell may respond to positional information in one of four ways: cell division, cell migration, cell differentiation, and cell death (**Figure 19.3**). First, positional information may stimulate a cell to divide. Second, positional information in animals may cause the migration of a cell or group of cells in a particular direction from one region of the embryo to another. Third, it may cause a cell to differentiate into a specific cell type such as a nerve cell. Finally, positional information may promote cell death. This process, known as **apoptosis**, is a necessary event during normal development.

As an example of how the coordination of these four processes is required for pattern formation, let's consider the development of the vertebrate limb. **Figure 19.4a** shows the growth and development of a human arm during the embryonic stage. Cell division with accompanying cell growth increases the size of the limb (**Figure 19.4b**). Cell migration is also important in this process. For example, embryonic cells that eventually form muscles in the arm and hand must migrate long distances to reach their correct location within the limb.

(a) Cell division   **(b) Cell migration**

(c) Cell differentiation

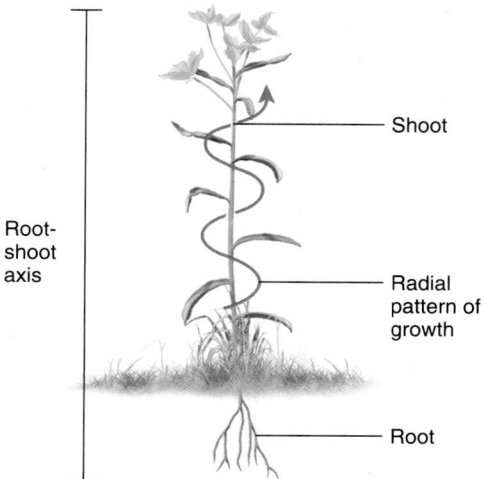

(b) Body plan found in many seed-bearing plants

Figure 19.2 Body plan axes in animals and plants.

(a) Body plan found in many animals

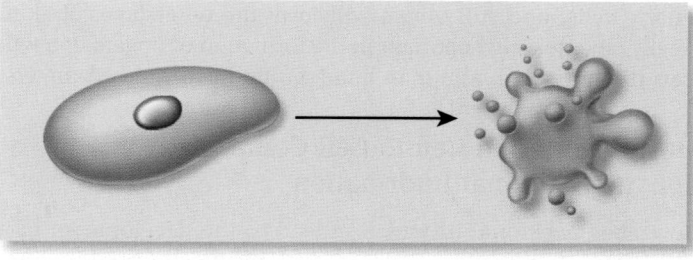

(d) Cell death (apoptosis)

Figure 19.3 Four types of cellular responses to positional information in animals.

Biological inquiry: Which of these four responses do you expect to be more prevalent in the early stages of development, and which would become more prevalent in later stages?

(a) Limb development in a human embryo

1 Cell division promotes the growth of the limb.

Limb bud

2 Cell migration causes cells to reach their correct location.

3 Cell differentiation produces specialized cells such as muscle cells.

4 Apoptosis eliminates cells in between the fingers.

(b) Four cellular processes that promote limb formation

Figure 19.4   Limb development in humans.   (a) Photographs of limb development in human embryos. The limb begins as a protrusion called a limb bud that eventually forms an arm and hand. **(b)** The development of a human limb from an embryonic limb bud.

As development proceeds, cell differentiation produces the various tissues that will eventually be found in the fully developed limb. Some cells will become nerve cells, others will be muscle cells, and still others will form the outer layer of skin. Finally, apoptosis is important in the formation of fingers. If apoptosis did not occur, a human hand would have webbed fingers.

Morphogens and Cell-to-Cell Contacts Convey Positional Information

How does positional information lead to the development of a body plan? Two main molecular mechanisms are used to communicate positional information. One of these mechanisms involves molecules called morphogens. **Morphogens** impart positional information and promote developmental changes at the cellular level. A morphogen influences the fate of a cell by promoting cell division, cell migration, cell differentiation, or apoptosis. A key feature of morphogens is that they act in a concentration-dependent manner. At a high concentration, a morphogen will restrict a cell into a particular developmental pathway, while at a lower concentration it will not. There is often a critical **threshold concentration** above which the morphogen will exert its effects but below which it is ineffective.

Morphogens typically are distributed asymmetrically along a concentration gradient. This may happen at one of two times during early development. First, morphogenic gradients may be established in the oocyte or egg cell precursor (**Figure 19.5a**). Second, a morphogenic gradient can be established in the embryo by secretion and transport (**Figure 19.5b**). A certain cell or group of cells may synthesize and secrete a morphogen at a specific stage of development. After secretion, the morphogen is transported to neighboring cells. The concentration of the morphogen is usually highest near the cells that secrete it. The morphogen may then influence the developmental fate of cells that are exposed to it. The process by which a cell or group of cells governs the developmental fate of neighboring cells is known as **induction**.

(a) Asymmetric distribution of morphogens in the oocyte

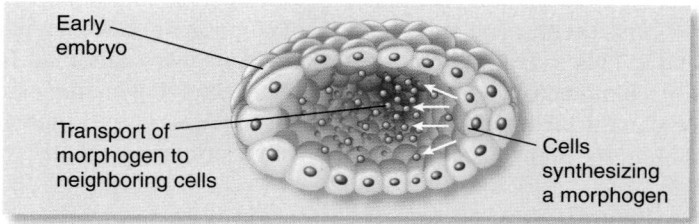

(b) Induction: Asymmetric synthesis and extracellular distribution of a morphogen

(c) Cell adhesion: Cell-to-cell contact conveys positional information

Figure 19.5 Molecular mechanisms that convey positional information. Asymmetric distribution of a morphogen in the **(a)** oocyte or **(b)** embryo. **(c)** Positional information may also be obtained by cell-to-cell contact.

Another mechanism used to convey positional information involves **cell adhesion** (**Figure 19.5c**). Each cell makes its own collection of surface receptors that enable it to adhere to other cells. Such receptors, which are known as **cell adhesion molecules (CAMs)**, are described in Chapter 10. The positioning of a cell within a multicellular organism is strongly influenced by the combination of contacts it makes with other cells and with the extracellular matrix.

The phenomenon of cell adhesion, and its role in multicellular development, was first recognized by H. V. Wilson in 1907. He took multicellular sponges and separated them into individual cells. Remarkably, the cells actively migrated until they adhered to one another to form a new sponge, complete with the chambers and canals that characterize a sponge's internal structure! When sponge cells from different species were mixed, they sorted themselves properly, adhering only to cells of the same species. Overall, these results indicate that cells possess specific CAMs, which are critical in cell-to-cell recognition. Cell adhesion plays an important role in governing the position that a cell will adopt during development. Together, the actions of morphogens and CAMs convey positional information to the cells in a developing organism.

A Hierarchy of Transcription Factors Plays a Key Role in Orchestrating a Program of Development in Plants and Animals

The formation of a body, both in plants and animals, occurs in a series of organizational phases. As an overview of this process, let's consider four general phases of pattern development in an animal (**Figure 19.6**). The first phase is to organize the body along major axes. The anteroposterior axis determines the organization from head to tail, the dorsoventral axis governs the structure from front (ventral) to back (dorsal), and the right-left axis provides organization from side to side. During the second phase, the body becomes organized into smaller regions that will eventually contain organs and other structures such as legs. In insects, these regions form well-defined segments. In mammals, some segmentation of the body is apparent during embryonic development but defined boundaries are lost as the embryo proceeds to the fetal and adult stages. After segmentation, the third phase occurs when the cells within the segments organize themselves in ways that will produce particular body parts. Finally, during the fourth phase, the cells themselves change their morphologies and become differentiated. This final phase of development produces an organism with many types of tissues, organs, and other body parts with specialized functions.

Certain genes are expressed at a specific phase of development in a particular cell type, a phenomenon called **differential gene regulation**. Developmental geneticists have discovered a parallel between the expression of specific transcription factors and the four major phases of animal development. Many morphogens, particularly those that act at an early phase of development, are transcription factors. Such transcription factors regulate the expression of genes in a way that controls the formation of the body axes. Next, these early transcription factors cause the expression of other transcription factors that direct the segmented body plan. After the body plan has been segmented, a third category of transcription factors controls what structures will be made within each segment. Finally, a fourth category of transcription factors controls the differentiation of each cell type. Overall, as diagrammed in Figure 19.6, development is largely determined by a hierarchy of transcription factors that controls a program of developmental changes.

When considering the events of animal development, the plan progresses from large to small. First, the axes of the entire animal are determined. Next, the entire animal is divided into regions or segments. Then, the organization within each region leads to the formation of tissues, organs, and other body structures. And finally, the cells within each region become differentiated. As these phases occur, the transcription factors controlling each phase regulate the production of proteins that convey positional information to cells, which alters their behavior and characteristics. In the early phases, this positional information mostly promotes cell division and cell migration. During the last phase, cell differentiation becomes a common event, and some cells die via apoptosis.

Hierarchy of transcription factors

Posterior
Right
Dorsal
(ventral is
underneath)
Anterior
Left

1 Phase 1:
Transcription factors
determine the formation of
the body axes and control
the transcription factors of
phase 2.

Evidence of
segmentation

2 Phase 2:
Transcription factors cause
the embryo to become
subdivided into regions that
have properties of
individual segments. They
also control transcription
factors of phase 3.

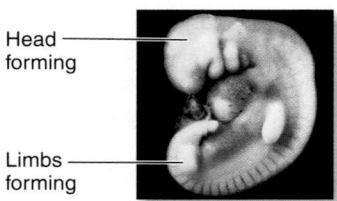

Head
forming

Limbs
forming

3 Phase 3:
Transcription factors cause
each segment and groups of
segments to develop
specific body parts and
control transcription factors
of phase 4.

4 Phase 4:
Transcription factors cause
cells to differentiate into
specific cell types such as
skin, nerve, and muscle
cells.

Figure 19.6   Pattern development in a human embryo.
As shown here, pattern formation in animals occurs in four
phases that are controlled by a hierarchy of transcription factors.
The example shown here involves human development, although
research suggests that pattern formation in all complex animals
follows a similar plan. The ideas in this scenario are based largely
on analogies between pattern formation in mammals and *Drosophila*.
Many of the transcription factors that are likely to control the early
phases of pattern formation in mammals have yet to be identified.
Also, note that the phases of development are overlapping. For
example, cell differentiation begins to occur as the cells are
adopting their correct locations.

19.2 Development in Animals

In the previous section, we considered the general features of development as they apply to animals and plants. In this section,
we will begin by examining the steps of *Drosophila* development, and then focus our attention on embryonic development.

During this stage, the overall body plan is determined. We will
see how the differential expression of particular genes and the
localization of positional information within the embryo control
the developmental processes of pattern formation and segmentation. The roles of genes in the organization of mammalian embryos are not as well understood as they are in *Drosophila*. Even
so, the analysis of the genomes of mammals and many other
species has revealed many interesting parallels in the developmental program of all animals.

This section will end with an examination of cell differentiation. This process is better understood in mammals than in
Drosophila because researchers have been studying mammalian
cells in the laboratory for many decades. To explore cell differentiation, we will consider mammals as our primary example.

Embryonic Development Determines the Pattern of Structures in the Adult: The Development of *Drosophila*

Figure 19.7 illustrates the general sequence of events in *Drosophila* development. Let's examine these steps before we consider the differential gene expression that causes them to happen.
The oocyte is critical to establishing the pattern of development
that will ultimately produce an adult organism. It is an elongated cell that contains positional information: As noted in Figure 19.7a, the oocyte already has anterior and posterior ends that
correspond to those found in the adult (compare Figure 19.7a
and f). After fertilization takes place, the zygote develops into a
blastoderm. First, the zygote goes through a series of nuclear
divisions that are not accompanied by cytoplasmic division,
producing many free nuclei. Initially, these nuclei are scattered
throughout the yolk (Figure 19.7b), but eventually they migrate
to the periphery, producing a stage called a syncytial blastoderm (Figure 19.7c). After the nuclei have lined up along the
cell membrane, individual cells are formed as portions of the
membrane surround each nucleus to create a cellular blastoderm (Figure 19.7d).

After blastoderm formation is complete, some dramatic
changes occur during **gastrulation**, which involves a well-
ordered rearrangement of cells in the embryo. During this process, some cells migrate into the interior. The result is three
cellular layers called the ectoderm, endoderm, and mesoderm,
producing a **gastrula** (Figure 19.7e). In general, the ectoderm
remains on the outside of the gastrula, the endoderm is on the
inside, and the mesoderm is wedged in the middle. These cell
layers are important because each of them leads to the development of specific cell types at a later stage. For example, the
mesoderm produces muscle cells.

A key process in *Drosophila* embryonic development is the
creation of a segmented body pattern. The embryo is subdivided into visible units. In *Drosophila* the segments can be
grouped into three general areas, the head, the thorax, and the
abdomen. Figure 19.7f shows the segmented pattern of a *Drosophila* embryo about 10 hours after fertilization. Later in this
section, we will explore how the coordination of gene expression underlies the formation of these segments.

An embryo then develops into a **larva**, which is a free-living organism that is morphologically very different from the adult (Figure 19.7g). *Drosophila* undergoes three larval stages. After the third larval stage, the organism becomes a **pupa** (Figure 19.7h) and proceeds through a process known as **metamorphosis**, in which it changes into a mature adult. Each segment in the adult develops its own characteristic structures; for example, the wings are on a thoracic segment. The adult fly then emerges from the pupal case (Figure 19.7i). From beginning to end, this process takes about 10 days.

Phase 1 Pattern Development: Maternal Effect Genes Promote the Formation of the Main Body Axes

The first phase in *Drosophila* embryonic pattern development is the establishment of the body axes, which occurs before the embryo becomes segmented (see Figure 19.6). The morphogens necessary to establish these axes are distributed prior to fertilization. In most invertebrates and some vertebrates, certain morphogens, which are important in early developmental stages, are deposited asymmetrically within the egg as it develops. Later, after the egg has been fertilized and development begins, these morphogens will initiate developmental programs that govern the formation of the body axes of the embryo. The products of several different genes ensure that proper development will occur.

As an example of one morphogen that plays a role in axis formation, let's consider the product of a gene in *Drosophila* called *bicoid*. Its name is derived from the observation that a mutation that inactivates the gene results in a larva with two posterior ends (**Figure 19.8**). During normal egg development, the *bicoid* gene product accumulates in the anterior region of the oocyte, causing the development of the anterior end of the embryo.

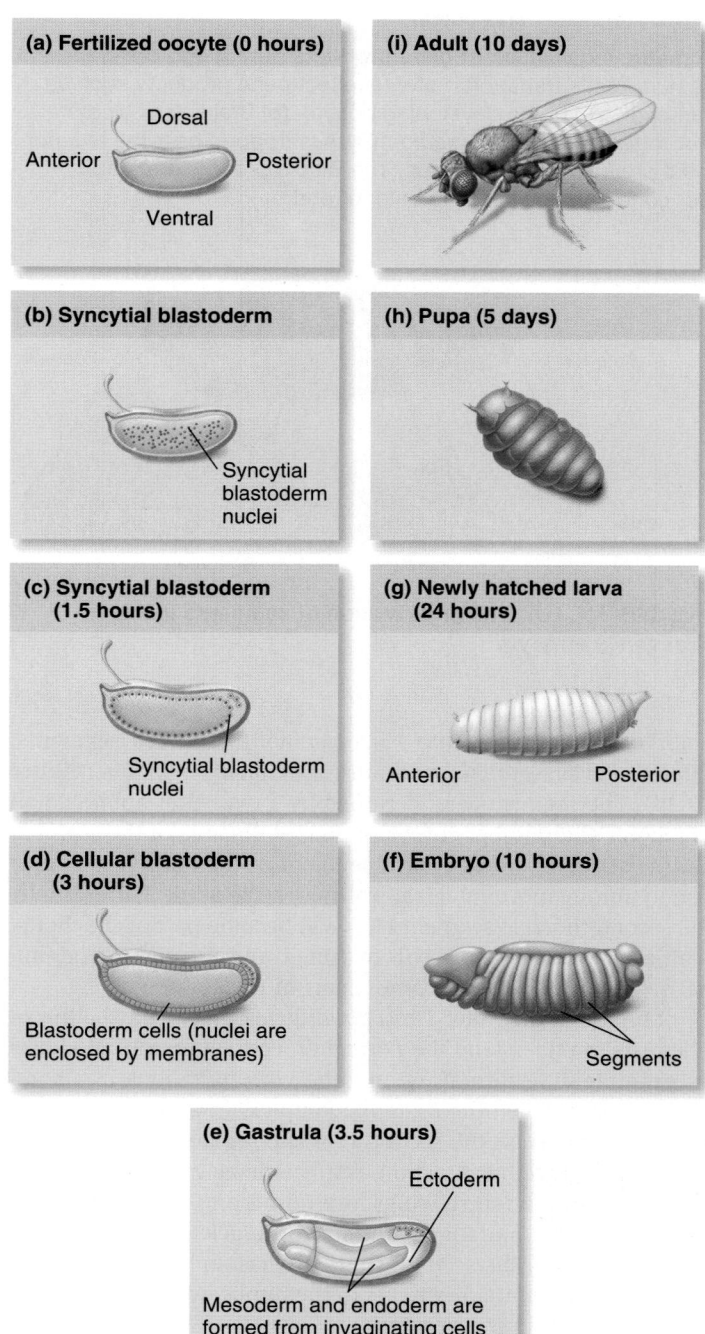

Figure 19.7   Developmental stages of the fruit fly *Drosophila*.

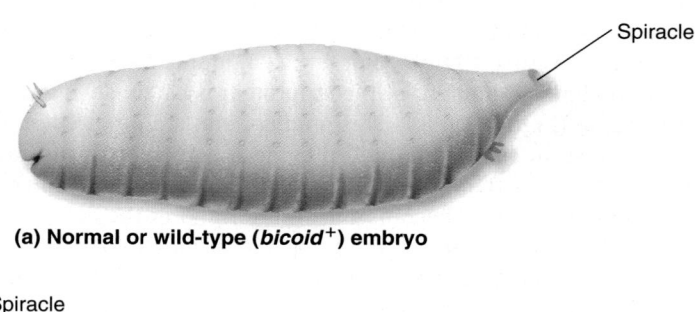

(a) Normal or wild-type (*bicoid*⁺) embryo

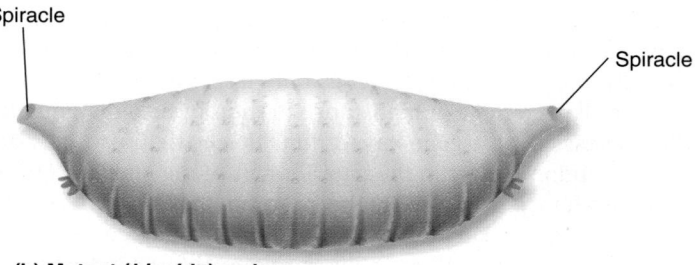

(b) Mutant (*bicoid*⁻) embryo

Figure 19.8   The *bicoid* mutation in *Drosophila*. **(a)** A normal *bicoid*⁺ embryo. **(b)** A *bicoid*⁻ embryo, in which both ends of the larva develop posterior structures. For example, both ends develop a spiracle, which normally is found only at the posterior end.

Biological inquiry: What would you expect the phenotype to be of a larva in which the bicoid *gene was expressed in both the anterior region and the posterior region of the oocyte?*

(a) Transport of maternal effect gene products (*bicoid* mRNA) into the oocyte

Nurse cell / Anterior side of oocyte / Oocyte
Follicle cell / *bicoid* mRNA

(b) Staining of *bicoid* mRNA in oocyte

(c) Staining of Bicoid protein in an embryonic blastoderm

Figure 19.9   Asymmetric localization of gene products during egg development in *Drosophila*.   **(a)** The nurse cells transport maternal effect gene products such as *bicoid* mRNA into the anterior end of the developing oocyte. **(b)** Staining of *bicoid* mRNA in an oocyte prior to fertilization. The *bicoid* mRNA is trapped near the anterior end. **(c)** Staining of Bicoid protein after fertilization. The Bicoid protein forms a gradient, with its highest concentration near the anterior end.

How does the *bicoid* gene product accumulate in the anterior region of the oocyte? The answer involves specialized nurse cells that are found next to the oocyte in the ovaries of female flies. As discussed in Chapter 17, nurse cells supply the products (for example, mRNA) of maternal effect genes. These genes cause an unusual pattern of inheritance called **maternal effect** (refer back to Figure 17.15). In *Drosophila*, the nurse cells are located toward the anterior end of the oocyte rather than the posterior end. The *bicoid* gene is transcribed in the nurse cells, and *bicoid* mRNA is then transported into the anterior end of the oocyte and trapped there (**Figure 19.9a**). Prior to fertilization, the *bicoid* mRNA is highly concentrated near the anterior side of the oocyte (**Figure 19.9b**). After fertilization, the *bicoid* mRNA is translated and a gradient of Bicoid protein is established across the zygote (**Figure 19.9c**). This gradient starts a progression of developmental events that will provide the positional information that causes the end of the zygote with a high Bicoid protein concentration to become the anterior end of the embryo.

The Bicoid protein is a morphogen that functions as a transcription factor to activate particular genes at specific times. The ability of Bicoid to activate a given gene is tuned exquisitely to its concentration. Due to its asymmetric distribution, the Bicoid protein will activate genes only in certain regions of the embryo. For example, a high concentration of Bicoid stimulates the expression of a gene called *hunchback* (that also encodes a transcription factor) in the anterior half of the embryo, but its concentration is too low in the posterior half to activate the *hunchback* gene. The ability of Bicoid to activate genes in certain regions but not others plays a role in the second phase of pattern formation, which is segmentation.

The Study of *Drosophila* Mutants Has Identified Genes That Control the Development of Segments and Their Characteristics

As described earlier in Figure 19.6, the second phase of pattern formation is the development of segments. The normal *Drosophila*

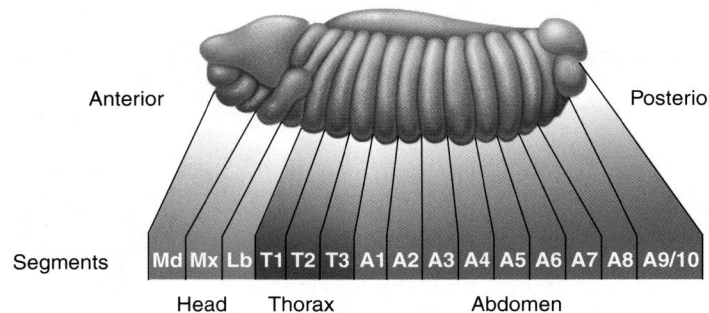

Anterior   Posterior

Segments | Md | Mx | Lb | T1 | T2 | T3 | A1 | A2 | A3 | A4 | A5 | A6 | A7 | A8 | A9/10 |

Head   Thorax   Abdomen

Figure 19.10   The organization of segments in the *Drosophila* embryo.

embryo is subdivided into 15 segments: three head segments, three thoracic segments, and nine abdominal segments (**Figure 19.10**). The segments are identified by a two-letter abbreviation or a letter and number corresponding to the area of the segment and its sequential order. Each segment of the embryo will give rise to unique morphological features in the adult. For example, the second thoracic segment (T2) will become part of the thorax with a pair of legs and a pair of wings, and the eighth abdominal segment (A8) will become a part of the abdomen.

Mutations that alter *Drosophila* development have allowed scientists to understand the normal process. **Figure 19.11** shows an example of a normal fly and one with mutations in a complex of genes called the *bithorax* complex. In a normal fly, two wings are found on the second thoracic segment, and two halteres, which together function as a balancing organ that resembles a pair of miniature wings, are found on the third thoracic segment. In this mutant fly, the third thoracic segment has the characteristics of the second, so the fly has no halteres and four wings. The term *bithorax* refers to the duplicated characteristics of the second thoracic segment.

Edward Lewis, a pioneer in the genetic study of development, became interested in the bithorax phenotype and began

(a)

(b)

Figure 19.11 The *bithorax* mutation in *Drosophila.* **(a)** A normal fly has two wings on the second thoracic segment, and two halteres on the third thoracic segment. **(b)** This mutant fly contains mutations in a complex of genes called the *bithorax* complex. In this fly, the third thoracic segment has the same characteristics as the second thoracic segment, thereby producing a fly with four wings instead of two.

(a) Gap gene

(b) Pair-rule gene

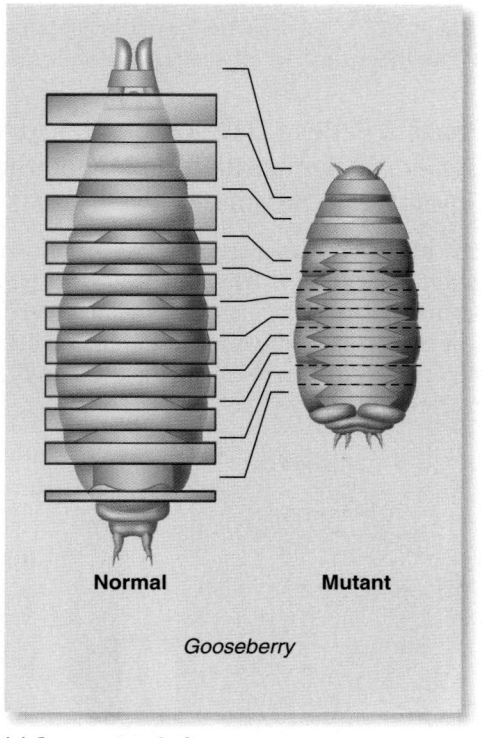

(c) Segment-polarity gene

Figure 19.12 Phenotypic effects in *Drosophila* larvae that have mutations in segmentation genes. In each pair, the larva on the *left* is normal. The *right* side shows a larva with a defect in **(a)** a gap gene called *Krüppel*, **(b)** a pair-rule gene called *Even-skipped*, and **(c)** a segment-polarity gene called *Gooseberry*.

investigating it in 1946. He discovered that the mutant chromosomal region actually contains a complex of three genes that play a role in the third phase of development, which will be described later in this chapter. During the 1960s and 1970s, as the study of molecular genetics blossomed, it became clear that the genomes of multicellular organisms contain groups of genes that initiate a program of development.

In the 1970s, Christiane Nüsslein-Volhard and Eric Wieschaus undertook a systematic search for *Drosophila* mutants with dis-

rupted development. In particular, they focused their search on genes that alter the segmentation pattern of the *Drosophila* embryo and larva, called **segmentation genes**. Their pioneering efforts identified most of the genes required for the embryo to develop a segmented pattern. Based on the characteristics of abnormal larva, they identified three classes of segmentation genes that they termed gap genes, pair-rule genes, and segment-polarity genes (**Figure 19.12**). When a mutation inactivates a **gap gene**, several adjacent segments are missing in the larva (Figure 19.12a).

A defect in a **pair-rule gene** may cause alternating segments or parts of segments to be deleted (Figure 19.12b). Finally, **segment-polarity gene** mutations cause portions of segments to be missing either an anterior or a posterior region and for adjacent regions to become mirror images of each other (Figure 19.12c).

An important advantage of isolating mutant fruit flies as well as other mutant organisms is that a mutation allows researchers to eventually identify, at the molecular level, both the mutant allele and normal allele. Using molecular techniques similar to those described in Chapter 20, the DNA from mutant and normal strains can be compared to determine which gene has become altered to cause the observed changes in development. Once the gene has been identified, the expression of the gene can be studied throughout development. For example, after the *bicoid* gene was identified at the molecular level, researchers could study when and where the Bicoid protein was made, as described earlier in Figure 19.9. As discussed next, this approach also made it possible to understand how gene expression leads to pattern formation.

Phase 2 Pattern Development: Segmentation Genes Act Sequentially to Divide the *Drosophila* Embryo into Segments

The study of segmentation genes has provided information regarding the process by which segments are formed. To create a segment, a group of genes acts sequentially to govern the fate of a given region of the body. A simplified scheme of gene expression that leads to a segmented pattern in the *Drosophila* embryo

is shown in **Figure 19.13**. Many more genes are actually involved in this process.

In general, maternal effect genes, which promote phase 1 pattern development, activate gap genes; this activation is seen as broad bands of gap gene expression in the embryo (Figure 19.13, step 2). The gap genes and maternal effect genes then activate the pair-rule genes in alternating stripes in the embryo (Figure 19.13, step 3). Once the pair-rule genes are activated, their gene products then regulate the segment-polarity genes. As you follow the progression from maternal effect genes to segment-polarity genes, notice that a body pattern is emerging in the embryo that matches the segmentation pattern found in the larva and adult animal. As you can see in step 4 of Figure 19.13, the expression of a segment-polarity gene corresponds to portions of segments in the adult fly. To appreciate this phenomenon, notice that the embryo at this stage is curled up and folded back on itself. If you imagine that the embryo was stretched out linearly, the 15 stripes seen in this embryo correspond to portions of the 15 segments of an adult fly.

Phase 3 Pattern Development: Homeotic Genes Control the Development of Segment Characteristics

Thus far, we have considered how the *Drosophila* embryo becomes organized along axes and then into a segmented body pattern. During the third phase of pattern development, each segment begins to develop its own unique characteristics (see Figure 19.6).

Phase 1	Phase 2		
Asymmetric localization of maternal effect gene products	Gap gene expression	Pair-rule gene expression	Segment-polarity gene expression

1 **Maternal effect genes** establish the anteroposterior and dorsoventral axes. They activate gap genes.

2 **Gap genes** produce broad bands of proteins that act as genetic regulators of pair rule genes. They bind to enhancers that are located adjacent to pair-rule genes.

3 **Pair-rule genes** are expressed in alternating stripes. Each stripe of protein is found in a portion of every other segment. Pair-rule genes control segment-polarity genes.

4 **Segment-polarity genes** are expressed in a portion of every segment. At this stage, all of the segments of the body are defined.

Figure 19.13 **Overview of segmentation in *Drosophila*.** The *left* micrograph shows an embryonic blastoderm, and the other three depict the progression during the first few hours of embryonic development. The micrographs show the expression of protein products of a maternal effect gene (step 1) or segmentation genes (steps 2–4). In step 1, the protein is stained brown and is found in the *left* side of the embryonic blastoderm. In step 2, three different proteins encoded by gap genes are stained green, yellow, and red. In step 3, a protein encoded by a pair-rule gene is stained in light blue. In step 4, a protein encoded by a segment-polarity gene is stained pink. When comparing steps 3 and 4, the embryo has undergone a 180° turn, folding back on itself.

Geneticists use the term **cell fate** to describe the ultimate morphological features that a cell or group of cells will adopt. For example, as mentioned previously, the fate of cells in segment T2 is to develop into a thoracic segment containing two legs and two wings. In *Drosophila*, the cells in each segment of the body have their fate determined at a very early stage of embryonic development, long before the morphological features become apparent.

Our understanding of developmental fate has been greatly aided by the identification of mutant genes that alter cell fates. In animals, the first mutant of this type was described by the German entomologist G. Kraatz in 1876. He observed a sawfly (*Climbex axillaris*) in which part of an antenna was replaced with a leg. During the late 19th century, the English zoologist William Bateson collected many of these types of observations and published them in 1894 in a book entitled *Materials for the Study of Variation Treated with Especial Regard to Discontinuity in the Origin of Species*. In this book, Bateson coined the term **homeotic** to describe changes in which one body part is replaced by another. We now know these are caused by mutant alleles of what we call **homeotic genes**.

The *bithorax* mutation discussed earlier that was studied by Edward Lewis is an example of a homeotic mutation (see Figure 19.11). Lewis also studied strains of *Drosophila* with other homeotic mutations. From these studies, he eventually discovered that each homeotic gene controls the fate of a particular region of the body. *Drosophila* contains two clusters of homeotic genes called the *bithorax* complex and the *Antennapedia* complex (**Figure 19.14**). The *Antennapedia* complex contains

five genes, designated *lab*, *pb*, *Dfd*, *Scr*, and *Antp*. The *bithorax* complex has three genes, *Ubx*, *abd-A*, and *Abd-B*. Both of these complexes are located on the same chromosome, but a long stretch of DNA separates them. Homeotic genes are controlled by many genes that are initially expressed during embryonic development, including segmentation genes.

Interestingly, the order of homeotic genes along the chromosome correlates with their expression along the anteroposterior axis of the body. This phenomenon is called the **colinearity rule**. For example, *lab* (for labial) is expressed in the anterior segment and governs the formation of mouth structures. The *Antp* (for antennapedia) gene is expressed in the thoracic region during embryonic development and controls the formation of thoracic structures. Transcription of the *Abd-B* (for abdominal B) gene occurs in the posterior region of the embryo and promotes the formation of the posteriormost abdominal segments.

The role of homeotic genes in determining the identity of particular segments has been revealed by mutations that alter their function. For example, a mutation in the *Antp* gene has been identified in which the gene is incorrectly expressed in an anterior segment (**Figure 19.15**). A fly with this mutation has the bizarre trait in which it develops legs where antennae are normally found!

Homeotic genes encode homeotic proteins that function as transcription factors. The coding sequence of homeotic genes contains a 180-bp sequence known as a **homeobox** (**Figure 19.16a**). This sequence was first discovered in the *Antp* and *Ubx* genes, and it has since been found in many *Drosophila* homeotic genes. The homeobox is also found in other genes affecting pattern development, such as *bicoid*, which is not a homeotic gene. The homeobox encodes a region of the protein called a **homeodomain**, which functions in binding to the DNA (**Figure 19.16b**). The arrangement of α helices in the homeodomain promotes the binding of the protein to the DNA.

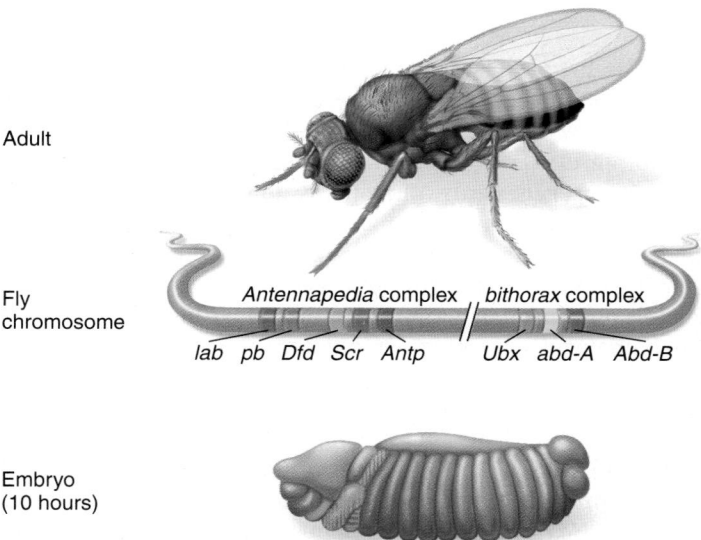

Adult

Fly chromosome

Antennapedia complex | *bithorax* complex

lab pb Dfd Scr Antp | *Ubx abd-A Abd-B*

Embryo (10 hours)

Figure 19.14 **Expression pattern of homeotic genes in *Drosophila*.** The order of homeotic genes, *labial* (*lab*), *proboscipedia* (*pb*), *deformed* (*Dfd*), *sex combs reduced* (*Scr*), *antennapedia* (*Antp*), *ultrabithorax* (*Ubx*), *abdominal A* (*abd-A*), and *abdominal B* (*Abd-B*), correlates with their spatial order of expression in the embryo.

(a) Normal fly | **(b) Mutant fly**

Figure 19.15 **The *Antennapedia* mutation in *Drosophila*.** **(a)** A normal fly with antennae. **(b)** This fly has a mutation in which the *Antp* gene is expressed in the embryonic segment that normally gives rise to antennae. The abnormal expression of *Antp* causes this region to have legs rather than antennae.

Biological inquiry: What phenotype would you expect if the Antp *gene was expressed in the last abdominal segment?*

(a) Homeotic gene

(b) Homeotic protein bound to DNA

Figure 19.16   Molecular features of homeotic genes and proteins.   (a) A homeotic gene (shown in blue) contains a 180-bp sequence called the homeobox (shown in dark blue). (b) Homeotic genes encode transcription factors. The homeobox encodes a region of the protein called a homeodomain, which functions in binding to the DNA at a regulatory site such as an enhancer that controls the expression of a gene.

Figure 19.17   A comparison of homeotic genes in *Drosophila* and the mouse.   The mouse and other mammals have four gene complexes, *HoxA–D*, that correspond to certain homeotic genes found in *Drosophila*. Thirteen different types of homeotic genes are found in the mouse, although each *Hox* complex does not contain all 13 genes. In this drawing, homologous genes are aligned in columns. For example, *lab* is the homologue to *HoxA-1*, *HoxB-1*, and *HoxD-1*; *Ubx* is homologous to *HoxA-7* and *HoxB-7*.

The primary function of homeotic proteins is to activate the transcription of specific genes that promote developmental changes in the animal. The homeodomain binds to enhancer sequences, which are described in Chapter 13. These enhancers are found in the vicinity of specific genes that control development. Most homeotic proteins also contain a transcriptional activation domain (Figure 19.16). After the homeodomain binds an enhancer, the transcriptional activation domain of the homeotic protein activates RNA polymerase to begin transcription. Some homeotic proteins also function as repressors of transcription.

GENOMES & PROTEOMES

A Homologous Group of Homeotic Genes Is Found in All Animals

Because most vertebrates have long generation times and produce relatively few offspring, it is not practical for researchers to screen large numbers of their embryos or offspring in search of mutant phenotypes with developmental defects. As an alternative,

the most successful way of identifying genes that affect vertebrate development has been the use of molecular techniques to identify vertebrate genes that are homologous to those that control development in simpler organisms such as *Drosophila*. Homologous genes are evolutionarily derived from the same ancestral gene and have similar DNA sequences. For this reason, a DNA strand from one gene (a probe) will bind to a complementary DNA strand of a homologous gene. As described in Chapter 20, this phenomenon, termed hybridization, can be used to identify homologous genes among different species.

The general approach of using *Drosophila* genes as probes to identify homologous vertebrate genes has been quite successful. Researchers have found complexes of homeotic genes in vertebrate species that bear striking similarities to those in the fruit fly. In the mouse, these groups of adjacent homeotic genes are called **Hox complexes**. The mouse has four *Hox* complexes, designated *HoxA*, *HoxB*, *HoxC*, and *HoxD*. Thirty-eight genes are found in the four complexes, which represent 13 different gene types. You can see from **Figure 19.17** that several of the homeotic genes in fruit flies and mammals are evolutionarily related. Among the first six types of *Hox* genes, five of them are homologous to genes found in the *Antennapedia* complex of

1 2 3 4 5 6   7 8 9 10 11 12 13

Mouse chromosome

Embryo

Adult mouse

Figure 19.18 **Expression pattern of *Hox* genes in the mouse.** A schematic illustration of *Hox* gene expression in the embryo and the corresponding regions in the adult. The order of *Hox* gene expression, from anterior to posterior, parallels the order of genes along the chromosome.

Drosophila. Among the last seven (genes numbered 7–13), three are homologous to the genes of the *bithorax* complex.

Like the *Antennapedia* and *bithorax* complexes in *Drosophila*, the arrangement of *Hox* genes along the mouse chromosome follows the colinearity rule, reflecting their pattern of expression from the anterior to the posterior end (**Figure 19.18**). Mutations in these mouse genes affect the formation of structures along this axis. These results indicate that the *Hox* genes play a role in determining the fates of regions along the anteroposterior axis. Nevertheless, additional research will be necessary to understand the individual role that each of the 38 *Hox* genes plays during embryonic development.

Hox genes are present in all animals. Sponges, which are a primitive type of animal, have a single *Hox* gene. The study of *Hox* genes in many different animal species has shown that the *Hox* cluster, with its colinear expression and its role of determining the anteroposterior axis, originated very early in the evolution of animals. At the level of genetics, fundamental similarities are observed in the ways that animals, such as worms, fruit flies, and mammals, undergo embryonic development. Researchers have suggested there is a "universal body plan" for animal development. A portion of the genome of all animals is devoted to the execution of this plan. The biological diversity that we see among animals is due to variation from this common plan.

Phase 4 Pattern Development: Stem Cells Can Divide and Differentiate into Specialized Cell Types

Thus far we have focused our attention on patterns of gene expression that occur during the early stages of development. These genes control the basic body plan of the organism. During the fourth phase of pattern development, the emphasis shifts to cell differentiation (see Figure 19.6). But how do cells differentiate into the necessary cell types? For example, how does an undifferentiated mesodermal cell differentiate into a specialized muscle cell, or how does an ectodermal cell differentiate into a nerve cell?

Although invertebrates have been instrumental in our understanding of pattern formation in animals, cell differentiation has been studied more extensively in mammals. One reason is because researchers have been able to grow mammalian cells in the laboratory for many decades. The availability of laboratory-grown cells makes it much easier to analyze the process of cell differentiation.

By studying mammalian cells in the laboratory, geneticists have determined that the profound morphological differences between two different types of differentiated cells, such as muscle and nerve cells, arise through gene regulation. Though nerve and muscle cells contain the same set of genes, they regulate the expression of their genes in very different ways. Certain genes that are transcriptionally active in muscle cells are inactive in nerve cells, and vice versa. Therefore, nerve and muscle cells express different proteins that affect the morphological and physiological characteristics of the respective cells in distinct ways. In this manner, differential gene expression underlies cell differentiation.

To understand the process of cell differentiation in a multicellular organism, we need to consider the special properties of **stem cells**, which divide and supply the cells that construct the bodies of all animals and plants. Stem cells have two common characteristics. First, they have the capacity to divide, and second, their daughter cells can differentiate into one or more specialized cell types. The two daughter cells that are produced from the division of a stem cell may have different fates (**Figure 19.19**). One of the cells may remain an undifferentiated stem cell, while the other daughter cell can differentiate into a specialized cell type. With this asymmetric division/differentiation pattern, the population of stem cells remains constant, yet stem cells provide a population of specialized cells. For example, in mammals, this mechanism is needed to replenish cells that have a finite life span such as skin cells and red blood cells.

In mammals, stem cells are commonly categorized according to their developmental stage and their ability to differentiate (**Figure 19.20**). The ultimate stem cell is the fertilized egg, which, via multiple cellular divisions, can give rise to an entire organism. A fertilized egg is considered to be **totipotent** because it can produce all of the cell types in the adult organism. The early mammalian embryo contains **embryonic stem cells (ES cells)**, which are initially found in the inner cell mass of the blastocyst.

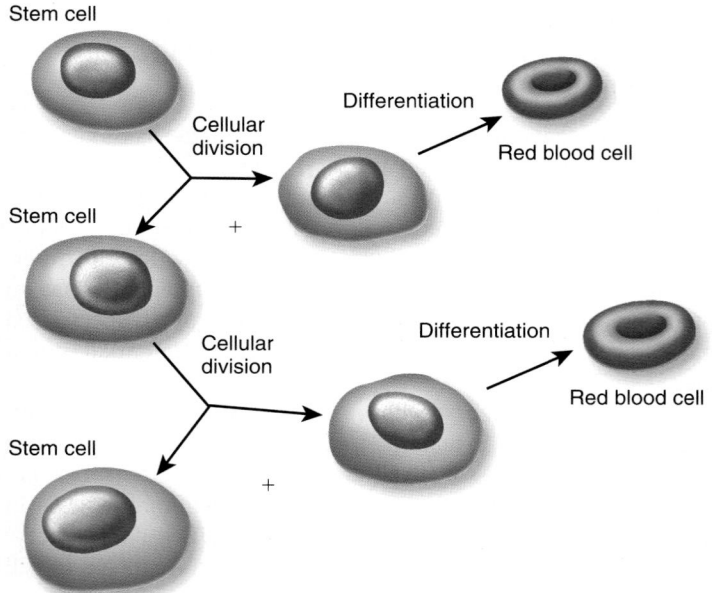

Figure 19.19 **Growth pattern of stem cells.**   When a stem cell divides, one of the two daughter cells may remain a stem cell, while the other cell can differentiate into a specialized cell type, such as the red blood cells shown here.

Biological inquiry: What are the two key features of stem cells?

The blastocyst stage of embryonic development occurs prior to uterine implantation. Embryonic stem cells are **pluripotent**, which means they can also differentiate into almost every cell type of the body. However, a single embryonic stem cell has lost the ability to produce an entire, intact individual. At the early fetal stage of development, the cells that later give rise to sperm or eggs cells, known as the **embryonic germ cells (EG cells)**, also are pluripotent.

During the embryonic and fetal stages of development, multipotent and unipotent stem cells are produced that give rise to the differentiated cells within the tissues and organs of an animal's body. Adults also have multipotent and unipotent stem cells. A **multipotent** stem cell can differentiate into several cell types, but far fewer than an embryonic stem cell can differentiate into. For example, hematopoietic stem cells (HSCs) found in the bone marrow can supply cells that populate two different tissues, namely, the blood and lymphoid tissues (**Figure 19.21**). Furthermore, each of these tissues contains several cell types. Multipotent hematopoietic stem cells can follow a pathway in which cell division produces a myeloid progenitor cell, which then differentiates into various cells of the blood and immune systems. Alternatively, an HSC can follow a path in which it becomes a lymphoid progenitor cell that differentiates into lymph cell types. Other stem cells found in the adult seem to be **unipotent**, or only able to produce daughter cells that differentiate into one cell type. For example, stem cells in the testis produce daughter cells that only differentiate into sperm.

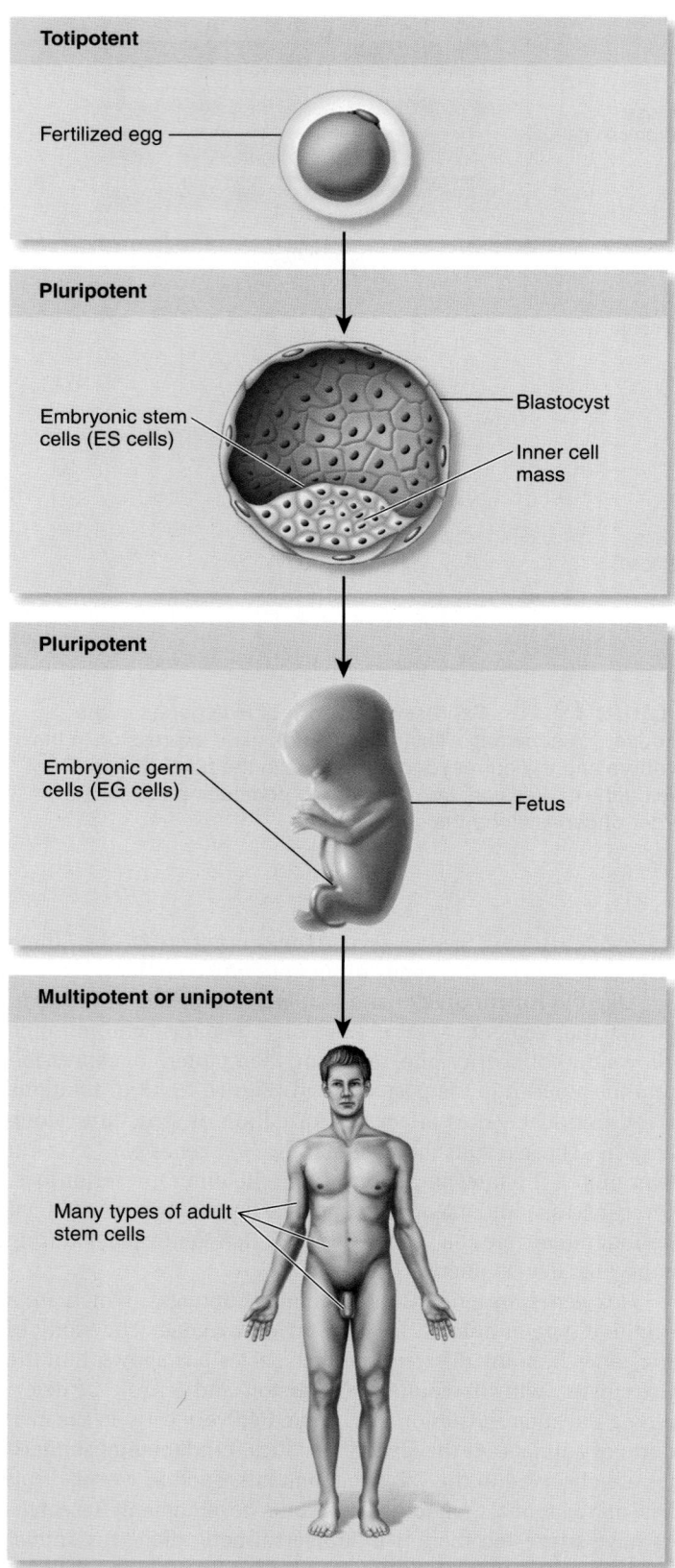

Figure 19.20   Occurrence of stem cells at different stages of mammalian development.

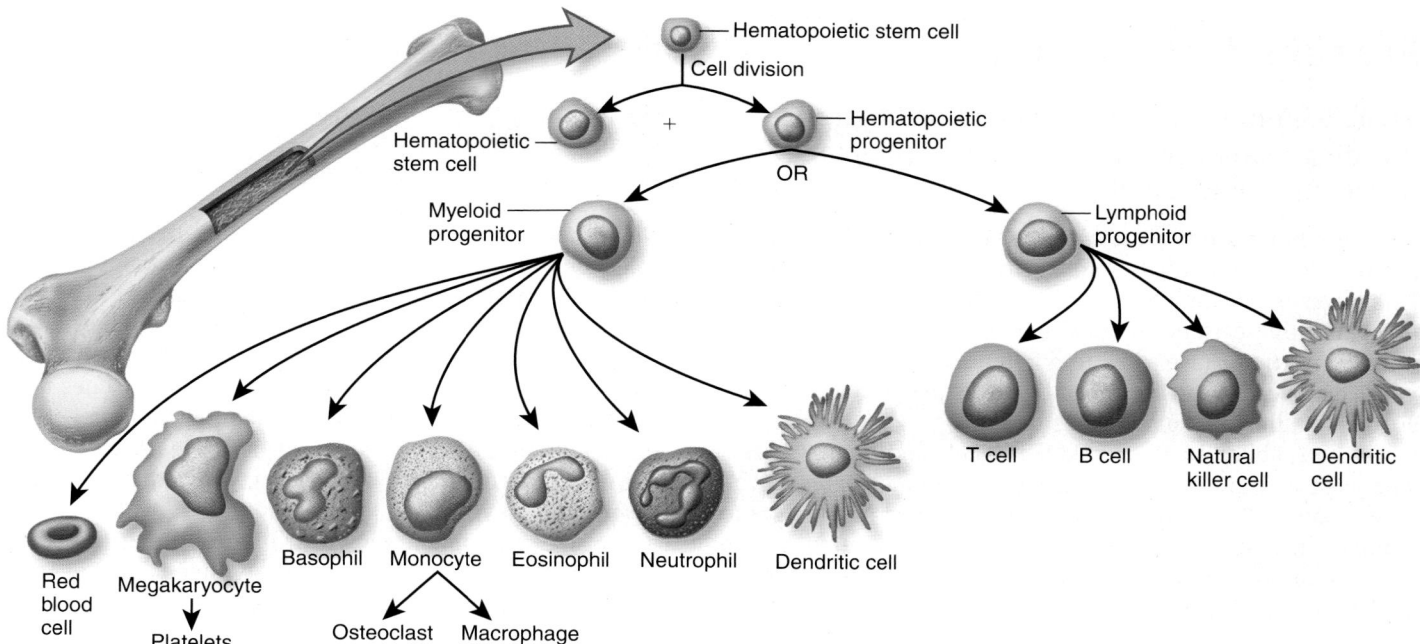

Figure 19.21 Fates of hematopoietic stem cells.

A compelling medical reason why researchers are interested in stem cells is the potential to treat human diseases or injuries that cause cell and tissue damage. This application has already become a reality in certain cases. For example, bone marrow transplants are used to treat patients with certain forms of cancer. When bone marrow from a healthy person is injected into the body of a patient who has had their immune system wiped out via radiation, the stem cells within the transplanted marrow have the ability to proliferate and differentiate into various types of blood cells within the body of the patient.

Renewed interest in the use of stem cells in the potential treatment of many other diseases has been fostered by studies in 1998 that showed researchers can obtain ES cells from blastocysts and EG cells from aborted fetuses and successfully propagate them in the laboratory. As mentioned, ES and EG cells are pluripotent. Therefore, they have the capacity to produce many different kinds of tissue. Embryonic cells could potentially be used to treat a wide variety of diseases associated with cell and tissue damage (**Table 19.1**). Much progress has been made in testing the use of stem cells in animal models. However, more research will need to be done before the use of stem cells to treat such diseases in humans is realized.

From an ethical perspective, the primary issue that raises debate is the source of stem cells for research and potential treatments. Most ES cells have been derived from human embryos that were produced from *in vitro* fertilization, and that were not used by the couples. Most EG cells are obtained from aborted fetuses, either those that were spontaneously aborted or those in which the decision to abort was not related to donating the fetal tissue to research. Some feel that it is morally wrong to use such tissue in research and/or the treatment of disease. Furthermore, some people fear that this technology could lead to intentional abortions for the sole purpose of obtaining fetal tissues for

Table 19.1	Some Potential Uses of Stem Cells to Treat Diseases
Cell/tissue type	**Disease treatment**
Nerve	Implantation of cells into the brain to treat Parkinson disease
	Treatment of injuries such as those to the spinal cord
Skin	Treatment of burns and skin disorders
Cardiac	Repair of heart damage associated with heart attacks
Cartilage	Repair of joints damaged by injury or arthritis
Bone	Repair of damaged bone or replacement with new bone
Liver	Repair or replacement of liver tissue damaged by injury or disease
Skeletal muscle	Repair or replacement of damaged muscle

transplantation. Alternatively, others feel that the embryos and fetuses that have been the sources of ES and EG cells were not going to become living individuals, and therefore it is beneficial to study these cells and to use them in a positive way to treat human diseases and injury. It is not clear whether these two opposing viewpoints can reach a common ground. Many governments are enacting laws that limit or prohibit the use of embryos or fetuses to obtain stem cells, yet permit the use of stem cell lines that are already available in research laboratories. In the U.S., for example, federal law prohibits the use of government funding for research projects that involve the destruction of embryos to obtain stem cells. However, government-sponsored research can be done on stem cell lines that were created prior to this legislation.

FEATURE INVESTIGATION

Davis, Weintraub, and Lasser Identified Genes Encoding Transcription Factors That Promote Muscle Cell Differentiation

A key question regarding the study of stem cells is, What causes a stem cell to differentiate into a particular cell type? Though the answer is not understood in great detail, researchers have discovered that certain proteins function as "master transcription factors" that cause a cell to differentiate into a specific cell type. The investigation described here was one of the first studies to reveal this phenomenon.

In 1987, Robert Davis, Harold Weintraub, and Andrew Lasser conducted a study to identify genes that promote muscle cell differentiation. The initial strategy for their experiments was to identify genes that are expressed only in differentiating muscle cells, not in nonmuscle cells. Though methods of gene cloning are described in Chapter 20, let's briefly consider these scientists'

cloning methods so we can understand their approach. They began with two different laboratory cell lines that could differentiate into muscle cells. From these two cell lines, they identified about 10,000 different genes that were transcribed into mRNA. Next, they compared the expressed genes in these two muscle cell lines with genes that were expressed in a nonmuscle cell line. Davis, Weintraub, and Lasser's comparison revealed 26 genes that were expressed only in the two muscle cell lines but not in the nonmuscle cell line. To narrow their search further, they compared these 26 genes with other nonmuscle cell lines they had available. Among the 26, only 3 of them were not expressed in other nonmuscle cell lines. Based on these criteria, they were left with three different cloned genes, which they termed *MyoA*, *MyoD*, and *MyoH*.

In the experiment shown in **Figure 19.22**, the scientists' goal was to determine if any of these three cloned genes could cause nonmuscle cells to differentiate into muscle cells. Using

Figure 19.22   Davis, Weintraub, and Lasser and the promotion of muscle cell differentiation in fibroblasts by the expression of *MyoD*.

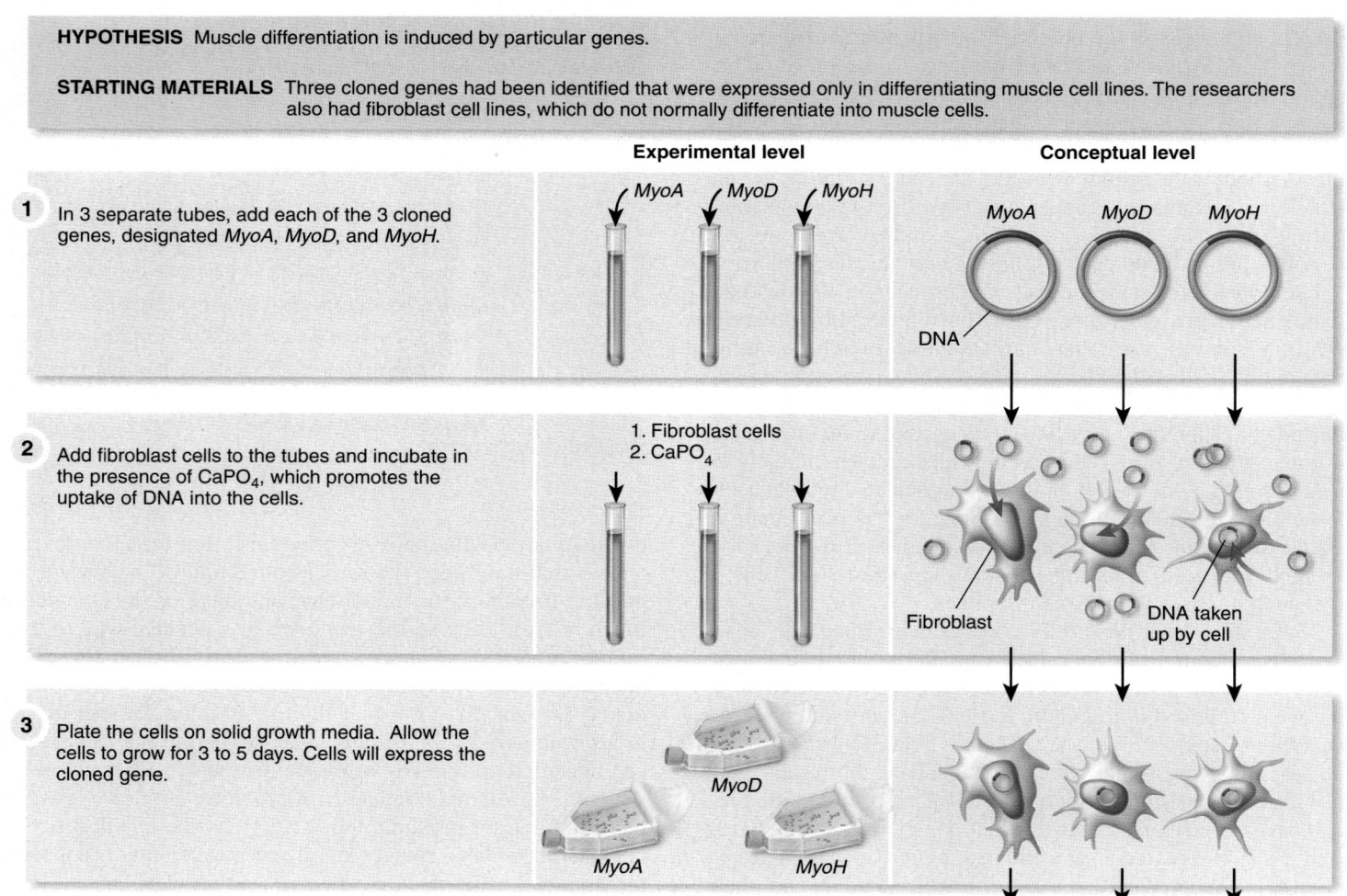

HYPOTHESIS Muscle differentiation is induced by particular genes.

STARTING MATERIALS Three cloned genes had been identified that were expressed only in differentiating muscle cell lines. The researchers also had fibroblast cell lines, which do not normally differentiate into muscle cells.

Experimental level | Conceptual level

1 In 3 separate tubes, add each of the 3 cloned genes, designated *MyoA*, *MyoD*, and *MyoH*.

2 Add fibroblast cells to the tubes and incubate in the presence of CaPO₄, which promotes the uptake of DNA into the cells.

3 Plate the cells on solid growth media. Allow the cells to grow for 3 to 5 days. Cells will express the cloned gene.

4 Examine the cells under a microscope to determine if they have the morphology of differentiating muscle cells.

Now looks like a muscle cell

Still looks like a fibroblast

5 Also, determine if the cells are synthesizing myosin, which is a protein that is abundantly made in muscle cells. This is done by adding a labeled antibody that recognizes myosin and determining the amounts of antibody that bind.

Colony labeled with myosin antibody

MyoD

MyoA   *MyoH*

Antibodies

6 THE DATA

Results from step 4:

DNA added	Microscopic morphology of cells
MyoA	Fibroblasts
MyoD	Muscle cells
MyoH	Fibroblasts

Results from step 5:

DNA added	Colonies labeled with antibody that binds to myosin?
MyoA	No
MyoD	Yes
MyoH	No

molecular techniques described in Chapter 20, the coding sequence of each cloned gene was placed next to an active promoter, and then the genes were introduced into fibroblasts, which are a type of cell that normally differentiates into osteoblasts (bone cells), chrondrocytes (cartilage cells), adipocytes (fat cells), and smooth muscle cells, but *in vivo* never differentiates into skeletal muscle cells. However, when the cloned *MyoD* gene was expressed in fibroblast cells in a laboratory, the fibroblasts differentiated into skeletal muscle cells. These cells contained large amounts of myosin, which is a protein that is expressed in muscle cells. The other two cloned genes (*MyoA* and *MyoH*) did not cause muscle cell differentiation or promote myosin production.

Since this initial discovery, researchers have found that *MyoD* belongs to a small group of genes termed **myogenic bHLH genes** that initiate muscle development. Myogenic bHLH genes encode transcription factors that contain a basic domain and a helix-loop-helix domain (bHLH). They are found in all vertebrates, and they have been identified in several invertebrates, such as *Drosophila* and *C. elegans*. In all cases, myogenic bHLH genes are activated during skeletal muscle cell development.

At the molecular level, two key features enable myogenic bHLH proteins to promote muscle cell differentiation. First, the basic domain binds specifically to an enhancer DNA sequence that is adjacent to genes that are expressed only in muscle cells. Second, the protein contains a transcriptional activation domain that stimulates the ability of RNA polymerase to initiate transcription. Therefore, bHLH proteins can bind to enhancers and activate the expression of many different muscle cell–specific genes. When the encoded proteins are synthesized, they change the characteristics of an undifferentiated cell into those of a highly specialized skeletal muscle cell.

The function of myogenic bHLH proteins relies on their interaction with other cellular proteins. Myogenic bHLH proteins can form a dimer with a protein called Id (for inhibitor of differentiation) or with an E protein, which is also expressed in cells that should differentiate into muscle cells. The helix-loop-helix domain is necessary for dimer formation. The Id protein is produced during early stages of development and prevents myogenic bHLH proteins from promoting muscle differentiation too soon. When a dimer forms between a myogenic bHLH protein and Id protein, the dimer cannot bind to DNA because the Id protein lacks a basic domain (**Figure 19.23a**). At later stages of development, the amount of Id protein falls, and myogenic bHLH proteins can then combine with E proteins to induce muscle differentiation. When a dimer forms between a myogenic bHLH protein and an E protein, the dimer binds to the DNA and activates gene expression (**Figure 19.23b**).

Muscle-cell-specific enhancer sequence

Muscle-cell-specific enhancer sequence

Dimer is unable to bind — gene is not activated.

Basic domain

Binding of dimer activates the gene.

Myogenic bHLH protein

Id protein

Transcriptional activation domain

E protein

Myogenic bHLH protein

(a) Action of bHLH-Id dimer

(b) Action of bHLH-E dimer

Figure 19.23 **Regulation of muscle-cell-specific genes by myogenic bHLH proteins.** **(a)** When a myogenic bHLH protein forms a dimer with an Id protein, it cannot bind to the DNA and therefore does not activate gene transcription. **(b)** A dimer formed from a myogenic bHLH protein and an E protein can bind to a muscle-cell-specific enhancer sequence and activate gene expression. This leads to muscle cell differentiation.

19.3 Development in Plants

As discussed at the beginning of the chapter, the morphology of more complex plants has two key features (see Figure 19.2b). The first is the root-shoot axis. Most plant growth occurs via cell division near the tips of the shoots and the bottoms of the roots. Second, this growth occurs in a well-defined radial pattern. For example, early in *Arabidopsis* growth, a rosette of leaves is produced from leaf buds that emanate in a spiral pattern directly from the main shoot (**Figure 19.24**). Later, the shoot generates branches that produce leaf buds as they grow. Overall, the radial pattern in which a plant shoot generates buds is an important mechanism that determines much of the general morphology of the plant.

At the cellular level too, plant development differs markedly from animal development. For example, cell migration does not occur during plant development. In addition, the development of a plant does not rely on morphogens that are deposited asymmetrically in the oocyte, as in complex animals. In plants, an entirely new individual can be regenerated from many types of somatic cells. Certain somatic cells of plants are totipotent, with the ability to produce an entire individual.

In spite of these apparent differences, the underlying molecular mechanisms of pattern formation in plants still share some similarities with those in animals. Like animals, plants use the mechanism of differential gene expression to coordinate the

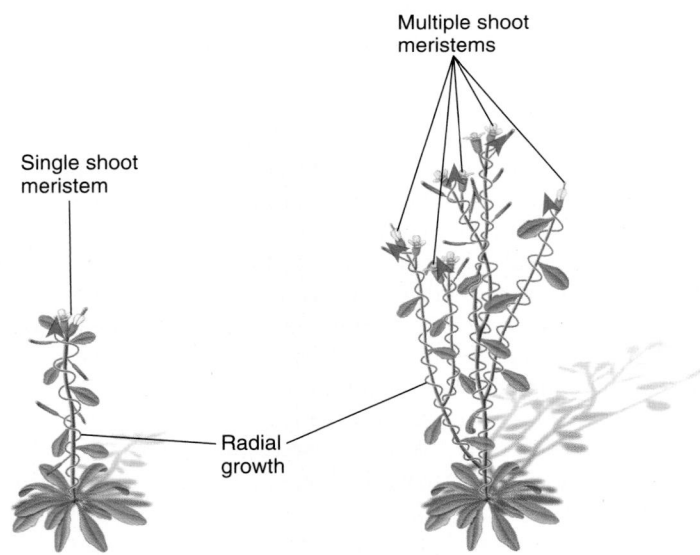

Multiple shoot meristems

Single shoot meristem

Radial growth

Figure 19.24 **Radial pattern of shoot growth in plants.** Early in development, as shown here in *Arabidopsis*, a single shoot promotes the formation of early leaves on the plant. Later, buds will form from this main shoot that will go on to form branches. The buds that produce the branches and the leaves that form on the branches are also produced in a radial manner.

development of a body plan. Like their animal counterparts, a plant's developmental program relies on the use of transcription factors, determining when and how much gene products are made. In this section, we will consider pattern formation in plants, and study how transcription factors play a key role in plant development.

Plant Development Occurs from Meristems That Are Formed in the Embryo

Before we discuss genes that control plant development, let's examine pattern formation in the early development of the plant embryo. **Figure 19.25** illustrates a common sequence of events that takes place in the embryonic development of seed plants such as *Arabidopsis*. After fertilization, the first cellular division is asymmetrical and produces a smaller apical cell and a larger basal cell (Figure 19.25a). The apical cell will give rise to most of the embryo and later develop into the shoot of the plant. In *Arabidopsis*, the basal cell will give rise to the root, along with a structure called the suspensor, which will become tissue that is required for seed formation (Figure 19.25b).

At the heart stage, which is composed of only about 100 cells, the basic organization of the plant has been established (Figure 19.25c). Plants have organized groups of actively dividing stem cells called **meristems**. As discussed earlier in this chapter, stem cells retain the ability to divide and their daughter cells differentiate into multiple cell types. The meristem produces offshoots of proliferating and differentiating cells. The **root meristem** gives rise only to the root, while the **shoot meristem** produces all aerial parts of the plant, which include the stem as well as lateral structures such as leaves and flowers.

The heart stage then progresses to the formation of a seedling that has two cotyledons, which are embryonic leaves that store nutrients for the developing embryo and seedling. In the seedling shown in Figure 19.25d, you can see three main regions. The **apical region** produces the leaves and flowers of the plant. The **central region** creates the stem. The radial pattern of cells in the central region causes the radial growth observed in plants. Finally, the **basal region** produces the roots. Each of these three regions develops differently, as indicated by their unique cell division patterns and distinct morphologies.

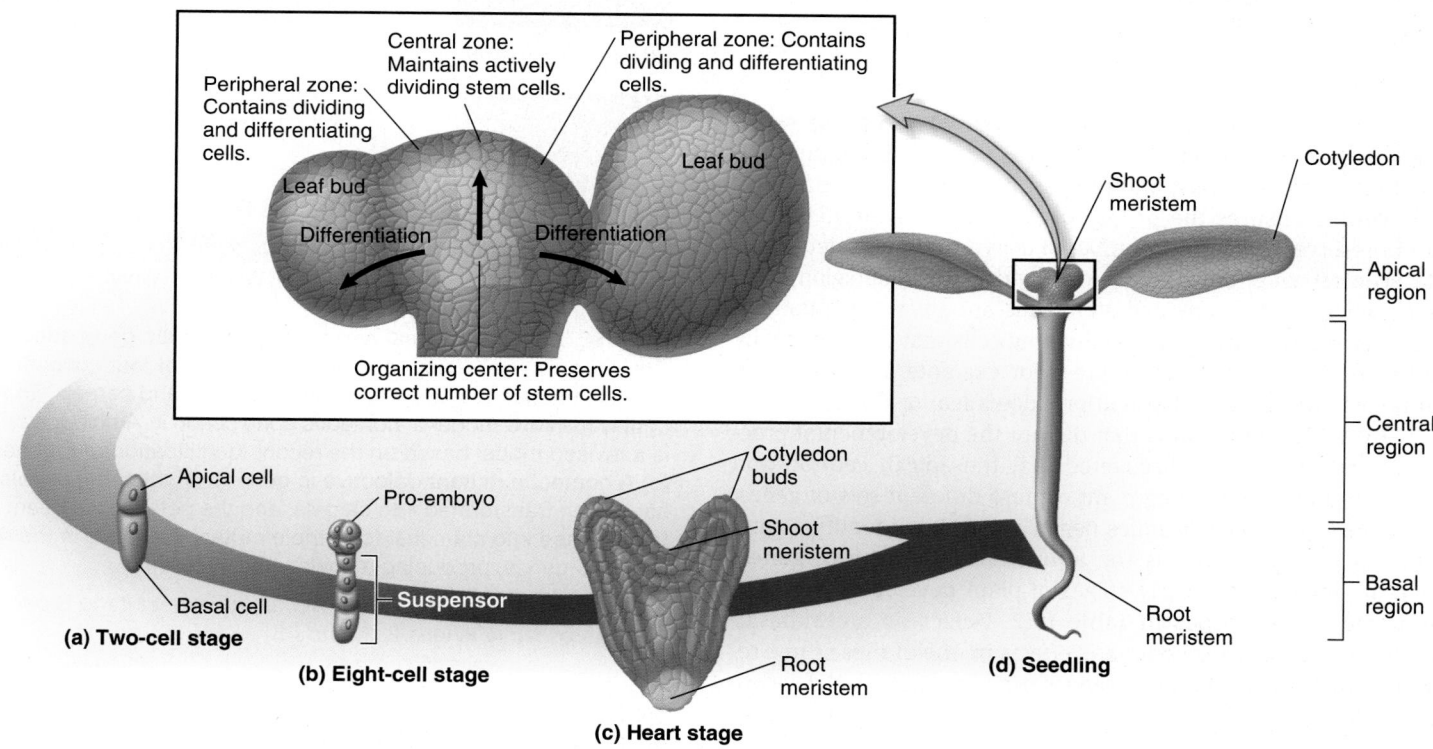

Figure 19.25 **Developmental steps in the formation of a plant embryo.** **(a)** The two-cell stage consists of the apical cell and basal cell. **(b)** The eight-cell stage consists of a pro-embryo and a suspensor. The suspensor gives rise to extra embryonic tissue that is needed for seed formation. **(c)** At the heart stage, all of the plant tissues have begun to form. Note that the shoot meristem is located between the future cotyledons, while the root meristem is on the opposite side. **(d)** A seedling showing basal, central, and apical regions. The inset shows the organization of the shoot meristem. Note: The steps shown in this figure occur during seed formation, and the embryo would be enclosed within a seed.

Table 19.2	Examples of *Arabidopsis* Apical-Basal-Patterning Genes
Region: *Gene*	**Description**
Apical:	
Gurke	Mutations that cause minor loss of function in the *Gurke* gene produce seedlings with highly reduced or no cotyledons. Complete loss of function eliminates the entire shoot.
Aintegumenta	Encodes a transcription factor that is also expressed in the peripheral zone. Its expression maintains the growth of lateral buds.
Central:	
Scarecrow	Encodes a transcription factor that plays a role in the asymmetric division that produces the radial pattern of growth in the stem. Note: The Scarecrow protein also affects cell division patterns in roots and plays a role in sensing gravity.
Basal:	
Monopterous	Encodes a transcription factor. When this gene is defective, the plant embryo cannot initiate the formation of root structures, but root structures can be formed post-embryonically under the correct growth conditions. This gene seems to be required for organizing root formation in the embryo, but is not required for root formation *per se*.

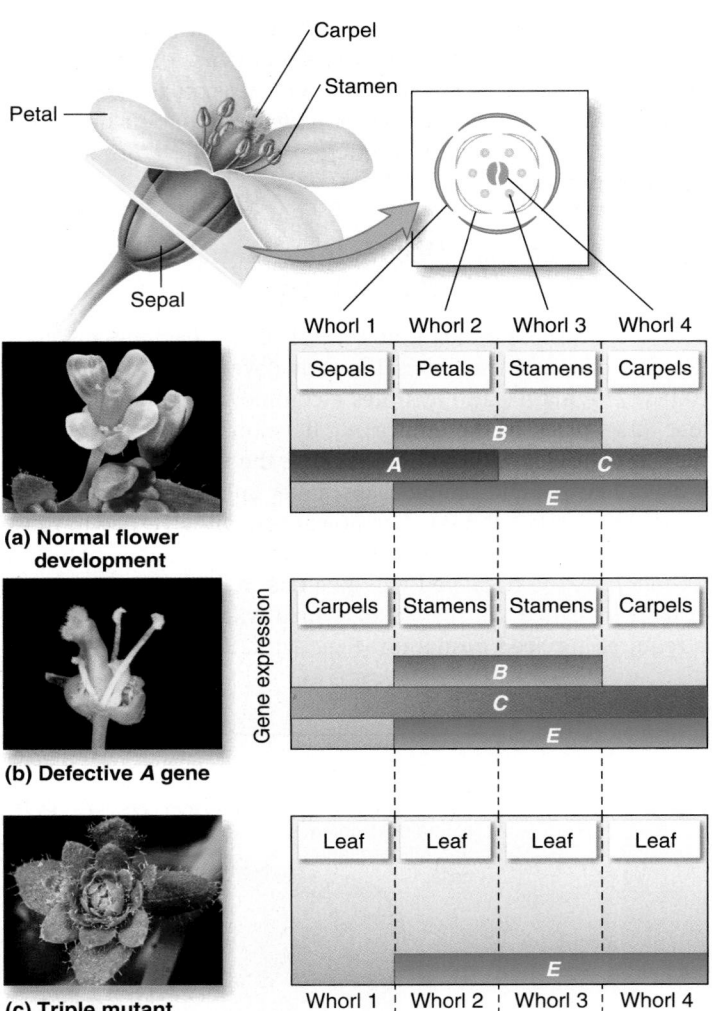

Figure 19.26 Normal and mutant homeotic gene action in *Arabidopsis.* **(a)** A normal flower composed of four concentric whorls of structures: sepals, petals, stamens, and carpels. To the right is the ABC model of homeotic gene action in *Arabidopsis*. This is a revised model based on the recent identification of *E* genes. **(b)** A homeotic mutant defective in gene *A* in which the sepals have been transformed into carpels, and the petals have been transformed into stamens. **(c)** A triple mutant defective in the *A*, *B*, and *C* genes, producing a flower with all leaves.

Biological inquiry: What pattern would you expect if the B gene was expressed in whorls 2, 3, and 4?

As you can see in the inset to Figure 19.25d, the shoot meristem is organized into three areas called the organizing center, the central zone, and the peripheral zone. The **organizing center** ensures the proper organization of the meristem and preserves the correct number of actively dividing stem cells. The **central zone** (not to be confused with the central region) is an area where undifferentiated stem cells are always maintained. The **peripheral zone** contains dividing cells that will eventually differentiate into plant structures. For example, the peripheral zone may form a bud that will produce a leaf or flower.

By analyzing mutants that disrupt the developmental process, researchers have discovered that the apical, central, and basal regions of a growing plant express different sets of genes. G. Jürgens and his colleagues began a search to identify a category of genes, known as the **apical-basal-patterning genes**, that are important in early stages of plant development. A few examples are described in **Table 19.2**. Defects in apical-basal-patterning genes cause dramatic effects in one of these three regions. For example, the *gurke* gene is necessary for apical development. When it is defective, the embryo lacks apical structures.

Plant Homeotic Genes Control Flower Development

Although William Bateson coined the term homeotic to describe such mutations in animals, the first known homeotic genes were described in plants. Naturalists in ancient Greece and Rome, for example, recorded their observations of double flowers in which stamens were replaced by petals. In current research, geneticists are studying these types of mutations to better understand devel-

opmental pathways in plants. Many homeotic mutations affecting flower development have been identified in *Arabidopsis* and also in the snapdragon (*Antirrhinum majus*).

A normal *Arabidopsis* flower is composed of four concentric whorls of structures (**Figure 19.26a**). The first, outer whorl contains four **sepals**, which protect the flower bud before it opens. The second whorl is composed of four **petals**, and the third whorl contains six stamens. The **stamens** are structures that make the male gametophyte, pollen. Finally, the fourth, innermost whorl contains two carpels that are fused together. The **carpel** produces the female gametophyte.

By analyzing the effects of many different homeotic mutations in *Arabidopsis*, Elliot Meyerowitz and his colleagues in 1994 proposed the ABC model for flower development. In this model, three classes of genes, called *A*, *B*, and *C*, govern the formation of sepals, petals, stamens, and carpels. More recently, a fourth category of genes called the *E* genes was found to also be required for this process. In *Arabidopsis*, Figure 19.26a illustrates how these genes affect normal flower development. In whorl 1, gene *A* product is made. This promotes sepal formation. In whorl 2, *A*, *B*, and *E* gene products are made, which promotes petal formation. In whorl 3, the expression of genes *B*, *C*, and *E* causes stamens to be made. Finally, in whorl 4, the products of *C* and *E* genes promote carpel formation.

Now let's consider what happens in certain homeotic mutants that undergo transformations of particular whorls. According to the original ABC model, genes *A* and *C* repress each other's expression, and gene *B* functions independently. In a mutant defective in gene *A* expression, gene *C* will also be expressed in whorls 1 and 2. This produces a carpel-stamen-stamen-carpel arrangement in which the sepals have been transformed into carpels and the petals into stamens (**Figure 19.26b**). When gene *B* is defective, a flower cannot make petals or stamens. Therefore, a gene *B* defect yields a flower with a sepal-sepal-carpel-carpel arrangement. When gene *C* is defective, gene *A* is expressed in all four whorls. This results in a sepal-petal-petal-sepal pattern. If the expression of *E* genes is defective, the flower consists entirely of sepals.

Overall, the genes described in Figure 19.26 promote a pattern of development that leads to sepal, petal, stamen, or carpel structures. But what happens if genes *A*, *B*, and *C* are all defective? This produces a flower that is composed entirely of leaves (**Figure 19.26c**). These results indicate that the leaf structure is the default pathway and that the *A*, *B*, *C*, and *E* genes cause development to deviate from a leaf structure in order to make something else. In this regard, the sepals, petals, stamens, and carpels can be viewed as modified leaves. With astonishing insight, Johann Goethe, a German poet, novelist, playwright, and natural philosopher, originally proposed this idea over 200 years ago!

Like the *Drosophila* homeotic genes, plant homeotic genes are part of a hierarchy of gene regulation. Genes that are expressed within the flower bud produce proteins that activate the expression of these homeotic genes. Once they are transcriptionally activated, the homeotic genes then regulate the expression of other genes, the products of which promote the formation of sepals, petals, stamens, or carpels.

CHAPTER SUMMARY

- Development refers to a series of changes in the state of a cell, tissue, organ, or organism.

19.1 General Themes in Development

- A cell that is determined has a particular cell fate. A cell that is differentiated has a specialized morphology and function.
- *Drosophila*, *C. elegans*, the mouse, zebrafish, and *Arabidopsis* are model organisms studied by developmental geneticists. (Figure 19.1)
- The process that gives rise to a plant or animal with a particular body structure is called pattern formation. (Figure 19.2)
- Four responses to positional information are cell division, cell migration, cell differentiation, and apoptosis. (Figures 19.3, 19.4)
- Morphogens and direct contacts between adjacent cells are two ways that cells obtain positional information. (Figure 19.5)
- Transcription factors control the program of development in animals and plants. In animals, this program occurs in four phases. (Figure 19.6)

19.2 Development in Animals

- Embryonic development in *Drosophila* occurs in a series of steps, starting with a zygote, then an embryo, larvae, pupa, and an adult. The basic body plan is established in the embryo. (Figure 19.7)
- Maternal effect genes control the formation of body axes. (Figures 19.6, phase 1, 19.8, 19.9)
- *Drosophila* is divided into many segments. The identification of *Drosophila* mutants has aided in our understanding of segmentation. (Figures 19.10, 19.11, 19.12)

- The sequential expression of three categories of segmentation genes divides the embryo into segments. (Figure 19.13)
- Homeotic genes control the development of structures in a particular segment or group of segments. (Figures 19.14, 19.15, 19.16)
- Invertebrates, such as *Drosophila*, and vertebrates, such as the mouse, both have a homologous set of homeotic genes. In mammals, these are called the *Hox* genes. (Figure 19.17, 19.18)
- Stem cells have the ability to divide and to differentiate. (Figure 19.19)
- In mammals, a fertilized egg is totipotent, certain embryonic and fetal cells are pluripotent, while stem cells in the adult are multipotent or unipotent. (Figures 19.20, 19.21)
- Stem cells have the potential to treat a variety of human disorders. (Table 19.1)
- Certain "master transcription factors" control the cell differentiation process. An example is *MyoD*, which Davis, Weintraub, and Lasser showed experimentally can cause a cell to differentiate into a muscle cell. (Figures 19.22, 19.23)

19.3 Development in Plants

- Plants grow in a radial pattern along a root and shoot axis. (Figure 19.24)
- Plant meristems contain dividing cells that promote the development of plant structures such as roots, stems, leaves, and flowers. (Figure 19.25)
- Several types of genes have been identified in plants that influence pattern development. (Table 19.2)
- Four classes of homeotic genes in plants, *A*, *B*, *C*, and *E*, control flower formation. (Figure 19.26)

TEST YOURSELF

1. The process whereby a cell's morphology and function has changed is called
   a. determination.
   b. cell fate.
   c. differentiation.
   d. genetic engineering.
   e. both a and c.

2. Pattern formation in plants is along the _____ axis.
   a. dorsoventral
   b. anteroposterior
   c. right-left
   d. root-shoot
   e. all of the above

3. Positional information is important in determining the destiny of a cell in a multicellular organism. Cells respond to positional information by
   a. dividing.
   b. migrating.
   c. differentiating.
   d. undergoing apoptosis.
   e. all of the above.

4. Morphogens are
   a. molecules that disrupt normal development.
   b. molecules that convey positional information.
   c. mutagenic agents that cause apoptosis.
   d. receptors that allow cells to adhere to the extracellular matrix.
   e. both a and c.

5. What group of molecules plays a key role in controlling the program of developmental changes?
   a. motor proteins
   b. transporters
   c. transcription factors
   d. restriction endonucleases
   e. cyclins

6. Using the following list of events, determine the proper sequence for the events of animal development:
   1. Formation of tissues, organs, and other body structures in each segment.
   2. Axes of the entire animal are determined.
   3. Cells become differentiated.
   4. The entire animal is divided into segments.

   a. 2, 3, 4, 1
   b. 1, 2, 4, 3
   c. 2, 4, 3, 1
   d. 3, 2, 4, 1
   e. 2, 4, 1, 3

7. The homeotic genes in *Drosophila*
   a. determine the structural and functional characteristics of different segments of the developing fly.
   b. encode motor proteins that transport morphogens throughout the embryo.
   c. are dispersed apparently randomly throughout the genome.
   d. are expressed in similar levels in all parts of the developing embryo.
   e. both a and c.

8. Which of the following genes do not play a role in the process whereby segments are formed in the fruit fly embryo?
   a. homeotic genes
   b. gap genes
   c. pair-rule genes
   d. segment-polarity genes
   e. All of the above play a role in segmentation.

9. A type of stem cell that can give rise to any type of cell of an adult organism but cannot produce an entire, intact individual is called
   a. totipotent.
   b. pluripotent.
   c. multipotent.
   d. unipotent.
   e. antipotent.

10. During plant development, the leaves and the flowers of the plant are derived from
    a. the central region.
    b. the basal region.
    c. the suspensor.
    d. the apical region.
    e. both a and d.

CONCEPTUAL QUESTIONS

1. Define development.
2. Explain the hierarchy of gene expression that controls segmentation in *Drosophila*.
3. What are two characteristics of the proteins that are encoded by homeotic genes that give clues to their function?

EXPERIMENTAL QUESTIONS

1. What was the goal of the research conducted by Davis, Weintraub, and Lasser?
2. How did Davis, Weintraub, and Lasser's research identify the candidate genes for muscle differentiation?
3. Once the researchers identified the candidate genes for muscle differentiation, how did they test the effect of each gene on cell differentiation? What were the results of the study?

COLLABORATIVE QUESTIONS

1. Discuss four types of cellular responses that cells can exhibit in response to positional information.
2. Discuss meristems in plants.

---

www.brookerbiology.com

This website includes answers to the Biological Inquiry questions found in the figure legends and all end-of-chapter questions.

20
GENETIC TECHNOLOGY

Copycat, the first cloned pet. In 2002, the cat shown here called Copycat or CC (for carbon copy) was produced by cloning, a procedure described later in this chapter.

The past several decades have witnessed amazing technical advances in the ability of biologists to study life at the molecular level. **Recombinant DNA technology** is the use of laboratory techniques to isolate and manipulate fragments of DNA. Such technology produces **recombinant DNA**, which is any DNA molecule that has been manipulated so that it contains DNA from two or more sources. In the early 1970s, the first successes in making recombinant DNA molecules were accomplished independently by two groups at Stanford University: David Jackson, Robert Symons, and Paul Berg; and Peter Lobban and A. Dale Kaiser. Both groups were able to isolate and purify pieces of DNA in a test tube, and then covalently link together two or more DNA fragments. Shortly thereafter, these recombinant DNA molecules were introduced into living cells. Once inside a host cell, the recombinant molecules are replicated to produce many identical copies, a process known as cloning. When copies contain a gene of interest, the technique is called **gene cloning**. Recombinant DNA technology and cloning have enabled geneticists to probe relationships between gene sequences and phenotypic consequences and have been fundamental to our understanding of gene structure and function.

Later in this chapter we will consider the topic of genomics, which involves the molecular analysis of the entire genome of a species. In recent years, molecular techniques have progressed to the point where researchers can study the structure and function of many genes as large integrated units. For example, the expression of all genes in a genome can be analyzed under different conditions, such as in normal versus cancer cells. This information can help us to understand how changes in gene expression can cause uncontrolled cell growth.

In the last section of this chapter, we will explore the topic of **biotechnology**, which is the use of living organisms or the products of living organisms for human benefit. We will learn that **genetic engineering**, the direct manipulation of genes for practical purposes, is playing an ever-increasing role in the creation of strains of microorganisms, plants, and animals that have characteristics that are useful to people. These include bacteria that make human insulin, plants that are resistant to herbicides, and farm animals that make human medicines.

20.1 Gene Cloning

Molecular biologists want to understand how the molecules within living cells contribute to cell structure and function. Because proteins are the workhorses of cells, many researchers focus their attention on the structure and function of proteins, or the genes that encode them. In any given laboratory, it is common to study just one or perhaps a few different genes or proteins. At the molecular level, this poses a daunting task. Any given cell can express thousands of different proteins, making the study of any single gene or protein akin to finding a needle in a haystack. To overcome this truly formidable obstacle, researchers frequently take the approach of cloning the genes that encode their proteins of interest.

As mentioned, the term gene cloning refers to procedures that lead to the formation of many copies of a particular gene. **Figure 20.1** provides an overview of the steps and goals of gene cloning. The process is usually done with one of two goals in mind.

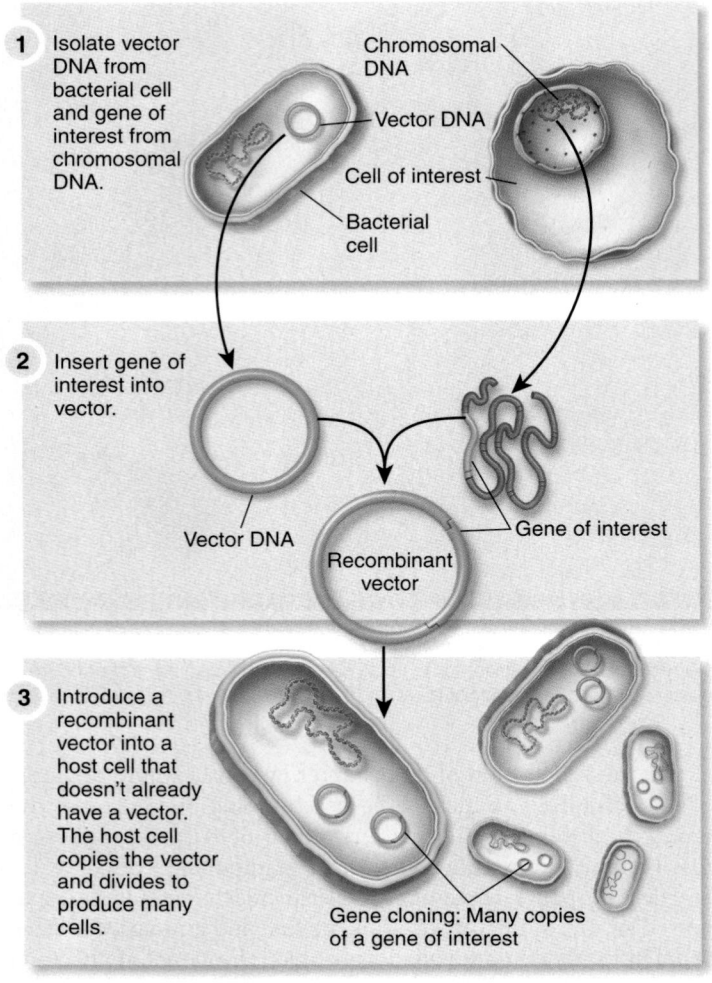

1 Isolate vector DNA from bacterial cell and gene of interest from chromosomal DNA.

Chromosomal DNA

Vector DNA

Cell of interest

Bacterial cell

2 Insert gene of interest into vector.

Vector DNA

Recombinant vector

Gene of interest

3 Introduce a recombinant vector into a host cell that doesn't already have a vector. The host cell copies the vector and divides to produce many cells.

Gene cloning: Many copies of a gene of interest

4 Gene cloning is done to achieve one of two main goals:

Producing large amounts of DNA of a specific gene	Expressing the cloned gene to produce the encoded protein
Examples	*Examples*
• Cloned genes provide enough DNA for DNA sequencing. The sequence of a gene can help us understand how a gene works and identify mutations that cause diseases.	• Large amounts of the protein can be purified to study its structure and function.
• Cloned DNA can be used as a probe to identify the same gene or similar genes in other organisms.	• Cloned genes can be introduced into bacteria or livestock to make pharmaceutical products such as insulin.
	• Cloned genes can be introduced into plants and animals to alter their traits.
	• Cloned genes can be used to treat diseases—a clinical approach called gene therapy.

Figure 20.1 **Gene cloning.** The process of gene cloning produces copies of a gene or copies of its protein product, which have many uses.

One goal would be that a researcher or clinician wants many copies of the gene, perhaps to study the DNA directly or to use the DNA as a tool. Alternatively, the goal may be to obtain lots of the gene product—mRNA or protein. For example, biochemists use gene cloning to obtain large amounts of proteins to study their structure and function. In modern molecular biology, the many uses for gene cloning are remarkable. Gene cloning has provided the foundation for critical technical advances in a variety of disciplines including molecular biology, genetics, cell biology, biochemistry, and medicine. In this section, we will examine the procedures that are used to copy genes.

Step 1: Vector DNA and Chromosomal DNA Are the Starting Materials to Clone a Gene

In the first step of gene cloning, a key material is a type of DNA known as a **vector** (Figure 20.1). The role of vector DNA is to act as a carrier of the DNA segment that is to be cloned. (The term vector comes from a Latin term meaning "carrier.") In cloning experiments, a vector may carry a small segment of chromosomal DNA, perhaps only a single gene. By comparison, a chromosome carries many more genes, perhaps a few thousand. When a vector is introduced into a living cell, it can replicate, and so the DNA that it carries is also replicated. This produces many identical copies of the inserted gene.

The vectors commonly used in gene cloning experiments were derived originally from two natural sources, plasmid or viral. Some vectors are **plasmids**, which are small circular pieces of DNA. As discussed in Chapter 18, plasmids are found naturally in many strains of bacteria and also occasionally in eukaryotic cells. Commercially available plasmids have been genetically engineered for effective use in cloning experiments. They contain unique sites into which geneticists can easily insert pieces of DNA. An alternative type of vector used in cloning experiments is a **viral vector**. Viruses can infect living cells and propagate themselves by taking control of the host cell's metabolic machinery. When a chromosomal gene is inserted into a viral vector, the gene will be replicated whenever the viral DNA is replicated. Therefore, viruses can be used as vectors to carry other pieces of DNA.

Another material that is needed to clone a gene is the gene itself, which we will call the "gene of interest." If a scientist wants to clone a particular gene, the source of the gene is the chromosomal DNA that carries the gene. To prepare chromosomal DNA, an experimenter first obtains some cells from the organism of interest. The preparation of chromosomal DNA involves breaking open cells and extracting and purifying the DNA using biochemical techniques such as chromatography and centrifugation.

Step 2: Cutting Chromosomal and Vector DNA into Pieces and Linking Them Together Produces Recombinant Vectors

The second step in a gene cloning experiment is the insertion of chromosomal DNA into a plasmid or viral vector (Figure 20.1).

Table 20.1	Examples of Restriction Enzymes Used in Gene Cloning	

Restriction enzyme*	Bacterial source	Sequence recognized†
*Bam*HI	*Bacillus amyloliquefaciens* H	↓ 5′–GGATCC–3′ 3′–CCTAGG–5′ ↑
*Eco*RI	*Escherichia coli* RY13	↓ 5′–GAATTC–3′ 3′–CTTAAG–5′ ↑
*Sac*I	*Streptomyces achromonogenes*	↓ 5′–GAGCTC–3′ 3′–CTCGAG–5′ ↑

* Restriction enzymes are named according to the species in which they are found. The first three letters are italicized because they indicate the genus and species names. Since a species may produce more than one restriction enzyme, the enzymes are designated I, II, III, etc., to indicate the order in which they were discovered in a given species.

† The arrows show the locations in the upper and lower DNA strands where the restriction enzymes cleave the DNA backbone.

DNA fragments must be cut and pasted to produce recombinant vectors. To cut DNA, researchers use enzymes known as **restriction enzymes** or restriction endonucleases. These enzymes, which were discovered by Werner Arber, Hamilton Smith, and Daniel Nathans in the 1960s and 1970s, are made naturally by many different species of bacteria. Restriction enzymes protect bacterial cells from invasion by foreign DNA, particularly that of bacteriophages. Currently, several hundred different restriction enzymes from various bacterial species have been identified and are available commercially to molecular biologists.

The restriction enzymes used in cloning experiments bind to a specific base sequence and then cleave the DNA backbone at two defined locations, one in each strand. The sequences recognized by restriction enzymes are called **restriction sites**. Most restriction enzymes recognize sequences that are palindromic, which means that the sequence is identical when read in the opposite direction in the complementary strand (**Table 20.1**). For example, the sequence recognized by the restriction enzyme *Eco*RI is 5′–GAATTC–3′ in the top strand. Read in the opposite direction in the bottom strand, this sequence is also 5′–GAATTC–3′.

Figure 20.2 shows the action of a restriction enzyme and the insertion of a gene into a vector. This vector also carries the *amp*^R and *lacZ* genes, whose functions will be discussed later. Certain restriction enzymes are useful in cloning because they digest DNA into fragments with **sticky ends**. This means that these DNA fragments have single-stranded ends that will hydrogen-bond to each other due to their complementary sequences.

Figure 20.2   Step 2 of gene cloning: The actions of a restriction enzyme and DNA ligase to produce a recombinant vector.   The restriction enzyme binds to a specific sequence in both the chromosomal and vector DNA. It then cleaves the DNA backbones, producing DNA fragments. The complementary single-stranded ends of the DNA fragments hydrogen-bond with each other. The enzyme DNA ligase then covalently links these fragments, producing a recombinant vector.

Biological inquiry: In the experiment shown in this figure, has the gene of interest been cloned?

The hydrogen bonding between the sticky ends of DNA fragments promotes temporary interactions between two DNA fragments. However, this interaction is not stable, because it involves only a few hydrogen bonds between complementary bases. To establish a permanent connection between two DNA fragments, the sugar-phosphate backbones within the DNA strands must be covalently linked together. This linkage is catalyzed by DNA ligase (Figure 20.2, step 3).

In some cases, the two ends of the vector will simply ligate back together, restoring it to its original circular structure; this forms what is called a recircularized vector. In other cases, a fragment of chromosomal DNA may become ligated to both ends of the vector. When this happens, a segment of chromosomal DNA has been inserted into the vector. The result is a vector containing a piece of chromosomal DNA, which is called a **recombinant vector** or a hybrid vector. We then have a recombinant vector that is ready to be cloned. A recombinant vector may contain the gene of interest or it may contain a different piece of chromosomal DNA.

Step 3: Putting Recombinant Vectors into Host Cells and Allowing Those Cells to Propagate Achieves Gene Cloning

The procedure we have considered in Figure 20.2 seeks to insert a chromosomal gene into a plasmid vector. The vector already carries an antibiotic-resistance gene called the amp^R gene. Such a gene is called a **selectable marker** because the presence of the antibiotic selects for the growth of cells expressing the amp^R gene. The amp^R gene encodes an enzyme known as β-lactamase that degrades the antibiotic ampicillin, which normally kills bacteria. Bacteria containing the amp^R gene can grow on media containing ampicillin because they can degrade it. In a cloning experiment where the amp^R gene is found within the plasmid, the growth of cells in the presence of ampicillin identifies bacteria that contain the plasmid.

The third step in gene cloning is the actual cloning of the gene of interest. In this step, the goal is for the recombinant vector carrying the desired gene to be taken up by bacterial cells treated with agents that render them permeable to DNA molecules (**Figure 20.3**). This procedure is called transformation, when a plasmid vector is used, or transfection, when a viral vector is introduced into a cell. During transformation, some cells take up a single plasmid while most cells fail to take up a plasmid. The bacteria are then streaked on petri plates containing a bacterial growth medium and ampicillin. In the experiment shown here, the bacterial cells were originally sensitive to ampicillin. Bacteria that have not taken up a plasmid will be killed by the antibiotic. In contrast, any bacterium that has taken up a plasmid carrying the amp^R gene will grow and divide many times to form a bacterial colony containing tens of millions of cells. Because each cell in a single colony is derived from the same original cell that took up a single plasmid, all cells within a colony contain the same type of plasmid DNA.

1   Mix plasmid DNA with many *E.coli* cells that have been treated with agents that make them permeable to DNA.

Gene of interest

Part of *lacZ*

Part of *lacZ*

amp^R

Origin of replication

In this example, the gene of interest was inserted into a plasmid. This separates the *lacZ* gene and renders it nonfunctional. It is also possible for any other chromosomal DNA fragment to be inserted into the plasmid, or the plasmid may recircularize without an insert.

This shows a bacterial cell that has taken up the plasmid carrying the gene of interest. Many bacterial cells fail to take up a plasmid.

2   Plate cells on media containing X-Gal and ampicillin. Incubate overnight. Note: The *ampR* gene allows bacteria to grow in the presence of ampicillin. The *lacZ* gene encodes β-galactosidase that degrades X-gal to produce a blue color.

Blue colony

White colony

Each bacterial colony is derived from a single cell; so all the cells in a colony are genetically identical.

Recircularized vector without an insert—*lacZ* gene functional and produces blue color

Recombinant vector with an insert—*lacZ* gene is nonfunctional

Figure 20.3   **Step 3 of gene cloning: The cloning of a recombinant vector.**   For cloning to occur, a recombinant vector is introduced into a host cell, which copies the vector and divides to produce many cells. This produces many copies of the gene of interest.

In the experiment shown in Figure 20.3, the experimenter can distinguish bacterial colonies that contain a recombinant vector from those containing a recircularized vector—and therefore no chromosomal DNA. In a recombinant vector, a piece of chromosomal DNA carrying a desired gene has been inserted into a region of the vector that contains the *lacZ* gene, which encodes the enzyme β-galactosidase. The insertion of chromosomal DNA into the vector disrupts the *lacZ* gene. By comparison, a recircularized vector has a functional *lacZ* gene. The functionality of *lacZ* can be determined by providing the growth medium with a colorless compound, X-Gal, which is cleaved by β-galactosidase into a blue dye. Bacteria grown in the presence of X-Gal will form blue colonies if they have a functional β-galactosidase enzyme, and white colonies if they do not. In this experiment, therefore, bacterial colonies containing recircularized vectors will form blue colonies, while colonies containing recombinant vectors will be white.

The net result of gene cloning is an enormous number of gene copies within a huge number of recombinant vectors. During transformation, a single bacterial cell usually takes up a single recombinant vector. Two subsequent events lead to the production of many copies of the cloned gene. First, because the vector has a highly active origin of replication, the bacterial host cell produces many copies of the recombinant vector per cell. Second, the bacterial cells divide approximately every 20 minutes. Following overnight growth, a population of many millions of bacteria will be obtained from a single cell. Each of these bacterial cells will contain many copies of the cloned gene. For example, a bacterial colony may comprise 10 million cells, with each cell containing 50 copies of the recombinant vector. Therefore, this bacterial colony would have 500 million copies of the cloned gene!

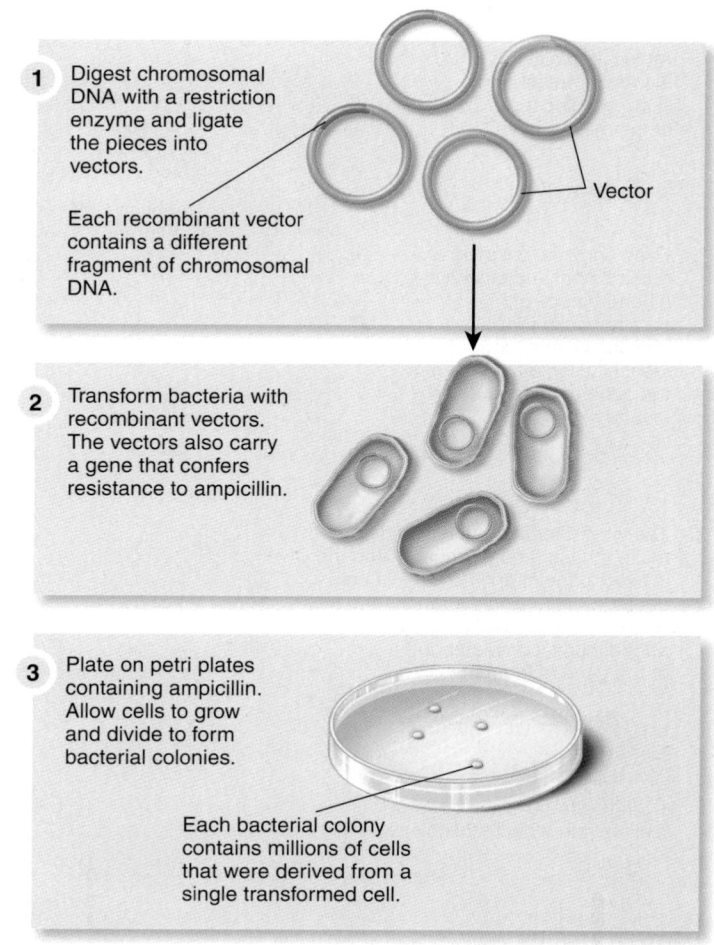

1 Digest chromosomal DNA with a restriction enzyme and ligate the pieces into vectors.

Each recombinant vector contains a different fragment of chromosomal DNA.

Vector

2 Transform bacteria with recombinant vectors. The vectors also carry a gene that confers resistance to ampicillin.

3 Plate on petri plates containing ampicillin. Allow cells to grow and divide to form bacterial colonies.

Each bacterial colony contains millions of cells that were derived from a single transformed cell.

Figure 20.4   A DNA library.   Each colony in a DNA library contains a vector with a different piece of chromosomal DNA.

A Cloned Gene of Interest Can Be Identified in a DNA Library with the Use of Colony Hybridization

In our previous discussion of gene cloning, we focused on the insertion of a particular gene into a vector, and its introduction into a host cell. But how do we know that a vector contains a gene of interest? In a typical cloning experiment, such as the one described in Figures 20.2 and 20.3, the treatment of chromosomal DNA with restriction enzymes actually yields tens of thousands of different DNA fragments. Therefore, after the DNA fragments are ligated individually to vectors, a researcher has a collection of many recombinant vectors, with each vector containing a particular fragment of chromosomal DNA. This collection of vectors is known as a **DNA library** (**Figure 20.4**). Researchers make DNA libraries using the methods shown in Figures 20.2 and 20.3 and then use those libraries to obtain clones of genes in which they are interested.

Two types of DNA libraries are commonly made. The library is called a **genomic library** when the inserts are derived from chromosomal DNA. Alternatively, researchers can isolate mRNA

and use the enzyme reverse transcriptase, which is described in Chapter 18, to make DNA molecules using mRNA as a starting material. Such DNA is called **complementary DNA** or **cDNA**. A **cDNA library** is a collection of recombinant vectors that have cDNA inserts. From a research perspective, an important advantage of cDNA is that it lacks introns. Because introns can be quite large, it is much simpler for researchers to insert cDNAs into vectors rather than work from chromosomal DNA if they want to focus their attention on the coding sequence of a gene.

In many cloning experiments, the ultimate goal is to clone a specific gene. After making a DNA library, how is it possible to identify those rare colonies that happen to contain the gene of interest? One method that can be used is **colony hybridization**, in which a researcher uses a probe to identify colonies that contain a desired gene. The master plate shown at the top of **Figure 20.5** has many bacterial colonies, each composed of bacterial cells containing a recombinant vector with a different piece of chromosomal DNA. In this example, the goal is to identify a colony that contains the gene of interest.

1 A filter is gently laid onto the master plate and lifted, yielding a replica of the master plate.

Master plate with a DNA library

Each bacterial colony has a recombinant vector with a different piece of chromosomal DNA. The goal in this experiment is to identify colonies that have the gene of interest as an insert.

Nitrocellulose filter

2 The lifted filter is treated with detergent to make the bacteria permeable and the DNA is fixed to the filter. The DNA is then denatured into single strands, and a radioactively labeled probe is added that is complementary to the gene of interest.

Radiolabeled probe
5' ⌐ 3'
3' ⌐⌐⌐⌐⌐⌐⌐⌐⌐⌐⌐⌐ 5'
Single strand of the gene of interest in a bacterial colony

3 After allowing time for the probe to hybridize with the desired gene, the filter is washed to remove unbound probe and then placed next to X-ray film.

X-ray film

4 Based on the orientation of the filter and X-ray film (the X on both mark where they match), the colonies containing the gene of interest are identified on the master plate.

Master plate (see above)

Colonies containing the cloned gene of interest

Figure 20.5 **Identifying a cloned gene by colony hybridization.** This method can be used to identify colonies containing recombinant vectors produced by a gene cloning experiment.

A nitrocellulose filter is laid gently onto the master plate containing many bacterial colonies. When the filter is lifted, some cells from each colony are attached to it. In this way, the filter paper contains a replica of the colonies on the master plate.

The cells attached to the filter are exposed to a detergent, which makes them permeable, and the DNA inside the cells is fixed to the filter. The DNA is denatured into single strands and the filter is then submerged in a solution containing a radiolabeled DNA probe. In this case, the probe is single-stranded DNA that is complementary to the gene of interest. A probe that is 15 to 20 nucleotides long is usually of sufficient length to specifically bind to a gene. The probe is given time to hybridize to the DNA on the filter. If a bacterial colony contains the desired gene, the probe will hybridize to the DNA in this colony. Most bacterial colonies are not expected to contain this gene. The unbound probe is then washed away, and the filter is placed next to X-ray film. If the DNA within a bacterial colony did hybridize to the probe, a dark spot will appear on the developed film in the corresponding location. Therefore, a dark spot identifies a bacterial colony that contains the desired gene. Following the identification of labeled colonies, the researcher can go back to the master plate to select and grow bacteria containing the cloned gene. The growth of cells will produce many copies of the cloned gene.

GENOMES & PROTEOMES

Blotting Methods Can Identify Gene Families

As we have just seen, blotting methods such as colony hybridization can allow us to detect a gene within a DNA library. Another blotting method, **Southern blotting**, can detect the presence of a particular gene within a mixture of many chromosomal DNA fragments separated on a gel. This method, developed by Edwin Southern in 1975, can be used to determine the number of copies of a gene within the genome of an organism. For example, Southern blotting has revealed that rRNA genes are found in multiple copies, whereas structural genes are usually unique. In the study of human genetic diseases, Southern blotting can also detect small gene deletions. Another common use of Southern blotting is to identify gene families, in which two or more genes are derived from the same ancestral gene. The members of a gene family are homologous, having similar but not identical DNA sequences. Southern blotting can distinguish the homologous members of a gene family within a single species or identify homologous genes among different species.

Before we discuss the Southern blotting method, we first need to examine **gel electrophoresis**, a technique that is used to separate macromolecules on a gel. In the example shown in **Figure 20.6**, gel electrophoresis was used to separate different fragments of chromosomal DNA based on their masses. It consists of a flat, semisolid gel called a slab gel with depressions at

1 Load samples of DNA fragments into wells.

Samples

2 Apply an electric field.

3 Wait additional time.

Higher-mass molecules

Each band is a group of DNA fragments with the same mass.

Lower-mass molecules

Figure 20.6 **Separation of molecules by gel electrophoresis.** In this example, samples containing many fragments of DNA are loaded into wells at the top of the gel and then subjected to an electric field that causes the fragments to move toward the bottom of the gel. This separates the fragments according to their masses, with the smaller DNA fragments near the bottom of the gel.

Biological inquiry: One DNA fragment contains 600 bp and another has 1,300 bp. Following electrophoresis, which would be closer to the bottom of a gel?

the top called wells where samples are added. An electric field is applied to the gel, causing charged molecules to migrate from the top of the gel toward the bottom—a process called electrophoresis. DNA is negatively charged and moves toward the positive end of the gel, which is at the bottom in this figure.

Smaller DNA fragments move more quickly through the gel polymer and therefore are located closer to the bottom of the gel compared to larger ones. After the slab gel has run, the fragments are separated into bands within the gel. The fragments in each band can then be stained with a dye for identification. Alternatively, in a Southern blotting experiment, specific bands in a gel are identified with a probe, as described next.

To conduct a Southern blotting experiment, a strand of DNA from a gene of interest is labeled (for example, radiolabeled) *in vitro*, and then the labeled strand is used as a probe to detect the presence of the gene within a mixture of many DNA fragments obtained from chromosomal DNA. The basis for a Southern blotting experiment, as with other hybridization experiments, is that two DNA fragments will bind to each other only if they have complementary sequences. In such an experiment, a labeled strand of the probe will pair specifically with a complementary DNA strand from a chromosomal fragment.

To begin a Southern blotting experiment, the chromosomal DNA is isolated and digested with a restriction enzyme (**Figure 20.7**). Because the restriction enzyme cuts the chromosomal DNA at many different sites within the chromosomes, this step produces thousands of DNA pieces of different sizes. The chromosomal pieces are loaded onto a gel that separates them according to their masses. The DNA pieces within the gel are denatured into single-stranded molecules and then transferred from the gel to a nitrocellulose filter, a process called blotting. At this point, the filter contains many unlabeled DNA strands that have been separated according to mass.

The next step is to determine if any of these unlabeled strands contain sequences complementary to a probe. The filter, which has the unlabeled chromosomal DNA attached to it, is submerged in a solution containing the radiolabeled probe. If the labeled probe and a strand of chromosomal DNA are complementary, they will hydrogen bond to each other. Any unbound radiolabeled probe is then washed away, and the filter is exposed to X-ray film. Locations where the labeled probe has bound appear as dark bands on the developed X-ray film.

Important variables in the Southern blotting procedure are temperature and the ionic strength of the hybridization and wash steps. If these steps are done at very high temperatures and/or at low salt concentrations, the probe and chromosomal DNA must be very complementary—nearly a perfect match—to hybridize. This condition is called high stringency and is used to detect a close match between the probe and a chromosomal DNA fragment. If the temperature is lower and/or the ionic strength is higher, DNA sequences that are similar but not necessarily identical may hybridize to the probe. This is called low stringency and can be used to detect homologous genes. In the results shown in Figure 20.7, conditions of high stringency reveal that the probe, which is a strand of DNA from a gene of interest, binds to a single chromosomal DNA fragment, indicating that the gene is found in a single copy in the genome. At low stringency, however, two other bands are detected. These results suggest that this gene is a member of a gene family composed of three distinct members.

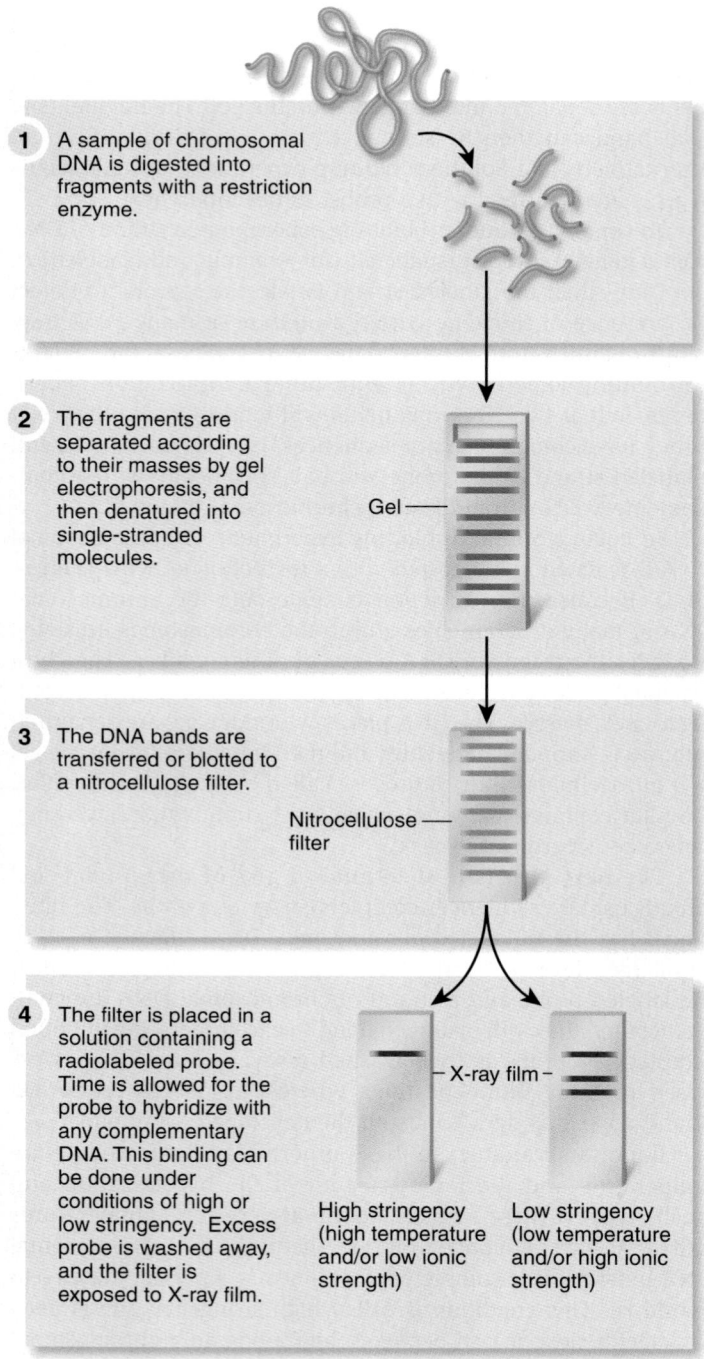

1 A sample of chromosomal DNA is digested into fragments with a restriction enzyme.

2 The fragments are separated according to their masses by gel electrophoresis, and then denatured into single-stranded molecules.

Gel

3 The DNA bands are transferred or blotted to a nitrocellulose filter.

Nitrocellulose filter

4 The filter is placed in a solution containing a radiolabeled probe. Time is allowed for the probe to hybridize with any complementary DNA. This binding can be done under conditions of high or low stringency. Excess probe is washed away, and the filter is exposed to X-ray film.

X-ray film

High stringency (high temperature and/or low ionic strength)

Low stringency (low temperature and/or high ionic strength)

Figure 20.7 **The technique of Southern blotting.** This technique is used to identify genes that are complementary to a labeled DNA probe from a known gene. Under conditions of low stringency, the probe may bind to genes that are similar but not identical to the probe, thereby identifying homologous genes that constitute a gene family.

Biological inquiry: The globin gene family is described in Chapter 21 (look ahead to Figure 21.8). Let's suppose you started with human chromosomal DNA and a probe that is complementary to the β-globin gene. How many bands would you see under conditions of high stringency? How many under low stringency?

Polymerase Chain Reaction (PCR) Can Also Be Used to Make Many Copies of DNA

Thus far we have examined cloning methods in which the gene of interest was inserted into a vector and then introduced into a host cell. Another way to copy DNA without the aid of vectors and host cells is a technique called **polymerase chain reaction (PCR)** that was developed by Kary Mullis in 1985 (**Figure 20.8**). The goal of PCR is to make many copies of DNA in a defined region, perhaps encompassing a gene or part of a gene. Several reagents are required for synthesis of the DNA. These include a high concentration of two primers that are complementary to sequences at the ends of the DNA region to be amplified, deoxynucleoside triphosphates (dNTPs), and a heat-stable form of DNA polymerase called *Taq* **polymerase** (isolated from the bacterium *Thermus aquaticus*, which lives in hot springs). A heat-stable form of DNA polymerase is necessary because PCR involves heating steps that would inactivate most other natural forms of DNA polymerase.

To make copies, a sample of chromosomal DNA, called the double-stranded template DNA, is denatured into single-stranded molecules by heat treatment and then the primers bind to the DNA as the temperature is lowered. The binding of the primers to the specific sites in the template DNA is called annealing. Once the primers have annealed, *Taq* polymerase will catalyze the synthesis of complementary DNA strands, thereby doubling the amount of DNA in the region that is flanked by the primers. The sequential process of denaturation—annealing—synthesis is then repeated to double the amount of template DNA many times over. This method is called a chain reaction because the products of each previous step are used as reactants in subsequent steps.

A thermal reactor that automates the timing of each cycle, known as a thermocycler, is used to carry out PCR. The experimenter mixes the DNA sample, an excess amount of primers, *Taq* polymerase, and deoxynucleotides together in a single tube. Then the experimenter places the tube in a thermocycler, and sets the machine to operate within a defined temperature range and number of cycles. During every cycle, the thermocycler increases the temperature to denature the DNA strands and then lowers the temperature to allow annealing and DNA synthesis to occur. After a few minutes, the cycle is repeated by increasing and then lowering the temperature. A typical PCR run is likely to involve 20 to 30 cycles of replication and takes a few hours to complete. The PCR technique can amplify the amount of DNA by a staggering amount. After 30 cycles of amplification, a DNA sample will increase 2^{30}-fold, which is approximately a billionfold!

The PCR reaction shown in Figure 20.8 seeks to amplify a particular DNA segment. The sequences of the PCR primers are complementary to two specific sequences within the DNA. Therefore, the two primers bind to these sites and the intervening region is replicated. To conduct this type of PCR experiment, the researcher must have prior knowledge about the sequence of the DNA in order to design primers that are complementary to the ends of the DNA sequence. When specific primers can be constructed, PCR can amplify a specific region of DNA from a

Primer annealing site

5′ ——————————— 3′

T G C A C C A G C A T C C G A T C
A C G T G G T C G T A G G C T A G

3′ ——————————— 5′

Primer

1 Start with double-stranded template DNA, present in low amounts, plus dNTPs, *Taq* polymerase, and 2 primers that are present in high amounts and complementary to the ends of the region that will be amplified.

Region that will be amplified

2 During each cycle, the template DNA strands are separated via heating. The temperature is then lowered to allow the primers to bind and a complementary strand to be made.

Cycle 1

3 With each succesive cycle, the relative amount of DNA fragments that end exactly at both primer sites (marked ∗) increases. Therefore, after many cycles, the vast majority of DNA fragments only contain the region that is flanked by the 2 primer sites.

Cycle 2   Cycle 3

Figure 20.8 **Polymerase chain reaction (PCR).** During each PCR cycle, short DNA sequences that are complementary to the ends of the targeted DNA sequence bind to the DNA and act as primers for the synthesis of this DNA region (see small illustrations next to arrows in step 2). The primers used in actual PCR experiments are usually 15 to 20 nucleotides in length. The region between the two primers is typically hundreds of nucleotides in length, not just several nucleotides as shown here. The net result of PCR is the synthesis of many copies of DNA in the region that is flanked by the two primers.

complex mixture of DNA or from a small amount of DNA. For example, PCR is routinely used in crime investigations to amplify DNA from small samples of cells that may have been left behind at a crime scene.

20.2 Genomics and Proteomics

As discussed throughout Unit III, the genome is the total genetic composition of an organism. As genetic technology has progressed over the past few decades, researchers have gained an increasing ability to analyze the composition of genomes as a whole unit. The term **genomics** refers to the molecular analysis of the entire genome of a species. The two main types of genomic research are structural and functional. **Structural genomics** is aimed at the direct analysis of the DNA itself. Segments of chromosomes are cloned and analyzed in progressively smaller pieces, the locations of which are known on the intact chromosomes. This is the mapping phase of structural genomics. The mapping of a genome ultimately progresses to the determination of the complete DNA sequence, which provides the most detailed description available of an organism's genome at the

molecular level. By comparison, **functional genomics** is aimed at studying the expression of a genome. For example, functional genomics can be used to analyze which genes are turned on or off in particular cell types, such as nerve and muscle cells. In this section, we will consider a few of the methods that are used in structural and functional genomics.

A long-term goal of researchers is to determine the roles of all cellular proteins, as well as how those proteins interact to produce the characteristics of particular cell types, and the traits of complete organisms. As discussed at the end of this section, **proteomics** involves techniques that are used to identify and study groups of proteins.

BAC Cloning Vectors Are Used to Make Contigs of Chromosomes to Map a Genome

A goal of structural genomics is to clone and analyze the entire genome of a species. For large eukaryotic genomes, cloning an entire genome is much easier when a cloning vector can accept very large chromosomal DNA inserts. In general, most plasmid and viral vectors can accommodate inserts only a few thousand to perhaps tens of thousands of nucleotides in length.

If a plasmid or viral vector has a DNA insert that is too large, it will have difficulty with DNA replication and is likely to suffer deletions in the insert. By comparison, a type of cloning vector known as a **bacterial artificial chromosome (BAC)** can reliably contain much larger inserted DNA fragments. BACs are derived from large plasmids called F factors (see Chapter 18). They can typically contain inserts up to 500,000 bp. BACs are used in genomic research in the same way as other types of vectors. Similarly, YACs, yeast artificial chromosomes, are used as vectors in yeast. An insert in a YAC can be several hundred thousand to perhaps 2 million nucleotides in length.

The term **mapping** refers to the process of determining the relative locations of genes or other DNA segments along a chromosome. After many large fragments of chromosomal DNA have been inserted into BACs or YACs, the first step of mapping is to determine the relative locations of the inserted chromosomal pieces as they would occur in an intact chromosome. This is called physical mapping. To obtain a complete physical map of a chromosome, researchers need a series of clones that contain overlapping pieces of chromosomal DNA. Such a collection of clones, known as a **contig**, contains a <u>contig</u>uous region of a chromosome that is found as overlapping regions within a group of vectors (**Figure 20.9**). These overlapping regions allow researchers to identify their order along the chromosome.

The Dideoxy Chain-Termination Method Is Used to Determine the Base Sequence of DNA

Once researchers have cloned DNA into vectors and obtained a physical map, the next phase of genomic research is **DNA sequencing**, which is a method to determine the base sequence of DNA. Scientists can learn a great deal of information about the function of a gene if its nucleotide sequence is known. For example, the investigation of genetic sequences has been vital in our understanding of the genetic basis of human diseases.

During the 1970s, two methods for DNA sequencing were devised. One method, developed by Alan Maxam and Walter Gilbert, involves the base-specific cleavage of DNA. Another method, developed by Frederick Sanger and colleagues, is known as the **dideoxy chain-termination method** or more simply, **dideoxy sequencing**. Because it has become the more popular method of DNA sequencing, we will consider the dideoxy method here.

The dideoxy procedure of DNA sequencing is based on our knowledge of DNA replication. As described in Chapter 11, DNA polymerase connects adjacent deoxynucleotides by catalyzing a covalent linkage between the 5′–phosphate on one nucleotide and the 3′—OH group on the previous nucleotide. Chemists, however, can synthesize nucleotides, called dideoxynucleoside triphosphates (ddNTPs), that are missing the —OH group at the 3′ position (**Figure 20.10**, step 1). Note: The prefix "dideoxy-" refers to the fact that there are two (di) removed (de) oxygens (oxy) compared with ribose, which has —OH groups at both the 2′ and 3′ positions. Sanger reasoned that if a dideoxynucleotide is added to a growing DNA strand, the strand can no longer grow because the 3′—OH group is missing, which results in chain termination.

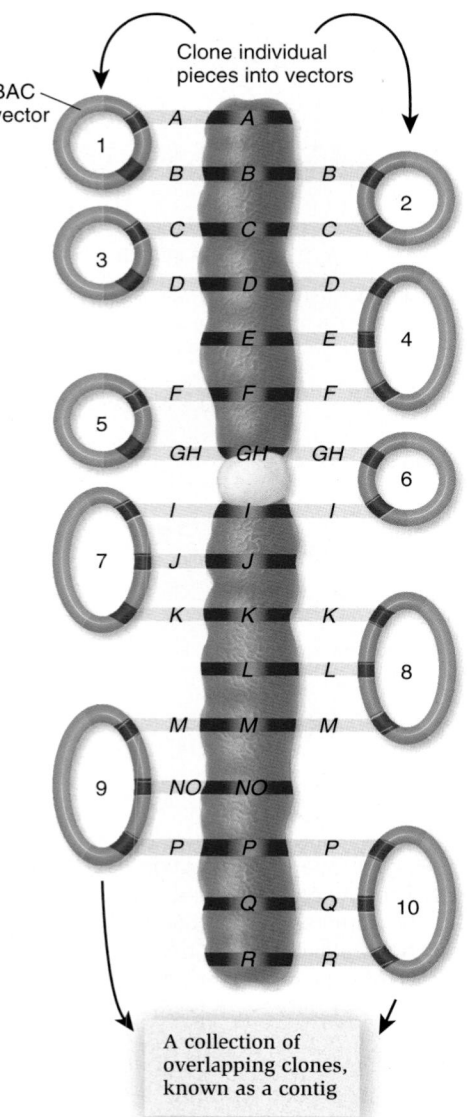

Figure 20.9 **A contig.** As shown here, a contig is a collection of clones that have overlapping pieces of DNA from a particular chromosome. The numbers denote the order of the members of the contig. The chromosome is labeled with letters that denote the locations of particular genes. The members of the contig have overlapping regions that have the same genes, which allows you to order them.

Biological inquiry: What does it mean when we say that two members of a contig have overlapping regions?

Before describing the steps of this DNA sequencing protocol, we need to become acquainted with the DNA segments that are used in a sequencing experiment. The segment of DNA to be sequenced must be obtained in large amounts by using the gene cloning or PCR techniques that were described earlier in this chapter. In Figure 20.10, the segment of DNA to be sequenced was inserted into a vector next to a primer-annealing site, the site where the primer will bind.

1 The following reagants are added to each of 4 tubes:
- Many copies of target DNA
- Primers
- DNA polymerase
- All 4 deoxynucleotides (radiolabeled)
- One of the 4 dideoxynucleotides

Guanine (G)

2′, 3′- dideoxyguanosine triphosphate

Sequence to be analyzed (target DNA)

Primer

3′ G T G G C A T T C C T G A C

5′

1 2 3 4 5 6 7 8 9 10 11 12 13 14

Annealing site

Vector

+ddGTP   +ddATP   +ddTTP   +ddCTP

2 Incubate samples to allow the synthesis of DNA strands. On occasion, a dideoxynucleotide is incorporated instead of the normal deoxynucleotide. This causes DNA synthesis to terminate at different positions. Run samples on a gel and expose gel to X-ray film.

Number of nucleotides attached to the primer

14 13 12 11 10 9 8 7 6 5 4 3 2 1

Sequence of newly made DNA, which is complementary to target DNA sequence in the vector

3′ G T C A G G A A T G C C A C 5′

(5) CACCddG
(9) CACCGTAAddG
(10) CACCGTAAGddG
(14) CACCGTAAGGACTddG

(2) CddA
(7) CACCGTddA
(8) CACCGTAddA
(11) CACCGTAAGGddA

(6) CACCGddT
(13) CACCGTAAGGACddT

(1) ddC
(3) CAddC
(4) CACddC
(12) CACCGTAAGGAddC

(a) The procedure used in traditional dideoxy sequencing

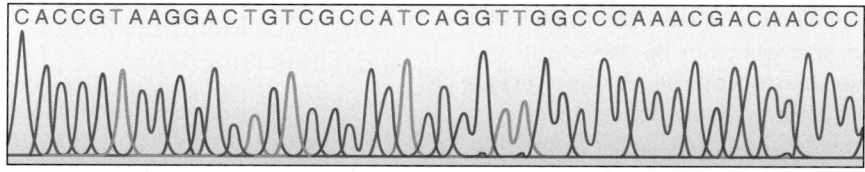

C A C C G T A A G G A C T G T C G C C A T C A G G T T G G C C C A A A C G A C A A C C C

(b) Output from automated dideoxy sequencing

Figure 20.10 **DNA sequencing by the dideoxy method.** (a) Traditional dideoxy sequencing used a mixture of radiolabeled deoxy and nonlabeled dideoxynucleoside triphosphates. Following electrophoresis, the bands were detected by autoradiography. Step 1 shows the structure of dideoxyguanosine triphosphate, abbreviated ddGTP. It has a hydrogen, shown in red, instead of a hydroxyl group. (b) Automated sequencing uses a fluorescence detector that measures the four kinds of dideoxynucleotides as they emerge from the gel.

Let's now examine the steps involved in DNA sequencing. The DNA containing the sequence to be analyzed is actually double stranded. Figure 20.10a, step 1, shows the DNA after it has been denatured into a single strand by heat treatment. Many copies of this single-stranded DNA are placed into four tubes and mixed with primers that bind to the primer-annealing site. All four types of regular deoxynucleotides and DNA polymerase are also added to each tube. Finally, each of the four tubes has a low concentration of just one of the four possible dideoxynucleotides (ddGTP, ddATP, ddTTP, or ddCTP). The tubes are then incubated to allow DNA polymerase to make strands complementary to one strand of the target DNA sequence. However, the dideoxynucleotides will occasionally cause DNA synthesis to terminate early. For example, let's consider the third tube, which contains ddTTP. Synthesis of new DNA strands will occasionally stop at the sixth or thirteenth position after the annealing site if a ddTTP, instead of a dTTP, is incorporated into the growing DNA strand. The target DNA has a complementary A at the sixth and thirteenth positions.

Within the four tubes, mixtures of DNA strands of different lengths are made. These DNA strands are separated according to their lengths by subjecting them to gel electrophoresis. The shorter strands move to the bottom of the gel more quickly than do the longer strands. To detect the newly made DNA strands, the deoxynucleotides that were added to each reaction are radiolabeled. This enables the strands to be visualized as bands when the gel is exposed to X-ray film. In Figure 20.10a, the DNA strands in the four tubes were run in separate lanes on a gel. Because we know which dideoxynucleotide was added to each tube, we also know which base is at the very end of each DNA strand separated on this gel, because dideoxynucleotides cause chain termination. Therefore, we can read the DNA sequence by reading which base is at the end of every DNA strand and matching this sequence with the length of the strand.

Dideoxy sequencing can now be done much more quickly with automated sequencing. Instead of having four separate tubes with a single type of dideoxynucleotide in each tube, automated sequencing uses one tube containing all four dideoxynucleotides. Each dideoxynucleotide (ddGTP, ddATP, ddTTP, and ddCTP) has a different-colored fluorescent label attached. After incubating the template DNA with deoxynucleotides, the four types of fluorescent dideoxynucleotides, and DNA polymerase, the sample is then loaded into a single lane of a gel and the fragments are separated by electrophoresis. Electrophoresis is continued until each band emerges from the bottom of the gel, where a laser excites the fluorescent dye. A fluorescence detector records the amount of fluorescence emission at four wavelengths, corresponding to the four dyes. An example of a printout from a fluorescence detector is shown in Figure 20.10b. The peaks of fluorescence correspond to the DNA sequence that is complementary to the target DNA. The intensity of the fluorescent peaks is not always the same because dideoxynucleotides get incorporated at certain sites more readily than at other sites.

A Microarray Can Identify Which Genes Are Transcribed by a Cell

Let's now turn our attention to functional genomics. Researchers have developed an exciting new technology, called **DNA microarrays** (or gene chips), that is used to monitor the expression of thousands of genes simultaneously. A DNA microarray is a small silica, glass, or plastic slide that is dotted with many different sequences of single-stranded DNA, each corresponding to a short sequence within a known gene. Each spot contains multiple copies of a specific DNA sequence. For example, one spot in a microarray may correspond to a sequence within the β-globin gene, while another spot could correspond to a different gene, such as a gene that encodes an iron transporter. A single slide contains tens of thousands of different spots in an area the size of a postage stamp. The DNA sequences of each spot is known. These microarrays are typically produced using spotting technologies that are quite similar to the way that an inkjet printer works.

A DNA microarray is used as a hybridization tool. In the experiment shown in **Figure 20.11**, mRNA was isolated from a sample of cells and then used to make fluorescently labeled cDNA. The labeled cDNAs were then incubated with a DNA microarray. The DNA in the microarray is single stranded and corresponds to the sense strand—the strand that has a sequence like mRNA. Those cDNAs that are complementary to the DNAs in the microarray will hybridize and thereby remain bound to the microarray. The array is then washed and placed in a microscope equipped with a computer that scans the fluorescence intensity for each spot. If the fluorescence intensity in a spot is high, there was a large amount of cDNA in the sample that hybridized to the DNA at this location. For example, if the β-globin gene was expressed in the cells being tested, a large amount of cDNA for this gene would be made, and the fluorescence intensity for that spot would be high. Because the DNA sequence of each spot is already known, a fluorescent spot identifies cDNAs that are complementary to those DNA sequences. Furthermore, because the cDNA was generated from mRNA, this technique identifies genes that have been transcribed in a particular cell type under a given set of conditions. Thus far, the most common use of microarrays is to study gene expression patterns. In addition, the technology of DNA microarrays has found many other important uses (**Table 20.2**).

Proteomics Identifies Cellular Proteins and Their Functions

Thus far we have considered ways to characterize the genome of a given species and to study its function. Because most genes encode proteins, a logical next step is to examine the functional roles of the proteins that a species can make. As previously mentioned, this technical approach is called proteomics, and the entire collection of a species' proteins is its proteome. As we move into the 21st century, a key challenge facing molecular biologists is the study of proteomes. Much like genomic research, this study

1. Isolate mRNA from cells of interest. Add reverse transcriptase along with fluorescent nucleotides.

In this example, the cells make 3 different mRNAs labeled A, D, and F.

DNA microarray

Each spot contains single-stranded DNA molecules that correspond to a short sequence of a particular gene.

This process produces fluorescently labeled cDNA that is complementary to the mRNA.

2. Hybridize cDNAs to the microarray, and wash away any unbound cDNAs.

3. View hybridized fluorescent DNA on microarray with a scanning fluorescence microscope.

Figure 20.11 **Identifying transcribed genes within a DNA microarray.** In this simplified example, only three cDNAs specifically hybridize to spots on the microarray. Those genes were expressed in the cells from which the mRNA was isolated. In an actual experiment, there are typically hundreds or thousands of different cDNAs and tens of thousands of different spots on the array.

Table 20.2	Applications of DNA Microarrays
Application	**Description**
Cell-specific gene expression	A comparison of microarray data using cDNAs derived from mRNA of different cell types can identify genes that are expressed in a cell-specific manner.
Gene regulation	Environmental conditions play an important role in gene regulation. A comparison of microarray data using cDNA derived from mRNA isolated from cells exposed to two different environmental conditions may reveal genes that are induced under one set of conditions and repressed under another set of conditions.
Elucidation of metabolic pathways	Genes that encode proteins that participate in a common metabolic pathway are often expressed in a parallel manner and can be revealed from a microarray analysis.
Tumor profiling	Different types of cancer cells exhibit striking differences in their profiles of gene expression, which can be revealed by a DNA microarray analysis. This approach is gaining widespread use to classify tumors that are sometimes morphologically indistinguishable.
Genetic variation	A mutant allele may not hybridize to a spot on a microarray as well as a wild-type allele. Therefore, microarrays are gaining widespread use as a tool to detect genetic variation. This application has been used to identify disease-causing alleles in humans, and to elucidate mutations that contribute to quantitative traits in plants and other species.
Microbial strain identification	Microarrays can distinguish between closely related bacterial species and subspecies.

One major goal of researchers in the field of proteomics is to identify and functionally characterize all the proteins that a cell type will make. Because cells produce thousands of different proteins, this is a daunting task. Nevertheless, along with genomic research, the past decade has seen important advances in our ability to isolate and identify cellular proteins.

An important technique in the field of proteomics is **two-dimensional (2D) gel electrophoresis**, which is a technique that can separate hundreds or even thousands of different proteins within a cell extract (**Figure 20.12**). To begin this procedure, an experimenter loads a sample of cellular proteins onto the top of a gel that is shaped like a tube. This tube gel separates proteins according to their net charge at a given pH. A protein migrates to the point in the gel where its net charge is zero, a process termed isoelectric focusing. After the tube gel has run, the experimenter places it on top of a slab gel that contains a detergent called sodium dodecyl sulfate (SDS). The SDS is negatively charged and coats each protein uniformly. During this process, the proteins lose their three-dimensional structure or become denatured. The proteins in the slab gel are separated according to their molecular masses. The end result is a collection of spots, each spot corresponding to a unique cellular protein.

will require the collective contributions of many research scientists, as well as improvements in technologies that are aimed at explaining the complexities of the proteome.

Due to gene regulation, any given cell will produce only a subset of the proteins that are found in the proteome of a species. For example, the human genome has approximately 20,000 to 25,000 different genes, yet a human red blood cell makes less than half that number of different proteins. The subset of proteins that a cell makes depends primarily on what type of cell it is, its stage of development, and its environmental conditions.

Figure 20.12 **Two-dimensional gel electrophoresis.**
The technique involves two electrophoresis steps. First, a mixture of proteins is separated on an isoelectric focusing gel that has the shape of a tube. Proteins migrate to the point where their net charge is zero. This tube gel is placed into a long well on top of an SDS-polyacrylamide gel. This second gel separates the proteins according to their masses. In this diagram, only a few spots are seen, but an actual experiment would involve a mixture of hundreds or thousands of different proteins.

Two-dimensional gel electrophoresis is used as a method to separate a mixture of cellular proteins. The next step is to identify those proteins, based on the concept that each protein has its own unique amino acid sequence. To identify a protein, a spot on a 2D gel can be cut out of the gel to obtain a tiny amount of the protein within the spot. In essence, the 2D gel procedure purifies a small amount of the cellular protein of interest. A protein in a given spot can then be identified via a technique called **mass spectrometry** that allows researchers to determine the amino acid sequence of short regions within a protein. The technique determines the mass of the peptide fragments produced by digesting a purified protein with an enzyme called a protease that cuts the protein into small peptide fragments. The peptides are mixed with an organic acid and dried onto a metal slide, and then the sample is struck with a laser. This causes the peptides to temporarily vaporize in the form of an ionized gas, in which the peptide contains one or more positive charges. The charged gaseous peptides are then accelerated by an electric field toward a detector. The time of flight to the detector is determined

Figure 20.13 **Tandem mass spectrometry.** This technique is used to determine the amino acid sequence of a peptide.

by their mass and net charge. A measurement of the flight time provides an extremely accurate way to measure the mass of a peptide.

To determine the amino acid sequence of a given peptide, two mass spectrometers are used, a method called tandem mass spectrometry (**Figure 20.13**). The first spectrometer measures the mass of a given peptide, and then this same peptide is analyzed by a second spectrometer after the peptide has been digested into smaller fragments. The differences in the masses of the peaks in the second spectrum reveal the amino acid sequence of the peptide, because the masses of all 20 amino acids are known. For example, suppose that a peptide had a mass of 1,652 Daltons. If one amino acid at the end were removed, and the smaller peptide had a mass that was 87 Daltons less (that is, 1,565 Daltons), this would indicate that a serine is at one end of the peptide because the mass of serine within a polypeptide chain is 87 Daltons. If two amino acids were removed at one end and the mass was 224 Daltons less, this would correspond to the removal of one serine (87 Daltons) and one histidine (137 Daltons). Thus, from these measurements, we would conclude that the amino acid sequence from one end of the peptide was serine-histidine.

After a researcher has obtained a few short peptide sequences from a given protein, genomic information can readily predict the entire amino acid sequence of the protein. For example, if a peptide had the sequence serine-histidine-leucine-asparagine-serine-asparagine, a researcher could determine the possible codon sequences that could encode such a peptide. More than one sequence is possible due to the degeneracy of the genetic code. Using computer software described in Chapter 21, the codon sequences would be used to scan the entire genomic sequence if it has been determined in a genome sequencing project. For example, the DNA sequence of the entire human genome has already been determined. Computer software can locate a match between a predicted codon sequence and a specific gene within the human genome. In this way, mass spectrometry makes it possible to identify the gene that encodes the entire protein. The gene sequence, in turn, can be used to predict the amino acid sequence of the whole protein.

20.3 Biotechnology

As mentioned at the beginning of this chapter, biotechnology is defined as technologies that involve the use of living organisms, or products from living organisms, as a way to benefit humans. Biotechnology is not a new topic. Its use began about 12,000 years ago when humans began to domesticate animals and plants for the production of food. Since that time, many species of microorganisms, plants, and animals have become routinely used by people. More recently, the term biotechnology has become associated with molecular genetics. Since the 1970s, molecular genetic tools have provided new ways to make use of living organisms to benefit humans.

In this section, we will consider genetic techniques to alter the characteristics of microorganisms, plants, and animals. Examples include bacteria that can degrade pollutants in the environment and livestock that make human hormones. We will also examine several topics that you often hear about in the news. We will explore the technical side of mammalian cloning and also touch upon the ethical issues. Another newsworthy topic is DNA fingerprinting, a common method used to analyze the DNA from an individual. Like conventional fingerprinting,

this technique can be used for identification purposes, as in a crime investigation. Finally, we will learn how certain inherited diseases may be treated via gene therapy. In this relatively new approach, cloned genes are introduced into individuals with genetic diseases in an attempt to compensate for mutant genes.

Many Important Medicines Are Produced by Recombinant Microorganisms

In 1982, the U.S. Food and Drug Administration approved the sale of human insulin made by recombinant bacteria. In healthy individuals, insulin is produced by the beta cells of the pancreas. Insulin functions to regulate several physiological processes, particularly the uptake of glucose into fat and muscle cells. Persons with insulin-dependent diabetes cannot synthesize an adequate amount of insulin due to a defect in their beta cells. Today, people with this disorder are usually treated with human insulin that is made by genetically engineered bacteria. Prior to 1982, insulin was isolated from pancreases removed from cattle. Unfortunately, in some cases, diabetic individuals became allergic to cow insulin. These allergic patients had to use expensive combinations of insulin from human cadavers and other animals. Now, of course, they can use human insulin made by recombinant bacteria.

As shown in **Figure 20.14**, insulin is a hormone composed of two polypeptides, called the A and B chains. To make this hormone using bacteria, the coding sequence of the A or B chains is inserted into a plasmid vector next to the coding sequence of the E. coli protein, β-galactosidase. After transforming bacteria with such vectors, the cells produce many copies of a fusion protein comprising β-galactosidase and the A or B chain. This step is necessary because the A and B chains are rapidly degraded when expressed in bacterial cells by themselves; the fusion proteins, however, are not. These fusion proteins are then extracted from the bacterial cells and treated with a chemical, cyanogen bromide (CNBr), that cleaves after methionine to separate β-galactosidase from the A or B chain. The A and B chains are purified and mixed together under conditions in which they will fold and associate with each other to form a functional insulin hormone molecule.

Microorganisms Can Reduce Environmental Pollutants

Bioremediation is the use of microorganisms to decrease pollutants in the environment. As its name suggests, it is a biological remedy for pollution. During bioremediation, enzymes produced by a microorganism modify a toxic pollutant by altering or transforming its structure, an event called biotransformation. In many cases, biotransformation results in biodegradation, in which the toxic pollutant is degraded, yielding less complex, nontoxic metabolites.

Since the early 1900s, people have intentionally used microorganisms in the treatment and degradation of sewage. More recently, the field of bioremediation has expanded into the treatment of hazardous and refractory chemical wastes (that is,

1 Transform recombinant plasmids containing insulin A or B chain sequences into *E.coli*.

Bacterial promoter
β-galactosidase
Bacterial promoter
β-galactosidase
Insulin A chain
Insulin B chain
amp^R
amp^R

2 Culture cells so many copies of fusion proteins are made.

3 Purify β-gal-insulin fusion proteins.

β-gal   A chain   β-gal   B chain
Met   Met

4 Treat with cyanogen bromide (CNBr) to separate A or B chain from β-gal.

CNBr cleaves the peptide bond after methionine.

β-gal   β-gal
A chain   B chain
Met   Met

5 Purify A and B chains.

6 Polypeptides fold to form functional hormone.

Active insulin with disulfide bonds
Disulfide bond

Figure 20.14 **The use of bacteria to make human insulin.**

chemicals that are difficult to degrade), usually associated with chemical and industrial activity. These pollutants include heavy metals, petroleum hydrocarbons, and halogenated organic compounds such as those with chlorine atoms, as well as pesticides, herbicides, and organic solvents. Many new applications that use microorganisms to degrade these pollutants are being tested. The field of bioremediation has been fostered by better knowledge of how pollutants are degraded by microorganisms, the identification of new and useful strains of microbes, and the ability to enhance bioremediation through genetic engineering. Molecular genetic technology is critical to identifying genes that

encode enzymes involved in bioremediation. The characterization of the relevant genes greatly enhances our understanding of how microbes can modify toxic pollutants. In addition, recombinant strains created in the laboratory can be more efficient at degrading certain types of pollutants.

In 1980, in a landmark case (*Diamond v. Chakrabarty*), the U.S. Supreme Court ruled that a live, recombinant microorganism is patentable as a "manufacture or composition of matter." The first recombinant microorganism to be patented was an "oil-eating" bacterium, which contained a laboratory-constructed plasmid. This strain can oxidize the hydrocarbons commonly found in petroleum. It grew faster on crude oil than did any of the natural strains that were tested. However, it has not been a commercial success because this recombinant strain metabolizes only a limited number of toxic compounds. Unfortunately, the strain did not degrade many higher-molecular-weight compounds that tend to persist in the environment.

Currently, bioremediation should be considered a developing industry. In the future, recombinant microorganisms may provide an effective way to decrease the levels of toxic chemicals within our environment. However, this approach will require careful studies to demonstrate that recombinant organisms are effective at reducing pollutants and safe when released into the environment.

Gene Replacements and Knockouts in Mice Can Be Used to Understand Gene Function and Human Disease

Let's now turn our attention to the genetic engineering of animals. Researchers can introduce a cloned gene into a fertilized egg or embryonic cells and create animals that carry the cloned gene. The term **transgenic** is used to describe an organism that carries genes that were introduced using molecular techniques such as gene cloning. Transgenics are also called **genetically modified organisms (GMOs)**. In some cases, the cloned gene will recombine with the normal gene on a chromosome, a phenomenon called **gene replacement**. For eukaryotic species that are diploid, only one of the two copies is initially replaced. In other words, the initial gene replacement creates a heterozygote carrying one normal copy of the gene and one copy that has been replaced with a cloned gene. Heterozygotes can be crossed to each other to obtain homozygotes, which carry both copies of the cloned gene. If the cloned gene carries a mutation that inactivates its function, such a homozygote is said to have undergone a **gene knockout**. The inactive cloned gene has replaced both copies of the normal gene. In this way, the function of the normal gene has been "knocked out." Gene replacements and gene knockouts have become powerful tools for understanding gene function.

A particularly exciting avenue of gene replacement research is its application in the study of human disease. As an example, let's consider the disease cystic fibrosis (CF), which is one of the most common and severe inherited human disorders. In humans, the defective gene that causes CF has been identified. Likewise, the homologous gene in mice was later identified. Using the technique of gene replacement, researchers produced mice that are homozygous for the same type of mutation that is found in humans with CF. Such mice exhibit disease symptoms resembling those found in humans, namely, respiratory and digestive abnormalities. Therefore, these mice can be used as model organisms to study this human disease. Furthermore, these mice models have been used to test the effects of various therapies in the treatment of the disease.

Biotechnology Holds Promise in Producing Transgenic Livestock

The technology of creating transgenic mice has been extended to other animals, and much research is under way to develop transgenic species of livestock including fish, sheep, pigs, goats, and cattle. A novel avenue of research involves the production of medically important proteins in the mammary glands of livestock. This approach is sometimes called **molecular pharming**. (The word pharming refers to the use of farm animals to make pharmaceuticals.) Several human proteins have been successfully produced in the milk of domestic livestock such as goats, sheep, and cattle. These include Factor IX to treat a certain type of hemophilia, tissue plasminogen activator to dissolve blood clots, and α-1-antitrypsin for the treatment of emphysema.

Compared with the production of proteins in bacteria, one advantage of molecular pharming is that certain proteins are more likely to function properly when expressed in mammals. This may be due to the post-translational modification of proteins that occurs in mammals but not in bacteria. In addition, certain proteins may be degraded rapidly or folded improperly when expressed in bacteria. Furthermore, the yield of recombinant proteins in milk can be quite large. Each dairy cow, for example, produces about 10,000 liters of milk per year. In most cases, a transgenic cow can produce approximately 1 g/L of the transgenic protein in its milk.

To introduce a human gene into a mammal so that the encoded protein will be secreted into its milk, the strategy is to clone the gene next to the promoter of a gene that is specifically expressed in mammary cells. As we learned in Chapter 13, gene regulation promotes the expression of genes in certain cell types and/or under certain conditions. Researchers have identified several genes that are transcribed only in mammary cells. One example is β-lactoglobulin, a protein that is specifically found in milk. To express a human hormone into a domestic animal's milk, the promoter for one of these milk-specific genes is linked to the coding sequence of a human hormone gene and inserted into a plasmid vector (**Figure 20.15**). In addition to the promoter, a short signal sequence may also be necessary so that the protein hormone will be secreted from the mammary cells and into the milk. Such a vector is then injected into an oocyte, such as a sheep oocyte, where it integrates into the genome. The egg is then fertilized by exposure to sperm, and then implanted into the uterus of a female sheep. The resulting offspring carries the cloned gene. If the offspring is a female, the protein encoded by the human gene will be expressed within the mammary gland and secreted into the milk. The milk can then be obtained from the animal, and the human protein isolated and purified.

1. Clone a human hormone gene into a plasmid vector next to a sheep β-lactoglobulin promoter. This promoter is functional only in mammary cells so that the protein product is secreted into the milk.

Human hormone gene

β-lactoglobulin promoter

Plasmid vector

2. Inject this recombinant plasmid into a sheep oocyte. The plasmid DNA will integrate into the chromosomal DNA, resulting in the addition of the hormone gene into the sheep's genome.

Sheep oocyte

3. The oocyte is fertilized and implanted into a female sheep, which then gives birth to a transgenic sheep offspring.

Transgenic sheep

4. Obtain milk from a female transgenic sheep. The milk contains a human hormone.

Milk containing human hormone

5. Purify the hormone from the milk.

Figure 20.15 **Molecular pharming.** This procedure is used for expressing human genes in a domestic animal's milk. The β-lactoglobulin promoter is normally expressed in mammary cells, whereas the human hormone gene is not. To express the human hormone gene in milk, the promoter from the β-lactoglobulin gene is linked to the coding sequence of the hormone gene. This causes the human hormone to be expressed in the milk of the animal, from which it can be purified.

Agrobacterium tumefaciens Can Be Used to Make Transgenic Plants

As we have just seen, the introduction of cloned genes into oocytes or embryonic cells can produce transgenic animals. The production of transgenic plants is somewhat easier because plant cells are totipotent, which means that an entire organism can be regenerated from somatic cells. Therefore, a transgenic plant can be made by the introduction of cloned genes into somatic tissue, such as the tissue of a leaf. After the cells of a leaf have become transgenic, an entire plant can be regenerated by treating the leaf with plant growth hormones that cause it to form roots and shoots.

Molecular biologists can use the bacterium *Agrobacterium tumefaciens*, which naturally infects plant cells and causes tumors, to produce transgenic plants. A plasmid from the bacterium is known as the **Ti plasmid**, for tumor-inducing plasmid. This plasmid contains a region called the T DNA (for transferred DNA) that is transferred from the bacterium to the plant cell. The T DNA from the Ti plasmid becomes integrated into the chromosomal DNA of the plant cell. After this occurs, genes within the T DNA that encode plant growth hormones cause uncontrolled plant cell growth and produce a crown gall tumor, a bulbous growth on the plant.

Researchers have modified the Ti plasmid to use it as a vector to introduce cloned genes into plant cells. In the modified Ti plasmid, the genes that cause a crown gall tumor to form have been removed. Fortunately for genetic engineers, the T DNA is still taken up into plant cells and integrated into the plant chromosomal DNA. In addition, a selectable marker gene has been inserted into the T DNA to allow selection of plant cells that have taken up the T DNA. A gene that provides resistance to the antibiotic kanamycin, called *Kan^R*, is a commonly used selectable marker. Finally, the Ti plasmid used in cloning experiments has been modified to contain unique restriction sites for the convenient insertion of any gene of interest.

Figure 20.16 shows the general strategy for producing transgenic plants via T DNA–mediated gene transfer. A desired gene is inserted into the T DNA of a genetically engineered Ti plasmid and then transformed into *A. tumefaciens*. Plant cells are exposed to the transformed *A. tumefaciens*. After allowing time for infection, the plant cells are grown on a solid medium that contains kanamycin and carbenicillin. Carbenicillin kills *A. tumefaciens*, and kanamycin kills any plant cells that have not taken up the T DNA. Therefore, the only surviving cells are those plant cells that have integrated the T DNA into their genome. Because the T DNA also contains the cloned gene of interest, the selected plant cells are expected to have received this cloned gene as well. The cells are then transferred to a medium that contains the plant growth hormones necessary for the regeneration of entire plants.

Transgenic plants are approved for human consumption. Their production has become routine practice for many agriculturally important plant species, including alfalfa, corn, cotton, soybean, tobacco, and tomato. Transgenic plants can be given characteristics that are agriculturally useful, such as those that improve plant quality and protection. In terms of quality, gene additions have been made to improve the nutritional value of some plants, such as making the canola grain produce more oil, and making the Brazil nut contain more of the amino acid methionine. Frequently, transgenic research has sought to produce plant strains resistant to insects, disease, and herbicides. For example, transgenic plants highly tolerant of particular herbicides have been made. The Monsanto Company has produced

1. Gene of interest is inserted into the T DNA of the Ti plasmid.

Ti plasmid
Gene of interest
T DNA
Kan^R
Site where restriction enzyme cuts
Recombinant Ti plasmid

2. The recombinant Ti plasmid is transformed into *A. tumefaciens*.

A. tumefaciens
Recombinant Ti plasmid

3. Plant cells are exposed to *A. tumefaciens*. The T DNA is transferred and incorporated into the plant cell chromosome.

Plant cell
Inserted T DNA carrying new gene

4. The plant cells are placed in a medium containing kanamycin and carbenicillin. Only plant cells that have taken up T DNA can grow. Carbenicillin kills *A. tumefaciens*. The cells are transferred to growth media that has plant hormones necessary for regenerating an entire plant.

Plant with cloned gene

Figure 20.16 Using the Ti plasmid and *Agrobacterium tumefaciens* to make transgenic plants.

transgenic plant strains tolerant of glyphosate, the active agent in the herbicide Roundup™. Compared with nontransgenics, these plants grow quite well in the presence of glyphosate-containing herbicides (**Figure 20.17**). Another important approach is to make plant strains that are disease resistant. In many cases, virus-resistant plants have been developed by introducing a gene that encodes a viral coat protein. When the plant

Figure 20.17 **Transgenic plants that are resistant to glyphosate.** This field of soybean plants has been treated with the herbicide glyphosate. The plants in the back rows have been genetically engineered to contain a herbicide-resistance gene, so they are resistant to killing by glyphosate. By comparison, the dead or stunted plants in the front row with the orange stick do not contain this gene.

cells express the viral coat protein, they become resistant to infection by that pathogenic virus.

Research into transgenic plants has also led to the development of new products such as biodegradable plastics and medical products. For example, transgenic plants have been modified to produce vaccines against many human and animal diseases, including hepatitis B, cholera, and malaria. Transgenic plants have been made that produce a variety of medicines, such as human epidermal growth factor for wound repair and human interferon to fight viral diseases and cancer.

Researchers Have Succeeded in Cloning Mammals from Somatic Cells

We now turn our attention to cloning as a way to genetically manipulate plants and animals. The term cloning has many different meanings. At the beginning of this chapter we discussed gene cloning, which involves methods that produce many copies of a gene or its protein product. The cloning of an entire organism is a different matter. By accident, this happens in nature. Identical twins are genetic clones that began from the same fertilized egg. Similarly, researchers can take mammalian embryos at an early stage of development (for example, the two-cell to eight-cell stage), separate the cells, implant them into the uterus of a female, and obtain multiple births of genetically identical individuals.

In the case of plants, the cloning of new individuals is relatively easy. Plants can be cloned from somatic cells. In most cases, it is easy to take a cutting from a plant, expose it to growth hormones, and obtain a separate plant that is genetically identical to the original. However, this approach has not been possible with mammals. For several decades, scientists believed that chromosomes within the somatic cells of mammals had incurred irreversible genetic changes that render them unsuitable for cloning. However, this hypothesis has proven to be incorrect.

1. Donor sheep's mammary cell is extracted and grown in a tissue culture flask. Another sheep's unfertilized egg is extracted, and the nucleus is removed.

Mammary cell

Unfertilized egg

Nucleus

2. The cells are fused together with electric pulses.

Egg with nucleus removed

Mammary cell

3. The donor nucleus from the mammary cell and the maternal proteins within the egg initiate development of the egg into an embryo.

4. The embryo is transferred into a surrogate ewe.

5. Pregnancy is allowed to proceed.

6. A lamb (almost) genetically identical to the donor sheep is then born.

Figure 20.18 **Protocol for the successful cloning of sheep.** In this protocol, the genetic material from a somatic cell is used to make a cloned mammal, in this case the sheep Dolly.

In 1996, Ian Wilmut and his colleagues at the Roslin Institute in Edinburgh, Scotland, created clones of sheep using the genetic material from somatic cells. As you may have heard, they named the first cloned lamb Dolly.

Figure 20.18 illustrates how Dolly was created. The researchers removed mammary cells from an adult female sheep and grew them in the laboratory. The researchers then extracted the nucleus from a sheep oocyte and fused a diploid mammary cell with the enucleated oocyte cell. Fusion was promoted by electric pulses. After fusion, the zygote was implanted into the uterus of an adult sheep. One hundred and forty-eight days later, Dolly was born.

Dolly was (almost) genetically identical to the sheep that donated a mammary cell to create her. Dolly and the donor sheep were (almost) genetically identical in the same way that identical twins are. They carry the same set of genes and look remarkably similar. However, Dolly and her somatic cell donor may have some minor genetic differences due to possible differences in their mitochondrial DNA and may exhibit some phenotypic differences due to maternal effect genes.

Mammalian cloning is still at an early stage of development. Nevertheless, creating Dolly was a breakthrough that showed that it is technically possible. In recent years, cloning using somatic cells has been achieved in several mammalian species, including sheep, cows, mice, goats, pigs, and cats. In 2002, the first pet was cloned, which was named Copycat (see chapter-opening photo). The cloning of mammals provides the potential for many practical applications. With regard to livestock, cloning would enable farmers to use the somatic cells from their best animals to create genetically homogeneous herds. This could be advantageous in terms of agricultural yield, although such a genetically homogeneous herd may be more susceptible to certain diseases.

Aside from the practical uses of cloning agricultural species, however, people have become greatly concerned with the possibility of human cloning. This prospect has raised serious ethical questions. Some people feel that it is morally wrong and threatens the basic fabric of parenthood and family. Others feel that it is a modern technology that offers a new avenue for reproduction, one that could be offered to infertile couples, for exam-

ple. Human cloning is a complex subject with many more viewpoints than these two. In the public sector, the sentiment toward human cloning has been generally negative. Indeed, many countries have issued a complete ban on human cloning, while others permit limited research in this area. In the future, our society will have to wrestle with the legal and ethical aspects of cloning as it applies not only to animals but also to people.

DNA Fingerprinting Is Used for Identification and Relationship Testing

DNA fingerprinting, also known as DNA profiling, is a technology that identifies particular individuals using properties of their DNA. Like the human fingerprint, the DNA of each individual is a distinctive characteristic that provides a means of identification. When subjected to DNA fingerprinting, the chromosomal DNA produces a series of bands on a gel (**Figure 20.19a**). The unique pattern of these bands is a distinguishing feature of each individual.

In the past two decades, the technique of DNA fingerprinting has become automated, much like the automation that changed the procedure of DNA sequencing described earlier in this chapter. DNA fingerprinting is now done using PCR, which amplifies **short tandem repeat sequences (STRs)**. STRs are found in multiple sites in the genome of humans and other species, and they vary in length among different individuals. Using primers, the STRs from a sample of DNA are amplified by PCR and then separated by gel electrophoresis according to their molecular masses. Like automated DNA sequencing, the amplified STR fragments are fluorescently labeled. A laser excites the fluorescent molecule within an STR, and a detector records the amount of fluorescence emission for each STR. The DNA finger-

(a) Traditional DNA fingerprinting

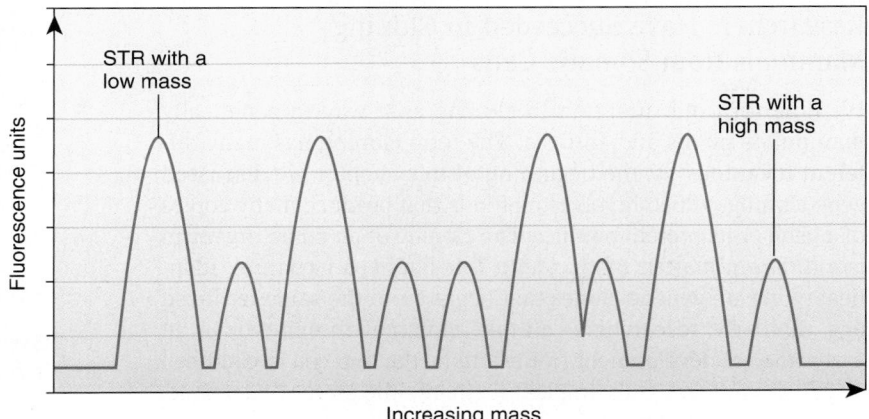

(b) Automated DNA fingerprinting

Figure 20.19 DNA fingerprinting. (a) Chromosomal DNA from two different individuals was subjected to traditional DNA fingerprinting. Their DNA appears as a series of bands on a gel. The dissimilarities in the patterns of these bands distinguish different individuals, much as the differences in physical fingerprint patterns can be used for identification. DNA evidence at a crime scene (E) matches suspect 2 (S2) but not suspect 1 (S1). (b) Automated DNA fingerprinting compares fluorescently labeled DNA sequences called short tandem repeat sequences (STRs), which differ in length among various individuals. A printout from the fluorescence detector is shown here.

print yields a series of peaks, each peak having a characteristic molecular mass (**Figure 20.19b**). In this automated approach, the pattern of peaks is an individual's DNA fingerprint.

Within the past decade or so, the uses of DNA fingerprinting have expanded in many ways. DNA fingerprinting has gained acceptance as a precise method of identification. In medicine, it is used to identify different species of bacteria and fungi, and it can even distinguish among closely related strains of the same species. This is useful so that clinicians can treat patients with the appropriate antibiotic or fungicide.

A second common use is forensics—providing evidence in a criminal case. DNA fingerprinting can be used as evidence that an individual was at a crime scene. Forensic DNA was first used in the U.S. court system in 1986. When a sample taken from a crime scene matches the DNA fingerprint of an individual, the probability that a match could occur simply by chance can be calculated. Each STR size is given a probability score based on its observed frequency within a reference human population (Caucasian, Asian, etc.). An automated DNA fingerprint contains many peaks, and the probability scores for each peak are multiplied together to arrive at the likelihood that a particular pattern of peaks would be observed. For example, if a DNA fingerprint contains 20 fluorescent peaks, and the probability of an individual having each peak is 1/4, then the likelihood of having that pattern would be $(1/4)^{20}$, or roughly 1 in 1 trillion. Therefore, a match between two samples is rarely a matter of random chance.

Another important application of DNA fingerprinting is relationship testing. Persons who are related genetically will have some peaks in common. The number they share depends on the closeness of their genetic relationship. For example, offspring are expected to receive half of their peaks from one parent and half from the other. Therefore, DNA fingerprinting can be used as evidence in paternity cases.

FEATURE INVESTIGATION

Blaese and Colleagues Performed the First Gene Therapy to Treat ADA Deficiency

Gene therapy is the introduction of cloned genes into living cells in an attempt to cure disease. It represents a new, potentially powerful means of treating a wide variety of illnesses. Many current research efforts in gene therapy are aimed at alleviating inherited human diseases. More than 4,000 human genetic diseases involve a single gene abnormality—common examples include cystic fibrosis and sickle-cell anemia. Many inherited diseases have been investigated as potential targets for gene therapy. These include metabolic diseases, such as phenylketonuria, and blood disorders, such as hemophilia and severe combined immunodeficiency. In addition, gene therapies have also been aimed at treating diseases, such as cancer and cardiovascular disease, that may occur later in life. Some researchers are even pursuing research that will use gene therapy to combat infectious diseases such as AIDS. Human gene therapy is still at an early stage of development. Relatively few patients have been successfully treated with gene therapy in spite of a large research effort. In addition, experimental gene therapy in humans has been associated with adverse reactions. In 1999, a patient even died from a reaction to a gene therapy treatment.

Adenosine deaminase (ADA) deficiency was the first inherited disease treated with gene therapy. When present, the adenosine deaminase enzyme deaminates the nucleoside deoxyadenosine. This is an important step in the proper metabolism of nucleosides. If both copies of the *ADA* gene are defective, however, deoxyadenosine will accumulate within the cells of the individual. At high concentration, deoxyadenosine is particularly toxic to lymphocytes in the immune system, namely, T cells and B cells. In affected individuals, the destruction of T and B cells leads to a severe combined immunodeficiency disease (SCID). If left untreated, SCID is typically fatal at an early age (generally 1 to 2 years old), because the compromised immune system of these individuals cannot fight infections.

Three approaches are used to treat ADA deficiency. In some cases, a bone marrow transplant is received from a compatible donor. A second method is to treat SCID patients with purified ADA enzyme that is coupled to polyethylene glycol (PEG). This PEG-ADA is taken up by lymphocytes and can correct the ADA deficiency. Unfortunately, these two approaches are not always available and/or successful. A third, more recent approach is to treat ADA patients with gene therapy.

On September 14, 1990, the first human gene therapy was approved for a young girl suffering from ADA deficiency. This work was carried out by a large team of researchers including R. Michael Blaese, Kenneth Culver, W. French Anderson, and several collaborators at the National Institutes of Health (NIH). Prior to this clinical trial, the normal gene for ADA had been cloned into a retroviral vector. The retroviral vector also contained mutations that prevented it from causing a viral disease, yet it still enabled the virus to infect human cells. The general aim of this therapy was to remove lymphocytes from the blood of the girl with SCID, introduce the normal *ADA* gene into the cells, and then return them to her bloodstream.

Figure 20.20 outlines the protocol for the experimental treatment. The researchers removed lymphocytes from the girl and cultured them in a laboratory. The lymphocytes were then transfected with a recombinant retrovirus that contained the normal *ADA* gene. During the life cycle of a retrovirus, the retroviral genetic material is inserted into the host cell's DNA.

Figure 20.20   The first human gene therapy for adenosine deaminase (ADA) deficiency by Blaese and colleagues.

HYPOTHESIS   Infecting lymphocytes with a retrovirus containing the *ADA* normal gene will correct the inherited deficiency of the mutant *ADA* gene in patients with ADA deficiency.

STARTING MATERIALS   A retrovirus with the normal *ADA* gene.

Experimental level   Conceptual level

1   Remove *ADA*-deficient lymphocytes from the patient with severe combined immunodeficiency disease (SCID).

Lymphocytes

Lymphocyte

Mutant *ADA* gene

2   Culture the cells in a laboratory.

3   Infect the cells with a retrovirus that contains the normal *ADA* gene. As described in Chapter 18, retroviruses insert their DNA into the host cell chromosome as part of their reproductive cycle.

Retrovirus with *ADA* gene

Retrovirus   Lymphocyte

4   Infuse the *ADA*-gene-corrected lymphocytes back into the SCID patient.

Normal *ADA* gene

Retroviral inserted DNA

5   **THE DATA**

Therefore, because this retrovirus contained the normal *ADA* gene, this gene also was inserted into the chromosomal DNA of the girl's lymphocytes. After this occurred in the laboratory, the cells were reintroduced back into the patient.

In this clinical trial, two U.S. patients were enrolled, and a third patient was later treated in Japan. The data in Figure 20.20 show the results of this trial for one of the three patients. Over the course of two years, this patient was given 11 infusions of lymphocytes that had been corrected with the normal *ADA* gene. Even after four years, this patient's lymphocytes were still making ADA. Therefore, these results suggest that this first gene therapy trial may offer some benefit. However, because the patients were also treated with PEG-ADA, the researchers could not determine whether gene transfer into T cells was of significant clinical benefit.

CHAPTER SUMMARY

20.1 Gene Cloning

- Recombinant DNA technology is the use of laboratory techniques to isolate and manipulate fragments of DNA.

- Gene cloning is the process of making multiple copies of a gene. The procedure is used to obtain large amounts of the DNA that encodes a particular gene, or to obtain large amounts of the gene product. (Figure 20.1)

- Plasmid and viral vectors are used in gene cloning. To obtain recombinant DNA, a vector and chromosomal DNA are cut with restriction enzymes. The DNA fragments bind to each other via their sticky ends, and the pieces are linked together via DNA ligase. (Table 20.1, Figure 20.2)

- When a recombinant vector is introduced into a bacterial cell, the cell replicates the vector and also divides to produce many cells. This achieves gene cloning. (Figure 20.3)

- A collection of recombinant vectors, each with a particular piece of chromosomal DNA, is introduced into bacterial cells to create a DNA library. If the DNA inserts are derived from cDNA, which is made from mRNA, this is a cDNA library. (Figure 20.4)

- Colony hybridization is a method that uses a labeled probe that recognizes a specific gene to identify that gene in a DNA library. (Figure 20.5)

- Gel electrophoresis is used to separate macromolecules by using an electric field that causes them to pass through a gel matrix. Gel electrophoresis typically separates molecules according to their masses. (Figure 20.6)

- In a Southern blotting experiment, a labeled probe, which is a strand of DNA from a specific gene, is used to identify that gene in a mixture of many chromosomal DNA fragments. (Figure 20.7)

- Polymerase chain reaction (PCR) is another technique to make many copies of a gene. Primers are used that flank the region of DNA to be amplified. (Figure 20.8)

20.2 Genomics and Proteomics

- Genomics is the study of genomes as whole units, while proteomics is the study of large groups of proteins made by a particular cell or organism.

- For structural genomics, large fragments of chromosomal DNA are cloned into vectors such as BACs. One goal of structural genomics is to make a contig, which is a collection of clones that cover a contiguous region of a chromosome. This is a type of mapping— determining the relative locations of genes or other DNA segments along a chromosome. (Figure 20.9)

- The dideoxy method of DNA sequencing uses dideoxynucleotides to determine the base sequence of a segment of DNA. (Figure 20.10)

- An important technique in functional genomics is a microarray that contains a group of spots, each with a specific type of DNA. It is used as a hybridization tool to study the expression of groups of genes. (Figure 20.11, Table 20.2)

- Two-dimensional gel electrophoresis is used to separate complex mixtures of proteins, such as all the proteins that are made by a particular cell type. The first dimension separates proteins according to their isoelectric points and the second dimension separates according to mass. (Figure 20.12)

- Tandem mass spectrometry is a method to determine short amino acid sequences within proteins. These sequences are then used to identify the protein, provided the genome sequence encoding the protein is already known. (Figure 20.13)

20.3 Biotechnology

- Microorganisms can be genetically engineered to produce human products such as insulin. (Figure 20.14)

- Microorganisms are also used to reduce pollutants in the environment, a phenomenon called bioremediation.

- Transgenic organisms, also called genetically modified organisms, are made via gene replacement or gene addition.

- Transgenic livestock can be genetically engineered to produce human hormones in their milk. (Figure 20.15)

- The Ti plasmid in *Agrobacterium tumefaciens* has been extensively used to produce transgenic plants, such as plants that are resistant to herbicides. (Figures 20.16, 20.17)

- The cloning of mammals can be achieved by fusing a somatic cell with an egg that has had its nucleus removed. The possibility of human cloning raises serious ethical questions. (Figure 20.18)

- DNA fingerprinting is a method of identification based on the occurrence of segments of DNA in the genomes of all individuals, such as STRs, that are variable in length among different individuals. (Figure 20.19)

- Gene therapy is a method to treat human diseases by the introduction of cloned genes into cells. The first gene therapy involved a disease called severe combined immunodeficiency syndrome (SCID). (Figure 20.20)

TEST YOURSELF

1. Vectors used to clone genes were derived originally from
   a. proteins.
   b. plasmids.
   c. viruses.
   d. all of the above.
   e. b and c only.

2. Restriction enzymes
   a. are used to cut DNA into pieces for gene cloning.
   b. are naturally produced by bacteria cells to prevent viral infection.
   c. produce sticky ends on DNA fragments.
   d. all of the above.
   e. a and c only.

3. A DNA library produced by isolating mRNA from a cell and using reverse transcriptase to make DNA molecules is called a _____ library.
   a. genomic
   b. mRNA
   c. proteonomic
   d. cDNA
   e. chromosomal

4. Researchers can identify the colonies that contain the vector with the gene of interest by
   a. screening the different colonies using a probe that is complementary to the gene of interest.
   b. keeping records of the particular colonies that were supposedly inoculated with the particular probe.
   c. using PCR to determine the gene sequence of the DNA in the different colonies.
   d. using DNA fingerprinting techniques to identify the particular gene of interest.
   e. none of the above.

5. A method used to detect a particular DNA sequence within a mixture of many DNA fragments is
    a. PCR.
    b. colony hybridization.
    c. DNA fingerprinting.
    d. DNA sequencing.
    e. Southern blotting.

6. Why is *Taq* polymerase used in PCR rather than other DNA polymerases?
    a. *Taq* polymerase is a synthetic enzyme that produces DNA strands at a faster rate than natural polymerases.
    b. *Taq* polymerase is a heat-stable form of DNA polymerase that can function after exposure to the high temperatures that are necessary for PCR.
    c. *Taq* polymerase is easier to isolate than other DNA polymerases.
    d. *Taq* polymerase is the DNA polymerase commonly produced by most eukaryotic cells.
    e. All of the above.

7. The method of determining the base sequence of DNA is
    a. PCR.
    b. gene cloning.
    c. DNA fingerprinting.
    d. DNA sequencing.
    e. gene mapping.

8. During bioremediation, microorganisms are used to
    a. clone genes from eukaryotic organisms.
    b. introduce correct genes into individuals with genetic diseases.
    c. decrease pollutants in the environment.
    d. produce useful products such as insulin.
    e. all of the above.

9. Organisms that carry genes that were introduced using molecular techniques are called
    a. transgenics.
    b. clones.
    c. mutants.
    d. genetically modified organisms.
    e. both a and d.

10. DNA fingerprinting is used
    a. to provide a means of precise identification of an organism, such as the identification of specific strains of bacteria.
    b. as a forensics tool to provide evidence in a criminal case.
    c. to determine genetic relationships between individuals.
    d. to determine the identity of an individual.
    e. all of the above.

CONCEPTUAL QUESTIONS

1. Define recombinant DNA technology and recombinant DNA.
2. Explain how using one restriction enzyme to cut both a plasmid and a gene of interest will allow the gene to be inserted into the plasmid.
3. Explain how gel electrophoresis separates DNA fragments.

EXPERIMENTAL QUESTIONS

1. What is gene therapy? What is ADA deficiency?
2. In the investigation of Figure 20.20, how did the researchers treat ADA deficiency?
3. How successful was the gene therapy for ADA deficiency?

COLLABORATIVE QUESTIONS

1. Discuss the use of microorganisms for bioremediation.
2. Discuss the process of molecular pharming.

www.brookerbiology.com
This website includes answers to the Biological Inquiry questions found in the figure legends and all end-of-chapter questions.

21

GENOMES, PROTEOMES, AND BIOINFORMATICS

Genomes and computer technology. The amount of data derived from the analyses of genomes is so staggering in size and complexity that researchers have turned to the use of computers to unravel the amazing information that genomes contain.

The unifying theme of biology is evolution. The genome of every living species is the product of over 3.5 billion years of evolution. We can understand the unity of modern organisms by realizing that all species evolved from an interrelated group of ancestors. Throughout this textbook, each chapter touches upon "Genomes & Proteomes" as a way to see the evolutionary connections among all forms of life, and also to understand how the genetic material produces the form and function of living organisms. By now, you may feel familiar with the concept of a **genome**, which is the complete genetic makeup of a cell or organism. The genome of each species is critical to its existence in several ways, including:

- The genome stores information in the form of genes, which provide a blueprint to produce the characteristics of organisms.

- The genome is copied and transmitted from generation to generation.

- The accumulation of genetic changes over the course of many generations produces the evolutionary changes that alter species and create new species.

An extension of genome analysis is the study of **proteomes**, which refers to the entire complement of proteins that a cell or organism can make. The function of most genes is to encode proteins, which are the key participants in maintaining cell structure and carrying out most cell functions. Analyzing the proteome of a single species and comparing the proteomes of different species allow biologists to understand the structure and function of cells, multicellular organisms, and the interactions between organisms and their environment.

The terms genomes and proteomes are sometimes confused with the terms genomics and proteomics. When we speak of

genomes and proteomes, we are talking about characteristics of living organisms. By comparison, genomics and proteomics are experimental approaches to studying genomes and proteomes, respectively. An exciting advance in biology has been the ability to analyze the DNA sequence of genomes and the expression of groups of genes. This technology is genomics. Likewise, other tools have been invented to study large groups of proteins; this technology is proteomics. The techniques of genomics and proteomics are discussed in Chapter 20.

In the first two sections of this chapter, our focus will be primarily on biology, not on experimental techniques. We will consider the sizes and compositions of the genomes and proteomes of prokaryotic and eukaryotic species. As you will learn, genomes and proteomes are full of surprises. For example, did you know that most of your DNA has no known function? From a molecular perspective, genomes and proteomes contain extensive and intriguing information, which has challenged researchers to study them using computer technology. In the last section of this chapter, we will consider how the field of **bioinformatics**, which uses computers to study biological information, has been critical in the study of genomes and proteomes.

21.1 Genome Sizes and Composition

The past decade has seen remarkable advances in our overall understanding of the entire genomes of several species. As genetic technology has progressed, researchers have gained an increasing ability to analyze the composition of genomes as a whole unit.

For many species, we now know their complete DNA sequence, which provides the most detailed description available of an organism's genome at the molecular level. In this section, we will survey the sizes of genomes in prokaryotic and eukaryotic species and examine their compositions. Genomes consist not only of genes but also of other types of noncoding sequences. For example, the genomes of all species have repetitive DNA—short repeated sequences. We will learn how certain types of repetitive DNA sequences are formed by a process called transposition. We will also examine how the duplication of genes can lead to families of related genes.

Prokaryotic Genomes Often Contain a Circular Chromosome with a Few Thousand Genes and Little Repetitive DNA

Geneticists have made great progress in the study of bacterial and archaeal genomes. Some of the key features of prokaryotic chromosomes are described in Chapter 18 (refer back to Figure 18.2). Researchers are interested in the genomes of prokaryotic species for three main reasons. First, bacteria cause many different diseases that affect people, as well as plants and animals. Studying the genomes of bacteria reveals important clues about the process of infection and also may help us find ways to combat bacterial infection. A second reason for studying prokaryotic genomes is that the information we learn about these tiny creatures often applies to more complex organisms. For example, basic genetic mechanisms such as DNA replication and gene regulation were first understood in *Escherichia coli*. That knowledge provided a critical foundation to understand how these processes work in humans and other eukaryotic species. A third reason is evolution. The origin of the first eukaryotic cell probably involved a union between an archaeal and a bacterial cell. The study of prokaryotic genomes helps us understand how all modern species evolved.

Because they are both relatively small and of great interest to us, the entire genomes of many prokaryotic species have been sequenced and analyzed. The chromosomes of prokaryotes are usually in the range of several hundred thousand to a few million base pairs in length. Genomic researchers refer to 1 million base pairs as 1 megabase pair, abbreviated Mbp or Mb. Most prokaryotes that have been studied contain a single type of chromosome, though multiple copies may be present in a single cell. However, some prokaryotes are known to have different chromosomes. For example, *Vibrio cholerae*, the bacterium that causes the diarrheal disease known as cholera, has two different chromosomes in each cell, one 2.9 Mb and the other 1.1 Mb. Bacterial chromosomes are often circular. For example, the two chromosomes in *V. cholerae* are circular, as is the single type of chromosome found in *E. coli*. However, linear chromosomes are found in some prokaryotic species, such as *Borrelia burgdorferi*, which is the bacterium that causes Lyme disease, the most common tick-borne disease in the U.S. Certain bacterial species may even contain both linear and circular chromosomes. *Agrobacterium tumefaciens*, which infects plants

Table 21.1	Examples of Prokaryotic Genomes That Have Been Sequenced*		
Species	**Genome size (Mb)†**	**Number of genes‡**	**Description**
Methanobacterium thermoautotrophicum	1.7	1,869	An archaeon that produces methane.
Haemophilus influenzae	1.8	1,743	One of several different bacterial species that causes respiratory illness and meningitis.
Sulfolobus solfataricus	3.0	3,032	An archaeon that metabolizes sulfur.
Lactobacillus plantarum	3.3	3,052	A type of lactic acid bacterium used in the production of cheese and yogurt.
Mycobacterium tuberculosis	4.4	4,033	The bacterium that causes the respiratory disease tuberculosis.
Escherichia coli	4.6	4,289	A common intestinal bacterium; certain strains can cause human illness.
Bacillus anthracis	5.2	5,439	The bacterium that causes the disease anthrax.

* Prokaryotic species often exist in different strains that may differ slightly in their genome size and number of genes. The data are from common strains of the indicated species. The species shown in this table have only one type of chromosome.

† Mb equals 1 million base pairs.

‡ The number of genes is an estimate based on the analysis of genome sequences.

and causes a disorder called crown gall disease, has one linear chromosome (2.1 Mb) and one circular chromosome (3.0 Mb).

Table 21.1 compares the sequenced genomes from several prokaryotic species. They range in size from 1.7 to 5.2 Mb. The total number of genes is correlated with the total genome size. Roughly 1,000 genes are found for every Mb of DNA. Compared to eukaryotic genomes described later, prokaryotic genomes are less complex. They lack centromeres and telomeres, and they have a single origin of replication. Also, compared to their eukaryotic counterparts, prokaryotic chromosomes have relatively little repetitive DNA.

In addition to one or more chromosomes, prokaryotes often have plasmids, usually circular pieces of DNA that exist independently of the bacterial chromosome. Plasmids are typically small, in the range of a few thousand to tens of thousands of base pairs in length, though some can be quite large, even hundreds of thousands of base pairs. The various functions of plasmids are described in Chapter 18. Their use as vectors in genetic engineering is discussed in Chapter 20.

FEATURE INVESTIGATION

Venter, Smith, and Colleagues Sequenced the First Complete Genome, That of *Haemophilus influenzae*

The first genome to be entirely sequenced was that of the bacterium *Haemophilus influenzae*. This bacterium causes a variety of diseases in humans, including respiratory illnesses and bacterial meningitis. *H. influenzae* has a relatively small genome size, approximately 1.8 Mb.

Scientists can follow different strategies when tackling a genome-sequencing project. One strategy, which has been used for larger eukaryotic genomes, requires extensive mapping. This means that the genome is cut into large pieces of DNA whose locations are known within a given chromosome. These large pieces are then cut into smaller and smaller pieces, whose relative locations are known within the larger pieces. Once small DNA pieces have been mapped throughout the whole genome, their DNA sequences are determined by the procedure known as dideoxynucleotide DNA sequencing, which is described in Chapter 20.

An alternative strategy for sequencing an entire genome is called **shotgun DNA sequencing**. In this approach, researchers use the technique of dideoxy sequencing to randomly sequence many DNA fragments from the genome. As a matter of chance, some of the fragments are overlapping, which allows researchers to order them as they are found in the intact chromosome (refer back to Figure 20.9). The advantage of shotgun DNA sequencing is that it does not require extensive genetic mapping, which can be time-consuming. A disadvantage is that researchers will waste time sequencing the same region of DNA more than once.

To obtain a complete sequence of a genome with the shotgun approach, how do researchers decide how many fragments to sequence? We can calculate the probability that a base will not be sequenced using this approach with the following equation:

$$P = e^{-m}$$

where P is the probability that a base will be left unsequenced, e is the base of the natural logarithm ($e = 2.72$), and m is the number of bases sequenced divided by the total genome size. For example, in the case of *H. influenzae*, with a genome size of 1.8 Mb, if researchers sequenced 9.0 Mb, $m = 5$ (i.e., 9.0 Mb divided by 1.8 Mb):

$$P = e^{-m} = e^{-5} = 0.0067, \text{ or } 0.67\%.$$

This means that if we randomly sequence 9.0 Mb, which is five times the length of a single genome, we are likely to miss only 0.67% of the genome. With a genome size of 1.8 Mb, we would miss about 12,060 nucleotides out of approximately 1.8 million. Such missed sequences are typically on small DNA fragments that—as a matter of random chance—did not happen to be sequenced. The missing links in the genome can be sequenced later using mapping methods.

The general protocol conducted by Craig Venter, Hamilton Smith, and colleagues in this discovery-based investigation is described in **Figure 21.1**. The researchers isolated chromosomal DNA from *H. influenzae* and broke the DNA into small fragments, approximately 2,000 bp in length. These fragments were randomly cloned into vectors that allow the DNA to be propagated in *E. coli*. Each *E. coli* clone carried a vector with a different piece of DNA from *H. influenzae*. As discussed in Chapter 20, this collection of clones is called a DNA library. The researchers then subjected many of these clones to the procedure of DNA sequencing. They sequenced a total of approximately 10.8 Mb of DNA.

Figure 21.1   Determination of the complete genome sequence of *Haemophilus influenzae* by Venter, Smith, and colleagues.

GOAL   The goal is to obtain the entire genome sequence of *Haemophilus influenzae*. This information will reveal its genome size and also which genes the organism has.

STARTING MATERIALS   A strain of *H. influenzae*.

Experimental level   Conceptual level

1   Purify DNA from a strain of *H. influenzae*. This involves breaking open cells and adding phenol. Most protein components go into the phenol phase, while DNA remains in the aqueous (water) phase.

DNA in aqueous (water) phase
Proteins in phenol phase

H. influenzae chromosomal DNA

2   Sonicate the DNA to break it into small fragments.

Sound waves

DNA fragments in aqueous phase

Sound waves

3 Clone the DNA fragments into vectors. The procedures for cloning are described in Chapter 20. This produces a DNA library.

Refer back to Figures 20.2 and 20.3.

Vector DNA
Piece of *H. influenzae* DNA
A DNA library

4 Subject many clones to the procedure of dideoxy DNA sequencing, also described in Chapter 20. A total of 10.8 Mb was sequenced.

Refer back to Figure 20.10.

Produces a large number of overlapping sequences.

5 Use tools of bioinformatics, described in the last section of this chapter, to identify various types of genes in the genome.

Explores the genome sequence and identifies and characterizes genes.

6 THE DATA

1,600,000
1,830,137 bp
~1,800 genes
400,000
1,200,000
800,000

Functions of Genes

% of genome		% of genome	
6.8	Amino acid biosynthesis	5.3	Metabolism of purines, pyrimidines, nucleosides, and nucleotides
5.4	Biosynthesis of cofactors, prosthetic groups, carriers	6.3	Regulatory functions
8.3	Cell envelope	8.6	Replication
5.3	Cellular processes	12.2	Transport and binding proteins
3.0	Central intermediary metabolism	14.0	Translation
10.4	Energy metabolism	2.7	Transcription
2.5	Fatty acid/phospholipid metabolism	9.2	Other categories

The outcome of this genome-sequencing project was a very long DNA sequence. In 1995, Venter, Smith, and colleagues published the entire DNA sequence of *H. influenzae*. The researchers then analyzed the genome sequence by computer to obtain information about the properties of the genome. Questions that they asked included, How many genes does the genome contain? and What are the likely functions of those genes? Later in this chapter, we will learn how scientists can answer such questions with the use of computers. The data in Figure 21.1 summarize the results that the researchers obtained. The *H. influenzae* genome is composed of 1,830,137 bp of DNA. The computer analysis predicted 1,743 genes. Based on their similarities to known genes in other species, the researchers also predicted the functions of nearly two-thirds of these genes. The diagram shown in the data of Figure 21.1 places genes in various categories based on their predicted function. These results gave the first complete "genome picture" of a living organism!

The Nuclear Genomes of Eukaryotes Are Sets of Linear Chromosomes That Vary Greatly in Size and Composition Among Different Species

As we learned in Chapter 15, the genome found in the nucleus of eukaryotic species is usually found in sets of linear chromosomes. In humans, for example, one set contains 23 chromosomes—22 autosomes and one sex chromosome, X or Y. In addition, certain organelles in eukaryotic cells contain a small amount of their own DNA. These include the mitochondrion, which plays a role in ATP synthesis, and the chloroplast found in plants and algae, which carries out photosynthesis. The DNA found in these organelles is referred to as extranuclear DNA to distinguish it from the DNA that is found in the cell nucleus. In this chapter, we will focus on the nuclear genome of eukaryotes.

Table 21.2	Examples of Eukaryotic Nuclear Genomes That Have Been Sequenced		
Species	**Nuclear genome size (Mb)**	**Number of genes**	**Description**
Saccharomyces cerevisiae (baker's yeast)	12.1	~6,200	One of the simplest eukaryotic species; it has been extensively studied by researchers to understand eukaryotic cell biology and other molecular mechanisms.
Caenorhabditis elegans (nematode worm)	100	~19,000	A model organism used to study animal development.
Drosophila melanogaster (fruit fly)	180	~14,000	A model organism used to study many genetic phenomena, including development.
Arabidopsis thaliana (thale cress)	120	~26,000	A model organism studied by plant biologists.
Oryza sativa (rice)	460	~40,000–50,000	A cereal grain with a relatively small genome; it is very important worldwide as a food crop.
Homo sapiens (humans)	2,900	~20,000–25,000	The sequencing of the human genome will help to elucidate our inherited traits and may aid in the identification and treatment of diseases.

(b) *Echinops bannaticus*

(c) *Echinops nanus*

Figure 21.2 **Genome sizes among selected groups of eukaryotes.** **(a)** Genome sizes among various groups of eukaryotes are shown on a log scale. As an example for comparison, two closely related species of globe thistle are pictured, *Echinops bannaticus* in **(b)** and *Echinops nanus* in **(c)**. These species have similar characteristics, but *E. bannaticus* has nearly double the amount of DNA that *E. nanus* does due to the accumulation of repetitive DNA sequences.

Biological inquiry: What are two reasons why the groups of species shown in (a) have variation in their total amount of DNA?

In the past decade or so, the DNA sequence of entire nuclear genomes has been determined for several eukaryotic species (**Table 21.2**). The genomes of simpler eukaryotes, such as yeast, carry a few thousand different genes, while the genomes of more complex eukaryotes contain tens of thousands of genes. Motivation to sequence these genomes comes from four main sources. First, the work of researchers can greatly benefit from genome sequences that allow them to identify and characterize the genes of model organisms. This has been the impetus for genome projects involving baker's yeast (*Saccharomyces cerevisiae*), the fruit fly (*Drosophila melanogaster*), a nematode worm (*Caenorhabditis elegans*), the simple plant called thale cress (*Arabidopsis thaliana*), and the mouse (*Mus musculus*). A second reason for genome sequencing is to gather more information in order to identify and treat human diseases, which is an important aim for sequencing the human genome. Researchers hope that the DNA sequence of the human genome will help to identify genes in which mutation plays a role in disease. Third, the genomes of agriculturally important species have also been

the subject of genome-sequencing projects. An understanding of a species' genome may help us to develop new strains of livestock and plant species that have improved traits from an agricultural perspective. Fourth and finally, evolutionary biologists are interested in genome sequences as a way to establish evolutionary relationships.

When we speak of genome size, this means the total amount of DNA, often measured in megabase pairs. Genome size is not the same as the number of genes. The relative sizes of nuclear genomes vary dramatically among different eukaryotic species (**Figure 21.2a**). In general, increases in the amount of DNA are correlated with increases in cell size, cell complexity, and body complexity. For example, yeast have smaller genomes than animals. However, major variations in genome sizes are observed within organisms that are similar in form and function. For example, the total amount of DNA found within different species of amphibians can vary over 100-fold. The DNA content of closely related species can also vary. **Figure 21.2b,c** compares two closely related species of the plant called the globe thistle, *Echinops bannaticus* and *Echinops nanus*. These species have similar numbers of chromosomes, but *E. bannaticus* has nearly double the amount of DNA. The larger genome of *E. bannaticus* is not likely to contain twice as many genes. Rather, the genome composition includes many **repetitive sequences**, which are short DNA sequences that are present in many copies. Repetitive sequences are often abundant in eukaryotic species.

Repetitive sequences fall into two broad categories, moderately and highly repetitive. **Moderately repetitive sequences** are found a few hundred to several thousand times in the genome.

In some cases, these sequences are multiple copies of the same gene. For example, the genes that encode ribosomal RNA (rRNA) are found in many copies. The cell needs a large amount of rRNA for its cellular ribosomes. This is accomplished by having and expressing multiple copies of the genes that encode rRNA. In addition, other types of functionally important sequences can be moderately repetitive. For example, multiple copies of origins of replication are found in eukaryotic chromosomes. Other moderately repetitive sequences may play a role in the regulation of gene transcription and translation.

Highly repetitive sequences are those that are found tens of thousands or even millions of times throughout the genome. Each copy of a highly repetitive sequence is relatively short, ranging from a few nucleotides to several hundred nucleotides in length. Most of these sequences have no known function, and whether they benefit the organism is a matter of debate. A widely studied example is the *Alu* family of sequences found in humans and other primates. The *Alu* sequence is approximately 300 bp long. This sequence derives its name from the observation that it contains a site for cleavage by a restriction endonuclease known as *Alu*I. The *Alu* sequence is present in about 1 million copies in the human genome. It represents about 10% of the total human DNA and occurs approximately every 5,000–6,000 bases. Evolutionary studies suggest that the *Alu* sequence arose 65 million years ago from a section of a single ancestral gene known as the 7SL RNA gene. Remarkably, over the course of 65 million years, the *Alu* sequence has been copied and inserted into the human genome so often that it now appears more than 1 million times. The mechanism for the proliferation of *Alu* sequences will be described later.

Some highly repetitive sequences, like the *Alu* family, are interspersed throughout the genome. However, other highly repetitive sequences are clustered together in a tandem array in which a very short nucleotide sequence is repeated many times in a row. In *Drosophila*, for example, 19% of the chromosomal DNA is highly repetitive DNA found in tandem arrays. An example is shown here:

A A T A T A A T A T A A T A T A A T A T A A T A T A T A A T
T T A T A T T A T A T T A T A T T A T A T T A T A T A T T A

In this particular tandem array, two related sequences, AATAT and AATATAT, are repeated many times. Highly repetitive sequences, which contain tandem arrays of short sequences, are commonly found in centromeric regions of chromosomes and can be quite long, sometimes more than 1 million bp in length!

Figure 21.3 shows the composition of the relative classes of DNA sequences that are found in the nuclear genome of humans. Surprisingly, the coding regions of genes make up only about 2% of our genome! The other 98% is composed of noncoding sequences. Though we often think of genomes as being the repository of sequences that code for proteins, most eukaryotic genomes are largely composed of other types of sequences. Intron DNA is the second most common category at 24%. Unique noncoding DNA, whose function is largely unknown, constitutes 15%. Repetitive DNA makes up 59% of the DNA in the genome. Much of the repetitive DNA is derived from transposable elements, which are described next.

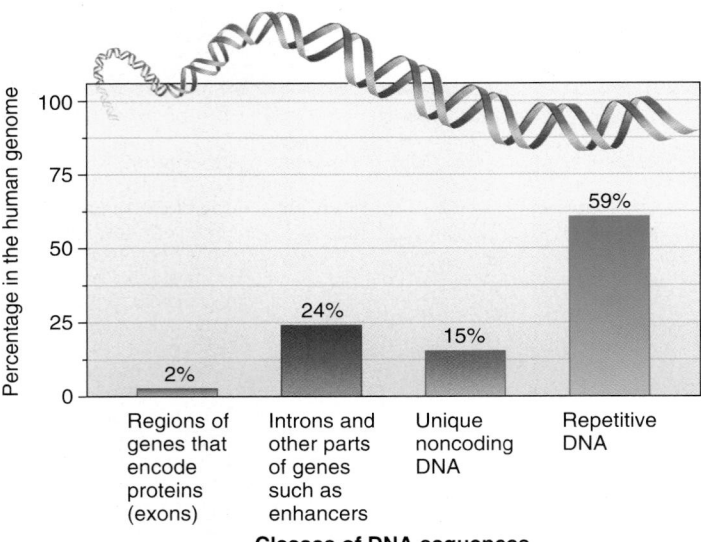

Figure 21.3   The composition of DNA sequences that are found in the nuclear genome of humans.   Note that only about 2% of our genome codes for proteins; most of our genome is made up of repetitive sequences.

Transposable Elements Can Move from One Chromosomal Location to Another

As we have seen, genomes are composed of several types of DNA sequences. Such sequences include not just the coding sequences of genes but also other sequences, including introns and repetitive sequences. Much of the moderately repetitive and highly repetitive DNA that is found in genomes is derived from a process called **transposition**, in which a short segment of DNA moves within a cell from its original site to a new site in the genome. The DNA segments that transpose themselves are known as **transposable elements (TEs)**. They range from a few hundred to several thousand base pairs in length. TEs have sometimes been referred to as "jumping genes," because they are inherently mobile. Barbara McClintock first identified transposable elements in the late 1940s from her studies with corn plants (**Figure 21.4**). She identified a segment of DNA that could move into and out of a gene that affected the color of corn kernels, producing a speckled appearance. Since that time, biologists have discovered many different types of TEs in prokaryotes, protists, fungi, plants, and animals—transposable elements have been found in all species examined.

Though Barbara McClintock identified TEs in corn in the late 1940s, her work was met with great skepticism because many researchers had trouble believing that DNA segments could be mobile. The advent of molecular technology in the 1960s and 1970s allowed scientists to understand more about the characteristics of TEs that enable their movement. Most notably, research involving bacterial TEs eventually progressed to a molecular understanding of the transposition process. In 1983, over 30 years after her initial discovery, McClintock was awarded the Nobel Prize.

(a) Organization of a simple transposable element

(a)   (b)

Figure 21.4 **Barbara McClintock, who discovered transposable elements.** As shown in part **(b)**, when a transposable element is found within a pigment gene in corn, its frequent movement causes the kernel color to be speckled.

As researchers have studied TEs from many species, they have found that DNA sequences within transposable elements are organized in several different ways, and they can move by different molecular mechanisms. Both ends of many TEs have inverted repeats (IRs), DNA sequences that are identical (or very similar) but run in opposite directions (**Figure 21.5a**), such as the following:

5′—CTGACTCTT—3′   and   5′—AAGAGTCAG—3′
3′—GACTGAGAA—5′   3′—TTCTCAGTC—5′

Depending on the particular transposable element, inverted repeats range from 9 to 40 base pairs in length. In addition, TEs may contain a central region that encodes **transposase**, an enzyme that facilitates transposition.

As shown in **Figure 21.5b**, one type of transposition occurs by a cut-and-paste mechanism. Transposase first recognizes the inverted repeats in the TE and then removes the sequence from its original site. Next, the transposase/TE complex moves to a new location, and transposase then inserts the sequence into the chromosome. Transposition may occur when a cell is in the process of DNA replication. If a transposable element is removed from a site that has already replicated, and is inserted into a chromosomal site that has not yet replicated, the TE will increase in number after DNA replication is complete. This is one way for TEs to become more prevalent in a genome.

Another category of transposable elements moves via an RNA intermediate. This form of transposition is very common but is found only in eukaryotic species. These types of elements are known as **retroelements** or **retrotransposons**. By comparison, TEs that move via DNA are called **transposons**. The *Alu* sequence in the human genome is an example of a retroelement. Some retroelements contain genes that encode the enzymes reverse transcriptase and integrase, which are needed in the transposition process (**Figure 21.6a**). Retroelements may also contain repeated sequences called terminal repeats at each end that facilitate their recognition. The mechanism of retroelement

(b) Cut-and-paste mechanism of transposition

Figure 21.5 **Simple transposable elements and their mechanism of transposition.** **(a)** Simple transposable elements (TEs) contain inverted repeat (IR) sequences at each end and a gene that encodes transposase in the middle. **(b)** Simple transposition occurs by a cut-and-paste mechanism.

movement is shown in **Figure 21.6b**. First, the element is transcribed into RNA by the enzyme RNA polymerase. Reverse transcriptase uses this RNA as a template to synthesize a double-stranded DNA molecule. The ends of the double-stranded DNA are then recognized by integrase, which catalyzes the insertion of the DNA into the host chromosomal DNA. The integration of retroelements can occur at many locations within the genome. Furthermore, because a single retroelement can be copied into many RNA transcripts, retroelements may accumulate rapidly within a genome. This explains how the *Alu* element in the human genome was able to proliferate and constitute 10% of our genome.

(a) Organization of a retroelement

Terminal repeat

Reverse transcriptase gene

Integrase gene

Terminal repeat

Retroelement

(b) Mechanism of movement of a retroelement

1   RNA polymerase transcribes the retroelement into RNA.

RNA polymerase

RE

RNA

2   Reverse transcriptase uses the RNA as a template to synthesize a double-stranded DNA molecule.

Reverse transcriptase

DNA

Integrase

3   Integrase inserts this retroelement DNA into the chromosome.

RE

4   The chromosome now contains 2 copies of the retroelement.

Figure 21.6   **Retroelements and their mechanism of transposition.**   Retroelements are found only in eukaryotic species. **(a)** Some retroelements contain terminal repeats and may contain genes that encode the enzymes reverse transcriptase and integrase, which are needed in the transposition process. **(b)** The process that adds a copy of a retroelement into a host chromosome.

Biological inquiry: Based on their mechanism of movement, which type of TEs do you think would proliferate more rapidly in a genome, simple TEs (see Figure 21.5b) or retroelements?

The biological significance of TEs remains a matter of debate. According to the **selfish DNA hypothesis**, TEs exist because they have the characteristics that allow them to insert themselves into the host cell DNA. In other words, they resemble parasites in the sense that they inhabit the host without offering any advantage. They can proliferate within the host as long as they do not harm the host to the extent that they significantly disrupt survival. However, TEs can do harm. For example, if they jump into the middle of an important gene and thereby disrupt its function, this may have a negative impact on the phenotype of an organism.

Alternatively, other biologists have argued that TEs may provide benefits to a given species. For example, bacterial TEs often carry an antibiotic-resistance gene that provides the organism with a survival advantage. In addition, TEs may cause greater genetic variability by promoting chromosomal rearrangements. As discussed next, such rearrangements can cause a misaligned crossover during meiosis, and promote the formation of a gene family.

Gene Duplications Provide Additional Material for Genome Evolution, Sometimes Leading to the Formation of Gene Families

Let's now turn our attention to a way that the number of genes in a genome can increase. Small chromosomal duplications are important because they provide raw material for the addition of more genes into a species' genome. Such duplications can create two or more copies of the same gene (**Figure 21.7a**). Over the course of many generations, each version of the gene accumulates different mutations, producing **homologous genes**—two or more similar genes that are derived from a single ancestral gene. During evolution, this type of event can occur sev-

eral times, creating a family of many similar genes called a **gene family**. Homologous genes within a single species are also called **paralogs**.

A mechanism that produces gene duplications is a misaligned crossover (**Figure 21.7b**). In this example, two homologous chromosomes have paired with each other during meiosis, but the homologs are misaligned. If a crossover occurs, this produces one chromosome with a gene duplication, one with a deletion, and two normal chromosomes. Each of these chromosomes will be segregated into different haploid cells. If a haploid cell carrying the chromosome with the gene duplication participates in fertilization with another gamete, an offspring with a gene duplication is produced. In this way, gene duplications can form and be transmitted to future generations. The presence of multiple copies of the same transposable element in a genome can foster this process, because the chromosomes may misalign while attempting to align TEs that are at different locations in the same chromosome.

A well-studied example of gene duplications is the globin gene family found in animals. The globin genes encode polypeptides that are subunits of proteins that function in oxygen binding. Hemoglobin, which is made in red blood cells, carries oxygen throughout the body. In humans, the globin gene family is composed of 14 homologous genes that were originally derived from a single ancestral globin gene (**Figure 21.8**). According to an evolutionary analysis, the ancestral globin gene duplicated between 500 and 600 million years ago. Since that time, additional duplication events and chromosomal rearrangements have occurred to produce the current number of 14 genes on three different human chromosomes. Four of these are pseudogenes, which are genes that have been produced by gene duplication but have accumulated mutations that make them nonfunctional so they are not transcribed into RNA.

(a) Gene duplication and the formation of homologous genes

(b) Mechanism of gene duplication

Figure 21.7   **Gene duplication and the evolution of homologous genes.**   (a) A gene duplication produces two copies of the same gene. Over time, these copies accumulate different random mutations, which results in homologous genes with similar but not identical DNA sequences. (b) Mechanism of gene duplication. If two homologous chromosomes misalign during meiosis, a crossover will produce a chromosome with a gene duplication.

Figure 21.8   **The evolution of the globin gene family in humans.**   The globin gene family evolved from a single ancestral globin gene. The first gene duplication produced two genes that accumulated mutations and became the genes encoding myoglobin and the family of hemoglobins. The modern myoglobin gene is found on chromosome 22. An ancestral hemoglobin gene duplicated to produce the α- and β-globins. Further duplications of ancestral α-globin and β-globin genes produced several homologous genes on chromosomes 16 and 11, respectively. The four genes shown in gray are nonfunctional pseudogenes.

Biological inquiry: What is the advantage of a gene family?

Gene families have been important in the evolution of complex traits. Even though all of the globin polypeptides are subunits of proteins that play a role in oxygen binding, the accumulation of different mutations in the various family members has created globins that are more specialized in their function. For example, myoglobin is better at binding and storing oxygen in muscle cells, whereas the hemoglobins are better at binding and transporting oxygen via the red blood cells. Also, different globin genes are expressed during different stages of human development. For instance, the epsilon (ε)- and zeta (ζ)-globin genes are expressed very early in embryonic life, while the gamma (γ)-globin genes are maximally expressed during the second and third trimesters of gestation. Following birth, the γ-globin genes are turned off and the β-globin gene is turned on. These differences in the expression of the globin genes reflect the differences in the oxygen transport needs of humans during the embryonic, fetal, and postpartum stages of life.

The Human Genome Project Has Stimulated Genomic Research

Before ending our discussion of genomes, let's consider what is now the largest genome project in history, the **Human Genome Project**. Scientists had been discussing how to undertake this project since the mid-1980s. In 1988, the National Institutes of Health in Bethesda, Maryland, established an Office of Human Genome Research with James Watson as its first director. The Human Genome Project, which officially began on October 1, 1990, was a 13-year effort coordinated by the U.S. Department of Energy and the National Institutes of Health. From its outset, the Human Genome Project had the following goals:

1. *To identify all human genes.* This involved mapping the locations of genes throughout the entire genome. The data from the Human Genome Project suggest that humans have about 20,000 to 25,000 different genes.
2. *To obtain the DNA sequence of the entire human genome.* The first draft of a nearly completed DNA sequence was published in February 2001, and a second draft was published in 2003. The entire genome is approximately 2.9 billion nucleotides in length. If the entire human genome were typed in a textbook like this, with about 3,000 letters per page, it would be nearly 1 million pages long!
3. *To develop technology for the generation and management of human genome information.* Some of the efforts of the Human Genome Project have involved improvements in molecular genetic technology such as gene cloning, DNA sequencing, and so forth. The Human Genome Project has also developed computer tools to allow scientists to easily access up-to-date information from the project, and analytical tools to interpret genomic information.
4. *To analyze the genomes of model organisms.* These include *E. coli*, *S. cerevisiae*, *D. melanogaster*, *C. elegans*, *A. thaliana*, and *M. musculus*.
5. *To develop programs focused on understanding and addressing the ethical, legal, and social implications of the results obtained from the Human Genome Project.*

An important question is: Who should have access to genetic information? Should employers, insurance companies, law enforcement agencies, and schools know our genetic makeup? The answer is complex and will require discussion among many groups.

A great benefit expected from the characterization of the human genome is the ability to identify our genes. A complete DNA sequence has made it profoundly easier for researchers to locate such genes. Furthermore, the DNA sequence of the human genome provides researchers with insight into the types of proteins encoded by these genes.

21.2 Proteomes

Thus far in this chapter, we have considered the genome characteristics of many different species, including humans. Because most genes encode proteins, a logical next step is to examine the functional roles of the proteins that a species can make. As mentioned, the entire collection of proteins that a cell or organism produces is called a proteome. As we move into the 21st century, a key challenge facing molecular biologists is the study of proteomes. Much like the study of genomes, this will require the collective contributions of many research scientists, as well as improvements in technologies that are aimed at explaining the complexities of the proteome. In this section, we will begin by considering the functional categories of proteins and examining their relative abundance in the proteome. Then, we will explore the molecular mechanisms that cause an organism's proteome to be much larger than its genome.

The Proteome Is a Diverse Array of Proteins with Many Kinds of Functions

The genomes of simple, unicellular organisms such as bacteria and yeast contain thousands of structural genes, while the genomes of complex, multicellular organisms contain tens of thousands. Such genome sizes can produce proteomes with thousands of different proteins. To bring some order to this large amount of complex information, researchers often organize proteins into different categories based on their functions.

Table 21.3 describes some general categories of protein function and provides examples of each type. Many approaches are used to categorize proteins. The categories listed in Table 21.3 are just one of the more general ways to evaluate protein function. For example, the data of Figure 21.1 categorize protein function in a different, more complicated way.

Dividing proteins into categories by function allows scientists to compare the amount of each type of protein, the protein abundance. The relative abundance of proteins can be viewed at two levels. First, we can consider abundance in the genome: the numbers of genes in the genome that encode a particular type or category of protein. For example, if a cell has genes that encode 10,000 different types of proteins, and 1,500 of these are different types of transporters, we would say that 15% of the genome is composed of transporters. However, such an analysis

Table 21.3	Categories of Proteins Found in the Proteome

Function	Examples
Metabolic enzymes	Hexokinase: Phosphorylates glucose during the first step in glycolysis. Glycogen synthetase: Uses glucose to synthesize a large carbohydrate known as glycogen.
Structural proteins	Tubulin: Forms cytoskeletal structures known as microtubules. Collagen: Found abundantly in the extracellular matrix of animals.
Motor proteins	Myosin: Involved in muscle cell contraction. Kinesin: Involved in the movement of chromosomes during cell division.
Cell-signaling proteins	Insulin: Influences target cell metabolism and growth. Insulin receptor: Recognizes insulin and initiates a cellular response.
Transport proteins	Lactose permease: Transports lactose across the bacterial cell membrane. Hemoglobin: Found in red blood cells, and transports oxygen throughout the body.
Gene expression and regulation proteins	Transcription factors: Regulate the expression of genes. Ribosomal proteins: Make up the structure of ribosomes, which are needed for the synthesis of new proteins.
Protective proteins	Antibodies: Fight viral and bacterial infections in vertebrate species. Antiviral proteins: Prevent viruses from binding to cells or successfully infecting cells in plants and animals.

Liver cell

Abundance in genome	
Genes for metabolic enzymes	25%
Genes for structural proteins	5%
Genes for motor proteins	< 2%

Abundance in cell	
Metabolic enzymes	> 50%
Structural proteins	< 10%
Motor proteins	< 5%

Skeletal muscle cell

Abundance in genome	
Genes for metabolic enzymes	25%
Genes for structural proteins	5%
Genes for motor proteins	< 2%

Abundance in cell	
Metabolic enzymes	< 10%
Structural proteins	20–30%
Motor proteins	25–40%

Figure 21.9 **A comparison of the proteomes in human liver and skeletal muscle cells.** Because all cells of the human body carry the same genome, the percentages of proteins that are encoded in the genome are the same in each cell type. However, the relative amounts of proteins that are made in different cell types can be vastly different, as is the case between liver and skeletal muscle cells.

Biological inquiry: What genetic process explains the differences in protein abundance in liver cells versus muscle cells?

ignores the phenomenon that genes are expressed at different levels—in other words, different types of proteins are made in different amounts. Therefore, a second way to view protein abundance is to consider abundance in the cell: the amount of a given protein or protein category that is actually made by a living cell. For example, less than 1% of human genes encode proteins, such as collagen, that are found in the extracellular matrix. Even so, these genes are highly expressed in certain cells, so that a large amount of this type of protein is made compared to other types.

Figure 21.9 is a general comparison of protein abundance in two cell types in humans—liver and muscle cells. The liver plays a key role in metabolism, while muscles are involved in bodily movements. Both liver and muscle cells have the same genes. Therefore, at the level of the genome, the percentages of the different protein categories are identical. However, at the cellular level, the relative abundance of certain protein categories is quite different. Liver cells make a large number of different enzymes that play a role in the metabolism of fats, proteins, and carbohydrates. By comparison, their level of structural and motor proteins is relatively small. In contrast, muscle cells have fairly low levels of enzymes, but they have a high percentage of structural and motor proteins.

The following discussion describes each category of protein, its relative abundance in both the genome and the cell, and trends in abundance that are observed among different groups of species and cell types.

Metabolic Enzymes Metabolic enzymes, which accelerate chemical reactions within the cell, are a key category of proteins. Some enzymes assist in the breakdown of molecules or macromolecules into smaller units. These are known as catabolic enzymes and are important in generating cellular energy. In contrast, anabolic enzymes function in the synthesis of molecules and macromolecules. For example, glycogen synthetase is required for the synthesis of glycogen from glucose building blocks. In terms of abundance in the genome, typically 20–30% of a cell's genes encode metabolic enzymes.

However, many of these genes are expressed at relatively low levels, so that the abundance of metabolic enzymes in most cell types may be less than 20–30% of the total protein.

Structural Proteins Structural proteins are involved in providing shape and form to cells and organisms. Examples include cytoskeletal proteins, such as actin or tubulin, and proteins of the extracellular matrix, such as collagen. Structural proteins are important in all species, but they tend to be more abundant in multicellular organisms such as plants and animals. In terms of abundance in the genome, only about 5% of a eukaryotic cell's genes encode structural proteins. However, certain structural proteins are expressed at very high levels, so these proteins can be much more abundant than 5% of the proteins made by particular cells. For example, in mammals, roughly 25% of the total protein in the body is a type of structural protein called collagen.

Motor Proteins Motor proteins use energy to facilitate intracellular movements and the movements of whole cells. Examples are dynein and myosin. These motor proteins provide the power to move eukaryotic flagella and cause muscles to contract, respectively. Overall, genes that encode motor proteins constitute a small percentage (that is, less than 2%) of the genome of a cell or organism. However, in certain cell types, motor proteins can be abundant. For example, myosin is very abundant in muscle cells. In skeletal or heart muscle cells, usually 25–40% of the total protein is myosin.

Cell-Signaling Proteins Many different types of proteins are needed so that cells can respond to environmental signals and send signals to each other. Proteins involved in cell signaling include hormones, hormone receptors, and intracellular proteins that form signal cascades, allowing a cell to respond to a signaling molecule. Cell signaling is needed by all species to respond to environmental changes. Among multicellular organisms, such as plants and animals, cell-to-cell signaling is more complex, so that the cells of the body can coordinate their activities. In terms of abundance in the human genome, roughly 12% of our genes encode proteins that are devoted to cell signaling. However, many proteins involved with cell signaling are only needed in small amounts, so their abundance in cells is considerably less than 12%.

Transport Proteins A diverse array of membrane proteins are involved in the transport of ions and molecules across membranes. Among all species, typically 10–15% of the genome is made up of genes that encode proteins involved in transport. The protein abundance in the cell is usually less than 10–15%, though certain specialized cells, such as nerve cells, have a greater abundance of transport proteins in their membranes. Most transport proteins are involved in the transmembrane transport of solutes. These include proteins such as sugar transporters, ion pumps, and ion channels. Other proteins are involved in intercellular transport. For example, hemoglobin plays a role in the transport of oxygen throughout the bodies of vertebrates.

Gene Expression and Regulation Proteins For proteins to be made, genes must be expressed. This process involves transcription, mRNA modification, and translation. In addition, genes are regulated so that their encoded product will be expressed at the correct times and in the correct amounts. Genes that encode proteins involved in gene expression and regulation constitute a large percentage of the genomes of all organisms, perhaps 25–30%. In terms of abundance in the cell, well over 25% of the different types of proteins made by cells are devoted to gene expression and regulation. Among more complex eukaryotes, the percentage tends to be higher. This increase can be attributed largely to the complexities of gene regulation and mRNA modification. In complex plants and animals, a substantial percentage of the genome is made up of genes that encode transcription factor proteins that regulate gene expression. For example, in both *Arabidopsis* and humans, roughly 5% of all genes encode proteins that function as transcription factors.

Protective Proteins All species make proteins that help them to survive environmental stress. These include proteins that help organisms withstand high temperatures and other protective proteins that prevent damage caused by infectious agents. Mammals, for example, produce proteins called antibodies that attack infectious agents such as viruses and bacteria and thereby help to ward off infection. Plants also produce a diverse array of proteins that prevent viral invasion.

From the perspective of abundance, this category of proteins is somewhat difficult to calculate. At the genome level, the percentage of genes that encode proteins involved in protection would be relatively small, less than 2%. However, in vertebrates, genes that encode antibodies undergo a specialized type of rearrangement in the B cells of the immune system (see Chapter 53). With regard to the whole organism, antibodies are by far the most diverse of all proteins; a single person can make millions of different types of antibodies, each with a slightly different amino acid sequence. Even so, each B cell makes only one type of antibody. The amount of protein made by a given cell can vary greatly depending on the cell's environment. For example, in the absence of an infection, the amounts of antiviral proteins that are synthesized by animal and plant cells are relatively small. In contrast, when confronted with a pathogen, the amounts of protective proteins produced by these cells greatly increase.

Proteomes Are Larger Than Genomes

From the sequencing and analysis of genomes, researchers can identify all or nearly all of the genes that a given species has. For example, the human genome is predicted to contain between 20,000 and 25,000 different genes that encode proteins. Even so, humans can make many more than 25,000 different types of proteins. The larger size of the proteome relative to the genome is primarily due to two types of cellular processes, as described next.

First, changes in pre-mRNA structure may ultimately affect the resulting amino acid sequence of a protein. The most important alteration that occurs commonly in eukaryotic species is **alternative splicing**, which is described in Chapter 13. For many genes, a single pre-mRNA can be spliced into more than one version (**Figure 21.10a**). The splicing is often cell specific, or it may be related to environmental conditions. Alternative splicing is widespread, particularly among more complex eukaryotes (refer back to Table 13.1). It can lead to the production of several or perhaps dozens of different polypeptide sequences from the same pre-mRNA. This greatly increases the number of potential proteins in the proteome.

A second process that greatly diversifies the composition of a proteome is the phenomenon of **post-translational covalent modification** (**Figure 21.10b**). Such modifications can be permanent or reversible. Permanent modifications are primarily involved with the assembly and construction of functional proteins. These alterations include proteolytic processing (the cleavage of a polypeptide to a smaller unit), disulfide bond formation, and the attachment of prosthetic groups, sugars, or lipids. In contrast, reversible modifications, such as phosphorylation, acetylation, and methylation, often transiently affect the function of a protein. The covalent bonds are formed and then later broken by cellular enzymes. Because a given type of protein may be subjected to several different types of modifications, this can greatly increase the forms of a particular protein that are found in a cell at any given time.

21.3 Bioinformatics

In the two previous sections, we have learned that the number of genes in a genome and the number of proteins that are made by a given cell type are extremely large. In the 1960s and 1970s, when the tools of molecular biology first became available, researchers tended to focus on the study of just one or a few genes and proteins at a time. While this is a useful approach, scientists came to realize that certain properties of life arise by complex interactions involving the expression of many genes and the functioning of many different proteins. This is the field of systems biology. Such an awareness challenged researchers to invent new tools to study many genes and many proteins at the same time. These tools involved experimental procedures—genomics and proteomics—that allowed researchers to study complex interactions. To manage the huge amounts of data produced by these studies, researchers turned to the use of computers.

As a very general definition, bioinformatics describes any use of computers to handle biological information. Scientists primarily use computers to store and analyze data. We often think of bioinformatics in the context of analyzing genetic data. Even so, bioinformatics can also be applied to information from various sources, such as patient statistics and scientific literature.

(a) Alternative splicing

(b) Post-translational covalent modification

Figure 21.10 Cellular mechanisms that increase protein diversity. (a) During alternative splicing, the pattern of exons that remains in a mature mRNA can be different, creating multiple types of transcripts from the same gene. (b) In post-translational covalent modification, after a protein is made, it can be modified in a variety of ways, some of which are permanent and some reversible.

This rapidly developing branch of biology is highly interdisciplinary, using principles from mathematics, statistics, biology, chemistry, and physics.

But why do we need bioinformatics? Simply put, the main issues are size and speed. Earlier in this chapter, we learned that the human genome has been sequenced and that it is approximately 2.9 billion base pairs long. A single person, or even a group of talented mathematicians, cannot, in a reasonable length of time, analyze such an enormous amount of data. Instead, the data are put into computers and then scientists devise computational procedures to study it. The ability of computers to analyze data at a rate of millions or even billions of operations per second has made it possible to solve biological problems that were thought intractable a few decades ago. Biologists have made many important discoveries using a bioinformatic approach. For example, we can compare the entire genome sequences of different organisms, which contain millions or even billions of base pairs, and identify genes that are homologous in two or more different species. We can also compare genes in the same species and discover how mutations in particular genes are correlated with the development of diseases such as cancer and diabetes.

In this section, we will consider the branch of bioinformatics that focuses on using molecular information to study biology. This area, also called **computational molecular biology**, uses computers to characterize the molecular components of living things. Molecular genetic data, which comes in the form of a DNA, RNA, or protein sequence, are particularly amenable to computer analysis. Such studies usually rely on three basic components: a computer (or hardware), a computer program (or software), and some type of data. In genetic research, the data are typically a particular genetic sequence or several sequences that a researcher or clinician wants to study. In this section, we will first survey the fundamental concepts that underlie the analysis of genetic sequences. We will then consider how these methods are used to provide knowledge about how biology works at the molecular level.

Sequence Files Are Stored and Analyzed by Computers

The first step in bioinformatics is to collect and store data in a computer. As an example, let's consider a gene sequence as a type of data. The gene sequence must first be determined experimentally using the technique of DNA sequencing. After the sequence is obtained, the next step is to put that data into a computer. To store data in a computer, a scientist creates a computer data file and enters the data into the file. For short genetic sequences, less than a few hundred nucleotides in length, this may be done using a keyboard to enter (that is, type) the DNA sequence into the file. For very long genetic sequences, however, like those obtained in genome-sequencing projects, keyboarding data would be a tedious and error-prone process. Typically, genetic sequence data are entered into a computer file by laboratory instruments that can read experimental data—such as data from a DNA-sequencing experiment—and enter the sequence directly into a computer.

Genetic sequence data in a computer data file can then be investigated in many different ways, corresponding to the many questions a researcher might ask about the sequence and its functional significance, including:

1. Does a sequence contain a gene?
2. Does a gene sequence contain a mutation that might cause a disease?
3. Where are functional sequences such as promoters, regulatory sites, and splice sites located within a particular gene?
4. From the sequence of a structural gene, what is the amino acid sequence of the polypeptide encoded by that gene?
5. Is a genetic sequence similar to any other known sequences?
6. What is the evolutionary relationship between two or more genetic sequences?

To answer these and many other questions, computer programs have been written to analyze genetic sequences in particular ways.

As an example, let's consider a computer program aimed at translating a DNA sequence into an amino acid sequence. **Figure 21.11** shows a short computer data file of a DNA sequence that is presumed to be part of the coding sequence of a structural gene. In this figure, only the coding strand of DNA is shown. A computer program can analyze this sequence and print out the possible amino acid sequences that this DNA sequence would encode. The program relies on the triplet genetic code (refer back to Table 12.1). In the example shown in Figure 21.11, the computer program shows the results for all three possible reading frames, beginning at nucleotide 1, 2, or 3, respectively. In a newly obtained DNA sequence, a researcher would not know the proper reading frame, so the computer program provides all three. If you look at the results, reading frames 1 and 2 include several stop codons, while reading frame 3 does not. From these results, reading frame 3 is likely to be the correct one. Also, for a new DNA sequence, a researcher may not know which DNA strand is the coding strand. Therefore, the sequence of the other DNA strand, which is not shown in this figure, would also be analyzed by this computer program.

The Scientific Community Has Collected Computer Data Files and Stored Them in Large Computer Databases

Over the past several decades, the amount of genetic information generated by researchers and clinicians has become enormous. The Human Genome Project, for example, has produced more data than any other undertaking in the history of biology. With these advances, scientists have realized that another critical use of computers is to store the staggering amount of data produced from genetic research.

When a large number of computer data files are collected, stored in a single location, and organized for rapid search and retrieval, this collection is called a **database**. The files within databases are often annotated, which means they contain a concise description of each gene sequence, the name of the organism from which the sequence was obtained, and the function of the encoded protein, if it is known. The file may also

Computer DNA sequence file

```
5' ATGTCCACGC GGTCCTGGAA AACCCAGGCT TGGGCAGGAA ACTCTCTGAC
   TTTGGACAGG AAACAAGCTA TATTGAAGAC AACTGCAATC AAAATGGTGC
   CATATCACTG ATCTTCTCAC TCAAAGAAGA AGTTGGTGCA TTGGCCAAAG
   TATTGCGCTT ATTTGAGGAG AATGATGTAA ACCTGACCCA CATTGAATCT
   AGACCTTCTC GTTTAAAGAA AGATGGTAT  GAATTTTTCA CCCATTTGGA
   TAAACGTAGC CTGCCTGCTC TGACAAACAT CATCAAGATC TTGAGGCATG
   ACATTGGTGC CACTGTCCAT GAGCTTTCAC GAGATAAGAA GAAAGACACA
   GTGCCCTGGT TTCCCAAG 3'
```

Run a computer program that translates this DNA sequence into an amino acid sequence in all 3 reading frames.

Possible amino acid sequences

5' ⟶ 3' Frame 1

Met Ser Thr Arg Ser Trp Lys Thr Gln Ala Trp Ala Gly Asn Ser Leu Thr Leu Asp Arg Lys Gln Ala Ile Leu Lys Thr Thr Ala Ile Lys Met Val Pro Tyr His **STOP** Ser Ser His Ser Lys Lys Lys Leu Val His Trp Pro Lys Tyr Cys Ala Tyr Leu Arg Arg Met Met **STOP** Thr **STOP** Pro Thr Leu Asn Leu Asp Leu Leu Val **STOP** Arg Lys Met Ser Met Asn Phe Ser Pro Ile Trp Ile Asn Val Ala Cys Leu Leu **STOP** Gln Thr Ser Ser Arg Ser **STOP** Gly Met Thr Leu Val Pro Leu Ser Met Ser Phe His Glu Ile Arg Arg Lys Thr Gln Cys Pro Gly Ser Gln

5' ⟶ 3' Frame 2

Cys Pro Arg Gly Pro Gly Lys Pro Arg Leu Gly Gln Glu Thr Leu **STOP** Leu Trp Thr Gly Asn Lys Leu Tyr **STOP** Arg Gln Leu Gln Ser Lys Trp Cys His Ile Thr Asp Leu Leu Thr Gln Arg Arg Ser Trp Cys Ile Gly Gln Ser Ile Ala Leu Ile **STOP** Gly Glu **STOP** Cys Lys Pro Asp Pro His **STOP** Ile **STOP** Thr Phe Ser Phe Lys Glu Arg **STOP** Val **STOP** Ile Phe His Pro Phe Gly **STOP** Thr **STOP** Pro Ala Cys Ser Asp Lys His His Gln Asp Leu Glu Ala **STOP** His Trp Cys His Cys Pro **STOP** Ala Phe Thr Arg **STOP** Glu Glu Arg His Ser Ala Leu Val Pro Lys

5' ⟶ 3' Frame 3

Val His Ala Val Leu Glu Asn Pro Gly Leu Gly Arg Lys Leu Ser Asp Phe Gly Gln Glu Thr Ser Tyr Ile Glu Asp Asn Cys Asn Gln Asn Gly Ala Ile Ser Leu Ile Phe Ser Leu Lys Glu Glu Val Gly Ala Leu Ala Lys Val Leu Arg Leu Phe Glu Glu Asn Asp Val Asn Leu Thr His Ile Glu Ser Arg Pro Ser Arg Leu Lys Lys Asp Glu Tyr Glu Phe Phe Thr His Leu Asp Lys Arg Ser Leu Pro Ala Leu Thr Asn Ile Ile Lys Ile Leu Arg His Asp Ile Gly Ala Thr Val His Glu Leu Ser Arg Asp Lys Lys Lys Asp Thr Val Pro Trp Phe Pro

Figure 21.11 The use of a computer program to translate a DNA sequence into an amino acid sequence. The top part of this figure shows the sequence of a segment of the coding strand of a structural gene (artificially divided into groups of 10 nucleotide bases for ease of reading). A computer program translates the DNA sequence into an amino acid sequence based on the genetic code. The program produces three different amino acid sequences, as shown at the bottom of the figure. In this example, reading frame 3 is likely to be the correct reading frame because it does not contain any stop codons.

describe other features of significance and provide a published reference that contains the sequence.

The research community has collected genetic information from thousands of research labs and created several large databases. **Table 21.4** describes some of the major genetic databases in use worldwide. These databases enable researchers to access and compare genetic sequences that are obtained by many laboratories. Later in this chapter, we will learn how researchers can use databases to analyze genetic sequences.

The databases described in Table 21.4 collect genetic information from many different species. Scientists have also created more specialized databases, called **genome databases**, that focus on the genetic characteristics of a single species. Genome databases have been created for most model organisms such as bacteria (*E. coli*), yeast (*S. cerevisiae*), worms (*C. elegans*), fruit flies (*D. melanogaster*), plants (*A. thaliana*), mice (*M. musculus*), and humans (*H. sapiens*). The primary aim of genome databases is to organize the information from sequencing and mapping projects for a single species. Genome databases identify the known genes within an organism and describe their map locations in the genome. In addition, a genome database may provide information concerning gene alleles, bibliographic information, a directory of researchers who study the species, and other pertinent information.

Table 21.4	Examples of Major Computer Databases
Type	**Description**
Nucleotide sequence	DNA sequence data are collected into three internationally collaborating databases: GenBank (a U.S. database), EMBL (European Molecular Biology Laboratory Nucleotide Sequence Database), and DDBJ (DNA Databank of Japan). These databases receive sequence and sequence annotation data from genome projects, sequencing centers, individual scientists, and patent offices. These databases are accessed via the Internet and on CD-ROM.
Amino acid sequence	Amino acid sequence data are collected into a few international databases, including Swissprot (Swiss protein database), PIR (Protein Information Resource), Genpept (translated peptide sequences from the GenBank database), and TrEMBL (Translated sequences from the EMBL database).
Three-dimensional structure	PDB (Protein Data Bank) collects the three-dimensional structures of biological macromolecules with an emphasis on protein structure. These are primarily structures that have been determined by X-ray crystallography and nuclear magnetic resonance (NMR), but some models are included in the database. These structures are stored in files that can be viewed on a computer screen.

From: Persson, Bengt (2000) Bioinformatics in protein analysis. EXS 88, 215–231.

Computer Programs Can Identify Homologous Sequences

Let's now turn our attention to the use of computer technology to identify genes that are evolutionarily related. Organisms that are closely related evolutionarily tend to have genes with similar DNA sequences. As an example, let's consider the gene that encodes β-globin. As discussed earlier, β-globin is a polypeptide found in hemoglobin, which carries oxygen in red blood cells. The β-globin gene is found in humans and other vertebrates. **Figure 21.12a** compares a short region of this gene from the laboratory mouse (*Mus musculus*) and rat (*Rattus norvegicus*). As you can see, the gene sequences are similar but not identical. In this 60-nucleotide sequence, five differences are observed. The reason for the sequence similarity is that the genes are derived from the same ancestral gene. This idea is shown schematically in **Figure 21.12b**. An ancestral gene was found in a rodent species that was a common ancestor to both mice and rats. During evolution, this ancestral species diverged into different species, which eventually gave rise to several modern rodent species, including mice and rats. Following divergence, the β-globin genes accumulated distinct mutations that produced somewhat different base sequences for this gene. Therefore, in mice and rats, the β-globin genes have homologous sequences—their sequences are similar because they are derived from the same ancestral gene, but they are not identical because each species has accumulated a few different random mutations. Homologous genes in different species are also called **orthologs**.

Analyzing genes that are homologous to each other helps biologists understand the evolutionary relationships among modern species. This topic is considered in Units IV and V. Later in this section, we will also see how the study of homology may provide important clues about gene function.

But how do researchers, with the aid of computers, determine if two genes are homologous to each other? To evaluate the similarity between two sequences, a matrix can be constructed. In a general way, **Figure 21.13** illustrates the use of a simple dot matrix. In Figure 21.13a, the word BIOLOGY is compared with itself. Each point in the grid corresponds to one position of each sequence. The matrix allows all such pairs to be compared simultaneously. Dots are placed where the same letter occurs at the two corresponding positions. Sequences that are alike produce a diagonal line on the matrix. In contrast, Figure 21.13b compares two similar but different sequences: BIOLOGY and ECOLOGY. This comparison produces only a partial diagonal line. Overall, the key observation is that regions of similarity are distinguished by the occurrence of many dots along a diagonal line within the matrix. This same concept holds true when genetic sequences are compared with each other.

To relate homologous genes in different species, researchers must compare relatively long DNA sequences. For such long sequences, a simple dot matrix approach is not adequate. Instead, dynamic computer programming methods are used to identify similarities between genetic sequences. This approach was

(a) A comparison of one DNA strand of the mouse and rat β-globin genes

(b) The formation of homologous β-globin genes during evolution of mice and rats

Figure 21.12 Structure and formation of the homologous β-globin genes in mice and rats.   **(a)** A comparison of a short region of the gene that encodes β-globin in laboratory mice (*Mus musculus*) and rats (*Rattus norvegicus*). Only one DNA strand is shown. Bases that are identical between the two sequences are connected by a vertical line. The β-globin genes are similar because they are derived from the same ancestral gene, but they are not identical because each species has accumulated a few different random mutations since the divergence of their latest common ancestor. **(b)** The formation of these homologous β-globin genes during evolution. An ancestral β-globin gene was found in a rodent species that was a common ancestor to both mice and rats. This ancestral species later diverged into different species, which gave rise to modern rodent species, such as mice and rats. During this process, the β-globin genes accumulated different random mutations, causing the DNA sequences of these two homologous genes to be slightly different from each other.

proposed originally by Saul Needleman and Christian Wunsch in 1970. Dynamic programming methods are theoretically similar to a dot matrix, but they involve mathematical operations that are beyond the scope of this textbook. In their original work, Needleman and Wunsch demonstrated that whale myoglobin and human β-globin genes have similar sequences.

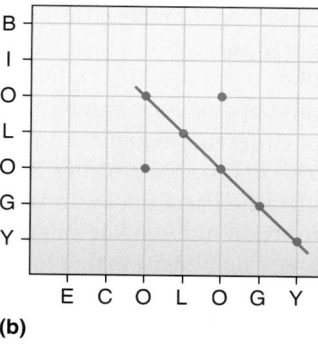

(a)   **(b)**

Figure 21.13 **The use of a simple dot matrix.** In these comparisons, a diagonal line indicates sequence similarity. **(a)** The word BIOLOGY is compared with itself. Dots are placed where the same letter occurs at the two corresponding positions. Notice the diagonal line that is formed. **(b)** Two similar but different sequences, BIOLOGY and ECOLOGY are compared with each other. Notice that only a partial line is formed by this comparison.

A Database Can Be Searched to Identify Homologous Sequences and Thereby Identify Gene Function

Because they are derived from the same ancestral gene, homologous genes usually carry out similar or identical functions. In many cases, the first indication of the function of a newly determined gene sequence is through homology to known sequences in a database. An example is the gene that is altered in cystic fibrosis patients. After this gene was identified in humans, bioinformatic methods revealed that it is homologous to several genes found in other species. Moreover, a few of the homologous genes were already known to encode proteins that function in the transport of ions and small molecules across the plasma membrane. This observation provided an important clue that cystic fibrosis involves a defect in membrane transport.

The ability of computer programs to identify homology between genetic sequences provides a powerful tool for predicting the function of genetic sequences. In 1990, Stephen Altschul, David Lipman, and their colleagues developed an approach called a **basic local alignment search tool**, or **BLAST**. The BLAST program has been described by many biologists as the single most important tool in computational molecular biology. This computer program can start with a particular genetic sequence—either a nucleotide or an amino acid sequence—and then locate homologous sequences within a large database.

As an example of how the BLAST program works, let's consider the human enzyme phenylalanine hydroxylase, which functions in the metabolism of phenylalanine, an amino acid. Recessive mutations in the gene that encodes this enzyme are responsible for the disease called phenylketonuria (PKU). The computational experiment shown in **Table 21.5** started with the amino acid sequence of this protein and used the BLAST program to search the Swissprot database, which contains hundreds

Table 21.5	Results from a BLAST Program Comparing Human Phenylalanine Hydroxylase with Database Sequences

Match	Percentage of identical* amino acids	Species	Function of of sequence†
1	100	Human (*Homo sapiens*)	Phenylalanine hydroxylase
2	99	Orangutan (*Pongo pygmaeus*)	Phenylalanine hydroxylase
3	95	Mouse (*Mus musculus*)	Phenylalanine hydroxylase
4	95	Rat (*Rattus norvegicus*)	Phenylalanine hydroxylase
5	89	Chicken (*Gallus gallus*)	Phenylalanine hydroxylase
6	82	Pipid frog (*Xenopus tropicalis*)	Phenylalanine hydroxylase
7	82	Green pufferfish (*Tetradon nigroviridis*)	Phenylalanine hydroxylase
8	82	Zebrafish (*Danio rerio*)	Phenylalanine hydroxylase
9	80	Japanese pufferfish (*Takifugu rubripes*)	Phenylalanine hydroxylase
10	75	Fruit fly (*Drosophila melanogaster*)	Phenylalanine hydroxylase

* The number indicates the percentage of amino acids that are identical with the amino acid sequence of human phenylalanine hydroxylase.

† In some cases, the function of the sequence was determined by biochemical assay. In other cases, the function was inferred due to the high degree of sequence similarity with other species.

of thousands of different protein sequences. The BLAST program can determine which sequences in the Swissprot database are the closest matches to the amino acid sequence of human phenylalanine hydroxylase. Table 21.5 shows the results—the 10 best matches to human phenylalanine hydroxylase that were identified by the program. Because this enzyme is found in nearly all eukaryotic species, the program identified phenylalanine hydroxylase from many different species. The column to the right of the match number shows the percentage of amino acids that are identical between the species indicated and the human sequence. Because the human phenylalanine hydroxylase sequence is already in the Swissprot database, the closest match of human phenylalanine hydroxylase is to itself. The next nine sequences are in order of similarity. The next most similar sequence is from the orangutan, a close relative of humans. This is followed by two mammals, the mouse and rat, and then five vertebrates that are not mammals. The tenth best match is from *Drosophila*, an invertebrate.

You can see two trends in Table 21.5. First, the order of the matches follows the evolutionary relatedness of the various species to humans. The similarity between any two sequences is related to the time that has passed since they diverged from a common ancestor. Among the species listed in this table, humans are most similar to themselves, followed by the orangutan, other mammals, other vertebrates, and finally invertebrates. A second trend you may have noticed is that several of the matches involve species that are important from a research, medical, or agricultural perspective. Currently, our databases are biased toward organisms that are of interest to humans, particularly model organisms such as mice and *Drosophila*. Over the next several decades, the sequencing of genomes from many different species will tend to lessen this bias.

Overall, Table 21.5 is an example of the remarkable computational abilities of current computer technology. In less than a minute, the amino acid sequence of human phenylalanine hydroxylase can be compared with hundreds of thousands of different sequences to yield the data shown in this table! The main power of the BLAST program is its use with newly identified sequences, in which a researcher does not know the function of a gene or an encoded protein. When the BLAST program identifies a match to a sequence whose function is already known, it is likely that the newly identified sequence has an identical or similar function.

CHAPTER SUMMARY

21.1 Genome Sizes and Composition

- The genome is the complete genetic makeup of a cell or organism.
- Prokaryotic genomes are typically a single circular chromosome that has several hundred thousand to a few million base pairs of DNA. Such genomes usually have a few thousand different genes. (Table 21.1)
- Venter, Smith, and colleagues used a shotgun DNA-sequencing strategy to determine the first sequence of a prokaryotic genome, that of *Haemophilus influenzae*. (Figure 21.1)
- The nuclear genomes of eukaryotic species are composed of sets of linear chromosomes with a total length of several million to billions of base pairs. They typically contain several thousand to tens of thousands of genes. (Table 21.2)
- Genomes sizes vary among eukaryotic species. In many cases, this variation is due to the accumulation of noncoding regions of DNA, particularly repetitive DNA sequences. (Figures 21.2, 21.3)
- Transposable elements are segments of DNA that can move from one site to another, a process called transposition. (Figure 21.4)
- The enzyme transposase mediates the movement of transposable elements by a cut-and-paste mechanism. (Figure 21.5)
- Retroelements move to new sites in the genome via RNA intermediates. (Figure 21.6)
- Gene duplication may occur by a misaligned crossover. This is one mechanism that creates gene families, which are composed of two or more homologous genes. The members of a gene family often have similar yet specialized functions. (Figures 21.7, 21.8)
- The Human Genome Project undertook the DNA sequencing of the entire human genome.

21.2 Proteomes

- A proteome is the collection of proteins that a given cell makes.
- Proteins are often placed into broad categories based on their functions. These include metabolic enzymes, structural proteins, motor proteins, cell-signaling proteins, transport proteins, proteins involved with gene expression and regulation, and those involved with protection. (Table 21.3)
- Protein abundance can refer to their relative abundance in the genome or in the cell. (Figure 21.9)
- Protein diversity can increase via alternative splicing and post-translational modifications. (Figure 21.10)

21.3 Bioinformatics

- Bioinformatics involves the use of computers to analyze biological information, particularly genetic data such as DNA and protein sequences.
- Genetic information is stored in data files that can be analyzed using computer programs. (Figure 21.11)
- The research community has collected genetic information and created several large databases. (Table 21.4)
- Homologous genes are derived from the same ancestral genes and have accumulated random mutations that make their sequences slightly different. (Figure 21.12)
- A simple dot matrix illustrates the approach of identifying regions of similarity between two sequences. (Figure 21.13)
- Computer programs, such as BLAST, can identify homologous genes that are found in a database. (Table 21.5)

TEST YOURSELF

1. The entire collection of proteins produced by a cell or organism is
   a. a genome.
   b. bioinformatics.
   c. a proteome.
   d. a gene family.
   e. a protein family.

2. Important reasons for studying the genomes of prokaryotes include all of the following except:
   a. It may provide information that helps us understand how prokaryotes infect other organisms.
   b. It may provide a basic understanding of cellular processes that allow us to determine eukaryotic cellular function.
   c. It may provide the means to understand evolutionary processes.
   d. It will reveal the approximate number of genes that an organism has in its genome.
   e. All of the above are important reasons.

3. The enzyme that allows short segments of DNA to move within a cell from one location in the genome to another is
   a. transposase.
   b. DNA polymerase.
   c. protease.
   d. restriction endonuclease.
   e. reverse transcriptase.

4. A gene family includes
   a. a specific gene found in several different species that has a similar DNA sequence.
   b. all of the genes on the same chromosome.
   c. two or more homologous genes found within a single species.
   d. genes that code for structural proteins.
   e. both a and c.

5. Which of the following is not a goal of the Human Genome Project?
   a. Identify all human genes.
   b. Sequence the entire human genome.
   c. Address the legal and ethical implications resulting from the project.
   d. Develop programs to manage the information gathered from the project.
   e. Clone a human.

6. Bioinformatics is
   a. the analysis of DNA by molecular techniques.
   b. the use of computers to analyze and store biological information.
   c. a collection of gene sequences from a single individual.
   d. cloning.
   e. all of the above.

7. Using bioinformatics, evolutionary relationships between species can be characterized by identifying and analyzing
   a. phenotypes of selected organisms.
   b. homologous DNA sequences.
   c. fossils of ancestral species.
   d. all of the above.
   e. a and b only.

8. The BLAST program is a tool for
   a. inserting many DNA fragments into a cell at the same time.
   b. translating a DNA sequence into an amino acid sequence.
   c. identifying homology between a selected sequence and genetic sequences in large databases.
   d. all of the above.
   e. both b and c.

9. Repetitive sequences
   a. are short DNA sequences that are found many times throughout the genome.
   b. may be multiple copies of the same gene found in the genome.
   c. are more common in eukaryotes.
   d. all of the above.
   e. a and c only.

10. Proteins that provide movement of material inside the cell and movement of the entire cell are
    a. enzymes.
    b. structural proteins.
    c. motor proteins.
    d. transport proteins.
    e. antibodies.

CONCEPTUAL QUESTIONS

1. Define bioinformatics.
2. Explain how homologous genes may have arisen.
3. What are seven general categories of proteins found in the proteome?

EXPERIMENTAL QUESTIONS

1. What was the goal of the experiment conducted by Venter, Smith, and their colleagues?
2. How does shotgun DNA sequencing differ from procedures that involve mapping? What are an advantage and a disadvantage of the shotgun DNA sequencing approach?
3. What were the results of the study described in Figure 21.1?

COLLABORATIVE QUESTIONS

1. Compare the genomes of prokaryotic and eukaryotic organisms.
2. Discuss the concept and importance of transposable elements.

www.brookerbiology.com

This website includes answers to the Biological Inquiry questions found in the figure legends and all end-of-chapter questions.

22
ORIGIN AND HISTORY OF LIFE

A fossil fish. This approximately 50-million-year-old fossil of a unicorn fish (*Naso rectifrons*) is an example of the many different kinds of organisms that have existed since the origin of life on Earth.

Astronomers now think that the universe began with an explosion called the Big Bang about 13.7 billion years ago, when the first clouds of the elements hydrogen and helium were formed. Gravitational forces collapsed these clouds to create stars that converted hydrogen and helium into heavier elements, including carbon, nitrogen, and oxygen, which are the building blocks of life on Earth. These elements were returned to interstellar space by exploding stars called supernovas, forming clouds in which simple molecules such as water, carbon monoxide, and hydrocarbons were formed. The clouds then collapsed to make a new generation of stars and solar systems.

Our solar system began about 4.6 billion years ago after one or more local supernova explosions. According to one widely accepted scenario, about 500 planetesimals (asteroids and comets) occupied the region where Venus, Earth, and Mars are now found. The Earth, which is estimated to be 4.55 billion years old, grew from the accumulation of planetesimals over a period of 100–200 million years. For the first half billion years or so after its formation, the Earth was too hot to allow water to accumulate on its surface. By 4 billion years ago, the Earth had cooled enough for the outer layers of the planet to solidify and for oceans to form.

The period between 4.0 and 3.5 billion years ago marked the emergence of life on our planet. Though scientists can never be certain how the first primitive life forms arose, plausible hypotheses have emerged from our understanding of modern life. The first forms of life that we know about produced well-preserved microscopic fossils that were found in western Australia. These fossils, which are 3.5 billion years old, resemble photosynthetic bacteria called cyanobacteria that live today (**Figure 22.1**).

The first section of this chapter will survey a variety of hypotheses regarding the potential origins of biological molecules and living cells. Keep in mind that many of these hypotheses are speculative, and are being changed as new information comes to light. Starting 3.5 billion years ago, the formation of fossils (such as the one shown in the chapter opening photo) has provided biologists with a history of life on Earth from its earliest beginnings to the present day. The last section of this chapter surveys a time line for the history of life. This chapter emphasizes when particular forms of life arose. Later chapters in this unit examine the mechanisms by which populations of organisms change over the course of many generations. This process, termed **biological evolution**, involves genetic changes that occur over the course of many generations. Such genetic modifications often lead to dramatic changes in traits and even the formation of new species.

84 µm

(a) Fossil prokaryote

112 µm

(b) Modern cyanobacteria

Figure 22.1   Earliest fossils and living cyanobacteria. **(a)** A fossilized prokaryote about 3.5 billion years old that is thought to be an early cyanobacterium. **(b)** A modern cyanobacterium, which has a similar morphology. Cyanobacteria are unicellular, but they aggregate into chains as shown here.

22.1 Origin of Life on Earth

As we learned in Units I, II, and III, living cells are complex collections of molecules and macromolecules. DNA stores the information for the amino acid sequence of proteins, RNA acts as an intermediary in the process of protein synthesis, and proteins form the foundation for the structure and activities of living cells. Life as we now know it requires this interplay between DNA, RNA, and proteins for its existence and perpetuation. On modern Earth, all living cells are made from pre-existing cells.

But how did life get started? As described in Chapter 1, living organisms have several characteristics that distinguish them from nonliving materials. Because DNA, RNA, and proteins are the central players in the enterprise of life, scientists who are interested in the origin of life have focused much of their attention on the formation of these macromolecules and their building blocks, namely nucleotides and amino acids. To understand the origin of life, we can view the process as occurring in four overlapping stages:

Stage 1: Nucleotides and amino acids were produced prior to the existence of cells.

Stage 2: Nucleotides and amino acids became polymerized to form DNA, RNA, and proteins.

Stage 3: Polymers became enclosed in membranes.

Stage 4: Polymers enclosed in membranes evolved cellular properties.

Obviously we cannot take a time machine back 4 billion years and determine with certainty how these events occurred. Instead, scientists study the existence of modern life, as well as geological processes and fossils, and speculate about the conditions that existed on primitive Earth. This approach has led researchers to a variety of hypotheses regarding the origin of life, none of which can be firmly verified. Nevertheless, certain possibilities are becoming more plausible, and perhaps even compelling. In this section, we will consider a few scientific viewpoints that wrestle with the question, How did life begin?

Stage 1: Several Scientific Hypotheses Have Been Proposed to Explain the Origin of Organic Molecules

Let's begin our inquiry into the first stage of the origin of life by considering how nucleotides and amino acids may have been generated prior to the existence of living cells. In the 1920s, the Russian biochemist Alexander Oparin and the Scottish biologist John Haldane independently proposed that organic molecules such as nucleotides and amino acids arose spontaneously under the conditions that occurred on primitive Earth. According to this hypothesis, the spontaneous appearance of organic molecules produced what they called a "primordial soup," which eventually gave rise to living cells.

The conditions on primitive Earth, which were much different than they are today, may have been more conducive than modern conditions to the spontaneous formation of organic molecules. In particular, scientists think that little oxygen was present, and the atmosphere instead contained inorganic carbon dioxide, nitrogen gas, and water vapor. Current hypotheses suggest that organic molecules, and eventually macromolecules, formed spontaneously; this is termed prebiotic (before life) or abiotic (without life) synthesis. These slowly forming organic molecules accumulated because there was little free oxygen, so they were not spontaneously oxidized, and there were as yet no living organisms, so they were also not metabolized. The slow accumulation of these molecules in the early oceans over a long period of time formed what is now called the **prebiotic soup**. The formation of this medium was a key event that preceded the origin of life.

Though most scientists agree that life originated from the assemblage of nonliving matter on primitive Earth, the mechanism of how and where these molecules originated is widely debated. Many intriguing hypotheses have been proposed. Keep in mind that these hypotheses are not mutually exclusive. Indeed, more than one mechanism may have contributed to the formation of a prebiotic soup. A few of the more widely debated ideas are described next.

Reducing Atmosphere Hypothesis Based largely on geological data, many scientists in the 1950s thought that the atmosphere on primitive Earth was rich in water vapor (H_2O), hydrogen gas (H_2), methane (CH_4), and ammonia (NH_3). These components, along with a lack of atmospheric oxygen, produce a reducing atmosphere because methane and ammonia readily give up electrons and thereby reduce other molecules. As described in Chapters 7 and 8, such oxidation-reduction reactions are required for the formation of complex organic molecules from simple inorganic molecules.

In 1953, Stanley Miller, a student in the laboratory of Harold Urey, was the first scientist to use experimentation to test whether the prebiotic synthesis of organic molecules is possible. His experimental apparatus was intended to simulate the conditions on primitive Earth that were postulated in the 1950s (**Figure 22.2**). Water vapor from a flask of boiling water rose into another chamber containing H_2, CH_4, and NH_3. Miller inserted two electrodes that sent electrical discharges into the chamber to simulate lightning bolts. A condenser jacket cooled some of the gases from the chamber, causing droplets to form that dropped into a trap. He then took samples from this trap for chemical analysis. In his first experiments, he observed the formation of hydrogen cyanide (HCN) and formaldehyde (CH_2O). Such molecules are precursors of more complex organic molecules. These precursors also combined to make larger molecules such as the amino acid glycine. Later experiments by Miller and others demonstrated the formation of sugars, many types of amino acids, and nitrogenous bases found in nucleic acids (for example, adenine).

The studies of Miller and Urey were particularly important because they were the first attempt to apply scientific experimentation to our quest to understand the origin of life. Their pioneering strategy was to determine if a scientific hypothesis is plausible, although it cannot prove that an event in the past really happened that way. In spite of the importance of these studies, critics of the reducing atmosphere hypothesis have argued that Miller and Urey were wrong about the composition of primitive Earth's environment.

Since the 1950s, ideas about the atmosphere on early Earth have changed. More recently, many scientists have suggested that the atmosphere on primitive Earth was not reducing, but instead was a neutral environment composed mostly of carbon monoxide (CO), carbon dioxide (CO_2), nitrogen gas (N_2), and H_2O. These newer ideas are derived from studies of volcanic gas, which has much more CO_2 and N_2 than CH_4 and NH_3, and from the observation that UV radiation destroys CH_4 and NH_3, so that these molecules would have been short-lived on primitive Earth. Nevertheless, since the experiments of Urey, many newer investigations have shown that organic molecules can be made under a variety of conditions. Using different combinations of gases based on these corrected assumptions of atmospheric conditions on primitive Earth, researchers have still achieved similar results. Namely, organic molecules can be made prebiotically from a neutral environment composed primarily of CO, CO_2, N_2, and H_2O.

Figure 22.2 Testing the reducing atmosphere hypothesis for the origin of life—the Miller and Urey experiment.

Biological inquiry: With regard to the origin of life, why are biologists interested in the abiotic synthesis of organic molecules?

At this time, the idea that organic molecules were made in the atmosphere of primitive Earth is plausible. Even so, other scientists have proposed alternative mechanisms for the production of organic molecules before the emergence of life.

Extraterrestrial Hypothesis Many scientists have argued that sufficient organic carbon would have been present in the asteroids and comets that reached the surface of early Earth to make them instrumental in stocking the prebiotic soup. Modern evidence in support of this idea comes from the study of these fallen bodies, called meteorites. A significant proportion of meteorites belong to a class known as carbonaceous chondrites. Such meteorites contain a substantial amount of organic carbon, including amino acids and nucleic acid bases. Based on this observation, scientists have postulated that such meteorites could have transported a significant amount of organic molecules to primitive Earth.

However, opponents of this hypothesis argue that most of this material would have been destroyed by the intense heating that accompanies the passage of large bodies through the atmosphere and their subsequent collision with the surface of the Earth. The degree to which heat would have destroyed the organic molecules remains a matter of controversy.

Deep-Sea Vent Hypothesis In 1988, the German organic chemist Günter Wächtershäuser proposed that key organic molecules may have originated in deep-sea vents, which are cracks in the Earth's surface where superheated water rich in metal ions and hydrogen sulfide (H_2S) mixes abruptly with cold sea water. These vents release hot gaseous substances from the interior of the Earth at temperatures in excess of 300°C (572°F). Supporters of this hypothesis propose that biologically important molecules may have been formed in the temperature gradient between the extremely hot vent water and the cold water that surrounds the vent at the bottom of the ocean (**Figure 22.3a**).

(a) Deep-sea vent hypothesis   **(b) A deep-sea vent community**

Figure 22.3 The deep-sea vent hypothesis for the origin of life. (a) Deep-sea vents are cracks in the Earth's surface that release hot gases such as hydrogen sulfide (H_2S). This heats the water near the vent and creates a gradient between the very hot water adjacent to the vent and the cold water that is farther away from the vent. The synthesis of organic molecules can occur in this gradient. (b) Photograph of a biological community near a deep-sea vent consisting of giant tube worms, clams, and crabs.

Experimentally, the temperatures within this gradient are known to be suitable for the synthesis of molecules that form components of biological molecules. For example, the reaction between iron and H_2S yields pyrites and H_2, and has been shown to provide the energy necessary for the reduction of N_2 to NH_3. Nitrogen is an essential ingredient of the molecular building blocks of life, amino acids and nucleic acids. But N_2, which is found abundantly on Earth, is chemically inert, so it is unlikely to have given rise to life. Most scientists believe instead that NH_3 was required to help life get started.

Interestingly, complex biological communities are found in the vicinity of modern deep sea vents. Various types of fish, worms, crabs, clams, shrimps, and bacteria are found in significant abundance in those areas (**Figure 22.3b**). Unlike most other forms of life on our planet, these organisms receive their energy from chemicals in the vent and not from the sun.

Stage 2: Organic Polymers May Have Formed on the Surface of Clay

The preceding three hypotheses provide reasonable mechanisms whereby small organic molecules could have accumulated on primitive Earth. Scientists hypothesize that the second stage in the origin of life was a period in which simple organic molecules polymerized to form more complex organic polymers such as DNA, RNA, or proteins. Most ideas regarding the origin of life assume that polymers with lengths of at least 30–60 monomers are needed to store enough information to make a viable genetic system. Experimentally, the prebiotic synthesis of such polymers is not possible in aqueous solutions because hydrolysis competes with polymerization. For this reason, many scientists have speculated that the synthesis of polymers did not occur in a prebiotic soup, but instead took place on a solid surface or in evaporating tidal pools.

In 1951, John Bernal first suggested that the prebiotic synthesis of polymers took place on clay. In his book *The Physical Basis of Life* he wrote that "clays, muds and inorganic crystals are powerful means to concentrate and polymerize organic molecules." Many clay minerals are known to bind organic molecules such as nucleotides and amino acids. In addition, negative charges within the clay itself attract metal divalent cations, such as Mg^{2+}, that can catalyze the chemical reactions that produce polymers.

Experimentally, many research groups have demonstrated the formation of nucleic acid polymers and polypeptides on the surface of clay, given the presence of monomer building blocks. During the prebiotic synthesis of RNA, the purine bases of the nucleotides interact with the silicate surfaces of the clay. Divalent cations, such as Mg^{2+}, bind the nucleotides to the negative surfaces of the clay, thereby positioning the nucleotides in a way that promotes bond formation between the phosphate of one nucleotide and the ribose sugar of an adjacent nucleotide. In this way, polymers such as RNA may have been formed.

Stage 3: Cell-Like Structures May Have Originated When Polymers Were Enclosed by a Boundary

The third stage in the origin of living cells is believed to be the formation of a boundary that separated the environment from internal polymers, such as RNA. The term **protobiont** (or prebiont) is used to describe the first nonliving structures that evolved into living cells. Protobionts had four characteristics that put them on the path to living cells:

1. A boundary, such as a membrane, separated the external environment from the internal contents of the protobiont.
2. Polymers inside the protobiont contained information.
3. Polymers inside the protobiont had enzymatic functions.
4. The protobionts were capable of self-replication.

Scientists envision protobionts as aggregates of prebiotically produced molecules and macromolecules that acquired a boundary, such as lipid bilayer, that allowed them to maintain an internal chemical environment distinct from that of their surroundings. Protobionts were not capable of precise self-reproduction like living cells, but probably could divide to increase in number. Such protobionts are thought to have exhibited basic metabolic pathways in which the structures of organic molecules were changed. In particular, the polymers inside protobionts must have gained the enzymatic ability to link together organic building blocks to create new polymers. This would have been a critical step in the process that eventually provided protobionts with the ability to self-replicate. According to this scenario, the ability of protobionts to self-replicate became more refined over time, and metabolic pathways became more complex. Eventually, these structures exhibited the characteristics that we attribute to living cells.

Different scenarios have been proposed to explain the formation of protobionts. Russian biologist Aleksandr Oparin hypothesized in 1924 that living cells evolved from **coacervates**, droplets that form spontaneously from the association of charged polymers such as proteins, carbohydrates, or nucleic acids. Their name derives from the Latin *coacervare*, meaning to assemble together or cluster. Coacervates measure 1–100 μm (micrometers) across, possess osmotic properties, and are surrounded by a tight skin of water molecules (**Figure 22.4a**). This boundary allows the selective absorption of simple molecules from the surrounding medium.

If enzymes are trapped within coacervates, they can perform primitive metabolic functions (**Figure 22.4b**). For example, researchers have made coacervates containing the enzyme glycogen phosphorylase. When glucose-1-phosphate was made available to the coacervates, it was taken up into them and starch was produced. The starch merged with the wall of the coacervates, which increased in size and eventually divided into two. When the enzyme amylase was included, the starch was broken down to maltose, which was released from the coacervates.

An alternative pathway to protobionts is the formation of **microspheres**, which are small water-filled vesicles surrounded by a macromolecular boundary (**Figure 22.4c**). If hot solutions

Figure 22.4 **Protobionts and their lifelike functions.**   Primitive cell-like structures like these could have given rise to living cells. **(a)** This micrograph shows coacervates made by Oparin from a mixture of gelatine (composed primarily of protein) and gum arabic (composed of protein and carbohydrate). **(b)** The illustration shows simple metabolism that can be performed by coacervates. **(c)** Micrograph and illustration of microspheres, which are water-filled spheres of macromolecules such as protein. **(d)** An electron micrograph and illustration of liposomes. Each liposome is made of a phospholipid bilayer surrounding an aqueous compartment.

of proteins are cooled under the correct conditions, they produce microspheres about 2 μm in diameter that are hollow and have an outer layer of protein.

Similar experiments have been done with lipids. When certain types of lipids are dissolved in water, they spontaneously form **liposomes**, which are vesicles surrounded by a lipid bilayer (**Figure 22.4d**). Interestingly, in 2003, Martin Hanczyc, Shelly Fujikawa, and Jack Szostak showed that clay can catalyze the formation of liposomes that grow and divide, a primitive form of self-replication. Furthermore, if RNA was on the surface of the clay, the researchers discovered that liposomes were formed that enclosed RNA. These experiments are exciting because they showed that the formation of membrane vesicles containing RNA molecules is a plausible route to the first living cells based on simple physical and chemical forces.

Stage 4: Cellular Characteristics May Have Evolved via Chemical Selection, Beginning with an RNA World

The majority of scientists favor RNA as the first macromolecule that was found in protobionts. Unlike other polymers, RNA exhibits three key functions. First, RNA has the ability to store

information in its nucleotide sequence. Second, due to base pairing, its nucleotide sequence has the capacity for replication. And third, RNA can perform a variety of enzymatic functions. The results of many experiments have shown that RNA molecules can function as **ribozymes**, acting as enzymes to synthesize the macromolecules found in living cells. By comparison, DNA and proteins are not known to have all three attributes. DNA is not known to have enzymatic activity, and proteins are not known to undergo self-replication. Thus, RNA appears to be the most self-sufficient substance of living matter. RNA can perform functions that are characteristic of proteins, and at the same time can serve as genetic material with replicative and informational functions.

But how did the RNA molecules that were first made prebiotically evolve into more complex molecules that produced cell-like characteristics? Researchers propose that a process called chemical selection was responsible for the increased complexity. **Chemical selection** occurs when a chemical within a mixture has special properties or advantages that cause it to increase in number compared to other chemicals in the mixture. Initially, scientists speculate that the special properties that enabled certain RNA molecules to undergo chemical selection were its ability to self-replicate and to perform other enzymatic functions.

First step of chemical selection   Second step of chemical selection

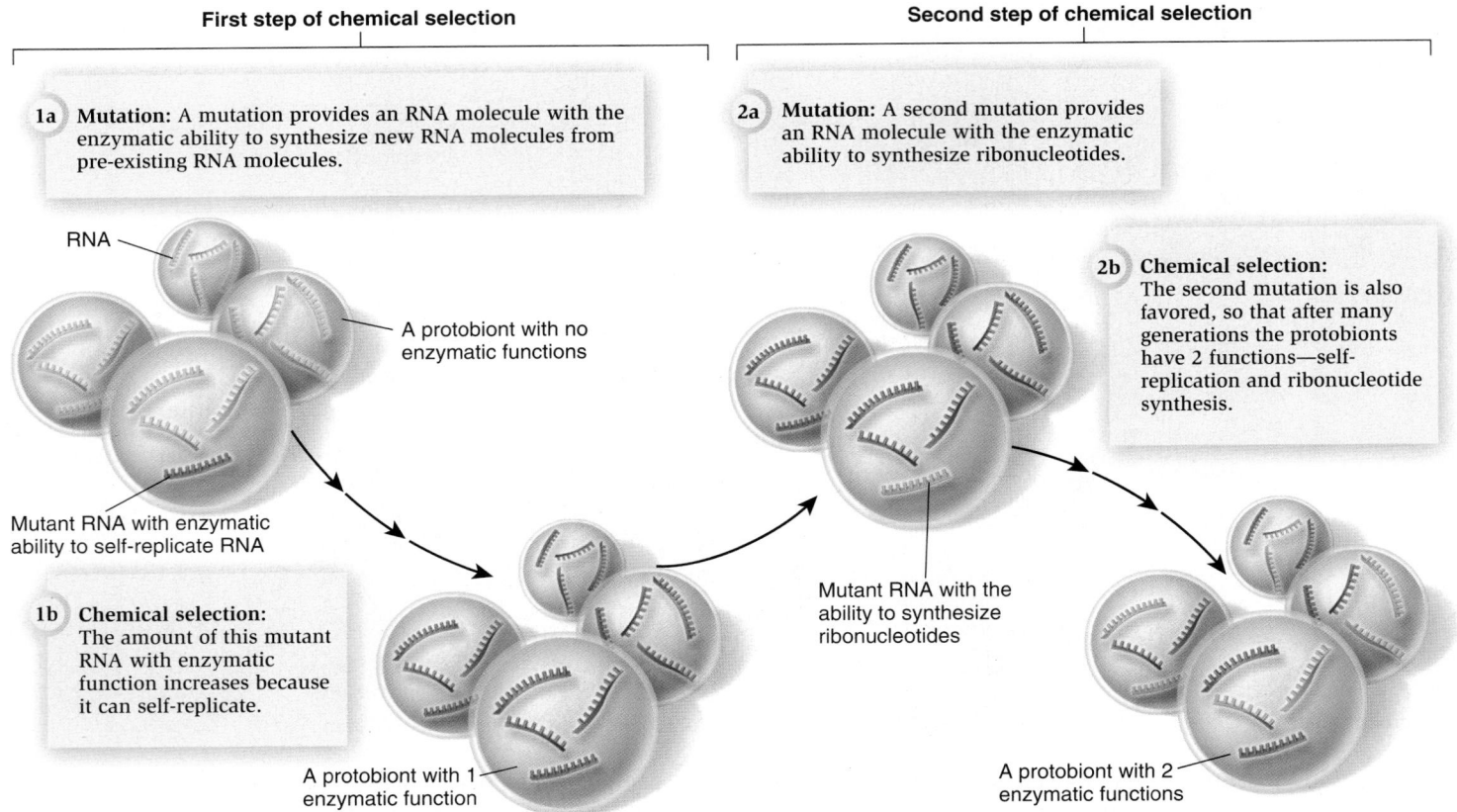

1a Mutation: A mutation provides an RNA molecule with the enzymatic ability to synthesize new RNA molecules from pre-existing RNA molecules.

2a Mutation: A second mutation provides an RNA molecule with the enzymatic ability to synthesize ribonucleotides.

RNA

A protobiont with no enzymatic functions

Mutant RNA with enzymatic ability to self-replicate RNA

2b Chemical selection: The second mutation is also favored, so that after many generations the protobionts have 2 functions—self-replication and ribonucleotide synthesis.

1b Chemical selection: The amount of this mutant RNA with enzymatic function increases because it can self-replicate.

Mutant RNA with the ability to synthesize ribonucleotides

A protobiont with 1 enzymatic function

A protobiont with 2 enzymatic functions

Figure 22.5   A hypothetical scenario illustrating the process of chemical selection.   This figure shows a two-step scenario. The first step selects for self-replication and the second step selects for the ability to synthesize ribonucleotides.

Biological inquiry: What is meant by the term chemical selection?

Because cells are a complex mixture of different macromolecules that carry out many enzymatic functions, researchers cannot establish with certainty the specific sequence of steps that produced modern cells. However, as a way to understand the concept of chemical selection, let's consider a hypothetical scenario showing two steps of chemical selection. **Figure 22.5** shows a group of protobionts that contain RNA molecules that were made prebiotically. RNA molecules inside these protobionts can be used as templates for the prebiotic synthesis of complementary RNA molecules. Such a process of self-replication, however, would be very slow because it would not be catalyzed by enzymes in the protobiont. In a first step of chemical selection, the sequence of one of the RNA molecules has undergone a mutation that gives it the enzymatic ability to attach nucleotides together, using RNA molecules as a template. This protobiont would have an advantage over the others because it would be capable of faster self-replication of its RNA molecules. Over time, due to its enhanced rate of replication, this type of protobiont would increase in number compared to the others. Eventually, the group of protobionts shown in the figure contains only this type of enzymatically functional RNA.

A second step of chemical selection is also shown in Figure 22.5 (right side). A second mutation in an RNA molecule could produce the enzymatic function that would promote the synthe-

sis of ribonucleotides, the building blocks of RNA. This protobiont would have the advantage of not having to rely on the prebiotic synthesis of ribonucleotides, which also is a very slow process. Therefore, the protobiont having the ability to both self-replicate its RNA molecules and synthesize ribonucleotides would have an advantage over a protobiont that could only self-replicate. Over time, the faster rate of ribonucleotide synthesis and self-replication would cause an increase in the numbers of the protobionts with both functions.

The **RNA world** is a hypothetical period on primitive Earth when both the information needed for life and the enzymatic activity of living cells were contained solely in RNA molecules. In this scenario, lipid membranes enclosing RNA exhibited the properties of life due to RNA genomes that were copied and maintained through the catalytic function of RNA molecules. Over time, scientists envision that mutations occurred in these RNA molecules, occasionally presenting new functional possibilities. Chemical selection for these new functions would have eventually produced an increase in complexity in these cells, with RNA molecules accruing activities such as peptide bond formation and other enzymatic functions.

But is an RNA world a plausible scenario? As described next in the Feature Investigation, chemical selection of RNA molecules can occur experimentally.

FEATURE INVESTIGATION

Bartel and Szostak Demonstrated Chemical Selection in the Laboratory

Remarkably, scientists have been able to perform experiments in the laboratory that can select for RNA molecules with a particular function. The first such study by David Bartel and Jack Szostak was conducted in 1993 (**Figure 22.6**). Using molecular techniques, they synthesized a mixture of 10^{15} RNA molecules that we will call the long RNA molecules. Each long RNA in this mixture contained two regions. The first region at the 5′ end was a constant region that formed a stem-loop structure. Its sequence was identical among all 10^{15} molecules. The constant region was next to a second region that was 220 nucleotides in length. A key feature of the second region is that its sequence was variable among the 10^{15} molecules. The researchers hypothesized that this variation could occasionally result in a long RNA molecule with the enzymatic ability to catalyze a phosphoester bond, which is a covalent bond between two adjacent nucleotides.

They also made another type of RNA molecule, which we will call the short RNA, with two important properties. First, the short RNA had a region that was complementary to a site near the constant region of the long RNA molecules. Second, the short RNA had a tag sequence that caused it to bind tightly to column packing material referred to as beads.

To begin this experiment, the researchers incubated a large number of the short and long RNA molecules together. During this incubation period, long and short RNA molecules would hydrogen bond to each other due to their complementary regions. Although hydrogen bonding is not permanent, this step allowed the long and short RNAs to recognize each other for a short period of time. The researchers reasoned that a long RNA with the enzymatic ability to form covalent phosphoester bonds may make this interaction more permanent by catalyzing a phosphoester bond between the long and short RNA molecules. Following this incubation, the mixture of RNAs was passed through a column with beads that specifically bound the short RNA. The aim of this approach was to select for longer RNA molecules that had covalently bonded to the short RNA molecule (see Conceptual Level of Figure 22.6).

The vast majority of RNAs would not have the enzymatic ability to catalyze a phosphoester bond. These would pass out of the column at step 2, because hydrogen bonding between the long and short RNAs is not sufficient to hold them together for very long. Such long RNAs would be discarded. Long RNAs with the ability to catalyze a phosphoester bond to the short RNA would remain bound to the column beads at step 2. These enzymatic RNAs were then flushed out at step 3 to generate a mixture of RNAs termed pool #1. The researchers expected this pool to contain several different long RNA molecules with varying abilities to catalyze a phosphoester bond.

To further the chemical selection process, the scientists used the first pool of long RNA molecules flushed out at step 3 to make more long RNA molecules. This was accomplished via PCR. This next batch also had the constant and variable regions.

Figure 22.6 Bartel and Szostak demonstrated chemical selection for RNA molecules that can catalyze phosphoester bond formation.

HYPOTHESIS Among a large pool of RNA molecules, some of them may contain the enzymatic ability to catalyze a phosphoester bond; these can be selected for in the laboratory.

STARTING MATERIALS Many copies of short RNA were synthesized that had a tag sequence that binds tightly to column packing material called beads. Also, a population of 10^{15} long RNA molecules was made that contained a constant region with a stem-loop structure and a 220-nucleotide variable region. Note: The variable regions of the long RNAs were made using a PCR step that caused mutations in this region.

1 Mix together the short RNAs with the 10^{15} long RNAs. Allow time for covalent connections to form if the long RNA happens to have the enzymatic activity for phosphoester bond formation.

Experimental level

Short RNAs   Long RNAs (10^{15})

Conceptual level

Short RNA   Tag sequence   Stem-loop structure   Long (10^{15}) RNA
3′ 5′
5′   Constant region
3′
Complementary region between long and short RNAs   220-nucleotide variable region

Covalent phosphoester bond
3′   5′

Of the 10^{15} long molecules, the variable region may rarely have the enzymatic ability to covalently connect the 3′ end of the short RNA to the 5′ end of the long RNA.

2 Pass the mixture through a column of beads that binds the tag sequence found on the short RNA. Add additional liquid to flush out long RNAs that are not covalently attached to short RNAs.

Column

Column bead

Tag sequence

Tag sequences promote the binding of the short RNA to the column beads. Long RNAs that are covalently attached to a short RNA will also be bound.

Discard these long RNAs.

This long RNA does not bind to the beads because the variable region does not possess the enzymatic ability to covalently attach to short RNA.

3 Add a low pH solution to prevent the tag sequence from binding to the beads. This causes the tightly bound RNAs to be flushed out of the column.

Low pH wash

4 The flushed-out RNAs are termed pool #1. Use pool #1 to make a second batch of long RNA molecules. This involved a PCR step using reverse transcriptase to make cDNA. The PCR primers recognized the beginning and end of the long RNA sequence and copied only this region. The cDNA was then used as a template to make long RNA via RNA polymerase.

Refer back to Figure 20.08 for a description of PCR.

Pool #1

5 Repeat procedure to generate 10 consecutive pools of RNA molecules.

Pool #1   Pool #2   Pool #3   Pool #4   Pool #5   Pool #6   Pool #7   Pool #8   Pool #9   Pool #10

6 Test a sample of the original population and each of the 10 pools for the enzymatic ability to catalyze a phosphoester bond.

Gap

Covalent bond

7 THE DATA

Rate of phosphoester bond formation

10^1
10^0
10^{-1}
10^{-2}
10^{-3}
10^{-4}
10^{-5}
10^{-6}

*   1   2   3   4   5   6   7   8   9   10

Pool

*Original 10^{15} molecules

However, the variable regions were derived from the variable regions of pool #1 RNA molecules that had enzymatic activity. The researchers reasoned that additional variation might occasionally produce an RNA molecule with improved enzymatic activity. This second batch of long RNA molecules was subjected to the same steps as was the first batch of 10^{15} molecules. In this case, the group of long molecules flushed out at step 3 was termed pool #2. This protocol was followed eight more times to generate 10 consecutive pools of RNA molecules. The researchers then analyzed the original random collection of 10^{15} RNA molecules and each of the 10 pools for the enzymatic ability to cata-lyze a phosphoester bond. As seen in the data, each successive pool became enriched for molecules with higher enzymatic activity. Pool #10 showed enzymatic activity that was approximately 3 million times higher than the original random pool of molecules!

Much like the work of Miller and Urey, these experiments demonstrated the feasibility of another phase of the prebiotic evolutionary process, in this case evolution via chemical selection. The results showed that chemical selection improves the functional characteristics of a group of RNA molecules over time by increasing the proportions of those molecules with enhanced function.

The RNA World Was Superseded by the Modern DNA/RNA/Protein World

Assuming that an RNA world was the origin of life, researchers have asked the question, Why and how did the RNA world evolve into the DNA/RNA/protein world we see today? Many potential pathways to a DNA/RNA/protein world are possible. Some researchers have argued that proteins evolved in parallel with RNA molecules. However, many proponents of the RNA world hypothesis suggest that DNA and proteins evolved later. If so, the RNA world may have been superseded by a DNA/RNA world or an RNA/protein world before the emergence of the modern DNA/RNA/protein world. Let's now consider the advantages of a DNA/RNA/protein world as opposed to the simpler RNA world, and how this modern biological world came into being.

Information Storage As we discussed previously, RNA can store information in its base sequence. So why did DNA take over that function, as is the case in modern cells? During the RNA world, RNA had to perform two roles, informational and catalytic. Scientists have speculated that the incorporation of DNA into cells would have relieved RNA of its informational role and thereby allowed RNA to perform a greater variety of other functions. For example, if DNA stored the information for the synthesis of RNA molecules, such RNA molecules could bind cofactors, have modified bases, or bind peptides that might enhance their catalytic function. Cells with both DNA and RNA would have had an advantage over those with just RNA, and so they would have been selected. Another advantage of DNA is stability. Compared to RNA, DNA is less likely to suffer mutations.

A second issue is, How did the DNA world come into being? Scientists have proposed that an ancestral RNA molecule had the ability to make DNA using RNA as a template. This function, known as reverse transcriptase activity, is described in Chapter 18. Interestingly, modern eukaryotic cells can use RNA as a template to make DNA. For example, telomerase, which is described in Chapter 11, copies the ends of chromosomes using an RNA template.

Metabolism and Other Cellular Functions Now let's consider the origin of proteins. The emergence of proteins as catalytic entities may have been a great advantage to early cells. Due to the many different chemistries of the 20 amino acids, proteins have vastly greater catalytic potential and efficiency than do RNA molecules, again providing a major advantage to cells that had both proteins and RNA. In modern cells, proteins have taken over most but not all catalytic functions. In addition, proteins can perform other important tasks. For example, cytoskeletal proteins carry out structural roles, and certain membrane proteins are responsible for the uptake of substances into living cells.

The answer to the question of how proteins came into being after an RNA world seems to be rooted in RNA function. Chemical selection experiments have shown that RNA molecules can catalyze the formation of peptide bonds and even attach amino acids to primitive tRNA molecules. Interestingly, modern protein synthesis still involves a central role for RNA. First, mRNA provides the information for a polypeptide sequence. Second, tRNA molecules act as adaptors for the formation of a polypeptide chain. And finally, ribosomes containing rRNA provide an arena for polypeptide synthesis. Furthermore, RNA within the ribosome acts as a ribozyme to catalyze peptide bond formation. Taken together, the analysis of translation in modern cells is consistent with an evolutionary history in which RNA molecules were instrumental in the emergence and formation of proteins.

22.2 History of Life on Earth

Thus far, we have considered how the first primitive cells may have come into existence. Recall that the first fossils of single-celled organisms were preserved approximately 3.5 billion years ago (see Figure 22.1). In this section, we will examine some of the major changes in life that have occurred since that time. As you will learn, the period from 3.5 billion years ago to the present has seen dramatic changes in the composition of life on Earth.

We will begin with a brief description of the geological changes that occurred on Earth that have impacted the emergence of new forms of life. Then we will examine how fossils are formed, and how they provide a fascinating yet incomplete journey through the history of life.

Many Environmental and Biological Changes Have Occurred Since the Origin of the Earth

The **geological timescale** is a time line of the Earth's history from its origin about 4.55 billion years ago to the present. This time line is subdivided into four eons, and then further subdivided into many eras. The first three eons are collectively known as the Precambrian. **Figure 22.7** provides the geological timescale and describes some of the major events that occurred during the history of life. The names of several eons and eras end in *-zoic* (meaning animal life) because we often recognize these time intervals on the basis of animal life. We will examine these time periods later in this chapter.

The changes that have occurred in living organisms over the past 4 billion years are the result of two interactive processes. First, as discussed in the next several chapters, genetic changes in organisms can affect their characteristics. Such changes often have an important impact on the ability of organisms to survive in their native environment. Second, the environment on Earth has undergone dramatic changes. Such environmental changes have profoundly influenced the types of organisms that have existed during different periods of time. As we will examine later, environmental influences can be both positive and negative. In some cases, a change can allow new types of organisms to come into being. Alternatively, environmental changes can cause the **extinction** of a species or group of species. A recurring pattern seen in the history of life is the emergence of new species and the extinction of other species. In some cases, biological changes and extinctions are correlated with major environmental changes, which include the following:

- *Climate/Temperature:* During the first 2.5 billion years of its existence, the surface of the Earth gradually cooled. However, during the last 2 billion years, the Earth has undergone major fluctuations in temperature, producing Ice Ages that alternate with warmer periods. Furthermore, the temperature on Earth is not uniform, which produces many different environments where the temperatures are quite different, such as a tropical rain forest and the North Pole.
- *Atmosphere:* The chemical composition of the gases surrounding the Earth has changed substantially over the past 4 billion years. One notable change involves oxygen. The emergence of organisms that were capable of photosynthesis added oxygen to the atmosphere. Prior to 2.5 billion years ago, relatively little oxygen was in the atmosphere. Levels of oxygen in the form of O_2 began to rise significantly at about this time. Our current atmosphere contains about 21% O_2.

- *Landmasses:* As the Earth cooled, landmasses formed that were surrounded by bodies of water. This created two different environments, terrestrial and aquatic. Furthermore, over the course of billions of years, the major landmasses, known as the continents, have shifted their positions, changed their shapes, and in some cases have become separated from each other. This phenomenon, called **continental drift**, is shown in **Figure 22.8**.
- *Floods:* Catastrophic floods have periodically had major impacts on the organisms in the flooded regions, sometimes causing extinction.
- *Glaciation:* On a periodic basis, glaciers have moved across continents and altered the composition of species on those landmasses. Glaciation also affects the water level of oceans.
- *Volcanic eruptions:* The eruptions of volcanoes can negatively impact the species in the vicinity of the eruption, sometimes causing extinctions. In addition, volcanic eruptions in the ocean can lead to the formation of new islands. Massive eruptions may also spew so much debris into the atmosphere that they can affect global temperatures and limit solar radiation, which limits photosynthetic production.
- *Meteorite impacts:* During its long history, the Earth has been struck by many meteorites. Large meteorites have had substantial impacts on the Earth's environment.

The effects of one or more of these changes have sometimes caused many species to go extinct at the same time. Such events are called **mass extinctions**. Five large mass extinctions occurred near the end of the Ordovician, Devonian, Permian, Triassic, and Cretaceous periods. Geological time periods are often based on the occurrences of these mass extinctions. The rapid extinction of many modern species due to human activities is sometimes referred to as the sixth mass extinction.

The study of the history of life involves an analysis of both environmental (geological) and biological changes that have occurred over the past 4 billion years. A key observation that provides us with a window into our biological past is the identification and examination of fossils, which are described next.

Fossils Provide a Glimpse into the History of Life

Fossils are the recognizable remains of past life on Earth. They can take many forms, including bones, shells, and leaves, the impression of cells, or other evidence, such as tracks or burrows. Scientists who study fossils are called **paleontologists** (*paleo* means ancient). The fossil record provides a record of the life forms that existed during particular geological periods. For example, rocks formed during the Proterozoic eon may have fossils of relatively simple organisms, such as bacteria, algae, and wormlike animals. By comparison, fossils formed during the Phanerozoic eon include complex animals and plants such as mammals and trees. Though the remains of dinosaurs often spark a childhood interest in fossils, the great bulk of the fossil record is dominated by the remnants of animals with shells and by the microscopic remains of plants and animals. These are the fossils studied by most paleontologists.

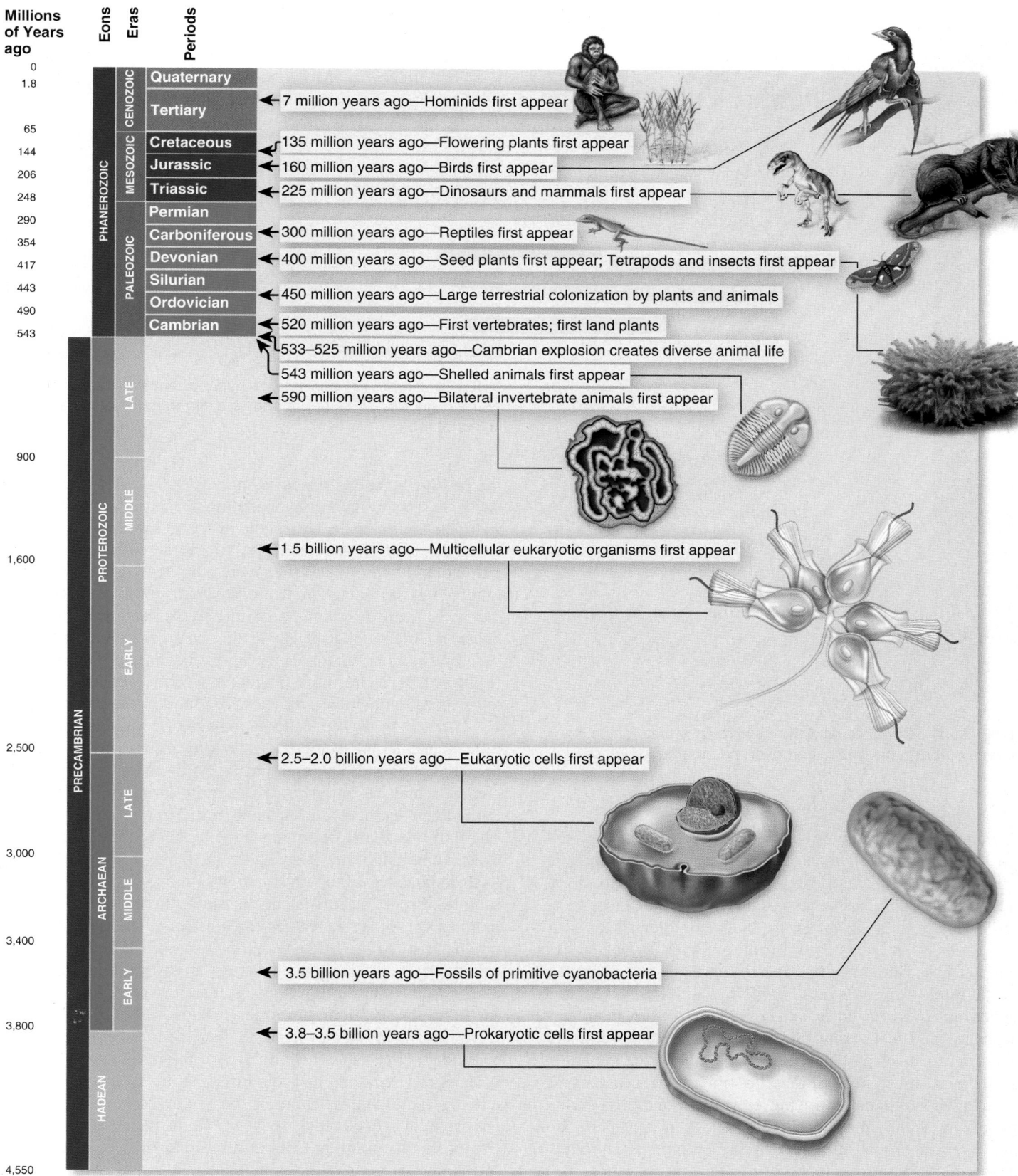

Millions of Years ago	Eons	Eras	Periods	
0			Quaternary	
1.8		CENOZOIC	Tertiary	← 7 million years ago—Hominids first appear
65				
144		MESOZOIC	Cretaceous	← 135 million years ago—Flowering plants first appear
206			Jurassic	← 160 million years ago—Birds first appear
248	PHANEROZOIC		Triassic	← 225 million years ago—Dinosaurs and mammals first appear
290			Permian	
354		PALEOZOIC	Carboniferous	← 300 million years ago—Reptiles first appear
417			Devonian	← 400 million years ago—Seed plants first appear; Tetrapods and insects first appear
443			Silurian	← 450 million years ago—Large terrestrial colonization by plants and animals
490			Ordovician	
543			Cambrian	← 520 million years ago—First vertebrates; first land plants

533–525 million years ago—Cambrian explosion creates diverse animal life

543 million years ago—Shelled animals first appear

← 590 million years ago—Bilateral invertebrate animals first appear

← 1.5 billion years ago—Multicellular eukaryotic organisms first appear

← 2.5–2.0 billion years ago—Eukaryotic cells first appear

← 3.5 billion years ago—Fossils of primitive cyanobacteria

← 3.8–3.5 billion years ago—Prokaryotic cells first appear

PROTEROZOIC — LATE, MIDDLE, EARLY
ARCHAEAN — LATE, MIDDLE, EARLY
HADEAN
PRECAMBRIAN

900
1,600
2,500
3,000
3,400
3,800
4,550

Figure 22.7   The geological timescale and an overview of the history of life on Earth.

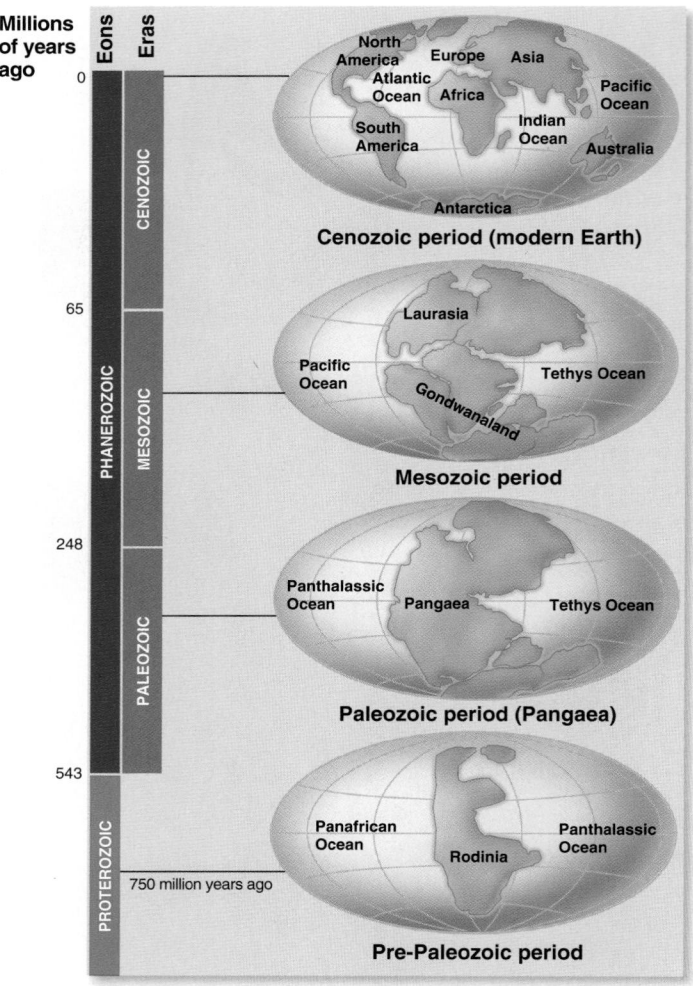

Figure 22.8   Continental drift.   The relative locations of the continents on Earth have changed over the past hundreds of millions of years.

Figure 22.9   An example of layers of sedimentary rock that contain fossils.   This photo shows a site that contains several layers of rock.

Because our understanding of the history of life is derived primarily from the fossil record, it is important to appreciate how fossils are formed and dated, and to understand some of the inherent biases in fossil analyses. Let's begin with fossil formation. Many of the rocks observed by paleontologists are sedimentary rocks that were formed from particles of older rocks broken apart by water or wind. These particles, such as gravel, sand, and mud, may settle and bury living and dead organisms at the bottoms of rivers, lakes, and oceans. Over time, more particles pile up, and sediments at the bottom of the pile eventually become rock. Gravel particles form rock called conglomerate, sand becomes sandstone, and mud becomes shale. Most fossils are formed when organisms are buried quickly, and then during the process of sedimentary rock formation, their hard parts are gradually replaced over millions of years by minerals, producing a recognizable representation of the original organism.

Information that helps to date fossils began to be collected in the mid-1600s, when Danish scientist Nicholas Steno studied the relative positions of sedimentary rocks. He noted that solid particles settle from a fluid according to their relative weight or size. The largest or heaviest settle first, and the smallest or lightest settle last. Because sedimentary rocks are formed particle by particle and bed by bed, the layers are piled one on top of the other. Thus, in any sequence of layered rocks, a given bed is older than any bed on top of it, and younger than any bed below it. Paleontologists often study changes in life-forms over time by studying the fossils in various beds from bottom to top (**Figure 22.9**). The more ancient life-forms are found in the lower beds, while newer species are found in the upper beds.

A common way to estimate the age of a fossil is by analyzing the elemental isotopes within the accompanying rock, a process called **radioisotope dating**. As discussed in Chapter 2, elements may be found in two or more forms called isotopes. Radioactive isotopes are unstable and decay at a specific rate. The **half-life** of a radioisotope is the length of time required for exactly one-half of the original isotope to decay (**Figure 22.10a**). Each radioactive isotope has its own unique half-life. Within a sample of rock, scientists can measure the amount of a given radioactive isotope as well as the amount of the isotope that is produced when the original isotope decays. For dating geological materials, several types of isotope decay patterns are particularly useful: carbon to nitrogen, potassium to argon, rubidium to strontium, and uranium to lead (**Figure 22.10b**). Except for recent fossils in which carbon-14 dating can be employed, fossil dating is not usually conducted on the fossil itself or on the sedimentary rock in which the fossil is found. Most commonly, igneous rock in the vicinity of the sedimentary rock is dated. For example, igneous rock derived from an ancient lava flow initially contains uranium-235 but no lead-207. By comparing the relative proportions of uranium-235 and lead-207, the age of the igneous rock can be accurately determined. Even so, paleontologists expect the fossil record to underestimate the actual date

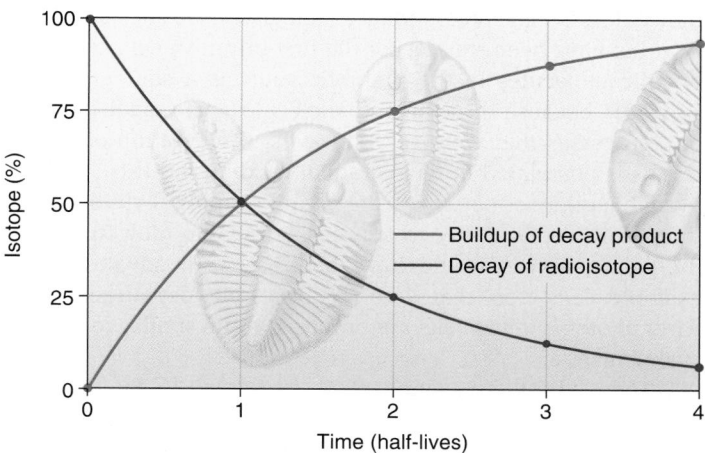

(a) Decay of a radioisotope

Radioisotope	Decay product	Half-life (years)	Useful dating range (years)
Carbon-14	Nitrogen-14	5,730	100–30,000
Potassium-40	Argon-40	1.3 billion	100,000–4.5 billion
Rubidium-87	Strontium-87	47 billion	10 million–4.5 billion
Uranium-235	Lead-207	710 million	10 million–4.5 billion
Uranium-238	Lead-206	4.5 billion	10 million–4.5 billion

(b) Radioisotopes that are useful for geological dating

Figure 22.10 Radioisotope dating of fossils. **(a)** Rocks can be dated by measuring the relative amounts of a radioisotope and its decay product that they contain. **(b)** These five isotopes are particularly useful for the dating of fossils.

Table 22.1	Biases That Occur in the Fossil Record
Factor	**Description**
Anatomy	Organisms with hard body parts, such as animals with a thick shell, are more likely to be preserved than are organisms composed of soft tissues.
Size	The remains of larger organisms are more likely than those of smaller organisms to be found as fossils.
Number	Species that existed in greater numbers are more likely to be preserved within the fossil record than are those that existed in smaller numbers.
Environment	Inland species are less likely to become fossilized than are those that lived in a marine environment or near the edge of water.
Geology	Due to the chemistry of fossilization, certain organisms are more likely to be preserved than are other organisms. In addition, parts of two or more different species may be subjected to mixing prior to fossilization. Such an event may distort the fossil record.
Time	Organisms that lived relatively recently are more likely to be found as fossils than are organisms that lived very long ago.
Paleontology	Certain types of fossils may be more interesting to paleontologists. In addition, a significant bias exists with regard to the locations where paleontologists search for fossils. For example, if a paleontologist were interested in dinosaurs, he or she would search in regions where other dinosaur fossils have already been found. Often, these are places where organized and academic paleontology has been going on most intensely, particularly in Europe and the U.S.

that a species came into existence because they are unlikely to find the first member of a particular species.

Before ending our discussion of fossils, we should consider some factors that impart a bias to the fossil record (**Table 22.1**). First, certain organisms are more likely to become fossilized than are others. Factors such as anatomy, size, number, and the environment and time in which they lived play important roles in determining the likelihood that an organism will be preserved in the fossil record. In addition, geological processes may create biases because they may favor the fossilization of certain types of organisms. Such processes can also create confusion because they may cause parts of different species to be mixed after death. Finally, unintentional biases arise that are related to the efforts of paleontologists. For example, scientific interests may favor searching for and analyzing certain species over others.

Though the fossil record should not be viewed as a comprehensive, balanced story of the history of life, it has provided a wealth of information regarding the types of life that existed in the distant past. The rest of this chapter will survey the emer-

gence of life-forms from 3.5 billion years ago to the present. In addition, the fossil record has provided compelling evidence for the theory of evolution. We will begin to examine this topic in Chapter 23.

Prokaryotic Cells Arose During the Archaean Eon

The Archaean (meaning "ancient") eon was a time when diverse microbial life flourished in the primordial oceans. As mentioned earlier in this chapter, the first known fossils of living cells were preserved in rocks that are dated 3.5 billion years old, though scientists postulate that cells arose many millions of years prior to this time. Based on the morphology of fossilized remains, these first cells were prokaryotic, lacking a true nucleus. During the more than 1 billion years of the Archaean eon, all life-forms were prokaryotic. At the time of their emergence, hardly any free oxygen (O_2) was in the Earth's atmosphere. Therefore, the single-celled microorganisms of this eon almost certainly used only **anaerobic** (without oxygen) metabolism.

As discussed in Chapter 4, prokaryotes are divided into two groups: bacteria and archaea. Bacteria (also called eubacteria) are the most prevalent prokaryotic organisms on modern Earth.

Many species of archaea (also called archaebacteria) have also been identified, though they are less common than bacteria, and tend to occupy extreme environments such as hot springs. Both bacteria and archaea share fundamental similarities, indicating that they are derived from a common ancestor. Even so, certain differences indicate that these two types of prokaryotes diverged from each other quite early in the history of life. In particular, bacteria and archaea show some interesting differences with regard to metabolism, lipid composition, and genetic pathways.

An important factor that greatly influenced the emergence of prokaryotic, and eventually eukaryotic species, is energy. As we learned in Unit II, living cells require energy to survive and reproduce. Organisms may follow two different strategies to obtain energy. Some are **heterotrophs**, which means that their energy is derived from eating other organisms or materials from other organisms. In this case, the energy of chemical bonds within the biological molecules of the food provides a source of energy. Alternatively, many organisms are **autotrophs** and have metabolic pathways that directly harness energy from either inorganic molecules or light. Among modern species, plants are an important example of autotrophs. As discussed in Chapter 8, plants can directly absorb light energy and use it (via metabolic pathways) to synthesize organic molecules such as glucose. On modern Earth, heterotrophs ultimately rely on autotrophs for the production of food.

Even though the first identified fossils are those of organisms that resembled photosynthetic cyanobacteria, scientists have hypothesized that the first living cells were actually heterotrophs. To be an autotroph, a cell must have the metabolic pathways that are needed to directly use energy to make organic molecules. Instead, evolutionary biologists have speculated that it would have been simpler for the first primitive cells to use the organic molecules in the prebiotic soup as a source of food. However, because these organic molecules were made by prebiotic processes that were very slow, the proliferation of heterotrophs is postulated to have gradually exhausted this supply of available organic matter. As the organic supply dwindled, cells that evolved the ability to synthesize organic molecules from inorganic sources would have had a growth advantage. For example, cells arose that were capable of photosynthesis. These early photosynthetic cells are proposed to be similar to modern cyanobacteria.

Then why were cyanobacteria preserved in fossils, while fossils of their heterotrophic ancestors were not? The most likely answer is related to their manner of growth. Interestingly, certain cyanobacteria promote the formation of a layered structure called a **stromatolite** (**Figure 22.11**). The aquatic environment where these cyanobacteria survive is rich in minerals such as calcium. The cyanobacteria, which grow in large mats, deplete the carbon dioxide in the surrounding water. This causes calcium carbonate to precipitate over the growing mat of bacterial cells, preserving those cells in the lower layers. The bacteria continue to grow upward to produce a new layer. Over time, many layers of sediment may be formed. As shown in Figure 22.11, this process still occurs today in places such as Shark Bay in western Australia, which is renowned for the stromatolite turfs rising along its beaches.

The emergence and proliferation of ancient cyanobacteria had two critical consequences. First, the autotrophic nature of these bacteria enabled them to produce organic molecules from carbon dioxide. This prevented the depletion of organic food-

(a) Fossil stromatolite

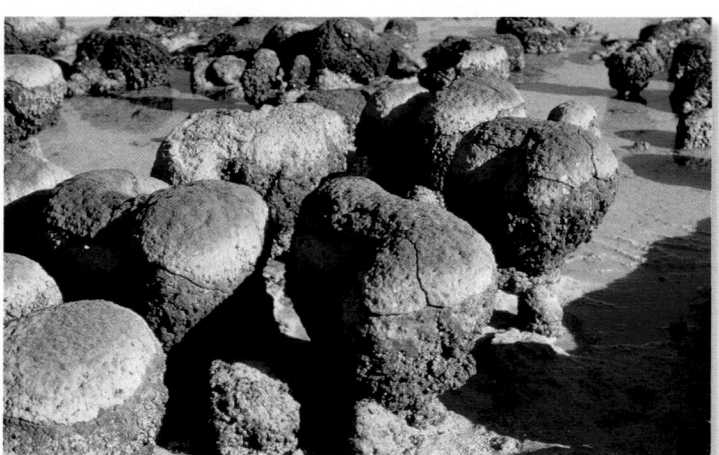

(b) Modern stromatolites

Figure 22.11 Fossil and modern stromatolites: evidence of autotrophic cyanobacteria. Each stromatolite is a rocklike structure, typically 1 meter in diameter. **(a)** Section of a fossilized stromatolite. These layers are mats of mineralized cyanobacteria, one layer on top of the other. The existence of fossil stromatolites provides evidence of early autotrophic organisms, which produced organic molecules and oxygen near the beginning of the history of life on Earth. **(b)** Modern stromatolites that have formed in western Australia.

stuffs that would have been exhausted if only heterotrophs existed. Second, cyanobacteria produce oxygen as a waste product of photosynthesis. During the Archaean and Proterozoic eons, the activity of cyanobacteria led to the gradual rise in atmospheric oxygen that we discussed earlier. The increase in atmospheric oxygen spelled doom for many prokaryotic groups. Anaerobic species became restricted to a few anoxic (without oxygen) environments such as deep within the soil. However, oxygen enabled the formation of new **aerobic** (with oxygen) prokaryotic species as well as the emergence and eventual explosion of eukaryotic life-forms, which are described next.

GENOMES & PROTEOMES

The Origin of Eukaryotic Cells During the Proterozoic Eon Involved a Union Between Bacterial and Archaeal Cells

Eukaryotic cells arose during the Proterozoic eon, which began 2.5 billion years ago and ended 543 million years ago (see Figure 22.7). The origin of the first eukaryotic cell is a matter of debate. In modern eukaryotic cells, genetic material is found in three distinct organelles. All eukaryotic cells contain DNA in the nucleus and mitochondria, and plant and algal cells also have DNA in their chloroplasts. To address the issue of the origin of eukaryotic species, scientists have paid great attention to the properties of the DNA found in these three organelles, and to how that DNA compares with DNA found in modern prokaryotic species. Many eukaryotic and prokaryotic genomes have been completely sequenced, as discussed in greater detail in Chapter 21. These DNA sequences are then compared to each other to determine evolutionary relationships. From such studies, the nuclear, mitochondrial, and chloroplast genomes appear to be derived from once-separate cells that came together.

Let's begin with the nuclear genome. From a genome perspective, both bacteria and archaea have contributed substantially to modern eukaryotic genomes. Eukaryotic genes encoding proteins involved in metabolic pathways and lipid biosynthesis appear to be derived from ancient bacteria, while genes involved with transcription and translation are derived from an archaeal ancestor. To account for this observation, several hypotheses have been proposed, the most widely accepted of which involve an association between ancient bacteria and archaea. Such relationships could have been symbiotic or endosymbiotic. A **symbiotic** relationship is one in which two different species live in direct contact with each other. For example, some scientists have postulated an ancient symbiotic relationship in which a bacterium and an archaeon formed a close association (**Figure 22.12a**). This eventually led to a fusion event that combined the genetic material of the two organisms. Over time, selection favored the retention of bacterial genes involved in metabolism and lipid biosynthesis and archaeal genes concerned with transcription and translation. A second possible scenario is an **endosymbiotic** relationship, in which one organism lives inside the other. According to this idea, one prokaryotic cell engulfed another, which became an endosymbiont (**Figure 22.12b**). For example, an ancient archaeon may have engulfed a bacterium, maintaining the bacterium in its cytoplasm as an endosymbiont. This process may have occurred via endocytosis, which is described in Chapter 5. Over time, genes were transferred to the archaeal host cell.

Let's now consider the origin of mitochondrial and chloroplast genomes. In 1905, a Russian botanist, Konstantin Mereschkowsky, was the first to suggest that such organelles may have an endosymbiotic origin. However, the question of endosymbiosis was largely ignored until researchers in the 1950s discovered that chloroplasts and mitochondria contain their own genetic material. The issue of endosymbiosis was revived and hotly debated when in 1981 Lynn Margulis published a book presenting evidence in support of this hypothesis entitled *Origin of Eukaryotic Cells*. During the 1970s and 1980s, the advent of molecular genetic techniques allowed researchers to analyze genes from mitochondria, chloroplasts, and prokaryotic species. The data are consistent with the endosymbiotic origin of these organelles.

Mitochondria found in eukaryotic cells are derived from a bacterial species that resembles modern α-proteobacteria, a diverse group of bacteria that can carry out oxidative phosphorylation that is used to make ATP. One possibility is that an endosymbiotic event involving this bacterial species created the first eukaryotic cell, and that the mitochondrion is a remnant of that event. Alternatively, symbiosis or endosymbiosis may have produced the first eukaryotic cell, and then a subsequent endosymbiosis resulted in mitochondria (see Figure 22.12). Over the next few years, the sequencing of many prokaryotic and mitochondrial genomes may help to resolve this controversy. DNA-sequencing data indicate that chloroplasts were derived from an endosymbiotic relationship between a primitive eukaryotic cell and a cyanobacterium.

Experimental support for an endosymbiotic origin of eukaryotic cells has also come from the study of proteobacteria. Curiously, an endosymbiotic relationship involving two different proteobacteria was recently reported, which demonstrates the feasibility of an endosymbiotic relationship among prokaryotic species. In mealy bugs, bacteria survive within the cytoplasm of large host cells of a specialized organ called a bacteriome. Recent analyses of these bacteria have shown that the bacteria inside the host cells share their own endosymbiotic relationship. In particular, γ-proteobacteria live inside β-proteobacteria. Such an observation is consistent with the idea that an endosymbiotic relationship could have given rise to the first eukaryotic species over 2 billion years ago.

1 A symbiotic association occurred between 2 prokaryotic species. The 2 cells fused with each other to make a single cell.

2 Over time, archaeal genes involved with metabolism and lipid biosynthesis were lost. Bacterial genes involved with transcription and translation were also lost.

3 Endosymbiosis of a bacterium resulted in the formation of mitochondria. The endomembrane system created the nucleus and other organelles such as the ER and Golgi.

4 A second endosymbiosis event involving cyanobacteria resulted in chloroplasts.

Archaeon   Bacterium

DNA

Eukaryotic cells: Animals, fungi, and some protists

Eukaryotic cells: Plants and algae

(a) Origin from symbiotic relationship

Archaeon

Smaller bacterium

DNA

1 A larger archaeal cell engulfed a smaller bacterial cell and established an endosymbiotic relationship.

2 Genes from endosymbiont were transferred to host cell.

3 This event may have resulted in mitochondria, or mitochondria may have arisen via a second endosymbiotic event. The endomembrane system created the nucleus and other organelles such as the ER and Golgi.

4 A subsequent endosymbiosis event involving cyanobacteria resulted in chloroplasts.

Eukaryotic cells: Plants and algae

Eukaryotic cells: Animals, fungi, and some protists

(b) Origin from endosymbiotic relationship

Figure 22.12   Possible symbiotic or endosymbiotic relationships that gave rise to the first eukaryotic cells.

Biological inquiry: What is the fundamental difference between the scenarios described in this figure?

Multicellular Eukaryotes and the Earliest Animals Arose During the Proterozoic Eon

Multicellular eukaryotes are thought to have first emerged about 1.5 billion years ago, in the middle of the Proterozoic eon, although the oldest fossils are dated at approximately 1.2 billion years old. Simple multicellular organisms could have originated in two different ways. One possibility is that several individual cells found each other and aggregated to form a colony. Cellular slime molds, discussed in Chapter 28, are examples of modern organisms in which groups of single-celled organisms can come together to form a small multicellular organism. According to the fossil record, such organisms have remained very simple for hundreds of millions of years. Alternatively, another way that multicellularity can occur is when a single cell divides, and the resulting cells stick together. This pattern occurs in many simple multicellular organisms, such as algae and fungi, as well as in species with more complex body plans, such as plants and animals. Biologists cannot be certain whether the first multicellular organisms arose by an aggregation process or by cell division and adhesion. However, the development of complex, multicellular organisms now occurs by cell division and adhesion.

An interesting example that compares unicellular organisms to more complex multicellular organisms is found among species of volvocine green algae that are related evolutionarily. These algae exist as unicellular species, as small clumps of cells of the same cell type, or as larger groups of cells with two distinct cell types. **Figure 22.13** compares four species of volvocine algae. *Chlamydomonas reinhardtii* is a unicellular alga (Figure 22.13a). It is called a biflagellate because the cells possess two flagella. *Gonium pectorale* is a multicellular organism composed of eight cells (Figure 22.13b). This simple multicellular organism is formed from a single cell by cell division and adhesion. All of the cells in this species are of the same cell type—biflagellate. Other volvocine algae have evolved into larger and more complex organisms. *Pleaodorina californica* has 64 to 128 cells (Figure 22.13c), while *Volvox aureus* has about 1,000 to 2,000 cells (Figure 22.13d). A new feature of these more complex organisms is that they have two cell types—somatic and reproductive cells. The somatic cells are biflagellate cells, while the reproductive cells are not. When comparing *P. californica* and *V. aureus*, *V. aureus* has a higher percentage of somatic cells than does *P. californica*.

Overall, an analysis of these four species of algae illustrates three important principles that are found among complex multicellular species. First, such organisms arise from a single cell that divides to produce daughter cells that adhere to one another. Second, the daughter cells can follow different fates, thereby producing multicellular organisms with different cell types. Third, as organisms get larger, more cells tend to be somatic cells. The somatic cells carry out the activities that are required for the survival of the multicellular organism. The reproductive cells are specialized for the sole purpose of producing offspring.

Toward the end of the Proterozoic eon, multicellular animals emerged. The first animals were **invertebrates**, soft-bodied animals without backbones. Among modern animals, most species, except for organisms such as sponges and corals, exhibit bilateral symmetry. This is a two-sided body plan with a left and right side that are mirror images of each other. Because each side of the body has appendages such as legs, one advantage of bilateral symmetry is that it facilitates locomotion. Bilateral animals also have anterior and posterior ends, with the mouth at the anterior end, and dorsal and ventral sides as described in Chapter 19. In 2004, Jun-Yuan Chen, David Bottjer, and colleagues discovered the earliest known ancestor of animals with bilateral symmetry. It was a minute creature barely visible to the naked eye that was shaped like a flattened helmet (**Figure 22.14**). This fossil was found in south China and is approximately 580–600 million years old.

   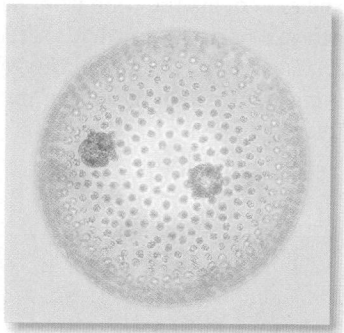

| 10 μm | 10 μm | 30 μm | 100 μm |

(a) *Chlamydomonas reinhardtii*, a unicellular alga

(b) *Gonium pectorale*, composed of 8 identical cells

(c) *Pleaodorina californica*, composed of 64 to 128 cells, has 2 cell types, somatic and reproductive

(d) *Volvox aureus*, composed of about 1,000 to 2,000 cells, has 2 cell types, somatic and reproductive

Figure 22.13 Variation in the level of multicellularity among volvocine algae.

Biological inquiry: Describe three changes that occur in these four types of algae.

Figure 22.14 **Fossil of an early invertebrate animal showing bilateral symmetry.** This fossil of an early animal, *Vernanimalcula guizouena*, is dated from 580 to 600 million years ago.

Phanerozoic Eon: The Paleozoic Era Saw the Diversification of Invertebrates and the Colonization of Land by Plants and Animals

The proliferation of multicellular eukaryotic life has been extensive during the Phanerozoic eon, which started 543 million years ago (mya) and extends to the present day. Phanerozoic means "well-displayed life," referring to the abundance of fossils of plants and animals that have been identified from this eon. As described earlier in Figure 22.7, the Phanerozoic eon is subdivided into three eras, the Paleozoic, Mesozoic, and Cenozoic. Because they are relatively recent and we have many fossils from these eras, each of them is further subdivided into periods. We will consider each era with its associated conditions and prevalent forms of life separately.

The term Paleozoic means "ancient animal life." The Paleozoic era covers the time from 543 to 248 mya, and is subdivided into six periods, known as the Cambrian, Ordovician, Silurian, Devonian, Carboniferous, and Permian. Periods are usually named after regions where rocks and fossils of that age were first discovered.

Cambrian Period (543 to 490 mya) The Cambrian climate was generally warm and wet, with no evidence of ice at the poles. During the Cambrian period, an event called the **Cambrian explosion** occurred in which there was an abrupt increase (on a geological scale) in the diversity of animal species. Many fossils telling the story of the Cambrian explosion were found in a rock bed in the Canadian Rockies called the Burgess Shale, which was discovered by Charles Walcott in 1909. At this site, both soft- and hard-bodied (shelled) invertebrates were buried in an underwater mudslide and preserved in water that was so deep and oxygen-free that decomposition was minimal (**Figure 22.15a**). The excellent preservation of these softer tissues is what makes this deposit unique (**Figure 22.15b**).

In the middle of the Cambrian period, all of the existing major types of marine invertebrates arose, plus many others that no longer exist. The Cambrian explosion generated over 100 major animal groups with significantly different body plans, but only about 30 of these occur in modern species. Examples of those that still exist include echinoderms (sea urchins and starfish), arthropods (insects, spiders, crustaceans), mollusks (clams and snails), and **chordates** (organisms with a spinal chord). Interestingly, although many new species of animals have arisen since this time, these later species have not shown a major reorganization of body plan, but instead exhibit variations on themes that were established during the Cambrian explosion. Approximately 520 million years ago, the first **vertebrates** (animals with backbones) appeared.

The cause of the Cambrian explosion is not understood. Because it occurred shortly after marine animals evolved shells, some scientists have speculated that the changes observed in animal species may have allowed them to exploit new environments and thereby evolve adaptations that would be beneficial in those environments. Alternatively, others have suggested that the increase in diversity may be related to atmospheric oxygen levels. During this period, oxygen levels were increasing, and perhaps more complex body plans became possible only after the atmospheric oxygen surpassed a certain threshold. In addition, as atmospheric oxygen reached its present levels, an ozone (O_3) layer was created that screens out ultraviolet radiation, thereby allowing complex life to live in shallow water and eventually on land.

(a) The Burgess Shale

(b) A fossilized arthropod, *Marella*

Figure 22.15 **The Cambrian explosion and the Burgess Shale.** (a) This photograph shows the original site in the Canadian Rockies discovered by Charles Walcott. Since its discovery, this site has been made into a quarry for the identification of fossils. (b) A fossil of an extinct arthropod, *Marella*, that was collected at this site.

Ordovician Period (490 to 443 mya) Like the Cambrian period, the climate of the early and middle parts of the Ordovician period was warm, and the atmosphere was very moist. This period had a diverse group of marine invertebrates, including trilobites and brachiopods (**Figure 22.16**). Marine communities consisted of invertebrates, algae, primitive jawless fish (a type of early vertebrate), mollusks, and corals. Fossil evidence also suggests that primitive land plants and arthropods may have first invaded the land during this period.

Toward the end of the Ordovician period, the climate changed rather dramatically. Large glaciers formed, which drained the relatively shallow oceans, causing the water levels to drop. This resulted in a mass extinction—estimates suggest that over 60% of the existing marine invertebrates became extinct during this period.

Silurian Period (443 to 417 mya) In contrast to the dramatic climate changes observed during the Ordovician period, the climate during the Silurian was relatively stable. The glaciers largely melted, which caused the ocean levels to rise.

No new major types of invertebrate animals appeared during this period, but significant changes were observed among existing vertebrate and plant species. Many new types of fish have been observed in the fossil record. In addition, the coral reefs made their first appearance during this period.

The Silurian marked a large colonization by terrestrial plants and animals. For this to occur, organisms evolved adaptations that prevented them from drying out, such as external cuticles. Ancestral relatives of spiders and centipedes became prevalent. Also, the earliest fossils of **vascular plants**, which can transport water, sugar, and salts throughout the plant body, were observed in this period.

Devonian Period (417 to 354 mya) In the Devonian, generally dry conditions occurred across much of the northern landmasses. However, the southern hemisphere was mostly covered by cool, temperate oceans.

The Devonian saw a major increase in the number of terrestrial species. At first, the vegetation consisted primarily of small plants, only a meter tall or less. Later, ferns, horsetails and **seed plants**, such as gymnosperms, also emerged. By the end of the Devonian, the first trees and forests were formed.

A major expansion of terrestrial animals also occurred. Insects emerged and other invertebrates became plentiful. In addition, **tetrapods**, which are vertebrates with four legs, came into existence. These first tetrapods were amphibians, living on land but requiring water to reproduce because of their jelly-coated eggs.

In the oceans, many types of invertebrates flourished, including brachiopods, echinoderms, and corals. This period is sometimes called the Age of Fishes, as many new types of fish emerged. During a period of approximately 20 million years near the end of the Devonian period, a prolonged series of extinctions eliminated many marine species. The cause of this mass extinction is not well understood.

Carboniferous Period (354 to 290 mya) The term Carboniferous refers to the rich coal deposits that were the result of the vegetation and climate of this period. The Carboniferous had the ideal conditions for the beginnings of coal. It was a cooler period and much of the land was covered by forest swamps. Coal was formed over many millions of years from compressed layers of rotting vegetation.

Plants and animals further diversified during the Carboniferous period. Very large plants and trees became prevalent. For example, tree ferns such as *Psaronius* grew to a height of 15 meters or more (**Figure 22.17**). The first flying insects emerged. For example, giant dragonflies with a wingspan of over two feet inhabited the forest swamps. Terrestrial vertebrates also became more diverse. Amphibians were very prevalent. One innovation that seemed particularly beneficial was the amniotic egg, which is covered with a leathery or hard shell. This prevented the desiccation of the embryo inside. This innovation was probably critical for the emergence of reptiles, which occurred during this period.

Permian Period (290 to 248 mya) At the beginning of the Permian, continental drift had brought much of the total land together, fused into a supercontinent known as Pangaea (see Figure 22.8). The interior regions of Pangaea were probably dry, with great seasonal fluctuations. The forests of fernlike plants shifted to gymnosperms. Species resembling modern conifers first appeared in the fossil record. During this period, amphibians were prevalent, but reptiles became the dominant vertebrate species. The first mammal-like reptiles also appeared.

(a) Trilobite

(b) Brachiopod

Figure 22.16 Shelled, invertebrate fossils of the Ordovician period.

Figure 22.17 **A giant tree fern, *Psaronius*, from the Carboniferous period.**   The illustration is a re-creation based on fossil evidence. The inset shows a fossilized section of the trunk, also known as petrified wood.

Figure 22.18 ***Megazostrodon*, the first known mammal of the Triassic period.**   The illustration is a re-creation based on fossilized skeletons.

At the end of the Permian period, the largest known mass extinction in the history of the Earth occurred. As a result of this event, 90–95% of marine species were eliminated. Similarly, a large proportion of terrestrial species became extinct. The cause of this extinction is the subject of much research and controversy. One possibility is that glaciation destroyed the habitats of terrestrial species and lowered ocean levels, which would have created greater competition among marine species. Another hypothesis is that enormous volcanic eruptions in Siberia produced large ash clouds that abruptly changed the climate on Earth.

Phanerozoic Eon: The Mesozoic Era Saw the Rise and Fall of the Dinosaurs

The Permian extinction marks the division between the Paleozoic and Mesozoic eras. Mesozoic means "middle animals." It was a time period that saw great changes in animal and plant species. This era is sometimes called the Age of Dinosaurs, as dinosaurs flourished during this time. The climate during the Mesozoic era was consistently hot, and terrestrial environments were relatively dry. Little if any ice was found at either pole. The Mesozoic is divided into three periods, the Triassic, Jurassic, and Cretaceous, which we will consider separately.

Triassic Period (248 to 206 mya) Reptiles were plentiful in this period, including new groups such as crocodiles and turtles. Some mammal-like reptiles continued to survive from the Permian period. The first dinosaurs emerged during the middle of the Triassic, as did the first true **mammals** such as the small Megazostrodon (**Figure 22.18**). Gymnosperms were the dominant land plant. Volcanic eruptions near the end of the Triassic are thought to have led to global warming, resulting in mass extinctions that eliminated many marine and terrestrial species.

Jurassic Period (206 to 144 mya) Gymnosperms, such as conifers, continued to be the dominant vegetation in this period. Dinosaurs became the dominant land vertebrate. Some dinosaurs attained enormous sizes, such as the massive *Brachiosaurus* that reached a length of 25 m (80 ft) and weighed up to 100 tons! The first known bird, *Archaeopteryx* (**Figure 22.19**), emerged in the Jurassic period, and mammals continued to exist although they were not prevalent.

Cretaceous Period (144 to 65 mya) On land, dinosaurs continued to be the dominant animals in this period. The earliest **flowering plants**, called angiosperms, which form seeds within a protective chamber, emerged and began to diversify.

The end of the Cretaceous also witnessed another mass extinction, which brought an end to many previously successful groups of organisms, such as dinosaurs. The dinosaurs and many other species abruptly died out. As with the Permian extinction, the cause of this mass extinction is also still debated. One plausible hypothesis suggests that a large meteorite or asteroid hit the Yucatán Peninsula of Mexico, lifting massive amounts of debris into the air and thereby blocking the sunlight from reaching the Earth's surface. Such a dense haze could have cooled the Earth's surface by 11–15°C (20–30°F). Evidence is also relatively strong that volcanic eruptions were the primary culprit for this mass extinction. Currently, both the volcanic and meteorite impact scenarios are reasonable hypotheses to explain the Cretaceous mass extinction, and both may have been contributing factors.

Phanerozoic Eon: Mammals and Flowering Plants Diversified During the Cenozoic Era

The Cenozoic era spans the most recent 65 million years. The Cenozoic is divided into two periods, the Tertiary and Quaternary.

Figure 22.19 A fossil of the first known bird, *Archaeopteryx*, which emerged in the Jurassic period.

In many parts of the world, tropical conditions were replaced by a colder, drier climate. The Cenozoic is sometimes called the Age of Mammals, because during this time mammals became the largest terrestrial animals. However, this phrase is perhaps misleading, because the Cenozoic era has also seen an amazing diversification of many types of organisms, including birds, fish, insects, and flowering plants.

Tertiary Period (65 to 1.8 mya) On land, the mammals, which survived from the Cretaceous period, began to diversify rapidly.

The diversification of mammals occurred during the early part of the Tertiary period. Whales emerged during this period. Likewise, birds and terrestrial insects also diversified. Angiosperms became the dominant land plant, and insects became important for their pollination. In the seas, fish also diversified, and sharks became abundant.

Toward the end of Tertiary period, about 7 million years ago, hominids came into existence. **Hominids** include modern humans, chimpanzees, gorillas, and orangutans plus all of their recent ancestors. By comparison, the term **hominin** refers to a subset of hominids, namely modern humans, extinct human species (for example, of the *Homo* genus), and our immediate ancestors. In 2002, a fossil of the earliest known hominid, *Sahelanthropus tchadensis*, was discovered in Central Africa. This fossil was dated at between 6 and 7 million years old. Another early hominid genus, called *Australopithecus*, first emerged in Africa about 4 million years ago. They walked upright, and had a protruding jaw, prominent eyebrow ridges, and a small braincase.

Quaternary Period (1.8 mya to Present) Periodic Ice Ages have been prevalent during the last 1.8 million years, covering much of Europe and North America. This period has witnessed the widespread extinction of many species of mammals, particularly larger species. Certain species of hominids became increasingly more human-like. Near the beginning of the Quaternary period, fossils were discovered of *Homo habilis*, who was called the "handy man" because tools have been found with the fossil remains. Fossils that are classified as *Homo sapiens* first appeared about 130,000 years ago. The evolution of hominids is discussed in more detail in Chapter 34.

CHAPTER SUMMARY

- Life began on Earth from nonliving material between 3.5 and 4.0 billion years ago. (Figure 22.1)

22.1 Origin of Life on Earth

- The first stage in the formation of life involved the synthesis of organic molecules to form a prebiotic soup. Possible scenarios of how this occurred are the reducing atmosphere, extraterrestrial, and deep-sea vent hypotheses. (Figures 22.2, 22.3)
- The second stage was the bonding of organic molecules to form polymers. This is thought to have occurred on the surface of clay.
- The third stage in the evolution of the first living cells occurred when polymers became enclosed in a structure that separated them from the external environment. Such structures, called protobionts, may have initially been coacervates, microspheres, or liposomes. (Figure 22.4)
- The fourth and final stage that led to the first living cells was chemical selection in which molecules with useful functional properties such as self-replication and other enzymatic functions increased in number. (Figure 22.5)
- The precursors of living cells as well as the first living cells themselves are thought to have used RNA for information storage and for carrying out enzymatic functions. This earliest phase of life is termed the RNA world.

- Bartel and Szostak demonstrated that chemical selection for RNA molecules that can catalyze phosphoester bond formation is possible experimentally. (Figure 22.6)
- The RNA world was later superseded by a DNA/RNA/protein world.

22.2 History of Life on Earth

- The geological timescale, which is divided into four eons and many eras and periods, charts the major events that occurred during the history of life. (Figure 22.7)
- The formation and extinction of new species, as well as mass extinctions, are often correlated with changes in temperature, atmosphere, and landmass locations, as well as floods, glaciation, volcanic eruptions, and meteorite impacts. (Figure 22.8)
- Fossils, which are preserved remnants of past life-forms, are formed in sedimentary rock. Radioisotope dating is one way to estimate the age of a fossil. The fossil record is incomplete and has several biases. (Figures 22.9, 22.10, Table 22.1)
- During the Archaean eon, the two groups of prokaryotes, bacteria and archaea, arose. The first prokaryotes were anaerobic heterotrophs. Later organisms, such as cyanobacteria, became phototrophs and produced oxygen. Cyanobacteria become preserved in structures called stromatolites. (Figure 22.11)
- Eukaryotic cells arose during the Proterozoic eon. This origin involved a union between bacterial and archaeal cells. The origin of mitochondria and chloroplasts was an endosymbiotic relationship. (Figure 22.12)

- Multicellular eukaryotes evolved during the Proterozoic eon, and first emerged about 1.5 billion years ago. Multicellularity now occurs via cell division and the adherence of the resulting cells to each other. A multicellular organism can produce multiple cell types. (Figure 22.13)

- The first bilateral animal emerged toward the end of the Proterozoic eon, approximately 580–600 million years ago. (Figure 22.14)

- The Phanerozoic eon is subdivided into the Paleozoic, Mesozoic, and Cenozoic eras. During the Paleozoic era, invertebrates greatly diversified, particularly during the Cambrian explosion, and the land became colonized by plants. Terrestrial vertebrates, including tetrapods, also became more diverse. (Figures 22.15, 22.16, 22.17)

- Dinosaurs were prevalent during the Mesozoic Era, particularly during the Jurassic period. Mammals and birds also emerged. (Figures 22.18, 22.19)

- During the Cenozoic Era, mammals diversified and flowering plants became the dominant species. The first hominids emerged approximately 7 million years ago.

TEST YOURSELF

1. The prebiotic soup was
   a. the assemblage of unicellular prokaryotes and eukaryotes that existed in the oceans of primitive Earth.
   b. the accumulation of organic molecules in the oceans of primitive Earth.
   c. the mixture of organic molecules that was found in the cytoplasm of the earliest cells on Earth.
   d. a pool of nucleic acids that contained the genetic information for the earliest organisms.
   e. none of the above.

2. Which of the following is not a characteristic of protobionts that was necessary for the evolution of living cells?
   a. a membrane-like boundary separating the external environment from an internal environment
   b. polymers capable of functioning in information storage
   c. polymers capable of enzymatic activity
   d. self-replication
   e. compartmentalization of metabolic activity

3. RNA is believed to be the first functional macromolecule in protobionts because it
   a. is easier to synthesize compared to other macromolecules.
   b. has the ability to store information, self-replicate, and perform enzymatic activity.
   c. is the simplest of the macromolecules commonly found in living cells.
   d. all of the above
   e. a and c only

4. The movement of landmasses that have changed their positions, shapes, and association with other landmasses is called
   a. glaciation.
   b. Pangaea.
   c. continental drift.
   d. biogeography.
   e. geological scale.

5. Paleontologists estimate the dates of fossils by
   a. the layer of rock in which the fossils are found.
   b. analysis of radioactive isotopes found in nearby rock.
   c. the complexity of the body plan of the organism.
   d. all of the above
   e. a and b only

6. The fossil record does not give us a complete picture of the history of life because
   a. not all past organisms have become fossilized.
   b. only organisms with hard skeletons can become fossilized.
   c. fossils of very small organisms have not been found.
   d. fossils of early organisms are located too deep in the crust of the Earth to be found.
   e. all of the above.

7. The endosymbiosis hypothesis explaining the evolution of eukaryotic cells is supported by
   a. DNA-sequencing analysis comparing bacterial genomes, mitochondrial genomes, and eukaryotic nuclear genomes.
   b. naturally occurring examples of endosymbiotic relationships between bacterial cells and eukaryotic cells.
   c. the presence of DNA in mitochondria and chloroplasts.
   d. all of the above
   e. a and b only

8. Which of the following explanations of the evolution of multicellularity in eukaryotes is seen in the development of complex, multicellular organisms today?
   a. endosymbiosis
   b. aggregation of cells to form a colony
   c. division of cells with the resulting cells sticking together
   d. multiple cell types aggregating to form a complex organism
   e. none of the above

9. The earliest fossils of vascular plants occurred during the _____ period.
   a. Ordovician
   b. Silurian
   c. Devonian
   d. Triassic
   e. Jurassic

10. The appearance of the first hominids date to the _____ period.
   a. Triassic
   b. Jurassic
   c. Cretaceous
   d. Tertiary
   e. Quaternary

CONCEPTUAL QUESTIONS

1. What are the four stages in which the origin of life occurred?
2. Define radioisotope dating and half-life.
3. Briefly discuss the Cambrian explosion.

EXPERIMENTAL QUESTIONS

1. What is chemical selection? What was the hypothesis tested by Bartel and Szostak?
2. In conducting the selection experiment among pools of long RNA molecules with various catalytic abilities, what was the purpose of using the short RNA molecules?
3. What were the results of the experiment conducted by Bartel and Szostak? What impact did this study have on our understanding of the evolution of life on Earth?

COLLABORATIVE QUESTIONS

1. Discuss three possible hypotheses of how organic molecules were first formed.
2. Discuss three periods in geological time.

---

www.brookerbiology.com

This website includes answers to the Biological Inquiry questions found in the figure legends and all end-of-chapter questions.

23

AN INTRODUCTION TO EVOLUTION

CHAPTER OUTLINE

Selective breeding. Many crop plants, which are produced after many generations of selective breeding, illustrate how the traits of organisms can dramatically change over time.

Organic life beneath the shoreless waves
Was born and nurs'd in Ocean's pearly caves
First forms minute, unseen by spheric glass,
Move on the mud, or pierce the watery mass;
These, as successive generations bloom,
New powers acquire, and larger limbs assume;
Whence countless groups of vegetation spring,
And breathing realms of fin, and feet, and wing.

From *The Temple of Nature* by Erasmus Darwin (1731–1802), grandfather of Charles Darwin. Published posthumously in 1803.

The word evolution is often associated with a process that involves change. **Biological evolution** is a heritable change in one or more characteristics of a population or species across many generations. Evolution can be viewed on a small scale as it relates to changes in a single gene in a population over time, or it can be viewed on a larger scale as it relates to the formation of new species or groups of related species. It is helpful to begin our discussion of evolution with a definition of a species. Unfortunately, as we will examine in Chapter 25, a precise definition of species is not always possible. As a working definition, biologists define **species** as a group of related organisms that share a distinctive form. Among species that reproduce sexually, such as plants and animals, members of the same species are capable of interbreeding to produce viable and fertile offspring. The term **population** refers to members of the same species that are likely to encounter each other, and so have the opportunity to interbreed. Some of the emphasis in the study of evolution is on understanding how populations change over the course of many generations to produce new species.

In the first part of this chapter, we will examine the history of evolutionary thought and some of the basic tenets of evolution, particularly those that were proposed by Charles Darwin in the mid-1800s. Though the theory of evolution has been refined over the past 150 years or so, the fundamental principle of evolution has remained unchanged, and has provided a cornerstone for our understanding of biology. Theodosius Dobzhansky, an influential evolutionary scientist of the 1900s, once said, "Nothing in biology makes sense except in the light of evolution." The extraordinarily diverse and seemingly bizarre array of species on our planet can be explained within the context of evolution. As is the case with all scientific theories, biological evolution is called a theory because it is supported by a substantial body of evidence and because it explains a wide range of observations. The theory of evolution provides answers to many questions related to the diversity of life.

In the second part of this chapter, we will survey the extensive data that show the results of evolution. These data not only support the theory of evolution but also allow us to understand the interrelatedness of different species, whose similarities are often related to descent from a common ancestor. Much of the early evidence supporting evolution came from visual observations and comparisons of modern and extinct species.

More recently, advances in molecular genetics, particularly those related to DNA sequencing and genomics, have revolutionized the study of evolution. Scientists now have information that allows us to understand how evolution involves changes in the DNA of a given species. These changes affect both a species' genes and the proteins they encode. The term **molecular evolution** refers to the molecular changes in genetic material that underlie the phenotypic changes associated with evolution.

A theme of this textbook, namely Genomes & Proteomes, is rooted in an understanding of these changes. In the last section of this chapter, we consider some of the exciting new information that helps us to appreciate evolutionary change at the molecular level. In the following chapters of this unit, we will examine how such changes are acted upon by evolutionary forces in ways that alter the traits of a given species and may eventually lead to the formation of new species.

23.1 The Theory of Evolution

Undoubtedly, the question "Where did we come from?" has been asked and debated by people for thousands, perhaps tens of thousands of years. Many of the earliest ideas regarding the existence of living organisms were strongly influenced by religion and philosophy, and these ideas suggested that all forms of life have remained the same since their creation. In the 1600s, however, scholars in Europe began a revolution that created the basis of empirical and scientific thought. **Empirical thought** relies on observation to form an idea or hypothesis, rather than trying to understand life from a nonphysical or spiritual point of view. The shift toward empirical thought encouraged scholars to look for the basic rationale behind a given object or phenomenon.

In the mid- to late-1600s, the first scientist to carry out a thorough study of the natural world was an Englishman named John Ray, who developed an early classification system for plants and animals based on anatomy and physiology. He established the modern concept of a species, noting that organisms of one species do not interbreed with members of another, and used it as the basic unit of his classification system. Ray's ideas on classification were later extended by the Swedish naturalist Carolus Linnaeus. Neither Ray nor Linnaeus proposed that evolutionary change promotes the formation of new species. However, their systematic classification of plants and animals helped scholars of this period perceive the similarities and differences among living organisms.

Late in the 1700s, a small number of European scientists began to quietly suggest that life-forms are not fixed. A French zoologist, George Buffon, actually said that living things do change through time. However, Buffon was careful to hide his views in a 44-volume natural history book series. While he was a quiet pioneer in asserting that species can change over generations, he publicly rejected the idea that one species could evolve into another species.

Around the same time, a French naturalist named Jean-Baptiste Lamarck suggested an intimate relationship between variation and evolution. By examining fossils, he realized that some species had remained the same over the millennia and others had changed. Lamarck hypothesized that species change over the course of many generations by adapting to new environments. He believed that living things evolved in a continuously upward direction, from dead matter, through simple to more complex forms, toward human "perfection." According to Lamarck, organisms altered their behavior in response to environmental change. He thought behavioral changes modified traits, and he hypothesized that such modified traits were inherited by offspring. He called this idea the **inheritance of acquired characteristics**. For example, according to Lamarck's hypothesis, giraffes developed their elongated necks and front legs by feeding on high tree leaves. The exercise of stretching up to the leaves altered the neck and legs, and Lamarck presumed that these acquired characteristics were transmitted to offspring. However, further research has rejected Lamarck's idea that acquired traits can be inherited.

Interestingly, Erasmus Darwin, the grandfather of Charles Darwin, who was a contemporary of Buffon and Lamarck, was an early advocate of evolutionary change. He was a physician, a plant biologist, and also a poet. He was aware that modern species were different from many fossil types and also saw how plant and animal breeders used breeding practices to change the traits of domesticated species (see chapter opener photo). He knew that offspring inherited features from their parents, and went so far as to say that life on Earth could be descended from a common ancestor.

Overall, Charles Darwin's many scientific predecessors set the stage for the theory of evolution. With this historical introduction, we will now consider Darwin's observations, and the tenets that provide the foundation for this theory.

Darwin Suggested That Species Are Derived from Pre-Existing Species

Charles Darwin, a British naturalist born in 1809, played a key role in developing the theory that existing species have evolved from pre-existing species. Darwin's unique perspective and his ability to formulate evolutionary theory were shaped by several different fields of study, including ideas of his time about physical and biological processes.

Two main geological hypotheses predominated in the early 19th century. Catastrophism was first proposed by French zoologist and paleontologist Baron Georges Cuvier to explain the age of the Earth. Cuvier suggested that the Earth was just 6,000 years old, and that only catastrophic events had changed its geological structure. This idea fit well with religious teachings. Alternatively, uniformitarianism, proposed by James Hutton and promoted by Sir Charles Lyell, suggested that changes in the Earth are directly caused by recurring events. For example, they suggested that geological processes such as erosion existed in the past, and that they happened then at the same gradual rate as they do now. For such slow geological processes to eventually lead to substantial changes in the Earth's characteristics, a great deal of time was required, so Hutton and Lyell were the first to propose that the age of the Earth was well beyond 6,000 years. The ideas of Hutton and Lyell helped to shape Darwin's view of the world.

Darwin's thinking was also influenced by a paper published in 1798 called *Essay on the Principle of Population* by Thomas Malthus, an English economist. Malthus asserted that the pop-

Figure 23.1 Charles Darwin and the voyage of the *Beagle*, 1831–1836. (a) A portrait of Charles Darwin (1809–1882) while in Ecuador. (b) Darwin's voyage on the *Beagle*, which took almost five years to circumnavigate the world.

(a) Charles Darwin

(b) The voyage of the *Beagle*

ulation size of humans can, at best, increase linearly due to increased land usage and improvements in agriculture, while our reproductive potential is exponential (for example, doubling with each generation). He argued that famine, war, and disease will limit population growth, especially among the poor. An important message from Malthus's work was that only a fraction of any population will survive and reproduce.

Darwin's evolutionary ideas were most influenced by his own experiences and observations. His famous voyage on the *Beagle*, which lasted from 1831 to 1836 (**Figure 23.1**), involved a careful examination of many different species. The main mission of the *Beagle* was to map the coastline of southern South America and take oceanographic measurements. Darwin's job was to record the weather, geological features, plants, animals, fossils, rocks, minerals, and indigenous people. He also collected many specimens of plants and animals, which had to be carefully packed and labeled.

Though Darwin made many interesting observations on his journey, he was particularly struck by the distinctive traits of island species that provided them with ways to better exploit their native environment. For example, Darwin observed several species of finches found on the Galápagos Islands. We now know that these finches all evolved from a single species similar to the blue-black grassquit finch (*Volatina jacarina*), commonly found along the Pacific Coast of South America. Once on the Galápagos Islands, the finches' ability to survive in their new habitat depended, in part, on changes in the size and shape of their bills over many generations. These specializations enabled succeeding generations to better obtain food (**Table 23.1**). For example, the ground and vegetarian finches have sturdy, crush-

ing bills that they use to crush various sizes of seeds or buds. The tree finches have grasping bills that they use to pick up insects from trees. The mangrove, woodpecker, warbler, and cactus finches have pointed, probing bills. The first three of these use their probing bills to search for insects in crevices; the cactus finches use their probing bills to open cactus fruits and eat the seeds. One species, the woodpecker finch, even uses twigs or cactus spines to extract insect larvae from holes in dead tree branches. Darwin clearly saw the similarities among these species, yet he noted the differences that provided them with specialized feeding strategies.

With an understanding of geology and population growth, and his observations from his voyage on the *Beagle*, Darwin had formulated his theory of evolution by the mid-1840s. However, he had cataloged and described all of the species he had collected on his *Beagle* voyage except for one type of barnacle. Some have speculated that Darwin may have felt that he should establish himself as an expert on one species before making generalizations about all of them. Therefore, he spent several additional years studying barnacles. During this time, the geologist Charles Lyell, who had greatly influenced Darwin's thinking, strongly encouraged Darwin to publish his theory of evolution. In 1856, Darwin began to write a long book to explain his ideas. In 1858, however, Alfred Wallace, a naturalist working in the East Indies, sent Darwin an unpublished manuscript to read prior to its publication. In it, Wallace proposed the same ideas concerning evolution. Darwin therefore quickly excerpted some of his own writings on this subject, and two papers, one by Darwin and one by Wallace, were published in the Proceedings of the Linnaean Society of London.

Table 23.1 | A Comparison of Beak Type and Diet Among the Galápagos Finches Darwin Studied

Type of finch	Species		Type of beak	Diet
Ground finches	Large ground finch (*Geospiza magnirostris*)		Crushing	Seeds—Ground finches have crushing beaks to crush various sizes of seeds; large beaks can crush large seeds, while smaller beaks are better for crushing small seeds.
	Medium ground finch (*G. fortis*)			
	Small ground finch (*G. fuliginosa*)			
	Sharp-billed ground finch (*G. difficilis*)			
Vegetarian finch	Vegetarian finch (*Platyspiza crassirostris*)		Crushing	Buds—Vegetarian finches have crushing beaks to pull buds from branches.
Tree finches	Large tree finch (*Camarhynchus psittacula*)		Grasping	Insects—Tree finches have grasping beaks to pick insects from trees. Those with heavier beaks can also break apart wood in search of insects.
	Medium tree finch (*Camarhynchus pauper*)			
	Small tree finch (*Camarhynchus parvulus*)			
	Mangrove finch (*Cactospiza heliobates*)		Probing	Insects—These finches have probing beaks to search for insects in crevices and then to pick them up. The woodpecker finch can also use a cactus spine for probing.
	Woodpecker finch (*Cactospiza pallidus*)			
Warbler finch	Warbler finch (*Certhidea olivacea*)		Probing	
Cactus finches	Large cactus finch (*G. conirostris*)		Probing	Seeds—Cactus finches have probing beaks to open cactus fruits and take out seeds.
	Cactus finch (*G. scandens*)			

These papers were not widely recognized. A short time later, however, Darwin finished his book, *The Origin of Species*, which described his ideas in greater detail and included observational support. This book, which received high praise from many scientists and scorn from others, started a great debate concerning evolution. Although some of his ideas were incomplete due to the fact that the genetic basis of traits was not understood at that time, Darwin's work remains one of the most important contributions to our understanding of biology.

The Theory of Evolution Explains How Natural Selection Acting on Genetic Variation Can Change Populations over Many Generations

The fundamental principle that underlies evolution is that biological species do not have a fixed, static existence but instead exhibit changing characteristics over the course of many generations. Darwin hypothesized that existing life-forms on our planet are the product of the modification of pre-existing life-forms. He expressed this concept of biological evolution as "the theory of descent with modification through variation and natural selection." As its name suggests, evolution is based on two factors: (1) variation within a given species and (2) forces of nature, which are termed natural selection. During the process of **natural selection,** certain individuals are less likely to survive and reproduce in a particular environment, while other individuals with traits that make them better suited to their native environment tend to flourish and reproduce. According to this idea, nature "selects" those individuals possessing certain traits that favor reproductive success. Over long periods of time, this process of natural selection eventually leads to **adaptation**, which is a form of evolutionary change in which a population's characteristics change to make its members better suited to their native environment. Note that natural selection affects the survival and/or reproduction of individuals, which over time affects the evolution of populations and ultimately species.

The genetic basis for variation within a species was not understood at the time Darwin proposed his theory of evolution by natural selection. In fact, Darwin's theory preceded, by a few years, Mendel's pioneering work in genetics. Even so, Darwin and many other people before him observed that offspring resemble their parents more than they do unrelated individuals. Therefore, he assumed that some traits are passed from parent to offspring.

Since the time of Darwin, the study of genetics has allowed scientists to understand the relationship between traits and inheritance. Genetic variation is a consistent feature of natural populations. Such variation may involve differences in genes, changes in chromosome structure, and alterations in chromosome number. In contrast to Lamarck's ideas, we now know that these genetic differences are not produced by an individual's behavior or other response to its environment. Rather, genetic variation is caused by random mutations that alter the genetic composition of particular individuals. In Chapter 24, we will consider how these random mutations are acted upon by evolutionary forces to change the genetic composition of populations over the course of many generations.

Based on Darwin's ideas regarding natural selection and a more modern understanding of genetics, biologists in the 1920s to 1940s were able to frame the theory of evolution in a modern perspective termed neo-Darwinism or the **modern synthesis of evolution**. Within a given population of interbreeding organisms, natural variation exists that is caused by random changes in the genetic material. Such genetic changes may affect the phenotype of an individual in a positive, negative, or neutral way. If a genetic change promotes an individual's survival and/or ability to successfully reproduce, natural selection may increase the prevalence of that trait in future generations.

For example, let's consider a population of finches that migrates from the South American mainland to a Galápagos island. The seeds produced on this island are larger than those produced on the mainland. Birds with larger beaks would be better able to feed on these seeds and so would be more likely to survive and pass on that trait to their offspring. Therefore, in succeeding generations, the population will tend to have a greater proportion of finches with larger beaks (**Figure 23.2**). Alternatively, if a genetic change happens to be detrimental to an individual's survival and/or reproduction, natural selection is likely to eliminate this type of variation. For example, if a mutation occurred that causes a finch in the same environment to have a slightly smaller beak, this bird would be less likely to survive and pass on this change to its offspring. Natural selection, which acts upon variation involving many different genes, may ultimately produce a new species with a combination of traits that are quite different from those of the original species, such as finches with larger beaks. In other words, the newer species has evolved from a pre-existing species. Now let's look at some other examples of change in organisms over time.

23.2 Observations of Evolutionary Change

Over the past 150 years, the research community has learned that no known concept other than descent with modification from a common ancestor can scientifically account for the diversity and unity of life on our planet. Observations regarding the theory of evolution can be gleaned from many sources (**Table 23.2**). Historically, the first descriptions of biological evolution came from studies of the fossil record, the distribution of living organisms on our planet, selective breeding experiments, and the comparison of similar anatomical features in different species. More recently, additional examples of the theory of evolution have been revealed at the molecular level. By comparing DNA sequences from many different species, evolutionary biologists have gained great insight into the relationship between the evolution of species and the associated changes in the genetic material. In this section, we will survey a variety of observations that show the process of evolutionary change.

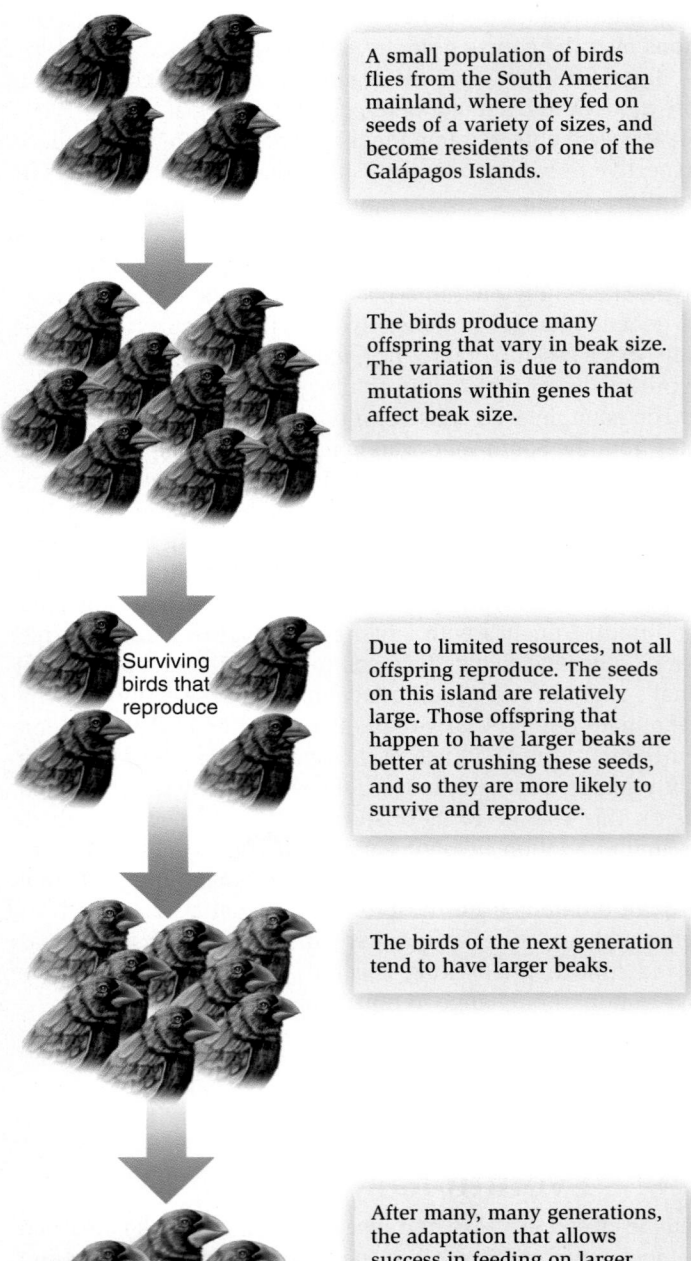

A small population of birds flies from the South American mainland, where they fed on seeds of a variety of sizes, and become residents of one of the Galápagos Islands.

The birds produce many offspring that vary in beak size. The variation is due to random mutations within genes that affect beak size.

Surviving birds that reproduce

Due to limited resources, not all offspring reproduce. The seeds on this island are relatively large. Those offspring that happen to have larger beaks are better at crushing these seeds, and so they are more likely to survive and reproduce.

The birds of the next generation tend to have larger beaks.

After many, many generations, the adaptation that allows success in feeding on larger seeds has created a new species with larger beaks, as well as other modified traits, such as changes in color, that are suited to the new environment.

Figure 23.2 **Evolutionary adaptation to a new environment via natural selection.**   The example shown here involves a species of finch adapting to a new environment on one of the Galápagos Islands. The plants on this island produce larger seeds than do the plants on the mainland from which the birds had originated. According to Darwin's theory of evolution by natural selection, the process of adaptation eventually led to the formation of a new species with larger beaks that were better suited to crushing the large seeds in its new environment.

Table 23.2	Observations of Biological Evolution

Type of observation	Description
Fossil record	When fossils are compared according to their age, from oldest to youngest, successive evolutionary change becomes apparent.
Biogeography	Unique species found on islands and other remote areas have arisen because the species in these locations have evolved in isolation from the rest of the world.
Convergent evolution	Two different species from different lineages sometimes become anatomically similar because they occupy similar environments. This indicates that natural selection promotes adaptation to a given environment.
Selective breeding	The traits in domesticated species have been profoundly modified by artificial selection practices.
Homologies:*	
Anatomical	Evolutionarily related species may possess homologous structures that have been modified in ways that allowed them to be used differently by each species. In some cases, such structures are no longer needed and degenerate to nonfunctional vestigial structures.
Developmental	An analysis of embryonic development often reveals similar anatomical features that point to past evolutionary relationships.
Molecular	At the molecular level, certain characteristics are found in all living cells, suggesting that all living species are derived from a common ancestor. In addition, species that are closely related evolutionarily tend to have DNA sequences that are more similar to each other than they are to those in distantly related organisms.

*Homology refers to fundamental similarities that occur due to descent from a common ancestor.

Fossils Show Successive Evolutionary Change

As discussed in Chapter 22, the fossil record reveals a history of life from its earliest beginnings some 3.5–4.0 billion years ago. Today, scientists have access to a far more extensive fossil record than was available to Darwin or other scientists of his time. Even though the fossil record is still incomplete, the many fossils that have been discovered often provide detailed information regarding evolutionary change in a series of related organisms. When fossils are compared according to their age, from oldest to youngest, successive evolutionary change becomes apparent.

Let's consider a few examples in which paleontologists have observed evolutionary change. In 2005, fossils of *Tiktaalik roseae*, nicknamed fishapod, were discovered by Edward Daeschler, Neil Shubin, and Farish Jenkins that illuminate the steps that led to the evolution of tetrapods, which are animals with four legs.

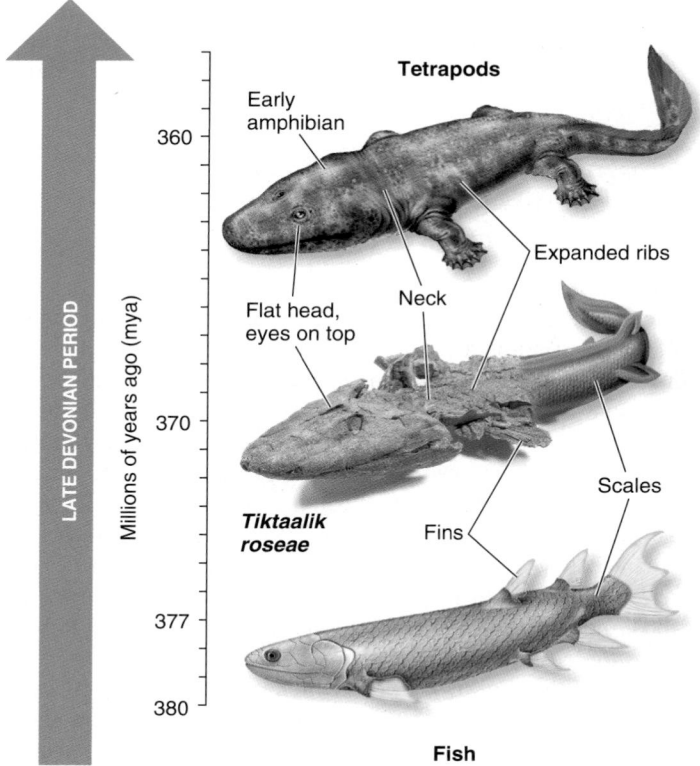

Tetrapods

Early amphibian

Expanded ribs

Neck

Flat head, eyes on top

Tiktaalik roseae

Fins

Scales

Fish

Figure 23.3 **Evolutionary change in the tetrapod lineage, showing a transitional form.** This figure shows two early tetrapod ancestors, a Devonian fish and the transitional form *Tiktaalik roseae*, as well as one of their descendents, an early amphibian. An analysis of the fossils shows that *T. roseae*, also known as a fishapod, had both fish and amphibian characteristics, so it probably was able to survive brief periods out of the water.

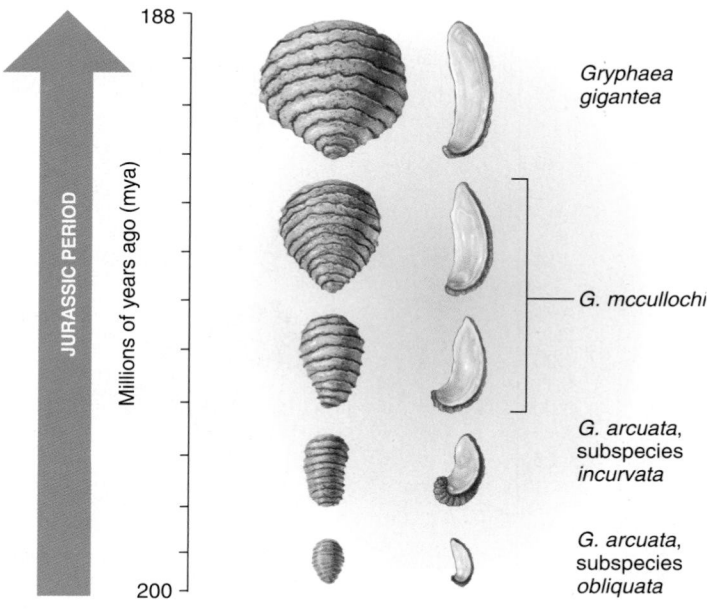

Gryphaea gigantea

G. mcculloch

G. arcuata, subspecies *incurvata*

G. arcuata, subspecies *obliquata*

Figure 23.4 **Evolutionary changes in the shape of oyster shells.** During a 12-million-year period in the early Jurassic, the shells of oysters of the extinct genus *Gryphaea* became larger and flatter. Scientists have hypothesized that such a change was an adaptation to disruptive water currents.

Biological inquiry: If a population of oysters was transferred to a shallow bay with calm water currents, how might the population change over the course of many generations?

T. roseae is called a **transitional form** because it provides a link between earlier species and many later species (**Figure 23.3**). In this case, the fishapod is a transitional form between fish, which are aquatic animals, and tetrapods, which are usually terrestrial animals. Unlike a true fish, *T. roseae* had a broad skull, a flexible neck, and eyes mounted on the top of its head like a crocodile. Its interlocking rib cage suggests it had lungs. Perhaps the most surprising discovery was that its pectoral fins (those on the side of the body) revealed the beginnings of a primitive wrist and five finger-like bones. These appendages would have been adequate for *T. roseae* to peek its head above the water from shallow river bottoms and look for prey. During the Devonian period, 417–354 million years ago, this could have been an important advantage in the marshy floodplains of large rivers.

As a second example, certain oysters began to undergo a change in shell structure about 200 million years ago. Oysters with smaller, curved shells were superseded by oysters with larger, flatter shells (**Figure 23.4**). This change was observed over the course of 12 million years during the early Jurassic period, when water currents became stronger. Scientists have hypothesized that larger, flatter shells are more stable in disrup-

tive water currents, so these shells were better adapted to the environmental change.

One of the best-studied examples of evolutionary change is our third example, that of the horse family. Modern members include horses, zebras, and donkeys. These species, which are large, long-legged animals adapted to living in open grasslands, are the remaining descendants of a long lineage that produced many extinct species since its origin approximately 55 million years ago. Examination of the horse lineage provides a particularly interesting case of how evolution involves adaptation to changing environments.

The earliest known fossils of the horse family (termed *Hyracotherium*) revealed that the animals were small with short legs and broad feet (**Figure 23.5**). Such early horses lived in wooded habitats and are thought to have eaten leaves and herbs. Between the time of these first members of the horse family and modern horses, the fossil record has revealed adaptive changes in size, foot anatomy, and tooth morphology. The first horses were the size of dogs, while modern horses typically weigh more than a half ton. *Hyracotherium* had four toes on its front feet and three on its hind feet. Instead of hooves, these toes were encased in fleshy pads. By comparison, the feet of modern horses have a single toe, enclosed in a tough, bony hoof. The fossil record shows an increase in the length of the central toe, the development of a bony hoof, and the loss of the other toes.

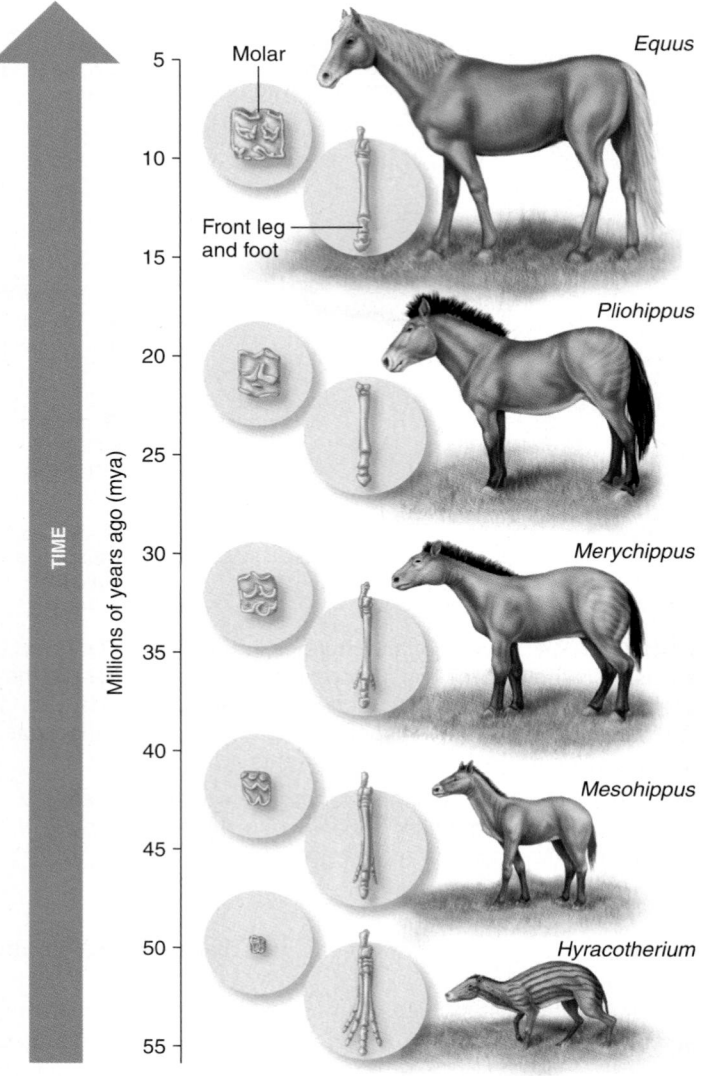

Molar

Front leg
and foot

TIME

Millions of years ago (mya)

5
10
15
20
25
30
35
40
45
50
55

Equus

Pliohippus

Merychippus

Mesohippus

Hyracotherium

Figure 23.5 **Evolutionary changes that led to the modern horse.** The major changes that occurred in horses' body size, foot anatomy, and tooth morphology were adaptations to the changing environment in which the horses and their ancestors lived over the last 55 million years of their evolution. Note: This figure is meant to emphasize changes that led to modern horses. The evolutionary pathway that produced modern horses involves several branches, and is described in Chapter 26.

Finally, the teeth of *Hyracotherium* were relatively small compared to those of modern horses. Over the course of millions of years, horse teeth have increased in size and developed a complex pattern of ridges on their molars.

These changes in horse characteristics can be attributed to natural selection producing adaptations to changing global climates. Over North America, where much of horse evolution occurred, large areas changed from dense forests to grasslands. Their increase in size and changes in their foot structure allowed horses to escape predators and travel great distances in search

of food. The changes seen in horse's teeth are consistent with a dietary shift from eating more tender leaves to eating grasses and other vegetation that are more abrasive and require more chewing.

Biogeography Indicates That Species in a Given Area Have Evolved from Pre-Existing Species

Biogeography is the study of the geographical distribution of extinct and modern species. Patterns of past evolution are often found in the natural geographical distribution of related species. From such studies, scientists have discovered that isolated continents and island groups have evolved their own distinct plant and animal communities. As mentioned earlier in this chapter, Darwin himself observed several species of finches found on the Galápagos Islands that had unique characteristics, such as beak shapes, when compared with similar finches found on the mainland. As we discussed, scientists now think that these island species evolved from mainland birds that had migrated to the islands and then became adapted to a variety of feeding habits. One example of this was shown in Figure 23.2.

Around the world, islands, which are often isolated from other landmasses, provide numerous examples in which geography has played a key role in the evolution of new species. Islands often have many species of plants and animals that are **endemic**, which means they are naturally found only in a particular location. Most endemic island species have closely related relatives on nearby islands or the mainland. For example, consider the island fox (*Urocyon littoralis*), which lives on the Channel Islands located off the coast of southern California between Los Angeles and Santa Barbara (**Figure 23.6**). This type of fox is found nowhere else in the world. It weighs about 3–6 pounds and feeds largely on insects, mice, and fruits. The island fox evolved from the mainland gray fox (*Urocyon cinereoargenteus*), which is much larger, usually 7–11 pounds. During the last Ice Age, about 16,000–18,000 years ago, the Santa Barbara channel was frozen and narrow enough for ancestors of the mainland gray fox to cross over to the Channel Islands. When the Ice Age ended, the ice melted and sea levels rose, causing the foxes to be cut off from the mainland. Over the last 16,000–20,000 years, these foxes evolved into the smaller island fox. The smaller size of the island fox is an example of island dwarfing, the phenomenon in which the size of large animals isolated on an island shrinks dramatically over many generations. It is a form of natural selection in which smaller size provides a survival advantage probably because of limited food.

The evolution of major animal groups is also correlated with known changes in the distribution of landmasses on the Earth. The first mammals arose approximately 225 million years ago. The fossils of the first placental mammals, which are those that have long internal gestation and give birth to offspring that are well developed, are only 80 million years old. The first mammals arose when the area that is now Australia was still connected to the other continents. However, the first placental mammals evolved after continental drift had separated Australia from the

Convergent Evolution Suggests Adaptation to the Environment

The process of natural selection is also evident in the study of plants and animals that have similar characteristics, even though they are not closely related evolutionarily. This similarity is due to **convergent evolution**, in which two different species from different lineages show similar characteristics because they occupy similar environments. For example, the long snout and tongue of both the giant anteater (*Myrmecophaga tridactyla*), found in South America, and the echidna (*Tachyglossus aculeatus*), found in Australia, are similar yet independently evolved adaptations that enable these animals to feed on ants (**Figure 23.7a**). The giant anteater is a placental mammal, while the echidna is an egg-laying mammal, so they are not closely related evolutionarily. Another example involves aerial rootlets found in vines such as English ivy (*Hedera helix*) and wintercreeper (*Euonymus fortunei*) (**Figure 23.7b**). Based on differences in their structures, these aerial rootlets appear to have developed independently as an effective means to cling to the support on which a vine attaches itself.

A third example of convergent evolution is revealed by the molecular analysis of fish that live in very cold water. Antifreeze proteins allow certain species of fish to survive the subfreezing temperatures of Arctic and Antarctic waters by preventing the formation of ice crystals in their blood. By studying these fish, researchers have determined that they are an interesting case of convergent evolution (**Figure 23.7c**). Among different species of fish, one of five different genes has independently evolved to produce antifreeze proteins. For example, in the sea raven (*Hemitripterus americanus*), the antifreeze protein is rich in the amino acid cysteine, and the secondary structure of the protein is in a β sheet conformation. In contrast, the antifreeze protein in the longhorn sculpin (*Trematomus nicolai*) evolved from an entirely different gene. The antifreeze protein in this species is rich in the amino acid glutamine, and the secondary structure of the protein is largely composed of α helices.

The similar characteristics in the examples shown in Figure 23.7, which are the result of convergent evolution, are called **analogous structures** or **convergent traits**. They represent cases in which structures have arisen independently, two or more times, because species have occupied similar types of environments on the Earth. By comparison, homologous structures have a single evolutionary origin.

Selective Breeding Is a Human-Driven Form of Natural Selection

The term **selective breeding** refers to programs and procedures designed to modify traits in domesticated species. This practice, also called **artificial selection**, is related to natural selection. In forming his theory of natural selection, Charles Darwin was influenced by his observations of selective breeding by pigeon breeders. The primary difference between natural and artificial selection is how the parents are chosen. Natural selection is due to natural variation in reproductive success.

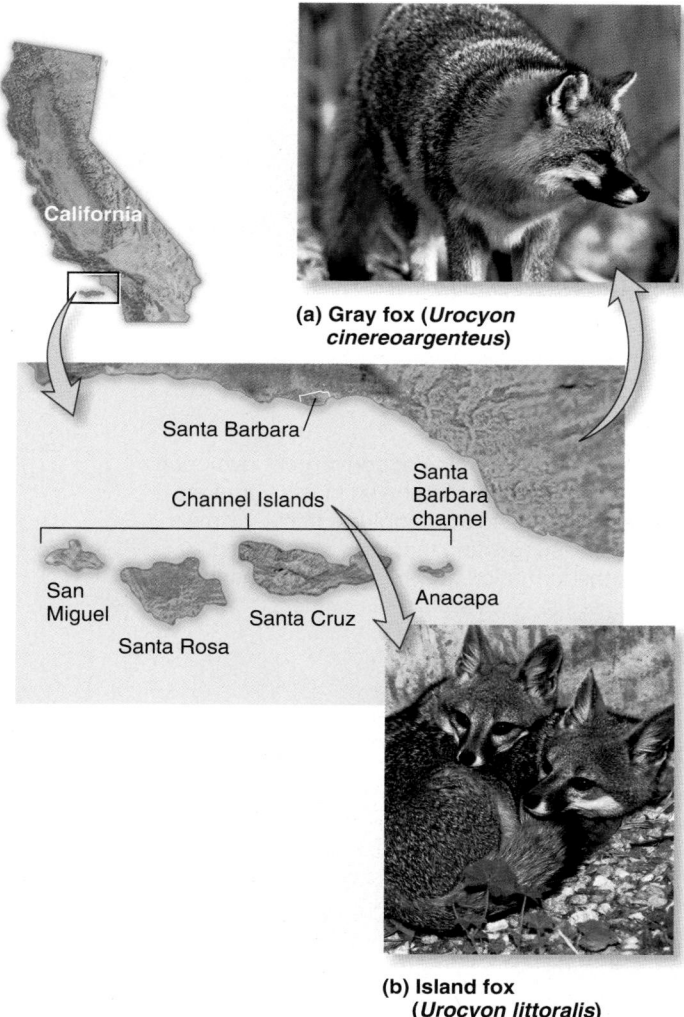

Figure 23.6 The evolution of an endemic island species from a mainland species. (a) The smaller island fox found on the Channel Islands evolved from (b) the gray fox found on the California mainland.

other continents (refer back to Figure 22.8). Except for a few species of bats and rodents that have migrated to Australia more recently, Australia lacks any of the larger, terrestrial placental mammals. This observation is consistent with the idea that placental mammals first arose somewhere other than Australia, and that the barrier of a large ocean prevented most terrestrial placental mammals from migrating there. Instead, Australia has more than 100 species of kangaroos, koalas, and other marsupials. Marsupials are a group of mammal species in which young are born in a very immature condition and then develop further in the mother's abdominal pouch, which covers the mammary glands. Most species of Australian marsupials are not found on other continents. Evolutionary theory is consistent with the idea that the existence of these unique Australian species is due to their having evolved in isolation from the rest of the world for millions of years.

(a) Their long snouts and tongues enable these species, the giant anteater (left) and the echidna (right), to feed on ants.

(b) Their aerial rootlets allow these vines, English ivy (left) and wintercreeper (right), to climb up supports.

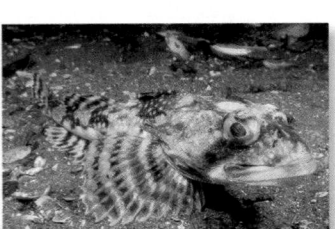

(c) Antifreeze proteins enable these fish, the sea raven (left) and the longhorn sculpin (right), to survive in frigid waters. The antifreeze proteins in these 2 species evolved from entirely different genes.

Figure 23.7 Examples of convergent evolution. All three pairs of organisms shown in this figure are not closely related evolutionarily, but occupy similar environments, suggesting that natural selection promotes the formation of similar adaptations that are well suited to a particular environment.

(a) Bulldog   **(b) Greyhound**   **(c) Dachshund**

Figure 23.8 Common breeds of dogs that have been obtained by selective breeding. By selecting parents carrying the alleles that influence traits desirable to humans, dog breeders have produced breeds with distinctive features. For example, the bulldog has alleles that give it short legs and a flat face. All the dogs shown in this figure carry the same kinds of genes (for example, genes that affect their size, shape, and fur color). However, the alleles for many of these genes are different among these dogs, thereby allowing humans to select for or against them and produce breeds with strikingly different phenotypes.

Organisms that are able to survive and reproduce are more likely to pass their genes on to future generations. Nature determines or "chooses" which individuals will be successful parents. In artificial selection, the breeder chooses as parents those individuals that possess traits desirable to humans.

The underlying phenomenon that makes selective breeding possible is genetic variation. Within a group of individuals of the same species, variation may exist in a trait of interest. For selective breeding to be successful, the underlying cause of the phenotypic variation must be related to differences in the **alleles**,

different forms of a particular gene, that determine the trait. The breeder will choose parents with desirable phenotypic characteristics. For centuries, humans have been practicing selective breeding to obtain domesticated species with interesting or agriculturally useful characteristics. For example, many common breeds of dog are the result of selective breeding strategies (**Figure 23.8**). All dogs are members of the same species, *Canis familiaris*, so they can be interbred to produce offspring. Selective breeding can dramatically modify the traits in a species. When you compare certain breeds of dogs (for example, a

Strain	Kohlrabi	Kale	Broccoli	Brussels sprouts	Cabbage	Cauliflower
Modified trait	Stem	Leaves	Flower buds and stem	Lateral leaf buds	Terminal leaf bud	Flower buds

Figure 23.9   **Crop plants developed by selective breeding of the wild mustard plant.**   Although these six agricultural plants look quite different from each other, they carry many of the same alleles as the wild mustard plant. However, they differ among each other in alleles that affect the formation of flowers, buds, stems, and leaves.

greyhound and a dachshund), they hardly look like members of the same species!

Likewise, most of the food we eat is obtained from species that have been profoundly modified by selective breeding strategies. This includes products such as grains, fruits, vegetables, meat, milk, and juices. For example, **Figure 23.9** illustrates how certain characteristics in the wild mustard plant (*Brassica oleracea*) have been modified by selective breeding to create several varieties of domesticated crops, including broccoli, Brussels sprouts, and cauliflower. This plant is native to Europe and Asia, and plant breeders began to modify its traits approximately 4,000 years ago. As seen here, certain traits in the domestic strains differ considerably from those of the original wild species. These varieties are all members of the same species. They can interbreed to produce viable offspring. For example, in the grocery store you may have seen brocciflower, which is produced from a cross between broccoli and cauliflower.

As a final example, **Figure 23.10** shows the results of an artificial selection experiment on corn begun at the Illinois Experiment Station in 1896, even before the rediscovery of Mendel's laws. This study began with 163 ears of corn with an oil content ranging from 4 to 6%. In each of 80 succeeding generations, corn plants were divided into two separate groups. In one group, members with the highest oil content in the kernels were chosen as parents of the next generation. In the other group, members with the lowest oil content were chosen. After many generations, the oil content in the first group rose to over 18%. In the other group, it dropped to less than 1%. These results show that selective breeding can modify a trait in a very directed manner.

Figure 23.10   **Results of selective breeding for oil content in corn plants.**   In this example, corn plants were selected for breeding based on high or low oil content of the kernels. Over the course of many generations, this had a major impact on the amount of corn oil—an agriculturally important product—that is made by the two groups of plants.

Biological inquiry: When comparing Figures 23.8, 23.9, and 23.10, what general effects of artificial selection do you observe?

FEATURE INVESTIGATION

The Grants Have Observed Natural Selection in Galápagos Finches

Since 1973, Peter Grant, Rosemary Grant, and their colleagues have studied the process of natural selection in finches found on the Galápagos Islands. For over 30 years, the Grants have focused much of their work on one of the Galápagos Islands known as Daphne Major (**Figure 23.11a**). This small island (0.34 km²) has a moderate degree of isolation (it is 8 km from the nearest island), an undisturbed habitat, and a resident population of the finch *Geospiza fortis*, the medium ground finch (**Figure 23.11b**).

To study natural selection, the Grants have observed various traits in finches over the course of many years. One example is beak size. The medium ground finch has a relatively small crushing beak, allowing it to more easily feed on small, tender seeds (see Table 23.1). The Grants quantified beak size among the medium ground finches of Daphne Major by carefully measuring beak depth (a measurement of the beak from top to bottom, at its base) on individual birds. During the course of their studies, they compared the beak sizes of parents and offspring by examining many broods over several years. The depth of the beak was transmitted from parents to offspring, regardless of environmental conditions, indicating that differences in beak sizes are due to genetic differences in the population. In other words, they found that beak depth was a heritable trait.

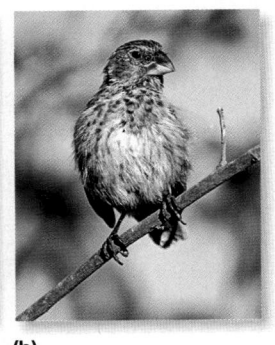

(a)   (b)

Figure 23.11 The Grants' investigation of natural selection in finches. (a) Daphne Major, one of the Galápagos Islands. (b) One of the medium ground finches (*Geospiza fortis*) that populate this island.

By measuring many birds every year, the Grants were able to assemble a detailed portrait of natural selection in action. In the study shown in **Figure 23.12**, they measured beak depth from 1976 to 1978. In the wet year of 1976, the plants of Daphne Major produced an abundance of the small seeds that these finches could easily eat. However, a drought occurred in 1977. During this year, the plants on Daphne Major tended to produce few of the smaller seeds, which the finches rapidly consumed.

Figure 23.12 The Grants and natural selection of beak size among the medium ground finch. The results are those recorded from 1976 and 1978.

HYPOTHESIS Dry conditions produce larger seeds and may result in larger beaks in succeeding generations of *Geospiza fortis* due to natural selection.	
STARTING MATERIALS A population of *G. fortis* on one of the Galápagos Islands called Daphne Major.	

		Experimental level	**Conceptual level**
1	In 1976, measure beak depth in parents and offspring of the species *G. fortis*.	Capture birds and measure beak depth.	This is a way to measure a trait that may be subject to natural selection.
2	Repeat the procedure on offspring that were born in 1978 and had reached mature size. A drought had occurred in 1977 that caused plants on the island to produce mostly larger seeds and relatively few small seeds.	Capture birds and measure beak depth.	This is a way to measure a trait that may be subject to natural selection.

3   **THE DATA**

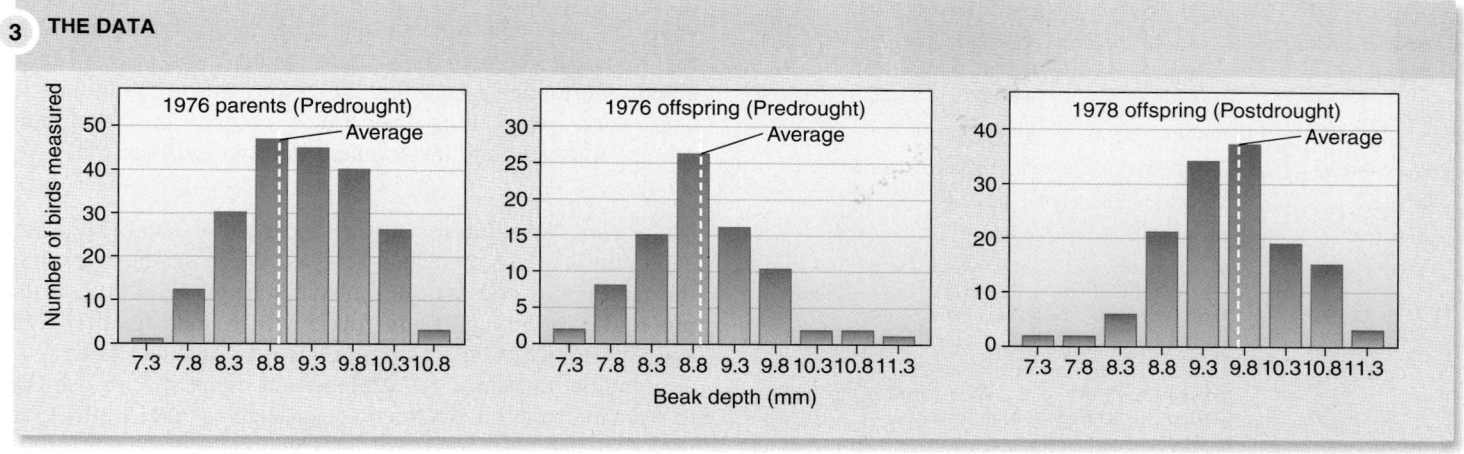

To survive, the finches resorted to eating larger, drier seeds, which are harder to crush. As a result, birds with larger beaks were more likely to survive because they were better at breaking open these large seeds. In the year after the drought, the average beak depth of birds in the population increased, because the surviving birds with larger beaks passed this trait on to their offspring. Overall, these results illustrate the power of natural selection to alter the nature of a trait, in this case beak depth, in a given population.

A Comparison of Anatomical, Developmental, and Molecular Homologies Shows Evolution of Related Species from a Common Ancestor

Let's now consider other widespread observations of the process of evolution among living organisms. In biology, the term **homology** refers to a fundamental similarity that occurs due to descent from a common ancestor. As a result of evolution, homology is often observed between different species. Two species may have a similar trait because the trait was originally found in a common ancestor. As described next, such homologies may involve anatomical, developmental, or molecular features.

Anatomical Homologies   As noted by Theodosius Dobzhansky, the theory of evolution provides a sensible framework for understanding the diversity of life. Many observations regarding anatomical features of plants and animals simply cannot be understood in any meaningful scientific way except as a result of evolution. A comparison of vertebrate anatomy is a case in point. An examination of the limbs of modern vertebrate species reveals similarities that indicate that the same set of bones has undergone evolutionary changes to become the bones used today for many different purposes. As seen in **Figure 23.13**, the forelimbs of vertebrates have a strikingly similar pattern of bone arrangements. These are termed **homologous structures** because they are considered to be derived from a common ancestor. The forearm has developed different uses among various vertebrates, including grasping, walking, flying, swimming, and climbing.

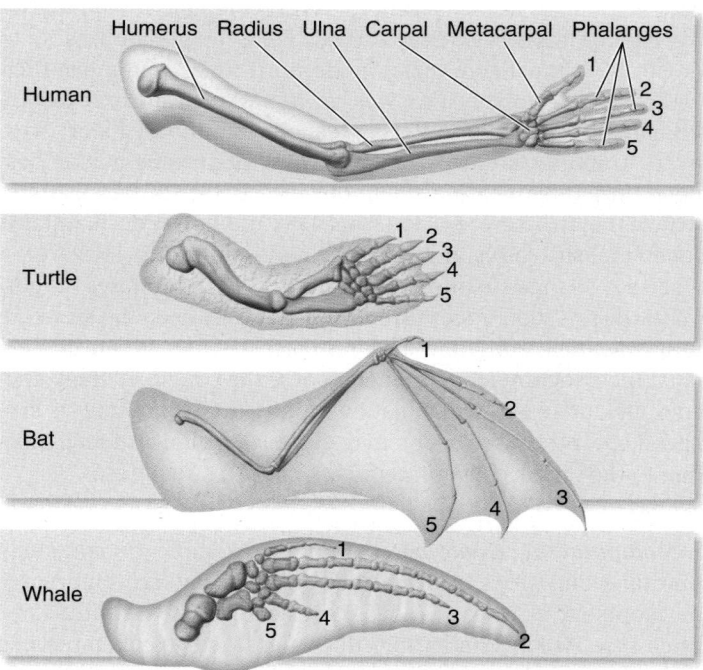

Figure 23.13   An example of anatomical homology: homologous structures found in vertebrates.   The same set of bones is found in the human arm, turtle arm, bat wing, and whale flipper, although these bones have been modified in ways that allow them to perform different functions. This homology suggests that all of these animals evolved from a common ancestor.

Table 23.3	Examples of Vestigial Structures
Organism	**Vestigial structure(s)**
Humans	Muscles to wiggle ears and tail bone in adult, rudimentary gill ridges in embryo
Boa constrictors	Skeletal remnants of hip bones and hind legs
Whales	Skeletal remnants of a pelvis
Manatees	Fingernails on the flippers
Dandelion	An asexually reproducing plant (*Taraxicum officinale*) that still makes flowers, even though it does not produce fertile seeds.
Hornbills and cuckoos	In certain families of birds, both of the common carotid arteries are nonfunctional. They are present as fibrous cords. Their vascular function has been assumed by other vessels.

The theory of evolution explains how these animals have descended from a common ancestor and how natural selection has modified the same initial pattern of bones in ways that ultimately allowed them to be used for several different purposes.

Another observation of evolution is the phenomenon of **vestigial structures**, which are anatomical features that have no apparent function but resemble structures of their presumed ancestors. **Table 23.3** describes several examples. An interesting case is found in humans. People have a complete set of muscles for moving their ears, even though most people are unable to do so. By comparison, many modern mammals can move their ears, and presumably this was an important trait in a distant human ancestor. Within the context of evolutionary theory, vestigial structures are evolutionary relics. Organisms having vestigial structures share a common ancestry with organisms in which the structure is functional. Natural selection maintains functional structures in a population of individuals. However, if a species changes its lifestyle so that the structure loses its purpose, the selection forces that would normally keep the structure in a functional condition are no longer present. When this occurs, the structure may degenerate over the course of many generations due to the accumulation of mutations that limit its size and shape. Natural selection may eventually eliminate such traits due to the inefficiency of producing unused structures.

Developmental Homologies Another example of homology due to evolution is the way that animals undergo embryonic development. Species that differ substantially at the adult stage often bear striking similarities during early stages of embryonic development. These temporary similarities are called developmental homologies. In addition, evolutionary history is revealed during development in certain organisms such as vertebrates. For example, if we consider human development, several features are seen in the embryo that are not present at birth. Human embryos possess rudimentary gill ridges like a fish embryo, even though human embryos receive oxygen via the umbilical cord. The presence of gill ridges indicates that humans evolved from

an aquatic animal with gill slits. A second observation is that every human embryo has a long bony tail. It is difficult to see the advantage of such a structure *in utero*, but easy to understand its presence assuming that an ancestor to the human lineage possessed a long tail. These observations, and many others, illustrate that development has evolved over time.

Molecular Homologies Our last example of homology due to evolution involves molecular studies. When scientists examine the features of cells at the molecular level, similarities called **molecular homologies** are found which indicate that living species evolved from a common ancestor or interrelated group of common ancestors. For example, all living species use DNA to store information. RNA molecules such as mRNA, tRNA, and rRNA are used to access that information, and proteins are the functional products of most genes. Furthermore, certain biochemical pathways are found in all or nearly all species, although minor changes in the structure and function of proteins involved in these pathways have occurred. For example, all species that use oxygen, which constitutes the great majority of species on our planet, have similar proteins that together make up an electron transport chain and an ATP synthase. In addition, nearly all living organisms can metabolize glucose via a glycolytic pathway that is described in Chapter 7. Taken together, these types of observations indicate that such molecular phenomena arose very early in the origin of life, and have been passed to all or nearly all modern forms.

The most compelling observation at the molecular level indicating that modern life-forms are derived from a common ancestor is revealed by analyzing genetic sequences and finding genetic homologies, or similar genes. The same type of gene is often found in diverse organisms. Furthermore, the degree to which a genetic sequence from different species is similar reflects the evolutionary relatedness of those species. As an example, let's consider a gene that encodes the p53 protein that plays a role in preventing cancer (see Chapter 14). **Figure 23.14** shows a short amino acid sequence that makes up part of the p53 protein from a variety of species, including five mammals, one bird, and three fish. The top sequence is the human p53 sequence, and the right column describes the percentages of amino acids within the entire sequence that are identical to the entire human sequence. Amino acids in the other species that are identical to humans are highlighted in orange. The sequences from the two monkeys are closest to humans, followed by the other two mammalian species (rabbit and dog). The three fish sequences are the least similar to the human sequence, but you may notice that the fish sequences tend to be similar to each other. Taken together, the data shown in Figure 23.14 illustrate two critical points regarding gene evolution. First, certain genes are found in a diverse array of species such as mammals, birds, and fish. Second, the sequences of closely related species tend to be more similar to each other than they are to distantly related species. The mechanism for this second observation is described in the next section.

	Short amino acid sequence within the p53 protein	Percentages of amino acids in the whole p53 protein that are identical to human p53
Human (*Homo sapiens*)	Val Pro Ser Gln Lys Thr Tyr Gln Gly Ser Tyr Gly Phe Arg Leu Gly Phe Leu His Ser Gly Thr	100
Rhesus monkey (*Macaca mulatta*)	Val Pro Ser Gln Lys Thr Tyr **His** Gly Ser Tyr Gly Phe Arg Leu Gly Phe Leu His Ser Gly Thr	95
Green monkey (*Cercopithecus aethiops*)	Val Pro Ser Gln Lys Thr Tyr **His** Gly Ser Tyr Gly Phe Arg Leu Gly Phe Leu His Ser Gly Thr	95
Rabbit (*Oryctolagus cuniculus*)	Val Pro Ser Gln Lys Thr Tyr **His** Gly **Asn** Tyr Gly Phe Arg Leu Gly Phe Leu His Ser Gly Thr	86
Dog (*Canis familiaris*)	Val Pro Ser **Pro** Lys Thr Tyr **Pro** Gly **Thr** Tyr Gly Phe Arg Leu Gly Phe Leu His Ser Gly Thr	80
Chicken (*Gallus gallus*)	Val Pro Ser **Thr Glu Asp** Tyr **Gly** Gly **Asp Phe Asp** Phe Arg **Val** Gly Phe **Val Glu Ala** Gly Thr	53
Channel catfish (*Ictalurus punctatus*)	Val Pro **Val Thr Ser Asp** Tyr **Pro** Gly **Leu Leu Asn** Phe **Thr** Leu **His** Phe **Gln Glu** Ser **Ser Gly**	48
European flounder (*Platichthys flesus*)	Val Pro **Val Val Thr Asp** Tyr **Pro** Gly **Glu** Tyr Gly Phe **Gln** Leu **Arg** Phe **Gln Lys** Ser Gly Thr	46
Congo puffer fish (*Tetraodon miurus*)	Val Pro **Val Thr Thr Asp** Tyr **Pro** Gly **Glu** Tyr Gly Phe **Lys** Leu **Arg** Phe **Gln Lys** Ser Gly Thr	41

Figure 23.14 **An example of genetic homology: a comparison of a short amino acid sequence within the p53 protein from nine different animals.** This figure compares a short region of the p53 protein, which plays a role in preventing cancer. Amino acids are represented by three-letter abbreviations. The orange-colored amino acids in the sequences are identical to those in the human sequence. The numbers in the right column indicate the percentage of amino acids within the whole p53 protein that is identical with the human p53 protein, which is 393 amino acids in length. For example, 95% of the amino acids, or 373 of 393, are identical between the p53 sequence found in humans and that in Rhesus monkeys.

Biological inquiry: In the sequence shown in this figure, how many amino acid differences are there between the following pairs: Rhesus and green monkeys, Congo puffer fish and European flounder, and Rhesus monkey and Congo puffer fish? What do these differences tell you about the evolutionary relationships among these four species?

23.3 The Molecular Processes That Underlie Evolution

Historically, the study of evolution was based on comparing the anatomies of extinct and modern species to identify similarities between related species. However, the advent of molecular approaches for analyzing DNA sequences has revolutionized the field of evolutionary biology. Now we can analyze how changes in the genetic material are associated with changes in phenotype, and how those changes have led to the formation of new species. In this section, we will examine some of the molecular changes in the genetic material that are associated with evolution.

Homologous Genes Are Derived from a Common Ancestral Gene

When two genes are derived from the same ancestral gene, they are called **homologous genes**. The analysis of homologous genes reveals the molecular details of evolutionary change. As an example, let's consider a gene in two different species of bacteria that encodes a transport protein involved in the uptake of metal ions into bacterial cells. Such genes, which are homologous yet from different species, are said to be **orthologs** of each other. Millions of years ago, these two species had a common ancestor (**Figure 23.15**). Over time, the common ancestor diverged

into additional species, eventually evolving into *Escherichia coli*, *Clostridium acetylbutylicum*, and many other species. Since this divergence, the metal transporter gene has accumulated mutations that alter its sequence, though the similarity between the *E. coli* and the *C. acetylbutylicum* genes remains striking. In this case, the two sequences are similar because they were derived from the same ancestral gene, but they are not identical due to the independent accumulation of different random mutations.

Gene Duplications Create Gene Families

Orthologs are examples of evolutionary change occurring in separate species. Demonstrations of evolutionary change can also be found within a single species. Two or more homologous genes found within a single species are termed **paralogs** of each other. Rare gene duplication events can produce multiple copies of a gene and ultimately lead to the formation of a gene family. A **gene family** consists of two or more copies of paralogous genes within the genome of a single organism. A well-studied example of a gene family is the globin gene family found in humans and many other animal species. The globin genes encode polypeptides that are subunits of proteins that function in oxygen binding. One such protein is hemoglobin that is found in red blood cells and carries oxygen throughout the body.

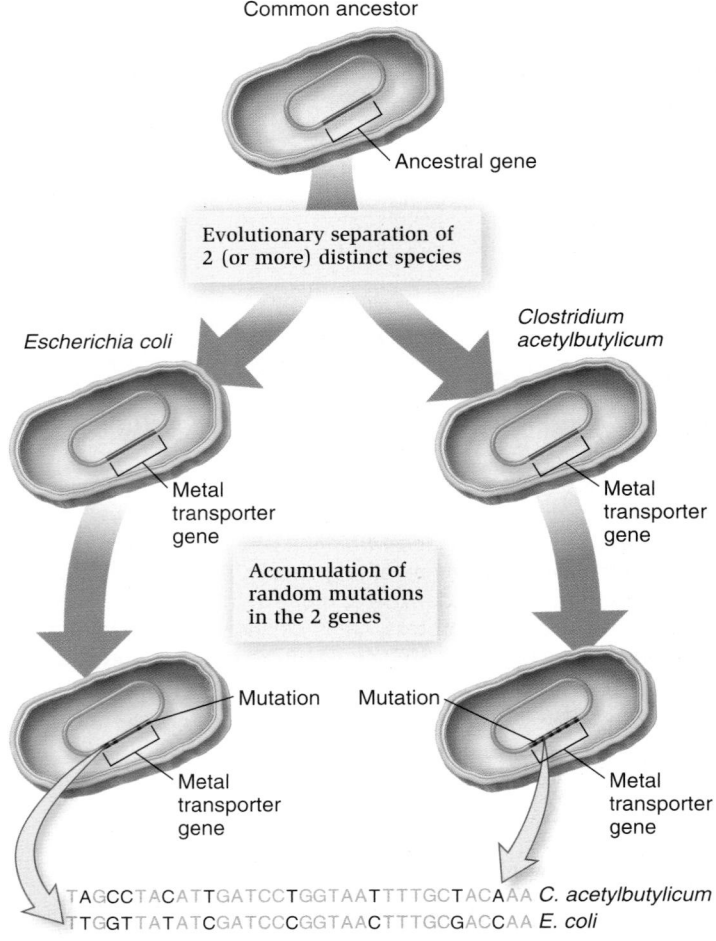

Common ancestor

Ancestral gene

Evolutionary separation of
2 (or more) distinct species

Escherichia coli

*Clostridium
acetylbutylicum*

Metal
transporter
gene

Metal
transporter
gene

Accumulation of
random mutations
in the 2 genes

Mutation   Mutation

Metal
transporter
gene

Metal
transporter
gene

TAGCCTACATTGATCCTGGTAATTTTGCTACAAA *C. acetylbutylicum*
TTGGTTATATCGATCCCGGTAACTTTGCGACCAA *E. coli*

Figure 23.15 The evolution of orthologs, homologous
genes from different species.   After two species diverged from
each other, the genes accumulated random mutations that
created similar but not identical gene sequences called orthologs.
These orthologs in *E. coli* and *C. acetylbutylicum* encode metal
transporters. Only one of the two DNA strands is shown from
each of the genes. Bases that are identical between the two
genes are shown in orange.

The globin gene family is composed of 14 genes that were originally derived from a single ancestral globin gene. According to
an evolutionary analysis, the ancestral globin gene first duplicated between 500 to 600 million years ago. Since that time,
additional duplication events and chromosomal rearrangements
have produced the current number of 14 genes on three different human chromosomes (refer back to Figure 21.8).

Gene families have been important in the evolution of
traits. Even though all of the globin polypeptides are subunits
of proteins that play a role in oxygen binding, the accumulation
of changes in the various family members has created globins
that are more specialized in their function. For example, myoglobin is better at binding and storing oxygen in muscle cells,
whereas the hemoglobins are better at binding and transporting
oxygen via the red blood cells. Also, different globin genes are

expressed during different stages of human development. The
epsilon (ε)- and zeta (ζ)-globin genes are expressed very early in
embryonic life, while the gamma (γ)-globin genes exhibit maximal expression during the second and third trimesters of gestation. Following birth, the γ-globin gene is turned off and the
β-globin gene is turned on. These differences in the expression
of the globin genes reflect the differences in the oxygen transport needs of humans during the embryonic, fetal, and postpartum stages of life.

What is the evolutionary significance of the globin gene
family regarding adaptation? Internal gestation is one way that
animals have adapted to a terrestrial environment. On land, egg
cells and small embryos are very susceptible to drying out if
they are not protected in some way. Species such as birds and
reptiles lay eggs with a protective shell around them. Most mammals, however, have adjusted to a terrestrial environment by
evolving the adaptation of internal gestation. The ability to develop young internally has been an important factor in the survival and proliferation of mammals. The embryonic and fetal
forms of hemoglobin allow the embryo and fetus to capture
oxygen from the bloodstream of the mother.

GENOMES & PROTEOMES

New Genes in Eukaryotes Have Evolved via Exon Shuffling

Thus far we have considered how evolutionary change results
in the formation of related genes, which are described as orthologs and paralogs. Evolutionary mechanisms are also revealed
when the parts of genes that encode protein domains are compared within a single species. Many proteins, particularly those
found in eukaryotic species, have a modular structure composed of two or more domains with different functions. For example, certain transcription factors have discrete domains involved with hormone binding, dimerization, and DNA binding.
As described in Chapter 13, the glucocorticoid receptor has a
domain that binds the hormone, a second domain that facilitates protein dimerization, and a third domain that allows the
glucocorticoid receptor to bind to glucocorticoid response elements (GREs) next to genes. By comparing the modular structure of eukaryotic proteins with the genes that encode them,
geneticists have discovered that each domain tends to be encoded by one exon, or by a series of two or more adjacent exons.
As we learned in Chapter 12, exons contain the coding sequences
of a gene, which are separated by noncoding introns.

During the evolution of eukaryotic species, many new genes
have been created by a type of mutation known as **exon shuffling**. During this process, an exon and the flanking introns are
inserted into a gene, thereby producing a new gene that encodes
a protein with an additional domain (**Figure 23.16**). This process
may also involve the duplication and rearrangement of exons.
Exon shuffling results in novel genes that express proteins with
diverse functional modules. Such proteins can then alter traits
in the organism that can be acted upon by natural selection.

Figure 23.16   **The process of exon shuffling.**   In this example, a segment of one gene containing an exon and the flanking introns has been inserted into another gene. A rare, abnormal crossing-over event called nonhomologous recombination may cause this to happen. This results in proteins that have new combinations of domains and possibly new combinations of functions.

Biological inquiry: What is the evolutionary advantage of exon shuffling?

Exon shuffling may occur by more than one mechanism. One possibility is that a double crossover could promote the insertion of an exon into another gene (see Figure 23.16). This is called nonhomologous or illegitimate recombination because the two regions involved in the crossover are not homologous to each other. Alternatively, transposable elements that are described in Chapter 21 may promote the movement of exons into other genes.

Horizontal Gene Transfer Also Contributes to the Evolution of Species

At the molecular level, the type of evolutionary change depicted in Figures 23.14 through 23.16 is called **vertical evolution**. In these cases, species evolve from pre-existing species by the accumulation of gene mutations, gene duplications, and exon shuffling. Vertical evolution involves genetic changes in a series of ancestors that form a lineage. In addition to vertical evolution, species accumulate genetic changes by another process called **horizontal gene transfer**, which involves the exchange of genetic material among different species.

Figure 23.17 illustrates one possible mechanism for horizontal gene transfer. In this example, a eukaryotic cell has engulfed a bacterial cell by endocytosis. During the degradation of the bacterium in an endocytotic vesicle, a bacterial gene happens to escape to the nucleus of the cell, where it is inserted into

Figure 23.17   **Horizontal gene transfer from a bacterium to a eukaryote.**   In this example, a bacterium is engulfed by a eukaryotic cell, and a bacterial gene is transferred to one of the eukaryotic chromosomes.

one of the chromosomes. In this way, a gene has been transferred from a bacterial species to a eukaryotic species. By analyzing gene sequences among many different species, researchers have discovered that horizontal gene transfer is a common phenomenon. This process can occur from prokaryotes to eukaryotes, from eukaryotes to prokaryotes, between different species of prokaryotes, and between different species of eukaryotes.

Therefore, when we view evolution, it is not simply a matter of one species evolving into one or more new species via the accumulation of random mutations. It also involves the horizontal transfer of genes among different species, enabling those species to acquire new traits that foster the evolutionary process.

Gene transfer among bacterial species is relatively widespread. As discussed in Chapter 18, bacterial species may carry out three natural mechanisms of gene transfer known as conjugation, transformation, and transduction. By analyzing the genomes of bacterial species, scientists have determined that many genes within a given bacterial genome are derived from horizontal gene transfer. Genome studies have suggested that as much as 20–30% of the variation in the genetic composition of modern prokaryotic species can be attributed to this process. For example, in *E. coli* and *Salmonella typhimurium*, roughly 17% of their genes have been acquired via horizontal gene transfer during the past 100 million years. The roles of these acquired genes are quite varied, though they commonly involve functions that are readily acted upon by natural selection. These include genes that confer antibiotic resistance, the ability to degrade toxic compounds, and pathogenicity (the ability to cause disease).

Evolution at the Genomic Level Involves Changes in Chromosome Structure and Number

Thus far, we have considered several ways that a species might acquire new genes. These include random mutations within pre-existing genes, gene duplications to create gene families, exon shuffling, and horizontal gene transfer. Evolution also occurs at the genomic level, involving changes in chromosome structure and number. When comparing the chromosomes of closely related species, changes in chromosome structure and/or number are common.

As an example, **Figure 23.18** compares the banding patterns of the three largest chromosomes in humans and the corresponding chromosomes in chimpanzees, gorillas, and orangutans. (See Chapter 15 for a description of chromosome banding.) The banding patterns are strikingly similar because these species are closely related evolutionarily. Even so, you can see some interesting differences. Humans have one large chromosome 2, but this chromosome is divided into two separate chromosomes in the other three species. This explains why human cells have 23 pairs of chromosomes while ape cells have 24. The fusion of the two smaller chromosomes during the development of the

Figure 23.18   **An example of genomic evolution.**   This figure is a comparison of banding patterns among the three largest human chromosomes and the corresponding chromosomes in apes. It is a schematic drawing of Giemsa-stained chromosomes. The differences between these chromosomes illustrate the changes that have occurred during the evolution of these related species.

human lineage may have caused this difference in chromosome pair numbers. Another interesting change in chromosome structure is seen in chromosome 3. The banding patterns among humans, chimpanzees, and gorillas are very similar, but the orangutan has a large inversion that flips the arrangement of bands in the centromeric region. As discussed in Chapter 25, changes in chromosome structure and number may affect the ability of two organisms to breed with one another. In this way, such changes have been important in the establishment of new species.

Chapter Summary

- Biological evolution is a heritable change in one or more characteristics of a population or species across many generations.

23.1 The Theory of Evolution

- Charles Darwin proposed the theory of evolution based on his understanding of geology and population growth and his observations of species in their natural settings. His voyage on the *Beagle*, during which he studied many species including finches on the Galápagos Islands, was particularly influential. (Figure 23.1, Table 23.1)
- Darwin's theory of evolution involves descent with modification that leads to better adaptation to environmental conditions. According to the modern synthesis, natural selection acts on existing genetic variation over the course of many generations to produce populations of organisms with traits that promote greater reproductive success (adaptations). (Figure 23.2)

23.2 Observations of Evolutionary Change

- Observations of evolution include the fossil record, biogeography, convergent evolution, selective breeding, and homologies. (Table 23.2)
- Fossils show successive evolutionary change over long periods of time. The fossil record often reveals transitional forms that link past ancestors to modern species. (Figures 23.3, 23.4, 23.5)
- The geographical distribution of species, biogeography, provides information on how certain species are evolutionarily related to each other. Often, when populations become isolated, they evolve into a new species. (Figure 23.6)
- Convergent evolution involves independent adaptations resulting in analogous structures that are similar because organisms have evolved in similar environments. (Figure 23.7)
- Selective breeding, also known as artificial selection, illustrates how changes in genetic variation over the course of many generations can dramatically change the traits of organisms. (Figures 23.8, 23.9, 23.10)
- The Grants showed that natural selection can promote changes in beak size in the medium ground finch. (Figures 23.11, 23.12)
- Homologous structures are similar because they are derived from the same ancestral structure. The set of bones in the arms of vertebrates is one example. (Figure 23.13)
- Vestigial structures are found in species because they are derived from structures that were once functional, but have degenerated because they no longer have use in a modern species. (Table 23.3)
- Homology also occurs during development and at the molecular level of gene and protein sequences. (Figure 23.14)

23.3 The Molecular Processes That Underlie Evolution

- Molecular evolution refers to the molecular changes in genetic material that underlie the phenotypic changes associated with evolution.

- Homologous genes are derived from the same ancestral gene. They accumulate random mutations that make their sequences similar but somewhat different. Orthologs are homologous genes in different species. (Figure 23.15)
- Paralogs are homologous genes in the same species, produced by gene duplication events. Paralogs constitute a gene family. An example is the globin gene family that promoted the evolutionary adaptation of internal gestation.
- Exon shuffling is a form of mutation in which exons are inserted into genes and thereby create proteins with additional functional domains. (Figure 23.16)
- Another mechanism that creates genetic variation is horizontal gene transfer, in which genetic material is transferred between different species. Such genetic changes are acted upon by natural selection. (Figure 23.17)
- Evolution is also associated with changes in chromosome structure and chromosome number. (Figure 23.18)

Test Yourself

1. The process involving changes in one or more characteristics of a population that are heritable and occur across many generations is called
   a. natural selection.
   b. sexual selection.
   c. population genetics.
   d. biological evolution.
   e. inheritance of acquired characteristics.

2. Lamarck's vision of evolution differed from Darwin's in that Lamarck believed
   a. living things evolved in an upward direction.
   b. behavioral changes modified heritable traits.
   c. genetic differences among individuals in the population allowed for evolution.
   d. a and b only
   e. none of the above.

3. Which of the following scientists influenced Darwin's views on the nature of population growth?
   a. Cuvier
   b. Malthus
   c. Lyell
   d. Hutton
   e. Wallace

4. An evolutionary change in which an organism's characteristics change in ways that make it better suited to its environment is
   a. natural selection.
   b. an adaptation.
   c. an acquired characteristic.
   d. evolution.
   e. both a and c.

5. Vestigial structures are anatomical structures
   a. that have more than one function.
   b. that have no function.
   c. that look similar in different species but have different functions.
   d. that have the same function in different species but have very different appearances.
   e. of the body wall.

6. Which of the following is an example of developmental homologies seen in human embryonic development?
   a. gill ridges
   b. umbilical cord
   c. tail
   d. both a and c
   e. all of the above

7. Two or more homologous genes found within a particular species are called
   a. homozygous.
   b. orthologs.
   c. paralogs.
   d. heterologs.
   e. duplicates.

8. The phenomenon of exon shuffling
   a. creates new gene products by changing the pattern of intron removal in a particular gene.
   b. creates new genes by inserting exons and flanking introns into a different gene sequence, thereby introducing a new domain in the gene product.
   c. rearranges the sequence of exons in a single gene.
   d. rearranges the introns in a particular gene, creating new gene products.
   e. both a and d.

9. Horizontal gene transfer is
   a. the transmission of genetic information from parent to offspring.
   b. the exchange of genetic material among individuals of the same species.
   c. the exchange of genetic material between mates.
   d. the exchange of genetic material among individuals of different species.
   e. none of the above.

10. Genetic variation can increase as a result of
    a. random mutations in genes.
    b. exon shuffling.
    c. gene duplication.
    d. horizontal gene transfer.
    e. all of the above.

CONCEPTUAL QUESTIONS

1. Briefly describe the various observations that support the theory of biological evolution.

2. Define convergent evolution and give an example.

3. Explain how homologous forelimbs of vertebrates support the idea of biological evolution.

EXPERIMENTAL QUESTIONS

1. What features of Daphne Major made it a suitable field site for studying the effects of natural selection?

2. Why is beak depth in finches a good trait for a study of natural selection? What environmental conditions were important to allow the Grants to collect information concerning natural selection?

3. What were the results of the Grants' study following the drought in 1977? What impact did these results have on the theory of evolution?

COLLABORATIVE QUESTIONS

1. Discuss evolution and how it occurs.

2. Discuss horizontal gene transfer.

www.brookerbiology.com

This website includes answers to the Biological Inquiry questions found in the figure legends and all end-of-chapter questions.

24

POPULATION GENETICS

Colorful African cichlids. The choice of mates among populations of cichlids may depend on color.

Population genetics is the study of genes and genotypes in a population. The central issue in population genetics is genetic variation. Population geneticists want to know the extent of genetic variation within populations, why it exists, and how it changes over the course of many generations. Population genetics helps us to understand how underlying genetic variation is related to phenotypic variation, and other issues such as mate preference (see chapter opening photo).

Population genetics emerged as a branch of genetics in the 1920s and 1930s. Its mathematical foundations were developed by theoreticians who extended the principles of Mendel and Darwin by deriving equations to explain the occurrence of genotypes within populations. These foundations can be largely attributed to British evolutionary biologists J. B. S. Haldane and Ronald Fisher, and American geneticist Sewall Wright. As we will see, several researchers who analyzed the genetic composition of natural and experimental populations provided support for their mathematical theories. More recently, population geneticists have used techniques to probe genetic variation at the molecular level. In addition, the staggering improvement in computer technology has aided population geneticists in the analysis of their genetic theories and data.

In this chapter, we will explore the extent of genetic variation that occurs in populations and how such variation is subject to change. In many cases, such changes are associated with adaptations, which are characteristics of a species that have evolved over a long period of time by the process of natural selection. The concept of adaptation was discussed in Chapter 23.

This chapter will examine the various ways that natural selection leads to adaptation.

24.1 Genes in Populations

Population genetics is an extension of our understanding of Darwin's theory of natural selection, Mendel's laws of inheritance, and newer studies in molecular genetics. All of the genes in a population make up its **gene pool**. Each member of the population receives its genes from its parents, which, in turn, are members of the gene pool. Individuals that reproduce contribute to the gene pool of the next generation. Population geneticists study the genetic variation within the gene pool, and how such variation changes from one generation to the next. The emphasis is often focused on an understanding of variation in alleles between members of a population. As discussed in Chapter 16, alleles are different forms of the same gene. In this section, we will examine some of the general features of populations and gene pools.

A Population Is a Group of Interbreeding Individuals

A **population** is a group of individuals of the same species that can interbreed with one another. Certain species occupy a wide geographic range and are divided into discrete populations. For example, distinct populations of a given species may be located on different continents. (A more detailed description of populations and their native environments is given in Chapter 56.)

Another common situation is that a large mountain or some other type of geographic barrier may separate two or more populations on the same continent.

Populations are dynamic units that change from one generation to the next. They may change in number, geographic location, and genetic composition. Natural populations may go through cycles of "feast or famine," during which the population gains or loses individuals. In addition, natural predators or disease may periodically decrease the size of a population significantly, and then later the population may rebound to its original size. Populations or individuals within populations may migrate to a new site and establish a distinct population at a new location that may differ in environment from the original site.

As population sizes and locations change, their genetic composition generally changes as well. Some of the genetic changes involve adaptive evolution, which means that a species is better adapted to its environment, making it more likely to survive and reproduce. For example, a population of mammals may move from a warmer to a colder geographic location. Over the course of many generations, adaptive evolution may change the population such that the fur of the animals is thicker and provides better insulation against the colder temperatures.

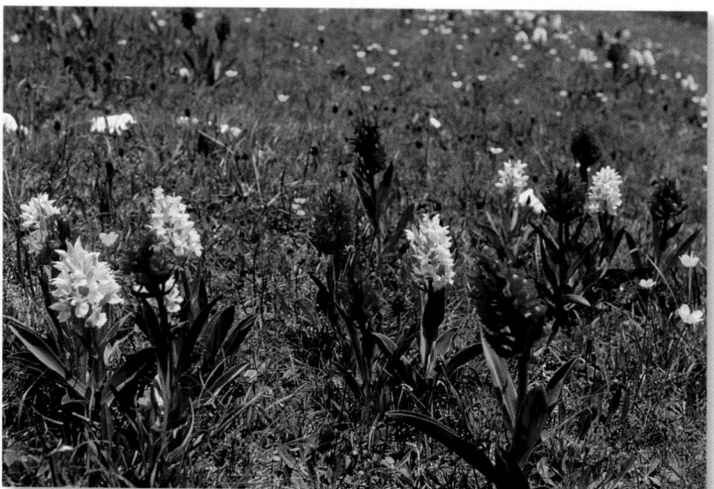

Figure 24.1   An example of polymorphism: the two color variations found in the orchid *Dactylorhiza sambucina*.

GENOMES & PROTEOMES

Genes in Natural Populations Are Usually Polymorphic

The term **polymorphism** (meaning many forms) refers to the phenomenon that many traits display variation within a population. Historically, polymorphism first referred to variation in phenotypes. Polymorphisms in color and pattern have long attracted the attention of population geneticists. **Figure 24.1** illustrates a striking example of polymorphism in the elder-flowered orchid (*Dactylorhiza sambucina*). Throughout the range of this species in Europe, both yellow- and red-flowered individuals are prevalent.

From a genetic perspective, polymorphism is due to two or more alleles that influence the phenotype of the individual that inherits them. In other words, it is due to genetic variation. Geneticists also use the term polymorphism to describe the variation in genes; this is sometimes called genetic polymorphism. A gene that commonly exists as two or more alleles in a population is described as a **polymorphic gene**. By comparison, a **monomorphic gene** exists predominantly as a single allele in a population. By convention, when 99% or more of the alleles of a given gene are identical, the gene is considered to be monomorphic. Said another way, a polymorphic gene must have one or more additional alleles that make up more than 1% of the alleles in the population.

At the level of a particular gene, a polymorphism may involve various types of changes such as a deletion of a significant region of the gene, a duplication of a region, or a change in a single nucleotide. This last phenomenon is called a single-

nucleotide polymorphism (SNP). SNPs ("snips") are the smallest type of genetic change that can occur within a given gene, and they are also the most common. In human populations, for example, SNPs represent 90% of all the variation in human DNA sequences that occurs among different people. Current estimates indicate that SNPs with a frequency of 1% or more are found very frequently in genes. In humans, a gene that is 2,000–3,000 bp in length will, on average, contain 10 different SNPs in the human population. The high frequency of SNPs indicates that polymorphism is the norm for most human genes. Likewise, relatively large, healthy populations of nearly all species exhibit a high level of genetic variation, as evidenced by the occurrence of SNPs within most genes. As discussed later in this chapter, genetic variation provides the raw material for populations to evolve over the course of many generations.

Population Genetics Is Concerned with Allele and Genotype Frequencies

To analyze genetic variation in populations, one approach is to consider the frequency of alleles in a quantitative way. Two fundamental calculations are central to population genetics: **allele frequencies** and **genotype frequencies**. Allele and genotype frequencies are defined as:

$$\text{Allele frequency} = \frac{\text{Number of copies of a specific allele in a population}}{\text{Total number of all alleles for that gene in a population}}$$

$$\text{Genotype frequency} = \frac{\text{Number of individuals with a particular genotype in a population}}{\text{Total number of individuals in a population}}$$

Though these two frequencies are related, make sure you keep in mind a clear distinction between them. As an example, let's consider a population of 100 four o'clock plants with the following genotypes:

49 red-flowered plants with the genotype *RR*

42 pink-flowered plants with the genotype *Rr*

9 white-flowered plants with the genotype *rr*

When calculating an allele frequency for diploid species, remember that homozygous individuals have two copies of an allele, whereas heterozygotes have only one. For example, in tallying the *r* allele, each of the 42 heterozygotes has one copy of the *r* allele, and each white-flowered plant has two copies. Therefore, the allele frequency for *r* equals

$$\text{Frequency of } r = \frac{(Rr) + 2(rr)}{2(RR) + 2(Rr) + 2(rr)}$$

$$\text{Frequency of } r = \frac{42 + (2)(9)}{(2)(49) + (2)(42) + (2)(9)}$$

$$= \frac{60}{200} = 0.3, \text{ or } 30\%$$

This result tells us that the allele frequency of *r* is 0.3. In other words, 30% of the alleles for this gene in the population are the *r* allele.

Let's now calculate the genotype frequency of *rr* (white-flowered) plants.

$$\text{Frequency of } rr = \frac{9}{49 + 42 + 9}$$

$$= \frac{9}{100} = 0.09, \text{ or } 9\%$$

We see that 9% of the individuals in this population have white flowers.

Allele and genotype frequencies are always less than or equal to one (that is, less than or equal to 100%). If a gene is monomorphic, the allele frequency for the single allele will equal or be close to a value of 1.0. For polymorphic genes, if we add up the frequencies for all the alleles in the population, we should obtain a value of 1.0. In our four o'clock example, the allele frequency of *r* equals 0.3. Therefore, we can calculate the frequency of the other allele, *R*, as equal to 1.0 − 0.3 = 0.7, because they must add up to 1.0.

The Hardy-Weinberg Equation Relates Allele and Genotype Frequencies in a Population

In 1908, Godfrey Harold Hardy, an English mathematician, and Wilhelm Weinberg, a German physician, independently derived a simple mathematical expression called the Hardy-Weinberg equation that relates allele and genotype frequencies when they

are not changing. Let's examine the Hardy-Weinberg equation using the population of four o'clock plants that we have just considered. If the allele frequency of *R* is denoted by the variable *p*, and the allele frequency of *r* by *q*, then

$$p + q = 1$$

For example, if $p = 0.7$, then q must be 0.3. In other words, if the allele frequency of *R* equals 70%, the remaining 30% of alleles must be *r*, because together they equal 100%.

For a gene that exists in two alleles, the Hardy-Weinberg equation states that

$(p + q)^2 = 1$ (Note: the number 2 in this equation reflects the fact that the genotype is due to the inheritance of two alleles, one from each parent.)

Therefore

$$p^2 + 2pq + q^2 = 1 \text{ (the \textbf{Hardy-Weinberg equation})}$$

If we apply this equation to our flower color gene, then

p^2 equals the genotype frequency of *RR*

$2pq$ equals the genotype frequency of *Rr*

q^2 equals the genotype frequency of *rr*

If $p = 0.7$ and $q = 0.3$, then

Frequency of $RR = p^2 = (0.7)^2 = 0.49$

Frequency of $Rr = 2pq = 2(0.7)(0.3) = 0.42$

Frequency of $rr = q^2 = (0.3)^2 = 0.09$

In other words, if the allele frequency of *R* is 70% and the allele frequency of *r* is 30%, the expected genotype frequency of *RR* is 49%, *Rr* is 42%, and *rr* is 9%.

To see the relationship between allele frequencies and genotypes in a population, **Figure 24.2** considers the relationship between allele frequencies and the way that gametes combine to produce genotypes. The Hardy-Weinberg equation reflects the way gametes combine randomly with each other to produce offspring. In a population, the frequency of a gamete carrying a particular allele is equal to the allele frequency in that population. For example, if the allele frequency of *R* equals 0.7, the frequency of a gamete carrying the *R* allele also equals 0.7. The frequency of producing an *RR* homozygote, which produces red flowers, is $0.7 \times 0.7 = 0.49$, or 49%. The probability of inheriting both *r* alleles, which produces white flowers, is $0.3 \times 0.3 = 0.09$, or 9%. In our Punnett square, two different gamete combinations can produce heterozygotes with pink flowers (Figure 24.2). An offspring could inherit the *R* allele from pollen and *r* from the egg, or *R* from the egg and *r* from pollen. Therefore, the frequency of heterozygotes is $pq + pq$, which equals $2pq$. In our example, this is $2(0.7)(0.3) = 0.42$, or 42%.

The Hardy-Weinberg equation predicts an **equilibrium** of unchanging allele and genotype frequencies in a population. If a population is in equilibrium, it is not adapting and evolution is not occurring. However, this prediction is valid only if certain conditions are met in a population. These conditions require that evolutionary mechanisms, those forces that can change allele and

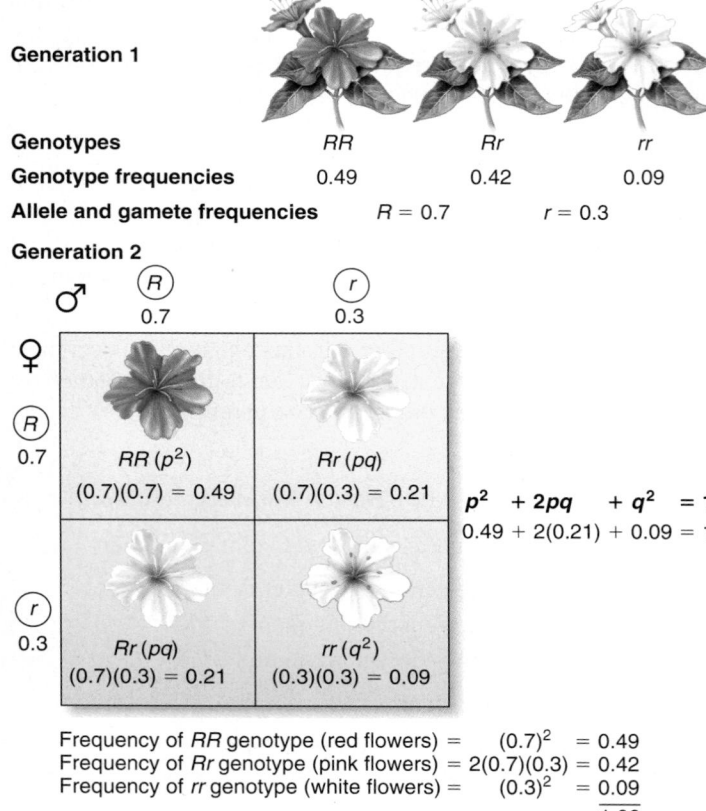

Generation 1

Genotypes	RR	Rr	rr
Genotype frequencies	0.49	0.42	0.09
Allele and gamete frequencies	R = 0.7		r = 0.3

Generation 2

♂ (R) 0.7   (r) 0.3

♀

(R) 0.7

RR (p^2)	Rr (pq)
(0.7)(0.7) = 0.49	(0.7)(0.3) = 0.21

(r) 0.3

Rr (pq)	rr (q^2)
(0.7)(0.3) = 0.21	(0.3)(0.3) = 0.09

$$p^2 + 2pq + q^2 = 1$$
$$0.49 + 2(0.21) + 0.09 = 1$$

Frequency of *RR* genotype (red flowers) = $(0.7)^2$ = 0.49
Frequency of *Rr* genotype (pink flowers) = 2(0.7)(0.3) = 0.42
Frequency of *rr* genotype (white flowers) = $(0.3)^2$ = 0.09
                                                                    ‾‾‾‾‾‾
                                                                    1.00

Figure 24.2   Comparing allele and genotype frequencies in a population with the Hardy-Weinberg equation and a Punnett square.   A population of four o'clock plants has allele and gamete frequencies of 0.7 for the *R* allele and 0.3 for the *r* allele. Knowing the allele frequencies allows us to calculate the genotype frequencies in the population.

Biological inquiry: What would be the frequency of pink flowers in a population where the allele frequency of R is 0.4 and the population is in Hardy-Weinberg equilibrium? Assume that R and r are the only two alleles.

genotype frequencies, are not acting on a population. With regard to a particular gene of interest, these conditions are:

- The population is so large that allele frequencies do not change due to random sampling error.

- The members of the population mate with each other without regard to their phenotypes and genotypes.

- No migration occurs between different populations.

- No survival or reproductive advantage exists for any of the genotypes—in other words, no natural selection occurs.

- No new mutations occur.

In reality, no population satisfies the Hardy-Weinberg equilibrium completely. Nevertheless, in large natural populations with little migration and negligible natural selection, the Hardy-Weinberg equilibrium may be nearly approximated for certain

genes. However, researchers often discover instead that allele and genotype frequencies for one or more genes in a given species are not in Hardy-Weinberg equilibrium. In such cases, we would say that the population is in disequilibrium—in other words, evolutionary mechanisms are affecting the population. When this occurs, population geneticists may wish to identify the reason(s) why disequilibrium has occurred because this may impact the future survival of the species.

24.2 Evolutionary Mechanisms and Their Effects on Populations

The genetic variation in all natural populations changes over the course of many generations. The term **microevolution** is used to describe changes in a population's gene pool from generation to generation. Such change is rooted in two related phenomena (**Table 24.1**). First, the introduction of new genetic variation into a population is one essential aspect of microevolution. As discussed in Chapter 23, genetic variation can originate by a variety of molecular mechanisms. New alleles of preexisting genes can arise by random mutation, and new genes can be introduced into a population by gene duplication, exon shuffling, and horizontal gene transfer. Such mutations, albeit rare, provide a continuous source of new variation to populations. In 1926, the Russian geneticist Sergei Tshetverikov was the first to suggest that random mutations are the raw material for evolution, but they do not constitute evolution itself. Mutations clearly supply new genetic variation to a population. However, due to their low rate of occurrence, mutations do not act as a major force in promoting widespread changes in a population. If mutations were the only type of change occurring in a population, that population would not evolve because mutations are so rare.

In this section we will discuss the second phenomenon that is required for microevolution, the action of evolutionary mechanisms that alter the prevalence of a given allele or genotype in a population. These mechanisms are natural selection, random genetic drift, migration, and nonrandom mating (Table 24.1). The collective contributions of these evolutionary mechanisms over the course of many generations have the potential to promote widespread genetic changes in a population.

To consider the effects of these evolutionary mechanisms, we will examine how they may affect the type of genetic variation that occurs when a gene exists in two alleles in a population. As you will learn, these mechanisms may cause one allele or the other allele to be favored, or they may create a balance where both alleles are maintained in a population. Although we will discuss the effects of these mechanisms on genetic variation involving alleles of a single gene caused by mutation, keep in mind that these same evolutionary mechanisms can also affect the frequencies of new genes that arise in a population by gene duplication, exon shuffling, and horizontal gene transfer.

Table 24.1	Factors That Govern Microevolution
Sources of new genetic variation*
New alleles	Random mutations within pre-existing genes introduce new alleles into populations, but at a very low rate. New mutations may be beneficial, neutral, or deleterious. For alleles to rise to a significant percentage in a population, evolutionary mechanisms, such as natural selection, random genetic drift, and migration, must operate on them.
Gene duplication	Abnormal crossover events and transposable elements may increase the number of copies of a gene. Over time, the additional copies can accumulate random mutations and create a gene family.
Exon shuffling	Abnormal crossover events and transposable elements may promote gene rearrangements in which one or more exons from one gene are inserted into another gene. The protein encoded by such a gene may display a novel function and can then be acted upon by evolutionary mechanisms.
Horizontal gene transfer	Genes from one species may be introduced into another species. Events such as endocytosis and interspecies mating may promote this phenomenon.
Evolutionary mechanisms that alter existing genetic variation
Natural selection	The phenomenon in which the environment selects for individuals that possess certain traits. Natural selection can favor the survival of members with beneficial traits or disfavor the survival of individuals with unfavorable traits. As a type of natural selection, sexual selection favors traits that increase the reproductive success of individuals.
Random genetic drift	This is a change in genetic variation from generation to generation due to random sampling error. Allele frequencies may change as a matter of chance from one generation to the next. This is much more likely to occur in a small population.
Migration	Migration can occur between two different populations that have different allele frequencies. The introduction of migrants into a recipient population may change the allele frequencies of that population.
Nonrandom mating	The phenomenon in which individuals select mates based on their phenotypes or genetic lineage. This can alter the relative proportion of homozygotes and heterozygotes that is predicted by the Hardy-Weinberg equation, but it will not change allele frequencies.

*Described in Chapter 23.

Natural Selection Favors Individuals with the Greatest Reproductive Success

As we discussed in Chapter 23, Charles Darwin and Alfred Wallace independently proposed the theory of evolution by natural selection. According to this theory, only a certain percentage of the offspring that a species produces will survive. This "struggle for existence" results in the selective survival of individuals that have inherited certain genotypes. Such genotypes confer greater **reproductive success**. In this regard, natural selection usually acts upon two aspects of reproductive success. First, certain characteristics make organisms better adapted to their environment and more likely to survive to reproductive age; such organisms have a greater chance to reproduce and contribute offspring to the next generation. Therefore, natural selection favors individuals with adaptations that provide a survival advantage. Second, natural selection favors individuals that produce viable offspring. As discussed later in this chapter, traits that enhance the ability of individuals to reproduce are often subject to natural selection.

Let's consider how natural selection can operate when a gene exists as two alleles in a population. Keep in mind that natural selection acts on individuals, while evolution occurs at the population level. A modern description of natural selection can relate our knowledge of molecular genetics to the phenotypes of individuals.

1. Within a population, allelic variation arises from random mutations that cause differences in DNA sequences. A mutation that creates a new allele may alter the amino acid sequence of the encoded protein. This in turn may alter the function of the protein.
2. Some alleles may encode proteins that enhance an individual's survival or reproductive capability compared to that of other members of the population. For example, an allele may produce a protein that is more efficient at a higher temperature, conferring on the individual a greater probability of survival in a hot climate.
3. Individuals with beneficial alleles are more likely to survive and contribute their alleles to the gene pool of the next generation.
4. Over the course of many generations, allele frequencies of many different genes may change through natural selection, thereby significantly altering the characteristics of a population. The net result of natural selection is a population that is better adapted to its environment and more successful at reproduction.

As mentioned earlier, Haldane, Fisher, and Wright developed mathematical relationships to explain the phenomenon of natural selection. To begin our quantitative discussion of natural selection, we need to consider the concept of **Darwinian fitness**, which is the relative likelihood that a genotype will contribute to the gene pool of the next generation as compared with other genotypes. Although this property often correlates with physical fitness, the two ideas should not be confused. Darwinian fitness is a measure of reproductive success. An extremely fertile individual may have a higher Darwinian fitness than a less fertile individual that appears more physically fit.

To examine Darwinian fitness, let's consider an example of a hypothetical gene existing in A and a alleles. We can assign fitness values to each of the three possible genotypes according to their relative reproductive success. For example, let's suppose that the average reproductive successes of the three genotypes are:

> *AA* produces five offspring
>
> *Aa* produces four offspring
>
> *aa* produces one offspring

By convention, the genotype with the highest reproductive success is given a fitness value of 1.0. Fitness values are denoted by the variable *W*. The fitness values of the other genotypes are assigned values relative to this 1.0 value.

> Fitness of *AA*: $W_{AA} = 1.0$
>
> Fitness of *Aa*: $W_{Aa} = 4/5 = 0.8$
>
> Fitness of *aa*: $W_{aa} = 1/5 = 0.2$

Variation in fitness occurs because individuals with certain genotypes have greater reproductive success. Such genotypes exhibit a higher fitness compared to others. Natural selection acts on phenotypes that are derived from an individual's genotype.

Likewise, the effects of natural selection can be viewed at the level of a population. The average reproductive success of members of a population is called the **mean fitness of the population**. Over many generations, as individuals with higher fitness values become more prevalent, natural selection also in-

creases the mean fitness of the population. In this way, the process of natural selection results in a population of organisms that is well adapted to its native environment and likely to be successful at reproduction.

Natural Selection Can Follow Different Patterns

By studying species in their native environments, population geneticists have discovered that natural selection can occur in several ways. In most of the examples described next, natural selection leads to adaptation so that a species is better able to survive to reproductive age.

Directional Selection **Directional selection** favors individuals at one extreme of a phenotypic distribution that have greater reproductive success in a particular environment. Different phenomena may initiate the process of directional selection. One way that directional selection may arise is that a new allele may be introduced into a population by mutation, and the new allele may confer a higher fitness in individuals that carry it (**Figure 24.3**). If the homozygote carrying the favored allele has the highest fitness value, directional selection may cause this favored allele to eventually become predominant in the population, perhaps even becoming a monomorphic allele.

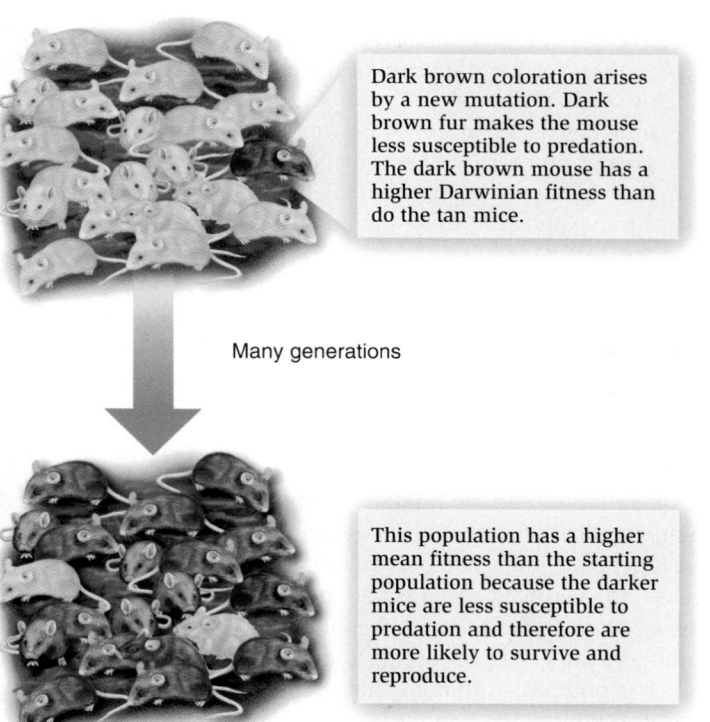

Dark brown coloration arises by a new mutation. Dark brown fur makes the mouse less susceptible to predation. The dark brown mouse has a higher Darwinian fitness than do the tan mice.

Many generations

This population has a higher mean fitness than the starting population because the darker mice are less susceptible to predation and therefore are more likely to survive and reproduce.

(a) An example of directional selection

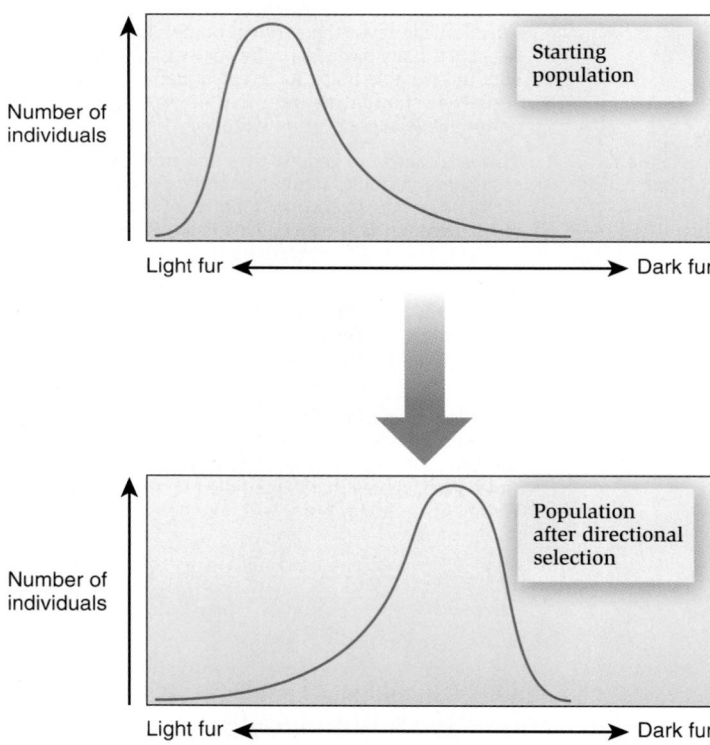

Number of individuals

Starting population

Light fur ←————————→ Dark fur

Number of individuals

Population after directional selection

Light fur ←————————→ Dark fur

(b) Graphical representation of directional selection

Figure 24.3 Directional selection.   This pattern of natural selection selects for one extreme of a phenotype that confers the highest fitness in the population's environment. **(a)** In this example, a mutation for darker fur arises in a population of mice. This new genotype confers higher Darwinian fitness, because mice with dark fur can evade predators and are more likely to survive and reproduce. Over many generations, directional selection will favor the prevalence of darker individuals. **(b)** These graphs show the change in fur color phenotypes in this mouse population before and after directional selection.

Biological inquiry: Over the short and long run, does directional selection favor the preservation of genetic diversity?

Another possibility is that a population may be exposed to a prolonged change in its living environment. Under the new environmental conditions, the relative fitness values may change to favor one genotype, and this will promote the elimination of other genotypes. As an example, let's suppose a population of finches on a mainland already has genetic variation that affects beak size (refer back to Figure 23.2). A small number of birds migrate to an island where the seeds are generally larger than they are on the mainland. In this new environment, birds with larger beaks would have a higher fitness because they would be better able to crack open the larger seeds, and thereby survive to reproductive age. Over the course of many generations, directional selection would produce a population of birds carrying alleles that promote larger beak size.

Stabilizing Selection **Stabilizing selection** favors the survival of individuals with intermediate phenotypes. The extreme values of a trait are selected against. Stabilizing selection tends to decrease genetic diversity. An example of stabilizing selection involves clutch size (number of eggs laid) in birds, which was first proposed by British biologist David Lack in 1947. Under stabilizing selection, birds that lay too many or too few eggs per nest have lower fitness values than do those that lay an intermediate number of eggs (**Figure 24.4**). Laying too many eggs has the disadvantage that many offspring will die due to inadequate

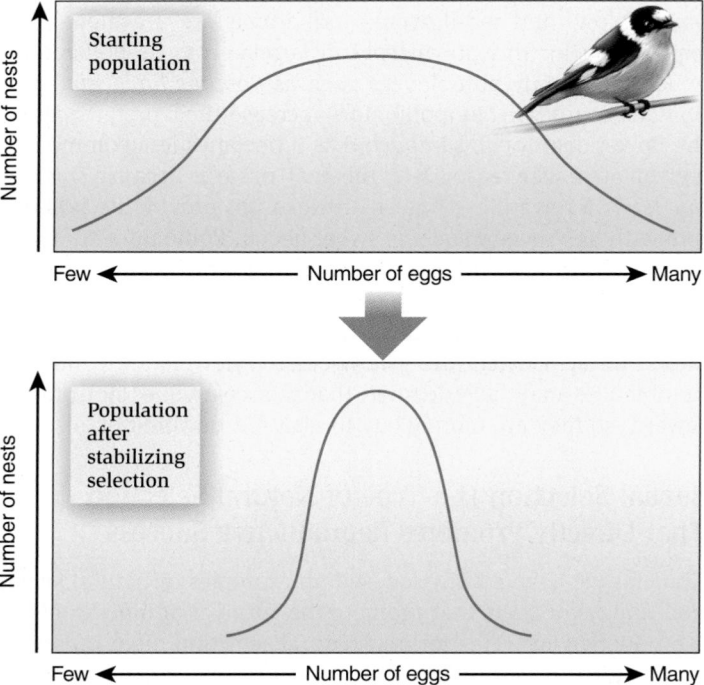

Figure 24.4 **Stabilizing selection.** In this pattern of natural selection, the extremes of a phenotypic distribution are selected against. Those individuals with intermediate traits have the highest fitness. These graphs show the results of stabilizing selection on clutch size in a population of collared flycatchers (*Ficedula albicollis*). This process results in a population with less diversity and more uniform traits.

parental care and food. In addition, the strain on the parents themselves may decrease their likelihood of survival and therefore their ability to produce more offspring. Having too few offspring, however, does not contribute many individuals to the next generation. Therefore, the most successful parents are those that produce an intermediate clutch size. In the 1980s, Swedish evolutionary biologist Lars Gustafsson and his colleagues examined the phenomenon of stabilizing selection in the collared flycatcher (*Ficedula albicollis*) on the island of Gotland south of Sweden. They discovered that Lack's hypothesis concerning an optimal clutch size appears to be true for this species.

Disruptive Selection **Disruptive selection** (also known as diversifying selection) favors the survival of two or more different genotypes that produce different phenotypes. In disruptive selection, the fitness values of a particular genotype are higher in one environment and lower in a different environment, while the fitness values of the other genotype vary in an opposite manner. Disruptive selection is likely to occur in populations that occupy diverse environments, so that some members of the species will survive in each type of environmental condition. An example involves colonial bentgrass (*Agrostis tenuis*). In certain locations where this grass is found, such as South Wales, isolated places occur where the soil is contaminated with high levels of heavy metals such as copper due to human activities such as mining. The relatively recent metal contamination has selected for the proliferation of mutant strains that show tolerance to the heavy metals. Such genetic changes enable the plants to grow on contaminated soil but tend to inhibit growth on normal, noncontaminated soil. These metal-resistant plants often grow on contaminated sites that are close to plants that grow on uncontaminated land and do not show metal tolerance (**Figure 24.5**).

In the case of metal-resistant and metal-sensitive grasses, the members of a population occupy heterogeneous environments that are geographically continuous; members of the populations can freely interbreed. In other cases, members of a single species may occupy two or more different environments that are geographically isolated from each other. Given enough time, disruptive selection due to heterogeneous environments can eventually lead to the evolution of two or more different species, a process that will be described in Chapter 25.

Balancing Selection Contrary to a popular misconception, natural selection does not always cause the elimination of "weaker" or less fit alleles. **Balancing selection** is a type of natural selection that maintains genetic diversity in a population. Over many generations, balancing selection can create a situation known as a **balanced polymorphism**, or a stable polymorphism, in which two or more alleles are kept in balance, and therefore are maintained in a population over the course of many generations.

Balancing selection does not favor one particular allele in the population. Population geneticists have identified two common ways that this pattern of selection can occur. First, for genetic variation involving a single gene, balancing selection favors the heterozygote rather than either corresponding homozygote.

(a) Growth of *Agrostis tenuis* on uncontaminated and contaminated soil

Figure 24.5   **Disruptive selection.**   This pattern of natural selection selects for two different phenotypes, each of which is most fit in a particular environment. **(a)** In this example, mutations have created metal-resistant alleles in colonial bentgrass (*Agrostis tenuis*), that allow it to grow on soil contaminated with high levels of heavy metals such as copper. These alleles provide high fitness where the soil is contaminated, but they confer low fitness where it is not contaminated. Because both metal-resistant and metal-sensitive alleles are maintained in the population, this situation is an example of disruptive selection due to heterogeneous environments. **(b)** These graphs show the change in phenotypes in this bentgrass population before and after disruptive selection.

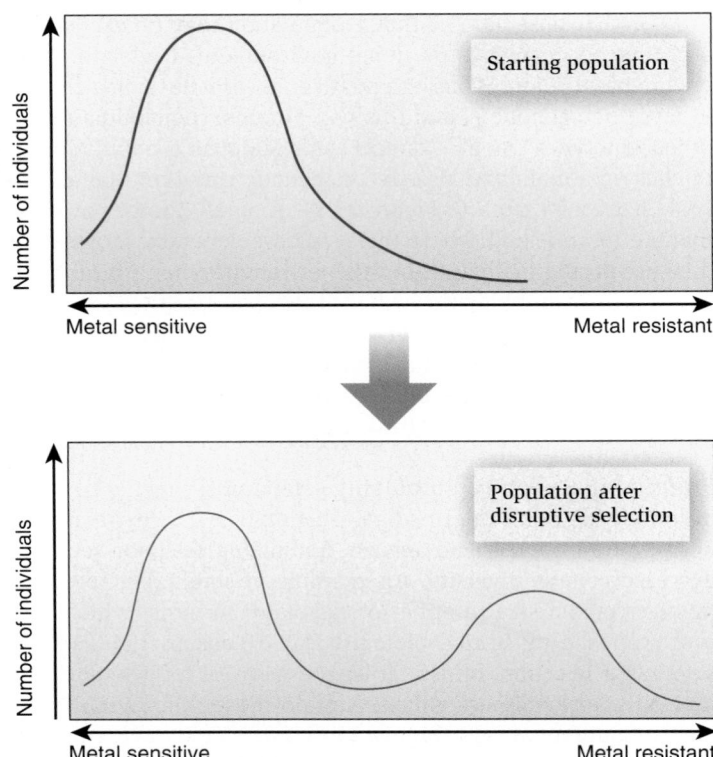

(b) Graphical representation of disruptive selection

This situation is called **heterozygote advantage**. Balanced polymorphisms can sometimes explain the high frequency of alleles that are deleterious in a homozygous condition. A classic example is the H^S allele of the human β-globin gene, which is described in Chapter 14. A homozygous $H^S H^S$ individual has sickle-cell anemia, a disease that leads to the sickling of the red blood cells. The $H^S H^S$ homozygote has a lower fitness than a homozygote with two copies of the more common β-globin allele, $H^A H^A$. However, the heterozygote, $H^A H^S$, has the highest level of fitness in areas where malaria is endemic. Compared with $H^A H^A$ homozygotes, heterozygotes have a 10–15% better chance of survival if infected by the malarial parasite, *Plasmodium falciparum*. Therefore, the H^S allele is maintained in populations where malaria is prevalent, even though the allele is detrimental in the homozygous state (**Figure 24.6**). The balanced polymorphism results in a higher mean fitness of the population. In areas where malaria is endemic, a population composed of all $H^A H^A$ individuals would have a lower mean fitness.

   Negative frequency-dependent selection is a second way that natural selection can produce a balanced polymorphism. In this pattern of natural selection, the fitness of a genotype decreases when its frequency becomes higher. In other words, rare individuals have a higher fitness and common individuals have a lower fitness. Therefore, rare individuals are more likely to reproduce while common individuals are less likely, thereby producing a balanced polymorphism in which no genotype becomes too rare or too common.

   An interesting example of negative frequency-dependent selection involves the elder-flowered orchid (*D. sambucina*), which was shown earlier in Figure 24.1. Throughout its range

both yellow- and red-flowered individuals are prevalent. The explanation for this polymorphism is related to its pollinators, which are mainly bumblebees such as *Bombus lapidarius* and *Bombus terrestris*. The pollinators increase their preference for the flower color of *D. sambucina* as it becomes less common in a given area. One reason why this may occur is because *D. sambucina* is a rewardless flower—it does not provide its pollinators with any reward such as sweet nectar. Pollinators are more likely to learn that the more common color of *D. sambucina* in a given area does not offer a reward, and this may explain their preference for the flower color that is less common. For example, in an area where the yellow-colored flowers are common, bumblebees may have learned that this color does not offer a reward, so they are more likely to visit red-flowered plants.

Sexual Selection Is a Type of Natural Selection That Directly Promotes Reproductive Success

Thus far we have mainly focused on examples of natural selection that favor traits that promote the survival of individuals to reproductive age. This form of natural selection often produces adaptations for survival in particular environments. Now let's turn our attention to a form of natural selection, called **sexual selection**, that is directed at certain traits of sexually reproducing species that make it more likely for individuals to find or choose a mate and/or engage in successful mating. Darwin originally described sexual selection as "the advantage that certain individuals have over others of the same sex and species solely with respect to reproduction." Within a species, members of the same sex (typically males) compete with each other for

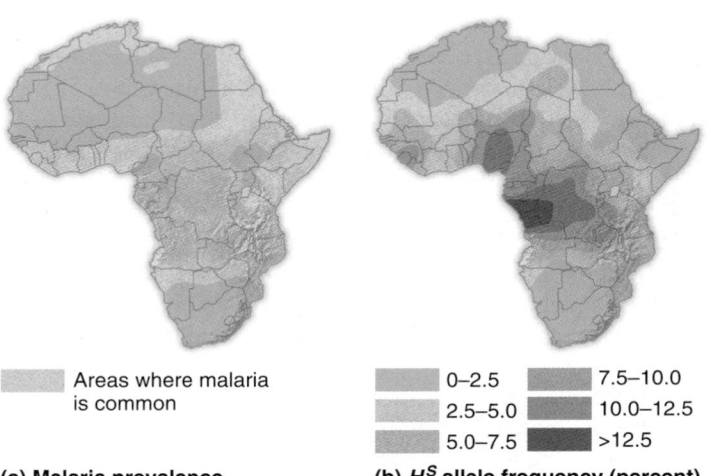

Areas where malaria is common	0–2.5	7.5–10.0
	2.5–5.0	10.0–12.5
	5.0–7.5	>12.5

(a) Malaria prevalence   **(b) H^S allele frequency (percent)**

Figure 24.6   **Balancing selection and heterozygote advantage.**   In this pattern of natural selection, genetic diversity is maintained in a population. This example shows balancing selection maintaining two alleles of the β-globin gene, designated H^A and H^S, in human populations in Africa. This situation occurs due to heterozygote advantage, because the heterozygous condition confers resistance to malaria. (a) The geographic prevalence of malaria in Africa. (b) The frequency of the H^S allele of the β-globin gene in the same area. In the homozygous condition, the H^S allele causes the disease sickle-cell anemia. However, this sickle-cell allele is maintained in human populations as a balanced polymorphism, because in areas where malaria is prevalent, the heterozygote carrying one copy of the H^S allele has a higher fitness than either of the corresponding homozygotes ($H^A H^A$ and $H^S H^S$).

the opportunity to mate with members of the opposite sex. Such competition results in sexual selection.

In many species of animals, sexual selection affects male characteristics more intensely than it does female. Unlike females, which tend to be fairly uniform in their reproductive success, male success tends to be more variable, with some males mating with many females and others not mating at all. (See Chapter 55 for a discussion of different mating strategies between the sexes.) Sexual selection results in the evolution of traits, called secondary sexual characteristics, that favor reproductive success. The result of this process is sometimes a significant difference between the appearances of the sexes in the same species, a situation called sexual dimorphism.

Sexual selection can be categorized as either **intrasexual selection**, between members of the same sex, or **intersexual selection**, between members of the opposite sex. Let's begin with intrasexual selection. Examples of traits that result from intrasexual selection in animals include horns in male sheep, antlers in male moose, and the enlarged claw of male fiddler crabs (**Figure 24.7a**). In many animal species, males directly compete with each other for the opportunity to mate with females, or they may battle for a particular territory. In fiddler crabs (*Uca paradussumieri*), males enter the burrows of females that are ready to mate. If another male attempts to enter the burrow, the male already inside the burrow stands in the burrow shaft and blocks the entrance with his enlarged claw.

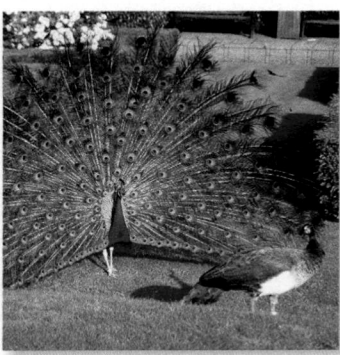

(a) Intrasexual selection   **(b) Intersexual selection**

Figure 24.7   **Examples of the results of sexual selection, a type of natural selection.**   (a) An example of intrasexual selection. The enlarged claw of the male fiddler crab is used in direct male-to-male competition. In this photograph, a male inside a burrow is extending its claw out of the burrow to prevent another male from entering. (b) An example of intersexual selection. Female peacocks choose males based on their colorful and long tail feathers, and the robustness of their display.

Now let's consider an example of intersexual selection, namely female choice. This type of sexual selection often results in showy characteristics in males. **Figure 24.7b** shows a classic example that involves the Indian peacock (*Pavo cristatus*), the national bird of India. Male peacocks have long and brightly colored tail feathers, which they fan out as a mating behavior. Females select among males based on feather color and pattern and physical prowess of the display.

A less obvious type of intersexual selection is cryptic female choice, in which the female reproductive system can influence the relative success of sperm. As an example of cryptic female choice, the female genital tract of certain animals selects for sperm that tend to be genetically unrelated to the female. Sperm from males closely related to the female, such as brothers or cousins, are less successful than are sperm from genetically unrelated males. The selection for sperm may occur over the journey through the reproductive tract. The egg itself may even have mechanisms to prevent fertilization by genetically related sperm. Cryptic female choice occurs in species in which females may mate with more than one male, such as many species of reptiles and ducks. A similar mechanism is found in many plant species in which pollen from genetically related plants, perhaps from the same flower, is unsuccessful at fertilization, while pollen from unrelated plants is successful. One possible advantage of cryptic female choice is that it inhibits inbreeding, which is described later in this chapter. At the population level, cryptic female choice may promote genetic diversity by favoring interbreeding among genetically unrelated individuals.

Sexual selection is sometimes a combination of both intrasexual and intersexual selection. During breeding season, male elk (*Cervus elaphus*) become aggressive and bugle loudly to challenge other male elk. Males spar with their antlers, which usually turns into a pushing match to determine which elk is stronger. Female elk then choose the strongest bulls as their mates.

Sexual selection can explain traits that may decrease an in- dividual's chances of survival but increase their chances of re- producing. For example, the male guppy (*Poecilia reticulata*) is brightly colored compared to the female. In nature, females pre- fer brightly colored males. Therefore, in places with few preda- tors, the males tend to be brightly colored. However, in places where predators are abundant, brightly colored males are less plentiful because they are subject to predation. In this case, the relative abundance of brightly and dully colored males depends on the balance between sexual selection, which favors bright coloring, and escape from predation, which favors dull coloring.

Many animals have secondary sexual characteristics, and evolutionary biologists generally agree that sexual selection is responsible for such traits. But why should males compete, and why should females be choosy? Researchers have proposed var- ious hypotheses to explain the underlying mechanisms. One possible reason is related to the different roles that males and females play in the nurturing of offspring. In some species, the female is the primary caregiver, while the male plays a minor role. In such species, the Darwinian fitness of males and females may be influenced by their mating behavior. Males increase their fitness by mating with multiple females. This increases their likelihood of passing their genes on to the next generation.

By comparison, females may produce relatively few offspring and their reproductive success may not be limited by the num- ber of available males. Females will have higher fitness if they choose males that are good defenders of their territory, and have alleles that confer a survival advantage to their offspring. One measure of alleles that confer higher fitness is age. Males that live to an older age are more likely to carry beneficial alleles. Many research studies involving female choice have shown that females tend to select traits that are more likely to be well de- veloped in older males than they are in immature males. In cer- tain species of birds, for example, females tend to choose males with a larger repertoire of songs, which is more likely to occur in older males.

Overall, sexual selection is a form of natural selection in which the evolution of certain traits occurs differently between the two sexes. Sexual selection is not some extra force in oppo- sition to natural selection. It is governed by the same processes involved in the evolution of traits that are not directly related to sex. Sexual selection can be directional, stabilizing, disruptive, or balancing. For example, directional selection probably played an important role in the evolution of the large and brightly col- ored tail of the male peacock. As described next, sexual selection can be diversifying if females select for males with different traits.

FEATURE INVESTIGATION

Seehausen and van Alphen Found That Male Coloration in African Cichlids Is Subject to Female Choice

Cichlids are tropical freshwater fish that are popular among aquarium enthusiasts. This family of fish (*Cichlidae*) is made up of more different species than is any other vertebrate family. The more than 3,000 species vary with regard to body shape, coloration, behavior, and feeding habits. By far the greatest diversity of these fish is found in Lake Victoria, Lake Malawi, and Lake Tanganyika in East Africa, where collectively more than 1,800 species are found. Lake Victoria, for example, has 500 species.

Cichlids have complex mating behavior and brood care. Females play an important role in choosing males with particu- lar characteristics. To study the importance of female choice, population geneticists Ole Seehausen and Jacques van Alphen investigated the effect of male coloration of *Pundamilia punda- milia* and *Pundamilia nyererei*. In some locations, *P. pundami- lia* and *P. nyererei* do not readily interbreed and behave like two distinct biological species, while in other places they behave like a single interbreeding species with two color morphs. They can interbreed to produce viable offspring, and both inhabit Lake Victoria. Males of both species have blackish underparts and blackish vertical bars on their sides (**Figure 24.8a**). *P. punda- milia* males are grayish white on top and on the sides, and they

Figure 24.8   Male coloration in African cichlids.   (a) Two males (*Pundamilia pundamilia*, top, and *Pundamilia nyererei*, bottom) under normal illumination. (b) The same species under orange monochromatic light, which obscures their color differences.

have a metallic blue and red dorsal fin, which is the uppermost fin. By comparison, *P. nyererei* males are orange on top and yel- low on their sides.

Seehausen and van Alphen hypothesized that females choose males for mates based on the males' coloration. The researchers took advantage of the observation that colors are obscured under orange monochromatic light. As seen in **Figure 24.8b**, males of

Figure 24.9   A study by Seehausen and van Alphen involving the effects of male coloration on female choice in African cichlids.

HYPOTHESIS Female African cichlids choose mates based on the males' coloration.

STARTING MATERIALS Two species of cichlid, *Pundamilia pundamilia* and *Pundamilia nyererei*, were chosen. The males differ with regard to their coloration. A total of 8 males and 8 females (4 males and 4 females from each species) were tested.

Experimental level	Conceptual level

1 Place 1 female and 2 males in an aquarium. Each male is within a separate glass enclosure. The enclosures contain 1 male from each species.

This is a method to evaluate sexual selection via female choice in 2 species of cichlid.

2 Observe potential courtship behavior for 1 hour. If a male exhibited lateral display (a courtship invitation) and then the female approached the enclosure that contained the male, this was scored as a positive encounter. This protocol was performed under normal illumination and under monochromatic illumination.

3 **THE DATA**

Female	Male	Light condition	Percentage of positive encounters*
P. pundamilia	*P. pundamilia*	Normal	16
P. pundamilia	*P. nyererei*	Normal	2
P. nyererei	*P. nyererei*	Normal	16
P. nyererei	*P. pundamilia*	Normal	5
P. pundamilia	*P. pundamilia*	Monochromatic	20
P. pundamilia	*P. nyererei*	Monochromatic	18
P. nyererei	*P. nyererei*	Monochromatic	13
P. nyererei	*P. pundamilia*	Monochromatic	18

*A positive encounter occurred when a male's lateral display was followed by the female approaching the male.

both species look similar under these conditions. As shown in **Figure 24.9**, a female of one species was placed in an aquarium that contained one male of each species within an enclosure. The males were within glass enclosures to avoid direct competition with each other, which would have likely affected female choice. The goal of the experiment was to determine which of the two males a female would prefer. Courtship between a male and female begins when a male swims toward a female and exhibits a lateral display (that is, shows the side of his body to the female). If the female is interested, she will approach the male, and then the male will display a quivering motion. Such courtship behavior was examined under normal light and under orange monochromatic light.

As seen in the data, Seehausen and van Alphen found that the females' preference for males was dramatically different depending on the illumination conditions. Under normal light, *P. pundamilia* females preferred *P. pundamilia* males, and *P. nyererei* females preferred *P. nyererei* males. However, such mating preference was lost under orange monochromatic light. If the light conditions in their native habitats are similar to the normal light used in this experiment, female choice would be expected to separate cichlids into two populations—*P. pundamilia* females mating with *P. pundamilia* males and *P. nyererei* females

mating with *P. nyererei* males. In this case, sexual selection appears to have followed a diversifying mechanism in which certain females prefer males with one color pattern while other females prefer males with a different color pattern. When this occurs, a possible outcome of such sexual selection is that it can separate one large population into smaller populations that selectively breed with each other, and eventually become distinct species. The topic of species formation is discussed in greater depth in Chapter 25.

In Small Populations, Allele Frequencies Can Be Altered by Random Genetic Drift

Thus far, we have focused on natural selection as an evolutionary mechanism that fosters genetic change. Let's now turn our attention to other ways that the gene pool of a population can change. In the 1930s, Sewall Wright played a large role in developing the concept of **random genetic drift**, which refers to changes in allele frequencies due to random sampling error. The term genetic drift is derived from the observation that allele frequencies may "drift" randomly from generation to generation as a matter of chance. Although the Darwinian fitness values of particular genotypes allow researchers to predict the allele frequencies of a population in future generations, random sampling error, or deviation between observed and predicted values, can arise due to random events that are unrelated to fitness. For example, an individual with a high fitness value may, as a matter of bad luck, not encounter a member of the opposite sex. Changes in allele frequencies due to genetic drift happen regardless of the fitness of individuals that carry those alleles. Likewise, random sampling error can influence which alleles happen to be found in the gametes that fuse with each other in a successful fertilization.

What are the effects of genetic drift? Over the long run, genetic drift favors either the loss or the fixation of an allele—when its frequency reaches 0% or 100% in a population, respectively. The rate at which an allele is either lost or fixed depends on the population size. **Figure 24.10** illustrates the potential consequences of genetic drift in one large (*N* = 1,000) and two small (*N* = 10) populations. This simulation involves the frequency of hypothetical *B* and *b* alleles of a gene for fur color in a population of mice—*B* is the black allele and *b* is the white allele. At the beginning of this hypothetical simulation, which runs for 50 generations, all of these populations had identical allele frequencies: *B* = 0.5 and *b* = 0.5. In the small populations, the allele frequencies fluctuated substantially from generation to generation. Eventually, in one simulation, the *B* allele was eliminated, while in the other, it was fixed at 100%. These small populations would then consist of only white mice or black mice, respectively. At this point, the allele has become monomorphic and cannot fluctuate any further. By comparison, the frequencies of *B* and *b* in the large population fluctu-

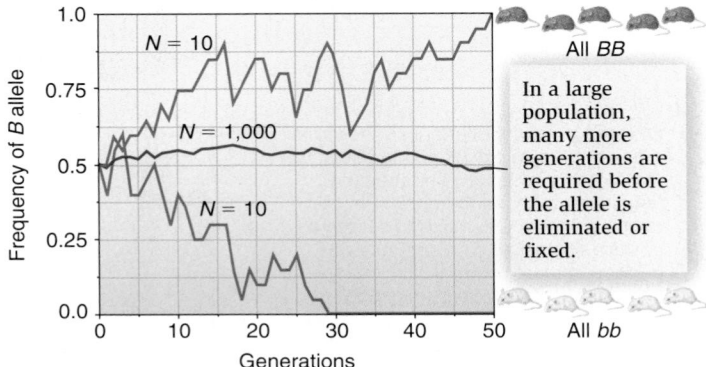

Figure 24.10   Genetic drift and population size.   This graph shows a hypothetical simulation of random genetic drift and its effects on small and large populations of black (*B* allele) and white (*b* allele) mice. In all cases, the starting allele frequencies are *B* = 0.5 and *b* = 0.5. The red lines illustrate two populations of mice in which *N* = 10. The blue line shows a population in which *N* = 1,000. Genetic drift leads to random changes in allele frequencies, eventually causing either the elimination or fixation of alleles. This happens much more quickly in small populations than it does in large ones. In this simulation, genetic drift has led to small populations of all-black (*BB*) or all-white (*bb*) mice in 50 generations or less.

ated much less. As discussed in Chapter 16, the relative effects of random sampling error are much less when the sample size is large. Nevertheless, genetic drift will eventually lead to allele loss or fixation even in large populations, but this will take many more generations to occur than it does in small populations.

In nature, genetic drift may rapidly alter allele frequencies when population sizes are small. One example is called the **bottleneck effect**. A population can be reduced dramatically in size by events such as earthquakes, floods, drought, and human destruction of habitat. Such occurrences may randomly eliminate most of the members of the population without regard to genetic composition. The period of the bottleneck, when the population size is very small, may be influenced by genetic drift. This may happen primarily for two reasons. First, the surviving members may have allele frequencies that differ from those of the original population. Second, allele frequencies are expected to drift substantially during the generations when the population size is small. In extreme cases, alleles may even be eliminated. Eventually, the bottlenecked population may regain its original size. However, the new population is likely to have less genetic variation than the original one. A hypothetical example of this process is shown with a population of frogs in **Figure 24.11**. In this example, a starting population of frogs is found in three phenotypes: yellow, dark green, and striped. Due to a bottleneck caused by a drought, the dark green variety is lost from the population.

As another example, the African cheetah has lost nearly all of its genetic variation. This was likely due to a bottleneck effect. An analysis by population geneticists has suggested that a severe bottleneck occurred in this species approximately 10,000–12,000 years ago, reducing it to near extinction. The species eventually rebounded in numbers, but the bottleneck apparently reduced its genetic variation to very low levels. The modern species is monomorphic for nearly all of its genes.

Another common phenomenon in which genetic drift may have a rapid impact is the **founder effect**. This occurs when a small group of individuals separates from a larger population and establishes a colony in a new location. For example, a few individuals may migrate from a large continental population and become the founders of an island population. The founder effect differs from a bottleneck in that it occurs in a new location, although both effects reduce the size of a population. The founder effect has two important consequences. First, the founding population, which is relatively small, is expected to have less genetic variation than the larger original population from which it was derived. Second, as a matter of chance, the allele frequencies in the founding population may differ markedly from those of the original population.

Population geneticists have studied many examples in which isolated populations were founded via colonization by members of another population. For example, in the 1960s, American geneticist Victor McKusick studied allele frequencies in the Old Order Amish of Lancaster County, Pennsylvania. At that time, this was a group of about 8,000 people, descended from just three couples that immigrated to the U.S. in 1770. Among this population of 8,000, a genetic disease known as the Ellis–van Creveld syndrome (a recessive form of dwarfism) was found at a frequency of 0.07, or 7%. By comparison, this disorder is extremely rare in other human populations, even the population from which the founding members had originated. The high frequency in the Lancaster County population is a chance occurrence due to the founder effect.

The Neutral Theory of Evolution Proposes That Genetic Drift Plays an Important Role in Promoting Genetic Change

In 1968, Japanese evolutionary biologist Motoo Kimura proposed that much of the variation seen in natural populations is caused by genetic drift. Because it is a random process, genetic drift does not preferentially select for any particular allele—it can eliminate both beneficial and deleterious alleles. Much of

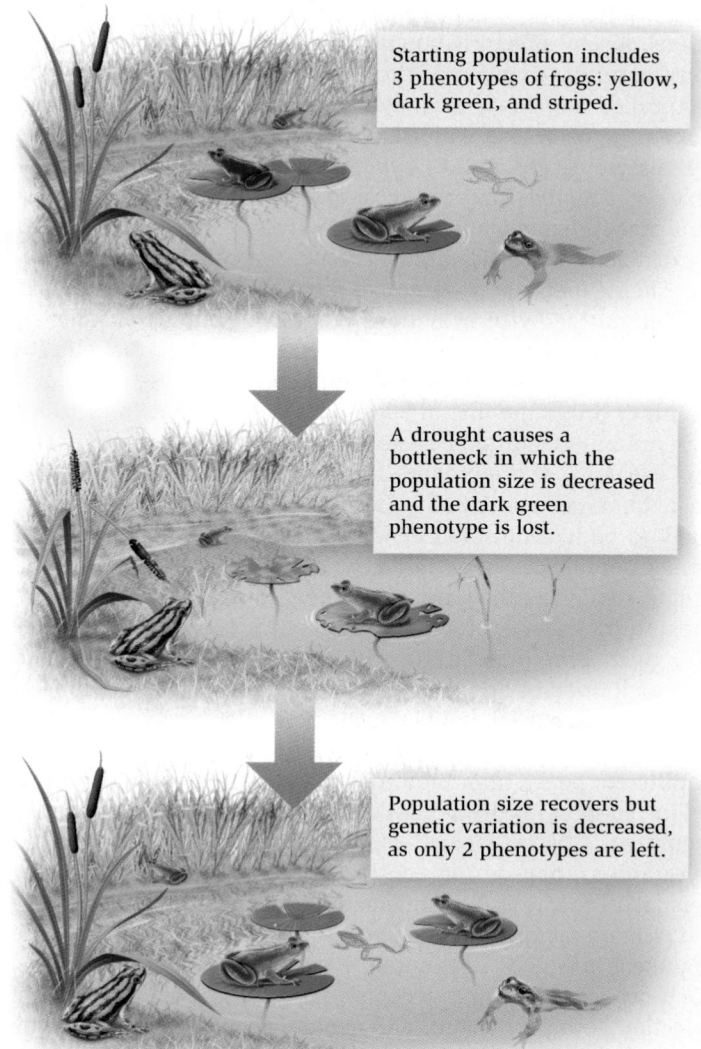

the time, genetic drift promotes **neutral variation**, which does not favor any particular genotype. According to Kimura's **neutral theory of evolution**, most genetic variation is due to the accumulation of neutral mutations that have attained high frequencies in a population via genetic drift.

Neutral mutations involve changes in genotypes that do not affect the phenotype of the organism, so they are not acted upon by natural selection. For example, a mutation within a structural gene that changes a glycine codon from GGG to GGC would not affect the amino acid sequence of the encoded protein.

Figure 24.11   **A hypothetical example of the bottleneck effect.**   This example involves a population of frogs in which a drought dramatically reduces population size, resulting in a bottleneck. The bottleneck reduces the genetic diversity in the population.

Biological inquiry: How does the bottleneck effect undermine the efforts of conservation biologists who are trying to save species nearing extinction?

Starting population includes 3 phenotypes of frogs: yellow, dark green, and striped.

A drought causes a bottleneck in which the population size is decreased and the dark green phenotype is lost.

Population size recovers but genetic variation is decreased, as only 2 phenotypes are left.

The resulting genotypes are equal in fitness. Because neutral mutations do not affect phenotype, they can spread throughout a population due to genetic drift (**Figure 24.12**). This theory has been called **non-Darwinian evolution** and also "survival of the luckiest" to contrast it with Darwin's "survival of the fittest" theory. Kimura agreed with Darwin that natural selection is responsible for adaptive changes in a species during evolution. His main idea is that much of the modern variation in gene sequences is explained by neutral variation rather than adaptive variation.

The sequencing of genomes from many species supports the neutral theory of evolution. When we examine changes of the coding sequence within structural genes, we find that nucleotide substitutions are more prevalent in the third base of a codon than they are in the first or second base. Mutations in the third base are often neutral because they may not change the amino acid sequence of the protein (refer back to Table 12.1). In contrast, random mutations at the first or second base are more likely to be harmful than beneficial and tend to be eliminated from a population. In addition, when mutations do change the coding sequence, they are more likely to involve conservative substitutions. For example, the difference between two alleles of a given gene may be the replacement of a nonpolar amino acid with another nonpolar amino acid. This change is conservative in the sense that it is less likely to affect protein function.

In general, the DNA sequencing of hundreds of thousands of different genes from hundreds of species has provided compelling support for the neutral theory of evolution. However, the argument is by no means resolved. Certain geneticists, called selectionists, oppose the neutralist theory. They often present persuasive theoretical arguments in favor of natural selection as the primary factor promoting genetic variation. In any case, the argument is largely a quantitative rather than a qualitative one. Each school of thought accepts that genetic drift and natural selection both play key roles in evolution. The neutralists argue that most genetic variation arises from neutral genetic mutations and genetic drift, whereas the selectionists argue that beneficial mutations and natural selection are primarily responsible.

Migration Between Two Populations Tends to Increase Genetic Variation

Earlier in this chapter, we considered how migration to a new location by a relatively small group can result in a founding population with an altered genetic composition due to genetic drift. In addition, migration between two different established populations can alter genetic variation. As a hypothetical example, let's consider two populations of a particular species of deer that are separated by a mountain range running north and south (**Figure 24.13**). On rare occasions, a few deer from the western population may travel through a narrow pass between the mountains and become members of the eastern population. If the two populations are different with regard to genetic variation, this migration will alter the frequencies of certain alleles in the eastern population. Of course, this migration could occur in the

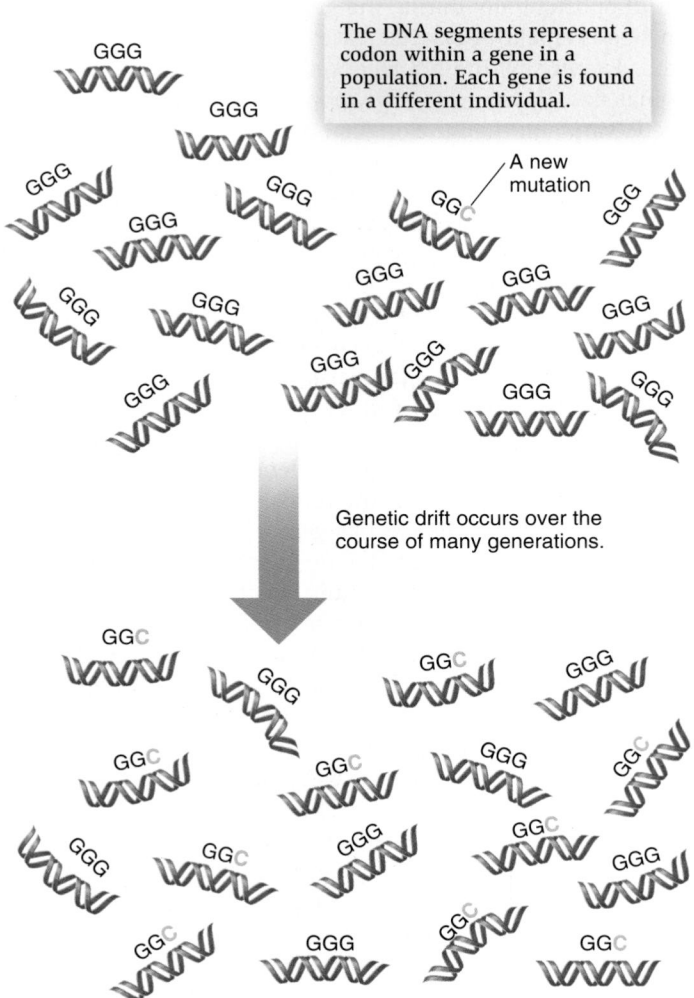

Figure 24.12 **Neutral evolution in a population.** In this example, a mutation within a gene (each gene shown represents a member of the population) changes a glycine codon from GGG to GGC, which does not affect the amino acid sequence of the encoded protein. Over the course of many generations, genetic drift may cause this neutral allele to become prevalent in the population, perhaps even monomorphic.

opposite direction as well and would then affect the western population. This phenomenon, called **gene flow**, occurs whenever individuals migrate between populations having different allele frequencies.

In nature, individuals commonly migrate in both directions. Such bidirectional migration has two important consequences. First, migration tends to reduce differences in allele frequencies between neighboring populations. In fact, population geneticists can evaluate the extent of migration between two populations by analyzing the similarities and differences between their allele frequencies. Populations that frequently mix their gene pools via migration tend to have similar allele frequencies,

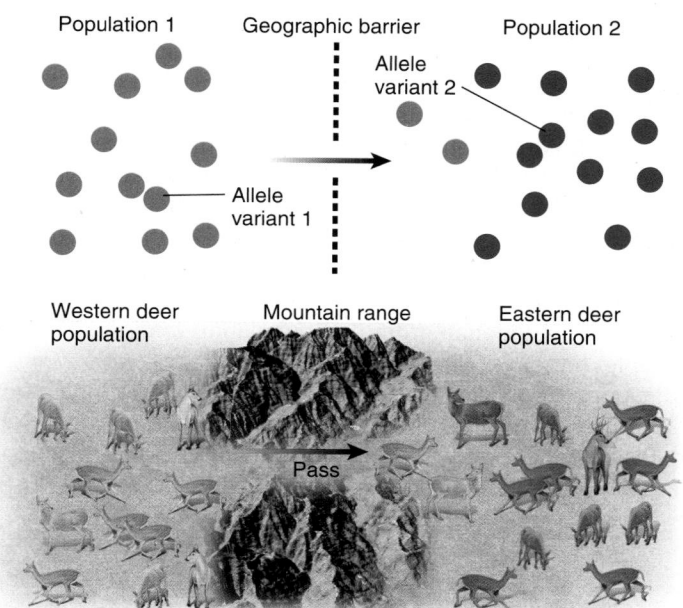

Figure 24.13 Migration and gene flow.   In this example, two populations of a deer species are separated by a mountain range. On rare occasions, a few deer from the western population travel through a narrow pass and become members of the eastern population, thereby changing some of the allele frequencies in the population and promoting gene flow.

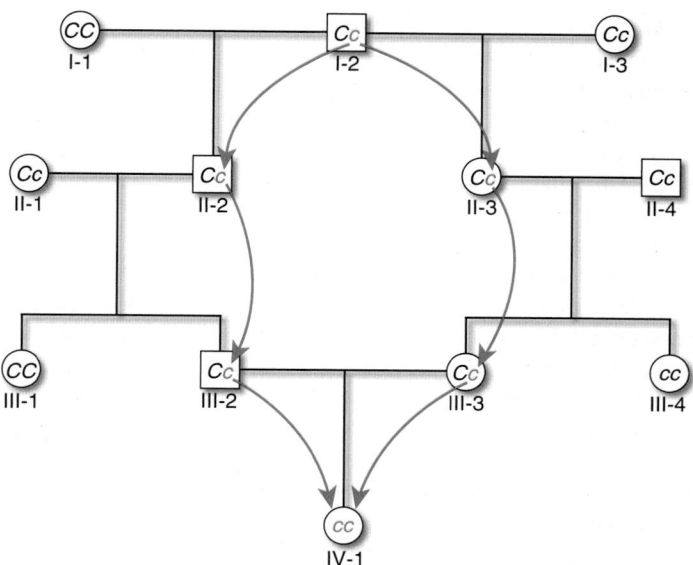

Figure 24.14 A human pedigree containing inbreeding. The parents of individual IV-1 are genetically related. Therefore, individual IV-1 has a higher probability of being homozygous for genes than do members of a noninbred population.

Biological inquiry: Although inbreeding by itself does not affect allele frequencies, how might inbreeding indirectly affect allele frequencies over the course of many generations if natural selection was also occurring?

whereas isolated populations are more disparate, due to the effects of natural selection and genetic drift. Second, migration tends to enhance genetic diversity within a population. As discussed earlier in this chapter, new mutations are relatively rare events. Therefore, a new mutation may arise in only one population. Migration may then introduce this new allele into a neighboring population.

Nonrandom Mating Affects the Relative Proportion of Homozygotes and Heterozygotes in a Population

As mentioned earlier in this chapter, one of the conditions required to establish the Hardy-Weinberg equilibrium is random mating. This means that individuals choose their mates irrespective of their genotypes. In many cases, particularly in human populations, this condition is frequently violated. Such **nonrandom mating** takes different forms. Assortative mating occurs when individuals with similar phenotypes are more likely to mate. If the similar phenotypes are due to similar genotypes, assortative mating tends to increase the proportion of homozygotes and decrease the proportion of heterozygotes in the population. The opposite situation, where dissimilar phenotypes mate preferentially, is called disassortative mating. This type of mating favors heterozygosity.

Another form of nonrandom mating involves the choice of mates based on their genetic history rather than their phenotypes. Individuals may choose a mate who is part of the same genetic lineage. The mating of two genetically related individuals, such as cousins, is called **inbreeding**. This sometimes occurs in human societies and is more likely to take place in nature when population size becomes very small.

In the absence of other evolutionary forces, nonrandom mating does not affect allele frequencies in a population. However, it will disrupt the balance of genotypes that is predicted by the Hardy-Weinberg equilibrium. As an example, let's consider inbreeding in a family pedigree. **Figure 24.14** illustrates a human pedigree involving a mating between cousins. Individuals III-2 and III-3 are cousins and have produced the daughter labeled IV-1. She is said to be inbred, because her parents are genetically related. The parents of an inbred individual have one or more common ancestors. In the pedigree of Figure 24.14, I-2 is the grandfather of both III-2 and III-3.

Inbreeding increases the relative proportions of homozygotes and decreases the likelihood of heterozygotes in a population. This is because an inbred individual has a higher chance of being homozygous for any given gene than does a noninbred individual, because the same allele for that gene could be inherited twice from a common ancestor. For example, let's suppose that individual I-2 is a heterozygote, *Cc* (see red lines in Figure 24.14). The *c* allele could pass from I-2 to II-2 to III-2, and finally to IV-1.

Likewise, the *c* allele could pass from I-2 to II-3 to III-3, and then to IV-1. Therefore, IV-1 has a chance of being homozygous because she inherited both copies of the *c* allele from a common ancestor of both of her parents. Inbreeding does not favor any particular allele—it does not favor *c* over *C*—but it does increase the likelihood that an individual will be homozygous for any given gene.

Although inbreeding by itself does not affect allele frequencies, it may have negative consequences with regard to recessive alleles. Rare recessive alleles that are harmful in the homozygous condition are found in all populations. Such alleles do not usually pose a problem because heterozygotes carrying a rare recessive allele are also rare, making it very unlikely that two heterozygotes will mate with each other. However, when inbreeding is practiced, homozygous offspring are more likely to be produced. For example, rare recessive diseases in humans are more frequent when inbreeding occurs.

In natural populations, inbreeding will lower the mean fitness of the population if homozygous offspring have a lower fitness value. This can be a serious problem as natural populations become smaller due to human habitat destruction. As the population shrinks, inbreeding becomes more likely because individuals have fewer potential mates from which to choose. The inbreeding, in turn, produces homozygotes that are less fit, thereby decreasing the reproductive success of the population. This phenomenon is called **inbreeding depression**. Conservation biologists sometimes try to circumvent this problem by introducing individuals from one population into another. For example, the endangered Florida panther (*Felis concolor coryi*) suffers from inbreeding-related defects, which include poor sperm quality and quantity, and morphological abnormalities. To alleviate these effects, individuals from Texas have been introduced into the Florida population of panthers.

CHAPTER SUMMARY

- Population genetics is the study of genes and genotypes in a population. The focus is on understanding genetic variation.

24.1 Genes in Populations

- All of the genes in a population constitute a gene pool.
- With regard to population genetics, a population is defined as a group of individuals of the same species that can interbreed with one another.
- Polymorphism refers to a genotype or phenotype that is found in two or more forms in a population. A monomorphic gene exists predominantly (>99%) as a single allele in a population. (Figure 24.1)
- An allele frequency is the number of copies of an allele divided by the total number of alleles in a population, while a genotype frequency is the number of individuals with a given genotype divided by the total number of individuals.
- The Hardy-Weinberg equation ($p^2 + 2pq + q^2 = 1$) relates allele and genotype frequencies. (Figure 24.2)
- The Hardy-Weinberg equation predicts an equilibrium if the population size is very large, mating is random, the populations do not migrate, no natural selection occurs, and no new mutations are formed.

24.2 Evolutionary Mechanisms and Their Effects on Populations

- Microevolution involves changes in a population's gene pool from one generation to the next.
- The sources of new genetic variation are random gene mutations, gene duplications, exon shuffling, and horizontal gene transfer.

This variation is acted upon by natural selection, genetic drift, migration, and nonrandom mating to alter allele and genotype frequencies, and ultimately cause a population to evolve over many generations. (Table 24.1)

- Natural selection favors individuals with the greatest reproductive success. Darwinian fitness is a measure of reproductive success. The mean fitness of a population is its average reproductive success.
- Directional selection is a form of natural selection that favors one extreme of a phenotypic distribution. (Figure 24.3)
- Stabilizing selection is a second form of natural selection that favors an intermediate phenotype. (Figure 24.4)
- Disruptive selection is a third pattern of natural selection that favors two or more genotypes. An example is when a population occupies a diverse environment. (Figure 24.5)
- Balancing selection maintains a balanced polymorphism in a population. Examples include heterozygote advantage and negative frequency-dependent selection. (Figure 24.6)
- Sexual selection is directed at traits that make it more likely for individuals to find or choose a mate and/or engage in successful mating. This can lead to traits described as secondary sexual characteristics. (Figure 24.7)
- Seehausen and van Alphen discovered that female choice of mates in cichlids is influenced by male coloration. This is an example of sexual selection. (Figures 24.8, 24.9)
- Genetic drift involves changes in allele frequencies due to random sampling error. It occurs more rapidly in small populations and leads to either the elimination or the fixation of alleles. (Figure 24.10)
- The bottleneck effect is a form of genetic drift in which a population size is dramatically reduced and then rebounds. During the bottleneck, genetic variation may be lost from a population. (Figure 24.11)

- The neutral theory of evolution by Kimura indicates that much of the genetic variation observed in populations is due to the accumulation of neutral genetic changes. (Figure 24.12)
- Gene flow occurs when individuals migrate between different populations and cause changes in the genetic composition of the resulting populations. (Figure 24.13)
- Inbreeding is a type of nonrandom mating in which genetically related individuals mate with each other. This tends to increase the proportion of homozygotes relative to heterozygotes. When homozygotes have lower fitness, this phenomenon is called inbreeding depression. (Figure 24.14)

TEST YOURSELF

1. Population geneticists are interested in the genetic variation in populations. The most common type of genetic change that can cause polymorphism in a population is
   a. a deletion of a gene sequence.
   b. a duplication of a region of a gene.
   c. a rearrangement of a gene sequence.
   d. a single-nucleotide substitution.
   e. an inversion of a segment of a chromosome.

2. The Hardy-Weinberg equation characterizes the genotype frequencies and allele frequencies
   a. of a population that is experiencing selection for mating success.
   b. of a population that is extremely small.
   c. of a population that is very large and not evolving.
   d. of a community of species that is not evolving.
   e. of a community of species that is experiencing selection.

3. Considering the Hardy-Weinberg equation, what portion of the equation would be used to calculate the frequency of individuals that do not exhibit a disease but are carriers of a recessive genetic disorder?
   a. q
   b. p^2
   c. $2pq$
   d. q^2
   e. both b and d

4. Which of the following does not alter allele frequencies?
   a. selection
   b. immigration
   c. mutation
   d. inbreeding
   e. emigration

5. Which of the following statements is correct regarding mutations?
   a. Mutations are not important in evolution.
   b. Mutations provide the source for genetic variation that other evolutionary forces may act upon.
   c. Mutations occur at such a high rate that they promote major changes in the gene pool from one generation to the next.
   d. Mutations are insignificant when considering evolution of a large population.
   e. Mutations are of greater importance in larger populations than in smaller populations.

6. In a population of fish, body coloration varies from a light shade, almost white, to a very dark shade of green. If changes in the environment resulted in decreased predation of individuals with the lightest coloration, this would be an example of _____ selection.
   a. disruptive
   b. stabilizing
   c. directional
   d. sexual
   e. artificial

7. Considering the same population of fish described in question 6, if the stream environment included several areas of sandy, light-colored bottom areas and lots of dark-colored vegetation, both the light- and dark-colored fish would have selective advantage and increased survival. This type of scenario could explain the occurrence of
   a. genetic drift.
   b. disruptive selection.
   c. mutation.
   d. stabilizing selection.
   e. sexual selection.

8. The microevolutionary force most sensitive to population size is
   a. mutation.
   b. migration.
   c. selection.
   d. genetic drift.
   e. all of the above.

9. The neutral theory of evolution differs primarily from Darwinian evolution in that
   a. neutral theory states natural selection does not exist.
   b. neutral theory states that most of the genetic variation in a population is due to neutral mutations, which do not alter phenotypes.
   c. neutral variation alters survival and reproductive success.
   d. neutral mutations are not affected by population size.
   e. both b and c.

10. Populations that experience inbreeding may also experience
    a. a decrease in fitness due to an increased frequency of recessive genetic diseases.
    b. an increase in fitness due to increases in heterozygosity.
    c. very little genetic drift.
    d. no apparent change.
    e. increased mutation rates.

CONCEPTUAL QUESTIONS

1. Explain the five conditions that are required for Hardy-Weinberg equilibrium.
2. List and define the four types of selection.
3. Define the founder effect.

EXPERIMENTAL QUESTIONS

1. What hypothesis is tested in the Seehausen and van Alphen experiment?

2. Describe the experimental design for this study, illustrated in Figure 24.9. What was the purpose of conducting the experiment under the two different light conditions?

3. What were the results of the experiment in Figure 24.9?

COLLABORATIVE QUESTIONS

1. Discuss four sources of new genetic variation in a population.

2. Discuss various patterns of natural selection that lead to environmental adaptation and also discuss sexual selection.

www.brookerbiology.com

This website includes answers to the Biological Inquiry questions found in the figure legends and all end-of-chapter questions.

25

ORIGIN OF SPECIES

CHAPTER OUTLINE

Two different species of zebras. Grey zebra (*Equus grevyi*) is shown here on the *left* and Grant's zebra (*E. quagga*), which has fewer and thicker stripes, is shown on the *right*. This chapter will examine how different species come into existence.

The origin of living organisms was described by several ancient philosophers as the great "mystery of mysteries." Perhaps that is why so many different views have been put forth to explain the existence of living species. At the time of Aristotle (4th century B.C.E.), most people believed that some living organisms could come into being by spontaneous generation, which is the idea that nonliving materials could give rise to living organisms. For example, it was commonly believed that worms and frogs could arise from mud. By comparison, many religious teachings contend that species were divinely made and have remained the same since their creation. In contrast to these ideas, the work of Charles Darwin provided the scientific theory of evolution by descent with modification. This theory helps us to understand the diversity of life, and in particular, it presents a logical explanation for how pre-existing species can evolve into new species.

This chapter provides an exciting way to build on the information that we have considered in previous chapters. In Chapter 22, we examined how the first primitive cells in an RNA world evolved into prokaryotic cells and eventually eukaryotes. Chapter 23 surveyed the tenets on which the theory of evolution is built, and in Chapter 24, we viewed evolution on a small scale as it relates to a single gene. In this chapter, we will consider evolution on a larger scale, as it relates to the formation of new species.

To biologists, the concept of a **species** has come to mean a group of organisms that maintains a distinctive set of attributes in nature. As a student, you may already have an intuitive sense of this concept. It is obvious that giraffes and mice are different species. However, as we will learn in the first section of this chapter, the distinction between different, closely related species is often blurred in natural environments, so that it may not be easy to definitively distinguish two species (see chapter-opening photo). Among other uses, species identification is important because it allows biologists to plan for the preservation and conservation of those species.

In this chapter we will also focus on the mechanisms that promote the formation of new species, a phenomenon called **speciation**. The term **macroevolution** refers to evolutionary changes that create new species and groups of species. It concerns the diversity of organisms established over long periods of time through the evolution and extinction of many species. Macroevolution occurs by the accumulation of microevolutionary changes, those that occur in a single gene (see Chapter 24). Natural selection results in the evolution of traits that promote environmental adaptation and reproductive success. In this chapter, we will learn how the same evolutionary mechanisms that account for microevolution also play a role in the formation of new species.

25.1 Species Concepts

The number of species on Earth is astounding. A study done by biologist E. O. Wilson and colleagues in 1990 estimated the known number of species at approximately 1.4 million. However, many species have yet to be identified. This is particularly true among prokaryotic organisms. Estimates of the number of unidentified species range from 2 to 100 million!

The difficulty of identifying whether certain groups constitute unique species is often rooted in the phenomenon that a single species may exist in two distinct populations that are in the slow process of evolving into two or more different species.

The amount of time that two populations are separated will have an important impact. If the time is short, the two populations are likely to be very similar, so they would be considered the same species. If the time is long, sufficient changes may have occurred so that the two populations would show unequivocal differences that allow them to maintain their distinctive set of features in nature. When studying natural populations, evolutionary biologists are often confronted with situations where some differences between two populations are apparent, but it is difficult to decide whether the two populations truly represent separate species. When two or more groups within the same species display one or more traits that are somewhat different but not enough to warrant their placement into different species, biologists sometimes classify such groups as subspecies.

In this section, we will begin by considering different attributes that biologists examine when deciding if two groups of organisms constitute different species. In later sections, we will consider mechanisms that explain how new species arise in nature.

The Members of a Species Have a Common Set of Characteristics That Distinguish Them from Other Species

Biologists adopt methods of species identification that are based on their own experience with the organisms they study. The characteristics that a biologist uses to identify a species depend, in large part, on the species in question. For example, the traits used to distinguish insect species would be quite different from those used to identify different bacterial species. The most commonly used characteristics are physical or morphological traits, the ability to interbreed, common evolutionary lineages, and ecological factors. These different ways of considering species have led to the use of different approaches for distinguishing species called **species concepts** (**Table 25.1**). A comparison of these concepts will help you to appreciate the various factors that are involved in categorizing the species on our planet.

Phylogenetic Species Concept According to the **phylogenetic species concept***, as advocated by Quentin Wheeler and Norman Platnick, the members of a single species are identified by having a unique combination of characteristics. Historically, the first way to categorize species was based on their physical characteristics. Organisms are classified as the same species if their anatomical traits appear to be very similar. Likewise, microorganisms can be classified according to morphological traits at the cellular level. In addition, molecular features such as DNA sequences can now be used to compare organisms. An advantage of the phylogenetic species concept is that it can be applied to all types of organisms.

---

*Although it is named the phylogenetic species concept, it is not based on phylogenetic trees, which are described in Chapter 26. The phylogenetic species concept is related to other species concepts not covered here called the morphological species concept and typological species concept.

Table 25.1	Species Concepts
Concept	**Description**
Phylogenetic species concept	Various physical characteristics can be analyzed to evaluate the identity of species. These often include morphological (anatomical) traits. In the case of unicellular organisms, characteristics such as cell-wall structure and other cellular traits may be examined. Molecular characteristics can also be compared.
Biological species concept	Two species are often judged to be separate species if they are unable to interbreed in nature to produce viable, fertile offspring.
Evolutionary species concept	An analysis of ancestry may help biologists to determine if two groups are members of the same species or represent evolutionarily distinct species.
Ecological species concept	The ability of organisms to successfully occupy their own ecological niche or habitat, including their use of resources and impact on the environment, may be used to distinguish species.

Although this concept is a common way for taxonomists to categorize species, it has a few drawbacks. First, it may be difficult to decide how many traits to consider when characterizing individuals. In addition, it is difficult to analyze quantitative traits that vary in a continuous way among members of the same species. Another drawback is that the degree of dissimilarity that separates different species may not be easy to decide upon. Researchers often disagree about how much morphological difference is necessary to separate different species. Another shortcoming of the phylogenetic species concept is that members of the same species sometimes look very different, and conversely, members of different species sometimes look remarkably similar to each other. For example, **Figure 25.1a** shows two different frogs of the species *Dendrobates tinctorius*, commonly called the dyeing poison frog. This species exists in many different-colored morphs, which are individuals of the same species that have noticeably dissimilar appearances. Alternatively, **Figure 25.1b** shows two different species of frog, the Northern leopard frog (*Rana pipiens*) and the Southern leopard frog (*Rana utricularia*), which look similar.

Biological Species Concept So why would biologists describe two species, such as the Northern leopard frog and Southern leopard frog, as being different if they are morphologically similar? One reason is that they are unable to breed with each other in nature. Therefore, a second way to define a species is by the ability to interbreed. In the late 1920s, geneticist Theodosius Dobzhansky proposed that each species is reproductively isolated from other species. Such **reproductive isolation** prevents one species from successfully interbreeding with other species.

(a)

(b)

Figure 25.1 Difficulties in applying the phylogenetic species concept.   In some cases, members of the same species can appear quite different. (a) Two color morphs of one species, the dyeing poison frog (*Dendrobates tinctorius*). (b) Two different species of frog, the Northern leopard frog (*Rana pipiens*, left) and the Southern leopard frog (*Rana utricularia*, right), which look similar.

In 1942, evolutionary biologist Ernst Mayr expanded on the ideas of Dobzhansky to provide a biological definition of a species. According to Mayr's **biological species concept**, a species is a group of individuals whose members have the potential to interbreed with one another in nature to produce viable, fertile offspring but cannot successfully interbreed with members of other species. As discussed later in this section, reproductive isolation among species of plants and animals can occur by an amazing variety of different mechanisms.

The biological species concept has been used to distinguish many plant and animal species, especially those that look alike but are reproductively isolated. Even so, it suffers from three main problems. First, in nature, it may be difficult to determine if two populations are reproductively isolated, particularly if they are large populations with overlapping geographical ranges. Second, taxonomists have noted many cases in which two different species can interbreed in nature yet consistently maintain themselves as separate species. For example, different species of Yucca plants, such as *Yucca pallida* and *Yucca constricta*, do interbreed in nature yet typically maintain populations with distinct characteristics. For this reason, they are viewed as distinct species based on the phylogenetic species concept. A third drawback of

the biological species concept is that it cannot be applied to asexual species such as bacteria, nor can it be applied to extinct species. Likewise, some species of plants and fungi are only known to reproduce asexually. Therefore, the biological species concept has been primarily used to distinguish closely related species of animals and plants that reproduce sexually.

Evolutionary Species Concept In 1961, American paleontologist George Gaylord Simpson proposed a species concept based on ancestry. According to the **evolutionary species concept**, a species is derived from a single lineage that is distinct from other lineages and has its own evolutionary tendencies and historical fate. A **lineage** is the genetic relationship between an individual or group of individuals and its ancestors. The evolutionary species concept is a theoretical viewpoint that is focused on the pathway that has led to the formation of each distinct species. In this regard, it can be applied to the formation of all species. We will consider the topic of evolutionary lineages in Chapter 26.

One drawback of the evolutionary species concept is that it does not provide an easy way to identify a unique species. In most cases, lineages are difficult to examine and evaluate quantitatively. The interpretation of lineages involving fossils may be particularly difficult to evaluate due to incomplete fossil remains as well as transitional forms that may be missing from the fossil record. Currently, the most common way for researchers to analyze lineages of modern species is by comparing the DNA sequences of particular genes. We can obtain samples of cells from different individuals and compare the genes within those cells to see how similar or different they are. Even so, it is difficult to decide where to draw the line when separating groups into different species. Is a 0.1% difference in their genome sequences sufficient to warrant placement into two different species, or do we need a 1% or 5% difference? A very tiny genetic change can have a dramatic effect on phenotype. Conversely, as discussed in Chapter 24, many genetic changes are neutral and have no discernable effect on phenotype. For these and other reasons, researchers cannot simply compare DNA sequences among different groups and always reach concrete conclusions as to whether or not the groups should be considered different species. Nevertheless, as we will examine in Chapter 26, sequence comparisons are very useful in evaluating whether multiple species—which are already considered distinct—are closely or distantly related evolutionarily.

Ecological Species Concept The **ecological species concept**, described by American evolutionary biologist Leigh Van Valen in 1976, is a viewpoint that considers a species within its native environment. Each species occupies an **ecological niche**, which is the unique set of habitat resources that a species requires, as well as its influence on the environment and other species. Within their own niche, members of a given species compete with each other for survival. If two organisms are very similar, their needs will overlap, which results in competition. Such competing individuals are likely to be of the same species.

According to this concept, species are formed because evolutionary mechanisms control how each type of species uses resources. This species concept is particularly useful in distinguishing bacterial species that do not reproduce sexually. Bacterial cells of the same species are likely to use the same types of resources (sugars, vitamins, etc.) and grow under the same types of conditions (temperature, pH, etc.).

Reproductive Isolating Mechanisms Help to Maintain the Distinctiveness of Each Species

Thus far we have considered four species concepts that provide biologists with approaches that are aimed at the identification of different species. In our discussion of these concepts, you may have realized that the identification of a species is not always a simple matter. With regard to plants and animals, the biological species concept proposed by Mayr has played a major role in the way that biologists study plant and animal species, partly because it identifies a possible mechanism for the process of forming new species—reproductive isolation. For this reason, much research has been done to try to understand the mechanisms that prevent interbreeding between different species, which are called **reproductive isolating mechanisms**. Keep in mind that populations do not intentionally erect these reproductive barriers. Rather, reproductive isolating mechanisms are merely a consequence of genetic changes that occur usually because a species becomes adapted to its own particular environment. The view of evolutionary biologists is that reproductive isolation typically evolves as a by-product of genetic divergence. Over time, as a species evolves its own unique characteristics, some of those traits are likely to prevent breeding with other species.

Biologists have discovered many mechanisms that prevent closely related species from interbreeding. These mechanisms fall into two categories, prezygotic and postzygotic. **Prezygotic mechanisms** prevent the formation of a zygote, while **postzygotic mechanisms** block the development of a viable and fertile individual after fertilization has taken place. **Figure 25.2** summarizes some of the more common ways that reproductive isolating mechanisms prevent reproduction between different species. When two species do produce offspring, such an offspring is called an **interspecies hybrid**.

Prezygotic Reproductive Isolating Mechanisms We will consider five types of prezygotic mechanisms. The first is an obvious way to prevent interbreeding, which is for members of different species to avoid contact with each other. This phenomenon, called habitat isolation, refers to a geographic barrier to interbreeding. For example, a large body of water may separate two different plant species that live on nearby islands. The second prezygotic isolating mechanism is temporal isolation, in which species reproduce at different times of the day or year. In the northeastern U.S., for example, the two most abundant field crickets, *Gryllus veletis* and *Gryllus pennsylvanicus* (spring and fall field crickets, respectively), do not differ in song or habitat and are morphologically very similar (**Figure 25.3**).

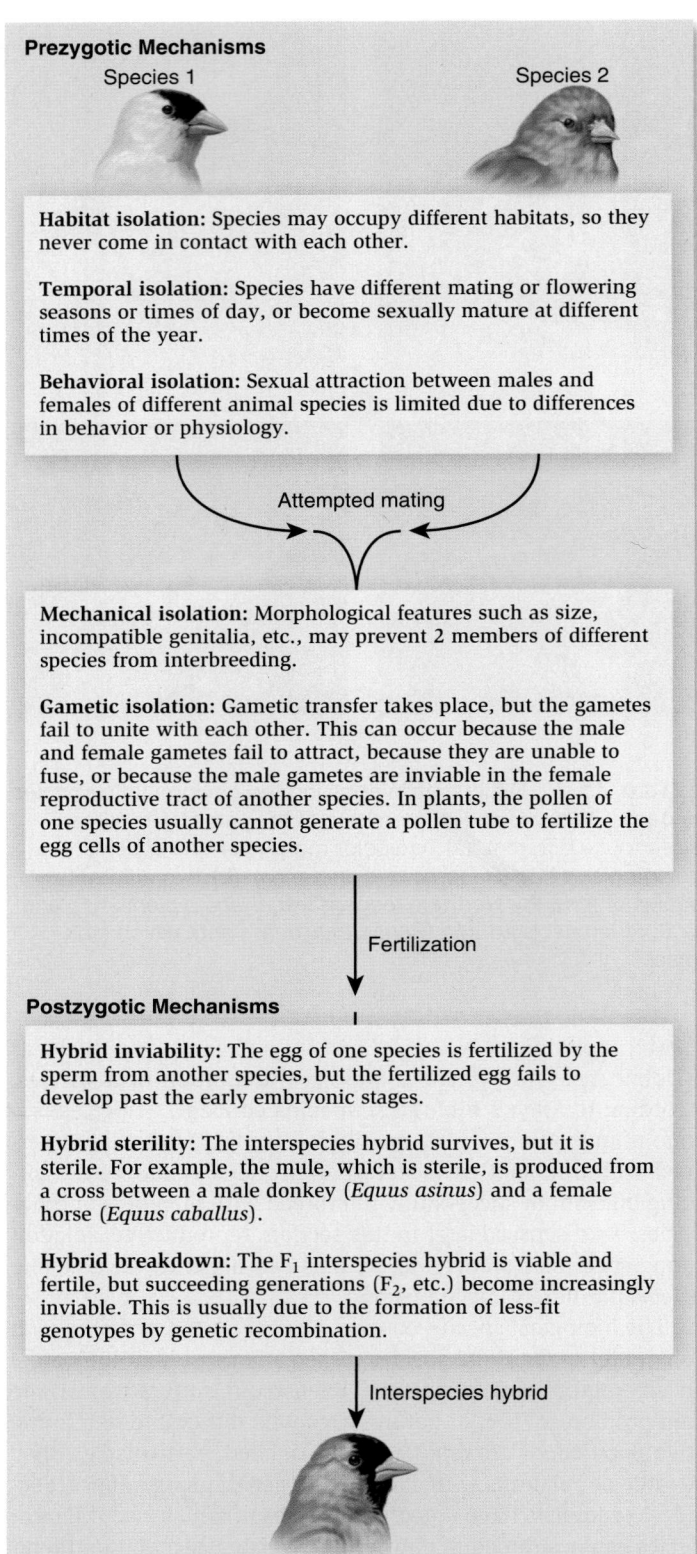

Figure 25.2   **Reproductive isolating mechanisms.**   These mechanisms prevent successful breeding between different species. They can occur prior to fertilization (prezygotic) or after fertilization (postzygotic).

(a) Spring field cricket (*Gryllus veletis*)

(b) Fall field cricket (*Gryllus pennsylvanicus*)

Figure 25.3 **An example of temporal isolation.** Interbreeding between these two species of crickets does not usually occur because *Gryllus veletis* matures in the spring, but *Gryllus pennsylvanicus* matures in the fall.

However, *G. veletis* matures in the spring, whereas *G. pennsylvanicus* matures in the fall. This minimizes interbreeding between the two species.

In the case of animals, mating behavior and anatomy often play key roles in promoting reproductive isolation. An example of the third type of isolation, behavioral isolation, is found between the eastern meadowlark (*Sturnella magna*) and western meadowlark (*Sturnella neglecta*). Both species are nearly identical in shape, coloration, and habitat, and their ranges overlap in the central U.S. (**Figure 25.4**). For many years, they were thought to be the same species. When biologists discovered that the western meadowlark is a separate species, it was given the species name *neglecta* to reflect the long delay in its recognition. In the zone of overlap, very little interspecies mating takes place between eastern and western meadowlarks, largely due to differences in their songs. The eastern meadowlark's song is a simple series of whistles, typically about four or five notes. By comparison, the song of the western meadowlark is a longer series of flutelike gurgling notes that go down the scale. These differences in songs enable meadowlarks to recognize potential mates as members of their own species.

A fourth type of isolation, called mechanical isolation, occurs when morphological features such as size or incompatible genitalia prevent two species from interbreeding. For example, male dragonflies use a pair of special appendages to grasp females during copulation. When a male tries to mate with a female of a different species, his grasping appendages do not fit her body shape.

A fifth type of prezygotic isolating mechanism can occur when two species attempt to interbreed, but the gametes fail to unite in a successful fertilization event. This phenomenon, called gametic isolation, is widespread among plant and animal species. In aquatic animals that release sperm and egg cells into the water, gametic isolation is important to prevent interspecies hybrids. For example, closely related species of sea urchins may release sperm and eggs into the water at the same time. Researchers have discovered that sea urchin sperm have a protein on their surface called bindin, which mediates sperm–egg attachment and membrane fusion. The structure of bindin is significantly different among different sea urchin species, and thereby

(a) Eastern meadowlark (*Sturnella magna*)

North America

(b) Western meadowlark (*Sturnella neglecta*)

Eastern meadowlark
Western meadowlark
Zone of overlap

Figure 25.4 **An example of behavioral isolation.** (a) The eastern meadowlark (*Sturnella magna*) and (b) western meadowlark (*Sturnella neglecta*) are very similar in appearance. The red region in this map shows where the two species' ranges overlap. However, very little interspecies mating takes place due to differences in their songs. Meadowlarks use their songs to recognize potential mates of their own species.

ensures that fertilization occurs between sperm and egg cells of the same species.

In flowering plants, gametic isolation is a particularly vital mechanism to achieve reproductive isolation. As discussed in Chapter 39, plant fertilization is initiated when a pollen grain lands on the stigma of a flower and sprouts a pollen tube that ultimately reaches an egg cell (look ahead to Figure 39.13). When pollen is released from a plant, it could be transferred to the stigma of many different plant species. In most cases, when a pollen grain lands on the stigma of a different species, it either fails to generate a pollen tube or the tube does not grow properly and reach the egg cell. The mechanism that controls pollen tube growth in plants has not been widely studied. The process is best understood in members of the mustard genus (*Brassica*), which includes turnips and cabbage. In these species, pollen tube growth is known to involve molecules released by pollen grains that activate receptors found in the cell wall of the stigma. If a similar mechanism is used to achieve gametic isolation between different species, the receptors that recognize molecules released by the pollen grain must be very specific for the molecules of their own species, and fail to recognize those molecules released by pollen grains of other species.

Postzygotic Reproductive Isolating Mechanisms Let's now turn to postzygotic mechanisms of reproductive isolation, of which we will discuss three types. These mechanisms tend to be less common in nature, because they are more costly in terms of energy and resources used. The first such mechanism is hybrid inviability, in which an egg of one species is fertilized by a sperm from another species, but the fertilized egg cannot develop past the early embryonic stages. A second mechanism is hybrid sterility, in which an interspecies hybrid may be viable but sterile. A classic example of hybrid sterility is the mule, which is produced by a mating between a male donkey (*Equus asinus*) and a female horse (*Equus caballus*) (**Figure 25.5**). Because a horse has 32 chromosomes per set and a donkey has 31, a mule inherits 63 chromosomes (32 + 31). When a mule tries to produce offspring, all of its chromosomes do not have pairs, so it is sterile. Note that the mule has no species name because it is not considered a species due to this sterility. Finally, interspecies hybrids may be viable and fertile, but subsequent generations may harbor genetic abnormalities that are detrimental. This third mechanism, called hybrid breakdown, can be caused by changes in chromosome structure. The chromosomes of closely related species may have structural differences from each other, such as inversions. In hybrids, a crossover may occur in the region that is inverted in one species but not the other. This will produce gametes with too little or too much genetic material. For this reason, such hybrids would often have offspring with developmental abnormalities.

Male donkey (*Equus asinus*)

×

Female horse (*Equus caballus*)

Mule

Figure 25.5 **An example of hybrid sterility.**   When a male donkey (*Equus asinus*) mates with a female horse (*Equus caballus*), their offspring is a mule, which is sterile.

25.2 Mechanisms of Speciation

The formation of a new species, or speciation, is caused by genetic changes in a particular group that make it different from the species from which it was derived. As discussed in Chapter 24, random mutations in genes can be acted upon by natural selection and other evolutionary mechanisms to alter the genetic composition of a population. New species commonly evolve in this manner. In addition, interspecies matings, changes in chromosome number, and horizontal gene transfer may also cause new species to arise. In all of these cases, the underlying cause of speciation is the accumulation of genetic changes that ultimately promote enough differences so that we judge a population to constitute a unique species.

Even though genetic changes account for the phenotypic differences we observe among living organisms, such changes do not fully explain the existence of many distinct species on our planet. Why does life diversify into the more or less discrete populations that we recognize as species? Two main explanations have been proposed:

1. Species are a consequence of adaptation to different ecological niches.
2. Species of sexually reproducing organisms arise via reproductive isolation.

Depending on the species involved, one or both factors may play a dominant role in the formation of new species. In this section, we will begin by considering two different patterns of speciation that occur over the course of many generations. We will then consider several examples of speciation occurring in nature. As you will learn, adaptation and/or reproductive isolating mechanisms are critical aspects of the speciation process.

Speciation Can Occur as a Linear or a Branching Process

Before we consider mechanisms of speciation, let's first examine the patterns of speciation that are commonly observed. By studying the fossil record and the diversity of modern species, evolutionary biologists have found two different patterns of speciation. During **anagenesis** (from the Greek, *ana*, up, and *genesis*, origin), a single species is transformed into a different species over the course of many generations (**Figure 25.6a**). In this process, evolutionary mechanisms cause the characteristics of the species to change. For example, a single species of fish in an isolated lake may evolve into a different species due to changes in available food sources.

By comparison, **cladogenesis** (from the Greek *clados*, branch) involves the division of a species into two or more species. In the case of the birds shown in **Figure 25.6b**, the original mainland species has remained relatively unchanged, while populations on neighboring islands have evolved substantially different traits by natural selection. Such cladogenesis is a more common form of speciation. One reason why is related to geographic isolation. If one or more individuals move to a new location

(a) Anagenesis   **(b) Cladogenesis**

Figure 25.6   **A comparison between anagenesis and cladogenesis, two patterns of speciation.**   **(a)** Anagenesis is the change of one species into another, whereas **(b)** cladogenesis occurs when one population diverges into two or more different species.

Biological inquiry: Explain why cladogenesis requires reproductive isolation and anagenesis does not.

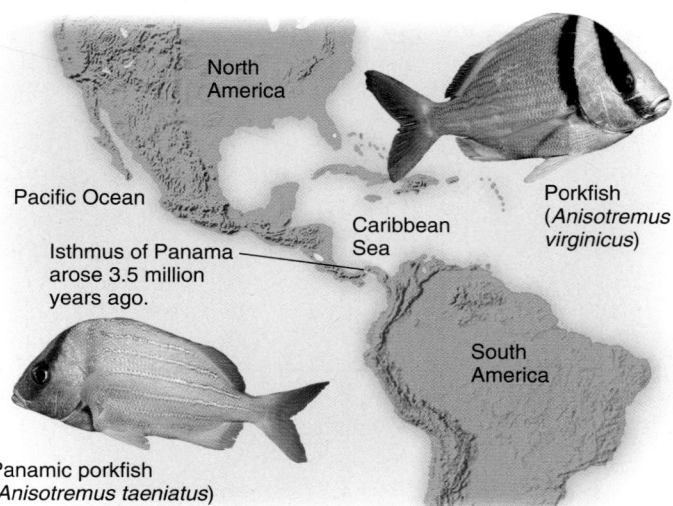

Figure 25.7   **An example of allopatric speciation.**   An ancestral fish population was split into two by the formation of the Isthmus of Panama about 3.5 million years ago. Since that time, different genetic changes occurred in the two populations because of their geographic isolation; these changes eventually led to the formation of different species. The porkfish (*Anisotremus virginicus*) is found in the Caribbean Sea, whereas the Panamic porkfish (*Anisotremus taeniatus*) is found in the Pacific Ocean.

that is geographically isolated from the original population, evolutionary mechanisms such as natural selection will operate independently on the two populations. This mechanism of speciation is described next.

Geographic Isolation Can Promote Allopatric Speciation

Although sexual reproduction is not a barrier to anagenesis, it is a barrier to cladogenesis. A population will evolve as a single unit if the members can successfully breed with one another, thereby preventing its divergence into two or more discrete species. The process of cladogenesis begins only when gene flow becomes limited between two or more populations. **Allopatric speciation** (from the Greek *allos*, other, and the Latin *patria*, homeland) is thought to be the most prevalent way for cladogenesis to occur. This form of speciation occurs when some members of a species become geographically separated from the other members. In some cases, the separation may be caused by slow, geological events that eventually produce quite large geographic barriers. For example, a mountain range may emerge and split a species that occupies the lowland regions. Or a creeping glacier may divide a population. **Figure 25.7** shows an interesting example in which geological separation promoted speciation. A fish called the porkfish (*Anisotremus virginicus*) is found in the Caribbean Sea, whereas the Panamic porkfish (*Anisotremus taeniatus*) is found in the Pacific Ocean. These two species were derived from an ancestral species that was split by the formation of the Isthmus of Panama about 3.5 million years ago. Since this event, the two populations have been geographically isolated and have evolved into distinct species via natural selection and other evolutionary mechanisms.

Allopatric speciation can also occur via a second mechanism related to the founder effect that was described in Chapter 24. This occurs when a small population moves to a new location that is geographically separated from the main population. For example, a storm may force a small group of birds from a mainland to a distant island. In this case, migration between the island and the mainland population is an infrequent event. In a relatively short period of time, the founding population on the island may evolve into a new species. Different evolutionary forces may contribute to this rapid evolution. First, as discussed in Chapter 24, genetic drift may quickly lead to the random fixation of certain alleles and the elimination of other alleles from the small population. Another factor is natural selection, because the environment on the island may differ significantly from the mainland environment.

The Hawaiian Islands are a showcase of the founder effect. The islands' extreme isolation coupled with their phenomenal array of ecological niches has enabled a small number of founding species to evolve into a vast assortment of different species. Biologists have investigated several examples of **adaptive radiation**, in which a single ancestral species has evolved into a wide array of descendant species that differ greatly in their habitat, form, or behavior. For example, approximately 1,000 species of *Drosophila* are found dispersed throughout the islands. Evolutionary studies suggest that these were derived from a single colonization by one species of fruit fly! Another striking example is seen with a family of birds called honeycreepers (*Drepanidinae*) (**Figure 25.8**). Researchers estimate that the honeycreepers' ancestor arrived in Hawaii 3–7 million years ago.

Figure 25.8   **An example of the founder effect with subsequent adaptive radiation.**   The honeycreepers' ancestor is believed to be related to a Eurasian rosefinch that arrived on the Hawaiian Islands approximately 3–7 million years ago. Since that time, at least 54 different species of honeycreepers (*Drepanidinae*) have evolved on the islands. A few selected examples are shown here. Adaptations to feeding have produced species with notable differences in beak morphology.

This ancestor was a single species of finch, possibly a Eurasian rosefinch (*Carpodacus* sp.) or, less likely, the North American house finch (*Carpodacus mexicanus*). At least 54 different species of honeycreepers, many of which are now extinct, have evolved from this founding event to fill available niches in the islands' habitats. Seed eaters developed stouter, stronger bills for cracking tough husks. Insect-eating honeycreepers developed thin, warbler-like bills for picking insects from foliage or strong, hooked bills to root out wood-boring insects. And nectar-feeding honeycreepers evolved curved bills for extracting nectar from the flowers of Hawaii's endemic plants.

FEATURE INVESTIGATION

Podos Found That an Adaptation to Feeding Also May Have Promoted Reproductive Isolation in Finches

To investigate how environmental adaptation may contribute to reproductive isolation, in 2001 American evolutionary biologist Jeffrey Podos analyzed the songs of Darwin's finches on the Galápagos Islands. Like the honeycreepers, the differences in beak sizes and shapes among the various species of finches are adaptations to different feeding strategies. Podos hypothe-

sized that changes in beak morphology could also impact the songs that the birds produce, thereby having the potential to affect mate choice. The components of the vocal tract of birds, including the trachea, syrinx, and beak, work collectively to produce a bird's song. Birds actively modify the shape of their vocal tracts during singing. Beak movements are normally very rapid and precise.

Podos focused on two aspects of a bird's song. The first feature is the frequency range, which is a measure of the minimum and maximum frequencies in a bird's song. The second feature is the trill rate. A trill is a series of notes or group of notes repeated in succession. **Figure 25.9** shows a graphic depiction of the songs of Darwin's finches. As you can see, the song patterns of these finches are quite different from each other.

To quantitatively study the relationship between beak size and song, Podos first captured male finches on one of the Galápagos Islands (Santa Cruz) and measured their beak sizes (**Figure 25.10**). The birds were banded and then released into the wild; the banding provided a way to identify the birds whose beaks had already been measured. After release, the songs of the banded birds were recorded on a tape recorder, and then their range of frequencies and trill rate were analyzed. Podos then compared the data for the Galápagos finches to a large body of data that had been collected on many other bird species. This comparison was used to evaluate whether beak size, in this case, beak depth—the measurement of the beak from top to bottom, at its base—constrained either the frequency range and/or the trill rate of the finches.

The results of this comparison are shown in the data of Figure 25.10. As seen here, the relative constraint on vocal performance became higher as the beak depth became larger. This means that birds with larger beaks had a more narrow frequency range and/or a slower trill rate. Podos proposed that as jaws and beaks become adapted for strength to crack open larger seeds, they will be less able to perform the rapid movements associated with certain types of songs. In contrast, the finches whose beaks were adapted to probe for insects or eat smaller seeds had less constraint on their vocal performance. From the perspective of evolution, the changes observed in song patterns for the Galápagos finches could have played an important role in promoting reproductive isolation, because song pattern is one factor involved in mate selection in birds. Therefore, a by-product of beak adaptation for feeding is that it also appears to have an effect on song pattern, possibly promoting reproductive isolation. In this way, populations of finches would have evolved into their own distinct species.

Figure 25.9   **Differences in the songs of Galápagos finches.**   These spectrograms depict the frequency of each bird's song over time, measured in kilohertz (kHz). The songs are produced in a series of trills that have a particular pattern and occur at regular intervals. Notice the differences in frequency and trill rate between different species of birds.

Figure 25.10   **The effects of beak depth on song among different species of Galápagos finches.**

HYPOTHESIS Changes in beak morphology that are an adaptation to feeding may also affect the songs of Galápagos finches and thereby lead to reproductive isolation between species.

STARTING MATERIALS This study was conducted on finch populations of the Galápagos Island of Santa Cruz.

	Experimental level	Conceptual level
1 Capture male finches and measure their beak depth. Beak depth is measured at the base of beak, from top to bottom.		This is a measurement of phenotypic variation in beak size.

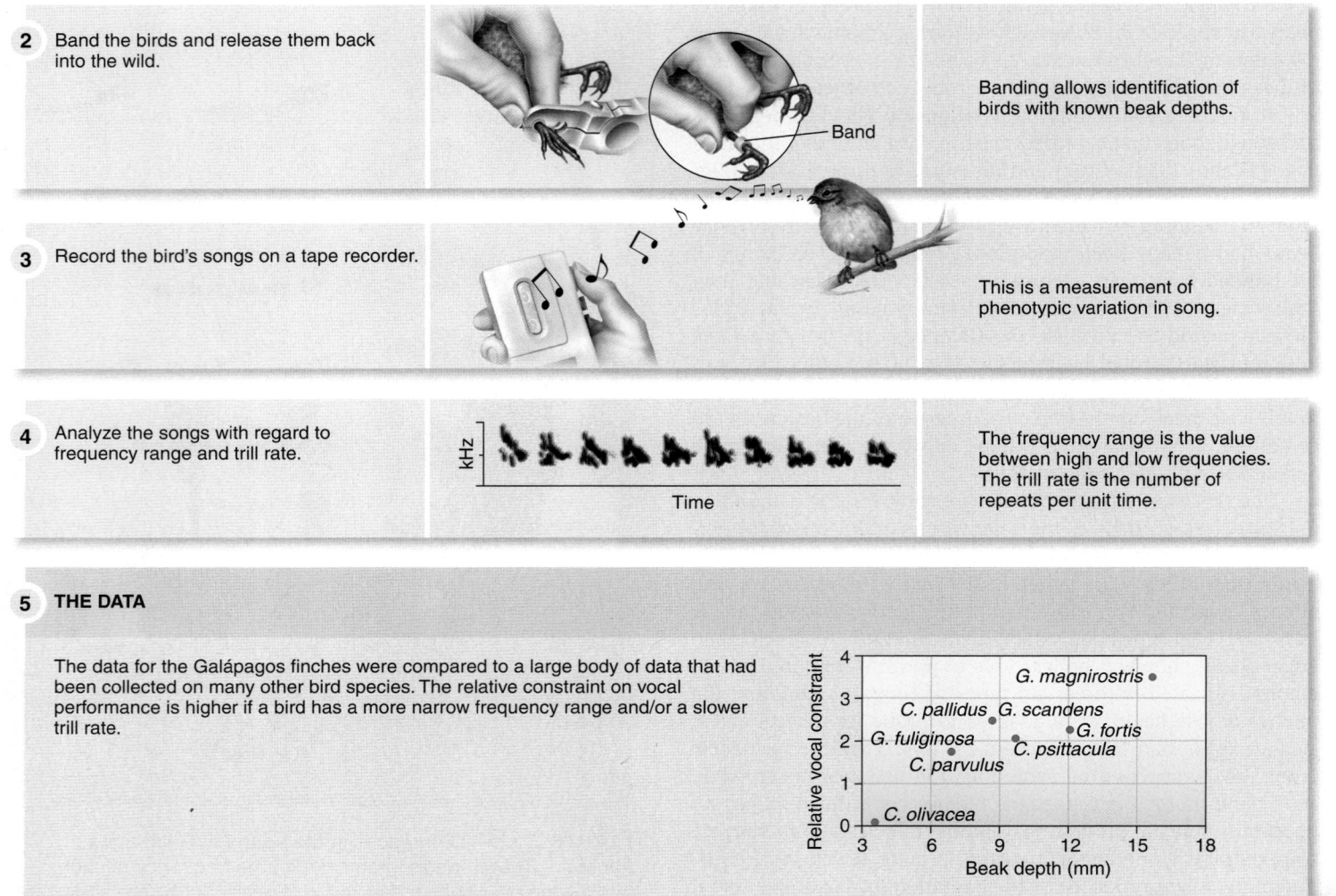

2 Band the birds and release them back into the wild.

Band

Banding allows identification of birds with known beak depths.

3 Record the bird's songs on a tape recorder.

This is a measurement of phenotypic variation in song.

4 Analyze the songs with regard to frequency range and trill rate.

kHz

Time

The frequency range is the value between high and low frequencies. The trill rate is the number of repeats per unit time.

5 **THE DATA**

The data for the Galápagos finches were compared to a large body of data that had been collected on many other bird species. The relative constraint on vocal performance is higher if a bird has a more narrow frequency range and/or a slower trill rate.

Relative vocal constraint

G. magnirostris
C. pallidus G. scandens
G. fuliginosa G. fortis
C. parvulus C. psittacula
C. olivacea

Beak depth (mm)

Hybridization Can Occur When Reproductive Isolation Between Two Populations Is Incomplete

Speciation sometimes occurs when members of a species are only partially separated or when a species lives in one small area or moves about very little. In these cases, the geographic separation is not complete. For example, a mountain range may divide a species into two populations, but there may be breaks in the range where the two groups are connected physically. In these zones of contact, the members of two populations can interbreed, although this tends to occur infrequently. Likewise, speciation may occur among species that live in one small area even though no large geographic isolation exists. Certain organisms move about so little that 100–1,000 meters may be sufficient to limit interbreeding between neighboring groups. Plants, terrestrial snails, rodents, grasshoppers, lizards, and many flightless insects may speciate in this manner.

Prior to complete reproductive isolation, the zones where two populations can interbreed are known as **hybrid zones**. **Figure 25.11** shows a hybrid zone along a mountain pass that connects two deer populations. For speciation to occur, the amount of gene flow within hybrid zones must become very limited (refer back to Figure 24.13). But how does this happen? As the two populations accumulate different genetic changes, this may decrease the ability of individuals from different populations to mate with each other in the hybrid zone. For example, natural selection in the western deer population may favor an increase in body size that is not favored in the eastern population. Over time, as this size difference between members of the two populations becomes greater, breeding in the hybrid zone may decrease. Larger individuals may not interbreed easily with smaller individuals due to mechanical isolation. Alternatively, larger individuals may prefer larger individuals as mates, while smaller individuals may also prefer each other. Once gene

Figure 25.11   **A hypothetical hybrid zone.**   A narrow mountain pass allows members of two deer populations to occasionally interbreed with each other, creating hybrids between the two populations.

flow through the hybrid zone is greatly diminished, the two populations are reproductively isolated. Over the course of many generations, such populations may evolve into distinct species.

Sympatric Speciation Occurs When Populations in Direct Contact Become Reproductively Isolated from Each Other

Thus far, we have considered how geographic isolation can result in allopatric speciation. In the early stages of this process, reproductive isolation occurs because geographic separation prevents interbreeding. After long separation periods, reproductive isolation may also involve the accumulation of genetic changes that directly affect reproduction, such as changes in breeding season or mating behavior. By comparison, **sympatric speciation** (from the Greek *sym*, together) occurs when members of a species that initially occupy the same habitat within the same range diverge into two or more different species. Sympatric speciation tends to involve abrupt genetic changes that quickly lead to the reproductive isolation of a group of individuals. Let's consider a few examples in which such sudden changes can quickly lead to the formation of a new species within the same geographic area as the original species.

One abrupt genetic change that may occur is a change in chromosome number. Plants tend to be more tolerant of changes in chromosome number than are animals, as discussed in Chapter 15. In particular, alterations in the number of sets of chromosomes, such as polyploidy, often produce healthy plants.

Such changes, which are relatively common in plants on an evolutionary timescale, can result in sympatric speciation. Polyploidy is so frequent in plants that it is a major mechanism of their speciation. In ferns and flowering plants, about 30–50% of the species are polyploid. By comparison, polyploidy is much less common in animals, but it can occur. For example, roughly 30 species of reptiles and amphibians have been identified that are polyploids derived from diploid ancestors.

As described in Chapter 15, complete nondisjunction of chromosomes during gamete formation can increase the number of chromosome sets in the same species (autopolyploidy) or in hybrids of different species (allopolyploidy). The formation of a polyploid can abruptly lead to reproductive isolation. As an example, let's consider the origin of a natural species of a plant called the common hemp nettle, *Galeopsis tetrahit*. This species is thought to be an allotetraploid derived from two diploid species, *Galeopsis pubescens* and *Galeopsis speciosa* (**Figure 25.12**). These two diploid species contain 16 chromosomes each ($2n = 16$), while *G. tetrahit* contains 32. Its origin is not known, but one possibility is that an interspecies mating between *G. pubescens* and *G. speciosa* produced an allodiploid with 16 chromosomes (one set from each species) and then the allodiploid underwent complete nondisjunction to become an allotetraploid carrying four sets of chromosomes—two from each species.

The allotetraploid is fertile, because all of its chromosomes occur in homologous pairs that can segregate evenly during meiosis. However, a cross between an allotetraploid and a diploid produces an offspring that is monoploid for one chromosome set and diploid for the other set. The chromosomes of the monoploid set cannot be evenly segregated during meiosis. These offspring are expected to be sterile, because they will produce gametes that have incomplete sets of chromosomes. This hybrid sterility, which is an example of a postzygotic isolating mechanism, causes the allotetraploid to be reproductively isolated from both diploid species. Therefore, this process could have led to the formation of a new species, *G. tetrahit*, by sympatric speciation.

Sympatric speciation may also occur when genetic changes enable members of a species to occupy a new niche within the same geographic range as the original species. An example in which this process may be occurring involves colonial bentgrass (*Agrostis tenuis*), which was discussed in Chapter 24. In certain locations such as South Wales, isolated places are found where the soil is contaminated with high levels of heavy metals such as copper. On these sites, natural selection has promoted the proliferation of plants carrying alleles that confer resistance to the heavy metals. These alleles enable *A. tenuis* to grow on contaminated soil but tend to inhibit growth on normal, noncontaminated soil. Metal-resistant populations are continuous with populations that grow on uncontaminated land and do not show metal tolerance (refer back to Figure 24.5). In recent years, the metal-tolerant plants are starting to show a change in their flowering season. Over time, if this process continues, the metal-tolerant population may evolve into a new species that cannot interbreed with the original (metal-sensitive) species.

Figure 25.12 **Polyploidy and sympatric speciation.** This example shows how polyploidy may have caused reproductive isolation between three natural species of hemp nettle. Research suggests that the diploid species *Galeopsis pubescens* and *Galeopsis speciosa* produced the allotetraploid *Galeopsis tetrahit*. If *G. tetrahit* is mated with either of the other two species, the resulting offspring would be monoploid for one chromosome set and diploid for the other set, making them sterile. Therefore, *G. tetrahit* is reproductively isolated from the diploid species, making it a new species.

Biological inquiry: Suppose that G. pubescens *was crossed to* G. tetrahit *to produce an interspecies hybrid as shown at the bottom of this figure. If this interspecies hybrid was crossed to* G. tetrahit, *how many chromosomes do you think an offspring would have? The answer you give should be a range, not a single number.*

25.3 The Pace of Speciation

Throughout the history of life on Earth, the rate of evolutionary change and speciation has not been constant, although the degree of inconstancy has been debated since the time of Darwin. Even Darwin himself suggested that evolution can be fast or slow. **Figure 25.13** illustrates contrasting views concerning the rate of evolutionary change. These ideas are not mutually exclusive but represent two different ways to consider the tempo of evolution. The concept of **gradualism** suggests that each new species evolves continuously over long spans of time (Figure 25.13a). The principal idea is that large phenotypic differences that produce new species are due to the accumulation of many small genetic changes. By comparison, the concept of **punctuated equilibrium**, advocated in the 1970s by American paleontologists and evolutionary biologists Niles Eldredge and Stephen Jay Gould, suggests that the tempo of evolution is more sporadic (Figure 25.13b). According to this idea, species exist

relatively unchanged for many generations. During this period, the species is in equilibrium with its environment. These long periods of equilibrium are punctuated by relatively short periods (that is, on a geological timescale) during which evolution occurs at a far more rapid rate, perhaps due to environmental changes via the founder effect or new mutations that are quickly acted upon by natural selection.

In reality, neither of the views presented in Figure 25.13 fully accounts for evolutionary change. The occurrence of punctuated equilibrium is often supported by the fossil record. Paleontologists rarely find a gradual transition of fossil forms. Instead, new species seem to arise rather suddenly in a layer of rocks, persist relatively unchanged for a very long period of time, and then become extinct. Scientists think that the transition period during which a previous species evolved into a new species was so short that few, if any, of the transitional members were preserved as fossils. Even so, these rapid periods of change were probably followed by long periods of equilibrium that likely

(a) Graphism

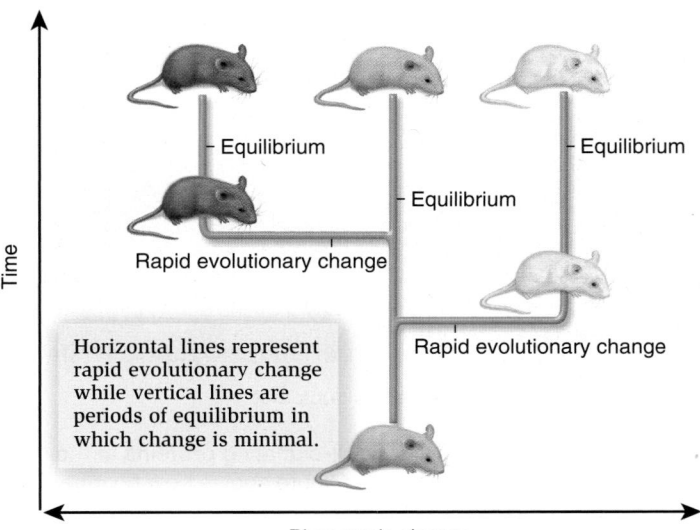

Change occurs gradually over a long time period.

Phenotypic change

(a) Gradualism

Equilibrium

Equilibrium

Equilibrium

Rapid evolutionary change

Rapid evolutionary change

Horizontal lines represent rapid evolutionary change while vertical lines are periods of equilibrium in which change is minimal.

Phenotypic change

(b) Punctuated equilibrium

Figure 25.13 **A comparison of gradualism and punctuated equilibrium.** (a) During gradualism, the phenotypic characteristics of a species gradually change due to the accumulation of small genetic changes. (b) During punctuated equilibrium, species exist essentially unchanged for long periods of time, during which they are in equilibrium with their environment. These equilibrium periods are punctuated by relatively short periods of evolutionary change during which phenotypic characteristics may change rapidly.

involved the additional accumulation of many small genetic changes, consistent with gradualism.

As discussed earlier, rapid evolutionary change can be explained by genetic phenomena. Single gene mutations can have dramatic effects on phenotypic characteristics. Therefore, only a small number of new mutations may be required to alter phe-

notypic characteristics, eventually producing a group of individuals that makes up a new species. Likewise, events such as changes in chromosome number and alloploidy may abruptly create individuals with new phenotypic traits. On an evolutionary timescale, these types of events can be rather rapid, because one or only a few genetic changes can have a major impact on the phenotype of the organism.

In conjunction with genetic changes, species may also be subjected to sudden environmental shifts that quickly drive the gene pool in a particular direction via natural selection. For example, a small group may migrate to a new environment in which different alleles provide better adaptation to the surroundings. Alternatively, a species may be subjected to a relatively sudden environmental event that has a major impact on survival. For example, the climate may change or a new predator may infiltrate the geographic range of the species. Natural selection may lead to a rapid evolution of the gene pool by favoring those genetic changes that allow members of the population to survive the climatic change or to have phenotypic characteristics that allow them to avoid the predator.

Finally, another issue associated with the speed of evolution is generation time. Species of large animals with long generation times tend to evolve much more slowly than do microbes with short generations. Many new species of bacteria will come into existence during our lifetime, while new species of large animals tend to arise on a much longer timescale. This is an important consideration for people because bacteria have great environmental impact; they are decomposers of organic materials and pollutants in the environment, and they also play a role in many diseases of plants, animals, and humans.

25.4 Evo-Devo: Evolutionary Developmental Biology and the Form of New Species

As we have learned in the preceding sections of this chapter, the origin of new species involves genetic changes that lead to adaptations to environmental niches and/or to reproductive isolating mechanisms that prevent closely related species from interbreeding. These genetic changes result in morphological and physiological differences that distinguish one species from another. In recent years, many evolutionary biologists have begun to investigate how genetic variation produces species and groups of species with novel shapes and forms. The underlying reasons for such changes are often rooted in the developmental pathways that control an organism's morphology. **Evolutionary developmental biology** (referred to as **evo-devo**) is an exciting and relatively new field of biology that compares the development of different organisms in an attempt to understand ancestral relationships between organisms and the developmental mechanisms that bring about evolutionary change. During the past few decades, developmental geneticists have gained a better understanding of biological development at the molecular level.

Much of this work has involved the discovery of genes that control development in experimental organisms. As more and more organisms have been analyzed, researchers have become interested in the similarities and differences that occur between closely related and distantly related species. Evolutionary developmental biology has arisen in response to this trend.

A central question in evo-devo studies is, How do new morphological forms come into being? For example, how does a nonwebbed foot evolve into a webbed foot? Or how does a new organ, such as an eye, come into existence? As we will learn, such novelty arises through genetic changes, also called genetic innovations. Certain types of genetic innovation have been so advantageous that they have resulted in groups of many new species. For example, the innovation of feathered wings resulted in the evolution of many different species of birds. In this section, we will learn that proteins that control developmental changes, such as cell-signaling proteins and transcription factors, often play a key role in promoting the morphological changes that occur during evolution.

The Spatial Expression of Genes That Affect Development Can Have a Dramatic Effect on Phenotype

In Chapter 19, we considered the role of genetics in the development of multicellular organisms. As we learned, genes that play a role in development may influence cell division and growth, cell differentiation, cell migration, and cell death. The interplay among these four processes creates an organism with a specific body form and function. As you might imagine, developmental genes are very important to the phenotypes of individuals. They affect traits such as the shape of a bird's beak, the length of a giraffe's neck, and the size of a plant's flower. In recent years, the study of development has indicated that developmental genes are key players in the evolution of many types of traits. Changes in such genes affect traits that can be acted on by natural selection. Furthermore, variation in the expression of such genes may be commonly involved in the acquisition of new traits that promote speciation.

As an example, let's compare the formation of a chicken's foot with that of a duck. Developmental biologists have discovered that the morphological differences between a webbed and a nonwebbed foot are due to the differential expression of two different cell-signaling proteins called bone morphogenetic protein 4 (BMP4) and gremlin. The *BMP4* gene is expressed throughout the developing limb; this is shown in **Figure 25.14a**, in which the BMP4 protein is stained blue. The BMP4 protein causes cells to undergo apoptosis and die. The gremlin protein, which is stained brown in **Figure 25.14b**, inhibits the function of BMP4 and thereby allows cells to survive. In the developing chicken limb, the *gremlin* gene is expressed throughout the limb, except in the regions between each digit. Therefore, these cells die, and a chicken develops a nonwebbed foot (**Figure 25.14c**). By comparison, in the duck, *gremlin* is expressed throughout the entire

Chicken · Duck

(a) BMP4 protein levels

Future interdigit regions

(b) Gremlin protein levels

(c) Comparison of a chicken and a duck foot

Figure 25.14   The role of cell-signaling proteins in the morphology of birds' feet.   This figure shows how changes in developmental gene expression can produce new traits.
(a) Expression of the *BMP4* gene in the developing limbs. BMP4 protein is stained blue here. **(b)** Expression of the *gremlin* gene in the developing limbs. Gremlin protein is stained brown here. Note that *gremlin* is expressed in the interdigit region only in the duck. Gremlin inhibits BMP4, which causes programmed cell death. **(c)** Because BMP4 is not inhibited in the interdigit regions in the chicken, the cells in this region die, and the foot is not webbed. By comparison, inhibition of BMP4 in the interdigit regions in the duck results in a webbed foot.

limb, including the interdigit regions, which results in a webbed foot. Interestingly, researchers have been able to introduce gremlin protein into the interdigit regions of developing chicken limbs. This produces a chicken with webbed feet!

During the evolution of birds, variation in the expression of these genes determined whether or not their feet were webbed. Mutations occurred that provided variation in the expression of the *BMP4* and *gremlin* genes. In terrestrial settings, having nonwebbed feet is an advantage because these are more effective at holding onto perches, running along the ground, and snatching prey. Therefore, natural selection would maintain nonwebbed feet in terrestrial environments. This process explains the occurrence of nonwebbed feet in chickens, hawks, crows, and many other terrestrial birds. In aquatic environments, webbed feet are an advantage because they act as paddles for swimming, so genetic variation that produced webbed feet would have been promoted by natural selection. Over the course of many generations, this gave rise to the webbed feet that are now found in ducks, geese, penguins, and other aquatic birds.

But how does having webbed or nonwebbed feet influence speciation? Perhaps this trait would not directly affect the ability of two individuals to mate. However, due to natural selection, birds with webbed feet would become more prevalent in aquatic environments, while birds with nonwebbed feet would be found in terrestrial locations. Therefore, reproductive isolation would occur because the populations would occupy different environments.

The *Hox* Genes Have Been Important in the Evolution of a Variety of Body Plans Found in Different Species of Animals

The study of developmental genes has revealed interesting trends among large groups of species. *Hox* genes, which are discussed in Chapter 19, are found in all animals. Developmental biologists have speculated that genetic variation in the *Hox* genes may have been a critical event that spawned the formation of many new body types, yielding many different species. As shown in **Figure 25.15**, the number and arrangement of *Hox* genes varies considerably among different types of animals. Sponges, the simplest of animals, have at least one *Hox* gene, whereas insects typically have nine or more. In most cases, multiple *Hox* genes occur in a cluster in which the genes are close to each other along a chromosome. In mammals, such *Hox* gene clusters have been duplicated twice during the course of evolution to form four clusters, all slightly different, with a total of 38 genes.

Figure 25.15   **Relationship between *Hox* gene number and body complexity in different types of animals.**   Researchers speculate that the duplication of *Hox* genes and *Hox* gene clusters played a key role in the evolution of more complex body plans in animals. A correlation seems to exist between increasing numbers of *Hox* genes and increasing complexity of body structure. The *Hox* genes are divided into four groups, called anterior, group 3, central, and posterior, based on their relative similarities. Each group is represented by a different color in this figure.

Researchers propose that increases in the number of *Hox* genes have been instrumental in the evolution of many animal species with greater complexity in body structure. To understand how, let's first consider *Hox* gene function. All *Hox* genes encode transcription factors that act as master control proteins to direct the formation of particular regions of the body. Each *Hox* gene controls a hierarchy of many genes that includes other control genes, which regulate the expression of target genes, as well as structural genes, which encode proteins that ultimately affect the morphology of the organism. The evolution of complex body plans is associated with an increase not only in the number of regulatory genes—as evidenced by the increase in *Hox* gene complexity during evolution—but also in structural genes, which ultimately affect an organism's form and function.

But how would an increase in *Hox* genes enable more complex body forms to evolve? Part of the answer lies in the spatial expression of the *Hox* genes. In fruit flies, for example, different *Hox* genes are expressed in different segments of the body along the anteroposterior axis (refer back to Figure 19.17). Therefore, an increase in the number of *Hox* genes allows each of these master control genes to become more specialized in the region that it controls. One segment in the middle of the fruit fly body can be controlled by a particular *Hox* gene and form wings and legs, while a segment in the head region can be controlled by a different *Hox* gene and develop antennae. Therefore, research suggests that one way for new, more complex body forms to evolve is by increasing the number of *Hox* genes, thereby making it possible to form many specialized parts of the body that are organized along a body axis.

Three lines of evidence support the idea that *Hox* gene complexity has been instrumental in the evolution and speciation of animals with different body patterns. First, as discussed in Chapter 19, *Hox* genes are known to control body development. Second, as described in Figure 25.15, a general trend is observed in which animals with a simpler body structure tend to have fewer *Hox* genes and fewer *Hox* clusters in their genomes than do the genomes of more complex animals. Third and finally, a comparison of *Hox* gene evolution and animal evolution bear striking parallels. Researchers can analyze *Hox* gene sequences among modern species and make estimates regarding the timing of past events. If the DNA sequences of homologous genes in two different species are fairly different, the modern genes evolved from an ancestral gene in the very distant past. In contrast, if two homologous genes are very similar, they arose from an ancestral gene that was more recent. Using this type of approach, geneticists can estimate when the first *Hox* gene arose by gene innovation; the date is difficult to pinpoint but is well over 600 million years ago. The single *Hox* gene found in the sponge has descended from this primordial *Hox* gene. In addition, gene duplications of this primordial gene produced clusters of *Hox* genes in other species. Clusters, such as those found in modern insects, were likely to be present approximately 600 million years ago. A duplication of that cluster is estimated to have occurred around 520 million years ago. Remarkably, these estimates of *Hox* gene origins correlate with major speciation events in the history of animals. As described in Chapter 22, the Cambrian explosion, which occurred from 533 to 525 million years ago, saw a phenomenal diversification in the body plan of invertebrate species. This diversification occurred after the *Hox* cluster was formed and was possibly undergoing its first duplication to create two *Hox* clusters. Also, approximately 420 million years ago, a second duplication produced species with four *Hox* clusters. This event precedes the proliferation of tetrapods—vertebrates with four limbs—that occurred during the Devonian period, approximately 417–354 million years ago. Modern tetrapods, such as mammals, have four *Hox* clusters. This second duplication may have been a critical event that led to the evolution of complex terrestrial vertebrates with four limbs, such as lizards, bears, and humans.

Developmental Genes That Affect Growth Rates Can Also Have a Dramatic Effect on Phenotype and Can Lead to the Formation of New Species

In our previous example in Figure 25.14, a difference in the spatial expression of the *gremlin* gene affected whether a bird has nonwebbed or webbed feet. In this case, the expression determined if apoptosis would occur in a defined region. Another way that genetic variation can influence morphology is by influencing the relative growth rates of different parts of the body. The pattern whereby different parts of the body grow at different rates with respect to each other is called **allometric growth**. As an example, **Figure 25.16a** compares the progressive growth of the head between a human and chimpanzee. At the fetal stage, the size and shape of the heads look fairly similar. However, after this stage, the relative growth rates of certain regions become markedly different, thereby affecting the shape and size of the head in the adult. In the chimpanzee, the jaw region grows faster, giving the chimpanzee a much larger and longer jaw. In the human, the jaw grows more slowly and the region of the skull that surrounds the brain grows slightly faster. Therefore, humans have smaller jaws but larger skulls.

Changes in growth rates can also affect the developmental stage at which one species reproduces compared to that of another species. This can occur in two ways. One possibility is that the parts of the body associated with reproduction could grow and mature faster than the rest of the body. Alternatively, reproduction could happen at the same absolute age, but the development of nonreproductive body parts could be slowed down in one species, causing it to reproduce at an earlier stage. In either case, the morphological result is the same—reproduction in the adult is observed at an earlier stage in one species than it is in another. The retention of juvenile traits in an adult organism is called **paedomorphosis** (from the Greek *paedo*, meaning "young" or "juvenile," and *morph*, relating to the form of an organism). It is particularly common among salamanders. Even though paedomorphic species have certain juvenile traits as adults, they still have the ability to reproduce successfully. For example, Cope's giant salamander (*Dicamptodon copei*) becomes mature and reproduces in the aquatic form, without changing into a terrestrial adult as do other salamander species (**Figure 25.16b**). The adult form of Cope's giant salamander has gills and a large paddle-shaped tail, features that resemble those of the tadpole stage of other salamander species.

Human	Chimpanzee
Fetus	
Infant	
Adult	

(a) The effects of allometric growth on head morphology in the human and chimpanzee

Figure 25.16 **Effects of growth rate on development.** **(a)** Allometric growth refers to the phenomenon in which one region of the body grows faster than another. Allometric growth explains why the heads of adult chimpanzees and humans have different shapes even though their fetal shapes are quite similar. **(b)** Paedomorphosis occurs when an adult species retains characteristics that are juvenile traits in another related species. Cope's giant salamander reproduces at the tadpole stage.

(b) Paedomorphosis in Cope's giant salamander

GENOMES & PROTEOMES

The Study of the *Pax6* Gene Indicates That Different Types of Eyes Evolved from a Simpler Form

Thus far in this section, we have focused on the roles of particular genes as they influence the development of the body. Explaining how a complex organ comes into existence is another major challenge for evolutionary biologists. While it is relatively easy to understand how a limb could undergo evolutionary modifications to become a wing, flipper, or arm, it is more difficult to understand how a body structure comes into being in the first place. In his book *The Origin of Species*, Charles Darwin addressed this question and admitted that the existence of complex organs was difficult to understand. His book had an entire chapter devoted to the question of eye development and evolution. As noted by Darwin, the eyes of vertebrate species are exceedingly complex, being able to adjust focus, let in different amounts of light, and detect a spectrum of colors. Darwin speculated that such complex eyes must have evolved from a simpler structure through the process of descent with modification. With amazing insight, he suggested that a very simple eye would be composed of two cell types, a photoreceptor cell and an adjacent pigment cell. The photoreceptor cell, which is a type

of nerve cell, is able to absorb light and respond to it. The function of the pigment cell is to stop the light from reaching one side of the photoreceptor cell. This primitive, two-cell arrangement would allow an organism to sense both light and the direction from which the light comes.

A primitive eye would provide an additional way for an organism to sense its environment, possibly allowing it to avoid predators or locate food. Vision is nearly universal among animals, which indicates that there must be a strong selective advantage to better eyesight. Over time, eyes could become more complex by enhancing the ability to absorb different amounts and wavelengths of light, and also by refinements in structures such as the addition of lenses that focus the incoming light.

Since the time of Darwin, many evolutionary biologists have wrestled with the question of eye evolution. From an anatomical point of view, researchers have discovered many different types of eyes. For example, the eyes of fruit flies, squid, and humans are quite different from each other. Furthermore, species that are closely related evolutionarily sometimes have different types of eyes. This observation led evolutionary biologists such as Luitfried von Salvini-Plawen and Ernst Mayr to propose that eyes may have independently arisen many different times during evolution. Based solely on morphology, such a hypothesis seemed reasonable and for many years was accepted by the scientific community.

(a) **Abnormal expression of *Drosophila eyeless* gene**

(b) **Abnormal expression of mouse *Pax6* gene in a fruit fly leg**

Figure 25.17 Formation of additional eyes in *Drosophila* due to the abnormal expression of a master control gene for eye morphogenesis. (a) When the *Drosophila eyeless* gene is expressed in the antenna region, eyes are formed where antennae should be located. (b) When the mouse *Pax6* gene is expressed in the leg region of *Drosophila*, a small eye is formed there.

Biological inquiry: What do you think would happen if a researcher used genetic engineering techniques to express the Drosophila eyeless *gene at the tip of a mouse's tail?*

The situation took a dramatic turn when geneticists began to study eye development. Researchers identified a master control gene, *Pax6*,* that controls the expression of many other genes and thereby influences eye development in both rodents and humans. In mice and rats, a mutation in this gene results in small eyes. A mutation in the human *Pax6* gene causes an eye disorder called aniridia, in which most of the eye is underdeveloped. Similarly, *Drosophila* has a homologous gene named *eyeless* that also causes a defect in eye development when mutant. The *eyeless* and *Pax6* genes are homologous; they were derived from the same ancestral gene.

In 1995, Swiss geneticist Walter Gehring and his colleagues were able to show experimentally that the abnormal expression of the *eyeless* gene in other parts of the fruit fly body could promote the formation of additional eyes. For example, using genetic engineering techniques, they were able to express the *eyeless* gene in the region where antennae should form. As seen in **Figure 25.17a**, this resulted in the formation of fruit fly eyes where antennae are normally found! Remarkably, the expression of the mouse *Pax6* gene in *Drosophila* can also cause the formation of eyes in unusual places. For example, **Figure 25.17b** shows the formation of an eye on the leg of a fruit fly.

Note that the mouse *Pax6* master control gene switches on eye formation in *Drosophila*, but the eye produced is a *Drosophila* eye, not a mouse eye. This happens because the genes that are activated by the *Pax6* master control gene are all from the *Drosophila* genome. In fact, in *Drosophila*, the *Pax6* homolog called *eyeless* switches on a cascade involving 2,500 different genes required for eye morphogenesis. In more primitive eyes, the *Pax6* gene would be expected to control a cascade of fewer genes.

Taken together, these results are consistent with two ideas. First, the *Pax6* gene and its *Drosophila* homolog are master control

*Pax is an acronym for p̲aired bo̲x̲. The protein encoded by this gene contains a domain called a paired box.

The ancestral *Pax6* gene controlled other genes that produced a primitive 2-celled eye.

Photoreceptor cell   Pigment cell   Simple eye

Ancestral *Pax6* gene

During evolution, species diverged from each other, but each species retained a *Pax6* homolog.

Drosophila eyeless gene   Mammal *Pax6* gene

Over time, gene duplications and other genetic changes produced many more genes that added to eye complexity. These additional genes remained under the control of the *Pax6* gene and its homologs.

Ommatidia

Retina   Cornea   Lens   Iris   Optic nerve

***Drosophila* eye**   **Mammal eye**

Figure 25.18 Genetic control of eye evolution. In this diagram, genetic changes, under control of the ancestral *Pax6* gene, led to the evolution of different types of eyes.

genes that promote the formation of an eye. The second idea is that the eyes of *Drosophila* and mammals are evolutionarily derived from the modification of an eye that arose once during evolution. If *Drosophila* and mammalian eyes had arisen independently, the *Pax6* gene from mice would not be expected to induce the formation of eyes in *Drosophila*.

Since the discovery of the *Pax6* and *eyeless* genes, homologs of this gene have been discovered in many different species. In all cases where it has been tested, this gene directs eye development. The *Pax6* gene and its homologs encode a transcription factor protein that controls the expression of many different genes. Gehring and colleagues have hypothesized that the eyes from many different species all evolved from a common ancestral form consisting of, as proposed by Darwin, one photoreceptor cell and one pigment cell (**Figure 25.18**). As mentioned, such a very simple eye can accomplish some rudimentary form of vision by detecting light and its direction. Eyes such as these are still found in modern species such as the larvae of certain types of mollusks. Over the course of evolution, simple eyes were transformed into more complex types of eyes by modifications that resulted in the addition of more types of cells such as lens cells and muscle cells.

CHAPTER SUMMARY

- A species is a group of organisms that maintains a distinctive set of attributes in nature. Speciation is the process whereby new species are formed. Macroevolution refers to the evolutionary changes that create new species and groups of species.

25.1 Species Concepts

- Four species concepts that are used to identify a species are phylogenetic, biological, evolutionary, and ecological. (Table 25.1)
- The phylogenetic species concept identifies species based on similarities in form, which sometimes can be misleading. (Figure 25.1)
- The biological species concept evaluates species based on their ability to interbreed.
- The evolutionary species concept considers the lineages that gave rise to species.
- The ecological species concept identifies species by their ability to occupy an ecological niche.
- By studying the biological species concept, researchers have identified many reproductive isolating mechanisms, both prezygotic and postzygotic, that prevent two different species from breeding with each other. (Figures 25.2, 25.3, 25.4, 25.5)

25.2 Mechanisms of Speciation

- Anagenesis is the evolution of one species into another species, while cladogenesis is the divergence of one species into two or more different species. (Figure 25.6)
- Allopatric speciation occurs when a population becomes geographically isolated from other populations and evolves into one or more new species. When speciation occurs multiple times due to the migration of populations into new environments, the evolutionary process is called adaptive radiation. (Figures 25.7, 25.8)
- Podos discovered that changes in beak depth, associated with evolutionary changes involved with feeding, may also promote reproductive isolation by altering the song pattern of finches. (Figures 25.9, 25.10)
- As allopatric speciation is occurring, limited reproduction between adjacent populations may take place in a hybrid zone, so that few interspecies hybrids are produced. (Figure 25.11)
- Sympatric speciation involves the formation of different species that are not geographically isolated from one another. (Figure 25.12)

25.3 The Pace of Speciation

- Gradualism refers to a pace of speciation that involves many small genetic changes, while punctuated equilibrium is a pattern of evolution in which new species arise rapidly, on an evolutionary timescale, and then remain unchanged for long periods of time. (Figure 25.13)

25.4 Evo-Devo: Evolutionary Developmental Biology and the Form of New Species

- Evolutionary developmental biology is concerned with changes in developmental patterns that have given rise to new species and groups of species. These changes often involve variation in the expression of transcription factors and cell-signaling proteins.

- The spatial expression of genes that affect development can change phenotypes dramatically, as shown by the expression of the *gremlin* and *BMP4* genes to form birds with webbed or nonwebbed feet. (Figure 25.14)
- The evolution and duplication of *Hox* genes played an important role in producing groups of animals with specific body plans that are organized along an anteroposterior axis. (Figure 25.15)
- Patterns that affect the relative growth rates of body parts are called allometric growth. Paedomorphosis occurs when an adult species retains characteristics that are juvenile traits in another related species. (Figure 25.16)
- The *Pax6* gene and its homolog in other species are master control genes that control eye development in animals. This result suggests that eyes arose only once during evolution. (Figures 25.17, 25.18)

TEST YOURSELF

1. Macroevolution refers to the evolutionary changes that
   a. occur in multicellular organisms.
   b. create new species and groups of species.
   c. occur over long periods of time.
   d. cause changes in allele frequencies.
   e. occur in large mammals.

2. The biological species concept classifies a species based on
   a. morphological characteristics.
   b. reproductive isolation.
   c. the niche the organism occupies in the environment.
   d. genetic relationships between an organism and its ancestor.
   e. both a and b.

3. Which of the following would be considered an example of a postzygotic isolating mechanism?
   a. incompatible genitalia
   b. different mating seasons
   c. incompatible gametes
   d. mountain range separating two populations
   e. fertilized egg fails to develop normally

4. Hybrid breakdown occurs when species hybrids
   a. do not develop past the early embryonic stages.
   b. have a reduced life span.
   c. are infertile.
   d. are fertile but produce offspring with reduced viability and fertility.
   e. produce offspring that only express the traits of one of the original species.

5. The evolution of one species into a single new species is
   a. hybridization.
   b. anagenesis.
   c. cladogenesis.
   d. adaptive radiation.
   e. microevolution.

6. Founder events may lead to rapid speciation because of
   a. differences in natural selection on the new population versus the original population.
   b. genetic differences due to genetic drift.
   c. enhanced gene flow between the new population and the original population.
   d. all of the above.
   e. a and b only.

7. A major mechanism of speciation in plants but not animals is
   a. anagenesis.
   b. polyploidy.
   c. hybrid breakdown.
   d. genetic changes that alter the organism's niche.
   e. both a and d.

8. The concept of punctuated equilibrium suggests that
   a. the rate of evolution is constant, with short time periods of no evolutionary change.
   b. evolution occurs gradually over time.
   c. small genetic changes accumulate over time to allow for phenotypic change and speciation.
   d. long periods of little evolutionary change are interrupted by short periods of major evolutionary change.
   e. both b and c.

9. Researchers suggest that an increase in the number of *Hox* genes
   a. would lead to reproductive isolation in all cases.
   b. could explain the evolution of color vision.
   c. allows for the evolution of more complex body forms.
   d. results in the decrease in the number of body segments in insects.
   e. all of the above.

10. The observation that the mammalian *Pax6* gene and the *Drosophila eyeless* gene are homologous genes that promote the formation of different types of eyes suggests that
    a. *Drosophila* eyes are more complex.
    b. mammalian eyes are more complex.
    c. eyes arose once during evolution.
    d. eyes arose at least twice during evolution.
    e. eye development is a simple process.

EXPERIMENTAL QUESTIONS

1. What did Podos hypothesize regarding the effects of beak size on a bird's song? How could changes in beak size and shape lead to reproductive isolation among the finches?

2. How did Podos test the hypothesis that beak morphology caused changes in the birds' songs?

3. Did the results of Podos's study support his original hypothesis? Explain. What is meant by the phrase "by-product of adaptation" and how does it apply to this particular study?

COLLABORATIVE QUESTIONS

1. Discuss how geographic isolation can lead to speciation.

2. Evolution can proceed at different rates. Discuss the two models that address the speed at which evolution occurs.

www.brookerbiology.com

This website includes answers to the Biological Inquiry questions found in the figure legends and all end-of-chapter questions.

CONCEPTUAL QUESTIONS

1. Define species.

2. Distinguish between prezygotic isolating mechanisms and postzygotic isolating mechanisms and give an example of each.

3. Define punctuated equilibrium.

26

TAXONOMY AND SYSTEMATICS

CHAPTER OUTLINE

26.1 Taxonomy
26.2 Systematics

A new species, the African forest elephant. In 2001, biologists discovered that this is a unique species of elephant, now named *Loxodonta cyclotis*. This is one example of how scientists have changed taxonomic groupings in response to recent discoveries in molecular systematics.

In Chapter 25, we considered the formation of new species and learned that species diversity is enormous. Biologists estimate that between 10 and 100 million species currently exist on our planet. To make some sense out of such amazing variation, and to be able to communicate with each other about it, biologists have categorized living organisms into groups of related species. **Taxonomy** (from the Greek *taxis*, order, and *nomos*, law) is the field of biology that is concerned with the theory, practice, and rules of classifying living and extinct organisms and viruses. Taxonomy results in the ordered division of species into groups based on similarities and dissimilarities in their characteristics. This task has been ongoing for over 300 years. As we discussed in Chapter 23, John Ray, who was the first scientist to carry out a thorough study of the natural world in the 1600s, developed an early classification system for plants and animals based on anatomy and physiology. Ray made the first attempt to broadly classify all known forms of life and therefore was a pioneer in the field of modern taxonomy. Ray's ideas regarding taxonomy were later extended by Carolus Linnaeus in the mid-1700s.

Systematics is the study of biological diversity and the evolutionary relationships among organisms, both extinct and modern. In 1950, the German entomologist Willi Hennig began classifying organisms in a new way. Hennig proposed that evolutionary relationships should be inferred from new features shared by descendants of a common ancestor. Since that time, biologists have

applied systematics to the field of taxonomy. In other words, taxonomic groups are now based on hypotheses regarding evolutionary relationships that are derived from systematics.

As scientists obtain additional information regarding genetic variation and evolutionary trends, taxonomic groupings have changed to accommodate the new and more accurate data. For example, until recently, biologists thought there were two species of elephant in the world—the African savanna elephant (*Loxodonta africana*) and the Asian elephant (*Elephas maximus*). However, by analyzing the DNA of African elephants, a third species, now called the African forest elephant (*Loxodonta cyclotis*), was discovered (see chapter-opening photo). This surprising finding was made somewhat by accident in 2001. A DNA identification system was set up to trace ivory poachers. By studying the DNA, researchers discovered that Africa has two different *Loxodonta* elephant species. The African forest elephant is found in the forests of central and western Africa. The African savanna elephant, which is larger and has longer tusks, lives on large, dry grasslands.

In this chapter, we will begin with a discussion of taxonomy from a traditional point of view. However, because recent information has dramatically changed our view of taxonomy, we will also consider newer ways to group organisms that are more logically based on evolutionary relationships. We will then examine how the analyses of morphological and molecular genetic data are used to place species into such groups.

26.1 Taxonomy

A hierarchy is a system of organization that involves successive levels. In biology, every species is placed into several different groups. Some groups are very large and encompass many species that have fundamental similarities and, at the same time, many differences. For example, a leopard and a fruit fly are both classified as animals, though they differ in many traits. Leopards and lions are placed into a smaller group called felines (more formally named Felidae), which are predatory cats. The felines are a subset of the animal group, which has species that share many similar traits. Different species that are found in small taxonomic groups are likely to share many of the same characteristics.

In this section, we will consider how biologists use a hierarchy to group similar species. We also survey the major groups of organisms and some of their common features. As you will learn later in this chapter, a few of our traditional ideas about taxonomy have been challenged as biologists have gathered new information regarding the genetic composition of modern species. For example, taxonomists now believe that fungi and animals are much more closely related than they once did.

Taxonomy Provides a Hierarchy of Groups of Organisms

Modern taxonomy places species into progressively smaller hierarchical groups. Each group is called a **taxon** (plural, *taxa*). The taxonomic group called the **kingdom** was originally the highest and most inclusive. Linnaeus classified all life into two kingdoms, plants and animals. In 1969, American ecologist Robert Whittaker proposed a five-kingdom system in which all life was classified into the kingdoms Monera, Protista, Fungi, Plantae, and Animalia. (Monera included all prokaryotes.) However, as biologists began to learn more about the evolutionary relationships among these groups, a subject we will discuss later in this chapter, they found that these groupings did not correctly reflect the relationships among them.

In the late 1970s, based on information in the sequences of genes, American biologist Carl Woese proposed the idea of creating a category called a **domain**. Under this system, all forms of life are grouped within three domains called **Bacteria**, **Archaea**, and **Eukarya** (also spelled Eucarya) (**Figure 26.1**). The original kingdom Monera was split into two domains, Bacteria and Archaea, due to major differences between these two types of prokaryotes. The terms Bacteria and Archaea are capitalized when referring to the domains. Alternatively, these terms can also refer to prokaryotic cells and species, in which case they are not capitalized. A single bacterial cell is called a bacterium, while a single archaeal cell is an archaeon. The domain Eukarya includes the traditional four eukaryotic kingdoms, **Protista**, **Fungi**, **Plantae**, and **Animalia**. However, as described later in this chapter, this method of grouping eukaryotic species has been revised.

Figure 26.1 **A classification system for living and extinct organisms.** All organisms can be grouped into three large domains, Bacteria, Archaea, or Eukarya. Eukaryotes have been traditionally divided into the four kingdoms Protista (protists), Fungi (fungi), Plantae (plants), and Animalia (animals). As discussed later, this traditional system has been revised.

The three domains of life contain millions of different species. Subdividing them into progressively smaller groups makes it easier for biologists to appreciate the relationships among living species. Proceeding down the hierarchical ladder, kingdoms are in turn subdivided into **phyla** (singular, *phylum*), each of which is divided into **classes**, then **orders**, **families**, and **genera** (singular, *genus*). Each of these taxa contains fewer species that are more similar to each other than they are to the members of the taxon above it in the hierarchy. For example, the taxon Animalia, which is at the kingdom level, has a larger number of fairly diverse species than does the class Mammalia, which contains species that are relatively similar to each other.

To further understand taxonomy, let's consider the classification of a species such as the gray wolf (*Canis lupus*) (**Figure 26.2**). The original range of *C. lupus* covered the majority of the Northern Hemisphere. Due primarily to habitat destruction, gray wolf populations are much less prevalent now, but they are still found in a few areas in North America and Eurasia. The gray wolf is placed in the domain Eukarya, and then within the kingdom Animalia, which includes the approximately 2 million species of all animals. Next, the gray wolf is classified in the phylum Chordata. The 50,000 species of animals in this group all have four common features at some stage of their development. These are a notochord (a cartilaginous rod that runs along the back of all chordates at some point in their life cycle), a tubular nerve or spinal cord located above the notochord, gill slits or arches, and a postanal tail. Examples of animals in the Chordata phylum include fishes, reptiles, and mammals.

The gray wolf is in the class Mammalia, which includes all 5,000 species of mammals. Two common features of animals in this group are hair and mammary glands. The mammary glands produce milk, which nourishes the young, and hair helps insulate the body to maintain a warm, constant body temperature. There are 26 orders of mammals; the order that includes the gray wolf is called Carnivora and has about 270 species. The gray wolf is next categorized in the family known as Canidae. This is

Taxonomic group	Gray wolf found in	Number of species	
Domain	Eukarya	~4–10 million	
Kingdom	Animalia	~2 million	
Phylum	Chordata	~50,000	
Class	Mammalia	~5,000	
Order	Carnivora	~270	
Family	Canidae	34	
Genus	*Canis*	7	
Species	*lupus*	1	

Figure 26.2 A taxonomic classification of the gray wolf (*Canis lupus*).

a relatively small family of 34 species including different types of wolves, jackals, foxes, wild dogs, and the coyote and domestic dog. All species in the family Canidae are doglike animals. The smallest grouping that contains the gray wolf is its genus, *Canis*, which includes only two types of wolves, four types of jackals, and the coyote.

As originally advocated by Linnaeus, **binomial nomenclature** is the standard method for naming species. The scientific name of every species has two Latinized names, which are its genus name and its unique species epithet. The genus name is always capitalized, while the species epithet is not. Both names are italicized or underlined. After the first mention, the genus name is abbreviated to a single letter. For example, we would write that *Canis lupus* is the gray wolf, and in subsequent sentences the species would be referred to as *C. lupus*.

The rules for naming animal species, such as *Canis lupus*, have been established by the International Commission on Zoological Nomenclature (ICZN). The ICZN provides and regulates a uniform system of nomenclature to ensure that every animal has a unique and universally accepted scientific name. ICZN publishes the International Code of Zoological Nomenclature containing the rules accepted as governing the application of scientific names to all animals. The ICZN also provides rulings on individual nomenclatural problems brought to its attention. As long as ICZN rules are followed, new species can be named by anyone, not only by scientists. When someone believes he or she has identified a new species, that person generally does research through various publications (journals, books, etc.) to determine if it is a new species. In addition, the person needs to provide specimens to a museum where they will be preserved so someone else can verify the description or do research using the original specimens. When naming a new species, genus names are always nouns or treated as nouns, while species epithets may be either nouns or adjectives. The names often have a Latin or Greek origin and refer to characteristics of the species or to features of its habitat. For example, the genus name of the newly discovered African forest elephant, *Loxodonta*, is from Greek *loxo*, meaning slanting, and *odonta*, meaning tooth. The species epithet *cyclotis* refers to the observation that the ears of this species are rounder compared to those of *L. africana*. However, sometimes the naming of a species is not taken very seriously. For example, there is a beetle named *Agra vation* and a spider named *Draculoides bramstokeri*. International organizations are also involved in the nomenclature of species other than animals. The rules for naming plants and fungi are governed by the International Association of Plant Taxonomy (IAPT), and the naming of prokaryotes is overseen by the International Committee on Systematics of Prokaryotes (ICSP).

Organisms of the Three Domains Have Basic Similarities Yet Distinctive Differences

As discussed in Chapter 22, scientists think that all life originated from primordial prokaryotic cells, sometime between 4.0 and 3.5 billion years ago. Soon thereafter, the two prokaryotic domains, Bacteria and Archaea, diverged from each other.

Between 2.5 and 2.0 billion years ago, the first unicellular eukaryotic species came into being via an endosymbiotic or symbiotic relationship between archaea and bacteria. Eventually, multicellular eukaryotic species arose approximately 1.5 billion years ago.

Due to the evolutionary relatedness of Bacteria, Archaea, and Eukarya, the three domains of life share striking similarities:

- DNA is used as the genetic material.
- All species use the same genetic code (with only a few rare codon exceptions).
- Messenger RNA encodes the information to produce proteins.
- Transfer RNA and ribosomes are needed to synthesize proteins, using mRNA as a source of genetic information.
- All living cells are surrounded by a plasma membrane.
- Certain metabolic pathways, such as glycolysis, are found in all three domains.

Biologists believe that these traits are universal because all three domains evolved from a common prokaryotic ancestor. However, when we compare the domains of life in detail, we find that some characteristics are not shared by all three. For example, the cytoplasm of eukaryotic cells is compartmentalized into various types of organelles, while those of bacterial and archaeal cells are not. Such dissimilarities exist because major evolutionary changes have occurred since the time that the three domains diverged from each other. **Table 26.1** compares a variety of molecular and cellular characteristics among Bacteria, Archaea, and Eukarya.

On modern Earth, the domain Bacteria is a diverse collection of many species that can live almost anywhere (**Figure 26.3a**). Bacteria are so widespread that we can make only general statements about their ecology. They are found on the tops of mountains, at the bottoms of oceans, in the guts of animals, and even in the frozen rocks of Antarctica. More bacteria are found in a person's digestive tract than there are people in the world! The human mouth is home to more than 500 different species of bacteria. A key reason why bacteria have been so successful in occupying various habitats is their metabolic diversity. Different bacterial species have evolved an amazing variety of ways to obtain energy and organic molecules from their environment. Most species of bacteria decompose existing organic materials, though some can carry out photosynthesis or break down inorganic molecules to create their own organic molecules. Bacterial species can survive on many different sources of energy. Many bacteria metabolize common organic molecules such as sugars and amino acids. However, some species can use organic molecules that are less common, and are not usually broken down by archaea or eukaryotes. These include compounds found in gasoline, crude oil, pesticides, and industrial solvents!

Bacterial species come in a myriad of shapes and sizes, but individual bacterial cells are usually quite small, in the range of 1 to 5 microns long. Under the microscope, bacterial cells and archaeal cells do not possess morphological differences that

Table 26.1	Distinguishing Cellular and Molecular Features of Domains Bacteria, Archaea, and Eukarya*		
Characteristic	**Bacteria**	**Archaea**	**Eukarya**
Chromosomes	Usually circular	Circular	Usually linear
Nucleosome structure	No	No	Yes
Chromosome segregation	Fission	Fission	Mitosis/ meiosis
Introns in genes	Rarely	Rarely	Commonly
Ribosomes	70S	70S	80S
Initiator tRNA	Formylmethionine	Methionine	Methionine
Operons	Yes	Yes	No
Capping of mRNA	No	No	Yes
RNA polymerases	One	Several	Three
Promoters of structural genes	–35 and –10 sequences	TATA box	TATA box
Cell compartmentalization	No	No	Yes
Membrane lipids	Ester-linked	Ether-linked	Ester-linked

* The descriptions in this table are meant to represent the general features of most species in each domain. Some exceptions are observed. For example, certain bacterial species have linear chromosomes.

would allow them to be distinguished. This is one reason why it took so long for researchers to realize that bacteria and archaea should be placed in different domains. Nevertheless, as described in Table 26.1, certain of their molecular features are different.

The domain Archaea is less diverse than Bacteria. However, this domain was discovered relatively recently, in the 1970s, and the list of archaeal species continues to expand. Thus far, many archaeal species have been found to live in extreme environments (**Figure 26.3b**). The name archaea, meaning ancient, refers to the observation that archaea tend to occupy environments that are thought to be similar to those found on the ancient Earth. Thus far, most archaea can be placed into one of three categories: extreme halophiles, methanogens, or hyperthermophiles. Extreme halophiles live in salty environments. Some species can live in water with salt concentrations above 15% (seawater averages 3.5% salinity). Methanogens release methane as a waste product of cellular metabolism. These species live at the bottom of lakes and swamps and in the intestinal tracts of animals. Methanogens living in your digestive tract cause intestinal gas. Hyperthermophiles live in extremely hot water, even temperatures over 100°C (212°F). Examples include archaea that live in hot springs and around deep-sea thermal vents. More recently, biologists have come to realize that archaea are not entirely restricted to extreme environments. For example, newer research has shown they are abundant in the open sea.

Bacillus thuringiensis, a soil bacterium (SEM) 2 μm

Escherichia coli, an intestinal bacterium (SEM) 3 μm

Anabaena torulosa, a freshwater bacterium (LM) 71 μm
(a)

Methanobrevibacter smithii, an intestinal archaeon 1 μm

Hot springs
(b)

Thermus aquaticus, an archaeon found in hot springs (SEM) 6 μm

Figure 26.3   **A comparison of selected species of bacteria and archaea.**   **(a)** Bacteria are found in nearly all environments, such as lakes (*Anabaena torulosa*), soil (*Bacillus thuringiensis*), and even within the digestive tracts of animals (*Escherichia coli*). Archaea are also found within the digestive tracts of animals (*Methanobrevibacter smithii*) and **(b)** often occupy extreme environments such as hot springs (*Thermus aquaticus*).

Though less diverse than Bacteria, Archaea is nevertheless a diverse domain of successful species.

The third domain, Eukarya, is so diverse that it has been traditionally divided into four kingdoms, based on a set of characteristics that is unique to each kingdom. We will consider these kingdoms next.

Table 26.2	A Comparison of the Four Traditional Kingdoms of Domain Eukarya			
Characteristic	**Protista**	**Fungi**	**Plantae**	**Animalia**
Multicellular	A few species	Most species	All species	All species
Cell wall	Some species, various types	Yes, chitin	Yes, cellulose	No
Multiple cell and tissue types	Some species	Some species	Yes	Yes
Capable of photosynthesis	Some species	No	Nearly all species	No
Use of organic molecules in the environment	Some species	All species, digestion begins externally	Rarely	All species, digestion is usually within a digestive tract

The Organisms of Each Traditional Eukaryotic Kingdom Have a Distinctive Set of Characteristics

The four traditional kingdoms of the Eukarya domain, Protista, Fungi, Plantae, and Animalia, provide a broad basis for biologists to appreciate the diversity of eukaryotic species. The members of each kingdom possess certain characteristics that tend to set them apart from the other kingdoms. **Table 26.2** compares some of the common features that are found among species of these four kingdoms.

Protists are species within the traditional kingdom Protista, which contains members that are considered to be the simplest eukaryotes (**Figure 26.4**). Most species are unicellular, but some are colonial, and others are simple multicellular organisms closely related to unicellular protist species. Certain protists, such as algae, are capable of photosynthesis, while the remaining species eat bacterial or other protistan cells, or organic material suspended in water. Protists live in aquatic habitats, and they represent a critical step in the early evolution of eukaryotes. They evolved from prokaryotes and eventually gave rise to the other three kingdoms of eukaryotes.

From the perspective of taxonomy, protists are an unusually diverse group of organisms that were put together because they did not seem to belong to any other group. In some ways, the kingdom Protista was a place for leftover eukaryotic species that could not be classified into any of the other three eukaryotic kingdoms. Over 250,000 species of protists are estimated to currently exist. Many biologists are studying the evolutionary relationships among the protists, and the categorization of protists into a single kingdom has been revised. Current models are discussed later in this chapter and in Chapter 28.

The kingdom Fungi contains mushroom-forming fungi, molds, and yeasts (**Figure 26.5**). Species of fungi are present all over the world, in aquatic as well as terrestrial environments. Scientists estimate their diversity at over 100,000 species. Many fungi have symbiotic relationships with plants. These organisms

(a) An amoeba (*Entamoeba histolytica*)

(b) A paramecium (*Paramecium caudatum*) (SEM)

(c) A green alga (*Acrosiphonia coalita*)

Figure 26.4   Examples of protists.

(a) A species that forms mushrooms, golden chanterelle (*Cantharellus cabaruis*)

(b) A common bread mold (*Neurospora crassa*)

(c) Baker's yeast (*Saccharomyces cerevisiae*) (SEM)

Figure 26.5   Examples of fungi.

typically secrete digestive enzymes into their environment and then transport the smaller breakdown products into their cytoplasm. Along with bacteria, fungi play an important role in the decomposition and recycling of organic materials on our planet. A distinguishing cellular feature is that fungal cells are surrounded by a cell wall that is made of the polysaccharide chitin. Though a few species such as yeast are unicellular, most fungi are multicellular, consisting of a mass of threadlike filaments called hyphae that combine to make up the fungal body called the mycelium. The mycelium usually grows underground. For many fungal species, the familiar mushrooms that we can see aboveground are actually the reproductive structures of the fungus. Fungi are discussed in more detail in Chapter 29.

The kingdom Plantae is composed of multicellular species that are nearly all capable of photosynthesis. These include mosses, ferns, conifers, and flowering plants (**Figure 26.6**). Approximately 300,000 species are thought to currently exist. Though many species are found in aquatic environments, plants are the foundation of all terrestrial habitats. Rather than using organic molecules from their environment, nearly all plants produce their own organic molecules via photosynthesis. They are therefore the Earth's primary producers of the organic molecules that sustain both themselves and most other living species. (Certain species of bacteria and algae are also capable of photosynthesis.) At the cellular level, a distinguishing feature of plant cells is a cell wall that is made primarily of the carbohydrate cellulose. Chapters 30 and 31 provide an introduction to plants, which are covered in more detail in Unit VI.

All species of the kingdom Animalia are multicellular and rely on other organisms for their nourishment. It is a particularly diverse kingdom composed of more than 1 million species. These include sponges, worms, insects, mollusks, fish, amphibians, reptiles, birds, and mammals (**Figure 26.7**). Most ingest food and digest it in an internal cavity. Except for sponges, the bodies of animals are composed of cells organized into several types of tissues; each tissue is composed of cells that are specialized to perform a particular function. Most animals are capable of complex and relatively rapid movement compared to other organisms.

(a) Juniper moss (*Polytrichum juniperinum*)

(b) Royal fern (*Osmunda regalis*)

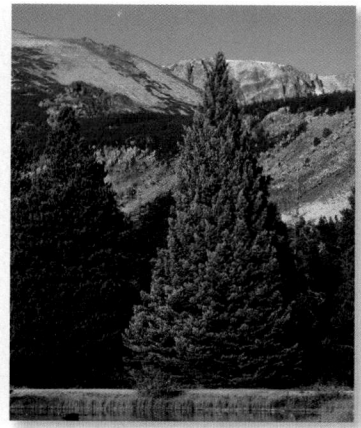

(c) A conifer, the Colorado blue spruce (*Picea pungens*)

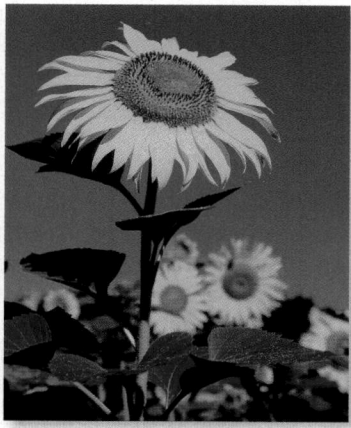

(d) A flowering plant, the common sunflower (*Helianthus annuus*)

Figure 26.6 Examples of plants.

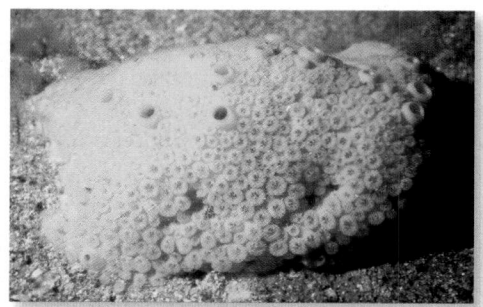

(a) Yellow sponge (*Cleona celata*)

(b) Firefly (*Photinus granulatus*)

(c) Zebra mussel (*Dreissena polymorpha*)

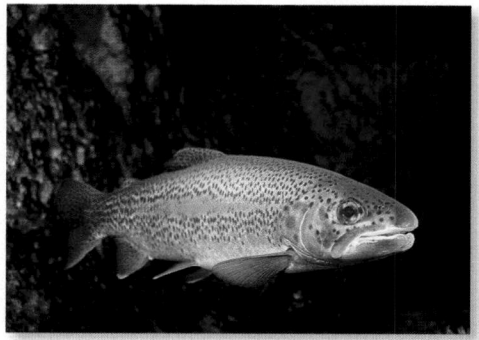

(d) Rainbow trout (*Oncorhynchus mykiss*)

(e) The lesser flamingo (*Phoeniconaias minor*)

(f) Red squirrel (*Tamiasciurus hudsonicus*)

Figure 26.7 Examples of animals.

A nervous system enables animals to receive environmental stimuli and respond with specialized movements. At the cellular level, a distinguishing feature is that animal cells lack a rigid cell wall such as that surrounding plant, fungal, and bacterial cells. Animal diversity is discussed in Chapters 32–34, and animal form and function are described in more detail in Unit VII.

Our traditional view of four eukaryotic kingdoms has recently changed based on a clearer understanding of the evolutionary relationships between the organisms involved. For example, protists are no longer considered to make up a single kingdom. Later in this chapter, we will examine the newer groupings of eukaryotic phyla. In the next section, we will see how the information generated by the study of systematics has been instrumental in reshaping the classification of organisms.

26.2 Systematics

Systematics is the study of biological diversity and evolutionary relationships. Biologists use systematics to develop the methods to construct taxonomic groups. For example, the classification of the gray wolf described earlier in Figure 26.2 is based on systematics. Therefore, taxonomy is one part of systematics. By studying the similarities and differences among species, biologists can gain information about **phylogeny**, which is the evolutionary history of a species or group of species. As you will learn, phylogeny helps us understand the relationships between ancestors and their descendants.

In this section, we will first consider how evolutionary biologists construct diagrams or trees that describe the evolutionary relationships among various species, both living and extinct. Biologists typically gather morphological or molecular genetic data, and then they use different mathematical strategies to analyze those data and construct evolutionary trees. Newer approaches using molecular genetic data have revolutionized our understanding of evolutionary relationships, causing biologists to revise the broad taxonomic groupings that have existed for the past few decades.

A Phylogenetic Tree Depicts the Evolutionary Relationships Between Past and Present Species

A systematic approach is followed to produce a **phylogenetic tree**—a diagram that describes a phylogeny. Such a tree is a hypothesis of the evolutionary relationships among various species, based on the information available to and gathered by systematists. Let's look at what information a phylogenetic tree contains and the form in which it is presented.

Figure 26.8 shows a hypothetical phylogenetic tree of the relationships between various bird species, in which the species are labeled A through K. The vertical axis represents time, with the oldest species at the bottom. As discussed in Chapter 25 (refer back to Figure 25.6), new species can be formed by anagenesis, in which a single species evolves into a different species, or more commonly by cladogenesis, in which a species diverges into two or more species. The nodes or branch points in a phylogenetic tree illustrate times when cladogenesis has occurred. For example, approximately 12 million years ago, species A diverged into species A and species B. Figure 26.8 also shows anagenesis. After species B split into species B and D, species D then evolved into species G by anagenesis. The tips of branches represent species that became extinct in the past, such as species B, C, and E, or modern species, such as F, I, G, J, H, and K, which are at the top of the tree. Species A and D are also extinct but gave rise to modern species.

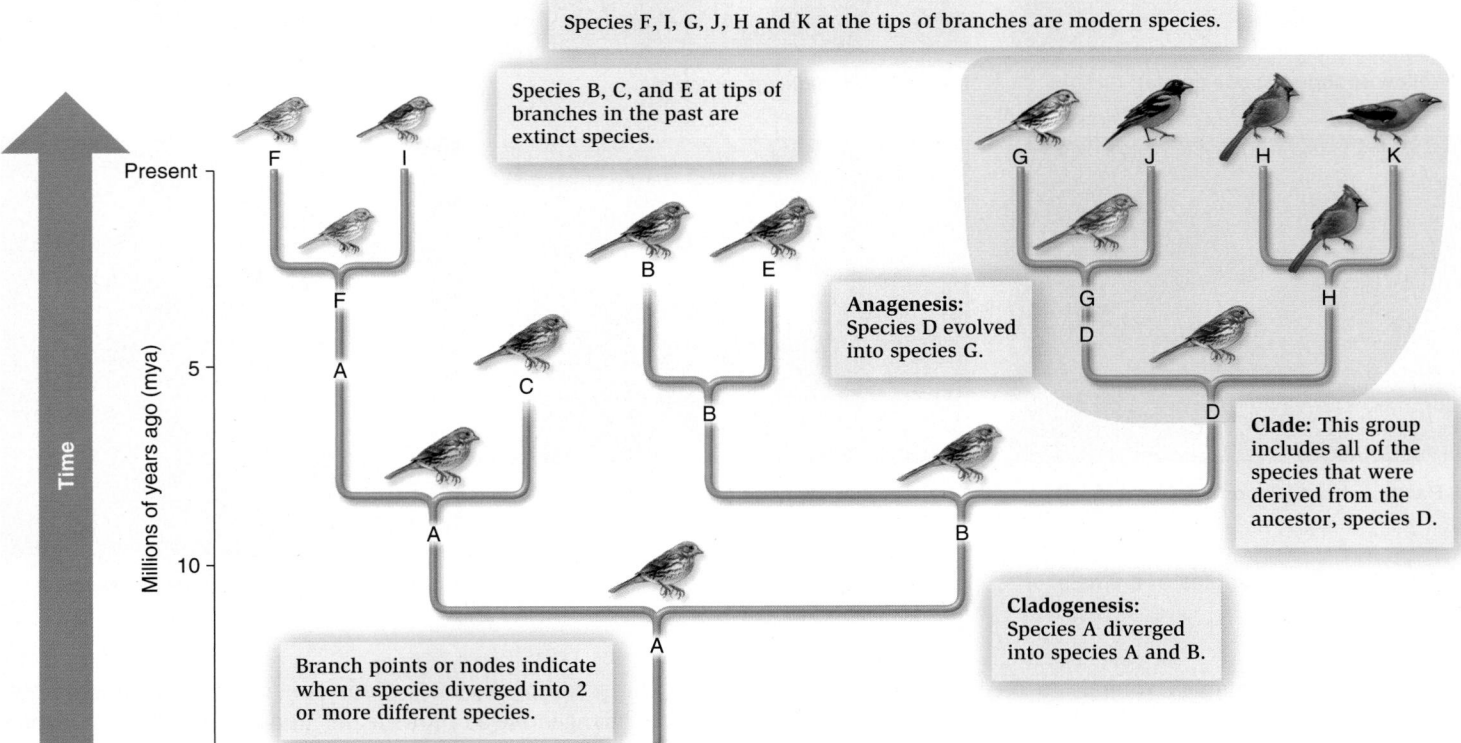

Figure 26.8 **How to read a phylogenetic tree.**   This hypothetical tree shows the proposed relationships between various bird species. Species are grouped into clades derived from a common ancestor.

Biological inquiry: Can two different species have more than one common ancestor?

By studying the branch points of a phylogenetic tree, researchers can group species according to common ancestry. A **monophyletic group**, also known as a **clade**, is a group of species, a taxon, consisting of the most recent common ancestor and all of its descendants. For example, the group highlighted in light green in Figure 26.8 is a clade derived from the common ancestor labeled D. The present-day descendants of a common ancestor can also be called a clade. In this case, species G, J, H, and K would form a modern clade. Likewise, the entire tree forms a clade, with species A as a common ancestor. Thus, smaller and more recent clades are subsets of larger clades.

The relationship between a clade and taxonomy depends on how far back we go to choose a common ancestor. For broader taxa, such as a kingdom, the common ancestor existed a very long time ago, on the order of hundreds of millions or even billions of years ago. For smaller taxa, such as a family or genus, the common ancestor occurred much more recently, on the order of millions or tens of millions of years ago. This idea is shown schematically in **Figure 26.9**. This small, hypothetical kingdom is a monophyletic group that contains 64 species. (Actual kingdoms are obviously larger and more complex.) The diagram emphasizes the taxa that contain the species designated number 43. The common ancestor that gave rise to this kingdom of organisms existed approximately 1 billion years ago. Over time, more recent species arose that became the common ancestors to the phylum, class, order, family, and genus that contain species number 43.

A goal of modern systematics is to create taxa according to evolutionary relationships. Systematics attempts to organize species into monophyletic groups, which means that each group includes an ancestral species and all of its descendants. Ideally, every taxon, whether it is a kingdom, phylum, class, order, family, or genus, should be a monophyletic group. However, as biologists gather more information about evolutionary relationships among extinct and modern species, we sometimes discover that previously established taxonomic groups are not monophyletic. When this occurs, efforts are made to revise taxonomy so it is consistent with evolutionary relationships.

The Study of Systematics Is Usually Based on Morphological or Genetic Homology

Systematics assumes that species that have diverged from each other in the very distant past have had a large amount of time to accumulate many genetic differences. Ultimately, such genetic changes are likely to cause differences in morphology and other attributes. By comparison, systematics also assumes that species that have diverged from each other more recently tend to be much more similar both genetically and morphologically.

As we discussed in Chapter 23, the term **homology** refers to similarities among various species that occur because the species are derived from a common ancestor. Attributes that are the result of homology are said to be homologous. For example, the wing of a bat, the arm of a human, and the front leg of a cat are homologous structures (refer back to Figure 23.13). Similarly, genes found in different species are homologous if they have been derived from the same ancestral gene (refer back to Figure 23.14).

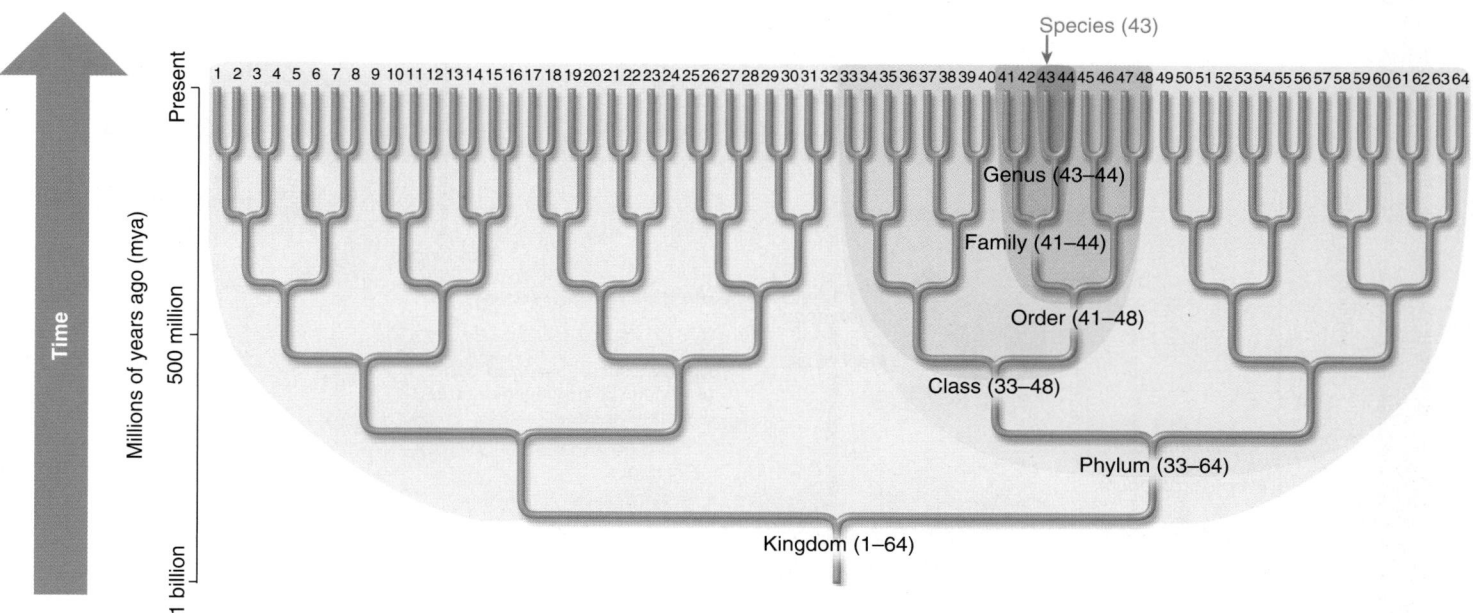

Figure 26.9 Schematic relationship between a phylogenetic tree and taxonomy. The shaded areas highlight the kingdom, phylum, class, order, family, and genus for species number 43. All of the taxa are monophyletic groups or clades. Broader taxa, such as phyla and classes, are derived from more ancient common ancestors. Smaller taxa, such as families and genera, are derived from more recent common ancestors. These smaller taxa are subsets of the broader taxa.

In systematics, researchers identify homologous features that are shared by some species but not by others. This allows them to group species based on their similarities. Researchers usually study homology at the level of morphological traits or at the level of genes. In addition, the data they gather are viewed in light of geographic data. Because most organisms do not migrate extremely long distances, species that are closely related evolutionarily are likely to inhabit neighboring or overlapping geographic regions. Let's now consider how researchers would analyze morphological and genetic homology to construct phylogenetic trees.

Morphological Analysis Traditionally, the first studies in systematics, which occurred prior to the advent of molecular genetic techniques, focused on morphological features of extinct and modern species. To establish evolutionary relationships based on morphological homology, many traits have to be analyzed to identify similarities and differences as a way to obtain a comprehensive picture of species' relatedness. Researchers use a variety of methods to combine information about different traits.

At times, the data can be complex because traits can change more than once during evolution, and the same trait can arise independently in different species. As described in Chapter 23, convergent evolution sometimes leads to convergent or analogous traits, those that arise independently in different species due to adaptation to similar environments. For example, the giant anteater (*Myrmecophaga tridactyla*), in South America, and the echidna (*Tachyglossus aculeatus*), found in Australia, have similar adaptations, such as long snouts and tongues, that enable these animals to feed on ants (refer back to Figure 23.7). These are convergent traits because they were not derived from a common ancestor. Rather, they arose twice during evolution due to adaptation to similar environments. In systematics, convergent evolution can cause errors if a researcher assumes that a particular trait arose only once, and that all species having the trait are derived from a common ancestor.

By studying morphological features of extinct species in the fossil record, paleontologists have constructed many phylogenetic trees that chart a series of organisms that led to the existence of modern species. In this approach, the tree is based on

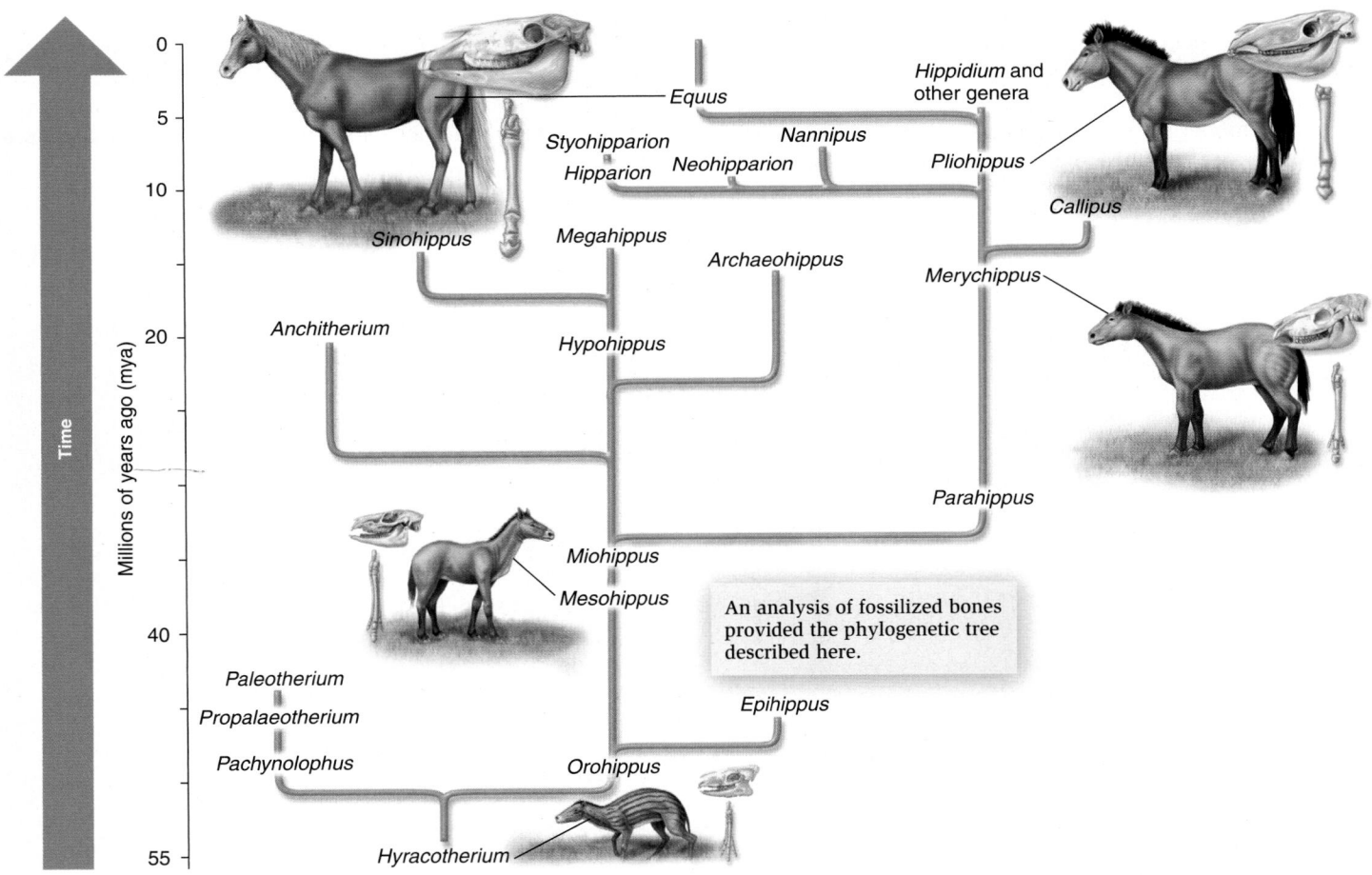

Figure 26.10   **Horse evolution.**   An analysis of morphological traits was used to produce this phylogenetic tree showing the evolutionary changes that led to the modern horse. Three important morphological changes that occurred in these genera were larger size, fewer toes, and a shift toward a jaw structure suited for grazing.

morphological features that change over the course of many generations. As an example, **Figure 26.10** depicts a current hypothesis of the evolutionary changes that led to the development of the modern horse. This figure shows genera, not individual species. Many morphological features were used to construct this tree. Because hard parts of the body are preserved in the fossil record, this tree is largely based on the analysis of changes in hoof structure, lengths and shapes of various leg bones, skull shape and size, and jaw and tooth morphology. Over an evolutionary time scale, the accumulation of many genetic changes has had a dramatic impact on species' characteristics. In the genera that are depicted in this figure, a variety of morphological changes occurred such as an increase in size, fewer toes, and a modified jaw structure suitable for grazing on fibrous grasses.

Molecular Systematics and Molecular Clocks The field of **molecular systematics** involves the analysis of genetic data, such as DNA sequences, to identify and study genetic homology and construct phylogenetic trees. In 1963, Austrian biologist Emile Zuckerkandl and American chemist and Nobel laureate Linus Pauling were the first to suggest that molecular data should be used to establish evolutionary relationships. Nucleotide base sequences in DNA and amino acid sequences of proteins are particularly well suited to studying relationships, because genetic sequences change over the course of many generations due to the accumulation of mutations. Therefore, when comparing homologous genes in different organisms, DNA sequences from closely related organisms are more similar to each other than they are to sequences from distantly related species.

As discussed in Chapter 24, researchers have speculated that most genetic variation that exists in populations is neutral, meaning that it is not acted upon by natural selection. The reasoning behind this concept is that favorable mutations are likely to be very rare, and detrimental mutations are likely to be eliminated from a population by natural selection. A large body of evidence supports the idea that much of the genetic variation observed in modern species is due to the accumulation of neutral mutations. From an evolutionary point of view, if neutral mutations occur at a relatively constant rate, they can act as a **molecular clock** on which to measure evolutionary time.

Figure 26.11 illustrates the concept of a molecular clock. The graph's y-axis is a measure of the number of nucleotide sequence differences between pairs of species. The x-axis plots the amount of time that has elapsed since a pair of species shared a common ancestor. As seen in this diagram, the number of sequence differences is higher when two species shared a common ancestor in the very distant past than it is in pairs that shared a more recent common ancestor. The explanation for this phenomenon is that the gene sequences of the species accumulate independent mutations after they have diverged from each other; a longer period of time since their divergence allows for a greater accumulation of mutations, which makes their sequences more different.

Figure 26.11 suggests a linear relationship between the number of sequence changes and the time of divergence. Such a

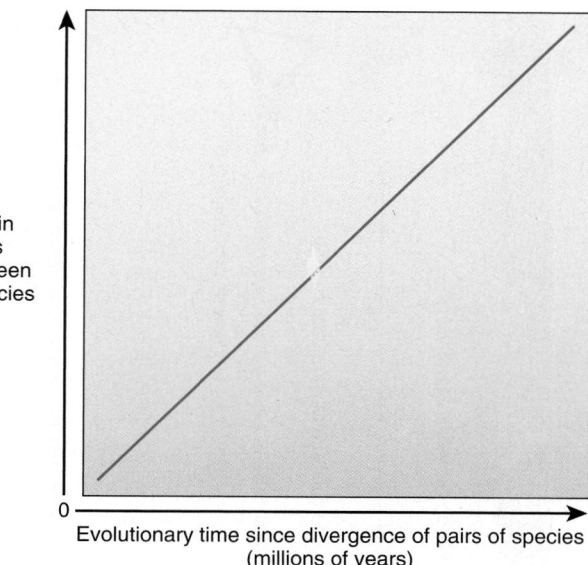

Nucleotide differences in homologous genes between pairs of species

0

Evolutionary time since divergence of pairs of species (millions of years)

Figure 26.11 **A molecular clock.** According to the concept of a molecular clock, neutral mutations accumulate over evolutionary time. When comparing homologous genes between species, those species that diverged more recently tend to have fewer differences than do those whose common ancestor occurred in the very distant past.

relationship indicates that the observed rate of neutral mutations remains constant over millions of years. For example, a linear relationship predicts that a pair of species that has 20 nucleotide differences in a given gene sequence would have a common ancestor that is roughly twice as old as that of a pair showing 10 nucleotide differences. While actual data sometimes show a relatively linear relationship over a defined time period, evolutionary biologists do not believe that molecular clocks are perfectly linear over very long periods of time. Several factors can contribute to nonlinearity of molecular clocks. These include differences in the generation times of the species being analyzed, the presence of mutations that are acted upon by natural selection, and variation in mutation rates between different species.

To obtain reliable data, researchers must calibrate their molecular clocks. How much time does it take to accumulate a certain percentage of nucleotide changes? To perform such a calibration, researchers must have information regarding the date when two species shared a common ancestor. Such information could come from the fossil record, for instance. The genetic differences between those species are then divided by the amount of time since their last common ancestor to calculate a rate of change. For example, research suggests that humans and chimpanzees shared a common ancestor approximately 6 million years ago. The percentage of nucleotide differences between mitochondrial DNA of humans and chimpanzees is 12%. From these data, the molecular clock for changes in mitochondrial DNA sequences of primates is calibrated at roughly 2% nucleotide changes per million years.

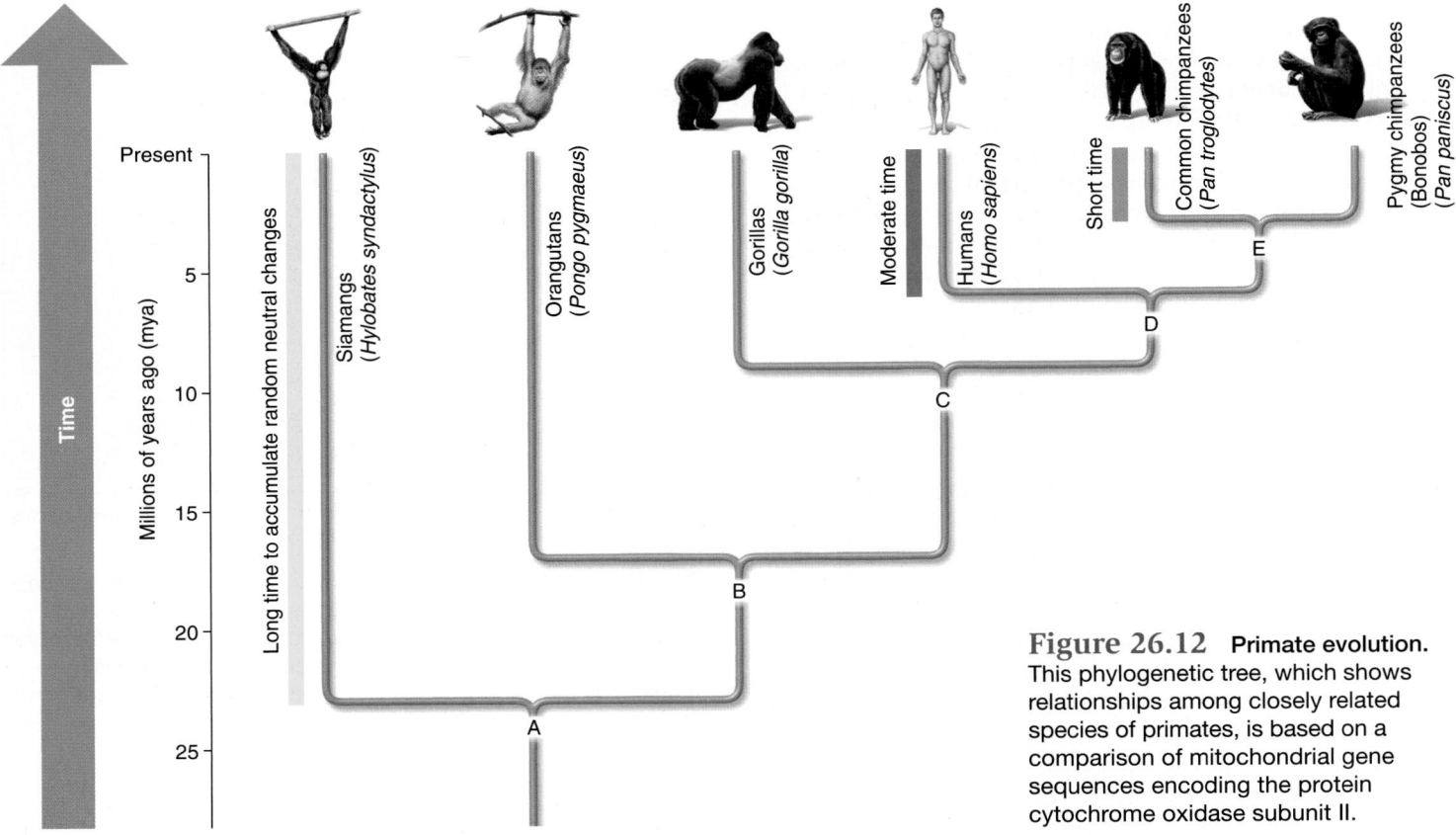

Figure 26.12 **Primate evolution.** This phylogenetic tree, which shows relationships among closely related species of primates, is based on a comparison of mitochondrial gene sequences encoding the protein cytochrome oxidase subunit II.

To understand the concept of a molecular clock, let's consider the evolution of some species of primates. **Figure 26.12** illustrates a simplified diagram of evolutionary relationships among several species that was derived by comparing DNA sequences in a mitochondrial gene. This gene encodes a protein called cytochrome oxidase subunit II, which is involved with cellular respiration. The vertical scale represents time, and the branch points that are labeled with letters represent common ancestors. Let's take a look at three branch points (labeled A, D, and E) and relate them to the concept of a molecular clock. The common ancestor labeled A diverged into two species that ultimately gave rise to siamangs and the other five species. Since this divergence, there has been a long time (approximately 23 million years) for the siamang genome to accumulate random neutral changes that would be different from the random changes that have occurred in the genomes of the other five species (see yellow bar in Figure 26.12). Therefore, the gene in the siamangs is more different than are the genes in the other seven species. Now let's compare humans and chimpanzees. The common ancestor that gave rise to these species is labeled D. This species diverged into two species that eventually gave rise to humans and chimpanzees. This divergence occurred a moderate time ago, approximately 6 million years ago, as illustrated by the red bar. Compared to humans and chimpanzees, humans and siamangs have more differences in their gene sequences because there has been more time for them to accumulate random neutral mutations. Finally, let's consider the two species of chimpanzees,

whose common ancestor is labeled E. Since the divergence of species E into two species, approximately 3 million years ago, the time for the molecular clock to tick (that is, accumulate random mutations) is relatively short, as depicted by the green bar in Figure 26.12. Therefore, the two modern species of chimpanzee have very similar gene sequences.

Genetic sequence information is primarily used for studying relationships between modern organisms. However, as discussed later in our Feature Investigation, sometimes DNA can be obtained from extinct organisms as well. For evolutionary comparisons, the DNA sequences of many genes have been obtained from a wide range of sources. Several genes have been used to construct phylogenetic trees. For example, the gene that encodes an RNA of the small subunit of the ribosome (SSU rRNA) is commonly analyzed to compare higher taxa such as phyla. Because rRNA is universal in all living organisms, its function must have been established at an early stage in the evolution of life on this planet, and its sequence has changed fairly slowly. The molecular clock for SSU rRNA has been calibrated at approximately 1% sequence change per 50 million years. Furthermore, SSU rRNA is a rather large molecule, so it contains a large amount of sequence information. This gene has been sequenced from thousands of different species.

Slowly changing genes such as the gene that encodes SSU rRNA are useful for evaluating distant evolutionary relationships, such as comparing plants and animals. Such genes may also be useful in assessing evolutionary relationships among bacterial species that evolve fairly rapidly due to their short generation times. Other genes have changed more rapidly because of a greater tolerance of neutral mutations. For example, the mitochondrial genome and DNA sequences within introns can more easily incur neutral mutations (compared to the coding sequences of genes), and so their sequences change frequently during evolution. More rapidly changing DNA sequences have been used to study recent evolutionary relationships, particularly among eukaryotic species such as large animals that have long generation times and tend to evolve more slowly. In these cases, slowly evolving genes may not be very useful for establishing evolutionary relationships because two closely related species are likely to have identical or nearly identical DNA sequences for such genes. Instead, sequence differences are more easily found among closely related species when the DNA sequences are more rapidly changing. Our example in Figure 26.12 of closely related species of primates uses sequence changes in the mitochondrial gene for cytochrome oxidase subunit II because this gene tends to change fairly rapidly on an evolutionary timescale.

A Cladistic Approach Is the Most Common Way to Make a Phylogenetic Tree

A **cladistic approach** reconstructs a phylogenetic tree by considering the various possible pathways of evolution and then choosing the most plausible tree. Cladistics is generally accepted as the best method available for the construction of phylogenetic trees, which are known as **cladograms.** A cladistic approach compares traits that are either shared or not shared by different species. A trait shared with a distant ancestor is called a **shared primitive character** or **symplesiomorphy**. Such traits are viewed as being older traits—ones that occurred earlier in evolution. In contrast, a **shared derived character**, or **synapomorphy**, is a trait that is shared by a group of organisms but not by a distant common ancestor. Compared to primitive characters, derived characters are more recent traits on an evolutionary timescale. For example, among mammals, only some species have flippers, such as whales and dolphins. In this case, flippers were derived from the two front limbs of an ancestral species. The word derived refers to the phenomenon that evolution involves the modification of traits in pre-existing species. In other words, newer populations of organisms are derived from changes in pre-existing populations. The basis of the cladistic approach is to analyze many shared derived characters among groups of species to deduce the pathway that gave rise to the species.

To understand the concept of primitive and shared derived characters, **Figure 26.13** shows a cladogram that compares several traits among five species of animals. Shared derived characters are used to establish a cladogram. The various colors in this tree emphasize when particular traits arose during the evolution of these species. A branch point, or node, is where two species differ in shared derived characters. The oldest common ancestor, which would now be extinct, had a notochord and gave rise to all five species. Vertebrae are a shared derived character of the lamprey, salmon, lizard, and rabbit, but not amphioxus. By com-

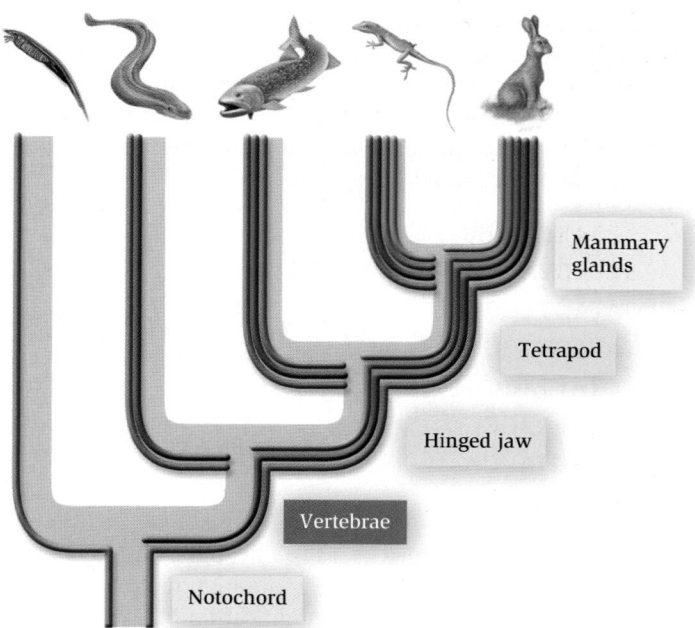

(a) Cladogram based on morphological traits

	Amphioxus	Lamprey	Salmon	Lizard	Rabbit
Notochord	Yes	Yes	Yes	Yes	Yes
Vertebrae	No	Yes	Yes	Yes	Yes
Hinged jaw	No	No	Yes	Yes	Yes
Tetrapod	No	No	No	Yes	Yes
Mammary glands	No	No	No	No	Yes

(b) Characteristics among species

Figure 26.13 **Primitive versus shared derived characters involving morphological traits.** **(a)** This phylogenetic tree illustrates both primitive and shared derived characters in a cladogram of relationships among five animal species. **(b)** A comparison of characteristics among these species.

Biological inquiry: What shared derived character is common to the salmon, lizard, and rabbit, but not the lamprey?

parison, a hinged jaw is a shared derived character of the salmon, lizard, and rabbit, but not of the lamprey or amphioxus. The table shown in Figure 26.13b compares the characters that are found in these five species.

In a cladogram, an **ingroup** is a monophyletic group in which we are interested. By comparison, an **outgroup** is a species or group of species that is most closely related to an ingroup. For example, let's suppose that the salmon, lizard, and rabbit are an ingroup. The outgroup would be the lamprey, because this species is the most closely related to the ingroup but lacks one of the shared derived characters, namely a hinged jaw. All traits that are shared by the outgroup and the ingroup must have arisen in a common ancestor that predates the divergence of the two groups.

Likewise, the concept of shared derived characters can apply to molecular data such as a sequence of a gene. Let's consider an example to illustrate this idea. Our example involves molecular data obtained from seven different hypothetical species called A–G. In these species, a homologous region of DNA was sequenced as shown here:

```
    12345678910
A: GATAGTACCC
B: GATAGTTCCC
C: GATAGTTCCG
D: GGTATTACCC
E: GGTATAACCC
F: GGTAGTACCA
G: GGTAGTACCC
```

The cladogram of **Figure 26.14** is a hypothesis of how these DNA sequences arose. In this case, a mutation that changes the DNA sequence is analogous to a modification of a characteristic. Species that share such genetic changes possess shared derived characters because the new genetic sequence was derived from a more primitive sequence.

Now that we have an understanding of primitive and shared derived characters, let's consider the steps a researcher would follow to construct a cladogram using a cladistics approach.

1. **Choose the species in whose evolutionary relationships you are interested**. In a simple cladogram, such as those described in this chapter, individual species are compared to each other. In more complex cladograms, species may be grouped into larger taxa (for example, families) and

compared with each other. If such grouping is done, the groups must be clades for the results to be reliable.

2. **Choose characters for comparing different species.** A character is a general feature of an organism. Characters may come in different versions called character states. For example, hair color is a character, while brown hair and red hair are character states.

3. **Determine the order of character states.** In other words, determine if a character state is primitive or derived. This information may be available by examining the fossil record, for example, but is usually done by comparing the ingroup with the outgroup. An assumption is made that the outgroup has the primitive state(s).

4. **Group species (or higher taxa) based on shared derived characters**.

5. **Build a cladogram based on the following principles**:
   - All species (or higher taxa) are placed on tips in the phylogenetic tree, not at branch points.
   - Each cladogram branch point should have a list of one or more shared derived characters that are common to all species above the branch point unless the character is later modified.
   - All shared derived characters appear together only once in a cladogram unless they arose independently during evolution more than once.

6. **Choose the most likely cladogram among possible options**. When grouping species or higher taxa, more than one cladogram may be possible. Therefore, analyzing the data and producing the most likely cladogram is a key aspect of this process. As described next, different theoretical approaches can be followed to achieve this goal.

Figure 26.14 **Primitive versus shared derived characters involving a molecular trait.** This phylogenetic tree illustrates a cladogram of relationships involving homologous gene sequences found in seven species. Mutations that alter a primitive DNA sequence are shared among certain species and thereby allow the construction of a cladogram.

Biological inquiry: What nucleotide change is a shared derived character for species A, B, and C, but not for species G?

Different Strategies Can Be Followed to Produce the Most Likely Cladogram

The challenge in a cladistic approach is to determine the correct order of events. It may not always be obvious which traits are ancestral and came earlier, and which are derived and came later in evolution. Different approaches can be used to deduce the correct order. First, for morphological traits, a common way to deduce the order of events is to analyze fossils and determine the relative dates that certain traits arose. A second strategy that can be used to deduce the correct order is to assume that the best hypothesis is the one that requires the fewest number of evolutionary changes. This concept, called the **principle of parsimony**, states that the preferred hypothesis is the one that is the simplest. For example, if two species possess a tail, we would initially assume that a tail arose once during evolution, and that both species have descended from a common ancestor with a tail. Such a hypothesis is simpler, and more likely to be correct, than assuming that tails arose twice during evolution, and that the tails in the two species are not due to descent from a common ancestor.

The principle of parsimony can also be applied to gene sequence data, as can other methods called maximum likelihood and Bayesian analysis. In these methods, a model for the most likely changes during gene evolution is applied to the data. Maximum likelihood, for example, will produce the tree that is most likely based on the data and the application of a particular evolutionary model. These models take into account various types of factors, such as whether mutations would affect the first, second, or third nucleotide in a codon (mutations affecting the third codon are often neutral) and whether the data conform to a molecular clock. When constructing their trees, researchers often have preferences about the kinds of strategies they employ. More than one strategy may be used to construct a cladogram.

Now that we have a better understanding of synapomorphies and the steps used to build a cladogram, let's look at a hypothetical example. We will use molecular data for four taxa (A–D), where A is presumed to be the outgroup and has all of the primitive states.

```
  12345
A: GTACA (outgroup)
B: GACAG
C: GTCAA
D: GACCG
```

Given this information, three potential trees are shown in **Figure 26.15**, although more are possible. In these examples, tree 1 requires seven mutations, and tree 2 requires six, while tree 3 requires only five. Therefore, tree 3 requires the fewest number of mutations and is considered the most parsimonious. Based on the principle of parsimony, it would be the more likely choice.

Figure 26.15 **The cladistic approach from molecular genetic data.** These are three possible evolutionary trees for the evolution of a short DNA sequence, but many more are possible. Changes in nucleotide sequence are shown along each tree. According to the principle of parsimony, tree number 3 is the more likely choice because it requires only five mutations. When constructing cladograms based on long genetic sequences, researchers use computers to generate trees with the fewest possible genetic changes.

FEATURE INVESTIGATION

Cooper and Colleagues Extracted DNA from Extinct Flightless Birds and Then Compared It with DNA from Modern Species to Reconstruct Their Phylogeny

Starting with small tissue samples from extinct species, scientists have discovered that it is occasionally possible to obtain DNA sequence information. This is called ancient DNA analysis or molecular paleontology. Since the mid-1980s, some researchers have become excited about the information derived from sequencing DNA of extinct specimens. Debate has centered on how long DNA can remain intact after an organism has died. Over time, the structure of DNA is degraded by hydrolysis and the loss of purines. Nevertheless, under certain conditions (cold temperature, low oxygen, etc.), DNA samples may be stable as long as 50,000–100,000 years. In most studies involving extinct specimens, the ancient DNA is extracted from bone, dried muscle, or preserved skin. These samples are often from museum specimens that have been gathered by paleontologists. In recent years, this approach has been used to study evolutionary relationships between modern and extinct species.

In the Feature Investigation shown in **Figure 26.16**, Alan Cooper, Cécile Mourer-Chauviré, Geoffrey Chambers, Arndt von Haeseler, Allan Wilson, and Svante Pääbo investigated the evolutionary relationships among some extinct and modern species of flightless birds. This is an example of discovery-based science. The researchers wanted to gather data to propose a hypothesis regarding the evolutionary relationships among certain species.

Figure 26.16   DNA analysis of phylogenetic relationships among modern and extinct flightless birds by Cooper and colleagues.

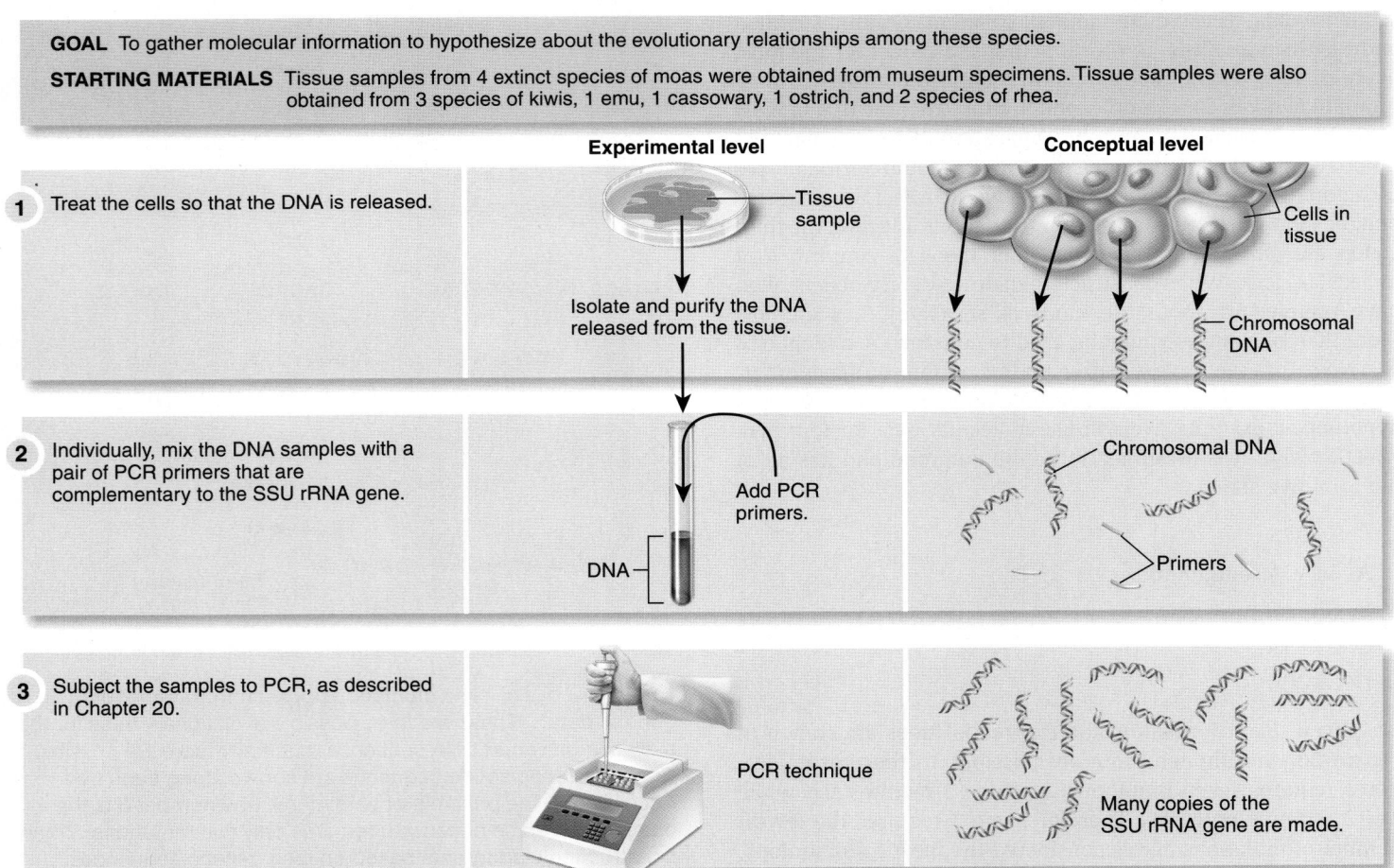

GOAL   To gather molecular information to hypothesize about the evolutionary relationships among these species.

STARTING MATERIALS   Tissue samples from 4 extinct species of moas were obtained from museum specimens. Tissue samples were also obtained from 3 species of kiwis, 1 emu, 1 cassowary, 1 ostrich, and 2 species of rhea.

Experimental level   **Conceptual level**

1 Treat the cells so that the DNA is released.

Tissue sample

Isolate and purify the DNA released from the tissue.

Cells in tissue

Chromosomal DNA

2 Individually, mix the DNA samples with a pair of PCR primers that are complementary to the SSU rRNA gene.

Add PCR primers.

DNA

Chromosomal DNA

Primers

3 Subject the samples to PCR, as described in Chapter 20.

PCR technique

Many copies of the SSU rRNA gene are made.

4 Subject the amplified DNA fragments to DNA sequencing, as described in Chapter 20.

ATTGAACCAT
GGATTAACGC
ACTGGTTGAC
TCCTGATGTC
ATTGAACCAT
GGATTAACC

Sequence the amplified DNA.

The amplification of the SSU rRNA gene allows it to be subjected to DNA sequencing.

5 Align the DNA sequences to each other, using computer techniques described in Chapter 21.

Align sequences using computer programs.

Align sequences to compare the degree of similarity.

6 THE DATA

```
Moa 1       GCTTAGCCCTAAATCCAGATACTTACCCTACACAAGTATCCGCCCGAGAACTACGAGCACAAACGCTTAAAACTCTAAGGACTTGGCGGTGCCCCAAACCCA
Kiwi 1      ··················T·G·····GT··CT····C································································T······
Emu         ···············TT·······C···T···CAG·C······T·······································T······
Cassowary   ···············TT·····CG·TA···CTG····································································T······
Ostrich     ········T·····AT·····C··CT·····································································T······
Rhea 1      ··············T········C··CT·······································································T······

Moa 1       CCTAGAGGAGCCTGTTCTATAATCGATAATCCACGATACACCCGACCATCCCTCGCCCGT-GCAGCCTACATACCGCCGTCCCCAGCCCGCCT--AATGAAA
Kiwi 1      ····················C·········A·····T·T···AAC-A·······T·········G····T·····AA·····G·
Emu         ····················C·········A·····T·T···AA·-A·················G·····························--·····
Cassowary   ····················C·········AG····T·T···AA·TA··················G·····················G··G·
Ostrich     ··············T····A···C···T···A--T·············G·············C---··G·
Rhea 1      ····················C·········T·T···A--··········G·······TA·G·

Moa 1       G-AACAATAGCGAGCACAACAGCCCTCCCCCGCTAACAAGACAGGTCAAGGTATAGCATATGAGATGGAAGAAATGGGCTACATTTTCTAACATAGAACACC
Kiwi 1      ·····C····A·········TA··-·A··············C··············A····T·T
Emu         -·····T····AC--TT··················G·········T·T
Cassowary   -····T·····AC--T·····················G·······················T··
Ostrich     -······T····A--·················GAG··················T·A
Rhea 1      -······C···AG··T··T··TA---··················G·····TC··········A·

Moa 1       C-------------ACGAAAGAGAAGGTGAAACCCTCCTCAAAAGGCGGATTTAGCAGTAAAATAGAACAAGAATGCCTATTTTAAGCCCGGCCCTGGGGC
Kiwi 1      -·············A·GGT····T·-C··T·G···················C··T···GA·T·········-··T·····A·····
Emu         -·············AG·T····T·AC·T··G···················C··T···GA·T·········A-·T···T··A·····
Cassowary   -·············A·G·T····T·A···T·G···················C··T···GA·T·········A--·T···T··A·····
Ostrich     -·············G·TA····T·A···G···················T···GA·T·········-T···T··A·····
Rhea 1      -············G·····GGCA······-AC···CG············G··G·TC····A··C·C·······-·····A·····
```

The kiwis and moas existed in New Zealand during the Pleistocene. Species of modern kiwis still exist, but the moas are now extinct. Eleven known species of moas formerly existed. In this study, the researchers investigated the phylogenetic relationships of four extinct species of moas, which were available as museum samples, kiwis of New Zealand, and several other living species of flightless birds. These included the emu and the cassowary (both found in Australia and New Guinea), the ostrich (found in Africa and formerly Asia), and two rheas (found in South America).

Samples from the various species were subjected to PCR to amplify a region of the SSU rRNA gene. This provided enough DNA for DNA sequencing. The data in Figure 26.16 illustrate a comparison of the sequences of a continuous region of the SSU rRNA gene from these species. The first line shows the DNA sequence for one of the four extinct moa species. Below it are the sequences of several of the other species they analyzed. When the other sequences are identical to the first sequence, a dot is placed in the corresponding position. When the sequences are different, the changed nucleotide base (A, T, G, or C) is placed there.

In a few regions, the genes are different lengths. In these cases, a dash is placed at the corresponding position.

As you can see from the large number of dots, the gene sequences among these flightless birds are very similar, though some differences occur. If you look carefully at the data, you will notice that the sequence from the kiwi (a New Zealand species) is actually more similar to the sequence from the ostrich (an African species) than it is to that of the moa, which was once found in New Zealand. Likewise, the kiwi is more similar to the emu and cassowary (found in Australia and New Guinea) than to the moa. Contrary to their original expectations, the researchers concluded that the kiwis are more closely related to Australian and African flightless birds than they are to the moas. From these results, they proposed that New Zealand was colonized twice by ancestors of flightless birds. First, New Zealand was colonized by the ancestor of moas, and then at a later date by an ancestor of the kiwis, which evolved independently of the moas in Australia and New Guinea. As shown in **Figure 26.17**, the researchers constructed a new evolutionary tree that illustrates the revised relationships among these modern and extinct species.

Figure 26.17 **A revised phylogenetic tree of flightless birds.** This phylogenetic tree was based on a comparison of DNA sequences from extinct and modern birds. The species included are moas, kiwis, emus, cassowaries, ostriches, and rheas. Moas are now extinct, but all other species still exist.

Biological inquiry: With regard to geography, why are the results in this figure surprising?

Molecular Systematics Is Changing Our View of Taxonomy

Taxonomy is a work in progress. As researchers gather new information, they sometimes discover that some of the current taxonomic groups are not monophyletic. **Figure 26.18** compares a monophyletic taxon with those that are not. As mentioned earlier, a monophyletic taxon contains a common ancestor and all of the species that are derived from that ancestor (Figure 26.18a). The ideal goal of taxonomy is to place organisms into monophyletic groups. A **paraphyletic taxon** is a group that contains a common ancestor and some, but not all, of its descendants (Figure 26.18b). In contrast, a **polyphyletic taxon** consists of members of several evolutionary lines and does not include the most recent common ancestor of the included lineages (Figure 26.18c). Over time, as we learn more about evolutionary relationships, taxonomic groups are being reorganized in an attempt to recognize monophyletic groups in biological classifications.

For higher eukaryotic taxa, such as kingdoms, molecular genetic data has shed new light regarding the placement of species into monophyletic groups. Many recent models propose several major groups, sometimes called **supergroups**, as a way to organize eukaryotes into monophyletic groups. **Figure 26.19** shows a diagram that hypothesizes eight supergroups. Of the four traditional kingdoms, both Fungi and Animalia are within the Opisthokonta supergroup, while Plantae is found within Archaeplastida. The remaining branches and supergroups used to be classified within the single kingdom Protista. Thus, molecular data and newer ways of building trees reveal that protists played a key role in the evolution of many diverse groups of eukaryotic species, producing several large monophyletic groups. A survey of the characteristics of these different types of protists will be described in Chapter 28.

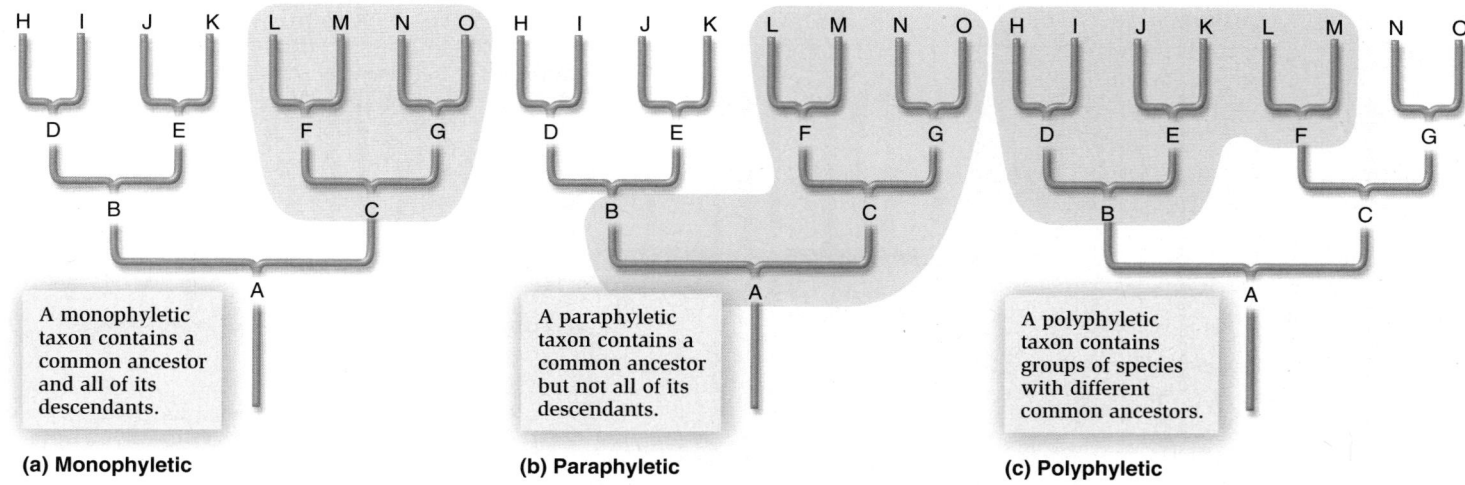

Figure 26.18 A comparison of monophyletic, paraphyletic, and polyphyletic taxonomic groups.

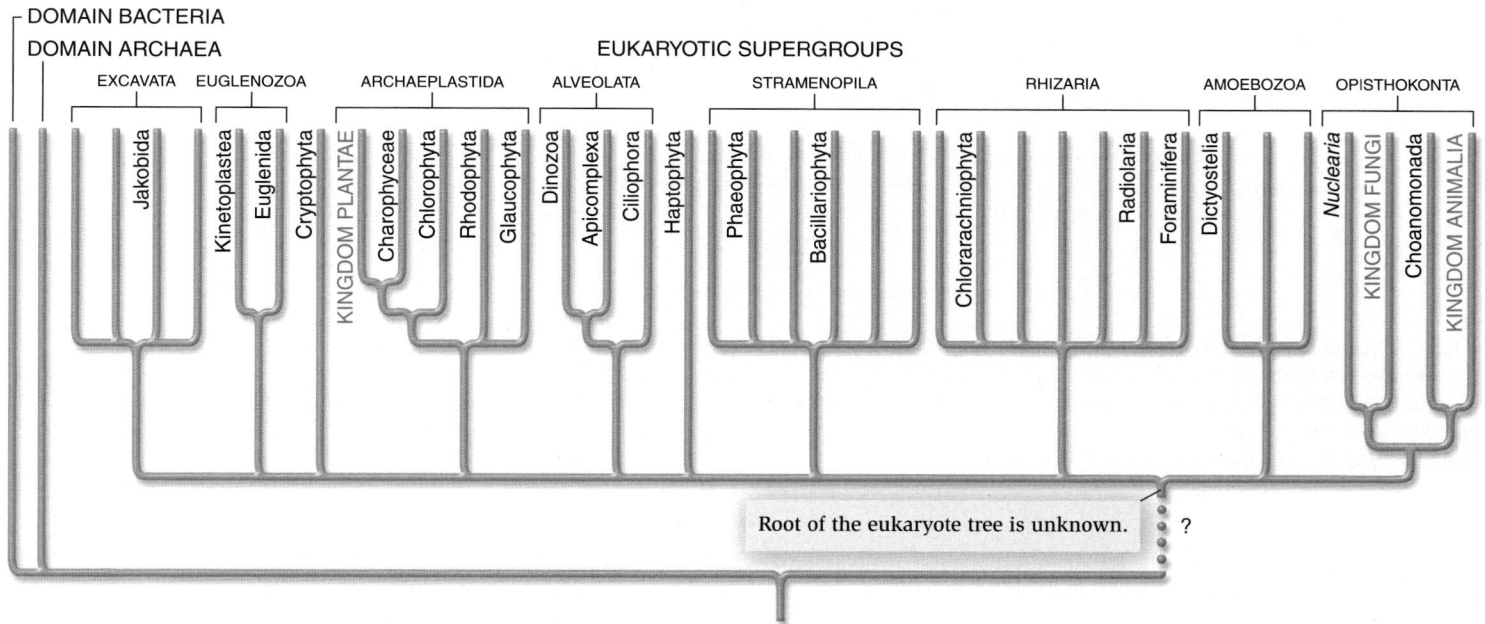

Figure 26.19 **A modern cladogram for eukaryotes.** This drawing should be considered a working hypothesis. The arrangement of these supergroups relative to each other is not entirely certain.

GENOMES & PROTEOMES

Due to Horizontal Gene Transfer, the Tree of Life Is Really a "Web of Life"

Thus far, we have considered various ways to construct phylogenetic trees, which describe the relationships between ancestors and their descendents. The type of evolution that is depicted in previous figures is vertical evolution, which involves changes in groups of species due to descent from a common ancestor. Since the time of Darwin, vertical evolution has been the tradi-

tional way that biologists view the evolutionary process. However, over the past couple of decades researchers have come to realize that evolution is not so simple. In addition to vertical evolution, horizontal gene transfer has also played a significant role in the phylogeny of all living species.

As discussed in Chapter 23, horizontal gene transfer is the transfer of genes between different species. An analysis of many bacterial genomes has shown that horizontal gene transfer has been a major force in the evolution of prokaryotes. Horizontal gene transfer continued after the divergence of the three major domains of life, including transfer between prokaryotic and

Figure 26.20   **A web of life.**   This phylogenetic tree shows not only the modern classification of all life on Earth but includes the contribution of horizontal gene transfer in the evolution of species on our planet. This phenomenon was prevalent during the early stages of evolution when all organisms were unicellular. Horizontal gene transfer continues to be a prominent factor in the speciation of Bacteria and Archaea. Note: This tree is meant to be schematic. For example, Figure 26.19 is a more realistic representation of the evolutionary relationships among modern eukaryotic species.

eukaryotic species and among eukaryotic species. With regard to modern organisms, horizontal gene transfer is still prevalent among prokaryotic species. By comparison, this process is less common among modern eukaryotes, though it does occur. Researchers have speculated that multicellularity and sexual reproduction have presented barriers to horizontal gene transfer in most eukaryotes. For a gene to be transmitted to eukaryotic offspring, it would have to be transferred into a eukaryotic cell that is a gamete or a cell that gives rise to gametes.

Recently, scientists have debated the role of horizontal gene transfer in the earliest stages of evolution, prior to the emergence of the two prokaryotic domains. The traditional viewpoint was that the three domains of life arose from a single type of prokaryotic (or preprokaryotic) cell called the universal ancestor. However, genomic research has suggested that horizontal gene transfer may have been particularly common during the early stages of evolution on Earth, when all species were unicellular. Rather than proposing that all life arose from a single type of prokaryotic cell, horizontal gene transfer may have

been so prevalent that the universal ancestor may have actually been an ancestral community of cell lineages that evolved as a whole. If that were case, the tree of life cannot be traced back to a single universal ancestor.

Figure 26.20 illustrates a schematic scenario for the evolution of life on Earth that includes the roles of both vertical evolution and horizontal gene transfer. This has been described as a "web of life" rather than a "tree of life." Instead of a universal ancestor, a web of life began with a community of primitive cells that transferred genetic material in a horizontal fashion. Horizontal gene transfer was also prevalent during the early evolution of bacteria and archaea, and when eukaryotes first emerged as unicellular species. In modern bacteria and archaea, it remains a prominent way to foster evolutionary change. By comparison, the region of the diagram that contains most eukaryotic species has a more treelike structure, because horizontal gene transfer has become less common in these species, though it does occur occasionally.

CHAPTER SUMMARY

26.1 Taxonomy

- Taxonomy is the field of biology that is concerned with the theory, practice, and rules of classifying living and extinct organisms and viruses.
- Taxonomy places all living organisms into groups called taxa. The broadest groups are the three domains called Bacteria, Archaea, and Eukarya. (Figure 26.1, Table 26.1)
- Domains are divided into a hierarchy composed of kingdoms, phyla, classes, orders, families, genera, and species. (Figure 26.2)
- Binomial nomenclature provides each species with a scientific name that refers to its genus and species epithet.
- The two prokaryotic domains, Bacteria and Archaea, possess distinctive characteristics at the molecular level and occupy particular environments. (Figure 26.3)
- Eukaryotes were traditionally divided into four kingdoms called Protista, Fungi, Plantae, and Animalia. (Figures 26.4, 26.5, 26.6, 26.7, Table 26.2)

26.2 Systematics

- Systematics is the study of biological diversity and evolutionary relationships. The evolutionary history of a species is its phylogeny.
- A phylogenetic tree is a hypothesis that describes the phylogeny of particular species. A monophyletic group or clade includes all of the species that are derived from a common ancestor. (Figure 26.8)
- The hierarchy of taxonomy is related to the timing of common ancestors. Smaller taxa, such as families and genera, are derived from more recent common ancestors than are broader taxa such as kingdoms and phyla. (Figure 26.9)
- Morphological and genetic data are used to construct phylogenetic trees. (Figure 26.10)
- Assuming that the rate of neutral mutation is relatively constant, genetic data provide a molecular clock on which to measure evolutionary time. (Figures 26.11, 26.12)
- A cladistic approach constructs a phylogenetic tree, also called a cladogram, by considering the various possible pathways of evolution. Species are grouped together according to shared derived characters that come from primitive characters. An ingroup is a clade of interest, while an outgroup is closely related but lacks one or more shared derived characters. (Figures 26.13, 26.14)
- The cladistic approach can produce many possible cladograms. The most likely cladogram is chosen by a variety of methods such as the principle of parsimony or maximum likelihood. (Figure 26.15)
- Cooper and colleagues analyzed DNA sequences from extinct and modern flightless birds and hypothesized a phylogeny in which New Zealand was colonized twice, once by moas and later by kiwis. (Figures 26.16, 26.17)
- Molecular systematics, which involves the analysis of genetic sequences, has led to major revisions in taxonomy. Ideally, all taxa should be monophyletic, though previously established taxa sometimes turn out to be paraphyletic or polyphyletic. (Figure 26.18)
- Modern eukaryotes are now divided into eight supergroups instead of the four traditional kingdoms. (Figure 26.19)
- Due to horizontal gene transfer, the tree of life should really be described as a web of life. (Figure 26.20)

TEST YOURSELF

1. The study of biological diversity based on evolutionary relationships is
   a. paleontology.
   b. evolution.
   c. systematics.
   d. ontogeny.
   e. both a and b.

2. Which of the following is the correct order of the taxa used to classify organisms?
   a. kingdom, domain, phylum, class, order, family, genus, species
   b. domain, kingdom, class, phylum, order, family, genus, species
   c. domain, kingdom, phylum, class, family, order, genus, species
   d. domain, kingdom, phylum, class, order, family, genus, species
   e. kingdom, domain, phylum, order, class, family, species, genus

3. When considering organisms within the same taxon, which level includes organisms with the greatest similarity?
   a. kingdom
   b. class
   c. order
   d. family
   e. genus

4. Which of the following characteristics is not shared by prokaryotes and eukaryotes?
   a. DNA is the genetic material.
   b. Messenger RNA encodes the information to produce proteins.
   c. All cells are surrounded by a plasma membrane.
   d. The cytoplasm is compartmentalized into organelles.
   e. Both a and d.

5. The characteristics that define organisms in the kingdom Plantae include
   a. unicellular.
   b. photosynthetic.
   c. multicellular.
   d. a and b only.
   e. b and c only.

6. The evolutionary history of a species is
   a. ontogeny.
   b. taxonomy.
   c. evolution.
   d. phylogeny.
   e. embryology.

7. A group composed of all species derived from a common ancestor is referred to as
   a. a phylum.
   b. a monophyletic group or clade.
   c. a phenogram.
   d. an outgroup.
   e. a taxon.

8. The goal of modern taxonomy is to
   a. classify all organisms based on morphological similarities.
   b. classify all organisms in monophyletic groups.
   c. classify all organisms based solely on genetic similarities.
   d. determine the evolutionary relationships between similar species.
   e. none of the above.

9. The concept that the preferred hypothesis is the one that is the simplest is
   a. phenetics.
   b. cladistics.
   c. the principle of parsimony.
   d. maximum likelihood.
   e. both b and d.

10. Researchers believe that horizontal gene transfer is less prevalent in eukaryotes because of
    a. the presence of organelles.
    b. multicellularity.
    c. sexual reproduction.
    d. all of the above.
    e. b and c only.

CONCEPTUAL QUESTIONS

1. Explain binomial nomenclature and give an example.
2. Explain both the usefulness and potential pitfalls of morphological analysis.
3. Explain the value of a molecular clock.

EXPERIMENTAL QUESTIONS

1. What is molecular paleontology? What was the purpose of the study conducted by Cooper and colleagues?
2. What birds were examined in the Cooper study and what are their geographic distributions? Why were the different species selected for this study?
3. What results did Cooper and colleagues obtain by comparing these DNA sequences? How did the results of this study impact the proposed phylogeny of flightless birds?

COLLABORATIVE QUESTIONS

1. Discuss taxonomy and its hierarchical organization.
2. Discuss systematics and the concept of a phylogenetic tree.

www.brookerbiology.com

This website includes answers to the Biological Inquiry questions found in the figure legends and all end-of-chapter questions.

27

THE BACTERIA AND ARCHAEA

CHAPTER OUTLINE

A visible cyanobacterial bloom gives a pea-soup appearance to lake water.

Among Earth's life-forms, bacteria and archaea are unique in several ways. These organisms have the simplest cell structure and include the smallest known cells. Bacteria and archaea are also the most abundant organisms on Earth. About half of Earth's total biomass consists of an estimated 10^{30} individuals; just a pinch of garden soil can contain 2 billion, and about a million occur in 1 ml of seawater. Bacteria and archaea live in nearly every conceivable habitat, including extremely hot or salty waters that support no other life, and they are also Earth's most ancient organisms, having originated more than 3.5 billion years ago. Their great age and varied habitats have fostered very high diversity. Today, many millions of species of bacteria and archaea collectively display more diverse metabolic processes than occur in any other group of organisms. Many of these metabolic processes are important on a global scale, influencing Earth's climate, atmosphere, soils, water quality, and human health and technology. In this chapter we will survey the diversity, structure, reproduction, metabolism, and ecology of bacteria and archaea. This survey will illustrate major principles of diversity, including descent with modification and horizontal gene transfer.

27.1 Diversity and Evolution

As we have noted, one of the prominent features of bacteria and archaea is their astounding diversity. In the past, microbiologists studied diversity by isolating these organisms from nature and growing cultures in the laboratory. Such cultures allowed microbiologists to observe variation in structure and metabolism, major features used to classify bacteria and archaea. Today, microbiologists also use molecular techniques to detect diverse bacteria and archaea in nature. By using these new techniques,

they have discovered that bacteria and archaea are vastly more diverse than previously realized, and many new species have been discovered. For example, in 2004, gene sequencing expert Craig Venter and associates found 148 new bacteria and archaea by sequencing DNA extracted from microbial organisms from the Sargasso Sea. Similar studies of other habitats have also revealed much new diversity, though only 1% of the newly discovered species have been cultured in the laboratory. Many species of bacteria and archaea are known only as a distinctive molecular sequence.

Though much remains to be learned about the diversity of Earth's microorganisms, extensive molecular analysis has supported the concept that prokaryotic microbes can be classified into two major domains of life: the **Archaea** and **Bacteria** (also called Eubacteria) (**Figure 27.1**). The terms prokaryote and prokaryotic are often used to refer to archaeal and bacterial cells because these organisms lack nuclei and other cellular features typical of eukaryotes. As we noted in Chapter 26, in the 1970s microbiologist Carl Woese and associates proposed splitting the kingdom Monera, which had included all prokaryotes, into these two domains, based on comparisons of ribosomal RNA sequences from diverse microorganisms. In this section we will first survey the major kingdoms and phyla of the domains Archaea and Bacteria and then explore how horizontal gene transfer—the transfer of genes between different species—has influenced their evolution.

Domain Archaea Includes Inhabitants of Extremely Harsh Environments

Organisms classified in the domain Archaea, commonly known as archaea, display some unique characteristics. First, archaea possess a number of features in common with the eukaryotic nucleus

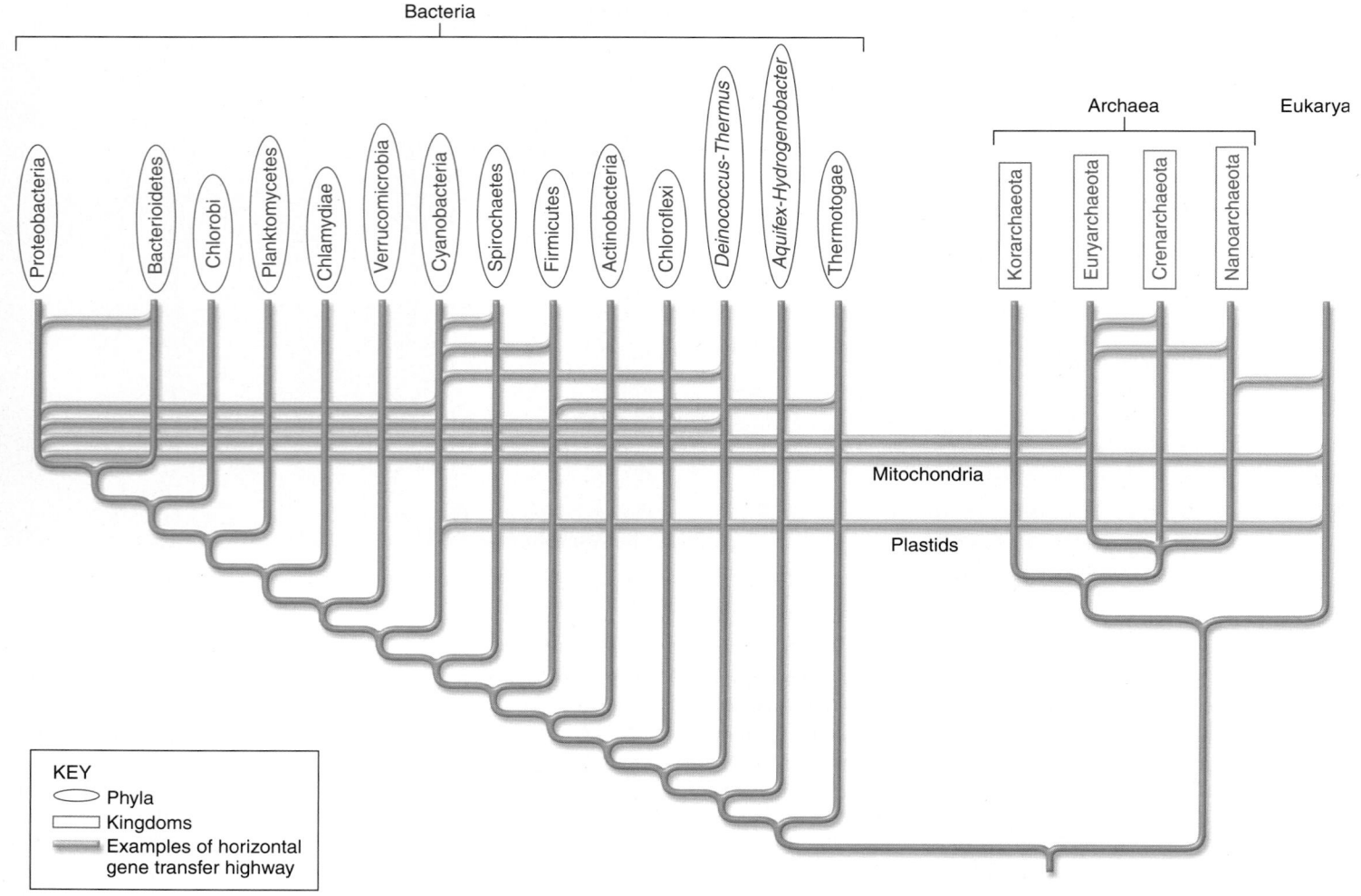

Figure 27.1 **Relationships and diversification of the Bacteria and Archaea.** Bacteria and Archaea are the two domains featuring prokaryotic cells. Eukarya is the domain of organisms composed of eukaryotic cells. Each domain has diversified into multiple kingdoms and phyla. Many cases of horizontal gene transfer among phyla, kingdoms, and domains are known. These cases include the acquisition of mitochondria and plastids by eukaryotes.

and cytoplasm, suggesting common ancestry. For example, histone proteins are typically associated with the DNA of both archaea and eukaryotes, but they are absent from most bacteria. Another distinctive feature of archaea is their membrane lipids, which are formed with ether linkages (in contrast, ester linkages characterize the membrane lipids of bacteria and eukaryotes). Ether-linked membranes are resistant to damage by heat and other extreme conditions, which helps explain why many archaea are able to grow in extremely harsh environments.

Though many archaea occur in soils and surface ocean waters of moderate conditions, diverse archaea occupy habitats of very high salt content, acidity, methane levels, or temperatures that would kill most bacteria and eukaryotes. Organisms that occur primarily in extreme habitats are known as **extremophiles**. One example is the methane producer *Methanopyrus*, which grows best at deep-sea thermal vent sites where the temperature is 98°C. In fact, *Methanopyrus* is so closely adapted to

its extremely hot environment that it will not grow when the temperature is less than 84°C. Such archaea are known as hyperthermophiles. Some archaea prefer habitats having both high temperatures and extremely low pH. For example, microbiologist Thomas Brock discovered the archaeal genus *Sulfolobus* in samples taken from sulfur hot springs having a pH of 3 or lower.

Extreme halophiles ("salt-lovers") occupy evaporation ponds used to produce salt from seawater, often growing so abundantly that they color the ponds red (**Figure 27.2**). Halophiles are often red because their plasma membranes contain large amounts of rhodopsins, proteins combined with the red light-sensitive pigment known as retinal. (Similar rhodopsins play important roles in eukaryote light sensing, and they are essential for animal vision.) In bacteria and archaea, rhodopsin functions as a proton pump, a protein that can move protons and other ions across the plasma membrane.

The domain Archaea includes the kingdoms Crenarchaeota, Euryarchaeota, Korarchaeota, and Nanoarchaeota. Crenarchaeota includes *Sulfolobus* and other organisms that grow in extremely hot or cold habitats. Euryarchaeota includes methane producers and extreme halophiles. Korarchaeota is primarily known from DNA sequences found in samples from hot springs. Nanoarchaeota includes the hyperthermophile *Nanoarchaeum equitans*, which appears to be a parasite of the thermal vent crenarchaeote *Ignicoccus*. Molecular biologist Elizabeth Waters and associates sequenced the exceptionally small genome of *N. equitans* and determined that it represents an early branching archaeal lineage.

Domain Bacteria Includes Proteobacteria, Cyanobacteria, and Many Other Phyla

Molecular studies suggest the existence of 50 or so bacterial phyla (considered kingdoms by some). However, the structural and metabolic features of about half of these are unknown. Though some members of domain Bacteria live in extreme environments, many more favor moderate conditions. Many bacteria form symbiotic associations with eukaryotes and are thus of concern in medicine and agriculture. The characteristics of 14 prominent bacterial phyla are briefly summarized in **Table 27.1**.

Table 27.1	Representative Bacterial Phyla
Phyla	**Characteristics**
Thermotogae	Hyperthermophiles; an ancient and slowly evolving lineage.
Aquifex-Hydrogenobacter	Hyperthermophiles; thought to be a very ancient lineage.
Deinococcus-Thermus	Extremophiles. The genus *Deinococcus* is known for high resistance to ionizing radiation, and the genus *Thermus* inhabits hot springs habitats. *Thermus aquaticus* has been used in commercial production of Taq polymerase enzyme used in the polymerase chain reaction (PCR), an important procedure in molecular biology laboratories.
Chloroflexi	Known as the green nonsulfur bacteria; conduct photosynthesis without releasing oxygen (anoxygenic photosynthesis).
Actinobacteria	Gram-positive bacteria producing branched filaments; many form spores. *Mycobacterium tuberculosis*, the agent of tuberculosis in humans, is an example. Actinobacteria are notable antibiotic producers; over 500 different antibiotics are known from this group. The pharmaceutical industry produces antibiotics from large-scale cultures of the actinobacterium *Streptomyces*. Some fix nitrogen in association with plants.
Firmicutes	Diverse Gram-positive bacteria, some of which produce endospores.
Spirochaetes	Motile bacteria having distinctive corkscrew shapes, with flagella held close to the body. They include the pathogens *Treponema pallidum*, the agent of syphilis, and *Borrelia burgdorferi*, which causes Lyme disease.
Cyanobacteria	The oxygen-producing photosynthetic bacteria (some are also capable of anoxygenic photosynthesis). Photosynthetic pigments include chlorophyll *a* and phycobilins, which often give cells a blue-green pigmentation. Occur as unicells, colonies, unbranched filaments, and branched filaments. Many of the filamentous species produce specialized cells: dormant akinetes and heterocysts in which nitrogen fixation occurs. In waters having excess nutrients, cyanobacteria produce blooms and may release toxins harmful to the health of humans and wild and domesticated animals.
Verrucomicrobia	Notable for producing tubulin-like proteins.
Chlamydiae	Notably tiny, obligate intracellular parasites. Some cause eye disease in newborns or sexually transmitted diseases.
Planktomycetes	Reproduce by budding; cell walls lack peptidoglycan; and cytoplasm contains nucleus-like bodies.
Chlorobi	Known as the green sulfur bacteria; conduct anoxygenic photosynthesis. Use H_2S or organic compounds as photosynthetic electron donors. Store elemental sulfur, which can be further oxidized to sulfate.
Bacteroidetes	Includes representatives of diverse metabolism types; some are common in the human intestinal tract, and others are primarily aquatic.
Proteobacteria	A very large group of Gram-negative bacteria, collectively having high metabolic diversity. Includes many species important in medicine, agriculture, and industry.

Figure 27.2 Hypersaline waters colored red by numerous halophilic archaea.

Biological inquiry: What material explains the red color of the water in this figure?

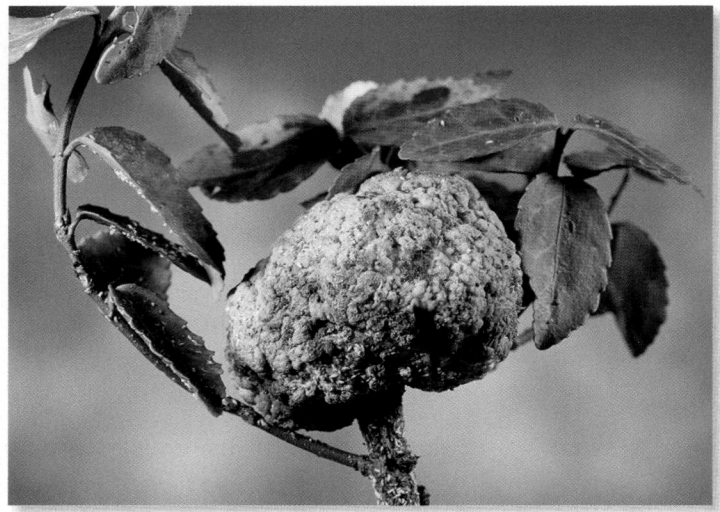

Figure 27.3 *Agrobacterium tumifaciens* infection causes cancer-like tumors to grow on plants.

Among these, the Proteobacteria and the Cyanobacteria are particularly diverse and relevant to eukaryotic cell evolution, global ecology, and human affairs.

Proteobacteria Though Proteobacteria share molecular and cell-wall features, this phylum displays amazing diversity of form and metabolism. Genera of this phylum are classified into five major subgroups: alpha (α), beta (β), gamma (γ), delta (δ), and epsilon (ε). As we saw in Chapter 22, the ancestry of mitochondria can be traced to the α-proteobacteria, which also include several genera noted for symbiotic relationships with animals and plants. For example, *Rhizobium* and related genera of α-proteobacteria form nutritionally beneficial associations with the roots of legume plants and are thus agriculturally important (see Chapter 37). Another α-proteobacterium, *Agrobacterium tumifaciens*, causes destructive cancer-like galls to develop on susceptible plants, including grapes and ornamental crops (**Figure 27.3**). *A. tumifaciens* induces gall formation by injecting DNA into plant cells, a property that has led to the use of the bacterium in the production of transgenic plants (see Chapter 20).

The genus *Nitrosomonas*, a soil inhabitant important in the global nitrogen cycle, represents the β-proteobacteria. *Neisseria gonorrhoeae*, the agent of the sexually transmitted disease gonorrhea, is another member of the β-proteobacteria. *Vibrio cholerae*, a γ-proteobacterium, causes cholera epidemics when drinking water becomes contaminated with animal waste during floods and other natural disasters. The γ-proteobacteria *Salmonella enterica* and *Escherichia coli* strain O157:H7 also cause human disease, and food and water are widely tested for their presence. The δ-proteobacteria include the colony-forming myxobacteria and predatory bdellovibrios, which drill their way through the cell walls of other bacteria in order to consume them. *Helicobacter pylori*, which causes stomach ulcers, belongs to the ε-proteobacteria.

Cyanobacteria The phylum Cyanobacteria contains photosynthetic bacteria that are abundant in fresh waters, oceans, and wetlands and on the surfaces of arid soils. Cyanobacteria are named for the typical blue-green (cyan) coloration of their cells (**Figure 27.4**). Blue-green pigmentation results from the presence of accessory phycobilin pigments that help chlorophyll absorb light energy. Cyanobacteria are the only prokaryotes that generate oxygen as a product of photosynthesis. Ancient cyanobacteria produced Earth's first oxygen-rich atmosphere, which allowed the rise of eukaryotes. The plastids of eukaryotic algae and plants arose from cyanobacteria.

Cyanobacteria display the greatest structural diversity found among bacterial phyla. Some occur as single cells, while others form colonies of cells held together by a thick gluey substance called mucilage (Figure 27.4a,b). Many cyanobacteria form filaments of cells that are attached end to end (Figure 27.4c). Some of the filamentous cyanobacteria produce specialized cells and display intercellular chemical communication, the hallmarks of multicellular organisms. Many cyanobacteria that grow in conditions of high light intensity produce protective brown sunscreen compounds at their surfaces (Figure 27.4d).

Cyanobacteria play essential ecological roles by producing organic carbon and fixed nitrogen (see Section 27.4). However, several kinds of cyanobacteria, notably the genera *Microcystis*, *Anabaena*, and *Cylindrospermopsis*, form nuisance growths in freshwater lakes during the warm season. Such growths, known as blooms, give the water a pea-soup appearance (see chapter-opening photo). Blooms develop when natural waters receive excess fertilizer from sewage discharges or agricultural runoff. Such blooms are becoming more common every year and are of serious concern because they may produce toxins in amounts sufficient to harm the health of humans and other animals. Consequently, it is inadvisable for people or pets to swim in or consume water that has a visible cyanobacterial bloom.

(a) *Chroococcus* spp.   0.1 mm

(b) *Merismopedia* spp.   60 μm

(c) *Oscillatoria* spp.   0.2 mm

(d) *Stigonema* spp.   50 μm

Figure 27.4 **The phylum Cyanobacteria illustrates major types of microbial cell aggregations.**   (a) The genus *Chroococcus* occurs as unicells. (b) The genus *Merismopedia* is a flat colony of cells held together by mucilage. (c) The genus *Oscillatoria* is an unbranched filament. (d) The genus *Stigonema* is a branched filament having a mucilage sheath, which has sunscreen compounds that are colored brown.

Now that we have learned something about the diversity of Archaea and Bacteria, let's consider the effects of gene exchanges within and between these domains.

Horizontal Gene Transfer Influences Diversity of Bacteria and Archaea

Horizontal gene transfer, also known as lateral gene transfer, is the movement of one or more genes from one species to another. This process contrasts with vertical gene transfer from parent to progeny. Horizontal gene transfer increases genetic diversity and influences the methods used to infer the phylogeny of bacteria and archaea.

Horizontal gene transfer is common among bacteria and archaea, and it can result in large genetic changes. For example, at least 17% of the genes present in the common human gut inhabitant *E. coli* came from other bacteria. In addition, genes move among the bacterial, archaeal, and eukaryotic domains. For example, about a third of the genes present in the archaeon *Methanosarcina mazei* originally came from bacteria, and there is genetic evidence for transfer of genes from Nanoarchaeota to protists. Viruses are probably important gene transfer vectors. This is illustrated by the fact that photosynthetic genes typical of oceanic cyanobacteria widely occur in marine viruses that infect these prokaryotes. These genes may evolve in the viral host before being transferred to another cyanobacterial cell.

To explore the extent of horizontal gene transfer, evolutionary microbiologists Robert Beiko, Timothy Harlow, and Mark Ragan examined the pattern of occurrence of more than 220,000 proteins encoded in 144 microbial genomes. These investigators discovered that major gene-sharing highways occur, particularly among close relatives or between microorganisms living in similar habitats (see Figure 27.1). As an example of the latter, Niels-Ulrik Frigaard, Edward DeLong, and their associates discovered a case of probable horizontal transfer between proteobacteria and euryarchaeota living in the same well-lit surface ocean waters. Proteobacteria occurring in such waters produce distinctive rhodopsins known as proteorhodopsins. As noted earlier, rhodopsins are pigment-protein complexes that enable cells to use light energy to drive proton pumps, a very useful trait for microbes occupying light-rich habitats. Because proteorhodopsins had not previously been found in euryarchaeota, the investigators were surprised to find that about 10% of euryarchaeota collected from the same area also possessed proteorhodopsin genes. This observation suggests that proteorhodopsin genes were horizontally transferred to the euryarchaeota from their proteobacterial neighbors. Acquisition of proteorhodopsin genes allowed some of the euryarchaeota to also benefit from light-driven proton pumping. This example and others have shown that horizontal gene transfer is an effective evolutionary process in bacteria and archaea, allowing them to acquire new metabolic processes despite lacking the sexual processes typical of eukaryotes. Horizontal transfer thereby increases the metabolic diversity of bacteria and archaea.

Horizontal gene transfer has the potential to interfere with human efforts to deduce evolutionary relationships. For example, if proteorhodopsin genes had been used in phylogenetic studies of ocean microorganisms, systematists might have falsely concluded that proteobacteria and euryarchaeota were closely related, because both groups possess these distinctive genes. Such a conclusion would be erroneous, because euryarchaeota did not inherit proteorhodopsin genes vertically. To understand prokaryote relationships, molecular systematists analyze ribosomal RNA (rRNA) genes and other sequences thought to move infrequently in a horizontal manner. Such analyses may more accurately reflect patterns of vertical inheritance. However, even ribosomal genes can be vulnerable to horizontal transfer. Scott Miller, Michelle Wood, and their associates discovered part of an rRNA gene from a proteobacterium within the genome of a cyanobacterium obtained from high-salinity water.

Phylogenetic trees displaying current understanding of relationships (see Figure 27.1) reveal several important concepts.

Bacteria and Archaea probably evolved from a common ancestor, and the eukaryotic nucleus and cytoplasm likely arose in an ancient archaeal organism. In addition, mitochondria and plastids originated from proteobacteria and cyanobacteria by endosymbiosis (see Chapter 22). In these cases, endosymbiosis resulted in the transfer of many genes from bacteria to eukaryotes. Finally, bacteria and archaea are amazingly diverse, but many phyla and species lack scientific names because microbiologists know so little about them. The next section surveys ways in which the better-known bacteria and archaea vary in structure and locomotion.

27.2 Structure and Motility

Bacteria and archaea have several features in common, including small size, rapid growth, and simple cellular structure. With few exceptions, bacteria and archaea are 1–5 μm in diameter and are thus known as microorganisms or microbes. (By contrast, most plant and animal cells are between 10 and 100 μm in diameter.) Small cell size limits the amount of materials that can be stored within cells but allows faster cell division. When nutrients are sufficient, many microorganisms can divide several times within a single day. This explains how bacteria can spoil food rapidly, and why infections can spread quickly within the human body. Despite these common features, prokaryotes differ in many ways. In this section we explore variation in cell structure and shape, in surface and cell wall features, and in movement. We will also learn how these characteristics influence the roles of bacteria and archaea in nature and human affairs.

Bacteria and Archaea Vary in Cellular Structure

Prokaryotic cells are much simpler than eukaryotic cells. Even so, many prokaryotes display surprising complexity of cellular structure resulting from adaptive evolution. For example, cyanobacteria and other photosynthetic bacteria are able to use light energy to produce organic compounds and typically contain large numbers of intracellular tubules known as thylakoids (Figure 27.5). The extensive membrane surface of the thylakoids has large amounts of chlorophyll and other components of the photosynthetic apparatus. Thylakoids thus enable photosynthetic bacteria to take maximum advantage of light energy in their environments.

Thylakoids develop by ingrowth of the plasma membrane, and in some bacteria, plasma membrane ingrowth has generated additional intriguing adaptations—magnetosomes and nucleus-like bodies—that are sometimes described as bacterial organelles. Magnetosomes are tiny crystals of an iron mineral known as magnetite, each surrounded by a membrane. These structures occur in the bacterium *Magnetospirillum* and related genera (Figure 27.6). In each cell, about 15 to 20 magnetosomes occur in a row, together acting as a compass needle to orient the bacteria within the Earth's magnetic field. This helps the bacteria to locate the low-oxygen habitats they prefer. Microbiologists Arash Komeili, Grant Jensen, and their colleagues used

Thylakoids provide a greater surface area for chlorophyll and other molecules involved in photosynthesis.

Thylakoids

Gas vesicles (cross sections)

Gas vesicles (long sections)

0.6 μm

The gas vesicles buoy this photosynthetic organism to the lighted water surface, where it often forms conspicuous scums.

Figure 27.5 Cells of the cyanobacterial genus *Microcystis* contain photosynthetic thylakoid membranes and numerous gas vesicles.

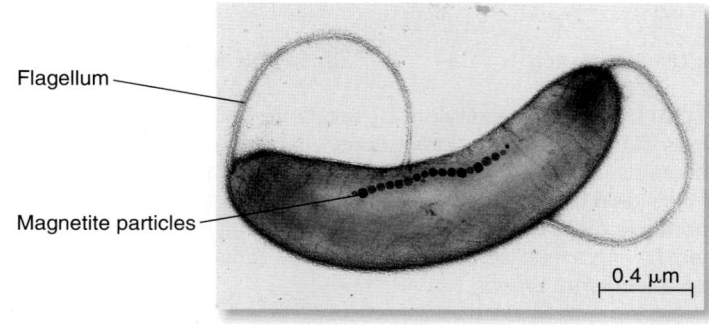

Flagellum

Magnetite particles

0.4 μm

Figure 27.6 The spirillum *Magnetospirillum magnetotacticum*. An internal row of iron-rich magnetite crystals, each enclosed by a membrane, functions like a compass needle, allowing this bacterium to detect the Earth's magnetic field. This feature allows *M. magnetotacticum* to locate its preferred habitat, low-oxygen subsurface waters. Cells use their flagellum to move from less favorable to more attractive locations.

rapid freezing techniques and a special type of transmission electron microscope to observe the development of magnetosomes. They found that the process begins with ingrowth of the plasma membrane to form a row of spherical vesicles. If *Magnetospirillum* cells are grown in media having low iron levels, the vesicles remain empty. But if iron is available, a magnetite crystal forms within each vesicle. The investigators also observed that fibrils of an actin-like protein keep the magnetosomes aligned in a row. (Recall from Chapter 4 that actin is a major cytoskeletal protein of eukaryotes.) Mutant bacteria lacking a functional form of this protein produce magnetosomes, but they do not remain aligned in a row. Instead, magnetosomes scatter around mutant cells, disrupting their ability to detect a magnetic field.

Plasma membrane invaginations also produce nucleus-like bodies in *Gemmata obscuriglobus* and other members of the bacterial phylum Planktomycetes. In *G. obscuriglobus*, an envelope composed of a double membrane encloses all cellular DNA and some ribosomes. Although this bacterial envelope lacks nuclear pores characteristic of the eukaryotic nuclear envelope, it likely plays a similar adaptive role in isolating DNA from other cellular influences.

A final example of surprising cell structure complexity occurs in a bacterial phylum known as the Verrucomicrobia. These bacteria produce cellular proteins that are remarkably similar to eukaryotic tubulins, proteins that self-assemble into microtubules (see Chapter 4). The existence of actin- and tubulin-like proteins in prokaryotic cells suggests that eukaryotic microfilaments and microtubules originated from ancestral proteins by descent with modification.

Archaea and Bacteria Vary in Cell Shape and Arrangement

Microbial cells occur in five major shapes (**Figure 27.7**): spheres (**cocci**), rods (**bacilli**), comma-shaped cells (**vibrios**), and spiral-shaped cells that are either flexible (**spirochaetes**) or rigid (**spirilli**; see Figure 27.6). In addition, such cells can occur in several types of arrangements. Some microorganisms occur only as single cells or pairs of cells resulting from recent division (Figure 27.7a). Others occur as cell aggregates or are attached end to end to form filaments.

The shapes of bacterial cells, and the degree of their arrangement in groups, can be important diagnostic features in water and medical testing. For example, the presence of abundant rod-shaped cells in samples suggests possible contamination by *E. coli* or other proteobacteria that normally occur in the human gut. The proteobacterial genus *Neisseria* occurs as paired cocci, a feature that aids rapid detection of this organism when disease is suspected. Microorganisms may also produce surface mucilage, which plays important roles in disease and ecology.

Slimy Mucilage Often Coats Microbial Surfaces

Many microorganisms produce a coat of slimy mucilage that varies in consistency and extent. Mucilage is composed of polysaccharides, protein, or both, which are secreted from cells and serves a variety of functions. One example is the mucilage coat (known as a capsule) that helps some disease bacteria to evade the defense system of their host. You may recall that Frederick Griffith discovered the transfer of genetic material while experimenting with capsule-producing pathogenic forms and capsule-less nonpathogenic strains of the bacterium *Streptococcus pneumoniae* (refer back to Figure 11.1). The immune system cells of mice are able to destroy this bacterium only if it lacks a capsule.

Mucilage also holds together the cells of colonial microorganisms (see Figure 27.4b), helps aquatic species to float in water, binds mineral nutrients, and repels predators. Slime sheaths (see Figure 27.4d) often coat bacterial filaments, where they may help to prevent drying. Mucilage is also critical to the formation of biofilms, which are environmentally and medically important. **Biofilms** are aggregations of microorganisms that secrete adhesive mucilage, thereby gluing themselves to surfaces (**Figure 27.8**). They help microbes to remain in favorable locations for growth; otherwise body or environmental fluids would wash them away. A process known as **quorum sensing** fosters biofilm formation. During quorum sensing, individual microbes secrete small molecules having the potential to influence the behavior of nearby microbes. If enough individuals are present (a quorum), the concentration of signaling molecules builds to a level that causes collective behavior. In the case of biofilms, populations of microbes respond to chemical signals by moving to a common location and producing mucilage.

From a human standpoint, biofilms have both beneficial and harmful consequences. In aquatic and terrestrial environments, they help to stabilize and enrich sand and soil surfaces. In contrast, biofilms that form on the surfaces of animal tissues can be harmful. Dental plaque is an example of a harmful biofilm (Figure 27.8); if allowed to remain, the bacterial community secretes acids that can damage tooth enamel.

(a) Lactococcus lactis   1 μm   **(b) Lactobacillus plantarum** 11.4 μm   **(c) Vibrio cholerae** 15 μm   **(d) Leptospira jaundice** 7.5 μm

Figure 27.7 **Major types of microbial cell shapes.** Scanning electron microscopic views. (a) Ball-shaped cocci. (b) Rod-shaped bacilli. (c) Comma-shaped vibrios. (d) Spiral-shaped spirochaetes.

(a)   (b)

Figure 27.8 A biofilm composed of a community of micro-organisms glued by mucilage to a surface. **(a)** This transmission electron micrograph shows a thin slice of dental plaque, with the tooth surface at the lower edge. **(b)** Close-up view showing chains of *Streptococcus sobrinus* bacterial cells, which are dark because they have been labeled with an identifying antibody. The plaque layer is about 50 μm deep.

Biofilms may also develop in industrial pipelines, where the attached microbes can contribute to corrosion by secreting enzymes that chemically degrade metal surfaces.

Archaea and Bacteria Differ in Cell-Wall Structure

Whether coated with mucilage or not, most prokaryotes possess a rigid cell wall outside the plasma membrane. Cell walls maintain cell shape and help protect against attack by viruses or predatory bacteria. Cell walls also help microbes avoid lysing in hypotonic conditions, when the solute concentration is higher inside the cell than outside. The structure and composition of cell walls vary in ways that can be important in medical and other contexts.

The cell walls of most archaea and certain bacteria are composed of protein or glycoprotein. In contrast, the polymer known as **peptidoglycan** is lacking from archaea but is an important component of most bacterial cell walls. Peptidoglycan is composed of carbohydrates that are cross-linked by peptides. Bacterial cell walls occur in two major forms that differ in their amount of peptidoglycan, staining properties, and response to antibiotics. The first type of bacterial cell wall features a relatively thick peptidoglycan layer (**Figure 27.9a**). The second bacterial cell-wall type has relatively less wall peptidoglycan and is enclosed by a thin, outer envelope whose outer leaflet is rich in **lipopolysaccharides** (**Figure 27.9b**). This outer layer envelope is a lipid bilayer, but is distinct from the plasma membrane. Lipopolysaccharides and peptidoglycan affect bacterial responses to antibiotics and sometimes disease symptoms.

Acidic polysaccharides

Thick peptidoglycan layer

Plasma membrane

(a) Thick cell wall, no outer envelope

Lipopolysaccharide rich outer envelope

Thin peptidoglycan layer

Plasma membrane

(b) Thinner cell wall, with outer envelope

Figure 27.9 Cell-wall structure of Gram-positive and Gram-negative bacteria. **(a)** The structure of the cell wall of Gram-positive bacteria. **(b)** The structure of the cell wall and lipopolysaccharide envelope typical of Gram-negative bacteria.

For example, part of the peptidoglycan covering of *Bordetella pertussis* is responsible for the extensive tissue damage associated with whooping cough. To identify the two major types of bacterial cell walls, microbiologists use a staining procedure known as the Gram stain.

The Gram Stain In the late 1800s, the Danish physician Hans Christian Gram developed the **Gram stain** procedure to more easily detect and distinguish bacteria. The Gram stain remains a useful tool to identify bacteria and predict their responses to antibiotics. To perform a Gram stain, a microbiologist starts by smearing bacteria onto a glass slide and heating it briefly to aid cell attachment. The microbiologist then floods the slide

(a) Gram-positive bacteria **(b) Gram-negative bacteria**

Figure 27.10 Gram-positive and Gram-negative bacteria. **(a)** *Streptococcus pneumoniae* stains positive (purple) with the Gram stain. **(b)** *Escherichia coli* stains negative (pink) when the Gram stain procedure is applied.

with crystal violet, a purple dye, followed by an iodine solution. The iodine binds the purple dye, forming an insoluble complex. Next, the microbiologist adds alcohol. The alcohol dehydrates peptidoglycan, thereby trapping the purple crystal violet-iodine complex. The alcohol is able to remove purple dye from thin peptidoglycan walls, but not dye bound in cell walls with thick peptidoglycan layers. Finally, safranin, a pink stain, is applied. At the end of the procedure, some bacteria will remain purple; these are known as Gram-positive bacteria (**Figure 27.10a**). Other types of bacteria will lose the purple stain at the alcohol step but retain the final pink stain; these are known as Gram-negative bacteria (**Figure 27.10b**). If a fluorescence microscope is available, a single-step fluorescent stain can be used to distinguish Gram-positive from Gram-negative bacteria (**Figure 27.11**). A closer look at Gram-positive and Gram-negative bacteria will reveal how these staining differences are useful.

Gram-Positive Bacteria Gram-positive bacteria occur in the phyla Firmicutes and Actinobacteria (see Figure 27.1, Table 27.1). Gram-positive bacteria typically have thick peptidoglycan cell walls lacking a lipopolysaccharide envelope. When treated by the Gram stain process, Gram-positive bacteria appear purple because their more abundant peptidoglycan traps a large amount of the crystal violet-iodine dye complex. Such bacteria are typically vulnerable to penicillin and related antibiotics, because these antibiotics interfere with peptidoglycan synthesis. An example of a Gram-positive bacterium is *S. pneumoniae* (see Figure 27.10a). Strains of this organism cause strep throat, streptococcal pneumonia, streptococcal meningitis (an infection of the spinal fluid), and eye infections. Streptococci also include the infamous "flesh-eating" bacteria that cause necrotizing fasciitis, a disease whose progression is notoriously difficult to control. The entire DNA sequence of *S. pneumoniae* has been determined, revealing the presence of many genes conferring the ability to break down several types of molecules in human tissues for use as food. The presence of these genes explains why *S. pneumoniae* can cause such varying disease symptoms. *Staphylococcus aureus* is another medically important Gram-positive bacterium, which is commonly found on human skin.

Gram-Negative Bacteria Gram-negative bacteria have thin peptidoglycan cell walls enclosed by a lipopolysaccharide envelope.

Figure 27.11 A commercial fluorescent stain can be used to distinguish Gram-positive and Gram-negative bacteria. This preparation shows green-fluorescent cells of the Gram-negative bacterium *E. coli* mixed with orange-yellow fluorescent cells of the Gram-positive bacterium *Staphylococcus aureus*.

Biological inquiry: What are the advantages and disadvantages of this fluorescence staining procedure as compared to the classical Gram stain process?

Diverse phyla of bacteria display Gram-negative staining. The outer lipopolysaccharide envelope and lower amount of cell-wall peptidoglycan enable Gram-negative bacteria to resist the effects of penicillin and chemically similar antibiotics. (Gram-negative bacteria would be treated with different antibiotics.) *E. coli* is an example of a Gram-negative bacterium that lives in the human lower intestine (see Figure 27.10b). A particularly virulent strain, *E. coli* O157:H7, is a significant cause of foodborne illness.

Microbes Display Diverse Types of Motility Structures

Many microorganisms have structures at the cell surface or within cells that enable them to change position in their environment. Motility allows microbes to move to favorable conditions within gradients of light, gases, or nutrients. In addition, motility structures allow them to respond to chemical signals emitted from other microbes during quorum sensing and mating.

Microbes move by swimming, twitching or gliding, adjusting their floatation in water, or pirating cytoskeletal proteins of invaded eukaryotic cells.

Structures known as **flagella** enable swimming behavior. Prokaryotic flagella differ from eukaryotic flagella in several ways. Prokaryotic flagella lack an internal cytoskeleton of microtubules, the motor protein dynein, and a plasma membrane covering, all features that characterize eukaryotic flagella (see Chapter 4). Unlike eukaryotic flagella, prokaryotic flagella do not repeatedly bend and straighten. Instead, prokaryotic flagella are propelled by molecular machines composed of a filament,

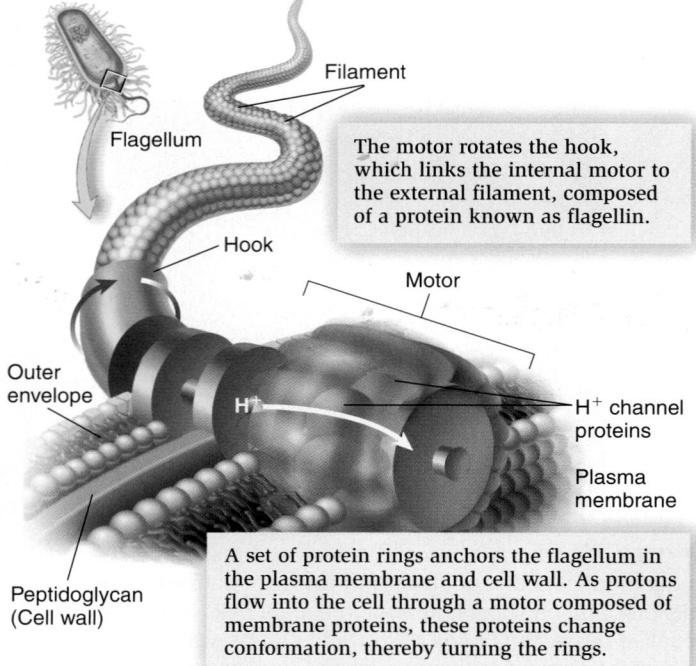

Filament

Flagellum

The motor rotates the hook, which links the internal motor to the external filament, composed of a protein known as flagellin.

Hook

Motor

Outer envelope

H⁺

H⁺ channel proteins

Plasma membrane

A set of protein rings anchors the flagellum in the plasma membrane and cell wall. As protons flow into the cell through a motor composed of membrane proteins, these proteins change conformation, thereby turning the rings.

Peptidoglycan (Cell wall)

Figure 27.12 Diagram of a prokaryotic flagellum, showing motor, hook, and filament.

hook and motor that work together somewhat like a boat's outboard motor (**Figure 27.12**). Lying outside the cell, the long, stiff, curved filament acts as a propeller. The hook links the filament with the motor, a set of protein rings at the cell surface. Hydrogen ions (protons), which have been pumped out of the cytoplasm, usually via the electron transport system, diffuse back into the cell through channel proteins. This proton movement powers the turning of the hook and filament.

Prokaryotic species differ in the number and location of flagella, which may occur singly, in clumps at one pole of a bacterial cell, or emerging from around the cell (**Figure 27.13**). Differences in flagellar number and location cause microorganisms to exhibit different modes of swimming. For example, spirochaete flagella are located outside the peptidoglycan cell wall but within the confines of an outer membrane that holds them close to the cell. Rotation of these flagella causes spirochaetes to display characteristic bending, flexing, and twirling motions.

Some prokaryotes twitch or glide across surfaces, using threadlike cell surface structures known as **pili** (**Figure 27.14**). *Myxococcus xanthus* cells, for example, move by alternately extending and retracting pili from one pole or the other. This process allows directional movement toward food materials. If nutrients are low, cells of these bacteria glide together to form tiny treelike colonies, which are part of a reproductive process. As we will see, pili can also play important roles in bacterial reproduction and disease processes.

Cyanobacteria and some other bacteria that live in aquatic habitats use cytoplasmic structures known as **gas vesicles** to adjust their buoyancy. This process allows them to move up or down in the water column, an advantage in finding nutrients and avoiding harmful conditions. Gas vesicles are hollow cylinders whose water-tight walls are made of protein (see Figure 27.5). These vesicles do not actually contain gas concentrations different from the rest of the cell, but their density is lower than that of surrounding cytoplasm, helping bacteria to float. Gas vesicles can repeatedly collapse and reassemble, depending on

1.6 μm

(a) *Vibrio parahaemoliticus*

1.4 μm

(b) *Salmonella enterica*

Figure 27.13 Microbial cells can produce one or more flagella at the poles, or numerous flagella around the periphery. **(a)** *Vibrio parahaemoliticus*, a bacterium that causes seafood poisoning, has a single short flagellum. **(b)** *Salmonella enterica*, another bacterium that causes food poisoning, has many flagella distributed around the cell periphery.

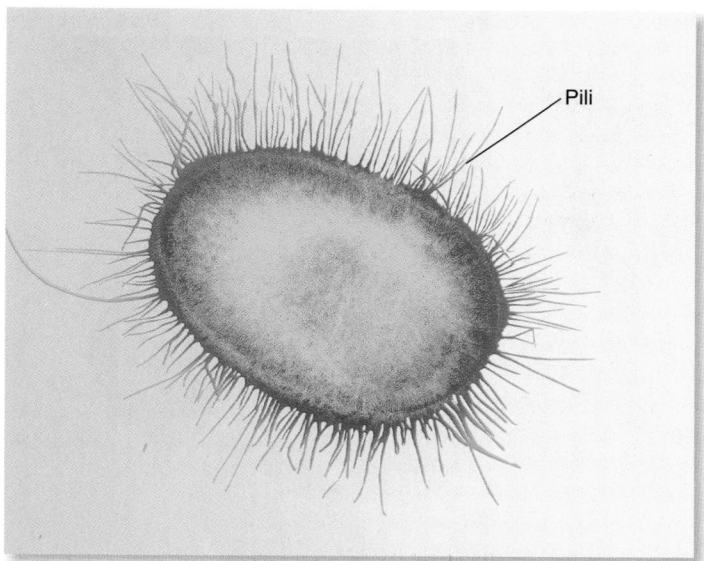

Figure 27.14   Pili extending from the surface of *Proteus mirabilis*.

the cell's solute content. When sugars and other solutes are abundant in the cytoplasm, the high osmotic pressure collapses the vesicles. Cells in which the gas vesicles have collapsed experience increased density, so that they tend to sink. Sinking removes bacteria from damaging radiation at the water's surface and allows them to harvest inorganic minerals from deeper water. But when organic solutes made via photosynthesis are used up during cell metabolism, the gas vesicles reassemble, providing floatation to the surface allowing photosynthesis to resume. The presence of gas vesicles explains why cyanobacteria often form buoyant scums at the surfaces of ponds and lakes (see chapter-opening photo).

27.3 Reproduction

Bacteria and archaea enlarge their populations by means of a process known as binary fission. In addition, some microbes produce tough cells that can withstand deleterious conditions for long periods in a dormant condition. Finally, although bacteria and archaea lack meiosis and other features typical of eukaryotic sexual reproduction, they are able to obtain additional genes by a variety of methods. Each of these aspects of reproduction helps to explain how microbes function in nature and human affairs.

Populations of Bacteria and Archaea Increase by Binary Fission

The cells of bacteria and archaea divide by splitting in two, a process known as **binary fission** (**Figure 27.15a**; also refer back to Figure 18.6). If sufficient nutrients are available, an entire population of identical cells can be produced from a single parental cell by repeated binary fission. This growth process allows microbes to become very numerous in water, food, or animal tissues, potentially causing harm.

Binary fission is the basis for a widely used method for detecting and counting bacteria in patient fluids, food, or water samples. Microbiologists who study the spread of disease need to quantify bacterial cells in samples taken from the environment. Medical technicians often need to count bacteria in body fluid samples to assess the likelihood of infection. But because bacterial cells are small and often unpigmented, they are difficult to count directly. One way that microbiologists count bacteria is to place a measured volume of sample into plastic dishes filled with a semisolid nutrient medium. Bacteria in the sample undergo repeated binary fission to form colonies of cells that are visible to the unaided eye (**Figure 27.15b**).

| 0.8 μm |

(a) Bacterium undergoing binary fission   **(b) Colonies developed from single cells**   **(c) Bacteria stained with fluorescent DNA-binding dye**

Figure 27.15   **Binary fission and counting microbes.**   (a) Division of a bacterial cell as viewed by scanning electron microscopy. (b) When samples are spread onto the surfaces of laboratory dishes containing nutrients, single cells of bacteria or archaea may divide repeatedly to form visible colonies, which can be easily counted. The number of colonies is an estimate of the number of culturable cells in the original sample. (c) If a fluorescence microscope is available, cells can be counted directly by applying a fluorescent stain that binds to cell DNA. Each cell glows brightly when illuminated with ultraviolet light.

Biological inquiry: Which procedure would you choose to count bacteria in a sample that is known to include many species that have not as yet been cultured?

Since each colony represents a single cell that was present in the original sample, the number of colonies in the dish reflects the number of living bacteria in the original sample.

Another way to detect and count bacteria is to treat samples with a stain that binds bacterial DNA, causing cells to glow brightly when illuminated with ultraviolet light. The glowing bacteria can be viewed and counted by the use of a fluorescence microscope (**Figure 27.15c**). The fluorescence method must be used when the microbes of interest cannot be cultured in the laboratory. Microorganisms can also be stained with fluorescent reagents that bind to specific DNA sequences, allowing microbiologists to detect and count bacteria and archaea of precise genetic types. This procedure, known as fluorescence *in situ* hybridization (FISH), is a powerful technique for finding particular types of microorganisms in mixed populations.

Some Bacteria Survive Harsh Conditions as Akinetes or Endospores

Some bacteria produce thick-walled cells known as akinetes or endospores that are able to survive unfavorable conditions in a dormant state. Akinetes and endospores develop when bacteria have experienced stress, such as low nutrients or unfavorable temperatures. Such cells are able to germinate into metabolically active cells when conditions improve again. For example, aquatic filamentous cyanobacteria often produce large, food-filled **akinetes** when winter approaches (**Figure 27.16a**). Akinetes are able to survive winter at the bottoms of lakes, and they produce new filaments in spring when they are carried by water currents to the brightly lit surface.

Endospores (**Figure 27.16b**) are cells having tough protein coats that are produced inside bacterial cells and then released when the enclosing cell dies and breaks down. Endospores can remain alive, though in a dormant state, for hundreds of years at least. Some studies suggest even longer lifetimes. For example, Raúl Cano and Monica Borucki cultured bacterial cells from *Bacillus sphaericus* endospores that were carefully isolated from the gut of a bee that had been trapped in amber (hardened tree resin) for more than 25 million years. These and other researchers determined that DNA obtained from the cultures was identical to DNA taken directly from fossil bees. They also found that similar bacteria and endospores commonly occur in modern bees of the same species.

The ability to produce endospores allows some Gram-positive bacteria to cause serious diseases. For example, *Bacillus anthracis* causes the disease anthrax, a potential agent in bioterrorism and germ warfare. Most cases of human anthrax result when endospores of *B. anthracis* enter wounds, causing skin infections that are relatively easily cured by antibiotic treatment. But sometimes the endospores are inhaled, or consumed in undercooked, contaminated meat, potentially causing more serious illness or death. *Clostridium botulinum* can contaminate the improperly canned food that has not been heated to temperatures high enough to destroy its tough endospores. When

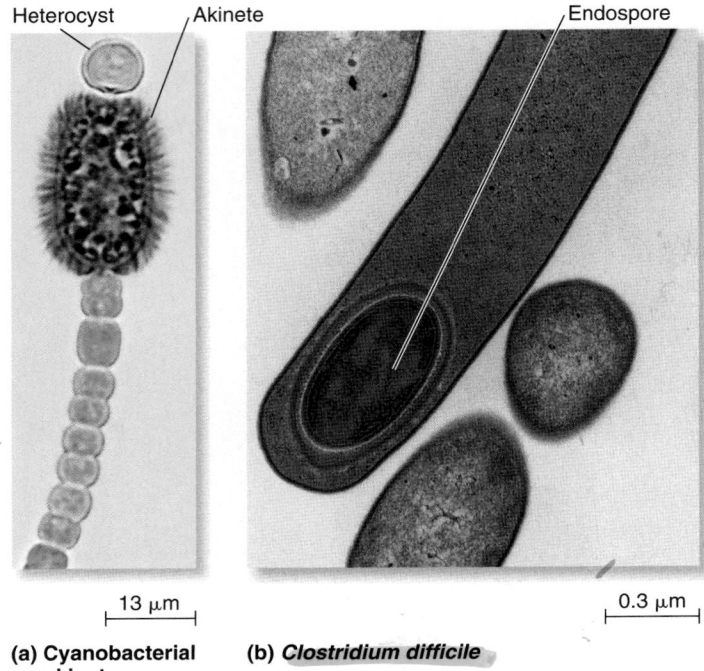

Heterocyst   Akinete   Endospore

|← 13 µm →| |← 0.3 µm →|

(a) Cyanobacterial akinete   **(b)** *Clostridium difficile*

Figure 27.16   Specialized cells capable of dormancy. **(a)** Akinetes are thick-walled, food-filled cells produced by some cyanobacteria. They are able to resist stressful conditions and generate new populations when conditions improve. As discussed later, the heterocyst is a specialized cell in which nitrogen fixation occurs. **(b)** An endospore with a resistant wall develops within the cytoplasm of the bacterium *Clostridium difficile*.

the endospores germinate and bacterial cells grow in the food, they produce a deadly toxin, as well as NH_3 and CO_2 gas, which causes can lids to bulge. If humans consume the food, the toxin causes botulism, a severe type of food poisoning that can lead to respiratory and muscular paralysis. The botulism toxin has been recently marketed commercially as Botox, which is injected into the skin, where it paralyzes facial muscles, thereby reducing the appearance of wrinkles. *Clostridium tetani* produces a nerve toxin that causes lockjaw, also known as tetanus, when bacterial cells or endospores enter wounds from soil. The ability of the genera *Bacillus* and *Clostridium* to produce resistant endospores helps to explain their widespread presence in nature and their potential danger to humans.

Bacteria and Archaea Obtain Genetic Material by Transduction, Transformation, and Conjugation

In Chapter 18, you were introduced to the varied ways in which microorganisms can acquire genetic material from other cells. In this chapter we have noted the ecological and evolutionary impacts of horizontal gene exchange. DNA may enter cells by means of viral vectors in the process known as **transduction**.

Microbes are able to take up DNA directly from their environments, in the process known as **transformation**. Some bacteria transmit DNA during a mating process known as **conjugation**. Pili (see Figure 27.14) help mating cells attach to each other, allowing conjugation to proceed in Gram-negative bacteria. Such gene exchange processes have contributed to the diversity of microbial nutrition and metabolism, our next topic.

27.4 Nutrition and Metabolism

Bacteria and archaea use a wide variety of materials and chemical transformations to obtain carbon and energy for growth (**Table 27.2**). These materials and transformations explain the nutritional diversity of microorganisms. Bacteria and archaea together display more diverse types of metabolism than other groups of organisms. Microbes can be classified according to type of nutrition, response to oxygen, and presence of specialized metabolic processes.

Bacteria and Archaea Display Diverse Types of Nutrition and Responses to Oxygen

Cyanobacteria and some other bacteria are **autotrophs** (meaning "self-feeders"), organisms that are able to produce all or most of their own organic compounds. Autotrophic microorganisms fall into two categories: photoautotrophs and chemoautotrophs. **Photoautotrophs** are able to use light as a source of energy for synthesis of organic compounds from CO_2 and H_2O, or H_2S. **Chemoautotrophs** are able to use energy obtained by chemical modifications of inorganic compounds to synthesize organic compounds. Such chemical modifications include nitrification (the conversion of ammonia to nitrate) and the oxidation of sulfur, iron, or hydrogen.

Heterotrophs (meaning "other feeders") are organisms that require at least one organic compound, and often more. Some microorganisms are **photoheterotrophs**, meaning that they are able to use light energy to generate ATP, but they must take in organic compounds from their environment. **Chemoheterotrophs** must obtain organic molecules for both energy and as a carbon source. Among the many types of bacterial chemoheterotrophs is *Propionibacterium acnes*, which causes acne, affecting up to 80% of adolescents in the U.S. The genome sequence of *P. acnes* has revealed numerous genes that allow it to break down skin cells and consume the products.

Microorganisms differ in their need for oxygen. **Obligate aerobes** require O_2. **Facultative aerobes** are more adaptable than obligate aerobes. They can use O_2 in aerobic respiration, obtain energy via anaerobic fermentation, or use inorganic chemical reactions to obtain energy. One fascinating example of a facultative aerobe is *Thiomargarita namibiensis*, a giant bacterium that lives in marine waters off the Namibian coast of Africa. This heterotroph obtains its energy in two ways: by oxidizing sulfide with oxygen when this is available or, when oxygen is low or unavailable, by oxidizing sulfide with nitrate. In either case, the cells convert sulfide to elemental sulfur, which is stored within the cells as large globules.

In contrast to obligate aerobes, **obligate anaerobes** such as the bacterial genus *Clostridium* are poisoned by O_2. People suffering from gas gangrene (caused by *Clostridium perfringens* and related species) are usually treated by placement in a chamber having a high oxygen content (called a hyperbaric chamber), which kills the organisms and deactivates the toxins. **Aerotolerant anaerobes** do not use O_2, but they are not poisoned by it either. These organisms obtain their energy by anaerobic respiration, which uses electron acceptors other than oxygen in electron transport processes. Anaerobic metabolic processes include denitrification (the conversion of nitrate into N_2 gas) and the reduction of manganese, iron, and sulfate, which are all important in the Earth's mineral cycles.

Bacteria Play Important Roles as Nitrogen Fixers

Many cyanobacteria and some other microbes are known as **diazotrophs** ("dinitrogen consumers") because they conduct a specialized metabolic process called **nitrogen fixation**. The removal of nitrogen from the gaseous phase is called fixation. During nitrogen fixation, the enzyme nitrogenase converts inert atmospheric gas (N_2) into ammonia (NH_3). Plants and eukaryotic algae can use ammonia (though not N_2) to produce proteins and other essential nitrogen-containing molecules. As a result, many plants have developed symbiotic relationships with diazotrophs, which provide ammonia to the plant partner. For example, nitrogen-fixing cyanobacteria typically live in leaf cavities of water ferns that float in rice paddies. The fern partner and the rice crop both take up ammonia produced by the diazotrophs. This natural fertilizer allows the rice plants to produce more protein. Many types of heterotrophic soil bacteria also fix nitrogen. Examples include *Rhizobium* and its relatives, which form close associations with the roots of legumes (see Chapter 37).

Table 27.2	Major Nutritional Types of Bacteria and Archaea		
Nutrition type	**Energy source**	**Carbon source**	**Example**
Autotroph			
Photoautotroph	Light	CO_2	Cyanobacteria
Chemoautotroph	Inorganic compounds	CO_2	*Sulfolobus*
Heterotroph			
Photoheterotroph	Light	Organic compounds	Chloroflexi
Chemoheterotroph	Organic compounds	Organic compounds	Many

Oxygen binds to nitrogenase, irreversibly disabling it. Thus, most diazotrophs are able to conduct nitrogen fixation only in low-oxygen habitats. Many cyanobacteria accomplish nitrogen fixation in specialized cells known as **heterocysts** (see Figure 27.16a). Heterocysts display many adaptations that reduce nitrogenase exposure to oxygen. These adaptations include thick walls, which reduce inward oxygen diffusion; absence of oxygen-producing photosystems; and increased respiration, which consumes oxygen.

GENOMES & PROTEOMES

Gene Expression Studies Revealed How Cyanobacteria Fix Nitrogen in Hot Springs

The microbial communities of hot springs in Yellowstone National Park and other thermal areas around the world have long fascinated microbiologists interested in the occurrence of life at high temperatures. Thermal pools characteristically display beautiful, multicolored microbial mats (**Figure 27.17**). Such mats are composed of diverse photoheterotrophs, chemoautotrophs, chemoheterotrophs, and photoautotrophs, including many types of cyanobacteria. However, in Yellowstone thermal pools where temperatures range from 50 to 70°C single-celled cyanobacteria of the genus *Synechococcus* are the only photoautotrophic organisms present. Heterocyst-producing cyanobacteria are absent from such hot waters, and few heterotrophic diazotrophs tolerate such temperatures, yet nitrogen fixation occurs in these pools. It was not clear which organisms could fix nitrogen until genomic information provided an essential clue.

Anne-Soisig Steunou, Arthur Grossman, and associates sequenced the genomes of *Synechococcus* cultures obtained from

Figure 27.18   Daily changes in photosynthesis and nitrogen fixation gene expression in *Synechococcus* cultures isolated from thermal pools.

Biological inquiry: At what time of the day does the photosynthesis gene psaB reach its lowest level of expression, and when does the nitrogen fixation gene nifK reach its highest expression level?

these hot springs and discovered that *nif* (<u>ni</u>trogen <u>f</u>ixation) genes were present. They also determined that *Synechococcus nif* genes were expressed in parts of the microbial mat having temperatures near 60°C. These results indicated that *Synechococcus* conducts nitrogen fixation in the mats, producing fixed nitrogen and other compounds used by other community members. But the investigators wondered how this single-celled bacterium managed to fix nitrogen without the oxygen arising from photosynthesis disabling the functioning of nitrogenase. They tracked the expression of *Synechococcus* genes over a 24-hour period and found that after nightfall the expression of a photosynthesis gene (*psaB*) fell dramatically, while expression of nitrogen fixation genes (*nifH*, *nifD*, and *nifK*) increased (**Figure 27.18**). They concluded that *Synechococcus* turns on nitrogen fixation at dusk, when oxygen production from photosynthesis drops. But nitrogen fixation is an energy-intensive process, so how does *Synechococcus* fuel it? Steunou and colleagues used differential gene expression analyses to establish that *Synechococcus* uses fermentation, an anaerobic metabolic pathway, to supply the ATP needed for nitrogen fixation. This study not only illustrates important features of microbial metabolism but also the essential ecological roles played by bacteria and archaea, a topic that we will explore in more detail next.

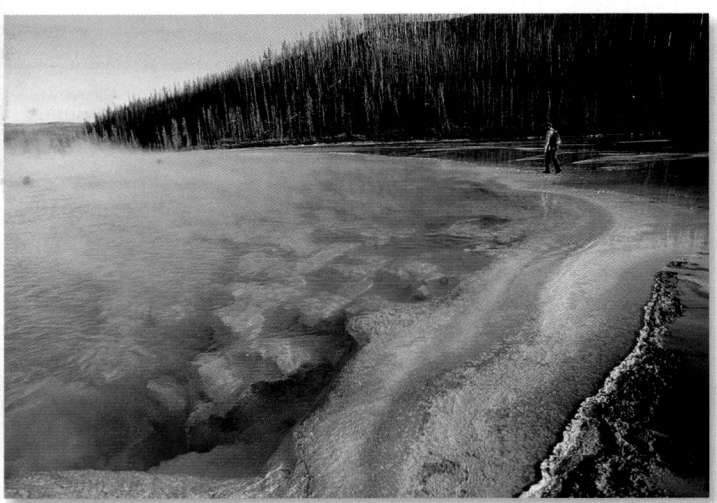

Figure 27.17   **A thermal pool in Yellowstone National Park.** Brightly colored mats at the pool edge are microbial communities that include thermophilic cyanobacteria.

27.5 Ecological Roles and Biotechnology Applications

Bacteria and archaea play several key ecological roles, including the production and cycling of carbon. Earth's carbon cycle depends on microorganisms that produce and degrade organic compounds, including methane. Bacteria also play fascinating roles as symbionts living in close associations with eukaryotes.

In this section, we focus on these diverse ecological roles and also provide examples of ways that humans use the metabolic capabilities of bacteria and archaea in biotechnology.

Bacteria and Archaea Play Important Roles in Earth's Carbon Cycle

Earth's carbon cycle is the sum of all the transformations that occur among compounds that contain carbon. (See Chapter 59 for a detailed discussion of the carbon cycle.) Bacteria and archaea are important in producing and degrading organic compounds. For example, cyanobacteria and other autotrophic bacteria are important **producers**. These bacteria, together with algae and plants, synthesize the organic compounds used by other organisms for food. **Decomposers**, also known as saprobes, include heterotrophic microorganisms (as well as fungi and animals). These organisms break down dead organisms and organic matter, releasing minerals for uptake by living things. Microbial decomposers can consume diverse materials, including oil, explosives, and pesticides, making them useful in cleaning up toxic spills and dumps.

Archaea and bacteria also influence Earth's carbon cycle by producing and consuming methane (CH_4). Methane—the major component of natural gas—is a powerful greenhouse gas, as are CO_2 and H_2O vapor. Atmospheric methane thus has the potential to alter the Earth's climate. Several groups of anaerobic archaea known as the **methanogens** convert CO_2, methyl groups, or acetate to methane and release it from their cells. Methanogens live in swampy wetlands, in deep-sea habitats, or in the digestive systems of ungulate animals, such as cattle. Marsh gas produced in wetlands is largely composed of methane, and large quantities of methane produced long ago are trapped in deep-sea and subsurface Arctic deposits.

Microbes also play a role in limiting atmospheric methane. The balance of methane in Earth's atmosphere is maintained by the activities of aerobic bacteria known as **methanotrophs**, which consume methane. Some methanotrophs live in symbiotic association with marine invertebrates in deep-sea thermal vents or hydrocarbon seep communities. Methane-consuming bacteria also live in close association with wetland plants, which release the oxygen needed by these bacteria to metabolize methane. In the absence of methanotrophs, Earth's atmosphere would be much richer in the greenhouse gas methane, which would substantially increase global temperatures.

Many Bacteria Live in Symbiotic Associations with Eukaryotes

As we have seen, many bacteria live in symbiotic associations with eukaryotic organisms, a relationship called **symbiosis**. If symbiotic associations are beneficial to both partners, the association is known as a **mutualism**. If one partner benefits at the expense of the other, the association is known as a **parasitism**, and the partner that benefits is termed a parasite or a pathogen. There are numerous examples of mutualistic and parasitic bacteria.

Mutualistic Microbes Bacteria are involved in many mutually beneficial partnerships in aquatic and terrestrial habitats. Many aquatic protists depend on bacterial partners for vitamins or other essential compounds. For example, the common green seaweed *Ulva* does not display its typical lettuce-leaf-like structure unless bacterial partners belonging to the phylum Bacteroidetes are present. The bacteria produce a compound that induces normal seaweed development. Bioluminescent bacteria form symbiotic relationships with squid and other marine animals. In deep-sea thermal vent communities, sulfur-oxidizing bacteria live within the tissues of tube worms and mussels, supplying these animals with carbon compounds used as food. Bacteriologist Cameron Curry and his associates have documented a complex land association involving four partners: ants, fungi that the ants cultivate for food, parasitic fungi that attack the food fungi, and mutualistic Actinobacteria, which produce antibiotics. These antibiotics control the growth of the fungal parasite, preventing it from destroying the ants' fungal food supply. The ants rear the useful bacteria in cavities on their body surfaces; glands near these cavities supply the bacteria with nutrients. Aphids also harbor mutualistic bacteria that help to defend against pathogens or parasites. Nitrogen-fixing cyanobacteria partner with a variety of plants (see Chapter 31) and, together with certain fungi, form many types of lichens (see Chapter 29).

Parasitic and Pathogenic Microbes **Parasites** are organisms that obtain organic compounds from living hosts. If parasitic microbes cause disease symptoms in their hosts, the microorganisms are known as **pathogens**. Cholera, leprosy, tetanus, pneumonia, whooping cough, diphtheria, Lyme disease, scarlet fever, rheumatic fever, typhoid fever, bacterial dysentery, and tooth decay are among the many examples of human diseases caused by bacterial pathogens. Bacteria also cause many plant diseases of importance in agriculture, including blights, soft rots, and wilts. How do microbiologists determine which bacteria cause these diseases? The pioneering research of the Nobel Prize–winning German physician Robert Koch provides the answer.

In the mid- to late 1800s, Koch established a series of steps to determine whether a particular organism causes a specific disease. First, the presence of the suspected pathogen must correlate with occurrence of symptoms. Next, the pathogen must be isolated from an infected host and grown in pure culture if possible, and cells from the pure culture should cause disease when inoculated into a healthy host. Finally, one should be able to isolate the same pathogen from the second-infected host. Using these steps, known as **Koch's postulates**, Koch discovered the bacterial causes of anthrax, cholera, and tuberculosis. Subsequent investigators have used Koch's postulates to establish the causes of many other infectious diseases. As recently as the 1980s, bacteriologists used Koch's postulates to establish that Legionnaires' disease is caused by a bacterial pathogen, *Legionella pneumophila*.

How Pathogenic Bacteria Attack Cells Modern research is providing new information about how bacteria attack cells. Such knowledge aids in developing strategies for disease prevention and treatment. Many pathogenic bacteria attack cells by binding to the target cell surfaces and injecting substances that help them utilize cell components. During their evolution, some pathogenic bacteria have transformed flagella into needle-like systems for injecting infection proteins into animal or plant cells as part of the infection process. Such modified flagella are known as type III secretion systems called injectisomes (**Figure 27.19a**). *Yersinia pestis* (the agent of bubonic plague) and *Salmonella enterica* (which causes salmonellosis) are examples of bacteria whose type III secretion systems allow them to attack human cells. These bacteria also induce the host cell to form a plasma membrane pocket that encloses the bacterial cell, bringing it into the target cell. Once within a host's cell, pathogenic bacteria use the cell's resources to reproduce and spread to nearby tissues.

Some other bacterial pathogens use a type IV secretion system to deliver toxins or transforming DNA into cells (**Figure 27.19b**). Examples of such bacteria that cause human disease include *Helicobacter pylori*, *Legionella pneumophila*, and *Bordetella pertussis*. The plant pathogen *Agrobacterium tumifaciens* uses a type IV secretion system to transfer DNA (T DNA) into plant cells. The bacterial T DNA encodes an enzyme that affects normal plant growth, with the result that cancer-like tumors develop (see Figure 27.3). Type IV systems evolved from pili and other components of bacterial mating. Thus, types III and IV attack systems are examples of descent with modification, the evolutionary process by which organisms acquire new features. Some experts propose that type III and IV systems may be useful in human gene therapy, to deliver DNA to target cells.

Antibiotic-Resistant Pathogens Antibiotic compounds are widely used to treat diseases caused by bacterial pathogens, but overuse of antibiotics can lead to production of resistant bacteria, a serious health problem. For example, some strains of *Staphylococcus aureus*, which causes "staph" infections, have developed resistance to penicillin and similar antibiotics. Livestock are often fed antibiotics to reduce the effects of minor bacterial infections or to prevent infections from spreading, thereby increasing the production profit margin. Unfortunately, this activity promotes the evolution of antibiotic-resistant bacterial strains. When humans become infected by such strains, antibiotic treatments are not effective, and cases have occurred in which patients' infections were resistant to all known antibiotics. The World Health Organization has advised eliminating the routine use of antibiotics in livestock.

Some Bacteria Are Useful in Industrial and Other Applications

Several industries have harnessed the metabolic capabilities of microbes obtained from nature. The food industry uses bacteria to produce chemical changes in food that improve consistency

(a) Type III secretion system

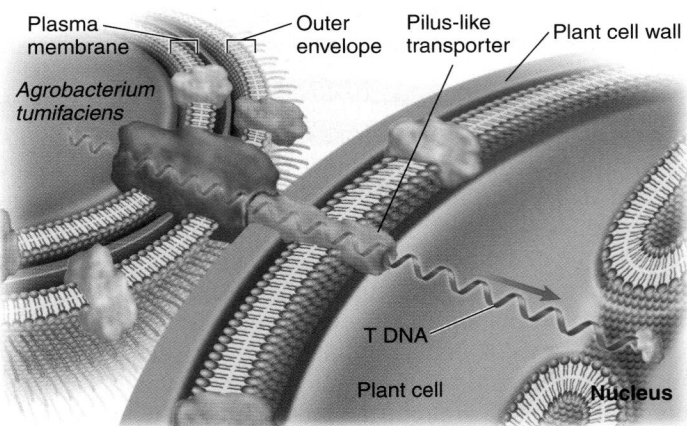

(b) Type IV secretion system

Figure 27.19 Attack systems of pathogenic bacteria. **(a)** The Type III secretion system functions like a syringe to inject proteins into host cells, thereby starting a disease process. **(b)** The Type IV secretion system forms a channel through which DNA can be transmitted from a pathogen to a host cell, in this case from the bacterium *Agrobacterium tumifaciens* into a plant cell.

or flavor—to make dairy products, including cheese and yogurt, for example. Cheese makers add pure cultures of certain bacteria to milk. The bacteria consume milk sugar (lactose) and produce lactic acid, which aids in curdling the milk.

The chemical industry produces materials such as vinegar, amino acids, enzymes, vitamins, insulin, vaccines, antibiotics, and other useful pharmaceuticals by growing particular bacteria in giant vats. For example, bacteria produce the antibiotics streptomycin, tetracycline, kanamycin, gentamycin, bacitracin, polymyxin-B, and neomycin.

The ability of microorganisms to live in harsh environments and break down organic compounds makes them very useful in treating wastewater, industrial discharges, and harmful substances such as explosives, pesticides, and oil spills. This process is known as bioremediation. *Geobacter sulfurreducens*, for example, is used to precipitate metals such as uranium from contaminated water, thereby purifying it. This bacterium uses metal ions and elemental sulfur in an electron transport process to oxidize acetate to CO_2, in the process producing ATP needed for growth and reproduction. Taking advantage of this fact, engineers drip acetate into the groundwater, thereby enriching the population of *G. sulfurreducens* to 85% of the microbial community at contaminated sites, which more efficiently precipitates uranium.

In addition to food production and waste treatment, bacteria are also used in agriculture. Several species of *Bacillus*, particularly *B. thuringiensis*, produce crystalline proteins known as Bt-toxins, which kill insects that ingest them. Tent caterpillars, potato beetles, gypsy moths, mosquitoes, and black flies are among the pests that can be controlled by Bt-toxin. For this reason, genes involved in Bt-toxin production have been engineered into some crop plants such as corn to limit pests, thereby increasing crop yields. *Deinococcus radiodurans* is another example of a bacterial species whose unusual ecological properties may prove useful to humans, as described next.

FEATURE INVESTIGATION

The Daly Experiments Revealed How Mn(II) Helps *Deinococcus radiodurans* to Avoid Radiation Damage

The bacterial phylum known as *Deinococcus-Thermus* includes *Deinococcus radiodurans*, which is unusually resistant to chemical mutagens and nuclear radiation (see Table 27.1). This bacterium can survive brief radiation doses greater than 10,000 Gray (Gy) and continuous radiation levels as high as 50 Gy/hour. (By contrast, a radiation dose of 5 Gy is lethal to humans. This trait evolved as an adaptation that aids *Deinococcus* survival in its natural arid desert habitats. Like nuclear radiation, drying and solar radiation cause chromosome breakage, and *Deinococcus* has acquired very effective methods for repairing damaged DNA.) M. J. Daly and associates were interested in learning more about how *D. radiodurans* survives treatment with high radiation, with the hope that such information might lead to better ways of treating victims of radiation sickness.

As the result of a series of experiments, Daly and colleagues learned that radiation-resistant bacteria tended to have higher levels of manganese—Mn(II)—ions than do radiation-sensitive bacteria. These facts suggested the hypothesis that *D. radiodurans* might protect itself by accumulating Mn(II) ions to high levels. In experiments reported in 2004 and described in **Figure 27.20**, Daly and colleagues first grew *D. radiodurans* in basic culture media containing three different levels of manganese ion (50, 100, or 250 nM Mn(II)). As they grew, the bacteria took up Mn(II) ions from the media. Researchers then inoculated bacteria from each type of medium into separate sectors of laboratory dishes containing media to which low or high levels of Mn(II) ions had been added. Next, they exposed some of the dishes to 50 Gy/hour of radiation, leaving one as a nonirradiated control. After a period of growth, irradiated bacteria grown in dishes containing high Mn(II) ion levels grew as well as the control, but irradiated bacteria grown in dishes containing low Mn(II) ion levels did not grow as well. Irradiated bacteria that had been pregrown with high Mn(II) ion levels fared particularly well.

This experiment revealed that *D. radiodurans* cells that had accumulated high levels of Mn(II) ions were better able to grow in the presence of ionizing radiation.

Figure 27.20   Manganese helps *Deinococcus radiodurans* to avoid radiation damage.

HYPOTHESIS High levels of intracellular Mn(II) help to protect *Deinococcus radiodurans* from the effects of ionizing radiation.

STARTING MATERIALS *Deinococcus radiodurans*.

Experimental level

1 Cultivate *Deinococcus radiodurans* on media having different Mn(II) ion content (50 nM, 100 nM, 250 nM).

50 nM   100 nM   250 nM

2 Inoculate the bacteria onto the surfaces of media differing in Mn(II) ion content.

(1) Control (25 nM)

(2) High Mn medium (2.5 M)

(3) Low Mn medium (2.5 nM)

3 Expose media plates to 50 Gy/hour of ionizing radiation.

Not Irradiated

4 Compare bacterial growth to control plates not irradiated.

5 THE DATA

Results from step 4: Even when exposed to ionizing radiation, *D. radiodurans* grew well when high levels of Mn(II) ions are present in its environment (compare plates 2 and 1). The higher the level of Mn accumulated by cells, the better the growth after ionizing radiation (plate 2). When environmental levels of Mn are low, cells are more readily damaged by ionizing radiation (compare plates 2 and 3).

CHAPTER SUMMARY

27.1 Diversity and Evolution

- Domains of life include Eukarya, Bacteria, and Archaea (known informally as eukaryotes, bacteria, and archaea). (Figure 27.1)
- Many representatives of the Archaea occur in extremely hot, salty, or acidic habitats. Ether-linked membranes are among the features of archaea that enable their survival in extreme habitats. (Figure 27.2)
- The domain Bacteria includes 50 or more phyla, including Proteobacteria and Cyanobacteria, which are particularly diverse and of great evolutionary and ecological importance. (Table 27.1)
- Mitochondria arose from a proteobacterial ancestor, and many modern proteobacteria occur in symbiotic relationships with animals or plants. (Figure 27.3)

- Algal and plant plastids arose from a cyanobacterial ancestor. Modern cyanobacteria play important ecological roles, including the production of harmful blooms in overfertilized waters. (Figure 27.4)
- Widespread horizontal DNA transfer has occurred among bacteria and archaea. Horizontal DNA transfer is particularly common among close relatives and inhabitants of the same environment, and it allows microorganisms to evolve rapidly.

27.2 Structure and Motility

- Bacteria and archaea are composed of prokaryotic cells that are smaller and simpler than those of eukaryotes. Even so, structures such as thylakoids, magnetosomes, and nucleus-like bodies are examples of surprising cell structure complexity. (Figures 27.5, 27.6)

- Cells of bacteria and archaea vary in shape. Major cell shape types are spherical cocci, rod-shaped bacilli, comma-shaped vibrios, and coiled spirilli and spirochaetes. Some cyanobacteria display specialized cells and intercellular communication, the hallmarks of multicellular organisms. (Figure 27.7)

- Many microbes occur within a coating of slimy mucilage, which may play a role in diseases or aid in the development of biofilms. Biofilm development is influenced by quorum sensing, a process in which group activity is coordinated by chemical communication. (Figure 27.8)

- Most prokaryotic cells possess a protective cell wall. Archaea and some bacteria have walls composed of proteins, but most bacterial cell walls contain peptidoglycan, which is composed of carbohydrates cross-linked by peptides. (Figure 27.9)

- Gram-positive bacterial cells have walls rich in peptidoglycan, while Gram-negative cells have less peptidoglycan in their walls and are enclosed by a lipopolysaccharide envelope. Gram-positive bacteria can be distinguished from Gram-negative bacteria by use of the Gram stain and other staining procedures. (Figures 27.10, 27.11)

- Motility enables microbes to change positions within their environment, which aids in locating favorable conditions for growth. Some microorganisms swim by means of flagella; others twitch or glide by the use of threadlike pili, or adjust their buoyancy in water by means of intracellular vesicles. (Figures 27.12, 27.13, 27.14)

27.3 Reproduction

- Populations of bacteria and archaea enlarge by binary fission, a simple type of cell division that provides a means by which culturable microbes can be counted. (Figure 27.15)

- Some bacteria are able to survive harsh conditions as dormant akinetes or endospores. (Figure 27.16)

- Many microorganisms obtain new DNA sequences directly from their environment (transformation), by means of viral vectors (transduction), or by a mating process (conjugation).

27.4 Nutrition and Metabolism

- Bacteria and archaea can be grouped according to nutritional type, response to oxygen, or presence of distinctive metabolic features. Major nutritional types are photoautotrophs, chemoautotrophs, photoheterotrophs, and chemoheterotrophs. (Table 27.2)

- Obligate aerobes require oxygen, while facultative aerobes are able to live with oxygen or without it by using different processes for obtaining energy. Obligate anaerobes are poisoned by oxygen, while aerotolerant anaerobes do not use oxygen but are not poisoned by it; both obtain their energy by anaerobic respiration.

- Nitrogen fixation is an example of distinctive metabolism displayed only by certain microorganisms. A number of plants display symbiotic associations with bacteria that fix nitrogen, called diazotrophs; many of these associations are ecologically or agriculturally important. (Figures 27.17, 27.18)

27.5 Ecological Roles and Biotechnology Applications

- Bacteria and archaea play key roles in Earth's carbon cycle as producers, decomposers, symbionts, or pathogens.

- Methane-producing methanogens and methane-using methanotrophs are important in the carbon cycle, and they influence the Earth's climate.

- Parasitic bacteria obtain organic compounds from living hosts, and if disease symptoms result, such bacteria are known as pathogens.

- Bacteria attack eukaryotic cells by means of modified flagella, known as Type III secretion systems, or Type IV secretion systems, which evolved from pili and other mating components. (Figure 27.19)

- Many bacteria and archaea are useful in industrial and other applications; others are used to make food products or antibiotics or to clean up polluted environments. The cellular adaptations of bacteria that are extremely resistant to ionizing radiation may suggest new ways to treat radiation sickness in humans. (Figure 27.20)

TEST YOURSELF

1. Which of the following features is common to prokaryotic cells?
   a. a nucleus, featuring a nuclear envelope with pores
   b. mitochondria
   c. plasma membranes
   d. mitotic spindle
   e. none of the above

2. The bacterial phylum that produces oxygen gas as the result of photosynthesis is
   a. the proteobacteria.
   b. the cyanobacteria.
   c. the Gram-positive bacteria.
   d. all of the above.
   e. none of the above.

3. The Gram stain is a procedure that microbiologists use to
   a. determine if a bacterial strain is a pathogen.
   b. determine if a bacterial sample can break down oil.
   c. infer the structure of a bacterial cell wall and bacterial response to antibiotics.
   d. count bacteria in medical or environmental samples.
   e. all of the above.

4. Place the following steps in the correct order, according to Koch's postulates:
   I. Determine if pure cultures of bacteria cause disease symptoms when introduced to a healthy host.
   II. Determine if disease symptoms correlate with presence of a suspected pathogen.
   III. Isolate the suspected pathogen and grow it in pure culture, free of other possible pathogens.
   IV. Attempt to isolate pathogen from second-infected hosts.
   a. II, III, IV, I
   b. II, IV, III, I
   c. III, II, I, IV
   d. II, III, I, IV
   e. I, II, III, IV

5. Cyanobacteria play what ecological role?
   a. producers
   b. consumers
   c. decomposers
   d. parasites
   e. none of the above

6. Bacterial structures that are produced by pathogenic bacteria for use in attacking host cells include
   a. Type III and IV secretion systems.
   b. magnetosomes.
   c. gas vesicles.
   d. thylakoids.
   e. none of the above.

7. The structures that enable some Gram-positive bacteria to remain dormant for extremely long periods of time are known as
   a. akinetes.
   b. endospores.
   c. biofilms.
   d. lipopolysaccharide envelopes.
   e. pili.

8. By means of what process do populations of bacteria or archaea increase their size?
   a. mitosis
   b. meiosis
   c. conjugation
   d. transduction
   e. none of the above

9. By what means do bacterial cells acquire new DNA?
   a. by conjugation, the mating of two cells of the same bacterial species
   b. by transduction, the injection of viral DNA into bacterial cells
   c. by transformation, the uptake of DNA from the environment
   d. all of the above
   e. none of the above

10. How do various types of bacteria move?
    a. by the use of flagella composed of motor, hook, and filament
    b. by means of pili, which help cells twitch or glide along a surface
    c. by using gas vesicles to regulate buoyancy in water bodies
    d. all of the above
    e. none of the above

CONCEPTUAL QUESTIONS

1. Explain why many microbial populations grow more rapidly than do eukaryotes, and how bacterial population growth influences the rate of food spoilage or infection.

2. Why does the overuse of antibiotics in medicine and agriculture result in widespread antibiotic resistance?

3. What organisms are responsible for the blue-green blooms that often occur in warm weather on lake surfaces?

EXPERIMENTAL QUESTIONS

1. What feature of *Deinococcus radiodurans* attracted the attention of researchers?

2. What hypothesis did Daly and associates develop to explain radiation resistance in *D. radiodurans*?

3. As shown in the Feature Investigation in Figure 27.20, bacterial cells were grown in media having various levels of manganese ion, which they absorbed. Later, some of these bacteria were exposed to high levels of ionizing radiation, while control bacteria were not exposed. What results of this experiment support the hypothesis that cellular manganese plays a role in radiation resistance?

COLLABORATIVE QUESTIONS

1. How would you go about cataloging the phyla of bacteria and archaea that occur in a particular place?

2. How would you go about developing a bacterial product that could be sold for remediation of a site contaminated with materials that are harmful to humans?

www.brookerbiology.com

This website includes answers to the Biological Inquiry questions found in the figure legends and all end-of-chapter questions.

28
PROTISTS

A population of green algal cells, each surrounded by a halo of protective mucilage.

Protists are eukaryotes that live in moist habitats and are mostly microscopic in size. Despite their small size, protists have a greater impact on global ecology and human affairs than most people realize. For example, the photosynthetic protists known as algae generate at least half of the oxygen in the Earth's atmosphere and produce organic compounds that feed marine and freshwater animals. The oil that fuels our cars and industry derives from pressure-cooked algae that accumulated on the ocean floor over millions of years. Today, algae are being engineered into systems for producing biofuels and cleaning pollutants from water.

Protists also include some parasites that cause serious human illnesses. For example, in 1993 the waterborne protist *Cryptosporidium parvum* sickened 400,000 people in Milwaukee, Wisconsin, costing $96 million in medical expenses and lost work time. The related protist *Plasmodium falciparum*, which is carried by mosquitoes in many warm regions of the world, causes the disease malaria. Every year, nearly 500 million people become ill with malaria and more than 2 million die of this disease. As we will see in this chapter, sequencing the genomes of these and other protist species has suggested new ways of battling such deadly pathogens.

In this chapter, we will survey protist diversity, including structural, nutritional, and ecological variations. We begin by exploring ways of informally classifying protists, by ecological roles, habitat, and motility. We then will focus on the defining features and evolutionary importance of the major protist phyla. Next, the nutritional modes and defensive adaptations of protists are discussed, and we conclude by looking at the reproductive adaptations that allow protists to exploit and thrive in a variety of environments. In the course of the discussion, you will also see the enormous role that protists play in ecosystem functioning and in human health. Our study of protists also illustrates important principles of diversity, particularly ways in which organisms modify their environments and provide essential biological services.

28.1 An Introduction to Protists

Protists are eukaryotes that are not classified in the plant, animal, or fungal kingdoms, though some protists are closely related to plants or animals or fungi (**Figure 28.1**). The term protist comes from the Greek word *protos*, meaning "first," reflecting the fact that protists were Earth's first eukaryotes. Protists display two common characteristics: They are most abundant in moist habitats, and most of them are microscopic in size. Despite these shared features, modern phylogenetic analyses based on comparative analysis of DNA sequences and cellular features reveal that protists do not form a monophyletic group. Instead, protist phyla are classified into several eukaryotic **supergroups** that each display distinctive features (**Table 28.1**). One of these supergroups contains several phyla of protists in addition to the plant kingdom, while another supergroup links certain protists to the animal and fungal kingdoms (see Figure 28.1). Such classifications reveal the great importance of protists in our understanding of eukaryotic evolution.

In addition to classification based on evolutionary relationships, protists are often informally classified according to diverse ecological roles, habitats, or type of motility. In this section, we will examine these informal classifications as a way of introducing protist diversity.

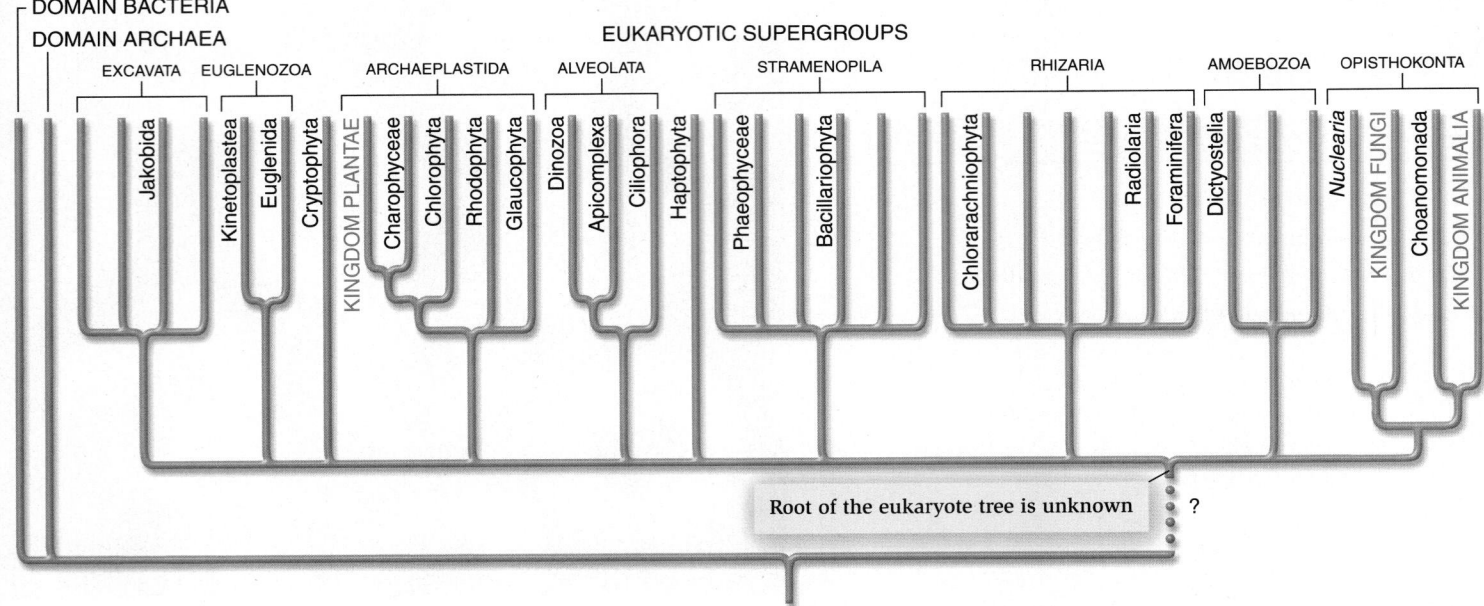

Figure 28.1 **A phylogenetic tree showing major eukaryotic supergroups.** Eight supergroup names are indicated above brackets, the names of eukaryotic kingdoms are shown in red, and major protist phyla and classes discussed in this chapter are also shown. The position of the root of the eukaryotic tree is uncertain, as are the branching patterns of supergroups and many phyla.

Protists Can Be Informally Classified According to Diverse Ecological Roles

Protists are often classified according to ecological roles into three major groups: algae, protozoa, and fungus-like protists. The term **algae** (Latin for "seaweeds") applies to about 10 phyla of protists that include both photosynthetic and nonphotosynthetic species (see Table 28.1). Photosynthetic algae produce organic compounds and oxygen, thereby providing substances used by heterotrophic organisms. For example, the single-celled algae known as diatoms often serve as food for heterotrophic protists (**Figure 28.2**). Despite the common feature of photosynthesis, algae do not form a monophyletic group descended from a single common ancestor (Figure 28.1).

Similarly, the term **protozoa** (Greek for "first life") is commonly used to describe diverse heterotrophic protists. Many protozoa are mobile by means of flagella or other specialized motility structures. Protozoa feed by absorbing small organic molecules or ingesting prey. For example, the protozoa known as ciliates move by means of many short motility structures known as cilia, and they munch on smaller cells such as diatoms (Figure 28.2). Like the algae, the protozoa do not form a monophyletic group. Instead, they are widely distributed among the major protist supergroups (Figure 28.1).

The heterotrophic **fungus-like protists** often resemble true fungi in having threadlike, filamentous bodies and absorbing nutrients from their environments (**Figure 28.3a**; see Chapter 29).

Figure 28.2 **Algal and protozoan cells.** A large, heterotrophic ciliate protozoan, whose surface is covered with motility cilia, has consumed several golden-pigmented, glass-walled algal cells known as diatoms. The algal cells were ingested by the process of phagotrophy and will be digested as food. Diatom cells that have avoided capture glide nearby.

Table 28.1	Eukaryotic Supergroups and Examples of Constituent Kingdoms, Phyla, Classes, or Species

Supergroup	KINGDOMS, **Phyla**, classes, or *species*	Distinguishing features
EXCAVATA	**Jakobida** *Giardia lamblia* *Trichomonas vaginalis*	Unicellular flagellates, often with feeding groove; mitochondria highly modified in specialized parasites
EUGLENOZOA	**Kinetoplastea** (kinetoplastids) *Trypanosoma brucei* **Euglenida** (euglenoids)	Unicellular flagellates; disk-shaped mitochondrial cristae; secondary plastids (when present) derived from endosymbiotic green algae
ARCHAEPLASTIDA	**Glaucophyta** (glaucophytes) **Rhodophyta** (red algae) **Chlorophyta** (green algae) Charophyceae (charophyceans) KINGDOM PLANTAE (land plants)	Primary plastids having only two envelope membranes
ALVEOLATA	**Dinozoa** (dinoflagellates) *Pfiesteria shumwayae* **Ciliophora** (ciliates) **Apicomplexa** (apicomplexans) *Cryptosporidium parvum* *Plasmodium falciparum*	Peripheral membrane sacs (alveoli); some Dinozoa have secondary plastids derived from red algae, some have secondary plastids derived from green algae, and some have tertiary plastids derived from diatoms or cryptomonads; Apicomplexa sometimes have secondary plastids derived from red or green algae
STRAMENOPILA	**Bacillariophyta** (diatoms) Phaeophyceae (brown algae) *Phytophthora infestans* (fungus-like)	Strawlike flagellar hairs; fucoxanthin accessory pigment common in autotrophic forms
RHIZARIA	**Chlorarachniophyta** **Radiolaria** **Foraminifera**	Thin, cytoplasmic projections; secondary plastids (when present) derived from endosymbiotic green algae
AMOEBOZOA	*Entamoeba histolytica* **Dictyostelia** (a slime mold phylum) *Dictyostelium discoideum*	Amoeboid movement by pseudopodia
OPISTHOKONTA	**Choanomonada** (choanoflagellates) KINGDOM ANIMALIA KINGDOM FUNGI *Nuclearia* spp.	Swimming cells possess a single posterior flagellum

Protist phyla listed in this chapter whose supergroup affiliations are controversial or unknown:

	Haptophyta (haptophytes)	Haptonema, derived from a flagellum that aids food-gathering and attachment; many produce calcium carbonate cell coverings; secondary plastids derived from red algae
	Cryptophyta (cryptomonads)	Flagellates; most have secondary plastids derived from red algae

While the structure and reproduction of fungus-like protists resemble those of a true fungus (**Figure 28.3b**), they are actually more closely related to diatoms than they are to fungi. These examples illustrate that the terms algae, protozoa, and fungus-like protists, while useful in describing ecological roles, lack taxonomic or evolutionary meaning. Next we will survey ways in which protists have been informally classified according to diverse habitats.

Protists Can Be Informally Classified According to Diverse Habitats

Although protists occupy nearly every type of moist habitat, they are particularly common and diverse in oceans, lakes, wetlands, and rivers. Even extreme aquatic environments such as Antarctic ice and acidic hot springs serve as habitats for some protists.

In such places protists may swim or float in open water, or live attached to surfaces such as rocks or beach sand. These different habitats influence protist structure and size.

Swimming or floating protists are members of an informal group of organisms known as plankton, a group that also includes bacteria, viruses, and small animals. The photosynthetic protists in plankton are called **phytoplankton** ("plantlike" plankton), while heterotrophic protists in plankton are described as protozoan plankton. Planktonic protists are necessarily quite small in size; otherwise they would readily sink to the bottom. Staying afloat is particularly important for phytoplankton, which need light for photosynthesis. For this reason, planktonic protists occur primarily as single cells, colonies of cells held together with mucilage, or short filaments of cells linked end to end (**Figure 28.4a–c**).

Many protists live within **periphyton**, communities of micro-organisms that are attached by mucilage to underwater surfaces such as rocks, sand, and plants. Because sinking is not a problem for attached protists, these often produce multicellular bodies, such as branched filaments (**Figure 28.4d**). Although some **seaweeds** are very large single cells, most are multicellular algae, which often produce large and complex bodies (**Figure 28.4e**). Seaweeds usually grow attached to underwater surfaces such as rocks, sand, or offshore oil platforms. Seaweeds require sunlight and carbon dioxide for photosynthesis and growth, so most grow along coastal shorelines, fairly near the water surface. Even so, some seaweeds occur at amazing depths. By using a submersible vehicle, algal ecologists Mark Littler, Diane Littler, and associates found seaweeds growing 210 m beneath the surface of ocean waters near San Salvador in the Bahama Islands, the deepest known occurrence of photosynthetic eukaryotes.

(a) Protist attacking a leaf

(b) True fungus

Figure 28.3   A fungus-like protist and a true fungus.   (a) A fungus-like protist, *Phytophthora infestans*, has absorbed enough nutrients from the plant to produce reproductive structures that emerge from pores on the lower leaf surface. The reproductive structures will eventually break off and be dispersed to other host leaves. (b) A true fungus (*Phragmidium* spp.) produces threadlike filaments that likewise grow through leaf tissues, absorbing nutrients. These nutrients will be used to produce reproductive structures that perpetuate the parasite.

Biological inquiry: How could a microbiologist determine whether unknown filaments growing in diseased plant leaves were fungus-like protists or true fungi?

(a) *Chlamydomonas* with flagella

(b) The colonial genus *Pediastrum*

(c) The filamentous protist genus *Desmidium*

(d) The branched filamentous protist genus *Cladophora*

(e) The seaweed genus *Acetabularia*

Figure 28.4   Diversity of algal body types.   (a) The single-celled flagellate genus *Chlamydomonas* occurs in the phytoplankton of lakes. (b) The colonial genus *Pediastrum* is composed of several cells arranged in a lacy star shape. This arrangement aids in keeping this alga afloat in water. (c) The filamentous genus *Desmidium* occurs as a twisted row of cells. (d) The branched filamentous genus *Cladophora*, which occurs in the attached periphyton, is large enough to see with the unaided eye. (e) The relatively large seaweed genus *Acetabularia* grows on rocks and coral rubble in shallow tropical oceans.

Protists Can Be Informally Classified According to Type of Motility

Microscopic protists have evolved diverse ways to propel themselves in moist environments. Swimming by means of eukaryotic flagella, cilia, amoeboid movement, and gliding are major types of protist movements.

Many types of photosynthetic and heterotrophic protists are able to swim because they produce one or more eukaryotic flagella, cellular extensions whose movement is based on interactions between microtubules and the motor protein dynein (eukaryotic flagella are described in Chapter 4). Eukaryotic flagella rapidly bend and straighten, thereby pulling or pushing cells through water. Protists that use flagella to move in water are commonly known as **flagellates** (see Figure 28.4a). Flagellates are typically composed of one or only a few cells and are small—usually within 2–20 μm long—because flagellar motion is not powerful enough to keep larger bodies from sinking. Some flagellate protists are sedentary, living attached to underwater surfaces. These protists use flagella to collect bacteria and other small particles for consumption as food. Seaweeds and other immobile protists often produce small, flagellate reproductive cells that have a greater capacity for mobility. Flagellate reproductive cells allow these protists to mate and disperse to new habitats.

An alternate type of protist motility relies on cilia, tiny hairlike extensions on the outsides of cells. Cilia are structurally similar to eukaryotic flagella but are shorter and more abundant on cells (see Figure 28.2). Protists that move by means of cilia are known as **ciliates** and are classified in the phylum Ciliophora (see Table 28.1). The presence of many cilia allows ciliates to achieve larger sizes than flagellates yet still remain buoyant in water. The coordinated movement of cilia enables some ciliates to scamper across underwater surfaces rich in food materials, and other ciliates to generate hurricane-like vortices of water that concentrate food particles.

A third type of motility is amoeboid movement. This kind of motion involves extending protist cytoplasm into lobes, known as pseudopodia ("false feet"). Once these pseudopodia move toward a food source or other stimulus, the rest of the cytoplasm may flow after them, thereby changing the shape of the entire organism as it creeps along. Protist cells that move by pseudopodia are described as **amoebae** (**Figure 28.5**). Finally, many diatoms, the malarial parasite *Plasmodium falciparum*, and some other protists glide along surfaces in a snail-like fashion by secreting protein or carbohydrate slime. With the exception of ciliates, motility classification does not correspond with the phylogenetic classification of protists, our next topic.

28.2 Evolution and Relationships

Modern biologists study protist diversity and infer their phylogenetic relationships by comparing cellular structure and gene sequences. Because the relationships of some protists are uncertain or disputed and new protist species are continuously

Figure 28.5 The amoebozoan genus *Pelomyxa*, showing pseudopodia.

being discovered, concepts of protist evolution and relationships are constantly changing. Even so, molecular and cellular data reveal that many protists can be classified into diverse eukaryotic supergroups (see Figure 28.1). In this section, we survey these supergroups, focusing on the defining features and evolutionary importance of the major protist phyla. We will also examine ways in which protists are important ecologically or in human affairs.

A Feeding Groove Characterizes Many Protists Classified in the Excavata

The protist supergroup known as the Excavata is related to some of Earth's earliest eukaryotes and is therefore important in understanding the early evolution of eukaryotes. The Excavata is named for a feeding groove "excavated" into the cells of many representatives, such as the genus *Jakoba* (phylum Jakobida) (**Figure 28.6**). The feeding groove is an important adaptation that allows these organisms to ingest small particles of food in their aquatic habitats. Once collected within the feeding groove, food particles are taken into cells by a type of endocytosis known as **phagotrophy** or phagocytosis. During phagotrophy, a vesicle of plasma membrane surrounds each food particle and pinches off within the cytoplasm. Enzymes within these food vesicles break the food particles down into small molecules that, upon their release into the cytoplasm, can be respired for energy. Phagotrophy is the evolutionary basis for the process of endosymbiosis, because particle ingestion provides a way for host cells to take in endosymbionts. Early in protist history, endosymbiotic proteobacterial cells gave rise to mitochondria, the organelle that is the major site of ATP synthesis. Consequently, most protists possess mitochondria, though these may be highly modified in parasitic species.

Some modern protists that are members of Excavata have become parasitic within animals, including humans. Instead of eating particles via feeding grooves, parasitic species attack host cells and absorb food molecules released from them.

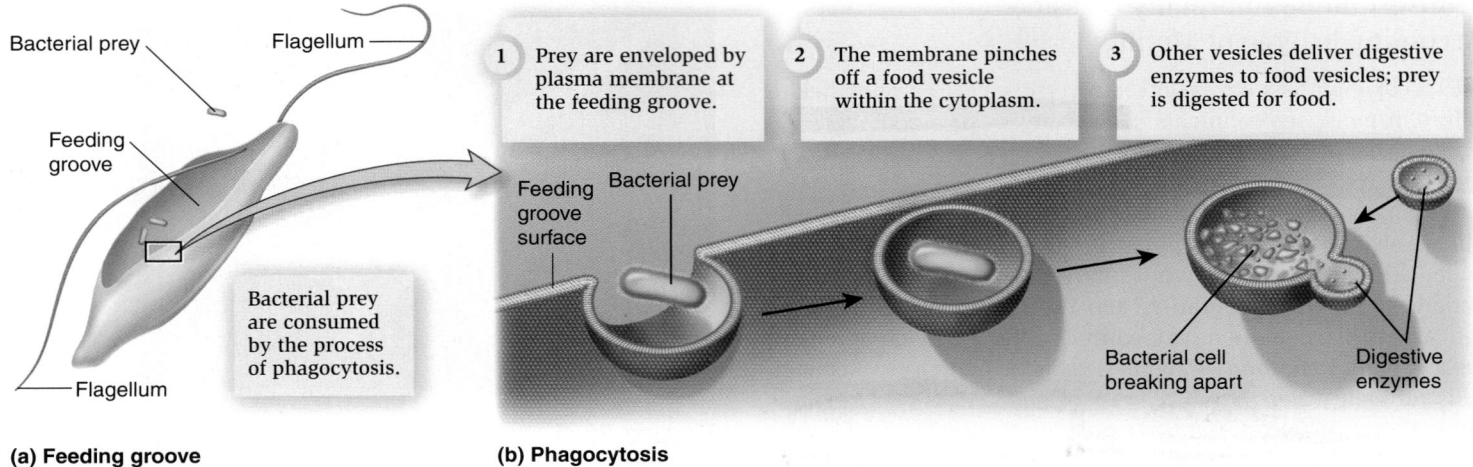

Bacterial prey

Flagellum

Feeding groove

Flagellum

Bacterial prey are consumed by the process of phagocytosis.

(a) Feeding groove

1. Prey are enveloped by plasma membrane at the feeding groove.

2. The membrane pinches off a food vesicle within the cytoplasm.

3. Other vesicles deliver digestive enzymes to food vesicles; prey is digested for food.

Feeding groove surface

Bacterial prey

Bacterial cell breaking apart

Digestive enzymes

(b) Phagocytosis

Figure 28.6   **Feeding groove and phagotrophy displayed by many species of supergroup Excavata.**   **(a)** Diagram of *Jakoba libera*, phylum Jakobida, showing flagella emerging from the feeding groove. **(b)** Diagram of phagocytosis, the process by which food particles are consumed at a feeding groove.

Biological inquiry: What happens to ingested particles after they enter feeding cells?

Trichomonas vaginalis, for example, causes a sexually transmitted infection of the human genitourinary tract. It has an undulating membrane and flagella that allow it to move over mucus-coated skin. *Giardia lamblia* causes giardiasis, an intestinal infection that can result from drinking untreated water or from unsanitary conditions in day-care centers (**Figure 28.7**). *T. vaginalis* and *G. lamblia* were once thought to lack mitochondria but are now known to possess structures that are highly modified mitochondria. In *T. vaginalis*, the modified mitochondria are called hydrogenosomes because they produce hydrogen gas. In *G. lamblia*, reduced mitochondria known as mitosomes do not themselves produce ATP, but they do generate iron-sulfur cofactors essential for ATP production in the cytosol by anaerobic processes. Microbiologist Pavel Dolezal and colleagues recently demonstrated that mitosomes and hydrogenosomes possess inner membrane protein import mechanisms otherwise found only in mitochondria. This evidence demonstrates the evolutionary origin of mitosomes and hydrogenosomes from mitochondria.

Disk-Shaped Mitochondrial Cristae Occur in Euglenozoa

The Euglenozoa is a supergroup of flagellates whose name originates from that of a common representative genus, *Euglena*. Members of the Euglenozoa feature mitochondrial cristae that are disk-shaped (**Figure 28.8**). (In contrast, the mitochondrial cristae of most eukaryotes are typically flattened or tube-shaped; refer back to Figure 4.24.) How or if these different types of cristae function differently is unknown. Euglenozoa includes the phyla informally known as kinetoplastids (Kinetoplastea) and euglenoids (Euglenida) (see Table 28.1).

Euglenoids possess unique, interlocking ribbon-like protein strips just beneath their plasma membranes (**Figure 28.9a**).

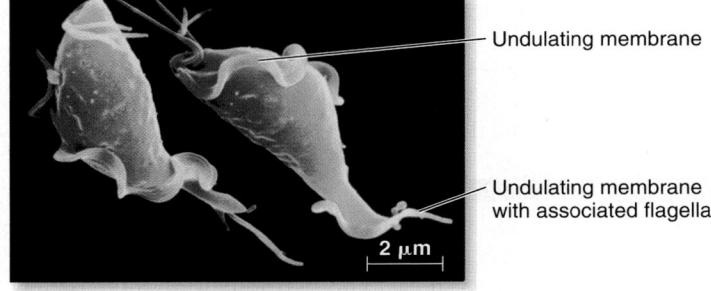

Undulating membrane

Undulating membrane with associated flagella

2 μm

(a) *Trichomonas vaginalis*

Flagella

2.5 μm

(b) *Giardia lamblia*

Figure 28.7   **Parasitic members of the supergroup Excavata.** **(a)** *Trichomonas vaginalis*. **(b)** *Giardia lamblia*. These specialized, parasitic flagellates absorb nutrients from living hosts.

These strips make the surface of some euglenoids so flexible that they can crawl through mud. This unique type of motility is known as euglenoid movement or metaboly. Many euglenoids are heterotrophic, but some possess green plastids and

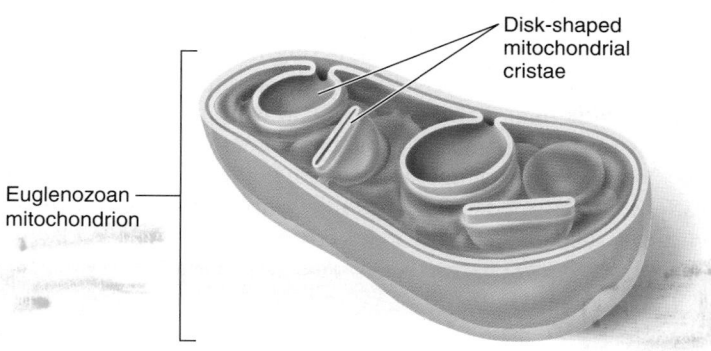

Figure 28.8 **A euglenozoan trait.** The disk-shaped mitochondrial cristae characteristic of the supergroup Euglenozoa.

(a) *Euglena*

(b) *Leishmania*

(c) *Trypanosoma*

Figure 28.9 *Euglenozoa.* (a) *Euglena* has helical protein ribbons near the surface, internal green plastids, a red eyespot, and white storage carbohydrate granules. (b) LM showing the kinetoplast DNA mass typical of kinetoplastid mitochondria. (c) In this SEM, several undulating kinetoplastids appear near disk-shaped red blood cells.

thus are photosynthetic. Plastids are organelles found in plant and algal cells that are distinguished by their synthetic abilities. Many euglenoids possess a light-sensing system that includes a conspicuous red structure known as an eyespot or stigma, and light-sensing molecules located in a swollen region at the base of a flagellum. Most euglenoids produce conspicuous storage carbohydrate particles known as paramylon. Euglenoids are particularly abundant and ecologically significant as photosynthesizers and phagotrophs in wetlands.

Kinetoplastids are named for an unusually large mass of DNA (known as a kinetoplast) that occurs in their single large mitochondrion (**Figure 28.9b**). These protists also feature an unusual modified peroxisome that contains glycolytic enzymes; in most eukaryotes, glycolysis occurs in the cytosol. Some kinetoplastids, such as *Trypanosoma brucei*, the causative agent of sleeping sickness, are serious pathogens of humans and other animals (**Figure 28.9c**).

Archaeplastida Are Distinguished by Plastids Arising from Primary Endosymbiosis

The ancestors of the supergroup Archaeplastida ("ancient plastids") most likely obtained plastids by the process of **primary endosymbiosis**. Such plastids are known as **primary plastids**; these can be identified by the presence of an envelope composed of two membranes (**Figure 28.10a**). Primary plastids arose when heterotrophic host cells captured cyanobacterial cells, thereby acquiring photosynthetic ability. These cyanobacterial endosymbionts evolved into plastids, in the process losing their cell walls and ability to live independently. Most species classified in Archaeplastida possess plastids, though some are heterotrophic because photosynthetic pigments have been lost.

The Archaeplastida encompasses some very diverse and ecologically important groups of organisms, including (1) the phylum Rhodophyta, informally known as red algae; (2) the phylum Chlorophyta, informally known as the green algae; (3) the kingdom Plantae, the land plants, which evolved from green algal ancestors; and (4) a small phylum of algae having blue-green plastids, the Glaucophyta, informally called glaucophytes (see Figure 28.1). Grouping the red algae, green algae, land plants,

and glaucophytes together within the supergroup Archaeplastida is based on the assumption that all primary plastids originated with a single endosymbiotic event, a notion that remains somewhat controversial. Evolutionary biologist John Stiller and several other experts have used molecular evidence to frame an alternate hypothesis that green and red primary plastids might have evolved independently as the result of separate endosymbiotic events.

(a) A primary plastid

(b) *Calliarthron*

(c) *Chondrus crispus*

Figure 28.10   Red algae and primary plastid of the supergroup Archaeplastida.   (a) A primary plastid, showing the two membranes of the primary plastid envelope. **(b)** The red algal genus *Calliarthron* has cell walls that are impregnated with calcium carbonate. This stony, white material makes the red alga appear pink. **(c)** *Chondrus crispus* is an edible red seaweed.

Red algae characteristically lack flagella, and most species are multicellular marine seaweeds (**Figure 28.10b,c**). Some red algae are cultivated in ocean waters for production of industrial and scientific materials or food. Sushi wrappers are composed of the sheetlike red algal genus *Porphyra*. Carrageenan is a complex red algal cell-wall polysaccharide that has numerous applications in the food industry, for example, keeping chocolate particles suspended evenly in ice cream or milk. Agar is a material extracted from red seaweeds that is widely used in research laboratories to solidify growth media used for cultivating microorganisms. Molecular biologists commonly use agarose, likewise obtained from red algae, to separate DNA molecules by gel electrophoresis.

Green algae of diverse structural types occur in fresh waters, the ocean, and on land (see Figure 28.4). A class of modern green algae informally known as charophyceans is closely related to the ancestor of land plants. Charophycean green algae resemble land plants in many aspects of cell structure, reproduction, and molecular biology (see Chapter 30). Thus, charophycean green algae are useful in deciphering the early evolutionary history of land plants.

Membrane Sacs Lie at the Cell Periphery of Protists Classified in Alveolata

The supergroup Alveolata includes three important phyla: (1) the Ciliophora, informally known as ciliates (see Figure 28.2); (2) the Dinozoa, informally known as dinoflagellates; and (3) the Apicomplexa, a medically important group of parasites. This supergroup is named for saclike membranous vesicles known as alveoli that are present at the cell periphery in all of these phyla (**Figure 28.11a**). The alveoli in some dinoflagellates seem empty,

(a)

(b)

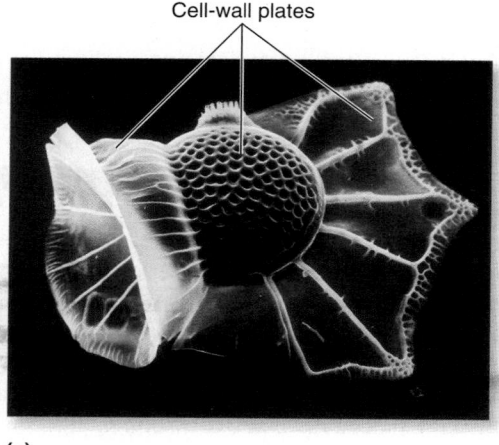

(c)

Figure 28.11   Dinoflagellates of the supergroup Alveolata and characteristic alveoli.   (a) Sac-shaped membranous vesicles known as alveoli lie beneath the plasma membrane of a dinoflagellate, along with trichocysts ready for discharge. **(b)** The surface of *Peridiniopsis berolinensis* appears smooth because the alveoli seem empty. Two types of flagella are seen on this freshwater dinoflagellate. One flagellum coils around a cellular groove; as it moves, this flagellum causes the cell to spin. By contrast, a straight flagellum extends from the cell, acting as a rudder to determine the direction of backward or forward movement. **(c)** The alveoli of the marine dinoflagellate genus *Ornithocercus* contain cellulose cell-wall plates.

Biological inquiry: Why do parts of the cell wall of Ornithocercus *resemble sails?*

so that the cell surface appears smooth (**Figure 28.11b**). By contrast, the alveoli of many dinoflagellates contain plates of cellulose, which taken together form an armor-like enclosure (**Figure 28.11c**). These plates are often modified in ways that provide adaptive advantage, such as protection from predators or increased ability to float.

About half of dinoflagellate species possess photosynthetic plastids of diverse types (see Table 28.1), and half lack plastids and are thus heterotrophic. Dinoflagellates play particularly important roles in coastal oceans as the result of their nutrition and defenses (see Section 28.3). Ciliates are notable for a complex mating behavior known as conjugation (see Section 28.4). Apicomplexans include the malarial agent *Plasmodium falciparum* (see Section 28.4), the related protist *Cryptosporidium parvum*, and other serious pathogens of humans and other animals.

Flagellar Hairs Distinguish Stramenopila

The supergroup Stramenopila (informally known as the stramenopiles) encompasses a wide range of algae, protozoa, and fungus-like protists that usually produce flagellate cells at some point in their lives. The Stramenopila (from the Greek *stramen*, straw + *pila*, hair) is named for distinctive strawlike hairs composed of glycoprotein that occur on the surfaces of flagella (**Figure 28.12**). These flagellar hairs function something like oars to greatly increase swimming efficiency. Biologists discovered this by using antibodies to clump the flagellar hairs together, which caused hairs to detach without otherwise harming the cells. As a result of this treatment, the flagellate cells could no longer swim as fast. Stramenopiles are also informally known as heterokonts (a term meaning "different flagella") because the two flagella often present on swimming cells have slightly different structures.

Fungus-like stramenopiles cause serious diseases of seaweeds, mollusks, fish, and terrestrial crop plants. For example, the fungus-like protist *Phytophthora infestans* (see Figure 28.3) caused the historic Irish potato crop failure that dramatically influenced immigration to the U.S. in the mid-19th century and remains a serious crop pest today. A related species, *Phytophthora ramorum*, causes sudden oak death, and *Phytophthora sojae* is a serious pest of soybeans. These fungus-like protists produce flagellate reproductive cells that enable them to spread on and between plants in films of water. The genomes of these crop parasites are currently being analyzed with the goal of finding better ways to diagnose the presence of the organisms and control their spread in the environment.

Stramenopiles also include many phyla or classes of algae having golden- or brown-colored plastids. Examples include diverse glass-enclosed diatoms (Bacillariophyta) (see Figure 28.2) and giant brown seaweeds (Phaeophyceae) that form extensive kelp forests in cold and temperate coastal oceans. Kelp forests are essential nurseries for fish and shellfish.

Molecular and biochemical evidence indicates that the plastids of stramenopile algae arose by the process of **secondary endosymbiosis** from ingested red algal cells (**Figure 28.13**).

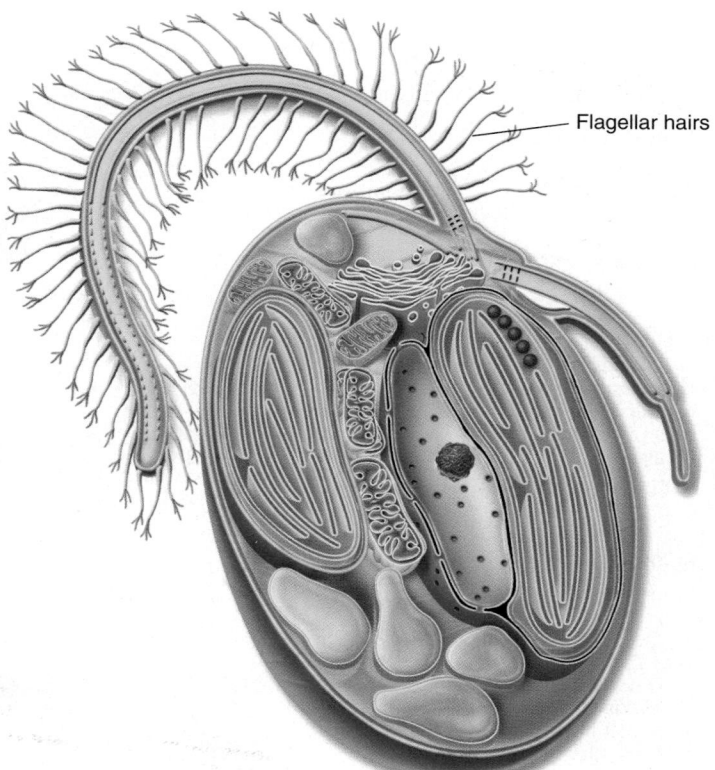

Flagellar hairs

Figure 28.12 Diagram of a flagellate stramenopile cell, showing characteristic flagellar hairs.

Biological inquiry: How do the flagellar hairs aid cell motion?

Secondary endosymbiosis occurs when the endosymbiont is a eukaryote having a primary plastid. Recall that red algal cells possess primary plastids. **Secondary plastids** originate with the endosymbiotic incorporation of a eukaryotic cell having a primary plastid. The host cell digests most of the endosymbiont but retains its plastid. Secondary plastids possess envelopes composed of more than two membranes.

Molecular evidence indicates that the plastids of most photosynthetic dinoflagellates also arose by secondary endosymbiosis from red algal cells, as did plastids of the algal phyla Cryptophyta (flagellates commonly known as cryptomonads) and Haptophyta (flagellate haptophytes) (see Table 28.1). Evolutionary protistologist Patrick Keeling and some other biologists hypothesize that alveolates, cryptomonads, haptophytes, and stramenopiles originated from a single common ancestor that had acquired a secondary plastid from a red algal endosymbiont. These scientists explain the occurrence of heterotrophic species in these phyla as the result of multiple plastid loss events, and they have proposed that these phyla should be classified into a supergroup named Chromalveolata.

By contrast, biogeochemist Paul Falkowski and colleagues argue that red algal cells possess molecular features that favored their incorporation into diverse types of host cells on multiple

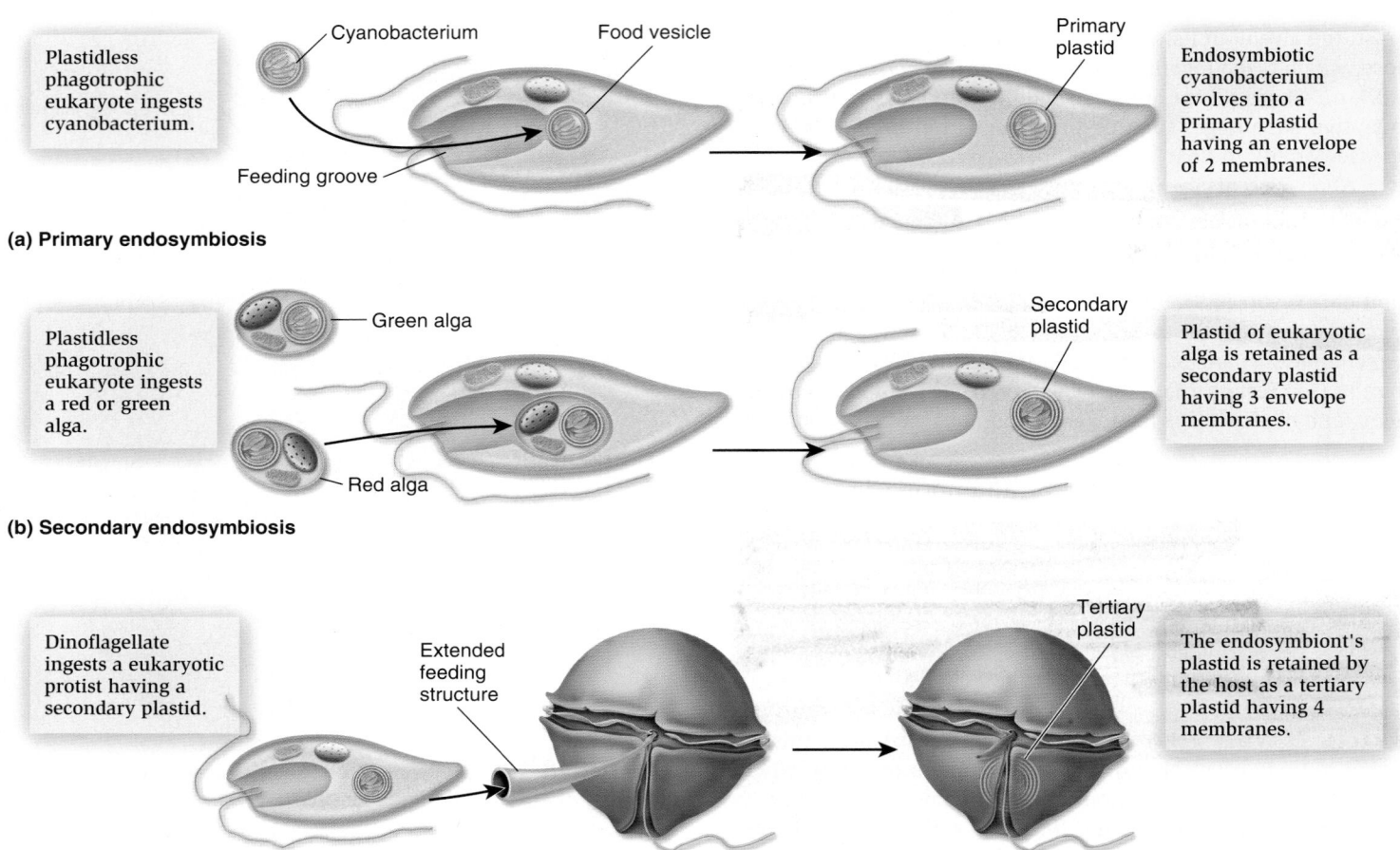

Plastidless phagotrophic eukaryote ingests cyanobacterium.

Cyanobacterium

Food vesicle

Feeding groove

Primary plastid

Endosymbiotic cyanobacterium evolves into a primary plastid having an envelope of 2 membranes.

(a) Primary endosymbiosis

Plastidless phagotrophic eukaryote ingests a red or green alga.

Green alga

Red alga

Secondary plastid

Plastid of eukaryotic alga is retained as a secondary plastid having 3 envelope membranes.

(b) Secondary endosymbiosis

Dinoflagellate ingests a eukaryotic protist having a secondary plastid.

Extended feeding structure

Tertiary plastid

The endosymbiont's plastid is retained by the host as a tertiary plastid having 4 membranes.

(c) Tertiary endosymbiosis

Figure 28.13   Primary, secondary, and tertiary endosymbiosis.

occasions. In other words, the similar plastids of photosynthetic stramenopiles, alveolates, cryptomonads, and haptophytes could have arisen by parallel evolution. If this explanation is correct, grouping these distinctive organisms into a single supergroup would not appropriately reflect phylogenetic relationships.

Relevant to this controversy is the fact that some dinoflagellates possess **tertiary plastids** acquired by engulfing diatoms or cryptomonads. Such plastids were obtained by independent occurrences of **tertiary endosymbiosis**—the acquisition by hosts of plastids from cells that already possessed secondary plastids (see Figure 28.13). In addition, many heterotrophic protists and aquatic animals have independently acquired diverse types of algal or plastid endosymbionts, because this adaptation is nutritionally valuable. The acquisition of plastids by endosymbiosis is an example of horizontal gene transfer. It is clear that endosymbiosis has been a powerful force in the evolution of protists.

Spiky Cytoplasmic Extensions Are Present on the Cells of Many Protists Classified in Rhizaria

Several groups of flagellates and amoebae that have thin, hair-like extensions of their cytoplasm, known as filose pseudopodia,

are classified into the supergroup Rhizaria (from the Greek *rhiza*, root). Rhizaria includes the phylum Chlorarachniophyta, whose spider-shaped cells possess secondary plastids obtained from endosymbiotic green algae (see Table 28.1). The Radiolaria (**Figure 28.14a**) and Foraminifera (**Figure 28.14b**) are two phyla of ocean plankton that produce exquisite mineral shells and frequently have symbiotic algal partners. The accumulated calcium carbonate shells of ancient foraminifera (and haptophyte algae) produced huge chalk deposits about 100 million years ago. Examples include notable geological formations such as the white cliffs of Dover, England (**Figure 28.14c**). Together, abundant ancient populations of foraminifera and haptophytes also produced the famous North Sea oil deposits.

Fossil foraminiferan shells are extensively used to infer past climatic conditions. For example, David Field, Mark Ohman, and fellow scientists used the abundances of different species of fossil foraminifera deposited in Southern California coastal ocean sediments over the past 1,400 years to demonstrate recent climate warming. These scientists observed that over the course of the 20th century, the number of foraminifera typical of warm climates rose dramatically, while the number of species that prefer cold climates fell.

(a) Radiolarian

0.5 mm
(b) Foraminiferan

(c) Fossil deposit containing rhizarians

Figure 28.14 Living representatives of supergroup Rhizaria and fossil deposits. (a) A radiolarian, *Acanthoplegma* spp., showing long filose pseudopodia. (b) A foraminiferan, showing calcium carbonate shell with long filose pseudopodia extending from pores in the shell. (c) Fossil carbonate remains of foraminifera and haptophyte algae deposited over millions of years form the White Cliffs of Dover in the United Kingdom.

Amoebozoa Includes Many Types of Amoebae with Pseudopodia

The supergroup Amoebozoa includes many types of amoebae that move by extension of pseudopodia (see Figure 28.5). One example is the human parasite *Entamoeba histolytica*, which causes a severe form of intestinal illness. Several types of protists known as slime molds are also classified in this supergroup (see Table 28.1). For example, the phylum Dictyostelia includes the slime mold *Dictyostelium discoideum*, widely used as a model laboratory system for understanding communication among cells. These slime molds produce amoebae that feed on small particles. Under starvation conditions, these amoebae aggregate and form reproductive structures known as "fruiting bodies" because they resemble the spore-producing structures of fungi.

A Single Flagellum Occurs on Swimming Cells of Opisthokonta

The supergroup Opisthokonta is named for the presence of a single posterior flagellum on swimming cells. Because their flagellate cells exhibit this feature, the animal and fungal kingdoms are included in Opisthokonta, as are the protists known as choanoflagellates (formally, the Choanomonada; see Table 28.1). The choanoflagellates are single-celled or colonial organisms that live in fresh water or marine periphyton, which are communities of attached microorganisms (**Figure 28.15**). Choanoflagellates feature a distinctive collar surrounding the single flagellum. The collar is made of cytoplasmic extensions known as tentacles, which filter bacterial food from water currents generated by flagellar motion. Molecular evidence suggests that choanoflagellates are the modern protists most closely related to the common ancestor of animals. Thus, evolutionary biologists interested in the origin of animals study choanoflagellates for molecular clues to this important event in our evolutionary history.

Flagellum

Collar of tentacles

Figure 28.15 A choanoflagellate, of the supergroup Opisthokonta.

Biological inquiry: How do choanoflagellates obtain food?

The preceding surveys of protist diversity and evolutionary relationships provide the foundation for a closer look at protist functions in nature. A focus on protist nutritional, defensive, and reproductive adaptations will help explain these roles, which often influence human affairs.

28.3 Nutritional and Defensive Adaptations

Wherever you look in moist places, you will find protists playing diverse and important ecological roles. In this section we will survey protist nutritional and defensive adaptations, which help to explain their ecological functions.

Protists Display Four Basic Types of Nutrition

Protist nutrition occurs in four basic types: phagotrophy, osmotrophy, autotrophy, and mixotrophy. All of these are important in ecological food webs that connect organisms of many types. Heterotrophic protists that specialize in phagotrophy (particle feeding) are known as **phagotrophs**, whereas those relying on osmotrophy (uptake of small organic molecules) are **osmotrophs**. Photosynthetic protists are **autotrophs**, organisms that can make their own organic nutrients. **Mixotrophs** are protists that are able to use autotrophy as well as phagotrophy or osmotrophy to obtain organic nutrients. The genus *Dinobryon* (**Figure 28.16**), a photosynthetic stramenopile that lives in phytoplankton lakes and consumes enormous numbers of bacteria, is an example of a mixotroph. Mixotrophic protists may switch back and forth between autotrophy and heterotrophy, depending on conditions in their environment. If sufficient light, carbon dioxide, and other minerals are available, mixotrophs produce their own organic food. If any of these resources limits photosynthesis, or organic food is especially abundant, mixotrophs can function as heterotrophs. Mixotrophs thus have remarkable nutritional flexibility.

Heterotrophic protists that feed on nonliving organic material function as decomposers, also known as saprobes. Decomposers, which also include many prokaryotes and fungi, are essential in breaking down wastes and releasing minerals for use by other organisms. Heterotrophic protists that feed on the living cells of other organisms are parasites. Some parasites are pathogens, which cause disease in other organisms. *Trichomonas vaginalis, Giardia lamblia, Entamoeba histolytica,* and *Phytophthora infestans* are examples of pathogenic protists that have previously been described in this chapter. Humans view such protists as pests when they harm us or our agricultural animals and crops, but such protists also play important roles in nature by controlling the population growth of other organisms. Such roles illustrate the principle that organisms modify their environments and provide essential biological services. The diversity of photosynthetic protists and their storage molecules further illustrate these principles.

Algal Protists Vary in Photosynthetic Pigments and Food Storage Molecules

As we have earlier noted, algae display a surprising diversity of coloration, from gold to brown, red, and green. Why do so many pigmentation types occur? The answer is related to light availability in these protists' watery environment.

If you dive more than a few feet into lakes or the ocean, your aquatic world will appear intensely blue-green. This occurs because water absorbs the longer red to yellow wavelengths of light to a greater degree than shorter blue and green wavelengths. Little red light penetrates far into natural waters, depriving chlorophyll *a*, the photosynthetic pigment universally present in photosynthetic eukaryotes, of much of the light it would ordinarily absorb. Photosynthetic protists living in water have coped with this filtering effect by adapting their photosynthetic systems so that they capture more of the available blue-green light. Seaweeds, for example, may be colored red, gold, or brown, in addition to green, because they produce diverse accessory pigments that are able to absorb more of the light that is available underwater and transfer the energy to chlorophyll *a* for photosynthesis. The red accessory pigment phycoerythrin is abundant in plastids of red algae, and golden fucoxanthin enriches the color of golden and brown algae. Carotene and lutein have similar accessory pigment functions in green algae and were inherited by their land plant descendants, thus playing important roles in human nutrition. These and other accessory pigment adaptations explain why algae occur in so many different colors.

Photosynthetic protists also vary in the types of molecules that serve as food storages. Starch stored in the plastids of green algae and oil droplets in diatom cells are two common examples. Algae use such food storage cells for energy when light or minerals are too low to allow photosynthesis; such cells are also desirable food sources for heterotrophs. As a consequence of photosynthesis and food storage, autotrophic and mixotrophic protists play essential roles in aquatic food webs as primary producers. Such protists generate the organic food and atmospheric oxygen consumed by many other organisms.

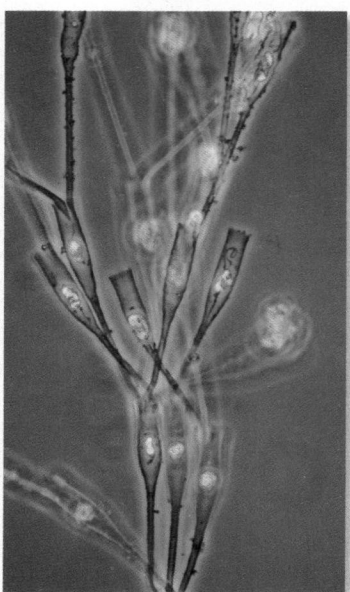

Figure 28.16
A mixotrophic protist. The genus *Dinobryon* is a colonial flagellate that occurs in the phytoplankton of freshwater lakes. The photosynthetic cells have golden plastids and also capture and consume bacterial cells.

Protists Defend Themselves in Diverse Ways

Protists use a wide variety of defensive adaptations to ward off attack. Major types of defenses are slimy, tough, or spiny cell coverings; sharp projectiles that can be explosively shot from cells; light flashes; and toxic compounds.

Many protists have cell coverings such as slimy mucilages (see chapter-opening photo) or cell walls that provide protection from attack by herbivores or pathogens. Cell walls may also aid in preventing osmotic damage or enhance floatation in water. Rigid cellulose walls are common in brown and green algae, while slimy polysaccharide polymers form a protective matrix around red algal cells. Calcium carbonate forms a stony coat for many protist cells, including foraminifera, haptophytes, and some marine seaweeds. Ornate glassy coatings of silica protect diatoms, while metallic iron and manganese crystals armor other protists.

Several types of protist cells contain compressed protein structures known as trichocysts (see Figure 28.11a). Upon attack, trichocysts rapidly elongate into spear-shaped projectiles that are shot from protist cells, thereby discouraging herbivores from feeding. Some species of ocean dinoflagellates emit flashes of blue light when disturbed, explaining why ocean waters teeming with these protists display bioluminescence. The light flashes may deter herbivores by startling them, but when ingested the dinoflagellates make the herbivores also glow, revealing them to hungry fish. Light flashes benefit dinoflagellates by helping to reduce populations of herbivores that consume the algae.

Various protist species produce **toxins**, compounds that inhibit animal physiology. Toxins probably originated as a means of protection from relatively small herbivores that might otherwise eat protists. Dinoflagellates are probably the most important protist toxin producers; they synthesize several types of toxins that affect humans and other animals. Why does this happen? Under natural conditions, small populations of dinoflagellates produce low amounts of toxin that do not harm large organisms. Dinoflagellate toxins become dangerous to humans when people contaminate natural waters with excess mineral nutrients such as nitrogen and potassium from untreated sewage, industrial discharges, or fertilizer that washes from agricultural fields. The excess nutrients foster explosive growth of algal populations called blooms, which then produce sufficient toxin to affect birds, aquatic mammals, fish, and humans. Toxins can concentrate in organisms, and humans who ingest shellfish that have accumulated dinoflagellate toxins can suffer poisoning.

Some dinoflagellates use toxins to attack fish, which they consume for food, and excessive growth of these dinoflagellates can also be harmful to people. In 1992, ecologist JoAnn Burkholder and colleagues reported that a new type of dinoflagellate had caused major fish kills in the overly fertile waters of the Chesapeake Bay. These investigators gave the new organism the generic name *Pfiesteria* to honor the late dinoflagellate expert Lois Pfiester. The scientists described the genus *Pfiesteria* as a "phantom predator" because populations of swimming cells seemed to appear rapidly in response to the presence of fish or their excretions. *Pfiesteria piscida* and *Pfiesteria shumwayae* tend to lie dormant on the ocean bottom until chemical signals emitted by their fish prey stimulate production of swimming cells. Burkholder and colleagues reported that the dinoflagellate toxin damages the fish skins, allowing the voracious dinoflagellate predators easier access to flesh. These biologists also discovered that *Pfiesteria*-associated toxin caused amnesia and other nervous system conditions in fishers and scientists who were exposed to it. The discovery of *Pfiesteria* excited the media, which dubbed it a "killer alga" and focused attention on nutrient pollution of Chesapeake Bay, the fundamental cause of *Pfiesteria*'s excessive growth and fish kills. Because of this organism's importance to the fishing industry and human health, teams of aquatic ecologists have continued to study the genus *Pfiesteria* and its toxin.

FEATURE INVESTIGATION

Burkholder and Colleagues Demonstrated That Strains of the Dinoflagellate Genus *Pfiesteria* Are Toxic to Mammalian Cells

A team of investigators led by JoAnn Burkholder performed an experiment to determine whether or not two strains of *P. shumwayae* were toxic to mammalian cells (**Figure 28.17**). One of these strains (CCMP 2089) had earlier been reported to be nontoxic, and neither strain had been tested for its impact on mammalian cells. The experiment was conducted in a biohazard containment facility, for the safety of the investigators. In a first step, the team grew the two *Pfiesteria* strains on different food sources, because the impact of food on dinoflagellate toxicity was unclear. Both strains were grown with algal cells as food and with juvenile fish as a food source.

In a second step, the dinoflagellates grown in step 1 were transferred to tanks with fish to elicit maximal toxin production, before treating mammalian cell cultures with the dinoflagellates. Dinoflagellates were not added to a control tank of fish. Toxin was detected, and fish deaths occurred, in all of the tanks except the control. In a third step, samples from the step 2 treatments were added to mammalian cell cultures. The investigators then determined the relative levels of toxicity to the mammalian cells. They found that both strains of *P. shumwayae* subjected to both feeding treatments were toxic to mammalian cells.

Figure 28.17 Burkholder and colleagues demonstrated that some strains of *Pfiesteria shumwayae* are toxic to fish and mammalian cells.

GOAL To determine the toxicity of two *Pfiesteria shumwayae* strains, grown on differing food regimes, to mammalian cells.

STARTING MATERIALS
1) *P. shumwayae* strain CCMP 2089
2) *P. shumwayae* strain CAAE 1024C
3) *Cryptomonas* sp.—algal food for dinoflagellates
4) Juvenile tilapia (*Oreochromis* spp.)—fish food for dinoflagellates
5) Mammalian pituitary GH_4C_1 cell line

Experimental level   **Conceptual level**

1 Grow strains CCMP 2089 and CAAE 1024C with algal food in culture flasks, or with fish as food in tanks. Use a biohazard containment facility to prevent toxins from harming scientists.

CCMP 2089 — Culture flask — CAAE 1024C
Algal food
CCMP 2089 — Fish as food — CAAE 1024C

Food source, algae or juvenile fish, might affect the amount of toxin produced by dinoflagellates.

The 2 dinoflagellate strains might differ in toxin production response.

2 Transfer dinoflagellates grown under step 1 conditions to new tanks containing juvenile fish.

(a) CCMP 2089 grown on algae
(b) CCMP 2089 grown on fish
(c) CAAE 1024C grown on algae
(d) CAAE 1024C grown on fish
(e) Control No dinoflagellates

Grow with fish to elicit maximal toxin production.

3 Expose mammalian cells to dinoflagellates from treatments a–d and control water from tank e in step 2.

Glass dishes with lids   Mammalian cells

(a) (b) (c) (d) (e)

Determine toxicity of dinoflagellates to mammalian cells.

4 **THE DATA**

Results from step 3:

28.4 Reproductive Adaptations

Diverse reproductive adaptations allow protists to thrive in an amazing variety of environments. These include specialized asexual reproductive cells, tough-walled dormant cells that allow protists to survive periods of environmental stress, and several types of sexual life cycles.

Protist Populations Increase by Means of Asexual Reproduction

All protists are able to reproduce themselves by asexual means that involve mitotic cell divisions of parental cells to produce progeny. When resources are plentiful, repeated mitotic divisions of single-celled protists will generate large protist populations, as in the case of Chesapeake Bay *Pfiesteria* species. By contrast, multicellular protists often generate specialized asexual cells that help disperse the organisms in their environment. For example, red seaweeds disperse single nonflagellate cells that drift with the currents until they encounter a suitable substrate for attachment and growth of new seaweeds. Many other multicellular or nonmotile protists produce flagellate cells called zoospores that swim through water or water films, thereby dispersing in the environment.

Many protists produce unicellular **cysts** as the result of asexual (and in some cases, sexual) reproduction (**Figure 28.18**). Cysts often have thick, protective walls and can remain dormant through periods of unfavorable climate or low food availability. Dinoflagellates commonly produce cysts. *Pfiesteria* species lie in wait on the ocean bottom as cysts until fish prey are present. Dinoflagellate cysts in ship ballast water can be transported from one port to another, a problem that has caused harmful

Figure 28.18 Cysts. The round cells are dormant, tough-walled cysts of the dinoflagellate *Peridinium limbatum*. The pointed cell is an actively growing cell of the same species. As cysts develop, the outer cellulose plates present on actively growing cells are cast off.

Cysts

Active cell

dinoflagellate blooms to appear in harbors around the world. Ship captains can help to prevent such ecological disasters by heating ballast water before it is discharged from ships.

Many protozoan pathogens spread from one host to another via cysts. For example, waterborne cysts are the form in which the apicomplexan pathogen *Cryptosporidium parvum* infects humans. Water supplies that are contaminated with large numbers of cysts can sicken many people. In the case of the outbreak in Milwaukee, Wisconsin, noted at the beginning of the chapter, *C. parvum* cysts probably entered drinking water supplies in water that has drained from cattle feedlots. *Entamoeba histolytica* is a pathogen that infects people who consume food or water that is contaminated with cysts. Once inside the human digestive system, *E. histolytica* attacks intestinal cells, causing amoebic dysentery, a worldwide problem.

Sexual Reproduction Provides Multiple Benefits to Protists

Eukaryotic sexual reproduction, featuring gametes, zygotes, and meiosis first arose among protists. Sexual reproduction has not been observed in some protist phyla such as kinetoplastids and euglenoids. However, it occurs widely among green algae, red algae, apicomplexans, ciliates, and dinoflagellates. Sexual reproduction is generally adaptive because it produces diverse genotypes, thus increasing the potential for faster evolutionary response to environmental change. Many protists reap additional benefits from sexual reproduction, illustrated by several types of sexual life cycles.

Zygotic Life Cycles Most unicellular protists that reproduce sexually display what is known as a **zygotic life cycle** (**Figure 28.19**). In this type of life cycle, haploid cells transform into gametes. Some protists produce nonmotile eggs and smaller flagellate sperm. However, many other protists have gametes that look similar to each other structurally but have distinctive biochemical features and hence are known as + and − mating strains. Gametes fuse to produce thick-walled diploid zygotes, which give this type of life cycle its name. Such zygotes often have tough cell walls and can survive stressful conditions, much like cysts. Some of the earliest fossils regarded as ancient protists are probably tough-walled zygotes.

Sporic Life Cycles Many multicellular green and brown seaweeds display a **sporic life cycle**, which is also known as alternation of generations (**Figure 28.20a**). Organisms having sporic life cycles produce two types of multicellular organisms: a haploid gametophyte generation that produces gametes (sperm or eggs), and a diploid sporophyte generation that produces spores by the process of meiosis. This type of life cycle takes its name from the characteristic production of spores as the result of meiosis. Each of the two types of multicellular organisms can adapt to distinct habitats or seasonal conditions, thus allowing protists to occupy more types of environments and for longer periods.

Figure 28.19 Zygotic life cycle, illustrated by the unicellular flagellate genus *Chlamydomonas.* In *Chlamydomonas,* mature cells are haploid.

Many red seaweeds display a variation of the sporic life cycle that involves alternation of three distinct multicellular generations (**Figure 28.20b**). Their unique type of sexual life cycle is an adaptation that allows red algae to cope with the lack of flagella on sperm. Because red algal sperm are unable to swim to eggs, fertilization occurs only when sperm carried by ocean currents happen to drift close to eggs. Fertilization can thus be rare. Many red algae compensate by making millions of spores that are produced by two distinct sporophyte generations.

Gametic Life Cycle and the Problem of Diatom Size Diatoms are examples of a relatively few protists known to display a **gametic life cycle** (**Figure 28.21**). In gametic life cycles, all cells except the gametes are diploid, and gametes are produced by meiosis. Diatoms use sexual reproduction not only to increase genetic variability but also to regenerate maximal species size. Many diatoms must do this as the result of a unique problem that develops during asexual reproduction by mitosis.

In many diatoms, one daughter cell arising from mitosis is smaller than the other, and it is also smaller than the parent cell. This happens because diatom cell walls are composed of two overlapping halves, much like two-part round glass laboratory dishes having lids that overlap bottoms. After each mitotic division, each daughter cell receives one-half of the parent cell wall. The daughter cell that inherits a larger, overlapping parental

"lid" then produces a new "bottom" that fits inside. This daughter cell will be the same size as its parent. However, the daughter cell that inherits the parental "bottom" uses this wall half as its lid and produces a new, even smaller "bottom." This cell will be smaller than its sibling or parent. Consequently, after many such mitotic divisions, the average cell size of diatom populations often declines over time. If diatom cells become too small, they lose the ability to survive. Sexual reproduction helps to solve this problem.

Sexual reproduction allows diatom species to recover maximal cell size, thereby preventing species extinction. Diatom cells mate within a blanket of mucilage, each partner undergoing meiotic divisions to produce gametes. The large, spherical diatom zygotes that result from fertilization (Figure 28.21b) later undergo a series of mitotic divisions to produce new diatom cells having the maximal size for the species.

Ciliate Sexual Reproduction Among protists, ciliates have one of the most complex sexual processes known. Ciliates are unusual in having two types of nuclei, a single large macronucleus and one or more smaller micronuclei. Macronuclei, which develop from micronuclei, serve as the source of information for cell function, and they divide when ciliates reproduce asexually by mitosis (**Figure 28.22a**). Sexual reproduction in ciliates begins when two cells pair and fuse longitudinally—a process known

KEY
- Haploid
- Diploid

1 When mature, the seaweed produces spores by meiosis.

Spores

Meiosis

Sperm-producing gametophyte (*n*)

Mitosis

2 One type of haploid (*n*) spore swims away and eventually attaches to a surface and grows into a small sperm-producing gametophyte.

Mitosis

Egg (*n*)

Sperm (*n*)

Egg-producing gametophyte (*n*)

3 Another type of haploid (*n*) spore swims away and eventually settles, growing into a small egg-producing gametophyte.

Zygote (2*n*)

Egg (*n*)

Sperm (*n*)

Mitosis

Mature sporophyte (2*n*)

Fertilization

4 Eggs secrete chemicals that attract sperm, and fertilization occurs. Zygotes grow into diploid sporophytes by mitosis.

(a) *Laminaria* life cycle – alternation of 2 generations

Male gametophyte (*n*)

Sperm (*n*)

2 Red algal gametophytes produce eggs that remain on the female gametophyte or sperm, which are dispersed.

Egg (*n*)

Fertilization

3 Sperm, lacking flagella, drift to egg cells, and fertilization produces a zygote.

Nourishing female gametophyte tissue (*n*)

Mitosis

Haploid spores (*n*)

Female gametophyte (*n*)

2*n* zygote nucleus

Diploid spores (2*n*)

Small sporophyte (2*n*)

Large sporophyte (2*n*)

1 Some cells of the large sporophyte undergo meiosis, producing haploid spores that grow into gametophytes.

Meiosis

Mitosis

4 The diploid zygote uses food from the female gametophyte to undergo many mitotic divisions, producing a small sporophyte that releases many diploid spores, which are dispersed.

5 Diploid spores attach to a surface and grow into a large sporophyte.

(b) *Polysiphonia* life cycle – alternation of 3 generations

Figure 28.20 **Sporic life cycles.** **(a)** Sporic life cycle as illustrated by the brown seaweed *Laminaria*. **(b)** Sporic life cycle involving three alternating generations, illustrated by the red seaweed *Polysiphonia*.

as **conjugation** (**Figure 28.22b**). In *Paramecium caudatum*, the single micronucleus undergoes meiosis early in conjugation. Of the four haploid micronuclei produced, all disintegrate except one, which divides by mitosis. Then the paired ciliates exchange one of each pair of micronuclei, after which the paired cells separate and the macronuclei disintegrate. The two genetically different micronuclei in each cell then fuse, and the new diploid nuclei divide mitotically to produce four macronuclei and multiple micronuclei. Subsequent mitotic divisions occur, distributing one macronucleus plus one or more micronuclei (depending on the species) to each cell. Conjugation seems to be essential for continued existence of ciliate species.

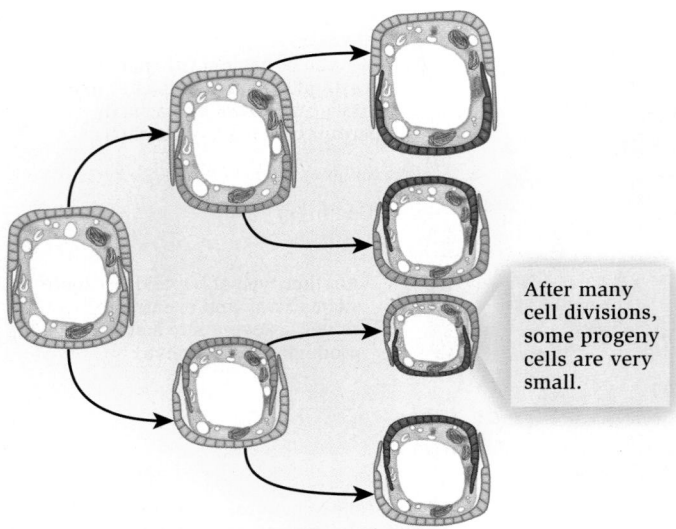

(a) Asexual reproduction in diatoms

Figure 28.21   Gametic life cycle as illustrated by diatoms. (a) Diatom asexual reproduction involves repeated mitotic division. Because a new lower cell-wall piece is always synthesized, asexual reproduction may eventually cause the mean cell size to decline in a diatom population. (b) Small cell size may trigger sexual reproduction, which regenerates maximal cell size.

After many cell divisions, some progeny cells are very small.

2 Blanketed by mucilage, each cell produces one or more haploid gametes by meiosis. The gametes may look alike or take the form of sperm and eggs.

Meiosis

Haploid gametes

3 The gametes fuse to form a diploid zygote that is larger and rounder than a typical diatom cell.

Fertilization

Mucilage

KEY
Haploid
Diploid

Lipid food storage

Plastids

1 When diatom cells reach a critical small size or are stimulated by environmental factors, they may begin the process of reproduction.

4 The 2n zygote undergoes mitotic divisions to produce diploid cells that have the normal shape and maximum size for that species.

(b) Sexual reproduction in diatoms

Parasitic Protists May Use Alternate Hosts for Different Life Stages

Parasitic protists are notable for often using more than one host organism, in which different life stages occur. The malarial parasite genus *Plasmodium* is a prominent example. About 40% of humans live in tropical regions of the world where malaria occurs, and as noted earlier, millions of infections and human deaths result each year. Malaria is particularly deadly for young children. The malarial parasite's alternate hosts are mosquitoes classified in the genus *Anopheles*. Though insecticides can be used to control mosquito populations, and antimalarial drugs exist, malarial parasites can develop drug resistance. Experts are concerned that cases may double in the next 20 years.

Plasmodium enters the human bloodstream via a mosquito bite as life stages known as sporozoites (**Figure 28.23**). These eventually reach the victim's liver and enter the liver cells. Following several cycles of cell division, life stages known as merozoites develop. Merozoites have protein complexes at their front ends, or apices, that allow them to invade human red blood cells. (The presence of these apical complexes gives rise to the phylum name Apicomplexa.) The merozoites consume the hemoglobin in red blood cells. While living within red blood cells, they form "ring stages," which can be visualized by staining and use of a microscope, allowing diagnosis. Merozoites reproduce asexually, generating large numbers of new merozoites that synchronously break out of red blood cells at intervals of 48 or 72 hours. These merozoite reproduction cycles correspond to cycles of chills and fever experienced by the infected person.

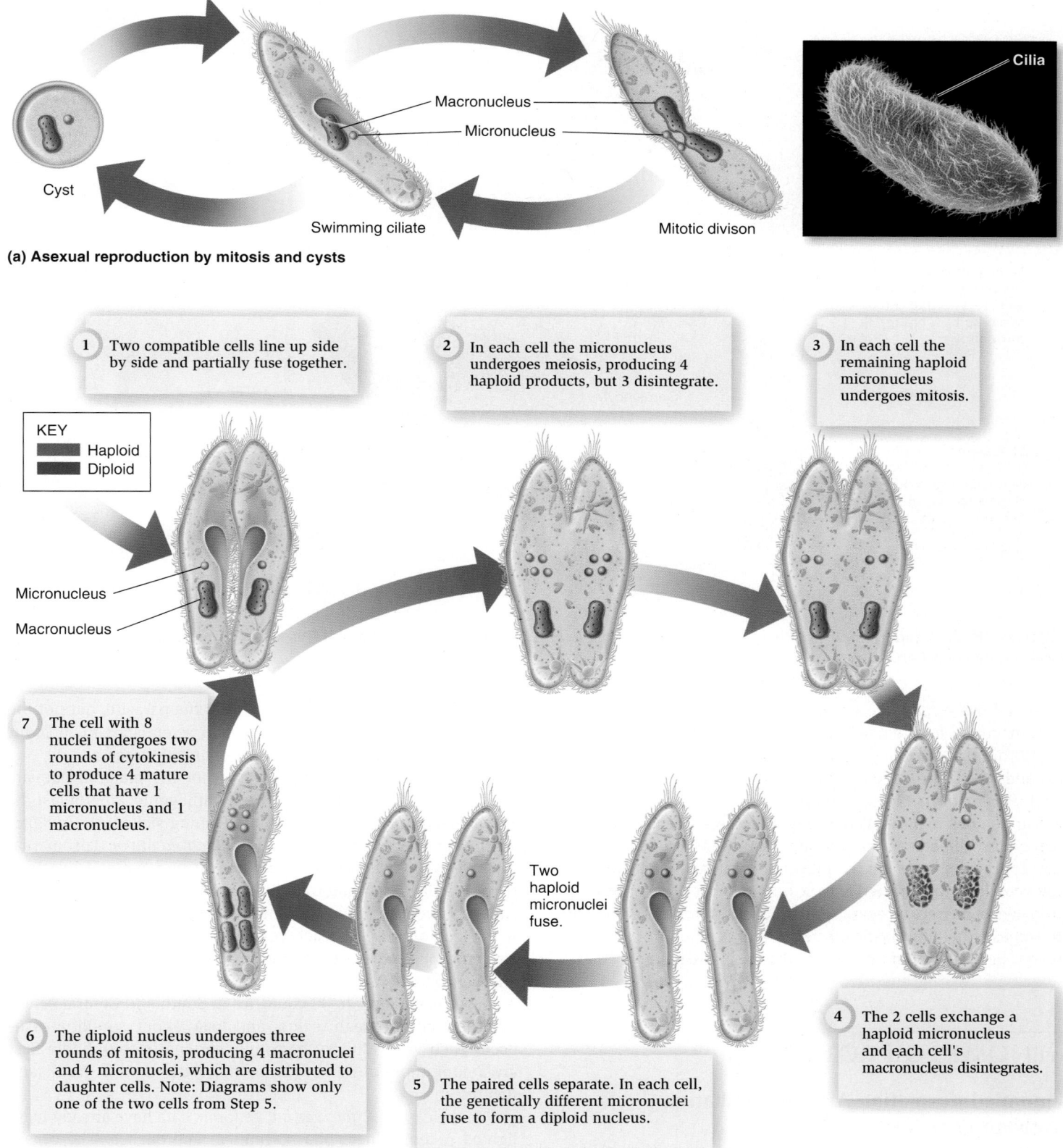

Cilia

Macronucleus
Micronucleus

Cyst

Swimming ciliate

Mitotic divison

(a) Asexual reproduction by mitosis and cysts

1 Two compatible cells line up side by side and partially fuse together.

2 In each cell the micronucleus undergoes meiosis, producing 4 haploid products, but 3 disintegrate.

3 In each cell the remaining haploid micronucleus undergoes mitosis.

KEY
Haploid
Diploid

Micronucleus
Macronucleus

7 The cell with 8 nuclei undergoes two rounds of cytokinesis to produce 4 mature cells that have 1 micronucleus and 1 macronucleus.

Two haploid micronuclei fuse.

4 The 2 cells exchange a haploid micronucleus and each cell's macronucleus disintegrates.

6 The diploid nucleus undergoes three rounds of mitosis, producing 4 macronuclei and 4 micronuclei, which are distributed to daughter cells. Note: Diagrams show only one of the two cells from Step 5.

5 The paired cells separate. In each cell, the genetically different micronuclei fuse to form a diploid nucleus.

(b) Sexual reproduction by conjugation

Figure 28.22   Ciliate reproduction.   (a) The asexual reproduction process in ciliates. **(b)** The sexual reproduction process in the ciliate *Paramecium caudatum*, which involves conjugation.

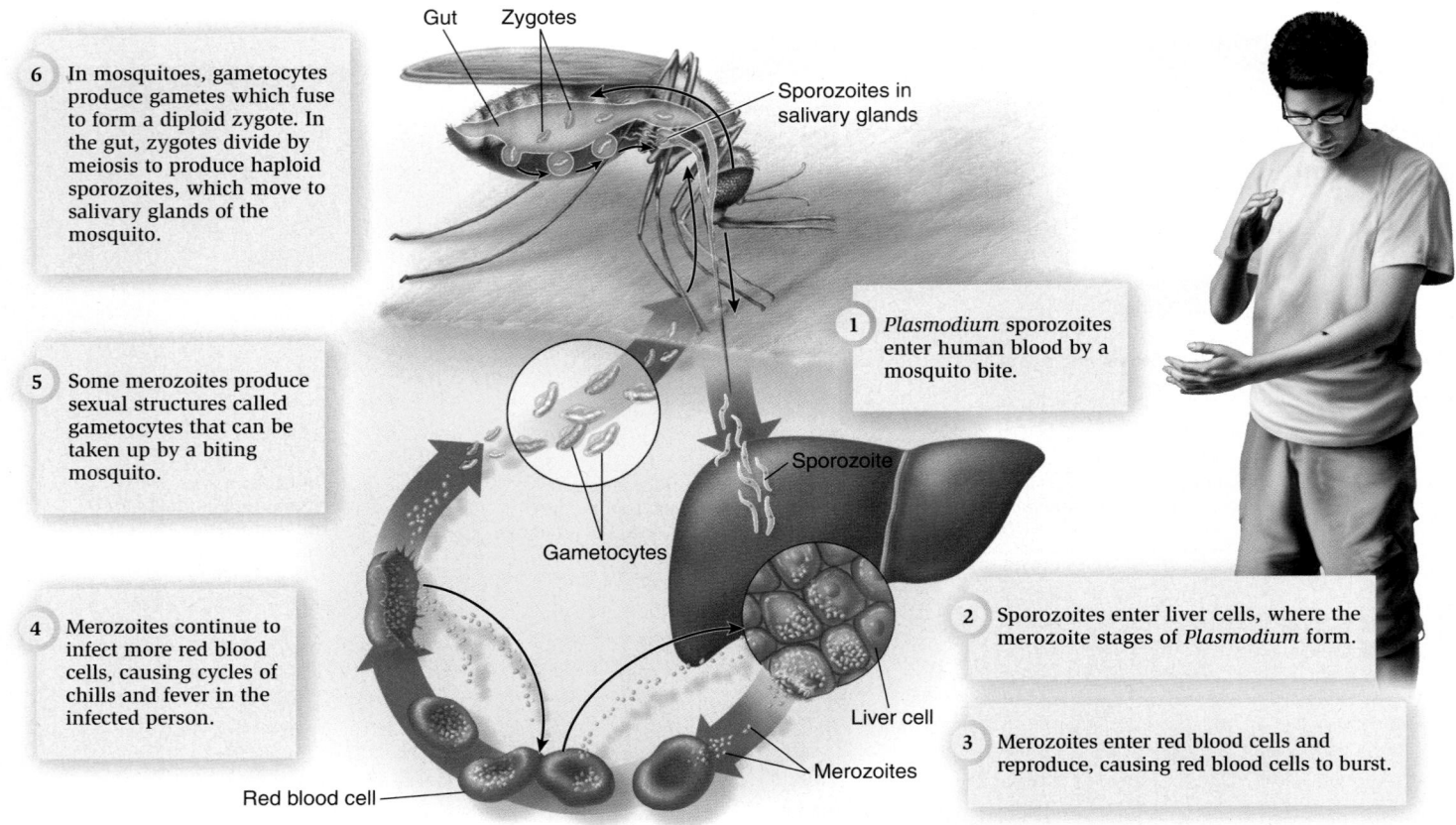

Gut   Zygotes

6 In mosquitoes, gametocytes produce gametes which fuse to form a diploid zygote. In the gut, zygotes divide by meiosis to produce haploid sporozoites, which move to salivary glands of the mosquito.

Sporozoites in salivary glands

1 *Plasmodium* sporozoites enter human blood by a mosquito bite.

5 Some merozoites produce sexual structures called gametocytes that can be taken up by a biting mosquito.

Sporozoite

Gametocytes

4 Merozoites continue to infect more red blood cells, causing cycles of chills and fever in the infected person.

2 Sporozoites enter liver cells, where the merozoite stages of *Plasmodium* form.

Liver cell

3 Merozoites enter red blood cells and reproduce, causing red blood cells to burst.

Red blood cell

Merozoites

Figure 28.23   Diagram of the life cycle of *Plasmodium falciparum*, the agent of malaria.   This life cycle requires two alternate hosts, humans and *Anopheles* mosquitoes.

Some merozoites produce sexual structures, gametocytes, which are transmitted to a female mosquito as she bites.

Within the mosquito's body, the gametocytes produce gametes and fertilization occurs, yielding a zygote, the only diploid cell in *Plasmodium*'s life cycle. Within the mosquito gut, the zygote undergoes meiosis, generating structures filled with many sporozoites, the stage that can be transmitted to a new human host. Sporozoites move to the mosquito's salivary glands, where they remain until injected into a human host when the mosquito feeds. In recent years, genomic information bearing on life stages is helping medical scientists to develop new ways to prevent or treat malaria and other diseases caused by protists.

GENOMES & PROTEOMES

Genomic Information Aids in Combating Protistan Parasites

Genome sequences have been determined for several protists that cause human disease. This has been done as a way to find new cellular targets for drug treatments that will kill the parasites without harming the human host. The goal is to identify metabolic processes that are present in the parasite, but not the host. With such information, medical scientists can look for drugs that affect only the distinctive parasite process.

Genomic data are now available for the malarial parasite *Plasmodium falciparum* and three pathogenic kinetoplastids: *Trypanosoma brucei* (the agent of sleeping sickness), *Trypanosoma cruzi* (Chagas disease), and *Leishmania major* (leishmaniasis). Sleeping sickness is transmitted to humans and livestock by tsetse flies in equatorial Africa as well as in parts of South America and Asia. Nearly half a million people become infected with sleeping sickness each year, and the disease is fatal if untreated. Chagas disease is primarily a hazard in South America and can be transmitted via blood and organ donations. Leishmaniasis is endemic to 88 countries in South and Central America, the Mediterranean region, and Asia; about 2 million cases occur each year. Military personnel serving in the Middle East are among those at risk of contracting leishmaniasis from sand flea bites.

In the case of *P. falciparum*, genomic data have already highlighted potential new pharmaceutical approaches. The organism's nuclear genome consists of 14 chromosomes and about 5,300 genes. A large proportion of *P. falciparum*'s genes are related to evasion of the host's immune system and interactions between host and parasite. Two-thirds of the predicted 5,268 proteins appear to be unique to this organism, an unusual situation among eukaryotes. About 550 (some 10%) of the nuclear-encoded proteins are likely imported into a modified plastid known as an apicoplast, where they are needed for fatty acid metabolism and other processes. Because plastids are not present in the cells of mammals, enzymes in apicoplast pathways are possible targets for development of drug therapy. Mammals also lack calcium-dependent protein kinases (CDPKs), enzymes

that are essential to *P. falciparum*'s sexual development, offering another potential drug target. By contrast, genomic data reveal that kinetoplastids lack plastids, so drug strategies focused on this organelle will not affect these parasites. However, comparative study of their genomes reveals that *T. brucei*, *T. cruzi*, and *L. major* share some distinctive DNA transcription features that might prove useful in developing drug treatments.

CHAPTER SUMMARY

28.1 An Introduction to Protists

- Protists are eukaryotes that are abundant in moist habitats, and most are microscopic in size. Modern phylogenetic analysis has revealed that protists do not form a monophyletic group; instead, they are classified into several eukaryotic supergroups. (Figure 28.1, Table 28.1)

- Protists are often classified according to ecological roles into three major groups. The term algae is used informally for protist phyla that include photosynthetic protists. The term protozoa describes diverse heterotrophic protists that are often mobile. The term fungus-like protists describes protists having a structure and nutrition that resemble those of true fungi. (Figures 28.2, 28.3)

- Protists are particularly diverse in aquatic environments. The floating or swimming plankton are relatively small and simple in structure; the phytoplankton are adapted to life in well-illuminated near-surface waters; and the periphyton are attached to underwater surfaces. The multicellular protists known as seaweeds often display complex structure. (Figure 28.4)

- Microscopic protists propel themselves in several ways. Diverse small protists that swim by means of eukaryotic flagella are known as flagellates. The ciliates are a phylum of protists that move by means of many short cilia on their surfaces. Amoebae are diverse protists that move by means of pseudopodia. Other protists are able to glide across surfaces by secreting slime. (Figure 28.5)

28.2 Evolution and Relationships

- New discoveries are constantly changing concepts of protist evolution and relationships, and the relationships of some protists are unknown or controversial.

- The supergroup Excavata includes flagellate protists characterized by a feeding groove (such as the genus *Jakoba*), as well as the specialized parasites *Trichomonas vaginalis* and *Giardia lamblia*, which have modified mitochondria. (Figures 28.6, 28.7)

- The supergroup Euglenozoa consists of flagellates having disk-shaped mitochondrial cristae. Phyla include the kinetoplastids, which include some human parasites, and euglenoids, some of which are photosynthetic. (Figures 28.8, 28.9)

- The supergroup Archaeplastida includes three algal phyla (green algae, red algae, and glaucophytes), as well as the plant kingdom, all of which display primary plastids with two envelope membranes. (Figure 28.10)

- The supergroup Alveolata includes the ciliates, dinoflagellates, and apicomplexans, all known for the presence of saclike membranous vesicles called alveoli. (Figure 28.11)

- The supergroup Stramenopila includes a diverse group of protists that produce flagella with strawlike hairs that aid swimming. Stramenopiles include the fungus-like plant parasite *Phytophthora*, as well as diatoms and diverse algal classes whose secondary

plastids originated from a red alga by secondary endosymbiosis. Such plastids also occur in species of three other algal phyla: Haptophyta, Cryptophyta, and Dinozoa. (Figure 28.12)

- Tertiary plastids arose in protists on more than one occasion, and many heterotrophic protists and aquatic animals contain endosymbiotic algal cells. (Figure 28.13)

- The supergroup Rhizaria consists of flagellates and amoebae having thin cellular processes known as filose pseudopodia. Three prominent phyla are Chlorarachniophyta, which have secondary green plastids; mineral-shelled Radiolaria; and Foraminifera, which have shells composed of calcium carbonate. (Figure 28.14)

- The supergroup Amoebozoa is composed of many types of amoebae that move by means of pseudopodia, and includes the parasite *Entamoeba histolytica* and slime molds such as *Dictyostelium discoideum*.

- The supergroup Opisthokonta includes organisms that produce swimming cells that bear a single posterior flagellum. It includes the fungal and animal kingdoms. The protists known as choanoflagellates are considered the modern protists most closely related to the common ancestor of animals. (Figure 28.15)

28.3 Nutritional and Defensive Adaptations

- Protists display four basic types of nutrition. Heterotroph protists called phagotrophs rely on particle feeding, whereas heterotroph protists called osmotrophs absorb small organic molecules. Autotroph protists make their own organic food, and protists that use both autotrophic and heterotrophic means to obtain nutrients are known as mixotrophs. (Figure 28.16)

- Protists that break down wastes and dead organisms are called decomposers, whereas those that feed on living cells are termed parasites.

- Protists defend themselves from attackers by means of defensive adaptations that include protective cell coverings, sharp projectiles, light flashes, and toxic compounds. Dinoflagellates are particularly important as toxin producers. Aquatic ecologists have established that populations of the unicellular dinoflagellate genus *Pfiesteria* kill fish by means of a toxin that can also harm human cells. (Figure 28.17)

28.4 Reproductive Adaptations

- Protist populations grow by means of asexual reproduction involving mitosis, and many are able to persist through unfavorable conditions in the form of tough-walled cysts. (Figure 28.18)

- Protists having zygotic life cycles often use tough-walled zygotes to persist through unfavorable conditions. (Figure 28.19)

- Protists displaying sporic life cycles are able to occupy more types of habitats because they produce two or more alternating life stages having differing environmental preferences. (Figure 28.20)

- Diatoms, which have a gametic life cycle, use sexual reproduction to solve a cell size problem that originates from their unique mode of asexual cell division. (Figure 28.21)
- Ciliates display a unique type of sexual reproduction (conjugation) that involves the exchange of genetic material between a mating pair of cells. (Figure 28.22)
- Parasitic protists may have life cycles involving alternate hosts. One example is *Plasmodium*, the malarial agent, whose alternate hosts are humans and mosquitoes. Genomic sequence data have been obtained for several parasitic protists, with the goal of developing new treatments that will kill the parasites without harming hosts. (Figure 28.23)

TEST YOURSELF

1. Which protist phylum is most closely related to the animal kingdom?
   a. Rhodophyta
   b. Euglenida
   c. Choanomonada
   d. Radiolaria
   e. None of the above

2. Which protist phylum is most closely related to the plant kingdom?
   a. Chlorophyta
   b. Chlorarachniophyta
   c. Choanomonada
   d. Bacillariophyta
   e. Radiolaria

3. Which informal ecological group of protists includes autotrophs?
   a. protozoa
   b. algae
   c. fungus-like protists
   d. all of the above
   e. none of the above

4. How would you recognize a secondary plastid?
   a. It would have one envelope membrane.
   b. It would have two envelope membranes.
   c. It would have more than two envelope membranes.
   d. It would have pigments characteristic of primary plastids of red or green algal cells.
   e. Both c and d are correct.

5. What organisms have tertiary plastids?
   a. certain stramenopiles
   b. certain euglenoids
   c. certain cryptomonads
   d. certain opisthokonts
   e. none of the above

6. What is unusual about mixotrophs?
   a. They have no plastids, but they occur mixed in communities with autotrophs.
   b. They have mixed heterotrophic and autotrophic nutrition.
   c. Their cells contain a mixture of red and green plastids.
   d. Their cells contain a mixture of mitochondria having differently shaped cristae.
   e. They consume a mixed diet of algae.

7. What advantages do diatoms obtain from sexual reproduction?
   a. increased genetic variability
   b. increased ability of populations to respond to environmental change
   c. evolutionary potential
   d. regeneration of maximal cell size for the species
   e. all of the above

8. What are trichocysts?
   a. hairs that occur on flagella
   b. membrane sacs that occur beneath the cell surface
   c. tough-walled asexual cells that are able to withstand unfavorable conditions
   d. spearlike defensive structures that are shot from cells under attack
   e. none of the above

9. How do autotrophic protists use accessory pigments?
   a. Accessory pigments provide camouflage, so herbivores cannot see algae.
   b. Accessory pigments are able to absorb underwater light and transfer the energy to chlorophyll *a* for use in photosynthesis.
   c. Accessory pigments attract aquatic animals that carry gametes from one seaweed to another.
   d. All of the above.
   e. None of the above.

10. What are the two alternate hosts of the malarial parasite *Plasmodium falciparum*?
    a. humans and ticks
    b. ticks and mosquitoes
    c. humans and *Anopheles* mosquitoes
    d. humans and all types of mosquitoes
    e. none of the above

CONCEPTUAL QUESTIONS

1. Explain why protists are classified into multiple supergroups, rather than a single kingdom or phylum.

2. Why have molecular biologists sequenced the genomes of several parasitic protists?

3. Why are protistan cysts important to epidemiologists, biologists who study the spread of disease?

EXPERIMENTAL QUESTIONS

1. Why did the Burkholder team test two different strains of *Pfiesteria shumwayae*?

2. Why did the Burkholder team grow *Pfiesteria shumwayae* with algae or fish as food?

3. Why did the Burkholder team use a biohazard containment system?

COLLABORATIVE QUESTIONS

1. Imagine that you are studying an insect species, and you discover that the insects are dying of a disease that results in the production of cysts of the type that protists often generate. Thinking that the cysts might have been produced by a parasitic protist that could be used as an insect control agent, how would you go about identifying the disease agent?

2. Imagine that you are part of a marine biology team seeking to catalogue the organisms inhabiting a threatened coral reef. The team has found two new seaweeds, each of which occurs during a particular time of the year when the water temperature differs. You suspect that the two seaweeds might be different generations of the same species that have differing optimal temperature conditions. How would you go about testing your hypothesis?

---

www.brookerbiology.com
This website includes answers to the Biological Inquiry questions found in the figure legends and all end-of-chapter questions.

29

THE KINGDOM FUNGI

The orange peel fungus, *Aleuria aurantia*.

Y ou might think that the largest organism in the world is a whale or perhaps a giant redwood tree. Amazingly, giant fungi would also be good candidates. For example, an individual of the fungus *Armillaria ostoyae* weighs hundreds of tons, is more than 2,000 years old, and spreads over 2,200 acres of Oregon forest soil! Scientists discovered the extent of this enormous fungus when they found identical DNA sequences in samples taken over this wide area. Other examples of such huge fungi have been found, and mycologists—scientists who study fungi—suspect that they may be fairly common, underfoot yet largely unseen.

Regardless of their size, fungi typically occur within soil or other materials, becoming conspicuous only when reproductive portions such as mushrooms extend above the surface. Even though fungi can be inconspicuous, they play essential roles in the Earth's environment, are associated in diverse ways with other organisms, and have many technological applications. In this chapter, we will explore the distinctive features of fungal structure, growth, nutrition, reproduction, and diversity. In the process you will learn how fungi are connected to forest growth, food production and food toxins, sick building syndrome, and other topics of great importance to humans.

29.1 Distinctive Features of Fungi

The eukaryotes known as fungi are so distinct from other organisms that they are placed in their own kingdom, the kingdom Fungi (**Figure 29.1**). The monophyletic kingdom Fungi includes early-diverging phyla such as chytrids, zygomycetes, and AM fungi, and higher fungi ascomycetes and basidiomycetes. Because fungi are closely related to the animal kingdom, fungi and animals display some common features. For example, both are **heterotrophic**,

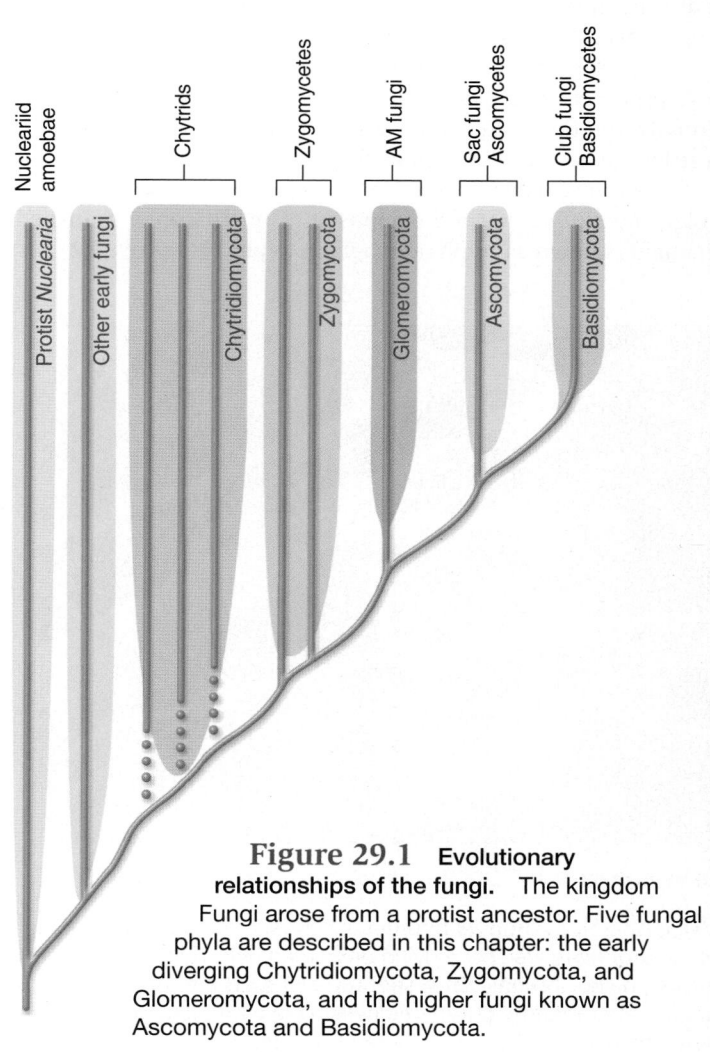

Figure 29.1   Evolutionary relationships of the fungi.   The kingdom Fungi arose from a protist ancestor. Five fungal phyla are described in this chapter: the early diverging Chytridiomycota, Zygomycota, and Glomeromycota, and the higher fungi known as Ascomycota and Basidiomycota.

meaning that they cannot produce their own food but must obtain it from the environment. Fungi use an amazing array of organic compounds as food, which is termed their **substrate**. The substrate could be the soil, a rotting log, a piece of bread, a living tissue, or a wide array of other materials. Fungi are also like animals in having **absorptive nutrition**. Both fungi and animals secrete enzymes that digest organic materials and absorb the resulting small organic food molecules, as well as minerals and water, into their cells. In addition, both fungi and animals store surplus food as the carbohydrate glycogen in their cells. Despite these nutritional commonalities fungal structure is quite distinctive.

Fungi Have a Unique Cell-Wall Chemistry and Body Form

Unlike animal cells, which lack rigid cell walls, fungal cells are enclosed by tough cell walls composed of **chitin**, a polysaccharide that contains nitrogen. Chitin, which also forms the exoskeletons of arthropods such as insects, resists bacterial attack, thereby helping to protect fungal cells. Fungal cell walls influence two other features that distinguish fungal cells from animal cells: nutrition and motility. Because they have cell walls, fungal cells cannot engulf food particles by phagotrophy. In contrast, some animal cells are capable of phagotrophy, a feature inherited from ancestral protists (see Chapter 28). Additionally, the rigid walls of fungal cells, like those of plant cells, restrict the mobility of nonflagellate cells.

Most fungi have distinctive bodies known as **mycelia** (singular, *mycelium*) that are composed of microscopic, branched filaments known as **hyphae** (singular, hypha) (**Figure 29.2**). As

discussed later, hyphae may be septate or aseptate, and cells may contain one nucleus or multiple nuclei. Most of the fungal mycelium is diffuse, or spread out, within food substrates and thus inconspicuous. In contrast, **fruiting bodies** are the visible fungal reproductive structures, which are composed of more densely packed hyphae that typically grow out of the substrate (**Figure 29.3**). Fruiting bodies are thus the only parts of fungal bodies that most people see. Mushrooms are one type of fungal fruiting body. Fungal fruiting bodies are specialized to produce the walled reproductive cells known as **spores**. Fungal fruiting bodies are amazingly diverse in form, color, and odor because they are adapted in ways that foster spore dispersal by wind, water, or animals. When spores settle in places where conditions are favorable for growth, they produce new mycelia. When the new mycelia undergo sexual reproduction, they produce new fruiting bodies.

If you remove the bark from rotting logs or look under dead, wet leaves in a forest, you may find what look like white or colored strings. The strings are **rhizomorphs**, fungal mycelia that have the shape of roots (**Figure 29.4**). Rhizomorphs transport water to other parts of the same mycelium living in low-moisture habitats.

Fungi Have Distinctive Growth Processes

If you have ever watched food become increasingly moldy over the course of several days, you have observed fungal growth. When a food source is plentiful, fungal mycelia can grow rapidly, adding as much as a kilometer of new hyphae per day. The mycelia grow at their edges as the fungal hyphae extend their tips through the undigested substrate. The narrow dimen-

(a) Hyphae on hair

(b) Unmated septate hypha

(c) Mated septate hypha

(d) Unmated aseptate hypha

(e) Mated aseptate hypha

Figure 29.2 **Fungal hyphae.** Fungal bodies are composed of delicate filaments known as hyphae. **(a)** Hyphae of *Trichophyton mentagrophites* are shown lying on human hair. This fungus produces enzymes that break down keratin, a protein that occurs abundantly in skin, nails, and hair. **(b)** A septate hypha prior to mating. Each cell contains only one nucleus. **(c)** A septate, dikaryotic hypha that occurs after mating. Dikaryotic cells have two nuclei of differing genetic types. **(d)** An aseptate hypha prior to mating lacks septa and contains only one genetic type of nuclei. **(e)** An aseptate hypha after mating contains two genetic types of nuclei.

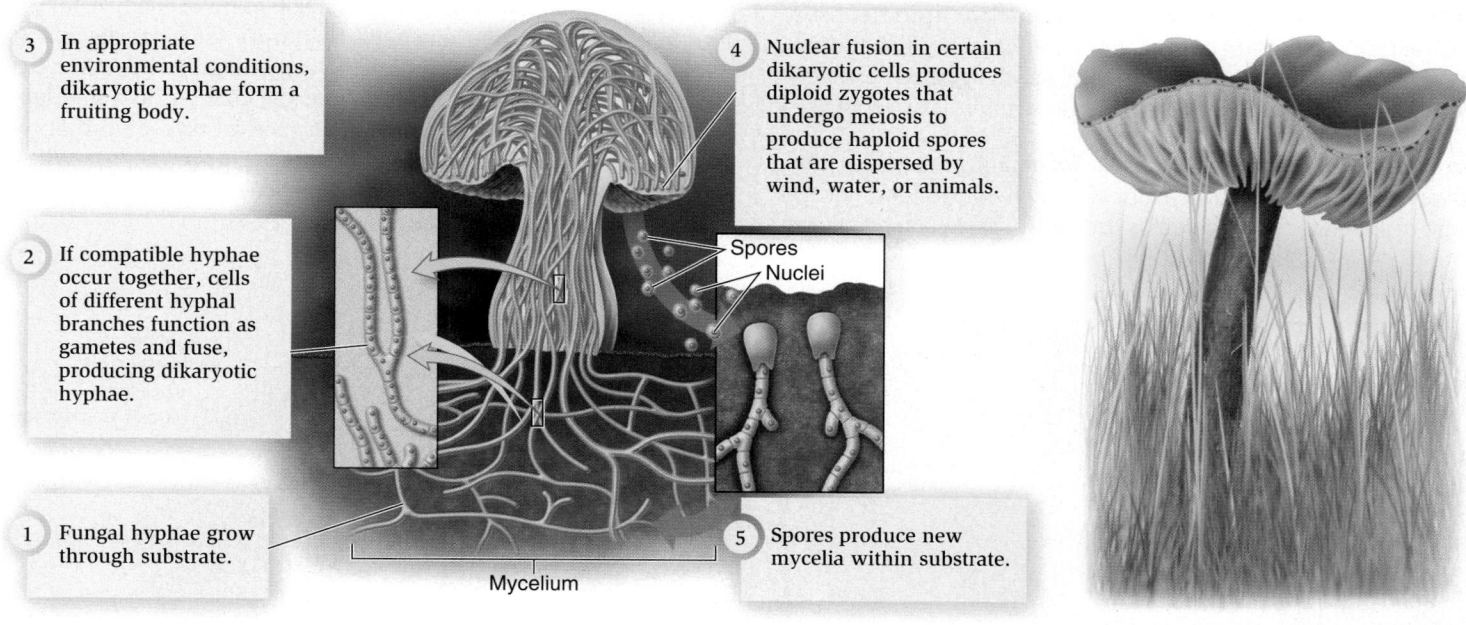

3 In appropriate environmental conditions, dikaryotic hyphae form a fruiting body.

2 If compatible hyphae occur together, cells of different hyphal branches function as gametes and fuse, producing dikaryotic hyphae.

1 Fungal hyphae grow through substrate.

4 Nuclear fusion in certain dikaryotic cells produces diploid zygotes that undergo meiosis to produce haploid spores that are dispersed by wind, water, or animals.

Spores
Nuclei

5 Spores produce new mycelia within substrate.

Mycelium

(a)

(b) A fruiting body of the genus *Laccaria*

Figure 29.3 **The reproductive fungal body.** **(a)** Most of a fungus consists of hyphae that grow and branch from a central point to form a diffuse mycelium within a food substrate, such as soil. After mating, fungal hyphae may aggregate to form fleshy masses known as fruiting bodies that extend above the substrate surface. Fruiting bodies produce spores that are dispersed by wind, water, or animals. In suitable sites, spores may germinate, producing new mycelia. **(b)** The aboveground portion of the genus *Laccaria* and other mushrooms are examples of fruiting bodies.

Rhizomorphs
Underside of log

Figure 29.4 **Rhizomorphs.** In moist forests, hyphae of fungi such as the honey fungus (*Armillaria mellea*) clump to form conspicuous string-shaped rhizomorphs.

sions and extensive branching of hyphae provide a very high surface area for absorption of organic molecules, water, and minerals.

Hyphal Tip Growth How do hyphae grow? Cytoplasmic streaming and osmosis are important cellular processes in hyphal growth. Osmosis (see Chapter 5) is the diffusion of water through a membrane from an area with a low solute concentra-

tion into an area with a high solute concentration. Water enters fungal hyphae by means of osmosis because their cytoplasm is rich in sugars, ions, and other solutes. Water entry swells the hyphal tip, producing the force necessary for tip extension. Masses of tiny vesicles carrying enzymes and cell-wall materials made in the Golgi apparatus collect in the hyphal tip (**Figure 29.5**). The vesicles then fuse with the plasma membrane. Some vesicles deliver cell-wall materials to the hyphal tip, allowing it to extend. Other vesicles release enzymes that digest materials in the environment, releasing small organic molecules that are absorbed as food.

Hyphal Structure The hyphae of most fungi are divided into many small cells by cross walls known as **septa** (see Figure 29.2b,c). Each of these cells has one or two nuclei. In such fungi, known as septate fungi, each round of nuclear division is followed by cross-wall formation. A central pore, which may be simple or more complex, perforates hyphal septa (**Figure 29.6**). Complex pores may have caps of modified endoplasmic reticulum. Septal pores are large enough to allow cytoplasmic structures and materials to pass through the hyphae. The hyphae of some fungi are not partitioned into smaller cells; rather, these hyphae are **aseptate** and multinucleate (see Figure 29.2d,e), a condition that results when nuclei repeatedly divide without intervening cytokinesis.

Fungal Nuclear Division When the nuclei of fungi divide, in most cases a spindle forms within the nuclear envelope, which does not break down. This **intranuclear spindle** distinguishes nuclear division of fungi from that of animals and plants. (By contrast, in land plants and animals, the nuclear envelope vesiculates during prometaphase and then re-forms at telophase.)

Variations in Mycelium Growth Form Fungal hyphae grow rapidly through a substrate from areas where the food has become depleted to food-rich areas. In nature, mycelia may take an irregular shape, depending on the availability of food. A fungal mycelium may extend into food-rich areas for great distances, even spreading over thousands of acres like the giant fungus *A. ostoyae*, noted at the beginning of the chapter. In liquid laboratory media, fungi will grow as a spherical mycelium that resembles a cotton ball floating in water (**Figure 29.7a**). When grown in flat laboratory dishes, the mycelium assumes a more two-dimensional growth form (**Figure 29.7b**).

(a)

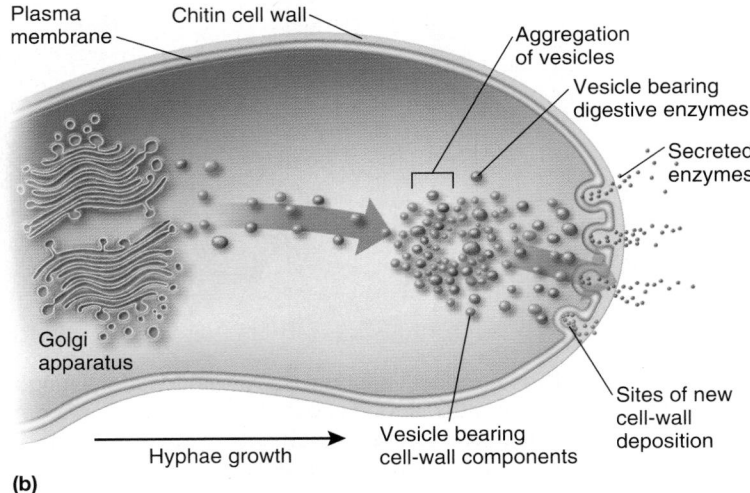

(b)

Figure 29.5   **Hyphal tip growth and absorptive nutrition.**   **(a)** TEM showing the hyphal tip of *Aspergillus nidulans*, a fungus commonly used as a genetic model system. **(b)** Diagram of a hyphal tip, showing vesicles of two types: those containing cell-wall materials, and those bearing digestive enzymes. Upon fusion of these vesicles with the plasma membrane, both types of vesicle contents are released to the cell exterior. Digestive enzymes degrade extracellular organic polymers into smaller components that can be taken into hyphae by means of plasma membrane transporter proteins. The uptake of minerals and organic molecules causes osmotic water uptake, with the result that cells enlarge and hyphal tips extend. New cell-wall materials are added to the hyphal tip as it extends.

Biological inquiry: What do you think would happen to fungal hyphae that begin to grow into a substrate having higher solute concentration? How might your answer be related to food preservation techniques such as drying or salting?

(a) Simple pore

(b) Complex pore

Figure 29.6   **Septa divide hyphae of higher fungi into compartments.**   **(a)** The septa of ascomycetes have simple pores at the centers. **(b)** More complex pores distinguish the septa of most types of basidiomycetes.

(a) **(b)**

Figure 29.7 **Fungal shape shifting.** **(a)** When a mycelium, such as that of *Rhizoctonia solani*, is surrounded by food substrate in a liquid medium, it will grow into a spherical form. **(b)** When the food supply is limited to a two-dimensional supply, as shown by *Neotestudina rosatii* in the laboratory dish shown here, the mycelium will form a disk. Likewise, distribution of the food substrate determines the mycelium shape in nature.

(a) **(b)**

Figure 29.8 **Sexual and asexual reproductive cells of fungi.** **(a)** SEM showing spores produced after a sexual mating, zygote formation, and subsequent meiosis in the forest mushroom *Coprinus disseminatus*. When mature, these genetically variable spores detach and are dispersed into the environment, where they may find suitable conditions for growth. **(b)** SEM of the asexual conidia of *Aspergillus versicolor*, which causes skin infections in burn victims and lung infections in AIDS patients. Each of these small cells is able to detach and grow into an individual that is genetically identical to the parent fungus and able to grow in similar conditions.

29.2 Fungal Sexual and Asexual Reproduction

Many fungi are able to reproduce both sexually and asexually by means of microscopic spores, each of which can grow into a new adult (**Figure 29.8**). Sexual reproduction generates new allele combinations that may allow fungi to colonize new types of habitats. In contrast, asexual reproduction is a natural cloning process; it produces genetically identical organisms. Production of asexual spores allows fungi that are well adapted to a particular environment to disperse to similar, favorable places.

Fungi Have Distinctive Sexual Reproductive Processes

As is typical for eukaryotes, the fungal sexual reproductive cycle involves the union of gametes, the formation of zygotes, and the process of meiosis. However, many aspects of fungal sexual reproduction are unique, including the function of hyphal branches as gametes and the development of fruiting bodies (see Figure 29.3).

Fungal Gametes and Mating The gametes of most fungi are inconspicuous hyphal branches. During sexual reproduction, these hyphal branches fuse with those of a different mycelium of compatible mating type. Particular genes control mating compatibility. In contrast, some primitive fungi that live in the water produce flagellate sperm and nonmotile eggs.

The actual mating process in fungi is also remarkable. In most sexual organisms, gametes undergo fusion of their cytoplasms—a process known as **plasmogamy** and then the nuclei fuse—a process known as **karyogamy**. In contrast, the gamete nuclei of many fungi behave differently. After plasmogamy

occurs, the haploid gamete nuclei may remain separate for a long time. During this time period, the gamete nuclei both divide at each cell division, producing a **dikaryotic** (meaning "two nuclei") mycelium. Each cell of a dikaryotic mycelium possesses two unfused gamete nuclei (see Figure 29.2c). Dikaryotic mycelia are also known as **heterokaryons** (meaning "different nuclei"), reflecting the fact that the two nuclei of each cell are genetically distinct. Some fungi normally persist as heterokaryons, producing clones that can live for hundreds of years. Although the nuclei of dikaryotic mycelia remain haploid, alternate copies of many alleles occur in the separate nuclei. Thus, dikaryotic mycelia are functionally diploid. Eventually, dikaryotic mycelia produce fruiting bodies, the next stage of reproduction.

Fruiting Bodies Under appropriate environmental conditions such as seasonal change, a heterokaryotic mycelium may produce a fleshy fruiting body that emerges from the substrate (see Figure 29.3). All the cells of the fruiting body are dikaryotic. When the fruiting body is mature, the two nuclei in cells at the surface undergo nuclear fusion. This process produces many zygotes, which are the only cells in the fungal life cycle that have a diploid nucleus. In most cases, the fungal zygotes soon undergo meiosis to produce haploid spores. Each spore acquires a tough wall that protects it from drying and other stresses. Wind, rain, or animals disperse the mature spores, which grow into haploid mycelia. If a haploid mycelium encounters hyphae of an appropriate mating type, hyphal branches will fuse and start the sexual cycle over again.

(a)   (b)

Figure 29.9 Fruiting body adaptations that foster spore dispersal. (a) When touched by wind gusts or animal movements, spores puff from fruiting bodies of the puffball fungus (*Lycoperdon perlatum*). (b) Insects visit fruiting bodies of stinkhorn fungi such as *Phallus impudicus*, attracted by foul odors and colors like those of dung or rotting meat. Spores, produced in a sticky matrix, attach to the insects and are thereby carried away.

Figure 29.10 Toxic fruiting body of *Amanita muscaria*. Common in conifer forests, *A. muscaria* is both toxic and hallucinogenic. Ancient people used this fungus to induce spiritual visions and to reduce fear during raids. This fungus produces a toxin, amanitin, which specifically inhibits RNA polymerase II of eukaryotes (but not prokaryotes).

Biological inquiry: What effect would the amanitin toxin have on human cells?

By now, it should be clear why fruiting bodies usually emerge from the substrate, while most of the fungal mycelium lies inconspicuously within it. Mycelium growth requires organic molecules, minerals, and water provided by the substrate, but spores are more easily dispersed if released outside of the substrate. The structure of fruiting bodies varies in ways that reflect different adaptations that foster spore dispersal by wind, rain, or animals. For example, mature puffballs have delicate surfaces upon which just a slight pressure causes the spores to puff out into wind currents (**Figure 29.9a**). Birds' nest fungi form characteristic egg-shaped spore clusters. Raindrops splash on these clusters and disperse the spores. The fruiting bodies of stinkhorn fungi smell and look like rotting meat, which attracts carrion flies (**Figure 29.9b**). The flies land on the fungi to investigate the potential meal and then fly away, in the process dispersing spores that stick to their bodies. The fruiting bodies of fungal truffles are also specialized for spore dispersal by animals. However, they are unusual in being produced underground. Mature truffles emit odors that attract wild pigs and dogs, which break up the fruiting structures while digging for them, thereby dispersing the spores. When collectors seek to harvest valuable truffles from forests for the market, they use trained leashed pigs or dogs to locate these fungi.

Though fungi such as truffles are edible, the fruiting bodies of many fungi produce toxic substances which may deter animals from consuming them (**Figure 29.10**). For example, several fungi that attack stored grains, fruits, and spices produce **aflatoxins**; these cause liver cancer and are a major health concern worldwide. When people consume the forest mushroom *Amanita virosa*, known as the "destroying angel," they are also ingesting a powerful toxin that may cause liver failure so severe that a transplant may be required. Each year, thousands of people in North America alone are poisoned when they consume similarly toxic mushrooms gathered in the wild. There is no reliable way for nonexperts to distinguish poisonous from nontoxic fungi; it is essential to receive instruction from an expert before foraging for mushrooms in the woods. Because several species of edible fungi are cultivated for market sale, many experts recommend that it is better to forage for mushrooms in the grocery store than in the wild (**Figure 29.11**).

Several types of fungal fruiting structures produce hallucinogenic or psychoactive substances. As in the case of fungal toxins, fungal hallucinogens may have evolved as antiherbivore adaptations. Humans have inadvertently experienced their effects. For example, *Claviceps purpurea*, which causes a disease of rye crops and other grasses known as ergot, produces a psychogenic compound related to LSD (lysergic acid diethylamide) (**Figure 29.12**). Some experts speculate that cases of hysteria, convulsions, infertility, and a burning sensation of the skin that occurred in Europe during the Middle Ages resulted from ergot-contaminated rye used in foods. Ergot poisoning may also have caused the behavioral changes that were at the heart of the Salem witch trials in the late 17th century. Another example of a hallucinogenic fungus is the "magic mushroom" (*Psilocybe*), which is used in traditional rituals in some cultures. Like ergot, the magic mushroom produces a compound similar to LSD, a controlled substance. Consuming hallucinogenic fungi is risky because the amount used to achieve psychoactive affects is dangerously close to a poisonous dose.

Asexual Reproduction Does Not Involve Mating and Meiosis

Asexual reproduction is particularly important to fungi, allowing them to spread rapidly through favorable environments by means of asexual spores. To reproduce asexually, fungi do not need to find compatible mates or expend resources on fruiting-body formation and meiosis. Though many fungi reproduce both sexually and asexually, more than 17,000 species reproduce primarily or exclusively by asexual means. DNA-sequencing studies have revealed that many types of modern fungi that reproduce only asexually have evolved from ancestors that had both sexual and asexual reproduction.

Figure 29.11 Several types of edible, delicious fungi are grown for the market.

Figure 29.13
An indoor mold fungus. The black mold *Stachybotrys chartarum* is one of several types of conidial fungi that grow on indoor paper and wood that have been wet for an extended period.

58 μm

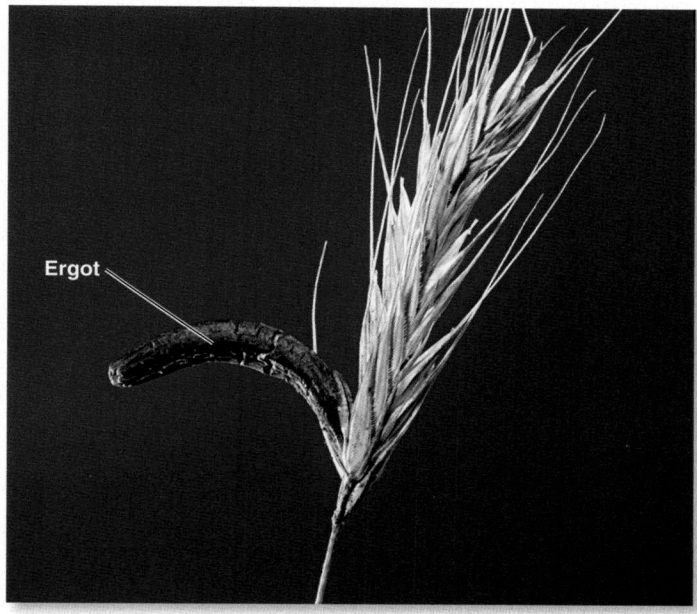

Ergot

Figure 29.12 **Ergot of rye.** The fungus *Claviceps purpurea* infects rye and other grasses, producing hard masses of mycelia known as ergots in place of some of the grains (fruits). Ergots produce alkaloids related to LSD and thus cause psychotic delusions in humans and animals that consume products made with infected rye. Ergots were used in folk medicine to treat migraine or hasten childbirth.

Most fungi reproduce asexually by generating chains of tiny spores from the tips of hyphae. Many fungi produce asexual spores known as **conidia** (from the Greek word for "dust") (see Figure 29.8b). When they land on a favorable substrate, conidia germinate into a new mycelium that produces many more conidia. The green molds that form on citrus fruits are familiar examples of conidial fungi. A single fungus can produce as many as 40 million conidia per hour over a period of 2 days.

Because they can spread so rapidly, asexual fungi are responsible for costly fungal food spoilage and allergies. *Aspergillus fumigatus* is a common conidial mold that causes a potentially fatal lung disease. Such harmful conidia-producing molds growing in poorly ventilated, moist places may be the cause of "sick building syndrome," a term used to describe situations in which occupants of a building experience acute health effects that appear to be linked to time spent inside, but where no specific cause can be identified. However, some black molds that also grow on moist building materials are blamed for human illness, even though there is little evidence that they are harmful to adults (**Figure 29.13**). Medically important fungi that reproduce primarily by asexual means include the athlete's foot fungus (*Epidermophyton floccosum*) and the infectious yeast (*Candida albicans*).

29.3 Fungal Ecology and Biotechnology

The ability of fungi to degrade diverse materials is important in ecology, medicine, and human biotechnology applications. Fungal decomposers, also known as saprobes, are able to decompose nonliving organic materials. Recycling of materials in ecosystems depends on the activities of fungal decomposers. Fungal predators consume whole, though tiny, living organisms. Fungi that attack living animal or plant tissues are pathogens. As hyphae of pathogenic fungi grow through the tissues of plants and animals, food and minerals are absorbed from the host, causing disease symptoms. Mutualistic fungi have formed partnerships with photosynthetic organisms that provide the fungi with organic food. In return, fungi contribute mineral nutrients and water to their photosynthetic partners. In this section, we focus more closely on the ecological roles of decomposer, predatory, and mutualistic fungi.

Figure 29.14 **A predatory fungus.** The fungus *Arthrobotrys anchonia* traps nematode worms in hyphal loops that suddenly swell in response to the animal's presence. Fungal hyphae then grow into the worm's body and digest it.

Decomposer and Predatory Fungi Play Important Ecological Roles

Decomposer fungi are essential components of the Earth's ecosystems. Together with bacteria, they decompose dead organisms and organic wastes, preventing litter buildup. For example, only certain bacteria and fungi can break down cellulose, and a few fungi are the major decomposers of lignin, the decay-resistant component of wood. In the absence of decomposers, ecosystems would become clogged with organic debris. Decomposers are needed to break down organic compounds into carbon dioxide, which is used by algae and plants for photosynthesis. Decomposer fungi and bacteria are also Earth's recycling engineers. They release minerals to the soil and water, where plants and algae take up the minerals for growth. In the absence of fungi and bacteria, such minerals would remain forever bound up in dead organisms, unavailable for the growth of forests, coral reefs, and other biotic communities.

Some soil fungi are predators—they trap tiny soil animals, such as nematodes, in nooselike hyphae and then proceed to digest their bodies (**Figure 29.14**). Such fungi help control populations of nematodes, some of which attack plant roots. Some fungi attack and kill insects, and certain of these species are used as biological control agents to kill black field crickets, red-legged earth mites, and other pests.

Some Fungal Species Cause Plant and Human Diseases

Five thousand fungal species cause serious crop diseases and recent results show that new diseases can arise by horizontal gene transfer. Wheat rust is an example of a common crop disease caused by fungi (**Figure 29.15**). Rusts are named for reddish spores that emerge from the surfaces of infected plants. Many types of plants can be attacked by rust fungi, but rusts are of particular concern when new strains attack crops. For example, in late 2004, agricultural scientists discovered that a devastating rust, *Phakopsora pachyrhizi*, had begun to spread in the U.S. soybean crop. This rust kills soybean plants by attacking the leaves, causing complete leaf drop in less than 2 weeks. The disease had apparently spread to U.S. farms by means of

Figure 29.15 **Wheat rust.** (a) The plant pathogenic fungus *Puccinia graminis* grows within the tissues of wheat plants, using plant nutrients to produce rusty streaks of red spores that erupt at the stem and leaf surface where spores can be dispersed. (b) Red spore production is but one stage of a complex life cycle involving several types of spores. Rusts infect many other crops, causing immense economic damage.

spores blown on hurricane winds from South America. To control the spread of fungal diseases, agricultural experts work to identify effective fungicidal chemicals and develop resistant crop varieties.

In addition to being spread by wind, the spores of crop-disease fungi can also be introduced on travelers' clothing and other belongings. To reduce the entry of new crop disease fungi—as well as crop viruses and insect pests—agricultural customs inspectors closely monitor the entry of plants, soil, foods, and other materials that might harbor these organisms.

Fungi cause several types of disease in humans. For example, athlete's foot and ringworm are common skin diseases caused by fungi that are known as dermatophytes because they colonize the human epidermis (see Figure 29.2a). *Pneumocystis carinii* is a fungal pathogen that infects individuals with weakened immune systems, such as AIDS patients, sometimes causing death by pneumonia. *Coccidioides immitis, Cryptococcus neoformans,* and *Histoplasma capsulatum* are examples of fungi that cause lung and other diseases. These diseases are of special concern when they occur in cancer patients whose immune systems are weakened.

Though fungal diseases that attack humans are of medical concern, in nature, fungal pathogens often help to control populations of other organisms, which is an important ecological role. Some fungi play important ecological roles as beneficial partners in symbiotic associations with other organisms.

Fungi Form Mutually Beneficial Ecological Associations

Fungi form several types of associations that appear to benefit both partners; these are known as mutualistic interactions (see Chapter 57). For example, leaf-cutting ants, certain termites and beetles, and the salt marsh snail (*Littoraria irrorata*) cultivate particular fungi for food, much as do human mushroom growers. Other fungi obtain organic food molecules from photosynthetic organisms: plants, green algae, or cyanobacteria. We focus next on three types of fungi—mycorrhizal fungi, endophytes, and lichen fungi—that are beneficially associated with other organisms.

Mycorrhizae   Associations between the hyphae of certain fungi and the roots of most seed plants are known as **mycorrhizae** (literally, "fungus roots") (see Chapter 37). Mycorrhizal associations are very important in ecology and agriculture; more than 80% of terrestrial plants form mycorrhizae. Plants that have mycorrhizal partners receive an increased supply of water and mineral nutrients, primarily phosphate, copper, and zinc. They do so because an extensive fungal mycelium is able to absorb minerals from a much larger volume of soil than can plant roots lacking fungal associates. Added together, all the branches of a fungal mycelium in 1 m^3 of soil can reach 20,000 km in total length. Experiments have shown that mycorrhizae greatly enhance plant growth in comparison to plants lacking fungal partners. In return, plants provide fungi with organic food molecules, sometimes contributing as much as 20% of their photosynthetic products—a very good investment.

By binding soils, fungal hyphae also reduce water loss and erosion, and they help protect plants against pathogens and toxic wastes. Fungi thereby help plants adapt to and thrive in new sites, thus playing an important role in plant succession (see Chapter 58). For this reason, ecologists increasingly incorporate mycorrhizal fungi into plant-community restoration projects.

The two most common types of mycorrhizae are ectomycorrhizae and endomycorrhizae. **Ectomycorrhizae** (from the Greek *ecto*, outside) are beneficial interactions between temperate forest trees and soil fungi (**Figure 29.16a**) whose hyphae coat tree-root surfaces (**Figure 29.16b**) and grow into the spaces between root cells (**Figure 29.16c**). Some species of oak, beech, pine, and spruce trees will not grow unless their ectomycor-

rhizal partners are also present. Mycorrhizae are thus essential to the success of commercial nursery tree production and reforestation projects.

Endomycorrhizae (from the Greek *endo*, inside) are partnerships between plants and fungi in which the fungal hyphae penetrate the spaces between root cell walls and plasma membranes and grow along the surfaces of these membranes (**Figure 29.17**). In such spaces, endomycorrhizal fungi often form highly branched, bushy arbuscules (from the word "arbor," referring to tree shape). As the arbuscules develop, the root plasma membrane also expands. Consequently, the arbuscules and the root plasma membranes surrounding them have very high surface areas that facilitate rapid and efficient exchange of materials: Minerals flow from fungal hyphae to root cells, and organic food molecules move from root cells to hyphae. These fungus-root associations are known as **arbuscular mycorrhizae**, abbreviated AM. AM fungi are associated with apple trees, coffee shrubs, and many herbaceous plants, including legumes, grasses, tomatoes, strawberries, and peaches.

Fungal Endophytes   Other mutualistic fungi, known as **endophytes**, live compatibly within the leaf and stem tissues of various types of plants. The endophytes obtain organic food molecules from plants, and in turn contribute toxins or antibiotics that deter foraging animals, insect pests, and microbial pathogens. Endophytic fungi also help some plants to tolerate higher temperatures. As a result, plants with endophytes often grow better than plants of the same species without endophytic fungi. For example, fungal endophytes grow throughout the leaves and stems of many cool-season grasses, such as tall fescue (*Festuca arundinacea*). When such grasses reproduce, fungal hyphae transmitted in seeds will grow in the progeny plants. Grasses with endophytic fungi are larger, more toxic to herbivores, and more drought resistant than those without. These advantages allow tall fescue to invade plant communities and displace other plant species, with deleterious effects on wildlife.

(a)

(b)

(c)

Figure 29.16   Ectomycorrhizae.   **(a)** The fruiting body of the common forest fungus *Boletus*. This is an ectomycorrhizal fungus that is associated with tree roots. **(b)** Ectomycorrhizal fungal hyphae of *Laccaria bicolor* cover the surfaces of young *Pinus resinosa* root tips. **(c)** Diagram showing that the hyphae of ectomycorrhizal fungi do not penetrate root cell walls but grow within intercellular spaces. In this location, fungal hyphae are able to obtain organic food molecules produced by plant photosynthesis.

Biological inquiry: What benefits do plants obtain from this association?

(a)
(b)

Figure 29.17   Endomycorrhizae.   (a) Light micrograph showing black-stained AM fungi within the roots of the forest herb *Asarum canadensis*. Endomycorrhizal fungal hyphae enter plant roots via root hair cells, and then branches grow in the spaces between plant cell walls and plasma membranes. **(b)** Diagram showing the position of highly branched arbuscules. Hyphal branches of arbuscules are coated with plant plasma membrane, with the result that both hyphae and plant membrane have very high surface areas.

Fungal endophytes also occur in many tropical plants, producing compounds that help their hosts to repel attacks by herbivores and disease microbes. These compounds offer promise for the development of new antibiotics and anticancer drugs for use in human medicine.

Lichens **Lichens** are partnerships of particular fungi and certain photosynthetic green algae or cyanobacteria, and sometimes both. There are at least 25,000 lichen species, but these did not all descend from a common ancestor. DNA-sequencing studies suggest that lichens evolved independently in at least five separate fungal lineages. Molecular studies also show that some fungi have lost their ancestral ability to form lichen associations.

Lichen bodies take one of three major forms: (1) crustose—flat bodies that are tightly adherent to an underlying surface (**Figure 29.18a**); (2) foliose—flat, leaflike bodies (**Figure 29.18b**); or (3) fruticose—bodies that grow upright (**Figure 29.18c**) or hang down from tree branches. The photosynthetic green algae or cyanobacteria typically occur in a distinct layer close to the lichen's surface (**Figure 29.18d**). Lichen structure differs dramatically from that of the fungal components grown separately, demonstrating that the photosynthetic components influence lichen form.

The photosynthetic partner provides lichen fungi with organic food molecules and oxygen, and in turn it receives carbon dioxide, water, and minerals from the fungal partner. Lichen fungi also protect their photosynthetic partners from environmental stress. For example, lichens that occupy exposed habitats of high light intensity often produce bright yellow, orange, or red-colored compounds that help prevent damage to the photosynthetic apparatus (see Figure 29.18a). Lichen fungi also produce distinctive organic acids and other compounds that deter animal and microbial attacks.

(a)
(b)
(c)
(d)

Figure 29.18   Lichen structure.   (a) An orange-colored crustose lichen grows tightly pressed to the substrate. **(b)** The flattened, leaf-shaped genus *Umbilicaria* is a common foliose lichen. **(c)** The highly branched genus *Cladonia* is a common fruticose lichen. **(d)** A handmade thin slice of *Umbilicaria* viewed with a compound microscope reveals that the photosynthetic algae occur in a thin upper layer. Fungal hyphae make up the rest of the lichen.

The fungal partners of many lichens can undergo sexual reproduction, producing fruiting bodies and sexual spores much like those of related fungi that do not form lichens. DNA studies have shown that some lichen fungi can self-fertilize, which is advantageous in harsh environments where potential mates may be lacking. To produce new lichens, hyphae that grow from sexual spores must acquire new photosynthetic partners, but only particular green algae or cyanobacteria are suitable. Even if compatible fungi and algae or cyanobacteria occur together, lichens may not form. In the laboratory, lichen formation has been accomplished only when the photosynthetic partner and fungus are exposed to stressful conditions. These results suggest that lichens form in nature under similarly harsh conditions.

The fungi associated with about one-third of lichen species can only reproduce asexually. Asexual reproductive structures include **soredia**, small clumps of hyphae surrounding a few algal cells that can disperse in wind currents (**Figure 29.19**). Soredia are lichen clones. By forming soredia, lichen fungi can disperse along with their appropriate photosynthetic partners.

Lichens often grow on rocks, buildings, tombstones, tree bark, soil, or other surfaces that easily become dry. When water is not available, the lichens are dormant until moisture returns. Thus lichens may spend much of their time in an inactive state, and for this reason they often grow very slowly. However, because they can persist for long periods, lichens can be very old; some are estimated to be more than 4,500 years old. Lichens occur in diverse types of habitats, and a number grow in some of the most extreme, forbidding terrestrial sites on Earth—deserts, mountaintops, and the Arctic and Antarctic—places where most plants cannot survive. In these locations, lichens serve as a food source for reindeer and other hardy organisms. Though unpalatable, lichens are not toxic to humans and have also served as survival foods for aboriginal peoples in times of shortages.

Lichens are recognized for their soil-building activities, which occur over very long timescales. Lichen acids help to break up the surfaces of rocks, beginning the process of soil

Figure 29.19   Lichen asexual reproduction.   Soredia are examples of lichen asexual structures that contain algal or cyanobacterial cells wrapped by fungal hyphae. SEM of a soredium of the lichen *Cladonia coniocraea* that will break off from the parent lichen for dispersal in the environment.

development. Lichens having cyanobacterial partners can also increase soil fertility by adding fixed nitrogen. One study showed that such lichens released 20% of the nitrogen they fixed into the environment, where it is available for uptake by plants.

Lichens are useful as air-quality monitors because they are particularly sensitive to air pollutants such as sulfur dioxide. Air pollutants severely injure the photosynthetic components, causing death of the lichens. The disappearance of lichens serves as an early warning system of air-pollution levels that are also likely to affect humans. Lichens can also be used to monitor atmospheric radiation levels because they accumulate radioactive substances from the air. After the Chernobyl nuclear power plant accident, lichens in nearby countries became so radioactive that reindeer, which consume lichens as a major food source, became unfit for use as human food or in milk production.

FEATURE INVESTIGATION

Piercey-Normore and DePriest Discovered That Some Lichens Readily Change Partners

One of the most interesting current issues regarding symbiotic relationships is the extent to which partners have influenced each other's evolution and diversification. When symbiotic partners have influenced each other's evolution, they are said to have cospeciated (coevolved). In the past it was unclear whether or not lichen fungi and their photosynthetic partners coevolved. The fairly recent development of molecular techniques and computer software for analyzing relationships among microorganisms has

allowed researchers to attempt to answer this and many other fundamental questions about lichen biology.

Researchers Michelle Piercey-Normore and Paula DePriest devised a way to look for evidence of cospeciation of the fungal and algal partners in a large and diverse lichen group known as the Cladoniaceae (**Figure 29.20**). This group of lichens includes the common "reindeer lichen" (*Cladonia cristatella*). The investigators wanted to know if the pattern of speciation for the fungal partners matched that for the algal partner. The investigators first extracted DNA from the fungal and algal partners in 33 lichens, including diverse species of the genus *Cladonia*.

Figure 29.20   Piercey-Normore and DePriest discovered that some lichens readily change partners.

GOAL   To determine if lichen fungal and algal phylogenies are consistent, as evidence for or against cospeciation.

STARTING MATERIALS   Fungal and algal components of 33 lichen species in the family Cladoniaceae.

Experimental level   **Conceptual level**

1   Isolate DNA from fungal and algal partners of 33 lichens.

Lichen   Fungal component

Algal component

Fungal DNA

Algal DNA

2   Amplify a region of the rRNA gene using the technique of polymerase chain reaction (PCR) (described in Chapter 20). Use DNA-sequencing techniques (described in Chapter 20) to determine DNA sequences of the rRNA gene region from the algal and fungal partners of 33 lichens.

Use PCR primers specific for fungal rRNA genes and green algal rRNA genes.

Collect DNA sequences from lichen fungi and lichen algae.

PCR

Thermocycler

rDNA

Fungal DNA

rDNA

Algal DNA

Fungal sequences

Amplified sequences

Algal sequences

3   Use DNA-sequence information and computer programs to infer the phylogenies of the algae and the fungi from 33 lichens.

4   **THE DATA**

Connect lines between partners from the same lichen. Look for cases of cospeciation—when phylogenies of the algal and fungal partners match.

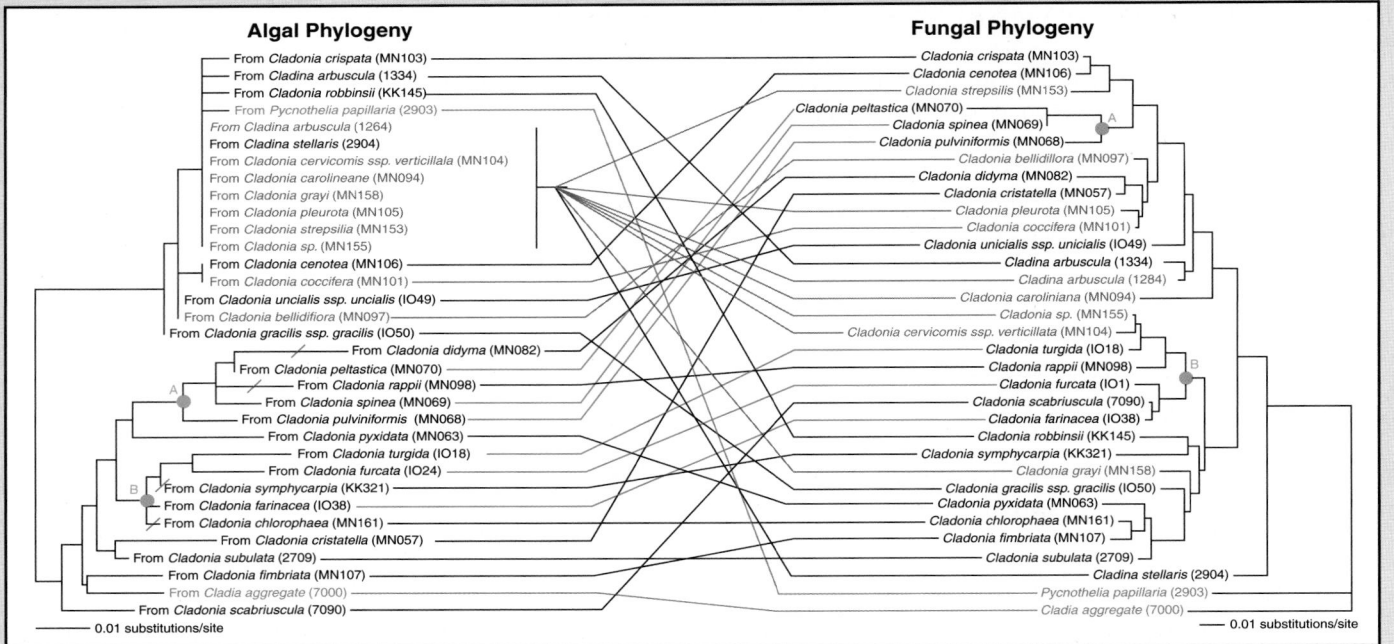

Results from step 4: The green dots marked A and B indicate lichen lineages showing some cospeciation. The partners of these lichens are connected by green lines. Other lineages show little evidence of cospeciation, shown by extensive crossed lines. These results imply a high degree of partner switching in this group of lichens. Partner switching even occurred in lineages A and B (marked with red slashes). These data show that in the lichen family Cladoniaceae, partner switching has been a more frequent occurrence than cospeciation.

Next, they amplified a region of the rRNA gene for all samples and determined the sequences of these DNAs. Finally, Piercey-Normore and DePriest used the DNA sequences to produce separate phylogenetic trees for the algal and fungal partners, and they used statistical procedures to test the possibility that the evolutionary divergence patterns of the fungi and algae were linked.

The results revealed that the fungal and algal phylogenies did not match, indicating that the patterns of their diversifica-tion were not congruent. These data indicated that lichen fungi and algae had not generally influenced each other's speciation patterns. If the lichen fungi and algae had undergone species diversification in tandem, their phylogenies should match, that is, be congruent. Instead, the phylogenetic evidence suggested that lichens often switch algal partners, "trading-up" for better algae. The results of this study, published in 2001, revealed that the lichen symbiosis is more complex than had previously been realized.

Fungi Have Many Applications in Biotechnology

The ability of fungi to grow on many types of substrates and produce many types of organic compounds represents their diverse ecological adaptations. Humans have harnessed fungal biochemistry in many types of biotechnology applications. Fungal biochemistry is a valuable asset to the chemical, food processing, and waste-treatment industries. A variety of industrial processes use fungi to convert inexpensive organic compounds into valuable materials such as citric acid used in the soft drink industry, glycerol, antibiotics such as penicillin, and cyclosporine, a drug widely used to prevent rejection of organ transplants. In the food industry, fungi are used to produce the distinctive flavors of blue cheese and other cheeses. Other fungi secrete enzymes that are used in the manufacture of protein-rich tempeh and other food products from soybeans. The baking industry depends on the yeast *Saccharomyces cerevisiae* for bread production, and the brewing and winemaking industries also find yeasts essential.

Fungi are increasingly being used in industrial processes to replace chemical procedures that generate harmful waste materials. For example, during the production of paper, wood pulp is chemically bleached to remove lignin, but this bleaching process generates harmful compounds, including dioxin, a carcinogenic compound. The forest fungus *Phanerochaete chrysosporium* produces several enzymes that allow it to break down lignin in wood (**Figure 29.21**). Removal of lignin allows the fungi to gain access to cell-wall polysaccharides, which they break down, thereby releasing sugars that the fungus absorbs as food. Fungal lignolytic ("lignin-breaking") enzymes such as those produced by *P. chrysosporium* can be used to bleach paper without producing dioxin, thereby reducing a harmful by-product of paper production. In addition, Khadar Valli, Hiroyuki Wariishi, and Michael Gold demonstrated that *P. chrysosporium* is able to decompose dioxin into nontoxic products, suggesting additional potential uses of the fungus in preventing pollution.

29.4 Evolution and Diversity of Fungi

The kingdom Fungi arose from a protist ancestor (see Figure 29.1). The modern fungi have been classified into five phyla, listed in **Table 29.1** by both their common and formal names:

Hyphae of *Phanerochaete chrysosporium*

Woody tissue; cell walls impregnated with lignin

├— 25 μm ┤

Figure 29.21   Ecological features of fungi can be harnessed in biotechnology.   The forest fungus *Phanerochaete chrysosporium* produces a cobweb-like mycelium on the surfaces of rotting wood. This fungus releases enzymes that break down lignin, thereby making plant cell-wall carbohydrates such as cellulose more accessible.

chytrids (Chytridiomycota), zygomycetes (Zygomycota), AM fungi (Glomeromycota), ascomycetes (Ascomycota), and basidiomycetes (Basidiomycota) (Table 29.1). (The suffixes *mycota* and *mycetes* derive from a Greek word meaning fungus.) DNA and other data strongly indicate that these phyla all descended from a single common ancestor. Several types of slime molds and fungus-like protists—though often studied with fungi—are classified with protists rather than true fungi (see Chapter 28). In this section, we will survey the characteristics of five phyla of true fungi and examine their distinctive reproductive cycles.

Chytrids Primarily Live in Aquatic Environments

Chytrids are the simplest fungi, and molecular evidence indicates that chytrids were among the earliest fungi to appear. Some chytrids occur as single, spherical cells that may produce hyphae (**Figure 29.22**), while others exist mainly as branched, aseptate hyphae. Chytrids are the only fungi that produce flagellate cells; these are used for spore or gamete dispersal.

Table 29.1	Distinguishing Features of Fungal Phyla			
Common name (Formal name)	**Habitat**	**Ecological role**	**Reproduction**	**Examples cited in this chapter**
Chytrids (Chytridiomycota)	Water and soil	Mostly decomposers; some pathogens	Flagellate spores	*Batrachochytrium dendrobatidis*
Zygomycetes (Zygomycota)	Mostly terrestrial	Decomposers and pathogens	Nonflagellate asexual spores produced in sporangia; resistant sexual zygospores	*Rhizopus stolonifer*
AM Fungi (Glomeromycota)	Terrestrial	Form mutually beneficial mycorrhizal associations with plants	Distinctively large, nonflagellate, multinucleate asexual spores	The genus *Glomus*
Ascomycetes (Ascomycota)	Mostly terrestrial	Many form lichens; some are mycorrhizal	Nonflagellate sexual spores (ascospores) in sacs (asci) on fruiting bodies (ascocarps); asexual conidia	*Venturia inaequalis, Aleuria aurantia, Saccharomyces cerevisiae*
Basidiomycetes (Basidiomycota)	Terrestrial	Less commonly form lichens; many are mycorrhizal; decomposers	Nonflagellate sexual spores (basidiospores) on club-shaped basidia on fruiting bodies (basidiocarps); several types of asexual spores	*Coprinus disseminatus, Rhizoctonia solani, Armillaria mellea, Puccinia graminis, Ustilago maydis, Phanerochaete chrysosporium, Laccaria bicolor, Amanita muscaria, Phallus impudicus, Lycoperdon perlatum*

The presence of a single, posterior flagellum on chytrid spores or gametes links fungi with the ancestry of choanoflagellates and animals (see Chapter 28). Since species within the phylum Chytridiomycota do not have a single, common ancestor, this phylum is not a monophyletic group. Using molecular data, systematists are now assigning some chytrids into new phyla.

Chytrids live in aquatic habitats or in moist soil. Most chytrids are decomposers, but some are parasites of protists (Figure 29.22), plants, or animals. For example, the chytrid *Batracho-*

Figure 29.22   **Chytrids growing on a freshwater protist.** The colorless chytrids produce hyphae that penetrate the cellulose cell walls of the dinoflagellate *Ceratium hirundinella*, absorbing organic materials from the alga. Chytrids use these materials to produce spherical flagellate spores that swim away to attack other algal cells.

chytrium dendrobatidis has been associated with declining harlequin frog populations (look ahead to Figure 54.1).

Zygomycetes Produce Distinctive Zygospores

The **zygomycetes** feature a mycelium that is mostly composed of aseptate hyphae (those lacking cross walls) and distinctive reproductive structures. For example, like most zygomycetes, the black bread mold *Rhizopus stolonifer* produces asexual spores in enclosures known as **sporangia** (singular, sporangium) (**Figure 29.23a**). Bread mold sporangia form at hyphal tips in such large numbers that they make moldy bread appear black. Zygomycete sporangia may each release up to 100,000 spores into the air! The great abundance of such spores means that bread easily molds unless retardant chemicals are added.

Zygomycetes are named for the **zygospore**, a distinctive feature of sexual reproduction (**Figure 29.23b**). Zygospore production begins with the development of **gametangia** ("gamete bearers"). In the zygomycete fungi, gametangia are hyphal branches whose cytoplasm is isolated from the rest of the mycelium by cross walls. These gametangia enclose gametes that are basically a mass of cytoplasm containing several haploid nuclei. When food supplies run low, and if compatible mating strains are present, the gametangia of compatible mating types fuse, as do the gamete cytoplasms. The resulting cell, known as a zygosporangium, contains many haploid parental nuclei that fuse, producing many diploid nuclei. Eventually, a dark-pigmented, thick-walled spore known as a **zygospore** matures within the zygosporangium. Each zygospore contains many diploid nuclei and is capable of surviving stressful conditions.

Figure 29.23   The life cycle of a zygomycete, the black bread mold *Rhizopus stolonifer*.

When the environment is suitable, the zygospore may undergo meiosis and germinate, dispersing many haploid spores. If the spores land in a suitable place, they germinate to form aseptate hyphae containing many haploid nuclei produced by mitosis.

Most zygomycetes are saprobes in soil, living on decaying materials, but some are parasites of plants or animals. The phylum Zygomycetes is not a monophyletic group and thus is likely to be split as more information becomes available.

AM Fungi Live with Plant Partners

The phylum Glomeromycota—commonly known as the **AM** (for a̲rbuscular m̲ycorrhizal) **fungi**—has only recently been defined as a distinct group. AM fungi were previously classified with the Zygomycota. Glomeromycota have aseptate hyphae and reproduce only asexually by means of unusually large spores with many nuclei (**Figure 29.24**). Many vascular plants depend on AM fungi, and these fungi are not known to grow separately from plants or cyanobacterial partners.

Molecular evidence suggests that Glomeromycota originated around 600 million years ago. Fossils having aseptate hyphae and large spores similar to those of modern Glomeromycota are known from the time when land plants first became common and widespread, about 460 million years ago (see Chapter 30). This and other fossil evidence suggests that the ability of early plants to live successfully on land may have depended on help from fungal associates, as is common today.

Ascomycetes Produce Sexual Spores in Saclike Asci

The **ascomycetes**, along with the basidiomycetes, are regarded as higher fungi. DNA evidence indicates that the ascomycetes originated more recently than the earliest chytrids, zygomycetes, and AM fungi.

The name *ascomycetes* derives from unique sporangia known as **asci** (from the Greek *asco*, bags or sacs), which produce sexual spores known as **ascospores** (**Figure 29.25**). The asci are produced on fruiting bodies known as **ascocarps**. Although many ascomycetes have lost the ability to reproduce sexually, hyphal septa with simple pores (see Figure 29.6a) and DNA data can be used to identify them as members of this phylum.

Ascomycetes occur in terrestrial and aquatic environments, and they include many decomposers as well as parasites.

Figure 29.24   **The genus *Glomus* is an example of the Glomeromycota.**   The hyphae of these endomycorrhizal (arbuscular mycorrhizal) fungi are found in roots of many types of plants, aiding them to acquire water and nutrients. AM fungi produce large, multinucleate spores, primarily by asexual processes.

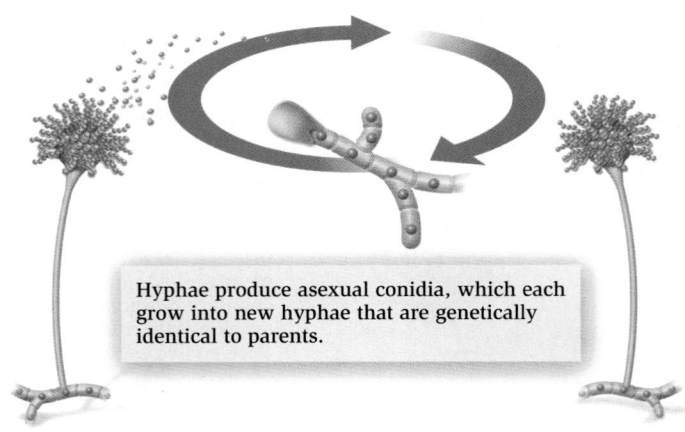

Hyphae produce asexual conidia, which each grow into new hyphae that are genetically identical to parents.

(a) Asexual reproduction

Figure 29.25   **The life cycles of ascomycete fungi.**

1 Compatible hyphae mate by plasmogamy of hyphal branches, combining nuclei of 2 genetic types.

2 The mated cell produces a dikaryotic mycelium by mitotic division of both types of nuclei.

3 The dikaryotic mycelium produces a fruiting body known as an ascocarp.

4 At the fruiting body surface, hyphae produce sac-shaped asci, each containing 2 haploid nuclei of distinct genotype.

5 The 2 nuclei fuse to form a diploid zygote nucleus in the process known as karyogamy.

6 The diploid nucleus undergoes meiosis, then each of the 4 daughter nuclei divide again by mitosis. The cytoplasm around each nucleus secretes a spore wall.

7 When ascospores are mature, they are explosively released from asci into the air.

8 Ascospores grow into hyphae having 1 nucleus per cell.

Hyphal branches
Dikaryotic mycelium
Dispersing spores
Eight ascospores
Four haploid nuclei
Diploid nucleus
Meiosis
Ascospores
Fruiting body

KEY
Haploid
Diploid
Heterokaryotic

(b) Sexual reproduction

Important ascomycete plant pathogens include powdery mildews, chestnut blight (*Cryphonectria parasitica*), Dutch elm disease (the genus *Ophiostoma*), and apple scab (*Venturia inaequalis*) (**Figure 29.26**). Cup fungi (see chapter-opening photo) are common examples of ascomycetes. Edible truffles and morels are the fruiting bodies of particular ascomycetes whose mycelia form mycorrhizal partnerships with plants. Ascomycetes are the most common fungal components of lichens. Members of the ascomycetes include most **yeasts**, fungi that can occur as unicells and that reproduce by budding (**Figure 29.27**). (Some yeasts are classified in the Zygomycota or the Basidiomycota.) Ascomycete yeasts are important to the baking and brewing industries, and some are medically significant, as agents of disease. For example, the ascomycete yeast *Candida albicans* causes yeast infections and thrush (oral candidiasis). Yeasts are also widely used for fundamental biological studies.

Figure 29.26   Apple infected with the ascomycete fungus *Venturia inaequalis*, which causes apple scab disease.   This fungus grows on leaves, flowers, and fruits, leaving harmless but unsightly scabs on their surfaces. Growers usually try to control apple scab by spraying trees with fungicides.

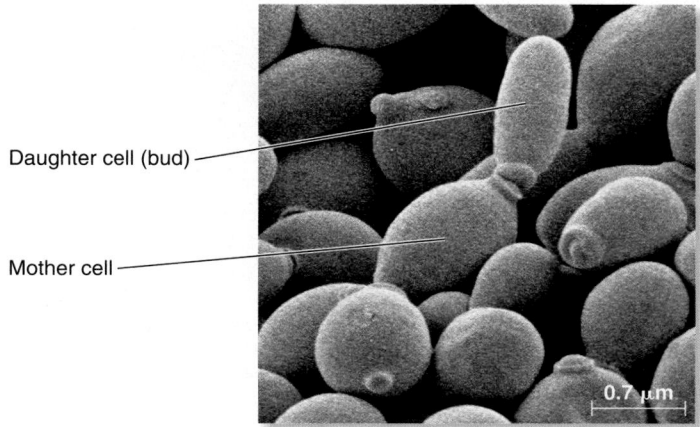

Daughter cell (bud)

Mother cell

0.7 μm

Figure 29.27   The budding yeast *Saccharomyces cerevisiae*.

Biological inquiry: Are all yeasts classified in the ascomycetes?

GENOMES & PROTEOMES

Yeast Is Used as a Model System in Genomics, Proteomics, and Metabolomics

Ascomycete yeasts such as *Saccharomyces cerevisiae* (Figure 29.27) have long served as model systems for cellular, biochemical, genetic, and molecular research. Because of *S. cerevisiae*'s scientific importance, this organism's genome was completely sequenced by an international coalition of more than 600 scientists in more than 100 laboratories, led by molecular biologist André Goffeau. This team found that *S. cerevisiae* has about 6,000 genes on 16 chromosomes and exhibits many cases of gene duplication and divergence, as do most organisms.

Genome sequences have recently been obtained for two of *S. cerevisiae*'s close relatives, the yeast *Schizosaccharomyces pombe* and *Ashbya gossypii*, a filamentous fungus used in industry for producing vitamin B_2. These data allow biologists to perform comparative studies that shed light on the evolution of entire genomes. Fred Dietrich, Peter Philippsen, and colleagues found that by comparison with other eukaryotes, *A. gossypii* has a comparatively small genome of only 9.2 million base pairs, which includes around 4,700 protein-coding genes on seven chromosomes. The genome of *A. gossypii* has no transposons, relatively few introns, and only a few gene duplications. A similarly "stripped-down" genome is present in *S. pombe*, with fewer than 5,000 protein-coding genes. These two fungi may illustrate the minimum genome size possible for free-living (nonparasitic) fungi and reveal the genetic mechanisms that result in small genome size.

Based on genomic data, the *S. cerevisiae* proteome has been estimated to include over 5,000 soluble proteins and more than 1,000 transmembrane proteins. In 2006, Anne-Claude Gavin, Giulio Superti-Furga, and many associates reported the existence of about 500 protein complexes in *S. cerevisiae*. These protein complexes combine in various ways with other proteins, in modular fashion, to form many types of cellular metabolic "machines."

Yeast research has also led the way in metabolomics, the study of all of the small molecules produced in an organism. For example, in 2005 Michael Snyder and his associates treated glass chips bearing an array of 1,300 yeast proteins with 87 different purified protein kinases plus radiolabeled ATP. The labeled ATP allowed researchers to determine which kinases phosphorylate particular proteins during signal transduction (see Chapter 9). This experiment revealed over 4,000 different signal transduction interactions. Such metabolomic studies have important applications in human medicine, because protein kinases are important drug targets.

Basidiomycetes Produce Diverse Fruiting Bodies

DNA-sequencing comparisons indicate that **basidiomycetes** are the most recently evolved group of fungi. Even so, they have a long fossil history, extending back nearly 300 million years.

Today, basidiomycetes are very important as decomposers and mycorrhizal partners in forests, producing diverse fruiting bodies commonly known as mushrooms, puffballs, stinkhorns, shelf fungi, rusts, and smuts (**Figure 29.28**; see also Figures 29.3b, 29.9, 29.10, and 29.16a).

The name given to the basidiomycetes derives from **basidia**, the club-shaped cells that produce sexual spores known as **basidiospores** on their surfaces (**Figure 29.29**). Basidia are typically located on the undersides of fruiting bodies, which are gener-ally known as **basidiocarps**. Though some basidiomycetes have lost sexual reproduction, they can be identified as members of this phylum by unique hyphal structures known as **clamp connections** that help distribute nuclei during cell division. Basidiomycetes can also be identified by distinctive septa having complex pores (see Figure 29.6b) and by DNA methods. Basidiomycetes reproduce asexually by various types of spores. Rust fungi, for example, can produce up to four different types of asexual spores during their life cycle.

(a)   **(b)**

Figure 29.28   **Basidiomycetes produce diverse fruiting bodies.**   **(a)** Corn smut (*Ustilago maydis*) produces dikaryotic mycelial masses within the kernels (fruits) of infected corn plants. These mycelia produce many dark spores in which karyogamy and meiosis occur. Masses of these dark spores cause the smutty appearance. When the spores germinate, they produce basidiospores that can infect other corn plants. **(b)** Shelf fungi such as this sulfur shelf fungus are the fruiting bodies of basidiomycete fungi that have infected trees.

KEY
- Haploid
- Diploid
- Heterokaryotic

3 To ensure that one of each nuclear type is regularly distributed to each daughter cell, hyphal branches known as clamp connections bridge recently divided cells.

Long-lived dikaryotic mycelium

Mitosis in tip cell

Hyphal branch carries 1 nucleus

Clamp connection forms

New septum forms

Nuclear distribution complete

2 The dikaryotic cell divides by mitosis to produce a dikaryotic mycelium, which can be very long-lived.

4 Under appropriate conditions, dikaryotic mycelium may form a fruiting body.

1 Hyphae of compatible genotypes mate by plasmogamy of hyphal branches.

9 Basidiospores grow into mycelia, the cells of which possess 1 haploid nucleus.

Basidium with haploid nuclei

Diploid nucleus

8 Basidia each produce 4 haploid basidiospores which are dispersed.

7 Basidia undergo meiosis to produce 4 haploid nuclei.

6 Nuclei in basidia fuse to form diploid nuclei.

5 Dikaryotic basidia occur at the surfaces of gills (or pores of some mushrooms).

Figure 29.29   **The sexual cycle of the basidiomycete fungus *Coprinus disseminatus*.**

CHAPTER SUMMARY

29.1 Distinctive Features of Fungi

- Fungi form a monophyletic kingdom of heterotrophs that, like animals, display absorptive nutrition. Fungi secrete enzymes into food substrates, breaking down complex organic molecules into small organic molecules that are absorbed as food. (Figure 29.1)

- Fungal cells possess cell walls composed of chitin, a polysaccharide resistant to microbial attack. Fungal bodies, known as mycelia, are composed of microscopic branched filaments known as hyphae. (Figure 29.2)

- During sexual reproduction, fungi produce fruiting bodies, formed by aggregated hyphae, that usually extend out of the substrate. Fruiting bodies produce sexual spores and aid their dispersal by means of wind, water, or animals. (Figure 29.3)

- In most cases, the fungal body is inconspicuously dispersed within the substrate, but sometimes fungal hyphae aggregate to form stringlike rhizomorphs that aid fungal water transport. (Figure 29.4)

- Fungal mycelia grow at their edges, by tip growth of hyphae. Septate hyphae are divided into cells by cross walls, or septa, perforated by simple or complex pores. Some fungi have aseptate hyphae that are not subdivided into cells. (Figures 29.5, 29.6)

- Mycelial shape depends on the location of nutrients in the environment, which determines the direction in which cell division and hyphal growth will occur. (Figure 29.7)

29.2 Fungal Sexual and Asexual Reproduction

- Fungi disperse in their environments by means of spores produced by asexual or sexual reproduction. (Figure 29.8)

- The gametes of most fungi are relatively unspecialized hyphal branches. During sexual reproduction, hyphal branches fuse with those of a different mycelium of compatible mating type. In many fungi, dikaryotic hyphae (having two nuclei per cell) that result from mating often persist for long periods before nuclear fusion occurs. Nuclear fusion generates zygotes. Fungal zygotes are the only cells in the life cycle that possess a diploid nucleus.

- Zygotes undergo meiosis to produce haploid spores, which disperse from fruiting bodies. Spores germinate to produce haploid fungal mycelia. Fungi produce diverse types of fruiting bodies that foster spore dispersal by wind, water, or animals. Many fungal fruiting bodies produce defensive toxins or hallucinogens. (Figures 29.9, 29.10, 29.11, 29.12)

- Asexual reproduction does not involve mating or meiosis, and it occurs by means of asexual spores such as conidia. Many common and destructive molds spread by means of conidia. (Figure 29.13)

29.3 Fungal Ecology and Biotechnology

- Fungi play important roles in nature as decomposers, predators, and pathogens. Several types of fungi cause disease. (Figures 29.14, 29.15)

- Fungi form important mutualistic symbioses with other organisms. Mycorrhizae are common associations between fungi and plant roots. Ectomycorrhizae coat root surfaces, extending into root intercellular spaces. Endomycorrhizae commonly form highly branched arbuscules in the spaces between root cell walls and plasma membranes. (Figures 29.16, 29.17)

- Lichens are partnerships between fungi and photosynthetic green algae and/or cyanobacteria. Lichens can reproduce asexually by means of structures such as soredia, which consist of fungal hyphae wrapped around a few algal cells. When lichen fungi reproduce sexually, the hyphae arising from spore germination must find new algal partners. (Figures 29.18, 29.19)

- Lichens occur in diverse habitats, including harsh environments, and often grow slowly and to great age. Lichens help to build soils and are useful air-quality monitors.

- In at least one lichen family, the Cladoniaceae, the fungal partners have not evolved in tandem with symbiotic algae, and data suggest that such lichen fungi commonly switch algal partners. (Figure 29.20)

- Fungal biochemistry is useful in chemical, food processing, and waste-treatment industries, and fungi are increasingly used to replace chemical procedures that generate harmful waste materials. (Figure 29.21)

29.4 Evolution and Diversity of Fungi

- Currently, fungi are classified into five phyla commonly known as chytrids, zygomycetes, AM fungi, ascomycetes, and basidiomycetes. (Table 29.1)

- Chytrids are among the simplest and earliest-divergent fungi. They commonly occur in aquatic habitats and moist soil, where they produce flagellate reproductive cells. (Figure 29.22)

- Zygomycetes are named for their distinctive, large zygospores, the result of sexual reproduction. Common black bread mold and other zygomycetes reproduce asexually by means of many small spores. (Figure 29.23)

- The AM fungi produce distinctive large, multinucleate spores, and they form mutualistic arbuscular mycorrhizal relationships with many types of plants. (Figure 29.24)

- Ascomycetes produce sexual ascospores in saclike asci located at the surfaces of fruiting bodies known as ascocarps. Many are lichen symbionts. The ascomycete yeast *Saccharomyces cerevisiae* is widely used as a laboratory model system for genomic, proteomic, and metabolomic studies. (Figures 29.25, 29.26, 29.27)

- Basidiomycetes produce sexual basidiospores on club-shaped basidia located on the surfaces of fruiting bodies known as basidiocarps. Such fruiting bodies take a wide variety of forms, including mushrooms, puffballs, stinkhorns, shelf fungi, rust, and smuts. The hyphae of basidiomycete fungi are characterized by clamp connections, structures that aid in distributing nuclei of two types after cell division occurs in dikaryotic hyphae. (Figures 29.28, 29.29)

TEST YOURSELF

1. Fungal cells differ from animal cells in that fungal cells
   a. lack ribosomes, though these are present in animal cells.
   b. lack mitochondria, though these occur in animal cells.
   c. have cell walls, whereas animal cells lack rigid walls.
   d. lack cell walls, whereas animal cells possess walls.
   e. none of the above.

2. Conidia are
   a. cells produced by some fungi as the result of sexual reproduction.
   b. fungal asexual reproductive cells produced by the process of mitosis.
   c. structures that occur in septal pores.
   d. the unspecialized gametes of fungi.
   e. none of the above.

3. What are mycorrhizae?
   a. the bodies of Fungi, composed of hyphae
   b. fungi that attack plant roots, causing disease
   c. fungal hyphae that are massed together into stringlike structures
   d. fungi that have symbiotic partnerships with algae or cyanobacteria
   e. mutually beneficial associations of particular fungi and plant roots

4. Where could you find diploid nuclei in an ascomycete or basidiomycete fungus?
   a. in spores
   b. in cells at the surfaces of fruiting bodies
   c. in conidia
   d. in soredia
   e. all of the above

5. Which fungi are examples of hallucinogens?
   a. *Claviceps* and *Psilocybe*
   b. *Epidermophyton* and *Candida*
   c. *Pneumocystis carinii* and *Histoplasma capsulatum*
   d. *Saccharomyces cerevisiae* and *Phanerochaete chrysosporium*
   e. *Cryphoenectria parasitica* and *Ventura inaequalis*

6. What role do fungal endophytes play in nature?
   a. They are decomposers.
   b. They are human parasites that cause skin diseases.
   c. They are plant parasites that cause serious crop diseases.
   d. They live within the tissues of grasses and other plants, helping to protect plants from herbivores and pathogens.
   e. All of the above.

7. What forms do lichens take?
   a. crustose, flat bodies
   b. foliose, leaf-shaped bodies
   c. fruticose, erect or dangling bodies
   d. single cells
   e. a, b, and c

8. Lichens consist of a partnership between fungi and what other organisms?
   a. red algae and green algae
   b. green algae and cyanobacteria
   c. heterotrophic bacteria and archaea
   d. choanoflagellates and *Nuclearia*
   e. none of the above

9. How can ascomycetes be distinguished from basidiomycetes?
   a. Ascomycete hyphae have simple pores in their septa and lack clamp connections, whereas basidiomycete hyphae display complex septal pores and clamp connections.
   b. Ascomycetes produce sexual spores in sacs, whereas basidiomycetes produce sexual spores on the surfaces of club-shaped structures.
   c. Ascomycetes are commonly found in lichens, whereas basidiomycetes are less commonly partners in lichen associations.
   d. Ascomycetes are not commonly mycorrhizal partners, but basidiomycetes are commonly present in mycorrhizal associations.
   e. All of the above are correct.

10. Which group of organisms listed is most closely related to the kingdom Fungi?
    a. the animal kingdom
    b. the green algae
    c. the land plants
    d. the bacteria
    e. the archaea

CONCEPTUAL QUESTIONS

1. Explain three ways that fungi are like animals, and two ways in which fungi resemble plants.

2. Explain why some fungi produce toxic or hallucinogenic compounds.

3. Explain three ways in which fungi function as mutualistic symbionts and what benefit the fungi receive from the partnerships.

EXPERIMENTAL QUESTIONS

1. What is meant by the term cospeciation?

2. How did Piercey-Normore and DePriest determine whether or not cospeciation had occurred in the Cladoniaceae?

3. What results did Piercey-Normore and DePriest find?

COLLABORATIVE QUESTIONS

1. Thinking about the natural habitats closest to you, where could you find fungi, and what roles do these fungi play?

2. Imagine that you are helping to restore the natural vegetation on a piece of land that had long been used to grow crops. You are placed in charge of planting pine seedlings (*Pinus resinosa*) and fostering their growth. In what way could you consider using fungi?

---

www.brookerbiology.com
This website includes answers to the Biological Inquiry questions found in the figure legends and all end-of-chapter questions.

30

PLANTS AND THE CONQUEST OF LAND

CHAPTER OUTLINE

A temperate rain forest in Olympic National Park, Washington State, containing diverse plant phyla.

W hen thinking about plants, people envision lush green lawns, shady street trees, garden flowers, or leafy fields of valuable crops. On a broader scale, they might imagine dense jungles, vast grassy plains, or tough desert vegetation. Shopping in the produce section of the local grocery store may remind us that plant photosynthesis is the basic source of our food. Just breathing crisp fresh air might bring to mind the role of plants as oxygen producers—the ultimate air fresheners. Do you start your day with a "wake-up" cup of coffee, tea, or hot chocolate? Then you may appreciate the plants that produce these and many other materials we use in daily life: medicines, cotton, linen, wood, bamboo, cork, and even the paper on which this textbook has been printed.

In addition to their importance to humans and modern ecosystems, plants have played dramatic roles in Earth's past. Throughout their evolutionary history, diverse plants have influenced Earth's atmospheric chemistry, climate, and soils and the evolution of many other groups of organisms, including humans. In this chapter, we will survey the diversity of modern plant phyla and their distinctive features. This chapter also explains how early plants adapted to land and how plants have continued to adapt to changing terrestrial environments. During this process, we will continue to gain insight into principles of diversity and evolution such as descent with modification.

30.1 Ancestry and Diversity of Modern Plants

Several hundred thousand modern species are formally classified into the kingdom Plantae, informally known as the plants or land plants (**Figure 30.1**). **Plants** can be defined as eukaryotic, primarily photosynthetic organisms that mostly live on land and display many adaptations to life in terrestrial habitats (**Table 30.1**). Molecular and other evidence indicate that land plants most likely evolved from aquatic algal ancestors. In this section, we will describe the modern algae that are most closely related to plants and survey the diverse phyla of living land plants.

Modern Green Algae Are Closely Related to the Ancestry of Land Plants

Molecular, biochemical, and structural data reveal that the plant kingdom is monophyletic, having probably originated from a single common protist ancestor (see Chapter 28). This common ancestor could probably be classified with modern freshwater green algae that are informally known as **charophyceans**. Such algae are named for the genus *Chara*, a multicellular alga that displays relatively complex structure and reproduction (**Figure 30.2a**), as does the related genus *Coleochaete* (**Figure 30.2b**). Either *Chara* and its close relatives or *Coleochaete* are the modern protists most closely related to the ancestry of land plants, depending on which genes are used in phylogenetic analyses. Thus, it is not surprising that these complex charophyceans share many features of biochemistry, cell structure, cell division, and sexual reproduction with land plants.

For example, complex charophyceans accomplish sexual reproduction by means of flagellate sperm and larger, nonmotile eggs, as do land plants. In contrast, more simply structured, early diverging charophyceans, such as the beautiful desmids (**Figure 30.2c**) and the single-celled flagellate *Mesostigma viride* (**Figure 30.2d**), reproduce differently. Such reproductive differences suggest that plant eggs and sperm evolved prior to divergence of the complex charophyceans. Diverse modern charophyceans are a vast reservoir of information about the origin of land plant sexual reproduction and other plant features.

Although charophyceans by themselves do not form a monophyletic group, what is known as a **clade**, some experts classify charophyceans together with land plants to form a clade known informally as streptophytes (see Figure 30.1). Charophycean algae have also been classified into larger groups in a variety of ways based on photosynthetic features. For example, charophyceans and other green algae are classified together with other protists having primary plastids, and plants into the eukaryote supergroup Archaeplastida (see Chapter 28). In addition, some experts classify plants and green algae together and name the combination in various ways (Viridiplantae, Chlorobionta, and Chloroplastida) that reflect the common feature of green plastids. The plastids of green algae and plants possess the same photosynthetic pigments: the main photosynthetic pigment chlorophyll *a*, the green accessory pigment chlorophyll *b*, and orange accessory pigments such as β-carotene. You may recall that accessory pigments absorb light energy and transfer it to chlorophyll *a* for use in photosynthesis. Because chlorophylls *a* and *b* are so abundant in plastids, the plastids of green algae and plants are generally green. By contrast, other algal groups have different colored plastids whose abundant amounts of red, gold, or brown accessory pigments hide chlorophyll *a*.

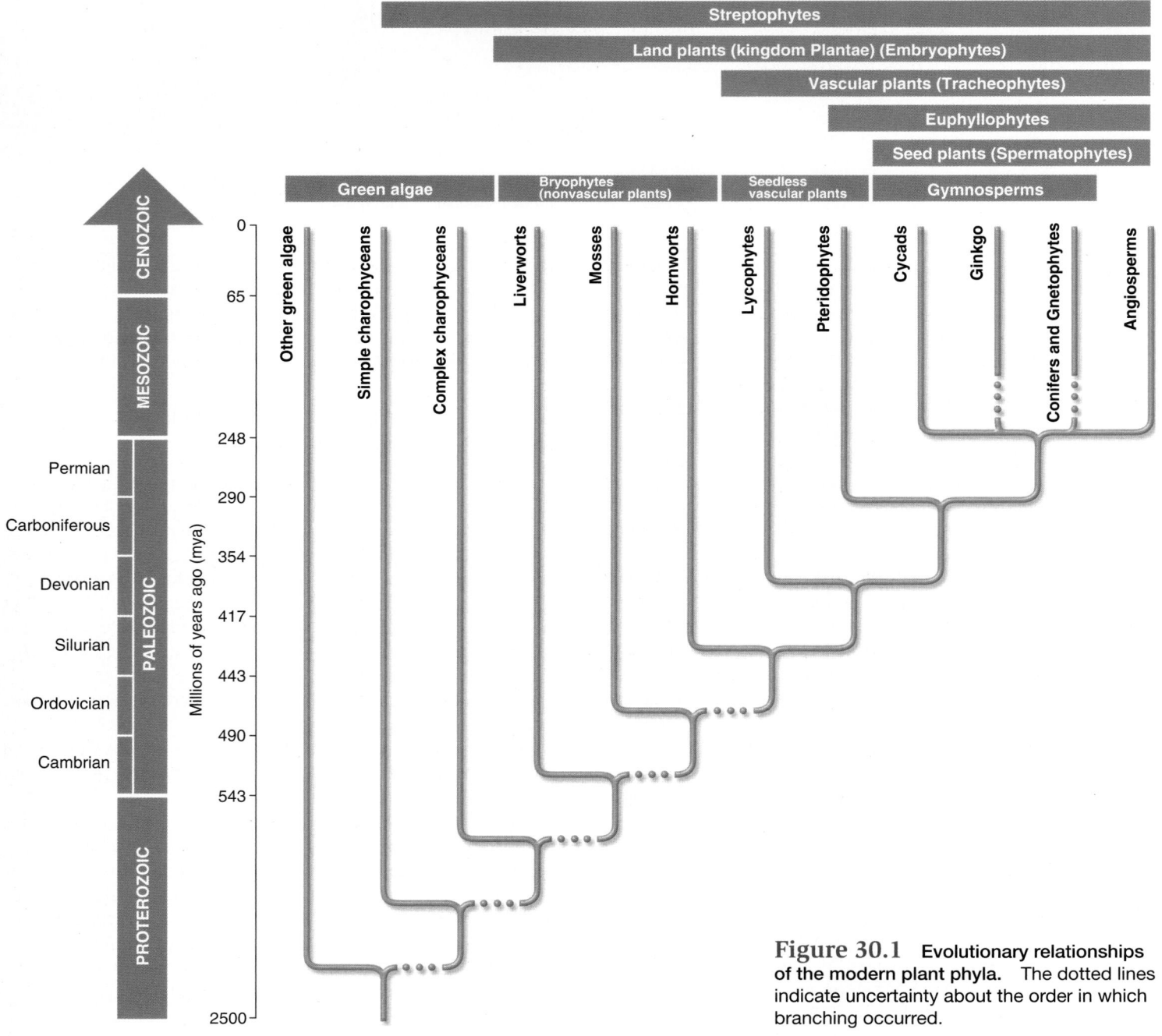

Figure 30.1 **Evolutionary relationships of the modern plant phyla.** The dotted lines indicate uncertainty about the order in which branching occurred.

Green algae and land plants share another photosynthetic feature, storage of the polysaccharide starch within plastids. By contrast, other photosynthetic eukaryotes store photosynthetic products in different locations or in other biochemical forms (see Chapter 28). Chlorophyll *b*, starch, and other features of

ancestral green algae were inherited by diverse phyla of fossil and living land plants. We next focus on the phyla of modern plants and their distinguishing features.

Modern Land Plants Can Be Classified into 10 Phyla

Plant systematists use molecular and structural information from living and fossil plants to classify plants into phyla. However, many controversies exist regarding the time of first appearance and species composition of some phyla, and the relationships among plant phyla are not completely clear. As a result of our incomplete understanding of plant relationships, biologists have classified modern plants in different ways, and classifications continue to change as new information becomes available.

Table 30.1	Distinguishing Features of Modern Charophyceans and Land Plants*
Charophyceans	Primarily aquatic habitat; zygotic life cycle; rosette cellulose synthesizing complexes; sporangia absent; sporophytes absent; cell-wall xyloglucans absent

LAND PLANTS (EMBRYOPHYTES)
Primarily terrestrial habitat; sporic life cycle consisting of alternation of two multicellular generations—diploid sporophyte and haploid gametophyte; multicellular embryos are nutritionally dependent for at least some time during development; spore-producing sporangia; gamete-producing gametangia; sporopollenin spore walls; cell-wall xyloglucans

Nonvascular plants (Bryophytes) (**liverworts**, **hornworts**, and **mosses**)
Dominant gametophyte generation; supportive, lignin-containing vascular tissue absent; true roots, stems, and leaves absent; sporophytes unbranched and unable to grow independently of gametophytes

VASCULAR PLANTS (TRACHEOPHYTES) (**lycophytes**, **pteridophytes**, spermatophytes)
Dominant sporophyte generation; lignified water conducting tissue—xylem; specialized organic food conducting tissue—phloem; sporophytes branched; sporophytes eventually become independent of gametophytes

SEEDLESS VASCULAR PLANTS (**lycophytes**, **pteridophytes**)
Lycophytes Leaves generally small with a single, unbranched vein (lycophylls or microphylls); sporangia borne on sides of stems

EUPHYLLOPHYTES (**pteridophytes**, SPERMATOPHYTES)
Pteridophytes Leaves relatively large with extensively branched vein system (euphylls or megaphylls); sporangia borne on leaves

SEED PLANTS (SPERMATOPHYTES) Seeds present; leaves are euphylls

Gymnosperms (**cycads, ginkgos, conifers, gnetophytes**)
Flowers and fruits absent; seed food stored before fertilization in female gametophyte, endosperm absent

Angiosperms (flowering plants)
Flowers and fruits present; seed food stored after fertilization in endosperm formed by double fertilization

*Key: **Phyla**; LARGER MONOPHYLETIC CLADES (synonyms). All other classification terms are not clades.

1 cm   52 μm

(a) *Chara zeylanica* **(b) *Coleochaete pulvinata***

 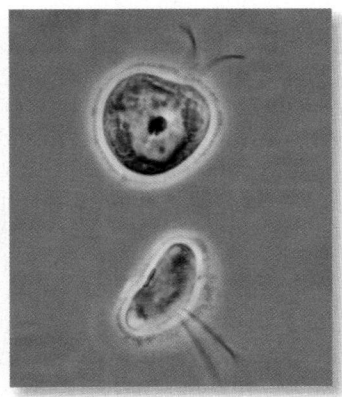

37 μm   10 μm

(c) *Micrasterias radiosa* **(d) *Mesostigma viride***

Figure 30.2 Charophycean green algal relatives of the land plants. **(a)** *Chara zeylanica*. **(b)** *Coleochaete pulvinata*. **(c)** The desmid *Micrasterias radiosa*. **(d)** *Mesostigma viride*, named for its central red spot and green plastid color. All of these algae inhabit freshwater lakes and ponds, and they display structural, reproductive, biochemical, and molecular features in common with land plants.

In this textbook, 10 phyla of living land plants are described: (1) the plants informally known as **liverworts** (formally called Hepatophyta), (2) **hornworts** (Anthocerophyta), (3) **mosses** (Bryophyta), (4) **lycophytes** (Lycopodiophyta), (5) **pteridophytes** (Pteridophyta), (6) **cycads** (Cycadophyta), (7) **ginkgos** (Ginkgophyta), (8) **conifers** (Coniferophyta), (9) **gnetophytes** (Gnetophyta), and (10) **angiosperms** (Anthophyta), also known as the **flowering plants** (see Table 30.1). Fossils reveal that additional plant phyla once lived but have since become extinct.

Phylogenetic information suggests that the modern plant phyla arose in a particular sequence. Liverworts, mosses, and hornworts originated relatively early; lycophytes and pteridophytes arose later; cycads, ginkgos, conifers, and gnetophytes came next; and flowering plants appeared most recently (see Figure 30.1). Each of these modern plant phyla displays features that reveal how plants became increasingly better adapted to life on land.

Bryophytes Include the Liverworts, Hornworts, and Mosses

Liverworts, hornworts, and mosses are Earth's simplest land plants (**Figures 30.3, 30.4,** and **30.5**). There are about 6,500 species of modern liverworts, about 100 species of hornworts, and 12,000 or more species of mosses. These monophyletic phyla are together known informally as **bryophytes**. Although the term bryophyte does not reflect a clade, it is useful for expressing common structural, reproductive, and ecological features of liverworts, mosses, and hornworts. For example, bryophytes possess simpler structure and reproduction than other plants, and bryophytes are most common and diverse in moist habitats. Because they diverged early in the evolutionary history of land plants (see Figure 30.1), bryophytes serve as models of the earliest terrestrial plants. They display several features that are absent from charophycean algae but are present in all other land plants. Such bryophyte features thus likely reflect early adaptations to life on land. The bryophyte life cycle, shared with other plants, is a good example. A comparison between the life cycle of aquatic charophyceans and that of bryophytes reveals the bryophyte life cycle's adaptive value on land.

The charophycean algae display a zygotic life cycle in which the diploid generation consists of only one cell, the zygote (**Figure 30.6a**). By contrast, sexual reproduction in bryophytes and all other plants involves a **sporic life cycle**, also known as **alternation of generations** (**Figure 30.6b**). A sporic life cycle originated independently in early plants and several other eukaryotic lineages that are not related to land plants (see Chapter 28). The plant life cycle is believed to have originated by a delay in meiosis, with the result that the diploid generation became multicellular before undergoing meiosis. For the earliest land plants, this multicellular diploid generation provided advantages in coping with terrestrial conditions. A closer look at the process of sexual reproduction in bryophytes reveals these advantages and also highlights ways in which bryophytes differ from other plants.

(a) *Marchantia polymorpha*

Asexual propagules

Upper surface of liverwort

(b) *Marchantia polymorpha*　　　　**(c) A species of leafy liverwort**

Figure 30.3 **Liverworts.** **(a)** The common liverwort *Marchantia polymorpha*, with raised, umbrella-shaped structures that bear sexual reproductive structures on the undersides. **(b)** A close-up of *M. polymorpha* showing surface cups that contain multicellular, frisbee-shaped asexual structures called propagules that are dispersed by wind and grow into new liverworts. **(c)** A species of liverwort having leaflike structures and thus known as a leafy liverwort.

Biological inquiry: Why do you think liverworts produce their spores on raised structures?

Spore dispersal

Sporophyte

Gametophyte

Figure 30.4 **Hornworts.** Sporophytes are the diploid, spore-producing structures that grow on gametophytes, the haploid, gamete-producing structures growing close to the ground. Hornwort sporophytes mature and open at the top, dispersing spores.

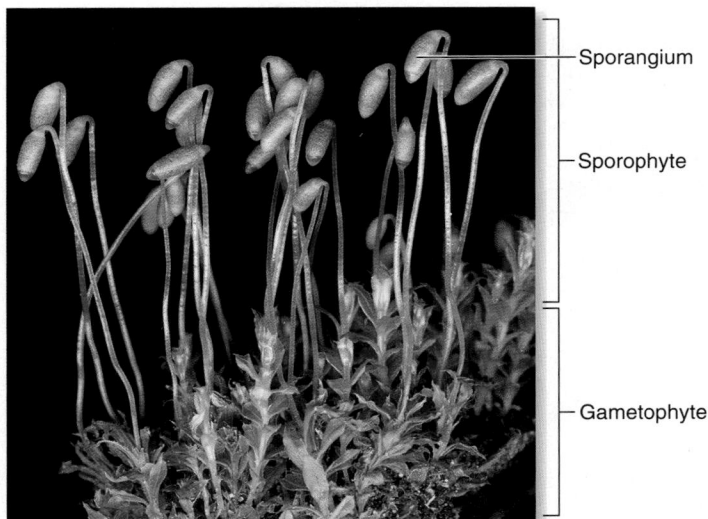

Figure 30.5 **Moss.** The common moss genus *Mnium* has leafy green gametophytes and unbranched, dependent sporophytes each bearing a sporangium at its tip. When mature, these sporophytes sprinkle spores into the wind.

Bryophyte Reproduction Illustrates Adaptation to Life on Land

Bryophytes—in common with other organisms having sporic life cycles—produce two multicellular life stages (**Figure 30.7**). These two stages are the sporophyte and the gametophyte, each stage having a distinctive reproductive role. The diploid **sporophyte** generation produces haploid spores by the process of meiosis, and multicellular plant sporophytes can produce many spores. Plant **spores** are single-celled reproductive structures that are dispersed into the air and are able to grow into gametophytes if they find suitable habitats. In contrast to the sporophyte generation, the haploid **gametophyte** generation produces gametes. Bryophytes may have separate male and female gametophytes. As we have earlier noted, plant **gametes** (from the Greek word meaning "wife") are nonflagellate eggs and smaller flagellate sperm that may fuse in pairs to form single-celled, diploid **zygotes**. Let's take a closer look at how sporophytes and gametophytes function together during the life cycle of a bryophyte, starting with gamete production.

Gametophytes The role of plant gametophytes is to produce haploid gametes, but meiosis is not involved in this process. The gametophytes of bryophytes and many other land plants produce gametes in specialized structures known as **gametangia** ("gamete containers"), in which developing gametes are protected by a jacket of tissue. The gametangial jacket protects delicate gametes from drying and microbial attack while they develop. Round or elongate gametangia that produce sperm are known as **antheridia**, whereas flask-shaped gametangia that enclose an egg cell are known as **archegonia** (Figure 30.7). When mature, if moist conditions exist, plant sperm are released from antheridia into

(a) Life cycle of charophyceans

KEY
■ Haploid
■ Diploid

(b) Life cycle of early plants

Figure 30.6 Origin of the plant life cycle.

films of water. Under the influence of sex-attractant molecules secreted from archegonia, sperm swim toward eggs, twisting their way down the tubular archegonial necks. Sperm then fuse with egg cells to form diploid zygotes, which grow into new sporophytes. However, fertilization cannot occur in bryophytes unless water is present. Conditions of uncertain moisture, common in the land habitat, can thus limit plant reproductive fitness. The plant life cycle, featuring several reproductive advantages, is an adaptive response to this environmental challenge.

Sporophytes One reproductive advantage of the plant life cycle is that zygotes remain enclosed within gametophyte tissues, where they are sheltered and fed (a process described in more detail in Section 30.3). This critical innovation gives zygotes a good start as they begin to grow into young sporophytes, which are

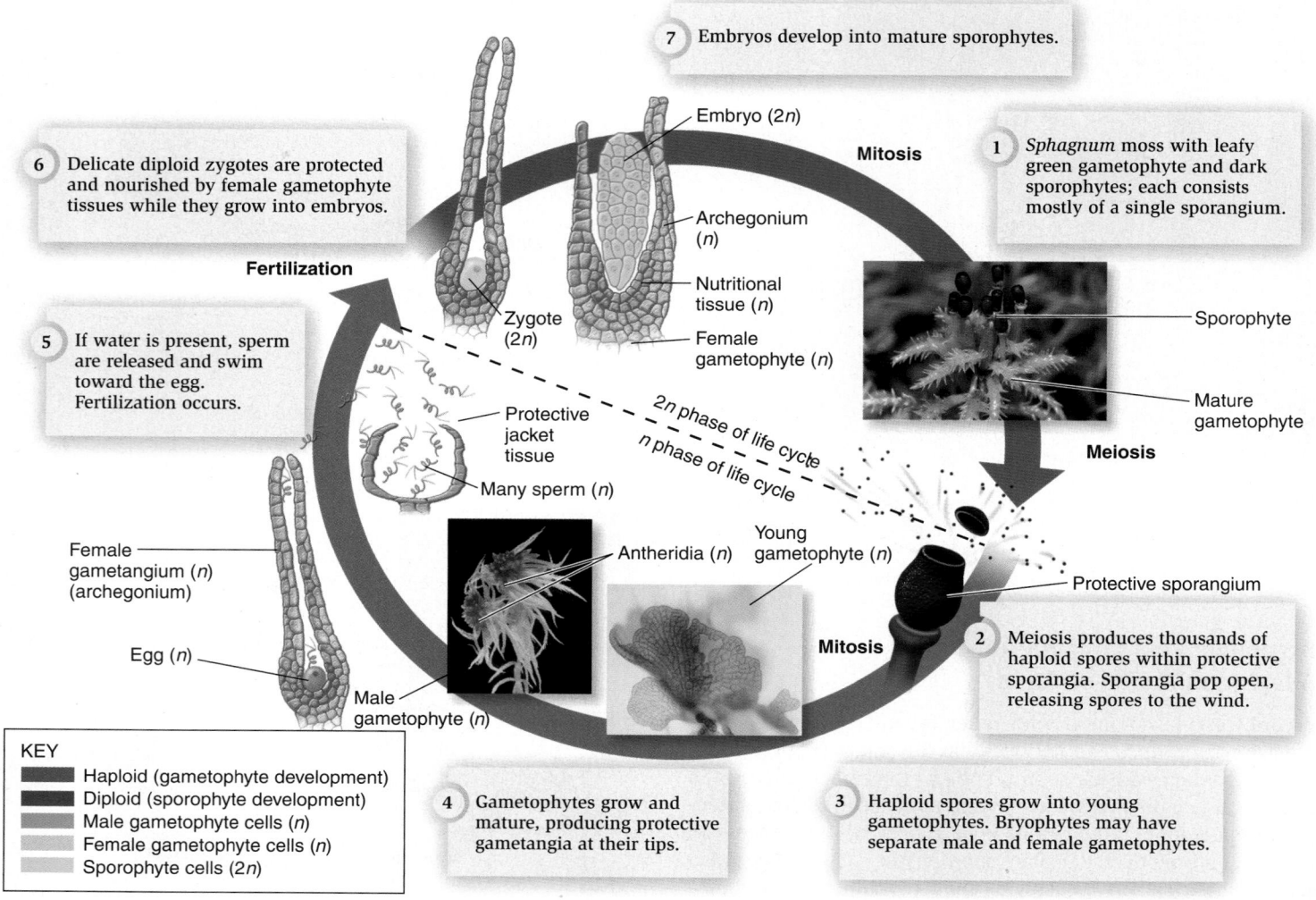

Figure 30.7 The life cycle of *Sphagnum* moss, illustrating reproductive adaptations that likely helped early plants to reproduce on land.

known as **embryos**. For this reason, all land plants are known as **embryophytes**. Sheltering and feeding embryos is particularly important when embryo production is limited by water availability, as is the case for bryophytes. Another reproductive advantage is that, when mature, the multicellular sporophytes produce many **spores** in protective enclosures known as **sporangia** ("spore containers"). Bryophyte sporangia open in specialized ways that foster dispersal of spores into the air, allowing spore transport by wind. A third reproductive advantage is that plant spores have cell walls containing a tough material known as **sporopollenin** that helps to prevent cellular damage during transport in air. If spores reach habitats favorable for growth, their walls crack open, and new gametophytes develop by mitotic divisions, completing the life cycle.

Spore production is a measure of plant fitness, because plants can better disperse progeny throughout the environment when they produce more spores. The larger the diploid generation, the more spores a plant can produce. As a result, during plant evolution, the plant sporophyte has become larger and more complex (**Figure 30.8**).

Bryophytes Display Several Distinguishing Features

Bryophytes can be distinguished from other land plants by several features. First, bryophyte gametophytes are more common in nature, larger, and longer-lived than bryophyte sporophytes. Patches of moss in the woods are primarily gametophytes. In order to see bryophyte sporophytes, you would have to look very closely, because this life stage is quite small and attached to female gametophytes throughout its short lifetime (see Figures 30.5 and 30.7). Plant biologists say that bryophyte gametophytes are the dominant generation in their life cycle. By contrast, other plants have dominant sporophyte generations (see Figure 30.8).

Bryophyte sporophytes are small because they remain attached to parental gametophytes, are unable to branch, and have short lifetimes. At their tips, bryophyte sporophytes produce only a single sporangium containing a limited number of spores. By contrast, the sporophytes of other land plants become independent, produce branches with many sporangia, and are often able to produce spores for many years (see Figure 30.8).

Figure 30.8 Relative sizes of the sporophyte generations of bryophytes, ferns, and seed plants.

In all land plants except bryophytes, the sporophyte generation is the larger, more complex, and longer-lived generation.

Yet another distinguishing feature of bryophytes is that they lack tissues that are specialized for both structural support and conduction. Plant tissues that provide both structural support and conduction of water, minerals, and organic compounds are known as **vascular tissues**. Although the gametophytes of some bryophytes display simple conducting tissues, these do not provide structural support. Thus, bryophytes are informally known as **nonvascular plants**. Other modern plant phyla are known as **vascular plants** because they produce vascular tissues that function in both conduction and support.

Lycophytes and Pteridophytes Are Vascular Plants That Do Not Produce Seeds

If you take a look outside, most of the plants in view are probably vascular plants. The presence of vascular tissue allows most vascular plants to grow much taller than bryophytes and to reproduce more effectively. As a result, vascular plants are more prominent than bryophytes in most modern plant communities. Vascular plants have been important to Earth's ecology for a long time. Fossils tell us that the first vascular plants appeared later than the earliest bryophytes (see Section 30.2), and that several early vascular plant lineages, such as the trimerophytes and psilophytes, once existed but became extinct. Molecular data indicate that the lycophytes are the oldest phylum of living vascular plants and that pteridophytes are the next oldest living plant phylum (see Figure 30.1). In the past, lycophytes were very diverse and included tall trees, but the tree lycophytes became extinct and now about 1,000 relatively small species exist (**Figure 30.9**). Pteridophytes have diversified more recently, and there are about 12,000 species of modern pteridophytes, including horsetails, whisk ferns, and other ferns (**Figure 30.10**).

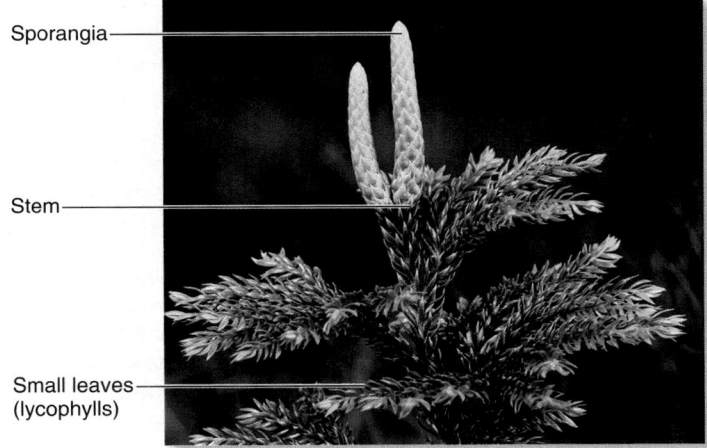

Figure 30.9 The lycophyte *Lycopodium obscurum* growing in a forest. The stems bear many tiny leaves, known as lycophylls, and sporangia generally occur in club-shaped clusters. For this reason, lycophytes are informally known as club mosses or spike mosses, though they are not true mosses. The gametophytes of lycophytes are small structures that often occur underground.

The lycophytes and pteridophytes diverged prior to the origin of seeds, and they are informally known as seedless vascular plants. In contrast to bryophytes, and in common with the seed plants, the sporophytes of lycophytes and pteridophytes possess vascular tissues providing both structural support and conduction. Consequently, lycophytes, pteridophytes, and seed-producing plants are together known as the vascular plants or **tracheophytes**. The latter term takes its name from **tracheids**, a type of specialized vascular cell that conducts water and minerals and provides structural support.

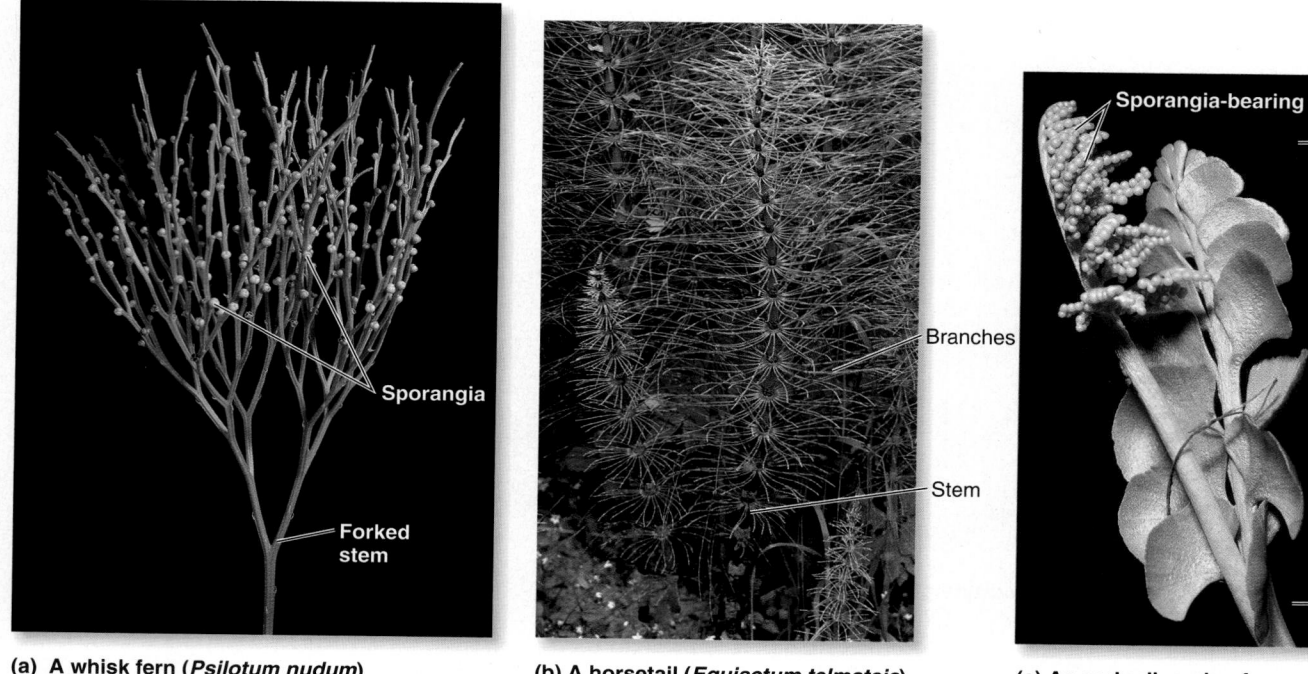

(a) A whisk fern (*Psilotum nudum*)

(b) A horsetail (*Equisetum telmateia*)

(c) An early-diverging fern
(*Botrychium lunaria*)

Figure 30.10   Pteridophyte diversity.   (a) The leafless, rootless green stems of the whisk fern (*Psilotum nudum*) branch by forking and bear many clusters of yellow sporangia that disperse spores via wind. The gametophyte of this plant is a tiny white structure that lives underground in a partnership with fungi. (b) The giant horsetail (*Equisetum telmateia*) displays branches in whorls around the green stems. The leaves of this plant are tiny, light brown structures that encircle branches at intervals. This plant produces spores in cone-shaped structures, and the wind-dispersed spores grow into small green gametophytes. (c) The fern *Botrychium lunaria*, showing a green photosynthetic leaf with leaflets and a modified leaf that bears many round sporangia. (d) The fern *Blechnum capense*, viewed from above, showing a whorl of young leaves that are in the process of unrolling from the bases to the tips. The surrounding older leaves have many leaflets. The stem of this fern grows parallel to the ground and is thus not illustrated. Most ferns produce spores in sporangia on the undersides of leaves. (e) The life cycle of a typical fern.

(d) A later-diverging fern (*Blechnum capense*)

Stems, Roots, and Leaves   In common with other tracheophytes, the sporophytes of lycophytes and pteridophytes produce specialized organs—stems, roots, and leaves—that contain vascular tissue. **Stems** are branching structures that contain vascular tissue and also produce leaves and reproductive structures. At their centers, stems contain the specialized conducting tissues known as **phloem** and **xylem**, the latter of which contain tracheids. Such conducting tissues enable vascular plants to conduct organic compounds, water, and minerals throughout the plant body. The xylem also provides structural support, allowing vascular plants to grow taller than nonvascular plants. This support function arises from the presence of a compression-

and decay-resistant waterproofing material known as **lignin** that occurs in the cell walls of tracheids and some other types of plant cells. Most vascular plants also produce **roots**—organs specialized for uptake of water and minerals from the soil—and **leaves**, which generally have a photosynthetic function. Roots and leaves possess vascular systems that connect to the vascular tissue of stems, forming a continuous conduction system in the plant body. Vascular plant roots efficiently take up water and minerals from the soil, conduct these materials to stems via xylem, and receive sugar from phloem. The leaves of vascular plants obtain water and minerals needed for photosynthesis from the xylem, and they export sugar via the phloem.

2 Sporangia are multicellular structures that develop on the undersides of the mature fern sporophyte leaves. Sporangia occur in clusters known as sori (singular, *sorus*).

1 The diploid sporophyte is the dominant generation in the life of ferns and other vascular plants.

Meiosis

Sporangium

3 Meiosis occurs in cells within sporangia to produce haploid spores, which are dispersed by the wind.

Spores

Spore (n)

Protective sporopollenin wall

8 The embryo matures into a sporophyte. After developing a root and leaf, fern sporophytes become independent of their gametophyte parent, which eventually rots away.

Sori

Mitosis

Gametophyte (n)
Young sporophyte (2n)

4 Under favorable conditions, spores undergo mitosis to produce gametophytes. These are often thumbnail-sized and heart-shaped, anchored by cells known as rhizoids.

Rhizoids

Mitosis

KEY
 Haploid
 Diploid

Diploid zygote (2n)

Blue-stained gametophyte

Female gametangium (archegonium)

Egg cell

Fertilization

Sperm (n)

7 The resulting diploid zygote is retained on the gametophyte, undergoes mitosis, and grows into a multicellular embryo that receives essential nutrients from the gametophyte.

Egg (n)

Male gametangia (antheridia)

6 When water is present, the male gametangia release the flagellate sperm, which swim to the female gametangia and fertilize the eggs.

5 Mature gametophytes produce eggs in female gametangia and sperm in male gametangia.

(e) The life cycle of a fern

Lycophyte roots and leaves differ from those of pterido-phytes. For example, lycophyte roots fork at their tips, whereas roots of pteridophytes branch from the inside like the roots of seed plants (see Chapter 35). Lycophyte leaves are relatively small and possess only one unbranched vein, whereas pterido-phyte leaves are larger and have branched veins, as do those of seed plants (compare Figures 30.9 and 30.10d). The evolutionary origin of pteridophyte leaves will be discussed in Section 30.4 of this chapter.

Adaptations That Foster Stable Internal Water Content Lyco-phytes, pteridophytes, and other vascular plants display several adaptations that increase their ability to maintain stable internal water content. For example, a protective **waxy cuticle** is present on most surfaces of vascular plant sporophytes (**Figure 30.11**). The plant cuticle contains a polyester polymer known as **cutin**, which helps to prevent attack by pathogens, and wax, which helps to prevent desiccation. The surface tissue of vascular plant stems and leaves contains **stomata** (singular, stomate), pores that are able to open and close (see Figure 30.11). Stomata allow vascular plants to take in carbon dioxide needed for photosynthesis and release oxygen to the air, while conserving water. When the soil is very dry, stomata close, which reduces water loss from plants via these pores. When the soil is moist, stomata open, allowing photosynthetic gas exchange to occur. These water-conserving features have allowed vascular plants to exploit a wide spectrum of land habitats.

Gymnosperms and Angiosperms Are the Modern Seed Plants

The modern phyla commonly known as cycads, ginkgos, conifers, and gnetophytes are collectively known as gymnosperms (**Figure 30.12**). **Gymnosperms** reproduce using both spores and seeds, as do the flowering plants or angiosperms (**Figure 30.13**). Gymnosperms and angiosperms are thus known informally as the **seed plants**. **Seeds** are complex structures having specialized tissues that protectively enclose embryos, that is, the young sporophytes. Seeds also contain stores of carbohydrate, lipid, and protein that can be used to provide energy for seed germination and seedling development.

Several additional seed plant phyla once existed and left fossils, but they are now extinct. Collectively, all of the living and fossil seed plant phyla are formally known as **spermatophytes** (see Figure 30.1). Modern and fossil seed plants are also formally known as **lignophytes**. Such plants are descended from seedless ancestors that produced wood, a specialized vascular tissue that is rich in lignin. Wood production enables plants to increase in girth and become tall. Many modern seed plants produce wood and are thus known informally as woody plants. By contrast, seedless plants lack wood, as do many modern seed plants. Plants that produce little or no wood are informally known as herbaceous plants. Woody and herbaceous plants do not represent clades.

The **angiosperms** are distinguished by the presence of flowers, fruits, and a specialized seed tissue known as endosperm.

Figure 30.11   A cross section through a stem of the pteridophyte *Psilotum nudum* shows tracheophyte adaptations for transporting and conserving water.   When viewed with fluorescence microscopy and illuminated with violet light, an internal core of xylem tracheids glows yellow, as does the surface cuticle. Surface pores known as stomata are also present.

— Vascular tissue

— Stomate

— Cuticle

120 μm

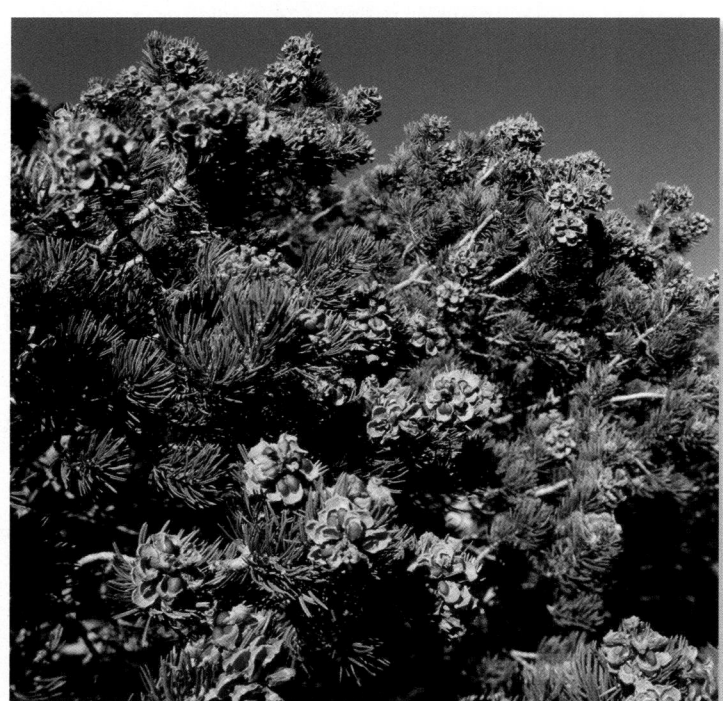

Figure 30.12   The pinyon pine (*Pinus edulis*) is an example of a gymnosperm.

Figure 30.13 This bleeding heart plant (the genus *Dicentra*) is an example of an angiosperm.

Flowers are short stems bearing organs that are specialized in ways that enhance seed production (see Figure 30.13). **Fruits** are structures that develop from flower organs, enclose seeds, and foster seed dispersal in the environment. The term angiosperm means "enclosed seeds," reflecting the fact that the flowering plants produce seeds within fruits. **Endosperm** is a nutritive seed tissue that increases the efficiency by which food is stored in the seeds of flowering plants (see Section 30.4). Flowers, fruits, and endosperm are defining features of the angiosperms, and they are integral components of animal nutrition.

Though gymnosperms produce seeds, and many are woody plants, they lack flowers, fruits, or endosperm. The term gymnosperm means "naked seeds," reflecting the fact that gymnosperm seeds are not enclosed within fruits (see Table 30.1). Despite their lack of flowers, fruits, and seed endosperm, the modern gymnosperms are diverse and abundant in many places. Like the other phyla of plants that we have just surveyed, gymnosperms played significant roles in the past evolutionary history of plants, our next topic.

30.2 An Evolutionary History of Land Plants

A billion years ago, Earth's terrestrial surface was comparatively bare of life. Some green or brown crusts of cyanobacteria most likely grew in moist places, but there would have been very little soil, no plants, and no animal life. The origin of the first land plants was essential to development of the first substantial soils, the evolution of modern plants, and the ability of animals to colonize land. It is safe to say that if land plants had not appeared hundreds of millions of years ago, and become so marvelously well adapted to life on land, humans and the modern ecosystems on which we depend would not have evolved in their present form.

How can we know about events such as the origin and diversification of land plants? Biologists are able to infer the past by comparing molecular and other features of modern plants and by examining plant remains in the fossil record. Thus, we will begin our survey of plant history by taking a closer look at these techniques.

Modern and Fossil Plants Are Helpful in Inferring Evolutionary History

The distinctive features of living plant phyla (summarized in Table 30.1) reveal the order in which they first appeared—bryophytes earliest, lycophytes and pteridophytes later, gymnosperms next, and angiosperms most recently (see Figure 30.1). Molecular and fossil data support and provide additional information for this inference.

Molecular Approaches to Plant Evolutionary History Plant evolutionary biologists often use computers and special software to compare gene sequences from diverse plants. Such analyses aid in understanding relationships among plants and the time when different clades originated. Hypotheses of organism relationships are represented as diagrams known as phylogenetic trees (see Figure 30.1). The arrangement of branches on phylogenetic trees often changes as new data become available.

Sequence data often reveal past changes in genome structure, such as chromosomal rearrangements and the addition of introns to genes. You may recall that many eukaryotic genes possess noncoding regions known as introns and that transcribed intron sequences are removed during mRNA processing before translation occurs. Introns can be inserted into genes, but such events are relatively rare. As a result, introns can reveal ancient phylogenetic divergences. For example, in 1998, evolutionary biologist Y. L. Qiu and associates reported that liverworts lack three introns generally present in mitochondrial genes of other plants. More recently, Milena Groth-Malonek and coworkers identified additional introns that are missing from liverworts but present in other living plant phyla. Plant evolutionary biologists suspect that these introns first appeared within plant genes after liverworts had diverged from the mainstream of plant evolution, indicating that liverworts are the most ancient modern plant phylum. This conclusion has been supported by an analysis of sequence data for several genes, reported in 2005 by bryophyte expert Barbara Crandall-Stotler and associates.

Another example of the use of sequence data to infer plant phylogeny is an analysis of pteridophyte relationships published in 2001 by Kathleen Pryer and associates. Based on sequences of small subunit (SSU) rRNA and three chloroplast genes, Pryer's group concluded that pteridophytes are monophyletic and are the closest modern relatives of seed plants. The Pryer phylogeny contained a surprise—that plants commonly known as horsetails and whisk ferns were part of a clade that also contained the ferns. Plant biologists had earlier used structural features to classify horsetails and whisk ferns into phyla distinct from the ferns.

The Use of Fossils to Infer Plant Evolutionary History Fossils are the remains of dead organisms that have survived decay and other destructive processes long enough to become buried in materials that harden into rocks. The enclosed organic remains become stacked in layers, with the oldest at the bottom and the most recent fossils closer to the rock surface. This layering, together with chemical features, allows biologists to infer the relative ages of fossil remains (refer back to Figure 22.9).

The tough plant compounds lignin, cutin, and sporopollenin help to preserve the structure of plants as they fossilize (**Figure 30.14**). Such preservation allows plant scientists to deduce how fossil plants looked when alive. Plant biologists compare fossils to other fossils and to living plants to deduce the kinds of evolutionary changes that occurred during plant history. For example, several types of fossil plants were noted to have different types of early vascular tissues, but their relationships to the vascular tissue of modern plants were uncertain. Evolutionary biologists Martha Cook and Ned Friedman helped to resolve this issue by comparing vascular tissues of a modern lycophyte with those of fossilized early vascular plants. These investigators treated the stems and roots of a modern lycophyte with enzymes that would degrade all but the most resistant plant materials—those most likely to fossilize. In 2004, Cook and Friedman reported that the treated stem conducting tissues closely resembled particular fossils, enabling them to better interpret the fossils and understand the evolution of plant vascular tissue.

Another example of the use of fossils to deduce plant history was reported by evolutionary biologists Gar Rothwell and Kevin Nixon in 2006. These investigators performed a phylogenetic analysis of the structural features of many fossils, as well as living pteridophytes. Their study suggested the possibility that plants currently classified within the pteridophytes might actually belong to more than one clade, thus questioning the strength of conclusions based primarily on molecular data. Rothwell and Nixon have emphasized that current concepts of plant evolutionary history are incomplete and that studies of plant fossils are needed to complement molecular analyses.

The study of fossils, as well as molecular and other features of modern plants, has revealed an amazing story—how plants conquered the land. This story can be conceptualized as three dramatic episodes: (1) aquatic charophycean algae give rise to the first land-adapted plants; (2) seedless plants transform Earth's atmosphere and climate; and (3) an ancient cataclysm marks the rise of angiosperms.

Aquatic Charophycean Algae Give Rise to the First Land-Adapted Plants

Land plants got their start in the water. As we have earlier noted, plants most likely evolved from an aquatic ancestor similar to modern, complex charophycean algae. During their evolution in the water, charophyceans acquired several cellular and reproductive adaptations that aided survival in aquatic habitats. Some of these traits also proved useful on land, helping the first land plants to survive there. As an example, let's consider cytokinesis, the process by which the cytoplasm divides after nuclear division.

A distinctive feature of plant cytokinesis, the **phragmoplast**, first appeared during the diversification of charophyceans. Phragmoplasts are constructed largely of microtubules arranged perpendicularly to a developing cell plate (refer back to Figure 15.14b) and microfilaments. Phragmoplasts also promote the development of intercellular connections known as **plasmodesmata**, which are important modes of plant cell-to-cell communication (see Chapter 10). The earliest land plants inherited phragmoplasts and plasmodesmata from complex charophyceans, making use of them to produce cohesive tissues in which component cells could effectively communicate. Subsequent land plants used these traits to build increasingly more complex bodies that were even better adapted to the stresses of terrestrial life.

In addition to traits inherited from charophycean algae, early land plants acquired other features in response to stresses present on land, but not in the water. This fact explains why all land plants possess several features in common that are absent from charophyceans (see Table 30.1). For example, plant biologists Zoe Popper and Stephen Fry discovered that the cell walls of all land plants possess xyloglucan carbohydrates, which cross-link cellulose microfibrils, but that charophycean algae lack this feature. Cell-wall xyloglucans are thus among the new features that appeared as early plants began to adapt to land, probably aiding in the development of more complex bodies. The terrestrial environment also influenced the evolution of genes that control another essential plant cell-wall polysaccharide, cellulose.

Figure 30.14 Plant fossils. Fossil of *Pseudosalix handleyi*, an angiosperm.

Biological inquiry: What biochemical components of plants favor the formation of fossils?

GENOMES & PROTEOMES

The Number of Genes That Controls Cellulose Production Increased During Plant Evolutionary History

Land plant cell walls possess an essential component—the polymer of glucose known as cellulose. In fact, cellulose-rich cell walls, which strengthen and protect plant cells, are a hallmark of the plants and many of their green algal relatives. In plant cell walls, cellulose occurs as 3.5- to 10-nm-diameter cylindrical microfibrils containing 36 to 90 closely packed glucan chains (refer back to Figure 10.5). Extensive hydrogen bonding among the glucan polymers confers great strength, explaining the role of cellulose in cell walls, and its many useful applications, such as the paper pages of this textbook. The manner in which plant cellulose polymers are produced is key to comprehending how plants achieve this high level of biochemical organization. Understanding how plants produce cellulose is also fundamental to human efforts to genetically engineer plant cellulose, which has many industrial and medical applications.

Cellulose is spun outward from highly organized arrays of proteins known as terminal complexes, which are located in the plasma membrane (**Figure 30.15**). The terminal complexes of charophycean algae and plants have a unique rose shape and are thus known as rosettes. Cell biologists have determined that the rosettes contain cellulose synthase, an enzyme that synthesizes cellulose from sugars, and geneticists have discovered that *CesA* genes encode cellulose synthase.

Plant molecular biologist Alison Roberts and her associates compared the *CesA* genes of charophycean algae, seedless plants, and seed plants. They found that *Mesotaenium caldariorum*, a single-celled charophycean desmid, has at least two *CesA* genes.

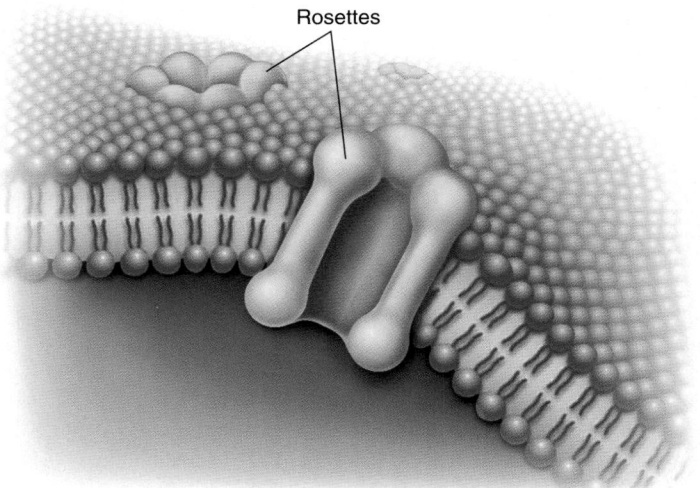

Figure 30.15   Rosette-shaped structures in plant plasma membranes contain cellulose synthase, which produces the cellulose-rich plant cell wall.

By contrast, the moss *Physcomitrella patens* has at least four *CesA* genes, and seed plants have numerous *CesA* genes. For example, the flowering plant *Arabidopsis thaliana* has a 10-member *CesA* gene family, and different *CesA* genes are expressed in particular tissues. Roberts and associates concluded that land plants had inherited cellulose synthase genes from ancestral charophycean algae and that the *CesA* gene family subsequently diversified by gene duplication and divergence. These investigators noted that cellulose synthase gene diversification is correlated with the evolution of greater plant structural complexity. As plants became larger and more complex, they exerted increasingly greater effects on their environments, as we will see next.

Seedless Plants Transformed Earth's Ecology

The spread of early plants dramatically changed the Earth's landscapes, fostering soil buildup, changes in atmospheric chemistry, and colonization of land by fungi and animals. For this reason, evolutionary biologists are interested in learning when land plants first appeared, and what ecological effects resulted.

The First Land Plants   When did the earliest land plants first appear? Some evolutionary biologists use molecular clock methods to attempt to answer this question, and others use fossils. Molecular clock methods use rates of molecular sequence changes in modern organisms to infer the point in time when lineages diverged from a common ancestor. Based on molecular clock data, Daniel Heckman and colleagues have suggested that land plants might have been present as early as 700 million years ago. Although fossils of such a great age have not been identified, Heckman and associates hypothesized that populations of the earliest land plants were very small, inhibited by cold conditions that prevailed 750–580 million years ago (see Chapter 22). If so, the earliest plants would not have left many fossils. Only when warmer conditions allowed plant populations to expand do fossils become more abundant.

An early fossil record of presumed land plants consists of tiny sporelike structures reported by paleobiologists Wilson Taylor, Paul Strother, and colleagues, who studied rocks of the Cambrian period that are more than 500 million years old. Early plants whose bodies were too delicate to fossilize well presumably produced these tougher structures. By about 450 million years ago, during the Ordovician period, spores similar to those of modern bryophytes had become abundant and geographically widespread. This fact led paleobiologist Jane Gray to propose that bryophyte-like land plants had become well established by this point in time, millions of years prior to the appearance of vascular plant fossils. In 2003, paleobiologist Charles Wellman and associates reported that they had found fossil sporangia with spores, similar to those of modern bryophytes, in 470-million-year-old rocks.

Ecological Effects of Ancient and Modern Bryophytes Later-diverging modern liverworts and mosses typify a later phase of plant evolution. In addition to possessing tough spores and sporangia, these plants produce decay-resistant body tissues similar to fossils left by ancient land plants. Such tissues help modern bryophytes to avoid attack by decay bacteria and fungi. Plant evolutionary biologists Linda Graham and David Hanson and colleagues determined the amount of decay-resistant mass produced by several modern bryophytes and used these data to estimate the ecological impact of early nonvascular plants. The results suggested that nonvascular plants likely contributed organic substances to early soils, thereby helping to enrich them. The results also indicated that ancient nonvascular plants could have begun a process by which atmospheric carbon dioxide (CO_2) was converted to organic carbon that was buried and thus not respired back to carbon dioxide. Burial of organic carbon is a process that helps to reduce the amount of the greenhouse gas CO_2 in Earth's atmosphere, thereby influencing temperature and precipitation. The investigators calculated that such effects on soil, atmospheric chemistry, and climate might have been significant because they could have occurred over large geographic areas, and for millions of years before vascular plants became dominant.

Modern bryophytes likewise play important roles by storing CO_2 as decay-resistant organic compounds. The abundant modern moss genus *Sphagnum* contains so much decay-resistant mass that dead moss has accumulated over thousand of years into deep peat deposits. By storing organic carbon for long periods, *Sphagnum* moss functions as a giant global thermostat that helps to keep Earth's climate steady, to the benefit of humans and other life-forms. How does this biological thermostat work? Under cooler than normal conditions, *Sphagnum* grows more slowly and thus absorbs less CO_2, allowing atmospheric CO_2 to rise a bit. Since atmospheric CO_2 helps to warm Earth's climate, increasing CO_2 warms the climate a little. When the climate warms sufficiently, *Sphagnum* grows faster, thereby sponging up more CO_2 as peat deposits. Reducing atmospheric CO_2 returns the climate to slightly cooler conditions. The world's peat deposits currently store an estimated 400 gigatons (billion tons) of organic carbon. Ecologists have expressed concern that modern global warming could increase the rate at which decay organism populations increase and break down organic peat into CO_2. Such processes have the potential to increase global temperatures sufficiently to harm human and other life.

Ecological Effects of Ancient Vascular Plants Vascular plants originated from extinct plants called protracheophytes such as the fossil *Aglaophyton major*. Protracheophytes had branched sporophytes and produced numerous sporangia, but their water-conducting cells lacked lignin and thus did not provide structural support, as does the xylem of vascular plants. Vascular plant fossils first appear in rocks deposited 420–430 million years ago and rapidly became diverse and abundant. These fossils reveal that the earliest vascular plants had no leaves or roots, but they did have stems with a central core of lignin-coated water-conducting cells, a tough outer cuticle, and stomata, much like modern pteridophytes (see Figure 30.11). These features suggest that early vascular plants had achieved the ability to maintain a stable internal water level. The presence of lignin and cutin also fostered the ability of vascular plant bodies to resist decaying long enough to fossilize, explaining why vascular plants have left a more extensive fossil record than did the earliest plants.

Fossils tell us that extensive forests dominated by tree-sized lycophytes, pteridophytes, and early lignophytes occurred in widespread swampy regions during the warm, moist Carboniferous period (354–290 million years ago) (**Figure 30.16**). Much

Giant lycophytes   Giant dragonfly   Giant horsetail (pteridophyte)

Figure 30.16   Reconstruction of a Carboniferous (Coal Age) forest, showing tree-sized lycophytes and pteridophytes.

Biological inquiry: Why did giant dragonflies occur during this time, but not now?

of today's coal derives from the abundant remains of these ancient plants, explaining why the Carboniferous is commonly known as the Coal Age. Carboniferous plants converted huge amounts of atmospheric CO_2 into decay-resistant organic materials such as lignin. Long-term burial of these materials, compressed into coal, together with chemical interactions between soil and the roots of vascular plants, dramatically changed Earth's atmosphere and climate.

Mathematical models of ancient atmospheric chemistry, supported by measurements of natural carbon isotopes, led paleoclimatologist Robert Berner to propose that the Carboniferous proliferation of vascular plants was correlated with a dramatic decrease in atmospheric carbon dioxide, which reached a his-

toric low about 300 million years ago (**Figure 30.17**). During this period of very low CO_2, atmospheric oxygen levels rose to historic high levels, because less O_2 was being used to break down organic carbon into CO_2. High atmospheric oxygen content has been invoked to explain the occurrence of giant Carboniferous dragonflies and other huge insects, which obtain their air by diffusion. Because atmospheric CO_2 is a greenhouse gas—allowing Earth's atmosphere to retain heat—the great Carboniferous decline in CO_2 level caused cool, dry conditions to prevail in the late Carboniferous and early Permian periods. As a result of this relatively abrupt global climate change, many of the giant seedless lycophytes and pteridophytes that had dominated Carboniferous forests became extinct, as did organisms such as the giant dragonflies. Cooler, drier Permian conditions favored extensive diversification of the first seed plants, the gymnosperms. Seed plants were better able than nonseed plants to reproduce in cooler, drier habitats (as we will see in Section 30.4). As a result, seed plants came to dominate Earth's terrestrial communities, as they continue to do.

An Ancient Cataclysm Marks the Rise of Angiosperms

Diverse phyla of gymnosperms dominated Earth's vegetation through the Mesozoic era (248–65 million years ago), what is sometimes called the Age of Dinosaurs (**Figure 30.18**). In fact, gymnosperms were probably a major source of food for terrestrial herbivorous dinosaurs, some of which grew to enormous sizes.

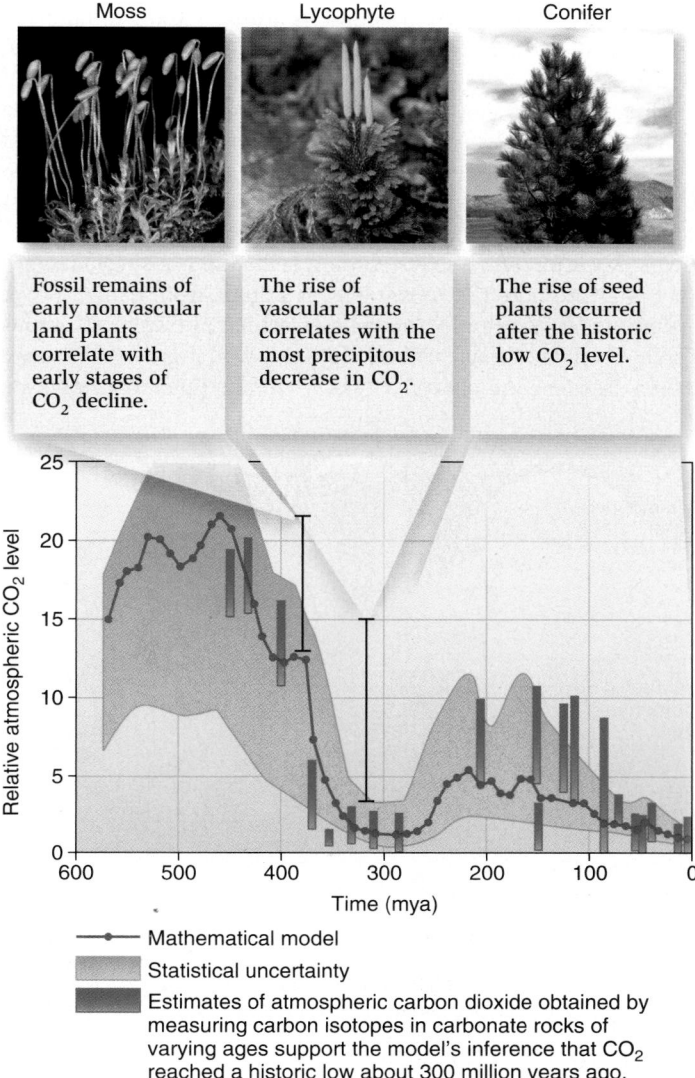

Figure 30.17 Graph showing changes in Earth's atmospheric carbon dioxide levels over geological time. Geological evidence indicates that carbon dioxide levels in Earth's atmosphere were once higher than they are now, but that the rise of land plants caused CO_2 to reach a historic low about 300 million years ago.

Figure 30.18 Gymnosperms dominated Earth's flora and dinosaurs dominated animal life during the Mesozoic era.

Though both early flowering plants and early mammals were present, they primarily existed in the shadows of gymnosperms and dinosaurs. One fateful day about 65 million years ago, disaster struck from the sky.

That day, at least one large meteorite or comet crashed into the Earth near the present-day Yucatan Peninsula in Mexico. This episode is known as the **K/T event** because it marks the end of the Cretaceous (sometimes spelled with a K) period and the beginning of the Tertiary (T) period. The impact, together with substantial volcanic activity that also occurred at this time, is thought to have produced huge amounts of ash, smoke, and haze that dimmed the sun's light long enough to kill many of the world's plants. Many groups of plants were forever lost, though some survived and persist to the present time. With a severely reduced food supply, the dinosaurs were also doomed, the only exceptions being their descendants, the birds.

After the K/T event, weedy ferns dominated long enough to leave huge numbers of fossil spores, and then surviving groups of flowering plants began to diversify into the space left by the demise of previous plants. The rise of angiosperms fostered the diversification of beetles (see Chapter 33) and other types of insects that associate with plants. For example, a large molecular analysis of ant relationships allowed Corrie Moreau, Naomi Pierce, and their associates to infer that the development of forests dominated by flowering plants had fostered ant diversification.

Our brief survey of plant evolutionary history reveals some important diversity principles. While environment certainly influenced the diversification of plants, plant diversification has also changed Earth's environment in ways that affected the evolution of other organisms. Plant evolutionary history also serves as essential background for a closer focus on the evolution of **critical innovations**, new features that foster the diversification of phyla. Among the critical innovations that appeared during plant evolutionary history, embryos, leaves, and seeds were particularly important, as we will discuss next.

30.3 The Origin and Evolutionary Importance of the Plant Embryo

The embryo, absent from charophyceans, was probably one of the first distinctive traits acquired by land plants. Recall that plant embryos are young sporophytes that develop from zygotes, enclosed by maternal tissues that provide sustenance. This feature is critical to plant reproduction in terrestrial environments. Drought, heat, ultraviolet light, and microbial attack could kill delicate plant egg cells, zygotes, and embryos if these were not protected and nourished by enclosing parental tissues. For land plants, the evolutionary origin of a dependent plant embryo was a critical innovation. The first embryo-producing plants diversified into hundreds of thousands of diverse modern species, as well as many species that have become extinct. The embryo is such a defining feature of plants that plant biologists often use

the term embryophytes as a synonym for plants (see Figure 30.1). A closer look at embryos reveals why their origin and evolution is so important to all land plants.

A plant embryo has several characteristic features. First, as we have previously noted, plant embryos are multicellular and diploid (see Section 30.1). Plant embryos develop by repeated mitosis from a single-celled zygote resulting from fertilization (see Figure 30.6b). In addition, we have also learned that plant eggs are fertilized while still attached to the maternal plant body, with the result that embryos begin their development within the protective confines of maternal tissues (see Figure 30.7). Plant biologists say that plants retain their zygotes and embryos. Third, plant embryo development depends on organic and mineral materials supplied by the mother plant. Nutritive tissues composed of specialized **placental transfer tissues** aid in the transfer of nutrients from parent to embryo. Let's take a closer look at the valuable role played by placental transfer tissues.

Placental transfer tissues function similarly to the placenta of mammals, which fosters nutrient movement from the mother's bloodstream to the developing fetus. Plant placental transfer tissues often occur in haploid gametophyte tissues that lie closest to embryos and in the diploid tissues of young embryos themselves. Such transfer tissues contain cells that are specialized in ways that promote the movement of solutes from gametophyte to embryo. For example, the cells of placental transport tissues display complex arrays of finger-like cell-wall ingrowths (**Figure 30.19**). Because the plant plasma membrane lines this elaborate

Figure 30.19 **TEM of placental transfer tissue in the liverwort genus *Monoclea*.** Placental transfer tissues contain specialized cells having extensive finger-shaped cell-wall ingrowths. Such cells help nutrients to move rapidly from parental gametophytes to embryonic sporophytes, a process that fosters plant reproductive success.

Biological inquiry: How does increasing plasma membrane surface area foster rapid movement of nutrients in placental transfer tissue?

plant cell wall, the ingrowths vastly increase the surface area of plasma membrane. This increase provides the space needed for abundant membrane transport proteins, which move solutes into and out of cells. With more transport proteins present, materials can move at a faster rate from one cell to another. Experiments have revealed that dissolved sugars, amino acids,

and minerals first move from maternal cells into the intercellular space between parent tissues and embryo. Then, transporter proteins in the membranes of nearby embryo cells efficiently import materials into the embryo. Classic experiments have revealed the adaptive value of this process in land plant reproduction.

FEATURE INVESTIGATION

Browning and Gunning Demonstrated That Placental Transfer Tissues Increase the Rate at Which Organic Molecules Move from Gametophytes to Sporophytes

In the 1970s, plant cell biologists Adrian Browning and Brian Gunning explored placental transfer tissue function. Using a simple moss experimental system, they determined that placental transfer tissues increase the rate at which radioactively labeled carbon moves through placental transfer tissues from green gametophytes into young sporophytes. This process is expected to increase the ability of mature sporophytes to produce progeny

spores, thereby increasing reproductive success. Recall that embryos are very young, few-celled sporophytes, and that in mosses and other bryophytes all stages of sporophyte development are nutritionally dependent on gametophyte tissues. Browning and Gunning investigated nutrient flow into young sporophytes because these slightly older and larger developmental stages were easier to manipulate in the laboratory than were tiny embryos.

In a first step, the investigators grew many gametophytes of the moss *Funaria hygrometrica* in a greenhouse until young sporophytes developed as the result of sexual reproduction (**Figure 30.20**). In a second step, they placed black glass sleeves over young sporophytes as a shade to prevent photosynthesis,

Figure 30.20 Browning and Gunning demonstrated that placental transfer tissues increase plant reproductive success.

HYPOTHESES 1) Placental transfer tissues allow organic nutrients to flow from plant gametophytes to sporophytes faster than such nutrients move through plant tissues lacking transfer cells.
2) The rate of organic nutrient transfer into larger sporophytes is faster than into smaller sporophytes.

STARTING MATERIALS Moss *Funaria hygrometrica*, $^{14}CO_2$ (radiolabeled carbon dioxide)

Experimental level   Conceptual level

1 Grow moss gametophytes until young sporophytes develop from embryos, and measure sporophyte size.

Young sporophytes receive organic nutrients in the same way as embryos but are easier to handle.

2 Darken young sporophytes with blackened glass tubing. Expose plants to $^{14}CO_2$ for 15 minutes.

Clear jar
Sporophyte
Dark tubing
Gametophyte
Threaded lid

$^{14}CO_2$

Photosynthesis, which requires light, will convert $^{14}CO_2$ into ^{14}C-sugar in gametophytes but not sporophytes.

Light
$^{14}CO_2$
$^{14}CO_2$
Labeled sugar

3 Expose plants to a large amount of nonradioactive CO_2. This is called a chase. Incubate up to 8 hours.

Non-radiolabeled CO_2

Removal of radioactive CO_2 gas is known as a "chase".

Light
Non-radiolabeled CO_2

$^{14}CO_2$ no longer taken up by plant.

4 Pluck young sporophytes from gametophytes. Assay ^{14}C in both sporophytes and gametophytes using a scintillation counter. This was done immediately following the chase, or 2 or 8 hours after the chase.

Scintillation counter

Determine how much organic carbon flowed into sporophytes during each chase time.

5 THE DATA I

Carbon transfer from gametophyte to sporophyte:

Mean ^{14}C content of 5 gametophytes at 0 chase time	Mean ^{14}C lost from gametophytes after 8-hour chase	Mean ^{14}C gained by sporophytes after 8-hour chase
228 units	145 units	51 units

6 THE DATA II

Sporophyte size effect:

Sporophyte size	Mean ^{14}C content of 8 sporophytes at 0-hour chase	Mean ^{14}C content of 8 sporophytes after 2-hour chase
5–7 mm	1.52 ± 0.48 units	8.47 ± 4.29 units
11–13 mm	0.22 ± 0.26 units	9.93 ± 3.94 units
23–25 mm	0.13 ± 0.05 units	24.97 ± 5.30 units

enclosed moss gametophytes and their attached sporophytes within transparent jars, and supplied the plants with radioactively labeled carbon dioxide for measured time periods known as pulses. Because the moss gametophytes were not shaded, their photosynthetic cells were able to convert the radioactively labeled carbon dioxide into labeled organic compounds, such as sugars and amino acids. Shading prevented the young sporophytes, which possess some photosynthetic tissue, from using labeled CO_2 to produce organic compounds.

In a third step, the researchers added an excess amount of nonradioactive CO_2 to prevent the further uptake of the radioactive CO_2 from their experimental system, a process known as a chase. This process stopped the radiolabeling of photosynthetic products. (Experiments such as these are known as pulse-chase experiments.) In a final step, Browning and Gunning plucked young sporophytes of different sizes (ages) from gametophytes and measured the amount of radioactive organic carbon present in the separated gametophyte and sporophyte tissues at various times following the chase.

From these data, they were able to calculate the relative amount of organic carbon that had moved from the photosynthetic moss gametophytes to sporophytes. Browning and Gunning discovered that about 22% of the organic carbon produced

by gametophyte photosynthesis was transferred to the young sporophytes during an 8-hour chase period. They also calculated the rate of nutrient transfer between generations and compared this rate to the rate at which organic carbon moves in several other plant tissues that lack specialized transfer cells (determined in other studies). By so doing, Browning and Gunning discovered that organic carbon moved from moss gametophytes to young sporophytes nine times faster than organic carbon moves within other plant tissues. These investigators inferred that the increased rate of nutrient movement could be attributed to placental transfer cell structure, namely, the fact that cell-wall ingrowths enhanced plasma membrane surface area. By comparing the amount of radioactive carbon accumulated by young sporophytes of differing ages, they also learned that larger sporophytes absorbed labeled carbon about three times as fast as smaller ones.

These data are consistent with the hypothesis that placental transfer tissues increase plant reproductive success by providing embryos and older sporophytes with more nutrients than they would otherwise receive. Supplied with these greater amounts of nutrients, sporophytes are better able to grow larger than they otherwise would, and eventually they produce more progeny spores.

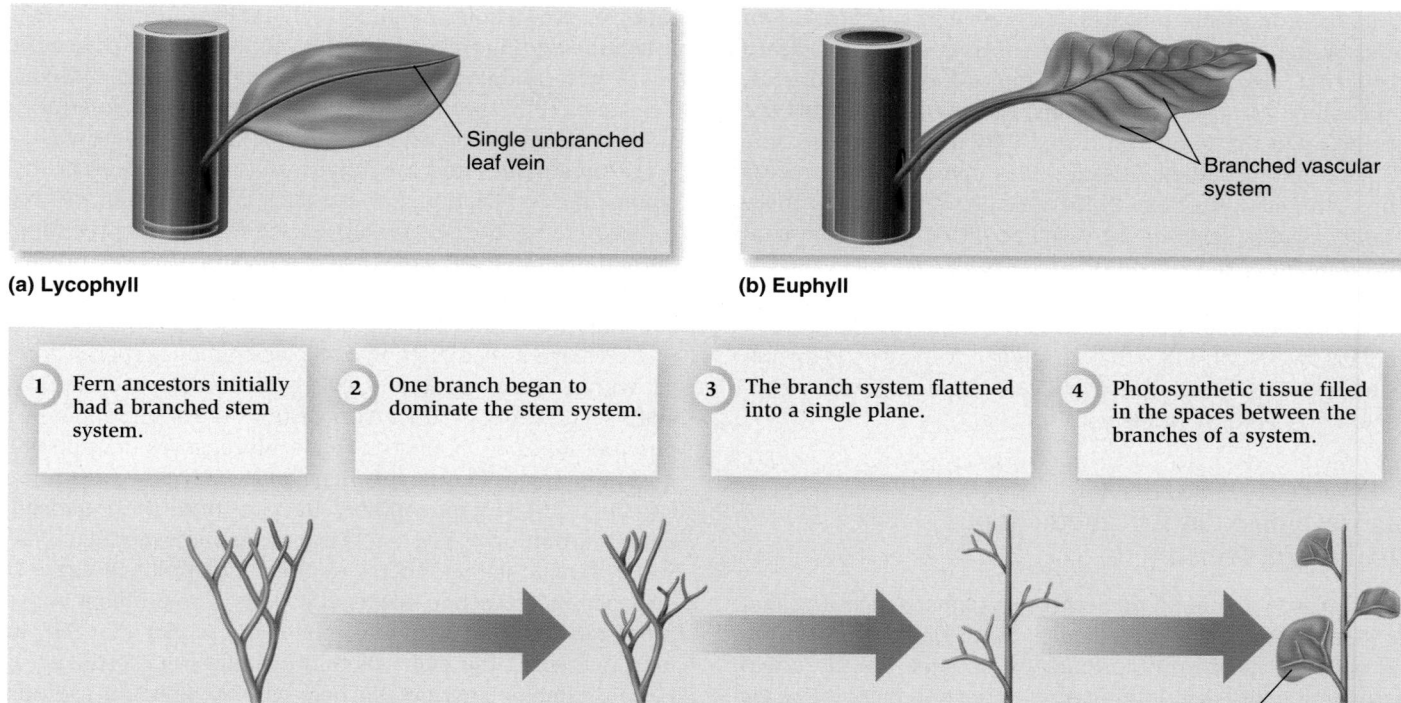

(a) Lycophyll

(b) Euphyll

1. Fern ancestors initially had a branched stem system.

2. One branch began to dominate the stem system.

3. The branch system flattened into a single plane.

4. Photosynthetic tissue filled in the spaces between the branches of a system.

(c) Euphyll evolution process

Figure 30.21   Lycophylls and euphylls.   (a) Most lycophylls possess only a single unbranched leaf vein having limited conduction capacity, explaining why lycophylls are generally quite small. **(b)** Euphylls possess branched vascular systems having greater conduction capacity, explaining why many euphylls are relatively large. **(c)** Fossil evidence suggests how pteridophyte euphylls might have evolved from branched stem systems.

30.4 The Origin and Evolutionary Importance of Leaves and Seeds

Like plant embryos, leaves and seeds are critical innovations that allowed plant phyla to diversify extensively. However, unlike the plant embryo, which likely originated just once at the birth of the plant kingdom, leaves and seeds probably evolved several times during plant evolutionary history. Comparative studies of diverse types of leaves and seeds in fossil and living plants suggest how these critical innovations might have originated.

The Leaves of Lycophytes Differ in Structure and Origin from Those of Other Vascular Plants

Leaves are the solar panels of the plant world. Their flat structure enables leaves to effectively capture sunlight for use in photosynthesis. This explains why leaflike structures evolved even on the gametophytes of leafy liverworts and mosses (see Figures 30.3 and 30.5), and also why they occur on the sporophytes of most vascular plants. Among the vascular plants, lycophytes produce the simplest and most ancient leaves. Modern lycophytes have tiny leaves, known as **lycophylls** or microphylls, that typically have only a single unbranched vein (**Figure 30.21a**). These small leaves are thought to have evolved from sporangia, which are borne along the stems of lycophytes.

In contrast, the leaves of other vascular plant phyla have extensively branched veins. Leaves with branched veins are known as **euphylls** (meaning "true leaves") (**Figure 30.21b**). This explains why the clade that includes pteridophytes and seed plants is known as the **euphyllophytes** (see Figure 30.1). The branched veins of euphylls are able to supply relatively large areas of photosynthetic tissue with water and minerals. Thus, euphylls are typically much larger than lycophylls, explaining why euphylls are also known as megaphylls (meaning "large leaves"). Euphylls provide considerable photosynthetic advantage to ferns and seed plants, because they provide more surface for solar energy capture than do small leaves. Hence, the evolution of relatively large leaves allowed plants to more effectively accomplish photosynthesis, enabling them to grow larger and produce more progeny.

Study of fern fossils indicates that euphylls likely arose from leafless, branched stem systems by a series of steps (**Figure 30.21c**). First, one branch assumed the role of the main axis, then the whole branch system became flattened, and finally the spaces between the branches of this flattened system became filled with photosynthetic tissue. This hypothetical process explains why euphylls have branched vascular systems; these apparently originated from the vascular system of an ancestral branched stem. Plant evolutionary biologists suspect that euphylls arose several times, and it is unclear whether or not the leaves of seed plants originated in the same way as those of ferns. Next, we'll consider what adaptive advantages seeds provide and how seeds might have evolved, beginning with a consideration of seed development.

Seeds Develop from the Interaction of Ovules and Pollen

The seed plants dominate modern ecosystems, suggesting that seeds offer reproductive advantages. Seed plants are also the plants with the greatest importance to humans. For these reasons, plant biologists are interested in understanding why seeds are so advantageous and how they evolved. To consider these questions, we must first take a closer look at seed structure and development.

Plants produce seeds by means of reproductive structures known as ovules and pollen, which are structures unique to seed plants. An **ovule** is a sporangium that contained only a single spore that developed into a very small egg-producing game-

tophyte, the whole enclosed by modified leaves known as **integuments** (**Figure 30.22a**). You can think of an ovule as being like a nesting doll with four increasingly smaller dolls inside. The smallest doll would correspond to an egg cell; intermediate-sized dolls would stand for the gametophyte, spore wall, and megasporangium; and the largest doll would represent the integuments. Fertilization converts such layered ovules into seeds. In seed plants, the sperm needed for fertilization are supplied by **pollen**, tiny male gametophytes enclosed by sporopollen in spore walls. A closer look at pollen and ovules will help in understanding how seeds develop.

We have earlier noted that all plants produce spores by meiosis within sporangia, and seed plants are no exception. However, seed plants produce two distinct types of spores in two different types of sporangia. Small **microspores** develop within microsporangia, and larger **megaspores** develop within megasporangia. Male gametophytes develop from the microspores, and the resulting pollen is released from microsporangia. Meanwhile, female gametophytes develop and produce eggs while enclosed by protective megaspore walls. The problem with this process is that female gametophytes are so tiny that they need help in feeding the embryos that develop from fertilized eggs. Female gametophytes get this help from the previous sporophyte generation by remaining attached to it. This is an advantage because the large sporophyte generation is thereby able to provide gametophytes with the nutrients needed for embryo development.

Embryos develop as the result of fertilization, which cannot occur until after **pollination**, the process by which pollen of the same species reaches ovules. Pollination typically occurs by means of wind or animal transport (see Chapter 31). Fertiliza-

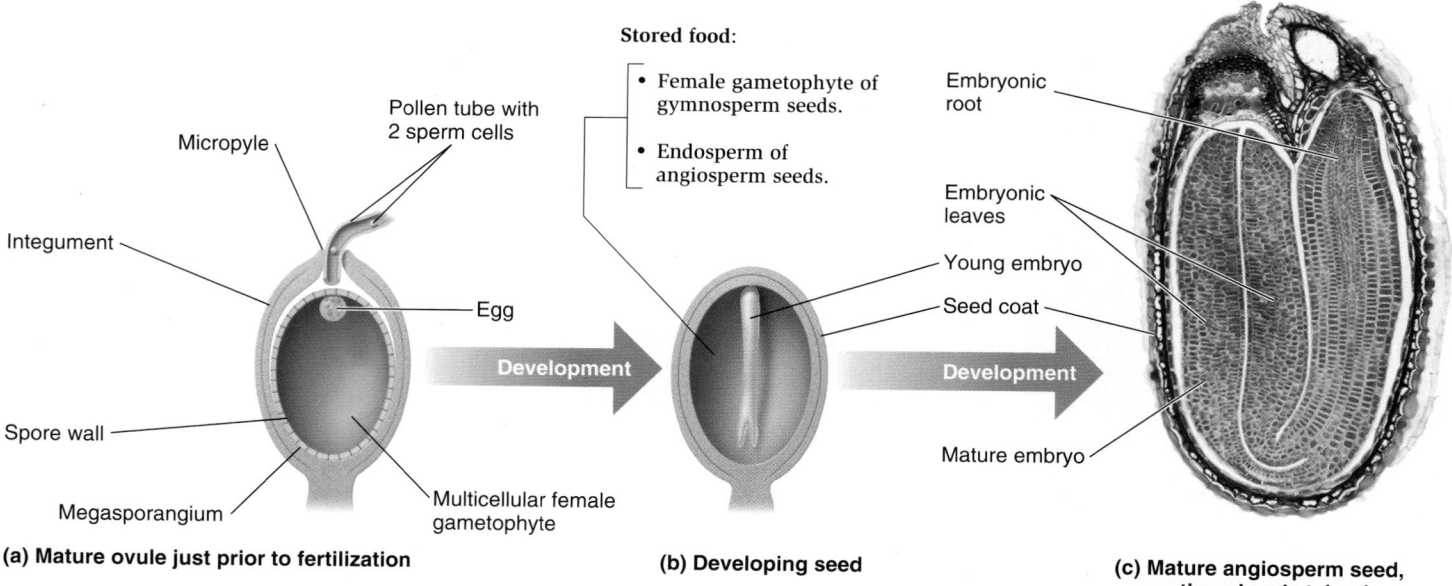

Figure 30.22   Ovule and seed structure.   (a) Ovule structure. (b) Seed structure. (c) Light micrograph of an angiosperm seed.

Biological inquiry: Why does the seed photo show little or no endosperm?

tion occurs when a male gametophyte extends a slender pollen tube that carries two sperm toward an egg. After a pollen tube enters the micropyle and releases the sperm, fertilization may occur. The fertilized egg becomes an embryo, and the ovule's integument develops into a protective, often hard and tough **seed coat** (**Figure 30.22b,c**).

Gymnosperm seeds contain female gametophyte tissue that has accumulated large amounts of protein, lipids, and carbohydrates prior to fertilization. These nutrients are used during seed germination to help support growth of the seedling. Angiosperm seeds also contain this useful food supply, but angiosperm ovules do not store food materials prior to fertilization. Instead, angiosperm seeds store food only after fertilization occurs, ensuring that the food is not wasted if an embryo does not form. How is this accomplished? The answer is a process known as **double fertilization**. This process produces both a zygote and a food storage tissue known as **endosperm**, which is a tissue unique to angiosperms. One of the two sperm delivered by each pollen tube fuses with the egg, producing a diploid zygote, as you might expect. The other sperm nucleus fuses with different gametophyte nuclei to form an unusual cell that has more than the diploid number of chromosomes; this cell generates the endosperm food tissue (see Chapter 39).

Seeds allow embryos access to food supplied by the older sporophyte generation, an option not available to seedless plants. The layered structure of ovules explains why seeds are also layered like nesting dolls, with a protective seed coat (developed from the ovule integuments) enclosing the embryo and also stored food, in many cases. These seed features improve the chances of embryo and seedling survival, thereby increasing seed plant fitness.

Seeds Confer Important Ecological Advantages

Seeds provide plants with numerous ecological advantages, which explains why seed plants have dominated Earth's ecosystems since the end of the Coal Age. First, many seeds are able to remain dormant in the soil for long periods, until conditions become favorable for germination and seedling growth. In contrast, most single-celled spores are unable to survive for long; if conditions are not suitable for germination, they die. In addition, seeds are larger and more complex than spores, which improves the ability of seeds to resist mechanical damage and pathogen attack. Further, seed coats have evolutionary adaptations that improve dispersal in diverse habitats. For example, many plants produce winged seeds that are effectively dispersed by wind. Other plants produce seeds with fleshy coverings that attract birds, which consume the seeds, digest their fleshy covering, and eliminate them at some distance from the originating plants.

Another advantage of seeds is that they can store considerable amounts of food, which helps plant seedlings grow large enough to compete for light, water, and minerals. This is especially important for seeds that must germinate in shady forests.

Finally, the sperm of seed plants can reach eggs without having to swim through water, because pollen tubes deliver sperm directly to ovules. Consequently, seed plant fertilization is not typically limited by lack of water, in contrast to that of seedless plants. Therefore, seed plants are better able to reproduce in arid and seasonally dry habitats. For these reasons, seeds are considered to be a particularly significant plant reproductive adaptation to reproduction in a land habitat.

Ovule and Seed Evolution Illustrate Descent with Modification

As we have seen, seed plants reproduce using both spores and seeds, and it is important to recognize that seed plants have not replaced spores with seeds. Instead, during seed plant evolution, ovules and seeds were added to an ancestral life history that begins with spores. Ovules and seeds thus originated by a sequence of changes to pre-existing structures and processes. Some clues about ovule and seed evolution arise from comparing reproduction in living lycophytes, pteridophytes, and spermatophytes. Fossils provide additional information.

Let's first consider reproduction of lycophytes and pteridophytes. Most modern lycophytes and pteridophytes release one type of spore and one type of gametophyte, which lives in the open environment and produces both male and female gametangia (see Figure 30.10e). However, some lycophytes and pteridophytes produce separate microspores and megaspores, which grow into separate male and female gametophytes, a process known as **heterospory** (meaning "different spores"). Such gametophytes also grow within the confines of microspore and megaspore walls and thus are known as **endosporic gametophytes**. The advantage of heterospory is that it has the potential to increase cross-fertilization, the fusion of eggs and sperm from gametophytes of distinct genotypes. Cross-fertilization thus increases the potential for genetic variation, which aids evolutionary flexibility.

The advantage of endosporic gametophytes is that they are protected to some degree by surrounding spore walls from environmental damage. As we have earlier noted, seed plants also produce distinct megaspores and microspores, which produce internal female and male gametophytes. From these observations, we can infer that heterospory and endosporic gametophytes were probably also features of seed plant ancestors, and constitute early steps toward seed evolution. Because seed plant ancestors are extinct, lycophytes and pteridophytes are useful sources of information about the origin of heterospory and endosporic gametophytes.

A next step in seed evolution may have been retention of megaspores within megasporangia, rather than releasing them. A further step may have been production of only one megaspore per sporangium rather than multiple spores per sporangium, which is common in nonseed plants. Reduction of megaspore numbers would have allowed plants to channel more nutrients into each megaspore. A final step might have been the retention of megasporangia on parental sporophytes. As we have noted, this adaptation would allow food materials to flow from photosynthetic plants to their dependent gametophytes and young embryos.

Fossils provide information about when the process of seed evolution first occurred. Very early fossil seeds such as *Elkinsia polymorpha* and *Archaeosperma arnoldii* were present 365 million years ago. Thus, paleobiologists think that the first seeds arose prior to the Coal Age, during the Devonian period (see Figure 30.1). A newly discovered Devonian fossil known as *Runcaria heinzelinii* is thought to represent a seed precursor (**Figure 30.23**). It had a lacy integument that did not completely enclose the megasporangium.

These hypothesized stages of seed evolution illustrate the stepwise way in which plants probably achieved an increased ability to cope with the stresses of life on land. The evolutionary journey illustrated by the transition from aquatic charophycean algae to bryophytes, to seedless plants, and finally to seed plants reveals how adaptation is related to environmental change, as well as ways in which plants themselves shaped Earth's ecosystems.

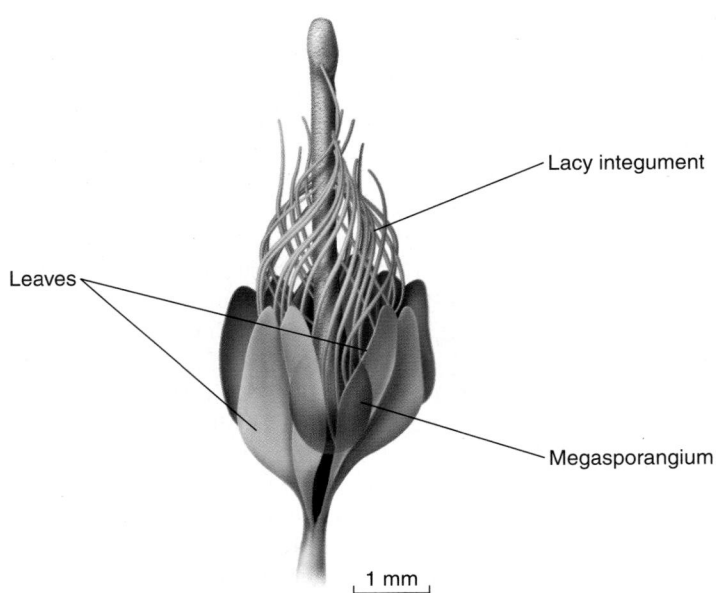

Figure 30.23   The fossil *Runcaria heinzelinii*.

CHAPTER SUMMARY

30.1 Ancestry and Diversity of Modern Plants

- Plants are multicellular organisms composed of eukaryotic cells having plastids, which display many adaptations to life on land, and primarily live on land. The modern plant kingdom consists of several hundred thousand species that can be classified into 10 phyla, informally known as the liverworts, mosses, hornworts, lycophytes, pteridophytes, cycads, ginkgos, conifers, gnetophytes, and angiosperms. (Figure 30.1, Table 30.1)

- The land plants evolved from ancestors that probably would be classified with modern, complex charophycean algae of freshwater habitats. (Figure 30.2)

- The monophyletic liverwort, hornwort, and moss phyla are together known informally as the bryophytes. Bryophytes illustrate early evolved features of land plants, such as a sporic life cycle involving embryos that develop within protective gametophytic tissues. (Figures 30.3, 30.4, 30.5, 30.6)

- Bryophytes are distinguished by a dominant gametophyte generation and a dependent, nonbranching, short-lived sporophyte generation. By contrast, other plant phyla possess dominant sporophyte generations that become independent, branch, and have increased reproductive capacity. Bryophytes also lack supportive vascular tissues, in contrast to other modern plant phyla, which are known as the vascular plants or tracheophytes. (Figures 30.7, 30.8)

- Lycophytes and pteridophytes possess stems, leaves, and roots having vascular tissues composed of xylem and phloem, but the roots and leaves of these phyla differ in distinctive ways. (Figures 30.9, 30.10, 30.11)

- Cycads, ginkgos, conifers, and gnetophytes are collectively known as gymnosperms. Gymnosperms inherited an ancestral capacity to produce wood. Angiosperms, also known as the flowering plants, produce seeds and many also produce wood. Flowers, fruits, and seed endosperm are distinctive features of the angiosperms. (Figures 30.12, 30.13)

30.2 An Evolutionary History of Land Plants

- Paleobiologists and plant evolutionary biologists infer the history of land plants by analyzing the molecular features of modern plants, and by comparing the structural features of fossil and modern plants. (Figures 30.14, 30.15)

- Seedless plants transformed Earth's ecology by fostering soil buildup and altering atmospheric chemistry and climate. Such plants also influenced the evolutionary history of terrestrial animal life. (Figures 30.16, 30.17, 30.18)

- The K/T meteorite or comet impact event that occurred 65 million years ago helped cause the extinction of previously dominant dinosaurs and many types of gymnosperms, leaving space into which angiosperms, insects, birds, and mammals diversified.

30.3 The Origin and Evolutionary Importance of the Plant Embryo

- Origin of the plant embryo was a critical innovation that fostered diversification of the land plants. Plant embryos are supported by nutrients supplied by female gametophytes with the aid of specialized placental transfer tissues. (Figure 30.19)

- In a classic experiment, Browning and Gunning inferred that placental transfer tissues were responsible for an enhanced flow rate of nutrients from parental gametophytes to embryos. (Figure 30.20)

30.4 The Origin and Evolutionary Importance of Leaves and Seeds

- Leaves are specialized photosynthetic organs that evolved more than once during plant evolutionary history. Lycophylls, which occur in lycophytes, are relatively small leaves having a single unbranched vein. Pteridophyte leaves are larger and have an extensively branched vascular system; they are known as euphylls. Fossils indicate that fern euphylls evolved from branched stem systems by dominance by one branch, flattening of the whole branch system, and development of photosynthetic tissue between the branches. (Figure 30.21)

- Seeds develop from ovules, megasporangia enclosed by leaflike integuments. Ovules develop into seeds after fertilization, following pollination. Pollen produces thin cellular tubes that deliver sperm cells to eggs produced by female gametophytes. Mature seeds contain an embryo sporophyte that develops from the zygote. Seeds also contain food stored within female gametophytes of gymnosperms or within endosperm tissues arising from double fertilization in angiosperms. (Figure 30.22)

- Seeds confer many reproductive advantages, including dormancy through unfavorable conditions, greater protection for embryos from mechanical and pathogen damage, seed coat modifications that enhance seed dispersal, and reduction of plant dependence on water for fertilization. Fossil seeds display stages in the evolution of seeds. (Figure 30.23)

TEST YOURSELF

1. The simplest and most ancient phylum of modern land plants is probably
   a. the pteridophytes.
   b. the cycads.
   c. the liverworts.
   d. the angiosperms.
   e. none of the above.

2. An important feature of land plants that originated during the diversification of charophycean algae is
   a. the sporophyte.
   b. spores, which are dispersed in air and coated with sporopollenin.
   c. tracheids.
   d. plasmodesmata.
   e. fruits.

3. A phylum whose members are also known as bryophytes is commonly known as
   a. liverworts.
   b. hornworts.
   c. mosses.
   d. all of the above.
   e. none of the above.

4. Plants possess a life cycle that involves alternation of two multicellular generations: the gametophyte and
   a. the lycophyte.
   b. the bryophyte.
   c. the pteridophyte.
   d. the lignophyte.
   e. the sporophyte.

5. The seed plants are also known as
   a. bryophytes.
   b. spermatophytes.
   c. pteridophytes.
   d. lycophytes.
   e. euphyllophytes.

6. A waxy cuticle is an adaptation that
   a. helps to prevent water loss from tracheophytes.
   b. helps to prevent water loss from charophyceans.
   c. helps to prevent water loss from bryophytes.
   d. aids in water transport within the bodies of vascular plants.
   e. all of the above.

7. Plant photosynthesis transformed a very large amount of carbon dioxide into decay-resistant organic compounds, thereby causing a historic low in atmospheric carbon dioxide levels during the geological period known as
   a. the Cambrian.
   b. the Ordovician.
   c. the Carboniferous.
   d. the Permian.
   e. none of the above.

8. Which phylum among the plants listed is likely to have the largest leaves?
   a. liverworts
   b. hornworts
   c. mosses
   d. lycophytes
   e. pteridophytes

9. Euphylls, also known as megaphylls, probably evolved from
   a. the leaves of mosses.
   b. lycophylls.
   c. branched stem systems.
   d. modified roots.
   e. none of the above.

10. A seed develops from
    a. a spore.
    b. a fertilized ovule.
    c. a microsporangium covered by integuments.
    d. endosperm.
    e. none of the above.

CONCEPTUAL QUESTIONS

1. Why do evolutionary biologists think that land plants evolved from ancestors related to modern charophycean algae?

2. Why have bryophytes such as mosses been able to diversify into so many species even though they have relatively small, dependent sporophytes?

3. What features help vascular plants to maintain stable internal water content?

EXPERIMENTAL QUESTIONS

1. What were the goals of the Browning and Gunning investigation?

2. How did Browning and Gunning prevent photosynthesis from occurring in moss sporophytes during the experiment (shown in the Feature Investigation in Figure 30.20), and why did they do this?

3. What measurements did Browning and Gunning make after adding an excess amount of unlabeled CO_2?

COLLABORATIVE QUESTIONS

1. Discuss at least one difference in environmental conditions experienced by early land plants and ancestral complex charophycean algae.

2. Discuss as many plant adaptations to land as you can.

www.brookerbiology.com
This website includes answers to the Biological Inquiry questions found in the figure legends and all end-of-chapter questions.

31

THE DIVERSITY OF MODERN GYMNOSPERMS AND ANGIOSPERMS

CHAPTER OUTLINE

The Madagascar periwinkle (*Catharanthus roseus*).

Humans have long recognized that plant extracts are useful in treating illnesses. For example, Native Americans used extracts of the flowering mayapple plant (*Podophyllum peltatum*) to treat skin tumors. Modern medicine has validated this practice. Compounds such as podophyllotoxin, obtained from the mayapple plant, are now used to treat leukemia and other medical conditions. Leukemia is also treated with vincristine, a drug extracted from another angiosperm, the beautiful Madagascar periwinkle (*Catharanthus roseus*). Vinblastine—another extract from *C. roseus*—is used to treat lymphatic cancers. Taxol, a compound used in the treatment of breast and ovarian cancers, was first discovered in extracts of the Pacific yew tree, a gymnosperm known as *Taxus brevifolia*.

Podophyllotoxin, vincristine, vinblastine, taxol, and many other plant-derived medicines are examples of plant secondary metabolites, which are distinct from the products of primary metabolism (carbohydrates, lipids, proteins, and nucleic acids). Secondary metabolites play essential roles in protecting plants from disease organisms and plant-eating animals, and they also aid plant growth and reproduction. Though all plants produce secondary metabolites, these natural products are exceptionally diverse in seed plants: the nonflowering gymnosperms and the flowering angiosperms.

In this chapter, we will learn how the hundreds of thousands of modern seed plants play many additional important roles in modern ecosystems and the lives of humans. This chapter also illustrates many diversity principles, including descent with modification, horizontal gene transfer, critical innovation, coevolution, biological services, and human impacts on biodiversity.

31.1 The Diversity of Modern Gymnosperms

Gymnosperms are plants that produce seeds that are exposed rather than enclosed in fruits, as is the case for angiosperms. The word gymnosperm, which means "naked seed," comes from the Greek *gymnos*, meaning naked (referring to the unclothed state of ancient athletes), and *sperma*, meaning seed. Gymnosperms inherited many adaptations to life on land from their ancestors (**Figure 31.1**) (see also Chapter 30). In this section, we will first consider some fossil plants that help to explain important gymnosperm traits. Then we will survey the structure, reproduction, and ecological roles of modern gymnosperm phyla.

Modern Gymnosperms Arose from Woody Ancestors

Most modern gymnosperms are woody shrubs or trees, such as the famous giant sequoias (*Sequoiadendron giganteum*) native to the Sierra Nevada of the western U.S. Giant sequoias are among Earth's largest organisms, weighing as much as 6,000 tons and reaching an amazing 100 m in height. The large size of sequoias and other trees is based on the presence of **wood**, a tissue composed of numerous pipelike arrays of empty, water-conducting cells whose walls are strengthened by an exceptionally tough secondary metabolite known as lignin. These properties enable woody tissues to transport water upward for great distances and also to provide the structural support needed for trees to grow tall and produce many branches.

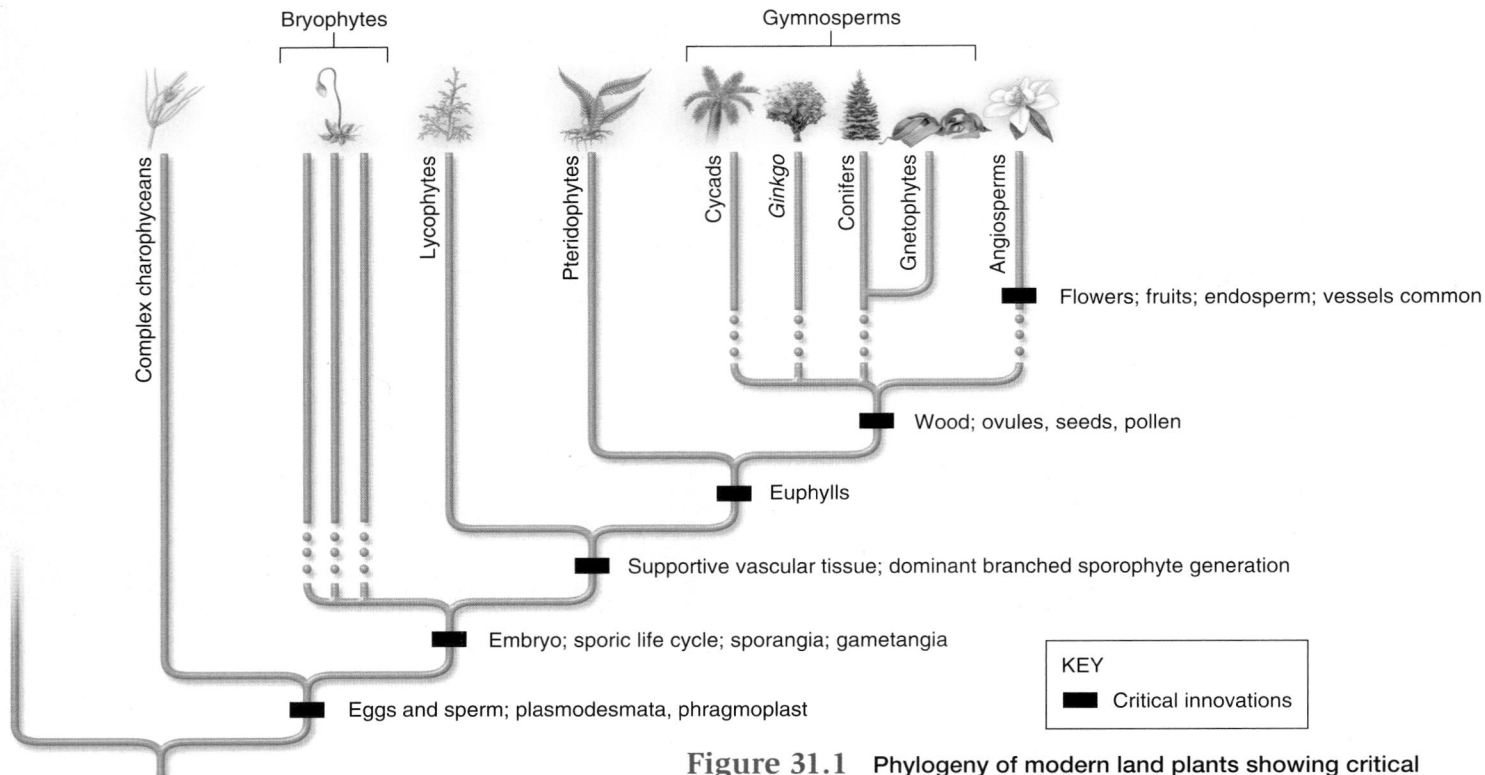

Figure 31.1   **Phylogeny of modern land plants showing critical innovations.**   The dotted lines indicate uncertainty about branching order.

In modern seed plants, a special tissue known as the **vascular cambium** produces both thick layers of wood and thinner layers of inner bark. The inner bark transports watery solutions of organic compounds. (The structure and function of the vascular cambium, wood, and bark are described in more detail in Chapter 35.) Vascular cambium, wood, and inner bark are critical innovations that helped gymnosperms and other seed plants to compete effectively for light and other resources needed for photosynthesis.

Wood first appeared in a group of plants known as the **progymnosperms** (meaning "before gymnosperms"). Woody progymnosperms, such as the fossil plant genus *Archaeopteris*, which lived 370 million years ago, were the first true trees (**Figure 31.2**). Progymnosperms were able to produce a vascular cambium and wood because their vascular tissue was arranged in a ring around a central pith of nonvascular tissue. (In contrast, the vascular tissue of earlier tracheophytes was arranged differently.) This ring of vascular tissue, known as a **eustele**, contained cells that were able to develop into the vascular cambium as seedlings grew into saplings. The vascular cambium then produced wood, allowing saplings to grow into tall trees. Modern seed plants inherited the eustele, explaining why many gymnosperms and angiosperms are also able to produce vascular cambia and wood. Despite the fact that progymnosperms were woody plants, they did not produce seeds. This fact reveals that wood originated before seeds evolved.

Figure 31.2   **The progymnosperm *Archaeopteris*, an early tree.**   This illustration was reconstructed from fossil data.

Table 31.1	Distinguishing Characteristics of Modern Seed Plant Phyla			
Seed plant group	**Vessels in wood**	**Flowers, fruit, seed endosperm**	**Flagellate sperm**	**Female gametangia**
Gymnosperms				
Cycads	No	No	Yes	Yes
Ginkgo biloba	No	No	Yes	Yes
Conifers	No	No	No	Yes
Gnetophytes	Yes	No	No	No
Angiosperms	Yes	Yes	No	No

(a) Emergent cycad stem   **(b) Submergent cycad stem**

Figure 31.3   Cycads.   Palmlike foliage and conspicuous seed-producing cones are features of most cycads. **(a)** The stems of some cycads emerge from the ground. **(b)** The stems of other cycads are submerged in the ground.

Progymnosperms and diverse early gymnosperms were the major vegetation present during the Mesozoic era, also known as the Age of Dinosaurs. Some groups of gymnosperms became extinct before or as a result of the K/T event at the end of the Cretaceous period 65 million years ago (see Chapter 30). Only a few gymnosperm phyla have survived to modern times: cycads (the Cycadophyta); *Ginkgo biloba*, the only surviving member of a once large phylum termed Ginkgophyta; conifers (the Coniferophyta); and gnetophytes (Gnetophyta). These phyla display distinctive reproductive features and play important roles in ecology and human affairs (**Table 31.1**).

Cycads Are Endangered in the Wild but Widely Used as Ornamentals

Nearly 300 cycad species occur today, primarily in tropical and subtropical regions. However, many species of cycads are rare, and their tropical forest homes are increasingly threatened by human activities. Consequently, all cycads are listed as endangered species, and commercial trade in cycads is regulated by CITES (Convention on International Trade in Endangered Species of Wild Fauna and Flora), a voluntary international agreement.

The structure of cycads is so interesting and attractive that many species are cultivated for use in outdoor plantings or as houseplants. The nonwoody stems of some cycads emerge from the ground much like tree trunks, some reaching 15 m in height, while other cycads produce subterranean stems (**Figure 31.3**). Cycads display spreading, palmlike leaves (*cycad* comes from a Greek word meaning "palm"). Mature leaves of the African cycad *Encephalartos laurentianus* can reach an astounding 8.8 m in length!

In addition to underground roots, which provide anchorage and take up water and minerals, many cycads produce coralloid roots. Such roots extend aboveground and have branching shapes resembling corals (**Figure 31.4**). Coralloid root tissues harbor a bright blue-green ring of symbiotic cyanobacteria (see Figure 31.4 inset). These cyanobacteria use light to produce ATP by cyclic electron flow. The ATP helps to fuel cyanobacterial nitrogen fixation, which produces nitrogen minerals used in host growth (see Chapters 27, 37).

Root surface

Cyanobacteria

(a) Coralloid roots

(b) Coralloid root cross section

Figure 31.4   Coralloid roots of cycads.   (a) Many cycads produce aboveground branching roots that resemble branched corals. **(b)** This magnified cross section of a coralloid root shows a ring of symbiotic blue-green cyanobacteria, which provide the plant with fixed nitrogen.

Biological inquiry: Why do the coralloid roots grow aboveground?

Recent studies have revealed that the cyanobacteria in *Cycas micronesica* produce an unusual amino acid that is distributed to the host plant's leaves and seeds. The distinctive amino acid, known as BMAA (β-N-methylamino-L-alanine), is harmful to the health of humans who consume flour made from cycad seeds or the meat of bats that have fed on *C. micronesica*. Botanist Paul Alan Cox and associates linked this toxin to the unusually high occurrence of a dementia resembling Alzheimer's disease among the Chamorro people of Guam in the Mariana Islands chain.

These researchers also found BMAA in the brain tissues of dementia patients who had not consumed foods originating from cycads. In an effort to understand these cases, Cox and associates examined many types of cyanobacteria from diverse habitats. In 2005, these investigators reported that most cyanobacteria produce BMAA in nature, suggesting that the toxic amino acid could be more widely present in the environment than previously thought. Studies of cycad toxicity have thus revealed a potential human health hazard of widespread concern. Because cycads generally produce a variety of toxins that likely deter herbivorous animals, experts recommend that humans should not consume food products made from these plants.

Cycad reproduction is distinctive in several ways. Individual cycad plants produce conspicuous conelike structures that bear either ovules and seeds, or pollen (see Figure 31.3). When mature, both types of reproductive structure emit odors that attract beetles. These beetles carry pollen to ovules, where the pollen produces tubes that deliver flagellate sperm to eggs.

Ginkgo Biloba Is the Last Survivor of a Once Diverse Group

The beautiful tree *Ginkgo biloba* is the single remaining species of a once diverse phylum that existed during the Age of Dinosaurs (**Figure 31.5a**). *G. biloba* takes its species name from the lobed shape of its leaves, which have unusual forked veins (**Figure 31.5b**). Today, *G. biloba* may be nearly extinct in the wild; widely cultivated modern *Ginkgo* trees are descended from seeds produced by a tree found in a remote Japanese temple garden and brought to Europe by 17th-century explorers.

G. biloba trees are widely planted along city streets because they are ornamental and also tolerate cold, heat, and pollution better than most trees. These trees are also long-lived. Individuals can live for more than a thousand years and grow to 30 m in height. Individual trees produce either ovules and seeds or pollen, based on a sex chromosome system much like that of humans. Ovule-producing trees have two X chromosomes; pollen-producing trees have one X and one Y chromosome. Wind disperses pollen to ovules, where pollen grains germinate to produce pollen tubes. These tubes grow through ovule tissues for several months, absorbing nutrients that are used for sperm development. Eventually the pollen tubes burst, delivering flagellate sperm to egg cells. After fertilization, zygotes develop into embryos, and the ovule integument develops into a fleshy, bad-smelling outer seed coat and a hard, inner seed coat. For streetside or garden plantings, people usually select pollen-producing trees to avoid the stinky seeds (**Figure 31.5c**).

Conifers Are the Most Diverse Modern Gymnosperm Lineage

The conifers are a lineage of trees named for their seed cones, of which pinecones are familiar examples (**Figure 31.6**). Conifers have been important components of terrestrial vegetation since the end of the Carboniferous period some 300 million years ago. Modern conifer families have existed for about 200 million years, survived the K/T event, and today include more than 500 species in 50 genera. Conifers are particularly common in mountain and high-latitude forests and are important sources of wood and paper pulp.

Conifers produce simple pollen cones and more complex ovule-bearing cones (**Figure 31.7**). The pollen cones of conifers bear many leaflike structures, each bearing a microsporangium in which meiosis occurs and pollen grains develop. By contrast, the ovule cones of conifers are composed of many short branch systems that bear ovules. No one could explain why conifer pollen and ovule cones were so different in structure until a new Coal Age fossil pollen cone came to light. In 2001, evolutionary botanist Genaro Hernandez-Castillo and colleagues reported that this early conifer pollen cone had the same complex structure as ovulate cones. This evidence suggests that modern, simple pollen cones probably evolved from more complex structures.

(a) *Ginkgo biloba* **tree**

(b) *Ginkgo biloba* **leaf**

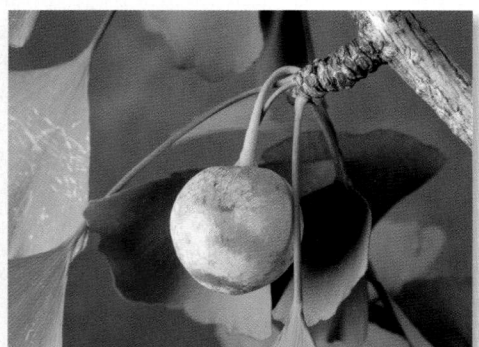
(c) *Ginkgo biloba* **seed**

Figure 31.5   *Ginkgo biloba.*   **(a)** A *Ginkgo biloba* tree; **(b)** fan-shaped leaves with forked veins; and **(c)** seeds having foul-smelling, fleshy seed coats.

Figure 31.6
Representative conifers.

(a) Pine (*Pinus ponderosa*)

(b) Dawn redwood (*Metasequoia glyptostroboides*)

1 Sporophytes produce 2 types of cones: pollen cones and ovule cones.

2 In ovule cones, megaspores are produced by meiosis, within megasporangia. In pollen cones, microspores are produced by meiosis within microsporangia.

3a In ovule cones, megaspores produce female gametophytes with eggs in archegonia. The entire structure, including the outer integuments, is an ovule.

Ovule cone

Megaspore

Egg (*n*)

Meiosis

Megasporangium

Integument

Female gametophyte (*n*)

Megasporangium (2*n*)

Archegonium (*n*)

Mature sporophyte (2*n*)

Pollen cone

Pollen grain (*n*)

Seed coat

Microspores

Microsporangium

KEY

Haploid

Diploid

Seedling

3b In pollen cones, microspores develop into pollen grains, which contain young male gametophytes.

Ovule

Seed

4 Pollen grains are dispersed into the wind and encounter ovules.

8 Seeds germinate and embryo sporophytes grow into seedlings.

Male gametophyte (*n*)

Embryo (2*n*)

Fertilization

Sperm

5 The pollen grain's male gametophyte matures, producing sperm cells in a pollen tube.

7 The zygote produces an embryo in a seed. Mature seeds are dispersed.

6 The pollen tube delivers sperm to eggs where fertilization occurs.

Figure 31.7   The life cycle of *Pinus* spp.

(a) *Pinus* **spp. seed**

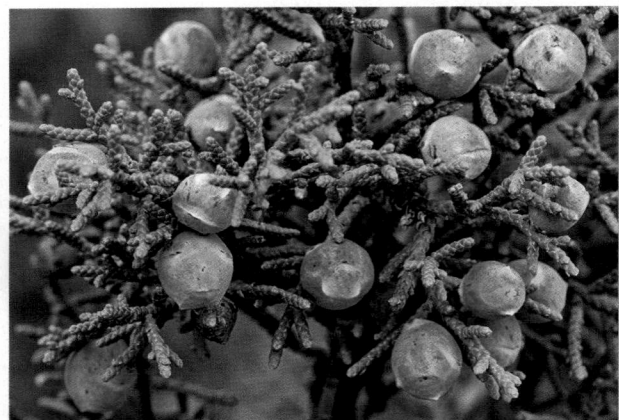

(c) *Juniperus scopularum* **cones**

(b) *Taxus baccata* **seeds**

Figure 31.8 Conifer seeds.
(a) Winged, wind-dispersed seed of *Pinus* spp. **(b)** Fleshy-coated, bird-dispersed seeds of yew (*Taxus* spp.). **(c)** Fleshy cones of juniper (*Juniperus* spp.) contain one or more seeds and are dispersed by birds. Juniper seeds are used in the production of gin.

When conifer pollen is mature, it is released into the wind, which transports pollen to ovules. All together, it takes more than two years for pine (the genus *Pinus*) to complete the processes of male and female gamete development, fertilization by means of nonflagellate sperm, and seed development (Figure 31.7). The seed coats of pine and some other conifers develop wings that aid in wind dispersal. Other conifers, such as yew and juniper, produce seeds with fleshy coatings that are attractive to birds, which help to disperse the seeds (**Figure 31.8**).

Conifer wood contains water transport cells that are adapted for efficient conduction even in dry conditions. These transport cells, known as tracheids, are devoid of cytoplasm and occur in long columns that function like plumbing pipelines (**Figure 31.9a**). Tracheid side and end walls possess many thin-walled, circular **pits** through which water moves both vertically and laterally from one tracheid to another. Conifer pits have a porous outer region that lets water flow through and a nonporous, flexible central region—the **torus**—that functions like a valve

(**Figure 31.9b**). If tracheids become dry and fill with air, they are no longer able to conduct water. In this case, the torus presses against the pit opening, sealing it (**Figure 31.9c**). The torus valve thereby prevents air bubbles from spreading to the next tracheid. This adaptation localizes air bubbles, preventing them from stopping water conduction in other tracheids. These specialized tracheids help to explain why conifers have been so successful for hundreds of millions of years. Conifer wood (and leaves) may also display conspicuous resin ducts, passageways for the flow of syrup-like resin that helps to prevent attack by pathogens and herbivores. Resin that exudes from tree surfaces and hardens in the air may form amber, which traps and preserves insects and other organisms.

Many conifers occur in cold climates and thus display numerous adaptations to such environments. Their conical shapes and flexible branches help conifer trees shed snow, preventing heavy snow accumulations from breaking branches. Conifer leaf shape and structure are adapted to resist damage from drought that occurs in both summer and winter, when liquid water is scarce.

Figure 31.9 **Tracheids in conifer wood.** **(a)** The lignin-rich cell walls of tracheids, the major type of water-conducting cell in gymnosperms and seedless vascular plants. **(b)** Detailed view of a portion of a tracheid that shows the thin-walled areas known as pits, each with a torus. **(c)** A water-filled tracheid with open pit and an air-filled tracheid with pit sealed by the flexed torus.

Conifer leaves are often scalelike (**Figure 31.10a**) or needle-shaped (**Figure 31.10b**); these shapes reduce the area of leaf surface from which water can evaporate. In addition, a thick, waxy cuticle coats conifer leaf surfaces (**Figure 31.10c**), retarding water loss and attack by disease organisms.

Many conifers are evergreen; that is, their leaves live for more than one year before being shed and are not all shed during the same season. Retaining leaves through winter helps conifers start up photosynthesis earlier than deciduous trees, which in spring must replace leaves lost during the previous autumn. Evergreen leaves thus provide an advantage in the short growth season of alpine or high-latitude environments. However, some conifers do lose all their leaves in the autumn. The bald cypress (*Taxodium distichum*) of southern U.S. floodplains, tamarack (*Larix laricina*) of northern bogs, and dawn redwood (*Metasequoia glyptostroboides*) are examples of deciduous conifers. Fossils indicate that dawn redwoods once grew abundantly across wide areas of the Northern Hemisphere until a few million years ago and then disappeared. However, in the 1940s, a forester found a living dawn redwood growing in a remote Chinese village, and subsequent expeditions located forests of the conifers. Since then, dawn redwood trees have been widely planted as ornamentals, prized for their attractive foliage and cones. As recently as 1994, botanists found a previously unknown conifer species, *Wollemia nobilis*, in an Australian national park. Like the dawn redwood, *Wollemia* is an attractive tree that is likely to become more widely distributed as the result of human cultivation.

Gnetophytes Are of Evolutionary Interest

The modern gnetophytes consist of three unusual genera, *Gnetum*, *Ephedra*, and *Welwitschia*. *Gnetum* is unusual among modern gymnosperms in having broad leaves similar to those of many tropical plants (**Figure 31.11a**). More than 30 species of the genus *Gnetum* occur as vines, shrubs, or trees in tropical Africa or Asia. *Ephedra*, a gnetophyte genus native to arid regions of the southwestern U.S., has tiny brown scale-like leaves and green, photosynthetic stems (**Figure 31.11b**). *Ephedra* produces secondary metabolites that aid in plant protection but also affect human physiology. Early settlers of the western U.S. used *Ephedra* to treat colds and other medical conditions. In fact, the modern decongestant drug pseudoephedrine is based on the chemical structure of ephedrine, which was named for and originally obtained from *Ephedra*. Pseudoephedrine sales are now restricted in many places because this compound can be used as a starting point for the synthesis of illegal drugs. Ephedrine has also been used to enhance sports performance, a practice that has elicited medical concern.

Welwitschia, the third gnetophyte genus, has only one living representative species. *Welwitschia mirabilis* is a strange-looking plant that grows in the coastal Namib Desert of southwestern Africa, one of the driest places on Earth (**Figure 31.11c**). A long taproot anchors a stubby stem that barely emerges from the ground. Two very long leaves grow from the stem but rapidly become wind-shredded into many strips. The plant is thought to obtain most of its water from coastal fog, explaining how *W. mirabilis* can grow and reproduce in such a dry place.

Evolutionary plant biologists are interested in gnetophytes because their relationships to other seed plants are unclear. Some morphological features suggest that gnetophytes might be closely related to the angiosperms. For example, female gametangia (archegonia; refer back to Figure 30.7) are present in the ovules of most gymnosperms but have been lost from gnetophytes and angiosperms. Gnetophytes also possess vessels—a type of large-diameter water-conducting feature—in their vascular tissue, as do most angiosperms but not other gymnosperms (see Table 31.1). However, the vessels of gnetophytes are thought to have evolved independently of angiosperm vessels. In addition, recent molecular evidence links gnetophytes more closely to the conifers or specifically with the conifer genus *Pinus* (pine). The identity of the gymnosperm group from which the angiosperms arose remains unresolved.

(a) Scale-shaped leaves

(b) Needle-shaped leaves of pine

Thick cuticle

(c) Stained cross section of pine needle, showing the thick cuticle

Figure 31.10 Conifer leaves.

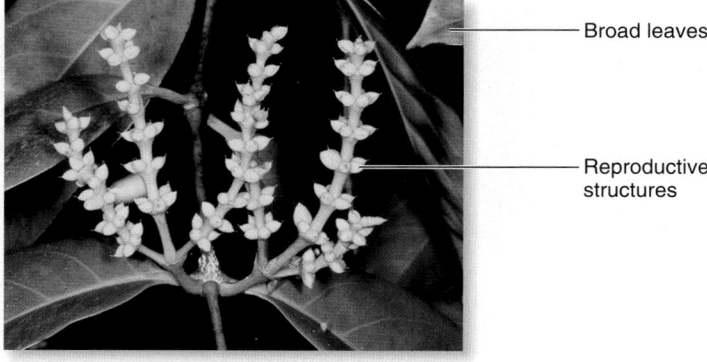

Broad leaves

Reproductive structures

(a) *Gnetum* spp.

Tiny-scale like leaves

Green stem

Reproductive structures

(b) *Ephedra californica*

Reproductive structures

Leaves

(c) *Welwitschia mirabilis*

Figure 31.11 Gnetophytes. **(a)** The tropical *Gnetum* spp., displaying broad leaves and reproductive structures. **(b)** *Ephedra* spp. growing in deserts of North America, showing miniscule brown leaves on green, photosynthetic stems and reproductive structures. **(c)** *Welwitschia mirabilis* growing in the Namib Desert of southwestern Africa, showing long, wind-shredded leaves and reproductive structures.

31.2 The Diversity of Modern Angiosperms

Angiosperms, the flowering plants, retained many structural and reproductive features from ancestral plants. In addition, flowering plants evolved several traits not found or seldom found among gymnosperms and other land plants. Flowers and fruits are two of the defining features of angiosperms (**Figure 31.12**), because these features do not occur in other modern plants (see Table 31.1). The term **angiosperm** means "enclosed seed," which reflects the presence of seeds within fruits. Seed endosperm is another defining feature of the flowering plants (see Chapters 30 and 39).

Although humans obtain wood, medicines, and other valuable products from gymnosperms, we depend even more on the angiosperms. Our food, beverages, and spices—flavored by an amazing variety of secondary metabolites—primarily come from flowering plants. People surround themselves with ornamental flowering plants and decorative items displaying flowers or fruit. We also commonly use flowers and fruit in ceremonies. In this section, we focus on how flowers, fruits, and secondary metabolites played key roles in angiosperm diversification. We will also learn that features of flowers, fruits, and secondary metabolites are used to classify and identify angiosperm species.

Diverse Flower Types Are Adaptations That Foster Seed Production

Flowers are complex reproductive structures that are specialized for the efficient production of pollen and seeds. The sexual reproduction process of angiosperms depends on flowers. Thus, as the flowering plants diversified, flowers of varied types evolved as reproductive adaptations to differing environmental conditions. To understand this process, let's start by considering the basic flower parts and their roles in reproduction.

Flower Parts and Their Reproductive Roles Flowers are produced at stem tips and contain four types of organs: **sepals**, **petals**, pollen-producing **stamens**, and ovule-producing **carpels**

(**Figure 31.13**). These flower organs are supported by tissue known as a **receptacle**, located at the tip of a flower stalk—a **peduncle**. The functioning of several genes that control flower organ development explains why carpels are the central-most flower organs, why stamens surround carpels, and why petals and sepals are the outermost flower organs (refer back to Figure 19.26).

Many flowers produce attractive petals that play a role in **pollination**, the transfer of pollen among flowers. Sepals of many flowers are green and form the outer, protective layer of flower buds. By contrast, the sepals of other flowers resemble attractive petals. All of a flower's petals and sepals are collectively known as the **perianth**. Most flowers produce one or more **stamens**, the structures that produce and disperse pollen. Most flowers also contain carpels, structures that produce ovules. Some flowers lack perianths, stamens, or carpels. Flowers that possess all four types of flower organs are known as **complete flowers**, while flowers lacking one or more organ types are known as **incomplete flowers**. Flowers that contain both stamens and carpels are described as **perfect flowers**, while flowers lacking either stamens or carpels are **imperfect flowers**.

Flowers also differ in the numbers of organs they produce. Some flowers produce only a single carpel, others display several separate carpels, and many possess several carpels that are fused together into a compound structure. Both a single carpel and compound carpels are referred to as a **pistil** (from the Latin word *pistillum*, pestle), because it resembles the device people use to grind materials to powder in a mortar (see Figure 31.13). Only one pistil is present in flowers that have only one carpel and in flowers with fused carpels. By contrast, flowers possessing several separate carpels display multiple pistils.

Pistils are usually differentiated into three regions having distinct functions. A topmost portion of the pistil, known as the **stigma**, receives and recognizes pollen of the appropriate species or genotype. The stigma allows pollen of appropriate genetic type to germinate, producing a long pollen tube that grows through the elongate **style**. The pollen tube thereby delivers nonflagellate sperm cells to ovules and the eggs inside, allowing fertilization (**Figure 31.14**). If fertilization occurs, the ovule

Figure 31.12 **Angiosperm flowers and fruits.** Citrus plants display the defining features of flowering plants: flowers and fruits shown here and seed endosperm.

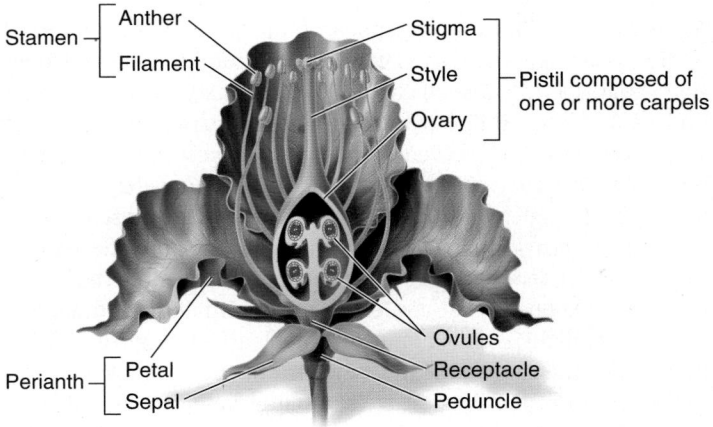

Figure 31.13 **Generalized flower structure.**

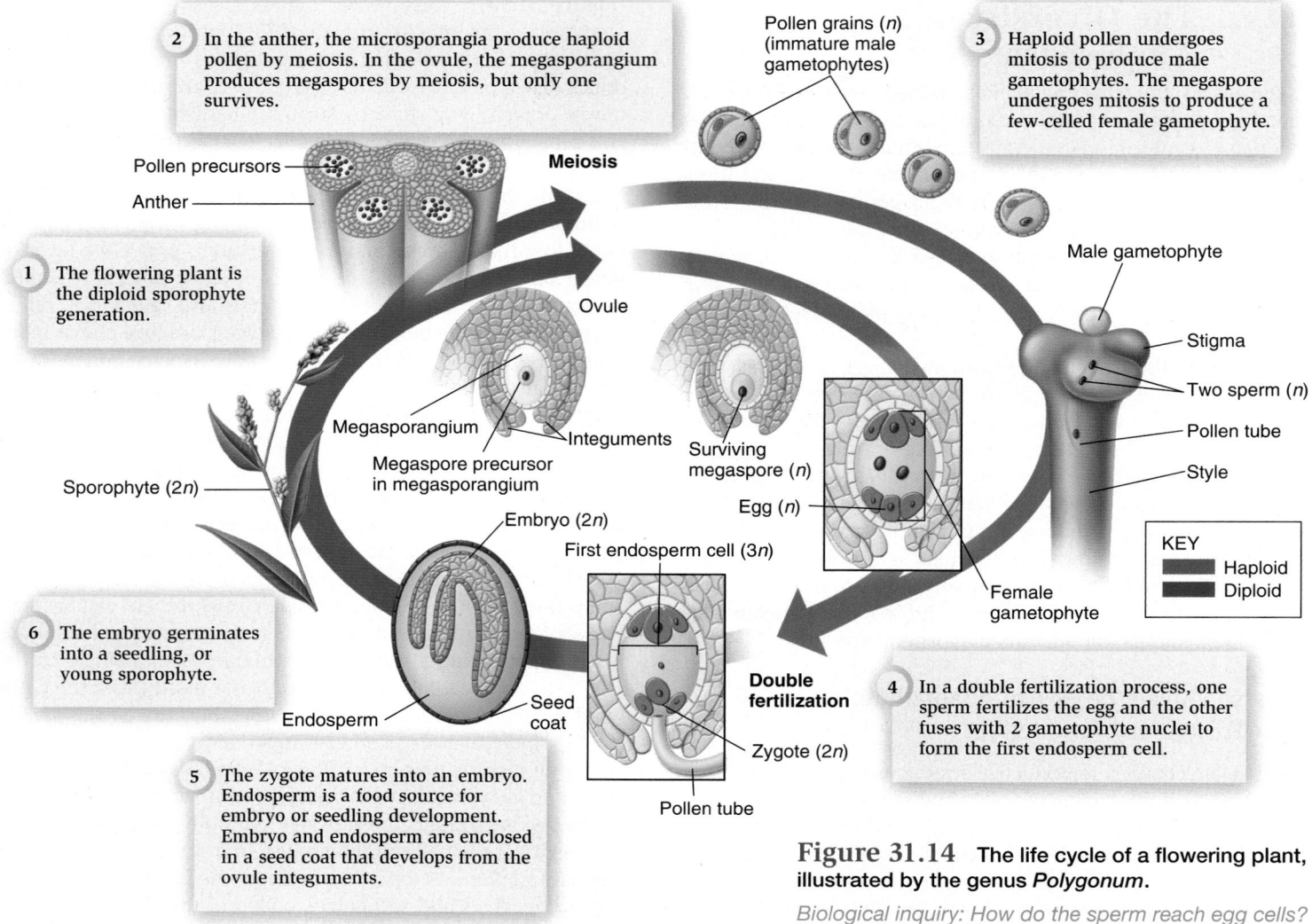

2 In the anther, the microsporangia produce haploid pollen by meiosis. In the ovule, the megasporangium produces megaspores by meiosis, but only one survives.

Pollen grains (*n*) (immature male gametophytes)

3 Haploid pollen undergoes mitosis to produce male gametophytes. The megaspore undergoes mitosis to produce a few-celled female gametophyte.

Pollen precursors

Anther

Meiosis

1 The flowering plant is the diploid sporophyte generation.

Ovule

Male gametophyte

Stigma

Two sperm (*n*)

Pollen tube

Style

Megasporangium

Integuments

Megaspore precursor in megasporangium

Surviving megaspore (*n*)

Egg (*n*)

Sporophyte (2*n*)

Embryo (2*n*)

First endosperm cell (3*n*)

KEY
Haploid
Diploid

Female gametophyte

6 The embryo germinates into a seedling, or young sporophyte.

Endosperm

Seed coat

Double fertilization

4 In a double fertilization process, one sperm fertilizes the egg and the other fuses with 2 gametophyte nuclei to form the first endosperm cell.

5 The zygote matures into an embryo. Endosperm is a food source for embryo or seedling development. Embryo and endosperm are enclosed in a seed coat that develops from the ovule integuments.

Zygote (2*n*)

Pollen tube

Figure 31.14 The life cycle of a flowering plant, illustrated by the genus *Polygonum*.

Biological inquiry: How do the sperm reach egg cells?

develops into a seed. The lowermost portion of the pistil is the **ovary**, which encloses and protects ovules. Ovaries (and sometimes additional flower parts) develop into fruits.

Early Flowers Flowers first appear in the fossil record about 150 million years ago and were a critical innovation that led to extensive angiosperm diversification. Evolutionary botanist Peter Endress has studied flower evolution by comparing the structures of modern and fossil flowers. His studies suggest that the earliest flowers were quite small—only about 1 mm in diameter—and had few parts, though both stamens and carpels were present. Comparative studies also suggest how modern stamens and carpels might have arisen. Early fossil flowers and some modern flowers have broad stamens that are leaf-shaped, with elongated, pollen-producing microsporangia on the stamen surface (**Figure 31.15a**). In contrast, the stamens of most modern plants have become narrowed to form **filaments**, or stalks, that elevate **anthers**, clusters of microsporangia that produce pollen and then open to release it (see Figure 31.13). Filaments and anthers are thus adaptations that foster pollen dispersal. Plant biologists hypothesize that

carpels evolved from leaflike structures bearing ovules on their surfaces and that such leaves folded over ovules, protecting them. In support of this hypothesis is the fact that the carpels of some early diverging modern plants are leaflike structures that fold over ovules, with the carpel edges stuck together by secretions (**Figure 31.15b**). In contrast, most modern flowers produce carpels whose edges have fused together into a tube whose lower portion (ovary) encloses ovules. Plant biologists hypothesize that such evolutionary change increased ovule protection, which would improve plant fitness.

Flowering Plants Diversified into Several Lineages

Amborella trichopoda, a shrub that occurs in cloud forests on the South Pacific island of New Caledonia, is an example of a modern angiosperm whose flowers display hypothesized ancient features. For example, the fairly small flowers display stamens with broad filaments and have several separate carpels (**Figure 31.16**). Gene sequencing studies indicate that *A. trichopoda* is a very early diverging flowering plant, as are water lilies and the

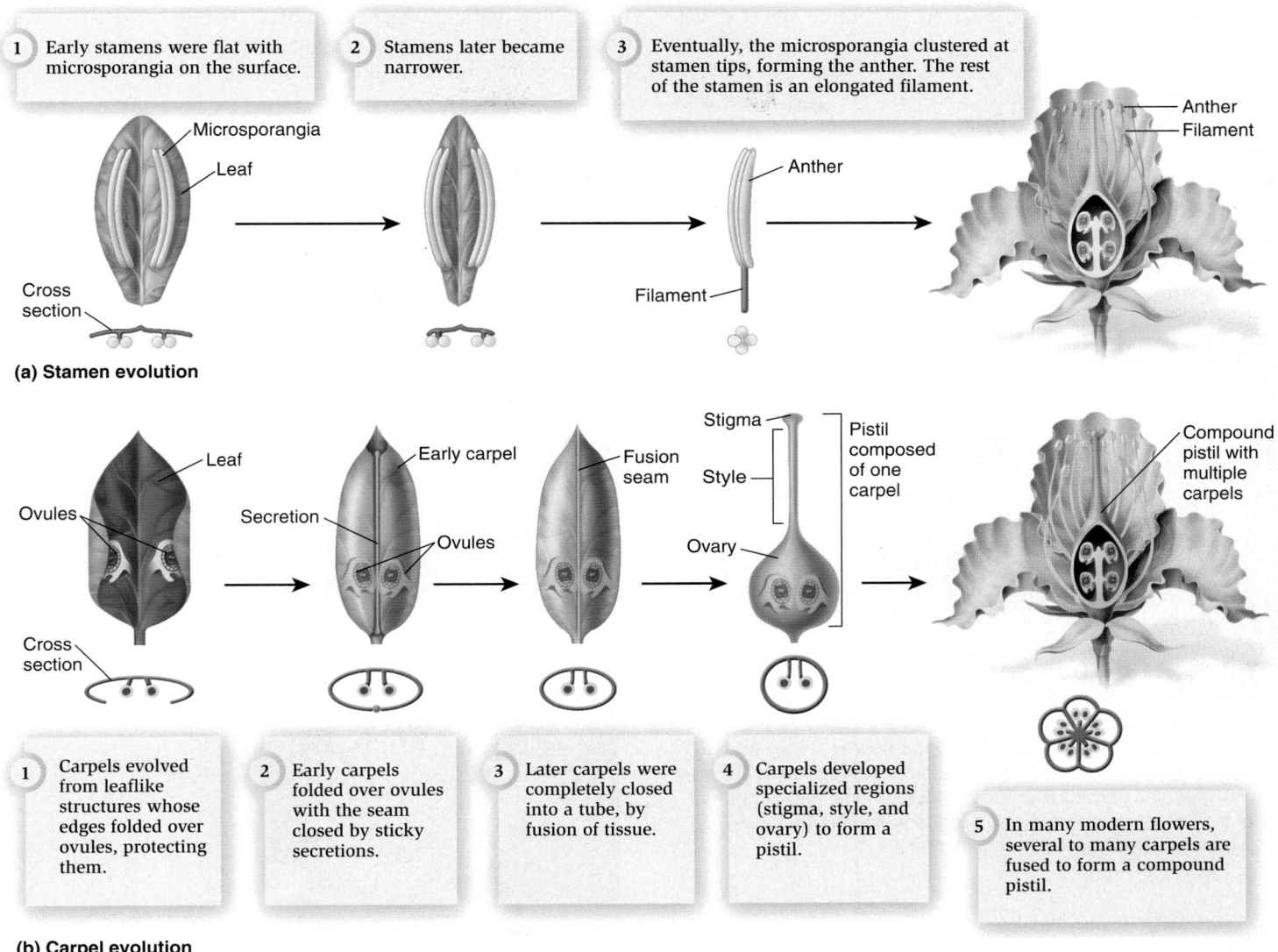

Figure 31.15 Hypothetical evolution of stamens, carpels, and pistil.

star anise plant and its close relatives (**Figure 31.17**). Magnoliids, represented by the genus *Magnolia*, are closely related to two very large and diverse angiosperm lineages, the **monocots** and the **eudicots**. Monocots and eudicots are named for differences in the number of leaves that occur on their embryos, located within seeds; monocot embryos possess one seed leaf, while eudicots possess two embryonic leaves. Monocots differ from eudicots in several additional ways (see Chapter 35). For example, monocots typically have flowers with parts numbering three or some multiple of three (**Figure 31.18a**). In contrast, eudicot flower parts often occur in fours, fives, or a multiple of four or five (**Figure 31.19a**).

Fossil pollen evidence has shown that monocots and eudicots diverged more than 120 million years ago. In 2004, plant evolutionary biologist Else Marie Friis and colleagues described monocot pollen grains from the 120-million-year-old fossil genus *Mayoa*. Evolutionary botanist Peter Crane and associates found

distinctive early eudicot fossil pollen in 125-million-year-old deposits. From this early start, monocots and eudicots have come to dominate modern plant communities.

Flowering Plant Lineages Display Horizontal Gene Transfer

Although the major lineages of flowering plants (see Figure 31.17) diverged more than a hundred million years ago, genes may sometimes be transferred between different plant species. You may recall that horizontal gene transfer is a common feature among prokaryotes (see Chapter 27), and that many protists obtained new genes from endosymbionts that evolved into mitochondria and chloroplasts (see Chapter 28). However, horizontal gene transfer was not known to occur among plants until 2003, when plant evolutionary biologists Ulfar Bergthorsson, Jeffrey Palmer, and their colleagues reported it in flowering plants.

These researchers discovered that mitochondrial genes encoding ribosomal and respiratory proteins move between flowering plants that are only distantly related. For example, the eudicot bloodroot plant (*Sanguinaria canadensis*) has at least one gene that is composed of parts typical of both monocots and eudicots. In this case, the monocot sequences were horizontally transferred to the eudicot. In another example, a gene encoding one of the proteins of ATPase was horizontally transferred from a eudicot to *Amborella trichopoda* (see Figure 31.17). These

surprising findings suggest that horizontal gene transfer might be an important factor in angiosperm diversification, as were changes in flower structure, our next topic.

Flowers Diversified by Undergoing Several Types of Changes

During the diversification of flowering plants, flowers evolved into many different types. Some of the processes by which flowers

Figure 31.16 *Amborella trichopoda* **flower.** This small flower is only about 3–4 mm in diameter. It displays several central, greenish carpels; nonfunctional stamens; and a pink perianth. This plant species also produces flowers that lack carpels but have many functional stamens.

Carpel

Perianth

Peduncle

Nonfunctional stamen

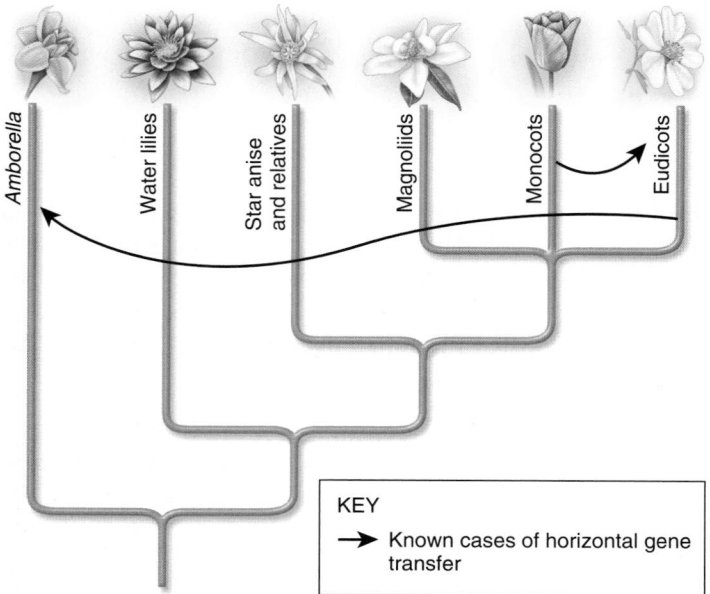

Figure 31.17 A phylogeny showing the major modern angiosperm lineages and some known cases of horizontal gene transfer.

KEY

→ Known cases of horizontal gene transfer

(a) *Lilium* spp.

(b) *Cattleya califlora*

Perianth parts

Fused pistil and stamens

Perianth parts

(c) *Triticum* spp.

Figure 31.18 **Representative monocots.** **(a)** Flowers and buds of lily (genus *Lilium*), displaying six perianth parts. **(b)** An orchid of the genus *Cattleya* has six perianth parts, one of which is specialized to form a lower lip. **(c)** Grass flowers of the genus *Triticum* lacking showy perianth.

evolved include differently shaped petals (**Figure 31.19b**), fusion of flower organs, changes in flower symmetry, loss of flower organs, change in organ position, and arrangement of flowers into bunches known as **inflorescences** (**Figure 31.19c**) These changes fostered the effective transfer of pollen by various means.

For example, during the evolution of both eudicots and monocots, the petals of some flowers fused into floral tubes. Such tubes hold sugar-rich **nectar** as a reward for **pollinators**, animals that carry pollen between flowers. The diameters of floral tubes vary among flowers and are evolutionarily tuned to the feeding structures of diverse animals, which range from the narrow tongues of butterflies to the wider bills of nectar-feeding birds (**Figure 31.20**). Nectar-feeding bats stick their heads into even larger tubular flowers to lap up nectar with their tongues. Floral tubes have thus allowed plants to diversify in association with pollinators, a process that we will describe later in this chapter.

(a) *Malus* spp. (b) *Pisum* spp. (c) *Helianthus* spp.

Figure 31.19 **Representative eudicots.** **(a)** A flower and buds of apple (genus *Malus*), showing five flower petals. Sepals are visible around the pink buds. **(b)** A flower of a pea plant (genus *Pisum*), illustrating differently shaped petals, which favor bee pollinators. **(c)** An inflorescence (head) of sunflower (genus *Helianthus*). This inflorescence includes a rim of flowers with conspicuous petals that attract pollinators and an inner disk of flowers that lack attractive perianths.

(a) Zinnia and butterfly (b) Hibiscus and hummingbird (c) Saguaro cactus and bat

Figure 31.20 **Floral tubes and coevolved pollinators.** **(a)** This zinnia is a type of inflorescence known as a head, which is composed of an outer rim of showy flowers and a central disk of narrow tubular flowers that produce nectar. Butterflies, but not other pollinators, are able to reach the nectar by means of narrow tongues. **(b)** The hibiscus flower produces nectar in a floral tube whose diameter corresponds to the dimensions of a hummingbird bill. **(c)** The saguaro cactus (*Carnegiea gigantea*) flower forms a floral tube that is wide enough for nectar-feeding bats to get their heads inside.

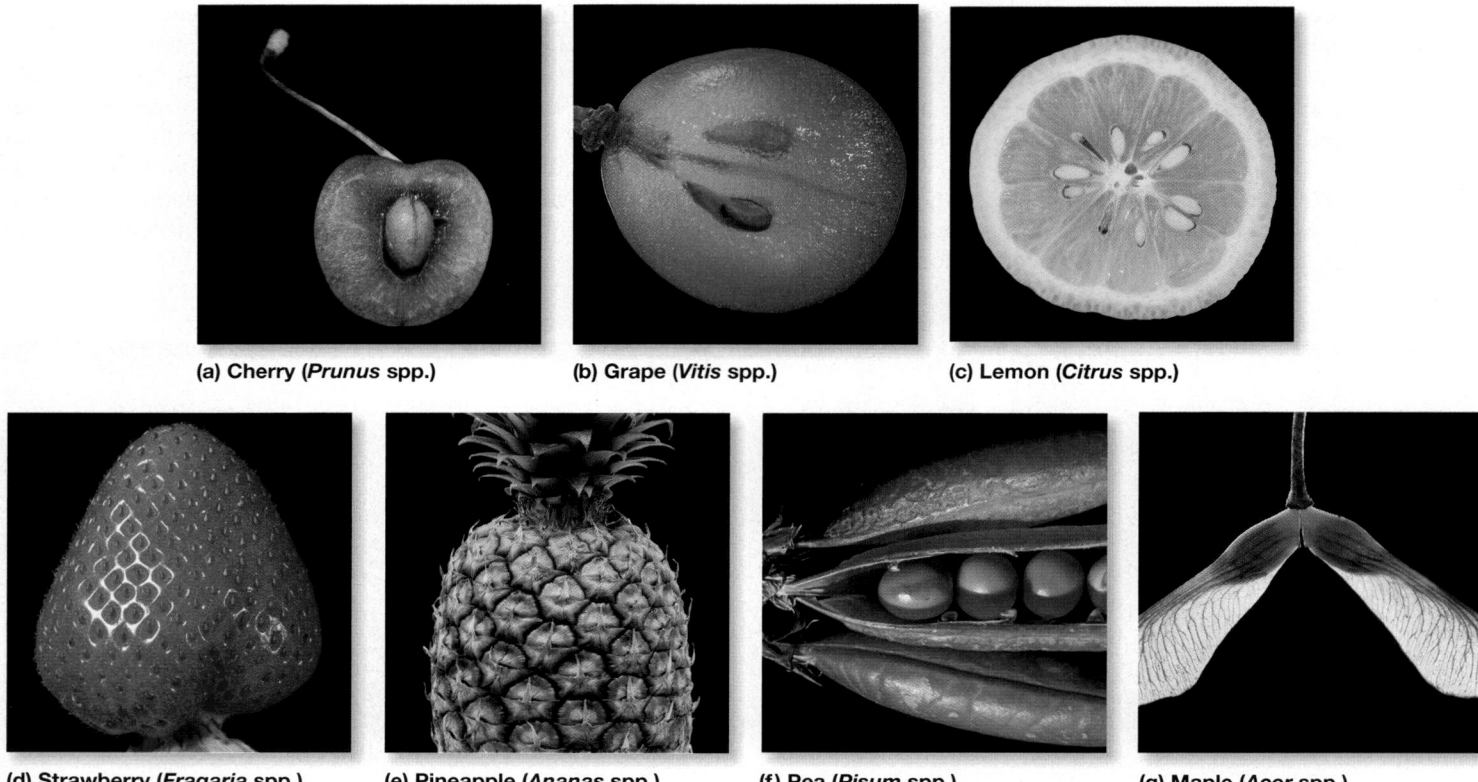

(a) Cherry (*Prunus* spp.)   **(b) Grape (*Vitis* spp.)**   **(c) Lemon (*Citrus* spp.)**

(d) Strawberry (*Fragaria* spp.)   **(e) Pineapple (*Ananas* spp.)**   **(f) Pea (*Pisum* spp.)**   **(g) Maple (*Acer* spp.)**

Figure 31.21   **Representative fruit types.**   **(a–c)** Cherry, grape, and lemon are fleshy, sweet fruits adapted to attract animals that consume the fruits and excrete the seeds. **(d)** Strawberry is an aggregate fruit, consisting of many tiny, single-seeded fruits produced by a single flower. The fruits are embedded in the surface of a fleshy receptacle that is adapted to attract animal seed dispersal agents. **(e)** Pineapple is a large multiple fruit formed by the aggregation of smaller fruits, each produced by one of the flowers in an inflorescence. **(f)** Peas produce legumes, fruits that open on two sides to release seeds. **(g)** Maple trees produce dry fruits with wings adapted for wind dispersal.

Flowers have also diversified by fusing different types of flower organs together, and orchids provide an interesting example. Orchid stamens and carpels are fused together into a single reproductive column that is surrounded by attractive petals and sepals (see Figure 31.18b). This arrangement of flower organs fosters orchid pollination by particular insects and is a distinctive feature of the orchid family. The sunflower family features unique inflorescences, groups of small flowers gathered into a head (see Figure 31.19c). Heads allow pollinators to transfer pollen among a large number of flowers at the same time (see Figure 31.20a). The grass family features flowers having few or no perianths, explaining why grass flowers are not showy (see Figure 31.18c). This adaptation fosters wind pollination, because petals would only get in the way of pollen transfer by wind. Diversification of fruit types, which we will consider next, has likewise contributed to the vast number of modern flowering plants.

Diverse Types of Fruits Function in Seed Dispersal

Fruits are structures that develop from ovary walls in diverse ways that aid the dispersal of enclosed seeds. Seed dispersal helps to prevent seedlings from competing with their larger par-

ents for scarce resources such as water and light. Dispersal of seeds also allows plants to colonize new habitats. Diverse fruit types illustrate many ways in which plants have become adapted for effective seed dispersal. Like flower types, fruit types are useful in classifying and identifying angiosperms.

Many mature angiosperm fruits, such as cherries, grapes, and citrus, are attractively colored, soft, juicy, and tasty (**Figure 31.21a–c**). Such fruits are adapted to attract animals that consume the fruits, digest the outer portion as food, and eliminate the seeds, thereby dispersing them. Hard seed coats prevent such seeds from being destroyed by the animal's digestive system. Strawberries are actually an aggregation of many fruits that all develop from a single flower having multiple pistils (**Figure 31.21d**). The ovaries of these pistils develop into the tiny, single-seeded yellow fruits on a strawberry surface; the fleshy, red, sweet portion of a strawberry develops from the flower receptacle. The strawberry system of aggregate fruits allows a single animal consumer, such as a bird, to disperse many seeds at the same time.

Pineapples (**Figure 31.21e**) are juicy multiple fruits that develop when many ovaries of an inflorescence fuse together. Multiple fruits have the advantage of being larger than individual fruits of the same species. Larger fruits attract relatively large animals that have the ability to disperse seeds for long distances.

Mulberries and figs are additional examples of multiple fruits that provide similar benefits.

The plant family informally known as **legumes** is named for its distinctive fruits, dry pods that open down both sides when seeds are mature, thereby releasing them (**Figure 31.21f**). Nuts and grains are additional examples of dry fruits, which have the advantage of rotting less rapidly than juicy fruits. **Grains** are the characteristic single-seeded fruits of cereal grasses such as rice, corn (maize), barley, and wheat. Maple trees produce dry and thus lightweight fruits having wings, features that foster effective wind dispersal (**Figure 31.21g**). Other plants produce dry fruits having surface burrs that attach to animal fur or possess floating fruits that disperse in water. Thus, flowering plants have diverse mechanisms for dispersing their seeds.

Angiosperms Produce Diverse Secondary Metabolites That Play Important Roles in Structure, Reproduction, and Protection

As we discussed in Chapter 7, secondary metabolism involves the synthesis of molecules that are not essential for cell structure and growth. These molecules, called **secondary metabolites**, are produced by various prokaryotes, fungi, protists, and plants but are most diverse in the angiosperms. About 100,000 different types of secondary metabolites are known, most of these produced by flowering plants. Because secondary metabolites play essential roles in plant structure, reproduction, and protection, diversification of these compounds has influenced flowering plant evolution. Three major classes of plant secondary metabolites occur: (1) terpenes and terpenoids; (2) phenolics, which include flavonoids and related compounds; and (3) alkaloids (**Figure 31.22**).

About 25,000 types of plant terpenes and terpenoids are derived from units of the hydrocarbon gas isoprene. Taxol, widely used in the treatment of cancer, is a terpene, as are citronella and a variety of other compounds that repel insects. Rubber, turpentine, rosin, and amber are complex terpenoids that likewise serve important roles in plant biology as well as having useful human applications.

Phenolic compounds are responsible for some flower and fruit colors as well as the distinctive flavors of cinnamon, nutmeg, ginger, cloves, chilies, and vanilla. Phenolics absorb ultraviolet radiation, thereby preventing damage to cellular DNA. They also help to defend plants against insects and disease microbes. Some phenolic compounds found in tea, red wine, grape juice, and blueberries have antioxidant properties that prevent the formation of damaging free radicals.

Alkaloids are nitrogen-containing secondary metabolites that often have potent effects on the animal nervous system. Plants produce at least 12,000 types of alkaloids, and some plants produce many alkaloids. Caffeine, nicotine, morphine, ephedrine, cocaine, heroin, and codeine are examples of alkaloids that influence the physiology and behavior of humans and are thus of societal concern. Like flower and fruit structure, secondary metabolites are useful in distinguishing among Earth's hundreds of thousands of flowering plant species.

(a) Natural rubber produced by *Hevea brasiliensis* is an example of a complex terpene.

(b) Natural vanillin, extracted from the orchid *Vanilla planifolia*, is an example of a phenolic compound.

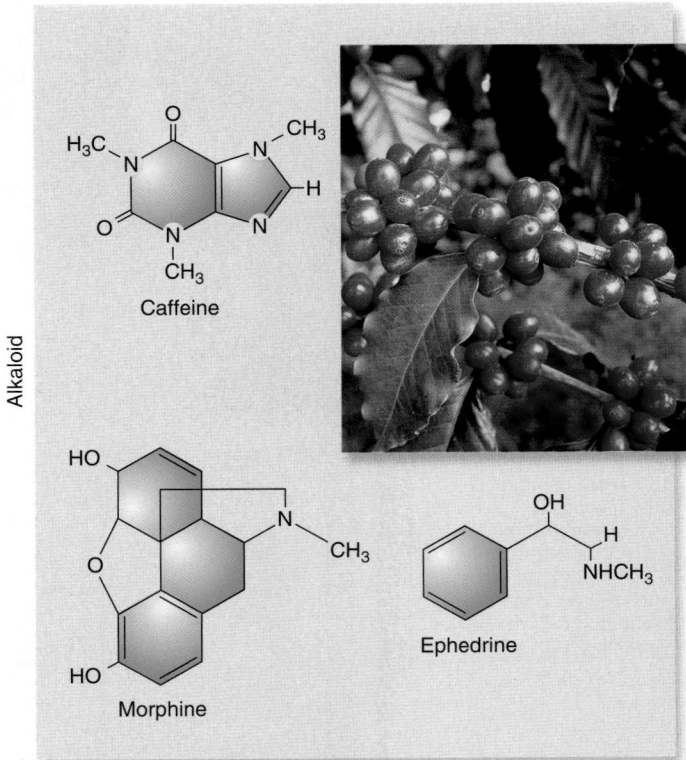

(c) Alkaloids include caffeine produced by *Coffea arabica*, morphine, and ephedrine.

Figure 31.22   Major types of plant secondary metabolites.

FEATURE INVESTIGATION

Hillig and Mahlberg Used Secondary Metabolites to Explore Species Diversification in the Genus *Cannabis*

The genus *Cannabis* has long been a source of hemp fiber used for ropes and fabric. People have also used *Cannabis* in traditional medicine and as a hallucinogenic drug. *Cannabis* produces THC (tetrahydrocannabinol), a secondary metabolite called a cannabinoid. THC mimics compounds known as endocannabinoids, which are naturally produced and act in the brain and elsewhere in the body. THC affects humans by binding to receptor proteins in plasma membranes in the same way as natural endocannabinoids. Today, cancer patients sometimes choose to use cannabis to stimulate their appetite, which can decline as a side effect of cancer treatment. Cannabinoids are produced in glandular hairs that cover most of the *Cannabis* plant's surface but are particularly rich in leaves located near the flowers.

Because humans have subjected cultivated *Cannabis* plants to artificial selection for so long, plant biologists have been uncertain how many species of *Cannabis* existed in cultivation and in the wild. In the past, plants cultivated for drug production were often identified as *Cannabis indica*, whereas those grown for hemp were typically known as *Cannabis sativa*. However, these species are difficult to distinguish on the basis of structural features, and the relevance of these names to wild cannabis was unknown. At the same time, species identification has become important for biodiversity studies, agriculture, and law enforcement. For these reasons, plant biologists Karl Hillig and Paul Mahlberg investigated the possibility that ratios of THC to another cannabinoid known as CBD (cannabidiol) might aid in defining *Cannabis* species and identifying plant samples to the species level, as shown in the Feature Investigation in **Figure 31.23**. They began by collecting *Cannabis* fruits from nearly a hundred diverse locations around the world and then growing

Figure 31.23   Hillig and Mahlberg's analysis of secondary metabolites in *Cannabis* spp.

GOAL   To determine if cannabinoids aid in distinguishing *Cannabis* species.

STARTING MATERIALS   Cannabis fruits obtained from nearly 100 different sources, worldwide.

	Experimental level	Conceptual level
1 Grow replicate *Cannabis* plants from seeds under standard conditions in a greenhouse.		Eliminates differential environmental effects on cannabinoid content.
2 Extract cannabinoids from leaves surrounding flowers.		Extracts were made from tissues richest in cannabinoids; reduces the chance that cannabinoids present in lower levels would be missed.
3 Analyze cannabinoids by gas-liquid chromatography. Determine ratios of THC (tetrahydrocannabinol) to CBD (cannabidiol) in about 200 *Cannabis* plants.		Previous data suggested that ratios of THC to CBD might be different in separate species.

4 **THE DATA**

Cannabis plants obtained from diverse sources worldwide formed 2 groups—those having relatively high THC to CBD ratios and those having lower THC to CBD ratios.

Plants having high THC to CBD ratios, often used as drug sources, corresponded to the species *C. indica*.

Plants having lower THC to CBD ratios, often used as hemp fiber sources, corresponded to the species *C. sativa*.

Cannabinoids support a concept of 2 *Cannabis* species.

these plants from seed under uniform conditions in a greenhouse. The investigators next extracted cannabinoids, analyzed them by means of gas-liquid chromatography, and determined the ratios of THC to CBD. The results, published in 2004, suggested that the wild and cultivated *Cannabis* samples evaluated in this study could be classified into two species: *C. sativa*, dis-playing relatively low THC levels, and *C. indica*, having relatively high THC levels. As a result of this work, ecologists, agricultural scientists, and forensic scientists can reliably use ratios of THC to CBD to classify samples. Similar studies of plant secondary metabolites offer the benefit of uncovering potential new medicinal compounds or other applications of significance to humans.

31.3 The Role of Coevolution in Angiosperm Diversification

In the previous section, we learned that flowering plants are commonly associated with animals in ways that strongly influence plant evolution. As a matter of fact, plants have likewise influenced animal evolution, in a diversity-generating process known as coevolution. Generally speaking, **coevolution** is the process by which two or more species of organisms influence each other's evolutionary pathway. During the diversification of flowering plants, coevolution with animals has been a major evolutionary force. Coevolution explains the diverse forms of most flowers and many fruits, and how plants accomplish effective pollen and seed dispersal. Human attraction to flowers and fruit also reflects coevolutionary associations. This is because human sensory systems are similar to those of various animals that have coevolved with angiosperms.

Pollination Coevolution Influences the Diversification of Flowers and Animals

Animal pollinators transfer pollen from the anthers of one flower to the stigmas of other flowers of the same species. Pollinators thereby foster genetic variability and plant potential for evolutionary change. Insects, birds, bats, and other pollinators learn the characteristics of particular flowers, visiting them preferentially. This animal behavior, known as constancy or fidelity, increases the odds that a flower stigma will receive pollen of the appropriate species. Animal pollinators thus offer precision of pollen transfer, which reduces the amount of pollen that plants must produce to achieve pollination. By contrast, wind-pollinated plants must produce much larger amounts of pollen because windblown pollen reaches appropriate flowers by chance.

Flowers attract the most appropriate pollinators by means of attractive colors, odors, shapes, and sizes. Secondary metabolites influence the colors and odors of many flowers. For example, flavonoids color many blue, purple, or pink flowers. More than 700 types of chemical compounds contribute to floral odors. Most flowers reward pollinators with appropriate food: sugar-rich nectar, lipid- and protein-rich pollen, or both. Thus, flowering plants provide an important biological service, providing food for many types of pollinator animals. But some flowers "trick" pollinators into visiting or trap pollinators temporarily, thereby achieving pollination without actually rewarding the pollinator. Examples include flowers that look and smell like dead meat, thereby attracting flies, which are fooled but accomplish pollination anyway.

Although many flowers are pollinated by a variety of animals, others have flowers that have become specialized for particular pollinators, and vice versa. These specializations, which have resulted from coevolution, are known as **pollination syndromes** (**Table 31.2**). For example, odorless red flowers, such as those of hibiscus (see Figure 31.20b), are attractive to birds, which can see the color red but lack a sense of smell. By contrast, bees are not typically attracted to red flowers because bee vision does not extend to red. Rather, bees are attracted to blue, purple, yellow, and white flowers having sweet odors. If you are allergic to bee stings or just want to reduce the possibility of being stung, do not dress in bee-attracting flower colors or wear fragrant perfumes when in locales frequented by bees. People who work in remote regions where bees occur and where medical care may not be readily available may choose to wear khaki or other colors that do not attract bees.

Table 31.2	Pollination Syndromes
Animal features	**Coevolved flower features**
Bees	
Color vision includes UV, not red	Often blue, purple, yellow, white (not red) colors
Good sense of smell	Fragrant
Require nectar and pollen	Provide nectar and abundant pollen
Butterflies	
Good color vision	Blue, purple, deep pink, orange, red colors
Sense odors with feet	Light floral scent
Need landing place	Provide landing place
Feed with long tubular tongue	Nectar in deep, narrow floral tubes
Moths	
Active at night	Open at night; white or bright colors
Good sense of smell	Heavy, musky odors
Feed with long, thin tongue	Nectar in deep, narrow floral tubes
Birds	
Color vision, includes red	Often colored red
Often require perch	Strong, damage-resistant structure
Poor sense of smell	No fragrance
Feed in daytime	Open in daytime
High nectar requirement	Copious nectar in floral tubes
Hover (hummingbirds)	Pendulous (dangling) flowers
Bats	
Color blind	Light, reflective colors
Good sense of smell	Strong odors that attract bats
Active at night	Open at night
High food requirements	Copious nectar and pollen provided
Navigate by echolocation	Pendulous or borne on tree trunks

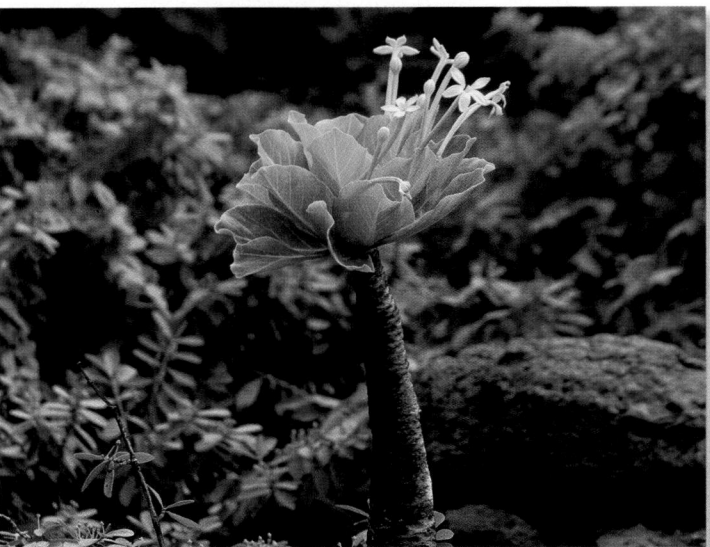

Figure 31.24 *Brighamia insignis*, an endangered plant. The pollinator that coevolved with *B. insignis* has become extinct, with the result that the plant is unable to produce seed unless artificially pollinated by humans.

Biological inquiry: What kind of animal likely pollinated B. insignis?

Pollination syndromes are also of practical importance in agriculture and in conservation biology. Fruit growers often import bees to pollinate flowers of fruit crops, thus increasing crop yields. Some plants have become so specialized to particular pollinators that if the pollinator becomes extinct, the plant becomes endangered. An example is the Hawaiian cliff-dwelling *Brighamia insignis* (**Figure 31.24**), whose presumed moth pollinator has become extinct. Humans that hand-pollinate *B. insignis* are all that stand between this plant and extinction.

Seed-Dispersal Coevolution Influences the Evolution of Fruits and Particular Animals

As in the case of pollination, coevolution between plants and their animal seed-dispersal agents has influenced both plant fruit characteristics and those of seed-dispersing animals. In addition, flowering plant fruits provide food for animals, an important biological service. For example, many of the plants of temperate forests produce fruits that are attractive to resident birds. Such juicy, sweet fruits have small seeds that readily pass

through bird guts. Many plants signal fruit ripeness by undergoing color changes from unripe green fruits to red, orange, yellow, blue, or black (**Figure 31.25**). Because birds have good color vision, they are able to detect the presence of ripe fruits and consume them before the fruits drop from plants and rot. Apples, strawberries, cherries, blueberries, and blackberries are examples of fruits whose seed dispersal adaptations have made them attractive food for humans as well. By contrast, the lipid-rich fruits of Virginia creeper (*Parthenocissus quinquefolia*) and some other autumn-fruiting plants energize migratory birds but are not tasty to humans. The Virginia creeper's leaves often turn fall colors earlier than surrounding plants, thereby signaling the availability of nutritious, ripe fruit to high-flying birds. It is important to such plants that lipid-rich fruits be consumed expeditiously because they rot easily, in which case seed dispersal cannot occur.

Primates of tropical Asia and Africa, but not generally those of Central and South America, have **trichromatic color vision**—the ability to distinguish blue, green, and red colors. Trichromatic color vision in Old World primates is the result of a gene duplication event that occurred early in their evolution, but after their divergence from New World primates. Primate specialists have speculated that trichromatic vision may enable Old World tree-dwelling primates to better detect the young, tender leaves of tropical trees, which are often slightly redder in pigmentation than mature leaves. Trichromatic vision also allows primates to more easily detect yellow, red, or orange ripe fruits against a background of leafy green foliage. Since humans evolved from primates native to Africa, we inherited their trichromatic color vision, helping to explain why humans find brightly colored flowers and fruits so attractive.

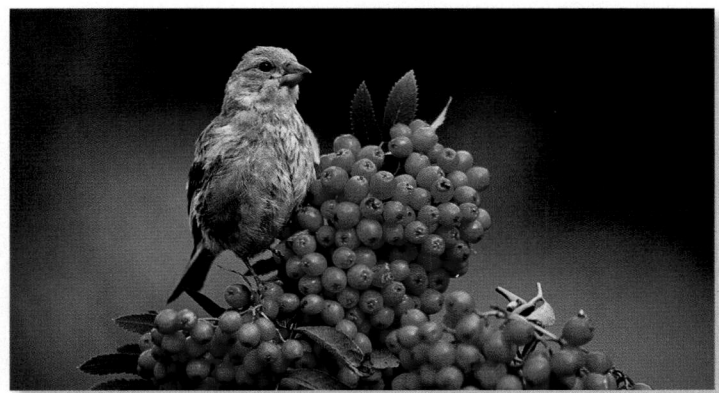

Figure 31.25 **Fruits attractive to animal seed dispersal agents.** Color and odor signals alert coevolved animal species that fruits are ripe, thus favoring the dispersal of mature seeds.

Figure 31.26 **Ears and grains of modern corn and its ancestor, teosinte.** Domesticated corn ears are larger than those of the ancestral grass teosinte. Corn fruits are softer and more edible than are grains of teosinte.

Biological inquiry: In what other way do corn ears differ from those of teosinte?

31.4 Human Influences on Angiosperm Diversification

By means of the process known as **domestication**, which involves artificial selection for traits desirable to humans, ancient humans transformed wild plant species into new crop species. More recently, human populations have increased so much that larger areas of natural habitat are being transformed for human use. Deforestation, for example, often results from the conversion of forests to agricultural land. Such habitat destruction is a leading cause of the extinction of plant and other species. In this section, we will consider how humans have produced new crop species and influenced the loss of wild species.

Humans Produced New Crop Plants by Domesticating Wild Species

Between 10,000 and 5,000 years ago, agriculture originated independently in eight different locations around the world. One of the earliest domesticated crops was an African plant commonly known as the bottle gourd (*Lagenaria siceraria*). The bottle gourd was grown for use as containers, musical instruments, and floats for fishing. For planting crops, humans selected seeds from gourds that had thicker rinds because these resisted breakage better than wild gourds. These differences in rind thickness can be detected in fossil remains that are about 10,000 years old. The fossils indicate that bottle gourds were grown as a crop in Asia and from there were transported to the Americas by ancient human colonists.

Cultivated bread wheat (*Triticum aestivum*) was probably among the earliest food crops, having originated more than 8,000 years ago, in what is now southeastern Turkey and northern Syria. Bread wheat originated by a series of steps from wild ancestors (*Triticum boeoticum* and *Triticum dicoccoides*). Among the earliest changes that occurred during wheat domestication was the loss of **shattering**, the process by which ears of wild wheat break apart and disperse the single-seeded fruits known

as grains. A mutation probably caused the ears of some wheat plants to remain intact, a trait that is disadvantageous in nature but beneficial to humans. Nonshattering ears would have been easier for humans to harvest than normal ears. Early farmers probably selected plants as seed stock having nonshattering ears and other favorable traits such as larger grains. These ancient human selection processes, together with modern breeding efforts, explain why cultivated wheat differs from its wild relatives in shattering and other properties. The accumulation of these trait differences explains why cultivated and wild wheat plants are classified as different species.

About 6,500 years ago, people living in what is now Mexico domesticated a native grass known as teosinte (*Zea* spp.), producing a new species, *Zea mays*, known as corn or maize. The evidence for this pivotal event includes ancient ears that were larger than wild ones and distinctive fossil pollen. Modern ears of corn are much larger than those of teosinte, corn grains are larger and softer, and modern corn ears do not shatter, as do those of ancestral teosinte (**Figure 31.26**). These and other trait changes reflect artificial selection accomplished by humans. An analysis of the corn genome, reported in 2005 by Stephen Wright, Brandon Gaut, and coworkers, suggests that human selection has influenced about 1,200 genes.

Molecular analyses indicate that domesticated rice (*Oryza sativa*) originated from ancestral wild species of grasses (*Oryza nivara* and/or *Oryza rifipogon*). As in the cases of wheat and corn, domestication of rice involved loss of ear shattering. In 2006, Changbao Li, Ailing Zhou, and Tao Sang reported that a key amino acid substitution was primarily responsible for the loss of ear shattering in rice. As a result of this mutation, rice ears remain intact. Ancient humans might have unconsciously selected for this mutation while gathering rice from wild populations, because the mutants would not so easily have shed grains during the harvesting process. Eventually, the nonshattering mutant became a widely planted crop throughout Asia, and today it is the food staple for millions of people.

Although humans generated these and other new plant species, in modern times humans have caused the extinction of plants and other species as the result of habitat destruction. Protecting biodiversity will continue to challenge humans as populations and demands on the Earth's resources increase. Techniques such as the use of molecular barcodes have helped in the process of identifying and cataloging Earth's biodiversity, a process that aids in habitat conservation and restoration efforts.

GENOMES & PROTEOMES

DNA Barcodes Help in Identifying and Cataloging Plants

In view of the fact that flowering plants provide many essential biological services, ecologists have become concerned about current and future extinctions caused by human alterations of the environment. Plant extinction is a tragedy because so many plants produce useful secondary metabolites (including those having medicinal value) or serve as important food sources for other organisms. Botanists have responded by attempting to catalog the world's 300,000 species of plants, most of which are angiosperms. A globally useful cataloging system can help biologists keep track of all these plants, identify new species, and spot threats to plant biodiversity. One such system is **molecular barcodes**, DNA sequences that are used to identify and catalog Earth's biodiversity.

Sequences of a short region of the mitochondrial DNA have proved useful in barcoding animal species. Because each species has a distinctive genetic code for this region, species identifications can be made from tissue samples. Biologists accomplish this by extracting DNA from the samples and using polymerase chain reaction (PCR) techniques to make many copies of the barcode region, which can then be sequenced (see Chapter 20). Biologists then compare barcode sequences to those of other organisms in genetic databases, using computer methods to determine the specimen's closest relatives. Future miniaturization of the DNA processing and sequencing steps could allow field biologists to transmit information to databases and receive results, thereby identifying known or new organisms without removing them from their natural habitats.

Barcoding plant species has been a problem, because plant mitochondrial DNA lacks sufficient variation to distinguish closely related plant species. In other words, two closely related plant species may have identical mitochondrial gene sequences. For this reason, plant scientist W. John Kress and colleagues looked for a plastid DNA sequence that would parallel the use of the mitochondrial barcode in animals. By checking the variability of several plastid genes from a wide range of angiosperm species, including pairs of close relatives, these investigators identified a specific plastid gene for cataloging flowering plants. This sequence region can be easily amplified, even from dried specimens as old as 20 years. This work, reported in 2005, provides the information needed to barcode even larger arrays of plants, such as the flowering plants of entire countries. Such extensive barcoding efforts are expected to foster conservation and restoration projects, thereby helping to protect Earth's biodiversity.

CHAPTER SUMMARY

31.1 The Diversity of Modern Gymnosperms

- The seed plants inherited features of ancestral nonseed plants but display distinctive adaptations. The major phyla of seed plants mainly differ in reproductive features. (Figure 31.1)

- Gymnosperms are plants that produce exposed seeds rather than seeds enclosed in fruits. Many gymnosperms produce wood by means of a special tissue called vascular cambium.

- Several phyla of gymnosperms once existed but have become extinct and are known only from fossils. The diversity of modern gymnosperms includes four modern phyla: cycads, *Ginkgo biloba*, the conifers, and the gnetophytes. (Figure 31.2, Table 31.1)

- Cycads primarily live in tropical and subtropical regions. Features of cycads include palmlike leaves, nonwoody stems, coralloid roots with cyanobacterial endosymbionts, toxins, and large conelike seed-producing structures. (Figures 31.3, 31.4)

- *Ginkgo biloba* is the last surviving species of a phylum that was diverse during the Age of Dinosaurs. (Figure 31.5)

- Conifers have been widespread and diverse members of plant communities for the past 300 million years, and they are important sources of wood and paper pulp. Reproduction involves simple pollen cones and complex ovule-producing cones. Conifer wood contains water-conducting tracheids with thin-walled pits that allow water to move from one tracheid to another. Many conifers display additional adaptations that help them to survive in cold climates. (Figures 31.6, 31.7, 31.8, 31.9, 31.10)

- The gnetophytes are of particular interest to plant evolutionary biologists because their relationships to other seed plants are unclear. (Figure 31.11)

31.2 The Diversity of Modern Angiosperms

- Angiosperms inherited seeds and other features from ancestors but display distinctive features, such as flowers and fruits. Diversification of flower structure, fruit structure, and secondary metabolites has played important roles in angiosperm evolution. (Figure 31.12)

- Flowers foster seed production and are adapted in various ways that aid pollination in varying circumstances. The major flower organs are sepals, petals, stamens, and carpels, but some flowers lack one or more of these organs. Pollination is the transfer of pollen from a stamen to a pistil. Pistils display regions of specialized function:

the stigma is a receptive surface for pollen, pollen tubes grow through the style, and ovules develop within the ovary. If pollen tubes successfully deposit sperm near eggs in ovules, and fertilization occurs, ovules develop into seeds, and ovaries develop into fruits. (Figures 31.13, 31.14)

- Stamens and carpels may have evolved from leaflike structures bearing sporangia. The two largest lineages of flowering plants, the monocots and eudicots, diverged more than 120 million years ago. (Figures 31.15, 31.16, 31.17, 31.18, 31.19)

- Horizontal gene transfer occurs between modern monocots and eudicots, and between these plants and earlier-diverging lineages, with potential effects on evolution and diversification.

- Flower diversification involved evolutionary changes such as fusion of parts, changes in symmetry, loss of parts, and aggregation into inflorescences. (Figure 31.20)

- Fruits are structures that enclose seeds and aid in their dispersal. Fruits occur in many types that foster seed dispersal in varying circumstances. (Figure 31.21)

- Angiosperms produce three main groups of secondary metabolites: (1) terpenes and terpenoids; (2) phenolics, flavonoids, and related compounds; and (3) alkaloids. Secondary metabolites play essential roles in plant structure, reproduction, and defense. (Figure 31.22)

- Hillig and Mahlberg demonstrated the use of particular secondary metabolites in distinguishing species of the genus *Cannabis*. (Figure 31.23)

31.3 The Role of Coevolution in Angiosperm Diversification

- Coevolutionary interactions with animals that serve as pollen- and seed-dispersal agents played a powerful role in the diversification of both flowering plants and animals. (Figures 31.24, 31.25)

- Some flowers form specialized relationships with specific pollinators. These interdependent relationships are known as pollination syndromes. (Table 31.2)

- Human appreciation of flowers and fruits is based on sensory systems similar to those present in the animals with which angiosperms coevolved.

31.4 Human Influences on Angiosperm Diversification

- Humans have produced new crop species by domesticating wild plants. The process of domestication involved artificial selection for traits such as nonshattering ears of wheat, corn, and rice. (Figure 31.26)

- Large human populations are causing habitat destruction leading to species extinction. Genetic barcodes allow biologists to identify and catalog plants, thus aiding biodiversity conservation and habitat restoration efforts.

TEST YOURSELF

1. What feature must be present for a plant to produce wood?
   a. a type of conducting system in which vascular bundles occur in a ring around pith
   b. a eustele
   c. a vascular cambium
   d. all of the above
   e. none of the above

2. What is the correct order of evolution for these critical adaptations?
   a. embryos, vascular tissue, wood, seeds, flowers
   b. vascular tissue, embryos, wood, flowers, seeds
   c. vascular tissue, wood, seeds, embryos, flowers
   d. wood, seeds, embryos, flowers, vascular tissue
   e. seeds, vascular tissue, wood, embryos, flowers

3. How long have gymnosperms been important members of plant communities?
   a. 10,000 years, since the dawn of agriculture
   b. 100,000 years
   c. 300,000 years
   d. 65 million years, since the K/T event
   e. 300 million years, since the Coal Age

4. What similar features do gnetophytes and angiosperms possess that differ from other modern seed plants?
   a. Gnetophytes and angiosperms both produce flagellate sperm.
   b. Gnetophytes and angiosperms both produce flowers.
   c. Gnetophytes and angiosperms both produce tracheids, but not vessels, in their vascular tissues.
   d. Gnetophytes and angiosperms both produce fruits.
   e. None of the above.

5. Which part of a flower receives pollen from the wind or a pollinating animal?
   a. perianth
   b. stigma
   c. filament
   d. peduncle
   e. ovary

6. The primary function of a fruit is to
   a. provide food for the developing seed.
   b. provide food for the developing seedling.
   c. disperse pollen.
   d. disperse seeds.
   e. none of the above.

7. What are some ways in which flowers have diversified?
   a. color
   b. symmetry
   c. fusion of organs
   d. aggregation into inflorescences
   e. all of the above

8. Flowers of the genus *Fuchsia* produce deep pink to red flowers that dangle from plants, produce nectar in floral tubes, and have no scent. Based on these features, which animal is most likely to be a coevolved pollinator?
   a. bee
   b. bat
   c. hummingbird
   d. butterfly
   e. moth

9. Which type of plant secondary metabolite is best known for antioxidant properties of human foods such as blueberries, tea, and grape juice?
   a. alkaloids
   b. cannabinoids
   c. carotenoids
   d. phenolics
   e. terpenoids

10. What features of domesticated grain crops might differ from those of wild ancestors?
    a. the degree to which ears shatter, allowing for seed dispersal
    b. grain size
    c. number of grains per ear
    d. softness and edibility of grains
    e. all of the above

CONCEPTUAL QUESTIONS

1. Explain why humans should not consume food products made from cycads.
2. Explain why fruits such as apples, strawberries, and cherries are attractive and harmless foods for humans.
3. Is a sunflower really a flower?

EXPERIMENTAL QUESTIONS

1. Why did the investigators in the Feature Investigation in Figure 31.23 obtain nearly a hundred *Cannabis* fruit samples from around the world?
2. Why did Hillig and Mahlberg grow plants in a greenhouse before conducting the cannabinoid analysis?
3. Why did Hillig and Mahlberg collect samples from the leaves growing nearest the flowers?

COLLABORATIVE QUESTIONS

1. Where in the world would you have to travel to find wild plants representing all of the gymnosperm phyla, including the three types of gnetophytes?
2. How would you go about trying to solve what Darwin called an "abominable mystery," that is, the identity of the seed plant group that was ancestral to the flowering plants?

www.brookerbiology.com

This website includes answers to the Biological Inquiry questions found in the figure legends and all end-of-chapter questions.

32

An Introduction to Animal Diversity

Chapter Outline

There is a staggering variety of animal species on Earth, including thousands of vertebrates and hundreds of thousands of invertebrates.

Animals constitute a very species-rich kingdom. Well over a million species have been found and described by biologists, with many more species likely awaiting discovery and classification. Beyond being members of this kingdom ourselves, humans are dependent on animals. We eat many different kinds of animals, use a diverse array of animal products, and employ animals such as horses and oxen as a source of labor. We enjoy many animal species as companions and are dependent on other species to test lifesaving drugs. On the other hand, we are in competition with animals such as insects that threaten our food supply and are in turn parasitized by others. With such a huge number and diversity of existing animals, and with animals featuring so prominently in our lives, understanding animal diversity is of paramount importance. Researchers have spent great effort in determining the unique characteristics of different taxonomic groups and identifying their evolutionary relationships.

Since the time of Carl Linnaeus in the 1700s, scientists have classified animals based on their morphology, that is, on their physical structure. Then, as now, a lively debate has surrounded the question of what constitutes the "correct" animal phylogeny. In the 1990s, animal classifications based on similarities in DNA and rRNA became more common. Quite often, classifications based on morphology and those based on molecular data were similar, but some important differences arose. In this chapter, we will begin by defining the key characteristics of animals and then take a look at the major features of the animal body plans that form the basis of the traditional view of animal classification. We will explore how new molecular evidence has made significant alterations to this categorization of the animal kingdom and examine some of the similarities and differences between the morphological and molecular-based phylogenies. As more molecular-based evidence becomes available, systematists will likely continue to redraw the tree of animal life. Thus, as you read this chapter, keep in mind that animal classification is not set in stone but rather is a work in progress.

32.1 Characteristics of Animals

The Earth contains over a million known animal species living in environments from the deep sea to the desert and exhibiting an amazing array of characteristics. Most animals move and eat multicellular prey, and thus they are loosely differentiated from species in other kingdoms. However, coming up with a firm definition of an animal is tricky, because animals are so diverse that biologists can find exceptions to nearly any given characteristic. Even so, a number of key features exist that can help us broadly characterize the group we call animals (Table 32.1).

In brief, animals are multicellular heterotrophs that lack cell walls, and most have nerves, muscles, the capacity to move at some point in their life cycle, and the ability to reproduce sexually. Unlike plants, animals cannot synthesize their essential organic molecules from inorganic sources and so must feed on other organisms. Many, if not most, animals are capable of some type of movement or locomotion in order to acquire food. This ability has led to the development of specialized systems of sensory structures and a nervous system to coordinate movement and prey capture. Sessile species, such as barnacles, use bristled appendages to obtain food. In many such sessile species, while adults are immobile, the larvae can swim.

Table 32.1	Common Characteristics of Animals
Characteristic	**Example**
Multicellularity	Even relatively simple types of animals such as sponges are multicellular, in contrast to the single-celled eukaryotic microorganisms called protists (see Chapter 28).
Heterotrophs	Animals obtain their food by eating other organisms or their products. This contrasts with plants and algae, which are autotrophs and essentially make their own food.
No cell walls	While plant, fungal, and bacterial cells are rigid because they possess a cell wall, animal cells lack a cell wall and are quite flexible.
Nervous tissue	The presence of a nervous system in most animals enables them to respond rapidly to environmental stimuli.
Movement	Most animals have a muscle system, which, combined with a nervous system, allows them to move in their environment.
Sexual reproduction	Most animals reproduce sexually, with small, mobile sperm uniting with a much larger egg to form a fertilized egg or zygote.
Extracellular matrix	Proteins such as collagen bind cells together to give them added support and strength.
Special clusters of *Hox* genes	All animals possess *Hox* genes, which function in patterning the body axis (see Chapter 19).
Similar rRNA	Animals have very similar genes that encode for RNA of the small subunit (SSU) rRNA.
Characteristic cell junctions	Animals have characteristic cell junctions called anchoring, tight, and gap junctions.

The lack of a rigid cell wall and the existence of nerves and muscles facilitate movement but reduce structural support. Instead, animal cells exist in an extensive extracellular matrix that forms strong fibers outside of the cell (see Chapter 10). Additionally, a group of unique cell junctions, called anchoring, tight, and gap junctions, plays an important role in holding animal cells in place and allowing cell communication. Finally, most animals reproduce sexually, with sperm fusing directly with eggs.

While many animals live in aquatic systems, most are terrestrial and therefore face the problem of drying out, or desiccation. This problem necessitated the development of features that enabled animals to venture onto land, in much the same way as seeds and other adaptations permitted plants to colonize terrestrial habitats (see Chapter 30). Such features include internal fertilization and the development of a tough, protective shell around the eggs of many species.

32.2 Traditional Classification of Animals

Although animals constitute an extremely diverse kingdom, most biologists agree that the kingdom is monophyletic, meaning that all taxa have evolved from a single common ancestor. We can appreciate this by comparing a portion of the sequence of the small subunit (SSU) rRNA genes of a sponge, seagull, and flatworm (**Figure 32.1**), in much the same way as we did in Chapter 12 (refer back to Figure 12.19). The three animal genomes are very similar compared to that of a protist, such as a paramecium.

With the monophyletic nature of the animal kingdom in mind, scientists have attempted to characterize the organism from which animals most likely evolved. According to research, the most likely ancestor is a colonial flagellated protist that is likely related to the present-day protists known as choanoflagellates.

GAGGTTCGAAGACGATCAGATACCGTCGTAGTTCCGACCATAAACGATG — Sponge

GAGGTTCGAAGACGATCAGATACCGTCGTAGTTCTGACCATAAACGATG — Seagull

GAGGTTCGAAGACGATCAGATACCGTCGTAGTTCCAACCATAAACGATG — Flatworm

GGGGATCAAAGACGATCAGATACCGTCGTAGTCTTAACTATAAACTATA — Paramecium

Figure 32.1   Comparison of small subunit (SSU) rRNA gene sequences from three animals and a protist.   Note the similarities between the animals, even though they are very different species, and the differences with the protist. This and other comparative studies of gene sequences underscore the likelihood that animals share a common ancestor. Key: Yellow, identical in all four species; green, identical in two or three animal species; gray, dissimilar in one animal species; and red, dissimilar in the paramecium.

Choanoflagellates are tiny single-celled organisms, each with a single flagellum surrounded by a collar composed of cytoplasmic tentacles (refer back to Figure 28.15). A number of species are colonial, usually taking the form of a cluster of undifferentiated cells on a single stalk. Scientists think that some of these cells may have gradually taken on specialized functions, for example, movement or nutrition, while still maintaining coordination with other cells and cell types. Over time, such coordinated groups of cells evolved into a more complex organism that we characterize as an animal. Molecular data also point to choano-flagellates as the common animal ancestor.

Today about 35 animal phyla are recognized. At first glance, many of these phyla seem so distantly related to one another (for example, chordates and jellyfish) that making sense of this diversity with a classification scheme seems very challenging. However, over the course of centuries, scientists have come to some basic conclusions about the evolutionary relationships among animals. In this section, we explore the major features of animal body plans and development that are the basis of the traditional view of animal phylogeny.

The Traditional Classification of Animals Is Based on Body Plans

Without the availability of modern molecular techniques, biologists traditionally classified animal diversity in terms of four main morphological and developmental features of animal body plans listed here.

1. Presence or absence of different tissue types
2. Type of body symmetry
3. Presence or absence of a true body cavity
4. Patterns of embryonic development

We will discuss each of these major features of animal body plans next.

Tissues Collectively, animals are known as **Metazoa**. Animals can be divided into two subgroups based on whether or not they have specialized types of tissues, that is, stable associations of cells that have a similar structure and function. The **Parazoa** ("alongside animals") are not generally thought to possess specialized tissue types or organs, although they may have several distinct types of cells. Those cells can change their shape and location, making any associations temporary. The Parazoa consist of a single phylum, Porifera (sponges) (**Figure 32.2a**). In contrast, the **Eumetazoa** ("true animals") have more than one type of tissue and, for the most part, have different types of organs.

Symmetry The Eumetazoa are divided according to their type of symmetry. Symmetry refers to the existence of balanced proportions of the body on either side of a median plane. Radially symmetric animals, the **Radiata**, can be divided equally by any longitudinal plane passing through the central axis (**Figure 32.2b**). Such animals are often circular or tubular in shape, with a mouth at one end, and include the animals called cnidarians and ctenophores.

Bilaterally symmetric animals, the **Bilateria**, can be divided along a vertical plane at the midline to create two halves (**Figure 32.2c**). Thus, a bilateral animal has a left side and a right side, which are mirror images, as well as a **dorsal** (upper) and a **ventral** (lower) side, which are not identical, and an **anterior** (head) and a **posterior** (tail) end. Note that a radial animal only has an **oral** (top) and an **aboral** (bottom) side. Bilateral symmetry is strongly correlated with both the ability to move through the environment and **cephalization**, the localization of sensory structures at the anterior end of the body. Such abilities permitted animals to encounter their environment initially with their head, which is best equipped to detect and consume prey and in turn detect and respond to predators and other dangers.

Another key difference between the Radiata and Bilateria is that radial animals have two layers of embryonic cell layers,

(a) Parazoa: no tissues

(b) Eumetazoa: tissues
Radiata: radial symmetry

(c) Eumetazoa: tissues
Bilateria: bilateral symmetry

Figure 32.2 **Early divisions in the animal phylogeny.**   Animals can be categorized based on **(a)** the absence of different tissue types (Parazoa; the sponges) or **(b,c)** the presence of tissues (Eumetazoa; all other animals). Further categorization is based on the presence of (b) radial symmetry (Radiata; the cnidarians and ctenophores) or (c) bilateral symmetry (Bilateria; all other animals).

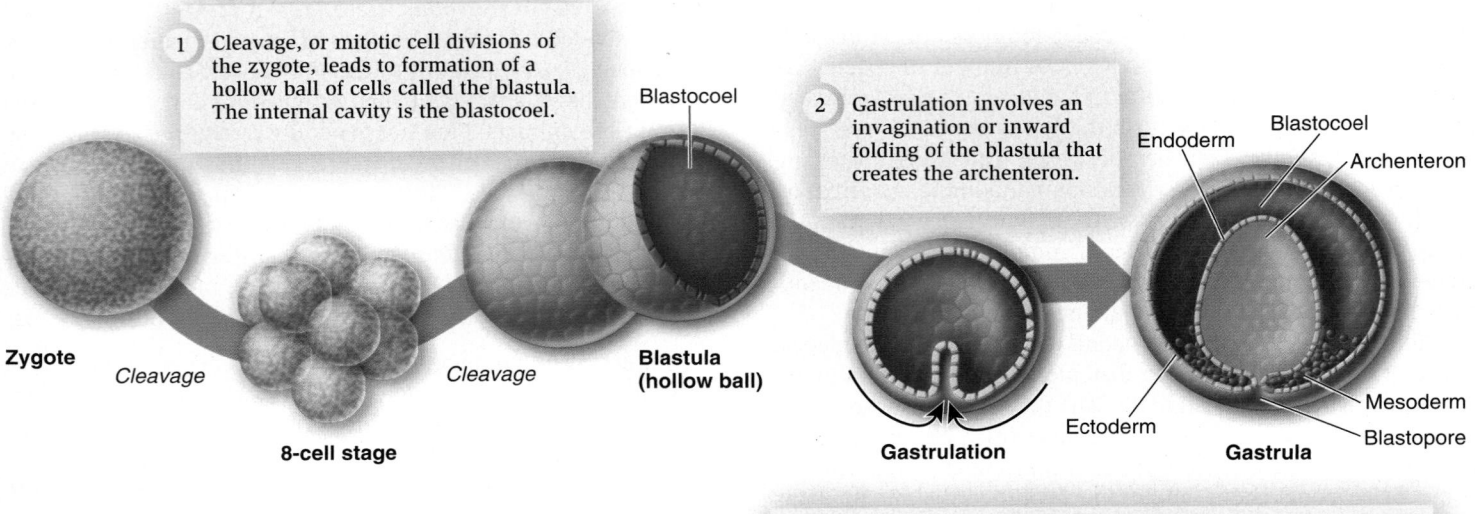

Figure 32.3 Embryonic development in animals.

called **germ layers**, while bilateral animals have three germ layers. In all animals except the sponges, the growing embryo develops different layers of cells through a process known as **gastrulation** to produce a **gastrula** (**Figure 32.3**). Fertilization of an egg by a sperm creates a diploid zygote. The zygote then undergoes **cleavage**, a succession of rapid cell divisions with no significant growth that produces a hollow sphere of cells called a **blastula**. In gastrulation, an area in the blastula invaginates and folds inward, creating in the process the primary germ layers. The inner layer of cells becomes the **endoderm**, which lines the **archenteron**, or primitive digestive tract. The outer layer, or **ectoderm**, covers the surface of the embryo and differentiates into the epidermis and nervous system.

The Bilateria develop a third layer of cells, termed the **mesoderm**, which develops between the ectoderm and endoderm. Mesoderm forms the muscles and most other organs between the digestive tract and the ectoderm. Because the Bilateria have these three distinct germ layers, they are often referred to as **triploblastic**, while the Radiata, which have only ectoderm and endoderm, are termed **diploblastic**.

Body Cavity The next three significant divisions in the classification of animals concern the development of a fluid-filled body cavity called a **coelom** (**Figure 32.4**). In many animals, the body cavity is completely lined with mesoderm and is called a true coelom. Animals with a true coelom are termed **coelomates**. If the coelom is not completely lined by tissue derived from mesoderm, it is known as a **pseudocoelom**. Animals with a pseudocoelom are termed **pseudocoelomates** and include rotifers and roundworms. Some animals, such as flatworms, lack a fluid-filled body cavity and are termed **acoelomates**. Instead of fluid, this region contains mesenchyme tissue. In some coelomate animals, such as mollusks and arthropods, the coelom is reduced to small pockets around the heart and excretory organs, and the main body cavity is known as the hemocoel. In these animals, a fluid called hemolymph bathes the tissues directly.

A body cavity has many important functions, perhaps the most important being that its fluid is relatively incompressible and thus cushions internal organs such as the heart and intestinal tract, helping to prevent injury from external forces. A body cavity also enables internal organs to move and grow independently of the outer body wall. Furthermore, in some soft-bodied invertebrates such as earthworms, the coelom functions as a **hydrostatic skeleton**, a fluid-filled body cavity surrounded by muscles that gives support and shape to the body of organisms. Because liquid is relatively incompressible, muscle contractions at one part of the body can push fluid toward another part of the body. This type of movement can best be observed in an earthworm (see Chapter 33). Finally, in some organisms, the fluid in the body cavity also acts as a simple circulatory system.

Embryonic Development In the developing zygote, cleavage may occur by two mechanisms (**Figure 32.5a**). In **spiral cleavage**, the planes of cell cleavage are oblique to the axis of the embryo, resulting in an arrangement in which newly formed upper cells lie centered between the underlying cells. Animals that exhibit spiral cleavage are called **protostomes** and include mollusks, annelid worms, and arthropods. In **radial cleavage**, the cleavage planes are either parallel or perpendicular to the

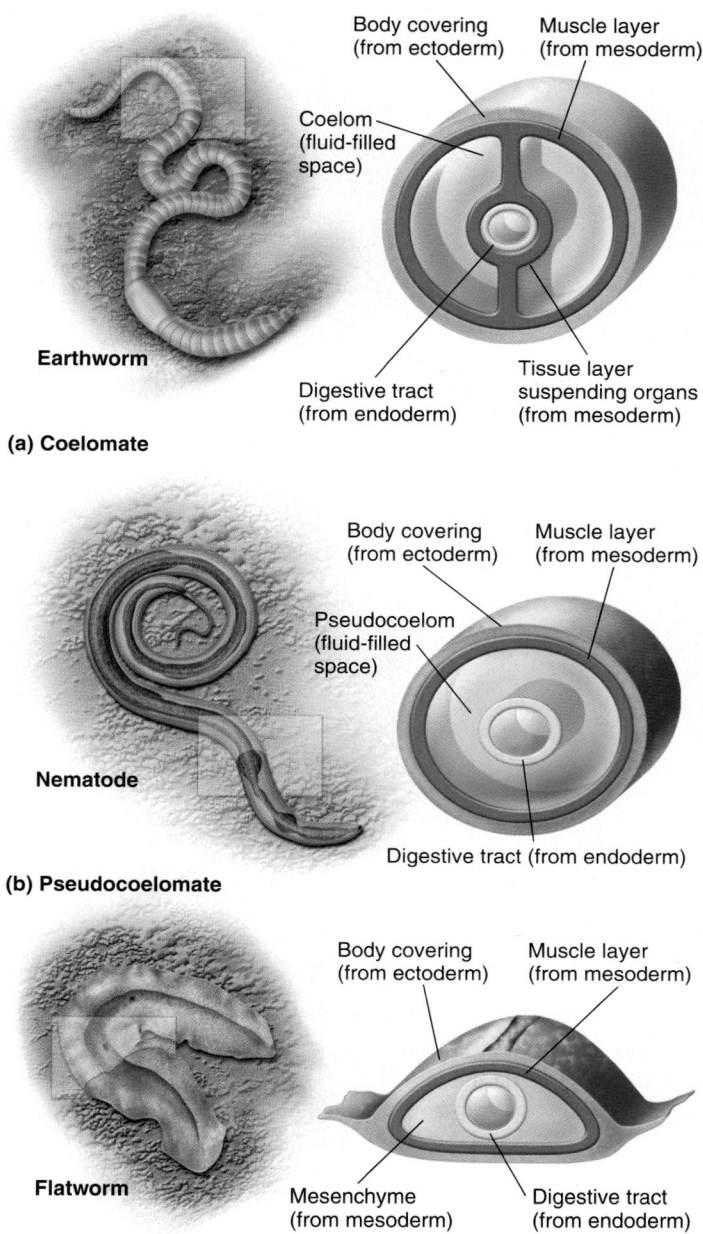

Body covering (from ectoderm)
Muscle layer (from mesoderm)
Coelom (fluid-filled space)
Earthworm
Digestive tract (from endoderm)
Tissue layer suspending organs (from mesoderm)

(a) Coelomate

Body covering (from ectoderm)
Muscle layer (from mesoderm)
Pseudocoelom (fluid-filled space)
Nematode
Digestive tract (from endoderm)

(b) Pseudocoelomate

Body covering (from ectoderm)
Muscle layer (from mesoderm)
Flatworm
Mesenchyme (from mesoderm)
Digestive tract (from endoderm)

(c) Acoelomate

Figure 32.4   **The three basic body plans of bilaterally symmetrical animals.**

vertical axis of the egg. This results in tiers of cells, one directly above the other. Animals exhibiting radial cleavage are called **deuterostomes** and include echinoderms and vertebrates.

Protostome development is also characterized by so-called **determinate cleavage**, in which the fate of each embryonic cell is determined very early (**Figure 32.5b**). If we remove one of the cells from a four-cell mollusk embryo, neither the single cell nor the remaining three-cell mass can form viable embryos, and

development is halted. In contrast, deuterostome development is characterized by **indeterminate cleavage**, in which each cell produced by early cleavage retains the ability to develop into a complete embryo. For example, when one cell is excised from a four-cell sea urchin embryo, both the single cell and the remaining three can go on to form viable embryos. Other embryonic cells compensate for the missing cells. In human embryos, if individual embryonic cells separate from one another early in development, identical twins can result. Indeterminate embryonic cells are also known as stem cells and are found not just in the four-cell embryo but also in later embryonic stages. They are called pluripotent, because they can develop into almost every cell type of the body.

In addition to differences in cleavage patterns, protostomes and deuterostomes differ in two other embryonic features. The most fundamental of these concerns the development of a mouth and anus (**Figure 32.5c**). In gastrulation, the endoderm forms an indentation, the **blastopore**, which is the opening of the archenteron to the outside. In protostomes (from the Greek *protos*, first, and *stoma*, mouth), the blastopore becomes the mouth. If an anus is formed in a protostome, it develops from a secondary opening. In contrast, in the deuterostomes (Greek *deuteros*, second), the blastopore becomes the anus, and the mouth is formed from the secondary opening.

The last difference between protostomes and deuterostomes concerns coelom formation. In protostomes, a solid mass of mesoderm cells splits to form the cavity that becomes the coelom. This pattern is called **schizocoelous** development (from the Greek *schizo*, to split). In contrast, in deuterostomes, a layer of mesoderm cells form outpockets that bud off from the developing gut to form the coelom. This pattern is known as **enterocoelous** development. The resultant coeloms in both protostomes and deuterostomes are similar; they are just formed in different ways.

Other Methods of Classification   In the traditional phylogenetic tree of animal life, further branches are based on features such as the possession of an exoskeleton (arthropods) or the development of a notochord (chordates). One other key feature of the animal body plan is the presence or absence of segmentation. In segmentation, also called **metamerism**, the body is divided into nearly identical subunits called segments. It is most obvious in the annelids, or segmented worms, in which each segment contains the same set of blood vessels, nerves, and muscles, but it is also evident in arthropods and chordates (**Figure 32.6**). Some segments may differ, such as those containing the brain or the sex organs, but most segments are very similar.

The advantage of segmentation is that it allows specialization of body regions. For example, in arthropods, some segments are specialized to form wings for flight. Recent studies have shown that changes in specialization among arthropod body segments can be traced to relatively simple changes in homeotic, or *Hox*, genes. In chordates, we can see segmentation in the backbone and muscles.

(a) Cleavage pattern (b) Fate of embryonic cells (c) Fate of blastopore

Figure 32.5 **Differences in embryonic development between protostomes and deuterostomes.** **(a)** Many protostomes have spiral cleavage, while most deuterostomes have radial cleavage. The dotted arrows indicate the direction of cleavage. **(b)** Protostomes have determinate cleavage, whereas deuterostomes have indeterminate cleavage. **(c)** In protostomes, the blastopore becomes the mouth. In deuterostomes, the blastopore becomes the anus.

In earthworms, each ring is a distinct segment.

Lobsters have developed specialized appendages on many segments.

Fish exhibit segmentation in their muscles and backbone.

Figure 32.6 **Segmentation.** Annelids, arthropods, and chordates all exhibit segmentation.

GENOMES & PROTEOMES

Changes in *Hox* Gene Expression Control Body Segment Specialization

As we will see in Chapter 33, arthropods exhibit a vast degree of specialization of their segments. For example, many insects have wings and only three pairs of legs, whereas centipedes have no wings and many legs. Crabs, lobsters, and shrimps have highly specialized thoracic appendages called maxillipeds that aid in feeding. In the 1990s, Michalis Averof and coworkers elegantly showed how relatively simple changes in the expression patterns of **Hox genes**, genes involved in patterning the body axis, can account for this large variation in appendage types. As described in Chapter 19, animals have several *Hox* genes that are expressed in particular regions of the body. Some are expressed in anterior segments, while others are expressed in posterior segments (refer back to Figure 19.18). The *Hox* genes are designated with numbers 1 through 13.

Shifts in patterns of expression of *Hox* genes in the embryo along the anteroposterior axis are prominent in evolution. In vertebrates, the transition from one type of vertebra to another, for example, from cervical (neck) to thoracic (chest) vertebrae, is also controlled by particular *Hox* genes (**Figure 32.7**). The site of the cervical/thoracic boundary appears to be influenced by the *HoxC-6* gene. Differences in its relative position of expression, which occurs prior to vertebrae development, control neck length in vertebrates. In mice, which have a relatively short neck, the expression of *HoxC-6* begins between vertebrae numbers 7 and 8. In chickens and geese, which have longer necks, the expression begins further back, between vertebrae 14 and 15, or 17 and 18, respectively. The forelimbs also arise at this boundary in all vertebrates. Interestingly, snakes, which essentially have no neck or forelimbs, do not exhibit this boundary, and *HoxC-6* expression occurs toward their heads. This in effect means that snakes got longer by losing their neck and lengthening their chest.

Evolutionary and development biologist Sean Carroll has remarked that it is very satisfying to find that the evolution of body forms and novel structures in two of the most successful and diverse animal phyla, arthropods and vertebrates, is shaped by the shifting of *Hox* genes. It also reminds us of one of the basic tenets of diversity, that modern organisms illustrate Darwin's concept of descent with modification. Much of the diversity in animal phyla can be seen as variations on a common theme.

32.3 Molecular Views of Animal Diversity

While what we have called the traditional view of animal phylogeny has been accepted by most biologists for over a century, biologists are now using new molecular techniques to classify animals by comparing similarities in the DNA and the ribosomal RNA of animals, especially sequences of nucleotides in the gene that encodes RNA of the small ribosomal subunit (SSU rRNA) (see Chapter 26). The advantage of the molecular approach is that the data are generally more objective and subject to more rigorous testing than morphological data. The DNA sequence contains four easily identified and mutually exclusive character states: A, T, G, and C (RNA has A, U, G, and C).

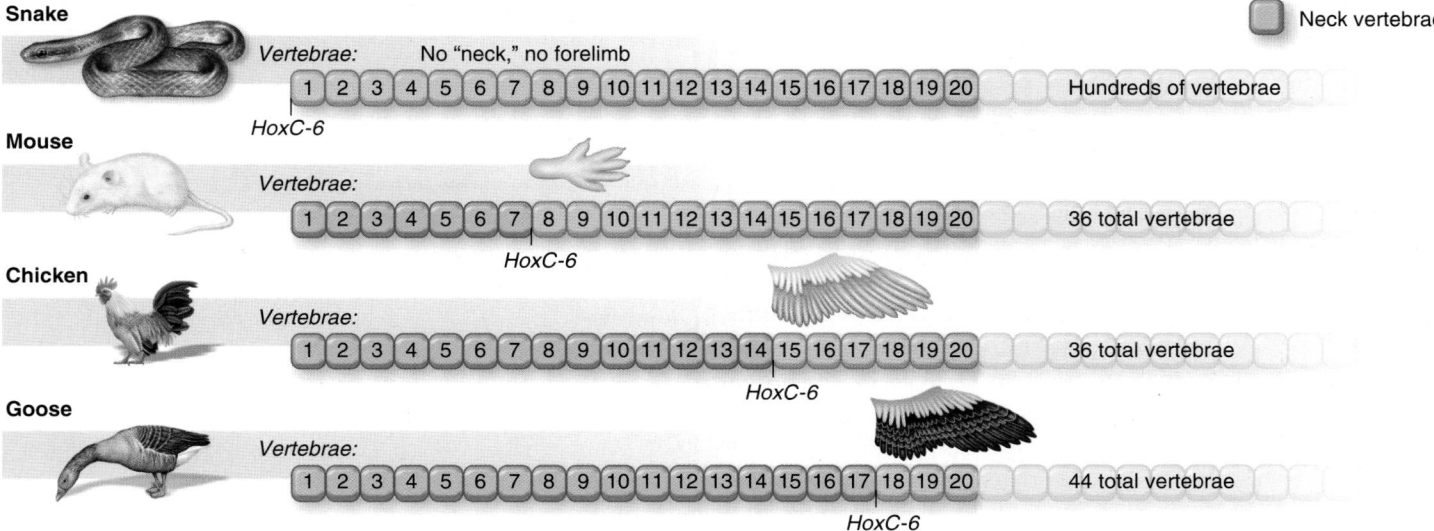

Figure 32.7 Shifting zones of *Hox* gene expression control segment specialization. In vertebrates, the transition between neck and trunk vertebrae is controlled by the position of the *HoxC-6* gene. In snakes, the expression of this gene is shifted so far forward that a neck does not develop.

Contrast this with morphological and embryological data, where characters are scored more subjectively, often based on the qualitative assessment of many traits. One of the most influential of the modern studies was the 1997 paper of Anna Marie Aguinaldo and colleagues, which established evidence for a new monophyletic group (clade) of molting animals, the Ecdysozoa (see Feature Investigation).

In addition, researchers have also studied the *Hox* genes that control early development and are present in all animals. *Hox* genes are important because many branches in the traditional phylogeny are based on early developmental differences such as cleavage, so examination of the genes that regulate these differences should provide insight into the evolution of animal development and to understanding how, when, and why animal body plans diversified.

While the use of molecular techniques in taxonomy is relatively new, the techniques have had a dramatic impact on traditional classification schemes. Phylogenies based on SSU rRNA and *Hox* genes are similar and, as we will explore, in many cases agree with the structure of the traditional phylogenetic tree (**Figure 32.8**). However, there are some important differences. Recently, the journal *Science* brought together numerous molecular taxonomists to discuss the current available data. In this section, we summarize their consensus on the new molecular phylogeny and examine the similarities and differences between its hypotheses and those of traditional phylogeny.

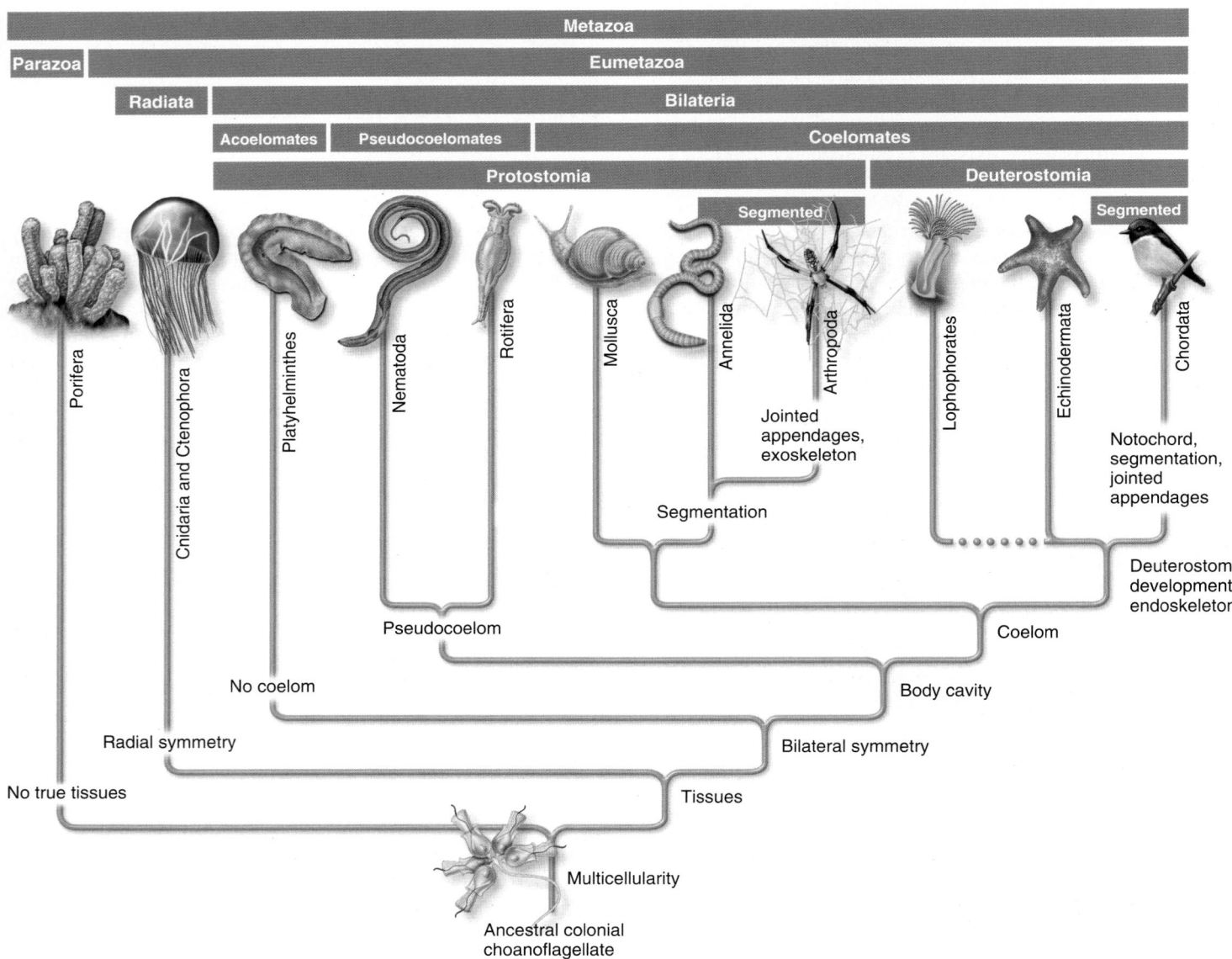

Figure 32.8 **A traditional animal phylogeny based on body plan.** Though there are about 35 different animal phyla, we will focus our discussions here and in the next two chapters on the 11 groups with the greatest numbers of species. The dotted line represents the uncertainty of including the lophophorates with the deuterostomes.

FEATURE INVESTIGATION

Aguinaldo and Colleagues Used 18S rRNA to Analyze the Taxonomic Relationships of Arthropods to Other Taxa

In 1997, Anna Marie Aguinaldo, James Lake, and colleagues analyzed the relationships of arthropods to other taxa by sequencing the complete gene that encodes SSU rRNA from a variety of representative taxa (**Figure 32.9**). Total genomic DNA was isolated using standard techniques and amplified by the polymerase chain reaction (PCR). PCR fragments were then subjected to DNA sequencing, a technique described in Chapter 20. Using approaches that we have described in Chapter 26, the evolutionary relationships among 50 species were determined. The data indicated the existence of a monophyletic clade—the Ecdysozoa—containing the arthropods and nematodes plus a number of smaller phyla (see data in Figure 32.9). The name Ecdysozoa means molting animals, referring to the fact that all organisms in the clade undergo shedding of their exoskeleton.

The hypothesis that nematodes are more closely related to arthropods than previously thought has important ramifications. In particular, it implies that two well-researched model organisms, *Caenorhabditis elegans* (a nematode) and the fruit fly, *Drosophila melanogaster* (an arthropod), are more closely related than had been believed (see their positions in the data of Figure 32.9). Traditional classification assumed that developmental features common to both of these organisms had arisen early in metazoan evolution and thus might be broadly relevant to development in other coelomates, and in particular, humans. This new classification brings into question the applicability of studies of these organisms to human biology, since commonalities between these organisms might have evolved after the Ecdysozoans diverged. In this case, both nematodes and arthropods would be more closely related to each other than to humans.

Figure 32.9 A molecular animal phylogeny based on sequencing of SSU rRNA.

Biological inquiry: What is the major difference between molecular phylogeny and traditional phylogeny?

GOAL To determine the evolutionary relationships among many animal species.

STARTING MATERIALS Cellular samples from about 50 animals in different taxa.

Experimental level	Conceptual level

1 Subject the samples to polymerase chain reaction (PCR) to obtain enough material for DNA sequencing. PCR is described in Chapter 20.

For more detail, refer back to figure 20.8.

The goal of PCR is to amplify a region in the SSU rRNA gene.

2 Sequence the amplified DNA by dideoxy sequencing, also described in Chapter 20.

For more detail, refer back to figure 20.10.

CACCGTA

Dideoxy sequencing, in which DNA strands are separated according to their lengths by subjecting them to gel electrophoresis, is used to determine the base sequence of DNA.

3 Compare the DNA sequences and infer phylogenetic relationships using the cladistic approach described in Chapter 26.

The cladistic approach compares traits that are either shared or not shared by different species.

4 THE DATA

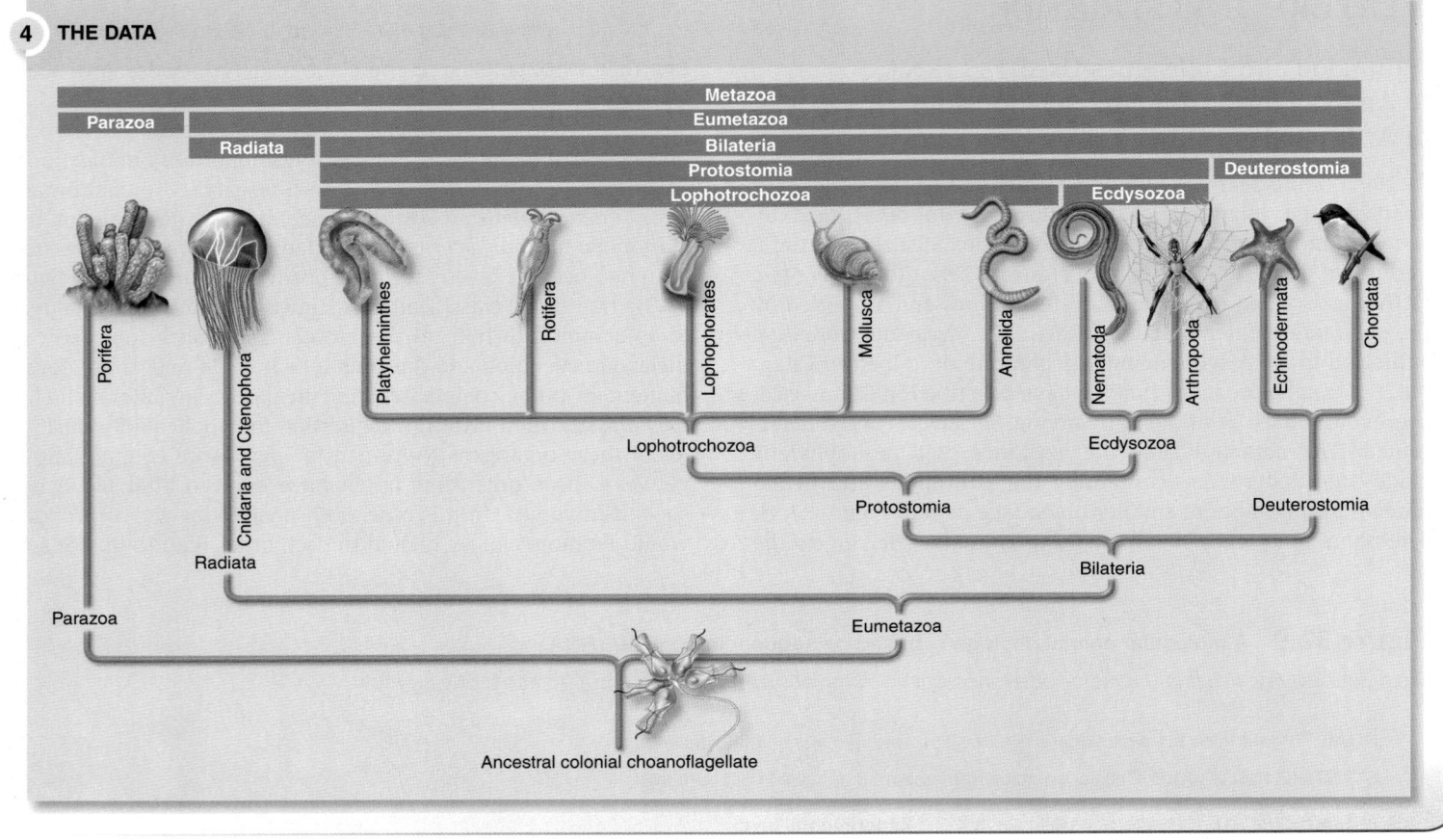

Molecular Phylogeny Shares Many Similarities with Traditional Phylogeny

Molecular phylogeny is in agreement with traditional phylogeny over the following features of the animal kingdom:

1. The clade called Metazoa is monophyletic, meaning all animals came from a single common ancestor.
2. At the earliest stages of evolution, molecular phylogeny supports the traditional view of the split between Parazoa and Eumetazoa.
3. There is also agreement about an early split between Radiata and Bilateria, with most animal phyla belonging to the Bilateria.
4. Molecular phylogeny also agrees that the echinoderms and chordates belong to a clade called the Deuterostomia.

Molecular Phylogeny Has Some Important Differences from Traditional Phylogeny

Recent molecular analyses have revealed two key differences between models of traditional evolutionary relationships among animals and newer models. The two differences concern first

the division of the protostomes into two separate clades and second, the presence or absence of a body cavity.

Protostomes: Lophotrochozoa and Ecdysozoa The most important difference between molecular and traditional phylogenies involves relationships among the Bilateria. In the traditional view of animal phylogeny, the bilaterally symmetrical animals are split into two clades, the Deuterostomia and the Protostomia, reflecting two basic modes of embryonic development. However, recent molecular studies, from James Lake and others, suggest a different grouping. The deuterostomes are still separate, but the protostomes are divided into two major clades: the **Lophotrochozoa**, which encompasses the annelids, mollusks, and several other phyla, and the **Ecdysozoa**, primarily the arthropods and nematodes (Figure 32.9).

While the clade was organized primarily through analysis of molecular data, the name Lophotrochozoa stems from two morphological features seen in organisms of this clade. The "lopho" part is derived from the **lophophore**, a horseshoe-shaped crown of tentacles used for feeding that is present on some members of the group (**Figure 32.10a**). The "trocho" part refers to the **trochophore larva**, a distinct larval stage of many of the phyla (**Figure 32.10b**).

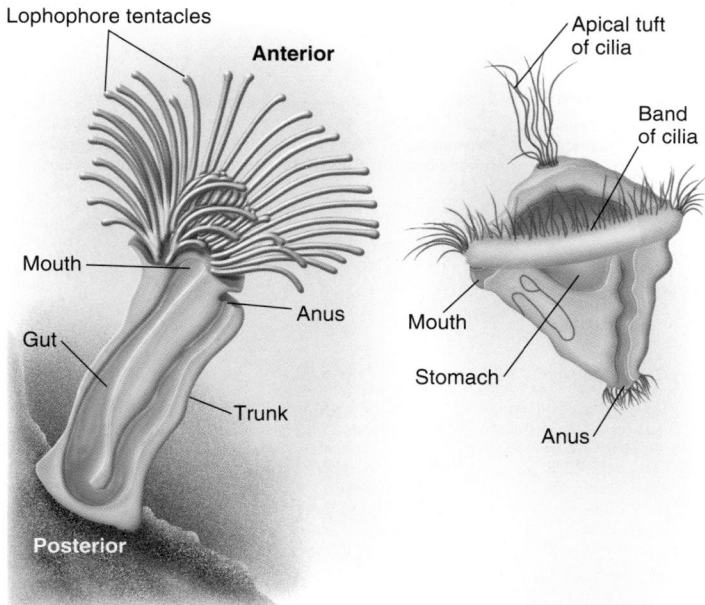

(a) Lophophore of a phoronid worm   **(b) Trochophore larva**

Figure 32.10   Characteristics of the Lophotrochozoa.
(a) A lophophore, a crown of ciliated tentacles, generates a current to bring food particles into the mouth. **(b)** The trochophore ("wheel-bearer") larval form is found in several animal lineages, such as polychaete worms and mollusks, indicating that both may have similar ancestry.

Figure 32.11   **Ecdysis.**   The Ecdysozoa are a clade of animals exhibiting ecdysis, the periodic shedding (molting) and re-formation of the exoskeleton.

Biological inquiry: What are the main members of the Ecdysozoa?

In traditional phylogeny, much debate surrounded the classification of the three minor phyla that possess lophophores—the Bryozoa, Phoronida, and Brachiopoda (collectively called the lophophorates). While these taxa exhibited some characteristics of protostomes and some of deuterostomes, they were often classified as the Deuterostomia (see Figure 32.8). However, molecular data support their inclusion within the Lophotrochozoa, along with annelids and mollusks.

The **Ecdysozoa** are so named because its members secrete a nonliving cuticle, typically an external skeleton (exoskeleton); think of the hard shell of a beetle or that of a crab. As these animals grow, the exoskeleton becomes too small and the animal molts, or breaks out of its old exoskeleton, and secretes a newer, larger one (**Figure 32.11**). This molting process is called **ecdysis**; hence the name Ecdysozoa. While named for this morphological characteristic, the clade is strongly supported by molecular evidence such as similarities in DNA.

Body Cavity   The second most important difference between molecular and traditional phylogenies involves the presence or absence of a body cavity. In the traditional view of animal phylogeny, the bilaterally symmetrical animals are divided into those lacking a coelom (acoelomates), those with a pseudocoelom (pseudocoelomates), and those possessing a coelom (coelomates) (see Figure 32.4). The Platyhelminthes, or flatworms,

are classified as acoelomate and are thus seen as separate from coelomate phyla. Molecular evidence, however, now suggests that the flatworms should be included with the Lophotrochozoa. In this view, flatworms are not primitive acoelomate animals. Rather, they evolved from an ancestor that possessed a coelom but lost it during evolutionary modification. Similarly, molecular-based phylogeny places the rotifers and nematodes, the pseudocoelomate phyla, within the Lophotrochozoa and Ecdysozoa, respectively. Thus, molecular data suggest that the presence or absence of a coelom, a distinction traditionally used in the construction of animal phylogenies, may not be a useful way to classify animals.

It is important to stress, however, that more similarities are found between the molecular and traditional phylogenies than differences. Most of the major branch points in the phylogenies are in agreement. As a reference, **Table 32.2** summarizes the characteristics of the major animal phyla.

In the following two chapters, the discussion of animal phylogeny is based primarily on findings of molecular data. In Chapter 33, we will discuss the diversity of the Parazoa, Radiata, Lophotrochozoa, and Ecdysozoa phyla, and indeed all animals without a backbone. In Chapter 34, we will turn our attention to one phylum, Chordata, and in particular the vertebrates: fish, amphibians, reptiles, birds, and mammals.

Table 32.2   Summary of Characteristics of the Major Animal Phyla

Feature	*Porifera* Sponges	*Cnidaria* and *Ctenophora* Hydra, anemones, jellyfish	*Platyhel-minthes* Flatworms	*Rotifera* Rotifers	*Lopho-phorates* Bryozoans and others	*Mollusca* Snails, clams, squid	*Annelida* Segmented worms	*Nematoda* Roundworms	*Arthropoda* Insects, arachnids, crustaceans	*Echino-dermata* Sea stars, sea urchins	*Chordata* Vertebrates and others
Estimated number of species	8,000	11,000	20,000	2,000	4,000+	110,000	15,000	20,000	1,000,000+	6,000+	47,000+
Level of organization	Cellular; lack tissues and organs	Tissue; lack organs	Organs	Organs	Organs	Organs	Organs	Organs	Organs	Organs	Organs
Symmetry	Absent	Radial	Bilateral	Bilateral	Bilateral	Bilateral	Bilateral	Bilateral	Bilateral	Bilateral larvae, radial adults	Bilateral
Cephalization	Absent	Absent	Present	Present	Reduced	Present	Present	Present	Present	Absent	Present
Germ layers	Absent	Two	Three	Three	Three	Three	Three	Three	Three	Three	Three
Body cavity	Absent	Absent	Absent	Pseudo-coelom	Coelom	Reduced coelom	Coelom	Pseudo-coelom	Reduced coelom	Coelom	Coelom
Segmentation	Absent	Absent	Absent	Absent	Absent	Absent	Present	Absent	Present	Absent	Present (but reduced)
Digestive system	Absent	Gastro-vascular cavity; Ctenophores have complete gut	Incomplete gut	Complete gut (usually)	Complete gut	Complete gut	Complete gut	Complete gut	Complete gut	Usually complete gut	Complete gut
Circulatory system	Absent	Absent	Absent	Absent	Absent, open, or closed	Open	Closed	Absent	Open	Absent	Closed
Respiratory system	Absent	Absent	Absent	Absent	Absent	Gills, lungs	Absent	Absent	Trachae, gills, or book lungs	Tube feet, gills, respiratory tree	Gills, lungs
Excretory system	Absent	Absent	Protonephridia with flame cells	Protonephridia	Meta-nephridia	Meta-nephridia	Meta-nephridia	Excretory gland cells	Excretory glands resembling metanephridia	Absent	Kidneys
Nervous system	Absent	Nerve net	Brain, nerve cords	Brain, nerve cords	No brain, nerve ring	Ganglia, nerve cords	Brain, ventral nerve cord	Brain, nerve cords	Brain, ventral nerve cord	No brain, nerve ring and radial nerves	Well-developed brain; dorsal hollow nerve cord
Reproduction	Sexual; asexual (budding)	Sexual; asexual (budding)	Sexual (most her-maphroditic); asexual (body splits)	Mostly partheno-genetic; males appear only rarely	Sexual (some her-maphroditic); asexual (budding)	Sexual (some her-maphroditic)	Sexual (some her-maphroditic)	Sexual (some her-maphroditic)	Usually sexual (some her-maphroditic)	Sexual (some her-maphroditic); partheno-genetic; asexual by regeneration (rare)	Sexual, rarely partheno-genetic
Support	Endo-skeleton of spicules and collagen	Hydrostatic skeleton	Hydrostatic skeleton	Hydrostatic skeleton	Exoskeleton	Hydrostatic skeleton and shell	Hydrostatic skeleton	Hydrostatic skeleton	Exoskeleton	Endoskeleton of plates beneath outer skin	Endoskeleton of cartilage or bone

CHAPTER SUMMARY

32.1 Characteristics of Animals

- Animals constitute a very species-rich kingdom. They share a number of key characteristics, including multicellularity, heterotrophic feeding, the possession of nervous and muscle tissues, and sexual reproduction. (Table 32.1)

32.2 Traditional Classification of Animals

- The animal kingdom is monophyletic, meaning that all taxa have evolved from a single common ancestor. This can be observed in comparing gene sequences from animals and a protist. (Figure 32.1)
- The traditional classification of animals is based on four morphological and developmental features of animal body plans. (Table 32.2)
- Animals can be categorized according to the absence of different types of tissues (the Parazoa or sponges) and the presence of tissues (Eumetazoa or all other animals). The Eumetazoa can also be divided according to their type of symmetry, whether radial (Radiata, the cnidarians and ctenophores) or bilateral (Bilateria, all other animals). (Figure 32.2)
- The Radiata have two layers of embryonic cell layers (germ layers) called the endoderm and the ectoderm. The Bilateria develop a third germ layer termed the mesoderm, which develops between the ectoderm and the endoderm. (Figure 32.3)
- Animals can be classified according to the presence or absence of a coelom, or true body cavity. Animals with a coelom are termed coelomates. Animals that possess a pseudocoelom, or coelom that is not completely lined by tissue derived from mesoderm, are called pseudocoelomates. Those animals lacking a fluid-filled body cavity are termed acoelomates. (Figure 32.4)
- Animals are also classified according to patterns of embryonic development. Animals with spiral cleavage are called protostomes, and those exhibiting radial cleavage are considered deuterostomes. In protostomes, the blastopore, or opening of the gut to the outside, becomes the mouth; in deuterostomes, the blastopore becomes the anus. (Figure 32.5)
- Metamerism, the division of the body into identical subunits called segments, is another key feature of the animal body plan. (Figure 32.6)
- Shifts in the pattern of expression of *Hox* genes are prominent in evolution. In vertebrates, the transition from one type of vertebra to another is controlled by certain *Hox* genes. (Figure 32.7)

32.3 Molecular Views of Animal Diversity

- New molecular techniques that compare similarities in DNA and ribosomal RNA of animals are having a dramatic effect on traditional classification schemes. (Figure 32.8)
- In many cases, phylogenies based on these techniques are similar to those of traditional approaches; however, some important differences exist. Recent molecular studies propose a division of the protostomes into two major clades: the Lophotrochozoa and the Ecdysozoa. (Figure 32.9)
- The Lophotrochozoa are grouped primarily through analysis of molecular data, but they are distinguished by two morphological features—the lophophore, a crown of tentacles used for feeding, and the trochophore larva, a distinct larval stage. The lophophorates were often classified as deuterostomes, but molecular data support their inclusion within the Lophotrochozoa. (Figure 32.10)

- The Ecdysozoa are so named because its members secrete a nonliving cuticle, typically an exoskeleton or external skeleton. Ecdysis is the periodic shedding and re-formation of the exoskeleton. (Figure 32.11)
- Molecular evidence suggests that flatworms evolved from a coelomate ancestor and should be considered to be within the Lophotrochozoa, and not as primitive acoelomate animals.

TEST YOURSELF

1. Which of the following is *not* a distinguishing characteristic of animals?
   a. the capacity to move at some point in their life cycle
   b. possession of cell walls
   c. multicellularity
   d. heterotrophy
   e. all of the above are characteristics of animals
2. Terrestrial adaptations seen in animals include
   a. internal fertilization.
   b. tough, protective shells around eggs.
   c. a waxy cuticle covering exposed tissue.
   d. a and b only.
   e. all of the above.
3. Eumetazoa are animals that have
   a. true tissues.
   b. more than one tissue type.
   c. only one tissue type.
   d. radial symmetry.
   e. spiral cleavage.
4. The localization of sensory structures at the anterior end of the body is
   a. bilateral symmetry.
   b. spiral cleavage.
   c. cephalization.
   d. radial symmetry.
   e. gastrulation.
5. The germ layer that is present in triploblastic animals but is absent in diploblastic animals is
   a. the ectoderm.
   b. the mesoderm.
   c. the endoderm.
   d. the pseudocoelom.
   e. the coelom.
6. Pseudocoelomates
   a. lack a fluid-filled cavity.
   b. have a fluid-filled cavity that is completely lined with mesoderm.
   c. have a fluid-filled cavity that is partially lined with mesoderm.
   d. have a fluid-filled cavity that is not lined with mesoderm.
   e. have an air-filled cavity that is partially lined with mesoderm.
7. Protostomes and deuterostomes can be classified based on
   a. cleavage pattern.
   b. destiny of the blastopore.
   c. whether the fate of the embryonic cells is fixed early during development.
   d. how the coelom is formed.
   e. all of the above.
8. Naturally occurring identical twins are possible only in animals that
   a. have spiral cleavage.
   b. have determinate cleavage.
   c. are protostomes.
   d. have indeterminate cleavage.
   e. a, b, and c.

9. Genes involved in the patterning of the body axis, that is, in determining characteristics such as neck length and appendage formation, are called
    a. small subunit (SSU) rRNA genes.
    b. *Hox* genes.
    c. metameric genes.
    d. determinate genes.
    e. none of the above.

10. A major difference between the molecular phylogeny of animals and traditional phylogeny of animals is that
    a. the presence or absence of the mesoderm is not important in molecular phylogeny.
    b. molecular phylogeny suggests that all animals do not share a single common ancestor.
    c. body symmetry, whether radial or bilateral, is not an important determinant in molecular phylogeny.
    d. molecular phylogeny does not include the echinoderms in the deuterostome clade.
    e. molecular phylogeny suggests that the presence or absence of a coelom is not important for classification.

CONCEPTUAL QUESTIONS

1. The traditional classification is based on what four features of animal body plans?
2. Distinguish between radial and bilateral symmetry.
3. Define cleavage and gastrulation.

EXPERIMENTAL QUESTIONS

1. What was the purpose of the study conducted by Aguinaldo and colleagues?
2. What was the major finding of this particular study?
3. What impact does the new view of nematode and arthropod phylogeny have on other areas of research?

COLLABORATIVE QUESTIONS

1. Discuss the different types of body cavities.
2. Discuss some of the characteristics of animal life.

www.brookerbiology.com

This website includes answers to the Biological Inquiry questions found in the figure legends and all end-of-chapter questions.

33
THE INVERTEBRATES

This plantlike organism is actually a feather star, an echinoderm. The feather star's tube feet grip the ocean substrate while its arms filter feed on drifting microorganisms.

As we saw in Chapter 22, the history of animal life on Earth has evolved over hundreds of millions of years. While many date the divergence of animals to the early Paleozoic era, over 500 million years ago, biologist Blair Hedges and others have shown that molecular data point to a divergence among animals that may be as long as 800–1,200 million years ago. Some scientists suggest that changing environmental conditions, such as a buildup of dissolved oxygen and minerals in the ocean or an increase in atmospheric oxygen, eventually permitted higher metabolic rates and increased the activity of a wide range of animals. Others suggest that with the development of sophisticated locomotor skills, a wide range of predators and prey evolved, leading to an evolutionary arms race in which predators evolved powerful weapons and prey evolved more powerful defenses against them. Such adaptations and counteradaptations would have led to a proliferation of different lifestyles and taxa.

Over the next two chapters, we will survey the wondrous array of animal life on Earth (see chapter-opening photo). In this chapter, we examine the **invertebrates**, or animals without a backbone, a category that makes up more than 95% of all animal species. We begin by exploring some of the earliest animal lineages, the Parazoa and Radiata. We then turn to the Lophotrochozoa and Ecdysozoa, the two sister groups of protostomes introduced in Chapter 32. Finally, we turn to the deuterostomes, focusing here on the echinoderms and the invertebrate members of the phylum Chordata.

While the more modern molecular classification outlined in Chapter 32 will serve as the basis of our discussion of animal lineages, we will not ignore the concept of body plans, because it still can provide clues about how different phyla have evolved. The newer molecular phylogeny is still in its infancy, and many refinements will undoubtedly be made as increasing numbers of genes from more species are sequenced and compared. For this reason, many biologists are not yet ready to totally set aside the older body plan–based phylogeny.

33.1 Parazoa: Sponges, the First Multicellular Animals

The Parazoa consist of one phylum, Porifera (from the Latin, "pore bearers"), whose members are commonly referred to as sponges. Sponges are loosely organized and lack true tissues, groups of cells that have a similar structure and function. However, sponges are multicellular and possess several types of cells that perform different functions. Biologists have identified approximately 8,000 species of sponges, the vast majority of which are marine. Sponges range in size from only a few millimeters across to more than 2 m in diameter. The smaller sponges may be radially symmetrical, but most have no apparent symmetry. Some sponges have a low encrusting growth form, while others grow tall and erect (**Figure 33.1a**). While adult sponges are sessile, that is, anchored in place, the larvae are free-swimming.

The body of a sponge looks similar to a vase pierced with small holes or pores (**Figure 33.1b**). Water is drawn through these pores (ostia, singular, ostium) into a central cavity, the **spongocoel**, and flows out through the large opening at the top called the **osculum**. The water enters the pores by the beating action of the flagella of the **choanocytes**, or collar cells, that line the spongocoel (**Figure 33.1c**). In the process, the choanocytes trap and eat small particulate matter and tiny plankton.

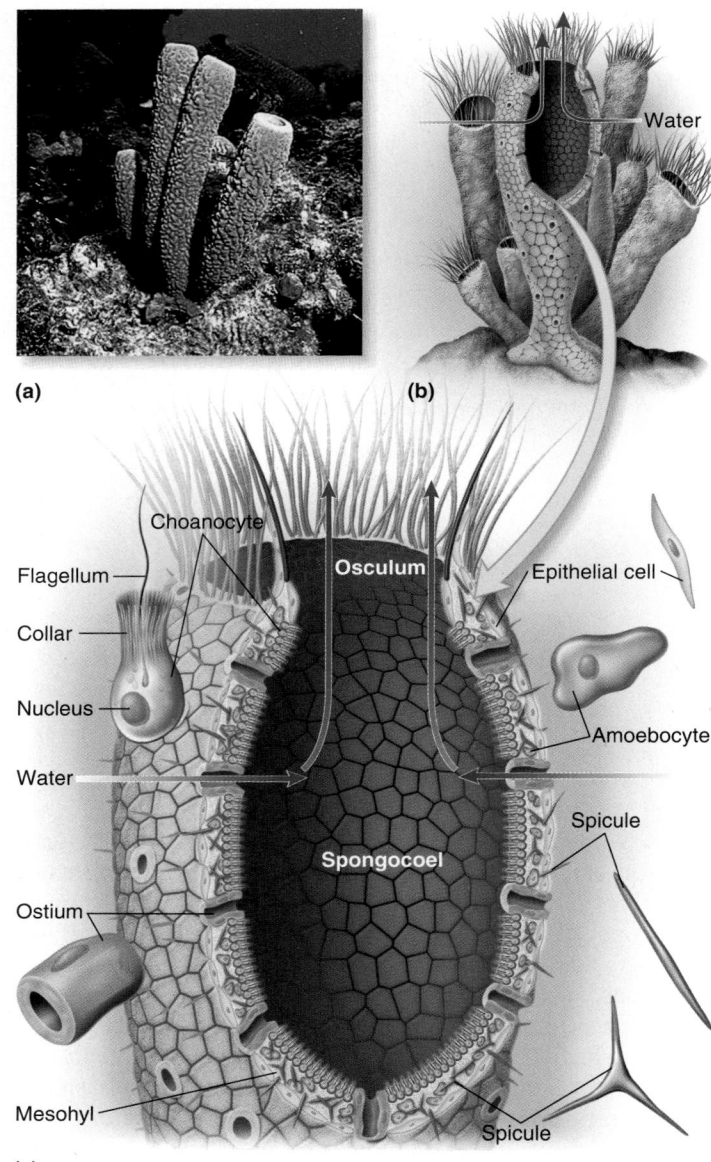

As we noted in Chapter 32, because of striking morphological and molecular similarities between choanocytes and choanoflagellates, a group of modern protists having a single flagellum, it is believed that sponges originated from a common choanoflagellate ancestor.

A layer of flattened epithelial cells similar to those making up the outer layer of other phyla protects the sponge body. In between the choanocytes and the epithelial cells lies a gelatinous, protein-rich matrix called the **mesohyl**. Within this matrix are mobile cells called **amoebocytes** that absorb food from choanocytes, digest it, and carry the nutrients to other cells. Thus, considerable cell-to-cell contact and communication exists in sponges.

Some amoebocytes can also form tough skeletal fibers that support the body. In many sponges, this skeleton consists of sharp spicules formed of calcium carbonate or silica. For example, some deep-ocean species, called glass sponges, are distinguished by needle-like silica spicules that form elaborate lattice-like skeletons. The presence of such tough spicules may help explain why there is not much predation of sponges. Sponge spicules come in a diverse array of shapes and sizes, and they are valuable taxonomic tools by which to distinguish different types of sponges. In a small family of carnivorous sponges, the spicules are sticky and capture small crustaceans. In these sponges, other cells migrate around the immobilized crustaceans and digest them extracellularly. Not all sponges have **spicules**, however. Others have fibers of a tough protein called **spongin** that lend skeletal support. Spongin skeletons are still commercially harvested and sold as bath sponges. Many species produce toxic defensive chemicals, some of which are thought to have possible antibiotic and anti-inflammatory effects in humans.

Figure 33.1 **Sponge morphology.** **(a)** The stovepipe sponge (*Aplysina archeri*) is a common sponge found on Caribbean reefs. **(b)** Many sponges have a vaselike shape. **(c)** A cross section reveals that sponges are truly multicellular animals, having various cell types but no distinct tissues.

Biological inquiry: If sponges are soft and sessile, why aren't they eaten by other organisms?

Around the turn of the 20th century, biologist Henry V. Wilson made the incredible discovery that if a sponge is dissociated into its individual cells after being passed through a sieve, its cells can reaggregate into a functional sponge within a short time. Wilson concluded that in order to do this, individual cells recognized and reaggregated with other cells of their own kind. Researchers have since discovered that the cells of other multicellular organisms also recognize cells of their own kind and tend to adhere when mixed with other cells. For example, in mammals, liver cells recognize and stick better to other liver cells, and brain cells recognize and adhere to brain cells.

Sponges reproduce through both sexual and asexual means. Most sponges are **hermaphrodites** (from the Greek god Hermes and the goddess Aphrodite), individuals that can produce both sperm and eggs. Gametes are formed in the mesohyl by amoebocytes or choanocytes. While eggs remain in the mesohyl, the sperm are released into the water and carried by water currents to fertilize the eggs of neighboring sponges. Zygotes develop into flagellated swimming larvae that eventually settle on a suitable substrate to become sessile adults. In asexual reproduction, a small fragment or bud may detach and form a new sponge.

33.2 Radiata: Jellyfish, Radially Symmetrical Animals

The Radiata consists of two closely related phyla: the Cnidaria (from the Greek *knide*, nettle, and *aria*, related to; pronounced nid-air'-e-ah) and the Ctenophora (from the Greek *ktenos*, comb, and *phora*, bearing; pronounced teen-o-for'-ah). Members of the Radiata phyla, or radiates, are mostly found in marine environments, although a few, primarily hydra, are freshwater species. The Cnidaria includes hydra, jellyfish, box jellies, sea anemones, and corals, and the Ctenophora consists of the comb jellies. The Radiata have only two embryonic germ layers: the ectoderm and the endoderm, which give rise to the epidermis and the gastrodermis, respectively. A gelatinous substance called the **mesoglea** connects the two layers. In jellyfish, the mesoglea is enlarged and forms the buoyant, transparent jelly, whereas in coral, the mesoglea is very thin.

Figure 33.2 Cnidarians exist as a polyp or medusa form. Both **(a)** polyp and **(b)** medusa forms have two layers of cells, an outer epidermis (from ectoderm) and an inner layer of gastrodermis (from endoderm). In between is a layer of mesoglea, which is thick in jellyfish and thin in corals.

Both cnidarians and ctenophores possess a **gastrovascular cavity**, where extracellular digestion takes place (**Figure 33.2**). This feature allows the ingestion of larger food particles and represents a major advance over the sponges, which utilize only intracellular digestion. Most radiates have tentacles around the mouth that aid in food detection and capture. Radiates also have true nerve cells arranged as a **nerve net** consisting of interconnected neurons with no central control organ. In this section, we will provide an overview of the biology and diversity of the cnidarians and ctenophores.

The Cnidarians Have Specialized Stinging Cells

Most cnidarians exist as two different body forms and associated lifestyles: the sessile **polyp** or the motile **medusa** (Figure 33.2). For example, jellyfish exist predominantly in the medusa form, and corals exhibit only the polyp form. Many cnidarians, such as *Obelia*, have a life cycle that prominently features both polyp and medusa stages (**Figure 33.3**).

The polyp form has a tubular body with an opening at the oral end that is surrounded by tentacles and functions as both mouth and anus. The aboral end is attached to the substrate. In the 18th century, the Swiss naturalist Abraham Trembley discovered that when a freshwater hydra was cut in two, each part not only survived but could also regenerate the missing half. Polyps exist colonially, as they do in corals, or alone, as in sea anemones. Corals take dissolved calcium and carbonate ions from seawater and precipitate them as limestone underneath their bodies. In some species, this leads to a buildup of limestone deposits. As each successive generation of polyps dies, the limestone remains in place and new polyps grow on top. Thus, huge underwater limestone deposits called coral reefs are formed (look ahead to Figure 54.27b). The largest of these is Australia's Great Barrier Reef, which stretches over 2,300 km.

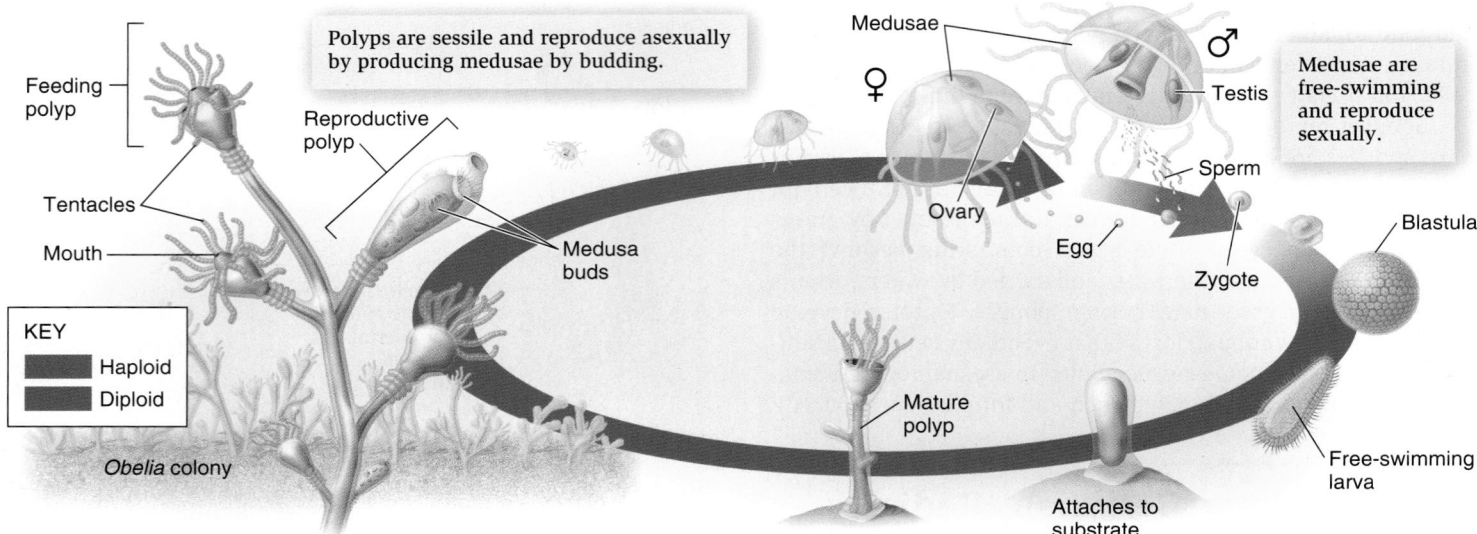

Figure 33.3 **Life cycle of *Obelia*, a colonial cnidarian.** This species exhibits clearly defined polyp and medusa stages.

Biological inquiry: What are the dominant life stages of the following types of cnidarians: jellyfish, sea anemone, and Portuguese Man-of-War?

Many other extensive coral reefs are known, including the reef system along the Florida Keys, all of which occur in warm water, generally between 18°C and 30°C.

The free-swimming medusa form has an umbrella-shaped body with a mouth on the concave underside that is surrounded by tentacles. More mobile medusae possess simple sense organs near the bell margin, including organs of equilibrium called **statocysts** and photosensitive organs known as **ocelli**. When one side of the bell tips upward, the statocysts on that side are stimulated and muscle contraction is initiated to right the medusa. The ocelli allow medusae to position themselves in particular light levels.

One of the unique and characteristic features of the cnidarians is the existence of stinging cells called **cnidocytes**, which function in defense or the capture of prey (**Figure 33.4a**). Cnidocytes contain **nematocysts**, powerful capsules with an inverted coiled and barbed thread. While several different types of nematocysts are recognized, they all have the same general structure and function. Each cnidocyte has a hairlike trigger called a **cnidocil** on its surface. When the cnidocil is touched or a chemical stimulus is detected, the nematocyst fires the thread, which penetrates the prey and injects a small amount of toxin. Small prey are immobilized and passed into the mouth by the tentacles. Alternatively, some nematocyst threads can be sticky rather than stinging. After discharge, the cnidocyte is absorbed and a new one grows to replace it. The nematocysts of most cnidarians are not harmful to humans, but those on the tentacles of the larger jellyfish and the Portuguese Man-of-War (**Figure 33.4b**) can be extremely painful and even fatal. Tentacles of the largest jellyfish, *Cyanea arctica*, may be over 40 m long.

Muscles and nerves exist in their simplest forms in cnidarians. Contractile muscle fibers are found in both the epidermis and gastrodermis. While not true muscles, which only arise

(a) Cnidocytes

(b) Portugese Man-of-War (*Physalia physalis*)

Figure 33.4 **Cnidarians have specialized stinging cells called cnidocytes.** (a) Cnidocytes, which contain stinging capsules called nematocysts, are situated in the tentacles. (b) Portuguese Man-of-War (*Physalia physalis*) is an example of the medusa body form and lifestyle. A gas-filled float allows the animal to stay on the surface of the water.

from the mesoderm and therefore do not appear in diploblastic animals, these muscle fibers can contract to change the shape of the animal. For example, in the presence of a predator, an anemone can expel water very quickly through its open mouth and shrink down to a very small body form. The muscle fibers work against the fluid contained in the body, which thus acts as a hydrostatic skeleton. A nerve net that conducts signals from

Table 33.1 | Main Classes and Characteristics of the Cnidaria

Class and examples (est. # of species)	Class characteristics
Hydrozoa: Portuguese Man-of-War, *Hydra*, *Obelia*, some corals (2,700)	Mostly marine; most have both polyp and medusa stages with polyp stage colonial
Scyphozoa: Jellyfish (200)	All marine; medusa stage dominant and large (up to 2 m); reduced polyp stage
Anthozoa: Sea anemones, sea fans, most corals (6,000)	All marine; polyp stage dominant; medusa stage absent; many are colonial
Cubozoa: Box jellies, sea wasps (20)	All marine; medusa stage dominant; box-shaped

Figure 33.5 A ctenophore. *Mnemiopsis leidyi*, commonly called the sea walnut, was accidentally introduced via the ballast water of ships to the Black and Caspian Seas, where it fed on plankton and decreased the food base for local fish, whose populations dramatically declined.

sensory nerves to muscle cells allows coordination of simple movements and shape changes.

The phylum Cnidaria consists of four classes—Hydrozoa (including *Obelia* and Portuguese Man-of-War), Scyphozoa (jellyfish), Anthozoa (sea anemones and corals), and Cubozoa (box jellies)—whose distinguishing characteristics are shown in **Table 33.1**.

Ctenophores Have a Complete Gut

Ctenophores, also known as comb jellies, are a small phylum of fewer than 100 species, all of which are marine and look very much like jellyfish (**Figure 33.5**). They have eight rows of cilia on their surface that resemble combs. The coordinated beating of the cilia, rather than muscular contractions, propels the ctenophores. Averaging about 1–10 cm in length, comb jellies are probably the largest animals to use cilia for locomotion. There are even a few ribbon-like species up to 1 m long.

Comb jellies possess two long tentacles but lack stinging cells. Instead, they have colloblasts, cells that secrete a sticky substance onto which small prey adhere. The tentacles are then drawn over the mouth. As with cnidarians, digestion occurs in the gastrovascular cavity, but waste and water are eliminated through two anal pores. Thus, the comb jellies possess the first complete gut. Prey are generally small and may include tiny crustaceans called copepods and small fish. Comb jellies are often transported around the world in ships' ballast water. *Mnemiopsis leidyi*, a ctenophore species native to the Atlantic coast of North and South America, was accidentally introduced into the Caspian and Black Seas in the 1980s. With a plentiful food supply and a lack of predators, *Mnemiopsis* underwent a population explosion and caused devastation to the local fishing industries.

All ctenophores are hermaphroditic, possessing both ovaries and testes, and gametes are shed into the water to eventually form a free-swimming larva that grows into an adult. There is no polyp stage. Nearly all ctenophores exhibit **bioluminescence**, a phenomenon that results from chemical reactions that give off light rather than heat. Thus, individuals can be particularly evident at night, and ctenophores that wash up onshore can make the sand or mud appear luminescent.

33.3 Lophotrochozoa: The Flatworms, Rotifers, Lophophorates, Mollusks, and Annelids

In the traditional view of animal phylogeny (refer back to Figure 32.8), the bilaterally symmetrical animals are split into those with no coelom (the platyhelminthes), those with a pseudocoelom (the nematodes and rotifers), and those with a coelom (the remaining phyla). However, as we explored in Chapter 32, molecular data suggest a different grouping in which the deuterostomes are still separate, but the protostomes are divided into two major lineages: the Lophotrochozoa and the Ecdysozoa (refer back to Figure 32.9). The Lophotrochozoa are a diverse group that generally includes taxa that possess either a lophophore (a crown of ciliated tentacles) or a distinct larval stage called a trochophore. In this grouping are seven major Lophotrochozoa phyla: the Platyhelminthes (flatworms), Rotifera (rotifers), Lophophorata (the lophophorates, a group of three phyla), Mollusca (mollusks), and Annelida (segmented worms). In this section, we explore the distinguishing characteristics of these phyla, beginning with some of the simplest lophotrochozoans, the Platyhelminthes.

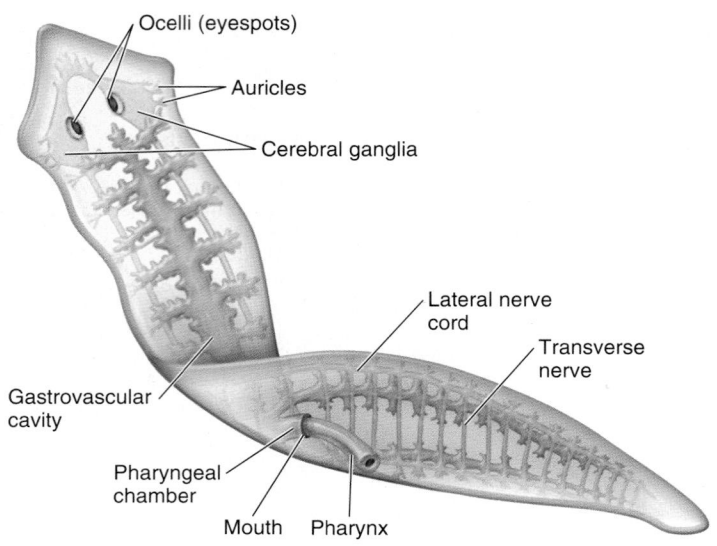

Figure 33.6 **Body plan of a flatworm.** Flatworm morphology as represented by a planarian, a member of the class Turbellaria.

The Phylum Platyhelminthes Consists of Flatworms with No Coelom

Platyhelminthes (from the Greek *platy*, flat, and *helminth*, worm), or flatworms, lack a specialized respiratory or circulatory system to transport gases. They must obtain oxygen by diffusion, which makes a flattened shape necessary, in that no cell can be too far from the surface. Flatworms were among the first animals to develop an active predatory lifestyle. Platyhelminthes, and indeed most animals, are bilaterally symmetrical, with a head bearing sensory appendages.

The flatworms are also believed to be the first animals to develop three embryonic germ layers—ectoderm, endoderm, and mesoderm—with mesoderm replacing the simpler gelatinous mesoglea of cnidarians. As such, they are said to be triploblastic. The muscles in flatworms, which are derived from mesoderm, are well developed. The development of mesoderm was thus a critical evolutionary innovation in animals, because it also led to the development of more sophisticated organs. Flatworms are sometimes regarded as the first animals to reach the organ-system level of organization. The mesoderm fills the body spaces apart from the gastrovascular cavity; thus, the flatworms are acoelomate, lacking a fluid-filled body cavity in which the gut is suspended (**Figure 33.6**).

The digestive system of flatworms is incomplete, with only one opening, which serves as both mouth and anus, as in cnidarians. Most flatworms possess a muscular pharynx that may be extended through the mouth. The pharynx opens to a gastrovascular cavity, where food is digested. In large flatworms, the gastrovascular cavity is branched enough to distribute nutrients to all parts of the body. Any undigested material is egested back through the pharynx. The incomplete digestive system of flatworms thus prevents continuous feeding. Some flatworms are predators, but many species have invaded other animals as parasites.

Flatworms have a distinct excretory system, consisting of **protonephridia**, two lateral canals with branches capped by **flame cells**. The flame cells, which are ciliated and waft water through the lateral canals to the outside (look ahead to Figure 49.8), exist primarily to maintain osmotic balance between the flatworm's body and the surrounding fluids. Simple though this system is, its development was key to permitting the invasion of freshwater habitats and even moist terrestrial areas.

Some free-living flatworms in the class Turbellaria possess light-sensitive eyespots or ocelli at the anterior end and chemoreceptive and sensory cells that are concentrated in organs called auricles. A pair of **cerebral ganglia** receives input from photoreceptors in eyespots and sensory cells. From the ganglia, a pair of lateral nerve cords running the length of the body allows rapid movement of information from anterior to posterior. In addition, transverse nerves form a nerve net on the ventral surface similar to that of cnidarians. Thus, the flatworms have retained the cnidarian-style nervous system, while possessing the beginnings of a more centralized type of nervous system seen throughout much of the rest of the animal kingdom. In all the Platyhelminthes, reproduction is either sexual or asexual. Most species are hermaphroditic but do not fertilize their own eggs. Flatworms can also reproduce asexually by splitting into two parts, with each half regenerating the missing fragment.

The four classes of flatworms are the Turbellaria, Monogenea, Trematoda (flukes), and Cestoda (tapeworms) (**Table 33.2**). Both cestodes and trematodes are internally parasitic and hence are of great medical and veterinary importance. They possess a variety of organs of attachment, such as hooks and suckers, that enable them to remain embedded within their hosts (**Figure 33.7**). Cestodes often require two separate vertebrate host

Table 33.2	Main Classes and Characteristics of Platyhelminthes	
	Class and examples (est. # of species)	**Class characteristics**
	Turbellaria: Planarian (3,000)	Free-living flatworms; mostly marine; predatory or scavengers
	Monogenea: Fish flukes (1,000)	Marine and freshwater; usually external parasites of fish; simple life cycle (no intermediate host)
	Trematoda: Flukes (11,000)	Internal parasites of vertebrates; complex life cycle with several intermediate hosts
	Cestoda: Tapeworms (5,000)	Internal parasites of vertebrates; no digestive system, nutrients absorbed across epidermis; complex life cycle, usually with one intermediate host

Figure 33.7 A tapeworm, *Taenia pisiformis*, a member of the class Cestoda.   Note the scolex, the organ of attachment at the head end, complete with tiny hooks and suckers.

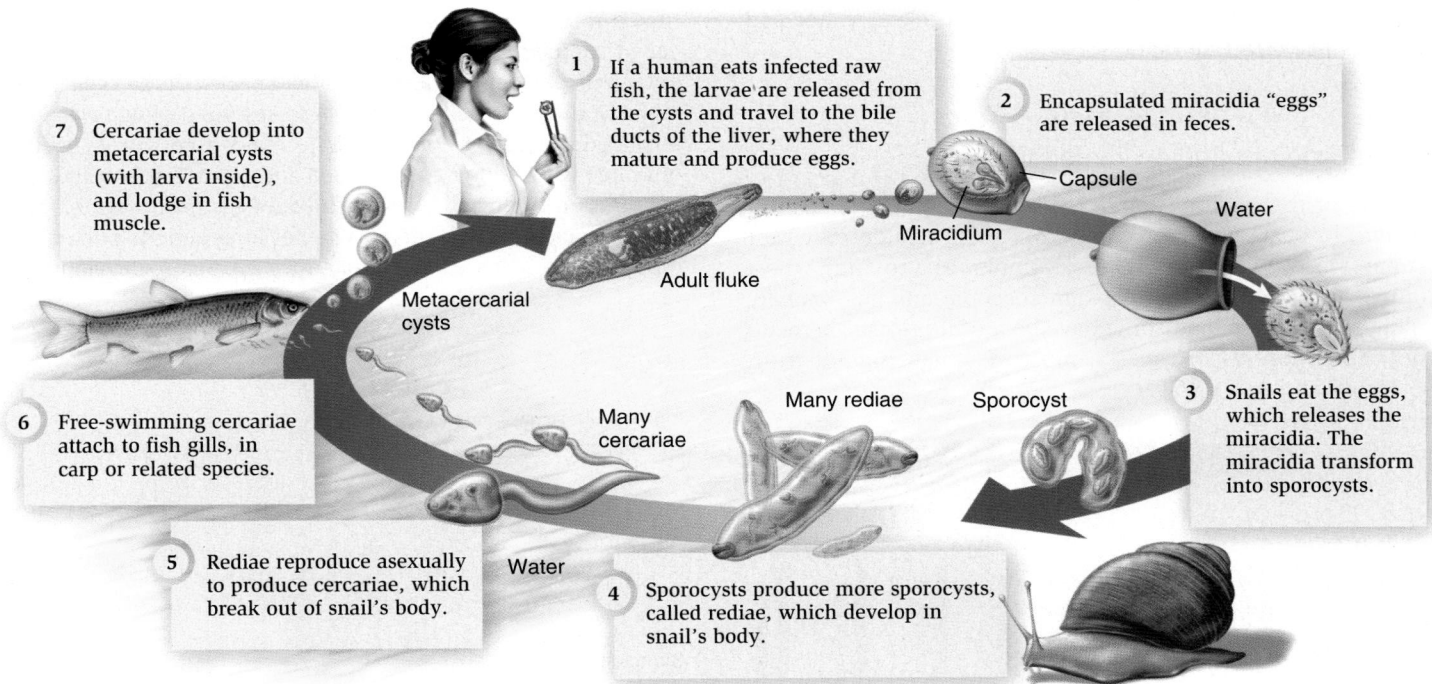

1 If a human eats infected raw fish, the larvae are released from the cysts and travel to the bile ducts of the liver, where they mature and produce eggs.

2 Encapsulated miracidia "eggs" are released in feces.

3 Snails eat the eggs, which releases the miracidia. The miracidia transform into sporocysts.

4 Sporocysts produce more sporocysts, called rediae, which develop in snail's body.

5 Rediae reproduce asexually to produce cercariae, which break out of snail's body.

6 Free-swimming cercariae attach to fish gills, in carp or related species.

7 Cercariae develop into metacercarial cysts (with larva inside), and lodge in fish muscle.

Adult fluke — Miracidium — Capsule — Water — Sporocyst — Many rediae — Many cercariae — Metacercarial cysts — Water

Figure 33.8 The complete life cycle of a trematode, as illustrated by the life cycle of the Chinese liver fluke (*Clonorchis sinensis*).

species, such as pigs or cattle to begin their life cycle and humans to complete their development. Many tapeworms can live inside humans who consume undercooked, infected meat—hence the value of thoroughly cooking meat to kill any parasites inside.

The life history of trematodes is even more complex than that of cestodes, involving multiple hosts. The first host, called the intermediate host, is usually a mollusk, and the final host, or definitive host, is usually a vertebrate, but often a second or even

a third intermediate host is involved. In the case of the Chinese liver fluke (*Clonorchis sinensis*), the adult parasite lives and reproduces in the definitive host, and the resultant "eggs" (encapsulated miracidia) pass from the host via the feces (**Figure 33.8**). An intermediate host, such as a snail, eats the eggs. The miracidia are released and transform into sporocysts. The sporocysts produce more sporocysts asexually called rediae. The rediae reproduce asexually to produce cercariae. Cercariae bore their way out of the snail and infect their definitive hosts directly by boring into their feet when in water or, for species with a second intermediate host (as in the Chinese liver fluke), by entering fish. Here, the cercariae develop into metacercarial cysts (juvenile flukes) and lodge in fish muscle, which the definitive host will eat. In the definitive host, the cyst protects the metacercaria from the host's gastric juices. In the small intestine, the metacercariae travel to the liver and grow into adult flukes and the life cycle begins anew. The life cycle of a trematode can thus involve at least seven stages: adult, egg (encapsulated embryo), miracidium, sporocyst, rediae, cercaria, and metacercaria. Because of the low probability of each larva reaching a suitable host, trematodes must produce large numbers of offspring to ensure that some survive.

Blood flukes, *Schistosoma* spp., are the most common parasitic trematodes infecting humans and cause the disease known as schistosomiasis. Over 200 million people worldwide, primarily in tropical Asia, Africa, and South America, are infected with schistosomiasis. The inch-long adult flukes can live for years in human hosts, and the release of eggs may cause chronic inflammation and blockage in many organs. Untreated schistosomiasis can lead to severe damage to the liver, intestines, and lungs and can eventually lead to death. Access to clean water can greatly reduce infection rates.

Members of the Phylum Rotifera Have a Pseudocoelom and a Ciliated Crown

Members of the phylum Rotifera (from the Latin *rota*, wheel, and *fera*, to bear) get their name from their ciliated crown or **corona**, which, when beating, looks similar to a rotating wheel (**Figure 33.9**). Most rotifers are microscopic animals, usually less than 1 mm long, and some have beautiful colors. There are about 2,000 species of rotifers, most of which inhabit fresh water, with a few marine or terrestrial species. Most often they are bottom-dwelling organisms, living on the pond floor or along lakeside vegetation.

Rotifers have an alimentary canal, a digestive tract with a separate mouth and anus, which means that they can feed continuously. The corona creates water currents that propel the animal through the water and that waft small planktonic organisms or decomposing organic material toward the mouth. The mouth opens into a circular muscular pharynx called a **mastax**, which has jaws for grasping and chewing. The mastax, which in some species can protrude through the mouth to seize small prey, is a structure unique to rotifers. The body of the rotifer bears a jointed foot with one to four toes. **Pedal glands** in the foot secrete a sticky substance that aids in attachment to the substrate. The internal organs lie within a pseudocoelom, a fluid-filled body cavity that is not completely lined with mesoderm. The pseudocoelom serves as a hydrostatic skeleton and as a medium for the internal transport of nutrients and wastes. Rotifers also have a pair of protonephridia with flame bulbs that collect excretory and digestive waste and drain into a cloacal bladder, which passes waste to the anus.

Reproduction in rotifers is unique. In some species, unfertilized diploid eggs that have not undergone meiotic division, called amictic eggs, develop into females through a process known as **parthenogenesis**. In other species, some eggs undergo meiosis and become haploid. These so-called mictic eggs, if unfertilized, develop into degenerate males that cannot feed and only live long enough to produce and release sperm that fertilize other mictic eggs. These fertilized eggs form zygotes, which have a thick shell and can survive for long periods of harsh conditions, such as if a water supply dries up, before developing into

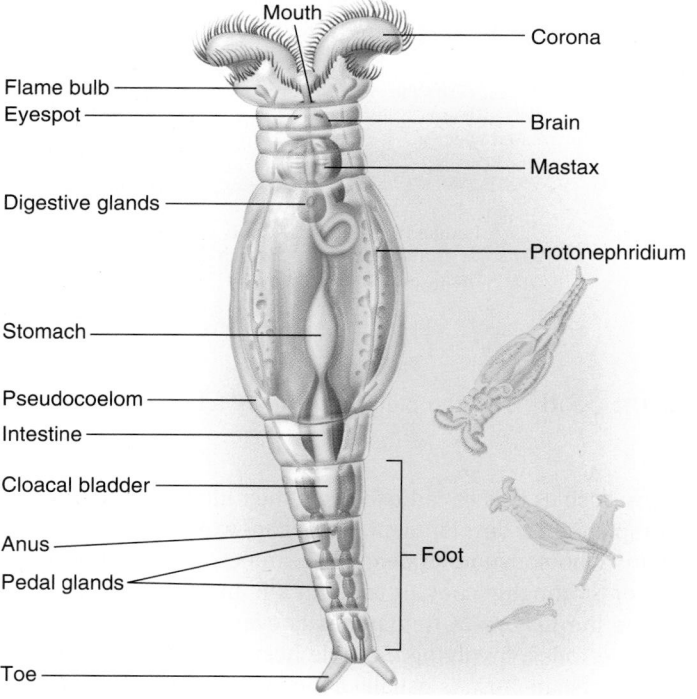

Figure 33.9 Body plan of a common rotifer, *Philodina* genus.

new females. Because the tiny zygotes are easily transported, rotifers show up in the smallest of aquatic environments, such as roof gutters or birdbaths.

The Lophophorata Includes Three Closely Related Phyla: Phoronida, Bryozoa, and Brachiopoda

The Lophophorates consist of three distinct phyla: the Phoronida, the Bryozoa, and the Brachiopoda. They all possess a lophophore, a ciliary feeding device (refer back to Figure 32.10a), and a true coelom (refer back to Figure 32.4a). The lophophore is a circular fold of the body wall that bears tentacles that draw water towards the mouth. Because a thin extension of the coelom penetrates each tentacle, the tentacles also serve as a respiratory device. Gases diffuse across the tentacles and into or out of the coelomic fluid and are carried throughout the body.

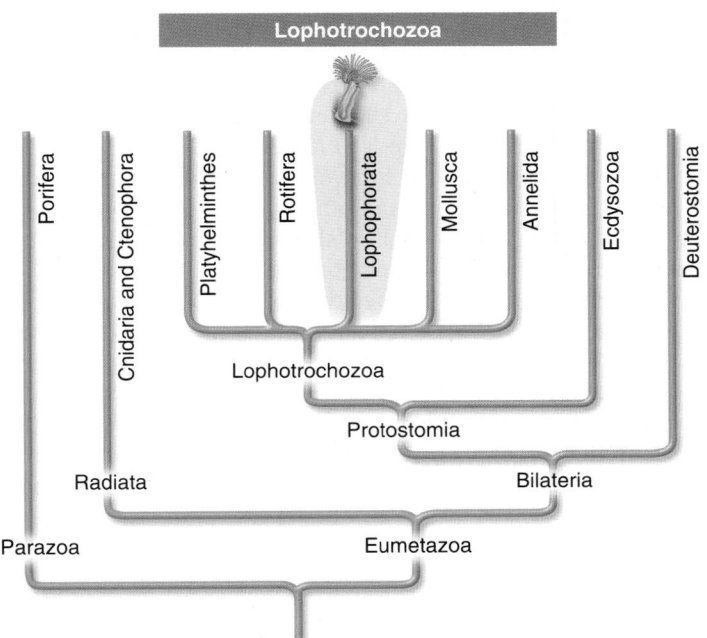

All three phyla have a U-shaped alimentary canal, with the anus located near the mouth but outside of the lophophore.

Phylum Phoronida Members of the Phoronida (from the Greek *phoros*, bearer, and the Latin *nidus*, nest) are elongated, tube-dwelling marine worms ranging in size from 1 mm to 50 cm in length. They live in a tough, leather-like chitinous tube that they secrete and that is often buried in the ground so only the lophophore sticks out (**Figure 33.10a**). The lophophore can be retracted quickly in the presence of danger. Only about 15 species of phoronids are found worldwide.

Phylum Bryozoa The bryozoans (from the Greek *bryon*, moss, and *zoon*, animal) are small colonial animals, most of which are less than 0.5 mm long, that can be found encrusted on rocks in shallow aquatic environments. They look very much like plants. There are about 4,000 species, many of which are fouling organisms that encrust boat hulls and have to be scraped off periodically. Within the colony, each animal secretes and lives inside a nonliving case called a **zoecium** (**Figure 33.10b**). The walls of the zoecium may be composed of chitin or calcium carbonate. For this reason, bryozoans have been important reef-builders, and since they date back to the Ordovician era, many fossil forms have been discovered and identified.

Phylum Brachiopoda Brachiopods (from the Greek *brachio*, arm, and *podos*, foot) are marine organisms with two shell halves, much like modern clams (**Figure 33.10c**). Unlike bivalve mollusks, however, which have a left and right valve (side) of the shell, brachiopods have a dorsal and ventral valve. Brachiopods are bottom-dwelling species that attach to the substrate via a muscular pedicle. While they are a relatively small group, with about 300 living species, brachiopods flourished in the Paleozoic and Mesozoic eras, and about 30,000 fossil species have been identified. Some of these fossil forms tell of organisms that reached 30 cm in length, although their current relatives are only 0.5–8.0 cm long.

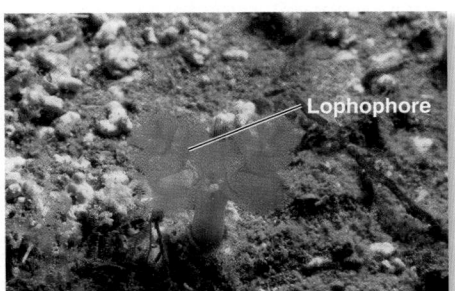

(a) A phoronid worm (*Phoronis californica*), buried in the sand with the lophophores extended.

(b) Bryozoans, such as this *Plumatella repens*, are colonial lophophorates that reside in a nonliving case called a zoecium.

(c) Brachiopods including this northern lamp shell (*Terebratulina septentrionalis*) have a dorsal and ventral shell.

Figure 33.10   Lophophorates.

Biological inquiry: What are the two main functions of the lophophore?

The Mollusca Is a Large Phylum Containing Snails, Slugs, Oysters, Clams, Octopuses, and Squids

Mollusks (from the Latin *mollis*, soft) constitute a very large phylum, with over 100,000 living species, including organisms as diverse as snails, clams and oysters, cephalopods, and chitons. They are an ancient group, as evidenced by the classification of about 35,000 fossil species. Mollusks have a considerable economic, aesthetic, and ecological importance to humans. Many serve as sources of food, including scallops, oysters, clams, and squid. A significant industry involves the farming of oysters to produce cultured pearls, and rare and beautiful mollusk shells are extremely valuable to collectors. Snails and slugs can damage vegetables and ornamental plants, and boring mollusks can penetrate wooden ships and wharfs. Mollusks are intermediate hosts to many parasites, and several exotic species have become serious pests. For example, populations of the zebra mussel (*Dreissena polymorpha*) were introduced into North America from Asia via ballast water from transoceanic ships. Since their introduction, they have spread rapidly throughout the Great Lakes and an increasing number of inland waterways, significantly impacting native organisms and clogging water intake valves to municipal water treatment plants around the lakes.

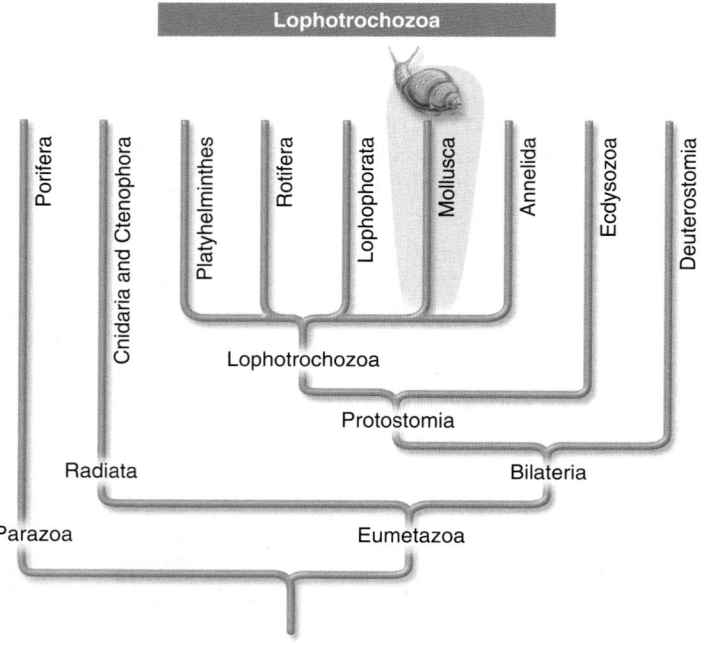

One common feature of the mollusks is their soft body, which exists, in many species, under a protective external shell. Most mollusks are marine, although some have colonized fresh water. Many snails and slugs have even moved onto land, but they survive only in humid areas and where the calcium necessary for shell formation is abundant in the soil. The ability to colonize freshwater and terrestrial habitats has led to a diversification of mollusk body plans. Thus, we again see how organismal diversity is related to environmental diversity.

While great variation in morphology occurs between classes, mollusks have a basic body plan consisting of three parts (**Figure 33.11**). A muscular **foot** is usually used for movement, and a **visceral mass** containing the internal organs rests atop the foot. The **mantle**, a fold of skin draped over the visceral mass, secretes a shell in those species that form shells. The mantle often extends beyond the visceral mass, creating a chamber called the **mantle cavity**, which houses delicate **gills**, specialized filamentous organs that are rich in blood vessels. A continuous current of water, often induced by cilia present on the gills or by muscular pumping, flushes out the wastes from the mantle cavity and brings in new oxygen-rich water.

Mollusks are coelomate organisms, but the coelom is confined to a small area around the heart. The mollusks' organs are served with oxygen and nutrients via a circulatory system. Mollusks have an **open circulatory system** with a heart that pumps body fluid called hemolymph through vessels and into sinuses, which are open, fluid-filled cavities. In comparison, in a closed circulatory system, the fluid called blood is always contained within vessels. The organs and tissues are thus continually bathed in hemolymph. From these sinuses, the hemolymph drains into vessels that take it to the gills and then back to the heart. The anus and pores from organs called **metanephridia**, which extract nitrogenous and other wastes, discharge into the mantle cavity. The metanephridial ducts may also serve to discharge sperm or eggs from the gonads.

The mollusk's mouth may contain a **radula**, a unique, protrusible, tonguelike organ that has many teeth and is used to eat plants, scrape food particles off rocks, or, if the mollusk is predatory, bore into shells of other species and tear flesh. In the cone shells (*Conus* spp.), the radula is reduced to a few poison-injecting teeth on the end of a long proboscis that is cast about in search of prey, such as a worm or even a fish (**Figure 33.12**). Some Indo-

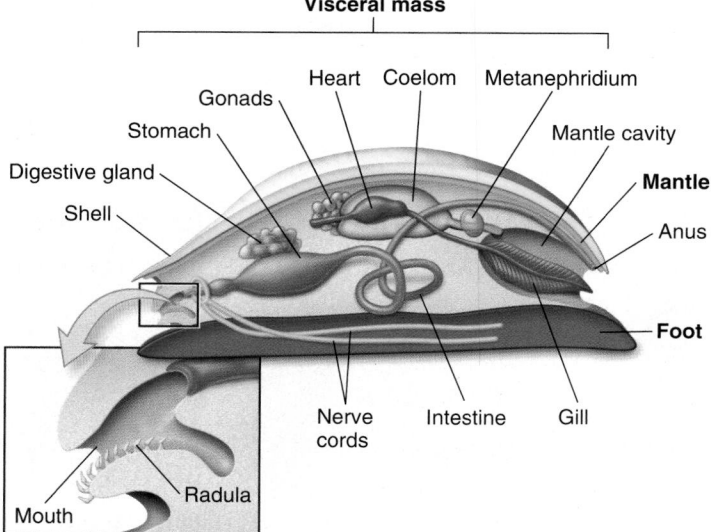

Figure 33.11 The mollusk body plan. The generalized body plan of a mollusk includes the characteristic foot, mantle, and visceral mass.

Figure 33.12 **Some gastropods are predators.** The cone shell, *Conus* spp., uses its long proboscis, equipped with poison-injecting teeth, to paralyze its prey—usually a fish.

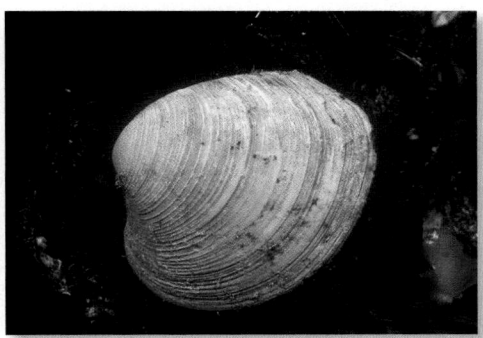

Figure 33.13 **Bivalve shells have growth rings.** Quahog clams (*Mercenaria mercenaria*), also known as cherrystones or littlenecks (depending on their size), can live over 20 years.

Figure 33.14 **A veliger larva.** The immature free-swimming stage of molluscans develops from the trochophore larva.

Pacific cone shell species produce a neuromuscular toxin that can kill humans. Other mollusks, particularly bivalves, are suspension feeders that filter water brought in by ciliary currents.

Most shells are complex three-layered structures secreted by the mantle that continue to grow as the mollusk grows. Shell growth is often seasonal, resulting in distinct growth lines on the shell, much the same as tree rings (**Figure 33.13**). Using shell growth patterns, biologists have discovered some bivalves that are over 100 years old. The innermost layer of the shells of oysters, mussels, abalone, and other mollusks is a smooth, iridescent lining called **nacre**, which is commonly known as mother-of-pearl and is often collected from abalone shells for jewelry. Actual pearl production in mollusks, primarily oysters, occurs when a foreign object, such as a grain of sand, becomes lodged between the shell and the mantle, and layers of nacre are laid down around it to reduce the irritation.

Most mollusks have separate sexes, although some are hermaphroditic. Gametes are usually released into the water, where they mix and fertilization occurs. In some snails, however, fertilization is internal, with the male inserting sperm directly into the female. Internal fertilization was a key evolutionary development, enabling some snails to colonize land, and can be considered a critical innovation that fostered extensive adaptive radiation. In many species, reproduction involves the production of a trochophore larva that develops into a **veliger**, a free-swimming larva that has a rudimentary foot, shell, and mantle (**Figure 33.14**).

Of the eight molluscan classes, the four most common are the polyplacophora (chitons), gastropoda (snails and slugs), bivalvia (clams and mussels), and cephalopoda (octopuses, squid, and nautiluses) (**Table 33.3**). The class Gastropoda (from the Greek *gaster*, stomach, and *podos*, foot) is the largest group of mollusks and encompasses about 75,000 living species, including snails, periwinkles, limpets, and other shelled members. The class also includes species such as slugs and nudibranchs, whose shells have been greatly reduced or completely lost during their evolution (**Figure 33.15**). Most are marine or freshwater species, but some species, including snails and slugs, have

Table 33.3	Main Classes and Characteristics of Mollusks	
	Class and examples (est. # of species)	**Class characteristics**
	Polyplacophora: Chitons (860)	Marine; eight-plated shell
	Gastropoda: Snails, slugs, nudibranchs (75,000)	Marine, freshwater, or terrestrial; most with coiled shell, but shell absent in slugs and nudibranchs; radula present
	Bivalvia: Clams, mussels, oysters (30,000)	Marine or freshwater; shell with two halves or valves; primarily filter feeders with siphons
	Cephalopoda: Octopuses, squids, nautiluses (780)	Marine; predatory, with tentacles around mouth, often with suckers; shell often absent or reduced; closed circulatory system; jet propulsion via siphon

also colonized land. Most gastropods are slow-moving animals that are weighed down by their shell. Unlike bivalves, gastropods have a one-piece shell, into which the animal can withdraw to escape predators.

The 780 species of Cephalopoda (from the Greek *kephale*, head, and *podos*, foot) are the most morphologically complex of the mollusks and indeed among the most complex of all invertebrates. They include the octopuses, squids, cuttlefish, and nautiluses. Most are fast-swimming marine predators that range from organisms just a few centimeters in size to the giant squid (*Architeuthis*) which is known to reach over 17 m in length and 2 tons in weight. A cephalopod's mouth is surrounded by many long tentacles commonly armed with suckers. Octopuses have eight arms with suckers, and squids and cuttlefish have ten arms—eight with suckers and two long tentacles with suckers limited to their ends. Nautiluses have between 60 to 90 tentacles around the mouth.

All cephalopods have a beaklike jaw that allows them to bite their prey, and some, such as the blue-ringed octopus (*Hapalochlaena lunulata*), deliver a deadly poison through their saliva (**Figure 33.16**). Only one group, the nautiluses, has retained its external shell. In octopuses, the shell is not present, and in squid and cuttlefish, it is greatly reduced and internal. However, the fossil record is full of shelled cephalopods, called ammonites, some of which were as big as truck tires (**Figure 33.17**). They became extinct at the end of the Cretaceous period, although the reasons for this are not well understood.

The foot of some cephalopods has become modified into a muscular siphon. Water drawn into the mantle cavity is quickly expelled through the siphon, propelling the organisms forward or backward in a kind of jet propulsion. Such vigorous movement requires powerful muscles and a very efficient circulatory system to deliver oxygen and nutrients to the muscles. Cephalopods are the only mollusks with a **closed circulatory system**, in which blood flows throughout an animal entirely within a series of vessels. One of the advantages of this type of system is that the heart can pump blood through the tissues rapidly. The blood of cephalopods contains the copper-rich protein hemocyanin for transporting oxygen. Less efficient than the iron-rich hemoglobin of vertebrates, hemocyanin gives the blood a blue color.

The nautiluses are impeded by a coiled, chambered shell and do not move as fast as the jet-propelled squids and octopuses (**Figure 33.18**). As it grows, the nautilus secretes a new chamber and seals off the old one with a **septum**. The older

Figure 33.17   A fossil ammonite.   These shelled cephalopods were abundant in the Cretaceous period.

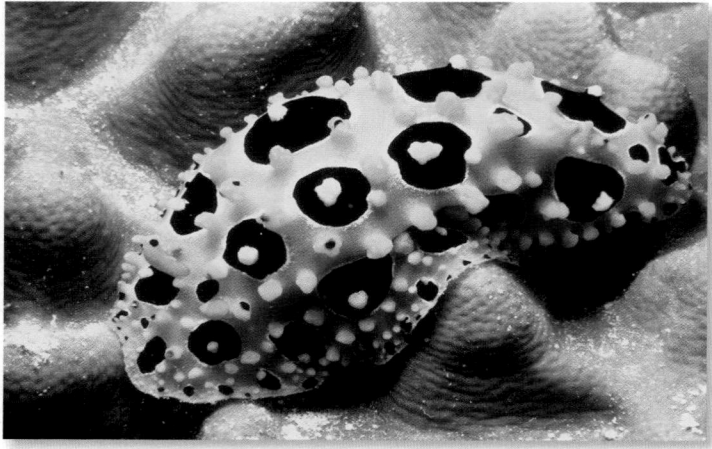

Figure 33.15   A sea slug (*Phyllidia ocellata*).   The sea slugs, or nudibranchs, are a gastropod subclass whose members have lost their shell altogether.

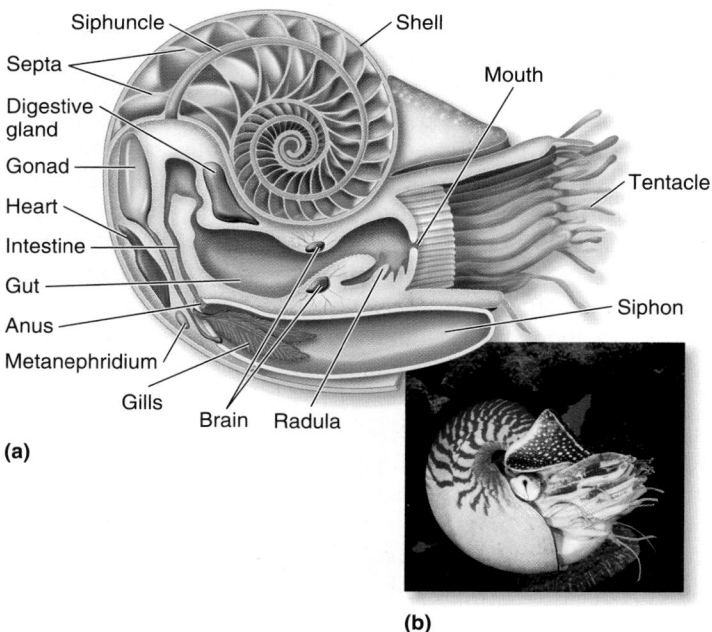

Siphuncle
Shell
Septa
Digestive gland
Gonad
Heart
Intestine
Gut
Anus
Metanephridium
Gills
Brain   Radula
Mouth
Tentacle
Siphon

(a)

(b)

Figure 33.18   The nautilus.   (a) A longitudinal section of a nautilus, showing the coiled shell with many chambers. The animal secretes a new chamber each year and lives only in the new one. (b) The chambered nautilus (*Nautilus pompilius*).

Figure 33.16   The blue-ringed octopus (*Hapalochlaena lunulata*) is highly poisonous.

chambers are gas filled and act as buoyancy chambers. A thin strip of living tissue called the siphuncle removes liquid from the old chamber and replaces it with gas. The gas pressure within the chambers is only 1 atmosphere, despite the fact that nautiluses may be swimming at 400 m depths at a pressure of about 40 atmospheres. The shell's structure is strong enough to withstand this amount of pressure differential.

Cephalopods have a well-developed nervous system and brain that support their active lifestyle. Their sense organs, especially their eyes, are also very well developed. Many cephalopods (with the exception of nautiluses) have an ink sac that contains the pigment melanin; the sac can be emptied to provide a "smokescreen" to confuse predators. In many species, melanin is also distributed in special pigment cells in the skin that can produce color changes. Octopuses often change color when alarmed or during courtship, and they can rapidly change color to blend in with their background and escape detection. The central nervous system of the octopus is among the most complex in the invertebrate world. Behavioral biologists have demonstrated that octopuses can behave in sophisticated ways, and scientists are currently debating to what degree they are capable of learning by observation.

FEATURE INVESTIGATION

Fiorito and Scotto's Experiments Showed Invertebrates Can Exhibit Sophisticated Observational Learning Behavior

We tend to think of the ability to learn from others as a vertebrate phenomenon, especially among species that live in social groups. However, in 1992, Italian researchers Graziano Fiorito and Pietro Scotto demonstrated that octopuses can learn by observing the behavior of other octopuses (**Figure 33.19**). This was a surprising finding, in part because *Octopus vulgaris*, the species they studied, lives a solitary existence for most of its life.

In their experiments, octopuses were trained to attack either a red ball or a white ball by use of a reward (a small piece of fish placed behind the ball that it could not see) and a punishment (a small electric shock for choosing the wrong ball). This type of learning is called classical conditioning (see Chapter 55). Because octopuses are color blind, they must have been distinguishing between the relative brightness of the balls. Octopuses were considered to be trained when they made no mistakes in five trials. Observer octopuses in adjacent tanks were then allowed to watch the trained octopuses attacking the balls. In the third part of the experiment, the observer octopuses were themselves tested. In these cases, observers nearly always attacked the same color ball as they had observed the demonstrators attacking. In addition, learning by observation was achieved more quickly than the original training. This remarkable behavior is considered by some as the precursor to more complex forms of learning, including problem solving.

Figure 33.19   Observational learning in octopuses.

HYPOTHESIS Octopuses can learn by observing another's behavior.

STUDY LOCATION Laboratory setting with *Octopus vulgaris* collected from the Bay of Naples, Italy.

	Experimental level		Conceptual level
1 Train 2 groups of octopuses, one to attack white balls, one to attack red.	Reward choice of correct ball (with fish) and punish choice of incorrect ball (with electric shock). Training complete when octopus makes no "mistakes" in 5 trials.		Conditions a demonstrator octopus to attack a particular color of ball.
2 In adjacent tanks, allow observer octopus to watch trained demonstrator octopus.	Observer Demonstrator		Observer octopus may be learning the correct ball by watching the demonstrator octopus.

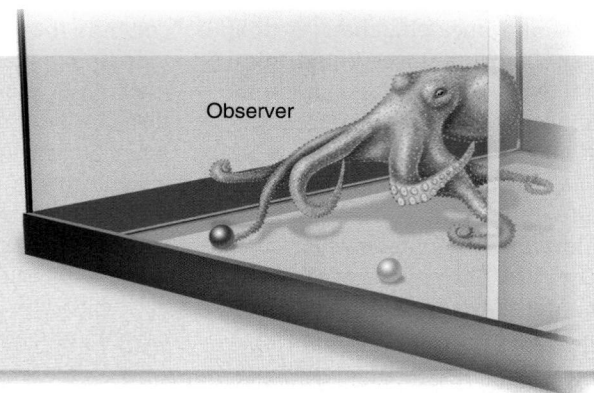

3 Drop balls into the tank of the observer octopuses. Test the observer octopus to see if it makes the same decisions as the demonstrator octopus.

If the observer octopus is learning from demonstator octopus, the observer octopus should attack the ball of the same color as the demonstrator octopus was trained to attack.

4 **THE DATA**

Participant	Color of ball chosen in 5 trials*	
	Red	**White**
Observers (watched demonstrator attack red)	4.31	0.31
Observers (watched demonstrator attack white)	0.40	4.10
Untrained (did not watch demonstrations)	2.11	1.94

*Average of 5 trials, data do not always sum to 5, because some trials resulted in no balls being chosen.

The Phylum Annelida Consists of the Segmented Worms

If you look at an earthworm, you will see little rings all down its body. Indeed, the phylum name Annelida is derived from the Latin, *annulus*, meaning little ring. Each ring is a distinct segment of the annelid's body, with each segment separated from the one in front and the one behind by a septum (**Figure 33.20a**). Segmentation, the division of the body into nearly identical subunits, is a critical evolutionary innovation in the annelids and confers at least three major advantages. First, many components of the body are repeated in each segment, including blood vessels, nerves, and excretory and reproductive organs. Excretion is accomplished by metanephridia, paired excretory organs in every segment that extract waste from the blood and coelomic fluid, emptying it to the exterior via pores in the skin (look ahead to Figure 49.9). If the excretory organs in one segment fail, the organs of another segment will still function.

Second, annelids possess a fluid-filled coelom that acts as a hydrostatic skeleton. In unsegmented coelomate animals, muscle contractions can distort the entire body during movement. However, such distortion is minimized in segmented animals, which allows for more effective locomotion over solid surfaces. In an earthworm, when the circular muscles around a segment contract against the hydrostatic skeleton, that segment becomes elongated. When the longitudinal muscles contract, the segment becomes compact. Waves of muscular contraction ripple down the segments, which elongate or contract independently.

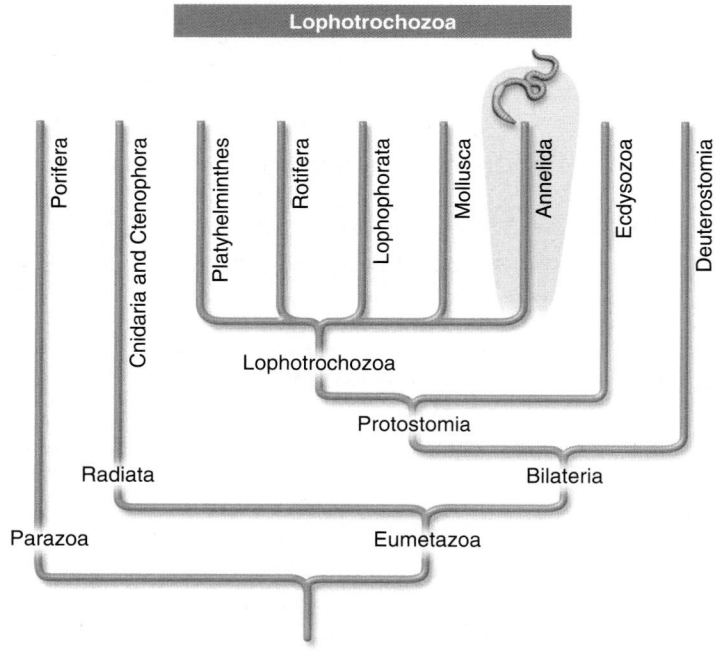

Third, segmentation also permits specialization of some segments, although such specialization is only minimally present at the annelid's anterior end. Annelids have a relatively sophisticated nervous system involving a pair of cerebral ganglia that connect to a subpharyngeal ganglion (**Figure 33.20b**).

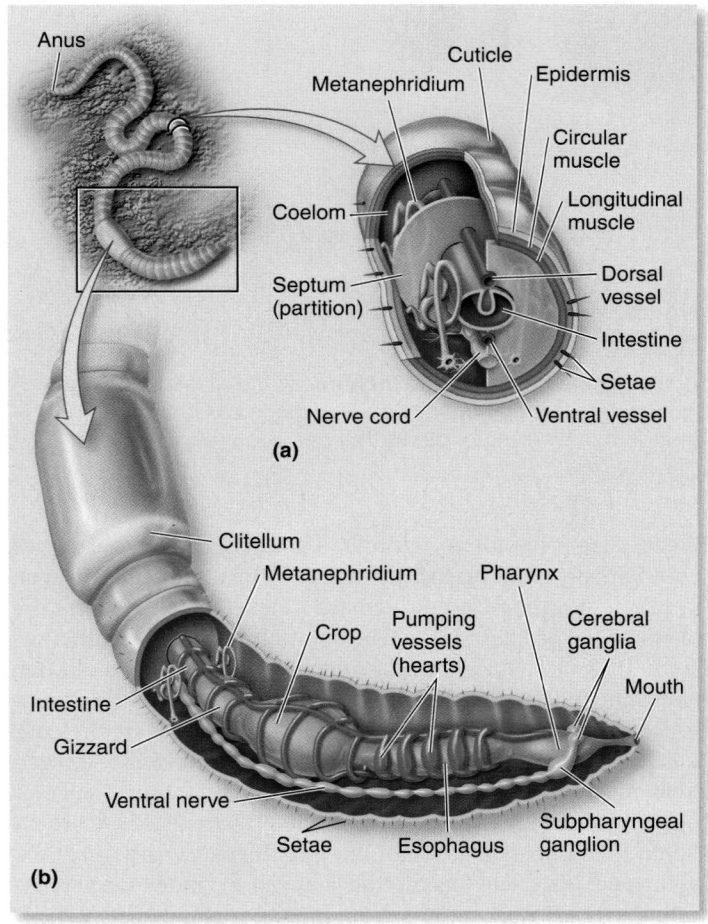

Anus
Cuticle
Metanephridium — Epidermis
Circular muscle
Coelom
Longitudinal muscle
Septum (partition)
Dorsal vessel
Intestine
Setae
Nerve cord
Ventral vessel
(a)

Clitellum
Metanephridium — Pharynx
Crop — Pumping vessels (hearts) — Cerebral ganglia
Mouth
Intestine
Gizzard
Ventral nerve
Subpharyngeal ganglion
Setae — Esophagus
(b)

Figure 33.20 **The segmented body plan of an annelid, as illustrated by an earthworm.** (a) Individual segments are separated by a septum. (b) The segmented nature of the worm is apparent externally as well as internally.

Biological inquiry: What are some of the advantages of segmentation?

Table 33.4	Main Classes and Characteristics of Annelids	
	Class and examples (est. # of species)	**Class characteristics**
	Polychaeta: Marine worms (10,000)	Well-developed head; usually free living; parapodia present
	Oligochaeta: Terrestrial and freshwater worms such as earthworms (3,500)	Undeveloped head; setae; no parapodia
	Hirudinea: Leeches (630)	Mostly ectoparasites; suckers present at both ends; flattened body; reduced coelom; no setae

From there, a large ventral nerve cord runs down the entire length of the body. The ventral nerve chord is unusual because it contains a few very large nerve cells called **giant axons** that facilitate high-speed nerve conduction and rapid responses to stimuli. Such axons are found in other invertebrates, the best known of which is the squid.

Annelids essentially have a double transport system, since both the circulatory system and the coelomic fluid carry nutrients, wastes, and respiratory gases, to some degree. The circulatory system is usually closed, with dorsal and ventral vessels connected by five pairs of pumping vessels that serve as muscular hearts. The blood of most annelid species contains the respiratory pigment hemoglobin, an iron-containing protein involved in oxygen transport in annelids as well as in vertebrates. Respiration occurs directly through the permeable skin surface, which restricts annelids to moist environments. The digestive system is complete and unsegmented, with many specialized regions: mouth, pharynx, esophagus, crop, gizzard, intestine, and

anus. Sexual reproduction involves two individuals, often of separate sexes, but sometimes hermaphrodites, which exchange sperm via internal fertilization. In some species, asexual reproduction by fission occurs, in which the posterior part of the body breaks off and forms a new individual.

Annelids are a large phylum with about 15,000 described species. Its members include the familiar earthworm, marine polychaete worms, and leeches, and range in size from less than 1 mm to enormous Australian earthworms that can reach a size of 3 m. All annelids except the leeches have chitinous bristles called **setae** on each segment. In one class, the polychaetes, these are situated on fleshy, footlike **parapodia** ("almost feet") that are pushed into the substrate to provide traction during movement. Many annelid species burrow into the Earth or into muddy marine sediments and extract nutrients from ingested soil or mud. Some annelids also feed on dead or living vegetation, while others are predatory or parasitic.

The phylum consists of three main classes: the Polychaeta, Oligochaeta, and Hirudinea (**Table 33.4**). Some biologists have recently suggested Oligochaeta and Hirudinea be combined into one larger class called the Clitellata, because they share a common structure called a clitellum, a glandular region of the body that has a role in reproduction. However, this newer taxonomic grouping is still being debated.

Class Polychaeta With over 10,000 species, the Polychaeta is the most species-rich class of the annelids. Many polychaetes are brightly colored, and all have many long setae bristling out of their body (polychaete means "many bristles"). Most of them are marine organisms, living in burrows in the mud or sand, or in rock crevices, and are often abundant in the intertidal mudflats. They are important prey for predators such as fish and crustaceans. The polychaete head is well developed and, in predatory species, may exhibit powerful jaws. Some species are filter feeders and have a crown of tentacles that sticks up out of the mud while the bulk of the worm remains hidden.

Class Oligochaeta The Oligochaeta (meaning "few bristles") includes the common earthworms and many species of freshwater worms. Earthworms play a unique and beneficial role in conditioning the soil, primarily due to the effects of their burrows and castings. Earthworms ingest soil and leaf tissue to extract nutrients and in the process create burrows in the Earth. As plant material and soil passes through the earthworm's digestive system, it is finely ground in the gizzard into smaller fragments. Once excreted, this material—called castings—enriches the soil. Because a worm can eat is own weight in soil every day, worm castings on the soil surface can be extensive. The biologist Charles Darwin was interested in earthworm activity, and his last work, *The Formation of Vegetable Mould, through the Actions of Worms, with Observations on Their Habits*, was the first detailed study of earthworm ecology. In it, he wrote, "All the fertile areas of this planet have at least once passed through the bodies of earthworms."

Class Hirudinea Leeches (class Hirudinea) are primarily found in freshwater environments, but there are also some marine species as well as terrestrial species that inhabit warm, moist areas such as tropical forests. Leeches have a fixed number of segments, usually 34, though the septa have disappeared in most species. All leeches feed on other organisms, often as blood-sucking parasites of vertebrates. They have powerful suckers at both ends of the body, and the anterior sucker is equipped with razor-sharp jaws that can bore or slice into the host's tissues. The salivary secretion of leeches (hirudin) acts as an anticoagulant to stop blood clotting. Leeches can suck up to several times their own weight in blood. As such, they were once used in the medical field in the practice of bloodletting, the withdrawal of often considerable quantities of blood from a patient in the belief that this would prevent or cure illness and disease. Even today, leeches may be used after surgeries, particularly those involving the reattachment of digits (**Figure 33.21**). In these cases, the blood vessels are not fully reconnected and much excess blood accumulates, causing swelling. This excess blood switches off the delivery of new blood and stops the formation of new arteries. If leeches remove the accumulated blood, new capillaries will be more likely to form, and the tissues will become healthy.

Unlike cestode and trematode flatworms, which are internally parasitic and quite host specific, leeches are generally external parasites that feed on a broad range of hosts, including fish, amphibians, and mammals. However, there are always exceptions. *Placobdelloides jaegerskioeldi* is a parasitic leech that lives only in the rectum of hippopotamuses.

33.4 Ecdysozoa: The Nematodes and Arthropods

The Ecdysozoa is the sister group to the Lophotrochozoa. While the separation is supported by molecular evidence, the Ecdysozoa is named for a morphological characteristic, the physical phenomenon of ecdysis, or molting (refer back to Figure 32.11).

Figure 33.21   A leech, a member of the class Hirudinea. This species, *Hirudo medicinalis*, is sucking blood from a hematoma, a swelling of blood that can occur after surgery.

All ecdysozoans possess a **cuticle**, a nonliving cover that serves to both support and protect the animal. Once formed, however, the cuticle typically cannot increase in size, which restricts the growth of the animal inside. The solution for growth is the formation of a new, softer cuticle under the old one. The old one then splits open and is sloughed off, allowing the new, soft cuticle to expand to a bigger size before it hardens. Where the cuticle is thick, as in arthropods, it impedes the diffusion of oxygen across the skin. Such species acquire oxygen by lungs, gills, or a set of branching, air-filled tubes called tracheae. There are no cilia on the cuticle for locomotion, and thus a variety of appendages specialized for locomotion evolved in many species, including legs for walking or swimming, and wings for flying.

The ability to shed the cuticle opened up developmental options for the ecdysozoans. For example, many species undergo a complete metamorphosis, changing from a wormlike larva into a winged adult. Animals with internal skeletons cannot do this because growth only occurs by adding more minerals to the existing skeleton. Another significant adaptation is the development of internal fertilization, in which the male deposits sperm directly into the female, where fertilization takes place. This trait, which allows animals to breed on dry land, evolved independently in the vertebrates.

Because of these innovations, ecdysozoans are an incredibly successful group. Of the eight ecdysozoan phyla, we will consider the most common two: the nematodes and arthropods. The grouping of nematodes and arthropods is a relatively new idea and implies that the process of molting arose only once in animal evolution. In support of this, certain hormones that stimulate molting have been discovered to exist in nematodes as well as arthropods.

The Phylum Nematoda Consists of Small Pseudocoelomate Worms Covered by a Tough Cuticle

The nematodes (from the Greek *nematos*, thread), also called roundworms, are small, thin worms that range from less than

1 mm to about 5 cm (**Figure 33.22**), although some parasitic species measuring 1 m or more have been found in the placenta of sperm whales. Nematodes are ubiquitous organisms that exist in nearly all habitats, from the poles to the tropics. They are found in the soil, in both freshwater and marine environments, and inside plants and animals as parasites. A shovelful of soil may contain a million nematodes. Over 20,000 species are known, but there are probably at least five times as many undiscovered species.

Figure 33.22   Scanning electron micrograph of a nematode within a plant leaf.

and John Sulston for their studies of the genetic regulation of development and programmed cell death in *C. elegans*. This nematode has 1,090 cells but 131 die, leaving exactly 959 cells. The cells die via a genetically controlled cell death. Many diseases in humans, including acquired immunodeficiency syndrome (AIDS), cause extensive cell death, while others, such as cancer and autoimmune diseases, reduce cell death so that cells that should die do not. Researchers are studying the process of programmed cell death in *C. elegans* in the hope of finding treatments for these and other human diseases.

A large number of nematodes are parasitic in humans and other vertebrates. The large roundworm (*Ascaris lumbricoides*) is a parasite of the small intestine that can reach up to 30 cm in length. Over a billion people worldwide carry this parasite. While infections are most prevalent in tropical or developing countries, the prevalence of *A. lumbricoides* is relatively high in rural areas of the southeastern U.S. Eggs pass out in feces and can remain viable in the soil for years. Eggs require ingestion before hatching into an infective stage. Hookworms (*Necator americanus*), so named because their anterior end curves dorsally like a hook, are also parasites of the human intestine. The eggs pass out in feces, and recently hatched hookworms can penetrate the skin of a host's foot to establish a new infection. In areas with modern plumbing, these diseases are uncommon.

Pinworms (*Enterobius vermicularis*), while a nuisance, have relatively benign effects on their hosts. The rate of infection in the U.S., however, is staggering: 30% of children and 16% of adults are believed to be hosts. Adult pinworms live in the large intestine and migrate to the anal region at night to lay their eggs, which causes intense itching. The resultant scratching spreads the eggs. In the tropics, some 250 million people are infected with *Wuchereria bancrofti*, a fairly large (100 mm) worm that lives in the lymphatic system, blocking the flow of lymph, and, in extreme cases, causing elephantiasis or extreme swelling of the legs and other body parts (**Figure 33.23**). Females release tiny, live young called microfilariae, which are transmitted to new hosts via mosquitoes.

Nematodes have several distinguishing characteristics. A tough cuticle covers the body. The cuticle is secreted by the epidermis and is made primarily of **collagen**, a structural protein also present in vertebrates. The cuticle is shed periodically as the nematode grows. Beneath the epidermis are longitudinal muscles but no circular muscles, which means that muscle contraction results in more thrashing of the body than smoother wormlike movement. Nematodes possess a pseudocoelom that acts as a hydrostatic skeleton and a circulatory system. Diffusion of gases occurs through the cuticle. Roundworms have a complete digestive tract composed of a mouth, pharynx, intestine, and anus. The mouth often contains sharp, piercing organs called **stylets**, and the muscular pharynx functions to suck in food.

Nematode reproduction is usually sexual, with separate males and females, and fertilization takes place internally. Females are generally larger than males and can produce prodigious numbers of eggs, in some cases over 100,000 per day. Development in nematodes is easily observed because the organism is transparent and the generation time is short. For these reasons, the small, free-living nematode *Caenorhabditis elegans* has become a model organism for researchers to study (refer back to Figure 19.1b). In fact, the 2002 Nobel Prize in Medicine or Physiology was shared by Sydney Brenner, Robert Horvitz,

Figure 33.23 **Elephantiasis in a human leg.** The disease is caused by the nematode parasite *Wuchereria bancrofti*, which lives in the lymphatic system and blocks the flow of lymph.

The Phylum Arthropoda Contains the Insects, Crustaceans, and Spiders, All with Jointed Appendages

The arthropods (from the Greek *arthron*, joint, and *podos*, foot) constitute perhaps the most successful phylum on Earth. About three-quarters of all described living species are arthropods, and scientists have estimated that they are also numerically common, with an estimated 10^{18}, or a billion billion, individual organisms present on Earth. The huge success of the arthropods, in terms of their sheer numbers and diversity, is related to a body plan that permitted conquest of the major biomes on Earth, from the poles to the tropics, and from marine and freshwater habitats to dry land.

The body of a typical arthropod is covered by a hard cuticle, an **exoskeleton** (external skeleton) made of layers of chitin and protein. The cuticle can be extremely tough in some parts, as in the shells of crabs, lobsters, and even beetles, yet be soft and flexible in other parts, between body segments and segments of appendages, to allow for movement. In the class of arthropods called crustaceans, the exoskeleton is reinforced with calcium carbonate to make it extra hard. The exoskeleton provides protection and also a point of attachment for muscles, all of which are internal. It is also relatively impermeable to water, a feature that may have enabled many arthropods to conserve water and colonize land, in much the same way as a tough seed coat allowed plants to colonize land (see Chapter 30). From this point of view, the development of a hard cuticle was a critical innovation. It also reminds us that the ability to adapt to diverse environmental conditions can itself lead to increased organismal diversity.

Arthropods are segmented, and many of the segments bear appendages for locomotion, food handling, or reproduction. In many orders, the body segments have become fused into functional units, or **tagmata**, such as the head, thorax, and abdo-

men of an insect (**Figure 33.24**). Cephalization is extensive, and arthropods have well-developed sensory organs, including organs of sight, touch, smell, hearing, and balance. Arthropods have compound eyes composed of many independent visual units called **ommatidia** (singular, ommatidium) (look ahead to Figure 45.12). Each ommatidium functions as a separate photoreceptor capable of forming an independent image. Together, these lenses render a mosaic-like image of the environment. Some species, particularly some insects, possess additional simple eyes, or ocelli, that are probably only capable of distinguishing light from dark.

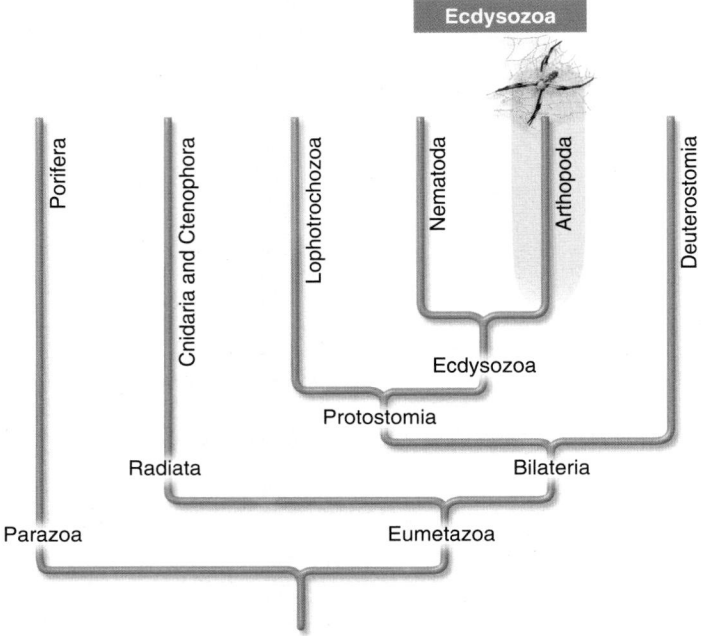

The arthropod brain is quite sophisticated, consisting of two or three ganglia connected to several smaller ventral nerve ganglia. Like most mollusks, arthropods have an open circulatory system (look ahead to Figure 47.2), in which hemolymph is pumped from the heart into the aorta or short arteries and then into sinuses, open fluid-filled cavities surrounding the major organs that coalesce to form the hemocoel. From the sinuses, gases and nutrients diffuse into tissues. The hemolymph flows back into the heart via pores, called ostia, that are equipped with valves.

Because the cuticle impedes the diffusion of gases through the body surface, arthropods possess special organs that permit gas exchange. In aquatic arthropods, these consist of feathery gills that have an extensive surface area in contact with the surrounding water. Terrestrial species have a highly developed **tracheal system**, a series of finely branched air tubes called tracheae that lead into the body from pores called **spiracles** (look ahead to Figure 48.9). The tracheal system delivers oxygen directly to tissues and cells. Some spiders have book lungs, consisting of a series of sheetlike structures extending into a hemolymph-filled chamber on the underside of the abdomen. Gases also diffuse across thin areas of the cuticle.

(a) External anatomy

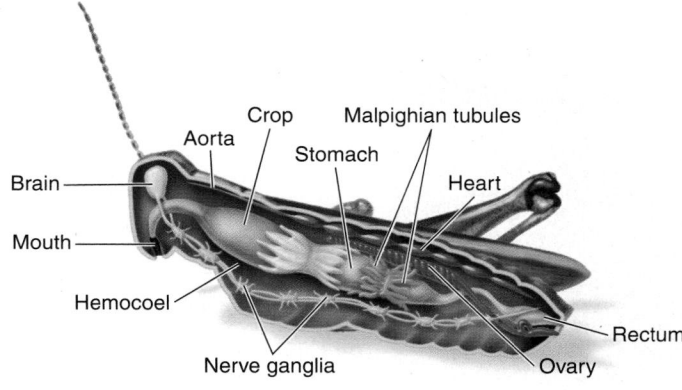

(b) Internal anatomy

Figure 33.24 Body plan of an arthropod, as represented by a grasshopper.

Table 33.5	Main Classes and Characteristics of Arthropods	
	Class and examples (est. # of species)	**Class characteristics**
	Arachnida: Spiders, scorpion, ticks, mites (74,000)	Body usually with cephalo-thorax and abdomen only; six pairs of appendages, including four pairs of legs, one pair of fangs, and one pair of pedipalps; terrestrial; predatory or parasitic
	Diplopoda: Millipedes (10,000)	Body with head and highly segmented trunk, each segment with two pairs of walking legs; herbivorous
	Chilopoda: Centipedes (3,000)	Body with head and highly segmented trunk; each segment with one pair of walking legs; predatory, poison jaws
	Insecta: Beetles, butterflies, flies, fleas, grasshoppers, ants, bees, wasps, termites (>1 million)	Body with head, thorax, and abdomen; mouthparts modified for biting, chewing, sucking, or lapping; usually with two pairs of wings and three pairs of legs; mostly terrestrial, some fresh-water; herbivorous, parasitic, or predatory
	Crustacea: Crabs, lobsters, shrimp (45,000)	Body of two to three parts; three or more pairs of legs; chewing mouthparts; usually marine

The digestive system is complex and often includes a mouth, crop, stomach, intestine, and rectum. Excretion is accomplished by a specialized metanephridia or, in insects and some other taxa, by **Malpighian tubules**, delicate projections from the digestive tract that protrude into the hemolymph (look ahead to Figure 49.10). Nitrogenous wastes are absorbed by the tubules and emptied into the gut. The intestine and rectum reabsorb water and salts. This excretory system, allowing the retention of water, was another critical innovation that permitted the colonization of land by arthropods.

There are six main classes of arthropods, one now-extinct class, Trilobita (the trilobites), and five living classes, Arachnida (spiders and scorpions), Diplopoda (millipedes), Chilopoda (centipedes), Insecta (insects), and Crustacea (crabs and relatives) (**Table 33.5**).

Class Trilobita: Extinct Early Arthropods The trilobites were among the earliest arthropods, flourishing in shallow seas of the Paleozoic era, some 500 million years ago, and dying out about 250 million years ago. Most trilobites were bottom feed-

ers and were generally 3–10 cm in size, although some reached almost 1 m in length (**Figure 33.25**). Like many arthropods, they had three main tagmata: the head, thorax, and abdomen. Trilobites also had two dorsal grooves that divided the body longitudinally into three lobes—a median lobe and two anterior lobes—a structural characteristic giving the class its name. Most of the body segments showed little specialization. In contrast, as we will explore, more advanced arthropods developed specialized appendages on many segments, including appendages for grasping, walking, and swimming.

Class Arachnida: The Spiders, Scorpions, Ticks, and Mites
The class Arachnida contains predatory spiders and scorpions as well as the ticks and mites, some of which are blood-sucking parasites that feed on vertebrates. All species have a body consisting of two tagmata: a fused head and thorax called a **cephalo-thorax**, and an abdomen (**Figure 33.26a**).

In spiders (order Araneae), the two body parts are joined by a **pedicel**, a narrow, waistlike point of attachment. Spiders have six pairs of appendages: the chelicerae, or fangs (**Figure 33.26b**); a pair of **pedipalps**, which have various sensory, predatory, or reproductive functions; and four pairs of walking legs. The fangs are supplied with venom from poison glands.

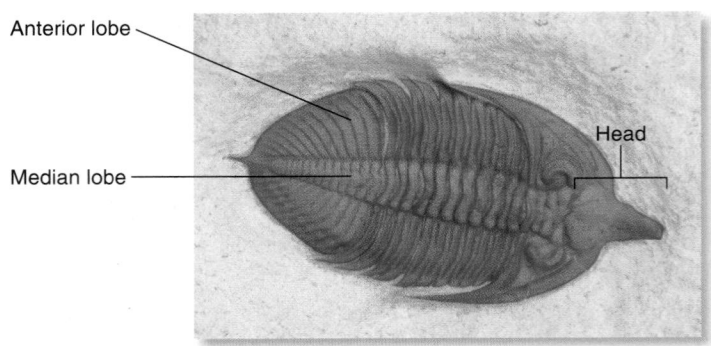

Figure 33.25 **A fossil trilobite.** These early arthropods were common in the shallow seas of the Paleozoic era but died out some 250 million years ago. About 4,000 fossil species, including *Huntonia huntonesis* shown here, have been described.

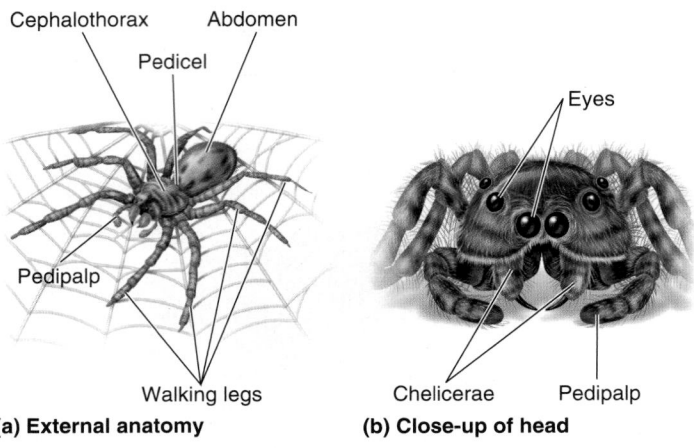

(a) **External anatomy**   (b) **Close-up of head**

Figure 33.26 Spider morphology.

(a)   (b)   (c)

Figure 33.27 Spider webs. (a) Normal web. (b) Web spun by spider fed with prey containing caffeine. (c) Web spun by spider fed with prey containing marijuana.

Most spider bites are harmless to humans, although they are very effective in immobilizing and/or killing their insect prey. Venom from some species, including the black widow (*Latrodectus mactans*) and the brown recluse (*Loxosceles reclusa*), are potentially, although rarely, fatal to humans. The toxin of the black widow is a neurotoxin, which interferes with the functioning of the nervous system, while that of the brown recluse is hemolytic, meaning it destroys tissue around the bite. After the spider has subdued its prey, it pumps digestive fluid into the tissues via the fangs and sucks out the partially digested meal.

Spiders have abdominal silk glands called **spinnerets**, and many spin webs to catch prey (**Figure 33.27a**). The silk is a protein that stiffens after extrusion from the body because the mechanical shearing causes a change in the organization of the amino acids. Silk is stronger than steel of the same diameter. Each species constructs a characteristic size and style of web and can do it perfectly on its first attempt, indicating that web spinning is an innate (inherited) behavior (see Chapter 55). Spiders also use silk to wrap up prey and to construct egg sacs. Interestingly, spiders that are fed drugged food (flies) spin their webs differently than undrugged spiders (**Figure 33.27b,c**). Some scientists have suggested that web-spinning spiders be

used to test substances for the presence of drugs or even to indicate environmental contamination. Not all spiders use silk extensively. Other spiders, including the wolf spider (**Figure 33.28a**), actively pursue their prey.

Scorpions (order Scorpionida) are generally tropical or subtropical animals that feed primarily on insects, though they may eat spiders and other arthropods as well as smaller reptiles and mice. Their pedipalps are modified into large claws, and their abdomen tapers into a stinger, which is used to inject venom. While the venom of most North American species is generally not fatal to humans, that of the *Centruroides* genus from deserts in the U.S. Southwest and Mexico can be deadly. Fatal species are also found in India, Africa, and other countries. Unlike spiders, which lay eggs, scorpions bear live young that the mother subsequently carries around on her back until they have their first molt (**Figure 33.28b**).

In mites and ticks (order Acari), the two main body segments (cephalothorax and abdomen) are fused and appear as one large segment. Many mite species are free-living scavengers that feed on dead plant or animal material. Other mites are serious pests on crops, and some, like chiggers (*Trombicula alfreddugesi*), are parasites of humans that can spread diseases such as

(a) (b) (c) (d)

Figure 33.28 **Common arachnids.** **(a)** This wolf spider (*Lycosa tarantula*) does not spin a web but instead runs after its prey. Note the pedipalps, which look like short legs. **(b)** The Cape thick-tailed scorpion (*Parabuthus capensis*) is highly venomous and carries its white young on its back. **(c)** The chigger mite (*Trombicula alfreddugesi*) can cause irritation to human skin. **(d)** These South African bont ticks (*Amblyomma hebraeum*) are feeding on a white rhinoceros.

Biological inquiry: What is one of the main characteristics distinguishing arachnids from insects?

typhus (**Figure 33.28c**). Chiggers are parasites only in their larval stage. It is a myth that chiggers bore into the skin. Rather, it is their bite and salivary secretions that cause skin irritation. *Demodex brevis* is a hair-follicle mite that is common in animals and humans. The mite is estimated to be present in over 90% of adult humans. While the mite causes no irritation in most humans, *Demodex canis* causes the skin disease known as mange in domestic animals, particularly dogs.

Ticks are larger organisms than mites, and all are ectoparasitic, feeding on the body surface, on vertebrates. Their life cycle includes attachment to a host, sucking blood until they are replete, and dropping off the host to molt (**Figure 33.28d**). Ticks can carry a huge variety of viral and bacterial diseases, including Lyme disease, a bacterial disease so named because it was first found in the town of Lyme, Connecticut, in the 1970s.

Classes Diplopoda and Chilopoda: The Millipedes and Centipedes The millipedes and centipedes are both wormlike arthropods that are among the earliest terrestrial phyla known. Millipedes (class Diplopoda) have two pairs of legs per segment, as their Latin name denotes (*diplo*, two, and *podos*, feet), not 1,000 legs, as their common name suggests (**Figure 33.29a**). They are slow-moving herbivorous creatures that eat decaying leaves and other plant material. When threatened, the millipede's response is to roll up into a protective coil. Many millipede species also have repugnatorial glands on their underside that can eject a variety of toxic, repellent secretions. Some millipedes are brightly colored, warning potential predators that they can protect themselves.

Class Chilopoda (from the Latin *chilo*, lip, and *podos*, feet), or centipedes, are fast-moving carnivores that have one pair of walking legs per segment (**Figure 33.29b**). The head has many sensory appendages, including a pair of antennae and three pairs of appendages modified as mouthparts, including powerful claws connected to poison glands. The toxin from venom of some of the larger species, such as *Scolopendra heros*, is powerful enough to cause pain in humans. Most species do not have a waxy waterproofing layer on their cuticle and are restricted to

(a) (b)

Figure 33.29 **Millipedes and centipedes.** **(a)** Millipedes have two pairs of legs per segment. **(b)** The venom of the giant centipede (*Scolopendra heros*) is known to produce significant swelling and pain to humans.

moist environments under leaf litter or in decaying logs, usually coming out at night to actively hunt their prey.

Class Insecta: A Diverse Array of Insects Insects are in a class by themselves, literally and figuratively. There are more species of insects than all other species of animal life combined. One million species of insects have been described and, according to best estimates, 9 million more species are awaiting description. At least 90,000 species of insects have been identified in the U.S. and Canada alone. It is believed that the explosion of diversity in the flowering plants (angiosperms) greatly promoted the huge diversification of insects relative to other classes.

GENOMES & PROTEOMES

Insect Diversity May Be Explained by the Appearance of the Angiosperms

More species are found in the class Insecta, over 1 million, than in any other class. Of these, over half a million are beetles, order Coleoptera. To understand what factors influence biodiversity, we need to understand why there are so many species of insects and beetles. In 1998, biologist Brian Farrell used

Table 33.6	Five Phylogenetic Comparisons of the Species Richness of Related Beetle Families and Subfamilies			
Comparison*	Gymnosperm-associated family/subfamily	# of Species	Angiosperm-associated family/subfamily	# of Species
1	*Nemonychidae*	85	*Curculionidae*	44,002
2	*Oxycoryninae*	30	*Belinae*	150
3	*Aseminae*	78	*Cerambycinae*	25,000
4	*Palophaginae*	3	*Megalopodinae*	400
5	*Orsodacnidae*	26	*Chrysomelidae*	33,400

* In each comparison, the families and subfamilies are closely related. For example, the *Nemonychidae* are closely related to the *Curculionidae*, but the latter has many more species.

a genomic study to show that the appearance of angiosperm plants led to an increased richness of insects in general and leaf-eating beetles in particular.

For about the first 50 million years of their order's existence, beetles fed on detritus, dead plant and animal material, and fungi. Plant feeding arose later in beetle history, during the Permian period, but relatively few fossilized beetle genera exist from that time. Jurassic fossils showed the existence of about 150 phytophagous (plant-feeding) beetle genera, which fed on gymnosperm plants: the conifers and cycads. However, many lineages of these ancestral beetles colonized angiosperms independently in the Cretaceous and Tertiary, after the diversification of angiosperms. In each case, the lineages that fed on angiosperms have become much more diverse than those that fed on gymnosperms.

In seeking to explain the huge diversity of beetles, Farrell analyzed DNA sequences for the entire 18S ribosomal subunit genes for 115 beetle species, drawn from all beetle subfamilies. In addition, he constructed a data set of 212 morphological characters of the same species. From both sets of data, Farrell constructed a phylogenetic tree. The oldest of the lineages for each of the subfamilies are gymnosperm-affiliated, while the more recently derived lineages are angiosperm-associated species.

Farrell also made five specific phylogenetic comparisons of the species richness of beetles in the same subfamilies but that either fed on gymnosperms or angiosperms (Table 33.6). In each case, the angiosperm-feeding group speciated frequently and became very diverse. In contrast to the gymnosperm-associated lineages, which generally feed on the pollen-rich reproductive parts of conifers and cycads, the angiosperm-associated subfamilies diversified in feeding habit to chew and mine leaves and feed on seeds and roots as well as reproductive parts. This illustrates how critical innovations can lead to extensive adaptive radiation, one of the principles of diversity noted at the beginning of this unit.

Insects are the subject of an entire field of scientific study, **entomology**. They are studied in large part because of their significance as pests of the world's agricultural crops. Insects live in all terrestrial habitats, and virtually all species of plants are fed upon by at least one, usually tens, and sometimes, in the case of large trees, hundreds of insect species. Because insects eat approximately one-quarter of the world's crops, we are constantly trying to find ways to reduce insect pest densities. Pest reduction often involves chemical control, the use of pesticides, or biological control, the use of living organisms, to reduce pest populations. Many species of insects are also important pests or parasites of humans and livestock, both by their own actions and as vectors of diseases such as malaria and sleeping sickness.

On the other hand, insects provide many types of essential biological services. We depend on insects such as honeybees to pollinate our crops. Bees also produce honey, and silkworms are the source of silk fiber. Despite the revulsion they provoke, fly larvae (maggots) are important in the decomposition process of both dead plants and animals. Insect parasites and predators are used in biological control to reduce densities of pest insects on crops.

Of paramount importance to the success of insects was the development of wings, a feature possessed by no other arthropod and indeed no other living animal except birds and bats. Unlike vertebrate wings, however, insect wings are actually outgrowths of the body wall cuticle and are not true segmental appendages. This means that insects still have all their walking legs. Insects are thus like the mythological horse Pegasus, which sprouted wings out of its back while retaining all four legs. In contrast, birds and bats have one pair of appendages (arms) modified for flight, which leaves them considerably less agile on the ground.

The great diversity of insects is illustrated by the fact that there are 35 different orders, some of which have over 100,000 species. The most common of the orders are discussed in **Table 33.7**. Different orders of insects have slightly different wing structures, and many of the orders are based on wing type (*pteron* is a Greek word meaning wing). Wasps and bees (Hymenoptera) have two pairs of wings hooked together that move as one wing. Butterflies (Lopidoptera) have wings that are covered in scales (from the Greek, *lepido*, scale), while other insects generally have clear, membranous wings. Flies (Diptera) possess only one pair of wings (the front pair); the back pair has been modified to a small pair of balancing organs, called halteres, that act like miniature gyroscopes. On the other hand, in beetles (Coleoptera) only the back pair of wings is functional, as the front wings have been hardened into protective shell-like coverings (elytra), under which the back pair folds when not in use. In ant and termite colonies, female individuals called workers have lost their wings, while the queen and the drones (males) have retained theirs. Other species, such as fleas and lice, are completely wingless.

Table 33.7 | Main Orders and Characteristics of Insects

Order and examples (est. # of species)		Order characteristics
Coleoptera: Beetles, weevils (500,000)		Two pairs of wings (front pair thick and leathery, acting as wing cases, back pair membranous); armored exoskeleton; biting and chewing mouthparts; complete metamorphosis; largest order of insects
Hymenoptera: Ants, bees, wasps (190,000)		Two pairs of membranous wings; chewing or sucking mouthparts; many have posterior stinging organ on females; complete metamorphosis; many species social; important pollinators
Diptera: Flies, mosquitoes (190,000)		One pair of wings with hindwings modified into halteres (balancing organs); sucking, piercing, or lapping mouthparts; complete metamorphosis; larvae are grublike maggots in various food sources; some adults are disease vectors
Lepidoptera: Butterflies, moths (140,000)		Two pairs of colorful wings covered with tiny scales; long tubelike tongue for sucking; complete metamorphosis; larvae are plant-feeding caterpillars
Hemiptera: True bugs; assassin bug, bedbug, chinch bug, cicada (100,000)		Two pairs of membranous wings; piercing or sucking mouthparts; incomplete metamorphosis; many plant feeders; some predatory or blood feeders; vectors of plant diseases
Orthoptera: Crickets, roaches, grasshoppers, mantids (30,000)		Two pairs of wings (front pair leathery, back pair membranous); chewing mouthparts; mostly herbivorous; incomplete metamorphosis; powerful hind legs for jumping
Odonata: Damselflies, dragonflies (6,000)		Two pairs of long, membranous wings; chewing mouthparts; large eyes; predatory on other insects; incomplete metamorphosis; larvae aquatic; considered primitive insects
Anoplura: Sucking lice (2,400)		Wingless ectoparasites; sucking mouthparts; flattened body; reduced eyes; legs with clawlike tarsi for clinging to skin; incomplete metamorphosis; very host specific; vectors of typhus
Siphonoptera: Fleas (2,000)		Wingless, laterally flattened; piercing and sucking mouthparts; adults are bloodsuckers on birds and mammals; jumping legs; complete metamorphosis; vectors of plague
Isoptera: Termites (2,000)		Two pairs of membranous wings when present; some stages wingless; chewing mouthparts; social species; incomplete metamorphosis

Insects in different orders have also evolved a variety of mouthparts (**Figure 33.30**). Grasshoppers, beetles, dragonflies, and many others have mouthparts adapted for chewing. Mosquitoes and many plant pests have mouthparts adapted for piercing and sucking. Butterflies and moths have a coiled tongue (**proboscis**) that can be uncoiled, enabling them to drink nectar from flowers. Finally, some flies have lapping, spongelike mouthparts that sop up liquid food. Their varied mouthparts allow insects to specialize their feeding on virtually anything: plant matter, decaying organic matter, and other living animals. The biological diversity of insects is thus related to environmental diversity, in this case, the variety of foods that insects eat. Parasitic insects attach themselves to other species, and there are even insect parasites that feed on parasites (called hyperparasites), proving, as the 18th-century English poet Jonathan Swift noted:

> Big fleas have little fleas
> upon their backs to bite 'em;
> little fleas have smaller fleas
> and so *ad infinitum*.

(a) Chewing (grasshopper)

(b) Blood sucking (mosquito)

Proboscis

(c) Nectar sucking (butterfly)

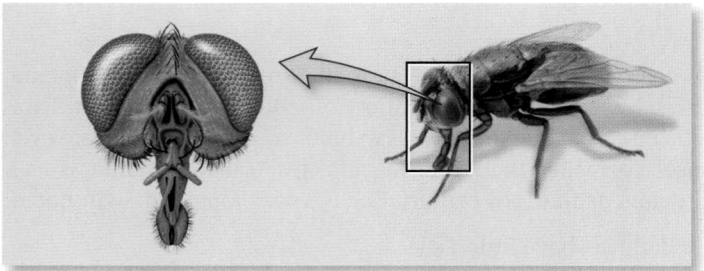

(d) Lapping up liquid (housefly)

Figure 33.30 **A variety of insect mouthparts.** Insect mouthparts can be modified to allow insects to feed in a variety of ways, including **(a)** chewing (Orthoptera, Coleoptera, and others), **(b)** blood sucking (Diptera), **(c)** nectar sucking (Lepidoptera), and **(d)** lapping up liquid (Diptera).

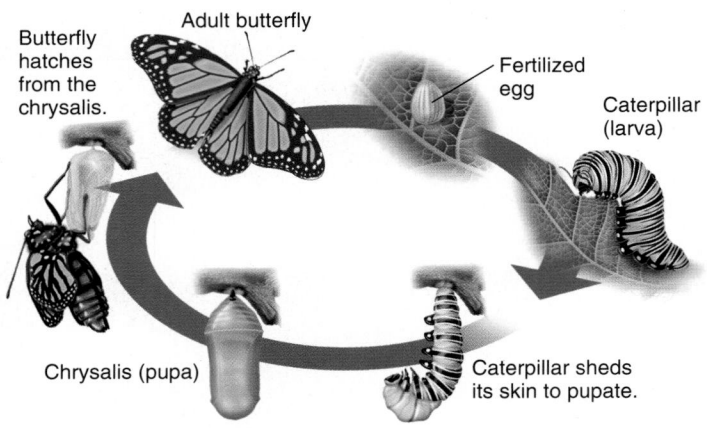

Butterfly hatches from the chrysalis.

Adult butterfly

Fertilized egg

Caterpillar (larva)

Chrysalis (pupa)

Caterpillar sheds its skin to pupate.

(a) Complete metamorphosis

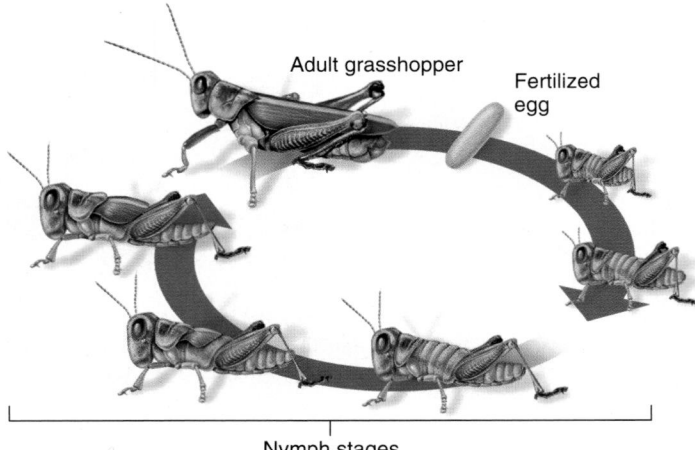

Adult grasshopper

Fertilized egg

Nymph stages

(b) Incomplete metamorphosis

Figure 33.31 **Metamorphosis.** **(a)** Complete metamorphosis, as illustrated by the life cycle of a monarch butterfly. The adult butterfly has a completely different appearance than the larval caterpillar. **(b)** Incomplete metamorphosis, as illustrated by the life cycle of a grasshopper. The eggs hatch into nymphs, essentially miniature versions of the adult.

All insects have separate sexes, and fertilization is internal. During development, the majority (approximately 85%) of insects undergo a change in body form known as **complete metamorphosis** (from the Greek *meta*, change, and *morph*, form) (**Figure 33.31a**). Complete metamorphosis has four stages: egg, larva, pupa, and adult. In these species, the larval stage is often spent in an entirely different habitat from that of the adult, and larval and adult forms utilize different food sources. They thus do not compete directly for the same resources. The dramatic body transformation from larva to adult occurs in the pupa stage. The remaining insects undergo **incomplete metamorphosis**, in which change is more gradual (**Figure 33.31b**). Incomplete metamorphosis has only three stages: egg, nymph, and adult. Young insects, called nymphs, look like miniature adults when they hatch from their eggs. As they grow and feed, they shed their skin several times, each time entering a new **instar**, or stage of growth.

(a)   (b)

Figure 33.32   The division of labor in insect societies. Individuals from the same insect colony may appear very different. Among these army ants (*Eciton burchelli*) from Paraguay, there are **(a)** workers that forage for the colony and soldiers that protect the colony from predators, and **(b)** the queen, which reproduces and lays eggs.

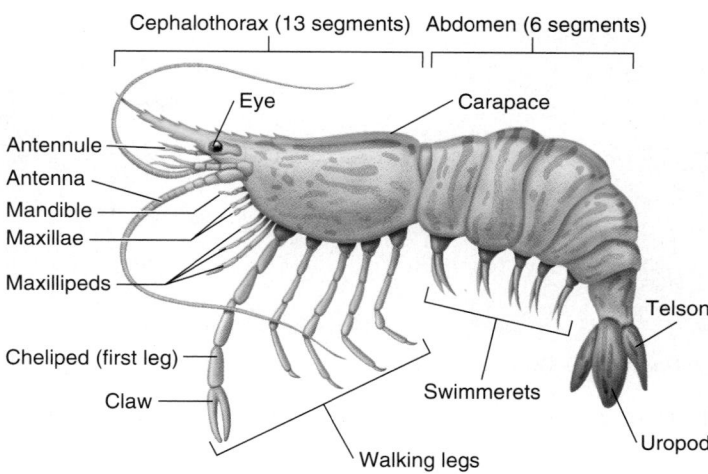

Figure 33.33   Body plan of a crustacean, as represented by a shrimp.

Finally, some insects, such as bees, wasps, ants, and termites, have developed complex social behavior and live cooperatively in underground or aboveground nests. Such colonies exhibit a division of labor, in that some individuals forage for food and care for the brood (workers), others protect the nest (soldiers), and some only reproduce (the queen and one or two males) (**Figure 33.32**).

Class Crustacea: Crabs, Lobsters, Barnacles, and Shrimp

The crustaceans are common inhabitants of marine environments, although some species live in fresh water and a few are terrestrial. Many are economically important food items for humans, including crabs, lobsters, crayfish, and shrimp, and smaller species are important food sources for other predators.

The crustaceans are unique among the arthropods in that they possess two pairs of antennae at the anterior end of the body—the antennule (first pair) and antenna (second pair) (**Figure 33.33**). In addition, they have three or more sensory and feeding appendages that are modified mouthparts, called mandibles, maxillae, and maxillipeds. These are followed by walking legs and, often, additional abdominal appendages called **swimmerets** and a powerful tail consisting of a telson and uropod. In some orders, the first pair of walking legs, or chelipeds, is modified to form powerful claws. A lost crustacean appendage can regrow. The head and thorax are often fused together, forming the cephalothorax. In many species, the cuticle covering the head extends over most of the cephalothorax, forming a hard protective fold called the **carapace**. For growth to occur, a crustacean must shed the entire exoskeleton.

Many crustaceans are predatory in nature, but others, such as barnacles, are suspension feeders. Gas exchange typically occurs via gills, and crustaceans, like all other arthropods, have an open circulatory system. Crustaceans possess two excretory organs: antennal glands and maxillary glands, both modified metanephridia, which open at the bases of the antennae and maxillae, respectively. Reproduction usually involves separate

Figure 33.34   Crustacean larvae.   The nauplius is a distinct larval type possessed by most crustaceans, which molt several times before reaching maturity.

sexes, and fertilization is internal. Most species carry their eggs in brood pouches under the female's body. Eggs of most species produce larvae that must go through many different molts prior to assuming adult form. The first of these larval stages, called a **nauplius**, is very different in appearance than the adult crustacean (**Figure 33.34**).

While there are many crustacean orders, most are small and obscure, although many feature prominently in marine food chains, a series of organisms in which each member of the chain feeds on and derives energy from the member below it. These include the Ostracoda, Copepoda, and Euphausiacea. Ostracods are tiny creatures that superficially resemble clams, and copepods are tiny and abundant planktonic crustaceans, both of which are a food source of filter-feeding organisms and small fish. Euphausiids are shrimplike krill that grow to about 3 cm and provide a large part of the diet of many whales.

(a) Barnacles—Order Cirripedia   **(b) Pill bug—Order Isopoda**   **(c) Coral crab—`Order Decapoda**

Figure 33.35   Common crustaceans.   **(a)** Barnacles on intertidal rocks. **(b)** Pill bug or wood louse. **(c)** Coral crab (*Carpilius maculates*).

The order Cirripedia is composed of the barnacles, crustaceans whose carapace forms calcified plates that cover most of the body (**Figure 33.35a**). Their legs are modified into feathery filter-feeding structures. The order Isopoda (**Figure 33.35b**) contains many small species that are parasitic on marine fish. There are also terrestrial isopods, better known as pill bugs or wood lice, that retain a strong connection to water and need to live in moist environments such as leaf litter or decaying logs. When threatened, they curl up into a tight ball, making it difficult for predators to get a grip on them.

The most famous order, however, is the Decapoda, which includes the crabs and lobsters, the largest crustacean species (**Figure 33.35c**). As their name suggests, these decapods have 10 walking legs (five pairs), although the first pair is invariably modified to support large claws. Most decapods are marine, but there are many freshwater species, such as crayfish, and in hot, moist tropical countries, even some terrestrial species called land crabs. The larvae of many larger crustaceans are planktonic and grow to about 3 cm. These are abundant in some oceans and are a staple food source for many species.

33.5 Deuterostomia: The Echinoderms and Chordates

As we explored in Chapter 32, the deuterostomes are grouped together because they share similarities in patterns of development (refer back to Figure 32.5). Molecular evidence also supports a deuterostome clade. All animals in the phylum Chordata, which includes the vertebrates, are deuterostomes. Interestingly, so is one invertebrate group, the phylum Echinodermata, which includes the sea stars, sea urchins, and sea cucumbers. While there are far fewer phyla and species of deuterostomes than ecdysozoans, the species are generally much more familiar to us. After all, we humans are deuterostomes. While the majority of the deuterostome clade will be discussed in Chapter 34, we will conclude our discussion of invertebrate biology by turning our attention to the invertebrate deuterostomes. In this section, we will explore the phylum Echinodermata and then introduce the phylum Chordata, looking in particular at its distinguishing characteristics and at its two invertebrate subphyla: the urochordates, also known as the tunicates, and the cephalochordates, commonly referred to as the lancelets.

The Phylum Echinodermata Includes Sea Stars and Sea Urchins, Species with a Water Vascular System

The phylum Echinodermata (from the Greek *echinos*, spiny, and *derma*, skin) consists of a unique grouping of deuterostomes. A striking feature of all echinoderms is their modified radial symmetry. The body of most species can be divided into five parts pointing out from the center. As a consequence, cephalization is absent in most classes, and echinoderms move only very slowly. There is no brain and only a simple nervous system. The radial symmetry of echinoderms is secondary, however, because the free-swimming larvae have bilateral symmetry and metamorphose into the radially symmetrical adult form.

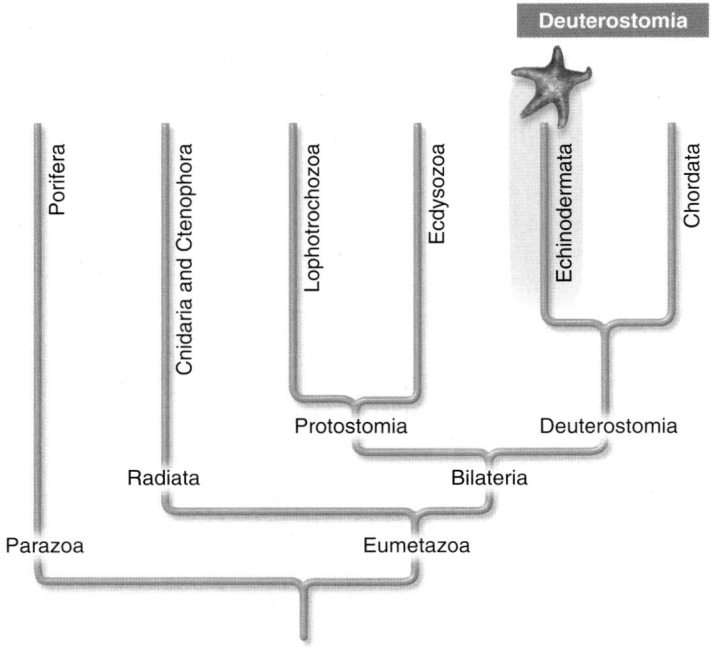

Most echinoderms have an **endoskeleton** (internal skeleton) composed of calcareous plates overlaid by a thin skin. The skeleton is covered with spines and jawlike pincers called **pedicellariae**, the primary purpose of which is to deter settling of animals such as barnacles (**Figure 33.36**). These structures can also possess poison glands.

A portion of the coelom has been adapted to serve as a unique **water vascular system**, a network of canals that branch into tiny **tube feet** that function in movement, gas exchange, and feeding. The water vascular system is powered by hydraulic power, that is, by water pressure generated by the contraction of muscles that enables the extension and contraction of the tube feet, allowing echinoderms to move slowly.

Water enters the water vascular system through the **madreporite**, a sieve-like plate on the animal's surface. From there it flows into a **ring canal** in the central disk, into five radial canals, and into the tube feet. At the base of each tube foot is a muscular sac called an **ampulla**, which stores water. Contractions of the ampullae force water into the tube feet, causing them to straighten and extend. When the foot contacts a solid surface, muscles in the foot contract, forcing water back into the ampulla. Sea stars also use their tube feet in feeding, where they can exert a constant and strong pressure on bivalves, whose adductor muscles eventually tire, allowing the shell to open slightly. At this stage, the sea star everts its stomach and inserts it into the space, and then digests its prey using juices secreted from extensive digestive glands. Sea stars also feed on sea urchins, brittle stars, and sand dollars, prey that cannot easily escape them.

Echinoderms cannot osmoregulate, so no species have entered freshwater environments. No excretory organs are present, and both respiration and excretion of nitrogenous waste take place by diffusion across the tube feet. Coelomic fluid circulates around the body.

Most echinoderms exhibit **autotomy**, the ability to intentionally detach a body part, such as a limb, that will later regenerate. In some species, a broken limb can even regenerate into a whole animal. Some sea stars regularly reproduce by breaking in two. Most echinoderms reproduce sexually and have separate sexes. Fertilization is usually external, with gametes shed into the water. Fertilized eggs develop into free-swimming bilaterally symmetrical larvae, which undergo metamorphosis into sedentary adults.

While over 20 classes of echinoderms have been described from the fossil record, only 5 main classes of echinoderms exist today: the Asteroidea (sea stars), Ophiuroidea (brittle stars), Echinoidea (sea urchins and sand dollars), Crinoidea (sea lilies and feather stars), and Holothuroidea (sea cucumbers). The key features of the echinoderms and their classes are listed in **Table 33.8**.

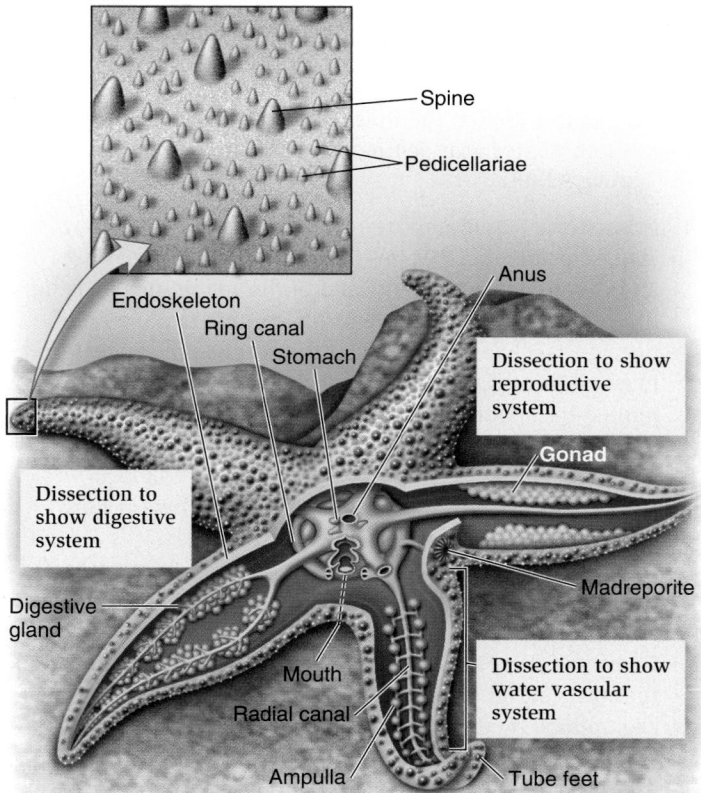

Figure 33.36 Body plan of an echinoderm, as represented by a sea star. The arms of this sea star have been dissected to different degrees to show the echinoderm's various organs.

Biological inquiry: Echinoderms and chordates are both deuterostomes. What are three defining features of deuterostomes?

Table 33.8	Main Classes and Characteristics of Echinoderms	
	Class and examples (est. # of species)	**Class characteristics**
	Asteroidea: Sea stars (1,600)	Five arms; tube feet; predatory on bivalves and other echinoderms; eversible stomach
	Ophiuroidea: Brittle stars (2,000)	Five long, slender arms; tube feet not used for locomotion; no pedicellaria; browse on sea bottom or filter feed
	Echinoidea: Sea urchins, sand dollars (1,900)	Spherical (sea urchins) or disc-shaped (sand dollars); no arms; tube feet and moveable spines; pedicellaria present; many feed on seaweeds
	Crinoidea: Sea lilies and feather stars (700)	Often attached to substrate via stalk; arms feathery and used in filter feeding; very abundant in fossil record
	Holothuroidea: Sea cucumbers (1,200)	Cucumber-shaped; no arms; spines absent; endoskeleton reduced; tube feet

Figure 33.37 Sea cucumber with pearl fish.

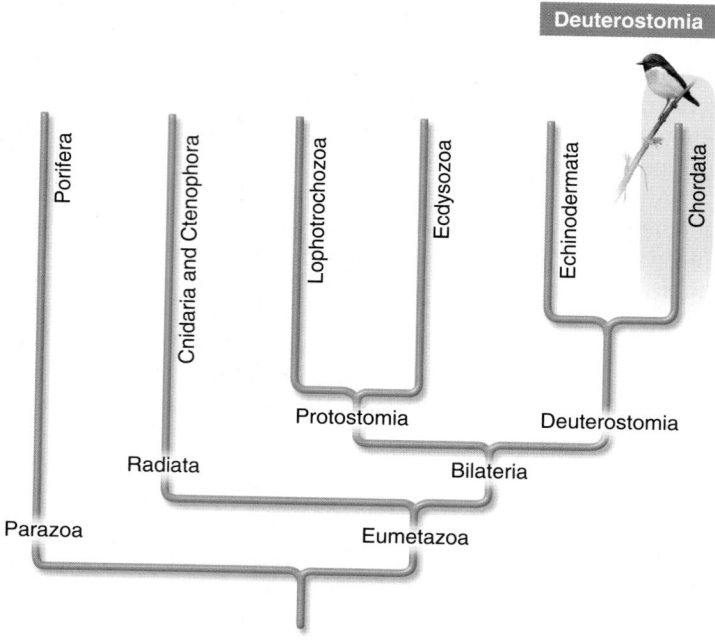

The most unusual of the echinoderms are members of the class Holothuroidea, the sea cucumbers. These animals really do look like a cucumber rather than a sea star or sea urchin (**Figure 33.37**). The hard plates of the endoskeleton are less extensive, so that the animal appears and feels fleshy. Sea cucumbers possess specialized respiratory structures called respiratory trees that pump water in and out of the anus. They are typically deposit feeders, ingesting sediment and extracting nutrients.

When threatened by a predator, a few tropical species of sea cucumber can eject sticky, toxic substances from their anus. If these do not serve to deter the predator, these species can undergo the process of evisceration, ejecting its digestive tract, respiratory structures, and gonads from the anus. If the sea cucumber survives, it can regenerate its organs later. The pearl fish has a particularly intriguing relationship with sea cucumbers. It takes advantage of the sea cucumber's defenses by darting inside its anus for protection. At first, the sea cucumber resists the pearl fish, but since it breathes through its anus, it is only a matter of time before an opportunity is afforded to the fish.

The Phylum Chordata Includes All the Vertebrates and Some Invertebrates

The deuterostomes consist of two major phyla, the echinoderms and the chordates (from the Greek *chorde*, string). As deuterostomes, both phyla share similar developmental traits. In addition, both have an endoskeleton, consisting in the echinoderm of calcareous plates and in chordates, for the most part, of bone. However, the endoskeleton of the echinoderms is usually covered only by a thin layer of skin and functions in much the same way as the arthropod exoskeleton, in that its primary function is providing protection. The chordate endoskeleton serves a very different purpose. In early divergent chordates, the endoskeleton is composed of a single flexible rod situated dorsally, deep inside the body. Muscles move this rod, and their contractions cause the back and tail end to move from side to side, permitting a swimming motion in water.

The endoskeleton becomes more complex in different lineages that develop limbs, as we will see in Chapter 34, but it is always internal, with muscles attached. This arrangement permits the possibility of complex movements, including the ability to move on land.

Let's take a look at the four critical innovations in the body design of chordates that distinguish them from all other animal life (**Figure 33.38**):

1. **Notochord.** Chordates are named for the **notochord**, a single flexible rod that lies between the digestive tract and the nerve cord. Composed of fibrous tissue encasing fluid-filled cells, the notochord is stiff yet flexible and provides skeletal support for all primitive chordates. In most chordates, such as vertebrates, a more complex jointed backbone usually replaces the notochord and only remnants exist as the soft material within the discs of vertebrae.

2. **Dorsal hollow nerve cord.** Many animals have a long nerve cord, but in nonchordate invertebrates, it is a solid tube that lies ventral to the alimentary canal. In contrast, the nerve cord in chordates is a hollow tube that develops dorsal to the alimentary canal. In 1822, the French naturalist Geoffroy Saint-Hilaire argued that this difference suggested that the ventral side of nonchordate invertebrates (as exemplified by the lobster) was homologous to the dorsal side of vertebrates. Some recent molecular work suggests that, as Hilaire first proposed, there was an inversion of the dorsoventral axis during animal evolution. In vertebrates, the dorsal hollow nerve cord develops into the brain and spinal cord.

3. **Pharyngeal slits.** Chordates, like many animals, have a complete gut, from mouth to anus. However, in chordates, slits develop in the pharyngeal region, close to the mouth,

that open to the outside. This permits water to enter through the mouth and exit via the slits, without having to go through the digestive tract. In early divergent chordates, **pharyngeal slits** function as a filter-feeding device, while in more advanced chordates, they develop into gills for gas exchange. In terrestrial chordates, the slits do not fully form and become modified for other purposes, such as the auditory (Eustachian) tubes in ears.

4. **Postanal tail.** Chordates possess a postanal tail of variable length that extends posterior to the anal opening. In aquatic chordates such as fish, the tail is used in locomotion. In terrestrial chordates, the tail may be used in a variety of functions or may be absent, as in humans. In virtually all other nonchordate phyla, the anus is terminal.

While few chordates apart from fishes possess all of these characteristics in their adult life, they all exhibit them at some time during development. For example, in adult humans the notochord becomes the spinal column and the dorsal hollow nerve cord becomes the central nervous system. However, humans only exhibit pharyngeal slits and a postanal tail during early embryonic development. All the pharyngeal slits, except one, which forms the Eustachian tubes in the ear, are eventually lost, and the postanal tail regresses to form the tailbone (the coccyx).

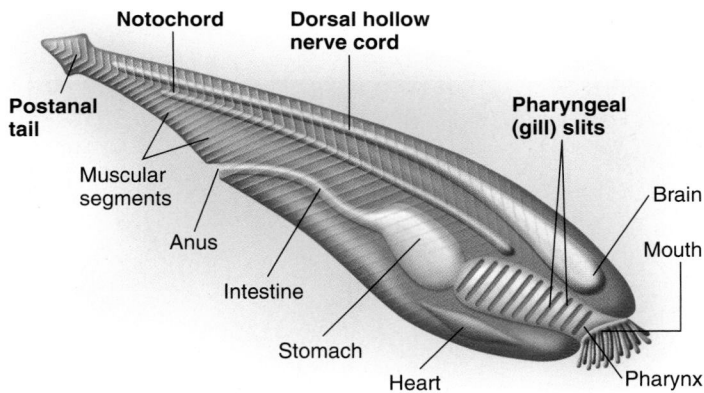

Figure 33.38   Chordate characteristics.   The generalized chordate body plan has four main features: notochord, dorsal hollow nerve cord, pharyngeal slits, and postanal tail.

The phylum Chordata consists of the invertebrate chordates—the subphylum Urochordata (tunicates) and the subphylum Cephalochordata (lancelets)—along with the subphylum Vertebrata. While the Vertebrata is by far the largest of these subphyla, biologists have focused on the Urochordata and Cephalochordata for clues as to how the chordate phylum may have evolved. Comparisons of gene sequences coding for 18S rRNA show that these two subphyla in general, and the cephalochordates in particular, are our closest nonvertebrate relatives (**Figure 33.39**).

Subphylum Urochordata: The Tunicates   The urochordates (from the Greek *oura*, tail) are a group of 3,000 marine species also known as tunicates. Looking at an adult tunicate, you might never guess that it is a relative of modern vertebrates. The only one of the four distinguishing chordate characteristics that it possesses is pharyngeal slits (**Figure 33.40a**). The larval tunicate, in contrast, looks like a tadpole and exhibits all four chordate hallmarks (**Figure 33.40b**). The larval tadpole swims for only a few days, usually without feeding, because the siphons are not functional in the larvae. Larvae settle on and attach to a rock surface via rootlike extensions called stolons. Here they metamorphose into adult tunicates and in the process lose most of their chordate characteristics. In 1928, the marine biologist Walter Garstang suggested modern vertebrates arose from a larval tunicate form that had somehow acquired the ability to reproduce. However, both the adult tunicate body plan and modern molecular data have led Billie Swalla and colleagues to propose that tunicates are so distantly related to other chordates that they should be considered a separate deuterostome phylum called Tunicata.

Adult tunicates are marine animals, some of which are colonial and others solitary, that superficially resemble sponges or cnidarians. Tunicates are filter feeders that draw water through the mouth through an **incurrent siphon**, using a ciliated pharynx, and filter it through extensive pharyngeal slits. The food is trapped on a mucus sheet secreted by what is known as an endostyle; passes via ciliary action to the stomach, intestine, and anus; and exits through the **excurrent siphon**. The whole animal is enclosed in a nonliving **tunic**, which it secretes, made of a protein and a cellulose-like material called tunicin.

Figure 33.39   Comparison of small subunit rRNA gene sequences of vertebrate, invertebrate chordate, and invertebrate species.   The similarities between the invertebrate chordates (represented by the lancelet) and the vertebrates (represented by a human) suggest they are indeed our closest relatives.

(a) Adult tunicate

(b) The larval form of the tunicate

(c) Typical tunicate

Figure 33.40   Tunicates.   **(a)** Body plan of the sessile, filter-feeding adult tunicate. **(b)** The larval form, which shows the four characteristic chordate features, has been proposed as a possible ancestor of modern vertebrates. **(c)** Blue tunicate, *Rhopalaea crassa*.

(a) Body plan of the lancelet

(b) Lancelet in the sand

Figure 33.41   Lancelets.   **(a)** The body plan of the lancelet clearly displays the four characteristic chordate features. **(b)** Lancelet in feeding position, showing the segmented body with chevron-shaped muscles in each segment.

Tunicates are also known as sea squirts for their ability to squirt out water from the excurrent siphon when disturbed. There is a rudimentary circulatory system with a heart and a simple nervous system of relatively few nerves connected to sensory tentacles around the incurrent siphon. The animals are mostly hermaphroditic.

Subphylum Cephalochordata: The Lancelets The cephalochordates (from the Greek *cephalo*, head) look a lot more chordate-like than do tunicates. They are commonly referred to as lancelets, in reference to their bladelike shape and size, about 5–7 cm in length (**Figure 33.41a**). Lancelets are a small subphylum of 26 species, all marine filter feeders, with 4 species occurring in North American waters. Most of them belong to the genus *Branchiostoma*.

The lancelets live mostly buried in sand, with only the anterior end protruding into the water so that they can filter feed through the mouth (**Figure 33.41b**). Lancelets have the four distinguishing chordate characteristics: a clearly discernible notochord (extending well into the head), dorsal hollow nerve cord, pharyngeal slits, and postanal tail. In a fashion similar to tunicates, water enters the mouth and moves into the pharynx, where it is filtered through the pharyngeal slits. A mucus net across the pharyngeal slits traps food particles, and ciliary action

takes the food into the digestive tract, while water exits via the excurrent siphon, here called the atriopore. Gas exchange generally takes place across the body surface. While the lancelet is usually sessile, it can leave its sandy burrow and swim to a new spot, using a series of serially arranged muscles that appear like chevrons ($<<<<$) along their sides. These muscles reflect the segmented nature of their body. With a body shape and a swimming motion characteristic of fish, it is not difficult to see how lancelets may have been the precursors to modern fish, especially the more ancient jawless fishes, which we will explore in Chapter 34.

CHAPTER SUMMARY

33.1 Parazoa: Sponges, the First Multicellular Animals

- Invertebrates, or animals without a backbone, make up more than 95% of all animal species. An early lineage—the Parazoa—consists of one phylum, the Porifera, or sponges. While sponges lack true tissues, they are multicellular animals possessing several types of cells. (Figure 33.1)

33.2 Radiata: Jellyfish, Radially Symmetrical Animals

- The Radiata consists of two phyla: the Cnidaria (hydra, jellyfish, sea anemones, corals, and box jellies) and the Ctenophora (comb jellies). Radiata have only two embryonic germ layers: the ectoderm and endoderm, with a gelatinous substance (mesoglea) connecting the two layers.

- Cnidarians exist as two forms: polyp and medusa. A characteristic feature of cnidarians is their stinging cells or cnidocytes, which function in defense and prey capture. Ctenophores possess the first complete gut and nearly all exhibit bioluminescence. (Figures 33.2, 33.3, 33.4, 33.5, Table 33.1)

33.3 Lophotrochozoa: The Flatworms, Rotifers, Lophophorates, Mollusks, and Annelids

- The Lophotrochozoa include taxa that possess either a lophophore or trochophore larva. Platyhelminthes, or flatworms, are regarded as the first animals to reach the organ-system level of organization. (Figure 33.6, Table 33.2)

- The four classes of flatworms are the Turbellaria, Monogenea, Trematoda (flukes), and Cestoda (tapeworms). Flukes and tapeworms are internally parasitic, with complex life cycles. (Figures 33.7, 33.8)

- Rotifers are microscopic animals that have a complete digestive tract with separate mouth and anus; the mastax, a muscular pharynx, is a structure unique to the rotifers. (Figure 33.9)

- The Lophophorates consist of the phoronids, bryozoa, and brachiopods, all of which possess a lophophore, a ciliary feeding structure. (Figure 33.10)

- The mollusks, which constitute a large phylum with over 100,000 diverse living species, have a basic body plan with three parts—a foot, a visceral mass, and a mantle—and an open circulatory system. (Figures 33.11, 33.12, 33.13, 33.14, 33.15)

- The four most common mollusk classes are the polyplacophora (chitons), gastropoda (snails and slugs), bivalvia (clams and

mussels), and cephalopoda (octopuses, squid, and nautiluses). (Table 33.3)

- Cephalopods are among the most complex of all invertebrates. They are the only mollusks with a closed circulatory system, that have a well-developed nervous system and brain, and are believed to exhibit learning by observation. (Figures 33.16, 33.17, 33.18, 33.19)

- Segmentation, in which the body is divided into nearly identical subunits, is a critical evolutionary innovation in the annelids, although specialization is only minimally present at the anterior end. (Figure 33.20)

- Annelids are a large phylum with three main classes: Polychaeta (the most species-rich class), Oligochaeta (which includes the earthworms), and Hirudinea (leeches). (Figure 33.21, Table 33.4)

33.4 Ecdysozoa: The Nematodes and Arthropods

- The ecdysozoans are so named for their ability to shed their cuticle, a nonliving cover providing support and protection. The two most common ecdysozoan phyla are the nematodes and the arthropods.

- Nematodes, which exist in nearly all habitats, have a cuticle made of collagen, a structural protein. The small, free-living nematode *Caenorhabditis elegans* is a model organism. Many nematodes are parasitic in humans, including *Wuchereria bancrofti*, which causes elephantiasis. (Figures 33.22, 33.23)

- Arthropods are perhaps the most successful phylum on Earth. The arthropod body is covered by a cuticle made of layers of chitin and protein, and it is segmented, with segments fused into functional units called tagmata. (Figure 33.24)

- The six main classes of arthropods are Trilobita (the trilobites; now extinct), Arachnida (spiders and scorpions), Diplopoda (millipedes), Chilopoda (centipedes), Insecta (insects), and Crustacea (crabs and relatives). (Figures 33.25, 33.26, 33.27, 33.28, 33.29, Table 33.5)

- More insect species are known than all other animal species combined. Over half a million insect species are beetles. Researcher Brian Farrell found that the appearance of angiosperm plants led to an increased richness of leaf-eating beetles. (Tables 33.6, 33.7)

- The development of a variety of wing structures and mouthparts was a key to the success of insects. (Figure 33.30)

- Insects undergo a change in body form during development, either complete metamorphosis or incomplete metamorphosis, and have developed complex social behaviors. (Figures 33.31, 33.32)

- Most crustacean orders are small and feature prominently in marine food chains. The most well-known order of crustaceans is the Decapoda, which includes the crabs and lobsters. (Figures 33.33, 33.34, 33.35)

33.5 Deuterostomia: The Echinoderms and Chordates

- The Deuterostomia includes the phylums Echinodermata and Chordata. A striking feature of the echinoderms is their radial symmetry, which is secondary; the free-swimming larvae are bilaterally symmetrical. Most echinoderms possess an internal skeleton called an endoskeleton. (Figure 33.36)
- Five main classes of echinoderms exist today: the Asteroidea (sea stars), Ophiuroidea (brittle stars), Echinoidea (sea urchins and sand dollars), Crinoidea (sea lilies and feather stars), and Holothuroidea (sea cucumbers). (Figure 33.37, Table 33.8)
- The phylum Chordata is distinguished by four critical innovations: the notochord, dorsal hollow nerve chord, pharyngeal slits, and postanal tail. (Figure 33.38)
- The subphylum Urochordata (tunicates) and subphylum Cephalochordata (lancelets) are invertebrate chordates. Genetic studies have shown that cephalochordates are the closest nonvertebrate relatives of the vertebrate chordates (subphylum Vertebrata). (Figures 33.39, 33.40, 33.41)

TEST YOURSELF

1. Choanocytes are
   a. a group of protists that are believed to have given rise to animals.
   b. specialized cells of sponges that function to trap and eat small particles.
   c. cells that make up the gelatinous layer in sponges.
   d. cells of sponges that function to transfer nutrients to other cells.
   e. cells that form spicules in sponges.

2. Which of the following is *not* a characteristic of ctenophores?
   a. complete gut
   b. bioluminescence
   c. stinging cells
   d. ciliary locomotion
   e. all of the above are characteristics of ctenophores

3. Which of the following organisms can produce female offspring through parthenogenesis?
   a. cnidarians
   b. flukes
   c. choanocytes
   d. rotifers
   e. annelids

4. A lophophore is
   a. a contractile structure that aids in movement through the water column.
   b. a ciliated structure that aids in movement through the water column.
   c. a ciliated structure that functions primarily in feeding.
   d. a contractile structure that moves food into the coelom.
   e. a contractile structure important for defense.

5. In the annelids, the metanephridia function in
   a. transport of nutrients.
   b. movement.
   c. cellular communication.
   d. waste removal.
   e. reproduction.

6. A defining feature of the Ecdysozoa is
   a. a segmented body.
   b. a closed circulatory system.
   c. a cuticle.
   d. a complete gut.
   e. a lophophore.

7. In arthropods, the tracheal system is
   a. a unique set of structures that function in ingestion and digestion of food.
   b. a series of branching tubes extending into the body that allow for gas exchange.
   c. a series of tubules that allow waste products in the blood to be released into the digestive tract.
   d. the series of ommatidia that form the compound eye.
   e. none of the above.

8. Characteristics of the class Arachnida include
   a. two tagmata.
   b. six walking legs.
   c. an aquatic lifestyle.
   d. a lobed body.
   e. both b and d.

9. Incomplete metamorphosis
   a. is characterized by distinct larval and adult stages that do not compete for resources.
   b. is typically seen in arachnids.
   c. involves gradual changes in life stages where young resemble the adult stage.
   d. is characteristic of the majority of insects.
   e. always includes a pupal stage.

10. In the phylum Echinodermata, the tube feet function in
    a. movement.
    b. gas exchange.
    c. feeding.
    d. excretion.
    e. all of the above.

CONCEPTUAL QUESTIONS

1. Define hermaphrodite.
2. Define nematocyst and explain its function.
3. Explain the difference between complete metamorphosis and incomplete metamorphosis.

EXPERIMENTAL QUESTIONS

1. What was the hypothesis tested by Fiorito and Scotto?
2. What were the results of the experiment? Did these results support the hypothesis?
3. What is the significance of performing the experiment on both observer and untrained octopuses?

COLLABORATIVE QUESTIONS

1. Discuss some of the characteristics of the phylum Mollusca.
2. Discuss the four defining characteristics of chordates.

---

www.brookerbiology.com

This website includes answers to the Biological Inquiry questions found in the figure legends and all end-of-chapter questions.

34

THE VERTEBRATES

CHAPTER OUTLINE

Giant pandas (*Ailuropoda melanoleuca*). This species, shown here in Sichuan, China, is one of many charismatic vertebrate species.

In Chapter 33, we discussed two chordate subphyla, the Urochordates (tunicates) and Cephalochordates (lancelets). The third subphylum of chordates, the Vertebrata, or vertebrates, with about 48,000 species, is by far the largest and most dominant group of chordates. Species range in size from tiny fish weighing 0.1 g to huge whales of over 100,000 kg. They occupy nearly all of Earth's habitats, from the deepest depths of the oceans to mountaintops and the sky beyond. Throughout history, humans have depended on many vertebrate species for their welfare and have domesticated species such as horses, cattle, pigs, sheep, and chickens and had countless species, including cats and dogs, as pets. Many other vertebrate species are the subjects of conservation efforts, as we will see in Chapter 60 (see also chapter-opening photo). For example, some vertebrate species require large areas for their survival and as such are considered to be umbrella species, in that their preservation will help ensure conservation of others.

In this chapter, we begin by outlining the characteristics of **craniates**, chordates that have a brain encased in a skull. We will take a brief look at the hagfish, a craniate that has some, but not all, vertebrate characteristics. In the remainder of the chapter, we will explore the characteristics of vertebrates (**Figure 34.1**), discussing in some depth the evolutionary development of the major vertebrate classes, including fishes, amphibians, reptiles, birds, and mammals.

34.1 The Craniates

Although the hagfish, an eel-like, eyeless scavenger, is often called a primitive fish, its resemblance to a fish is misleading (**Figure 34.2**). The hagfish is not a true vertebrate and thus cannot be considered a true fish. At the same time, it is different from the tunicates and lancelets, the nonvertebrate chordates we discussed in Chapter 33. Although hagfish do not have a vertebral column—one of the vertebrate characteristics—they do have a skull, in this case, one made of cartilage and not bone. Paleontologists in China have recently discovered similar fishlike fossils that date to around 530 million years ago, the time of the Cambrian explosion. These are believed to be the earliest known craniates, of which the hagfish is the only living relative.

Craniates Are Distinguished by a Cranium and Neural Crest

Craniates have two defining characteristics that distinguish them from other nonvertebrate chordates:

1. **Cranium.** In craniates, the anterior end of the nerve cord elaborates to form a more developed brain that is encased in a protective bony or cartilaginous housing called the **cranium**. This continues the trend of cephalization—the development of the head end in animals.
2. **Neural crest.** The **neural crest** is a group of embryonic cells found on either side of the neural tube as it develops. The cells disperse throughout the embryo, where they contribute to the development of the skeleton, especially the cranium and other structures, including nerves, jaws, and teeth.

While these are the main distinguishing characteristics of craniates, there are others. For example, craniates have at least two clusters of *Hox* genes, compared to the single cluster of *Hox* genes in tunicates and lancelets. This additional gene cluster is believed to have permitted increasingly complex morphologies than those possessed by other nonvertebrate chordates.

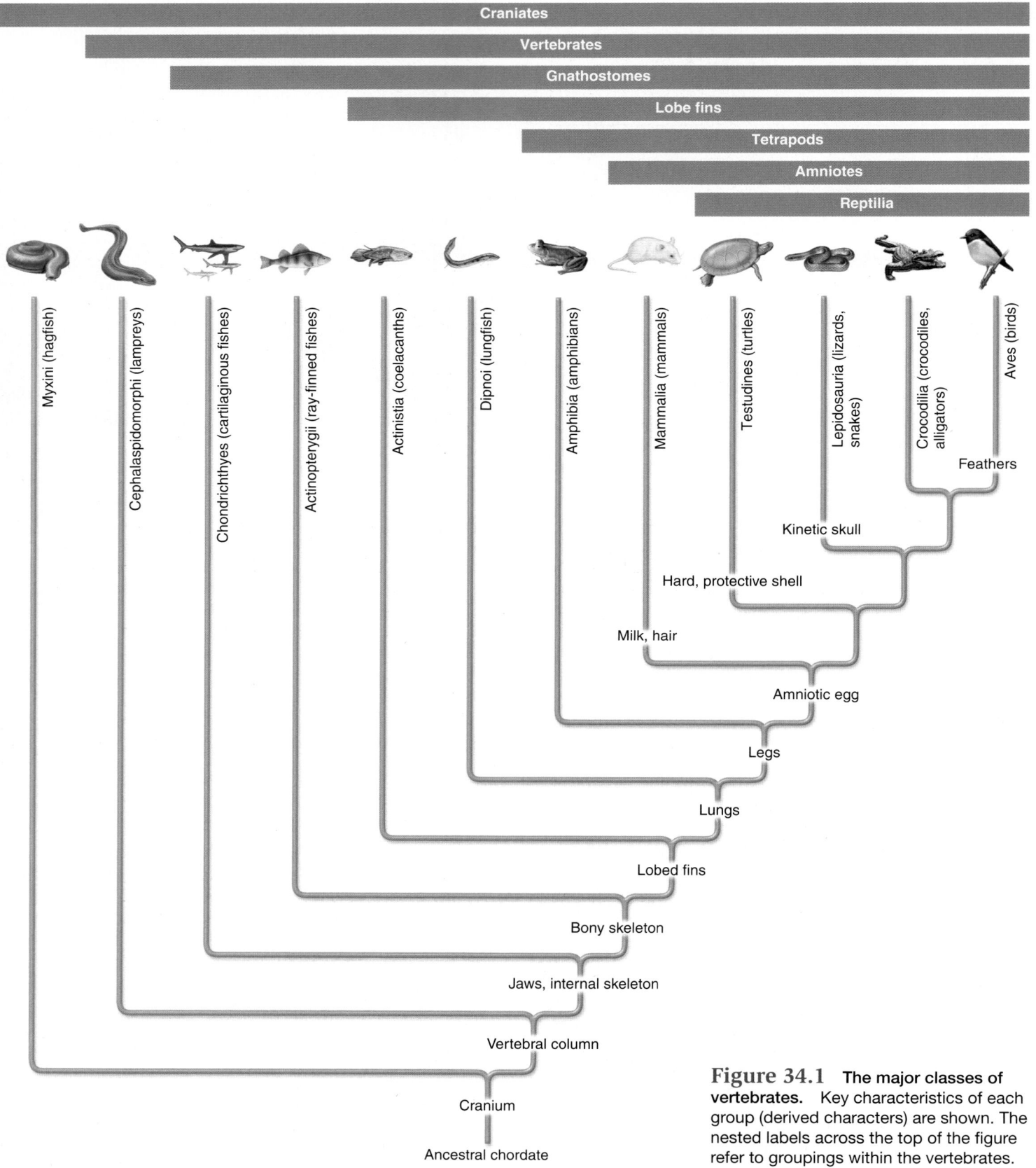

Figure 34.1 **The major classes of vertebrates.** Key characteristics of each group (derived characters) are shown. The nested labels across the top of the figure refer to groupings within the vertebrates.

Figure 34.2 Craniates. The hagfish possesses craniate characteristics (a skull and neural crest) but is not a true fish or vertebrate.

Biological inquiry: Why isn't the hagfish a true fish?

The Hagfish, Class Myxini, Are the Most Primitive Living Craniates

The hagfish are entirely marine, jawless, finless craniates that lack vertebrae. The hagfish skeleton consists largely of a skull that encloses the brain and a cartilaginous notochord. The lack of a vertebral column leads to extensive flexibility. Hagfish live in

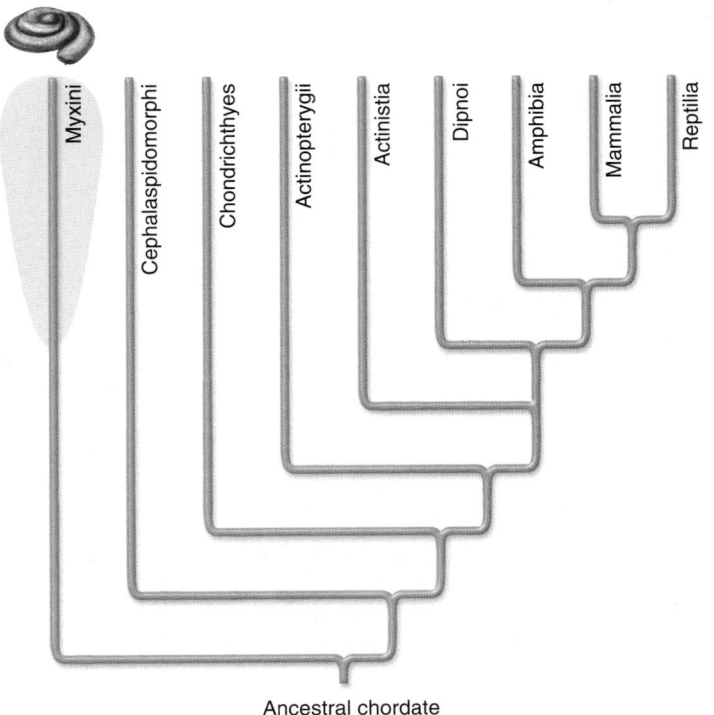

Ancestral chordate

the cold waters of northern oceans, close to the muddy bottom. Essentially blind, hagfish have a very keen sense of smell and are attracted to dead and dying fish, which they attach themselves to via toothed plates on the mouth. The powerful tongue then rasps off pieces of tissue. However, most of their diet consists of dead or disabled marine worms and other invertebrates. Though the hagfish cannot see approaching predators, they have special glands that produce copious amounts of slime, which may deter predatory attacks. When provoked, the hagfish's slime production increases dramatically, enough to potentially distract predators or coat their gills and interfere with breathing. Hagfish can sneeze to free their nostrils of their own slime.

34.2 Characteristics of Vertebrates

The **vertebrates** (from the Latin, *vertebratus*, back boned) retain all chordate characteristics that we outlined in Chapter 33 and all craniate characteristics noted previously, as well as possessing several additional traits, including:

1. **Vertebral column.** During development in vertebrates, the notochord is replaced by a bony or cartilaginous column of interlocking **vertebrae** that provides support and also protects the nerve cord, which lies within its tubelike structure.
2. **Endoskeleton of cartilage or bone.** The cranium and vertebral column are parts of the endoskeleton, the living skeleton of vertebrates that forms within the animal's body. Most vertebrates also have two pairs of appendages, whether fins, legs, or arms. The endoskeleton is composed of either bone or cartilage, both of which are very strong materials, yet they are more flexible than the chitin found in insects and other arthropods. The endoskeleton also contains living cells that secrete the skeleton, which grows with the animal, unlike the nonliving exoskeleton of arthropods.
3. **Internal organs.** Vertebrates possess a great diversity of internal organs, including a liver, kidneys, endocrine glands, and a heart with at least two chambers. The liver is unique to vertebrates, while the vertebrate heart, kidneys, and endocrine system are more complex than analogous structures in other taxa.

While these features are exhibited in all vertebrate classes, some classes developed innovations that helped them succeed in specific environments such as on land or in the air. For example, birds developed feathers and wings, structures that enable most species to fly. In fact, each of the vertebrate classes is distinctly different from one another, as outlined in **Table 34.1**. Of the 11 different classes of vertebrates, 5 of them are classes of fish, reflecting the dominance of fish within the vertebrates.

Table 34.1 | The Main Classes and Characteristics of Living Vertebrates

Class	Examples (approx. # of species)	Main characteristics
Cephalaspidomorphi	Lampreys (41)	Early divergent fish with no appendages, that is, fins; jawless sucking mouth; parasitic on other fish
Chondrichthyes	Sharks, skates, rays (850)	Fish with cartilaginous skeleton; teeth not fused to jaw; no swim bladder; well-developed fins; internal fertilization; single blood circulation
Actinopterygii	Ray-finned fish, most bony fish (24,600)	Fish with ossified skeleton; single gill opening covered by operculum; fins supported by rays, fin muscles within body; swim bladder often present; mucus glands in skin
Actinistia	Lobe-finned fish, of which coelacanth is only living member (1)	Fish with ossified skeleton; bony extensions, together with muscles, project into pectoral and pelvic fins; swim bladder filled with oil
Dipnoi	Lungfish (6)	Fish with ossified skeleton; rudimentary lungs allow fish to come to the surface to gulp air; limblike appendages
Amphibia	Frogs, toads, salamanders (4,000)	Tetrapods; adults able to live on land; fresh water needed for reproduction; development usually involving metamorphosis from tadpoles; adults with lungs and double blood circulation; moist skin; shell-less eggs
Mammalia	Mammals (5,500)	Mammary glands, hair, specialized teeth, enlarged skull, external ears, endothermic, highly developed brains, diversity of body forms
Testudines	Turtles (330)	Body encased in hard shell; no teeth; head and neck retractable into shell; eggs laid on land
Lepidosauria	Lizards, snakes (7,800)	Lower jaw not attached to skull; skin covered in scales
Crocodilia	Crocodiles, alligators (23)	Four-chambered heart; large aquatic predators; parental care of young
Aves	Birds (9,600)	Feathers, hollow bones, air sacs, reduced internal organs, endothermic, four-chambered heart

34.3 The Fishes

A critical innovation in vertebrate evolution is the hinged jaw, which first developed in fish. However, not all fish developed hinged jaws. The **ostracoderms**, an umbrella term for several classes of early divergent, heavily armored fish, now extinct, lacked a jaw (**Figure 34.3**). These fish most likely swam slowly above the substrate, using a strong pharyngeal pump as a way of vacuuming up particles. The ostracoderms were common in the Silurian and Devonian periods and many had fins, suggesting they were accomplished swimmers. However, their heavy armor also suggests that predatory fish, with hinged jaws, were also common at that time. A class of jawless fish—the lampreys—exists today, although these evolved separately from the ostracoderms.

Figure 34.3 Ancient jawless fish. *Cephalaspis* is one of an extinct group of primitive fish called ostracoderms that were partially armored and lacked a hinged jaw. Jawless species like this were particularly common in the Silurian and Devonian periods.

The jawed mouth was a significant evolutionary development. It enabled an animal to grip its prey more firmly, which may have increased its rate of capture, and to attack larger prey species, thus increasing its potential food supply. All vertebrate species that possess jaws are called **gnathostomes** (meaning "jaw mouth") (see Figure 34.1). In contrast, the jawless species are termed agnathans (meaning "without jaws"). Accompanying the jawed mouth was the development of more sophisticated head and body structures, including two pairs of appendages called fins. Gnathostomes also possess two additional *Hox* gene clusters (bringing their total to four), which permitted increased morphological complexity.

Biologists have identified about 25,000 species of living fish, more than all other species of vertebrates combined. All are aquatic, gill-breathing ectotherms that possess fins and a scaly skin. There are five separate classes of fish, each of which has distinguishing characteristics (see Table 34.1). We will begin by examining the most primitive class, the Cephalaspidomorphi (lampreys), which lack jaws and appendages. Most of our attention, however, will focus on the jawed fish: the Chondrichthyes (cartilaginous fish), Actinopterygii (ray-finned fish), Actinistia (coelacanths), and Dipnoi (lungfish).

The Lampreys, Class Cephalaspidomorphi, Are Eel-like Fish That Lack Jaws

Lampreys are unlike members of other classes of fish because they lack both a hinged jaw and true appendages. However, lampreys do possess a notochord surrounded by a cartilaginous rod that represents a rudimentary vertebral column, and thus they represent one of the most early divergent groups of vertebrates. Lampreys can be found in both marine and freshwater environments. Marine lampreys are parasitic as adults. They grasp other fish with their circular mouth (**Figure 34.4a**) and rasp a hole in the fish's side, sucking blood, tissue, and fluids until they are full (**Figure 34.4b**). Reproduction of all species is similar, whether they live in marine or freshwater environments. Males and females spawn in freshwater streams, and the resultant larval lampreys bury into the sand or mud, much like lancelets (refer back to Figure 33.41b), emerging to feed on small invertebrates or detritus at night. This stage can last for three to seven years, at which time the larvae metamorphose into an adult.

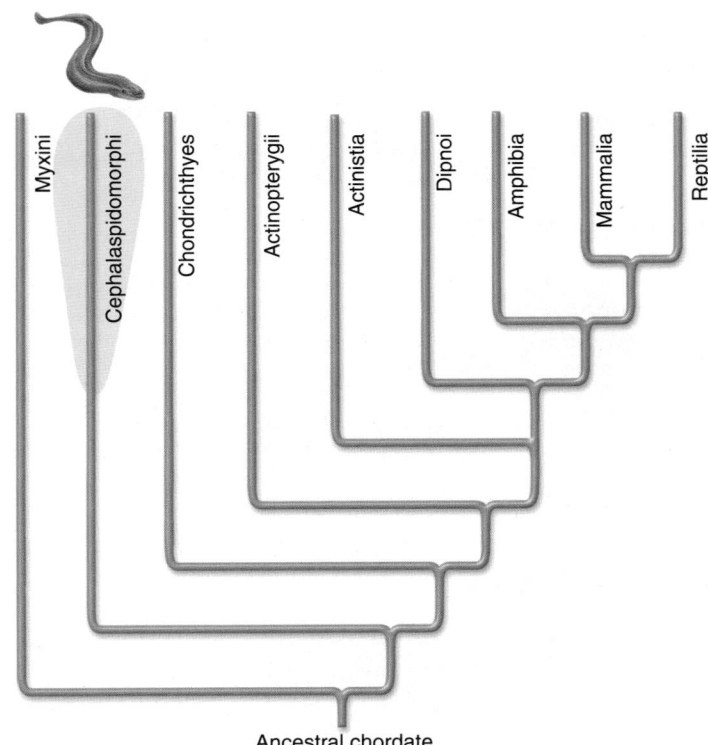

Ancestral chordate

Sometimes, in strictly freshwater species, the adults do not feed at all but quickly mate and die. Marine species migrate from freshwater back to the ocean for the remainder of their life. For parasitic species, the adults then begin their parasitic lifestyle.

Jawed Fish Include the Sharks and Rays and the Bony Fish

As we noted previously, one of the main innovations in fish was the development of the hinged jaw. The hinged jaw developed from the pharyngeal arches, cartilaginous rods that function to help support the respiratory tissue. Primitive jawless fish had nine gill arches surrounding the eight gill slits (**Figure 34.5a**). During the late Silurian period, about 417 million years ago, some of these gill arches became modified in jawless fish. The first and second gill arches were lost, while the third and fourth pairs became modified to form the jaws (**Figure 34.5b,c**).

(a) The sea lamprey (*Petromyzon marinus*) has a circular, jawless mouth.

(b) A sea lamprey feeding on a fish.

Figure 34.4 Modern jawless fish.

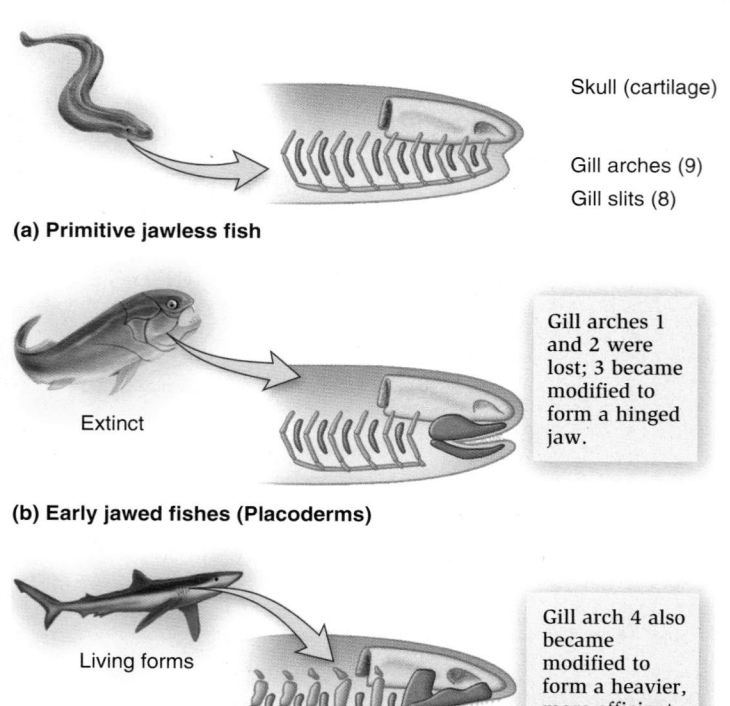

(a) Primitive jawless fish

Skull (cartilage)

Gill arches (9)
Gill slits (8)

Gill arches 1 and 2 were lost; 3 became modified to form a hinged jaw.

(b) Early jawed fishes (Placoderms)

Extinct

Gill arch 4 also became modified to form a heavier, more efficient jaw.

Living forms

(c) Modern jawed fishes (cartilaginous and bony fishes)

Figure 34.5   The evolution of the vertebrate jaw. (a) Primitive fish and extant jawless fish such as lampreys have nine cartilaginous gill arches that support eight gill slits. (b) In early jawed fish such as the placoderms, the first two pairs of gill arches were lost, and the third pair became modified to form a hinged jaw. This left six gill arches to support the remaining five gill slits, which were still used in breathing. (c) In modern jawed fish, the fourth gill arch is also incorporated into the jaw, allowing stronger, more powerful bites to be delivered.

This is how evolution usually works; body features do not appear *de novo* but instead become modified to serve other functions.

By the mid-Devonian period, two classes of jawed fish, the Acanthodii (spiny fish) and Placodermi (armored fish) were common. Some of the placoderms were huge individuals, over 9 m long. Both classes died out by the end of the Devonian as part of one of several mass extinctions that occurred in the Earth's geological and biological history. The reasons for this extinction are not well understood, but other types of jawed fish—the cartilaginous and bony fish—did not go extinct. We will discuss the fish in these classes next.

Chondrichthyans: The Cartilaginous Fish Members of the class Chondrichthyes (the **chondrichthyans**)—sharks, skates, and rays—are also called cartilaginous fish because their skeleton is composed of flexible cartilage rather than bone. The cartilaginous skeleton is not considered an ancestral character but rather a derived character. This means that the ancestors of the chondrichthyans had bony skeletons, but that members of this class subsequently lost this feature. This hypothesis

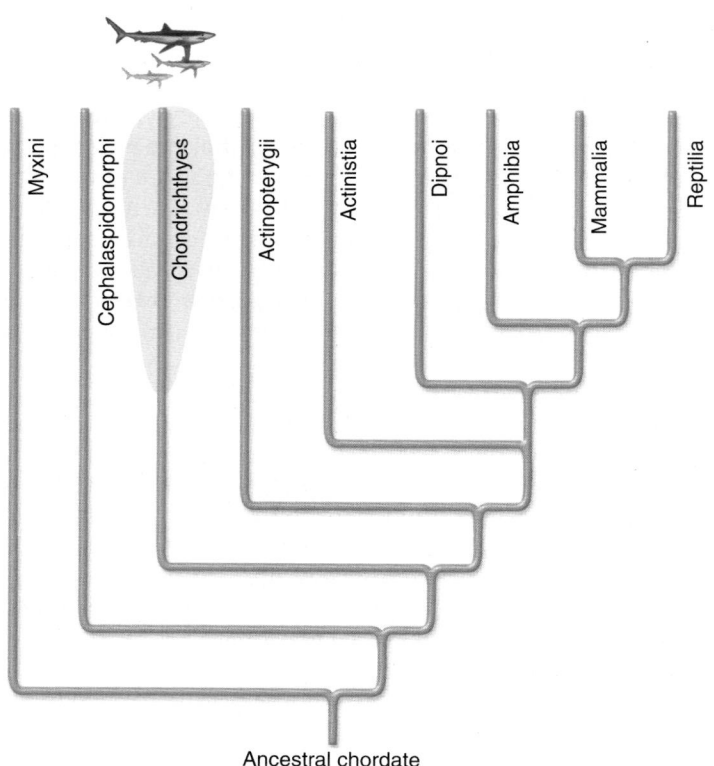

Myxini

Cephalaspidomorphi

Chondrichthyes

Actinopterygii

Actinistia

Dipnoi

Amphibia

Mammalia

Reptilia

Ancestral chordate

is reinforced by the fact that during development, the skeleton of most vertebrates is cartilaginous, and then it becomes bony (ossified) as a hard calcium-phosphate matrix replaces the softer cartilage. A change in the developmental sequence of the cartilaginous fishes is believed to prevent the ossification process.

In the Carboniferous period, 354–290 million years ago, sharks were the great predators of the ocean. Aided by appendages called fins, sharks became fast, extremely efficient swimmers (**Figure 34.6a**). Perhaps the most important fin for propulsion is the large and powerful caudal fin, or tail fin, which, when swept from side to side, thrusts the fish forward at great speed. For example, great white sharks (*Carcharodon carcharias*) can swim at over 40 km per hour, and Mako sharks (*Isurus oxyrinchus*) have been clocked at nearly 50 km per hour. The paired pelvic fins (at the back) and pectoral fins (at the front) act like flaps on airplane wings, allowing the shark to dive deeper or rise to the surface. They also aid in steering. In addition, the dorsal fin (on the back) acts as a stabilizer to prevent the shark from rolling in the water as the tail fin pushes it forward.

Sharks were among the earliest fish to develop teeth. Their teeth evolved from rough scales on the skin that also contain dentin and enamel. While shark's teeth are very sharp and hard, they are not set into the jaw, as are human teeth, so they break off easily. To offset this, the teeth are continually replaced, row by row (**Figure 34.6b**). Sharks may have 20 rows of teeth, with the front pair in active use and the ones behind ready to grow as replacements when needed. Tooth replacement time varies from 9 days in the cookie-cutter shark (named for its

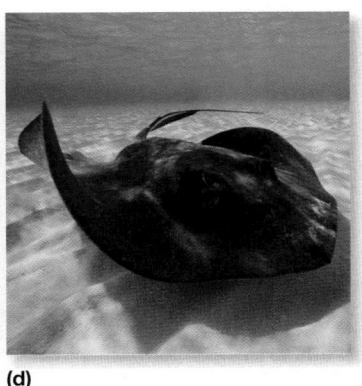

(a)   **(b)**   **(c)**   **(d)**

Figure 34.6   Cartilaginous fish.   (a) The silvertip shark (*Carcharhinus albimarginatus*) is one of the ocean's most powerful predators. **(b)** Close-up of the mouth of a sand tiger shark (*Carcharias taurus*), showing rows of teeth. **(c)** This mermaid's purse (egg pouch) of a dogfish shark (*Scyliorhinus canicula*) is entwined in vegetation to keep it stationary. **(d)** Stingrays are essentially flattened sharks with very large pectoral fins.

characteristic of biting round plugs of flesh from its prey) to 242 days in great whites. Some experts estimate that certain sharks can use up to 20,000 teeth in a lifetime.

All chondrichthyans are denser than water, which theoretically means that they would sink if they stopped swimming. Many sharks never stop swimming and maintain buoyancy via the use of their fins and a large oil-filled liver. Another advantage of swimming is that water continually enters the mouth and is forced over the gills, allowing the sharks to extract oxygen and breathe. How then do skates and rays breathe when they rest on the ocean floor? These species, and a few sharks like the nurse shark, use a muscular pharynx and jaw muscles to pump water over the gills. In these and indeed all species of fish, the heart consists of two chambers, an atrium and a ventricle, that contract in sequence. They employ what is known as a single circulation, in which blood is pumped from the heart to capillaries in the gills to collect oxygen, and then it flows through arteries to the tissues of the body, before returning to the heart (look ahead to Figure 47.4a).

Many active predators possess a variety of acute senses, and sharks are no exception. They have a powerful sense of smell, facilitated by sense organs in the nostrils (sharks and other fish do not use nostrils for breathing). They can see well but cannot distinguish colors. While sharks have no eardrum, they can detect pressure waves generated by moving objects. All jawed fish have a row of microscopic organs in the skin, arranged in a line that runs laterally down each side of the body, that can detect movements in the surrounding water. This system of sense organs, known as the **lateral line**, picks up pressure waves and sends nervous signals to the inner ear and then on to the brain.

Fertilization is internal in chondrichthyans, with the male transferring sperm to the female via a pair of **claspers**, extensions of the pelvic fins. Some shark species are **oviparous**, that is, they lay eggs, often inside a protective pouch called a mermaid's purse (**Figure 34.6c**). The eggs, which are not guarded by either parent, then hatch into tiny sharks. In **ovoviparous** species, the eggs are retained within the female's body, but there

is no placenta to nourish the young. A few species are **viviparous**; the eggs develop within the uterus, receiving nourishment from the mother via a placenta. Both ovoviparous and viviparous sharks give birth to live young.

The sharks have been a very successful vertebrate group, with many species identified in the fossil record. While many species died out in the mass extinction at the end of the Permian period (290–248 million years ago), there was a further period of speciation of the survivors in the Mesozoic era, when most of the 375 modern-day species appeared. Skates and rays are essentially flattened sharks that cruise along the ocean floor using hugely expanded pectoral fins. In addition, their thin and whiplike tails are often equipped with a venomous barb that is used in defense (**Figure 34.6d**). Most of the 475 or so species of skates and rays feed on bottom-dwelling crustaceans and mollusks.

The Bony Fish   Bony fish are the most numerous types of fish, with more individuals and more species (about 24,600) than any other. Most authorities now recognize three living classes: the Actinopterygii (ray-finned fish), the Actinistia, (coelacanths), and the Dipnoi (lungfish). Fish in all three classes possess a bony skeleton and scale-covered skin. The skin of bony fish, unlike the rough skin of sharks, is slippery and slimy because of glands that produce mucus, an adaptation that reduces drag during swimming. Just as in the cartilaginous fish, water is drawn over the gills for breathing, but in bony fish, a protective flap called an **operculum** covers the gills (**Figure 34.7**). Some early bony fish lived in shallow, oxygen-poor waters and developed lungs as an embryological offshoot of the pharynx. These fish could rise to the water surface and gulp air. As we will see, modern lungfish operate in much the same fashion. In most bony fish, these lungs evolved into a **swim bladder**, a gas-filled, balloon-like structure that helps the fish remain buoyant in the water even when it is completely stationary. The swim bladder is connected to the circulatory system, and gases can diffuse in and out of the blood, allowing the fish to change the volume of the swim bladder and to rise and sink.

Ancestral chordate

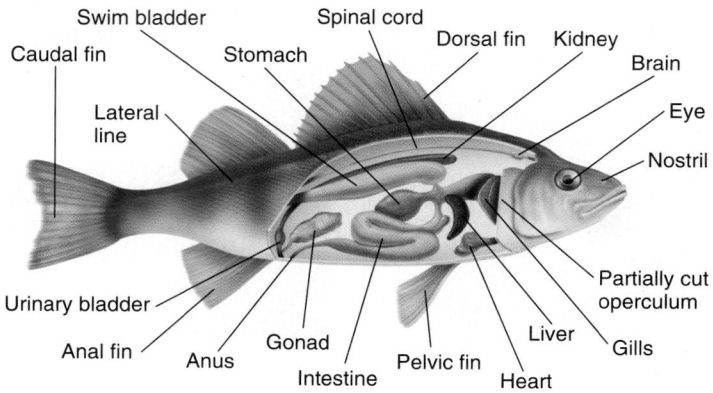

Figure 34.7 Generalized body plan of a bony fish.

logical adaptations to the different osmotic problems seawater presents compared to fresh water (see Chapter 49).

The most species-rich class of bony fish is the Actinopterygii, or **ray-finned fish**, which includes all bony fish except the coelacanths and lungfish. In Actinopterygii, the fins are supported by thin, bony, flexible rays and are moved by muscles on the interior of the body. The class has a diversity of forms, from large predatory moray eels to delicate sea dragons (**Figure 34.8**). Whole fisheries are built around the harvest of species such as cod, anchovies, and salmon.

The Actinistia (coelacanths) and Dipnoi (lungfish) are both considered Sarcopterygii or **lobe fins**. The name Sarcopterygii used to refer solely to the lobe-finned fish, but since it has become clear that terrestrial vertebrates (tetrapods) evolved from such fish, the definition of the group has been expanded to include both lobe-finned fish and tetrapods (see Figure 34.1). In the **lobe-finned fish**, the fins are part of the body, and they are supported by skeletal extensions of the pectoral and pelvic areas that are moved by muscles residing in the fins.

The fossil record revealed that the Actinistia, or coelacanths, were a very successful group in the Devonian period, but all fish of the class were believed to have died off at the end of the Mesozoic era, some 65 million years ago. You can therefore imagine the scientific excitement in 1938, when a modern coelacanth was

Thus, unlike the sharks, many bony fish can remain motionless and utilize a "sit-and-wait" ambush style. These three features—bony skeleton, operculum, and swim bladder—distinguish bony fish from cartilaginous fish.

Reproductive strategies of bony fish vary tremendously, but most species reproduce via external fertilization, with the female shedding her eggs and the male depositing sperm on top of them. While adult bony fish can maintain their buoyancy, their eggs tend to sink. This is why many species spawn in shallow, more oxygen- and food-rich waters, and why coastal areas are such important fish nurseries.

Bony fish have colonized all aquatic habitats. Most cartilaginous fish are marine, but bony fish probably evolved in freshwater habitats and secondarily returned to marine environments. This, of course, required the redevelopment of physio-

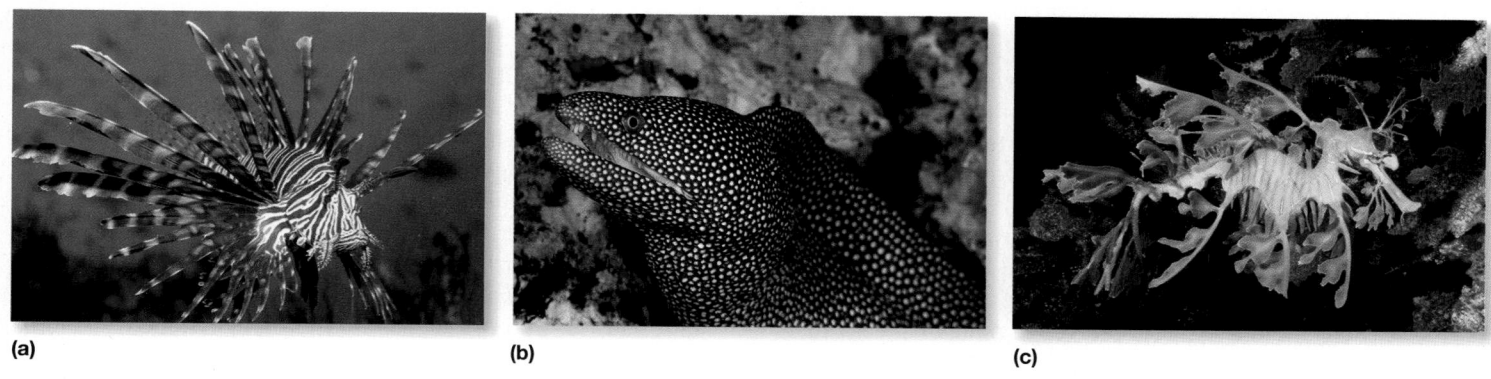

(a) (b) (c)

Figure 34.8 **The diversity of ray-finned fish.** (a) Lionfish (*Pterois volitans*). (b) Whitemouth moray eel (*Gymnothorax meleagris*). (c) Leafy sea dragon (*Phycodurus eques*).

Figure 34.9 A lobefin fish, the coelacanth (*Latimeria chalumnae*).

Figure 34.10 An Australian lungfish (*Neoceratodus forsteri*).

Biological inquiry: How are lungfish similar to coelacanths?

discovered as part of the catch of a boat fishing near the Chalumna River in South Africa (**Figure 34.9**). Intensive searches in the area revealed that coelacanths were living off the southern African coast and especially off a group of islands near the coast of Madagascar called the Comoros Islands. Another species was more recently found in Indonesian waters.

Early diverging lobe-finned fish probably evolved in fresh water and had lungs, but the coelacanth lost them and returned to the sea. One distinctive feature is a special joint in the skull that allows the jaws to open extremely wide and gives the coelacanth a powerful bite. As further evidence of the coelacanth's unusual body plan, its swim bladder is filled with oil rather than gas, although it serves a similar purpose—to increase buoyancy.

The Dipnoi, or **lungfish**, like the coelacanths, are also not currently a very species-rich class, having just three genera and six species (**Figure 34.10**). Lungfish live in oxygen-poor freshwater swamps and ponds. They have both gills and lungs, the latter of which enable them to come to the surface and gulp air. In fact, lungfish will drown if they are unable to breathe air. When ponds dry out, some species of lungfish can dig a burrow and survive in it until the next rain. Because they also have muscular lobe fins, they are often able to successfully traverse quite long distances over shallow-bottomed lakes that may be drying out.

The morphological features of coelacanths, lungfish, and early divergent land animals, together with the similarity of coelacanth and lungfish nuclear genes to those of primitive terrestrial vertebrates, suggest to many scientists that lobe-fin ancestors gave rise to three lineages: the coelacanths, the lungfish, and the tetrapods.

34.4 Amphibia: The First Tetrapods

During the Devonian period, from about 417 to 354 million years ago, a diversity of plants colonized the land. The presence of plants served as both a source of oxygen and an extensive food source for animals that ventured out of the aquatic environment.

Terrestrial arthropods, especially insects, had evolved to feed on these plants and provided an additional food source for any vertebrate that could colonize the land.

The transition to life on land involved a large number of adaptations. Paramount among these were adaptations preventing desiccation and making locomotion and reproduction on land possible. We have seen that the lungfish evolved the ability to breathe air. In this section, we begin by outlining the development of the **tetrapods**, vertebrate animals having four legs or leglike appendages. We will discuss the first terrestrial vertebrates and their immediate descendants, the amphibians. We will then explore the characteristic features and diversity of modern amphibians.

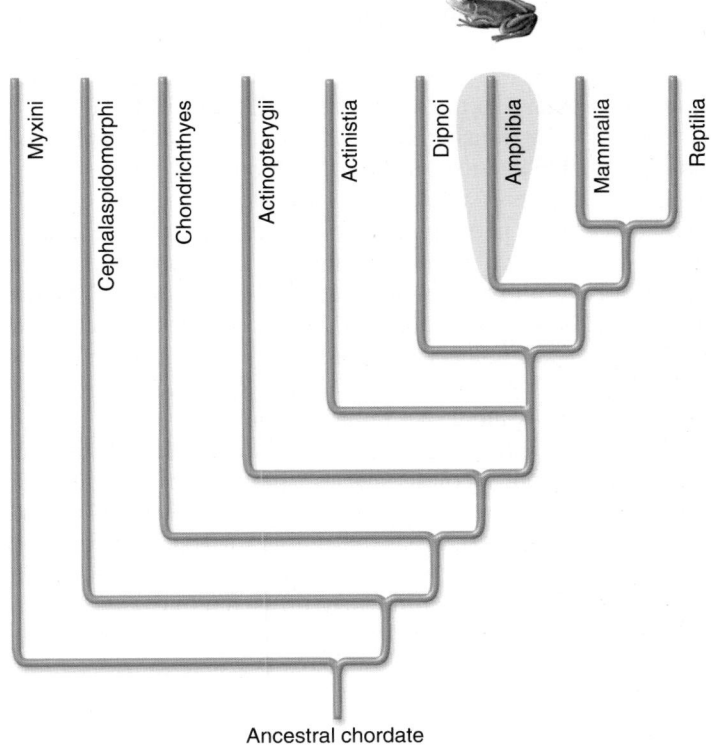

Myxini
Cephalaspidomorphi
Chondrichthyes
Actinopterygii
Actinistia
Dipnoi
Amphibia
Mammalia
Reptilia

Ancestral chordate

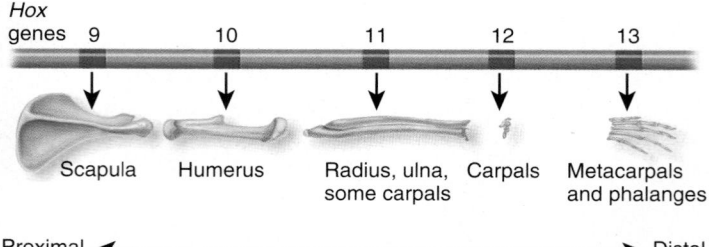

Figure 34.11 *Hox* genes 9–13 work together to specify limb formation from the proximal to distal direction. The axes of limb development in mice are shown, together with their associated genes.

(a) *Acanthostega*

(b) *Cacops*

Figure 34.12 The development of primitive tetrapods. **(a)** *Acanthostega* represents a transitional form between a lobe-finned fish and a true amphibian. It possessed gills and was aquatic but had the tetrapod skeletal structure. **(b)** *Cacops* was a large, early amphibian of the Permian period.

The Origin of Tetrapods Involved the Development of Four Limbs

Over the Devonian period, fossils record the evolution of sturdy lobe-finned fishes that became fishes with four limbs. The abundance of light and nutrients in shallow waters encouraged a profusion of plant life and the invertebrates that fed on them. The development of lungs enabled lungfish to colonize these productive yet often oxygen-poor waters. Here, the ability to move in shallow water chock full of plants and debris was more vital than the ability to swim swiftly through open water and may have favored the progressive development of sturdy limbs. As an animal's weight began to be borne more by the limbs, the vertebral column strengthened, and hip bones and shoulder bones were braced against the backbone for added strength. Such modifications, while seemingly dramatic, are believed to be caused by relatively simple changes in the expression of genes, especially *Hox* genes (see Feature Investigation). In particular, *Hox* genes 9–13 work together to specify limb formation from the proximal to the distal direction (**Figure 34.11**).

Some of the transitional taxa between fish and amphibians possessed gills, like fish, but they had four limbs and free digits, characteristics of amphibians. *Acanthostega*, a very early diverging tetrapod, had forelimbs and hindlimbs, and eight digits on each hand, but it also retained many adaptations for aquatic life because it probably only rarely ventured onto land (**Figure 34.12a**). This is an important species, for it may represent a so-called stem species, a common tetrapod ancestor located at the beginning of the tetrapod branch on the tree of life. Eventually, however, species more like modern amphibians evolved, species that were still tied to water for reproduction but increasingly fed on land. In these species, the vertebral column, hip bones, and shoulder bones grew even sturdier (**Figure 34.12b**). Such changes were needed as the animal's weight was no longer supported by water but was borne entirely on the limbs. The evolution of a rib cage provided protection for the internal organs, especially the lungs and heart.

By the middle of the Carboniferous period, about 320 million years ago, species similar to modern amphibians had become common in the terrestrial environment. For example, *Cacops* was a large amphibian, as big as a pony (Figure 34.12b). Its skin was heavy and tough, an adaptation that helped prevent water loss; its breathing was accomplished more by lungs than by skin; and it possessed **pentadactyl limbs** (limbs ending in five digits). With a bonanza of terrestrial arthropods to feast on, the amphibians became very numerous and species rich, and the mid-Permian period, some 260 million years ago, is sometimes known as the Age of Amphibians. However, most of the large amphibians became extinct at the end of the Permian period, coincident with the radiation of the reptiles, though whether the rise of reptiles was responsible for the extinction of amphibians is not known. Most surviving amphibians were smaller organisms resembling modern-day species.

FEATURE INVESTIGATION

Davis, Capecchi, and Colleagues Showed a Genetic-Developmental Explanation for Limb Length in Tetrapods

The development of limbs in tetrapods was a vital step in enabling animals to colonize land. The diversity of vertebrate limb types is amazing, from fins in fish and marine mammals to legs and arms in primates to different wing types in bats and birds.

In 1995, PhD student A. P. Davis, professor M. R. Capecchi, and colleagues analyzed the effects of mutations in specific *Hox* genes that are responsible for determining limb formation in mice. As described in **Figure 34.13**, they began with strains of mice carrying loss-of-function mutations in *HoxA-11* or *HoxD-11*. They bred the mice to obtain offspring carrying one, two, three, or four mutations.

Figure 34.13   Relatively simple changes in *Hox* genes control limb formation in tetrapods.

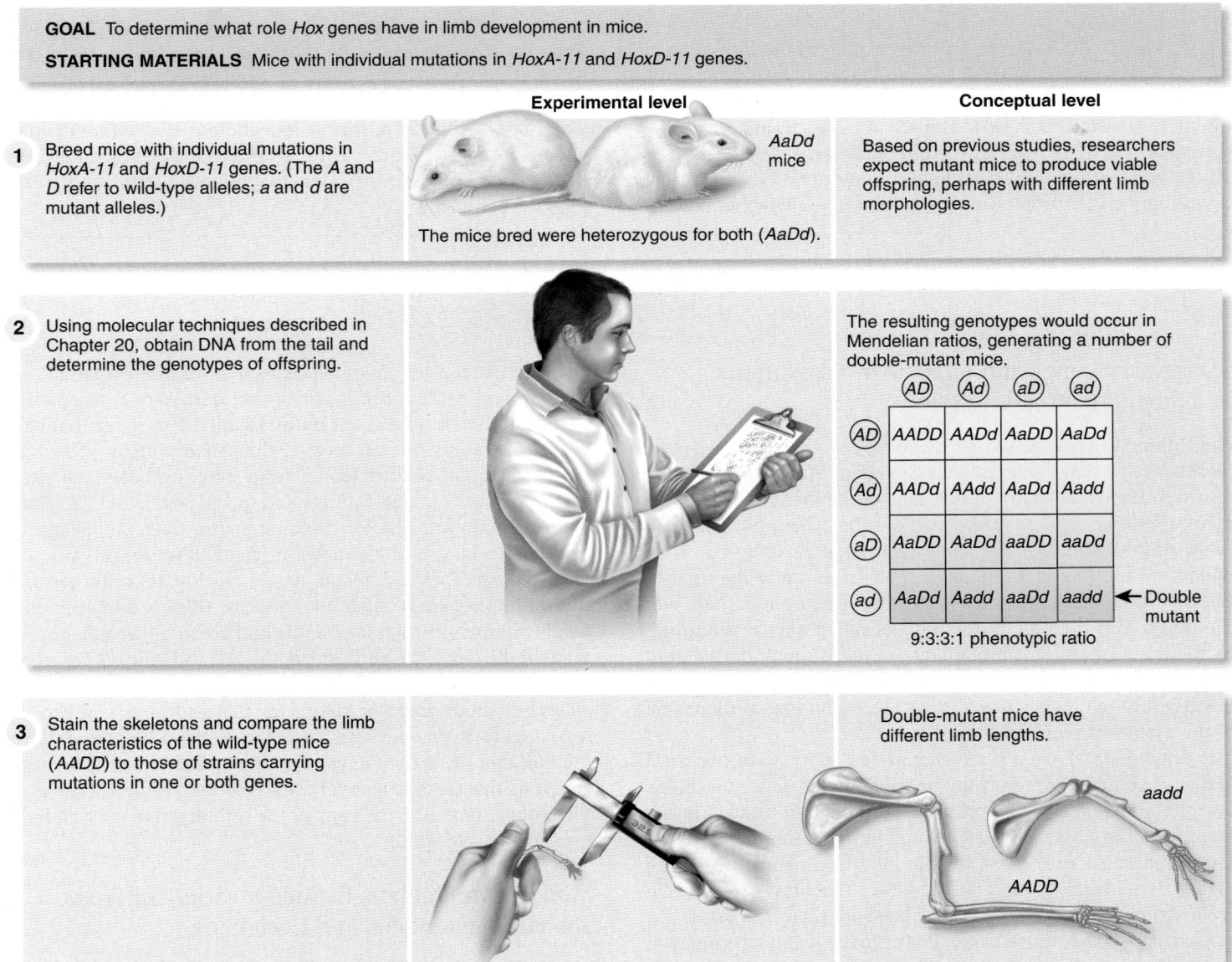

GOAL To determine what role *Hox* genes have in limb development in mice.

STARTING MATERIALS Mice with individual mutations in *HoxA-11* and *HoxD-11* genes.

Experimental level / **Conceptual level**

1 Breed mice with individual mutations in *HoxA-11* and *HoxD-11* genes. (The *A* and *D* refer to wild-type alleles; *a* and *d* are mutant alleles.)

AaDd mice

The mice bred were heterozygous for both (*AaDd*).

Based on previous studies, researchers expect mutant mice to produce viable offspring, perhaps with different limb morphologies.

2 Using molecular techniques described in Chapter 20, obtain DNA from the tail and determine the genotypes of offspring.

The resulting genotypes would occur in Mendelian ratios, generating a number of double-mutant mice.

	(AD)	(Ad)	(aD)	(ad)
(AD)	AADD	AADd	AaDD	AaDd
(Ad)	AADd	AAdd	AaDd	Aadd
(aD)	AaDD	AaDd	aaDD	aaDd
(ad)	AaDd	Aadd	aaDd	aadd ← Double mutant

9:3:3:1 phenotypic ratio

3 Stain the skeletons and compare the limb characteristics of the wild-type mice (*AADD*) to those of strains carrying mutations in one or both genes.

Double-mutant mice have different limb lengths.

aadd

AADD

4 THE DATA

Genotype	Carpal bone fusions (% of mice showing the fusion)			
	Normal (none fused)	NL-T	T-P	NL-T-P
AADD	100			
AaDD	100			
aaDD	33	17	50	
AADd	100			
AAdd		17	17	67
AaDd	17	17	33	33

As seen in the data, the mutations affected the formation of limbs. For example, the wrist contains seven bones: three proximal carpals—called pisiform (P), triangular (T), and navicular lunate (NL)—and four distal carpals (d1–d4). In mice with the genotypes *aaDD* and *AAdd*, the proximal carpal bones are usually fused together. Individual heterozygotes (*AADd* and *AaDD*) do not show this defect, but compound heterozygotes (*AaDd*) often do. Therefore, any two mutant alleles (either from both *HoxA-11* and *HoxD-11* or one from each locus) will cause carpal fusions. Deformities became even more severe with three mutant alleles (*Aadd* or *aaDd*) or four mutant alleles (*aadd*) (data not shown in the figure). Thus, scientists have shown that relatively simple mutations can control relatively large changes in limb development.

Amphibian Lungs and Limbs Are Adaptations to a Semiterrestrial Lifestyle

Amphibians (from the Greek, meaning "two lives") live in two worlds: They have successfully invaded the land but must return to the water to reproduce. One of the first challenges terrestrial animals had to overcome was breathing air when on land. Amphibians use the same technique as lungfish: They open their mouths to let in air then close and raise the floor of the mouth creating a positive pressure, thus pumping air into the lungs. This method of breathing is called **buccal pumping**. In addition, the skin of amphibians is much thinner than that of fish, and many amphibians can absorb oxygen from the air directly through their moist skin or the skin lining of the inside of the mouth or pharynx.

Amphibians have a three-chambered heart, with two atria and one ventricle. One atrium receives blood from the body, and the other receives blood from the lungs. Both atria pump blood into the single ventricle, which pumps some blood to the lungs and some to the rest of the body (look ahead to Figure 47.4b). This form of circulation allows the tissues to receive well-oxygenated blood at a higher pressure than is possible via single circulation because blood returns to the heart to be pumped to the tissues and is not slowed down by passage through the lung capillaries. This development enhances the delivery of nutrients and oxygen to the tissues, but it is still not maximally efficient because some oxygenated and deoxygenated blood are mixed in the ventricle.

Because the skin of amphibians is so thin, the animals face the problem of desiccation, or drying out. As a consequence, even amphibian adults are more abundant in damp habitats, such as swamps or rain forests, than in dry areas. Also, amphibians cannot venture too far from water because their larval stages are still aquatic. In frogs and toads, fertilization is generally external, with males shedding sperm over the gelatinous egg masses laid by the females in water (**Figure 34.14a**). The fertilized eggs lack a shell and would quickly dry out if exposed to the air. They soon hatch into tadpoles (**Figure 34.14b**), small fishlike herbivores that lack limbs and breathe through gills. As the tadpole nears the adult stage, the tail and gills are resorbed, and limbs and lungs appear (**Figure 34.14c**). Such a dramatic change in body form is known as metamorphosis, a process regulated by hormones from the thyroid gland. Reproduction is the link that ties amphibians to water, though a few species circumvent this need. Relatively few species are ovoviviparous or viviparous, retaining the eggs in the reproductive tract and giving birth to live young.

Modern Amphibians Include a Variety of Frogs, Toads, Salamanders, and Caecilians

Approximately 4,800 living amphibian species are known, and the vast majority of these, nearly 90%, are frogs and toads of the order Anura, meaning tail-less ones (**Figure 34.15a**). The other two orders are the Caudata ("tail visible"), the salamanders, and Gymnophiona ("naked snake"), the wormlike caecilians.

(a)   (b)   (c)

Figure 34.14 Amphibian development in the wood frog (*Rana sylvatica*).   (a) Amphibian eggs are laid in gelatinous masses in water. (b) The eggs develop into tadpoles, aquatic herbivores with a fishlike tail that breathe through gills. (c) During metamorphosis, the tadpole loses its gills and tail and develops limbs and lungs.

Adult anurans are carnivores, eating a variety of invertebrates by catching them on a long, sticky tongue. In contrast, the aquatic larvae (tadpoles) are primarily herbivores. Frogs generally have smooth, moist skin and long hind legs, making them excellent jumpers and swimmers. In addition to secreting mucus, which keeps their skin moist, some frogs can also secrete poisonous chemicals that deter would-be predators. Some amphibians advertise the poisonous nature of their skin with warning coloration (look ahead to Figure 57.10b). Others use camouflage as a way of avoiding detection by predators. Toads have a drier, bumpier skin and shorter legs than frogs. While they are less impressive leapers, toads can tolerate drier conditions than can frogs.

The salamanders (order Caudata) possess a tail and have a more elongate body than anurans (**Figure 34.15b**). During locomotion, they seem to sway from side to side, perhaps reminiscent of how the earliest tetrapods may have walked. Like frogs, salamanders often have colorful skin patterns that advertise their distastefulness to predators. Salamanders retain their moist skin by living in damp areas under leaves or logs or beneath lush vegetation. They generally range in size from 10 to 30 cm. Fertilization is usually internal, with females using their cloaca to pick up sperm packets deposited by males. A very few salamander species do not undergo metamorphosis, and the newly hatched young resemble tiny adults. On the other hand, some species, such as the axolotl, retain the gills and tail fins characteristic of the larval stage into adulthood, a phenomenon known as paedomorphosis.

Caecilians (order Gymnophiona) are a small order of about 160 species of legless, nearly blind amphibians (**Figure 34.15c**). Most are tropical and burrow in forest soils, while a few live in ponds and streams. They are secondarily legless, which means that they developed from legged ancestors. Caecilians eat worms and other soil invertebrates and have tiny jaws equipped with teeth. In this order, fertilization is internal, and females usually bear live young. The young are nourished inside the mother's body by a thick, creamy secretion known as uterine milk. In most caecilian species, the young grow into adults about 30 cm long, though some species up to 1.3 m in length are known.

(a)

(b)   (c)

Figure 34.15 Amphibians.   (a) Most amphibians are frogs and toads of the order Anura, including this red-eyed tree frog (*Agalychnis callidryas*). (b) The order Gymnophiona includes wormlike caecilians such as this species from Colombia, *Caecilia nigricans*. (c) The order Caudata includes species such as this mud salamander (*Pseudotriton montanus*).

Biological inquiry: Do all amphibians produce tadpoles?

34.5 Amniotes: The Appearance of Reptiles

While successfully living in a terrestrial environment, adult amphibians remain tied to water for reproduction. Amphibians lay their eggs in water or in a very moist place, so the shell-less eggs do not dry out on exposure to air. Thus, a critical innovation in animal evolution was the development of a shelled egg that sheltered the embryo from desiccating conditions on land.

Ancestral chordate

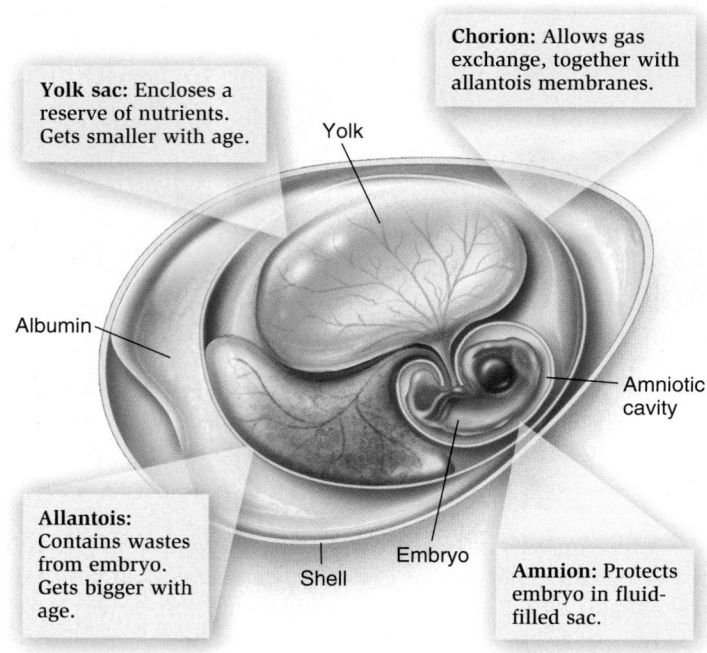

Yolk sac: Encloses a reserve of nutrients. Gets smaller with age.

Chorion: Allows gas exchange, together with allantois membranes.

Yolk

Albumin

Amniotic cavity

Allantois: Contains wastes from embryo. Gets bigger with age.

Shell

Embryo

Amnion: Protects embryo in fluid-filled sac.

Figure 34.16   The amniotic egg.

A shelled egg containing fluids was like a personal enclosed pond for each developing individual. Such an egg evolved in the common ancestor of turtles, lizards, snakes, crocodiles, birds, and mammals—a group of tetrapods collectively known as the **amniotes**. The amniotic egg permitted animals to lay their eggs in a dry place so that reproduction was no longer tied to water. It was truly a critical innovation. The amniotic egg untethered animals from water in a similar way as the development of seeds liberated plants from water (see Chapter 31).

In time, the amniotes came to dominate the Earth. Mammals are considered amniotes, too, because even though most of them do not lay eggs, they retain other features of amniotic reproduction. In this section, we begin by discussing in detail the morphology of the amniotic egg and other adaptations that permitted species to become fully terrestrial. We then discuss the biology of the reptiles, the first group of vertebrates to fully exploit land.

The Amniotic Egg and Other Innovations Permitted Life on Land

The **amniotic egg (Figure 34.16)** contains the developing embryo and the four separate extraembryonic membranes that it produces:

1. The innermost membrane is the **amnion**, which protects the developing embryo in a fluid-filled sac called the amniotic cavity.
2. The **yolk sac** encloses a stockpile of nutrients, in the form of yolk, for the developing embryo.

3. The **allantois** functions as a disposal sac for metabolic wastes.
4. The **chorion**, along with the allantois, provides gas exchange between the embryo and the surrounding air.

Surrounding the chorion is the albumin, or egg white, which also stores nutrients. The **shell** provides a tough, protective covering that is impermeable to water and prevents the embryo from drying out. However, the shell remains permeable to oxygen and carbon dioxide, so the embryo can breathe. In birds, this shell is hard and calcareous, while in reptiles and primitive mammals like the platypus and echidna, it is soft and leathery. Most mammals, however, do not have shelled eggs. Instead, the embryos embed into the wall of the uterus and receive their nutrients directly from the mother.

Along with the amniotic egg, other critical innovations that enabled the conquest of land included:

- **Desiccation-resistant skin.** While the skin of amphibians is moist and aids in respiration, the skin of amniotes is thicker and watertight and contains keratin, a tough protein. This requires that gas exchange take place through the lungs.
- **Thoracic breathing.** Amphibians breathe by contracting the mouth to force air into the lungs. In contrast, amniotes use **thoracic breathing**, in which coordinated contractions of muscles expand the rib cage, creating a negative pressure to suck air in and then forcing it out later. This results in a greater volume of air being displaced with each breath.
- **Water-conserving kidneys.** The ability to concentrate wastes prior to elimination and thus conserve water is an important role of the amniote kidneys.

- **Internal fertilization**. Because sperm cannot penetrate a shelled egg, fertilization occurs internally, within the female's body before the shell is secreted. In this process, the male of the species often uses a copulatory organ (penis) to transfer sperm into the female reproductive tract. However, birds transfer sperm from cloaca to cloaca.

Classification of the Reptiles Is Currently Under Revision

Early amniote ancestors gave rise to all modern amniotes we know today, from lizards and snakes to birds and mammals. The traditional view of amniotes involved three living classes: the reptiles (turtles, lizards, snakes, and crocodilians), birds, and mammals. As we will see later in the chapter, modern systematists have argued that enough similarities exist between birds and the classic reptiles that birds should be considered part of the reptilian lineage. Other systematists believe that enough differences are found between the living reptilian taxa that each should be considered a distinct class: the Testudines (turtles), Lepidosauria (lizards and snakes), Crocodilia (crocodiles), and Aves (birds), as illustrated in Figure 34.1. This is the classification scheme that we will follow in this chapter. The fossil record includes other reptilian classes, all of which are extinct, including two classes of dinosaurs (ornithischian and saurischian dinosaurs), flying reptiles (pterosaurs), and two classes of ancient aquatic reptiles (icthyosaurs and plesiosaurs).

Class Testudines: The Turtles Turtles is an umbrella term for terrestrial species, also called tortoises, and aquatic species, also known as terrapins. The turtle lineage is ancient and has remained virtually unchanged for 200 million years. The major distinguishing characteristic of the turtle is a hard protective shell into which the animal can withdraw its head and limbs. In most species, the vertebrae and ribs are fused to this shell. All turtles lack teeth but have sharp beaks for biting.

The majority of turtles are aquatic and have webbed feet. The forelimbs of marine species have evolved to become large flippers. All turtles, even the aquatic species, lay their eggs on land, usually in soft sand. The gender of hatchlings is temperature dependent, with high temperatures producing more females. Marine species often make long migrations to sandy beaches to lay their eggs (**Figure 34.17a**). Most land tortoises are quite slow movers, possibly due to a low metabolic rate and a heavy shell. On the other hand, they are very long-lived species, often surviving for 120 years or more. Many turtle species are in danger of extinction, due to egg hunting, destruction of habitat and nesting sites, and death from entanglement in fishing nets.

Class Lepidosauria: Lizards and Snakes The class Lepidosauria is the largest class within the traditional reptiles, with about 3,000 species of snakes (order Serpentes) and 4,800 species of lizards (order Sauria). Many species have an elongated body form. One of the defining characteristics of the orders is a **kinetic skull**, in which the joints between various parts of the skull are extremely mobile. The lower jaw does not join directly to the skull but rather is connected by a multijointed hinge, and the upper jaw is hinged and moveable from the rest of the head. This allows the jaws to open relatively wider than other vertebrate jaws, with the result that lizards, and especially snakes, can swallow large prey (**Figure 34.18**). Nearly all species are carnivores.

A main difference between lizards and snakes is that lizards generally have limbs, whereas snakes do not. Also, snakes may be venomous, whereas lizards are generally not. However, there are exceptions to these general rules. Many legless lizard species exist (see **Figure 34.17b**), and two lizards are poisonous: the Gila monster (*Heloderma suspectum*) of the U.S. Southwest (see **Figure 34.17c**) and the Mexican beaded lizard (*Heloderma horridum*). A more reliable distinguishing characteristic is that lizards have moveable eyelids and external ears (at least ear canals), whereas snakes do not.

(a)

(b)

(c)

Figure 34.17 **A variety of reptiles.** (a) A green turtle (*Chelonia mydas*) laying eggs in the sand in Malaysia. (b) The Florida worm lizard (*Rhineura floridana*) is an amphisbaenian, a type of legless soil-burrowing lizard. (c) The Gila monster (*Heloderma suspectum*), one of only two poisonous lizards, is an inhabitant of the desert Southwest of the U.S. and of Mexico.

Class Crocodilia: The Crocodiles and Alligators The Crocodilia is a small class of large aquatic animals that have remained essentially unchanged for nearly 200 million years (**Figure 34.19**). Indeed, these animals existed at the same time as the dinosaurs. Most of the 23 recognized species live in tropical or subtropical regions. There are only two extant species of alligators: one living in the southeastern U.S. and one found in China.

While the class is small, it is evolutionarily very important. Crocodiles have a four-chambered heart, a feature they share with birds and mammals (look ahead to Figure 47.4c). In this regard, crocodiles are more closely related to birds than any other living reptile class. Their teeth are set in sockets, a feature typical of the dinosaurs and the earliest birds. Similarly, crocodiles care for their young, another trait they have in common with birds. These and other features suggest that crocodiles and birds are more closely related than crocodiles and lizards. As with turtles, nest temperature influences the sex ratio of offspring. With crocodiles, however, when temperatures are warm, more males are produced. Biologists are concerned that a scenario of global warming may reduce the number of breeding females, leading to extinction of some species.

The Dinosaurs: Classes Ornithischia and Saurischia In 1841, the English paleontologist Richard Owen coined the term **dinosaur**, meaning terrible lizard, to describe some of the wondrous fossil animals discovered in the 19th century. About 215 million years ago, dinosaurs were the dominant tetrapods on Earth and remained so for 150 million years, far longer than any other vertebrate. The two main classes were the ornithischian or bird-hipped dinosaurs, which were herbivores such as *Stegosaurus*, and the saurischian or lizard-hipped dinosaurs, which were fast, bipedal carnivores such as *Tyrannosaurus* (**Figure 34.20**). In contrast to the limbs of lizards, amphibians, and crocodiles, which splayed out to the side, the legs of dinosaurs were positioned directly under the body, like pillars, a position that could help support their heavy body. Because less energy was devoted in lifting the body from the ground, some dinosaurs are believed to have been fast runners. Members of different but closely related classes—the pterosaurs (the first vertebrates to fly), and plesiosaurs and ichthyosaurs (marine reptiles)—were also common at this time.

Figure 34.18 **The kinetic skull.** In snakes and lizards, the top of the jaw, as well as the bottom of the jaw, is hinged on the skull, thus permitting large prey to be swallowed. This Halloween snake (*Pliocercus euryzonus*) is swallowing a Costa Rican rain frog.

(a)

(b)

Figure 34.19 **Crocodilians.** The Crocodilia is an ancient class that has existed unchanged for millions of years. **(a)** Alligators, such as this American alligator (*Alligator mississippiensis*), have a broad snout, and the lower jaw teeth close on the inside of the upper jaw (and thus are almost completely hidden when the mouth is closed). **(b)** Crocodiles, including this American crocodile (*Crocodylus acutus*), have a longer, thinner snout, and the lower jaw teeth close on the outside of the upper jaw (and thus are visible when the mouth is closed).

Biological inquiry: In what ways are crocodilians similar to birds?

(a) Ornithischian (*Stegosaurus*)   **(b) Saurischian (*Tyrannosaurus*)**

Figure 34.20 Dominant terrestrial dinosaur classes.
(a) Ornithischians included *Stegosaurus*, and (b) saurischians included bipedal species such as *Tyrannosaurus*.

Dinosaurs were the biggest animals ever to walk on the planet, with some animals weighing up to 50 tonnes (metric tons) or over 100,000 pounds. The variety of the thousands of dinosaur species found in fossil form around the world is staggering. However, perhaps not surprisingly for such long-extinct species, the details of their lives are still being hotly debated. For example, an issue still unresolved is whether some dinosaur species were **endothermic**, that is, capable of generating and retaining body heat through their metabolism, just as birds and mammals are. Another issue is whether dinosaurs exhibited parental care of their young.

All dinosaurs, and many other animals, went extinct quite abruptly at the end of the Cretaceous period, about 65 million years ago. Although widely attributed to climatic change brought about by the impact of an asteroid, there is ongoing debate over the cause of this mass extinction. It is not known why all the dinosaurs died out, while many other animals, including small mammals, survived.

34.6 Birds

The defining characteristics of birds (class Aves, plural of the Latin *avis*, bird) are that they have feathers and nearly all species can fly. As we will see, the ability to fly has shaped nearly every feature of the bird body. The other vertebrates that have evolved the ability to fly, the bats and the now-extinct pterosaurs, used skin stretched tight over elongated limbs to fly. Such a surface can be irreparably damaged, though some holes may heal remarkably quickly. In contrast, birds utilize feathers, epidermal outgrowths that can be replaced if damaged. Recent studies in cell signaling show that the developmental changes necessary for animals to grow feathers instead of scales may not be that complex at the molecular level.

In this section, we will begin by discussing the likely evolution of birds from dinosaur ancestors, outline the key characteristics of birds, and provide a brief overview of the various bird orders.

GENOMES & PROTEOMES

The Differentiation of Scales, Feathers, and Fur May Be Caused by Simple Changes in Developmental Pathways

The skin of reptiles is protected by tough scales containing keratin, a tough waterproof protein. In birds, the skin is covered in feathers, and in mammals, it is covered by hair or fur. These epithelial appendages look very different but share similarities in initial development. All originate as **epithelial placodes**, regions of slightly thickened epithelial cells. Their spatial distribution on the body is determined by many developmental signaling pathways, including those controlled by proteins designated Wnt, Notch, and BMP that regulate the properties of the epithelial cells (or epidermis) and the dermis. The timing and level of expression of these regulatory molecules determine the actual appendage that develops in the dermal layer. Wnt and Notch promote cell proliferation, while BMP programs cell death. Scales are formed as an outgrowth of the skin and are turned on by low levels of Wnt and Notch and high levels of BMP, while feathers develop with inverse levels of such regulatory molecules (**Figure 34.21**).

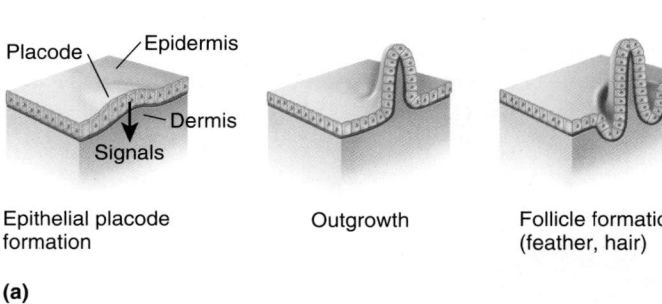

(a)

Epithelial placode formation   Outgrowth   Follicle formation (feather, hair)

Morphology	Scale	Feather
Wnt levels (cell proliferation)	Low	High
Notch levels (cell proliferation)	Low	High
BMP levels (cell death)	High	Low

(b)

Figure 34.21 Development of epithelial appendages.
(a) Epithelial appendages begin as an outgrowth of the epidermis and dermis. In the development of feathers and hair, this outgrowth gives way to the formation of a follicle. (b) The difference between the identities of different epithelial appendages, such as scales and feathers, are caused, in part, by the timing and level of early signaling events.

Different regulatory genes acting in later stages of development can alter the properties of the scales, feathers, or hair. For example, the mouse *HoxC-13* gene is necessary for normal development of mouse hair. Mice with a mutation in this gene have fragile, brittle hair that breaks easily. Other features of the body that are hardened by keratin, the structural component of hair, may also be affected, such as nails and even filiform papillae, which are small projections on the surface of the tongue. The seemingly radical differences in body coverings of different vertebrate classes are actually caused by relatively simple changes in the levels of regulatory molecules.

Modern Birds Probably Evolved from Small, Feather-Covered Dinosaurs

To trace the evolution of birds, we must look for the earliest type of animals that had feathers. One of the first known fossils exhibiting the faint impression of feathers was *Archaeopteryx lithographica* (meaning "ancient wings" and "stone picture"), found in a limestone quarry in Germany in 1861. The fossil was dated at 150 million years old. Except for the presence of feathers, *Archaeopteryx* appears to have had features similar to those of dinosaurs (**Figure 34.22a**). First, the fossil had an impression of a long tail with many vertebrae, a dinosaur feature. Some modern birds have long tails, but they are made of feathers, with the actual tailbone being much reduced. Second, the wings had claws

halfway down the leading edge, another dinosaur-like character. Among modern birds, only the hoatzin, a South American swamp-inhabiting bird, has claws on its wings, which enable the chicks to climb back into the nest if they fall out. A third dinosaur-like feature is *Archaeopteryx*'s toothed beak. Fourth, the fossils show that *Archaeopteryx* lacked an enlarged breastbone, a feature that modern birds possess to anchor their large flight muscles, so it likely could not fly.

Similarities between the structure of the skull, feet, and hind leg bones have led scientists to conclude that *Archaeopteryx* is closely related to **theropods**, a group of bipedal saurischian dinosaurs. The wings and feathers of *Archaeopteryx* may have enabled it to glide from tree to tree, to help it keep warm, or to fold over its head when hunting, in much the same way as some herons fold their wings over their heads when they are fishing. Later, the wings and feathers may have taken on functions of flight. We often find that evolution proceeds when previously evolved structures are co-opted into different uses, a diversity principle known as descent with modification.

In 1997, paleontologists unearthed fossils of about the same age of *Archaeopteryx* in China that similarly suggest a close kinship between dinosaurs and modern birds. *Caudipteryx zoui* was a dinosaur-like animal with feathers on its wings and tail and a toothed beak (**Figure 34.22b**). *Confuciusornis sanctus* was a small, flightless but completely feathered dinosaur lacking the long bony tail and toothed jaw found in other theropod dinosaurs. Its large tail feathers may have functioned in courtship displays (**Figure 34.22c**).

Avian features:
1. Feathered wing and tail

Reptilian features:
1. Long bony tail
2. Claws on wing
3. Toothed beak

(a) *Archaeopteryx lithographica*

Avian features:
1. Feathered wing and tail

Reptilian features:
1. Short bony tail
2. Claws on wing
3. Toothed beak

(b) *Caudipteryx zoui*

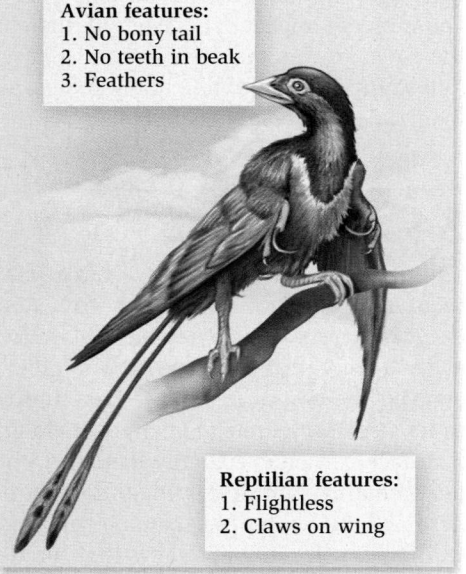

Avian features:
1. No bony tail
2. No teeth in beak
3. Feathers

Reptilian features:
1. Flightless
2. Claws on wing

(c) *Confuciusornis sanctus*

Figure 34.22   **Missing links between dinosaurs and birds.**   (a) *Archaeopteryx lithographica* was a Jurassic animal with dinosaur-like features as well as wings and feathers. (b) *Caudipteryx zoui* was a dinosaur with feathers on its tail and wings. (c) *Confuciusornis sanctus* was a birdlike animal with a horny bill.

These three species—*Archaeopteryx*, *Caudipteryx*, and *Confuciusornis*—help trace a lineage from dinosaurs to birds. By the early Cretaceous period, and only a relatively short period after *Archaeopteryx* evolved, the fossil record shows the existence of a huge array of bird types resembling modern species. These were to share the skies with pterosaurs for 70 million years, before eventually having the airways to themselves.

Birds Display Four Key Characteristics

Modern birds possess many characteristics that reveal their reptilian ancestry. For example, they have scales on their feet and legs, and they lay shelled eggs. In addition, however, among living animals they have four features unique to birds, all of which are associated with flight.

1. **Feathers.** Feathers are modified scales that keep birds warm, enable flight, and are supported on a modified forelimb (**Figure 34.23a**). Soft, downy feathers maintain heat, while stiffer contour feathers give the wing the airfoil shape it needs to generate lift. Each contour feather develops from a follicle, a tiny pit in the skin. If a feather is lost, a new one can be regrown. The contour feathers consist of many paired barbs, each of which supports barbules that hook together with barbules from neighboring barbs to give the feather its shape (**Figure 34.23b**).

2. **Lightweight skeleton.** Most bird bones are thin and hollow and are crisscrossed internally by tiny pieces of bone to give them a honeycomb structure (**Figure 34.23c**). The keeled breastbone, or **sternum**, provides an anchor on which a bird's powerful flight muscles attach. These muscles may contribute up to 30% of the bird's body weight. The bird skull lacks teeth, but the keratin of bird beaks is tough and malleable, and it takes very different shapes in different species (**Figure 34.24**).

3. **Air sacs.** Flight requires a great deal of energy generated from an active metabolism that requires abundant oxygen. Birds have nine air sacs, large, hollow sacs that may extend into the bones (look ahead to Figure 48.13). On inhalation, air passes into the posterior air sacs. On exhalation, half of this air is forced back out through the lungs, where oxygen is extracted, and then goes into the anterior air sacs. On the next inhalation, air is sucked from the posterior air sacs into the anterior air sacs via the lungs. Air is therefore being constantly moved across the lungs during inhalation and exhalation. While making bird breathing very efficient, this process also makes birds especially susceptible to airborne toxins (hence, the utility of the canary in the coal mine; the bird's death signaled the presence of harmful carbon dioxide or methane gas that was otherwise unnoticed by miners).

4. **Reduction of organs.** To decrease the total mass the bird must carry, some organs are reduced in size or are lacking altogether. For example, birds have only one ovary and can carry relatively few eggs. As a result, they lay fewer eggs than most other reptile species. In fact, the gonads of both males and females are reduced, except during the breeding season, when they increase in size. Most birds also lack a urinary bladder.

(a) Wing

(b) Feather structure

(c) Pelican bone (*Pelicanus occidentalis*)

Figure 34.23 **Features of the bird wing and feather.** (a) The wing is supported by an elongated and modified forelimb with extended fingers. (b) Each feather has a hollow shaft that supports many barbs, which in turn support barbules that interlock with hooks to give the feather its form. (c) The bones of a pelican (*Pelicanus occidentalis*) are hollow but crisscrossed with a honeycomb structure that provides added strength.

Biological inquiry: What adaptations in birds help reduce their body weight to enable flight?

Birds also have other distinct features, though mammals also possess some of these. For example, birds have a warm body temperature, which ensures rapid metabolism and the quick production of ATP that these active organisms need to fuel flight and other activities. In fact, birds' body temperatures are generally 40–42°C, considerably warmer than the human body's average of 37°C. Birds have a double circulation and a four-chambered heart to ensure rapid blood circulation. Rapid flight also requires acute vision, and bird vision is among the best in the vertebrate world. Generally, plant material is not energy-rich enough to supply the dietary needs of birds, so most birds are carnivores, eating insects and other invertebrates, though some birds like parrots specialize on more nutrient-rich fruits and seeds. Bird eggs also have to be kept warm for successful development, which entails brooding by an adult bird. Often, the males and females take turns brooding so that one parent can feed and maintain its strength. Picking successful partners is therefore an important task, and birds often engage in complex courtship rituals.

Birds Have Many Orders, All with the Same Body Plan

Birds are the most species-rich class of terrestrial vertebrates, with 28 orders, 166 families, and about 9,600 species (**Table 34.2**). Despite this diversity, birds lack the variety of body shapes that exist in the other endothermic class of vertebrates, the mammals, where some species swim, others fly, others walk on four legs, and yet others walk only on two legs. Most birds fly and thus most have the same general body shape. The biggest departures from this body shape are the flightless birds, including the kiwis, ostriches, cassowaries, emus, and rheas. These birds have smaller wing bones and the keel on the breastbone is greatly reduced or absent. Penguins are also flightless birds whose upper limbs are modified as flippers for swimming.

Figure 34.24   A variety of bird beaks.   Birds have evolved a variety of beak shapes for different types of food gathering. **(a)** Hyacinthe macaw (*Anodorhynchus hyacinthinus*)—cracking. **(b)** White pelican (*Pelecanus onocrotalus*)—scooping. **(c)** Verreaux's eagle (*Aquila verreauxii*)—tearing. **(d)** American avocet (*Recurvirostra americana*)—probing. **(e)** Lucifer hummingbird (*Calothorax lucifer*)—nectar feeding. **(f)** Roseate spoonbill (*Ajaia ajaia*)—sieving.

34.7 Mammals

Mammals evolved from amniote ancestors earlier than birds. About 225 million years ago, the first mammals appeared in the mid-Triassic period. They evolved from small mammal-like lizards that went extinct about 170 million years ago. Mammals survived, although most are believed to have been small, insect-eating species that lived in the shadows of dinosaurs. However, in January 2005, two fossils of a 130-million-year-old mammalian species called *Repenomamus* were discovered that challenge the notion of mammals as small insect eaters. One fossil was of an animal estimated to weigh about 13 kg (30 lb), while the other had the remains of a baby dinosaur in its stomach.

After extinction of the dinosaurs in the Cretaceous period, some 65 million years ago, mammals flourished. Today, biologists have identified about 5,500 species of mammals with a diverse array of lifestyles, from fishlike dolphins to birdlike bats, and from small insectivores such as shrews to large herbivores such as giraffes and elephants. The range of sizes and body forms of mammals is unmatched by any other group, and mammals are prime illustrations of the concept that organismal diversity is related to environmental diversity. In this section, we will outline the features that distinguish mammals from other taxa. We will also examine the diversity of mammals that exists on Earth and will end by turning our attention to primates and in particular to the evolution of modern humans.

Table 34.2	The Main Orders of Birds, in Order of Species Richness	
Order	**Examples (approx. # of species)**	**Main characteristics**
Passeriformes	Robins, starlings, sparrows, warblers (5,300)	Perching birds with perching feet; songbirds
Apodiformes	Hummingbirds, swifts (430)	Fast fliers with rapidly beating wings; small bodies
Piciformes	Woodpeckers, toucans (380)	Large with specialized bills; two toes pointing forward and two backward
Psittaciformes	Parrots, cockatoos (340)	Large, powerful beaks
Chadradriiformes	Seagulls, wading birds (330)	Shorebirds
Columbiformes	Doves, pigeons (300)	Round bodies; short legs
Falconiformes	Eagles, hawks, kestrels, vultures (290)	Diurnal carnivores; birds of prey; powerful talons, and strong beaks
Galliformes	Chickens, pheasants, quail (270)	Often large birds; weak flyers; ground nesters
Coraciiformes	Hornbills, kingfishers (200)	Large bills; cavity nesters
Anseriformes	Ducks, swans, geese (150)	Able to swim; webbed feet; broad bills
Strigiformes	Owls (150)	Nocturnal carnivores; powerful talons, and bills
Pelecaniformes	Pelicans, frigate birds, cormorants (55)	Large, colonial fish eaters; often tropical
Sphenisciformes	Penguins (18)	Flightless, wings modified into flippers for swimming; marine; Southern Hemisphere
Casuariformes	Cassowaries, emus (3)	Large, flightless; Australia and New Guinea
Struthioniformes	Ostrich (1)	Large, flightless; only two toes; Africa only

Mammals Have Mammary Glands, Hair, Specialized Teeth, and an Enlarged Skull

The distinguishing characteristics of mammals are the possession of mammary glands, hair, specialized teeth, and an enlarged skull.

- **Mammary glands.** Mammals, or the class Mammalia (from the Latin, *mamma*, breast), are named after the female's distinctive **mammary glands**, which secrete milk.

Milk is a fluid rich in fat, sugar, protein, and vital minerals, especially calcium. Newborn mammals suckle this fluid, which helps promote rapid growth.

- **Hair.** All mammals have hair, although some have more than others. Whales retain only a few hairs on their snout, and humans are relatively hairless. In some animals, the hair is dense and referred to as fur. In some aquatic species such as beavers, the fur is so dense it cannot be thoroughly wetted, so the hair underneath remains dry.

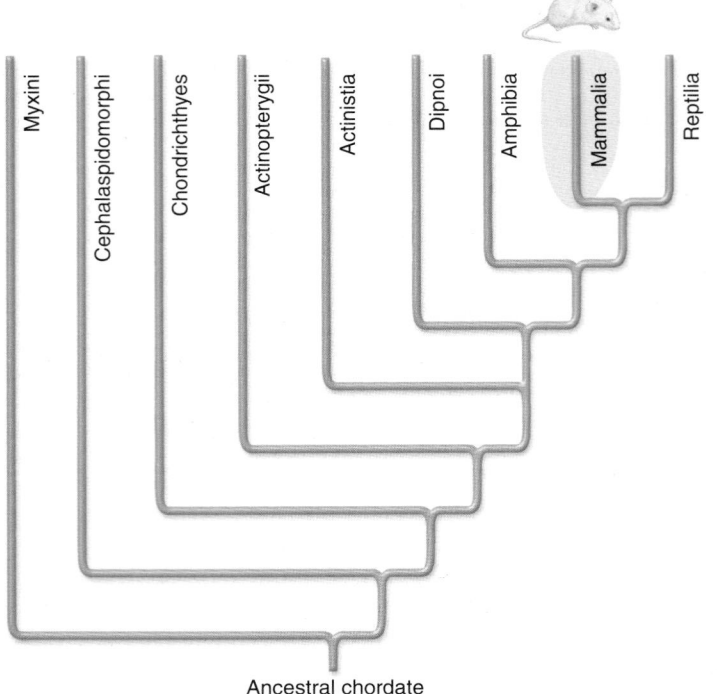

Ancestral chordate

Mammals are endothermic, and their fur is an efficient insulator. Hair can also take on functions other than insulation. Many animals, including cats, dogs, walruses, and whales, have sensory hairs called vibrissae (**Figure 34.25a**). Hair can be of many colors, to allow the mammals to blend into their background (**Figure 34.25b**). In some cases, as in porcupines and hedgehogs, the hairs become long, stiffened, and sharp (quills) and serve as a defense mechanism (**Figure 34.25c**).

- **Specialized teeth.** Mammals are the only vertebrates with teeth that are adapted for different types of diets (**Figure 34.26**). While teeth are generally present in all species, different teeth are larger, smaller, lost, or reduced, depending on diet. Of particular importance to carnivores are the piercing canine teeth, while herbivorous species like antelopes depend on their chisel-like incisors to snip off vegetation and on their many molars to grind plant material. Only mammals chew their food in this fashion. Rodent incisors grow continuously throughout life, and species such as beavers wear them down by gnawing tough plant material such as wood.

- **Enlarged skull.** The mammalian skull differs from other amniote skulls in several ways. First, mammals have a single lower jawbone—the dentary—unlike reptiles, whose lower jaw is composed of multiple bones. Second, mammals have three bones in the inner ear, as opposed to reptiles, which have one bone in the inner ear. Third, mammals have pinnae, or external ears. Fourth, the brain is enlarged and is contained within a relatively large skull.

In addition to those uniquely mammalian characteristics, some but not all mammals possess these additional features:

- **The ability to digest plants.** Apart from tortoises and marine iguanas, certain species of mammals are the only large vertebrates that can exist on a steady diet of grasses or tree leaves; indeed, most large mammals are herbivores. Though mammals cannot digest cellulose, the principal constituent of the cell wall of many plants, some species have a large four-chambered stomach containing cellulose-digesting bacteria. These bacteria can break down the cellulose and make the plant cell contents available to the animal.

- **Horns and antlers.** Mammals are the only living class of vertebrates to possess horns or antlers. Many mammals, especially antelopes, cattle, and sheep, have horns, typically consisting of a bony core that is a permanent outgrowth of the skull surrounded by a hairlike keratin sheath (**Figure 34.27a**). Rhinoceros horns are outgrowths of the epidermis consisting of very tightly matted hair (**Figure 34.27b**).

Figure 34.25 **Mammalian hair.** **(a)** The sensory hairs (vibrissae) of the walrus (*Odobenus rosmarus*). **(b)** The camouflaged coats of Burchell's zebra (*Equus burchelli*). **(c)** The defensive quills of the crested porcupine (*Hystrix africaeaustralis*).

(a)

(b)

(c)

In contrast, deer antlers are made entirely out of bone (**Figure 34.27c**). Deer grow a new set of antlers each year and shed them after the mating season. Hooves and claws are also made of keratin and protect an animal's toes from the impact of its feet striking the ground and aid in prey capture, respectively.

Mammals Are the Most Diverse Group of Vertebrates Living on Earth

Modern mammals are incredibly diverse (**Table 34.3**). They vary in size from tiny insect-eating bats, weighing in at only 2 g, to leviathans such as the blue whale, the largest animal ever known, which tips the scales at 100 tonnes (over 200,000 lbs.). The 26 different mammalian orders are divided into three dis-

tinct subclasses. The subclass Prototheria contains only the order Monotremata, or **monotremes**, which are found in Australia and New Guinea. There are only three species: the duck-billed platypus (**Figure 34.28a**) and two species of echidna, a spiny animal resembling a hedgehog. Monotremes are considered primitive mammals because they lay eggs rather than bear live young, lack a placenta, and have mammary glands with poorly developed nipples. The mothers incubate the eggs, and upon hatching, the young simply lap up the milk as it oozes onto the fur.

The subclass Metatheria, or the **marsupials**, is a group of seven orders, with about 280 species (**Figure 34.28b**). Once widespread, members of this order are now largely confined to Australia, although some marsupials exist in South America and one species—the opossum—is found in North America.

(a)   (b)   (c)

(d)   (e)

Figure 34.26 Mammalian teeth. Mammals have different types of teeth, according to their diet. **(a)** The wolf has long canine teeth for biting its prey. **(b)** The deer has a long row of front molars for grinding plant material. **(c)** The beaver, a rodent, has long, continually growing incisors. **(d)** The elephant's incisors are extremely modified into tusks. **(e)** Dolphins and other fish or plankton feeders have numerous small teeth for prey capture.

(a)   (b)   (c)

Figure 34.27 Horns and antlers. Mammals have a variety of outgrowths that are used for defense or by males as weapons in contests over females. **(a)** The horns of this male kudu (*Tragelaphus strepsiceros*) are bony outgrowths of the skull covered in a keratin sheath. **(b)** The horns of the black rhinoceros (*Diceros bicornis*) are outgrowths of the epidermis and made of tightly matted hair. **(c)** The antlers of the caribou (*Rangifer tarandus*) are grown and shed each year.

Table 34.3 — The Main Orders of Mammals, in Order of Species Richness

Order		Examples (approx. # of species)	Main characteristics
Rodentia		Mice, rats, squirrels, beaver, porcupine (2,277)	Plant eating, gnawing habit, with two pairs of continually growing incisor teeth
Chiroptera		Bats (1,116)	Insect or fruit eating; small; have ability to fly; navigate by sonar; nocturnal
Soricomorpha		Shrews, moles (428)	Insect eaters; primitive placental mammals
Primates		Monkeys, apes, humans (404)	Opposable thumb; binocular vision; large brains
Carnivora		Cats, dogs, weasels, bears, seals, sea lions (286)	Flesh-eating mammals; canine teeth
Artiodactyla		Deer, antelopes, cattle, sheep, goats, camels, pigs (240)	Herbivorous hoofed mammals, usually with two toes, hippopotamus and others with four toes; many with horns or antlers
Diprotodontia		Kangaroos, koalas, opossums, wombats (143)	Pouched mammals mainly found in Australia
Lagomorpha		Rabbits, hares (92)	Powerful hind legs; rodent-like teeth
Cetacea		Whales, dolphins (84)	Marine fish or plankton feeders; front limbs modified into flippers; no hind limbs; little hair except on snout
Erinaceomorpha		Hedgehogs, moonrats (24)	Insect eaters; nocturnal; hedgehogs with stiff spines
Perissodactyla		Horses, zebras, tapirs, rhinoceroses (18)	Hoofed herbivorous mammals with odd number of toes, one (horses) or three (rhinoceroses)
Monotremata		Duck-billed platypus, echidna (3)	Egg-laying mammals only found in Australia and New Guinea
Proboscidea		Elephants (3)	Long trunk; large, upper incisors modified as tusks

Unlike other mammals, marsupials are extremely small when they are born (often only 1 or 2 cm) and make their way to a ventral pouch called a marsupium for further development. Initially, they are too feeble to suck on the mother's nipple so she simply oozes milk for them to lick up. However, as they grow, marsupial young eventually attach to the nipple, often for a long time.

All the other mammalian orders are members of the subclass Eutheria and considered **eutherians** or placental mammals (**Figure 34.28c**). Even though marsupials have a placenta, it is not as long-lived or complex as that of eutherians. In eutherians, fertilization is internal and reproduction is viviparous, as in marsupials, but the developmental period, or gestation, of the young is prolonged, and the mother's placenta provides the fetus with nourishment during development.

Of all the world's animals, the diversity of vertebrates, and especially mammals, is the most threatened by human activities. Many species are hunted for food. Others, such as wild cats and whales, are hunted for their products (fur and oil), and still others, like the oryx, have simply been shot for sport.

Primates Are Mammals with Opposable Thumbs and a Large Brain

Of the mammalian orders, the primates, and specifically humans, have had the greatest impact on the world. Primates are primarily tree-dwelling species that are believed to have evolved from a group of small, arboreal insect-eating mammals about 85 million years ago, before dinosaurs went extinct. Primates have

(a)

(c)

(b)

Figure 34.28 **Diversity among mammals.**
(a) Prototherians, like the duck-billed platypus (*Ornithorhynchus anatinus*) are considered primitive because they lay eggs and possess mammary glands with poorly developed nipples. **(b)** Metatherians, or marsupials, such as this rock wallaby (*Petrogale assimilis*) feed and carry their developing young, or "joeys," in a ventral pouch. **(c)** Gestation lasts longer in eutherians, and their young are more developed at birth, as illustrated by this young orangutan (*Pongo pygmaeus*).

several defining characteristics, mostly relating to their tree-dwelling nature:

- **Grasping hands.** All primates have grasping hands, a characteristic that enables them to hold onto branches (see Figure 34.30a,b). Most primate species also possess an **opposable thumb**, a thumb that can be placed opposite the fingers of the same hand, which gives them a precision grip and enables the manipulation of small objects. All primates, except humans, also have an opposable big toe.

- **Large brain.** Acute vision and other senses enhancing the ability to move quickly through the trees require the efficient processing of large amounts of information. As a result, the primate brain is large and well developed. In turn, this has facilitated complex social behaviors.

- **At least some digits with flat nails instead of claws.** This feature is believed to aid in the manipulation of objects.

- **Binocular vision.** Primates have forward-facing eyes that are positioned close together on a flattened face, though other mammals share this characteristic. Jumping from branch to branch requires accurate judgment of distances. This is facilitated by binocular vision, in which the field of vision for both eyes overlaps, producing a single image.

There are several ways of classifying primates, but many taxonomists divide them into two groups: the prosimians and the anthropoids (**Figure 34.29**). The **prosimians** contain the smaller species such as bush babies, lemurs, pottos, and tarsiers. These are generally nocturnal and smaller-brained primates (**Figure 34.30a**).

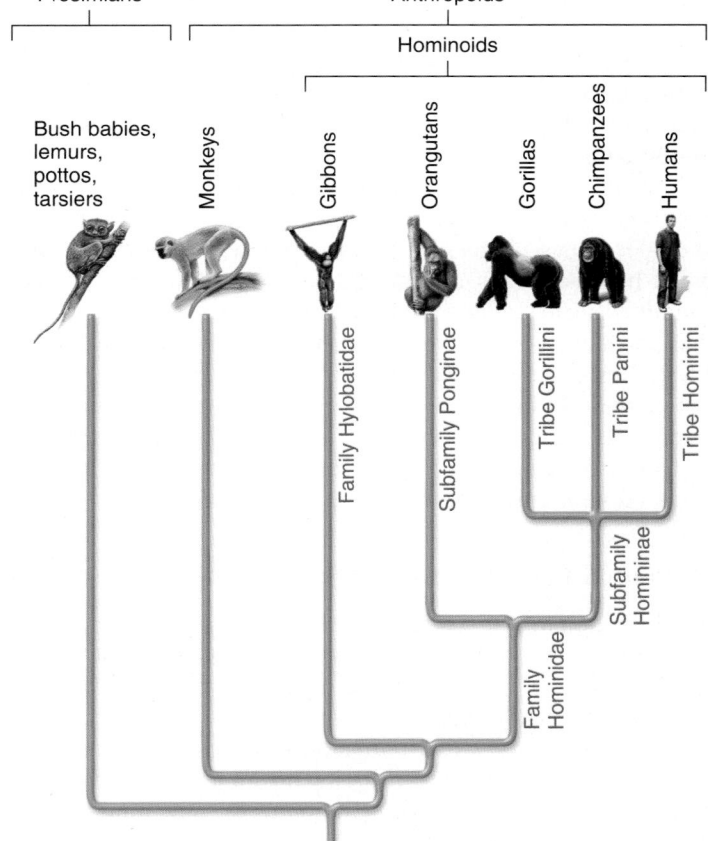

Figure 34.29 Evolutionary tree for the primates.

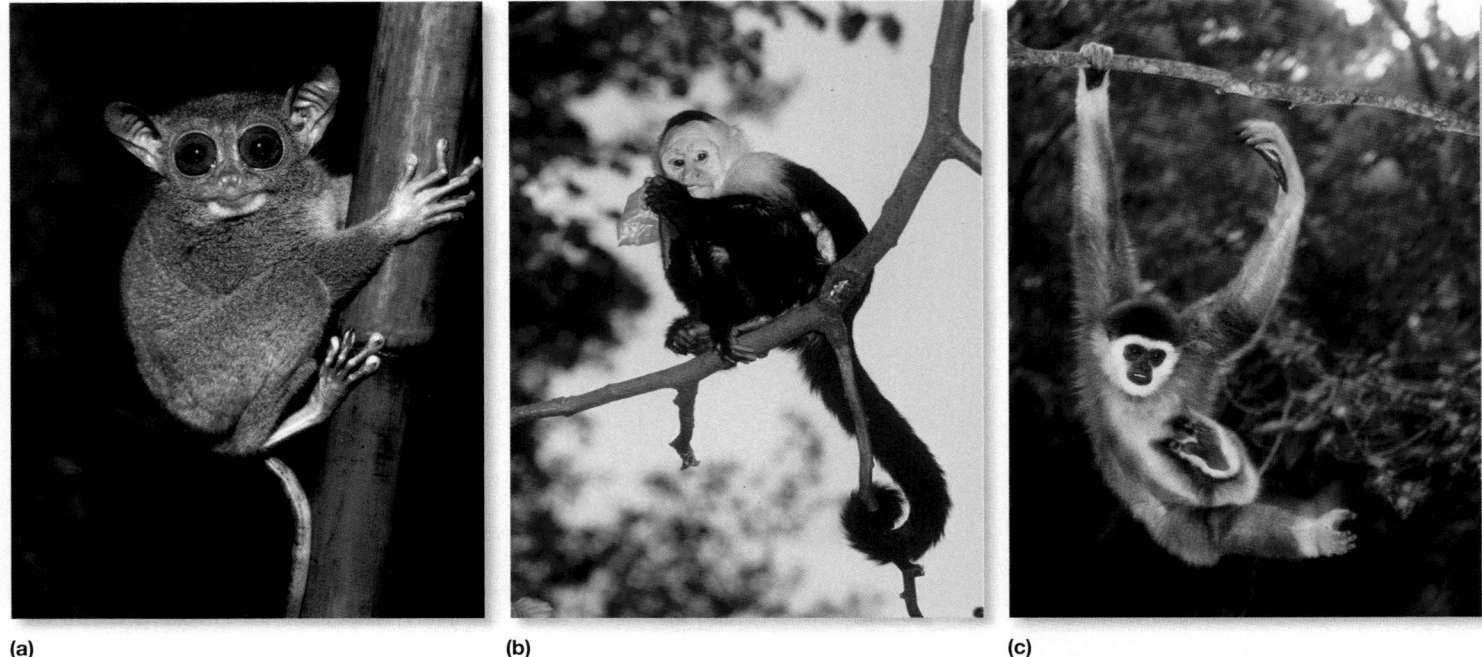

(a)   (b)   (c)

Figure 34.30   Primate classification.   Many authorities divide the primates into two groups: **(a)** the prosimians, smaller, nocturnal species such as this tarsier (*Tarsius syrichta*), and the anthropoids, larger diurnal species. **(b)** Anthropoids comprise the monkeys, such as this Capuchin monkey, *Cebus capucinus*, and **(c)** the hominoids, species such as this white-handed gibbon (*Hylobates lar*).

Biological inquiry: What are the defining features of primates?

The second group consists of the larger-brained and diurnal **anthropoids**: the monkeys (**Figure 34.30b**) and the **hominoids** (gibbons, orangutans, gorillas, chimpanzees, and humans) (**Figure 34.30c**).

What differentiates monkeys from hominoids? Most monkeys have tails, whereas hominoids do not. In addition, the shoulder bones in the two groups have a different structure; while hominoids can swing from branch to branch, monkeys cannot. Instead, monkeys run along the tops of branches and their movements are more like other four-footed mammals. The 13 species of hominoids are split into two groups: the lesser apes (family Hylobatidae), or the gibbons; and the greater apes (family Hominidae), or the orangutans, gorillas, chimpanzees, and humans. The lesser apes are strictly arboreal, while the greater apes often descend to the ground to feed.

Humans Evolved from Ancestral Primates

Although humans are closely related to chimpanzees and gorillas, they did not evolve directly from them. Rather, all hominoid species shared a common ancestor. Recent molecular studies show that gorillas, chimpanzees, and humans are closer to one another than gibbons and orangutans, so scientists have split the hominoids into groups, including the Ponginae (orangutans) and the Homininae (gorillas, chimpanzees, and humans and their ancestors). In turn, the Homininae are split into three tribes: the Gorillini (gorillas), the Panini (chimpanzees), and the Hominini (humans and their ancestors).

About 6 million years ago in Africa, a lineage that led to humans began to separate from other primate lineages. The evolution of humans should not be viewed as a neat, stepwise progression from one species to another. Rather, human evolution, like the evolution of most species, can be visualized more like a tree, with one or two **hominin** species (extinct and modern forms of humans) likely coexisting at the same point in time, with some eventually going extinct and some giving rise to other species (**Figure 34.31**).

One of the key characteristics differentiating hominins from other apes is that hominins walked on two feet, that is, they are **bipedal**. At about the time when hominins diverged from other ape lineages, the Earth's climate had cooled and the forests of Africa had given way to grassy savannas. Now that the hands were not engaged in climbing, bipedalism could have been selected for, as it is a more energetically economical form of walking. A bipedal method of locomotion and upright stance may also have been advantageous in allowing hominins to peer over the tall grass of the savanna.

In any event, bipedalism resulted in many anatomical changes in hominins. First, the opening of the skull where the spinal cord enters shifted forward, allowing the spine to be more directly underneath the head. Second, the hominin pelvis became broader to support the additional weight, and third, the lower limbs, used for walking, became relatively larger than those in other apes. These are also the types of anatomical changes paleontologists look for in the fossil record to help determine whether fossil remains are hominin. The earliest group

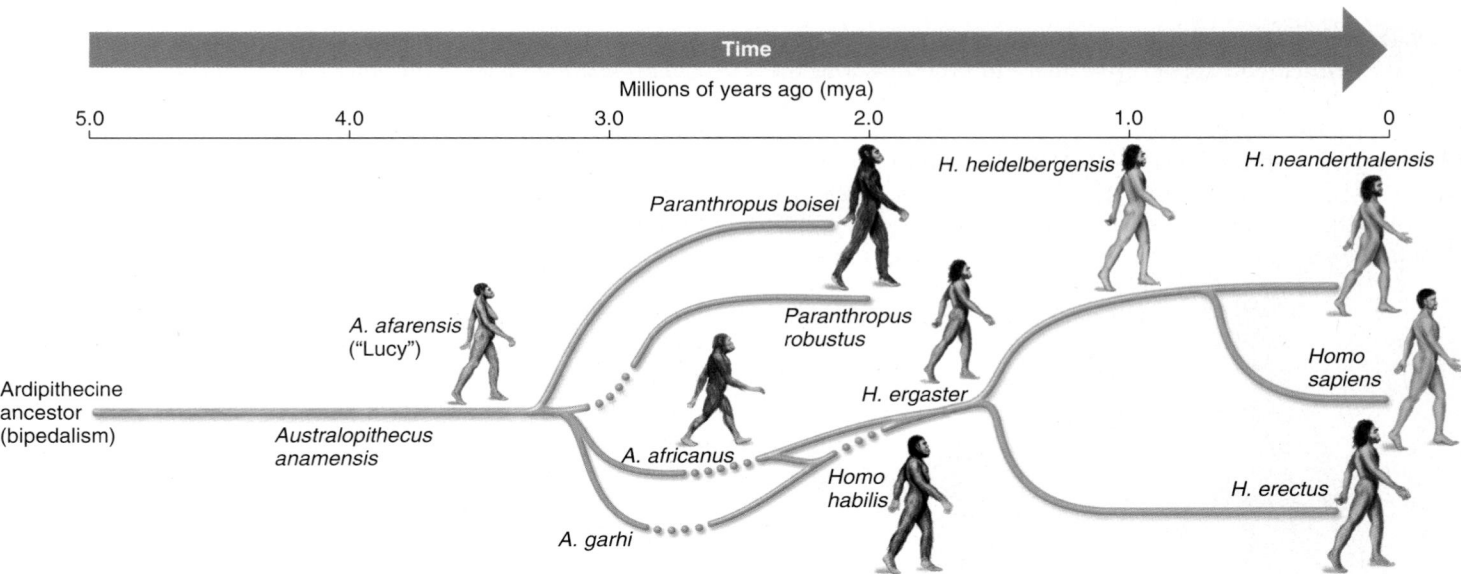

Figure 34.31   Human evolution.   In this human family tree (based on the ongoing work of Donald Johanson), several hominid species lived contemporaneously with one another, but only one lineage gave rise to modern humans (*Homo sapiens*).

of hominins included several species of a smaller-brained genus, *Australopithecus*, one of which gave rise to the genus *Homo*, the humans.

Australopithecines   Since 1924, when the first fossil australopithecine ("southern ape") was found in South Africa, hundreds of fossils of this group have been unearthed all over southern and eastern Africa, the areas where fossil deposits are best exposed to paleontologists. This was a widespread group, with at least six species. Compared to modern humans, all were relatively small, about 1–1.5 m in height, and possessed a facial structure and brain size similar to those of a chimp. Females were much smaller than males, a condition known as sexual dimorphism. In 1974, paleontologist Donald Johanson unearthed the skeleton of a female *Australopithecus afarensis*, dubbed Lucy (the Beatles' song "Lucy in the Sky with Diamonds" was playing in the camp the night when Johanson was sorting the unearthed bones). Over 40% of the skeleton had been preserved, enough to provide a good idea of the physical appearance of australopithecines. When compared to modern humans, it is striking how small australopithecines were (**Figure 34.32**). Examination of the bones revealed that *A. afarensis* walked on two legs. Two of the larger species now considered to be a separate genus, *Paranthropus*, weighed about 40 kg and lived contemporaneously with australopithecines and members of *Homo* species. Both *Paranthropus* species died out rather suddenly about 1.5 million years ago.

***Modern Humans and the Genus* Homo**   In the 1960s, paleontologist Louis Leakey found hominin fossils estimated to be about 2 million years old in Olduvai Gorge, Tanzania. Two particularly interesting facts stand out about these fossils. First, reconstruction of the skull showed an increased brain size, about 680 cubic centimeters (cc) compared to the 500 cc of *Australopithecus*.

Second, the fossils were found with a wealth of stone tools. As a result, Leakey assigned the fossils to a new species, *Homo habilis*, meaning "handyman." The discovery of several more *Homo* fossils followed, but there have been no extensive finds, as there were with Lucy. This makes it difficult to determine which *Australopithecus* lineage gave rise to the *Homo* lineage (see Figure 34.31), and scientists remain divided on this point.

One of the most important species of *Homo* is believed to be *Homo ergaster*, a hominin with a "human-looking" face and skull, with downward-facing nostrils. *H. ergaster* also evolved in Africa and is thought to have given rise to many species, including *Homo erectus*, *Homo heidelbergensis*, *Homo neanderthalensis*, and *Homo sapiens*. *H. ergaster* is believed to be a direct ancestor of modern humans, with *Homo heidelbergensis* viewed as an intermediary step.

H. erectus was a large hominin, as large as a modern human but with heavier bones and a smaller brain capacity of between 750 and 1,225 cc. Fossil evidence shows that *H. erectus* was a social species that used tools, hunted animals, and cooked over fires. *H. erectus* spread out of Africa soon after it appeared, over a million years ago, and fossils have been found as far away as eastern Asia.

Homo heidelbergensis gave rise to two species, *H. neanderthalensis* and *H. sapiens*. *H. neanderthalensis* was named for the Neander Valley of Germany, where the first fossils of its type were found. Neanderthals were a shorter, stockier species than modern humans, with a more massive skull and large brain size of about 1,450 cc, perhaps associated with their bulk. Males were about 168 cm (5 ft 6 in) in height and would have been very strong by modern standards. However, about 30,000 years ago, this species was replaced by another hominin species, *H. sapiens* ("wise man"), our own species. *H. sapiens* was a taller, lighter-weight species with a slightly smaller brain capacity of about 1,350 cc.

Paleontologists remain divided as to whether *H. sapiens* evolved in Africa and spread from there to other areas of the world, or whether premodern humans such as *H. ergaster* migrated from Africa and evolved to become modern humans in different parts of the world. The first model, the Out of Africa hypothesis, posits that the migration of hominins, after evolution in Africa, happened at least twice, once for *H. ergaster* and once for *H. sapiens*, with *H. sapiens* gradually replacing

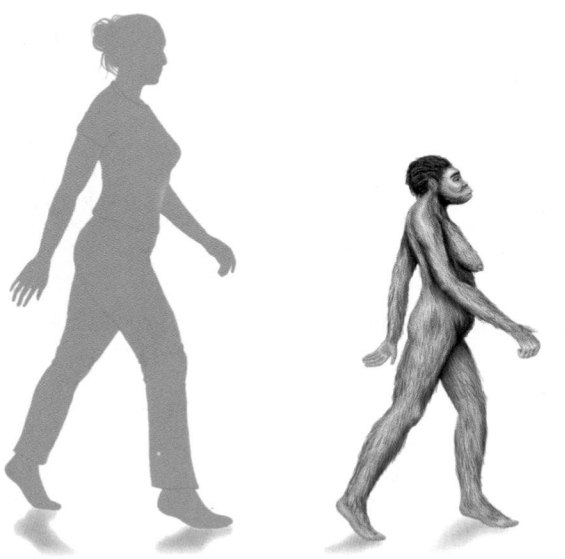

Figure 34.32 Australopithecines. These ancestors of modern humans, as illustrated by this reconstruction of the famous fossil "Lucy," were only about 1 m tall and weighed around 18 kg.

H. ergaster and other intermediate species such as *H. erectus* and *H. neanderthalensis* in other parts of the world. Some scientists find this difficult to accept and suggest that human groups have evolved from *H. ergaster* populations in a number of different parts of the world, a model known as the multiregional hypothesis. According to this hypothesis, gene flow between neighboring populations prevented the formation of several different species.

Studies of human mitochondrial DNA, which occurs only in the cellular organelles called mitochondria, show that all modern people share a common ancestor that dates to about 170,000 years ago. This evidence is consistent with the Out of Africa model, since the common ancestor would have to be much older than that to support the multiregional hypothesis. Furthermore, recent analyses of DNA from Neanderthal bones show it to be quite distinct from the DNA of *H. sapiens*. This also suggests that there was no interbreeding of Neanderthals with *H. sapiens* that migrated into Europe.

Evidence overall appears to support the Out of Africa hypothesis, in which *H. ergaster* evolved in Africa and spread to Europe and Asia. *H. erectus* evolved in Asia and *H. neanderthalensis* evolved in Europe. Both species shared the same fate, extinction at the hands of another species, the "wise man," *H. sapiens*. Many, but not all researchers think that humans evolved in Africa about 170,000 years ago from *H. heidelbergensis* and then spread into the rest of the world. Recently, scientists have suggested that a genetic mutation in modern humans some 50,000 years ago enabled increased problem-solving skills and cognitive abilities, which may have allowed them to succeed at the expense of Neanderthals and others.

CHAPTER SUMMARY

34.1 The Craniates

- The subphylum Vertebrata is the largest and most dominant group of chordates, occupying nearly all of Earth's major habitats. (Figure 34.1)

- Craniates have two defining characteristics that distinguish them from other nonvertebrate chordates: a cranium and a neural crest. The hagfish (class Myxini) is considered the most primitive living craniate. (Figure 34.2)

34.2 Characteristics of Vertebrates

- Vertebrates have several characteristic features, including a vertebral column, endoskeleton of cartilage or bone, and internal organs. Of the 11 different vertebrate classes, 5 are classes of fish, reflecting the dominance of fish. (Table 34.1)

34.3 The Fishes

- A critical innovation in vertebrate evolution is the hinged jaw, which first developed in fish. Ostracoderms, a name for several extinct classes of fish, and lampreys, eel-like fish, lack a hinged jaw. The hinged jaw evolved from cartilaginous gill arches. (Figures 34.3, 34.4, 34.5)

- The chondrichthyans (sharks, skates, and rays) have a skeleton composed of flexible cartilage and powerful appendages called fins. They are active predators with acute senses and were among the earliest fish to develop teeth. (Figure 34.6)

- Bony fish consist of the Actinopterygii (ray-finned fish, the most species-rich class), Actinistia (coelacanths), and the Dipnoi (lungfish). In Actinopterygii, the fins are supported by thin, flexible rays and moved by muscles. (Figures 34.7, 34.8)

- The lobe fins comprise the lobe-finned fish (Actinistia and Dipnoi) and the tetrapods. In the lobe-finned fish, the fins are part of the body. (Figures 34.9, 34.10)

34.4 Amphibia: The First Tetrapods

- Fossils record the evolution of lobe-finned fishes to fishes with four limbs (*Acanthostega*). Recent research has shown that relatively simple mutations control large changes in limb development. (Figures 34.11, 34.12, 34.13)

- Amphibians live on land but return to the water to reproduce. The larval stage undergoes metamorphosis, losing gills and tail for lungs and limbs. (Figure 34.14)

- The majority of amphibians belong to the order Anura (frogs and toads). Other orders are the Caudata (salamanders) and Gymnophiona (caecilians). (Figure 34.15)

34.5 Amniotes: The Appearance of Reptiles

- The amniotic egg permitted animals to become fully terrestrial. Other critical innovations included desiccation-resistant skin, thoracic breathing, water-conserving kidneys, and internal fertilization. (Figure 34.16)
- Living reptilian classes include the Testudines (turtles), Lepidosauria (lizards and snakes), Crocodilia (crocodiles), and Aves (birds). Other reptilian classes, all of which are extinct, include two classes of dinosaurs (Ornithischia and Saurischia). (Figures 34.17, 34.18, 34.19, 34.20)

34.6 Birds

- The differences in body coverings of different vertebrate classes (scales, feathers, hair) are caused by relatively simple changes in the levels of regulatory molecules. (Figure 34.21)
- Three species—*Archaeopteryx*, *Caudipteryx*, and *Confuciusornis*—help trace a lineage from dinosaurs to birds. (Figure 34.22)
- The four key characteristics of birds are feathers, a lightweight skeleton, air sacs, and reduced organs. Birds are the most species-rich class of terrestrial vertebrates. (Figures 34.23, 34.24, Table 34.2)

34.7 Mammals

- The distinguishing characteristics of mammals are the possession of mammary glands, hair, specialized teeth, and an enlarged skull. Other unique characteristics of some mammals are the ability to digest plants and possession of horns or antlers (Figures 34.25, 34.26, 34.27)
- Three subclasses exist: the Prototheria (monotremes), Metatheria (marsupials), and Eutheria (placental mammals). (Figure 34.28, Table 34.3)
- Many defining characteristics of primates relate to their tree-dwelling nature and include grasping hands, large brain, nails instead of claws, and binocular vision. (Figures 34.29, 34.30)
- About 6 million years ago in Africa, a lineage that led to humans began to separate from other primate lineages. A key characteristic of hominins (extinct and modern humans) is bipedalism.
- Human evolution can be visualized like a tree, with one or two hominin species coexisting at the same point in time, with some eventually going extinct and some giving rise to other species. (Figures 34.31, 34.32)
- The Out of Africa hypothesis posits that the migration of hominins from Africa happened at least twice, with *Homo sapiens* gradually replacing other hominin species in other parts of the world. The multiregional hypothesis posits that human groups evolved in a number of different parts of the world, a model known as the multiregional hypothesis.

TEST YOURSELF

1. Which of the following is *not* a defining characteristic of craniates?
   a. cranium
   b. neural crest
   c. two clusters of *Hox* genes
   d. protective housing around the brain
   e. cephalization

2. The hinged jaw is an important adaptation in vertebrates that allowed for
   a. movement onto land.
   b. evolution of more complex body development.
   c. improvement in prey capture.
   d. evolution of speech.
   e. all of the above.

3. The presence of a bony skeleton, an operculum, and a swim bladder are all defining characteristics of
   a. Myxini.
   b. lampreys.
   c. Chondrichthyes.
   d. bony fishes.
   e. amphibians.

4. Organisms that lay eggs are said to be
   a. oviparous.
   b. ovoviparous.
   c. viviparous.
   d. placental.
   e. none of the above.

5. Adaptations in animals that allow for life on land include
   a. a waxy cuticle.
   b. strong limbs that provide movement.
   c. internal fertilization.
   d. b and c only.
   e. all of the above.

6. In some amphibians, the adult retains certain larval characteristics, which is known as
   a. metamorphosis.
   b. parthenogenesis.
   c. cephalization.
   d. paedomorphosis.
   e. hermaphrodism.

7. The membrane of the amniotic egg that serves as a site for waste storage is
   a. the amnion.
   b. the yolk sac.
   c. the allantois.
   d. the chorion.
   e. the albumin.

8. Which of the following is *not* a distinguishing characteristic of birds?
   a. amniotic egg
   b. feathers
   c. air sacs
   d. lack of certain organs
   e. lightweight skeletons

9. Placental mammals that gestate their young for a prolonged time are
   a. monotremes.
   b. marsupials.
   c. eutherians.
   d. therapsids.
   e. all of the above.

10. *Homo sapiens* are believed to have evolved most recently from
    a. *Homo erectus.*
    b. *Homo neanderthalensis.*
    c. *Homo habilis.*
    d. *Homo ergaster.*
    e. *Australopithecus africanus.*

CONCEPTUAL QUESTIONS

1. What are the two defining characteristics that distinguish craniates from other nonvertebrate chordates?

2. Explain the function of the lateral line and the operculum.

3. List the four extraembryonic membranes in the amniotic egg and explain the function of each.

EXPERIMENTAL QUESTIONS

1. What was the purpose of the study conducted by A. P. Davis and colleagues?

2. How were the researchers able to study the effects of individual genes?

3. Explain the results of the experiment shown in Figure 34.13 and how this relates to limb development in vertebrates.

COLLABORATIVE QUESTIONS

1. Discuss the characteristics of the vertebrates.

2. Discuss the three different ways mammals bring their offspring into the world.

www.brookerbiology.com

This website includes answers to the Biological Inquiry questions found in the figure legends and all end-of-chapter questions.

35

AN INTRODUCTION TO FLOWERING PLANTS

CHAPTER OUTLINE

Representatives of two major groups of flowering plants. These corn and bean plants display similar architecture, common to flowering plants, yet differ in other structural and reproductive features.

Corn, more formally known as *Zea mays*, and the common bean (*Phaseolus vulgaris*) are among the world's most important agricultural crop plants. Farmers grow corn for food and fuel production. The seeds of beans and related plants are extraordinarily rich sources of protein in the human diet, and therefore of particular importance to vegetarians. Corn and bean plants share many features that are common to flowering plants, including stems, roots, leaves, flowers, fruits, and seeds. However, corn and bean plants also display differences in these features, as you can see in the chapter-opening photo. In fact, the hundreds of thousands of modern flowering plants display an amazing variety of forms that represent adaptations to diverse habitats.

This chapter provides an introduction to the flowering plants, focusing on fundamental principles of structure and function. These principles will help to explain why flowering plants resemble each other in many ways, yet differ in features that influence plant roles in nature and agriculture. We will begin by describing the general life of a flowering plant—from seed germination to reproductive maturity. A survey of major plant structural features and growth processes will follow. Finally, we will consider diverse stem, leaf, and root adaptations. These topics provide essential background for subsequent chapters in this unit, which focus on flowering plant behavior, nutrition, transport, and reproduction.

35.1 From Seed to Seed—The Life of a Flowering Plant

Several major events punctuate the lives of flowering plants, also known as the **angiosperms**. When seeds germinate, dormant embryos wake to metabolic activity and begin the process of seedling development. Seedlings grow and develop into mature plants capable of reproduction. Finally, flowers and fruits produce and disperse the next generation of seeds. In this section we will briefly survey the life of flowering plants, focusing on the basic structural features of each life stage.

Seedlings Develop from Embryos in Seeds

Seeds are reproductive structures produced by flowering plants and other seed plants, usually as the result of sexual reproduction. Seeds contain embryos that develop into young plants—seedlings—when seeds germinate. The embryo is an essential stage in the sexual cycle of a flowering plant. The plant sexual cycle explains how embryos arise.

Sexual reproduction in plants requires two multicellular stages: a gamete-producing **gametophyte** and a spore-producing **sporophyte**. In the life cycle of flowering plants, these two life stages alternate with one another, a process called **alternation of generations** (**Figure 35.1**). Flowering plants produce relatively large sporophytes and microscopic gametophytes that grow and develop within flowers. Diploid sporophytes produce haploid spores by the process of meiosis. These spores grow into gametophytes that produce plant gametes—eggs and sperm. Fusion of egg and sperm in the process of fertilization generates a diploid zygote, which undergoes repeated mitotic divisions to form the plant **embryo**.

The plant embryo is a very young sporophyte that lies dormant within seeds, accompanied by a supply of stored food and enclosed by a tough, protective seed coat (**Figure 35.2a**). The seed coat protects the delicate embryo during the dispersal of seeds from parent plants into the environment. Dispersed seeds may remain dormant in the soil—sometimes for long periods—but germinate when temperature, moisture, and light conditions are favorable. Such conditions activate embryo metabolism, causing embryos to enlarge and break the seed coat. Once free of the seed coat, embryos grow into seedlings (**Figure 35.2b**), and if sufficient resources such as water and minerals are available, seedlings develop into mature plants (**Figure 35.2c**).

Growth is an increase in weight or size. **Development** is an increase in number of organs, accompanied by differentiation of tissues and specialized cells having distinctive structure and function. One or more types of specialized cells make up **tissues**, groups of adherent cells that have particular functions.

Figure 35.1   **The sexual cycle of flowering plants involves alternation of sporophyte and gametophyte generations.** The sporophyte is the dominant, conspicuous generation, whereas the tiny gametophytes are mostly hidden within flowers. The bean plant shown here produces both male and female gametophytes.

Seedlings and mature plants produce new tissues in areas called meristems. A **meristem** (from the Greek *merizein*, meaning "to divide") is a region of undifferentiated cells that produces new tissues by cell division (see Figure 35.2b). Tissues of several types can be found in roots, stems, and leaves, the basic **organs** found in plant embryos, seedlings, and mature plants (**Figure 35.3**).

Roots provide anchorage in the soil and also foster efficient uptake of water and minerals. **Stems** produce leaves and branches and bear the reproductive structures of mature plants. **Leaves** are flattened structures that emerge from stems; foliage leaves are specialized in ways that enable photosynthesis.

The embryonic root, known as the radicle, is the first organ to emerge from a germinating seed and provides the water and minerals needed for seedling growth (see Figure 35.2a). A short, stemlike region of the embryo, known as a hypocotyl, produces embryonic leaves known as seed leaves or **cotyledons**. The embryos of bean plants and relatives possess two (*di*) seed leaves and are thus known as **eudicots** (**Table 35.1**). Other flowering plants (including corn) are known as **monocots** because their embryos have only one (*mono*) seed leaf. Although seed leaves are not green, the leaves of seedlings and mature plants are usually green and function in photosynthesis.

Embryos in mature seeds are not green or capable of photosynthesis. Thus, seedlings obtain the energy needed to fuel early growth and development by respiring organic food stored in seeds. Corn seeds, for example, contain the starch-rich tissue known as **endosperm**. Bean seeds are particularly rich in protein, which is stored in the cotyledons. Some seeds, such as those of the oil palm, store large amounts of lipids. These organic food storage compounds explain why seeds are often used as food sources by animals, including humans.

Figure 35.2   **The seed-to-seed life of flowering plants.**   This example depicts the life of the eudicot *Arabidopsis thaliana*. **(a)** Seed embryos possess embryonic leaves, known as cotyledons, a dormant shoot meristem, and an embryonic root, the radicle, with a dormant root meristem. **(b)** When seeds germinate, the radicle emerges and produces the seedling root, and the young shoot of the seedling begins to produce leaves. **(c)** Reproductively mature plants have branched shoot and root systems, and bear flowers and fruits that disperse seeds.

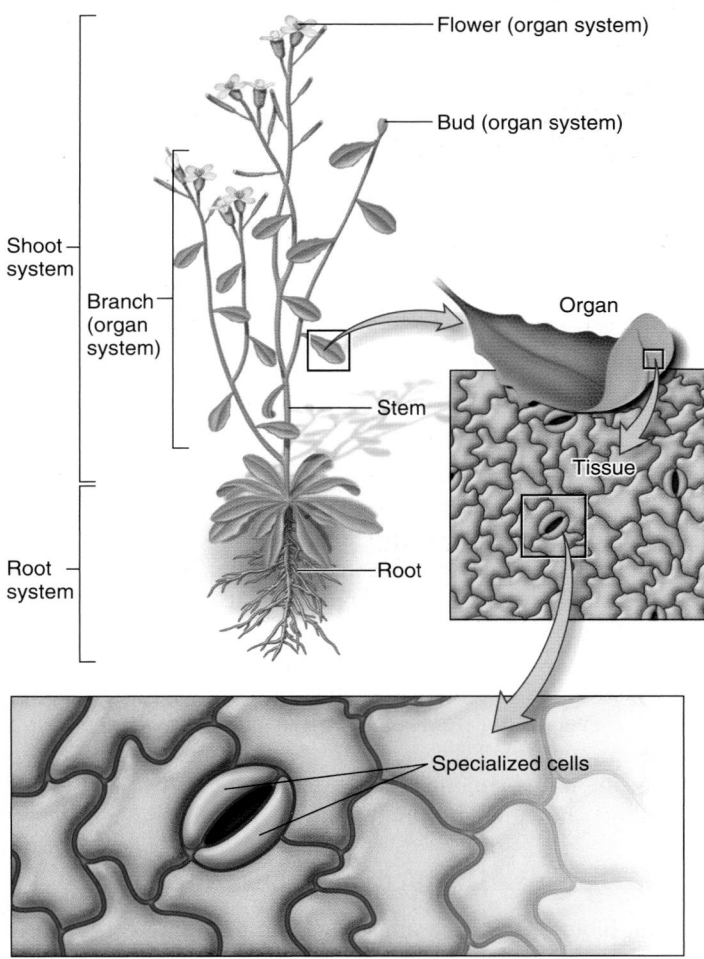

Figure 35.3 **Levels of biological organization in a plant.** Flowering plant sporophyte bodies consist of a root system and a shoot system, each composed of organ systems, organs, tissues, and specialized cells.

Table 35.1	Distinguishing Features of Eudicots and Monocots, Two Major Groups of Flowering Plants	
Feature	**Eudicots**	**Monocots**
Number of seed leaves (cotyledons)	Two	One
Number of flower parts	Usually four, five, or multiples of these	Usually three or multiple of three
Stem vascular bundles	Arranged in a ring	Scattered
Root system	Branched taproot	Fibrous; adventitious
Leaf venation	Netted or branched	Often parallel
Pollen	Three pores or slits	One pore or slit

Mature Sporophytes Develop from Seedlings

As seedlings develop into mature sporophytes, the aboveground shoot typically becomes green and photosynthetic and thus able to produce organic food. Photosynthesis powers the transformation of seedlings into mature plants. Seedlings produce new shoot tissues at shoot tips (apices), in a meristem. These regions of rapidly dividing cells at shoot apices are known as **shoot apical meristems**, or **SAMs**. The collection of organs produced by the SAMs is known as the **shoot system**. The shoot system includes stems and leaves, and structures composed of more than one of these organs, which are **organ systems**. Roots also grow at their tips, where new root tissues are produced by the **root apical meristems**—the **RAMs**. The collection of roots and root branches produced by the RAMs is the **root system**. The production of new tissues by the SAMs and RAMs during seedling development and growth of mature plants is known as **vegetative growth**. Mature shoot meristems may also undergo reproductive development.

Vegetative Growth During their growth, plant shoots produce vegetative **buds**, miniature shoots each having a dormant SAM. Tough leaflike scales protect bud contents. Under favorable conditions, the bud scales fall off and the vegetative buds open. Newly opened buds display young leaves on a short shoot. The shoot SAM then becomes active, producing new stem tissue and leaves. In this way, buds generate leafy branches. Buds are examples of organ systems, because they contain more than one type of organ.

Vegetative shoots often display **indeterminate growth**, meaning that SAMs continuously produce new stem tissues and leaves as long as conditions remain favorable. This process explains how very large plants can develop from seedlings.

However, plant size is also under genetic control, so some plants remain small even when they are mature.

Reproductive Development Under favorable conditions, mature plants produce reproductive structures: flowers, seeds, and fruits. Flowers open from floral buds. Floral buds and **flowers** are reproductive shoots that develop when shoot apical meristems produce flower parts instead of leaves and new stem tissues. Flower development occurs under the control of several genes whose roles are well understood (see Chapter 19). In contrast to vegetative shoots (which often show indeterminate growth), flowers are produced by **determinate growth**, which is growth of limited duration. A floral shoot no longer produces new stem growth or leaves. Thus, flowering and shoot growth are alternative processes. In order to flower, a plant must give up some of its potential to continue vegetative growth. Eudicot flowering plants often produce flowers having parts in four, five, or a multiple of those numbers, whereas monocot flower parts usually occur in threes or a multiple of three (Table 35.1).

Flower tissues enclose and protect tiny male and female gametophytes during their growth and development. Female gametophytes produce eggs within structures known as ovules, produced in the ovary of a flower pistil. Male gametophytes begin their development within pollen grains produced in the anthers of a flower stamen (see Figure 35.1). Pollen is dispersed to the flower pistil, whereupon pollen grains may germinate, producing a tube that delivers sperm to eggs (refer back to Figure 31.14). Fertilization generates zygotes, which develop into embryos, and also triggers the process by which ovules develop into seeds and flower parts develop into fruits. **Fruits** thus enclose seeds and function in seed dispersal. Together with vegetative buds, flower buds, flowers, fruits, and seeds are organ systems because they consist of more than one organ.

Flowering Plants Vary in Seed-to-Seed Lifetime

The seed-to-seed lifetime of a flowering plant can vary from a few weeks to many years. Plants that die after producing seed during their first year of life are known as **annuals**. Corn and the common bean are examples of annual crops whose nutrient-rich seeds are harvested within a few months after planting, and must be replanted at the beginning of each new growing season. Plants that do not reproduce during the first year of life but may reproduce within the following year are known as **biennials**. Many plants with fleshy roots that we use for food are biennials. Trees are examples of **perennials**, plants that live for more than two years, often producing seed each year after they reach reproductive maturity. Many flowering plants use environmental signals to time flowering in ways that enhance seed production. Temperature and day length are examples of environmental factors that determine flowering time. Plant seed-to-seed lifetimes are also influenced by the longevity of their seeds. Seeds of some plants are able to germinate after more than a thousand years of dormancy, whereas other plant seeds are unable to remain alive for long periods.

We have seen that the life of a flowering plant includes seed germination, seedling development into mature plants, and reproduction. We have also learned that during reproduction, a plant accomplishes its sexual cycle—alternation of generations. During the plant sexual cycle, a parental sporophyte produces many gametophytes, from which develop the next generation of young sporophytes—embryos in seeds. In this way, a single plant can produce many progeny. We have also surveyed the basic components of embryo, seedling, and adult plant structure. These topics provide useful background as we next focus more closely on principles of plant growth and development.

35.2 How Plants Grow and Develop

Four processes are essential to plant growth and development: cell division, growth, cell specialization, and apoptosis—controlled cell death. However, in contrast to animals, cell migration does not occur in plants. Additional principles of plant growth and development include (1) development and maintenance of a distinctive architecture throughout life, (2) increase in length by the activity of primary meristems, (3) maintenance of a population of ever-young stem cells in meristems, and (4) expansion of cells in controlled directions, by water uptake. All of these features are under genetic control and can be studied by the use of mutants showing abnormal structure and development.

Plants Display a Distinctive Architecture

In plant biology, the term apical has two distinct meanings. As we have seen, apical refers to the tips of shoots and roots, as in shoot apical meristems or root apical meristems. A second meaning for apical is the part of a plant that projects upward, which is the top of the shoot. By contrast, the bottom of a root is at the basal region of a plant. Thus, the shoot apical meristem occurs at the apical pole, and the root apical meristem occurs at the basal pole. This property, known as **apical-basal polarity**, explains why plants produce shoots at their tops and roots at their lower regions. Apical-basal polarity originates during embryo development (**Figure 35.4a**). As seedlings and maturing plants grow in length by the activity of shoot and root meristems, apical-basal polarity is maintained.

A second architectural feature common to plants is **radial symmetry**. Plant embryos display a cylindrical shape, also known as an axis, which is retained in the stems and roots of seedlings and mature plants. A thin slice or cross section of an embryo, stem, or root is typically circular in shape. Another component of radial symmetry is that plants produce new leaves or flower parts in circular whorls, or spirals, around shoot tips (Figure 35.4a). Buds and branches likewise emerge from stems in radial patterns, as do branch roots from a central root axis.

Apical-basal polarity and radial symmetry are essential properties of normal plant development, a concept that is illustrated by genetic mutations that abolish apical/basal polarity.

Shoot apical meristem

Radial symmetry

Apical-basal polarity

Root apical meristem

(a) Normal seedling

(b) Abnormal GNOM mutants

Figure 35.4 **Plant architecture is based on apical-basal polarity and radial symmetry.**   **(a)** Normal plants exhibit apical-basal polarity and radial symmetry, as shown by this seedling. Such growth occurs at two meristems, one at the shoot (SAM) and one at the root (RAM). **(b)** *GNOM* mutants of *Arabidopsis thaliana* lack apical-basal polarity and thus produce abnormal embryos and seedlings.

For example, mutations in the gene *GNOM* result in plant embryos that are spherical, and thus lack apical-basal polarity (**Figure 35.4b**). Such embryos produce abnormal seedlings. Together, apical-basal polarity and radial symmetry explain why diverse plant species have a fundamentally similar architecture. Next, we will learn more about how primary meristems allow plants to grow in length from both shoot and root tips.

Primary Meristems Increase Plant Length and Produce Plant Organs

We have previously noted that plant embryos grow into seedlings by adding new cells from only two growth points, the shoot apical meristem (SAM) and the root apical meristem (RAM). During plant development, the SAM and RAM of the embryo give rise to many apical meristems located in the buds of shoots and at the tips of roots. Each apical meristem, in turn, produces additional meristematic tissues that increase plant length and produce new organs. Collectively, such meristems are known as the **primary meristems**, and they ultimately produce **primary tissues** and organs of diverse types (**Table 35.2**). Tissues and organs are composed of cells of diverse types. Simple primary tissues are those composed of only one or two cell types, while complex primary tissues are made up of more than one cell type.

Apical meristems produce three primary tissue meristems called protoderm, procambium, and ground meristem that are

Table 35.2 | **Example Tissues and Specialized Cells Found in Flowering Plants***

Simple primary tissues (composed of one or two cell types)		Plant cell types found in those tissues
Parenchyma (G)		Parenchyma cells
Collenchyma (G)		Collenchyma cells
Sclerenchyma (G)		Fibers and sclereids
Root endodermis (G)		Endodermal cells
Root pericycle (G)		Pericycle cells

Complex primary tissues (composed of at least two cell types)	Plant cell types found in those tissues
Leaf or stem epidermis (D)	Flattened epidermal cells, trichomes, stomatal guard cells
Root epidermis (D)	Flattened epidermal cells, root hairs
Leaf mesophyll (G)	Spongy parenchyma cells, palisade parenchyma cells
Leaf, stem, or root xylem (V)	Tracheids, vessel elements, fibers, parenchyma cells
Leaf, stem, or root phloem (V)	Sieve-tube elements, companion cells, fibers, parenchyma cells

Complex secondary tissues	Plant cell types found in those tissues
Secondary xylem (V) (wood)	Tracheids, vessel elements, fibers, parenchyma
Secondary phloem (V) (inner bark)	Sieve-tube elements, companion cells, fibers, parenchyma
Outer bark	Cork cells

*This list does not include all of the tissues and cell types found in flowering plants. Some of these examples will be described later in this chapter.

D = dermal, G = ground, V = vascular

present in young stems, leaves, and roots (**Figure 35.5**). The **protoderm** generates the outermost **dermal tissue**. The **procambium** produces **vascular tissues**, which make up a vascular system that conducts materials within the plant body and also provide support. The **ground meristem** gives rise to **ground tissues** defined by their locations—between dermal and conducting tissues, or at the center of mature stems where it is called pith. However, note that plant cell specialization and tissue development do not depend much on the lineage (the parentage) of a cell or tissue. Chemical influences are much more important in determining the type of specialized tissue that will be produced by unspecialized plant cells.

Shoot apical meristem
D
G
V

D G
V

V
G
D

V
G
D

Procambium (V)
Ground meristem (G)
Protoderm (D)
Root apical meristem

Figure 35.5   Plant apical meristems and the primary tissues they produce.   Shoot and root apical meristems generate three primary meristems: Protoderm produces the dermal tissues (shown in tan color), ground meristem produces ground tissues (shown in green or brown color), and procambium yields vascular tissues (shown in blue color). The vascular systems of leaves, stem, and root are connected.

We next consider the role of primary meristems in the development of stems, leaves, and roots and briefly describe the basic features of these organs. A closer focus on stem and leaf structural adaptations can be found in Section 35.3, and a deeper view of the root system is provided in Section 35.4 of this chapter.

Development and Structure of Primary Stem Tissues New primary stem tissues arise by the cell division activities of primary meristems located near the bases of SAMs. A layer of dermal tissue known as the **epidermis** develops at the stem surface. The epidermis produces a waxy surface coating known as the **cuticle**, which helps to reduce water loss from the plant surface. The epidermis and cuticle also help to protect plants from damage by insects and disease microorganisms. Beneath the epidermis lies the stem **cortex**, which is largely composed of a ground tissue known as **parenchyma tissue**. This tissue is composed of only one cell type, thin-walled cells known as **parenchyma cells**. These cells store starch in plastids and thus serve

as an organic food reserve. Stem parenchyma also has the ability to undergo cell division (meristematic capacity), which aids wound healing when stems are damaged. The cell division capability of stem parenchyma also explains how people are able to grow new plants from stem cuttings. Stems also contain **collenchyma tissue** composed of flexible **collenchyma cells**, and rigid **sclerenchyma tissue** composed of tough-walled **fibers** and **sclereids**. Collenchyma tissue and fibers provide support, while sclereids often aid in defense. New water and food conducting tissues develop at the core of a young shoot. These conducting tissues, also known as primary vascular tissues, are composed of several types of specialized tissues and cells (see Section 35.3 and Table 35.2). Newly formed stem vascular tissues connect with older conducting tissues that extend throughout the stem system and are linked with vascular tissues of the root system (see Figure 35.5). Vascular tissues, made up of **xylem** and **phloem**, supply water, minerals, and organic food resources needed for continued cell division in shoot meristems, and for stem growth and development.

Primary vascular tissues are arranged in elongate clusters known as **vascular bundles** that appear round or oval when cross-cut. In the primary stems of beans and other eudicots, the vascular bundles are arranged in a ring, which is easily seen in thin slices made across a stem (see Table 35.1). Celery strings are examples of eudicot vascular bundles. In contrast, vascular bundles are scattered in the stems of corn and other monocots.

Leaf Structure and Development Young leaves are produced at the sides of a SAM as small bumps known as **leaf primordia** (**Figure 35.6a**). As leaves develop, they become flattened, a process that expands the area of leaf surface available for light collection during photosynthesis. Many leaves are also thin, an adaptation that helps them to shed excess heat. Leaves also become bilaterally symmetrical, meaning that they can be divided into two equal halves in only one direction, from the leaf tip to its base (**Figure 35.6b**). In addition, upper (adaxial—facing toward the stem) and lower (abaxial—facing away from the stem) leaf tissues develop differently in several ways that foster photosynthesis (**Figure 35.6c**). For example, the more shaded abaxial leaf epidermis displays more pores, known as **stomata**, than the sunnier adaxial leaf surface. When open, stomatal pores allow CO_2 to enter and O_2 to escape leaf tissues; when closed, these pores help to prevent excessive loss of water from plant surfaces. Another example of differentiation between the abaxial and adaxial regions of a leaf is revealed by the positions of different types of photosynthetic ground tissue. The adaxial **palisade parenchyma** consists of closely packed, elongate cells adapted to absorb sunlight efficiently, whereas abaxial **spongy parenchyma** contains rounder cells separated by abundant air spaces. These air spaces, located near stomata, foster CO_2 absorption and O_2 release by leaves. Together the palisade and spongy parenchyma are known as the leaf **mesophyll**.

Veins composed of vascular tissue commonly occur at the junction of palisade and spongy parenchyma, or within the spongy parenchyma (Figure 35.6c). During leaf development,

(a) SAM producing leaf primordia   **(b) Mature leaf**   **(c) Leaf internal structure**

Figure 35.6   Leaf development and structure.   (a) Leaves develop from primordia produced at the sides of shoot apical meristems, as shown in this thinly sliced, stained shoot tip. **(b)** Mature leaves are typically thin and flat, and show bilateral symmetry. **(c)** An internal view of a thinly sliced and stained leaf reveals top to bottom (adaxial to abaxial) tissue differentiation. A layer of palisade parenchyma lies just beneath the upper epidermis capped with waxy cuticle. Veins of conducting tissue (xylem and phloem) are embedded in the photosynthetic mesophyll. Spongy parenchyma lies above the lower epidermal leaf surface, which displays stomatal pores. These structural features of mature leaves facilitate photosynthesis.

leaf vascular tissues connect with those of the stem. This process allows water and minerals to reach leaves, fostering photosynthesis, and also allows leaves to export the sugar products of photosynthesis to other parts of the plant.

Root System Structure and Development In beans and other eudicots, a main root develops from the embryonic root and then produces branch roots. The root system of eudicots is known as a **taproot system**; this kind of root system has one main root with many branch roots. In contrast, the embryonic root of monocots dies soon after seed germination, and it is replaced by a **fibrous root system** consisting of multiple roots that grow from the stem base (see Table 35.1). Fibrous roots are examples of **adventitious roots**, structures that are produced on the surfaces of stems (and sometimes leaves) of both monocots and eudicots. Roots that develop at the bases of stem cuttings are also adventitious roots.

As we have earlier noted, the tips of roots and their branches each possess an apical meristem that adds new cells. Expansion of these new cells allows roots to grow into the soil. As they lengthen, roots produce branches, also known as lateral roots. In contrast to stems, roots do not produce buds. As a result, lateral roots develop from meristematic tissues located within the root (see Section 35.4). The root system both anchors plants in the soil and plays an essential role in harvesting water and mineral nutrients. Root tissues are usually not green and photosynthetic, and thus they rely on organic compounds conducted from the shoot. The plant root system and shoot system thus depend on each other.

Plant Meristems Contain Ever-Young Stem Cells

Plant meristems include undifferentiated, forever-youthful cells known as initials or **stem cells**. In the late 19th century, the biologist Alexander Maximow coined the term *Stammzelle*, which is derived from the German words *Stamm*, meaning "stem,"

such as a plant stem, or "family," and *Zelle*, meaning "cell." Maximow used the term "stem cell" to describe animal cells that remain undifferentiated but are also able to generate specialized tissues. Animal stem cells are currently much in the news because of their potential for use in the treatment of damaged human tissues. The term stem cell is now also widely used for cells of the plant meristem that likewise remain undifferentiated but also can produce new tissues. In the context of plant development, the term does not mean any cell located in a plant stem, only the undifferentiated cells located within the meristems of the shoot and root.

When plant stem cells divide, they produce two cells, one that remains young and unspecialized plus another cell. This second cell may differentiate into various types of specialized cells, but it often retains the ability to divide. As a result of these properties, stem cell numbers influence the size of a meristem, which in turn affects plant growth. Recent molecular studies have shed new light on how stem cell populations and meristem size are controlled.

GENOMES & PROTEOMES

Plant Shoot Apical Meristem Size Is Genetically Controlled

Much of what is known about the function of stem cells in plant meristems comes from study of the model plant system *Arabidopsis thaliana*. This plant is particularly useful for genetic studies because its entire genome has been sequenced. The normal *Arabidopsis* SAM consists of several hundred stem cells organized into at least three distinct cell regions having different functions: a central zone, a peripheral zone, and an area called the rib meristem (**Figure 35.7**). The central zone consists of stem cells that divide but remain undifferentiated.

Figure 35.7   **The shoot apical meristem and stem cells.**   In a shoot apical meristem, a central zone of stem cells is surrounded by a peripheral zone of cells in the process of differentiating into dermal tissue. An underlying region of rib meristem cells divides to produce new ground and vascular tissues of the stem.

The peripheral zone is made up of cells that originate from the edges of the central zone and produce leaf and flower primordia at the shoot surface. The rib meristem consists of a group of meristematic cells lying beneath the central zone that generates new ground and vascular tissues. Normal growth depends on a plant's ability to maintain normal sizes of the central zone and SAM. *Arabidopsis* mutants have been used to discover how plants accomplish this process.

Previous studies had established that stem cells in the central zone express and secrete a small peptide known as CLAVATA3. This peptide specifically binds to proteins located in the membranes of underlying meristem cells, causing them to produce substances that control the size of the central zone. In 2005, plant molecular biologists G. Venugopala Reddy and Elliot Meyerowitz reported the results of experiments that added more information about the influence of CLAVATA3 on SAM size. These investigators observed that plants having mutant *CLAVATA3* genes produced very large SAMs with enlarged central zones. There were several possible explanations for this result. First, the central zone cells of mutant plants might divide more rapidly than is normal. Alternatively, cells at the edges of the central zone might not develop into peripheral zone cells, as is usual. Finally, peripheral zone cells might abnormally develop into central zone cells.

To determine which of these possibilities was most likely, the investigators disabled the *CLAVATA3* gene in experimental plants. They also labeled the central zone with a glowing marker molecule in order to observe changes in its size by microscopic examination. They discovered that loss of *CLAVATA3* function caused peripheral zone cells to change their identity to central zone cells. This had the effect of increasing the size of the central zone. In addition, the investigators observed that cell division increased in other parts of the SAM. These results suggest that CLAVATA3 peptide normally prevents peripheral zone cells from differentiating into central zone cells, and it controls the overall cell division rate in the SAM. Together with CLAVATA3 effects that had earlier been observed, these newly discovered processes maintain the central zone at a steady size, which is necessary for normal shoot growth.

Plant Cells Expand in a Controlled Way by Absorbing Water

As we have observed, meristem production of new cells is an important component of plant growth. In addition, plant growth involves cell expansion. The diameters of newly formed stem and root cells are usually equal in all dimensions, but many soon begin to extend lengthwise, thereby helping shoots and roots to grow longer. Cell extension occurs when water enters the central vacuole by osmosis (**Figure 35.8**). As the central vacuole expands, cell volume also increases. By taking up water, plant cells can enlarge quickly, allowing rapid plant growth. Bamboo, for instance, can grow taller by 2 m within a week, and can grow up to 30 m in less than three months! The importance of water uptake in cell expansion helps to explain why plant growth is so dependent on water supply.

Plant cell walls contain cellulose microfibrils that are held together by cross-linking polysaccharides. When plant cells and their vacuoles absorb water, pressure builds on cell walls. In response to this pressure and under acidic conditions, proteins unique to plants—known as expansins—are produced. **Expansins** unzip cross-linking cell-wall polysaccharides from cellulose microfibrils, so that the cell wall can stretch (**Figure 35.9**). As a result, cells enlarge, often by elongating in a particular direction, which is important to plant form. Some plant cells are able to elongate up to 20 times their original length.

The direction in which a plant cell expands depends on the arrangement of cellulose microfibrils in its cell wall, which is determined by the orientation of cytoplasmic microtubules. These microtubules are thought to influence the positions of cellulose-synthesizing protein complexes located in the plant plasma membrane. The protein complexes convert sugars into cellulose polymers, spinning cellulose microfibrils onto the cell surface to form the cell wall (refer back to Figure 30.15). As a result, cell-wall cellulose microfibrils encircle cells in the same orientation as underlying cytoplasmic microtubules (**Figure 35.10**). Because cellulose microfibrils do not extend lengthwise, plant cell walls expand more easily in a direction perpendicular to them. To visualize this process, imagine encircling a spherical balloon with parallel bands of tape before it has been completely inflated.

Figure 35.8 Plant cells expand by taking up water into their vacuoles.

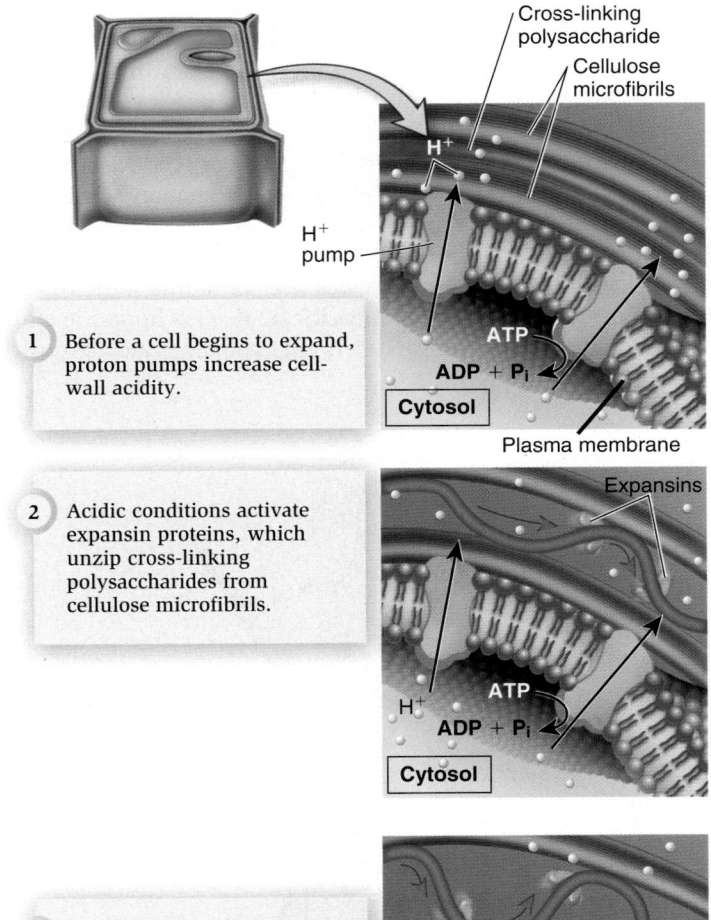

1 Before a cell begins to expand, proton pumps increase cell-wall acidity.

2 Acidic conditions activate expansin proteins, which unzip cross-linking polysaccharides from cellulose microfibrils.

3 The cellulose microfibrils are free to glide apart. As the cell takes up water, the cytoplasm exerts pressure on the cell wall, causing it to expand.

Figure 35.9 A hypothetical model of the process of cell-wall expansion.

KEY
Cytoplasmic microtubules
Cell-wall cellulose microfibrils

Figure 35.10 The direction of plant cell expansion is controlled by microfibrils and microtubules. Plant cells enlarge in the direction perpendicular to encircling cell-wall cellulose microfibrils, which run parallel to the orientation of underlying cytoplasmic microtubules.

The tape bands, which also do not extend lengthwise, represent encircling cellulose microfibrils. As you add more air, the balloon will tend to extend in the direction perpendicular to the bands of tape.

Microtubules control not only the direction of cell expansion but also the plane of cell division, which is also critical to plant form. The *FASS* gene mutation in *Arabidopsis* illustrates the importance of microtubule orientation to plant structure. In cells of these mutants, microtubules are randomly arranged, resulting in cell division planes that are abnormal, cells that do not elongate, and mature plants with abnormal, stubby organs.

35.3 The Shoot System: Stem and Leaf Adaptations

As we have seen, the shoot system includes all of a plant's stems, branches, and leaves. It also produces flowers and fruits when the plant has reached reproductive maturity. Thus, the shoot system is essential to plant growth, photosynthesis, and reproduction. In this section, we examine stem and leaf structure and development in more detail. We will observe that features of shoot stems and leaves vary among plants in ways that explain plant ecological function in nature, and that these features are very useful to us in distinguishing plant species.

Shoot Systems Have a Modular Structure

More than 200 years ago, the German author, politician, and scientist Johann Wolfgang von Goethe realized that plants are modular organisms, composed of repeated units. Shoots are notably modular (**Figure 35.11**). Each shoot module, often known as a phytomere, consists of four parts: a stem node and an internode, a leaf, and an axillary meristem or bud. A **node** is the stem region from which one or more leaves emerge. An **internode** is the region of stem between adjacent nodes. Differences in numbers and lengths of internodes help to explain why plants differ in height. A young leaf with axillary meristem is a derivative of the shoot apical meristem. The axillary meristem occurs at the upper angle formed where a leaf or branch emerges from the stem. This angle is known as an axil (from the Greek *axilla*, armpit). Axillary meristems generate **axillary buds**, which can produce branches, also known as lateral shoots. New leaves are produced by the shoot apical meristem located at the apices of main and lateral shoots. What causes new leaves to arise? The answer involves chemical messengers known as hormones.

Hormones Influence Leaf Development

As we have noted, the very youngest leaves, known as leaf primordia, originate as surface bumps at the sides of a shoot apical meristem (see Figure 35.6a). Production of such leaf primordia is under the control of a hormone known as **auxin**. In general, **hormones** are signaling molecules that exert their effects at a site distant from the place where such compounds are produced.

Plant hormones are important in coordinating both plant development and plant responses to environmental conditions (see Chapter 36). The outermost epidermal layer of cells at shoot tips produces auxin, which moves from cell to cell by means of specific membrane transport proteins. Auxin accumulates in particular locations because cells of the shoot apex differ in their ability to import and export auxin. When auxin accumulates in a particular apical region, the hormone causes expansin gene expression to increase. When expansin loosens their cell walls (see Figure 35.9), cells expand by taking up water, thereby forming a tissue bulge—a leaf primordium. The development of leaf primordia depletes auxin from nearby tissue, with the result that the next leaf primordium will develop in a different place on the shoot apex, where the auxin level is higher. Such changes in auxin concentration on the surface of the shoot explain why leaf (and flower) primordia develop in spiral or whorled patterns around the shoot tip. The youngest leaf primordia occur closest to the shoot tip, and successively older leaf primordia occur on the sides of the shoot tip (see Figure 35.6a).

Hormones also influence the transformation of primordia into leaves. The cells of leaf primordia do not produce a protein known as KNOX (short for <u>Kn</u>otted-like Home<u>ob</u>ox), which is produced by other shoot meristematic cells. KNOX is a transcription factor, a protein that regulates gene transcription. The absence of KNOX proteins induces leaf primordia to produce a plant hormone known as **gibberellic acid**. This hormone stimulates both cell division and cell enlargement, causing young leaves to grow larger.

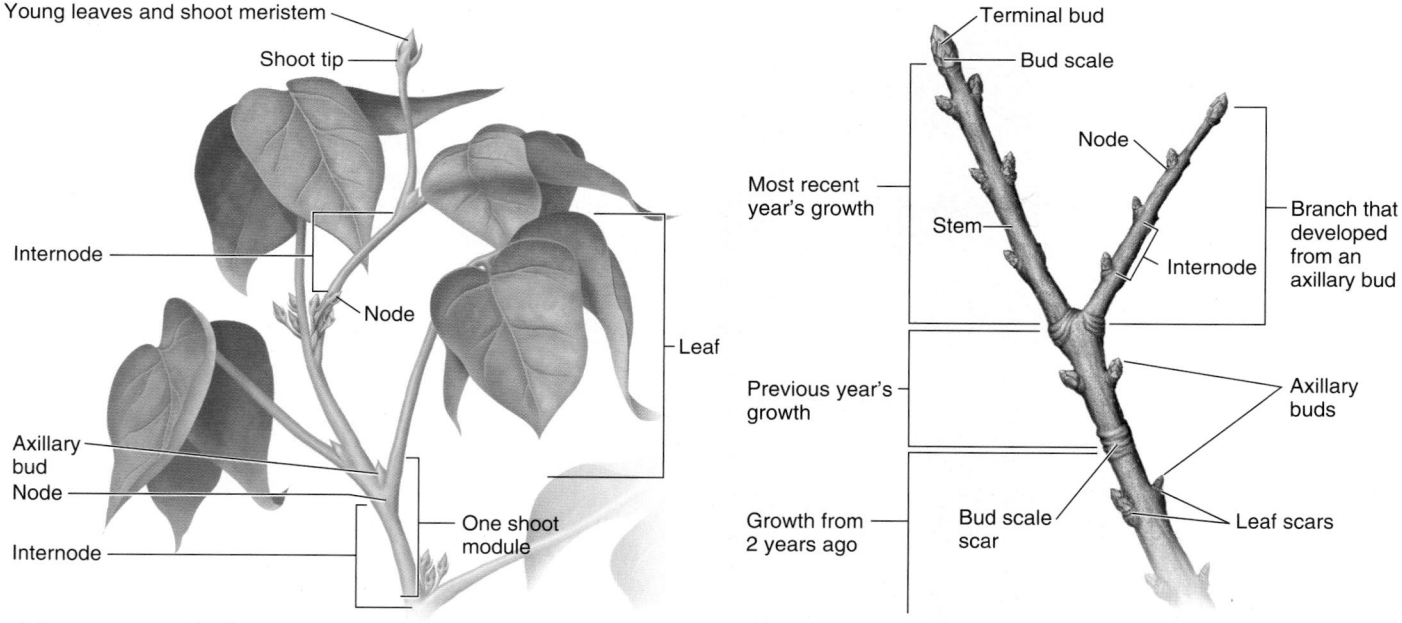

(a) Modular structure of herbaceous shoot

(b) Modular structure of woody shoot in winter

Figure 35.11 **Plant shoots have a modular organization.** (a) The top end of an herbaceous stem showing the shoot modules. Each module consists of a node with its associated leaf and axillary meristem or bud and an internode. (b) The modular organization shown by a woody stem as it appears during winter. Axillary buds lie above the scars left by leaf fall. Regions between successive sets of bud scale scars mark each year's growth.

Other molecules produced by a SAM direct leaf flattening and differentiation of upper adaxial and lower abaxial leaf tissues (see Figure 35.6c). These signaling effects were first demonstrated in 1955 by plant developmental biologist Ian Sussex, who made surgical cuts around leaf primordia, thereby isolating them from chemical substances produced by the meristem. The isolated primordia retained their initial cylindrical, radially symmetrical structure rather than becoming flat and bilaterally symmetrical—the normal process. Later investigators discovered that a particular set of genes (*PHABULOSA, PHAVOLUTA,* and *REVOLUTA*) expressed in young primordia causes leaf adaxial tissues to develop, and that abaxial development is specified by a different set of genes (*KANADI* and *YABBY*). What causes this differential gene expression? In 2004, plant biologists Catherine Kidner and Robert Martienssen reported that a specific type of microRNA (miRNA; described in Chapter 13) accumulates in the abaxial regions of a developing leaf. In this location, the miRNA interrupts expression of the set of genes that specify adaxial tissue development, helping to explain why leaf adaxial and abaxial tissues differ. Other types of miRNA molecules influence leaf shape, explaining the diversity of leaf shapes produced by different plant species.

Leaf Shape and Surface Features Reflect Adaptation to Environmental Stress

As we have noted, leaf flatness facilitates solar energy collection, and thinness helps leaves to avoid overheating. Leaf shape and surface features also reflect adaptation to stressful environmental conditions.

Leaf Form The flattened portion of a leaf is known as the leaf blade. In beans and most other dicots, blades are attached to the stem by means of a stalk known as a petiole (**Figure 35.12a**). An axillary bud occurs in the leaf axil. Leaf shape can be simple or complex, each having particular advantages. Simple leaves have only one blade, though the edges may be smooth (Figure 35.12a), toothed, or lobed. Simple leaves are advantageous in shady environments because they provide maximal light absorption surface, but they can overheat in sunny environments. As an evolutionary solution to the heating problem, the blades of some leaves have become highly dissected into leaflets. Such leaves are known as complex leaves or compound leaves (**Figure 35.12b**). You can distinguish leaflets from leaves by the fact that leaflets lack axillary buds at their bases. Complex leaves are common in hot environments because leaflets foster heat dissipation. During the development of complex leaves, the transcription factor *KNOX* becomes active shortly after the leaf primordia form, causing these primordia to produce multiple growth points that generate the leaflets. In contrast, during simple leaf development *KNOX* is not active, because a protein known as *PHANTASTICA* suppresses expression of the *KNOX* gene.

Leaf Vein Patterns Leaf vein patterns are known as venation. Eudicot leaves have either pinnate (feather-like) or palmate (spreading like the fingers of your hand) venation, with smaller veins connected in a net or branching pattern (see Figure 35.12a and Table 35.1). In contrast, monocot leaves such as those of corn often have distinctive parallel venation (see **Figure 35.12c** and Table 35.1). However, evolutionary biologists Thomas Givnish, Chris Pires, and their colleagues discovered that netted veins have evolved among monocots at least 26 different times over the past 90 million years, probably as an adaptation to shady conditions. Plants growing in shady conditions typically have broad leaves, which increase the amount of solar collection surface but require structural support, and netted veins provide more support than parallel veins. Species of the genus *Trillium*, which grows on the floors of shady temperate forests, and the vanilla orchid (*Vanilla planifolia*) of dense tropical forests are examples of monocots that have leaves with net venation.

  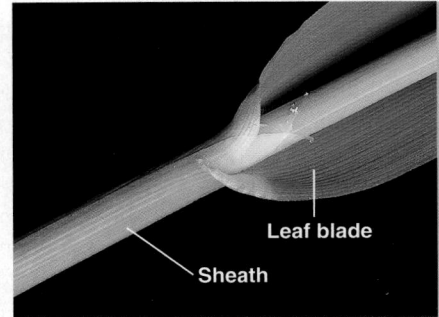

(a) Eudicot stem with simple leaf   **(b) A complex leaf**   **(c) Monocot stem and leaf**

Figure 35.12 **Examples of variation in leaf form.** (a) A simple eudicot leaf, showing blade, petiole, and axillary bud. This leaf has a pinnate venation pattern. (b) A complex (compound) leaf divided into leaflets. (c) Leaf of a monocot, showing how the base of the leaf encircles the stem and parallel veins on the blade.

Biological inquiry: Is the leaf shown in (b) from a eudicot or a monocot plant?

Leaf Surface Features Leaf surfaces also show adaptive features. As we have previously noted, a layer of epidermal tissue occurs at upper and lower leaf surfaces (see Figure 35.6c). These epidermal cells secrete a cuticle composed of protective wax and polyester compounds (**Figure 35.13**). The cuticle helps

Figure labels:
- Cuticular wax
- Trichome
- Closed stomatal pores with guard cells

27 μm

Figure 35.13 Leaf surface features.   These features, viewed by SEM, include cuticular wax, stomatal pores with guard cells, and trichomes.

plants to avoid drying in the same way that enclosure in waxed paper keeps food moist. Plants that grow in very arid climates often have thick cuticles, whereas plants native to moist habitats typically have thinner cuticles. Cuticles also filter damaging UV radiation, reduce attack by microbes and animals, and foster a self-cleaning process by which water droplets and debris wash from the leaf surface.

Some leaf epidermal cells are specialized to form guard cells of stomatal pores (Figure 35.13). These **guard cells** allow pores to close when conditions are too dry, and open under moist conditions, allowing the entry of CO_2 needed for photosynthesis. Guard cell development begins with an unequal cell division, a common early event in plant cell differentiation. The smaller epidermal cell secretes a protein that inhibits division by adjacent cells but does not affect cells farther away. This process distributes stomata evenly and prevents too many of them from forming, which could increase the loss of water from plant surfaces.

Other leaf epidermal cells differentiate into spiky or hairlike projections known as **trichomes** (Figure 35.13). Blankets of trichomes offer protection from excessive light, ultraviolet radiation, extreme air temperature, or attack by herbivores—animals that consume plant tissues. Broken trichomes of the stinging nettle, for example, release a caustic substance that irritates animals' skin, causing them to avoid these plants. Producing specialized cells such as trichomes can be costly to plants, but it may be worth the cost if such cells aid plant survival.

FEATURE INVESTIGATION

Elle and Associates Investigated the Cost to *Datura wrightii* of Producing Sticky Leaf Trichomes

The jimsonweed plant (*Datura wrightii*) grows in arid regions of the southwestern U.S. and Mexico. This plant produces leaves whose epidermal surfaces feature hair-shaped trichomes. Some *D. wrightii* plants produce leaf trichomes that secrete sticky sugar solutions at their surfaces and are thus known as glandular trichomes; such plants are described as "sticky" plants. In contrast, other *D. wrightii* plants produce similarly shaped, nonglandular leaf trichomes that do not secrete sugars. The leaves of these plants are therefore smooth rather than sticky, and they are described as "velvety." In *D. wrightii*, trichome phenotype is controlled by variation in a single gene, and sticky is dominant over velvety. If the phenotypes were of equal advantage in nature, sticky and velvety plants should occur in a 1:1 proportion. However, the actual frequency of sticky plants in nature ranges from 0 to 93%, depending on the environment. This variation and other data suggest that sticky trichomes deter certain herbivores, thereby providing a growth advantage. However, sticky trichomes may be excessively costly in terms of resources when those herbivores are not present, because sugar secretion uses up photosynthetic products.

Evolutionary ecologists Elizabeth Elle, Nicole van Dam, and Daniel Hare devised an experiment to determine the relative cost of producing sticky trichomes, which is shown in the Feature Investigation in **Figure 35.14**. The team hypothesized that sticky plants might produce fewer seeds than velvety plants, because plant photosynthetic products are diverted from reproduction. The investigators grew sticky and velvety *D. wrightii* plants together, protecting some from insect herbivores but not protecting others. Then the investigators compared the numbers of seeds capable of germinating that were produced by sticky and velvety plants. (In this experiment, seed production was a measure of Darwinian fitness.) The team found that velvety plants sustained more insect damage than sticky plants, verifying the protective function of glandular trichomes. When insect herbivores were present, there was no significant fitness difference between sticky and velvety plants. However, when herbivores were absent, though both types of plants flowered similarly, the sticky plants produced 45% fewer viable seeds than did the velvety plants. These data supported the initial hypothesis that sticky trichomes were more costly than smooth trichomes. This and other studies suggest that the investment required to produce sticky trichomes may pay off by reducing the destructive effects of insects.

Figure 35.14 Elle and associates investigated the cost to *Datura wrightii* of producing sticky leaf trichomes.

HYPOTHESIS Although sticky *Datura wrightii* plants are better protected from certain herbivores, they may produce fewer viable seeds than velvety plants.

STARTING MATERIALS Seeds from natural populations of *Datura wrightii* that produce sticky or velvety leaves.

	Experimental level	Conceptual level

1 Grow sticky and velvety plants from seeds in a greenhouse, protecting all seedlings from herbivores.

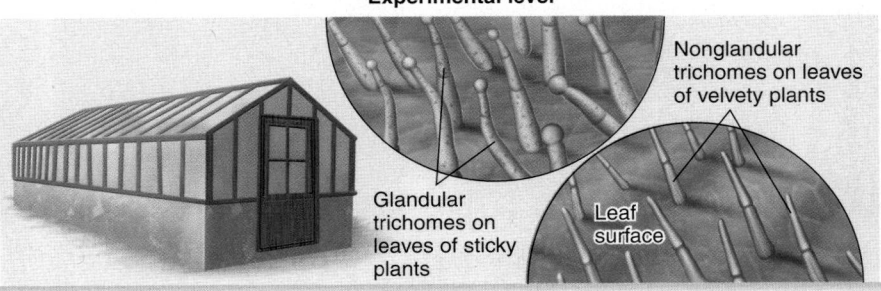

Nonglandular trichomes on leaves of velvety plants

Glandular trichomes on leaves of sticky plants

Leaf surface

Growing plants from seed in a protected environment helps to produce a fairly uniform population of experimental plants, and protects seedlings during the most vulnerable stage of plant growth.

2 Transplant 3-week-old plants to an outdoor experimental garden. Treat some adult plants of each type (sticky and velvety) with insecticide to deter herbivores; leave other plants exposed.

Treated with insecticide

Not treated with insecticide

Sticky plants   Velvety plants   Sticky plants   Velvety plants

Moving mature plants outdoors exposes them to insect herbivores. Treating half with herbicide allows comparison of trichome effects on the level of insect damage.

3 Monitor plant growth and flower, fruit, and seed production.

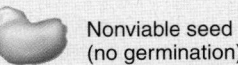

Plant growth   Flower production   Fruit production   Seed production

Monitoring growth and reproduction allows observation of different impacts on plants of trichome production.

4 Compare numbers of seeds capable of germinating (viable seeds).

Viable seed (germination)

Nonviable seed (no germination)

Seed viability is a measure of Darwinian fitness.

5 **THE DATA**

(a) American vetch
(*Vicia americana*)

(b) Sycamore

(c) Poinsettia

(d) Giant saguaro cactus

Figure 35.15 **Examples of modified leaves.** **(a)** The tendrils of an American vetch plant are modified leaves that help the plant attach to a trellis. **(b)** Bud scales such as those on this sycamore bud are modified leaves that protect buds from winter damage. **(c)** The attractive red bracts of poinsettia are modified leaves that function like flower petals to attract pollinator insects to the small flowers. **(d)** Cactus spines such as these on this giant saguaro are modified leaves that function in defense.

Biological inquiry: Since cactus leaves are so highly modified for defense that they cannot effectively accomplish photosynthesis, how do cacti obtain organic compounds?

Modified Leaves Perform Diverse Functions

Though most leaves function primarily as photosynthetic organs, some plants produce leaves that are modified in ways that allow them to play other roles. For example, threadlike tendrils that help some plants such as the American vetch attach to a supporting structure are modified leaves or leaflets (**Figure 35.15a**). The tough scales that protect buds on plants such as the sycamore from winter damage are modified leaves (**Figure 35.15b**). Poinsettia "petals" are actually modified leaves known as bracts, which are larger and more brightly colored than the flowers they surround (**Figure 35.15c**). These conspicuous bracts help attract pollinators to the flowers. Cactus spines such as those on the giant saguaro are actually modified leaves that have taken on a defensive role, leaving photosynthesis to the cactus stem (**Figure 35.15d**).

Stems May Contain Primary and Secondary Vascular Systems

Stems, leaves, roots, buds, flowers, and fruits all contain vascular systems composed of xylem and phloem tissues that conduct water, minerals, and organic compounds. **Herbaceous plants** such as corn and bean produce mostly primary vascular tissues. In contrast, **woody plants** produce both primary and secondary vascular tissues. A comparison of primary and secondary vascular tissues will aid in understanding their roles.

Primary Vascular Tissues **Primary vascular tissues** are composed of primary xylem and phloem. Let's begin with primary xylem, a complex tissue containing several cell types (see Table 35.2). These include unspecialized parenchyma cells; stiff fibers

that provide structural support; and two types of cells that have differentiated in ways that facilitate water transport: tracheids and vessel elements. Arranged in pipeline-like arrays, **tracheids** and **vessel elements** conduct water, along with dissolved minerals and certain organic compounds. Mature tracheids and vessel elements are no longer living cells and lack cytoplasm, which facilitates water flow. During development, these cells lose their cytoplasm by the process of controlled cell death (apoptosis). However, the cell walls of tracheids and vessel elements do not break down or collapse because they are impregnated with a tough polymer known as **lignin** (**Figure 35.16**). The rigid cell walls of tracheids and vessel elements not only foster water conduction but also help support the plant body.

In contrast to xylem, living phloem tissues transport organic compounds such as sugars and certain minerals in a watery solution. Phloem tissues include **sieve-tube elements**, thin-walled living cells that are arranged end to end to form pipelines (**Figure 35.17**). Pores in the end walls of sieve-tube elements allow solutions to move from one cell to another. Phloem tissue also includes companion cells that aid sieve-tube element metabolism, parenchyma cells, and supportive fibers (see Table 35.2).

Secondary Vascular Tissues Trees are woody plants that typically have only one major stem, shrubs are woody plants with multiple stems, and woody vines are known as lianas. Woody plants begin life as herbaceous seedlings that possess only primary vascular systems. But as these plants mature, they produce secondary vascular tissues and bark. Secondary vascular tissues are composed of secondary xylem and phloem. **Secondary xylem** is also known as **wood**, while **secondary phloem** is the inner bark. **Outer bark** is protective layers of mostly dead cork cells that cover the outside of woody stems and roots.

Figure 35.16   **Water-conducting cells of the xylem.**   In this thinly sliced portion of a stem, the stained, lignin-impregnated walls of narrow tracheids and wider vessel elements can be distinguished.

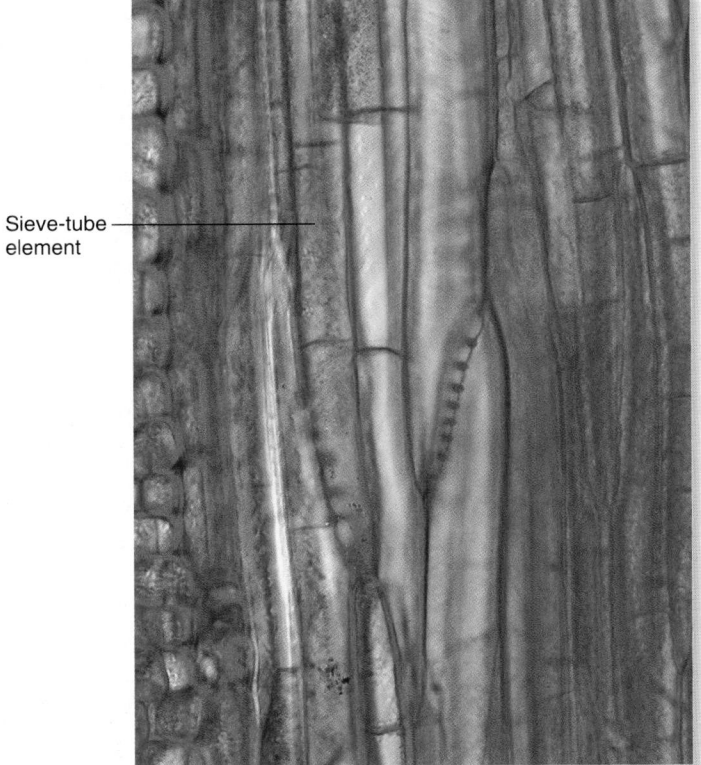

81 μm

Figure 35.17   **Food-conducting cells of the phloem.**   This thinly sliced portion of a stem shows stained, thin-walled sieve elements, which conduct watery solutions of organic compounds such as sugar and certain minerals.

Therefore, bark includes both inner bark (secondary phloem) and outer bark (cork). Woody plants produce secondary vascular tissues, wood and bark, by means of two types of secondary meristem: vascular cambium and cork cambium (**Figure 35.18**).

The meristem known as **vascular cambium** is a ring of dividing cells that produces secondary xylem to its interior and secondary phloem to its exterior. Secondary xylem conducts most of a woody plant's water and minerals. Cell divisions that occur in secondary meristems increase the girth of woody stems. During each new growing season, the vascular cambium produces new cylinders of secondary xylem (wood) and secondary phloem (**Figure 35.19**). Each year's addition of new secondary xylem forms growth rings that can be observed on the cut surfaces of woody stems (**Figure 35.20**). If environmental conditions favor plant growth, the growth rings formed at that time will be wider than those formed during stressful conditions. Climatologists use growth ring widths in samples of old wood to deduce past climatic conditions, and archeologists use growth ring data to determine the age of wood constructions and artifacts left by ancient cultures.

Secondary xylem may transport water for several years, but usually only the current year's production of secondary phloem is active in food transport. This is because thin-walled sieve elements typically live for only a year, after which they are crushed by the addition of new wood. Thus, only a thin layer of phloem, the **inner bark**, conducts most of the sugar transport in a large tree. Deep abrasion of tree bark may damage this thin phloem layer, disrupting a tree's food transport. In fact, if a groove is cut all the way around a tree trunk—a process known as girdling—the tree will die because all of its functional phloem transport routes will have been interrupted.

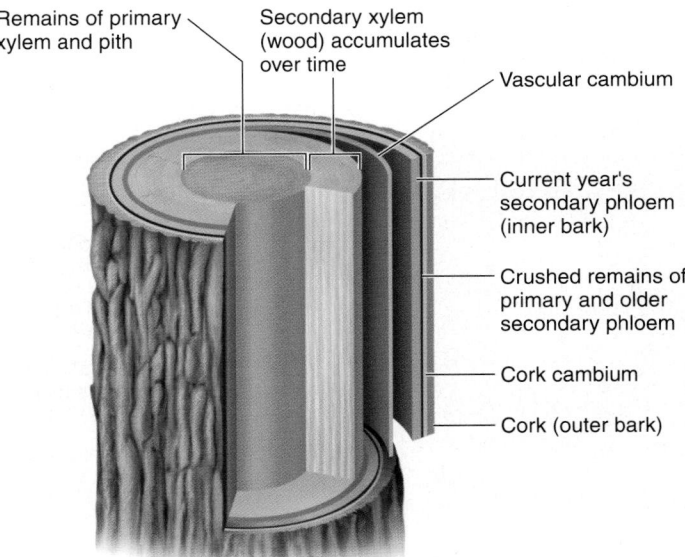

Figure 35.18   **Secondary meristems generate wood and bark.**   The vascular cambium is a thin cylinder of tissue that produces a thick cylinder of wood (secondary xylem) toward the inside of the stem, and a thinner cylinder of inner bark (secondary phloem) toward the outside of the stem. The cork cambium forms an outer coating of protective cork (outer bark).

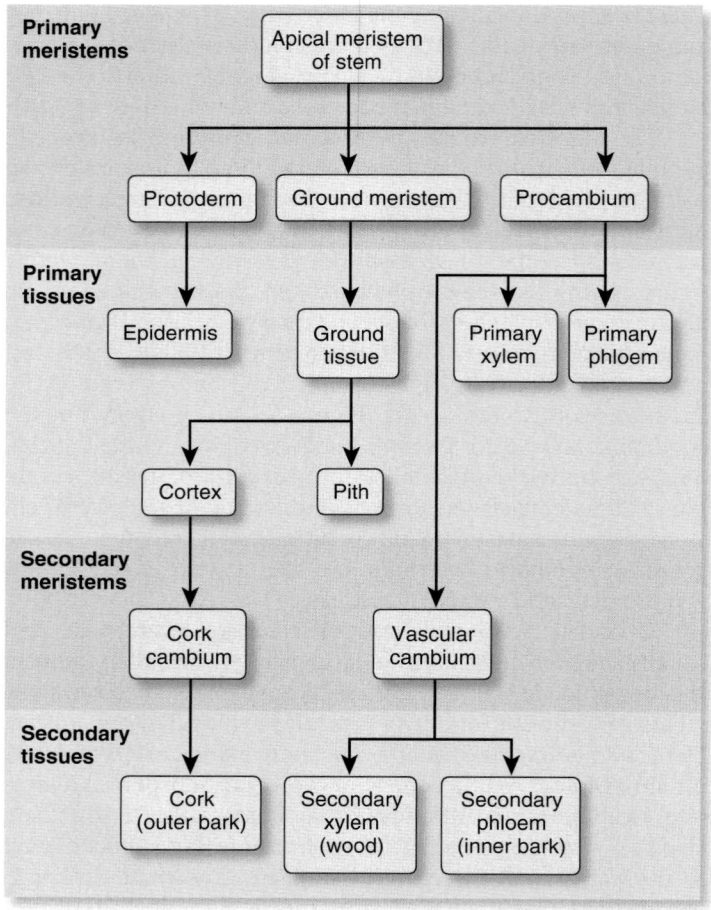

Primary
meristems

Apical meristem
of stem

Protoderm   Ground meristem   Procambium

Primary
tissues

Epidermis   Ground
tissue   Primary
xylem   Primary
phloem

Cortex   Pith

Secondary
meristems

Cork
cambium   Vascular
cambium

Secondary
tissues

Cork
(outer bark)   Secondary
xylem
(wood)   Secondary
phloem
(inner bark)

Figure 35.19   An overview of the process of primary and secondary growth in a woody stem.

Pith   Remains of
primary xylem

Annual ring of
secondary
xylem

Secondary
xylem (wood)

Vascular
cambium

Secondary
phloem (inner
bark)

Cork cambium
and outer bark
(cork)

0.5 mm

Figure 35.20   **The anatomy of a tree trunk.**   Each year a new cylinder of wood is produced; this yearly wood production appears as annual rings on the cut surface of a woody stem.

Biological inquiry: Why do tree trunks have a thicker layer of wood (secondary xylem) than of inner bark (secondary phloem)?

As a young woody stem begins to increase in diameter, its thin epidermis eventually ruptures and is replaced by outer bark, which is composed of protective cork tissues. Cork is produced by a secondary meristem called the **cork cambium**, which is another ring of actively dividing cells. The cork cambium surrounds the secondary phloem (see Figure 35.19). Cork cells are dead when mature, and their walls are layered with tough lignin and suberin, materials that help to prevent both attack by microbial pathogens and water loss from the stem surface. Cork tissues also produce tannins, compounds that protect against pathogens by inactivating their proteins, and protective gums and rubbery latexes. The cracked surfaces of tree trunks are dead cork tissues of the outer bark. Commercial cork is harvested from the cork oak tree (*Quercus suber*) for production of flooring material, bottle stoppers, and other items. Additional information about the structure and function of primary and secondary xylem and phloem can be found in Chapter 38.

Modified Stems Display Diverse Forms and Functions

Stems mostly grow upright because light is required for photosynthesis. But some stems, known as rhizomes, occur underground and grow horizontally. For example, potato tubers are the swollen, food-storing tips of rhizomes. Grass stems also grow horizontally, as either rhizomes just beneath the soil surface or stolons, which grow along the soil surface. The leaves and reproductive shoots of grasses grow upward from the point where they are attached to these horizontal stems. Grass blades continue to elongate from their bases even if you cut their tips off, explaining why lawns must be mowed repeatedly during the growing season. The horizontal stems of grasses are adaptations that help to protect vulnerable shoot apical meristems against natural hazards such as fire and grazing animals.

35.4 Root System Adaptations

Roots play the essential roles of absorbing water and minerals, and anchoring plants in soil. The external form of roots varies among flowering plants, reflecting adaptation to particular life spans or habitats. In contrast, root internal structure is more uniform. In this section, we first consider variation in root external structure and then focus on root internal structure and development.

Modified Roots Display Diverse Forms and Functions

As we have observed, the common bean and other eudicots display an underground taproot system, whereas corn and other monocots have a fibrous root system (see Table 35.1). Plants produce several other types of roots that provide adaptive

(a) *Pterocarpus hayesii* **buttress roots**

Pneumatophores

(b) Pneumatophores

Figure 35.21 Examples of modified aboveground roots. **(a)** Buttress roots help to keep tropical trees such as this *Pterocarpus hayesii* from toppling in windstorms. **(b)** Pneumatophores produced by mangroves are roots that extend upward into the air. These roots take up air and then transmit it to underwater roots that grow in oxygen-poor sediments.

advantages. For example, corn and many other plants produce supportive prop roots from the lower portions of their stems. Many tropical trees grow in such thin soils that the trees are vulnerable to being blown down in windstorms. Such trees often produce dramatic aboveground buttress roots that help keep trees upright (**Figure 35.21a**). Many mangrove trees that grow along tropical coasts produce pneumatophores ("breath bearers"), roots that grow upward into the air (**Figure 35.21b**). Functioning like snorkels, pneumatophores absorb oxygen-rich air, which diffuses to submerged roots growing in oxygen-poor sediments. This is necessary because all roots require a supply of oxygen in order to produce ATP, which is needed to power the uptake of mineral nutrients (see Chapter 37).

The external form of roots may also reflect adaptation to variations in plant life span or habitat. For example, biennial plants—which require two years' growth to produce seed—store carbohydrates in fleshy storage roots during their first year of growth. This stored food is used to help fuel the production of flowers, fruits, and seeds in the biennial plant's second year of life. Carrots, sugar beets, parsnips, and rutabagas are examples of biennial plants with food storage roots that are also useful as human food. *D. wrightii*, whose leaf trichomes were earlier described, also grows each year from a fleshy storage root.

Root Growth and Tissue Specialization Occur in Distinct Zones

In their study of gene expression in *Arabidopsis* roots, plant molecular biologists Kenneth Birnbaum, Philip Benfrey, and associates identified 15 distinct regions of cellular specialization. Such studies reveal that roots are amazingly complex in structure. However, for our purposes, a simpler microscopic examination of root internal structure reveals three major zones: (1) an apical meristem, which produces the root and root cap, (2) a zone of root elongation, and (3) a zone of maturation in which specialized cells can be observed (**Figure 35.22**).

Root Apical Meristem and Root Cap As we have earlier observed, an apical meristem occurs at the tips of roots and their branches. The root apical meristem contains stem cells, and the same types of primary meristems that produce shoot tissues. These are protoderm, which generates epidermal tissues; ground meristem, which produces ground tissue; and procambium, which makes vascular tissues (Figure 35.22a). The root apical meristem also produces a protective **root cap**.

The root cap senses gravity and touch, which helps roots extend downward in the soil and around obstacles such as rocks. Root cap cells live for only a few days and then slough off and are replaced by the new cap cells continuously produced by the root meristem. Root tips are embedded in a blanket of gooey **mucigel** (Figure 35.22), which is composed of water bound by sticky polysaccharides secreted by root-tip epidermal cells. Mucigel lubricates roots, aiding in their passage through the soil. In addition, mucigel helps in water and mineral absorption, prevents root drying, and provides an environment hospitable to beneficial microbes.

Zones of Elongation and Maturation Above the root apical meristem lies the **zone of elongation**, in which cells extend by water uptake, thereby dramatically increasing root length (Figure 35.22b). Above the zone of elongation lies the **zone of maturation**, where most root cell differentiation and tissue specialization occur (Figure 35.22c). Specialized tissues include mature vascular tissues at the root core, epidermal cells, and starch-storing cortex tissue located between the epidermis and vascular tissues. The hormone auxin influences the development of root vascular tissue, as also occurs in shoots.

The zone of maturation can be identified by the presence of numerous microscopic hairs that emerge from the root epidermis. **Root hairs** are specialized epidermal cells that can be as long as 1.3 cm, about the width of your little finger, but are only 10 μm in diameter. Their small diameter allows root hairs to obtain water and minerals from soil pores that are too narrow for even the smallest roots to enter. Root hair plasma membranes are rich in transport proteins that use ATP to selectively absorb materials from the soil (see Chapter 37).

The production of hairs from root epidermal cells is controlled by the activity of the gene *GLABRA-2* (from the Latin *glaber*, meaning "bald") (**Figure 35.23**). Whether this gene is expressed or not depends on the position of an epidermal cell.

Figure 35.22 Three zones of root growth. A longitudinal view of a typical root reveals three major zones: **(a)** a root apical meristem where cell division occurs protected by a root cap, **(b)** a zone of elongation, and **(c)** a zone of maturation, characterized by epidermal root hairs and other specialized cells. Phloem matures sooner than xylem as a way of supplying the energy needs of the growing root tip with organic compounds from the shoot.

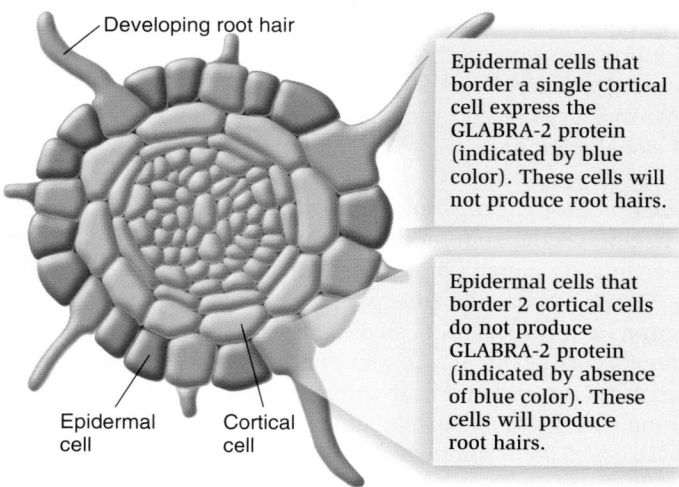

Figure 35.23 Genetic control of root hair development. The *Arabidopsis* gene *GLABRA-2* affects root hair development.

If an epidermal cell contacts two cortical cells, *GLABRA-2* is repressed, causing a root hair to develop. In contrast, no hair will develop if an epidermal cell lies in contact with only one cortical cell. A mutation that disables *GLABRA-2* in every epidermal cell causes all of them to produce root hairs.

Root hairs are so delicate that they are easily damaged by abrasion as roots grow through the soil, and they live for only 4 or 5 days. As a result, root hairs are absent from older regions above the zone of maturation. To compensate, roots must continually produce new root hairs. The average rate of root hair production has been estimated at more than 100 million per day for some plants. It is important to use care when transplanting seedlings to prevent extensive damage to the root hair zone.

The epidermis of mature roots encloses a region of ground parenchyma known as the **root cortex** (**Figure 35.24**). Much like the stem cortex, root cortex cells are often rich in starch and thus serve as a food storage site for plants. The root cortex of some plants contains intercellular air spaces that arise from apoptosis and provide routes for oxygen diffusion within the root. Water and dissolved minerals also diffuse from the environment into roots through spaces between cortex cells, stopping only when they reach a specialized tissue known as **endodermis**, an important component of the mechanism by which roots absorb selected minerals (see Chapter 37).

The primary vascular system in a mature root includes a core of xylem enclosed by strands of phloem. The xylem exports water and minerals upward to shoots, and the phloem imports a watery solution of organic compounds from the shoots. A cylinder of tissue having cell division (meristematic) capacity, known as the **pericycle**, encloses the root vascular tissue (Figure 35.24). The pericycle produces lateral (branch) roots that force their way through the cortex to the surface. In 2006 plant developmental biologists Judith Coates, Laurent Laplaze, and Jim Haseloff reported that the *Arabidopsis* gene *ARABIDILLO* promotes lateral root development. In some roots the pericycle generates a vascular cambium that produces wood—secondary xylem. Such woody roots also possess a cork cambium that makes a protective covering of suberin-coated cork tissue. Like woody stems, woody roots produce primary vascular tissues in their youth, and secondary vascular tissues at maturity.

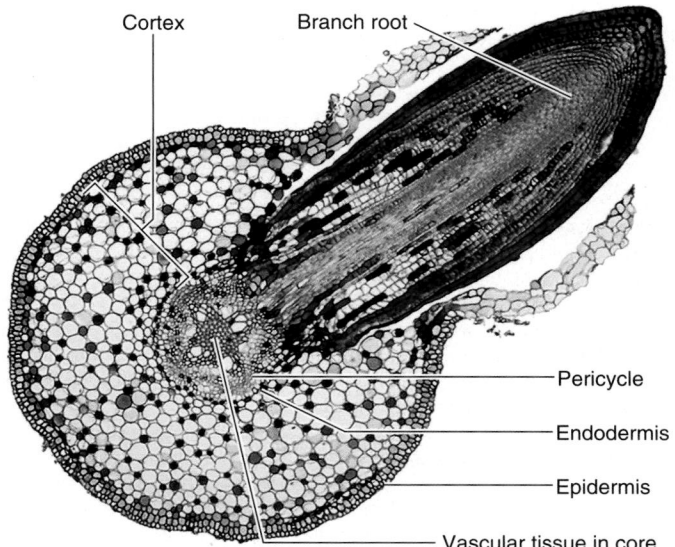

Cortex   Branch root

Pericycle

Endodermis

Epidermis

Vascular tissue in core

Figure 35.24   **Cross-section of a mature root.**   This stained light micrograph shows the epidermis and cortex of a root surrounding a central core of vascular tissue. An inner cortex layer is the endodermis, which surrounds a cylinder of meristematic tissue called pericycle. The pericycle has produced a young branch (lateral) root that has grown through the cortex and the epidermis.

Biological inquiry: Why must lateral roots be produced in this way?

CHAPTER SUMMARY

35.1 From Seed to Seed—The Life of a Flowering Plant

- Seed embryos, seedlings, and mature plants are components of the sporophyte generation in the plant sexual cycle; tiny gametophytes develop and grow within flowers. (Figure 35.1)

- Plant organs are composed of two or more tissues that contain specialized cells. The basic plant organs are roots, stems, and leaves. Shoot systems include stems and stem branches, and stems produce leaves, buds, flowers, and fruits. Root systems include one or more main roots with branches. Buds, flowers, fruits, and seeds are organ systems, composed of more than one organ. (Figures 35.2, 35.3)

- The two major groups of flowering plants, eudicots and monocots, differ in the structure of their roots, stems, leaves, flowers, and seed embryos and pollen. (Table 35.1)

35.2 How Plants Grow and Develop

- The principles of plant growth and development include the presence of a fundamental architecture featuring apical-basal polarity and radial symmetry throughout the life of a plant. (Figure 35.4)

- Plants grow by producing new cells at meristems and controlled cell enlargement involving water uptake.

- Shoot and root apical meristems produce primary meristems that increase plant length and produce organs. The primary meristems are (1) protoderm, which generates dermal tissues; (2) ground meristem, which produces ground tissues; and (3) procambium, which makes the primary vascular tissues. Thus the primary meristems produce primary tissues. Primary vascular tissue occurs in vascular bundles that include xylem and phloem tissues. Leaves develop from primordia produced at shoot tips. (Table 35.2, Figures 35.5, 35.6)

- Meristems include ever-young stem cells, whose numbers influence plant growth and structure and are thus genetically controlled. (Figure 35.7)

- Plant cells are able to expand under conditions that result in loosening of cell-wall components, and by water uptake into vacuoles. The direction in which plant cells expand is determined by the arrangement of wall cellulose microfibrils, which is influenced by the orientation of microtubules in the nearby cytoplasm. (Figures 35.8, 35.9, 35.10)

35.3 The Shoot System: Stem and Leaf Adaptations

- Shoots are modular systems; each module includes a node, internode, leaf, and axillary meristem. An axillary bud develops in leaf axils; such buds may grow into new branches. (Figure 35.11)

- Chemical hormones such as auxin and gibberellic acid influence leaf development, other aspects of plant development, and environmental responses.

- Variations in leaf structure reflect adaptations that aid photosynthesis or protect against stress. For example, Elle and associates demonstrated the value and cost of sticky, sugar-secreting leaf trichomes when destructive insects were present or absent. (Figures 35.12, 35.13, 35.14)

- Leaves function not only in photosynthesis but they also play other roles, including attachment, attraction, and protection. (Figure 35.15)

- Herbaceous plants are those whose stems produce little or no wood, and are mostly composed of tissues. The primary vascular tissues are primary xylem and primary phloem. (Figures 35.16, 35.17)

- In addition to primary tissues, woody plants—trees, shrubs, and lianas—possess secondary meristems that produce wood and bark. The vascular cambium produces secondary xylem (wood) and secondary phloem (inner bark). The cork cambium produces cork tissues that form outer bark. (Figures 35.18, 35.19, 35.20)

- Stems occur in diverse forms that reflect adaptation to environmental conditions. Examples include grass rhizomes, which grow horizontally underground and are thereby better protected from fire and grazing animals (as well as lawnmowers).

35.4 Root System Adaptations

- Roots occur in multiple forms that reflect adaptation to environmental conditions. Examples of aboveground roots include prop roots, buttress roots, and pneumatophores. (Figure 35.21)
- The internal organization of roots is comparatively uniform, and three major zones can be recognized with the use of a microscope: the root apical meristem and root cap, a zone of cell and root elongation, and a zone of tissue maturation. Features of the mature root include epidermal root hairs that aid nutrient uptake, a food-storing cortex, an endodermis that functions in mineral selection, a pericycle that produces lateral (branch) roots (and vascular cambium in the case of woody roots), and an inner core of vascular tissue. (Figures 35.22, 35.23, 35.24)

TEST YOURSELF

1. Where would you look to find the gametophyte generation of a flowering plant?
   a. at the shoot apical meristem
   b. at the root apical meristem
   c. in seeds
   d. in flower parts
   e. none of the above

2. What is a radicle?
   a. an embryonic leaf
   b. an embryonic stem
   c. an embryonic root
   d. a mature root system of a monocot
   e. an organism that has extreme political views

3. What is the correct term for a plant that does not reproduce in its first year of life, but often produces flowers and seeds within the following year?
   a. an annual
   b. a biennial
   c. a perennial
   d. a centennial
   e. none of the above

4. Which of the following terms best describe the distinctive architecture of plants?
   a. radial symmetry and apical-basal polarity
   b. bilateral symmetry and apical-basal polarity
   c. radial symmetry and absence of apical-basal polarity
   d. bilateral symmetry and absence of apical-basal polarity
   e. absence of symmetry and absence of apical-basal polarity

5. Which is the most accurate description of how plants grow?
   a. by the addition of new cells at meristems that include stem cells
   b. by cell enlargement as the result of water uptake
   c. by both the addition of new cells and cell expansion
   d. by addition of fat cells
   e. none of the above

6. Where would you look for leaf primordia?
   a. at a vegetative shoot tip
   b. at the root apical meristem
   c. at the vascular cambium
   d. at the cork cambium
   e. in a floral bud

7. Which leaf tissues display the greatest amount of air space?
   a. the adaxial epidermis
   b. the abaxial epidermis
   c. the palisade parenchyma
   d. the spongy parenchyma
   e. the vascular tissues

8. What are adventitious roots?
   a. roots that develop on plant cuttings that have been placed in water
   b. buttress roots that grow from tree trunks
   c. the only kinds of roots produced by monocots, because their embryonic root dies soon after seed germination
   d. any root that is produced by stem (or sometimes leaf) tissue, rather than developing directly from the embryonic root
   e. all of the above

9. During its development, a tracheid will elongate in a direction parallel to the shoot or root axis. Based on this information, what can you say about the orientation of cellulose cell-wall microfibrils and cytoplasmic microtubules in this developing tracheid?
   a. The microfibrils will be oriented perpendicularly (at right angles) to the long axis of the developing tracheid, encircling it, but the cytoplasmic microtubules will be oriented parallel to the direction in which the tracheid is elongating.
   b. Microfibrils and microtubules will both be oriented perpendicularly (at right angles) to the elongating axis of the tracheid.
   c. Microfibrils and microtubules will both be oriented parallel to the direction of tracheid elongation.
   d. Microfibrils will be oriented parallel to the direction of tracheid elongation, but microtubules will be perpendicular (at right angles) to both the microfibrils and the elongating tracheid.
   e. None of the above.

10. What are examples of woody plants?
    a. trees
    b. shrubs
    c. woody vines, known as lianas
    d. all of the above
    e. none of the above

CONCEPTUAL QUESTIONS

1. What would be the consequences if overall plant architecture were bilaterally symmetrical?

2. What would be the consequences if leaves were radially symmetrical (shaped like spheres or cylinders)?

3. Why are most tall plants also woody, rather than herbaceous?

EXPERIMENTAL QUESTIONS

1. Why did Elle and colleagues begin their investigation by growing *Datura wrightii* plants from seeds in a greenhouse?

2. Why did Elle and colleagues move the mature plants outdoors but treat only half of them with herbicide?

3. Why did the investigators keep track of plant growth as well as plant reproduction during the experiment?

COLLABORATIVE QUESTIONS

1. Find a tree stump or a large limb that has recently been cut from a tree (or imagine doing so). Which of the following features could you locate with the unaided eye: the outer bark, the inner bark, the secondary xylem, the vascular cambium, annual rings?

2. Which physical factors would you expect influence shoot growth most strongly? Which physical factors would you expect most strongly influence underground root growth?

---

www.brookerbiology.com

This website includes answers to the Biological Inquiry questions found in the figure legends and all end-of-chapter questions.

36

FLOWERING PLANTS: BEHAVIOR

CHAPTER OUTLINE

Behavior of the snow buttercup. The snow buttercup (*Ranunculus adoneus*) holds its flowers above the surface of the snow. The flowers move so that they always face the sun during the day, a behavior known as sun tracking.

T he snow buttercup (*Ranunculus adoneus*) grows in deep snowbanks in the high Rocky Mountains, with flower stems protruding above the snow surface toward the sun, as shown in the chapter-opening photo. Amazingly, snow buttercup flowers change their position so that they face the sun throughout the day, a movement process known as sun tracking. Experiments have demonstrated that sun tracking warms snow buttercup flowers, thereby favoring pollen development and germination. Thus, sun tracking is an adaptation that increases snow buttercup reproductive fitness. Sun tracking by leaves of alfalfa, lupine, soybean, common bean, cotton, and some other plants aids in photosynthesis. Sun tracking is but one example of the many ways in which plants display behavior—that is, responses to stimuli. Although sun tracking and some other plant responses are relatively rapid, plants often display behavior as changes in growth and development that occur more slowly.

We begin this chapter by surveying the diverse types of stimuli that induce plant behavior, and consider how plant cells perceive and respond to stimuli by means of signal transduction processes. Next we will learn about two types of internal, mobile molecules that influence plant behavior—plant hormones and RNA. Finally, we consider how plant responses to environmental stimuli foster survival and reproduction.

36.1 Plant Behavioral Responses

As mentioned, **behavior** is defined as response to stimuli. Thus, plants display behavior because they exhibit many types of responses to stimuli. Examples include plant movements, some types of which were described in 1880 by Charles Darwin and his son Francis in their book *The Power of Movement in Plants*.

Modern time-lapse photography reveals that most plants are constantly in motion, bending, twisting, or rotating in dancelike movements known as nutation (**Figure 36.1a**). Some plants display quite rapid movements, illustrated by the sensitive plant (*Mimosa pudica*), whose leaves quickly fold when touched (**Figure 36.1b**). Plants also display behavior when they grow and develop in response to stimuli, though such responses are slower.

Time 0

< 1 sec

(a) Nutation movements   **(b) Leaf folding**

Figure 36.1   Examples of plant movements.   (a) Sixteen superimposed photographs of a shoot of the honeysuckle vine *Lonicera japonica*, taken over a period of 2 hours, reveal the circular movement known as nutation. **(b)** Photographs of the sensitive plant (*Mimosa pudica*) made before and shortly after a touch reveal the rapid process of leaf folding. Even if only one leaflet is touched, electrical signals travel throughout the complex leaf, causing the entire organ to fold. The leaves will eventually unfold.

For example, plant shoots grow toward light and against the pull of gravity, and roots grow toward water and in the same direction as the gravitational force. Seeds germinate when they detect the presence of sufficient light and moisture for successful seedling growth. Producing flowers, fruits, and seeds only at the season most favorable for reproductive success is a behavioral response to environmental change. Plants also sense attack by disease microbes or hungry animals and take protective actions.

The reason that humans are often unaware of plant behavior is that it typically occurs on a longer timescale than our sensory systems are adapted to notice. Despite this fact, plant behavior is important to humans because it influences agricultural productivity and plant roles in nature. To gain a more complete understanding of plant behavior, we will begin by surveying the types of stimuli that cause plant responses.

Plant Behavior Involves Responses to Internal and External Stimuli

Most people are aware that both internal chemical hormones and environmental factors influence animal behavior. Bird nesting behavior in spring, for example, involves hormonal changes triggered by seasonal conditions. Plants likewise respond to both internal and environmental influences. As **Figure 36.2** shows, each of these can be physical or biological in nature.

Internal Stimuli Plants respond to internal electrical impulses and mobile internal chemical signals. Electrical impulses—known as action potentials—can be transmitted along the plasma membrane from one cell to another in plants, as is also the case in animal nervous tissue. Such impulses are particularly important in rapid plant movements such as leaf-folding by the sensitive plant.

Chemical signals that travel from one location in the plant body to another include **hormones** (also known as plant growth regulators) and RNA molecules, including those known as **microRNAs (miRNAs)**. These mobile chemical signals control plant cell, tissue, and organ development and allow plants to respond to environmental stimuli. Like animal hormones, diverse types of plant hormones bind specific cellular receptor molecules that trigger gene expression responses. Only cells that express the appropriate receptors can bind plant hormones and thus perceive the chemical signal. MiRNAs, which are discussed in chapter 13, cause the destruction of mRNA that would otherwise be translated into transcription factors or other proteins. In this way, miRNAs influence gene expression and thus plant behavior.

Environmental Stimuli Plants sense and respond to many types of external physical and biological stimuli (Figure 36.2). Physical stimuli in natural plant environments include light, atmospheric gases such as CO_2 and water vapor, temperature, touch, wind, gravity, water, rocks and other barriers to root growth, and soil minerals. Biological stimuli include herbivores (animals that consume plant parts), pathogens (disease-causing microbes) dispersed in air, organic chemicals emitted from neighboring plants, and beneficial or harmful soil organisms.

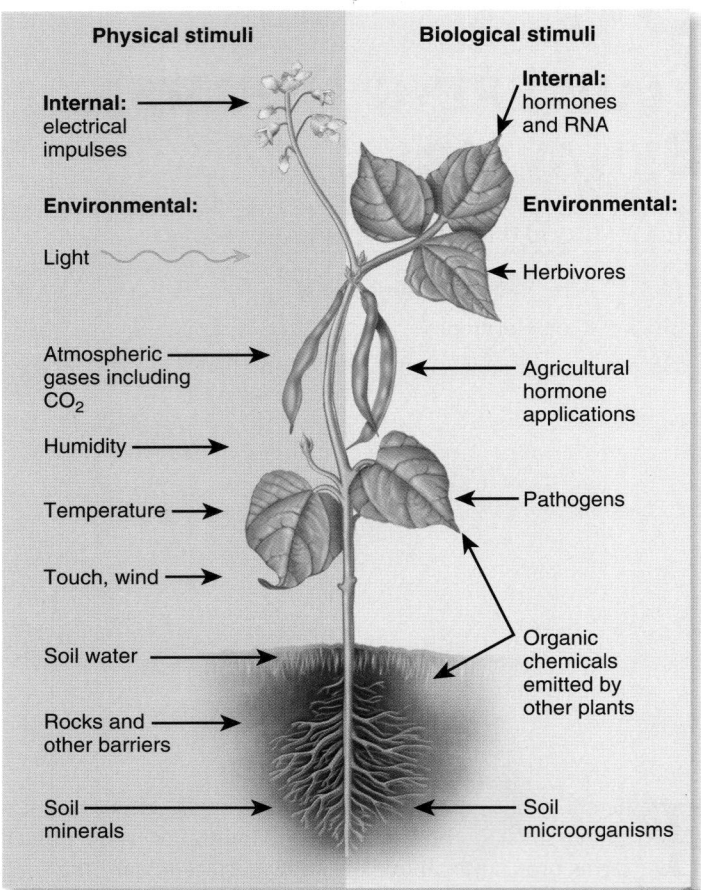

Figure 36.2 **Types of plant stimuli.** Plants respond to both physical and biological stimuli. These stimuli either come from the environment or are internal to the plant.

Crop plants also respond to applications of agricultural chemicals, which may include hormones. That plants have evolved such a broad array of sensory capacity is not surprising, because all of the environmental influences listed here affect plant survival and reproduction.

Plant Responses to Environmental Stimuli Though plants lack the specialized sense organs typical of animals, receptor molecules located in plant cells sense stimuli and cause responses. When many cells of a tissue receive and respond to the same chemical or physical signal, whole organs or plant bodies display behavior. For example, houseplants tend to grow toward a light source such as a window. This process, known as **phototropism**, involves both a cellular perception of light and a growth response of stem tissue to an internal chemical signal. How can an environmental signal such as directional light cause such stem movements? In the case of phototropism, the plant senses the direction of light and responds by changing the location of a plant hormone known as auxin. We will next consider how plant cells receive signals and transmit them intracellularly, a process known as signal transduction.

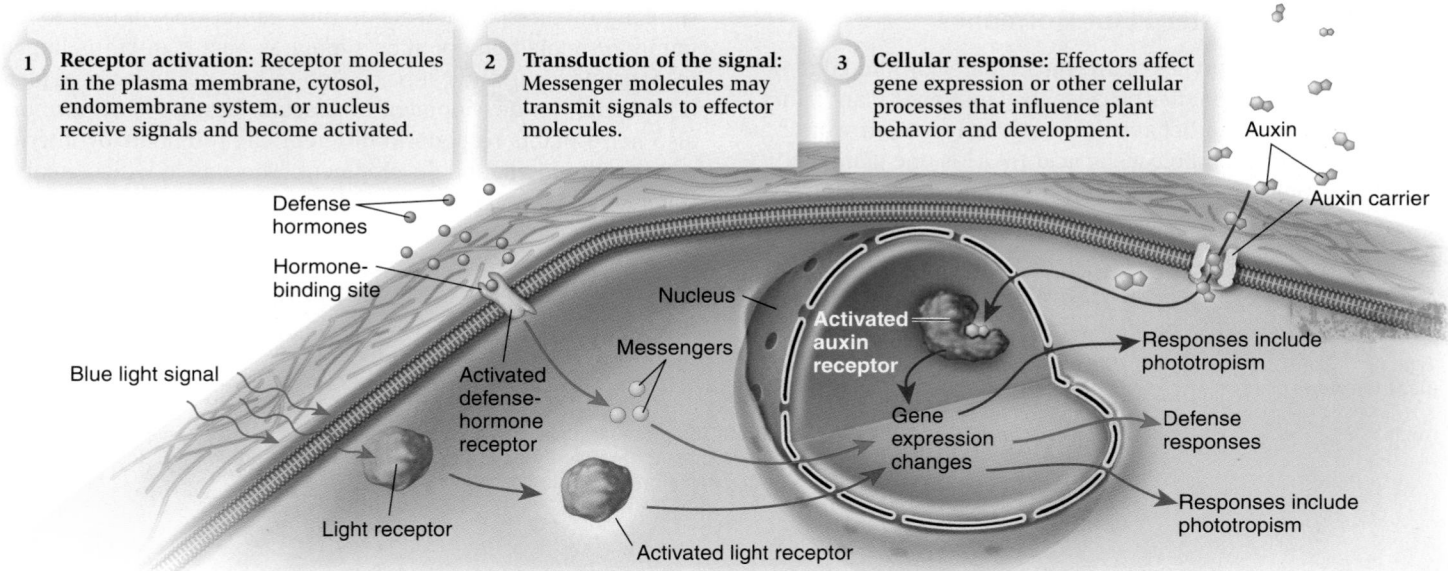

1 **Receptor activation:** Receptor molecules in the plasma membrane, cytosol, endomembrane system, or nucleus receive signals and become activated.

2 **Transduction of the signal:** Messenger molecules may transmit signals to effector molecules.

3 **Cellular response:** Effectors affect gene expression or other cellular processes that influence plant behavior and development.

Figure 36.3   **An overview of plant signal transduction.**   Plant cells respond to hormonal and RNA signals produced within the plant body, as well as to environmental stimuli. Three different signal transduction processes are shown here, those started by blue light, a defense hormone, and auxin.

Plant Signal Transduction Involves Receptors, Messengers, and Effectors

Signal transduction is the process in which a cell perceives a physical or chemical signal, thereby switching on an intercellular pathway that leads to a cellular response (**Figure 36.3**). The process of signal transduction occurs in three stages: receptor activation, transduction of the signal, and a cellular response (see Chapter 9). Signal transduction involves receptor, messenger, and effector molecules.

Receptors (also known as sensors) are proteins that become activated when they receive a specific type of signal (Figure 36.3). Receptors occur in diverse cellular locations. Whereas some plant hormone receptors are located within the plasma membrane, light receptors occur in the cytosol, and auxin binds to specific receptors located in the nucleus (see Chapter 9) and endomembrane system. Some activated receptors directly generate a response, such as increased flow of ions across a membrane. In contrast, many activated receptors first bind to signaling molecules and then initiate an intracellular signaling pathway. For example, the binding of defense hormones to plasma membrane receptors results in the intracellular production of signaling molecules called second messengers.

Second messengers transmit messages from many types of activated receptors to effector molecules. Cyclic AMP, inositol trisphosphate (IP_3), and calcium ions are major types of messengers in animal and plant cells. Calcium ions are particularly common messengers in plant cells. Touch and various other stimuli cause calcium ions to flow from storage sites in the ER lumen into the cytosol. Calmodulin or other calcium-binding proteins then bind the calcium ions. Calcium binding alters the structure of such proteins, causing them to interact with other cell proteins or alter enzymatic function.

Effectors are molecules that directly influence cellular responses. In plants, calcium-dependent protein kinases (CDPKs) are particularly important effector molecules. Signal transduction ends when an effector causes a cellular response such as opening or closing an ion channel or switching the transcription of particular genes on or off. A single activated receptor can dispatch many second messenger molecules, which, in turn, can activate scores of effectors, leading to many molecular responses within a single cell. Now that we have considered how plant cells perceive and respond to stimuli, we are prepared to focus more closely on the roles of plant hormones.

36.2 Plant Hormones

As we have seen, chemical signals transported within the plant body include plant hormones, which bind to receptors, causing cellular responses. Plant hormones include about a dozen small molecules synthesized in metabolic pathways that also make amino acids, sterols, or secondary metabolites. Of these, auxins, cytokinins, gibberellins, ethylene, abscisic acid, and brassinosteroids are considered to be the major plant hormones (**Table 36.1**). Individual plant hormones often have multiple effects, and different concentrations or combinations of hormones can produce distinct growth or developmental responses. For example, individual members of the same plant species that experience different environmental conditions may display considerable variation in structure or behavior, a feature known as **phenotypic plasticity**. A closer look at the major types of plant hormones reveals their multifaceted roles.

Auxins Are the Master Plant Hormones

Plants produce several types of **auxins** that are considered to be the "master" plant hormones because they influence plant structure, development, and behavior in many ways, often working with other hormones. Indoleacetic acid (IAA) is one plant auxin (Table 36.1), but other natural and artificial compounds have similar structures and effects. In this section we will refer to this family of related compounds simply as auxin. Auxin is largely responsible for the apical-basal (top-to-bottom) polarity of plant bodies and the development of vascular tissues. This hormone also helps plants respond to different environments by altering patterns of growth and development. The way in which auxin is transported is integral to these processes.

Table 36.1	The Major Types of Plant Hormones	
Type of plant hormone	**Chemical structure of an example**	**Functions**
Auxins	Indoleacetic acid (IAA)	Establish apical-basal polarity, mediate phototropism, induce vascular tissue development, promote formation of adventitious roots, inhibit leaf and fruit drop, and stimulate fruit development
Cytokinins	Zeatin	Promote cell division, influence cell specialization and plant aging, activate secondary meristem development, promote adventitious root growth, and promote shoot development on callus
Gibberellins	Gibberellic acid	Stimulate cell division and cell enlargement, promote seed germination, stimulate stem elongation and flowering, and retard leaf and fruit aging
Ethylene	Ethylene $H_2C = CH_2$	Promotes seed germination, induces leaf and flower aging and fruit ripening, promotes leaf and fruit drop, functions in plant responses to flooding, and coordinates plant development and stress responses
Abscisic acid	Abscisic acid	Slows or stops metabolism in stress responses, prevents seed germination in unfavorable conditions, induces bud and seed dormancy, and influences stomatal opening and closing
Brassinosteroids	Brassinolide	Promote cell elongation, stimulate shoot elongation, stimulate xylem development, retard leaf drop, and promote stress responses

Auxin Transport Auxin is produced in apical shoot tips and young leaves, and it is directionally transported from one living parenchyma cell to another. Auxin in an uncharged form (IAAH) may enter cells by means of diffusion while the negatively charged form (IAA⁻) requires the aid of AUX1, a plasma membrane protein known as the **auxin influx carrier**. Several types of PIN proteins transport auxin out of cells; they are named for the pin-shaped shoot apices of plants having mutations in *PIN* genes. PIN proteins are **auxin efflux carriers**. In shoots, AUX1 is located at the apical ends of cells, whereas PIN proteins often occur at the basal ends (**Figure 36.4a**). This polar distribution of auxin carriers explains why auxin primarily flows downward in shoots and into roots, a process called **polar transport** (**Figure 36.4b**). However, the locations of auxin carriers can also change within cell plasma membranes, allowing lateral or upward transport of auxin. Differences in the presence and positions of auxin carrier proteins explain variations in auxin concentration within plants. By measuring the local auxin concentration, plant cells determine their position within the plant body and respond by dividing, expanding, or specializing.

Auxin Effects Many of auxin's effects are of practical importance to humans. For example, embryos within developing seeds produce auxin that stimulates flower ovaries to mature into fruits. This explains why some types of seedless fruits are produced with the aid of auxin applications. Auxin also retards premature fruit drop, explaining why apple or pear growers may spray trees with auxin. This hormone also fosters development of adventitious roots—those developing from stems. Houseplant stem cuttings will more readily develop roots if you dip the stem ends into commercial rooting compounds, which contain auxin. Gardeners know that if they remove the topmost portion of a plant shoot, nearby lateral buds will begin to grow and produce new branches, allowing plants to become bushier. Such decapitation works by disrupting the flow of auxin from the shoot apex. Recent research suggests that when less auxin than usual is present near lateral buds, a lipid molecule produced by roots and transported to buds stimulates their growth. Auxin produced by intact shoot tips inhibits lateral bud growth, a process known as apical dominance. Some experts speculate that such auxin works by inhibiting the newly discovered lipid. Although this example reveals that there is still much to learn about auxin function, auxin's role in phototropism has been elucidated by a series of elegant experiments.

The Role of Auxin Phototropism In the 1880s, Charles Darwin and his son Francis were the first to publish results of experiments on plant phototropism. In a simple but elegant experiment, they covered either the tips or lower shoot portions of some grass seedlings with shading materials such as blackened

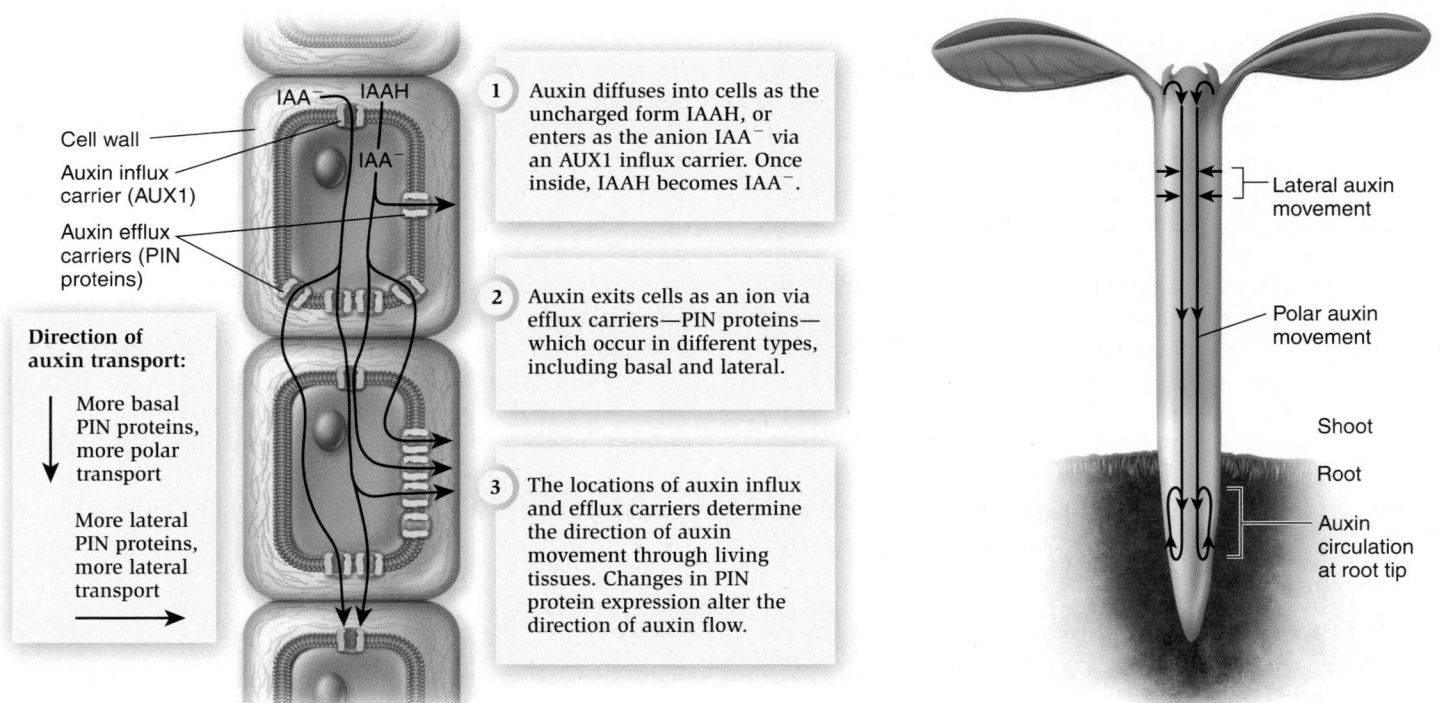

(a) Cellular mechanism of auxin transport

1 Auxin diffuses into cells as the uncharged form IAAH, or enters as the anion IAA⁻ via an AUX1 influx carrier. Once inside, IAAH becomes IAA⁻.

2 Auxin exits cells as an ion via efflux carriers—PIN proteins—which occur in different types, including basal and lateral.

3 The locations of auxin influx and efflux carriers determine the direction of auxin movement through living tissues. Changes in PIN protein expression alter the direction of auxin flow.

Cell wall
Auxin influx carrier (AUX1)
Auxin efflux carriers (PIN proteins)

Direction of auxin transport:

More basal PIN proteins, more polar transport

More lateral PIN proteins, more lateral transport

(b) Auxin transport throughout a plant

Lateral auxin movement
Polar auxin movement
Shoot
Root
Auxin circulation at root tip

Figure 36.4 **Auxin transport.** (a) Polar and lateral auxin transport is controlled by the distribution of auxin influx and efflux carriers located in the plasma membrane. When efflux carriers primarily occur at the basal ends of cells, auxin will flow downward. Auxin may flow laterally when auxin efflux carriers occur at the sides of cells. (b) In a whole plant, auxin primarily flows downward from shoot tips to root tips, where it then flows upward for a short distance.

Biological inquiry: How could auxin carriers be organized to allow auxin to move upward in roots?

glass tubes, left other seedlings uncovered, and removed the tips of some seedlings. They then compared how those seedlings responded to illumination from the side. The shoots of seedlings whose tips were left uncovered grew toward the light, whereas shoots of seedlings whose tips were covered or removed did not. The Darwins concluded that seedling shoot tips transmit some "influence" to lower shoot portions, causing them to bend toward the light. You can probably guess what this influence was, but technology available at the time did not allow the Darwins to determine this.

Three decades later, in the 1910s, Danish botanist Peter Boysen-Jensen confirmed the Darwins' results and demonstrated that the "influence" was a chemical substance that diffused from the tips of the seedlings to other parts of the shoots. To do this, Boysen-Jensen cut off the tips of oat seedlings and placed either a porous layer of gelatin or a nonporous material such as a sheet of the mineral mica on the cut shoot surface. Then he replaced the shoot tips. Shoots of the oat seedlings layered with porous gelatin displayed a normal phototropic response, bending toward the light, but those layered with nonporous mica did not. Boysen-Jensen's experiment demonstrated that the phototropic substance was a diffusible chemical, but exactly which one, and how it worked, remained unknown. A series of additional experiments provided some answers.

FEATURE INVESTIGATION

Experiments Performed by Went and Briggs Revealed the Role of Auxin in Phototropism

In the 1920s, the Dutch plant physiologist Frits Went named the substance discovered by Boysen-Jensen auxin (from the Greek word *auxein*, meaning "to increase") and performed experiments that helped explain how auxin works, as shown in the Feature Investigation in **Figure 36.5a**. In a first step, Went cut the shoot tips off of oat seedlings and placed these tips onto agar blocks. Agar, a complex polysaccharide derived from red algae, forms a mesh capable of holding considerable water and dissolved compounds. Agar's permeability to auxin is similar to that of the protein gelatin used by Boysen-Jensen, but agar is much more stable to heat and microbial breakdown and thus is easier to use in laboratory experiments. In Went's experiment, the auxin diffused from cut shoot tips into these agar blocks. In the next steps, he treated decapitated shoots in one of four ways: (A) placed auxin-laden agar blocks off-center on some, (B) placed auxin-laden blocks evenly onto others, (C) placed plain agar blocks off-center on some, and (D) left some uncapped. All shoots were then kept in darkness throughout the experiment. Only shoots capped to one side with agar containing the material later identified as auxin grew in the direction away from the agar block. This experiment demonstrated that auxin application

could substitute for the directional light stimulus, and it suggested that asymmetric auxin distribution is the mechanism by which light causes shoots to bend.

Went and N. O. Cholodny independently proposed that light causes auxin to move to the unlit side of shoot tips, causing cells on that side to elongate more, which results in shoot bending. But other scientists argued that bending could result if auxin on the lit side of a shoot were destroyed by light, which would result in greater elongation of cells on the unlit side. In the 1950s, American plant biologist Winslow Briggs designed two elegant experiments to test these alternate hypotheses.

In a first experiment, to test the hypothesis that auxin might be destroyed by light, Briggs first grew corn seedlings in the dark (**Figure 36.5b**). Then he cut off their shoot tips, put the tips on agar blocks, and exposed some to light on one side, leaving other tips in the dark. During this process, auxin from tips diffused evenly into the agar blocks. If auxin were destroyed by light, agar blocks under lighted tips should receive less auxin than blocks under tips kept in the dark. The auxin-destruction hypothesis also predicts that when agar blocks from lighted tips are placed on one side of decapitated shoots, they should cause less bending than would blocks from tips kept in the dark. However, Briggs discovered that both types of agar blocks caused the same amount of shoot bending.

Figure 36.5 Went and Briggs demonstrated the relationship between light perception and auxin function. (a) The Went experiment, (b) Briggs experiment 1, and (c) Briggs experiment 2.

(a) Went experiment

HYPOTHESIS A chemical substance produced by shoot tips causes the shoots to bend.

STARTING MATERIALS Oat seedlings.

Experimental level   Conceptual level

1 Collect diffusible substance from cut seedling tips into agar blocks. Note: Once auxin enters the agar block, it can diffuse throughout the whole block.

Scalpel
Seedling
Seedling tip
Agar block

Diffusible substance in tip of seedling
Diffusible substance in agar block

2 Cut more shoot tips from seedlings. Discard tips.

Discard tips

To prepare shoots for experimentation.

3 A Cap some shoots off-center with agar blocks from step 1.
    B Cap some shoots evenly with agar blocks from step 1.
    C Cap some shoots off-center with plain agar blocks (control).
    D Leave some shoots uncapped (control).

A   B   C   D

Diffusion from agar block into shoot

4 Place all shoots in darkness.

A   B   C   D

Seedling growth will be affected by caps, not by light.

5 **THE DATA**

Degree of bending among treated shoots:

| Bending | No bending | No bending | No bending |

6 **CONCLUSION**

Uneven distribution of diffusible tip substance, named auxin, causes bending. Auxin can substitute for a directional light stimulus.

Went's new hypothesis:
Light causes auxin to move to unlit side of shoots, causing that side to elongate more.

(b) Briggs experiment 1

HYPOTHESIS Light destroys auxin on lit side of shoot tips, causing unequal auxin distribution. Unlit side should grow more than lit side.

STARTING MATERIALS Corn seedlings.

Experimental level   **Conceptual level**

1 Collect auxin into agar blocks from:
    A dark-grown tips
    B tips grown with directional light

Dark-grown tip   Auxin diffusion   Directional light-grown tip

A   B

Light

If light destroys auxin on one side, less auxin will enter the block.

2 Place agar blocks on right side of decapitated shoots.

Agar block   Dark-grown   Shoot   Light-grown

A   B

If the block on the right side has less auxin, it will cause less bending.

3 **THE DATA**

Dark-grown 26°   Light-grown 26°

A   B

4 **CONCLUSION**

Similar bending demonstrates that light did not destroy auxin in the directionally lit shoot tip. If it had, less auxin would have been present in the agar block in B, and the degree of bending would have been less. The hypothesis described under conceptual level (above) is incorrect.

(c) Briggs experiment 2

HYPOTHESIS Directional light causes auxin to move to the shaded shoot tip side.

STARTING MATERIALS Corn seedlings.

	Experimental level	Conceptual level

1 Place shoot tips on agar blocks.

Shoot tip/block combinations will be used together with decapitated shoots as a way to compare auxin concentration.

2 Divide some tip/block combinations completely with a mica sheet, which prevents diffusion between the 2 halves of the tip and agar block. Divide some tip/block combinations only partially with a mica sheet. This allows auxin diffusion across the tip, but not across the agar block. Expose both to directional light.

Mica sheet A   Mica sheet B

If directional light causes auxins to move to shaded side, agar block in B will contain more auxin on right side.

3 Remove agar block halves from tips. Place agar halves onto right sides of shoots, which have their tips removed.

If directional light causes auxins to move laterally, the block half beneath the left side of the partially divided tip shown in B should cause the least shoot bending, whereas the block half beneath the right side of B should cause the greatest amount of bending.

4 **THE DATA**

11°   11°   8°   15°

5 **CONCLUSION**

In A, the mica sheets prevented auxin from moving to the shaded side of tips. In B, auxin was able to move in response to directional light. Agar block pieces from the shaded side of tips contained more auxin and thus caused greater shoot bending.

This result is consistent with the hypothesis that light does not destroy auxin.

In a second experiment, Briggs tested the Cholodny-Went hypothesis that light causes auxin to move to the shaded side of shoots (**Figure 36.5c**). Briggs set shoot tips onto agar blocks (step 1) and then used a mica sheet (which is impervious to auxin) to completely divide some tips and blocks into halves (step 2A). In other cases he divided blocks completely but left tips incompletely divided, allowing auxin to diffuse across tips but not the block halves (step 2B). Then Briggs exposed all sets of tips and blocks to directional light. He predicted that auxin would not be able to move across tips having complete mica barricades, but that auxin would be able to move across tips that had been only partially divided. Auxin diffused from tips into blocks, but it could not diffuse evenly across blocks divided by mica sheets. When Briggs later placed the agar block halves on decapitated shoots (step 3), those receiving auxin from completely divided tips were bent by the same amount. By contrast,

agar block halves from the unlit side of partially divided tips induced more bending, while halves from the lit sides of partially divided tips caused the least bending (see the data in Figure 36.5c). These experimental results support the hypothesis that unidirectional light causes auxin to accumulate on the shaded side of a shoot. Modern plant scientists would explain such auxin movement as the result of lateral transport involving PIN proteins.

How might auxin accumulation cause bending? One widely held hypothesis is that auxin accumulation on the shaded side of a shoot causes plasma membrane proton pumps located there to work at a faster rate. As protons are pumped into the cell wall, expansins and cell-wall-loosening enzymes increase their activities, allowing cells to extend (refer back to Figure 35.9). This process might explain how auxin accumulation in cells located on the shaded side of shoot tips causes them to extend more than do cells on the sunny side, causing the tip to bend toward the light.

GENOMES & PROTEOMES

Genomic Analyses Support the Cholodny-Went Hypothesis

The Cholodny-Went hypothesis predicts that auxin movement should result in corresponding gene expression changes. More specifically, the transcription of genes related to phototropism should differ in cells taken from the lighted and shaded sides of a shoot. In 2006, C. Alex Esmon, Emmanuel Liscum, and their coworkers reported the use of genomic techniques to identify genes from *Brassica oleracea* that were expressed differently as the result of phototropism. These investigators worked with *B. oleracea* rather than the genetic model plant *Arabidopsis thaliana* because *B. oleracea* shoots are larger. This choice allowed the scientists to easily cut tissue from lit and shaded portions of shoots. *B. oleracea* has the additional advantage of being closely related to *A. thaliana*. Such a close relationship allowed the investigators to use commercially available *A. thaliana* genomic microarrays (gene chips; see Chapter 20) to study differential gene expression in *B. oleracea*. Microarrays are thin glass plates onto which are deposited dots that each contain a different portion of the DNA making up a genome. These features allowed the researchers to compare the gene expression patterns of shaded and lit shoot tissues.

Esmon and coworkers extracted mRNA from both shaded and lit shoot tissues and used it to produce nucleic acid labeled with a fluorescent marker. They applied this nucleic acid to separate gene chips, where it bound complementary DNA, causing particular regions of the gene chip to light up. By this process, the investigators discovered that at least eight genes were expressed at their highest level in shaded, elongating tissues. The investigators inferred that these eight genes were involved in phototropism, and they determined that some encode enzymes involved in cell-wall extension. This genomic study provides strong molecular support for the Cholodny-Went hypothesis.

Cytokinins Stimulate Cell Division

Like auxins, the plant hormones known as **cytokinins** play varied and important roles in plants. The name of these hormones reflects their major effect—an increase in the rate of plant cytokinesis, or cell division. Cytokinins also influence cell differentiation and senescence (plant aging). Root tips are the major sites of cytokinin production, but shoots and seeds also make this plant hormone. Cytokinins are transported in the xylem to meristems, seeds, leaves, and fruit, where they induce cell cycle progression to mitosis. Cytokinins also work together with auxin to activate secondary meristems and foster adventitious root growth.

Plant Tissue Culture In the laboratory, cytokinin and auxin are essential to cloning plants. This involves a process, known as **plant tissue culture**, that is used commercially to produce thousands of identical plants having the same desirable charac-

teristics. Plant tissue culture begins with pieces of stem, leaf, or root that have been removed from a plant, and their surfaces are sterilized to prevent growth of microbes (**Figure 36.6**, step 1). The cleaned plant pieces are then placed into dishes containing nutrients (minerals, vitamins, and sugar) and various proportions of auxin and cytokinin. If the proportions of auxin and cytokinin are about the same (1:1), plant cells undergo division, forming a mass of white tissue known as a callus (step 2). If the callus is then transferred to a new dish containing the same nutrients, but auxin-to-cytokinin proportions greater than 10:1, the callus will form roots (step 3). Auxin-to-cytokinin proportions of less than 10:1 cause the callus to develop green shoots (step 4). Thus, by altering the ratios of auxin and cytokinin, entire plants can be regenerated from a callus. A single callus can be divided into many pieces and each piece treated with these hormones, thereby producing many hundreds of identical new plants.

The Discovery of Cytokinin Plant tissue culture would not be possible without the understanding of cytokinin biochemistry that started with the pioneering work of plant physiologists Carlos Miller and Folke Skoog in the 1950s. These investigators discovered that the hormones now known as cytokinins are derivatives of adenine, a compound also found in ATP, DNA, and RNA. They made this major discovery during a laborious search for compounds that would foster growth and division of plant cells cultivated in the lab. Miller found that extracts from a particular yeast sample and a particular batch of herring sperm both stimulated plant cultures. Furthermore, the active substance in the yeast and sperm extracts could be precipitated by silver nitrate under lowered pH conditions, suggesting the presence of nucleotides. Suspecting that the active substance was similar to adenine, Miller experimented with autoclaved DNA, purifying from it white crystals that had the expected properties and activity. With the aid of biochemists, Miller and Skoog determined the chemical structure of these crystals, naming the substance kinetin. Kinetin was the first cytokinin to be discovered. Later, Miller and Skoog purified the cytokinin known as zeatin (see Table 36.1) from corn (*Zea mays*). These investigators also demonstrated the effects of different auxin-to-cytokinin ratios on the development of shoots and roots from a cultured plant callus (see Figure 36.6). Their discoveries paved the way for many basic discoveries about the roles of cytokinin in plant function and the widespread use of tissue culture to clone plants in the horticulture industry.

Gibberellins Stimulate Cell Division and Elongation

Another family of plant hormones, the **gibberellins** (also known as gibberellic acids or GA; see Table 36.1), also act as plant hormones. Gibberellins are produced in apical buds, roots, young leaves, and seed embryos. Gibberellins stimulate both cell division and cell elongation, thereby fostering growth. They accomplish this by promoting the destruction of certain nuclear proteins—known as DELLA proteins—that function like brakes to restrain cell division and expansion. Gibberellins promote stem elongation and flowering, and they retard leaf and fruit aging.

1. A block of tissue is removed from a plant and the surfaces are sterilized.

2. Tissue is cultivated in dishes on nutrient media. Treatment with equal proportions of auxin and cytokinin causes formation of an undifferentiated callus.

3. Treatment with auxin-to-cytokinin ratios greater than 10:1 causes root development on many replicate plantlets.

4. Treatment with auxin-to-cytokinin ratios less than 10:1 induces shoot development on many replicate plantlets.

Plant tissue

Callus

1 : 1
auxin : cytokinin

>10 : 1
auxin : cytokinin

<10 : 1
auxin : cytokinin

Figure 36.6   **The process of plant tissue culture.**   Plant tissue culture illustrates the impact of different proportions of auxin and cytokinin on plant organ development.

Biological inquiry: How do commercial growers use this process to produce many identical plants?

They also play roles in seed germination, a topic covered in Chapter 39.

The plant pathologist Eiichi Kurosawa discovered gibberellins when he observed that excessively tall, spindly rice plants were infected with a fungus called *Gibberella*. The fungus produced a substance, later identified chemically and named gibberellin, that caused the rice seedlings to grow taller than normal. Later, it was discovered that plants also produce gibberellins. More than a hundred different forms of gibberellin have been found. Many kinds of dwarf plants are short because they produce less gibberellin than taller varieties of the same species. The dwarf strain of pea plants Mendel used in some of his breeding experiments is an example. When dwarf varieties of plants are experimentally sprayed with gibberellin, their stems grow to normal heights. However, dwarf wheat and rice crops are valued in agriculture because they can be more productive and less vulnerable to storm damage than taller varieties.

Ethylene Influences Cell Expansion

The plant hormone **ethylene** is particularly important in coordinating plant developmental and stress responses. Ethylene is a simple hydrocarbon gas produced during seedling growth, flower development, and fruit ripening. This hormone also plays important roles in leaf and petal aging and drop, and defense against osmotic stress and pathogen attack. As a gas, ethylene is able to diffuse through the plasma membrane and cytosol to ethylene receptors localized in the endoplasmic reticulum. When activated by ethylene binding, these receptors inactivate a protein kinase known as CTR1. This action ultimately enables transcription factors to induce the transcription of various genes.

People first noticed the effects of ethylene gas on plants in the 1800s when they observed that street-side trees exposed to leaking street lanterns unexpectedly lost their leaves. A 17-year-old student in St. Petersburg, Dimitry Neljubov, performed the first experiments to explore the effects of illumination gas on plants. He exposed pea seedlings grown in the laboratory to illumination gas and noticed that the pea seedlings grew sideways rather than upward. So he tested the individual chemical components of illumination gas for the same effect. After conducting many experiments, in 1901 Neljubov reported that ethylene was the only component of illumination gas that caused the seedlings to grow horizontally, and that ethylene was effective in very low concentrations (as low as 0.06 parts per million in air). Later, scientists established that ethylene influences cell expansion, often in association with auxin. Ethylene does this by increasing the disorder of microtubules within cells, thereby causing random orientation of cell-wall microfibrils. As a result, cells exposed to ethylene tend to expand in all directions rather than elongating.

Ethylene's effects on cell expansion explain this hormone's important role in seedling germination. The tender apical meristems of seedlings could be easily damaged during their growth through crusty soil. Ethylene helps seedlings avoid such damage by inducing what is known as the "triple response" (**Figure 36.7**). First, ethylene prevents the seedling stem and root from elongating. Second, the hormone induces the stem and root to swell radially, thereby increasing in thickness. Together, these responses strengthen the seedling stem and root. A third ethylene response is that the seedling stem bends so that embryonic leaves and the delicate meristem grow horizontally rather than vertically, the sideways growth response that Neljubov first observed. The bent portion of the stem, known as a hook, then pushes up through the soil (look ahead to Figure 39.22). The hook forms as the result of an imbalance of auxin across the stem axis, which causes cells on one side of the stem to elongate faster than cells on the other side. Ethylene drives this auxin imbalance.

0.00   0.10   0.20   0.40   0.80
Ethylene concentration (parts per million)

Figure 36.7 **Seedling growth showing the triple response to ethylene.** When applied at levels above a particular concentration (0.80 ppm), ethylene causes seedling stems to cease elongation, swell radially, and bend to form a hook that can push upward through the soil. Ethylene produced naturally within seedlings causes the same response.

Biological inquiry: What adaptive advantage does this seedling behavior provide?

Knowledge of the effects of ethylene on fruits has been very useful commercially. Ripe fruits are easily damaged during transit, but fruits such as tomatoes and apples can be picked before they ripen for transport with minimal damage. At their destination, these fruits can be ripened by treatment with ethylene. However, fruit that becomes overripe may exude ethylene that hastens ripening in nearby, unripe fruit. For this reason, fruit that must be stored for extended periods is kept in ethylene-free environments.

Several Hormones Help Plants Cope with Environmental Stress

Several plant hormones share the property of helping plants respond to environmental stresses such as flooding, drought, high salinity, cold, heat, and attack by disease microorganisms and animal herbivores. These protective hormones include the major plant hormones known as **abscisic acid** and **brassinosteroids** (see Table 36.1). Additional protective hormones are salicylic acid (SA), whose chemical structure is similar to that of aspirin, and a peptide known as systemin. The fragrant compound jasmonic acid, whose structure is similar to that of mammalian prostaglandins, and the gas nitric oxide (NO) are also protective plant hormones.

Abscisic Acid Abscisic acid, abbreviated as ABA, was named at a time when plant biologists thought that it played a role in leaf

or fruit drop, also known as abscission. Later they discovered that ethylene actually causes leaf and fruit abscission, whereas abscisic acid slows or stops plant metabolism when growing conditions are poor. For example, ABA may induce bud and seed dormancy. Dormant buds and seeds resume growth only when specific environmental signals reveal the onset of conditions suitable for survival. In preparation for winter, ABA stimulates the formation of tough, protective scales around the buds of perennial plants. Seed coats of apple, cherry, and other plants also accumulate ABA, which prevents seeds from germinating unless temperature and moisture conditions are favorable for seedling growth. Water-stressed roots also produce ABA, which is then transported to shoots, where it helps to prevent water loss from leaf surfaces by inducing leaf pores (stomata) to close.

Brassinosteroids Brassinosteroids are named after the plant genus *Brassica* (which includes cruciferous plants such as cabbage and broccoli), from which they were first identified. However, seeds, fruits, shoots, leaves, and flower buds of all types of plants contain brassinosteroids. These hormones induce vacuole water uptake and influence enzymes that alter cell-wall carbohydrates, thereby fostering cell expansion. Mutations that affect brassinosteroid synthesis cause plants to exhibit dwarfism. Such plants have small, dark green cells because their tissues are unable to expand. Brassinosteroids also impede leaf drop, help grass leaves to unroll, and stimulate xylem development. They can be applied to crops to help protect plants from heat, cold, high salinity, and herbicide injury.

Brassinosteroids are chemically related to animal steroid hormones, such as human sex hormones. However, unlike animal steroid hormone receptors, which are located at the nucleus, brassinosteroids receptors are plasma membrane kinases. Upon activation, these kinases inhibit a protein that otherwise phosphorylates two transcription factors, targeting them for destruction. Brassinosteroids allow these transcription factors to accumulate and influence gene expression.

In this section, we have surveyed the roles of major plant hormones produced within plants. While plant hormones coordinate many genetically determined developmental processes, they also help plants to respond to environmental stimuli, our next topic.

36.3 Plant Responses to Environmental Stimuli

Plants encounter many types of environmental challenges and behave accordingly. Take seed germination, for example. If buried seeds were to germinate beneath soil layers too deep for light to penetrate, or beneath a cover of established plants, seedlings would not be able to obtain sufficient light for photosynthesis and would die. A related reproductive challenge for plants is to flower at times of the year that are most beneficial for achieving pollination or seed dispersal. How do plants determine if there is enough light for seeds to germinate and for seedlings to grow? How do plants determine when to flower?

The answer is that plants possess cellular systems for measuring light and determining the seasonal time of year. Using these systems, plant seeds germinate only when there is sufficient light for seedling growth, and flowering occurs during the most advantageous season. Therefore, plants can sense and respond to their light environments. Plants are also able to respond to other physical and biological stimuli. In this section, we survey plant responses to these external stimuli, beginning with light.

Plants Detect Light and Measure Day Length

A plant's ability to measure and respond to light amounts and day length, a process called **photoperiodism**, is based on the presence of light receptors within cells. Such light sensors, known as **photoreceptors**, are distinct from the light-absorbing pigments that function in photosynthesis. Each type of photoreceptor has a light-absorbing component as well as other regions that respond to light absorption by switching on signal transduction. Responses by many cells in a tissue or organ cumulatively result in behaviors such as sun tracking, phototropism, flowering, and seed germination.

Blue-Light Receptors Cryptochromes and phototropins are two types of blue-light receptors—molecules that absorb and respond to blue light. Experiments suggest that **cryptochromes** help young seedlings determine if their light environment is bright enough to allow photosynthesis. If not, seedlings continue to elongate through the soil, toward the light. How does this process work? In the dark, seedlings produce a nuclear protein known as COP1 (constitutive photomorphogenesis1) that prevents the expression of genes for chlorophyll synthesis and photosynthesis. The result is that seedlings continue to elongate. When seedlings reach the lighted surface, COP1 moves to the

cytosol and associates with activated cryptochrome. This prevents COP1 from inhibiting photosynthesis genes. As a result, seedlings are able to produce chlorophyll.

Phototropin is the main blue-light sensor involved in phototropism. This sensor has two components: a protein that has a kinase domain and a flavin pigment molecule that can absorb blue light. In the dark, the flavin is not covalently bound to the protein. However, when the flavin absorbs blue light, it changes conformation and becomes able to covalently bind phototropin. Flavin binding, in turn, changes the conformation of the phototropin protein, allowing it to phosphorylate itself by means of the protein kinase domain. When a plant organ is exposed to directional blue light, phototropin becomes phosphorylated. In this way, a light signal is converted into a chemical signal. However, the events that connect phosphorylation to changes in auxin movement are so far unknown.

Phytochrome, the Red- and Far-Red-Light Receptor Many plant growth and developmental processes are influenced by **phytochrome**, a red- and far-red-light receptor. Phytochrome operates much like a light switch, flipping back and forth between two conformations (**Figure 36.8**). When red light is abundant, as in full sunlight, phytochrome absorbs red light and changes to a conformation that only absorbs far-red light (light having a wavelength longer than that of red light). This form of phytochrome, known as P_{fr}, activates cellular responses (which are described later). When left in the dark for a long period, P_{fr} slowly transforms into the inactive red light–absorbing form, known as P_r. Far-red light is more abundant than red light in situations such as deep within a canopy of leaves, because chlorophyll absorbs most of the available red light. In this environment, P_{fr} rapidly switches to P_r. P_r can only absorb red light, and it does not activate cellular responses.

Red light

Far-red light

Protein

Light-sensitive region

Inactive P_r (cytosol)

Active P_{fr} (cytosol)

Active P_{fr} (nucleus)

1 P_r, the inactive form of phytochrome, occurs in the cytosol and is a receptor for red light.

2 Red light activates phytochrome, converting it to P_{fr}, a receptor for far-red light.

3 Activated P_{fr} moves into the nucleus, where it interacts with specific proteins, thereby causing responses such as seed germination.

Figure 36.8   How phytochrome acts as a molecular light switch.

The role of phytochrome as a plant "light switch" has been shown experimentally in studies of lettuce-seed germination. Researchers have found that water-soaked lettuce seeds will not germinate in darkness, but they will germinate if exposed to as little as 1 minute of red light (**Figure 36.9**). This amount of light exposure is sufficient to transform a critical amount of P_r to the active P_{fr} form, which stimulates germination. However, if this brief red-light treatment is followed by a few minutes of treatment with far-red light, the lettuce seeds will not germinate. This short period of far-red illumination is enough to convert seed P_{fr} back to the inactive P_r form. The most recent light exposure determines whether the phytochrome occurs in the active or inactive form. In nature, if seeds are close enough to the surface that their phytochrome is switched on by red light, the seeds will germinate. But if seeds are buried too deeply for red light to penetrate, they will not germinate. In this way, seeds can sense if they are close enough to the surface to begin the germination process. More than 50 years ago, plant physiologist Harry Borthwick and colleagues performed the classic experiments with lettuce seeds that first revealed the existence of phytochrome, and these researchers purified and named the phytochrome molecule. Since that time, molecular genetic approaches have revealed more about how phytochrome works at the cellular and molecular levels.

Most of our understanding of phytochrome's effect on gene expression comes from the study of the model plant *Arabidopsis*. This plant has five phytochrome genes (*PHYA* to *PHYE*). Each of the five types of phytochrome is composed of two proteins, each having a chromophore—a light-sensitive region (see Figure 36.8). In the dark, phytochrome molecules in the P_r state reside in the cytosol. After exposure to red light, activated phytochrome (P_{fr}) molecules typically move from the cytosol to the nucleus. Within the nucleus, P_{fr} interacts with a protein known as PIF3 (phytochrome interacting factor 3). PIF3 binds to the regulatory elements of several phytochrome-responsive genes, functioning as a positive regulator of some genes and as a negative regulator of others.

Photoperiodism Phytochromes play a critical role in photoperiodism, which influences the timing of dormancy and flowering. Flowering plants can be classified as long-day, short-day, or day-neutral plants according to the way their flowering responds to night length. When scientists named these groups they thought that plants measured the amount of daylight. However, later researchers discovered that plants actually measure night length.

Lettuce, spinach, radish, beet, clover, gladiolus, and iris are examples of **long-day plants** because they flower in spring or early summer, when the night period is shorter (and thus the day length is longer) than a defined period (**Figure 36.10a**). In contrast, asters, strawberries, dahlias, poinsettias, potatoes, soybeans, and goldenrods are examples of **short-day plants** because they flower only when the night length is longer than a defined period. Such night lengths occur in late summer, fall, or winter, when days are short (**Figure 36.10b**). Roses, snapdragons, cotton, carnations, dandelions, sunflowers, tomatoes, and

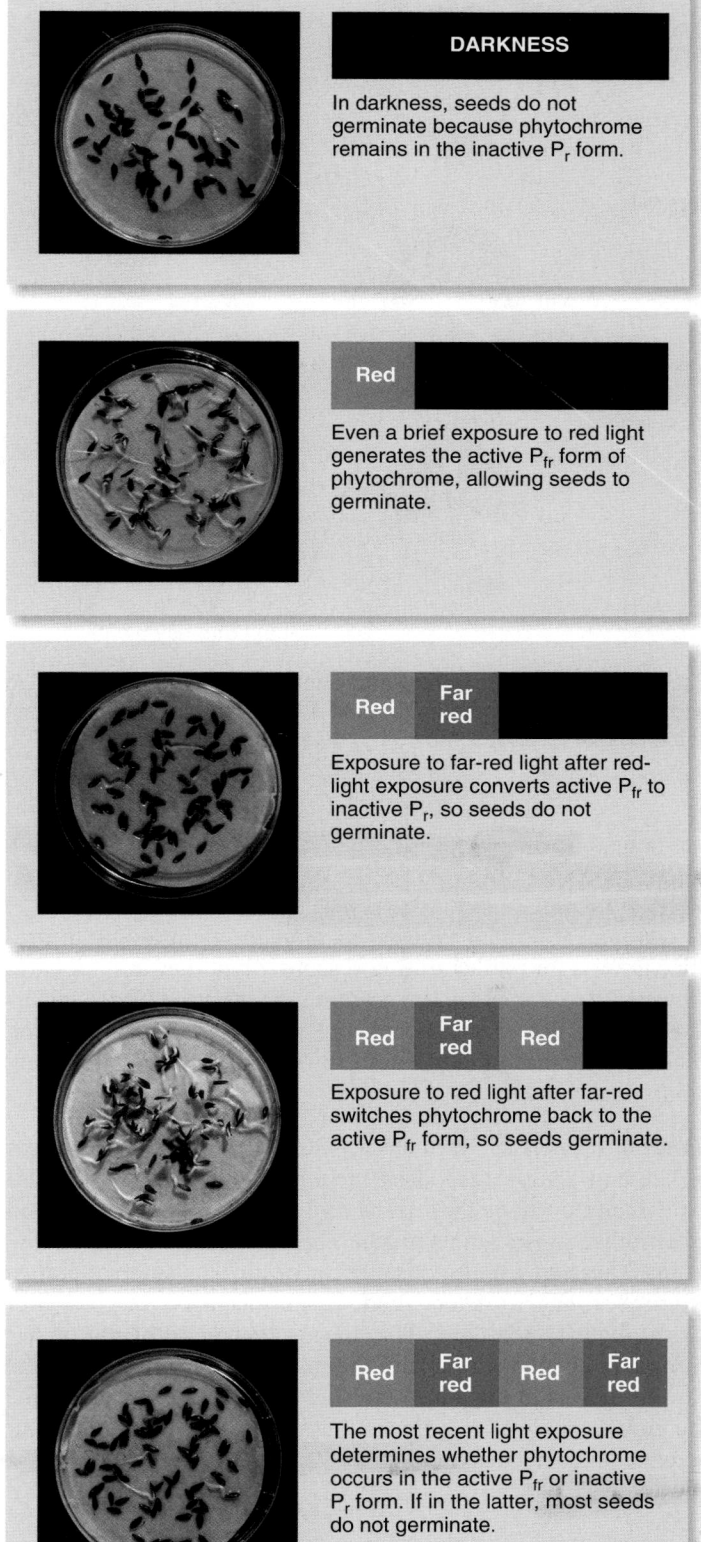

Figure 36.9 **How phytochrome influences seed germination.**

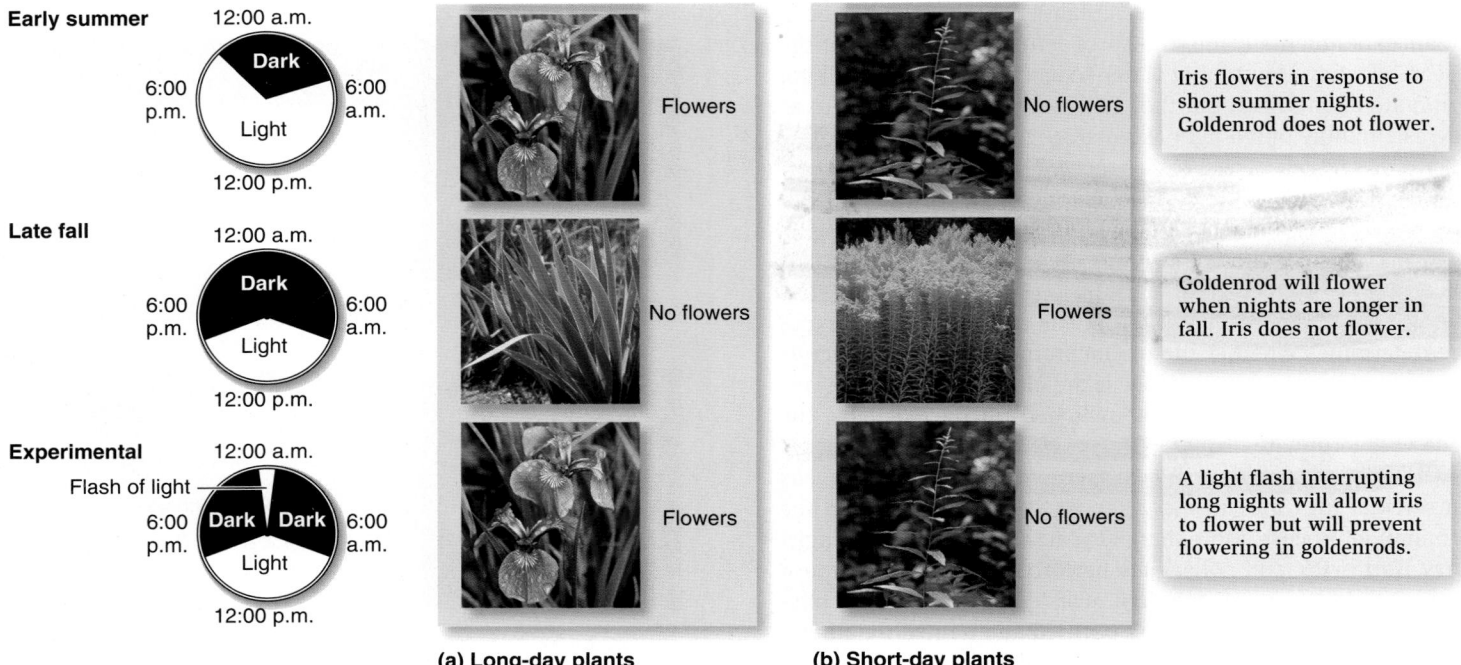

Figure 36.10   **Flowering and photoperiodism.**   (a) Iris is a long-day plant that flowers in response to the short nights of late spring and early summer, whereas (b) goldenrod is a short-day plant that flowers in response to the longer nights of autumn. The length of night is the critical factor, as shown by the effects of light flashes.

Biological inquiry: What would happen if you gave these plants a brief exposure to darkness in the middle of the daytime?

cucumbers flower regardless of the night length, as long as day length meets the minimal requirements for plant growth, and are thus known as **day-neutral plants**.

Ornamental plant growers manipulate night length to produce flowers for market during seasons when they are not naturally available. For example, chrysanthemums are short-day plants that usually flower in the fall, but growers use light-blocking shades to increase night length in order to produce flowering plants at any season.

Shading Responses   Phytochrome also mediates plant responses to shading. These responses include the extension of leaves from shady portions of a dense tree canopy into the light, and growth that allows plants to avoid being shaded by neighboring plants. These growth responses occur by the elongation of branch internodes. Leaves detect shade as an increased proportion of far-red light to red light. This means that more of the phytochrome in shaded leaves is in the inactive (P_r) state than is the case for leaves in the sun. Activated phytochrome (P_{fr}) inhibits the growth of shoot internodes, but phytochrome in the inactivated state does not, so branches bearing shaded leaves extend toward sunlight.

Plants Respond to Gravity and Touch

Now that we have explored how plants behave under changing light conditions, let's consider how plants respond to other environmental factors, such as gravity and touch. The upward growth of shoots and the downward growth of roots are behaviors known

as **gravitropism**, growth in response to the force of gravity. Shoots are said to be negatively gravitropic because they grow in the direction opposite to the gravitational force. If a potted plant is turned over on its side, the shoot will eventually bend and begin to grow vertically again (**Figure 36.11**). Most roots are said to be positively gravitropic because they grow in the same direction as increased gravity.

Both roots and shoots detect gravity by means of starch-heavy plastids, known as **statoliths**. Specialized gravity-sensing cells that contain statoliths are known as **statocytes**. In shoots, statocytes are located in a layer of cells known as the endodermis or starch sheath, which is located near vascular tissues. In roots, gravity-sensing cells occur in the center of the root cap. Gravity causes the relatively heavy statoliths to sink, which causes changes in calcium ion messengers that affect the direction of auxin transport. This process induces changes in the direction of shoot or root growth. For example, in a root that becomes oriented horizontally, the statoliths are pulled by gravity to the lower sides of statocytes (**Figure 36.12**). The change in statolith positions causes auxin to move to cells on the lower sides of roots. In roots, auxin inhibits cell elongation (in contrast to its action in shoots). Thus, root growth slows on the lower side, while cell elongation continues normally on the upper side. This process causes the root to bend, so that it eventually grows downward again.

Recent studies suggest that gravity responses are related to touch responses, known as **thigmotropism** (from the Greek *thigma*, touch). For example, when roots encounter rocks or other barriers to their downward growth in the soil, they display a touch response that temporarily supersedes their response to gravity. Such roots grow horizontally until they get around the barrier, whereupon downward growth in response to gravity resumes. Plant shoots also respond to touch; examples include vines with tendrils that wind around or clasp supporting structures. Wind also induces touch responses. In very windy places, trees tend to be shorter than normal, giving them the advantage of being less likely to blow over than are taller trees. In the laboratory, plant scientists have simulated natural touch responses by rubbing plant stems and found that this treatment can result in shorter plants. Touch causes the release of calcium ion messengers that influence gene expression.

More rapid responses to touch, such as leaf folding by the sensitive plant (see Figure 36.1b), are based on changes in the water content of cells within structures known as pulvini. A pulvinus consists of a thick layer of parenchyma cells that surrounds a core of vascular tissue (**Figure 36.13**). In the complex leaves

of sensitive plants, a pulvinus occurs at the base of each leaflet pair. When touched, an action potential opens ion channels in parenchyma cells near the top (adaxial) surfaces of the leaflets. These cells lose potassium and chloride ions, causing water to flow out and the cells to become limp. This bends the leaflets downward and together, starting the leaf-folding process. An electrical impulse generated at the touch site also flows through the leaf, causing many or all of the leaflets to also bend, with the result that the entire leaf folds. Reversal of this process allows the leaf to unfold.

The action of pulvini also explains some plant movements that are unrelated to touch: sleep movements and sun tracking. Sleep movements are changes in leaf position that occur in response to day–night cycles. Sun tracking, as mentioned, is the movement of leaves or flowers in response to the sun's position.

Plants Respond to Flooding, Drought, High Salinity, Heat, and Cold

Plants display many types of adaptations that help them cope with unfavorable growth conditions, such as flooding, drought, high salinity, heat, and cold. The following examples of plant responses reveal important roles of hormones and miRNA.

Flooding The major harmful effect of flooding is that too much water makes roots unable to obtain sufficient oxygen to fuel respiratory processes. Without oxygen from the air, roots cannot produce the ATP needed to absorb minerals from soil. Many plants can reduce the effects of flooding by producing internal, snorkel-like airways that allow more oxygen to flow from shoots to roots. These airways occur in tissue known as **aerenchyma** (**Figure 36.14**). Aerenchyma formation involves the action of ethylene, which leads to programmed cell death, followed by cell collapse. In plants such as corn, aerenchyma forms only when roots are subjected to flooding. However, aerenchyma occurs in the roots of many plants native to wetland habitats even when the soil is not wet.

Figure 36.11 **Shoot growth displays negative gravitropism.** This tomato shoot system has resumed upward growth after being placed on its side. Upward growth started about 4 hours after the plant was turned sideways; this photo was taken 20 hours later. Shoots sense gravity by means of starchy statoliths present in stem tissue near the central vascular tissue.

Zone of elongating cells

Direction of growth

Auxin flow

Root cap

Statocytes

Statoliths

When roots are vertical, statoliths settle onto the bottom ends of statocytes in the root cap. Auxin moves upward equally on all sides. Cells elongate equally on all sides, so the root grows downward.

Cells elongate

Less auxin

More auxin: cell elongation inhibited

When roots are horizontal, statoliths settle to the sides of statocytes that now face downward. This causes auxin to be transported to the lower side of the root, inhibiting cell elongation there. Cells on the upper side continue to elongate, causing the root to bend downward.

Figure 36.12 **Root growth displays positive gravitropism.** Root-tip cells sense gravity by means of starchy statoliths present in cells at the center of the root cap.

Figure 36.13 **Leaf folding in the sensitive plant: how pulvini change the positions of leaflets and leaves.** The electrical signals known as action potentials result from the rapid flow of ions through membrane ion channels. Electrical signals spread from one cell to another through plasmodesmata, intercellular connections. Cells respond to ion flow by losing or taking up water, causing them to shrink or swell.

Drought, High Salinity, Heat, and Cold Drought, high salinity, heat, and cold are related environmental stresses because they all reduce the amount of liquid water present in plant cells. Most plants that lose half or more of their water are unable to recover. Thus, plants possess diverse adaptations that reduce water loss, and the hormone abscisic acid often coordinates these responses.

One way in which plants cope with drought stress is by regulating aquaporins, proteins that form water channels in plasma membranes. Aquaporins allow plant cells to take up or lose water much faster than could be accomplished by diffusion alone. Because water uptake is such an important component of plant growth responses and movements, plants regulate water channel opening and closing. In 2006, Susanna Törnroth-Horsefield, Per Kjellbom, and coworkers described the X-ray crystallographic structure of spinach plasma membrane aquaporin from cells experiencing different levels of water stress. These investigators found that under drought conditions, particular aquaporin amino acids lose phosphate groups, causing a loop of the protein to stopper the membrane channel. This helps to prevent plant cells from losing water via these pores. When water becomes available again, phosphorylation of aquaporin causes the loop to lift off of the pore, thereby opening it for water passage. Drought-stressed plants also close their epidermal pores—the stomata. Stomatal closure and other plant responses to water stress are described in Chapter 38.

Saline soils stress plants by restricting root growth and the amount of water that roots can absorb. When too much salt enters plant tissues, it slows their growth by causing gibberellic acid levels to decline. One way that plants respond to harmful levels of salt is by producing miRNA that degrades specific mRNAs. This process eventually alters the levels of particular

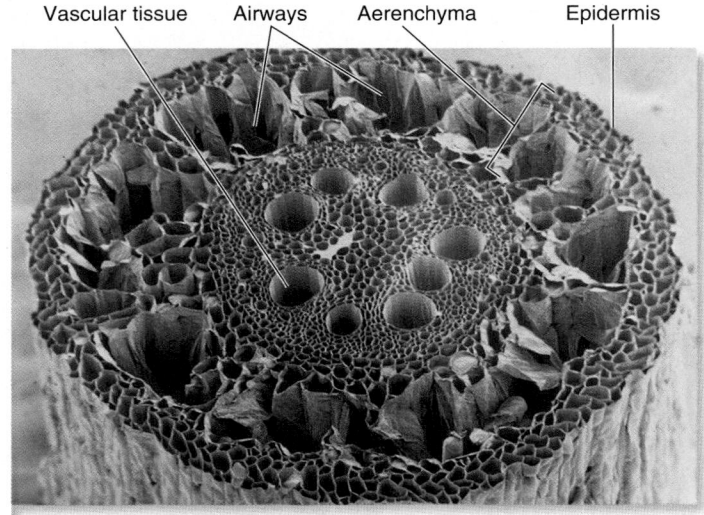

Figure 36.14 **A plant response to flooding.** A slice of a root is shown in this micrograph. Air channels in aerenchyma tissue allow air to readily flow to roots from shoots even when the plant is partially submerged.

enzymes in the cells and thereby leads to the buildup of cellular proline, an amino acid. Together with some other organic compounds, proline stabilizes plasma membranes against damage.

Heat stress causes proteins to fold abnormally and thus lose normal function. Plants cope with heat stress by producing increased amounts of several types of **heat shock proteins**. These proteins help to protect proteins from heat damage and refold proteins to their functional state.

An example of plant adaptation to cold is **vernalization** (from the Latin *vernalis*, spring). This process allows plants

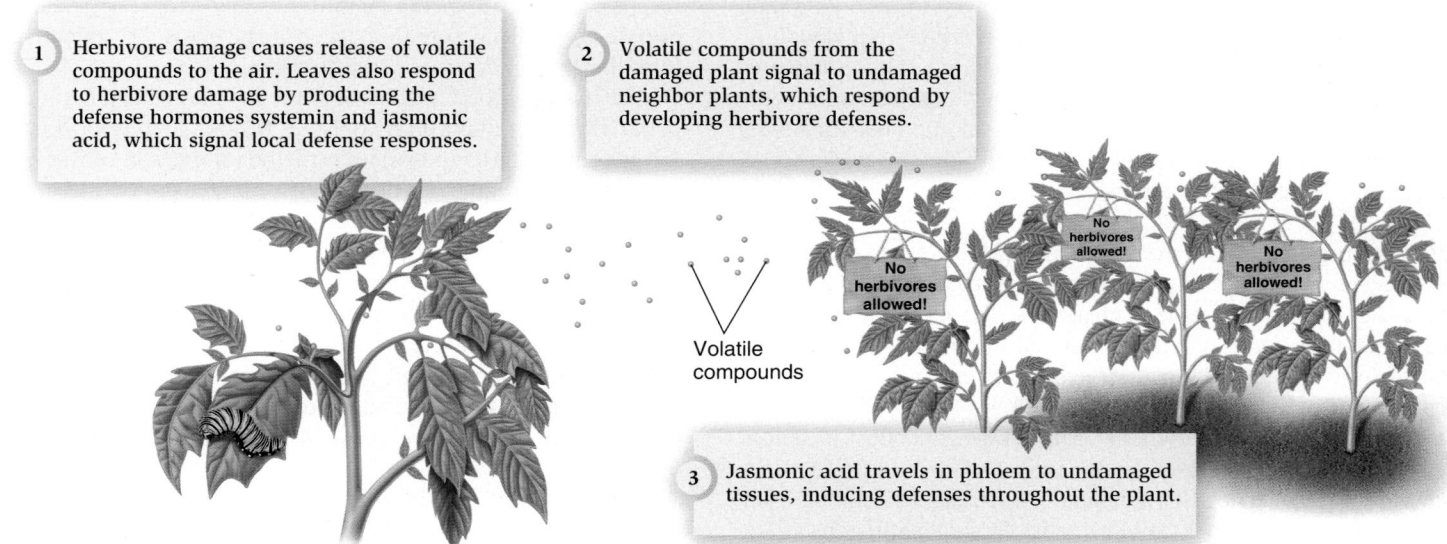

1. Herbivore damage causes release of volatile compounds to the air. Leaves also respond to herbivore damage by producing the defense hormones systemin and jasmonic acid, which signal local defense responses.

2. Volatile compounds from the damaged plant signal to undamaged neighbor plants, which respond by developing herbivore defenses.

No herbivores allowed!

No herbivores allowed!

No herbivores allowed!

Volatile compounds

3. Jasmonic acid travels in phloem to undamaged tissues, inducing defenses throughout the plant.

Figure 36.15 **Plant responses to herbivore attack.** Wounding releases volatile compounds that diffuse through the air, fostering the production of chemical defenses by neighboring plants, attracting enemies of the attacker, or inducing defensive responses in other parts of the plant.

that grow best in the cool, moist autumn to flower only in the warm season, after unfavorable winter conditions have passed. Wheat, barley, and other cereals native to regions having cold climates often display vernalization behavior. For example, winter wheat varieties require a long period of low temperature exposure before they can flower, explaining why farmers sow winter wheat in the fall. Such plants measure the length of winter, flowering only in the warm season. By contrast, farmers plant spring wheat varieties in the spring because these do not require a period of cold exposure to flower.

A team of agricultural scientists led by plant molecular biologist Jorge Dubcovsky recently distinguished the vernalization genes *VRN1* and *VRN2*, whose expression explains these variations. VRN1 protein promotes spring flowering and is expressed in both spring and winter wheat. By contrast, VRN2 protein, which is expressed only in winter wheat varieties, represses flowering, probably by silencing certain flowering genes. A long cold period is necessary to destroy VRN2 protein and allow flowering. When Dubcovsky's team manipulated winter wheat plants to prevent the production of VRN2 protein, the plants flowered in the absence of a cold treatment.

Another plant response to cold is the production of antifreeze compounds that protect plant cells from damage by ice-crystal formation. Some of these antifreeze proteins also function in protection against attack by biological agents, our next topic.

Plants Defend Themselves Against Herbivore and Pathogen Attack

Plants are vulnerable to attack by animal herbivores and pathogens—disease-causing viruses, bacteria, and fungi. Structural barriers such as cuticles, epidermal trichomes, and outer bark help to reduce infection and herbivore attack (see Chapter 35).

Like other organisms, plants use nucleic-acid–degrading enzymes as defenses against attacking viruses. Plants also use many types of chemical defenses to deter herbivores and respond to pathogens. Such defenses explain why remarkably little natural vegetation is lost to herbivore or pathogen attack. However, agricultural crops can be more vulnerable to attack than their wild counterparts. This is because some protective adaptations have been lost during crop domestication as the result of genetic changes that increase edibility. For this reason, crop scientists are particularly interested in understanding plant defense behavior, with the goal of being able to breed or genetically engineer crop plants that are better protected from pests.

Plant Responses to Herbivores Plants use a wide variety of chemical defenses against herbivores. Defense compounds include the secondary metabolites alkaloids, terpenes and terpenoids, tannins, and phytoalexins. Some of these substances act directly on herbivores, making plants taste bad so that herbivores learn to avoid them. Other chemical compounds function indirectly. For example, when attacked by insect caterpillars, cruciferous plants release terpenoids that attract the "bodyguard" wasp (*Cotesia rubecula*), which attacks the caterpillars. Peptides or fatty acid–amino acid combinations present in the mouth secretions of attacking insects also mobilize indirect defenses. Corn and other plants respond to such mouth secretions by producing volatile leaf compounds: alcohols, aldehydes, and esters (**Figure 36.15**). As these compounds diffuse through the air, they attract natural enemies of insect attackers. Furthermore, neighboring plants may detect the volatile compounds and respond by producing defensive compounds that deter herbivores (see Figure 36.15). Plants that can detect nearby battles are thus able to take preemptive action to arm themselves against attack.

Defenses against herbivores may involve the action of plant hormones (see Figure 36.15). For example, when insects wound tomato plants, damaged cells release systemin. This protective hormone induces undamaged cells to produce jasmonic acid, which functions as an alarm system. Jasmonic acid travels in the phloem to other parts of the plant, inducing widespread production of defensive compounds.

Plant Responses to Pathogen Attack Bacterial and fungal pathogens produce many types of compounds known as **elicitors**. Such elicitors promote **virulence**, the infection of in this case plant tissues. A peptide produced by the plant pathogenic bacterium *Pseudomonas syringae* and the fungal cell-wall compound chitin are examples of elicitors. Despite their name, **avirulence genes**—known as *Avr* genes—encode virulence-enhancing elicitors (**Figure 36.16, step 1**). Some bacteria inject elicitors into plant cells by means of syringe-like systems that are also used to attack animal cells. Fungal pathogens often deliver elicitors by means of penetrating structures known as haustoria. These haustoria (whose name derives from a Latin word meaning "to drink") are also the means by which fungi absorb food materials from plant cells.

In response to pathogen attack, plants have evolved defense systems based on about 20 types of **resistance genes (R genes)** (**Figure 36.16, step 2**). Many alleles of *R* genes occur in plant populations, providing plants with a large capacity to cope with many types of pathogens. Most resistance genes encode proteins that function as receptors for elicitors. Some elicitor receptors occur in plasma membranes, allowing early detection of pathogens. Receptors that occur in the cytosol recognize elicitors that have been injected into cells.

In 1955, based on studies of rust fungi and flax plants (*Linum usitatissimum*), plant pathologist H. H. Flor proposed the gene-for-gene hypothesis to explain how the interaction of *Avr* and *R* gene products influences disease. If a plant is genetically unable to produce a receptor that can recognize a pathogen's elicitor, disease will result. By contrast, plants successfully resist disease when the product of a dominant *R* gene (a receptor) recognizes a pathogen's dominant *Avr* gene product (elicitor). Receptor-elicitor binding stimulates signal transduction processes that induce plant defensive responses. As described next, the hypersensitive response is a local defensive response, whereas systemic acquired resistance occurs throughout plant body.

The Hypersensitive Response to Pathogen Attack The plant **hypersensitive response (HR)** occurs when a plant recognizes the elicitors released by a pathogen and responds in such a way that the disease symptoms are limited (**Figure 36.16, steps 3 and 4**). This plant defensive response has several components. One of these is increased production of hydrogen peroxide (H_2O_2), which can kill infectious agents, help to strengthen the cell wall by cross-binding polymers, and induce programmed cell death. Programmed cell death deprives pathogens of food and helps to prevent their spread. The hormone NO is also produced as part of the hypersensitive response, and together with H_2O_2 it stimulates synthesis of hydrolytic enzymes, defensive

Figure 36.16 Pathogen/plant interactions and the hypersensitive response to pathogen attack.

secondary metabolites, the hormone salicylic acid, and tough lignin in cell walls of nearby tissues. Necrotic spots are brown patches on plant organs that reveal where disease pathogens have attacked plants, and plant tissues have battled back. Salicylic acid or the related compound methyl salicylate signals other parts of the plant, which respond by preparing defenses.

Systemic Acquired Resistance A localized hypersensitive response can result in the production of alarm signals that travel to noninfected regions of a plant and induce widespread resistance to diverse pathogens. This response of the whole plant, known as **systemic acquired resistance (SAR)**, is a type of plant immune system (**Figure 36.17**). The SAR system uses the same type of long-distance signaling process that is associated with herbivore attack (see Figure 36.15). At or near a wound, the production of systemin induces the production of jasmonic acid in nearby vascular tissues. Jasmonic acid is then transported throughout the plant. In addition, the hormone salicylic acid can be converted to volatile methyl salicylate, which perfumes the air surrounding a plant, inducing resistance in noninfected tissues. In response, plant tissues may produce defensive enzymes, which can break down pathogen cell walls, or they may generate tannins, which are toxic to microorganisms. Experimental infection of *Arabidopsis* leaves with the bacterial pathogen *Pseudomonas syringae* led to induction of SAR within 6 hours. HR and SAR are of great importance to agricultural scientists as they endeavor to find new ways to protect crop plants from attack by herbivores and pathogens.

Figure 36.17 Systemic acquired resistance to pathogen attack.

CHAPTER SUMMARY

36.1 Plant Behavioral Responses

- Plants sense and respond to diverse external stimuli and internal signals, and thus display behavior. (Figures 36.1, 36.2)
- During the process of signal transduction, cellular receptors respond to environmental stimuli as well as to internal hormonal signals and mobile RNA molecules. The process involves receptor activation, transduction of the signal by messengers, and cellular response due to effectors. (Figure 36.3)
- Cellular responses include changes in gene expression and ion channels that influence plant growth, development, reproduction, chemistry, and movements such as sun tracking and leaf folding.

36.2 Plant Hormones

- RNA molecules and plant hormones interact with environmental stimuli to control plant development, growth, and behavior. (Table 36.1)
- Auxin plays an important role in many aspects of plant behavior, including phototropism—as demonstrated by the classic experiments of Went, Briggs, and others. Auxin can be transported downward or upward (polar transport) and sideways (lateral transport) in the plant. The position of auxin efflux carriers determines the direction of auxin transport. (Figures 36.4, 36.5)
- Other major plant hormones are the cytokinins, gibberellins, ethylene, abscisic acid, and brassinosteroids. Auxin and cytokinin

are used in plant tissue culture. Several additional hormones (systemin, salicylic acid, jasmonic acid, and nitric oxide) play important roles in stress responses. (Figures 36.6, 36.7)

36.3 Plant Responses to Environmental Stimuli

- Light-sensitive pigments such as cryptochrome, phototropin, and phytochrome allow plants to respond to light stimuli: Sun tracking, seed germination, and photoperiodic control of flowering are examples. (Figures 36.8, 36.9, 36.10)
- Plant shoots and roots respond to gravity (in the process of gravitropism) by means of starch-heavy statoliths located within statocytes. Sun tracking and rapid touch responses (thigmotropism) such as leaf folding in the sensitive plant depend on changes in the water content of cells in motor structures known as pulvini. (Figures 36.11, 36.12, 36.13)
- Plants can cope with stresses such as flooding, drought, high salinity, heat, and cold with the involvement of hormones and miRNA. (Figure 36.14)
- Plants defend themselves against attack by herbivores and pathogens by means of structural and chemical adaptations. A wide range of secondary metabolites provide chemical defenses against herbivores. Defense against pathogens starts when a pathogen produces an elicitor encoded by an *Avr* gene. If the plant can produce a receptor for that elicitor, encoded by an *R* (resistance) gene, then the plant can resist infection by that pathogen. The hypersensitive response (HR) is a local defensive response to pathogen attack, and systemic acquired resistance (SAR) is a whole-plant defensive response. (Figures 36.15, 36.16, 36.17)

TEST YOURSELF

1. What kinds of molecules move within plants from one place to another, influencing plant development and behavior?
   a. plant hormones and RNA molecules such as microRNAs (miRNAs)
   b. secondary messengers
   c. effectors
   d. receptors
   e. none of the above

2. Phototropism is
   a. the production of flowers in response to a particular day length.
   b. the production of flowers in response to a particular night length.
   c. the growth response of a plant, organ system, or organ to directional light.
   d. illustrated by sun tracking.
   e. none of the above.

3. What is the most accurate order of events during signal transduction?
   a. first, receptor activation, then messenger signaling, and last, an effector response
   b. first, an effector response, then messenger signaling, and last, receptor activation
   c. first, messenger signaling, then receptor activation, and last, an effector response
   d. first, an effector response, then receptor activation, and last, messenger signaling
   e. none of the above

4. Which of the plant hormones is known as the "master hormone"?
   a. cytokinin
   b. gibberellin
   c. abscisic acid
   d. brassinolide
   e. none of the above

5. Which of the major plant hormones is a diffusible gas?
   a. auxin
   b. gibberellin
   c. cytokinin
   d. ethylene
   e. abscisic acid

6. Which of these molecules is considered to be a plant photoreceptor?
   a. cryptochrome
   b. phototropin
   c. phytochrome
   d. all of the above
   e. none of the above

7. Thigmotropism is a plant response to
   a. light.
   b. cold.
   c. touch.
   d. gravity.
   e. drought.

8. Which response is an adaptation to flooding?
   a. production of heat shock proteins
   b. stomatal closure
   c. vernalization
   d. production of aerenchyma
   e. all of the above

9. What are avirulence genes?
   a. plant genes that encode proteins that prevent infection (virulence)
   b. plant genes that cause infection when the proteins they encode bind to pathogen elicitors
   c. pathogen genes that prevent the pathogens from causing plant disease
   d. pathogen genes that encode elicitors that foster disease in plants
   e. none of the above

10. How do plants defend themselves against pathogens?
    a. Plants produce resistance molecules (usually proteins) that bind pathogen elicitors, thereby preventing disease.
    b. Plants display a hypersensitive response that limits the ability of pathogens to survive and spread.
    c. Plants display systemic acquired resistance, whereby an infection induces immunity to diverse pathogens in other parts of a plant.
    d. All of the above are correct.
    e. None of the above is correct.

CONCEPTUAL QUESTIONS

1. Why can plants be said to display behavior?
2. Why do plants produce so many types of resistance (*R*) genes?
3. Because diverse plants exude volatile compounds in response to herbivore or pathogen attack, some experts have written about "talking trees." Is there any such thing?

EXPERIMENTAL QUESTIONS

1. Boysen-Jensen used gelatin to collect diffusible auxin from shoot tips, but Went switched to agar, and Briggs followed suit. Why is agar better for this purpose than gelatin?
2. How did Briggs determine that light does not destroy auxin?
3. What is the current hypothesized mechanism by which auxin accumulation causes shoot bending in response to directional light?

COLLABORATIVE QUESTIONS

1. Why are most wild plants distasteful, and some even poisonous, to people?
2. How could you increase the resistance of a particular crop plant species to particular types of herbivores?

---

www.brookerbiology.com
This website includes answers to the Biological Inquiry questions found in the figure legends and all end-of-chapter questions.

37

FLOWERING PLANTS: NUTRITION

The leaves of the sundew (*Drosera rotundifolia*), shown with a trapped fly, are a plant adaptation for acquisition of nutrients.

Many types of fascinating carnivorous ("meat-eating") plants grow abundantly in wetlands around the world, even though the soils in these places are infertile. How is this possible? In common with most plants, carnivorous plants are photosynthetic and thus produce their own organic food from water and carbon dioxide, using sunlight as an energy source. These resources are abundant in wetlands, but wetland soils are low in other minerals such as nitrogen that are needed for plant growth. Carnivorous plants such as the sundew pictured in the chapter-opening photo have adapted by obtaining minerals from the bodies of trapped insects and other small animals. Carnivorous plants lure animals with enticing fragrances, brightly colored leaves, or sugar-rich glistening drops of nectar. The unsuspecting prey fall into deep, water-filled pitchers, become ensnared by gluelike mucilage, or are trapped within the walls of leafy jails whose doors suddenly snap shut. Decomposition of the animal bodies releases minerals that plant leaves quickly absorb. Other wild and cultivated plants face similar nutritional challenges and likewise display adaptations that help them acquire sufficient resources for growth and reproduction.

This chapter focuses on plant nutrition, the processes by which plants obtain essential resources. We will begin by describing plant nutrients, the resources needed by plants for completion of their seed-to-seed life cycle in good health. Next, we will explore the role of soil as an essential resource for plants. Last, we will examine the biological sources of plant nutrients, focusing on nutritional associations between plants and microorganisms, sources of nutrients for carnivorous plants, and how some plants obtain nutrients from other plants. An understanding of these topics is crucial for those who seek ways to grow more plant-derived food for humans without causing harmful mineral pollution of Earth's waters.

Plant nutritional information is also useful to people who tend gardens or houseplants.

37.1 Plant Nutritional Requirements

Let's begin our consideration of plant nutritional requirements by defining nutrients more specifically. A **nutrient** is a substance that is metabolized by or incorporated into an organism. Two additional features are characteristic of nutrients. Deficiency symptoms develop in plants that receive too little of these substances, and the scarcity of nutrients selects for adaptations that help plants to acquire them.

Like other species, plants have a specific set of nutritional requirements. **Essential nutrients** are defined as those substances needed by plants in order to complete their reproductive cycle, while avoiding the symptoms of nutrient deficiency. For healthy growth and reproduction, green plants require light and certain **minerals**, defined here as inorganic substances (**Figure 37.1**). Carbon dioxide is primarily absorbed from air, while water and more than a dozen elements—such as potassium, nitrogen, and calcium—are primarily taken up from soil.

Elements required by plants play many roles in plant metabolism, often functioning as enzyme cofactors (**Table 37.1**). Elements that are required in amounts of at least 1 g/kg of plant dry matter are known as **macronutrients**. In contrast, elements that are needed in amounts at or less than 0.1 g/kg of plant dry mass are known as **micronutrients** or trace elements. Because insufficient amounts of light, carbon dioxide, water, and other mineral nutrients can limit the extent of green plant growth, these resources are known as **limiting factors**.

Figure 37.1 The major types of plant nutrients and their sources.

Table 37.1	Plant Essential Elements			
Element (chemical symbol)	**Percent of plant dry mass**	**Major source**	**Form taken up by plants**	**Function(s)**
Macronutrients				
Carbon (C)	45	Air	CO_2	Component of all organic molecules
Oxygen (O)	45	Air, soil, water	CO_2, O_2 H_2O	Component of all organic molecules
Hydrogen (H)	6	Water	H_2O	Component of all organic molecules, protons used in chemiosmosis and cotransport
Nitrogen (N)	1.5	Soil	NO_3^-, NH_4^+	Component of proteins, nucleic acids, chlorophyll, coenzymes, alkaloids
Potassium (K)	1.0	Soil	K^+	Has essential role in cell ionic balance
Calcium (Ca)	0.5	Soil	Ca^{2+}	Component of cell walls, messenger in signal transduction
Magnesium (Mg)	0.2	Soil	Mg^{2+}	Component of chlorophyll, activates some enzymes
Phosphorus (P)	0.2	Soil	HPO_4^{2-}	Component of nucleic acids, ATP, phospholipids, some coenzymes
Sulfur (S)	0.1	Soil	SO_4^{2-}	Component of proteins, some coenzymes, defense compounds
Micronutrients				
Chlorine (Cl)	0.01	Soil	Cl^-	Required for water splitting in photosystem II, cell ion balance
Iron (Fe)	0.01	Soil	Fe^{3+}, Fe^{2+}	Enzyme cofactor, component of cytochromes
Manganese (Mn)	0.005	Soil	Mn^{2+}	Enzyme cofactor
Boron (B)	0.002	Soil	$B(OH)_3$	Enzyme cofactor, component of cell walls
Zinc (Zn)	0.002	Soil	Zn^{2+}	Enzyme cofactor
Sodium (Na)	0.001	Soil	Na^+	Required to generate PEP in C_4 and CAM plants
Copper (Cu)	0.0006	Soil	Cu^+, Cu^{2+}	Enzyme cofactor
Molybdenum (Mo)	0.00001	Soil	MoO_4^{2-}	Enzyme cofactor
Nickel (Ni)	0.000005	Soil	Ni^{2+}	Enzyme cofactor

In this section, we will take a closer look at the nutritional factors that foster or limit plant growth.

Light Is an Essential Resource for the Growth of Green Plants

All photosynthetic plants require light, which can be regarded as a plant nutrient for several reasons. First, light energy is needed for the formation of the covalent bonds of organic compounds that make up the plant body. In addition, green plants' use of light energy as an essential resource parallels animals' nutritional requirements for organic food as a source of chemical energy. For the several hundred species of plants that have lost their photosynthetic capacity, such as the Indian pipe (*Monotropa uniflora*), essential nutrients include organic compounds that replace light as a source of energy (**Figure 37.2**).

Yet another reason for considering light to be a plant nutrient is that light-deprived plants display deficiency symptoms, as do plants that receive too little carbon dioxide, water, or other minerals. For example, if houseplants receive too little light, they must use stored carbohydrates for cellular respiration. As a result, the leaves may die and fall off, a symptom of light deficiency. When reserves are exhausted, the plant will die. Even if a plant is able to obtain enough light to survive, it may not absorb enough light to reproduce. This explains why many plants cannot live or flower inside the relatively dim interiors of human homes.

In nature, plants must adapt to environments with varying amounts of light and shade. For example, in forests, light availability limits the growth of tree seedlings and other small plants that are shaded by the leafy tree canopy overhead. But if a canopy tree dies from disease or is blown over by wind, creating a hole in the canopy called a light gap, the plants growing beneath it receive more sunlight and are able to grow taller. Conversely, plants growing in deserts or on mountains often experience light that is so intense that it can damage the photosynthetic components. Let's now consider some of the adaptations that help plants cope with environments that have too little or too much light.

Figure 37.2 The heterotrophic flowering plant, Indian pipe (*Monotropa uniflora*). This plant lacks photosynthetic capacity and thus must absorb organic compounds for use as an energy source. In contrast, nearby autotrophic green plants use sunlight as a source of energy.

(a) Shade leaf

(b) Sun leaf

Figure 37.3 Shade and sun leaves. These scanning electron micrographs of cut leaves reveal the thinner mesophyll and greater amount of air spaces in (a) shade leaves as compared to (b) sun leaves.

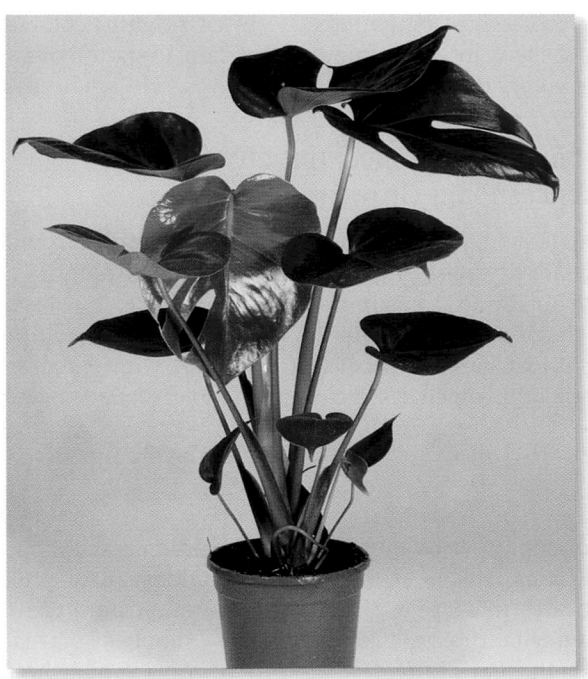

Figure 37.4 Houseplants can survive indoors because they have evolved in dimly lit natural habitats. *Monstera deliciosa* makes an attractive houseplant because its large, deep green leaves are adapted to the moist and shady conditions present in the interiors of tropical forests.

leaves have thinner mesophyll layers, with more air spaces than do sun leaves. Because they have a thicker layer of chlorophyll-containing mesophyll, sun leaves are able to harvest more of the bright sunlight that penetrates deeply into the leaf. In contrast, the thinner shade leaves avoid wasting energy by not producing photosynthetic tissues that dim light cannot reach.

Most epidermal cells at leaf surfaces are transparent, allowing light to penetrate to green mesophyll, and in some plants, these epidermal cells are shaped like lenses that focus more light into the green tissue. The arrangement of leaves on plants and stem branching patterns can also reflect adaptations that reduce shading. For example, many tropical forest trees that must compete for light with closely crowded neighbors are extremely tall, and they produce branches and leaves only at their very tops. Shorter plants native to the shady interiors of tropical rain forests are so well adapted to these moist but dim conditions that they make excellent houseplants (**Figure 37.4**).

Adaptations to Excessive Light Too much light can damage the photosynthetic machinery of plants by destroying an essential photosynthetic protein, called D1 protein. As an adaptation to excess light, specific xanthophyll pigments in the chloroplast absorb some of the light energy and dissipate it as harmless heat. These protective pigments can be produced from a precursor xanthophyll within minutes of a plant's exposure to excess light. Other adaptations prevent UV damage. Harmful amounts of UV radiation are absorbed by the plant's surface cuticle, as well as by carotenoid and flavonoid compounds located within leaf cells.

Adaptations to Shade One way in which plants adapt to shading is by producing thin, translucent leaves that allow some light to pass through to other leaves. In a shady environment, a plant or leaf may also produce more total chlorophyll, thereby maximizing the amount of light absorbed. In addition, many plants produce distinctive shade and sun leaves (**Figure 37.3**). Shade

These protective compounds are often brightly colored and, when present in large amounts, explain the attractive red or purple colors of some leaves.

Carbon Dioxide Concentration Influences Plant Growth

Although light provides the energy for plant photosynthesis, most plant dry mass originates from carbon dioxide (CO_2). Most plants obtain CO_2 gas from the atmosphere, by absorption through stomatal pores. As CO_2 dissolves in water on the surfaces of leaf mesophyll cells, it combines with water to form carbonic acid, which can dissociate into bicarbonate ion and H^+.

$$H_2O + CO_2 \longleftrightarrow H_2CO_3 \longleftrightarrow HCO_3^- + H^+$$
$$\text{Carbonic acid} \quad \text{Bicarbonate ion}$$

Leaf mesophyll cells then absorb the carbonic acid and bicarbonate ions. Under experimental conditions, plant photosynthetic rates increase with CO_2 concentration until the Calvin cycle enzyme rubisco has become fully supplied, that is, saturated with CO_2. (To review the role of rubisco and the Calvin cycle in photosynthesis, see Chapter 8.)

CO$_2$ Limitation In nature, plants are often unable to obtain enough CO_2 for maximal photosynthesis. As a result, CO_2 limits agricultural crop productivity. This is partly because the modern atmospheric concentration of CO_2 is only 350 $\mu L/L$ (0.035% of atmospheric content), whereas 1,000 $\mu L/L$ of CO_2 would be required to saturate photosynthesis in most plants. In addition, oxygen can compete with carbon dioxide for the active site in rubisco, causing a reduction in photosynthetic productivity known as photorespiration (see Chapter 8).

Evidence that plant photosynthesis can be limited by CO_2 availability is provided by studies of crops grown in greenhouses where atmospheric gas content can be controlled. When supplied with air enriched with CO_2, tomatoes, cucumbers, leafy vegetables, and some other crops can double their growth rate. Experimental addition of CO_2 to natural environments also results in increased plant productivity (defined as grams of organic carbon produced per square meter of land per year). For example, a recent report authored by Richard Norby and other environmental biologists revealed that the productivity of several types of temperate forests increased by 23% under elevated CO_2 conditions.

Plants that experience hot, dry conditions are particularly vulnerable to low CO_2 levels when their stomata close to conserve water. When stomata are closed, plants cannot absorb CO_2, which limits photosynthesis. Many plants possess structural and biochemical adaptations that help them cope with CO_2 limitation by improving CO_2 absorption. Such adaptations may be helpful in genetic engineering efforts to improve crop productivity.

Engineering CO$_2$-Absorption Adaptations into Crop Plants
Many plants that evolved in arid or hot environments display an adaptation known as C_4 photosynthesis, which improves productivity by aiding CO_2 absorption. An amazing 30,000 plant species utilize C_4 photosynthesis. In addition, 7,500 other plant species possess a variation of C_4 photosynthesis known as CAM (crassulacean acid metabolism; see Chapter 8). However, many other plants that grow in hot environments lack C_4 photosynthesis. Rice, a staple crop for much of the world's population, is a prominent example. Agricultural scientists envision using genetic engineering techniques to endow rice with C_4 photosynthesis with the goal of increasing this crop's productivity.

To understand how this valuable crop modification might be accomplished, let's briefly review C_4 photosynthesis. Both C_4 and CAM photosynthesis rely on the leaf mesophyll enzyme phosphoenolpyruvate (PEP) carboxylase. PEP carboxylase binds CO_2 more readily than does rubisco, thereby effectively trapping scarce CO_2 within four-carbon organic acids. You may recall that this process is the basis for the term C_4 photosynthesis. In the leaves of most C_4 plants such as maize (corn), the four-carbon organic acids move through intracellular connections from mesophyll cells into specialized bundle-sheath cells that surround leaf veins. Within the bundle-sheath cells, enzymes release CO_2 from the organic acids, and the CO_2 is bound by rubisco (see Chapter 8).

Molecular biologists have succeeded in transferring maize genes encoding PEP carboxylase and other C_4 enzymes into the rice genome. However, this procedure primarily improved plant stress tolerance, rather than photosynthetic performance, which was the primary goal. Because bundle-sheath cells are often central to C_4 photosynthesis, an alternative strategy is to transform rice with the genes that control development of these specialized cells. Important progress toward this goal was made by plant biologist Jane Langdale and colleagues, who discovered a family of genes known as the *G* (for "golden") genes. In maize leaves, which display C_4 photosynthesis, a particular *G* gene family member is expressed primarily in bundle-sheath cells, whereas a different gene of the same family is expressed in mesophyll cells. These results will help to identify more precisely the genes essential to bundle-sheath cell specialization, which are potentially useful in genetic engineering efforts.

Some other plants accomplish C_4 photosynthesis without specialized bundle-sheath cells, offering additional strategies for crop genetic engineering. For example, PEP carboxylase and rubisco occupy the same mesophyll cells in the aquatic plant *Hydrilla verticillata*. In this case, PEP carboxylase is localized within the cytoplasm, while rubisco occurs within plastids. *Borszczowia aralocaspica*, native to salty, near-desert regions of central Asia, and *Bienertia cycloptera*, native to Iranian deserts, are likewise remarkable for single-cell C_4 photosynthesis. Plant biologists Elena Voznesenskaya, Gerald Edwards, and colleagues discovered that *B. aralocaspica* and *B. cycloptera* achieve C_4 photosynthesis within the same photosynthetic cells by producing two types of plastids, one type containing PEP carboxylase and the other containing rubisco (**Figure 37.5**). These and other recent discoveries reveal that wild plants have acquired diverse adaptations that improve CO_2 absorption and show that both basic and applied research are essential to efforts to improve food crops.

Figure 37.5 A CO$_2$-acquisition adaptation.   In contrast to most C$_4$ plants, *Bienertia cycloptera* is able to conduct C$_4$ photosynthesis within the confines of a single cell. In this fluorescence photograph of a *B. cycloptera* cell, the red plastids at the cell periphery function like the mesophyll cells of most C$_4$ plants. In contrast, the red plastids clustered near the green nucleus function much like the bundle-sheath cells of most C$_4$ plants. The plastids appear red because chlorophyll emits red light when it fluoresces.

Biological inquiry: Which plastids should have more rubisco enzyme in them, plastids at the cell periphery or those clustered near the nucleus?

Water Is an Essential Plant Resource

Water is essential to plants for several reasons. As a nutrient, water is the source of most of the hydrogen atoms and some of the oxygen atoms in organic compounds (see Table 37.1). You may recall that hydrogen is incorporated during the process of photosynthesis, and oxygen is incorporated during hydrolysis reactions. Water is also the solvent for other mineral nutrients and is the main transport medium in plants, allowing movement of minerals and other solutes throughout the plant body. Cytoplasmic and vacuolar water also help to support plants by maintaining hydrostatic pressure on the cell wall.

Though plants vary in water content, water typically makes up about 90% of the weight of living plants. Most plants die when their water content falls below half of the amount normal for that particular species. While many types of desiccation-resistant plants display adaptations that allow them to survive for extended periods in nearly dry conditions, all plants require an adequate supply of water for active metabolism and growth. Water is also a source of dissolved mineral nutrients for aquatic plants, but for terrestrial plants, soil is the main source of dissolved nutrients.

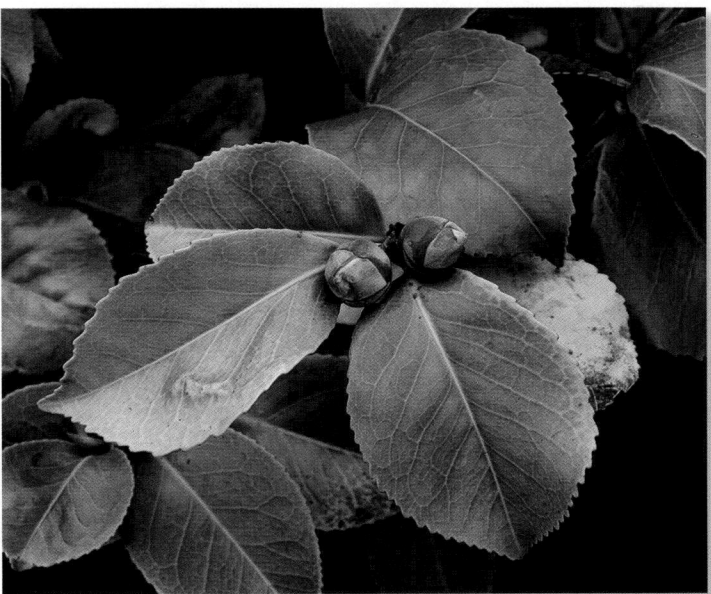

Figure 37.6 Chlorosis as a symptom of mineral deficiency. This camellia plant is suffering from an iron deficiency, as revealed by the yellow leaves, a symptom known as chlorosis.

Biological inquiry: Does chlorosis always indicate iron deficiency?

Soil and Water Provide Additional Plant Nutrients

In addition to light, CO$_2$, and water, plants require additional elements. Macronutrients—the essential elements required in relatively large amounts—include nitrogen, potassium, calcium, magnesium, phosphorus, and sulfur. Nitrogen and phosphorus play many important roles in plants (see Table 37.1), and potassium helps to maintain or alter plant cell water content. Calcium is used in the construction of cell walls and as a cellular signal, while magnesium is an essential component of chlorophyll. Sulfur occurs in the amino acids cysteine and methionine, as well as other cell constituents. In addition, the distinctive flavors of onions, garlic, broccoli, mustard, and horseradish result from plant defense compounds that contain sulfur. Micronutrients—essential elements that plants require in relatively small amounts—include chlorine, iron, manganese, boron, zinc, sodium, copper, molybdenum, and nickel (see Table 37.1). Terrestrial plants obtain these essential elements in the form of ions dissolved in soil water (see Figure 37.1).

Plant biologists have quantitatively analyzed the mineral ion requirements of plants by growing them hydroponically, that is, by bathing plant roots in a water solution to which minerals are added in various combinations and amounts. Hydroponic studies reveal that when plants lack an adequate supply of an essential mineral nutrient, they display characteristic deficiency symptoms. Such symptoms include failure to reproduce, tissue death, and changes in leaf color. Yellowing of leaves, known as **chlorosis**, is a common mineral deficiency symptom, because many nutrients are needed for chlorophyll production (**Figure 37.6**). Deficiency symptoms are often specific to particular plants. For example, zinc deficiency causes chlorosis in corn, but in fruit trees zinc deficiency causes smaller than normal leaves.

Such deficiency symptoms provide clues that crop, garden, or houseplant soils are deficient in one or more essential minerals. Farmers commonly have their soil analyzed for mineral content, and if the soil is deficient, they may amend it with **fertilizers**, soil additions that enhance plant growth by providing essential elements. Fertilizers are also sold for use in enriching garden soil or growing houseplants.

37.2 The Role of Soil in Plant Nutrition

Soil is an essential resource for most wild and cultivated plants, providing water and other essential nutrients. For this reason, extensive loss of soil by wind and water erosion is of wide concern. Soils vary greatly in fertility, that is, their ability to support plant growth. Thus, plants of many types have had to adapt to the challenges of obtaining nutrients from poor soils. In this section, we explore soil structure and chemistry from the perspective of plant growth and describe how plants take up nutrients from the soil.

The Physical Structure of Soils Affects Their Water-Holding Capacity and Fertility

Natural soils display layers, known as **soil horizons** (**Figure 37.7a**). The remains of dead plants and other organisms form a layer of litter above the topsoil, the topmost layer of soil. Many of the inorganic minerals and organic materials that enrich high-quality topsoil arise from the activities of microorganisms that decompose the litter. In this way, the minerals contained in living plants are eventually recycled to subsequent generations. Beneath the topsoil lie layers called subsoil and soil base, which are largely composed of mineral materials, and bedrock. Plant roots play an important role in conveying deep-lying minerals to the surface, thereby helping to enrich topsoil.

Soil horizons vary in composition and thickness, depending on various factors—including climate, vegetation, and bedrock type—and the history of human impact. For example, natural grasslands produce deep, rich topsoil that is used for cropland in many regions of the world. In dramatic contrast, tropical rain forests often have only thin layers of topsoil; their soils are composed mostly of inorganic materials that are not useful to plants (**Figure 37.7b,c**). However, certain Amazonian soils, known as "terra preta," are rich and black with organic materials, because ancient inhabitants fertilized extensive gardens with large amounts of decayed plant material dredged from nearby riverbanks. In other parts of the world, however, human activities have increased topsoil loss by wind or water erosion, a worldwide agricultural concern.

These examples illustrate that soils are composed of both inorganic minerals and organic materials. The sizes of inorganic particles and the proportions of organic to inorganic materials are used to classify soils into different types. Soils also display variation in amount of aeration, water-holding capacity, pH, and mineral content. All of these soil properties affect plant growth. For this reason, we will take a closer look at the organic and inorganic constituents of soils, and the role of fertilizers.

Organic Soil Constituents The organic constituents of soils are collectively known as **humus**. Humus is largely derived from plant detritus, the partially decomposed remains of plants, although animal wastes and decayed animal bodies also contribute to the organic content of soils. Organic soil constituents derived from plants include chemically stable lignin and cutin, as well as organic acids containing phenolic groups that are known as humic acids. These organic compounds give many soils their characteristic brown to black coloration. Soils that contain less than 1% organic materials are said to be humus-poor, whereas soils containing more than 8% are considered to be humus-rich. Though some plants are adapted to humus-poor soils, optimal growth generally requires soil that has an organic content close to 8%, because humus helps soils hold more water and ions. Humus particles accomplish this by electrostatically binding positively charged ions, cations such as K^+, to negative charges at their surfaces. In this way, humus helps to

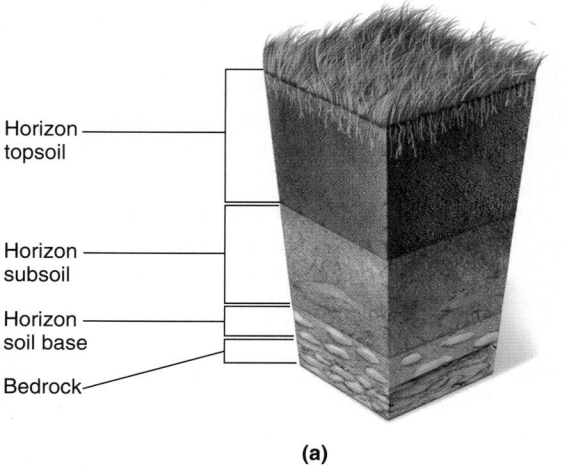

Horizon topsoil

Horizon subsoil

Horizon soil base

Bedrock

(a)   (b)   (c)

Figure 37.7 Soil horizons, the structural layers of soil.   (a) Diagram showing general soil structure. **(b)** A vertical view of an agricultural soil, showing a relatively deep layer of dark topsoil. **(c)** A vertical view of a tropical rain forest soil, showing a thin layer of dark topsoil.

prevent ions from washing out of soils, thereby maintaining soil fertility. Humus also gives most soils their soft consistency, a property that fosters root growth.

The benefits of humus explain why farmers often plow manure and plant residues left after crop harvesting into agricultural soils. Garden soils that are low in organic materials can be amended with peat moss or with composted (partially decomposed) manure, which is available at garden stores. You can produce your own humus-rich compost by layering vegetable waste from the kitchen and yard with soil, turning the pile occasionally to introduce the oxygen necessary for decomposition (**Figure 37.8**).

Inorganic Soil Constituents Inorganic materials in soil are derived from the physical and chemical breakdown of rock, a process known as **weathering**. Rock is physically weathered by changes such as cycles of freezing and thawing. Lichens and plant roots may produce organic acids that contribute to chemical weathering of rocks. During chemical weathering, soluble salts are washed out and minerals are hydrolyzed or oxidized. **Leaching** is the dissolution and removal of inorganic ions as water percolates through materials. Heavy rainfall can reduce the fertility of soils by leaching large amounts of nutrients from them. The leached minerals often end up in natural bodies of water, where they can foster the growth of cyanobacteria, algae, and aquatic plants.

Inorganic soil materials occur as particles that can be categorized according to their size as gravel, coarse sand, fine sand, silt, or clay (**Figure 37.9**). Coarse sand grains range from 200 to 2,000 μm (micrometers) in diameter, whereas fine sand particles can be as small as 20 μm. Particles of silt range from 2 to 20 μm in diameter, and clay particles are less than 2 μm in size. Soils can be classified according to their relative content of coarse and fine materials. For example, soils that contain 45% or more sand and 35% or less clay are classified as sandy soils. The other main types are silty, clay, and loam soils. Loam soil contains a mixture of sand, silt, and clay.

Because of their size differences, sand, silt, and clay particles confer different properties on soils. The relatively large size and irregular shapes of sand particles allow air and water to move rapidly through sandy soils, which are said to be porous (**Figure 37.10**). Sandy soils are well aerated, which is favorable to the growth of plant roots, because they require oxygen for cellular respiration. However, rapid percolation of water through sandy soils reduces the amount of water available to the roots and leaches minerals from the soil.

Figure 37.9   The relative sizes of inorganic soil components.

Figure 37.8   Composting.   Compost can be produced by layering small amounts of soil with vegetable waste from the kitchen or garden and periodically turning the pile to introduce the oxygen needed for decomposition. Compost can be used to increase the humus and mineral content of garden soils.

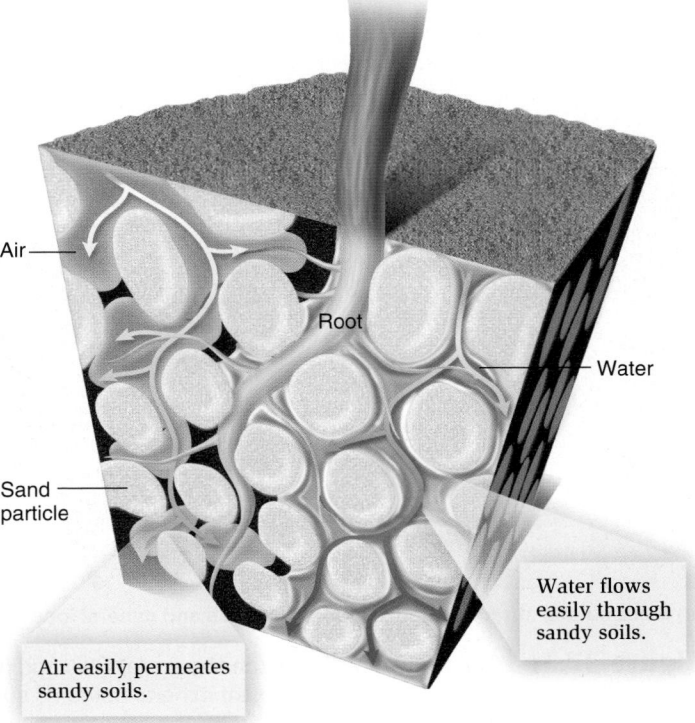

Figure 37.10   The movement of air and water in sandy soil.

Biological inquiry: In which type of soil is nutrient leaching a more common problem, sandy or clay soils?

Silt and clay particles fit closely together, so soils containing larger amounts of these materials are less porous than sandy soils. Water percolates less easily through silty and clay soils, which therefore retain more ionic mineral nutrients than do sandy soils. Like organic humus, clay particles have negative charges on their surfaces that electrostatically bind cations (such as NH_4^+, Ca^{2+}, and Fe^{2+}) (**Figure 37.11a**). Cations having higher valence numbers (such as Fe^{2+}) are bound more tightly than ions having lower valence numbers (NH_4^+). Hydrogen ions (H^+, protons) are able to replace mineral cations on the surfaces of humus or clay particles, a process known as **cation exchange** (**Figure 37.11b**). Cation exchange releases cations to soil water, making them more available for uptake by plant roots. However, free ions are also more easily washed out of soil. If the H^+ concentration becomes too high, large numbers of mineral ions are released and can be leached from soil by heavy rainfall. Such leached minerals may include heavy metals such as aluminum that would otherwise be bound in soil. Cation exchange is also the mechanism by which acid rain, which adds H^+ to soil, causes loss of soil fertility and pollution of streams with toxic substances that can harm human health.

Despite their water- and mineral-retention features, silt- and clay-rich soils may be poorly aerated and therefore unfavorable to root growth. Gardeners often mix organic materials and sand into silt or clay-rich soils to improve their aeration properties. A loam soil is the preferred soil for agriculture because it combines the aeration provided by sand with the mineral and water retention capacity of clay and silt.

The Role of Fertilizers The addition of fertilizer to soils can compensate for deficiencies in soil humus or mineral content and thus improve soil fertility. Fertilizers occur in organic and inorganic forms. Organic fertilizers are those in which most of the minerals are bound to organic molecules and are thus released relatively slowly. Organic fertilizers are often produced on the same farm where they will subsequently be used to enrich soils. They play an important role in **organic farming**, a term referring to the production of crops without the use of inorganic fertilizers, growth substances, and pesticides. Manure and compost are examples of organic fertilizers.

In contrast, inorganic fertilizers consist of inorganic minerals, which are immediately useful to plants but are more easily leached from soils by heavy rainfall. Nitrogen (N), phosphorus (P), and potassium (K) are the mineral nutrients that most frequently limit crop growth. For this reason, these minerals are the main components of the most common type of commercial inorganic fertilizers, known as NPK fertilizers. Such fertilizers are available in different ratios of minerals, which are optimal for different types of plants.

Excessive application of fertilizers to farm fields and lawns is undesirable because minerals not taken up by plant roots are easily washed by rain into waterways. In aquatic habitats, excess mineral nutrients fuel large growths of cyanobacteria, algae, and aquatic plants that harm other aquatic life-forms. For example, large areas of the Gulf of Mexico and other coastal regions are known as "dead zones" because microbial decomposition of large algal populations has depleted oxygen from the water, suffocating the animal life. These large populations of algae are fostered by fertilizers that wash from farm soils into rivers, such as the Mississippi River, that drain into coastal oceans. More careful application of fertilizers, and planting stream and river edges with vegetation that helps to absorb mineral nutrients, are actions that can reduce or prevent dead zones.

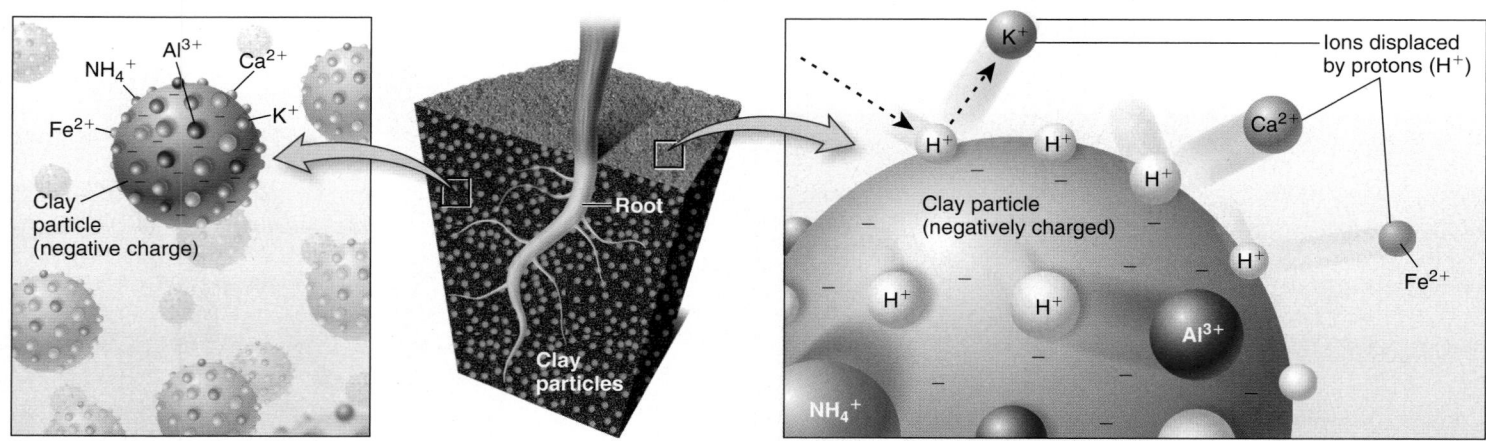

(a) Electrostatic attraction between clay particles and mineral ions

(b) Cation exchange

Figure 37.11 Cation binding and exchange. **(a)** Clay and humus particles display negative electrostatic surface charges that bind cations. Bound cations include not only plant mineral nutrients such as NH_4^+ but also cations that are not plant nutrients, such as Al^{3+}. **(b)** Cation exchange occurs when protons displace cations, releasing them to soil water. This process makes cations more available for uptake by plant roots, but it also increases the potential for cations to leach away during rains or floods. Protons in acid soils or arising from acid rain displace so many cations from soil that its fertility decreases, and toxic ions such as Al^{3+} more readily damage plants.

Plants Require Fixed Nitrogen

Nitrogen is frequently limiting to plant growth in nature and in crop fields, because large amounts of it are required for plant synthesis of amino acids, nucleotides, and alkaloids, among many other cellular constituents. Nitrogen is the largest component of plants by mass after carbon, oxygen, and hydrogen. Although the Earth's atmosphere is 78% nitrogen gas (N_2), plants cannot utilize nitrogen in this form.

To be of use to plants, soil nitrogen must occur in a combined form, such as ammonia (NH_3), ammonium ion (NH_4^+), or nitrate ion (NO_3^-). Combined forms of nitrogen are also known as **fixed nitrogen**. Ammonia and its dissolved form—ammonium ion—can be used directly for amino acid production by plants, explaining why ammonia is often applied as a fertilizer to farm fields in springtime. However, in oxygen-rich soils, microorganisms oxidize much of the ammonium to nitrate, so nitrate may be the form in which fixed nitrogen enters most plants.

Nitrate is imported into root cells by means of plasma membrane transporter proteins (**Figure 37.12**). Once inside the plants, the nitrate can be transported in the xylem for use elsewhere, stored for later use, or assimilated for immediate use.

Cells store nitrate by transporting it into vacuoles. To assimilate nitrate, plant cells must reduce it to ammonium ion in a two-step process. First, the cytosolic enzyme nitrate reductase reduces nitrate (NO_3^-) to nitrite (NO_2^-). Second, in plastids, nitrite is reduced to ammonium ion by nitrite reductase. Another plastid enzyme, glutamine synthetase, uses ammonium ion to produce the amino acid glutamine.

Atmospheric Nitrogen Is Fixed by Natural and Industrial Processes

Much of the fixed nitrogen in soils has been recycled from compounds previously utilized by other organisms. New fixed nitrogen can be added to soils by the action of lightning, fire, and air pollution, as well as biological and industrial nitrogen fixation. **Nitrogen fixation** is the process by which atmospheric nitrogen gas is combined with hydrogen to produce ammonia. Most of the fixed nitrogen in soils is produced by **biological nitrogen fixation**, which is performed in nature only by certain prokaryotes. Nitrogen fertilizers applied to crops are produced by **industrial nitrogen fixation**, a human activity.

1 **Uptake:** Protons (H^+) pumped from plant cells provide the driving force for the cotransport of protons (H^+) and nitrate (NO_3^-) into the cytosol.

2 **Storage:** Nitrate can be transported into a vacuole for storage.

OR

3a **Assimilation:** Cytoplasmic nitrate reductase converts nitrate (NO_3^-) to nitrite (NO_2^-), which enters plastids.

H^+
ADP + P_i
ATP
Root cell plasma membrane
$2H^+$
NO_3^-
Vacuole
NO_3^-
Nitrate reductase
NO_2^-
Nitrite reductase
Plastid
Glutamine synthetase
NH_4^+
Glutamine

3b In plastids, nitrite reductase converts nitrite to ammonium ion (NH_4^+) and glutamine synthetase uses NH_4^+ to synthesize the amino acid glutamine.

Figure 37.12 How plant cells take up and transform nitrate.

Biological Nitrogen Fixation by Bacteria Nitrogen-fixing pro-karyotes include many types of cyanobacteria, which are photosynthetic and occur in oceans, lakes, and other aquatic systems, as well as in surface soil crusts (**Figure 37.13**). Various types of nonphotosynthetic bacteria living in water and soil are also able to fix nitrogen; actinobacteria and the genera *Clostridium*, *Klebsiella*, and *Azotobacter* are examples. Nitrogen-fixing prokaryotes often excrete a substantial amount of fixed nitrogen, and their death makes still more fixed nitrogen available to plants. Many plants have nitrogen-fixing, prokaryotic symbionts that transfer fixed nitrogen directly to plant cells. Nitrogen-fixation

Figure 37.13 **A soil surface crust that includes nitrogen-fixing, soil-enriching cyanobacteria.** Such crusts are widespread in grasslands and other arid regions.

symbioses are so important in nature and in agriculture that they are discussed in more detail in a later section of this chapter.

All nitrogen-fixing prokaryotes utilize relatively large amounts of ATP and an enzyme known as **nitrogenase** to fix nitrogen (**Figure 37.14**). This process occurs in three steps. The first step is that nitrogenase binds to a molecule of nitrogen gas (N_2). In the second step, the bound nitrogen is reduced by the addition of two hydrogen atoms (2 H), a reaction powered by the breakdown of ATP. Such a reduction occurs three times, with the addition of a total of three hydrogen atoms to each nitrogen atom. In a third and final step, two molecules of ammonia (NH_3) are released and dissolve in cell water to form ammonium ions. The nitrogenase enzyme is then free to bind more nitrogen gas.

Because the O_2 molecule resembles N_2, oxygen can bind to the active site of nitrogenase. Oxygen-binding disables nitrogenase, thereby stopping nitrogen fixation. Many of the genes involved in prokaryotic nitrogen fixation are known, and crop scientists are working to genetically engineer nitrogen fixation capacity into crop plants such as rice and maize (corn). However, the vulnerability of nitrogenase to oxygen means that this enzyme may need to be altered so that it binds oxygen less readily, or plants must be engineered with some mechanism that protects nitrogenase from oxygen.

Industrial Nitrogen Fixation Worldwide, farmers apply more than 80 million metric tons of nitrogen fertilizer per year. The fixed nitrogen found in fertilizer is produced industrially from nitrogen gas by means of a procedure invented by German chemists Fritz Haber and Carl Bosch in 1909. The reduction of N_2 gas to NH_3 is energetically favorable at room temperature but the activation energy is very high, so the reaction occurs

Figure 37.14 **The biological process of nitrogen fixation.**

Biological inquiry: What common substance inactivates nitrogenase enzyme by binding to its active site?

extremely slowly. Using an iron catalyst, temperatures of 400–650°C (752–1,202°F), and high pressures (150–400 atmospheres), the Haber-Bosch process generates NH_3 rapidly. However, because of its high energy requirements, industrial nitrogen fixation can be costly from the perspective of many of the world's farmers. The high cost of fertilizers helps to explain why agricultural scientists are so interested in the possibility of genetically engineering nitrogen fixation into crop plants.

Plants Display Adaptations for Acquiring Phosphate

Phosphorus (P) is another soil mineral that often limits plant growth. Plants obtain phosphorus from the ion known as phosphate, which occurs in the soil in three dissolved forms, H_3PO_4, $H_2PO_4^-$, and HPO_4^{2-}. The HPO_4^{2-} ion (called hydrogen phosphate ion) is the form most commonly absorbed by plants. Phosphate uptake involves ATP-requiring proton cotransport, as is the case for other anions such as nitrate and sulfate. The uptake of minerals at the root hair surface is discussed further in Chapter 38.

Although phosphate is abundant in soil, it is often unavailable to plants. One reason is that phosphate forms tightly bound complexes with clay, iron and aluminum oxides, and calcium carbonate in soils. In addition, soil microbes convert phosphate ions into organic compounds that are not taken up by plants.

Because plants need copious supplies of phosphate for a variety of cell processes, they have evolved various adaptations that increase their ability to obtain phosphorus from soil. A common adaptation for acquiring phosphorus is the symbiotic association of plant roots with various types of fungi (see Section 37.3). In addition, plants that grow in soils having low phosphorous content may produce more highly branched roots and more and longer root hairs. Plant roots also secrete protons and organic acids such as citrate and malate into the soil; these help release phosphorus from inorganic complexes. For example, the plant *Lupinus alba* releases as much as 25% of its total photosynthetic carbon into the soil as organic acids—a high price to pay, but apparently one that is essential for the plant to obtain sufficient phosphorus. Plants may also secrete phosphatase enzymes from roots; these enzymes release phosphorus from organic compounds in the soil. The general process by which phosphorus, nitrogen, CO_2, and other minerals are released from organic compounds is called **mineralization**.

Farmers and gardeners apply phosphate-rich fertilizers to crop fields and gardens as a way of preventing phosphorous deficiencies, which reduce yields. Phosphorous fertilizers are obtained from phosphate-rich mineral deposits, but experts have warned that inexpensive sources of phosphate will be exhausted within the next 90 years. Consequently, there is much interest in devising ways to maximize the efficiency by which plants are able to take up and use phosphorus. Genetic engineering to produce "smart plants" may offer some options.

FEATURE INVESTIGATION

Hammond and Colleagues Engineered "Smart Plants" That Can Communicate Their Phosphate Needs

If farmers could apply fertilizer to crops in the precise amounts needed by plants, they would not only save money, but less fertilizer would run off fields into aquatic habitats, where it can lead to harmful ecological effects. Plant biologists have used genetic engineering to produce "smart plants" that signal impending nutrient deficiency. Such plants could serve as sentinels, reflecting the conditions of an entire field. With this information, farmers could apply just enough mineral nutrients to prevent deficiency, thereby avoiding overapplication of fertilizers. Recent research has provided proof that this concept can work, as shown in **Figure 37.15**.

Working with the model plant *Arabidopsis*, John Hammond, Philip White, and their associates used microarray technology (see Chapter 20) to identify some of the genes whose expression changes when plants are transferred from hydroponic nutrient solutions containing sufficient phosphorus to solutions lacking phosphorus. They found that only four genes were expressed differently within 24 hours after plants were transferred to solutions lacking phosphorus, but that an additional 61 genes were affected 100 hours after phosphorus was no longer available. This timing is important because genes expressed between 24 and 72 hours after phosphate removal are considered useful as phosphorous monitors. During this window of time, plant tissue levels of phosphate decreased but had not yet affected plant growth. One gene that met this timing criterion was *SQD1*, which is required for the synthesis of sulfur-containing lipids. Expression of *SQD1* allows plants to respond to low phosphorous levels by replacing plastid phospholipids with sulfur-containing lipids, thereby reducing their phosphorous requirement. This evidence suggested that smart plants could be engineered to communicate impending phosphate deficiency when they express *SQD1*.

To make smart plants, the researchers first placed the reporter gene *GUS* under the control of the *SQD1* promoter and transformed this gene into *Arabidopsis* plants. The researchers then grew these genetically engineered plants for various time periods in hydroponic solutions of differing phosphate levels. After different time periods, they removed leaves and chemically treated them with a compound that produces a blue color when the *GUS* gene is expressed. Because the *GUS* gene was under the control of the *SQD1* promoter, leaves from plants that were developing phosphate deficiency turned blue! These smart plants were able to communicate impending phosphorous deficiency in time for a farmer to apply fertilizer.

Figure 37.15   The experiment of Hammond and colleagues showed that plants can be engineered to communicate changes in the level of nutrients.

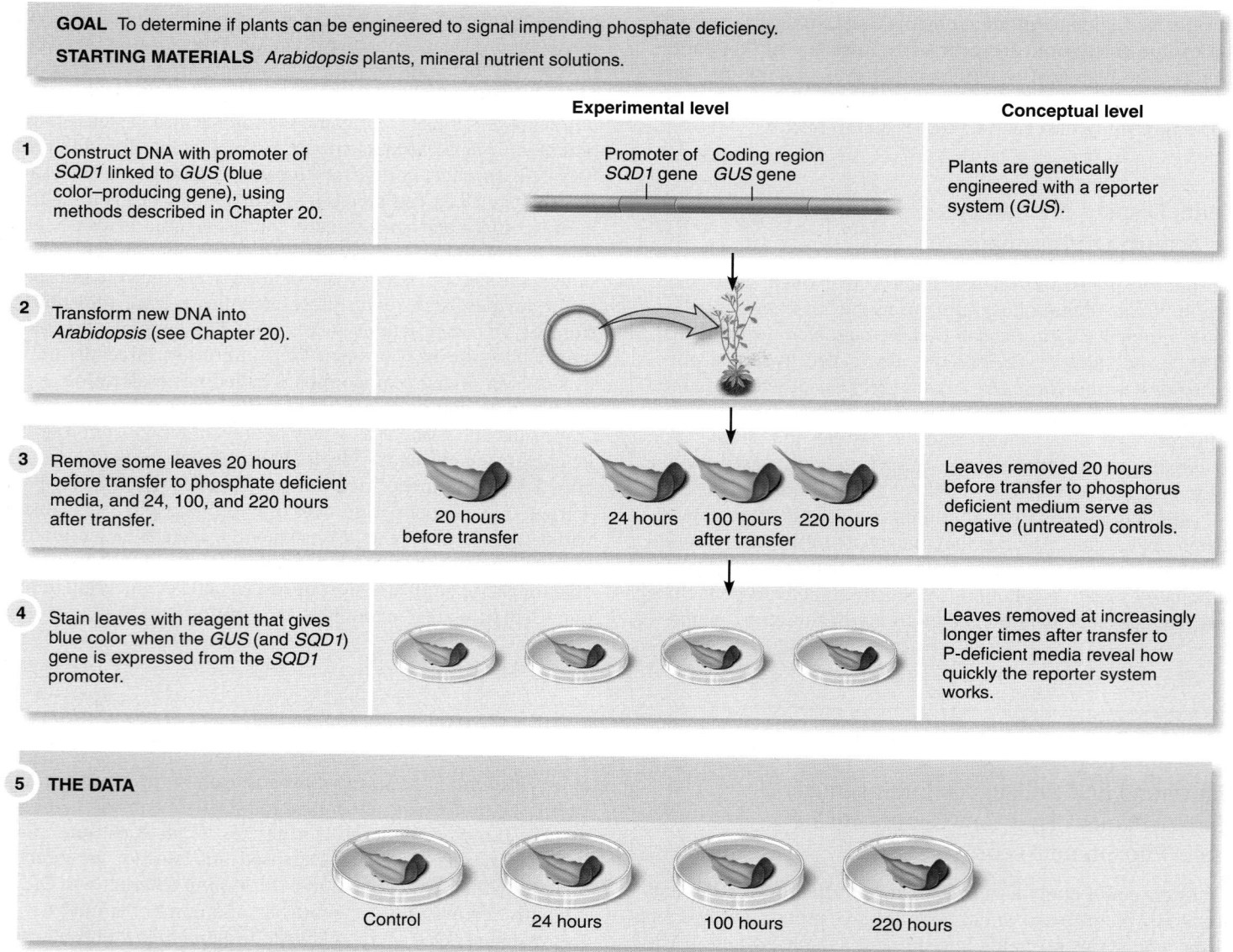

GOAL   To determine if plants can be engineered to signal impending phosphate deficiency.

STARTING MATERIALS   *Arabidopsis* plants, mineral nutrient solutions.

Experimental level

Conceptual level

1   Construct DNA with promoter of *SQD1* linked to *GUS* (blue color–producing gene), using methods described in Chapter 20.

Promoter of *SQD1* gene   Coding region *GUS* gene

Plants are genetically engineered with a reporter system (*GUS*).

2   Transform new DNA into *Arabidopsis* (see Chapter 20).

3   Remove some leaves 20 hours before transfer to phosphate deficient media, and 24, 100, and 220 hours after transfer.

20 hours before transfer

24 hours   100 hours   220 hours after transfer

Leaves removed 20 hours before transfer to phosphorus deficient medium serve as negative (untreated) controls.

4   Stain leaves with reagent that gives blue color when the *GUS* (and *SQD1*) gene is expressed from the *SQD1* promoter.

Leaves removed at increasingly longer times after transfer to P-deficient media reveal how quickly the reporter system works.

5   **THE DATA**

Control   24 hours   100 hours   220 hours

In a later and more extensive study of gene expression in *Arabidopsis*, molecular biologists Julie Misson, Marie-Christine Thibaud, and associates discovered that phosphate induces 612 genes and represses 254 genes. Some of these genes may encode proteins useful in monitoring plant phosphorous status. If smart plant technology can be developed for crop plants, farmers may be able to monitor and fertilize fields with much greater precision.

Some Common Soil Minerals Can Be Toxic to Plants

Some soil minerals, including micronutrients, inhibit plant growth when present at high concentrations and thus are considered toxic to plants. Minerals that may occur in toxic concentrations in soils include aluminum, lead, cadmium, chromium, mercury, and uranium, and the micronutrients boron, copper, nickel, and zinc. These minerals enter plant roots through water-soaked cell walls and intercellular spaces. Boron is an example of an element that functions as a micronutrient when present in low levels but that is toxic at higher concentrations. Aluminum is not known to be an essential element, but has important ecological effects on plants even in very low concentrations.

Boron Toxicity   At concentrations less than 10 μg/g of plant tissue, boron (B) is an essential micronutrient, required for cross-linking polysaccharides in plant cell walls (see Table 37.1).

However, boron is toxic to wheat plants, for example, in concentrations of 100 µg/g or higher. Boron is easily taken up by plant tissues because this mineral occurs as uncharged boric acid, $B(OH)_3$, which readily crosses membranes. When applying fertilizer to relieve boron deficiency, farmers must take care not to add too much.

Plant scientists Junpei Takano, Toru Fujiwara, and their associates recently reported that the model flowering plant *Arabidopsis* has cellular adaptations that regulate the concentration of boron. These investigators demonstrated that when boron is present in small amounts, a boron transporter protein (BOR1) is expressed in the plasma membranes of root pericycle tissue. In this case, BOR1 moves boric acid into root xylem for transport to shoots, where boron can be used in producing new cell walls. However, when potentially harmful amounts of boric acid are applied to roots, endocytotic vesicles move the BOR1 protein to the cell vacuole for degradation. This process prevents too much boric acid from moving into shoots. Together, these processes protect plants from boron deficiency and toxicity.

Aluminum Toxicity   Aluminum, in the form of Al^{3+}, is one of the most common minerals in soil, so small amounts may seep into plant tissues. However, aluminum is not known to be an essential element and is toxic to plants even at micromolar concentrations. Aluminum binds to cellular carbohydrates, phosphates such as ATP, and sulfates, thereby inhibiting root elongation and uptake of minerals and water. Aluminum toxicity is primarily a problem for plants rooted in acidic soils, which are common in the tropics and subtropics, including at least 12% of cultivated land. In addition, aluminum toxicity can occur in temperate regions that receive acid rain. In the cation exchange process, the abundant protons in acidic soils or acid rain replace aluminum ions bound to soil particles (see Figure 37.11) and allow the released aluminum ions to enter plant roots more easily.

Some plants defend themselves against aluminum toxicity by secreting organic acids such as citrate, oxalate, and malate from their root tips. In the soil, these acids form complexes with Al^{3+}, which helps to prevent aluminum from entering roots. Within plant cells, certain proteins also bind aluminum and other metals, thereby protecting cell components. Some plants accumulate aluminum or other toxic metals in their tissues, often sequestering the metals in cell vacuoles, where they do little harm. For example, when hydrangea plants are grown in acid soil, aluminum accumulations cause the showy sepals of the flowers to change from pink to blue (**Figure 37.16**). Many other kinds of plants protect themselves from toxic metals by storing them in plant tissues.

Hyperaccumulators and Phytoremediation   Some plants are known as hyperaccumulators because they accumulate and safely bind high concentrations of toxic metal minerals within their tissues, an adaptation that allows such plants to grow on metal-rich soils. Some 400 different plant species are hyperaccumulators. Examples include the alpine pennycress (*Thlaspi caerulescens*), which accumulates zinc and cadmium, and the brake fern (*Pteris vittata*), which accumulates arsenic. These

Figure 37.16   **An adaptation protecting against aluminum toxicity.**   The showy sepals of hydrangea (*Hydrangea macrophylla*) flowers turn from pink to blue when soil conditions become more acidic. Acidic conditions cause soil aluminum ions to become more mobile and thus to enter plants more easily. Hydrangea stores aluminum ions in the cell vacuole, a protective adaptation that prevents aluminum from damaging other cell components. The aluminum causes pigments that are also located in the vacuoles of sepals to change from pink to blue color.

plant adaptations can be usefully applied to remove harmful metals from soils, a process known as **phytoremediation**. In this process, hyperaccumulator plants are grown on metal-contaminated soils, harvested, and burned to ashes for disposal and/or metal recovery.

37.3   Biological Sources of Plant Nutrients

This section focuses on several fascinating ways in which plants use other organisms as sources of nutrients. Biological sources of plant nutrients include symbiotic fungi or bacteria, the animal prey of carnivorous plants, and green plants that serve as hosts for nonphotosynthetic plant parasites.

Mycorrhizal Associations Help Most Plants Obtain Mineral Nutrients

About 90% of seed plants have symbiotic associations with fungi that live within the tissues of plant roots or that envelop root surfaces (see Chapter 29). These associations are termed **mycorrhizae**; the prefix *myco* refers to fungi, and *rhiza* means root, so the term literally means "fungus root."

In mycorrhizal associations, soil fungi obtain organic food from the roots of a photosynthetic plant host, while the fungal partner supplies the plant with water and mineral nutrients.

Figure 37.17 **Mycorrhizae aid nutrient acquisition.** In all forests, but particularly those of tropical regions, mycorrhizal fungi rapidly collect soil minerals released from decaying organisms and transport them directly to plant roots. Such efficient nutrient cycling bypasses the soil, from which mineral ions can be easily leached by heavy rainfall. This process explains how lush forests can grow on thin, infertile soils.

Figure 37.18 *Gunnera* growing on nitrogen-poor soil. Cyanobacteria that live within small cavities in this plant's stems or leaf petioles provide the plant with fixed nitrogen, which explains how such a large plant can grow on infertile soils.

These fungal-root associations provide an exceptionally efficient way for plants to harvest water and minerals, especially phosphorus, from a much larger volume of soil than is available to roots by themselves. The presence of lush vegetation on tropical rain forest soils of low mineral content is largely due to the ability of mycorrhizae to rapidly absorb mineral nutrients released by decay, and transmit the nutrients directly to plant roots (**Figure 37.17**). In tropical rain forests, mineral nutrients occur within the bodies of living organisms, rather than accumulating in the soil where they could easily be leached away by heavy, frequent rains.

Various species of ghostly pale plants have lost their photosynthetic pigments (see Figure 37.2) and have thus become dependent on organic compounds supplied by fungi that form mycorrhizal associations with a photosynthetic host, such as a nearby tree. In this process, known as mycoheterotrophy, the fungus serves as an underground conduit for the flow of organic nutrients from a green, photosynthetic plant to a heterotrophic plant. Many plant seedlings that grow in the shade of taller plants also use mycoheterotrophy to survive until they are able to obtain enough light for photosynthesis.

Plant-Prokaryote Symbioses Provide Some Plants with Fixed Nitrogen

Some kinds of plants have symbiotic relationships with bacteria that provide them with fixed nitrogen. Though many nitrogen-fixing bacteria live freely within the soil, some form nitrogen-

fixing partnerships with plants, actually living within plant cells or tissues. Such symbioses are advantageous to both partners. The plants provide organic nutrients to the bacteria, while the bacteria supply the plants with a much higher supply of fixed nitrogen than the plants could obtain from most soils. Representatives of three types of prokaryotes—cyanobacteria, actinobacteria, and proteobacteria—are symbiotically associated with specific types of plants. (For more information about the characteristics of these bacterial groups, see Chapter 27.)

Cyanobacterial-Plant Symbioses Although cyanobacteria are themselves photosynthetic, organic compounds supplied by plant partners subsidize the high energy costs of nitrogen fixation. This allows the cyanobacteria to fix more nitrogen than they require, secreting the excess to plant partners. Nitrogen-fixing cyanobacteria form symbioses with some bryophytes, ferns, and gymnosperms, as well as the flowering plant *Gunnera*. This plant, commonly known as the giant or prickly rhubarb, can produce leaves almost 3 m across on stems as long (**Figure 37.18**). Nitrogen-fixing symbionts are advantageous to *Gunnera* because this large plant grows in nitrogen-poor habitats, such as on volcanic slopes in Hawaii. *Gunnera* harbors cyanobacteria within stems and leaf petioles. In these locations, the cyanobacteria can use cyclic electron flow to transform light energy into ATP, needed to produce fixed nitrogen. The presence of nitrogen-fixing cyanobacteria helps to explain why *Gunnera* can grow to dramatic size on poor soils.

Woody Plant—Actinobacteria Symbioses In contrast to cyanobacteria, actinobacteria are heterotrophic, nitrogen-fixing bacteria. Actinobacteria known as *Frankia* occur in nodules formed on the underground roots of certain shrubs or trees, such as alder (*Alnus*) and myrtle (*Myrica*). These plants receive fixed nitrogen from their bacterial partners, which in turn obtain organic nutrients. Woody plants, such as *Ceanothus* shrubs, that have *Frankia* symbionts are able to grow abundantly even in places where soil nitrogen is low. This symbiosis helps to explain why *Ceanothus* covers extensive areas in mountainous chaparral regions of the western U.S.

Legume-Rhizobia Symbioses The nitrogen-fixation symbioses most important in nature and to agriculture involve certain proteobacteria that are collectively known as **rhizobia** (from the Greek *rhiza*, root). Rhizobia live within root cells of wild and cultivated legumes, forming legume-rhizobia symbioses. In nature, legume plants are important sources of fixed nitrogen for other plants. When legumes die, they generate humus that is enriched in fixed nitrogen. Consequently, wild legumes are regarded as particularly valuable members of natural plant communities. For example, plant conservationists are concerned about the survival of a rare Appalachian legume known as running buffalo clover (*Trifolium stoloniferum*), because it is a valuable source of environmental fixed nitrogen.

Important legume crops include soybean, pea, beans, peanuts, clover, and alfalfa. Foods produced from soybeans, peas, beans, and peanuts are valued for their high protein content. Clover and alfalfa are used for animal food and to enrich fields with the fixed nitrogen needed by subsequent food crops. The value of these crops arises from their nitrogen-fixation abilities. The amount of ammonia produced by legume-rhizobia symbioses nearly equals the world's entire industrial production.

Rhizobia can live independently in the soil, but they fix nitrogen only when they occur within lumpy **nodules** that form on legume roots (**Figure 37.19**). Different species of rhizobia preferentially form symbioses with particular plant species. Because of their agricultural importance, these legume-rhizobial symbioses have been extensively studied, and a great deal is now known about the molecular basis of their development. This information is potentially useful in efforts to genetically engineer nitrogen-fixation capacity into nonlegume crops.

Nodule development involves a series of chemical signals sent back and forth between rhizobia and their host plants (**Figure 37.20**). Legumes start this exchange by secreting particular flavonoid compounds from their roots. Recall that flavonoids are phenolic secondary metabolites that play essential roles in plant structure, reproduction, and protection (see Chapter 31). These flavonoids bind to receptors in the plasma membranes of compatible soil rhizobia (Figure 37.20, step 1). In response, the

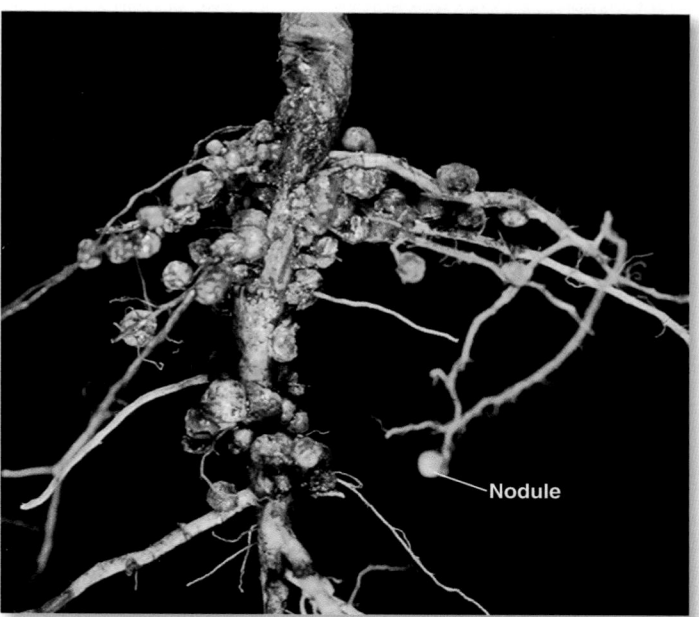

Figure 37.19 **Legume root nodules.** The cells of nodules on the roots of this soybean plant (*Glycine max*) and other legumes contain nitrogen-fixing bacteria known as rhizobia.

rhizobia secrete **Nod factors** (nodulation factors). Each rhizobial species produces Nod factors with distinctive structural variations that can be recognized by the preferred host species. These Nod factors function something like keys that unlock doors, allowing bacteria to enter roots via root hairs. These factors bind to receptors in the membranes of root hair cells in the host plant (step 2). Within minutes after its membrane receptors bind Nod factors, the root hair plasma membrane allows an influx of calcium ions and a few minutes later, root hair calcium concentrations start oscillating rapidly. Root hairs respond to these calcium changes by swelling at their tips and curling around the rhizobia (step 3). The rhizobia then inject infection proteins into root hairs. In response, the root hair cell wall degrades and the plasma membrane forms a tubular thread that the rhizobia enter. Rhizobia move through this infection thread into the root cortex and the tip of the infection thread fuses with the plasma membrane of a cortex cell. The rhizobia are then released into the cortex cell cytoplasm, enclosed by the host membrane (step 4).

Meanwhile, plants produce proteins known as **nodulins** that foster nodule development. Within 18–30 hours after the initial infection, root cortex cells start to divide to form root nodules. Environmental conditions in developing nodules cause rhizobia to undergo changes in their structure and gene expression patterns. The modified rhizobia present in mature nodules are known as **bacteroids** (step 5). Mature nodules also produce vascular tissue that moves nitrogen fixed by bacteroids to the root vascular system for transport throughout the plant. These nodule vascular tissues also supply organic food produced by the legume to their bacteroid partners (step 6).

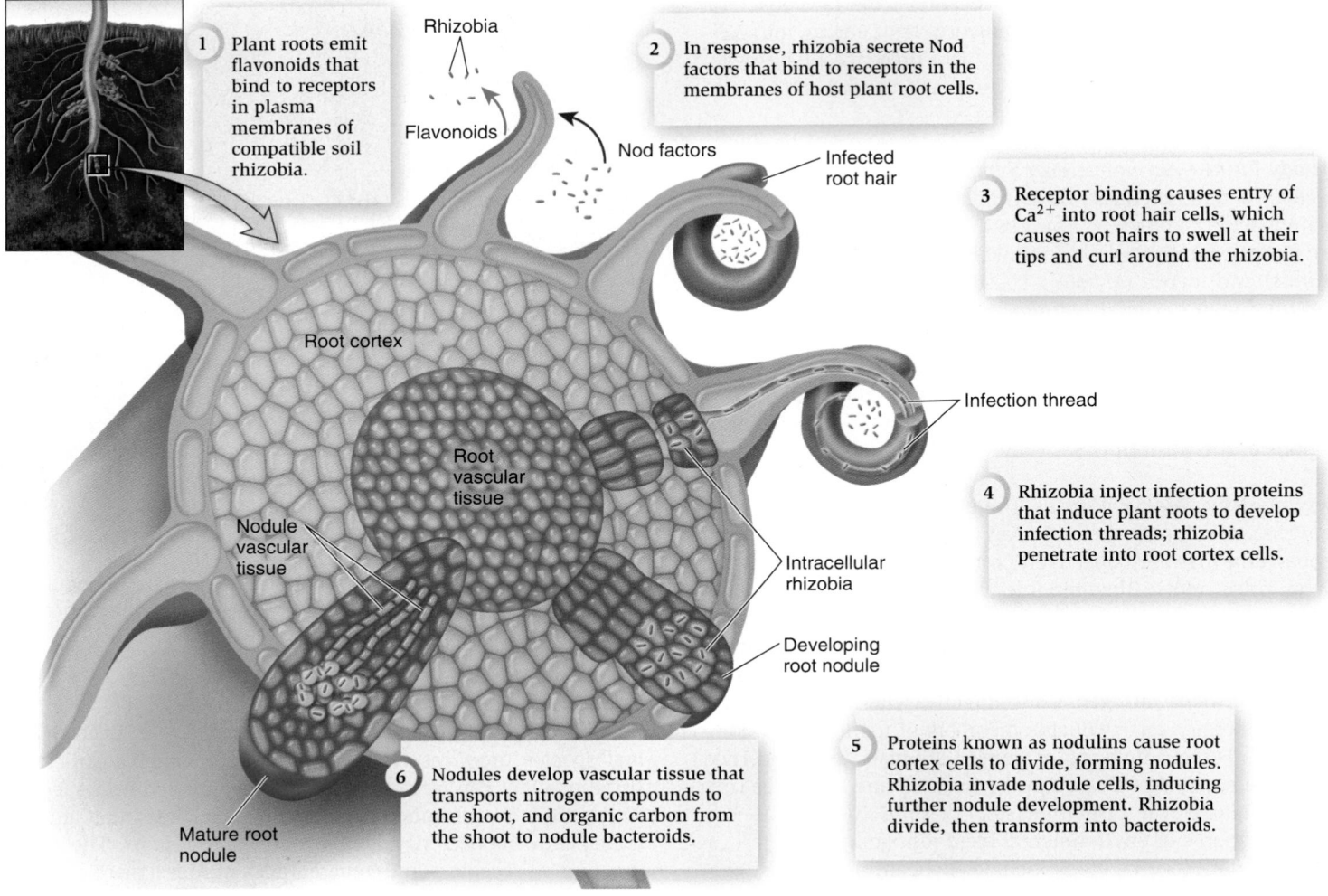

Figure 37.20   Root nodule development.

Image labels:

1. Plant roots emit flavonoids that bind to receptors in plasma membranes of compatible soil rhizobia.

Rhizobia

Flavonoids

2. In response, rhizobia secrete Nod factors that bind to receptors in the membranes of host plant root cells.

Nod factors

Infected root hair

3. Receptor binding causes entry of Ca^{2+} into root hair cells, which causes root hairs to swell at their tips and curl around the rhizobia.

Root cortex

Infection thread

Root vascular tissue

4. Rhizobia inject infection proteins that induce plant roots to develop infection threads; rhizobia penetrate into root cortex cells.

Nodule vascular tissue

Intracellular rhizobia

Developing root nodule

6. Nodules develop vascular tissue that transports nitrogen compounds to the shoot, and organic carbon from the shoot to nodule bacteroids.

5. Proteins known as nodulins cause root cortex cells to divide, forming nodules. Rhizobia invade nodule cells, inducing further nodule development. Rhizobia divide, then transform into bacteroids.

Mature root nodule

Plant molecular biologists Melanie Barnett, Sharon Long, and their colleagues recently produced a microarray chip called a "symbiosis chip" (see Chapter 20) that contains the genomes of both a legume plant host, *Medicago truncatula*, and its rhizobial symbiont, *Sinorhizobium meliloti*. This chip allows investigators to study gene expression in both partners simultaneously, revealing that nearly 5,000 gene expression changes are associated with the legume-rhizobia symbiosis.

GENOMES & PROTEOMES

Early Stages in the Development of Mycorrhizae and Legume-Rhizobia Associations Involve Similar Genes and Proteins

Even though fungi and bacteria are distantly related, some of the genes and proteins involved in early stages of the establishment of mycorrhizal and legume-rhizobial symbioses are amaz-ingly similar. Experts regard these similarities as evidence that legume-rhizobial symbioses evolved by modification of more ancient mycorrhizal associations known from the fossil record.

Symbiosis experts Julien Levy, Frederic Debelle, and their colleagues discovered these genetic similarities by studying the model legume *Medicago truncatula*, which can form symbiotic associations with both mycorrhizal fungi and rhizobia. These investigators found that expression of the *M. truncatula* gene *DMI3* (an abbreviation for "Does Not Make Infections") is required for the establishment of both types of symbioses. When root hair plasma membrane receptors bind either rhizobial Nod factors or compounds released by mycorrhizal fungi, calcium ions that function as signal transduction messengers bind to the DMI protein. This causes the DMI protein to transfer phosphate groups to other proteins, thereby promoting the development of nodules or mycorrhizae. This key role of DMI was discovered when investigators observed that mutant plants that do not produce functional DMI protein displayed normal calcium changes in response to the presence of rhizobia or fungi, but did not produce nodules or mycorrhizae.

Carnivorous Plants Are Autotrophs That Obtain Mineral Nutrients from Animals

About 600 species of flowering plants have adapted to low-nitrogen environments by evolving mechanisms for trapping and digesting animals, and are thus known as carnivorous plants. Their leaves are modified in ways that allow them to capture animal prey, primarily insects, though larger animals are sometimes snared as well. (Despite the popular play and movie *Little Shop of Horrors*, there are no wild or cultivated carnivorous plants that, like Audrey II, are large enough to consume humans!) Carnivorous plants are photosynthetic autotrophs that supply their own organic compounds; prey animals are primarily sources of mineral nutrients. The experimental use of radioactively labeled prey insects has revealed that carnivorous plants obtain as much as 87% of their nitrogen from animals.

The trapping mechanisms used by carnivorous plants are classified as passive or active. Plants with passive trapping mechanisms depend on the prey to fall or wander into the trap. For example, tropical pitcher plants (*Nepenthes* spp.) have leaves that are folded and partially fused to form tubes that collect rainwater (**Figure 37.21a**). The interior walls of these pitchers are slippery and have downward-pointing hairs. Insects and other small animals such as lizards and frogs that fall into the pitchers are unable to climb out. Eventually, the trapped animals drown and are digested by microbes living within the pitchers.

Plants with active mechanisms, such as Venus flytraps and sundews, have traps that are stimulated by touch. Charles Darwin, who was fascinated by carnivorous plants, was one of the first to study the trapping mechanisms of Venus flytraps and sundews. The Venus flytrap (*Dionaea muscipula*) has an active trap formed by two-lobed leaves that are edged with lance-shaped teeth (**Figure 37.21b**). The leaf surface has glands that secrete carbohydrates, which lure prey, as well as glands that secrete digestive enzymes after prey has been trapped. Also present on leaf surfaces are modified hairs, usually three per leaf lobe. If a single hair is touched—perhaps by wind or rain, or debris—and another touch does not occur soon thereafter, nothing happens. But when a fly or similar prey lands on the leaf and brushes against the same hair twice, or touches a second hair within 20–40 seconds, the leaf lobes snap shut around it. Experimental studies indicate that an action potential develops in the stimulated hairs and then travels along plasma membranes, via plasmodesmata, at about 10 cm per second. This signal causes leaf cells to take up ions and water so that the leaf enlarges and changes shape, springing the trap. Digestion of the prey is typically finished within 10 days, whereupon the trap may reopen. Trap leaves can go through three or four digestive cycles during their lifetime, but if they trap too large an animal, they may blacken and die prematurely.

Sundews (such as *Drosera votundifolia*) have leaves bearing glandular hairs whose sticky tips glisten in the sunlight. Insects that land on sundew leaves get mired in the sticky mucilage exuded by these hairs, as shown in the chapter-opening photo. As the insects struggle to get away, they become covered with more mucilage and eventually smother as their breathing pores become clogged. Darwin discovered that sundew leaves bend after being touched, and that glandular hairs not originally in contact with the insects also bend, folding over the prey as you would fold your fingers over an object in your palm. Later investigators discovered that this bending involves the plant hormone auxin. In response to touch, auxin accumulates in sundew leaf tips and then flows downward, stimulating the cell expansion that causes the leaf bending. The glandular hairs produce enzymes that digest the prey.

Parasitic Plants Are Heterotrophs That Obtain Nutrients from Photosynthetic Plants

More than 4,500 species of plants live as complete or partial **parasites**, obtaining all or much of their water, minerals, and organic compounds from photosynthetic plant hosts. Dodder and witchweed are prominent examples of plants that are completely parasitic.

(a) This pitcher plant (*Nepenthes* spp.) passively captures animals that accidentally fall into its water-filled passive trap.

(b) The Venus flytrap (*Dionaea muscipula*) has an active trap that is stimulated by the touch of its prey, in this case a fly.

Figure 37.21   Carnivorous plants.

Dodder (*Cuscuta pentagona*) lacks roots and does not grow from the soil. Instead, all of the 150 species of this parasite live aboveground (**Figure 37.22**). These parasites twine their yellow or orange stems around green plant hosts, into which they sink peg-shaped, absorptive structures known as haustoria. These haustoria tap into the host plant's vascular system, stealing water, minerals, and sugar, which the parasite uses for growth and reproduction. The long, flexible stems of dodder often loop from one plant to another, such that an individual dodder plant can tap into many different plants at the same time. In fact, ecologists Colleen Kelly and Kevin Horning discovered that individual dodder plants grew larger when they infected two hosts of different species rather than two hosts of the same species. Dodder reproduces very rapidly by means of broken-off stem fragments and seeds. A single dodder plant can produce more than 16,000 seeds. In consequence, dodder is a widespread agricultural pest that attacks citrus, tomatoes, and many other fruit, vegetable, forage, and flower crops.

Another group of parasitic plants, the witchweeds (*Striga* spp.), are serious problems for African agriculture, because these parasites attack the major cereal crops: corn, sorghum, and millet. Witchweed seeds lie dormant in soil until secretions from host plant roots stimulate their germination. Genetic engineers are working to find ways to protect crops from the debilitating effects of these crop parasites.

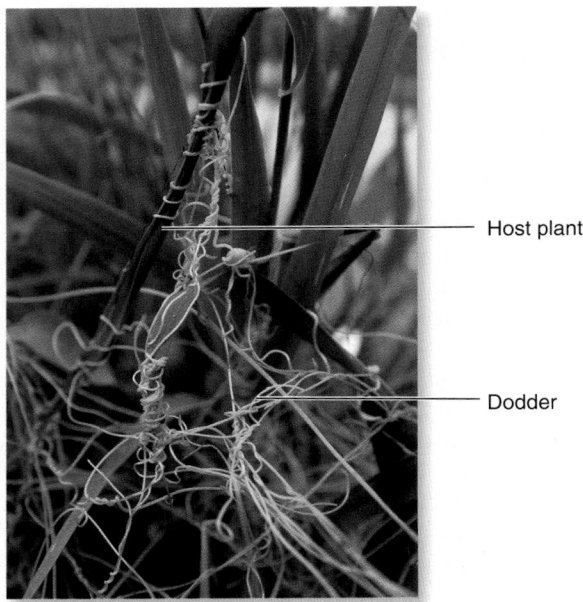

Host plant

Dodder

Figure 37.22   A parasitic plant. Dodder (*Cuscuta pentagona*) is an example of a parasitic plant that obtains all of its water, minerals, and organic compounds from one or more green plant hosts.

CHAPTER SUMMARY

37.1 Plant Nutritional Requirements

- Green plants are autotrophs that use light as the energy source for producing organic compounds. However, some flowering plants have lost photosynthetic capacity; these heterotrophs must obtain energy by metabolizing organic compounds absorbed from their environment. (Figures 37.1, 37.2)

- Plants display many adaptations that allow them to cope with insufficient or excess amounts of light and inadequate CO_2. Some of these adaptations may be usefully engineered into crop plants as a way of improving food production. (Figures 37.3, 37.4, 37.5)

- In addition to light, green plants require minerals, including CO_2 absorbed from the air, and water and several types of ions usually absorbed from the soil. All of these nutrients can limit plant growth and cause nutrient-deficiency symptoms. (Table 37.1, Figure 37.6)

37.2 The Role of Soil in Plant Nutrition

- Soil structure affects water-holding capacity, aeration, and fertility, factors that influence root and plant growth. Soil structure depends largely on particle sizes of inorganic soil components (sand, clay, and silt) and the amount of organic material that is present. (Figures 37.7, 37.8, 37.9, 37.10)

- Inorganic soil components include mineral nutrients dissolved in soil water. Clay particles electrostatically bind cations, which can be released by ion exchange. (Figure 37.11)

- Organic soil components, collectively known as humus, help soil hold water and mineral nutrients and provide the soil's soft consistency.

- The addition of organic fertilizers, which release minerals relatively slowly, can compensate for humus and mineral deficiencies in the soil. Inorganic fertilizers are washed more rapidly from soils and can contaminate natural waters when excessively applied. In aquatic habitats, excess mineral nutrients washed from agricultural fields can cause excessive growths of cyanobacteria, algae, and plants that harm other organisms.

- Plants take up fixed nitrogen in the form of ammonium or nitrate. To utilize nitrate, plant cells must convert it to ammonium ion. (Figure 37.12)

- Biological or industrial processes can convert atmospheric nitrogen gas into fixed nitrogen. Biological nitrogen fixation can be performed only by certain prokaryotes. (Figures 37.13, 37.14)

- Plants display several types of adaptations to cope with phosphorous deficiency. Genetically modified "smart plants" can signal impending phosphorous deficiency. (Figure 37.15)

- Some soil minerals inhibit plant growth, and some plants, called hyperaccumulators, are adapted to cope with high concentrations of toxic soil minerals. Such plants can be used to remove harmful minerals from soil, in the process known as phytoremediation. (Figure 37.16)

37.3 Biological Sources of Plant Nutrients

- Mycorrhizal fungi, which are associated with the roots of most plants, provide plants with water, phosphorus, and other minerals. (Figure 37.17)

- Nitrogen-fixing prokaryotes living within the tissues of some plants provide them with fixed nitrogen. Legume-rhizobia associations are particularly important in nature and in agriculture. (Figures 37.18, 37.19, 37.20)
- Carnivorous plants obtain mineral nutrients from the digested bodies of trapped animals. (Figure 37.21)
- Parasitic plants obtain water, mineral ions, and organic compounds from green plant hosts. (Figure 37.22)

TEST YOURSELF

1. Which of the following substances can limit plant growth in nature?
   a. sunlight
   b. water
   c. carbon dioxide
   d. fixed nitrogen
   e. all of the above

2. In what form do plants take up most soil minerals?
   a. as ions dissolved in water
   b. as neutral salts
   c. as mineral-clay complexes
   d. linked to humus particles
   e. none of the above

3. Why do plants need sulfur?
   a. in the construction of cell walls
   b. as an essential component of chlorophyll
   c. to produce the amino acids cysteine and methionine
   d. all of the above
   e. none of the above

4. Humus is
   a. a food produced from garbanzo beans.
   b. the organic constituents of soils.
   c. the inorganic constituents of soils.
   d. the bedrock layer of soils.
   e. none of the above.

5. Which environments are most conducive to heavy leaching of minerals from soils?
   a. those having soils that are composed primarily of sand particles
   b. those having acidic soils
   c. those impacted by acid rain
   d. regions characterized by heavy rainfall
   e. all of the above

6. Which property is not characteristic of clay-rich soils?
   a. high mineral nutrient retention
   b. high water retention
   c. high aeration
   d. lower amounts of sand than clay
   e. all of the above

7. Which of the substances listed is toxic to plants?
   a. carbon dioxide
   b. oxygen
   c. atmospheric nitrogen gas
   d. aluminum
   e. none of the above

8. What kinds of organisms occur in nitrogen-fixing symbioses with plants?
   a. cyanobacteria
   b. actinobacteria
   c. rhizobia bacteria
   d. all of the above
   e. none of the above

9. How do legume roots attract rhizobia?
   a. They secrete flavonoids.
   b. They secrete carotenoids.
   c. They secrete alkaloids.
   d. They secrete Nod factors.
   e. None of the above.

10. Which plant uses a passive trap to obtain animal prey as a source of mineral nutrients?
    a. the Indian pipe (*Monotropa uniflora*)
    b. the tropical pitcher plant (*Nepenthes* spp.)
    c. the Venus flytrap (*Dionaea muscipula*)
    d. dodder (*Cuscuta* spp.)
    e. all of the above

CONCEPTUAL QUESTIONS

1. Why are agricultural experts and ecologists alike concerned about overfertilization of crop fields?

2. How do plant roots acquire enough boron to serve essential needs without transporting toxic amounts of this mineral to shoots?

3. What are the evolutionary implications of the finding that early steps in the establishment of mycorrhizal and rhizobia symbioses by the legume *Medicago trunculata* involve similar genes, proteins, and signal transduction processes?

EXPERIMENTAL QUESTIONS

1. Why did Hammond and colleagues seek to identify genes whose expression changed between 24 and 72 hours after plants experience phosphorous limitation?

2. What advantage do plants obtain when the *SQD1* gene is expressed?

3. How were the investigators able to identify plants that were starting to experience phosphorous deficiency?

COLLABORATIVE QUESTIONS

1. Imagine that you have bought a farm and want to start growing a crop to sell at a local market. How could you go about determining if the soil needs to be fertilized and with what mineral nutrients?

2. Imagine that you own a large farm through which runs a trout stream. How would you protect the water quality of the stream?

---

www.brookerbiology.com
This website includes answers to the Biological Inquiry questions found in the figure legends and all end-of-chapter questions.

38

FLOWERING PLANTS: TRANSPORT

CHAPTER OUTLINE

38.1 Movement of Water, Minerals, and Organic Compounds at the Cellular Level

38.2 Tissue-Level Transport

38.3 Long-Distance Transport

A shade tree. The evaporation of water from plant leaves cools them and us, and even affects local and global climate.

On hot days, people naturally gravitate to the cool shade beneath trees, as shown in the chapter-opening photo. But most people do not realize that trees are not just sun umbrellas. Plants actually cool the air around them as water evaporates from their surfaces. That's why grass feels cool when you walk barefoot on it, even on a hot day.

Plants benefit from this evaporation process—known as transpiration—because it cools their surfaces and powers the movement of a continuous stream of water from the soil, through roots and stems, to leaves. This evaporation pump not only distributes water throughout the plant body, it also is essential for moving dissolved minerals and organic compounds such as sugars and hormones over long distances within plants. Transport is thus crucial for the functions of plant growth, behavior, and nutrition, which were described in the preceding three chapters.

In addition, plant transport plays a crucial role in Earth's global climate. On a worldwide basis, plant transpiration moves huge amounts of soil water into the atmosphere as water vapor. Along with water evaporated from the surfaces of oceans and freshwater bodies, plant-produced water vapor is a source of rain. Along with other atmospheric gases (carbon dioxide and methane), water vapor works as a greenhouse gas that helps to warm Earth's climate by absorbing the sun's heat. Plant transport processes are also relevant to agriculture, as humans seek to improve crop productivity and the efficiency of water and nutrient use.

To comprehend plant transport more fully, in this chapter we will first consider how water and dissolved materials are transported by plant cells. We will then examine how these materials are moved within plants over short and long distances, and explore some of the plant adaptations that allow such transport to be as efficient as possible in a variety of environments. In the process, we will learn why plants are so cool!

38.1 Movement of Water, Minerals, and Organic Compounds at the Cellular Level

Our previous studies of general angiosperm plant structure and nutrition described the interdependence of plant root and shoot systems (see Chapters 35 and 37). The root system absorbs water and dissolved minerals from the soil. The shoot system uses these materials together with carbon dioxide during photosynthesis to produce sugar needed by roots and for overall plant growth and reproduction. The exchange of water, dissolved minerals, and sugar between root and shoot systems, as well as the long-distance transport of materials throughout the plant body, occurs within a continuous system of conducting tissues. Recall that the complex vascular tissues of plants that conduct water and dissolved minerals are known as the xylem, and those that conduct organic substances in a watery sap are termed the phloem. These long-distance transport tissues allow some trees to achieve heights well over 100 m and are equally important to shorter plants.

To fully understand how plants accomplish long-distance transport, it is useful to begin by reviewing the processes by which minerals, organic compounds, and water move at the cellular level. Cells use both passive and active processes to import or export materials. Let's explore how these processes differ.

Passive Transport Does Not Require the Input of Energy

Recall from your earlier studies of plasma membranes that water, gases, and certain other small, uncharged compounds can passively diffuse across membranes in the direction of their concentration gradients. **Passive transport** is the movement of materials into or out of cells down a concentration gradient without the expenditure of energy in the form of ATP. Passive transport across plasma membranes occurs in two ways: by passive diffusion or facilitated diffusion. **Passive diffusion** is the movement of a solute through a phospholipid bilayer down a gradient. **Facilitated diffusion** is the transport of molecules across plasma membranes down a concentration gradient with the aid of membrane transport proteins (**Figure 38.1a**; see also Chapter 5).

The two main types of membrane transport proteins that function in facilitated diffusion are transporters and channels. **Transporters** are proteins that transport molecules by binding them on one side of the membrane and then changing conformation so that the molecule is released to the other side of the membrane (refer back to Figure 5.19). Transporters increase the rate at which specific mineral ions and organic molecules are able to enter plant cells and vacuoles. **Channels** are membrane pores formed by proteins that allow movement of ions and molecules across membranes (refer back to Figure 5.17).

You may also remember that osmosis is the diffusion of water across a selectively permeable membrane in response to differences in solute concentrations. Osmotic water uptake into living plant cells is essential to photosynthesis, as well as to cell

expansion and structural support. However, the passive diffusion of water does not occur rapidly enough to supply the water needs of rapidly expanding plant cells. In this case, facilitated diffusion of water occurs through channels formed by proteins known as aquaporins, which occur widely in living things (refer back to Figure 5.16). Aquaporins are so important to plants that at least 35 distinct aquaporin genes have been identified in the genome of the model plant *Arabidopsis*. In response to the blue light component of sunlight, plants produce aquaporins as well as abscisic acid and gibberellin, hormones that signal the onset of water stress (see Chapter 36). Aquaporins increase the rate at which water flows into expanding plant cells and their vacuoles. In the same way, many other types of plasma membrane protein channels and transporters facilitate the diffusion of specific mineral ions and organic molecules into plant cells and vacuoles. If a substance must be transported across a plasma membrane against its concentration gradient, however, work must be performed in the process known as active transport, our next topic.

ATP Hydrolysis Powers Active Transport

During **active transport**, membrane transporter proteins use energy to move substances against their concentration gradients. An example is the H^+-ATPase proton pump, found in the plasma membranes of plant cells, which uses ATP to pump H^+, which are protons, against a gradient (**Figure 38.1b**). This proton gradient generates an electrical difference, which is known as a **membrane potential**. Energy is released when protons pass back across the plasma membrane, in the direction of their

(a) Passive transport: facilitated diffusion

(b) Active transport

Figure 38.1 Passive and active transport.

electrochemical gradient. This energy can then be used to power other active transport processes. For example, it might be used to open or close ion channels or in proton **cotransport** or **symport** (transport of two substances in the same direction across a membrane) of organic materials such as sugars, amino acids, and nucleotide bases.

Proton pumps occur in vast numbers in the vacuolar membrane as well as in the plasma membrane. The large vacuole present in many plant cells serves as a storage site for ions and other substances that move in or out via active transport. For example, vacuoles store calcium ions until they are needed in signal transduction processes and they store reserves of nitrate (see Chapter 37). Active transport proteins are particularly abundant in root cell membranes (**Figure 38.2**), allowing root cells to concentrate dissolved mineral nutrients to more than 75 times their abundance in soil. As a result, soil water flows into root cells by osmosis. We next take a closer look at osmotic water movement into and out of plant cells.

Cellular Water Movement Is Influenced by Osmosis and Cell-Wall Pressure

The water content of plant cells depends on two factors: osmosis and turgor pressure. Turgor is Latin for "swelling" and refers to the pressure within the cell that builds when water moves into a plant cell. **Turgor pressure** is the hydrostatic pressure that increases as water enters plant cells, because their cell walls restrict the extent to which the cells can swell. Because animal cells lack cell walls, they do not experience turgor pressure.

A **turgid** plant cell is one whose cytosol is so full of water that the plasma membrane presses right up against the cell wall (**Figure 38.3a**). As a result, turgid cells are firm or swollen. The pressure relationship between the cytosol and cell wall recalls the way that a soccer ball's leather skin presses inward upon the air within, while at the same time the internal air presses on the ball's cover. If you add more air to a limp ball, the ball will stiffen. In the same way, if a nonturgid, flaccid cell absorbs more water, it will become more rigid as the water exerts pressure on the cell wall.

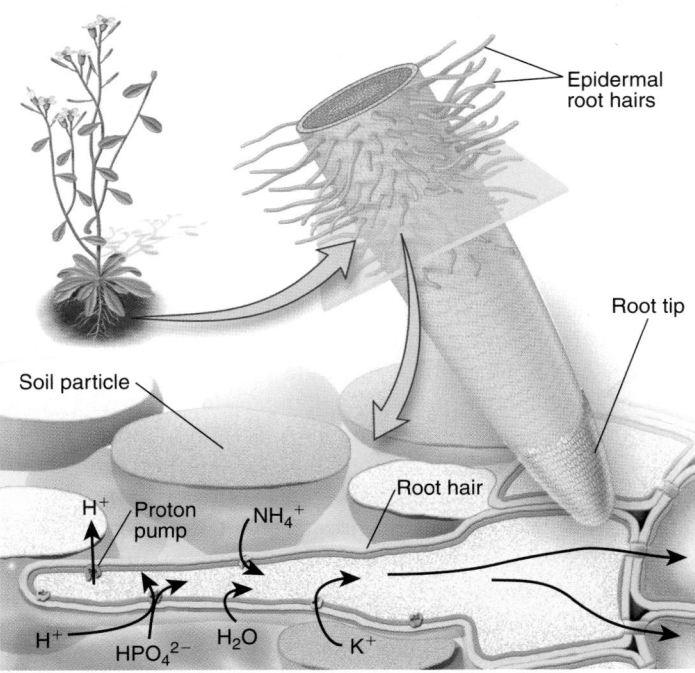

Figure 38.2 **Ion uptake at root hair membranes.** When soil mineral ion concentrations are lower than those within cells, root hair plasma membranes take up nutrient ions by active transport. The H^+-ATPase establishes an electrochemical gradient that drives the active uptake of solutes. The resulting increase in intracellular solute concentration also drives the osmotic diffusion of water into the cell.

Figure 38.3 **Turgid, plasmolyzed, and flaccid plant cells.** **(a)** When the concentration of solutes inside a cell is higher than that outside (the cell is surrounded by a hypotonic solution), more water will enter the cell than will leave it. As a result, a plant cell will become turgid. **(b)** When the concentration of solutes outside a cell is greater than within it (the cell is surrounded by a hypertonic solution), more water will leave the cell than will enter it. As a result, a plant cell will become plasmolyzed. **(c)** When the concentration of solutes is the same outside and inside a plant cell (the cell is bathed in an isotonic solution), it will be flaccid.

By contrast, a plasmolyzed cell is one that has lost so much water by osmosis that turgor pressure has also been lost. **Plasmolysis** is the condition in which the plasma membrane no longer presses on the cell wall (**Figure 38.3b**). A cell having a water content between these two extremes is termed a flaccid cell (**Figure 38.3c**). Together, turgor pressure and osmosis influence an important plant cell property known as water potential.

Recall that potential energy is the stored energy of a material or system, with the capacity to perform work. **Water potential** is the potential energy of water. A waterfall analogy illustrates that water moves from a region of higher water potential to a region of lower water potential, in this case, as the result of gravity. Pressure also influences water potential; a waterfall would flow upward if pressure greater than the force of gravity were applied. Water potential is measured in pressure units known as megapascals (MPa) (a pascal is equal to 1 newton per square meter). One MPa is equal to 10 times the average air pressure at sea level. As another reference point, 1 MPa is several times the pressure in typical home plumbing pipes, which you experience when turning on a water faucet.

In the study of plants, the concept of water potential is used in two ways: to understand the movement of water into and out of cells, or cellular water potential, and to understand the water status of entire plants or organs, or relative water content. Let's first consider the water potential of cells.

Cellular Water Potential Two major factors influence cellular water potential: the concentration of solutes inside and outside of cells, and the turgor pressure within cells. A water potential equation can be used to predict the direction of water movement, given information about the solute concentrations inside and outside of plant cells, and a measure of pressure at the cell wall–membrane interface (**Figure 38.4**). In this equation, water potential is symbolized by the Greek letter psi (ψ) with the subscript "W" for water: ψ_W. In its simplest form, total ψ_W is calculated as the sum of the **solute potential** (ψ_S), also known as the osmotic potential, and the **pressure potential** (ψ_P), also known as the wall potential. A simple form of the water potential equation is thus:

$$\psi_W = \psi_S + \psi_P$$

Solute potential is the component of water potential due to the presence of solute molecules. As you might expect, solute potential is proportional to the concentration of solutes in a solution. The solute potential of pure water open to the air, at sea level and room temperature, is defined as zero. When solutes are added, they form bonds with water molecules, dilute the water, and may increase its disorder. As a result, fewer free water molecules are present, which reduces the potential energy of water. Thus, in the absence of a pressure potential, water that contains solutes always has a negative solute potential. The higher the concentration of dissolved solutes, the lower (more negative) the solute potential.

Pressure potential is the component of water potential due to hydrostatic pressure. In plant cells, the hydrostatic pressure is determined in part by the resistance provided by the cell wall,

Figure 38.4   Plant cell water potential.   The water potential of a plant cell, which predicts the direction of water movement into or out of a cell, is the sum of the solute potential and the pressure potential resulting from pressure exerted by the cell wall–plasma membrane complex. Examples of water potential calculations are shown for a turgid cell, a plasmolyzed cell, and a flaccid cell.

Biological inquiry: If a cell of each of these types is placed into a solution of pure water (whose water potential, ψ_W, is defined as 0 MPa), in which direction will water move in each case?

the turgor pressure. Because of this resistance, the value for pressure potential can be either positive or negative. For example, a turgid cell has a positive pressure potential, which typically measures about 1 MPa. This high pressure inside turgid plant cells is a testimony to the strength of their cellulose-rich plant cell walls. In contrast to turgid cells, flaccid or plasmolyzed cells will have pressure potentials lower than 1 MPa. Plants or plant organs having many cells with low turgor pressure will appear wilted. If your houseplants become wilted, watering will enable the cells to increase their pressure potential, restoring cell turgor and normal plant appearance. Thus, the water content of an entire organ or plant is influenced by the water potential of its component cells.

Relative Water Content The property known as **relative water content** (**RWC**) is often used to gauge the water content of a plant organ or entire plant. RWC integrates the water potential of all cells within an organ or plant and is thus a measure of relative turgidity. Measurements of RWC can be used to predict a plant's ability to recover from the wilted condition. An RWC of less than 50% spells death for most plants, but some plants can tolerate lower water content for substantial time periods.

A standard method for determining RWC was developed by plant biologists H. D. Barrs and Paul E. Weatherley in 1962. This process involves three simple weight measurements: fresh weight, turgid weight, and dry weight. Sample tissue taken from a plant is first weighed to obtain the fresh weight, and then it is completely hydrated in water within an enclosed, lighted chamber until constant turgid weight is achieved. Finally, the sample is dried to a constant dry weight. Researchers use these measurements to calculate RWC using the equation:

$$RWC = \frac{(fresh\ weight\ -\ dry\ weight)}{(turgid\ weight\ -\ dry\ weight)} \times 100$$

RWC measurements have been very useful in ecological studies of natural plant adaptation to cold, drought, or salt stress, and in agricultural research for developing drought-tolerant crops. Agricultural research reported by R. Chandra Babu, Henry Nguyen, and colleagues in 2004 provides an example. These investigators used measurements of leaf RWC to evaluate rice crops that had been genetically engineered with a barley gene that confers dehydration tolerance. Developing new crops that are better able to withstand water stress requires an understanding of not only water potential but also how plant cells cope with cellular osmotic stress, which leads to water stress.

Plant Adaptations to Cellular Osmotic Stress Plants native to cold, dry, or saline environments have evolved many different adaptations that allow them to cope with low water content. For example, in 2005 Chilean plant physiologist Nélida Olave-Concha and associates reported their use of RWC to study the stress responses of Antarctic hair grass (*Deschampsia antarctica*), one of the few flowering plants able to inhabit Antarctica (**Figure 38.5a**). These researchers determined that proteins known as dehydrins accumulate in *D. antarctica* at the same time that RWC decreases. These investigators concluded that dehydrins, which protectively bind to membranes, help this plant survive extreme conditions.

In addition to producing dehydrins, such plants often increase the solute concentrations of their cell cytosol, a process known as **osmotic adjustment**. Increased amounts of the amino acid proline, monomeric sugars such as glucose and fructose, or sugar alcohols such as mannitol decrease the cells' water potential, thus drawing water into cells. By increasing the concentration of solutes inside cells, cold-resistant plants prevent water from moving out of their cells when ice crystal formation in intercellular spaces lowers the water potential outside cells. The additional solutes also lower the freezing point of the cytosol, for the same reason that you might add antifreeze to your car's radiator in winter if you live in a cold climate.

Plants of arid lands often possess adaptations that help them survive water stress. Many can survive in a nearly dry state for as much as 10 months of the year, growing and reproducing only after the rains come. The cytosol of such desiccation-tolerant plants is typically rich in sugars that bind to phospholipids to form a glasslike structure. This helps to stabilize the cellular membranes, preventing them from becoming damaged

(a) A cold-tolerant plant, *Deschampsia antarctica*

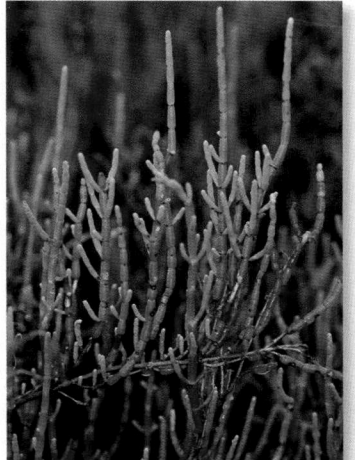

(b) A halophyte, *Salicornia perennis*

Figure 38.5 **Plant adaptations to osmotic stress.** **(a)** Osmotic adjustment and dehydrin proteins are adaptations that help the Antarctic hair grass plant (*Deschampsia antarctica*) to survive in its harsh environment. **(b)** Perennial glasswort (*Salicornia perennis*), a salt marsh plant of the Everglades National Park, is an example of a halophyte, a plant that is adapted to withstand higher than usual external salt concentrations.

during plasmolysis (see Figure 38.3b). As previously noted, plant cells under water stress may also increase the number of plasma membrane aquaporins. These additional protein channels increase the rate of water uptake, allowing cells to recover turgor more quickly when water becomes available.

Halophytes are plants that are able to grow in salty habitats, including coastal salt marshes. The perennial glasswort (*Salicornia perennis*) is an example of a halophyte (**Figure 38.5b**). Plant cells cannot readily absorb salty water because of its highly negative water potential. Halophytes have thus acquired adaptations to help them cope with osmotic stress. Some halophytes accumulate inorganic salts in their vacuoles and organic solutes in their cytosol. Such storages balance solute concentrations inside and outside cells, preventing osmotic water loss. Halophytes also display adaptations that help rid their bodies of excess salt. Some halophytes excrete salt from root surfaces, while others transport excess salt to their leaf epidermal cells (look ahead to Figure 54.16). Some halophytes accumulate salt within multicellular surface glands that break off or burst harmlessly. The water potential adaptations of desiccation-tolerant plants and halophytes are of special interest to agricultural scientists seeking to develop crops that are able to grow in arid lands or in places where the soil or water used for irrigation is salty.

38.2 Tissue-Level Transport

Now that we have learned how water, dissolved minerals, and organic compounds enter or leave plant cells, we are prepared to consider short-distance transport within and among nearby tissues. Tissue-level transport occurs in three forms: transmembrane transport, symplastic transport, and apoplastic transport (**Figure 38.6**).

Transmembrane transport involves the export of a material from one cell into the intercellular space, followed by import of the same substance by an adjacent cell (Figure 38.6a). One prominent example of transmembrane transport is movement of the plant hormone auxin downward, from shoots to roots, by moving from one phloem parenchyma cell to another in a linear series. This process explains how auxin produced in one part of the plant body can influence more distant tissues.

Symplastic transport is the movement of a substance from the cytosol of one cell to the cytosol of an adjacent cell via membrane-lined channels called plasmodesmata (Figure 38.6b). Plasmodesmata also have a central tubule called a desmotubule that connects endoplasmic reticulum of adjoining cells (see Chapter 10). Plasmodesmata are large enough in diameter to allow transport of proteins and nucleic acids, as well as smaller molecules. These molecules primarily move by diffusion, though in some cases special movement proteins facilitate transport through plasmodesmata. Together, all of a plant's protoplasts (the cell contents without the cell walls) and plasmodesmata form the **symplast**. Symplastic transport has the potential to move molecules among the cells and tissues of the entire plant body. However, diffusion of molecules through water is much slower (on the order of millimeters per day) than through air. Although perfumes and other odors diffuse rapidly in air from one place to another, a glucose molecule would take more than 32 years to diffuse across a single meter of water. Low rates of molecular diffusion in water explain why symplastic transport is most important at the level of tissues. Glucose molecules can diffuse across a cell and into another cell through plasmodesmata within a few seconds.

In contrast to the symplast, the **apoplast** refers to the continuum of water-soaked cell walls and intercellular spaces (Figure 38.6c). **Apoplastic transport** is thus the movement of solutes along cell walls and the spaces between cells. Water and dissolved minerals often move through plant tissues for short distances by apoplastic transport.

Both symplastic and apoplastic transport play important roles in mineral nutrient transport through the outer tissues of roots (**Figure 38.7**). As we have noted, the plasma membrane

(a) Transmembrane
Molecules repeatedly cross plasma membranes and cell walls.

(b) Symplastic
Molecules move through the symplast.

(c) Apoplastic
Molecules move through the apoplast.

Figure 38.6 Three routes of tissue-level transport in plants: transmembrane, symplastic, and apoplastic.

Figure 38.7 Symplastic and apoplastic transport of mineral ions in roots.

of epidermal root hair cells is rich in channels and transporters that selectively absorb essential mineral ions from soil water. Absorbed ions can move symplastically from the cytosol of root hairs, root cortex, and endodermal cells directly to xylem parenchyma cells. Plasmodesmata make such cell-to-cell, tissue-level transport possible.

Apoplastic transport moves soil water and dissolved minerals through root epidermal and cortex tissues. However, apoplastic movement of water and minerals stops at the root endodermis, a term meaning "inside skin." The endodermis is a thin cylinder of root tissue whose close-fitting cells and specialized cell walls form a barrier between the root cortex and the central core of vascular tissue. Materials in the root apoplast cannot penetrate farther into the root unless endodermal cells transport them into their cytosol, a process that requires specific transporter proteins.

Endodermal cell walls possess ribbon-like strips of waterproof suberin, composed of wax and phenolic polymers. These suberin ribbons, known as **Casparian strips**, prevent apoplastic transport through endodermal cell walls and into the root vascular tissues (**Figure 38.8**). The endodermal cell walls work like filter paper to prevent small soil particles from entering vascular tissues, where such particles might clog narrow transport pipes. The endodermis also prevents harmful solutes (such as toxic metal ions) from moving through the apoplast to vascular tissues, and being transported to the shoot. For example, recall that aluminum ions (Al^{3+}) are commonly dissolved in soil water, but they are not plant nutrients and are highly toxic to

plants (see Chapter 37). Aluminum ions can penetrate the root epidermis and cortex by moving through the apoplast, but they stop at the endodermis because they are unable to enter the cytosol of endodermal cells. Keep in mind, however, that while the endodermis is effective in reducing the entry of toxic metal ions, these substances can enter roots at root tips, which lack a mature endodermis, and in places where a branch root has broken through the endodermis (refer back to Figure 35.24).

Endodermal plasma membranes possess specific channels and transporters for essential mineral nutrients (such as K^+), which are thereby able to enter the cytosol of endodermal cells. By moving through endodermal cytosol, symplastically transported essential minerals are able to bypass the endodermal barrier that limits apoplastic transport. Thus, the root endodermis functions as a molecular filter that allows the passage of beneficial solutes that have entered from the symplast or have been specifically transported into endodermal cytosol. Once solutes have passed the endodermal barrier, they are actively transported out of xylem parenchyma cells and into the apoplast of the vascular system, which includes xylem conducting cells (see Figure 38.8). The endodermis prevents solutes from returning to outer root tissues or the soil, so the solute concentrations of vascular tissues rise. As a result, water flows into vascular tissues from outer root tissues and the soil. In the final stage of the process known as **xylem loading**, large amounts of water enter the long-distance conducting cells of the xylem, carrying solutes along. In the next section we will see how water and solutes are transported for long distances through the plant.

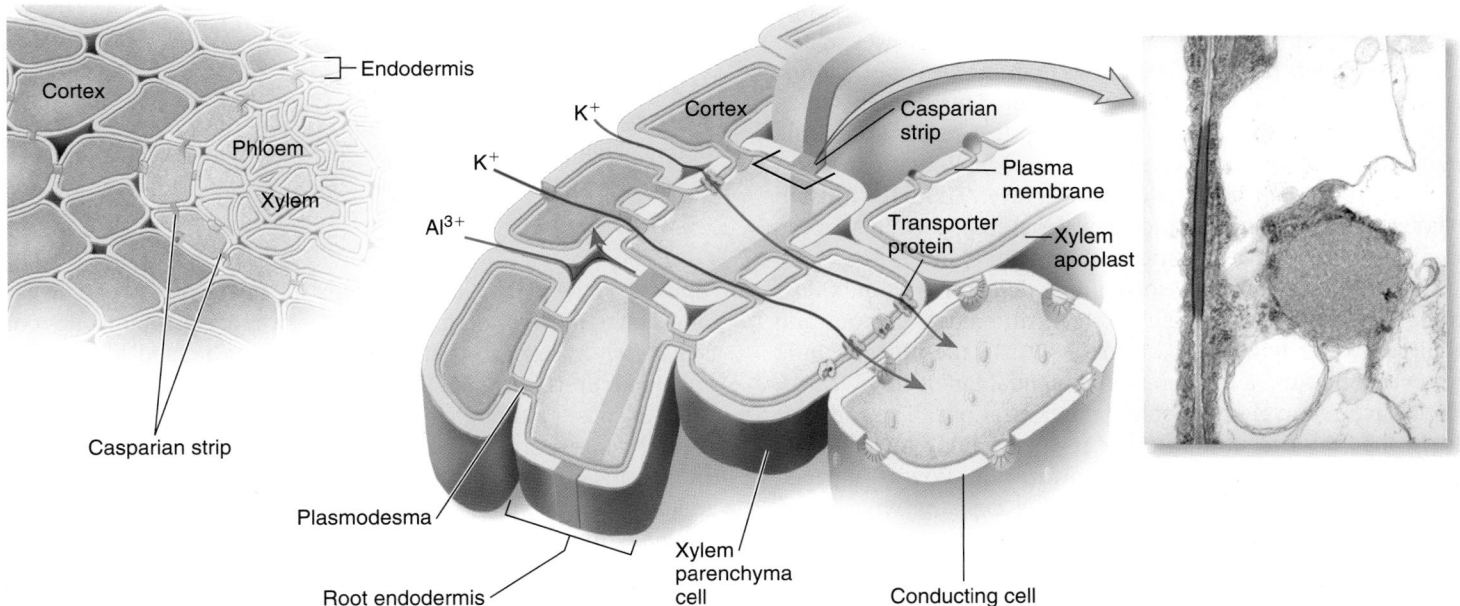

Figure 38.8 **Ion transport pathways across the root endodermis.** Casparian strips in endodermal cell walls prevent apoplastic transport across the root endodermis, limiting entry of harmful soil minerals such as Al^{3+}. Mineral nutrients that are transported into the cytosol of endodermal cells are able to pass through the endodermal barrier to xylem parenchyma cells via plasmodesmata. Once past the endodermis, nutrient ions such as K^+ are moved across plasma membranes to the apoplast of the vascular tissue and are thus able to enter xylem. Inset shows a transmission electron micrograph (TEM) of a Casparian strip in the wall of an endodermal cell.

Biological inquiry: Where in most plants would you expect to find the highest concentration of Al^{3+}?

38.3 Long-Distance Transport

Plants rely on long-distance transport to move water and dissolved materials from roots to shoots and among organs. Tall trees are able to transport water and minerals to astounding heights, more than 110 m in some cases. This is possible because plants possess an extensive, branched, long-distance vascular system composed of xylem and phloem tissues. Watery solutions move through these tissues by **bulk flow**, the mass movement of liquid caused by pressure, gravity, or both. Plant conducting tissues are specialized in ways that foster bulk flow and aid plants in adapting to water stress. The evaporation of water at plant surfaces, the cohesion of water molecules to each other, the adhesion of water to the walls of conducting cells, and the tension that pulls water through those cells all are important physical forces that influence long-distance water transport. Gravity, though of negligible effect for shorter plants, must also be considered in the case of water transport in tall trees. In this section we will take a closer look at bulk flow and the major factors involved in long-distance transport by this process.

Bulk Flow Is Water Movement Under the Influence of Pressure and Gravity

Bulk flow (also known as mass flow) occurs when molecules of liquid all move together from one place to another as the result of differences in potential energy. One example of bulk flow is leaching, the movement of water and dissolved minerals downward through soil layers as the result of gravity. Bulk flow is thus one way in which mineral ions can move through soil toward plant roots; diffusion is the other. However, liquids move much faster by bulk flow than by diffusion. Diffusing molecules move only millimeters per day through water, while phloem sap moves by bulk flow up to 1 m per hour. Bulk flow within xylem moves water from the roots to the uppermost leaves of 100-m-tall trees in about 10 days. Though diffusion operates effectively in cell- and tissue-level transport, it is far too slow for efficient long- distance transport in plants.

In plants, water and dissolved materials move by bulk flow through xylem tissues as the result of differences in root and shoot cell water potentials (**Figure 38.9**). In the phloem, bulk flow occurs because a high pressure potential builds up in leaves, which have relatively high solute concentrations. The pressure causes the water to flow to other regions having lower solute concentrations, such as developing fruit. Bulk flow is impeded by obstructions such as cytoplasm. This explains why xylem transport cells are devoid of cytoplasm and why phloem transport cells have reduced amounts of cytoplasm. Xylem and phloem share additional transport features related to bulk flow, which we will discuss later in this chapter.

Although xylem generally serves as the transport system for water and minerals, and phloem for organic compounds dissolved in water, the transport functions of xylem and phloem overlap. For example, xylem can transport organic compounds just as phloem does. In early spring, trees convert starch stored in stem parenchyma cells into sugars, which are transported

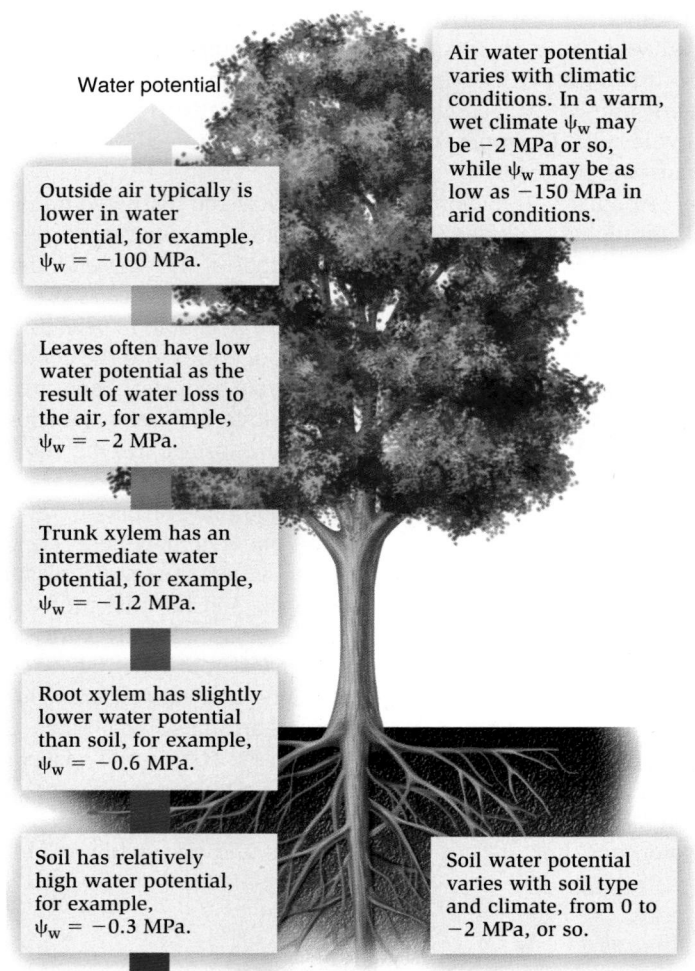

Figure 38.9 Upward transport of water in xylem.   Water potential differences between root and shoot surfaces drive the upward movement of water in plants.

upward in the xylem sap, supplying the needs of bud expansion and flower development. Maple trees produce a copious flow of sugar-rich xylem sap that people have long tapped for making maple syrup (**Figure 38.10**).

Likewise, although phloem tissues are usually associated with the transport of organic compounds, mineral ions are often more concentrated in phloem sap than in xylem sap. Whereas xylem sap is the main source of minerals for leaves, phloem sap provides most of the mineral ions needed by developing fruits and seeds. Minerals are transferred from xylem to phloem at particular locations, including **leaf gaps**, the points where leaf veins separate from the stem vascular system. At leaf gaps, **vascular transfer cells** with extensive inward cell-wall protrusions are abundant (**Figure 38.11**). These protrusions increase the surface area of the plasma membrane, where transporter proteins are found. Vascular transfer cells allow minerals to move rapidly from one transport system to another, much as travelers might transfer from one bus or plane route to another.

Figure 38.10 Maple sugar tapping. In early spring, xylem transports sugar from storage sites to the shoot buds of woody plants such as this sugar maple. We use the xylem sap of sugar maples to produce maple syrup.

Cell-wall ingrowths lined by plasma membranes that contain numerous transport proteins

3.6 μm

Figure 38.11 A vascular transfer cell. Plant leaf gaps contain many vascular transfer cells, such as the one shown here in the common sow thistle (*Sonchus deraceus*). This TEM of a vascular transfer cell shows the characteristic wall ingrowths that increase plasma membrane surface area. Numerous transport proteins in these membranes facilitate movement of materials between xylem and phloem.

The plant vascular system is thus highly integrated in ways that foster bulk flow. Next we will consider additional features of xylem and phloem structure and function that foster long-distance transport.

Long-Distance Transport of Water and Minerals Occurs in the Xylem

Xylem structure plays an essential role in its transport function. The xylem of flowering plants contains several types of special-

End wall

Pits

(a) Tracheids

8.5 μm

(b) Extensible tracheids

Figure 38.12 Tracheid cells in xylem tissue. **(a)** Tracheids are long, tubular cells with slanted end walls. Water and ions move from cell to cell through the pits. **(b)** Light micrograph of extensible tracheids from the xylem of pumpkin.

ized cells, some of which remain alive at maturity, and some of which are dead when they are fully functional. Xylem parenchyma cells are alive, while thick-walled supportive fibers may be alive or dead at maturity. Two types of specialized water-conducting cells are always dead and empty of cytosol when mature: tracheids and vessel elements.

Tracheids Long and narrow in shape, tracheids typically have slanted end walls that fit together to form long files (**Figure 38.12a**). During tracheid development, a tough, lignin-containing secondary wall is deposited in patterns on the inside of the primary cell wall. Although lignin deposition can render much of the tracheid wall impervious to water, tracheid walls also have nonlignified areas that allow water to flow from one tracheid to another. Tracheids whose walls are extensively lignified display numerous small, lignin-free cell wall regions known as **pits**. At such pits, the thin primary wall of the tracheid remains readily permeable to water. Water moves both laterally and vertically through pits.

(a) **Vessels made up of vessel elements**

(b) **Vessel elements in a walnut tree**

43 μm

(c) **Perforations in a vessel element end wall**

6 μm

Figure 38.13   Vessels composed of vessel elements in xylem tissue.   **(a)** This illustration shows the wide diameter of vessel elements with many pits and end-wall perforations. **(b)** SEM of vessels in the wood of the walnut tree (*Juglans* spp.). **(c)** SEM of a perforated vessel element end wall from the tulip tree (*Liriodendron tulipifera*).

Biological inquiry: Which structural features of a vessel element explain the vulnerability of vessels to embolism, that is, blockage by air bubbles?

Because lignin is resistant to compression, microbial decay, and water infiltration, this polymer confers strength, durability, and waterproofing. Like the plumbing pipes of a building, tracheids do not readily collapse as water moves through them under tension. These characteristics explain why tracheids contribute to structural support of the plant body as well as transport.

Tracheids that develop in regions that are still growing by cell expansion are extensible because they have rings or spirals of lignin. These allow tracheids to continue to elongate, much as a coiled appliance cord can stretch (**Figure 38.12b**). In contrast, tracheids that develop in tissues that have already expanded have more lignin, which makes them rigid. Such tracheids cannot elongate any more.

Vessels and Vessel Elements   Mature **vessel elements** are a second type of water-conducting cell present in xylem tissue. Vessel elements are aligned in pipeline-like files known as **vessels** (**Figure 38.13a**). Flowering plants are distinguished from other plant groups by the abundance of vessels; nonflowering plants primarily rely on tracheids for water conduction. Vessel elements are larger in diameter than tracheids, conferring greater capacity for bulk flow, and thus represent one of the many ways in which flowering plants are particularly well adapted to life on land.

Development of vessel elements resembles that of tracheids in some ways. For example, lignified secondary walls are deposited in spirals or sheets that cover most of a vessel element's inner cell wall. Vessel elements also have numerous pits in their side walls. In contrast to tracheids, the end walls and some side walls of vessel elements are extensively perfo-

rated, meaning that all cell-wall material is removed from some areas (**Figure 38.13b,c**). This allows water to flow faster from one vessel element to another than it can flow from one tracheid to another.

Because the perforated end walls and large diameter of vessels allow them to transport more water at a faster rate than tracheids, you might ask why vessels have not completely replaced tracheids in flowering plants. The answer is that vessels are more vulnerable than tracheids to embolism, meaning blockage by air bubbles. Once an embolism forms in a vessel element, it can extend through the large end-wall perforations into many elements, thereby blocking a vessel. Just as air bubbles can cause disruption of blood circulation in people, sometimes leading to death, such bubbles also disrupt water transport in plants, sometimes severely. Emboli can form within vessels as the result of physical damage, drought, or repeated cycles of freezing and thawing. Air bubbles form frequently during winter, because air does not dissolve in ice. In fact, by the end of winter, the functional vessels of many woody plants have become almost completely blocked by air. Blocked vessels in trees often cease to function in water transport and must be replaced by new growth in the spring. Fortunately, even if vessels become blocked, water conduction can still occur via tracheids. This is because tracheid pits are so small that they do not allow air bubbles to move to other tracheids. Thus, an air bubble is confined to the single tracheid in which it first formed, and water continues to flow through nearby tracheids. Tracheids thus provide a fail-safe conduction route when vessels have become disabled by air blockages.

Some plants are able to refill embolized vessels by means of a process known as **root pressure**. At night, the xylem of roots may accumulate high concentrations of ions that are not immediately transported upward to shoots. In this case, the root acts much like a cell rich in solutes, with the result that water gushes in so rapidly that it pushes upward to leaves. Evidence of this process can be observed in the early morning as droplets of water at the edges of leaves, a phenomenon known as **guttation** (**Figure 38.14**). As the water rushes upward, it can dislodge air bubbles or dissolve them, thereby reversing embolism. Root pressure refilling has been observed to occur in nonwoody plants such as corn (*Zea mays*) and in some woody plants, including the sugar maple (*Acer saccharum*).

Figure 38.14 Guttation, the result of root pressure.

FEATURE INVESTIGATION

Holbrook and Associates Revealed the Dynamic Role of Xylem in Transport

Plant biologists once thought that xylem sap moved from one transport cell to another at a uniform rate, unless blocked by an embolism. After all, botanists hypothesized, the cells are dead and thus lack cytoplasmic components that might influence sap flow rates. However, recent experiments have revealed that xylem sap actually moves more readily from one vessel to another through pits when the solute concentration of xylem sap is relatively high. These studies suggest that this variation results from dynamic changes in pit-wall microstructure.

More than 20 years ago, plant biologists noticed that tap water moved more rapidly through cut pieces of stem than did pure water. This observation suggested that ions present in the tap water might have influenced bulk flow rates. But this hypothesis remained untested until 2001, when N. Michelle Holbrook's lab reported the results of experiments designed to explore this phenomenon, as shown in **Figure 38.15**. These investigators compared the flow rate of pure water with that of artificial sap, pure water to which they had added various levels of potassium chloride (KCl) or other salts that normally occur in xylem sap. The artificial sap and pure water were introduced into the same experimental tobacco plant through separate flaps cut into the stem. (Using different flaps on the same plant helps to remove experimental variation that might have resulted if investigators had compared results from separate plants.)

Figure 38.15 The Holbrook team's dynamic xylem experiment.

HYPOTHESIS Dissolved minerals increase water flow through microchannels in xylem vessel pit walls.

STARTING MATERIALS Plants with split stems.

Experimental level

Conceptual level

1 Initially, supply both stem flaps with pure water. After 500 seconds, supply flap A with artificial sap and flap B with pure water. Measure the rate of flow into the stem with a flowmeter.

The flow of A is 2 times that of control in B.

Tobacco plant

Artificial sap   A

B   Pure water (control)

A Width ← Pectin   B

Cellulose microfibrils

With artificial sap, pectins shrink, micropores widen.

With pure water, pectins swell, micropores narrow.

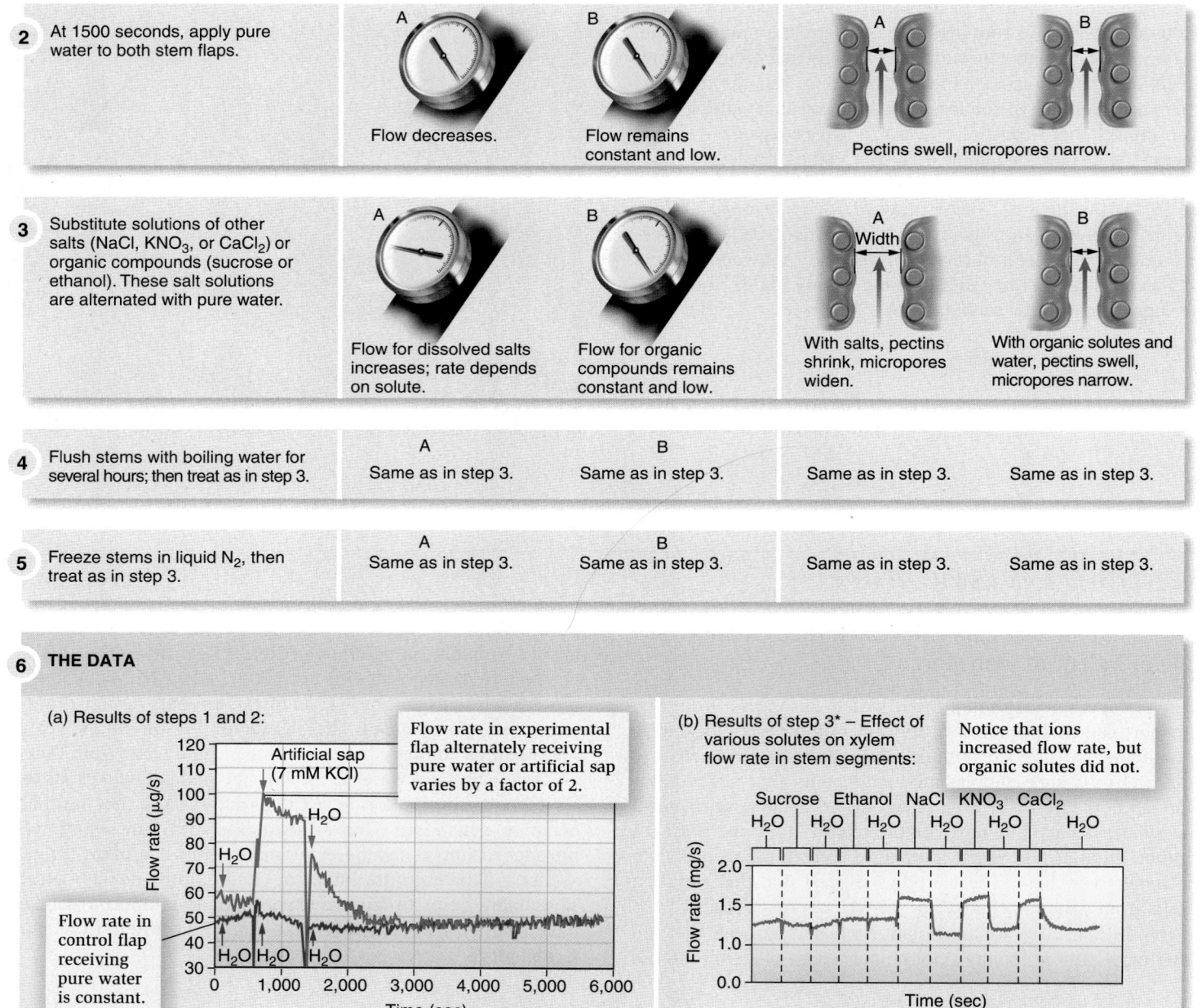

2 At 1500 seconds, apply pure water to both stem flaps.

A — Flow decreases.

B — Flow remains constant and low.

A / B — Pectins swell, micropores narrow.

3 Substitute solutions of other salts (NaCl, KNO$_3$, or CaCl$_2$) or organic compounds (sucrose or ethanol). These salt solutions are alternated with pure water.

A — Flow for dissolved salts increases; rate depends on solute.

B — Flow for organic compounds remains constant and low.

A — Width — With salts, pectins shrink, micropores widen.

B — With organic solutes and water, pectins swell, micropores narrow.

4 Flush stems with boiling water for several hours; then treat as in step 3.

A — Same as in step 3.
B — Same as in step 3.
Same as in step 3.
Same as in step 3.

5 Freeze stems in liquid N$_2$, then treat as in step 3.

A — Same as in step 3.
B — Same as in step 3.
Same as in step 3.
Same as in step 3.

6 THE DATA

(a) Results of steps 1 and 2:

Flow rate in experimental flap alternately receiving pure water or artificial sap varies by a factor of 2.

Flow rate in control flap receiving pure water is constant.

Artificial sap (7 mM KCl)

H$_2$O

Flow rate (μg/s) — 120, 110, 100, 90, 80, 70, 60, 50, 40, 30

Time (sec) — 0, 1,000, 2,000, 3,000, 4,000, 5,000, 6,000

(b) Results of step 3* – Effect of various solutes on xylem flow rate in stem segments:

Notice that ions increased flow rate, but organic solutes did not.

Sucrose Ethanol NaCl KNO$_3$ CaCl$_2$
H$_2$O H$_2$O H$_2$O H$_2$O H$_2$O H$_2$O

Flow rate (mg/s) — 2.0, 1.5, 1.0, 0.0

Time (sec)

*Similar results were observed for step 4 and 5.

These researchers observed that the xylem flow rate was higher when the flap received artificial sap than when it was receiving pure water. Then the investigators supplied pure water to both stem flaps, with the result that the flow rate equalized (see step 2 in Figure 38.15 and also data, part a). The investigators then provided artificial sap containing various types of solutes to one flap, along with alternating additions of pure water. The flow rate increased for artificial sap containing dissolved salts, but not for artificial sap containing the organic compounds sucrose or ethanol (see data, part b).

Next, the scientists performed experiments designed to determine if living cells in the xylem tissue were causing these effects. They flushed the stems with boiling water or froze the stems in liquid nitrogen before treating them with water or artificial sap as in step 3. These treatments kill living cells. They found that flow rates through the boiled or frozen stems yielded the same results: The flow rate was higher when the flap received sap containing dissolved salts rather than pure water or organic solutes. These results indicated that living cells did not cause the effect.

Investigators suspected that ions change vessel element structure in some way that allows greater flow through them. In later experiments not shown here, the scientists devised an apparatus for measuring the flow of sap through individual xylem vessels versus sap flow through groups of vessels. They found that flow through an isolated vessel did not respond to changes in sap ion content, whereas flow between adjacent vessels responded. The researchers interpreted these results as evidence that small pores called micropores in the thin pit walls between adjacent vessel elements play a role in the change in flow rates. These micropores occur within a mesh of cellulose microfibrils and pectins. The pectins, which are water-rich gels known as hydrogels, shrink in response to sap ions and swell in response to pure water. As a result, micropores widen in response to sap ions, increasing flow rates, but narrow in response to pure water (see Figure 38.15, step 1, Conceptual level). Researchers speculated that this process hastens the delivery of ion-rich sap to transpiring tissues. Investigators also proposed that this process, which increases sap flow, may help plants compensate for partial loss of xylem sap flow when an air embolism occurs.

The Cohesion-Tension Theory Explains the Role of Transpiration in Long-Distance Water Transport

Now that we have a better understanding of the structure of cells that form xylem, let's reexamine the phenomenon of bulk flow, which was described earlier in Figure 38.9. Plants expend no energy to transport water and minerals by bulk flow in the xylem. Instead, the sun's energy indirectly powers this process by generating a water potential difference between moist soil and drier air. In warm dry air, water evaporates from plant surfaces. This evaporation process is known as **transpiration** (from the French *transpirer*, to perspire) (**Figure 38.16a**).

How does evaporation at plant surfaces influence long-distance water transport in the xylem? To answer this question, we must consider the unique physical properties of water. Liquid water molecules are linked by hydrogen bonds (see Chapter 2). As a result, liquid water is amazingly cohesive, explaining why water tends to form continuous streams (**Figure 38.16b**). Consequently, when water molecules evaporate from plants, water films present in the intercellular spaces of leaves display high surface tension, causing a curved water surface known as a meniscus to form (**Figure 38.16c**). This force pulls on neighboring liquid water molecules and eventually on water in the nearest vein, which is connected to the plant's entire water supply. As the result of water's cohesion, and the tension exerted on water by evaporation at the plant's surface, a continuous stream of water can be pulled by bulk flow through the plant body from the soil, into roots, through stems, and into leaves. This explanation for long-distance water movement in plants is known as the **cohesion-tension theory**. Although considerable experimental data support the cohesion-tension concept, some experts think that additional factors might be important in long-distance water transport. These factors include adhesion of water (its tendency to stick) to the lignified walls of xylem conducting cells (see Figure 38.16b), and a squeezing effect exerted by nonconducting tissues on conducting tissues.

Even though plant biologists are still seeking a more complete understanding of the forces acting in long-distance water transport, measurements of xylem pressure clearly reveal that a water potential gradient occurs in the soil-plant-atmosphere continuum. The water potentials of hydrated soil and plant roots are relatively high, while the water potentials of leaves and the atmosphere are relatively low (see Figure 38.9). This water potential gradient explains the upward movement of water in plants.

(a) **Transpiration occurs when leaf water is exposed to drier air.**

(b) **Cohesion in xylem causes water to form a continuous stream.**

(c) **Surface tension increases in the intercellular spaces of cells, pulling on water stream in xylem.**

Figure 38.16   The roles of transpiration, cohesion, adhesion, and tension in long-distance water transport.

Figure 38.17   Plant-transpired water vapor mist rising from a tropical rain forest.   This mist visually illustrates the enormous amount of water that is transpired from the surfaces of plants into the atmosphere. Water vapor derived from plant transpiration is an important source of rainfall, and the process of evaporation cools plant surfaces as well as local and global climate.

Biological inquiry: Why does evaporation of water have such a powerful cooling effect?

Along the route from roots to leaves, water is distributed to tissues and cells, depending on their relative water potential.

Plant transpiration moves huge amounts of water from the soil to the atmosphere. About 99% of the water that enters plants via roots is generally lost as water vapor during transpiration. Each crop season, a single corn plant (*Zea mays*) loses more than 200 L of water, which is more than 100 times the corn plant's mass. A typical tree loses 400 L of water per day! On a regional and global basis, plant transpiration has enormous climate effects. For example, an estimated one-half to three-quarters of rainfall received by the Amazon tropical rain forest actually originates from plant-transpired water vapor, often visible as mist (**Figure 38.17**). Furthermore, about half of the solar heat received by Amazonian plants is dispersed to the atmosphere during transpiration. This heat dispersal has a cooling effect on regional ground temperature, which would be much higher in the absence of plant transpiration. Such cooling effects result from water's unusually high heat of vaporization, the amount of heat needed to isolate water molecules from the liquid phase and move them to the vapor phase. Most of this energy is needed to break the large numbers of hydrogen bonds that occur in liquid water. The evaporation of large amounts of water from plant surfaces effectively dissipates heat, explaining how plants cool themselves and their environments.

Although evaporation of water from plant surfaces plays an essential role in bulk flow through xylem, if plants lose too much water, they will die. Consequently, plant surfaces, including those of leaves, are typically coated with a cuticle, a wax-containing layer that retards water loss. Only about 5% of water evaporated from plant surfaces emerges through the cuticle. More than 90% of the water that evaporates from plants is lost through **stomata**, surface pores that can be closed to retain water or opened to allow the entry of CO_2 needed for photosynthesis and the exit of oxygen and water vapor. Stomata are often abundantly located on the lower surfaces of leaves. Tobacco leaves, for example, possess an estimated 12,000 stomata per square centimeter of leaf surface! Plants face a constant dilemma: whether to open their stomata for CO_2 intake and suffer the impact of reduced water content or to close stomata to retain water, thereby preventing CO_2 uptake. For this reason, plants have acquired many adaptations to cope with transpirational water loss, our next topic.

Plant Adaptations Help to Reduce Transpirational Water Loss

Under some conditions, almost all plants experience water stress, which is an inadequate amount of water. Water stress is common for plants of the world's arid regions, and their growth is often limited by water availability. Even plants of moist, forested regions of the world experience water stress during drier or colder seasons or under windy conditions. The leaves at the tops of tall trees are generally under considerable water stress because gravity has a substantial impact on their water potential. Earlier, we considered examples of plant cellular adaptations to deal with water stress, also called osmotic stress (see Figure 38.5). Plants have evolved two additional ways to prevent excessive loss of water by transpiration: stomatal movements and leaf drop.

Stomatal Movements   Plant stomata close to conserve water under conditions of water stress and open when the stress has been relieved, allowing air exchange with the leaf's spongy mesophyll. Stomata are bordered by a pair of **guard cells**, which are sausage-shaped chloroplast-containing cells attached at their ends (**Figure 38.18a**). The distinctive structural features of guard cells explain how they are able to open and close a pore. As guard cells become fully turgid, their volume expands by 40–100%. This expansion does not occur evenly, however, because the innermost cell walls are thicker and less extensible than are other parts of the guard cell walls. In addition, the cells expand primarily in the lengthwise direction because bands of radially oriented cellulose microfibrils prevent the guard cells from expanding laterally (**Figure 38.18b**). Thus, when guard cells are turgid, a stomatal pore opens between them, allowing air exchange with the leaf's spongy mesophyll. Conversely, when the guard cells lose their turgor, their volume decreases and the stomatal pore becomes closed.

What causes the change in turgor? Stomata often open early in the morning, in response to sunlight. This response makes

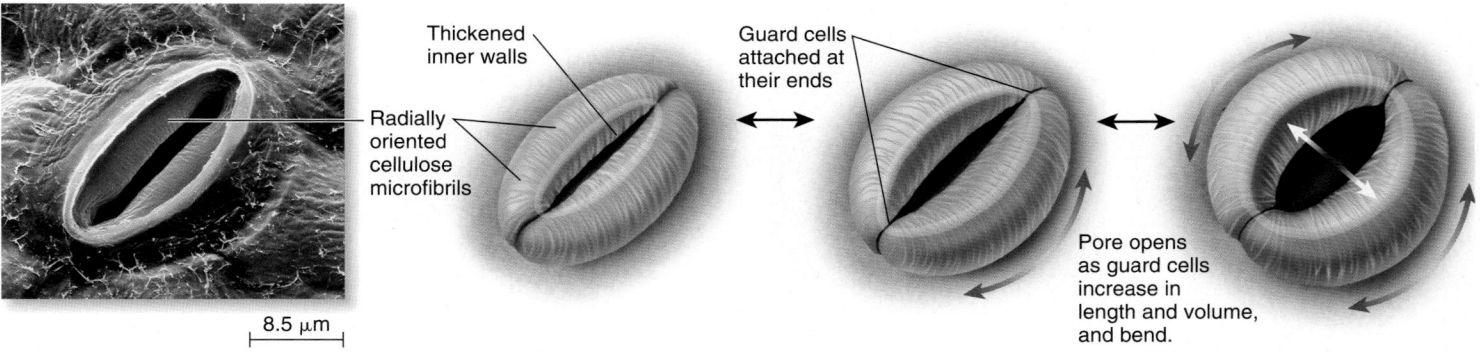

(a) Stomatal guard cells

(b) The roles of radial orientation of cellulose microfibrils and thickened inner walls in opening or closing guard cells

Figure 38.18   The structure of stomatal guard cells affects their function. When flaccid, guard cells close stomatal pores. Turgid guard cells produce a stomatal opening. **(a)** An SEM of a stoma in a rose leaf, showing the two guard cells bordering a partly open pore. **(b)** Thickened inner cell walls and radial orientation of cellulose microfibrils in the guard cell walls explain why they separate when turgid, forming a pore.

Blue light stimulates H^+-ATPase proton pumps, providing the membrane potential needed for guard cells to import K^+ and sugar. As a result, water enters by osmosis via aquaporin channels. The water-swollen guard cells separate, opening the pore.

At night or under conditions of water stress that stimulate ABA production, K^+ and other solutes are pumped out of guard cells. This causes water to leave, deflating guard cells and closing the pore.

(a) A closed stomate in the process of opening

(b) An open stomate in the process of closing

Figure 38.19   How stomatal pores open and close. **(a)** Stomata usually open in response to the blue light of sunlight. **(b)** Stomata usually close in response to lack of sunlight. They can also close during the day under conditions of water stress, which induces plants to produce more of the hormone abscisic acid (ABA). Stomatal guard cell plasma membranes possess ABA receptors, which receive the drought signal.

sense, given that light, water, and carbon dioxide are all required for photosynthesis. Blue light stimulates H^+-ATPase proton pumps, leading to guard cell uptake of ions, especially potassium (K^+), and other solutes. As a result of increases in solute concentrations inside guard cells, osmotic water uptake occurs via plasma membrane aquaporins, resulting in cell expansion and stomatal opening (**Figure 38.19a**).

At night, the reverse process closes stomata. Potassium and other solutes are pumped out of guard cells, causing water to exit and deflate guard cells, resulting in pore closure. Plants also close stomata during the daytime under conditions of water stress, a process that is mediated by the stress hormone abscisic acid (ABA). Water stress causes a 50-fold increase in ABA, which

can be produced by any cell having chloroplasts or starch-rich amyloplasts. ABA is transported in the xylem sap to guard cells. ABA then binds to a receptor, which elicits a Ca^{2+} second messenger, causing the guard cells to lose solutes and deflate (**Figure 38.19b**).

Some plants of arid habitats keep their stomata closed during the hot daytime hours and open them during the cooler night, allowing carbon dioxide to enter without losing too much water. These plants, known as CAM plants, bind the CO_2 taken in at night into an organic acid and release it when the sun rises, thus allowing photosynthesis to occur. Other forms of C_4 photosynthesis are also adaptations that foster plant water retention in hot, dry conditions (see Chapters 8 and 37).

(a) Ocotillo with leaves   **(b) Ocotillo without leaves**

Figure 38.20   **Leaf abscission as a drought adaptation.**
The ocotillo (*Fouquieria splendens*), a plant native to North American deserts, is known for its ability to respond to intermittent rain and drought by producing and dropping leaves multiple times within a year.

Sometimes stomatal movements are not completely under the plant's control. Certain pathogens cause plant wilt diseases by influencing stomata. For example, the fungal genus *Fusicoccum*, a pathogen of fruit trees, produces a toxin known as fusicoccin that stimulates plant plasma membrane proton pumps. This fungal pathogen thereby causes guard cell turgor to increase. As a result, the stomata open under conditions when they would normally be closed. The subsequent water loss causes leaves of infected trees to wilt.

Leaf Abscission   Plants that are under water stress may drop their leaves, a process called leaf drop or **leaf abscission**. In some plants, leaf abscission occurs normally to prevent water stress. This process lets plants avoid very low leaf water potentials and the consequent danger of xylem embolism. Leaf abscission also reduces the amount of root mass that plants must produce to obtain water under arid conditions. The process is a particularly valuable adaptation for desert plants and angiosperm trees of seasonally cold habitats. For example, the ocotillo (*Fouquieria splendens*) of North American deserts can produce leaves after sporadic rains and then drop all of its leaves as many as six times a year (**Figure 38.20**). The sugar maple (*Acer saccharum*) is an example of the many types of temperate forest trees that drop their leaves each autumn and are thus known as deciduous trees. Such trees would otherwise experience water stress in winter, because transpiration would continue while roots would be unable to take up water frozen in the soil.

Leaf abscission is a highly coordinated developmental process. The hormone ethylene stimulates an abscission zone to develop at the bases of leaf petioles (**Figure 38.21a**). The abscission zone contains two types of tissues: a separation layer

Leaf petiole

Conducting tissue

Abscission zone

Stem

25 μm

(a) LM of leaf abscission zone at the junction of a petiole and stem, stained with dyes

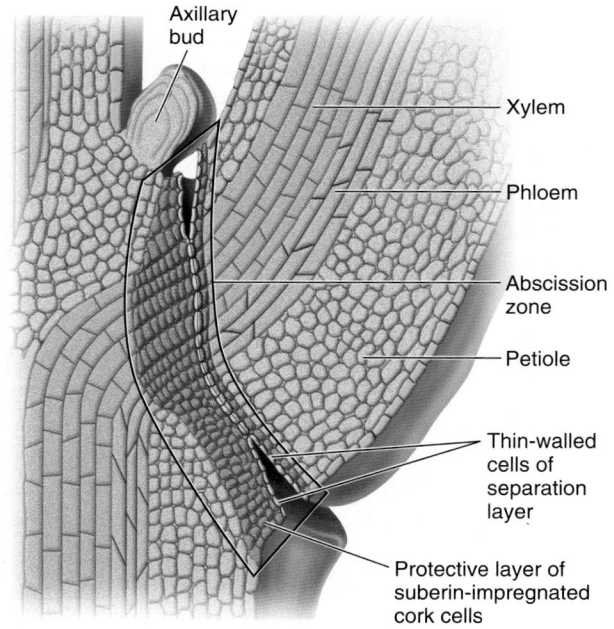

Axillary bud

Xylem

Phloem

Abscission zone

Petiole

Thin-walled cells of separation layer

Protective layer of suberin-impregnated cork cells

(b) Tissues involved in leaf abscission

Figure 38.21   **Leaf abscission.**

of short, thin-walled cells and an underlying protective layer of suberin-impregnated cork cells (**Figure 38.21b**). Suberin contains both waterproofing wax and phenolic polymers that retard microbial attack. As the abscission layer develops across the vein linking the petiole with the stem, it eventually cuts off the water supply to the leaf. Chlorophyll degrades, revealing colorful orange and yellow carotenoid and xanthophyll pigments that were hidden beneath. In addition, some plants synthesize red and reddish blue pigments in response to changing environmental conditions. The presence of these pigments explains colorful autumn vegetation. Enzymes eventually break down the cell-wall components of the separation layer, causing the petiole to break off the stem. The underlying protective layer forms a leaf scar that seals the wound, helping to protect the plant from water loss and pathogen attack.

The Pressure-Flow Hypothesis Explains Transport in Phloem Tissues

Our examination of xylem structure and function provides the background needed to understand phloem structure and function. As we have noted, phloem plays an essential role in long-distance transport of dissolved organic compounds and ions in the plant body. Phloem often transports sugars from where they are produced to where they will be used. Recall that primary phloem occurs in the vascular bundles of herbaceous plants, and secondary phloem of woody plants occurs as the inner portion of bark, a thin cylinder external to the wood (secondary xylem; see Chapter 35). In contrast to xylem, whose transport tissues are dead and empty of cytoplasm at maturity, mature phloem tissues remain alive and retain at least some cytoplasmic components. In addition, phloem works under pressure, unlike xylem, which is under tension. A closer look at phloem structure and function will help to illuminate these differences.

Phloem Structure Phloem tissues of flowering plants include supporting fibers, parenchyma cells, sieve-tube elements, and adjacent companion cells. **Sieve-tube elements** are arranged end to end to form transport pipes (**Figure 38.22**), analogous to the way that the xylem's vessel elements are aligned to form longitudinal vessels. Together, the sieve-tube elements and companion cells form a transport system whose structure explains its function.

Each pair of sieve-tube elements and companion cells has a common origin. They are produced by an unequal division of a single precursor cell and are thus linked by plasmodesmata formed at cytokinesis. The smaller of the two cells develops into a companion cell, whose name reflects its life-support function. The larger of the two cells develops into a sieve-tube element, which loses its nucleus and most of its cytoplasm as an adaptation that reduces obstruction to bulk flow. Mature sieve-tube elements retain only a thin film of peripheral cytoplasm that includes some endoplasmic reticulum, plastids, and mitochondria. The end walls of developing sieve-tube elements become perforated by the action of wall-digesting enzymes that enlarge existing plasmodesmata. The perforated end walls of mature sieve-tube elements are known as **sieve plates**, and the numerous perforations are known as **sieve plate pores**. Phloem sap passes through these plates from one sieve-tube element to another.

Mature sieve-tube elements are not dead. However, because they lack a nucleus, they are dependent on their neighboring companion cell for messenger RNA (mRNA) and proteins, which are supplied via plasmodesmata. For example, when a plant's conducting system is damaged, a short-term wound response occurs that involves production of a protein known as **P protein** (for phloem protein). Masses of this protein congeal along sieve plates, preventing loss of phloem sap (**Figure 38.23**). This "slime plug" functions much like a clot that helps reduce blood loss from wounded animals. P protein also binds to the cell walls of disease microorganisms, thereby helping to prevent infection at wounds.

Sieve plate pore
Sieve plate
Sieve-tube element
Companion cell

(a) Sieve-tube elements and companion cells

20 μm

(b) Light micrograph of phloem stained with blue dye, showing sieve-tube elements

Figure 38.22 Sieve-tube elements and companion cells in phloem tissue.

5 μm

Sieve plate
P protein
Companion cell

Figure 38.23 Phloem wound response. P protein, stained red in this LM, collects along sieve plates of wounded sieve elements, thereby helping to prevent infection and leakage of solutes.

In some plants, sugar moves from sugar-producing cells into companion cells and sieve-tube elements via plasmodesmata.

In some plants, sugar moves from sugar-producing cells into the apoplast. ATP is required for active transport into companion cells. Sugar moves into sieve-tube elements via membrane sugar transporter proteins and via plasmodesmata.

(a) Symplastic phloem loading

(b) Partially apoplastic phloem loading

Figure 38.24 Symplastic and partly apoplastic phloem loading.

However, sieve-tube elements cannot produce P protein by themselves. Their companion cells provide either P protein mRNA or the protein itself to sieve-tube elements. In the longer term, plants plug wounded phloem with the carbohydrate callose.

Phloem Loading Companion cells also play an essential role in conveying sugars to sieve-tube elements for long-distance transport, a process known as **phloem loading**. Although glucose and some other sugars can occur in phloem, the disaccharide sucrose is the main form in which most plants transport sugar over long distances. Plant biologists think that sucrose is less vulnerable to metabolic breakdown en route than are monosaccharides.

Two types of phloem loading occur, symplastic and partly apoplastic. Many woody plants transport sucrose from sugar-producing cells of the leaf mesophyll to companion cells and then to sieve-tube elements via plasmodesmata, a process known as symplastic phloem loading (**Figure 38.24a**). The advantage of symplastic loading is that it does not require ATP; by moving through plasmodesmata, sugar does not have to cross plasma membranes.

In contrast, most herbaceous plants, including important crop plants and the model plant *Arabidopsis*, load sugar into sieve-tube elements or companion cells from intercellular spaces, often against a concentration gradient. ATP must be used to move the sugar across a plasma membrane into a companion cell or sieve-tube element (**Figure 38.24b**). Thus, this second type of phloem loading is partly apoplastic and partly a transmembrane process.

Translocation of Phloem Sap We have learned that the sun's energy drives transpirational water movement in plant xylem by a process of cohesion, adhesion, and tension. Once sugar has been loaded into phloem sieve-tube elements, how does it move within the plant? The most common explanation is that phloem transport is driven by differences in turgor pressure that occur between cells of a **sugar source**, where sugar is produced, and those of a **sugar sink**, where sugar is consumed. Photosynthetic leaf mesophyll is the main sugar source. Roots and developing leaves, seeds, and fruits are examples of sugar sinks. In the process known as **translocation**, phloem transports substances from source to sink. The direction of phloem movement can be horizontal as well as vertical, depending on the relative positions of the sources and sinks.

Because sieve-tube elements near source tissues have comparatively high solute contents, water tends to rush into them from nearby xylem, thereby building turgor pressure. In contrast, sieve-tube elements near sink tissues have lower solute concentration, so water that contains solutes moves from phloem into sink tissues. The resulting water pressure difference drives the bulk flow of phloem sap from source to sink tissues. This explanation for phloem transport is known as the **pressure-flow hypothesis** (**Figure 38.25**). German plant physiologist Ernst Münch first proposed this hypothesis in 1927, though later investigators have modified it. The pressure-flow hypothesis is supported by studies that show relatively high solute concentrations in phloem samples taken from tree leaves (source) and relatively low solute concentrations in samples taken from a

Figure 38.25 Pressure-flow hypothesis for phloem transport.

tree trunk (sink). The production of turgor pressure requires an intact plasma membrane. This explains why mature phloem sieve-tube elements must be alive in order to function, in contrast to xylem tracheids and vessel elements.

At sink tissues, sugar is typically unloaded through plasmodesmata (Figure 38.25). Because plasmodesmata are very narrow in comparison to sieve-tube elements, they slow the flow of phloem sap from sieve-tube elements into sink tissues. This reduction in flow rate helps to equalize the distribution of phloem sap, preventing delivery of too much sap to any single sink. When the solute concentration of phloem sap has been sufficiently reduced, water flows from the phloem back into the xylem, where upward transport occurs. The reliance of phloem bulk flow on water supplied by the xylem explains the close proximity of phloem and xylem tissues in vascular bundles and woody stems.

GENOMES & PROTEOMES

Microarray Studies of Gene Transcription Reveal Xylem- and Phloem-Specific Genes

Xylem and phloem tissues have important economic, agricultural, and ecological roles. For example, secondary xylem produced by the vascular cambium and otherwise known as wood forms the basis of our forest products industry. The ability of plants to respond to environmental factors also relies on vascular tissues, which transport hormones and RNAs. Seeking increased fundamental understanding and enhanced ability to engineer plants, molecular biologists have tried to identify the genes involved in xylem and phloem development. The use of microarrays, also known as gene chips, is providing a genome-wide view of gene expression related to xylem and phloem.

Chengsong Zhao, Eric Beers, and their coworkers recently compared gene expression in xylem (X), phloem plus vascular cambium (PC), and nonvascular (NV) tissues dissected from young *Arabidopsis* plants. Recall that the vascular cambium is the meristematic tissue that produces secondary phloem (inner bark) and secondary xylem (wood; see Chapter 35). The investigators used a dissecting microscope to view 1-cm-long pieces of tissue and a razor blade to make a longitudinal cut into nonvascular tissues and phloem, isolating the xylem. Then they peeled the nonvascular tissue away from the phloem. They ensured that their tissue samples were pure and authentic by checking for the expression of marker genes already known to be tissue specific. The investigators ascertained that their X and NV tissue samples had very low expression of certain PC genes and that their PC samples had very low expression of certain X and NV genes. They then extracted RNA from the three separate tissues and used it to make labeled nucleic acid for hybridization to gene microarrays containing about 90% of the *Arabidopsis* genome (refer back to Figure 20.11).

Binding of labeled nucleic acid revealed which genes were expressed. By comparing the binding patterns on different chips hybridized with nucleic acid from distinct tissues, the researchers were able to determine the genes that were preferentially expressed in X, PC, and/or NV tissues. These differences are revealed by a "triangle plot" that reflects the proportions of genes expressed in the three tested tissues (**Figure 38.26**). Genes expressed primarily in only one tissue type cluster at the triangle apices, and genes expressed in two tissues but not the third are plotted at the triangle edges. Genes expressed in all three tissues fall into the center. The researchers identified 319 X-biased genes, 211 PC-biased genes, and 154 NV-biased genes. Not surprisingly, 17% of the X-expressed genes were related to cell-wall biosynthesis, including lignin production. The investigators then confirmed, via microscopy, that several proteins predicted by microarray data to be located in X, PC, or NV tissue were actually made in these tissues. This study is an essential first step toward determining the function, timing of expression, and cellular locations of xylem- and phloem-specific proteins.

Figure 38.26 Triangle plot showing gene expression related to xylem, phloem, and nonvascular tissue in *Arabidopsis*.

CHAPTER SUMMARY

38.1 Movement of Water, Minerals, and Organic Compounds at the Cellular Level

- Passive transport of substances along a concentration gradient does not require ATP. Passive diffusion of water, gases, and other small, uncharged molecules across plasma membranes may occur in the absence of transport proteins. Facilitated diffusion involves plasma membrane and vacuolar protein channels and transporters. (Figure 38.1)

- Transport of materials across plasma membranes against concentration gradients is known as active transport and usually requires ATP. Plasma membrane proton pumps use the energy released by ATP hydrolysis to move protons from the cytosol into the intercellular space, thereby generating a membrane potential. Potential energy released by the flow of protons back into the cell can be coupled to the transport of ions and organic compounds into cells, a process known as cotransport. (Figure 38.2)

- Turgor pressure arises when the cell wall restricts the extent to which plant cells can swell when water enters as the result of osmosis. A cell that is so full of water that the plasma membrane presses closely against the cell wall is turgid. By contrast, a cell that is not swollen with water is flaccid, and a cell that contains so little water that the plasma membrane pulls away from the cell wall is plasmolyzed. (Figure 38.3)

- Solute potential and pressure potential arising from the presence of a cell wall generate cellular water potential. The integrated water potential of all cells in a plant can be estimated by a measure known as relative water content. (Figure 38.4)

- Plants display a variety of adaptations that help them cope with osmotic stress. In osmotic adjustment, high solute concentrations accumulate in the cytosol, a process that helps cells retain their water. Halophytes are plants adapted to very saline conditions, which impose extreme osmotic stress. (Figure 38.5)

38.2 Tissue-Level Transport

- Transmembrane transport involves the movement of materials from one cell to another from intercellular spaces, across plasma membranes, and into the cytosol. In apoplastic transport, water and solutes move through the apoplast, the water-filled cell walls and intercellular spaces of tissues. The symplast is the continuum of all plant cytosolic compartments linked by plasmodesmata. Symplastic transport allows materials to move from one cell to another without crossing plasma membranes. (Figures 38.6, 38.7)

- In roots, waxy Casparian strips on cell walls of endodermal tissue act as filters that reduce the movement of toxic ions into the plant vascular system. (Figure 38.8)

38.3 Long-Distance Transport

- Water and solutes move for long distances by bulk flow within the xylem and phloem. Plant vascular tissues are adapted in ways that reduce resistance to bulk flow. (Figure 38.9)

- Xylem is the main conduit for water and dissolved mineral nutrients, but it may also transport certain organic compounds. At maturity, tracheids and vessel elements are dead, empty of cytoplasm, and have lignified cell walls. Pits in tracheid and vessel walls allow water entry and exit. Experiments reveal that pits narrow or constrict in response to xylem sap solute content. Narrow tracheids fit together in long cell files. Vessel elements are wider but are more vulnerable than tracheids to blockage by air bubbles, or embolism. Root pressure, whose effects include the morning water drops on leaf tips known as guttation, helps some plants to refill embolized vessels. (Figures 38.10, 38.11, 38.12, 38.13, 38.14, 38.15)

- Transpiration is the evaporative loss of water from plant surfaces into sun-heated air. Plant transpiration has extensive local, regional, and global effects on climate. The cohesion-tension theory explains long-distance water transport as the combined effect of the cohesive forces of water and evaporative tension. (Figures 38.16, 38.17)

- Expansion of guard cells causes stomatal pores to open, allowing CO_2 intake. Guard cell deflation causes pores to close, an adaptation that helps to prevent excess water loss. Plants under water stress often drop their leaves in a process known as abscission, an adaptive response that lets plants avoid very low water potentials and the consequent threat of embolism. (Figures 38.18, 38.19, 38.20, 38.21)

- Organic solutes and minerals are transported in phloem sap as the result of osmosis. Phloem sap moves within sieve-tube elements that are living when mature, but lack a nucleus and are thus dependent on companion cells. Phloem loading occurs by symplastic or partly apoplastic transport. The pressure-flow hypothesis explains sugar translocation as a process driven by differences in turgor pressure that occur between a sugar source (for example, leaves) and a sugar sink (for example, developing fruit). (Figures 38.22, 38.23, 38.24, 38.25, 38.26)

TEST YOURSELF

1. An aquaporin is
   a. a channel protein that allows the influx of K^+ into cells, causing water to also flow in between the phospholipids of a plasma membrane.
   b. a type of blue-colored pore in the epidermal surfaces of plants.
   c. a protein channel in plasma membranes that facilitates the diffusion of water.
   d. a protein transporter in plasma membranes that uses protons to cotransport water.
   e. none of the above.

2. Why is turgor pressure a property of plant cells?
   a. Plant cells possess the necessary chloroplasts.
   b. Plant cells possess a cell wall, necessary for formation of turgor.
   c. Plant cells possess mitochondria, which provide the ATP needed for turgor.
   d. All of the above.
   e. None of the above.

3. How might plant cells avoid losing too much water in very cold, dry, or saline habitats?
   a. They may balance the osmotic condition of their cytosol with that of the environment.
   b. They may drop their leaves.
   c. They may stabilize their membranes with sugars or dehydrin proteins.
   d. They may produce more aquaporin water channels to take maximum advantage of available moisture.
   e. All of the above are possible.

4. What are ways in which plants accomplish tissue-level transport?
   a. transmembrane transport of solutes from one cell to another
   b. symplastic transport of materials from one cell to another via plasmodesmata
   c. apoplastic transport of water and dissolved solutes through cell walls and intercellular spaces
   d. all of the above
   e. none of the above

5. A root endodermis is
   a. an innermost layer of cortex cells that each display characteristic Casparian strips.
   b. a layer of cells just inside the epidermis of a root.
   c. a layer of cells just above the epidermis of a root.
   d. a group of cells that occur within the root epidermis.
   e. none of the above.

6. Xylem loading is
   a. the process by which water from the air enters vascular tissues of the leaf.
   b. the process by which sugar is transported via plasmodesmata directly into vessel elements.
   c. the process by which sugar is transported directly into sieve-tube elements.
   d. the process in which ions are transported across the membranes of root xylem parenchyma into the xylem apoplast, followed by water.
   e. none of the above.

7. What features of water explain how it can be drawn up a tall tree from roots to leaves?
   a. cohesion, the result of extensive hydrogen bonding
   b. adhesion, water's tendency to stick to surfaces such as the inner walls of tracheid and vessels
   c. high surface tension that develops when water evaporates from intercellular leaf spaces
   d. all of the above
   e. none of the above

8. What feature of terrestrial plants contributes to their ability to maintain relatively stable internal water content?
   a. a waxy surface cuticle
   b. an extensive root system that mines water from soil
   c. specialized water-conducting tissues composed of dead cells
   d. epidermal pores that open and close
   e. all of the above

9. What structural features of stomatal guard cells foster their ability to form an open pore in plant epidermal surfaces?
   a. thickened inner cell walls and radially oriented microfibrils
   b. thickened outer cell walls and radially oriented microfibrils
   c. thickened inner cell walls and longitudinal microfibrils
   d. thickened outer cell walls and longitudinal microfibrils
   e. uniform thickness of cell walls and randomly arranged microfibrils

10. What substances plug wounded sieve-tube elements, thereby preventing the leakage of phloem sap?
    a. X protein and callose
    b. C protein and callose
    c. P protein and callose
    d. P protein and sucrose
    e. none of the above

CONCEPTUAL QUESTIONS

1. Why is it a bad idea to overfertilize your houseplants? If the amount recommended on the package is good, wouldn't more be better?

2. Why is it a bad idea for subsistence farmers (those barely able to grow enough crops to feed themselves) to allow livestock to graze natural vegetation to the point that it disappears?

3. In the desert southwestern U.S., the ocotillo plant is often used as a landscaping plant, thanks to its interesting shape and beautiful floral displays. Imagine that you are responsible for property landscaped with ocotillo but find the plants bare of leaves. Can you assume that the plants are dead and need to be replaced?

EXPERIMENTAL QUESTIONS

1. Why did the Holbrook team use the same plant to examine the effect of a pure water control and an experimental, artificial xylem sap?

2. What caused water to flow from the source of artificial sap (or water) into the xylem of the plants used in the experiment conducted by Holbrook and associates?

3. Why were sap flow rates the same in living and dead plants in the experiment conducted by the Holbrook team?

COLLABORATIVE QUESTIONS

1. Imagine that you are part of a team that is assigned to determine what environmental conditions best suit a new crop, so that the crop can be recommended to farmers in appropriate climate regions. What features of the crop plants might you investigate?

2. Take a look outside or imagine a forest or grassland. What can you deduce about the availability of soil water from the types of plants that occur?

www.brookerbiology.com

This website includes answers to the Biological Inquiry questions found in the figure legends and all end-of-chapter questions.

39

FLOWERING PLANTS: REPRODUCTION AND DEVELOPMENT

CHAPTER OUTLINE

The reproductive success of dandelions.

Dandelions sometimes seem to be taking over the world, growing abundantly in open, sunny areas. Cheerful bright yellow flower heads, as shown in the chapter-opening photo, are one of the secrets of dandelions' success. If you pull a dandelion flower head apart, you can see that it is actually a bouquet of 200 or so small flowers. Each flower produces a tiny, one-seeded fruit equipped with a "parachute" for effective long-distance dispersal by wind. Each dandelion plant can produce up to 5,000 fruits during its lifetime, which explains how dandelions can spread so rapidly across a landscape. Flowers, fruits, and seeds are reproductive features of flowering plants that help to explain why hundreds of thousands of flowering plant species dominate most modern terrestrial ecosystems.

This chapter begins with an overview of the reproductive cycle of flowering plants that describes how flowers, fruits, and seeds function. This overview provides essential background for a closer look at flower structure and development, and some of the genes that control flower production and appearance. We will also examine the sexual reproductive processes by which plants produce gametes and accomplish fertilization, thereby producing zygotes and embryos. We will explore how seeds, fruits, and seedlings develop. Finally, we will consider ways in which some plants reproduce without using the sexual process. Throughout this chapter you will find examples of ways in which flowering plant reproduction plays an essential role in ecology and agriculture.

39.1 An Overview of Flowering Plant Reproduction

Most flowering plants display **sexual reproduction**. This is the process by which new individuals arise by the fusion of gametes (fertilization) to form zygotes that develop into mature adults.

Flowering plants, also known as angiosperms, inherited their sexual life cycle, known as alternation of generations, from ancestors extending back to the earliest land plants (as we explored in Chapter 30). Though all plants share the same basic life cycle, flowering plants display unique reproductive features. In this section, we first review the general features of alternation of generations and then consider more specific features of the angiosperm life cycle.

Flowering Plants Display Alternation of Generations

All groups of land plants produce two multicellular life cycle stages, in essence, two distinct plants. These two life cycle stages are the diploid, spore-producing **sporophyte**, and the haploid, gamete-producing **gametophyte**. In all groups of plants, haploid spores are typically produced by diploid sporophytes as the result of meiosis. These spores undergo mitotic cell divisions to produce multicellular gametophytes. Certain cells within the gametophytes differentiate into gametes. Thus, meiosis does not directly generate the gametes of plants. The processes of meiosis and fertilization form the transitions between the sporophyte and gametophyte life stages and link them in a cycle (**Figure 39.1**). The land plant life cycle is known as **alternation of generations** because it involves the cycling between distinct sporophyte and gametophyte generations.

During the evolutionary diversification of land plants, the sporophyte generation has become larger and more complex. By contrast, the gametophyte generation has become smaller and less complex. To illustrate this, let's compare the life cycle stages of mosses to those of angiosperms. Mosses diverged early in the history of land plants, whereas angiosperms first appeared much later. During the intervening time, the relative sizes and dependence of the sporophyte and gametophyte generations changed dramatically. Moss sporophytes are small structures that always grow attached to larger, photosynthetic gametophytes,

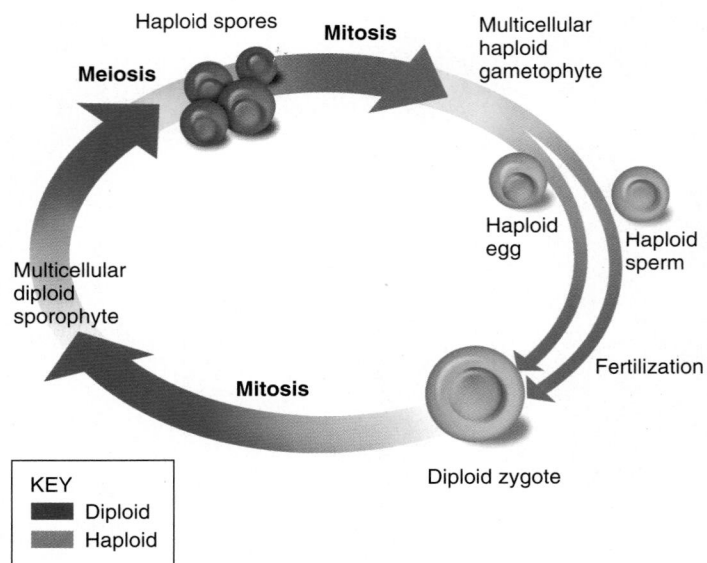

Figure 39.1 Alternation of generations, the plant life cycle.

(a) Gametophyte-dominant bryophyte (moss)

(b) Sporophyte-dominant flowering plant (oak)

Figure 39.2 Evolutionary shift in plant life cycle stage dominance. (a) In mosses, the gametophyte is the dominant life cycle stage, and the sporophyte is dependent on the gametophyte for resources. (b) In flowering plants such as oak, the sporophyte life cycle stage is dominant. Flowering plant gametophytes develop and grow within sporophytic flower tissues, and are completely dependent on sporophytes.

because moss sporophytes are incapable of independent life (**Figure 39.2a**). Moss sporophytes are dependent on gametophytes, because the gametophytes supply the sporophytes with essential nutrients.

In contrast, flowering plant sporophytes are notably larger and more complex than gametophytes. A tall oak tree, for example, is a single sporophyte. However, oak gametophytes are few-celled structures that develop and grow within flowers (**Figure 39.2b**). In addition, photosynthetic oak seedlings and trees grow independently, but nonphotosynthetic oak gametophytes are completely dependent on the sporophyte generation for their nutrition. A closer look at flower structure will help us to gain a more complete view of angiosperm gametophyte structure and function.

Flowers Produce and Nurture Male and Female Gametophytes

The literary wit Gertrude Stein famously paraphrased Shakespeare when she wrote, "A rose is a rose is a rose." Everyone knows that a rose is a flower, but what, exactly, is a flower? A **flower** is defined as a reproductive shoot, a stem branch that produces reproductive organs instead of leaves. Flower organs are thought to have evolved from leaflike structures by descent with modification (refer back to Figure 31.15). Flower organs are produced by shoot apical meristems much like those that generate leaves.

A flower shoot generally produces four types of organs: sepals, petals, stamens, and carpels (**Figure 39.3**). **Sepals** often function to protect the unopened flower bud. **Petals** usually serve to attract insects or other animals for pollen transport (refer back to Figures 31.19 and 31.20). **Stamens** and **carpels** each produce distinctive types of spores by the process of meiosis. From these spores, multicellular gametophytes develop

and certain gametophytic cells become specialized gametes. Nutritive tissues of stamens and carpels channel organic food, mineral ions, and water from the adult sporophyte into gametophytes. A closer look at the ways in which sporophytic stamens and carpels provide nutritional support to dependent gametophytes will help us to understand how flowering plants produce seeds.

Stamens Stamens produce **male gametophytes** and foster their early development. Most stamens display an elongate stalk, known as a **filament**, which is topped by an anther (see Figure 39.3). Filaments contain vascular tissue that delivers nutrients from parental sporophytes to anthers. Each **anther** is a group of four sporangia, structures in which spores are produced. Within the anther's sporangia, many diploid cells undergo meiosis, each producing four tiny, haploid spores. Because they are so small, generally 25–50 μm in diameter, the spores produced within anthers are known as **microspores**. Immature male gametophytes, otherwise known as **pollen grains**, develop from microspores. The term "pollen" comes from a Latin word meaning "fine flour," reflecting the small size of pollen grains. Pollen grains are eventually dispersed through pores or slits in the anthers. At the time of dispersal, the pollen grain is a two- or three-celled male gametophyte produced by mitotic division. During a later phase of development, a mature male gametophyte produces **sperm cells**.

(a) Flower parts

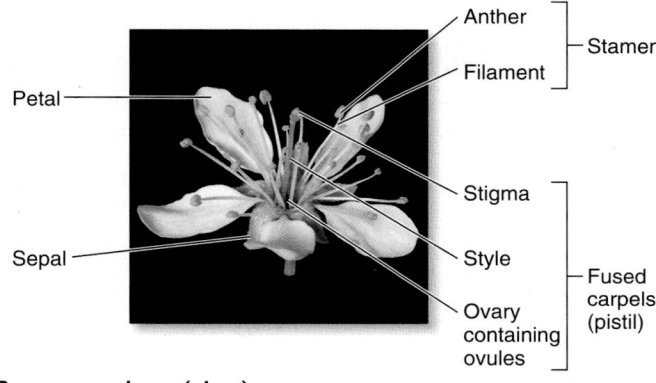

(b) *Prunus americana* **(plum)**

Figure 39.3   The structure of a typical flower.

Carpels   Carpels are vase-shaped structures that produce, enclose, and nurture **female gametophytes** and maturing male gametophytes. Carpels contain veins of vascular tissue that deliver nutrients from the parent sporophyte to the developing gametophytes. A flower contains one or more carpels that form a **pistil** (named for its resemblance to the pestle used to grind materials to a powder). Most flowers produce a single pistil composed of several fused carpels (see Figure 39.3), but some flowers have separate pistils that are each composed of a carpel. The topmost portion of a pistil, known as a **stigma** (meaning "mark"), receives pollen grains. The **style** is the middle portion of the pistil, and an ovary is at the bottom of the pistil. The **ovary** produces and nourishes one or more ovules. An **ovule** consists of a spore-producing structure (a sporangium) and enclosing tissues known as **integuments**. A small opening in these integuments is called a **micropyle**. Within an ovule, a diploid cell produces four **megaspores** by meiosis, but three of them die. The surviving megaspore generates a female gametophyte by mitosis. The female gametophytes of flowering plants typically consist of seven cells, one of which is the female gamete, the **egg cell**. This basic information about male and female gametophytes will help us to understand how they function to produce a young sporophyte within a seed.

Fertilization Triggers the Development of Embryonic Sporophytes, Seeds, and Fruits

In flowering plants, fertilization leads to the production of a young sporophyte that lies within a seed, completing the life cycle (**Figure 39.4**). However, fertilization cannot occur until pollen grains released from anthers first find their way to the stigma of a compatible flower, a process known as **pollination**. Many flowers are attractive to insects or other animals that transport pollen, but many angiosperm flowers are adapted for pollen transport by wind, and a few move pollen by means of water currents. Some plants display **self-pollination**, in which pollen from the anthers of a flower is transferred to the stigma of the same flower, or between flowers of the same plant. **Cross-pollination**, which occurs when a stigma receives pollen from a different plant of the same species, is also common.

Development of a Mature Male Gametophyte   When pollen grains land on the stigma, the stigma functions as a gatekeeper, allowing only pollen of appropriate genotype to germinate. During germination, a pollen grain produces a long, thin **pollen tube**. A mature male gametophyte is composed of the pollen grain plus the pollen tube that contains two sperm cells. Supplied with nutrients from the style, the pollen tube grows through it toward the ovary. Upon reaching the ovules, the pollen tube grows through the micropyle and delivers sperm to the female gametophyte (Figure 39.4). These sperm unite with haploid cells of the female gametophyte in the process of **fertilization**. Note that pollination and fertilization are distinct processes in flowering plants.

Double Fertilization   Angiosperms display an amazing phenomenon known as **double fertilization**. In this process, two different fertilization events occur. One of the two sperm cells delivered by a pollen tube fertilizes the egg cell, thereby forming a diploid **zygote**. This zygote may develop by mitotic division into a young sporophyte, known as an **embryo**. Fertilization thus begins a new cycle of alternation between sporophyte and gametophyte generations. The other sperm delivered by the same pollen tube fuses with two nuclei located in a central cell of the female gametophyte. The cell formed by this second fertilization undergoes mitosis, eventually producing a nutritive tissue known as the **endosperm**. The embryo and the endosperm are essential parts of maturing seeds. Fertilization not only starts the development of zygotes into embryos but also triggers the transformation of ovules into seeds and ovaries into fruits. Embryo, seed, and fruit development occur at the same time. Let's consider each of these.

Embryos and Seeds   An embryo is a young, multicellular, diploid sporophyte that develops from a single-celled zygote by mitosis. Because they are not yet capable of photosynthesis, embryos depend on organic food and other materials supplied by sporophytes. Thus, embryo development occurs within developing seeds located in a flower ovary. **Seeds** develop from

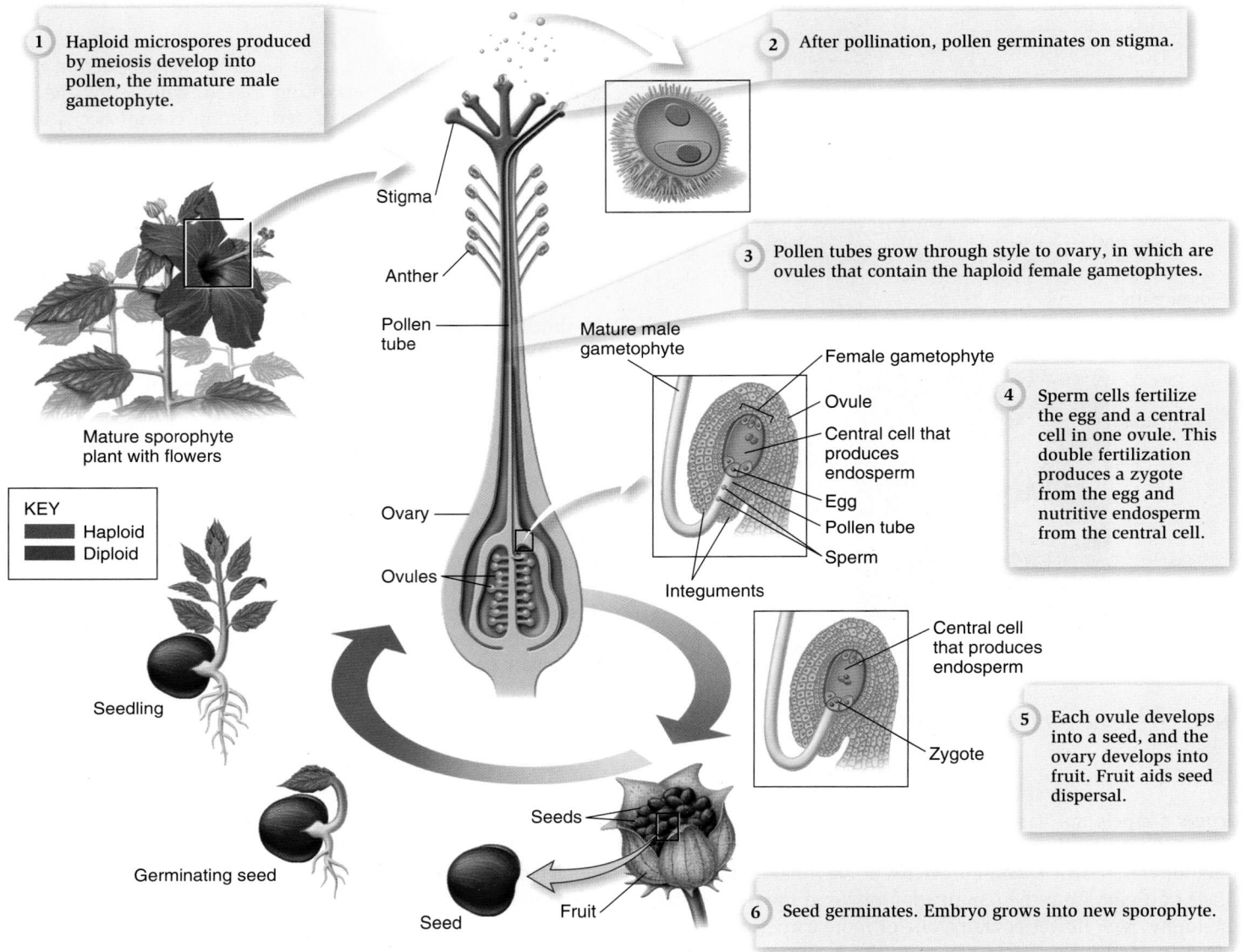

1 Haploid microspores produced by meiosis develop into pollen, the immature male gametophyte.

2 After pollination, pollen germinates on stigma.

3 Pollen tubes grow through style to ovary, in which are ovules that contain the haploid female gametophytes.

4 Sperm cells fertilize the egg and a central cell in one ovule. This double fertilization produces a zygote from the egg and nutritive endosperm from the central cell.

5 Each ovule develops into a seed, and the ovary develops into fruit. Fruit aids seed dispersal.

6 Seed germinates. Embryo grows into new sporophyte.

Stigma

Anther

Pollen tube

Mature sporophyte plant with flowers

KEY
Haploid
Diploid

Ovary

Ovules

Seedling

Germinating seed

Seed

Fruit

Seeds

Mature male gametophyte

Female gametophyte

Ovule

Central cell that produces endosperm

Egg

Pollen tube

Sperm

Integuments

Central cell that produces endosperm

Zygote

Figure 39.4   **The life cycle of a flowering plant.**   The plant reproductive cycle is illustrated here by hibiscus.

Biological inquiry: What advantage does the hibiscus flower gain by clustering its stamens around the pistil?

fertilized ovules. Each developing seed contains an embryo and nutritive endosperm tissue. Seeds also possess a tough **seed coat** that develops from the ovule integuments, which are from the sporophytic parent. When embryos and the seed coat have fully matured, they undergo drying and the seed enters a phase of metabolic slowdown known as **dormancy**. Fully mature, dormant seeds are ready to be dispersed.

Fruit and Seed Dispersal A **fruit** is a structure that encloses and helps to disperse seeds (Figure 39.4). Seed dispersal benefits plants by reducing competition for resources among seedlings and parental plants, and it allows plants to colonize new sites.

Fruits develop from the flower's ovary and sometimes include other flower parts. Young fruits bearing immature seeds are typically small and green. During the time that embryos and seeds are developing, the fruit also matures. The ovary wall changes into a fruit wall known as a **pericarp** (from the Greek word meaning "surrounding the fruit"). Mature fruits vary greatly among plant species in size, shape, color, and water content. These variations represent adaptations for seed dispersal in different ways. For example, single-seeded dandelion fruits are dry, lightweight, and bear a fluffy "parachute" derived from the flower's sepals (see the chapter-opening photo). These features foster dispersal by wind. In contrast, coconut fruits feature

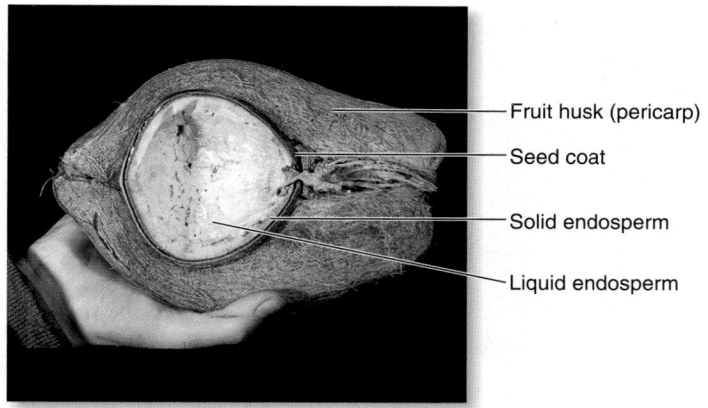

- Fruit husk (pericarp)
- Seed coat
- Solid endosperm
- Liquid endosperm

(a) Coconut fruit and seed

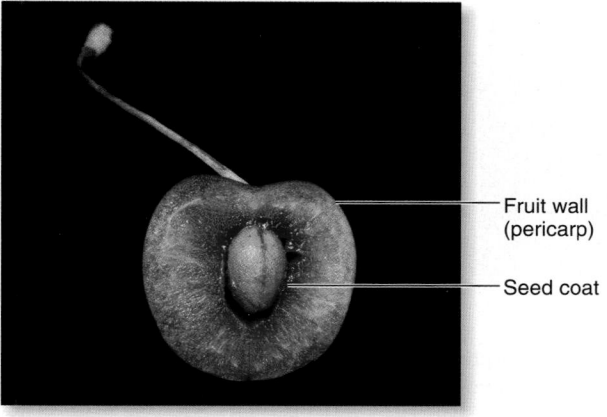

- Fruit wall (pericarp)
- Seed coat

(b) Cherry fruit and seed

Figure 39.5 **The structure of fruits as adaptations for seed dispersal.** (a) The coconut's outer fruit wall, called a husk, allows the fruit to float and thus disperse its seed among tropical shores. (b) The cherry is a juicy, sweet fruit that attracts animals to consume it, thereby helping to disperse its seed.

an airy husk (the outer pericarp) that keeps them afloat in ocean currents, allowing coconuts to disperse from one tropical shore to another. Inside the coconut fruit is a single large seed loaded with liquid and solid endosperm, which people use as coconut milk and coconut meat (**Figure 39.5a**). These large amounts of endosperm provide nutrients that sustain coconut seedling growth on infertile, sandy shores. Fruit variation is also extremely important to wild animals and in human agriculture. For example, most fruit crops, such as cherry, are juicy and sweet, with relatively small seeds (**Figure 39.5b**). In nature, these features foster dispersal by birds and other animals that feed on such fruits.

Seed Germination and Seedlings If a dispersed seed encounters favorable conditions, including sufficient sunlight and water, it will undergo seed **germination**. During seed germination, the embryo absorbs water, becomes metabolically active, and grows out of the seed coat, producing a seedling. If the seedling obtains

sufficient nutrients from the environment, it grows into a mature sporophyte capable of producing flowers. In the next section, we focus on flower production, structure, and development.

39.2 Flower Production, Structure, and Development

Many flowers are essential sources of food for diverse animals that help to disperse pollen. As the result of coevolutionary relationships, flowers occur in a spectacular array of colors and forms that attract particular pollinators (refer back to Table 31.2). Flowers also attract humans. We give bouquets to show love and appreciation; decorate homes, workplaces, and objects with flowers; display flower arrangements on ceremonial occasions; and make perfume from flowers. Consequently, many types of flowers are grown for the florist and perfume industries. Flowers are also integral to the production of grain and other food crops. Thus, both crop scientists and evolutionary biologists study flower development and the genes that control it.

Flowers develop when a leaf-producing shoot apical meristem becomes transformed into a reproductive meristem that produces flower organs. In this section, we focus on flower structure and how it varies as the result of genetic influences on development. We begin by describing recent research that helps to explain how the environment and genes work together to induce flower development.

GENOMES & PROTEOMES

Environmental Signals and Developmental Genes Control Flower Production

How do plants determine when to flower, and how do plants transform leaf-producing shoots into flowering shoots? Studies of the genes and proteins of *Arabidopsis*, rice (*Oryza*), and other plants whose genomes have been completely sequenced have provided answers to these questions.

Although temperature often influences flowering, day length is the main flowering signal for many plants. For example, crop plants such as rice and soybean are known as short-day plants because they flower when day length is decreasing. In contrast, pea and other plants are known as long-day plants because they flower when day length has increased beyond a critical level (refer back to Figure 36.10). The model plant *Arabidopsis* can flower during periods of long days or short days, but flowering occurs much faster in response to long days. How do plants detect the day-length signal?

At sunrise each day, the light-sensing pigment phytochrome in leaves is stimulated by red light and resets a molecular clock that occurs within plant cells. Each day, this clock starts the accumulation of cellular proteins that influence flowering. In short-day plants such as rice, a protein that represses flowering fails to accumulate to critical levels in short days, so flowering occurs. In long-day plants, including *Arabidopsis*, the cellular

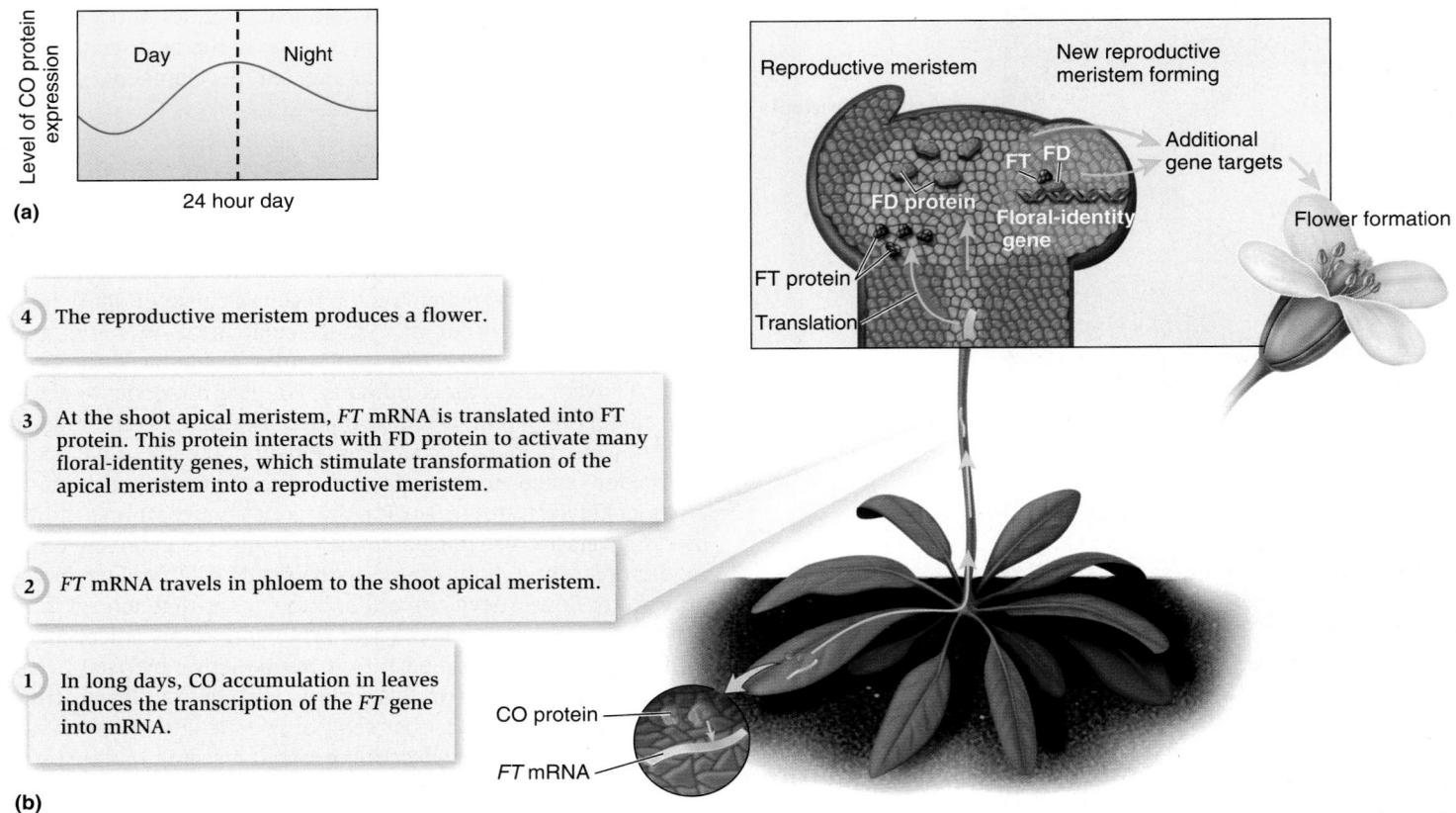

(a)

(b)

4 The reproductive meristem produces a flower.

3 At the shoot apical meristem, *FT* mRNA is translated into FT protein. This protein interacts with FD protein to activate many floral-identity genes, which stimulate transformation of the apical meristem into a reproductive meristem.

2 *FT* mRNA travels in phloem to the shoot apical meristem.

1 In long days, CO accumulation in leaves induces the transcription of the *FT* gene into mRNA.

Figure 39.6 **Long-day stimulation of flower development in *Arabidopsis*.** **(a)** In long days, the transcription factor CONSTANS (CO) accumulates to relatively high levels. **(b)** The accumulation of CO induces transcription of the *FT* gene into mRNA, which travels in phloem to shoot meristems and induces flower formation.

clock starts the production of a nuclear protein known as CON-STANS (CO) that accumulates during the day but degrades at night. In long days, enough CO accumulates to induce flowering in *Arabidopsis* (**Figure 39.6a**).

Let's consider how CO exerts its effects. Experiments performed in the 1970s revealed that leaves transmit a flowering signal to the shoot apical meristem via phloem. Based on these results, plant scientists hypothesized the existence of a flowering hormone, which they named **florigen**, meaning "flower generator." However, the chemical nature of florigen remained unknown until recently. In 2005, three groups of plant molecular biologists—Mitsutomo Abe, Takashi Araki, and colleagues; Philip Wigge, Detlef Weigel, and associates; and Tao Huang and Ove Nilsson and coworkers—working with *Arabidopsis* made critical discoveries that revealed florigen's identity. In long days, CO accumulation induces the transcription of a flowering time gene (*FT*) in leaf phloem tissues (**Figure 39.6b**). Messenger RNA transcribed from *FT* then travels in the phloem to the shoot apical meristem. At the meristem, this mRNA is translated into FT protein, which then binds to a transcription factor called FD. This interaction activates many floral-identity genes, which in turn stimulate the transformation of a leaf-producing apical meristem into a reproductive meristem. FT is probably the long-sought florigen molecule.

Developmental Genes Control Flower Structure

In this section, we survey the major ways in which flowers vary and consider some of the genes that influence flower structure by altering development. We begin with the genetic specification of the four basic flower organs—sepals, petals, stamens, and carpels.

The Genetic Basis of Flower Organ Identity Sepals, petals, stamens, and carpels occur in four concentric rings known as **whorls**. Sepals (collectively known as the **calyx**) form the outermost whorl, while petals (together known as the **corolla**) occur in an adjacent whorl, stamens (the **androecium**) in a third whorl, and carpels (the **gynoecium**) form the innermost whorl (**Figure 39.7**). A **perianth** is the calyx plus the corolla.

You may recall that *A*, *B*, *C*, and *E* genes encode transcription factors that control the production and arrangement of these whorls (refer back to Figure 19.26). Class A genes control the development of sepals and petals, class B genes specify petal and stamen formation, and class C genes determine the identities of the stamen and carpels. *E* gene expression distinguishes the sepals from the other whorls. However, not all flowers produce all four types of organs.

Variation in Number of Whorls Flowers that possess all four types of flower whorls—calyx, corolla, androecium, and gynoecium—are known as **complete flowers**. In contrast, flowers that lack one or more flower whorls are described as **incomplete flowers**. Flowers having both stamens and carpels are said to be **perfect**, whereas flowers lacking stamens or carpels are described as **imperfect**. An imperfect flower that produces carpels is known as a carpellate flower (or pistillate flower). Imperfect flowers that produce only stamens are described as staminate flowers.

Corn is a plant that produces both imperfect carpellate and staminate flowers (**Figure 39.8**). The flowers of corn start to develop as perfect flowers, but in carpellate flowers, the stamens stop developing. After pollination and fertilization, each carpellate flower produces one of the kernels on a cob of corn. In contrast, staminate flowers of corn, which are found in corn tassels, produce the pollen. Holly and willow also produce staminate and carpellate flowers, though on separate plants, and are thus described as **dioecious** (meaning "two houses"). In contrast, corn is **monoecious** (meaning "one house") because it produces carpellate and staminate flowers on the same plant.

Variation in Flower Organ Number In addition to variation in whorls, flowers vary in number of organs. You may recall that eudicot flower organs often occur in fours or fives or a multiple of these numbers. By contrast, monocot flower organs often occur in threes or a multiple of three. Many plants sold for use in gardens have been bred so that the flowers produce multiple organs. Garden roses, for example, typically have many more petals than do wild roses. This change results from a mutation that causes organs that would have become stamens to instead develop into additional petals. Other flowers possess relatively few organs. For example, the miniscule flowers of the tiny aquatic flowering plant *Lemna gibba* produce no perianths and only two stamens.

Variation in Flower Color Different flower parts may vary in color. The calyx and corolla of monocot flowers such as tulip are often similar in appearance and attractive function. In contrast, eudicot flowers tend to have green, leaflike sepals that are quite distinct from petals, which are often colorful and fragrant. For example, the sepals of petunia (*Petunia hybrida*) resemble leaves in having numerous stomata on both epidermal surfaces, and internal parenchyma tissue rich in chloroplasts. Petunia petal epidermis lacks stomata, and the parenchyma tissue usually lacks chlorophyll. Instead, the cell vacuoles of petals are often colored by **flavonoids**, a type of secondary metabolite (see Chapter 31). Among the pigments responsible for pink, red, blue, violet, purple, or yellow colors of petals and other flower parts, flavonoids are the most important. Color variations arise from differences in gene action that influence pigment biosynthesis pathways. Petunia has become an important model system for understanding the evolution of genes and proteins that control petal color.

Interestingly, research involving flower color has led to important discoveries in genetics. In 1990, plant molecular biologist Rich Jorgensen and colleagues reported gene silencing in petunia. In an attempt to produce flowers of deeper color, the

Figure 39.7   **The occurrence of flower parts in whorls.**

Calyx (sepals) ⎤
Corolla (petals) ⎦ Perianth
Androecium (stamens)
Gynoecium (carpels)

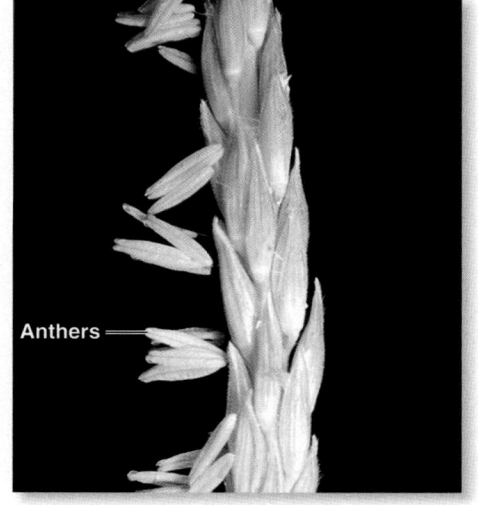

(a) Staminate flowers of Zea mays (corn)

Anthers

(b) Carpellate flowers of Zea mays (corn)

Ovary
Stigma

Figure 39.8   **Imperfect flowers of corn, a monoecious plant.**   Staminate flowers lack carpels, and carpellate flowers lack stamens, but both types of flowers occur on a single corn plant. In contrast, dioecious plants produce staminate and carpellate flowers on separate plants.

Biological inquiry: What inference can you make from the observation that corn flowers lack showy petals?

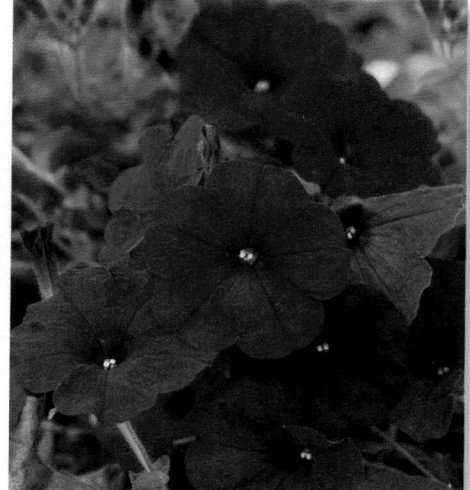

(a) Normal purple petunia flowers

(b) Petunia flower affected by microRNA

Figure 39.9   Gene silencing influences flower color in petunia.   (a) Normal purple petunia (*Petunia hybrida*) flowers produced by the expression of all genes involved in flavonoid synthesis. (b) Flower of a genetically engineered plant displays petal tissues in which one of the genes needed for production of purple pigment production has been silenced (suppressed) by microRNA.

researchers introduced an extra copy of a pigment-producing gene into petunia plants. To their surprise, they sometimes obtained flowers whose petals had white patches or were completely white (**Figure 39.9**). Adding the extra gene caused the production of microRNA (see Chapter 13). This RNA not only silenced the extra gene, but also the pigment-producing gene, explaining the white patches on the petals.

Variation in Flower Fragrance   The fragrances of flowers result from secondary metabolites that diffuse into the air from petals and other flower organs. One example is the terpene known as geraniol, which is alluded to in Shakespeare's famous phrase, "A rose by any other name would smell as sweet." Diverse fragrances function to attract particular types of animal pollinators. Because humans possess sensory systems similar to those of many pollinators, we are also attracted to many of these fragrances. Diverse genetically controlled biochemical pathways are responsible for the many types of flower scents.

Flower Shape Variation Resulting from Organ Fusion   During their development, many flowers undergo genetically controlled fusion of the organs within a whorl or fusion between whorls. For example, stamen filaments often partially fuse with the corolla or form a tube surrounding the pistil, a feature displayed by hibiscus flowers (see Figure 39.4). As we have seen, pistils themselves are often compound, composed of two or more fused carpels. The rosy periwinkle (*Cantharanthus roseus*), the original source of widely used leukemia drugs, provides an example. This plant's flowers each have two carpels, whose surfaces fuse together during flower development. In 1995, plant developmental biologist Judy Verbeke and colleagues reported that when the two young carpels first come into contact with each other, epidermal cells change their pattern of differentia-

tion. This re-differentiation is necessary for carpel fusion. When Verbeke experimentally removed one of the carpels, or inserted a nonporous barrier between the two carpels, cells did not re-differentiate and carpels did not fuse. If she inserted a porous barrier, however, normal differentiation and carpel fusion occurred. These experiments indicate that carpels communicate with each other by means of diffusible chemical signals that foster the cellular changes necessary for carpel fusion.

Mutants of the *HOTHEAD* (*HTH*) gene of *Arabidopsis* display similar floral organ fusions that likewise arise from epidermal interactions. In 2005, plant molecular biologists Susan Lolle, Robert Pruitt, and colleagues reported their surprising discovery that about 10% of *HOTHEAD* plants with two copies of a mutant gene could revert to producing normal flowers whose organs are not fused. According to Mendel's laws of inheritance, all of such offspring should have displayed the same mutation. These investigators propose that the *HOTHEAD* reversions result from a previously unknown genetic repair mechanism involving a store of nongenomic DNA. These results, like Mendel's use of pea crosses to uncover basic laws of inheritance and Jorgensen's discovery of gene silencing in petunia, illustrate the importance of flowering plant systems to basic biological research.

Flower Symmetry Variations   Flower shape variation can also result from changes in symmetry. Flowers that possess radial symmetry are described as regular (actinomorphic) flowers. Flowers having radial symmetry can be divided into two equal parts by more than one plane inserted through the center of the flower. In contrast, flowers that display bilateral symmetry are known as irregular (zygomorphic) flowers. Flowers having bilateral symmetry can be divided into two equal parts by only a single plane inserted through the center. The evolution of bilateral symmetry in flowers can often be linked to bee pollination.

(a) **(b)**

Figure 39.10 Genetic control of flower symmetry. **(a)** Normal snapdragon flowers, with functioning *CYCLOIDEA* genes, are bilaterally symmetrical. **(b)** Snapdragon plants carrying mutations in the *CYCLOIDEA* gene produce flowers that have radial symmetry.

This is because the development of bilateral symmetry typically generates a complex flower through which bees, but not other insect pollinators, are strong enough to push. Symmetry, like other flower features, is under genetic control. For example, snapdragon (*Antirrhinum majus*) flowers are normally bilaterally symmetrical, but mutation in the *CYCLOIDEA* gene causes these flowers to display radial symmetry (**Figure 39.10**). The *CYCLOIDEA* gene also specifies bilateral symmetry in another eudicot, the legume *Lotus japonicus*, which is only distantly related to snapdragon.

Variation in Flower Arrangements Flowers can occur singly or arranged in clusters of various types known as **inflorescences**. Diverse inflorescence types represent adaptations to particular pollination circumstances. For example, cereals such as corn and other grasses are wind-pollinated and thus produce inflorescences whose component flowers lack showy perianth parts (see Figure 39.8). Pollen is readily shed from such flowers into the wind. For example, a single staminate flower of corn disperses more than 10,000 pollen grains. Hairy grass stigmas effectively catch pollen from the wind, and clustering flowers into inflorescences also aids in pollen capture.

In corn, an inflorescence of staminate flowers is called a tassel, and an inflorescence of carpellate flowers is the ear (see Figure 39.8). Both types of inflorescence result from branching of the flowering meristem, with the result that a single corn shoot meristem produces many flowers rather than just one. One gene that plays an essential role in corn tassel and ear development is *barren inflorescence2* (*bif2*). In 2001, plant molecular biologists Paula McSteen and Sarah Hake reported that nonfunctional *bif2* mutant corn plants produce few or no inflorescence branches and thus make few or no flowers. Such corn plants are reproductively barren.

Many flowers that are pollinated by animals also occur in inflorescences. Daisies, marigolds, zinnias, thistles, sunflowers,

Figure 39.11 An inflorescence adaptation for pollination. The clustering of flowers into a type of inflorescence known as a head, displayed here by a cut dandelion, is an adaptation that fosters pollination by insects.

and dandelions are examples of plants that produce flowers in inflorescences known as heads. In the case of dandelions, each of the 200 perfect yellow flowers produces just one strap-shaped petal (**Figure 39.11**). Such individual flowers would be too small to attract pollinators from very far, but inflorescences are visible for much greater distances. Insects can easily walk around on the surfaces of heads, delivering pollen to many flowers during a single visit, thereby increasing pollination efficiency.

Now that we have surveyed aspects of flower development that influence overall flower structure, let's turn our attention to the development of male and female gametophytes within stamens and carpels.

39.3 Male and Female Gametophytes and Double Fertilization

Though flowering plant gametophytes are microscopic in size, they are essential to produce gametes. Agricultural scientists are interested in understanding gametophyte development in order to manipulate it as a way of improving crop plants.

Recall that mature male gametophytes are pollen tubes that deliver sperm cells, the male gametes. Mature female gametophytes are located within ovules and produce egg cells, the female gametes. In this section, we will begin by examining male and female gametophyte development, and how stamens and carpels aid this process. These gametes participate in fertilization, a process we consider next. We will also take an in-depth look at how scientists first achieved the technique of *in vitro* fertilization, which has applications in producing transgenic crops and allowing plant biologists to gain a better understanding of plant fertilization processes.

Pollen Grains Are Immature Male Gametophytes

As we have earlier noted, pollen grains are immature male gametophytes that are adapted for transport through air from one flower to another. Pollen grains develop within the sporangia of anthers. Inside these sporangia, diploid cells undergo meiosis and cytokinesis. As a result, each diploid cell gives rise to a cluster of four haploid microspores, each having a thin cellulose cell wall. The development of microspores into pollen grains involves two processes that occur at the same time. First, the microspore divides within its cell wall, producing a two- or three-celled young male gametophyte. Second, each male gametophyte develops a tough pollen wall that protects the gametophyte during pollen transport. Both of these processes are completed before anthers release mature pollen.

Early Male Gametophyte Development Each microspore nucleus undergoes one or two mitotic divisions to form a young male gametophyte. The first division gives rise to two specialized cells: a tube cell and a generative cell suspended within the tube cell (**Figure 39.12a**). The **generative cell** divides to produce two sperm cells, either before or (more commonly) after anthers release mature pollen. The **tube cell** stores proteins and lipids that may be used during later stages of male gametophyte development, after pollen has reached carpels. After pollination, this cell forms the pollen tube that grows to reach the female gametophyte.

Pollen Wall Development A mature pollen grain has a tough wall, and each plant species produces pollen whose wall has a distinctive sculptural shape (**Figure 39.12b**). The **pollen wall**, which surrounds the plasma membrane, is composed largely of a nearly indestructible polymer known as **sporopollenin**. In fact,

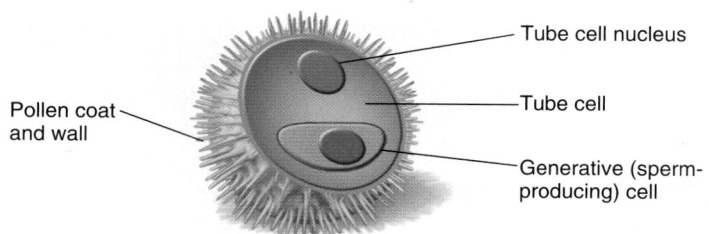

(a) A cut pollen grain showing immature male gametophyte

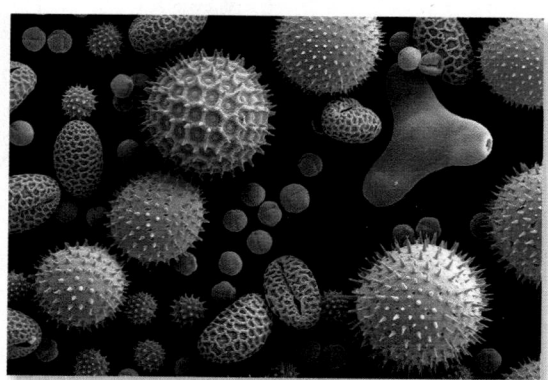

(b) SEM of whole pollen grains showing distinctive pollen wall ornamentation

Figure 39.12 Pollen grains.

Biological inquiry: What is the maximum number of cells in a mature male gametophyte of a flowering plant?

sporopollenin is so chemically inert that its biochemical composition is still uncertain. Named for its presence on the surfaces of mature spores and pollen, sporopollenin confers physical strength, chemical inertness, and resistance to microbial attack. Sporopollenin protects spores and pollen from damage. It also is responsible for the occurrence of intact fossil pollen grains in deposits that are millions of years old.

Development of the pollen wall starts with deposition of a blanket of the carbohydrate **callose** around each cluster of four microspores after they form by meiosis. Callose plays several crucial roles in male gametophyte development. First, the callose blanket seals microspores off from the influences of adjacent sporophyte tissues, thereby aiding pollen differentiation. Callose also provides a surface pattern for sporopollenin deposition and holds microspores together until an anther enzyme degrades the callose, freeing developing pollen grains from each other. Finally, as we shall later see, callose deposits help the pollen tubes of mature male gametophytes to grow.

As pollen grains mature, anther cells secrete a **pollen coat**, a layer of material that covers the sporopollenin-rich pollen wall. Coat materials include additional sporopollenin and pigments that give pollen its typically yellow, orange, or brown coloration, and lipids and proteins that aid in pollen attachment to carpels. Certain of these pollen coat compounds are responsible for allergic reactions in people exposed to particular types of airborne pollen. It is no wonder that pollen allergies are common, because about 10% of flowering plants are wind-pollinated, and such plants produce copious amounts of pollen. For example, ragweed plants (*Ambrosia* spp.), which are commonly associated with allergies, each produce an estimated 1 billion pollen grains during a year.

In some plants, pollen release can be explosive. In 2005, plant biologist Joan Edwards and coworkers discovered that flower buds of the bunchberry (*Cornus canadensis*) forcefully open in less than 0.5 millisecond. This sudden flower opening can occur by itself, thereby catapulting pollen into wind currents. It can also occur when an insect pollinator touches a sensitive hair on the bud. As a result, the stamens spring up and spray the insect with pollen. The insect may then carry this pollen to the stigma of the next bunchberry flower it visits. Such adaptations foster rapid transport of pollen to stigmas, which is necessary because pollen grains dry very quickly after their release, surviving for periods ranging only from a few hours to a few days.

The production and dispersal of pollen grains is just the first phase of male gametophyte development. We will learn more about the second phase of male gametophyte development, the production of a pollen tube, after focusing on female gametophyte development.

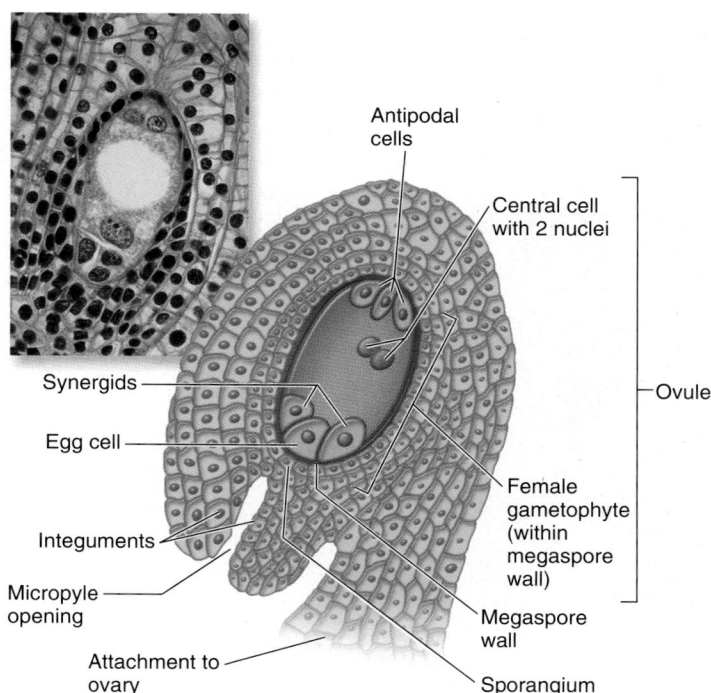

Figure 39.13   Female gametophyte within an ovule.

A Female Gametophyte Develops Within Each Ovule

Each ovule produces a single female gametophyte (**Figure 39.13**). Although flowering plant female gametophytes vary somewhat in size and structure, many possess seven cells and eight nuclei. One of these cells is an egg cell, which lies wedged between two cells known as **synergids**. These synergids have characteristic cell-wall ingrowths that increase the area of the plasma membrane, thereby helping to move nutrients from the larger sporophyte to the nonphotosynthetic female gametophyte. The other four cells of the angiosperm female gametophyte consist of three antipodal cells, whose functions are not well understood, and a large **central cell**. The central cell contains two nuclei, which accounts for the extra nucleus in the female gametophyte.

After Pollination, the Pistil Controls Pollen Germination

Pollen grains adhere to the surfaces of the stigma, which provides a receptive surface. The stigma and the style determine whether or not pollen grains germinate and pollen tubes grow. Pistils usually prevent the germination of pollen from different species and pollen that is genetically too similar, such as pollen from the same flower. Let's look at how the latter event is accomplished.

Rejection of pollen that is genetically too similar to the pistil is a phenomenon known as **self-incompatibility (SI)**. This feature helps to decrease the likelihood of recessive disorders in offspring. In the absence of self-incompatibility, self-fertilization and matings between closely related parent plants are more likely to produce progeny that are homozygous for many deleterious alleles, which would then be expressed. How then do pistils recognize compatible pollen, grains of the appropriate species that are genetically distinct? Recognition involves interactions between proteins of pollen and pistil cells. These interactions influence the ability of pollen to take up water, a process known as rehydration. Pollen is nearly dry when it reaches the stigma, and incompatible pollen will not rehydrate and will eventually die. In contrast, compatible pollen rehydrates, a process that activates pollen metabolism and allows it to germinate. Pollen-stigma interactions thus form an important part of a flowering plant's mate recognition and selection system.

Let's look at SI systems a bit more closely. In many plants, SI involves the *S* gene locus, which encodes SI proteins. Each locus contains two genes, one that determines pollen compatibility traits and another that determines pistil compatibility traits. Multiple *S* alleles for both genes occur in plant populations. Two major types of SI are known—gametophytic SI and sporophytic SI. Gametophytic SI occurs when an *S* gene within a pollen grain itself determines compatibility. Because pollen is a haploid gametophyte, the pollen genome contains only one pollen-compatibility allele. This allele encodes a specific S protein that is located in the pollen cytosol. When the gametophyte controls compatibility, tubes may start to grow from incompatible pollen, but an S protein encoded by the same locus in pistil cells enters the tubes and destroys RNA. This halts tube growth.

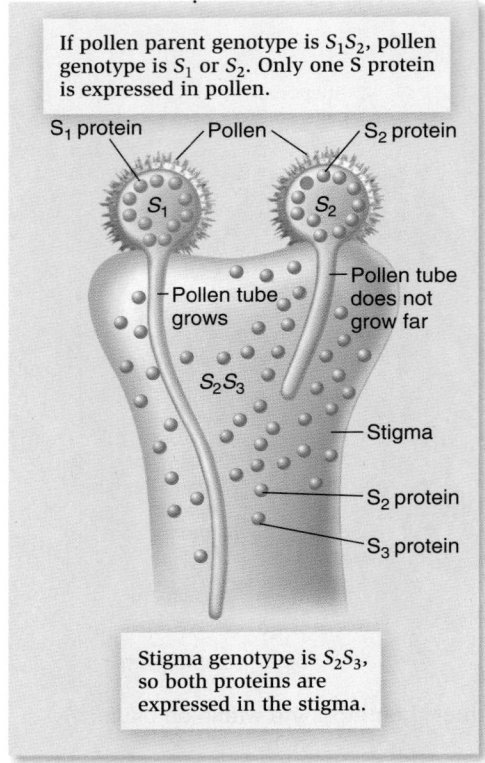

If pollen parent genotype is S_1S_2, pollen genotype is S_1 or S_2. Only one S protein is expressed in pollen.

S_1 protein — Pollen — S_2 protein

Pollen tube grows

Pollen tube does not grow far

Stigma

S_2 protein

S_3 protein

Stigma genotype is S_2S_3, so both proteins are expressed in the stigma.

(a) In gametophytic SI, if pollen *S* allele does not match either stigma allele, pollen will germinate.

If pollen parent genotype is S_1S_2, the pollen genotype is S_1 or S_2. However, the pollen parent makes both types of proteins, which are placed into the pollen coat.

Pollen tubes do not grow

S_1 protein

S_3 protein

If stigma genotype is S_1S_3 or S_2S_3, pollen will not germinate.

Pollen tubes grow

S_3 protein

S_4 protein

If stigma genotype is S_3S_4, S_1S_2 pollen will germinate.

(b) In sporophytic SI, if pollen coat S proteins do not match either stigma S protein, pollen tubes will grow.

Figure 39.14   Self-incompatibility.   Self-incompatibility helps plants to avoid the combination of gametes that are too genetically similar, and it is controlled by interactions between proteins of pollen and the pistil. **(a)** In gametophytic self-incompatibility, compatibility between pollen and pistil is determined by the haploid genotype of the pollen. In this case, S protein is located in the pollen cytosol. **(b)** In sporophytic self-incompatibility, compatibility between pollen and pistil is determined by the sporophyte that produced the pollen and contributed S proteins to its coat.

In contrast, S protein within genetically different pollen binds the pistil-produced protein, thereby preventing destruction of pollen tube RNA and allowing tube growth to continue (**Figure 39.14a**).

Sporophytic SI occurs when pollen compatibility is determined by the sporophyte that produces the pollen. This control is exerted when anthers deposit proteins into the pollen coat. Because anther cells belong to the sporophyte generation, they possess two *S* genes and thus heterozygotes will produce two different pollen coat S proteins. When the sporophyte controls compatibility, pollen cannot germinate when S proteins in the plasma membranes of stigma cells recognize (bind) the S proteins of incompatible pollen. In this case, protein binding leads to signal transduction processes that prevent pollen germination. However, pollen can germinate if stigma proteins are unable to bind genetically distinct S proteins in its coat (**Figure 39.14b**).

If the SI system becomes inactivated by mutation, a plant species can become self-pollinating. This has occurred during the evolution of self-pollinating lines of the model plant *Arabidopsis* from an ancestor that possessed SI. Self-pollination is useful to scientists performing genetic experiments with *Arabidopsis*.

Pea plants are also self-pollinating, a feature that was useful to Mendel in performing his classic experiments on inheritance (see Chapter 16).

During Pollen Germination, the Pollen Tube Grows, and Double Fertilization Occurs

In the second phase of male gametophyte development, a pollen grain germinates by taking up water and producing a pollen tube. In addition, the pollen generative nucleus usually divides by mitosis to produce two sperm cells. Upon rehydration, it takes a few minutes to an hour or so for a pollen tube to appear. Pollen tubes extend from a thin region in the grain's wall into the spaces between cells of the style. To deliver sperm to egg cells, the tube must grow from the stigma, through the style, to the ovule (**Figure 39.15**).

The Role of the Style   In some flowers, the style plays a role in mate selection by nourishing or inhibiting pollen tube growth. The style also guides pollen tube growth. For example, in 2005,

Figure 39.15 Pollen tubes delivering sperm to ovules.
This fluorescence microscopic view shows pollen grains (the light objects) germinating on the stigma surface (SG), and pollen tubes (PT) growing through the style (ST) toward ovules.

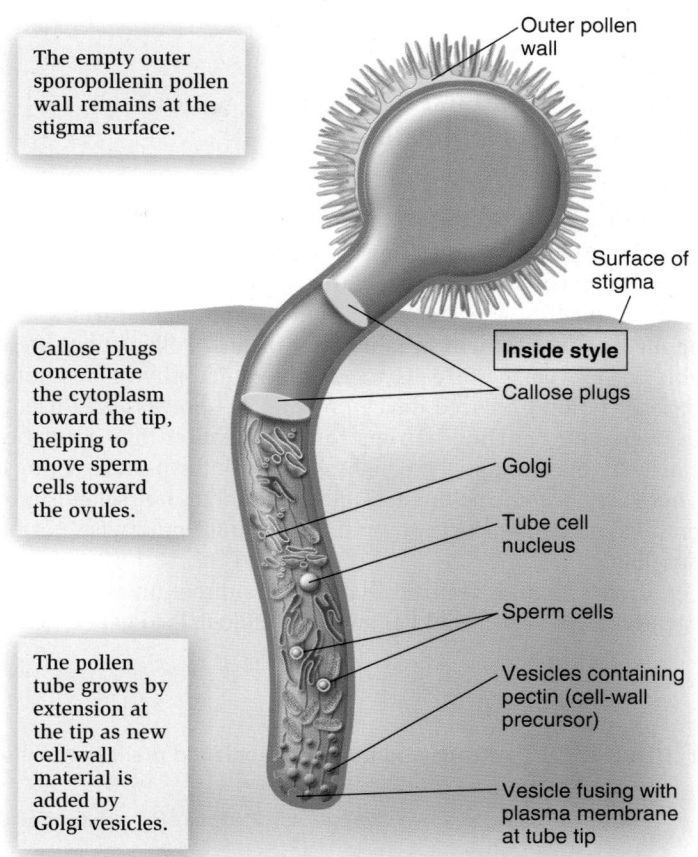

Figure 39.16 Tip growth by a pollen tube.

Elizabeth Lord and associates reported that the *Arabidopsis* style produces a blue copper-containing protein known as plantacyanin. These investigators also noted that the concentration of plantacyanin is normally low at the stigma but gradually increases in the style, with highest levels occurring near the ovary. Their observation suggested that plantacyanin might guide pollen tube growth toward ovules. Lord and associates tested this hypothesis by producing transgenic plants that expressed abnormally high levels of plantacyanin in the stigma. This change disrupted normal pollen tube behavior, causing pollen tubes to grow in circles around stigma cells or, in rare cases, up into the air, away from the style! This experimental evidence supports the concept that a gradient of plantacyanin in the style likely guides *Arabidopsis* pollen tubes toward ovules.

Tip Growth and Sperm Delivery A pollen tube transports two sperm cells to a female gametophyte in an ovule. A pollen tube does this by a process called tip growth, which is controlled by the tube cell nucleus. During tip growth, new cytoplasm and cell-wall material are added to the tip of an elongating cell (**Figure 39.16**). Golgi vesicles continuously deliver the carbohydrate pectin to the tube tip, whose thin cell wall is largely composed of pectin. As a pollen tube grows, callose plugs are commonly deposited in the older parts of the tubes. These plugs concentrate the tube cytoplasm at the tip, which helps to maintain the turgor pressure necessary for continued tip growth.

A pollen tube grows toward an ovule at about 0.5 mm per hour, taking from 1 hour to 2 days to reach its destination. When a pollen tube encounters an ovule, it enters through the micropyle. Attracted by secretions from a nearby synergid, the pollen tube penetrates the synergid. The thin tube tip wall then bursts, releasing the sperm into the female gametophyte. Plant biologists used to think that these two sperm cells were identical, but now it is known that at least some plants produce sperm that differ in structure and role. For example, plant reproductive biologist Scott Russell reported in 1985 that pollen tubes of leadwort (*Plumbago zeylanica*) contain a smaller sperm cell that is destined to fuse with the egg and a larger sperm that participates in endosperm formation. The two sperm shed their cytoplasm before double fertilization occurs.

Double Fertilization During the process of double fertilization, one sperm nucleus fuses with the egg cell to produce a zygote, the first cell of a new sporophyte generation (see steps 4 and 5 in Figure 39.4). The other sperm fuses with the two nuclei of the central cell to form the first endosperm cell. As the zygote develops into an embryo, the endosperm develops into a nutritive tissue that is usually triploid. We'll next see how plant biologists first achieved plant fertilization in the laboratory.

FEATURE INVESTIGATION

Kranz and Lörz First Achieved Plant *in Vitro* Fertilization

In vitro fertilization (IVF) is a widely practiced technology in human medicine and in agriculture. The term *in vitro* is Latin, meaning "in glass," referring to lab dishes. In past decades, plant biologists had only been able to perform *in vitro* pollination. To do this, they removed ovules from plant carpels and applied pollen to them in laboratory dishes. Pollen tubes were able to deliver sperm cells to female gametophytes, resulting in fertilization and seed development. This process was useful in overcoming mating barriers to hybridization that occur because stigmas and styles usually prevent interspecies matings. Therefore, *in vitro* pollination allowed agricultural scientists to produce new interspecies hybrids that would otherwise not be possible.

Crop scientists realized that IVF, the direct union between egg and sperm cell, would allow plant biologists to gain a better understanding of plant fertilization processes and might ultimately be useful in producing hybrid crops. However, plant gametes are so deeply buried within gametophytic and sporophytic tissues that scientists found it difficult to isolate and manipulate them. As a result, plant IVF was not accomplished until 1993. In that year, German researchers Erhard Kranz and Horst Lörz reported that they had used isolated gametes to produce mature hybrid corn (maize) plants, as shown in the Feature Investigation in **Figure 39.17**.

The development of a method for plant IVF required several research steps. First, by conducting many experiments, Kranz and Lörz developed procedures for isolating and handling single eggs and sperm. They also designed a process for using electrical current to induce gamete fusion. Further, they defined the optimal chemical and physical conditions necessary for *in vitro* development of zygotes into mature plants. For example, the researchers discovered that "nurse" cells grown from excised natural embryos were critical to the survival of zygotes and

Figure 39.17 Kranz and Lörz first achieved plant *in vitro* fertilization.

GOAL To demonstrate *in vitro* plant fertilization using isolated gametes.

STARTING MATERIALS Maize egg cell donor plants with green stigmas and hairs, and maize sperm cell donor plants with red stigmas and hairs.

	Experimental level	Conceptual level
1 Isolate sperm cells from pollen ruptured by osmotic shock. Isolate egg cells from ovules by microdissection of embryo sacs with a glass needle, using an enzyme solution that degrades cell walls.	Sperm Egg	
2 Under a light microscope, use a micromanipulator to line up egg and sperm on electrode surface.	Micromanipulator	Sperm and egg are aligned.
3 Fuse gametes with electrical pulses.	Pt (platinum) Electrode Electrical pulse	
4 Culture zygotes on porous membrane over a suspension of maize "nurse" cells in a solution of minerals and glucose at pH 5.5, on a rotary shaker, at 25°C, with 16-hour light/day.	Liquid medium — Zygotes and embryos — Porous membrane — Nurse cells	Maize nurse cells allow a sustained growth of zygotes into embryos.

5 Transfer embryos to solid media containing various concentrations of sucrose and plant hormones.

Embryo

Solid medium

Embryos develop into young plants.

6 Transplant young plants to soil. When plants flower, examine plants to see if their stigmas and stigma hairs were those expected of a hybrid.

7 THE DATA

Plants produced by *in vitro* fertilization (step 6) had green stigma, red hairs.

Plants donating the egg cells (step 1) had green stigma and hairs.

Plants donating the sperm cells (step 1) had red stigma and hairs.

---

embryos produced *in vitro*. The IFV corn embryos displayed normal development and grew into normal seedlings and reproductive adults. These investigators demonstrated their accomplishment by showing the hybrid origin of traits in progeny plants (see the data in Figure 39.17). The hybrid progeny plants had flower pistils with green stigmas with red hairs on them.

These traits were inherited from different parents: the green stigmas were inherited from the female parent, while the red hairs were inherited from the male parent. This work provided a foundation for additional IVF work with plants and also enabled more detailed investigations of plant embryo development, our next subject.

39.4 Embryo, Seed, Fruit, and Seedling Development

Seeds and fruits are major components of plant reproduction. Seeds contain dormant plant embryos that may develop into seedlings under favorable conditions. As earlier noted, fruits aid seed dispersal, which allows plants to colonize new sites. Embryos, seeds, and fruits mature simultaneously, and their development is coordinated by hormonal signals. Seedling development is also hormonally regulated. In this section, we will examine embryo, seed, fruit, and seedling development.

Endosperm Provides Nutrients for Developing Embryos Within Seeds

Endosperm is a tissue present in developing angiosperm seeds and also occurs in the mature seeds of many plants. Rich in protein, lipid, carbohydrate, vitamins, and minerals, endosperm supplies the nutritional needs of developing embryos and often seedlings as well. We have already noted how humans use the coconut's liquid and solid endosperm as food (see Figure 39.5a).

In fact, a large percentage of human and animal food protein comes from seed endosperm of grain crops. Corn, wheat, rice, and other grain crops generate more than 380 billion pounds of endosperm per year in the U.S. alone. How does this valuable tissue develop?

Recall that the first cell of the endosperm tissue arises by the fusion of a sperm nucleus with the two nuclei of the female gametophyte's central cell. The resulting nucleus is typically triploid, combining one set of chromosomes from the male parent with two sets from the female parent. Mitotic division of the triploid cell generates endosperm tissue. The food stored in endosperm comes from the parent sporophyte via apoplastic transport through moist cell walls and intercellular spaces.

As the embryo develops, it uses food stored in the endosperm. By the time that most eudicot seeds are mature, they contain little or no endosperm. The eudicot embryo stores organic food in two embryonic leaves, the cotyledons. The seeds of legumes, such as beans, peas, and peanuts, are valued as food because of their protein-rich cotyledons. In contrast, many mature monocot seeds retain considerable endosperm, explaining the food value of grain crops and coconuts.

Embryos Develop from Zygotes Within Seeds

Mature plant embryos are young sporophytes that possess several types of organs and lie dormant within seeds. Embryos develop from single-celled zygotes by mitotic divisions, in the process known as **embryogenesis**. This process begins with an unequal cell division and proceeds through several distinctive stages, as illustrated by the model eudicot plant *Arabidopsis* (**Figure 39.18**).

Early Embryogenesis Sometime within a period of days to several weeks following fertilization, a zygote begins to divide. At this point, a zygote is blanketed with a layer of callose, which helps to seal it off from the environment, thereby fostering embryo-specific gene expression. The zygote's first cell division is unequal, producing a smaller cell and a larger cell. These two cells differ in cytoplasmic contents, which Kranz and Lörz observed in IVF corn embryos. This unequal division helps to establish the apical-basal (top-bottom) polarity of the embryo, which persists through the life of the plant. The smaller cell develops into the embryo, whose radial symmetry is established at this point and continues in adult plants. The larger cell

develops into a **suspensor**, a short chain of cells anchored near the micropyle at the ovule entrance. The suspensor channels nutrients and hormones from the parent sporophyte into the young embryo, which absorbs them at its surfaces. Researchers have found that the functional loss of a gene called *TWN* (twin) transforms the suspensor into a second, twin embryo. This indicates that when TWN protein is active, it prevents the suspensor from developing into an embryo, thereby maintaining normal suspensor attachment and nutrient transport functions. The suspensor usually disappears by controlled cell death during embryo development. Older embryos rely on nutrients supplied by the endosperm.

Later Embryogenesis Young eudicot embryos are spherical, but they soon become heart-shaped as the seedling leaves, called **cotyledons**, start to develop. At this point, the young shoot and root are established at the apical and basal poles, respectively. This process is controlled by the expression of the gene *TOPLESS*, which represses genes that promote root development at the apical pole. (Mutants that have lost normal *TOPLESS* function have roots at both apical and basal poles, but no shoots, and are thus topless.) Eudicot embryos such as *Arabidopsis* then become

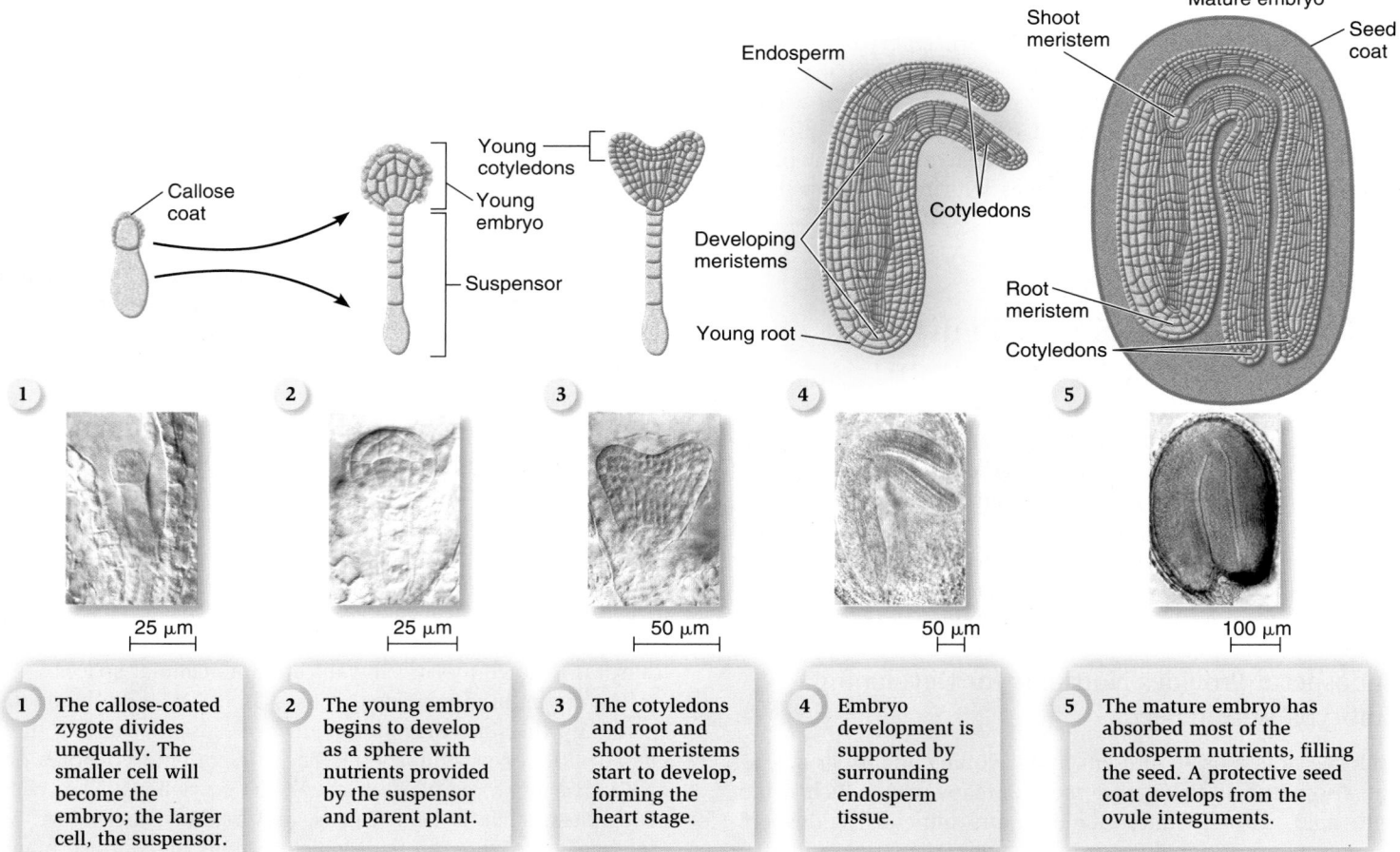

1 The callose-coated zygote divides unequally. The smaller cell will become the embryo; the larger cell, the suspensor.

2 The young embryo begins to develop as a sphere with nutrients provided by the suspensor and parent plant.

3 The cotyledons and root and shoot meristems start to develop, forming the heart stage.

4 Embryo development is supported by surrounding endosperm tissue.

5 The mature embryo has absorbed most of the endosperm nutrients, filling the seed. A protective seed coat develops from the ovule integuments.

Figure 39.18   **Embryogenesis in the eudicot *Arabidopsis*.**

Biological inquiry: How would this embryo differ if its TOPLESS genes were nonfunctional?

torpedo-shaped, and as the cotyledons grow, they often curl to fit within the developing seed (Figure 39.18). In contrast, mature monocot embryos are cylindrical, with a single cotyledon and a side notch where the apical meristem forms. Mature embryos then become dormant as seeds mature.

Mature Seeds Contain Dormant Embryos

The structure of mature monocot and eudicot seeds differs. Within eudicot seeds, mature embryos often display an **epicotyl**, the portion of an embryonic stem with two tiny leaves in a first bud that is located above the point of attachment of the cotyledons (**Figure 39.19a**). The **hypocotyl** is the portion of an embryonic stem located below the point of attachment of the cotyledons. An embryonic root, the **radicle**, extends from the hypocotyl. Much of the endosperm has been absorbed into the large cotyledons.

In contrast, mature monocot embryos, such as those of corn, feature an epicotyl with a first bud enclosed in a protective sheath known as the **coleoptile**. The young monocot root is enclosed within a protective envelope known as the **coleorhiza** (**Figure 39.19b**).

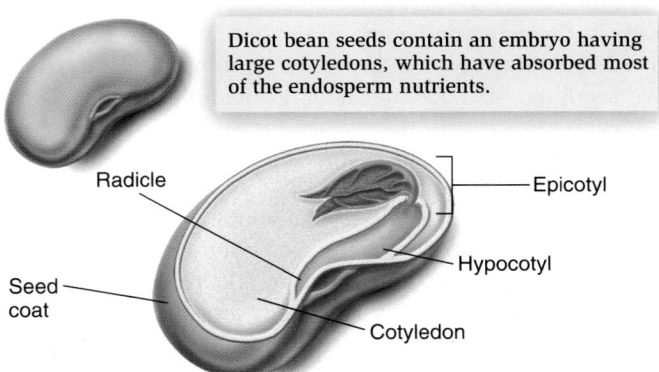

Dicot bean seeds contain an embryo having large cotyledons, which have absorbed most of the endosperm nutrients.

Radicle

Epicotyl

Hypocotyl

Seed coat

Cotyledon

(a) Eudicot bean seed, showing embryo with epicotyl, hypocotyl, and radicle

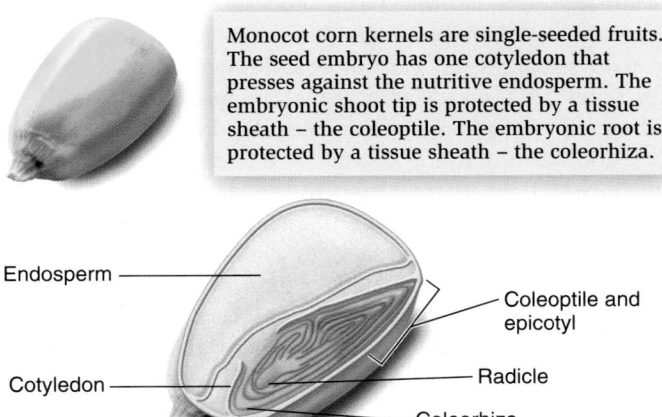

Monocot corn kernels are single-seeded fruits. The seed embryo has one cotyledon that presses against the nutritive endosperm. The embryonic shoot tip is protected by a tissue sheath – the coleoptile. The embryonic root is protected by a tissue sheath – the coleorhiza.

Endosperm

Coleoptile and epicotyl

Cotyledon

Radicle

Coleorhiza

(b) Monocot corn seed, showing an embryo protected by coleoptile and coleorhiza

Figure 39.19   Structure of mature seeds and embryos.

As seeds mature, they undergo changes leading to dormancy, an adaptation that prevents them from germinating when environmental conditions are not suitable for seedling growth. Seed maturation includes transformation of the ovule's integuments into a tough seed coat (see Figure 39.19a). The seed coat restrains seedlings from growing and prevents the entry of water and oxygen, which maintains low seed metabolism. In addition, the coats of some seeds are darkly colored with pigments that may help to prevent damage by UV radiation or microbial attack. Another change leading to seed dormancy is gradual, controlled loss of water from the embryo and other seed tissues. As the result, the water content of dispersed seeds is only 5 to 15%. Abscisic acid (ABA) is a hormone that induces the activity of genes that help embryo tissues to survive the drying process. Some of these desiccation-tolerance genes encode proteins that form loose coils enclosing cell contents, thereby preventing damage as the cytoplasm becomes almost completely dry. When the seeds of flowering plants are dry and ready for dispersal, they are released from the plant while enclosed in a fruit or released when the fruit breaks open.

Fruits Develop from Ovaries and Other Flower Parts

Though all fruits develop from ovaries (and sometimes other flower parts), fruits occur in diverse forms that aid seed dispersal. Some fruits are dry, whereas others are moist and juicy; some fruits open to release seeds and others do not. Fruits also display a wide variety of sizes, colors, and fragrances. These variations result from differences in the process of fruit development. Plant hormones, including auxin, gibberellic acid, and cytokinin, control this transformation. Abscisic acid stimulates cell expansion, and ethylene influences fruit ripening. For instance, ethylene helps to ripen nuts, a type of dry fruit, by inducing plasma membranes to rupture, causing water loss. Under the influence of plant hormones, the pericarp (ripened ovary wall) of peaches, plum, and related fruits swells and softens, while orange or red chromoplasts replace green chloroplasts. As fruits mature, the outer protective cuticle often becomes very thick, contributing to peel toughness, which helps to prevent microbe attack. In addition, many maturing fruits increase their sugar and acid content, which produces the distinctive tastes of ripe fruit. Many fruits also produce fragrant volatile compounds.

Differences in the fragrance, color, shape, and moisture content of wild fruits reflect evolutionary adaptation for effective seed dispersal. Though many fruits and seeds are dispersed by wind or water, or by attaching to animal fur, others are consumed by fruit-eating animals that are attracted by fruit color and fragrance. Blackberries provide a good example of fruits adapted for animal dispersal. Blackberry flowers produce many separate pistils, each containing a single ovule (**Figure 39.20a**). Following pollination and fertilization, the ovary of each pistil develops into a sweet, juicy fruitlet containing a single seed. As the individual fruitlets develop, they fuse together at the sides.

(a) *Rubus allegheniensis* (common blackberry) flower

(b) Blackberry fruit

Figure 39.20   Blackberry flower and fruit.   (a) Each of the many separate pistils in a blackberry flower is able to produce a single one-seed fruit (called a fruitlet) if fertilization occurs. (b) Together, the individual fruitlets of the blackberry compose an aggregate fruit. A shriveled style with a stigma is attached to each fruitlet.

Consequently, the many fruitlets produced by a single blackberry flower are dispersed together, in a structure known as an aggregate fruit (**Figure 39.20b**). Attracted by the color, birds consume the whole aggregate and excrete the seeds, thereby dispersing many at a time. Many other types of fruits occur and these likewise represent adaptations that foster seed dispersal (refer back to Figure 31.21). Although a fruit is usually defined as a mature ovary containing seeds, commercial seedless fruits are produced by genetic modification or treatment with artificial auxin.

Environmental and Internal Factors Influence Seed Germination

Seeds vary greatly in their ability to germinate after dispersal. Small seeds such as those of dandelions and lettuces germinate quickly if light is available. Other seeds require a period of dormancy before germination occurs. Some seeds can remain dormant for amazingly long time periods. For example, a lotus (*Nelumbo nucifera*) seed collected from a lake bed in China germinated at the age of 1,450 years, as determined by radiocarbon dating. In 2005, plant scientists germinated a 2,000-year-old date seed found in Israel. Arctic lupine (*Lupinus arcticus*) seeds germinated after having been collected among Yukon animal remains that were dated at more than 10,000 years of age!

Water is generally required to rehydrate seeds so that embryos can resume their metabolic activity. Water absorption also swells seeds, helping to break the seed coat and allowing embryonic organs to emerge. In some cases, rainfall of sufficient duration to leach germination-inhibiting compounds out of seeds is required. The optimal temperature for germination of most seeds lies between 25 and 30°C (77 and 86°F). This explains why gardeners wait until the soil is warm before planting seeds outdoors in spring. However, some seeds need a period of cold treatment or seed coat abrasion before they will germinate.

Such physical stimuli induce the activity of more than 2,000 genes associated with seed germination. In 2005, British plant physiologists Nompumelelo Masubelele, James Murray, and colleagues reported that several phases of cell division occur after dry *Arabidopsis* seeds have been moistened. Cell division in the radicle, the embryonic root, occurs first. As a result, the radicle is the first organ to emerge from a germinating seed, which allows seedlings to take up water even more rapidly. Next, cell division rates rise in the cotyledons and then in the shoot apical meristem, which increases the size of the embryonic shoot.

When grass seeds rehydrate, the young shoot secretes the hormone gibberellic acid from the seed cotyledon into the outermost endosperm layer, known as the aleurone. In response, the aleurone secretes digestive enzymes into the central endosperm layer, releasing sugars from stored starch (**Figure 39.21**). The seedling uses these sugars for growth. This highly coordinated process allows grass seeds to quickly germinate when it rains, an advantage in arid grassland habitats.

Humans also use this basic process to make beer. Beer brewers apply gibberellic acid to barley seeds to induce them to germinate simultaneously, a process known as malting. The barley seeds are then baked at a high temperature to stop germination, a process that produces malt. Baking also caramelizes sugars, contributing to the distinctive flavors and color of beer. To make beer, brewers treat malt with water and heat, add the dried flowers of the hop plant (the genus *Humulus*), and add yeasts to ferment the plant sugars to alcohol.

Once seeds have germinated, plants vary in the process by which the embryonic shoot emerges. When bean and onion seeds germinate, the embryo forms a hook that first breaches the soil surface and then straightens, thereby pulling the rest of the embryo and cotyledons aboveground (**Figure 39.22a,b**). In contrast, when pea seeds germinate, the epicotyl forms a hook that pulls the shoot tip out of the ground, leaving the cotyledons beneath the soil surface (**Figure 39.22c**). In such cases, the

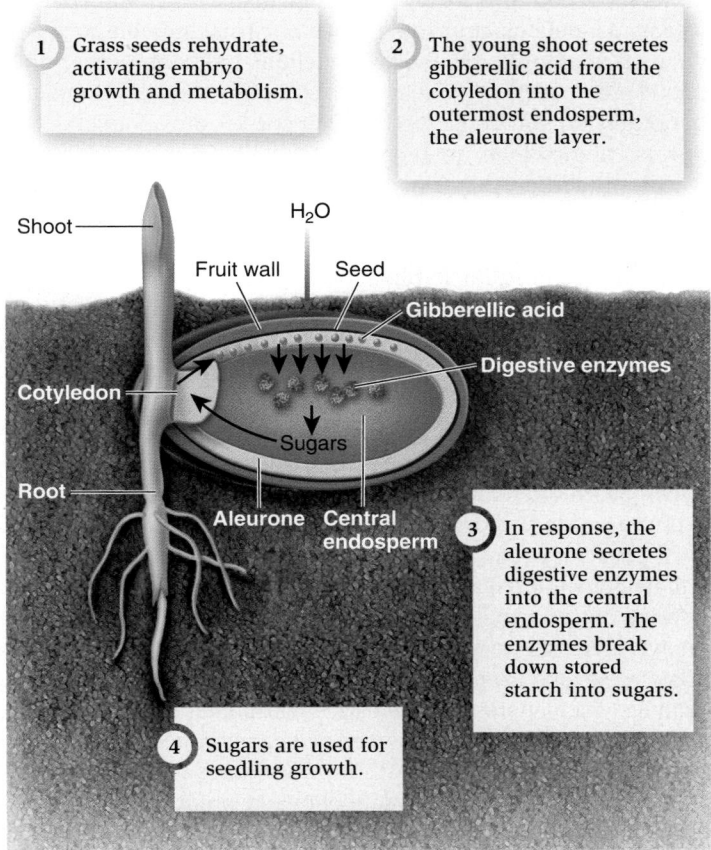

1. Grass seeds rehydrate, activating embryo growth and metabolism.

2. The young shoot secretes gibberellic acid from the cotyledon into the outermost endosperm, the aleurone layer.

Shoot

H_2O

Fruit wall Seed

Gibberellic acid

Cotyledon

Digestive enzymes

Sugars

Root

Aleurone Central endosperm

3. In response, the aleurone secretes digestive enzymes into the central endosperm. The enzymes break down stored starch into sugars.

4. Sugars are used for seedling growth.

Figure 39.21 Germination of grass seeds is mediated by hormones.

tough hook cells bear the brunt of passage through hard surface soil crusts, thereby protecting the delicate shoot tips. The plant hormone ethylene controls seedling hook formation. As they grow through the soil, the shoot tips of corn seedlings and those of other grasses are protected by the coleoptile, which functions as a protective tube for the first foliage leaves (**Figure 39.22d**).

39.5 Asexual Reproduction in Flowering Plants

Many plants rely on sexual reproduction. By contrast, most dandelions and a wide variety of other angiosperms reproduce primarily by asexual means, and other plants commonly utilize both sexual and asexual reproduction. **Asexual reproduction** is the production of new individuals from a single parent without the occurrence of fertilization. Although sexual reproduction provides beneficial genetic variation, asexual reproduction can be advantageous in other ways. For example, asexual reproduction maintains favorable gene combinations that allow faster population growth in stable environments.

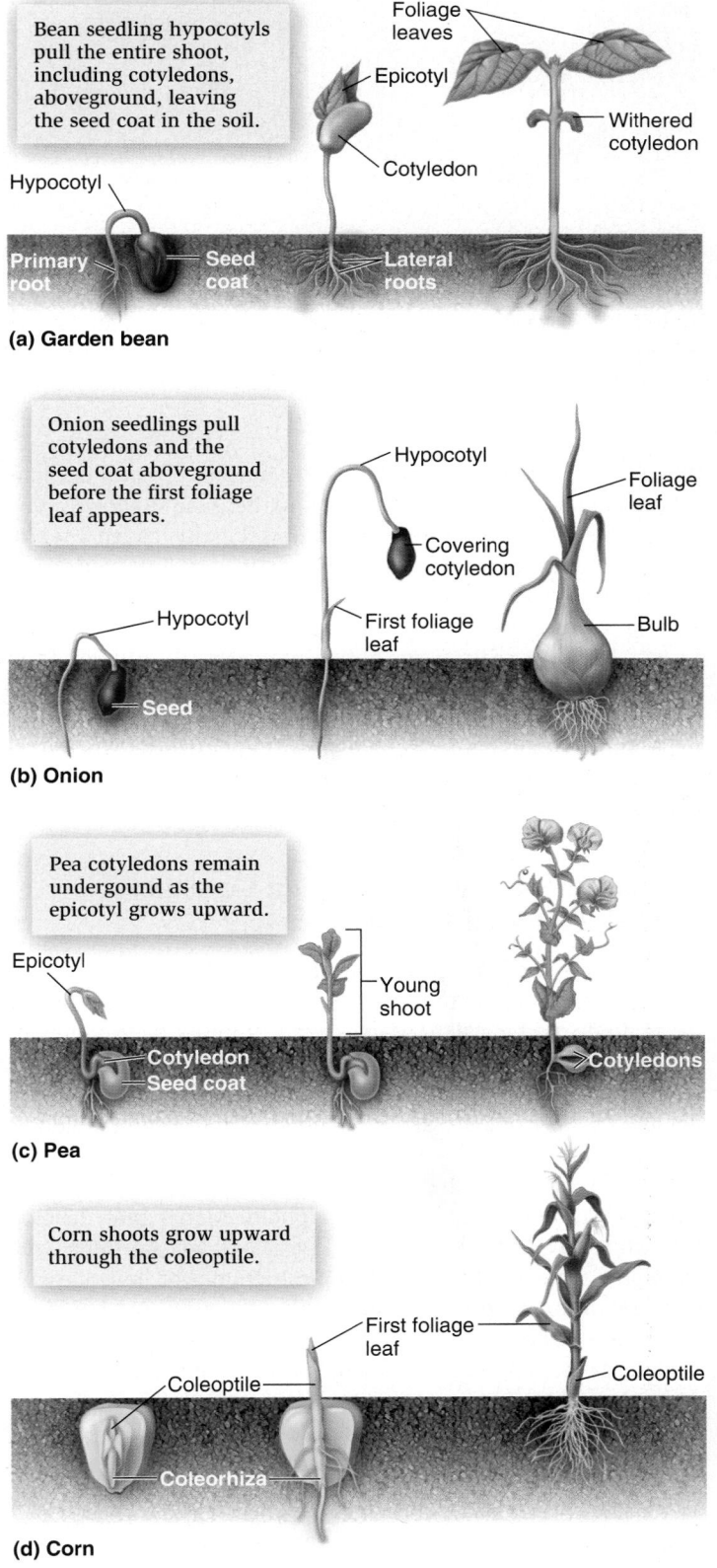

Bean seedling hypocotyls pull the entire shoot, including cotyledons, aboveground, leaving the seed coat in the soil.

Foliage leaves

Epicotyl

Hypocotyl

Withered cotyledon

Cotyledon

Primary root

Seed coat

Lateral roots

(a) Garden bean

Onion seedlings pull cotyledons and the seed coat aboveground before the first foliage leaf appears.

Hypocotyl

Foliage leaf

Covering cotyledon

Hypocotyl

First foliage leaf

Bulb

Seed

(b) Onion

Pea cotyledons remain undergound as the epicotyl grows upward.

Epicotyl

Young shoot

Cotyledons

Cotyledon

Seed coat

(c) Pea

Corn shoots grow upward through the coleoptile.

First foliage leaf

Coleoptile

Coleoptile

Coleorhiza

(d) Corn

Figure 39.22 Variations in seed germination and seedling growth patterns.

Figure 39.23   **An example of asexual reproduction.**   The leaves of this *Kalanchoë* plant bear small plantlets around the edges. When mature, these plantlets drop off, and under the right conditions, grow into new plants.

Asexual reproduction is also advantageous in stressful habitats where pollinators or mates can be rare, because it allows a single individual to start a new population. Finally, asexual reproduction allows some plants to persist for very long periods of time. The oldest known plants are creosote bushes that are asexual clones of a parent that grew from a seed about 12,000 years ago! The artificial vegetative propagation of plants from cuttings is a form of asexual reproduction that is widely used commercially and by home gardeners. In this section, we will explore the three main mechanisms of plant asexual reproduction: specialized vegetative structures, apomixis, and somatic embryogenesis.

Vegetative Asexual Reproduction Generates Plant Clones from Organs

Roots, stems, and leaves are vegetative plant organs that can function as asexual reproductive structures. Root sprouts, such as those produced by aspens, can generate entire groves of genetically identical trees. The common houseplant *Kalanchoë* has leaves bearing many tiny plantlets at their edges. When these detach, they are able to take root and grow into new individuals (**Figure 39.23**).

Sucker shoots such as those appearing at the bases of banana plants and date palms, and potato plants that develop from the pieces of tuber-bearing "eyes" (which are buds), are examples of vegetative asexual reproduction that have agricultural importance. Attractive horticultural varieties of African violets and other plants can be propagated from leaf cuttings. Such asexual offspring grow into adult plants that have the same valued properties as their parent plant. In contrast, these crops grown from seed would produce diverse progeny, not all of which would have economically prized properties.

Apomixis Is Seed Production Without Fertilization

Apomixis, which means "away from mixing" (that is, genetic mixing), is a natural asexual reproductive process in which fruits and seeds are produced in the absence of fertilization.

More than 300 species of flowering plants, including hawkweeds, dandelions, and some types of citrus, are able to reproduce asexually by apomixis. Dandelions and some other apomictic plants require pollination to stimulate seed development, but others do not. Some weedy plant species are apomictic; this process allows them to reproduce quickly, a characteristic of weeds. In addition to their concern about weeds, agricultural scientists are interested in apomixis as a potential method for producing genetically uniform seeds, propagating hybrids, and removing the need for fertilization.

Most studies of apomixis have been carried out with dandelions, because they are widespread and locally abundant. Dandelions are herbaceous perennials that flower in their first year. Although some European dandelion populations contain individuals that reproduce sexually, many dandelion populations are composed mainly of individuals that reproduce by apomixis. In apomictic plants, meiosis produces diploid megaspores because homologous chromosomes do not pair and meiosis II does not occur. These diploid megaspores produce diploid female gametophytes, which generate diploid eggs. These eggs develop into normal embryos without fertilization, a process known as parthenogenesis. In apomictic dandelions, the endosperm develops from diploid central cells. Though such plants produce microspores via meiosis, most pollen grains have abnormal numbers of chromosomes and can produce pollen tubes, but not sperm cells. Hence, fertilization does not occur. However, pollination stimulates dandelions to produce their single-seeded fruits. Fruits that develop in the absence of fertilization are known as parthenocarpic fruit.

Somatic Embryogenesis Produces Embryos from Body Cells

Somatic embryogenesis is the production of plant embryos from body (somatic) cells. Embryos can develop from many types of cells, including microspores. Somatic embryos develop normally to the torpedo stage, but do not dehydrate and become dormant, as is normal for zygotic embryos. Rather, somatic embryos produce root and shoot systems and develop into mature plants. Somatic embryogenesis occurs naturally in citrus, mango, onion, and tobacco plants, but agricultural scientists have made use of it as well.

In 1958, plant biologist F. C. Steward and associates were the first researchers to successfully clone a complex organism, carrot plants, by means of somatic embryogenesis. They used differentiated cells from carrot roots, grew the cells in conditions that caused some cells to lose their specialized properties and develop into embryos, and then cultivated each embryo in conditions that favored development into a mature carrot plant. Many types of plants are now cloned by somatic embryogenesis, which allows commercial growers to produce large numbers of genetically identical individuals.

CHAPTER SUMMARY

39.1 An Overview of Flowering Plant Reproduction

- Flowering plants display a sexual life cycle known as alternation of generations, as do all plant groups. The gamete-producing male and female gametophytes of flowering plants are very small and entirely dependent on nurturing sporophytic tissues. Plant gametes arise by the process of mitosis. (Figures 39.1, 39.2)

- Flowers are reproductive shoots that develop from a shoot apical meristem. The role of flowers is to promote seed production. A flower shoot generally produces four types of organs: sepals, petals, carpels, and stamens. Stamens produce pollen, otherwise known as immature male gametophytes. Carpels produce ovules that contain female gametophytes. (Figure 39.3)

- Mature male gametophytes, or pollen tubes, each produce two sperm and deliver them to ovules. Flowering plants display double fertilization: One of the two sperm released from a pollen tube combines with an egg cell to form a zygote, while the other fuses with two nuclei located in a central cell of the female gametophyte, producing the first cell of endosperm tissue. Endosperm is a nutritive tissue that supports development of an embryonic sporophyte. (Figure 39.4)

- Seeds are reproductive structures that contain a dormant embryo enclosed by a protective seed coat that develops from ovule integuments. Fruits are structures that contain seeds and foster seed dispersal. Like flowers and endosperm, fruits are unique features of flowering plants. (Figure 39.5)

39.2 Flower Production, Structure, and Development

- Plants flower in response to environmental stimuli, such as temperature and day length, by the conversion of a leaf-producing shoot into a flowering shoot. In *Arabidopsis*, phytochrome in leaves interacts with clock genes to induce expression of the transcription factor protein CO. The buildup of CO in long days induces the production of *FT* mRNA in phloem, which conducts the mRNA to shoot meristems, where translation occurs. The resulting FT protein binds to FD protein, which triggers expression of the floral identity genes that produce flower organs. (Figure 39.6)

- Flowers vary in the type of whorls present, the number of flower organs, color, fragrance, organ fusion, symmetry, and arrangement (whether single or in inflorescences). These variations are related to pollination mechanisms and are genetically controlled. Studies of genes controlling flower development led to the discovery of gene silencing and previously unknown genetic repair mechanisms. (Figures 39.7, 39.8, 39.9, 39.10, 39.11)

39.3 Male and Female Gametophytes and Double Fertilization

- Pollen grains are immature male gametophytes protected by a tough sporopollenin wall. Female gametophyte development occurs within an ovule. Mature female gametophytes include an egg, a central cell with two nuclei, and a few additional cells. (Figures 39.12, 39.13)

- After pollination, interactions between proteins of pistil cells and those of pollen determine pollen germination. (Figure 39.14)

- Germinated pollen delivers two sperm to female gametophytes by means of a long pollen tube. The style plays a role in the guidance, nutrition, and fate of the pollen tube. Pollen tubes that enter female gametophytes burst, releasing sperm. One sperm nucleus fuses with the egg to produce a zygote, the first cell of a new sporophyte generation; the other sperm nucleus fuses with the two nuclei of the central cell, generating the first cell of the nutritive endosperm tissue. (Figures 39.15, 39.16, 39.17)

39.4 Embryo, Seed, Fruit, and Seedling Development

- Unequal division of a zygote leads to development of a nutritive suspensor and an embryo. Young eudicot embryos are heart-shaped as two embryonic leaves (cotyledons) develop and assume a torpedo shape as the embryonic root forms. Monocot embryos produce only a single cotyledon. (Figures 39.18, 39.19)

- Mature seeds contain embryos that become dry and are protected by desiccation-resistance proteins and a tough seed coat. These adaptations enable seeds to withstand long periods of dormancy, germinating only when conditions are favorable for seedling survival. Mature fruits develop from ovaries and aid in seed dispersal. (Figure 39.20)

- Seed germination is influenced by environmental and internal factors. The embryonic root (radicle) is the first organ to emerge, an adaptation that allows rapid water uptake, essential for seedling development. (Figures 39.21, 39.22)

39.5 Asexual Reproduction in Flowering Plants

- Asexual reproduction is the production of new individuals from a single parent without the occurrence of fertilization. Vegetative reproduction is the development of whole plants from nonreproductive structures. Apomixis is a mechanism by which some plants produce seeds from flowers without fertilization. Somatic embryogenesis is the production of embryos from body cells. (Figure 39.23)

TEST YOURSELF

1. Where do immature male gametophytes of flowering plants, otherwise known as pollen grains, develop?
   a. in the anthers of a flower
   b. in the carpels of a flower
   c. while being dispersed by wind, water, or animals
   d. within ovules
   e. none of the above

2. Where do mature male gametophytes of flowering plants primarily develop?
   a. in the anthers of a flower
   b. in the carpels of a flower
   c. while being dispersed in wind, water, or by animals
   d. within ovules
   e. none of the above

3. Where would you find female gametophytes of a flowering plant?
   a. in the anthers of a flower
   b. at the stigma of a pistil
   c. in the style
   d. within ovules in a flower's ovary
   e. none of the above

4. How does double fertilization occur in flowering plants?
   a. The two sperm in a pollen tube fertilize the two egg cells present in each female gametophyte.
   b. One of the two sperm in a pollen tube fertilizes the single egg in a female gametophyte, while the other fuses with the two nuclei present in the central cell.
   c. Two sperm, one contributed by each of two different pollen tubes, fertilize the two egg cells in a single female gametophyte.
   d. Two sperm contributed by separate pollen tubes enter a single female gametophyte; one of the sperm fertilizes the egg cell, while the other fertilizes the central cell.
   e. None of the above.

5. A seed is
   a. an embryo produced by the fertilization of an egg, which is protected by a seed coat.
   b. a structure that germinates to form a seedling under the right conditions.
   c. an embryo produced by parthenogenesis that is enclosed by a seed coat.
   d. all of the above.
   e. none of the above.

6. What is the likely chemical composition of florigen, the long-sought chemical stimulus of flowering?
   a. the hormone auxin
   b. the protein CO
   c. the carbohydrate callose
   d. the mineral ion K^+
   e. none of the above

7. How many whorls of organs occur in complete flowers?
   a. two
   b. four
   c. six
   d. eight
   e. none of the above

8. If an ovary contains eight ovules, how many seeds could potentially result if pollen tubes reach all eight ovules?
   a. one
   b. four
   c. eight
   d. more than 20
   e. none of the above

9. What function(s) does the polysaccharide callose have in the reproduction of flowering plants?
   a. Callose forms a coat that isolates young embryos during their early development.
   b. Callose forms a coat that isolates groups of four microspores during their early development into pollen grains.

c. Callose helps to pattern the sculptured sporopollenin walls of pollen grains.
   d. Callose forms plugs that concentrate pollen tube cytoplasm at the tips, thereby aiding tip growth.
   e. All the above.

10. From what structure does a fruit pericarp primarily develop?
    a. the style
    b. a stamen filament
    c. the ovary wall
    d. a group of fused sepals
    e. the stigma

CONCEPTUAL QUESTIONS

1. Why are pollen grain walls composed of sporopollenin?

2. Why are seed coats often tough?

3. Why do flowers occur in such a diversity of shapes and colors?

EXPERIMENTAL QUESTIONS

1. Why had *in vitro* plant fertilization been so difficult for plant biologists to accomplish before the work of Kranz and Lörz?

2. What procedure did Kranz and Lörz use to accomplish fertilization of isolated eggs by sperm cells?

3. How did Kranz and Lörz demonstrate that their plants were actually hybrids arising from *in vitro* fertilization involving specific genetic parents?

COLLABORATIVE QUESTIONS

1. Observe or think of orchid flowers. Are these flowers bilaterally symmetrical or radially symmetrical, are these flowers more likely to be wind-pollinated or pollinated by animals, and what gene might be involved in the production of orchid flower shape?

2. How do plants prevent the production of many offspring expressing deleterious recessive traits?

www.brookerbiology.com
This website includes answers to the Biological Inquiry questions found in the figure legends and all end-of-chapter questions.

40

INTRODUCTION TO ANIMAL FORM AND FUNCTION

CHAPTER OUTLINE

Archer fish moving between environments to feed on a spider.

In Chapters 32–34, we considered the diversity of animal life. Yet, despite the obvious differences between a worm and a human, or an insect and a fish, fundamental similarities link the millions of animal species together. We will introduce some of these similarities in this chapter and explore them throughout this unit. For example, all animals must obtain energy to survive, and they must have the ability to cope with or adapt to changes in their environment. Furthermore, all animals are composed of similar types of cells. In this chapter, we will look at some of the characteristics common to animals and the link between structure and function. We will also see how animals maintain internal conditions within normal limits despite fluctuations in their external environments. **Anatomy** is the study of animal structure, while **physiology** is the study of how those structures function. Collectively, anatomy and physiology form the foundation of animal biology. We begin, therefore, with an examination of the physical structure and organization of animal bodies.

40.1 Organization of Animal Bodies

All animal cells share similarities in the ways in which they exchange materials with their surroundings, obtain energy from organic nutrients, synthesize complex molecules, duplicate themselves, and detect and respond to signals in their immediate environment. Animals usually begin life as a single cell—most commonly a fertilized egg—which divides to create two cells, each of which divides in turn, resulting in four cells, and so on. If cell multiplication were the only event occurring, the end result would be a spherical mass of identical cells.

As we will see in Chapter 52, however, cells become specialized during development to perform a particular function (that is, they differentiate), such as muscle or nerve cells. Cells also migrate to new locations within the developing organism and form clusters with other cells. In this way, the cells of an animal's body are arranged in various combinations to form organized, multicellular structures (**Figure 40.1**). Cells with similar properties group to form tissues (for example, muscle tissue), which combine with other types of tissues to form organs (for example, a kidney), which are anatomically or functionally linked together to form organ systems (for example, the urinary system).

Specialized Cells Are Organized into Tissues

Specialized cells of a given type often cluster together to form **tissues.** The tissues in a typical animal's body can be classified into four categories, according to the types of functions they perform: muscle, nervous, epithelial, and connective tissues. Most structures in an animal's body contain at least two types of tissues and often contain all four. Within each of these functional categories, subtypes of tissues perform variations of that function. For example, as described next, there are three subtypes of muscle tissue, all of which generate force, but only one of which produces animal locomotion.

Muscle Tissues **Muscle tissues** consist of cells that are specialized to contract, generating the mechanical forces that produce body movement, exert pressure on a fluid-filled cavity, or decrease the diameter of a tube. Complex animals have three types of muscle tissue: skeletal, smooth, and cardiac (**Figure 40.2**).

Skeletal muscles are generally attached to bones in vertebrates and to the exoskeleton of invertebrates. Contraction of these muscles is under voluntary control and can produce the types of movements required for locomotion, such as extending limbs or flapping wings. Skeletal muscles may also attach to skin, such as the muscles producing facial expressions. **Smooth muscles** surround hollow tubes and cavities inside the body's organs, such that their contraction can propel the contents of those organs. For example, the contraction of smooth muscle in the stomach wall propels partially digested food into the intestines, where it can be digested fully. Smooth muscle also forms part of small blood vessels and airway tubes (bronchioles). Contraction in those regions reduces blood flow or movement of air. Contraction of all smooth muscle is involuntary and occurs automatically. In the third type, **cardiac muscle**, physical and electrical connections between individual cells enable many of the cells to contract simultaneously. Like smooth muscle, cardiac muscle is involuntary. It is found only in the heart, however, where it provides the force that generates sufficient pressure to pump blood through an animal's body.

Figure 40.1   The internal organization of cells, tissues, organs, and organ systems in a mammal.   Most animals share the same four tissue types.

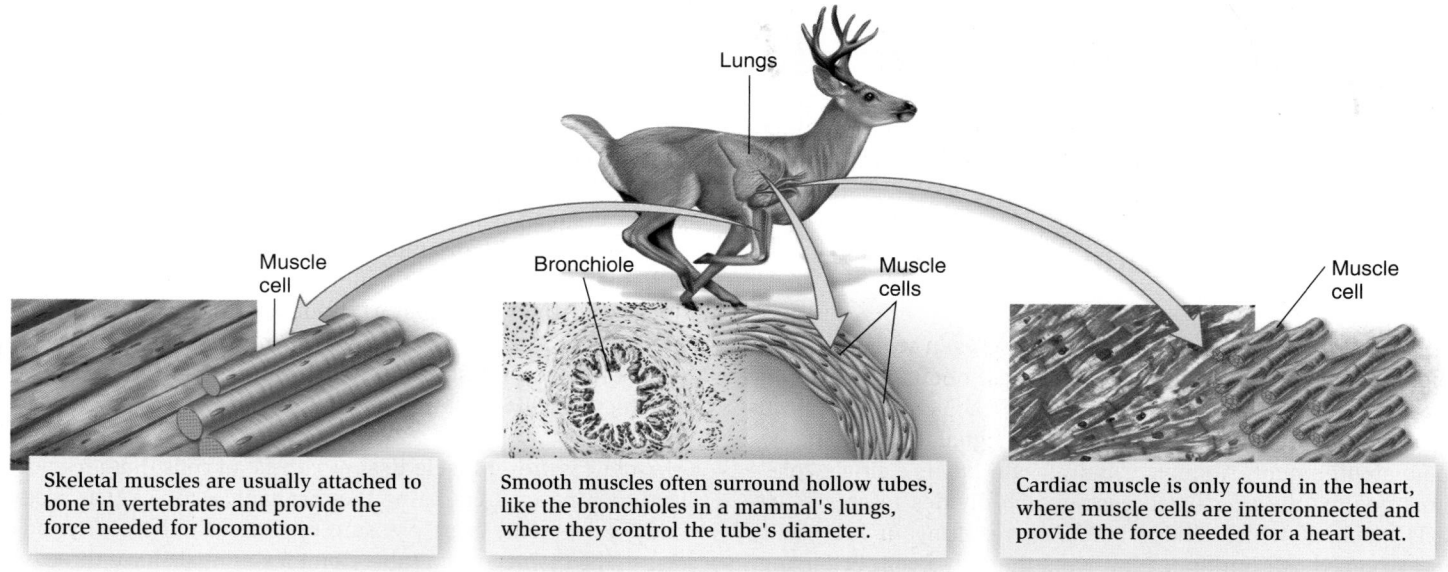

Skeletal muscles are usually attached to bone in vertebrates and provide the force needed for locomotion.

Smooth muscles often surround hollow tubes, like the bronchioles in a mammal's lungs, where they control the tube's diameter.

Cardiac muscle is only found in the heart, where muscle cells are interconnected and provide the force needed for a heart beat.

Figure 40.2   Three types of muscle tissue: skeletal, smooth, and cardiac muscle.   All three types produce force, but they differ in their appearance and in their locations within animals' bodies.

Biological inquiry: All muscles produce movement, but only skeletal muscle produces locomotion. What is meant by this statement?

Nervous Tissues **Nervous tissues** initiate and conduct electrical signals from one part of an animal's body to another part (**Figure 40.3**). A single nerve cell may connect two or more other nerve cells and be only a few microns long. Such nerve cells are found widely throughout the vertebrate brain, for example. By sharp contrast, a nerve cell leaving the spinal cord of a giraffe and extending to its hoof can extend for 2 meters! Depending on where it is located in an animal's body, an electrical signal produced in one nerve cell may stimulate or inhibit other nerve cells to initiate new electrical signals, stimulate muscle tissue to contract, or stimulate glandular cells to release chemicals into the animal's blood. Thus, nervous tissue provides a critical means of controlling many diverse activities of body cells.

Epithelial Tissues **Epithelial tissues** are sheets of densely packed cells that cover the body or individual organs or line the walls of various cavities inside the body. Epithelial cells (see Chapter 10) are specialized to protect structures and secrete and absorb ions and organic molecules. For example, epithelial tissue can invaginate (fold inward) to form sweat glands that secrete water and ions onto the surface of an animal's skin. Like muscle and nerve cells, epithelial cells come in a variety of shapes and sizes (**Figure 40.4**). Epithelial cells rest on an extracellular matrix called the basal lamina or basement membrane (see Chapter 10), and they form boundaries between different body compartments (we will say more about compartments later in this chapter).

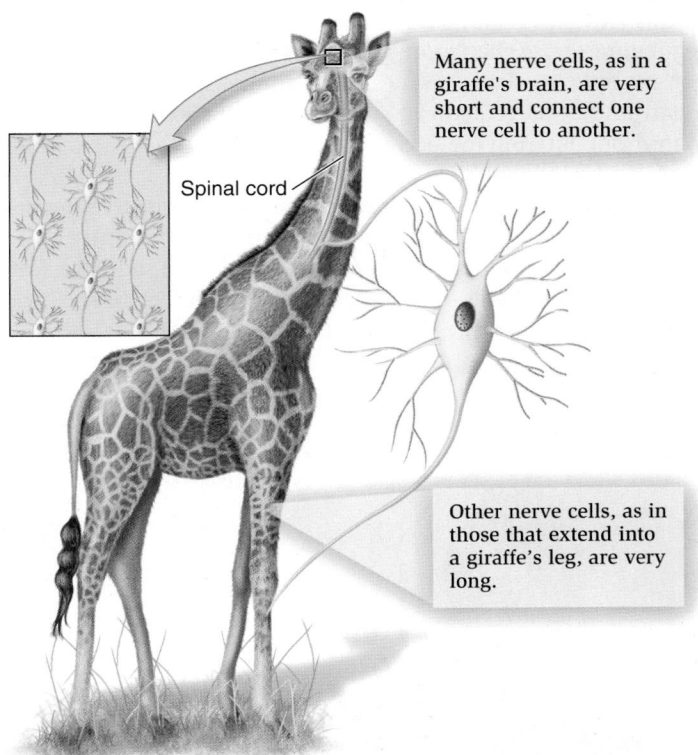

Many nerve cells, as in a giraffe's brain, are very short and connect one nerve cell to another.

Spinal cord

Other nerve cells, as in those that extend into a giraffe's leg, are very long.

Figure 40.3   Variation in shape and length of nerve cells. Most nerve cells in the brain of an animal are only microns long, while others, like those that send signals to an animal's extremities, can be much longer.

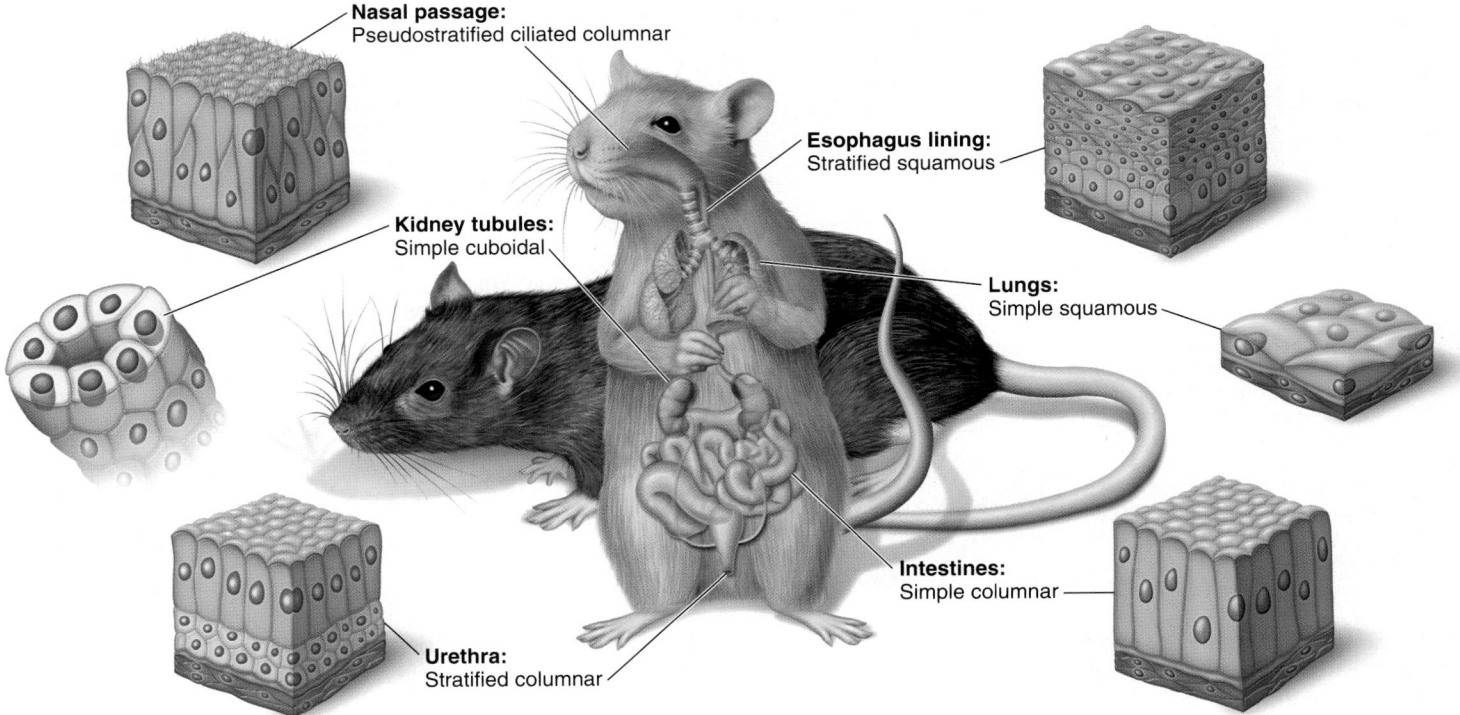

Nasal passage:
Pseudostratified ciliated columnar

Kidney tubules:
Simple cuboidal

Esophagus lining:
Stratified squamous

Lungs:
Simple squamous

Intestines:
Simple columnar

Urethra:
Stratified columnar

Figure 40.4   **Examples of epithelial tissue.**   There are several types of epithelial tissue, distinguished by their appearance. Epithelial tissue is used to construct body coverings and the protective sheets that line and cover hollow tubes and cavities.

In this way, epithelial tissues can function as selective barriers regulating the exchange of molecules between compartments. For example, epithelial tissues in an animal's skin help to form a barrier that prevents most substances in the external environment from entering the body.

Connective Tissues As their name implies, **connective tissues** connect, surround, anchor, and support the structures of an animal's body. Connective tissues include blood, adipose (fat-storing) tissue, bone, cartilage, loose connective tissue, and dense connective tissue (**Figure 40.5**).

An important function of some types of connective tissue is to form the extracellular matrix around cells (see Chapter 10) by secreting a mixture of fibrous proteins and carbohydrates, such as glycosaminoglycans. These carbohydrates may attach to proteins to form proteoglycans. In some cases, the extracellular matrix is rich in minerals. The final characteristics of any type of connective tissue are determined in part by the relative proportions and types of proteins, proteoglycans, and minerals secreted into the extracellular matrix. The matrix serves several general functions, which include (1) providing a scaffold to which cells attach, (2) protecting and cushioning parts of the

Blood is composed of red and white blood cells and cell fragments called platelets, all three of which are suspended in a watery fluid called plasma. The cells are formed in spaces within certain bones.

Adipose tissue is composed of fat-filled cells, which provide a layer of protection and insulation around internal organs and under the skin. Adipose tissue is also a major energy store.

Adipose tissue

Blood

Dense connective tissue has tightly packed layers of collagen fibers in parallel arrays, giving the tissue great strength but very little flexibility, as in tendons and ligaments.

Bone is composed of bone-forming cells that secrete the protein collagen. The collagen is embedded in a hard casing composed of calcium and phosphorus, which gives bone its inflexible, tough characteristics suitable for support and protection.

Bone

Cartilage

Dense connective tissue

Loose connective tissue

Loose connective tissue is abundant throughout animals' bodies, where it holds internal organs in place and provides much of the internal framework of the body. It is composed of loosely arranged collagen fibers mixed with elastin fibers, which allows it to be flexible.

Cartilage is formed by collagen-secreting cells. Cartilage is not mineralized and is therefore softer and more flexible than bone, providing flexibility of movement and cushioning of joints in animals with bony skeletons.

Figure 40.5 **Examples of connective tissue.** Connective tissue connects, surrounds, anchors, and supports other tissues and may exist as isolated cells (blood), clumps of cells (fat), or tough, rigid material like that which forms bone and cartilage. The samples have been stained or the micrographs have been computer-colorized to reveal connective tissue.

body, (3) providing mechanical strength, and (4) transmitting information to the cells that helps regulate their activity, migration, growth, and differentiation.

The proteins of the extracellular matrix of a tissue consist of insoluble fibrous proteins such as **collagen** and the rubberband-like protein **elastin**; these proteins are often referred to as fibers. A second category are adhesive proteins that serve to organize the protein and carbohydrate components of the extracellular matrix.

Different Tissue Types Combine to Form Organs and Organ Systems

An **organ** is composed of two or more kinds of tissues arranged in various proportions and patterns, such as sheets, tubes, layers, bundles, or strips. For example, the vertebrate stomach (**Figure 40.6**) consists of;

- a connective tissue layer covering and cementing the organ together;
- layers of smooth muscle tissues whose contraction propels food through the stomach;
- nervous tissue that comes in close contact with the smooth muscle tissues and helps regulate their activity;
- an inner lining of epithelial tissue that secretes molecules important in the digestive process.

In an **organ system**, different organs work together to perform an overall function. In the example just described, the stomach is part of the digestive system, along with other structures like the intestines, mouth, and anus. In another familiar example, the kidneys, the urinary bladder, the tubes leading from the kidneys to the bladder, and the tube leading from the bladder to the exterior of the body constitute the urinary system in mammals, which helps remove waste products from the blood in the form of urine (see Figure 40.1). The organ systems found in animals are listed in **Table 40.1**. Some organ systems may be rudimentary in some animals (such as the simple nervous systems of worms and other ancestral animals).

Organ systems frequently work together. For example, signals from the nervous system and endocrine system strongly influence how much water the mammalian kidney retains as it forms urine, an adaptation that can be lifesaving under certain circumstances. Likewise, nerve signals from the bladder relay information to the cells of the brain when the bladder is full and needs to be emptied.

The spatial arrangement of organs into organ systems is part of the overall body plan of animals. Organ systems and other structures, such as an animal's limbs, tentacles, and other appendages, develop at specific times and locations within the body and along the anteroposterior body axis. Scientists have long wondered how the layout of animal bodies is determined during the period when an embryo is developing. Remarkably, the body plan of animals appears to be under the control of a highly conserved family of genes with homologs in all animals, as described next.

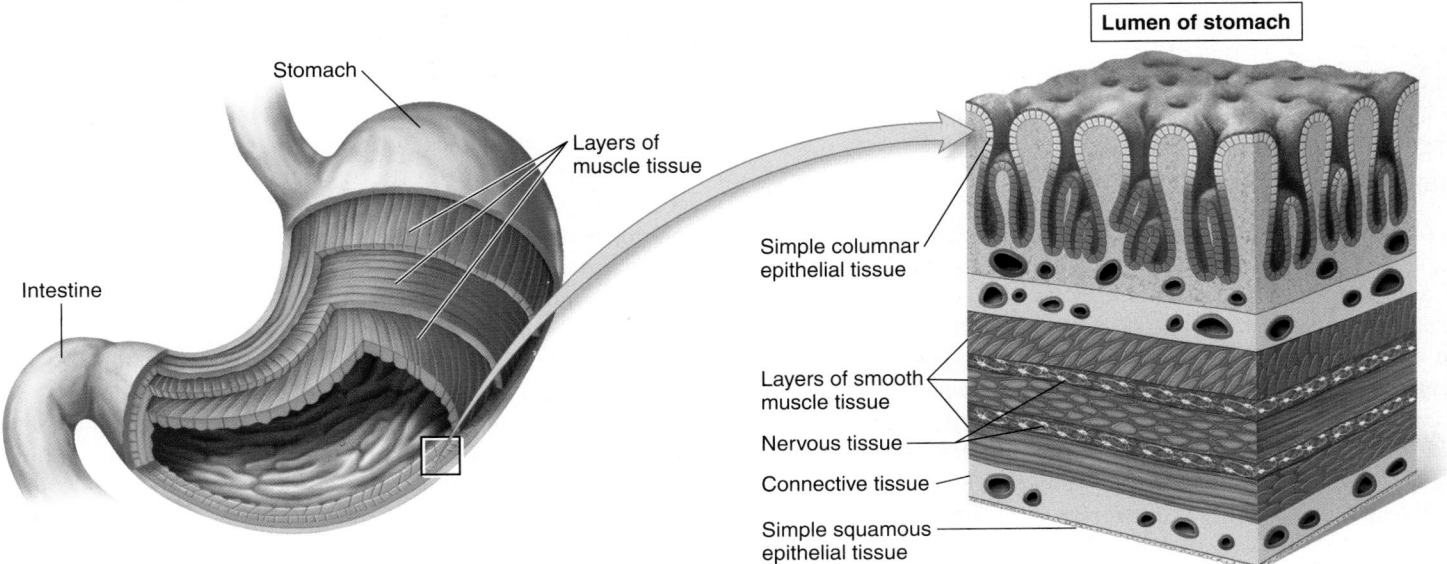

Figure 40.6 The vertebrate stomach as an example of an organ composed of all four tissue types.

Table 40.1	Organ Systems Found in Animals	
Organ system	**Major components***	**Major functions**
Circulatory	Contractile element (heart or vessel); distribution network (blood vessels); blood or hemolymph	Transport and distribute solutes (nutrients, gases, wastes, etc.) to and from all parts of an animal's body
Digestive	Ingestion structures (mouth or mouthparts); storage structures (crop, stomach); digestive and absorptive structures (intestines); elimination structures (rectum, anus); accessory structures (pancreas, gall bladder)	Break complex foods into smaller, absorbable units; absorb salts and water; eliminate solid wastes
Endocrine	All glands, organs, or tissues that secrete hormones; examples include the brain, pituitary and thyroid glands	Regulate and coordinate processes like growth, development, metabolism, mineral balance, and reproduction
Excretory	Kidneys or comparable structures; storage sites for soluble wastes (bladder); tubes connecting kidneys and bladder, and bladder to external environment	Eliminate soluble metabolic wastes; regulate body fluid volume and solute concentrations
Immune and lymphatic	Circulating white blood cells; lymph vessels and nodes	Defend against pathogens
Integumentary	Body surfaces (skin)	Protect from dehydration and injury; defend against pathogens; regulate body temperature; in some animals gas and fluid transport
Muscular-skeletal	Muscles, bones, cartilage, tendons, ligaments, chitin	Produce locomotion; generate force; propel materials through body organs; support body
Nervous	Brain, spinal cord, peripheral nerves and ganglia, sense organs	Regulate and coordinate many body activities, such as movement, learning, and sensation
Reproductive	Gonads and associated structures	Produce gametes (sperm and egg); in some animals, provide nutritive environment for embryo and fetus, and nourish offspring
Respiratory	Gills, skin, lungs, tracheae (insects)	Exchange oxygen and carbon dioxide with environment; regulate blood pH

*Selected examples only; these do not necessarily pertain to all animals.

GENOMES & PROTEOMES

Organ Development and Function Are Controlled by Homeotic Genes

In previous chapters, you learned about a family of ancient, highly conserved genes called homeotic genes that are found in all animals. These genes determine the timing and spatial patterning of the anteroposterior body axis during development. For example, we saw in Chapter 19 how homeotic genes determine the number and position of legs and wings in *Drosophila*. In vertebrates, the homeotic genes are known as *Hox* genes, and they play a similar role in determining the spatial patterning of the vertebrate body and appendages. Recently, scientists have begun exploring the role of *Hox* genes in the development and spatial patterning of internal organs that constitute animals' organ systems.

By generating homozygous, null mutant ("knockout") mice that fail to express one or more *Hox* genes, researchers have discovered the important role these genes play in determining where within the vertebrate body particular organs form. Recall from Chapter 19 (refer back to Figure 19.18) that *Hox* genes are arranged in four clusters, designated A–D, with 13 genes per cluster. Paralogous genes (for example, *HoxA-3*, *HoxB-3*, *HoxD-3*) typically act in concert to regulate similar developmental processes. For example, *HoxA-3* is important for development of anterior parts of the body, including the neck. When this gene is knocked out, mouse embryos show defects in neck structure, such as abnormal blood vessels. Also, the organs within the neck—including the thymus, thyroid, and parathyroid glands—do not develop normally. When two or more paralogs of this group are knocked out, certain neck organs fail to form at all. Experimental deletion of *Hox* genes that are associated with other body segments does not affect the development and function of neck glands and organs. Likewise, investigators have uncovered vital roles of different *Hox* genes in lung development within the thorax, and the proper positioning and development of the vertebrate kidneys in the abdomen.

Of particular interest is the discovery that *Hox* genes are important not only for spatial patterning of organs but also for their growth, development, and function. Genes in the *Hox* 1 and 3 groups, for instance, help determine the final branching patterns of the airways of the lungs, the final size of the lungs, and the ability of the lungs to produce secretions that are important for breathing air after birth. Other *Hox* genes have been shown to control cell proliferation, shape changes, apoptosis, cell migration, and cell-cell adhesion within various organs. This is also true in invertebrates, such as the leech, where homeotic genes are first expressed during organ formation, and in *Drosophila*, where the final shape and size of the heart is partly controlled by homeotic genes.

Chapter 52 will describe how organs form in greater detail. It is clear, however, that *Hox* genes in vertebrates and their homologs in invertebrates are important in organ positioning, development, and function. This explains in part why the thyroid gland forms in the neck region of vertebrates and not in the legs or tail, for example. Interestingly, *Hox* genes continue to be expressed long after embryonic development has ceased, and it is believed that they play roles in organ function even in adult animals. In fact, expression of abnormal *Hox* genes in a mature organ has been hypothesized to be a contributing factor in organ disease, such as lung disorders and leukemia (cancer of the white blood cells).

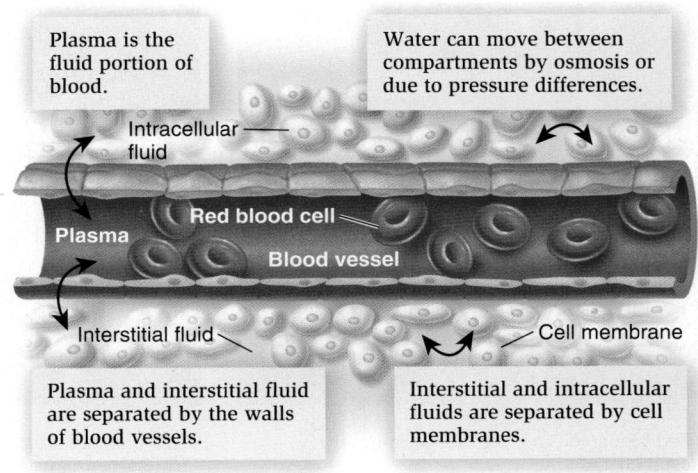

Figure 40.7   Fluid compartments in a typical vertebrate. Most of the fluid within an animal's body exists within cells (intracellular fluid). Extracellular fluid is that portion of the body's fluid that lies outside cells (interstitial fluid) and within blood vessels (plasma). Arrows indicate directions of water movement between adjacent compartments.

Body Fluids Are Distributed into Compartments

In addition to various tissues and organs, all animal bodies are composed primarily of water. Indeed, even terrestrial animals can be thought of as "living in water," since all the cells of an animal's body are filled with and surrounded by water, as we will now see.

Most of the water in an animal's body is contained within the **intracellular fluid** inside its cells (from the Latin, *intra*, inside of). The rest of the water in its body exists outside of the cells and is therefore called **extracellular fluid** (Latin *extra*, outside of). Extracellular fluid is composed of the fluid part of blood, called **plasma**, and the fluid-filled spaces that surround cells, called **interstitial** fluid (Latin *inter*, between) (**Figure 40.7**). In vertebrates and some invertebrates, plasma and interstitial fluid are kept separate within a closed circulatory system, while in many invertebrates, plasma and interstitial fluid are intermingled into a single fluid called hemolymph (see Chapter 47).

The solute composition of the extracellular fluid may be very different from that of the intracellular fluid. Maintaining differences in fluid composition across the plasma membrane is an important way in which animal cells regulate their own activity. For example, many different proteins are confined to intracellular fluid, because they are important in regulating cellular events such as mitosis, cytokinesis, and metabolism.

The fluids depicted in Figure 40.7 are said to be enclosed in compartments. The sum of the water in the three compartments (intracellular fluid, plasma, and interstitial fluid) accounts for about two-thirds of the body weight of a typical vertebrate. Normally, up to two-thirds of total body water is intracellular, and the remaining one-third is extracellular, the majority of which is located in the interstitial compartment.

Movement of Water Between Compartments Barriers separating adjacent fluid compartments determine which solutes can move between them. Solute movement in turn accounts for the differences in composition of the different compartments. In the case of the body fluid compartments, plasma membranes separate the intracellular fluid from the extracellular fluid. The two components of extracellular fluid in vertebrates and many invertebrates—the interstitial fluid and the plasma—are separated by the walls of the blood vessels (for example, arteries, capillaries, and veins).

Fluid can readily move between adjacent compartments in an animal's body, because barriers like plasma membranes tend to be highly permeable to water. This movement depends on osmosis (see Chapter 5), in which water moves from a region of lower solute concentration to a region of higher solute concentration. For cells to function properly, they require a relatively stable internal composition, including ion and protein concentrations, cellular volume, and pH. An increase in solute concentration outside a cell, for example, would cause water to move by osmosis from inside the cell to outside. In other words, osmosis redistributes fluid from the intracellular to interstitial compartments. This would cause a cell to become deformed as it shrinks due to loss of fluid. A deformed animal cell rarely operates as well as a normal cell, and in fact it may die if its membrane ruptures. **Figure 40.8** shows examples of mammalian red blood cells in which intracellular fluid levels have been altered. This could occur, for example, if the blood cells were exposed to extracellular fluids that were either more dilute (hypoosmotic) or more concentrated (hyperosmotic) than the fluid inside the blood cell. Such dramatic deformities can rupture blood cells and cause them to release their contents and die.

Red blood cell in a solution of normal osmolarity.

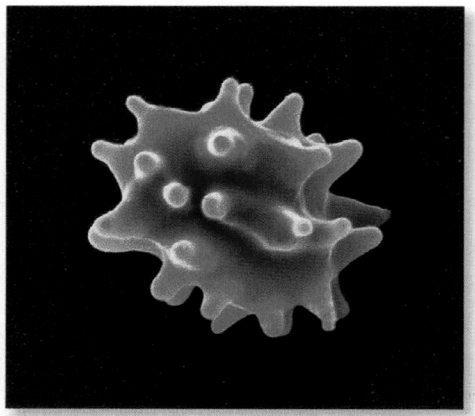

Red blood cell that has lost intracellular fluid when placed in a solution of higher than normal osmolarity.

Red blood cell that has gained intracellular fluid when placed in a solution of lower than normal osmolarity.

Figure 40.8   Changes in cell shape due to alterations in intracellular fluid volume.   Alterations in intracellular fluid volume can have drastic effects on cell shape, as shown by these scanning electron micrographs of red blood cells. Each cell is approximately 7 µm in diameter. Dramatic changes in shape like these are usually lethal for cells.

Biological inquiry: What effect would changes in intracellular fluid volume have on intracellular solute concentration?

Pressure differences may also cause fluid to move between compartments in an animal's body. As we will see in Chapter 47, the pressure generated by the beating of the heart can force plasma through tiny openings that are found between the endothelial cells that make up the walls of the smallest blood vessels. This allows water and solutes such as nutrient molecules to be transferred from the blood to the interstitial fluid, from which they may gain access to cells.

Movement of Solutes Between Compartments Like water, solutes must also move between body fluid compartments. In this way, cells in the body maintain normal levels of ions, nutrients, and gases such as oxygen. We discussed the mechanisms by which solutes move in Chapter 5. Let's summarize those mechanisms, which apply to all animal cells.

Passive diffusion, a major form of solute transport, is the movement of a solute down its concentration gradient without the aid of a transport protein or hydrolysis of ATP. Passive diffusion is one way in which cells gain and lose solutes. Molecules capable of crossing lipid bilayers are able to passively diffuse into or out of a cell. Examples include nonpolar solutes such as lipids and extremely small polar solutes such as alcohol. Most polar molecules and ions can only diffuse through a plasma membrane if the membrane has pores or channels of some type that permit the ion to pass through the bilayer. The rate of diffusion of any solute depends on several factors, notably the concentration gradient of the solute and the area across which it is diffusing. The rate of diffusion can be calculated by the Fick diffusion equation shown here in simplified form:

$$J = DA\,(C_1 - C_2)$$

where J is the rate of diffusion, D is a constant, A is the cross-sectional area of the barrier across which diffusion is occurring, and C_1 and C_2 are the concentrations of the solute at two locations (for example, inside and outside a cell).

The way in which most solutes move between compartments or across plasma membranes is mediated by transport proteins via facilitated diffusion and the various types of active transport (see Chapter 5). Each of these processes is critical to animal regulation of intracellular and extracellular fluid composition. In Chapter 42, we will discuss one example, the mechanism by which cells obtain their most important energy source, glucose. Because glucose is a relatively large and polar molecule, it cannot diffuse through lipid bilayers. Instead, it is transported from the interstitial fluid into cells by membrane-bound proteins.

40.2 The Relationship Between Structure and Function

A key theme throughout this textbook is that structure determines function. The organization of an animal's body into tissues, organs, organ systems, and fluid compartments can often help us predict the function of each of those structures. For example, the respiratory systems of animals exchange oxygen from the environment with carbon dioxide generated by the body. Let's compare the respiratory systems of an insect and a mammal (**Figure 40.9**). Although many important differences will be covered in Chapter 48, notably the absence of lungs in insects, certain structural similarities suggest that both systems

(a) Insect

(b) Human

Figure 40.9 Comparison of the branching air tubes in (a) an insect and (b) a mammal. Note the similar features of highly branching, internalized hollow tubules that connect to the outside air, suggesting that these systems perform similar functions.

serve similar functions. In both cases, for example, a series of internal branching tubes composed of epithelial and connective tissues arises from one or more openings that connect with the outside environment (the mouth and nose in the mammal, and the spiracles in the insect). These tubes become smaller and smaller as they continue to branch, eventually terminating in narrow structures that are only one cell thick.

Without knowing anything else about the respiratory systems of these two animals, we can surmise that in both cases these branching tubes serve as conduits for air to flow back and forth between the environment and the internal spaces of the animal. In the insect, the ends of the branching tubes called tracheoles are where oxygen diffuses from the air to the fluid around individual cells (and from there to intracellular fluid). In the mammal, the ends of the tubes form saclike structures through which oxygen diffuses into the bloodstream.

If we examine the mammalian lung in greater detail (Figure 40.9b), we see that the air sacs are composed of extremely thin, simple squamous epithelial cells. The shape of the cells provides a clue to their function. Their thinness permits diffusion of gases across the cells as rapidly and efficiently as possible. Imagine the resistance to oxygen diffusion if the cells were thick or scaly, like the cells of the body surface of many animals, for example. Thus, both the gross and microscopic anatomy of the gas-exchange surfaces of respiratory systems facilitates their functions. An additional structural similarity found in essentially all respiratory surfaces, including gills, is an extensive sur-

face area. In fact, we can expand our discussion to include all cells, tissues, and organs that mediate diffusion or absorption of a solute from one compartment to another. Consider, for instance, the finger-like projections of the small intestine of a human, the skin folds of a frog, the cellular extensions on the surface of nerve cells of a turtle, and the feathery antennae of a moth (**Figure 40.10**). What do these structures have in common? They all display a large surface area, which maximizes the ability of a tissue or organ to absorb solutes (intestine), exchange oxygen and carbon dioxide with the environment (frog skin), communicate with other cells (nerve cells), or sense airborne molecules (moth antennae). However, increased surface area could also mean that the structure requires more space inside or outside the body. In some cases, this is avoided by changes in shape (see, for example, the way the inner surface of the intestines in Figure 40.10a fold inward). The ratio between a structure's surface area and the volume in which the structure is contained is called the **surface area/volume (SA/V) ratio**. This concept will appear throughout this unit as we explore the ways in which animals obtain energy, regulate their body temperature, eliminate wastes, obtain oxygen, and regulate their metabolism.

When all of an animal's organ systems operate normally and body fluid levels and solute composition are maintained within normal limits, the animal is generally considered to be in a healthy condition. In the rest of this chapter, we will examine the process of achieving and maintaining this condition.

(a) Human intestine   **(b) Frog skin**   **(c) Nerve cells in turtle brain**   **(d) Moth antennae**

Figure 40.10 Examples of structures in which extensive surface area is important for function. A large surface area allows **(a)** high rates of transport of digested foods across the intestine of a human, **(b)** rapid diffusion of oxygen across the skin of a frog, **(c)** extensive communication between nerve cells in a turtle's brain, and **(d)** detection of airborne chemicals by moth antennae.

40.3 Homeostasis

The environmental conditions in which animals live are rarely if ever constant. Animals are exposed to fluctuations in air and water temperatures, food and water supplies, pH, and in some cases oxygen availability. Any one of these environmental changes could be harmful or even fatal if an organism is unable to respond appropriately. However, as you might expect from the incredible diversity of environments in which they exist, animals can adjust in many ways to their surroundings and thrive.

The process of adjusting to the external environment and maintaining a stable internal environment is known as **homeostasis** (from the Greek *homoios*, similar, and *stasis*, standing still). The term was coined in the 20th century by the American physician and physiologist Walter Cannon, but the concept itself originated in the 19th century with the French physician and physiologist Claude Bernard, who postulated that a constant *milieu interieur* (internal environment) was a prerequisite for good health. The ability to maintain homeostasis imparts a clear selection advantage for animals, because many of the fundamental processes of life—for example, enzyme activity—operate most efficiently within narrow limits of such variables as intracellular pH.

Some Animals Conform to External Environments; Others Regulate Their Internal Environments

Generally speaking, animals use two mechanisms to maintain homeostasis: conforming and regulating. Some animals conform to their environments so that some feature of their internal body composition matches their external surroundings. A marine crab, for example, maintains about the same solute concentration in its body fluids as is found in seawater. Likewise, the body temperatures of many fish and aquatic invertebrates match the temperatures of the surrounding waters. Energetically speaking, conforming is a cheap strategy for survival. It would take a great deal of energy for a small fish to maintain its body temperature at, say, 37°C when swimming in the waters of Antarctica.

Other animals regulate the internal composition of their fluids and solutes at levels that are different from those of the external environment. Most vertebrates are regulators, but as just noted for fish, exceptions exist. A single animal may be both a conformer and a regulator with respect to different physiological variables. A fish may conform its body temperature to the environment but regulate its internal solute concentrations at levels different from that of fresh or salty water. Regulating the internal environment requires considerable energy in the form of ATP. For this reason, homeostasis for regulators comes with a high price tag.

Vertebrates Maintain Most Physiological Variables Within a Narrow Range

The common physiological variables found in healthy organisms—levels of bloodborne factors like minerals, glucose, and oxygen, for example—are maintained within a certain range despite variable external environmental conditions (**Table 40.2**). In its simplest form, homeostasis may be defined as a state of stable balance of physiological variables. This simple definition cannot capture the scope of homeostasis, however. For example, no physiological function is constant for very long, which is why we call them "variables." Some variables may fluctuate around an average value during the course of a single day yet still be considered in balance. That is because homeostasis is a dynamic process, not a static one.

Consider an example in your own body. Normally, blood sugar (glucose) remains at fairly steady and predictable levels in any healthy individual. After a meal, however, the level of glucose in your blood can increase quickly, especially if you have just eaten something sweet. Conversely, if you skip a few meals, your blood sugar level may drop slightly (**Figure 40.11**). Such fluctuations above and below the normal value would suggest that blood glucose levels are not homeostatic, but this is incorrect. The important fact is that once blood glucose increases or decreases, homeostatic mechanisms restore glucose levels back toward normal. In the case of glucose, the endocrine system is primarily responsible for this quick adjustment, but in other examples, a wide variety of control systems may be initiated. In later chapters, we will see how every organ and tissue of an animal's body contributes to homeostasis, sometimes in multiple ways, and usually in concert with each other.

Table 40.2	Examples of Homeostatic Variables in Animals	
Variable	**Factors that influence homeostasis**	**Examples of functions**
Minerals	Eating food; excreting wastes	
Na^+ and K^+		Establish electrical differences across plasma membranes (for example, in muscle, nerve, and heart cells)
Ca^{2+}		Muscle contraction; nerve function; skeleton and shell formation
Fe^{2+}		Bind and transport oxygen in blood or body fluids (some invertebrates use copper instead of iron)
Fuel sources	Eating food; expending energy	
Glucose		Broken down to provide energy for use by all cells, especially brain cells
Fats		Provide an alternate source of energy, particularly for cells not in the nervous system; major component of plasma membranes
ATP		Provides energy to drive most chemical reactions and body functions; modifies function of many proteins by transferring terminal phosphate group to protein
Body temperature	Changing rates of energy expenditure; behavioral mechanisms (see Chapter 42)	Determines the rate of chemical reactions in an animal's body
pH of body fluids	Hydrogen ion pumps in cells; buffers in body fluids; rates of energy expenditure	Affects enzymatic activity in all cells
Other variables		
Oxygen	Movement of air or water across respiratory surfaces (for example, lungs and gills); rates of energy expenditure	Circulates in body fluids and enters cells, where it is used to make ATP
Water	Drinking, eating, excretion of wastes, perspiration, osmosis across body surface (for example, skin or gills)	Numerous biological functions (see Chapter 2)

Thus, homeostasis does not imply that a given physiological function is rigidly constant. Instead, homeostasis means that a variable fluctuates within a certain range, and that once disturbed from that range, compensatory mechanisms restore the variable toward normal levels.

Homeostatic Control Systems Maintain the Internal Environment

The activities of cells, tissues, and organs must be regulated and integrated with each other so that any change in the extracellular fluid—the internal environment—initiates a reaction to correct the change. These compensating regulatory responses are performed by homeostatic control systems. A homeostatic control system must have several components, including:

- a sensor, which monitors a particular variable;
- an integrator, which compares signals from the sensor to a baseline **set point** (the normal value);
- an effector, which compensates for any deviation between the actual value and the set point.

Figure 40.12 shows an example of a homeostatic control system that regulates body temperature in mammals. This system

Figure 40.11   An example of a homeostatically controlled variable, blood levels of glucose in a human.   Note that glucose levels in the blood may rise or fall depending on whether an animal has recently eaten. However, even after a sugary meal or a prolonged fast, homeostatic mechanisms either return glucose levels to normal or enable those levels to remain within the range required for survival.

is somewhat analogous to the heating system of a home. In that case, a sensor and integrator within the thermostat compare the actual room temperature to the set point temperature that was determined by turning the thermostat dial to a given temperature. If the room temperature becomes cooler than the thermostat setting, the effector (furnace) is activated and adds heat to the room. In a mammal, the sensors are temperature-sensitive nerve endings in the skin and elsewhere, while the integrator is within the brain. Signals from the brain are sent along nerves to the effectors, which include skeletal muscles. The muscles contract vigorously in response to these signals, resulting in shivering—a key way in which mammals' bodies generate heat. We will discuss other heat-conserving and heat-generating mechanisms that contribute to this homeostatic control system in Chapter 42.

When a homeostatic control system operates normally, changes in a physiological variable are kept to a minimum. These changes are limited by the process of feedback, as we will now see.

Feedback Is a Key Feature of Homeostasis

Feedback is a fundamental feature of homeostasis and a major way in which disturbances to a physiological variable are minimized. The temperature regulation system just described is an example of a **negative feedback loop**, in which a change in the variable being regulated brings about responses that move the variable in the opposite direction. Thus, a decrease in body temperature leads to responses that increase body temperature— that is, move it back toward its original value.

Negative feedback also prevents homeostatic responses from overcompensating. When the blood pressure of a bleeding animal falls, for example, pressure sensors in certain blood vessels detect the change and send the information to the brain, where the signal is compared to a set point (**Figure 40.13**). The brain stimulates the endocrine system to release hormones whenever blood pressure decreases for any reason. These hormones act in conjunction with the nervous system on the effectors of the compensatory response, such as the heart and blood vessels, to raise blood pressure back toward normal. The increase in pressure removes the stimulus from the sensor (negative feedback). The hormones also feed back to inhibit the brain's ability to continue stimulating the endocrine system. If feedback inhibition did not occur, the blood pressure would not only rebound back to baseline but might rise to abnormally high and possibly dangerous levels.

Negative feedback may occur at the organ, cellular, or molecular level. For instance, feedback mechanisms regulate many enzymatic processes. In one example, ATP regulates the rate of its own formation in cells by inhibiting certain of the intracellular enzymes that catalyze the breakdown of glucose molecules, a key event in the production of ATP.

Figure 40.12 An example of a homeostatic control system. The sensor and effector for responding to a decrease in body temperature are shown. Different homeostatic control systems have different sensors and effectors.

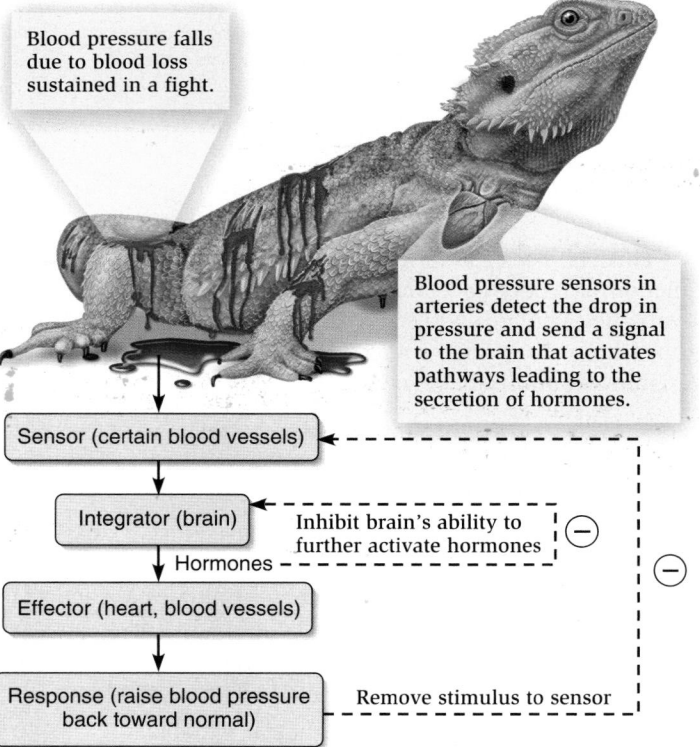

Figure 40.13 Negative feedback as a mechanism by which homeostatic control systems operate. In this example, loss of blood results in a drop in blood pressure, which could be life-threatening if not corrected. Effectors such as the heart and blood vessels help restore blood pressure toward normal. They do not raise blood pressure above normal, however, because of negative feedback (as denoted by the minus sign).

Not all forms of feedback contribute to homeostasis. In some cases, a **positive feedback loop** may accelerate a process, leading to what is sometimes called an explosive system (think of an avalanche that begins with a small snowball rolling down a steep hill). This is contrary to the principle of homeostasis, because positive feedback has no obvious means of being stopped. Not surprisingly, perhaps, positive feedback is far less common in animals than is negative feedback. Nonetheless, examples are found in animal biology where positive feedback is crucial. One is the process of birth in mammals (**Figure 40.14**), which is triggered by a positive feedback loop between nerve signals arising from smooth muscle cells of the birth canal and the mother's brain and pituitary gland (a component of the endocrine system). As the uterus pushes the baby's head against the birth canal, nerve signals from the canal send information to the brain, which triggers hormone release from the pituitary gland. The hormones stimulate the uterus to contract with more force, which causes additional nerve signals to the brain, and so on. Eventually, the uterus contracts strongly enough to deliver the fetus.

Feedforward Regulation Prepares for an Upcoming Challenge to Homeostasis

Built into the homeostatic mechanisms of many animals, particularly those with well-developed nervous systems, is another feature designed to minimize large swings in physiological variables. In **feedforward regulation**, an animal's body begins preparing for a change in some variable (for example, blood glucose levels) before it even occurs. Consider the anticipatory changes that occur when a hungry dog smells or sees food. First, the animal starts to salivate and its stomach begins to churn. Salivation and movements of the stomach wall are important components of the digestive process, yet at this stage the animal has not actually eaten any food. Instead, its digestive system is already preparing for the arrival of food in order to maximize digestive efficiency, speed the flow of nutrients into the blood, and minimize the time required for active cells to replenish energy stores. Thus, feedforward regulation speeds up the body's homeostatic responses and minimizes fluctuations in the level of the variable being regulated—that is, it reduces the amount of deviation from the set point.

In the preceding example, feedforward control utilizes sensory detectors that recognize odors and sights. Many examples of feedforward control, however, result from or are modified by the phenomenon called learning. The first times they occur, early in life, changes in the external environment might trigger relatively large fluctuations in regulated internal environmental factors. In responding to these changes, the nervous system learns to anticipate and resist them more effectively. Familiar examples are the increased heart rate and breathing rate that occur just before an athletic competition, a phenomenon not unique to human athletes (**Figure 40.15**). The most famous example, however, was first demonstrated in the early 20th century by Russian physiologist Ivan Pavlov.

> The brain triggers the release of hormones from the pituitary gland, which enter the blood and increase the strength of uterine contractions. The strong contractions stimulate the brain to make even more hormones.

Pituitary gland

> As the uterus begins to contract, nerve impulses travel from the birth canal to the brain.

Uterus

Figure 40.14 **The human birth process as an example of positive feedback.**   Nerve signals from the birth canal activate the brain to stimulate hormone release from the posterior pituitary gland. (The plus sign indicates stimulation.) These hormones act on muscle cells of the uterus to produce stronger uterine contractions, which in turn activate the release of still more hormones, and so on. Eventually the contractions become strong enough to deliver the fetus, at which time the positive feedback cycle ceases.

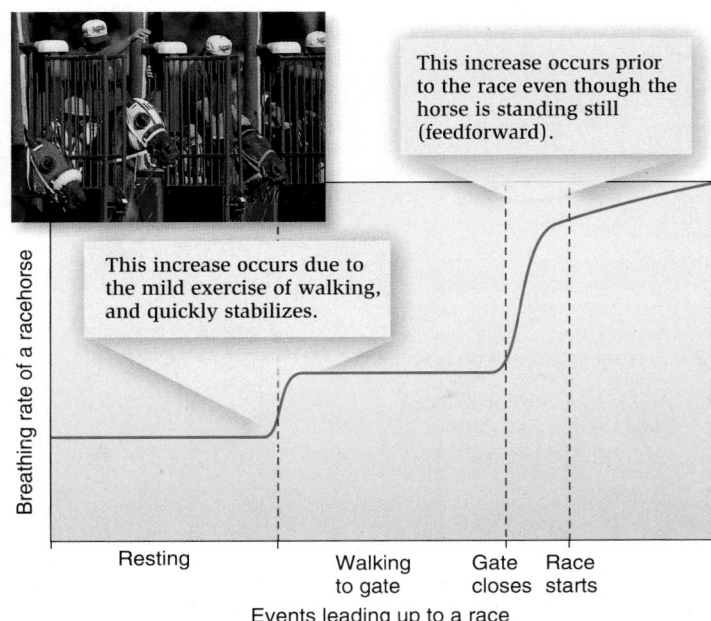

> This increase occurs prior to the race even though the horse is standing still (feedforward).

> This increase occurs due to the mild exercise of walking, and quickly stabilizes.

Breathing rate of a racehorse

Resting   Walking   Gate   Race
                 to gate   closes   starts

Events leading up to a race

Figure 40.15 **Feedforward control of breathing rate in an animal trained for athletic exercise.**   Feedforward processes prepare the body for an ensuing challenge or event, such as the one shown here.

FEATURE INVESTIGATION

Pavlov Demonstrated the Relationship Between Learning and Feedforward Processes

Ivan Pavlov made numerous contributions to our understanding of the digestive processes in mammals (see Chapter 41), for which he earned a Nobel Prize in Physiology and Medicine in 1904. Today, however, Pavlov is best remembered for work he did later in his career, when he demonstrated that feedforward processes associated with digestion could be conditioned to an irrelevant stimulus.

Pavlov was interested in the factors that increase production of saliva in a hungry animal. Saliva is an important secretion made by glands in the mouth, because it aids in swallowing and contains enzymes that begin the process of digesting certain foods. We are all familiar with the sight of a hungry dog or other animal salivating when it sees or smells food; indeed, the same phenomenon happens in humans. Pavlov discovered that dogs that were accustomed to being fed by the same researcher each day would begin to salivate when they saw the researcher approaching, even before receiving, seeing, or smelling any food. This did not happen in dogs that were not yet accustomed to the regular feeding schedule of the laboratory. Pavlov questioned whether any stimulus could elicit the feedforward process of salivation, even one that is not normally associated with feeding, so long as that stimulus was somehow paired with feeding.

To test this hypothesis, Pavlov presented two groups of dogs with either food or food and an irrelevant stimulus together, as shown in the Feature Investigation in **Figure 40.16**. Each dog was isolated in a room so that it could not see, hear, or smell the researcher or any other cues that might interfere with the experiment. In fact, the researcher observed the experiments from behind a wall at the rear of the animal. Contrary to popular belief, the stimulus Pavlov used was not the ringing of a bell but the ticking of a metronome, as well as various other sounds.

Figure 40.16 Feedforward events can be conditioned to external stimuli.

HYPOTHESIS It is possible to condition a feedforward response to an irrelevant stimulus through learning.

STARTING MATERIALS Experimental animals (dogs).

Experimental level   **Conceptual level**

1. Divide dogs into 2 groups. The control dogs are given food only. The conditioned dogs are exposed to a metronome when food is given.

Food only (control dogs)

Food dish

Separate soundproof testing chambers

Start metronome simultaneously with food presentation for several days (conditioned dogs)

Ticking metronome

Food dish

The dogs exposed to metronome become conditioned to the sound.

2. Surgically prepare dogs for saliva collection. Collect saliva from the control dogs and the conditioned dogs at various times after giving food or exposing them to a metronome.

Soundproof walls

Testing room

Harness

Salivary gland duct exteriorized to empty into funnel and collection tube

Researcher behind wall observing through small glass window

Graduated cylinder

Salivary gland

Funnel glued to bottom of chin

Plastic tube from salivary gland

3 THE DATA

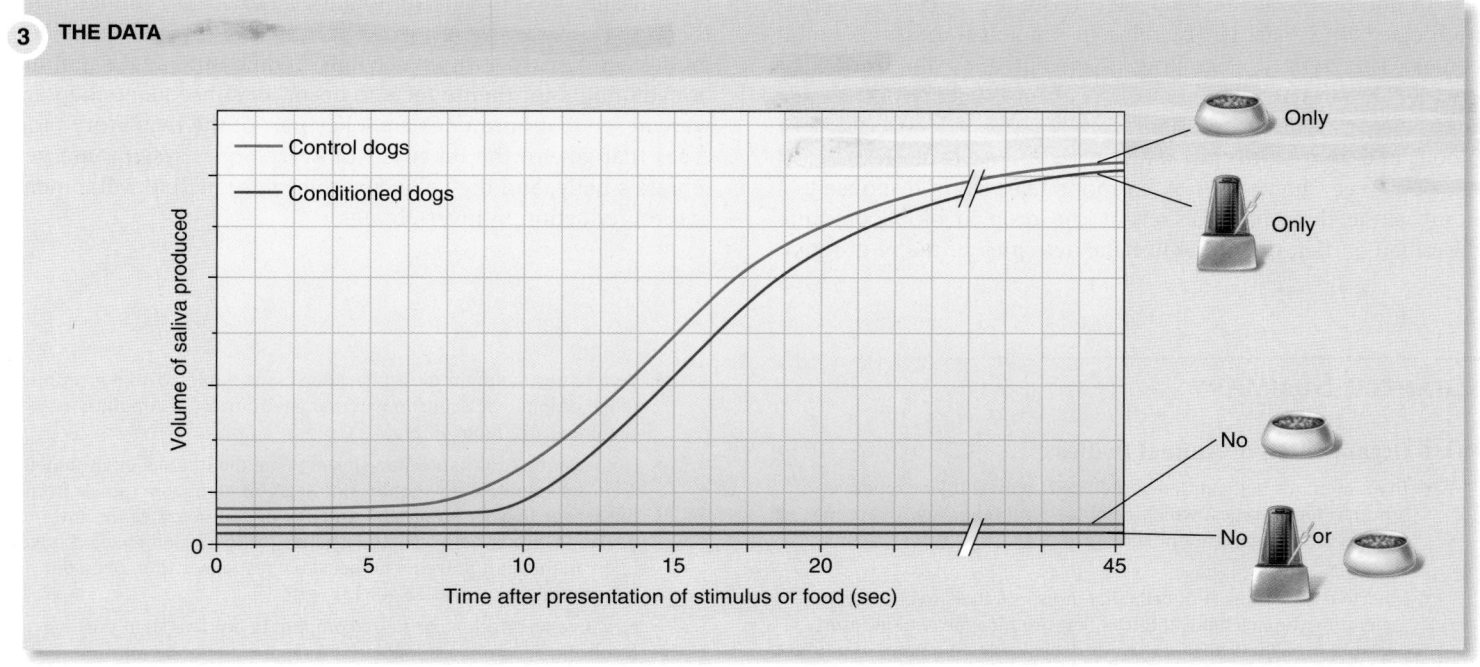

Pavlov reasoned that after some period of time, the dog would become "conditioned" to the auditory stimulus, which he called the conditioned stimulus, and would learn to associate it with the arrival of food. If this was correct, the metronome by itself should eventually be sufficient to elicit the feedforward process of salivation.

To measure salivary production rates, Pavlov surgically inserted a plastic tube into the ducts leading from the salivary glands located under the dogs' tongues. The tube exited through a small opening created in the cheek and emptied the saliva into a funnel glued beneath the chin. Pavlov's hypothesis was confirmed when, after a conditioning period of several days, the conditioned stimulus by itself elicited increased salivation in the dog in less than 10 seconds. The amount of saliva produced

was comparable to that produced in control dogs that had not been exposed to conditioning trials, and who were then presented with food in the usual way.

In later experiments, Pavlov demonstrated that other stimuli, such as touch, produced the same phenomenon he observed with sound. He also showed that the conditioned response was not permanent. If a dog that was already conditioned to the sound of a metronome was then repeatedly exposed to that sound without the simultaneous presentation of food, the salivary response to the sound would eventually become extinguished. These experiments demonstrated that certain feedforward processes can be modulated by experience and learning, and suggested that animals have a much greater ability to recognize and adapt to changes in their environment than previously thought.

Local Homeostatic Responses Fine-Tune the Activities of Tissues and Organs

A common thread that links all homeostatic processes together is communication between cells, whether cells are close to each other or in different parts of an animal's body. Some homeostatic responses may be highly localized, occurring only in the area of a disturbance. For example, damage to an area of skin causes cells in the injured area to release molecules that help contain the injury, prevent infections, and promote tissue repair in the immediate vicinity (see Chapter 53). Local responses provide areas of the body with mechanisms for local self-regulation. It would be of no benefit to an animal to promote tissue repair in regions of the body that are not injured. This type of cellular communication—in which molecules are released into the inter-

stitial fluid and act on nearby cells—is called **paracrine** signaling (see Chapter 9).

Another example of extremely localized paracrine signaling occurs between nerve cells. Most nerve cells communicate through the release of **neurotransmitters**, small signaling molecules that are synthesized and stored in nerve cells. Neurotransmitters are released from one nerve cell where they diffuse to combine with receptor proteins on a single, adjacent nerve cell, which alters the activity of that cell. This type of cell-to-cell communication is typically very rapid, finishing within milliseconds. Consequently, neurotransmitter responses are capable of making immediate homeostatic adjustments, like those associated with reflexes. These are just two of the many types of paracrine signaling that occur in animals' bodies, and that will be described in subsequent chapters in this unit.

In addition to localized paracrine signaling, cells can communicate "long distance" by releasing chemical messenger molecules. This type of signaling is mediated by the endocrine systems of animals. A **hormone** is a chemical messenger that is produced in a gland or other structure, is secreted into the blood, and acts on distant target cells in one or more parts of the body (see Chapter 50). A hormone released in response to a homeostatic disturbance, such as the drop in blood pressure described earlier, can influence the activities of many different cells, tissues, and organs simultaneously because the hormone is carried throughout the entire blood circulation. Some hormones act quickly—within seconds—while others take minutes or even hours for their effects to occur. In subsequent chapters, we will see that hormones are a key part of the regulatory processes that govern the functions of every organ system in a vertebrate's body, and they play key roles in growth, development, and reproduction in invertebrates.

CHAPTER SUMMARY

40.1 Organization of Animal Bodies

- In an animal's body, differentiated cells with similar properties group to form tissues, which combine with other types of tissues to form organs. Organs are linked together to form organ systems. (Figure 40.1)

- Muscle tissues consist of cells that are specialized to contract, generating the mechanical forces that produce movement, exert pressure on a fluid-filled cavity, or decrease the diameter of a tube. There are three categories of muscle tissue: skeletal, smooth, and cardiac. Nervous tissues initiate and conduct electrical signals from one part of an animal's body to another part. Epithelial tissues are specialized to protect structures, and to secrete and absorb ions and organic molecules. Connective tissues connect, surround, anchor, and support the structures of an animal's body. Connective tissues include bone, cartilage, blood, adipose (fat-storing) tissue, loose connective tissue, and dense connective tissue. (Figures 40.2, 40.3, 40.4, 40.5)

- An organ is composed of two or more kinds of tissues. In an organ system, different organs work together to perform an overall function. (Figure 40.6, Table 40.1)

- The development, spatial positioning, and function of many body organs is under the control of homeotic genes, such as *Hox* genes in vertebrates.

- An animal's body fluids are distributed into three compartments (plasma, interstitial fluid, and intracellular fluid). The properties of the barriers between the compartments determine which substances can move between them. (Figures 40.7, 40.8)

40.2 The Relationship Between Structure and Function

- Similarities in structure are often associated with similarities in function. (Figure 40.9)

- Extensive surface area maximizes the ability of a tissue or organ to absorb solutes, exchange oxygen and carbon dioxide with the environment, communicate with other cells, and receive sensory information from the environment. The high surface area/volume (SA/V) ratio principle applies to many structures and across all animal taxa. (Figure 40.10)

40.3 Homeostasis

- Homeostasis is the dynamic process of adapting to the external environment and maintaining a relatively stable internal environment. Some animals are conformers; others are regulators.

- Vertebrates maintain most physiological variables within a certain range despite variations in external environmental conditions. (Figure 40.11, Table 40.2)

- Homeostatic control mechanisms regulate the internal environment. A sensor monitors a variable, an integrator compares signals from the sensor to a set point, and an effector compensates for any deviations from the set point. Negative feedback minimizes changes in a variable and prevents homeostatic responses from overcompensating. (Figures 40.12, 40.13)

- Negative feedback is far more common in animals than positive feedback. Feedforward regulation prepares the body for an upcoming challenge to homeostasis. (Figures 40.14, 40.15)

- Pavlov's experiments with dogs demonstrated that certain feedforward responses such as salivation could be conditioned to irrelevant stimuli, like the ticking of a metronome. (Figure 40.16)

- Chemical communication between cells is essential to homeostasis. Types of chemical messengers include paracrine signals, such as neurotransmitters, and long-distance signals, such as hormones.

TEST YOURSELF

1. Tissue that is specialized to conduct electrical signals from one structure in the body to another structure is _____ tissue.
   a. epithelial
   b. connective
   c. nervous
   d. adipose

2. Structures that are composed of two or more tissue types arranged in various proportions and patterns are
   a. cells.
   b. tissues.
   c. organs.
   d. organ systems.
   e. organisms.

3. The extracellular matrix of connective tissue
   a. contains different proteins that provide structural support to cells.
   b. provides a scaffolding for the cells of the tissue.
   c. plays a role in cellular communication.
   d. all of the above.
   e. a and b only.

4. When examining the structure of many animal organs
   a. it is apparent that all organ systems are fully functional in all animals.
   b. the function can be predicted based on the structural adaptations.
   c. the function of the structure is difficult to determine because of the lack of variation among the different structures of the body.
   d. all four tissues are equally represented in all organs.
   e. none of the above.

5. Most of the water in an animal's body
   a. lacks any type of ions or other solutes.
   b. is found in the spaces between cells.
   c. is contained inside the cells.
   d. is located in the extracellular fluid.
   e. is unable to move between body compartments.

6. The folds, convolutions, or extensions found in many structures of animals results in
   a. decreased level of activity in that particular structure.
   b. interruption in the normal functioning of the structure.
   c. increased surface area for absorption, communication, or exchange.
   d. higher susceptibility to infection.
   e. none of the above.

7. Adapting to changes in the external environment and maintaining a relatively stable internal environment is
   a. natural selection.
   b. evolution.
   c. positive feedback.
   d. homeostasis.
   e. both c and d.

8. Which of the following statements regarding negative feedback is *not* correct?
   a. It helps regulate variables like body temperature and blood pressure.
   b. It is part of the process by which birth occurs in mammals.
   c. It is a major feature of homeostatic control systems.
   d. It prevents homeostatic responses from overcompensating.
   e. It may occur at the organ, cellular, or molecular level.

9. The ability of an animal's body to prepare for a change in some variable before the change occurs is called
   a. positive feedback.
   b. negative feedback.
   c. homeostasis.
   d. feedforward regulation.
   e. autoregulation.

10. A hormone differs from a neurotransmitter in that
    a. hormones act extracellularly; neurotransmitters act within the cell that synthesized them.
    b. hormones are only released by nerve cells; neurotransmitters are released by many different types of cells.
    c. hormones cause fast responses (seconds or less) to stimuli; neurotransmitters cause slow responses (minutes to hours) to stimuli.
    d. hormones only affect epithelial cells; neurotransmitters only affect muscle cells.
    e. hormones are released into the bloodstream and can activate many cells in many parts of the body; neurotransmitters are released by nerve cells and affect adjacent cells.

CONCEPTUAL QUESTIONS

1. Define anatomy and physiology.
2. Briefly describe adipose tissue.
3. Define homeostasis.

EXPERIMENTAL QUESTIONS

1. What process did Pavlov study in his experimental animals, and what was his hypothesis?
2. How did Pavlov control for the possibility that other stimuli, such as the appearance of the investigator, were somehow causing the feedforward response?
3. What did Pavlov measure and what was his major finding?

COLLABORATIVE QUESTIONS

1. Discuss the terms homeostasis, negative feedback loop, and positive feedback loop.
2. Discuss the organization of animal bodies.

www.brookerbiology.com

This website includes answers to the Biological Inquiry questions found in the figure legends and all end-of-chapter questions.

41

NUTRITION, DIGESTION, AND ABSORPTION

Obtaining and utilizing nutrients is a fundamental process in all animals, such as this orangutan.

A **nutrient** is any substance consumed by an animal that is needed for survival, growth, development, tissue repair, or reproduction. All animals require nutrients to survive. Because animals cannot utilize the energy of sunlight to make nutrients, as plants do, they must have ways of obtaining food that contains these vital substances. If the nutrients in food are in a form that cannot be directly used by cells, they must be broken down into smaller molecules, a process known as **digestion**. This is followed by the process of **absorption**, in which these small molecules are transported from the digestive cavity into an animal's circulatory system. Once delivered to the cells of the body, these molecules have several possible fates. Enzymes act on some organic nutrients to transfer the energy in the nutrients' chemical bonds into high-energy bonds of ATP. Alternatively, organic molecules in food can serve as building blocks to synthesize new cellular organelles, provide the needed components for cells to divide and multiply, and produce new proteins that carry out cellular activities. Other nutrients are inorganic—for example, iron, copper, zinc, and other minerals—and function as cofactors in enzymatic reactions and other processes. In this chapter, we look at the diverse ways in which animals obtain, digest, and absorb nutrients. We begin by examining the types of nutrients they consume.

41.1 Animal Nutrition

The organic nutrients required by animals fall into five categories: carbohydrates, lipids, proteins, nucleic acids, and vitamins. In addition, animals require inorganic nutrients in the form of water and minerals such as Ca^{2+}, K^+, and Fe^{3+}. In many cases, the same nutrients serve comparable functions in different animals. For example, we have seen that the plasma membranes of every animal are made of proteins, carbohydrates, and lipids. As another example, all animals that have muscles use amino acids to make the same specialized proteins that allow their muscles to contract. All animal cells require calcium ions to carry out common activities such as exocytosis and cell division. Similarly, all animal cells require a regular source of energy to replace the high-energy bonds of ATP that are used in phosphorylation and other biochemical reactions.

Despite these basic similarities, however, nutritional demands differ depending upon a given animal's physiology. The digestive systems of **herbivores**, animals that eat only plants, contain microbes that assist in the digestion of cellulose, and so these animals are well adapted to subsist on plants and algae. By contrast, **carnivores** are primarily adapted to consume animal flesh or fluids, while **omnivores** have the ability to eat and survive on both plant and animal products. In this section, we will examine the various organic and inorganic nutrients consumed by animals, and some of the major functions of these molecules.

Animals Require Nutrients for Macromolecule Synthesis, Energy, and Chemical Reactions

As stated earlier, all animals must obtain nutrients from the environment. Nutrients common to animals include those organic and mineral compounds that provide the substrates for building and replenishing cellular components such as proteins and nucleic acids (DNA and RNA). The chemical bonds in organic compounds, such as sugars and fats, are also used as a source of energy from which ATP can be formed. Indirectly, therefore, sugars and fats provide the energy needed for most of the chemical reactions that occur in animals' bodies.

Table 41.1	Major Nutrients in Animals, Their Dietary Sources, and Some of Their Functions in Vertebrates		
Class of nutrient	**Dietary sources**	**Functions in vertebrates**	**Symptoms of deficiency**
Food source			
Carbohydrates	All food sources	Energy source; components of some proteins; source of carbon	Muscle weakness, weight loss, ketone formation (decreases pH of blood)
Lipids	All food sources, especially fatty meats, dairy products, plant oils	Major component of cell membranes; energy source; thermal insulator	Hair loss, dry skin, weight loss, hormonal and reproductive disorders
Proteins	All food sources, especially meat, legumes, cereals, roots	Provides amino acids to make new proteins; builds muscle; some amino acids used as energy source	Weight loss, muscle loss, weakness, weakened immune system, increased likelihood of infections
Nucleic acids	All food sources	Provides nucleotides to make DNA, RNA, and ATP	
Minerals			
Calcium (Ca)	Dairy products, cereals, legumes, whole grains, green leafy vegetables, bones (eaten by some animals)	Bone and tooth formation; exocytosis of stored secretions in nerves and other cells; muscle contraction; blood clotting	Muscular disorders, loss of bone, reduced growth in children
Chlorine (Cl)	Meats, dairy foods, blood, natural deposits of salt	Participates in electrical, acid-base, and osmotic balance across cell membranes, notably nerve and heart cells	Muscular and nerve disorders
Chromium (Cr)	Liver, seafood, some nuts, meats, mushrooms, some vegetables	Needed for proper glucose metabolism, possibly by aiding insulin action	Disorders of lipid and glucose balance in blood
Copper (Cu)	Fish, shellfish, nuts, legumes, liver, and other organs	Required for hemoglobin production and melanin synthesis; needed for connective tissue formation; serves as oxygen-binding molecule in some invertebrates	Anemia (abnormally low number of red blood cells), bone changes
Iodine (I)	Seaweed, seafood, milk, iodized salt	Required for formation of thyroid hormones	Inability to make thyroid hormones, resulting in enlarged thyroid gland
Iron (Fe)	Liver and other organs, some meats, eggs, legumes, leafy green vegetables	Oxygen-binding component of hemoglobin	Anemia
Magnesium (Mg)	Hay, grasses, whole grains, green leafy vegetables	Cofactor for ATP and many enzymes	Changes in nervous system function
Manganese (Mn) and Molybdenum (Mo)	Nuts, whole grains, legumes, vegetables, liver	Cofactors for many enzymes	Poor growth, abnormal skeletal formation, nervous system disorders (convulsions)
Phosphorus (P)	Dairy foods, grains, legumes, nuts, meats	Bone and tooth formation; component of DNA, RNA, and ATP	Bone loss, muscle weakness
Potassium (K)	Meats, fruits, vegetables, dairy foods, grains	See Chlorine	Muscle weakness, serious heart irregularities, gastrointestinal symptoms
Selenium (Se)	Seafood, eggs, chicken, soybeans, grains	Antioxidant; cofactor for some enzymes	Keshan disease (damage to and loss of heart muscle)
Sulfur (S)	Proteins from any source	Component of two amino acids (methionine and cysteine)	Inability to synthesize many proteins
Sodium (Na)	Many fruits, vegetables, meats, and natural salt deposits	See Chlorine	Muscle cramps, changes in nerve activities
Zinc (Zn)	Widely found in meats, fish, shellfish (oysters), grains	Many functions related to tissue repair; sperm development; cofactor for many metabolic enzymes; needed for certain transcription factors to bind to DNA	Stunted growth, loss of certain sensations like taste, impaired immune function, skin lesions

Minerals such as iron and zinc are needed as cofactors or constituents of enzymes and other proteins. Other minerals such as calcium are needed for bone, muscle, and nerve function, while still others—sodium and potassium—establish ion gradients across plasma membranes, and therefore are critical for heart, skeletal muscle, and nerve activity. **Table 41.1** summarizes the major nutrients, their dietary sources, some of their important functions, and some health consequences associated with their deficiencies.

Table 41.1	Major Nutrients in Animals, Their Dietary Sources, and Some of Their Functions in Vertebrates, *continued*		
Class of nutrient	**Dietary sources**	**Functions in vertebrates**	**Symptoms of deficiency**
Water-soluble vitamins			
Biotin	Liver, legumes, soybeans, eggs, nuts, mushrooms, some green vegetables	Coenzyme for gluconeogenesis and fatty acid synthesis; needed to catabolize certain amino acids	Skin rash, nausea, loss of appetite, mental disorders (depression or hallucinations)
Folic acid	Green vegetables, nuts, legumes, whole grains, organ meats (especially liver, kidney, heart)	Coenzyme required for synthesis of nucleic acids	Anemia, depression, birth defects
Niacin	Legumes, nuts, milk, eggs, meat	Involved in many oxidation/reduction reactions	Skin rashes, diarrhea, mental confusion, memory loss
Pantothenic acid	Nearly all foods	Part of coenzyme A, which is involved in numerous synthetic reactions, including formation of cholesterol	Burning sensation in hands and feet, gastrointestinal symptoms, depression
Vitamin B$_1$ (thiamine)	Meats, legumes, whole grains	Required for certain steps of intermediary metabolism, leading to energy production from some amino acids	Beriberi: Muscular weakness, anemia, heart problems, loss of weight
Vitamin B$_2$ (riboflavin)	Dairy foods, meats, organ meats, cereals, some vegetables	Respiratory coenzyme; required for metabolism of fats, carbohydrates, and proteins	Seborrhea (excessive oil secretion from skin glands resulting in skin lesions)
Vitamin B$_6$ (pyridoxine)	Meats, liver, fish, nuts, whole grains, legumes	Cofactor for over 100 enzymes that participate in amino acid metabolism, lipid metabolism, and heme synthesis	Seborrhea, nerve disorders, depression, confusion, muscle spasms
Vitamin B$_{12}$	Meats, liver, eggs, some shellfish, dairy foods	Required for red blood cell formation	Anemia, nervous system disorders leading to sensory problems, balance and gait problems, loss of bladder and bowel control
Vitamin C (ascorbic acid)	Citrus fruits, green vegetables, tomatoes, potatoes	Antioxidant and free radical scavenger; aids in iron absorption; helps maintain healthy connective tissue and gums	Scurvy (connective tissue disease associated with skin lesions, weakness, poor wound healing, tooth decay), bleeding gums
Fat-soluble vitamins			
Vitamin A (retinol)	Liver, green and yellow vegetables, some fruits in small amounts	Component of visual pigments; regulatory molecule affecting transcription; important for reproduction and immunity	Night blindness due to loss of visual ability, skin lesions, impaired immunity
Vitamin D	Fish oils, fish, egg yolk, liver, synthesized in skin via sunlight	Required for calcium and phosphorus absorption from gut; bone growth	Rickets (weakened, deformed bones) in children, osteomalacia (weak bones) in adults
Vitamin E	Meats, vegetable oils, cereals, small amounts in some fruits and vegetables	Antioxidant; inhibits prostaglandin synthesis	Unknown, possibly skeletal muscle atrophy, peripheral nerve disorders
Vitamin K	Legumes, green vegetables, some fruits, some vegetable oils (olive oil, soybean oil), liver, also synthesized by hindgut bacteria	Component of blood-clotting mechanism	Reduced blood clotting ability

Although different species consume a wide variety of foods, all animals require the same fundamental organic nutrients. These include carbohydrates to supply energy-yielding glucose; lipids to supply fatty acids, cholesterol, and triglycerides for energy and membrane-building; and proteins to supply amino acids, which are used to make new proteins and the purines and pyrimidines needed for DNA and RNA synthesis (**Figure 41.1**). The amount of each nutrient needed by an animal depends on its activity level, or metabolic rate, which we will discuss in Chapter 42. Generally, highly active and energetic animals, such as most birds and mammals, require a greater amount of nutrients each day than do relatively inactive or sedentary animals like nonmotile invertebrates. However, exceptions to this general rule exist. If a normally sedentary animal temporarily expends more energy than usual, it will require additional nutrients to replenish that energy. Conversely, some mammals that are active at one time of year may hibernate for the winter, during which time they use very little energy and exist on nutrients that were previously consumed and stored (for example, as fat).

	Carbohydrates	**Lipids**	**Proteins**
Polymer	Polysaccharides	Triglycerides	Polypeptides
Monomer	Monosaccharides	Fatty acids	Amino acids

Figure 41.1   The three classes of organic nutrients that are used to provide energy, make new molecules, or build cellular structures.   Many nutrients are ingested as polymers, which are broken down during digestion to monomers. However, some nutrients are ingested in monomer form, such as fructose (a monosaccharide).

In many cases, an animal may consume monomers, polymers, or both. The polymers are typically too large to be absorbed across the epithelial cells that line the digestive tract. Therefore, these molecules are first broken down by enzymes into monomers and then transported into the blood and cells of the body. Once in the cells, metabolic enzymes can synthesize new polymers from these monomers. In this way, cells can obtain the polysaccharides, proteins, triglycerides, and nucleic acids they require. Thus, even though organic nutrients may be consumed as polymers, they typically are first broken down into monomers that can more easily exit the digestive tract and enter the blood circulation.

Essential Nutrients Must Be Obtained from the Diet

Certain compounds cannot be synthesized from any ingested or stored precursor molecule. These **essential nutrients** must be obtained in the diet in their complete form. The word essential refers to the fact that they must come from food. It does not mean that other nutrients are less important, since many nutrients are essential for an animal's survival. The essential nutrients can be classified into four groups: essential amino acids, essential fatty acids, minerals, and vitamins.

Essential Amino Acids Eight **essential amino acids** are required in the diet of humans and many other animals—isoleucine, leucine, lysine, methionine, phenylalanine, threonine, tryptophan, and valine. These amino acids are required for building proteins but cannot be synthesized by the animal's cells. Also, animal cells do not store amino acids. Without a recurring supply of these eight amino acids, therefore, protein synthesis in each cell in an animal's body would slow down or stop

completely. Carnivores and omnivores readily obtain all of the essential amino acids, because meat (animal muscle) contains all 20 amino acids. People who follow strict vegetarian diets must find ways to balance the types of plant matter they eat. Unlike animal meat, most plants do not contain every essential amino acid in sufficient quantities to supply a human's nutritive needs. By contrast, some herbivores such as cows have evolved the capacity to synthesize the essential amino acids, which allows them to subsist entirely on a plant diet.

Essential Fatty Acids The **essential fatty acids** are certain polyunsaturated fatty acids such as linoleic acid that cannot be synthesized by animal cells. Saturated fatty acids, on the other hand, can be synthesized from carbon sources like glucose. Linoleic acid is vital to an animal's health because it is the building block that is converted in cells to arachidonic acid. This fatty acid is the precursor for production of several compounds known to be important in many aspects of animal physiology. Some animals—such as felines—cannot synthesize arachidonic acid from linoleic acid, and so arachidonic acid is an essential fatty acid in those species. Unsaturated fats are found primarily in plants, which provide a dietary source for both herbivores and omnivores. Strict carnivores such as cats, however, obtain their essential fatty acids from fish, or from the adipose tissue of birds and mammals.

Minerals **Minerals** are inorganic ions required to build skeletons, maintain the balance of salts and water in the body, provide a source of electric current across plasma membranes, and provide a mechanism for exocytosis and muscle contraction, among other functions. Table 41.1 summarizes some of the most important minerals and their functions. Many minerals are required in only trace amounts, far less than 1 mg/day in a relatively large mammal like ourselves. Nonetheless, without regular consumption of these small amounts, serious health problems may arise.

Some minerals can be stored in a typical vertebrate animal's body, reducing the risk of deficiency when the mineral is not available in the diet. Iodine, for example, is stored in large amounts in the thyroid gland, where it is required for proper thyroid function. If a vertebrate's diet is deficient in iodine, which is common in areas where iodine periodically leaches from the soil as a result of considerable rainfall, there is usually enough stored iodine to permit normal thyroid gland function for days or even months, depending on the species. As another example, calcium is stored in huge quantities in bone in vertebrates and in the shells of some invertebrates. If the dietary intake of this mineral falls, calcium is released from these storage sites and secreted into the blood.

Not all minerals are used the same way or at the same rate by all animals. For instance, copper helps to bind and transport oxygen in the fluids of some invertebrates such as horseshoe crabs, while in all vertebrates and in most other invertebrates iron serves this function. Factors that affect mineral usage are an animal's species, age, weight, overall health status, and the types of food it eats.

Vitamins **Vitamins** are important organic nutrients that serve as coenzymes for many metabolic and biosynthetic reactions (see Table 41.1). The two categories of vitamins are fat-soluble and water-soluble. Fat-soluble vitamins, such as vitamin A, are stored in adipose tissue, whereas water-soluble vitamins, for example, vitamin C, are not stored in the body and must be regularly ingested. Not all animals require the same vitamins in their diet, however. Among vertebrates, for example, only primates and guinea pigs cannot synthesize their own vitamin C and must therefore consume it in the diet.

Inadequate vitamin intake can have serious consequences (see Table 41.1), as shown by the health problems of British sailors in the 18th century. Sailors would often remain at sea for months or even years, subsisting on a poor diet with few fruits or vegetables. Many developed **scurvy**, a potentially fatal vitamin C deficiency characterized by weakness, bleeding gums, and tooth loss. The symptoms of scurvy had been recognized for centuries. Only in 1795, when British physician James Lind performed a crucial controlled experiment, was its cause finally understood. Lind was aware of earlier suggestions that citrus fruits were important for overall health. During one voyage, he divided 12 sailors who were already sick with scurvy into six groups, each of which received a different supplement to their meager diet. The two sailors who were given oranges and lemons to eat showed a remarkable recovery from the disease, while the other five groups failed to recover. Lind had shown that scurvy could be prevented with the simple procedure of including citrus foods in the diet, a discovery that revolutionized thinking about the link between nutrition and disease. Later studies revealed that vitamin C in citrus fruits is responsible for this outcome.

41.2 Ways in Which Animals Obtain Nutrients

Several factors determine how and when an animal obtains food, including its energetic demands, whether or not it is capable of locomotion, the local environment, and the structure of its digestive system. In this section, we will see that despite the enormous number of animal species and the highly varied environments in which they live, the major ways in which animals obtain food can be classified into just a few categories.

Animals May Eat Plants, Other Animals, or Both

Unlike plants, which make their own food, all animals are heterotrophs, which means they must consume their food. As stated earlier, most animals can be grouped into one of three dietary categories: herbivores, carnivores, or omnivores (**Figure 41.2**).

(a) Herbivores eat plants or algae.

(b) Carnivores eat animals or animal products.

(c) Omnivores eat animals and plants.

Figure 41.2   **Dietary categories among animals.**   One way of classifying animals is by the types of food they consume: **(a)** plants, **(b)** other animals, or **(c)** both plants and animals.

Biological inquiry: The dietary preferences of one animal often influence its behaviors as well as the behaviors of other animals. How does this occur?

The gastrointestinal (GI) tracts, or **gut**, of carnivores and omnivores resemble each other much more than either resembles those of herbivores. This reflects differences in the enzymatic processing and the energetic quality of the foods they eat. Generally the nutritive value of meat is much greater than that of plants. Consequently, carnivores usually need to eat less frequently than herbivores, many of whom eat almost continuously to supply sufficient nutrients for energy and growth.

Although useful, the three dietary categories are limited when describing the diversity of animal species. For example, some animals are almost strictly carnivores at one time of year but herbivores at other times. Many nonmigratory birds, for example, feed on insects and worms during the summer but switch to eating whatever vegetation, buds, or seeds they can find during the winter. Similarly, a coyote prefers to eat only meat but will consume plants and fruits if hungry enough. A hungry fruit-eating bat will eat leaves, nectar, or pollen if fruit is not available. Animals like these are said to be **opportunistic**. They have a strong preference for one type of food but can adjust their diet if the need arises.

An animal's life stage may also influence its diet. Mammals begin life as milk-drinkers and later switch to consuming plants, animals, or both. Caterpillars eat leaves, but after metamorphosis, most species of moth or butterfly are strictly fluid-drinkers, typically the nectar in flowers. Other animals that undergo metamorphosis also change their diet as adults. Tadpoles, for example, are mostly herbivores, while frogs eat insects and so are carnivores.

The three major categories also do not indicate what type of plant or animal is consumed. Grasses, cereals, and fruits, for example, each have different energy and nutrient contents. Some herbivores eat primarily fruits, and some carnivores only drink the blood of other animals. To better describe the nutrition of animals, therefore, we need to investigate the strategies by which they obtain food.

Animals Have Evolved Multiple Strategies for Obtaining Food

The ways in which an animal obtains its food are related to its environment. Not surprisingly, a sessile (nonmotile) marine invertebrate and a highly mobile terrestrial vertebrate face different challenges and opportunities.

Aquatic animals that are sedentary or nonmotile may trap food that floats or swims their way using sticky surfaces or tentacles. Sea squirts are coated with a layer of sticky mucus that traps suspended food particles, while sea anemones use tentacles to capture and kill prey that swims within reach.

Other aquatic animals, both motile and nonmotile, include **suspension feeders**, which sift water, filtering out the organic matter and expelling the rest (**Figure 41.3**). Shelled mollusks filter seawater and capture floating bits of organic material on mucus-covered cilia located in their gills, which move the material into the animal's mouth. Baleen whales have specialized plates suspended from the roofs of their mouths like piano wire. The whale engulfs a mouthful of seawater and squeezes it back out through the plates, which screen for small animals and protozoa.

Carnivores, which inhabit aquatic and terrestrial environments, are generally **predators** (they kill live prey) or **scavengers**

Figure 41.3 Suspension feeding. Suspension feeders include species of both invertebrates and vertebrates. These animals, like the clam and baleen whale illustrated here, filter out tiny organisms or bits of organic matter from the water.

(they eat the remains of dead animals). Carnivores' teeth are well developed for biting and tearing. They normally do not chew food to any large degree but rather swallow large chunks at once. This does not hinder their ability to digest their food, because meat is still fully digested if swallowed whole, although not as quickly as it would be if it were chewed. Swallowing food whole allows a carnivore to quickly get another bite when competing with other hungry animals. The strategy of consuming entire chunks of food without chewing is seen in species from all the vertebrate classes. Birds of prey, for example, use their sharp beaks to tear away pieces of flesh that can be engulfed all at once. Chewing, when or if it occurs in a carnivore, is mostly a means of breaking food into pieces small enough to swallow.

Herbivores, on the other hand, have powerful jaw muscles and large, broad teeth that are highly adapted for grinding tough, fibrous plants. **Grazers**—herbivores that feed almost constantly on grasses—must chew to facilitate the digestion of plant material. Some herbivores such as the ruminants (for example, sheep, goats, llamas, cows) will even regurgitate partially digested food and chew it a second time. Other herbivores called **frugivores** are adapted primarily to feed on fruits (**Figure 41.4**).

Fluid-feeders lick or suck fluid from plants or animals and so do not need teeth except, perhaps, to puncture an animal's skin. Fluid-feeding has evolved independently in many types of animals, including insects, worms, birds, and mammals. When the fluid is obtained from an animal, the fluid-feeder usually has a specialized mouthpart, such as the piercing needle-like extension of a mosquito's mouth or the tiny jaws and teeth of a blood-sucking leech (**Figure 41.5a**). Fluid-feeders that consume the blood of other animals have developed a fascinating set of strategies to ensure a full meal. Leeches, for example, secrete a local anesthetic at the site of the bite to dull the pain. An enzyme digests the host's connective tissue to allow the leech to firmly embed its mouth into its host's flesh, and a locally acting chemical is secreted to keep the host's blood vessels open.

Finally, to ensure that the blood does not clot before the leech has drunk its fill, the saliva of leeches and other blood-sucking animals often contains an anticoagulant. Scientists are studying these anticoagulants to develop anticlotting drugs for people with certain forms of heart and blood vessel disease. Recently some physicians have applied leeches to surgical sites to prevent blood clots, such as when surgically reattaching a severed finger.

Not all fluid-feeders consume blood. Many birds and bats drink nectar from plants and have specialized beaks, tongues, or mouths that enable them to reach the nectar of a particular flower (**Figure 41.5b**).

Regardless of what an animal eats, the useful parts of the eaten material must be digested into molecules that its cells can absorb. Next, we turn to a discussion of how animals digest their food and absorb the nutrients.

Blood-sucking animals like this leech use their teeth to prick the skin of their host.

(a)

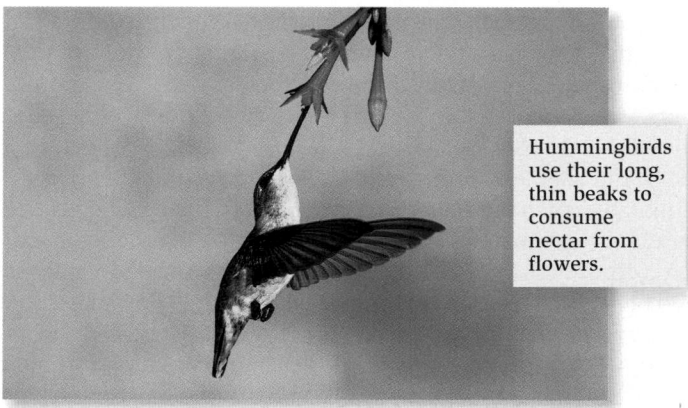

Hummingbirds use their long, thin beaks to consume nectar from flowers.

(b)

Figure 41.5   Strategies of fluid-feeders.   Fluid-feeders use their specialized mouths to suck (a) the blood of animals or (b) the nutrient-rich nectar of plants.

Biological inquiry: To obtain sufficient nutrients, blood-sucking animals like mosquitoes, leeches, and some bats must consume large amounts of blood with each meal. For mosquitoes and bats, would this hinder their ability to take flight after a meal?

Figure 41.4   An example of a frugivore.   Frugivores like this bat (*Pteropus hypomelanus*) feed exclusively or primarily on fruit when it is available.

41.3 Principles of Digestion and Absorption of Food

Digestion requires enzymes capable of hydrolyzing the chemical bonds in carbohydrates, lipids, proteins, and nucleic acids. These enzymes are named after their substrates. Thus, we have polysaccharidases, lipases, proteases, and nucleases. Once the enzymes digest the food, the products of digestion must be absorbed by diffusion or by energy-dependent transport across plasma membranes. Not all of the food that is eaten, however, is digested. Minerals, vitamins, and monomers of macromolecules do not require digestion and are absorbed intact. Digestion is only needed to convert polymers into smaller units that can be absorbed across plasma membranes. In this section, we will examine some of the major principles of digestion and absorption in animals, beginning with where digestion takes place.

Digestion Can Occur Intracellularly or Extracellularly

Intracellular digestion occurs only in some very simple invertebrates such as sponges and single-celled organisms. It involves the phagocytosis of food particles directly into a cell, where the food is segregated from the rest of the cytoplasm in food vacuoles. Once inside the vacuoles, hydrolytic enzymes digest the food into monomers, which exit the vacuole and are used directly by that cell. Intracellular digestion cannot support the metabolic demands of an active animal for long, because only tiny bits of food can be phagocytosed at one time. It also does not provide a mechanism for storing large quantities of food so that an animal can digest it slowly while going about its other activities.

Most animals digest food via extracellular digestion in a cavity of some sort. Extracellular digestion protects the interior of the cells from the actions of hydrolytic enzymes and allows animals to consume large prey or large pieces of plants. Food enters the digestive cavity, where it is stored, slowly digested, and absorbed gradually over long periods of time ranging from hours (for example, after a human eats a pizza) to weeks (after a python eats a gazelle).

In the simplest form of extracellular digestion—seen in invertebrates such as flatworms and cnidarians—the digestive cavity has one opening that serves as both an entry and an exit port (**Figure 41.6**). The digestive cavity of these animals is called a **gastrovascular cavity**, because not only does digestion occur within it, but fluid movements in the cavity serve as a circulatory—or vascular—system to distribute digested nutrients throughout the animal's body. Food within a gastrovascular cavity is partially digested by enzymes that are secreted into the cavity by the cells lining the cavity. As the food particles become small enough, they are phagocytosed by the lining cells and further digested intracellularly. Nondigested material that remains in the gastrovascular cavity is expelled.

Most Digestive Cavities Are Tubes with Specialized Regions and Openings at Opposite Ends

In contrast to the gastrovascular cavities of simple invertebrates, all other animals possess digestive systems that consist of a

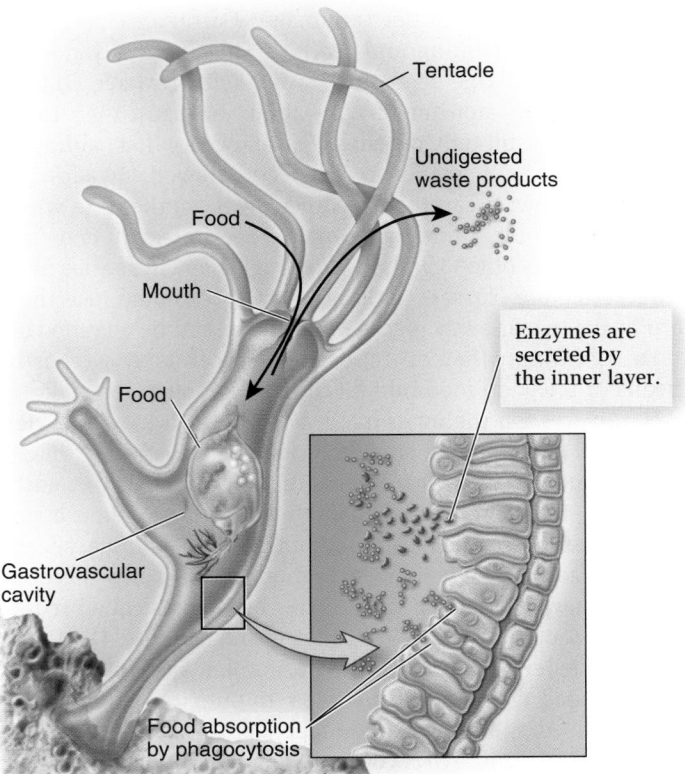

Figure 41.6 Extracellular digestion. In animals with gastrovascular cavities, such as the cnidarian Hydra, digestion occurs extracellularly. Food sources are trapped by tentacles and delivered to the mouth. Digestion occurs within the gastrovascular cavity. Digested food products are phagocytosed directly into the cells that line the cavity, and wastes are excreted out the same opening in which food entered.

single elongated tube with an opening at both ends, through which food passes from one end to the other. Along its length, the tube, or **alimentary canal**, is lined on its interior surface by a layer of epithelial cells. These cells synthesize and secrete digestive enzymes and other factors into the alimentary canal, and secrete certain hormones that help regulate digestive processes. The cells are also involved with transporting digested material out of the canal.

The alimentary canal has several specialized regions along its length. Because of these specializations, digestive processes requiring acidic conditions can be segregated from those requiring higher pH, and undigested food can be stored in one region while digestion continues in another area. The ability to store food in the stomach, for example, allows some animals to reduce the number of times they must feed each day.

Absorption of Food May Be Passive or Active

Once food is digested, the nutrients must be absorbed by the epithelial cells that line specialized portions of the alimentary canal. This occurs in different ways, either by passive or facilitated diffusion or by active transport. Small, hydrophobic molecules such as fats passively diffuse down concentration gradients across the epithelium, while ions and other molecules are transported by facilitated diffusion or active transport. Minerals are ions and therefore do not readily cross plasma membranes. Instead, they are usually actively transported across the membranes of the epithelial cells of the canal by ATP-driven ion pumps. In other cases, small, hydrophilic organic nutrients are transported by secondary active transport, usually with sodium ions.

After nutrients enter the epithelial cells of the alimentary canal, the cells use some of the nutrients for their own needs. Most of the nutrients, however, are transported out of the cells and into nearby blood vessels, where they enter the blood and circulate to the other cells of the body. Thus, nutrients enter the alimentary canal in food, are digested within the canal into monomers that can be transported into epithelial cells, and from there are released into the blood, where they can reach all of the body's cells. In the special case of water, osmotic gradients established by the transport of ions and other nutrients out of the epithelial cells draw water by osmosis from the canal, across the epithelial cells, and from there into the blood.

The mechanisms that activate and control the digestive and absorptive functions of the alimentary canal have been extensively studied in vertebrates and have great importance for human health. In the rest of this chapter, we will explore the structure and function of the digestive systems of vertebrates.

41.4 Overview of Vertebrate Digestive Systems

The vertebrate **digestive system** consists of the alimentary canal—also known as the gastrointestinal, or GI, tract—plus several associated structures, not all of which are found in all vertebrates: the tongue, teeth, salivary glands, liver, gallbladder, and pancreas (**Figure 41.7a**). The differences between the digestive systems of various vertebrates reveal much about their respective feeding strategies. In this section, we will look at some of the most important differences as we discuss the form and function of each part of the vertebrate digestive system. We begin with an overview of the form, or anatomy, of a typical vertebrate digestive system.

Alimentary Canals Are Divided into Functional Regions

The alimentary canal is one continuous tube that changes in appearance and function along its length (**Figure 41.7b**). The anterior end, which functions primarily in the ingestion of food, contains the oral cavity, salivary glands, pharynx (throat), and esophagus. The middle portion, which functions in storage and initial digestion of food, contains one or more food storage or digestive organs, including the crop, gizzard, and stomach depending on species. The upper part of the small intestine, and accessory structures of the gastrointestinal system that connect with the intestine, including the pancreas, liver, and gallbladder,

follow next. The posterior part of the canal functions in final digestion and absorption, and the elimination of nondigestible wastes. It consists of the remainder of the small intestine, the large intestine, the rectum, and the anus.

From the midesophagus to the anus, the gastrointestinal tract has the same general structure, with a hollow cavity called a lumen that is lined by a layer of epithelial cells. Included in the epithelial layer are secretory cells that release a protective coating of mucus into the lumen of the tract, and other cells that release hormones into the blood in response to the presence of food. Connected to the epithelium are secretory glands that release acid, enzymes, water, and ions into the lumen. From the stomach onward, the epithelial cells are linked together along the edges of their luminal surfaces by tight junctions that prevent digestive enzymes and undigested food from moving across and out of the canal. In some places, like the small intestine, the luminal surface is highly convoluted, a feature that increases the surface area available for digestion and absorption.

The epithelial cell layer is surrounded by layers of tissue made up of smooth muscles, nerves, and connective tissue. The nerves are activated by the sight and smell of food, and by the presence of food in the gut. Contraction of the muscles is controlled by these nerves and results in mixing of the contents of the stomach and intestine. This helps speed up digestion and also brings digested foods into contact with the epithelium to facilitate absorption.

Food Processing and Polysaccharide Digestion Begin in the Mouth

As stated earlier, herbivores chew their food extensively while many carnivores do not chew at all. Chewing tough, fibrous plant material breaks up the fibers into more easily digestible pieces. Many herbivores have large, flat molars and well-developed jaw muscles. Incisors and canine teeth are poorly developed or absent in many herbivores. When present, they are usually for defense (as in hippos and some primates) or for nipping grass (as in horses). Chewing occurs by a rotary motion, which grinds up the tough plant cell walls and releases the digestible intracellular contents. Carnivores, by contrast, chew in a vertical motion and usually have jagged molars and sharp incisors and canine teeth, all of which are adaptations for slicing and tearing animal flesh. Omnivores often have a combination of grinding molars and sharp canines and incisors.

Saliva Saliva is a watery fluid containing proteins, mucus, and antibacterial agents. Salivary glands within the mouth produce a constant flow of saliva, which keeps the mouth moist and clean. In response to food, chemical and pressure detectors in the walls of the mouth and tongue increase its secretion. Saliva production can also be increased simply by the smell or sight of food.

Most herbivores salivate constantly because they eat almost continuously while awake. Increased secretion of saliva can be produced in most animals, however, by a large increase in blood flow to the salivary glands, which occurs in response to signals sent from the nervous system. In mammals, the volume of

(a)

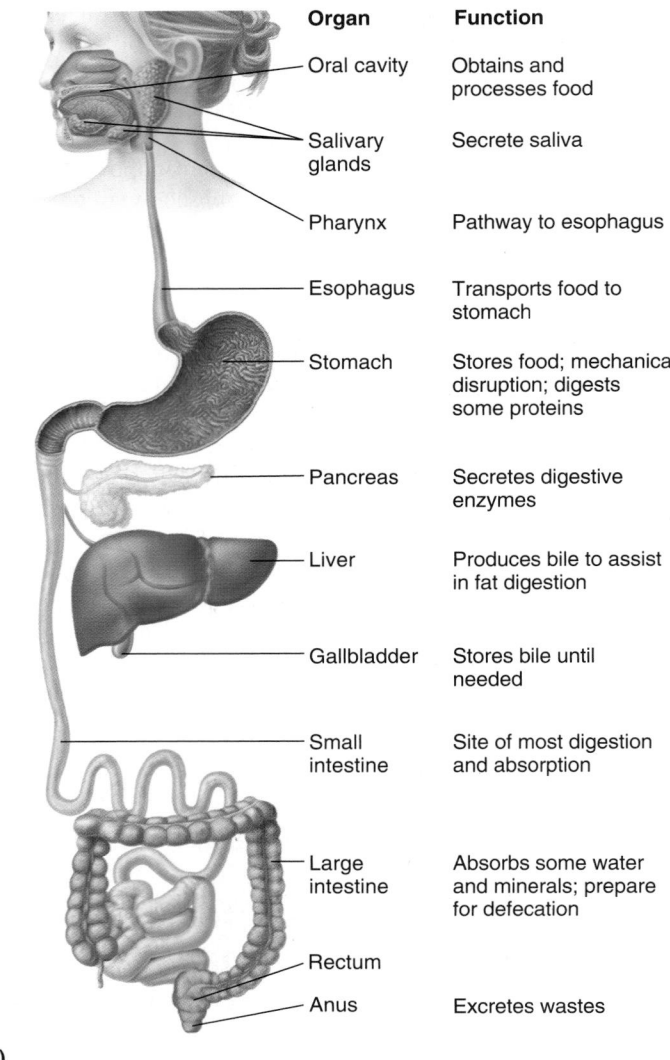

Organ	Function
Oral cavity	Obtains and processes food
Salivary glands	Secrete saliva
Pharynx	Pathway to esophagus
Esophagus	Transports food to stomach
Stomach	Stores food; mechanical disruption; digests some proteins
Pancreas	Secretes digestive enzymes
Liver	Produces bile to assist in fat digestion
Gallbladder	Stores bile until needed
Small intestine	Site of most digestion and absorption
Large intestine	Absorbs some water and minerals; prepare for defecation
Rectum	
Anus	Excretes wastes

(b)

Figure 41.7 Two views of a vertebrate digestive system, as shown in the human. (a) The components of the human digestive system. Not all vertebrates share identical features of the digestive system. (b) An extended view of the digestive system making it easier to visualize the various regions.

saliva secreted per gram of tissue is among the largest secretion of any of the glands in the animal. Saliva has several functions, notably to:

1. moisten and lubricate food to facilitate swallowing;
2. dissolve food particles to facilitate the ability of specialized chemical-sensing structures called taste buds to taste food;
3. kill ingested bacteria with a variety of antibacterial compounds, including antibodies;
4. initiate digestion of polysaccharides through the action of a secreted enzyme called **amylase**.

Digestion is the least important of these functions, accounting for only a few percent of total polysaccharide digestion. Nonetheless, it is possible to detect the action of amylase in your own saliva with a simple test. If you chew on a starchy soda cracker and leave it in your mouth for a while, you may notice it begins to taste sweet. That is because some of the polymers in the cracker have been digested to sweet-tasting disaccharides.

The other functions of saliva, however, are very important. For example, imagine trying to swallow a large piece of unchewed food with a perfectly dry mouth. Also, the antibiotic properties of saliva help cleanse the mouth. In people who have had their salivary glands removed due, for example, to cancer of the glands, the teeth and gums often become diseased to the point of losing teeth.

Pharynx, Esophagus, and Crop The next segments of the alimentary canal, the **pharynx** (throat) and **esophagus**, do not contribute to digestion but serve as a pathway to the stomach. The muscles in the walls of these segments control swallowing. In the pharynx, swallowing begins as a voluntary action but continues in the esophagus by the process of **peristalsis**—rhythmic, spontaneous waves of muscle contraction that begin near the mouth and end at the stomach. When most vertebrates eat, the mouth and stomach are roughly horizontal with respect to each other. In fact, the head of a grazing animal is often lower than its stomach when eating. Peristalsis allows propulsion of food toward the stomach without the aid of (or even against) gravity. The wavelike action of peristalsis ensures that food will be pushed toward the stomach and not merely sit in

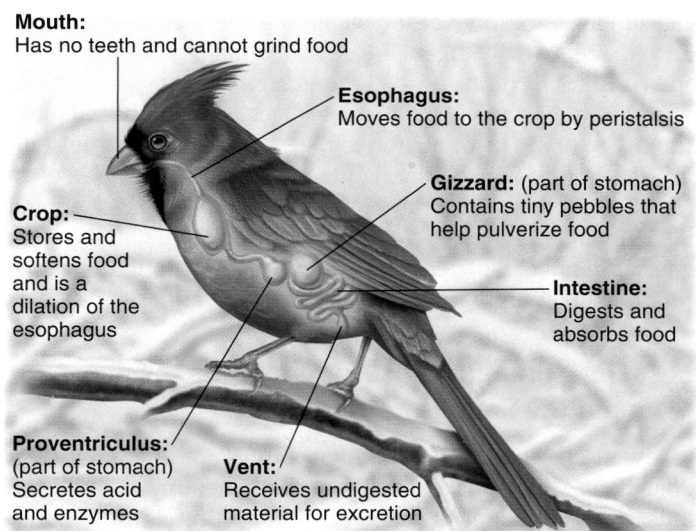

Mouth:
Has no teeth and cannot grind food

Esophagus:
Moves food to the crop by peristalsis

Crop:
Stores and
softens food
and is a
dilation of the
esophagus

Gizzard: (part of stomach)
Contains tiny pebbles that
help pulverize food

Intestine:
Digests and
absorbs food

Proventriculus:
(part of stomach)
Secretes acid
and enzymes

Vent:
Receives undigested
material for excretion

Figure 41.8   **The alimentary canal of birds.**   The avian alimentary canal contains specialized regions for storing and softening food (the crop) and pulverizing food (the gizzard). The gizzard and proventriculus constitute the stomach. Undigested material is excreted through the vent.

the esophagus or even move backward into the mouth if the head is lowered.

In some animals, food moves directly from the esophagus to a storage organ called the **crop**, which is a dilation of the lower esophagus (**Figure 41.8**). Crops are found in most birds and many invertebrates, including insects and some worms. Food is stored and softened by watery secretions in the crop, but little or no digestion occurs there. Because they need to process large amounts of tough food, birds that eat primarily grains and seeds have larger crops than do birds that eat insects and worms. The material that birds regurgitate to their young comes from the crop. In some species, like pigeons and doves, the cells that line the crop wall secrete a lipid-rich watery solution called crop or pigeon milk into the material to be regurgitated.

Despite the small amount of polysaccharide digestion that occurs in the mouth, no absorption of carbohydrates or any other nutrients into the blood occurs up to this point, except for a small amount of water.

Food Processing Continues and Protein Digestion Begins in the Stomach

The **stomach** is a saclike organ that most likely evolved as a means of storing food. In addition to its storage function, however, the stomach partially digests some of the macromolecules in food and regulates the rate at which the contents empty into the small intestine. Glands within the stomach wall secrete hydrochloric acid and an inactive molecule called pepsinogen into the stomach lumen. One function of the acid is to convert pepsinogen into the active enzyme **pepsin**, which begins the digestion of protein. Stomach cells secrete pepsinogen, because

if the cells produced active pepsin they would digest their own cellular proteins.

Within the stomach lumen, hydrochloric acid kills many of the microbes that may have been ingested with food. The acid also dissolves the particulate matter in food. The acid environment in the stomach (or gastric) lumen alters the ionization of polar molecules, especially proteins. This disrupts the structural framework of the tissues in food and makes the proteins in food more accessible to pepsin. The proteins released by the dissolving action of hydrochloric acid are partially digested in the stomach by pepsin. By contrast, no significant digestion of carbohydrates or fat occurs in the stomach.

In birds, the stomach is divided into two parts, the proventriculus and the gizzard (Figure 41.8). The **proventriculus** is the glandular portion of the stomach, and it secretes the acid and pepsinogen. Partially digested and acidified food then moves to the **gizzard**, a muscular structure with a rough inner lining capable of grinding food into smaller fragments. The gizzard contains sand or tiny stones swallowed by the bird, which take the place of teeth and help mash and grind ingested food. It is believed that this function of the gizzard evolved partly as an adaptation for flight, since large jaws and associated chewing muscles and teeth would result in a larger, heavier, and less aerodynamic head.

Eventually, the tiny pebbles in the gizzard become smaller as they are worn away, and they are excreted. Thus, birds need to occasionally restock the gizzard with new grinding stones. Grain-eating birds, particularly chickens and other fowl, many passerines, and pigeons and doves, generally have more muscular gizzards than do insectivorous birds, because of the difficulty in breaking down plant cell walls. In other birds, such as owls, gizzards help compress the nondigestible parts of their meals (bones, teeth, fur, feathers) into a pellet that can be regurgitated. Gizzards, incidentally, are not unique to birds. Certain reptiles that are closely related to birds, such as crocodiles, also contain muscular gizzards.

Digestive actions of the stomach reduce food particles to **chyme**, a solution that contains water, salts, molecular fragments of proteins, nucleic acids, polysaccharides, droplets of fat, and various other small molecules. Virtually none of these molecules, except water, can cross the epithelium of the stomach wall, and thus little or no absorption of organic nutrients occurs in the stomach.

Some Herbivores Use Microbes to Aid Digestion in the Stomach

Herbivores face a special challenge because they must digest cellulose, but they lack the enzyme cellulase required for the job. Instead, herbivores rely on microbes living within their digestive tracts that do have the capacity to digest cellulose. The microbes break down the cellulose into monosaccharides that can be absorbed along with other by-products of microbial digestion such as fatty acids and some vitamins. In this way, bacteria and protists predigest the food for the animal.

In animals with simple stomachs, such as horses, the microbes exist in the large intestine and the region that connects the small intestine to the large intestine is called the cecum. Other herbivores, such as ruminants, have complex stomachs consisting of several chambers, beginning with three outpouchings of the lower esophagus collectively referred to as the forestomach (**Figure 41.9**). The forestomach is composed of the rumen, reticulum, and omasum, in sequence. The rumen and reticulum contain microbes that digest the cellulose, while the

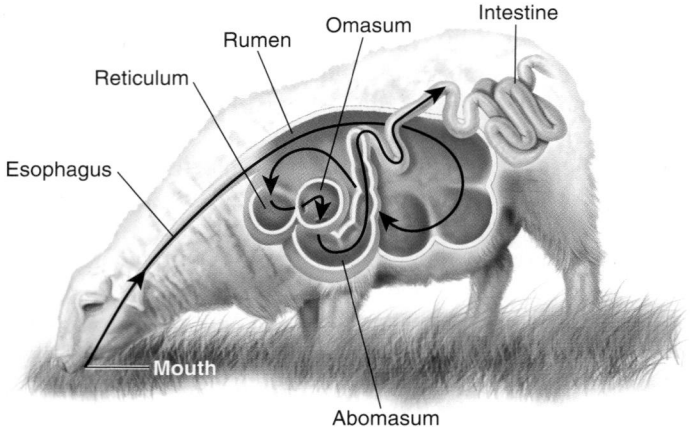

Figure 41.9 **Digestive tract of a ruminant.** Ruminants have a complex arrangement of three modified pouches arising from the esophagus (the rumen, reticulum, and omasum) connected to the true stomach, or abomasum. The rumen and reticulum act as storage and processing sites, and the omasum absorbs some water and salts. Digestion by acid and pepsin takes place in the abomasum, which connects with the small intestine. In large ruminants, the rumen may store up to 95 liters of undigested food.

omasum absorbs some of the water and salts released from the chewed and partially digested food. The tough, partially digested food (the cud) is occasionally regurgitated, rechewed, and swallowed again. Eventually the partially digested food, the microbes, and the by-products of microbial digestion reach the true stomach, the abomasum, which contains the acid and proteolytic enzymes typical of other vertebrate stomachs. From the abomasum, the material passes to the intestines, where digestion and absorption are completed. Some microbes remain in the rumen and quickly multiply to replenish their populations.

Most Digestion and Absorption Occurs in the Small Intestine

Nearly all digestion of food, and absorption of food and water, occur in the first quarter of the **small intestine**, a tube that leads from the stomach to the large intestine. Hydrolytic enzymes break down molecules of organic nutrients into monosaccharides, fatty acids, monoglycerides, nucleotides, and amino acids. Some of these hydrolytic enzymes are on the luminal surface of the intestinal lining cells, while others are secreted by the pancreas and enter the intestinal lumen. The products of digestion are absorbed across the epithelial cells and enter the blood. Vitamins and minerals, which do not require enzymatic digestion, are also absorbed in the small intestine. Water is absorbed by osmosis from the small intestine in response to the movement of nutrients across the intestinal epithelium.

The ability of the small intestine to carry out the bulk of digestion and absorption is aided by infoldings and specializations along its length. Extending from the luminal surface into the lumen of the small intestine are finger-like projections known as **villi** (**Figure 41.10**). The surface of each villus is cov-

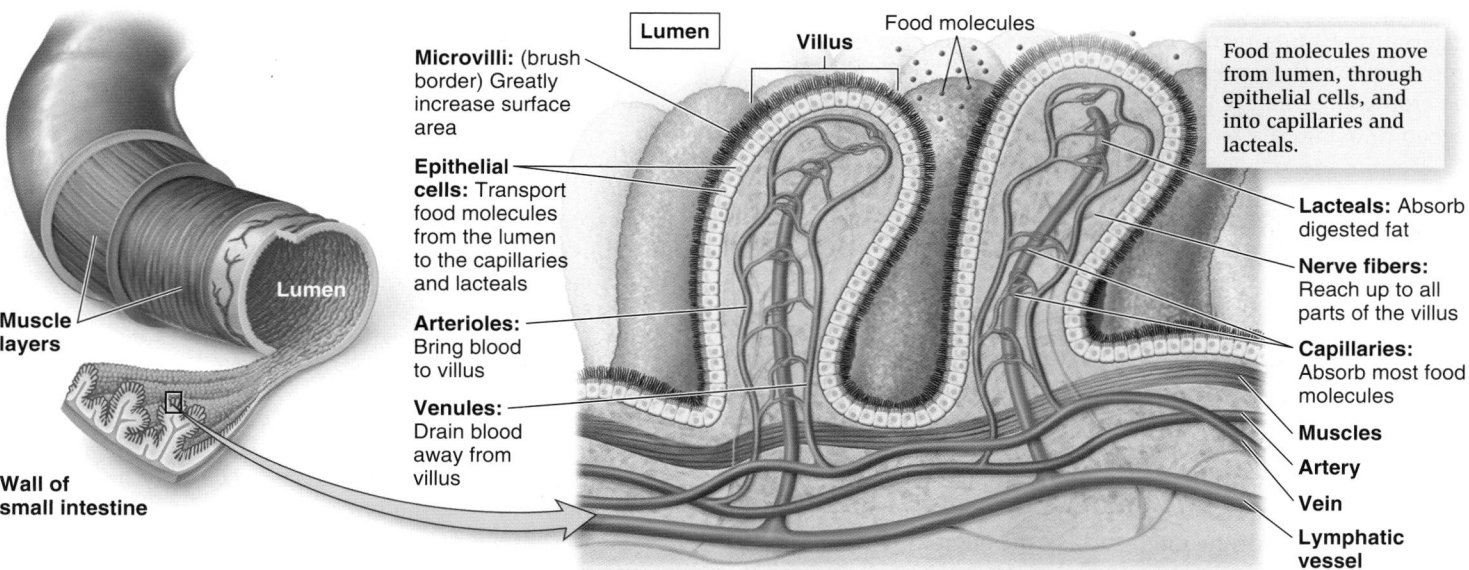

Figure 41.10 **The specialized arrangement of tissues in the small intestine.** The small intestine is folded into numerous villi, which increase the surface area for digestion and absorption. Within each villus are capillaries and lymphatic vessels (lacteals) into which absorbed nutrients are transported. The epithelial cells of the villi have extensions from their surface called microvilli that constitute the brush border of the intestine and greatly add to the total surface area.

ered with a layer of epithelial cells whose surface membranes form small projections called **microvilli**, known collectively as the **brush border**. The combination of folded mucosa, villi, and microvilli increases the small intestine's surface area about 600-fold over that of a flat-surfaced tube having the same length and diameter. This brings the total surface area of the human small intestine to about 300 m^2—roughly the size of a tennis court! This enormous surface area means that the likelihood of an ingested food particle encountering a digestive enzyme and being absorbed across the epithelium is very high, so digestion and absorption proceed much more rapidly.

The center of each intestinal villus is occupied by a special type of vessel called a **lacteal**, and by capillaries, the smallest blood vessels in the body (Figure 41.10). Most of the fat absorbed in the small intestine exists as bulky protein-bound particles that are too large to enter capillaries. Consequently, absorbed fat enters the larger, wider lacteals, which are part of the lymphatic system. Material absorbed by the lacteals eventually empties into the circulatory system. Other nutrients are absorbed directly into the capillaries and from there into veins.

The length of the small intestine varies among species. Herbivores generally have much longer small intestines than do carnivores, which allows added time for plant material to be digested and absorbed (**Figure 41.11**). Even within a single ani-

mal, the length of the small intestine can change. In a bird that switches from eating insects and worms in summer to buds and other nutrient-poor vegetation in winter, the small intestine grows and elongates considerably to meet the digestive challenges associated with an herbivorous diet.

The Pancreas and Liver Secrete Substances That Aid Digestion

Two major organs—the pancreas and liver—secrete substances that flow via ducts into the small intestine. The **pancreas**, an elongated gland located behind the stomach, has several functions, but in this chapter we will focus on those that are directly involved in digestion (**Figure 41.12**). The gland secretes digestive enzymes and a fluid rich in bicarbonate ions (HCO_3^-). The bicarbonate neutralizes the acidic chyme coming from the stomach, which would otherwise inactivate the pancreatic enzymes in the small intestine.

The **liver** is the site of bile production. **Bile** contains bicarbonate ions, cholesterol, phospholipids, a number of organic wastes, and a group of substances collectively termed **bile salts**. The bicarbonate ions, like those from the pancreas, help neutralize acid from the stomach, while the bile salts solubilize dietary fat and increase its accessibility to digestive enzymes.

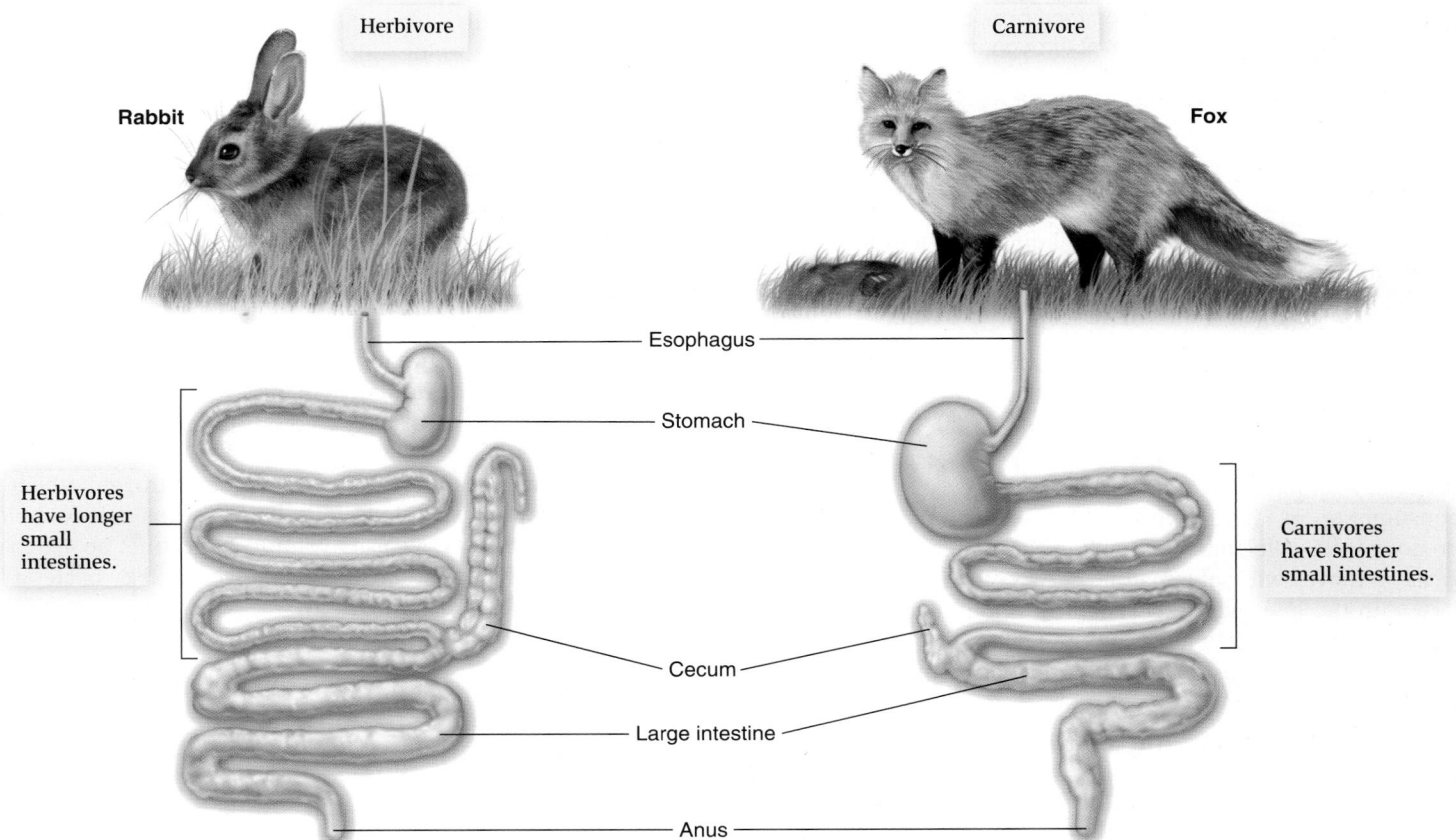

Figure 41.11   **The alimentary canals of a nonruminant herbivore and a small carnivore.**   Note that the small intestine and cecum are considerably longer in the herbivore, adaptations that facilitate the digestion of plant matter.

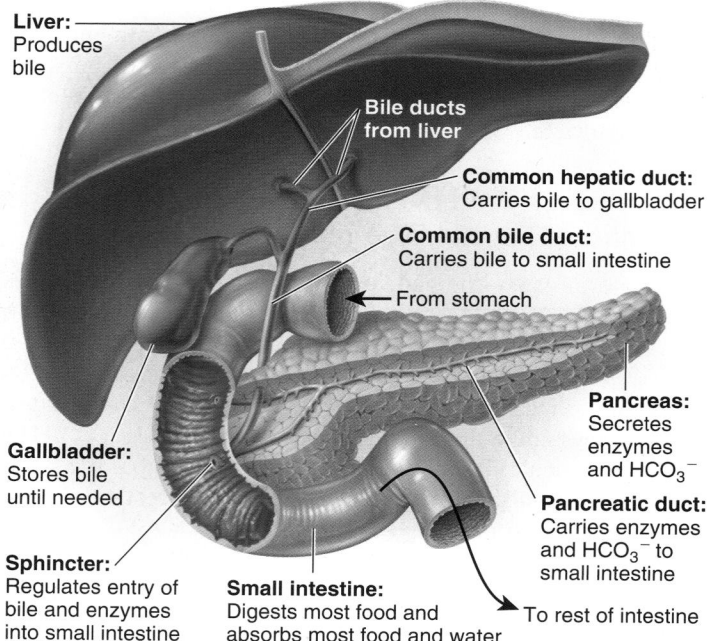

Liver:
Produces bile

Bile ducts from liver

Common hepatic duct:
Carries bile to gallbladder

Common bile duct:
Carries bile to small intestine

From stomach

Gallbladder:
Stores bile until needed

Pancreas:
Secretes enzymes and HCO_3^-

Pancreatic duct:
Carries enzymes and HCO_3^- to small intestine

Sphincter:
Regulates entry of bile and enzymes into small intestine

Small intestine:
Digests most food and absorbs most food and water

To rest of intestine

Figure 41.12 **The arrangement of the vertebrate liver, gallbladder, pancreas, and small intestine.** Bile drains from the liver into the gallbladder through the common hepatic duct. During a meal, bile is secreted from the gallbladder and enters the small intestine through the common bile duct. Simultaneously, secretions from the pancreas travel through another duct that joins with that from the gallbladder and empties into the small intestine. A muscular sphincter controls the entrance to the small intestine.

The liver secretes bile into small ducts that join to form the common hepatic duct (from the Greek *hepar*, liver). Between meals, secreted bile is stored in the **gallbladder**, a small sac underneath the liver. During a meal, the smooth muscles in the gallbladder wall contract, injecting the bile solution into a connecting duct called the common bile duct (Figure 41.12). The opening of a sphincter allows entry of the bile into the small intestine. The gallbladder, therefore, is a storage site that allows the release of large amounts of bile to be precisely timed to the consumption of fats. However, many animals that secrete bile do not have a gallbladder. In humans, the gallbladder can be surgically removed without impairing bile secretion by the liver or its flow into the intestinal tract. People without a gallbladder can still digest fat, but they must limit the amount of fat they eat at one time since bile secretion can no longer be timed to a meal.

The Large Intestine Concentrates Undigested Material

The digested nutrients, along with minerals, vitamins, and water, are absorbed across the plasma membranes of the brush border cells. Peristalsis in the walls of the small intestine slowly propels the remaining luminal contents toward the large intestine.

In humans and other similarly sized animals, the large intestine is a tube about 6 cm in diameter and a little over a meter long. Its first portion, the **cecum**, forms a small pouch from which extends the **appendix**, a finger-like projection having no known essential function but that may at one time have been an important part of the body's defense mechanisms. However, the cecum of many herbivores is much more developed and serves as a storage organ that permits resident microbes sufficient time to further digest cellulose (see Figure 41.11). The next part of the large intestine, the **colon**, consists of three relatively straight segments—the ascending, transverse, and descending portions. The terminal portion of the descending colon is S-shaped, forming the sigmoid colon, which empties into the rectum, a short segment of the large intestine that ends at the anus.

Chyme enters the cecum through a sphincter. This sphincter is normally closed, but it periodically relaxes after a meal, allowing chyme to enter. Although the large intestine has a greater diameter than the small intestine, its epithelial surface area is far less—particularly in carnivores and omnivores—because it lacks the convolutions and villi of the small intestine.

The primary functions of the large intestine are to store and concentrate fecal material before defecation, and to absorb some of the remaining salts and water that were not absorbed in the small intestine. Because most substances are absorbed in the small intestine, only a small volume of water, salts, and undigested material is passed on to the large intestine. The large intestine temporarily stores the undigested material and concentrates it by absorbing salts and water. **Defecation** occurs when contractions of the rectum and relaxation of associated sphincter muscles expel the feces through the final portion of the canal, the anus (see Figure 41.7).

The large intestine also absorbs some of the products formed by bacteria inhabiting this region, including small amounts of vitamins. Although this source of vitamins generally provides only a small part of the normal daily requirement, it may make a significant contribution when dietary vitamin intake is low. Sometimes people develop a vitamin deficiency if treated with antibiotics that inhibit these species of bacteria. Other bacterial products include gas (**flatus**), which is a mixture of nitrogen and carbon dioxide, with small amounts of hydrogen, methane, and hydrogen sulfide. Bacterial processing of undigested polysaccharides produces these gases, except for nitrogen, which is derived from swallowed air. Certain foods (beans, for example) contain large amounts of carbohydrates that cannot be digested by intestinal enzymes but are readily metabolized by bacteria in the large intestine, producing flatus.

41.5 Mechanisms of Digestion and Absorption in Vertebrates

The preceding sections provided an overview of nutrition and the basic features of digestive systems. We turn now to a more detailed description of how each of the three major foodstuffs are processed in the vertebrate digestive system, and how the

Figure 41.13 **Digestion of disaccharides and absorption of monosaccharides.**   Enzymes on the luminal surface of the small intestine epithelial cells digest disaccharides such as maltose into monosaccharides, which are absorbed by secondary active transport with sodium ions or, in the case of fructose, by facilitated diffusion. Digestion and absorption occur in the same cells but are shown separately here for clarity. The absorbed monosaccharides enter the body by facilitated diffusion from the epithelial cells. For simplicity, the microvilli (brush border) are not shown.

Biological inquiry: The absorption of many nutrients requires secondary active transport. What can we conclude from that about the energetic cost of absorption?

end products of digestion are absorbed across intestinal cells. In this section, we will focus on how digestion and absorption occur at the cellular and molecular levels.

Carbohydrates Are Digested and Absorbed in the Small Intestine

In a typical omnivore like ourselves, most of the ingested carbohydrates are the polysaccharides starch and cellulose from plants, and glycogen from animals. The remainder consists of simple carbohydrates such as the monosaccharides fructose and glucose in fruit, and disaccharides such as lactose in milk. Humans, as is well known, also add the disaccharide sucrose (table sugar) to their diets, although sucrose also occurs naturally, in maple sap and sugarcane, for example, and is consumed by other animals.

Starch digestion by salivary amylase begins in the mouth, but the acid in the stomach destroys the amylase and prevents further starch digestion there. Starch digestion resumes in the small intestine by amylase secreted into the intestine by the pancreas. The products of starch digestion via amylase are molecules of maltose (**Figure 41.13**). Maltose, along with ingested sucrose and lactose, are broken down into monosaccharides—fructose, glucose, and galactose—by enzymes located on the brush border of the small intestine epithelial cells. The monosaccharides are then transported across the intestinal epithelium into the blood. Fructose enters the epithelial cells by facilitated diffusion, while glucose and galactose undergo secondary active transport coupled to sodium. Monosaccharides then leave the epithelial cells and enter the blood by way of facilitated diffusion transporters located in the basolateral membranes of

the epithelial cells. The transport of substances from the lumen to the blood is called **transepithelial transport** because it occurs across a layer of epithelial cells.

Do you feel ill after drinking milk or eating dairy products? If so, you are among the majority of people who cannot digest lactose, the chief disaccharide in milk. A small percentage of humans, however, retain the ability to digest lactose throughout life. Next, we'll examine these phenomena.

GENOMES & PROTEOMES

Genetics Explains Lactose Intolerance

With rare exceptions, the milk of all mammals contains lactose. A human with lactose intolerance cannot digest lactose because the intestinal enzyme lactase, which cleaves lactose into glucose and galactose, is either inactive, absent, or present only in very small amounts. Normally, symptoms of the condition do not appear until a child is several years old.

Milk is the sole food of most mammals shortly after birth, and the primary food for various lengths of time thereafter until weaning, the transition from consuming mother's milk to eating a diet of solid foods. Once weaned, mammals never again drink milk, except, of course, for humans. Since the only dietary source for lactose is milk, it is not surprising that older mammals lose the ability to digest this disaccharide. The developmental mechanisms that turn off lactase production and activity are not firmly established, but they are known to involve decreased transcription of the lactase gene.

If an adult mammal were to drink milk, the undigested lactose would remain in the gut. As a result, water that normally would be absorbed by osmosis with the digested monosaccharides formed from lactose would also remain in the gut. Farther down the alimentary canal, however, microbes in the large intestine digest some of the lactose for their own use, and in the process release by-products such as hydrogen and other gases. The combination of water retention and bacterial action results in gastrointestinal symptoms such as diarrhea, gas, and cramps. Adult housepets and farm animals suffer the same symptoms and will usually refuse milk if it is offered. The popular notion of cats lapping up milk from a bowl is a misconception. Some visits to the veterinarian are in fact related to gastrointestinal symptoms caused by well-meaning owners who fed milk to their pets.

If the ability to digest lactose is normally lost upon weaning, why are some people able to consume dairy products without getting ill? In fact, it is estimated that at least 90% of the world's human population cannot fully digest lactose after early childhood. In other words, lactose intolerance is not a disorder, it is a normal condition for humans just as it is for all other mammals.

The ability of human adults to digest lactose is clearly linked to genetic background (**Table 41.2**). Lactose intolerance is an example of a human polymorphism, a genetic trait that varies among people. Other than individuals who trace their ancestry to northern Europe and a few isolated regions in western Africa, nearly every population of humans shows a considerable degree of lactose intolerance, reaching essentially 100%

in most of south and east Asia. One hypothesis for this phenomenon appears to be a behavioral and cultural change that occurred in Neolithic times, when certain populations domesticated cattle and added cow's milk to their diet. Adults who were able to digest lactose—presumably due to some mutation affecting the expression of the lactase gene—enjoyed a selective advantage and tended to thrive and pass on their genes more frequently than those whose digestive systems could not handle milk. In other words, natural selection resulted in certain populations carrying mutations that caused the lactase gene to be expressed after weaning. Eventually, nearly everyone in those regions of the world retained sufficient intestinal lactase to use milk as a staple food.

To understand how this trait was passed on, let's examine the gene that codes for lactase. Surprisingly, the coding regions of the lactase gene and the lactase core promoter are identical whether individuals are lactose intolerant or lactose tolerant. Recently, however, Finnish investigators have uncovered two single nucleotide changes located in presumed regulatory sites that control the expression of the lactase gene. These changes are associated with prolonged lactase expression, allowing it to occur after weaning. People carrying these mutations are lactose tolerant and can consume milk products through adulthood. By comparison, adults who lack these mutations—and are therefore lactose intolerant—can consume dairy products only in small amounts or not at all. However, their ability to do so is greatly improved if the product has been commercially treated with purified lactase to predigest the lactose.

Table 41.2	Incidence of Lactose Intolerance in the Human Population	
Population	**Percentage of people with lactose intolerance***
South and East Asia	>99%
Sub-Saharan Africa (average)	75–95%
Fulani tribe (West Africa)	15–20% (traditional cattle herders)
Yoruba tribe (S.W. Nigeria)	99% (no history of cattle herding)
Greenland Inuit	80%
Southern India	70–90%
North-central India	15–50%
Ashkenazi Jews	70–80%
Israeli Jews	55–60%
Southern Europe (for example, southern Italy)	70–75%
Northern Europe (for example, Denmark, U.K.)	1–15%
U.S. (average)	25%
African American	70–77%
Mexican American	50–55%

* Percentages are estimates obtained from a variety of sources, including U.S. Department of Health and Human Services.

Proteins Are Digested in the Stomach and Small Intestine, and Absorbed in the Small Intestine

Proteins are broken down to peptide fragments in the stomach by pepsin, and in the small intestine by the proteases **trypsin** and **chymotrypsin**. The pancreas secretes the latter two enzymes as zymogens, which are inactive precursors of enzymes. This prevents the active enzymes from digesting the pancreas itself. Once the zymogen form of trypsin enters the small intestine, it is enzymatically cleaved into the active molecule by another enzyme whose active site is located on the luminal membranes of the intestinal cells. Trypsin then activates the zymogen form of chymotrypsin.

The peptide fragments produced by the cleavage of ingested proteins by trypsin and chymotrypsin are further digested into individual amino acids by the actions of specific proteases located on the luminal membranes of the small intestine epithelial cells. These proteases, called aminopeptidases and carboxypeptidases, cleave off one amino acid at a time from the N-terminus and C-terminus of peptides, respectively. Individual amino acids enter the epithelial cells by secondary active transport coupled to sodium. Amino acids leave the intestinal cell and enter the blood through a facilitated diffusion carrier in the plasma membranes on the opposite side of the intestinal lumen.

As with carbohydrates, protein digestion and absorption are largely completed in the upper portion of the small intestine.

Fat Digestion and Absorption Occur in the Small Intestine

Most ingested fat is in the form of triglycerides. Fat digestion occurs entirely in the small intestine. The major digestive enzyme in this process is pancreatic **lipase**, which catalyzes the splitting of bonds linking fatty acids to the first and third carbon atoms of glycerol, producing two free fatty acids and a monoglyceride as products.

Lipase
Triglyceride $\rightarrow$ 2 fatty acids + Monoglyceride

Fats are poorly soluble in water and aggregate into large lipid droplets, much like a mixture of oil and vinegar after shaking. Because pancreatic lipase is a water-soluble enzyme, its digestive action in the small intestine can take place only at the surface layer of a lipid droplet. Therefore, if most of the ingested fat remained in large droplets, lipid digestion would be very slow. The rate of digestion is substantially increased by the process of **emulsification**, which disrupts the large lipid droplets into many tiny droplets, each about 1 mm in diameter, thereby increasing their total surface area and exposure to lipase action. The resulting suspension of small lipid droplets is called an emulsion.

Emulsification of fat requires mechanical disruption of the large fat droplets into smaller droplets (**Figure 41.14a**) and an emulsifying agent, which prevents them from recombining back into larger droplets (**Figure 41.14b**). The muscular contractions of the stomach and small intestine provide mechanical disruption to mix the luminal contents. Phospholipids and bile salts, both secreted in the bile, serve as emulsifying agents because they are amphipathic. The nonpolar portions of the phospholipids and bile salts associate with the nonpolar interior of the lipid droplets, and the polar portions remain exposed at the water surface outside the droplet. When two lipid droplets come together, the polar regions of the phospholipids and bile salts coating the droplets repel each other. This prevents the reaggregation of smaller lipid droplets into larger ones.

Although emulsification speeds up digestion, absorption of the poorly soluble products of the lipase reaction is facilitated by a second action of the bile salts, the formation of **micelles**, which are similar in structure to emulsion droplets but much smaller at 4–7 nm in diameter (**Figure 41.14c**). Micelles consist of bile salts, phospholipids, fatty acids, and monoglycerides clustered together, with the polar ends of each molecule oriented toward the micelle's surface and the nonpolar portions forming the micelle's core. Also in the core of the micelle are small amounts of fat-soluble vitamins and cholesterol. Micelles continuously break down and re-form near the epithelium of the intestine. When a micelle breaks down, individual molecules are released into the solution and diffuse across the intestinal epithelium. As the lipids diffuse into epithelial cells, micelles release more lipids into the aqueous phase. Thus, the micelles keep most of the insoluble fat digestion products in small soluble aggregates, while gradually releasing very small quantities of lipids to diffuse into the intestinal epithelium. Note that it is not the micelle that is absorbed but rather the individual lipid molecules that are released from the micelle.

Although fatty acids and monoglycerides enter epithelial cells from the intestinal lumen, it is triglycerides that are released on the other side of the cell into the interstitial fluid. In other words, during their passage through the epithelial cells, fatty acids and monoglycerides are resynthesized into triglycerides. This occurs in the smooth endoplasmic reticulum, where the enzymes for triglyceride synthesis are located. This process lowers the concentration of cytosolic free fatty acids and monoglycerides in the epithelial cells, and thus maintains a diffusion gradient for these molecules from the lumen into the cell. The resynthesized triglycerides aggregate into **chylomicrons**, large droplets coated with amphipathic proteins that perform an emulsifying function similar to that of bile salts (**Figure 41.14d**). In addition to triglycerides, chylomicrons contain phospholipids, cholesterol, and fat-soluble vitamins that have been absorbed by the same process that led to fatty acid and monoglyceride movement into the epithelial cells of the small intestine.

Chylomicrons are released by exocytosis from the epithelial cells and pass into the lacteals within the intestinal villi. Chylomicrons are too large to diffuse into blood capillaries. However, lacteals have large, slit pores in their walls through which the chylomicrons can pass. The fluid from the lacteals eventually empties into a large vein and from there into the general blood circulation.

Vitamins, Minerals, and Water Are Not Digested but Must Be Absorbed

As stated earlier, vitamins, minerals, and water do not require digestion, and they are absorbed in their complete form. Most water-soluble vitamins are absorbed by diffusion or active transport. The fat-soluble vitamins—A, D, E, and K—follow the pathway for fat absorption described in the previous section. Any interference with the secretion of bile or the action of bile salts in the intestine decreases the absorption of fat-soluble vitamins, a condition known as **malabsorption**. Malabsorption syndromes can lead to vitamin and mineral deficiencies. For example, a common genetic disorder in humans called **celiac sprue** results in a loss of intestinal surface area due to a sensitivity to the wheat protein gluten. In addition to symptoms such as diarrhea, the decreased absorptive area can reduce absorption of fat-soluble vitamins like vitamin D. Because vitamin D promotes calcium absorption from the small intestine, celiac sprue may lead to decreased calcium absorption and imbalances in blood calcium levels.

Water is the most abundant substance in chyme. Small amounts of ingested water are absorbed in the stomach, but the stomach has a much smaller surface area available for diffusion and lacks the solute-absorbing mechanisms that create the osmotic gradients necessary for water absorption. The epithelial membranes of the small intestine are very permeable to water,

The grinding action of the stomach breaks fat into small droplets.

Stomach

Chyme enters the small intestine.

(a)

Fat droplets

Bile enters the intestine from the gallbladder.

Lipases from the pancreas begin fat digestion.

Emulsification occurs as droplets are coated with bile salts and phospholipids that are present in bile.

Small intestine

(b)

Emulsified fat droplet

Cholesterol

Small intestine epithelial cell

Bile salt

Fat-soluble vitamins

Triglyceride

Phospholipid

Lipase (digestion)

Triglycerides are broken down by lipase into fatty acids and monoglycerides. These small lipids along with cholesterol and vitamins form micelles.

Small lipids

Diffusion (absorption)

Bile salt polar surface

The small lipids gradually leave the micelles and diffuse into the epithelial cells of the intestine.

Micelle (enlarged)

(c)

Figure 41.14   Digestion and absorption of fat.

(a) The churning of the stomach helps break fat into many small droplets, which then enter the intestine. **(b)** The fat droplets are emulsified with a coating of amphipathic bile salts and phospholipids, which attach their nonpolar surface to the fat droplet and extend their polar surface to the aqueous environment in the intestinal lumen. This prevents the droplets from joining into larger droplets, which would decrease the surface area available for digestion. Lipases are released from the pancreas into the small intestine and begin the process of digesting fats into smaller molecules. **(c)** Monoglycerides and fatty acids formed by the action of lipases, along with small, fat-soluble compounds such as cholesterol and certain vitamins, form micelles. These small lipids gradually leave the micelle and diffuse into the epithelial cells of the intestine. **(d)** Inside the epithelial cells, triglycerides are re-formed in the smooth endoplasmic reticulum, packaged with other fats, and coated with protein to form chylomicrons, which exit the cell and enter the lacteals of the intestinal villus and are eventually returned to the blood.

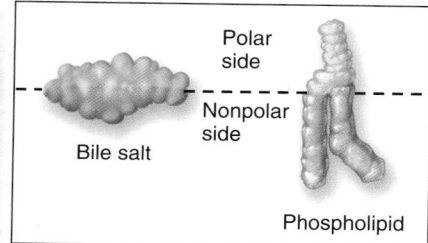

Bile is formed in the liver. It contains bile salts and phospholipids that have polar and nonpolar surfaces (they are amphipathic).

Polar side

Nonpolar side

Bile salt

Phospholipid

In epithelial cells, triglycerides are reformed and along with other fats are enclosed by a membrane from the ER. They are coated with proteins to form chylomicrons and enter lacteals to be transported to the blood.

Lacteal

Fatty acids and monoglycerides

Triglycerides and other fats

Diffusion

Proteins

Chylomicron

Free molecules of fatty acids and monoglycerides

Epithelial cell

To blood

(d)

and water diffusion occurs across the epithelium whenever an osmotic gradient is established by the active absorption of solutes, particularly Na^+, Cl^-, and HCO_3^-. Other minerals present in smaller concentrations, such as potassium, magnesium, and calcium, are also absorbed, as are trace elements such as zinc and iodide. The mechanisms of absorption of these molecules generally involve transport proteins and/or ion pumps.

41.6 Regulation of Digestion

The digestive systems of animals are under complex control regulated in part by other organ systems, especially the nervous and endocrine systems. Secretions from nerve endings and endocrine glands control the volume of saliva produced, the amount of acid produced in the stomach, the timing and amount of secretions from the gallbladder and pancreas, and the rate and strength of muscle contractions along the walls of the alimentary canal.

In this section, we examine the major mechanisms by which the nervous and endocrine systems control the activity of the digestive system.

The Nervous System Controls Muscle and Secretory Activity

The nervous system can affect the activities of the digestive system in two major ways: (1) control of muscle and glandular activity by the local nerves within the alimentary canal, and (2) long-distance regulation by the brain.

Within the walls along the length of the gastrointestinal tract is a highly branched, interconnected collection of nerve cells. These nerve cells interact with nearby smooth muscles, glands, and epithelial cells located within the tract. Stimulation of nerve activity at one point along the alimentary canal can lead to impulses that are transmitted up and down the canal.

When food enters the small intestine, for example, the intestine is stretched. This directly activates the nerve cells in the intestinal wall. Impulses are sent from these nerve cells to the muscles of the stomach, where they decrease the contractions of the stomach. This slows the rate at which chyme moves from the stomach into the small intestine, giving the intestine sufficient time for digestion and absorption. In this way, the alimentary canal can regulate its own function independent of the brain. However, the brain can communicate with nerve cells in the walls of the stomach and intestines, and thereby influence the movement and secretory activity of the gastrointestinal tract. For example, emotional stress, a brain-related event, can affect digestive processes. Likewise, the sight, smell, and taste of food activate digestive functions even before food reaches the stomach. These stimuli act via nerves from the brain to "get the ball rolling" so digestion can begin as soon as food is swallowed.

Hormones Regulate the Rate of Digestion

Hormones are chemical messengers secreted by specialized cells into the blood, where they travel to all parts of the body and act on various target cells (see Chapter 50). Hormones that control the digestive system are secreted mainly by cells scattered throughout the epithelium of the stomach and small intestine. One surface of each hormone-producing cell is exposed to the lumen of the gastrointestinal tract. At this surface, chemical substances in chyme stimulate the cell to release its hormones into the blood, by which they travel to their target cells. These include the cells of the pancreas, which responds to the presence of the hormones by secreting digestive enzymes into the small intestine, and the gallbladder, which responds by secreting bile. One such hormone released by the small intestine is secretin, a polypeptide whose discovery not only initiated the field of endocrinology—the study of hormones—but provided the first clear understanding of how the gut and its accessory structures communicate, as described next.

FEATURE INVESTIGATION

Starling and Bayliss Discovered a Mechanism by Which the Small Intestine Communicates with the Pancreas

The dependence of digestive function on the nervous system and the brain was established by the pioneering work of Ivan Pavlov. Pavlov's work suggested that nerves conveyed signals between the gut and other structures, such as the brain and pancreas, and in this way the gut could communicate with these structures. Such communication is essential for the synchronized release of pancreatic enzymes and bicarbonate ions

with the arrival of food in the intestine. Two English scientists, Ernest Starling and his brother-in-law William Bayliss, hypothesized that nerves were not the only structures controlling gut activity.

To test this hypothesis, Starling and Bayliss carefully dissected away the nerves in the small intestine of an anesthetized dog. Next, they directly injected a small quantity of acid into the intestinal lumen, because it was known at the time that the acidic contents arriving from the stomach somehow triggered secretion of digestive enzymes and bicarbonate ions from the pancreas, as shown in the Feature Investigation in **Figure 41.15**.

Figure 41.15   Bayliss and Starling discovered the mechanism by which the small intestine and pancreas work together in digestion.

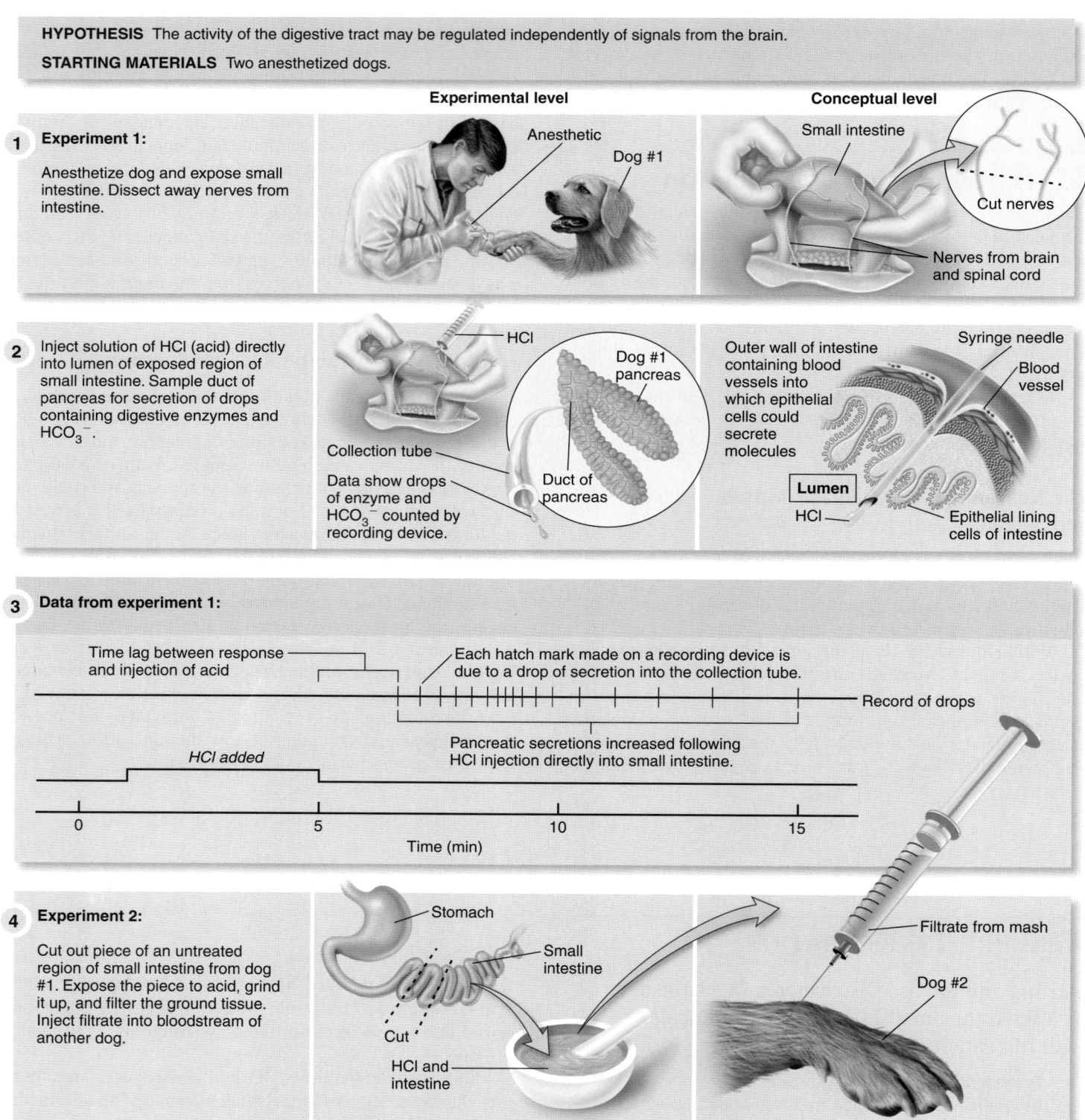

HYPOTHESIS The activity of the digestive tract may be regulated independently of signals from the brain.

STARTING MATERIALS Two anesthetized dogs.

Experimental level   **Conceptual level**

1 **Experiment 1:**

Anesthetize dog and expose small intestine. Dissect away nerves from intestine.

Anesthetic
Dog #1

Small intestine
Cut nerves
Nerves from brain and spinal cord

2 Inject solution of HCl (acid) directly into lumen of exposed region of small intestine. Sample duct of pancreas for secretion of drops containing digestive enzymes and HCO_3^-.

HCl
Dog #1 pancreas
Collection tube
Data show drops of enzyme and HCO_3^- counted by recording device.
Duct of pancreas

Outer wall of intestine containing blood vessels into which epithelial cells could secrete molecules
Syringe needle
Blood vessel
Lumen
HCl
Epithelial lining cells of intestine

3 **Data from experiment 1:**

Time lag between response and injection of acid

Each hatch mark made on a recording device is due to a drop of secretion into the collection tube.

Record of drops

Pancreatic secretions increased following HCl injection directly into small intestine.

HCl added

0   5   10   15
Time (min)

4 **Experiment 2:**

Cut out piece of an untreated region of small intestine from dog #1. Expose the piece to acid, grind it up, and filter the ground tissue. Inject filtrate into bloodstream of another dog.

Stomach
Small intestine
Cut
HCl and intestine
Filtrate from mash
Dog #2

5 Test for pancreatic secretions into small intestine of second dog.

Pancreas of dog #2

Pancreatic secretions present?

If the filtrate from dog #1 contained chemical signals released upon exposure to acid, those signals should stimulate the pancreas of dog #2 to secrete enzymes and HCO_3^-.

6 Data from experiment 2:

Prior to the injection of the filtrate into dog #2, the pancreas did not produce any secretion.

After the injection, pancreatic secretion increased rapidly. Each hatch mark corresponds to a drop of fluid collected from the duct of the pancreas.

Filtrate from dog #1 injected

Time (min)

Conclusions: The small intestine contains molecules whose secretion into the blood is stimulated by acid, and that travel to the pancreas through the blood, causing secretion of pancreatic enzymes and HCO_3^-.

They placed a tube into the duct that carries secretions from the pancreas to the small intestine, so they could collect the pancreatic secretions. Each drop of secretion was recorded as a hatch mark on a chart recorder. They discovered that even in the absence of intact nerves, simply adding acid into the intestinal lumen resulted in increased pancreatic secretions after a short lag period. Starling and Bayliss postulated that the lag was due to the time required for the intestinal cells to produce and secrete a factor that entered the blood and traveled to the pancreas.

To confirm their results, and to counter the suggestion made by Pavlov and others that their first experiment merely failed to sever all the intestinal nerves, Starling and Bayliss performed a follow-up experiment. First, they removed a section of a dog's small intestine, added acid to it, and then ground it up in the acid. The mashed tissue was then filtered to produce an extract of any secretions that may have been produced by the intestine in response to the acid. Next, they injected the filtered extract into the bloodstream of another anesthetized dog whose pancreatic duct was opened as in the first experiment. What they discovered was remarkable: The extract derived from the acid-stimulated intestine of one dog caused the almost immediate secretion of pancreatic enzymes and bicarbonate from the second dog. This demonstrated conclusively that the small intestine secreted some chemical or chemicals in response to the presence of acid, and that these secretions could reach the pancreas—presumably through the bloodstream—and stimulate the gland to release its contents. Starling and Bayliss called these chemicals hormones, from a Greek verb meaning "to excite or arouse to action," and they named the hormone that activated the pancreas secretin. Thus was born not only the concept of a hormone but the recognition that these chemical messengers play a vital role in regulating the digestive system.

In the rest of this chapter, we will examine the impact of digestive system disorders on human health.

41.7 Impact on Public Health

As we have seen, the functioning of the vertebrate digestive system is extraordinarily fine-tuned by the brain, nerves, and hormones. When people are well nourished, therefore, you might assume that digestive problems would be rare. However, each year in the U.S. alone, gastrointestinal complaints account for approximately 40 million visits to doctors and 10 million visits to hospitals and emergency rooms. Among the most common GI problems registered by hospitals and physicians are heartburn, ulcers, and diarrhea.

Excess Stomach Acid Production Can Lead to Heartburn

Approximately one in four people in the U.S. suffers at some time from heartburn. The word heartburn is a misnomer, because the painful burning sensation is caused by stomach acid and arises within the esophagus, not the heart. Normally, the stomach contents do not move backward into the esophagus, largely because of the muscular sphincter at the esophageal/stomach juncture. However, under some conditions, the sphincter either does not close entirely or is forced open by the pressure of material in the stomach. When this happens, the acid in the stomach lumen enters the esophagus and irritates nerve endings there.

Many circumstances may contribute to heartburn. Overeating, for example, enlarges the volume of the stomach to the point of forcing its contents through the esophageal sphincter. Lying down after a big meal removes the effect of gravity on food in the stomach and may allow some acid to leak backward. Heartburn is also associated with smoking and consumption of alcohol, citrus fruits (which are acidic), chocolate (which contains caffeine, a known inducer of heartburn), and fatty foods (which take longer to digest than other foods and therefore remain in the stomach longer). One of the most common causes is pregnancy. Towards the middle of pregnancy, the growing fetus pushes up on the abdominal contents, which tends to force material from the stomach into the esophagus.

Common antacids contain carbonate ions, which buffer the acid in the stomach and esophagus. In severe cases, heartburn can damage the walls of the esophagus enough to cause a chronic cough and pain, or even perforate the esophagus. Antacids may not be sufficient to treat these patients, who are instead given drugs that inhibit the stomach's ability to produce acid. These drugs are among the most widely prescribed medications in the United States.

Erosion of the Walls of the Alimentary Canal Causes an Ulcer

Although heartburn is usually treated before the acid begins eroding away the lining of the esophagus or stomach, occasionally this does occur. Erosion of any portion of the alimentary canal for any reason is called an ulcer. Most ulcers occur in the stomach, lower esophagus, and the part of the small intestine that connects to the stomach, because these sites have the greatest acid concentrations. Ulcers are typically less than an inch wide, and they may allow luminal contents to leak into the surrounding body cavity where the enzymes and acids from the stomach can do considerable damage. As many as 20 million Americans (6–7% of all adults in the U.S.) have an ulcer. Each year, 40,000 patients require surgery to repair tissue damaged by ulcers, and about 6,000 die due to complications from the disease.

Acid is essential for ulcer formation, but we now know that it is not usually the primary cause. In fact, many patients with ulcers have perfectly normal rates of acid production. Contrary to popular belief, stress is probably not a major cause of ulcers either.

In the 1980s, two Australian scientists, Barry Marshall and J. Robin Warren, proposed that most stomach ulcers arise from a bacterial infection. This idea did not gain quick acceptance because many scientists believed that bacteria could not survive the acidic conditions in the stomach. However, Marshall and Warren demonstrated the existence of *Helicobacter pylori* in the stomachs of a majority of patients with ulcers, and killing the bacteria with antibiotics proved to be an effective treatment. Once the bacteria are gone, the normal body repair mechanisms heal the wound created by their secretions. Marshall and Warren received the 2005 Nobel Prize in Medicine for their revolutionary discovery, which has improved the quality of life for millions of people. Nowadays, a combined treatment of antibiotics and the same drugs used to neutralize or inhibit stomach acid production in heartburn patients is effective in reducing or eliminating the symptoms of ulcers.

Diarrhea Is the Most Common Gastrointestinal Disorder Worldwide

There are over 1 billion cases of diarrhea—loose, watery stools occurring at least three times per day—in the U.S. every year. Nearly all episodes of diarrhea run their course in a day or two. Typically they result from a viral or bacterial infection but may also be caused by food sensitivities (such as lactose intolerance, discussed earlier), reactions to medications, stress-related disorders, or parasites.

One cause of diarrhea stands apart as particularly dangerous. Cholera is a disease caused by infection by the bacterium *Vibrio cholerae*, usually ingested by consuming contaminated food or water. According to the World Health Organization, at least 2,000 people die of cholera each year, with another 100,000 people contracting the disease but surviving. Nearly all cases of cholera occur in Africa, parts of Asia (notably China), and India, although scattered outbreaks have occurred nearly everywhere except the U.S., with the most recent worldwide pandemic lasting from 1961 to 1971. *V. cholerae* releases a toxin that alters the permeability of salts and water in the large intestine, resulting in a massive flow of these substances into the intestinal lumen. As in all cases of diarrhea, the chief concern with cholera is the loss of nutrients and water and the dehydration that ensues. In addition to killing the bacteria with antibiotics, therefore, the major treatment of cholera is the same as for any cause of diarrhea, which includes drinking solutions of salts and water to replace those that were lost in the feces.

CHAPTER SUMMARY

41.1 Animal Nutrition

- Animals require organic nutrients—carbohydrates, proteins, and lipids—and inorganic nutrients in the form of water and minerals. Differences in nutritional demands reflect an animal's physiology and environment. (Figure 41.1)

- Essential nutrients (essential amino acids, essential fatty acids, vitamins, and minerals) must be obtained from the diet. There are eight essential amino acids in humans and many other animals that are required for building proteins. The essential fatty acids are polyunsaturated fatty acids, which cannot be synthesized by animal cells. Vitamins are organic nutrients that serve as coenzymes for metabolic and biosynthetic reactions. Minerals are inorganic ions that serve many functions. (Table 41.1)

41.2 Ways in Which Animals Obtain Nutrients

- Herbivores eat only plants, carnivores primarily consume animal flesh or fluids, and omnivores eat both plant and animal products. (Figure 41.2)

- An animal's environment may determine how it obtains food. Suspension feeders sift water to filter out organic matter. (Figure 41.3)

- Carnivores, which have teeth developed for biting and tearing, are generally predators or scavengers. The teeth and jaw muscles of herbivores are adapted for grinding plants. Fluid-feeders do not need teeth except to puncture an animal's skin. (Figures 41.4, 41.5)

41.3 Principles of Digestion and Absorption of Food

- Intracellular digestion occurs in simple invertebrates and single-celled organisms.

- Most animals digest food via extracellular digestion. Flatworms and cnidarians have a gastrovascular cavity with one opening that serves as both entry and exit port. (Figure 41.6)

- All other animals have an alimentary canal, open at the mouth and anus and segregated into specialized regions, through which food passes from one end to the other. Once food has been digested, the nutrients must be absorbed via passive diffusion, facilitated diffusion, or primary or secondary active transport.

41.4 Overview of Vertebrate Digestive Systems

- The vertebrate digestive system consists of the alimentary canal plus associated structures, not all of which are found in all vertebrates. The anterior portion of the canal contains the oral cavity, salivary glands, pharynx (throat), and esophagus. The middle portion contains food storage or digestive organs (crop, gizzard, or stomach, depending on species), and the upper part of the small intestine. The final part of the alimentary canal contains the remainder of the small intestine, the large intestine, the rectum, and the anus. (Figures 41.7, 41.8)

- Saliva helps animals lubricate, chew, and swallow food. Peristalsis moves food through the pharynx and esophagus to the stomach. Some animals have a crop where food is stored and softened. The stomach stores food, partially digests proteins, and regulates the rate at which chyme empties into the small intestine.

- Herbivores rely on microbes living within their digestive tracts to digest cellulose. (Figure 41.9)

- Nearly all digestion of food and absorption of food and water occur in the small intestine. The combination of villi and microvilli increases the small intestine's surface area and maximizes the efficiency of digestion and absorption. Herbivores generally have longer intestines than carnivores, which allows added time for plant material to be digested and absorbed. (Figures 41.10, 41.11)

- The pancreas secretes digestive enzymes and a bicarbonate-rich fluid that neutralizes chyme. The liver secretes bile, which aids in the digestion of fat. (Figure 41.12)

41.5 Mechanisms of Digestion and Absorption in Vertebrates

- Starch digestion begins in the mouth and is completed in the small intestine. Starch is digested by amylases into molecules of the disaccharide maltose. Maltose and other disaccharides are digested into monomers by cells of the brush border. (Figure 41.13)

- Humans with lactose intolerance cannot digest lactose because the intestinal enzyme lactase is inactive or not present, a trait that is linked to genetic background. (Table 41.2)

- Proteins are broken down to peptide fragments in the stomach by pepsin, and they are further broken down into amino acids in the small intestine.

- Fat digestion occurs entirely in the small intestine. The major digestive enzyme in this process is pancreatic lipase. (Figure 41.14)

- Most water-soluble vitamins are absorbed by diffusion or active transport. Fat-soluble vitamins follow the pathway for fat absorption.

- The large intestine stores and concentrates fecal material before defecation, and it absorbs some of the remaining salts and water that were not absorbed in the small intestine.

41.6 Regulation of Digestion

- The digestive systems of animals are regulated in part by other organ systems, especially the nervous and endocrine systems. The nervous system can affect the digestive system in two major ways: (1) control of muscular and glandular activity by the local nerves in the alimentary canal, and (2) long-distance regulation by the brain.

- Starling and Bayliss demonstrated that hormones play a vital role in gastrointestinal function. (Figure 41.15)

41.7 Impact on Public Health

- Common GI complaints registered by hospitals and physicians include heartburn, ulcers, and diarrhea.

Test Yourself

1. The process of breaking down large molecules into smaller molecules that can be used by the cell is
   a. absorption.
   b. anabolism.
   c. ingestion.
   d. digestion.
   e. secretion.

2. Essential nutrients are
   a. all the carbohydrates, proteins, and lipids ingested by an organism.
   b. nutrients that an animal cannot manufacture.
   c. nutrients that must be obtained from the diet in their complete form.
   d. all of the above.
   e. b and c only.

3. An animal that has a strong preference for a particular food but can adjust its diet if necessary is said to be
   a. herbivorous.
   b. omnivorous.
   c. carnivorous.
   d. opportunistic.
   e. both b and d.

4. Which of the following statements is *not* correct?
   a. Fluid-feeders include animals that consume animal and plant fluids.
   b. Suspension-feeding only occurs in invertebrates.
   c. Predators are animals that kill and consume live prey.
   d. Animals that eat primarily fruits are called frugivores.
   e. Blood-sucking animals may contain anticoagulants in their saliva.

5. The pancreas connects to which part of the alimentary canal?
   a. esophagus
   b. stomach
   c. small intestine
   d. cecum
   e. large intestine

6. Which of the following statements regarding the vertebrate stomach is *not* correct?
   a. Its cells secrete the protease enzyme pepsin.
   b. It is a saclike organ that evolved to store food.
   c. Its cells secrete hydrochloric acid.
   d. It is the initial site of protein digestion.
   e. Little or no absorption of nutrients occurs there.

7. Absorption in the small intestine is increased by
   a. the many villi that are present on the inner surface of the small intestine.
   b. the brush border formed by microvilli on the cells of the villi.
   c. the presence of numerous transporter molecules on the epithelial cells.
   d. all of the above.
   e. a and b only.

8. In birds, the portion of the stomach responsible for the secretion of acid and pepsinogen is
   a. the crop.
   b. the gizzard.
   c. the proventriculus.
   d. the vent.
   e. the gallbladder.

9. Bile is secreted by
   a. the liver.
   b. the gallbladder.
   c. the pancreas.
   d. the small intestine.
   e. the cecum.

10. Which of the following is a function of the large intestine?
    a. It participates in cellulose digestion by microbes that exist in the cecum of herbivores.
    b. It stores and concentrates fecal material.
    c. Its cells absorb salts and water that remain in chyme after it leaves the small intestine.
    d. Its cells absorb certain vitamins produced by bacteria.
    e. All of the above.

Conceptual Questions

1. Distinguish between digestion and absorption.

2. Explain the difference between the alimentary canals of herbivores and carnivores.

3. Explain the functions of the crop and gizzard in birds.

Experimental Questions

1. Explain the first experiment conducted by Starling and Bayliss, which indicated that other factors besides signals from nerves may influence digestive gland secretion.

2. What criticism did Starling and Bayliss need to address to provide more conclusive evidence that the secretion was not due to neural regulation, and how did they address it?

3. What conclusions did the investigators draw from their second experiment?

Collaborative Questions

1. Discuss several strategies that animals use to obtain food.

2. Discuss some of the nutrients that animals must obtain through their diet.

---

www.brookerbiology.com
This website includes answers to the Biological Inquiry questions found in the figure legends and all end-of-chapter questions.

42

CONTROL OF ENERGY BALANCE, METABOLIC RATE, AND BODY TEMPERATURE

CHAPTER OUTLINE

A dormouse, *Muscardinus avellanarius*, hibernates in its nest.

I n Chapter 41, we saw that all animals require sources of carbon, nitrogen, and other nutrients to assemble the macromolecules that make up body tissues. Some of the ingested nutrients, such as sugars, lipids, and proteins, also represent a form of fuel, or energy, that can be used to generate ATP within cells.

Many organisms, however, do not have the luxury of a constant supply of nutrients. Most animals, for instance, sleep or rest for part of the day, during which time they do not eat. In addition, environmental changes may reduce the food supply, leading to long fasts. As a consequence of the irregular and sometimes unpredictable flow of nutrients into the body, animals have evolved an array of mechanisms to adequately maintain levels of important fuel molecules even during a fast. In the first part of this chapter, we will explore how ingested nutrients are stored in the body for such times of need, and the mechanisms by which these stores are tapped.

Metabolism refers to all the activities and chemical reactions in an organism's body, and **metabolic rate** describes the rate at which an organism uses fuel to supply ATP for these reactions. Different animals may have widely different metabolic rates, which determine the amount of nutrients they need. The greater an animal's metabolic rate, the more heat it generates as a by-product of breaking down nutrients and using the energy of their chemical bonds to synthesize ATP. Some of this heat escapes to the environment, and some is used to warm an animal's body. Metabolism and body temperature are therefore closely related, and we will examine this relationship in the second part of the chapter.

42.1 Nutrient Utilization and Storage

The utilization of nutrients can be divided into two alternating phases. The **absorptive state** occurs when ingested nutrients enter the blood from the gastrointestinal tract. The **postabsorptive state** occurs when the gastrointestinal tract is empty of nutrients and the body's own stores must supply energy. An average meal in a human requires approximately 4 hours for complete absorption, and therefore our usual three-meal-a-day pattern places us in the postabsorptive state during the late morning and afternoon and part of the night.

During the absorptive period, some of the ingested nutrients supply the immediate energy needs of the body. The rest are added to the body's energy stores to be called upon during the next postabsorptive period. Total-body energy stores are adequate for the average human to withstand a fast of several weeks, provided that water is available. By contrast, some animals can barely survive a single skipped meal—particularly if they have low energy reserves—because their metabolic needs are much higher than our own. In this section, we will focus on nutrient absorption, utilization, and storage in vertebrates, with a closer look at mammals.

Absorbed Carbohydrates Are Consumed for Energy or Stored for Future Use

The chief carbohydrate monomers absorbed from the gastrointestinal tract of vertebrates are glucose, galactose, and fructose.

Carbohydrates,
fats, and proteins

Glucose

Blood
vessel

Triglycerides

Amino acids

All cells:

Glucose or
fatty acids

ATP

Excess glucose is stored as
glycogen in liver and muscle
and as triglycerides in
adipose cells.

Triglycerides are stored
in adipose cells.

Amino acids are transported
into all cells, including
muscle, where they form
proteins. Excess amino acids
are converted to triglycerides
and glucose.

A portion of absorbed
glucose and fats is used
to supply energy needs.

Figure 42.1   **Events of the absorptive state.**   The products of digestion are used for energy, or they are deposited in cells (see text for details about mechanisms) as energy stores or as macromolecules important in cell function.

In mammals, the latter two sugars are either converted to glucose by the liver or enter the same metabolic pathways as does glucose.

Glucose is one of the body's two major energy sources during the absorptive state, fats being the other. Much of the absorbed glucose enters cells and is enzymatically broken down to carbon dioxide and water, in the process releasing the energy required to form ATP (**Figure 42.1**). Because skeletal muscle makes up a large fraction of body mass in most vertebrates, it is a major consumer of glucose, particularly when an animal is active. Skeletal muscle also incorporates some of the glucose into the polymer glycogen, which is stored in the muscles to be used later during fasting periods. If more glucose is absorbed than is needed for immediate energy, a portion of the excess is incorporated into glycogen in the liver, and the remainder into triglycerides in fat cells. The structures of glycogen and triglycerides are described in Chapter 3.

Most Absorbed Triglycerides Are Stored for Future Use

Triglycerides are too large to diffuse across the plasma membranes of the intestinal epithelial cells. As described in Chapter 41, they are digested into monoglycerides and fatty acids in the lumen of the small intestine, and then resynthesized into triglycerides once they diffuse into the intestinal epithelial cells. The triglycerides and other ingested lipids (for example, choles-

terol) are packaged into chylomicrons, which enter the blood circulation. As blood moves through adipose tissue, a blood vessel enzyme called lipoprotein lipase releases the fatty acids in the chylomicrons' triglycerides. The released fatty acids then enter fat cells and combine with glycerol to form triglycerides in adipose cells (Figure 42.1). These triglycerides are stored in fat cells until the body requires additional fuel.

As with glucose, some of the ingested fat is not stored but is used during the absorptive state by most organs to provide energy. The relative amounts of carbohydrate and fat used for energy during the absorptive period depend largely on the composition of a meal.

Absorbed Amino Acids Are Primarily Used to Build Proteins

Amino acids are taken up by all body cells, where they are used to synthesize proteins (Figure 42.1). All cells require a constant supply of amino acids, because proteins are constantly being synthesized and degraded. Any excess amino acids that are ingested are not stored as protein in the sense that glucose is stored as glycogen or that both glucose and fat are stored as triglycerides. Instead, they are converted by liver cells into glucose or triglycerides. Therefore, eating large amounts of protein does not normally increase stores of body protein. An exception would be a young, rapidly growing animal with its continuous increase in body protein.

Glycogen and Noncarbohydrate Precursors Provide Glucose

As the absorptive period ends, synthesis of glycogen and fat slows and the breakdown of these substances begins. No glucose is absorbed from the intestinal tract during this time, yet the blood glucose concentration must be maintained because the brain normally utilizes only glucose for energy. If blood glucose levels fall too low, brain function can be affected, ranging from subtle impairment of mental function to seizures, coma, or even death.

The events that maintain blood glucose concentration fall into two categories: (1) reactions that provide glucose to the blood and (2) cellular utilization of fat for energy, thus sparing glucose for the brain and nervous system.

Vertebrates can increase their blood glucose levels during the postabsorptive period in two major ways (**Figure 42.2**). First, the stored glycogen can be broken back down into molecules of glucose by hydrolysis, a process known as **glycogenolysis**. This occurs primarily in the liver, following which the glucose is released into the blood, where it can travel to all cells.

The amount of glycogen available to provide glucose during the postabsorptive period varies among animals, but it is generally not sufficient to maintain blood glucose levels for more than a brief time, such as an overnight fast in a human. Therefore, a second mechanism for maintaining blood glucose levels is required if the postabsorptive period continues longer. In the process of **gluconeogenesis** (creation of new glucose), enzymes in the liver convert noncarbohydrate precursors into glucose, which is then secreted into the blood.

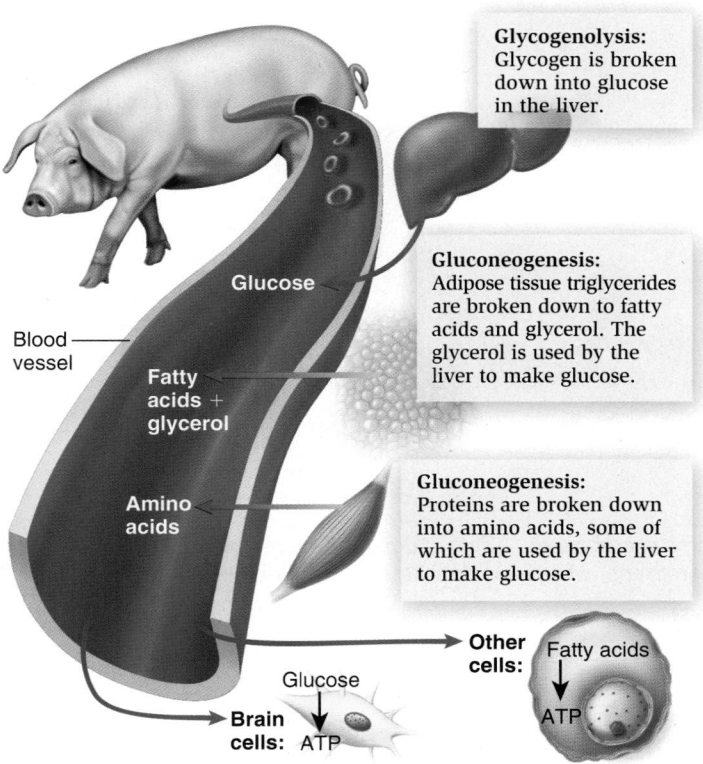

Glycogenolysis: Glycogen is broken down into glucose in the liver.

Gluconeogenesis: Adipose tissue triglycerides are broken down to fatty acids and glycerol. The glycerol is used by the liver to make glucose.

Gluconeogenesis: Proteins are broken down into amino acids, some of which are used by the liver to make glucose.

Blood vessel

Glucose

Fatty acids + glycerol

Amino acids

Other cells: Fatty acids → ATP

Glucose

Brain cells: ATP

Figure 42.2   Events of the postabsorptive state. Macromolecules formed during the absorptive state are broken down to supply monomers that can be used for energy. This begins with glycogenolysis—the breakdown of glycogen into glucose. The breakdown products of triglycerides and proteins can be used for energy directly in the case of fatty acids or indirectly by their conversion in the liver to glucose by gluconeogenesis.

The primary precursor for gluconeogenesis is the sugar-alcohol glycerol, which is released from triglycerides in adipose tissue by the process of lipolysis. In lipolysis, enzymes within fat cells hydrolyze triglycerides into fatty acids and glycerol, both of which enter the bloodstream. The fatty acids diffuse into cells, where they are used as an alternate energy source to glucose (except for parts of the nervous system). The glycerol is taken up by the liver, where enzymes convert it into glucose, which is then released back into the bloodstream.

If the postabsorptive period continues for a rather long time—as when an animal fails to find food—protein becomes an important source of blood glucose. Large quantities of protein in muscle and other tissues can be broken down to amino acids without serious cellular malfunction. The amino acids

enter the blood and are taken up by the liver, which removes the amino groups and converts the remaining molecule into glucose. This process, however, has limits. Continued protein loss can result in the death of cells throughout the body because they depend on proteins for such vital processes as plasma membrane function, enzymatic activity, and formation of organelles.

Fat Metabolism by Other Tissues Reserves Glucose for the Nervous System

Organs and tissues that need glucose the most—such as the brain—must get a major share of the available glucose. This is accomplished by having most organs and tissues reduce their glucose utilization and increase their fat utilization during the postabsorptive period. This metabolic adjustment, termed **glucose sparing**, reserves the glucose produced by the liver for use by the nervous system.

The essential step in glucose sparing is lipolysis, the breakdown of adipose tissue triglycerides, which as stated earlier liberates glycerol and fatty acids into the blood. In vertebrates, the circulating fatty acids are taken up and used to provide energy by almost all tissues, excluding the nervous system, whose cells do not express the enzymes required to break down fatty acids for energy.

Of the body's tissues and organs, the liver is unique in that most of the fatty acids entering it during the postabsorptive state are not used by the liver for energy. Instead they are processed into three small compounds collectively called **ketones** or ketone bodies. Ketones are released into the blood and provide an important energy source during prolonged fasting for the many tissues, including the brain, capable of oxidizing them via the citric acid cycle.

The utilization of fatty acids and ketones during fasting provides energy for the body, sparing the available glucose for the brain. Moreover, as just mentioned, the brain can utilize ketones as energy, and it does so increasingly as ketones build up in the blood during the first few days of a fast. The survival value of this phenomenon is significant. When the brain reduces its glucose requirement by utilizing ketones, much less protein breakdown is required to supply amino acids for gluconeogenesis. Protein stores last longer, enabling the animal to survive a long fast without serious tissue damage.

The combined effects of glycogenolysis, gluconeogenesis, and glucose sparing are so efficient that, after several days of complete fasting, human blood levels of glucose fall by only a few percent. **Table 42.1** summarizes the relative changes in these and other variables during the absorptive and postabsorptive periods.

Table 42.1	Relative Changes in Fuel Usage and Generation During the Absorptive and Postabsorptive Periods					
	Glucose absorption from gut	**Glucose utilization by cells**	**Synthesis of triglycerides**	**Fat utilization by cells**	**Breakdown of glycogen**	**Blood levels of glucose**
Absorptive period	High	High	High	Low	Low	Normal
Postabsorptive period	Absent	Highest in brain	Low	High	High	Normal

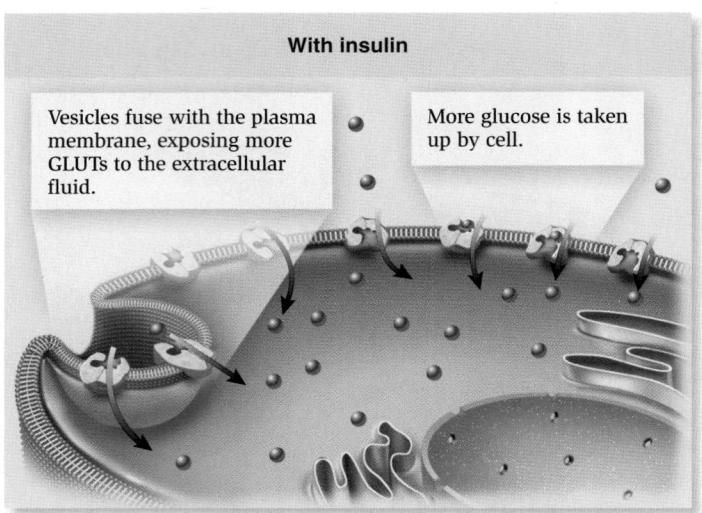

Figure 42.3 Insulin promotes the transport of glucose across plasma membranes. This is done by the recruitment of intracellular vesicles containing glucose transporter (GLUT) proteins to the plasma membrane, where they facilitate glucose diffusion into the cell.

Biological inquiry: What benefit is there to having GLUT proteins premanufactured and stored in intracellular vesicles?

42.2 Regulation of the Absorptive and Postabsorptive States

Tight control mechanisms are required to maintain homeostatic levels of fuel molecules in the blood. These controls come in the form of several hormones produced by cells of the endocrine system, and from nerve signals arising from cells of the nervous system. In this section, we will learn that one common function of the endocrine and nervous systems is to regulate the processes of glycogenolysis and gluconeogenesis, so that glucose is made available to cells at all times.

Insulin Is a Key Regulator of Metabolism

The blood concentration of insulin, a hormone made by the pancreas, increases during the absorptive state and decreases during the postabsorptive state. Insulin binds to a cell surface receptor and stimulates a signaling pathway that acts in cells to facilitate the diffusion of glucose from blood into the cytosol of cells. This is necessary because as a relatively large and polar molecule, glucose cannot diffuse by itself through plasma membranes. Insulin increases glucose uptake by promoting the movement of glucose transporters (GLUTs) located within intracellular vesicles to the plasma membrane (**Figure 42.3**). Thus, insulin assists the glucose in entering cells, which lowers blood glucose levels. Insulin also inhibits glycogenolysis and gluconeogenesis in the liver, because these processes are not required during the absorptive state.

Insulin exerts its effects mainly on muscle cells and adipose tissue cells, because these cells have insulin receptors in their plasma membranes. However, not all GLUTs are identical, and not all require insulin for their activity, as we see next.

GENOMES & PROTEOMES

GLUT Proteins Transport Glucose in Animal Cells

All animal cells require transporters to move glucose across their plasma membranes. GLUTs in mammals make up a family of at least 14 related proteins that share similar structures but that are expressed in different tissues (**Table 42.2**). All GLUTs have 12 membrane-spanning helical domains that anchor the protein in plasma membranes. The structures of GLUTs are very similar across phyla. For example, the sequence of one *Drosophila* GLUT mRNA is nearly 70% identical to that of one of the human GLUTs. In more closely related taxa, such as birds and mammals, GLUTs are even more similar, with up to 95% of their

Table 42.2	Distribution and Affinities of Common GLUTs in Mammals		
Transporter	**Tissue distribution**	**Affinity for glucose**	**Insulin-sensitive**
GLUT1	Cells that make up blood vessels of the brain, red blood cells, embryo cells	Very high	No
GLUT2	Liver, kidney, pancreas, small intestine	Low	No
GLUT3	Brain and nervous system cells, testes, placenta, fibroblasts	Very high	No
GLUT4	Skeletal and cardiac muscle cells, adipose cells	Low	Yes

amino acid sequences being the same. These and other considerations indicate that the GLUTs arose by duplication of a common ancestral gene.

The different GLUTs vary in their ability to bind glucose. For example, some GLUTs have high affinity for glucose and others have low affinity. High affinity means the protein binds glucose even at very low concentrations of the glucose. Muscle and fat express the protein GLUT4, which has a low affinity for glucose that matches the concentration of glucose normally found in blood. This is also the only GLUT protein whose movement to the plasma membrane requires the cell to be stimulated by insulin. Thus, as glucose levels rise in the blood after a meal, insulin recruits more GLUT4 molecules from the cell cytosol to the plasma membrane of muscle and fat cells.

By comparison, GLUT1 and GLUT3 are found in the brain, where they act in concert to mediate transport of glucose from blood vessels to the interstitial fluid of the brain, and from there into brain cells. GLUT1 and GLUT3 have much higher affinity for glucose than other GLUT proteins, and unlike GLUT4, insulin is not required for their expression. This means that nerve cells of the brain can transport glucose into their cytosol even if the blood level of glucose is very low, and even in the absence of insulin. Thus, if an animal's blood glucose level plummets, perhaps due to disease or starvation, the most important cells in the body—those of the brain and nervous system—would still receive adequate energy for survival.

Expressing multiple types of the same functional class of protein means that different parts of an animal's body can meet their own particular metabolic demands. Moreover, these needs may change during development. High-affinity GLUTs such as GLUT1 and GLUT3, therefore, are often found in cells with very high glucose requirements, such as in a developing embryo. In another example, one intestinal GLUT protein (GLUT5) can also transport another sugar, fructose. However, this GLUT is not expressed in mammals until after weaning (the time of transition from mother's milk to solid food), because milk does not contain fructose.

If insulin is the key regulatory molecule that controls blood glucose levels through its actions on GLUT4, what factors regulate the production of insulin? This is described next.

Blood Glucose Levels Control Insulin Secretion

What stimulates the pancreas to secrete insulin into the blood? The answer is the blood glucose concentration. An increase in blood glucose levels, such as after a meal, stimulates the cells of the pancreas to secrete insulin. After glucose is taken up by cells of the body, a decrease in blood glucose levels removes the signal for secretion (**Figure 42.4**). Thus, rising glucose levels stimulate insulin secretion, which lowers blood glucose levels, which in turn removes the signal for insulin secretion. This is an example of negative feedback, as described in Chapter 40.

In addition to blood glucose levels, inputs from nerve cells to the pancreas also play a role in regulation of insulin secre-

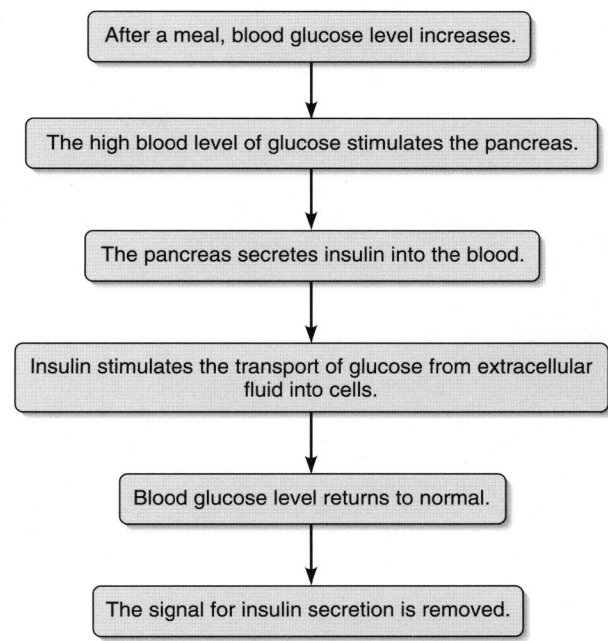

Figure 42.4 **The relationship between levels of glucose and insulin in the blood.** Glucose stimulates insulin secretion from the pancreas. Insulin stimulates glucose transport into cells, thus lowering blood glucose levels, which removes the stimulatory signal for the pancreas.

tion. During a meal, nerve signals from the brain stimulate the secretion of insulin. Other nerve signals can inhibit insulin, for example, during acutely stressful situations where an animal's blood glucose increases to supply the active brain.

Blood Glucose Levels Are Maintained within a Normal Range

Like all important variables in an animal's body, blood glucose levels are controlled by a system of checks and balances. Several factors act in concert to prevent blood glucose from falling below the normal homeostatic range, even when an animal is fasting. Otherwise, glucose could drop so low that despite the high-affinity GLUT1 and GLUT3 proteins in brain cells, there would not be enough glucose to keep brain cells alive.

If blood glucose levels fall below the normal homeostatic range for an animal, this activates brain cells that are specialized for monitoring glucose concentrations in extracellular fluid. Signals from the brain then stimulate the production of glucose-elevating factors. These include numerous hormones, notably another one produced by the pancreas called glucagon. Glucagon stimulates the processes of glycogenolysis, gluconeogenesis, and the synthesis of ketones in the liver (**Figure 42.5**). In addition, other factors released by nerve cells and endocrine glands stimulate adipose tissue to release fatty acids into the blood. The fatty acids diffuse across plasma membranes and provide another source of energy for cellular synthesis of ATP.

Figure 42.5   **The role of the liver in fuel supply to the blood.** If glucose levels decrease, glucose-monitoring regions in the brain initiate a series of responses that result in the secretion of hormones, such as glucagon, that stimulate the breakdown of glycogen, glycerol, amino acids, and fatty acids in the liver. The glucose and ketones that form as a result of this breakdown enter the bloodstream, where they travel to all the body's cells and supply energy.

The overall effects are to increase the blood concentrations of glucose, fatty acids, and ketones during the postabsorptive period or during a prolonged fast.

More Fuel Is Required During Exercise

We think of exercise as something humans do for fun or fitness, but in its broadest sense, **exercise** can be defined as any physical activity that increases an animal's metabolic rate. Generally, an animal becomes active to seek something, such as food, shelter, or a mate, or to elude something, such as a predator or a storm. The type of exercise animals engage in, therefore, can be quite varied. When a cheetah sprints after a zebra, for example, the activity of both predator and prey is brief and intense, perhaps lasting only a few seconds. By contrast, a tuna may spend its entire life swimming at a moderate pace, while a migrating bird may fly a hundred miles a day or more over a span of weeks.

For any type of exercise, fuel in the form of nutrients must be available to provide the energy required for skeletal muscle contraction, increased heart and lung activity, and increased activity of the nervous system, which directs the activities of all the other organs in the body. These fuels include glucose and fatty acids as well as the muscle's own glycogen.

The liver supplies the blood with the glucose used during exercise by breaking down its glycogen stores and by gluconeogenesis. This occurs even in the absorptive state, and thus blood glucose levels increase above normal when an animal exercises at such times. In addition, an increase in adipose tissue lipolysis releases fatty acids into the blood, which provides a source of energy for the exercising muscle. These events are mediated by the same hormones and nerves that are responsible for the regulation of the postabsorptive state. Thus, the body taps stored forms of energy in similar ways in response to both fasting and exercise.

42.3 Energy Expenditure

Animals have a wide range of metabolic needs that depend on numerous factors. Active animals such as migrating birds burn fuel more rapidly than inactive animals such as hibernating mammals. Juveniles typically burn fuel more quickly than older animals. An animal's **energy expenditure** refers to the amount of fuel it uses in a given period of time to power all of its metabolic requirements.

A fundamental feature of energy is that it can be neither created nor destroyed, but it can be converted from one form to another. The breakdown of organic molecules liberates the energy locked in their molecular bonds and transfers it to the bonds in ATP. This is the energy that cells use to perform various biological activities like muscle contraction, active transport, and molecular synthesis. We refer to these functions as work. Not all of the energy is used to do work, however. Some of it appears as heat, which contributes to an animal's body temperature. In this section, we examine how energy expenditure is measured in animals, and the factors that control energy consumption and expenditure in order to achieve a balance between the two.

Energy Expenditure Is Determined by Measuring Metabolic Rate

Biologists quantify the energy of metabolism in calories. A **calorie** is the amount of heat required to raise the temperature of 1 gram of water 1 degree Celsius. Most biological activities, however, require much greater amounts of energy than a calorie, and thus the more common unit of measurement is the kilocalorie (1,000 calories, or **kcal**). (In food labeling, a Calorie with a capital C is also the same as kcal.) The **metabolic rate** is the total energy expenditure of an animal per unit of time. Biologists often compare the metabolic rates of different animals to learn, for example, how some animals are capable of hibernating, how an animal's body temperature influences its metabolic rate, and how hormones and other factors alter an animal's metabolism.

The most common method for comparing metabolic rates of different species is to obtain the **basal metabolic rate (BMR)**. In the basal condition, the animal is at rest in the postabsorptive state, and at a standard temperature. For **endotherms**, animals that generate their own internal heat and maintain relatively constant body temperatures, the temperature is within the range that causes the animal to neither gain heat (for example, by shivering) nor lose heat (for example, by perspiration). This is called the animal's thermoneutral zone. The BMR of **ectotherms**, animals whose body temperature changes with the environmental temperature, is measured at a standard temperature for each species—one that approximates the average temperature that a species normally encounters. In this case, the term **standard metabolic rate (SMR)** is used instead of BMR, since the basal condition in ectotherms is harder to define than for endotherms. Because BMR and SMR apply only to resting, postabsorptive animals at a standard temperature, any animal that has recently eaten or been active has a higher metabolic rate than its basal metabolic rate.

The BMR is often called the metabolic cost of living, and most of it can be attributed to the routine functions of the heart, liver, kidneys, and brain. Two methods are used to obtain a good estimate of BMR. **Direct calorimetry** was invented by Antoine Lavoisier in 1780 (**Figure 42.6**). In Lavoisier's experiment, an animal was placed in an enclosed, insulated chamber that was surrounded by ice. As the animal metabolized fuel, it generated heat, which dissipated into the chamber and melted the ice. The amount of water collected from an opening at the bottom of the chamber could be used to estimate the amount of heat generated by the animal. Today, direct calorimetry is measured using more sophisticated instrumentation, but the principle remains the same. It provides an ideal measure of metabolism, because energy expenditure and heat production are directly related. However, the method is not very practical, particularly with large animals.

The second and more practical method of measuring BMR, **indirect calorimetry**, is based on the principle that animals require oxygen to metabolize fuel. The more fuel being metabolized—that is, the greater the BMR—the more oxygen must be consumed by the animal. By measuring the rate at which an animal uses oxygen, therefore, we can obtain a good estimate of BMR. Indirect calorimetry can also be used to compare the metabolic rates of an animal at rest and when active, when oxygen consumption increases (**Figure 42.7**). One limitation to this method is that a small percentage of fuel is metabolized without oxygen consumption, and thus indirect calorimetry underestimates actual metabolic rate.

BMR and SMR are indicators of total body metabolic rates. However, not all tissues in the body use oxygen and produce heat at the same rate. Some structures, such as skin, consume very little oxygen under resting conditions, while others, such as the brain, heart, or liver, have high rates of metabolism even when an animal is sleeping. Moreover, metabolism is not always basal, and the metabolic rates of different tissues can vary independently of each other. For example, the metabolism of the gut increases when food is being digested, and that of skeletal muscle increases during exercise (**Figure 42.8**).

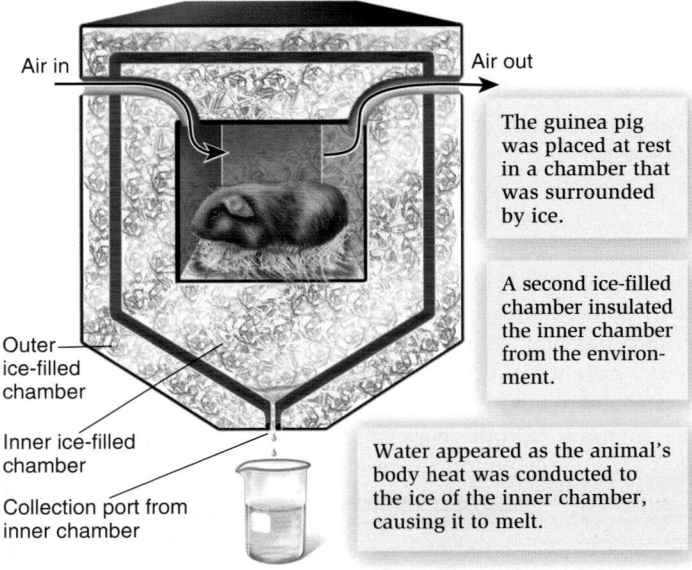

Figure 42.6 Lavoisier's experiment, a pioneering example of direct calorimetry.   Direct calorimetry determines an animal's basal metabolic rate by determining how much the heat given off by its body raises the temperature of a surrounding ice-filled chamber. The heat from the guinea pig's body melts the ice in the surrounding chamber. The volume of melted ice is proportional to the amount of heat produced.

Figure 42.7 Indirect calorimetry as a tool to measure oxygen consumption.   Many animals, like this goose, can be trained to walk on a treadmill, which allows scientists to compare metabolism during rest and exercise. Oxygen consumption can be determined by sampling the air exhaled into a tightly fitting mask. One-way valves prevent inhaled and exhaled air from mixing.

Biological inquiry: If the air had been sampled from the mask just before starting the treadmill while the goose was resting, would this have been a good estimate of BMR?

Figure 42.8 **Changes in metabolic rate of selected structures in a mammal's body during different activities.** Red areas are regions of high metabolism; blue areas are regions of low metabolism. Note that during exercise, skeletal muscle becomes more active and areas associated with food digestion and absorption become less active. At all times, however, the heart, liver, and brain are highly active.

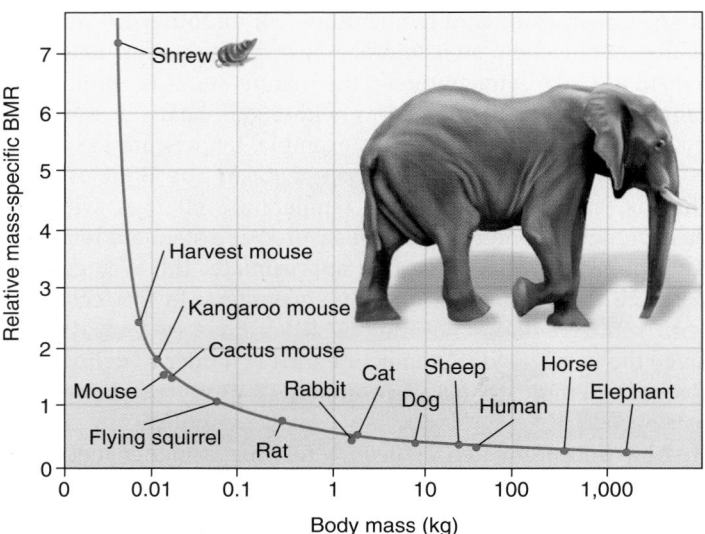

Figure 42.9 **Metabolic rates of animals that differ in size.** Metabolism can be scaled to body mass by measuring oxygen consumption and normalizing it to the animal's body mass (mass-specific BMR). Note that when expressed in this way, the mass-specific BMR of a shrew is higher than that of an elephant, even though the total oxygen consumption and heat output of the elephant would be much greater. The values on the y-axis are relative units of metabolism.

Biological inquiry: Can the relationship between body size and metabolic rate be used to make hypotheses about metabolic rates of extinct animals?

Metabolic Rate Depends on Many Factors

As just stated, metabolic rate is not always basal. Many factors have an impact on metabolism. These include skeletal muscle activity, whether an animal has recently eaten, body size, and whether or not an animal is hibernating.

The primary factor that increases metabolic rate is altered skeletal muscle activity. Even minimal increases in muscle contraction significantly increase metabolic rate, and strenuous activity even more so. For example, the total daily expenditure of kilocalories may vary for a healthy young adult human from approximately 1,350 kcal for a small person at rest to more than 7,000 kcal for a cyclist competing in the Tour de France. Changes in muscle activity also affect metabolic rate during sleep due to decreased muscle activity, and during exposure to cold temperatures due to increased muscle activity from shivering.

Eating and digesting food increase the metabolic rate. Particularly in mammals that eat meat, this may raise metabolic rate by 10% to nearly 50% for a few hours after eating. You may have noticed this **food-induced thermogenesis** after consuming a large meal, such as Thanksgiving dinner. Ingested protein—for example, turkey—produces the greatest effect, while carbohydrate and fat produce less. The increased heat production is believed to result at least partly from the processing of the absorbed nutrients by the liver, and the energy expended by the gastrointestinal tract in digestion and absorption. Food-

induced thermogenesis is observed in nearly all vertebrates, but it is most notable in certain reptiles that eat large and infrequent meals. Because of food-induced thermogenesis, BMR tests must be performed in the postabsorptive state.

A third factor affecting metabolic rate is body size. In general, a large animal uses greater amounts of energy than does a small animal because the large animal has more mass and more cells, all of which consume fuel and generate heat. The total energy expenditure and heat generation of an elephant is clearly greater than that of a mouse, for instance. However, when the energy expenditures of an elephant and a mouse are scaled to their respective body masses, we find that the energy expenditure per gram of body mass in a mouse is much higher than the comparable calculation in an elephant. **Mass-specific BMR** is the amount of energy expended per gram of body mass. Mass-specific BMR is a relative term that allows comparisons to be made of metabolic rates among animals of different sizes. Research has shown that the relationship between mass-specific BMR and body mass follows an exponential curve (**Figure 42.9**). One possible explanation for this phenomenon is that because the ratio of an animal's surface area to body mass is greater in smaller animals than in larger animals, smaller animals lose heat more rapidly than do larger ones. According to this hypothesis, smaller animals must generate more heat per gram of body mass than larger animals to replace their heat loss. However, while this hypothesis appears to provide an explanation for the relationship between metabolism and body size in endotherms, it does not explain the fact that the same relationship exists in almost all animals, including ectotherms.

The smallest endotherms—hummingbirds, shrews, mice and other rodents, and some bats—face the special challenge of fueling their very high mass-specific metabolic rates. This becomes difficult and even impossible during cold months, or any time when food is unavailable. Many animals have evolved a strategy of lowering their internal body temperature to just a few degrees above that of the environment, a process called **torpor**. Torpor may occur on a nightly basis, while the animal continues to be active during the daylight hours, or it can extend for months, in which case it is called **hibernation**.

A reduction in body temperature reduces the metabolic demands of all body cells. BMR in a small hibernating rodent, for example, may drop to less than 1% of what it would be at the normal body temperature of around 38°C. This allows the animal to conserve energy for remarkably long times. Ground squirrels, for example, may hibernate for up to eight months.

Total Body Energy Stores Are Balanced Between Consumption and Expenditure

When the daily amount of energy consumed is equal to the amount of energy expended, an animal's body weight remains stable. Tipping the balance in either direction causes weight gain or loss, that is, the total-body energy content increases or decreases. Normally, energy is stored in the form of fat in adipose tissue.

Body weight in an adult animal usually hovers around a stable set point that differs among species. Body weight is maintained by adjusting caloric intake and energy expenditure in response to changes in body weight. This mechanism usually works very precisely in those animals in which it has been studied. For example, a mammal that eats less one day will eat enough the next day to compensate for the previous day's deficit. Similarly, if an animal is overfed one day, it may eat less the next day, and its metabolic rate will increase to burn off the extra fuel in its body. The hypothesis accepted today by scientists is that appetite and metabolism change when food intake is more or less than the amount needed to maintain the body's set point. This phenomenon explains why some dieters initially lose weight easily and then become stuck at a plateau, because appetite increases and metabolic rate decreases to compensate for the weight loss. Conversely, it also helps explain why some very thin people have difficulty gaining weight.

Hormones and the Brain Control Food Intake

Short-term control of feeding generally involves a feeling of **satiety**, that is, fullness. As an animal's stomach and small intestine stretch to accommodate food, nerves send signals from these structures to the brain. At the same time, the stomach and small intestine release hormones into the blood that suppress appetite. These **satiety signals** remove the sensation of hunger and set the time period before hunger returns again (**Figure 42.10**).

1 As the stomach and small intestine stretch, nerve signals tell the brain to reduce appetite.

3 Satiety center in the brain reduces appetite after a meal.

Small intestine — Stomach — Electrical signal — Brain — Food

2 Hormones secreted into the blood by the gut also signal the brain when to stop eating.

Figure 42.10 **Brain signals that control hunger.** Appetite is controlled by a satiety center in the brain that receives nerve and hormonal signals from the stretched stomach and intestines after a meal. When digestion and absorption are complete, the stomach and intestines return to their original size and the brain no longer senses that an animal feels "full."

Over the long term, such as weeks, months, or years, total energy consumption tends to remain fairly constant. An animal may go through a period of fasting and then make up for it when food becomes available. In other words, meals may vary from day to day, or during different seasons of the year, but the average amount of food consumed does not change much. Long-term control of food intake is mediated by many different brain molecules, by hormones, and by emotional state, particularly in humans. One molecule that has received considerable attention in recent years for its ability to control appetite and metabolic rate is **leptin** (Greek *leptos*, thin). Leptin is produced by adipose cells in proportion to fat mass. As more fat is stored in the body, more leptin is secreted into the blood. Leptin acts on brain centers to reduce appetite and increase metabolic rate. In this way, the brain is made aware of how much fat is stored in the body at all times, and it can adjust appetite and energy expenditure appropriately if fat stores decline or increase.

If an animal fasts for a period of time, its adipose cells shrink as they release their stored fat into the blood. The decrease in leptin secretion resulting from the decreased fat mass results in a decrease in BMR and an increase in appetite. This may be the true evolutionary significance of leptin, namely that its disappearance from the blood lowers the BMR, thus prolonging life during periods of starvation (**Figure 42.11**). Leptin was not discovered until 1994, but its existence was postulated decades before that by the pioneering work of Douglas Coleman, who investigated the nature of mutations in mice that result in obesity.

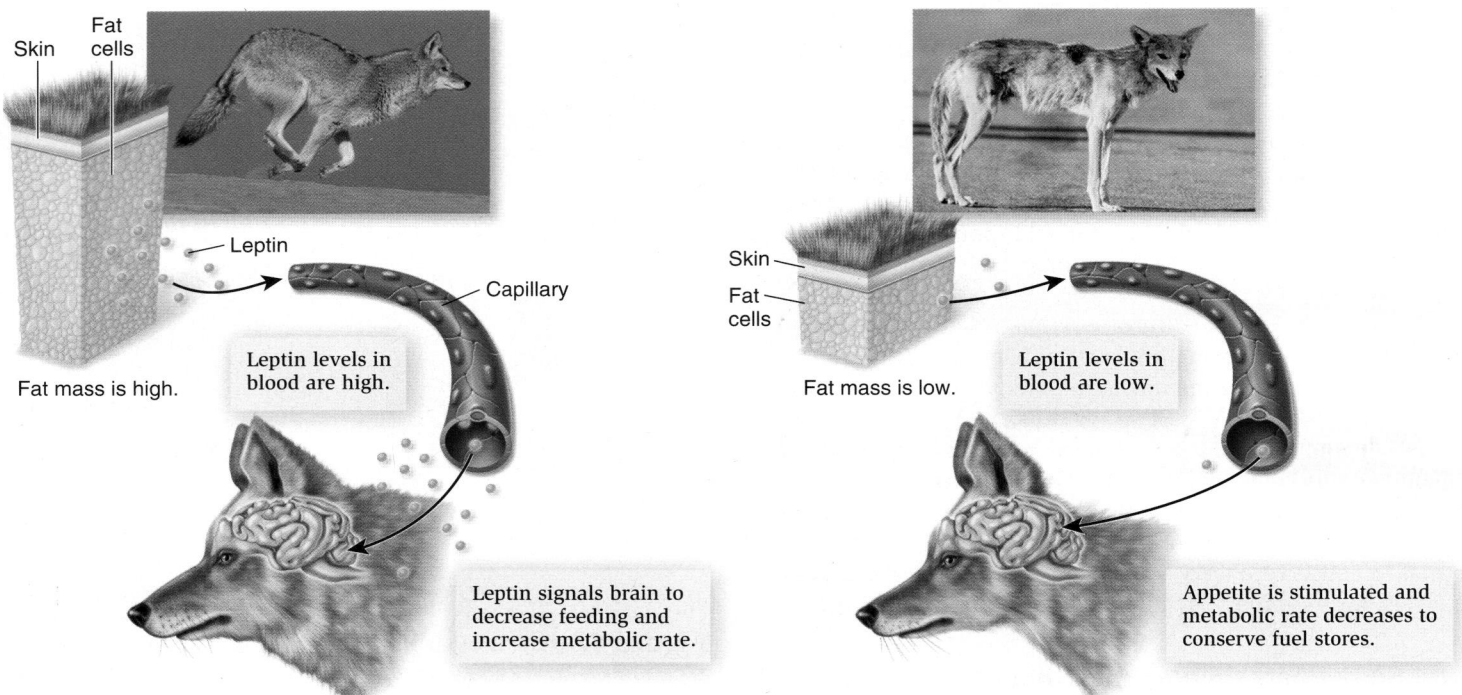

Figure 42.11 **The role of leptin in regulating appetite and metabolic rate.** In animals, like this coyote, changes in blood leptin levels result directly from changes in fat mass. Animals with more fat make more leptin.

FEATURE INVESTIGATION

Coleman Revealed a Satiety Factor in Mammals

For many years, scientists wondered how most animals appeared to regulate their body mass around a predetermined set point, despite fluctuations in the food supply. They postulated that other parts of the body somehow communicated with the brain to signal when energy stores were above or below normal. In the 1970s, researcher Douglas Coleman tested this hypothesis in an experiment involving parabiosis, the surgical connection of the bloodstreams of two animals, such that the blood from one animal mingles with that of the other.

Coleman used two strains of mice called ob and db mice, carrying different mutations that resulted in inherited forms of obesity. Coleman first connected the bloodstream of a wild-type mouse, one that lacked these mutations, with the bloodstream of either an ob mouse, or a db mouse, as shown in the Feature Investigation in **Figure 42.12**. He discovered that when the circulatory system of the ob mouse was in contact with the circulation of the wild-type mouse, the ob mouse ate less and gained less weight than usual. This suggested that the blood of the wild-type mouse contained a circulating factor that signals the brain when an animal has sufficient fat stored in its body, and

adjusts appetite accordingly. The ob animal apparently lacked this factor, but when exposed to it via the wild-type mouse's circulation, it responded in the correct way. The wild-type mouse of the parabiosis pair apparently retained a sufficient amount of the factor in its blood, because it maintained its body weight at a normal level.

Coleman noticed, however, that a db mouse continued to gain weight at an abnormally high rate even when parabiosed with a wild-type mouse. Instead, in this case the wild-type animal lost weight while the db animal remained obese. Coleman concluded that the db mouse must produce the same factor as the wild-type mouse, but for some reason it was unable to respond to it. The wild-type mouse that was parabiosed to the db mouse lost weight because it received the factor from the db mouse, in addition to having its own supply of the factor. Thus, whether the factor was absent as in the ob mouse, or present but unable to act as in the db mouse, the resulting phenotype was the same—obesity.

In 1994 Jeffrey Friedman and coworkers at Rockefeller University identified this circulating factor as the protein leptin. The ob mice were found to be homozygous for a mutation in the leptin gene, which produced an inactive leptin molecule,

Figure 42.12 Coleman's parabiosis experiments with genetically obese and normal (wild-type, wt) mice.

HYPOTHESIS Body weight is controlled by a factor that circulates in the blood. This factor is absent in strains of mice that have an inherited form of obesity.

STARTING MATERIALS Two different strains of genetically obese mice, and normal (wild-type) mice.

Experimental level

Conceptual level

1 Surgically connect the blood circulations of an obese and normal (wt; wild-type) mouse. Monitor changes in body weight. Note: 2 different strains of obese mice were tested, called ob and db mice.

ob or db   wt

Parabiosis

Blood flow

Blood moves between both circulations

Blood vessel from ob or db mouse

Blood vessel from wt mouse

2 Allow mice a normal diet for several weeks, then visually inspect and weigh each pair.

55g

Obese mouse vessel

wt vessel

Weight-reducing factor

3 **THE DATA**

ob wt

db wt

Weight gain per pair per week (grams)

ob–wt   db–wt

while db mice produced leptin but did not respond to it. In fact, the db mice were found to produce even greater amounts of leptin than wild-type mice, which explained why the wild-type mouse in Coleman's experiments lost weight when parabiosed with a db mouse. Friedman and others later showed that adipose cells produce leptin in direct proportion to the total fat mass of an animal. The brain responds to increased amounts of leptin by reducing appetite and increasing metabolic rate, both of which result in weight loss. Conversely, a decrease in blood levels of leptin, which arises when fat stores decline, causes an increase in appetite and a decrease in metabolic rate, which helps

prevent further weight loss. It is now known that all classes of vertebrates produce leptin.

At first, the work of Coleman and Friedman generated considerable excitement that leptin might be useful to treat obesity in humans, but this has thus far proven difficult. Why? Recent research has revealed that most obese humans are more like the db mice than the ob mice. That is, they produce leptin but apparently fail to respond adequately to it, and therefore simply increasing blood levels of leptin may not have a significant effect on body weight.

42.4 Regulation of Body Temperature

As we have seen, energy expenditure is linked to body temperature. Next, we discuss why body temperature is important for the health and survival of all animals, and consider mechanisms by which the body gains or loses heat.

Temperature Affects Chemical Reactions, Plasma Membrane Structure, and Protein Function

Most animals can survive only in a relatively narrow range of temperatures, and there are several reasons why. First, chemical reactions depend on temperature. Heat accelerates the motion of molecules, so as an animal's temperature rises, the rates at which the molecules in its body move and react with each other also increase. In addition, enzymes have an optimal temperature range for their maximal catalytic function. Compared to temperatures below the optimal range, most chemical reactions in an animal's body are observed to double or even triple in rate for every 10°C increase in body temperature. Low temperatures slow down chemical reactions, making it harder for an animal to remain active.

A second effect of temperature is that heat alters the structures of plasma and intracellular membranes. At low temperatures, membranes become less fluid and more rigid. Rigid membranes are less able to perform biological functions, such as transporting ions and fuels and binding extracellular molecules to receptors on the membrane surface. Alternatively, if the temperature becomes too high, membranes can become leaky.

Likewise, unusually high temperature can also inhibit enzyme function and slow chemical reactions. At very high temperatures many proteins are denatured, that is, they lose the three-dimensional structure that is crucial to their ability to function properly. This occurs because the bonds that form tertiary and quaternary protein structures result from weak interactions, such as hydrogen bonds, and can be disrupted by heat. Denaturation of enzymes is especially serious, because enzymes catalyze many reactions in the body, including those involved in metabolism. Most animals have an upper limit of body temperature at which they can survive. In humans, for example, a body temperature of 41°C (106°F) causes loss of protein function and breakdown of the nervous system, and a body temperature of 42–43°C (107–109°F) is fatal. Birds, which have slightly higher resting body temperatures than mammals (approximately 40–41°C compared with 35–38°C for most mammals), cannot survive at body temperatures above 46–47°C. At environmental temperatures greater than 50°C, nearly all animals die.

At the other end of the temperature spectrum, though, extreme cold is better tolerated. For example, some animals can freeze and survive after thawing. Freezing/thawing is normally dangerous because ice crystals form inside cells and rupture membranes. Also, the ice forms from water in the cells' cytosol, which dehydrates the cells. However, many insects, such as the wooly caterpillar, a few species of amphibia such as the wood

Figure 42.13   **Adaptation to extreme cold in an ectotherm.** Several species of insects, reptiles, and amphibia—like the wood frog shown here—can reduce the risk of freezing by increasing glucose levels, which decreases the freezing point of their body fluids. Furthermore, some species can tolerate being partially frozen because they can prevent ice crystals from forming inside their cells.

frog (**Figure 42.13**), and a very small number of reptiles such as the painted turtle, can block crystal formation in their cells. They do this by responding to ice on their skin surfaces with an enormous outpouring of glucose from the liver. The glucose enters the blood and the cells, lowering the freezing point so that the cells do not freeze solid. Other regions of the body that are less critical, such as the lumens of the stomach and bladder, do freeze. These animals can have 65% or more of their bodies completely frozen for long periods, only to thaw during warm periods without harmful effects. The glucose gets reabsorbed by the liver at that time.

Ectotherms and Endotherms May Have Fluctuating or Stable Body Temperatures

In the past, animals were classified into two categories. Cold-blooded animals require an external heat source such as sunlight to warm themselves. By contrast, warm-blooded animals use internal heat to maintain their body temperature. These terms are misleading, however, because many cold-blooded animals can generate considerable heat by exercising their skeletal muscles. Indeed, many have a body temperature during daylight hours that is at least as warm as that of warm-blooded animals like birds and mammals.

Biologists now classify animals according to both their source of heat and their ability to maintain body temperature. Recall that ectotherms depend on external heat sources to warm their bodies, while endotherms use their own metabolically generated heat to warm themselves. **Homeotherms** maintain their body temperature within a narrow range, while **heterotherms** have body temperatures that vary with the environment

Figure 42.14 **Body temperature and environmental temperature in homeotherms versus heterotherms.** Homeotherms maintain stable body temperatures across a wide range of environmental temperatures. Heterotherms, by contrast, have body temperatures that depend in part on the temperature of their surroundings.

Figure 42.15 **Types of heat exchange.** The four ways in which animals exchange heat with the environment are radiation, conduction, convection, and evaporation.

(**Figure 42.14**). Most animals fall into two categories. Birds and mammals are endothermic and homeothermic, while other vertebrates and most invertebrates are ectothermic and heterothermic.

Not all animals, however, can be neatly classified into two categories at all times. Hibernating mammals, for example, are endotherms. They are homeothermic, but during the winter their body temperature drops dramatically as their metabolism slows to conserve energy for the winter. Hibernators behave like heterotherms during the transition from fall to winter and again from winter to spring. During the winter, however, they are homeothermic except for brief periods of arousal, but at a lower body temperature than at other times of year. Similarly, a fish swimming in deep ocean waters is an ectotherm but also homeothermic because the temperature of the water—and therefore of its body—is essentially constant. Fish that live in waters with fluctuating temperatures, by contrast, are ectothermic and heterothermic.

Even endothermic homeotherms do not have truly constant body temperatures. They have a narrow range of body temperatures that increases or decreases slightly in extreme climates, during exercise, or even during sleep. The important feature is that birds and mammals can quickly adjust the body's mechanisms for retaining or releasing heat such that body temperature remains relatively stable. This provides the advantage that the body's chemical reactions are at optimal levels even when the environment imposes extreme challenges. The metabolic rate of a resting mammal, for example, is roughly six times greater than that of a comparably sized reptile. A suddenly awakened mammal is instantly capable of intense activity even on a winter day, but an icy-cold reptile could be at the mercy of a predator because of the time required to warm itself and flee.

Endothermy does have two major disadvantages, however. First, to produce sufficient heat by metabolic processes, endotherms must consume larger amounts of food to provide substrates for ATP production. Small endotherms with high BMRs, such as shrews, must eat almost continuously and may die if deprived of food for as little as a day. By contrast, many ectotherms, such as snakes, can go for weeks without eating. Second, endotherms run the risk of overheating during periods of intense activity, even in cold weather.

Animals Exchange Heat with the Environment in Four Ways

The surface of an animal's body can lose or gain heat from the external environment via four mechanisms. These include radiation, conduction, convection, and evaporation (**Figure 42.15**).

Radiation is the emission of electromagnetic waves by the surfaces of objects. The rate of emission is determined by the temperature of the radiating surface. Thus, if the body surface is warmer than the environment, the body loses heat at a rate that depends on the temperature difference. If the outside temperature is warmer than body temperature, the body gains heat, for instance from sunlight. We can observe radiated heat from

Exposed to cold   °F   After exercising

Figure 42.16 **Visualization of heat exchange during different conditions.** Thermal-imaging cameras can detect radiated heat from an animal's body. The photo on the left shows a woman who is outside in the cold, while the photo on the right shows the same person after exercising. Note the differences in heat radiation.

an animal's body with imaging devices that detect infrared light (**Figure 42.16**).

In **conduction**, the body surface loses or gains heat through direct contact with cooler or warmer substances. The greater the temperature difference, the greater the rate of heat transfer. Different materials have different abilities to absorb heat, however. As we saw in Chapter 2, water has a higher heat capacity than air, meaning that at any temperature water will retain greater amounts of heat than will air. Thus, aquatic animals in water that is 10°C lose considerably more heat in a short time than terrestrial animals in air that is 10°C. Indeed, on a hot day, terrestrial animals can lose heat efficiently by immersing themselves in cooler water. An animal's body surface area plays an important role in the rate of heat conduction across its body surface. In some animals, certain body regions are particularly good heat conductors, such as the ears of an elephant or the wings of a bat (**Figure 42.17**).

Convection is the transfer of heat by the movement of air or water next to the body. For example, the air close to an endotherm's body is heated by conduction. Since warm air is less dense than cold air, the warm air near the body rises and carries away heat by convection. Convection is aided by creating currents of air around an animal's body. Humans do this by sitting near fans, but other animals can create cooling air currents by other means such as when a bat flaps its wings or an elephant waves its ears.

Animals can also lose body heat through **evaporation** of water from the skin and membranes lining the respiratory tract, including the surface of the tongue. A large amount of energy in the form of heat is required to transform water from the liquid to the gaseous state. Thus, whenever water vaporizes from the body's surface, the heat required to drive the process is conducted from the surface, thereby cooling the animal.

The four processes of heat transfer just described can be regulated in animals, such that heat is retained within the body at some times and lost at other times, as we see next.

Several Mechanisms Can Alter Rates of Heat Gain or Loss

For purposes of temperature control, think of the body as a central core surrounded by a shell consisting of skin and subcutaneous (just below the skin) tissue. The temperature of the central core of endotherms is regulated at approximately 37–40°C, but the temperature of the outer surface of the skin varies considerably. If the skin were a perfect insulator, the body would never lose or gain heat by conduction. The skin does not insulate completely, however, so the temperature of its outer surface generally is somewhere between that of the external environment and the core. Only in animals that store large amounts of subcutaneous fat (blubber) does the body surface provide considerable insulation. In endotherms without blubber, the main form of insulation is a covering of hair, fur, or feathers, which traps heat from the body in a layer of warm air near the skin, reducing heat loss.

Changes in Skin Blood Flow Rather than acting as an insulator in most animals, the skin functions as a heat exchanger that can be adjusted to increase or decrease heat loss from the body.

(a)

(b)

Figure 42.17 Striking examples where high surface area releases heat. Structures with large surface area, such as (a) the elephant's ears and (b) the bat's wings, aid the conduction of heat from an animal's body into the surrounding air.

As more blood reaches the skin from the core, the skin's temperature approaches that of the core. In effect, blood vessels deliver heat to the skin surface, some of which is lost by conduction to the external environment (**Figure 42.18**). Temperature-sensitive nerves and smooth muscles largely control the opening and closing of these blood vessels. Diving birds and mammals are good examples of this mechanism. Ducks, seals, and walruses dramatically reduce the amount of blood flowing to the skin when they dive in cold waters. This allows them to retain body heat that would otherwise be conducted into the water. In many terrestrial endotherms, certain areas of skin play a more prominent role in heat exchange than others—recall the elephant ears and bat wings mentioned earlier—so skin temperature varies with location.

Countercurrent Heat Exchange Both endotherms and ectotherms regulate heat loss to the environment through **countercurrent heat exchange**, which conserves heat by returning it to the body's core and keeping the core much warmer than the extremities. In endotherms, countercurrent heat exchange occurs primarily in the extremities—the flippers of dolphins, for example, or the legs of birds and certain other terrestrial animals (**Figure 42.19a**). As warm blood travels through arteries down a bird's leg, heat moves by conduction from the artery to adjacent veins carrying cooler blood in the other direction (**Figure 42.19b**). By the time the arterial blood reaches the tip of the leg, its temperature has dropped considerably, reducing the amount of heat lost to the environment, and returning the heat to the body's core.

Ectotherms such as many fish use countercurrent heat exchange to warm their muscles. As the swimming muscles become active, they generate heat from metabolism that warms the blood in the veins leaving the muscles. The veins are in close contact with nearby arteries bringing fresh, oxygen-rich blood from the gills. In tuna, as in other ectotherms, arterial blood is cold. Blood entering the gills comes in contact with cold seawater, and its temperature rapidly adjusts to that of the seawater.

Section of skin

Outer layer of skin

Cold day

These vessels constrict to reduce heat loss.

These vessels are a bypass for blood.

Warm blood enters from arteries.

Blood leaves in veins, retaining most of its heat.

Hot day

Heat is lost when these vessels dilate.

These vessels constrict to shunt blood to the skin surface.

Warm blood enters from arteries.

Blood leaves in veins, having lost some of its heat.

Figure 42.18 Regulation of heat exchange in the skin. The skin functions as a variable heat exchanger. Skin surface blood vessels constrict (get narrower to reduce blood flow) on cold days to retain body heat, and they dilate (widen to increase blood flow) on hot days to dissipate heat to the environment. The arrows in the blood vessels indicate direction of blood flow.

Warm arterial
blood
40°C   38°C

5°C

10°C   5°C
Cold venous
blood

Heat is transferred from arteries to nearby veins, thus reducing heat loss to the water.

(a) Countercurrent heat exchange in the leg of an endotherm

Artery

Veins   Artery

Veins

(b) Cross section and surface view of veins covering an artery

Figure 42.19   Countercurrent heat exchange.   (a) Countercurrent exchange retains heat in the leg of an endotherm. Arrows in vessels indicate direction of blood flow. **(b)** An SEM of the arrangement of veins surrounding an artery in a bird's leg. The artery is almost completely covered by overlying veins, allowing efficient heat exchange between the vessels.

As the cold blood reaches the muscles, however, heat from the warm veins leaving the muscles is conducted from the veins into the nearby arteries entering the muscle. In this way, the heat generated by the muscles is returned to the muscles rather than being lost from the gills. As warm temperatures stimulate the rate of chemical reactions, the muscles are able to operate more efficiently.

Evaporative Heat Loss Recall that animals can lose body heat through evaporation of water from the skin and membranes lining the respiratory tract. Heat exchange can be regulated by changing the rate of water evaporation via perspiration. Nerves to the sweat glands stimulate the production of sweat, a dilute solution containing sodium chloride. The most important factor determining evaporation rate—and therefore heat loss—is the water vapor concentration, or humidity, of the air. Your discomfort on a humid day is due to the failure of evaporation. Your sweat glands continue to secrete, but the sweat simply remains on your skin. In endotherms that lack sweat glands, panting (short, rapid breaths with the mouth open) helps evaporate water from the tongue surface. Panting has advantages

over sweating, because no salt is lost, and panting provides the air current that promotes heat exchange by convection. However, the surface area of the mouth and tongue is relatively small, which limits the rate at which heat can be eliminated.

Animals that neither sweat nor pant can still benefit from evaporative heat loss. Licking the skin or splashing the skin with water also draws heat from the body.

Behavioral Adaptations Behavioral mechanisms can also alter heat loss by radiation and conduction. Two such behaviors involve changing exposed surface area and changing surroundings. Curling up into a ball, huddling in groups, burying the head and feet in feathers, hunching the shoulders, and similar maneuvers reduce the surface area exposed to a cold environment and decrease heat loss by radiation and conduction (**Figure 42.20**). For example, Emperor penguins huddle together to survive months of inactivity in Antarctica at environmental temperatures lower than −30°C. Conduction occurs only when two objects such as the body surface and the air have different temperatures, so huddling surrounds the penguins with air that is almost as warm as their own skin and greatly reduces heat loss.

People adjust their exposed surface area by wearing clothing, our substitute for the insulating feathers of birds and fur of other mammals. The outer surface of the clothing becomes the exterior of the body. The skin loses heat to the air space trapped by the clothes, which in turn pick up heat from the inner air layer and transfer it to the external environment. The thickness of the trapped air layer determines the insulating ability of the garments.

The other major behavioral mechanism for altering heat exchange between the body and the environment is to seek more comfortable surroundings, such as moving from the shade into the sunlight on a cold day. Although endotherms do this to prevent small swings in body temperature, it is the critical way in which ectotherms obtain or lose heat. Many terrestrial ectotherms will seek shade or burrow into the ground when the sun is high, or they will bask on hot objects to conduct heat into their bodies when they need to warm up.

Muscle Activity and Brown Fat Metabolism Increase Heat Production

We have discussed how heat is gained or lost to the environment, and how heat can be retained by reducing blood flow to the skin on a cold day. Body temperature, however, is a balance between these factors and heat production. Changes in muscle activity constitute the major control of heat production for temperature regulation in endotherms.

When an endotherm is in its thermoneutral zone, no significant adjustments are necessary to maintain core body temperature. When exposed to conditions below the thermoneutral zone, however, core body temperature begins to fall. One of the first responses to this change is a general increase in skeletal muscle contraction. This may lead to shivering, which consists of rapid muscle contractions without any locomotion.

(a) (b) (c)

(d) (e)

Figure 42.20 Examples of behavioral mechanisms by which animals cope with extreme temperatures. Common behaviors include (a) burrowing, (b) huddling together, (c) and (d) postural changes, and (e) seeking more comfortable surroundings.

Biological inquiry: Clothing serves as an effective insulator against heat loss in humans, because humans are endothermic and their bodies are relatively free of insulating hair or fur. In very hot environments, however, humans, like all endotherms, must reduce heat gain from the environment. Under such conditions, which would be a better way to reduce heat gain from the sun, wearing loose fitting clothing or going nude?

Virtually all of the energy liberated by the contracting muscles appears as internal heat, known as **shivering thermogenesis**. Many birds that remain in cold climates during the winter shiver almost continuously.

In many mammals, chronic cold exposure induces **nonshivering thermogenesis**, an increase in the metabolic rate and therefore heat production that is not due to increased muscle activity. Nonshivering thermogenesis occurs primarily in **brown fat**, a specialized tissue in small mammals such as hibernating bats, small rodents living in cold environments, and many newborn mammals including humans. Brown fat is responsive to hormones and signals from the nervous system, which are activated when body temperature falls. The mitochondria in the brown fat contain uncoupling proteins, a class of proteins that uncouple oxidative phosphorylation, described in Chapter 7. The H^+ gradient becomes uncoupled from ATP synthesis. Instead, the H^+ gradient is used to generate heat, which helps maintain body temperature.

Endotherms Can Adapt to Chronic Changes in Their Environments

Although some animals spend their lives in relatively constant environmental temperatures, most do not. Such animals must adapt to environmental changes using the mechanisms we have discussed in this section. In some cases, however, long-term exposure to a challenging environment, either very hot or very cold, can fine-tune the adaptive mechanisms that persist as long as the animal lives in that environment. This process of **acclimatization** occurs particularly well in humans, who have more sweat glands than any known mammal. Increases in the amount and speed of sweat production, and a decrease in the temperature threshold for initiating sweating, for example, can acclimatize an animal to chronic high temperatures. A person newly arrived in a hot climate has trouble coping. Body temperature rises, and too much activity can lead to a breakdown in the body's temperature-regulating systems. After several days to weeks, however, acclimatization begins. Body temperature stabilizes and it becomes much easier to function. Body temperature does not rise as much because sweating begins sooner and the volume of sweat produced is greater. Long-term exposure to hot conditions also increases the dilation of skin blood vessels and blood flow to the skin, helping to dissipate heat by conduction.

Cold acclimatization has been less studied than heat acclimatization, but seasonal changes occur in many endotherms that live in variable climates. Birds grow an extra layer of insulating feathers, while mammals may grow additional fur in the winter. Feathers and fur are shed in the summer. The additional insulation reduces heat loss up to 50% in some animals. Ectotherms, too, can acclimatize to changes in temperature. For example, marine animals that migrate from warm to very cold waters undergo a change in the lipid composition of their cellular membranes such that they contain more unsaturated fats and therefore remain fluid even at extremely cold temperatures. In addition, many cold-acclimated ectotherms have cellular enzymes with a wider range of temperature tolerance than those that are found in animals living in mild climates.

42.5 Impact on Public Health

As we have seen, most animals, if given adequate nutrients, maintain their body mass around a set point that is normal for that species. Healthy animals may overeat if given the opportunity, but they will never choose to undereat if given sufficient food. It is rare to observe healthy animals in nature that are overweight. Generally, only domesticated animals become sufficiently sedentary that they gain weight (think of an overweight housecat). Humans, too, are prone to weight gain, particularly when living sedentary lives. Many people maintain a healthy body weight but, as discussed in this section, an increasing number are unable to meet this goal.

Obesity Is a Global Health Issue

Excess body fat—in humans and in other species, too—increases the risk of many diseases, including high blood pressure, cancer, heart disease, and diabetes. At what point does fat accumulation in humans start to pose a health risk? Historically, this question has been evaluated by research studies that investigate possible correlations between disease rates with some measure of body fat.

One of the currently preferred methods for assessing body fat and health risks is the **body mass index (BMI)**, a ratio of weight compared to height. A person's BMI is calculated by dividing the weight in kilograms by the square of the height in meters. For example, a 70-kg human with a height of 180 cm would have a BMI of 21.6 ($70/1.8^2$).

Current National Institutes of Health guidelines categorize BMIs of 25 or more as **overweight**, that is, as having some increased health risk because of excess fat. BMIs of 30 or greater are considered **obese**, with a highly increased health risk. The Centers for Disease Control and Prevention in Atlanta, Georgia,

estimates that 65% of U.S. adults age 20 and older are now overweight, and 25–30% are obese. Since the early 1990s, the CDC estimates that the average body weight of Americans has risen by 10 pounds. This is a worldwide trend. According to the World Health Organization, more than 1 billion adults globally are overweight and 300 million are obese. Even more troubling, the rate of childhood obesity is also rising.

The increased health risk associated with obesity may be at least partly due to a lack of physical activity, not body fat *per se*. Humans today are more sedentary than past generations, with the advent of television, computer games, and the Internet and a move from farming to urban lifestyles. Lack of exercise is associated with fat accumulation because it reduces total daily energy expenditure.

Growing evidence suggests that the location of body fat has important health consequences, too. For reasons that are not yet clear, people with mostly abdominal fat ("apples") are at greater risk for developing serious conditions such as diabetes and cardiovascular disease than are people whose fat is mainly in the lower body, on the buttocks, hips, and thighs ("pears") (**Figure 42.21**).

Some studies indicate that genetic factors play an important role in obesity. Identical twins separated soon after birth and raised in different households have strikingly similar body weights as adults. Researchers hypothesize that natural selection favored the evolution of so-called **thrifty genes**, which boosted our ancestors' ability to store fat from each feast in order to sustain them through the next famine. Given today's abundance of high-fat foods in many countries, what was once a survival mechanism could now be a liability.

The methods and goals of treating obesity are undergoing extensive rethinking. An increase in body fat must be due to an excess of food intake over energy expenditure, and overweight people have traditionally been advised to follow a low-calorie diet. However, such diets alone have limited effectiveness, because over 90% of obese people regain most or all of their lost weight within five years. This disturbing phenomenon may

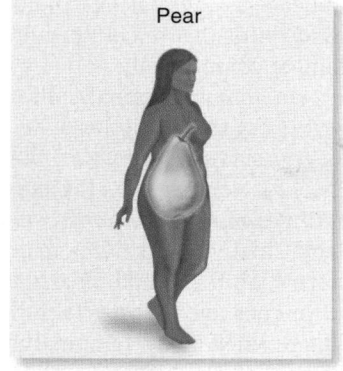

Figure 42.21   Distribution patterns of fat in overweight humans. Men tend to be of the "apple" type, while women are more likely to be "pears." The apple pattern of abdominal obesity is more closely associated with adverse health consequences.

be related to the observation that metabolic rate decreases as leptin levels decrease, sometimes falling low enough to prevent further weight loss on as little as 1,000 calories per day.

Research indicates that crash diets are not an effective long-term method for controlling weight. Instead, caloric intake should be set at a realistic level that can be maintained for the rest of one's life. This should lead to a slow, steady weight loss of no more than 1 pound per week until body weight stabilizes at a new, lower level. Most important, any program of weight loss should include increased physical activity. The exercise itself burns calories, but more importantly it partially offsets the tendency, described earlier, for the metabolic rate to decrease during long-term weight loss. As a bonus, the combination of exercise and caloric restriction causes the person to lose more fat and less protein than with caloric restriction alone.

The impact of obesity on public health is enormous, accounting for many illnesses requiring hospitalization and chronic drug therapy, and nearly 400,000 premature deaths per year. Its impact is far-reaching on the economy as well. The toll of obesity-related illnesses is felt in the loss of worker-hours in the workplace, for example. Obesity can also impact society in unexpected ways. As an example, the increasing weight load of passengers forces airplanes to burn 350 million additional gallons of fuel each year, compared to just 10 years ago! This translates into nearly 4 million tons of additional pollution released into the atmosphere every year.

Anorexia Nervosa and Bulimia Nervosa Can Have Serious Health Consequences

Being obese can be dangerous, but so can being extremely underweight. One eating disorder, anorexia nervosa, is characterized by weight loss and is found primarily in adolescent girls and young women.

People with anorexia nervosa become pathologically obsessed with weight and body image, and they reduce their food intake to the point of starving. Anorexia nervosa may have both biological and psychological causes. Many other abnormalities are associated with the disorder, such as loss of menstrual periods, low blood pressure, low body temperature, and altered secretion of many hormones. These may be simply signs of starvation, although they may also result from malfunction of the part of the brain that normally controls appetite.

Another eating disorder, called bulimia nervosa but often simply called bulimia, involves recurrent episodes of fasting and overeating. Weight loss may occur because of regular self-induced vomiting and the use of laxatives to stimulate bowel movements or diuretics to stimulate urination, as well as strict dieting, fasting, or vigorous exercise. As with individuals with anorexia nervosa, people with bulimia are obsessed with body weight, although they are usually within 10% of their ideal weight. Successful treatments of anorexia and bulimia rely primarily on counseling, nutritional education, and medications.

CHAPTER SUMMARY

42.1 Nutrient Utilization and Storage

- From an energetic point of view, an animal's life can be divided into two phases: the absorptive state, during which ingested nutrients are entering the blood from the gastrointestinal tract, and the postabsorptive state, during which the GI tract is empty of nutrients and the body's own stores must supply energy.

- Glucose and fats are the body's two major energy sources during the absorptive state. Much of the absorbed glucose immediately enters cells and is enzymatically broken down to carbon dioxide and water, in the process providing the energy required to synthesize ATP. Most absorbed triglycerides are stored in fat cells until the body requires additional energy. Amino acids are taken up by all body cells, where they are used to synthesize proteins. (Figure 42.1)

- The events that maintain blood glucose concentration fall into two categories: (1) reactions that provide glucose to the blood and (2) cellular utilization of fat for energy, which spares glucose for use by the nervous system. (Figure 42.2)

- An animal can increase its blood glucose levels during the postabsorptive period via glycogenolysis or gluconeogenesis. Most organs and tissues reduce their glucose utilization and increase their fat utilization during this time. (Table 42.1)

42.2 Regulation of the Absorptive and Postabsorptive States

- Tight control mechanisms, in the form of several hormones and the nervous system, maintain homeostatic levels of fuel in the blood. The hormone insulin acts on cells to facilitate the diffusion of glucose from blood into the cell cytosol. (Figure 42.3)

- All animal cells utilize glucose transporter proteins (GLUTs) to move glucose across their plasma membranes. In vertebrates, an increase in blood glucose levels stimulates the cells of the pancreas to secrete insulin, and a decrease in glucose removes the signal for secretion. (Table 42.2, Figures 42.4, 42.5)

- Exercise is any type of physical activity that increases an animal's metabolic rate. Exercise increases an animal's requirement for nutrients to provide energy. The source of these nutrients includes glucose and fatty acids.

42.3 Energy Expenditure

- An animal's energy expenditure refers to the amount of energy it uses in a given period of time to power all of its metabolic needs.

- The most common method for comparing metabolic rates of different species is to compare the basal metabolic rate (BMR), most of which is due to the routine functions of the heart, liver, kidneys, and brain. (Figures 42.6, 42.7)

- Many factors have an impact on metabolism, including skeletal muscle activity, whether an animal has recently eaten, and body size. (Figures 42.8, 42.9)

- When the daily amount of energy consumed is equal to the amount of energy expended, an animal's body weight remains stable. Tipping the balance in either direction causes weight gain or loss by increasing or decreasing total-body energy content.

- Short-term control of feeding generally involves satiety signals that remove the sensation of hunger and set the time period before hunger returns again. Experiments by Coleman and Friedman investigated leptin as a satiety factor in mammals. Leptin has since been found in all classes of vertebrates. (Figures 42.10, 42.11, 42.12)

- Torpor and hibernation allow animals to conserve energy during winter.

42.4 Regulation of Body Temperature

- Most animals can survive only in a relatively narrow temperature range that allows molecules in the body to move and react with one another, maintains the structures of membranes, and avoids denaturing proteins. (Figure 42.13)

- Animals can be classified according to their source of heat and their ability to maintain body temperature. Ectotherms depend on external heat sources to warm their bodies, while endotherms use their own metabolically generated heat to warm themselves. Homeotherms maintain their body temperature within a narrow range, while heterotherms have body temperatures that vary with the environment. All animals fall into two of these categories. (Figure 42.14)

- The surface of an animal's body can lose or gain heat from the external environment via four mechanisms: radiation, conduction, convection, and evaporation. (Figures 42.15, 42.16, 42.17)

- Blood vessels deliver heat to the skin surface to be lost by conduction to the external environment. Both endotherms and ectotherms regulate heat loss through countercurrent heat exchange, which conserves heat by returning it to the body's core and keeping it warmer than the extremities. Heat exchange can also be regulated by changing the rate of water evaporation via perspiration. Behavioral mechanisms can alter heat loss by radiation, conduction, and convection. (Figures 42.18, 42.19, 42.20)

- Muscle activity and brown fat metabolism increase the production of heat.

- Acclimatization can fine-tune an animal's adaptive mechanisms to a changing environment.

42.5 Impact on Public Health

- Excess body fat increases the risk of many diseases. A body mass index (BMI) of 25 or more is generally considered overweight, while a BMI of 30 or greater is considered obese. (Figure 42.21)

- Being underweight is also unhealthy. People with anorexia nervosa become pathologically obsessed with weight and body image and reduce their food intake to the point of starving. Bulimia nervosa involves recurrent episodes of fasting and overeating, plus self-induced vomiting of ingested food. Both eating disorders can be treated with counseling, nutritional education, and medications.

TEST YOURSELF

1. During the absorptive phase, an animal is
   a. fasting.
   b. relying on stored molecules for energy.
   c. absorbing nutrients from a recently ingested meal.
   d. metabolizing lipids stored in adipose tissue to supply ATP to its cells.
   e. both a and b.

2. Gluconeogenesis occurs when
   a. the liver synthesizes glucose from noncarbohydrate precursors.
   b. glycogen is broken down to glucose.
   c. an animal is in the absorptive phase.
   d. triglycerides are being stored in adipose cells.
   e. none of the above.

3. During _____, most tissues of the body increase metabolism of fat to ensure that _____ tissue has an adequate supply of glucose.
   a. gluconeogenesis, muscle
   b. glucose sparing, epithelial
   c. gluconeogenesis, nervous system
   d. glucose sparing, nervous system
   e. gluconeogenesis, epithelial

4. Ketones are compounds derived from
   a. glucose.
   b. glycogen.
   c. fatty acids.
   d. amino acids.
   e. proteins.

5. Insulin primarily regulates blood glucose levels by
   a. stimulating the recruitment of glucose transporter proteins from the cytosol to the plasma membrane for transport of glucose from extracellular to intracellular fluid.
   b. stimulating gluconeogenesis.
   c. suppressing glucose uptake by muscle tissue.
   d. stimulating the release of glucose from glycogen reserves.
   e. none of the above.

6. The rate at which the body uses fuel to synthesize ATP, which is used for all the activities and chemical reactions occurring in an animal's body, is
   a. body mass index.
   b. an animal's energy consumption.
   c. metabolic rate.
   d. an animal's energy expenditure.
   e. both c and d.

7. Which factor or factors may affect metabolic rate?
   a. muscle activity
   b. food-induced thermogenesis
   c. an animal's body size
   d. fasting
   e. all of the above

8. The molecule that has an effect on brain centers in reducing appetite and increasing metabolic rate is
   a. GLUT.
   b. insulin.
   c. leptin.
   d. glycogen.
   e. ketones.

9. Animals that have body temperatures that are maintained within a very narrow range are
    a. endotherms.
    b. ectotherms.
    c. homeotherms.
    d. heterotherms.
    e. both b and d.

10. The rate of heat loss is regulated by
    a. the degree of blood flow at the surface of the skin.
    b. the level of perspiration.
    c. behavioral adaptations.
    d. all of the above.
    e. a and b only.

CONCEPTUAL QUESTIONS

1. Explain the function of insulin.
2. Explain how appetite is controlled.
3. Explain countercurrent heat exchange.

EXPERIMENTAL QUESTIONS

1. What observation led to the experiments conducted by Coleman?
2. What was the hypothesis tested by Coleman, and how did he test it?
3. How did the experimental linking of the bloodstreams of the wild-type mice and the mutant mice affect the body weight of both strains?

COLLABORATIVE QUESTIONS

1. Discuss the differences between being ectothermic and endothermic, and heterothermic and homeothermic.
2. Discuss four ways animals exchange heat with their environment.

---

www.brookerbiology.com

This website includes answers to the Biological Inquiry questions found in the figure legends and all end-of-chapter questions.

43

NEUROSCIENCE I: CELLS OF THE NERVOUS SYSTEM

CHAPTER OUTLINE

A human neuron. These types of cells are responsible for sending signals throughout the nervous systems of animals.

When we speak of someone being "nervous," we imagine a person who fidgets a lot and has trouble maintaining self-control. Ironically, such imagery contradicts the amazing function of **nervous systems**, which are groups of cells that sense internal and environmental changes and transmit signals that enable an animal to respond in an appropriate way. Nervous systems help us to exert control over our bodies. **Neuroscience** is the scientific study of nervous systems. Neuroscientists are interested in topics like the structure and function of the brain, and the biological basis of consciousness, memory, and learning. Neuroscience is now at an unprecedented level of new discoveries and rapid growth. It interfaces with other disciplines such as cell biology, psychology, and behavioral biology.

In this chapter, our focus on neuroscience will be at the cellular level. Nervous systems are composed of circuits of **neurons**, highly specialized cells that communicate with each other and with other types of cells by electrical or chemical signals (see chapter-opening photo). In complex animals, neurons become organized into **nuclei** (singular, *nucleus*), clusters of cells that carry out a particular function. Nuclei typically form in a central processing area of the nervous system called a **brain**. The brain sends commands to various parts of the body through **nerves**—bundles of neuronal cell projections running to various tissues and organs of the body. Nerves control an animal's movements, its ability to respond to changes in its environment, and its ability to maintain internal homeostasis.

Before you continue reading, stop and think. Can you describe everything that your nervous system is doing right now? Nervous systems allow us to sense what is going on in the outside world, initiate actions that influence events and respond to demands, and regulate internal processes—all while maintaining homeostasis. You are conscious of some of these functions, such as reading this text, listening to music, or feeling hungry. However, many others, such as maintaining your body temperature and controlling your heart rate, occur without your awareness.

In this chapter we will investigate the special features of neurons that make them suited for rapid communication between cells. The next two chapters will explore how nervous systems are organized, how the brain functions, and how animals use their nervous systems to sense the world around them.

43.1 Cellular Components of Nervous Systems

Nervous systems are organized to provide extremely rapid responses to changes in an animal's external or internal environment, and to maintain homeostasis. In complex animals, the **central nervous system (CNS)** consists of a brain and a nerve cord, which in vertebrates extends from the brain through the vertebral column and is called the **spinal cord**. The **peripheral nervous system (PNS)** consists of all neurons and projections of their plasma membranes that are outside of and connect with the CNS, such as projections that end on muscle and gland cells. In certain invertebrates with simple nervous systems, the distinction between central and peripheral nervous systems is less clear or not present.

The evolution of nervous systems has allowed animals to receive information about the environment via their PNS, interpret that information in a central nervous system, and, if necessary, initiate a behavioral response via their PNS (**Figure 43.1**).

Figure 43.1 **Roles of the central and peripheral nervous systems.** In this example, a hungry hyena senses a smell, which the brain interprets as a potential food source. This initiates a biological response (salivation) that prepares the hyena for eating.

For example, if a hungry hyena receives a stimulus such as the smell of food, odor-sensing cells in the nose act as receptors for the stimulus and then send signals to the brain, which interprets it and thereby recognizes the smell. The brain then sends a signal to nerves that stimulate gland cells in the mouth, which respond by producing saliva in preparation for the arrival of food. In this section, we will survey the general properties of the cells of the CNS and PNS.

Cells of the Nervous System Are Specialized to Transfer Signals

Nervous systems transfer signals from one part of the body to another and direct the activities of cells, tissues, and organs. While these are complex tasks, nervous systems have only two unique classes of cells, neurons and glia.

Neurons Neurons are the cells in the nervous system that send and receive electrical and chemical signals to and from other neurons throughout the body (**Figure 43.2**). All animals except sponges have neurons. The number of neurons in the nervous systems of different species varies widely, partly as a function of the size of an animal's head and brain, but also as a function of the complexity of its behavior. Thus, the tiny, short-lived

(a) Micrograph of a neuron

(b) A myelinated neuron

Figure 43.2 **Structure of a typical neuron and associated glial cells.** (a) A neuron seen at high magnification (confocal microscopy). (b) A diagrammatic representation of a neuron with glial cells, also known as Schwann cells. The Schwann cells wrap their membranes around the axon at regular intervals, creating a myelin sheath.

nematode *C. elegans* has only 302 neurons in its nervous system, compared with several thousand in a wasp, several hundred thousand in a salamander, 300 million in an octopus, and over 100 billion neurons in a human!

Regardless of the total number, neurons in one animal species look and act much like neurons from any other species. A neuron is composed of a **cell body** or **soma**, which contains the cell nucleus and other organelles. Two types of extensions or projections that arise from the cell body are the dendrites and the axon. **Dendrites** (from the Greek word *dendron*, meaning tree) may be single projections of the plasma membrane but more commonly are elaborate treelike structures with numerous branching extensions. Chemical and electrical messages from other neurons are received by the dendrites, and electrical signals move toward the cell body. An **axon** is an extension of the plasma membrane that is involved in sending signals from a neuron to neighboring cells. An axon may be less than a millimeter long in an insect or as long as 2 m in a giraffe's neck. A typical neuron has a single axon, which may have branches. The part of the axon closest to the cell body is named the **axon hillock**, which, as we will see later, is important in nerve cell communication. At the other end of the axon are terminal branches or nerve terminals that send electrical or chemical messages to other cells, such as neurons or muscle cells.

Within an animal's body, many axons tend to run in parallel bundles to form nerves. Each axon within a nerve is surrounded by glial cells, and the entire bundle is covered by a protective layer of connective tissue. Nerves enter and leave the CNS, and they make contacts with structures outside the CNS. Along the way, the terminal branches of axons communicate with particular cells of the body. Nerves transmit signals between the CNS and PNS.

Glia Surrounding the neurons are cells called the **glia** (from the Greek word meaning "glue"). Glial cells perform various functions and are many times more numerous than neurons, depending on the species. In the human brain, for example, glial cells outnumber neurons by about 50 to 1. In vertebrates, specialized glial cells wrap around the axons to form an insulating layer called a **myelin sheath** (Figure 43.2). In the human brain and spinal cord, these cells are called **oligodendrocytes**, while **Schwann cells** are the glial cells that form myelin on axons that travel outside the brain and spinal cord. As we will see later, myelin increases the speed with which electrical signals pass down the axon.

Glial cells also perform other roles. Astrocytes, which are a type of glial cell, provide metabolic support for neurons and also are involved in forming the blood-brain barrier, which is a physical barrier between blood vessels and most parts of the central nervous system. This barrier protects the CNS by preventing the passage of toxins and other damaging chemicals from the blood into the CNS. Astrocytes also help to maintain a constant concentration of ions in the extracellular fluid. Other glia, called microglia, remove cellular debris produced by damaged or dying cells. In developing embryos, glia form tracks along which neurons migrate to form the nervous system. In addition, some glia function as stem cells to produce more glial cells and neurons.

Sensory, Motor, and Interneurons Form Pathways in a Nervous System

Neurons can be categorized into three main types: sensory neurons, motor neurons, and interneurons. The structures of these types of neurons reflect their specialized functions (**Figure 43.3**).

Sensory Neurons As their name suggests, the role of **sensory neurons** is to detect or sense information from the outside world, such as light, sound, touch, and heat. In addition, sensory neurons detect internal body conditions such as blood pressure and body temperature. Sensory neurons are also called afferent neurons because they transmit information to the CNS. They often have a very long axon, which is involved in rapidly

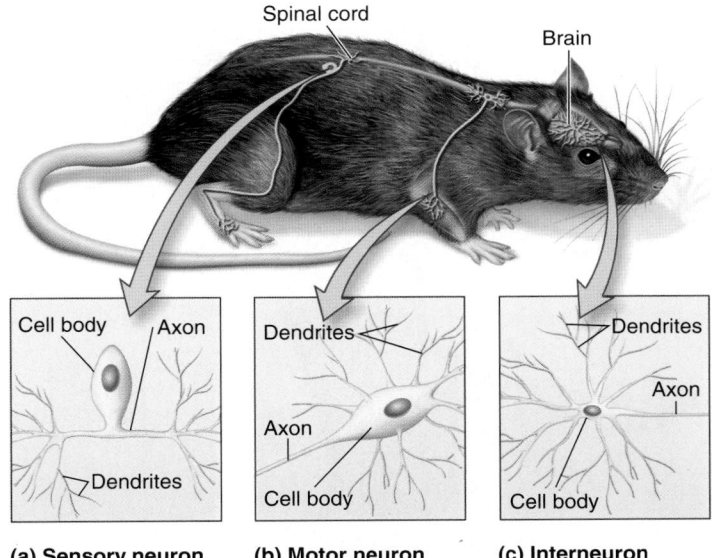

(a) Sensory neuron   **(b) Motor neuron**   **(c) Interneuron**

Figure 43.3 **Types of neurons.** **(a)** The vertebrate sensory neurons are afferent neurons with an axon that bypasses the cell body and projects to the CNS. **(b)** Motor neurons are efferent neurons that leave the CNS and usually have long axons that enable them to act on distant cells. **(c)** Interneurons are usually short neurons that connect two or more other neurons within the CNS. Although short, the axons and dendrites may have extensive branches, allowing them to receive many inputs and transmit signals to many neurons.

Biological inquiry: In which direction do signals move along the axons of motor neurons?

communicating signals. Sensory neurons are unusual because the axon bypasses the cell body and goes directly to the CNS (Figure 43.3a). This arrangement allows for the rapid transmission of a sensory signal to the CNS.

Motor Neurons The role of **motor neurons** is to send signals away from the CNS and elicit some type of response. They are so named because one type of response they cause is movement. However, motor neurons may cause other effects such as the secretion of hormones from endocrine glands. Because they send signals away from the brain, motor neurons are also called efferent neurons. Like sensory neurons, motor neurons tend to have long axons (Figure 43.3b). However, the axons of motor neurons originate from an axon hillock at the cell body.

Interneurons A third type of neuron, called the **interneuron**, forms interconnections between other neurons in the CNS. The greatest complexity of nervous systems occurs among interneurons, such as those found in the brain. The signals that are sent between interneurons are critical in the interpretation of information that the CNS receives, as well as the response that it may elicit. Interneurons tend to have many dendrites, and their axons are typically short and highly branched (Figure 43.3c). This arrangement allows interneurons to form complex connections with other cells.

Figure 43.4 **A reflex arc.** The knee-jerk response is an example of a reflex arc. A tap on the kneecap acts as a stimulus, initiating a reflex arc that causes the extensor muscle on top of the thigh to contract. At the same time, an interneuron inhibits the motor neuron of the flexor muscle, causing it to relax.

Reflex Circuits As a way to understand the interplay between sensory neurons, interneurons, and motor neurons, let's consider a simple example in which these types of neurons form interconnections with each other. Neurons transmit information to each other through a series of connections that form a circuit. An example of a simple circuit is a **reflex arc**, which allows an organism to respond rapidly to inputs from sensory neurons and consists of only a few neurons (**Figure 43.4**). The stimulus from sensory neurons is sent to the CNS, but there is little or no interpretation of the signal; few, if any, interneurons are involved. The signal is then transmitted to motor neurons, which elicit a response, such as a knee jerk. Such a response is very quick and automatic. Reflexes are among the oldest and most important features of nervous systems, because they allow animals to respond quickly to potentially dangerous events. For instance, many vertebrates will immediately cringe, jump, leap, or take flight in response to a loud noise, which could represent sudden danger. Animals that live in the water will reflexively dive deeper in response to a shadow overhead, which could represent a passing shark or other predator. Infant primates have strong grasping reflexes that help them hold onto their mothers as they move about. Countless examples of useful reflexes are found in animals, and their importance is evident from the observation that they exist in simple animals, such as flatworms, as well as complex mammals, such as humans.

43.2 Electrical Properties of Neurons

In the late 18th century, Italian scientists Luigi Galvani and Alessandro Volta experimented with ways to stimulate the contraction of frog leg muscles placed in salt solutions. They dis-covered that the stimulation of the nerve that led to the muscle, or stimulation of the muscle itself with any source of electric current, caused the muscle to contract. Eventually, Galvani postulated that electric current could somehow be generated by the tissue itself, something he called "animal electricity."

Today, we know that Galvani's animal electricity comes from neurons, which use electrical signals to communicate with other neurons, muscles, or glands. These signals, or **nerve impulses**, involve changes in the amount of electric charge across a cell's plasma membrane. In this section, we examine the concept of chemical and electrical gradients across the plasma membrane of neurons. Later in this chapter, we will explore how such gradients provide a way for neurons to conduct signals.

Neurons Establish Differences in Ion Concentration and Electric Charge Across Their Membranes

Like cellular membranes throughout the body, the plasma membrane of a neuron acts as a barrier that separates charges. Ion concentrations differ between the interior and exterior of the cell. Such differences in charge act as an electrical force measured in **volts**, named after Alessandro Volta. Analogous to a battery, there are positive and negative poles, but these are the outside and inside of the plasma membrane. For this reason, a neuron is said to be electrically **polarized**. The difference between the electric charges inside and outside the cell is called a potential difference or **membrane potential**. The **resting potential** refers to the membrane potential of a cell that is not sending nerve impulses.

How do scientists measure electrical changes in a structure as tiny as a neuron? Several invertebrates, such as squid, lobsters, and earthworms, have large neurons that have been used successfully to measure membrane potentials. The squid giant

Section of a squid giant axon in a bathing medium. Forceps are used to remove surrounding muscle.

(a)

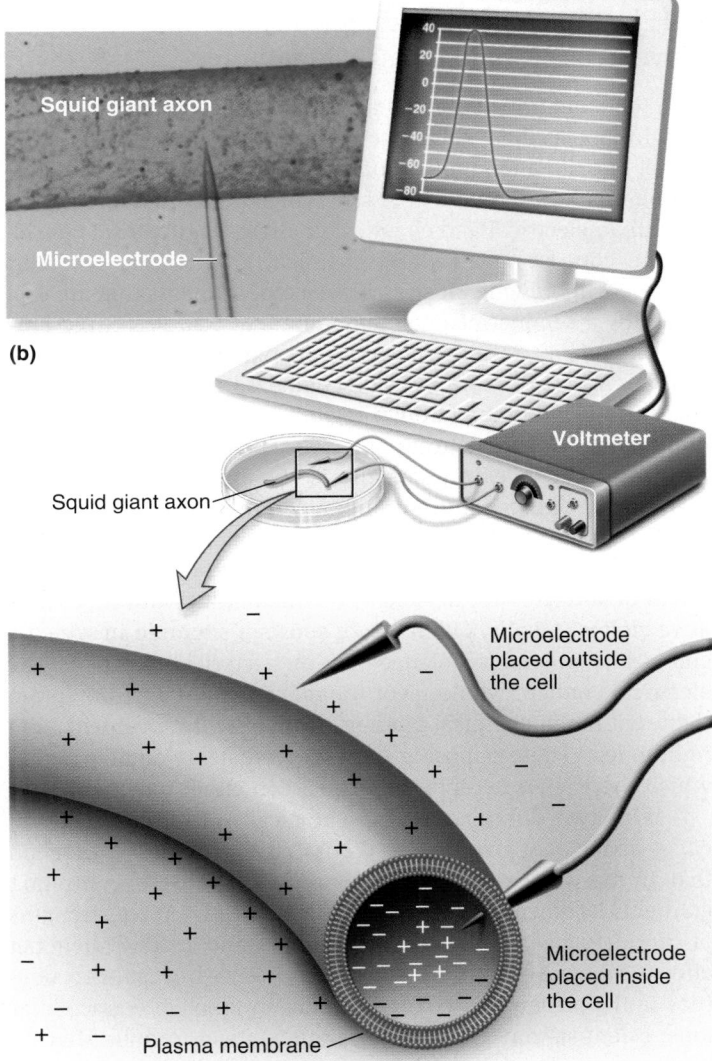

Squid giant axon

Microelectrode

(b)

Voltmeter

Squid giant axon

Microelectrode placed outside the cell

Microelectrode placed inside the cell

Plasma membrane

(c)

Figure 43.5 **Recording the membrane potential of neurons.** (a) The giant axon from a squid is removed from the animal and placed in a bathing medium. (b) A microelectrode, seen here under a microscope, impales the axon. (c) The potential difference across the membrane of the axon is recorded by comparing the charge inside and outside the axon. The membrane potential is steady at times and fluctuates at other times, depending on experimental conditions.

axon, for example, has a diameter up to 1 mm, making it relatively easy to impale with a microelectrode, which is a recording instrument constructed of a glass pipette with an extremely thin tip of less than 1 μm in diameter. Within the microelectrode is a salt solution that can conduct ions. The squid giant axon has been utilized extensively since the mid-20th century to determine the mechanisms by which nerve cells generate electrical signals (**Figure 43.5**). A neuron is dissected from the squid and placed in a solution that has similar ionic concentrations as normal extracellular fluid. A microelectrode is pushed through the axon membrane to record from inside the cell, and another microelectrode is placed within the solution bathing the neuron outside the cell. A voltmeter records the voltage difference between the two microelectrodes, which is a measure of the membrane potential. This measurement is made as a function of time and displayed on a computer screen.

Let's begin our discussion of electrical signaling by examining how the resting potential is established and maintained. The plasma membrane is selectively permeable to cations and anions, so it separates charge by keeping different ions largely inside or outside the cell. When investigators first measured the resting potential of a squid giant axon, they registered a voltage that read about −70 mV (millivolts) inside the cell with respect to the outside bathing medium. This means that the interior of the cell had a more negative charge than the exterior, which turns out to be true of all animal cells in their resting state. A resting potential of −70 mV is tiny compared to the voltages used to provide electric current in a home (approximately 120 V), or even that of a small 1.5-V battery. Nonetheless, this tiny difference in charge across the membrane of a neuron is sufficient to generate a nerve impulse that can travel from one end of a neuron to the other, as we will see later in this chapter.

The resting potential is determined by the ions located along the inner and outer surfaces of the plasma membrane (**Figure 43.6a**). Ions of opposite charges align on either side of the membrane because they are drawn to each other due to electrical forces. Negative ions within the cell are drawn to the positive ions arrayed on the outer surface of the plasma membrane. The actual number of ions that contribute to the resting membrane potential is extremely small compared to the rest of the ions within and outside the cell. **Table 43.1** lists the ions that are important in maintaining the resting potential and their concentrations inside and outside the cell. The ions that are most critical for establishing the resting potential are Na⁺, K⁺, and intracellular anions such as negatively charged proteins and Cl⁻.

Three factors are primarily responsible for the resting membrane potential (**Figure 43.6b**). First, the sodium-potassium pump (Na^+/K^+-ATPase) within the plasma membrane continually moves sodium ions out of the cell and potassium ions into the cytosol (see Chapter 5). The sodium-potassium pump utilizes ATP to transport three Na⁺ out of the cell for every two K⁺ it moves into the cell. Thus, the pump contributes to a charge difference across the plasma membrane and establishes gradients for Na⁺ and K⁺. Second, the plasma membrane contains ion-specific channels that affect the permeability of ions across the membrane. Channels that are specific for Na⁺ or K⁺ influence the resting potential by allowing the passive movement of ions.

(a) Distribution of charges across the neural plasma membrane

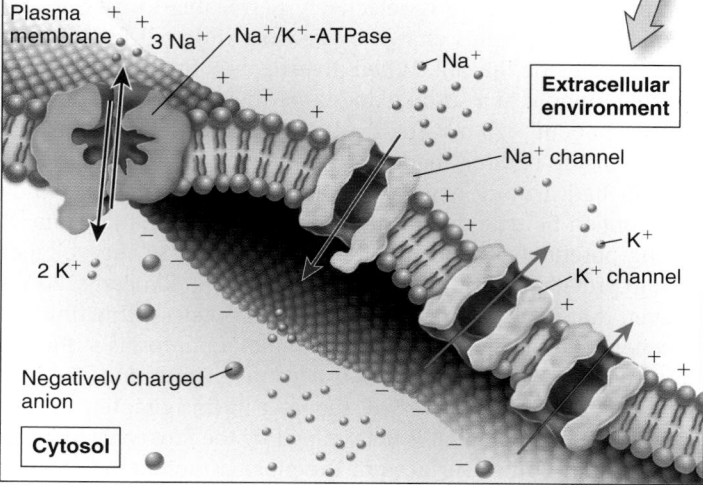

(b) Factors that influence the resting potential

Figure 43.6   **The resting membrane potential.**   **(a)** Positive charges are more numerous on the outside of a neuron, while negative charges predominate inside the cell. The positive and negative charges tend to line up along the membrane. **(b)** Three major factors influence the resting potential. The Na⁺/K⁺-ATPase is largely responsible for establishing and maintaining the ion gradients. Also, this transporter pumps 3 Na⁺ out of the cell and 2 K⁺ into the cytosol. The resting potential is also influenced by Na⁺ and particularly K⁺ channels that allow the passive movement of these ions. Finally, certain anions tend to be more abundant in the cytosol (for example, negatively charged proteins) and contribute to the internal negative charge.

In particular, K⁺ channels tend to be open more frequently at the resting potential. In most neurons, there are about 50 times more potassium channels than sodium channels, so the membrane is more permeable to potassium than to sodium. Third, negatively charged molecules such as proteins are more abundant inside the cell. These anions do not readily move through the plasma membrane so they contribute more negative electric charge to the interior of the cell.

Table 43.1	Extracellular and Intracellular Concentrations of Ions for a Typical Mammalian Neuron	

	Concentration (mM/L)	
Ion	**Extracellular**	**Intracellular**
Na⁺	145	15
K⁺	5	150
Anions found in macromolecules	<1	65
Cl⁻	110	7

An Electrochemical Gradient Governs the Movement of Ions Across a Membrane

Both the membrane potential and the chemical concentration of ions influence the direction of ion movement across a membrane. The direction that an ion will move depends on the **electrochemical gradient** for that ion, which is the combined effect of both an electrical and chemical gradient. **Figure 43.7** considers the concept of an electrochemical gradient for K⁺. This hypothetical drawing illustrates two compartments that are separated by a semipermeable membrane that can permit the flow of K⁺. Figure 43.7a shows a chemical gradient in which K⁺ concentration is higher on one side compared to the other. In this scenario, K⁺ will move from a region of high to low concentration. Figure 43.7b illustrates an electrical gradient. In this case, the chemical concentration of K⁺ is equal on both sides of the membrane, but the concentrations of other ions such as Na⁺ and Cl⁻ are unequal on both sides of the membrane and thereby produce an electrical gradient. Because K⁺ is positively charged, it will be attracted to the side of the membrane with more negative charge. Finally, Figure 43.7c shows a situation in which a chemical gradient is balanced by an electrical gradient. The electrical gradient would favor the movement of K⁺ from left to right, while the chemical gradient would favor movement from right to left. These opposing forces can create an equilibrium in which there is no net movement of K⁺ in either direction.

With two different forces—chemical and electrical gradients—acting on a given ion, is it possible to predict the direction that an ion will move across a membrane at any concentration gradient? In other words, can we compare the relative strengths of the chemical and electrical gradients and predict their net effect? By measuring the membrane potential of isolated neurons in the presence of changing concentrations of extracellular ions, scientists have deduced a mathematical formula that relates chemical and electrical gradients to each other. This formula, named the **Nernst equation** after the Nobel laureate Walther Nernst, gives the **equilibrium potential** for an ion at any given concentration gradient. For monovalent cations such as Na⁺ and K⁺, the Nernst equation can be expressed as:

$$E = 60 \text{ mV } \log_{10} ([X_{\text{extracellular}}]/[X_{\text{intracellular}}])$$

where, E is the equilibrium potential, inside relative to outside, [X] is the concentration of an ion, outside or inside the cell, and 60 mV is a constant that depends on temperature, valence of the ion, and other factors.

(a) Chemical gradient for K⁺, no electrical gradient

(b) Electrical gradient, no chemical gradient for K⁺

(c) K⁺ equilibrium, chemical gradient for K⁺ is balanced by the electrical gradient

Figure 43.7 **Chemical and electrical gradients.** This hypothetical example depicts two chambers separated by a membrane that is permeable to K⁺. **(a)** Initially, there is no electrical gradient across the membrane, and the left compartment contains a lower concentration of KCl compared to the right compartment. In water, the KCl dissociates to K⁺ and Cl⁻. Under these conditions, K⁺ diffuses down its chemical concentration gradient from right to left. **(b)** In this example, the compartments initially contain equal concentrations of K⁺, but an electrical gradient exists due to an unequal distribution of Na⁺ and Cl⁻. Potassium ions are attracted to the higher amount of negative charge on the right side of the membrane. **(c)** This example illustrates opposing chemical and electrical gradients. The right compartment contains a higher chemical concentration of K⁺ while the left side has a higher amount of positive charge. These gradients balance each other, so that no net movement of K⁺ occurs.

Biological inquiry: In part (a), does the diffusion of K⁺ down its chemical gradient result in an electrical gradient? What will eventually stop the net diffusion of K⁺?

The Nernst equation allows neuroscientists to predict when an ion is in equilibrium and when it is not. To understand the usefulness of this equation, let's consider two examples. First, let's suppose the membrane potential is −88.6 mV and the K⁺ concentration is 5 mM outside and 150 mM inside. If we plug these chemical concentrations into the Nernst equation:

$$E = 60 \text{ mV} \log_{10} (5/150) = 60 \text{ mV} (-1.48)$$
$$E = -88.6 \text{ mV}$$

Under these conditions, K⁺ is in equilibrium and no net flow of K⁺ will occur, even when many K⁺ channels are open.

As a second example, let's suppose that the membrane potential is at the typical resting potential of −70 mV and the Na⁺ concentration is 100 mM outside and 10 mM inside. If we plug these chemical concentrations into the Nernst equation:

$$E = 60 \text{ mV} \log_{10} (100/10) = 60 \text{ mV} (1)$$
$$E = 60 \text{ mV}$$

At a resting potential of −70 mV, the value of +60 mV tells us that Na⁺ is not in equilibrium. Na⁺ will tend to flow into the cell. Both the chemical and electrical gradients favor an inward flow. However, at Na⁺ concentrations of 100 mM outside and 10 mM inside, Na⁺ would not tend to flow—it would be in equilibrium—if the membrane potential was at 60 mV.

By establishing electrochemical gradients and maintaining them with the sodium-potassium pump, neurons have the ability to suddenly allow ions to move across the plasma membrane by opening additional channels that were previously closed. The movement of a charged ion down its gradient results in an electric current. This current is different from the type of current that runs through electrical wires and is much smaller in amplitude. Even so, this small current provides the electrical signal that neurons use to communicate with one another, as we will see next.

43.3 Communication Between Neurons

Communication between neurons begins when one cell receives a stimulus and sends an electrical signal along its plasma membrane via currents generated by ion movements. This signal will then influence the next neuron in a circuit. Each signal is brief (only a few milliseconds), but a neuron may receive and transmit millions of signals in its lifetime. In this section, we will survey the ability of neurons to send brief signals with amazing speed.

Signaling by a Neuron Occurs Through Changes in the Membrane Potential

Recall that a cell is polarized because of the separation of charge across the membrane. The resting potential is more negative inside. Changes in the membrane potential, therefore, are changes in the degree of polarization. **Depolarization** occurs when the cell becomes less polarized, that is, less negative relative to the surrounding fluid. As described later, when a neuron is stimulated by an electrical or chemical stimulus, one or more types of gated channels open and sodium ions move into the cell, bringing with them their positive charge. This increases the membrane potential to a level somewhat more positive than the resting membrane potential. The membrane potential is said to be depolarized. In contrast, **hyperpolarization** occurs when the cell becomes more polarized. For example, increased movement of K^+ out of the cell would make the internal charge of the cell more negative than the resting potential.

Though all cells in an animal's body have a membrane potential, neurons and muscle cells are called **excitable cells** because they have the capacity to generate electrical signals. For a cell to communicate using electrical signals, it must be able to change its membrane potential very rapidly. This is accomplished by gated ion channels, so called because they open and close like a gate in a fence (**Figure 43.8**). **Voltage-gated ion channels** open and close in response to changes in voltage across the membrane. **Ligand-gated ion channels**, also known as chemically gated ion channels, open or close when ligands, such as small molecules, bind to them. The opening and closing of voltage-gated and ligand-gated channels are responsible for two types of changes in the neuron's membrane potential called graded potentials and action potentials.

Graded Potentials A **graded potential** is a depolarization or hyperpolarization that varies depending on the strength of the stimulus. A large change in membrane potential will occur when a strong stimulus opens many channels, while a small stimulus causes a small change because only a few channels are opened (**Figure 43.9**). Graded potentials occur locally on a particular area of the plasma membrane, such as dendrites and the cell body, where a chemical or electrical stimulus opens ion channels. From this area, a graded potential spreads a small distance across a region of the plasma membrane. In a short time, the membrane potential returns to the resting potential.

Graded potentials occur on all neurons and are particularly important for the function of sensory neurons, which must distinguish between strong and weak stimuli coming into the organism from the environment. When a graded potential is sufficiently large, this stimulates the second type of electrical signal, the action potential.

Action Potentials **Action potentials** are the nerve impulses that carry an electrical signal along an axon. In contrast to a graded potential, an action potential is always a large depolarization, and all action potentials in a given neuron have the same amplitude, which is the degree to which an action potential changes the membrane potential away from its resting state.

(a) Voltage-gated ion channel

(b) Ligand-gated ion channel

Figure 43.8 **Examples of gated ion channels.** **(a)** A voltage-gated ion channel allows ions to diffuse into the cell. These channels open or close depending on changes in charge (voltage) across the membrane. **(b)** A ligand-gated ion channel opens or closes in response to ligand binding. In the example here, the binding of neurotransmitter opens the channel.

Figure 43.9 **Graded membrane potentials.** Within a limited range, the change in the membrane potential occurs in proportion to the intensity of the stimulus. If the membrane potential becomes more positive, the resulting change is called a depolarization. Hyperpolarization results from the membrane potential becoming more negative following a stimulus.

Once an action potential has been triggered, it is expressed in an all-or-none fashion. In other words, it cannot be graded. Unlike a graded potential, an action potential is actively propagated along the axon, renewing itself as it travels. In this sense, an action potential is said to be regenerative. An action potential travels rapidly down the axon to the axon terminal, where it initiates a response at the junction with the next cell.

Figure 43.10a shows the electrical changes that occur in a localized region of an axon when an action potential is occurring. Voltage-gated sodium channels and potassium channels are present in very high numbers from the axon hillock to the axon terminus. An action potential begins when a graded potential is large enough to spread to the axon hillock and depolarizes the membrane to a value called the threshold potential. The **threshold potential** is the membrane potential, typically around −50 mV, which is sufficient to open voltage-gated Na$^+$ channels and to trigger an action potential.

Sodium channels open rapidly when the membrane potential changes from the resting potential (−70 mV) to the threshold potential (−50 mV). The opening of sodium channels involves a change in the conformation of the membrane-spanning region of the channel (**Figure 43.10b**). When the protein changes shape, the central pore opens and Na$^+$ rapidly diffuses into the cell down its electrochemical gradient. This further depolarizes the cell, causing even more voltage-gated Na$^+$ channels to open and resulting in the spike in membrane potential that characterizes the action potential. This process is so rapid that, within 1 msec (millisecond), the membrane potential reaches its peak positive value. In a typical action potential, approximately 7,000 sodium ions move through each voltage-gated Na$^+$ channel!

When the membrane potential becomes very polarized, a conformational change in the sodium channel blocks the continued flow of Na$^+$ into the cell. The conformational change involves the **inactivation gate**, a string of amino acids that juts out from the channel protein into the cytosol (Figure 43.10b). The inactivation gate swings into the channel, thereby preventing the movement of Na$^+$ into the cell. Under these conditions, the Na$^+$ channel is said to be inactivated. The inactivation gate

(a)

(b)

Figure 43.10   Changes that occur during an action potential.   (a) The electrical gradient across the plasma membrane of an axon first depolarizes and then hyperpolarizes. **(b)** These changes in the membrane potential are caused by the opening and closing of Na$^+$ and K$^+$ channels.

Biological inquiry: How would the curve in part (a) be affected if the Na$^+$ channels were missing their inactivation gate?

does not swing out of the channel until the membrane returns to its resting potential.

Voltage-gated potassium channels are also triggered to open by the change in voltage to the threshold potential, but they open about 1 msec later than Na$^+$ channels. When K$^+$ channels open, potassium ions leave the cell down their electrochemical gradient and the membrane potential becomes more negative again. The membrane shows a brief period of hyperpolarization because the K$^+$ channels remain open and the electrical gradient approaches the equilibrium potential for K$^+$, which is slightly more negative than the resting potential. When the voltage-gated K$^+$ channels eventually close, the membrane returns to the resting potential. At this stage, both the Na$^+$ and K$^+$ channels are closed, and they have the ability to reopen if the resting potential rises to the threshold potential.

The evolution of K$^+$ channels with a slightly slower opening time than Na$^+$ channels was a key event that led to the formation of nervous systems. Imagine what would happen if the voltage-gated Na$^+$ and K$^+$ channels opened simultaneously: As Na$^+$ ions entered the cell down their electrochemical gradient, K$^+$ ions would leave the cell down their electrochemical gradient, and they would negate each other's effects on the membrane potential.

While the inactivation gate of the Na$^+$ channel is closed, the cell is **refractory**, or unresponsive to another stimulus. A change in voltage will not open the Na$^+$ channels while they are inactivated. Therefore, this refractory period places limits on the frequency at which a neuron can generate action potentials. As we will see, this also ensures that the action potential does not "retrace its steps" by moving backward toward the cell body.

Many natural toxins work by blocking the actions of sodium and potassium channels. Some animals use these toxins as defensive mechanisms or for capturing prey, and they can be deadly. For instance, tetrodotoxin, a chemical produced by pufferfish, blocks voltage-gated Na$^+$ channels and therefore nerve activity. An animal that eats a pufferfish dies because the poison paralyzes muscles, including those that depend on nerve stimulation to control breathing and movement. Eating even a small amount can paralyze or kill a human. Other toxins from wasps, bees, and scorpions can block the action of voltage-gated K$^+$ channels, resulting in abnormal action potentials and loss of normal muscle responses.

Our present understanding of action potentials began with experiments carried out in the 1940s by two British physiologists, Alan Hodgkin and Andrew Huxley. Using the squid giant axon and measuring ion movements into and out of the cell, they showed that the action potential depends on voltage-gated Na$^+$ and K$^+$ channels. They also showed that the timing with which these channels open and close is responsible for shaping the action potential. They noted that Na$^+$ channels open and close rapidly, while voltage-gated K$^+$ channels open later than Na$^+$ channels and close only toward the end of the action potential. This results in Na$^+$ channels being involved in the initiation and early phase of the action potential, while K$^+$ channels control the duration and termination of the action potential.

Since those early studies, we have learned a great deal about the structure and function of voltage-gated Na$^+$ and K$^+$ channels. These studies have largely involved selectively mutating amino acids within these proteins and observing whether changes occur in the function of the channels. The Na$^+$ channel consists of one long protein that crosses or spans the membrane 24 times, while the K$^+$ channel consists of four separate protein subunits that each span the membrane 6 times. The pores of the Na$^+$ and K$^+$ channels are composed of different amino acids and have different diameters, and this results in each ion being able to pass only through its own channel. Researchers have identified 10 different Na$^+$ channel genes in mammals alone and as many as 100 types of K$^+$ channel genes, each encoding channels with slightly different properties such as how quickly they open or close.

Action Potentials Are Conducted Down the Axon with Great Speed

Thus far we have considered the electrical changes that happen when an action potential is initiated. We will now consider how this event occurs in the direction from the axon hillock to the axon terminus (**Figure 43.11**). Let's begin at the axon hillock. Because the neuron receives stimuli from other cells, this results in a graded potential at the axon hillock that changes the resting potential to the threshold potential. This triggers the abrupt opening of several voltage-gated Na$^+$ channels just beyond the axon hillock, where many Na$^+$ channels are found. This event triggers the opening of more Na$^+$ channels, which allow Na$^+$ to depolarize a region closer to the axon terminus. In this way, the sequential opening of Na$^+$ channels along the axon membrane conducts a wave of depolarization from the axon hillock to the axon terminus.

But why doesn't the action potential move backwards from the terminus to the hillock? The key is the inactivation state of the Na$^+$ channel. Again, let's begin at the hillock. When these Na$^+$ channels open for 1 msec, they allow Na$^+$ to rapidly enter the cell, and then they become inactivated. This inactivation prevents the action potential from moving backwards (see Figure 43.11). While they are inactivated, K$^+$ channels open and the resting potential is restored. At the resting potential, the Na$^+$ channels switch from the inactivated to the closed state. The opening of K$^+$ channels also travels from the axon hillock to the axon terminus, but it causes a wave of hyperpolarization that re-establishes the resting potential.

The speed with which the action potential is conducted down the axon can be as fast as 100 m/sec or as slow as a centimeter or two per second. The speed is determined by two factors: the axon diameter and the presence or absence of myelin.

The axon diameter influences the rate at which incoming ions can spread along the inner surface of the plasma membrane. In a broad axon, the flow of ions meets less resistance compared to a thin axon, just as water moves more easily through a wide hose than a narrow one. Therefore, in a broad axon the action potential will move faster. The large axons of

Low density of Na⁺ channels

Axon hillock

Unstimulated nerve (Na⁺ channels are closed)

High density of Na⁺ channels along the axon

1 Threshold potential is reached at axon hillock. Na⁺ channels open.

Na⁺ channel open

2 Entry of Na⁺ depolarizes membrane and opens channels to the right. Previously opened Na⁺ channels are inactivated.

Inactivation gate closed

3 Continued entry of Na⁺ depolarizes the membrane and opens channels farther to the right. Previously opened Na⁺ channels are inactivated. Previously inactivated Na⁺ channels switch to a closed state, when K⁺ channels open and restore the resting potential.

4 Process continues.

KEY

Closed Na⁺ channels

Open Na⁺ channels

Inactivated Na⁺ channels

Figure 43.11 Conduction of the action potential along an axon. For simplicity, only Na⁺ channels are shown.

the squid or lobster conduct action potentials very rapidly, allowing the animals to move quickly when threatened by a predator.

Myelination is the second factor that influences the speed of an action potential. Myelinated axons conduct action potentials at a faster rate than unmyelinated axons. Invertebrate neurons lack myelin, while vertebrate neurons may be myelinated or unmyelinated. Recall that glial cells (oligodendrocytes and Schwann cells) wrap around axons to form an insulating sheath of membrane. However, this myelin sheath is not continuous (**Figure 43.12**). At regular intervals, the axons of myelinated neurons have exposed areas known as the **nodes of Ranvier**, which contain many voltage-gated Na⁺ channels. The nodes of Ranvier are the only areas of myelinated axons that have enough Na⁺ channels to elicit an action potential. When sodium ions move into the cell, the charge moves rapidly through the cytosol to the next node, where the action potential continues. This type of conduction is called **saltatory conduction** (from the Latin, *saltare*, to leap) because the action potential seems to "jump" from one node to the next. Saltatory conduction speeds up the conduction process because it takes less time for the action potential to travel from node to node rather than continuously along the length of the axon.

Neurons Communicate Electrically or Chemically at Synapses

Neurons communicate with other cells at a **synapse**, which is a junction where a nerve terminal meets a target neuron, muscle cell, or gland. At a synapse, an electrical or chemical signal passes from the nerve terminal to the next cell. A synapse includes the axon terminal of the neuron that is sending the message, the nearby plasma membrane of the receiving cell, and the **synaptic cleft**, or extracellular space between the two cells. The **presynaptic cell** sends the signal and the **postsynaptic cell** receives it (**Figure 43.13**).

By studying neurons from both invertebrates and vertebrates, researchers have identified two types of synapses. The first type, called an **electrical synapse**, directly passes electric current from the presynaptic to the postsynaptic cell. The electrical signal passes through this type of synapse extremely rapidly, because the plasma membranes of adjacent cells are connected by gap junctions that can move electric charge freely from one cell to the other (see Chapter 10). An example of cells that use electrical synapses are cardiac muscle cells. In some aquatic invertebrates like leeches, electrical synapses are known to occur where a group of neurons must fire rapidly and synchronously, such as when an animal must coordinate a number of muscles to swim or escape danger. A second type of synapse is a **chemical synapse** in which a chemical called a **neurotransmitter** is released from the nerve terminal and acts as a signal from the presynaptic to the postsynaptic cell. Chemical synapses are more common, particularly in vertebrates. Most neurons release only one type of neurotransmitter, though some neurons can release two or more different neurotransmitters.

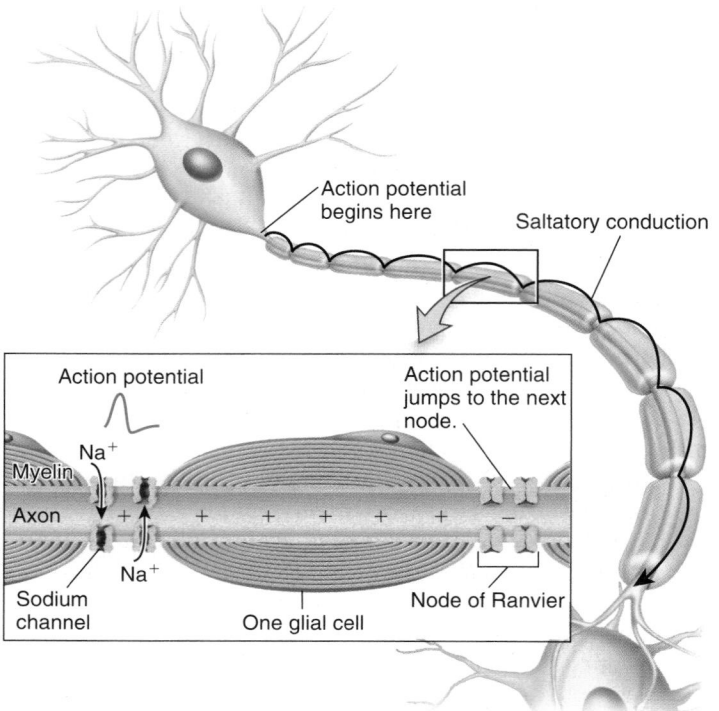

Figure 43.12 **Saltatory conduction along a myelinated axon.** Action potentials are generated only at the nodes of Ranvier, which lack a surrounding sheath of myelin.

Biological inquiry: What is the main advantage of saltatory conduction?

Figure 43.13 **Presynaptic and postsynaptic cells.** The arrows show the direction of signal transmission from one neuron to the next. Note that neuron 2 is postsynaptic with respect to neuron 1, and presynaptic with respect to neuron 3.

Figure 43.14 describes the steps that occur when two cells communicate via a chemical synapse. The presynaptic nerve terminal of a chemical synapse contains vesicles—small, membrane-enclosed packets, each containing thousands of molecules of neurotransmitter. When an action potential arrives at the nerve terminal, voltage-gated calcium channels open, which allows calcium ions to diffuse down their electrochemical gradient into the cell. Calcium is known to bind to a protein called synaptotagmin in the vesicle membrane. This causes exocytosis, in which the vesicle fuses with the presynaptic membrane, thereby releasing neurotransmitter molecules into the synaptic cleft. The neurotransmitter molecules diffuse across the 10–20-nm-wide synaptic cleft between the presynaptic and postsynaptic cells and bind to channels or receptors in the postsynaptic membrane.

The binding of neurotransmitter molecules opens or closes ion channels and thereby changes the membrane potential of the postsynaptic cell. An excitatory neurotransmitter depolarizes the postsynaptic membrane, and the response is an **excitatory postsynaptic potential (EPSP)**. The response is called excitatory because the depolarization of the membrane brings the membrane potential closer to the threshold potential that would trigger an action potential. Conversely, an inhibitory neurotransmitter usually hyperpolarizes the postsynaptic membrane, a so-called **inhibitory postsynaptic potential (IPSP)**, which reduces

the likelihood of an action potential. To end the synaptic signal, neurotransmitter molecules in the synaptic cleft can be broken down by enzymes or taken back up into the presynaptic terminal and repackaged into vesicles for reuse.

We can appreciate the power of neurotransmitters when we consider the effects of toxins that are known to affect neurotransmitter release. For example, the venom of the black widow spider, α-latrotoxin, acts on presynaptic proteins to cause massive release of neurotransmitter from motor neurons. Correspondingly, the symptoms of black widow spider bites in humans are painful muscle cramping and rigidity. Researchers have also demonstrated the function of several proteins involved in vesicle membrane fusion and exocytosis by experimenting with certain toxins that decrease neurotransmitter release. For instance, the bacterial botulinum toxin prevents exocytosis of vesicles containing neurotransmitter. If ingested in food, the toxin causes botulism, a sometimes fatal form of food poisoning characterized by weak muscles, trouble focusing the eyes, and difficulty swallowing. Surprisingly, researchers have discovered a commercial use for botulinum toxin—when injected into a muscle, it blocks contraction of the muscle by preventing the motor neurons from releasing neurotransmitter. Clinicians administer a synthetic form of the toxin (called BOTOX) to patients who wish to reduce facial wrinkles. The effect is to reduce the contractions of the facial muscles, which cause frown lines.

Neurons Respond to Multiple Synaptic Inputs

A neuron may receive inputs from many synapses (**Figure 43.15**), and some of these synapses may release their neurotransmitter onto the neuron at the same time. Certain neurotransmitters are excitatory, while others are inhibitory. How is the decision made to initiate an action potential? The effect of a

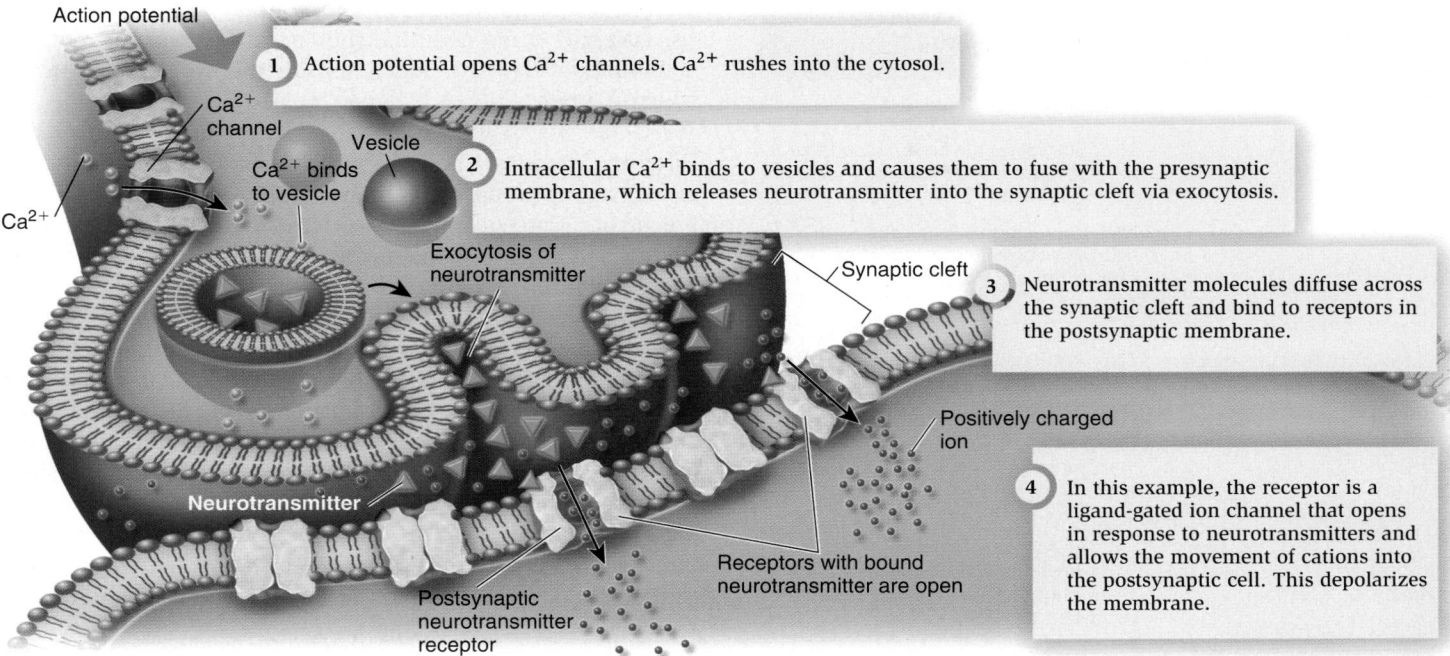

1 Action potential opens Ca^{2+} channels. Ca^{2+} rushes into the cytosol.

2 Intracellular Ca^{2+} binds to vesicles and causes them to fuse with the presynaptic membrane, which releases neurotransmitter into the synaptic cleft via exocytosis.

3 Neurotransmitter molecules diffuse across the synaptic cleft and bind to receptors in the postsynaptic membrane.

4 In this example, the receptor is a ligand-gated ion channel that opens in response to neurotransmitters and allows the movement of cations into the postsynaptic cell. This depolarizes the membrane.

Figure 43.14 **Structure and function of a chemical synapse.**   Calcium enters the presynaptic neuron terminal in response to an action potential and results in vesicle fusion with the plasma membrane, which releases neurotransmitter molecules into the synapse. The neurotransmitter molecules then bind to receptors in the plasma membrane of the postsynaptic cell.

single synapse is usually far too weak to elicit an action potential in the postsynaptic neuron (though a single synapse can cause a muscle cell to contract). However, many EPSPs generated at one time can sum together and bring the membrane potential to the threshold potential at the axon hillock for action potential firing. This summation is a process known as **synaptic integration**.

In addition to the number of synapses that stimulate the postsynaptic membrane, the location of the synapses is also important. Synapses initially cause a graded potential in the postsynaptic membrane that spreads only a short distance. If two excitatory synapses are close together and activated at the same time, the depolarization will be larger and will spread farther. This spread increases the chances that the depolarization will reach the axon hillock, where the high concentration of Na^+ channels can trigger an action potential. Synapses that occur far from the axon hillock, for instance at the end of a long dendrite, are less effective than synapses on the cell body nearer the axon hillock.

Several Classes of Neurotransmitters Stimulate Responses in Postsynaptic Cells

Neuroscientists have identified more than 100 different neurotransmitters in animals. Generally, neurotransmitters are categorized by size or molecular structure (**Table 43.2**). Both excitatory and inhibitory neurotransmission occurs in nervous systems. The changing balance between excitation and inhibition controls the state of nervous system circuits at any one time.

(a)

KEY

■ Synapses that are excitatory

■ Synapses that are inhibitory

0.4 μm

(b)

Figure 43.15 **Integration of synaptic inputs.**   **(a)** Many synapses on a neuron can be activated at the same time. **(b)** SEM of a cell body with multiple presynaptic terminals making contact.

Table 43.2	Classes of Neurotransmitters and Their Functions	
Transmitter	**Some major functions**	
Acetylcholine	CNS: Stimulates the brain; important in memory, motor control, and many other functions	
	PNS: Stimulates skeletal muscle at neuromuscular junction, inhibits cardiac muscle, promotes digestion	
Biogenic amines		
Catecholamines: dopamine, norepinephrine, epinephrine	CNS: Regulate mood, attention, learning, and many other functions	
Serotonin	PNS: Catecholamines stimulate cardiac muscle, improve lung function, help animals respond to stressful situations	
Histamine		
Amino acids		
Excitatory amino acids: glutamate, aspartate	Widespread mediators of activity in all areas of CNS; the major on and off signals of the CNS	
Inhibitory amino acids: gamma-aminobutyric acid (GABA), glycine		
Neuropeptides		
Opiate peptides: endorphin, enkephalin	CNS: Modulate postsynaptic cell response to neurotransmitters; play a role in mood, behavior, appetite, pain perception, and many others	
Oxytocin		
Gases		
Nitric oxide, carbon monoxide	CNS: Possible role in memory and odor sensation	
	PNS: Relaxes smooth muscle, especially in blood vessels	

To understand how neurotransmitters work, imagine driving a car with one foot on the accelerator and one foot on the brake. To speed up, you could press down harder on the accelerator, ease up on the brake, or both, while to slow down, you could do the opposite. All nervous systems operate in this way, with combined excitatory and inhibitory actions of neurotransmitters.

Next, we highlight major features of the different chemical classes of neurotransmitters found in animals, which include acetylcholine, biogenic amines, amino acids, neuropeptides, and gaseous neurotransmitters. With the exception of acetylcholine, all of these classes contain several different neurotransmitters that are similar in chemical structure but may have different functions in nervous systems.

Acetylcholine Acetylcholine, one of the most widespread neurotransmitters in animals, is released at the synapses of **neuromuscular junctions** in vertebrates, where a neuron contacts skeletal or cardiac muscle. It is also released at synapses within the brain. Acetylcholine acts as an excitatory neurotransmitter in the brain and skeletal muscle cells, but it is inhibitory to neurons that control cardiac muscle contraction. Therefore, it stimulates skeletal muscle but inhibits cardiac muscle. As we will see later, the same neurotransmitter can exert both excitatory and inhibitory effects depending on the type of receptor to which it binds.

Biogenic Amines The biogenic amines are compounds containing amine groups that are formed from amino acids or their precursors. Catecholamines are formed from tyrosine and include dopamine, norepinephrine, and epinephrine. Dopamine and epinephrine are primarily active in the brain, while norepinephrine acts in the brain and at synapses in the peripheral nervous system. Serotonin is formed from the amino acid tryptophan. In addition to widespread physiological effects such as control of heart and lung function, these neurotransmitters are psychoactive, that is, they affect mood, attention, and learning. In humans, abnormally high or low levels of the biogenic amines have been associated with a variety of disorders, including schizophrenia, Parkinson's disease, and depression. Histamine is formed from the amino acid histidine and plays various roles in the body. As a neurotransmitter, it helps to maintain the awake state.

Amino Acids In addition to their role in the formation of proteins, the amino acids glutamate, aspartate, and glycine function as neurotransmitters. Glutamate is the most widespread excitatory neurotransmitter found in animal nervous systems. Though not involved in protein formation, another amino acid, gamma-aminobutyric acid (GABA), is the most common inhibitory neurotransmitter. GABA hyperpolarizes the postsynaptic membrane by opening chloride channels, allowing chloride ions to move into the cell (see Table 43.1) and acting as the major "brake" on the central nervous system.

Neuropeptides Neuropeptides are short chains of 2 to about 15 amino acids. They remained largely unknown until the 1970s, when it began to be apparent that perhaps 100 or so peptides can act as neurotransmitters. Like the other neurotransmitters discussed, neuropeptides can be excitatory or inhibitory. They are produced within the cell body, packaged in vesicles in the Golgi apparatus, and transported to the axon terminal, where they are released. Neuropeptides are often called **neuromodulators**, because they can alter or modulate the response of the postsynaptic neuron to other neurotransmitters. For example, a cell exposed to a neuropeptide may increase the number of receptors for another neurotransmitter, which makes it more responsive to that neurotransmitter. One group of neuropeptides is called the opiate peptides because opium-like drugs (for example, morphine) bind to their receptors. Opiate peptides include the endorphins, a group of peptides that decrease pain and cause natural feelings of euphoria, and that may be important in helping animals deal with stress. A second example is enkephalin, which acts as a sedative in the body and appears to affect mood and motivation. Oxytocin is a neuropeptide that acts as both a hormone and a neurotransmitter. In the brain, oxytocin seems to play a role in bonding and social recognition, and it might even be involved in the formation of trust between people.

Gaseous Neurotransmitters Gaseous molecules such as nitric oxide (NO) and carbon monoxide (CO) act locally in many tissues and sometimes function as neurotransmitters. Unlike other neurotransmitters, they are not sequestered into vesicles and are produced locally as needed. Gaseous neurotransmitters are short-acting and influence other cells by diffusion. In humans, NO is responsible for relaxing the smooth muscle surrounding blood vessels, including those in the penis. When a male becomes sexually aroused, NO levels increase in this tissue, dilating the vessels and increasing blood flow into the penis, producing an erection. Several drugs used for male sexual dysfunction enhance erections by increasing or mimicking the action of NO on smooth muscle. The functions of CO are still uncertain, but it is believed to act as a neurotransmitter in the sense of smell in some animals, such as the terrestrial mollusk *Limax maximus.*

FEATURE INVESTIGATION

Otto Loewi Discovered Acetylcholine

German physiologist Otto Loewi was interested in how neurons communicate with muscle. He already knew that the electrical stimulation of a nerve in a frog's leg would result in muscle contraction, so it appeared that neurons communicate with the muscle by electrical signals. In the early 1920s, he turned his studies to the heart. If he removed a frog's heart and put it in a bath containing saline, it continued to beat for several hours. As we will see in Chapter 47, all vertebrate hearts receive both stimulatory and inhibitory nerve impulses that regulate the rhythm and intensity of the heartbeat. When dissected from an animal, these nerves remain attached to the heart and their function can be studied in a bath solution.

When Loewi electrically stimulated a certain nerve attached to the frog's heart, the rate at which the heart contracted increased. However, if he stimulated a different nerve, known as the vagus nerve, the heart rate decreased. Loewi wondered how electrical activity in different neurons could have opposite effects. He hypothesized that a stimulated neuron released a chemical substance that could influence muscles and other neurons in different ways, but he did not know how to test the hypothesis. As he described later, the idea for the definitive experiment came to him in the middle of the night: "The night before the Easter Sunday of that year I awoke, turned on the light, and jotted down a few notes on a tiny slip of paper. Then I fell asleep again. It occurred to me at six o'clock in the morning that during the night I had written down something most important, but I was unable to decipher the scrawl. The next night, at three o'clock, the idea returned. . . . I got up immediately, went to the laboratory, and performed a simple experiment on a frog heart according to the nocturnal design."

As shown in **Figure 43.16**, Loewi placed two frog hearts in separate, but connected, chambers containing saline. He used an electrode to stimulate the vagus nerve that was still attached to the first heart and recorded the heart rate. As he expected from an earlier experiment, the heart rate slowed. Within a few minutes, the heart rate of the second heart also slowed even though he had not stimulated the nerve to that heart at all!

Figure 43.16   Loewi's experimental discovery of chemical neurotransmission.

HYPOTHESIS Neurons release chemical substances that influence the rate of the heartbeat.

STARTING MATERIALS Two frog hearts.

Experimental level

Conceptual level

1 Dissect hearts from 2 frogs and place in chambers with a saline solution. Heart 1 still has its vagus nerve attached. Chambers are interconnected.

Vagus nerve

Heart 1 in chamber 1

Saline solution

Heart 2 in chamber 2

2 Electrically stimulate vagus nerve of heart 1. Do not stimulate any nerves that are attached to heart 2.

Stimulating electrode

Heart 1

Action potential

Action potentials travel along vagus nerve to heart 1.

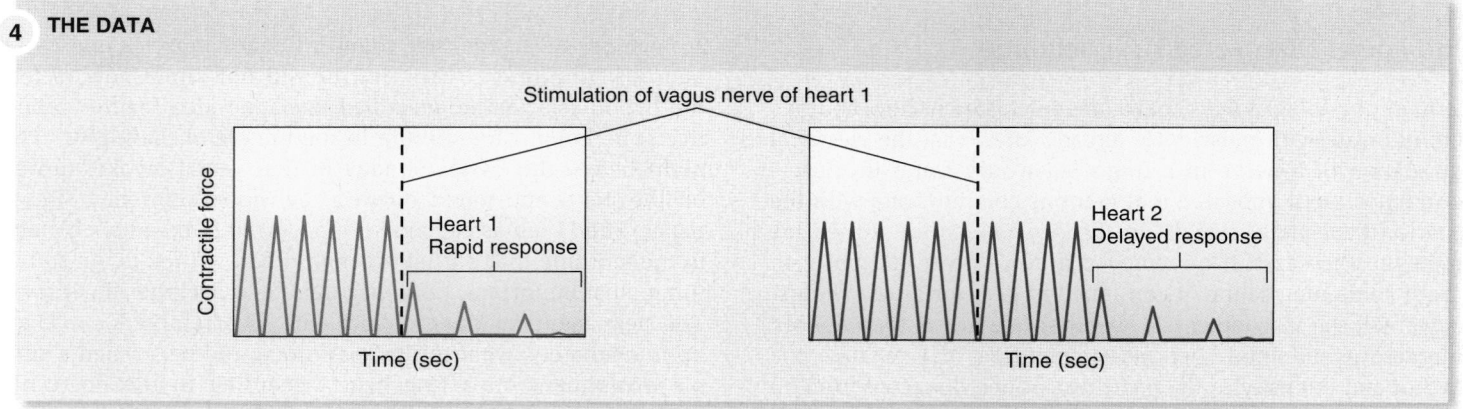

3 Record strength and number of beats in heart 1 before and after electrical stimulation of vagus nerve. Likewise, record heartbeats in heart 2. This was done using mercury manometers that were connected to each heart. The manometers measure pressure, which is due to the contractile force of the heart beating.

Contractile force measured with manometers.

Action potentials in vagus nerve caused release of chemicals, which slowed heart rate.

Chemicals from the vagus nerve leaked into the bath solution of chamber 1 and from there diffused via tubing to chamber 2.

The delayed response of heart 2 was due to the time required for chemicals to diffuse from chamber 1 to chamber 2.

4 **THE DATA**

Stimulation of vagus nerve of heart 1

Contractile force

Heart 1 Rapid response

Time (sec)

Heart 2 Delayed response

Time (sec)

He concluded that a chemical substance was released from the nerve of the first heart into the surrounding fluid and diffused into the second chamber, where it exerted its effects on the second heart.

Loewi initially named this substance *Vagusstoff* (vagus substance) after the vagus nerve he stimulated, but it was later renamed acetylcholine when its chemical nature (acetic acid bonded to choline) was determined. Acetylcholine was the first neurotransmitter discovered, which was a major achievement in neuroscience. Loewi's research opened the door for what we now know about chemical transmission at synapses, and the enormous pharmaceutical industry, which builds on this knowledge to treat neurological disorders.

Postsynaptic Receptors Determine the Response to Neurotransmitters

As we have seen, the same neurotransmitter can have both excitatory and inhibitory effects. The response of the postsynaptic neuron depends on the type of receptor that is present in the postsynaptic membrane. The two major types of postsynaptic receptors are ionotropic and metabotropic, and many neurotransmitters act on both (**Figure 43.17**). Neurotransmitter molecules bind to the extracellular portion of these receptors.

Ionotropic receptors are ligand-gated ion channels that open in response to binding of a neurotransmitter molecule. When neurotransmitter molecules bind to these receptors, ions flow through the channels to cause an EPSP or IPSP. Acetylcholine, the biogenic amines, and amino acids bind to ionotropic receptors. Ionotropic receptors are composed of multiple sub-units. Like K^+ channels, the subunits associate in a ring to form the receptor's channel. For instance, the ionotropic acetylcholine receptor has five subunits, named alpha, beta, etc. The α subunit is the only one that binds acetylcholine.

Metabotropic receptors are G-protein-coupled receptors (GPCRs), which are described in Chapter 9. They do not form a channel but instead are coupled to an intracellular signaling pathway that initiates changes in the postsynaptic cell. Metabotropic cascades produce second messengers as part of signal transduction pathways that produce a postsynaptic response. A common type of response is the phosphorylation of ion channels for sodium, potassium, or calcium ions, which are present in the plasma membrane. In Chapter 45, we will see that metabotropic receptors are important in activating sensory cells that respond to visual and other stimuli.

Figure 43.17   **The two major categories of postsynaptic receptors.**   **(a)** Ionotropic receptors have several subunits. Neurotransmitters bind to ionotropic receptors and directly open ion channels in the membrane. **(b)** Metabotropic receptors are GPCRs, which are discussed in Chapter 9. Neurotransmitters bind to metabotropic receptors and open ion channels indirectly through G proteins.

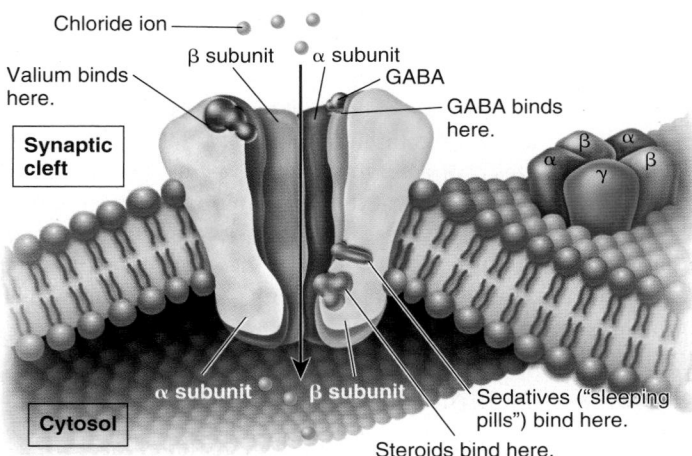

Figure 43.18   **The structure and function of the GABA receptor.**   This receptor has five subunits that form an anion channel that allows the passage of Cl⁻ ions. When GABA binds to the receptor, chloride ions move through the open channel and hyperpolarize the cell. Various other molecules bind to different sites on each subunit. These include naturally occurring molecules in animals' bodies, such as steroid hormones, and drugs, such as Valium, used to treat anxiety and other disorders.

GENOMES & PROTEOMES

Neurotransmitter Receptors Have Varied Subunit Compositions

As mentioned earlier, gamma-aminobutyric acid (GABA) is an inhibitory neurotransmitter that opens Cl⁻ channels. Though cells can possess different types of GABA receptors, a major category functions as a ligand-gated ion channel. This ionotropic receptor, which we will simply refer to as the GABA receptor, binds GABA and thereby opens the channel. This event allows Cl⁻ to diffuse into the cell, causing a hyperpolarization of the plasma membrane or clamping the membrane at the equilibrium potential for Cl⁻ (usually around −90 mV) (**Figure 43.18**). In this way, GABA binding to this receptor decreases the likelihood that a neuron will generate an action potential.

The GABA receptor is a good example of how receptor subunits can influence a postsynaptic response to a neurotransmitter.

GABA receptor proteins are usually composed of five subunits (designated α, β, and so on). The genomes of humans and other mammals have a group of homologous genes that encode at least 16 different GABA receptor subunits. In addition, subunit variation can be further increased by alternative splicing, which is described in Chapter 13. This amazing variety in subunits allows cells to express many different kinds of GABA receptors.

What are the benefits of having so many different GABA subunits? Though the answer is not entirely understood, each type of subunit has its own unique properties that can fine-tune the function of the GABA receptor so that it works optimally in the neuron in which it is expressed. The various subunits may differ in their affinity for GABA and the rate of Cl⁻ movement through the channel. In addition, neuroscientists have been particularly interested in whether the various subunits differ in their ability to recognize molecules other than GABA (Figure 43.18). This work has shown that the subunits of the GABA receptor bind a variety of other molecules. Presumably, the binding of these molecules enhances or reduces GABA's effectiveness in activating the receptor. This knowledge has proven beneficial in understanding how certain drugs exert their actions. For example, ethanol—found in alcoholic drinks—binds to one of the GABA receptor subunits expressed in brain and motor neurons, and enhances the actions of GABA. This explains why alcohol depresses the activity of the brain and impairs motor coordination, among other effects. Other subunits of the GABA receptor bind drugs, such as Valium, that are used to treat certain psychological problems like chronic anxiety. Apparently, the inhibitory effects of GABA are part of the mechanism for achieving a balance between alertness, anxiety, and calmness.

43.4 Impact on Public Health

When neurons fail to develop properly or their function is impaired, the consequences can be severe, affecting mood, behavior, and even the ability to think or move. Over 100 neurological disorders (diseases of the nervous system) have been identified in humans, and drugs to treat them are among the most widely prescribed medicines today (**Table 43.3**). Many recreational drugs also exert their effects by altering neurotransmission. The use of these drugs, therefore, results in symptoms similar to those of neurological disorders.

Disorders of Neurotransmission Can Impact Mood

Several neurological disorders result from disrupted neurotransmission between cells. Genetic processes involved in the production of neurotransmitters or malfunction of synaptic events can increase or decrease activity at synapses, which in turn affects emotions and behavior. One mood disorder is manic depression, also called bipolar disorder, in which individuals experience shifts between periods of euphoria (extreme happiness) and despair.

The most common mood disorder is major depression (unipolar depression), which results in prolonged periods of sadness and despair without alternating episodes of euphoria. Major depression affects 5–12% of men and 10–25% of women at some point in their lifetime. This condition is thought to result from decreased activity of synapses that release biogenic amines, such as serotonin, which changes neuronal activity within specific areas in the brain involved in processing emotion. Drugs used to treat major depression include the serotonin reuptake inhibitors, such as Prozac, Zoloft, and Paxil, which prevent the reuptake of serotonin into the presynaptic terminal after it is released. This allows serotonin to accumulate in the synaptic cleft, counteracting the deficit that causes the alteration in mood.

In many respects, the use of such drugs to treat depression reflects a profound change in the public's attitude toward mental illness. Older attitudes toward mood disorders held that individuals who were depressed lacked the ability to cope with stressful events in their lives. Only recently has it become accepted that mood disorders can be caused by changes in the balance of neurotransmitters in the brain. Perhaps the best evidence of a physiological basis for depression is that mood disorders occur more frequently in certain families, suggesting a genetic component. Drugs can be very effective in treating these disorders. Patients taking serotonin reuptake inhibitors often report decreased sadness and increased energy.

Recreational Drugs Disrupt Normal Neurotransmission

Most so-called recreational drugs work at the synapse to either enhance or interfere with the normal mechanisms of neurotransmission (see Table 43.3). In the presynaptic terminal, drugs can decrease neurotransmitter release by reducing calcium movement into the cell or preventing the exocytosis of vesicles containing stored neurotransmitters. In the synaptic cleft, drugs can slow the rate at which the neurotransmitter is broken down into an inactive form or taken back up into the presynaptic neuron, thereby prolonging the action of the neurotransmitter in the synaptic cleft. Some substances act on the postsynaptic membrane by either preventing the neurotransmitter from binding to its receptor, or acting as a substitute for the neurotransmitter by stimulating the receptor.

In effect, these drugs produce changes or imbalances in neurotransmission similar to those observed in some neurological disorders. These substances can induce euphoria, increase activity, alter mood, and produce hallucinations. They can also have serious, sometimes life-threatening effects and may be highly addictive.

Some drugs, such as cocaine, block the removal of dopamine from the synaptic cleft by preventing its reuptake into the presynaptic terminal. Morphine and marijuana mimic the actions of biological substances already in the brain, binding to receptors on the postsynaptic membrane. For these drugs, the resulting effects are much stronger than the effects of natural neurotransmitters. It is no surprise that recreational drugs are mind altering. They do, after all, change the ways in which neurons communicate with each other.

Disorders of Conduction May Result in Motor Problems and Abnormal Neuronal Development

Some human diseases are caused by the inability of the axon to properly conduct an action potential. This occurs most commonly because an axon fails to become myelinated, or a myelinated axon becomes demyelinated.

In **cretinism**, axons fail to become wrapped with myelin during fetal life, which leads to slow conduction speeds and abnormal connections between brain neurons. This results in profound mental defects that cannot be reversed unless treatment begins immediately after birth. Cretinism is caused by insufficient levels of thyroid hormone in the fetus, usually because the mother's dietary intake of iodine (which is required for synthesis of thyroid hormone) was too low. Cretinism is rare in the U.S. since the advent of iodized table salt, but it is not uncommon in many parts of the world.

Unlike cretinism, **multiple sclerosis (MS)** begins in adulthood—usually between the ages of 20 and 50—in individuals with apparently healthy nervous systems. With MS, the patient's own body, for reasons that are unknown, attacks and destroys myelin as if it were a foreign substance. Eventually, these repeated attacks leave multiple scarred (sclerotic) areas of tissue in the nervous system and impair the function of myelinated neurons that control movement, speech, memory, and emotion. MS is a serious and unpredictable disease, characterized by flare-ups followed by periods of remission in which symptoms are reduced or absent. No cure is currently available, but certain drugs may slow its progression and reduce the severity of symptoms. MS affects roughly 2.5 million people worldwide, 60–75% percent of them women.

Table 43.3	Effects of Common Drugs on Neurotransmitter Action and Mood		
Name of drug	**Actions on neurotransmission**	**Effects on mood**	**Effects of abuse**
Alcohol (ethanol)	Enhances inhibitory GABA transmission, increases dopamine transmission, inhibits glutamate transmission	Relaxation, euphoria, sleepiness	Liver damage, brain damage
Amphetamines *"uppers," "crystal meth," "pep pills"*	Stimulate the release of dopamine and norepinephrine; prevents their reuptake from synapse	Euphoria, increased activity	High blood pressure, psychosis
Antidepressants *Tricyclic antidepressants, for example, Elavil, Anafranil*	Block the reuptake of norepinephrine from synapses	Relieve depression and obsessive-compulsive disorder	Drowsiness, confusion
Selective serotonin reuptake inhibitors for example, *Prozac, Zoloft, Paxil, Lexapro*	Block the reuptake of serotonin from synapses	Relieve depression and obsessive-compulsive disorder	Insomnia, anxiety, headache
Monoamine oxidase inhibitors, for example, *Parnate, Nardil*	Block the breakdown of biogenic amine neurotransmitters	Relieve depression	Liver damage, hyperexcitability
Antianxiety drugs *Benzodiazepines,* for example, *Xanax, Valium, Librium, Rohypnol,* ("date rape" drug, "roofies," "roach")	Bind to GABA receptors and increase inhibitory neurotransmission	Sedation, sleepiness, amnesia	Drowsiness
Antipsychotic drugs *Phenothiazines,* for example, *Thorazine, Mellaril, Stelazine, Abilify, Risperdal*	Block dopamine receptors	Ease schizophrenic symptoms	Decreased movement
Cocaine	Causes release of dopamine and norepinephrine; blocks dopamine reuptake	Intense euphoria followed by depression	Convulsions, hallucinations, death from overdose
LSD (lysergic acid diethylamine)	Binds to serotonin receptors	Hallucinations, sensory distortions	Unpredictable and irrational behavior
Marijuana (tetrahydrocannibinol)	Binds to receptors for natural cannabinoids	Increased sense of well-being, decreased short-term memory, decreased goal-directed behavior, increased appetite	Delusions, paranoia, confusion
Narcotics *Heroin, morphine, Demerol, codeine*	Bind to opiate receptors	Pain relief, euphoria, sedation	Slowed breathing, death from overdose
Nicotine	Initially stimulates but then depresses activity in adrenal medulla and neurons in the peripheral nervous system; increases dopamine in brain	Increased attention, decreased irritability	Heart disease and lung disease
PCP (phencyclidine) *"Angel dust," "ozone"*	Blocks channel for excitatory amino acid neurotransmitters; increases dopamine activity	Violent behavior, feelings of power, numbness, disorganized thoughts	Psychosis, convulsions, coma, death

CHAPTER SUMMARY

43.1 Cellular Components of Nervous Systems

- The central nervous system (CNS) is composed of a brain and a nerve cord. The peripheral nervous system (PNS) consists of all neurons and their projections that are outside of and connect with the CNS. (Figure 43.1)
- The two major classes of cells in nervous systems are neurons and glia. (Figure 43.2)
- Types of neurons include sensory neurons, motor neurons, and interneurons. A neuron's structure is a reflection of its function. (Figure 43.3)
- The most basic circuit is a reflex arc, which occurs rapidly in response to inputs from sensory neurons and consists of only one or a few afferent and efferent cells. (Figure 43.4)

43.2 Electrical Properties of Neurons

- Neurons use electrical signals to communicate with other neurons, muscles, or glands. These nerve impulses involve changes in the amount of electric charge across a cell's plasma membrane. (Figures 43.5, 43.6, Table 43.1)
- Diffusion of ions through membrane channels occurs as a result of the relative concentration of an ion on both sides of the membrane and the electric charge across the membrane. Ions move in response to an electrochemical gradient. (Figure 43.7)
- The Nernst equation gives the equilibrium potential for an ion at any given concentration gradient.

43.3 Communication Between Neurons

- Gated ion channels enable a cell to communicate by changing its membrane potential rapidly. The opening and closing of voltage-gated and ligand-gated ion channels cause two types of changes in the neuron's membrane potential—graded potentials and action potentials. (Figures 43.8, 43.9)
- When a graded potential is large enough to spread to the axon hillock and depolarize the membrane to its threshold potential, this causes an action potential that carries an electrical signal from the axon hillock to the axon terminal. (Figures 43.10, 43.11)
- Axon diameter and myelination influence the speed of an action potential. (Figure 43.12)
- Electrical synapses are connected via gap junctions. Most vertebrate neurons communicate via chemical synapses in which a neurotransmitter carries the signal from the presynaptic to the postsynaptic cell. Many EPSPs generated at one time can sum together and bring the membrane potential to the threshold potential for action potential firing. (Figures 43.13, 43.14, 43.15)
- Chemical classes of neurotransmitters found in animals include acetylcholine, biogenic amines, amino acids, neuropeptides, and gaseous neurotransmitters. (Figure 43.16, Table 43.2)
- The response of the postsynaptic neuron determines the types of signals that pass from one neuron to the other. The two major types of postsynaptic receptors are ionotropic and metabotropic. (Figures 43.17, 43.18)

43.4 Impact on Public Health

- Most neurological conditions can be classified as disorders of either neurotransmission or conduction. Mood disorders caused by disrupted neurotransmission include major depression (unipolar depression). Drugs used in the treatment of neurological disorders and recreational drugs usually alter neurotransmission. (Table 43.3)
- Some neurological conditions are caused by the inability of the axon to conduct an action potential. This occurs most commonly because axons fail to become myelinated (cretinism), or myelinated axons become demyelinated (multiple sclerosis).

TEST YOURSELF

1. In vertebrates, the brain and the spinal cord are parts of
   a. the peripheral nervous system.
   b. the enteric nervous system.
   c. the central nervous system.
   d. the autonomic nervous system.
   e. the endocrine system.

2. The structures of a neuron that function mainly in receiving signals from other neurons are
   a. the myelin sheaths.
   b. the axons.
   c. the nerve termini.
   d. the dendrites.
   e. the K^+ channels.

3. The glial cells that form the myelin sheath in the peripheral nervous system are called
   a. astrocytes.
   b. oligodendrocytes.
   c. microglia.
   d. neurons.
   e. Schwann cells.

4. Neurons that function mainly in connecting other neurons in the central nervous system are
   a. sensory neurons.
   b. efferent neurons.
   c. motor neurons.
   d. afferent neurons.
   e. interneurons.

5. The difference in charges across the plasma membrane of an unstimulated cell is called
   a. the membrane potential.
   b. the resting potential.
   c. homeostasis.
   d. the graded potential.
   e. the action potential.

6. Which of the following influences the resting potential of a membrane?
   a. negatively charged molecules or ions inside the cell
   b. active transport of ions across the membrane
   c. concentration of Na^+ and K^+ inside and outside of the cell
   d. all of the above
   e. b and c only

7. A cell has reached a threshold potential when it has depolarized to the point where voltage-gated
   a. K^+ channels open.
   b. Na^+ channels open.
   c. K^+ channels close.
   d. Na^+ channels close.
   e. both b and d.

8. The speed of transmission of an action potential along an axon is influenced by
   a. the presence of myelin.
   b. an increased concentration of Ca^{2+}.
   c. the diameter of the axon.
   d. all of the above.
   e. a and c only.

9. Gap junctions are characteristic of
   a. electrical synapses.
   b. chemical synapses.
   c. acetylcholine synapses.
   d. GABA synapses.
   e. c and d only.

10. The response of the postsynaptic cell is determined by
   a. the type of neurotransmitter released at the synapse.
   b. the type of receptors the postsynaptic cell has.
   c. the number of Na^+ channels in the postsynaptic membrane.
   d. the number of K^+ channels in the postsynaptic membrane.
   e. all of the above.

CONCEPTUAL QUESTIONS

1. Distinguish between neurons and glial cells and give an example of a glial cell.

2. Describe the difference between a graded potential and an action potential.

3. Explain how the transmission of an impulse occurs across a chemical synapse.

EXPERIMENTAL QUESTIONS

1. What observations led Loewi to develop his hypothesis of how nerves stimulate muscle contractions?

2. What was Loewi's hypothesis to explain the observations of changes in the rate of contraction of the frog heart muscle? Describe his experimental design to test his hypothesis. Why did he use two hearts?

3. What were the results of Loewi's experiment? Did the results support his hypothesis?

COLLABORATIVE QUESTIONS

1. Discuss the organization of nervous systems in animals.

2. Discuss the parts of a neuron.

www.brookerbiology.com

This website includes answers to the Biological Inquiry questions found in the figure legends and all end-of-chapter questions.

44

NEUROSCIENCE II: EVOLUTION AND FUNCTION OF THE BRAIN AND NERVOUS SYSTEMS

CHAPTER OUTLINE

Connections between neurons in the brain of a mouse embryo. Nervous systems are constructed from many interconnected neurons that receive sensory information, transmit and interpret that information, and then elicit a response.

The brain of an adult person weighs only about 3 pounds, yet it consumes 20–30% of the body's energy! This amazing fact emphasizes the enormous level of metabolic activity that goes on continually in the brain, even during sleep. The beauty of the brain lies in its incredible complexity. The human brain, for example, has several thousand miles of interconnected neurons, which number around 100 billion. The brain allows us to move, think, and experience sensation and emotion. Groups of neurons also coordinate homeostatic functions such as breathing, blood circulation, and body temperature. When we examine the way that groups of neurons communicate, we begin to understand the complex mental functions of nervous systems, including learning, memory, and motivation.

Neuroscience—the study of nervous systems—is an area of intense research activity worldwide. For example, at one recent annual scientific meeting, more than 30,000 neuroscientists from dozens of countries met to present their research and learn about the latest developments in this dynamic field. The challenge of neuroscience is to transform the mind-boggling complexity of nervous systems into manageable proportions. For this reason, most neuroscientists do not study the human nervous system, because of its great complexity and the difficulties inherent in doing research on humans. Many animals, however—including the nematode *Caenorhabditis elegans*, the sea slug *Aplysia californica*, the fruit fly *Drosophila melanogaster*, the zebrafish *Danio rerio*, and the mouse *Mus musculus*—provide excellent opportunities to study how neurons work and how groups of neurons cooperate to produce animal behavior. The genomes of most of these animals have been sequenced, and neuroscientists are identifying genes that are critical for the structure and function of nervous systems. Many of the relevant genes are homologous to human genes, so the study of the molecular control of the nervous system in these model animals has the potential to reveal new treatments for many genetically inherited neurological diseases in humans.

In this chapter, we will first survey a variety of nervous systems, which allow animals to sense and respond to environmental changes (see chapter-opening photo). We will then examine the complex nervous system of humans. However, keep in mind that we still have much to learn about the organization, connectivity, and functions of nervous system structures. Our nervous system is fascinating and mysterious, and the study of how it functions will ultimately tell us much about what makes us human.

44.1 The Evolution and Development of Nervous Systems

The nervous systems of animals are the products of hundreds of millions of years of evolution. They provide advantages to animals that promote reproductive success. For example, nervous systems allow animals to keenly sense their environment and respond to changes in an appropriate way. In addition, nervous systems form connections with muscle systems and facilitate movements, which has allowed animals to travel short and long distances to obtain food. Likewise, nervous systems help animals to avoid predation.

Studying the evolution and development of nervous systems helps us understand how particular nervous systems are adapted to different functions. At the structural level, the organization of nervous systems ranges from a relatively simple network of a few cells to the marvelous complexity of the human brain. The characteristics of an animal's nervous system determine the behaviors that it displays. In this section, we will survey the nervous systems of invertebrates and vertebrates, and also examine the brains of vertebrates in greater detail.

Nervous Systems Evolved to Sense and Respond to Changes in the Environment

Except for sponges, all animals have a nervous system. The simplest nervous system is the **nerve net** of the cnidarians (jellyfish, hydras, and anemones) (**Figure 44.1a**). In these organisms, no single group of neurons controls all the others. Instead, the neurons connect to each other in a network that can be activated all at once. This allows the organism to contract and move large areas of its tentacles and body simultaneously. Some cnidarians, such as the jellyfish, have two nerve nets, one for moving tentacles and one for swimming. Sea stars and other echinoderms have a slightly more sophisticated nervous system with a nerve ring around the mouth that is connected to larger radial

nerves extending into the arms (**Figure 44.1b**). This arrangement allows the mouth and arms to operate independently.

More complex invertebrate nervous systems usually consist of one or two **nerve cords** that extend from the anterior end of the animal to the tail. In the flatworm (*Planaria*), collections of neurons in the animal's head form **ganglia** (singular, *ganglion*), which are groups of neuronal cell bodies that perform basic functions of integrating inputs from sense organs, such as the eyes, and controlling motor outputs like those involved in swimming (**Figure 44.1c**). Two nerve cords extend along the ventral surface of the animal into the tail and are connected to each other by transverse nerves. In annelids (segmented worms), the basic structure is similar, except that more neurons are present and the ventral nerve cord has ganglia located in each body segment (**Figure 44.1d**). In the head, a ganglion, known as an integrative center, functions as a rudimentary brain to control body movements. The other ventral ganglia along the nerve cord receive sensory information from a particular body segment and control local movements. In the simpler forms of mollusks, such as the snail, the nervous system is very similar to that of the annelids. The brain is composed of a pair of anterior ganglia; paired nerve cords extend from these ganglia to the eyes, muscular foot, and gut.

During the evolution of animals, more complex body types have been associated with **cephalization** (from the Greek, *cephalo,*

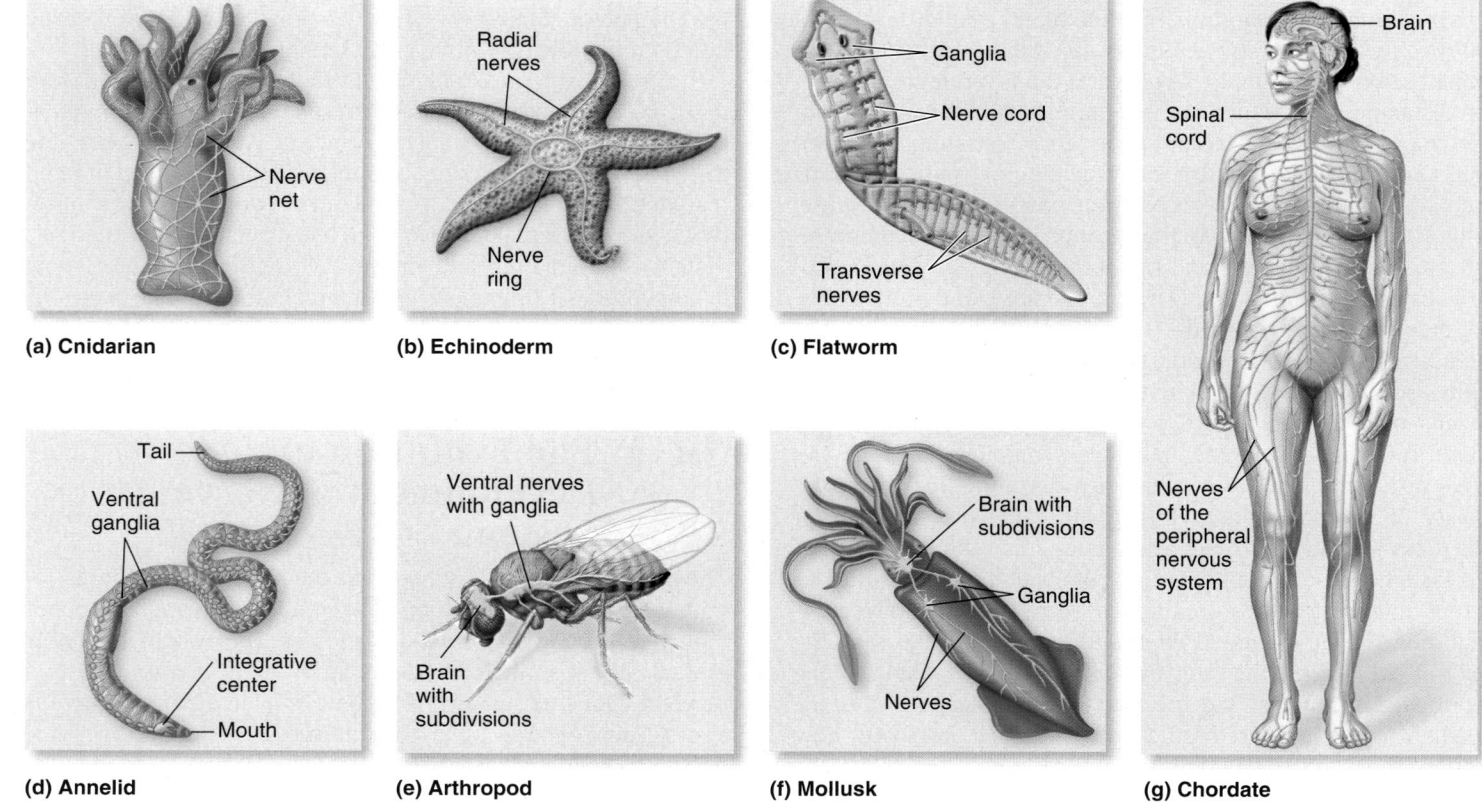

(a) Cnidarian
(b) Echinoderm
(c) Flatworm
(d) Annelid
(e) Arthropod
(f) Mollusk
(g) Chordate

Figure 44.1 Representative nervous systems throughout the animal kingdom.

head), which is the formation of an increasingly complex **brain** in the head that controls sensory and motor functions of the entire body. Within the brain, neuronal pathways provide integrative functions necessary for an animal to make more sophisticated responses to its environment. Brains are found in all vertebrates and most invertebrates, and they are usually composed of more than one anatomical and functional region. For instance, in *Drosophila* (**Figure 44.1e**), the brain has several subdivisions with separate functions, such as a region devoted to learning and memory. More advanced mollusks, such as the squid and octopus, have brains with well-developed subdivisions that allow these animals to coordinate complex visual and motor behaviors necessary for their predatory lifestyle (**Figure 44.1f**). In vertebrates and simpler chordates, the brain is connected to the **spinal cord** and constitutes the **central nervous system** (**Figure 44.1g**). As discussed later, nerves from the **peripheral nervous system** route information into and out of the central nervous system at separate regions along the spinal cord. The vertebrate spinal cord and the bony vertebrae have an organization that shows similarities to the segmented nervous system of invertebrates.

Brains of Vertebrates Have Three Basic Divisions

Development of the vertebrate brain begins with the formation of a central fold in the embryo called the **neural tube**. This hollow tube is the structure from which the entire nervous system develops. Bending and folding of the neural tube during embryonic development results in bulges that become separate divisions of the nervous system. The anterior end develops into the brain, while the posterior portion becomes the spinal cord.

In vertebrates, the brain has a triad structure. The three major divisions are the **hindbrain**, **midbrain**, and **forebrain** (**Figure 44.2a**). In fish, the brain remains a three-part structure.

Fossils of agnathan fishes that lived 500 million years ago show that their brains were already organized into the three basic divisions that have been retained in all modern vertebrates. In vertebrates with more complex brains, the hindbrain and forebrain each form two separate divisions (**Figure 44.2b**). The hindbrain subdivides into the metencephalon and the myelencephalon. The forebrain subdivides into the telencephalon and the diencephalon. The midbrain, by contrast, does not subdivide. Its single division is termed the mesencephalon. The development of brain subdivisions provides additional specialization and increases the capacity of the brain to perform complex functions.

Hindbrain The hindbrain includes the **medulla oblongata**, **pons**, and **cerebellum** (**Figure 44.2c**). The medulla oblongata coordinates many basic reflexes and bodily functions, such as breathing, that maintain the normal homeostatic processes of the animal. The cerebellum and pons are responsible for monitoring and coordinating body movements.

Midbrain The midbrain processes several types of sensory inputs, including vision, smell, and hearing. It controls sophisticated tasks such as coordinating eye movement with visual inspection of the environment.

Forebrain The forebrain also processes sensory inputs and consists of a group of structures that are responsible for the higher functions of conscious thought, planning, and emotion. Many of these functions are attributed to the **cerebrum**. The surface layer is called the **cerebral cortex**, which is only a few millimeters thick. Other structures of the forebrain are beneath the cerebrum. These include the thalamus and hypothalamus, which we will discuss later in this chapter.

(a) Embryo (4 weeks)

(b) Embryo (5 weeks)

(c) Adult

Figure 44.2 **Development of the human brain.** The structures shown here occur during embryonic development at **(a)** 4 weeks and **(b)** 5 weeks, or **(c)** in the adult. Most structures beneath the cerebrum are not shown.

The relative sizes of the brain divisions vary in different animals. Usually, a larger area plays a greater role in an animal's life. In addition, the functions performed by these divisions have changed over the course of evolution. For instance, the midbrain in fishes is relatively large compared to the other two areas, and it plays a key role in vision. By comparison, the forebrain is the largest division in birds and mammals and is largely responsible for vision.

Increased Brain Complexity Involves a Larger, Highly Folded Cerebrum

As evolution produced animals with more complex nervous systems, the size of the cerebrum also increased, making up a greater proportion of the brain. As discussed later in this chapter, many of the important functions of the cerebrum are carried out by neurons along its outer surface called the cortex. Therefore, increased complexity of the brain is also correlated with an increased surface area of the cerebral cortex. During the evolution of more complex mammals, this increase in surface area is thought to have occurred more rapidly than an expansion in the size of the skull. But how could this occur? The answer is that the external surface of the cerebrum in complex animals is highly convoluted, forming many folds. Compare the relatively smooth-looking surface of the cerebrum of a rat with the highly folded one of a human in **Figure 44.3**. If the surface of the human brain were spread out, it would be about 2.5 square feet!

As body size increases across the animal kingdom, you might expect that brain mass would increase proportionately—that the brain of an elephant would be larger than that of a monkey, for instance. That is generally the case, with a few important exceptions (**Figure 44.4**). In particular, the masses of the human and dolphin brains are greater than would be expected on the basis of body mass.

Brain mass and the amount of folding are correlated with more complex behaviors. Why is this so? The outer surface of the brain, the cerebral cortex, plays a key role in conscious thought, reasoning, and learning. Greater size and folding provide more surface area, which allows greater processing and interpretation of information. Even so, evidence does not suggest that people with small differences in brain size differ in intelligence. Also, it would be wrong to assume that an animal with a small brain is profoundly limited in its behavioral repertoire. A bat with a 0.9-g brain and an elephant with a 2,500-g brain can both perform a great variety of interesting and complex behaviors, such as navigating across great distances and interacting with fellow members of their species.

Figure 44.3   **The degree of cerebral cortex folding in different mammalian species.**   The brains are not shown to scale.

Biological inquiry: Based on size and cerebral cortex folding, pick two animals capable of more complex behaviors versus two that would be less capable.

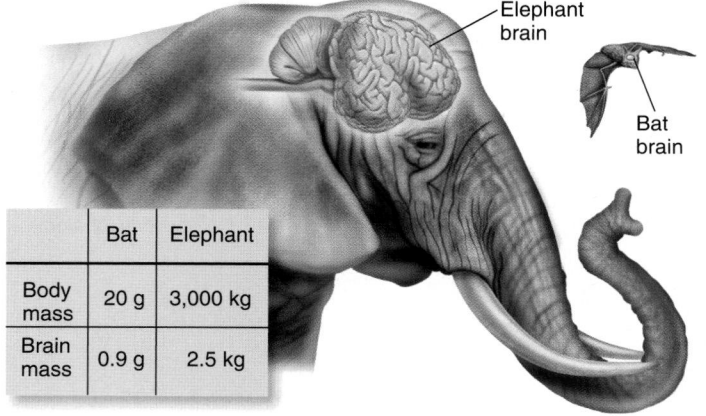

	Bat	Elephant
Body mass	20 g	3,000 kg
Brain mass	0.9 g	2.5 kg

Figure 44.4   **Brain mass as a function of body mass in mammals.**   For most mammals, brain weight is in proportion to body mass. However, humans and dolphins have a much higher brain mass relative to their body mass than do other mammals.

44.2 Structure and Function of the Human Nervous System

The nervous system of the human is wonderfully complex—the brain alone has over 100 billion neurons and has the highest ratio of glia/neurons of any vertebrate. Moreover, complexity is defined by more than just numbers of cells. Within the human brain, for example, are enormous numbers of connections between neurons—a single neuron in the cerebellum may have as many as 10^5 synapses with other cells! In this section, we will examine the human nervous system, with an emphasis on the functions of the major parts of the brain and spinal cord.

The Nervous System Is Composed of the Central and Peripheral Nervous Systems

As mentioned, the brain and spinal cord constitute the central nervous system (CNS) of vertebrates (**Figure 44.5**). The peripheral nervous system (PNS) is composed of neurons and axons of neurons that are outside the CNS and communicate with it. The CNS and PNS are connected anatomically and functionally. The CNS receives information about the internal or external environment from the PNS. The CNS interprets that information and may initiate a response that is then carried out by the PNS. For example, suppose you accidentally lean against a newly painted fence. Neuronal endings in your skin, which are part of

the PNS, would transmit information about the sense of touch through axons that bring information directly into the spinal cord. From there, the information travels to your brain, where the sensation is analyzed and identified as something wet. Signals are sent from your brain, down your spinal cord, and through the neurons of the PNS to your muscles, causing you to move away.

Within the nervous system, groups of neurons may associate with each other and perform a particular function. In the CNS, the cell bodies of neurons that are involved in a similar function may be grouped into a structure called a **nucleus** (plural, *nuclei*), which may include thousands of cells. For instance, cell bodies that regulate body temperature and those that recognize visual information are located in separate nuclei in the brain. As mentioned earlier, the term ganglion is also used to refer to a group of neuronal cell bodies but usually in the PNS (Figure 44.5). Within an animal's body, many myelinated axons may run in parallel bundles. (Myelination is described in Chapter 43.) Such a structure is called a **tract**, when it is found in the CNS. Tracts convey information within the brain and between the brain and the spinal cord. Bundles of myelinated axons also form **nerves**, which are found in the PNS. Nerves make contacts with structures outside the CNS and transmit signals that enter or leave the CNS. Nerve connections between the PNS and the CNS occur at the brain or spinal cord. **Cranial nerves** are directly connected to the brain. They are located in the head and transmit incoming and outgoing information between the PNS and the brain. By comparison, **spinal nerves** are connections between the PNS and spinal cord (Figure 44.5).

One of the most obvious characteristics of the CNS becomes noticeable in nonliving tissue. After death, some parts look white, while others appear gray (**Figure 44.6**). The **white matter** consists of myelinated axons that are bundled together in large numbers to form tracts. The **gray matter** consists of neuronal cell bodies, dendrites, and some unmyelinated axons. The cerebral cortex is composed of gray matter that sits on top of a large collection of white matter pathways. In the spinal cord, the gray matter is located in the center and forms two dorsal horns and two ventral horns. Each dorsal horn connects to a dorsal root, which is part of a spinal nerve. Dorsal roots receive incoming information from sensory (afferent) nerves of the PNS. The ventral horn connects to the ventral root, which is also part of a spinal nerve that transmits outgoing information to motor (efferent) nerves.

Unlike the PNS, the CNS is encased in protective structures including bone (the skull and backbone) and three layers of sheathlike membranes called **meninges** (**Figure 44.7**). The outermost membrane, the dura mater (from the Latin, meaning "hard mother"), is a thick protective layer that lies just inside the skull and vertebrae. The middle membrane is called the arachnoid mater (Latin, meaning "spidery mother") because it has numerous weblike tissue connections to the innermost membrane, the pia mater (Latin, meaning "thin mother"). The pia mater is a very thin membrane that lies on the surface of the brain and spinal cord, folding with the brain's surface.

Between the arachnoid mater and pia mater is the subarachnoid space. This space is filled with **cerebrospinal fluid,**

Figure 44.5 **Organization of the human nervous system.** The central nervous system consists of the brain and spinal cord, both of which are encased in bone (not shown). The peripheral nervous system includes cranial nerves, ganglia, and spinal nerves, which carry information to and from the CNS, and many other neurons throughout the body.

Central nervous system

Brain

Spinal cord

Peripheral nervous system

Cranial nerves

Ganglia

Spinal nerves

Figure 44.6 Gray matter and white matter in the CNS. The gray matter is composed of groups of cell bodies, dendrites, and unmyelinated axons, while the white matter consists of tracks of myelinated axons.

Biological inquiry: Is a spinal nerve composed of afferent or efferent nerves, or both?

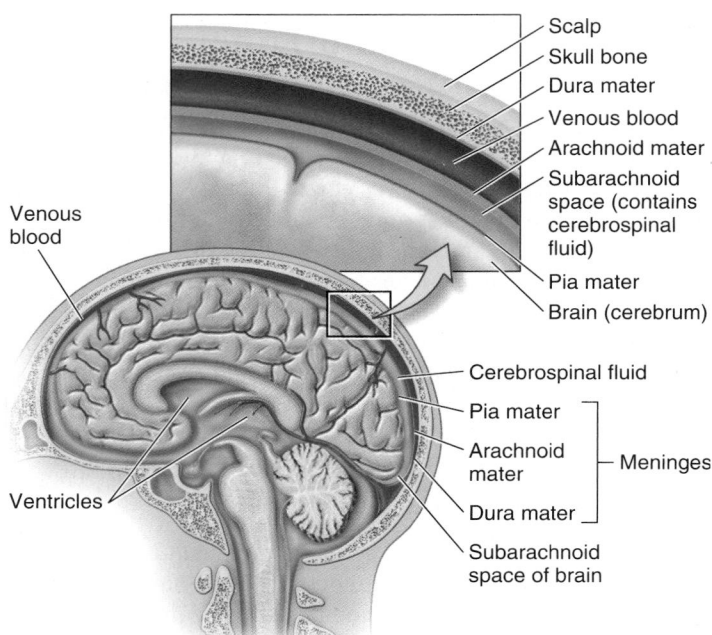

Figure 44.7 The meninges and ventricles of the CNS. The thickness of the meninges are exaggerated for illustration purposes. Note that the cerebrospinal fluid encases the entire CNS and also fills the ventricles.

which surrounds the exterior of the brain and spinal cord and absorbs physical shocks to the brain resulting from sudden movements or blows to the head. The cerebrospinal fluid contains nutrients, hormones, and other substances that are taken up by cells of the brain, and the fluid is also a reservoir for waste products that are carried away by the circulatory system. In addition to the subarachnoid space, the cerebrospinal fluid also fills a series of connected cavities called the ventricles that lie deep within the brain and run along a central canal that extends the length of the spinal cord. These fluid-filled structures provide a cushion of support and protection for the CNS.

The PNS Carries Information to and from the CNS

The PNS of vertebrates is subdivided into two major functional and anatomical components, the somatic nervous system and the autonomic nervous system. Both divisions have sensory (afferent) nerves and motor (efferent) nerves.

Somatic Nervous System The primary function of the **somatic nervous system** is to sense the external environmental conditions and control skeletal muscles. The sensory neurons of the somatic nervous system receive stimuli, such as heat, vision, smell, taste, hearing, and touch, and transmit signals to the CNS. The motor neurons of the somatic nervous system control skeletal muscles. The cell bodies of these motor neurons are actually located within the CNS. The axons from these cells leave the spinal cord and project directly onto skeletal muscle without making any intermediary synapses along the way.

The somatic nervous system is said to be voluntary because the responses can usually be controlled consciously. For example, we use our somatic nervous system to walk and hold a pencil. However, not all responses are voluntary. An example is a reflex arc, such as the knee-jerk response, which is automatic (see Chapter 43).

Autonomic Nervous System The **autonomic nervous system** (also called the visceral nervous system) regulates homeostasis and organ function. For example, it is involved in regulating the rate of heart contractions, blood pressure, and the amount of stomach acid secreted. Though the autonomic nervous system is predominantly composed of motor neurons, certain of its sensory neurons detect internal body conditions. For example, baroreceptors are sensory neurons that detect blood pressure. For the most part, the autonomic nervous system is not subject to voluntary control. We usually cannot consciously change our heart rate or blood pressure.

The efferent pathways of the autonomic nervous system involves two motor neurons. The cell body of the first neuron is within the CNS and synapses on a second neuron in a ganglion of the PNS. This second neuron sends its axon to an effector cell, where it alters that cell's function. These neurons control smooth muscles, cardiac muscle, and glands.

The efferent nerves of the autonomic system are subdivided into the sympathetic and parasympathetic divisions (**Figure 44.8**). Both divisions of the autonomic system act on the same organs and usually have opposing actions. The **sympathetic division** is responsible for rapidly activating systems that

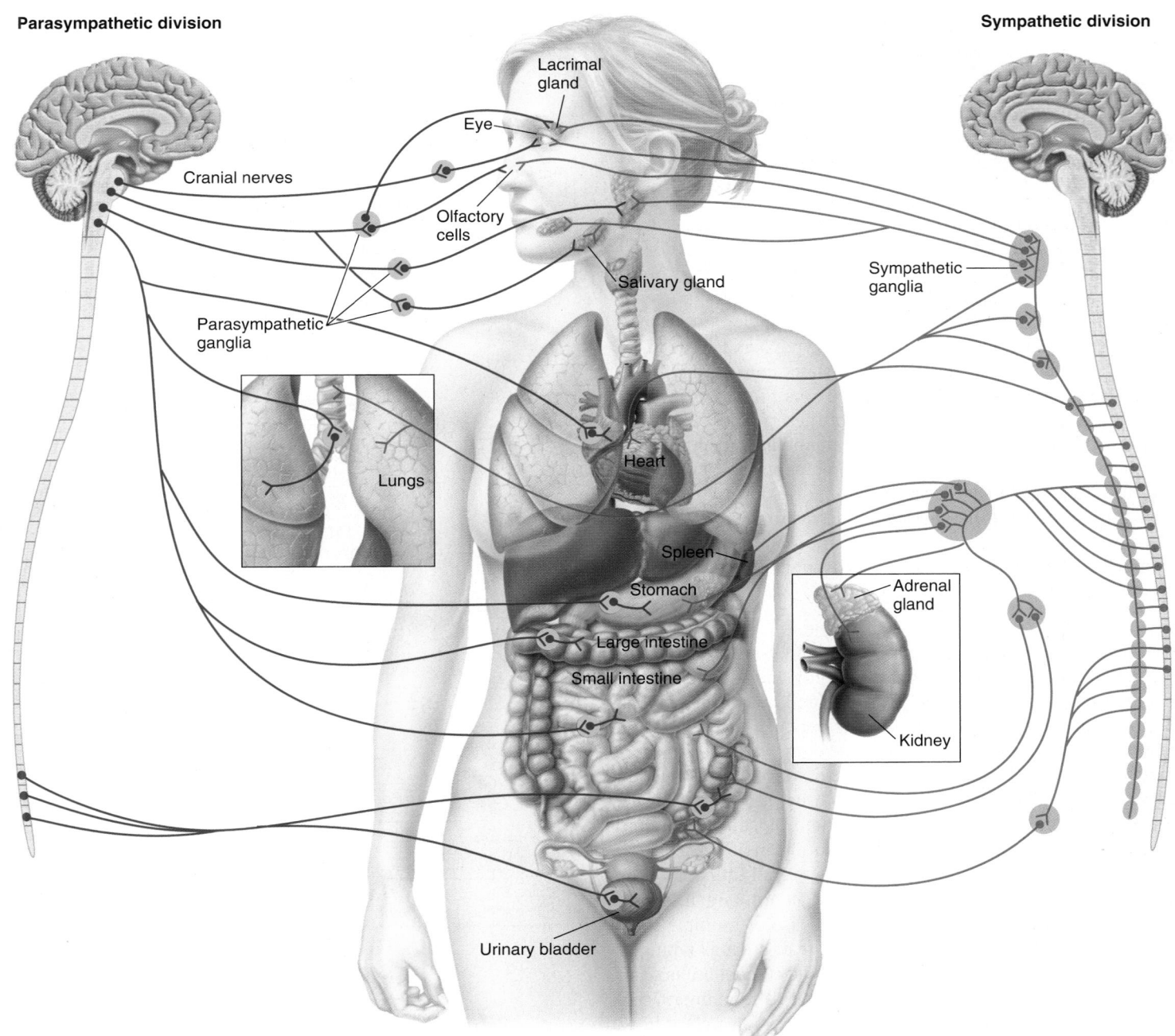

Figure 44.8 **Parasympathetic and sympathetic divisions of the autonomic nervous system.** For simplicity, only some of the major functions of each division are shown in this figure. The parasympathetic and sympathetic systems tend to have opposite effects, and most parts of the body receive inputs from both divisions. Nerves from the sympathetic division make connections with ganglia, most of which are alongside the spinal cord, while nerves from the parasympathetic division make connections in ganglia near or in their targets.

prepare the body for danger or stress. This is the "fight-or-flight" response, which is characterized by increased heart rate, stronger pumping action of the heart, faster breathing, additional blood flow to skeletal muscles, and increased secretion of glucose and fats into the blood by the liver and adipose tissue. These features prepare us to confront (fight) or avoid (flight) a perceived threat.

The **parasympathetic division** of the autonomic nervous system is involved in maintaining and restoring body functions. It is active during restful periods or after a meal, which is why it is sometimes called the "rest-or-digest" system. Neurons of the parasympathetic division promote digestion and absorption of food from the gut, slow the heart rate, and decrease the amount of fuel supplied to the blood from the liver and adipose tissue.

The Hindbrain Is Important for Homeostasis and Coordination

Let's now turn our attention to the structure and function of the human brain (**Figure 44.9**). We will begin with the evolutionarily oldest structures of the brain, some of which are located in the hindbrain and control the basic processes that sustain life.

Medulla Oblongata The medulla oblongata coordinates many basic reflexes and bodily functions that maintain the normal homeostatic processes of a person. It is involved in the control of heart rate, breathing, cardiovascular function, digestion, swallowing, and vomiting.

The medulla oblongata as well as the pons and areas of the midbrain, collectively called the brainstem, contain additional nuclei (groups of cell bodies) whose axons project dorsally to many other parts of the brain. Each of these nuclei consists of neurons that use a single biogenic amine neurotransmitter (see Chapter 43), such as dopamine or norepinephrine, and project into specific target areas. Some of these nuclei form the reticular formation, a network-like pathway that maintains and controls alertness and sleep. Because of the importance of the brainstem's functions, damage to it is catastrophic and may result in coma or death.

Cerebellum and Pons The cerebellum and pons are responsible for monitoring and coordinating body movements. Major nerve tracts pass through the pons into and out of the cerebellum so that the pons serves as a relay between the cerebellum and other areas of the brain. The cerebellum, a large structure that sits dorsal to the pons and medulla oblongata, receives sensory inputs from the cerebral cortex and the hearing and visual areas of the brain. It also receives inputs from the spinal cord that convey information about the position of joints and the contraction or relaxation of muscles. The overall function of the cerebellum is to maintain balance and coordinate hand-eye movements. The cerebellum controls the use of multiple muscles at one time and synchronizes motor feats such as typing or making a jump shot in basketball. When the cerebellum is damaged, as in an accident, people find it difficult to maintain balance and fine-tune motor functions. Reaching for a glass filled with water, for example, might result in spilling some of the water while bringing the glass to the mouth.

The Midbrain Processes Sensory Inputs

In addition to forming part of the reticular formation, the midbrain processes several types of sensory inputs, including vision, smell, and hearing. It has tracts that pass this information to other parts of the brain for further processing and interpretation. In vertebrates with less developed brains such as fish, the midbrain is the only area that interprets vision, but in humans this function is largely performed by the forebrain, to which we now turn.

(a)

(b)

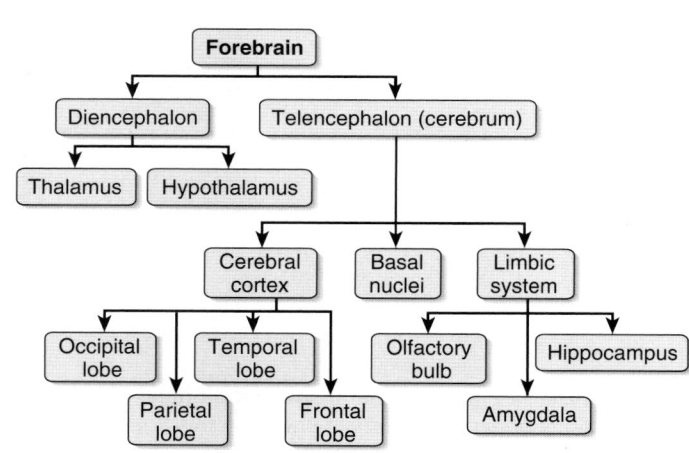

(c)

Figure 44.9 **Major structures of the human brain.** (a) A cross section of the brain, showing many internal structures (not all structures are visible in this plane, such as the basal nuclei). The limbic system consists of the olfactory bulbs, amygdala, hippocampus and parts of the thalamus and hypothalamus. (b) The four lobes of the cerebral cortex. (c) Relationships of the various parts of the forebrain.

Biological inquiry: If a person received an injury to the cerebellum, how would that affect his or her behavior?

The Forebrain Is Responsible for the Higher Functions of Thought, Learning, and Emotion

The forebrain comprises the thalamus and hypothalamus (diencephalon) and the cerebrum (telencephalon), which consists of the basal nuclei, limbic system, and cerebral cortex (Figure 44.9).

Thalamus and Hypothalamus In vertebrates, the **thalamus** plays a major role in relaying sensory information to appropriate parts of the cerebrum and, in turn, sending outputs from the cerebrum to other parts of the brain. It receives input from all sensory systems. The thalamus is organized according to the type of sensation. For example, certain parts of the thalamus process visual inputs while other parts process sounds. It also directs outgoing motor instructions that it receives from the cortex, sending directions to the spinal cord to generate voluntary movements. The thalamus is involved in the perception of pain and the degree of mental arousal in the cortex.

The **hypothalamus**, located below the thalamus at the floor of the forebrain, controls functions of the gastrointestinal and reproductive systems, and it regulates many basic behaviors such as eating and drinking. This area has great importance for homeostasis of the body and the control of behavior. Though small in size, it is composed of many nuclei, each with its own function. A major role of the hypothalamus is the production of hormones, which travel to the pituitary gland located just beneath the brain. The pituitary gland, in turn, regulates hormone secretion from other glands in the body, including the thyroid, gonads, and adrenal glands. In addition to producing hormones, the hypothalamus is sensitive to their actions. For example, certain hormones produced by cells in the stomach, intestine, adipose tissue, gonads, and elsewhere act within the hypothalamus to facilitate the expression of feeding, drinking, sexual, and aggressive behaviors. Finally, a small pair of hypothalamic nuclei called the suprachiasmatic nuclei are the "master clocks" of the CNS, establishing circadian rhythms, which control the expression of behavioral, physiological, and hormonal rhythms over the 24-hour day.

Basal Nuclei The **basal nuclei** are also called the basal ganglia, though the term ganglia usually refers to structures in the PNS. The basal nuclei are a group of nuclei that surrounds the thalamus on both sides and lies beneath the cerebral cortex. Like the cerebellum, the basal nuclei are involved in planning and learning movements. They also function via a complex circuitry to initiate or inhibit movements.

Parkinson's disease, a common neurological disorder that affects the basal nuclei, provides a good illustration of the function of these nuclei. People with Parkinson's disease have trouble initiating movement, such as beginning to move their legs when they wish to walk. They are capable of walking once movement has begun, but they move slowly with muscle tremors and a shuffling, jerky gait. These symptoms result from the gradual deterioration of dopamine-releasing neurons in an area of basal nuclei called the substantia nigra. People in the early stages of Parkinson's disease can be treated with L-dopa, a molecule that enters the blood and travels to the basal nuclei, which are capable of converting it into dopamine. L-dopa, therefore, substitutes for dopamine and reduces the symptoms.

Limbic System The **limbic system** refers to certain areas of the telencephalon and parts of the diencephalon. It is made of evolutionarily older structures that form an inner layer at the base of the forebrain. These include structures such as the **olfactory bulbs** (which process information about smells), amygdala, and hippocampus, as well as parts of the thalamus and hypothalamus. However, not all neuroscientists agree on which brain structures constitute the limbic system. The limbic system is primarily involved in the formation and expression of emotions, and it also plays a role in learning, memory, and the perception of smells. The expression of emotions occurs early in childhood before the more advanced functions of the cortex are evident. Thus, even very young babies can express fear, distress, and anger as well as bond emotionally with their parents.

The **amygdala** is one of the areas known to be critical for understanding and remembering emotional situations. This area also is involved in the ability to recognize emotional expression in others. Emotions are not unique to humans, however, and are clearly present in other primates and mammals. Being able to express and detect emotions imparts a selective advantage by enabling animals to establish and maintain relationships. Emotions such as fear help an animal defend itself against danger by avoiding conflict. Likewise, anger is associated with aggression, a key behavior by which many animals defend their territories.

The **hippocampus** is composed of several layers of cells that are connected together in a circuit. Its main function appears to be establishing memories for spatial locations, facts, and sequences of events. Damage to certain parts of the hippocampus in humans results in an inability to form new memories, a devastating condition that prevents recognition of other people or even an awareness of daily events. Experiments with laboratory animals have also demonstrated the importance of the hippocampus for memory and learning in other mammals. In a particularly well-studied example, rats are placed into a pool of milky water containing a hidden platform. The animals swim until they find the platform, on which they can safely stand. The time it takes to find the platform in subsequent trials is shorter as they learn and remember its whereabouts. This type of spatial learning depends on activity in the hippocampus. Rats with parts of their hippocampus destroyed fail to improve their times with repeated trials. The hippocampus also receives extensive inputs from the olfactory bulbs, which may explain why smells are such potent triggers of memory in humans, and why many animals use their sense of smell as a major way to learn and remember aspects of their environments.

Cerebral Cortex As mentioned earlier, the cerebral cortex is the surface layer of gray matter that covers the cerebrum (see Figure 44.6). The **neocortex** (literally, new cortex) is the layer that evolved most recently in mammals and has only six layers of cells.

Older parts of the cortex, including areas involved in processing the sense of smell, have fewer layers. Although the cerebral cortex is only a few millimeters thick, it contains about 10% of all the neurons in the brain. Sensory and motor information are integrated in the cortex, which is responsible for the initiation of voluntary acts, generation of speech, learning and memory, and the production of emotions.

The cerebral cortex is divided into four lobes (see Figure 44.9b). Each lobe has a number of functions, many of which are still being actively investigated by researchers. The frontal lobe is important for conscious thought and social awareness. The parietal lobe is involved in attention and making associations between events and incoming information. The occipital lobe controls many aspects of vision, while the temporal lobe is necessary for language, hearing, and some types of memory.

One particularly interesting aspect of temporal lobe activity that illustrates the subtle functions of the various lobes is the recognition of objects and people. One area of the temporal lobe contains so-called face cells that are activated specifically in response to seeing faces. Electrical activity of individual cells within the temporal lobe of rhesus monkeys increases when the monkey views human and monkey faces, though not when looking at other images. Humans who have damage to that area are unable to recognize other people by their faces even if they knew them before the brain damage occurred. Fortunately, these patients can recognize other traits such as gait or tone of voice, which they use to identify the people in their lives. It is not known whether face cells are active in nonprimate mammalian species.

An amazing finding is that sensory inputs come into the cortex and motor outputs leave the cortex in areas that stretch out like a map of the body (**Figure 44.10**). The amount of space in the cortex assigned to a particular part of the body is proportional to either the degree of sensory sensitivity of that body part or the number of muscles required for its movement. For instance, a larger part of the cortex is devoted to the lips than to other areas of the face. The lips have more nerve endings and are more sensitive to touch than these other areas. The organization of the cortex may not be permanent, however, because the map may change depending on the amount of use or disuse of a given part of the body, as discussed in the Feature Investigation later in this chapter.

Cerebral Hemispheres One of the most visible features of the surface of the cerebrum is its division into two halves, or **hemispheres**. Each hemisphere is connected to the other by a major tract called the **corpus callosum** (**Figure 44.11a**). In the 1950s, American neuroscientists Roger Sperry and Ronald Meyers began examining the separate functions of the hemispheres in laboratory animals by performing split-brain surgeries, in which the corpus callosum was surgically severed. The animals that underwent split-brain surgery maintained their overall health and functioning, and thus the surgery was considered safe for humans. This became important in 1961, when the procedure was used for the first time to treat patients with severe epilepsy,

Figure 44.10   Mapping of body parts along the cortex that control body movements (motor function) or interpret sensory information.   The relative sizes of body parts reflect the relative amount of cortex devoted to them.

a disorder characterized by uncontrolled electrical activity that begins in one place in the brain and can spread via the corpus callosum to the other side. Cutting the connection between the hemispheres reduced the severity of epileptic flare-ups.

Split-brain surgery also enabled the researchers to make critical observations that could not be made in laboratory animals. Split-brain humans generally show normal behavior and intellectual function, because both hemispheres can function fairly independently. However, psychological tests revealed that the two sides of the brain process different types of information. One study demonstrated that the left hemisphere produces a descriptive word for an object but does not identify certain

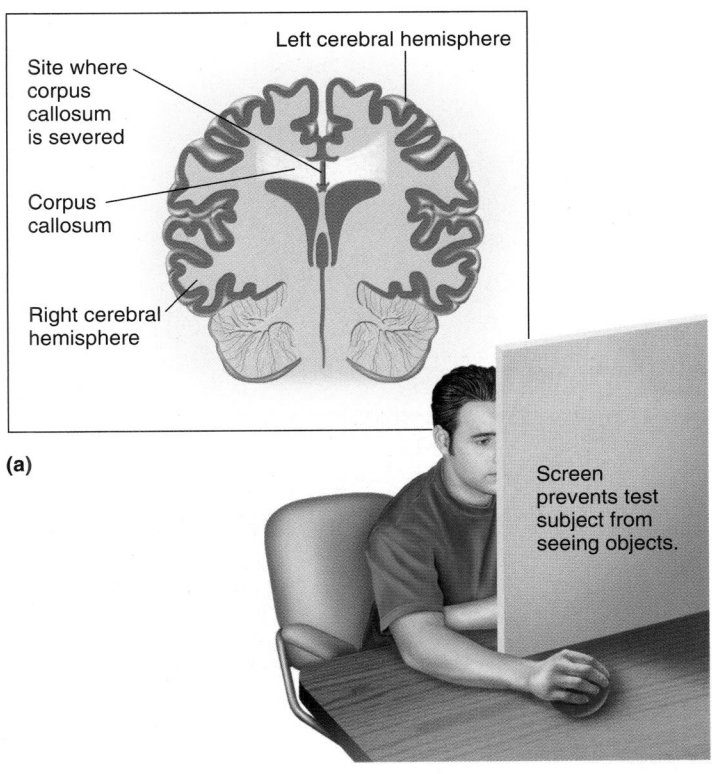

(a)

(b)

Figure 44.11 **The hemispheres of the brain.** **(a)** The cerebral hemispheres and their connection by the corpus callosum. (Note: the right hemisphere controls the left side of the body, while the left hemisphere controls the right side.) **(b)** Split-brain patient being tested for hemispheric dominance. By using this apparatus, Roger Sperry and his collaborators showed that the left and right cerebral hemispheres have different capabilities. When a split-brain patient held an object in his right hand but could not see it or touch it with his left hand, he could give it a name (for example, a ball). When he held another object in his left hand, he could describe it (for example, rough or smooth), but could not name it.

Biological inquiry: With her eyes closed, a split-brain patient was given a rock to hold, and she described it as a rock. Which hand was it in?

characteristics of that object, such as shape and texture (**Figure 44.11b**). The right hemisphere, in contrast, cannot use words to name the object but can identify other qualities. The studies of Sperry, Meyers, and other neuroscientists have concluded that the left hemisphere is involved in understanding language and producing speech in most people. Therefore, the left hemisphere is said to be dominant for those functions. The right hemisphere is dominant for nonverbal memories, recognizing faces, and interpreting emotions. In 1981, Sperry received the Nobel Prize for his insight regarding specialization in each hemisphere of the brain.

GENOMES & PROTEOMES

Several Genes Have Been Important in the Evolution of the Cerebral Cortex

A number of genes are known to be involved in the development of the cerebral cortex. Some have been identified by examining genetic mutations in developmentally disabled individuals, and others by comparing human genes with genes that are known to be involved in brain development in other species such as *Drosophila*. Researchers have also compared these genes in many species that show notable differences in cerebral structure. This last approach can determine whether a relationship exists between the expression of a particular gene and the organization of the cerebral cortex.

One inherited disorder that involves abnormal development of the cerebral cortex is polymicrogyria (meaning "many small folds"), which results in mental impairment as well as disrupted gait and language production. In people with this disorder, the cortex is characterized by multiple abnormal surface folds and fewer layers of cells. One type of polymicrogyria is a recessively inherited condition for which eight different mutations of a single gene are known. This gene, called GPR56, encodes a G-protein-coupled receptor (described in Chapter 9), which has large extracellular loops. All eight mutations that produce polymicrogyria alter these extracellular loops of the receptor, and it is thought that this alters the ability of the G-protein-coupled receptor to bind its ligand.

Two other genes, called microcephalin (meaning "small head") and ASPM (a̲bnormal s̲pindle-like m̲icrocephaly-associated gene), have been shown to be determinants of brain size and thus possibly involved in the evolution of brain size. For example, mutations of these genes in the human population produce individuals with a much smaller frontal cortex. Interestingly, the sequences of these genes in several primates, including humans, as well as in other mammals such as dogs and sheep, have shown that the proteins produced by the normal microcephalin and ASPM genes have undergone greater changes in humans and great apes than in other species. Therefore, these genes may have been under greater selective pressure in animals with larger cortices, suggesting that the genes play a key role in cortex development.

44.3 Cellular Basis of Learning and Memory

In the past few decades, an exciting advance in neuroscience is that researchers have begun to understand complex behaviors, such as learning and memory, at the cellular level. Though it is difficult to separate the two concepts, **learning** can be defined as the process by which new information is acquired, while **memory** involves the retention of that information over time. Memory is the "glue" that connects our experiences throughout life.

Our behavior is largely controlled by what we have learned and remember from past experiences. Neuroscientists want to understand how the brain learns, and how it captures memories. In this section, we will examine some current ideas about how this may be achieved at the cellular level, and consider experimental approaches that neuroscientists follow when trying to investigate such complicated phenomena.

Learning and Memory Occur via Changes Within Neurons and Their Connections with Each Other

Beginning in the 1960s, research along two fronts led to key insights regarding the cellular basis of memory. Norwegian neuroscientist Terje Lomo and British researcher Timothy Bliss focused their efforts on the hippocampus. As described earlier, this is a key region of the brain involved with learning and memory. Lomo and Bliss conducted experiments on anesthetized rabbits in which they monitored signal transmission across particular regions of the hippocampus. Their key discovery involved the effects of multiple stimuli. Experimentally, a series of short, electrical stimulations to a neuron was shown to strengthen, or potentiate, its synapse with an adjacent cell for minutes or hours. Such multiple stimuli caused neurons to communicate more readily; responses were stronger and more prolonged. This phenomenon was termed **long-term potentiation (LTP)**. LTP is the long-lasting strengthening of the connection between neurons. Later work showed that LTP occurs naturally in the hippocampus and can last from hours to days, and even years.

Austrian neuroscientist Eric Kandel also was interested in learning and memory, and some of his early studies involved the hippocampus. In the 1960s, however, he took a reductionist approach by studying learning and memory in a simpler organism called the California sea slug (*Aplysia californica*). He chose this organism for several key reasons. First, it has only about 20,000 nerve cells, making it easier to identify pathways that are involved in specific types of behavior. Second, some of the neurons in this organism are extremely large, allowing the study of action potentials via microelectrodes (as described in Chapter 43). In addition, the large size made it possible to inject substances into the neurons to study their effects. Finally, another major advantage is that Kandel and colleagues could isolate proteins and mRNA from these large neurons, and identify the biochemical and genetic changes that occur when the animal learns about its outside world and retains memory.

Much of Kandel's work focused on one type of learning involving the gill-withdrawal reflex, which is thought to involve less than 100 neurons in the CNS. The gill and siphon are organs involved in respiration. When the siphon is gently touched with a fine probe, the sea slug closes the siphon and retracts its gills into the mantle for protection (**Figure 44.12**). Though called a reflex, this behavior is subject to learning. It can be modified to occur at a faster or slower rate. For example, if the touching of the siphon is accompanied with a shock in the tail, the sea slug

(a) Gill-withdrawal reflex

(b) *Aplysia californica*

Figure 44.12 The gill-withdrawal reflex in the sea slug. (a) When touched with a probe, the siphon closes and the gill retracts. In this drawing, the mantle is moved apart for a better view of the gill. (b) Photo of *A. californica*.

can learn to withdraw its gill more quickly. The sea slug can learn that the two events occur at the same time.

Over the course of many years, the work of Kandel and colleagues revealed many clues regarding the cellular basis of learning and memory. Like vertebrates, these processes in the sea slug occur in two phases (**Figure 44.13**). Short-term memory lasts for minutes or hours. This type of memory is typically caused by a single stimulus. Kandel found that short-term memory does not require the synthesis of new proteins. Rather, a single stimulus activates intracellular second messenger pathways (described in Chapter 9) that make it easier for neurons involved in a particular behavior to communicate with each other. For example, as shown in Figure 44.13a, a single stimulus may lead to the activation of protein kinases in the presynaptic (sensory) cell that phosphorylate proteins such as ion channels and proteins involved with the release of neurotransmitter. These changes enhance the transmission of a signal between the presynaptic and postsynaptic cells.

Kandel and colleagues also discovered that repeated stimuli result in long-term memory, which lasts days or weeks. Such repeated stimuli require the synthesis of new proteins (Figure 44.13b). Long-term memory involves the activation of genes in the presynaptic cell, which leads to the synthesis of mRNA and the translation of the encoded proteins. Once made, such proteins cause the formation of additional synaptic connections.

(a) Short-term memory

Presynaptic cell (unstimulated)   Ion channel   Presynaptic vesicles   Postsynaptic cell

PKA (inactive)

A single stimulus activates PKA (protein kinase A) in the presynaptic cell. PKA phosphorylates proteins such as ion channels and proteins in presynaptic vesicles. This enhances the transmission of a signal between the presynaptic and postsynaptic cells.

PKA (active)   Phosphate

For the short term, the communication between these 2 cells is stronger.

(b) Long-term memory

Presynaptic cell (unstimulated)

Repeated stimuli activates genes in the nucleus. The mRNAs are translated into proteins that cause the cell to form more synaptic connections. This enhances transmission between the presynaptic and postsynaptic cell.

Postsynaptic cell

Transcription
Translation
Proteins involved with forming new synapses

For the long term, the additional synaptic connections cause the communication between the 2 cells to be stronger. Note the additional synapses on the postsynaptic cell.

Figure 44.13 Cellular changes associated with short-term and long-term memory in the sea slug.

These connections also allow the presynaptic and postsynaptic cells to communicate with each other more readily. Such a change in synapses, which occurs as a result of learning, is termed **synaptic plasticity**.

The work of Kandel provides a foundation for our ability to understand how learning and memory may occur at the cellular level. Short-term memory may involve changes in pre-existing cellular proteins that make it easier for neurons to communicate, while long-term memory results in protein synthesis that causes physical changes in the synapse itself, also affecting communication. In the case of the sea slug, learning does not change the pathways of neurons but instead changes the strength of the signals along those pathways. When a light touch on the siphon was coupled with an unpleasant shock, the animal became sensitized and could react more strongly to the light touch because the set of connections had strengthened. Kandel also studied how learning can occur in the opposite way. A sea slug can learn to ignore a harmless touch. In that case, the connections between neurons became weaker. Later studies by Kandel and others showed that such changes also occur in vertebrates such as the mouse. For his work on learning and memory, Kandel was awarded the Nobel Prize in Physiology or Medicine in 2000.

Neurogenesis May Also Contribute to Learning and Memory

Until fairly recently, neuroscientists had thought the adult brain of vertebrates is incapable of **neurogenesis**—the production of new neurons by cell division. In 1983, an early study showing neurogenesis in an adult vertebrate was carried out by Fernando Nottebohm and colleagues. Their research showed an increase in the number of neurons in certain brain areas of the canary in conjunction with mating season. In the late 1990s, evidence also showed that the primate and human CNS, like other parts of the body, contain stem cells (see Chapter 19). American researchers Elizabeth Gould and Bruce McEwen have demonstrated the appearance of new neurons in the hippocampus and olfactory bulbs of adult marmosets and rhesus monkeys. In 1998, American researcher Fred Gage and Swedish physician Peter Eriksson made a key discovery: They found new hippocampal cells in the brains of recently deceased adult cancer patients. Those patients had previously been treated with bromodeoxyuridine (BrdU) to combat their cancer. BrdU is also taken up by noncancerous, newly formed cells, and its presence can be detected with special stains on sections of brain and other tissue. Gage conducted such staining procedures on the brains of the deceased patients and observed the presence of BrdU in the hippocampus, demonstrating recent neurogenesis.

A key question is whether the neurogenesis observed in adult brains is involved in learning and memory. This question is hotly debated and not resolved. However, some evidence suggests that it could play a role. For example, McEwen and co-workers have shown that the hippocampus of adult rhesus monkeys grows new neurons when the animals are placed in enriching environments, and the formation of these neurons slows when animals are chronically stressed. Also, studies in the rat by Gould and colleagues suggested that new neurons are retained in the hippocampus in response to training on particular tasks that require hippocampus function.

(a)

(b)

Figure 44.14 Exploring the functional activity of brain regions using fMRI scans. Part **(a)** shows the brain activity of a person who is thinking about hammering a nail. Part **(b)** shows the brain activity when the same person is actually performing this task.

Brain Images Are Used to Assess Brain Structure and Function

Because of the enormous number of neurons and connections between neurons in the vertebrate brain, a key challenge in neuroscience is to understand how such complexity results in sophisticated forms of learning, memory, and responses to environmental conditions. Several imaging techniques allow doctors and researchers to examine the structure and activity level of the brain without anesthesia or surgery. The earliest technique to be developed was computerized tomography (from the Greek word *tomas*, meaning "a cutting"). This **CT scan** involves the use of an X-ray beam and a series of detectors that rotate around the head, producing slices of images that are reconstructed into three-dimensional images based on differences in the density of brain tissue. CT scans can easily visualize the ventricles and differences between white and gray matter, but they do not have the capacity to examine the brain in great detail.

A more sensitive method, **magnetic resonance imaging (MRI)**, was developed in the 1980s. The patient's head is placed in a device that contains a magnet powerful enough to generate a magnetic field many thousands of times greater than that of the Earth. This stabilizes the spinning, or resonance, of atomic nuclei (usually hydrogen atoms in water molecules) so that most of the nuclei align with the magnetic field. When body tissue is stimulated with a beam of radio waves, its atoms absorb the energy of the waves and the resonance of their nuclei changes, thereby altering their alignment with the magnetic field. When the radio wave pulse stops, the atoms release their energy, which is recorded by a detector. This information is analyzed by a computer and an image is produced. MRI images allow detection of structures as small as one-tenth of a millimeter. For example, they can provide information about abnormal tissue, such as brain tumors, which respond to magnetic and radio frequency pulses differently than normal tissue. MRIs are widely used in medicine to check for injured tissue, cancers, and other abnormalities throughout the body. In 2003, Paul Lauterbur and Peter Mansfield received the Nobel Prize in Medicine for their work in developing this method.

With certain modifications, MRI can be used to assess the functional activity of areas within the brain. This technique, which is widely used by neuroscientists, is called functional MRI (fMRI). It takes advantage of the fact that blood flow increases to areas where neurons are active. To supply more oxygen to such regions, the amount of oxygenated hemoglobin is increased relative to deoxygenated hemoglobin. This increase is detected via fMRI. In this way, fMRI determines which neurons in particular areas of the brain are active when an individual performs certain intellectual or motor tasks (**Figure 44.14**). The principle is similar to the standard MRI, except that the higher oxygen content of active tissue alters the resonance of local hydrogen atoms.

FEATURE INVESTIGATION

Gaser and Schlaug Showed That the Sizes of Certain Brain Structures Differ Between Musicians and Nonmusicians

MRI and fMRI have been extremely useful in revealing which brain areas are involved in a particular function. They have also shown that the human brain is surprisingly adaptable. A number of studies have been carried out on musicians, because they practice extensively throughout their lives, enabling researchers to study the effects of repeated use on brain function.

American neuroscientist Christian Gaser and German neuroscientist Gottfried Schlaug used MRI to examine the sizes of brain structures in three groups of people—professional musicians, amateur musicians, and nonmusicians (**Figure 44.15**).

Individuals were assigned to each group based on their reported history of musical training: professional musicians with over 2 hours of musical practice time each day, amateur musicians who played a musical instrument regularly but not professionally (practicing about 1 hour/day), and those who never played a musical instrument regularly. The researchers hypothesized that repeated exposure to musical training would increase the size of brain areas associated with visual, motor, and auditory (hearing) skills, because each of these activities is used to read, make, and interpret music. The results showed that brain areas involved in hearing, moving the fingers, and coordinating movements with vision and hearing were larger in professional musicians than in amateur musicians, and larger in amateurs than in nonmusicians. The region of the brain that controls finger movements appeared

Figure 44.15   Gaser and Schlaug's study of the size of visual, motor, and auditory nuclei in the brains of musicians and nonmusicians.

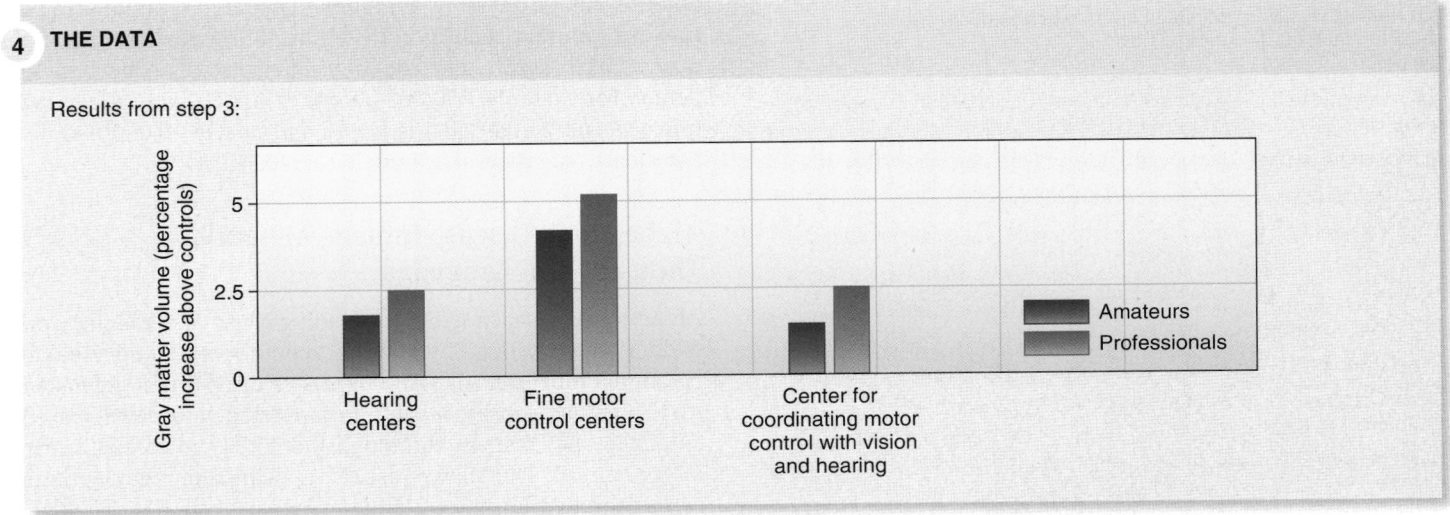

HYPOTHESIS Musical training is associated with structural differences in the brain.

STARTING MATERIALS Volunteer subjects with different degrees of musical training.

Experimental level   **Conceptual level**

1 Establish 3 groups of subjects with different musical backgrounds.

Interview subjects for musical history and assign to 1 of 3 groups.

Control (nonmusicians)
No musical training.

Amateur musicians
Play an instrument about 1 hr/day and are not employed as musicians.

Professional musicians
Employed as musicians and practice their instrument > 2 hr/day.

2 Control for possible factors that might affect results.

Make group assignments in a way that matches subjects' age and other characteristics.

	Controls	Amateurs	Professionals
Age (average)	23	26	27
Type of instrument	None	Keyboard	Keyboard
Mental IQ	119	123	118
Sex	Male	Male	Male

MRI scan

Resulting images

3 Perform MRI and calculate volume of gray matter in different brain regions.

Region of brain involved with a particular skill

Posterior ——————————————————→ Anterior

4 **THE DATA**

Results from step 3:

Gray matter volume (percentage increase above controls)

5

2.5

0

Hearing centers   Fine motor control centers   Center for coordinating motor control with vision and hearing

Amateurs
Professionals

particularly well developed in the professional musicians, an interesting finding because all of the musicians in this study played keyboard instruments such as the piano.

In another study, American researchers Vincent Schmithorst and Scott Holland used fMRI to determine if musicians' brains were activated differently than nonmusicians' brains when they heard music. They found that one area of the cortex was selectively activated by melodies only in musicians. This study differed from that of Gaser and Schlaug, because Schmithorst and Holland examined the activity of brain areas as well as their sizes. Their results showed that listening to music activates certain neurons and pathways in the brains of musicians but not in nonmusicians.

The human studies of Gaser, Schlaug, Schmithorst, and Holland have not determined the underlying reason(s) for increased brain size. One possibility is that people with increased brain size in these regions are more likely to become musicians. Alternatively, musical training may actually cause certain regions of the brain to grow larger and alter their neuronal pathways. In other research studies involving experimental animals, groups of animals have been randomly separated into those learning a task versus controls, which do not learn the task. Such experiments have shown increases in brain size that are associated with learning and memory. The increased size may result from the formation of new synapses, the growth of blood vessels to the region, and/or the production of more glial cells.

44.4 Impact on Public Health

Most neurological disorders can be classified into several broad groups (Table 44.1). In addition to the conditions already discussed in this chapter and Chapter 43, we will consider two of these disorders that result from very different causes and impact millions of individuals worldwide.

Meningitis Is an Infectious Disease That Attacks the Meninges

Meningitis is caused by a viral or bacterial infection of the meninges, which were described earlier in Figure 44.7. The result is an inflammation in which fluid accumulates in the affected area. The excess fluid in the subarachnoid space raises pressure within the skull, which can lead to severe headaches,

Table 44.1	Categories of Diseases Affecting the Human Central Nervous System
Category	**Examples**
Infectious diseases (inflammation of the brain)	Meningitis (discussed in this section), encephalitis
Neurodegenerative disorders	Alzheimer's disease (discussed in this section)
Movement disorders	Parkinson's disease (described in Section 44.2)
Seizure disorders	Epilepsy (described in Section 44.2)
Sleep disorders	Sleep apnea (brain fails to regulate breathing during sleep)
Tumors	Glioma (a tumor arising from glial cells)
Headache disorders	Migraines
Mood disorders	Unipolar and bipolar depression (see Chapter 43)
Demyelinating disorders	Multiple sclerosis (see Chapter 43)
Injury-related disorders	Brain and spinal cord injuries due to accidents

fever, seizures, unconsciousness, and even death within hours if untreated. Many patients with meningitis develop a stiff neck because the inflammation proceeds down the spinal cord.

Several different viruses or bacterial species can cause meningitis. It usually results from another infection typically in neighboring regions, such as the sinuses behind the eyes, nose, or ears, that has gone untreated. Meningitis can be confirmed by using a long needle to sample the cerebrospinal fluid (CSF) in the spinal cord and analyzing the pressure and contents of the CSF. Large numbers of white blood cells, which are the body's chief infection-fighting cells, indicate infection in the CSF and meninges. If the infection is the result of bacterial invasion, meningitis can be treated with bacterial-killing agents like antibiotics. Antibiotics do not kill viruses, but fortunately the viral form of meningitis is usually less serious than the bacterial form and runs its course after several days or weeks.

Meningitis strikes roughly 25,000 people a year in the U.S. and can affect people of any age. Its incidence in children has greatly declined since the widespread use of a vaccine against the bacterium *Haemophilus influenzae* began in the U.S. in the early 1990s. Despite the vaccine, meningitis is still a dangerous and prevalent disease, and it tends to occur in individuals living in close quarters such as college dormitories, where infections may spread rapidly. Occasionally, meningitis can become epidemic. For example, 250,000 people in sub-Saharan Africa were infected and 25,000 died in 1996, and nearly 75,000 people in Southeast Asia died of meningitis in 2004.

Alzheimer's Disease Impairs Memory, Thought, and Language

Alzheimer's disease (AD), the leading cause of dementia worldwide, is characterized by a loss of memory and intellectual and emotional function. AD is a progressive disease that begins with small memory lapses, leading in later stages to problems with language and abstract thinking, and finally to loss of normal motor control. The disease usually appears after age 65, though some inherited forms can strike people in their 30s and 40s.

Although psychological testing can help to diagnose AD, a definitive diagnosis can be made only after death when the brain is examined microscopically. Brains of Alzheimer's patients show two noticeable changes: senile plaques and neurofibrillary tangles (**Figure 44.16**). Senile plaques are extracellular deposits of an abnormal protein, β-amyloid, that forms large, sticky aggregates. These plaques were first noted in 1906 by German physician Alois Alzheimer, after whom the disease was named. Neurofibrillary tangles are intracellular, twisted accumulations of cytoskeletal fibers. It is not clear how these changes influence intellectual function and memory. AD is also associated with the degeneration and death of neurons, particularly in the hippocampus and parietal lobes, which is why it is considered a neurodegenerative disease.

Researchers have identified variation in a few genes whose products are associated with the likelihood of developing AD later in life, but the underlying changes that result in the expression of these and other possible AD-related genes are still the subject of considerable research. Although genetics undoubtedly plays a role in AD, it is not the only possible cause. For example, when one identical twin develops AD, the other appears to be at increased risk but does not always develop the disease, even if he or she survives to very old age. Moreover, evidence suggests that severe head injuries, metabolic diseases like diabetes, and heart and blood vessel disease may predispose a person to AD in later life.

Currently, AD cannot be prevented or cured. However, three major clinical approaches are currently being tested to prevent or slow down its progression. These approaches are designed to (1) induce a person's immune system to destroy β-amyloid as soon as it is formed, (2) prevent the formation of β-amyloid with drugs that block its synthesis, or (3) prevent the accumulation of β-amyloid into large aggregates using antiaggrega-

Figure 44.16   **The brain of a person who died from Alzheimer's disease.**   Note the senile plaques and neurofibrillary tangles.

tion drugs. Each of these approaches holds great promise but is still unproven.

Until a cure for AD is found, its impact on public health remains enormous. Currently, about 4–5 million Americans have AD, and this number is expected to grow to nearly 16 million by 2050. The prevalence of the disease is about 3% for people between the ages of 65 and 74, and 25–50% for people over 85. Estimated costs associated with providing health care and housing for AD patients (30% of whom live in nursing homes), as well as lost productivity in the workplace, total a staggering $100 billion in the U.S. per year. This number will rise substantially as the population spike known as the baby boomer generation approaches age 65.

CHAPTER SUMMARY

44.1 The Evolution and Development of Nervous Systems

- All multicellular animals except sponges have a nervous system. Simpler nervous systems include the nerve net of cnidarians and ganglia, which integrate inputs from sense organs. As animal bodies become more complex, the integrative center in the head becomes a brain that is larger and capable of more functions. (Figure 44.1)

- In all vertebrates, the three major divisions of the brain are the hindbrain, midbrain, and forebrain. (Figure 44.2)

- Additional folding of the brain and increased mass allows for expansion of regions associated with conscious thought, reasoning, and learning. (Figures 44.3, 44.4)

44.2 Structure and Function of the Human Nervous System

- In humans and other vertebrates, the brain and spinal cord are the central nervous system (CNS), while the neurons and all axons outside the CNS, including the cranial and spinal nerves, constitute the peripheral nervous system (PNS). The CNS cannot learn anything about the internal or external environment without the PNS, and the PNS relies on commands from the CNS. (Figure 44.5)

- The gray matter of the CNS is composed of dendrites, cell bodies, and unmyelinated axons, while the white matter consists of tracts of myelinated axons. (Figure 44.6)

- The meninges are protective coverings of the CNS. Cerebrospinal fluid fills the subarachnoid space and ventricles. (Figure 44.7)

- The PNS can be subdivided into the somatic and autonomic nervous systems. The somatic nervous system senses external environmental conditions and controls skeletal muscles and skin. The autonomic nervous system senses internal body conditions and controls homeostasis.

- The efferent part of the autonomic nervous system is divided into two components: sympathetic (fight or flight) and parasympathetic (rest or digest). (Figure 44.8)

- The evolutionarily oldest structures of the brain, some of which are located in the hindbrain, control the basic processes that sustain life. These include the medulla oblongata, cerebellum, and pons.

- In addition to forming part of the reticular formation, the midbrain processes several types of sensory inputs, including vision, smell, and hearing.

- The forebrain is made of the thalamus and hypothalamus (diencephalon) and the cerebrum (telencephalon), which consists of the basal nuclei, limbic system, and cerebral cortex. The human cerebral cortex is divided into four lobes, each of which has a number of functions. (Figures 44.9, 44.10)

- The cerebrum is divided into two hemispheres. Each hemisphere is specialized to perform certain aspects of behavior and can operate independently. (Figure 44.11)

44.3 Cellular Basis of Learning and Memory

- Learning is the process by which new information is acquired, while memory involves the retention of that information over time.
- Repeated stimuli result in long-term potentiation in which the connections between adjacent neurons are stronger.
- Studies of the sea slug indicate that short-term memory is caused by a single stimulus that activates second messenger pathways, while long-term memory is caused by repeated stimuli that activate genes, which results in more synaptic connections, a phenomenon called synaptic plasticity. (Figures 44.12, 44.13)
- Imaging techniques such as CT scans, MRI, and fMRI allow us to examine the structure and activity of the brain. (Figures 44.14, 44.15)

44.4 Impact on Public Health

- Disorders of the human central nervous system can be placed into several broad categories. (Table 44.1)
- Meningitis is a potentially life-threatening infectious disease in which the meninges become inflamed.
- Alzheimer's disease, the leading cause of dementia worldwide, is a progressive disorder characterized by the formation of senile plaques and neurofibrillary tangles in brain tissue. (Figure 44.16)

TEST YOURSELF

1. A nerve net consists of
   a. bilateral neurons that extend from the head of the animal to the tail.
   b. a group of neurons that are interconnected and are activated all at once.
   c. a single nerve cord with ganglia in each body segment.
   d. a central nervous system with peripheral nerves associated with different body structures.
   e. none of the above.

2. The division of the vertebrate brain that includes the cerebellum is
   a. the hindbrain.   d. the forebrain.
   b. the telencephalon. e. the diencephalon.
   c. the midbrain.

3. In general, the brains of more complex vertebrates
   a. are larger.
   b. have fewer neurons.
   c. have more folds in the cerebral cortex.
   d. use less oxygen.
   e. both a and c.

4. The white matter of the CNS is composed of
   a. dendrites.
   b. unmyelinated axons.
   c. myelinated axons.
   d. cell bodies.
   e. a and b only.

5. The division of the nervous system that controls voluntary muscle movement is
   a. the autonomic nervous system.
   b. the sensory division.
   c. the somatic nervous system.
   d. the parasympathetic division.
   e. the sympathetic division.

6. Which of the following is *not* a characteristic response to the activation of the sympathetic division of the autonomic nervous system?
   a. increased breathing rate
   b. decreased heart rate
   c. increased blood flow to the skeletal muscles
   d. increased blood glucose levels
   e. all of the above are characteristic responses to activation of the sympathetic division of the autonomic nervous system

7. The ___ acts as a relay for the cerebrum.
   a. medulla   d. midbrain
   b. pons   e. thalamus
   c. hypothalamus

8. The _____ is a portion of the limbic system that is important for memory formation.
   a. amygdala   d. thalamus
   b. hippocampus   e. mesencephalon
   c. pons

9. In humans, the _____ hemisphere of the cerebrum is dominant in nonverbal processing.
   a. right
   b. left
   c. both a and b

10. _____ is a progressive disease that causes a loss of memory and intellectual and emotional function.
    a. Meningitis   d. Alzheimer's disease
    b. Parkinson disease e. Stroke
    c. Amnesia

CONCEPTUAL QUESTIONS

1. List the three major divisions of the brain of vertebrates and briefly describe the function of each.

2. Explain what makes white matter white.

3. What are the two subdivisions of the efferent pathways of the autonomic nervous system, and what are their functions?

EXPERIMENTAL QUESTIONS

1. What was the hypothesis proposed by Gaser and Schlaug?

2. How did Gaser and Schlaug test this hypothesis? What were the results of their experiment?

3. How did the research of Schmithorst and Holland impact the findings of Gaser and Schlaug? How did their experiments differ from that of Gaser and Schlaug?

COLLABORATIVE QUESTIONS

1. Discuss two different types of nervous systems found in animals.

2. Discuss the parts of the human brain.

---

www.brookerbiology.com
This website includes answers to the Biological Inquiry questions found in the figure legends and all end-of-chapter questions.

45

NEUROSCIENCE III: SENSORY SYSTEMS

Vision in the dark. As with many nocturnal animals, a cat's eye contains a reflective layer of tissue called a tapetum lucidum. This increases their ability to see in the dark.

Senses allow living organisms to perceive their environment. Animals with nervous systems often have acute senses, such as the ability to perceive light, touch, and sound. In neuroscience, a broad definition of a **sense** would be a system that consists of sensory cells that respond to a specific type of chemical or physical stimulus and send signals to the central nervous system, where the signals are received and interpreted. The senses allow animals to detect subtle and complex aspects of their environment. They are the windows through which animals experience the world around them. The nervous systems of many animals are also designed to sense signals arising from within an animal's body, such as hunger or pain. However, neuroscientists do not always agree on the actual number of different senses. Common examples include sight, hearing, touch, taste, smell, balance, and the ability to sense temperature and pain.

Of course, some organisms have a greater ability to detect stimuli than others. Hawks can see mice from 100 m aboveground, while polar bears can smell a seal several kilometers away. Humans do not see ultraviolet light but some bees do, and apes cannot hear extremely high-pitched sounds although rodents can.

We can never know exactly how an animal perceives its external and internal environments. Biologists can, however, determine experimentally the capabilities of an animal's sensory systems. For instance, we can examine the structure of a cat's eye, do behavioral studies to see if they can discriminate between different colors, and measure the electrical responses of neurons in the feline visual system to different visual stimuli. From these studies, we know that cats sense color very poorly, if at all, but can see exceptionally well even in the dark.

In this chapter, we will examine how nervous systems collect incoming sensory information and how membrane potentials of specialized neurons change in response to sensory inputs. We will learn that other structures of the nervous system may modify or enhance this neural activity before sending it to the brain, where it is interpreted.

45.1 An Introduction to Sensory Receptors

Sensory systems convert chemical or physical stimuli from an animal's body and the external environment into a signal that causes a change in the membrane potential of sensory neurons. **Sensory transduction** is the process by which incoming stimuli are converted into neural signals. Sensory transduction involves cellular changes, such as opening of ion channels, which cause either graded potentials or action potentials in neurons.

Perception is an awareness of the sensations that are experienced. For instance, touching a hot stove generates a thermal sensation, which initiates a neuronal response, giving us the perception that this stimulus is hot. Perception requires integration of incoming stimuli by the CNS. Not all sensations are consciously perceived by an organism. Most of the time, for example, we are not aware of the touch of our clothing. The brain also processes sensory information in areas that do not generate conscious thought. For instance, certain neurons constantly monitor blood pressure, blood levels of oxygen and glucose, and heart rate, but we are not aware that this is occurring.

We will begin our study of sensory systems by examining the specialized cells—called sensory receptors—whose function it is to receive sensory inputs. A **sensory receptor** recognizes an internal or environmental stimulus and initiates signal transduction by creating graded potentials (described in Chapter 43) in the same cell or an adjacent cell. Sensory receptor is a term that neuroscientists use to describe certain types of cells, which are either neurons or specialized epithelial cells that respond to internal or environmental stimuli (**Figure 45.1**). When a response is strong enough, sensory receptors can initiate electrical responses to stimuli, such as chemicals, light, touch, heat, and sound, which leads to action potentials that are sent to the CNS.

Figure 45.1 **Sensory receptors.** **(a)** Many sensory receptors are neurons that directly sense stimuli. **(b)** Others are specialized epithelial cells that sense stimuli and secrete neurotransmitter that stimulates nearby sensory neurons. When stimulated, the neurons send action potentials to the CNS, where the signals are interpreted.

Biological inquiry: What is the difference between the sensory receptor described in this chapter, and the membrane receptors described in Chapter 9?

An Intense Stimulus Generates More Frequent Action Potentials

Sensory transduction begins when the specialized endings of a sensory receptor, such as dendrites, respond to a stimulus. In some cases, sensory receptors are depolarized when a stimulus opens ion channels that allow sodium ions into the cell. The amount of depolarization is directly related to the intensity of the stimulus, because a stronger stimulus will open more ion channels. When a stimulus is strong enough, it will depolarize the membrane to the threshold potential at the axon hillock and produce an action potential in a sensory neuron (see Chapter 43).

The first response of a sensory receptor is usually a graded change in the membrane potential that is proportional to the intensity of the stimulus (**Figure 45.2**). This example involves the sense of touch. The membrane potential, known as the **receptor potential** in these cells, becomes more and more positive as the strength of the stimulus increases. A higher membrane potential results in more action potentials being generated by the sensory receptor itself, which is a neuron, as described in Figure 45.1a. These action potentials are transmitted to the CNS, where the stimulus is perceived.

Recall from Chapter 43 that action potentials proceed in an all-or-none fashion, regardless of the nature of the stimulus that elicits them. So, how can action potentials provide information about the intensity of a stimulus? The answer is that the strength of the stimulus is indicated by the frequency of action potentials generated. A particularly strong stimulus will generate many action potentials in a short period of time. As a result, the frequency of action potentials is higher when the stimulus is strong than when it is weak. The action potentials are transmitted into the CNS and carried to the brain for interpretation. The brain interprets a higher frequency of action potentials as a more intense stimulus.

The CNS Processes Each Sense Within Its Own Pathway

Different stimuli produce different sensations because they activate specific neural pathways that are dedicated to processing only that type of stimulus. We know that we are seeing light because the signals generated by visual sensory receptors are transmitted along a neural pathway that sends action potentials into areas of the brain that are devoted to processing vision. For this reason, the brain interprets such signals as visual stimuli.

We can classify sensory receptors into general classes, based on the type of stimulus to which they respond. Each type uses a different mechanism to respond to signals and transmits the response to different regions of the CNS. **Mechanoreceptors** transduce mechanical energy such as pressure, touch, stretch, movement, and sound. **Thermoreceptors** respond to cold and heat. **Nociceptors**, or pain receptors, respond to extreme heat, cold, and pressure, as well as to certain molecules such as acids. **Electromagnetic receptors** detect radiation within a wide range of the electromagnetic spectrum, including visible, ultraviolet, and infrared light, as well as electrical and magnetic fields in

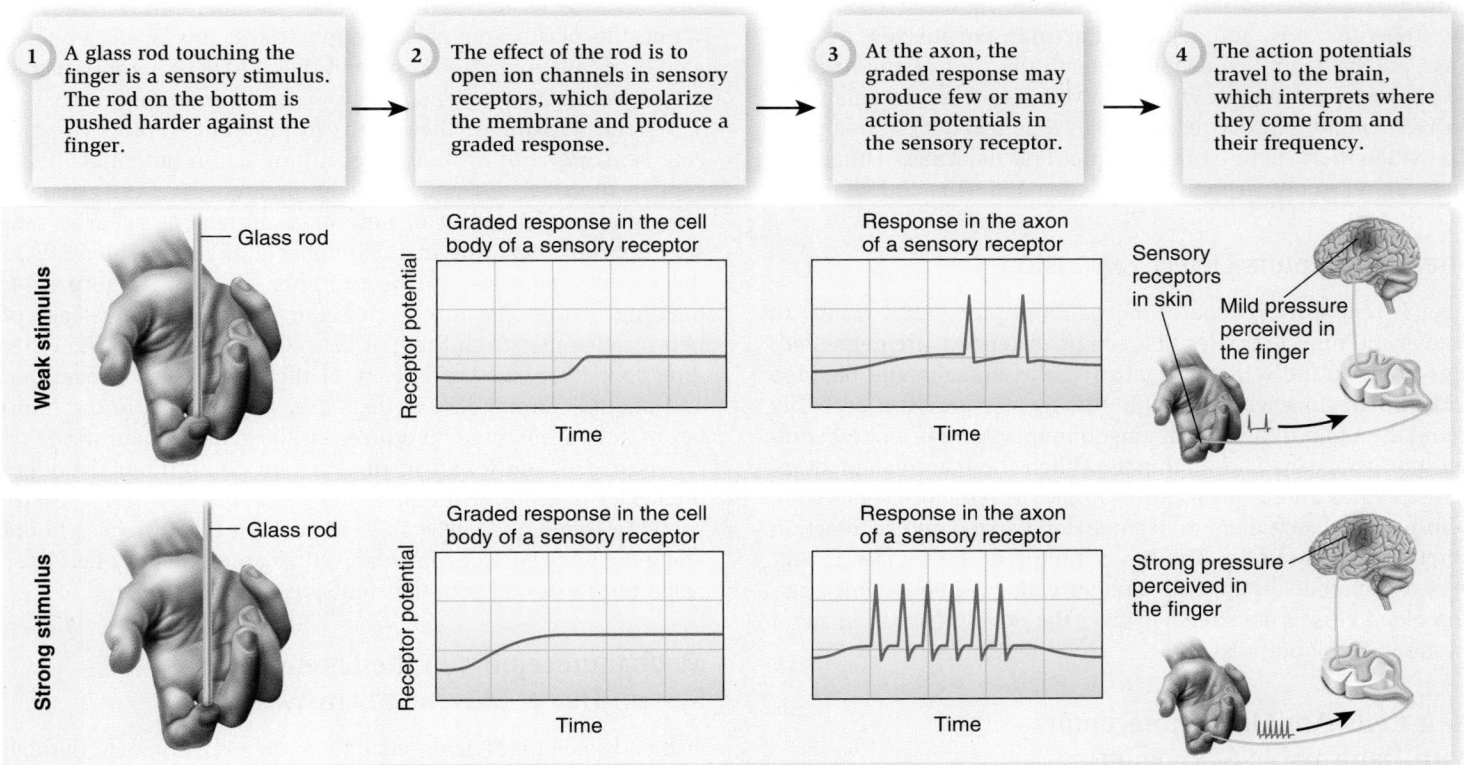

1 A glass rod touching the finger is a sensory stimulus. The rod on the bottom is pushed harder against the finger.

2 The effect of the rod is to open ion channels in sensory receptors, which depolarize the membrane and produce a graded response.

3 At the axon, the graded response may produce few or many action potentials in the sensory receptor.

4 The action potentials travel to the brain, which interprets where they come from and their frequency.

Weak stimulus — Glass rod

Graded response in the cell body of a sensory receptor — Receptor potential / Time

Response in the axon of a sensory receptor — Time

Sensory receptors in skin / Mild pressure perceived in the finger

Strong stimulus — Glass rod

Graded response in the cell body of a sensory receptor — Receptor potential / Time

Response in the axon of a sensory receptor — Time

Strong pressure perceived in the finger

Figure 45.2 **Transduction of a sensory stimulus of two different intensities.** In this example, the sensory receptor is a neuron.

some animals. **Photoreceptors** are electromagnetic receptors that respond to visible light energy. **Chemoreceptors** respond to specific chemical compounds. The remaining sections of this chapter will examine the structures and functions of these types of sensory receptors, and the organs in which they are found.

45.2 Mechanoreception

As mentioned, mechanoreceptors are cells that detect physical stimuli such as pressure, touch, stretch, movement, and sound. Physically touching or deforming a mechanoreceptor cell opens ion channels in the plasma membrane. As discussed in this section, some mechanoreceptors are neurons that send action potentials to the CNS in response to physical stimuli. Other mechanoreceptors are specialized epithelial cells that contain cilia that bend in response to mechanical forces.

Skin Receptors Respond to Touch and Pressure

Several specialized receptors in the skin respond to touch, deep pressure, or the bending of hairs on the skin. In mammals, these receptors are located at different depths below the surface of the skin, which makes them suitable for responding to different types of stimuli (**Figure 45.3**). For example, **Meissner's corpuscles**, which are structures that sense touch and light pressure, lie just beneath the skin surface. They are found throughout the skin

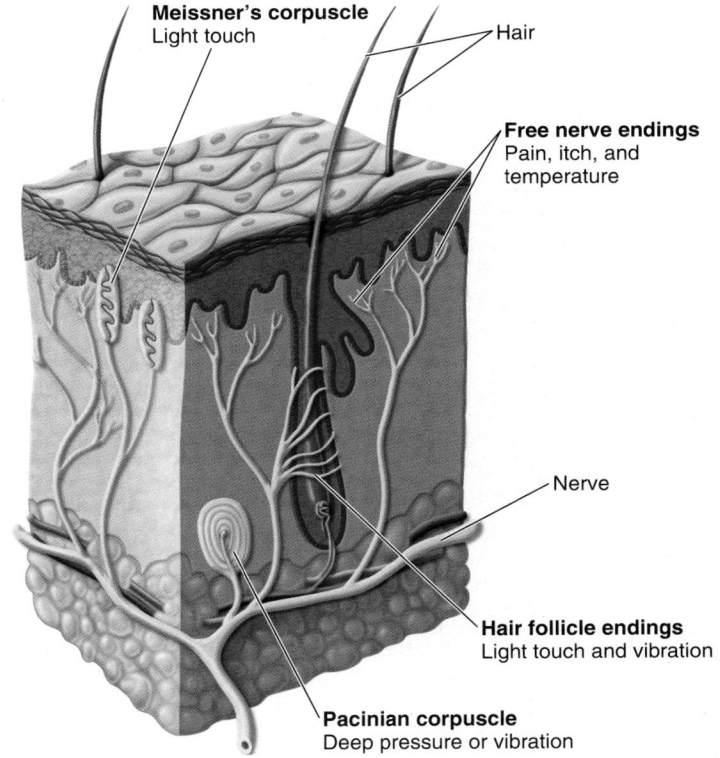

Meissner's corpuscle Light touch

Hair

Free nerve endings Pain, itch, and temperature

Nerve

Hair follicle endings Light touch and vibration

Pacinian corpuscle Deep pressure or vibration

Figure 45.3 **Examples of sensory receptors in the skin of mammals.**

but are concentrated in areas sensitive to light touch, such as the fingertips, lips, and genitals. **Pacinian corpuscles**, by contrast, are located much deeper beneath the surface and respond best to deep pressure or vibration. Meissner's and Pacinian corpuscles contain sensory receptor cells that generate action potentials when the structure of the corpuscle is deformed. Other skin mechanoreceptors respond to movements of hairs and whiskers.

Stretch Receptors Detect Expansion

Stretch receptors, a type of mechanoreceptor found widely in organs and muscle tissues that can be distended, are nerve endings found in the walls of organs like the stomach and bladder, and also in muscles. Although stretch receptors are probably found throughout the animal kingdom, they have been best studied in crustaceans and mammals. When the mammalian stomach stretches after a meal, for example, its stretch receptors are deformed, causing them to become depolarized and send action potentials to the brain. The brain interprets the signals as fullness, which reduces appetite. Similarly, the degree to which certain blood vessels are stretched gives the brain information about an animal's blood pressure.

Hair Cells Are Mechanoreceptors with Deformable Stereocilia

Thus far, we have considered skin and stretch receptors, which are neurons that detect physical stimuli. Other mechanoreceptors are specialized epithelial cells, called **hair cells**, which have deformable projections, called stereocilia, that resemble hairs under the microscope. These stereocilia are bent by water currents or other physical stimuli (**Figure 45.4**). Many types of hair cells contain ion channels that open or close when the

stereocilia bend and thereby change the membrane potential. When the plasma membrane depolarizes, this opens voltage-gated Ca^{2+} channels and results in the release of neurotransmitter molecules from the hair cells (refer back to Figure 43.14). The neurotransmitter then binds to protein receptors in adjacent sensory neurons and can result in action potentials being sent to the CNS. Even when unstimulated, hair cells usually release a small amount of neurotransmitter onto nearby sensory neurons, resulting in a resting level of action potentials in the sensory neurons. In the example shown in Figure 45.4, bending of the cilia in one direction increases the release of neurotransmitter from the hair cell, while bending in the other direction decreases the release of the same neurotransmitter. The result is an increase or decrease, respectively, in the number of action potentials produced in the sensory neurons.

Hair cells are found in the ear and equilibrium (balance) organs of many invertebrates and vertebrates, where they respond to sound or changes in head position. They are also found along the body surface of fish and some amphibians, where they respond to external water currents, as described next.

Mechanoreceptors in the Lateral Line System Detect Movements in Water

Fish and some toads detect changes in their environment through a **lateral line system** (**Figure 45.5**). This sensory system has hair cells that detect changes in water currents brought about by waves, nearby moving objects, and low-frequency sounds traveling through the water. The lateral line organ runs along both sides and the head of the animal. Small pores let water enter into a lateral line canal. The stereocilia of hair cells protrude into a gelatinous structure called a **cupula** within the lateral line organ. When the cupula is moved by the water, the

At rest, a small amount of neurotransmitter is released at all times, resulting in a steady number of action potentials being generated in the sensory neuron.

Fluid moving in one direction causes the release of more neurotransmitter, which results in more action potentials in the sensory neuron.

Fluid moving in the opposite direction inhibits the release of neurotransmitter, which results in fewer action potentials in the sensory neuron.

Figure 45.4   The response of hair cells to mechanical stimulation.

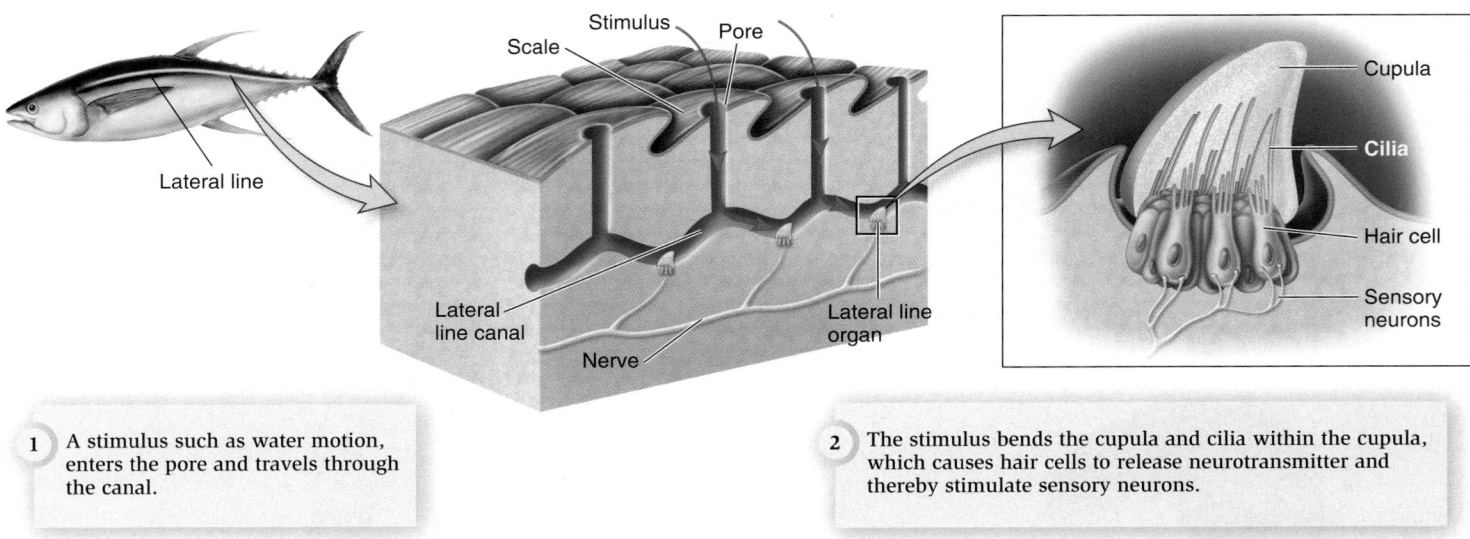

1. A stimulus such as water motion, enters the pore and travels through the canal.

2. The stimulus bends the cupula and cilia within the cupula, which causes hair cells to release neurotransmitter and thereby stimulate sensory neurons.

Figure 45.5   Mechanoreceptors in the lateral line system that detect changes in water movement.

cilia bend, causing the release of neurotransmitter from the hair cell, which stimulates a response in a neuron at the base of the hair cell. The response provides information to the brain about changes in water movement, such as the approach of a predator.

Hearing Involves the Reception of Sound Waves

The sense of hearing, **audition**, is the ability to detect and interpret sound waves. It is a critical sense for the survival of many types of animals. For example, a mother seal locates her pup by hearing its calls, and a male bird sings an elaborate song to attract a mate. Hearing is also important for detecting the approach of danger—a predator, a thunderstorm, an automobile—and locating its source.

Sound travels through air or water in waves. The distance from the peak of one wave to the next is a **wavelength**. The number of complete wavelengths that occur in 1 second is called the **frequency** of the sound, measured in number of waves per second, or Hertz (Hz), after the German physicist and pioneer of radio wave research, Heinrich Hertz. The length and frequency of sound waves impart certain characteristics to the stimulus. Short wavelengths have high frequencies that are perceived as a high **pitch** or tone, while long wavelengths have lower frequencies and a lower pitch.

The sense of hearing is present in vertebrates and arthropods, but not in other phyla. Arthropods do not appear to have more than a general sensitivity to sound, although some exceptions exist. For example, some species of moths have sound-sensitive membranes that detect the frequencies emitted by their chief predators, bats. The sense of hearing, however, is especially well developed in vertebrates (notably birds and mammals), and we turn therefore to a detailed discussion of the mammalian ear and the mechanism by which it detects sound.

Structure of the Mammalian Ear The mammalian ear has three main compartments (**Figure 45.6**). The **outer ear** consists of the external ear, or pinna, and the auditory canal. Mammals have a wide variety of shapes and sizes of the external ear, which reflects their different abilities to capture sound waves. The outer ear is divided from the **middle ear** by the tympanic membrane (eardrum); the middle ear contains three small bones called ossicles that connect the eardrum with the oval window. These are named the malleus (or hammer), the incus (or anvil), and the stapes (or stirrup). The oval window is another membrane similar to the eardrum that separates the middle ear from the **inner ear**. The inner ear is composed of the bony **cochlea** (from the Latin, meaning "snail")—a coiled structure containing the hair cells and other structures that generate the responses that travel via the auditory nerve to the brain—and the vestibular system, which plays a role in balance as described later.

Both the tympanic membrane and the oval window must be able to vibrate when sound waves meet the eardrum. For this to occur properly, the air pressure in the outer and middle ear compartments must be equal. The pressure in the outer ear is the atmospheric pressure. The **Eustachian tube**, a connection from the middle ear to the pharynx, maintains the pressure in the middle ear at atmospheric pressure as well. If you change altitude quickly, as during a plane takeoff, the pressure in the outer ear changes because atmospheric pressure decreases with altitude. After awhile, the pressures in the outer and middle ear equalize. However, if the outer ear pressure becomes greater than the middle ear pressure, which would occur when the plane descends, the eardrum will be pushed or bowed inwards. The eardrum is less able to vibrate in response to sounds when it is deformed under pressure, and thus hearing becomes muffled. Swallowing or yawning will open the Eustachian tube and equalize the pressure in the middle ear with atmospheric pressure.

Figure 45.6   **The structure of the human ear.**   The three main compartments are the outer, middle, and inner ear.

The popping sound you hear when the plane ascends or descends is the sudden return of the eardrum to its normal position.

Generation of Electrical Signals in the Mammalian Ear To understand how we hear, let's first consider how mechanical forces move through the ear. Sound waves entering the outer ear cause the tympanic membrane to vibrate back and forth. The ossicles transfer the vibration of the tympanic membrane to the oval window, causing it to vibrate against the cochlea (**Figure 45.7a**). This sends pressure waves, which also travel in a back-and-forth manner, through a fluid called perilymph, which is found within the vestibular canal and tympanic canal of the cochlea. The waves travel from the vestibular canal to the tympanic canal and eventually strike the round window, where they dissipate. Sounds of very low frequency create pressure waves that take the complete route through the vestibular and tympanic canals (see green arrows in Figure 45.7a). We would not be able to hear such sounds. Sound of higher frequency, however, produces pressure waves that take a shortcut. For example, the shortcut shown by the blue arrows in Figure 45.7a represents sound of 8,000 Hz. These waves pass through the cochlear duct, and then through the basilar membrane, before reaching the tympanic canal. As this happens, the basilar membrane vibrates—that is, moves up and down.

Mechanical forces are transduced into electrical signals within the organ of Corti (named after the Italian anatomist Alfonso Corti), which rests on top of the basilar membrane.

To understand how this works, we need to look at a cross section through the cochlea (**Figure 45.7b**). The organ of Corti contains supporting cells and rows of hair cells. The stereocilia of the hair cells are embedded in a gelatinous tectorial membrane. The back-and-forth vibration of the basilar membrane bends the hairs in one direction and then the other. When bent in one direction, the hair cells depolarize and release neurotransmitter, which activates adjacent sensory neurons that then send action potentials to the CNS via the auditory nerve. When bent in the other direction, the hair cells hyperpolarize and shut off the release of neurotransmitter. In this way, the frequency of action potentials by the sensory neurons is determined by the up-and-down vibration of the basilar membrane.

The basilar membrane is lined with protein fibers that span its width. These fibers function much like the strings of a guitar. The fibers near the oval and round windows at the base of the cochlea are short and rigid, and they vibrate in response to high-frequency waves. Longer and more resilient fibers are near the other end of the cochlea and vibrate to lower-frequency waves. For this reason, hair cells closer to the oval and round windows respond to high-pitched sounds, while those at the opposite end are triggered by lower-pitched sounds. When we hear a great number of sound frequencies at once, such as at a musical concert, the sound waves move hair cells all along the basilar membrane in a physical representation of the music! These cells stimulate sensory neurons, which send multiple action potentials to the auditory areas of the brain for processing.

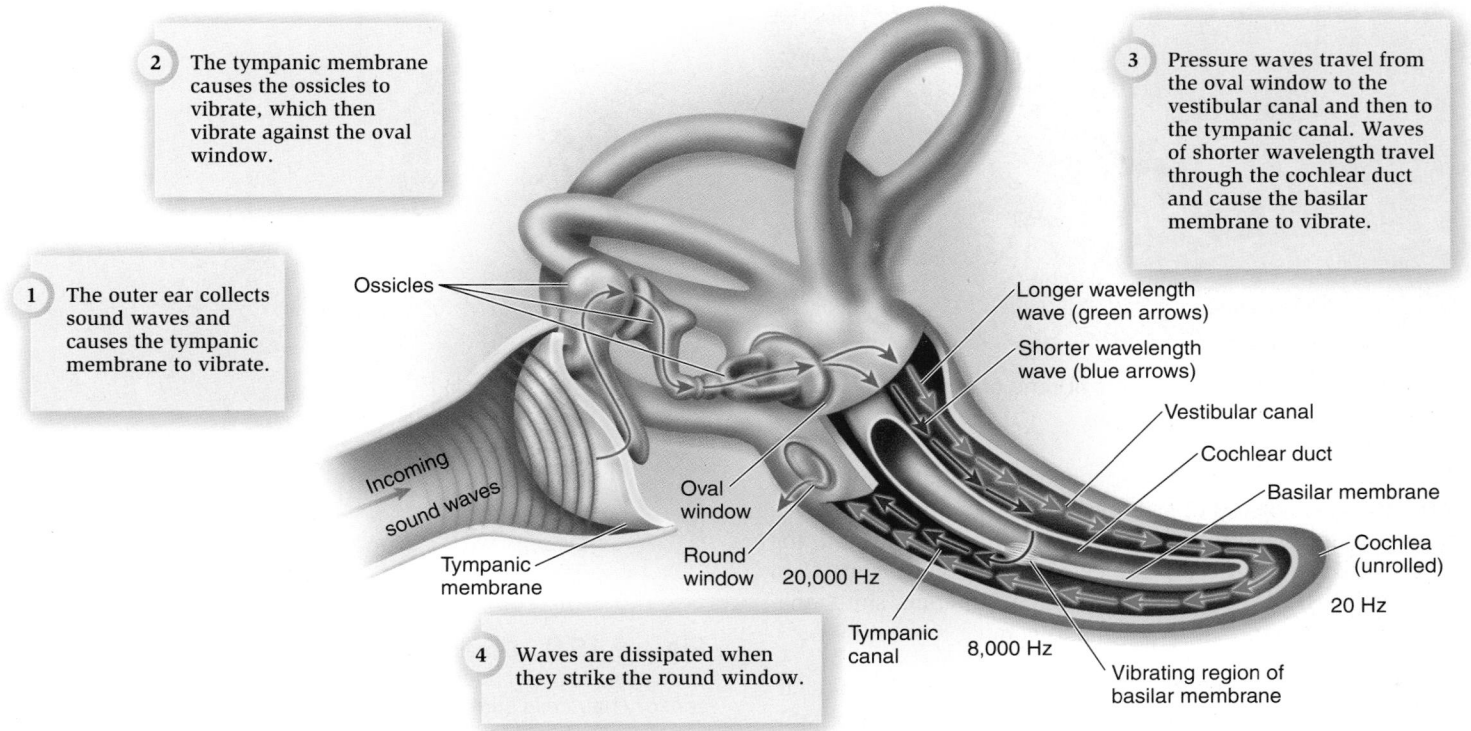

2 The tympanic membrane causes the ossicles to vibrate, which then vibrate against the oval window.

1 The outer ear collects sound waves and causes the tympanic membrane to vibrate.

3 Pressure waves travel from the oval window to the vestibular canal and then to the tympanic canal. Waves of shorter wavelength travel through the cochlear duct and cause the basilar membrane to vibrate.

4 Waves are dissipated when they strike the round window.

Ossicles

Incoming sound waves

Tympanic membrane

Oval window

Round window

20,000 Hz

8,000 Hz

Tympanic canal

Longer wavelength wave (green arrows)

Shorter wavelength wave (blue arrows)

Vestibular canal

Cochlear duct

Basilar membrane

Cochlea (unrolled)

20 Hz

Vibrating region of basilar membrane

(a) Movement of waves through the ear

Figure 45.7 Mechanism of hearing in the human ear.

The most incredible feature of this process, however, is that—unlike a radio, which can only be tuned to one frequency at a time—the mammalian ear and brain can "tune in" to all of these frequencies simultaneously.

Adaptations for Hearing Allow Animals to Live in a Wide Range of Habitats

Different animals are capable of hearing sounds of different pitches. For example, humans can hear between 20 and 20,000 Hz (conversation averages 90–300 Hz, while a human scream is 3,000 Hz). Bats and toothed whales probably can hear the highest-frequency sound waves, while baleen whales and elephants hear the lowest-pitched sounds. These adaptations increase the animals' ability to communicate and survive. For instance, high-pitched sounds are useful for locating small prey, such as flying insects, while low-pitched sounds carry great distances through water or air and are ideal for animals with large territories.

Locating a Sound A vital feature of hearing is for organisms to be able to locate the origin of a sound. For example, this ability makes the difference between a successful and an unsuccessful hunter. How does an animal locate a sound? Under most circumstances, sound does not arrive at both ears simultaneously. Sound waves coming from the right will excite the sensory receptors in the right ear first and the left ear some milliseconds

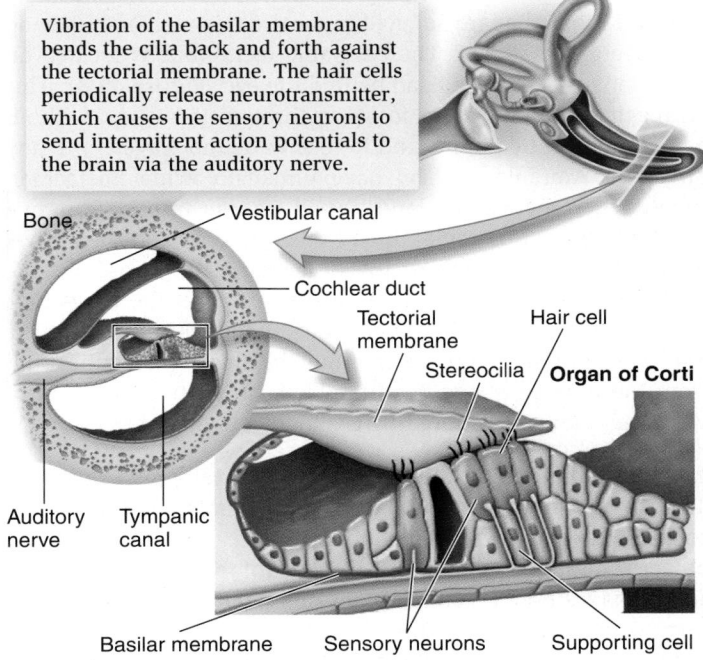

Vibration of the basilar membrane bends the cilia back and forth against the tectorial membrane. The hair cells periodically release neurotransmitter, which causes the sensory neurons to send intermittent action potentials to the brain via the auditory nerve.

Bone

Vestibular canal

Cochlear duct

Tectorial membrane

Hair cell

Stereocilia

Organ of Corti

Auditory nerve

Tympanic canal

Basilar membrane

Sensory neurons

Supporting cell

(b) Signal transduction in the organ of Corti

later, and therefore the brain receives action potentials from the auditory nerves of each ear at slightly different times. The brain interprets the time difference to determine the origin of the sound.

Animals such as owls that rely on hearing to locate prey tend to be extremely good at localizing the source of a sound. An interesting experiment demonstrated this by outfitting owls in a dark room with small headphones. Just as in a human hearing test, sounds could be sent to either headphone or to both.

Figure 45.8   The tympanic membrane of a frog.

If the investigator sent a high-pitched noise that mimicked a scurrying mouse first to the left headphone, and then a millisecond later to the right headphone, the owl turned its head to the left, because the owl's brain perceived the sound to come from that direction. If the noises reached both headphones simultaneously, the owl interpreted the signal as coming from in front.

Hearing in Air and Water Amphibians have the special challenge of hearing both on land and underwater, their two natural environments. Ears of amphibians have several interesting specializations that are adapted to these demands. First, they do not have external ears. The tympanic membrane is located on the outer surface of the head behind the eye (**Figure 45.8**). This arrangement allows them to swim through the water without being impeded by pinnae and to receive sound waves from any direction. Amphibians also have unusually wide Eustachian tubes. When they are on land, sound waves can pass from one ear through the Eustachian tube to the pharynx and then travel up the Eustachian tube of the other ear, where they hit the backside of the tympanic membrane. Air from the lungs also can pass into each ear in this manner. When a bullfrog emits its loud call, pressure coming through the Eustachian tubes and from the outside of the head prevents his own eardrum from vibrating and being injured by the loud sound.

Echolocation Bats in the air, whales and dolphins in the sea, and shrews in underground tunnels generate high-frequency sound waves to determine the distance and location of an object. This phenomenon is called **echolocation**, because the sound waves bounce off a distant object, like an echo, and return to the animal. The time it takes for the sound to return indicates the distance of the object. Echolocation can be useful in situations where vision is limited, such as in the dark.

The Sense of Balance Is Mediated by Statocysts in Invertebrates and the Vestibular System in Vertebrates

Let's now turn our attention to the sense of balance, also called equilibrium. Balance is part of a broader sense called proprioception, which is the ability to sense the position, orientation, and movement of the body. Being able to sense body position

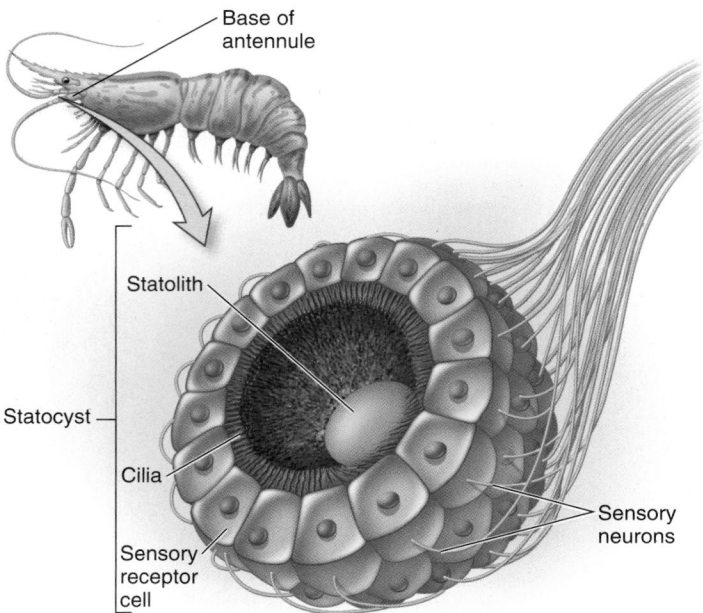

Figure 45.9   Sensing of balance in aquatic invertebrates. Statocysts located near the antennae consist of a sphere of sensory hair cells surrounding a stony statolith. When the animal moves, gravity shifts the statolith and stimulates the hair cells beneath it.

is vital for the survival of animals. This is how a lobster, for example, rights itself when flipped over by a predator, or a bird maintains its balance while flying rapidly through space. Many aquatic invertebrates have sensory organs called **statocysts** that send information to the brain about the position of the animal in space (**Figure 45.9**). Statocysts are small round structures made of an outer sphere of hair cells and **statoliths**, which are tiny granules of sand or other dense objects. When the animal moves, gravity alters the statoliths' position. If the animal turns on its left side, for example, the movement of statoliths stimulates a new set of hair cells to release neurotransmitter, generating action potentials in sensory neurons that inform the brain of the change in body position.

Several experiments have demonstrated the importance of the statoliths. In one particularly dramatic example, researchers replaced the statoliths of crayfish with iron filings. Moving a magnet to different positions around the animal displaced the filings, causing the animal to change its position, and even to swim upside down when the magnet was placed directly above its head.

The organ of balance in vertebrates, known as the **vestibular system**, is located in the inner ear next to the cochlea (**Figure 45.10**). The vestibular system is composed of a series of fluid-filled sacs and tubules. The utricle and saccule, two sacs nearest the cochlea, detect linear movements of the head as an animal runs and jumps. The hair cells within these structures are embedded in a gelatinous substance that contains granules of calcium carbonate called **otoliths** (from the Latin, meaning "ear stones"). When the head moves forward, the heavy otoliths move forward more slowly, and the inertia bends the stereocilia

Figure 45.10 **The vertebrate vestibular system.** In the cupula of the semicircular canals, hair cells move when fluid within the canal shifts. In the utricle and saccule, hair cells are stimulated by otoliths that bend the cilia in response to forward movement.

Biological inquiry: Note the orientation of the three semicircular canals with respect to each other. Why are the canals oriented in three different planes?

of the hair cells. This changes the membrane potential of the hair cells and alters the electrical responses of nearby sensory neurons.

Three **semicircular canals** open out of the utricle. The function of the semicircular canals is to detect rotational motions of the head. The hair cells in the semicircular canals are embedded in a gelatinous cone called the cupula (similar to the cupula of the fish lateral line). When the head moves, the fluid in the canal shifts in the opposite direction because of inertia, pushing on the cupula and bending the cilia of the hair cells. The three canals are oriented at right angles to each other, and each canal is maximally sensitive to motion in its own plane. For example, the canal that is oriented horizontally would respond greatest to rotations such as shaking the head "no." Overall, by comparing the signals from the three canals, the brain can interpret the motion in three dimensions.

The vestibular system of vertebrates provides conscious information about body position and movement. It also supplies unconscious information for reflexes that maintain normal posture, control head and eye movements, and assist in locomotion. Motion sickness in humans results when the vestibular system has not adapted to unfamiliar patterns of movement, such as spinning around or sailing aboard a ship. In these cases, the sense from your vestibular system conflicts with information coming from your vision. Vertigo, an illusion of movement or spinning, can result from malfunction of the vestibular system.

45.3 Thermoreception and Nociception

The perception of temperature and pain enables animals to respond effectively to their environments. As described in this section, these sensory stimuli are related in that their receptors are located in some of the same areas (skin), share similar physical features, and under certain conditions result in similar perceptions.

Thermoreceptors Detect Temperature

Sensing the outside temperature is important because body temperature is affected by the external temperature, particularly in ectotherms. Animals can survive at body temperatures only within certain limits, because the proteins in their cells function optimally within a particular temperature range. Thermoreceptors respond to cold or hot temperatures by activating or inhibiting enzymes within their plasma membranes, which alter membrane channels. Two types of thermoreceptors respond to either hot or cold. The endings of cold receptors are located nearer the surface of the skin than are hot receptors. Thermoreceptors are often linked with reflexive behaviors, such as when an animal steps on a hot surface and pulls its foot away.

In addition to skin receptors that sense the outside temperature, thermoreceptors in the brain also detect changes in core body temperature. Activation of skin or brain thermoreceptors triggers physiological and behavioral adjustments that help maintain body temperature. These changes, described in Chapter 42, include changes in blood flow, shivering, and behavioral events such as seeking shade or sunlight.

Nociceptors Warn of Pain

The sense of pain is one of the most important of all the senses. It tells an animal whether it has been injured, and triggers behavioral responses that protect it from further danger. Although in many cases we cannot know whether or how animals perceive pain, it is thought that all classes of animals have the ability to discern some type of painful sensation.

Nociceptors, the receptors for pain, are nerve endings in the skin and internal organs. They respond to tissue damage or to stimuli that are about to cause tissue damage. Nociceptors are unusual because they can respond not only to external stimuli, such as extreme temperatures, but also to internal stimuli, such as molecules released into the extracellular space from injured cells. Damaged cells release a number of substances, including acids and prostaglandins, that cause inflammation and make nociceptors more sensitive to painful stimuli. Aspirin and ibuprofen reduce pain by preventing the production of prostaglandins.

Signals arising from nociceptors travel to the CNS and reach the cerebrum, where the type or cause of the pain is interpreted. The signals are also sent to the limbic system, which holds memories and emotions associated with pain, and to the reticular formation, which increases alertness and arousal—an important response to a painful stimulus.

45.4 Electromagnetic Sensing

As mentioned earlier, electromagnetic receptors detect radiation within a wide range of the electromagnetic spectrum, including those wavelengths that correspond to visible light, ultraviolet light, and infrared light, as well as electrical and magnetic stimuli. Photoreceptors are specialized electromagnetic receptors that respond to light and are covered in a later section. Here, we will examine the ability of some animals to sense electrical and magnetic fields and also heat in the form of infrared radiation.

The ability to detect the presence of nearby prey or predators is an important adaptation in certain animals inhabiting regions where light intensity is low. The more ways in which an animal can detect other animals, the better it can avoid danger or obtain a meal. Many fish living in dark waters can detect weak electrical signals given off by other fish, and sharks and rays can even detect the tiny electrical signals generated by the heartbeats of their prey. Similarly, receptors on the skin of the platypus's bill detect very small electrical currents produced by their prey that might otherwise be undetectable in the murky waters of streams and ponds.

Homing pigeons utilize one type of electromagnetic sensing to return to their starting points from as far away as 1,500 km. This navigational feat is made possible by small particles of magnetite (iron oxide) in their beaks that act as a compass to tell direction. The magnetic particles respond to the Earth's magnetic field and alter neurotransmission in neurons that project to the brain. In one experiment, pigeons were placed individually in large tubes and trained that food was present in only one end of the tube. When the tube was placed in a changeable magnetic field, pigeons readily learned which end contained food based solely on the magnetic polarity of the tube. In another experiment, the pigeons lost this ability when their beaks were anesthetized or cooled down, procedures that block action potentials from being sent to the brain. This demonstrates that their magnetic sensing ability is located in the beak and communicated by nerves to the brain.

Magnetic field sensing is not unique to birds. Magnetite has also been found in the heads of migratory fish like rainbow trout. It is unlikely, however, that this entirely explains the extraordinary ability of migratory animals to navigate great distances, because other cues, such as smell and visual recognition of landmarks, also play a role.

Venomous snakes known as pit vipers (for example, copperheads and rattlesnakes) can localize prey in the dark with detectors that sense the heat emitted from animals as infrared radiation. These detectors are located in pits on each side of the head between the eyes and nostrils (**Figure 45.11**). Within the pit a thin, nerve-rich, temperature-sensitive membrane becomes activated in response to infrared waves emitted by live animals. When the snake detects the heat of the animal, it localizes its prey by moving its head back and forth until both pits detect the same intensity of radiation. This indicates that the prey is centered in front of the snake, and the snake is ready to strike.

Electrical, magnetic, and infrared sensing are adaptations for long-distance migration or low-light environments. When light is available, however, photoreception becomes a dominant sensory ability in many animals, as described next.

Pit—contains heat sensors

Figure 45.11   **Infrared sensing.**   Sensory pits enable a white-lipped pit viper to detect the heat given off by its prey.

45.5 Photoreception

Although it is a form of electromagnetic reception, photoreception is such an important and widespread sense that we will cover it separately here. Visual systems employ **photoreceptors**, which detect photons of light arriving from the sun or other light sources, or reflecting off an object. A photon is the fundamental unit of electromagnetic radiation and has the properties of both a particle and a wave. The properties of light are described in Chapter 8. In this section, we will examine the organs found in animals, usually called **eyes**, that detect light and send signals to the brain. The amazing features of these organs reflect the importance of vision in the animal world.

Eyecups and Compound Eyes Are Found in Certain Invertebrates

One of the simplest visual organs is found in the flatworm (*Planaria*); it is a concave **eyecup** containing the endings of photoreceptor cells (**Figure 45.12**). The eyecup, also called an eyespot, has a layer of pigment cells that shields the photoreceptors from one side. The left and right eyecups receive light from different directions. This allows the eyecups to detect the direction

as well as the presence or absence of light. The nervous system compares the amount of light detected by each eyecup, and the flatworm moves toward darkness, a behavior that protects it from predators. This type of photoreceptor is designed to sense only the presence or absence of light, not to form visual images of the environment.

In contrast, arthropods and some annelids have image-forming **compound eyes**, consisting of several hundred to several thousand light detectors called ommatidia (**Figure 45.13**). Each ommatidium makes up one facet of the eye. The **lens** and crystalline cone focus light onto a long central structure called a **rhabdom**. The rhabdom is a column of photosensitive microvilli that project from the cell membranes of retinula cells, which are the photoreceptor cells of the ommatidium. Each ommatidium senses the intensity and color of light. Combined with the different inputs from neighboring ommatidia, the compound eye forms an image that the brain interprets. Animals such as bees and fruit flies, with large numbers of ommatidia, presumably have sharper vision than those with fewer sensory cells, such as grasshoppers.

As anyone who has tried to swat a fly knows, the compound eye is extremely sensitive to movement and helps flying insects evade birds and other predators. Behavioral studies have shown, however, that the resolving power of even the best compound eye is considerably less than that of the single-lens eye, which we will consider next.

Figure 45.12 **The eyecup of the flatworm.** The orientation of the eyecup allows light to stimulate photoreceptors from only one direction. This type of eye senses only the presence or absence of light and does not form visual images.

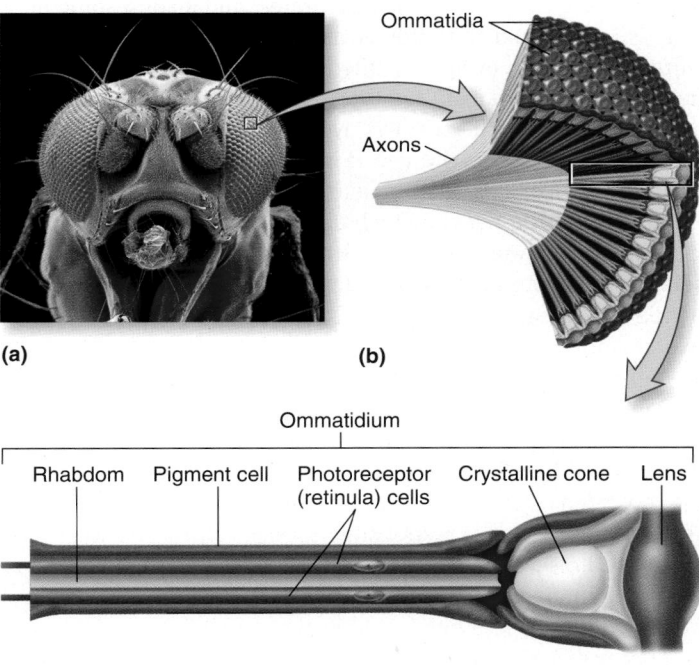

Figure 45.13 **The compound eye of insects.** (a) Close-up of the eyes of a fruit fly (*Drosophila melanogaster*). (b) Each eye has approximately 1,000 ommatidia, which form a sheet on the surface of the eye. (c) Each ommatidium has a lens that directs light to the photosensitive rhabdom.

Figure 45.14   **The vertebrate single-lens eye.**   **(a)** The structure of the human eye. **(b)** Changes in lens shape during accommodation. When an object is near, the ciliary muscles contract and the lens becomes rounder, causing light to bend more. When the object is far away, the ciliary muscles relax and the lens flattens.

Vertebrates and Some Invertebrates Have a More Complex Single-Lens Eye

Single-lens eyes are found in vertebrates, and also in some mollusks, such as squid and octopus, and in some snails and annelids. Different patterns of light emitted from images in the animal's field of view are transmitted through a small opening or **pupil**, through the lens, to a sheetlike layer of photoreceptors called the **retina** at the back of the eye (**Figure 45.14a**). Photoreceptors trigger electrical changes in neurons that pass out of the eye through the **optic nerve**, which carries the signals to the brain.

The vertebrate eye has a strong outer sheath called the **sclera** (the white of the eye), which in the front is continuous with a thin, clear layer known as the **cornea**. Within the eye are two chambers filled with liquid. At the front of the eye immediately behind the cornea is the smaller anterior chamber, which contains a thin liquid called the **aqueous humor**. The larger posterior chamber contains the thicker **vitreous humor**, which helps maintain the shape of the eye. Between the anterior and the posterior cavities are the **iris**—the circle of pigmented smooth muscle that is responsible for eye color—and the single lens. The hole in the center of the iris is the pupil. The size of the pupil changes when the muscles of the iris relax or contract to allow more or less light to enter the eye.

Because light radiates in all directions from a light source, light must be bent inward toward the photoreceptors at the back of the eye. This is accomplished by the cornea and the lens. Whenever light passes from one medium to another medium of a different density, light waves will bend (try looking at a pencil in a glass partly filled with water). The cornea is at the interface between the air and the aqueous humor, and it is the cornea

that initially bends the light. The light then passes through the thick lens, where it is focused onto the layer of photoreceptors, the retina, at the back of the posterior cavity. The bending of the incoming light results in an upside-down image on the retina, but the brain adjusts for this and the image is perceived correctly.

The lens is adjusted to focus light that comes from nearby or far away. In fish and amphibians, the lens is moved forward or backward. In the avian and mammalian eye, the lens remains stationary but changes shape to become more or less convex. When the lens is stretched, it flattens, and light passing through it bends less than when it is round. Contraction and relaxation of the ciliary muscles adjust the lens according to the angle at which light enters the eye, a process called **accommodation** (**Figure 45.14b**).

The single-lens eye is often compared to a film camera. Like the aperture mechanism of a camera, the iris regulates the amount of light that enters the eye. Just as the lens of a camera bends light and directs it toward the light-sensitive film, the lens of the eye bends and directs light toward the photoreceptor cells on the surface of the retina. Both the retina and a piece of film can be thought of as sheetlike structures across which different intensities and colors of light imprint an image. In the eye, the image initiates signals that travel from the retina to the brain through neurons that exit the eye in the optic nerve. In vertebrates, the point on the retina where the optic nerve leaves the eye does not have any photoreceptors, forming a "blind spot" where light does not activate a response. Invertebrates with single-lens eyes do not have a blind spot, because the optic nerve does not pass through the layer of photoreceptors before leaving the eye.

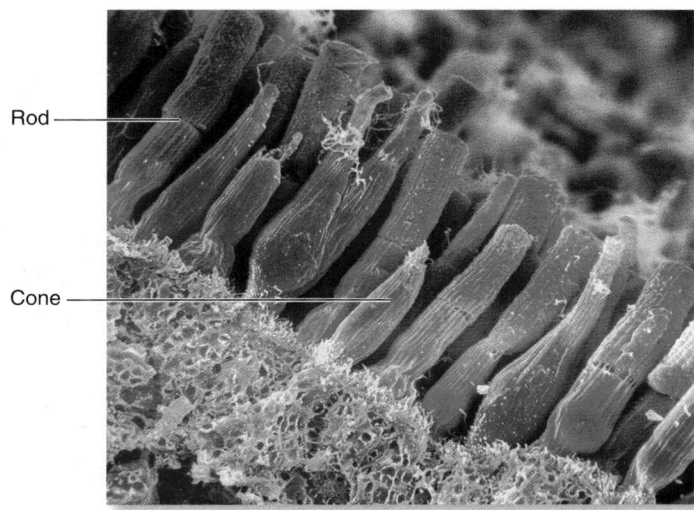

Figure 45.15   Rod and cone photoreceptors (SEM).

The **fovea**, a small area on the retina directly behind the lens where the image is focused, has numerous photoreceptors for color. The fovea is responsible for the sharpness with which we and other animals see in daylight.

Rods and Cones Are Photoreceptor Cells

Two types of photoreceptors have names that are derived from their shape (**Figure 45.15**). **Rods** are very sensitive to low-intensity light and can respond to as little as one photon, but they do not discriminate different colors. Rods are utilized mostly at night, and they send signals to the brain that generate a black-and-white visual image. **Cones** are less sensitive to low levels of light but unlike rods can detect color. Cones are used in daylight by most diurnal vertebrate species and by some insects such as the honeybee, which can detect the yellow color of pollen. Compared to rods, the human retina has fewer cones, which are clustered in and around the fovea. Cones provide sharp images because of their density at the fovea. Although they are less sensitive to light than rods, this is less critical in daylight because the amount of light reaching the eyes in daytime far exceeds what is needed to stimulate any photoreceptor cell. Because the two types of photoreceptor are specialized for either night or day vision, neither rods nor cones function at peak efficiency at twilight. This accounts for our relatively poor vision at this time.

Rods and cones have highly convoluted membranes that form the **outer segment** of the cell (**Figure 45.16**). The folds of the membrane form stacks of membranous discs. These discs contain the pigment molecules that absorb light. The **inner segment** of the cell contains the cell nucleus and other cytoplasmic organelles. The cell does not have an axon but has a synaptic terminal with neurotransmitter-containing vesicles, and which synapses with other neurons within the retina.

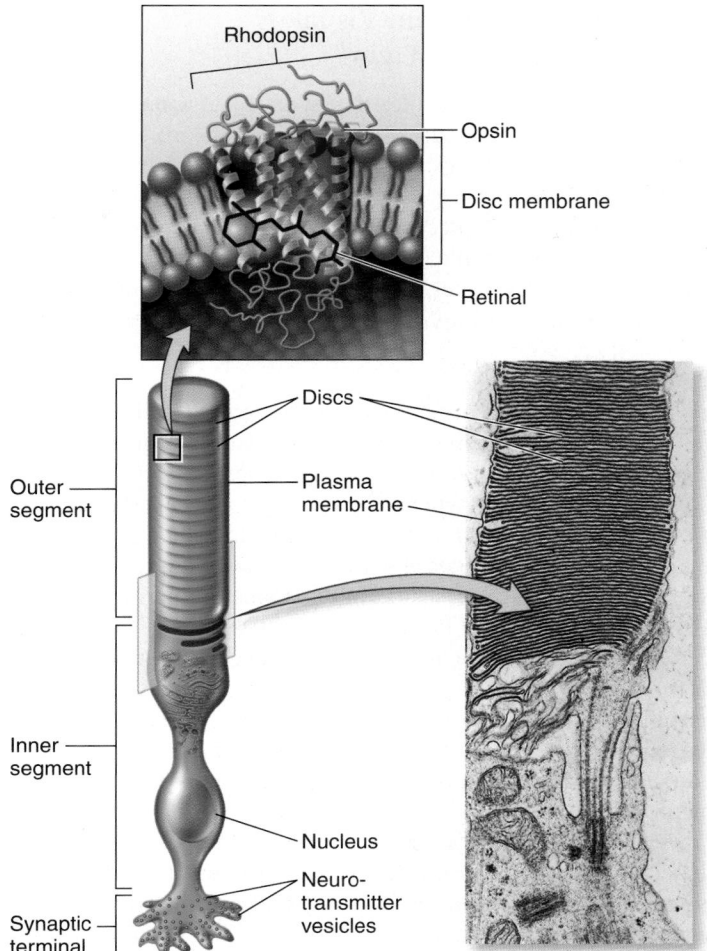

Figure 45.16   Structure of rod cells and rhodopsin. The illustration shows the structure of a rod photoreceptor and its appearance in an electron micrograph. Note the multiple stacks of membranous discs in the outer segment of the cell. Rhodopsin is found within the membrane of the photoreceptor discs. It is composed of a transmembrane protein, opsin, that is bonded to a molecule of retinal.

Nocturnal animals rely predominantly on rod vision, though some have limited color vision, too. In diurnal animals with both rods and cones, like humans, the rods are located around the periphery of the retina away from the fovea. Therefore, it is easiest to see low-intensity light if it comes into the eye at an angle. You can test this with a simple experiment. In early evening, before many stars are visible, look at the sky until you notice a star out of the corner of your eye. Now shift your gaze to where you thought you saw the star. You will probably not be able to locate it anymore. When you look away again so that light from the dim star enters your eye at an angle, it will reappear. This demonstrates that under low-light conditions, your vision is better when the light is directed to the part of the retina that contains only rods.

Rods and Cones Contain Visual Pigments That Detect Light

Visual pigments are molecules that span the disc membranes of the outer segment of rods and cones and absorb light. In the mid-20th century, American biologist and Nobel Prize winner George Wald discovered that these pigments consist of two components bonded together. The first is **retinal**, a derivative of vitamin A that is capable of absorbing light energy. The discovery of retinal in the visual pigment explains the need for vitamin A in the diet and its importance in vision. The second component of visual pigments is a protein, called **opsin**, of which there are several types.

Rods and cones contain visual pigments that differ with regard to the type of opsin protein they contain. In rods, the visual pigment is named **rhodopsin** (from "rod-opsin") (Figure 45.16). Cones contain several types of visual pigments called **cone pigments**. Opsins, found in both rhodopsin and cone pigments, are examples of G-protein-coupled receptors (see Chapter 9), which trigger a biochemical cascade that changes the permeability of membrane channels to ions.

In humans, cone pigments are composed of retinal plus one of three possible opsin proteins. Each type of opsin protein determines the wavelength of light that retinal can absorb. For this reason, each cone pigment can respond to either red, green, or blue light. Any given cone cell makes only one type of cone pigment. Many different shades of these colors can be perceived, however, because the brain uses information about the proportion of each type of cone that was stimulated to generate all other colors. Red, green, and blue opsins are not present in all species, and many species have only one or two opsins. Presumably, the more types of cones an animal has, the more shades of color it perceives. It is intriguing, therefore, to imagine how birds see the world, because they have up to five cones of overlapping wavelength sensitivity.

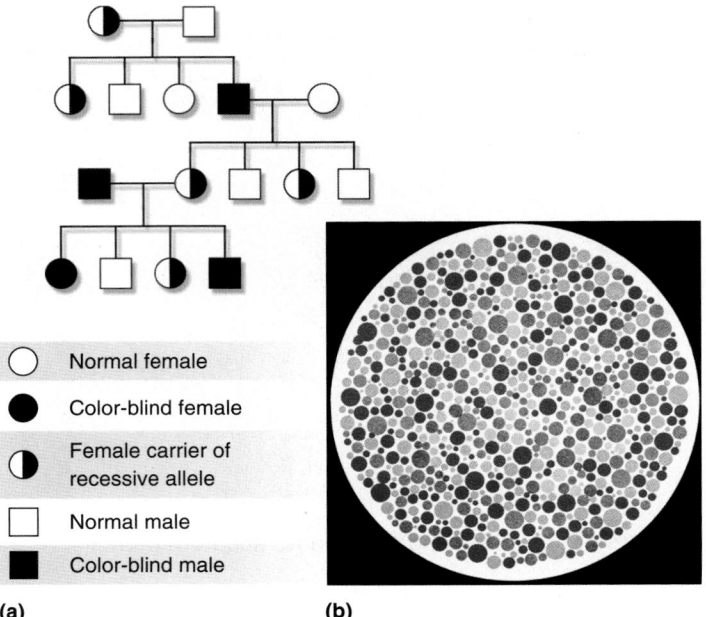

Normal female
Color-blind female
Female carrier of recessive allele
Normal male
Color-blind male

(a) **(b)**

Figure 45.17 Color blindness. **(a)** A pedigree for red-green color blindness showing all possible offspring. **(b)** A standard eye test to screen for red-green color blindness. People with red-green color blindness will not see the number 74 hidden in this picture.

Biological inquiry: Why is red-green color blindness rare in females?

GENOMES & PROTEOMES

Mutations in Cone Pigments Cause Color Blindness

In daylight, about 92% of human males and over 99% of females have normal color vision. However, problems in color vision may result from defects in the cone pigments arising from mutations in the opsin genes. The most common is red-green color blindness, which occurs predominantly in men (1 in 12 males compared with 1 in 200 females). Individuals with red-green color blindness either lack the red or green cone pigments entirely or, more commonly, have them in an abnormal form. In one form, for example, an abnormal green pigment responds to red light as well as green, making it difficult to discriminate between the two colors.

Color blindness was first described in the scientific literature in 1794 by John Dalton, the chemist mentioned in Chapter 2 after whom units of molecular mass are named. Dalton was himself color blind and willed that his eyes be preserved so that some-

day it might be possible to determine the cause of the defect. He hypothesized that it arose because his aqueous humor was filled with a blue-colored medium of some type, but this was proven wrong after his death when one of his eyes was carefully examined. In fact, color blindness results from a recessive mutation in one or more genes encoding the opsins. Genes encoding the red and green opsins are located very close to each other on the X chromosome, while the gene encoding the blue opsin is located on a different chromosome. In males, the presence of only one X chromosome means that a single recessive allele from the mother will result in red-green color blindness, even though the mother herself is not color blind (**Figure 45.17**).

Incidentally, in 1994, DNA testing of John Dalton's retina confirmed that he had classic red-green color blindness. For more than 100 years after his death, color blindness was known as Daltonism.

Photons Change Photoreceptor Activity by Altering the Conformation of Visual Pigments

Photoreceptors differ from other sensory receptor cells because their membrane potential is in a slightly depolarized state when the cell is at rest, and it is hyperpolarized rather than depolarized in response to a stimulus (**Figure 45.18a**). The membranes of the resting cell are highly permeable to sodium ions, which depolarizes the membrane potential. Sodium ions flow into the cell

Figure 45.18 **Mechanism of photoreception.** **(a)** Photoreceptor cells hyperpolarize in response to light and stop releasing the neurotransmitter glutamate. **(b)** Molecular changes in photoreceptor (cone) cell in response to light.

through open channels in the membrane of the outer segments and are pumped back out of the cell by Na^+/K^+-ATPase pumps in the membrane of the inner segment. In the dark, sodium channels remain open due to the binding of cGMP. The depolarization that is present when the photoreceptor is unstimulated (no light) results in a continuous release of the neurotransmitter glutamate from the synaptic terminal of the photoreceptor.

When the photoreceptor is exposed to light, the retinal within the visual pigment absorbs a photon. The energy of the photon alters the retinal from *cis*-retinal to *trans*-retinal, an isomer with a slightly different conformational state due to a rotation at one of the molecule's double bonds (**Figure 45.18b**). This change results in a brief dissociation of retinal from the opsin protein, causing the opsin to change shape and activate a G protein called transducin. The activated transducin, in turn, activates the enzyme phosphodiesterase. Phosphodiesterase reduces the amount of cGMP by converting it to GMP. GMP molecules cannot hold the sodium channels open, so the channels close, and sodium movement into the cell stops. The membrane potential of the cell becomes less positive than it was in the dark. Therefore, the response of the cell is a hyperpolarization that is proportional to the intensity of the light. The final result is a decrease in glutamate release from the photoreceptor (see Figure 45.18a).

The Visual Image Is Refined in the Retina

Thus far, we have considered the structure of the eye and how photoreceptors transduce light. We will now turn our attention to the neural pathways through which the visual signal travels

to reach the brain. To do so, we must consider the cellular organization of the vertebrate retina, which has three layers of cells (**Figure 45.19**). The rods and cones form the deepest layer, closest to the inside of the sclera along with a layer of pigmented epithelial cells that absorb scattered light. Light must pass through two other layers of cells, which are transparent, before it reaches the photoreceptors. The middle layer contains **bipolar cells**, so named because they make synapses with the photoreceptors and relay responses to the third layer of cells, the **ganglion cells**. The ganglion cells send their axons out of the eye into the optic nerve. In addition to these three layers, two types of cells, horizontal and amacrine cells, are interspersed across the retina.

The pathway for light reception begins at the photoreceptor cells (rods and cones). The photoreceptor cells release neurotransmitter molecules that affect the membrane potential of bipolar cells. The membrane potential of the bipolar cells determines the amount of neurotransmitter that bipolar cells release, which, in turn, controls the membrane potential of ganglion cells. When a threshold potential is reached in ganglion cells, action potentials are sent out of the eye via the optic nerve to the occipital lobe of the cerebral cortex, where visual information is further refined and interpreted. The visual cortex responds to such characteristics of the visual scene as whether something is moving, how far away it is, how one color compares to another, and the nature of the image (for example, a face).

Horizontal and amacrine cells modify electrical signals as they pass from the photoreceptors to the ganglion cells. These cells adjust the signal significantly, enhancing an animal's ability to visualize a scene by emphasizing the differences between images.

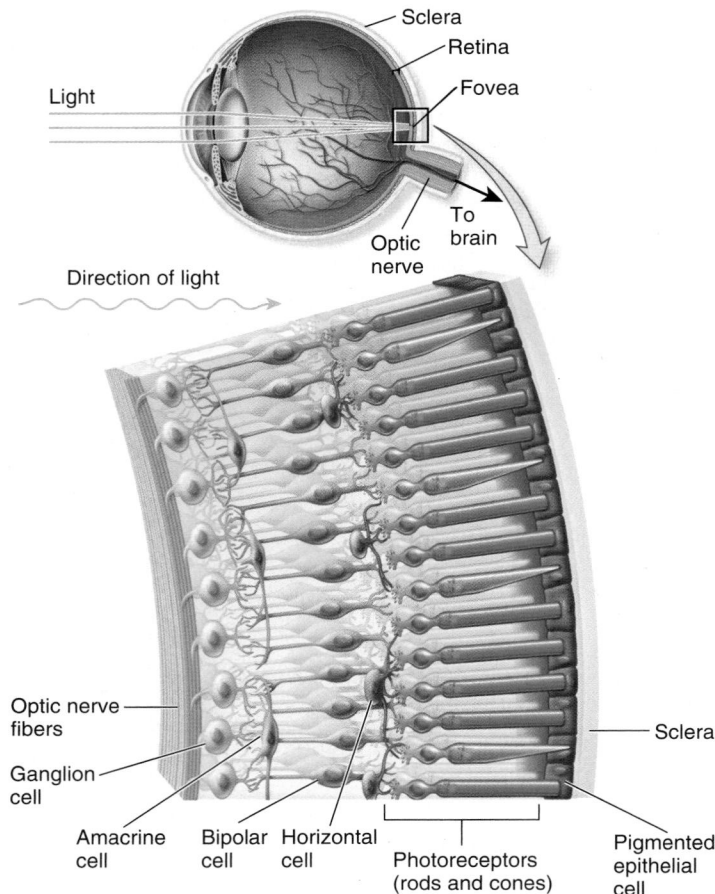

Figure 45.19   The arrangement of cells in the retina.
Light passes through two transparent cell layers before it reaches the photoreceptors. Amacrine and horizontal cells integrate the responses of the bipolar and ganglion cells. The ganglion cells generate action potentials to carry the information to the brain.

Horizontal cells make connections between photoreceptors and help to define the boundaries of an image, while amacrine cells are important in light adaptation and increasing the sensitivity of the eye to moving images. The ability of the retina to refine the image is especially well developed in birds and reptiles, which have complex retinas that process the image extensively before it is interpreted in the brain.

Vertebrate Eyes Are Adapted to Environmental Conditions and Life Histories

Many vertebrates show unique modifications of their visual systems that are the result of evolutionary adaptations to environmental conditions. Other adaptations have occurred as a result of behavioral requirements for obtaining food or attracting a mate.

Light Intensity Over the course of a day, wide variation occurs in the amount of light that stimulates the eye. The vertebrate eye can adjust to the differences in illumination in part by adjust-

ing the diameter of the pupil. In addition, the eye can adapt to light differences by changing the relative amounts of *cis-* and *trans*-retinal. Such changes can alter the light sensitivity of the eye by as much as a million-fold!

In a dark environment, a large amount of *cis*-retinal is available for light absorption. When light stimulates *cis*-retinal to change into *trans*-retinal, the visual pigment no longer responds to additional light and is said to be bleached. Such bleaching makes the eye less sensitive to low levels of light. When we enter a dark theater on a bright day, most of our visual pigments have been bleached by the sunlight and are temporarily unavailable for use. For a few minutes, we are unable to detect even low levels of light. This process of adjusting to dim light, called dark adaptation, involves the gradual reconversion of *trans*-retinal to *cis*-retinal.

Conversely, when we move from a dark room into bright sunlight, the large amount of *cis*-retinal makes our eyes extremely sensitive to light and temporarily overwhelms our ability to see colors and shapes. In light adaptation, the bright light stimulates bleaching of the visual pigments, which reduces the response to bright light.

Eye Placement Animals with both eyes located at the front of the head have **binocular vision**, because the overlapping images coming into both eyes are processed together in the brain to form one perception. Binocular vision provides excellent depth perception because the images come into each eye from slightly different angles. The brain processes those tiny differences to determine where an object is relative to other objects in its environment. Predators benefit from binocular vision because it helps them judge distance and determine the location of their prey. In contrast, animals with eyes on the sides of the head, such as most fish, have **monocular vision**. Monocular vision allows an animal to see a wide area at one time, though depth perception is reduced. Many prey species have monocular vision, perhaps because it helps them scan for predators across a wide field of vision.

Vision in the Deep Sea Fish and other deep-sea vertebrates have color vision that is limited primarily to the color blue. Light with longer wavelengths that would be seen as red or orange does not usually penetrate more than 6 m into the water, while the higher-energy short-wavelength light, which is seen as blue, can penetrate to greater depths. Aquatic animals that live in the deep sea are usually capable of seeing only blue, because they generally have only one opsin, which is responsive to blue light. Deep-dwelling fish tend to be drab colors, because most wavelengths of light do not penetrate that far and could not, therefore, reflect off the surface of the fish. In those deep-dwelling fish with more than one type of opsin, the additional visual pigments detect the bioluminescence (self-generated light, like that of a firefly) produced by their own or other species.

By contrast, fish that live near the water's surface sometimes have four or five different opsins, giving them excellent color vision. Not surprisingly, shallow water and surface-dwelling fish are often very colorful, because light of all wavelengths penetrates shallow water. These fish have adapted by using coloration for protection (camouflage) or for identification.

Vision with Speed Peregrine falcons, like other raptors, have exceptionally good eyesight. They dive at speeds that can reach 150 to 200 miles per hour to kill prey, and their visual system is specialized to maintain great sensitivity during the dive. The retina of a falcon has two foveas. A deep fovea has a central pit and provides the highest visual accuracy when objects are far away and to the side. However, when the falcon uses the deep fovea it does not have good depth perception. A shallow fovea provides the best images of nearby objects and good depth perception. When the falcon dives, it first uses the deep fovea to locate the prey and steer toward it. When the falcon gets closer, it switches to the shallow fovea. Some details of the image are sacrificed, but the improved depth perception allows it to gauge when it will reach its prey.

Tapetum Lucidum Did you ever wonder why cats' eyes seem to glow in photographs? What you see is light being reflected off the **tapetum lucidum**, a reflective layer of tissue located beneath the photoreceptors at the back of the eye (see chapter-opening photo). When light enters the retina, some of it passes into the layer containing the photoreceptors, but some light also passes through this layer. Normally, the pigmented epithelium at the back of the eye absorbs this light. However, in animals with a tapetum lucidum, the light reflects off this structure and back to the photoreceptors. This enhances the amount of light that the photoreceptors receive and makes it easier to detect low levels of light. Many nocturnal animals, both vertebrate and invertebrate, have a tapetum lucidum, which increases their ability to see in the dark.

45.6 Chemoreception: Smell and Taste

Chemoreception includes the senses of smell (**olfaction**) and taste (**gustation**), which involve detecting chemicals in air, water, and food. These chemicals bind to chemoreceptors, which initiate electrical responses in other neurons that pass into the brain. Amazingly, the binding of a single molecule to a receptor cell can sometimes be perceived as an odor! Airborne molecules that bind to olfactory receptors must be small enough to be carried in the air and into the nose. Taste molecules can be heavier because they are conveyed in food and liquid.

A close relationship exists between taste and smell. In fact, the distinction is largely meaningless for aquatic animals, because for them all chemoreception comes through the water. Even in terrestrial animals, about 80% of the perception of taste is actually due to activation of olfactory receptors. This is why food loses its flavor when olfaction is impaired, such as when you have a cold.

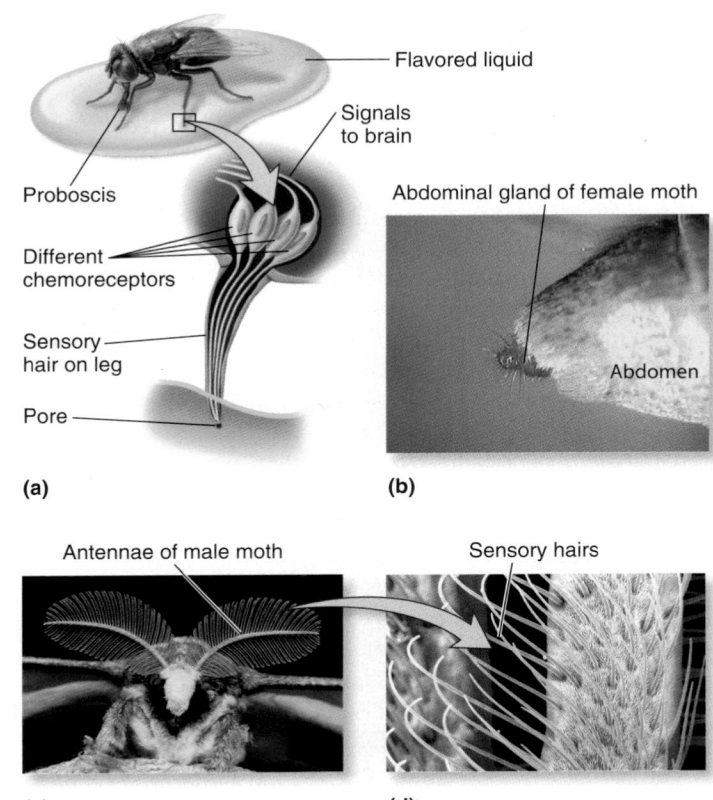

Figure 45.20 **Chemoreception in insects.** **(a)** Chemosensory cells of the blowfly located on the proboscis, legs, and feet sense different chemicals. **(b)** Female moths secrete a pheromone from abdominal glands, which is detected by **(c)** chemoreceptors in a male's antennae. **(d)** Sensory hairs on the antennae bind odor molecules.

Olfaction and Taste in Insects Involve Chemoreceptors in Sensory Hairs

Insects are highly dependent on odor and taste for finding food and mates. Chemoreceptors are located on sensory hairs on the proboscis, legs, feet, and antennae. Each sensory hair on the proboscis and feet has a pore at the tip through which the substance passes (**Figure 45.20a**). The blowfly has four separate chemoreceptors within each hair, and each of these neurons responds to different molecules. Dendrites of the chemoreceptor cells inside the pore bind to the molecules and initiate a sensory transduction pathway that opens ion channels in the membrane. This depolarizes the chemoreceptor cell and generates action potentials.

In certain moths, males have elaborate antennae that can sense pheromones, extremely potent signaling molecules given off by a female. The female secretes a sexual-attractant pheromone into the air from an abdominal gland (**Figure 45.20b**), and the chemosensory hairs on the male's antennae (**Figure 45.20c,d**) can detect extremely low concentrations of it from several kilometers away. This highly sensitive detection system enables the male to locate the female in the dark.

Mammalian Olfactory Receptors Respond to the Binding of Odor Molecules

The olfactory sensitivity of mammals varies widely depending on their supply of olfactory receptor cells, which ranges from 5 or 6 million in humans to 100 million in rabbits and 220 million in dogs. Olfactory receptors are located in the epithelial tissue at the upper part of the nasal cavity in mammals (**Figure 45.21**). These cells are surrounded by two additional cell types: supporting cells, which are located between the receptor cells and provide physical support for the olfactory receptors, and basal cells. The basal cells differentiate into new olfactory receptors every 30–60 days, replacing those that have died after prolonged exposure of their cell endings.

Olfactory receptors have long cilia that emerge into a mucus layer that covers the epithelium. Despite the similarity in structure, these cells do not function like the mechanoreceptor hair cells of the auditory and vestibular systems. Instead, olfactory receptors have receptor proteins within the plasma membrane of the cilia. Airborne molecules dissolve in the mucus and bind to these receptor proteins. When an odor molecule binds to its protein receptor, this initiates a signal transduction pathway that ultimately opens sodium channels in the plasma membrane. The subsequent depolarization results in action potentials being transmitted to the next series of cells in the olfactory bulbs, which are located at the base of the brain. From there, the information is sent to the cerebral cortex and limbic system for processing and interpretation.

The relative size of the olfactory bulbs indicates the importance of olfaction to an animal. In humans, the olfactory bulbs make up only about 5% of the weight of the brain, while in nocturnal animals like rats and mice they can comprise as much as 20%. Even with their limited olfactory sensitivity, however, humans have the capacity to detect up to 10,000 different odors, and other mammals are believed to respond to many more. Exactly how this is possible remained a mystery until 1991, when two scientists uncovered the molecular basis of olfaction.

Figure 45.21   **Olfactory structures in the human nose.** Odor molecules dissolve in a layer of mucus that coats the olfactory receptor cells. The molecules bind to protein receptors on the membrane of cilia that extend from the olfactory receptor cells. Action potentials in the olfactory receptor cells are conducted to cells in the olfactory bulb, and from there to the brain for interpretation. Basal cells periodically replace dead or damaged olfactory receptor cells.

FEATURE INVESTIGATION

Buck and Axel Discovered a Family of Olfactory Receptor Proteins That Bind Specific Odor Molecules

How does the olfactory system discriminate between thousands of different odors? American neuroscientists Linda Buck and Richard Axel set out to study this question. When they began, two hypotheses were proposed to explain this phenomenon. One possibility is that many different types of odor molecules might bind to one or just a few types of receptor proteins, with the brain responding differently depending on the number or distribution of the activated receptors. Alternatively, odors might be distinguished at the level of receptor proteins. The second hypothesis is that olfactory receptor cells can make many different types of receptor proteins, each type binding a particular odor molecule or group of odor molecules.

To begin their study, Buck and Axel assumed that olfactory receptor proteins would be highly expressed in the olfactory receptor cells, but not in other parts of the body. Based on previous work, they also postulated that the receptor proteins would

Figure 45.22   Buck and Axel identified olfactory receptor proteins in olfactory receptor cells.

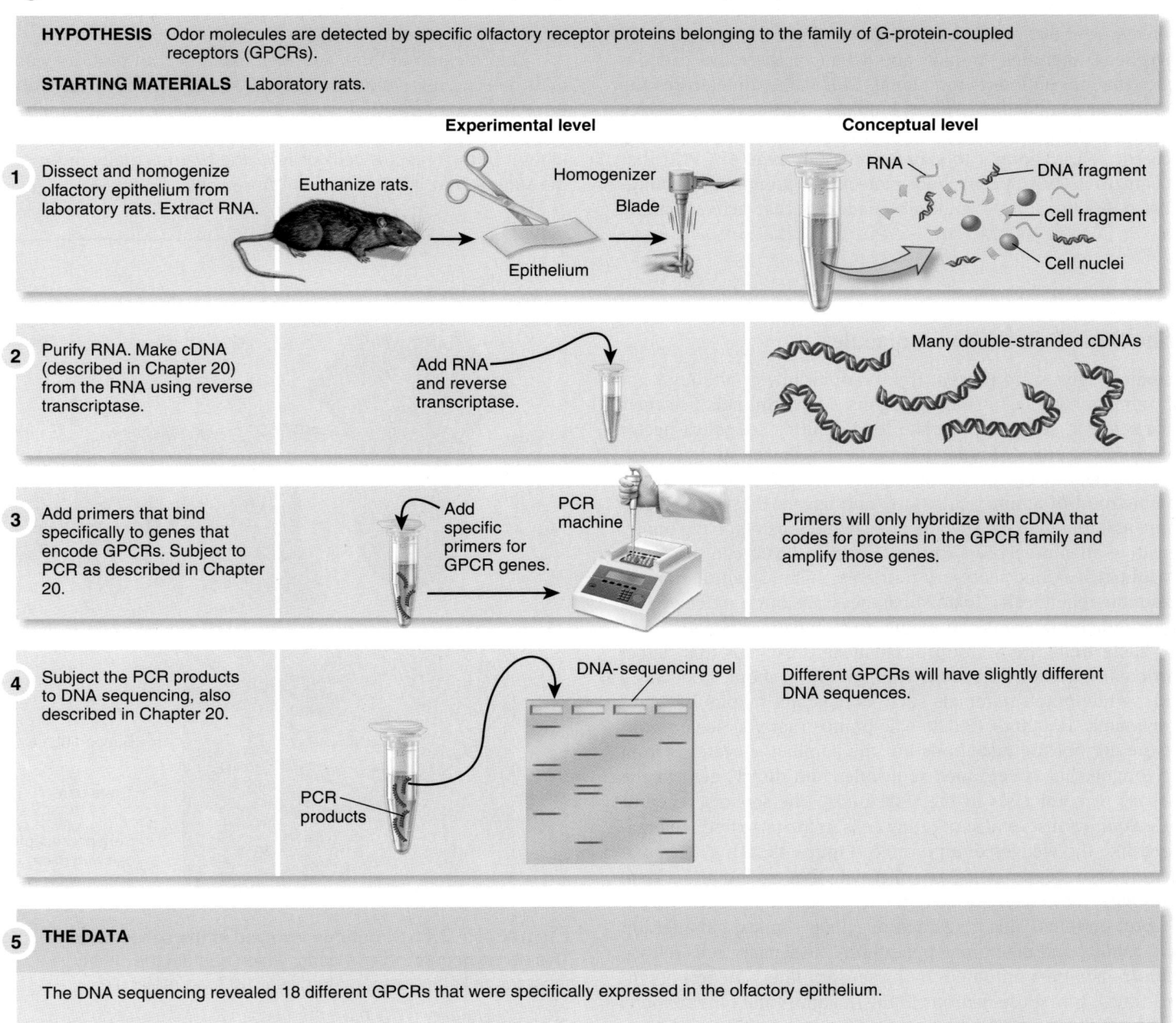

HYPOTHESIS   Odor molecules are detected by specific olfactory receptor proteins belonging to the family of G-protein-coupled receptors (GPCRs).

STARTING MATERIALS   Laboratory rats.

Experimental level   **Conceptual level**

1   Dissect and homogenize olfactory epithelium from laboratory rats. Extract RNA.

Euthanize rats.   Homogenizer   Blade   Epithelium

RNA   DNA fragment   Cell fragment   Cell nuclei

2   Purify RNA. Make cDNA (described in Chapter 20) from the RNA using reverse transcriptase.

Add RNA and reverse transcriptase.

Many double-stranded cDNAs

3   Add primers that bind specifically to genes that encode GPCRs. Subject to PCR as described in Chapter 20.

Add specific primers for GPCR genes.   PCR machine

Primers will only hybridize with cDNA that codes for proteins in the GPCR family and amplify those genes.

4   Subject the PCR products to DNA sequencing, also described in Chapter 20.

DNA-sequencing gel   PCR products

Different GPCRs will have slightly different DNA sequences.

5   **THE DATA**

The DNA sequencing revealed 18 different GPCRs that were specifically expressed in the olfactory epithelium.

be members of the large family of G-protein-coupled receptors (GPCRs), which are described in Chapter 9. As shown in the Feature Investigation in **Figure 45.22**, they isolated olfactory receptor cells from rats and then broke open the cells to release the RNA. The purified RNA was then used to make cDNA via reverse transcriptase. This would generate a large pool of cDNAs, which would represent all of the genes that are expressed in olfactory receptor cells. To determine if any of these cDNAs encoded GPCRs, they used primers that recognized regions within previously known GPCR genes that are highly conserved. A highly conserved region is a DNA sequence that rarely changes among different family members. The primers were used in the technique of PCR to amplify cDNAs that encoded GPCRs. This produced many PCR products that were then subjected to DNA sequencing. As shown in the data in Figure 45.22, Buck and Axel identified 18 different genes, each encoding a GPCR with a slightly different amino acid sequence. Further research showed that these 18 genes were expressed in nasal epithelia, but not in other parts of the rat's body. These results were consistent with the second hypothesis stated earlier, namely that organisms make a large number of receptor proteins, each type binding a particular odor molecule or group of odor molecules.

Since these studies, researchers have determined that this family of olfactory genes in mammals is surprisingly large. In humans, over 600 genes that encode olfactory receptor proteins have been identified, though about half of these are pseudogenes that are no longer functional. This value underscores the importance of olfaction even in humans, who have a relatively poor sense of smell compared to certain animals. Each olfactory receptor cell is thought to express only one type of GPCR that recognizes its own specific odor molecule or group of molecules. Most odors are due to multiple chemicals that activate many different types of odor receptors at the same time. We perceive odors based on the combination of receptors that become activated, and then send signals to the brain.

The research of Buck and Axel explained, in part, how animals detect a myriad of odors. In 2004, they received the Nobel Prize for this pioneering work. Even so, many intriguing questions remain. We do not yet understand the pattern of gene regulation in the receptor cells or how the brain actually interprets the signals that it receives from those cells.

Taste Buds Detect Food Molecules

Chemical senses are present in all animal phyla, although not all animals have taste-sensing organs. Many animals use their taste sense to select appropriate foods. Butterflies select nectar based on the sugars found in a particular flower, and carnivorous animals detect the taste of different meats based on the combination of amino acids, fats, and sugars that are present. Some aquatic animals, such as catfish and lobsters, have highly sensitive chemoreceptors for amino acids. Taste can also help an animal seek out necessary nutrients, such as salt, and avoid poisonous chemicals. Toxic substances are often perceived as bitter, which may cause an animal to stop eating immediately.

Taste buds are a group of chemosensory cells that detect particular molecules in food. Humans have about 9,000 taste buds, while other mammals such as pigs and rabbits have approximately 15,000 to 18,000. The bumps that you see on your tongue are not the taste buds but the papillae, elevations from the tongue that collect food molecules and direct them to the sensory receptor cells in the taste buds. The sensory receptor cells along with several supporting cells in the taste bud are organized like the wedges of an orange (**Figure 45.23**). The tips of the sensory receptor cells have microvilli that extend into a taste pore. Here, molecules in food that have dissolved in saliva bind to receptor proteins. This triggers intracellular signals that alter ion permeability and membrane potentials. The sensory cells release neurotransmitters onto underlying sensory fibers. Action potentials travel from these neurons to the thalamus and other regions of the cerebrum, where the taste is perceived and recognized.

The senses of taste and smell are enhanced when we are hungry, a phenomenon that most likely occurs in other animals as well. Once we have eaten, we are less aware of the smell and taste of food. The importance of this for survival is clear: A hungry animal needs to eat. An improved sense of smell aids in locating food, and a heightened sense of taste encourages an animal to eat. Afterward, these senses become temporarily dampened so as not to distract the animal from its other needs. The mechanism by which these changes occur is uncertain, but it may involve a temporary alteration in the number of smell and taste receptors, or in their ability to bind ligands.

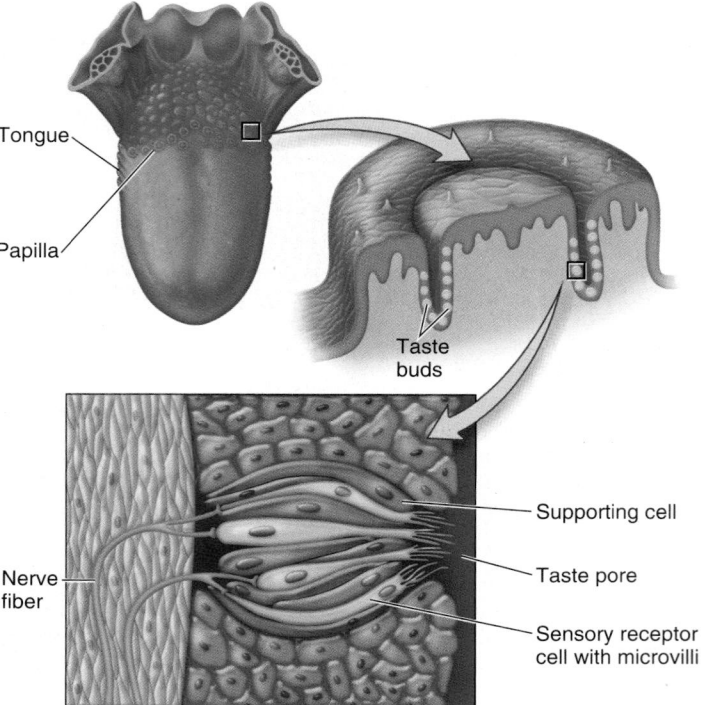

Figure 45.23   Structures involved in the sense of taste. This sense occurs in taste buds, which contain the sensory receptor cells that respond to dissolved food molecules.

Biological inquiry: Why have many animals, including humans, evolved the ability to sense salty, sweet, sour, and bitter?

45.7 Impact on Public Health

Sensory disorders are among the most common neurological problems found in humans and range from mild (for example, needing eyeglasses) to severe (for example, blindness). In this section, we present an overview of a few representative sensory disorders that have a major impact on the human population.

Visual Disorders Include Glaucoma, Macular Degeneration, and Cataracts

Visual disorders affect an enormous fraction of the world's population. In the U.S. alone, more than 10 million people have severe eye problems that cannot be corrected by eyeglasses, and more than 1 million people are blind (42 million worldwide). The costs to the U.S. government associated with blindness and severe visual loss amounts to nearly $4 billion yearly. Although vision loss has many causes, three account for over half of all cases: glaucoma, macular degeneration, and cataracts.

Normally, the fluid that makes up the aqueous humor is produced and reabsorbed (drained) in a circulation that keeps the fluid levels constant. In **glaucoma**, drainage of aqueous humor in the eye becomes blocked and the pressure inside the eye increases as the fluid level rises. If untreated, this eventually damages cells in the retina and leads to irreversible loss of vision. The cause of glaucoma is not always known, but in some cases it is due to trauma to the eye, chronic use of certain medicines, and diseases such as diabetes. Some forms of glaucoma may have a genetic component. Glaucoma can be treated with eye drops that reduce fluid production in the eye, or with laser surgery to reshape the structures in the eye. Incidentally, rubbing your eyes increases the intraocular pressure, and habitual rubbing can also damage vision. The American Foundation for the Blind estimates that 400,000 new cases of glaucoma are diagnosed each year in the U.S., and up to 10 million individuals have elevated intraocular pressure. Glaucoma accounts for roughly 10% of all cases of blindness in the U.S.

In **macular degeneration**, photoreceptor cells in and around the fovea of the retina are lost. Because this is the region where cones are densely packed, this condition is associated with loss of sharpness and color vision. It does not usually occur before age 60, but some cases are hereditary and can occur at any age. Macular degeneration is the leading cause of blindness in the U.S., accounting for roughly 25% of all cases, but its causes remain obscure.

Cataracts, the accumulation of protein in the lens, cloud the lens and cause blurring, poor night vision, and difficulty focusing on nearby objects. By age 65, as many as 50% of individuals have one or more cataracts in either eye and this jumps to 70% by age 75. Many cataracts are small enough not to affect vision. In fact, many people do not even realize they have cataracts until they undergo an eye exam. The causes of cataracts are not all known but include trauma, medicinal drugs, diabetes, and heredity. The treatment, when needed, is to wear a powerful contact lens to help do the job formerly done by the person's own lens, or to have the affected lens surgically removed and replaced with a plastic lens. Nearly 1.4 million such surgeries are performed each year in the U.S. Without the lens, which helps protect the retina by absorbing some of the high-energy ultraviolet light from the sun, the retina must be protected during the day by wearing dark sunglasses.

Damaged Hair Cells Can Cause Deafness

Deafness (hearing loss) is usually caused by damage to the hair cells within the cochlea, although some cases result from functional problems in brain areas that process sound or in nerves that carry information to the brain from the hair cells. When the hair cells are damaged, noises have to be louder to be detected, and an affected person may require the use of hearing aids to amplify incoming sounds.

Hearing loss may be mild or severe and may result from many causes, including injury to the ear or head, hereditary defects of the inner ear, and exposure to certain diseases (for example, rubella) or toxins during fetal life. By far, however, the most significant cause of hearing loss is repeated, long-term exposure to loud noise. When the amplitude of a sound wave (that is, the distance between the peak and the trough of the wave) is high, the loudness of the sound is great. The loudness of sound is measured in decibels (**Table 45.1**). The decibel (dB) scale is logarithmic; every increase of 10 dB is actually 10 times louder than the previous level. Noise-induced hearing loss, which usually results from job-related activities, accounts for 7–21% of all cases of hearing loss worldwide and is the leading occupational disorder in the U.S. It is, therefore, one of the most significant disabilities in the U.S. in terms of numbers of people affected and costs to society. It is estimated that nearly 40 million people—about 1 in 7 Americans—are exposed daily to dangerous noise levels as a result of their occupation.

Table 45.1	Intensity (Loudness) of Some Common Sounds	
Sound		**Intensity (dB)**
Mosquito buzzing		8
Leaves rustling		10–20
Watch ticking		20
Whisper		20–30
Bird chirping		35
Normal conversation		60
Department store or noisy office		70
Kitchen blender		90
Human scream		100–120
Chainsaw		108
Ambulance siren		120
Loud thunder		120
Loud rock concert		120
Jack hammer		130
Jet plane taking off		150

Recent research has led to a better understanding of the mechanism by which noise impairs hearing. Chronic exposure to loud sounds appears to produce a state of metabolic exhaustion in the hair cells of the cochlea. As a result, the cells become fatigued and are unable to maintain normal biochemical processes. One consequence of this is a buildup of free radicals. These compounds oxidize lipids in cellular membranes, damaging the membranes in the process. Mitochondrial membranes appear to be particularly susceptible to these free radicals. Once mitochondria are destroyed, a cell's ability to produce the ATP needed to fulfill its energy demands is compromised,

and the cell dies. As hair cells die, the ear becomes less sensitive to sound.

Researchers are investigating drugs that might prevent the formation of free radicals in the cells of the ear, but such drugs are not yet available. If the cochlea is severely damaged, artificial cochlear implants can be surgically implanted. These devices generate electrical signals in response to sound waves that can stimulate the auditory nerve, which communicates with the brain. Cochlear implants cannot restore hearing to normal, but they can make it possible to hear conversations.

CHAPTER SUMMARY

45.1 An Introduction to Sensory Receptors

- Sensory transduction is the process by which incoming stimuli are converted to neural signals. Perception is an awareness of a sense.
- Sensory receptors are either neurons or specialized epithelial cells that respond to stimuli and begin the process of sensory transduction. (Figure 45.1)
- A sensory receptor often responds to a stimulus by eliciting a graded response that is proportional to the intensity of the stimulus. (Figure 45.2)

45.2 Mechanoreception

- Mechanoreceptors respond to physical stimuli such as pressure, touch, stretch, movement, and sound.
- Mechanoreceptors in the skin are neurons that sense stimuli such as light touch and pressure. (Figure 45.3)
- Hair cells have stereocilia that respond to specific types of stimuli, and then release neurotransmitter that may result in action potentials in an adjacent sensory neuron. (Figure 45.4)
- The lateral line system of fish detects water movements. (Figure 45.5)
- Hearing is the ability to sense sound waves. Sound waves that travel at a high frequency are perceived as having a high pitch.
- The ear has three main compartments called the outer, middle, and inner ear. (Figure 45.6)
- As pressure waves move through the cochlea of the ear, they may cause the basilar membrane to vibrate. This causes the stereocilia of hair cells to bend back and forth, which results in intermittent action potentials being sent to the brain. (Figure 45.7)
- Many adaptations for hearing improve an animal's ability to sense sound. (Figure 45.8)
- Statocysts in certain invertebrates allow an animal to sense its body position or equilibrium. (Figure 45.9)
- The vestibular system in vertebrates allows animals to sense linear and circular motions. (Figure 45.10)

45.3 Thermoreception and Nociception

- Thermoreceptors in the skin and brain allow an animal to sense external and internal temperatures, respectively.
- Nociceptors in the skin and internal organs sense pain.

45.4 Electromagnetic Sensing

- Some animals can detect electrical and magnetic stimuli. (Figure 45.11)

45.5 Photoreception

- The eyecup in planaria is a primitive eye that detects light and its direction. (Figure 45.12)
- The compound eye found in many invertebrates consists of many ommatidia that focus light. (Figure 45.13)
- The eye found in vertebrates and certain invertebrates is a single-lens eye. Stretching and flattening of the lens in mammals aids in focusing on objects at different distances away. (Figure 45.14)
- Rods and cones are photoreceptors found in the vertebrate eye. The visual pigment consists of retinal and a protein called opsin. (Figures 45.15, 45.16)
- Red-green color blindness is due to a defect in a type of opsin found in cone pigments. (Figure 45.17)
- The retina is composed of layers of cells that receive light input and send electrical signals to the brain via the optic nerve. (Figures 45.18, 45.19)
- The vertebrate eye has several adaptations that aid in proper vision for particular animals.

45.6 Chemoreception: Smell and Taste

- Invertebrates have taste receptors on their proboscis, legs, feet, and antennae. Some insects can detect pheromones. (Figure 45.20)
- Olfactory receptors in vertebrates detect odor. They contain cilia with protein receptors that bind specific odor molecules. Buck and Axel discovered that olfactory receptors have many different types of protein receptors for odor molecules. (Figures 45.21, 45.22)
- Taste buds contain chemoreceptors that sense molecules in food. (Figure 45.23)

45.7 Impact on Public Health

- Common visual disorders include glaucoma, macular degeneration, and cataracts.
- Prolonged exposure to intense sound can damage hair cells and lead to deafness. (Table 45.1)

TEST YOURSELF

1. The process whereby incoming sensory stimulation is converted to neural signals is
   a. an action potential.
   b. a threshold potential.
   c. perception.
   d. sensory transduction.
   e. reception.

2. Photoreceptors are examples of
   a. nociceptors.
   b. baroreceptors.
   c. mechanoreceptors.
   d. electromagnetic receptors.
   e. thermoreceptors.

3. The sensory receptors for hearing are located in
   a. the organ of Corti.
   b. the Eustachian tube.
   c. the retina.
   d. the tympanic membrane.
   e. a, b, and c.

4. Statocysts are sensory organs for
   a. hearing found in many invertebrates.
   b. equilibrium found in mammals.
   c. equilibrium found in many invertebrates.
   d. water current changes found in fish.
   e. hearing found in amphibians.

5. The eyecups of planaria can
   a. focus light to form an image.
   b. detect light.
   c. detect the direction of light.
   d. all of the above.
   e. b and c only.

6. The light detectors of a compound eye are called
   a. retinas.
   b. opsins.
   c. cones.
   d. ommatidia.
   e. statocysts.

7. In the mammalian eye, light is focused on the retina when
   a. the lens moves forward or backward.
   b. the lens changes shape.
   c. the eyeball changes shape.
   d. the pupil dilates.
   e. none of the above.

8. The level of glutamate release from photoreceptors would be highest when
   a. a person is standing in full sunlight.
   b. a person is in a completely dark room.
   c. a person is in a dimly lit room.
   d. Na^+ channels of the photoreceptor are closed.
   e. both a and d.

9. Cone pigments detect different wavelengths of light due to
   a. their location in the retina.
   b. the amount of light they absorb.
   c. the type of retinal they have.
   d. the type of opsin protein they have.
   e. interactions with bipolar cells.

10. The stimulation for olfaction involves odorant molecules
    a. bending the cilia of olfactory receptor cells.
    b. binding to receptors of olfactory receptor cells.
    c. entering the cytoplasm of olfactory receptor cells.
    d. closing K^+ channels of olfactory receptor cells.
    e. all of the above.

CONCEPTUAL QUESTIONS

1. Define sensory transduction and perception.

2. Explain how the mammalian ear is adapted to distinguish sounds of different frequencies.

3. Explain how eye placement affects binocular vision.

EXPERIMENTAL QUESTIONS

1. What were the two major hypotheses to explain how animals discriminate between different odors? How did Buck and Axel test the hypothesis of multiple olfactory receptor proteins?

2. What were the results of Buck and Axel's study?

3. Considering the two hypotheses explaining how animals discriminate between different odors (see Experimental Question 1), which one was supported by the results of this experiment? With the evidence presented by Buck and Axel, what is the current hypothesis explaining the discrimination of odors in animals?

COLLABORATIVE QUESTIONS

1. Discuss two different types of mechanoreceptors.

2. Discuss different types of eyes found in animals.

---

www.brookerbiology.com
This website includes answers to the Biological Inquiry questions found in the figure legends and all end-of-chapter questions.

46

THE MUSCULAR-SKELETAL SYSTEM AND LOCOMOTION

European tree frog (*Hyla arborea*) leaping from leaf.

In the previous three chapters, you learned how the nervous system gives animals the ability to respond to external and internal cues. Another key feature of the nervous system of animals is its ability to control muscle function. Muscles are composed of highly specialized cells that have the ability to contract in response to stimuli. The three types of muscle—skeletal, smooth, and cardiac—were introduced in Chapter 40. In this chapter, you will learn about the structure and function of skeletal muscle, and how skeletal muscle controls **locomotion**, the movement of an animal from place to place.

For skeletal muscles to produce locomotion, they must exert their actions on an animal's skeleton. We begin the chapter, therefore, with a description of the types of animal skeletons. Then we examine the structure and function of skeletal muscle. Next, we consider the various modes of locomotion in animals. We conclude with a discussion of a few important bone and muscle diseases in humans.

46.1 Types of Animal Skeletons

When we think of the word skeleton, an image of the vertebrate system of bones usually comes to mind. However, invertebrates possess a skeleton as well, but their skeletons are not made of bone. It is better to think of a **skeleton** in a broader sense, therefore, as a structure or structures that serve one or more functions related to support, protection, and locomotion. Using this definition, the three types of skeletons found in animals are hydrostatic skeletons, exoskeletons, and endoskeletons. The first two are found in invertebrates, while endoskeletons are found in echinoderms, some sponges, and all vertebrates.

Hydrostatic Skeletons Consist of Fluid-Filled Body Compartments

Many soft-bodied invertebrates use hydrostatic pressure to support their bodies and generate movements. The combination of muscles and water in the body constitutes what is known as a **hydrostatic skeleton**. Water is nearly incompressible, which means that if an animal exerts a force on the water that fills its body cavities, it can use the hydrostatic pressure created to move the body, much like a balloon partly filled with water can be deformed by squeezing it along its length. In cnidarians such as hydras, for example, contractile cells in the body wall can exert a pressure on the water in the gastrovascular cavity. Depending on the direction of the force, a hydra's body and tentacles will either be elongated or shortened by the movement of water into and out of different regions of the cavity.

Segmented worms, such as earthworms, move forward by passing a wave of muscular contractions along the length of their body, segment by segment (**Figure 46.1a**). The muscles are arranged in two orientations: circular and longitudinal. Contraction of circular muscles squeezes and elongates the body, while contraction of the longitudinal muscles shortens and widens it. The worm is "squeezed" forward by hydrostatic pressures created along its body by the muscles. Bristles along its body surface help the animal grip the ground to prevent backsliding.

Exoskeletons Are on the Outside of an Animal's Body

Arthropods have an **exoskeleton**, an external skeleton that surrounds and protects most of the body surface (**Figure 46.1b**).

Exoskeletons provide support for the body, protection from the environment and predators, and protection for internal organs. The skeleton is made of a polysaccharide called chitin, and in crustaceans it is sometimes strengthened with calcium and other minerals. Exoskeletons are often tough, durable, and segmented to allow movement, but to allow growth they must be shed and then regrown (and strengthened again). Consequently, one disadvantage of exoskeletons is that when an animal must shed its exoskeleton to grow, it is temporarily more vulnerable to predators and the environment.

Exoskeletons vary enormously in their complexity, thickness, and durability. The differences in exoskeletons are usually adaptations that enhance an animal's survival. Think, for example, of the difference between the wing of a butterfly and the outer skeleton of a lobster. A butterfly's exoskeleton must be light enough for the animal to fly, while the thick, tough exoskeleton of the lobster provides a very effective defense against

predators in the sea. Exoskeletons may seem primitive compared to the endoskeletons of vertebrates (discussed next), particularly because they must be completely shed during growth periods. Even so, insects are among the most successful of all classes of animals living today, having survived for hundreds of millions of years and inhabiting nearly every possible ecological niche on the planet. Clearly, exoskeletons have been sufficient for one of the planet's greatest success stories.

Endoskeletons Are Internal Support Structures

Endoskeletons are functionally similar to exoskeletons in that both types of skeletons provide support and protection. Unlike exoskeletons, however, endoskeletons are internal structures and do not protect the body surface, only the internal organs and other structures.

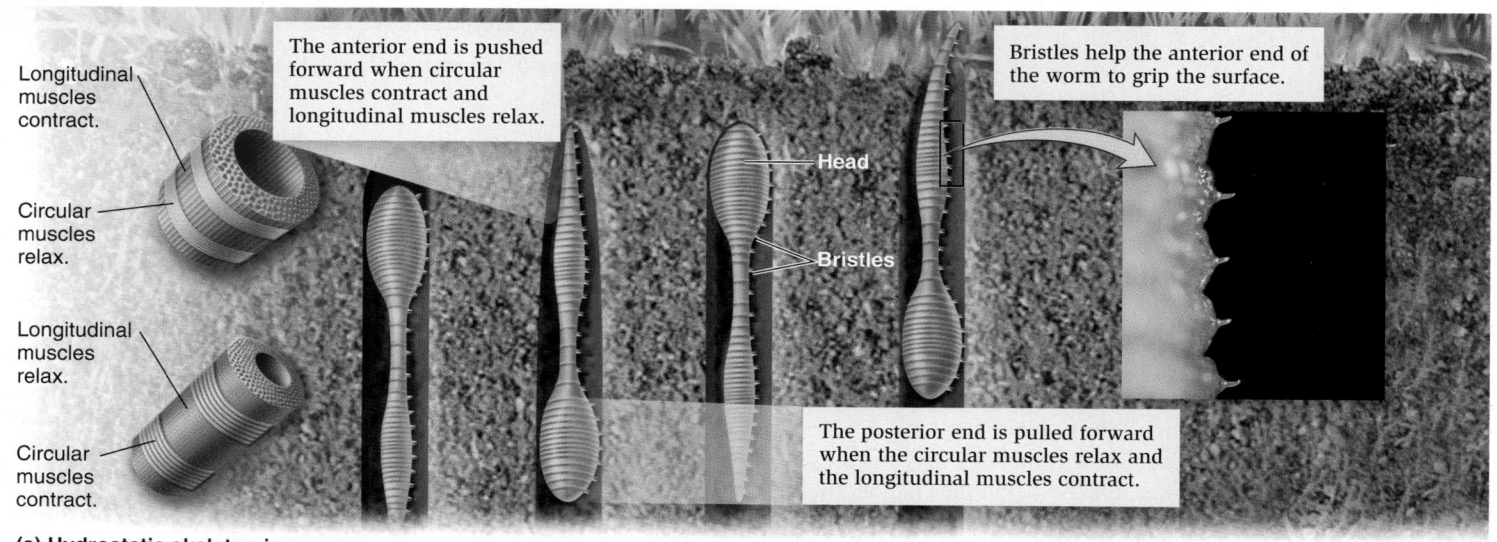

Longitudinal muscles contract.

Circular muscles relax.

Longitudinal muscles relax.

Circular muscles contract.

The anterior end is pushed forward when circular muscles contract and longitudinal muscles relax.

Head

Bristles

Bristles help the anterior end of the worm to grip the surface.

The posterior end is pulled forward when the circular muscles relax and the longitudinal muscles contract.

(a) Hydrostatic skeleton in a worm

(b) Exoskeleton in an arthropod

(c) Endoskeleton in an echinoderm

Figure 46.1 **Types of skeletons.** (a) Hydrostatic skeleton. By alternately contracting and relaxing circular and longitudinal muscles, earthworms use hydrostatic pressure to achieve locomotion. Clinging bristles along the body surface help prevent backsliding. (b) Exoskeleton. An arthropod's skeleton covers and protects its body, but it must be periodically shed and replaced to allow growth. (c) Endoskeleton. Echinoderms such as the sea star (starfish) have endoskeletons of bony plates made of calcium carbonate (shown in inset).

Biological inquiry: Endoskeletons do not provide protection for the body surface of animals. Can the lack of an external skeleton also be an advantage for animals?

Endoskeletons are found in some species of sponges, echinoderms, and all vertebrates. Minerals such as calcium, magnesium, phosphate, and carbonate supply the hardening material that gives the skeleton its firm structure. The endoskeletons of sponges and echinoderms consist of spiky networks of proteins and minerals or mineralized platelike structures. Beneath the body surface of echinoderms, for example, arrays of mineralized plates made largely of calcium carbonate extend into the spines and arms that radiate from the main body (**Figure 46.1c**).

Vertebrate skeletons, by contrast, are composed of either cartilage or bone. Next we consider the structure and function of the vertebrate skeleton in greater detail.

46.2 The Vertebrate Skeleton

Vertebrate skeletons are composed of either cartilage (see Chapter 40)—as in elasmobranchs—or both cartilage and bone, as in bony fish, amphibians, reptiles, birds, and mammals. First we consider the skeleton as a whole before examining bones in greater detail.

The Vertebrate Skeleton Performs Several Important Functions

In the vertebrate skeleton, bones are connected in ways that allow for support and movement. The vertebrate endoskeleton is often considered in two parts (**Figure 46.2**). The axial skeleton is composed of the bones that form the main longitudinal axis of an animal's body. The appendicular skeleton consists of the limb bones and the bones that connect the limb bones to the axial skeleton. A **joint** is formed where two or more bones come together. Some joints permit free movement (for example, the shoulders), while others allow no movement (fixed joints like those of the skull bones) or only limited movement (such as those of the vertebral column). Figure 46.2 illustrates the three types of freely movable joints: pivot, hinge, and ball-and-socket joints. Each of these allows a different type of movement.

The skeleton of vertebrates serves several functions in addition to support and locomotion (**Table 46.1**). For example, blood cells and platelets, the latter of which help blood to clot (see Chapter 47), are formed within the inner part of certain bones. In addition, calcium and phosphate homeostasis is achieved in large part through exchanges of these ions between bone and blood.

Figure 46.2   **The vertebrate endoskeleton.**   The axial skeleton is shown in green, and the appendicular skeleton in tan. The human skeleton consists of 206 separate bones. Examples of movable joints are shown.

Table 46.1	Functions of the Vertebrate Skeleton

Provide structural framework to support the body

Protect delicate internal organs like the heart and lungs

Provide a lever-like system for muscle attachment, thus enabling locomotion and movement of body parts and external objects

House and nurture immature blood cells

Provide a reservoir of calcium and phosphate ions that can be exchanged with the blood to maintain the circulating concentrations of these ions within a narrow, homeostatic range

For example, if dietary intake of calcium is low, calcium is removed from bone and added to the blood, so that all of the vital cellular activities that depend on calcium, such as nerve and muscle function, can continue to operate normally. When dietary calcium is restored to normal, calcium is redeposited in bone. In fact, about 99% of all the calcium in a typical vertebrate's body exists in bone. This represents a huge reservoir of calcium ions for the blood.

Bone Consists of a Mixture of Organic and Mineral Components

Bone is a living, dynamic tissue with both organic and mineral components. Organic materials include cells that form bone and cells that break it down, and fibers made from the protein collagen, whose unique triple helical structure gives bone both strength and flexibility. The mineral component is composed of a crystalline mixture of Ca^{2+}, PO_4^{2-}, and other ions that provides bone its rigidity. These ions must be obtained in an animal's diet, absorbed into the blood, and deposited in bone.

A proper ratio of the organic and mineral components is required for normal bone function. Too little mineral, for example, makes bone too easily fractured. Bone is formed at high rates during an animal's growth periods, but even in adulthood bone continues to be continuously formed, broken down, and re-formed. In fact, the skeleton is continually changing. The one in your body right now is not identical to the one in your body a few years ago, or the one you will have a few years from now.

Bones cannot move by themselves but instead provide the scaffold upon which muscles act to cause body movement. We turn now to a discussion of skeletal muscle and the mechanism by which it generates force.

46.3 Skeletal Muscle Structure and the Mechanism of Force Generation

As introduced in Chapter 40, vertebrates have three types of muscle that are classified according to their structure, function,

and control mechanisms. These include cardiac muscle, smooth muscle, and skeletal muscle. Cardiac muscle is found only in the heart and provides the force required for the heart to pump blood. Cardiac muscle will be covered in Chapter 47. Smooth muscle surrounds and forms the outer lining of hollow organs and tubes, including those of the gastrointestinal system, urinary bladder, uterus, blood vessels, and airways. Contraction of the smooth muscle in hollow organs may propel forward or help churn up the luminal contents of the organ, as when the stomach contracts after a meal. In other cases, smooth muscle regulates the flow of substances by changing the tube diameter, as in the widening or narrowing of small blood vessels. Smooth muscle contraction is not under voluntary control. Instead, it is controlled by the autonomic nervous system, hormones, and local chemical signals. Some smooth muscles contract on their own, even in the absence of such signals.

Skeletal muscle is found throughout the body and is directly involved in locomotion. In vertebrates, but not invertebrates, skeletal muscle is excitable. In vertebrates it can generate propagating action potentials in response to a stimulus. The muscle action potentials result in increases in cytosolic Ca^{2+}, which triggers force generation. Before understanding how that is possible, however, let's begin with an overview of skeletal muscle structure and function.

A Skeletal Muscle Is a Contractile Organ That Supports and Moves Bones

A **muscle** is a grouping of cells—called **muscle fibers**—bound together by a succession of connective tissue layers (**Figure 46.3a**). Skeletal muscles are usually linked to bones by bundles of collagen fibers known as tendons. The transmission of force from contracting muscle to bone can be likened to a number of people pulling on a rope attached to a heavy object. Each person corresponds to a single muscle fiber, the rope corresponds to the tendons, and the bone is the heavy object.

Some tendons are very long, with the site of tendon attachment to bone far removed from the end of the muscle. For example, some of the muscles that move the fingers are in the forearm. Wiggle your fingers and feel the movement of the muscles in your lower arm. These muscles are connected to the fingers by long tendons.

Skeletal muscle fibers increase in size during growth from infancy to adulthood, but no new fibers are formed during that time. As described later, enlargement of skeletal muscles in adults, as from weight lifting, is also a function of enlarged fibers, not the formation of new fibers.

Muscle Fibers Contain Myofibrils Composed of Arrays of Filaments

The most striking microscopic feature of skeletal muscle is a series of light and dark bands perpendicular to the muscle's long axis. Because of this characteristic striped pattern, it is also

Figure 46.3   Skeletal muscle structure.   (a) Skeletal muscles attach to bone by tendons, which are bundles of collagen fibers. Each muscle consists of bundles of muscle fibers (cells) bound together by connective tissue. **(b)** Each fiber consists of numerous myofibrils containing thick and thin filaments, the arrangement of which produces the striated banding pattern. Each repeating unit of the arrayed filaments constitutes a sarcomere.

known as **striated muscle**. This pattern results from the presence of cylindrical bundles known as **myofibrils**, each of which contains numerous thick and thin filaments (**Figure 46.3b**). Most of the cytosol of a muscle fiber is filled with myofibrils, each of which extends from one end of the cell to the other and is linked to the tendons at the ends of the fiber.

Within each myofibril, thick and thin filaments are arranged in a repeating pattern running the length of the myofibril. One complete unit of this repeating pattern is known as a **sarcomere** (from the Greek, *sarco*, muscle, and *mer*, part). The **thick filaments** are composed almost entirely of the motor protein **myosin**.

The **thin filaments**, which are about half the diameter of the thick filaments, contain the cytoskeletal protein **actin**, as well as two other proteins—troponin and tropomyosin—that play important roles in regulating contraction.

The thick filaments are located in the middle of each sarcomere, where their orderly parallel arrangement produces a wide, dark band known as the **A band** (Figure 46.3b). Each sarcomere contains two sets of thin filaments, each set anchored to a network of proteins at the **Z line**. The thin filaments overlap a portion of the thick filaments. Two successive Z lines define the limits of one sarcomere.

A light band known as the **I band** lies between the A bands of two adjacent sarcomeres. Each I band contains those portions of the thin filaments that do not overlap the thick filaments, and each I band is bisected by a Z line.

Two additional features are present in the A band of each sarcomere. The **H zone** is a narrow, light region in the center of the A band. It corresponds to the space between the two sets of thin filaments in each sarcomere. A narrow, dark band in the center of the H zone, known as the **M line**, corresponds to proteins that link together the central regions of adjacent thick filaments.

The spaces between overlapping thick and thin filaments are bridged by projections known as **cross-bridges**, which are regions of myosin molecules that extend from the surface of the thick filaments toward the thin filaments (Figure 46.3b). As we see next, during muscle contraction the cross-bridges make contact with the thin filaments and exert force on them.

Skeletal Muscle Shortens When Thin Filaments Slide over Thick Filaments

As used in muscle physiology, the term contraction does not necessarily mean shortening. Instead it refers to activation of the cross-bridges within muscle cells. Holding a heavy weight at a constant position, for example, requires muscle contraction, but muscle shortening does not occur. When the mechanisms that initiate force generation are turned off, contractions end, allowing relaxation of the muscle fiber.

The way in which a muscle fiber actually shortens is known as the **sliding filament mechanism** of muscle contraction (**Figure 46.4**). In muscle contraction, the sarcomeres shorten, but there is no change in the lengths of either the thick or thin filaments themselves. Instead, the thick filament remains stationary while the thin filaments actually slide. This movement is propelled by the cross-bridges.

During shortening, each myosin cross-bridge attaches to a thin filament actin molecule and moves in a motion much like an oar on a boat. This motion of many cross-bridges forces the thin filaments toward the center of the sarcomere, thereby shortening the sarcomere. One stroke of a cross-bridge produces only a very small movement of a thin filament relative to a thick filament. As long as a muscle fiber is stimulated to contract, however, each cross-bridge repeats its motion many times, resulting in continued sliding of the thin filaments. Thus, the ability of a muscle fiber to generate force and movement depends on the amount of interaction of actin and myosin.

Each actin molecule of a thin filament contains a binding site for myosin. Actin molecules form polymers and are arranged into two intertwined helical chains that are closely associated with molecules of troponin and tropomyosin (**Figure 46.5a**). Myosin, by contrast, is composed of six protein subunits: two

(a)

(b)

Figure 46.5 **Structure and function of the thick and thin filaments.** (a) Thin filaments are composed of two intertwined actin molecules and associated proteins tropomyosin and troponin. Thick filaments are made of the protein myosin, which has two intertwined heavy chains, each with two associated light chains, an actin-binding site, and an ATP-binding site. The end of a myosin molecule that contains the actin-binding sites is bent at an angle to form the globular heads. (b) When the cross-bridges on myosin molecules bind to actin filaments, the thin filaments are pulled toward each other, shortening the sarcomere.

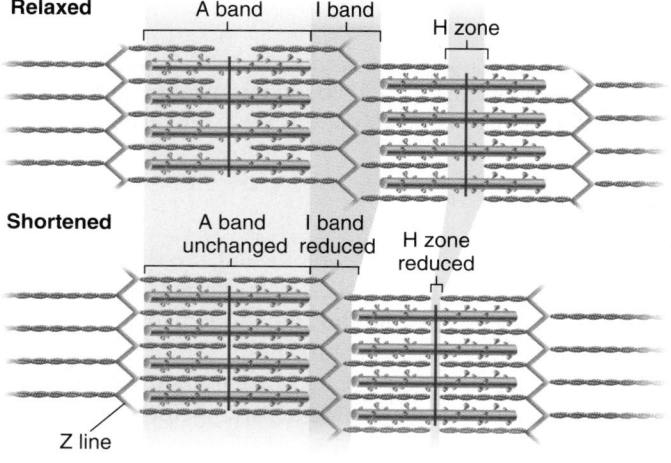

Figure 46.4 **Sliding filament mechanism of muscle contraction.** The sliding of thin filaments past the overlapping thick filaments shortens the sarcomere but does not change the lengths of the filaments themselves.

large, intertwined heavy chains and four smaller light chains. These polypeptides combine to form a protein that consists of two heads and a long tail formed by the two intertwined heavy chains. The tail of each myosin molecule lies along the axis of the thick filament, and the two globular heads extend out to the sides, forming the cross-bridges. Each globular head contains two binding sites, one for actin and one for ATP. The hydrolysis of ATP provides the energy for the cross-bridge (that is, the myosin head) to move.

The myosin proteins at the two ends of each thick filament are oriented in opposite directions, such that their tail ends are directed toward the center of the filament (**Figure 46.5b**). Because of this arrangement, the movements of the cross-bridges bring the attached thin filaments at the two ends of the sarcomere toward the center of the sarcomere during shortening.

The protein myosin was first discovered in extracts of frog leg muscle in the 1860s by the German physiologist Willi Kühne. We now know that eukaryotic genomes encode a family of related myosin proteins, as described next.

GENOMES & PROTEOMES

Did an Ancient Mutation in Myosin Play a Role in the Development of the Human Brain?

Myosins are among the most ancient of eukaryotic proteins, and also among the most highly conserved in amino acid sequence. Nonetheless, small differences in the sequences of genes that arose from a single primordial myosin gene have led to tissue-specific expression of various myosin proteins, each with a characteristic ability to bind actin and hydrolyze ATP. One recently discovered member of this gene family, called *MYH16*, codes for a heavy chain of a myosin molecule that is expressed mainly in the muscles of the jaw in primates. In 2004, H. Stedman and colleagues discovered that this gene, although present in humans, did not code for a functional protein because of a deletion of two base pairs in the gene. This mutation was present in 100% of people tested worldwide but absent from eight nonhuman primate species tested, including chimpanzees.

Based on genetic comparisons and estimates of genetic divergence among species, the researchers estimated that the mutation occurred approximately 2.4 million years ago. Significantly, the genus *Homo*, with its smaller jaw and larger braincase, is also thought to have first appeared about 2.4 million years ago, leading Stedman to suggest that the loss of the MYH16 protein led to the smaller, less-muscular jaws characteristic of humans (**Figure 46.6**). Because jaw muscles are anchored to skull bones, massive jaw muscles may have placed a mechanical constraint on skull growth in early hominids (as well as in modern nonhuman primates). Smaller, less-muscular jaws would place less strain on the forming skull and allow the brain to grow larger, expanding the braincase to the larger size seen in modern humans compared to our early ancestors.

Macaque   **Gorilla**   **Human**

Figure 46.6   Comparison of jaw size in modern primates. The jaws of modern primates, such as the macaque (*left*) and the gorilla (*center*), are much more massive than the jaw of humans (*right*). Stedman and colleagues proposed that the reduction in jaw size in humans resulted from a mutation in a gene coding for one form of myosin expressed in jaw muscles, and that jaw reduction subsequently led to the development of the enlarged braincase of humans. Note: Skulls are not to scale.

Stedman's hypothesis is compelling given that the mutation occurs in a myosin protein expressed predominantly in jaw muscles and is present only in humans and not in other primates tested thus far. Recently, though, another research group using complex statistical analyses of gene sequences in chimps and humans concluded that the mutation may have arisen earlier—perhaps 4.0 to 5.4 million years ago, prior to the appearance of small-jawed species of *Homo*. Stedman's hypothesis will require confirmation by future research but is at present an intriguing piece of the puzzle of how modern humans evolved.

The Cross-Bridge Cycle Requires ATP and Calcium Ions

Let's now turn our attention to how actin and myosin interact to promote contraction. The sequence of events that occurs between the time when a cross-bridge binds to a thin filament and when it is set to repeat the process is known as a **cross-bridge cycle**.

The chemical and physical events during the four steps of the cross-bridge cycle are illustrated in **Figure 46.7**. When ATP is hydrolyzed, the myosin cross-bridges are said to be in an energized state in which the hydrolysis products ADP and inorganic phosphate remain bound to myosin. This storage of energy in myosin is analogous to the storage of potential energy in a stretched spring. In this conformation, the myosin head is cocked backward.

Cross-bridge cycling begins whenever the Ca^{2+} level increases inside the cell, usually when neural input results in the release of calcium from intracellular storage sites (described in detail shortly). The four-step cycle begins with the binding of an energized myosin cross-bridge, with its associated ADP and P_i, to an actin molecule on a thin filament.

Step 1: Cross-bridge binds to actin.

$$\text{Actin} + \text{Energized Myosin} \cdot \text{ADP} \cdot \text{P}_i$$
$$\longrightarrow \text{Actin} \cdot \text{Myosin} \cdot \text{ADP} \cdot \text{P}_i \text{ complex}$$

Step 2: Cross-bridge moves and the thin filaments move.

The binding of an energized myosin cross-bridge to actin triggers the movement of the bound cross-bridge (sometimes called the **power stroke**) and the release of P_i. The release of P_i causes the cross-bridge to rotate forward, which moves the actin filament and releases ADP.

$$\text{Actin} \cdot \text{Myosin F} \cdot \text{ADP} \cdot \text{P}_i \text{ complex}$$
$$\longrightarrow \text{Actin} \cdot \text{Myosin complex} + \text{ADP} + \text{P}_i$$

This sequence of storage and release of energy by myosin is analogous to the operation of a mousetrap: Energy is first stored in the trap by cocking the spring (ATP hydrolysis, step 4) and then released by the springing of the trap (binding to actin and power stroke, steps 1 and 2). During the power stroke, the actin filament moves.

Step 3: ATP binds to myosin, causing the cross-bridge to detach.

During the cross-bridge movement in step 2, myosin is bound very firmly to actin, and this linkage must be broken to allow the cross-bridge to be re-energized and repeat the cycle. The binding of a new molecule of ATP to myosin breaks the link between actin and myosin.

$$\text{Actin} \cdot \text{Myosin complex} + \text{ATP} \longrightarrow \text{Actin} + \text{Myosin} \cdot \text{ATP}$$

The dissociation of actin and myosin by ATP is an example of allosteric regulation (see Chapter 7) of protein activity. The binding of ATP at one site on myosin decreases myosin's affinity for actin bound at another site. Note that ATP is not hydrolyzed in this step. Instead, ATP acts here only as an allosteric modulator of the myosin head that weakens the binding of myosin to actin.

Following the dissociation of actin and myosin, the ATP bound to myosin is then hydrolyzed (step 4), thereby re-forming the energized state of myosin.

Step 4: Hydrolysis of ATP re-energizes the cross-bridge.

$$\text{Actin} + \text{Myosin} \cdot \text{ATP} \longrightarrow \text{Actin} + \text{Energized Myosin} \cdot \text{ADP} \cdot \text{P}_i$$

If calcium is still present at this time, the cross-bridge can reattach to a new actin molecule in the thin filament, and the cross-bridge cycle repeats, causing further shortening of the muscle fiber.

As we have seen, ATP performs two different roles in the cross-bridge cycle. The energy released from ATP hydrolysis provides the energy for cross-bridge movement, and ATP binding breaks the link formed between actin and myosin during the cycle, allowing the cycle to be repeated. Although the precise mechanisms may differ slightly between vertebrates and invertebrates, all skeletal muscle is believed to function according to steps similar to those just described, and to require calcium ions, the actions of which we describe next.

Figure 46.7   **The four stages of cross-bridge cycling in skeletal muscle.**

Biological inquiry: Rigor mortis is a condition of contracted skeletal muscle that occurs shortly after death. Why would the muscle remain contracted after death, and why does such contraction eventually stop?

The Regulation of Muscle Contraction by Calcium Is Mediated by Tropomyosin and Troponin

How does the presence of calcium in the muscle fiber cytosol regulate the cycling of cross-bridges? The answer requires a closer look at the two additional thin filament proteins mentioned earlier, tropomyosin and troponin.

Tropomyosin is a rod-shaped molecule composed of two intertwined proteins, with a length equal to that of about seven actin molecules (**Figure 46.8a**). Chains of tropomyosin molecules are arranged end to end along the actin thin filament. In the absence of calcium ions, they partially cover the myosin-binding site on each actin molecule, thereby preventing cross-bridges from making contact with actin. Each tropomyosin molecule is held in this blocking position by **troponin**, a smaller, globular-shaped protein that is bound to both tropomyosin and actin. One molecule of troponin binds to each molecule of tropomyosin. In this way, troponin and tropomyosin cooperate to block access to myosin-binding sites on actin molecules in the relaxed muscle fiber.

Troponin is capable of binding calcium ions. The binding of Ca^{2+} produces a change in the shape of troponin, which—through troponin's linkage to tropomyosin—allows tropomyosin to move away from the myosin-binding site on each actin molecule (**Figure 46.8b**). This permits cross-bridge cycling to occur. Conversely, removal of Ca^{2+} from troponin reverses the process, turning off contractile activity.

Thus, the concentration of cytosolic Ca^{2+} determines the number of ions bound to troponin molecules, which in turn determines the number of actin sites available for cross-bridge binding. The cytosolic concentration of Ca^{2+}, however, is very low in resting muscle. How is the Ca^{2+} concentration increased so that contraction can occur? This topic is discussed next.

Contraction of Skeletal Muscle Is Coupled with Electrical Excitation

Like neurons, skeletal muscle cells are capable of generating and propagating action potentials in response to an appropriate stimulus. This causes a rise in the concentration of cytosolic Ca^{2+}, which triggers contraction of a muscle fiber. The sequence of events by which an action potential in the plasma membrane of a muscle fiber leads to cross-bridge activity by the mechanisms just described is called **excitation-contraction coupling**. The electrical activity in the plasma membrane does not directly act upon the contractile proteins but instead is the stimulus that increases cytosolic calcium concentration, which continues to activate the contractile apparatus long after electrical activity in the membrane has ceased.

The source of the increased cytosolic calcium that occurs following a muscle action potential is the fiber's **sarcoplasmic reticulum**, which acts as a Ca^{2+} reservoir. The sarcoplasmic reticulum, which is a specialized form of the endoplasmic reticulum found in most cells, is composed of interconnected sleeve-like compartments around each myofibril (**Figure 46.9**). Separate tubular structures, the **transverse tubules (T-tubules)**, are invaginations of the plasma membrane that open to the extracellular fluid. The T-tubules run around each myofibril and conduct action potentials from the outer surface of the muscle fiber to inner regions where the myofibrils are found. The action potential causes the opening of calcium channels in the sarcoplasmic reticulum, which allows Ca^{2+} to flow into the cytosol and bind to troponin.

A contraction continues until calcium is removed from troponin. This is achieved by ion pumps in the sarcoplasmic reticulum that lower the calcium concentration in the cytosol back to its resting level.

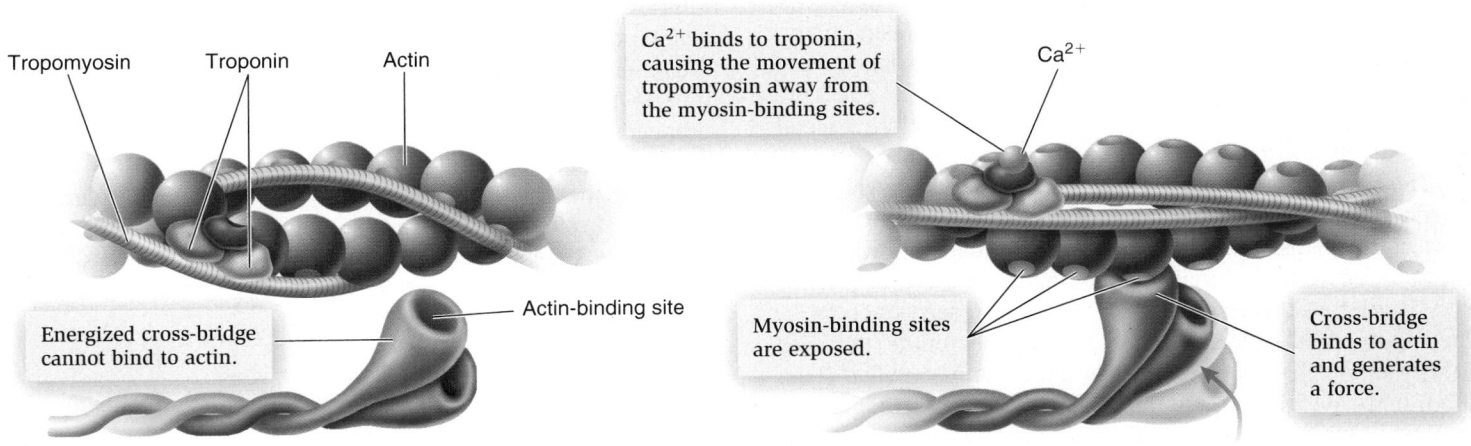

(a) Low cytosolic calcium, relaxed muscle

Tropomyosin   Troponin   Actin

Energized cross-bridge cannot bind to actin.

Actin-binding site

(b) High cytosolic calcium, activated muscle

Ca^{2+} binds to troponin, causing the movement of tropomyosin away from the myosin-binding sites.

Ca^{2+}

Myosin-binding sites are exposed.

Cross-bridge binds to actin and generates a force.

Figure 46.8 Role of calcium, tropomyosin, and troponin in cross-bridge cycling. (a) When cytosolic Ca^{2+} is low, the cross-bridge myosin-binding sites on actin are blocked by tropomyosin. (b) When cytosolic Ca^{2+} increases, Ca^{2+} binds to troponin, which in turn allows tropomyosin to move away from the myosin-binding sites on actin.

Opening of transverse tubule to extracellular fluid

Muscle fiber plasma membrane

1 Action potentials are propagated through T tubules.

Muscle fiber plasma membrane

Sarcoplasmic reticulum

2 Ca²⁺ channels open in response to depolarization.

T tubule

Ca²⁺

Ca²⁺

Cytosol

ATP   ADP + Pi

3 Ca²⁺ binds to troponin.

4 Ca²⁺ is returned to sarcoplasmic reticulum by Ca²⁺ pumps.

Ca²⁺

Myofibrils

Cytosol

Mitochondrion

Transverse tubules

Sarcoplasmic reticulum

Figure 46.9 **Arrangement of the sarcoplasmic reticulum, transverse tubules, and myofibrils in a single skeletal muscle fiber.** Action potentials propagate along the plasma membrane and down the transverse tubules. Inset: The depolarization produced by the action potentials opens Ca^{2+} channels in the membranes of the sarcoplasmic reticulum, through which Ca^{2+} diffuses into the cell cytosol. Ca^{2+} then binds to troponin, initiating muscle contraction. Ca^{2+} is then pumped back into the sarcoplasmic reticulum by ATP-dependent ion pumps. This results in muscle relaxation.

Electrical Stimulation of Skeletal Muscle Occurs at the Neuromuscular Junction

We have just seen that an action potential in the plasma membrane of a skeletal muscle fiber is the signal that triggers contraction. How are these action potentials initiated? The only mechanism by which action potentials are initiated in a skeletal muscle involves stimulation by a motor neuron.

The junction of a motor neuron's axon and the muscle fiber is known as a **neuromuscular junction** (**Figure 46.10**). Near the surface of the muscle fiber, the axon divides into several short processes, or terminals, containing synaptic vesicles filled with the neurotransmitter acetylcholine (ACh; see Chapter 43). The region of the muscle fiber plasma membrane that lies directly under the axon terminal is folded into so-called junctional folds, which increase the total surface area available for the membrane to respond to ACh.

When an action potential in a motor neuron arrives at the axon terminal, it releases its stored ACh, which traverses the synapse and binds to receptors in the junctional folds of the muscle fiber. The ACh receptor is a ligand-gated ion channel (see

Chapter 5). The result of ACh binding is an influx of Na^+ into the muscle cell through a channel created by the receptor. This influx causes depolarization of the muscle cell, resulting in an action potential that spreads over the surface of the muscle fiber. Most neuromuscular junctions are located near the middle of a muscle fiber, and newly generated muscle action potentials propagate from this region in both directions toward the ends of the fiber and throughout the T-tubule network.

Now that we have learned about the structural characteristics of skeletal muscle, and the events that initiate and produce force generation, we turn to a discussion of how skeletal muscle is adapted to meet the varied functional demands of vertebrates.

46.4 Skeletal Muscle Function

Animals use skeletal muscle for a wide variety of activities, such as locomotion, stretching, chewing, breathing, and maintaining balance, to name a few. Thus it is not surprising that not all

(a)

(b)

Myelin

Motor neuron

Axon terminal

Axon terminal

Synaptic cleft

Synaptic vesicle containing ACh

ACh receptor

ACh

Plasma membrane of muscle fiber

Junctional folds with ACh receptors

Figure 46.10   The neuromuscular junction.   The structure of a neuromuscular junction **(a)** as seen in a scanning electron micrograph, and **(b)** as depicted schematically. Action potentials in the motor neuron cause exocytosis of ACh-containing synaptic vesicles. ACh binds to receptors in the plasma membrane of the junctional folds of the skeletal muscle cell. This initiates an action potential in the muscle cell.

The second means of classifying skeletal muscle fibers relates to the type of metabolic pathways available for synthesizing ATP. (See Chapter 7 for a review of the biochemical processes associated with ATP production.) Fibers that contain numerous mitochondria and have a high capacity for oxidative phosphorylation are classified as **oxidative fibers**. Most of the ATP production by such fibers depends on blood flow to deliver oxygen and nutrients to the muscle. Not surprisingly, therefore, these fibers are surrounded by many small blood vessels. They also contain large amounts of the oxygen-binding protein **myoglobin**, which increases the availability of oxygen in the fiber by providing an intracellular reservoir of oxygen. The large amounts of myoglobin present in oxidative fibers give these fibers a dark-red color. Oxidative fibers are often referred to as red muscle fibers.

By contrast, **glycolytic fibers** have few mitochondria but possess both a high concentration of glycolytic enzymes (see Chapter 7 for a discussion of glycolysis) and large stores of glycogen, the storage form of glucose. Corresponding to their limited use of oxygen, these fibers are surrounded by relatively few blood vessels and contain little myoglobin. The lack of myoglobin is responsible for the pale color of glycolytic fibers and their designation as white muscle fibers.

On the basis of these two characteristics, three major types of skeletal muscle fibers have been distinguished:

1. **Slow-oxidative fibers** have low rates of myosin ATP hydrolysis but have the ability to make large amounts of ATP. These fibers are used for prolonged, regular activity, such as flight, long-distance swimming, or the maintenance of posture. These muscles, for example, are what give the red color to the dark meat of ducks, which use the muscles for flight. Long-distance runners have a high proportion of these fibers in their leg muscles. These types of activities require muscles that do not fatigue easily.
2. **Fast-oxidative fibers** have high myosin ATPase activity and can make large amounts of ATP. Like slow-oxidative fibers, these fibers do not fatigue quickly and can be used for long-term activities. They are particularly suited for rapid actions, such as the rapid trilling sounds made by the throat muscles in songbirds or the clicking sounds of a rattlesnake's shaking tail.
3. **Fast-glycolytic fibers** have high myosin ATPase activity but cannot make as much ATP as oxidative fibers, because their source of ATP is glycolysis. These fibers are best suited for rapid, intense actions, such as a short sprint at maximum speed. Fast-glycolytic fibers fatigue more rapidly than oxidative fibers. The breast meat of chickens, for example, appears white because unlike ducks, chickens do not fly except for very short distances, and therefore do not require oxidative pectoral muscles.

Different muscle groups within the body have different proportions of each fiber type interspersed with one another, and many activities require the action of all three types of fibers at once. This is important when you consider the wide range of animal activities related to locomotion alone, including walking,

skeletal muscle fibers share the same mechanical and metabolic characteristics. In this section, we will consider how different types of fibers can be classified on the basis of their rates of shortening (as either fast or slow), and the way in which they produce the ATP needed for contraction (as oxidative or glycolytic).

Skeletal Muscle Fibers Are Adapted for Different Activities

Different muscle fibers contain forms of myosin that differ in the maximal rates at which they can hydrolyze ATP. This, in turn, determines the maximal rates of cross-bridge cycling and muscle shortening. Fibers containing myosin with high ATPase activity are classified as **fast fibers**. Those containing myosin with lower ATPase activity are considered **slow fibers**. Although the rate of cross-bridge cycling is about four times faster in fast fibers than in slow fibers, the maximal force produced by both types of cross-bridges is approximately the same.

climbing, running, crawling, crouching, jumping, swimming, flying, and maintaining balance and posture. Depending on the needs of an animal at any given moment, the motor nerve inputs can be adjusted to stimulate different ratios of fiber types. When a person lifts a heavy weight for a brief time, the motor units containing fast-glycolytic fibers in the arm muscles are activated in large numbers. When a crab uses its pincers to grab prey, fast-glycolytic muscles snap the claws closed quickly, but then slow-oxidative fibers maintain a tight, relentless grip for as long as needed. The characteristics of the three types of skeletal muscle fibers are summarized in **Table 46.2**.

Muscles Adapt to Exercise

The regularity with which a muscle is used, as well as the duration and intensity of its activity, affects the properties of the muscle. Increased amounts of contractile activity—in other

words, exercise—can produce an increase in the size of muscle fibers as well as changes in their capacity for ATP production. Because the number of fibers in a muscle remains essentially constant throughout adult life, the increases in muscle size that occur with exercise do not result from increases in the number of muscle fibers, but instead from increases in the size of each fiber.

Exercise of relatively low intensity but long duration—popularly called aerobic exercise, including running and swimming—produces increases in the number of mitochondria in the fibers that are needed in this type of activity. In addition, the number of blood vessels around these fibers increases to supply the greater energy demands of active muscle. All of these changes increase endurance with a minimum of fatigue.

By contrast, short-duration, high-intensity exercise, such as weight lifting, primarily affects fast-glycolytic fibers, which are used during strong contractions. These fibers undergo an increase in fiber diameter due to the increased synthesis of actin and myosin filaments, which form more myofibrils. In addition, glycolytic activity is enhanced by elevated synthesis of glycolytic enzymes. The results of such high-intensity exercise are the increased strength and bulging muscles of a conditioned weight lifter. Such muscles, although very powerful, have little capacity for endurance and therefore fatigue rapidly.

A decline or cessation of muscular activity results in the condition called **atrophy**, a reduction in the size of the muscle. Likewise, if the neurons to a skeletal muscle are destroyed or the neuromuscular junctions become nonfunctional, the denervated muscle fibers will become progressively smaller in diameter. This condition is known as denervation atrophy. Even with an intact nerve supply, a muscle can atrophy if it is not used for a long period of time, as when a broken limb is immobilized in a cast.

The mechanism by which changes occur in skeletal muscle during exercise is an active area of research in biology, but a recent discovery has provided an intriguing clue, as described next.

Table 46.2	Characteristics of the Three Types of Skeletal Muscle Fibers		
	Slow-oxidative	Fast-oxidative	Fast-glycolytic
Primary source of ATP production	Oxidative phosphorylation	Oxidative phosphorylation	Glycolysis
Mitochondria	Many	Many	Few
Blood supply	High	High	Moderate
Myoglobin content	High (red muscle)	High (red muscle)	Low (white muscle)
Rate of fatigue	Slow	Intermediate	Fast
Myosin-ATPase activity	Low	High	High
Rate of contraction	Slow	Fast	Fast

FEATURE INVESTIGATION

Evans and Colleagues Activated a Gene to Produce "Marathon Mice"

In the course of investigating possible ways of reversing or preventing obesity in humans and other mammals, Ron Evans and his colleagues at the Salk Institute in California discovered one way in which the ratios of oxidative and glycolytic fibers change in skeletal muscle. Evans was interested in a gene that codes for a transcription factor called PPAR-delta. Activation of this protein results in the expression of genes that enable cells

to more efficiently burn fat instead of glucose for energy. Evans hypothesized that mice in which PPAR-delta was chronically activated at high levels would lose weight due to increased fat burning, as shown in **Figure 46.11**.

To test this hypothesis, Evans created transgenic mice (see Chapter 20) in which the *PPAR-delta* gene was introduced in a modified form. The modified gene had a skeletal muscle-specific promoter so that it would be expressed in skeletal muscle cells only. The coding sequence of the gene that encoded PPAR-delta was linked to a viral protein domain called VP16.

Figure 46.11 Activating a gene to produce "marathon mice."

HYPOTHESIS Upregulation of fat oxidation in cells will prevent obesity in mice.

STARTING MATERIALS Mice.

	Experimental level	**Conceptual level**

1 Experiment 1:

Prepare a modified gene containing a skeletal muscle promoter and a coding sequence that links *VP16* and *PPAR-delta*. See Chapter 20 for gene cloning methods

Skeletal muscle promoter

VP16   PPAR-delta

Skeletal muscle-specific promoter ensures gene is turned on only in skeletal muscle.

VP16 domain is a domain that always activates transcription.

2 Make transgenic mice expressing the *VP16–PPAR-delta* gene. See Chapter 20 for a discussion of gene addition.

All of the cells will carry this gene, but only skeletal muscle cells will express the gene.

3 Perform the following tests:

(a) Feed wild-type control mice and transgenic mice a normal-fat diet (4%), and then switch to a high-fat diet (35%). Weigh mice weekly.

(b) Examine the muscle fibers in the mice.

(c) Test their endurance on a treadmill.

Scale

Microscope

Treadmills

(a) Eating a high-fat diet is known to cause obesity in mice and other mammals.

(b) The appearance of skeletal muscle can be examined by light and electron microscopy.

(c) Treadmill is motorized to keep mice moving until they become exhausted, at which time treadmill is stopped.

4 THE DATA

(a) Weight gain

Characteristics of skeletal muscle in transgenic mice:

– redder than wild type
– more myoglobin
– more mitochondria
– more slow-oxidative fibers

(b) Difference in skeletal muscle

(c) Muscle endurance

This domain also facilitates gene activation. The combination of PPAR-delta and VP16 was expected to strongly activate genes that enable cells to more efficiently burn fat instead of glucose for energy. The modified gene was used to make transgenic mice that carried the gene in all of their cells and could therefore transmit the gene to their offspring. Evans then monitored the body weights of the mice after they reached adulthood. Transgenic mice gained significantly less weight compared to wild-type mice when fed high-fat diets (which normally cause mice—like humans—to gain weight), confirming Evans's hypothesis. As is often the case in scientific discovery, an unexpected finding arose from this study when Evans examined several tissues in these mice. Under the microscope, it was observed that the skeletal muscle of the transgenic mice showed a dramatic shift from glycolytic fibers to slow-oxidative fibers. The muscle in transgenic mice appeared redder than it did in wild-type mice. It contained more myoglobin and mitochondria and had higher levels of oxidative enzymes capable of providing the cells with sustained levels of ATP. These changes occurred even though the mice had not been subjected to exercise training.

Based on these observations, Evans also tested the hypothesis that the transgenic mice would have a greater capacity for prolonged exercise than wild-type mice. When the transgenic mice were challenged with an endurance exercise test, Evans discovered that they outperformed age- and weight-matched wild-type mice by a factor of nearly twofold! They could sustain a high level of activity on a miniature treadmill for nearly two times longer than wild-type mice (hence the nickname "marathon mice"). This effect occurred in transgenic mice even in the absence of prior exercise training. In other words, simply increasing the ratio of oxidative to glycolytic fibers gave the mice greater ability to sustain aerobic activity.

Elevating the levels of activated PPAR-delta facilitates an oxidative state that somehow signals skeletal muscle to convert fibers to types that are best suited for oxidative metabolism. This study suggested that the switch in fiber type that occurs normally in exercise training does not require exercise *per se*, and it may be mediated in part by proteins that activate or induce *PPAR-delta* expression. These results may have important implications for exercise endurance in humans, as well as for possible treatments for various muscle diseases.

Skeletal Muscles and Bones Create Lever Systems

A contracting muscle exerts a force on bones through its connecting tendons. When the force is great enough, the bone moves as the muscle shortens. A contracting muscle exerts only a pulling force, so that as the muscle shortens, the bones to which it is attached are pulled toward or away from each other. Muscles that bend a limb at a joint are called **flexors**, whereas muscles that straighten a limb are called **extensors**. Groups of muscles that produce oppositely directed movements at a joint are known as **antagonists**. For example, in **Figure 46.12** we can see that contraction of the hamstrings flexes the leg at the knee, whereas contraction of the antagonistic muscle, the quadriceps, causes the leg to extend. Both antagonistic muscles exert only a pulling force when they contract.

The muscles, bones, and joints in the body are arranged in lever systems. The basic principle of a lever is illustrated by the flexion of the arm by the biceps muscle (**Figure 46.13**), which exerts an upward pulling force on the forearm about 5 cm from the elbow joint. In this example, a weight held in the hand exerts a downward force about 35 cm from the elbow. When the biceps shortens 1 cm, the hand moves a distance of 7 cm. Since the muscle shortens 1 cm in the same amount of time that the hand moves 7 cm, the velocity at which the hand moves is seven times greater than the rate of muscle shortening. This lever system amplifies the velocity of muscle shortening so that short, relatively slow movements of the muscle produce faster movements of the hand. In a similar way, a pitcher can throw a baseball at 90–100 mph even though the arm muscles shorten at only a small fraction of this velocity.

Figure 46.12   Actions of flexors and extensors.   The figure shows how muscles cause flexion or extension of a limb in an animal with an endoskeleton. When the flexor muscle contracts, the extensor relaxes, and vice versa.

This concludes our discussion of the structure and function of skeletal muscle. We next consider the various strategies that different animals use to move through their environments.

46.5 Animal Locomotion

Animal locomotion occurs in only a limited number of forms, including swimming, flying, walking/running, crawling/sliding, or jumping, as in the frog shown in the chapter-opening photo. In all cases, animals experience certain constraints to locomotion. For example, all animals must overcome frictional forces (drag) generated by the air, water, or surface of the Earth. In addition, all forms of locomotion require energy to provide thrust, defined as the forward motion of an animal in any environment, and/or lift, which is movement against gravity.

Although the precise mechanism may differ among animals, locomotion always results from muscular contractions that exert force on one of the three types of skeletons discussed at the beginning of this chapter. In this section, we will examine the similarities and differences concerning locomotion in water, on land, or in air.

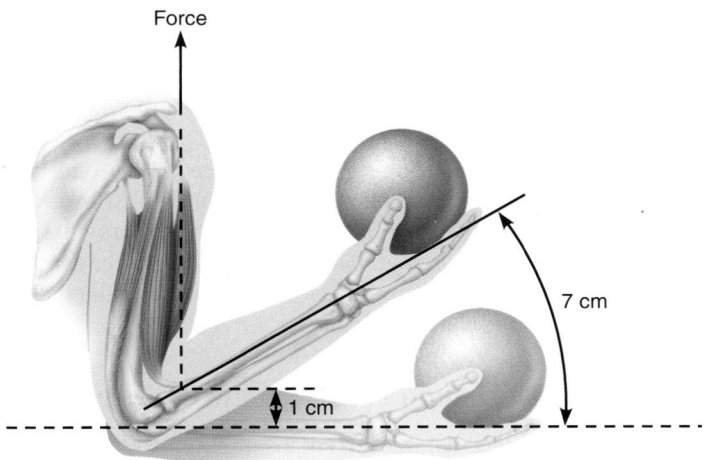

Figure 46.13 **The lever action of bones and skeletal muscles.**   Due to lever action, an object can be moved a greater distance than the distance the muscle shortens.

Aquatic Animals Must Overcome the Resistance of Water

The greatest challenge to locomotion in water is the density of water, which is much greater than that of air. This is apparent when you compare waving your hand through the air and underwater. The resistance to movement posed by the density of water increases exponentially as the speed of locomotion increases, which is one reason why many fish swim at relatively slow speeds. Overcoming this resistance requires considerable muscular effort. Most swimming animals, including fish, amphibians, reptiles, and marine mammals, have evolved streamlined bodies that reduce drag and make swimming more efficient (**Figure 46.14**). Other animals that swim only occasionally or spend time on the surface of the water have evolved adaptations, such as the webbed feet of ducks, that assist their muscles in generating greater thrust through the water.

An energetic advantage to swimming is that fish and other swimmers do not need to provide lift to overcome gravity. Because the density of water is similar to that of an animal's body, water provides buoyancy, which helps support the animal's weight.

The mechanism of swimming is similar among many different vertebrates. Most fish, for example, contract posterior skeletal muscles to move the tail end of the animal back and forth. This pushes water backward and propels the fish forward. Other fins provide additional thrust and enable changes in direction. Likewise, amphibians and marine reptiles rely predominantly on their hind legs for propulsion through the water. Confining most of the swimming muscles to the rear of an animal's body has other advantages. With the rear end devoted to movement, the front end is free to explore the environment, fight off aggressors, or find food.

(a) A barracuda

(b) Salamander

(c) Marine iguana

(d) Penguin

(e) Dolphin

Figure 46.14 **Adaptations for swimming.**   Streamlined bodies that greatly reduce frictional drag underwater are found in all classes of swimming vertebrates.

Biological inquiry: Is streamlining unique to swimming animals?

In contrast to swimming vertebrates, many aquatic invertebrates move through water by means other than swimming. Cephalopods such as squids use propulsion provided by the ejection of water to give them brief but rapid bursts of speed, while many other aquatic invertebrates move passively on water currents or crawl along rocks and other underwater surfaces.

Due to streamlining, the relatively slow speed of most swimmers, and the buoyancy of water, swimming is energetically the cheapest form of locomotion. By contrast, terrestrial animals face considerable energetic costs to locomotion, as we examine next.

Walking and Running on Land Are Energetically Costly

In contrast to swimming, locomotion on land is, on average, the energetically costliest means of locomotion (**Figure 46.15**). Flying may seem costlier, but it is not. Many migratory birds can travel hundreds of miles daily for many days. No terrestrial animal could possibly match such a feat by walking or running.

Figure 46.15 **Energy costs of locomotion.** The energy costs of different modes of locomotion for animals of different sizes are shown. The y-axis gives the relative energy costs, determined by measuring how much oxygen an animal consumes during locomotion. Energy costs are highest for runners compared to similarly sized fliers and swimmers.

Whereas gravity is not an important factor for locomotion in swimming animals, terrestrial animals must overcome gravity each time they take a step. Of even greater importance to walking and running animals, though, is the necessity of accelerating and decelerating the limbs with every step. In essence, each step is like starting a movement from scratch, without the luxury of occasionally gliding through water or air as fish and birds do. This challenge is even greater when an animal moves uphill or over rough terrain.

Apart from mollusks, which move along the surface of the Earth on a layer of secreted mucus, and snakes, which undulate along the ground on a portion of their ventral body surface, most terrestrial animals limit the amount of contact with the ground while moving, thereby minimizing the amount of friction they encounter. Tetrapods usually have only two feet on the ground at any time when walking, and for brief moments a horse galloping at full speed has all four feet off the ground (**Figure 46.16**).

Having fewer legs on the ground surface at any one time helps increase speed but can compromise stability. Arthropods, for example, have at least six legs. This apparently provides excellent stability but reduces maximal speed. At the other extreme are animals that move by jumping—fleas, certain spiders, click beetles, grasshoppers, and kangaroos.

Flying Has Evolved on Four Separate Occasions

Flying is a highly successful means of locomotion, having evolved on four occasions: in pterosaurs (extinct reptiles), insects, birds, and mammals (bats). The advantages to flying are numerous. Animals can escape land-based predators, scan their surroundings over great distances, and inhabit environments such as high cliffs that may be inaccessible to nonflying animals. The mechanics of flying, however, require animals to overcome gravity and air resistance, which makes flying energetically costlier than swimming but still cheaper than running on land (see Figure 46.15). As with swimming, resistance to flight is decreased by streamlined bodies. However, earthbound animals have one advantage over their flying cousins—they can grow to much larger

(a)

(b)

(c)

(d)

Figure 46.16 **Walking and running in a tetrapod.** By minimizing the number of feet in contact with the ground as they accelerate to running speed, many tetrapods reduce the energy cost of locomotion, as shown in this horse (a) walking, (b) trotting, (c) cantering, and (d) galloping.

sizes than animals that fly. The vast majority of flying animals have a mass between about 1 mg and 1 kg. Only a few large birds have masses exceeding 10 kg. Although this represents an enormous range, it falls far short of the sizes achieved by earth-bound or aquatic animals.

In flying vertebrates, lift and thrust are provided by pectoral and back muscles that move the wings. The pectoral muscles are so powerful and massive that they constitute as much as 15–20% of a bird's total body mass and up to 30% in humming-birds, which use their wings not only to fly but also to hover. The requirement for large, strong pectoral muscles is one reason why the remaining body mass of flying vertebrates is limited. The extinct pterosaurs would seem to be an exception because some species were known to have had wingspans of nearly 10m. However, it is believed that these large animals glided off of trees or cliffs and were unable to provide the force required to lift their massive bodies off the Earth.

In birds and bats, the wings are modifications of the fore-limbs. In general, bat wings are far more maneuverable than bird wings, because unlike birds, bats have fingers at the end of their forelimbs/wings. This allows bats to precisely alter the shape of their wings and provide for rapid, fine-tuned changes in direction, even at high speeds. Bird wings are more similar to those of a fixed-wing aircraft. The largest birds such as hawks and eagles are able to glide because of the great surface area of their large wings. By using a bird's momentum to propel it forward, gliding provides a considerable energy savings. Bats and small birds, however, can glide for only very brief moments.

In this chapter, we have seen how skeletal muscles and (in vertebrates) bones work together to provide animals with pro-tection and to enable them to move around in their environ-ments. Unfortunately, muscles and bone are frequently the sites of significant diseases with great impacts on human health. We conclude the chapter with a discussion of a few of these dis-eases in humans, each of which is known to result from a dif-ferent cause.

46.6 Impact on Public Health

A number of diseases can affect bone structure and function. Bone disease may involve defects in either the mineral or organic components of bone. Poor bone formation and structure may result from inadequate nutrition, hormonal imbalances, aging, or skeletal muscle atrophy, to name just a few of the common causes.

In addition, many diseases or disorders directly affect the contraction of skeletal muscle. Some of them are temporary and not serious, such as muscle cramps, whereas others are chronic and severe, such as the disease muscular dystrophy. Also, some diseases result from defects in parts of the nervous system that control contraction of the muscle fibers rather than from defects in the fibers themselves. One example is poliomyelitis (polio), a viral disease in which the destruction of motor neurons leads to skeletal muscle paralysis that may result in death from respira-tory failure. Polio, once thought to be nearly eliminated from the human population, is still a threat in parts of the world.

Rickets and Osteoporosis Affect the Bones of Millions of People

Bone diseases are fairly common, particularly among individu-als over age 50. Two major abnormalities can occur in bone. The first is improper mineral deposition in bone, usually due to inadequate dietary calcium intake or inadequate absorption of calcium from the small intestine. Without adequate minerals, bone becomes soft and easily deformed, as occurs in the weight-bearing bones of the legs of children with rickets (or osteomala-cia, as it is called in adults). These disorders are best prevented or treated with vitamin D, because this vitamin is the most important factor in promoting absorption of calcium from the small intestine.

Much more common is a disease called osteoporosis, in which both the mineral and organic portions of bone are re-duced. This disease, which affects four times as many women as men, occurs when the normal balance between bone forma-tion and bone breakdown is disrupted.

One way in which osteoporosis occurs is as a result of pro-longed disuse of muscles. In ways that are not completely clear, the force produced by active skeletal muscle contractions helps maintain bone mass. When muscles are not or cannot be used— due to paralysis or long-term immobilizing illnesses, or even during prolonged space flight in which the lack of gravity reduces the need for muscles to work hard—bone mass declines.

Osteoporosis may also result from hormonal imbalances. Some hormones—for example, estrogen—stimulate bone forma-tion. When estrogen levels decline after menopause (the period of time when a woman's reproductive cycles cease), bone den-sity may decline, increasing the risk of bone fractures. By con-trast, some hormones act to demineralize bone as part of the way in which the body normally maintains mineral homeosta-sis in the blood. If such hormones are present in excess, how-ever, they can cause enough demineralization of bone to cause osteoporosis. This may happen in rare cases when the glands that make these hormones malfunction and overproduce the hormones.

Osteoporosis can be minimized with adequate calcium in-take, weight-bearing exercise programs, and adequate vitamin D intake. In some cases, postmenopausal women may be given estrogen to replace what their bodies are no longer producing. Osteoporosis is the most prevalent bone disease in the U.S., affecting up to 15 million individuals, with another 30 million or so with mild to moderate bone loss. It results in annual national expenditures of approximately $15–20 billion in hospi-tal and other costs.

Myasthenia Gravis Is an Autoimmune Disease

Myasthenia gravis is a disease characterized by skeletal muscle fatigue and weakness. It ranges in severity, but at its worst it can make chewing, swallowing, talking, and even breathing difficult. It affects an estimated 10,000 to 30,000 Americans, usually not becoming apparent until adolescence.

In myasthenia gravis, skeletal muscle function is reduced because the body's immune system produces antibodies—proteins that attack foreign matter—that bind to and inactivate ACh receptors on skeletal muscle cells. Because the destruction of ACh receptors is brought about by the body's own defense mechanisms gone awry, myasthenia gravis is considered an autoimmune disease (see Chapter 53). Thus, although motor nerve signals to the muscles are normal, the muscles cannot fully respond to the ACh released from the nerves because the receptors are not functional.

A number of approaches are currently used to treat the disease. One is to administer inhibitors of an enzyme that normally breaks down ACh at the neuromuscular junction, a normal function that prevents overstimulation of the muscle. The inhibition of this enzyme prolongs the time that acetylcholine is available at the synapse, which to some degree compensates for reduced ACh receptor abundance. Other therapies are directed at suppressing the immune system to prevent destruction of the ACh receptors by antibodies. This is accomplished either by surgically removing certain glands that constitute part of the immune system or with drug therapy, but it has the disadvantage of making the person more susceptible to infections. Another therapy, called plasmapheresis, involves removing the liquid fraction of blood (plasma), which contains antibodies, and re-infusing the blood after the antibodies have been removed. A combination of these treatments has greatly reduced the mortality rate for myasthenia gravis.

Muscular Dystrophy Is a Rare Genetic Disease

The group of diseases collectively called muscular dystrophy affects 1 of every 3,500 American males. It is much less common in females. Muscular dystrophy is associated with the progressive degeneration of skeletal (and cardiac) muscle fibers, weakening the muscles and leading ultimately to death from lung or heart failure. The signs and symptoms become evident at about 2 to 6 years of age, and most affected individuals do not survive beyond the age of 30.

The most common form of muscular dystrophy, called Duchenne muscular dystrophy, is a sex-linked recessive disease resulting from a defective gene on the X chromosome. Females have two X chromosomes, and males only one. Therefore, a girl with one abnormal gene and one normal one will not generally develop the disease. In Duchenne muscular dystrophy, the affected gene codes for a protein known as dystrophin, which is absent in patients with the disease. Dystrophin is a large protein that links cytoskeletal proteins to the plasma membrane. It is believed to be involved in maintaining the structural integrity of the plasma membrane or of elements within the membrane, such as ion channels, in muscle fibers. In the absence of dystrophin, the plasma membrane of muscle fibers is disrupted, causing extracellular fluid to enter the cell. Eventually, the cell ruptures and dies.

CHAPTER SUMMARY

- Locomotion is the movement of an animal from place to place.

46.1 Types of Animal Skeletons

- Skeletons are structures that provide support and protection and also function in locomotion.

- Three types of skeletons are found in animals. In hydrostatic skeletons, found in many soft-bodied invertebrates, muscle contractions acting on fluid-filled body cavities create hydrostatic pressure that supports the body and generates movements. Exoskeletons, which are found in arthropods, are protective external structures that must be shed to accommodate growth of the animal. Endoskeletons are internal structures that grow with the animal but do not protect its body surface; endoskeletons are found in some species of sponges, echinoderms, and all vertebrates. (Figure 46.1)

46.2 The Vertebrate Skeleton

- A joint is formed where two or more bones of a vertebrate endoskeleton come together. (Figure 46.2)

- In addition to the functions of support, protection, and locomotion, the vertebrate skeleton produces blood cells and constitutes a reservoir for ions that are crucial to homeostasis. (Table 46.1)

46.3 Skeletal Muscle Structure and the Mechanism of Force Generation

- A skeletal muscle is a grouping of cells, called muscle fibers, bound together by a succession of connective tissue layers. The striations of skeletal muscle result from the presence within muscle fibers of cylindrical bundles known as myofibrils, each of which contains thick and thin filaments arranged in repeating units called sarcomeres. The arrangements of thick and thin filaments within each sarcomere create various light and dark regions designated the I and A bands, the Z and M lines, and the H zone. Thick filaments are composed almost entirely of the motor protein myosin, whereas thin filaments contain the cytoskeletal protein actin plus two other proteins called troponin and tropomyosin. Portions of the thick filaments are called cross-bridges. (Figure 46.3)

- During muscle contraction, the sarcomeres shorten by a process known as the sliding filament mechanism. In muscle contraction, the thick filaments remain stationary while the thin filaments slide past them propelled by action of the cross-bridges. (Figure 46.4)

- Sarcomeres are composed of thick (myosin) and thin (actin) filaments. (Figure 46.5)

- Mutations in a myosin gene may have allowed the human brain to become larger. (Figure 46.6)

- The repeated interaction of cross-bridges on the thick filaments and the protein components of the thin filaments occurs according to a process called the cross-bridge cycle. (Figures 46.7)

- Calcium plays an important role in excitation-contraction coupling, the sequence of events by which an action potential in the plasma membrane of a muscle fiber leads to cross-bridge activity. (Figure 46.8)

- The source of the cytosolic calcium involved in a muscle action potential is the muscle fiber's sarcoplasmic reticulum. Tubular structures called the transverse tubules (T-tubules) conduct action potentials from the outer surface of the muscle fiber to the myofibrils. (Figure 46.9)

- Electrical stimulation of skeletal muscle occurs at a neuromuscular junction, in which a motor neuron's axon and a muscle fiber are in close proximity. (Figure 46.10)

46.4 Skeletal Muscle Function

- On the basis of the rates at which muscle fibers shorten and the way in which they produce the ATP needed for contraction, three major types of skeletal muscle fibers have been distinguished. Slow-oxidative fibers have low rates of myosin ATP hydrolysis; they do not fatigue easily and are used for prolonged, regular activities. Fast-oxidative fibers have high myosin activity, do not fatigue quickly, and are particularly suited for rapid actions. Fast-glycolytic fibers have high myosin activity but cannot make as much ATP as oxidative fibers; they are best suited for rapid, intense actions. (Table 46.2)

- Increased expression of *PPAR-delta* in mice results in increased slow-oxidative muscle, greater exercise endurance, and weight loss. (Figure 46.11)

- Groups of muscles that produce oppositely directed movements at a joint are known as antagonists. Muscles that bend a limb at a joint are called flexors, whereas muscles that straighten a limb are called extensors. (Figure 46.12)

- To produce movement, muscles, bones, and joints are arranged in lever systems. (Figure 46.13)

46.5 Animal Locomotion

- Due to streamlining, the relatively slow speed of most swimmers, and the buoyancy of water, swimming is energetically the most efficient form of locomotion. Locomotion on land is, on average, the energetically costliest means of locomotion; the energy expenditure required for flight is intermediate between those for swimming and land-based locomotion. (Figures 46.14, 46.15, 46.16)

46.6 Impact on Public Health

- Several health conditions affect muscular-skeletal structures. Rickets (osteomalacia in adults) is characterized by soft, deformed bones, usually resulting from insufficient dietary intake or absorption of calcium. In osteoporosis, bone density is reduced when bone formation fails to keep pace with normal bone breakdown. Myasthenia gravis is an autoimmune disease characterized by skeletal muscle fatigue and weakness. Muscular dystrophy is an ultimately fatal genetic disease associated with the progressive degeneration of skeletal and cardiac muscle fibers.

TEST YOURSELF

1. The hydrostatic skeleton common in soft-bodied invertebrates is composed of
   a. muscle and cartilage.
   b. cartilage and a neural net.
   c. muscle and water.
   d. cartilage and water.
   e. bone and muscle.

2. Which of the following is *not* a function of the vertebrate skeleton?
   a. structural support
   b. protection
   c. calcium reserve
   d. blood cell production
   e. all of the above are functions of the vertebrate skeleton

3. The protein that provides strength and flexibility to bone is
   a. actin.
   b. myosin.
   c. myoglobin.
   d. collagen.
   e. elastin.

4. In a sarcomere, the _____ contains thin filaments and no thick filaments.
   a. A band
   b. M line
   c. I band
   d. H zone
   e. both a and d

5. The function of ATP during muscle contraction is to
   a. cause an allosteric change in myosin so it detaches from actin.
   b. provide the energy necessary for the movement of the cross-bridge.
   c. expose the myosin-binding sites on the thin filaments.
   d. all of the above.
   e. a and b only.

6. The function of calcium ions in skeletal muscle contraction is to
   a. cause an allosteric change in myosin so it detaches from actin.
   b. provide the energy necessary for the movement of the cross-bridge.
   c. expose the myosin-binding sites on the thin filaments.
   d. bind to tropomyosin.
   e. a and c only.

7. Stimulation of a muscle fiber by a motor neuron occurs at
   a. the neuromuscular junction.
   b. the transverse tubules.
   c. the myofibril.
   d. the sarcoplasmic reticulum.
   e. none of the above.

8. Muscle fibers that have a high number of mitochondria, contain large amounts of myoglobin and exhibit low rates of ATP hydrolysis are called _____ fibers.
   a. slow-glycolytic
   b. fast-glycolytic
   c. intermediate
   d. fast-oxidative
   e. slow-oxidative

9. Which of the following statements about movement and locomotion is incorrect?
   a. Terrestrial animals and flying animals expend energy to provide lift.
   b. Swimming animals expend energy to provide thrust but not lift.
   c. Flexors and extensors are examples of muscles called agonists.
   d. Flexors cause bending at a joint.
   e. Extensors cause straightening of a limb.

10. Swimming is energetically the cheapest type of locomotion because of
    a. streamlined body forms in aquatic organisms.
    b. slow speed of movement.
    c. buoyancy.
    d. a and c only.
    e. a, b, and c.

CONCEPTUAL QUESTIONS

1. Define locomotion, and list some of the various types that occur in animals.

2. Distinguish between exoskeletons and endoskeletons.

3. List and briefly discuss the steps in the cross-bridge cycle.

EXPERIMENTAL QUESTIONS

1. What is the normal function of the PPAR-delta protein in mice?

2. What was the hypothesis proposed by Evans in relation to PPAR-delta and obesity?

3. How did Evans and his colleagues test this hypothesis, and what did they observe?

COLLABORATIVE QUESTIONS

1. Discuss the structure and function of the three different types of skeletons found in animals.

2. Discuss the three types of muscle tissues found in vertebrates.

---

www.brookerbiology.com

This website includes answers to the Biological Inquiry questions found in the figure legends and all end-of-chapter questions.

47
CIRCULATORY SYSTEMS

Image from cardiac angiography (dye contrast injection to visualize coronary blood vessels).

In Chapters 41 and 42, we learned how animals obtain and use nutrients. In this chapter, we will examine how nutrients and other substances are transported throughout an animal's body to individual cells. Single-celled organisms are small enough that dissolved substances are able to diffuse into and out of their bodies. In larger animals, however, dissolved substances must move greater distances between cells, as well as between the internal and external environments. Transporting materials over long distances requires a circulatory system.

The primary function of **circulatory systems** is to transport necessary materials to all the cells of an animal's body, and to transport waste products away from the cells where they can be released into the environment. Necessary materials include organic and inorganic nutrients and gases such as oxygen. Waste products that must be eliminated include carbon dioxide and other breakdown products of metabolism. Most circulatory systems are capable of adjusting their activities to an animal's changing metabolic demands. For instance, we will see how exercise increases the ability of the vertebrate heart to pump blood and directs blood to the skeletal muscles where it is most needed.

The three basic types of circulatory systems are gastrovascular cavities, open systems, and closed systems. We begin with the simplest form of circulation, that provided by the water in which certain invertebrates live.

47.1 Gastrovascular Cavities

In cnidarians (jellyfish and hydras), the surrounding water not only contains the nutrients and oxygen needed to sustain life but also serves the functions of the animals' circulatory system.

These animals rely on water currents to bring a steady supply of water into contact with their body surfaces. Water enters the **gastrovascular cavity**, a body cavity with a single opening to the outside. Food is digested within the cavity and absorbed by body cells, and wastes are excreted into the cavity (see Chapter 41). Digested foods reach all the cells because water circulates throughout the entire cavity. All of the animal's body cells are located near the gastrovascular cavity or slender extensions that branch from it (**Figure 47.1**).

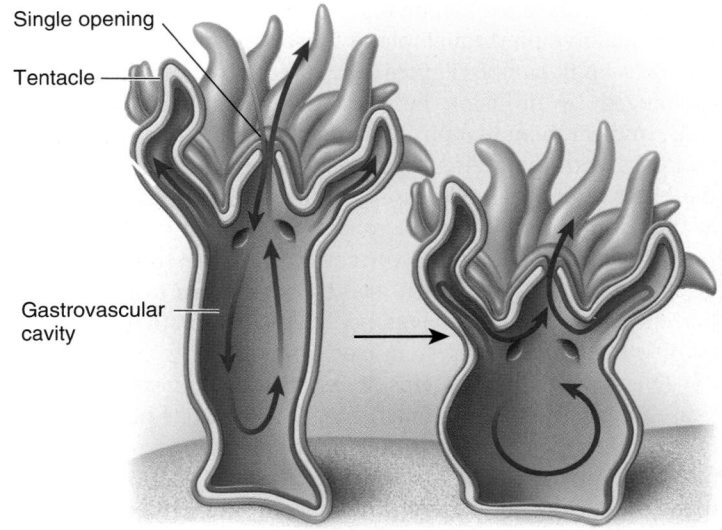

Figure 47.1 **Gastrovascular cavity of a hydra.** The animal's movements assist the flow of water throughout the cavity, which extends into the tentacles.

In some cnidarians, cilia that line the opening of the gastrovascular cavity propel water through these extensions. In addition, muscular efforts of the body wall—as a hydra stretches and relaxes, for instance, or a jellyfish propels itself by contracting its bell—help move water throughout the cavity. The harder the muscles work, the more effectively they propel water, thus providing a form of circulation. In this way, the muscles of cnidarians serve some of the same functions as a beating heart.

The muscles of the jellyfish bell are controlled by nerve cells that are scattered throughout the bell. This, too, is comparable to what occurs in animals with hearts, because nerves regulate the beating of a heart. Thus, animals that possess a gastrovascular cavity already show several elements of more complex circulatory systems.

47.2 Open Circulatory Systems

Arthropods and some mollusks have a true circulatory system with three basic components:

1. **blood** or **hemolymph**, an internal body fluid containing dissolved solutes;
2. **blood vessels**, a system of hollow tubes within the body through which blood travels;
3. one or more **hearts**, muscular structures that pump blood through the blood vessels.

Circulatory systems containing these three components are also known as a **cardiovascular system** (from the Greek *kardia*, heart, and Latin *vas*, vessel). In an **open circulatory system**, the vessels open into the animal's body cavity. Therefore, the fluid in the vessels and the interstitial fluid that surrounds cells mingle in one large, mixed compartment, rather than being located in separate body compartments. The blood and interstitial fluids form one fluid, called hemolymph. One or more muscular hearts pump hemolymph through blood vessels, where it exits into the animal's body cavity (**Figure 47.2**). Nutrients and wastes are exchanged by diffusion between the hemolymph and body cells, and the hemolymph is recirculated back to the heart. In mollusks with open circulatory systems, hemolymph re-enters the heart through blood vessels, while in arthropods it returns to the heart through small openings in the heart called **ostia**.

In insects, hemolymph primarily transports nutrients and wastes, while in mollusks it also brings oxygen to body cells. Insects have a unique oxygen-delivery system that does not involve the circulatory system (see Chapter 48).

Open circulatory systems are energetically inexpensive because they do not require much energy to circulate hemolymph. In addition, open systems can adapt to changes in an animal's metabolic demands. As an insect takes flight, for example, its flight muscles contract more forcefully and rapidly, which acts to expand and compress the animal's thorax. This helps propel hemolymph more rapidly throughout the body cavity, and into and out of the heart. In other words, as the animal's activity increases, its circulation becomes more efficient, recharging

Figure 47.2   **Open circulatory system.**   In open systems, one or more muscular, tubelike hearts pump hemolymph through open-ended vessels, where it percolates through the body. In arthropods, hemolymph re-enters the heart through ostia. The arrows show the movement of hemolymph.

metabolically active cells with nutrients. As noted earlier, the ability to adjust circulation to meet an animal's needs is one of the most important features of any circulatory system. Open systems, therefore, are highly effective, particularly for smaller animals.

Despite the phenomenal success of species with open systems, certain limitations are associated with this type of circulation. For instance, because the hemolymph empties in bulk into the general body cavity, it cannot be selectively delivered to body cells whose metabolic activity may have increased compared to other cells. By contrast, during exercise in vertebrates, larger amounts of blood are delivered to leg muscles and away from less metabolically active structures, an adaptation that results in greater endurance. This is possible because these animals have closed circulatory systems, as we see next.

47.3 Closed Circulatory Systems

Open circulatory systems are not ideal for larger, more active animals in which higher pressure is required to pump blood to all body cells at a rate sufficient to meet energetic demands. These animals have evolved a closed circulatory system, which we will examine in this section.

Closed Circulatory Systems Separate Blood from Interstitial Fluid

In a **closed circulatory system**, blood and interstitial fluid are physically separated by blood vessel walls, and they differ in their components and chemical composition. Closed circulatory systems are found in earthworms, cephalopods (squids and octopi), and all vertebrates (**Figure 47.3**). Despite differences in structure, closed circulatory systems share certain key features.

- Blood remains within tubelike vessels that distribute it throughout the body.

In the figure, labels read:
"Tubular hearts with ostia" and "Open-ended vessels"

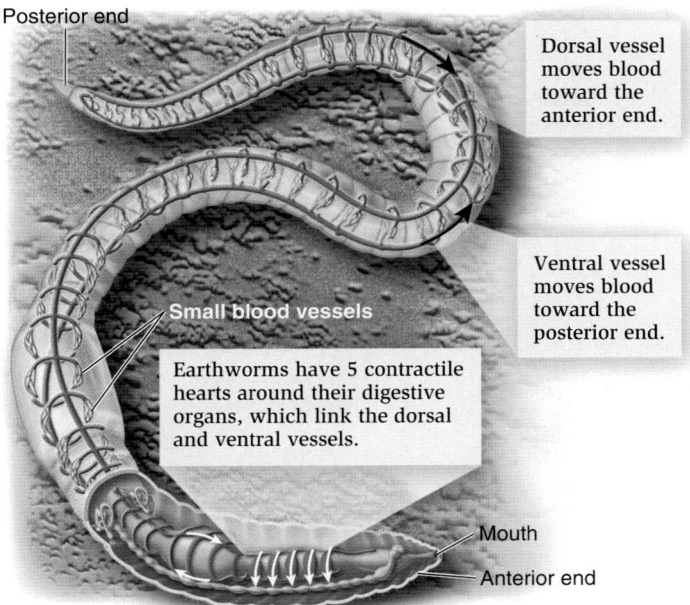

Posterior end

Dorsal vessel moves blood toward the anterior end.

Small blood vessels

Ventral vessel moves blood toward the posterior end.

Earthworms have 5 contractile hearts around their digestive organs, which link the dorsal and ventral vessels.

Mouth

Anterior end

Figure 47.3 **Closed circulatory system.** In closed systems, like that of the earthworm, blood remains in vessels or hearts and recirculates without emptying into the body cavity. Arrows indicate direction of blood flow.

- Blood is pumped under pressure by one or more contractile, muscular hearts.
- Blood contains dissolved solutes that are transported throughout the body, and that can be exchanged with the environment and the body's cells.
- In most cases, blood contains disease-fighting cells and molecules.
- The activity of the circulatory system can be adjusted to match the animal's metabolic demands.
- Closed circulatory systems generally have the capacity to heal themselves when wounded, by forming clots at the site of injury.
- Closed systems grow in size as an animal grows.

A closed circulatory system offers several advantages. First, animals can grow to a larger size, because blood can be directed to every cell of an animal's body, no matter how large. Nearly all body cells are within one or two cell-widths of a blood vessel. Second, blood flow can be selectively increased or decreased to supply different parts of the body with the precise amount of blood needed at any given moment. After a meal, for example, more blood can be directed to the gut to absorb nutrients, and on a hot day additional blood can be routed to the skin to dissipate heat. Animals with open systems cannot make these adjustments.

The closed circulatory systems of vertebrates can be divided into two major groups, single and double circulations. In a single circulation, blood is pumped under low pressure from the heart to the respiratory surface (for example, gills), where it picks up oxygen and drops off carbon dioxide. From there, blood circulates to the tissues of the body, where it releases oxygen and picks up carbon dioxide. The blood then circulates back to the heart. In a double circulation, blood is pumped under low pressure from the heart to the lungs, from where it returns to the heart, to be pumped under high pressure to the tissues and finally returned again to the heart. Single circulations are seen in fish, and double circulations are seen in crocodiles, birds, and mammals. Amphibians and most reptiles have a circulation that combines features of both.

In the single circulation of fish, the heart has a single filling chamber—an **atrium**—to collect blood from the tissues, and another chamber—a **ventricle**—to pump blood out of the heart (**Figure 47.4a**). Blood vessels called **arteries** carry blood away from the heart to the gills, where the fish picks up oxygen from the water in which it swims and unloads carbon dioxide into the water. The freshly oxygenated blood then circulates via other arteries to the rest of the body. There, oxygen and nutrients are delivered to cells and carbon dioxide diffuses from cells into the blood. Finally, the deoxygenated blood is returned to the heart via blood vessels called **veins**, where it is pumped back to the gills for another load of oxygen.

An important feature of all respiratory surfaces is that they function best when the blood flowing through them is maintained at a low pressure. Thus, the fish heart does not generate high pressure when it pumps blood to the gills. This means that blood leaving the gills will also be under low pressure, thus limiting the rate at which oxygenated blood can be delivered to the body's cells.

Unlike fish, most adult amphibians, including frogs, rely on lungs and their highly permeable skin to obtain oxygen and rid themselves of carbon dioxide. While amphibians are on land, their deoxygenated blood is pumped from the heart to the lungs and, to a lesser extent, the skin. Oxygen diffuses from the air into the blood vessels within the lungs and beneath the skin, and carbon dioxide diffuses in the opposite direction. While amphibians are underwater, deoxygenated blood from the heart bypasses the lungs and is directed almost entirely to the skin, so that oxygen in the water can diffuse across the skin and into the circulation. Oxygenated blood from the lungs and skin does not travel directly to the rest of the body but instead returns to the heart, where it is pumped out to the tissues of the body.

The amphibian heart, therefore, pumps oxygenated and deoxygenated blood to two separate locations. Blood is routed from the heart through different vessels to either the respiratory surfaces (the **pulmocutaneous circulation**) or the body tissues (the **systemic circulation**). This would not be possible if an amphibian's heart and circulation were the same as that of a fish, where blood travels through a single circulation. Instead, the amphibian heart has two separate atria for collecting blood (**Figure 47.4b**). The right atrium receives blood that has passed through all of the body except the lungs and is therefore depleted of much of its oxygen. However, some oxygen-rich blood returning from the skin also enters the right atrium and mixes with the blood returning from the systemic circulation. This is

(a) Circulation in fish

(b) Circulation in most amphibians

(c) Circulation in birds and mammals

Figure 47.4   Representative vertebrate circulatory systems.   (a) Fish have a single circulation in which blood is pumped from the heart to the gills, from where it circulates to the rest of the body tissues, and finally returns to the heart. (b) Amphibians such as frogs have a heart with a single ventricle that nevertheless is able to separate most of the oxygenated and deoxygenated blood entering it from the two atria. Some mixing does occur, however, as represented in this figure by the colors of the blood. A significant advantage of this type of circulation is the ability to redirect blood from the lungs to the skin when underwater. Because the skin is somewhat permeable to oxygen, this partly compensates for the lack of air-breathing. (c) Birds and mammals have a double circulation, in which oxygenated blood is pumped under high pressure to the body's tissues, and deoxygenated blood is pumped under low pressure to the lungs.

an important adaptation for amphibians when they are underwater, because as stated, the circulation through the lungs is decreased at such times and therefore diffusion across the skin becomes the sole means of obtaining oxygen.

When an amphibian breathes air, oxygen-rich blood from the lungs is delivered to the left atrium. Each atrium of the amphibian heart empties its blood into a single large ventricle. The internal structure of the single ventricle keeps the oxygenated and deoxygenated blood from the two atria mostly separate. Some mixing of the two streams of blood, however, does occur in the ventricle. This would not occur if each atrium emptied into a

separate ventricle. Because there is only a single ventricle, however, the blood traveling to the tissues is not fully oxygenated, and this imposes limits on the metabolic activity of amphibians.

Like amphibians, all of the noncrocodilian reptiles have hearts with two atria and a single ventricle, although the ventricle is further subdivided into two partially but incompletely separated chambers. These chambers allow more efficient separation of oxygenated and deoxygenated blood entering the ventricle from the two atria, which contributes in part to the higher activity levels of reptiles compared to amphibians. Despite this, because reptiles and amphibians have only a single ventricle that must pump blood to both the tissues and respiratory surfaces, blood must be pumped under low or moderate pressure in these animals to minimize the pressure of blood flowing through the lung tissue.

Crocodiles, Birds, and Mammals Have a Double Circulation with Four-Chambered Hearts

In crocodiles, birds, and mammals, oxygenated and deoxygenated blood are completely separated into two distinct circuits, the systemic circulation and the **pulmonary circulation** (**Figure 47.4c**). This is made possible by a heart that has a second ventricle. The crocodilian circulatory system is somewhat different from that of birds and mammals, because crocodiles spend part of their lives underwater and are considerably less active than birds and mammals. When resting or underwater, crocodiles can divert blood between the two circuits in ways that birds and mammals cannot. Nonetheless, all animals with double circulations have a left and right atrium, and a left and right ventricle. The left ventricle receives oxygenated blood from the left atrium, and the right ventricle receives deoxygenated blood from the right atrium.

A major advantage of a double circulation is that the two ventricles can function as if they were, in effect, two hearts, each with its own ability to pump blood under different pressures. This means that blood from the right ventricle can be pumped under low pressure to the lungs. Once the blood picks up oxygen from the lungs, it can be returned to the left side of the heart. The left ventricle is more muscular than the right ventricle and therefore can generate much higher pressures when pumping blood. Thus, the oxygenated blood leaving the left side of the heart has sufficiently high pressure to reach all the cells of the animal's body and deliver oxygen and nutrients at a high rate, even to regions above the heart that must contend with gravity. This is no small feat considering the distance blood must travel in some large mammals. The left ventricle of a giraffe, for example, is particularly muscular compared to that of other animals.

Blood Is Composed of Cells and Water with Dissolved Solutes

Blood is a fluid connective tissue consisting of cells and, in mammals, cell fragments suspended in a solution of water containing dissolved nutrients, proteins, gases, and other molecules. Blood has four components: plasma, leukocytes, erythrocytes, and platelets or thrombocytes.

Plasma If we collect a blood sample and spin it in a device called a centrifuge, the blood separates into distinct layers (**Figure 47.5**). The top layer is **plasma**, a yellowish solution of water and solutes. Plasma typically makes up about 35–60% of the total volume of blood in vertebrates. It contains water and the

Obtain a blood sample.

Centrifuge

Plasma

Leukocytes (white blood cells)

Erythrocytes (red blood cells)

Figure 47.5 **Components of blood.** When a blood sample is centrifuged, it forms three visible layers. Plasma is composed of water and dissolved solutes, leukocytes are the white blood cells that make up part of the body's immune system, and erythrocytes are red blood cells, which carry oxygen. An additional component of blood is not visible under these conditions because of small numbers and size. These are the thrombocytes (platelets in mammals) that participate in blood clotting reactions.

dissolved organic and inorganic nutrients that were absorbed from the digestive tract or secreted from cells. Plasma also contains dissolved oxygen, waste products of metabolism like carbon dioxide, and other molecules released by cells, such as hormones. In addition to transporting molecules throughout the body, plasma has other functions. For example, plasma contains buffers that help keep the body's pH within its normal range. It is also important in maintaining the fluid balance of cells. Changes in plasma salt concentrations can affect the movement of water between intracellular and extracellular fluid compartments. Plasma also transports cells of the immune system and cells involved in oxygen transport, as well as proteins that serve several important functions such as forming blood clots, which seal off wounds to blood vessels.

Leukocytes Beneath the plasma in our sample is a narrow white layer of **leukocytes**, also known as **white blood cells**. Leukocytes develop from the inner parts (the marrow) of certain bones in vertebrates. While there are several types—which we describe further in Chapter 53—all leukocytes perform vital functions that defend the body against infection and disease.

Figure 47.6 **Characteristic biconcave shape of mammalian erythrocytes.** The shape of these cells—seen here in a scanning electron micrograph magnified over 1,600 times—increases their surface area and is believed to facilitate diffusion of oxygen from the cells to extracellular fluid.

Figure 47.7 **Hemoglobin.** Erythrocytes contain large amounts of the protein hemoglobin. Oxygen binds reversibly to iron atoms at the core of each subunit of hemoglobin.

Erythrocytes The bottom visible layer of our blood sample consists of **erythrocytes**, also called **red blood cells** because of their color. **Hematocrit** refers to the volume of blood that is composed of red blood cells, usually between 40 and 65% among vertebrates. Erythrocytes serve the critical function of transporting oxygen throughout the body. There are approximately a thousand times more red blood cells than white blood cells in the circulation. Like leukocytes, red blood cells are derived from cells in the bone marrow. In most vertebrates, mature red blood cells retain their nuclei and other cellular organelles, but in all mammals (and a few species of fish and amphibia) the nuclei are lost upon maturation. The lack of a nucleus and many other organelles in the mammalian red blood cell increases the cell's oxygen-carrying capacity and contributes to its characteristic biconcave shape (**Figure 47.6**). The biconcave shape of the mammalian red blood cell increases its surface area compared to the flattened disk or oval shape seen in most other vertebrates. This is believed to make gas exchange between the red blood cell and the surrounding body fluids more efficient.

Oxygen is poorly soluble in plasma. Consequently, the amount of oxygen that dissolves in plasma usually cannot support a vertebrate's basal metabolic rate, let alone more strenuous activity. Within the cytosol of red blood cells are large amounts of the protein **hemoglobin**. Each molecule of hemoglobin contains four protein subunits, each with an atom of iron at its core that reversibly binds to a molecule of oxygen (**Figure 47.7**). Thus, each hemoglobin molecule has four iron atoms and can reversibly bind four oxygen molecules. The oxygen attached to hemoglobin greatly increases the reservoir of oxygen in the blood

and enables animals to be active. Chapter 48 describes the mechanisms by which hemoglobin binds and releases oxygen.

Anemia refers to lower than normal levels of hemoglobin, which reduces the amount of oxygen that can be stored in the blood. Among possible causes are loss of blood, impaired production of erythrocytes or increased breakdown of erythrocytes, each of which reduces hematocrit. By contrast, **polycythemia** is a condition of increased hemoglobin due to increased hematocrit. Polycythemia occurs when terrestrial animals living at low altitudes ascend to higher altitudes, where the air pressure is lower and it is harder to bring oxygen into the blood. The decreased blood oxygen levels trigger release from the liver and kidneys of a protein that stimulates the production of red blood cells. Polycythemia can also occur rapidly to increase the blood's oxygen-carrying capacity, such as when a seal dives underwater or a horse sprints. In this case, preformed red blood cells are released from the spleen, which acts like an erythrocyte storehouse.

Platelets Vertebrate blood has a fourth component—called **platelets** in mammals and **thrombocytes** in other vertebrates—which would not be visible in our centrifuge sample because of their low numbers. Thrombocytes are intact cells while platelets are cell fragments that lack a nucleus. Like leukocytes and erythrocytes, thrombocytes and platelets are formed in the bone marrow. They play a crucial role in the formation of blood clots, which limit blood loss after injury.

Although clot formation in mammals is a multistep process, two essential steps involve platelets (**Figure 47.8a**). First, platelets

(a)

(b)

Figure 47.8 **The process of blood clot formation.** (a) A blood clot forms in two major steps—a platelet plug reduces initial blood loss, and a fibrin clot then seals the wound. (b) Formation of the clot requires several clotting factors working together. These are enzymes, which catalyze a cascade of reactions that leads to the production of fibrin fibers. Several steps in the process are omitted here for simplicity.

secrete substances that cause them to stick in clumps to each other and to collagen fibers in the surrounding connective tissue at the wound site to form a plug that prevents continued blood loss. Second, other platelet secretions interact with plasma proteins to cause the precipitation from solution of the fibrous protein **fibrin**. Fibrin forms a meshwork of threadlike fibers that wrap around and between platelets and red blood cells, enlarging and thickening the clot. Blood clotting begins within seconds and helps prevent injured animals from bleeding to death. Eventually, the body absorbs the clot as the injured vessel heals. The importance of clotting is most evident when it fails to occur normally, as we see next.

GENOMES & PROTEOMES

Hemophilia Is a Genetic Defect in Clotting Factors

The process of blood clotting is not a simple one—at least a dozen different substances must be present in the right amounts for a clot to form. These substances are known as clotting factors and include proteins like fibrin and ions such as Ca^{2+}. Platelets secrete some of these substances, while others circulate in the plasma in an inactive form until activated by platelet and other cellular secretions.

The clotting process is a chain reaction that begins when an injury to a blood vessel exposes blood to the surrounding extracellular matrix (**Figure 47.8b**). When one particular clotting factor in blood interacts with proteins in the matrix, the factor is activated. Most of the activated clotting factors are enzymes, which activate a cascade of additional clotting factors, finally producing fibrin.

In the disease hemophilia, an inherited deficiency of a specific clotting factor inhibits the clotting process. The absence of the clotting factor interrupts the process leading to fibrin formation. Consequently, injured blood vessels take longer to heal and blood loss is excessive. Minor cuts are not necessarily serious, but uncontrolled bleeding in internal tissues and joints can be life-threatening.

In the most common form of hemophilia, an X-linked recessive mutation results in the inability to make an enzyme that is one of the clotting factors. Because the mutation is X-linked and recessive, it is much more common in males than females. Females who are heterozygous for the mutation are carriers and have a 50% chance of passing the disease genes to their sons.

Treating hemophilia requires transfusions of clotting factors purified from great amounts of human blood or from genetically engineered organisms. These procedures are extremely expensive, and transfusions run the risk of introducing infection or disease from the donor blood. Currently, researchers are attempting to treat hemophilia with gene therapy by using a viral vector (see Chapter 20) to introduce the normal gene into liver cells, which then secrete the missing factor into the blood.

Even if the introduced gene is not extremely efficient at forming new clotting factor, it could still improve patients' symptoms. This is because the missing clotting factor is an enzyme. Like all enzymes, it can act over and over again without being destroyed or chemically altered by the reaction it catalyzes. In addition, this enzyme is known to remain in the circulation for long periods without being degraded or excreted. Recent estimates suggest that restoring the clotting factor to only 2–5% of normal levels should prevent most symptoms of hemophilia. Gene therapy has proven to be effective in curing laboratory rodents and dogs of hemophilia, and testing has recently begun in humans. The results thus far are mixed but hopeful. One problem that needs to be overcome is that human immune systems appear to be more likely than those of experimental animals to destroy cells that incorporate the viral vector.

The Vertebrate Heart Has Multiple Chambers Separated by Valves

As described earlier, all vertebrate hearts have at least one upper or anterior chamber called an atrium, and at least one lower or posterior chamber called a ventricle. In hearts with more than one atrium or ventricle, like the mammalian heart, which has two of each chamber, the chambers are physically separated by a thick strip of connective tissue called a septum. **Figure 47.9** depicts the route blood takes through the mammalian heart. Blood enters the atria from systemic or pulmonary veins, moves

down a pressure gradient through the one-way **atrioventricular (AV) valves** into the ventricles, and is pumped out through the one-way **semilunar valves** into the systemic and pulmonary arteries.

The heart beats (contracts) with a steady rhythm that can increase during exercise. Each beat of the heart pumps out a volume of blood that depends on the size of the animal's heart (**Table 47.1**). Typically, heart size varies in proportion to body mass within a given class of vertebrates, with birds and mammals having larger hearts than similarly sized fish, amphibians, or reptiles. Birds and mammals, being endotherms, tend to be more active for longer periods of time, and thus they burn more fuel than ectotherms. A more powerful heart delivers oxygen and nutrients throughout the body at a faster rate, thus sustaining higher metabolic activity.

Vertebrates Have Myogenic Hearts Capable of Beating on Their Own

Animals cannot consciously initiate heart contractions. The beating of the heart is initiated either by nerves or by intrinsic activity of the heart muscle cells themselves. Many arthropods and decapod crustaceans have a **neurogenic heart** that will not beat unless it receives regular electrical impulses from the nervous system. All vertebrates, however, have a **myogenic heart**; that is, the signaling mechanism that initiates contraction resides within the cardiac muscle itself.

Figure 47.9   The mammalian heart.   The figure shows the major blood vessels entering and leaving the heart, the locations of the valves, and the direction of blood flow through the chambers of the heart. The aorta leads to the systemic circulation. The pulmonary arteries lead to the right and left lung, and the pulmonary veins return blood from the lungs. Oxygenated blood is shown in red, deoxygenated in blue. Note that the pulmonary veins are red because they return blood from the lungs, while veins from the systemic circulation are blue.

Myogenic hearts are electrically excitable and generate their own action potentials. The electrical signals produced by the cells of these hearts initiate the cellular events that trigger muscle contraction (see Chapter 46). Although nerves from the nervous system terminate on the vertebrate heart and can regulate the rate of the heart beat, myogenic hearts will continue to beat on their own if dissected out of an animal and placed in a nutrient bath, even without any nerves present.

Excitation of the Vertebrate Heart Begins in the SA Node

The electrical signals of myogenic hearts are generated either at the junction of the veins and the single atrium in fish, or within the wall of the right atrium in other vertebrates, at the **sinoatrial (SA) node**, or **pacemaker** (**Figure 47.10a**). The SA node is a collection of modified cardiac cells that spontaneously and rhythmically generate action potentials that spread across the entire atria.

| Table 47.1 | Comparative Features of Representative Mammalian Hearts |

Animal	Body mass* (kg)	Heart mass* (kg)	Stroke volume* (L)	Heart rate* (bpm)†
Shrew‡	0.0024 (2.4 g)	0.000035 (35 mg)	0.000008 (8 μl)	835
Rat	0.20	0.001	0.0018	360
Dog	23	0.12	0.025	95
Human	75	0.38	0.075	70
Elephant	4,000	25	4.0	25
Blue whale	100,000	600	100	10

*Values are based on average body masses and resting conditions. In some cases, stroke volumes are estimates based on heart size. Note that heart rate decreases as animals get larger, but heart mass and stroke volume increase roughly in proportion with body mass. Similar relationships are observed in other vertebrates, notably birds.

†bpm = beats per minute.

‡The shrew reported here is the Etruscan shrew, one of the smallest known mammals. Its heart is somewhat larger than would be predicted for its body mass. Note its heart rate; at 835 bpm, the heart beats 14 times per second!

	Atrial excitation		Ventricular excitation	
	Begins	Complete	Begins	Complete
AV valves	OPEN	OPEN	Beginning to close	CLOSED
Semilunar valves	CLOSED	CLOSED	CLOSED	OPEN
Phase of cardiac cycle	Diastole	Diastole	Systole beginning	Systole
Atria	Relaxed	Contracted	Beginning to relax	Relaxed
Ventricles	Relaxed	Relaxed	Beginning to contract	Contracted
Site of highest pressure	Atria	Atria	Ventricles	Ventricles

(a) Cycle of atrial and ventricular excitation

(b) Semilunar valves in nearly closed position

(c) Semilunar valves in opened position

Figure 47.10 **Coupling of electrical and mechanical activity in the mammalian heart.** (a) Electrical activity begins in the SA node and quickly spreads through the atria to the AV node. Branches from the AV node transmit electrical activity throughout the ventricles. The atria and ventricles do not contract until they have become electrically excited. Changes in pressure gradients between the atria and ventricles, and between the ventricles and the aorta and pulmonary arteries, cause the four heart valves to open and close. Semilunar valves in (b) closed and (c) opened positions.

The spread occurs because cardiac muscle forms a structure called a syncytium—a group of cells, all of which are electrically coupled by gap junctions (see Chapter 10). The action potentials trigger an influx of Ca^{2+} into the muscle cell cytosol, which activates contraction. Because the gap junctions allow electrical charge to flow freely from one cardiac cell to the next, the impulses spread very rapidly across the atria, so that both atria contract almost as if they were one large muscle cell. Atrial contraction pumps blood through the AV valves into the ventricles.

Electrical impulses initiated in the SA node reach another node of specialized cardiac cells, the **atrioventricular (AV) node**. The AV node sits near the junction of the atria and ventricles and conducts the electrical events from the atria to the ventricles. Like the SA node, the AV node is electrically excitable, but its cells require a longer time to become excited than do the cells of the atria. This allows time for the atria to contract before the ventricles do. Branches from the AV node spread electrical impulses throughout the walls of the ventricles (Figure 47.10). Once the ventricles are electrically excited, they contract.

Because the ventricular walls are thicker and stronger than the atrial walls, ventricular contractions generate much greater pressure. The pressure exerted by the ventricles on the blood within their cavities causes the AV valves to shut, preventing blood from flowing backward into the atria. This is why the electrical delay that is built into the AV node is important. If the atria and ventricles were excited and contracted at the same time, the AV valves would close too soon and the ventricles would not receive their normal volume of blood.

The Cardiac Cycle Has Two Phases, Diastole and Systole

Each beat of the vertebrate heart requires the coordinated activities of the atria and ventricles. The events that produce a single heartbeat are known as the **cardiac cycle**, which can be divided into two phases (Figure 47.10a). In the first phase, **diastole**, the ventricles fill with blood coming from the atria through the open AV valves. In the second phase, **systole**, the ventricles contract and eject the blood through the open semilunar valves.

The valves open and close as a result of pressure gradients established between the atria and ventricles, and the ventricles and arteries. During diastole, pressure in the ventricles is lower than in the atria and the arteries, and therefore the AV valves are open but the semilunar valves are closed (**Figure 47.10b**). Once the ventricles are electrically excited, they begin contracting and the pressure within the ventricles rapidly increases. This marks the beginning of systole. When the pressure in the ventricles exceeds that in the atria, the AV valves are forced closed. The closure of the AV valves makes the "lub" sound of the familiar "lub-dub" heard through a stethoscope. Ventricular pressure continues to rise and finally exceeds the pressure in the arteries, which opens the semilunar valves (**Figure 47.10c**). Blood is ejected out the semilunar valves into the arteries.

The pressure in the arteries increases during systole. As this is happening, the ventricles begin to return to their resting,

unexcited state, and the pressure in the ventricles falls below that in the arteries. The higher pressure in the arteries closes the semilunar valves, which prevents blood from flowing back into the heart. Closure of the semilunar valves creates the second heart sound, "dub," heard through a stethoscope. Throughout systole, meanwhile, the atria continue to fill with blood, which raises the pressure in the atria. Soon the pressure in the atria exceeds that in the relaxed ventricles, and the AV valves open again, bringing a new volume of blood into the ventricles and starting a new diastole.

Blood pressure in the arteries is highest during systole and lowest during diastole. For this reason, blood pressure is measured with two numbers, the systolic and diastolic pressures. For historical reasons, the units are usually given as mmHg (millimeters of mercury). Blood pressures are highest in mammals and birds, and lowest in fish and invertebrates. For example, a typical blood pressure in humans is around 120/80 mmHg (systolic/diastolic).

Unlike skeletal muscle, cardiac muscle does not fatigue. The vertebrate heart begins beating while an animal is still an embryo, and it never stops beating throughout life. The human heart beats about 70 times per minute. To envision how remarkable your heart is, try making a tight fist and relaxing it once every second for a few minutes. The muscles of your wrist and hand tire quickly, but the vertebrate heart does not. It continues beating steadily throughout life, approximately 2–3 billion beats in an average lifetime!

ECG Tracks Electrical Events During the Cardiac Cycle

An **electrocardiogram** (**ECG** or **EKG**) is a record of the electrical impulses generated during the cardiac cycle (**Figure 47.11**). Sensitive electrodes are placed on the surface of the body to monitor the wave of electricity initiated by the SA node, which travels through the atria, AV node, and ventricles. This procedure works because the body fluids that surround the heart conduct electricity, even the very weak impulses generated by a beating heart. The trace on an ECG reveals several waves of electrical excitation. The first is the P wave, which begins when the SA node fires and ends when the two atria completely depolarize. The next wave of excitation is a cluster of three waves, called the QRS complex. It begins when the branches from the AV node excite the ventricles and ends when both ventricles depolarize completely. The final wave is the T wave, which results from the repolarization of the ventricles back to their resting state. No wave is visible for atrial repolarization because it occurs simultaneously with the large QRS complex.

The ECG monitor displays both the amplitude (strength) of the electrical signal and the direction that the signal is moving in the chest. From this information, doctors can determine whether a person's heart is generating signals with a normal frequency, strength, duration, and direction.

Now that we have discussed the vertebrate heart, let's examine the vessels that transport blood to and from the heart and throughout the body.

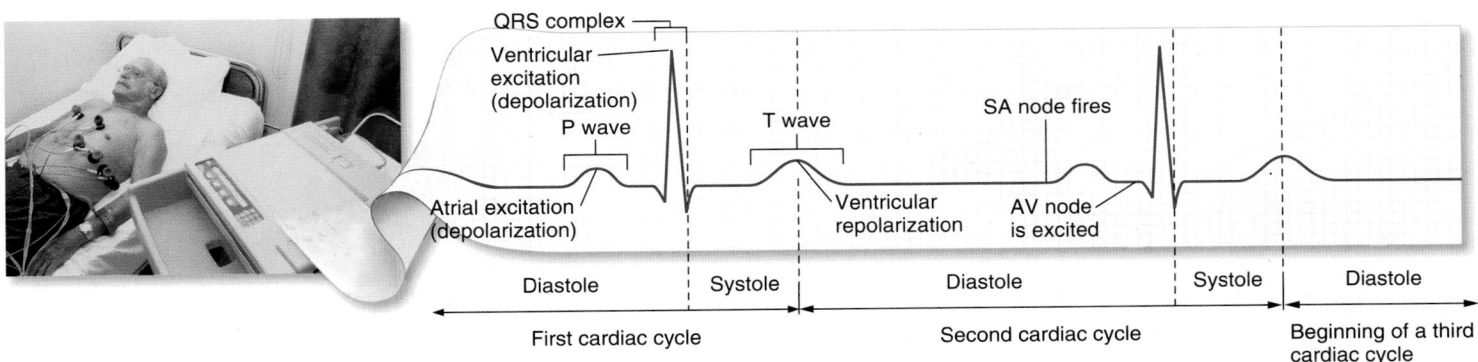

Figure 47.11   **An electrocardiogram (ECG).**   Electrodes placed on the skin detect electrical impulses occurring in the heart, and an electrocardiogram visualizes them. The resultant waveform is a useful indicator of cardiac health. Note that the ventricular wave (QRS complex) is taller than the atrial wave (P wave). This is because the ventricles are larger than the atria and generate more electrical activity.

Biological inquiry: Why is it possible to detect electrical changes in the heart using electrodes placed on the surface of the skin?

Arteries Distribute Blood to Organs and Tissues

Figure 47.12 illustrates the route that blood follows in a closed circulatory system. Recall that arteries are thick-walled vessels that carry blood away from the heart. They consist of layers of smooth muscle and connective tissue wrapped around a single-celled inner layer, the **endothelium**, which forms a smooth lining in contact with the blood (**Figure 47.13**). Because of the thick layers of tissue surrounding the endothelium, most dissolved substances cannot diffuse across artery walls. Instead, arteries act as conducting tubes that distribute blood leaving the heart to all the organs and tissues of an animal's body.

Figure 47.12   **Overview of blood flow through vessels in a closed circulatory system.**   Regions of gas exchange have been omitted for clarity.

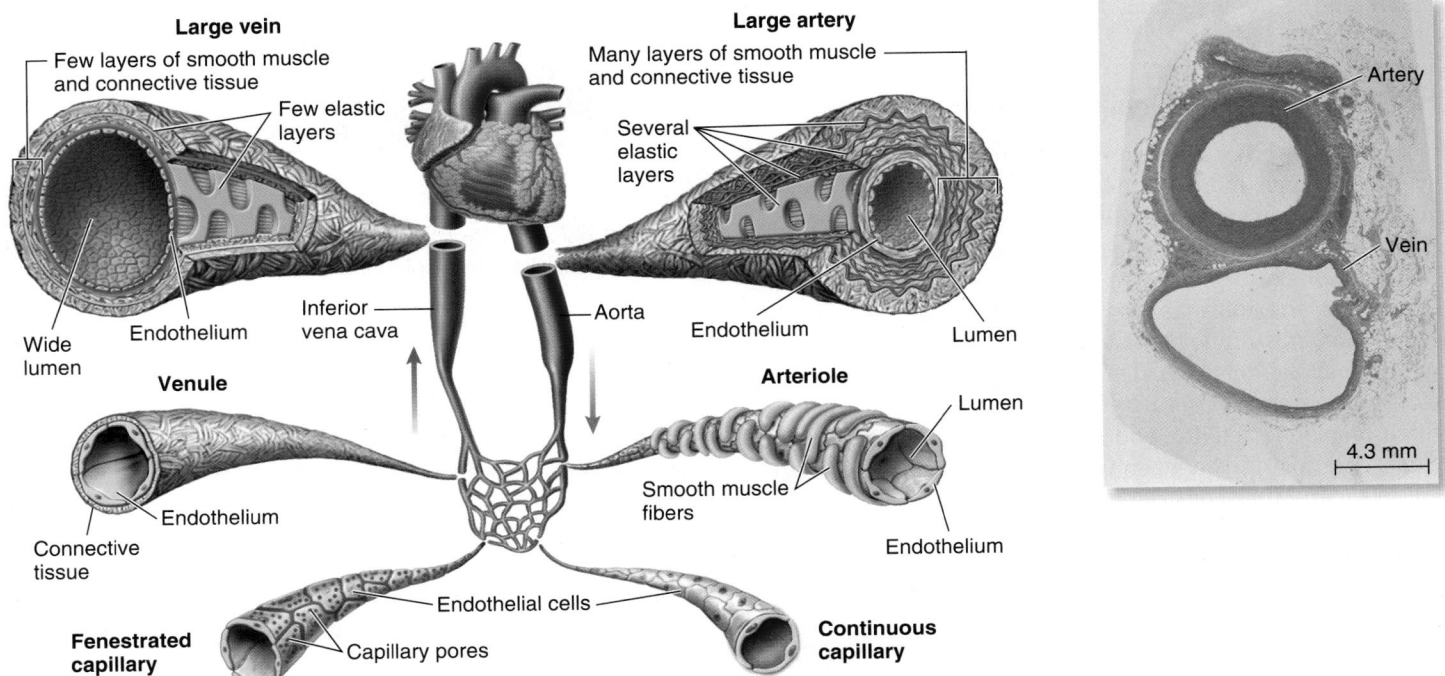

Figure 47.13   **Comparative features of blood vessels.**   Sizes are not drawn to scale. Inset: Light micrograph (enlarged four times) of a medium-sized artery near a vein. Note the difference between the two vessels in wall thickness and lumen diameter.

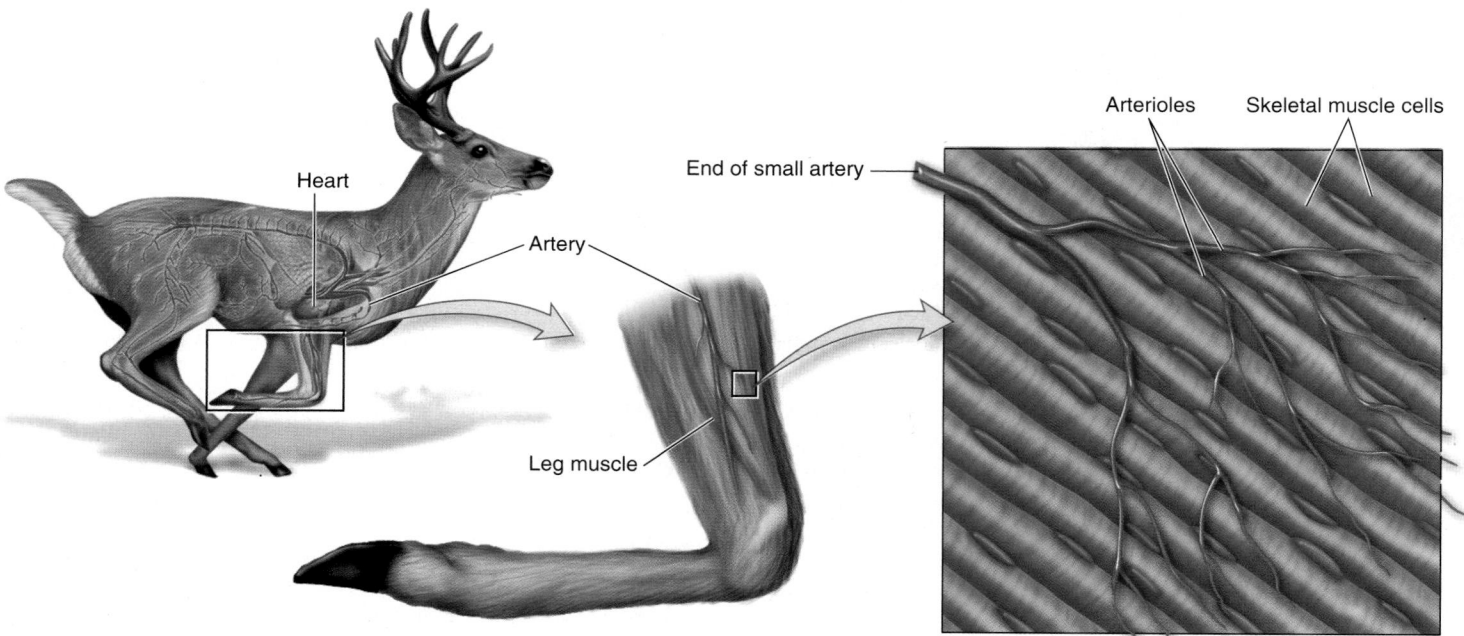

Figure 47.14 **Arteries and arterioles.** Arteries branch into smaller arteries, and into even smaller arterioles, in order to penetrate to all the cells of tissues and organs.

In vertebrates, the walls of the largest arteries, such as the aorta, also contain one or more layers of proteins that have elastic properties (Figure 47.13). As the aorta stretches to accommodate blood entering from the heart, the elastic proteins also stretch. When the heart relaxes as it readies for another beat, the elastic proteins in the aorta recoil to their original state, something like releasing a stretched rubber band. The elasticity of the large arteries stabilizes blood pressure by minimizing the spike in pressure during the beginning of systole and helping maintain pressure during diastole.

Arterioles Distribute Blood to Capillaries

As arteries carry blood away from the heart, they branch repeatedly and become more narrow to penetrate to the smallest reaches of an organ or tissue (**Figure 47.14**). Eventually, the vessels are little more than a single-celled layer of endothelium surrounded by one or two layers of smooth muscle and connective tissue. These **arterioles**, as they are called, deliver blood to the capillaries and distribute blood to regions of the body in proportion to metabolic demands. This is accomplished by changing the diameter of arterioles, such that they widen or dilate in areas of high metabolic activity, and narrow or constrict in inactive regions. Arterioles dilate when the smooth muscle cells around them relax, and they constrict when these cells contract.

Capillaries Are the Site of Gas and Nutrient Exchange

Arterioles branch into tiny thin-walled vessels called **capillaries**. At the capillaries, materials in the blood are delivered to the other cells of the body, and waste products and secretions from

Figure 47.15 **Erythrocytes (red blood cells, RBCs) moving through a capillary in single file.**

cells are delivered to blood. Capillaries are tubes composed of a single-celled layer of endothelium resting on an extracellular matrix called a basement membrane. Continuous capillaries have smooth walls while fenestrated capillaries contain tiny openings called fenestrations (see Figure 47.13). Capillaries are the smallest and narrowest blood vessels in the body, and they branch so extensively that essentially every cell in an animal's body is near one. The diameter of a capillary is about the same as the width of a red blood cell, so erythrocytes move through them in single file (**Figure 47.15**).

Blood enters capillaries under pressure that is created by the beating of the heart as blood is pumped into the arteries.

This pressure forces some of the water in blood out through the fenestrations and other tiny openings in capillary walls into the interstitial space. These openings are wide enough to permit water and small solutes, but not red blood cells and most proteins, to move out of the capillary. Once in the interstitial space, the solutes diffuse into body cells. Conversely, body cells release secretions and metabolic waste products into interstitial fluid, where they diffuse into capillaries and enter the circulation to be carried away.

If the water that leaves a capillary were to remain in the interstitial space, the volume of plasma in the blood would decrease and the interstitial fluid would swell. Most of the water that leaves at the beginning of a capillary, however, is recaptured at the capillary's end, for two reasons (**Figure 47.16**). First, near the arteriole the blood pressure inside the capillary is higher than the fluid pressure of the surrounding interstitial space, which creates a force that moves water out of the capillary. However, by the time blood reaches the end of a capillary, its pressure has decreased considerably, partly as a result of the loss of water. Second, proteins are trapped within the capillary

because they are too large to leak out through the capillary walls. This creates an osmotic force that draws water into the capillary from the interstitial fluid. Near the end of a capillary, the osmotic force drawing water into the capillary is greater than the pressure forcing water out, and thus water and dissolved solutes re-enter the capillary.

Despite the ability of capillaries to recapture their own fluid, the process is not 100% effective. Another set of vessels, those of the **lymphatic system**, collects excess fluid and returns it to the blood (see Chapter 53).

Venules and Veins Return Blood to the Heart

Once blood travels through the capillaries, picking up any substances secreted from the cells of the body, it then enters the **venules**, which are small, thin-walled extensions of the capillaries. The venules empty into larger vessels called **veins** that return blood to the heart for another trip around the circulation.

The walls of veins are much thinner, less muscular, and more easily distended or stretched than those of arteries (see Figure 47.13). Veins can fill with considerable volumes of blood, particularly veins in the lower parts of an animal's body, such as the legs, where gravity tends to cause blood to pool. In this way, veins serve as a reservoir of blood. When an animal is active, this reservoir can be tapped to provide the heart with sufficient blood to meet the metabolic requirements of the animal. At such times, the sympathetic nervous system stimulates the smooth muscles in the large veins to contract. This compresses the veins and helps force blood back to the heart.

By the time blood has traveled through the capillaries and reached the veins, the pressure of the blood is very low. Consequently, several factors assist the blood on its way toward the heart when flowing against gravity. One is the stimulation of the smooth muscle in leg veins, just described, by the sympathetic nervous system. In addition, the activity of skeletal muscles in the limbs assists the return of venous blood to the heart. For example, each time the leg skeletal muscles contract, they squeeze the veins passing through them (**Figure 47.17a**). This alone would not move blood upward toward the heart, though, because each time a vein is squeezed blood could be forced both up and down (**Figure 47.17b**). This does not happen because one-way valves inside veins ensure that blood returning from below the heart moves in only one direction, toward the heart. By contrast, veins located above the heart, like those in the necks of bipeds and some quadrupeds, lack valves because gravity pulls blood toward the heart.

You can easily observe the effects of gravity on venous blood flow. Hold your arms down by your side and look at the backs of your hands. The engorged blood vessels you see under the skin surface are veins. Now raise your arms above your head, and notice how the bulging veins quickly lose blood and become less visible. When blood from the veins returns to the heart, it must travel against gravity when your arms are at your side, and with gravity when your arms are elevated. As you just saw, blood drains from veins much more efficiently when gravity works in its favor.

Figure 47.16 **Water movement between capillaries and interstitial fluid.**   Water (and dissolved solutes other than proteins) exit capillaries near the arteriolar end because the capillary pressure is much greater than that of the interstitial fluid. As the volume of the water in the capillary decreases, however, the pressure within the capillary also decreases (but remains greater than interstitial fluid). Proteins remaining in the capillary contribute an osmotic force that tends to draw water back into the capillary. The combination of decreased pressure within the capillary and the osmotic force due to proteins leads to the recapture of much of the water that left the capillary. Lymph vessels drain any excess fluid from the interstitial space. For clarity, the system shown here contains a single capillary connecting an arteriole and venule, instead of the typical situation where capillaries divide into numerous branching vessels.

Leg muscle

Closed valve in a leg vein

0.4 mm

(a)

Blood flows toward heart.

Open valve

Contracting skeletal muscles of leg

When the leg muscle contracts, the lower valve stays closed while the upper valve opens. This causes blood flow toward the heart.

(b) Vein with valves

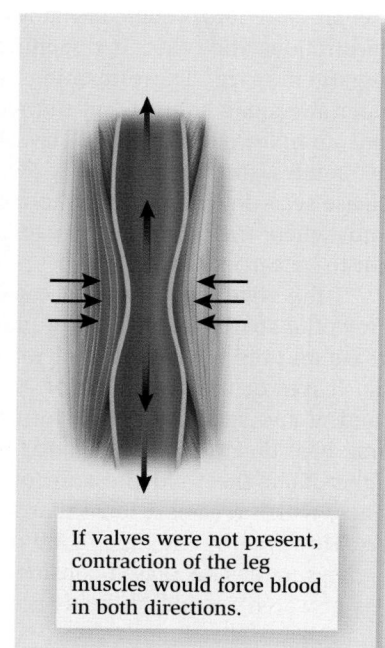

If valves were not present, contraction of the leg muscles would force blood in both directions.

Vein without valves

Figure 47.17   One-way valves.   (a) In some veins, one-way valves assist the return of blood to the heart against the force of gravity. Valves are typically present in the limbs, as shown in this light micrograph of a leg vein with a closed valve directed towards the heart. **(b)** Comparison of what would happen to blood flow through a vein with and without valves.

Biological inquiry: Giraffes have one-way valves in the veins of their long necks. Why?

47.4   Forces That Determine Blood Pressure and Flow

Blood pressure is defined as the force exerted by blood on the walls of blood vessels, and it is responsible for moving blood through the vessels. It is not the same in all regions of an animal's body. As we saw earlier, for example, the blood pressure in arteries is higher than that in veins. For this reason, we generally speak only of arterial pressure when discussing blood pressure. In this section, we will examine how blood pressure is regulated, how blood flow is distributed through the body, and the relationship between pressure and flow.

Resistance Determines Blood Pressure and Blood Flow

Resistance (R) refers to the tendency of blood vessels to slow down the flow of blood through their lumens, and it is a function of three variables, including vessel radius and length and blood viscosity. Wide, short tubes provide less resistance to flow than narrow, long tubes. Resistance is increased by the blood's viscosity, which is a measure of the blood's hematocrit. A high hematocrit increases the blood's viscosity—it makes the blood more sludgelike—and hinders its smooth flow through vessels.

The relationship between blood pressure, flow, and resistance is stated by Poiseuille's law, which is simplified here:

$$\text{Flow (F)} = \Delta\text{Pressure (P)}/\text{Resistance (R)}$$

Stated mathematically, blood flow through a blood vessel is directly proportional to the difference (Δ) in pressure of the blood at the beginning and end of the vessel, and inversely proportional to the resistance created by that vessel. Changes in any of these three variables determine how much blood flows through different body regions at any moment. The equation can be rearranged as $\Delta P = F \times R$, which demonstrates that blood pressure depends on both blood flow and resistance. Poiseuille's law applies to blood flow through a single vessel, an organ, or the entire body.

Arteriole Radius Is the Major Factor That Regulates Resistance

As stated earlier, selective and changeable distribution of blood throughout the body is a key adaptation of closed circulatory systems. Changes in arteriolar resistance are the major mechanism for increasing or decreasing blood flow to a region. In the short term, the length of an arteriole and the viscosity of blood do not normally change. Therefore, the radii of arterioles become the most important factor in determining minute-to-minute resistance.

The relationship between radius and resistance is not linear. Resistance is inversely proportional to the radius of the vessel raised to the fourth power: R α $1/r^4$, where α means "proportional to," and r is the radius of the arteriolar lumen. Let's consider an arteriole with a radius that increases by a factor of two. This would occur if the smooth muscles around the arteriolar wall relax sufficiently to allow the vessel to dilate and double its original radius. Because resistance is inversely proportional to the fourth power of vessel radius, an increase in radius by a factor of two will result in a decrease in resistance of 2^4 ($2 \times 2 \times 2 \times 2$), or 16-fold.

Vasodilation refers to an increase in blood vessel radius. **Vasoconstriction** refers to a decrease in blood vessel radius. Vasodilation and vasoconstriction are the most important mechanisms for directing blood flow to or away from specific regions of the body. The signals that control arteriolar radius come from three sources, including locally produced substances, hormones, and nervous system inputs. Locally, metabolic by-products such as carbon dioxide and lactic acid, as well as potassium ions and other substances secreted by metabolically active tissues, cause nearby arterioles to vasodilate immediately. According to Poiseuille's law, this permits more blood flow to the active region, facilitating oxygen and nutrient delivery and waste removal. Hormones secreted by glands throughout the body can also regulate arteriolar radius. For example, some hormones cause arterioles that deliver blood to the small intestine to vasoconstrict during fight-or-flight responses, routing blood away from the intestine and to areas of more immediate need such as the heart and skeletal muscles. Finally, smooth muscle cells that make up the outer wall of arterioles receive inputs from nerves that can stimulate the muscles to contract or relax. One of the most important regulators of blood vessel diameters, however, is a gas, as we see next.

FEATURE INVESTIGATION

Furchgott Discovered a Vasodilatory Factor Produced by Endothelial Cells

As we have seen, the cells surrounding blood vessels can produce ions, acids, and gases that diffuse to arteriolar smooth muscle cells and cause these cells to relax and vasodilate. Beginning in the 1970s, American scientist Robert Furchgott provided evidence that an unidentified substance released from within an artery—that is, from the endothelial lining of the vessel—also played a key role in regulating blood vessel radius.

For many years, Furchgott had used flattened strips of rabbit aorta in a culture bath to test whether various compounds stimulated or inhibited contraction of the smooth muscle in the artery. Although large arteries like the aorta do not show the dramatic vasodilation and vasoconstriction that arterioles do, they were more useful in experiments because of their larger size. Normally, when ACh is injected into an animal it causes vasodilation by relaxation of smooth muscle in arteriolar walls. However, in Furchgott's *in vitro* preparations, ACh caused the flattened strips of artery to contract. To determine if this apparent paradox could be the result of the method in which the tissues were prepared for *in vitro* tests, Furchgott compared the effects of ACh in different types of preparations. He discovered that ACh produced muscle relaxation when applied to circular rings of aorta, as described in step 1 of **Figure 47.18**.

Figure 47.18 The discovery that endothelial cells produce a vasodilating substance.

HYPOTHESIS The endothelium of arteries produces a vasodilatory substance.

STARTING MATERIALS Strips or rings of rabbit aorta.

Experimental level   **Conceptual level**

1   Attach a ring of aorta to a pressure gauge. Add acetylcholine (ACh) to bath medium. Record change in force (increased force = increased muscle contraction; decreased force = decreased muscle contraction).

ACh added to medium

Force transducer

Intact aorta with endothelium

In the presence of endothelium, ACh relaxes the muscles of the aorta, causing the vessel to widen.

2 Denude ring of its endothelium. The muscle layer is left behind. Retest effect of ACh.

Wooden stick is used to scrape away endothelium.

Endothelial cells

Muscle layer

The denuded ring is then placed in recording apparatus as before.

In absence of endothelium, ACh stimulates the muscles to contract, which narrows the vessel.

3 Perform "sandwich" experiment by combining denuded strip of aorta with an intact strip. Record effect of ACh on strip 1 alone, strip 2 alone, and strip 1 while part of a sandwich with strip 2.

Muscle layer only

Endothelium

Muscle layer

Strip 1

Strip 1   Strip 2   Sandwich   Strip 2

+ ACh →

Endothelium produces a vasodilating substance following ACh treatment that diffuses from strip 2 to strip 1, and causes strip 1 to relax.

4 **THE DATA**

Results of sandwich experiment

Force recorded on force transducer

Contraction — Relaxation

Time
Strip 1 alone: contraction

ACh

Time
Strip 2 alone: relaxation

ACh

Time
Strip 1 recorded while attached to strip 2: relaxation

How did Furchgott explain the results of muscle contraction when ACh was applied to the flattened strips? After examining the two preparations of tissue, Furchgott realized that the lining of endothelial cells along the inner wall of the vessel had been scraped away during preparation of the strips, but the endothelial lining was still present in the rings. He hypothesized that the endothelial cells must produce a factor whose secretion was stimulated by ACh, and that not only prevented contraction but even caused smooth muscles surrounding the blood vessel to relax. If so, this would explain the observation that when ACh was injected into rabbits it resulted in vasodilation, because the intact vessels of the animals contained their endothelial lining and thus resembled Furchgott's aortic rings.

To confirm his hypothesis, Furchgott performed two experiments. As shown in step 2 of Figure 47.18, he used a wooden rod to scrape away the endothelium from circular rings of aorta

and discovered that the relaxation effect of ACh was lost, as predicted. In step 3, he used a sandwich technique, in which he tested the effects of ACh on muscle contraction in a "denuded" strip of artery with the endothelium removed, and then again on the same strip, which was attached ("sandwiched") to another strip in which care was taken to ensure that the endothelium remained intact. When ACh was added to the water bath before the sandwich was formed, the denuded strip contracted, as expected. When the denuded strip was attached such that its muscle layer was exposed to the intact strip's endothelial surface, the denuded strip relaxed after ACh was added, demonstrating that the endothelium released a vasodilatory factor that diffused to the denuded strip and relaxed its muscles.

Two other scientists, Louis Ignarro and Ferid Murad, later determined that this factor was the gas nitric oxide (NO). The discovery of NO revealed a new category of signaling molecules.

Ignarro and Murad also revealed that nitroglycerin, which had for many years been used to treat patients with cardiovascular disease, acted by generating NO in the blood to produce vasodilation and lower blood pressure. Today NO is considered one of the most potent and important naturally occurring vasodilators in animals. Knowledge of the NO signaling pathway has led to new therapies for treating a wide variety of health disorders associated with blood vessels, including high blood pressure, glaucoma (an eye disease), and erectile dysfunction (see Chapter 51). For their efforts, Furchgott, Ignarro, and Murad shared the 1998 Nobel Prize.

Heart Rate and Size Determine Cardiac Output

Because blood vessels provide resistance to blood flow, the heart must beat forcefully enough to overcome that resistance. **Cardiac output (CO)** is the amount of blood the heart pumps per unit time, usually expressed in units of L/min. The CO depends on the size of an animal's heart and how often it beats each minute. Each beat, or stroke, of the heart ejects an amount of blood known as the **stroke volume (SV)** that is roughly proportional to the size of the heart. Thus, if we know the stroke volume of a heart, and can measure the heart rate (HR: the number of beats per minute), we can determine the CO. Simply put, CO = SV × HR.

Naturally, the CO of an elephant is far greater than that of a mouse, a sparrow, or a salmon. Smaller animals have smaller hearts and therefore smaller stroke volumes (see Table 47.1). The heart of the little brown bat, for example, is the size of a small pea, while the heart of a blue whale is as large as a cow. Their stroke volumes are similarly proportioned. However, small animals have faster heart rates than large animals. A hummingbird's or shrew's heart may beat over 1,100 times per minute during intense activity, while a whale's heart may beat only 10 times per minute, although the amount of blood ejected with each of those beats is enormous! Recall from Chapter 42 that metabolic rate is relatively greater in smaller animals. The higher heart rates of small animals give them a greater cardiac output than would be predicted for the size of their hearts, which helps them meet the extraordinary oxygen and nutrient demands of such highly metabolic organisms.

Cardiac Output and Resistance Determine Blood Pressure

Poiseuille's law can be adapted to the whole body. In this case, pressure refers to arterial blood pressure (BP) recorded using a blood pressure cuff, flow refers to CO, and resistance refers to the sum of all the resistance in all the arterioles (**total peripheral resistance**, or **TPR**). Thus, we have BP = CO × TPR. The CO and TPR determine the pressure the blood exerts in the arteries of a closed circulatory system. The greater the cardiac output and the higher the resistance, the higher the blood pressure. Imagine that the circulatory system is like a faucet (the heart) connected to a garden hose (the arteries) (**Figure 47.19**). If the faucet is fully open, analogous to maximal cardiac output,

Figure 47.19 Cardiac output, resistance, and blood pressure. A hose analogy shows the way in which cardiac output (the faucet) and resistance (constriction of the hose) impact blood pressure before and after the site of increased resistance.

the amount of water rushing into the hose will increase and so will the water pressure. If the faucet is only partially open, the water pressure will be lower. However, if the faucet is partially open but the end of the hose (that is, the arterioles) is constricted, the pressure of the water in the hose will increase between the faucet and the point where the hose is squeezed.

Arterial blood pressure, therefore, is a function of how hard the heart is working, and how constricted or dilated the various arterioles are. Blood pressure must be high enough for blood to perfuse through all body tissues even at the farthest extremities, but not high enough to burst blood vessels or force excess plasma out of capillaries. Blood pressure is roughly similar among most mammals. Although species differences exist, they are not as diverse as, for example, species differences in heart rates. Blood pressure is relatively low in nonmammalian vertebrates and in invertebrates, partly because of their smaller hearts.

47.5 Adaptive Functions of Closed Circulatory Systems

It is not enough for a circulatory system to pump blood and return it to the heart. The system must also adapt to changing conditions, including sleep, sudden activity, and emergencies such as blood loss or dehydration. In this section, we will examine how changes in the circulatory system help maintain cardiovascular activity when animals exercise or encounter changes in blood pressure arising from such events as blood loss following injury.

Circulatory Function Adapts to Exercise

Nearly all animals have periods of rest and activity. Therefore, the activity of the heart must adjust to fluctuating metabolic needs. For this reason, blood must be routed to different areas in proportion to their needs for oxygen and nutrients. Exercise provides a dramatic illustration of changes in circulatory function.

We often think of exercise as voluntary. In nature, exercise means an increase in locomotion for any reason. Typically, an animal exercises when it needs something, like food or shelter, or when it is escaping something, such as a predator. The increased activity may be momentary and moderate, as when a fish briefly swims faster to avoid the tentacles of an anemone, or it may last longer and be more intense, as when a cheetah sprints after a gazelle. Exercise may last for long periods of time, as when a salmon migrates from the sea to its freshwater spawning grounds or a migratory goose flies thousands of miles over several weeks. The circulatory systems of animals must quickly adapt to these increased metabolic demands and adjust to the intensity, duration, and type of exercise.

We have already noted some adjustments, such as vasodilation and vasoconstriction. These are mediated by changes in arteriolar radius, and therefore resistance. In addition, cardiac output increases during exercise, by increasing either stroke volume or heart rate, or both. Fish tend to increase stroke volume more than heart rate, while birds and mammals do the opposite. In humans, for example, stroke volume increases from approximately 75 mL (milliliters) at rest to 110 mL during exercise, a change of 1.5-fold, while heart rate may almost triple from 70 bpm to nearly 200 bpm. Thus, the combined effects of increasing stroke volume and heart rate raise the cardiac output by about 4.5-fold. Regardless of species, however, the result is the same: During exercise, cardiac output increases to supply more blood to the tissues that need it. As we will see in Chapter 48, the increased cardiac output is matched by an increase in the amount of oxygen obtained from the environment.

Whether cardiac output increases by changing stroke volume, heart rate, or both, one of the major mechanisms involved is an increase in blood levels of **epinephrine**, also known as adrenaline. Epinephrine is a hormone secreted by the adrenal glands, which receive input from the sympathetic nervous system during exercise. Epinephrine binds to receptors on heart muscle cells, making them contract more vigorously and thereby increasing stroke volume. In addition, epinephrine binds to receptors in the SA node cells of the atria, which stimulates the SA node to initiate impulses at a faster rate, thereby increasing heart rate. The latter effect of epinephrine is enhanced by the release of the neurotransmitter **norepinephrine** (or noradrenaline) directly from sympathetic nerve endings onto the SA node cells.

At rest, the heart does not pump out its entire content of blood. A reservoir of blood remains in each ventricle at the end of a beat. During exercise, this reservoir is tapped when epinephrine stimulates the heart muscles. In addition, the mechanical process of exercise itself enhances the return of blood to the heart, as skeletal muscles squeeze the veins within them and open their one-way valves. The combination of more blood getting returned to the heart and emptying the built-in reservoir of blood from the ventricles allows for increased stroke volume during exercise.

Because BP equals CO × TPR, you might imagine that increased cardiac output would raise blood pressure to possibly harmful levels during exercise. Although blood pressure does increase somewhat, it does not rise as much as might be predicted. This is because resistance decreases in the blood vessels of skeletal muscles to allow more blood to enter and, in terrestrial animals, in vessels of the skin to dissipate heat and prevent overheating. Therefore, CO increases but the total peripheral resistance in the body decreases, negating some of the effect on blood pressure.

Baroreceptors Maintain Blood Pressure

Within the walls of certain arteries, notably the aorta and the carotid arteries, are pressure-sensitive regions that contain the endings of nerve cells. These regions, known as **baroreceptors** (*baro*, pressure), are in constant communication with the brain (**Figure 47.20**). When the walls of these arteries stretch, the nerve cells stretch too, which increases the rate at which the nerves send action potentials to the brain. The brain interprets this increase as a signal that blood pressure is high—the arteries are being expanded more than normal. The brain responds by reducing the amount of norepinephrine released onto the cardiac SA node and inhibiting the glands that make epinephrine. The result is a decreased heart rate and stroke volume, resulting in lower cardiac output. In addition, nerves throughout

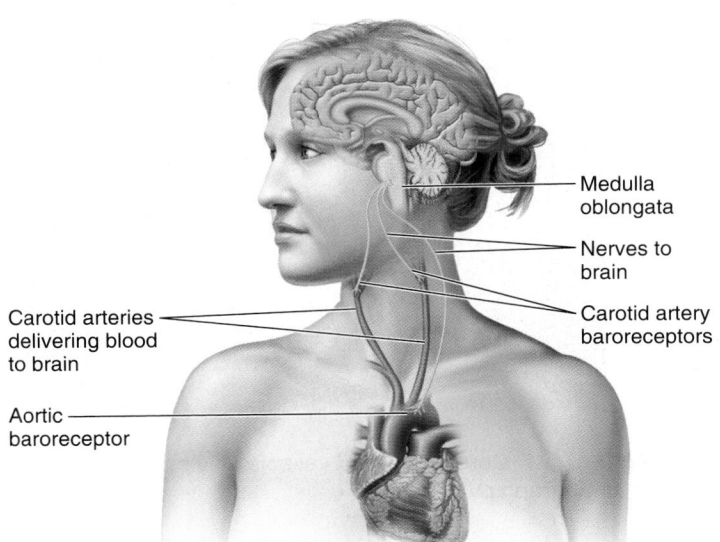

Figure 47.20 Location of the major baroreceptors. Increased blood pressure stretches arteries. This activates baroreceptors in the arteries, which send action potentials to the medulla oblongata. Nerve signals from the brain then initiate changes in heart function and total peripheral resistance that help restore blood pressure to normal.

Biological inquiry: In Chapters 43–45, we discussed the different types of ion channels found in neurons, and the ways in which such channels open and close. What type of channels would you expect to be activated when baroreceptors are stimulated?

the body release less vasoconstricting neurotransmitters onto the smooth muscle cells of arterioles, reducing total peripheral resistance. The combination of decreased cardiac output and resistance lowers blood pressure toward normal.

By contrast, if blood pressure falls below normal, the walls of these arteries would be less stretched than they normally are, and the baroreceptors would send fewer impulses to the brain. This results in a sequence of events that are opposite to those just described. More norepinephrine and epinephrine are secreted, thereby increasing cardiac output and vasoconstriction, thus increasing blood pressure toward normal. One situation in which blood pressure might drop is **dehydration**, which is a reduction in the amount of water in the body. Because blood is mostly water, dehydration reduces blood volume, and arteries—including those that contain baroreceptors—carry less blood than normal. Blood pressure could also drop due to **hemorrhage**, loss of blood from a ruptured blood vessel. Dehydration is primarily a problem for terrestrial animals, but hemorrhage may occur in any animal that has been injured. In a hemorrhage, there is a loss of both plasma and blood cells, and therefore hemorrhage may be more serious than dehydration. Like dehydration, though, hemorrhage results in reduced blood volume and therefore less baroreceptor activity.

Without baroreceptors, the circulatory system would have no means of relaying information about its status to the brain and the brain would be unable to initiate compensatory actions to maintain normal pressure. Baroreceptors even mediate nearly instantaneous adjustments to blood pressure, such as when you jump up quickly from a reclining position. Upon standing suddenly, gravity causes blood to pool a bit in the lower legs, temporarily reducing the volume and pressure of blood reaching the brain. This is why you may sometimes feel woozy when you stand up quickly. Baroreceptors immediately detect the change in pressure in your neck arteries, however, and cardiac output is rapidly increased to bring pressure back to normal until the system stabilizes.

47.6 Impact on Public Health

Cardiovascular disease—conditions affecting the heart and blood vessels—accounts for more deaths each year in the U.S. than any other cause, making it one of the most intensely researched areas of human health. Why is cardiovascular disease so devastating? One reason is that damage to these structures often occurs slowly, over many years, and without symptoms until the disease has reached late stages. In this section, we will consider several common cardiovascular disorders and their causes and treatments.

Systemic Hypertension and Atherosclerosis May Cause Heart and Blood Vessel Disease

Systemic hypertension, often called hypertension or high blood pressure, refers to an arterial blood pressure above normal. The normal range in humans varies from systolic/diastolic pressures of about 90/60 to 120/80 mmHg. A resting blood pressure above 140/90 mmHg defines hypertension, and values between 140/90 and 120/80 mmHg are considered borderline. Hypertension can have many causes, including obesity, smoking, aging, kidney disease, excess male hormones, and genetic factors. It can often be treated with diet, exercise, and drugs that cause vasodilation, thereby reducing arterial resistance.

Hypertension may also occur in the pulmonary side of the circulation. Pulmonary hypertension usually results from a diseased or damaged left ventricle that fails to pump out the usual amount of blood with each beat. This causes blood to "back up" in the pulmonary vessels, raising their pressure. Pulmonary hypertension is dangerous because the increased pressure forces more water out of the capillaries in the lung, and this accumulation of water hinders the ability of oxygen to diffuse from lung cells into the bloodstream. One of the chief symptoms of pulmonary hypertension, therefore, is shortness of breath. The water in the lungs gives them a congested sound in a stethoscope, and when the cause is failure of the heart to pump blood normally, this condition is called **congestive heart failure**.

Unlike pulmonary hypertension, systemic hypertension rarely has any symptoms. For this reason, it is important to have your

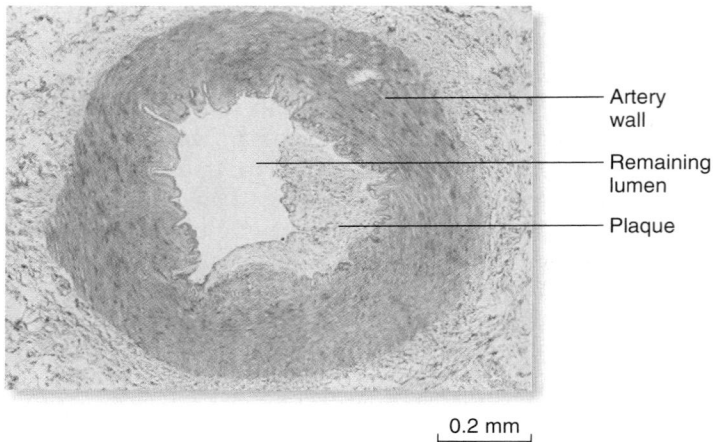

Figure 47.21   An atherosclerotic plaque in an artery.

0.2 mm

Figure 47.22   Coronary blood vessels as visualized by cardiac angiography.

blood pressure checked regularly. Systemic hypertension is dangerous because it can damage arteries, contributing to the formation of **plaques**—deposits of lipids, fibrous tissue, and smooth muscle cells—inside arterial walls and leading to a condition known as **atherosclerosis** ("hardening of the arteries") (**Figure 47.21**). Large plaques may occlude (block) the lumen of the artery entirely. In addition to hypertension, plaques are known to arise from a variety of factors, including calcium and fat deposits, and are correlated with obesity, high blood cholesterol levels, and smoking.

If plaques form in any artery, the regions of the body supplied with blood by that artery receive less oxygen and nutrients. Although atherosclerosis is dangerous anywhere, it is especially significant if it affects the **coronary arteries**, which carry oxygen and nutrients to the thick heart muscle. **Coronary artery disease** occurs when plaques form in the coronary vessels, and it can be life-threatening. One warning sign of coronary artery disease is **angina pectoris**, chest pain during exertion due to the heart being deprived of oxygen.

Myocardial Infarction Results in Death of Cardiac Muscle Cells

If a region of the heart is deprived of blood for an extended time, the result may be a **myocardial infarction (MI)** or **heart attack**. Heart attacks occur at localized regions of heart muscle that have died due to oxygen and fuel deprivation. Dead cardiac muscle tissue does not regenerate, diminishing the heart's ability to pump. If enough heart muscle dies, a heart attack can be deadly; each year in the U.S., more than 1 million people suffer a heart attack, and 40% of them are fatal. People often do survive heart attacks, however. In fact, the discomfort of a small heart attack may not even alarm someone enough to seek medical attention. Procedures are available that allow physicians to monitor the status of the coronary vessels in people suspected of having heart disease. The coronary vessels can be visualized by injecting a dye into a person's veins and then taking an X-ray

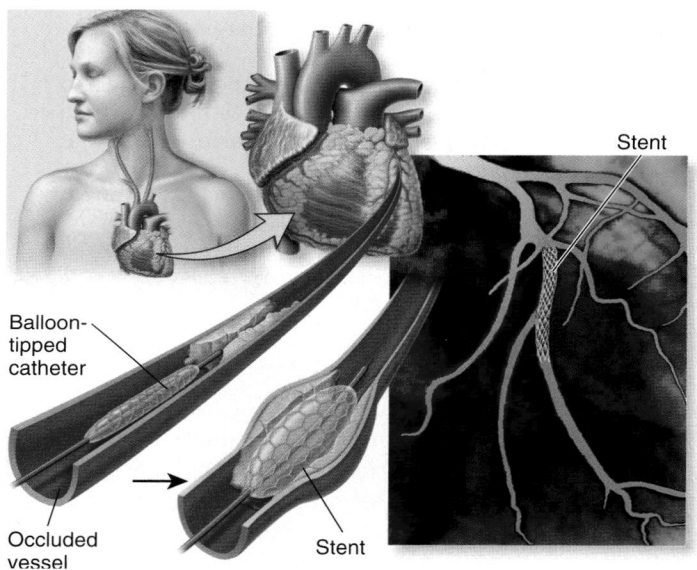

Figure 47.23   Treatments for blocked blood vessels. Balloon angioplasty can widen diseased arteries, followed by insertion of a stent. The inset shows a stent placed in a coronary artery of a human patient.

image of the chest, allowing a physician to determine if the vessels are narrowed by disease. This procedure is called cardiac angiography (**Figure 47.22**).

Several common treatments can restore blood flow through a blood vessel. One is **balloon angioplasty**, in which a thin tube with a tiny, inflatable balloon at its tip is threaded through the artery to the diseased area. Inflating the balloon compresses the plaque against the arterial wall, widening the lumen. In most cases, a wire-mesh device called a stent is inserted into the diseased artery after angioplasty has expanded it, providing a sort of lattice to hold the artery open (**Figure 47.23**). Another treatment is a **coronary artery bypass**, in which a small piece of healthy blood vessel is removed from one part of the body and surgically grafted onto the coronary circulation in such a way that blood bypasses the diseased artery.

Strokes Occur When Blood Flow to Part of the Brain Is Disrupted

If a blood vessel in the brain ruptures or is damaged by atherosclerosis, the result may be a **stroke**. In a stroke, the region of the brain normally receiving blood from the damaged vessel may die, resulting in permanent impairment. Depending on the region of the brain affected, this may reduce sensory function such as vision or hearing, or motor function. Some strokes arise when a blood clot forms elsewhere within the circulatory system, breaks free, and lodges in an artery or arteriole in the brain. Clots can form, for example, during heart attacks when heart muscle is destroyed. To minimize the risk of a stroke, patients who have suffered a heart attack may take blood-thinning drugs known as anticlotting factors.

CHAPTER SUMMARY

47.1 Gastrovascular Cavities

- Circulatory systems transport necessary materials to all cells of an animal's body, and transport waste products away from cells. The three basic types of circulatory systems are gastrovascular cavities, open systems, and closed systems.

- Cnidarians have a gastrovascular cavity that serves two functions: digestion of food and circulation of water containing the digested food. (Figure 47.1)

47.2 Open Circulatory Systems

- A cardiovascular system contains three components: blood or hemolymph, blood vessels, and one or more hearts. In an open circulatory system, the vessels open into the animal's body cavity. (Figure 47.2)

47.3 Closed Circulatory Systems

- In a closed circulatory system—found in earthworms, cephalopods (squids and octopi), and all vertebrates—blood and interstitial fluid are physically separated by blood vessel walls and differ in their components and chemical composition. (Figure 47.3)

- The closed circulatory systems of vertebrates can be divided into two major groups, single and double circulations. In crocodiles, birds, and mammals, oxygenated and deoxygenated blood are completely separated into the systemic circulation and pulmonary circulation. (Figure 47.4)

- Blood is a fluid connective tissue consisting of cells and (in mammals) cell fragments suspended in a solution of water containing dissolved nutrients, proteins, gases, and other molecules. Blood has four components: plasma, leukocytes, erythrocytes, and thrombocytes or platelets. (Figures 47.5, 47.6, 47.7, 47.8)

- All vertebrate hearts have at least one atrium and one ventricle. In hearts with more than one atrium or ventricle, connective tissue separates the atria and ventricles. Blood enters the atria from systemic or pulmonary veins, moves down a pressure gradient through the AV valves into the ventricles, and is pumped out through the semilunar valves into the systemic and pulmonary arteries. (Figure 47.9, Table 47.1)

- Many arthropods and decapod crustaceans have a neurogenic heart; all vertebrates have a myogenic heart.

- The electrical signals of myogenic hearts are generated either at the junction of the veins and the single atrium (in fish) or at the sinoatrial (SA) node (in other vertebrates). The electrical impulses spread through the atria and ventricles and stimulate the heart muscle to contract. The cardiac cycle has two phases, diastole and systole. (Figures 47.10, 47.11)

- Arteries are thick-walled vessels that carry blood away from the heart. Arterioles distribute blood to capillaries, and capillaries are the site of gas and nutrient exchange. (Figures 47.12, 47.13, 47.14, 47.15, 47.16)

- Blood leaving the capillaries empties into venules and then into veins, which return it to the heart. One-way valves help to return blood to the heart. (Figure 47.17)

47.4 Forces That Determine Blood Pressure and Flow

- Blood pressure is the force exerted by blood on the walls of blood vessels, and it is responsible for moving blood through the vessels.

- The endothelium of arteries produces a vasodilatory substance. (Figure 47.18)

- Blood flow through a vessel is directly proportional to the pressure of the blood entering the vessel, and inversely proportional to the resistance created by that vessel: Flow (F) = Pressure (P)/ Resistance (R). Arteriole radius is the major factor that regulates resistance.

- Cardiac output depends on the size of an animal's heart and how often it beats each minute.

- Cardiac output and total peripheral resistance determine blood pressure: BP = CO × TPR. Blood pressure must be high enough for blood to perfuse through all body tissues, but not high enough to burst vessels. (Figure 47.19)

47.5 Adaptive Functions of Closed Circulatory Systems

- The circulatory system adjusts to fluctuating metabolic needs through vasodilation, vasoconstriction, and increases in cardiac output. Baroreceptors maintain blood pressure by relaying information about circulatory system status to the brain, which initiates compensatory mechanisms. (Figure 47.20)

47.6 Impact on Public Health

- Cardiovascular disease accounts for more deaths each year in the U.S. than any other cause. Cardiovascular conditions include systemic hypertension, atherosclerosis, congestive heart failure, coronary artery disease, angina pectoris, myocardial infarction (MI, or heart attack), and stroke. (Figure 47.21)

- Cardiovascular diagnostic techniques and treatments include cardiac angiography, balloon angioplasty, and coronary artery bypass. (Figures 47.22, 47.23)

TEST YOURSELF

1. Hemolymph differs from blood in that it
   a. does not contain blood cells.
   b. is a mixture of blood and interstitial fluid.
   c. circulates through closed circulatory systems only.
   d. functions only in defense of the body and not transport.
   e. does not pass through the heart.

2. The heart chamber that functions as the receiving chamber is
   a. the atrium.
   b. the ventricle.
   c. the aorta.
   d. the pulmonary artery.
   e. the coronary artery.

3. Amphibians are unique in that
   a. they are the only vertebrates with an open circulatory system.
   b. oxygen diffuses into their blood through gills.
   c. oxygen diffuses into the blood from the lungs or through the skin.
   d. they have a four-chambered heart.
   e. none of the above.

4. A major advantage of a double circulation is that
   a. blood can be pumped to the upper portions of the body by one circuit and to the lower portions of the body by the other circuit.
   b. each circuit can pump blood with differing pressures to optimize the function of each.
   c. the oxygenated blood can mix with the deoxygenated blood before being pumped to the tissues of the body.
   d. less energy is needed to provide nutrients and oxygen to the tissues of the body.
   e. all of the above.

5. The function of erythrocytes is to
   a. transport oxygen throughout the body.
   b. defend the body against infection and disease.
   c. transport chemical signals throughout the body.
   d. secrete the proteins that form blood clots.
   e. all of the above.

6. The mammalian heart is a myogenic heart, meaning
   a. it is composed of four chambers.
   b. it acts as two pumps for two different circulations.
   c. contraction is regulated solely by the nervous system.
   d. contraction is regulated by specialized cells of the heart.
   e. it contains valves that maintain a one-way flow of blood.

7. During systole of the cardiac cycle
   a. the ventricles of the heart are relaxed.
   b. the ventricles of the heart are filling.
   c. the ventricles of the heart are contracting.
   d. the atria of the heart are contracting.
   e. both a and b.

8. Considering blood flow through a closed circulation, which of the following is the correct sequence of vessels beginning at the heart?
   a. arteriole, artery, capillary, vein, venule
   b. artery, capillary, arteriole, venule, vein
   c. vein, venule, capillary, arteriole, artery
   d. artery, arteriole, capillary, venule, vein
   e. artery, arteriole, capillary, vein, venule

9. Gas and nutrient exchange between the circulation and the tissues occurs at
   a. the arteries.
   b. the arterioles.
   c. the capillaries.
   d. the venules.
   e. the veins.

10. Which of the following factors determines blood pressure?
    a. cardiac output
    b. resistance
    c. arteriole diameter
    d. all of the above
    e. a and b only

CONCEPTUAL QUESTIONS

1. What are the three basic components of a true circulatory system?
2. List and briefly explain the components of blood.
3. Explain the cardiac cycle.

EXPERIMENTAL QUESTIONS

1. What observation did Furchgott make when testing the effects of acetylcholine on different treatments of rabbit aorta? How did he explain the results?

2. Based on his observations of the effects of acetylcholine on different preparations of rabbit aorta, what hypothesis did Furchgott propose to explain the relationship of acetylcholine and vasodilation? How did he test this hypothesis?

3. What did Furchgott conclude based on his experimental results?

COLLABORATIVE QUESTIONS

1. Discuss the difference between a closed and an open circulatory system.
2. Discuss the types of closed circulatory systems found in vertebrates.

---

www.brookerbiology.com
This website includes answers to the Biological Inquiry questions found in the figure legends and all end-of-chapter questions.

48

RESPIRATORY SYSTEMS

Llama (*Lama glama*) in the Andes Mountains.

I n the previous chapter we saw how circulatory systems function to transport blood or hemolymph throughout an animal's body. These fluids contain the nutrients that are absorbed from the digestive tract and must be delivered to all body cells. In addition, animals require oxygen, O_2. The oxygen is taken into the body by one of four major types of respiratory (breathing) structures, including the body surface, gills, lungs, or tracheae (insects). It then dissolves in plasma, binds to a carrier protein, and diffuses into cells. As described in Chapter 7, O_2 is used by mitochondria in the formation of ATP. One of the waste products of that reaction is carbon dioxide—CO_2—which diffuses out of cells. On the return trip to the respiratory organ, CO_2 is released into the environment, either the air or water. The process of moving oxygen and carbon dioxide in opposite directions between cells and the environment is called **gas exchange (Figure 48.1)**. This chapter deals with the mechanisms by which animals obtain and transport oxygen, rid themselves of carbon dioxide, and cope with the challenges imposed by their environments and their changing metabolic needs. We begin with an overview of the properties of gases in air and water, and why cells need oxygen.

Figure 48.1 **Overview of gas exchange between the environment, blood, and cells.** In vertebrates, oxygen diffuses from the environment across the cells of a respiratory organ, into the blood. From there, oxygen diffuses into tissue cells, where it is used to form ATP. Cells generate carbon dioxide as a waste product, which diffuses out of cells and into blood, and then across the respiratory organ into the environment.

48.1 Physical Properties of Gases

In 1774, the British physiologist Joseph Priestley demonstrated that heating mercuric oxide produced a pure gas that was "five or six times as good as common air," evidenced by its ability to sustain a burning candle or prevent a mouse from suffocating in an enclosed bell jar. Today, we know that the life-sustaining component of air is oxygen. Air is composed of about 21% oxygen, 78% nitrogen, and less than 1% carbon dioxide and other gases.

From a respiratory standpoint, nitrogen gas can usually be ignored because it plays no role in energy production, nor is it created as a waste product of metabolism. In this section, we begin by examining some of the basic properties of the gases important in respiration.

Gas Pressure Depends on Altitude

The gases in air exert pressure on the body surfaces of animals, although the pressure is not perceptible unless it changes suddenly, like when your ears "pop" in a descending airplane. This pressure is called **atmospheric** (or **barometric**) **pressure**. It can be measured by noting how high a column of mercury is forced upward by the air pressure in a device called a mercury manometer. The traditional unit of gas pressure, therefore, is the same as for blood pressure—mmHg—although there are several other ways of expressing pressure. For example, U.S. weather stations report atmospheric pressure in inches of Hg instead of mmHg, and most researchers today record gas pressures in kilopascals (kPa; 1 kPa = 7.5 mmHg).

At sea level, atmospheric pressure is 760 mmHg, and it decreases as we ascend to higher elevations (**Figure 48.2**). To visualize this, think how hydrostatic pressure increases at lower depths in an ocean. Now, imagine that you are standing at the bottom of an "ocean" of air. The closer you get to the "ocean" surface (the top of the atmosphere), the lower the pressure exerted on your body. At an elevation of 1700 m, as in Denver, the pressure is only about 640 mmHg, while at the highest elevations continuously inhabited by humans—in the Andes Mountains at 5,500 m—the pressure falls to 375 mmHg.

Atmospheric pressure is the sum of the pressures exerted by each gas in air, in exact proportion to their amounts. The individual pressure of each gas is its **partial pressure**, symbolized by a capital P and a subscript depending on the gas. Thus, the partial pressure of oxygen (P_{O_2}) in the air we breathe is 21% of the atmospheric pressure at sea level, or 160 mmHg (0.21×760 mmHg).

The percentage of oxygen and other gases in air remains the same regardless of altitude, but the lower the atmospheric pressure, the lower the partial pressure of oxygen in air. The partial pressure of oxygen in the environment provides the driving force for its diffusion from air or water across the respiratory surface and into the blood. All gases diffuse along pressure gradients, from regions of high pressure to regions of lower pressure. This is analogous to the diffusion of dissolved solids from regions of high to low concentration, or the way heat moves from hotter to cooler regions. Consequently, the rate of oxygen diffusion into the bloodstream of a terrestrial animal at sea level decreases when the animal moves to a higher altitude, where the partial pressure of oxygen is lower.

Pressure, Temperature, and Other Solutes Influence the Solubility of Gases

Gases dissolve in solution, including fresh water, seawater, and all body fluids. It is while in solution that gases like oxygen exert their biological effects. However, most gases dissolve rather poorly in water. There is less oxygen in a given volume of water than in air, for example, which limits how active many aquatic organisms can be. Among the factors that influence the solubility of a gas in water, three are particularly important: the pressure of the gas, the temperature of the water, and the presence of other solutes.

The higher the pressure of a gas that comes into contact with water, the more of that gas that will dissolve, up to a limit that is specific for each gas at a given temperature. Not all gases dissolve equally. Oxygen is less soluble in water than carbon dioxide, but more soluble than nitrogen. The partial pressure of gas in water is given in the same units as atmospheric pressure—mmHg. For example, if a solution of water is in contact with air that contains a P_{O_2} of 160 mmHg, the P_{O_2} of the water is also 160 mmHg.

Solubility also depends on the water's temperature, with more O_2 and CO_2 dissolved in 1 L of cold water than in 1 L of warm water. You can observe the effect of temperature with the following experiment. Open two bottles of a carbonated beverage and keep one on ice and the other at room temperature. Note that the "fizz" (actually carbon dioxide gas that has been added to the drinks) of the room-temperature beverage disappears sooner as CO_2 escapes due to its lower solubility at that temperature. Why? At higher temperatures, gases in solution have more thermal energy and are therefore more likely to escape the liquid. In terms of biological systems, the effect of water temperature means that animals inhabiting very warm waters have less oxygen available to them than animals living, for example, in the waters off Antarctica. In addition, animals that live in shallow water like ponds or tide pools, where the water temperature can fluctuate widely over a single day and night, must be able to adapt to dramatic daily swings in oxygen availability.

Finally, the presence of other solutes, namely salts, reduces the amount of gas that dissolves in water. Thus, warm salty blood dissolves less oxygen than does cold water. Likewise, less oxygen dissolves in seawater than fresh water at any given temperature and pressure. Animals living in cold freshwater lakes, therefore, typically have more oxygen available to them than do animals living in warm salty seas (although other factors can also influence the amount of oxygen, such as water depth and the amount of aquatic plant life).

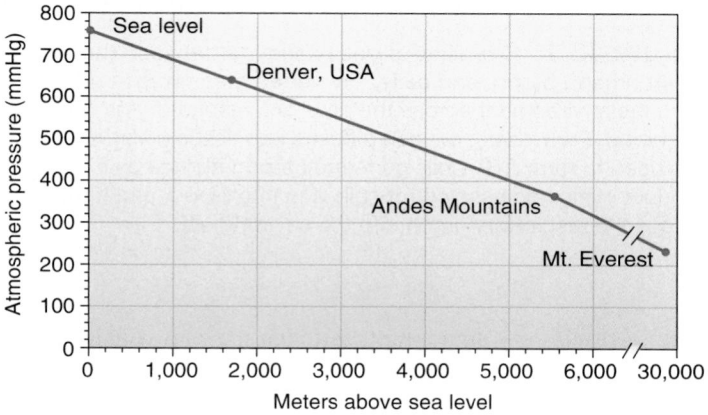

Figure 48.2   Atmospheric pressure decreases as altitude increases.

48.2 Types of Respiratory Organs

Despite the wide array of animal shapes, sizes, and habitats, animals obtain oxygen from their surroundings by only a few mechanisms. These include diffusion of oxygen across the body surface or across specialized respiratory or gas-exchange organs, including gills, tracheae, and lungs. **Ventilation** is the process of bringing oxygenated water or air into contact with a gas-exchange surface. In this section, we will examine the mechanisms of ventilation that are used by different animals.

The Structures of Respiratory Organs Are Adapted for Gas Exchange

Whether an animal utilizes respiratory organs or its body surface to obtain oxygen and expel carbon dioxide, all respiratory organs share certain common features. All of them have moist surfaces in which the gases can dissolve and diffuse. Further, all have adaptations that increase the amount of surface area available for gas exchange. As an example, if the gas-exchange regions of the human lungs were fully flattened, their total surface area would cover a tennis court!

The extensive surface area of respiratory organs is coupled with equally extensive blood flow, except in the special case of insects, described later in this chapter. The density of capillaries surrounding gills and lungs, for example, is among the highest found anywhere in an animal's body. The greater the amount of blood flowing to a respiratory surface, the more efficient will be the delivery and removal of gases.

Thick barriers greatly slow the diffusion of a gas. As a result, respiratory organs tend to be thin, delicate structures, which also means that they are easily damaged. For instance, the inner structures of fish gills are so thin that they cannot support their own weight out of water, and they collapse when exposed to air.

Terrestrial and Aquatic Animals Face Different Challenges

Terrestrial and aquatic animals face different challenges to gas exchange. First, as noted earlier, aquatic animals generally have less available oxygen than do terrestrial animals, which means that an active aquatic animal requires a very efficient means of extracting oxygen from water. Second, unlike terrestrial animals, aquatic animals must adapt to fluctuating oxygen availability when the temperature changes. Terrestrial animals, by contrast, must cope with the dryness of air. Passing dry air over the lung surfaces runs the risk of drying out the lungs, which would damage them, dehydrate the animal, and reduce the ability of gases to dissolve.

Aquatic animals bring water, not air, over their respiratory surfaces and therefore do not usually risk dehydration in the way that terrestrial animals do. However, moving water over a gas-exchange surface creates its own challenges. First, because water is denser than air, more muscular work is required to move water than air. Second, cold seawater or fresh water moving over the richly vascular gills removes considerable heat from

an animal's body, due to the high heat capacity of water. Third, moving fresh or salt water over a gas-exchange surface can create osmotic movement of water across the surface, potentially resulting in water imbalances in an animal's body.

Some Animals Use Their Body Surface to Exchange Gases with the Environment

In those invertebrates that are only one or a few cell layers thick, oxygen and carbon dioxide can diffuse directly across the body surface. In this way, oxygen reaches all the interior cells, in some cases without any specialized transport mechanism or circulatory system. Even in some large, complex animals, the body surface may be permeable to gases (**Figure 48.3**).

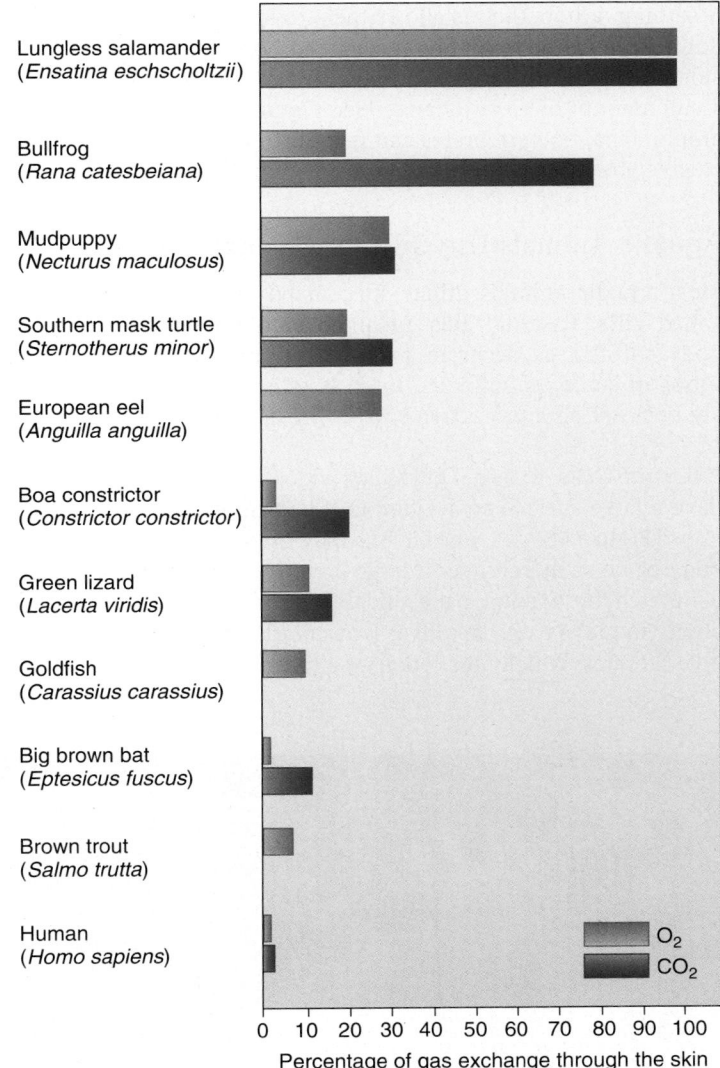

Figure 48.3 Gas exchange across body surfaces. This occurs readily in amphibians and some fish. Among mammals, gas exchange occurs to a very limited extent in species with body features that have high surface area, such as bat wings.

Amphibians, and a few species of fish including eels, have unusually permeable skin. On land, amphibians rely on their lungs and their moist skin to obtain oxygen and release carbon dioxide. Underwater, however, the lungs are no longer useful because they are not suited for extracting oxygen from water. Although oxygen diffusion across the skin is less efficient than in the lungs, this secondary ability to exchange gas permits prolonged times spent underwater for amphibians.

Many vertebrates such as seals are amphibious, in that they spend part of their lives on land and part in water, but only amphibians have skin that is highly permeable to gases. With some exceptions, very little gas diffuses across the body surfaces of fish, reptiles, birds, and mammals. Some reptiles, however, such as marine iguanas, can obtain oxygen from the water across the highly permeable mucous membranes of the nose, mouth, and anus. Although this provides less oxygen than air breathing, it nonetheless allows them to spend time underwater. Some animals have highly specialized body regions with considerable surface area—like the wings of a bat—that permit a small amount of gas exchange (see Figure 48.3), but except for amphibians, no vertebrates can rely for long periods of time exclusively on their skin to obtain oxygen and remain active.

Aquatic Animals Use Gills for Gas Exchange

Most aquatic animals utilize specialized respiratory structures called **gills**. External gills are uncovered extensions from the body surface, as occur in many invertebrates and the larval forms of some amphibians. Internal gills, which occur in fish, are enclosed in a protective cavity.

External Gills External gills vary widely in appearance, but all have a large surface area, often in the form of extensive projections (**Figure 48.4**). External gills may exist in one region of the body or be scattered over a large area. In many cases, they are ventilated by waving back and forth through the water. The ability to move external gills is particularly important for sessile invertebrates, which must otherwise rely on sporadic local water

currents or muscular efforts of their bodies to create local currents for ventilation.

Despite the success of marine invertebrates, external gills have several limitations. First, they are unprotected and therefore susceptible to damage from the environment. Second, because water is much more dense than air, considerable energy is required to continually wave the gills back and forth through the water (think of the difference between waving your hands through air or water). Finally, their appearance and motion may draw the attention of predators.

Internal Gills By contrast, fish have internal gills, which are covered by a bony plate called an operculum (**Figure 48.5a**). Fish gills have a more uniform appearance than external gills and are confined within the opercular cavity, which protects the gills and helps streamline the fish body.

The internal gills of fish contain numerous rows of **lamellae**, platelike structures that branch from structures called filaments. The filaments, in turn, arise from the main support structure of the gills, the gill arches (**Figure 48.5b**). Blood vessels run the length of the filaments, with oxygen-poor blood traveling through the afferent vessel along one side of the filament, and oxygen-rich blood returning through the efferent vessel along the other side. Within the lamellae are numerous capillaries, all oriented with blood flowing in the same direction, from the oxygen-poor vessel to the oxygen-rich one.

Water enters the fish's mouth and flows between the lamellae in the opposite direction to blood flowing through the lamellar capillaries. This arrangement of water and blood flow is another example of a **countercurrent exchange mechanism** (see Chapters 40 and 42). As oxygenated water encounters the lamellae, it comes into contact with deoxygenated blood in the gill capillaries. Recall that oxygen diffuses along pressure gradients from a region of high oxygen pressure to low oxygen pressure. Thus, oxygen diffuses from the water into the capillaries of the lamellae. As water continues to flow across the lamellar surface, it encounters regions of capillaries that have not yet picked up oxygen—in other words, even as oxygen begins to diffuse

(a) Nudibranch with clustered gill tufts

(b) Larval salamander

Figure 48.4   External gills. The gills of these animals have extensive projections to increase surface area, which facilitates oxygen diffusion from the surrounding water.

from the water into the gill capillaries, a sufficient pressure gradient remains along the lamellae to permit diffusion of more of the remaining oxygen from the water. This is an extremely efficient way to remove as much oxygen from the water as possible before the water passes out of the operculum.

Fish use one of two mechanisms to ventilate their gills. The first method, **buccal pumping** (the buccal cavity refers to the mouth), makes use of the muscles of the mouth and operculum to create a hydrostatic pressure gradient for water to follow in one direction (**Figure 48.6a**). First, the jaw is lowered, which enlarges the buccal cavity and lowers the pressure in the mouth, something like a suction pump. This draws water into the mouth, raising the water pressure again. At roughly the same time, the operculum begins to swing out from the body, enlarging the opercular cavity and lowering the water pressure there. This second suction pump is stronger than the buccal pump, so water flows down its hydrostatic pressure gradient from the outside to

the mouth, through the opercular cavity. Next, the mouth closes and the buccal cavity constricts. This creates positive pressure that forces water across the gills and out through the operculum, which is now open. A flap of tissue at the back of the mouth helps prevent accidental flow of water down the esophagus. Thus, fish do not swallow the water they inhale but instead send it on a one-way journey across their gills. In buccal pumping, a fish can remain stationary and still ventilate even in stagnant water by drawing water into its mouth.

The second way in which some fish ventilate their gills is by swimming with their mouths open, in essence using their large swimming muscles to bring water into their buccal cavity and from there across their gills. This method of ventilation, called **ram ventilation** (**Figure 48.6b**), is more energy efficient than buccal pumping. Ram ventilation still requires energy expenditure, but in this case the energy is used for swimming or for using muscles to remain stationary while facing upstream in moving water.

Figure 48.5 Structure of fish gills. (a) The gills are protected beneath the operculum, which has been pulled open in this photo to reveal the gills underneath. (b) The gills are composed of gill arches, from which numerous pairs of filaments arise. Platelike lamellae with large amounts of surface area are arrayed along the filaments. Blood vessels from the filaments form capillaries in the lamellae. Blood flows through the capillaries in the opposite direction of water flowing between lamellae, a process called countercurrent exchange. (c) Lamellae as revealed in a scanning electron micrograph.

Biological inquiry: Why do fish die when out of water?

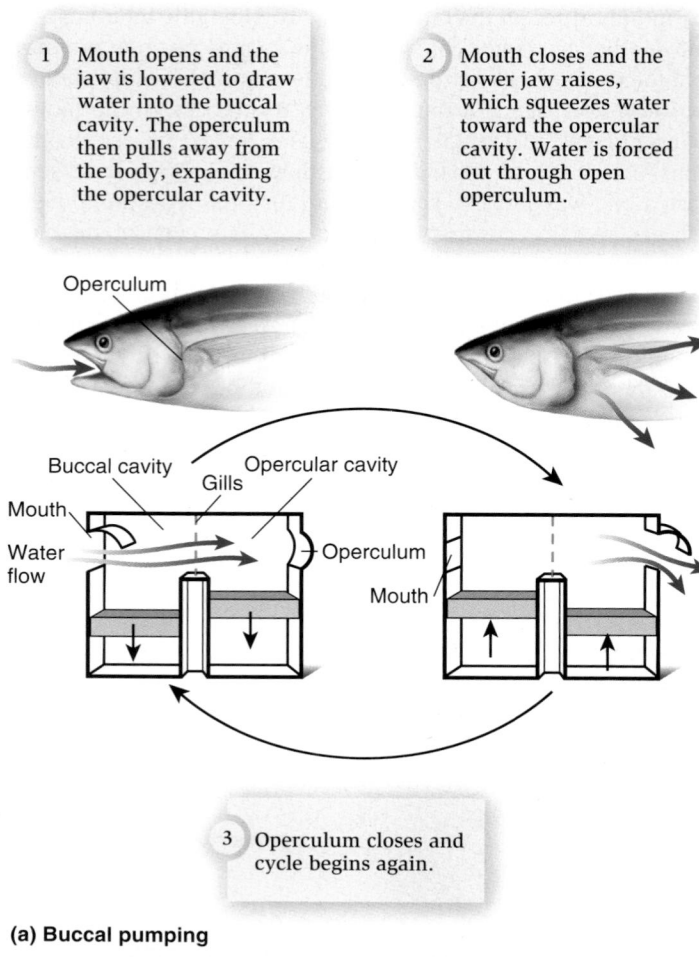

1 Mouth opens and the jaw is lowered to draw water into the buccal cavity. The operculum then pulls away from the body, expanding the opercular cavity.

2 Mouth closes and the lower jaw raises, which squeezes water toward the opercular cavity. Water is forced out through open operculum.

Operculum

Buccal cavity   Opercular cavity

Gills

Mouth

Water flow

Operculum

Mouth

3 Operculum closes and cycle begins again.

(a) Buccal pumping

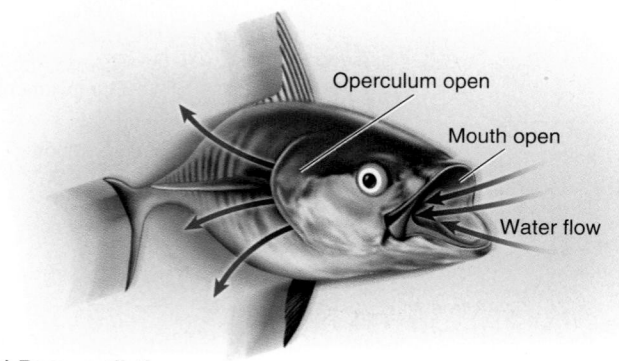

Operculum open

Mouth open

Water flow

(b) Ram ventilation

Figure 48.6 Mechanisms of gill ventilation. **(a)** Buccal pumping. Water is drawn into the mouth by lowering the jaw. Water exits via the operculum, which moves away from the body to draw water from the mouth. Flaps of tissue in the mouth, throat, and operculum help prevent water from moving in the reverse direction. The expansion of the buccal and opercular cavities acts like a suction pump, and closing the cavities acts like a pressure pump. **(b)** Ram ventilation. Many fish have the ability to ventilate by facing upstream in moving water with their mouths open. The water moves through the mouth and operculum. In some species, like this tuna, ram ventilation occurs as the animal swims forward with its mouth open. There is no buccal pump in this case.

Many fish employ both methods of ventilation, using buccal pumping when swimming slowly or in stagnant water, and switching to ram ventilation when swimming quickly or facing upstream. Some fish—like tuna—can only ram ventilate and therefore rarely stop moving.

Both buccal pumping and ram ventilation are **flow-through systems**—water moves unidirectionally such that the gills are constantly in contact with fresh, oxygenated water. As we will see, many air-breathing animals use a less efficient method of ventilating their lungs, called tidal ventilation, in which fresh air is breathed in and stale air is breathed out through the same route. Despite the efficiency of flow-through systems for maximizing oxygen extraction from water, they are energetically costly because of the work required to overcome the density of water. Fish may devote up to 10–20% of their resting metabolic rate simply to ventilating their gills, while a typical air-breather may allocate only 1–2% of its total energy usage to ventilating the lungs at rest.

Insects Use Tracheal Systems to Exchange Gases with the Environment

Except for terrestrial isopods like the pill bug (*Armadillidium vulgare*), which breathe air through very small, moist gills, the delicate nature of gill lamellae makes them unsuitable for gas exchange in air. Air-breathing probably evolved as an adaptation in aquatic animals inhabiting regions that were subject to periodic drought. One of the major mechanisms that animals evolved to breathe air is the tracheal system found in insects.

Running along the surface of both sides of an insect's body are tiny paired openings to the outside, called spiracles. Arising from the spiracles are **tracheae** (singular, trachea; **Figure 48.7**), sturdy tubes that are reinforced with chitin to keep them open. Tracheae branch extensively into ever-smaller tubes called tracheoles, eventually becoming small enough to bring their tips into contact with virtually every cell in the body. At their tips, tracheoles are filled with a small amount of fluid. Air moving down the tracheoles comes into contact with this fluid. Oxygen from the air dissolves in the fluid, and from there it diffuses across the tracheole wall and into nearby cells. Carbon dioxide diffuses in the opposite direction, from cells into the tracheoles, and from there to the environment.

When an insect's oxygen demands increase, muscular movements of its abdomen and thorax draw air into and out of the tracheae like a bellows, a phenomenon first demonstrated in the early 20th century by Danish physiologist and Nobel Laureate August Krogh, who studied the mechanics of ventilation in locusts. Krogh showed that an insect's muscles and tracheal system match ventilation with the animal's exercise intensity and oxygen requirements. This is particularly important in flying insects, whose metabolic demands are high. The faster an insect flies, the more oxygen it uses. Likewise, the more intensely and rapidly the body wall muscles move, the more air is moved in and out of the tracheae.

As discussed in Chapter 47, the open circulatory system of insects does not participate in gas exchange. Oxygen diffuses directly from air to trachea to tracheoles and finally to body cells.

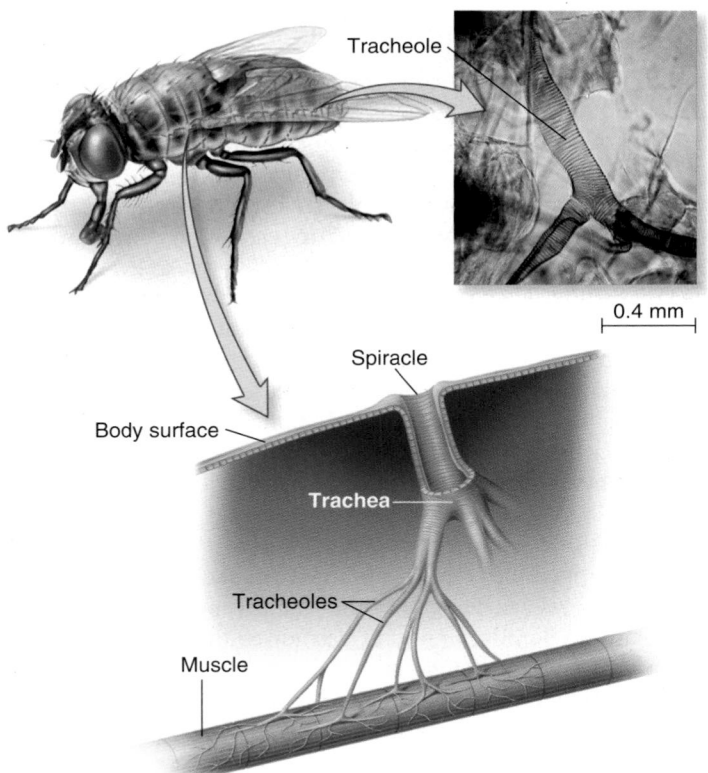

Tracheole

Spiracle

Body surface

Trachea

Tracheoles

Muscle

0.4 mm

Figure 48.7 **The tracheal system of insects.** Oxygen diffuses directly from the fluid-filled tracheole tips to cells that come into contact with the tips. The circulatory system plays no role in gas exchange. The micrograph illustrates a few branching tracheoles.

Biological inquiry: How might the design of the respiratory system of insects be related to why insects tend to be smaller than vertebrates?

This mechanism of ventilation and oxygen delivery is very efficient. The metabolic rate of insect flight muscles is among the highest known of any tissue in any animal, and the tracheal system supplies enough oxygen to meet those enormous demands.

Air-Breathing Vertebrates Use Lungs to Exchange Gases

Except for some rare amphibians such as the lungless salamander, all air-breathing terrestrial vertebrates use **lungs** to bring oxygen into the circulatory system and remove carbon dioxide. Scorpions and some spiders are the only invertebrates to have lungs, although those structures—called "book lungs"— actually resemble gills more than lungs. Lungs are internal paired structures that arise from the pharynx (throat). All lungs receive deoxygenated blood from the heart and return oxygenated blood back to the heart. Depending on the class of vertebrate, lungs may be filled using positive or negative pressure, and be ventilated by a tidal system or flow-through system.

Most amphibians have lungs that are simple sacs with relatively little surface area, and thus they are less effective surfaces for oxygen diffusion than are the lungs of other vertebrates. A 30-g mouse's lung, for example, has nearly 50 times more gas-exchange surface per cubic centimeter of lung tissue than a 30-g frog's lung.

The method by which amphibians ventilate their lungs is similar in some ways to the buccal pumping of fish. A frog, for example, lowers its jaw, which decreases the air pressure inside the mouth cavity. According to Boyle's law, gas pressure and volume are inversely related (**Figure 48.8**). For example, when the volume in which a gas is contained increases, the pressure of the gas in that container decreases. By expanding its mouth cavity, therefore, a frog creates a pressure gradient for air to move from the atmosphere (higher pressure) into its mouth cavity

Rest

Pressure gauge

Piston

Increased volume (analogous to inhaling)

Reduced volume (analogous to exhaling)

Gas molecules have thermal energy and exert a pressure on the walls of the cylinder.

Fewer collisions of gas molecules occur with cylinder wall (decreased pressure).

Greater number of collisions of gas molecules occur with cylinder wall (increased pressure).

Figure 48.8 **Boyle's law.** At a constant temperature, the volume and pressure of a gas are inversely related. This relationship creates the gas pressure gradients needed to ventilate the vertebrate lungs.

(lower pressure). The mouth and nostrils are then closed, and the mouth cavity is constricted by raising the bottom jaw. This raises the air pressure in the mouth, forcing air through a series of valves down into the lungs. Thus, **positive pressure filling** means that frogs and most other amphibians gulp air and force it under pressure into the lungs, as if inflating a balloon. They may do this many times in a row before finally exhaling.

Except for a few species of reptiles that also depend on positive pressure filling, all other terrestrial vertebrates use a different method of ventilation, negative pressure filling. To understand how negative pressure filling works, we will first examine the anatomy of the respiratory tract in vertebrates, using the mammalian and avian systems as models.

48.3 Design and Function of Mammalian and Avian Respiratory Systems

The **respiratory system** includes all components of the body that contribute to the exchange of gas between the external environment and the blood. In mammals, this includes the nose, mouth, airways, lungs, and muscles and connective tissues that encase these structures within the thoracic (chest) cavity. In this section, we begin by examining the mechanisms by which mammals ventilate their lungs. We then contrast this with the highly specialized and unique way in which birds ventilate their lungs.

Air Is Warmed, Humidified, and Filtered in the Nose and Mouth

In humans and other mammals, air first enters the nose and mouth (**Figure 48.9a**), where it is warmed and humidified. These processes protect the lungs from drying out. While in the nose, the air is cleansed as it flows over a coating of sticky mucus in the nasal cavity. The mucus and hairs in the nasal cavity trap some of the larger dust and other particles that are inhaled with air. These are then removed by the body's defense cells or swallowed.

The inhaled air from the mouth and nose converges at the back of the throat, or **pharynx**. From there, air passes through the **larynx**. Within the larynx are the vocal cords, folds of tissue through which air passes to create sound. Air flows from the larynx to the **trachea**, which leads to the lungs. At the opening to the trachea, known as the glottis, a flap of tissue called the epiglottis prevents food from entering the trachea by closing when food is swallowed.

The trachea is ringed by cartilage that provides rigidity and ensures that the trachea always remains open. The inner walls of the trachea are lined with cilia and cells called goblet cells. These cells secrete mucus into the lumen of the trachea, coating the cilia (**Figure 48.9b**). This mucous layer captures potentially harmful or irritating particles that escaped the nasal cleaning mechanism. The cilia beat constantly toward the mouth and move the mucus and its trapped particles into the mouth, where it can be expelled or swallowed. Without these active cilia, it would require repeated coughing to force the mucus out of the trachea. This is why humans who smoke tobacco develop a chronic "smoker's cough." The smoke and its components paralyze the cilia, causing mucus to pool in the trachea.

The trachea branches into two smaller tubes, called **bronchi** (singular, *bronchus*), which lead to each lung. The bronchi branch repeatedly into smaller and smaller tubes, eventually becoming thin-walled **bronchioles** surrounded by circular rings of smooth muscle (**Figure 48.9c**). Bronchioles can dilate or constrict in a manner analogous to that of arterioles (see Chapter 47). They may partially constrict when a damaging particle—such as a small bit of inhaled pollutant or dust—gets past the mucous layers of the upper airways. Constriction of a bronchiole prevents foreign particles from reaching delicate lung tissue. Once the body's defense cells have removed the particles, the bronchiole reopens to its normal diameter.

The bronchioles empty into the final, saclike regions of the lungs where gas exchange occurs—the **alveoli** (singular, alveolus; Figure 48.9c). Until now, air has flowed through the air tubes without any gas exchange taking place. The alveoli are highly adapted for gas exchange. They are only one cell thick and resemble extremely thin sacs, looking a little like bunches of grapes on a stem. Many capillaries containing deoxygenated blood pumped from the right ventricle of the heart surround the alveoli. Oxygen diffuses from the lumen of each alveolus across the alveolar cells, through the interstitial space outside the cells, and into the capillaries. Carbon dioxide diffuses in the opposite direction. The oxygenated blood from the lungs then flows to the left atrium of the heart and from there enters the left ventricle, where it is pumped out through the aorta to the rest of the body.

Like all cells, those that make up the lining of the alveoli are coated with extracellular fluid. This fluid layer is where gases dissolve. Unlike other internal body cells, however, alveolar cells come into contact with air, creating an air/liquid interface along the inner surface of the alveoli. This results in surface tension within the alveoli, which would tend to make them collapse. If many or all the alveoli collapsed, however, the amount of surface area available for gas exchange in the lungs would be greatly reduced. What prevents them from collapsing? Certain alveolar cells produce **surfactant**, a mixture of proteins and amphipathic lipids, and secrete it into the alveolar lumen. There, it forms a barrier between the air and the fluid layer inside the alveoli. This barrier reduces surface tension in the alveolar walls, allowing them to remain open.

Surface tension is particularly important in the transition from fetal to postnatal life in mammals. Most mammalian fetuses are encased in fluid within the uterus. Consequently, their lungs do not have an air/liquid interface and they do not start producing surfactant until the final stages of pregnancy. If a human baby is born more than 4–6 weeks prematurely, before sufficient surfactant is produced, many alveoli will quickly collapse after birth. This condition, known as **respiratory distress syndrome of the newborn**, can be partially alleviated by inserting a tube in the trachea and injecting synthetic surfactant.

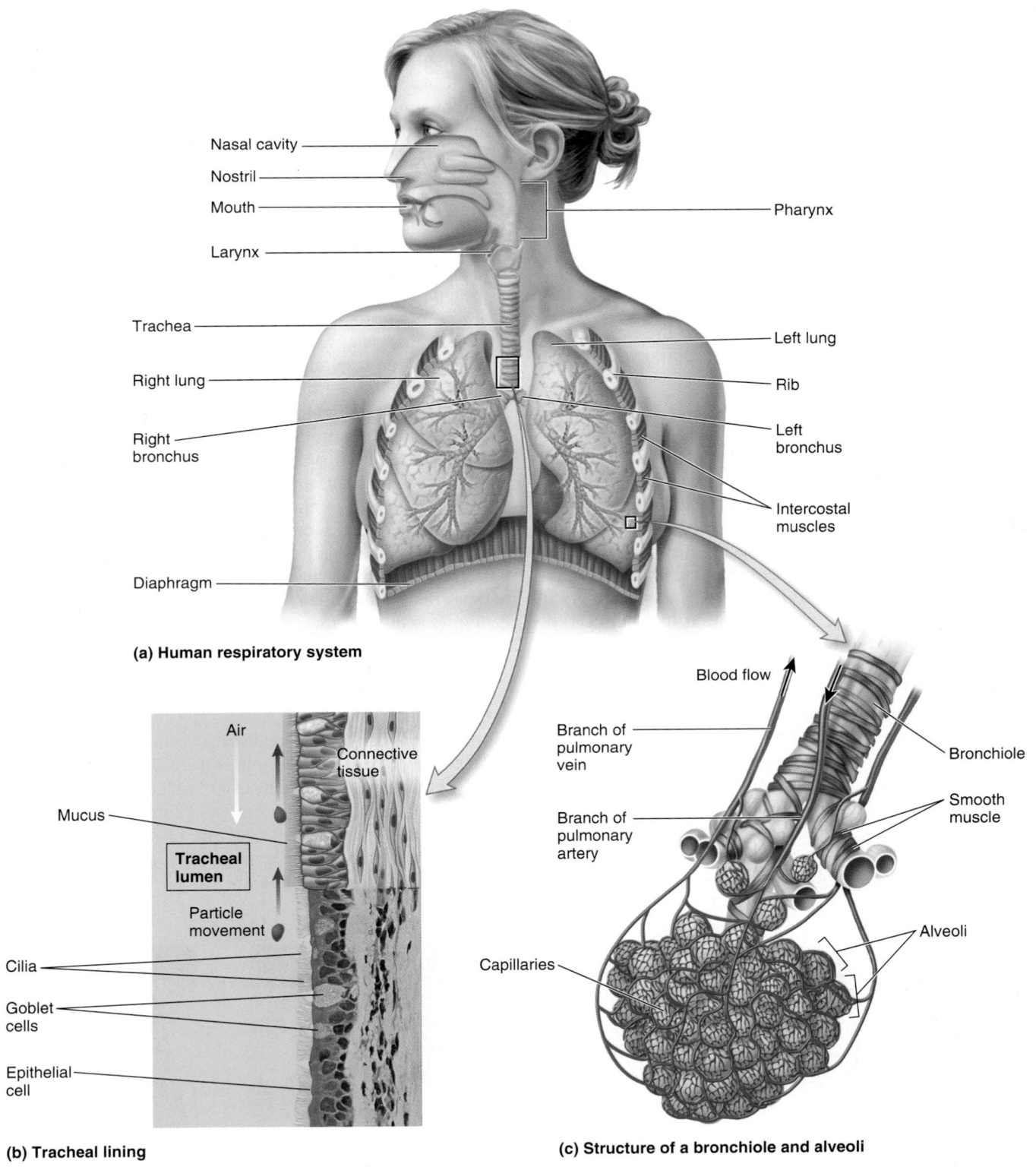

(a) **Human respiratory system**

Nasal cavity

Nostril

Mouth

Larynx

Pharynx

Trachea

Right lung

Right bronchus

Left lung

Rib

Left bronchus

Intercostal muscles

Diaphragm

(b) **Tracheal lining**

Air

Mucus

Tracheal lumen

Particle movement

Cilia

Goblet cells

Epithelial cell

Connective tissue

(c) **Structure of a bronchiole and alveoli**

Blood flow

Branch of pulmonary vein

Branch of pulmonary artery

Capillaries

Bronchiole

Smooth muscle

Alveoli

Figure 48.9   The mammalian respiratory system.   (a) In this overview, the ribs have been removed in front, and the major airways have been drawn in the lungs. The thoracic cavity is bounded by the ribs and intercostal muscles, and the muscular diaphragm. (b) Ciliated epithelial cells and mucus-producing goblet cells line the trachea. The mucus traps inhaled particles, and the cilia help move the mucus toward the mouth, where it can be swallowed. The top half of this illustration is a drawing; the bottom half is a light micrograph. (c) The bronchioles deliver air to clusters of alveoli. Note the smooth muscle cells around the bronchioles, which can cause the bronchioles to constrict or dilate. Capillaries surround the alveoli. Red represents oxygenated blood; blue represents deoxygenated blood.

Membranes Protect the Lungs

The lungs are soft, delicate tissues that could easily be damaged by the surrounding bone, muscle, and connective tissue of the thoracic cavity if not protected. Each lung is encased in a **pleural sac**, a double layer of sheathlike moist membranes. Between the two layers of membrane is a microscopically thin layer of water that acts as a lubricant and makes the two membranes adhere to each other.

In addition to protecting the lungs, the inner pleural sac adheres to its lung, and the outer pleural sac adheres to the chest wall. In this way, movements of the chest wall result in similar movements of the lungs. This is important because the lungs themselves are not muscular and cannot inflate themselves. Instead, as we will see, the lungs are inflated by the expansion of the thoracic cavity, which results from the contraction of muscles in the chest and abdomen.

Negative Pressure Ventilates Mammalian Lungs

The way in which you inflate a balloon, by forcing air from your mouth into the balloon, is called positive pressure filling. It is reminiscent of how amphibians ventilate their lungs. **Negative pressure filling** by contrast, is the mechanism by which reptiles, birds, and mammals ventilate their lungs (**Figure 48.10**). In this process, the volume of the lungs expands, creating a negative pressure that draws air into the lungs. The process differs in some ways among classes of vertebrates, but in mammals the work is provided by the **intercostal muscles**, which surround and connect the ribs in the chest, and a large muscle called the **diaphragm** (see Figure 48.9a), which subdivides the thoracic cavity from the abdomen. At the start of a breath, the diaphragm contracts, pulling downward and enlarging the thoracic cavity. Simultaneously, the intercostal muscles contract, moving the chest upward and outward, which also helps to enlarge the thoracic cavity. Recall that the pleural sacs adhere the lungs to the chest wall, so that as the chest expands, the lungs expand with it. According to Boyle's law, as the volume of the lungs increases, the pressures of the gases within them must decrease. In other words, the pressure in the lungs becomes negative with respect to the outside air. Air, therefore, flows down its pressure gradient from outside the mouth and nose, into the lungs.

Once the lungs are inflated with air, the chest muscles and diaphragm relax and recoil back to their original positions as an animal exhales. This compresses the lungs and forces air out of the airways. Whereas inhaling requires the expenditure of significant amounts of energy, exhaling is mostly passive and does not require much energy. During exhalation, air leaves via the same route that it entered during inhalation, and no new oxygen is delivered to the airways at that time. This type of breathing is called **tidal ventilation** (think of air flowing in and out of the lungs like the tides of an ocean). Tidal ventilation is less efficient than the unidirectional, flow-through system of fish in which gills are always exposed to oxygenated water during all phases of the respiratory cycle.

As you know from experience, the lungs are neither fully inflated nor deflated at rest. For example, you could easily take a larger breath than normal if you wished, or exhale more than the usual amount of air. The volume of air that is normally breathed in and out at rest is the **tidal volume**, about 0.5 L in an average-sized human. Tidal volume is proportional to body size among humans and between species. A 6-foot-tall adult, for example, has a larger tidal volume than a 4-foot-tall child because the adult has larger lungs. Similarly, horses have larger tidal volumes than humans, and humans have larger tidal volumes than dogs (see the first four columns of **Table 48.1**). During exertion, the lungs can be inflated further than the resting tidal volume to provide additional oxygen. Likewise, the lungs can be deflated beyond their normal limits at rest, by exerting a strong effort during exhalation. The lungs never fully deflate, however, partly because they are held open by their adherence to the chest wall. This is important for a simple reason. Think again of our analogy of a balloon. It is much easier to fill a balloon that is already partly inflated than it is to inflate a completely empty balloon. The same is true of the lungs. The most difficult breath is the very first one that a newborn mammal takes—the only time its lungs are ever completely empty of air.

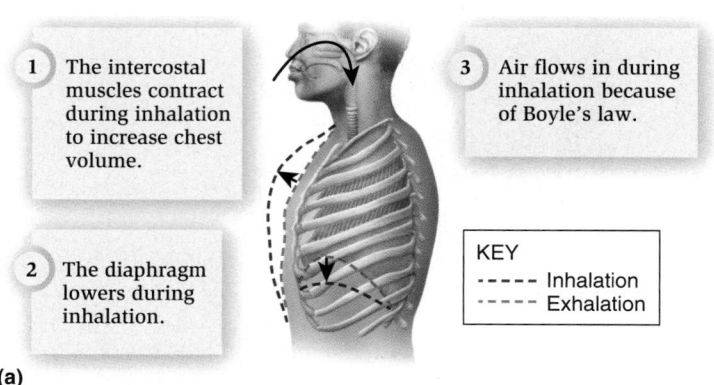

1 The intercostal muscles contract during inhalation to increase chest volume.

2 The diaphragm lowers during inhalation.

3 Air flows in during inhalation because of Boyle's law.

KEY
- - - - - Inhalation
- - - - - Exhalation

(a)

After exhaling

After inhaling

(b)

Figure 48.10   Ventilation of the mammalian lung.   (a) The intercostal muscles contract, which expands the chest cavity by moving the ribs up and out. The diaphragm also contracts, causing it to pull downwards, further expanding the cavity. The muscular efforts require energy, whereas the return to the resting state by exhaling is primarily by recoil. **(b)** X-ray images of the chest after exhaling and inhaling. The volume of the lungs after exhaling is superimposed in the latter case to illustrate the relative change in lung volume.

Table 48.1	Respiratory Characteristics of Different Mammals			
Animal	**Body mass**	**Breaths/min**	**Tidal volume**	**P_{50} value**
Shrew	0.0024 kg	700	30 μL	35 mmHg
Dog	25 kg	20	0.27 L	29 mmHg
Human	75 kg	12	0.50 L	26 mmHg
Horse	465 kg	9	6.50 L	24 mmHg

Note: All values are averages from resting animals. Tidal volume is the volume of air breathed in with each breath. Note that tidal volume increases as an animal's mass and therefore lung size increases. By contrast, breathing rate is higher in smaller animals. Similar relationships occur in other vertebrates, notably birds. As described later in the chapter, the P_{50} value is the oxygen pressure at which an animal's hemoglobin is 50% saturated with oxygen. Higher P_{50} values correspond to lower affinities of hemoglobin for oxygen (in other words, animals with high P_{50} values unload their oxygen from hemoglobin more readily).

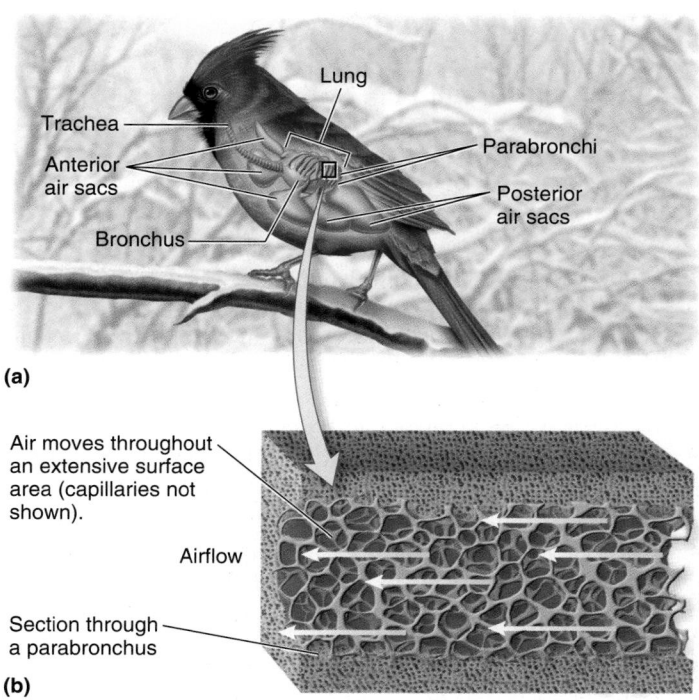

(a)

(b)

Figure 48.11 **Respiratory system of a bird.** (a) Air enters through the trachea, where it flows into the bronchi. Each lung is connected to several air sacs; this figure shows only one lung and its associated air sacs. The air sacs do not participate in gas exchange, but create a unidirectional flow of air through the respiratory system. (b) The lungs are composed of parabronchi, regions of high surface area where gas exchange takes place.

The Avian Lung Is a Flow-Through System

Unlike in mammals, ventilation in birds is a flow-through system. The avian respiratory system is unique among vertebrates in that it is supplemented with numerous **air sacs** (**Figure 48.11a**), which for simplicity we can lump together as those at the anterior or posterior regions of the body. The air sacs—not the lungs—expand when a bird inhales and shrink when it exhales. They expand by movements of the chest muscles, not by a diaphragm. The air sacs do not participate in gas exchange, as first demonstrated in a simple but elegant experiment by French physiologist J. Soum in 1896. Soum surgically plugged the openings of one of a pigeon's air sacs and then injected poisonous carbon monoxide (CO) gas into the sac. Birds are very sensitive to the toxic effects of CO, but in this experiment, the pigeon showed no signs of being poisoned. Soum concluded that gases do not diffuse from the air sacs into the blood.

Air enters the trachea of the avian respiratory system, and from there into two bronchi. Instead of branching into bronchioles and then alveoli, however, avian bronchi branch into a series of parallel air tubes called **parabronchi** (**Figure 48.11b**) that make up the lungs. The parabronchi are the regions of gas exchange. They have enormous surface area and an extensive network of blood capillaries. Blood flows through the lungs in a

crosscurrent direction with respect to movement of air (**Figure 48.12**). This is less efficient in extracting oxygen than the countercurrent flow arrangement in fish gills but more efficient than the tidal ventilation system of mammalian lungs.

The way in which air moves through avian lungs and the function of the air sacs remained a mystery to physiologists for many years, until the pioneering work of American physiologist Knut Schmidt-Nielsen, as described next.

Figure 48.12 Relationships between direction of flow of blood and water/air in different vertebrates. Blood flows through the avian lungs in a crosscurrent manner, which is more efficient at extracting oxygen from the lungs than the tidal flow in mammals but less efficient than the countercurrent flow in fish.

FEATURE INVESTIGATION

Schmidt-Nielsen Mapped Airflow in the Avian Respiratory System

In 1967, Schmidt-Nielsen and coworkers conducted a series of experiments using ostriches because of their large lung volumes. The researchers used O_2- and CO_2-sensitive probes to determine the concentrations of these gases in the birds' air sacs. In their first experiment, described in **Figure 48.13**, the investigators allowed the birds to breathe ordinary room air, which contained 21% oxygen. They used special syringes to extract a gas sample from the large air sacs of the birds and had the gas composition analyzed with the probes. They discovered that the O_2 content was high and the CO_2 content low in the posterior air sacs, with the reverse pattern in the anterior sacs. They concluded that fresh air must move through the trachea and into the posterior air sacs, while "stale" air, which had previously given up much of its oxygen in the lungs, must have entered the anterior sacs.

To test this hypothesis, they developed a method to track the movement of oxygen through the ostrich respiratory system, as shown in the second experiment of Figure 48.13. In this experiment, the scientists fitted ostriches with face masks, through which the birds inhaled a single breath of pure (100%)

oxygen. O_2-sensitive probes were surgically inserted into the posterior and anterior air sacs to monitor when the pure oxygen gas appeared in each sac. The researchers discovered that during the first inhalation, high levels of oxygen appeared in the posterior air sac but not the anterior sac. As the ostrich exhaled, some oxygen remained in the posterior air sac, but still no oxygen appeared in the anterior one. When the animal inhaled its next breath (this one of plain air, not pure oxygen), oxygen began appearing in the anterior sacs and disappearing from the posterior sacs. Finally, as the ostrich exhaled a second time, the oxygen in the anterior sacs decreased.

Schmidt-Nielsen concluded from this study that air flows through the trachea, down the bronchi, and into the posterior air sacs during inhalation. As a bird exhales, air exits the posterior sacs and flows through the parabronchi from the back to the front of the lungs. During its passage through the lungs, some of the oxygen diffuses into the capillaries, while CO_2 diffuses out of the capillaries and into the lungs. During the next inhalation, air at the anterior part of the lungs flows into the anterior air sacs, which serve as a sort of holding chamber. Finally, a second exhalation causes this air from the previous inhalation to exit the anterior sacs and leave the animal's body through the trachea.

Figure 48.13 Schmidt-Nielsen and coworkers demonstrated the route that air travels in the avian (ostrich) lung.

3 **Experiment 2:**
Give ostriches a single breath of pure O_2, which can be tracked as it moves through the respiratory system. Then remove mask and allow bird to breathe normal air.

Tight-fitting mask

100% O_2

Gas analyzer

O_2 probe

100% O_2

Air from previous breath

Gas analyzer

O_2 probe

4 **THE DATA**

Results from experiment 2: O_2 appears first in the posterior sacs, and does not appear in anterior sacs until the next breath.

Volume of air in lungs

Inhalation | Exhalation | Inhalation | Exhalation | Inhalation | Exhalation

Time

O_2 content* in posterior sacs

Time

O_2 content* in anterior sacs

Time

*The O_2 shown in these graphs is from the pure O_2 inhaled on the first breath, and is above the normal levels of O_2.

5 **CONCLUSION**

Proposed pattern of avian ventilation*:

First inhalation

Sacs expand

First exhalation

Sacs compress

Second inhalation

Second exhalation

*The blue color indicates the location of gas taken on the first breath.

Thus, it takes two complete breaths for air to move from the environment into the lungs, and back out again to the environment. Air flows through the lungs during inhalation, as it is drawn by negative pressure through the parabronchi and into the expanding anterior sacs. Air is also forced through the parabronchi by positive pressure during exhalation, as the posterior air sacs are compressed. The efficiency of the avian flow-through system, together with its crosscurrent blood supply, is a major reason why many birds are able to fly at altitudes with extremely low atmospheric pressures (even over Mt. Everest).

48.4 Control of Ventilation in Mammalian Lungs

Lungs are not muscles or electrically excitable tissue. Therefore, they are not capable of initiating and regulating their own expansion. Nonetheless, lungs require a mechanism to rhythmically expand and recoil, because animals cannot always consciously control breathing, such as during sleep. In this section, we examine the ways in which the brain and other structures control ventilation in mammals, and how these control mechanisms are linked with metabolism.

The Nervous System Contains the Pacemaker for Ventilation

The pacemaker that initiates expansion of the lungs is a collection of nuclei in the nervous system. In mammals, these **respiratory centers** constitute several regions of the brainstem (**Figure 48.14**). Electrical impulses from the brainstem travel through two sets of nerves. The first set, called the intercostal nerves, stimulates the intercostal muscles, and the second set, the phrenic nerves, stimulates the diaphragm. When the lungs expand in response to the contraction of these muscles, stretch-sensitive nerves in the walls of the lungs and chest send signals to the respiratory centers, informing them that the lungs are inflated. This temporarily turns off the stimulating signal until the animal exhales, whereupon a new signal is sent to the breathing muscles.

Although the brainstem controls breathing automatically, it can be overridden. Animals that dive underwater—including humans when we go swimming—can hold their breath temporarily. We can also breathe faster than normal even while resting, at least for short periods of time. Normally, however, increased breathing occurs in response to exercise or stress.

Factors that increase the respiratory rate:

Conscious effort

Exercise

Stress

Large decreases in blood levels of O_2

An increase in blood levels of CO_2 or H^+

Factors that inhibit the respiratory rate:

Stretching of the lungs during inhalation

Conscious effort (holding one's breath as in diving)

Sleep

Figure 48.14   Respiratory centers in the mammalian brain control breathing. Neurons in the brainstem send action potentials via nerves that stimulate the intercostal muscles and diaphragm. The factors listed here can modulate the rate of action potential generation and thus the respiratory rate.

During exercise, a variety of factors converge on the respiratory centers to increase the rate and strength of signals to the breathing muscles, resulting in faster and deeper breaths.

Chemoreceptors Modulate the Activity of the Respiratory Centers

The concentrations of hydrogen ions (pH) in blood, and the partial pressures of carbon dioxide and oxygen, influence the respiratory centers. **Chemoreceptors**—special cells located in the aorta, carotid arteries, and the brainstem—detect the circulating levels of these substances and relay that information through nerves or interneurons to the respiratory centers.

Normally, the oxygen level is not the most important variable monitored by chemoreceptors. However, if the oxygen level falls far below normal, as might occur at high altitude or in certain respiratory diseases, the chemoreceptors signal the respiratory centers to increase the rate and depth of breathing to increase ventilation of the lungs. Similarly, a buildup of carbon dioxide in the blood, which would occur if an animal's ventilation were lower than normal (again, often the result of respiratory disease), signals the respiratory centers to stimulate

breathing. Finally, hydrogen ions in the blood activate chemoreceptors, which signal the brain when the blood is too acidic. This leads to an increase in the rate of breathing.

One source of acid in the blood results from a by-product of anaerobic respiration, which occurs during intense exercise and is usually a sign that an animal needs more oxygen than it is getting. In addition, the concentration of hydrogen ions in the fluid bathing the brainstem chemoreceptors reflects the amount of carbon dioxide produced by cells during metabolism. About two-thirds of the carbon dioxide produced during metabolism is converted into a less toxic and more soluble form, bicarbonate ions. In the process, hydrogen ions are formed according to the following reaction, where H_2CO_3 is a short-lived compound called carbonic acid that immediately dissociates to a hydrogen ion and a bicarbonate ion:

$$CO_2 + H_2O \rightleftharpoons H_2CO_3 \rightleftharpoons H^+ + HCO_3^-$$

The left side of this reaction is readily reversible and is catalyzed in both directions by the enzyme carbonic anhydrase, which is present in high amounts in red blood cells. If CO_2 levels increase, the reactions move from left to right. Conversely, if the concentration of hydrogen ions rises or the levels of CO_2 decrease, the reactions move from right to left.

In body tissues, CO_2 levels are high because metabolism is generating the gas. Thus, the reactions move to the right, and CO_2 is converted to H^+ and HCO_3^-. As blood flows through the lungs, however, dissolved CO_2 diffuses into the alveoli and is exhaled. Thus, CO_2 levels in the lung capillaries decrease, which makes the reactions proceed in the opposite direction, right to left so that HCO_3^- is converted back to CO_2 and exhaled.

In addition to being transported as bicarbonate ions, about 25 % of the CO_2 in blood is bound to hemoglobin. The remainder—about 7–10 %—exists dissolved in solution in the plasma and within red blood cells.

Respiratory Center Activity Varies Among Species with Different Metabolic Rates

The respiratory centers are sensitive to a variety of other factors besides gases and hydrogen ions, such as sleep, stress, hormones, and body size. Recall from Chapter 42 that small animals tend to have higher metabolic rates than larger animals. Thus, small animals would be expected to require more oxygen per unit mass than would larger animals. This is not achieved by having disproportionately large lungs in small animals. Instead, small animals have higher breathing rates than do large animals. The respiratory centers are set at a higher frequency in smaller, active animals. For example, a typical adult human takes about 12 to 14 breaths per minute at rest, while hummingbirds, small rodents, and shrews have astonishingly high rates of breathing, up to many hundreds of breaths per minute (see Table 48.1)! Thus, the respiratory activity of small, highly active animals is matched with their metabolism by increasing the resting breathing rate. This is similar to the differences in cardiac function among species. Small animals have hearts that are roughly proportional to their body size, but with higher heart rates than are found in larger animals (see Chapter 47). Parallel changes in the heart and breathing rates are adaptations that allow small animals to have small hearts and lungs, yet achieve high metabolic rates.

48.5 Mechanisms of Oxygen Transport in Blood

Gases such as oxygen are soluble in the plasma component of blood, but as with all solutes, there are limits to how much of a gas can dissolve in a volume of water. The amount of gas dissolved in the body fluids is not sufficient to sustain life in most animals. In nearly all animals, therefore, the amount of oxygen in the body fluid compartments must be increased above that which can be physically dissolved. This is made possible because of the widespread occurrence in animals of oxygen-binding proteins, which increase the total amount of oxygen available to cells. In this section, we examine the nature, function, and evolution of these molecules.

Oxygen Binds to Respiratory Pigments

The amount of oxygen dissolved in blood under normal conditions is too low to provide the energy needed for the basal metabolic rate of most animals. Even the simple act of sleeping requires more oxygen than the amount dissolved in plasma. Therefore, animals have evolved a way to carry a reservoir of oxygen on plasma proteins. These proteins are called **respiratory pigments** because they have a color (blue or red). In vertebrates, the pigments are contained within red blood cells, but in many invertebrates the pigments circulate in the plasma. Enclosing the pigments within blood cells is an efficient way to package them, because the pigments are in close proximity to various cellular enzymes that assist in binding and unbinding the gases to the pigments.

Respiratory pigments are proteins containing one or more metal atoms in their cores. The metal atom binds to oxygen. Typically, the metal is iron (Fe^{2+}) in vertebrates and many marine invertebrates, but in decapod crustaceans, arachnids, and many mollusks (including cephalopods and some gastropods), it is copper (Cu^{2+}). The copper-containing pigment, **hemocyanin**, gives the blood or hemolymph a bluish tint. **Hemoglobin** is the major iron-containing pigment and gives blood a red color when oxygen is bound.

Hemoglobin gets its name because it is a globular (soluble) protein, and it contains a chemical group called a heme in its core. An atom of iron is bound within the heme group. In vertebrates, hemoglobin consists of four protein subunits, each with its own heme group and iron atom. Thus, a single molecule of hemoglobin can bind up to four molecules of oxygen.

Respiratory pigments share certain characteristics that make them ideal for transporting oxygen in blood. First, they all have a high affinity for binding oxygen. Second, the binding between the pigment and oxygen is noncovalent and reversible. It would not do much good for a protein to bind oxygen if it could not

unload the oxygen to cells that need it. Thus, the binding is reversible, as shown here:

$$O_2 + Hb \rightleftharpoons HbO_2$$

where Hb is hemoglobin, HbO_2 is called oxyhemoglobin, and the double arrows indicate that the reaction is reversible.

The amount of pigment present in the blood is great enough to provide sufficient oxygen to meet all but the most strenuous exertion. In humans, for example, the presence of hemoglobin gives blood about 45 times more oxygen-carrying capacity compared to plasma alone.

Differences exist among respiratory pigments, however, at the molecular level. For example, the amino acid sequences of hemoglobins from different species vary, in some cases considerably. Human hemoglobin is similar but not identical to that of other mammals, and mammalian hemoglobins have numerous amino acid differences from other vertebrate hemoglobins. The differences in amino acid sequences cause slight changes in the overall three-dimensional shape of hemoglobin, which alters its affinity for oxygen. As a general rule, animals with the highest metabolic rates tend to have hemoglobins with lower oxygen affinity, while animals with lower metabolic rates have higher affinities. That may sound paradoxical at first, but consider the preceding reactions again. Because the hemoglobin of small animals with high metabolic rates has lower oxygen affinity, the reverse reaction ($HbO_2 \longrightarrow Hb + O_2$) will be more likely to occur when oxygen is diffusing from the blood to active cells. In other words, hemoglobin unloads its oxygen more readily in those animals whose cells use oxygen at the greatest rates.

The Amount of Oxygen Bound to Hemoglobin Depends on the P_{O_2} of Blood

Recall that partial pressure of O_2 (P_{O_2}) is a measure of its dissolved concentration. When P_{O_2} is high, more O_2 binds to hemoglobin, while fewer O_2 molecules will be bound when P_{O_2} is low. **Figure 48.15** shows the relationship between O_2 binding and P_{O_2}, known as an **oxygen-hemoglobin dissociation curve**. At a P_{O_2} of 100 mmHg, which is found in the blood leaving the lungs, each hemoglobin protein binds four O_2 molecules. It is 100% saturated with oxygen. The P_{O_2} of blood leaving the tissue capillaries of other parts of the body is lower and depends on metabolic activity and exercise. At rest, the P_{O_2} of blood capillaries in these other parts of the body is typically around 40 mmHg. At this P_{O_2}, hemoglobin releases some O_2 molecules, becoming less saturated with O_2. At high levels of exercise, P_{O_2} drops even further (as low as 20 mmHg) and hemoglobin releases even more O_2. In this way, hemoglobin performs its role of oxygen delivery. In the lungs it binds O_2, and in other tissues of the body it releases some of the O_2 as needed.

You may have noticed that the curve in Figure 48.15 is not linear but S-shaped (sigmoidal). This is because the subunits of hemoglobin are said to cooperate with each other in binding oxygen. Once a molecule of oxygen binds to one subunit's iron atom, the shape of the entire hemoglobin molecule changes, making it easier for a second oxygen to bind to the next subunit,

Figure 48.15 **The human oxygen-hemoglobin dissociation curve.** Depending on the partial pressure of oxygen, oxygen is either loaded onto hemoglobin as in the lungs, or unloaded from hemoglobin, as in the rest of the body tissues. When P_{O_2} is high, more Fe^{2+} atoms are bound to O_2 and the hemoglobin is more saturated with O_2.

Biological inquiry: Members of the Channichthyidae family of Antarctic icefish are the only vertebrates that do not have red blood cells and hemoglobin. How do they survive without the oxygen reservoir provided by hemoglobin?

and so on. Thus, the relationship between oxygen pressure and the amount of oxygen bound to hemoglobin becomes very steep in the midrange of the curve, which are those pressures that occur in the tissue capillaries throughout the body except in the lungs. This steepness allows O_2 release to be very sensitive to changes in P_{O_2}, as occurs in exercise, for example.

One of the remarkable features of the oxygen-hemoglobin binding relationship is that it can be influenced by metabolic waste products like CO_2 and H^+, and also by temperature. **Figure 48.16a** shows three curves, one obtained under normal resting conditions and the others in the presence of low and high levels of CO_2. CO_2 binds to amino acids in the hemoglobin molecule (not to the iron, like O_2), and when it does, the ability of hemoglobin to continue binding oxygen decreases. Note how increased CO_2 shifts the curve to the right, such that at any P_{O_2}, less O_2 is bound to hemoglobin. Another way of saying this is that at any P_{O_2}, more O_2 has been released from hemoglobin, thus becoming available to cells. Similar shifts in the curve occur in the presence of acid and increased temperature. Cells generate each of these products—CO_2, acid, and heat—when they are actively metabolizing fuel. The metabolic products enter the surrounding blood vessels, where they disrupt the normal shape of hemoglobin, causing it to release more of its oxygen than would normally occur at that oxygen pressure. This phenomenon is a way in which individual body tissues can obtain more oxygen from the blood to match their changing metabolic demands. Thus, when an animal exercises, the skeletal muscles generate more waste products and heat than does, say, the skin, and therefore the oxyhemoglobin in muscle blood vessels releases more oxygen to the cells where it is immediately needed.

(a) Shift in the oxygen-hemoglobin dissociation curve

(b) Oxygen-hemoglobin dissociation curves of different animals

Figure 48.16 **Differences in oxygen-hemoglobin dissociation curves under different physiological conditions and among different species.** **(a)** Increasing or decreasing the amounts of carbon dioxide or hydrogen ions (acid), or the temperature of the blood, shifts the oxygen-hemoglobin dissociation curve. Actively metabolic tissues generate these products. By shifting the curve, different tissues obtain oxygen from hemoglobin in proportion to their metabolic needs. **(b)** Oxygen-hemoglobin dissociation curves for three mammals with low (elephant), moderate (human), or high (mouse) relative metabolic rates. For any P_{O_2}, such as the one shown (40 mmHg), which is typical of the P_{O_2} of tissue capillaries, less oxygen is bound to mouse hemoglobin than to human or elephant, and less is bound to human than to elephant hemoglobin. Thus, smaller animals unload O_2 from hemoglobin more readily.

The shift in the oxygen-hemoglobin dissociation curve occurs in all classes of vertebrates (although not in all species), but it has different magnitudes in different species. Not surprisingly, perhaps, the most metabolically active animals, such as mice, show a greater shift for a given increase in CO_2, H^+, or heat than do less active animals. Recall from Chapter 47 that these same waste products of metabolism also cause local vasodilation of arterioles. Thus, the more metabolically active a tissue is, the more blood flow it receives, which means more oxygen-bound hemoglobin. Moreover, the oxygen is unloaded from hemoglobin more readily due to the shift in the curve. This is another example of how adaptive changes in circulatory and respiratory functions often occur in parallel.

The hemoglobins of metabolically active animals also have a lower affinity for oxygen (**Figure 48.16b**). Small, active animals have curves that are displaced to the right of the human curve, which itself is to the right of the curve for an elephant. At high oxygen pressures, like those that occur in the lungs, these shifts have little relevance because nearly all the hemoglobin is bound to oxygen at those pressures. At lower oxygen pressures, however, like those that would occur in the blood vessels of actively metabolic tissues, the difference in the curves becomes significant. For example, look at the three curves at an oxygen pressure of 40 mmHg, a typical value found in tissues that are using oxygen at a resting rate. The mouse hemoglobin has less oxygen bound (its hemoglobin is less saturated) at that pressure than the human hemoglobin, which in turn has less oxygen bound than the elephant hemoglobin. In other words, the mouse hemoglobin has released more of its oxygen to the hungry tissues than have the other animal hemoglobins.

A useful way to compare the relative affinities of hemoglobin for oxygen in different species is to compare the oxygen pressure required to half-saturate hemoglobin—the so-called P_{50} value of a species' hemoglobin (see the final column of Table 48.1).

GENOMES & PROTEOMES

Hemoglobin Evolved over 500 Million Years Ago

Oxygen-carrying molecules are among the most ancient proteins found in animals. They appear to have begun as simple, single-subunit proteins such as muscle myoglobin, which stores oxygen in muscle cells. Around 500 million years ago, the gene for this ancient molecule duplicated, resulting in one gene that coded for myoglobin and the other for an early form of hemoglobin. Thus, animals gained the ability to not only store oxygen in tissues but also to transport it throughout the body via the circulation.

Modern hemoglobin in adult mammals contains four subunits, designated α1, α2, β1, and β2. The two alpha and two beta subunits are identical. Early in animal evolution, hemoglobin existed only as monomers (one subunit) or dimers (two subunits), which is still observed in lampreys. Later, gene duplication created the second set of subunits, leading to the four-subunit form, which first appeared in sharks and bony fishes.

As we have seen in this chapter, the subunit structure of hemoglobin enables one hemoglobin molecule to bind four oxygen molecules. Equally important, it allows the sigmoidal shape of the oxygen-hemoglobin dissociation curve.

As noted earlier, modern hemoglobins differ among species within the same class of vertebrates, and even more so between classes. This suggests that evolution of the hemoglobin genes continued even after the four-subunit structure appeared. The mutations that arose in the hemoglobin genes affected the structure of the hemoglobin molecule. One consequence is that the affinity of hemoglobin for oxygen differs among animals, as discussed earlier. The differences in affinity may result, for example, in better oxygen-capturing ability of animals living in regions of low oxygen pressure, for example, at high altitude. They also account, at least in part, for the ability of metabolically active animals to unload oxygen from hemoglobin more readily than less metabolically active organisms.

Despite the high frequency of mutations in the hemoglobin genes in vertebrates, certain regions of the molecule have remained well conserved. Not surprisingly, these are regions that are critically important for the function of the molecule. For instance, the amino acids that encase the heme group and iron atom of each subunit have changed little if at all during evolution. Similarly, the amino acid sequences associated with sites of subunit interactions have also resisted evolutionary change. Mutations that arise in these regions are typically not adaptive and may even be lethal.

Some mutations in the hemoglobin genes may appear to be detrimental because they negatively impact hemoglobin's function, and yet they provide a selective advantage under certain conditions. Consider the disease **sickle-cell anemia**. In this case, substitution of a thymine for an adenine at one position in the hemoglobin gene results in a single amino acid substitution. This produces an abnormal form of hemoglobin that tends to form polymers and precipitate under low oxygen conditions, particularly in capillaries and veins. The protein forms long fibrous strands that may permanently deform the red blood cell, making it sickle-shaped (**Figure 48.17**). Sickled cells are less able to

Figure 48.17 **Sickle-cell anemia.** A mutation in the hemoglobin gene can result in hemoglobin molecules that deform red blood cells, giving them a sickled appearance.

move smoothly through capillaries and can block blood flow, resulting in severe pain and cell death of the surrounding tissue.

The sickled red blood cells are also fragile and easily destroyed. The loss of those cells and their hemoglobin results in anemia, meaning that people with the disease have less oxygen-carrying capacity in their blood. Only individuals who are homozygous for this mutation show the dramatic phenotype of the disease. Heterozygotes are relatively unaffected.

The sickle-cell gene mutation is present in up to a third of individuals living in malaria-prone regions of Africa, to lesser extents in the Middle East and Eastern Europe, and in roughly 8% of African Americans. Why should such an obviously harmful mutation have persisted in the human population?

Malaria, a mosquito-borne disease, is a global health threat that is responsible for over 300 million severe illnesses and 1 million deaths every year. When an infected mosquito bites, it injects a parasitic protozoan (*Plasmodium falciparum*) into the circulation, where it spends part of its life cycle growing and multiplying within the host's red blood cells. The sickle-cell trait protects individuals from developing full-blown malaria, although the reason for this is uncertain. One favored hypothesis states that the parasite lowers the oxygen pressure in red blood cells by consuming oxygen to support its own metabolism. This may render the red blood cell susceptible to sickling even in heterozygous individuals. The damaged cells are destroyed and removed from the circulation, which also removes the parasite and limits its ability to multiply. Another hypothesis states that the heterozygotic condition results in a more efficient immune response of the body to the presence of the parasite. Thus, although both malaria and homozygous sickle-cell anemia are life-threatening, heterozygotes for the sickle-cell trait neither develop pronounced anemia nor severe malaria, and therefore have an advantage in areas where malaria is present. This heterozygote advantage explains why the sickle-cell mutation persists among certain human populations.

48.6 Adaptations to Extreme Conditions

Many animals are able to live permanently or temporarily in low-oxygen environments. Many humans, for instance, live in mountainous regions, and llamas and mountain goats spend most of their lives at extremely high altitudes. Other animals may only transiently encounter periods of oxygen deprivation, like some reptiles, birds, and mammals that dive underwater to forage for food. As described in this section, several adaptations allow animals to exploit these environments, but all of them include changes in both cardiovascular and respiratory activities.

Life at High Altitudes Requires More Hemoglobin

Animals that live at high altitudes must have special adaptations that permit them to obtain the oxygen they need at such low atmospheric pressures. Llamas, like the one shown in the chapter-opening photo, may live at altitudes up to 4,800 m, where the P_{O_2} is only about 85 mmHg (compared with 160 mmHg at sea level). Llama hemoglobin is quite different in amino acid sequence from that of other mammals, giving it an extraordinarily high affinity for binding oxygen even at very low atmospheric pressure. In other words, their hemoglobin curves are well to the left of a human's. In addition, llamas have larger hearts and lungs than would be predicted for their body size, and a higher number of red blood cells in a given volume of blood. These adaptations provide the oxygen-carrying capacity and cardiac output needed to deliver sufficient oxygen to the tissues, even at low partial pressures of oxygen.

When animals that normally inhabit lowland areas temporarily move to higher altitudes, they develop some of these same features. In humans, for example, the number of red blood cells increases from about 5.1×10^{12} cells/L to about 6.4×10^{12} cells/L. This is stimulated by the hormone erythropoeitin, which is secreted by the kidneys when the arterial P_{O_2} is low. This hormone acts on bone marrow to stimulate maturation of new red blood cells. Moving to higher elevations also increases ventilation due to a higher rate of breathing. The density of capillaries around the alveoli increases at high altitudes, an adaptation that facilitates oxygen diffusion into the blood. Finally, myoglobin content increases in muscle cells, expanding the reservoir of oxygen in the cytosol of those cells. Thus, after several days or weeks at high altitude, the cardiovascular and respiratory systems adapt together to maximize oxygen uptake, diffusion of oxygen into the blood, and the oxygen-carrying capacity of blood.

Diving Animals Rely on Anaerobic Respiration Once Their Oxygen Is Depleted

Many birds, reptiles, and a few mammals spend time underwater, foraging for food or escaping predators. During that time, the animal cannot breathe. In short dives, this is not a problem. We are all familiar with our own ability to stay underwater for a minute or so. After that time, the oxygen from our last breath is used up, blood levels of carbon dioxide begin to rise, and we must surface for air. Some marine mammals, however, have an astonishing ability to remain underwater for very long periods—up to nearly 2 hours in some species. During that time, they continue to be active and search for food. How do they do it?

Like high-altitude animals, many diving animals have unusually high numbers of red blood cells, allowing them to store more oxygen in their blood than nondiving animals. In some cases, like seals, the extra blood cells are stored in the spleen until they dive, at which time the spleen contracts like wringing a wet washcloth and ejects the blood cells into the circulation. When seals resurface, the blood cells are sequestered again in the spleen until needed. In addition to having more erythrocytes, diving mammals typically have larger blood volumes than comparably sized mammals that live exclusively on land.

The muscles of diving mammals usually contain large quantities of myoglobin and its bound oxygen. This means that the muscles do not need to consume the precious stores of oxygen circulating in blood. Instead, the blood and its oxygen can be routed to other critical structures that lack myoglobin, such as the eyes, certain glands, the brain, and the placenta if the animal is pregnant.

Eventually, even with these adaptations, the oxygen in muscles and blood can become so depleted that the only way to prolong the dive is for cells to begin respiring anaerobically. Most of a long dive, in fact, occurs under anaerobic conditions.

This concludes our examination of the ways in which animals obtain oxygen and eliminate carbon dioxide. We now address the importance of healthy lung function in humans, and the ways in which humans are affected by respiratory diseases.

48.7 Impact on Public Health

Respiratory diseases afflict over 10% of the U.S. population and cause an estimated 500,000 deaths per year, making lung disease among the top three causes of death in the U.S. As many as 25 million Americans have impaired lung function. The economic impact of respiratory disorders on the U.S. economy is staggering, with recent estimates of $40–150 billion per year in health-related costs and lost work time. Most of these diseases are chronic—once they appear, they last for the rest of a person's lifetime. Lifestyle factors, such as smoking tobacco and exposure to air pollution, can cause some respiratory disorders or make existing conditions worse. In this section, we examine a few of the most common respiratory disorders, as well as some of their causes and treatments.

Asthma Is a Disease of Hyperreactive Bronchioles

You learned earlier that the bronchioles deliver fresh air to the alveoli. Bronchioles are thin structures surrounded by smooth muscle cells that can contract in the presence of airborne pollutants or other potentially damaging substances. In the disease **asthma**, however, the muscles around the bronchioles are hyperexcitable and contract more than usual. Contraction of the smooth muscles narrows the bronchioles, a process called bronchoconstriction. This makes it difficult to move air in and out of the lungs, because resistance to airflow increases when the diameter of the airways decreases. Often, the resistance to airflow can be so great that the movement of air creates a characteristic wheezing sound.

Asthma tends to run in families and therefore may have a genetic basis. Several known triggers can elicit wheezing, including exercise, cold air, and allergic reactions. The latter is of interest because asthma is believed to be partly the result of an imbalance in the immune system (see Chapter 53), which controls inflammation and other allergic responses. In fact, during flare-ups of asthma, a viscous, mucus-like fluid may clog the airways and make symptoms worse.

The symptoms of asthma can be alleviated by having the patient inhale an aerosol mist containing **bronchodilators**, compounds that bind to the muscles around bronchioles and cause them to relax and widen. To help reduce the inflammation of the lungs, patients may inhale a mist containing hormones with anti-inflammatory actions. Currently there is no cure for asthma, but with regular treatment and the avoidance of known triggers, most asthmatics can lead perfectly normal lives.

Tobacco Smoke Can Cause Respiratory Health Problems and Cancer

Smoking tobacco products is one of the leading global causes of death, contributing to about 430,000 deaths each year in the U.S. alone and over 5 million per year worldwide. According to the Centers for Disease Control and Prevention and the American Lung Association, people who smoke up to one pack of cigarettes each day live seven years less on average than nonsmokers, while heavy smokers lose 15–25 years of life on average. Pregnant women who smoke run a high risk of their babies being born underweight, a potentially serious condition that may affect the newborn's long-term health.

Up to 85% of all new cases of lung cancer diagnosed each year are attributable to smoking, making lung cancer the leading cause of preventable death. Equally important, however, is that smoking is estimated to be responsible for nearly 30% of all cancers, including cancer of the mouth and throat, esophagus, bladder, pancreas, and ovaries. Smoking is also a leading cause of cardiovascular disease, high blood pressure, atherosclerosis, and stroke. Smoking as few as three to five cigarettes per day raises the risk of heart disease.

Because the products of tobacco smoke are inhaled directly into the lungs, the chemicals in smoke can do considerable damage to lung tissue. Even adolescents who have only recently started smoking have increased mucus (phlegm) production in their airways (which inhibits the flow of air in and out of the airways), shortness of breath, and reduced lung growth. Thousands of chemicals, including over 40 known cancer-causing compounds, have been identified in cigarette smoke. Some of these chemicals—such as formaldehyde—are toxic to all cells. Others, like the odorless gas carbon monoxide (CO), have harmful effects on lung function in particular. CO competes with oxygen for binding sites in hemoglobin, thereby reducing hemoglobin saturation. Heavy smokers who smoke more than a pack of cigarettes each day may have as much as 15% less oxygen-carrying capacity in their blood.

In addition to its effects on cancer, cardiovascular disease, and lung function, long-term smoking is the major cause of the serious and irreversible disease emphysema.

Emphysema Causes Permanent Lung Damage

Unlike asthma, in which the major problems are inflamed airways and hyperreactive bronchioles, emphysema involves extensive lung damage (**Figure 48.18**). The disease reduces the elastic quality of the lungs and the total surface area of the alveoli, which cuts the rate of oxygen diffusion from the lungs into the circulation. Consequently, one sign of emphysema is a lower than normal partial pressure of oxygen in the arteries.

Broken alveoli

Note large areas without gas-exchange surfaces

0.8 mm

Normal lung

0.5 mm

Diseased lung

Figure 48.18 **Emphysema.** The destruction of alveoli reduces the surface area for gas exchange in the lungs. These light micrographs compare a section of a normal lung with a lung from a person who died of emphysema.

It is also physically harder to exhale because of the loss of elasticity, and therefore arterial CO_2 levels increase. Finally, the terminal ends of the bronchioles are often damaged, which increases resistance to airflow and creates asthma-like symptoms and shortness of breath.

Reduced blood oxygen and poor lung function limit the patient's ability to function, and in its late stages emphysema results in a person being essentially bedridden. Oxygen therapy, in which the person breathes a mixture of air and pure oxygen from a portable gas tank, can provide some help. The extra oxygen increases the pressure gradient for oxygen from the alveoli to the lung capillaries, promoting oxygen diffusion into the blood.

It is difficult to estimate how many people have emphysema because the symptoms appear gradually, but more than 3 million people have severe cases of the disease and about 120,000 people die from it each year in the U.S. Emphysema is a progressive disease that does not go away. As the years go by, the disease usually worsens, although medical care can slow the rate at which it worsens. In some cases, emphysema results from an enzyme deficiency in the lungs that destroys the protein that provides the recoil during exhalation, or it may result from chronic exposure to air or industrial pollution. However, the overwhelming majority of cases, 85%, are due to smoking. Toxins in cigarettes and other tobacco products damage the

Region of inflammation

Normal lung

Figure 48.19 **Pneumonia.** This X-ray clearly shows the region and extent of fluid accumulation due to infection. One lung is normal and the other is inflamed.

lungs by stimulating white blood cells to release proteolytic enzymes that degrade lung tissue. The likelihood of developing emphysema is tightly correlated with the quantity of cigarettes smoked during a person's lifetime.

Pneumonia Is an Inflammation of the Lungs Caused by Infection

Unlike asthma, which has a genetic component, and emphysema, which is commonly due to smoking, pneumonia is an infectious disease. It is most often caused by viruses, bacteria, or other microorganisms that enter the lungs and multiply. Symptoms of pneumonia resemble those of the flu or common cold, and they may be moderate or severe. Up to 5 million people develop pneumonia each year in the U.S., and roughly 65,000 will die from the disease.

The body's immune system fights the infection, but one result of this is inflammation accompanied by fluid buildup (**Figure 48.19**). Inflammation that might occur in your elbow or knee when you hurt yourself is painful but normally not a medical concern. Inflammation in the lungs, however, can be serious, because excess fluid interferes with the ability of the airways to ventilate and exchange gases.

A vaccine is available for preventing the bacterial form of pneumonia. In addition, antibiotics are used to treat bacterial pneumonia. Antibiotics are not effective against viruses, however. There is no cure for viral pneumonia and the disease must run its course. In healthy people, pneumonia is not usually serious because the body's immune system does a good job of fighting the infection. In individuals with weakened immune systems, however, such as the elderly or chronically ill, pneumonia can be life-threatening.

CHAPTER SUMMARY

48.1 Physical Properties of Gases

- Gas exchange is the process of moving oxygen and carbon dioxide in opposite directions such as between cells and blood, and between blood and the environment. The partial pressure of oxygen in the environment provides the driving force for its diffusion from air or water across a respiratory organ and into the blood. (Figures 48.1, 48.2)

- Three factors—the pressure of the gas, temperature of the water, and presence of any other solutes—are particularly important for affecting the solubility of a gas in water.

48.2 Types of Respiratory Organs

- Ventilation is the process of bringing oxygenated water or air into contact with a respiratory organ. All respiratory organs have moist surfaces in which gases can dissolve and diffuse, as well as other adaptations that increase the amount of surface area and blood flow. Terrestrial and aquatic animals face different challenges to gas exchange.

- The body surface may be permeable to gases in some invertebrates and in amphibians, eels, and a few other species of fish. (Figure 48.3)

- Most aquatic animals utilize external or internal gills for gas exchange. (Figures 48.4, 48.5, 48.6)

- Air-breathing animals have evolved two mechanisms to exchange gas with the environment: tracheal systems in insects, and lungs in many arachnids, some land snails, and all terrestrial vertebrates. In insects, air moving down the tracheoles comes into contact with fluid at the tracheole tips. Oxygen from the air dissolves in this fluid and diffuses across the tracheole wall and into nearby cells. (Figure 48.7)

- Except for some rare amphibians, all air-breathing terrestrial vertebrates use lungs to bring oxygen into the circulatory system and remove carbon dioxide. All lungs receive deoxygenated blood from the heart and return oxygenated blood to the heart.

- Frogs and most other amphibians ventilate their lungs with positive pressure filling. (Figure 48.8)

48.3 Design and Function of Mammalian and Avian Respiratory Systems

- The mammalian respiratory system refers to all components of the body that contribute to the exchange of gas between the external environment and the blood. (Figure 48.9)

- Most reptiles and all birds and mammals ventilate their lungs by negative pressure filling. In mammals, the work is provided by the intercostal muscles and diaphragm. Tidal ventilation is less efficient than the unidirectional flow-through system of fish gills. Tidal volume is proportional to body size among humans and between species. (Figure 48.10, Table 48.1)

- The avian respiratory system, a flow-through system, is unique among vertebrates because it is supplemented with numerous air sacs. Blood flows through the lungs in a crosscurrent direction with respect to the movement of oxygen. This is less efficient in extracting oxygen than the countercurrent flow in fish gills but more efficient than the tidal ventilation system of mammals. (Figures 48.11, 48.12, 48.13)

48.4 Control of Ventilation in Mammalian Lungs

- In mammals, respiratory centers in the brainstem initiate the rhythmic expansion of the lungs. (Figure 48.14)

- Chemoreceptors detect blood levels of hydrogen ions and the partial pressures of carbon dioxide and oxygen, and they relay this information through nerves to the respiratory centers. The respiratory centers are also sensitive to other factors, such as sleep, stress, hormones, and body size.

48.5 Mechanisms of Oxygen Transport in Blood

- There are limits to how much of a gas can dissolve in water. These limits are overcome in nearly all animals by either transporting a gas reversibly bound to a protein carrier or by transforming the gas into a more soluble form.

- Animals have evolved a way to carry a reservoir of oxygen on respiratory pigments, large proteins containing one or more metal atoms that bind to oxygen. Typically, the metal is iron in vertebrates and many marine invertebrates, and copper in decapod crustaceans, arachnids, and many mollusks including cephalopods and some gastropods.

- The amount of oxygen bound to hemoglobin depends on the partial pressure of oxygen in the blood. Metabolic waste products can influence the oxygen-hemoglobin binding relationship. (Figures 48.15, 48.16, 48.17)

- The evolution of the globin gene family has resulted in several specialized hemoglobin proteins. Heterozygotes for the sickle cell allele are resistant to malaria. (Figure 48.17)

48.6 Adaptations to Extreme Conditions

- Several adaptations in cardiovascular and respiratory activities allow animals to exploit low-oxygen environments. Animals that inhabit high altitudes have larger hearts and lungs, and hemoglobin with a high affinity for binding oxygen. Many diving animals have unusually high hematocrits and also muscles with large quantities of myoglobin.

48.7 Impact on Public Health

- In asthma, the muscles around the bronchioles contract more than usual, increasing resistance to airflow.

- Smoking tobacco products is one of the leading global causes of death. Smoking is strongly linked to cancer, cardiovascular disease, stroke, and emphysema. (Figure 48.18)

- Pneumonia is an infectious disease involving inflammation and fluid buildup in the lungs. (Figure 48.19)

TEST YOURSELF

1. The driving force for diffusion of oxygen across the cells of a respiratory organ is
   a. the difference in the partial pressure of oxygen in the environment and in the blood.
   b. the humidity.
   c. the partial pressure of carbon dioxide in the blood.
   d. the air temperature.
   e. all of the above.

2. Carbon dioxide is considered a harmful by-product of cellular respiration because it
   a. lowers the pH of the blood.
   b. leads to damage of plasma membranes.
   c. competes with oxygen for transport in the blood.
   d. all of the above.
   e. a and b only.

3. The process of bringing oxygenated water or air into contact with a gas-exchange surface is
   a. respiration.
   b. inspiration.
   c. ventilation.
   d. resuscitation.
   e. exhalation.

4. The group of vertebrates that relies on gas exchange across the skin as well as at the lungs to maintain sufficient blood oxygen levels is
   a. the fishes.
   b. the reptiles.
   c. the amphibians.
   d. the birds.
   e. the mammals.

5. The countercurrent exchange mechanism seen in the gills of fishes
   a. maximizes oxygen uptake by the bloodstream.
   b. is a less efficient mechanism for gas exchange compared to mammalian lungs.
   c. occurs because the flow of blood is in the same direction as water flowing across the gills.
   d. increases the amount of energy necessary for respiration.
   e. requires that the fish swallow water.

6. The tracheal system of insects
   a. consists of several tracheae that connect to multiple lungs within the different segments of the body.
   b. consists of extensively branching tubes that are in close contact with all the cells of the body.
   c. allows oxygen to diffuse across the thin exoskeleton of the insect to the bloodstream.
   d. requires constant movement of the wings to move air into and out of the body.
   e. none of the above.

7. _____ is secreted by alveolar cells in the mammalian lung to prevent the collapse of alveoli due to surface tension at the interface of air and extracellular fluid.
   a. Hemoglobin
   b. Myoglobin
   c. Mucus
   d. Hemolymph
   e. Surfactant

8. In negative pressure filling, air moves into the lungs when
   a. the volume of the thoracic cavity increases.
   b. the pressure in the thoracic cavity decreases.
   c. air is forced down the trachea by muscular contractions of the mouth and pharynx.
   d. all of the above.
   e. a and b only.

9. Which of the following factors does *not* alter the rate of breathing by influencing the respiratory centers of the brain?
   a. carbon dioxide partial pressures in the blood
   b. oxygen partial pressures in the blood
   c. blood pH
   d. blood glucose levels
   e. hydrogen ion concentration in the blood

10. With rare exceptions, the majority of oxygen is transported in the blood of vertebrates
    a. by binding to plasma proteins.
    b. by binding to hemoglobin in erythrocytes.
    c. as a component of large organic molecules that are broken down by the cells.
    d. as dissolved gas in the cytoplasm in the erythrocytes.
    e. by binding to myoglobin.

CONCEPTUAL QUESTIONS

1. Define countercurrent exchange in the context of gas exchange.
2. Explain the special adaptations of the avian respiratory system.
3. Explain the special adaptations for life at high altitudes.

EXPERIMENTAL QUESTIONS

1. What was the purpose of the study conducted by Schmidt-Neilsen?
2. Considering the first experiment only (see Figure 48.13), what were the results and what conclusions did the researchers make based on these results?
3. Explain the purpose and procedure of the second experiment conducted by Schmidt-Neilsen and the conclusions he drew from his results.

COLLABORATIVE QUESTIONS

1. Discuss two ways animals exchange gases in an aqueous environment.
2. Discuss the components of the mammalian respiratory system.

www.brookerbiology.com

This website includes answers to the Biological Inquiry questions found in the figure legends and all end-of-chapter questions.

49

EXCRETORY SYSTEMS AND SALT AND WATER BALANCE

CHAPTER OUTLINE

The kangaroo rat (*Dipodomys panamintensis*), a desert-dweller that must conserve body stores of water.

Homeostasis has been a common theme of the previous several chapters. Animals maintain a variety of physiological processes—including energy intake and usage, blood pressure, body temperature, and blood oxygen levels—within normal and often narrow ranges. Homeostasis is also critical in the regulation of salt and water levels in body fluid compartments. Changes in the concentrations of salts in the extracellular and intracellular fluids have the potential to disrupt proper cellular function, for example, by altering the potential difference across plasma membranes in the heart and brain. Salt concentrations and water volumes in the different body compartments are related to each other, because a major way in which water moves between compartments is by osmosis, which in turn depends on the dissolved solutes in water. Consequently, changes in salt concentrations in body fluids may also affect cell volume, which in turn can cause cell death.

As we have learned, homeostasis is an energy-requiring process. A significant portion of an animal's daily energy expenditure goes toward maintaining salt and water homeostasis. The ability to do so is complicated by many factors, such as the environment and climate in which an animal lives, and its access to sufficient supplies of drinking water. In addition, animals use water from their bodies to eliminate metabolic waste products through their excretory system, and that water needs to be regularly replenished.

In this chapter, we examine why salt and water balance is vital for survival, how it is affected by the requirement to eliminate metabolic wastes, and how different excretory organs participate in these processes. We then highlight some of the major features of the vertebrate and mammalian kidney, examine how the kidney eliminates wastes and regulates salt and water balance, and consider how kidney disease affects human health.

49.1 Principles of Homeostasis of Internal Fluids

As we saw in Chapter 47, an animal's internal fluids exist in compartments within the body. In invertebrates, these fluids include the hemolymph and the intracellular fluid, whereas in vertebrates it includes the plasma component of blood, the interstitial fluid surrounding cells, and the intracellular fluid. The major salts found in animals' bodies, and some of their chief functions, were summarized in Chapter 41 (refer back to Table 41.1). In solution, salts dissociate to form charged ions. Because ions are electrically charged, such salts are referred to as electrolytes. In this section, we will examine why this is so important and how an animal's electrolyte concentration and water volume are maintained within normal ranges.

Balance of Water and Electrolytes Is Critical for Survival

Maintenance of normal body water levels is of great importance for all animals. Not only is water the major portion of an animal's body mass, it is also the solvent that permits dissolved solutes to interact in chemical reactions. Water is the transport vehicle that brings oxygen and nutrients to cells, and removes wastes generated by metabolism.

When an animal's water volume is reduced below the normal range, we say the animal is dehydrated. In terrestrial animals, dehydration may occur if sufficient drinking water is not available, or when water is lost by evaporation (perspiring or panting). Regardless of the cause, dehydration can be a serious, potentially life-threatening condition. For example, because blood is roughly 50% plasma (water), blood volume may decrease in dehydrated animals. Reduced fluid volume compromises the ability of the circulatory system to move nutrients and wastes throughout the body, and to help regulate body temperature on hot days (see Chapters 42 and 47).

Electrolyte balance is also very important for animals. A change of only a few percent in the extracellular fluid concentrations of K^+, for example, can trigger changes in nerve, heart, and skeletal muscle function by altering membrane potential (see Chapter 43). Other electrolytes, such as Ca^{2+}, Mg^{2+}, PO_4^{3-}, and SO_4^{2-}, participate in various biological activities. These include serving as cofactors for enzyme activation, participating in bone formation, forming part of the extracellular matrices around cells, and activating cellular events such as exocytosis and muscle contraction. A serious imbalance in any of these ions can disrupt cellular activities and in some cases may result in death of the cells.

Water and Salts Move Between Fluid Compartments by Different Mechanisms

Water moves between adjacent body compartments by osmosis down an osmotic gradient (see Chapter 5). Changes in the salt concentration in one compartment will lead to changes in fluid distribution between the compartments. These changes can cause cells to shrink or swell. When, for example, the salt concentration of extracellular fluid increases, water moves by osmosis from inside cells to the extracellular fluid, causing the cells to shrink. Shrinking or swelling of cells in the brain, heart, and other vital organs can rupture plasma membranes, leading to cell death.

Whereas water moves between fluid compartments by osmosis, dissolved salts such as sodium and potassium ions move by different mechanisms. All ions have very limited ability to diffuse across plasma membranes, because of their high water solubility and low lipid solubility. As described in Chapter 5, ions may cross membranes through channels formed by proteins that create a pore in the membrane. Alternatively, ions may be actively transported across epithelial cells that line tubelike structures such as those in the kidneys. Still other epithelia, such as those in gills, can actively transport salts between the surrounding salt or fresh water and the animal's body fluids. As this type of transport is an active process, it requires energy stored in the chemical bonds of ATP. Animals that face exceptional challenges to maintaining salt balance, such as marine fish, must expend a considerable share of their daily energy budget to transport ions across epithelial cells.

Salt and Water Balance Occurs Despite Obligatory Exchanges with the Environment

Many vital processes—obtaining oxygen and eliminating carbon dioxide, consuming and metabolizing food, regulating body temperature, and eliminating nitrogenous wastes—have the potential to disturb salt and water balance. Therefore, these processes require additional energy expenditure to minimize or reverse the disturbance. Exchanges of salt and water with the environment that occur as a consequence of vital processes, such as respiration or the elimination of wastes, are called obligatory exchanges because the animal is "obliged" to make them (**Figure 49.1**). Next we will examine a few of these obligatory exchanges and how they are related to an animal's environment and life history.

Elimination of Nitrogenous Wastes Proteins and RNA molecules exist for a finite period of time before they are broken down to smaller components. Among the degradation products of proteins and nucleic acids are **nitrogenous wastes** (**Figure 49.2**), molecules that include nitrogen from amino groups (NH_2). These wastes are toxic at high concentrations and must be eliminated from the body. Nitrogenous wastes are usually found in three forms—ammonia and ammonium ions, urea, or uric acid—depending on the species and the environment in which it lives. The elimination of nitrogenous wastes occurs via excretory organs such as the kidneys, or via other specialized structures such as gills. Only certain species, including sharks, rays, and skates, can survive with high levels of nitrogenous wastes in their blood and other body fluids.

Cells produce H_2O during metabolism.

Water vapor exits during breathing.

Salt and H_2O are lost in feces and urine.

Food introduces salt and H_2O.

H_2O is lost by evaporation.

Figure 49.1 Types of obligatory salt and water exchanges in a terrestrial animal. Obligatory exchanges with the environment occur as the result of necessary life processes.

Biological inquiry: Can animals completely eliminate all the losses resulting from obligatory exchanges?

Ammonia (NH₃) and ammonium ions (NH₄⁺) are the most highly toxic of the nitrogenous wastes because they disrupt pH, ion electrochemical gradients, and many chemical reactions that involve oxidations and reductions. Animals that excrete wastes in this form typically live in water. In marine invertebrates, ammonia and ammonium ions are continually excreted across the skin, whereas in freshwater fish these wastes are excreted via the gills and kidneys. Because ammonia is so toxic, aquatic animals excrete it as quickly as it is formed. However, some terrestrial snails and crustaceans can excrete ammonia in its gaseous form, which is less toxic and does not require water for elimination. The chief advantage of excreting nitrogenous wastes as NH₃ or NH₄⁺ is that energy is not required for their production, which is not the case for urea and uric acid, as described next.

All mammals, most amphibians, some marine fishes, some reptiles, and some terrestrial invertebrates convert ammonia into **urea**, which is then excreted. Urea is less toxic than ammonia, and it does not require large volumes of water for its excretion. Animals can tolerate some accumulation of urea in their blood, tissues, and storage organs such as the urinary bladder. This conserves water, removes the necessity for constant excretion, and reduces the likelihood of toxicity. One drawback of producing urea is that the enzymatic conversion of ammonia into urea requires a moderate expenditure of ATP, and thus uses up part of an animal's total daily energy budget.

Birds, insects, and most reptiles produce **uric acid** or other nitrogenous compounds called purines. Like urea, these compounds are less toxic than ammonia, but they are even more energetically costly to synthesize from ammonia. However, because they are poorly soluble in water, they are not excreted in a watery urine but instead are packaged with salts and other waste products into a semisolid, partly dried precipitate that is excreted. The energy investment required to produce uric acid, therefore, is balanced against the water conserved by excreting nitrogenous wastes in this form.

Water and Electrolyte Exchanges During Respiration To ventilate its lungs, a terrestrial animal passes air in and out of its airways. Water in the form of water vapor in the mouth, nasal cavity, and upper airways exits the body with each exhalation. As an animal becomes more active, it requires more oxygen and produces more carbon dioxide. These changes are met by an increase in respiratory activity. Breathing becomes deeper and more rapid, which in turn increases the rate of water loss from the body. Thus, respiration in animals with lungs is associated with significant water loss, as can be observed in cold weather when we can "see our breath."

As described in Chapter 48, small, active animals with high metabolic rates have higher breathing rates than do larger, less active animals. Consequently, the potential for water loss due to respiration is considerably greater in small animals, particularly in endotherms. Hummingbirds, for example, may have 15 to 20 times the water loss per gram of body mass than a large goose.

In aquatic animals, the osmoregulatory challenges are more complex, because fish and other aquatic animals move water, not air, over their respiratory organ (gills). Recall from Chapter 48 that gills, like all respiratory organs, are thin structures with large amounts of surface area and an extensive network of capillaries. These features make gills ideal for gas exchange by diffusion between the capillaries and the surrounding water. However, these same features also make gills ideal for salt and water movement by diffusion and osmosis, respectively.

Animals	Most aquatic animals	Mammals, most amphibians, some marine fishes, some reptiles and terrestrial invertebrates	Birds, insects, most reptiles
Nitrogenous waste	Ammonia or NH₃ Ammonium ions NH₄⁺	Urea $O=C\begin{smallmatrix}NH_2\\NH_2\end{smallmatrix}$	Uric acid
Energy required for production	None	Moderate	High
Amount of water required for excretion	High	Moderate	Low
Toxicity of waste	High	Low	Low

Figure 49.2 **Nitrogenous wastes produced by different animal groups.** The three forms of nitrogenous wastes, which are derived from the breakdown of proteins or nucleic acids, have different properties.

Does not drink.

H₂O enters gill capillaries by osmosis.

Osmolarity of body fluids: ~300 mOsm/L

Some salt and H₂O from food

Active uptake of Na⁺ and Cl⁻ across gill epithelia into capillaries

Very dilute, copious urine containing very little salts

Osmolarity of fresh water: 0–50 mOsm/L

(a) Freshwater fish

Drinks seawater.

H₂O leaves gill capillaries and enters seawater by osmosis.

Osmolarity of body fluids: ~300 mOsm/L

Some salt and H₂O from food

Active excretion of Na⁺ and Cl⁻ across gill epithelia

Small amount of concentrated urine containing large amounts of salts

Osmolarity of seawater: 1,000 mOsm/L

(b) Saltwater fish

Figure 49.3 **Salt and water balance in water-breathers.** Water-breathing creates osmoregulatory challenges due to diffusion of salts and osmosis of water across gills. These challenges differ between **(a)** freshwater and **(b)** saltwater fish and are met by alterations in urine output, by drinking or not drinking water, and by active transport of salts across the gills.

When differences occur in salt concentration between an aquatic animal's body fluids and the surrounding salt or fresh water, respiration via the gills has the potential to disrupt salt and water balance. The solute concentration of a solution of water is known as the solution's **osmolarity**, expressed as milliosmoles/liter (mOsm/L). The number of dissolved solute molecules determines a solution's osmolarity. For example, a 150-mM NaCl solution has an osmolarity of 300 mOsm/L, because each NaCl molecule dissociates into two molecules, one Na⁺ and one Cl⁻ ($2 \times 150 = 300$). The internal fluid osmolarity of most fish is usually within the range of 225–400 mOsm/L, similar to that of most other vertebrates. However, because freshwater lakes and rivers have very little salt content (usually <25 mOsm/L), a strong concentration gradient for salts could promote the loss of salts from a fish's body into the fresh water. Likewise, a strong osmotic gradient favors the movement of water into a freshwater fish. Water flows across the gills and into the underlying capillaries and bloodstream.

Freshwater fish, therefore, gain water and lose salt when ventilating their gills (**Figure 49.3a**). If left uncorrected, this would cause a dangerous decrease in blood salt concentrations. Freshwater fish avoid this problem using two different strategies. First, their kidneys are adapted to producing copious amounts of dilute urine—up to 30% of their body mass per day (an amount that would be equivalent to about 25 L per day in an average-sized man!). Second, specialized gill epithelial cells actively transport Na⁺ and Cl⁻ from the surrounding water into the fish's capillaries. Thus, these two important ions are recaptured from the water. As the preceding discussion suggests, freshwater fish rarely if ever drink water, except for any that might be swallowed with food.

Other freshwater animals, such as frogs and other adult amphibians, have body surfaces that are permeable to water and used in gas exchange. Like freshwater fish, therefore, they tend to gain water by osmosis and compensate by excreting a copious dilute urine. Epithelial cells of the skin actively transport necessary electrolytes from the water into the blood.

Saltwater fish have the opposite problem. They tend to gain salts and lose water across their gills, because seawater has a much higher osmolarity (about 1,000 mOsm/L) than that of their body fluids (**Figure 49.3b**). The gain of salts and the loss of water from the body are only partly offset by the kidneys, which in marine fish produce very little urine so that water can be retained in the body. The urine that is produced has a higher salt concentration than that of freshwater fish.

To prevent dehydration from occurring, marine fish must drink. However, the only water available to them to drink is seawater, which has a very high salt content. Paradoxically, therefore, marine fish drink seawater to replenish the water lost by osmosis through their gills. This creates a new problem: What does the fish do with all of the salt it ingested?

The ingested salt must be eliminated, and this process is accomplished by gill epithelial cells. In contrast to the gills of freshwater fish, which pump salt from the water into the fluids of the fish, gills of marine fish pump salt out of the fish and into the ocean. Thus, marine fish drink seawater to replace the water lost through their gills by osmosis, and then expend energy to transport the excess salt out of the body.

Ingestion-Related Fluid and Electrolyte Balance Because foods contain minerals and water, eating also involves an obligatory exchange of salt and water. Some plant products are over 95% water by weight, while other foods may contain high amounts of sodium or other minerals. Therefore, the type of diet an animal consumes determines how much salt and water it ingests.

Once food has been digested and absorbed, the unusable parts of food are excreted as solid wastes. Some water and salt are lost by this route in most animals, but exceptions exist. Desert-dwelling kangaroo rats (*Dipodomys panamintensis*), like the one shown in the chapter-opening photo, produce fecal pellets that are almost completely dry, which helps these animals conserve water.

As noted earlier, marine fish absorb some salt water. Other animals besides marine fish may also consume salt water, either because fresh water is unavailable or because they ingest some with the food they eat. Many marine reptiles and birds, for example, ingest seawater when consuming prey or, in some cases, when they spend prolonged periods at sea and thus have no access to fresh water for drinking. These animals have specialized epithelial cells that line structures called salt glands, located in groups around the nostrils, mouth, and eyes (**Figure 49.4**). Salts move from the blood into the interstitial fluid, from where they are actively transported by the epithelial cells of the salt glands into the tubules of the gland. The salts and a small amount of fluid then collect into a central duct and get excreted as highly concentrated solutions. Vertebrates without salt glands cannot survive by drinking seawater, because they have no means of creating and excreting a highly concentrated salt solution.

Use of Water to Regulate Body Temperature Endotherms use body water to cool off (see Chapter 42). For example, sweating and panting are used to cool the body. These behaviors use the evaporation of water to draw heat out of the body. In the process, however, the animal loses salt and water. You know from tasting sweat that it is salty, but the saltiness of sweat and of blood are not the same. Sweat is a hypoosmotic (more dilute) solution compared to blood, so the fluid left behind in the body after perspiration has both a lower volume and a higher salt concentration.

Other than perspiration and panting, very little water is gained or lost directly across the body surface of terrestrial vertebrates, because their skin is impermeable to water. An exception is the amphibians. Also, in invertebrates, the rate of water loss across the body surface depends on whether the animal is soft-bodied, like worms, or covered in a waxy, water-impermeable cuticle, like most insects.

Effects of Metabolism on Water Balance When food molecules are metabolized to provide the energy to synthesize ATP, water is generated as oxygen captures electrons and combines with hydrogen ions in mitochondria. This water is sometimes called "metabolic water" to indicate its origin. Some animals—especially desert dwellers, which lack ready access to drinking water—depend on this water to provide all or nearly all of their water requirements. In other animals, the production of metabolic water may result in more water than is needed at that time. This excess water is eliminated by the excretory organs or through other routes. As metabolism is ongoing and required for survival, the excretion or retention of metabolic water can be considered a type of obligatory exchange.

As described next, the significance of obligatory exchanges and their effects on homeostasis was dramatically illustrated by a long-term investigation by a research team at the University of Florida. Their discovery would lead to a revolution in our understanding of exercise physiology in humans.

This viscous solution is the salty secretion from the turtle's salt glands, which has given some observers the impression that the turtles are "crying."

NaCl diffuses from blood to interstitial fluid.

NaCl is actively transported from interstitial fluid to lumen of salt gland tubule.

Figure 49.4   **Salt glands as an adaptation for marine life.** Many marine birds and reptiles have salt glands, which contain a network of secretory tubules that actively transport NaCl from the interstitial fluid into the tubule lumen. The viscous solution then moves through a central duct and to the outside environment through pores in the nose, around the eyes, and in other locations. The black arrows indicate direction of flow of blood or salt gland secretions.

FEATURE INVESTIGATION

Cade and Colleagues at the University of Florida Discovered Why Athletes' Performance Wanes on Hot Days

On a typical hot summer day in the early 1960s in Gainesville, Florida, the University of Florida football team was practicing in full equipment. The players were rapidly becoming dehydrated and, unbeknownst to them, the osmolarity of their body fluids was increasing as their bodies produced copious amounts of dilute sweat in an effort to maintain body temperature. The athletes became aware of two things. First, they discovered that they did not need to urinate for long periods after a tough practice session, and second, their performance on the field suffered as they became increasingly fatigued and more susceptible to muscle cramps. In fact, athletes exercising in these conditions have been known to occasionally develop seizures—uncontrolled activity of neurons in the brain. This situation did not escape the notice of the team physicians and, notably, university faculty member and kidney specialist Robert Cade.

Many of the symptoms experienced by the players could be readily explained. The fatigue was directly related to body water loss, which put a strain on the circulatory system and reduced blood flow to muscles and other organs. The muscle cramps and even occasional seizures arose from an imbalance in extracellular electrolytes—notably sodium and potassium—which are secreted in perspiration. The resulting change in extracellular salt concentrations caused a change in the electrical potential across muscle and nerve cell membranes, which triggered the spasms. Lastly, the decreased urine production is one of the body's mechanisms for reducing fluid loss when body water is decreasing.

The key question was, How can these effects of extreme exercise best be reversed or prevented? The answer was simple and clever. Cade and his colleagues rejected the prevailing view that drinking any fluids during heavy exercise somehow contributed to cramps and other problems. Instead, they hypothesized that the best way to maintain salt and water homeostasis in a profusely sweating person is to restore to the body exactly what was lost; that is, the person should drink a solution of water and solutes that resembles sweat!

The first thing Cade needed to do was analyze precisely how much sodium, potassium, and other ions are actually present in sweat. Fortunately, he had an abundance of human sweat at his disposal to analyze. Once the players left the field, their jerseys were wrung out and the composition of the sweat was determined with an ion analyzer such as one called a flame spectrophotometer, as shown in the Feature Investigation in **Figure 49.5**. The concentrations could then be compared with known values of ion concentrations in human blood. The composition of human sweat is now known to change under certain conditions and to vary among people, but Cade's results were typical.

Figure 49.5 Cade and colleagues discovered a way to improve athletic performance and prevent salt and water imbalance during strenuous exercise.

HYPOTHESIS Athletic performance can be enhanced by maintaining the body's salt and H_2O balance during exercise.

STARTING MATERIALS Supply of human sweat for analysis.

Experimental level   Conceptual level

1 Obtain human sweat from exercising athletes.

Sweat

Dilute solution of salts of unknown composition

2 Analyze composition of sweat using a flame spectrophotometer, which measures ion concentrations. Prepare artificial solution that mimics composition of sweat. Compare composition to known ion concentration in human blood.

Flame spectrophotometer

Salts

Sweat   Artificial solution

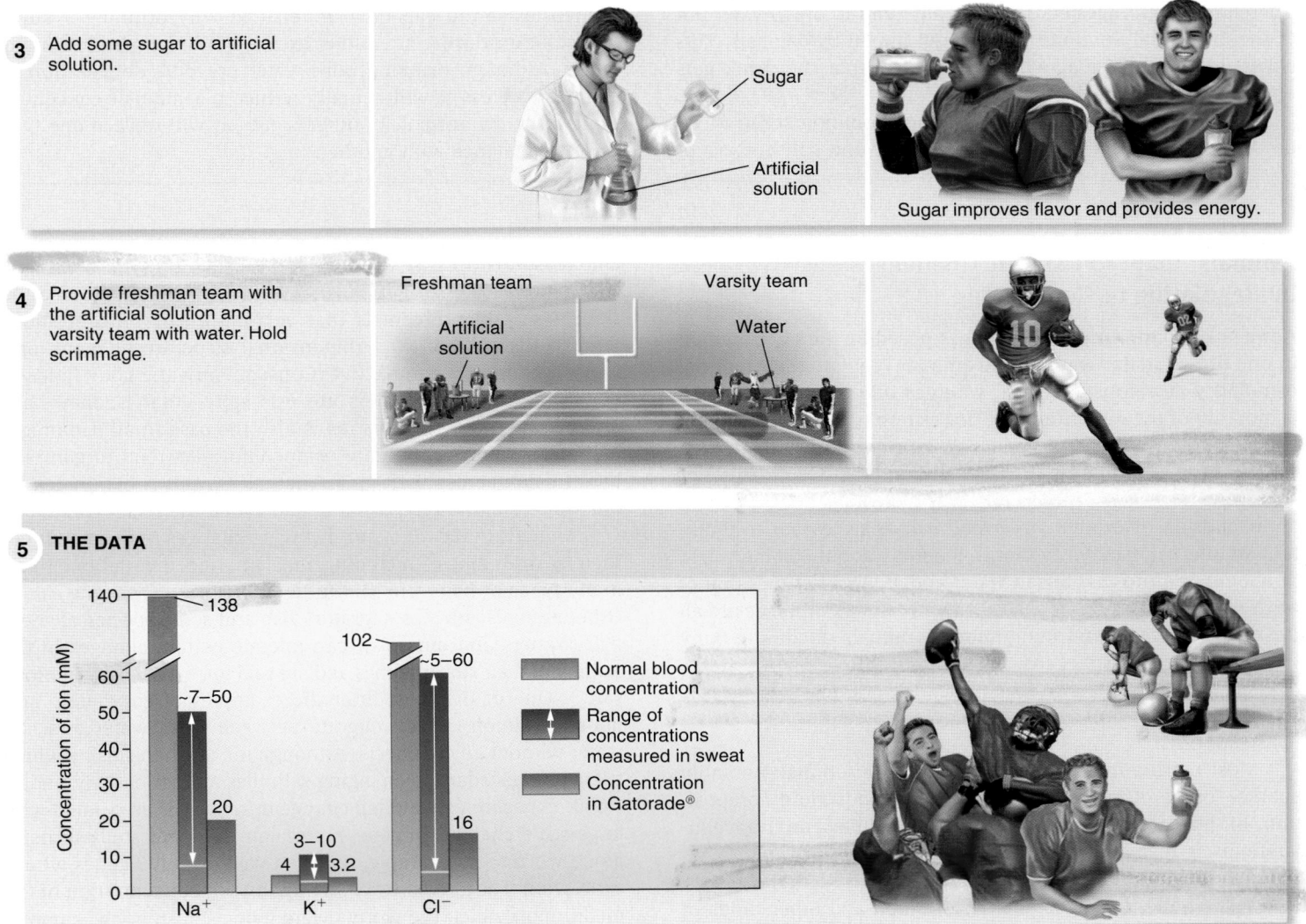

3 Add some sugar to artificial solution.

Sugar

Artificial solution

Sugar improves flavor and provides energy.

4 Provide freshman team with the artificial solution and varsity team with water. Hold scrimmage.

Freshman team

Artificial solution

Varsity team

Water

5 **THE DATA**

Concentration of ion (mM)

138

102

~7–50

~5–60

20

16

3–10

4

3.2

Na+

K+

Cl⁻

Normal blood concentration

Range of concentrations measured in sweat

Concentration in Gatorade®

The sweat of the athletes contained mostly Na⁺, K⁺, and Cl⁻ at concentrations that produced a dilute solution compared to blood. Once Cade completed this analysis, it was a relatively straightforward procedure to make a watery solution of a composition similar to human sweat. The next step was to give the solution to the players prior to and during the practice sessions and games. Improving its taste—adding some lemon flavoring and some sugar—removed any inhibitions the players may have had about drinking it, while also providing an energy boost.

For the first trial, Cade gave the solution to the freshman players during an intrasquad scrimmage against the more experienced varsity team, whose members received only pure water to drink as a control. Surprisingly, the freshman team outperformed the more experienced players and did not suffer the characteristic late-game fatigue the control subjects (the varsity) experienced. In subsequent years, Cade and other researchers would conduct carefully controlled experiments with humans and laboratory animals to confirm that a balanced solution of salts similar to that present in sweat effectively improves exercise performance and reduces the possibility of dehydration and its consequences.

Because the solution was envisioned as "aid" for the team known as the University of Florida "Gators," the drink eventually came to be called Gatorade®. Very shortly after its introduction, the Gators enjoyed their most successful season. In 1965 the Kansas City Chiefs of the former American Football League became the first professional sports team to try the drink, and shortly thereafter the team enjoyed its greatest success. Nowadays Gatorade is used at sporting events around the world, and for good scientific reason.

The effectiveness of a solution like Gatorade is due to its ability to restore the correct amounts of both water and ions lost during exercise. Importantly, it is very rapidly absorbed because its osmolarity is close to that of body fluids. Many of the other "sports drinks" subsequently invented contain additional solutes, such as vitamins and other minerals, and many contain higher amounts of sugar. Because of the presence of these other solutes, these drinks may be very hyperosmotic relative to body fluids.

Therefore, when ingested, they initially tend to draw water out of the interstitial fluid and into the gut lumen by osmosis. This slows down the rate at which the water from the drink gets absorbed into the blood.

The story of Gatorade is one of good common sense based on solid scientific principles of osmolarity and salt and water homeostasis. You can now understand why drinking a dilute salt solution during strenuous exercise is better than drinking water. Although drinking pure water prevents dehydration, if drunk in excess it will actually reduce plasma salt concentrations to below normal. In other words, it will replace one type of salt imbalance with another.

Animals Adapt to Osmotic Challenges by Regulating or Conforming

Animals adapt to osmotic challenges posed by the environment in one of two major ways. Some animals regulate their internal osmolarity at a constant level, whereas others conform to the osmolarity of their environment (for example, the sea). Animals that maintain constant internal salt concentrations and osmolarities, even when living in water with very different osmolarities than their body fluids, are called **osmoregulators**. Such animals drink or excrete water and salt as necessary to maintain an internal osmolarity that is generally about 300 mOsm/L, or about one-third that of salt water and at least 10 times that of fresh water. All terrestrial animals are osmoregulators, as are all freshwater animals and many marine animals, including bony fish and some crustaceans. Osmoregulators maintain stable cellular levels of ions and fluid, but this requires considerable expenditure of energy, primarily to pump ions into and out of epithelial cells.

Most marine invertebrates and some vertebrates—notably sharks—use a different means to control body fluid composition. In this case, the osmolarity of extracellular and intracellular fluids is matched with seawater. These animals are called **osmoconformers**, because their osmolarity conforms to that of their environment. The osmolarity of blood and other fluids of marine osmoconformers is like seawater, around 1,000 mOsm/L. An advantage of having body fluids conform to the osmolarity of the surrounding seawater is that there is much less tendency to gain or lose water across the skin or gills. Thus, sharks and other osmoconformers expend less energy to compensate for water gain or loss than do other aquatic animals. However, osmoconformers are generally limited to the marine environment, whereas many species of osmoregulators such as salmon can migrate between waters of different salinities.

Vertebrate osmoconformers employ a strategy in which they have a high concentration of uncharged molecules in their extracellular fluids. This allows the extracellular fluids and seawater to have similar osmolarities, but it prevents the excessive accumulation of ions in the body. The body fluids of sharks and other osmoconformers contain sugars, amino acids, and metabolic waste products—notably urea and an organic compound called trimethylamine oxide (TMAO). The total amount of salt and organic compounds in a shark's extracellular fluids produces an osmolarity very similar to that of seawater, even though the salt concentration is similar to that of osmoregulators.

Vertebrate osmoconformers cannot tolerate high ion concentrations in their body fluids any better than osmoregulators.

One reason is because a proper ion balance is required for normal electrical signaling in their nerve and muscle cells. In addition, a second problem is that very high salt concentrations tend to disrupt the three-dimensional structure of many proteins, rendering them inactive. Consequently, the body fluids of vertebrate osmoconformers are less salty—that is, they have fewer ions—than seawater, as is also the case in all osmoregulators. Therefore, vertebrate osmoconformers tend to gain salt by diffusion across their gills. That excess salt is eliminated by the kidneys and a type of salt gland called the rectal gland.

Animals that cannot survive wide changes in the salinity of their surroundings—including most marine invertebrates and many types of fish—are said to be stenohaline ("narrow salt"). Other animals, such as migratory fish and some species of shellfish (oysters and mussels), can tolerate certain changes in the salinity of their surroundings and are said to be euryhaline ("broad salt"). One of the most dramatic examples of euryhaline animals is salmon, which migrate between fresh water and salt water without any ill effects. Although the mechanisms by which such animals adapt to changing salinities are not entirely understood, exposure to a sudden change in salinity is associated with large-scale changes in gene expression. A recent study in nematode worms, for example, reported well over 100 genes whose expression was altered by changing the salt concentration of the medium on which the worms were living. Presumably, some of the genes are associated with such functions as ion transport in skin and other tissue.

Next we will examine the excretory organs found in different animals, and how such organs not only function to eliminate wastes but also are involved in salt and water homeostasis.

49.2 A Survey of Excretory Organs

Although the mammalian kidney, and in particular that of humans, has been especially well studied, enough is known about other classes of animals to make general statements regarding the regulation of salt, water, and waste levels in the body's fluid compartments. This is an ancient and important function that arose early in evolution.

As just described, animals make use of one or more different organs to rid themselves of metabolic wastes, excess water and salts, and toxins from their environment. Most excretory organs contain tubular structures lined with epithelial cells that have the capacity to actively transport ions across their membranes. Wastes are secreted out of the body by means of these tubes. In some cases, animals may have considerable ability to regulate the rate at which waste is secreted and how much water is lost in the process. For example, even though a thirsty mammal on a hot, sunny day must continue to rid its body of soluble waste products, it must also conserve water. In this section, we examine general principles common to most excretory systems. Next, we consider invertebrate and vertebrate excretory organs.

Urine Production Involves the Processes of Filtration, Reabsorption, and Secretion

Most excretory organs operate by using one or more of the following processes: filtration, reabsorption, and secretion. In **filtration**, an organ acts like a sieve or filter, removing some of the water and its small solutes from the blood, interstitial fluid, or hemolymph, while leaving behind blood cells and large solutes like proteins. A typical filtration system is that seen in the mammalian kidney, in which the plasma component of blood is forced under pressure through leaky capillaries and into the kidney tubules (**Figure 49.6**). The material that passes through the filter and enters the excretory organ for either further processing or excretion is called a **filtrate**.

Some of the material in the filtrate can be recaptured and returned to the blood. This is an important feature of many excretory organs, because the formation of a filtrate is not selective, apart from the exclusion of proteins and blood cells. In other words, in order to filter the blood and remove soluble wastes, necessary molecules such as salts, sugars, and amino acids also get filtered in the process. Recapturing those useful solutes requires active transport pumps or other transport systems and is known as **reabsorption**. The remaining part of the filtrate that gets excreted is called **urine**.

In some cases, solutes may get excreted from the body in quantities greater than those found in the filtrate. How is this possible? Some solutes are actively transported from the interstitial fluid surrounding the epithelial cells of the tubules, into the tubule lumens. This process, called **secretion**, supplements the amount of a solute that would normally be removed by filtration alone. This is often a way in which excretory organs eliminate particularly toxic compounds from an animal's body, and it can be very effective. Some marine fish, for example, make use of secretion as the sole means of cleansing the blood. These animals do not form a filtrate at all.

Most Invertebrates Use a Filtration Mechanism to Cleanse the Blood

The simplest filtration mechanism in invertebrates is the **protonephridia** system of flatworms (**Figure 49.7**). A series of branching tubules filters fluids from the body cavity into the tubule lumens by means of ciliated cells that cap the ends of the tubules. The beating of the cilia bears some resemblance to a flickering flame, which is why these cells are known as flame cells.

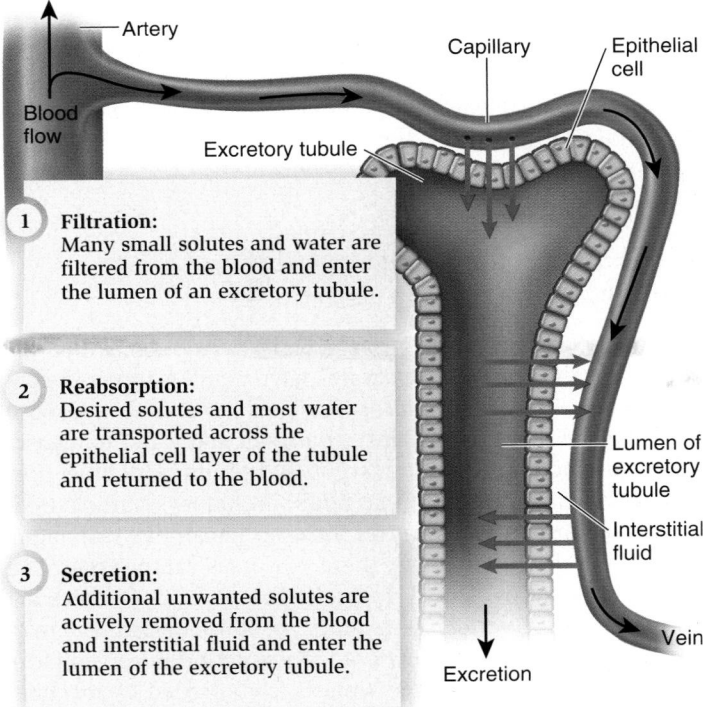

Figure 49.6 **Basic features of the function of many excretory systems.** A filtrate containing water and small solutes is formed from body fluids (in this case, blood) and enters an excretory tubule. Most of the water and the useful solutes in the filtrate get reabsorbed from the tubule into nearby capillaries. Some solutes get secreted from the capillaries into the tubule along its length. A portion of the filtrate, including wastes, gets excreted.

Figure 49.7 **The protonephridial system of flatworms.** As the filtrate moves along the tubules, the composition of the filtrate can be changed by the action of the tubule epithelial cells. The excreted fluid is typically hypoosmotic and exits through pores in the body wall called nephridiopores.

As fluid is drawn through slitlike openings of the flame cells and into the lumen by the beating of the cilia, it percolates through the tubule, where most solutes are reabsorbed back into the interstitial fluid. Excess water and some wastes are emptied through tiny openings in the body wall called nephridiopores. Much of the nitrogenous waste in flatworms actually diffuses across the body surface into the surrounding water, and therefore the protonephridia are primarily osmoregulatory organs. The urine is generally hypoosmotic compared to the rest of the body fluids, an adaptation for life in fresh water.

Annelids, by contrast, use a different filtration mechanism, called a **metanephridial system** (**Figure 49.8**). In annelids, pairs of metanephridia are located in each body segment. They consist of a tubular network that begins with a funnel-like structure called a nephrostome. The nephrostomes are open to the body cavity and collect coelomic fluid, which contains nitrogenous wastes, through tiny pores that exclude large solutes. Na^+, Cl^-, and other solutes are reabsorbed by active transport along the length of the tubules that extend from the nephrostomes, and from there diffuse into nearby capillaries. The nitrogenous wastes remain behind in the tubules and get excreted through nephridiopores in the body wall. Many annelids live in watery environments and thus, like flatworms, excrete a hypoosmotic urine.

Filtration mechanisms also operate in the kidney-like structures of many mollusks, and in the crustacean excretory organs called antennal glands. As in worms, reabsorption of useful solutes occurs along the length of the excretory system. Research has clearly indicated that mollusks can reabsorb organic nutrients well enough that under most conditions these nutrients do not appear at all in the urine. Like the protonephridial and metanephridial systems just discussed, urine is excreted via nephridiopores in these other animals.

Insects Excrete Wastes by Means of Secretory Organs Rather Than Filtration

The insect excretory system is quite different from other invertebrates, because it does not involve filtration of body fluids. Instead, a series of narrow, extensive tubes called **Malpighian tubules** arises from the midgut and extends into the surrounding hemolymph (**Figure 49.9**). The cells lining the tubules actively transport potassium ions, along with uric acid and other purines from the hemolymph, into the tubule lumen. This secretion process creates an osmotic gradient that draws water into the tubules. The fluid moves from the tubules into the hindgut—the intestine and rectum—where much of the useful salt and water is reabsorbed. The nitrogenous wastes and other waste compounds are excreted together with the feces through the anus.

Unlike other invertebrates, most terrestrial insects, apart from blood-sucking ones, excrete urine that is either isoosmotic or hyperosmotic to body fluids. This is a testament to the efficiency with which the insect hindgut reabsorbs water, and reflects the fact that life in dry environments is associated with a risk of dehydration.

The Kidney Is the Major Excretory and Filtration Organ in Vertebrates

The major excretory organ found in all vertebrates is the kidney. The kidneys of all vertebrates have many features in common. They typically contain specialized tubules composed of epithelial cells that participate in both salt and water homeostasis by promoting active transport of sodium and other ions across their membranes. In addition, all kidneys participate in the excretion of wastes. In response to an animal's changing water and salt requirements, these processes can be controlled, that is, sped up or slowed down, by the actions of nerves and hormones. Most vertebrate kidneys are filtration kidneys, with the exception of purely secretory kidneys found in some marine fish. Finally, filtration in the kidneys is controlled by mechanical forces, such as the hydrostatic pressure exerted by blood entering the capillaries of the kidneys. The mammalian kidney has been especially well studied and is examined in detail later in this chapter.

The need to eliminate waste products while simultaneously maintaining salt and water homeostasis occurs in all animals. How animals achieve this balance in highly disparate environments, and with different life histories, is the subject of our next discussion.

2 The filtrate moves through the tubules, where desired solutes are reabsorbed into capillaries.

Metanephridium in capillary network

Capillaries

1 Metanephridia filter coelomic fluid through nephrostomes.

Nephridiopore   Nephrostome

3 A dilute urine is excreted via nephridiopores.

Figure 49.8 **The metanephridial system of annelids.** Most internal body structures have been omitted for clarity. Only one of the two metanephridia in each segment is shown.

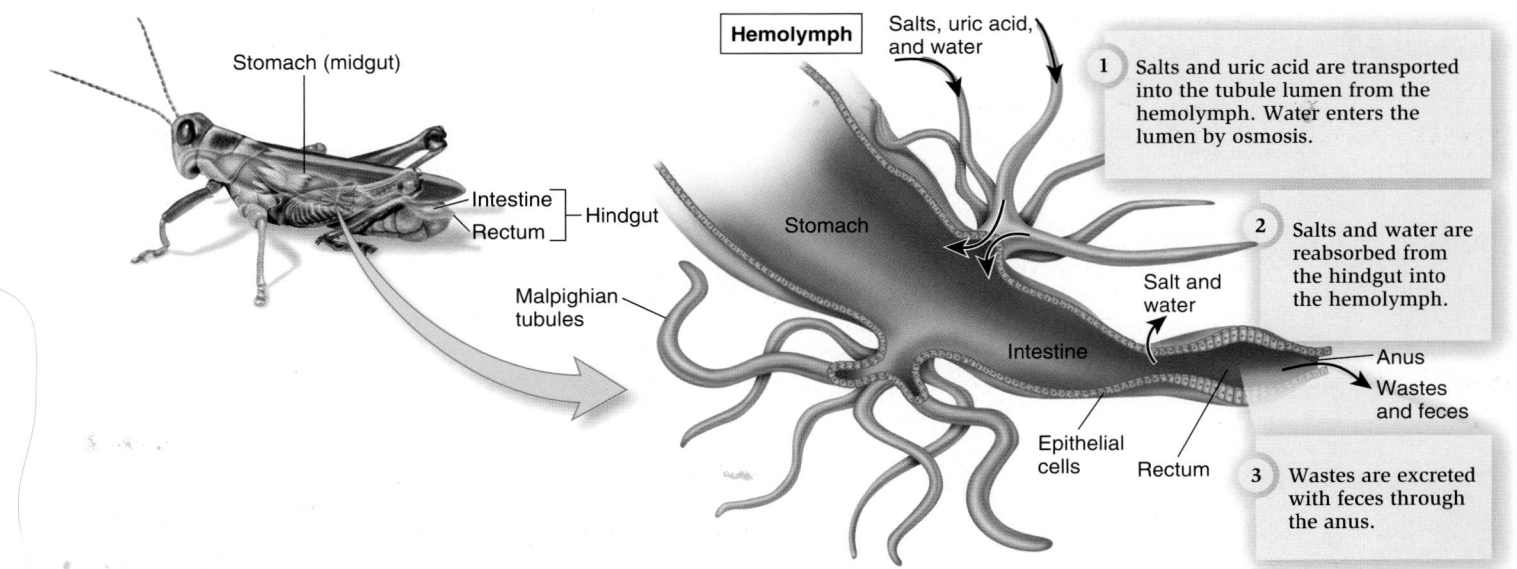

Figure 49.9   Malpighian tubules of an insect.   The tubules, which are longer and more convoluted than shown in this simplified illustration, extend into the body cavity, where they are surrounded by hemolymph. Epithelial cells of the tubules transport salts and uric acid into the tubule lumen. Water follows these solutes by osmosis. The tubular fluid—which is not a filtrate—enters the hindgut, where most of the water and useful salts are reabsorbed. The remaining nitrogenous waste is then excreted.

The Diet and Life History of Vertebrates May Influence Renal Function

As previously mentioned, not all animals have similar requirements for salt and water. **Table 49.1** summarizes some of the special needs of different vertebrates, and some major behavioral and physiological adaptations that allow them to maintain homeostasis. Note that when we refer to kidneys, we use the adjective renal, meaning "pertaining to the kidneys." For example, we refer to renal physiology and renal functions.

The demands placed on an animal by its environment can often be predicted by examining the activities of that animal's kidneys. For example, the kidneys of freshwater fish are specially adapted for rapid, large-scale filtering of the blood and producing a very dilute urine. Amphibians, because of their permeable skin, absorb fresh water from their environments, and consequently they have kidneys whose appearance and activity resemble those of freshwater fish. By contrast, the kidneys of marine fish and desert mammals are adapted to produce urine that is more concentrated than that of freshwater fish and amphibians. This is an important adaptation that helps conserve as much water as possible.

Some animals have a diet that is extreme in one way or another, and this is often associated with differences in renal structure and function. Many mammals and insectivores, for example, subsist entirely or almost entirely on sporadic, high-protein meals. These animals consume so much protein at a single meal that the rate at which they generate nitrogenous waste is very high. Not surprisingly, perhaps, the rate at which blood is filtered through the kidneys of such animals tends also to be very high, an adaptation that helps eliminate the toxic waste.

At the other extreme, polar bears may go four to five months without feeding. During the nonfeeding periods, polar bears neither defecate nor urinate and subsist by breaking down stores of fat and protein in their bodies. Thus, even though they are not eating, they generate nitrogenous wastes like urea. Despite the fact that they do not urinate during this time, however, the concentration of urea in their blood may actually drop! The explanation for this surprising phenomenon appears to be that polar bears have the extraordinary ability to recycle nitrogenous waste into synthesis of new protein, an ability that is of great interest to researchers who study the health consequences of kidney disease and consequent accumulation of urea in the bloodstream of humans.

Animals that subsist primarily on the blood of other animals, such as the vampire bat (*Desmodus rotundus*), face a special challenge because ingesting blood brings large amounts of water into the animal. The body mass of one of these bats may increase from around 30 g before a blood meal to as much as 50 g afterwards. This is roughly equivalent to a 125-pound woman weighing over 200 pounds after a meal! Vampire bats engorged with blood can even become too heavy to fly. Consequently, their kidneys have become adapted to rapidly excrete a dilute urine. Most of the water portion of the blood they consume is quickly excreted, while the nutrients are retained. Increased urine production begins within a few minutes of feeding and can be produced at rates as high as 4 mL/kg body mass/minute. This is roughly 400 times higher than the rate at which urine is produced in a human being. This only occurs during and after a blood meal. At other times, the kidneys of vampire bats can produce a concentrated urine to conserve water when necessary.

Table 49.1	A Comparison of the Vertebrate Mechanisms of Osmoregulation in Different Environments				
Type of animal	**Blood concentration relative to environment**	**Urine concentration relative to blood**	**Main nitrogenous waste**		**Osmoregulatory mechanisms and special features**
Marine fish	Hypoosmotic	Isoosmotic or hyperosmotic	Ammonia		Drinks seawater; secretes salt from gills
Freshwater fish	Hyperosmotic	Strongly hypoosmotic	Ammonia		Drinks no water; transports salt across gills
Amphibian (freshwater)	Hyperosmotic	Strongly hypoosmotic	Most secrete urea		Absorbs water through skin
Marine reptile	Hypoosmotic	Isoosmotic	Ammonia, urea, uric acid		Drinks seawater; hyperosmotic salt-gland secretion
Terrestrial reptile	—	Isoosmotic	Uric acid		Drinks fresh water
Marine bird	Hypoosmotic	Weakly hyperosmotic	Uric acid		Drinks seawater; hyperosmotic salt-gland secretion
Terrestrial bird	—	Weakly hyperosmotic	Uric acid		Drinks fresh water
Desert mammal	—	Strongly hyperosmotic	Urea		Drinks no water; very long loops of Henle; depends on water generated by cellular metabolism
Marine mammal	Hypoosmotic	Strongly hyperosmotic	Urea		Drinks no water
Other mammals	—	Strongly hyperosmotic	Urea		Drinks fresh water

The ability to respond to a blood meal by rapidly producing dilute urine has evolved on at least two separate occasions. Blood-sucking insects such as mosquitoes can also eliminate the water portion of their meal, even before they have finished ingesting a blood meal.

We turn now to an examination of one of the best understood excretory organs: the mammalian kidney.

49.3 Structure and Function of the Mammalian Kidney

All mammals have kidneys that are similar in structure and function, although some are more adapted to one type of life history and environment than another. Much of what we know today about the human kidney arose from observations made on the kidneys of rodents, dogs, and other mammals.

The two kidneys lie in the abdomen (**Figure 49.10a**). The urine formed in each kidney collects in the renal pelvis and from there flows through the **ureters** into the **urinary bladder**. Urine is eliminated via the **urethra**. Collectively, the kidneys, ureters, urinary bladder, and urethra constitute the **urinary system** in humans.

Each kidney is composed of an outer portion called the renal cortex, and an inner portion called the renal medulla, which in turn is composed of outer and inner regions (**Figure 49.10b**). The cortex is the primary site of blood filtration. In the medulla, urine becomes concentrated by the reabsorption of water back into the blood. In this section, we will examine the structural features of the kidney that allow it to function as a filtration system.

The Functional Units of the Kidney Are Called Nephrons

Depending on its size, the mammalian kidney contains as many as several million similar, single-cell-thick structures called **nephrons**. (It is from the word nephron that we get the name of the medical specialty nephrology.) As shown in **Figure 49.10c**, each nephron consists of (1) an initial filtering component called the **renal corpuscle** and (2) a narrow tubule that extends out from the renal corpuscle. The tubule empties into a larger tubule called a **collecting duct**.

The renal corpuscle forms a filtrate from blood that is free of cells and proteins. This filtrate then leaves the renal corpuscle and enters the lumen of the tubule. As it flows through different regions along the length of the tubule, substances are secreted into it or reabsorbed from it. Ultimately the urine remaining at the end of each nephron combines into the collecting ducts.

The Renal Corpuscle Each renal corpuscle contains a cluster of interconnected, fenestrated capillaries called the **glomerulus** (plural, glomeruli), or glomerular capillaries (**Figure 49.10d**). A fenestrated capillary has tiny holes in its walls that permit rapid flow of plasma out of the capillary (**Figure 49.10e**). Each glomerulus is supplied with blood under pressure by an **afferent arteriole**. The glomerulus protrudes into a fluid-filled space called Bowman's space, which lies within a capsule called **Bowman's capsule**. The combination of a glomerulus and a Bowman's capsule constitutes a renal corpuscle.

The Tubular Part of the Nephron The tubule of the nephron, which is continuous with Bowman's capsule, is made of a single layer of epithelial cells resting on a basement membrane. The epithelial cells differ in structure and function along the tubule's length, and 10 to 12 distinct segments are now recognized. For simplicity, however, it is possible to group two or more contiguous tubular segments when discussing function. Thus, the segment of the tubule that drains Bowman's capsule is the **proximal convoluted tubule** ("proximal" means adjacent to, with respect to the beginning of the nephron) (see Figure 49.10c). The next portion of the tubule is the **loop of Henle**, which is a long, hairpin-like loop consisting of a descending limb coming from the proximal tubule and an ascending limb leading to the next tubular segment, the **distal** ("away from") **convoluted tubule**. Fluid flows from the distal convoluted tubule into one of the many collecting ducts in the kidney.

All along its length, each tubule is surrounded by capillaries. These include the **peritubular capillaries** in the cortex, and the **vasa recta capillaries** in the medulla (see Figure 49.10c). These capillaries carry away reabsorbed solutes and water and return them to the bloodstream. Some of them also participate in the process of secretion in the nephron tubules.

Filtration: Blood Is Filtered by the Renal Corpuscle

Now that we have learned about the structural units of the mammalian kidney, we turn our attention to the mechanisms by which nephrons filter blood and produce urine for excretion, and how these activities are controlled under different conditions.

As blood flows through the glomerulus, about 20% of the plasma leaves the glomerular capillaries and filters into Bowman's space. The remaining blood then leaves the glomerulus by an **efferent arteriole** (see Figure 49.10d). Proteins and blood cells are prevented from leaving the glomerular capillaries because the fenestrations are exceedingly tiny. In addition, the glomerular capillaries are surrounded by cells called podocytes (Figure 49.10e) that form filtration slits that act as a physical barrier to movement of large solutes and cells.

The fluid that leaves the glomerular capillaries and enters Bowman's space is called the glomerular filtrate. The rate at which the filtrate is formed is called the **glomerular filtration rate (GFR)**. GFR can be increased by dilation of the afferent arteriole. When the afferent arteriole widens, more blood enters the glomerulus, increasing the hydrostatic pressure in those capillaries and forcing more plasma through the fenestrations in the glomerular capillaries and into Bowman's space. This might happen, for example, when there is an excess of water in the body that must be excreted in the urine. By contrast, constriction of

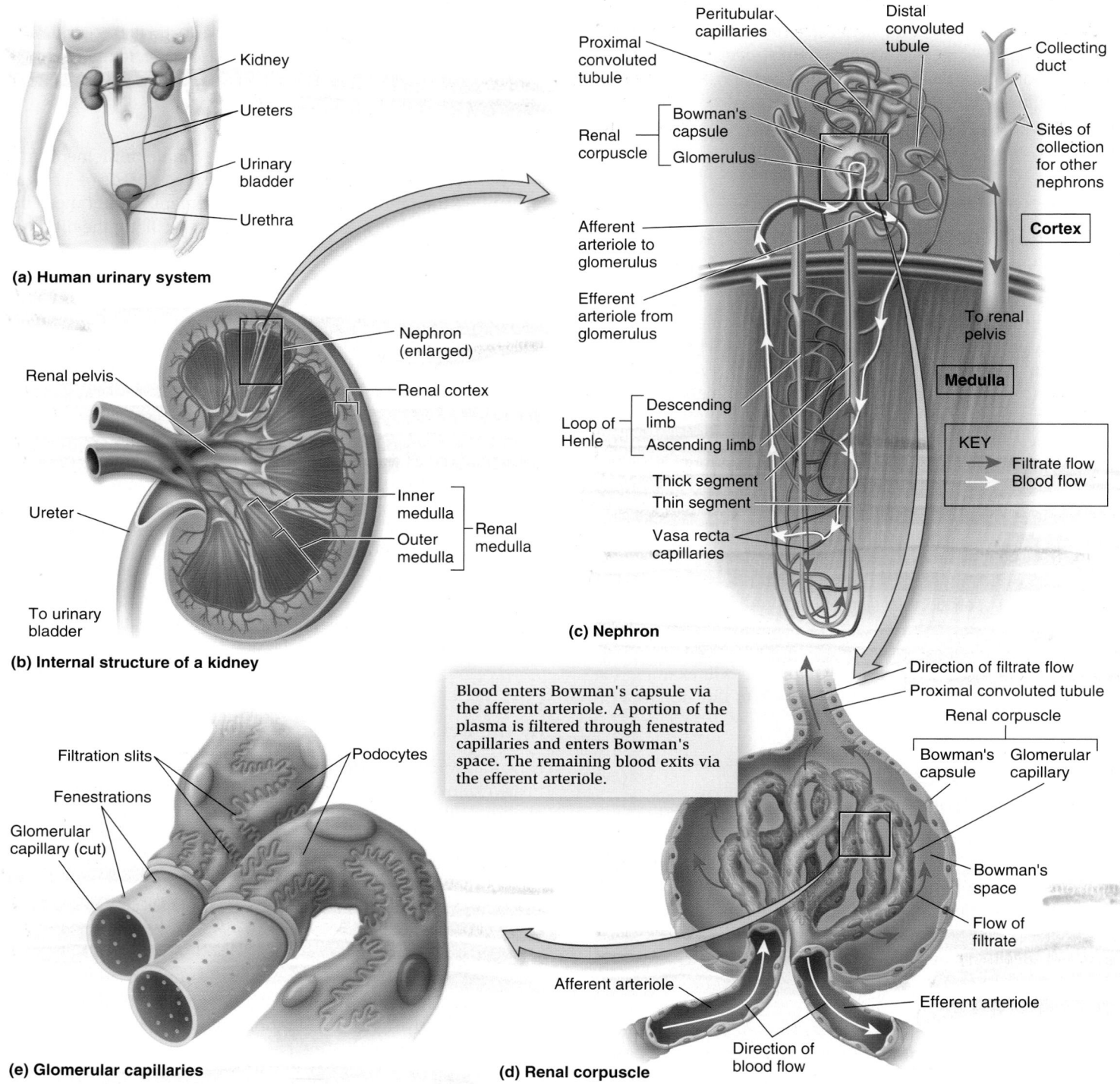

(a) Human urinary system

- Kidney
- Ureters
- Urinary bladder
- Urethra

(b) Internal structure of a kidney

- Renal pelvis
- Ureter
- To urinary bladder
- Nephron (enlarged)
- Renal cortex
- Inner medulla
- Outer medulla
- Renal medulla

(c) Nephron

- Peritubular capillaries
- Proximal convoluted tubule
- Distal convoluted tubule
- Collecting duct
- Renal corpuscle
  - Bowman's capsule
  - Glomerulus
- Sites of collection for other nephrons
- Afferent arteriole to glomerulus
- Efferent arteriole from glomerulus
- **Cortex**
- To renal pelvis
- **Medulla**
- Loop of Henle
  - Descending limb
  - Ascending limb
- Thick segment
- Thin segment
- Vasa recta capillaries
- **KEY**
  - Filtrate flow
  - Blood flow

(d) Renal corpuscle

Blood enters Bowman's capsule via the afferent arteriole. A portion of the plasma is filtered through fenestrated capillaries and enters Bowman's space. The remaining blood exits via the efferent arteriole.

- Direction of filtrate flow
- Proximal convoluted tubule
- Renal corpuscle
  - Bowman's capsule
  - Glomerular capillary
- Bowman's space
- Flow of filtrate
- Afferent arteriole
- Efferent arteriole
- Direction of blood flow

(e) Glomerular capillaries

- Filtration slits
- Fenestrations
- Glomerular capillary (cut)
- Podocytes

Figure 49.10 The anatomy of the mammalian urinary system. **(a)** The human urinary system in a woman. In the male, the urethra passes through the penis. **(b)** Enlarged view of a section through a kidney, showing the locations of the major internal structures and a single nephron (grossly enlarged; at the scale of this illustration, nephrons would be microscopic). **(c)** Structure of a nephron. Filtration occurs in the cortex. Urine concentration occurs in the medulla. The nephron begins at Bowman's capsule and terminates at a collecting duct. Many nephrons empty into a given collecting duct. Surrounding the nephron are capillaries called peritubular capillaries in the cortex, and the vasa recta capillaries in the medulla. Peritubular capillaries secrete solutes into the nephron, while both sets of capillaries carry reabsorbed solutes from the nephron to the bloodstream. **(d)** Anatomy of a renal corpuscle. Blood enters Bowman's capsule via the afferent arteriole. A portion of the plasma is filtered through fenestrated capillaries and enters the nephron. The remaining blood exits the corpuscle via the efferent arteriole. **(e)** Podocytes and glomerular capillaries. By forming filtration slits, podocytes act as a physical barrier to movement of large solutes such as proteins out of the glomerular capillaries.

constriction

the afferent arteriole would reduce the amount of blood entering the glomerular capillaries, and therefore would reduce GFR. This might occur following a loss of blood due to a severe injury. In such a scenario, reducing GFR results in less urine production, which in turn minimizes how much water is lost from the body and helps compensate for the blood lost due to the injury.

Reabsorption: Most Reabsorption of Useful Solutes Occurs in the Proximal Tubule

Anywhere from two-thirds to 100% of a particular solute is reabsorbed from the filtrate in the proximal tubule. This includes Na^+, K^+, Cl^-, HCO_3^-, and organic molecules such as glucose and amino acids. Some ions diffuse through channels in the membranes of the epithelial cells that form the single-cell-thick proximal tubule. Others are actively transported across the tubule. Organic molecules generally are reabsorbed by transporters such as the Na^+/glucose symporter.

Most of the water in the filtrate leaves by osmosis as the ions and organic molecules are transported from the lumen of the proximal convoluted tubule to the interstitial fluid outside the tubules. The amount of water reabsorbed is proportional to the amount of solutes reabsorbed. From there, the water and solutes enter peritubular capillaries to return to the blood. By the time the filtrate leaves the proximal tubule, its volume and composition have changed considerably. The amount of water and salts is much reduced now, and the organic molecules have all been removed.

Reabsorption: Water and Salt Reabsorption Continues Along the Loop of Henle

In the loop of Henle in the mammalian kidney, fluid moves down the descending limb of the loop, makes a U-turn at the bottom, and then moves back up the ascending limb of the loop. The permeabilities and transport characteristics of the epithelial cells lining the loop change over its length as it descends from the cortex into the medulla and then ascends to the cortex again (**Figure 49.11**).

Let's begin upstream of the loop of Henle, in the proximal convoluted tubule. As we have seen, solutes and water are reabsorbed here; the osmolarity of the initial filtrate is about 300 mOsm/L. The descending limb of the loop of Henle is very permeable to water but not to solutes. Water leaves the tubular fluid by osmosis because the surrounding interstitial fluid is hyperosmotic compared to the tubule contents. The hyperosmolarity of the interstitial fluid originates from three sources. First, the initial upturn of the thin segment of the ascending limb of the loop of Henle is very permeable to sodium and chloride but not to water. Therefore, these ions diffuse at high rates out of the loop into the interstitial fluid, significantly increasing the osmolarity of the inner medulla. Second, the epithelial cells of the thick segment of the ascending limb of the loop of Henle actively transport some of the remaining sodium and chloride ions out of the lumen of the loop and into the interstitial fluid of the outer medulla. Third, although urea is a waste product that gets excreted, some of the urea that is present in the filtrate does not get excreted in the urine, but

1 Fluid enters the loop of Henle at the same osmolarity as blood and becomes concentrated as water diffuses out of the water-permeable descending limb of the loop of Henle.

2 The thin segment of the ascending limb is permeable to Na^+ and Cl^-, but not water, and Na^+ and Cl^- diffuse out of the tubule. The thick segment is not permeable to water, but epithelial cells in this segment actively transport Na^+ and Cl^- ions out of the tubule.

3 By the time fluid enters the distal tubule, the total amounts of water and ions have greatly decreased, and the solution has become dilute.

4 The tubule fluid becomes concentrated again during passage through the collecting duct, which is permeable to water but not to ions. The lower ends of the collecting ducts are also permeable to urea, which diffuses out of the tubule and helps create the osmotic gradient in the inner medulla.

KEY
··· Active transport
— Passive diffusion
H_2O
Na^+,Cl^-
Urea

Figure 49.11 Tubule permeabilities and concentrations of the filtrate in the loop of Henle and collecting duct.

instead it diffuses out of the lower ends of the collecting ducts and into the inner medulla interstitial fluid. Collectively, these solutes create the osmotic force that draws water out of the descending limb. The water then enters local capillaries and rejoins the circulation.

As water diffuses out of the descending limb of the loop of Henle, the osmolarity of the fluid in the tubule increases from 300 to about 1,200 (or higher in some species) mOsm/L. During its passage up the ascending limb, however, the osmolarity decreases as ions diffuse and are transported out of the tubule. Finally, the osmolarity of the tubule fluid increases a second time during passage along the collecting ducts, which are permeable to water but not to ions, allowing water to diffuse out of the tubules by osmosis.

Note that the osmolarity of the kidney interstitial fluid increases from cortex to inner medulla. This extracellular osmolarity gradient allows water to diffuse by osmosis from the descending limb of the loop of Henle all along its length. In other words, this is an example of a countercurrent exchange system, like those that operate in heat and gas exchange in some animals (see Chapters 42 and 48). A major difference between those countercurrent exchange systems and the one in the loop of Henle, however, is that the latter requires energy-dependent ion pumps to maintain the necessary gradient. Energy-requiring systems like that in the kidney are referred to as countercurrent multiplication systems, because energy is used to increase—or multiply—the gradient. Energy is not used to establish a heat or oxygen gradient in those other countercurrent exchange systems.

The chief advantage to an animal provided by the loops of Henle is that urine volume can be reduced and concentrated by the recapture of water along the osmotic gradient. Animals in which total body water stores are regularly in danger of being depleted, such as desert mammals, tend to have very long loops of Henle with very large osmotic gradients in the medulla. By contrast, freshwater fish have no loops of Henle at all. In their case, it is advantageous to excrete as much water as possible to compensate for water entering the body by osmosis across the gills.

Reabsorption: The Medullary Blood Vessels Called the Vasa Recta Minimize the Loss of Solutes

Why does the blood flowing through the vasa recta capillaries of the medulla not eliminate the countercurrent osmotic gradient set up by the loops of Henle? One would think that as plasma having the usual osmolarity of ~300 mOsm/L enters the highly concentrated environment of the medulla, two types of massive net diffusion would occur: of sodium and chloride into the capillaries, and of water out of them. Thus, the interstitial gradient would be "washed away." However, this does not happen because the vasa recta form hairpin loops that run parallel to the loops of Henle and medullary collecting ducts. Near the top of the loop of Henle, blood in the vasa recta has a normal osmolarity. As the blood flows down the loop deeper and deeper into the inner medulla, sodium and chloride do indeed diffuse into, and water diffuses out of, the vasa recta. However, after the bend in the loop is reached, the blood then flows up the ascending vessel loop, where the process is almost completely reversed.

Thus, the hairpin-loop structure of the vasa recta minimizes excessive loss of solutes from the interstitial fluid by diffusion. At the same time, both the salt and water being reabsorbed from the loops of Henle and collecting ducts are carried away in equivalent amounts by the passive movement of water and solutes between fluid compartments, and the steady-state countercurrent gradient set up by the loops of Henle is maintained.

Fine-Tuning of Urine Composition Occurs in the Final Portions of the Nephron

The tubule does not become highly permeable to water again until the distal tubule. By this point, most of the reabsorbed water and salt have already been restored to the blood. However, fine-tuning of the concentrations of sodium, chloride, and potassium, along with water, is mediated by the actions of antidiuretic hormone (ADH) and a hormone called aldosterone. Aldosterone acts on the basolateral surface of epithelial cells of the distal convoluted tubule to stimulate the active transport of three molecules of Na^+ out of the tubule into the interstitial fluid (reabsorption) for every two molecules of K^+ it pumps into cells from the interstitial fluid (secretion) (**Figure 49.12**). Water from the tubule lumen follows the Na^+ by osmosis into the interstitial fluid and blood. By contrast, ADH acts to remove water from the urine, as described next.

The Final Solute Concentration of Urine Is Largely Determined by Events Within the Collecting Ducts

The final place where urine composition can be altered is within the collecting ducts. As previously described, cells along the collecting duct in the inner medulla are permeable to urea, which contributes to the osmotic gradient there. In addition, however, the permeability of the epithelial cells of the collecting ducts to water can be regulated, depending on the body's need at that moment for retaining or excreting water. This happens under the influence of ADH. When ADH is present in blood, it acts to increase the number of water channels in the membranes of the collecting duct cells. Because the collecting ducts travel through the hyperosmotic medulla of the kidney, an osmotic gradient moves water from the duct lumen into the surrounding interstitial fluid, from where it enters the vasa recta. In this way, as it travels through the collecting ducts, the urine becomes greatly concentrated (hyperosmotic) compared to blood—as much as four to five times more concentrated. The ability of the kidneys to produce hyperosmotic urine is a major determinant of an animal's ability to survive in conditions where water availability is limited. When the body's stores of water are plentiful, the level of ADH in the blood decreases, and this in turn decreases the water permeability of the collecting ducts. In such a case, urine increases in volume and becomes more dilute because water does not diffuse out of the collecting ducts.

Biological membranes like the plasma membranes of animal cells are composed of a lipid bilayer that inhibits the movement of water. The mystery of how water moves rapidly across membranes like those of the collecting ducts and loop of Henle

Figure 49.12 **Action of aldosterone on distal convoluted tubule epithelial cells.** Aldosterone stimulates the Na^+/K^+-ATPase activity of distal tubule cells on the side facing the interstitial fluid (the basolateral membrane). This activity creates an osmotic gradient as three Na^+ ions are reabsorbed for every two K^+ ions secreted into the tubule. Water leaves the tubule by osmosis. The net effect is reabsorption of Na^+ ions and water, and secretion of K^+ ions, which gets excreted in the urine. Na^+ ions and water enter the peritubular capillaries and are carried away in the bloodstream.

Biological inquiry: What special property of epithelial cells is demonstrated by the action of aldosterone?

was solved by the discovery in the early 1990s of a family of molecules that came to be known as aquaporins ("water pores").

GENOMES & PROTEOMES

Agre and Colleagues Discovered the Mechanism of Water Movement Across Biological Membranes

In the early 1990s, the American physiologist Peter Agre and colleagues discovered the first of what would eventually be recognized as a new family of proteins, called aquaporins, a discovery that earned Agre the Nobel Prize in Chemistry in 2003. Aquaporins, it is now known, are a subfamily of an even larger

(a) Secondary structure of aquaporin

(b) Three-dimensional (tertiary) structure of aquaporin

Figure 49.13 **Aquaporin structure.** All proteins of the aquaporin family share a similar structure, with six membrane-spanning domains (represented as cylinders) and two loops that come together to form a water channel. **(a)** Secondary structure of aquaporin. This representation highlights the two separate regions of the molecule that come together to form the water pore. **(b)** Tertiary structure of aquaporin showing pore formation.

family of proteins with membrane transport capabilities, found in all kingdoms of living organisms. The functions of all the members of this family are not yet known, but some aquaporins can transport other small molecules, such as glycerol and urea. There are currently 11 known members of the aquaporin family in the human genome.

Aquaporins are proteins with six transmembrane helical domains and two short loops in the membrane (**Figure 49.13**). In animals, the two short loops come together to form the three-dimensional core of the water channel. The importance of the loops is reflected in the fact that these portions are the most highly conserved sequences of the channel among different species. Water must pass through a zone of constriction, created by the loops, that reduces the channel opening to about 30 pm, or just about the width of a water molecule. Scattered along the inner part of the channel are arginine amino acids, which are positively charged. The charged arginines participate in hydrogen bonding with water molecules, facilitating their single-file movement through each channel at rates that have been estimated to be up to billions of water molecules per second!

Within the various extracellular and intracellular domains of aquaporins are sites that can be modified by enzymes, such as kinases. This suggests that the opening and closing of these channels may be gated by stimuli, like the way ion channels are gated in neurons and other cells. In addition, the promoter region of certain aquaporin genes contains a site that is recognized by transcriptional activator proteins that are responsive to the presence of cAMP, a common intracellular signaling molecule and one that is generated by cells stimulated by ADH. Thus, one mechanism by which ADH promotes osmosis of water out of the renal collecting ducts appears to be by stimulating the transcription of one or more aquaporin genes.

Our understanding of aquaporin function has allowed us to explain the molecular basis of one form of an inherited human disease in which patients are unable to produce a concentrated urine, and who consequently lose large amounts of water. A mutation in one aquaporin gene results in a form of the protein expressing any of several abnormalities: improper folding into its correct shape, impaired ability of the molecule to enter the plasma membrane, or impaired ability of the molecule to form a channel core. In addition, Agre's discovery may have widespread implications for other areas of biology and human health. For example, certain types of plant disease are associated with abnormal aquaporin expression in roots. In addition, some scientists are investigating the possibility that drugs that inhibit bacterial aquaporins may someday be useful antibiotics. The use of drugs to inhibit one class of aquaporins has even been suggested as a possible antiperspirant!

The Reabsorption of Solutes Has Limits

Many of the reabsorptive mechanisms in the renal tubule have a limit to the amounts of material they can transport in a given amount of time—called the transport maximum (T_m). This occurs because the binding sites on the membrane transport proteins become saturated with their ligands. Occasionally, certain molecules may reach such high concentrations in the blood that they exceed the T_m for that molecule in the renal tubule, and consequently get excreted in the urine. For example, most water-soluble vitamins are filtered in large amounts each day, but almost all of these filtered molecules are reabsorbed by the proximal tubule. If the plasma concentration becomes high enough, however, reabsorption of the filtered material will be less complete, and the substance will appear in the urine. Thus, people who ingest very large quantities of vitamin C, for example, have increased plasma concentrations of vitamin C. Eventually, the filtered load may exceed the tubular reabsorptive T_m for this substance, and any additional ingested vitamin C is excreted in the urine.

The mammalian kidney is a remarkably efficient organ serving many functions, most of them related to various aspects of homeostasis. Because of its central role in homeostasis, it should not be surprising that kidney damage or disease can have devastating consequences. In the final section we consider some causes of kidney diseases in humans, and their effects on the human condition.

49.4 Impact on Public Health

Diseases and disorders of the kidney are a major cause of illness in the human population. According to statistics released by the Centers for Disease Control and Prevention, up to 4.5% of the U.S. population—or about 13 million people—suffers from one form of kidney disease or another, with approximately 15,000 individuals requiring kidney transplants each year.

Kidney Damage and Disease Are Caused by Many Factors and Result in Disruption of Homeostasis

Many diseases affect the kidneys. Diabetes, bacterial infections, allergies, congenital defects, kidney stones (accumulation of mineral deposits in nephron tubules), tumors, and toxic chemicals are some possible sources of kidney damage or disease. A buildup of pressure due to obstruction of the urethra or a ureter may damage one or both kidneys and increase the likelihood of a bacterial infection.

The symptoms of renal malfunction are similar regardless of the cause of the disease, and all stem from the condition known as **uremia**—literally, "urine in the blood." Assuming that a person with diseased kidneys continues to ingest a normal diet containing the usual quantities of nutrients and electrolytes, what problems might arise? Potentially toxic waste products that would normally enter the tubules by filtration instead build up in the blood because kidney destruction significantly reduces the number of functioning nephrons. In addition, the excretion of potassium is impaired because too few nephrons remain capable of normal tubular secretion of this ion. Increased K^+ in the blood is an extremely serious condition, because of the importance of stable extracellular concentrations of K^+ ions in the control of heart and nerve function.

The kidneys are still able to perform their regulatory function quite well as long as 10% of the nephrons are functioning normally. The remaining nephrons undergo alterations in function—filtration, reabsorption, and secretion—to compensate for the missing nephrons. For example, each remaining nephron increases its rate of potassium secretion so that the total amount of potassium excreted by the kidneys can be maintained at normal levels. The kidneys' regulatory abilities are limited, however. If, for example, someone with severe renal disease were to eat a diet high in potassium, the remaining nephrons might not be able to secrete enough potassium to prevent its concentrations from increasing in the extracellular fluid.

Kidney Disease May Be Treated with Hemodialysis and Transplantation

Diseased kidneys may eventually reach a point when they can no longer excrete and reabsorb water and ions at rates that maintain salt and water homeostasis, nor excrete waste products as fast as they are produced. Adjusting a person's diet can help reduce the severity of these problems. For example, lowering potassium intake reduces the amount of potassium to be excreted, but such alterations may not eliminate the problems. In that case,

Figure 49.14   Simplified diagram of hemodialysis.   The dialyzer is composed of many strands of very thin, sievelike tubing. In the dialyzer, blood within the dialysis tubing and dialysis fluid bathing the tubing move in opposite directions (a countercurrent). The dialyzer thus provides a large surface area for diffusion of waste products out of the blood and into the dialysis fluid.

doctors must use various procedures to artificially perform the kidneys' excretory functions. The most important of these procedures is **hemodialysis**. The general term dialysis means to separate substances in solution using a porous membrane.

An artificial kidney is an apparatus that removes excess substances from the blood by hemodialysis (**Figure 49.14**). During this procedure, blood from one of the patient's arteries flows to the dialyzer. Within the dialyzer, blood flows through cellophane tubing that is surrounded by a special dialysis fluid. The tubing is highly permeable to most solutes but relatively impermeable to protein and completely impermeable to blood cells. In fact, these characteristics are designed to be quite similar to those of the body's own capillaries. The dialyzer is filled with a solution with ion concentrations similar to those in normal plasma, but with no urea or other substances that are to be completely removed from the plasma.

As blood flows through the tubing, small solutes diffuse out into the surrounding solution in the dialyzer until an equilibrium is reached. If, for example, the patient's plasma potassium concentration is above normal, potassium diffuses out of the blood across the cellophane tubing and into the dialysis fluid. Similarly, waste products and excess amounts of other substances also diffuse into the dialysis fluid and thus are eliminated from the body. Note in Figure 49.14 that blood and dialysis fluid flow in opposite directions through the dialyzer. This establishes an artificial countercurrent exchange system that increases the efficiency by which the blood is cleansed. The dialyzed, puri-

fied blood is then returned to one of the patient's veins through another type of tubing that leaves the dialyzer.

Some patients with reversible, temporary forms of kidney disease may require hemodialysis for only days or weeks. However, patients with chronic, irreversible kidney disease require treatment for the rest of their lives, unless they receive a kidney transplant. Such patients undergo hemodialysis several times a week. Each year nearly 300,000 Americans undergo some type of dialysis.

The treatment of choice for most patients with permanent kidney disease is kidney transplantation. Rejection of the transplanted kidney by the recipient's body is a potential problem with transplants, but great strides have been made in reducing the frequency of rejection. Many people who might benefit from a transplant, however, do not receive one, because the number of people needing a transplant exceeds the number of donors. Currently, the major source of kidneys for transplanting is from recently deceased persons, and improved public understanding may lead many more individuals to give permission to have their kidneys and other organs used following their death. Recently, donation from a living, related donor has become more common, particularly with improved methods for preventing rejection. As noted earlier in this chapter, the mammalian kidney can perform its functions with only a small fraction of its nephrons intact. Due to this large safety factor, a person who donates one of his or her kidneys can function quite normally with only one kidney.

CHAPTER SUMMARY

49.1 Principles of Homeostasis of Internal Fluids

- Exchanges of salt and water with the environment resulting from vital processes, such as respiration or the elimination of wastes, are called obligatory exchanges. (Figure 49.1)

- Among the important products of the breakdown of proteins and nucleic acids are nitrogenous wastes, molecules that include nitrogen from amino groups (NH_2). Most aquatic animals produce ammonia (NH_3) and ammonium ions (NH_4^+), which are the most highly toxic nitrogenous wastes but require no energy to produce. Many animals, including all mammals, convert ammonia into urea, which is less toxic than ammonia and requires moderate expenditures of water and energy. Birds, insects, and most reptiles produce uric acid or other nitrogenous compounds called purines. These nitrogenous wastes conserve water and are less toxic but energetically costlier than ammonia. (Figure 49.2)

- The solute concentration of a solution of water is known as the solution's osmolarity, expressed as milliosmoles/liter (mOsm/L). (Figure 49.3)

- Salt glands are adaptations found in marine birds and reptiles that permit these animals to eliminate excess salt from the body. (Figure 49.4)

- Robert Cade and coworkers discovered that fluid replacement during exercise is particularly beneficial if the fluid contains solutes at concentrations resembling those in sweat. (Figure 49.5)

- Animals that maintain constant internal salt concentrations and osmolarities, even when living in water with very different osmolarities than their body fluids, are called osmoregulators. Animals in which internal osmolarity conforms to the osmolarity of the environment are called osmoconformers.

49.2 A Survey of Excretory Organs

- Most excretory organs operate by one or more of three processes: (1) filtration, the passive removal of water and small solutes from the blood; (2) reabsorption, in which useful filtered solutes are returned to the blood via transport systems; and (3) secretion, in which unwanted solutes are actively transported from the blood and interstitial fluid and into the excretory tubule. (Figure 49.6)

- In the protonephridial system of flatworms, a series of branching tubules filters fluids from the body cavity into the tubule lumens via the actions of ciliated flame cells. (Figure 49.7)

- In the metanephridial system of annelids, pairs of metanephridia located in each body segment filter interstitial fluid through funnel-like structures called nephrostomes. A dilute urine is excreted via nephridiopores in the body wall. (Figure 49.8)

- In insects, cells of Malpighian tubules transport salts, uric acid, and water (by osmosis) from hemolymph into the tubule lumen. After useful salts and water are reabsorbed into the hemolymph, wastes are excreted from the body. (Figure 49.9)

- The mechanisms by which vertebrates osmoregulate vary depending on the animal's environment. (Table 49.1)

49.3 Structure and Function of the Mammalian Kidney

- The urinary system in humans consists of the kidneys, ureters, urinary bladder, and urethra. Each kidney is composed of an outer renal cortex and an inner renal medulla. The functional units of the kidney, called nephrons, are composed of a filtering component called the renal corpuscle, and a tubule that empties into a collecting duct. Each renal corpuscle contains a cluster of interconnected, fenestrated capillaries called the glomerulus within a structure called Bowman's capsule. Each glomerulus is supplied with blood under pressure by an afferent arteriole. Filtration occurs in the renal corpuscle. Each tubule of a nephron is composed of a proximal convoluted tubule, a loop of Henle, and a distal convoluted tubule. Tubules are surrounded by peritubular capillaries in the cortex and by the vasa recta capillaries in the medulla. (Figure 49.10)

- Different portions of the tubule have different permeabilities to solutes and water. Most reabsorption of useful solutes occurs in the proximal tubule. Water and ion reabsorption continues along the loop of Henle. The reabsorption of solutes by the vasa recta minimizes the loss of solutes from the renal medulla. Fine-tuning of urine composition occurs in the distal tubule, and the final concentration of urine is formed within the collecting ducts. (Figures 49.11, 49.12)

- Agre and coworkers discovered that water moves through plasma membranes through protein channels. Agre called these channels aquaporins. Aquaporins regulate water reabsorption in the kidneys. (Figure 49.13)

49.4 Impact on Public Health

- Regardless of the cause of renal malfunction, the symptoms are similar and stem from the condition known as uremia, which literally means "urine in the blood."

- One important treatment for kidney disease is hemodialysis, in which wastes in blood diffuse across a selectively permeable artificial membrane into a dialysis fluid. Hemodialysis uses a countercurrent exchange system. (Figure 49.14)

TEST YOURSELF

1. A change in electrolyte levels in the body may result in
   a. altered membrane potentials that disrupt normal cell function.
   b. disruption of certain biochemical processes that occur in the cell.
   c. cell death.
   d. a and b only.
   e. a, b, and c.

2. Nitrogenous wastes are the by-products of the metabolism of
   a. carbohydrates.
   b. lipids.
   c. nucleic acids.
   d. proteins.
   e. both c and d.

3. Marine fishes avoid water balance problems by
   a. producing a large volume of dilute urine.
   b. not drinking water.
   c. having gill epithelial cells that recapture lost salts from the environment.
   d. producing small volumes of concentrated urine.
   e. a, b, and c.

4. Metabolic water is water
   a. necessary to stimulate the process of cellular respiration.
   b. found within cells.
   c. produced during cellular respiration.
   d. produced by sweat glands.
   e. used by cells during the uptake of glucose.

5. Animals that maintain a constant water balance despite changes in water concentrations in the environment are
   a. osmoregulators.
   b. osmoconformers.

6. The excretory system found in insects is composed of
    a. protonephridia.
    b. metanephridia.
    c. Malpighian tubules.
    d. a filtration kidney.
    e. a secretory kidney.

7. In the mammalian kidney, filtration is driven by
    a. solute concentration in the tubular filtrate.
    b. solute concentration in the blood.
    c. water concentration in the blood.
    d. water concentration in the tubular filtrate.
    e. hydrostatic pressure in the blood vessels of the glomerulus.

8. In the mammalian urinary system, the urine formed in the kidneys is carried to the urinary bladder by
    a. the collecting duct.
    b. the renal tubule.
    c. the renal pelvis.
    d. the ureters.
    e. the urethra.

9. Which of the following causes an increase in sodium reabsorption in the distal convoluted tubule?
    a. an increase in aldosterone levels
    b. an increase in antidiuretic hormone levels
    c. a decrease in aldosterone levels
    d. a decrease in antidiuretic hormone levels
    e. none of the above

10. Aquaporins are
    a. ion channels.
    b. water channels.
    c. receptors for aldosterone.
    d. small pores in the fenestrated capillaries of the glomerulus.
    e. both a and c.

CONCEPTUAL QUESTIONS

1. Define nitrogenous wastes, and list four types.
2. Explain how salt glands are an adaptation for marine life.
3. List and define the three processes involved in urine production.

EXPERIMENTAL QUESTIONS

1. What symptoms are commonly seen in athletes after strenuous exercise? How are these symptoms related to water loss during exercise, and what did Cade and his colleagues hypothesize about this?
2. How did the researchers test their hypothesis?
3. What was the result of consuming the drink during exercise?

COLLABORATIVE QUESTIONS

1. Discuss two different types of filtration mechanisms found in invertebrates.
2. Briefly discuss the parts and functions of the nephron in the mammalian kidney.

www.brookerbiology.com
This website includes answers to the Biological Inquiry questions found in the figure legends and all end-of-chapter questions.

50
ENDOCRINE SYSTEMS

A three-dimensional MRI of a section through a human brain, highlighting the pituitary gland (white).

Communication takes several forms in nature and involves all of the major sensory systems. It may occur over long distances, as when sound waves propagate through the ocean from one whale to another over a kilometer or more, or when olfactory cues travel from one moth to another across hundreds of meters. At the other end of the scale, as you learned with neuronal synapses, communication can occur between individual cells across minute distances.

Between these two scales of communication lies yet another, that of long-distance cell-to-cell communication. **Endocrine glands** contain epithelial cells that secrete molecules called **hormones** into the bloodstream, where they circulate throughout the body. The hormones gain access to all of the body's cells in this way and influence many different cellular activities. A single hormone can influence many parts of the body at the same time. Typically, hormones are produced in response to a homeostatic challenge, such as a change in an animal's blood pressure or body temperature. It is a chief function of hormones to counter these challenges and maintain the body's homeostasis. In this chapter, we will learn about the chemical nature of hormones and their mechanisms of action, how the endocrine and nervous systems interact, and the ways in which hormones influence such diverse functions as metabolism, growth, and reproduction. We conclude with a discussion of how hormones impact the human condition.

50.1 Mechanisms of Hormone Action and Control

In this section, we will examine some of the general characteristics of hormones, how they act, and how they are controlled.

Long-Distance Signaling Connects Remote Organs and Tissues

Suppose an animal is injured and bleeding badly, such that the total amount of fluid in its body decreases significantly. The animal's heart would not fill adequately with blood. As a consequence, blood pressure would decrease, creating a potentially life-threatening situation. In vertebrates, the major organs capable of minimizing the loss of fluid from the body during such times are the kidneys. After a loss of blood, the heart communicates with the kidneys to improve their ability to retain water that would otherwise be lost in the urine. This helps prevent the fluid level of the body from dropping further. In this type of communication between organs, signals must be released from the heart—which detects the loss of blood—into the bloodstream, where they can be carried to the kidneys to exert their effects. As mentioned earlier, these signals, which are known as hormones, are made by cells in nearly all the body's organs, including the heart. In many cases, though, hormone-producing cells are packaged into discrete glands of their own. In those cases, the glands are called endocrine glands. Collectively, all the endocrine glands and other organs with hormone-secreting cells constitute the **endocrine system**, as depicted for a vertebrate in **Figure 50.1**. Unlike some other organ systems such as the urinary or respiratory system, the endocrine system does not consist of a small number of physically connected tissues and organs.

Hormones Can Have Short-Term or Long-Term Effects

The effects of a given hormone may occur within seconds or require several hours to develop, and they may last for as short as a few minutes or as long as days. This is a key difference between the communication processes in the nervous and endocrine systems.

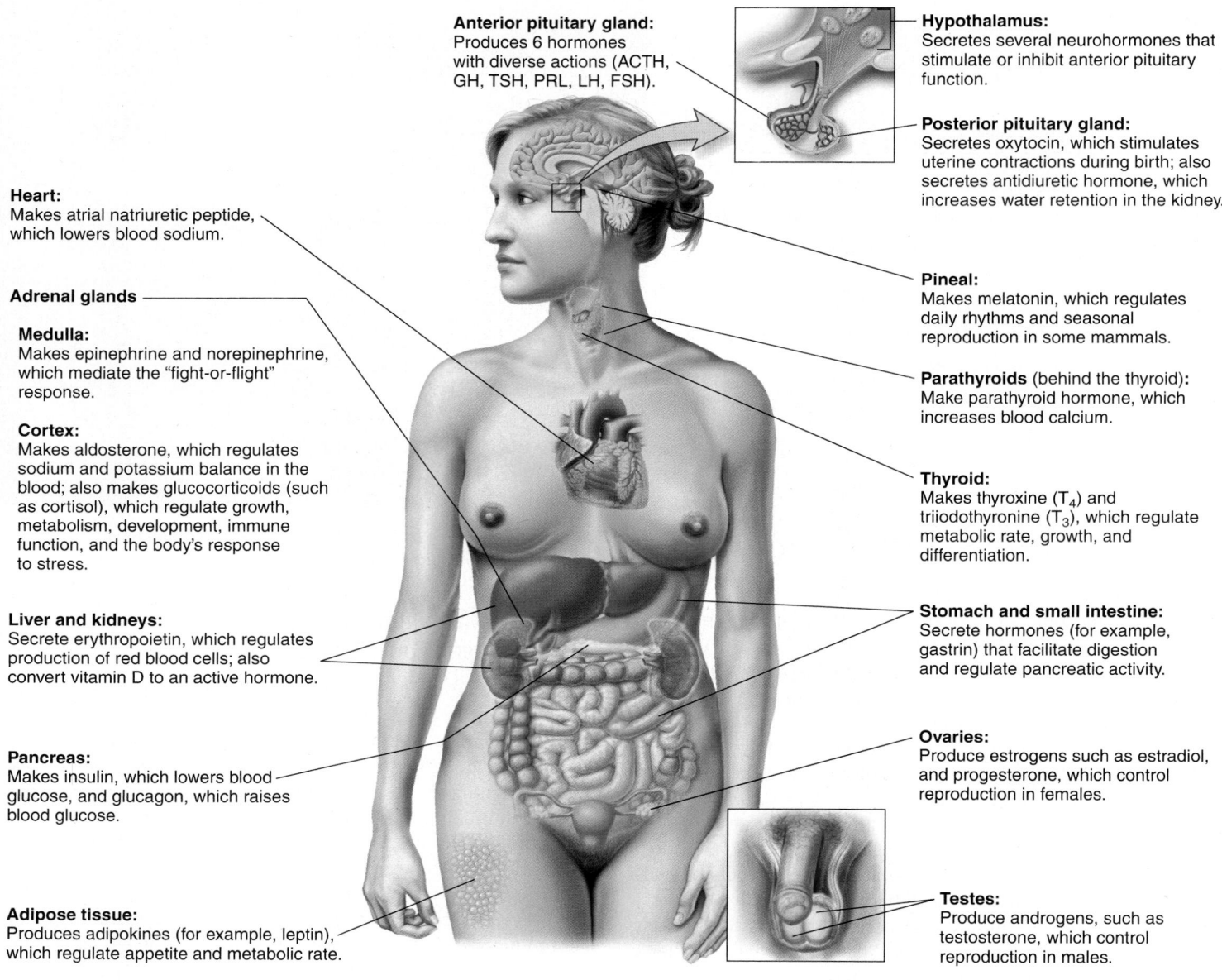

Anterior pituitary gland:
Produces 6 hormones with diverse actions (ACTH, GH, TSH, PRL, LH, FSH).

Hypothalamus:
Secretes several neurohormones that stimulate or inhibit anterior pituitary function.

Posterior pituitary gland:
Secretes oxytocin, which stimulates uterine contractions during birth; also secretes antidiuretic hormone, which increases water retention in the kidney.

Heart:
Makes atrial natriuretic peptide, which lowers blood sodium.

Pineal:
Makes melatonin, which regulates daily rhythms and seasonal reproduction in some mammals.

Adrenal glands

Medulla:
Makes epinephrine and norepinephrine, which mediate the "fight-or-flight" response.

Parathyroids (behind the thyroid)**:**
Make parathyroid hormone, which increases blood calcium.

Cortex:
Makes aldosterone, which regulates sodium and potassium balance in the blood; also makes glucocorticoids (such as cortisol), which regulate growth, metabolism, development, immune function, and the body's response to stress.

Thyroid:
Makes thyroxine (T_4) and triiodothyronine (T_3), which regulate metabolic rate, growth, and differentiation.

Liver and kidneys:
Secrete erythropoietin, which regulates production of red blood cells; also convert vitamin D to an active hormone.

Stomach and small intestine:
Secrete hormones (for example, gastrin) that facilitate digestion and regulate pancreatic activity.

Pancreas:
Makes insulin, which lowers blood glucose, and glucagon, which raises blood glucose.

Ovaries:
Produce estrogens such as estradiol, and progesterone, which control reproduction in females.

Adipose tissue:
Produces adipokines (for example, leptin), which regulate appetite and metabolic rate.

Testes:
Produce androgens, such as testosterone, which control reproduction in males.

Figure 50.1 Overview of many of the major endocrine glands that constitute the endocrine system in humans, and their hormones.

In the nervous system, signals are transmitted from one cell to another within milliseconds, and the effect on the postsynaptic cell occurs immediately. In the endocrine system, one hormone may act on a cell quickly and for a very short time, while another may act very slowly and have a lingering action. The explanation for these differences lies in the mechanisms by which different hormones act. To understand this, we first examine the different chemical types of hormones, because it is the chemical nature of hormones that determines their mechanisms of action.

The Three Classes of Hormones Differ in Composition and Modes of Action

Hormones fall into three broad classes, which include the amines, proteins/peptides, and steroids (**Table 50.1**). The amines and the proteins/peptides generally share similar chemical properties and modes of action, while the steroid hormones act very differently from the other classes.

The amine hormones are derived from an amino acid, either tyrosine or tryptophan. Tyrosine is the precursor for the hormones epinephrine and norepinephrine, which are produced in the adrenal medulla and are important in the body's response to stress (the so-called fight-or-flight reaction), and for dopamine, a hormone made by the brain. Tyrosine is also the chemical backbone of the hormones made by the thyroid gland, which are important regulators of metabolic rate, growth, and differentiation. The only hormone derived from tryptophan is melatonin, which is produced within an endocrine gland called the pineal gland located in the brain. Melatonin is important for controlling seasonal reproduction in certain mammals like sheep and hamsters.

Table 50.1	Chemical Classes of Hormones			
Class	Chemical properties	Location of target cell receptor	Mechanism of action	Examples
Amines	Small, water-soluble, derived from tyrosine or tryptophan	Plasma membrane (except thyroid hormones)	Second messengers (except thyroid hormones)	Epinephrine, norepinephrine, dopamine, melatonin, thyroid hormones
Proteins/Peptides	Water-soluble	Plasma membrane	Second messengers	Insulin, glucagon, leptin
Steroids	Derived from cholesterol, mostly lipid-soluble	Cytosol or nucleus	Usually stimulates gene transcription	Cortisol, testosterone, estradiol

Most hormones are peptides (small, water-soluble polypeptides) or proteins. These hormones participate in numerous body functions such as reproduction, metabolism, growth, and mineral balance.

The final group of hormones are the steroids, which are synthesized from the lipid cholesterol. All steroid hormones are lipids, unlike the other two classes of hormones. Steroids are less soluble in water than amines or protein and peptide hormones. Due to their limited solubility, steroids are usually bound to large, soluble proteins in the blood that serve as carriers. By combining with these proteins, steroid hormones can reach high concentrations in the blood. The major steroid hormones found in vertebrates include cortisol and aldosterone made by the adrenal cortex, androgens made by the testis, and estrogen and progesterone made by the ovary. In insects, ecdysone is produced in the prothoracic gland.

Water-Soluble Hormones Act on Membrane Receptors

A hallmark of cell signaling by water-soluble molecules is that a ligand binds to a receptor protein located in the plasma membrane. This is the mechanism by which neurotransmitters act, for example. Similarly, all the amine and protein/peptide hormones, with one exception (thyroid hormones), act in this way. As these hormones diffuse from the blood into the interstitial fluid, they must bind to a specific receptor protein located on the cell surface. Only cells having the proper receptors on their surfaces can respond to the hormone. Thus, although a hormone is released into the entire circulatory system, it activates only specific cells. The receptor interacts noncovalently and reversibly with the hormone. The reversibility of the binding between hormone and receptor is one way in which cells are prevented from being permanently stimulated.

Among cells throughout the body, different receptor proteins may be made that recognize the same hormone. These different receptors, called subtypes or isoforms, may be the product of different genes or produced by alternative splicing, which is described in Chapter 13. In this way, one hormone can perform more than one action in the body by binding to receptor isoforms that differ slightly from each other. These receptors may have different affinities for the hormone, such that low concentrations of a hormone may cause one type of effect by binding to a high-affinity receptor on one cell, and higher concentrations may cause entirely new actions on different cells by binding to a low-affinity receptor on another cell. In this way, most hormones have evolved to serve more than one function.

The binding of a water-soluble hormone to its plasma membrane receptor initiates intracellular signaling (see Chapter 9). Three major signaling pathways that are activated by water-soluble hormones are the cyclic AMP pathway, the calcium/phosphatidylinositol pathway, and the tyrosine kinase pathway. These signal transduction processes may be rapid, occurring in some cases within seconds, and involve the activity of enzymes. Both of these features are important, since occasionally the rapidity of the cell response to a hormone can mean the difference between life and death, as may occur, for example, under conditions of extreme danger. The presence of enzymes in the signaling process ensures that a small amount of hormone has its signal amplified greatly, because a single signaling molecule results in the production of many intracellular messengers.

Activation of signaling pathways may also lead to changes in cellular activities that occur more slowly. These slower changes usually require activation or inhibition of various genes in the nucleus.

Lipid-Soluble Hormones Act on Intracellular Receptors

Because steroid hormones are lipids, they are soluble in plasma membranes and diffuse across the membrane into the cell cytosol. Unlike the case for water-soluble hormones, receptors for steroid hormones are located within the cell, either within the cytosol or more commonly the nucleus. This is also true for the two thyroid hormones, which like steroids are poorly soluble in water.

The complex of a steroid hormone bound to its intracellular receptor functions as a transcriptional activator by binding to enhancers of particular genes (see Chapter 13). Once bound, transcription of a gene is greatly increased, which eventually increases the amount of that gene's protein product. These proteins may be important in any of a wide variety of cellular activities.

Steroid hormones can influence a number of genes within a single cell or in different cells. In this way, a single type of steroid hormone can exert a variety of actions throughout the body. The physical changes that accompany puberty result from the actions of two steroid hormones—androgens in males and estrogens in females—and are perhaps the most striking example of this widespread action.

We have learned that hormones act in all cases by binding to a receptor protein located either on the cell surface, in the cytosol, or in the nucleus. This raises an interesting question: Which evolved first, a given hormone or its receptor? It would appear that there would be no selection pressure to evolve one without the other. However, recent research has generated intriguing new hypotheses regarding the evolution of these signaling systems, as described next.

GENOMES & PROTEOMES

Hormones and Receptors Evolved as Tightly Integrated Molecular Systems

All hormones act by binding to a receptor. Without a receptor, a hormone has no function, and without a ligand, a receptor has no function. Hormones and their receptors function as tightly integrated molecular systems. Each molecule depends on the other for biological activity. How such systems could have evolved, or whether they evolved together, has long been a major puzzle.

American scientist Joseph Thornton and colleagues recently addressed this puzzle by examining how two structurally related steroid hormones evolved separate activities and distinct receptors. **Aldosterone**, as noted, is a steroid made by the adrenal cortex. Because it regulates the balance of certain minerals in the body, it is known as a **mineralocorticoid**. Cortisol is another steroid made by the adrenal cortex and is known as a **glucocorticoid** because one of its major functions is to regulate glucose balance. The actions of these hormones are mediated by intracellular receptors known as the mineralocorticoid receptor (MR) and the glucocorticoid receptor (GR), respectively. Phylogenetic analysis of the sequences of these two receptors suggests that they arose at least 450 million years ago by duplication of an ancient corticoid receptor (CR) gene.

The researchers analyzed the known sequences of the genes for the MR and GR of many vertebrate species, including the most ancient vertebrates, to deduce the theoretical sequence of an ancestral CR. They then synthesized this CR and tested its ability to bind aldosterone and cortisol. They discovered that the CR was capable of binding both hormones, particularly aldosterone. This was surprising, because aldosterone is only present in tetrapods, which arose long after the proposed gene duplication event that created the MR. Thus, it appears that a receptor with high affinity for aldosterone was present long before animals acquired the capacity to synthesize aldosterone. The receptor must have evolved to bind other steroids. Around 450 million years ago, the gene for the CR duplicated, and then the two resulting genes must have evolved such that one receptor gained high affinity primarily for mineralocorticoids and the other for glucocorticoids.

These studies suggest that the ability of an animal to respond to aldosterone evolved long before aldosterone did, because a receptor with high affinity was already in place in animals. When aldosterone evolved in tetrapods, the MR derivative of the ancestral CR was co-opted by aldosterone for its own use. In this one example, therefore, the answer to the "which came first" question appears to be that the receptor evolved first, and the hormone later.

Hormone Levels in Blood Depend on Rates of Synthesis and Removal

Although most hormones circulate in the blood at all times, their concentrations can increase or decrease dramatically when necessary. This can be accomplished in two major ways, by changing the rate of hormone production by an endocrine cell, and by changing the rates at which hormones are removed or inactivated.

Synthesis of amine hormones is mediated by enzymatic conversion of either tyrosine or tryptophan into their respective hormones. These reactions require several enzymes that add or remove chemical side groups and achieve the desired product (**Figure 50.2**). The synthesis of the hormone depends on the amounts and activities of the synthesizing enzymes. These enzymes are always present in the cell, but the amounts and activities of the enzymes can be greatly stimulated when additional hormone is required.

The protein and peptide hormones are synthesized at a steady rate in an unstimulated cell, until additional hormone is needed. In that case, transcription factors within the cell direct the increased transcription of the gene coding for that hormone. Conversely, when less hormone is required, gene transcription is slowed or stopped. These hormones are too large to diffuse across the plasma membrane from the cytosol where they are synthesized, out to the extracellular fluid. Instead, they are packaged into secretory vesicles much the same way as neurotransmitters are packaged in neuron terminals. This packaging provides a ready means of secreting the hormones by exocytosis and a reservoir of stored hormone available for immediate release when needed. When a cell is stimulated to secrete its stored protein or peptide hormones, it also is typically stimulated to synthesize new hormone molecules to replace them.

The synthesis of steroid hormones is quite different from that of other classes of hormones, because as noted earlier they are derived from cholesterol (**Figure 50.3**). To form steroid hormones, a cell must express enzymes capable of adding hydroxyl groups to specific carbons in the cholesterol skeleton. Other enzymes carry out reduction reactions, and still others can both hydroxylate a carbon and split carbon-carbon bonds. The activities and amounts of many of these enzymes change if more or less hormone is required.

Figure 50.2 Synthesis of the amine hormones dopamine, norepinephrine, epinephrine, and melatonin.

Steroid hormones, unlike water-soluble hormones, are not packaged into secretory vesicles because the steroid would diffuse out across the lipid membrane of the vesicle. Instead, steroid hormones are made on demand, and no significant storage exists for them.

Although hormones carry out vital functions, excessive stimulation of cells by hormones may produce detrimental effects. For example, hormones that are activated if an animal loses blood typically help restore blood volume and pressure.

Cortisol:
Helps regulate the body's response to stress.

Aldosterone:
Regulates osmotic balance.

Adrenal cells express enzymes that convert cholesterol to steroids such as cortisol and aldosterone.

Androgens:
"Sex steroids" responsible for the development and maintenance of male secondary sexual characteristics and reproduction.

(testosterone)

Testes make mostly androgens.

Estrogens:
"Sex steroids" responsible for the development and maintenance of female secondary sexual characteristics and reproduction.

(estradiol)

Cholesterol:
All steroid hormones are synthesized from the precursor cholesterol.

Progesterone:
"Sex steroid" required for reproduction.

Ovaries make mostly estrogens and progesterone.

Ecdysone:
Stimulates molting and pupa formation.

Prothoracic cells of insects make steroids such as ecdysone.

Figure 50.3 Synthesis and major functions of the chief steroid hormones in animals.

This cannot continue indefinitely, however, because once blood volume and pressure are restored, it would be harmful if they continued to increase until their levels were above normal. Therefore, once a hormone enters the blood and performs its functions, it must usually be prevented from exerting its effects indefinitely. This is accomplished in one or more ways:

- Hormones that bind to plasma membrane receptors may be engulfed by endocytosis into a cell, where lysosomal enzymes degrade the hormones.

- Small, water-soluble hormones are excreted in the urine.

- The liver chemically modifies many hormones to render them inactive and more easily excretable via the kidneys.

- Negative feedback processes (see Chapter 40) turn off the signals that were responsible for stimulating the synthesis and secretion of the hormone.

Generally, these processes ensure that hormone levels in the blood remain within a normal range under most circumstances, but have the capacity to be increased or decreased beyond that range if required. One of the ways in which changes in hormone levels are initiated is through sensory input to an animal's brain. As we see next, the nervous system and endocrine system are functionally linked in many animals, including all vertebrates.

50.2 Links Between the Endocrine and Nervous Systems

A key feature of the endocrine system in most animals is that the levels of many hormones in blood rise and fall with changes in an animal's environment. For this to happen, the endocrine system must somehow be made aware of environmental changes. This often occurs when sensory input is received by an animal's nervous system, which in turn modulates the activity of one or more endocrine glands. In this section, we explore how this happens in vertebrates and how the hypothalamus and pituitary gland play the major roles in linking the nervous and endocrine systems.

The Hypothalamus and Anterior Pituitary Gland Have Integrated Functions

Sensory stimuli detected by the nervous system can activate the endocrine system. For example, when an antelope detects the presence of a nearby lion, visual and olfactory sense information is relayed to the antelope's brain. The brain initiates changes in certain endocrine glands that release hormones to prepare the antelope for the possibility of an attack. As another example, the levels of certain hormones fluctuate in the blood of migrating fish as they move back and forth between feeding grounds in the sea and freshwater spawning sites. These hormones are activated by different salinities in the environment, which in turn are detected by sensory cells of the fish's nervous system. The hormones act to prepare the gills of the fish to handle the large swings in salinity of the water.

The common feature of these examples and many others is that a sensory cue, such as a predator or the salinity of water, must be perceived by a sensory receptor and converted into an endocrine response. Electrical signals are transmitted from the sensory receptors to different parts of the brain, including the hypothalamus. The hypothalamus sits at the bottom surface of the vertebrate brain and is vital for such diverse functions as reproduction, bodily rhythms, appetite and metabolism, and responses to stress (**Figure 50.4**). Within the hypothalamus are numerous neurons that synthesize a class of hormones called

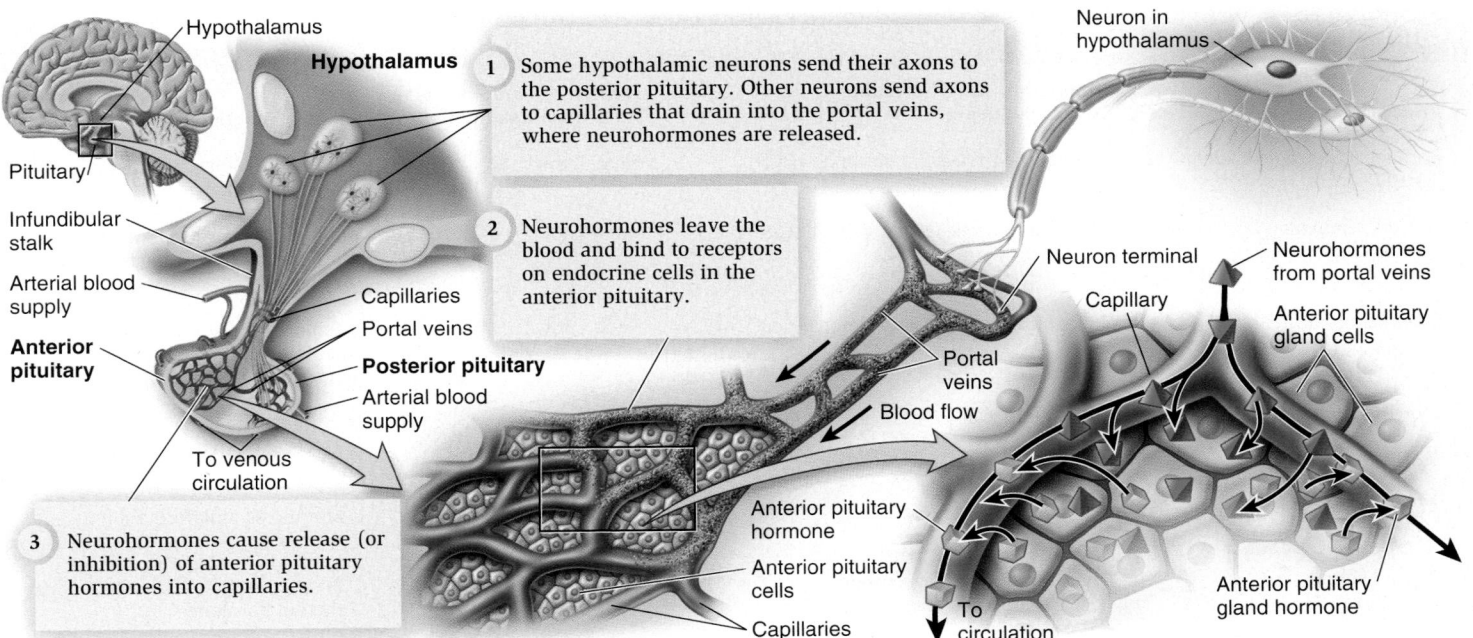

Figure 50.4 Stimulation of the anterior pituitary gland by the hypothalamus.

neurohormones, or hypothalamic releasing hormones, which are hormones made in and secreted by neurons. All neurohormones are either amines or peptides. Although they are produced within neurons, these molecular signals are not referred to as neurotransmitters, because the endings of the neurons do not terminate in a synapse with another cell. Instead, the neuron terminals end close to capillaries at the top of a structure known as the **infundibular stalk**. Here, the neurons secrete their neurohormones into the capillaries, which in turn collect into portal veins.

Portal veins differ from ordinary veins because not only do they collect blood from capillaries—like all veins—but they then also form another set of capillaries, as opposed to returning the blood directly to the heart like other veins. The portal veins extend through the length of the infundibular stalk before forming new capillaries. The stalk is physically connected to a multilobed endocrine gland sitting directly below the hypothalamus, called the **pituitary gland**, the highlighted structure in the chapter-opening photo. It is within the anterior lobe of the pituitary gland—often called simply the anterior pituitary gland—that the portal veins empty into a second set of capillaries. This arrangement of blood vessels bypasses the general circulation and allows for neurohormones to be delivered directly from the hypothalamus to the cells of the anterior pituitary gland in a quick, efficient manner.

The anterior pituitary gland synthesizes six different hormones that respond to the presence of the hypothalamic neurohormones in the capillaries arising from the portal veins. **Table 50.2** lists the major anterior pituitary hormones in vertebrates and the neurohormones that regulate them. As shown in the table, five of the anterior pituitary gland hormones are activated by neurohormones, one is inhibited by dopamine, and one is under both stimulatory and inhibitory control by the hypothalamus.

The six hormones of the anterior pituitary gland are secreted into the general blood circulation, where they reach their target organs.

Adrenocorticotropic hormone (ACTH) stimulates the cells of the adrenal cortex to synthesize and release cortisol or related steroid hormones. Follicle-stimulating hormone (FSH) and luteinizing hormone (LH) act on the ovaries or testes to stimulate egg or sperm development, as well as the synthesis and release of estrogens and androgens from the gonads. Growth hormone (GH) stimulates growth of immature animals and also participates in control of energy balance. Prolactin (PRL) levels are increased in mammals during late pregnancy and during lactation. This hormone stimulates milk production in mammals but affects salt and water balance in other vertebrates. Finally, thyroid-stimulating hormone (TSH) acts on cells of the thyroid gland to stimulate growth of the gland and production of thyroid hormones.

In addition to the anterior lobe, the pituitary gland contains a posterior lobe and, in some species, an intermediate lobe that is believed to be vestigial in primates. These lobes carry out numerous functions in the body, as described next for two small peptide hormones.

The Posterior Pituitary Gland Secretes Oxytocin and Antidiuretic Hormone

The posterior pituitary gland has a blood supply but, in contrast to the anterior pituitary gland, it is not connected to the hypothalamus by portal veins, and does not respond to neurohormones made in the hypothalamus. Instead, the posterior pituitary gland is an extension from the hypothalamus that lies in close contact with the anterior pituitary gland (Figure 50.4). In mammals, the neuron terminals in the posterior pituitary contain one of two

Table 50.2	Hormones of the Anterior Pituitary and Hypothalamus		
Pituitary hormone	**Stimulatory neurohormone from hypothalamus**	**Inhibitory neurohormone from hypothalamus**	**Major functions**
Adrenocorticotropic hormone (ACTH)	Corticotropin-releasing hormone (CRH)	None known	Stimulates adrenal cortex to make glucocorticoids
Follicle-stimulating hormone (FSH)	Gonadotropin-releasing hormone (GnRH)	None known	Stimulates germ cell development and sex steroid production in gonads
Luteinizing hormone (LH)	Gonadotropin-releasing hormone (GnRH)	None known	Stimulates release of eggs in females; stimulates sex steroid production from gonads
Growth hormone (GH)	Growth hormone–releasing hormone (GHRH)	Somatostatin	Promotes linear growth; regulates glucose and fatty acid balance in blood
Prolactin (PRL)	None known	Dopamine	Stimulates milk formation in mammals
Thyroid-stimulating hormone (TSH)	Thyrotropin-releasing hormone (TRH)	None known	Stimulates thyroid gland to make thyroid hormones

hormones, **oxytocin** or **a̱nti̱di̱uretic ho̱rmone** (**ADH**). When needed, these hormones are released directly from the neuron terminals into the bloodstream.

Oxytocin increases in the blood of pregnant mammals just prior to birth. It stimulates contractions of the smooth muscles in the uterus, which facilitates the birth process. After the offspring are born, oxytocin becomes important in the let-down of milk. When the mother's nipples are stimulated by the suckling of a newborn, neurons transmit a signal from the breast to the mother's hypothalamus, which stimulates the cells that produce oxytocin. Oxytocin, which is stored in the neuron terminals of the posterior pituitary gland is released into the blood where it travels to smooth muscle cells surrounding the milk glands and ducts of the breast. This stimulates the secretion of milk. Thus, oxytocin has two important and different functions, one during birth and one during lactation. In both cases the hormone stimulates contraction of muscle cells.

Antidiuretic hormone gets its name because it acts on kidney cells to decrease urine production—a process known as antidiuresis. A diuresis is an increased loss of water in the urine, as happens for example when you drink large amounts of coffee or tea. If the fluid content of the body is low, such as during dehydration or after blood loss, ADH is secreted into the blood from the posterior pituitary gland. It acts to increase the number of water-channel proteins called aquaporins present in the membranes of kidney tubule cells (see Chapter 49). It is through the aquaporins that water is recaptured from the forming urine. Minimizing the volume of water used to form urine is an adaptation that conserves body water when necessary.

At high concentrations, ADH also increases blood pressure by stimulating vasoconstriction of blood vessels. Like oxytocin, therefore, ADH has more than one function. Both of its major functions are related in that they contribute to maintaining blood pressure and fluid levels in the body. The two actions of ADH are mediated by different receptor subtypes, a feature that has proven to be of medical value in humans. Researchers have synthesized versions of ADH that have been chemically modified to activate ADH receptors in the kidney but not those in blood vessels. One such compound is sometimes used to treat bed-wetting (**nocturnal enuresis**) in children by decreasing urine production without the unwanted side effect of increasing blood pressure.

Oxytocin and ADH are well-studied examples of the evolution of hormones. These two hormones are found only in mammals. However, many invertebrates and all nonmammalian vertebrates secrete one or more peptides that share chemical similarities with oxytocin and ADH, but are not identical to the mammalian hormones. One of these is **vasotocin**, which combines some of the chemical structure of both oxytocin and ADH. It is believed that an ancestral gene duplicated at some point, and then the two new genes mutated slightly to produce two different and new hormones with new functions in mammals. Only mammals lactate, so the role of vasotocin must be different in birds, fish, and other vertebrates than that of oxytocin, for example. Research has shown that vasotocin is responsible for regulating salt and water balance in the blood of nonmammalian vertebrates.

The fact that members of the vasotocin/oxytocin/ADH gene family arose early in animal evolution also suggests that oxytocin and ADH may have additional, unrecognized actions that have been retained in mammals. For example, human males have oxytocin in their blood, but its role cannot be the same as in females because of course only females give birth. The key roles of oxytocin in male humans and other animals have not been unequivocally identified.

50.3 Control of Metabolism and Energy Balance by Hormones

An important function of the endocrine system is to control metabolic rate and energy balance. Hormones are partly responsible for regulating energy utilization by cells by modulating appetite, digestion, absorption of nutrients, and the levels of energy sources like glucose in the blood and its transport into cells. Although many hormones are involved in these processes, two from the thyroid gland, two from the pancreas, and one from adipose tissue play particularly important roles, as described in this section.

Thyroid Hormones Contain Iodine and Regulate Metabolic Rate

The thyroid gland lies within the neck of vertebrates. In mammals, it straddles the trachea just below the larynx (**Figure 50.5a**). It consists of many small, spherical structures called follicles that contain in their core a gel-like substance called the **colloid**. The colloid consists primarily of large amounts of the protein **thyroglobulin**.

Thyroid hormones are produced when the neurohormone thyrotropin-releasing hormone (TRH; see Table 50.2) stimulates the secretion into the blood of thyroid-stimulating hormone (TSH) from the anterior pituitary gland. TSH in turn stimulates the follicle cells of the thyroid gland to begin the process of making thyroid hormones. Production of TRH and TSH is prevented from getting too high by negative feedback actions of thyroid hormones on the hypothalamus and anterior pituitary.

Figure 50.5b shows the pathway leading to thyroid hormone synthesis. First, iodide—converted by the gut from dietary iodine—is transported from the extracellular fluid into the thyroid follicular cells. The iodide then enters the colloid, where it is oxidized and bonds to tyrosine side chains in thyroglobulin. When the thyroid follicle cells are stimulated by TSH, these tyrosines are enzymatically modified to form two active thyroid hormones, called **thyroxine** and **triiodothyronine**, which are cleaved from thyroglobulin and secreted from the cell. These molecules are unique among hormones in that they contain iodine molecules: four in the case of thyroxine, or T_4, and three in triiodothyronine, or T_3.

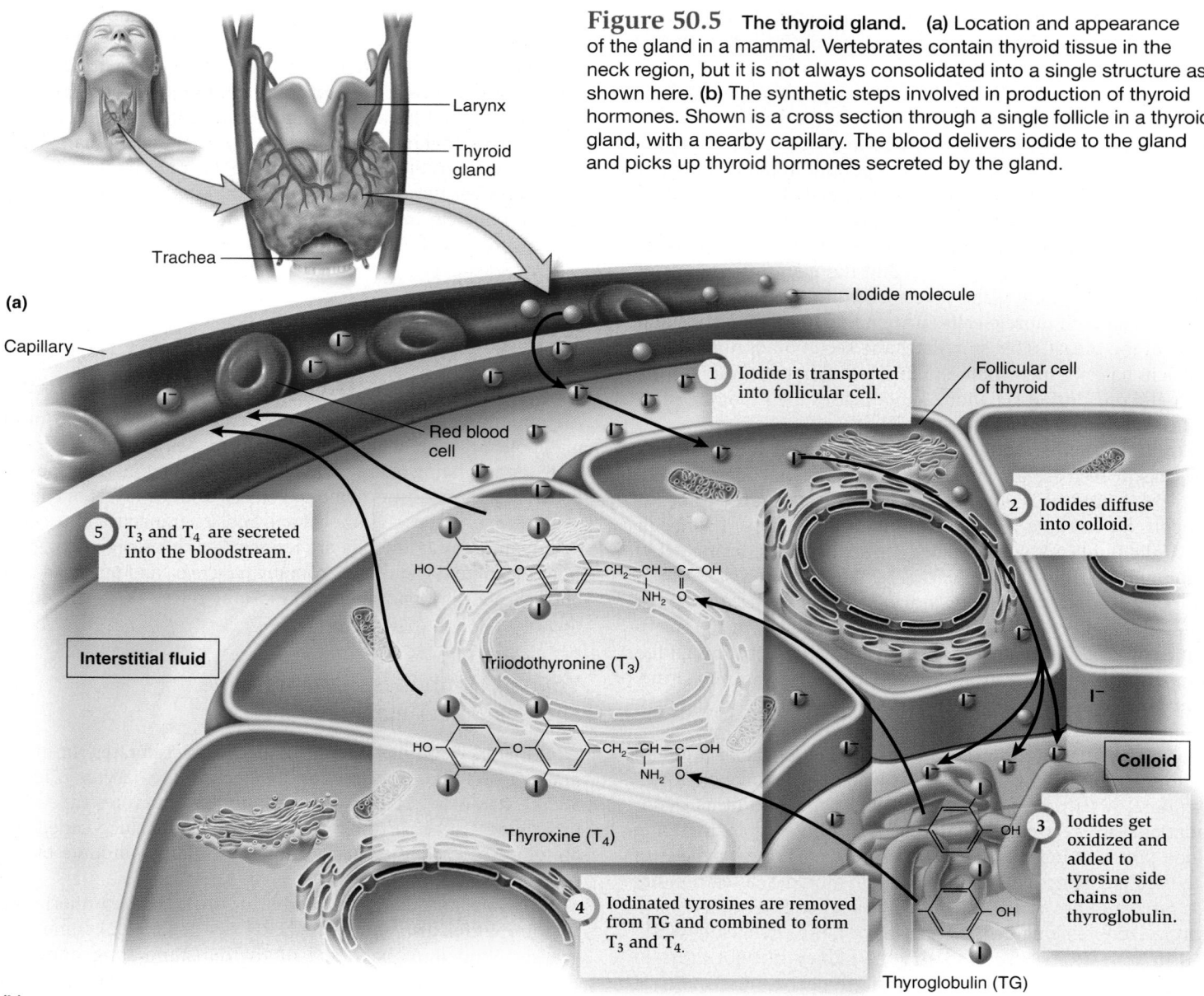

Figure 50.5 **The thyroid gland.** (a) Location and appearance of the gland in a mammal. Vertebrates contain thyroid tissue in the neck region, but it is not always consolidated into a single structure as shown here. (b) The synthetic steps involved in production of thyroid hormones. Shown is a cross section through a single follicle in a thyroid gland, with a nearby capillary. The blood delivers iodide to the gland and picks up thyroid hormones secreted by the gland.

Larynx

Thyroid gland

Trachea

(a)

Iodide molecule

Capillary

1 Iodide is transported into follicular cell.

Follicular cell of thyroid

Red blood cell

2 Iodides diffuse into colloid.

5 T_3 and T_4 are secreted into the bloodstream.

Triiodothyronine (T_3)

Interstitial fluid

Thyroxine (T_4)

Colloid

3 Iodides get oxidized and added to tyrosine side chains on thyroglobulin.

4 Iodinated tyrosines are removed from TG and combined to form T_3 and T_4.

Thyroglobulin (TG)

(b)

A major action of thyroid hormones in adult animals is to stimulate energy consumption by many different cell types. This occurs in large part by increasing the number and activity of the Na^+/K^+-ATPase pumps in plasma membranes. As these pumps hydrolyze ATP, cellular levels of ATP decrease. This decrease is compensated for by increasing the cell's metabolism of glucose. Whenever metabolism is increased, heat production is increased. Consequently, a person with a hyperactive thyroid gland (**hyperthyroidism**; *hyper*, meaning "excessive") generally feels warm, while the opposite condition (**hypothyroidism**; *hypo*, meaning "less than normal") results in a sensation of coldness. It is estimated that up to 70% of the heat produced by some homeotherms is attributable solely to the actions of thyroid hormones on metabolic rate.

The fact that thyroid hormones cannot be made without iodine presents some interesting and unique features of these hormones. The availability of iodine in the diet of most animals is variable. As a consequence, the thyroid evolved the ability to store large amounts of thyroglobulin in the colloid. In this way, during times when iodine ingestion is high, many thyroglobulin molecules have their tyrosine amino acids bound to iodines, one of the first steps in forming T_4 and T_3. During times of low iodine availability, this reservoir of iodinated tyrosines in thyroglobulin molecules can be tapped. Humans, for example, have at least a two-month supply of thyroid hormones even in the absence of ingested iodine. In most industrialized countries, iodine deficiency is rarely a problem since the introduction of iodized salt in the mid-20th century. However, in some regions of the world this is still a major health problem.

The right side of **Figure 50.6a** shows the consequences that follow when thyroid hormones are not produced in normal amounts, for example, due to a lack of iodine in the diet.

Normal iodine intake

Normally, T₃ and T₄ prevent overstimulation of the thyroid gland by TSH.

Hypothalamus

TRH   Excess TRH

Anterior pituitary gland

TSH   Excess TSH

Thyroid

Normal thyroid

5.1 cm

Normal negative feedback

T₃ and T₄

Reduced iodine intake

In iodine deficiency, T₃ and T₄ levels decrease, allowing TSH to increase and overstimulate the thyroid.

Enlarged thyroid

5.1 cm

Reduced negative feedback

Low T₃ and T₄

(a)   **(b)**

Figure 50.6 **Consequences of inadequate iodine in the diet.**   (a) The thyroid enlarges whenever TSH levels increase. In this example, a decrease in iodine intake leads to decreased synthesis of thyroid hormones. This removes the usual negative feedback inhibition of TSH secretion, and thus TSH levels increase, stimulating the growth of the gland. (b) An extreme example of an enlarged thyroid gland, or goiter, due to iodine deficiency.

The decreased T_4/T_3 levels remove the source of negative feedback on the hypothalamus and anterior pituitary gland, resulting in elevated TSH levels. The thyroid gland responds to the increased TSH by increasing the cellular machinery needed to produce more and more thyroglobulin, even though in the absence of iodine no additional T_4 or T_3 will get synthesized. What results is an overgrown gland that is nonetheless incapable of making thyroid hormone. This condition is known as an **iodine-deficient goiter** (**Figure 50.6b**). In humans, the problem can be alleviated either by adding iodine to the diet or by taking T_4 pills. Goiters are not unique to humans. Iodine deficiency is relatively common among vertebrates, and goiters are found frequently in reptiles and birds, particularly those that subsist on all-seed diets, which are generally low in iodine.

Hormones of the Pancreas and Adrenal Glands Regulate Fuel Levels in the Blood

Thyroid hormones regulate an animal's metabolism. For metabolism to proceed normally, however, body cells must have adequate energy available in the form of glucose, fatty acids, and amino acids. The brain, in particular, must have a constant supply of nutrients because brain cells have relatively limited storage capacity for fuel. Regulation of energy availability to cells is in large part the job of the hormones of the pancreas and the adrenal glands.

The pancreas is a complex organ consisting of two parts (**Figure 50.7**). Most of the mass of the pancreas comprises exocrine glands, which release their contents into the small intestine, where they aid digestion. The other part of the pancreas, the endocrine pancreas, produces the peptide hormones insulin and glucagon.

Spherical clusters of cells called **islets of Langerhans** are scattered in great numbers throughout the endocrine pancreas.

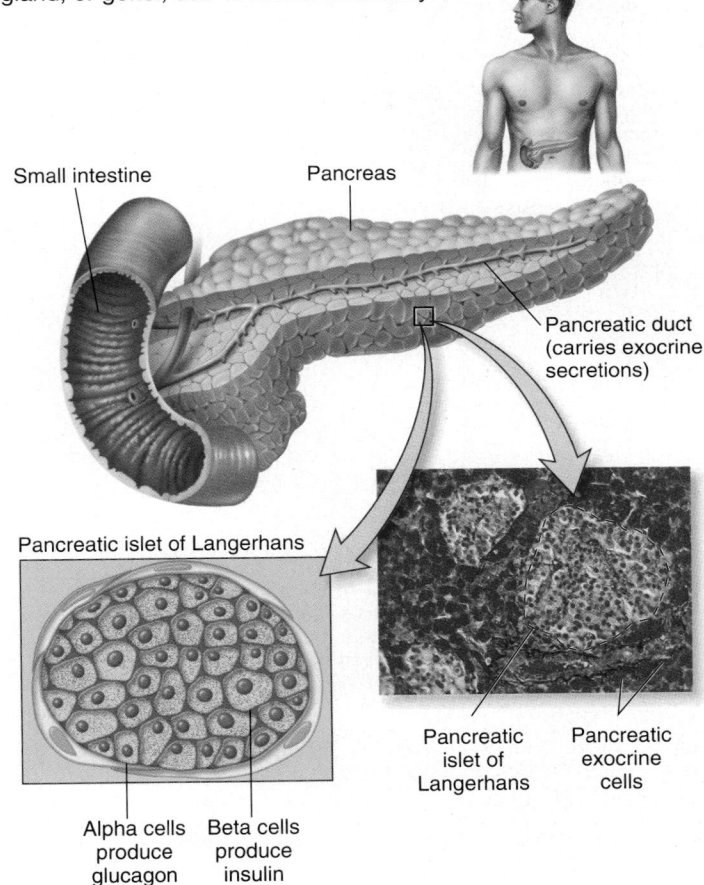

Small intestine   Pancreas

Pancreatic duct (carries exocrine secretions)

Pancreatic islet of Langerhans

Alpha cells produce glucagon   Beta cells produce insulin

Pancreatic islet of Langerhans   Pancreatic exocrine cells

Figure 50.7 **Location, appearance, and internal structure of the mammalian pancreas.**   The ducts of the pancreas empty into the small intestine. Amidst the exocrine pancreas are scattered islets of Langerhans, which are endocrine tissue.

Biological inquiry: The pancreas contains both exocrine and endocrine tissue. Is this property ever observed in other organs?

Within the islets are alpha cells, which make glucagon, and beta cells, which make insulin. These two hormones are antagonistic with respect to each other. For example, insulin lowers and glucagon raises blood sugar concentrations. When glucose levels in the blood need to be raised, glucagon is released, whereas insulin is released when glucose levels are high.

Maintaining normal glucose and other nutrient levels in the blood is a vital process that keeps cells functioning optimally. When an animal has not fed for some time such that its energy stores are becoming depleted and blood glucose levels are falling, glucagon is secreted into the blood, where it acts on the liver to stimulate the production of glucose (**Figure 50.8a**). The liver contains a limited supply of glucose in the form of stored glycogen, which is quickly broken down into many molecules of glucose, which are then secreted into the blood. This process, known as glycogenolysis (see Chapter 42), is stimulated within seconds by glucagon.

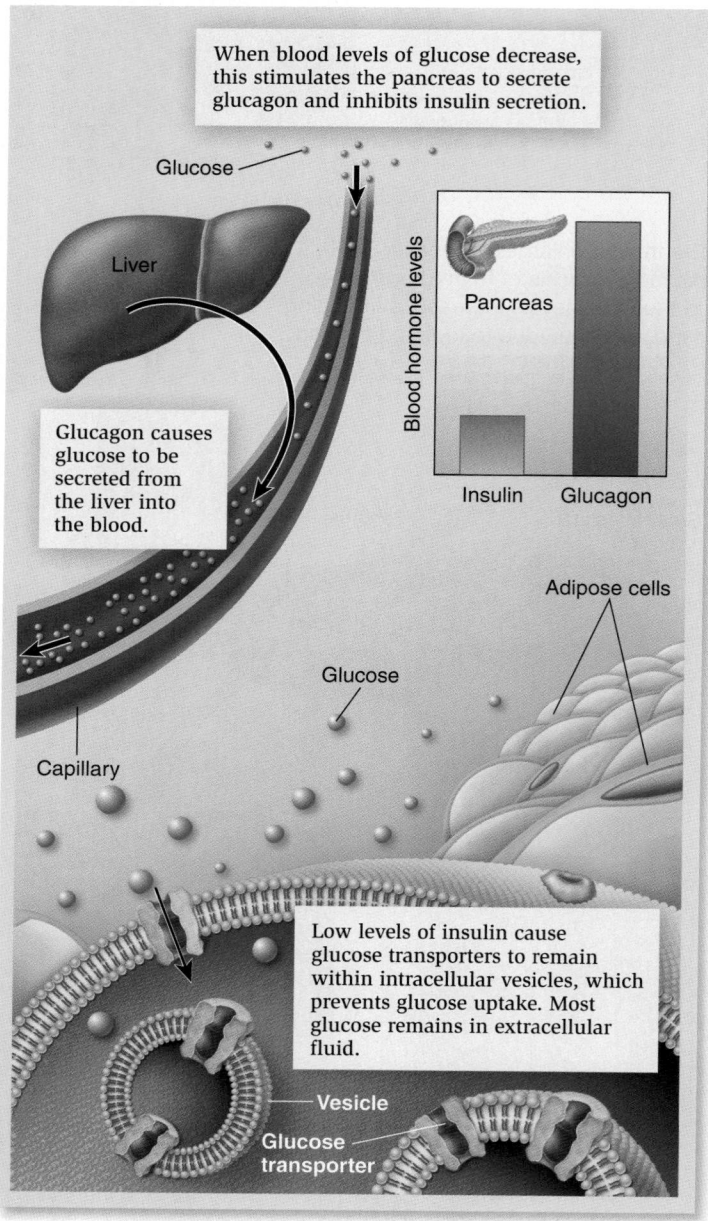

(a) Low blood glucose

When blood levels of glucose decrease, this stimulates the pancreas to secrete glucagon and inhibits insulin secretion.

Glucose

Liver

Glucagon causes glucose to be secreted from the liver into the blood.

Capillary

Glucose

Adipose cells

Blood hormone levels

Pancreas

Insulin   Glucagon

Low levels of insulin cause glucose transporters to remain within intracellular vesicles, which prevents glucose uptake. Most glucose remains in extracellular fluid.

Vesicle

Glucose transporter

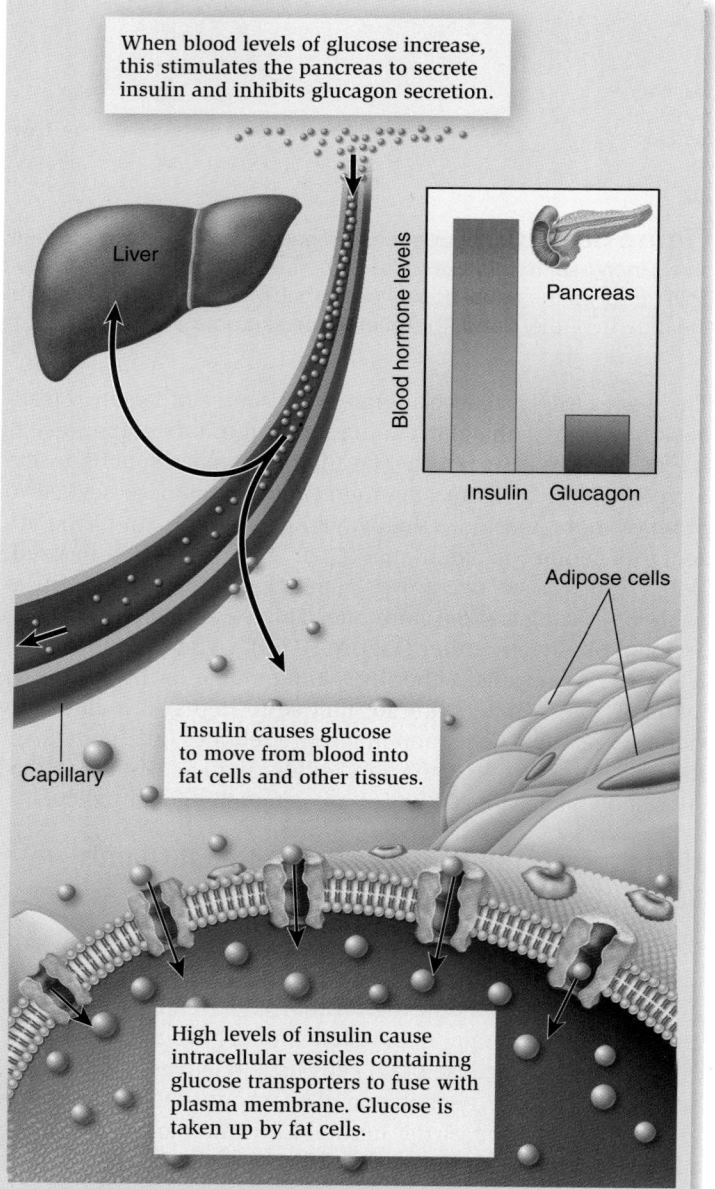

(b) High blood glucose

When blood levels of glucose increase, this stimulates the pancreas to secrete insulin and inhibits glucagon secretion.

Liver

Capillary

Insulin causes glucose to move from blood into fat cells and other tissues.

Adipose cells

Blood hormone levels

Pancreas

Insulin   Glucagon

High levels of insulin cause intracellular vesicles containing glucose transporters to fuse with plasma membrane. Glucose is taken up by fat cells.

Figure 50.8 **Compensatory mechanisms that help maintain glucose homeostasis in the blood.** (a) The liver releases glucose into the blood when plasma glucose levels decrease. This is stimulated by glucagon. Glucose transporters remain within the cytosol of adipose cells. (b) When plasma glucose concentrations increase, glucagon levels decrease and insulin increases. This causes an increase in glucose transporters in the plasma membranes of adipose and muscle cells, thus lowering plasma glucose to normal. Note that glucagon and insulin respond in opposite ways to changes in plasma glucose concentrations, as shown in the insets.

The adrenal glands, which sit atop the kidneys, also play a role in glucose metabolism by producing amine and steroid hormones. In addition to glucagon produced by the pancreas, epinephrine is secreted by the adrenal medulla and also stimulates glycogenolysis. Once the liver's glycogen is depleted, however, glucose must be produced in some other way if fasting continues or glucose levels continue to fall. In that case, cortisol is released from the adrenal cortex. Once in the blood, cortisol stimulates the liver to synthesize glucose from amino acids and glycerol. This process is vital to long-term survival in the absence of food. Because of the combined short-term and long-term actions of glucagon, epinephrine, and cortisol, blood glucose levels rarely fall or remain significantly below normal except in extreme circumstances.

In contrast to fasting, after an animal eats a meal the levels of glucose and other nutrients in the blood become elevated. Restoring the normal blood concentrations of glucose, fats, and amino acids is almost exclusively under the control of insulin, one of the very few hormones that is absolutely essential for survival in animals. The secretion of insulin is directly stimulated by increased concentrations of glucose in the blood (**Figure 50.8b**). Once in the blood, insulin acts on plasma membrane receptors primarily in adipose and muscle tissues to facilitate the transport of glucose across the plasma membrane into the cytosol. Once in the cell, glucose is used for cellular functions or converted to stored energy forms such as fat. As discussed in Chapter 42, the process of glucose transport involves the actions of proteins called glucose transporters. These molecules exist in membrane vesicles in the cytosol. The major function of insulin is to stimulate the movement of the vesicles with their glucose transporters to the plasma membrane. Once the vesicle fuses with the plasma membrane, its transporter molecules begin transporting glucose into the cell.

When blood glucose levels return to normal, the stimulus for insulin secretion disappears, and insulin levels in the blood decrease. This in turn results in a decrease in the number of glucose transporters in plasma membranes, because the glucose transporters are subjected to endocytosis and thereby return to membrane vesicles in the cytosol. The actions of insulin are not limited to glucose transport, however, because the hormone also stimulates movement of amino acids into cells and promotes fat deposition in adipose tissue. In other words, the broader role of insulin in the body is to facilitate the transfer of energy from the extracellular fluid into storage sites primarily in muscle and fat.

In the absence of sufficient insulin, as occurs in the disease **type 1 diabetes mellitus (T1DM)**, extracellular glucose cannot cross plasma membranes, and consequently glucose accumulates to very high concentrations in the blood.

T1DM occurs in many mammals and other vertebrates. It is caused when the body's immune system mistakenly attacks and destroys the insulin-producing cells of the islets of Langerhans. One consequence of the disease is that muscle and fat cells cannot receive their normal amount of glucose to provide the ATP they need. In addition, the unusually large amount of glucose in the blood overwhelms the kidneys' ability to reabsorb it from the kidney filtrate, and glucose appears in the urine. Fortunately, this form of the disease is treatable in humans with regular monitoring of blood glucose levels and daily administration of insulin. Similar methods are used to treat the disease in common house pets.

In humans, the more common form of diabetes, which accounts for roughly 90–95% of all cases (approximately 10–15 million people in the U.S. alone), is **type 2 diabetes mellitus (T2DM)**. In T2DM, the pancreas functions normally and is not attacked by the immune system. However, the cells of the body lose much of their ability to respond to insulin for reasons that are still unclear. T2DM is linked to obesity and can often be prevented or reversed with weight control. Drugs are also available that improve the body's ability to respond to insulin. Formerly, T2DM was known as "adult-onset" diabetes, because it usually appeared in middle age. This name is no longer used, however, because the rising tide of childhood obesity in developed countries like the U.S. has sparked a dramatic increase in the number of children and adolescents with the disease.

The discovery of insulin and its application to the human condition was one of the greatest and most influential achievements in the history of medical research. Next, we describe how this came to be.

FEATURE INVESTIGATION

Banting, Best, MacLeod, and Collip Isolated Active Insulin for the First Time

In the 19th century, scientists discovered that in addition to the exocrine part of the pancreas, there existed spherical clusters of cells that were not associated with the exocrine digestive functions of this organ. These clusters were called islets by their discoverer, German physiologist Paul Langerhans. By the early 20th century, German scientists had discovered that in dogs, complete removal of the pancreas resulted immediately in T1DM.

Researchers assumed, therefore, that cells of the islets of Langerhans produced a factor of some kind that prevented diabetes—in other words, it helped maintain glucose homeostasis by preventing blood glucose levels from getting too high. The factor, however, proved impossible to isolate using the chemical purification methods available at the time. It was assumed that during the process of grinding up a pancreas to produce an extract, the factor was destroyed by the digestive enzymes of the exocrine pancreas.

This problem was eventually solved in 1921 by a team of Canadian scientists working at the University of Toronto: surgeon Frederick Banting, medical student Charles Best, and biochemist James Bertram Collip. The work was performed in the laboratory of John MacLeod, a renowned expert in carbohydrate metabolism at the time. Banting had read a paper in a medical journal that described a deceased patient in whom the pancreatic duct, which carries digestive juices to the small intestine, had become clogged due to calcium deposits. The closed duct caused pressure to build up behind the blockage, which eventually caused the exocrine part of the pancreas to atrophy and die. The islets of Langerhans, however, survived intact. Banting hypothesized that if he were to tie off, or ligate, the pancreatic ducts of an animal, and then wait a sufficient time for the exocrine pancreas to die, he could more easily obtain an active glucose-lowering factor from the remaining islets without the problem of contamination by digestive enzymes.

Banting and Best proceeded to ligate the pancreatic ducts of several dogs as shown in the Feature Investigation in **Figure 50.9**. After waiting 7 weeks, an amount of time they previously had determined was sufficient for the exocrine part of the pancreas to atrophy, they prepared extracts of the remaining parts of the pancreas including its islets of Langerhans. This was done by removing the atrophied pancreas from each dog and grinding it up with a mortar and pestle in an acid solution. The extract was then injected into a second group of dogs in which diabetes had previously been induced by surgically removing the pancreas. The researchers discovered that the extract was

Figure 50.9   The isolation of insulin by Banting, Best, McLeod, and Collip.

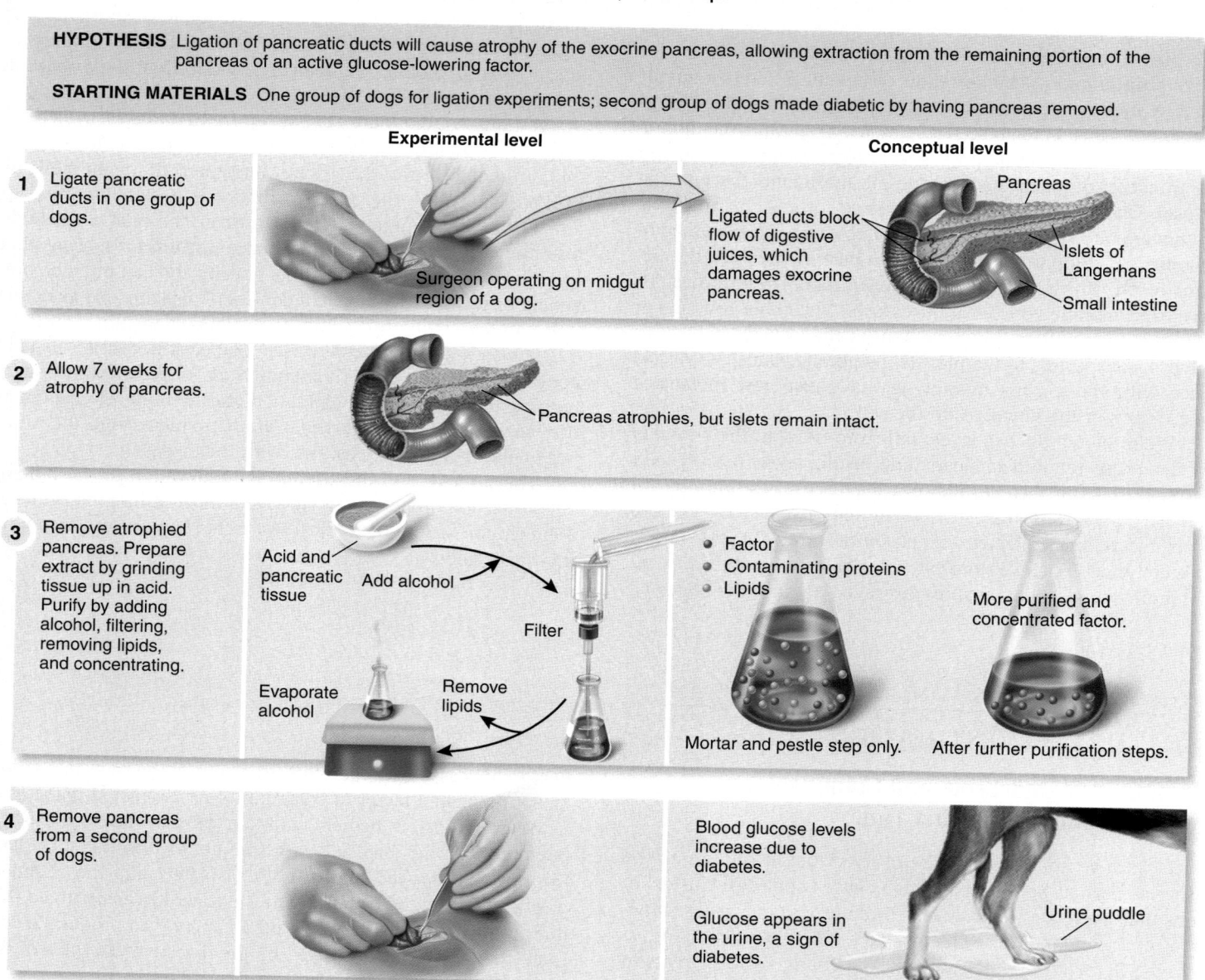

HYPOTHESIS Ligation of pancreatic ducts will cause atrophy of the exocrine pancreas, allowing extraction from the remaining portion of the pancreas of an active glucose-lowering factor.

STARTING MATERIALS One group of dogs for ligation experiments; second group of dogs made diabetic by having pancreas removed.

Experimental level — Conceptual level

1 Ligate pancreatic ducts in one group of dogs.

Surgeon operating on midgut region of a dog.

Ligated ducts block flow of digestive juices, which damages exocrine pancreas.

Pancreas

Islets of Langerhans

Small intestine

2 Allow 7 weeks for atrophy of pancreas.

Pancreas atrophies, but islets remain intact.

3 Remove atrophied pancreas. Prepare extract by grinding tissue up in acid. Purify by adding alcohol, filtering, removing lipids, and concentrating.

Acid and pancreatic tissue

Add alcohol

Filter

Evaporate alcohol

Remove lipids

• Factor
• Contaminating proteins
• Lipids

More purified and concentrated factor.

Mortar and pestle step only.

After further purification steps.

4 Remove pancreas from a second group of dogs.

Blood glucose levels increase due to diabetes.

Glucose appears in the urine, a sign of diabetes.

Urine puddle

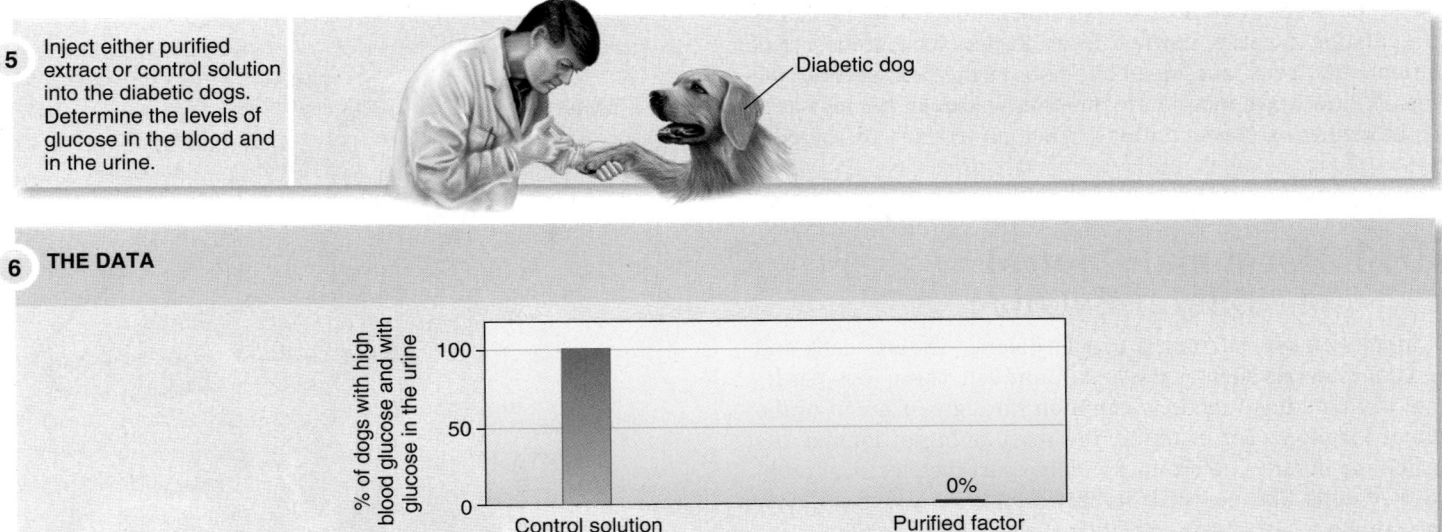

5 Inject either purified extract or control solution into the diabetic dogs. Determine the levels of glucose in the blood and in the urine.

Diabetic dog

6 **THE DATA**

% of dogs with high blood glucose and with glucose in the urine

0%

Control solution   Purified factor

capable of keeping the diabetic dogs alive for a brief time. However, they were unable to isolate sufficient quantities of the active factor in the extract to keep the experiments going longer than a day or so, nor were they able to obtain sufficiently pure factor to prevent side effects such as infection and fever. This is where Collip's expertise came in. Collip developed a method to precipitate contaminating proteins from the extract by adding alcohol to the acid (see step 3, Figure 50.9). This removed most of the contaminating proteins, leaving behind a much more purified and safe extract. The extract was then filtered to remove debris, and further extracted with hydrophobic solvents to remove lipids. This was done because the researchers believed that the factor was either a peptide or a protein and by removing lipids they would obtain a more purified preparation. Finally, the purified extract was concentrated by evaporating the alcohol, which increased its potency. When this factor was injected into diabetic dogs, their blood glucose levels were normalized and there was no longer any glucose in their urine. Dogs that only received a control solution that did not contain the factor continued to show high glucose levels in the blood and urine.

In later experiments, two subsequent innovations enabled the researchers to obtain larger amounts of the factor and more accurately assess its potency even without the ligation procedure. First, the researchers chose the very large pancreases of cows, obtained at a local slaughterhouse, for their starting material from which to prepare the extracts. Second, Collip developed a highly sensitive assay to more precisely measure the concentration of glucose in the blood of an animal before and after injection of the purified factor. The combination of the improved chemical purification steps, large amounts of starting material, and improved assays for testing the extract proved to be the keys that enabled the team to test the factor, which they eventually named insulin, on human patients. The first successful test came in 1922 on a 14-year-old boy in Toronto, who was seriously ill from T1DM. In 1982, recombinant human insulin first became available for widespread use.

The success of the team in rapidly isolating insulin and proving its effectiveness was so significant that Banting and MacLeod were awarded the Nobel Prize in 1923. The Nobel committee apparently felt that Banting and MacLeod were the leaders of the project and that it was they who deserved the prize, but the two scientists disagreed. Banting shared the monetary portion of his award with Best, and MacLeod shared his with Collip.

Adipose Tissue Secretes the Hormone Leptin to Regulate Appetite

Hormones also contribute to metabolism by exerting effects on appetite and, as a consequence, on food consumption. A chief source of appetite-regulating hormones is adipose tissue. One such hormone introduced in Chapter 42 is the protein leptin, which has been observed in species from all vertebrate classes. Leptin is released by adipose cells into the blood in direct proportion to the amount of adipose tissue in the body, and it acts on the hypothalamus to inhibit appetite. When adipose stores are low, however, leptin secretion from adipose cells decreases, resulting in reduced blood levels of leptin. This removes the inhibitory effect of leptin on appetite, causing an animal to consume more food. In these ways, the amount of energy stored in an animal's body in the form of fat is communicated to the brain to regulate how hungry an animal feels.

We have seen that hormones influence energy homeostasis by regulating appetite, nutrient levels in the blood, transport of nutrients into cells, and metabolic rate. In the next section, we consider how hormones control another feature of homeostasis, that of regulating concentrations of key minerals in the blood.

50.4 Hormonal Control of Mineral Balance

All animals must maintain a proper balance in their cells and fluids of minerals such as Ca^{2+}, Na^+, and K^+. These ions participate in numerous functions common throughout much of the animal kingdom. For example, the partitioning of ions across plasma membranes determines in large part the electrical properties of cells like neurons. In this section, we will see how maintaining a homeostatic balance of these ions in an animal's body is coordinated in large part by hormones.

Vitamin D and Parathyroid Hormone Regulate Ca^{2+} Levels in Blood

Because of the important roles calcium ions play in neuronal transmission, heart function, muscle contraction, and numerous other events, the level of Ca^{2+} in the blood is among the most tightly regulated variables in an animal's body.

Calcium is obtained from the diet and absorbed through the small intestine. Like all charged molecules, Ca^{2+} cannot readily cross plasma membranes, including those of the epithelial cells of the small intestine, and thus its transport must be facilitated. This process is controlled by the action of a derivative of **vitamin D**. In most mammals that receive regular exposure to the sun, skin cells produce vitamin D from a precursor called 7-dehydrocholesterol, a reaction that requires the energy of ultraviolet light (**Figure 50.10**). In addition, many animals obtain vitamin D from food or, in people today, from milk supplemented with the vitamin. This is especially important for people who live at latitudes that receive little sunlight for much or part of the year.

Before it can act, vitamin D must first be modified by two enzymes that both add one hydroxyl group to specific carbon atoms, first in the liver and then in the kidney. The final active product is called 1,25-dihydroxyvitamin D. This molecule is a hormone because it is secreted into the blood by one organ—the kidneys—and acts on a distant target tissue—the small intestine—where it stimulates the absorption of Ca^{2+}. The calcium ions then enter the blood, where they can be delivered to tissues for such activities as building bone and maintaining nerve, muscle, and heart functions. If the active form of vitamin D is not present in the blood, the bones lose calcium and become weakened. In children, this results in the weight-bearing bones of the legs becoming deformed, a condition called rickets.

When calcium is not present in the diet, or when 1,25-dihydroxyvitamin D is not formed in normal amounts because of insufficient exposure to sunlight, the blood levels of Ca^{2+}

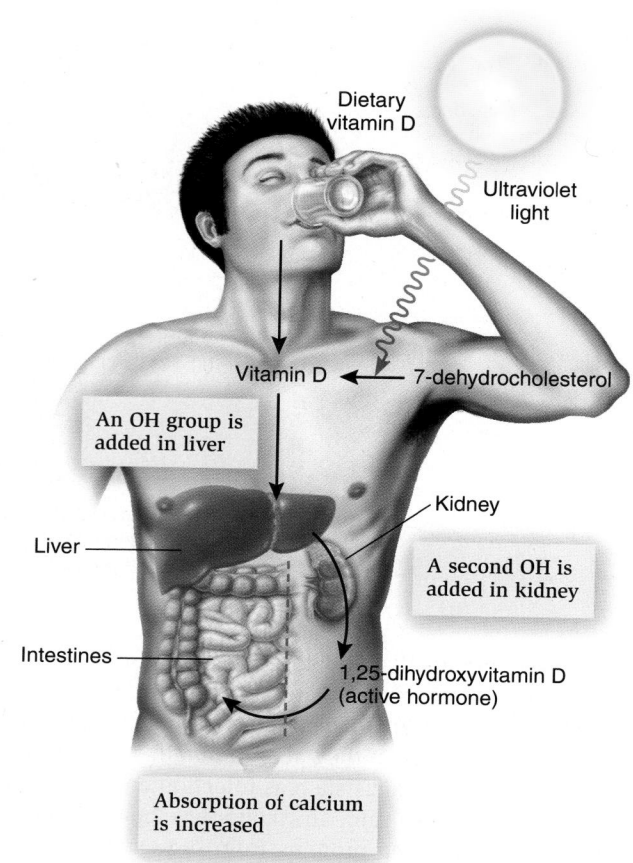

Figure 50.10   **Synthesis of the active hormone formed from vitamin D.**

Biological inquiry: Do you suppose that all mammals synthesize the active form of vitamin D using the energy of sunlight?

may decrease dramatically. This does not normally occur, however, because a hormone called **parathyroid hormone** (**PTH**) is secreted from several small glands called parathyroid glands located near the thyroid gland (**Figure 50.11a**). PTH acts on bone to stimulate the activity of cells that dissolve the mineral part of bone. This releases Ca^{2+}, which then enters the blood (**Figure 50.11b**). Thus, besides providing a skeletal framework for the vertebrate body, bone also serves as an important reservoir of Ca^{2+}. Without PTH, calcium homeostasis is not possible. The total absence of PTH is fatal in humans and other mammals.

Several Hormones Regulate Sodium and Potassium Ions in Vertebrates

Like calcium, concentrations of sodium and potassium ions in the body fluids of most animals are tightly regulated, because these ions play crucial roles in membrane potential formation and action potential generation (see Chapter 43), among other functions. Like calcium, sodium and potassium are ingested in the diet and excreted in the urine in vertebrates. Unlike calcium, however, no large reservoirs exist in the body for sodium and potassium.

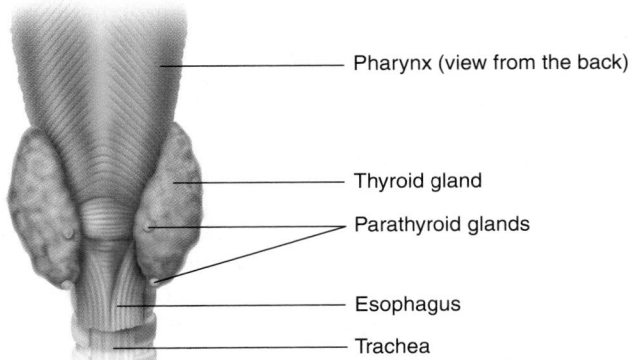

(a) Location of the parathyroid glands

Pharynx (view from the back)

Thyroid gland

Parathyroid glands

Esophagus

Trachea

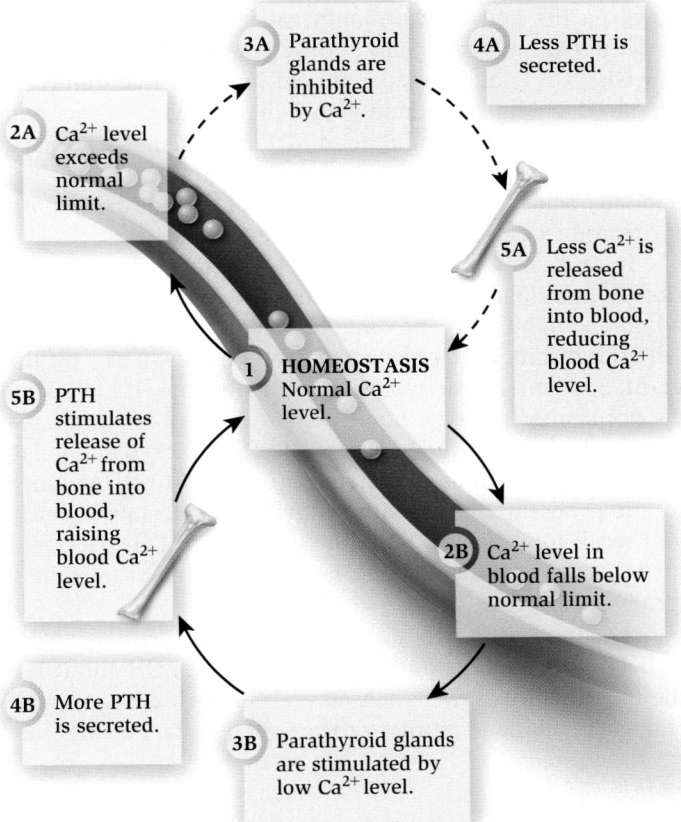

3A Parathyroid glands are inhibited by Ca^{2+}.

4A Less PTH is secreted.

2A Ca^{2+} level exceeds normal limit.

5A Less Ca^{2+} is released from bone into blood, reducing blood Ca^{2+} level.

1 HOMEOSTASIS Normal Ca^{2+} level.

5B PTH stimulates release of Ca^{2+} from bone into blood, raising blood Ca^{2+} level.

2B Ca^{2+} level in blood falls below normal limit.

4B More PTH is secreted.

3B Parathyroid glands are stimulated by low Ca^{2+} level.

(b) Homeostatic control of blood Ca^{2+} levels

Figure 50.11 Location of the parathyroid glands (a) and the homeostatic control of blood Ca^{2+} levels (b). Steps 2A–5A occur when Ca^{2+} is in excess, while steps 2B–5B occur when Ca^{2+} is below normal.

One of the key mechanisms that regulates the concentrations of these ions in the blood is altering the rate of sodium, potassium, and water reabsorption from the urine as it is being formed in the kidneys. This is accomplished in large part by the actions of three hormones: ADH, aldosterone, and atrial natriuretic peptide (**Figure 50.12**).

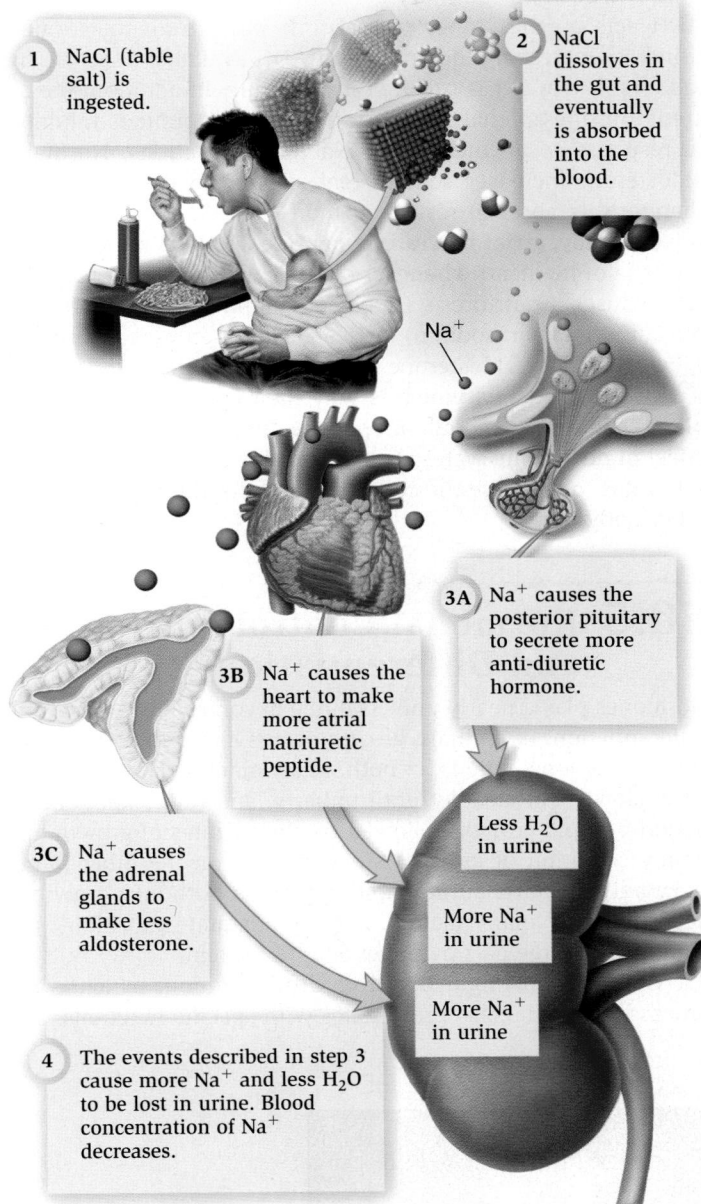

1 NaCl (table salt) is ingested.

2 NaCl dissolves in the gut and eventually is absorbed into the blood.

Na^+

3A Na^+ causes the posterior pituitary to secrete more anti-diuretic hormone.

3B Na^+ causes the heart to make more atrial natriuretic peptide.

3C Na^+ causes the adrenal glands to make less aldosterone.

Less H_2O in urine

More Na^+ in urine

More Na^+ in urine

4 The events described in step 3 cause more Na^+ and less H_2O to be lost in urine. Blood concentration of Na^+ decreases.

Figure 50.12 An example of sodium balance achieved by the coordinated actions of three hormones.

Biological inquiry: Why is there more than one hormone to regulate sodium and potassium balance?

The vertebrate kidney normally reabsorbs most sodium and potassium from the fluid filtered through the kidney glomerulus (see Chapter 49). However, dietary and other changes can alter the concentrations of these ions in the blood. When this happens, it is the role of the kidney to restore the ions to their normal concentrations. For example, if blood sodium levels increase above normal, the osmolarity of the blood will increase. Osmoreceptors in the brain detect this and stimulate ADH secretion from the posterior pituitary gland. Recall from Chapter 49 that ADH acts on the kidney to reabsorb water from the forming urine. By increasing the amount of ADH available to act on the kidneys, less water is excreted in the urine. In addition, the secretion of the adrenal steroid hormone aldosterone is inhibited in response to an increase in Na$^+$ concentrations. Normally, aldosterone increases sodium reabsorption in the kidney, and therefore its absence results in more sodium excretion in the urine. Finally, **a**trial **n**atriuretic **p**eptide (**ANP**) is secreted from the atria of the heart whenever blood levels of sodium increase. ANP causes a natriuresis (a loss of sodium in the urine; from the Latin *natrium*, sodium) by decreasing sodium reabsorption. Thus, ANP and aldosterone have opposite effects on sodium balance in the body, which is why their concentrations in the blood tend to rise and fall in opposite directions. The combined effect of decreasing water loss in the urine and reabsorbing less sodium is to decrease the concentration of sodium in blood and other body fluids.

50.5 Hormonal Control of Growth and Differentiation

Hormones play a nearly universal role in controlling growth and differentiation. In animals, growth occurs either in brief spurts, slowly over long periods, or both. Many mammals, for example, grow slowly but steadily until puberty, and then experience a period of rapid growth, followed by slower rates of growth and finally cessation of growth. Many insects, on the other hand, grow only in spurts during molting periods. Although growth is determined by many factors, notably adequate nutrition, it is regulated in large part by the endocrine system.

Growth is distinguished from differentiation, which is the process by which cells form tissues with specific functions, and

tissues form larger and more complex structures. In this section we describe how both processes depend in part on the endocrine system.

Vertebrates Require a Balance of Several Hormones for Normal Growth

In vertebrates, the anterior pituitary gland produces **growth hormone** (**GH**), which is under the control of the hypothalamus (see Table 50.2). GH acts on the liver to produce another hormone, called **insulin-like growth factor-1** (**IGF-1**). In mammals, IGF-1 stimulates the elongation of bones, especially during puberty, when mammals become reproductively mature. This growth is further accelerated by the steroid hormones of the gonads, leading to the rapid pubertal growth spurt. Eventually, however, the gonadal hormones cause the growth regions of bone to seal, preventing any further bone elongation.

GH, IGF-1, and gonadal steroid hormones continue to be produced in adult vertebrates—including humans—even though growth has ceased. In adulthood, GH in particular serves metabolic functions, such as helping regulate the levels of glucose and fatty acids in the blood. The gonadal steroids are important for reproduction in adults (see Chapter 51).

In rare cases, a tumor of the GH-secreting cells of the anterior pituitary gland produces excess GH during childhood and adulthood in humans and other mammals. In humans, a person with this disorder can grow very tall and is known as a **pituitary giant**. Soon after puberty, though, growth stops. If high GH levels occur after puberty, however, the excess GH causes many bones, like those of the hands and feet, to thicken and enlarge, a condition known as **acromegaly** (**Figure 50.13**). Acromegalics are generally treated with synthetic somatostatin, which as you recall from Table 50.2 is the inhibitor of GH made by the hypothalamus. In many cases, however, the tumor must be surgically removed.

By contrast, if the pituitary fails to make adequate amounts of GH during childhood, the concentrations of GH and IGF-1 in the blood will be lower than normal. In such cases, growth is stunted, resulting in **pituitary dwarfism**. Pituitary dwarfs can be

Figure 50.13   Acromegaly in one individual from a pair of identical twins.   Note the enlarged bones in the hands and feet.

treated with GH and will grow to relatively normal height, as long as treatment begins before puberty ends so that the bones may still elongate.

Hormones are also required during fetal life, not for growth but for development of the brain, lungs, and other tissues into fully mature, functional structures. For example, cortisol from the fetal adrenal gland is vital for proper lung formation. Premature babies born before the adrenal glands have matured have lungs that are not capable of expanding normally and therefore the infant must be hospitalized to survive.

As another example, thyroid hormones impact differentiation among all vertebrates. In amphibians, thyroid hormones play a critical role in metamorphosis, notably in tadpoles, where they promote the resorption of the tail and development of the legs (**Figure 50.14**). This effect can be dramatically demonstrated by experimentally decreasing or increasing the tadpole's thyroid hormone levels, which results in a tadpole that does not transform into a frog, or a tadpole in which metamorphosis occurs sooner than normal resulting in a tiny frog. In fish, thyroid hormones play an equally critical role in differentiation. For example, among the hundreds of species of flatfish that live on the ocean floor, thyroid hormones are responsible for the characteristic change in appearance that occurs in these species as they settle into a sedentary existence on the ocean bottom. The fins and gill covers migrate to the dorsal surface facing the water, the dorsal body surface becomes pigmented, and most remarkably the eyes migrate to the same side of the head, so that one eye is not constantly unused on the side facing the ocean floor. Each of these metamorphic events is under the direct control of thyroid hormones.

Invertebrates Grow in Spurts Under the Control of Three Major Hormones

Like vertebrate endocrine pathways, hormonal control systems in insects and other invertebrates often involve multiple glands and neural structures acting in concert. In insects, for example, the endocrine system is critical for the growth of larvae and their eventual differentiation into pupae (**Figure 50.15**). In the case of larval growth and metamorphosis, specialized nerve cells periodically secrete a hormone called **brain hormone**, which stimulates a pair of endocrine glands called the prothoracic glands. These glands synthesize and secrete a steroid hormone, **ecdysone**, not found in vertebrates. Ecdysone is secreted only periodically. In response to each burst of ecdysone, the larva molts—sheds its cuticle—and begins a new growth period until it must shed its cuticle again in response to another episode of ecdysone secretion.

Throughout larval development, a structure behind the brain called the corpus allatum secretes another hormone, this one a protein called **juvenile hormone**. Ecdysone and juvenile hormone have opposing actions. Ecdysone stimulates growth and differentiation of the larva, while juvenile hormone inhibits maturation of the larva into a pupa. As the larva ages, the amount of

A tiny froglet develops sooner than normal if thyroxine is experimentally administered here.

If the thyroid glands are surgically removed at this time, a permanent tadpole develops.

The red line indicates the normal rate of thyroxine secretion.

Thyroxine secretion rate

−35   −30   −25   −20   −15   −10   −5   0   +5   +10
Days from emergence of forelimb

Rapid growth

Reduced growth, rapid differentiation

Rapid differentiation

Fully developed frog

Stages of normal development

Figure 50.14 **The effect of thyroid hormones on tadpole development.**   Experimental manipulation of thyroid hormone levels can retard or accelerate development.

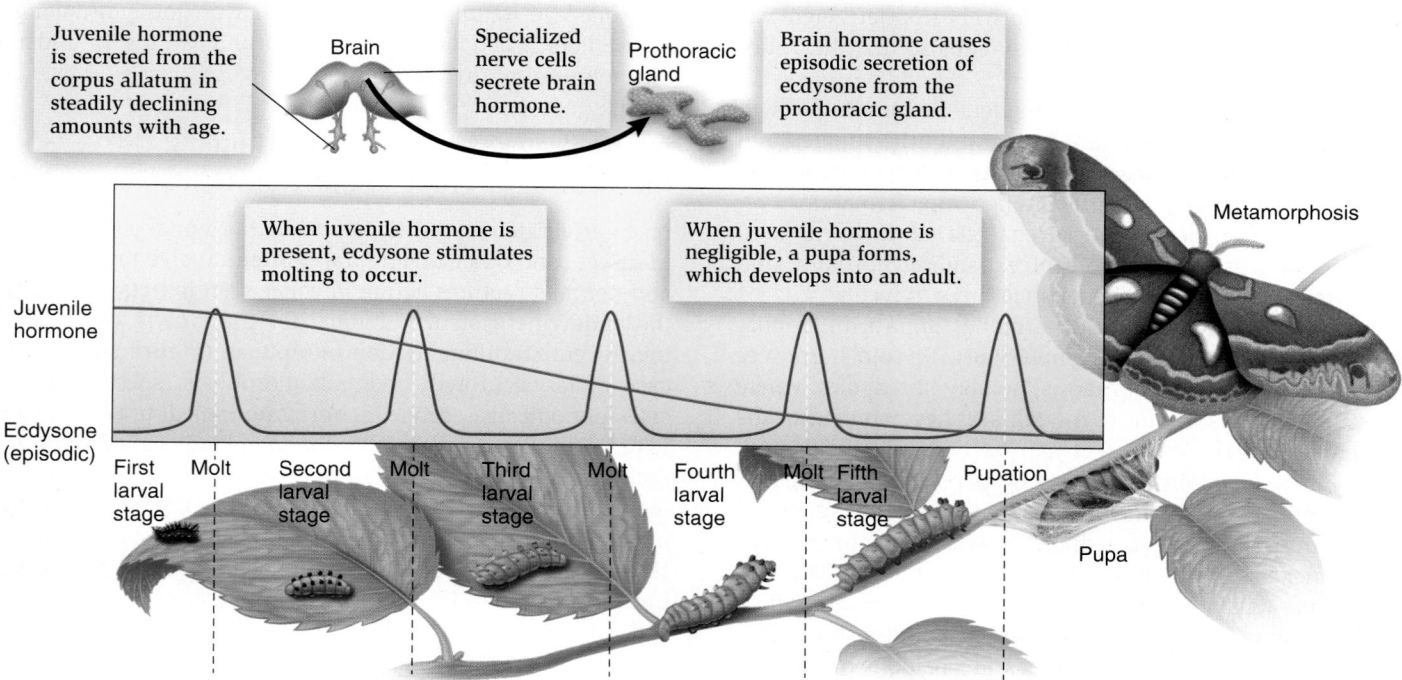

Juvenile hormone is secreted from the corpus allatum in steadily declining amounts with age.

Brain

Specialized nerve cells secrete brain hormone.

Prothoracic gland

Brain hormone causes episodic secretion of ecdysone from the prothoracic gland.

Metamorphosis

When juvenile hormone is present, ecdysone stimulates molting to occur.

When juvenile hormone is negligible, a pupa forms, which develops into an adult.

Juvenile hormone

Ecdysone (episodic)

First larval stage | Molt | Second larval stage | Molt | Third larval stage | Molt | Fourth larval stage | Molt | Fifth larval stage | Pupation

Pupa

Figure 50.15   **Hormonal control of insect development.**   Development of insects requires the coordinated actions of three hormones. Juvenile hormone suppresses pupa formation. Ecdysone stimulates molting when juvenile hormone is high, and it stimulates pupation and metamorphosis when juvenile hormone is low. Brain hormone usually causes the episodic secretion of ecdysone, but the levels of ecdysone and juvenile hormone in this figure are schematic and not representative of all insects.

juvenile hormone it produces gradually declines until its levels are nearly zero. During this time, however, ecdysone continues to be periodically secreted. This imbalance eventually tips the scale in ecdysone's favor, and the larva progresses to the pupa stage and finally transforms into an adult.

50.6 Hormonal Control of Reproduction

The topic of reproduction is covered in Chapter 51, but it is worth noting here that in all vertebrates and probably in most invertebrates, reproduction is closely linked with endocrine function. In this section, we describe the most common reproductive hormones and their actions in male and female animals.

The Gonads Secrete Sex Steroids That Influence Most Aspects of Reproduction

Hormones produced by the gonads of animals play vital roles in nearly all aspects of reproduction. These include sex-specific reproductive behaviors like courtship displays and mating rituals, development of an animal's sex-specific physical characteristics, its transition from a reproductively immature to a mature state, and its ability to produce offspring. In males, the gonads are called testes. They house the gametes, called sperm cells. In addition, though, the testes produce several related steroid hormones

collectively termed **androgens**. In many vertebrates, including humans, the primary androgen is **testosterone**. The gonads of females are called the ovaries, which contain the female gametes, or egg cells. Other cells within the ovaries secrete two major steroid hormones, **progesterone** and a family of related hormones called **estrogens**. The major estrogen in many animals including humans is **estradiol**. Collectively, the male and female gonadal steroid hormones are referred to as the sex steroids.

The sex steroids are responsible for male- or female-specific reproductive behaviors associated with courtship and mating. In vertebrates, androgens from the developing testes are required for the development of the male phenotype. Exposure of fetuses to abnormal levels of male hormones or female hormones can lead to ambiguous phenotypes and sexual behavior after birth. These steroids are also chiefly responsible for development of the secondary sex characteristics for each sex, which in humans include such things as growth of the appropriate external genitals, growth of the breasts, development of facial hair, and distribution and amount of fat and muscle in the body (see Chapter 51). Finally, the sex steroids are required for maturation of gametes, the transition of a young animal to a reproductively mature one, and the ability of a female to produce young.

The ability of the testes and ovaries to produce the sex steroids depends on the presence of the gonadotropins, which are the same in both sexes and include **follicle-stimulating hormone (FSH)** and **luteinizing hormone (LH)** (see Table 50.2). Thus, the way in which the gonads are controlled in male and female animals is through identical anterior pituitary gland hormones.

Nutrition and Reproduction Are Linked Through Hormones

An interesting feature of the endocrine control of reproduction is the long-noted observation that puberty is delayed in mammals, including people, that are very undernourished. Similarly, fertility—the ability to produce offspring—is reduced in women and other adult female mammals under such conditions. This makes sense, because supporting the nutritional demands of a growing fetus would be difficult without good nutrition and energy stores. In such a case, it would be more advantageous for an animal to delay reproduction until sufficient food is available. How does the brain of a female mammal determine when sufficient energy is stored in her body to support a pregnancy? The answer appears to be partly the result of hormonal signals. For example, if the amount of fat in a woman's body decreases, so does the concentration of leptin in her blood, as described earlier. Leptin has been demonstrated to stimulate the synthesis and secretion of key reproductive hormones like LH and FSH. Consequently, undernourishment causes a decrease in leptin, which in turn results in reduced production of reproductive hormones, and this contributes to a loss of fertility. Thus, leptin serves as a link between adipose tissue and the reproductive system.

50.7 Hormonal Responses to Stress

Stress in animal biology is often defined as any real or perceived threat to an animal's survival. This can take many forms depending on the species. To a crab, a passing shadow may indicate a hungry shark passing overhead, but to a human, a shadow may simply mean that a cloud has passed in front of the sun. Similarly, a severe storm can be extremely stressful for birds in exposed nests, but not to animals that live in sheltered environments. Overcrowding is adaptive for penguins trying to keep warm on ice floes, but it is very stressful to mice and rats.

In the 1920s, Canadian physiologist Hans Selye made the remarkable discovery that regardless of the nature of the stress in any given species, a mammal's adaptive responses to stress were highly similar. In this section, we describe how hormones made by the vertebrate adrenal glands are the common denominator of the vertebrate response to stress, and how these hormones act to prepare an animal to confront a challenge or escape it—the "fight-or-flight" response.

The Inner Core of the Adrenal Glands Makes Epinephrine

The adrenal glands, so named from the Latin (*ad*, toward, and *renis*, kidney) because they sit atop the kidneys, are multifunctional glands containing an inner region called the adrenal medulla and an outer region called the adrenal cortex (**Figure 50.16**). The cells of the adrenal medulla secrete the amine hormones norepinephrine and epinephrine. Together, these two

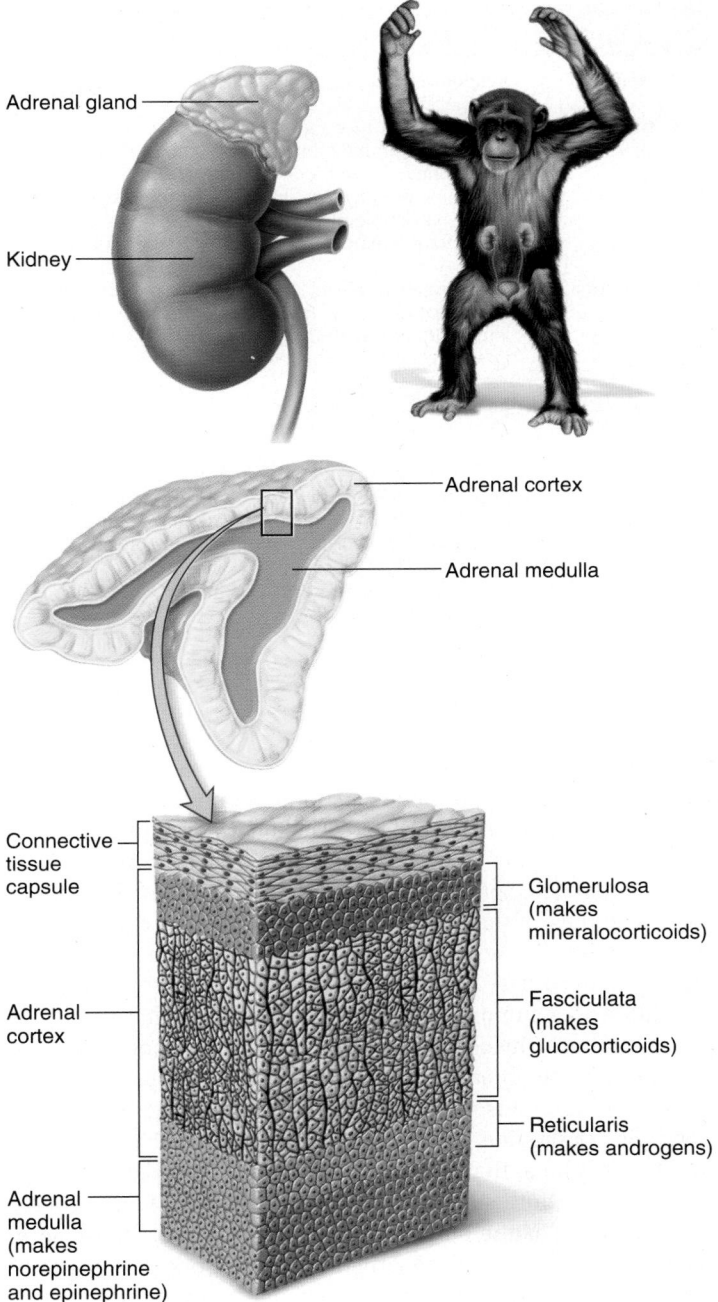

Figure 50.16 Location and cellular arrangement of the adrenal glands.

hormones are responsible for most of the fight-or-flight reactions that were described earlier in Chapter 44 and are summarized here in **Table 50.3**.

The Outer Regions of the Adrenal Glands Make Steroid Hormones

Norepinephrine and epinephrine are only a part of the adrenal gland's response to stress, however. The outer part of the adrenal, the cortex, is itself subdivided into three zones (Figure 50.16).

Table 50.3	The Fight-or-Flight Response to Stress
System	**Responses**
Cardiovascular	Increase heart rate and strength of heart contractions to maximize pumping of blood to all parts of the body; dilate blood vessels entering tissues needing more oxygen—like skeletal muscle—and constrict blood vessels to regions of less immediate importance—like the gut and kidney
Respiratory	Dilate small airways (bronchioles) to reduce resistance to airflow in mammals; increase rate and depth of breathing to maximize oxygen intake and carbon dioxide elimination
Metabolic	Increase glycogenolysis in muscle to provide glucose for muscle cells, and in liver to provide glucose to the blood, where it can reach all body cells; increase breakdown of adipose triglycerides into usable fuel (fatty acids) that can then enter the bloodstream; stimulate glucagon secretion, which acts on liver to promote gluconeogenesis
Nervous	Increase arousal and alertness; inhibit nonessential functions like appetite

Table 50.4	The Major Actions of Glucocorticoids in Stress
Site	**Actions**
Liver	Stimulate gluconeogenesis, thus providing glucose to the blood
Adipose tissue	Stimulate breakdown of triglycerides into fatty acids and glycerol for fuel
Muscle and adipose tissue	Inhibit sensitivity to insulin, making more glucose available to brain cells, which do not require insulin to move glucose across their plasma membranes
Bone	Inhibit bone growth and formation, because such processes require large amounts of nutrients that could be used to combat stress instead
Lungs	Stimulate lung maturation in the fetus
Immune	Suppress immune system function and reduce inflammation
Other	Regulate sodium and chloride balance in migratory fish; stimulate nervous system development in most vertebrates; stimulate protein breakdown to provide amino acids to the liver for gluconeogenesis; inhibit reproduction

The outer zone, the glomerulosa, is the region that makes aldosterone. Because this hormone acts to maintain mineral balance, it is known as a mineralocorticoid. The innermost cortical zone is known as the reticularis, which functions in humans to make certain androgens, but whose function in other animals is not as clear. The bulk of the cortex is the middle zone, the fasciculata. It is here that the so-called glucocorticoid hormones are made, among them cortisol and other structurally related steroids.

Table 50.4 summarizes the major actions of glucocorticoids in vertebrates. Glucocorticoids are catabolic hormones, that is, they promote the breakdown of tissue. For example, they act on bone, immune, muscle, and fat tissue to break down proteins and lipids to provide energy for the body's cells. This is important for an animal that is facing an acute stress, because feeding and digestion both stop during fight-or-flight reactions, leaving internal stores as the only source of energy. In fact, these adrenal steroids are called glucocorticoids because one of their major actions is to promote gluconeogenesis in the liver during times of stress, thus providing glucose to the blood.

Responding to acute stress is an important function of glucocorticoids, but excessive production of these steroids can create serious problems. Such a situation might arise if an animal were chronically stressed for some reason, for example, due to habitat destruction, intraspecies fighting and aggression, or chronic infection. Glucocorticoids and the amines increase the activity of the heart and can raise blood pressure. Therefore, prolonged stress can lead to disorders of the heart and blood vessels. In addition, chronic increases in glucocorticoids can suppress the immune system because of the catabolic actions of glucocorticoids on immune tissue. This may compromise an animal's ability to fight infection. For example, the prolonged stress a salmon experiences when it makes its exhausting swim upriver to its spawning grounds causes massive suppression of its immune system, and many mature salmon that die after spawning are found to be riddled with infections. People who are chronically stressed also tend to be susceptible to infections, or to having dormant viruses suddenly flare up, such as the herpes virus that causes cold sores.

Even an animal's ability to reproduce and grow is negatively affected by chronic stress, as noted most dramatically in medical situations involving humans. In children, for example, chronic exposure to high levels of glucocorticoids may stunt growth, at least temporarily. In women, chronic stress may result in the loss of monthly reproductive cycles. The explanation for this may be understood from an evolutionary perspective. Chronic stress generally suggests that an animal's fitness or survival is in jeopardy. Therefore, it is advantageous to have all energetically demanding activities—such as growth, reproduction, and even immune activity—come to a halt if these activities are not immediately required for staying alive.

The preceding discussion illustrates only a few ways in which the endocrine system can affect health and fitness in many animals, including humans. In the rest of this chapter, we highlight some aspects that are exerting a large impact on public health in humans.

50.8 Impact on Public Health

Hormones have become very much a part of the news. The use of hormones to promote more efficient food production in dairy animals, for example, has created considerable controversy around the world. On the one hand, injecting cows with bovine growth hormone makes for leaner, larger animals, which is of economic benefit to the dairy industry. On the other hand, many people are unsettled by the possibility that by-products of the hormone may find their way into our bodies, with unknown consequences. Let's look at two other ways in which factors that alter the function of the human endocrine system have important health implications, and then consider how hormones are used therapeutically.

Synthetic Compounds May Act as Endocrine Disruptors

A recent phenomenon—disturbing because of its potential to impact reproduction in animals and people—is the growing amount of so-called **endocrine disruptors** found in lakes, streams, ocean water, and soil exposed to pollution runoff. These chemicals are derived in many cases from industrial waste and have molecular structures that in some cases resemble estrogen sufficiently to bind to estrogen receptors. If these compounds make their way into drinking water or food, they can exert estrogen-like actions or, in some cases, inhibit the actions of the body's own estrogen. This can lead to dramatic consequences on fertility and on development of embryos and fetuses. The extent of the risk from these **xenoestrogens** (from the Greek *xeno*, foreign) is hotly debated, but the number of mature, functional germ cells produced in animals as diverse as mollusks and human males has declined dramatically during the past 50 years in the U.S. In addition, researchers throughout the world have noted feminization of freshwater fish downstream of wastewater facilities. For example, male fish that were exposed to such conditions during development show increased production of proteins normally made by females bearing eggs. They also show changes in gonadal structures that resemble the female appearance.

Hormone Misuse Can Be Disastrous

Some individuals, typically those in competitive sports, self-administer hormones such as androgens. This practice may increase muscle mass and improve athletic performance. However, the price is steep. First, androgens exert negative feedback actions on FSH and LH secreted by the anterior pituitary gland. As a consequence, the anterior pituitary gland stops producing and secreting FSH and LH while the user is taking supplements containing androgens. In males, this causes the testes to shrink, as they no longer are making sperm, and the man be-

comes infertile. Androgen administration has also been linked to extreme aggressive behavior (**"roid" rage**), cardiovascular disease and heart attacks, skin problems, and certain cancers. Women using androgens run similar health risks as men but in addition develop masculinizing traits, including growth of body hair and thinning scalp hair.

Another example of hormone misuse is to boost the number of red blood cells in the circulation to increase the oxygen-carrying capacity of the blood (often called **blood-doping**). This has become common among athletes participating in long-distance aerobic activities, such as cycling and cross-country skiing. It has replaced the former practice of actually transfusing suspensions of red blood cells into a person's circulation. The hormone that is used is **erythropoietin** (**EPO**), which acts by stimulating the maturation of red blood cells in the bone marrow and their release into the blood. EPO is normally made by the liver and kidneys in response to any situation where additional blood cells are required, such as following blood loss or when a person lives at high altitudes, where the oxygen pressure is low. When used abusively, however, the concentration of red blood cells reaches such high levels that the blood becomes much thicker than normal. This puts a serious strain on the heart, which must work harder to pump the thickened blood. Since the 1990s, the international cycling community has been rocked by a bizarre trend of world-class European cyclists who died of heart attacks in the prime of their lives. These individuals had been using EPO to gain an unfair and, as it turns out, unwise advantage over their peers. Testing continues to detect injected EPO in the blood of some cyclists and other endurance athletes.

Hormones Are Used Therapeutically to Treat Millions of People

The list of endocrine-related diseases in the human population is a long one, ranging from relatively common disorders like diabetes and thyroid disease to rare ones that may affect only 1 in every 100,000 or more individuals. Since the advent of large-scale hormone production, it has become routine to provide individuals with hormones that their own bodies fail to produce. For example, millions of people self-administer insulin each day, or take thyroxine to supplement their insufficient levels of thyroid hormones. Many women undergo hormone therapy to help induce pregnancy or control the symptoms of menopause. A growing number of men, too, take gonadal hormones if the levels produced by their bodies decline significantly in later life. Other examples include recombinant human growth hormone to treat abnormally short stature in prepubertal children, epinephrine inhalers to treat asthma, and glucocorticoids to treat inflammation, lung disease, and skin disorders, to name just a few.

CHAPTER SUMMARY

50.1 Mechanisms of Hormone Action and Control

- Endocrine glands contain epithelial cells that secrete hormones into the bloodstream, where they circulate throughout the body. Although slower than electrical signaling, chemical signaling has the advantages of being highly regulated and, in the case of some hormones, occurring at more than one site at once.

- The endocrine glands and other organs with hormone-secreting cells constitute the endocrine system. (Figure 50.1)

- The effects of a hormone may occur within seconds or require several hours to develop, and it may last for as short as a few minutes or as long as days.

- Hormones fall into three broad classes that include the amines, proteins/peptides, and steroids. The amines and the proteins/peptides generally share similar chemical properties and modes of action, while the steroid hormones act very differently from the other classes. (Table 50.1)

- Water-soluble hormones act on receptor molecules located in the plasma membrane, while lipid-soluble hormones act on intracellular receptors.

- Blood levels of hormones can increase or decrease through two mechanisms: changing the rate of hormone production by an endocrine cell, and changing the rates at which hormones are removed or inactivated. (Figures 50.2, 50.3)

50.2 Links Between the Endocrine and Nervous Systems

- Sensory input from an animal's nervous system modulates the activity of certain endocrine glands and influences blood levels of many hormones.

- Within the hypothalamus are numerous neurons that synthesize neurohormones and stimulate the anterior pituitary. (Figure 50.4)

- The anterior pituitary gland synthesizes six different hormones that respond to the presence of hypothalamic neurohormones. They are adrenocorticotropic hormone, follicle-stimulating hormone, luteinizing hormone, growth hormone, prolactin, and thyroid-stimulating hormone. (Table 50.2)

- In mammals, the neuron terminals in the posterior pituitary gland secrete one of two hormones, oxytocin or antidiuretic hormone (ADH).

50.3 Control of Metabolism and Energy Balance by Hormones

- Hormones are partly responsible for regulating energy utilization by cells, that is, for modulating appetite, digestion, absorption of nutrients, and blood levels of energy sources such as glucose. Although many hormones are involved in these processes, two from the thyroid gland (thyroxine and triiodothyronine), two from the pancreas (insulin and glucagon), and one from adipose tissue (leptin) play especially important roles.

- A major action of thyroid hormones in adult animals is to stimulate energy consumption by many different cell types. (Figures 50.5, 50.6)

- The endocrine pancreas produces the peptide hormones insulin and glucagon, which have opposite effects on blood glucose levels. The adrenal glands produce steroid hormones known as glucocorticoids, which increase blood glucose levels. (Figure 50.7)

- Maintaining normal glucose and other nutrient levels in the blood is a vital process that keeps cells functioning optimally. The combined short-term and long-term actions of glucagon, epinephrine, and cortisol help maintain normal blood glucose levels during fasting. (Figure 50.8)

- Groundbreaking research by Banting, Best, MacLeod, and Collip isolated insulin for therapeutic use in treating diabetes mellitus. (Figure 50.9)

- Adipose tissue is an important source of appetite-regulating hormones, including the protein leptin, which acts on the hypothalamus to inhibit appetite. Adipose cells release leptin into the blood in direct proportion to the amount of adipose tissue in the body.

50.4 Hormonal Control of Mineral Balance

- Because of the important roles that calcium plays in neuronal transmission, heart function, muscle contraction, and numerous other events, the concentration of Ca^{2+} in the blood is among the most tightly regulated variables in an animal's body. Vitamin D and parathyroid hormone regulate blood concentrations of Ca^{2+}. (Figures 50.10, 50.11)

- Sodium and potassium play crucial roles in membrane potential formation, action potential generation, and other functions. A key mechanism that regulates blood concentrations of these ions is to alter the rate of sodium, potassium, and water reabsorption from the urine as it is being formed in the kidneys. This is accomplished in large part by the actions of ADH, aldosterone, and atrial natriuretic peptide (ANP). (Figure 50.12)

50.5 Hormonal Control of Growth and Differentiation

- Hormones play a crucial role in regulating growth and differentiation. In vertebrates, normal growth depends on a balance between growth hormone, insulin-like growth factor-1 (IGF-1), and gonadal hormones. Thyroid hormones affect differentiation among all vertebrates. (Figures 50.13, 50.14)

- In insects, brain hormone, ecdysone, and juvenile hormone control the growth of larvae and their differentiation into pupae. (Figure 50.15)

50.6 Hormonal Control of Reproduction

- Hormones produced by the gonads play vital roles in nearly all aspects of reproduction. The ability of the testes and ovaries to produce the sex steroids depends on the presence of the gonadotropins, which are the same in both sexes and include luteinizing hormone (LH) and follicle-stimulating hormone (FSH).

- Leptin is a hormone produced by adipose cells that is required for fertility in mammals.

50.7 Hormonal Responses to Stress

- Regardless of the nature of the stress, animals' bodies respond to stress in similar ways. In vertebrates, hormones produced by the adrenal glands are the common denominator of the response to stress. Norepinephrine and epinephrine are responsible for most fight-or-flight reactions. (Figure 50.16, Table 50.3)

- Glucocorticoids promote the breakdown of storage compounds to provide energy for the body's cells during stress. (Table 50.4)

50.8 Impact on Public Health

- Endocrine disruptors, such as chemicals derived from industrial waste, may bind to estrogen receptors in animals' bodies. They may exert estrogen-like actions or inhibit the actions of the body's own estrogen.

- Androgen misuse can disrupt normal hormone levels and cause health risks such as aggression, cardiovascular disease, skin problems, cancer, and, in women, masculinizing traits. Blood-doping with erythropoietin can make blood dangerously thick.
- Hormones are also used therapeutically to treat a variety of human disorders.

TEST YOURSELF

1. Which is the defining feature of hormones?
   a. They are only produced in endocrine glands.
   b. They are secreted by one type of cell into the blood, where they may simultaneously reach many distant target cells and thereby alter cell function throughout the body.
   c. They are only released by neurons.
   d. They are never released by neurons.
   e. They are secreted into ducts, where they are carried to another nearby gland or other structure.

2. Steroid hormones are derived from
   a. proteins and bind membrane receptors.
   b. cholesterol and bind membrane receptors.
   c. tyrosine and bind intracellular receptors.
   d. proteins and bind intracellular receptors.
   e. cholesterol and bind intracellular receptors.

3. Protein hormones are unable to diffuse across the membrane due to
   a. their large size.
   b. their hydrophilic nature.
   c. their electric charge.
   d. all of the above.
   e. b and c only.

4. Which is *not* correct about hormone control?
   a. Many are regulated by negative feedback.
   b. They may be controlled by changes in blood levels of certain nutrients like glucose.
   c. They may be controlled by changes in blood levels of certain minerals.
   d. Their blood levels are controlled by their rates of synthesis and degradation.
   e. The rate of degradation of a hormone may go up or down, but synthesis is always constant.

5. The hypothalamus and the pituitary gland are physically connected by
   a. arteries.
   b. the infundibular stalk and portal vein system.
   c. the adrenal medulla.
   d. the spinal cord.
   e. the intermediate lobe.

6. The function of antidiuretic hormone is to
   a. increase water reabsorption at the kidneys.
   b. regulate blood pressure by constricting arterioles.
   c. decrease the volume of urine produced by the kidneys.
   d. increase blood pressure during times of high blood loss.
   e. all of the above.

7. Which of the following pairs of hormones are involved in the regulation of blood calcium levels in vertebrates?
   a. aldosterone and ANP
   b. insulin and glucagon
   c. parathyroid hormone and 1,25-dihydroxyvitamin D
   d. prolactin and oxytocin
   e. thyroxine and TSH

8. In invertebrates, growth of larvae is stimulated by
   a. growth hormone.
   b. cortisol.
   c. juvenile hormone.
   d. ecdysone.
   e. aldosterone.

9. Which of the following is *not* true of glucocorticoids?
   a. They stimulate maturation of the fetal lungs.
   b. They promote a decrease in blood glucose levels.
   c. They inhibit the sensitivity of cells to insulin.
   d. They are lipophilic.
   e. They reduce inflammation.

10. Endocrine disruptors are
    a. chemicals released by the nervous system to override the endocrine system.
    b. chemicals released by the male of a species to decrease the fertility of other males.
    c. drugs used to treat overactive endocrine structures.
    d. chemicals derived from industrial waste that may alter endocrine function.
    e. all of the above.

CONCEPTUAL QUESTIONS

1. List the hormones of the posterior pituitary gland and explain the function of each.
2. Explain the functions of insulin and glucagon.
3. Distinguish between type 1 diabetes mellitus and type 2 diabetes mellitus.

EXPERIMENTAL QUESTIONS

1. How did Banting and Best propose to obtain the glucose-lowering factor produced by the pancreas?
2. What was Collip's contribution to the isolation of insulin?
3. What subsequent innovations led to large-scale production of insulin for human patients?

COLLABORATIVE QUESTIONS

1. Discuss how hormones help control invertebrate development.
2. Discuss the role of steroid hormones and where they are produced in the vertebrate body.

www.brookerbiology.com
This website includes answers to the Biological Inquiry questions found in the figure legends and all end-of-chapter questions.

51

ANIMAL REPRODUCTION

Most animals reproduce by sexual reproduction. These mud daubers, shown here mating, reproduce by sexual reproduction, which favors genetic variation in a species.

When we examine all forms of life, we discover that individuals of every species are part of an unbroken cycle of life and death. Perpetuation of all life requires **reproduction**—the processes by which organisms replicate themselves and multiply. Because reproduction is the only way in which a species can perpetuate itself, enormous evolutionary pressure has been exerted on the processes of reproduction, and demands made on organisms by their environments have influenced the means by which reproduction occurs.

The biological mechanisms that favor successful reproduction in the animal kingdom are extraordinarily diverse. Many of the observable differences in animal behavior and anatomy are the result of adaptations to environmental pressures that increase an animal's chances of reproducing. Both the behaviors and the anatomical specializations that promote reproduction are under the control of a variety of factors, particularly hormones.

In this chapter, we will initially focus on the diverse means of reproduction that occur throughout the animal kingdom, including asexual and sexual reproduction. Later, we will highlight some of the anatomical, hormonal, and behavioral aspects of reproduction in mammals, including humans. We will conclude with a discussion of some key issues related to fertility (the ability to reproduce) in the human population today. We begin by examining how some animals reproduce without combining genetic material from two different sexes.

51.1 Asexual Reproduction

Asexual reproduction occurs when offspring are produced from a single parent, without the fusion of gametes from two parents. The offspring are therefore clones of the parent. Although many kinds of animals use this type of reproductive strategy, it is more prevalent in species that live in very stable environments, with little selection pressure for genetic diversity in a population.

Asexual reproduction has numerous advantages. First, an animal can reproduce asexually even if it is isolated from others of its own species, either because the animal is sessile (nonmotile) or rarely encounters another member of its species. Another advantage is that individuals can reproduce rapidly because they need not seek out, attract, and mate with the opposite sex. In addition, all asexual individuals can produce offspring, whereas in sexual reproduction involving separate sexes, two parents are needed to produce offspring. Asexual reproduction, therefore, can be a very efficient way of generating large numbers of offspring.

Animals use one of three major forms of asexual reproduction: budding, regeneration, or parthenogenesis. **Budding**, which is seen in cnidarians, occurs when a portion of the parent organism pinches off to form a complete new individual. In this process, cells from the parent undergo mitosis and differentiate into specific types of structures before the new individual breaks away from the parent. At any one time, a parent organism may have one, two, or multiple buds forming simultaneously (**Figure 51.1a**). Each new animal is fully formed but continues to grow. Budding continues throughout an animal's lifetime.

Some animals, including certain species of sponges, echinoderms, and worms, reproduce by the **regeneration** of a complete organism from small fragments of their body. In the sea star, for example, an arm removed by injury or predation can result in regrowth of a new individual (**Figure 51.1b**). Similarly, two new individuals regenerate from a flatworm bisected into two pieces.

(a) Budding

(b) Regeneration

Figure 51.1   Asexual reproduction.   Illustrated here are (a) budding, in a hydra, and **(b)** regeneration, shown in a sea star growing a complete new body from a single arm.

As in budding, the ability to regenerate a new individual from a body part, or to regenerate a severed body part (as also occurs in many echinoderms and cnidarians), continues throughout an animal's lifetime. This suggests that the molecular mechanisms responsible for differentiation into a complete animal remain active over the course of such animals' lives. Many of the genes responsible for these events have been found to be homologous to genes expressed during the early developmental stages of vertebrates. Why these genes are expressed only during a brief window of time in most vertebrates, but remain active throughout life in many invertebrates, remains a key question in developmental animal biology (see Chapter 52).

The asexual process called **parthenogenesis** (from the Greek, meaning "virgin birth") is the development of offspring from an unfertilized egg. It occurs in several invertebrate classes and in a few species of fish and reptiles. Animals produced by parthenogenesis are usually haploid. Some species—such as rotifers, many social insects, and the freshwater crustacean *Daphnia*—may reproduce either parthenogenically or sexually, depending on the time of year or environmental conditions. In honeybees, ants, and wasps, for instance, haploid males (drones) are produced by parthenogenesis, and diploid females (workers and queens) by sexual reproduction. Production of drones, whose major function is to mate with a queen, usually occurs in late spring. The queen stores sperm cells from several drones for up to two years, during which time she may lay hundreds of thousands of eggs, only some of which will be fertilized. As stored sperm runs out, the unfertilized eggs become new drones.

Parthenogenesis is of current interest to scientists who are using stem cells to develop new ways to treat degenerative diseases such as Parkinson's disease (see Chapter 44). Under specific conditions, certain chemical or mechanical stimuli can induce human eggs to undergo parthenogenesis. The cell divisions resulting from this type of parthenogenesis produce stem cells with the potential to develop into tissues that potentially could be used to replace damaged or diseased tissues. Such parthenogenically produced human cells cannot develop into viable embryos.

51.2 Sexual Reproduction

Sexual reproduction is the production of a new individual by the joining of two haploid **gametes**. The gametes are spermatozoa (usually shortened to **sperm**) from the male, and **eggs** or **ova** (singular, **ovum**) from the female. When a sperm unites with an egg—a process called **fertilization**—each haploid gamete contributes its set of chromosomes to produce a diploid cell called a **zygote**. As the zygote undergoes cell divisions and begins to develop, it is called an **embryo**. In this section, we will consider the advantages of sexual reproduction, how gametes are formed, and how two gametes join to form a new organism.

Sexual Reproduction Is Evolutionarily More Adaptive Than Asexual Reproduction

Whereas asexual reproduction provides a relatively straightforward way for an organism to produce many copies of itself, sexual reproduction requires the joining of gametes from two individuals to produce offspring. From an evolutionary perspective, it would seem to be more efficient to reproduce asexually than sexually, because more offspring can usually be produced per individual and reproduction can occur with only one individual. Yet given that most species reproduce sexually, a question has intrigued biologists since the time of Darwin: What is the advantage of sexual reproduction for the perpetuation of a species?

In the context of species survival, the major difference between asexual and sexual reproduction is that sexual reproduction allows for greater genetic variation due to genetic recombination. Only certain alleles from each parent are passed on, and when genes from one parent mix with a different set from the other parent, the offspring are never exactly like either of their parents. Thus a hallmark of sexual reproduction is increased genetic variation between successive generations. One prevalent hypothesis about why sex evolved is that sexual reproduction allows more rapid adaptation to environmental changes than asexual reproduction. Alleles that confer a selective advantage arise independently in members of a population. Sexual reproduction provides the opportunity for two or more beneficial alleles to be inherited by the same offspring. Also, as described next, sexual reproduction may make it easier for populations to eliminate harmful alleles.

FEATURE INVESTIGATION

Paland and Lynch Provided Evidence That Sexual Reproduction May Promote the Elimination of Harmful Mutations in Populations

Compared with asexual reproduction, sexual reproduction expends a lot more energy. Two types of gametes (sperm and eggs) must be made, and males and females need specialized body parts to mate with each other. Evolutionary biologists have speculated that these costs of sexual reproduction must somehow be outweighed by corresponding benefits.

In particular, sexual reproduction allows a species to redistribute alleles via crossing over and independent assortment across many generations. As a result, some offspring carry combinations of alleles that promote survival and reproduction, while other offspring may carry less favorable combinations. As described in Chapter 24, natural selection can favor those combinations that promote greater reproductive success while eliminating offspring with lower fitness. By comparison, asexual organisms cannot reassort their alleles from generation to generation. As a result, it is more difficult to accumulate beneficial alleles within individuals, and to eliminate harmful alleles from generation to generation. Evolutionary biologists have suggested that the inability of asexual species to reassort alleles may be a key disadvantage compared to sexually reproducing species.

To investigate this question, Susanne Paland and Michael Lynch studied the persistence of mutations in populations of *Daphnia pulex*, a freshwater organism known as the water flea. The researchers chose this organism because some natural populations reproduce asexually while others reproduce sexually.

In their experiment, shown in the Feature Investigation in **Figure 51.2**, Paland and Lynch studied the sequences of several mitochondrial genes in 14 sexually reproducing and 14 asexually

Figure 51.2   Paland and Lynch demonstrated the importance of sexual reproduction in reducing the frequency of maladaptive genetic mutations.

HYPOTHESIS   Sexual reproduction allows for greater mixing of alleles of different genes, and thereby may prevent the accumulation of detrimental alleles in a population.

STARTING MATERIALS   The researchers collected samples of *Daphnia pulex* from many natural populations. A total of 14 sexual strains and 14 asexual strains were studied.

Experimental level   **Conceptual level**

1   Isolate mitochondrial DNA from members of 28 strains of *D. pulex*. This involves breaking open cells, and extracting the DNA (refer back to Figure 21.1).

Daphnia   DNA   Segments of mitochondrial DNA

2   Amplify regions of mitochondrial genes using PCR. Subject the regions to DNA sequencing. The techniques of PCR and DNA sequencing are described in Chapter 20.

PCR (refer back to Figure 20.8)   DNA sequencing (refer back to Figure 20.10)

3   Using computer technology, align the sequences (refer back to Figure 21.12a) and determine the number of DNA changes that would cause amino acid substitutions. These amino acid changes were categorized as those that would be highly deleterious, moderately deleterious, mildly deleterious, or neutral for protein function. Compare the sexual and asexual strains for the persistence of these types of changes.

GGCACCTGACCC
GGCACCTAACCC
Stop codon

This change would be highly detrimental because it would put a stop codon into the gene.

4   THE DATA

Results from step 3:

Types of amino acid substitutions. The amino acid substitutions were due to rare mutations that occurred in the natural populations of *D. pulex*.	% of total amino acid substitutions	Allowed to persist in:	
		Sexual populations	Asexual populations
Highly deleterious	73.2	No	No
Moderately deleterious	13.3	No	Yes
Mildly deleterious	4.4	Yes	Yes
Neutral	9.1	Yes	Yes

reproducing populations of *D. pulex*. The researchers hypothesized that asexual species would have more trouble eliminating harmful mutations. As discussed in Chapter 24, random mutations that change the amino acid sequence of the encoded protein are much more likely to be harmful than beneficial. Sexually reproducing species can reassort their alleles from generation to generation, thereby producing succeeding generations in which the detrimental alleles are lost from the population. As seen in the data, the researchers discovered that both the sexual and asexual strains could eliminate highly deleterious muta-

tions from their populations. Organisms harboring such mutations probably died rather easily. In addition, the sexual and asexual strains both retained mildly deleterious and neutral mutations. However, moderately deleterious mutations were eliminated from the sexual strains but not from the asexual strains. One interpretation of these data is that sexual reproduction allowed for the reassortment of beneficial and detrimental alleles, making it easier for sexual strains to eliminate those mutations that are moderately detrimental.

Hermaphroditism Represents a Unique Form of Sexual Reproduction

In some species, individuals have both male and female reproductive systems. This is called **hermaphroditism** (after the male and female Greek gods, Hermes and Aphrodite). In some hermaphroditic species, individuals can fertilize their own eggs with their own sperm, but in most hermaphroditic species individuals exchange sperm with another individual. The latter situation has the selective advantage of creating additional genetic diversity in the population. Because sperm and egg join to produce offspring, hermaphroditism is a form of sexual reproduction, and it is the one case in which all individuals of a sexually reproducing species can produce offspring.

Hermaphrodites are often sessile animals or burrowing animals such as earthworms, which may live for long periods without encountering sexual partners. A single earthworm can fertilize its own eggs, or two earthworms can join together for several hours during which sperm from each worm fertilize the eggs of the other. In the latter case, individual worms act both as females (receiving sperm to fertilize their eggs) and males (giving sperm to fertilize another worm's eggs), so that the offspring carry genes from both individuals. This type of hermaphroditism is known as synchronous hermaphroditism, because individuals are simultaneously both male and female.

Another type of hermaphroditism, called sequential hermaphroditism, involves sex reversal, in which a female may change

into a male or vice versa. In such animals, individuals of both sexes have reproductive structures capable of producing sperm or eggs, and the particular gamete an individual produces at any given time depends on which genes are being expressed. Sex reversal occurs in some species of animals with strong social hierarchies. In some reef-dwelling species of fish, for example, a single dominant male defends a harem of several females within a specific territory. When that dominant male dies, the largest of the females reverses sex and becomes a male. Thus, these fish are protogynous—that is, female first but capable of becoming males later during their life cycle. Oysters and clownfish are also sequential hermaphrodites, but they are protandrous; that is, they are males first and only later become females. The advantage of protandrous hermaphroditism is that males change into females when they are older and larger—when they are likely to be capable of producing a greater number of eggs.

Sperm and Ova Are Produced During the Process of Gametogenesis

Male and female gametes are formed within the **gonads**—the **testes** (singular, **testis**) in males and the **ovaries** (singular, **ovary**) in females. Some similarities are found in the ways gametes develop in the testes and the ovaries, as well as some differences.

Gametogenesis—the formation of gametes—begins with primordial cells called germ cells, which multiply by mitosis,

(a) Spermatogenesis

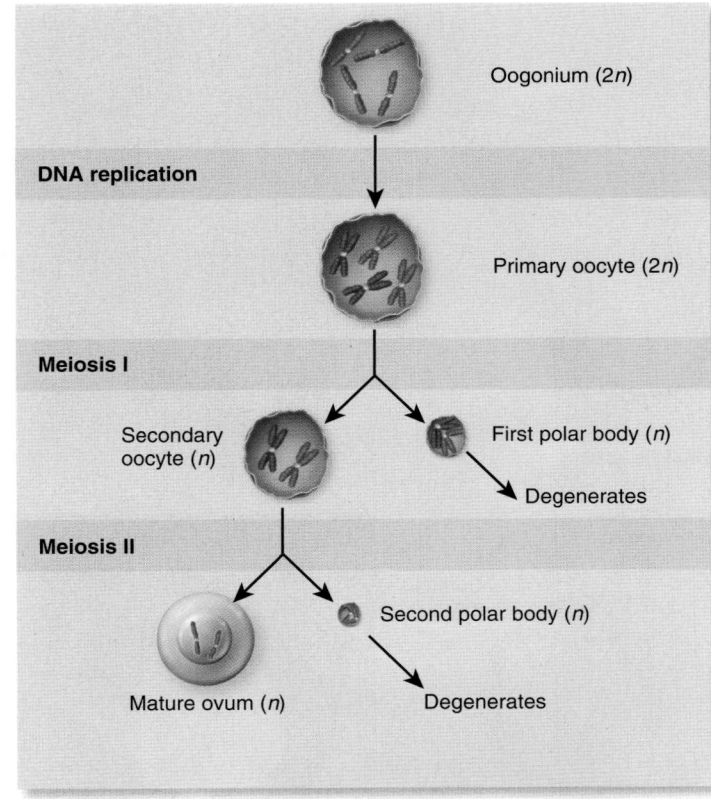

(b) Oogenesis

Figure 51.3   Gametogenesis in (a) males and (b) females.   Both male and female germ cells are diploid (2n) cells that undergo two meiotic divisions to produce mature haploid (n) gametes.

resulting in diploid cells (denoted as 2n) called **spermatogonia** (singular, **spermatogonium**) in males and **oogonia** (singular, **oogonium**) in females (**Figure 51.3**). Some of these cells become **primary spermatocytes** or **primary oocytes** that may begin the process of meiosis. Until this point, the development of sperm and eggs is similar. From then on, gametogenesis differs between the two types of gamete.

Spermatogenesis The formation of haploid sperm from the original diploid-germ cell is called **spermatogenesis**. As shown in Figure 51.3a, primary spermatocytes begin production of sperm by undergoing the first of two meiotic divisions (meiosis I). (For a review of the processes of mitosis and meiosis, see Chapter 15.) Meiosis I in the primary spermatocyte produces two haploid (n) cells called **secondary spermatocytes**. These cells also undergo meiosis (meiosis II), producing four haploid **spermatids** that eventually differentiate into sperm cells. Gametogenesis in males, therefore, results in four gametes from each spermatogonium.

The most striking change in each spermatid as it differentiates into a sperm is the formation of a flagellum or tail (**Figure 51.4a**), the movements of which require cellular energy and make the sperm motile. The chromosomes are in the head of the sperm, which is separated from the tail by a midpiece containing one or more mitochondria that produce the ATP required for

tail movements. At the tip of the head is a special structure called the **acrosome** that (as we will discuss shortly) contains proteolytic enzymes that help break down the protective outer layers that surround the ovum.

Oogenesis Whereas spermatogenesis produces four gametes from each primary spermatocyte, gametogenesis in the female, called **oogenesis**, results in the production of a single gamete from each primary oocyte (Figure 51.3b). Each meiotic division in oogenesis results in one large cell—either a **secondary oocyte** (meiosis I) or an ovum (meiosis II)—plus a smaller cell, called a polar body, that eventually degenerates. Only the larger of the two daughter cells resulting from meiosis II (that is, the ovum) contains the cellular machinery needed for development after fertilization.

Depending on the species, one or many ova can develop at a time. Within the ovaries, each ovum undergoes growth and development within a structure called a **follicle** before it leaves the ovary in the process of **ovulation**. Once ovulated, an egg cell can become fertilized if sperm are available. By the time ovulation occurs, the ovum is surrounded by two layers: an inner membrane-like structure called the zona pellucida, which surrounds the surface of the ovum, and an outer layer of cells called the cumulus mass (**Figure 51.4b**). These two layers play important roles in fertilization of the egg.

(a) Mature sperm

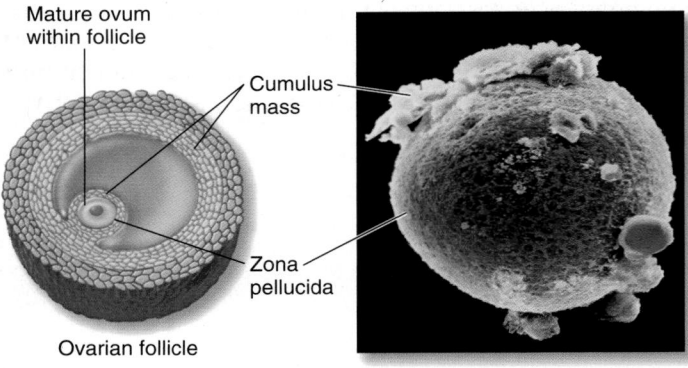

(b) Mature ovum

Figure 51.4 **The structure of gametes.** **(a)** Mature sperm have a characteristic head containing the nucleus and a structure called the acrosome, as seen in a drawing and the accompanying scanning electron micrograph (SEM) of a human sperm. The midpiece contains one or more mitochondria, which provide energy for movement of the tail. **(b)** Mature ovum. The drawing depicts a mature egg within its follicle; the SEM shows an isolated human egg covered by its zona pellucida and remnants of the cumulus mass.

In mammals, oogenesis begins in the female fetus before birth: Germ cells develop into primary oocytes and enter meiosis I, which is arrested partway through the process. Meiosis I does not resume until puberty, the time of a mammal's life when it first becomes capable of reproducing. In certain selected oocytes, meiosis I is completed, producing secondary oocytes. These cells then begin meiosis II and eventually develop into mature ova.

In a Given Species, Fertilization Occurs Either Inside or Outside the Female

For sperm to fertilize ova, the two gametes must physically come into contact. In some species, a sperm fertilizes the egg outside

Figure 51.5 **External fertilization.** The male frog (*Rana palustris*) clasps the female, which stimulates her to release her eggs. He then releases his sperm (not visible here) over the eggs. This process occurs in aquatic environments, which protect gametes from drying out.

Biological inquiry: What kinds of aquatic environments are most suitable for external fertilization?

of the female's body. This process, called **external fertilization**, occurs in aquatic environments, when eggs and sperm are released into the water in close enough proximity for fertilization to occur. The aqueous environment protects the gametes from drying out.

Animals that reproduce by external fertilization show species-specific behaviors that bring the eggs and sperm together. For instance, very soon after a female fish lays her eggs, a male deposits his sperm such that they spread over the clump of eggs. When frogs mate, the clasping behavior of the male stimulates the female to release her eggs into the water (**Figure 51.5**); the male then releases sperm onto the eggs. The fertilized ova then develop outside the parent's body.

Although the aqueous environment protects against desiccation, fertilized eggs can be eaten by predators, washed downstream by currents, or subjected to potentially lethal changes in water temperature. To counteract such problems, many aquatic or amphibious animals release very large numbers of eggs at once.

In contrast to external fertilization, terrestrial animals utilize **internal fertilization**, in which sperm are deposited within the reproductive tract of the female during the act called **copulation**, as seen in the mud daubers in the chapter-opening photo. Internal fertilization protects the delicate gametes from environmental hazards and predation, and also guarantees that sperm are placed and remain in very close proximity to ova. Once fertilization occurs within the female, the fertilized eggs then develop into offspring.

The behaviors and anatomical structures involved in achieving internal fertilization are extremely varied among species.

Typically, mating involves accessory sex organs, which are reproductive structures other than the gonads. The external accessory sex organs involved in copulation are the genitalia (for example, the **penis** and the **vagina**), which are used to physically join the male and female so that sperm can be deposited directly into the female's reproductive tract. A penis or analogous structure is present in most insects, reptiles, some species of birds (rattites), and all mammals. However, males of other vertebrate species that reproduce by internal fertilization lack a structure that can be inserted into the female, so they deposit sperm in the female by cloacal contact. The cloaca is a common opening for the reproductive, digestive, and excretory systems in these animals.

Another form of internal fertilization involves an indirect means of depositing sperm. In this circumstance, males produce small packets of sperm, called spermatophores, that are deposited externally and subsequently inserted into the female's reproductive tract by either the male or the female. During copulation in squid and octopus, the male uses a tentacle to transfer a spermatophore into the mantle cavity of the female. In spiders, the male places a droplet of sperm on a web and then uses a foreleg to insert the droplet into the female's reproductive tract. Subsequently the eggs are fertilized internally and deposited on the web or elsewhere in the environment.

Fertilization Involves Union of Sperm with Egg

Fertilization is a complex series of events by which the haploid male and female gametes unite and become a diploid zygote. Several important cellular and molecular processes must occur before the nuclei of the gametes can fuse. First, the egg and sperm must come into contact. The mechanism by which this occurs has been studied extensively in sea urchins, and some evidence suggests that a similar process occurs in humans. When chemical attractant molecules emitted by sea urchins eggs bind to nearby sperm, cellular respiration (that is, the breakdown of nutrients to synthesize ATP) within the sperm increases, which helps increase sperm motility. The sperm then swim toward the egg by following the attractant's concentration gradient. The attractants are species-specific; that is, sperm will only respond to the attractants produced by ova of their own species.

For a sperm to reach the ovum, it must first penetrate the cellular layers surrounding the plasma membrane of the egg. When the head of the sperm contacts the layers surrounding the ovum, chemicals in the cumulus mass stimulate the breakdown of the membrane covering the acrosome of the sperm. Proteolytic enzymes released from the acrosome digest a local area of the zona pellucida, allowing the sperm to contact the plasma membrane of the ovum. The plasma membrane of the ovum then fuses with the sperm head, facilitating its movement into the egg cell cytoplasm.

Once a sperm fuses with the egg, it is important that other sperm be prevented from also fusing with the egg, because in that case the zygote will be polyploid, a condition that few zygotes survive. In mammals, the penetration of one sperm induces metabolic changes within the egg that prevent additional sperm from penetrating the zona pellucida and entering the egg (see Chapter 52).

Fertilized Eggs Are Laid or Retained Within the Mother

As we have seen, internal fertilization occurs within the female, and the fertilized eggs then continue their development into offspring. When the embryo develops within the mother, the process is called **viviparity**. Alternatively, fertilized eggs can be laid in a well-protected site in the environment. Development of the embryo within an egg is called **oviparity**. Some animals develop by a process called **ovoviviparity**, which features aspects of both modes of development. In that case, eggs covered with a shell that is little more than a thin sheath hatch inside the mother's body, but the offspring receive no nourishment from the mother. Ovoviviparity occurs in sharks, lizards, some snakes, and invertebrates.

Oviparity is the rule in avian species and is also common in reptiles, fish, and amphibians. Among mammals, only the echidna and the platypus lay eggs. Although some animals lay only a single egg at one time, snakes and turtles lay between several and hundreds of eggs at a time, frogs and fish can lay thousands of eggs, and honeybee queens produce an astounding 250,000 eggs during a nearly continuous laying cycle over 1–2 years!

Although oviparity reduces the female's metabolic investment in the young, it increases the incidence of predation. In many amphibian and reptile species, the large number of eggs laid increases the chances of some young surviving even if predation occurs. The energetic cost to the female of producing numerous eggs is high, but usually little or no parental care is involved, freeing the parent to devote energy to other activities. In birds, however, incubation of the eggs after laying requires additional energy expenditure in the form of parental care, which typically continues after the eggs have hatched, and thus relatively few eggs are laid.

All eggs are single cells. The ostrich egg is the largest single cell among all living species. Eggs include the yolk (which contains nutrients for the developing embryo), a surrounding membrane, and the cytoplasm. The nucleus of the egg is located within the yolk, usually close to the surface at one end. In general, the larger the egg, the longer it takes the young to develop, and more yolk is available to provide nutrition.

Terrestrial animals lay eggs with shells that protect the future offspring from desiccation. Eggs may have leathery shells, as in snakes, lizards, some turtles, and insects, or hard shells made from calcium carbonate, as in birds and other turtles (**Figure 51.6**). These eggs contain all the nutrients necessary for the development of the young. Oxygen enters, and carbon dioxide exits, through tiny pores in the shell. The eggs of aquatic species lack shells so that external fertilization may occur. Such eggs are protected from drying out by their environment.

Figure 51.6 **Oviparity in terrestrial animals.** Terrestrial animals that lay eggs produce a wide range of egg types and sizes. **(a)** Snakes produce multiple eggs with leathery shells; such eggs can lose water and thus must be protected from desiccation. **(b)** Ostriches produce hard-shelled eggs that weigh up to 1.5 kg, including the shell.

(a) Eggs with leathery shells   **(b) Eggs with hard shells**

51.3 Mammalian Reproductive Structure and Function

We turn now to a detailed look at the mammalian reproductive system, including the processes by which gametes develop, and the establishment and maintenance of pregnancy. For both sexes, we will begin with a description of the anatomy of the reproductive system, including the gonads and the accessory sex structures. We will then examine the hormones that control the production of the gametes and the sexual behaviors that are important for internal fertilization.

The Male Reproductive Tract Is Specialized for Production and Ejaculation of Sperm

The external structures of the male reproductive tract—the genitalia—consist of the penis and the scrotum, the sac that contains the testes and holds them outside the body cavity (**Figure 51.7a**). The testes develop within the body cavity, and just before birth in human males they descend into the scrotum, where the temperature is approximately 2°C lower than core body temperature. The lower temperature is optimal for spermatogenesis.

Each testis is composed of tightly packed **seminiferous tubules** encased in connective tissue and surrounded by scattered Leydig cells—glandular cells that lie between the tubules and secrete the steroid hormone testosterone (**Figure 51.7b**). Spermatogenesis takes place within the walls of the seminiferous tubules. Cells at the earliest stages of spermatogenesis, the spermatogonia, are located nearest the wall. Cells of more advanced stages are located progressively inward, such that the sperm are released into the tubule lumen. Cells within the seminiferous tubules are continuously developing from spermatogonia into spermatocytes and eventually to sperm, so at any one time, all types of cells are present along the seminiferous tubule. Support cells, called Sertoli cells, surround the developing spermatogonia and spermatocytes, providing them with nutrients and protection and playing a role in their maturation into sperm.

Sperm moving out of the seminiferous tubules are emptied into the **epididymis**, a coiled, tubular structure located on the surface of the testis (Figure 51.7b). The epididymis is very long—approximately 6 m in humans—and it takes about 20 days for new sperm to reach its end. It is here that the sperm complete their differentiation by becoming motile and gaining the capacity to fertilize ova.

Sperm leave the epididymis through the **vas deferens**, a muscular tube leading to the **ejaculatory duct**, which then connects to the urethra (see Figure 51.7a). As noted in Chapter 49, the urethra originates at the bladder and extends to the end of the penis. In males, the urethra not only conducts urine but also carries **semen**, a mixture containing fluid and sperm, during **ejaculation**—the movement of semen through the urethra by contraction of muscles at the base of the penis. These contractions during ejaculation induce the pleasurable sensation of orgasm.

The liquid components of semen are important for the survival and movement of sperm through the female reproductive tract. This liquid is formed by three paired accessory glands that secrete substances into the urethra to mix with the sperm. The **seminal vesicles** secrete fructose, the main nutrient for sperm. The **bulbourethral glands** secrete an alkaline mucus that protects sperm by neutralizing the acidity in the urethra. The **prostate gland** secretes a thin fluid that protects sperm once they are deposited within the female reproductive tract. Secretions constitute about 95% of semen, while sperm make up only about 5% of total semen volume. The volume of semen released at ejaculation in humans is about 2–5 ml and contains 20–130 million sperm per milliliter. Although this seems like a huge excess of sperm to fertilize a single ovum, relatively few sperm actually reach the egg.

Introduction of sperm into the female reproductive system during copulation is made possible by erection of the penis. Erection occurs when blood fills spongy erectile tissue located along the length of the penis (see Figure 51.7a). In a number of mammals (for example, whales), the firmness of the erection is aided by the presence of a bone in the penis called a baculum.

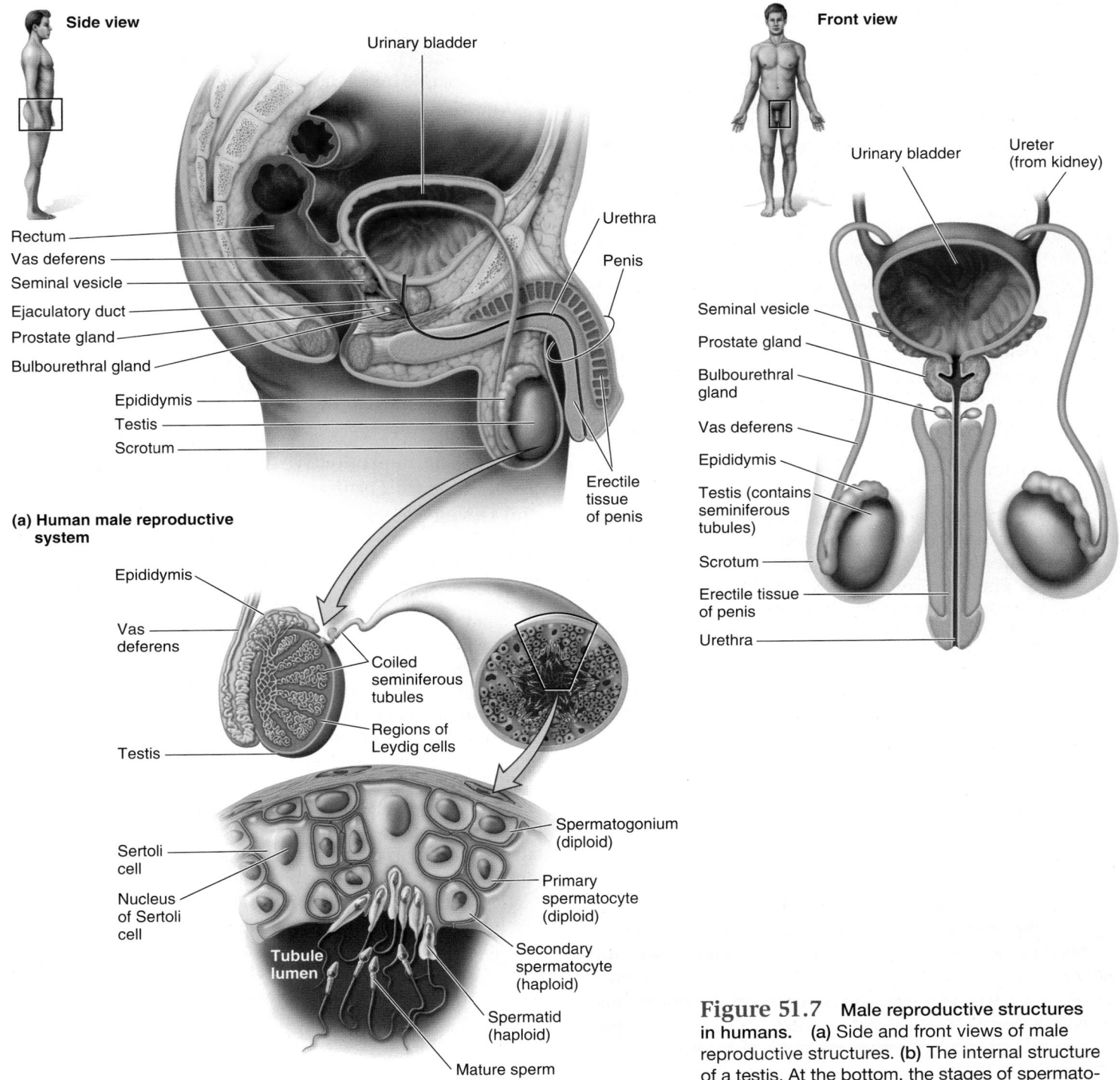

Side view

Urinary bladder

Urethra

Penis

Rectum

Vas deferens

Seminal vesicle

Ejaculatory duct

Prostate gland

Bulbourethral gland

Epididymis

Testis

Scrotum

Erectile tissue of penis

(a) Human male reproductive system

Front view

Urinary bladder

Ureter (from kidney)

Seminal vesicle

Prostate gland

Bulbourethral gland

Vas deferens

Epididymis

Testis (contains seminiferous tubules)

Scrotum

Erectile tissue of penis

Urethra

Epididymis

Vas deferens

Coiled seminiferous tubules

Regions of Leydig cells

Testis

Sertoli cell

Nucleus of Sertoli cell

Tubule lumen

Spermatogonium (diploid)

Primary spermatocyte (diploid)

Secondary spermatocyte (haploid)

Spermatid (haploid)

Mature sperm (haploid)

(b) Sperm formation

Figure 51.7 Male reproductive structures in humans. (a) Side and front views of male reproductive structures. (b) The internal structure of a testis. At the bottom, the stages of spermatogenesis within the seminiferous tubule are shown.

Sexual arousal stimulates release of the gaseous neurotransmitter nitric oxide (NO) in the penis, causing vasodilation of arteries. The pressure of the blood flowing into the penis constricts nearby veins, causing a reduction in venous drainage from the penis, engorging it with blood. After ejaculation, NO release is reduced, causing a reversal of the vascular changes responsible for erection. Both physiological and psychological factors can result in an inability to achieve an erection, a condition known as erectile dysfunction (also called impotence). Orally administered drugs are available that increase the occurrence of erections by stimulating the same intracellular signaling events as NO in the smooth muscle cells of penile blood vessels.

Male Reproductive Function Requires the Actions of Testosterone

The role of hormones in male reproductive function is diagrammed in **Figure 51.8**. The hypothalamus produces gonadotropin-releasing hormone (GnRH), which stimulates the anterior pituitary gland to release two gonadotropic hormones: luteinizing hormone (LH) and follicle-stimulating hormone (FSH).

LH stimulates the Leydig cells of the testes to produce androgens such as testosterone. In turn, testosterone acts on Sertoli cells and germ cells to stimulate spermatogenesis. Testosterone also stimulates the growth of the male reproductive tract and the genitalia during development and puberty, and the development of male secondary sex characteristics. In humans, these include facial hair growth, increased muscle size, and deepening of the voice. In other male mammals, they include such things as growth of the horns of a bull, enlargement of the nose of proboscis monkeys, and formation of cheek pads in some apes. Testosterone-dependent secondary sex characteristics are not unique to mammals, however. One familiar example is the bright coloration of extensive plumage in male birds such as the peacock.

The other pituitary gonadotropin, FSH, functions along with testosterone to stimulate spermatogenesis. FSH does this by stimulating the activity of the Sertoli cells within the seminiferous tubules. The Sertoli cells provide the nutritional and structural support necessary for development of the sperm. They also respond to testosterone produced by the Leydig cells, by stimulating mitosis and meiosis of the germ cells associated with them in the tubules.

Figure 51.8   The endocrine control of spermatogenesis. In response to LH, Leydig cells secrete testosterone, which, along with FSH, act on Sertoli cells to facilitate spermatogenesis. Negative signs indicate inhibitory effects via negative feedback.

Biological inquiry: What would happen to spermatogenesis in a man taking testosterone to increase muscle mass or improve athletic performance?

Production of sperm and testosterone is kept in check by negative feedback mechanisms that control the amount of gonadotropins produced (Figure 51.8). Testosterone inhibits the secretion of GnRH from the hypothalamus, so that both LH and FSH are inhibited when blood levels of testosterone are high. Testosterone also directly inhibits LH secretion by the pituitary gland. These feedback mechanisms maintain homeostatic levels of FSH and LH in the blood.

Before puberty, LH is not released in sufficient amounts to stimulate testicular production of testosterone, and the reproductive system is quiescent. Although the mechanisms that initiate puberty in mammals are still not completely understood, we know that increased GnRH production at that time initiates increased LH and FSH secretion from the pituitary. The testosterone induced by LH stimulates development of the secondary sexual characteristics, development of the internal reproductive organs, and stimulation of spermatogenesis. Testosterone is also responsible for an increased sex drive ("libido") at this time.

The Female Reproductive Tract Is Specialized for Fertilization of the Egg and Development of the Embryo

The female genitalia differentiate from the same embryonic tissues as the male genitalia. The female genitalia are composed of the large outer folds called the **labia majora** (major lips), which surround the external opening of the reproductive tract, plus the smaller, inner folds, called the **labia minora** (minor lips; **Figure 51.9a**). The labia majora originate from the same embryonic tissue as the scrotum of the male, whereas the labia minora originate from urethral primordial tissue. At the anterior part of the labia minora is the **clitoris**, which is erectile tissue of the same origin as the penis. Like the penis, the clitoris becomes engorged with blood during sexual arousal (as are the labia minora) and is very sensitive to sexual stimulation. Unlike males, however, the openings of the reproductive tract and the urethra are separate in females. The opening of the urethra is beneath the clitoris and just anterior to the opening of the reproductive tract.

In mammals alone, the external opening of the reproductive tract leads to the **vagina**, a tubular, smooth muscle structure into which sperm are deposited. At the end of the vagina is a fibrous structure called the **cervix**, which forms the opening to the large, conical-shaped **uterus**. Sperm pass through the cervix into the uterus, which is specialized for carrying the developing fetus. It consists of an inner lining of glandular and secretory cells called the endometrium, and a thick muscular layer called the myometrium. We will discuss the functions of the uterus later in the chapter.

Eggs develop within one of the two bilateral ovaries (Figure 51.9a), which are suspended within the abdominal cavity by connective tissue. In humans, each ovary is approximately the size of a ping-pong ball. Typically, one egg leaves the ovary and briefly enters the abdominal cavity. From there, the egg is quickly drawn into a thin tube, the **oviduct** (also called the fallopian tube), by the actions of undulating fimbriae (fingers) of the oviduct that extend out to the ovary.

The egg is moved down the length of the oviduct by cilia on the oviduct's inner surface. Fertilization of the ovum by a sperm typically occurs within the oviduct. Upon fertilization, the ovum completes meiosis II, and the union of sperm and egg creates a zygote. The zygote undergoes several cell divisions to become a **blastocyst**, a ball of approximately 32 to 150 cells that enters the uterus, where it will develop into an embryo (see Chapter 52).

Gametogenesis in Females Is a Cyclical Process Within the Ovaries

The process of oogenesis occurs within the ovaries (**Figure 51.9b**). In contrast to spermatogenesis, which continues throughout life in the testes in males, most female mammals appear to be born with all the primary oocytes they will ever have, although recent evidence in mice suggests that new oocytes may form later in life.

(a) **Human female reproductive system**

(b) **Ovarian structures**

1. Primary oocytes begin to mature within growing follicles.
2. A primary oocyte completes meiosis I to become a secondary oocyte.
3. The secondary oocyte begins meiosis II and matures into an ovum, which is released during ovulation when the follicle ruptures.
4. Corpus luteum develops from remnants of follicle.
5. Corpus luteum degenerates if fertilization of egg does not occur.

Figure 51.9   **Female reproductive structure and function in humans.**   (a) Side and front views of the female reproductive system. An ovum moves from the ovary into the oviduct (also called the Fallopian tube), where it is fertilized. Subsequently the blastocyst enters the uterus, where it will implant in the endometrium, the inner lining of the uterus. (b) Development of an ovum within the ovary, illustrating the events during a single ovarian cycle.

At birth, each ovary in a human female has about 1 million primary oocytes, which are arrested in prophase of meiosis I. Most of these degenerate before the onset of puberty, when each ovary contains about 200,000 primary oocytes. Other than this degeneration, the ovaries are quiescent until puberty, at which time they begin to show cyclical activity.

The process of producing ova occurs as a cycle that may be as brief as a few days in small rodents, or as lengthy as 15–16 weeks in elephants. In humans, a typical cycle lasts approximately 28 days, during which time several oocytes in each ovary begin to mature. However, all but one of these oocytes usually degenerates, and only a single ovum fully matures and gets released (ovulated) from the ovary each cycle. Thus, only 300 to 500 eggs are ovulated over a woman's 30- to 40-year reproductive lifetime.

Ovulation and degeneration of additional primary oocytes continues throughout adulthood. Eventually, the oocytes become nearly depleted, and a woman stops having ovarian cycles, an event called **menopause**. The average onset of menopause in the U.S. is approximately 51 years of age. After that time, a woman is no longer capable of becoming pregnant. Until 1900, the average life expectancy of a woman in the U.S. was approximately 47 years, and thus menopause was once much less common than it is today. At one time it was thought that menopause was unique to humans, but it appears that other female mammals, if they survive long enough, also become incapable of ovulation at some point.

The **ovarian cycle** involves the development of an ovarian follicle, the release of an ovum, and the formation and regression of a corpus luteum (Figure 51.9b). During the first week of the ovarian cycle in humans, several primary oocytes begin to mature, each within a follicle. At the beginning of the second week, all but one of the growing follicles and its primary oocyte degenerate, and the single remaining follicle continues to develop and enlarge. During that time, the primary oocyte of that follicle completes meiosis I, becomes a secondary oocyte, and begins meiosis II. The developing secondary oocyte is surrounded by cells of the cumulus mass, which both protect and nurture it, and which secrete estradiol, the major form of estrogen in humans. The estradiol is secreted into the blood, where it functions to control the secretion of LH and FSH from the anterior pituitary gland. Some estradiol is also secreted into the follicle, where it stimulates fluid secretion into the inner core of the follicle, called the antrum. As the follicle grows in response to continued stimulation by LH and FSH, the fluid pressure inside the antrum increases, until the follicle begins to form a bulge. Eventually, ovulation occurs as the follicle ruptures and the egg, zona pellucida, and some surrounding supportive cells of the cumulus mass are released from the ovary.

Cells in the empty follicle subsequently proliferate and develop into a structure called the **corpus luteum**. In humans, the corpus luteum is active for approximately the second half of the ovarian cycle. It is responsible for secreting hormones that stimulate the development of the uterus needed for sustaining the embryo in the event of a pregnancy. If pregnancy does not occur, the corpus luteum degenerates, and subsequently

a new group of follicles with their primary oocytes undergoes development.

The Ovarian Cycle Results from Changes in Hormone Secretion

The major estrogen produced by the mammalian ovaries—estradiol—not only is required for development of the follicle during the ovarian cycle but also influences the secondary sexual characteristics of females. Secondary sexual characteristics include development of breasts, widening of the pelvis in preparation for giving birth, and a particular pattern of fat deposition.

We saw that in males, testosterone produced in the testes exerts a negative feedback on secretion of GnRH and LH. In females, however, the situation is more complicated. Although GnRH also stimulates release of LH and FSH, estradiol can have both negative and positive feedback effects on the gonadotropic hormones. To understand these events, let's examine hormone changes during the ovarian cycle (**Figure 51.10**).

Figure 51.10   **The ovarian and uterine cycles in a human female.**   The ovarian cycle is divided into the follicular and luteal phases, while the uterine cycle is divided into the menstruation and the proliferative and secretory phases.

The first half of the ovarian cycle is called the follicular phase of the cycle, because this is when the growth and differentiation of the follicle are occurring. The fairly low levels of LH that occur during follicular development stimulate the cells of the follicle to make estradiol. The estradiol that is produced is important for enlargement and growth of the oocyte, and it also is secreted into the blood, where it can influence the secretion of LH and FSH.

As the follicle develops, estradiol (and to a lesser extent, progesterone) production continues, and consequently levels of estradiol in the blood slowly but steadily increase. Initially, estradiol exerts a negative feedback action on LH and FSH secretion, preventing their blood levels from rising significantly until the follicle is nearly ready to ovulate. When the follicle is fully developed and ready for ovulation, its production of estradiol increases, such that the blood concentrations of estradiol increase sharply. At that time, the feedback action of estradiol on LH and FSH switches from negative to positive, by mechanisms that are still being investigated but involve increased GnRH secretion from the hypothalamus. This results in a sudden, sharp surge in gonadotropin levels in the blood, particularly LH.

The LH released from the pituitary as a result of positive feedback by estradiol induces rupture of the follicle and ovulation. This type of ovulation is known as spontaneous ovulation, because it happens regularly on a cyclical basis without requiring any external stimulus. Some mammals, including rabbits, cats, and camels, undergo ovarian cycles that turn off unless mating occurs. If mating occurs at a time in the ovarian cycle when eggs are ready, the mating act itself triggers ovulation. This mechanism, called induced ovulation, helps ensure that eggs are not wasted by being ovulated when the female has not mated with a male.

Ovulation marks the end of the follicular phase and the beginning of the luteal phase of the ovarian cycle, named after the corpus luteum. Estradiol production decreases, and the continuing high LH levels initiate development of the corpus luteum. The corpus luteum secretes progesterone, which is the dominant ovarian hormone of the luteal phase, plus some estradiol. Progesterone inhibits LH and FSH secretion, and it further prepares the uterus for receiving and nourishing the embryo. If fertilization of the ovum does not occur, the corpus luteum degenerates after 2 weeks, allowing LH and FSH to initiate development of a new set of oocytes. However, if fertilization does occur, the blastocyst develops a surrounding layer of cells that secrete an LH-like hormone, called **chorionic gonadotropin (CG)**, which maintains the corpus luteum.

Maternal Hormones Prepare the Uterus to Accept the Embryo

In humans, the ovarian cycle occurs in parallel with changes in the lining of the uterus called the **uterine cycle** or **menstrual cycle**. The hormones produced by the ovarian follicle influence the development of the endometrium, the glandular inner layer of the uterus. As depicted at the bottom left of Figure 51.10, a period of bleeding called **menstruation** (from the Latin *mensis*, month) marks the beginning of the uterine cycle and the follicular phase of the ovarian cycle. During menstruation, the well-developed uterine wall, including the blood vessels that grew during the previous cycle, is sloughed off and released from the body.

Menstruation is initiated when progesterone and estrogen levels decrease substantially as the corpus luteum degenerates near the end of a luteal phase in which pregnancy did not occur. Menstrual cycles are found in many primates, including humans. Other mammals also have uterine cycles, but without the bleeding associated with menstruation. These cycles are called estrous cycles and are usually associated with a period of sexual receptivity in females that is timed to coincide with the preovulatory period. Cyclical changes in female sexual receptivity may be present in some primates with menstrual cycles but does not occur in humans.

By about the end of the first week of the menstrual cycle in humans, the endometrium is ready to grow in response to newly increasing levels of estrogen secreted by a developing follicle. This phase of the menstrual cycle, which corresponds to the latter part of the ovarian follicular phase, is called the proliferative phase (Figure 51.10). During this time, the endometrium becomes thicker and more vascularized. During the subsequent luteal phase of the ovarian cycle, progesterone from the corpus luteum initiates further endometrial growth, including the development of glands that secrete nutritive substances that sustain the embryo during its first 2 weeks in the uterus. This part of the menstrual cycle is called the secretory phase. If fertilization does not occur, degeneration of the corpus luteum and the associated decrease in progesterone secretion initiates menstruation, and the beginning of the next uterine cycle. If fertilization does occur, however, the blastocyst becomes embedded in the endometrium and pregnancy begins, as described in the next section.

51.4 Pregnancy in Mammals

Pregnancy, or gestation, is the time during which a developing embryo and fetus grows within the uterus of the mother. Physiologically, pregnancy is said to begin not at fertilization but when the embryo is established in the uterine lining. This occurs within days of fertilization in animals with short gestation lengths but may take weeks in large animals with long gestations.

In mammals, gestation length varies widely and is roughly related to the size of adults in a particular species. Thus, small animals like hamsters and mice have gestation periods of 16–21 days, canines have longer pregnancies of about 60–65 days, humans average about 268 days, and the Asian elephant carries its fetus up to 660 days! The advantages of prolonging prenatal development are twofold: The embryo is protected while it is developing in the uterus, and the offspring can be more fully developed at birth. This is especially important for animals whose survival depends on mobility shortly after birth, such as horses and ruminants.

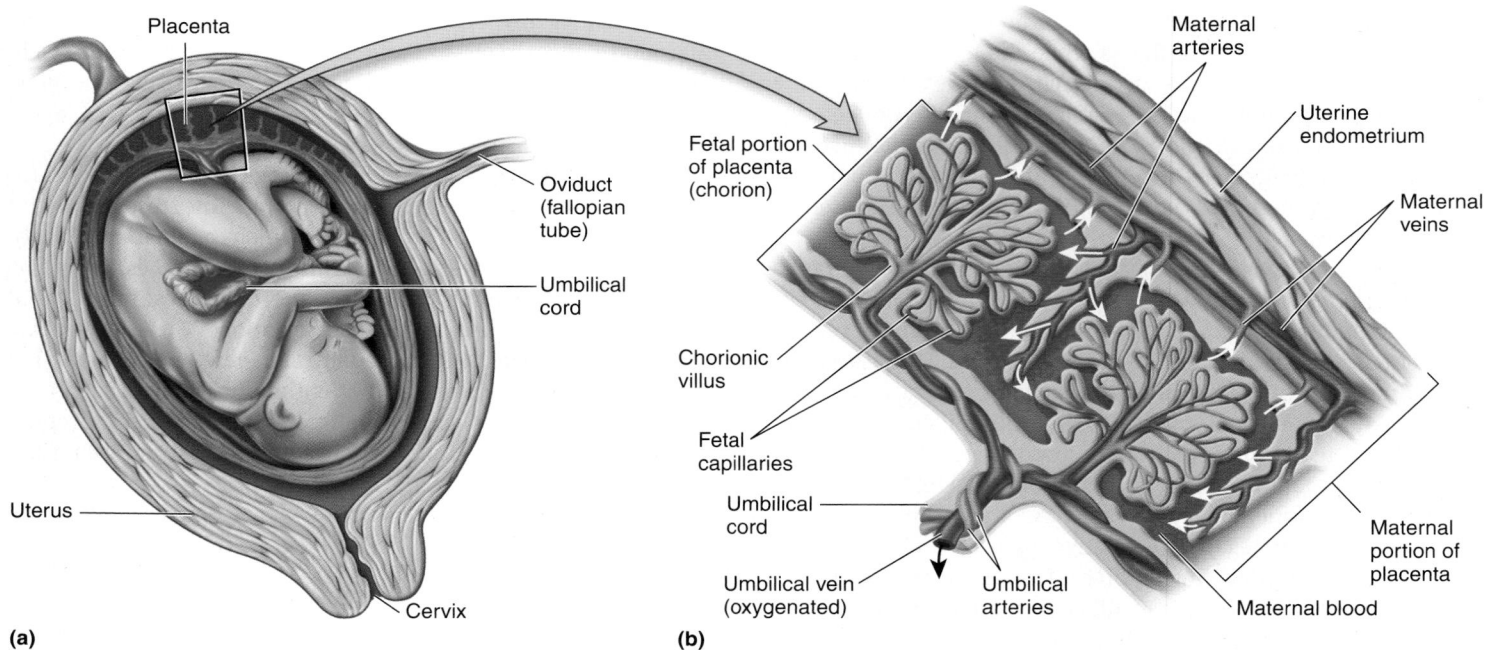

Figure 51.11   **The structure of the placenta.**   In all mammals, the placenta is composed of both fetal and maternal tissues. **(a)** Overview of placental structure in the human. **(b)** Enlarged view of the placenta showing the relationship between fetal and maternal structures. Note that in humans, blood in the fetal and maternal circulations does not mix.

Gestation length is influenced not only by adult body size but also by the number of offspring in a single pregnancy. Rats, for example, which bear up to 12 or so offspring per litter, have a short gestation period and produce young that are relatively undeveloped at birth and are totally dependent on the mother. In this section, we will examine how mammals have evolved to retain their young in the uterus for extended periods, the structure and function of the placenta, and the role of hormones in pregnancy.

Most Mammals Retain Their Young in the Uterus and Nourish Them via a Placenta

Three types of pregnancies are found in mammals. Monotremes, such as the platypus, are the only mammals that lay eggs. In marsupials such as the kangaroo, the young are born while still extremely undeveloped. They then crawl up the mother's abdomen to her pouch, where they attach to a nipple to suckle and obtain nourishment. They mature within the pouch. Compared to marsupials, humans and other eutherian mammals retain their young within the uterus a longer period of time, and nourish them via transfer of nutrients and gases through a structure called the **placenta**.

During pregnancy, many physiological changes occur in both the embryo and the mother. The first event of pregnancy in non–egg-laying mammals is **implantation**, when the blastocyst embeds within the uterine endometrium, which occurs in humans within 1–2 weeks, typically around 8–10 days after fertilization. Initially, the implanted blastocyst receives nutrients directly from the endometrial glands. However, shortly after implantation,

newly developing embryonic tissues merge with the endometrium to form the placenta (**Figure 51.11**), which remains in place and grows larger as the embryo matures into a **fetus**. (In humans, an embryo becomes a fetus after the eighth week of gestation.) The placenta, therefore, has a maternal portion and a fetal portion.

The placenta is rich in blood vessels from both the mother and fetus. The maternal and fetal sets of vessels lie in close proximity. The fetal portion of the placenta, called the chorion, contains convoluted structures called chorionic villi that provide a large surface area containing capillaries for exchange of nutrients, gases, and other solutes. Nutrients and oxygen from the mother are carried through maternal arteries, where her blood pools in large areas of the fetal placenta surrounding the fetal capillaries. Solutes diffuse from the maternal blood into fetal capillaries, and from there flow into the umbilical vein, a portion of the fetal circulation. In turn, carbon dioxide and other waste products from the fetus are carried through the umbilical artery to the placenta, where they diffuse into the mother's circulation, from which they can be excreted. Because of this placental organization, the blood of the mother and the fetus do not mix together.

Prenatal development in humans is generally divided into three trimesters, each of which lasts about three months (**Figure 51.12**). During the first two months of pregnancy, the organs of the embryo develop. At the end of the first trimester, the rudiments of the organs are present, and the developing fetus is about an inch long. The second trimester is an extremely rapid phase of growth. During the third trimester, the lungs of the fetus mature so that they are ready to function as gas-exchange organs, and the fetus positions itself with its head above the uterine cervix in preparation for birth.

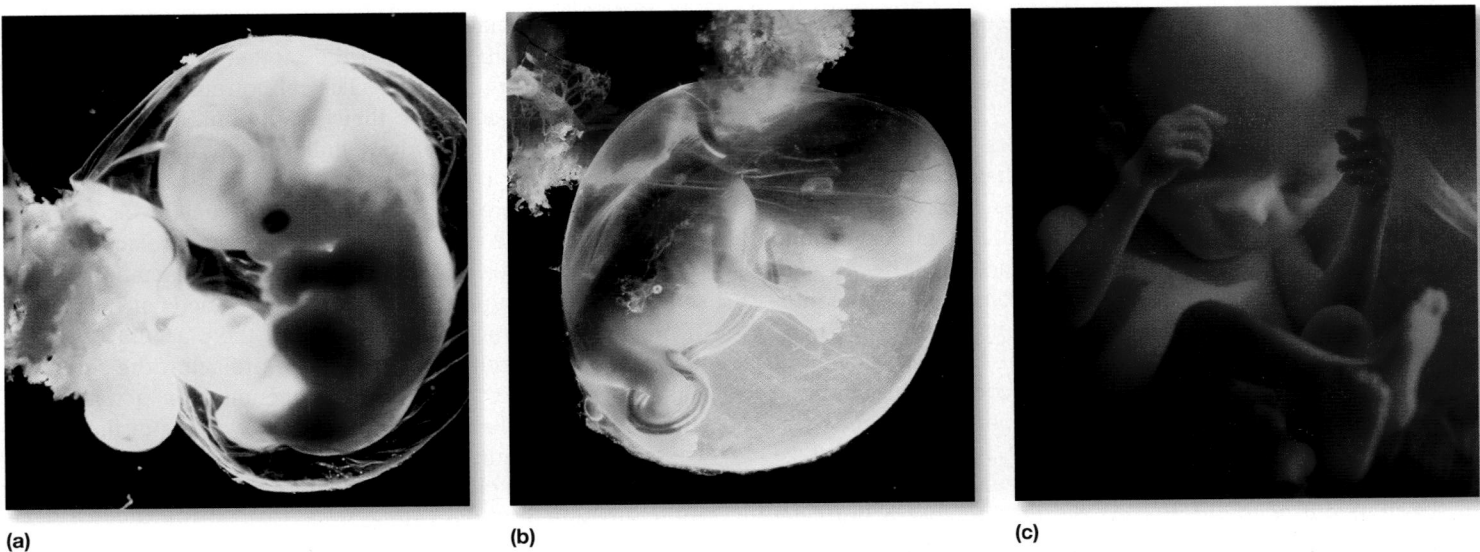

(a)   (b)   (c)

Figure 51.12   **Prenatal development in humans.**   The photos show the developing fetus **(a)** in the first trimester, **(b)** in the second trimester, and **(c)** at the start of the third trimester.

2 The push of the fetus against the cervix activates stretch-sensitive sensory neurons that send signals to the mother's hypothalamus.

(Sensory nerves to hypothalamus)

3 The hypothalamus stimulates secretion of oxytocin from the posterior pituitary gland.

1 Estradiol readies the uterus for a response to oxytocin.

⊕ Prostaglandins

Oxytocin

5 Prostaglandins secreted by the uterus also enhance contractions. Sensory input to the hypothalamus is further enhanced by contractions.

4 Oxytocin stimulates stronger uterine contractions.

Figure 51.13   **Hormone control of parturition.**   Birth relies on maternal hormones that act on the uterus, and neural signals from the uterus. In response to sensory neural input arising from the push of the fetus on the cervix, the maternal posterior pituitary gland releases oxytocin, which stimulates uterine smooth muscle contractions. The secretion of prostaglandins by the uterus also increases the strength of the contractions. Sensory receptors in the uterus detect the more forceful contractions and signal the mother's posterior pituitary gland to secrete more oxytocin, thus completing a positive feedback loop that further strengthens the contractions.

Birth Is Dependent on Maternal and Fetal Hormones

Birth—also called **parturition**—is initiated by the actions of several hormones and other factors secreted by the mother (**Figure 51.13**). In humans and many other mammals, the initial event of parturition is an increase in estradiol and oxytocin levels in the mother's blood. Recall from Chapter 50 that oxytocin is a posterior pituitary gland hormone that stimulates contraction of uterine muscle. The high levels of estradiol stimulate the production of oxytocin receptors within the smooth muscle layer of the uterus, making the uterus more sensitive to oxytocin. At the same time, the pressure of the fetus's head pressing on the cervix stretches the smooth muscle of the uterus and cervix.

This stretch is detected by neurons in these structures. Signals from the stretch-sensitive neurons are sent to the mother's hypothalamus, triggering the release of still more oxytocin from the posterior pituitary gland.

The binding of oxytocin to its receptors initiates the strong uterine muscle contractions that are the hallmark of **labor**. In addition to its direct action on uterine muscle, oxytocin stimulates uterine secretion of prostaglandins that act with oxytocin to increase the strength of the muscle contractions. The stronger contractions elicit more oxytocin release from the mother's pituitary, which causes yet stronger contractions, setting up a positive feedback cycle that continues until the baby is pushed out of the uterus and through the vagina. The contractions continue for a short while after birth, helping to expel the placenta.

Labor occurs in three stages (**Figure 51.14**). The initial stage induces dilation and thinning of the cervix to allow passage of the fetus out of the uterus. As the uterine contractions get stronger and more frequent toward the end of labor, the fetus moves, usually headfirst, through the cervix and the vagina and out into the world; this is the second stage of labor. In the third and final stage, blood vessels within the placenta and umbilical cord contract and block further blood flow, making the newborn independent from the mother. The placenta detaches from the uterine wall and is delivered a few minutes after the birth of the baby.

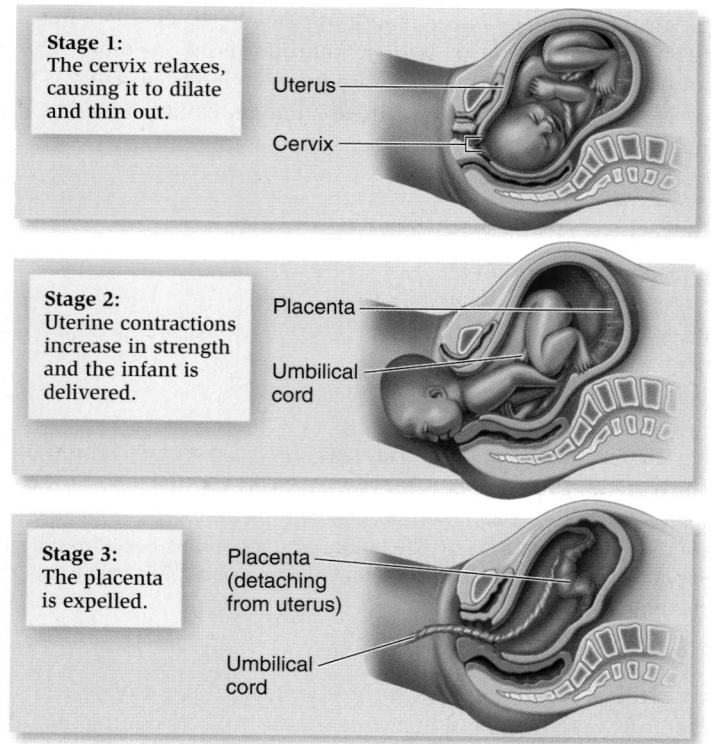

Figure 51.14   The three stages of labor.

Biological inquiry: Why do many female mammals consume the placenta after giving birth?

In most mammals, the young are nurtured for a period after birth by milk produced within the mother's mammary glands and secreted via the nipples. The monotremes, in which the young of some species use their pliable bills to attach to the breast of the mother and draw milk directly through the skin, are an exception. The production of milk is called **lactation**. During pregnancy, elevated blood progesterone levels suppress the secretion of the anterior pituitary hormone prolactin, which is required for milk production. In humans, lactation begins shortly after birth in response to declining blood concentrations of progesterone previously provided by the placenta. Whereas the formation of milk is dependent on prolactin, the actual release of milk from the breast is dependent on activation of smooth muscle cells surrounding secretory ducts in the breast. These cells are stimulated to contract by the presence of oxytocin. Thus oxytocin plays a key role in two major processes, birth and nourishment of the young.

GENOMES & PROTEOMES

The Evolution of the Globin Gene Family Has Been Important for Internal Gestation in Mammals

As discussed in Chapter 21, genes can become duplicated to create gene families. Gene families have been important in the evolution of complex traits because the various members of a gene family can enable the expression of complex, specialized forms and functions.

An interesting example is the globin gene family in animals. Globin genes encode polypeptides that are subunits of proteins that function in oxygen binding. Hemoglobin, which is made in erythrocytes, carries oxygen throughout the body in all vertebrates and many invertebrates, delivering oxygen to all of the body's cells (see Chapter 48). In humans, the globin gene family is composed of several homologous genes that were originally derived from a single ancestral globin gene (refer back to Figure 21.8).

Although all of the globin polypeptides are subunits of proteins that play a role in oxygen binding, the various family members tend to have specialized functions. For example, certain globin genes are expressed only during particular stages of embryonic development. This has particular importance in placental mammals, in which prenatal development occurs in the uterus of females, because the oxygen demands of a rapidly growing embryo and fetus are quite different from the demands of its mother. These different demands are met by the differential expression of hemoglobin genes during prenatal development.

The hemoglobin protein of adult mammals is composed of four globin polypeptides—two encoded by the α-globin gene and two encoded by the β-globin gene. Altogether, five globin genes, designated α, β, γ, ζ, and ϵ encode the major subunits that are found in hemoglobin proteins at different developmental stages.

Table 51.1	Globin Gene Expression During Mammalian Development		
Stage of development	Globin genes expressed	Hemoglobin composition	Oxygen affinity (P_{50})*
Embryo	ϵ-globin and ζ-globin	Two ϵ-globin and two ζ-globin subunits	5–13.5 mmHg
Fetus	γ-globin and α-globin	Two γ-globin and two α-globin subunits	19.5 mmHg
Birth to adult	β-globin and α-globin	Two β-globin and two α-globin subunits	26.5 mmHg

*P_{50} values represent the partial pressure of oxygen required to half-saturate hemoglobin (see Chapter 48): A lower P_{50} indicates a higher affinity of hemoglobin for oxygen. The value for embryos is an estimate based on *in vitro* experiments. All values are for human hemoglobins.

During embryonic development, the ϵ-globin and ζ-globin genes are turned on, resulting in embryonic hemoglobin with a very high affinity for oxygen (**Table 51.1**). At the fetal stage, these genes are turned off, and the α-globin and γ-globin genes are turned on, producing fetal hemoglobin with slightly less (but still high) affinity for oxygen. Finally, just before birth, expression of the γ-globin gene decreases, and the β-globin gene is turned on, resulting in adult hemoglobin, which has a lower affinity for oxygen than either the embryonic or fetal forms. The higher affinities of embryonic and fetal hemoglobins enable the embryo and fetus to remove oxygen from the mother's bloodstream and use that oxygen to meet their own metabolic demands. Thus the expression of different globin genes at particular stages of development enables placental mammals to develop in the uterus without either breathing on their own or being continually exposed to atmospheric oxygen (as occurs for vertebrate embryos that develop externally within eggs).

51.5 Timing of Reproduction with Favorable Times of Year

A species' survival depends on successful production of offspring, and the likelihood that those offspring will survive to adulthood and reproduce is influenced to a large degree by the environment into which they are born. For that reason, reproductive cycles often occur at times of the year when the likelihood of reproductive success is greatest—when the young will have sufficient nutrients to sustain them during their period of rapid growth. In general, animals that live under more uniformly favorable conditions—stable temperatures, ample rainfall, and abundant available food—have less dramatic cycles of reproductive activity. Many tropical species, for example, reproduce several times each year. By contrast, many temperate zone animals have seasonal reproductive cycles that reflect large fluctuations in environmental conditions. For example, insects such as the mayfly live as underwater larvae for a year before emerging into the adult (reproductively active) stage, but the emergence occurs only during one or two favorable summer months.

Animals synchronize the production of offspring with favorable environmental conditions by several means. In **sperm storage**, females store and nourish sperm in their reproductive tract for long periods of time, as long as two years in honeybees and four years in some turtles. Certain insectivorous bats such as *Myotis lucifugus* use a different strategy. These bats mate in the fall, but the ovarian cycle in females is halted before ovulation, and sperm are stored and nourished in the female's uterus over the winter. Upon arousal from hibernation in the spring, the female ovulates one or more eggs, which are fertilized by the stored sperm. This type of reproductive cycle, called **delayed ovulation**, ensures that bats mate when they are in prime condition, and that young are born when temperatures and food supplies are optimal.

Other animals have **delayed implantation**, in which a fertilized egg reaches the uterus but does not implant until later, when environmental conditions are more favorable for the newly produced young. This type of reproduction is common among carnivores, notably the mustelids (weasels) and some bears.

Finally, many animals (cats, some hamsters, sheep, and many birds) have seasonal periods of mating followed immediately by implantation and pregnancy. Such seasonal breeding results from neuroendocrine changes in the hypothalamus in response to changes in the length of daylight. As day length increases near the end of winter, neurohormone secretion and gamete development by the reproductive glands of some seasonal breeders increase, and these animals exhibit behavioral changes indicating their increased readiness to mate. Other seasonal breeders—sheep, for example—reproduce in response to shorter day length. Ewes are impregnated by rams in the fall, as the days grow shorter, and they carry the pregnancy through the winter. Lambs are born in the spring, providing a relatively long growth period before onset of the next winter. If sheep were to mate in summer, lambs would be born in late fall or winter, when conditions are much less favorable.

Unlike the examples just described, humans neither have a seasonal breeding cycle nor alter the timing of implantation or ovulation. Moreover, sperm can survive in a woman's reproductive tract for only a short time—2 or 3 days. Human reproduction, therefore, is far less responsive to environmental changes, day length, or other factors. However, human reproductive success can be seriously curtailed by sickness and other factors, as we see next.

51.6 Impact on Public Health

Approximately 5–10% of U.S. adults are not fertile; that is, they cannot reproduce. In men and women alike, fertility can be compromised by a variety of factors. In this section, we discuss some of the common causes of **infertility**—the inability of a man to produce sufficient numbers or quality of sperm to impregnate a woman, or the inability of a woman to become pregnant or maintain a pregnancy. We then conclude by examining the methods in use today to prevent pregnancy in fertile women.

Infertility May Result from Disease, Developmental Disorders, Inadequate Nutrition, and Stress

As many as 75% of infertility cases have some identifiable cause, and among the more prominent causes is disease. Primary among the diseases that affect fertility are sexually transmitted diseases (STDs). For example, some STDs may cause blockage in the ducts of the testes, preventing normal sperm transport.

A much rarer group of diseases that cause infertility are tumors. Tumors that form in the pituitary gland may prevent its normal functioning such that LH and FSH secretion is interrupted, resulting in decreased gamete development. Still other tumors may form in a reproductive organ—for example, the uterus—where they disrupt that organ's normal function.

Developmental disorders are conditions that are either present at birth or arise during childhood and adolescence. In some developmental disorders that affect fertility, inherited mutations of genes that code for enzymes involved in the biosynthesis of reproductive hormones cause abnormal expression of those genes. The result is either too much or too little of one or more of these hormones, notably estradiol or testosterone. Other developmental disorders that compromise fertility include malformations of the cervix or fallopian tubes.

Adequate nutrition is required for normal growth and development of all parts of the body, including the reproductive system. Because the reproductive system is not essential for the survival of any single organism, it often becomes inactive when nutrients are chronically scarce, such as during starvation. In this way, precious stores of energy in the body are preserved for vital functions, such as the operation of the brain and heart.

Nutrition can also affect reproduction before adulthood. Undernourished children may enter puberty several years later than normal. In light of recent evidence suggesting that the brains of mammals contain a center that monitors the body's fat stores, one of the triggers that initiate puberty in girls may be a signal from adipose tissue to the brain. Very low fat levels in undernourished girls somehow signal the brain that her body does not contain sufficient fat stores to support the energetic demands of pregnancy, and puberty is delayed.

Stress—any real or perceived threat to an animal's homeostasis—can affect fertility. In the short term, stress can produce hormonal changes that are adaptive in meeting a crisis. However, long-term stress is damaging to many aspects of health, including reproductive health. Because reproduction is not necessary for an individual's survival, many nonessential functions—including the maintenance of menstrual cycles in women—are suppressed by chronic stress. The reproductive effects of stress, like those of starvation, appear to be much greater in females than in males, perhaps because only females bear the energetic

cost of pregnancy. Interestingly, the human body responds to long-term strenuous exercise in a way that is similar to its response to long-term stress. This is why many young ballerinas and gymnasts experience delayed puberty, and why in female marathon runners menstrual cycles may be abnormal or absent.

When the causes of infertility cannot be ascertained, a variety of factors come under suspicion. Among these possible causes of infertility are ingestion of toxins (for example, certain heavy metals such as cadmium), tobacco smoking, marijuana use, injuries to the gonads, and aging. Recall that as women age, they experience a loss of fertility, an event called menopause. Although reproductive function declines with age in men, they do not experience complete cessation of gamete production even at very advanced ages.

Among several currently available treatments for increasing the likelihood of pregnancy in infertile women are hormone therapy to increase egg production, and a collection of procedures known as assisted reproductive technologies (ART). In the most common ART procedure, called *in vitro* fertilization, sperm and eggs collected from a man and a woman are placed together in culture dishes. Once the sperm have fertilized the eggs, and the eggs have undergone several cell divisions, one or more embryos are inserted into a woman's uterus in the hope that one will implant. When this procedure was first used in 1978, the children born as a result came to be known as "test-tube babies." Since then, over 200,000 children have been born using this technology, which is typically effective about 30–35% of the time.

We turn now to a discussion of the other side of fertility issues—namely, the ways in which individuals attempt to prevent pregnancy.

Contraception Prevents Pregnancy

The voluntary use of procedures to suppress fertility is commonly termed birth control. The use of such procedures to prevent fertilization or implantation of a fertilized egg is termed **contraception**. By contrast, procedures or circumstances that cause the death of an embryo or fetus after implantation produce an **abortion**. Abortions can occur either spontaneously—usually when the embryo is in some way defective—or be induced by substances or surgical procedures. A substance that induces an abortion is called an abortifacient.

Some forms of contraception prevent the transport of gametes through the reproductive tract. **Vasectomy** is a surgical procedure in men that severs the vas deferens, thereby preventing the release of sperm at ejaculation (however, semen is still released). In women, tubal ligation involves the cutting and sealing of the fallopian tubes. This procedure prevents the movement of the ovum from the oviduct into the uterus. Both of these procedures are essentially permanent, because it is difficult—sometimes impossible—to reverse the surgery.

Temporary methods of preventing fertilization are barrier methods, which prevent sperm from reaching an egg. Barrier methods include **vaginal diaphragms**, which are placed in the upper part of the vagina just prior to intercourse and block movement of sperm to the cervix, and **condoms**, which are sheathlike membranes worn over the penis that collect the ejaculate. In addition to their contraceptive function, condoms significantly reduce the risk of STDs such as HIV infection, syphilis, gonorrhea, chlamydia, and herpes.

Oral contraceptives (birth control pills) are synthetic forms of estradiol and progesterone, taken by mouth, that prevent ovulation in women by inhibiting pituitary LH and FSH release. The hormones in these pills also affect the composition of cervical mucus such that sperm cannot pass through it into the uterus. In addition to the oral route, hormones can be administered by injections and skin patches.

Although no chemical contraceptive agents are widely used by males, a recent experimental formulation of testosterone and progesterone given by injection has proven effective in suppressing sperm production in men. In another approach, American researcher Michael O'Rand and colleagues demonstrated in 2004 that long-term administration of antibodies against a protein expressed specifically in the testis and epididymis of male primates completely suppressed fertility in 100% of the monkeys tested. Whether or not this novel approach to controlling fertility will be effective in humans remains to be determined.

Another method of contraception involves placement in the uterus of an intrauterine device (IUD), a small object that interferes with the endometrial preparation required for acceptance of the blastocyst. Unlike the other forms of contraception described here, an IUD works after fertilization—by preventing implantation.

In addition to the contraceptive methods used before or during intercourse, within 72 hours after intercourse women can take a variety of drugs that typically interfere with ovulation, transport of the blastocyst to the uterus, or implantation. Approaches include a high dose of estrogen, or two large doses (12 hours apart) of a combined estrogen-progestin oral contraceptive. More effective and producing fewer side effects is the drug RU486 (mifepristone), which antagonizes progesterone's effects on the endometrium, causing it to erode. RU486 can also be used as an abortifacient if pregnancy has occurred.

Used prior to the advent of modern contraception, and still in use today by individuals who prefer not to use contraceptives, is the rhythm method, which involves abstaining from sexual intercourse near the time of ovulation. Its main drawback is the difficulty in precisely pinpointing the time of ovulation, even using laboratory techniques. One problem with predicting the time of ovulation is that several of the detectable changes characteristic of the midpoint of the ovarian cycle—including a small rise in body temperature and changes in the cellular characteristics of the vaginal epithelium—occur only after ovulation. This problem, plus the fact that among women ovulation can occur any time between days 5 and 15, explains why the rhythm method has such a relatively high failure rate.

Chapter Summary

51.1 Asexual Reproduction

- Asexual reproduction occurs when offspring are produced from a single parent, without the fusion of gametes from two parents. Budding occurs when part of the parent organism pinches off to form a complete new individual. Some animals reproduce by the regeneration of a complete organism from small body fragments. Parthenogenesis is the development of offspring from an unfertilized egg. (Figure 51.1)

51.2 Sexual Reproduction

- Sexual reproduction is the production of a new individual by the joining of two haploid gametes: sperm and eggs. The union of a sperm and an egg—fertilization—produces a zygote, which develops into an embryo.

- Sexual reproduction was shown by Paland and Lynch to be more effective in eliminating deleterious mutations from a population of *Daphnia* than asexual reproduction. (Figure 51.2)

- In hermaphroditism, individuals can fertilize their own eggs with their own sperm, but in most hermaphroditic species individuals exchange sperm with another individual.

- Male and female gametes are formed within the gonads—the testes in males and the ovaries in females.

- Gametogenesis—the formation of gametes—begins with diploid primordial cells called germ cells. In spermatogenesis, one spermatogonium becomes a primary spermatocyte, which gives rise to two secondary spermatocytes, which yield four spermatids

that mature into four sperm cells. In oogenesis, one oogonium becomes a primary oocyte, which produces a secondary oocyte that yields a mature egg (ovum). (Figures 51.3, 51.4)

- At the tip of a sperm's head is a special structure called the acrosome, which contains proteolytic enzymes that help break down the outer layers of the ovum.

- Within the ovaries, each ovum undergoes growth and development within a structure called a follicle before it leaves the ovary in a process called ovulation.

- In external fertilization, sperm and eggs are released into an aquatic environment where they unite and avoid desiccation. (Figure 51.5)

- Terrestrial animals utilize internal fertilization, in which sperm are deposited within the reproductive tract of the female during the act called copulation.

- External accessory sex organs, such as the penis, are used to physically join the male and female during copulation so that sperm can be deposited directly into the female's reproductive tract.

- Development of an embryo within an egg is called oviparity. When an embryo develops within the mother, the process is called viviparity. In ovoviviparity, eggs covered with a thin shell are produced and hatch inside the mother's body, but the offspring receive no nourishment from the mother. (Figure 51.6)

51.3 Mammalian Reproductive Structure and Function

- Sperm produced within each testis move out of the seminiferous tubules and into the epididymis, which leads into the vas deferens, a muscular tube leading to the ejaculatory duct. The urethra conducts semen, a mixture containing fluid and sperm, during ejaculation. The fluid components of semen are produced in the seminal vesicles, the bulbourethral glands, and the prostate gland. (Figures 51.7, 51.8)

- The female genitalia are composed of large outer folds called the labia majora, plus smaller, inner folds called the labia minora. At the anterior part of the labia minora is the clitoris, which is erectile tissue of the same origin as the penis. In mammals the external opening of the reproductive tract leads to the vagina, a tubular structure into which sperm are deposited. Sperm then move through a fibrous structure called the cervix, which forms the opening to the large, conical-shaped uterus. Sperm typically fertilize eggs in the oviduct, a tube extending between the uterus and an ovary. The fertilized egg undergoes several cell divisions to become a blastocyst, a ball of cells that enters the uterus, where it will develop into an embryo. (Figure 51.9)

- In females, changes in hormone secretion produce the ovarian cycle, and also control the uterine cycle or menstrual cycle. The latter name refers to menstruation, a period of bleeding at the beginning of the uterine cycle. (Figure 51.10)

- Following the release of an egg during ovulation, cells in the empty follicle proliferate and develop into a structure called the corpus luteum, which degenerates if fertilization does not occur. If fertilization does occur, the blastocyst develops a surrounding layer of cells that secrete an LH-like hormone, called chorionic gonadotropin (CG), which maintains the corpus luteum.

- The event during which a woman stops having ovarian cycles is called menopause.

51.4 Pregnancy in Mammals

- The time during which a developing embryo grows within the uterus of the mother is termed pregnancy or gestation. Humans and other eutherian mammals retain and nourish their young within the uterus via transfer of nutrients and gases through a structure called the placenta. (Figure 51.11)

- The first event of pregnancy is implantation, when the blastocyst imbeds within the uterine endometrium. Shortly after implantation, newly developing embryonic tissues merge with the endometrium to form the placenta, which remains in place and grows larger as the embryo matures into a fetus. (Figure 51.12)

- Birth, or parturition, is initiated by the actions of hormones. The hormone oxytocin stimulates the strong uterine muscle contractions that are the hallmark of the three-stage process called labor. (Figures 51.13, 51.14)

- In most mammals, the young are nurtured for a period after birth by milk produced in the process called lactation.

- The evolution of the globin gene family contributed to the ability of placental mammals to develop inside the mother's uterus. (Table 51.1)

51.5 Timing of Reproduction with Favorable Times of Year

- Animals synchronize the production of offspring with favorable environmental conditions by several means. In sperm storage, females store and nourish sperm in their reproductive tract for long periods of time. In delayed ovulation, the ovarian cycle in females is halted before ovulation, and sperm are stored and nourished in the female's uterus until a more favorable time for birth of the young. In delayed implantation, a fertilized egg reaches the uterus but does not implant until later, when environmental conditions are more favorable for the newly produced young. In addition, seasonal mating is practiced by many species.

51.6 Impact on Public Health

- Infertility is the inability of a man to produce sufficient numbers or quality of sperm to impregnate a woman, or the inability of a woman to become pregnant or maintain a pregnancy. A primary cause is STDs.

- The use of procedures to prevent fertilization or implantation of a fertilized egg is termed contraception. Procedures or circumstances that cause the death of an embryo or fetus after implantation produce an abortion.

- Vasectomy is a form of contraception in which the transport of gametes through the male reproductive tract is prevented by a surgical procedure that severs the vas deferens. Tubal ligation involves cutting and sealing of the fallopian tubes.

- Two type of barrier contraceptives are vaginal diaphragms, which are placed in the upper part of the vagina just prior to intercourse and block movement of sperm to the cervix, and condoms, which are sheathlike membranes worn over the penis that collect the ejaculate. Oral contraceptives are used to prevent ovulation.

TEST YOURSELF

1. The development of offspring from unfertilized eggs is
   a. budding.
   b. cloning.
   c. fragmentation.
   d. parthenogenesis.
   e. implantation.

2. Which of the following is considered an advantage of sexual reproduction?
   a. necessity to locate a mate
   b. increased energy expenditure in producing gametes that may not be used in reproduction
   c. increased genetic variation
   d. decreased genetic variation
   e. both a and b

3. Spermatogonia are
   a. germ cells.
   b. diploid cells.
   c. male gametes.
   d. all of the above.
   e. a and b only.

4. Advantages of internal fertilization over external fertilization include
   a. ensuring that male gametes come into close proximity to female gametes.
   b. protection of gametes from predation or other harmful environmental factors.
   c. decreased likelihood of desiccation of gametes.
   d. all of the above.
   e. a and b only.

5. Which of the following is an example of ovoviviparity?
   a. Honeybees lay soft eggs within the hive.
   b. Sharks hatch from shells held within the female body.
   c. Birds hatch from eggs laid within a nest.
   d. Kittens obtain nourishment from the mother through a placenta.
   e. Eggs are released into the water column, where they may be fertilized.

6. The fructose in semen is secreted by
   a. the epididymis.
   b. the seminiferous tubules.
   c. the seminal vesicles.
   d. the prostate gland.
   e. the bulbourethral glands.

7. The function of FSH in males is to
   a. stimulate the development of the gonads during early development.
   b. stimulate spermatogenesis.
   c. increase the secretion of testosterone by the testes.
   d. regulate the secretion of the bulbourethral glands.
   e. both a and b.

8. During the human ovarian cycle, ovulation is stimulated by
   a. a decrease in FSH secretion.
   b. an increase in progesterone secretion.
   c. an increase in LH secretion.
   d. the presence of semen in the vagina.
   e. a decrease in estradiol levels in the bloodstream.

9. During the secretory phase of the menstrual cycle, endometrial glands secrete
   a. hormones that increase the likelihood of pregnancy.
   b. nutritive substances that sustain the embryo during the first 2 weeks of development.
   c. hormones that prevent ovulation.
   d. waste products into the lumen of the uterus.
   e. both a and c.

10. During the _____ stage of labor in mammals, the fetus is expelled from the uterus.
    a. first
    b. second
    c. third

CONCEPTUAL QUESTIONS

1. Define asexual reproduction and give three examples.
2. Define hermaphroditism.
3. Explain how GnRH controls sexual functioning in males.

EXPERIMENTAL QUESTIONS

1. How did Paland and Lynch propose to test the hypothesis that sexual reproduction allowed for the reduction in deleterious mutations?
2. What did they discover?
3. What is the proposed evolutionary benefit for sexual reproduction?

COLLABORATIVE QUESTIONS

1. What are some of the costs of sexual reproduction?
2. Compare and contrast internal fertilization and external fertilization.

www.brookerbiology.com

This website includes answers to the Biological Inquiry questions found in the figure legends and all end-of-chapter questions.

52

ANIMAL DEVELOPMENT

CHAPTER OUTLINE

A chick embryo at 33 hours shows development of body segments.

I n the previous chapter, we learned how animals reproduce. In this chapter, we will learn about the cellular and molecular processes that lead to the formation of an embryo. The biological information that controls embryonic development resides in both the organism's genetic material—its DNA—and in the cytoplasm of the egg. A fertilized egg first becomes transformed into a cluster of cells without specialized functions, and ultimately into a complex organism containing organs with specific and evolutionarily conserved functions. These events occur during a period of time known as an animal's development. The process by which a fertilized egg is transformed into an organism with distinct physiological systems and body parts is called **embryonic development**. The process by which different cells within a developing organism acquire specialized forms and functions, due to the expression of cell-specific genes, is called **cellular differentiation**.

In this chapter, we will first describe the five different stages of embryonic development. Next, we will examine the cellular and molecular mechanisms that control development. We conclude with an overview of how abnormal development impacts the human condition.

52.1 Stages of Embryonic Development

Even though the adult forms of animals vary immensely in size and morphology (shape), embryonic developmental processes in most animals have similar properties. As described in Chapter 32, most modern animals develop from embryos with three cell layers. Such triploblasts include vertebrates, arthropods, echinoderms, and mollusks. Development in these animals can be categorized into five general stages: fertilization, cleavage, gastrulation, neurulation, and organogenesis (**Figure 52.1**). Many

species also go through a phase called metamorphosis, which is a transition from a feeding larval form to an adult. Metamorphosis occurs after organogenesis and facilitates the rapid growth of young organisms into mature ones. Examples of metamorphosis include the transformation of a caterpillar into a butterfly, and that of a tadpole into a frog. In this section, we will examine the key aspects of each of the five general stages of animal development.

Stage 1: Fertilization Activates Cell Division in the Embryo

In all species of triploblasts, the events in **fertilization** are quite similar (**Figure 52.2**). The description that follows summarizes some of the hallmark events following fertilization in sea urchins, a well-studied echinoderm. As mentioned in Chapter 51, sperm can contact the egg either outside or inside the female, depending on the species. In all cases, the sperm must penetrate a jelly-like layer before contacting the plasma membrane of the egg.

The initial contact is followed by the **acrosomal reaction**, in which hydrolytic enzymes are released from the acrosome in the tip of the sperm head onto the jelly coat of the egg (Figure 52.2a). These enzymes dissolve a localized region of the jelly coat, allowing the sperm head to bind to proteins in the egg's plasma membrane. Binding is followed by fusion of the sperm head with the egg membrane, and shortly thereafter by penetration of the sperm head with its nucleus into the egg. Additional sperm are prevented from fusing with the egg because the fusion of sperm and egg depolarizes the egg. This depolarization blocks other sperm from binding to egg membrane proteins and is known as the **fast block to polyspermy**. Without this block, a single egg could receive chromosomes from two or more sperm, resulting in zygotes that fail to develop normally or at all.

Figure 52.1 Overview of developmental stages in an amphibian.

The acrosomal reaction is followed by the **cortical reaction** (Figure 52.2b). Normally, cytosolic calcium concentration in eggs, as in most cells, is kept at a low level by several mechanisms, some of which pump Ca^{2+} out of the cytosol and into the endoplasmic reticulum. Upon binding of the sperm to the egg, inositol trisphosphate (IP_3) (see Chapter 9) is released from the region of the plasma membrane nearest to the sperm entry point. IP_3 then binds to nearby sites on the endoplasmic reticulum and opens Ca^{2+} channels. Within 10 seconds after a sperm cell binds to an egg, Ca^{2+} is released from the lumen of the endoplasmic reticulum and into the cytosol. This signal is propagated across the entire endoplasmic reticulum, resulting in the transmission of a calcium wave across the egg over a period of about 30 seconds. This calcium wave can be visualized by injecting the cytosol of an unfertilized egg with a calcium-sensitive fluorescent dye that becomes highly fluorescent when Ca^{2+} is released from the endoplasmic reticulum (Figure 52.2c).

The release of Ca^{2+} in the cortical reaction has several important effects. First, membrane-bound vesicles in the egg's cytosol, called cortical granules, release enzymes and other substances that inactivate the sperm-binding proteins on the plasma membrane. In addition, the outer coating of the egg cell, known as the vitelline layer in sea urchins, or the zona pellucida in vertebrates, becomes hardened and begins to separate from the plasma membrane. These events create another barrier to more sperm fusing with the egg, a process called the **slow block to polyspermy**. Additionally, the burst of cytosolic Ca^{2+} both leads to the activation of molecular signaling pathways that initiate the first cell cycle of the embryo and triggers an increase in protein synthesis and metabolism within the egg cell.

Shortly afterwards, the nucleus of the sperm fuses with the nucleus of the egg, creating a diploid fertilized egg that is now called a zygote. The first cell division of the zygote occurs approx-

imately 90 minutes after fertilization in sea urchins and amphibians, but it can take up to 24 hours after fertilization in mammals.

Stage 2: Cleavage Involves Cell Divisions Without Cell Growth

The initial cell cycles of embryos are unique because they involve repeated cell divisions without cell growth. The developmental phase during which these cell cycles occur is called **cleavage**. The embryonic cells repeatedly split in two, resulting in several generations of daughter cells that are roughly half the size of the cells that gave rise to them. These early cell cycles that lack cell growth are characterized as "biphasic" because they alternate only between the mitotic (M) phase and DNA synthesis (S) phase of the cell cycle. Neither the G_1 nor G_2 phase occurs (see Chapter 9).

In most species in which development occurs outside the mother, and thus eggs can be eaten by predators, cell division during cleavage represents some of the fastest cell cycles found in nature. The cell cycle during cleavage in amphibians, for example, requires only 20 minutes. During each 20-minute cell cycle, complete genome replication, mitosis, and duplication of the nuclear envelope are followed by cytokinesis. In eutherian mammals, however, whose development occurs within the protective environment of the mother's body, biphasic cell divisions during cleavage are relatively slow, requiring about 12 hours to complete.

The two half-size daughter cells produced by each cell division during cleavage are known as **blastomeres**. Individual blastomeres are bound together by cadherin molecules, and the outer single-cell layer of blastomeres forms a sheet of epithelial cells that separates the embryo from its environment. (Cadherins and epithelial cells are described in Chapter 10.) After formation of the outer epithelial layer, the embryos of many animals take up

(a) Acrosomal reaction

1. When a sperm cell contacts an egg, it releases hydrolytic enzymes that dissolve the jelly coat.

2. Sperm-binding proteins are exposed on the egg cell plasma membrane and bind to the sperm.

3. The sperm and egg plasma membranes fuse, and the sperm's nucleus enters the egg.

Vitelline layer

Sperm head

Sperm nucleus

Acrosome

Sperm-binding proteins

Jelly coat

Hydrolytic enzymes

Cortical granules

Egg plasma membrane

Egg cell cytoplasm

(b) Cortical reaction

1. IP$_3$ is released from the plasma membrane near the site of sperm fusion.

2. IP$_3$ stimulates Ca^{2+} release from the endoplasmic reticulum.

3. Ca^{2+} stimulates exocytosis of cortical granules.

4. The contents of the granules cause the vitelline layer and plasma membrane to separate, and also destroy sperm-binding proteins. The vitelline layer of the egg hardens.

Egg cell cytoplasm

IP$_3$

Ca^{2+}

Endoplasmic reticulum

(c) The Ca^{2+} wave in a sea urchin egg

Site of sperm entry

Time after sperm entry

15 seconds   25 seconds   31 seconds   36 seconds

Figure 52.2 **Fertilization of an egg.** **(a)** Acrosomal reaction.   The contact of a sperm with an egg initiates a series of events that permits the head of the sperm to bind to the plasma membrane of the egg. **(b)** Cortical reaction. Increased cellular levels of IP$_3$ near the site of sperm fusion cause increased cytosolic concentrations of Ca^{2+}, which stimulate cortical granules to release their contents. **(c)** Calcium wave in a sea urchin egg. The increase in cytosolic Ca^{2+} begins near the site of sperm entry and propagates throughout the egg. The highest concentration of cytosolic calcium released from the endoplasmic reticulum is indicated with brightest color.

water and form a cavity called a **blastocoel**. The embryo at this stage is called a **blastula**. The blastocoel provides a space into which cells will migrate to form the digestive tract and other structures of the embryo, as described later in stage 3.

Among triploblasts, cleavage-stage embryos can vary dramatically in size and appearance. This variation is in part related to the location and amounts of yolk that was deposited in the egg. Yolk is most concentrated toward one end—or pole—of the egg, called the **vegetal pole**. Much less yolk, and much more cytoplasm, is concentrated at the opposite pole, called the **animal pole** (**Figure 52.3**). These poles determine in part the future anteroposterior (head-tail) and dorsoventral (back-front or top-bottom, depending on the species) axes of the embryo.

Cells at the animal pole contain little yolk and much cytoplasm.

Cleavage-stage embryo

Cells at the vegetal pole contain much yolk and little cytoplasm.

Figure 52.3 **Polarity in an amphibian embryo.** The cleavage-stage embryo contains an animal pole with little yolk and much cytoplasm, and a vegetal pole with large amounts of yolk and little cytoplasm.

Figure 52.4   Meroblastic and holoblastic cleavage.   Early embryos of birds and many fishes undergo incomplete (meroblastic) cleavage, whereas most amphibian and mammalian embryos undergo complete (holoblastic) cleavage. The amount of yolk in the egg (not visible in these images) contributes to many of these morphological differences observed in various species.

In some animals, cleavage of the zygote is called incomplete because only the region of the zygote and embryo containing cytoplasm at the animal pole undergoes cell division. This incomplete type of cell cleavage, called **meroblastic cleavage**, occurs in birds, some fishes, and some other vertebrates whose eggs contain large amounts of yolk (**Figure 52.4**). Instead of forming a ball of cells (a blastula), a flattened disk of blastomeres known as a **blastoderm** develops on top of the yolk mass.

In animals whose eggs have smaller amounts of yolk, cleavage during the first cell division is complete and bisects the entire zygote into two equal-sized blastomeres. Such **holoblastic cleavage** occurs in amphibians and mammals (Figure 52.4). In amphibians, cleavage-stage embryos form a blastula, as previously noted. In mammals, however, cleavage-stage embryos undergo a process called compaction, in which the amount of physical contact between cells is maximized. At this stage, the embryo in these species is called a **morula**. The resulting blastomeres in mammals then proceed to form a **blastocyst**, the mammalian counterpart of a blastula.

In mammals, the events of fertilization and cleavage occur in the oviduct (**Figure 52.5**). The blastocyst has a different morphological appearance than the blastula or blastoderm embryos in nonmammalian species, and no animal-vegetal polarity that is analogous to that of other chordates exists. The blastocyst consists of an outer epithelial layer called the trophectoderm, which gives rise to the placenta, and an inner layer called the inner cell mass, which develops into the embryo. After the blastocyst forms, the embryo hatches from the zona pellucida and becomes embedded in the wall of the mother's uterus, a process known as implantation (Figure 52.5; also see Chapter 51).

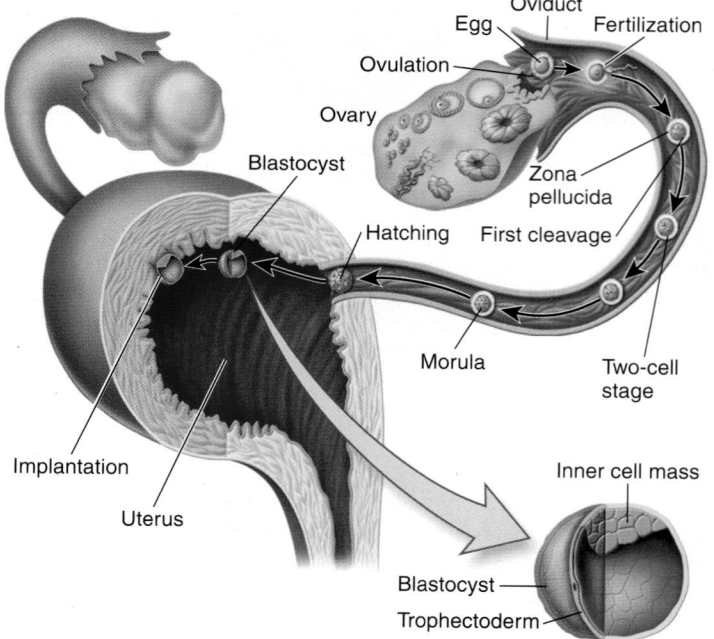

Figure 52.5   The sites of embryonic development in mammals.   After an ovulated egg is fertilized, initial cleavage and development occur in the oviduct of the mother. The blastocyst hatches from the zona pellucida before implanting into the inner lining of the uterus. A blastocyst is composed of an outer epithelial layer called the trophectoderm, which gives rise to extraembryonic tissues such as the placenta, and an inner cell mass, which develops into the embryo.

This entire process takes 4 days in mice and about 8–10 days in humans.

Because cell cycles can be extremely rapid during early cleavage, the DNA is often maintained in a conformation that is incompatible with transcription, so little or no transcription occurs during early cleavage. Therefore, nearly all cellular division and differentiation processes during the cleavage stage are regulated by cytoplasmic components that resided in the egg prior to fertilization. Subcellular components that were synthesized prior to fertilization are called maternal factors and include many mRNAs and proteins that are stockpiled in the egg during oogenesis to facilitate cleavage in the absence of transcription.

Toward the end of the cleavage stage of embryonic development, cell cycles become less synchronous, and the embryo begins to express its own genes. The embryo's shift from existing exclusively on maternal factors to developing in response to products derived from its own genome begins 6–24 hours after fertilization in vertebrates. This is followed by the next stage of development, called gastrulation.

Stage 3: Gastrulation Establishes the Three Germ Layers in the Embryo

Following cleavage is **gastrulation**, one of the most dramatic stages of embryonic development in animals. During gastrulation, the hollow ball of cells that makes up the blastula is developed into a highly organized structure in which the three germ layers—**ectoderm**, **mesoderm**, and **endoderm**—and the primordial germ cells are established. These distinct germ layers are partially differentiated tissues that are easily recognized by their appearance under a light microscope and ultimately give rise to specialized cells, tissues, and organs (**Figure 52.6**).

An embryo that has undergone gastrulation is called a **gastrula**. In the gastrula embryo, the three germ layers become clearly established and occupy discrete regions of the embryo, with an outer ectoderm, a middle mesoderm, and an inner endoderm layer. Each type of germ layer eventually gives rise to different structures, including the notochord, which is a key feature of all chordates described in a later section of this chapter. The organization that emerges during gastrulation is most evident by the clear establishment of the digestive tube and both body axes. Gastrulation is the first time when both the anteroposterior and dorsoventral body axes are clearly evident in the embryo.

The ultimate fate of the three germ layers has been established by an experimental procedure called **fate mapping**. In this technique, a single cell or a small population of cells within an embryo is specifically labeled with a harmless dye, and the fate of these labeled cells is followed to a later stage of embryonic development (**Figure 52.7**). Use of this technique in a variety of vertebrates has shown that the ectoderm in the gastrula forms the epidermis and nervous system in the later embryo. The mesoderm gives rise to heart, limbs, muscles, kidneys,

Figure 52.6   Examples of cell types derived from ectoderm, mesoderm, and endoderm.

Figure 52.7   Fate mapping during embryonic development. Fate mapping of single, labeled embryonic cells (blastomeres) shows that different cells in the embryo (in this case, a urochordate blastula) form different structures in the tadpole-like larva, as shown by the dark regions.

blood, and connective tissues. The endoderm becomes the epithelial lining of the gut, liver, pancreas, lungs, and bladder.

Some of our most detailed descriptions of the events in gastrulation come from the study of frog embryos, which are easy to observe under a dissecting microscope because they develop rapidly outside the mother and are relatively large (~1.5 mm in diameter). The major events in gastrulation are depicted in **Figure 52.8** and described next.

Invagination and Involution Prior to gastrulation, the blastula is enclosed in a simple, spherical epithelial cell layer. Gastrulation begins when a band of tissue extending perpendicularly to the animal-vegetal axis at the widest part of the embryo invaginates (pinches in), creating a small opening called the **blastopore** (Figure 52.8a). Invagination begins when a few epithelial cells located at the vegetal side of the blastula—called bottle cells—undergo a drastic change in their morphology through a process called **apical constriction**. As discussed in Chapter 10, epithelial cells may have a ring of actin filaments that underlie anchoring junctions between neighboring cells (refer back to Figure 10.19). In apical constriction, a reduction in the diameter of the actin rings connected to the adherens junctions causes the bottle cells to elongate toward their basal end, forcing the cells toward the center of the embryo (Figure 52.8a).

The initiating site of invagination becomes the dorsal lip of the blastopore. This change in morphology of only a few key cells in the embryo initiates the gastrulation process in amphibians. Once the bottle cells change their shape and push into the

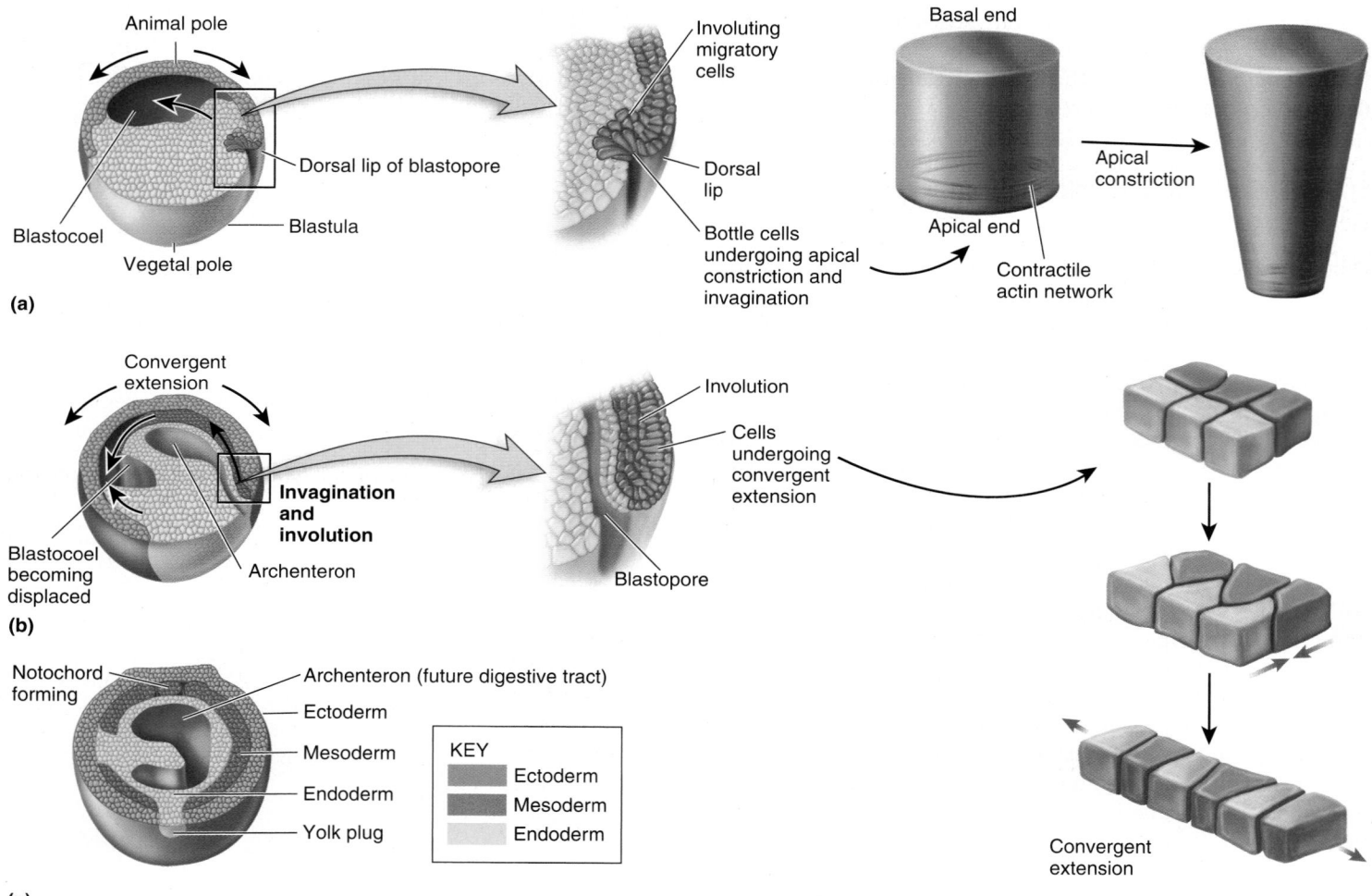

Figure 52.8   The events in gastrulation in amphibians. **(a)** Formation of the blastopore by apical constriction of bottle cells. Gastrulation is initiated by the formation of bottle cells. Invagination of bottle cells forces cells behind them to involute toward the future anterior end of the embryo. The curved arrows indicate directions of cell movements. **(b)** Formation of the archenteron by invagination and involution. Ectoderm spreads by convergent extension over the entire embryo after formation of the blastopore. **(c)** Completion of gastrulation with beginning of notochord formation. The blastocoel becomes displaced by the archenteron, which arises from the blastopore formed at the dorsal lip. By the end of gastrulation, the archenteron has displaced the blastocoel and becomes closed by a yolk plug. Involution continues by convergent extension. Some of the involuting cells become the mesoderm layer of the gastrula. The dorsal surface of the gastrula begins to thicken, and the dorsal mesoderm forms the notochord.

interior of the embryo, other cell movements occur, and together these orchestrated movements establish the mesoderm and future digestive tract of the organism.

Just after invagination occurs, cells of the animal pole begin to spread out and are pushed toward the bottle cells at the dorsal lip of the blastopore. When cells from the animal pole reach the bottle cells, they are forced back toward the animal pole along the roof of the blastocoel. This folding back into the interior of the embryo is called involution (Figure 52.8b). Both endodermal and mesodermal cells involute from the blastopore toward the opposite end of the embryo. After involution, dorsal mesodermal cells migrate toward the animal pole by crawling along the roof of the blastocoel, with endoderm following closely behind.

The spreading of ectoderm in the animal pole toward the vegetal pole is mediated by a cellular process called **convergent extension**, in which two rows of cells merge to form a single elongated layer (Figure 52.8b). Convergent extension leads to spreading of ectoderm from the animal pole to surround the entire vegetal pole. The result of these cellular rearrangements is an embryo with three distinct germ layers.

In chordates and echinoderms, the opening formed by the blastopore ultimately becomes the anus of the organism. As the opening from the blastopore extends into the embryo, a new cylindrically shaped cavity called the **archenteron** displaces the existing blastocoel (Figure 52.8c). The archenteron will become the organism's digestive tract. The blastopore opening remains sealed with a yolk-rich piece of tissue called the yolk plug until later in development.

How are the different classes of cells held together within the three germ layers of a gastrula embryo? Cells within ectoderm, mesoderm, and endoderm express different genes that encode distinct cadherin proteins, and only cadherins of the same type can bind to one another (**Figure 52.9**). As a result, mesodermal cells tend to adhere to each other, but not to endoderm cells or to ectoderm cells.

This property of cells within distinct germ layers is best illustrated by cell dissociation and mixing experiments. When the cells of a gastrula are completely dissociated in a calcium-free medium and then mixed with each other and allowed to reassociate, a fairly well organized ball of cells forms. When ectoderm, mesoderm, and endoderm are identified biochemically, it can be seen that the cells sort themselves out within the ball such that they usually bind only to cells of the same type, and that mesodermal cells tend to reside between the ectodermal and endodermal cells. This experiment indicates that cells within each germ layer can find each other and self-associate via cell-to-cell interactions. These interactions are mediated by distinct cadherin molecules.

Formation of the Notochord A distinguishing anatomical feature that forms at the end of gastrulation in all chordates is the **notochord**—a mesodermal structure that provides rigidity along the anteroposterior axis in the dorsal side of the gastrula. The presence of a notochord defines the phylum Chordates, which also includes the urochordates. Interestingly, urochordates such as the tunicates form a tadpole-like larva but develop into

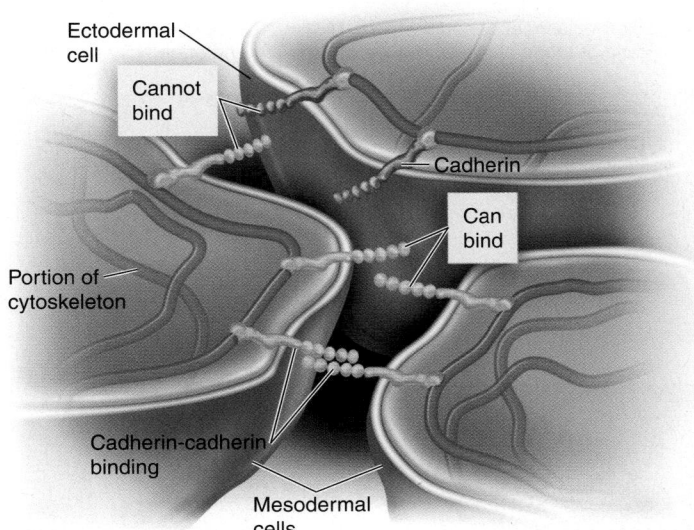

Figure 52.9   Role of cadherins in cell-to-cell contact. Developing cells express one of a family of proteins called cadherins. Similar cadherins bind with each other to promote cell-to-cell contacts. Cells expressing dissimilar cadherins cannot bind and thus do not form associations.

adults that lack vertebrae. The similarities between urochordate and vertebrate larva—including the formation of a notochord—have led many to believe that urochordates may be quite similar to the ancestor species of current-day vertebrates.

The notochord produces many signaling proteins that help establish tissue patterns in the embryo. During gastrulation, the notochord elongates through convergent extension, and proteins produced within it along the anteroposterior axis help further pattern the embryo and induce segment-specific expression of the *Hox* genes in subsequent stages of development. As described in Chapters 19 and 40, *Hox* genes are important for establishing structures along the anteroposterior axis. The notochord persists in the trunk and tail of fishes and amphibians, but it is somewhat constricted within each vertebra. In birds and mammals, the notochord disappears by the time vertebrae have formed.

Although the developmental steps vary among vertebrate species, gastrulation usually begins with invagination initiated by a small number of cells within an external epithelial cell layer that undergoes apical constriction. This leads to a morphologically distinct structure, such as the blastopore, that clearly defines the anteroposterior axis of the animal. The notochord then emerges from this structure. Since many of the genes responsible for specifying gastrulation processes are conserved from fishes to humans, it appears that basic genetic mechanisms for gastrulation are evolutionarily conserved in vertebrates.

By the time the notochord has formed, the dorsal ectoderm overlying the notochord begins to thicken, which initiates the next stage in development, called neurulation. Before we discuss that stage, however, we consider another important event that occurs during gastrulation: the establishment of germ cells.

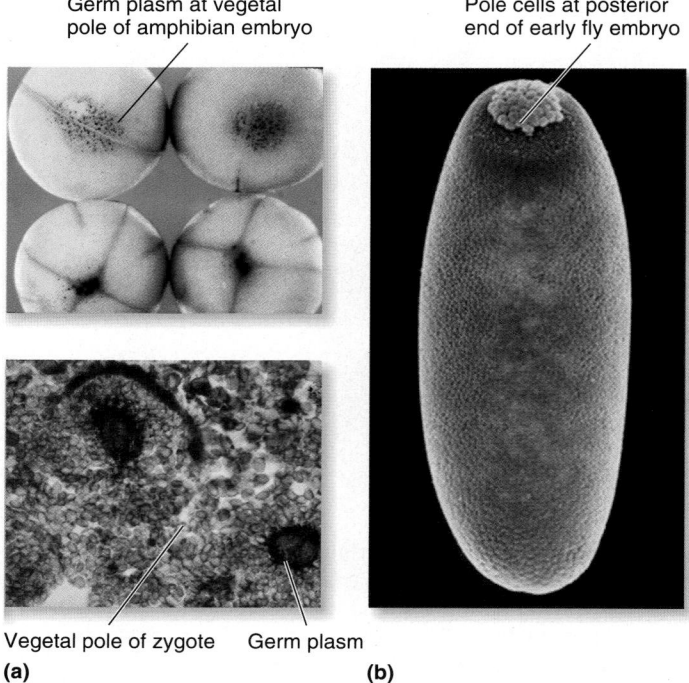

Germ plasm at vegetal pole of amphibian embryo

Pole cells at posterior end of early fly embryo

Vegetal pole of zygote   Germ plasm
(a)   **(b)**

Figure 52.10 Germ plasm and primordial germ cells (PGC) during gastrulation.   (a) Germ plasm in an amphibian embryo (*top*) visualized by labeling an mRNA that is specifically expressed in PGCs. Germ plasm is also easily identified by electron microscopy as dense granules (*bottom*). (b) In fly embryos, the PGCs are called pole cells and are the first cells to form.

Primordial Germ Cells During gastrulation, a specialized group of cells called primordial germ cells (PGCs) migrates to the future site of gonad development in the embryo. These PGCs are highly specialized cells that often arise independently of the three germ layers in the embryo. This cell lineage has two primary functions: (1) to protect and propagate the genetic content of the species, and (2) to undergo meiosis and differentiate into gametes—sperm or eggs—in the adult organism. PGCs are stem cells that can divide through mitosis to make copies of themselves, and some of the resulting daughter cells differentiate into gametes.

The eggs and resulting embryos of some species contain certain cytoplasmic determinants—called the **germ plasm** (**Figure 52.10a**)—that help define and specify the PGCs in the gastrula stage. In amphibian eggs, for example, the germ plasm occupies a small region of cytoplasm at the vegetal pole and contains a specific subset of maternal mRNAs. These cytoplasmic determinants are inherited by a subpopulation of blastomeres during cleavage. At the beginning of gastrulation, blastomeres that inherit these cytoplasmic determinants differentiate into PGCs and migrate from the posterior end of the late gastrula to a region called the genital ridge near the primordial kidneys. The genital ridge is where primordial gonads and PGCs will form testes or

ovaries. In flies, by contrast, PGCs are the first cells to form at the posterior end of the embryo and are called **pole cells** (**Figure 52.10b**).

In mammals, neither pole cells nor germ plasm have been identified. Instead, a few mesoderm-like cells begin to express PGC-specific genes early during gastrulation. These cells migrate along the hindgut during gastrulation and, as in amphibians, intermix with the gonad primordial cells at the genital ridge. Thus, the establishment of PGCs occurs during the earliest phases of animal development and presumably signifies the broad importance of this cell lineage for the propagation of species.

Stage 4: Formation of the Central Nervous System and Body Segmentation Occur During Neurulation

By studying development in several different vertebrate species, researchers are beginning to understand some of the fundamental steps in the formation of the central nervous system (CNS)—the brain and spinal cord—in vertebrates. The multistep embryological process responsible for initiating CNS formation is called **neurulation** (**Figure 52.11**). Neurulation occurs just after gastrulation and involves the formation of the **neural tube** from ectoderm located dorsal to the notochord. All neurons and their supporting cells in the CNS originate from neural precursor cells derived from the neural tube.

In the first step of neurulation shown in Figure 52.11, ectoderm overlaying the notochord thickens by the elongation of cells in the dorsal region to form the neural plate, with adjacent regions that will eventually form a structure called the neural crest (discussed shortly) and the epidermis. The neural plate then elongates by convergent extension, resulting in formation of a single, dorsal elongated epithelial cell layer that is aligned with the anteroposterior axis.

The neural plate forms the neural tube through a series of apical constrictions (steps 2 and 3, Figure 52.11). First, a column of cells along the midline of the neural plate—the medial hinge point—undergoes apical constriction. This initiates the folding phase of neurulation and leads to the formation of the neural groove characteristic of early neurula embryos. Two bilateral columns of cells in the dorsal lateral hinge points then undergo apical constriction after folding, leading to convergence of the two sides of the neural groove and generation of a tubelike structure that is not yet sealed on the dorsal side.

In a fourth step of neurulation, called fusion, the dorsalmost cells on either side of the neural tube are released from adjacent ectoderm and fuse with each other, culminating in the closure of the neural tube. At the same time, ectoderm on either side of the neural tube moves toward the centerline, then up and over the neural tube, where it fuses and forms the dorsal epidermis of the embryo.

Neural Crest Formation Another important cell lineage that arises during neurulation is the **neural crest**, which is unique

Figure 52.11 **Neurulation in vertebrates.** The steps of neurulation include: thickening and elongation; folding, which creates the neural groove; convergence, in which the neural tube begins to take shape; and fusion, in which the neural tube is completed. Apical constriction of three columns of cells—the medial hinge point along the centerline of the neural plate, and two dorsal lateral hinge points flanking the neural plate—facilitates folding and convergence. The neural plate is flanked on each side by cells that will become the neural crest and epidermis. The neural crest cells migrate away from the neural tube to form several other structures.

to vertebrates. It consists of cells that originate from the ectoderm overlaying the dorsal side of the newly formed neural tube and that migrate to other regions of the embryo (step 5, Figure 52.11). Once these cells reach their final destination in the embryo, they differentiate into a variety of cell types different from those that arise from the rest of the ectoderm. All neurons and supporting cells of the peripheral nervous system in vertebrates are derived from neural crest cells. In addition, the neural crest gives rise to melanocytes—specialized cells that provide pigmentation to the skin of vertebrates—and to cells that form facial cartilage and parts of the adrenal glands. The neural crest is so important that some refer to it as the "fourth germ layer" of vertebrates.

Segmentation and the Formation of Somites As embryos develop, distinct tissues acquire recognizable shapes and patterns, and the body often becomes segmented along the anteroposterior axis of the embryo. Segmentation allows individual body segments to have more specialized functions. In some species, such as insects, body segments are found in the adult animal. In vertebrates, the segmentation process helps define repeated structures such as vertebrae and ribs, which form later during development. Segmentation of the body plan along the anteroposterior axis becomes apparent during neurulation.

During neurulation, the mesoderm becomes segmented from the anterior end first, giving rise to blocklike structures of mesoderm called **somites** (**Figure 52.12**). The segmentation of somitic mesoderm continues toward the posterior end of the animal and into the tail, as can be seen in the chick embryo in the chapter-opening photo. The number of somite pairs that forms in a given species is relatively constant, but the number can vary considerably among species (50 in chicks, 65 in mice, 500 in some snakes, and 32 to 34 in humans). Because the rate of embryonic development within a given species often varies with temperature and other environmental factors, stages of embryo development are often standardized according to the number of somites that have formed.

Somites arise from a group of loosely packed mesodermal cell aggregates that condense into epithelial somites, which consist of a hollow ball or vesicle enclosed by an epithelial layer.

Epithelial somite

Neural tube

Figure 52.12   Segmentation of mesoderm during
neurulation.   This scanning electron micrograph shows early
events in the segmentation process: the formation of epithelial
somites at the anterior end of a chick embryo.

The scanning electron micrograph in Figure 52.12 shows a dorsal view of a chick embryo undergoing somite formation. These epithelial somites are located toward the anterior end, and unsegmented mesoderm can be seen extending toward the posterior end. The epithelial somites are transient structures, and the cells within them soon undergo transformation into two morphologically distinct structures that eventually form ribs/skeletal muscle and vertebrae, respectively.

Stage 5: Organs Are Formed During Organogenesis

As described in Chapter 40, organs are specialized structures that consist of arrangements of two or more tissue types. Most organs, such as the kidney, contain all four tissue types: nervous, muscle, epithelial, and connective tissue (see Chapters 10 and 40). The developmental stage during which cells and tissues form organs is called **organogenesis**. Each germ layer gives rise to different organs (see Figure 52.6).

Many organs begin to form during or just after neurulation. However, these organs become functional at different times during development. For example, the lungs of mammals do not acquire the ability to function until shortly before birth. By contrast, the heart is the first functional organ to form in the vertebrate embryo. It begins to beat and pump blood before all the embryo's somites have formed (by 2.5 days after fertilization in chicks, 9 days in mice, and about 22 days in humans).

As we saw in Chapter 40, the development of different organs in animals is controlled by genes in the embryo, notably the *Hox* genes. Many of the genes controlling the processes of gastrulation, neurulation, and organogenesis encode secreted proteins or growth factors that induce cells in their local vicinity to differentiate along a specific developmental pathway. Next, we will see how such growth factors can be studied in the context of embryonic development.

52.2 Control of Cell Differentiation and Morphogenesis During Animal Development

Thus far, we have examined the five stages of embryonic development. In this section, we will turn our attention to certain molecular mechanisms that are crucial for cell differentiation and morphogenesis.

Cytosolic Factors and Extracellular Signals Drive Cell Differentiation During Embryonic Development

The process of embryonic development requires that cells move to specific sites and acquire distinct functional properties (that is, differentiate). After the unicellular fertilized egg (zygote) divides into two cells, those cells then divide into four, then into eight, and soon the rates of cell divisions become asynchronous and more uniquely specified. The fate of each cell within this ever-growing collection of cells is determined through the cell's response to two primary mechanisms of cell differentiation. The first mechanism is the differential acquisition of various cytoplasmic factors during cell division, a process known as autonomous specification (**Figure 52.13a**). The second is the acquisition of properties through a variety of cell-to-cell signaling mechanisms, a process called conditional specification (**Figure 52.13b**). These two mechanisms provide embryonic cells with a continuously changing internal and external environment, in which cells ultimately fulfill their unique functional and spatial fates. This differentiation process involves alterations of gene expression such that specific cells may express a unique set of genes required for a particular function.

Throughout development, autonomous and conditional specification work together such that distinct cells differentiate into each of the numerous diverse cell types with unique functions in the developing organism. Such differentiation involves changes in both gene expression and subcellular organization. Identifying the precise conditional and autonomous signals that specify each cell lineage in the embryo is one of the biggest challenges facing developmental biologists in the 21st century.

Concentration Gradients of Morphogens Control Cell Differentiation in Vertebrate Embryos

The substances that elicit different cellular responses at distinct concentrations in an embryo are proteins called **morphogens**.

1 Cytoplasmic factors are asymmetrically distributed in the egg.

Cytoplasmic factors

2 Following fertilization and cell division, the factors are asymmetrically distributed to daughter cells.

(a) Autonomous specification

Signal

Second messenger

Cell-to-cell signaling by exocytosis of stored chemical messengers

Signal

Second messenger

Binding of membrane proteins to each other

Movement of cytoplasmic molecules through gap junctions

(b) Conditional specification

Figure 52.13 **Mechanisms of cell differentiation during embryonic development.** **(a)** In autonomous specification, cells differentiate through the unequal segregation of subcellular components during mitosis. **(b)** In conditional specification, cells respond to one of at least three types of chemical signals generated by neighboring embryonic cells.

For the biochemical identification of morphogens, researchers have often focused on amphibians. Three features of amphibians make them ideal for analyzing embryonic development using nongenetic mechanisms. First, fertilization and embryonic development occur outside the mother, making it easy to observe and manipulate embryos as they develop. Second, eggs are abundant—adult females carry thousands of eggs and can be induced to lay them by injecting a specific hormone—and are easily fertilized, usually by simply placing a male in an aquarium with an ovulating female. Third, amphibian embryos develop rapidly. In frogs, for example, development from fertilized egg to tadpole formation—including cleavage, gastrulation, neurulation, and complete organogenesis—takes only about 35 hours. Several types of experiments performed with amphibian embryos demonstrate how conditional specification directs cell differentiation in vertebrate embryos.

Conditional specification often involves the synthesis by a cell or group of cells of a signaling protein, a morphogen, that induces a response in another group of cells in the embryo. The idea of cellular induction during embryonic development arose from experiments in which cells isolated from one part of an embryo were analyzed in the presence or absence of cells isolated from a different region of the embryo.

An example of cellular induction was observed in experiments performed by Pieter Nieuwkoop during the 1950s. Nieuwkoop isolated cells from either the animal pole or the vegetal pole of late blastula (cleavage-stage) amphibian embryos. When cells from the animal pole, called an animal cap, were cultured in an appropriate growth medium, they formed undifferentiated spherical structures composed primarily of ectodermal cells. When cells isolated from the vegetal pole were cultured, undifferentiated clumps of endodermal cells resulted. However, when animal and vegetal cells were cultured together, Nieuwkoop observed the formation of mesodermal derivatives from the animal pole cells. This suggested that factors released by vegetal pole cells could induce differentiation of animal pole cells. This type of experiment, called an **animal cap assay**, has been used extensively to identify proteins secreted by embryonic cells that induce cells in the animal pole to differentiate into mesoderm. Nieuwkoop went on to show that different vegetal cells isolated from various positions along the dorsoventral axis induced specific types of mesoderm.

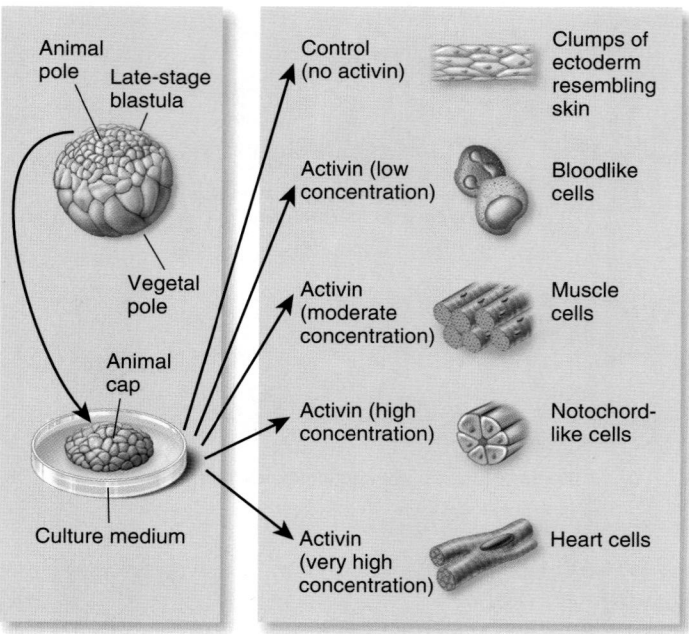

Figure 52.14 **Use of an animal cap assay to demonstrate morphogen activity.** Activin, a member of the TGFß protein family, functions as a morphogen in an animal cap assay by inducing different types of mesoderm at different concentrations. In the absence of activin, no mesodermal derivatives form and the animal caps form cells resembling skin cells. Low concentrations of activin induce mesodermal derivatives resembling blood cells, whereas moderate concentrations induce muscle cells, and higher concentrations induce such mesodermal derivatives as notochord and heart cells.

Biological inquiry: How might it be possible for a single molecule to exert different effects on cells at different concentrations?

One family of proteins—named transforming growth factor betas (TGFßs)—soon emerged as important factors involved in the induction of mesoderm. By purifying these proteins and then adding them to a culture medium containing animal caps, researchers demonstrated that TGFß proteins have mesoderm-inducing activity. In the absence of TGFßs, animal caps grow in culture to form clumps of ectodermal tissue that resemble skin cells. However, when purified TGFßs are added to these cells, they differentiate into mesoderm. TGFß proteins bind to a specific receptor molecule expressed on the surface of animal cap cells, inducing them to differentiate into mesoderm.

A key question was whether different TGFß proteins specify different types of mesoderm, or whether a single protein at different concentrations induced all types of mesoderm. The morphogenic activity of one TGFß protein—activin—in an animal cap assay is shown in **Figure 52.14**. At low concentrations, activin induces mesoderm consisting of bloodlike cells. However, at incrementally higher concentrations, muscle cells, mesoderm resembling the notochord, and fully differentiated heart cells are induced. These results show that certain TGFß proteins exert different effects depending on their concentrations.

Morphogenetic Fields Control the Formation of Certain Body Structures

One of the more interesting ideas to emerge from the study of developmental biology is the concept of the **morphogenetic field**, a group of embryonic cells that ultimately produce a specific body structure. Long before genes and their encoded proteins were identified, embryologists working with amphibians discovered that particular groups of embryonic cells have a striking characteristic: they form complete body structures when transplanted to another site in the embryo. This is true for cells that form the eyes, limbs, and heart in vertebrates, and the eyes, antennae, legs, and wings in insects. In insects, these morphogenetic fields are called imaginal discs.

One of the first morphogenetic fields discovered in vertebrates was the limb field. Between 1910 and 1930, two important observations were made concerning the limb field. First, removing this group of cells from either side of an early developing embryo led to an embryo that lacked a limb at the corresponding position. Second, transplanting a limb field to a new location within prelimb embryos led to the development of an additional limb at the new location. Interestingly, fate-mapping studies showed that certain cells within the field give rise to specific regions of the limb. These types of experiments revealed that cells within morphogenetic fields are uniquely specified to become particular embryonic structures before any physical evidence for the structure itself can be observed in the embryo.

It is thought that morphogenetic fields in amphibians are particularly sensitive to environmental pollutants. Over the past decade, a growing number of amphibians with abnormal numbers of eyes or limbs have been observed in polluted areas of the U.S., leading many scientists to consider these species to be reliable bioindicators of environmental pollution.

In the 1920s, the German zoologists Hans Spemann and Hilde Mangold discovered in an early gastrula an extremely important morphogenetic field with amazing properties. This morphogenetic field is now known as **Spemann's organizer**. Their experiments involved dissecting a small piece of tissue from the dorsal lip of the blastopore in an early gastrula of a newt, and transplanting it to the opposite side of another gastrula (**Figure 52.15**). The result was the formation of a second notochord and neural tube during gastrulation and neurulation in the host embryo, and ultimately the formation of an entirely new body axis—the resulting embryo developed two bodies!

Though technically difficult because the gastrulas were only about 2 mm wide, these experiments yielded clear results because the researchers used two very closely related species of newts. *Triton taeniatus*, which is pigmented, served as the tissue donor, and *Triton cristatus*, which is nonpigmented, served as the host embryo. By transplanting pigmented tissue from the donor embryo into a nonpigmented host embryo, Spemann and Mangold could visually track the origin of the newly developed tissue. The results showed that the secondary notochord and neural tube were composed in large part of host (nonpigmented) cells, with some pigmented cells remaining from the transplanted tissue. This indicated that the transplanted tissue—which they

1 **Transplantation**

Pigmented gastrula of donor embryo (*Triton taeniatus*)

Dorsal lip of blastopore

Nonpigmented gastrula of host embryo (*Triton cristatus*)

2 **Completion of embryonic development**

Primary structures (nonpigmented cells)
— Neural tube
— Notochord

Secondary structures (mostly nonpigmented cells)
— Notochord
— Neural tube

Primary embryo

Secondary (induced) nonpigmented embryo

Figure 52.15 **Experiment that led to the discovery of Spemann's organizer.** Note that the secondary body axis formed by the transplanted organizer is composed of host (nonpigmented) tissue, indicating that the organizer (donor) tissue induced the host tissue to differentiate.

named the "organizer"—had induced cells in the host to differentiate into neural tissue on the transplanted side of the embryo. More recent work has shown that the organizer secretes morphogens—unknown at the time of Spemann's experiments—responsible for inducing the formation of a new embryonic axis. In 1935 Spemann was acknowledged for his important discovery of cellular induction by receiving the first Nobel Prize for studies in developmental biology.

In the past two decades, scientists have identified several genes expressed specifically in Spemann's organizer. Strikingly, many of these genes, which were first discovered in amphibians, are evolutionarily conserved in all vertebrates. These organizer-specific genes are expressed in very small regions of early gastrula embryos, allowing researchers to identify the equivalent of Spemann's organizer in these other species. The names given

the organizer in different species vary. In chicks, it is called Hensen's node, whereas in mice it is simply referred to as the node.

With advances in molecular genetics during the 1960s and 1970s, molecular biologists began to re-examine the work of Hans Spemann and his contemporaries. A major quest of this new generation of molecular biologists was to identify the specific genes transcribed within Spemann's organizer that give this morphogenetic field the ability to form an entirely new body structure. Identifying the genes expressed exclusively in Spemann's organizer, it was thought, would give key insights into the genetic control systems that govern embryonic patterning during gastrulation in vertebrates. However, identifying unknown genes and their protein products was a daunting challenge. Eventually, in 1992, the first secreted morphogen expressed specifically in the organizer was isolated, as we see next.

FEATURE INVESTIGATION

Richard Harland and Coworkers Identified Genes Expressed Specifically in the Organizer

Richard Harland and his colleagues hypothesized in the early 1990s that morphogens expressed in the organizer should promote the formation of dorsal structures such as those found in the head. To search for genes that promote the formation of dorsal structures and are expressed in the organizer during gastrulation, Harland's group took advantage of the fact that unfertilized *Xenopus* (frog) eggs exposed to UV light at their vegetal pole fail to form tadpoles. However, embryos derived from UV-treated eggs do form primitive mesoderm, but only ventral

mesoderm forms. Dorsal mesoderm, which normally gives rise to the notochord and somites, does not form in UV-treated eggs, and the resulting "ventralized" embryos soon die. The embryos appear to be lacking one or more morphogens needed for proper dorsal development.

To identify genes in the organizer that encode morphogens that promote the formation of structures such as the notochord, Harland and colleagues used a strategy called expression cloning, as shown in the Feature Investigation in **Figure 52.16**. First, after dissecting a frog embryo they isolated and purified mRNA from dorsal-lip tissue containing the organizer. They next constructed a cDNA library (see Chapter 20) from the purified mRNA.

Figure 52.16   The discovery of the morphogen noggin using expression cloning.

HYPOTHESIS Cells within Spemann's organizer express specific genes that encode proteins that regulate the development of dorsal structures during gastrulation.

STARTING MATERIALS *Xenopus* gastrulas and eggs.

Experimental level   **Conceptual level**

1 Isolate mRNA from dorsal lip tissue, which contains the organizer.

Surgical scissors

Extract and purify mRNA.

Dorsal lip

Gastrula   Blastopore

mRNA

Contains many different mRNAs expressed in the organizer (~15,000 different mRNAs)

2 Create cDNA library. Use cDNA library to make 15,000 different mRNAs. See Chapter 20 for a description of cDNA libraries.

Plasmid

cDNA ligated into vector

Bacterium

Each bacterium takes up 1 plasmid.

Clones of bacteria, each containing a different cDNA

DNA is isolated and transcribed into mRNA *in vitro*.

3 Expose unfertilized *Xenopus* eggs to UV light, then inject with an mRNA. Fertilize the eggs to determine which if any develop. Note: each egg was injected with a single type of mRNA.

UV

Xenopus eggs

Vegetal pole

mRNA injected into egg

UV light inactivates a morphogen and prevents dorsal mesoderm formation and the embryos usually die.

Most fertilized eggs formed only ventral mesoderm and died. One, however, developed into a tadpole (rescue).

4 Inject into UV-treated eggs, increasing amounts of the mRNA identified from step 3 that rescued the embryo.

mRNA from step 3

Fertilize

Egg

Fertilized egg   mRNA   Protein product of mRNA

Few copies of mRNA

Moderate number of copies of mRNA

Many copies of mRNA

Low number of proteins formed

Moderate number of proteins formed

Many proteins formed

5   **THE DATA**

Embryo from UV-treated egg   Partial rescue   Normal tadpole   Abnormal tadpoles with excess dorsal and anterior development

Increasing amounts of dorsal mesoderm-promoting mRNA and its protein product

This resulted in approximately 15,000 different cDNAs, each of which corresponded to a single gene. Their next task was to transcribe each of these cDNAs back into mRNA *in vitro*. In a third step, the researchers injected these different mRNAs into thousands of unfertilized UV-treated eggs, which they then fertilized. Each egg was injected with a single type of mRNA. Their idea was that any mRNA that produced a protein with dorsal mesoderm-promoting activity might "rescue" the embryos and result in free-swimming tadpoles.

Using this procedure, the researchers identified one mRNA whose protein product rescued the embryo. They then injected increasing amounts of this mRNA into UV-treated eggs, fertilized the eggs, and then examined the morphology of the resultant tadpoles. They discovered that the protein product from this mRNA acted as a morphogen because it induced different embryonic structures at different concentrations. At low concentrations, it partially rescued the embryos. At moderate concentrations, it resulted in the formation of normal embryos. However, at very high concentrations, the embryos developed too much dorsal mesoderm and not enough ventral mesoderm. These abnormal embryos developed large heads and extremely small trunks. Further studies identified the gene (which they

called *noggin*) that coded for this mRNA, and verified that it was transcribed in Spemann's organizer.

Later work revealed that noggin protein promotes dorsal development by inhibiting ventral development. Noggin protein inhibits at least two other morphogens that are known to induce ventral but not dorsal mesoderm. These studies showed that antagonistically acting proteins expressed and secreted from certain cells in the organizer can specify precise structures in a concentration-dependent fashion. Further work in mice revealed that *noggin* is expressed in the node of mammals. Moreover, specific deletion of the *noggin* gene in mice leads to defects in dorsal structures. This indicates that *noggin* plays a fundamental role during gastrulation in mammals as well as amphibians.

The discovery of *noggin* and its mechanism of action revealed that the normal development of embryos requires a balance between stimulatory and inhibitory factors. The inhibition of developmental processes—like the suppression of ventral mesoderm by *noggin*—appears to be a normal part of the mechanism by which development of an organism occurs. This important control mechanism has since been observed in other developmental pathways, one example of which is described next.

GENOMES & PROTEOMES

The Frzb Protein Can Inhibit Developmental Pathways Activated by Other Morphogens

Our understanding of morphogens and morphogenetic fields has benefited tremendously by experiments designed to identify regulatory genes that affect key steps in developmental processes. As described in the Feature Investigation, one general principle that has emerged is that there are both activators and repressors of the genetic processes that control embryonic development, and a fine balance between the activities of positively and negatively acting proteins must be established for proper development to occur. Researchers continue to discover more genes that affect embryonic development. It is often possible to deduce the mechanism by which the protein products of these genes function during development, by comparing the amino acid sequences of proteins that are related to each other. A classic example involves a protein called Frzb, which was discovered simultaneously by two different laboratories, one of which was using novel methods to identify proteins expressed specifically in Spemann's organizer of frogs.

Frzb is a protein containing a region of about 300 amino acids that is 50% identical to a region of another, larger protein known as the Frizzled receptor (**Figure 52.17a**). The Frizzled receptor is a component of a signaling pathway called the Wnt pathway, which controls many aspects of embryonic development in animal species as diverse as flies and humans. At the

amino terminus of the Frizzled receptor is a signal sequence that ensures that this protein will be translated on the endoplasmic reticulum and sorted to the cell surface. Following the signal sequence in the protein is a cysteine-rich domain or CRD, which contains many cysteine residues. At the carboxyl terminus of the Frizzled receptor are seven transmembrane domains that integrate the Frizzled receptor into the cell's plasma membrane such that the CRD is outside the cell (**Figure 52.17b**).

During normal embryonic development, the CRD of the Frizzled receptor is bound by secreted protein morphogens of the Wnt family, activating the Wnt pathway and causing cell differentiation. Like the Frizzled receptor, Frzb has a signal sequence and a CRD domain, but it lacks the seven transmembrane domains (see Figure 52.17a). Because Frzb lacks the seven transmembrane domains, it has been postulated that Frzb is secreted from embryonic cells rather than becoming integrated into the plasma membrane. Secreted Frzb protein molecules are thought to inhibit the Wnt signaling pathway near the Spemann organizer by binding free Wnt proteins (see Figure 52.17b), thereby preventing the Wnt proteins from acting as morphogens (that is, preventing them from binding to and activating Frizzled receptors).

Experiments have shown that Wnt proteins can bind to Frzb proteins. In addition, the experimental overexpression of Frzb in embryos blocks the Wnt signaling pathway, leading to developmental defects associated with defective Wnt signaling.

(a) **A comparison of protein domains in Frizzled and Frzb**

(b) **Interactions between Wnt and the Frizzled receptor or Frzb**

Figure 52.17 **The actions of Frzb and the Frizzled receptor protein show how embryonic development represents a balance between activating and inhibiting factors.** (a) Structures of the Frizzled receptor protein and Frzb protein. The Frizzled receptor contains both cysteine-rich domains required for binding the morphogen Wnt and transmembrane domains required for localizing in the cell's plasma membrane. Frzb contains the cysteine-rich domains but lacks the transmembrane domains. (b) The binding of Wnt to the Frizzled receptor in the plasma membrane induces mesoderm development. When Frzb, which is expressed in the Spemann organizer of amphibians or node of mammals, binds to molecules of free Wnt, the effects of Wnt on mesoderm induction are prevented. It is the ratios in which Frzb and Frizzled receptor proteins are expressed during development that fine-tune the inductive activity of the Wnt morphogen.

Biological inquiry: Are there other examples in animals in which one protein binds another to prevent its action?

Because Frzb resembles the Frizzled receptor protein sufficiently to bind Wnt proteins and prevent them from acting on cells, it has been called a "molecular decoy." Thus, by identifying and characterizing both positive and negative regulators of signaling pathways that affect key developmental steps, we can generate a more complete picture of the way this amazing process is carried out.

Given the complexity and precision of the five developmental stages we have described in this chapter, it may not be surprising that on occasion these processes fail to work properly. When this happens in humans, its impact can be devastating, as we see next.

52.3 Impact on Public Health

Many health problems in humans are caused by genetic or environmental factors that disrupt embryonic development. As we have seen, the early development of an embryo is among the most complex processes found in nature. Because embryonic development is affected by genetic as well as environmental factors, a number of diseases and disorders in humans stem from problems that arise during this process.

An example of a condition involving defective embryonic development is spina bifida, caused by the failure of the neural tube to close at either the anterior or posterior end during neurulation. Signs and symptoms of spina bifida, which occurs in about 1 of every 1,000 births in the U.S., vary depending on the degree to which the neural tube fails to close. In minor forms there may be gaps in a few vertebrae at the bottom of the back, and impairment of motor or sensory function may not be apparent at birth. However, neurological deterioration often becomes evident during childhood or adulthood. Surgical procedures soon after birth can improve the long-term quality of life for patients with spina bifida, but prevention of this condition—by increasing the amount of folic acid the embryo gets during pregnancy—is much more effective. Up to 75% of spina bifida cases may be prevented if women in the first trimester of pregnancy increase their dietary intake of folic acid, a common B vitamin needed during the rapid growth that occurs during embryonic development.

Whereas spina bifida appears to be caused by deficient folic acid levels during neurulation, other embryonic defects are caused by foreign chemicals introduced during pregnancy. One of the more tragic examples involved a drug called thalidomide, which originally was developed in Europe during the late 1950s and early 1960s as an anticonvulsive drug for epilepsy and was later prescribed as an antihistamine for allergies. Although further testing revealed no beneficial effects on these conditions, it was noted that thalidomide seemed to qualm nausea and help people sleep. In 1961 thalidomide was the most widely used sleeping medication in Europe, where it was also prescribed to combat morning sickness (nausea) in pregnant women.

Soon afterward, however, physicians in Europe and Australia noticed that babies born to mothers who had taken thalidomide sometimes had severely malformed limbs, with hands and feet often emanating directly from the body. It became clear that cells within the limb fields in human embryos are selectively and adversely affected by this toxic compound. Ultimately, thousands of babies worldwide were deformed due to the ingestion of thalidomide. The drug was quickly eliminated from pharmaceutical markets, and the U.S. was largely spared from the effects of thalidomide due to prompt legislative action. Such tragedies underscore the importance of continued research into the mechanisms that regulate embryonic development in mammals.

Given the deleterious effects of certain chemical compounds on specific phases of embryonic development in humans, it is not surprising that ingested or inhaled compounds such as tobacco, alcohol, and other drugs can also have severe and devastating consequences on a developing fetus. Indeed, the leading overall cause of mental retardation worldwide is fetal alcohol syndrome (FAS), which occurs in babies whose mothers drink excessive amounts of alcohol during pregnancy. In addition to cognitive disorders, children born with FAS also show malformed facial features and joints, and altered overall growth characteristics. These morphological features are likely caused by the generally detrimental effects that ethanol has on cell division. The cognitive deficits are thought to result from the death of developing CNS neurons, which are particularly susceptible to the toxic effects of ethanol.

Unlike the previous conditions, which are caused by environmental conditions surrounding the embryo, other embryonic defects are caused by genetic defects. The most common genetic disorder affecting humans is Down syndrome, which arises from an egg containing two copies of chromosome 21. The effects in an embryo that ultimately acquires three copies of chromosome

21—two from the egg and one from the sperm— are complex, such that multiple physical and neurological disorders are associated with this syndrome.

The World Health Organization has compiled the number of birth defects around the world by geographic region in order to assess the extent of the impact of developmental disorders on public health. The resulting document, entitled *The World Atlas of Birth Defects*, indicates that regional differences in the incidences of a given birth defect can be dramatic. For example, the rate of spina bifida varies from 1 in every 10,000 births in France to about 15 in every 10,000 births in Mexico. The occurrence of Down syndrome, by contrast, varies from about 7 cases per 10,000 births in Cuba to 22 per 10,000 births in Chile. Together, 27 different birth defects affect 2–3% of all babies born in the world today, an astonishing number when one considers the world's population of roughly 6.5 billion, with annual worldwide births of nearly 150 million.

CHAPTER SUMMARY

- The process by which a fertilized egg is transformed into an organism with distinct physiological systems and body parts is called embryonic development.
- The process by which different cells within a developing organism acquire specialized forms and functions, due to the expression of cell-specific genes, is called cellular differentiation.

52.1 Stages of Embryonic Development

- Development in many animals, including vertebrates, involves five general stages: fertilization, cleavage, gastrulation, neurulation, and organogenesis. (Figure 52.1)
- Major events in fertilization include the acrosomal reaction and the cortical reaction. During the acrosomal reaction, the binding of a sperm with the egg membrane triggers a series of events producing the fast block to polyspermy, which prevents other sperm from binding to the egg. During the cortical reaction, events involving IP$_3$ and calcium produce additional barriers to more sperm, a process called the slow block to polyspermy. (Figure 52.2)
- During cleavage, which involves cell divisions without cell growth, the daughter cells are called blastomeres. When the embryo forms an outer epithelial layer and an inner cavity, it is called a blastula. The mammalian counterpart of a blastula is called a blastocyst.
- Cleavage-stage embryos in triploblast organisms have an animal pole and a vegetal pole. The poles determine in part the future anteroposterior and dorsoventral axes of the embryo. (Figure 52.3)
- Incomplete or meroblastic cleavage occurs in birds, some fishes, and some other vertebrates whose eggs contain large amounts of yolk. Complete or holoblastic cleavage occurs in amphibians and mammals, whose eggs have smaller amounts of yolk; the cleavage-stage embryo in mammals is called a morula. In mammals, cleavage occurs in the oviduct. (Figures 52.4, 52.5)
- During gastrulation, the hollow ball of cells that makes up the blastula is converted into a highly organized structure in which the three germ layers—endoderm, mesoderm, and ectoderm—and the primordial germ cells are established. Each germ layer gives rise to specific structures. An embryo that has undergone gastrulation is called a gastrula. (Figure 52.6)

- In fate mapping, a small population of embryonic cells is specifically labeled with a harmless dye, and the fate of these labeled cells is followed to a later stage of embryonic development. (Figure 52.7)
- Gastrulation begins when a band of tissue invaginates (pinches in), creating a small opening called the blastopore. During gastrulation, the embryo forms a cavity called the archenteron, which will become the organism's digestive tract. Cadherins are critical in promoting the correct contact between cells. (Figures 52.8, 52.9)
- During gastrulation, two cellular processes are crucial to development: apical constriction, in which a reduction in the diameter of the actin rings connected to the adherens junctions causes the cells to elongate toward their basal end, and convergent extension, in which two rows of cells merge to form a single elongated layer.
- A distinguishing anatomical feature that forms at the end of gastrulation in all chordates is the notochord, which produces many signaling proteins that help establish tissue patterns in the embryo.
- During gastrulation, primordial germ cells (PGCs) become established. Amphibian eggs contain certain cytoplasmic determinants, called the germ plasm, that help define and specify the PGCs in the gastrula stage. In flies, PGCs are the first cells to form at the posterior end of the embryo and are called pole cells. (Figure 52.10)
- Neurulation is the multistep embryological process responsible for initiating CNS formation. Neurulation occurs just after gastrulation and involves the formation of the neural tube from ectoderm located dorsal to the notochord. (Figure 52.11)
- Another important cell lineage that arises during neurulation is the neural crest, which gives rise to all neurons and supporting cells of the peripheral nervous system in vertebrates.
- During neurulation, the mesoderm becomes segmented from the anterior end first, giving rise to blocklike structures of mesoderm called somites. (Figure 52.12)
- The developmental stage during which cells and tissues form organs is called organogenesis.

52.2 Control of Cell Differentiation and Morphogenesis During Animal Development

- The process of cell differentiation during embryonic development is driven by two mechanisms: autonomous specification, which involves cytosolic factors, and conditional specification, which involves cell-to-cell signals. (Figure 52.13)

- The substances that elicit different cellular responses at distinct concentrations in an embryo are proteins called morphogens.

- A type of experiment called an animal cap assay has been used extensively to identify factors (proteins) secreted by embryonic cells that induce cells in the animal pole to differentiate into mesoderm. (Figure 52.14)

- A group of embryonic cells called a morphogenetic field controls the formation of certain body structures.

- An extremely important morphogenetic field in the early gastrula is known as Spemann's organizer. The organizer secretes morphogens responsible for inducing the formation of a new embryonic axis. (Figure 52.15)

- Richard Harland and coworkers discovered the *noggin* gene in frog embryos and demonstrated that it acted to promote dorsal development by inhibiting ventral development. (Figure 52.16)

- The Frzb protein and Wnt signaling pathway are examples of how embryonic development depends on a balance between stimulating and inhibiting factors. (Figure 52.17)

52.3 Impact on Public Health

- An important example of a condition involving defective embryonic development is spina bifida, caused by the failure of the neural tube to close completely during neurulation.

- Thalidomide is toxic to cells within the limb fields in human embryos; some babies born to mothers who had taken thalidomide to treat the symptoms of morning sickness had severely malformed limbs.

- The leading overall cause of mental retardation worldwide is fetal alcohol syndrome (FAS), which occurs in babies whose mothers drink excessive amounts of alcohol during pregnancy.

- In Down syndrome, the most common genetic disorder affecting humans, the embryo has three copies of chromosome 21. The effects are complex and include multiple physical and neurological disorders.

Test Yourself

1. The acrosomal and cortical reactions occur during which stage of development?
   a. fertilization
   b. cleavage
   c. gastrulation
   d. neurulation
   e. organogenesis

2. Cadherins are
   a. adhesion protein molecules that attach cells to extracellular material.
   b. genes necessary for proper germ layer formation.
   c. adhesion protein molecules that allow cells of a given germ layer to adhere to each other.
   d. diploblastic germ layers.
   e. proteins that provide structural support to the cytoplasm of cells.

3. Cell differentiation that results from cell-to-cell signaling is
   a. autonomous specification.
   b. conditional specification.
   c. communicative specification.
   d. gastrulation.
   e. embryogenesis.

4. The three germ layers of triploblasts are established during
   a. fertilization.
   b. cleavage.
   c. blastula formation.
   d. gastrulation.
   e. neurulation.

5. The limbs of vertebrates are formed from
   a. the endoderm.
   b. the mesoderm.
   c. the ectoderm.
   d. all of the above.

6. In vertebrates, the digestive tract forms from
   a. the blastopore.
   b. the dorsal lip.
   c. the archenteron.
   d. the mesoderm.
   e. both a and d.

7. The cells that give rise to gametes
   a. are derived from the mesoderm.
   b. often arise independently from the three germ layers.
   c. originate directly from the tissue that will develop into the gonads.
   d. are some of the last cells of the embryo to differentiate.
   e. are formed the same way in all vertebrates.

8. Cells of the neural crest
   a. form the central nervous system.
   b. originate from the ectoderm.
   c. migrate to different areas of the body and differentiate into a variety of cells, including neurons of the peripheral nervous system.
   d. all of the above.
   e. b and c only.

9. The first functional organ(s) to form in an embryo is/are
   a. the lungs.
   b. the brain.
   c. the heart.
   d. the kidneys.
   e. the liver.

10. Morphogens
    a. are proteins that stimulate responses in the cells of an embryo.
    b. are important signaling proteins in the process of embryonic development.
    c. can elicit different cell responses at different concentrations.
    d. all of the above.
    e. a and c only.

Conceptual Questions

1. Define embryonic development.

2. Define cell differentiation.

3. Define gastrulation.

Experimental Questions

1. Why were scientists interested in identifying genes expressed exclusively in Spemann's organizer?

2. What hypothesis did Harland and colleagues test?

3. How did the scientists identify possible genes that are important for the development of dorsal structures, and how did they test the gene products to determine their activity?

Collaborative Questions

1. Discuss the process of gastrulation.

2. Discuss the process of neurulation.

www.brookerbiology.com

This website includes answers to the Biological Inquiry questions found in the figure legends and all end-of-chapter questions.

53

DEFENSE MECHANISMS OF THE BODY

CHAPTER OUTLINE

Electron micrograph of flu viruses.

All animals must defend themselves against environmental factors that threaten their survival, including predation, intraspecific fighting, accidents, and hazardous substances. In addition, animals must contend with various threats within their internal environment, including the invasion of potentially harmful microorganisms such as bacteria, the presence of foreign molecules such as the products of microorganisms, and the presence of abnormal cells such as cancer cells. The ability of an animal to ward off these internal threats—an animal's **immunity**, or immune defenses—is the subject of this chapter. The cells and organs within an animal's body that contribute to immune defenses collectively constitute an animal's **immune system**.

The study of immunity is called immunology. Immunologists examine the processes by which the immune system protects an animal from foreign matter, whether living or nonliving. In these processes, the body's immune defenses recognize the body's own molecules as "self" and attack anything it recognizes as foreign, or "nonself."

Immune defenses are often divided into two types. In **nonspecific (innate) immunity**, the body's defenses are present at birth and act against foreign materials in much the same way regardless of the specific identity of the invading material. Nonspecific immunity includes the body's external barriers (skin and mucous membranes), plus a set of cellular and chemical defenses that oppose substances that breach those barriers. All animals have nonspecific immune defenses. In contrast, **specific (acquired) immunity** develops only after the body is exposed to foreign substances. The specificity of this type of immunity results from the ability of certain cells of the immune system to recognize a foreign substance and initiate a response that targets that substance specifically. Such immune defenses have been identified in all vertebrates except for the jawless fishes, but they have not been unequivocally identified in invertebrates.

We begin the chapter with a brief overview of the different microorganisms that cause disease in animals. We then consider the mechanisms that provide animals with nonspecific and specific defenses against harmful microorganisms, and conclude with a discussion of the public health implications of some selected immunity-related conditions in humans.

53.1 Types of Pathogens

As just noted, the body's immune defenses must protect against a variety of foreign materials, but most important among them are disease-causing microorganisms or pathogens. Pathogens exist in nearly every possible ecological niche on Earth. Both terrestrial and aquatic animals, including invertebrates and vertebrates, encounter each of the three major types of pathogens: bacteria, viruses, and parasites.

Bacteria can either damage tissues at open wound sites or release toxins that enter the bloodstream and disrupt functions in other parts of the body. Bacteria are responsible for many diseases and infections, including typhoid fever, strep throat, skin infections, middle ear infections, and food poisoning. The major ways in which bacteria gain entry into an animal's body are through direct bodily contact, open wounds, inhalation through the respiratory tract, and ingestion via fecal contamination of food or water. The latter situation may arise because many infectious bacteria enter the intestines and are excreted in feces, which may be deposited near food or water sources for some animals.

Viruses—including the flu virus shown in the chapter-opening photo—are nucleic acids enclosed within a protein coat. They are much smaller than bacteria and lack the metabolic machinery to synthesize the proteins they need to replicate themselves. Instead, viruses must infect a host cell and use its biochemical and genetic apparatus, including nucleotides and energy sources, to make more viruses. The viral nucleic acid directs the host cell to synthesize the proteins required for viral replication. After entering a cell, some viruses, such as the common cold virus, multiply rapidly, kill the cell, and then infect other cells. Other viruses can lie dormant within host cells before suddenly undergoing rapid replication, which causes cell damage. Finally, certain viruses can transform their host cells into cancer cells. Viruses are responsible for a great variety of illnesses, including some sexually transmitted diseases. Like bacterial infections, viral infections can spread rapidly among animals and can be lethal. Viruses typically enter the body through the respiratory tract or through open wounds in skin and mucous membranes.

Parasites—whether protozoa, fungi, or multicellular eukaryotic organisms like worms—damage a host by using the host's nutrients for their own growth and reproduction or by secreting toxic chemicals. In humans, parasites account for an enormous number of cases of disease annually. For example, several hundred million people have been infected with the mosquito-borne protozoan that causes malaria. Parasitic infections may enter a host through the bite of an infected insect, as in malaria, by ingestion of food or water containing parasitic organisms (many roundworms), or in some cases by penetrating the skin, as in blood flukes.

53.2 Nonspecific (Innate) Immunity

Nonspecific immune defenses protect against foreign cells or matter without having to recognize the invaders' specific identities. This type of defense mechanism is also called innate because animals inherit the ability to perform these protective functions, and because this type of immunity does not require prior exposure to invaders. Instead of distinguishing among foreign materials, nonspecific defenses recognize some general, conserved property marking the invader as foreign, such as particular classes of carbohydrates or lipids present in the cell walls of many different kinds of microbes.

In this section, we will consider the nonspecific immune defenses, which include defenses at the body surfaces, the actions of phagocytic cells, the response to injury known as inflammation, and various antimicrobial proteins.

The Body Surface Represents an Initial Line of Defense

An animal's initial defenses against pathogens are the barriers provided by surfaces exposed to the external environment. Very few microorganisms can penetrate the intact skin or body surface of most animals, particularly the tough, thick, or scaly skin characteristic of many vertebrates or the rigid exoskeleton of many arthropods. Moreover, glands in the body surfaces of both invertebrates and vertebrates secrete a variety of antimicrobial molecules, including mild acids and enzymes such as lysozyme that destroy bacterial cell walls.

The mucus secreted by cells in the mucous membranes lining the respiratory and upper gastrointestinal tracts of vertebrates

also contains antimicrobial molecules, but more important, mucus is sticky. Microbes that become stuck in it are prevented from penetrating the mucous membrane barrier. They are either swept up into the pharynx and then swallowed, or they are engulfed by cells that are present in both tracts. Pathogens ingested with food are often destroyed by the acidic environment of an animal's midgut.

If a pathogen is able to penetrate a barrier and gain entry into an animal's internal tissues and fluids, other nonspecific defense mechanisms are activated. These mechanisms are mediated by several types of cells that reside in the body fluids and tissues, as described next.

Phagocytic Cells Provide Nonspecific Defense Against Pathogens That Enter the Body

Several different types of cells in vertebrates play key roles in nonspecific immunity (**Figure 53.1**). Many of these cells are **phagocytes**—cells capable of **phagocytosis**. Phagocytosis is a type of endocytosis in which the cell engulfs particulate matter, which usually is then destroyed by proteases or oxidizing agents such as hydrogen peroxide. Phagocytes are found in the body fluids, such as hemolymph and blood, and also within various tissues and organs. Phagocytes are present in all classes of animals and are among the most fundamental and ancestral forms of body defenses.

In vertebrates, most phagocytes belong to the type of blood cells called white blood cells, or **leukocytes** (Figure 53.1). All leukocytes are derived from a common type of stem cell (see Chapter 19), which in mammals and birds resides in the bone marrow. These stem cells give rise to several types of leukocytes and other cells that have specialized functions. The leukocytes involved in nonspecific immunity include neutrophils, eosinophils, monocytes, macrophages, and basophils.

Neutrophils are phagocytes and the most abundant leukocytes. They are found in blood and within tissues. Neutrophils engulf bacteria by endocytosis. The bacteria are destroyed within endocytotic vacuoles by proteases, oxidizing agents, and antibacterial proteins called defensins. The production and release of neutrophils from bone marrow is greatly stimulated during the course of an infection. **Eosinophils** are phagocytes found in large numbers in mucosal surfaces lining the gastrointestinal, respiratory, and urinary tracts, where they fight off parasitic invasions. **Monocytes** are phagocytes that circulate in the blood for only a few days, after which they take up permanent residence as **macrophages** in virtually all organs and tissues. Macrophages are strategically located where they will encounter invaders, including epithelia in contact with the external environment, such as skin and the linings of respiratory and digestive tracts, and in many organs at sites through which blood or lymph flows. Macrophages are large phagocytes capable of engulfing viruses and bacteria (**Figure 53.2**).

In contrast to these other leukocytes, **basophils** are secretory cells. They secrete the anticlotting factor heparin at the site of an infection, which helps the circulation flush out the infected site. Basophils also secrete histamine, which attracts infection-fighting cells and proteins to the site.

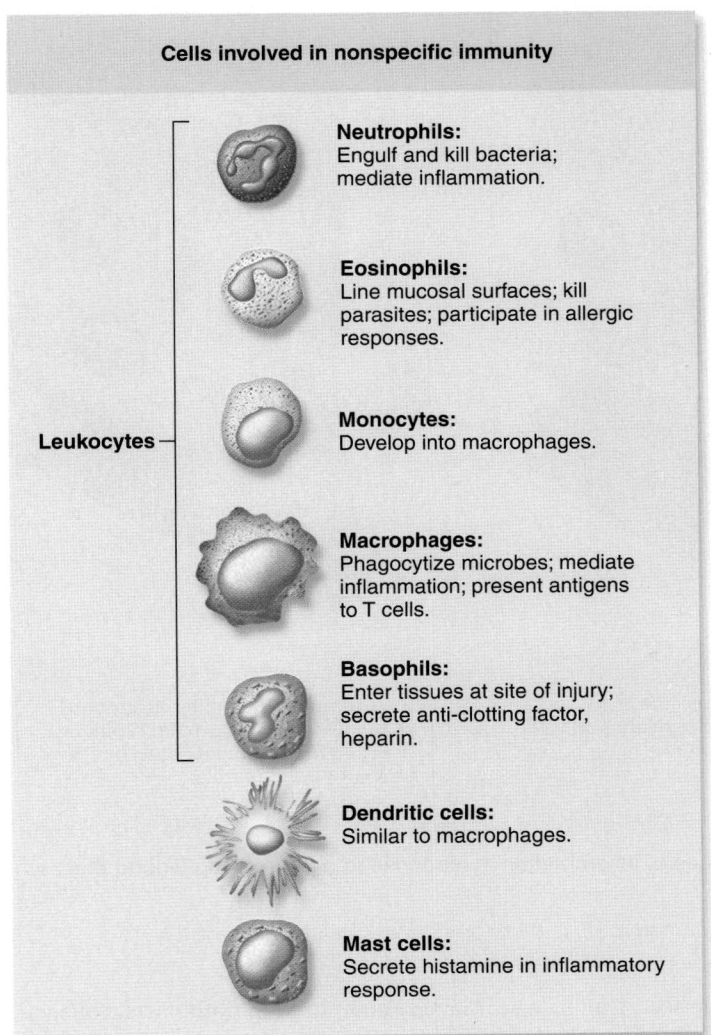

Cells involved in nonspecific immunity

Leukocytes {

Neutrophils:
Engulf and kill bacteria; mediate inflammation.

Eosinophils:
Line mucosal surfaces; kill parasites; participate in allergic responses.

Monocytes:
Develop into macrophages.

Macrophages:
Phagocytize microbes; mediate inflammation; present antigens to T cells.

Basophils:
Enter tissues at site of injury; secrete anti-clotting factor, heparin.

Dendritic cells:
Similar to macrophages.

Mast cells:
Secrete histamine in inflammatory response.

Figure 53.1 Cells involved in nonspecific immunity in vertebrates. Note that five of these types of cells are leukocytes.

Figure 53.2 Scanning electron micrograph of a macrophage engulfing numerous bacteria.

Inflammation Is a Nonspecific Response to Infection or Injury

Inflammation is an innate local response to infection or injury. The functions of inflammation are to destroy or inactivate foreign invaders, to clear the infected region of dead cells and other debris, and to set the stage for tissue repair. The key cellular mediators are phagocytes, primarily neutrophils, macrophages, dendritic cells, and mast cells.

The sequence of local events in a typical inflammatory response to a bacterial infection is summarized in **Figure 53.3**. The familiar signs and symptoms of tissue injury and inflammation are local redness, swelling, heat, and pain.

The events of inflammation are induced and regulated by chemical mediators. These include a family of proteins called **cytokines** that function in both nonspecific and specific immune defenses. Cytokines provide a chemical communication network that synchronizes the components of the immune response. Most cytokines are secreted by more than one type of immune system cell, and also by nonimmune cells such as endothelial cells and fibroblasts. Cytokine-like molecules have been identified in all vertebrates and in several evolutionarily distant invertebrate orders, in which they serve functions similar to some of those seen in vertebrates.

In addition to cytokines, substances secreted into the extracellular fluid from mast cells, injured tissue cells, and neutrophils also contribute to the inflammatory response (Figure 53.3, step 1). For example, histamine from mast cells and nitric oxide from endothelial cells cause dilation of the small blood vessels in an infected or damaged area, and the vessels become leaky (Figure 53.3, step 2).

These vascular changes provide two benefits. First, the increased blood flow to the inflamed area, which accounts for the redness and heat, increases the delivery of beneficial proteins and leukocytes. Second, the increased vascular permeability ensures that the plasma proteins that participate in inflammation can

Natural killer (NK) cells are another type of leukocyte that participate in both nonspecific immunity and specific immunity. These cells are part of the body's nonspecific defenses because they recognize general features on the surface of cancer cells or any virus-infected cells. NK cells arise in the bone marrow, but their life history remains unclear. They act by releasing chemicals into the vicinity of cancer or virus-infected cells and thereby kill those cells.

In addition to leukocytes, two other types of cells derived from bone marrow stem cells play important roles in nonspecific immunity. **Dendritic cells** are scattered throughout most tissues, where they perform various macrophage functions. **Mast cells** are found throughout connective tissues, particularly beneath the epithelial surfaces of the body. Mast cells secrete many locally acting molecules, including histamine. Histamine and other substances are involved in inflammation, an important component of the nonspecific defense mechanism to which we now turn.

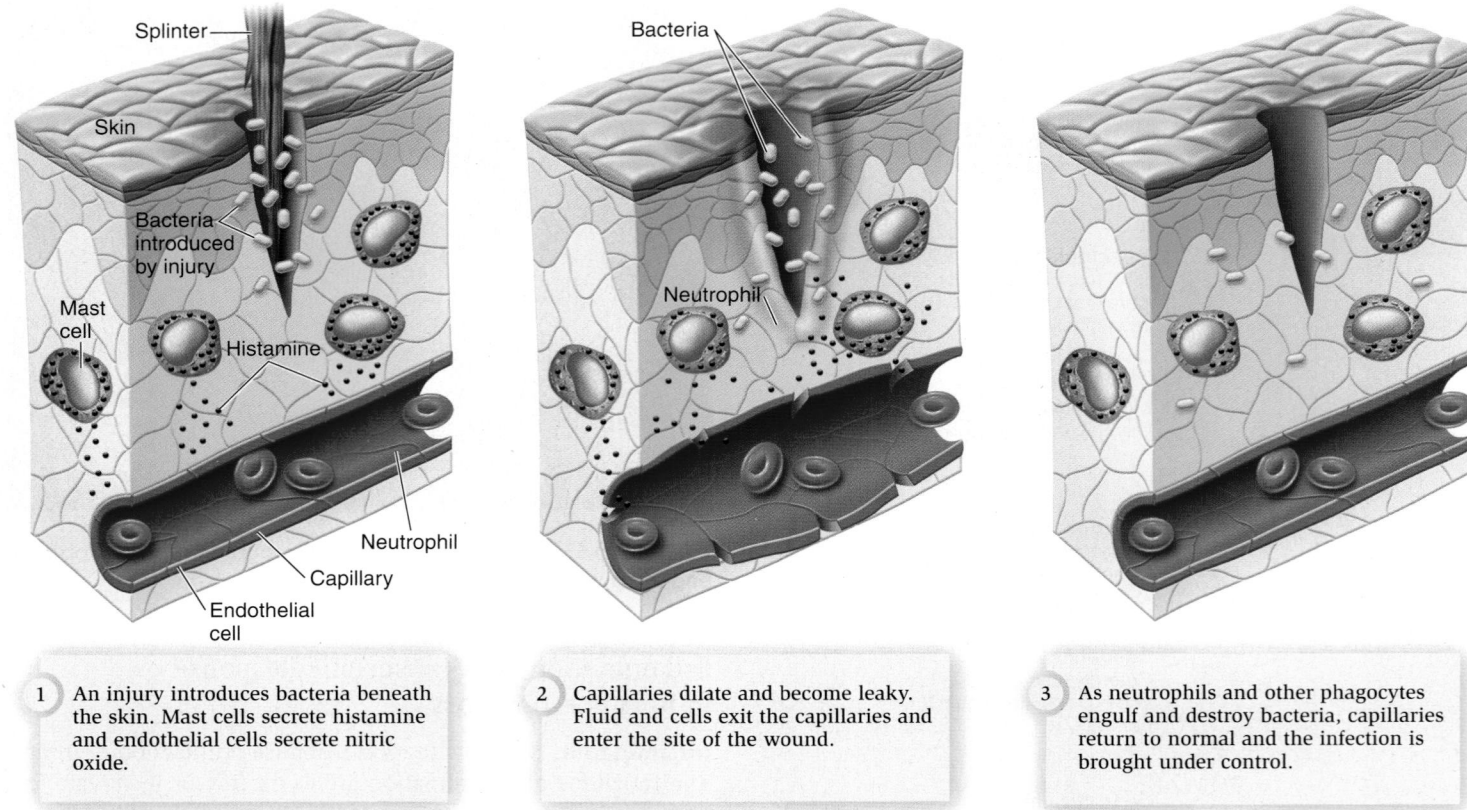

Splinter

Skin

Bacteria
introduced
by injury

Mast
cell

Histamine

Neutrophil

Capillary

Endothelial
cell

Bacteria

Neutrophil

1 An injury introduces bacteria beneath
the skin. Mast cells secrete histamine
and endothelial cells secrete nitric
oxide.

2 Capillaries dilate and become leaky.
Fluid and cells exit the capillaries and
enter the site of the wound.

3 As neutrophils and other phagocytes
engulf and destroy bacteria, capillaries
return to normal and the infection is
brought under control.

Figure 53.3   **The events in inflammation.**   Shown are the initial stages of inflammation in response to a penetrating wound that
introduces bacteria beneath the skin.

Biological inquiry: Inflammation often is associated with swelling of the inflamed area. Is there an adaptive value of swelling?

gain entry to the interstitial fluid. The swelling in an inflamed area also results from this increased leakiness of blood vessels, but it has no adaptive value of its own.

Once neutrophils and other leukocytes arrive at the site of an injury, they begin the process of phagocytizing and destroying invading microbes (Figure 53.3, step 3). The initial step in phagocytosis involves the interaction of phagocyte surface receptors with certain carbohydrates or lipids in the microbial cell walls. Subsequently, the neutrophils and other phagocytes also release into the extracellular fluid antimicrobial substances that can destroy microbes even before phagocytosis occurs. Other secreted substances like nitric oxide function as inflammatory mediators. The result is positive feedback: Once phagocytes enter the area and encounter microbes, they release inflammatory mediators that bring even more phagocytes into the area.

Inflammation sets the stage for tissue repair. In some tissues, undamaged cells surrounding an injury can multiply to replace destroyed cells and thereby restore tissue function, while in others they cannot. Liver cells, for example, multiply following injury or damage, but skeletal muscle cells do not. In either case, nearby connective tissue cells called fibroblasts divide rapidly and secrete large quantities of collagen, the major component of extracellular matrices, and nearby blood vessel cells proliferate to restore blood supply. Ultimately, tissue repair may be im-

perfect, leaving a scar largely composed of fibroblasts, collagen, and other proteins.

Antimicrobial Proteins Include Interferons and Complement Proteins

Interferons are proteins that generally inhibit viral replication inside host cells. In response to viral infection, most cell types produce interferons and secrete them into the extracellular fluid. When the interferons bind to plasma membrane receptors on the secreting cell and on other cells, the cell synthesizes a variety of proteins that interfere with the ability of the viruses to replicate. Interferons are not specific. Many kinds of viruses induce interferon synthesis, and the same interferons in turn can inhibit the multiplication of many different kinds of viruses.

The family of plasma proteins known as **complement** provides another means for extracellular killing of microbes without prior phagocytosis. Inactive complement proteins normally circulate in the blood at all times. When they contact the surface of microbes, a cascade of events results in the formation of active complement proteins. Five of the active proteins generated in the complement cascade form a multiunit protein called the **membrane attack complex (MAC)**, which by embedding itself in the microbial plasma membrane creates porelike channels.

Water and salts enter the microbe through the channels, and the microbe bursts.

We turn now to the type of immunity that is present in all jawed vertebrates, and that is distinguished by an ability to recognize great numbers of specific foreign molecules.

53.3 Specific (Acquired) Immunity

In specific immune defense mechanisms, cells of the immune system first encounter and later recognize a specific foreign cell or protein to be attacked, as opposed to recognizing some general feature of pathogens. Because this type of immunity requires exposure to foreign substances, it is also called acquired immunity. Any foreign molecule that can trigger a specific immune response is called an **antigen**. An antigen is any molecule that the host does not recognize as self. Most antigens are either proteins or very large polysaccharides. Antigens include the protein coats of viruses, specific macromolecules on pollens and other allergens, cancer cells, transplanted cells, and toxins.

This type of immunity is found in all classes of vertebrates except jawless fishes and was once thought to be absent in invertebrates. Recently, however, scientists who study the evolution of immune systems have uncovered several interesting features of invertebrate immune function that suggest that some invertebrates have a limited ability to adapt immune activity to a specific invader. Despite this intriguing finding, however, we will consider specific immune responses in the context in which they are best understood, the jawed vertebrates.

Lymphocytes Provide Specific Immunity Against Antigens

Lymphocytes are a type of leukocyte that are responsible for specific immunity (**Figure 53.4**). The two types of lymphocytes are **B cells** and **T cells**. B cells were first observed to mature in an avian organ called the bursa of Fabricius—thus the name B cells. In mammals, B cells mature within bone marrow. Some B cells differentiate further into **plasma cells**, which synthesize and secrete antibodies, proteins that bind to and help destroy foreign molecules, as described later. T cells are so named because they mature within the thymus gland.

Most T cells directly kill infected, mutated, or transplanted cells. T cells include two distinct types of lymphocytes, designated cytotoxic T cells and helper T cells. **Cytotoxic T cells** travel to the location of their targets, bind to these targets by combining with an antigen on them, and directly kill those targets via secreted chemicals. Responses mediated by cytotoxic T cells are directed against body cells that have become cancerous or infected by pathogens. Natural killer cells also destroy such cells by secreting toxic chemicals.

As the name **helper T cells** implies, these cells do not themselves function as "attack" cells. Instead, they assist in the activation and function of B cells and cytotoxic T cells. With only a few exceptions, B cells and cytotoxic T cells cannot function adequately unless they are stimulated by cytokines secreted from

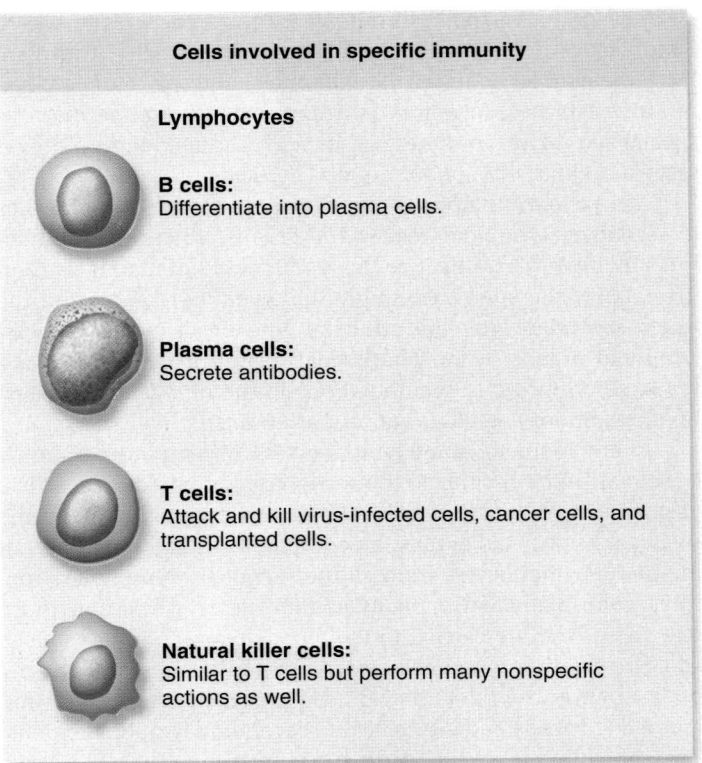

Figure 53.4   Cells involved in specific immunity in vertebrates.   Some important functions of each are included.

helper T cells. Helper T cells can be identified by the presence of a unique protein on their surface called CD4, which plays a role in the cell's ability to interact with other lymphocytes. Cytotoxic T cells are identified by a different membrane protein, CD8.

Lymphocytes carry out their role by recognizing antigens such as those found on viruses, bacteria, and the surface of cancer cells. The ability of lymphocytes to distinguish one antigen from another plays a central role in specific immunity. Immunologists recognize two types of specific immunity. In **humoral immunity**, plasma cells secrete antibodies that bind to antigens. In **cell-mediated immunity**, cytotoxic T cells directly attack and destroy infected body cells, cancer cells, or transplanted cells.

Most Lymphocytes Exist in Lymphoid Organs and Tissues

Now that we have a general understanding of the roles of lymphocytes, let's consider where they are found in the body. Like all leukocytes, lymphocytes circulate in the blood, but most of them reside in a group of organs and tissues that constitute the **lymphatic system** (**Figure 53.5a**). The system is composed primarily of a network of lymphatic vessels. These vessels drain the fluid known as lymph, which filters out of capillaries into the interstitial spaces. The lymph is eventually returned to the circulatory system. Various lymphoid organs and tissues are located throughout the lymphatic system. They are grouped into primary and secondary lymphoid organs.

The primary lymphoid organs are the structures in which lymphocytes differentiate into mature immune cells. These are the bone marrow in birds and mammals, and the thymus gland in all vertebrates. In animals without extensive bone marrow, specialized regions of other organs such as the kidney and liver serve as primary lymphoid organs.

The primary lymphoid organs supply mature lymphocytes to secondary lymphoid tissues and organs, where the lymphocytes multiply. These include the lymph nodes of mammals (**Figure 53.5b**), the spleen (found in all jawed vertebrates and the largest secondary lymphoid structure), the tonsils (small, rounded lymphoid organs in the pharyngeal region of mammals), and scattered lymphocyte accumulations in the linings of the intestinal, respiratory, genital, and urinary tracts.

At any moment, some lymphocytes are circulating from the bone marrow or thymus to the secondary lymphoid organs. The vast majority of lymphocytes, though, are in transit between the secondary lymphoid organs, blood, lymph, and all the tissues of the body. Lymphocytes from all the secondary lymphoid structures continually enter the lymphatic vessels draining those structures, and are carried to the bloodstream. Simultaneously, some circulating lymphocytes are exiting venules all over the body to enter the interstitial fluid. From there, they enter into lymphatic vessels and are carried to secondary lymphoid organs. They may then leave the lymphatic vessels to take up residence in a lymph node or other structure.

This recirculation is going on all the time, but during an infection the migration of lymphocytes into an inflamed area is greatly increased. Lymphocyte recirculation increases the likelihood that any given lymphocyte will encounter an antigen it is specifically programmed to recognize.

Specific Immune Responses Occur in Three Stages

Thus far, we have considered the types of lymphocytes involved in a specific immune response and how they circulate through the body. A specific immune response can usually be divided into three stages (**Figure 53.6**): (1) antigen encounter and recognition by lymphocytes, (2) lymphocyte activation and cell division to produce effector cells and memory cells, and (3) the attack launched by the activated lymphocytes and/or their secretions.

During its development, each lymphocyte synthesizes a type of membrane receptor that can bind to a specific antigen. If subsequently the lymphocyte encounters that antigen, the antigen becomes bound to the receptor. This specific binding is the meaning of the word "recognize" in immunology. Antigens that bind to a lymphocyte receptor are said to be recognized by the lymphocyte. The ability of lymphocytes to distinguish one antigen from another, therefore, is determined by the nature of their plasma membrane receptors. Each lymphocyte is specific for just one type of antigen.

The binding of antigen to a receptor must occur for lymphocyte activation to take place. Upon binding to an antigen, the lymphocyte undergoes multiple cycles of cell division.

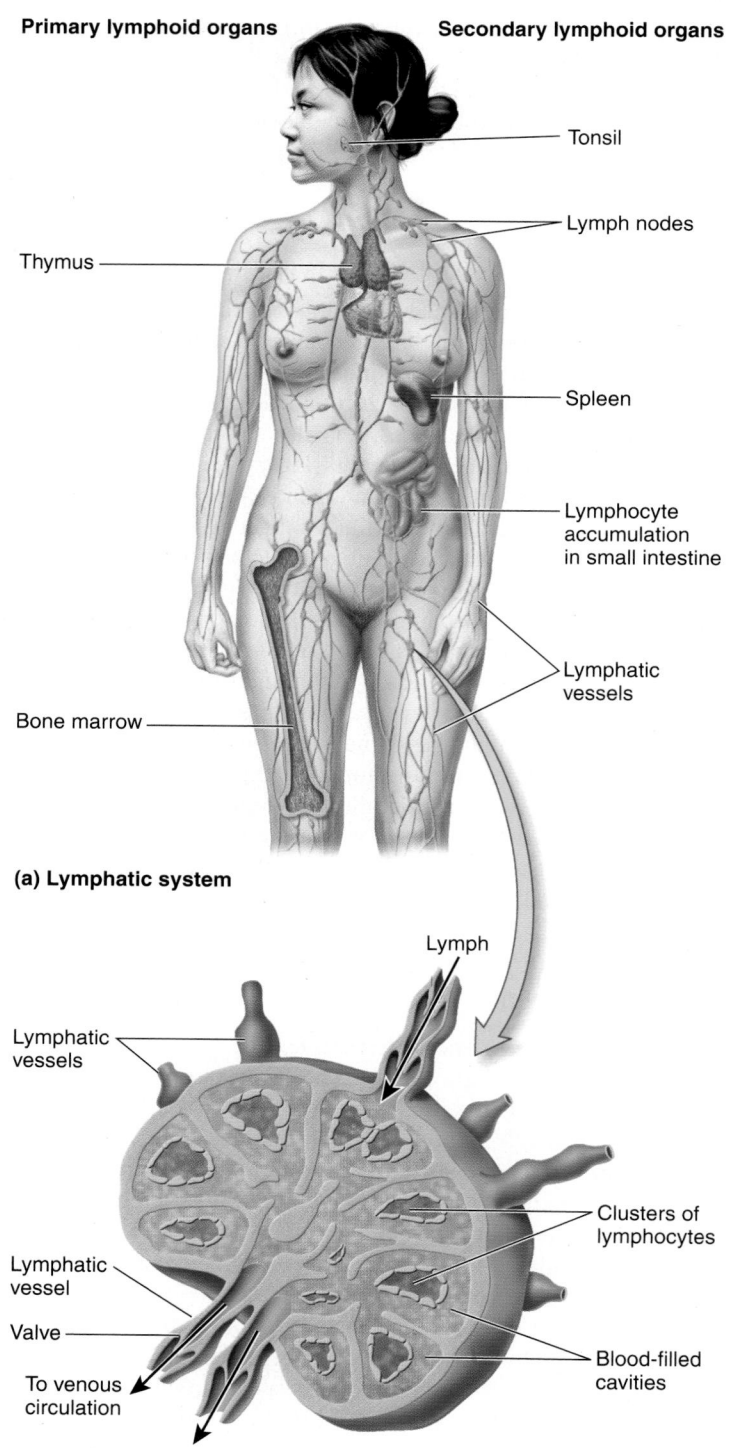

Primary lymphoid organs   **Secondary lymphoid organs**

Tonsil

Lymph nodes

Thymus

Spleen

Lymphocyte accumulation in small intestine

Lymphatic vessels

Bone marrow

(a) Lymphatic system

Lymph

Lymphatic vessels

Clusters of lymphocytes

Lymphatic vessel

Valve

Blood-filled cavities

To venous circulation

(b) Lymph node

Figure 53.5   The lymphatic system.   (a) The major components of the lymphatic system in humans. Primary lymphoid organs are shown in red, and secondary lymphoid structures are shown in green. **(b)** The structure of a lymph node. Lymph nodes occur along the course of lymphatic vessels, which drain interstitial fluid from tissues and return it to the venous circulation. Within a lymph node, lymph percolates through open cavities containing clusters of lymphocytes.

Stages of a specific immune response

Figure 53.6 **The three stages of a specific immune response.** Note the central role of helper T cells in specific immunity. The + symbol indicates a stimulatory action of cytokines. Both B cells and T cells undergo cell division to form clones when activated, and in both cases a portion of the cells are set aside as memory cells to fight off a future infection of the same type.

The result is the formation of many lymphocytes that express receptors identical to the receptor that first recognized the antigen. When an antigen-stimulated lymphocyte divides and replicates itself, the progeny of this lymphocyte—all of which express the same receptor—are called clones, and the process by which these clones are formed is called **clonal selection**. This term emphasizes that lymphocyte proliferation is selected by exposure to an antigen, an idea first established by Niels Jerne, Sir Frank MacFarlane, and Joshua Lederberg in the 1950s. The chemical nature of the antigen determines which individual lymphocytes will be activated to form clones. In a typical person, the size of the lymphocyte population is staggering. Over 100 million different lymphocytes, each with the ability to recognize a unique antigen, are found in a person's immune system. This vast population explains why our bodies are able to recognize so many different antigens as foreign, and eventually destroy them. The process of clonal selection requires the function of helper T cells that secrete cytokines—chemicals that promote cell division.

After activation, some of the cloned lymphocytes become **effector cells**, which carry out the attack response. Other cells function as **memory cells**, which remain poised to recognize the antigen if it returns in the future.

The effector plasma cells and cytotoxic T cells launch an attack against all antigens of the kind that initiated the immune response. Plasma cells carry out a humoral response by secreting antibodies into the blood. These antibodies then recruit and guide other molecules and cells that perform the actual attack. Activated cytotoxic T cells, by contrast, carry out cell-mediated immunity. They directly attack and kill the cells bearing the antigens.

Once the attack is successfully completed, the great majority of B cells, plasma cells, and cytotoxic T cells that participated in it die by apoptosis. The timely death of these effector cells prevents the immune defense from becoming excessive and possibly destroying the body's own tissues. However, memory cells persist even after the immune response has been successfully completed.

B Cells Mediate Humoral Immune Responses by Producing Antibodies

When B cells are activated by an antigen and by cytokines produced by helper T cells, they proliferate and differentiate into plasma cells, which secrete **antibodies**. These are proteins that travel all over the body to reach antigens identical to those that stimulated their production. In the extracellular body fluids, antibodies combine with these antigens and guide an attack that eliminates the antigens or the cells bearing them. As mentioned, such antibody-mediated responses are also called humoral immune responses, the adjective "humoral" denoting communication by way of soluble chemical messengers. Antibody-mediated responses are the major defense against bacteria, viruses, and other microbes in the extracellular fluid, and against toxin molecules.

Antibodies are members of a family of proteins called **immunoglobulins**. Some immunoglobulins, such as antibodies, are soluble proteins that are secreted from plasma cells (**Figure 53.7a**). Other immunoglobulins have a transmembrane domain that anchors them in the membrane. These immunoglobulins function as receptors for antigens **Figure 53.7b**).

Each immunoglobulin molecule is composed of four interlinked polypeptides: two long **heavy chains** and two short **light chains** (Figure 53.7a). A hinge region that provides the molecule with flexibility separates the light chains and upper parts of the heavy chains from the lower parts of the heavy chains. A key feature of immunoglobulins is their variable region, that varies among different B cells. The variable region is the site that specifically recognizes an antigen. The mechanism for variation in this region of the immunoglobulin will be described later.

Prior to activation, B cells produce B-cell receptors, but do not secrete antibodies. Upon activation, the resulting plasma cells begin to synthesize massive amounts of antibodies, which are secreted from the cell. Interestingly, the B-cell receptor and the antibodies made by plasma cells are encoded by the same genes. Upon B-cell activation, the pre-mRNA for the heavy chain is alternatively spliced, a phenomenon described in Chapter 13, so that the transmembrane domain is not present in the protein. For this reason, the B-cell receptor and secreted antibodies from

a particular B cell and its resulting plasma cells recognize the exact same antigen.

Mammals have five classes of immunoglobulins designated IgM, IgG, IgE, IgA, and IgD. All vertebrates have IgM molecules. Some vertebrates have only some of the other five classes and also express unique immunoglobulins not found in mammals. By contrast, invertebrates lack immunoglobulins. However, they have proteins containing regions called Ig-domains (or Ig-folds) with sequences that are similar to those of immunoglobulins. These proteins may be ancestral to immunoglobulins and in some cases have been shown to carry out immune activities.

The most abundant immunoglobulins in mammals are IgM and IgG, commonly called gamma globulin. Together these two immunoglobulin classes provide the bulk of specific immunity against bacteria and viruses in the extracellular fluid.

IgE antibodies participate in defenses against multicellular parasites and also mediate allergic responses. Although present in blood, they also exist in mast cell membranes. When mast cell IgE molecules bind antigen, the mast cell secretes its histamine into the extracellular fluid, causing vasodilation and contributing to the allergic response. In people who are particularly sensitive to allergens, this response is easily demonstrated by a pinprick injection of an antigen, such as proteins associated with hay fever, into a small region under the skin. The resultant local inflammation and reddening of the skin is mediated in large part by IgE molecules.

IgA antibodies are secreted by plasma cells in the linings of the gastrointestinal, respiratory, and genitourinary tracts, and in tear ducts and salivary glands. They act locally in the linings of these structures or on their surfaces. For example, IgA molecules secreted into saliva help keep animals' mouths relatively free of

(a) Secreted antibody   **(b) B-cell receptor**

Figure 53.7   **Immunoglobulins. (a) Secreted antibody, and (b) B-cell receptor.**   Immunoglobulins are composed of the two heavy chains and two light chains. Disulfide bonds hold the chains together. Within each immunoglobulin class, the constant regions, including the Fc regions (stems), have identical amino acid sequences. In contrast, the antigen-binding sites formed by the light- and heavy-chain variable regions have unique amino acid sequences and give each antibody its specificity for a particular antigen.

pathogens. IgA molecules are also secreted by the mammary glands of mammals shortly after birth and therefore are the major antibodies in milk. Although the functions of IgD are still unclear, IgD molecules are present both in blood and on the surface of B cells, and they are known to bind antigen on B cells, thus possibly contributing to B-cell activation.

The immunoglobulin's two heavy chains have a stem called the Fc region (Figure 53.7a), which contains two or three domains depending on the class of immunoglobulin. The amino acid sequences of the Fc domains are identical for all immunoglobulins of a given class and are therefore known as **constant regions**.

Another part of each heavy chain and its associated light chain contain a constant domain, plus a domain called the **variable region** that serves as the antigen-binding site. In contrast to the constant domains of the heavy and light chains, the amino acid sequences of the variable domains vary widely from immunoglobulin to immunoglobulin in a given Ig class. The enormous number of variable sequences results in countless unique structures of immunoglobulins within each class. Thus, each of the five classes of antibodies contains up to millions of unique immunoglobulins, each capable of combining with only one specific antigen or, in some cases, with several antigens whose structures are very similar. The genetic basis for this remarkable array of immunoglobulins was first identified in the 1970s, as we see next.

GENOMES & PROTEOMES

Recombination and Hypermutation Produce an Enormous Number of Different Immunoglobulin Proteins

The existence of so many different types of immunoglobulins raises an interesting question, because the human genome contains only about 200 genes that code for immunoglobulins: How can the body produce immunoglobulins having millions of different antigen-binding sites, when each immunoglobulin protein requires coding by a distinct gene? The answer is that immunoglobulin diversity is in part the result of a genetic process unique to lymphocytes.

Beginning in the 1970s, work by Susumu Tonegawa and others demonstrated that segments of genes that code for antigen binding sites of immunoglobulin light and heavy chains are cut by lymphocyte-specific enzymes, a segment is removed, and then the ends of the genes are rejoined to form new genes (**Figure 53.8**). The sites within the immunoglobulin genes where cutting and rejoining occur vary randomly from B cell to B cell, resulting in great diversity of the genes coding for the immunoglobulins that are produced by different B cells. In this way, immunoglobulin diversity is generated in somatic cells, not by inheritance through gametes.

Along the length of a typical immunoglobulin gene are numerous domains that code for a piece of the final immunoglobulin molecule. In light chains, these domains are of three types, called the variable, joining, and constant domains. A total of

Figure 53.8 The mechanism of immunoglobulin diversity. Events similar to those depicted here also occur in the heavy chains.

about 300 variable (V) domains code for the N-terminal region of the antigen binding site. These are next to four joining (J) domains and a single constant (C) domain. Each domain along the length of the gene is flanked by recognition sequences that bind two enzymes, called RAG-1 and RAG-2 (for recombination-activating gene). These enzymes, which are only expressed in developing lymphocytes, cut a random V domain and paste it to a random J domain. The result is a new, permanent immunoglobulin gene for that B cell. Because any V domain can be linked with any J domain, the number of possible final genes among different B cells is huge. Additionally, heavy chains have multiple domains that are spliced together in this way, except that they have yet another domain (designated D) and more V domains, yielding an even greater number of possible heavy-chain genes. Because any heavy chain can combine with any light chain in a given B cell, the number of possible combinations of immunoglobulins is immense.

The number of possible immunoglobulins is increased even further in two more important ways within B cells. First, the joining process of the V, D, and J domains is not always precise. Occasionally, a few nucleotides may be lost at a joining end, resulting in a different amino acid sequence in the immunoglobulin protein. Second, in a subset of activated B cells, the DNA coding for the variable antigen-binding sites of immunoglobulins undergoes a process known as **hypermutation**, which primarily produces point mutations. During this process, cytosines are deaminated into uracils, leading to numerous C to T mutations during DNA replication. In addition, the presence of numerous uracils within the DNA recruits error-prone DNA polymerases, leading to several additional types of mutations. The result is a hypervariable region of the light and heavy chains of all immunoglobulins. Hypermutation in lymphocytes appears to arise from lymphocyte-specific expression of a novel enzyme capable of deaminating cytosine.

The three processes of gene recombination, imprecise joining of domains, and hypermutation cause lymphocytes to produce an incredibly diverse array of antibodies capable of recognizing many different antigens. Because the body makes hundreds of millions of different types of lymphocytes, nearly any foreign antigen that is taken into the body will be recognized by some lymphocytes in this large population.

T-Cell Receptors Are Present in Plasma Membranes and Are Not Secreted

Like immunoglobulins, T-cell receptors for antigens have specific regions that differ from one T cell clone to another. As shown in **Figure 53.9a**, they are composed of two polypeptides, each with a variable and constant region, along with a transmembrane domain. The variable regions recognize an antigen. As in B-cell development, multiple DNA rearrangements occur during T-cell maturation, leading to millions of distinct clones of T cells, each possessing receptors of a unique specificity. For T cells, this maturation occurs during their development in the thymus.

In addition to their general structural differences, the B-cell receptors and T-cell receptors differ in a much more important way. T-cell receptors cannot combine with antigen unless the antigen is first complexed with certain of the body's own plasma membrane proteins. The T-cell receptor then combines with the entire self protein–antigen complex.

The plasma membrane self proteins that must be complexed with the antigen in order for T-cell recognition to occur are encoded by a gene family known as the **major histocompatibility complex (MHC)**, and thus the proteins are called MHC proteins. Other than identical twins, no two individuals have the same sets of MHC genes, so no two individuals have the same MHC proteins on the plasma membranes of their cells. MHC proteins are cellular "identity tags" that serve as genetic markers of self.

Two classes of MHC proteins are known. Class I MHC proteins are found on the surface of all human body cells except erythrocytes. Class II MHC proteins are found only on the surface of macrophages, B cells, and dendritic cells.

The two different types of T cells have different MHC requirements. Cytotoxic T cells require antigen to be associated with class I MHC proteins, whereas helper T cells require class II MHC proteins. One reason for this difference stems from the presence, as described earlier, of CD4 proteins on helper T cells and CD8 proteins on cytotoxic T cells; CD4 binds to class II MHC proteins, whereas CD8 binds to class I MHC proteins.

How do antigens, which are foreign, end up complexed with MHC proteins on the surface of the body's own cells? The answer involves the process known as antigen presentation, described next.

Antigen Presentation to Helper T Cells As previously noted, T cells can bind antigen only when the antigen appears on the plasma membrane of a host cell complexed with the cell's MHC proteins. Cells bearing these complexes, therefore, function as **antigen-presenting cells (APCs)**. Helper T cells require class II MHC proteins to function. Because only macrophages, B cells, and dendritic cells express class II MHC proteins, only these cells can function as APCs for helper T cells.

Let's consider the function of macrophages as APCs for helper T cells (**Figure 53.9b**). After a microbe or noncellular antigen has been phagocytized by a macrophage in a nonspecific response, antigens, such as proteins, are partially broken down into smaller peptide fragments by the macrophage's proteolytic enzymes within intracellular vesicles called endosomes. The resulting digested fragments then bind in the endosome to class II MHC proteins synthesized by the macrophage. Each fragment-MHC complex is then transported to the cell surface, where it is displayed in the plasma membrane. It is to this entire complex on the cell surface of the macrophage that a specific helper T-cell receptor binds. The CD4 protein helps link the two cells together. What is complexed to MHC proteins and presented to the helper T cells are not the intact antigens but instead the peptide fragments of the antigen—called antigenic determinants, or **epitopes**. Even so, it is customary to call this antigen presentation rather than epitope presentation.

B cells process antigen and present it to helper T cells in essentially the same way as occurs in macrophages. The ability of B cells to present antigen to helper T cells is a second function

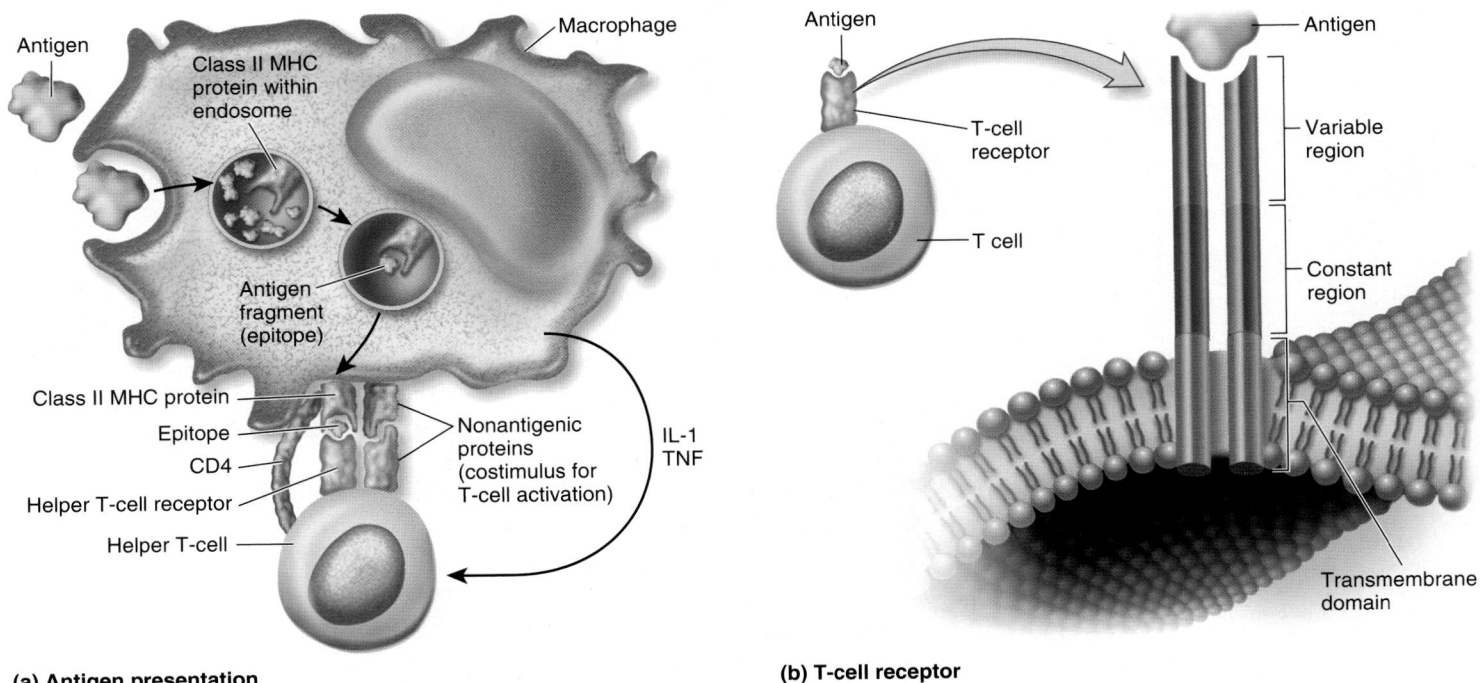

(a) Antigen presentation

(b) T-cell receptor

Figure 53.9 **The functions of antigen-presenting cells in helper T-cell activation.** In the initial events in helper T-cell activation, antigen fragments are complexed with a class II MHC protein within an antigen-presenting cell (a). The complex is then displayed on the cell surface and binds to a helper T-cell receptor (b). Also required for T-cell activation are the binding of nonantigenic proteins between the APC and the attached helper T cell, and the actions of the cytokines IL-1 and TNF.

of B cells in response to antigenic stimulation, in addition to their differentiation into antibody-secreting plasma cells.

The binding between the helper T-cell receptor and antigen bound to class II MHC proteins on an APC is the essential antigen-specific event in helper T-cell activation. However, by itself this specific binding will not result in T-cell activation. In addition, interactions occur between nonantigenic pairs of proteins on the surfaces of the attached helper T cell and the APC, and these interactions provide a necessary costimulus for T-cell activation (Figure 53.9).

Finally, the antigenic binding of the APC to the T cell plus the costimulus induce the APC to secrete large amounts of two cytokines—interleukin 1 (IL-1) and tumor necrosis factor (TNF). These molecules also stimulate the attached helper T cell.

Thus, the APC participates in the activation of a helper T cell in three ways: (1) antigen presentation, (2) provision of a co-stimulus in the form of a matching nonantigenic plasma membrane protein, and (3) secretion of IL-1 and TNF. As discussed later, activated T cells secrete molecules that stimulate other cells, such as B cells.

Antigen Presentation to Cytotoxic T Cells Unlike helper T cells, cytotoxic T cells require class I MHC proteins for activation. Class I MHC proteins are synthesized by all nucleated cells, and therefore any such cell can act as an APC for a cytotoxic T cell. This distinction helps explain the major function of cytotoxic T cells—destruction of any of the body's own cells that have become cancerous or infected with viruses. The crucial point is that the antigens that complex with class I MHC pro-

teins arise within body cells. They are endogenous antigens, synthesized by a body cell.

How do such antigens arise? In viral infections, once a virus has entered a host cell, the viral nucleic acid instructs the host cell to manufacture viral proteins, which are foreign to the cell. In cancerous cells, one or more of the cell's genes have become altered by chemicals, radiation, or other factors. The altered genes, called oncogenes, code for proteins that are not normally found in the body. Such abnormal proteins act as antigens.

In both virus-infected cells and cancerous cells, cytosolic enzymes hydrolyze some of the endogenously produced antigenic proteins into peptide fragments, which are transported into the endoplasmic reticulum. There the fragments are complexed with the host cell's class I MHC proteins and then shuttled by exocytosis to the plasma membrane, where a cytotoxic T cell specific for the antigen/MHC protein complex can bind to it. Once binding occurs, cytotoxic T cells release into the extracellular fluid chemicals that kill the infected or cancerous cell. This process will be discussed later.

Humoral Immune Responses Involve B Cells and T Cells Working Together

One classic humoral immune response results in the destruction of bacteria. The sequence of events, which is quite similar to the response to a virus in the extracellular fluid, is summarized in **Figure 53.10**. For this example, we consider the response in mammals, in which lymph nodes are present. Many features of the response, however, are similar in other vertebrates.

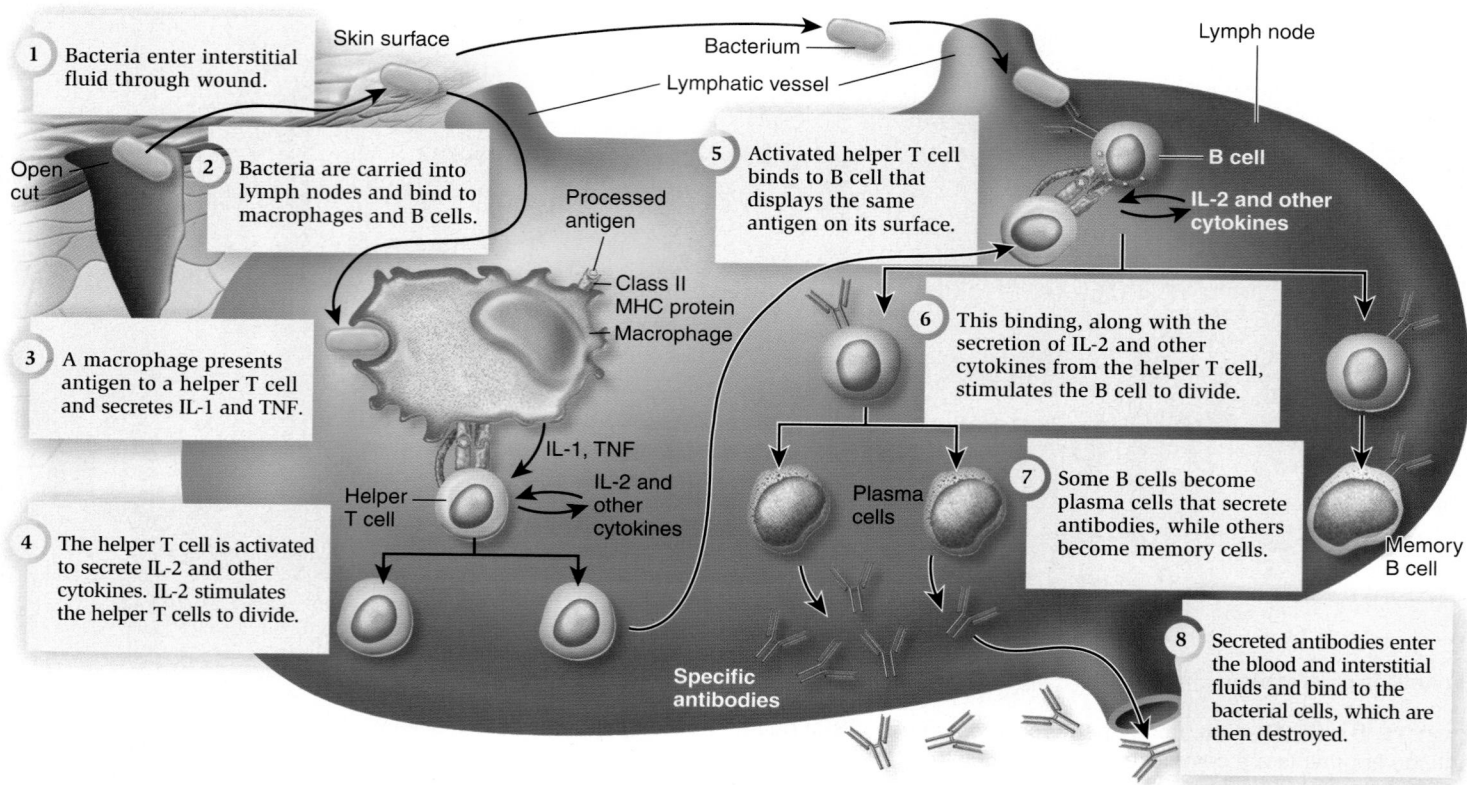

1 Bacteria enter interstitial fluid through wound.

Skin surface

Bacterium

Lymphatic vessel

Lymph node

2 Bacteria are carried into lymph nodes and bind to macrophages and B cells.

Open cut

Processed antigen

5 Activated helper T cell binds to B cell that displays the same antigen on its surface.

B cell

IL-2 and other cytokines

Class II MHC protein

Macrophage

3 A macrophage presents antigen to a helper T cell and secretes IL-1 and TNF.

6 This binding, along with the secretion of IL-2 and other cytokines from the helper T cell, stimulates the B cell to divide.

IL-1, TNF

Helper T cell

IL-2 and other cytokines

Plasma cells

7 Some B cells become plasma cells that secrete antibodies, while others become memory cells.

Memory B cell

4 The helper T cell is activated to secrete IL-2 and other cytokines. IL-2 stimulates the helper T cells to divide.

Specific antibodies

8 Secreted antibodies enter the blood and interstitial fluids and bind to the bacterial cells, which are then destroyed.

Figure 53.10 Events in a typical humoral immune response. Most of the events depicted occur within a lymph node.

This process starts the same way as for nonspecific responses, with the bacteria penetrating one of the body's linings and entering the interstitial fluid (Figure 53.10, step 1). The bacteria then move with lymph into the lymphatic system and are carried to lymph nodes (step 2). Within the lymph node, a macrophage and a B cell recognize the bacteria as a foreign substance and bind to the bacteria.

The process of B-cell activation usually requires the activation of helper T cells, which was depicted in Figure 53.9 and is also shown in step 3 of Figure 53.10. The helper T cell binds to a complex of antigen and class II MHC protein on an APC. In this case, the APC is a macrophage that has phagocytized one of the bacteria, hydrolyzed its proteins into peptide fragments, complexed the fragments with class II MHC proteins, and displayed the complexes on its surface. Once a helper T cell specific for the complex binds to it, the helper T cell becomes activated. The macrophage helps this activation process in two other ways: It provides a costimulus via nonantigenic plasma membrane proteins, and it secretes IL-1 and TNF.

IL-1 and TNF stimulate the helper T cell to secrete another cytokine, named interleukin 2 (IL-2). IL-2 stimulates the activated helper T cell to divide. This leads eventually to the formation of a clone of activated helper T cells (step 4), which bind to B cells and also secrete IL-2 and other cytokines. Certain of these cytokines provide the additional signals (step 5) that are usually required to activate nearby antigen-bound B cells to proliferate and differentiate into plasma cells (step 6), which then secrete specific antibodies (step 7). Some B cells are set aside to become memory cells to help ward off possible future attacks by the same antigen. Thus helper T cells are so named because their secretions help activate B cells that have bound antigen.

The secreted antibodies circulate through the lymphatic system and then the bloodstream, through which they reach the site of bacterial infection and inflammation. There the antibodies leave the blood and combine with the bacterial surface antigen that initiated the immune response. These antibodies then direct the attack against the bacteria to which they are now bound. Thus, immunoglobulins play two distinct roles in humoral immune responses. First, during antigen recognition, immunoglobulins on the surface of B cells (B-cell receptors) bind to antigen brought to them. Second, immunoglobulins secreted by the resulting plasma cells (antibodies) bind to bacteria bearing the same antigens, marking them as the targets to be attacked.

Instead of directly killing the microbes, antibodies bound to antigen on the microbial surface inactivate the microbes in various ways. Antibodies may physically link the microbes to phagocytes (neutrophils and macrophages), complement proteins, or NK cells. This linkage—called opsonization—triggers the attack mechanism and ensures that only the microbes and not nearby body cells are killed.

In a second mechanism, antibodies directed against bacterial toxins in the extracellular fluid bind to the toxins, thereby preventing them from harming susceptible body cells. The antibody-antigen complexes that are formed are then destroyed by phagocytes.

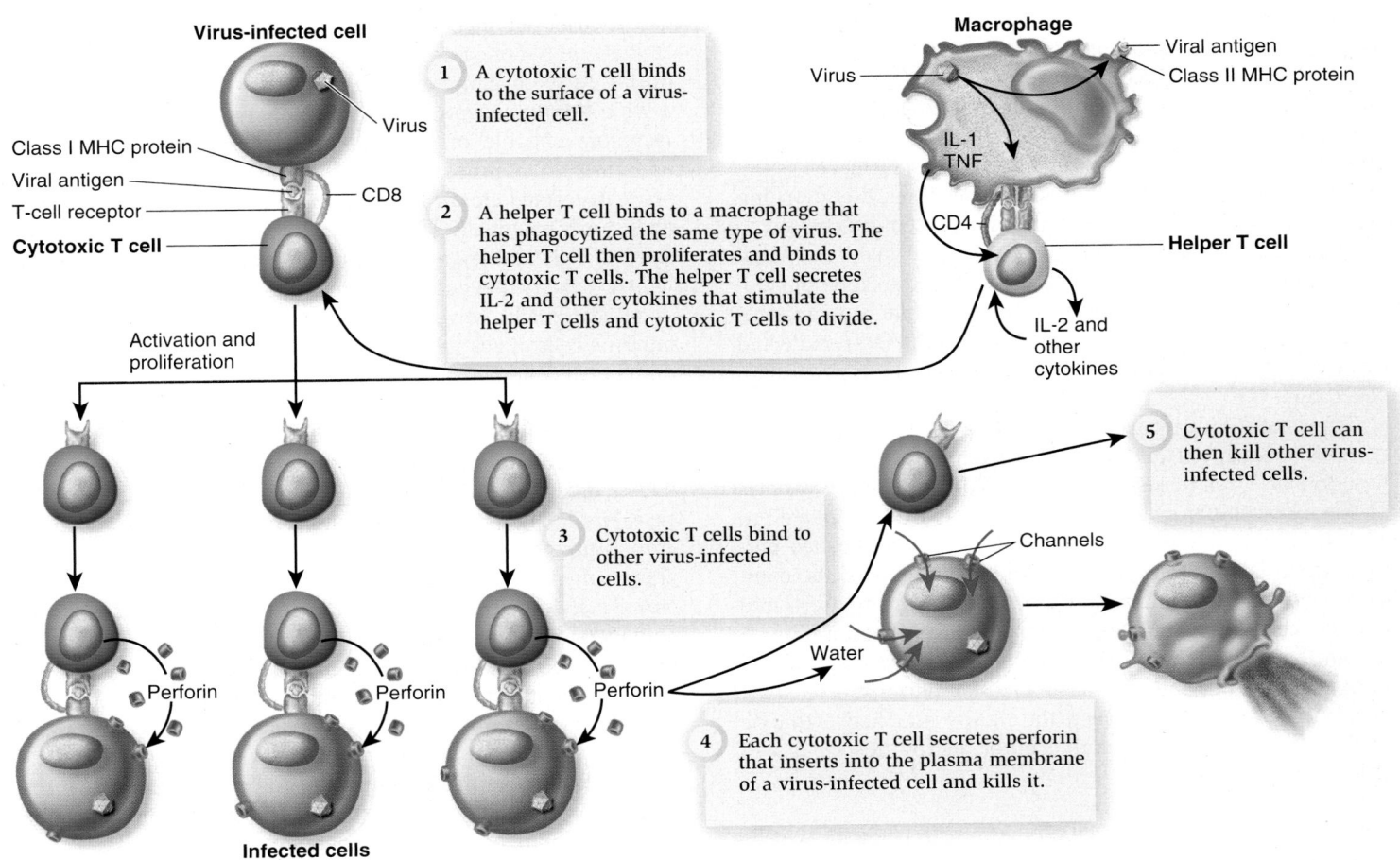

Figure 53.11   Summary of events in the killing of virus-infected cells by cytotoxic T cells.   The sequence is similar for cancer cells attacked by a cytotoxic T cell.

In a similar way, antibodies produced against certain viral surface proteins bind to the viruses in the extracellular fluid, preventing them from attaching to the plasma membranes of potential host cells. As with bacterial toxins, the antibody-virus complexes that are formed are subsequently phagocytized. Thus the body can eliminate viruses in two ways: through the humoral actions of antibodies in body fluids, and through the cell-mediated killing of virus-infected cells by cytotoxic T cells.

In Cell-Mediated Immunity, Cytotoxic T Cells Kill Infected or Cancer Cells

The previous sections described how immune responses provide long-term defenses against bacteria, viruses, and individual foreign molecules that enter the body's extracellular fluid. We now describe in detail how the body's own cells that have become infected by viruses or transformed into cancer cells are destroyed by cell-mediated immune responses.

What is the value of destroying virus-infected host cells? First, such destruction prevents cells from making more viruses. Second, for cells that already are making mature viruses, it

results in the release of the viruses into the extracellular fluid, where they can then be directly neutralized by circulating antibody, as just described. Generally, only a few host cells are sacrificed in this way. Once viruses have had a chance to replicate and spread from cell to cell, however, so many virus-infected host cells may be killed by the body's own defenses that organ malfunction may occur.

Role of Cytotoxic T Cells A typical cytotoxic T-cell response triggered by viral infection of a vertebrate's body cells is summarized in **Figure 53.11**. The response triggered by a cancer cell would be similar. A virus-infected cell produces foreign proteins, viral antigens that are processed and presented on the plasma membrane of the cell complexed with class I MHC proteins. Cytotoxic T cells specific for the particular antigen bind to the complex. As with B cells, binding to antigen alone does not cause activation of the cytotoxic T cell. Cytokines from adjacent activated helper T cells are also needed.

How are the helper T cells brought into play in these cases? Macrophages phagocytize extracellular viruses (or, in the case of cancer, antigens released from the surface of cancerous cells) and then process and present antigen, in association with class

II MHC proteins, to the helper T cells. In addition, the macrophages provide a costimulus and also secrete IL-1 and TNF. The activated helper T cell releases IL-2 and other cytokines, which stimulate proliferation of the helper T cell.

IL-2 and other cytokines also act on the cytotoxic T cell bound to the surface of the virus-infected or cancer cell, stimulating this attack cell to proliferate. Why is proliferation important if a cytotoxic T cell has already located and bound to its target? The answer is that there is rarely just one virus-infected cell or one cancer cell. By expanding the clone of cytotoxic T cells capable of recognizing the particular antigen, proliferation increases the likelihood that the other virus-infected or cancer cells will be encountered by the specific type of cytotoxic T cell.

Target cell killing by activated cytotoxic T cells occurs by several mechanisms, but one of the most important is as follows. The cytotoxic T cell releases the contents of its secretory vesicles into the extracellular space between itself and the target cell to which it is bound. These vesicles contain a protein called perforin, which is similar in structure to the proteins of the complement system's membrane attack complex. Perforin inserts into the target cell's membrane and forms channels ("perforations") through the membrane. In this manner, it causes the attacked cell to take in water and burst. The fact that perforin is released directly into the space between the tightly attached cytotoxic T cell and the target ensures that other nearby host cells will not be killed.

Role of Natural Killer (NK) Cells

Although cytotoxic T cells are very important attack cells against virus-infected and cancer cells, they are not the only ones. NK cells also destroy such cells by secreting toxic chemicals.

As mentioned earlier, NK cells can recognize general features on the surface of virus-infected cells or cancer cells, and participate in nonspecific immunity. In addition, NK cells can be linked to target cells by antibodies, and this constitutes another potential mechanism for bringing them into play against virus-infected or cancer cells.

B Cells and T Cells That Recognize Self Molecules Must Be Killed or Inhibited

As we have seen, the lymphocytes responsible for the specific immune response in vertebrates are very capable killers of pathogens—so capable, in fact, that it raises a question: Why don't these cells attack and kill normal self cells? In other words, how does the body distinguish between self and nonself components and develop what is called **immune tolerance**?

Recall that the huge diversity of lymphocyte receptors is ultimately the result of multiple random DNA cutting/recombination processes. It is virtually certain, therefore, that every animal possessing specific immune defenses would have clones of lymphocytes with receptors that could bind to that individual's own proteins. The continued existence and functioning of such lymphocytes would be disastrous because such binding would launch an immune attack against all body cells expressing these proteins.

At least two mechanisms explain why individuals normally lack active lymphocytes that respond to self components. First, during early development in vertebrates, T cells are exposed to a wide mix of self proteins in the thymus. Those T cells with receptors capable of binding self proteins are destroyed by apoptosis in a process termed **clonal deletion**. The second process, termed **clonal inactivation**, occurs outside the thymus and causes potentially self-reacting T cells to become nonresponsive. B cells undergo similar processes. The mechanisms by which these two events occur are still under investigation.

Occasionally, however, these mechanisms fail, and the body's immune cells attack the body's own cells. When this happens in humans, it is known as **autoimmune disease**. Autoimmune diseases are conditions in which the body's normal state of immune tolerance somehow breaks down, with the result that both humoral and cell-mediated attacks are directed against the body's own cells and tissues. A growing number of human diseases are being recognized as autoimmune in origin. Examples are multiple sclerosis, in which myelin is attacked; myasthenia gravis, in which the receptors for acetylcholine on skeletal muscle cells are the targets; rheumatoid arthritis, in which joints are damaged; and type 1 diabetes mellitus, in which the insulin-producing cells of the pancreas are destroyed.

Treatments for autoimmune disease range from treating the symptoms (for example, administering insulin to individuals with diabetes), to suppressing the immune system with drugs. Currently, researchers are investigating the possible use of transplanted hematopoietic stem cells that do not recognize the target protein as being foreign and thus antigenic.

Autoimmune diseases affect a large number of people, but there are many other immune disorders that have far greater impacts on public health, as described in Section 53.4.

Immunological Memory Is an Important Feature of Acquired Immunity

As we have learned, the acquired immune response to a given antigen depends on whether or not the body has previously been exposed to that antigen. Consider, for example, the humoral immune response. In mammals, antibody production in response to first contact with an antigen occurs slowly, over a few weeks (**Figure 53.12**). This response to an initial exposure to an antigen is termed a **primary immune response**. Any subsequent infection by the same pathogen elicits an immediate and heightened production of additional specific antibodies against that particular antigen, a reaction termed a **secondary immune response**.

In the case of humoral immunity, this secondary response occurs more quickly, is stronger, and lasts longer because memory B cells that had been produced in response to initial antigen exposure are quickly stimulated to multiply and differentiate into thousands of plasma cells. These cells then produce large amounts of specific antibodies. The immune system's ability to produce this secondary response is called **immunological memory**.

Immunological memory explains why we and other animals are able to fight off many illnesses to which we have been

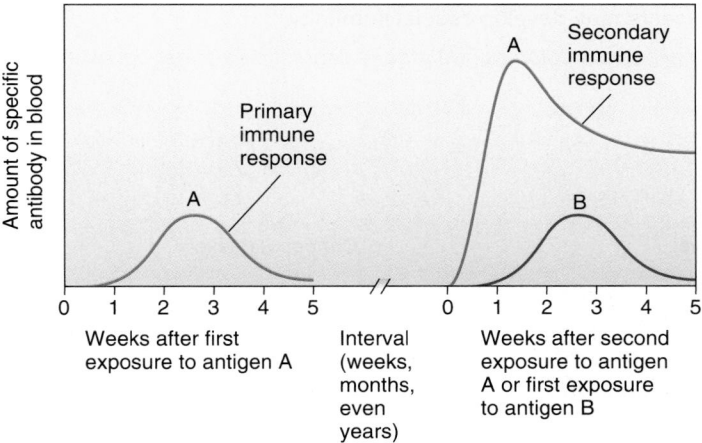

Figure 53.12 **Primary and secondary immune responses.** In a primary response, an initial exposure to an antigen produces modest levels of specific antibody over a period of weeks. In a secondary response, subsequent exposure to the same antigen results in increased antibody production that occurs more rapidly and lasts longer than a primary response. The secondary response is specific for that antigen. Exposure at that time to other antigens for the first time produces the usual primary response.

previously exposed, like many common childhood diseases. The acquired response to exposure to any type of antigen is known as **active immunity**. Active immunity not only results from natural exposure to antigens, but it is also the basis for the artificial exposures to antigen that occur in vaccinations and immunizations. In **vaccinations** and **immunizations**, small quantities of

living or dead microbes, small quantities of toxins, or harmless antigenic molecules derived from a microorganism or its toxin are injected into the body, resulting in a primary immune response including the production of memory cells. Subsequent natural exposure to the immunizing antigen results in a rapid, effective response that can prevent or reduce the severity of disease.

In contrast to active immunity is another type of acquired immunity called **passive immunity**, which confers protection against disease through the direct transfer of antibodies from one individual to another. Passive immunity can occur naturally, as when IgG molecules cross the mammalian placenta to protect a fetus from various pathogens, or when a newborn mammal receives antibodies from breast milk. It can also occur artificially, as when a human patient is given an injection of IgG molecules against hepatitis shortly after exposure to hepatitis viruses. Because antibodies are proteins with a limited life span, the protection afforded by the transfer of antibodies in passive immunity is relatively short-lived, usually lasting only a few weeks or months.

Immunizations are very important in preventing disease among humans and domesticated animals. This is especially important when humans or other animals live in dense populations. A good example is seen in certain invertebrates such as social insects, which live in populations with enormous numbers of individuals in confined areas. Do such populations have any naturally occurring protection against widespread disease that could otherwise ravage an entire colony of animals? This question has been explored recently by biologists studying social insects like ants and termites, as described next.

FEATURE INVESTIGATION

Traniello and Colleagues Demonstrated That Social Insects May Develop "Social Immunity"

Biologists have questioned how highly social animals living in dense colonies protect themselves from disease transmission. Such social animals include wood-eating termites that exist in very large numbers in complex nests. Termites may encounter a range of pathogens because they nest in moist, decaying wood, which provides a favorable environment for bacteria, fungi, and other parasites. A single infected termite could rapidly spread disease from insect to insect in a densely populated colony, but rarely are entire colonies wiped out by infection. In part, this occurs because termites kill microbes by producing antimicrobial secretions that are used to line their nests. In addition, behavioral adaptations such as grooming to remove pathogens from nest mates' cuticles, and the removal of sick or dead animals from the nest, help reduce disease transmission.

Recently, however, James Traniello and coworkers identified in termites another possible mechanism of protection from disease. "Social immunity," which involves the transfer of infection resistance from immunized to naive (susceptible) nest mates, is shown in the Feature Investigation in **Figure 53.13**. These investigators first discovered that *Zootermopsis angusticollis* ter-

mites survive a pathogen challenge better when they are living in a group than when they are isolated as individuals. This indicates that an individual's immunity is enhanced when they are surrounded by nest mates. How might this occur? Individuals of this species are highly susceptible to parasitic fungi. Traniello and his colleagues divided animals into two groups. To distinguish the groups from each other, one group was fed on paper that contained a dye that marked the animals with a visible color. The investigators then exposed this group of termites to a sublethal dose of fungal spores to immunize them against the fungus. The other (uncolored) group was exposed only to a control solution. One week following exposure to spores or control solutions, half of the control group of animals was combined into a single nest with the immunized animals, while the other half of the control group remained in a separate nest.

One week later, the animals in each group were exposed to a potentially lethal dose of fungal spores, and their survival was recorded after six days. Uncolored termites that had not experienced an initial exposure to spores but had later nested with animals that were exposed and thus immunized, had a significantly higher survival rate following exposure to the lethal dose of spores, compared to animals that did not nest with immunized animals. Thus, the investigators concluded that social

Figure 53.13 Traniello and colleagues demonstrated that social insects may develop "social immunity."

Biological inquiry: What other animals besides social insects live in extremely dense colonies and require protection against devastating illnesses sweeping through the entire population?

HYPOTHESIS Sociality promotes disease resistance in termites.

STARTING MATERIALS Termites (*Zootermopsis angusticollis*), pathogenic fungus spores.

	Experimental level	**Conceptual level**

1 Feed termites colored or uncolored filter paper.

Termite
Uncolored paper
Filter paper

These termites were fed uncolored paper.   These termites were fed colored paper and turned blue.

2 Expose colored termites to a suspension of pathogenic fungal spores at a sublethal concentration. Uncolored termites receive a control solution.

6,500 spores/mL (sublethal)   Control solution (no spores)

Petri dish

Colored termites only   Uncolored termites only (control)

Only the colored termites have been exposed to the pathogen, and will have an immune response.

3 One week after exposure, some of the unexposed (uncolored) termites were grouped with some of the exposed (colored) termites. This is called group 1. Group 2 consists of only uncolored termites.

Group 1   Group 2

Allowed to coexist for 1 week.

Close contact between exposed and unexposed termites may transfer immunity.

4 Expose group 1 and group 2 to a potentially lethal concentration of fungal spores. Measure the percentage of uncolored termites that survive, after 6 days.

65,000 spores/mL

A higher percentage of immunized animals should survive exposure to the spores.

5 **THE DATA**

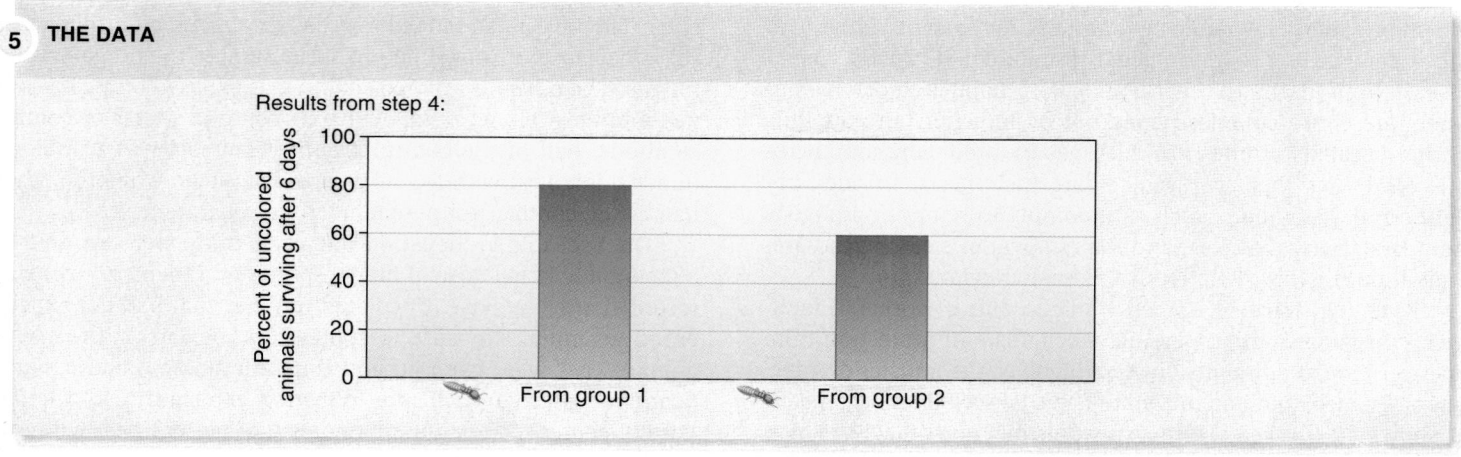

Results from step 4:

contact between nonimmunized and immunized animals resulted in a transfer of immunity to the more susceptible members of the group.

The mechanism of social immunization is not known but could be related to either the oral exchange of secretions or contact with inactivated spores remaining on the bodies of immunized animals. Whatever the mechanism may prove to be, the consequence is striking: The close physical contact that is characteristic of social insect behavior helps propagate colonywide immunization against potentially deadly pathogens.

53.4 Impact on Public Health

In this section, we describe a few ways in which health is affected by lifestyle, medical interventions, allergies, and destruction of immune cells. Collectively, the effects of these problems have an almost immeasurable impact on society in terms of worker productivity, health care resources, and the economy.

Lifestyle and Nutrition Have an Important Influence on Health

Protein-calorie malnutrition is the single greatest contributor to decreased resistance to infection worldwide. When inadequate amino acids are available to synthesize essential proteins, immune function is impaired. Deficits of specific nonprotein nutrients can also lower resistance to infection.

Both stress and state of mind can affect resistance to infection and to cancer. The immune system can alter neural and endocrine function, and in turn neural and endocrine activity modifies immune function. Lymphoid tissue is connected to nerves, and immune cells have receptors for many neurotransmitters and hormones. Conversely, immune cells release cytokines that have important effects on the brain and endocrine system. Moreover, lymphocytes secrete several hormones that are produced by endocrine glands. The multiple "mind-body" interactions that affect disease resistance are the subject of a field called psychoneuroimmunology.

Of the hormones associated with stress, the adrenal hormone cortisol has received the most attention due to its powerful suppressive activity on inflammation and specific immunity. Among other things, cortisol inhibits production of inflammatory mediators, reduces capillary permeability in injured areas, and suppresses the growth and activity of certain types of leukocytes. In this way, it acts as a sort of brake on the immune system. During chronic stress or when cortisol is used to treat certain illnesses for long periods of time, it may cause immunosuppression. This is a key link between stress and health. Chronic stress may lead to high cortisol levels that, by suppressing the body's immune responses, lower resistance to infection.

A third feature of a person's lifestyle that appears to affect immune function is exercise. The influence of physical exercise on the body's resistance to infection and cancer has been debated for decades. Present evidence indicates that the intensity, duration, regularity, and psychological stress of the exercise all have important influences, both negative and positive, on a variety of immune functions such as the numbers of circulating NK cells. Despite these complexities, most experts currently consider moderate exercise and physical conditioning to have net beneficial effects on the immune system and on disease resistance. A 2005 study suggested that exercise may be particularly beneficial in warding off the onset of breast cancer, one of the most common types of cancer in women.

Organ Transplants and Blood Transfusions Are Medical Procedures That Can Cause Significant Health Problems

Since the mid-20th century, organ transplants from a healthy or recently deceased donor to a recipient have become widespread. The United Network for Organ Sharing reports that approximately 27,000 organs are transplanted in the U.S. each year, with kidney (16,000), liver (6,000), heart (2,000), and lung (1,200) accounting for most of the transplants. The major obstacle to successful transplantation of tissues and organs is a reaction called graft rejection, in which the immune system recognizes the transplants (also called grafts) as foreign and attacks them as it would any other foreign cells. Although B cells and macrophages play some role in graft rejection, cytotoxic T cells and helper T cells are mainly responsible. To minimize this possibility, patients are given drugs that suppress immune function.

Except for identical twins, the class I MHC proteins on graft cells differ from those on the recipient's cells, as do the class II MHC proteins present on macrophages in the graft. Consequently, the recipient's T cells recognize the MHC proteins in the graft as foreign, and cytotoxic T cells (with the aid of helper T cells) destroy the graft cells.

Blood transfusion reactions, in which erythrocytes are destroyed after the transfusion, are a special example of tissue rejection in that antibodies rather than cytotoxic T cells are the major factor in rejection. Although erythrocytes lack MHC proteins, they have plasma membrane components that can function as antigens when transfused into another person's blood. Of the more than 400 erythrocyte antigens, the ABO system (see Chapter 16) of carbohydrates is the most important for transfusion reactions.

The four possible blood types—A, B, AB, and O—are genetically determined. Individuals with type A blood have erythrocytes bearing the A antigen. Those whose erythrocytes have the B antigen are type B, those that produce both antigens are type AB, and those that produce neither antigen are considered type O. Genes cannot code for the carbohydrates that function as antigens; instead, the genes code for the particular enzymes that catalyze the attachment of the carbohydrates onto cell surface proteins. Additionally, the plasma of type A individuals contains anti-B antibodies. Type B individuals have anti-A antibodies,

AB individuals have neither antibody, and type O individuals have both. How it is that these antibodies occur without prior exposure to the appropriate antigen is not clear.

What happens if a type A person is given type B blood? There are two incompatibilities. The recipient's anti-B antibodies attack the transfused erythrocytes, and the anti-A antibodies in the transfused plasma attack the recipient's erythrocytes. The latter reaction is generally of little consequence, because the transfused antibodies become too diluted in the recipient's plasma to cause damage. It is the destruction of the transfused cells by the recipient's antibodies that produces the problems. Thus a type A person cannot receive blood containing B antigens but can receive blood of types A and O.

By similar logic, type B individuals can receive blood of types B and O only. Because AB individuals have neither anti-A nor anti-B antibodies, they can receive blood of any type and are considered "universal recipients." By contrast, type O individuals, who have both anti-A and anti-B antibodies, can receive only type O blood. Because type O blood has neither A nor B antigens, it can be transfused into individuals of all types, and thus type O people are considered "universal donors."

Allergies Affect the Quality of Life of Millions of People

An allergy (also known as hypersensitivity) is a condition in which immune responses to environmental antigens cause inflammation and damage to body cells. Antigens that induce allergic reactions are called allergens. Common examples of allergens include ragweed pollen and poison ivy. Most allergens themselves are relatively or completely harmless. It is the immune responses to them that cause the damage. In essence, then, allergy is immunity gone awry, for the response is of inappropriate strength and duration for the stimulus. In the U.S. alone, as many as 40 million people (about 13.5% of the population) suffer from allergies.

For any allergy to develop, a genetically predisposed person must first be exposed to the allergen—a process called sensitization. Subsequent exposures elicit the damaging immune responses we recognize as an allergy.

Allergies can be broadly classified according to the speed of the reaction. Allergies that take up to several days to develop are considered delayed hypersensitivities. The skin rash that appears after contact with poison ivy is an example. More common are reactions considered immediate hypersensitivities, which can develop in minutes or up to a few hours. These allergies are also called IgE-mediated hypersensitivities because they involve IgE antibodies.

In immediate hypersensitivity, sensitization to the allergen leads to the production of specific antibodies and a clone of memory B cells. In individuals who are genetically susceptible to allergies, antigens that elicit immediate hypersensitivity reactions stimulate the production of IgE antibodies. Upon their release from plasma cells, these IgE molecules circulate throughout the body and attach themselves to mast cells in connective tissue. When the same antigen subsequently enters the body at some future time and binds with IgE bound to mast cells,

Figure 53.14 **Mast cells.** An unstimulated mast cell contains many secretory vesicles.

the mast cell is stimulated to secrete many inflammatory mediators, including histamine, that then initiate an inflammatory response.

The signs and symptoms of IgE-mediated allergy reflect both the effects of the inflammatory mediators and the body site in which the antigen–IgE–mast cell binding occurs. When, for example, a previously sensitized person inhales ragweed pollen, the antigen combines with IgE on mast cells in the respiratory passages (**Figure 53.14**). The mast cells release their contents, which induce increased mucus secretion, increased blood flow, swelling of the epithelial lining, and contraction of the smooth muscle surrounding airways. These effects produce the congestion, runny nose, sneezing, and difficulty in breathing characteristic of hay fever.

Acquired Immune Deficiency Syndrome (AIDS) Is a Growing Pandemic

AIDS is caused by the **human immunodeficiency virus (HIV)**, which incapacitates the immune response by preferentially infecting helper T cells. HIV is a retrovirus; such viruses have a nucleic acid core of RNA rather than DNA. Once inside a helper T cell, HIV uses the enzyme reverse transcriptase to transcribe the virus's RNA into DNA, which is then integrated into the T cell's chromosomes. Viral replication within the T cell results in the death of the cell (see Chapter 18).

HIV infects helper T cells because the CD4 protein in their plasma membranes acts as a receptor for an HIV capsid protein. But the binding to CD4 is not sufficient to enable HIV to enter the helper T cell. Another T-cell surface protein, which normally acts as a receptor for certain cytokines, must serve as a coreceptor. Interestingly, individuals possessing a mutation in this cytokine receptor are highly resistant to HIV infection, so much research is now focused on the possible therapeutic use of chemicals that can bind to and block this coreceptor.

HIV not only directly kills helper T cells, but it also indirectly causes additional helper T cell death by inducing cytotoxic T cells to kill HIV-infected helper T cells. In addition, by still poorly understood mechanisms, HIV causes the death of many uninfected helper T cells by apoptosis. Without adequate numbers of helper T cells, neither B cells nor cytotoxic T cells can function normally. Both humoral and cell-mediated immunity are compromised. Many individuals with AIDS die from infections and cancers that ordinarily would be readily handled by a fully functional immune system.

AIDS, first described in 1981, has since reached epidemic proportions. About 40 million people worldwide are infected with HIV, and nearly 15,000 new infections occur each day. The great majority of individuals now infected with HIV show no signs of AIDS. Their infections are diagnosed by the presence of anti-HIV antibodies or HIV RNA in the blood. However, most individuals infected with HIV will eventually develop AIDS.

If left untreated, HIV infection commonly develops into AIDS in about 10 years. During the first five years, killed helper T cells are typically replaced by new cells, so T-cell levels remain normal, and the individual remains asymptomatic. Over the next five years T-cell levels begin to decline, until at some point AIDS reveals itself in the form of opportunistic viral, bacterial, and fungal infections. Certain unusual cancers, such as Kaposi's sarcoma, also occur with high frequency. In untreated individuals, death usually ensues within two years after the onset of AIDS symptoms.

The major routes of HIV transmission are (1) unprotected sexual intercourse with an infected partner; (2) transfer of contaminated blood or blood products between individuals, such as during a blood transfusion or the sharing of needles among intravenous drug users; (3) transfer from an infected mother to her child across the placenta or during delivery; or (4) transfer via breast milk during nursing.

Treatment for HIV-infected individuals has two components: one directed against the virus itself to delay progression of the disease, and one to prevent or treat the opportunistic infections and cancers that ultimately cause death. One current antiviral approach involves administering a "cocktail" of four drugs, known as HAART (highly active anti-retroviral therapy). Two of the drugs inhibit the action of reverse transcriptase in converting viral RNA into DNA within the host cell, a third drug inhibits an HIV enzyme required for assembling new viruses, and a recently developed fourth class of drugs called fusion inhibitors prevents the virus from entering T cells. These treatments have proven to be effective in slowing the rate at which infection with HIV leads to AIDS.

CHAPTER SUMMARY

- An animal's cells and organs that collectively contribute to its immune defenses, or immunity, constitute the animal's immune system.
- In nonspecific (innate) immunity, the body's defenses are present at birth and act against foreign materials in much the same way regardless of the specific identity of the invading material. Specific (acquired) immunity develops only after the body is exposed to foreign substances and targets those foreign substances specifically.

53.1 Types of Pathogens
- Pathogens are of three major types: bacteria, viruses, and parasites.

53.2 Nonspecific (Innate) Immunity
- An important nonspecific defense is composed of phagocytes—cells capable of phagocytosis. In vertebrates, most phagocytes belong to the blood cells called leukocytes. The leukocytes involved in nonspecific immunity include neutrophils, eosinophils, monocytes, macrophages, basophils, and natural killer (NK) cells. (Figures 53.1, 53.2)
- Two other types of cells play important roles in nonspecific immunity: dendritic cells and mast cells.
- Inflammation is an innate local response to infection or injury characterized by local redness, swelling, heat, and pain. The events of inflammation are induced and regulated by chemical mediators called cytokines. (Figure 53.3)
- Antimicrobial proteins include interferons, which inhibit viral replication, and complement proteins, which kill microbes without prior phagocytosis. Activation of the complement proteins results in the formation of a membrane attack complex (MAC), which creates water channels in the microbial plasma membrane and causes the microbe to swell and burst.

53.3 Specific (Acquired) Immunity
- Any foreign molecule that the host does not recognize as self and that triggers a specific immune response is called an antigen.
- The lymphocytes responsible for specific immunity are B cells, which differentiate into antibody-producing cells called plasma cells, and T cells, which directly kill infected, mutated, or transplanted cells. T cells include cytotoxic T cells, which directly kill target cells via secreted chemicals, and helper T cells, which assist in the activation and function of B cells and cytotoxic T cells. (Figure 53.4)
- Immunologists recognize two types of specific immunity. In humoral immunity, plasma cells secrete antibodies that bind to antigens. In cell-mediated immunity, cytotoxic T cells directly attack and destroy infected body cells, cancer cells, or transplanted cells.
- Although lymphocytes circulate in the blood, most of them reside in a group of organs and tissues that constitute the lymphatic system. (Figure 53.5)
- When a lymphocyte is stimulated by an antigen, it divides and produces clones of itself, all of which express the same receptor. The process by which these clones are formed is called clonal selection. After activation, some of the cloned lymphocytes function as effector cells, which carry out the attack response; other cells function as memory cells, which remain poised to recognize the antigen if it returns in the future. (Figure 53.6)
- B cells that are activated by an antigen and by cytokines produced by helper T cells differentiate into plasma cells, which secrete antibodies—proteins that combine with the antigen and guide an attack that eliminates the antigen or the cells bearing it.
- Immunoglobulins (including antibodies and B-cell receptors) are Y-shaped proteins that include two heavy chains and two light chains. The lower halves of the heavy chains contain constant regions, which are identical for all immunoglobulins of a given Ig class. At the other end of the Ig molecule, the two chains form a variable region that serves as the antigen binding site. (Figures 53.7, 53.8)
- The process known as hypermutation, which primarily involves numerous C to T point mutations, is crucial to enabling lymphocytes to produce a diverse array of antibodies capable of recognizing many different antigens.
- Major histocompatibility complex (MHC) proteins are cellular "identity tags" that serve as genetic markers of self. Class I MHC proteins are found on the surface of all human body cells except

erythrocytes. Class II MHC proteins are found only on the surface of macrophages, B cells, and dendritic cells.

- Antigen-presenting cells are cells (APCs) bearing fragments of antigen, called antigenic determinants or epitopes, complexed with the cell's MHC proteins. (Figure 53.9)

- The process by which the body distinguishes between self and nonself components is called immune tolerance. Individuals normally lack active lymphocytes that respond to self components because of two mechanisms. T cells with receptors capable of binding self proteins are destroyed by apoptosis in a process termed clonal deletion. Clonal inactivation causes potentially self-reacting lymphocytes to become nonresponsive.

- Humoral immune responses are mediated by B cells and plasma cells, while cell-mediated immune responses are mediated by cytotoxic T cells. In both types of responses, helper T cells are required. (Figures 53.10, 53.11)

- When the body's immune cells attack the body's own cells, the result is called autoimmune disease.

- Upon initial exposure to an antigen, the body produces a primary immune response. Any subsequent exposure to the same antigen elicits an immediate and heightened response termed a secondary immune response. The immune system's ability to produce this secondary response is called immunological memory. (Figure 53.12)

- The acquired response to exposure to any type of antigen is known as active immunity. The artificial exposures to antigen that occur in vaccinations and immunizations also induce active immunity. In contrast, passive immunity confers protection against disease through the direct transfer of antibodies from one individual to another.

- Some animals, such as social insects, can confer immunity on each other through social contact, called "social immunity." (Figure 53.13)

53.4 Impact on Public Health

- Factors that affect the body's defense mechanisms include lifestyle and nutrition, organ transplants and blood transfusions, allergies, and AIDS, caused by the human immunodeficiency virus (HIV). AIDS reduces the body's immunity by killing helper T cells. (Figure 53.14)

TEST YOURSELF

1. Which of the following is *not* an example of a barrier defense in animals?
   a. skin   d. mucus
   b. secretions from skin glands   e. antibodies
   c. exoskeleton

2. The leukocytes that are found in mucosal surfaces and play a role in defending the body against parasitic infections are
   a. neutrophils.   d. monocytes.
   b. eosinophils.   e. NK cells.
   c. basophils.

3. The vascular changes of inflammation
   a. lead to an increase in bacterial cells at the injury site.
   b. decrease the number of leukocytes at the injury site.
   c. allow plasma proteins to move easily from the bloodstream to the injury site.
   d. increase the number of antibodies at the injury site.
   e. activate lymphocytes.

4. Which of the following statements is correct regarding specific immunity?
   a. Specific immunity only requires the presence of helper T cells to function properly.
   b. Specific immunity does not require exposure to a foreign substance.
   c. Specific immunity is triggered by contact with a particular antigen.
   d. Specific immunity includes inflammation.
   e. All of the above are correct.

5. Memory B cells are
   a. cloned lymphocytes that are active in subsequent infections.
   b. cloned lymphocytes that are active during a primary infection.
   c. NK cells that recognize cancer cells and destroy them.
   d. cells that produce antibodies.
   e. macrophages that have recognized self antigens.

6. The immunoglobulin that is passed from mother to fetus across the placenta is
   a. IgA.   d. IgG.
   b. IgD.   e. IgM.
   c. IgE.

7. The region of an antibody that serves as the antigen binding site is
   a. the constant region.   d. the light chain.
   b. the variable region.   e. the hinge point.
   c. the heavy chain.

8. A major difference between the activation of B cells and T cells is that
   a. T cells must interact with antigens bound to plasma membranes.
   b. B cells only interact with free antigens.
   c. B cells are not regulated by helper T cells.
   d. T cells produce antibodies.
   e. none of the above.

9. Cells that process foreign proteins and complex them with their MHC proteins are called
   a. cytotoxic T cells.   d. antigen presenting cells.
   b. plasma cells.   e. helper T cells.
   c. NK cells.

10. HIV causes immune deficiency because the virus
    a. destroys all the cytotoxic T cells.
    b. preferentially destroys helper T cells that regulate the immune system.
    c. directly inactivates plasma cells.
    d. causes mutations that lead to autoimmune diseases.
    e. all of the above.

CONCEPTUAL QUESTIONS

1. Distinguish between nonspecific immunity and specific immunity.

2. Explain the function of cytotoxic T cells.

3. Describe the basic structure of an immunoglobulin.

EXPERIMENTAL QUESTIONS

1. What features of termite physiology and/or behavior reduce the chance that an infection will spread throughout the colony?

2. What hypothesis was tested by Traniello and colleagues?

3. How did the researchers test this hypothesis, and what were the results of the study?

COLLABORATIVE QUESTIONS

1. Discuss three types of pathogens that affect the health of animals.

2. Describe the major difference between invertebrate and vertebrate immune systems.

www.brookerbiology.com
This website includes answers to the Biological Inquiry questions found in the figure legends and all end-of-chapter questions.

54

AN INTRODUCTION TO ECOLOGY AND BIOMES

CHAPTER OUTLINE

What controls the densities of these flowering plants at Glacier National Park, Montana? Is it temperature or rainfall? Or is it the availability of pollinators, herbivory by insects or vertebrates, or competition with other plant species for resources? Ecology seeks to answer questions such as these.

In 2006, a study led by J. Alan Pounds of the Monteverde Cloud Forest Preserve in Costa Rica reported that fully two thirds of the 110 species of harlequin frogs in mountainous areas of Central and South America had become extinct over the previous 20 years. The researchers noted that populations of other species, such the Panamanian golden frog (*Atelopus zeteki*), had been greatly reduced (**Figure 54.1**). The question was why. The culprit was identified as a disease-causing fungus, *Batrachochytrium dendrobatidis*, but this study implicated global warming as the agent causing outbreaks of the fungus.

Figure 54.1 **Diminishing and disappearing populations.** Population sizes of the Panamanian golden frog (*Atelopus zeteki*) have diminished greatly over the past 20 years, while populations of many other species of harlequin frogs have disappeared entirely. Ecologists are investigating the reasons for this decline.

One effect of global warming is to increase the cloud cover, which reduces daytime temperatures and raises nighttime temperatures. Researchers believe that this combination has created favorable conditions for the spread of *B. dendrobatidis*, which thrives in cooler temperatures. Pounds, the team's lead researcher and an ecologist, was quoted as saying, "Disease is the bullet killing frogs, but climate change is pulling the trigger."

Ecology is the study of interactions among organisms and between organisms and their environments. The interactions among organisms are called **biotic** interactions, while those between organisms and their nonliving environment are termed **abiotic** interactions. These interactions in turn govern the population densities of plants and animals and the numbers of species in an area. In this first chapter of the ecology unit, we will introduce you to the four broad areas of ecology: organismal, population, community, and ecosystems ecology. Next, we will explore how ecologists approach and conduct their work. We will then turn our focus to abiotic interactions and examine the effects of factors such as temperature, water, light, pH, and salt concentrations on the distributions of organisms. We conclude with a consideration of climate and its large influence on the major types of habitats, called biomes, where organisms are found.

Ecological studies have important implications in the real world, as will be amply illustrated by examples discussed throughout the unit. However, a distinction must be made between ecology and **environmental science**, the application of ecology to real-world problems. To use an analogy: Ecology is to environmental science as physics is to engineering. Both physics and ecology provide the theoretical framework on which to pursue more applied studies. Engineers rely on the principles of physics to build bridges. Environmental scientists rely on the principles of ecology to solve environmental problems.

Ecology describes the necessary framework for understanding how populations are affected by features of the physical environment, like temperature and moisture, and by other organisms. In this unit, we'll also learn how plants compete with one another, how herbivores affect plant abundance, and how natural enemies impact prey populations. We'll examine the effects of humans on the environment, including pollution, global warming, and the introduction of exotic species of plants and animals.

Ecologists are among the best-equipped scientists to study these phenomena. Before 1960, the field of ecology was dominated by taxonomy, natural history, and speculation about observed patterns. An ecologist's tools of the trade might have included sweep nets, quadrats (small, measured plots of land used to sample living things), and specimen jars. Since that time, an explosion in the number of ecological studies has occurred, and ecologists have become active in investigating environmental change on regional and global scales. Ecologists have embraced experimentation and adapted concepts and methods derived from agriculture, physiology, biochemistry, genetics, physics, chemistry, and mathematics. Now an ecologist's equipment is just as likely to include portable computers, satellite-generated images, and chemical autoanalyzers.

54.1 The Scale of Ecology

Ecology ranges in scale from the study of an individual organism through the study of populations to the study of communities and ecosystems (**Figure 54.2**). In this section, we introduce each of the broad areas of organismal, population, community, and ecosystems ecology. We will provide an investigation that helps illuminate the field of population ecology

and conclude with an exploration of how ecologists conduct their experiments.

Organismal Ecology Investigates How Adaptations and Choices by Individuals Affect Their Reproduction and Survival

Organismal ecology can be divided into two subdisciplines. The first, **physiological ecology**, investigates how organisms are physiologically adapted to their environment and how the environment impacts the distribution of species. Much of this chapter discusses physiological ecology.

The second area, **behavioral ecology**, focuses on how the behavior of an individual organism contributes to its survival and reproductive success, which in turn eventually affects the population density of the species. If we see individuals constantly competing, we may view nature as involving a fierce struggle for survival. If we see individuals cooperating with one another and acting in selfless, even altruistic ways, we tend to believe nature is neat and harmonious.

Population Ecology Describes How Populations Grow and Interact with Other Species

Population ecology focuses on groups of interbreeding individuals, called populations. A primary goal is to understand the factors that affect a population's growth and determine its size and density. Although the attention of a population ecologist may be aimed at studying the population of a particular species, the relative abundance of that species is often influenced by its interactions with other species. Thus, population ecology includes the study of **species interactions** such as predation, competition, and parasitism. Much of the theory of ecology is built upon the ecology of populations. Knowing what factors impact populations can help us lessen species endangerment, stop extinctions, and control invasive species.

(a)   **(b)**   **(c)**   **(d)**

Figure 54.2 **The scales of ecology.** **(a)** Organismal ecology. What is the temperature tolerance of this desert locust, *Schistocerca gregaria*? **(b)** Population ecology. What factors influence the growth of desert locust populations in Mauritania, Africa? **(c)** Community ecology. What factors influence the interaction of species in functional communities? These desert locusts were killed by fungal sprays developed in England. **(d)** Ecosystems ecology. How does energy flow among organisms, and how do nutrients cycle between organisms and the environment in the desert ecosystem? These desert locusts have died over the ocean and have been washed ashore.

FEATURE INVESTIGATION

Callaway and Aschehoug's Experiments Showed That the Secretion of Chemicals Gives Exotic Plants a Competitive Edge over Native Species

One important topic in the area of population ecology concerns **exotic species**, species moved from a native location to another location, usually by humans. Such species sometimes spread or invade so aggressively that they crowd out native species. Of the 300 most invasive exotic plants in the U.S., over half were brought in for gardening, horticulture, or landscaping purposes. Invasive exotic plants have traditionally been thought to succeed because they have escaped their natural enemies, primarily insects that remained in the country of origin and were not transported to the new locale. One way of controlling exotic species,

therefore, has been to import the plant's natural enemies. This is known as **biological control**. However, new research on the population ecology of diffuse knapweed (*Centaurea diffusa*), an invasive Eurasian plant that has established itself in many areas of North America, suggests a different reason for their success.

Researchers Ragan Callaway and Erik Aschehoug hypothesized that this particular species secretes powerful root exudates called **allelochemicals** that kill the roots of other species, allowing *Centaurea* to proliferate. To test their hypothesis, Callaway and Aschehoug collected seeds of three native Montana grasses, *Koeleria cristata*, *Festuca idahoensis*, and *Agropyron spicata*, and grew them both without other species and with the exotic *Centaurea* species (**Figure 54.3**). As hypothesized, *Centaurea* depressed the biomass of the native grasses.

Figure 54.3 Experimental evidence of the effect of allelochemicals on plant production.

HYPOTHESIS Exotic plants from Eurasia outcompete native Montana grasses by secreting chemicals from their roots.

STARTING LOCATION *Centaurea diffusa*, a Eurasian plant, is invading Montana grasslands because it outcompetes 3 native grasses: *Koeleria cristata*, *Festuca idahoensis*, and *Agropyron spicata*.

Experimental level / **Conceptual level**

1 Collect seeds of native Montana grasses and plant with and without seeds of invasive *C. diffusa* from Eurasia. Three months after sowing seeds, the plants are harvested, dried, and weighed.

C. diffusa significantly reduces biomass of native Montana grasses.

2 Collect seeds of grasses from Eurasia and plant with and without *C. diffusa*. Three months after sowing seeds, the plants are harvested, dried, and weighed.

C. diffusa doesn't depress the biomass of grasses native to Eurasia as much.

3 **THE DATA**

When the experiments were repeated with grasses native to Eurasia, *Koeleria laerssenii*, *Festuca ovina*, and *Agropyron cristatum*, the species were affected, but to a significantly lesser degree than the Montana species were.

In other experiments not described in Figure 54.3, Callaway and Aschehoug added activated carbon, which absorbs the chemical excreted by the *Centaurea* roots. With activated carbon added to the soil, the Montana grass species increased in biomass compared to the previous experiments. The researchers concluded that *C. diffusa* outcompetes Montana grasses by secreting an allelochemical and that Eurasian grasses are not as susceptible to the chemical's effect, since they coevolved with it. If the reason for the success of invasive plants can be attributed to the chemicals they secrete, this calls into question the effectiveness of biological control of exotic weeds by importation of their natural enemies. This study on the population biology of an exotic plant has changed the way we think about why invasive exotic species succeed and could affect the way we attempt to control them in the future.

Community Ecology Focuses on What Factors Influence the Number of Species in a Given Area

We've just seen how a population of an exotic grass may produce allelochemicals that give it a growth advantage over populations of other species. On a larger scale, **community ecology** studies how populations of species interact and form functional communities. In a forest, there are many populations of trees, herbs, shrubs, grasses, the herbivores that eat them, and the carnivores that in turn prey on the herbivores. Community ecology focuses on why certain areas have high numbers of species (that is, are species rich), while other areas have low numbers of species (that is, are species poor).

While ecologists are interested in species richness for its own sake, a link also exists between species richness and community function. Ecologists generally believe that species-rich communities perform better than do species-poor communities. It has also been posited that more species make a community more stable, that is, more resistant to disturbances such as introduced species. Community ecology also considers how species composition and community structure change over time and in particular after a disturbance, a process called succession.

Ecosystems Ecology Describes the Passage of Energy and Nutrients Through Communities

Ecosystems ecology deals with the flow of energy and cycling of nutrients among organisms within a community and between organisms and the environment. As such, it is concerned with both the biotic and abiotic components of the environment. Following this flow of energy and nutrients necessitates an understanding of feeding relationships between species, called food chains. In food chains, each level is called a trophic level and many food chains interconnect to form complex food webs.

The second law of thermodynamics states that in every energy transformation, free energy is reduced because heat energy is lost from the ecosystem in the process, and the entropy of the universe increases. There is, therefore, a unidirectional flow of energy through an ecosystem, with energy dissipated at every step. An ecosystem needs a recurring input of energy from an external source—in most cases, the sun—to sustain itself.

In contrast, chemicals such as nitrogen do not dissipate and constantly cycle between abiotic and biotic components of the environment, often becoming more concentrated in organisms in higher trophic levels.

Ecological Methods Focus on Observation and Experimentation

How do ecologists go about studying their subject? Let's suppose you are employed by the United Nations' Food and Agricultural Organization (FAO), which operates internationally out of Rome, Italy. As an ecologist, you are charged with finding out what causes outbreaks of locusts, a type of grasshopper that periodically erupts in Africa and other parts of the world, destroying crops and other vegetation. First of all, you might draw up a possible web of interaction between the factors that could affect locust population size (**Figure 54.4**). These interactions are many and varied, and they include

- abiotic factors such as temperature, rainfall, wind, and soil pH;
- host plants, including increases or decreases in either the quality or quantity of the plants;
- predators, including bird predators, insect parasites, and bacterial parasites;
- competitors, including other insects and larger vertebrate grazers.

With such a vast array of factors to be investigated, where is the best place to start? As discussed in Chapter 1, hypothesis testing involves a five-stage process: (1) observations, (2) hypothesis formation, (3) experimentation, (4) data analysis, and (5) acceptance or rejection of the hypothesis. In our study of locusts, we begin by careful observation of the organism in its native environment. We can analyze the fluctuations of locusts and determine if the populations vary with fluctuations in the other phenomena such as levels of parasitism, numbers of predators, or food supply. Imagine we found that locust numbers are affected by bird predation levels, and that an inverse relationship exists between predation levels and locust numbers.

Figure 54.4 Interaction web of factors that might influence locust population size.

As predation levels increase, locust numbers decrease. If we plotted this relationship graphically, the resulting graph would look like that depicted in **Figure 54.5a**. This result would give us some confidence that predation levels determined locust numbers and this would be our hypothesis. In fact, we would have so much confidence that we could create a "line of best fit" to represent a summary of the relationship between these two variables, as shown in **Figure 54.5b**.

However, if the points were not highly clustered, as in **Figure 54.5c**, we would have little confidence that predation affects locust density. Many statistical tests are used to determine whether or not two variables are significantly correlated. In the studies in this unit, unless otherwise stated, most graphs like Figure 54.5b imply that a meaningful relationship exists between the two variables. We call this type of relationship a significant **correlation**. If locust density shows a linear relationship with predation, we say that locust density is correlated with predation.

We have to be cautious when forming conclusions based on correlations. For example, large numbers of locusts could be associated with large, dense plants. We might conclude from this that food availability controls locust density. However, an alternative conclusion would be that large plants provide locusts refuge from bird predators, which cannot attack them in the dense interior. While it would appear that biomass affects locust density by providing abundant food, in actuality, predation would still be the most important factor affecting locust density.

Thus, correlation does not always mean causation. For this reason, after conducting observations, ecologists usually turn to experiments to test their hypotheses.

Continuing with our example, an experiment might involve removing predators from locust populations. If predators are having a significant effect, then removing them should cause an increase in locust numbers. Thus, we would have two groups: a group of locusts with predators removed (the experimental group) and a group of locusts with predators still intact (the control group), with equal numbers of locusts in both groups at the start of the experiment. Any differences in locust population density would be due solely to differences in predation. Reduced predation might be achieved by putting a cage made of chicken wire over and around bushes containing locusts, so that birds are denied access. We could look at locust survivorship over the course of one generation of locust and predator.

Performing the experiments several times is called **replication**. We might replicate the experiment 5 times, 10 times, or even more. At the end of the replications, we would add up the total number of living locusts and calculate the mean. In the experimental group, let's suppose that the surviving numbers of locusts in each replicate are 5, 4, 7, 8, 12, 15, 13, 6, 8, and 10; the mean number surviving would be 8.8. In the control group, which still allows predator access, the numbers surviving might be 2, 4, 7, 5, 3, 6, 11, 4, 1, and 3, with a mean of 4.6. Without predators, the mean number of locusts surviving would therefore be almost double the average number surviving with predators.

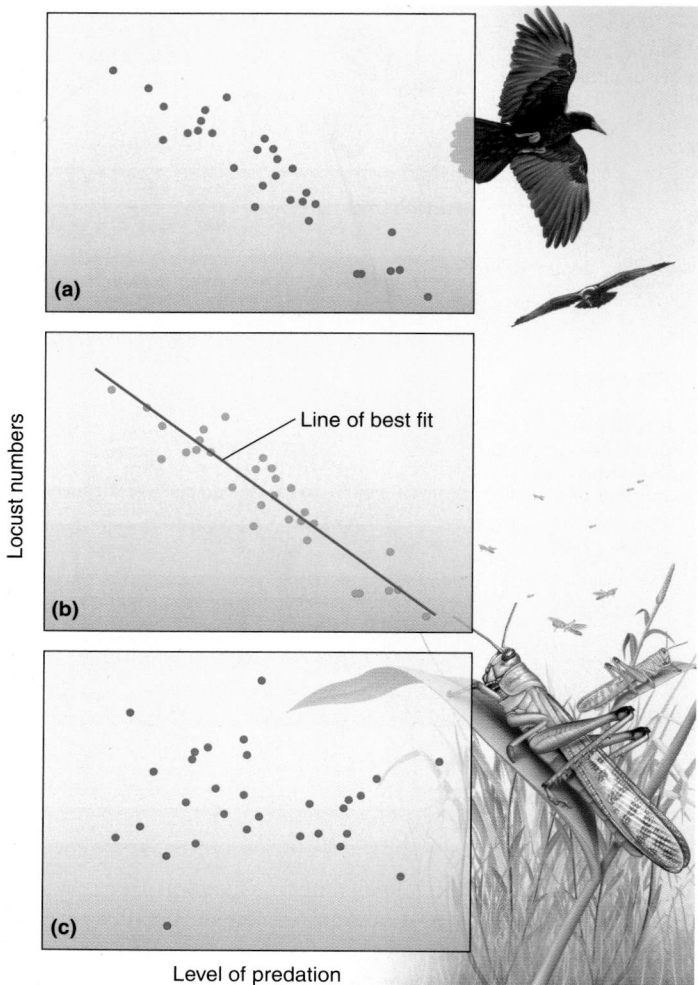

Figure 54.5 **How locust numbers might be correlated with predation.** **(a)** Higher locust numbers are found in nature where predation levels are lowest. We can draw a line of best fit **(b)** to represent this relationship. In **(c)**, the relationship between locust numbers and predation levels might be so weak that we would not have much confidence in a linear relationship between the variables.

Biological inquiry: What would it mean if the line of best fit sloped in the opposite direction?

Our data analysis would give us confidence that predators were indeed the cause of our changes in locust numbers. The results of such experiments can be illustrated graphically by a bar graph (**Figure 54.6**). Ecologists can use a variety of tests to see if these differences are statistically significant. We won't look at the mechanics of these tests, but in this unit, when experimental and control groups are presented as differing, these are considered to be statistically significant differences unless stated otherwise.

By the way, it turns out that predation is not really the primary factor that controls locust populations. The results we have been discussing were hypothetical. Weather, in particular, rain, is the most important feature governing locust population size.

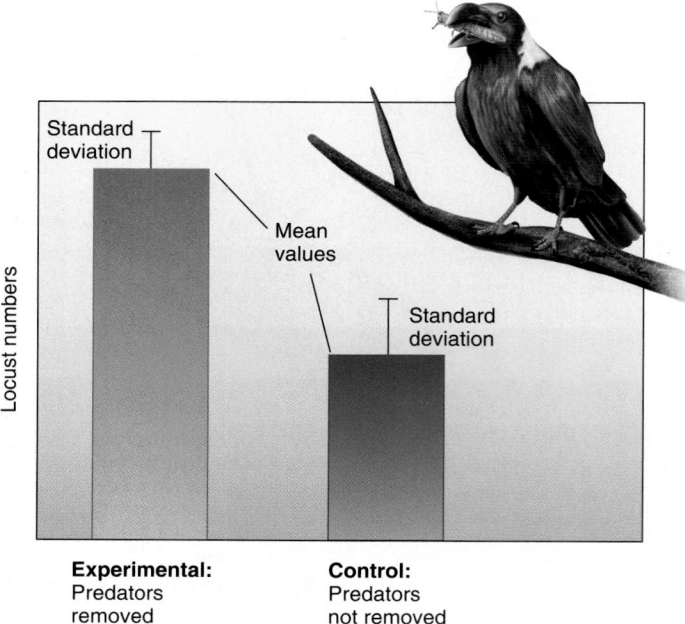

Figure 54.6 **Graphic display of hypothetical results of a predator removal experiment.** The two bars represent the average number of locusts where predators are removed (experimental) and where predators are not removed (control). The vertical lines (the standard deviations) give an indication of how tightly the individual replicate results are clustered around the mean. The shorter the lines, the tighter the cluster, and the more confidence we have in the result.

Moist soil allows eggs to hatch and provides water for germinating plants, allowing a ready source of food for the hatchling locusts. In fact, physical or abiotic factors such as amount of moisture usually have powerful effects in most ecological systems. In the next part of the chapter, we turn our attention to an examination of the effects of the physical environment on the distribution patterns of organisms.

54.2 The Environment's Impact on the Distribution of Organisms

Both the distribution patterns of organisms and their abundance are limited by physical features of the environment such as temperature, wind, availability of water and light, salinity, pH, and water currents. Some species can tolerate a relatively wide range of environmental conditions, and others only a narrow range, but each species usually functions best over only a limited part of the range known as a species' optimal range or **fundamental niche**. In this section, we will examine these features of the physical environment, focusing on the impact of each on the distribution of organisms.

(a)

Coral reef

(b)

Figure 54.7 **Locations of coral reefs.** (a) Coral reef formation is limited to waters bounded by the 20°C isotherm (dashed line), a line where the average daily temperature is 20°C during the coldest month of the year. (b) Coral reef from the Pacific Ocean.

Temperature Has an Important Effect on the Distribution of Plants and Animals

Temperature is perhaps the most important factor in the distribution of organisms because of its effect on biological processes and because of the inability of most organisms to regulate their body temperature precisely. For example, the organisms that form coral reefs secrete a calcium carbonate shell. Shell formation and coral deposition are accelerated at high temperatures but are suppressed in cold water. Coral reefs are therefore abundant only in warm water, and a close correspondence is observed between the 20°C isotherm for the average daily temperature during the coldest month of the year and the limits of the distribution of coral reefs (**Figure 54.7**). An isotherm is a line on a map connecting points of equal temperature. Coral reefs are located between the two 20°C isotherm lines that are formed above and below the equator.

In plants, cold temperature can be lethal because cells may rupture if the water they contain freezes. Frost is probably the single most important factor limiting the geographic distribution of tropical and subtropical plants. In the Sonoran Desert in Arizona, saguaro cacti can easily withstand frost for one night as long as temperatures rise above freezing the following day, but they are killed when temperatures remain below freezing for 36 hours. This means that the cactus's distribution is limited to places where the temperature does not remain below freezing for more than one night (**Figure 54.8**).

The geographic range limits of endothermic animals are also affected by temperature. For example, the eastern phoebe (*Sayornis phoebe*), a small bird, has a northern winter range that coincides with an average minimum January temperature of above −4°C. Such limits are probably related to the energy demands associated with cold temperatures. Cold temperatures mean higher metabolic costs, which are in turn dependent on high feeding rates. Below −4°C, the eastern phoebe cannot feed fast enough or, more likely, find enough food to keep warm.

- - - Boundary of saguaro cactus range

● Temperatures remain below freezing for 1 or more days/year

● Temperatures remain below freezing for less than 0.5 days/year

● No days below freezing on record

Figure 54.8 **Saguaro cacti in freeze-free zones.** A close correspondence is seen between the range of the saguaro cactus (dashed line) and the area in which temperatures do not go below freezing (0°C).

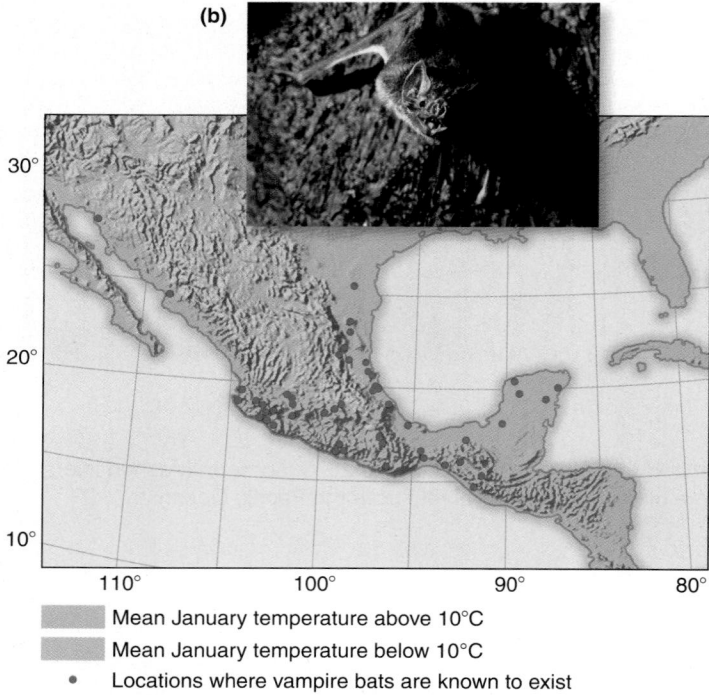

(a)

Mean January temperature above 10°C

Mean January temperature below 10°C

• Locations where vampire bats are known to exist

Figure 54.9 **Geographic range of vampire bats.** **(a)** The northern distribution of the vampire bat in Mexico is limited to areas warmer than the 10°C isotherm in January because of the animal's poor capacity to thermoregulate. Its southern distribution in Argentina and Chile is limited by the same isotherm. **(b)** The vampire bat, *Desmodus rotundus*.

Biological inquiry: Should customs officials in the U.S. be concerned about the entry of vampire bats into the U.S.?

Similarly, cold temperatures limit the distribution of the vampire bat (*Desmodus rotundus*). The vampire bat is found in an area from central Mexico to northern Argentina. Its range in Mexico is limited to that area where the average minimum temperature in January is above 10°C. Because of the bat's poor capacity for thermal regulation, it cannot survive in areas below that temperature (**Figure 54.9**).

High temperatures are also limiting for many plants and animals because relatively few species can survive internal temperatures more than a few degrees above their metabolic optimum. While we have discussed how corals are sensitive to low temperatures, they are sensitive to high temperatures as well. When temperatures are too high, the symbiotic algae that live within coral die and are expelled, causing a phenomenon known as coral bleaching. Once bleaching occurs, the coral tissue loses its color and turns a pale white. El Niño is a weather phenomenon characterized by a major increase in the water temperature of the equatorial Pacific Ocean. In the winter of 1982–1983, an influx of warm water from the eastern Pacific raised temperatures just 2–3°C for six months, which was enough to kill many of the reef-building corals on the coast of Panama. By May 1983, just a

Figure 54.10 **Giant sequoia.** A park ranger uses a drip torch to ignite a fire at Sequoia National Park in California. Periodic, controlled human-made fires mimic the sporadic wildfires that normally burn natural areas. Such fires are vital to the health of giant sequoia populations, since they serve to open the pine cones and release the seeds.

few individuals of one species, *Millepora intricata*, were alive. Scientists are concerned that future climate changes may increase the frequency of coral bleaching.

The ultimate high temperatures that many terrestrial organisms face are brought about by fire. However, some species depend on frequent, low-intensity fires for their reproductive success. The longleaf pine (*Pinus palustris*) of the southeast U.S. produces serotinous cones, which remain sealed by pine resin until the heat of a fire melts them open and releases the seeds. In the west, giant sequoia trees are similarly dependent on periodic low-intensity fires for germination of their seeds. Such fires both enhance the release of seeds and clear out competing vegetation at the base of the tree so that seeds can germinate (**Figure 54.10**). Fire-suppression practices that attempt to protect forests from fires actually can have the opposite result by preventing the regeneration of fire-dependent species. Furthermore, fire prevention can result in more growth of vegetation beneath the canopy (the understory) that may later fuel hotter and more damaging fires.

Keep in mind that it is not mean average temperatures that usually limit the range of species but rather the frequency of occasional extremes, such as freezes for the saguaro cacti. Farmers know this only too well. The frequency and strength of periodic freezes limit the northern distribution of oranges in Florida and the southern distribution of coffee in Brazil, not average temperatures for the coldest months. Experimentally moving organisms outside their normal range and monitoring survivorship is a useful way to establish what factors control the natural limits of species. Of course, such movements should be done only in a carefully controlled manner and when there is no risk of species becoming invasive in their new habitats. Again, physical extremes, such as fires or severe freezes, may be apparent only in isolated years, so ecologists often have to wait many years to determine whether extremes are limiting the distribution and abundance of plants and animals in the field.

Despite the obvious relationships between species distributions and temperature, we need to be cautious about solely relating the two. The temperatures measured for constructing isotherm maps are not always the temperatures that the organisms experience. In nature, an organism may choose to lie in the sun or hide in the shade, both of which affect the temperatures it experiences. Such local variations of the climate within a given area, or **microclimate**, can be important for a particular species. For example, the rufous grasshopper (*Gomphocerippus rufus*) is distributed widely in Europe, but in Great Britain it reaches its northern limit only 150 m from the south coast, where it is restricted to steep, south-facing, and therefore relatively sun-drenched and warm grassy slopes.

Because so many species are limited in their distribution patterns by global temperatures, ecologists are concerned that if global temperatures rise, many species will be driven to extinction or that their geographic ranges will shrink and the location of centers of agriculture and forestry will be altered. The increase in the average temperature of Earth's atmosphere and oceans is called global warming, and it is caused by a process known as the greenhouse effect.

The Greenhouse Effect The Earth is warmed by the **greenhouse effect**. In a greenhouse, sunlight penetrates the glass and raises temperatures, with the glass acting to trap the resultant heat inside. Similarly, solar radiation in the form of short-wave energy passes through the atmosphere to heat the surface of the Earth. At night this energy is radiated from the Earth's warmed surface back into the atmosphere, but in the form of long-wave infrared radiation. Instead of letting it escape back into space, however, atmospheric gases absorb much of this infrared energy and reradiate it to the Earth's surface, causing its temperature to rise further (**Figure 54.11**). Without some type of greenhouse effect, life on Earth would not exist. Global temperatures would be much lower than they are, perhaps averaging only $-17°C$ compared with the existing average of $+15°C$.

The greenhouse effect is caused by a small group of gases, mainly water vapor, that together make up less than 1% of the total volume of the atmosphere. After water vapor, the four most significant greenhouse gases are carbon dioxide, methane, nitrous oxide, and chlorofluorocarbons (**Table 54.1**).

Global Warming Ecologists are concerned that human activities are increasing the greenhouse effect, causing **global warm-**

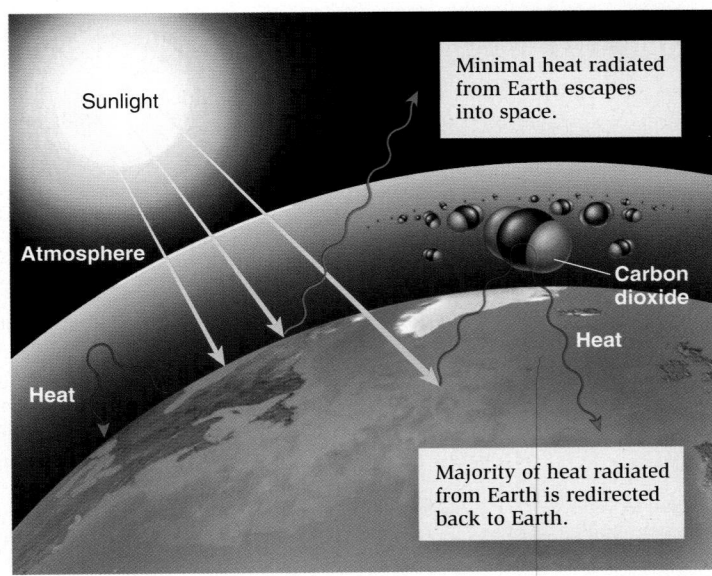

Figure 54.11   The greenhouse effect.   Solar radiation, in the form of short-wave energy, passes through the atmosphere to heat the Earth's surface. Long-wave infrared energy is radiated back into the atmosphere. Most infrared energy is reflected by atmospheric gasses, including carbon dioxide molecules, back to Earth, causing global temperatures to rise.

ing, a gradual elevation of the Earth's surface temperature. All greenhouse gases have increased in atmospheric concentration since industrial times. The most important of these gases is carbon dioxide (CO_2). As Table 54.1 shows, CO_2 has a lower global warming potential per unit of gas (relative absorption) than any of the other major greenhouse gases, but its concentration in the atmosphere is much higher. Concentrations of atmospheric CO_2 have increased from about 280 ppm (parts per million) in the preindustrial 18th century to 379 ppm in 2005.

To predict the effect of global warming, most scientists focus on a future point, about 2100, when the concentration of atmospheric CO_2 will have doubled—that is, increased to about 700 ppm compared with the late-20th-century level of 350 ppm. Ecologists argue that at that time, average global temperatures will be about $1-6°C$ (about $2-10°F$) warmer than present and will increase an additional $0.5°C$ each decade. This increase in heat might not seem like much, but it is comparable to the warming that ended the last ice age.

Assuming this scenario of gradual global warming is accurate, we need to consider what the consequences on natural and human-made ecosystems might be. Although many species can adapt to slight changes in their environment, the anticipated changes in global climate are expected to occur too rapidly to be compensated for by normal evolutionary processes such as natural selection. Plant species cannot simply disperse and move north or south into the newly created climatic regions that will be suitable for them. Many tree species take hundreds, even thousands, of years for seed dispersal.

Table 54.1	Greenhouse Gases and Their Contribution to Global Warming			
	Carbon dioxide (CO$_2$)	Methane (CH$_4$)	Nitrous oxide (N$_2$O)	Chlorofluorocarbons (CFCs)
Relative absorption per ppm of increase*	1	32	150	>10,000
Atmospheric concentration†	379 ppm	1.78 ppm	315 ppb	0.28–0.54 ppb
Contribution to global warming	50%	19%	4%	15%
Percent from natural sources; type of source	20–30%; volcanoes	70–90%; swamps, gas from termites and ruminants	90–100%; soils	0%
Major human-made sources	Fossil fuel use, deforestation	Rice paddies, landfills, biomass burning, coal and gas exploitation	Cultivated soil, fossil-fuel use, automobiles, industry	Previously industrially manufactured products (for example, as aerosol propellants) but banned in U.S. and the E.U.

*Relative absorption is the warming potential per unit of gas.

†ppm = parts per million, ppb = parts per billion.

Figure 54.12 **The range of sugar maples could be reduced by global warming.** The present geographic range of the sugar maple (blue shading) and its potential range under doubled CO$_2$ levels (red shading) in North America. Purple shading indicates the region of overlap, which is the only area where the sugar maple would be found before it spread into its new potential range.

Paleobotanist Margaret Davis predicted that in the event of a CO$_2$ doubling, the sugar maple (*Acer saccharum*), which is presently distributed throughout the midwestern and northeastern U.S. and southeastern Canada, would die back in all areas except in northern Maine, northern New Brunswick, and southern Quebec (**Figure 54.12**). Of course, this contraction in the tree's distribution could be offset by the creation of new favorable habitats in central Quebec. However, most scientists believe that the climatic zones would shift toward the poles faster than trees could migrate via seed dispersal, hence extinctions would occur. Interestingly, scientists are beginning to be able to genetically modify organisms to change their temperature tolerances.

GENOMES & PROTEOMES

Temperature Tolerance May Be Manipulated by Genetic Engineering

Below-freezing temperatures can be very damaging to plant tissue, either killing the plant or greatly reducing its productivity. Frost injury causes losses to agriculture of more than $1 billion annually in the U.S. While frost has been considered an unavoidable result of subfreezing temperatures, genetic engineering is beginning to change this view.

Between 0°C and −40°C, pure water will be a liquid unless provided with an ice nucleus or template on which an ice crystal can be built. Researchers discovered that some bacteria commonly found on leaf surfaces act as ice nuclei, triggering the formation of ice crystals and eventually causing frost damage. The genes that confer ice nucleation have been identified, isolated, and prevented from working in an engineered strain of the bacteria *Pseudomonas syringae*. When this strain is allowed to colonize strawberries, frost damage is greatly reduced, and plants can withstand an additional 5°C drop in temperature before frost forms. The promise of this technique for increasing agricultural yields and altering normal plant-distribution patterns is staggering.

At the other end of the temperature spectrum, heat shock proteins (HSPs) function as "molecular chaperones" to help organisms cope with the stress of high temperatures. At high temperatures, proteins may either unfold or bind to other proteins to form misfolded protein aggregations. HSPs act to prevent this type of event from taking place or to help aggregated proteins regain their normal configuration. HSPs normally constitute only about 2% of the cell's soluble protein content, but this can increase to 20% when a cell is stressed, whether by heat, cold, drought, or other stressors. In fact, HSPs are extremely common and are found in all organisms, from bacteria to plants and animals.

In the tropics, high temperatures can substantially decrease the growth rates and productivity of many crop species. There is now substantial interest in identifying crop strains with naturally high HSP levels and identifying thermally tolerant varieties for use in crop-breeding programs. Given the seemingly inevitable prospect of global warming, such research seems particularly timely.

Wind Can Amplify the Effects of Temperature

Wind is created by temperature gradients. As air heats up, it becomes less dense and rises. As hot air rises, cooler air rushes in to take its place. Hot air rising in the tropics is replaced by cooler air flowing in from more temperate regions, thereby creating northerly or southerly winds.

Wind affects living organisms in a variety of ways. It increases heat loss by evaporation and convection (the windchill factor). Wind also contributes to water loss in organisms by increasing the rate of evaporation in animals and transpiration in plants. For example, the tree line in alpine areas is often determined by a combination of low temperatures and high winds such that transpiration exceeds water uptake.

Winds can also intensify oceanic wave action, with resulting effects for organisms. On the ocean's rocky shore, seaweeds survive heavy surf by a combination of holdfasts and flexible structures. The animals of this zone have powerful organic glues and muscular feet to hold them in place (**Figure 54.13**).

The Availability of Water Has Important Effects on the Abundance of Organisms

Water has an important effect on the distribution of organisms. Cytoplasm is 85–90% water, and without moisture there can be no life. As noted in Chapter 3, water performs crucial functions in all living organisms. It acts as a solvent for chemical reac-

(a)

(b)

Figure 54.13   Animals and plants of the intertidal zone adhering to their rocky surface.   (a) The brown alga (*Laminaria digitata*) has a holdfast that enables it to cling to the rock surface. (b) The mussel (*Mytilus edulis*) attaches to the surface of a rock by proteinaceous threads (byssal threads) that extend from the animal's muscular foot.

tions, takes part in the reactions of hydrolysis, is the means by which animals eliminate wastes, and is used for support in plants and in some invertebrates as part of a hydrostatic skeleton.

The distribution patterns of many plants are limited by available water. Some plants, such as the water tupelo tree (*Nyssa aquatica*) in the southeast U.S., do best when completely flooded and are thus found predominantly in swamps. Others, for example, coastal plants that grow on sand dunes, experience very little fresh water. Their roots penetrate deep into the sand to extract moisture. In cold climates, water can be present but locked up as permafrost and, therefore, unavailable—this is termed a frost-drought situation. Alpine trees can be affected by frost drought. The trees stop growing at a point on the mountainside where they cannot take up enough moisture to offset transpiration losses. This point, known as the timberline, is readily apparent on many mountainsides. Not surprisingly, the density of many plants is limited by the availability of water. For example, a significant correlation is observed between increased rainfall and increased creosote bush density in the Mojave Desert.

Animals face problems of water balance, too, and their distribution and population density can be strongly affected by water availability. Because most animals depend ultimately on plants as food, their distribution is intrinsically linked to those of their food sources. Such a phenomenon regulates the number of buffalo (*Syncerus caffer*) in the Serengeti area of Africa. In this area, grass productivity is related to the amount of rainfall in the previous month. Buffalo density is governed by food availability, so a significant correlation is found between buffalo density and rainfall (**Figure 54.14**). The only exception occurs in the vicinity of Lake Manyara, where groundwater promotes plant growth.

The importance of water in limiting animal population density was underscored by an extraordinary event caused by El Niño. During the 1982–1983 rainy season, rainfall on Isla Genovesa in the Galápagos Islands off of Ecuador increased from its normal 100–150 mm to 2,400 mm. Plants responded with prodigious growth, and Darwin's finches (*Geospiza* spp.) bred up to eight times rather than their normal maximum of three, probably because of increased abundance of fruits and seeds.

Light Can Be a Limiting Resource for Plants and Algae

Because light is necessary for photosynthesis, it can be a limiting resource for plants. However, what may be sufficient light to support the growth of one plant species may be insufficient for another. Many plant species grow best in shady conditions, such as eastern hemlock (*Tsuga canadensis*). Its saplings grow in the understory below the forest canopy, reaching maximal photosynthesis at one-quarter of full sunlight. Other plants, like sugarcane (*Saccharum officinarum*) or the desert shrub *Larrea*, continue to increase their photosynthetic rate as light intensity increases.

One reason photosynthetic rates vary among plants is related to three different biochemical pathways by which carbon

fixation can occur: C_3, C_4, and CAM (see Chapter 8). Because C_4 plants such as sugarcane grow faster in areas with high daytime temperatures and intense sunlight, those species outcompete C_3 plants and are more common in tropical areas than in temperate areas. On the other hand, in cooler, cloudier temperate areas, C_3 species can tolerate lower light and live in areas where C_4 plants cannot. Thus, C_3 plants are much more common in areas outside the tropics. CAM plants, such as some desert succulents, are the opposite of typical plants in that they open their stomata to take up CO_2 at night, presumably as an adaptation to minimize water loss in the day. The absorbed CO_2 is stored as malic acid, which is then used to complete photosynthesis during the day. These plants are adapted to live in very dry desert areas where little else can grow.

In aquatic environments, light may be an even more limiting factor because water absorbs light, preventing photosynthesis at depths greater than 100 m. Most aquatic plants and algae are limited to a fairly narrow zone close to the surface, where light is sufficient to allow photosynthesis to exceed respiration. This zone is known as the **euphotic zone**. In marine environments, seaweeds at greater depths have wider thalli (leaflike light-gathering structures) than those nearer the surface, because wide thalli can collect more light. In addition, in aquatic environments, plant color changes with depth. At the surface, plants and algae appear green, as they are in terrestrial conditions, because they absorb red and blue light, but not green. At greater depths, red light is mostly absorbed by water, leaving predominantly blue-green light. Red algae occur in deeper water because these possess pigments that enable them to utilize blue-green light efficiently (**Figure 54.15**).

The Concentration of Salts in Soil or Water Can Be Critical

Salt concentrations vary widely in aquatic environments and have a great impact on osmotic balance in animals. Oceans contain considerably more dissolved minerals than rivers. Oceans continually receive the nutrient-rich waters of rivers, and the sun evaporates pure water from ocean surfaces, making concentrations of minerals such as salt even higher.

The phenomenon of osmosis influences how living organisms cope with different environments. Freshwater fish cannot live in salt water, and saltwater fish cannot live in fresh water. Each employs different mechanisms to maintain an osmotic balance with their environment. Freshwater fish are hyperosmotic (having a higher concentration of ions) to their environment and tend to gain water by osmosis as it passes over the thin tissue of the gills and mouth. To counter this, the fish continually eliminate water in the urine. However, to avoid losing all dissolved ions, many ions are reabsorbed into the bloodstream at the kidneys. Many marine fish are hypoosmotic (having a lower concentration of ions) to their environment and tend to lose water as seawater passes over the mouth and gills. They drink more water to compensate for this loss, but the water contains a higher concentration of salts, which must then be excreted at the gills and kidneys (look back to Chapter 49).

Figure 54.14   **The relationship between the amount of rainfall and the density of buffalo.**   In the Serengeti area of Africa, buffalo density is very much dependent on grass availability, which itself is dependent on annual rainfall. The main exception is where there is permanent water, such as Lake Manyara, where greater water availability leads to greater grass growth and buffalo densities.

(a)

(b)

Figure 54.15 **Green algae of surface marine waters compared to red algae of deeper marine waters.** (a) In the eastern Pacific Ocean, off the coast of California, these giant kelp floating at the ocean surface are green, just like terrestrial plants. (b) In contrast, at 75-m depth, in the McGrail Bank off of the Gulf of Mexico, most seaweeds are pink and red because the pigments can absorb the blue-green light that occurs at such depths.

Figure 54.16 **Plant adaptations for salty conditions.** Special salt glands in *Spartina* leaves exude salt, enabling this grass to exist in saline intertidal conditions.

(a)   (b)

Figure 54.17 **Species-rich floras of chalk grassland compared to species-poor floras of acid soils.** (a) At Mount Caburn, in the lime-rich chalk hills of Sussex County, England, there is a much greater variety of plant and animal species than at (b) a heathland site in England. Heathlands are a product of thousands of years of human clearance of natural forest areas and are characterized by acidic, nutrient-poor soils.

Salt in the soil also affects the growth of plants. In arid terrestrial regions, crystalline salt accumulates in soil where water evaporates. This can be of great significance in agriculture, where continued watering in arid environments, together with the addition of salt-based fertilizers, greatly increases salt concentration in soil and reduces crop yields. A very few terrestrial plants are adapted to live in saline soil along seacoasts. Here the vegetation consists largely of **halophytes**, species that can tolerate higher salt concentrations in their cell sap than regular plants. Species such as mangroves and *Spartina* grasses have salt glands that excrete salt to the surface of the leaves, where it forms tiny white salt crystals (**Figure 54.16**).

The pH of Soil or Water Can Limit the Distribution of Organisms

As discussed in Chapter 2, the pH of water can be acidic, alkaline, or neutral. Variation in pH can have a major impact on the distribution of organisms. Normal rainwater has a pH of about

5.6, which is slightly acidic because the absorption of atmospheric CO_2 and SO_2 into rain droplets forms carbonic and sulphuric acids. However, most plants grow best at a soil water pH of about 6.5, a value at which soil nutrients are most readily available to plants. Only a few genera, such as rhododendrons (*Rhododendron*) and azaleas (*Azalea*), can live in soils with a pH of 4.0 or less. Furthermore, at a pH of 5.2 or less, nitrifying bacteria do not function properly, which prevents organic matter from breaking down. In general, soils containing chalk and limestone, the so-called high-lime soils, have a higher pH and sustain a much richer flora (and associated fauna) than do acidic soils (**Figure 54.17**).

Generally, the number of fish and other species also decreases in acidic waters. The optimal pH for most freshwater fish and bottom-dwelling invertebrates is between 6.0 and 9.0. Acidity in lakes increases the amount of toxic metals, such as aluminum, mercury, and lead, which can leach into the water from surrounding soil and rock. Both too much mercury and too much aluminum can interfere with gill function, causing fish to suffocate. Interestingly, Florida has a great number of acidic lakes, more than any other region in the U.S., but most of these are naturally acidic, and most fish found in these lakes are tolerant of the relatively low pH levels.

The susceptibility of both aquatic and terrestrial organisms to changes in pH explains why ecologists are so concerned about **acid rain**, precipitation with a pH of less than 5.6. Acid rain results from the burning of fossil fuels such as coal, natural gas, and oil, which releases sulfur dioxide and nitrogen oxide into the atmosphere. These react with oxygen in the air to form sulfuric acid and nitric acid, which falls to the surface in rain or snow. When this precipitation falls on rivers and especially lakes, it can turn them more acidic and they lose their ability to sustain fish and other aquatic life. For example, lake trout disappear from lakes in Ontario and the eastern U.S. when the pH falls below about 5.2. Although this low pH does not affect survival of the adult fish, it affects the survival of juveniles.

Acid rain is important in terrestrial systems, too. For example, acid rain can directly affect forests by killing leaves or pine needles, as has happened on some of the higher mountaintops in the Great Smoky Mountains. It can also greatly depress soil pH, which can result in a loss of essential nutrients such as calcium and nitrogen. Low calcium in the soil results in calcium deficiencies in plants, in the snails that consume the plants, and in the birds that eat the snails, ultimately causing weak eggshells that break before hatching. Decreased soil pH also kills certain soil microorganisms, preventing decomposition and recycling of nitrogen in the soil. Decreases in soil calcium and nitrogen weaken trees and other plants, and may make them more susceptible to insect attack.

Acid rain is a common problem in the northeastern U.S. and Scandinavia, where sulfur-rich air drifts over from the Midwest and the industrial areas of Britain, respectively, causing the deposition of highly acidic rain. The problem was particularly acute during the 1960s and 1970s, but decreased manufacturing and the use of low-sulfur coal and the introduction of sulfur-absorbing scrubbers on the smokestacks of coal-burning power plants have reduced the problem somewhat in recent years. Acid rain is clearly a problem with a wide-ranging impact on ecological systems.

54.3 Climate and Its Relationship to Biological Communities

Temperature, water, wind, and light are components of **climate**, the prevailing weather pattern in a given region. Because the distribution and abundance of organisms are often limited by the abiotic environment, to understand the patterns of abundance of life on Earth, ecologists need to study the global climate. We begin this section by examining global climate patterns, focusing on how temperature variation drives atmospheric circulation and how features such as elevation and landmass can alter these patterns. Knowing how and why climate changes around the world enables us to understand and predict the occurrence of different **biomes**, the major community types on Earth such as tropical forests and hot deserts. In this section, we also explore the main characteristics of Earth's major terrestrial and aquatic biomes.

Atmospheric Circulation Is Driven by Global Temperature Differentials

Substantial differences in temperature occur over the Earth, mainly due to variations in the incoming solar radiation. In higher latitudes, such as northern Canada and Russia, the sun's rays hit the Earth obliquely and are spread out over more of the planet's surface than they are in equatorial areas (**Figure 54.18**). More heat is also lost in the atmosphere of higher altitudes because the sun's rays travel a greater distance through the atmosphere, allowing more heat to be dissipated by cloud cover. The result is that a much smaller amount of solar energy (40% less) strikes polar latitudes than equatorial areas. Generally, temperatures increase as the amount of solar radiation increases (**Figure 54.19**). However, at the tropics both cloudiness and rain reduce average temperature, so that temperatures do not continue to increase toward the equator.

The English meteorologist George Hadley made the initial contribution to a model of general atmospheric circulation in 1735. Hadley proposed that solar energy drives winds, which in

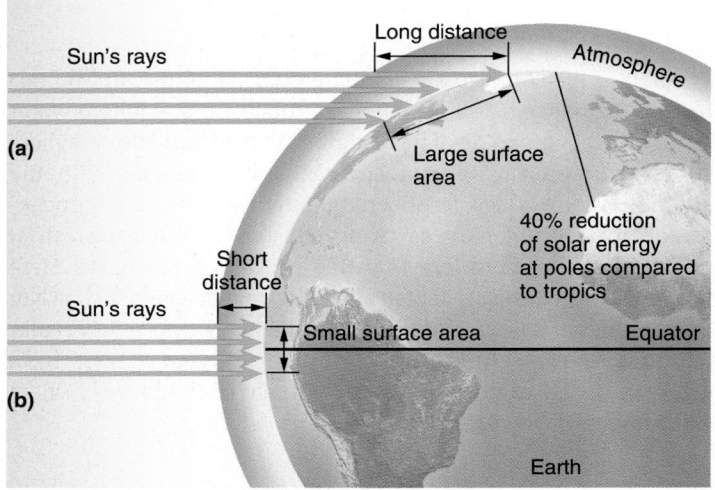

Figure 54.18   The intensity of solar radiation at different latitudes.   In polar areas **(a)**, the sun's rays strike the Earth at an oblique angle and deliver less energy than at tropical locations. In tropical areas **(b)**, the energy is concentrated over a smaller surface and travels a shorter distance through the atmosphere.

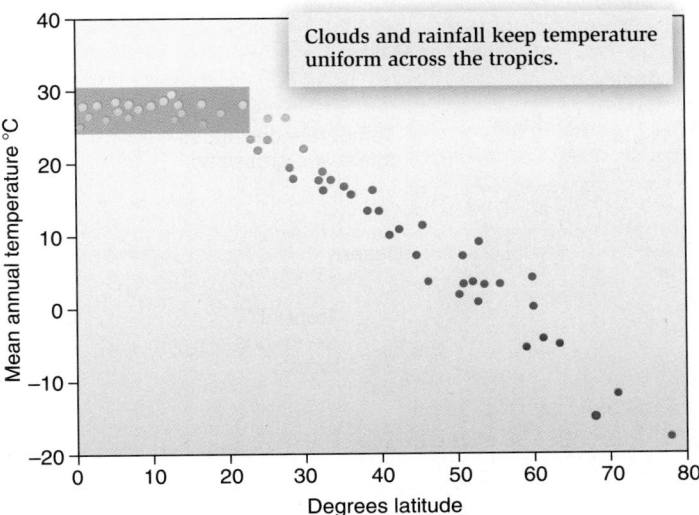

Figure 54.19 **Variation of the Earth's temperature.** The temperatures shown in this figure were measured at moderately moist continental locations of low elevation. Note the wide band of similar temperatures at the tropics.

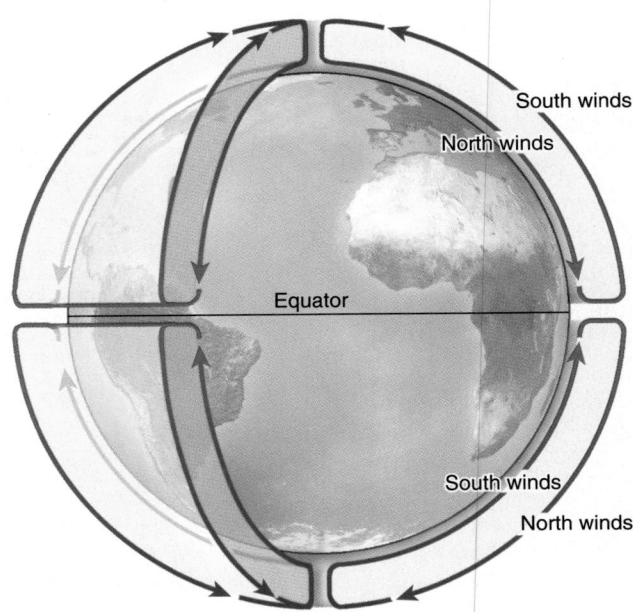

Figure 54.20 **George Hadley's 1735 model of atmospheric circulation.** In Hadley's model, simple convective circulation of air on a uniform, nonrotating Earth, heated at the equator and cooled at the poles, took the form of one large convection cell in each hemisphere. Winds are named according to the direction from which they blow, so the south wind blows from south to north.

turn influence the global circulation of the atmosphere. In his model, the warmth at the equator causes the surface equatorial air to heat up and rise vertically into the atmosphere. As the warm air rises away from its source of heat, it cools and becomes less buoyant, but the cool air does not sink back to the surface because of the warm air behind it. Instead, the rising air spreads north and south away from the equator, eventually returning to the surface at the poles. From there it flows back toward the equator to close the circulation loop. Hadley suggested that on a nonrotating Earth, this air movement would take the form of one large convection cell in each hemisphere, as shown in **Figure 54.20**.

When the effect of the Earth's rotation is added, however, the surface flow is deflected toward the west. This consequence is known as the **Coriolis effect**. Hadley's one-cell circulation has since been modified to account for the Coriolis effect and other more modern data. In the 1920s, a three-cell circulation in each hemisphere was proposed to fit the Earth's heat balance (**Figure 54.21**). The contribution of George Hadley is still recognized, in that the most prominent of the three cells, the one nearest the equator, is called the **Hadley cell**. In the Hadley cell, the warm air rising near the equator forms towers of cumulus clouds that provide rainfall, which in turn maintains the lush vegetation of the equatorial rain forests. As the upper flow in this cell moves toward the poles, it begins to subside, or fall back to Earth, at about 30° north and south of the equator. These **subsidence zones** are areas of high pressure and are the sites of the world's tropical deserts, because the subsiding air is relatively dry, having released all of its moisture over the equator. Winds are generally weak and variable near the center of this zone of descending air. Subsidence zones have popularly been called the horse latitudes. The name is said to have been coined by Spanish sailors crossing the Atlantic, whose ships were

sometimes rendered motionless in these waters and who reportedly were forced to throw horses overboard, or eat them, as they could no longer water or feed them.

From the center of the subsidence zones, the surface flow splits into the westerlies that flow toward the poles, and the equatorial flow, which is deflected by the Coriolis effect and forms the reliable trade winds. In the Northern Hemisphere, the trades are from the northeast, the direction from which they provided the sail power to explore the New World; in the Southern Hemisphere, the trades are from the southeast. The trade winds from both hemispheres meet near the equator in a region called the intertropical convergence zone (ITCZ), also known as the doldrums. Here the light winds and humid conditions provide the monotonous weather that may be the basis for the expression "in the doldrums."

In the three-cell model, the circulation between 30° and 60° latitude, called the **Ferrell cell**, is opposite that of the Hadley cell. The net surface flow is poleward, and because of the Coriolis effect, the winds have a strong westerly (flowing from west to east) component. These prevailing westerlies were known to Benjamin Franklin, perhaps the first American weather forecaster, who noted that storms migrated eastward across the colonies. In fact, the secondary zones of high precipitation can come anywhere from about 45° to 60°, with between 45° and 55° being most common. The final circulation cell is known as the **polar cell**. At the poles, the air has cooled and descends, but it has little moisture left, explaining why many high-latitude regions are actually desert-like in condition.

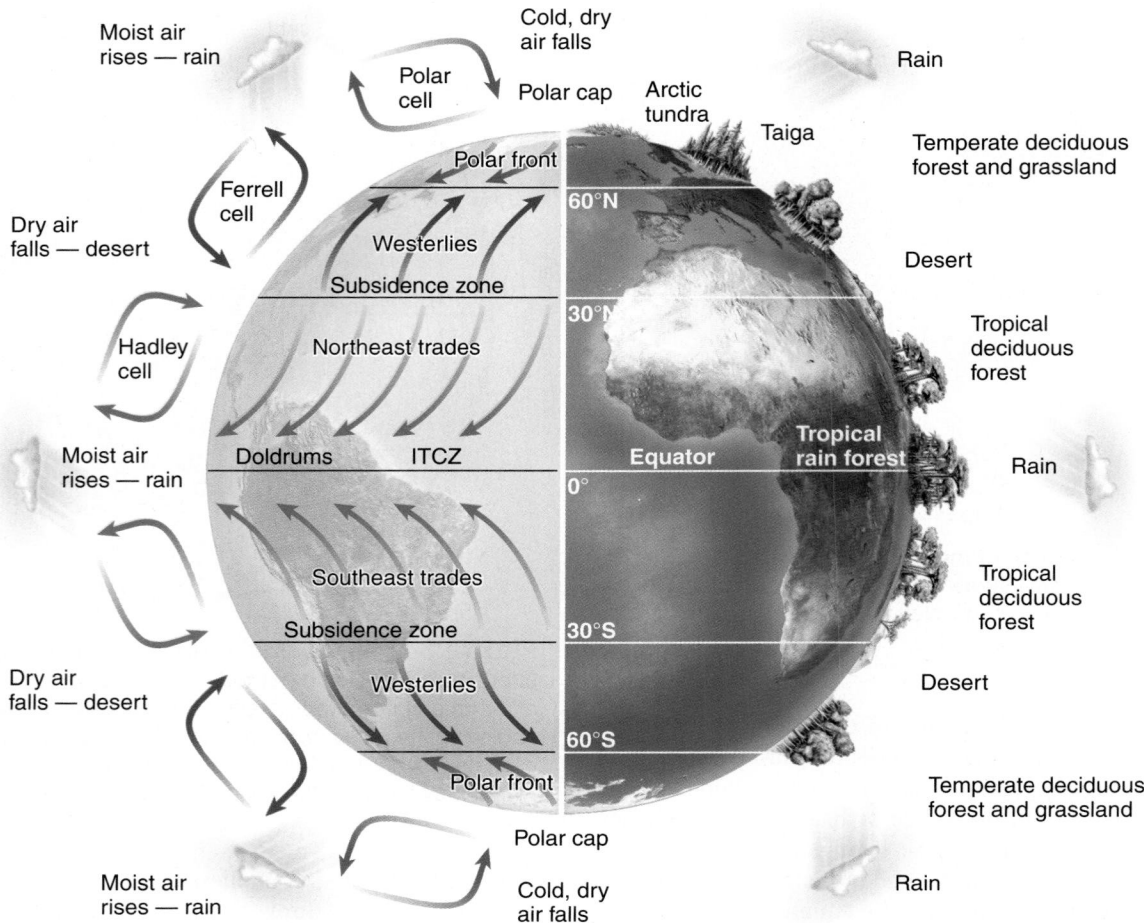

**Figure 54.21   Patterns of atmospheric circulation and biome positions.   **Three-cell model of the atmospheric circulation on a uniform, rotating Earth heated at the equator and cooled at the poles. Tropical forests exist mainly in a band around the equator, where it is hot and rainy. At around 30° north and south, the air is hot and dry, and deserts exist. A secondary zone of precipitation exists at around 45° to 55° north and south, where temperate forests are located. The polar regions are generally cold and dry.

Thus, the distributions of the major biomes are determined by temperature differences and the wind patterns they generate. Hot, tropical forest blankets the tropics, where rainfall is high. At about 30° latitude, the air cools and descends, but it is without moisture, so the hot deserts occur around that latitude. The middle cell of the circulation model shows us that at about 45° to 55° latitude the air has warmed and gained moisture, so it ascends, dropping rainfall over the wet, temperature forests of this region in the Pacific Northwest and Western Europe in the Northern Hemisphere and over New Zealand and Chile in the Southern Hemisphere.

Elevation and Other Features of a Landmass Can Also Affect Climate

Thus far, we have considered how global temperatures and wind patterns affect climate. The geographic features of a landmass can also have an important impact. For example, the elevation of a region greatly influences its temperature range. On mountains, temperatures decrease with increasing elevation.

This decrease is a result of a process known as **adiabatic cooling**, in which increasing elevation leads to a decrease in air pressure. When air is blown across the Earth's surface and up over mountains, it expands because of the reduced pressure. As it expands, it cools at a rate of about 10°C for every 1,000 m in elevation, as long as no water vapor or cloud formation occurs. (Adiabatic cooling is also the principle behind the function of a refrigerator, in which Freon gas cools as it expands coming out of the compressor.) A vertical ascent of 600 m produces a temperature change roughly equivalent to that brought about by an increase in latitude of 1,000 km. This explains why mountain-top vegetation, even in tropical areas, can have the characteristics of tundra.

Mountains can also influence patterns of precipitation. For example, when warm, moist air encounters the windward side of a mountain, it flows upward and cools, releasing precipitation in the form of rain or snow. On the side of the mountain sheltered from the wind (the leeward side), drier air descends, producing what is called a **rain shadow**, an area where precipitation is noticeably less (**Figure 54.22a**). In this way, the

As moist air blows across the windward side of a mountain, it rises, cools, and precipitation falls as rain or snow.

Moist air

Dry air

On the leeward side of a mountain, the cooler air descends and becomes warmer; little precipitation occurs.

Warm air

Cool air

During the day, as warm air rises, cooler air rushes in from the ocean to replace it.

(a) Rain shadow

(b) Sea breezes

Figure 54.22   The influence of elevation and proximity to water on climate.

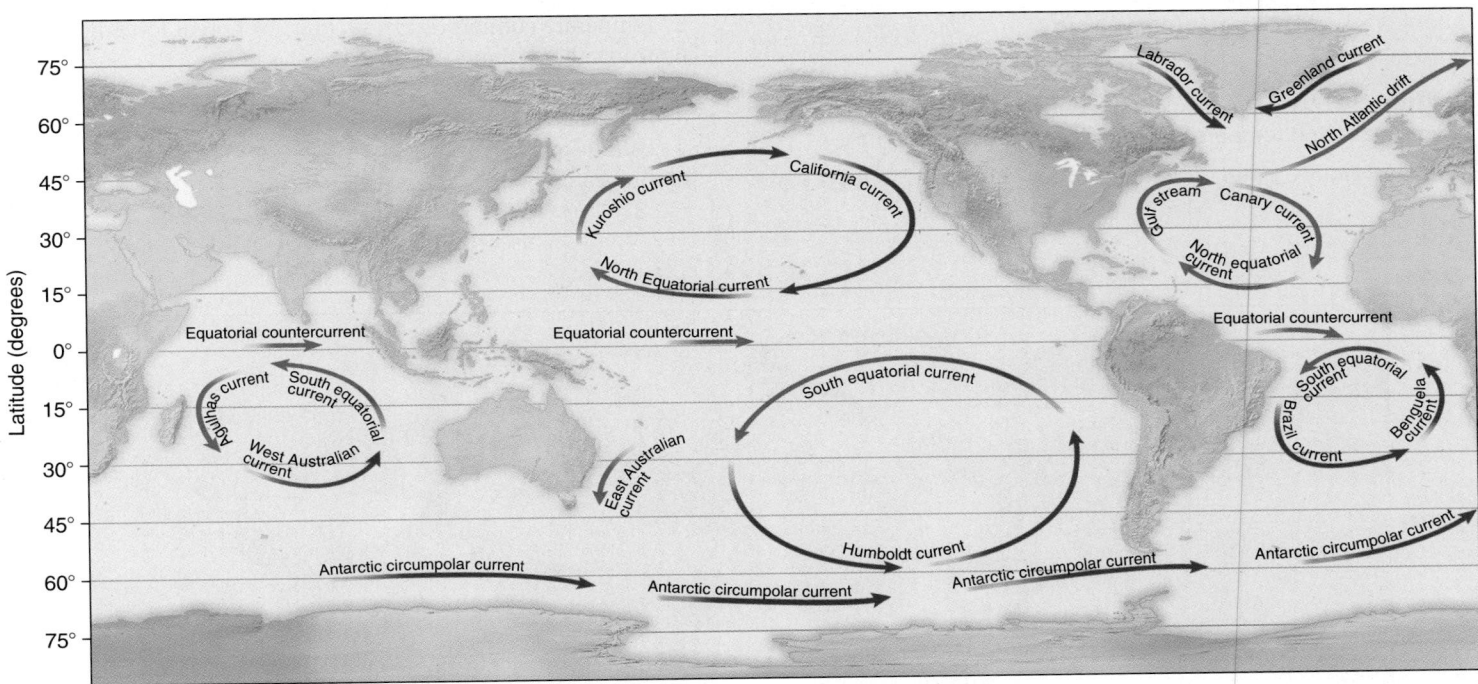

Figure 54.23   Ocean currents of the world.   The red arrows represent warm water; the blue arrows, cold water.

western side of the Cascade Range in Washington receives more than 500 cm of annual precipitation, whereas the eastern side receives only 50 cm.

The proximity of a landmass to a large body of water can affect climate because land heats and cools more quickly than the sea does. The specific heat capacity of the land is much lower than that of the water, allowing the land to warm more quickly in the day. The warmed air rises and cooler air flows in to replace it. This pattern creates the familiar onshore sea breezes in coastal areas (**Figure 54.22b**). At night, the land cools quicker than the sea, and so the pattern is reversed, creating offshore breezes. The sea, therefore, has a moderating effect on the temperatures of coastal regions and especially islands. The climates of coastal regions may differ markedly from those of their climatic zones. Many never experience frost, and fog is often evident. Thus, along coastal areas, different vegetation patterns may occur compared to those in areas farther inland. In fact, some areas of the U.S., including Florida, would be deserts were it not for the warm water of the sea and the moisture-laden clouds that form above them.

Together with the rotation of the Earth, winds also create ocean currents. The major ocean currents act as "pinwheels" between continents, running clockwise in the ocean basins of the Northern Hemisphere and counterclockwise in those of the Southern Hemisphere (**Figure 54.23**). Thus, the Gulf Stream, equivalent in flow to 50 times the world's major rivers combined, brings warm water from the Caribbean and the U.S. coasts across to Europe, the climate of which is correspondingly moderated. The Humboldt Current brings cool conditions almost to the equator along the western coast of South America.

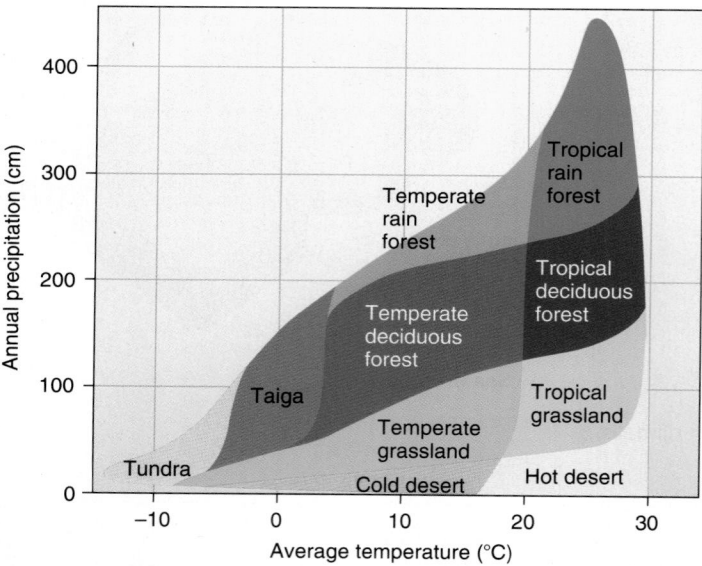

Figure 54.24 The relationship between the world's terrestrial biome types and temperature and precipitation patterns.

Biological inquiry: What other factors may influence biome types?

Terrestrial Biome Types Are Dictated by Climate Patterns

Differences in climate on Earth help to define its different terrestrial biomes. Many types of classification schemes are used for mapping the geographic extent of biomes, but one of the most useful was developed by the American ecologist Robert Whittaker, who classified biomes according to the physical factors of average annual precipitation and temperature (**Figure 54.24**). In this scheme, we can recognize 10 terrestrial biomes (**Figure 54.25**):

1. tropical rain forest;
2. tropical deciduous forest;
3. temperate rain forest;
4. temperate deciduous forest;
5. temperate coniferous forest (taiga or boreal forest);
6. tropical grassland (savanna);
7. temperate grassland (prairie);
8. hot desert;
9. cold desert;
10. tundra.

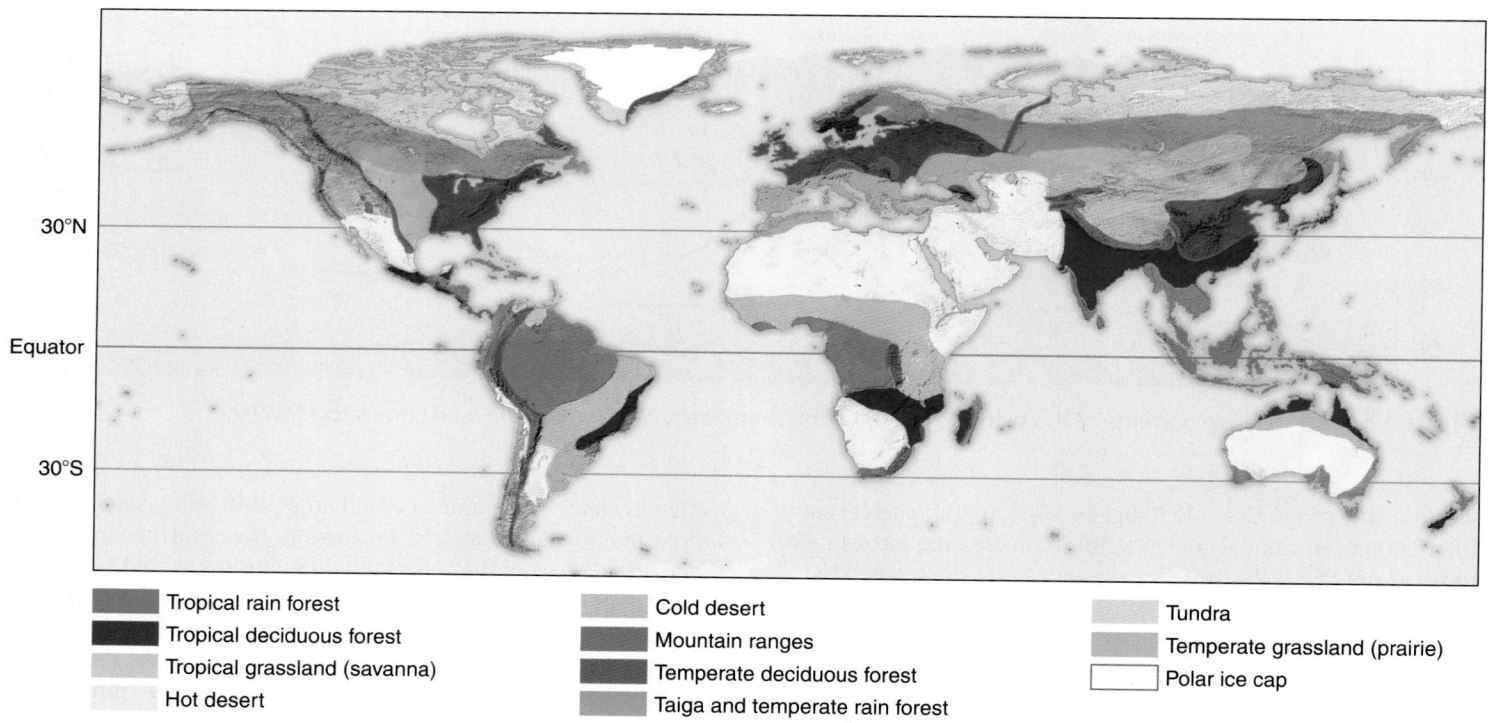

Tropical rain forest
Tropical deciduous forest
Tropical grassland (savanna)
Hot desert

Cold desert
Mountain ranges
Temperate deciduous forest
Taiga and temperate rain forest

Tundra
Temperate grassland (prairie)
Polar ice cap

Figure 54.25 Geographic location of terrestrial biomes. The distribution patterns of taiga and temperate rain forest are combined because of their similarity in tree species and because temperate rain forest is actually limited to a very small area.

Figure 54.26a–k illustrates these 10 terrestrial biomes and identify their main characteristics.

While these broad terrestrial biomes are a useful way of defining the main types of communities on Earth, ecologists acknowledge that not all communities fit neatly into one of these 10 major biome types. Also, one biome type often grades into another, as seen on mountain ranges (Figure 54.26k). Soil conditions can also influence biome type. In California, serpentine soils, which are dry and nutrient-poor, support only sparse vegetation. In the eastern U.S., most of New Jersey's coastal plain, called the Pine Barrens, consists of sandy, nutrient-poor soil that cannot support the surrounding deciduous forest and instead contains grasses and low shrubs growing among open stands of pygmy pitch pine and oak trees.

Some ecologists recognize chaparral, a type of temperate grassland, as a distinct biome type. Winters are cool and wet, and summers warm and dry. Summer conditions are so dry that fires occur normally and frequently at this time. Such conditions are most commonly seen at around 30° latitude, where cool ocean waters moderate the climate, as along the coasts of California, South Africa, Chile, and southwest Australia and in countries surrounding the Mediterranean Sea.

TROPICAL RAIN FOREST

Figure 54.26a

Tropical rain forest in Fiji

Physical Environment: Rainfall exceeding 230 cm per year, temperature hot year round, averaging 25–29°C, soils often shallow and nutrient-poor.

Location: Equatorial, between the Tropics of Cancer and Capricorn. Tropical forests cover much of northern South America, Central America, western and central Africa, Southeast Asia, and various islands in the Indian and Pacific oceans.

Plant Life: The numbers of plant species found in tropical forests can be staggering, often reaching as many as 100 tree species per square kilometer. Leaves often narrow to "drip-tips" at the apex so that rainwater drains quickly. Many trees have large buttresses that help support their shallow root systems. Little light penetrates the **canopy**, the uppermost layer of tree foliage, and the ground cover is often sparse. Epiphytes, plants that live perched on trees and are not rooted in the ground, are common. Bromeliads are common epiphytes in North and South American forests. Lianas, or climbing vines, are also common.

Animal Life: Animal life in the tropical rain forests is diverse; insects, reptiles, amphibians, and mammals are well represented. Large mammals, however, are not common. Because many of the plant species are widely scattered in tropical forests, it is risky for plants to rely on wind for pollination or to disperse their seed. This means that animals are important in pollinating flowers and dispersing fruits and seeds. Mimicry and bright protective coloration, warning of bad taste or the existence of toxins, are rampant.

Effects of Humans: Humans are impacting tropical forests greatly by logging and by clearing the land for agriculture. Also, many South American tropical forests are cleared to create grasslands for cattle.

TROPICAL DECIDUOUS FOREST

Figure 54.26b

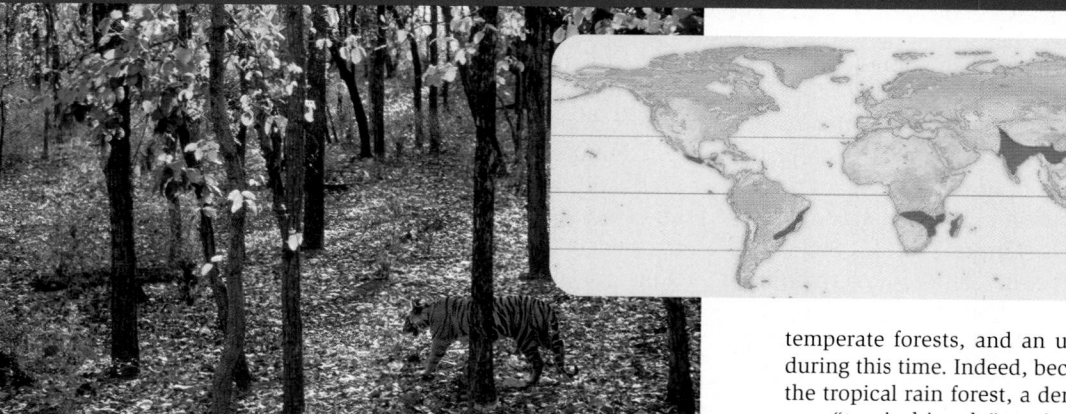

Tropical deciduous forest in Bandhavgarh National Park, India

Physical Environment: Rainfall is substantial, at around 130–280 cm a year, but the dry season is distinct, often two to three months or longer. Soil water shortages can occur in the dry season. Temperatures are hot year round, averaging 25–39°C.

Location: Equatorial, where rainfall is more seasonal. Much of India consists of tropical deciduous forest, containing teak trees. Brazil, Thailand, and Mexico also contain tropical deciduous forest. At the wet edges of this biome, it may grade into tropical forests; at the dry end, it may grade into tropical grasslands or savannas.

Plant Life: Because of the distinct dry season, many of the trees in tropical deciduous forests shed their leaves, just as they do in temperate forests, and an understory of herbs and grasses may grow during this time. Indeed, because the canopy is often more open than in the tropical rain forest, a denser closed forest—what we might think of as a "tropical jungle"—exists at the forest floor. Where the dry season is six to seven months long, tropical deciduous forests may contain shorter, thorny plants such as acacia trees and the forest is referred to as a tropical thorn forest.

Animal Life: The diversity of animal life is high, and species such as monkeys, antelopes, wild pigs, and tigers are present. However, as with plant diversity, animal diversity is less than in tropical rain forests. Tropical thorn forests may contain more browsing mammals; hence, the development of plant thorns as a defense.

Effects of Humans: Logging and agriculture have had a large impact on tropical deciduous forests. The litter layer of dead, decaying leaves is much thicker than in tropical rain forests and renders the soil more fertile.

TEMPERATE RAIN FOREST

Figure 54.26c

Hoh Rain Forest in Olympic National Park, Washington

Physical Environment: Temperatures are a little cooler, 5–25°C on average, and winters are mild, but there is abundant rainfall, usually exceeding 200 cm a year. The condensation of water from dense coastal fogs augments the normal rainfall.

Location: The coverage of this biome type is small, consisting of a thin strip along the northwest coast of North America from northern California through Washington State, British Columbia, and into southeast Alaska (here called tongass). It also exists in southwestern South America along the Chilean coast. Indeed, it is found only in coastal situations because of the moderating influence of the ocean on air temperature.

Plant Life: The dominant vegetation type, especially in North America, consists of large evergreen trees such as western hemlock, Douglas fir, and Sitka spruce. The high moisture content allows epiphytes to thrive. Cool temperatures slow the activity of decomposers, so that the litter layer is thick and spongy.

Animal Life: In North America, the temperate rain forest is rich in species such as mule deer, elk, squirrels, and numerous birds such as jays and nuthatches. Because of the abundant moisture and moderate temperatures, reptiles and amphibians are also common.

Effects of Humans: This biome is a prolific producer of wood and supplies much timber, although logging threatens the survival of the forest in some areas.

TEMPERATE DECIDUOUS FOREST

Figure 54.26d

Temperate deciduous forest in Minnesota

Physical Environment: Temperatures fall below freezing each winter but not usually below –12°C, and annual rainfall is generally between 75 and 200 cm.

Location: Large tracts of temperate deciduous forest are evident in the eastern U.S., eastern Asia, and western Europe. In the Southern Hemisphere, eucalyptus forests occur in Australia, and stands of southern beech are found in southern South America, New Zealand, and Australia.

Plant Life: Species diversity is much lower in temperate deciduous forests than in the tropical forests, with about only three to four species per square kilometer, and several tree genera may be dominant in a given locality—for example, oaks, hickories, and maples are usually dominant in the eastern U.S. Commonly, leaves are shed in the fall and reappear in the spring. Many herbaceous plants flower in spring before the trees leaf out and block the light, though even in the summer, the forest is not as dense as in tropical forests, so there is abundant ground cover. There are few epiphytes and lianas.

Animal Life: Animals are adapted to the vagaries of the climate; many mammals hibernate during the cold months, birds migrate, and insects enter diapause, a condition of dormancy passed usually as a pupa. Reptiles, which are dependent on solar radiation for heat, are relatively uncommon. Mammals include squirrels, wolves, bobcats, foxes, bears, and mountain lions.

Effects of Humans: Logging has eliminated much temperate forest from heavily populated portions of Europe and North America. Soils are rich because the annual leaf drop promotes high soil nutrient levels. With careful agricultural practices, soil richness can be conserved, and as a result agriculture can flourish.

TEMPERATE CONIFEROUS FOREST OR TAIGA

Figure 54.26e

Temperate coniferous forest in Canada

Physical Environment: Precipitation is generally between 30 and 70 cm, often occurring in the form of snow. Temperatures are very cold, below freezing for long periods of time.

Location: The biome of coniferous forests, known commonly by its Russian name, taiga, lies north of the temperate-zone forests and grasslands. Vast tracts of taiga exist in North America and Russia, and mountain taiga exists on mountainous areas. In the Southern Hemisphere, little land area occurs at latitudes at which one would expect extensive taiga to exist.

Plant Life: Most of the trees are evergreens or conifers with tough needles, hence its similarity to temperate rain forest. In this biome, spruces, firs, and pines generally dominate, and the number of tree species is relatively low. Many of the conifers have conical shapes to reduce bough breakage from heavy loads of snow. As in tropical forests, the understory is sparse because the dense year-round canopies prevent sunlight from penetrating. Soils are poor because the fallen needles decay so slowly in the cold temperatures that a layer of needles builds up. This layer of needles acidifies the soil, further reducing the numbers of understory species.

Animal Life: Reptiles and amphibians are rare because of the low temperatures. Insects are strongly periodic but may often reach outbreak proportions in times of warm temperatures. Mammals that inhabit this biome such as bears, lynxes, moose, beavers, and squirrels are heavily furred.

Effects of Humans: Humans have not extensively settled these areas but they have been quite heavily logged.

TROPICAL GRASSLAND

Figure 54.26f

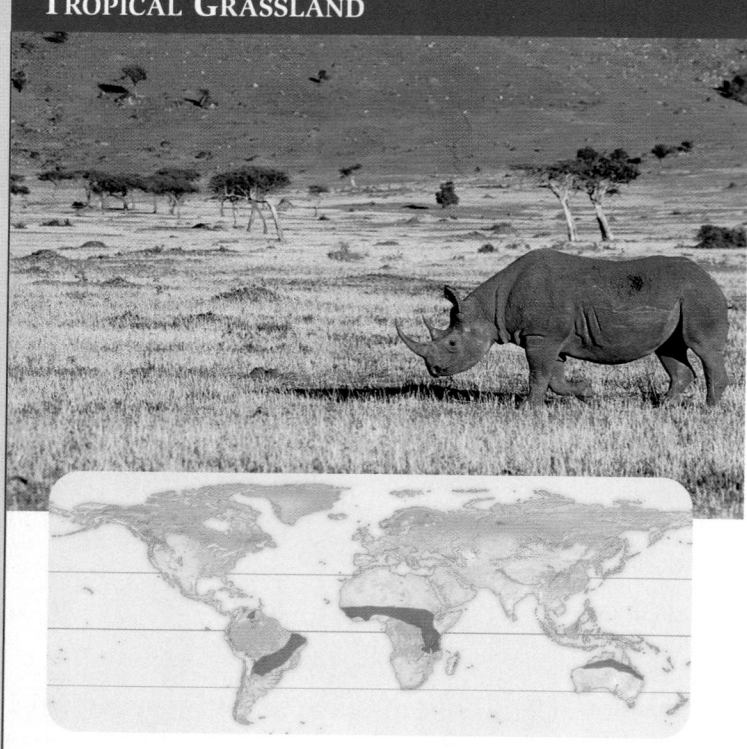

Tropical grassland of the Masai Mara Game Reserve in Kenya

Physical Environment: Hot, tropical areas, averaging 24–29°C with a low or seasonal rainfall of between 50 and 130 cm per year. There is often an extensive dry season.

Location: Extensive savannas occur in Africa, South America, and northern Australia.

Plant Life: Wide expanses of grasses dominate savannas but occasional thorny trees, such as acacias, may occur. Fire is prevalent in this biome, so most plants have well-developed root systems that enable them to resprout quickly after a fire.

Animal Life: The world's greatest assemblages of large mammals occur in the savanna biome. Herds of antelope, zebras, and wildebeest are found, together with their associated predators: cheetah, lion, leopard, and hyena. Termite mounds dot the landscape in some areas. The extensive herbivory of large grazers, together with frequent fires, may help maintain savannas and prevent their development into forests.

Effects of Humans: Savanna soils are often poor because the occasional rain leaches nutrients out. Nevertheless, conversion of this biome to agricultural land is rampant, especially in Africa. Overstocking of land for domestic animals can greatly reduce grass coverage through overgrazing, turning the area more desert-like. This process is known as **desertification**.

TEMPERATE GRASSLAND

Figure 54.26g

Temperate grassland in Wyoming State

Physical Environment: Annual rainfall generally between 25 and 100 cm, too low to support a forest but higher than that in deserts. Temperatures in the winter often fall below −10°C, while summers may be very hot, approaching 30°C.

Location: Temperate grasslands include the prairies of North America, the steppes of Russia, the pampas of Argentina, and the veldt of South Africa. In addition to the limiting amounts of rain, fire and grazing animals may also prevent the establishment of trees in the temperate grasslands. Where temperatures rarely fall below freezing and most of the rain falls in the winter, chaparral, a fire-adapted community featuring shrubs and small trees, occurs.

Plant Life: From east to west in North America and from north to south in Asia, grasslands show differentiation along moisture gradients. In Illinois, with an annual rainfall of 80 cm, tall prairie grasses such as big bluestem and switchgrass grow to about 2 m high. Along the eastern base of the Rockies, 1,300 km to the west, where rainfall is only 40 cm, prairie grasses such as buffalo grass and blue grama rarely exceed 0.5 m in height. Similar gradients occur in South Africa and Argentina.

Animal Life: Where the grasslands remain, large mammals are the most prominent members of the fauna: bison (buffalo) and pronghorn (antelope) in North America, wild horses in Eurasia, and large kangaroos in Australia. Burrowing animals such as North American gophers and African mole rats are also common.

Effects of Humans: Prairie soil is among the richest in the world, having 12 times the humus layer of a typical forest soil. Worldwide, most prairies have been converted to agricultural cropland, and original grassland habitats are among the rarest biomes in the world.

HOT DESERT

Figure 54.26h

Saguaro National Park in Arizona, part of the Sonoran Desert

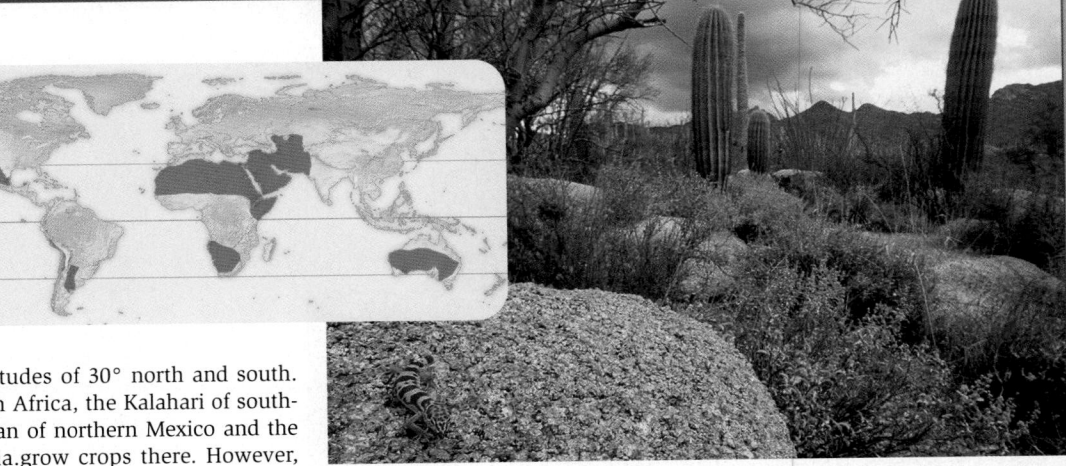

Physical Environment: Temperatures are variable from below freezing at night to as much as 50°C in the day. Rainfall is less than 30 cm per year.

Location: Hot deserts are found around latitudes of 30° north and south. Prominent deserts include the Sahara of North Africa, the Kalahari of southern Africa, the Atacama of Chile, the Sonoran of northern Mexico and the southwest U.S., and the Simpson of Australia.grow crops there. However, salinization of soils is prevalent.

Plant Life: Three forms of plant life are adapted to deserts: annuals, succulents, and desert shrubs. Annuals circumvent drought by growing only when there is rain. Succulents, such as the saguaro cactus and other barrel cacti of the southwestern deserts, store water. Desert shrubs, such as the spraylike ocotillo, have short trunks, numerous branches, and small, thick leaves that can be shed in prolonged dry periods. In many plants, spines or volatile chemical compounds serve as a defense against water-seeking herbivores.

Animal Life: To conserve water, desert plants produce many small seeds, and animals that eat those seeds, such as ants, birds, and rodents, are common. Reptiles are numerous because high temperatures permit these ectothermic animals to maintain a warm body temperature. Lizards and snakes are important predators of seed-eating mammals.

Effects of Humans: Ambitious irrigation schemes and the prolific use of underground water have allowed humans to colonize deserts and grow crops there. However, salinization of soils is prevalent. Off-road vehicles can disturb the fragile desert communities.

COLD DESERT

Figure 54.26i

The Gobi Desert of Mongolia

Physical Environment: Precipitation is less than 25 cm a year, often in the form of snow. Rainfall usually comes in the spring. In the daytime, temperatures can be high in the summer, 21–26°C, but average around freezing, −2 to 4°C, in the winter.

Location: Cold deserts are found in dry regions at middle to high latitudes, especially in the interiors of continents and in the rain shadows of mountains. Cold deserts are found in North America (the Great Basin Desert), in eastern Argentina (the Patagonian Desert), and in central Asia (the Gobi Desert).

Plant Life: Cold deserts are relatively poor in terms of numbers of species of plants. Most plants are small in stature, being only between 15 and 120 cm tall. Many species are deciduous and spiny. The Great Basin Desert in Nevada, Utah, and bordering states is a cold desert dominated by sagebrush.

Animal Life: As in hot deserts, large numbers of plants produce small seeds on which numerous ants, birds, and rodents feed. Many species live in burrows to escape cold and to keep warm. In the Great Basin Desert, pocket mice, jackrabbits, kit fox, and coyote are common.

Effects of Humans: Agriculture is hampered because of low temperatures and low rainfall, and human populations are not extensive. If the top layer of soil is disturbed by human intrusions such as off-road vehicles, erosion occurs rapidly and even less vegetation is able to exist.

TUNDRA

Figure 54.26j

Denali National Park in Alaska

Physical Environment: Precipitation is generally less than 25 cm per year and is often locked up as snow and unavailable for plants. Deeper water can be locked away for a large part of the year in **permafrost**, a layer of permanently frozen soil. The growing season here is short, only 50–60 days. Summer temperatures are only 3–12°C, and even during the long summer days, the ground thaws to less than 1 m in depth. Midwinter temperatures average −32°C.

Location: Tundra (from the Finnish *tunturia*, treeless plain) exists mainly in the Northern Hemisphere, north of the taiga, because there is very little land area in the Southern Hemisphere at the latitude where tundra would occur.

Plant Life: With so little available water, trees cannot grow. Vegetation occurs in the form of fragile, slow-growing lichens, mosses, grasses, sedges, and occasional shrubs, which grow close to the ground. Plant diversity is very low. In some places desert conditions prevail because so little moisture falls.

Animal Life: Animals of the arctic tundra have adapted to the cold by having good insulation. Many birds, especially shorebirds and waterfowl, migrate. The fauna is much richer in summer than in winter. Many insects spend the winter at immature stages of growth, which are more resistant to cold than the adult forms. The larger animals include such herbivores as musk oxen and caribou in North America, called reindeer in Europe and Asia, as well as the smaller hares and lemmings. Common predators include arctic fox, wolves, and snowy owls, and polar bears near the coast.

Effects of Humans: Though this area is sparsely populated, mineral extraction, especially of oil, has the potential to significantly impact this biome. Ecosystem recovery from such damage would be very slow.

MOUNTAIN RANGES

Figure 54.26k

Rocky Mountains of Colorado

Physical Environment: Mountain ranges must be treated differently than other biomes. Biome type relies predominantly on climate. On mountains, temperature decreases with increasing elevation through adiabatic cooling, as discussed in the text. Thus, precipitation and temperature may change dramatically, depending on elevation and whether the mountainside is on the windward or leeward side.

Location: Mountain ranges exist in many areas of the world, but among the largest are the Himalayas in Asia, the Rockies in North America, and the Andes in South America.

Plant Life: Biome type may change from temperate forest through taiga and into tundra on an elevation gradient in the Rocky Mountains, and even from tropical forest to tundra on the highest peaks of the Andes in tropical South America. In tropical regions, daylight varies little from the 12 hours per day throughout the year. Instead of an intense period of productivity, vegetation in the tropical alpine tundra exhibits slow but steady rates of photosynthesis and growth all year.

Animal Life: The animals of this biome are as varied as the number of habitats they contain. Generally, more species of plants and animals are found at lower elevations than at higher ones. At higher elevations, animals such as bighorn sheep and mountain goats have to be very sure-footed to climb the craggy slopes and have skidproof pads on their hooves. Despite the often-strong winds, birds of prey, such as eagles, are frequent predators of the furry rodents found at higher elevations, including guinea pigs and marmots.

Effects of Humans: Logging and agriculture at lower elevations can cause habitat degradation. Because of the steep slopes, mountain soils are often well drained, thin, and especially susceptible to erosion.

Aquatic Biomes Consist of Marine and Freshwater Regions

Within aquatic environments, several different biome types are also recognized, including marine aquatic biomes (the intertidal zone, coral reef, and open ocean) and freshwater lakes, rivers, and wetlands. These biomes are distinguished by differences in parameters such as salinity, oxygen content, depth, current strength, and availability of light (**Figures 54.27a–f**).

INTERTIDAL ZONE   Figure 54.27a

Olympic Coast National Marine Sanctuary in Washington State

Physical Environment: The **intertidal zone**, the area where the land meets the sea, is alternately submerged and exposed by the daily cycle of tides. The resident organisms are subject to huge daily variation in temperature, light intensity, and availability of seawater, which makes life difficult.

Location: Throughout the world, the area where the land meets the sea consists of sandy shore, mudflats, or rocky shore. Commonly, there is a vertical zonation consisting of three broad zones, most evident on rocky shores. The upper littoral zone is submerged only during the highest tides. The midlittoral zone is submerged during the highest regular tide and exposed during the lowest tide each day. The lower littoral zone is exposed only during the lowest tide.

Plant Life: Plant life may be quite limited because the sand or mud is constantly shifted by the tide. Mangroves may colonize mudflats in tropical areas, and salt marsh grasses may colonize mudflats in temperate locations. On the rocky shore, plant life consists of green algae and seaweeds.

Animal Life: Animal life may be quite diverse. On the rocky shore, sea anemones, snails, hermit crabs, and small fishes live in tide pools. On the rock face, there may be a variety of limpets, mussels, sea stars, sea urchins, snails, sponges, tube worms, whelks, isopods, and chitons. At low tides, organisms may be dry and vulnerable to predation by a variety of animals, including birds and mammals. High tides bring predatory fish. In sandy or muddy shores, the biome may contain burrowing marine worms, crabs, and small isopods.

Effects of Humans: Urban development has greatly reduced the beach area available to breeding turtles and shorebirds. Oil spills have greatly impacted some rocky intertidal areas.

CORAL REEF

Figure 54.27b

Caribbean Coral Reef

Physical Environment: Corals need warm water of at least 20°C but less than 30°C. They are also limited to the euphotic zone, where light penetrates. Sunlight is important because many corals harbor symbiotic algae, or dinoflagellates, that contribute nutrients to the animals and that require light to live.

Location: Coral reefs exist in warm tropical waters where there are solid substrates for attachment and water clarity is good. The largest coral reef in the world is the Great Barrier Reef off the Australian coastline, but other coral reefs are found throughout the Caribbean Sea and Pacific Ocean.

Plant Life: Dinoflagellate algae live within the coral tissue, and a variety of red and green algae live on the coral reef surface.

Animal Life: An immense variety of microorganisms, invertebrates, and fish live among the coral, making the coral reef one of the most interesting and species rich biomes on Earth. Probably 30–40% of all fish species on Earth are found on coral reefs. Prominent herbivores include snails, sea urchins, and fish. These are in turn consumed by octopus, sea stars, and carnivorous fish. Many species are brightly colored, warning predators of their toxic nature.

Effects of Humans: Collectors have removed many corals and fish for the aquarium trade, and marine pollution threatens water clarity in some areas. Perhaps the greatest threat is from global warming and climate change. Water temperatures that are too high (over 30°C) can cause coral bleaching.

THE OPEN OCEAN

Figure 54.27c

Manta Ray in the open ocean

Physical Environment: In the open ocean, sometimes called the **pelagic zone**, water depth averages 4,000 m and nutrient concentrations are typically low, though the waters may be periodically enriched by ocean **upwellings**, which carry mineral nutrients from the bottom waters to the surface. Pelagic waters are mostly cold, only warming near the surface.

Location: Across the globe, covering 70% of the Earth's surface.

Plant Life: Where light levels are high at the surface, many microscopic, photosynthetic organisms (**phytoplankton**) grow and reproduce. Phytoplankton account for nearly half the photosynthetic activity on Earth and produce much of the world's oxygen.

Animal Life: Open-ocean organisms include **zooplankton**, minute animal organisms consisting of some worms, copepods (tiny shrimplike creatures), tiny jellyfish, and the small larvae of invertebrates and fish that graze on the phytoplankton. The open ocean also includes free-swimming animals collectively called **nekton**, which can swim against the currents to locate food. The nekton include large squids, fish, sea turtles, and marine mammals that feed on phytoplankton, zooplankton, or each other. Only a few of these organisms live at any great depth. In some areas, there exists a unique assemblage of animals associated with deep-sea hydrothermal vents that spew out hot (350°C) water rich in hydrogen sulfide. In this dark, oxygen-poor environment, large polychaete worms exist together with other organisms that are chemoautotrophs (see Figure 59.3).

Effects of Humans: Oil spills and a long history of garbage disposal have polluted the ocean floors of many areas. Overfishing has caused many fish populations to crash, and the whaling industry has greatly reduced the numbers of most species of whales.

FRESHWATER LENTIC HABITATS

Figure 54.27d

Everglades National Park, Florida

Physical Environment: Freshwater habitats are traditionally divided into **lentic**, or standing-water habitats (from the Latin *lenis*, calm), and **lotic**, or running-water habitats (from the Latin *lotus*, washed). The lentic habitat consists of still, often deep water. Its physical characteristics depend greatly on the surrounding land, which dictates what nutrients collect in the lake. Young lakes often start off clear and with little plant life. Such lakes are called **oligotrophic**. With age, the phytoplankton bloom and algae spread, reducing the water clarity. Such lakes are termed **eutrophic**. The process of eutrophication is discussed more thoroughly in Chapter 59.

Location: Throughout all the continents of the world.

Plant Life: In addition to free-floating phytoplankton and algae, lentic habitats may have rooted vegetation, which often emerges above the water surface (emergent vegetation), such as cattails, plus deeper dwelling aquatic plants and algae.

Animal Life: Animals include fish, frogs, turtles, crayfish, insect larvae, and many species of insects. In tropical and subtropical lakes, alligators and crocodiles commonly are seen.

Effects of Humans: Agricultural runoff, including fertilizers and sewage, can greatly increase lake nutrient levels and speed up the process of eutrophication. This may result in algal blooms and fish kills. In some areas, exotic species of invertebrates and fish are outcompeting native species.

FRESHWATER LOTIC HABITATS

Figure 54.27e

Fast-flowing river in the Pacific Northwest

Physical Environment: In lotic habitats, flowing water prevents nutrient accumulations and phytoplankton blooms. The current also mixes water thoroughly, providing a well-aerated habitat of relatively uniform temperature. However, current, oxygen level, and clarity are greater in headwaters than in the lower reaches of rivers. Nutrient levels are generally less in headwaters.

Location: On all continents except Antarctica.

Plant Life: In slow-moving streams and rivers, algae and rooted plants may be present; in swifter moving rivers, leaves from surrounding forests are the primary food source for animals.

Animal Life: Lotic habitats have a fauna completely different from that of lentic waters. Animals are adapted to stay in place despite an often-strong current. Many of the smaller organisms are flat and attach themselves to rocks to avoid being swept away. Others live on the underside of large boulders, where the current is much reduced. Fish such as trout may be present in rivers with cool temperatures, high oxygen, and clear water. In warmer, murkier waters, catfish and carp may be abundant.

Effects of Humans: Animals of lotic systems are not well adapted for low-oxygen environments and are particularly susceptible to oxygen-reducing pollutants such as sewage. Dams across rivers have prevented the passage of migratory species such as salmon.

WETLANDS

Figure 54.27f

Yellow Waters River, Kakadu National Park, Northern Territory, Australia

Physical Environment: At the margins of both lentic and lotic habitats, wetlands may develop. Wetlands are areas regularly saturated by surface water or groundwater. They can range from marshes and swamps to bogs. Many are seasonally flooded when rivers overflow their banks or lake levels rise. Some wetlands also develop along estuaries, where rivers merge with the ocean and high tides can flood the land. Here, salt marshes or, in tropical areas, stands of mangroves can develop. Because of generally high nutrient levels, oxygen levels are fairly low. Temperatures vary substantially with location.

Location: Worldwide, except in Antarctica.

Plant Life: Wetlands are among the most productive and species-rich areas in the world. In North America, floating plants such as lilies and rooted species such as sedges, cattails, cypress, and gum trees predominate.

Animal Life: Most wetlands are rich in animal species. Wetlands are a prime habitat for wading and diving birds. In addition, they are home to a profusion of insects, from mosquitoes to dragonflies. Vertebrate predators include many amphibians, reptiles, otters, and alligators.

Effects of Humans: Long mistakenly regarded as wasteland by humans, many wetlands have been drained and developed for subdivisions and industry.

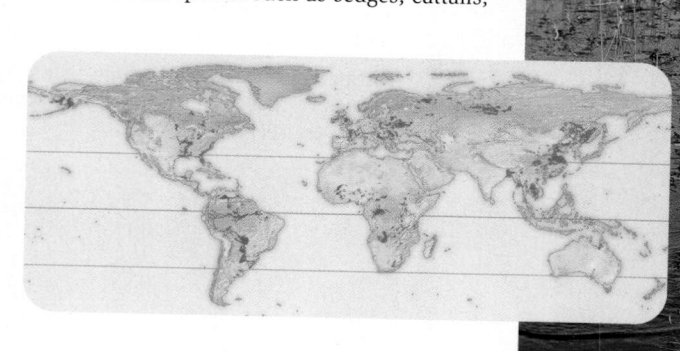

In this chapter we have discussed how the physical environment profoundly influences the distribution and abundance of life on Earth and the existence of different biome types. In subsequent chapters, we will learn how other factors influence the distribution of plants and animals. On a smaller scale, the presence of predators, parasites, or competitors can also influence where organisms are found. Part of an organism's fundamental niche may be occupied by a competitively superior species. We will examine the influences of such factors in Chapter 57. For animals, social interactions with other members of the same species such as fights over territory or mates can also influence population distributions. To discuss these issues, the next chapter examines animal behavior.

CHAPTER SUMMARY

54.1 The Scale of Ecology

- Ecologists study the interactions among organisms and between organisms and their environments. (Figure 54.1)
- The field of ecology can be subdivided into broad areas of organismal, population, community, and ecosystems ecology. (Figure 54.2)
- Organismal ecology considers how individuals are adapted to their environment and how the behavior of an individual organism contributes to its survival and reproductive success and the population density of the species.
- Population ecology explores those factors that influence a population's growth, size, and density, and community ecology studies how populations of species interact and form functional communities. (Figure 54.3)

- Ecosystems ecology examines the flow of energy and cycling of nutrients among organisms within a community and between organisms and the environment.
- Ecological methods focus on observation and experimentation. A variety of statistical tests exist that help determine whether two variables are related. (Figures 54.4, 54.5, 54.6)

54.2 The Environment's Impact on the Distribution of Organisms

- Abiotic factors such as temperature, wind, water, light, salinity, pH, and water currents can have powerful effects on ecological systems.
- Temperature exerts important effects on the distribution of organisms because of its effect on biological processes and the inability of most organisms to regulate their body temperature. (Figures 54.7, 54.8, 54.9, 54.10)

- Life on Earth is made possible by the greenhouse effect, in which short-wave solar radiation passes through the atmosphere to warm the Earth but is radiated back to space as long-wave infrared radiation. Much of this radiation is absorbed by atmospheric gases and reradiated back to Earth's surface, causing its temperature to rise. (Figure 54.11)

- The major atmospheric gases causing the greenhouse effect are water, carbon dioxide, methane, nitrous oxide, and chlorofluorocarbons. (Table 54.1)

- An increase in atmospheric gases is increasing the greenhouse effect, causing global warming, a gradual elevation of the Earth's surface temperature. Ecologists expect that global warming will have a large effect on the distribution of the world's organisms. (Figure 54.12)

- Wind can amplify the effects of temperature and modify wave action. (Figure 54.13)

- The availability of water has an important effect on the abundance of organisms. (Figure 54.14)

- Light can be a limiting resource for plants in both terrestrial and aquatic environments. (Figure 54.15)

- The concentration of salts and the pH of soil and water can limit the distribution of organisms. (Figures 54.16, 54.17)

54.3 Climate and Its Relationship to Biological Communities

- Global temperature differentials are caused by variations in incoming solar radiation and patterns of atmospheric circulation. (Figures 54.18, 54.19, 54.20, 54.21)

- Elevation and the proximity between a landmass and large bodies of water can similarly affect climate. (Figures 54.22, 54.23)

- Climate has a large effect on biomes, major types of habitats characterized by distinctive plant and animal life. (Figures 54.24, 54.25)

- Terrestrial biomes are generally named for their climate and vegetation type and include tropical rain forest, tropical deciduous forest, temperate rain forest, temperate deciduous forest, temperate coniferous forest (taiga), tropical grassland (savanna), temperate grassland (prairie), hot and cold deserts, and tundra. In mountain ranges, biome type may change on an elevation gradient. (Figure 54.26)

- Within aquatic environments, biomes include marine aquatic biomes (the intertidal zone, coral reef, and open ocean) and freshwater lakes, rivers, and wetlands. These are distinguished by differences in parameters including salinity, oxygen content, depth, current strength (lentic versus lotic), and availability of light. (Figure 54.27)

TEST YOURSELF

1. Which of the following is probably the most important factor in the distribution of organisms in the environment?
   a. light
   b. temperature
   c. salinity
   d. water availability
   e. pH

2. The greenhouse effect is
   a. a new phenomenon resulting from industrialization.
   b. due to the absorption of solar radiation by atmospheric gases.
   c. responsible for the natural warming of the Earth.
   d. all of the above.
   e. b and c only.

3. Which of the following is not a component of climate?
   a. temperature
   b. vegetation
   c. water
   d. wind
   e. light

4. The study of how groups of different species interact in their common environment is
   a. ecology.
   b. organismal ecology.
   c. community ecology.
   d. population ecology.
   e. environmental ecology.

5. In aquatic environments, plants and algae are usually found in the _____ zone near the surface of the water, where light is able to penetrate.
   a. aphotic
   b. littoral
   c. euphotic
   d. limnetic
   e. photosynthetic

6. What is the driving force that determines the circulation of the atmospheric air?
   a. temperature differences of the Earth
   b. winds
   c. ocean currents
   d. mountain ridges
   e. all of the above

7. The distribution of the major biomes of the Earth is determined by
   a. temperature differences.
   b. mountain ridges.
   c. ocean currents.
   d. temperature and moisture differences.
   e. all of the above.

8. What characteristics are commonly used to identify the biomes of the Earth?
   a. temperature
   b. precipitation
   c. vegetation
   d. all of the above
   e. a and b only

9. Terrestrial areas that are regularly saturated by surface water or groundwater are called
   a. chaparral.
   b. lakes.
   c. wetlands.
   d. deserts.
   e. grasslands.

10. Which is the most important contribution to human-caused global warming?
    a. carbon dioxide
    b. nitrous oxide
    c. sulfur dioxide
    d. methane
    e. chlorofluorocarbons

CONCEPTUAL QUESTIONS

1. Define ecology.
2. Explain the greenhouse effect.
3. Define biome.

EXPERIMENTAL QUESTIONS

1. Prior to Callaway and Aschehoug's study, what was the prevailing hypothesis of why invasive species succeed in new environments?
2. Briefly describe the evidence collected to support the allelochemical hypothesis.
3. What was the function of the activated charcoal used in a subsequent test of the hypothesis?

COLLABORATIVE QUESTIONS

1. Discuss several ways in which the physical environment affects the distribution of organisms.
2. Based on your knowledge of biomes, identify the biome in which you live. In your discussion, list and describe the organisms that you have observed in your biome.

www.brookerbiology.com

This website includes answers to the Biological Inquiry questions found in the figure legends and all end-of-chapter questions.

55

BEHAVIORAL ECOLOGY

CHAPTER OUTLINE

Chimpanzee (*Pan troglodytes*) using a stick to catch food.

Behavior is the observable response of organisms to external or internal stimuli. Defined as such, behavior takes into account a very broad range of activities. In this chapter, we focus our attention on the field of **behavioral ecology**, the study of how behavior contributes to the differential survival and reproduction of organisms. For example, nesting black-headed gulls (*Larus ridibundus*) always pick up broken eggshells after a chick has hatched and carry them away from the nest. One might think that they are being neat and tidy, or are minimizing the risk of bacterial infection to the chicks, but there is more to the behavior than this. The chicks and unhatched eggs are well camouflaged in the nest, but the white color of the empty eggshell quickly attracts the attention of predators such as crows that would kill and eat the chicks or remaining eggs. By removing the old eggshells, the gull parents are increasing the chances that their offspring—and thus their genes—will survive.

Behavioral ecology builds upon earlier work that focused primarily on how organisms behave. In the early 20th century, scientific studies of animal behavior, termed **ethology** (from the Greek *ethos,* habit, manner), focused on the specific genetic and physiological mechanisms of behavior. These factors are called **proximate causes**. For example, we could hypothesize that male deer rut, or fight with other males, in the fall because a change in day length stimulates the eyes, brain, and pituitary gland and triggers hormonal changes in their bodies. The founders of ethology, Karl von Frisch, Konrad Lorenz, and Niko Tinbergen, shared the 1973 Nobel Prize in Physiology or Medicine for their pioneering discoveries concerning the proximate causes of behavior.

However, we could also hypothesize that male deer fight to determine which deer get to mate with a female deer and pass on their genes. This hypothesis leads to a different answer than the one that is concerned with changes in day length. This answer focuses on the adaptive significance of fighting to the deer, that is, on why a particular behavior evolved, in terms of its effect on reproductive success. These factors are called **ultimate causes** of behavior. Since the 1970s, behavioral ecologists have focused more on understanding the ultimate causes of behavior.

In this chapter, we will explore the role of both proximate and ultimate causes of behavior. We begin the chapter by investigating how behavior is achieved, that is, we look at the role of both genetics and the environment. In doing so, we will examine the important contributions of ethologists von Frisch, Lorenz, and Tinbergen. We consider how different behaviors are involved in movement, gathering food, and communication. Later, we investigate how organisms interact in groups, whether an organism can truly behave in a way that benefits others at a cost to itself, and how behavior shapes different mating systems. You will note that much of the chapter focuses on animal behavior, because the behavior of other organisms is more limited and less well understood.

55.1 The Impact of Genetics and Learning on Behavior

Behavior is rarely due solely to genes (nature) or solely to environmental influences (nurture), but usually to both. Determining to what degree a behavior is influenced by genes versus the environment will depend on the particular genes and environment examined. However, in a few cases, changes in behavior may be caused by variation in just one gene. Even if a given behavior is influenced by many genes, if one gene is altered, it is possible that the entire behavior can change. To use the analogy of baking a cake, a change in one ingredient of the recipe may change the whole taste of the cake, but that does not mean that the one ingredient is responsible for the entire cake.

In this section, we begin by examining the effect that a single gene can exert on behavior and consider several examples of simple, genetically programmed behaviors. Later, we explore several types of learned behavior, including classical and operant conditioning and cognitive learning, and conclude the section by exploring an example of the interaction of genetics and learning on behavior.

GENOMES & PROTEOMES

Some Behavior Results from Simple Genetic Influences

An excellent example of the effect of a single gene on complex behavior was demonstrated in W. C. Rothenbuhler's 1964 work on honeybees. Some strains of bees are termed hygienic, that is, they detect and remove diseased larvae from the nest. This behavior involves two distinct maneuvers: uncapping the wax cells and then discarding the dead larvae. Other strains are not hygienic and do not exhibit such behavior. Rothenbuhler demonstrated, by genetic crosses, that one gene (u) controlled cell uncapping and another gene (r) controlled larval removal. Double recessives ($uurr$) were hygienic strains, and double dominants ($UURR$) were nonhygienic strains. When the two strains were crossed, all the F_1 hybrids were nonhygienic ($UuRr$). When the F_1 hybrids were backcrossed with the pure hygienic strain ($uurr$), four different genotypes were produced, as Mendel's law of independent assortment predicts (Chapter 16): one-quarter of the offspring were hygienic ($uurr$), one-quarter were nonhygienic and showed neither behavior ($UuRr$), one-quarter uncapped the cells but failed to remove the larvae ($uuRr$), and one-quarter removed the larvae but only if the cells were uncapped for them ($Uurr$).

In 1996, Jennifer Brown and her colleagues showed how a single gene, $fosB$, controlled nurturing behavior in mice. Normal mice, which carry the gene, clean their newborn pups, nurse them, and crouch over them to keep them warm. Brown's group created mutant mice lacking the $fosB$ gene. Despite fully functioning mammary glands, mutant mothers paid little attention to their newborn pups, which remained scattered around the cage and died within a few days. If the pups of the $fosB$ mutants were placed with normal females, they thrived, but if normal pups were given to mutant mothers, they too were neglected.

Genes for behavior actually act on the development of nervous systems and musculature, physical traits that evolve through natural selection. Many genes are needed for the proper development and function of the nervous system and musculature. Even so, as described by Rothenbuhler's and Brown's work, variation in a single gene can have a dramatic impact on behavior.

Fixed Action Patterns Are Genetically Programmed

Behaviors that seem to be genetically programmed are referred to as **innate** (also called instinctual). While we recognize that expression of genes varies, often in response to environmental stimuli, some behavior patterns evidently are genetically quite fixed. Most individuals will exhibit the same behavior regardless of the environment. A spider will spin a specific web without ever seeing a member of its own species build one. The courtship behaviors of many bird species are so stereotyped as to be virtually identical.

A classic example of innate behavior is the egg-rolling response in geese (**Figure 55.1**). If an incubating goose notices an egg out of the nest, she will extend her neck toward the egg, get up, and then roll the egg back to the nest using her beak. Such behavior functions to improve fitness because it increases the survival of offspring. Eggs that roll out of the nest get cold and fail to hatch. Geese that fail to exhibit the egg-rolling response would pass on fewer of their genes to future generations. Egg-rolling behavior is an example of what ethologists term a **fixed action pattern (FAP)**, a behavior that, once initiated, will continue until completed. For example, if the egg is removed while the goose is in the process of rolling it back toward the nest, the goose still completes the FAP, as though she were rolling back

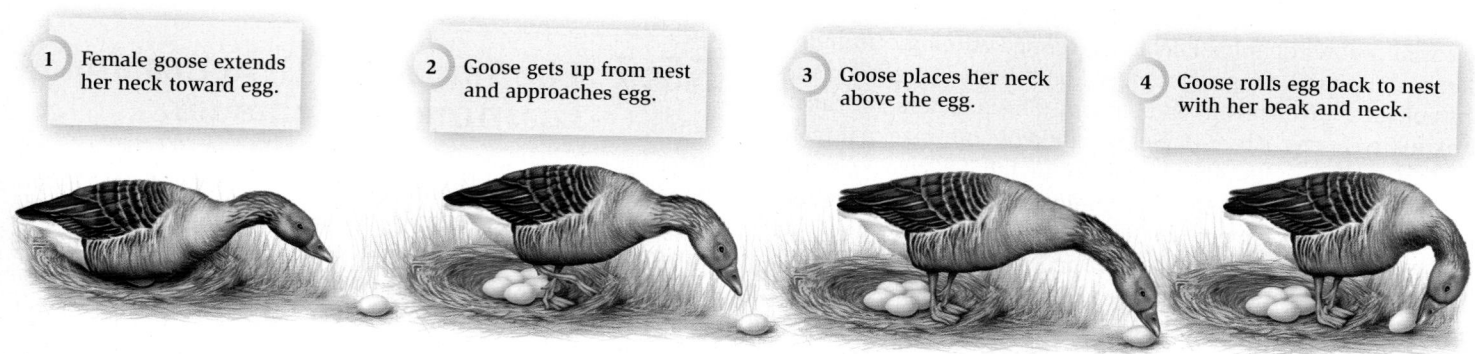

| 1 Female goose extends her neck toward egg. | 2 Goose gets up from nest and approaches egg. | 3 Goose places her neck above the egg. | 4 Goose rolls egg back to nest with her beak and neck. |

Figure 55.1   **A fixed action pattern as an example of innate behavior.**   Female geese retrieve eggs that have rolled outside the nest through a set sequence of movements. The goose will complete this entire sequence even if a researcher takes the egg away before the goose has rolled it back to the nest.

the now-absent egg to the nest. The stimulus to initiate this behavior is obviously a strong one, which ethologists term a **sign stimulus**. The sign stimulus for the goose is that an egg had rolled out of the nest. According to ethologists, this stimulus acts on the goose's central nervous system, which provides a neural stimulus to initiate the motor program or FAP. Interestingly, any round object will elicit the egg-rolling response, from a wooden egg to a volleyball. While sign stimuli usually have certain key components, they are not necessarily very specific.

Niko Tinbergen's study of male stickleback fish provides another classic example of FAPs and sign stimuli. Male sticklebacks, which have a characteristic red belly, will attack other male sticklebacks that invade their territory. Tinbergen found that sticklebacks attacked small, unrealistic model fish having a red ventral surface, while ignoring a realistic male stickleback model that lacked a red underside (**Figure 55.2**).

Conditioning Occurs When a Relationship Between a Stimulus and a Response Is Learned

Although many of the behavioral patterns exhibited by animals are largely innate, sometimes animals can make modifications to their behavior based on previous experience, a process called **learning**. Perhaps the simplest form of learning is **habituation**, in which an organism learns to ignore a repeated stimulus. For example, animals in African safari parks become habituated to the presence of vehicles containing tourists; these vehicles are neither a threat nor a benefit. After a while, birds can become habituated to the presence of a scarecrow, resulting in damage to crops. Habituation can be a problem at airfields, where birds eventually ignore the alarm calls designed to scare them away from the runways.

Habituation is a form of nonassociative learning in which there is a decrease in response to a stimulus due to repetition. Alternatively, an association may gradually develop between a stimulus and a response. Such a change in behavior is termed **associative learning**. In associative learning, a behavior is changed or conditioned through the association. The two main types of associative learning are termed classical conditioning and operant conditioning.

In **classical conditioning**, an involuntary response comes to be associated positively or negatively with a stimulus that did not originally elicit the response. This type of learning is generally associated with the Russian psychologist Ivan Pavlov. In his original experiments in the 1920s, Pavlov restrained a hungry dog in a harness and presented small portions of food at regular intervals. The dog would salivate whenever it smelled the food. Pavlov then began to sound a metronome when presenting the food. Eventually the dog would salivate at the sound of the metronome, whether or not the food was present. Pavlov termed the food the **unconditioned stimulus** for salivation, while the metronome was the **conditioned stimulus**. Likewise, he termed salivation in response to food as an **unconditioned response** and salivation in response to the metronome as a **conditioned response**. Classical conditioning is pervasive in human society and is widely observed in other animals. For example, many insects quickly learn to associate certain flower odors with nectar rewards and other flower odors with no rewards.

In **operant conditioning**, an animal's behavior is reinforced by a consequence, either a reward or a punishment. The classic example of operant conditioning is generally associated with the American psychologist B. F. Skinner, who placed laboratory animals, usually rats, in a specially devised cage that came to be known as a Skinner box. As the rat explored its cage, it could press on a small lever that would result in food being released. At the beginning of the experiment, the rat would often bump into the lever by accident, eat the food, and continue exploring its cage. Later, it would learn to associate the lever with getting food. Eventually, if it was hungry, the rat would almost continually press the lever. Operant conditioning, also called trial-and-error learning, is common in animals. Often it is associated with negative rather than positive reinforcement. For example, toads will eventually refuse to strike at insects that sting, such as wasps and bees, and birds will learn to avoid bad-tasting butterflies (**Figure 55.3**).

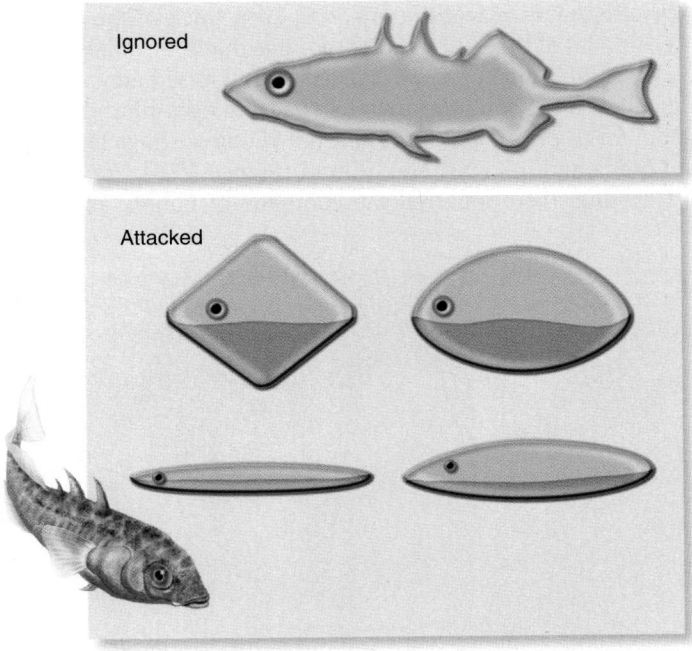

Figure 55.2 **A fixed action pattern elicited by a sign stimulus.** The sign stimulus for male sticklebacks to attack other males entering their territory is a red ventral surface. In experiments, male sticklebacks attacked all models that had a red underside, while ignoring a realistic model of a stickleback that lacked the red belly.

Cognitive Learning Involves Conscious Thought

Cognitive learning refers to the ability to solve problems with conscious thought and without direct environmental feedback.

(a)   **(b)**

Figure 55.3 Operant conditioning, also known as trial-and-error learning. (a) A young blue jay will eat a monarch butterfly, not knowing that it is noxious. (b) After the first experience of vomiting after eating a monarch, a blue jay will avoid the insects in the future.

In the 1920s, psychologist Wolfgang Kohler conducted a series of classic experiments with chimpanzees that suggested that animals could exhibit cognitive learning. In the experiments, a chimpanzee was left in a room with a banana hanging from the ceiling and out of reach (**Figure 55.4**). Also present in the room were several wooden boxes. At first, the chimp tried in vain to jump up and grab the bananas. After a while, however, it began to arrange the boxes one on top of another underneath the fruit. Eventually, the chimp climbed the boxes and retrieved the fruit. This clearly looks like an example of problem solving that involves cognitive learning.

In fact, many other examples of such behavior exist. Chimps strip leaves off twigs and use the twigs to poke into ant nests, withdrawing the twig and licking the ants off (see chapter-opening photo). Captive ravens have been shown to retrieve meat suspended from a branch by a string, even though they have never encountered the problem before. They pull up on the string, step on it, and then pull up on the string again, repeating the process until the meat is within reach.

The Interaction of Genetics and Learning on Behavior Often Involves Imprinting

Much of the behavior we have discussed so far has been presented as either innate or learned, but the behavior we observe in nature is often a mixture of both. Bird songs present a good example. Many birds learn their songs as juveniles, when they hear their parents sing. If juvenile white-crowned sparrows are raised in isolation, their adult songs do not resemble the typical species-specific song (**Figure 55.5**). If they hear only the song of a different species, such as the song sparrow, they again sing a poorly developed adult song. However, if they hear the song of the white-crowned sparrow, they will learn to sing a fully developed white-crowned sparrow song. The birds are genetically programmed to learn, but they will sing the correct song only if the appropriate instructive program is in place to guide learning.

Another example of how innate behavior interacts with learning can occur during a limited time period of development, called a **critical period**. At this time, many animals develop species-specific patterns of behavior. This process is called **imprinting**. One of the best examples of imprinting was demonstrated by the Austrian ethologist Konrad Lorenz in the 1930s. Lorenz noted that young birds of some species imprint on their mother during a critical period, usually within a few hours after hatching. This behavior serves them well, because in many species of ducks and geese it would be hard for the mother to keep track of all her offspring as they walk or swim. After imprinting takes place, the offspring keep track of the mother. The survival of the young ducks requires that they quickly learn to follow their mother's movements. Lorenz raised greylag geese from eggs, and soon after they hatched, he used himself as the model for imprinting. As a result, the young goslings imprinted on Lorenz and followed him around (**Figure 55.6**). For the rest of their life, they preferred the company of Lorenz and other

Figure 55.4 Cognitive behavior involves problem-solving ability. This chimp has devised a solution to the problem of retrieving bananas that were initially out of its reach.

Song heard by juvenile   **Song sung by juvenile**

No song heard   Abnormal song

Song of song sparrow   Abnormal song

Song of white-crowned sparrow   Normal song

Figure 55.5   The interaction between genetics and learning.   The juvenile white-crowned sparrow will sing an abnormal song if it is kept in isolation or hears only the song of a different species. However, the juvenile will sing the normal white-crowned sparrow song if exposed to it.

Biological inquiry: Cuckoos lay their eggs in other birds' nests, so their young are reared by parent birds of a different species. However, unlike the white-crowned sparrow, adult cuckoos always sing their own distinctive song, not that of the host species they hear as juveniles. How is this possible?

humans to geese. Studies have shown that even an object as foreign as a black box, watering can, or flashing light will be imprinted on if it is the first moving object the chick sees during the critical period. In nature, if the young geese failed to be provided with any stimulus during the critical period, they would fail to imprint on anything, and without parental care, they would almost certainly die.

Other animals imprint in different ways. Newborn shrews imprint on the scent of their mother. Mothers also can imprint on their own young within a few hours. For example, if sheep mothers are kept apart from their offspring for only a few hours after birth, they will reject them. In these situations, the innate behavior is the ability to imprint soon after birth, and the factors in the environment are the stimulus to which the imprinting is directed.

Finally, innate behavior can interact with learning during animal migration. Inexperienced juvenile birds will migrate in a particular direction but will fail to correct for deviations if they are blown off course. Experienced adult birds, on the other hand, can often correct for storm-induced displacement, indicating they have more complex navigational skills. There are in fact many complex behaviors involved in movement, as we will explore in the next section.

Figure 55.6   Konrad Lorenz being followed by his imprinted geese.   Newborn geese follow the first object that they see on hatching and later will follow that particular object only. They normally will follow their mother but can be induced to imprint on humans. The first thing these young geese saw on hatching was ethologist Konrad Lorenz.

55.2 Local Movement and Long-Range Migration

Organisms need to find their way, both locally and over what can be extremely long distances. Locally, organisms continually need to locate sources of food, water, and perhaps nesting sites. Migration involves the longer-distance seasonal movement of animals between overwintering areas and summer breeding sites; these are often hundreds or even thousands of kilometers apart. Several different types of behavior may be involved in these movements. In this section, we begin by exploring local movement and, in particular, how one species uses landmarks to guide its movements. We then consider the longer-range seasonal movement called migration and examine three possible mechanisms used by migrating animals to find their way.

Local Movement Can Involve Kinesis, Taxis, and Memory

The simplest forms of movement are mere responses to stimuli. A **kinesis** is a movement in response to a stimulus, but one that is not directed toward or away from the source of the stimulus. A simple experiment often done in classrooms is to observe the activity levels of woodlice, sometimes called sow bugs or pill bugs, in dry areas and moist areas. The woodlice move faster in drier areas, and they slow down when they reach moist environments. This behavior tends to keep them in damper areas, which they prefer in order to avoid desiccation.

A **taxis** is a more directed type of response either toward or away from a stimulus. Cockroaches exhibit negative phototaxis, meaning they tend to move away from light. Under low-light conditions, the photosynthetic unicellular flagellate *Euglena gracilis* shows positive phototaxis and moves toward a light source.

Sea turtle hatchings are also strongly attracted to light. On emerging from their nests, they crawl toward the brightest location, traditionally the reflected moonlight on the ocean's surface. Lighted houses on the shore can confuse the hatchlings, however, and cause them to lose their way, with sometimes disastrous results. Male silk moths orient themselves in relation to wind direction (anemotaxis). If the air current carries the scent of a female moth, they will move upwind to locate it. Some freshwater fish orient themselves to the currents of streams. Many fish exhibit positive rheotaxis (from the Greek *rheos*, current), in that they swim against the water current to prevent being washed downstream.

Sometimes memory and landmarks may be used to aid in local movements. The prominent Dutch-born ethologist Niko Tinbergen showed how the female digger wasp uses landmarks to relocate her nests as described next.

FEATURE INVESTIGATION

Tinbergen's Experiments Show That Digger Wasps Use Landmarks to Find Their Nests

In the sandy, dry soils of Europe, the solitary female digger wasp (*Philanthus triangulum*) digs four to five nests in which to lay her eggs. Each nest stretches obliquely down into the ground for 40–80 cm. The wasp, sometimes called a bee wolf, follows this by performing a sequence of apparently genetically programmed events. She catches and stings a honeybee, which paralyzes it; returns to the nest; drags the bee into the nest; and lays an egg on it. The egg hatches into a larva, which feeds on the paralyzed bee. However, the larva needs to ingest five to six bees before it is fully developed. This means the wasp must catch and sting four to five more bees for each larva. She can only carry one bee at a time. After each visit, the wasp must seal the nest, find a new bee, relocate the nest, open it, and add the bee. How does the wasp relocate the nest after spending considerable time away? Niko Tinbergen observed the wasps hover and

fly around the nest each time they took off. He hypothesized that they were learning the nest position by creating a mental map of the landmarks in the area.

To test his hypothesis, Tinbergen experimentally adjusted the landmarks around the burrow that the wasps might be using as cues (**Figure 55.7**). First, he put a ring of pinecones around the nest entrance to train the wasp to associate the pinecones with the nest. Then, when the wasp was out hunting, he moved the circle of pinecones a distance from the real nest and constructed a sham nest, making a slight depression in the sand and mimicking the covered entrance of the burrow. On returning, the wasp flew straight to the sham nest and tried to locate the entrance. Tinbergen chased it away. When it returned, it again flew to the sham nest. Tinbergen repeated this nine times, and every time the wasp chose the sham nest. Tinbergen got the same result with 16 other wasps and not once did they choose the real nest.

Figure 55.7   How Niko Tinbergen discovered the digger wasp's nest-locating behavior.

Biological inquiry: How would you test what type of spatial landmarks are used by female digger wasps?

HYPOTHESIS Digger wasps (*Philanthus triangulum*) use visual landmarks to locate their nests.

STARTING LOCATION The female digger wasp excavates an underground nest, to which she returns daily, bringing food to the larvae located inside.

Experimental level

1 Place ring of pinecones around nest to train the wasp to associate pinecones with the nest.

Pinecones

Digger wasp

2 After wasp leaves nest to hunt, move the pinecones 30 cm from the real nest. The wasp returns and flies to the center of the pinecone circle instead of the real nest. Repeated experiments yield similar results, indicating that the wasp uses landmarks as visual cues.

Move pinecones 30 cm from nest

3 To test whether it is the shape or the smell of the pinecones, perform same experiment as above, except add 2 small pieces of cardboard coated with pine oil.

Pine oil

Cardboard

4 After wasp leaves nest, move the pinecones 30 cm from the nest, but leave the scented cardboard at the nest. The wasps again fly to the pinecone nest (see data), indicating that it is the arrangement of cones, not their smell, that elicits the learning.

Move pinecones 30 cm from nest

5 **THE DATA**

Wasp #	Number of return visits to real nest with scented cardboard	Number of return visits to sham nest with pinecones
18	0	5
19	0	5
20	0	6
21	0	8
22	0	5

*Seventeen wasps were studied as described in steps 1 and 2. Five wasps, numbered 18–22, were studied as described in steps 3 and 4.

Next Tinbergen experimented with the type of stimulus that might be eliciting the learning. He hypothesized that the wasps could be responding to the distinctive scent of the pinecones rather than their appearance. He trained the wasps by placing a circle of pinecones that had no scent and two small pieces of cardboard coated in pinecone oil around the real nest.

He then moved the cones to surround a sham nest and left the scented cardboard around the real nest. The returning wasps again ignored the real nest with the scented cardboard and flew to the sham. He concluded that for the wasps, sight was apparently more important than smell in determining landmarks.

Migration Involves Long-Range Movement and More Complex Spatial Navigation

As well as learning to navigate over short-range distances, many animals undergo **migration**, long-range seasonal movement. Migrations occur in many species and usually involve a movement away from a birth area to feed and a return to the birth area to breed, with the movement generally being linked to seasonal availability of food. For example, nearly half the breeding birds of North America migrate to South America to escape the cold winters and feed, and then they return to North America in the spring to breed. Arctic terns that breed in Arctic Canada and Asia migrate to the Antarctic to feed in the winter and then return to breed. This staggering journey involves a 40,000-km (24,800-mi) round-trip, most of it over the open ocean, during which the birds must stay airborne for days at a time!

Figure 55.8 **Monarch butterfly migration.** Many monarch butterflies east of the Rocky Mountains migrate to a small area in Mexico to avoid the cold northern weather. Here they roost together in large numbers in fir trees (inset). Some butterflies may stay in Florida and Cuba. Butterflies west of the Rockies overwinter in mild coastal California locations.

Figure 55.9 **Orientation versus navigation.** Starlings normally migrate from breeding grounds in Scandinavia and northeastern Europe through the Netherlands and northern Germany to overwintering sites in northern France and southeast England. This involves a southwest flight. When juveniles were captured in the Netherlands and moved to Switzerland, they continued on in a southwestern direction and ended up in Spain. When experienced birds were captured and moved, they flew to the normal overwintering areas.

Many mammals, including wildebeest and caribou, make migrations that track the appearance of new vegetation on which they feed. The monarch butterfly of North America migrates to overwinter in California, Mexico, and probably south Florida and Cuba (**Figure 55.8**). This migration is unique in that none of the individuals has ever been to the migration sites before. An interesting point about the return journey of the monarch is that it involves several generations of butterflies to complete. On their way back to the northern U.S. and Canada, they lay eggs and die. The caterpillars develop on milkweed plants, and the resultant adults continue to journey farther north. This happens several times in the course of the return journey.

How do migrating animals find their way? Three mechanisms have been proposed: piloting, orientation, and navigation. In **piloting**, an animal moves from one familiar landmark to the next. For example, many whale species migrate between overwintering areas and summer calving grounds. Grey whales migrate between the Bering Sea near Alaska to coastal areas of Mexico. Features of the coastline, including mountain ranges, and rivers, may aid in navigation. In **orientation**, animals have the ability to follow a compass bearing and travel in a straight line. **Navigation** involves the ability not only to follow a compass bearing but also to set or adjust it.

An experiment with starlings helps illuminate the difference between orientation and navigation (**Figure 55.9**). European starlings breed in Scandinavia and northeastern Europe and migrate in a southwest direction toward coastal France and southern England to spend the winter. Migrating starlings were captured and tagged in the Netherlands and then transported south to Switzerland and released. Juvenile birds, which had never made the trip before, flew southwest in their migration and were later recaptured in Spain. Adult birds, with more experience, returned to their normal wintering range by adjusting their course by approximately 90°. This implies that the adult birds can actually navigate, whereas the juveniles rely on orientation.

Many species use a combination of navigational reference points, including the position of the sun, the stars (for nighttime travel), and the Earth's magnetic field. Homing pigeons have magnetite in their beaks that acts as a compass to tell direction (look back to Section 45.4, Electromagnetic Sensing). Navigation by the sun or the stars also requires the use of a timing device to compensate for the ever-changing position of these reference points. Many migrants, therefore, possess the equivalent of an internal clock. Pigeons integrate their internal clock with the position of the sun. Researchers have altered the internal clock of pigeons by keeping them under artificial lights for certain periods of time. When the pigeons are released, they display predictable deviations in their flight. For every hour that their internal clock is shifted, the orientation of the birds shifts about 15°.

Not all examples of animal migration are well understood. Green sea turtles feed off the coast of Brazil yet swim east for 2,300 km (1,429 mi) to Ascension Island, an 8-km-wide island in the center of the Atlantic Ocean between Brazil and Africa, to lay their eggs. It is not known why the turtles lay their eggs on this speck of an island or how they succeed in finding it. Perhaps fewer predators exist on Ascension than on other beaches. Thus, while scientists have made many discoveries about animal navigation, much remains to be learned about how animals acquire a "map sense."

To a large extent, local and long-distance movement involves searching for food. Organisms are constantly faced with a decision of how long to stay in a food patch and when to look for a new source of food. In the next section, we will investigate how such foraging decisions are made.

55.3 Foraging Behavior

Food gathering, or foraging, often involves decisions about whether to remain at a resource patch and look for more food or look for a completely new patch. The analysis of these decisions is often performed in terms of **optimality theory**, which predicts that an animal should behave in a way that maximizes the benefits of a behavior minus its costs. In this case, the benefits are the nutritional or calorific value of the food items, and the costs are the energetic or calorific costs of movement. When the difference between the energetic benefits of food gathering and the energetic costs of food gathering is maximized, an organism is said to be optimizing its foraging behavior. We should note that optimality theory can also be used to investigate other behavioral issues such as how large a territory to defend. Too small a territory would contain insufficient food, and too large a territory would be too energetically costly to defend. Theoretically, then, there is an optimal territory size for a given species.

Optimal Foraging Entails Maximizing the Benefits and Minimizing the Costs of Food Gathering

Optimal foraging proposes that in a given circumstance, an animal seeks to obtain the most energy possible with the least expenditure of energy. The underlying assumption of optimal foraging is that natural selection favors animals that are maximally efficient at propagating their genes and at performing all other functions that serve this purpose. In this model, the more net energy an individual gains in a limited time, the greater the reproductive success.

Studies show that animals tend to modify their behavior in a way that maximizes the difference between their energy uptake and their energy expenditure. For example, shore crabs (*Carcinus maenas*) will eat many different-sized mussels but tend to feed preferentially on intermediate-sized mussels, which give them the highest rate of energy return (**Figure 55.10**). Very

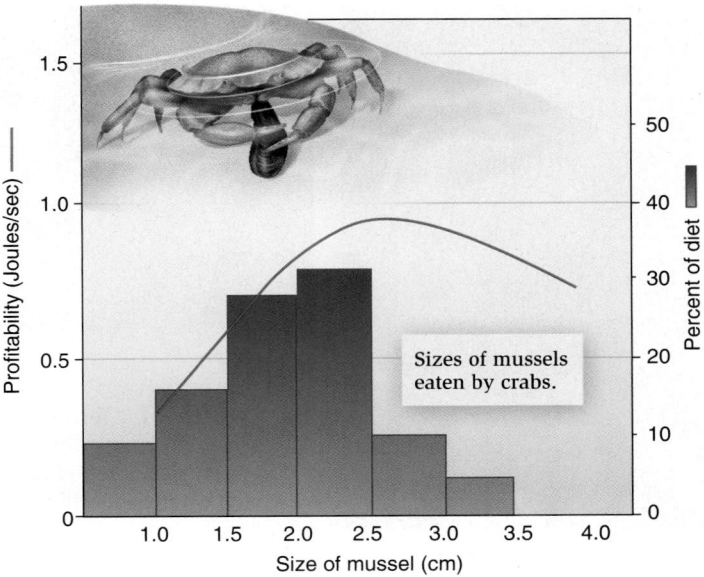

Figure 55.10 **Optimal foraging behavior in shore crabs.** In an aquarium setting, when offered a choice of equal numbers of each size mussel, the shore crab (*Carcinus maenas*) prefers intermediate-sized mussels that provides the highest rate of energy return. Profitability is the energy yield per second of time used in breaking open the shell.

large mussels yield more energy, but they take so long for the crab to open that they are actually less profitable, in terms of energy yield per unit time spent, than smaller sizes. Very small mussels are easy to crack open but contain so little flesh that they are not worth the effort. This leaves intermediate-sized mussels as the preferred size. Of course, the intermediate-sized mussels may take a longer time to locate, because more crabs are looking for them, so crabs eat some less profitable but more frequently encountered sizes. The result is that the diet consists of mussels in a range of sizes around the preferred optimal size.

In some cases, animals do not always forage optimally. For example, animals seek not only to maximize food intake but also to minimize the risk of predation. Some species may only dart out to take food from time to time. The risk of predation thus has an influence on foraging behavior. Many animals also maintain territories to minimize competition with other individuals and control resources, whether food, nesting sites, or mates. However, as we will see, defending these territories also has an energetic cost.

Defending Territories Has Costs and Benefits

Many animals, or groups of animals, such as a pride of lions, actively defend a **territory**, a fixed area in which an individual or group excludes other members of its own species, and sometimes other species, by aggressive behavior or territory marking.

(a) Golden-winged sunbird

(b) Cheetah

(c) Nesting gannets

Figure 55.11 **Territory sizes differ among animals.** **(a)** The golden-winged sunbird of East Africa (*Nectarinia reichenowi*) has a medium territory size that is dependent on the number of flowers it can obtain resources from and defend. **(b)** Cheetahs (*Acinonyx jubatus*) hunt over large areas and can have extensive territories. This male is urine-marking part of his territory in the southern Serengeti, near Ndutu, Tanzania. **(c)** Nesting gannets (*Morus bassanus*) have much smaller territories, in which each bird is just beyond the pecking range of its neighbor.

Territory owners tend to optimize territory size according to the costs and benefits involved. The primary benefit of a territory is that it provides exclusive access to a particular resource, whether it be food, mates, or sheltered nesting sites. Large territories may provide more of a resource but may be costly to defend, while small territories that are less costly to defend may not provide enough of a resource.

In studies of the territorial behavior of the golden-winged sunbird (*Nectarinia reichenowi*) in East Africa, researchers Frank Gill and Larry Wolf measured the energy content of nectar as the benefit of maintaining a territory while the energy costs included activities such as perching, foraging, and defending (**Figure 55.11a**). Defending the territory ensured that other sunbirds did not take nectar from available flowers, thus increasing the amount of nectar in each flower. In defending a territory, the sunbird saved 780 calories a day in reduced foraging activity. However, the sunbird also spent 728 calories in defense of the territory, yielding a net gain of 52 calories a day.

Territory size differs considerably among species, and optimality theory predicts that it should evolve to maximize the difference between benefits and costs, thus maximizing the fitness "profit" to the territory holder. Because cheetahs need large areas to be able to hunt successfully, they establish large territories relative to their size (**Figure 55.11b**). In contrast, territories set up solely to defend areas for mating or nesting are often relatively small. For example, male sea lions defend small areas of beach. The preferred areas contain the largest amount of females and are controlled by the largest breeding bulls. The size of the territory of some nesting birds, such as gannets, is determined by how far the bird can reach to peck its neighbor without leaving its nest (**Figure 55.11c**).

Territories may be held for a season, a year, or the entire lifetime of the individual. Ownership of a territory needs to be periodically proclaimed, and thus communication between individuals is necessary for territory owners. This may involve various types of signaling, which we discuss next.

55.4 Communication

Communication is the use of specially designed signals or displays to modify the behavior of others. It may be used for many purposes, including defining territories, maintaining contact with offspring, courtship, and contests between males. The use of different forms of communication between organisms depends on the environment in which they live. For example, visual communication plays little role in the signals of nocturnal animals. Similarly, animals in dense forests often cannot see each other, so sounds are of prime importance in mapping out territories. Sound, however, is a temporary signal. Scent can last longer and is often used to mark the large territories of some mammals. In this section, we outline the various types of communication—chemical, auditory, visual, and tactile—that occur among animals.

Chemical Communication Is Often Used to Mark Territories or Attract Mates

The chemical marking of territories is common among animals, especially among members of the canine and feline families (see Figure 55.11b). Scent trails are often used by social insects to recruit workers to help bring prey to the nest. Fire ants (*Solenopsis* spp.) attack large, living prey, and many ants are needed to drag the prey back to the nest. In this case, the scout that finds the prey lays down a scent trail from the prey back to the nest. The scent excites other workers, which follow the trail to the prey. The scent marker is very volatile, and the trail effectively disappears in a few minutes to avoid mass confusion over old trails.

Animals frequently use chemicals to attract mates. Female moths attract males by powerful chemical attractants called **pheromones**. Among social organisms, some individuals use pheromones to manipulate the behavior of others. A queen bee releases pheromones that suppress the reproductive system of workers, which ensures that she is the only reproductive female in the hive.

Auditory Communication Is Often Used to Attract Mates and to Deter Competitors

Many organisms communicate by making sound. Because the Earth itself can absorb sound waves, sound travels farther in the air, which is why many birds and insects use perches when singing. Air is on average 14 times less turbulent at dawn and dusk than during the rest of the day, which helps explain the preference of most animals for calling at these times. Some insects utilize the very plants on which they feed as a medium of song transmission. Many male leafhopper and planthopper insects vibrate their abdomens on leaves and create species-specific courtship songs that are transmitted by adjacent and touching vegetation and are picked up by nearby females of the same species.

While many males use auditory communication to attract females, some females use calls to attract the attention of males. Female elephant seals use this behavior to their advantage. When a nondominant male attempts to mate with a female, she screams loudly, attracting the attention of the dominant male, which drives the nondominant male away. In this way, she is guaranteed a mating with the strongest male. Sound production can attract predators as well as mates. Some bats listen for the mating calls of male frogs to find their prey. Parasitic flies detect and locate chirping male crickets and then deposit tiny maggots on or near them. The maggots latch onto and penetrate the cricket, and eventually kill it. Sound may also be used by males during competition over females. In many animals, lower-pitched sounds come from bigger males, so by calling to one another, males can gauge the size of their opponents and save fighting energy.

Visual Communication Is Often Used in Courtship and Aggressive Displays

In courtship, animals use a vast number of visual signals to identify and select potential mates. Competition among males for the most impressive displays to attract females has led to elaborate coloration and extensive ornamentation in some species. For example, peacocks and males of many bird species have developed elaborate plumage to attract females.

Male fireflies have developed light flashes that are species specific with regard to number and duration of flashes (**Figure 55.12**). Females respond with a flash of their own. Such bright flashes are also bound to attract predators. Some female fireflies use mimicry to their advantage. Female *Photuris versicolor* fireflies mimic the flashing responses normally given by females of other species, such as *Photinus tamytoxus*, in order to lure the males of those species close enough to eat them.

Visual signals are also used to resolve disputes over territories or mates. Deer and antelope have antlers or horns that they use to display and spar over territory and females. Most of these matches never develop into outright fights, since the males gauge their opponent's strength by the size of these ornaments. Among insects, the "horns" of rhinoceros beetles and the eye stems of stalk-eyed flies send similar signals.

Tactile Communication Is Used to Strengthen Social Bonds and to Convey Information About Food

Animals often use tactile communication to establish bonds between group members. Primates frequently groom one another, while canines and felines may nuzzle and lick each other. Many insects use tactile communication to convey information on the whereabouts of food. Members of the ant genus *Leptothorax* feed on immobile prey such as dead insects. When a scouting ant encounters such prey, it usually needs an additional worker to help bring it back to the nest.

(a)

(b)

(c)

Figure 55.12 **Visual communication in fireflies.** (a) Communication between fireflies is conducted by species-specific light flashes emitted by organs located on the underside of the abdomen. (b) At dusk, many different light flashes can be seen. (c) Sometimes females, such as this large *Photuris versicolor*, mimic the displays of other species, luring an unwitting male, such as this *Photinus tamytoxus*, and then eating him.

(a) Bees clustering around a recently returned worker.

(b) Round dance

(c) Waggle dance: The angle of the waggle to the vertical orientation of the honeycomb corresponds to the angle of the food source from the sun.

Figure 55.13 Tactile communication among honeybees regarding food sources. (a) Bees gather around a newly returned scout to receive information about nearby food sources. (b) If the food is less than 50 m away, the scout performs a round dance. (c) If the food is more than 50 m away, the scout performs a waggle dance, which conveys information about its location. If the dance is performed at a 30° angle to the right of the hive's vertical plane, then the food source is located at a 30° angle to the right of the sun.

Rather than laying a scent trail, which is energetically costly, the scout ant recruits a helper and physically leads it to the food source. The helper runs in tandem with the scout, its antennae touching the scout's abdomen.

Perhaps the most fascinating example of tactile communication among animals is the dance of the honeybee, elegantly studied by German ethologist Karl von Frisch in the 1940s. Bees commonly live in large hives; in the case of the European honeybee (*Apis mellifera*), the hive consists of 30,000–40,000 individuals. The flowering plants on which the bees forage are sometimes located miles from the hive and are distributed in a patchy manner, with any given patch usually containing many flowers that store more nectar and pollen than an individual bee can carry back to the nest. The scout bee that locates the resource patch returns to the hive and recruits more workers to join it (**Figure 55.13a**). Because it is dark inside the hive, the bee uses a tactile signal. The scout dances on the vertical side of a honeycomb, and the dance is monitored by other bees, which follow and touch her to interpret the message. If the food is relatively close to the hive, less than 50 m away, the scout performs a round dance, rapidly moving in a circle, first in one direction and then the other. The other bees know the food is relatively close at hand, and the smell of the scout tells them what flower species to look for (**Figure 55.13b**).

If the food is more than 50 m away, the scout will perform a different type of dance, called a "waggle dance." In this dance, the scout traces a figure 8, in the middle of which she waggles her abdomen and produces bursts of sound. Again, the other bees maintain contact with her. Occasionally, the scout will regurgitate a small sample of nectar so the bees know the type of food source they are looking for. The truly amazing part of the waggle dance is that the angle at which the central part of the figure 8 deviates from the vertical direction of the comb represents the same angle at which the food source deviates from the

sun (**Figure 55.13c**). The direction is always up-to-date, because the bee adjusts the dance as the sun moves across the sky.

As we have seen, much communication occurs not only to defend territories but also to communicate information to other individuals in the population, including potential mates. While living on your own and maintaining a territory has advantages, living in a group also has its benefits, including ready availability of mates and increased protection from predators. In the following section, we examine group living and the behavior it engenders.

55.5 Living in Groups

Much animal behavior is directed at other animals. As such, some of the more complex behavior occurs when animals live together in groups such as flocks or herds. If a central concern of ecology is to explain the distribution patterns of organisms, then one of our most important tasks is to identify and understand the behavior that results from group living. While congregations promote competition for food, there are also benefits of group living that compensate for the costs involved. As we discuss in this section, many of these benefits relate to group defense against predators. Group living can reduce predator success in at least two ways: through increased vigilance and through protection in numbers.

Living in Large Groups May Reduce the Risk of Predation Because of Increased Vigilance

For many predators, success depends on surprise. If an individual is alerted to an attack, the predator's chance of success is lowered. A woodpigeon (*Columba palumbus*) will take to the air when it spots a goshawk (*Accipiter gentilis*). Once one pigeon

Figure 55.14 Living in groups and the many eyes hypothesis. The larger the number of woodpigeons, the less likely an attack will be successful.

takes flight, the other members of the flock are alerted and follow suit. If each pigeon occasionally looks up to scan for a hawk, the bigger the group, the more likely that one bird will spot the hawk early enough for the flock to take flight. This is referred to as the **many eyes hypothesis (Figure 55.14)**. By living in groups, individuals may decrease the amount of time scanning for predators and increase the time they have to feed. Of course, cheating is a possibility, because some birds might never look up, relying on others to keep watch while they keep feeding. However, the individual that happens to be scanning when a predator approaches is most likely to escape, a fact that tends to discourage cheating.

Living in Groups Offers Protection by the "Selfish Herd"

Group living also provides protection in sheer numbers. Typically, predators take one prey animal per attack. In any given attack, an individual antelope in a herd of 100 has a 1 in 100 chance of being selected, whereas a single individual has a 1 in 1 chance. Large herds may be attacked more frequently than a solitary individual, but a herd is unlikely to attract 100 times more attacks than an individual, often because of the territorial nature of predators. Furthermore, large numbers of prey are able to defend themselves better than single individuals, which usually choose to flee. For example, groups of nesting black-headed gulls will mob a crow relentlessly, thereby reducing the crow's ability to steal the gulls' eggs.

Research has shown that within a group, each individual can minimize the danger to itself by choosing the location that is as close to the center of the group as possible. This was the subject of a famous paper, "The Geometry of the Selfish Herd," by the British evolutionary biologist W. D. Hamilton. The explanation of this type of defense is that predators are likely to attack prey on the periphery because they are easier to isolate visually. Many animals in herds tend to bunch close together when they are under attack, making it physically difficult for the predator to get to the center of the herd.

In the end, group size may be the result of a trade-off between the costs and benefits of group living. Although much group behavior serves to reduce predation, other complex behavior occurs in groups, including grooming behavior and behavior that appears to benefit the group at the expense of the individual. For example, a honeybee will sting a potential hive predator to discourage it. The bee's stinger is barbed, and once it has penetrated the predator's skin, the bee cannot withdraw it. The bee's only means of escape is to tear away part of its abdomen, leaving the stinger behind and dying in the process. In the next section, we explore the reasons for altruistic behavior, in which an individual risks its life for the benefit of others.

55.6 Altruism

In Chapter 23, we learned that a primary goal of an organism is to pass on its genes, yet we see many instances in which some individuals forego reproducing altogether, apparently to benefit the group. How do ecologists explain **altruism**, behavior that appears to benefit others at a cost to oneself. In this section, we begin by discussing whether such behavior evolved for the good of the group or for the good of the individual. As we will see, most altruistic acts serve to benefit the individual's close relatives. We explore the concept of kin selection, which argues that acts of self-sacrifice indirectly promote the spread of an organism's genes, and see how this plays out in an extreme form in the genetics of social insect colonies. We conclude by examining reciprocal altruism as an attempt to explain the evolution of altruism among nonkin.

In Nature, Individual Selfish Behavior Is More Likely Than Altruism

One of the first attempts to explain the existence of altruism was called **group selection**, the premise that natural selection produces outcomes beneficial for the whole group or species. In 1962, the British ecologist V. C. Wynne-Edwards argued that a group containing altruists, each willing to subordinate its interests for the good of the group, would have a survival advantage over a group composed of selfish individuals. In concept, the idea of group selection seemed straightforward and logical: a group that consisted of selfish individuals would overexploit its resources and die out, while the fitness of a group with altruists would be enhanced.

In the late 1960s, the idea of group selection came under severe attack. Leading the charge was the biologist G. C. Williams, who argued that evolution acts through **individual selection**, which proposes that adaptive traits generally are selected for because they benefit the survival and reproduction of the individual rather than the group. Williams' arguments against group selection follow.

Mutation Mutant individuals that readily use resources for themselves or their offspring will have an advantage in a population where individuals limit their resource use. Consider a species of bird in which a pair lays only two eggs, that is, it has a clutch size of two and there is no overexploitation of resources. Two eggs would ensure a replacement of the parent birds but would prevent a population explosion. Imagine a mutant bird arises that lays three eggs. If the population is not overexploiting its resources, sufficient food may be available for all three young to survive. If this happens, the three-egg genotype will eventually become more common than the two-egg genotype. This process would work for even larger brood sizes, such as four eggs or five eggs, and brood sizes would tend to increase until they became so large that the parents could not look after all their young. Field studies of great tits in Wytham Woods, England, show a median clutch size of eight to nine eggs, not because females couldn't incubate more, but because adult birds cannot reliably supply sufficient food for more than eight or nine chicks to survive.

Immigration Even in a population in which all pairs laid two eggs and no mutations occurred to increase clutch size, selfish individuals that laid more could still immigrate from other areas. In nature, populations are rarely sufficiently isolated to prevent immigration of "selfish" mutants from other populations.

Individual Selection For group selection to work, some groups must die out faster than others. In practice, groups do not become extinct very frequently. Individuals die off more frequently than groups, so individual selection will be the more powerful evolutionary force.

Resource Prediction Group selection assumes that individuals are able to assess and predict future food availability and population density within their own habitat. There is little evidence that they can. For example, it is difficult to imagine that songbirds would be able to predict the future supply of the caterpillars that they feed to their young and adjust their clutch size accordingly.

Most ecologists accept individual gain as a more plausible result of natural selection than group selection. Population size is more often controlled by competition in which individuals strive to command as much of a resource as they can. Such selfishness can cause some seemingly surprising behaviors. For example, male Hanuman langurs (*Semnopithecus entellus*) kill infants when they take over groups of females from other males

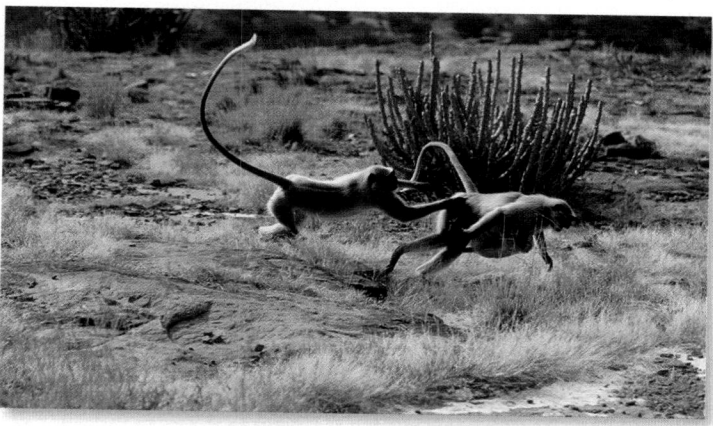

Figure 55.15   Infanticide as selfish behavior.   Male Hanuman langurs (*Semnopithecus entellus*) can act aggressively toward the young of another male, even killing them, hastening the day the females come into estrus and the time when the males can father their own offspring. Note that the mother is running with the infant.

(**Figure 55.15**). The reason for the behavior is that when they are not nursing their young, females become sexually receptive much sooner, hastening the day when the male can father his own offspring. Infanticide ensures that the male will father more offspring, and the genes governing this tendency spread by natural selection.

If individual selfishness is more common than group selection, how do we account for what appear to be examples of altruism in nature?

Apparent Altruistic Behavior in Nature Is Often Associated with Kin Selection

All offspring have copies of their parents' genes, so parents taking care of their young are actually caring for copies of their own genes. Genes for altruism toward one's young are favored by natural selection and will become more numerous in the next generation, because offspring have copies of those same genes.

The probability that any two individuals will share a copy of a particular gene is a quantity, r, called the **coefficient of relatedness**. During meiosis in a bi-allelic system, any given copy of a gene has a 50% chance of going into an egg or sperm. A mother and father are on average related to their children by an amount $r = 0.5$, because half of a child's genes come from its mother and half from its father. By similar reasoning, brothers or sisters are related by an amount $r = 0.5$ (they share half their mother's genes and half their father's); grandchildren and grandparents, by 0.25; and cousins, by 0.125 (**Figure 55.16**). In 1964, ecologist W. D. Hamilton realized the implication of the coefficient of relatedness for the evolution of altruism. An organism not only can pass on its genes through having offspring, it also can pass them on through ensuring the survival of siblings, nieces, nephews, and cousins. This means an organ-

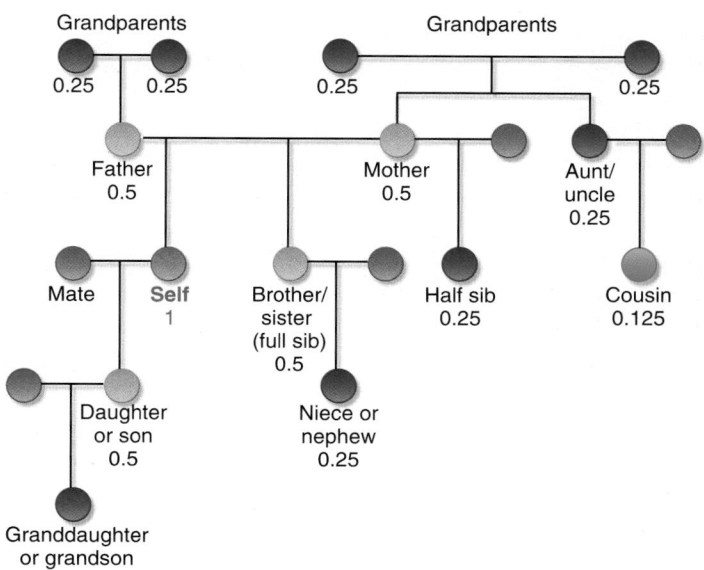

Figure 55.16 **Degree of genetic relatedness to "self" in a diploid organism.** Pink circles represent completely unrelated individuals.

Biological inquiry: In theory, should you sacrifice your life to save two sisters or nine cousins?

Figure 55.17 **Altruistic behavior or kin selection?** *Datana ministra* caterpillars exhibit a bright, striped warning pattern to advertise their bad taste to predators. All the larvae in the group are likely to be the progeny of one egg mass from one adult female moth. The death of the one caterpillar it takes to teach a predator to avoid the pattern benefits the caterpillar's close kin.

ism has a vested interest in protecting its brothers and sisters, and even their offspring.

The term **inclusive fitness** is used to designate the total number of copies of genes passed on through one's relatives, as well as one's own reproductive output. Selection for behavior that lowers an individual's own fitness but enhances the reproductive success of a relative is known as **kin selection**. Hamilton proposed that an altruistic gene will be favored by natural selection when

$$rB > C$$

where *r* is the coefficient of relatedness of donor (the altruist) to the recipient, *B* is the benefit received by the recipients of the altruism, and *C* is the cost incurred by the donor. This is known as **Hamilton's rule**.

Imagine two sisters who are not yet mothers. One has a rare kidney disease and needs a transplant from her sister. Let's assume both sisters will have two children of their own. The risk of the transplant to the donor involves a 1% chance of dying, but the benefit to the recipient involves a 90% chance of living and having children. In this example, $r = 0.5$, $B = 0.9 \times 2 = 1.8$, and $C = 0.01 \times 2 = 0.02$. Since the genetic benefit (*rB*) of 0.9 is much greater than the genetic cost (*C*) of 0.02, it makes evolutionary sense to proceed with the transplant. While humans are unlikely to do this type of calculation before deciding whether to risk their lives to save their brothers or sisters from a life-threatening event, this example shows how such behavior could arise and spread in nature.

Let's examine a situation involving altruism within a group of animals. Many insect larvae, especially caterpillars, are soft-bodied creatures. They rely on a bad taste or toxin to deter predators and advertise this condition with bright warning colors. For example, noxious *Datana ministra* caterpillars, which feed on oaks and other trees, have bright red and yellow stripes and adopt a specific posture with head and tail ends upturned when threatened (**Figure 55.17**). Unless it is born with an innate avoidance of this prey type, a predator has to kill and eat one of the caterpillars in order to learn to avoid similar individuals in the future. It is of no personal use to the unlucky caterpillar to be killed. However, animals with warning colors often aggregate in kin groups because they hatch from the same egg mass. In this case, the death of one individual is likely to benefit its siblings, which are less likely to be attacked in the future, and thus its genes will be preserved. This explains why the genes for bright color and a warning posture are successfully passed on from generation to generation. In a case where $r = 0.5$, *B* might be 50, and $C = 1$, the benefit of $25 > 1$, so the genes for this behavior will spread.

A common example of altruism in social animals occurs when a sentry raises an alarm call in the presence of a predator. This behavior has been observed in Belding's ground squirrels (*Spermophilus beldingi*) (**Figure 55.18**). The squirrels feed in groups, with certain individuals acting as sentries and watching for predators. As a predator approaches, the sentry typically gives an alarm call and the group members retreat into their burrows. In drawing attention to itself, the caller is at a higher risk of being attacked by the predator. However, in many groups, those closest to the sentry are most likely to be offspring or brothers or sisters; thus, the altruistic act of alarm calling is reasoned to be favored by kin selection. Supporting this is the fact that most alarm calling is done by females, because they are more likely to stay in the colony where they were born and thus have kin nearby, whereas the males are more apt to disperse far from the colony.

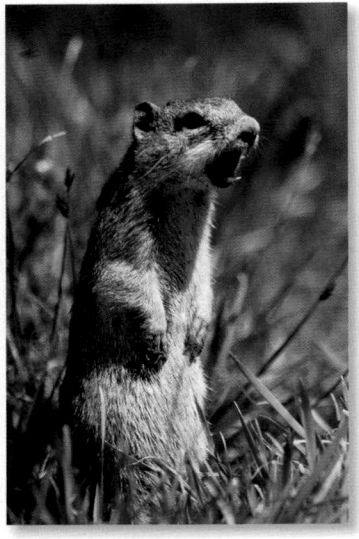

Figure 55.18 **Alarm calling, a possible example of kin selection.** This Belding's ground squirrel sentry is emitting an alarm call to warn other individuals, which are often close kin, of the presence of a predator. It is believed that by doing so, the sentry draws attention away from the others but becomes an easier target itself.

Figure 55.19 **A naked mole rat colony.** In this mammal species, females do not reproduce and only the queen (shown resting on workers) has offspring.

Altruism in Social Insects Arises Partly from Genetics and Partly from Lifestyle

Perhaps the most extreme form of altruism is the evolution of sterile castes in social insects, in which the vast majority of females, known as workers, rarely reproduce themselves but instead help one reproductive female (the queen) to raise offspring, a phenomenon called **eusociality**. The explanation of eusociality lies partly in the particular genetics of most social insect reproduction. Females develop from fertilized eggs and are diploid, the product of fertilization of an egg by a sperm. Males develop from unfertilized eggs and are haploid.

Such a genetic system is called **haplodiploidy**. If they have the same parents, each daughter receives an identical set of genes from her haploid father. The other half of a female's genes come from her diploid mother, so the coefficient of relatedness (r) of sisters is 0.50 (from father) + 0.25 (from mother) = 0.75. The result is that females are more related to their sisters (0.75) than they would be to their own offspring (0.50). This suggests it is evolutionarily advantageous for females to stay in the nest or hive and care for other female offspring of the queen, which are their full sisters.

Elegant though these types of explanations are, they do not provide the whole picture. Large eusocial colonies of termites exist, but termites are diploid, not haplodiploid. In this case, how do we account for the existence of eusociality?

In the 1970s, Richard Alexander suggested that it was the particular lifestyle of these animals, rather than genetics, that promoted eusociality. He argued that in a normal diploid organism, females are related to their daughters by 0.5 and to their sisters by 0.5, so it should matter little to them whether they rear siblings or daughters of their own. He predicted, well before eusociality was discovered to occur in mammals, that a eusocial mammalian species could exist when certain conditions were met, including that the nests or burrows be enclosed and subterranean, and that the colony have a food supply such

as tubules and roots. In addition, the soil would need to be hard, dry clay to keep the colony safe from digging predators. He proposed that the colony would be defended by a few members of the colony willing to give their lives in defense of others, and he posited the existence of mechanisms by which a queen could manipulate other individuals.

At the time, Alexander had no idea that a mammal with these characteristics existed. Surprisingly, subsequent discoveries confirmed the existence of a eusocial mammal that satisfied all of the predictions of Alexander's model: the naked mole rat. Naked mole rats are diploid species that live in arid areas of Africa in large underground colonies where only one female, the queen, produces offspring (**Figure 55.19**). A renewable food supply is present in the form of tubers of the plant *Pyrenacantha kaurabassana*. These weigh up to 50 kg and provide food for a whole colony, though the food would be insufficient if all the mole rats reproduced. Because the burrows are as hard as cement, there are few ways to attack them, and a heroic effort by a mole rat blocking the entrance can effectively stop a predator (commonly a rufous-beaked snake). The queen mole rat does indeed manipulate the colony members; she suppresses reproduction in other females by producing a pheromone in her urine that is passed around the colony by grooming. Hence, the mole rats seem to have evolved the appropriate behavior to exploit this ecological niche. As Alexander argued, lifestyle characteristics can provide an explanation for the evolution of eusociality in species such as naked mole rats, in which both sexes are diploid.

Unrelated Individuals May Engage in Altruistic Acts if the Altruism Is Likely to Be Reciprocated

Even though we have argued that kin selection can explain instances of apparent altruism, cases of altruism are known to exist between unrelated individuals. What drives this type of behavior appears to be a "You scratch my back, I'll scratch yours" type of reciprocal altruism, in which the cost to the animal of behaving altruistically is offset by the likelihood of a return benefit. This occurs in nature, for example, when unrelated chimps groom each other.

Researcher Gerald Wilkinson has noted that female vampire bats exhibit reciprocal altruism via food sharing. Vampire bats can die after 60 hours without a blood meal, because they can no longer maintain their correct body temperature. Adult females will share their food with their young, the young of other females, and other unrelated females that have not fed. The females roost together in groups of 8 to 12 and their dependent young. A hungry female will solicit food from another female by approaching and grooming her. The female being groomed then regurgitates part of her blood meal for the other. The roles of blood donor and recipient are often reversed, and Wilkinson showed that unrelated females are more likely to share with those that had recently shared with them. The probability of a female getting a free lunch is increased because the roost consists of individuals that remain associated with each other for long periods of time.

55.7 Mating Systems

In nature, most males seem superfluous because one male could mate with many females in a local area. If one male can mate with many females, why are there few species with a sex ratio of, say, 1 male : 20 females? The sex ratio is more often about 1:1. The answer lies with natural selection. Let's consider a hypothetical population that contains 20 females for every male; each male mates, on average, with 20 females. A parent whose children were exclusively sons could expect to have 20 times the number of grandchildren compared to a parent with the same number of daughters. Under such conditions, natural selection would favor the production of sons, and males would become prevalent in the population. However, if the population were mainly males, females would be at a premium, and natural selection would favor their production. Such constraints operate on the numbers of both male and female offspring, keeping the sex ratio at about 1:1.

Even though the sex ratio is fairly even in most species, that doesn't mean that one female always mates with one male or vice versa. Several different types of mating systems occur among animals. In monogamy, each individual mates exclusively with one partner over at least a single breeding cycle and sometimes for longer. In contrast, polygamy, a system in which individuals mate with more than one partner in a breeding season, is much more common among animals. There are two types of polygamy. In polygyny ("many females"), one male mates with more than one female, but females mate only with one male. In polyandry ("many males"), one female mates with several males, but males mate with only one female.

In this section, we examine the characteristics of these mating systems and explore the role of sexual selection, a type of natural selection in which competition for mates drives the evolution of certain traits.

Sexual Selection Involves Mate Choice and Mate Competition

As we learned in Chapter 24, sexual selection promotes traits that will increase an organism's mating success. You will recall that sexual selection can take two forms. In intersexual selection, members of one sex, usually females, choose mates based on particular characteristics such as color of plumage or sound of courtship song. In intrasexual selection, members of one sex, usually males, compete over partners, and the winner performs most of the matings. Let's explore each of these in a little more detail.

Female Mate Choice Females have many different ways to choose their prospective partners. Female hangingflies (*Hylobittacus* spp.) demand a nuptial gift of a food package, an insect prey item that the male has caught (**Figure 55.20**). Such a nutrient-rich gift may permit females to produce more eggs. The bigger the gift, the longer it takes the female to eat it and the longer the male can copulate with her. Females will not mate with males that do not offer such a package. Female spiders and mantids will sometimes eat their mate during or after copulation, with the male's body constituting the ultimate nuptial gift.

Males may also have parenting skills that females desire. Among 15-spined sticklebacks (*Spinachia spinachia*), males perform nest fanning, cleaning, and guarding of the offspring. Females prefer to mate with males that shake their bodies most during courtship, apparently using this cue to assess the quality of the male as a potential father.

Often, females choose mates without the offering of obvious material benefits and make their choices based on plumage color or courtship display. The male African long-tailed widowbird (*Euplectes progne*) has long tail feathers that he displays to females via aerial flights. Researcher Malte Andersson experimentally shortened the tails of some birds by clipping their tail feathers, and he lengthened the tails of others by taking the clippings and sticking them onto other birds with superglue.

Figure 55.20 **Female choice of males based on nuptial gifts.** A male hangingfly presents a nuptial gift, a small moth, to a female.

Males with experimentally lengthened tails attracted four times as many females as males with shortened tails and they fathered more clutches of eggs (**Figure 55.21**).

Some researchers have suggested that males with excessively long tail feathers must be very healthy to bear this "handicap" and that these ornaments therefore advertise the male's genetic quality. However, other important benefits may be associated with plumage quality. Bright colors are often caused by pigments called carotenoids that help stimulate the immune system. In zebra finches and red jungle fowl, colorful plumage has been associated with heightened resistance to disease, suggesting that females that choose such males are choosing genetically healthier fathers.

On rare occasions, a sex role reversal occurs and the male discriminates among females. In the Mormon cricket (*Anabrus simplex*), males only mate once because they provide a nutrient-rich nuptial gift of a spermatophore to females, which is energetically costly to produce. Here, males choose heavier females to mate with because these females have more eggs and the males can father more offspring.

Mate Competition Between Individuals In many species, females do not actively choose their preferred mate; instead, they mate with competitively superior males. In such cases, dominance is determined by fighting or by ritualized sparring. Outcomes may be dictated by the size of weapons such as antlers or horns or by body size. In the southern elephant seal (*Mirounga leonina*), females haul up onto the beach to give birth and gain safe haven for their pups from marine predators. Following birth, they are ready to mate. In this situation, dominant males are able to command a substantial group and constantly lumber across the beach to fight other males and defend their harem. Such competition tends to promote increased body size, and in species with male-male competition, males are often substantially larger than females (**Figure 55.22**).

Large body size does not always guarantee paternity. Smaller male elephant seals may intercept females in the ocean and attempt to mate with them there, rather than on the beach, where the competitively dominant males patrol. Such "satellite" males, which move around the edge of the mating arena, are also visible in other species. For example, small male frogs hang around ponds waiting to intercept females headed toward the call of dominant males. Thus, even though competitively dominant males father most offspring, smaller males can have reproductive success.

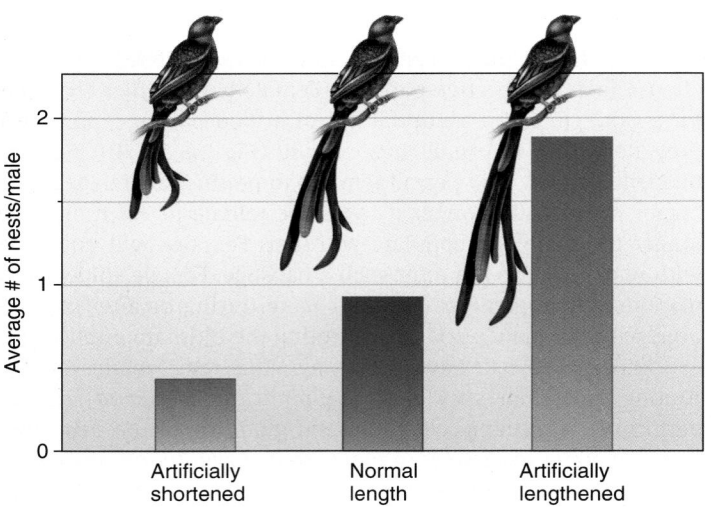

Figure 55.21 **Female choice based on male appearance.** Males with artificially lengthened tails mate with more females, and therefore have more nests, than males with a normal or artificially shortened tail.

Biological inquiry: Why do you think it is rarer for female birds or mammals to have more colorful plumage or elaborate adornments than males of the same species?

(a)

(b)

Figure 55.22 **Large male size and mating success.** (a) In southern elephant seals, the larger the male, in relation to female size, the greater the number of females that can be mated with and monopolized. (b) These male elephant seals are fighting to maintain control of their female harems.

In Monogamous Mating Systems, Males and Females Are Paired for at Least One Reproductive Season

Mating success is also dependent on the type of the particular species' mating system. In **monogamy**, one male mates with one female, and most individuals have mates. Males and females are generally similar in body size and appearance (**Figure 55.23a**). Several hypotheses explain the existence of monogamy. The first is the **mate-guarding hypothesis**, which suggests that males stay with a female to protect her from being fertilized by other males. Such a strategy may be advantageous when receptive females are widely scattered and difficult to find.

The **male assistance hypothesis** maintains that males remain with females to help them rear their offspring. Monogamy is common among birds, about 70% of which are socially monogamous, that is, the pairings remain intact during at least one breeding season. According to this hypothesis, monogamy is prevalent in birds because eggs and chicks take a considerable amount of parental care. Most eggs need to be incubated continuously if they are to hatch, and chicks require almost continual feeding. It is therefore in the male's best interest to help fledge his young, because he would have few surviving offspring if he did not.

The **female-enforced monogamy hypothesis** suggests that females stop their male partners from being polygynous. Male and female burying beetles (*Nicrophorus defodiens*) work together to bury small, dead animals, which will provide a food resource for their developing offspring. Males will release pheromones to attract other females to the site. However, while an additional female might increase the male's fitness, the additional developing offspring might compete with the offspring of the first female, decreasing her fitness. As a result, on smelling these pheromones, the first female will interfere with the male's attempts at signaling, preserving the monogamous relationship.

Recent research by Larry Young and Elizabeth Hammock has shown that fidelity may have a genetic basis. These researchers found that fidelity of male voles depends on the length of repeating, short genetic sequences, once considered junk DNA, in a gene that codes for a key hormone receptor. Adult male voles with the long version of the sequence were more apt to form pair bonds with female partners and nurture their offspring than were voles with the short version.

In Polygynous Mating Systems, One Male Mates with Many Females

In **polygyny**, one male mates with more than one female in a single breeding season, but females mate only with one male. Physiological constraints often dictate that female organisms must care for the young, because they are the ones most often left "holding the baby." Because of these constraints, at least in many organisms with internal fertilization, such as mammals and some fish, males are able to desert and mate with more females. Polygynous systems are therefore associated with uniparental care of young, with males contributing little. Sexual dimorphism is typical in polygynous mating systems, particularly when males engage in competition over mates (**Figure 55.23b**). Sexual maturity is often delayed in males that fight because of the considerable time it takes to reach a sufficiently large size to compete for females.

Polygyny is influenced by the spatial or temporal distribution of breeding females. In cases where all females are sexually receptive within the same narrow period of time, little opportunity exists for a male to garner all the females for himself. Where female reproductive receptivity is spread out over weeks or months, there is much more opportunity for males to mate with more than one female. For example, females of the common toad (*Bufo bufo*) all lay their eggs within a week, and males generally have time to mate only with one female. In contrast, female bullfrogs (*Rana catesbeiana*) have a breeding season of several weeks, and males may mate with as many as six females in a season.

(a)   (b)   (c)

Figure 55.23   **Sexual dimorphism in body size and mating system.**   **(a)** In monogamous species, such as these Manchurian cranes, *Grus japonensis*, males and females appear very similar. **(b)** In polygynous species, like white-tailed deer, *Odocoileus virginianus*, males are bigger than females and have large horns with which they engage in combat over females. **(c)** In polyandrous species, females are usually bigger, as with these golden silk spiders, *Nephila clavipes*.

Resource-Based Polygyny Where some critical resource is patchily distributed and in short supply, certain males may dominate the resource and breed with more than one visiting female. In the lark bunting (*Calamospiza melanocorys*), which mates in North American grasslands, males arrive at the grasslands first, compete for territories, and then display with special courtship flight patterns and songs to attract females. The major source of nestling death in this species is overheating from too much exposure to the sun. Prime territories are therefore those with abundant shade, and some males with shaded territories attract two females, even though the second female can expect no help from the male in the process of rearing young. Males in some exposed territories remain bachelors for the season. From the dominant male's point of view, resource-based polygyny is advantageous; from the female's point of view, there may be costs. Although by choosing dominant males, a female may be gaining access to good resources, she may also have to share these resources with other females.

Harem Mating Structures Sometimes males defend a group of females without commanding a resource-based territory. This pattern is more common when females naturally congregate in groups or herds, perhaps to avoid predation, as with southern elephant seals (see Figure 55.22). Usually the largest and strongest males command most of the matings, but being a harem master is usually so exhausting that males may only manage to remain the dominant male for a year or two.

Communal Courting Polygynous mating can occur where neither resources nor harems are defended. In some instances, particularly in birds and mammals, males display in designated communal courting areas called **leks** (**Figure 55.24**). Females come to these areas specifically to find a mate, and they choose a prospective mate after the males have performed elaborate displays. Most females seek to mate with the best male, so a few successful males perform the vast majority of the matings. At a lek of the white-bearded manakin (*Manacus manacus*) of South America, one male accounted for 75% of the 438 matings where there were as many as 10 males. A second male mated 56 times (13% of matings), while six others mated only a total of 10 times.

In Polyandrous Mating Systems, One Female Mates with Many Males

In most systems in which one individual mates with more than one individual of the opposite sex, the polygamous sex is the male. The opposite condition, **polyandry**, in which one female mates with several males, is much more rare. Nevertheless, it is practiced by some species of birds, fish, and insects. Sexual dimorphism is present, with the females being the larger of the sexes (see **Figure 55.23c**). In the Arctic tundra, the summer season is short but very productive, providing a bonanza of insect

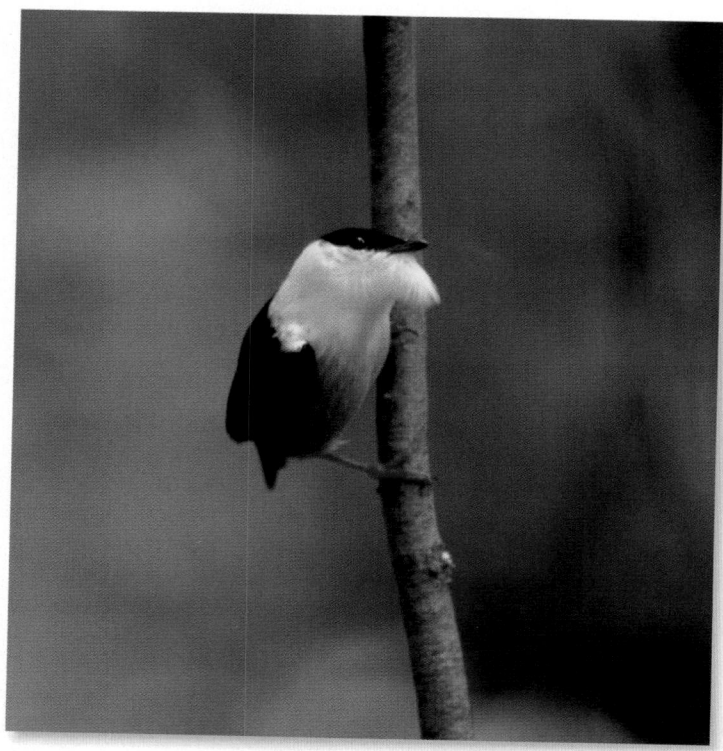

Figure 55.24 **Some male birds and mammals congregate in communal courting grounds called leks.** A male white-bearded manakin waits at a lek, where males have pulled the leaves off a branch to create a clear display area. Females visit the leks and males display to them, puffing up their white "beard" feathers.

food for two months. The productivity of the breeding grounds of the spotted sandpiper (*Actitis macularia*) is so high that the female becomes rather like an egg factory, laying up to five clutches of four eggs each in 40 days. Her reproductive success is limited not by food but by the number of males she can find to incubate the eggs, and females compete for males, defending territories where the males sit.

Polyandry is also seen in some species where egg predation is high and males are needed to guard the nests. For example, in the pipefish (*Syngnathus typhle*), males have brood pouches that provide eggs with safety and a supply of oxygen- and nutrient-rich water. Females produce enough eggs to fill the brood pouches of two males, and may mate with more than one male.

Ultimately, as we have seen, most behaviors have evolved to maximize an individual's reproductive output. In a successful group of individuals, this leads to population growth. But we are not knee-deep in sandpipers or elephant seals, so there must be some constraints on reproductive output. In Chapter 56, we next turn to the realm of population ecology to explore how populations grow and what factors limit their growth.

CHAPTER SUMMARY

55.1 The Impact of Genetics and Learning on Behavior

- Behavior is usually due to the interaction of an organism's genes and the environment.
- Genetically programmed behaviors are termed innate and often involve a sign stimulus that initiates a fixed action pattern. (Figures 55.1, 55.2)
- Organisms can often make modifications to their behavior based on previous experience, a process called learning. Some forms of learning include habituation, classical conditioning, operant conditioning, and cognitive learning. (Figures 55.3, 55.4)
- Much behavior is a mixture of innate and learned behaviors. A good example of this occurs in a process called imprinting, in which animals develop strong attachments that influence subsequent behavior. (Figures 55.5, 55.6)

55.2 Local Movement and Long-Range Migration

- The simplest forms of local movement involve kinesis, taxis, and memory. (Figure 55.7)
- Many animals undergo long-range seasonal movement called migration in order to feed or breed. They do this using three proposed mechanisms: piloting, the ability to move from one landmark to the next; orientation, the ability to follow a compass bearing; and navigation, the ability to set, follow, and adjust a compass bearing. (Figures 55.8, 55.9)

55.3 Foraging Behavior

- Animals use complex behavior in food gathering or foraging. Optimality theory views foraging behavior as a compromise between the costs and benefits involved.
- The theory of optimal foraging assumes that animals modify their behavior to keep the ratio of their energy uptake to energy expenditure high. (Figure 55.10)
- The size of a territory, a fixed area in which an individual or group excludes other members of its own species, tends to be optimized according to the costs and benefits involved. (Figure 55.11)

55.4 Communication

- Communication is a form of behavior. The use of different forms of communication between organisms depends on the environment in which they live.
- Chemical communication often involves marking territories; auditory and visual forms of communication are often used to attract mates. A fascinating form of tactile communication involves the dance of the honeybee. (Figures 55.12, 55.13)

55.5 Living in Groups

- Many benefits of group living relate to defense against predators, offering protection through sheer numbers and through what is called the many eyes hypothesis or the geometry of the selfish herd. (Figure 55.14)

55.6 Altruism

- Infanticide is an example of selfish behavior. (Figure 55.15)
- Altruism is behavior that benefits others at a cost to oneself. One of the first explanations of altruism, called group selection, posited that natural selection produced outcomes beneficial for the group. Biologists now believe that most apparently altruistic acts are often associated with outcomes beneficial to those most closely related to the individual, a concept termed kin selection. (Figures 55.16, 55.17, 55.18)
- Altruism among eusocial animals may arise partly from the unique genetics of the animals and partly from lifestyle. (Figure 55.19)
- Altruism is known to exist among nonrelated individuals that live in close proximity for long periods of time.

55.7 Mating Systems

- Sexual selection takes two forms: intersexual selection, in which the female chooses a mate based on particular characteristics, or intrasexual selection, in which males compete with one another for the opportunity to mate with a female. (Figures 55.20, 55.21, 55.22)
- Several types of mating systems are found among animals, including monogamy, polygyny, and polyandry. (Figure 55.23)
- Polygynous mating can often occur in situations where males dominate a resource, defend groups of females (harems), or display in common courting areas called leks. (Figure 55.24)

TEST YOURSELF

1. Behavioral ecology is the study of
   a. how organisms interact with their environment.
   b. courtship behavior.
   c. how an individual's behavior affects reproductive success.
   d. how an individual's behavior affects survival.
   e. both c and d.

2. Geotaxis is a response to the force of gravity. Fruit flies placed in a vial will move to the top of the vial. This is an example of _____ geotaxis.
   a. positive
   b. neutral
   c. innate
   d. negative
   e. learned

3. Certain behaviors seem to have very little environmental influence. Such behaviors are the same in all individuals regardless of the environment and are referred to as _____ behaviors.
   a. genetic
   b. instinctual
   c. innate
   d. b and c only
   e. all of the above

4. Patrick has decided to teach his new puppy a few new tricks. Each time the puppy responds correctly to Patrick's command, the puppy is given a treat. This is an example of
   a. habituation.
   b. classic conditioning.
   c. operant conditioning.
   d. imprinting.
   e. orientation.

5. An animal using landmarks to move from one area to another is exhibiting
   a. navigation.
   b. classic conditioning.
   c. migration.
   d. piloting.
   e. orientation.

6. For group living to evolve, the benefits of living in a group must be greater than the cost of group living. Which of the following is an example of a benefit of living in a group?
   a. reduced spread of disease and/or parasites
   b. increased food availability
   c. reduced competition for mates
   d. decreased risk of predation
   e. all of the above

7. The modification of behavior based on prior experience is called
   a. a fixed action pattern.
   b. learning.
   c. navigation.
   d. adjustment behavior.
   e. innate.

8. When an individual behaves in a way that reduces its own fitness but increases the fitness of others, the organism is exhibiting
   a. kin selection.
   b. group selection.
   c. altruism.
   d. selfishness.
   e. ignorance.

9. Hamilton's theory of kin selection suggests that altruism could evolve in a population if the altruistic behavior of one individual increased the reproductive behavior of
   a. all members of the group.
   b. only the females of the group.
   c. relatives.
   d. nonrelatives.
   e. the youngest individuals of the group.

10. When each female in the population mates with several males, but each male mates with only one female, the mating system is referred to as
    a. polygamy.
    b. polyandry.
    c. polygyny.
    d. monogamy.
    e. harem mating.

CONCEPTUAL QUESTIONS

1. Define ethology.
2. Define a fixed action pattern.
3. Describe the distinguishing features of the different mating systems.

EXPERIMENTAL QUESTIONS

1. What observations were important for the development of Niko Tinbergen's hypothesis explaining how digger wasps located their nests?
2. How did Tinbergen test the hypothesis that the wasps were using landmarks to relocate the nest? What were the results?
3. Did the Tinbergen experiment rule out any other cue the wasps may have been using besides the sight of pinecones?

COLLABORATIVE QUESTIONS

1. Discuss the two main types of associative learning and give an example of each.
2. Discuss several ways in which organisms communicate with each other.

www.brookerbiology.com

This website includes answers to the Biological Inquiry questions found in the figure legends and all end-of-chapter questions.

56

POPULATION ECOLOGY

CHAPTER OUTLINE

A population of king penguins (*Aptenodytes patagonicus*) in Salisbury Plain on South Georgia Island, Antarctica.

A population can be defined as a group of interbreeding individuals occupying the same area at the same time. In this way, we can think of a population of water lilies in a particular lake, the lion population in the Ngorogoro crater in Africa, or the human population of New York City. However, the boundaries of a population can often be difficult to define, though they may correspond to geographic features such as the boundaries of a lake or forest or be contained within a mountain valley or a certain island. Individuals may enter or leave a population, such as the human population of New York City or the deer population in North Carolina. Thus, populations are often fluid entities, with individuals moving into (immigrating) or out of (emigrating) an area. For the purposes of simplicity, in our discussion we will assume that immigration and emigration cancel each other out as factors.

This chapter explores **population ecology**, the study of how populations grow and what factors promote and limit growth. To study populations, we need to employ some of the tools of **demography**, the study of birth rates, death rates, age distributions, and the sizes of populations. We begin our discussion by exploring characteristics of populations, including density and how it is quantified, dispersion, reproductive strategies, and age classes. We will consider how life tables and survivorship curves help summarize demographic information such as birth and death rates. Similarly, growth rates will be examined by determining how many reproductive individuals are in the population and their fertility rate. The data are then used to construct simple mathematical models that allow us to analyze and predict population growth. We will also look at the factors that limit the growth of populations, and conclude the chapter by using the population concepts and models to explore the growth of human populations.

56.1 Understanding Populations

Within their areas of distribution, organisms occur in varying numbers. We recognize this pattern by saying a plant or animal is "rare" in one place and "common" in another. For more precision, ecologists quantify commonness further and talk in terms of population **density**, the numbers of organisms in a given unit area. Population growth affects population density, and knowledge of both can help us make decisions about the management of species. How long will it take for a population of an endangered species to recover to a healthy level if we protect it from its most serious threats? For example, in Florida, boat propellers kill dozens of manatees a year. A knowledge of manatee population growth rates and population densities would allow us to determine at what point populations can no longer recover from such losses and could help determine when and where to set speed limits for boaters. Since 1994, several parts of Georges Bank, once one of the world's richest fishing grounds, have been closed to commercial fishing because of overfishing. How many fish can we reasonably trawl from the sea and still ensure that an adequate population will exist for future use? Such information is vital in making determinations of size limits, catch quotas, and length of season for fisheries to ensure an adequate future population size.

In this section, we discuss density and other characteristics of populations within their habitats. We will also discuss the different reproductive strategies organisms use and assign individuals to different groups called age classes. Population growth can be predicted using knowledge of age classes, reproductive age, and reproductive strategies. We then analyze survivorship and fertility data, which can tell us at what rates populations may grow. Let's begin our exploration of populations by considering density and the various ways that ecologists attempt to quantify it.

Ecologists Use Many Different Methods to Quantify Population Density

The simplest method to measure population density is to visually count the number of organisms in a given area. We can reasonably do this only if the area is small and the organisms are relatively large. For example, we can determine the number of gumbo limbo trees (*Bursera simaruba*) on an island in the Florida Keys. Normally, however, population ecologists calculate the density of plants or animals in a small area and use this figure to estimate the total abundance over a larger area. Several different sampling methods exist for quantifying density in this way, including the use of traps to catch animals, from insects to mammals. Suction traps, like giant aerial vacuum cleaners, can suck flying insects from the sky. Pitfall traps set into the ground can catch species wandering over the surface, such as spiders, lizards, or beetles. Mist nets, consisting of very fine netting spread between trees, can entangle flying birds and bats. Simple baited snap traps, like mouse traps, or live traps, can catch small mammals. Population density can thus be estimated as the number of animals caught per unit area where the traps were set, for example, per 100 m² of habitat.

Sometimes population biologists will capture animals and then tag and release them (**Figure 56.1**). The rationale behind the **mark-recapture technique** is that after the tagged animals are released, they mix freely with unmarked individuals and within a short time are randomly mixed within the population. The population is resampled and the numbers of marked and unmarked individuals are recorded. We assume that the ratio of marked to unmarked individuals in the second sample is the same as the ratio of marked individuals in the first sample to the total population size. Thus:

$$\frac{\text{Number of marked individuals in first catch}}{\text{Total population size, } N} = \frac{\text{Number of marked recaptures in second catch}}{\text{Total number of second catch}}$$

Let's say we catch 50 largemouth bass in a lake and mark them with colored fin tags. A week later we return to the lake and catch 40 fish and 5 of them are previously tagged fish. If we assume that no immigration or emigration has occurred, which is quite likely in a closed system like a lake, and we assume there have been no births or deaths of fish, then the total population size is given by rearranging the equation:

$$\text{Total population size, } N = \frac{\text{Number of marked individuals in first catch} \times \text{Total number of second catch}}{\text{Number of marked recaptures in second catch}}$$

Using our data:

$$N = \frac{50 \times 40}{5} = \frac{2,000}{5} = 400$$

Figure 56.1   **The mark-recapture technique is often used to estimate population size.**   An ear tag identifies this Rocky Mountain goat (*Oreamnos americanus*) in Olympic National Park, Washington. Recapture of such marked animals permits accurate estimates of population size.

Biological inquiry: If we mark 110 Rocky Mountain goats and recapture 100 goats, 20 of which have ear tags, what is the estimate of the total population size?

From this equation, we estimate that the lake has a total population size of 400 largemouth bass. This could be useful information for game and fish personnel who wish to know the total size of a fish population in order to set catch limits. However, the mark-recapture technique can have drawbacks. Some animals that have been marked may learn to avoid the traps. Recapture rates will then be low, resulting in an overestimate of population size. Imagine that instead of 5 tagged fish out of 40 recaptured fish, we only get 2 tagged fish. Now our population size estimate is 2,000/2 = 1,000, a dramatic increase in our population size estimate. On the other hand, some animals can become trap-happy, particularly if the traps are baited with food. This would result in an underestimate of the population size.

Because of the limitations of the mark-recapture technique, ecologists also use other, more novel methods to estimate population density. For many species with valuable pelts we can track population densities through time by examining pelt records taken from trading stations. We can also estimate relative population density by examining catch per unit effort, which is especially valuable in commercial fisheries. We can't easily expect to count the number of fish in an area of ocean, but we can count the number caught, say, per 100 hours of trawling. For some species that leave easily recognizable fecal pellets, like rabbits or deer, we can count pellet numbers. For frogs or birds, we can count chorusing or singing individuals. Many plant individuals are clonal, that is, they grow in patches of genetically identical individuals, so that rather than count individuals we can use the amount of ground covered by plants as an estimate of vegetation density. We can also count leaf scars or chewed leaves as an estimate of the density of the animals that damage them.

Patterns of Spacing Individuals within a population can show different patterns of spatial **dispersion**; that is, they can be clustered together or spread out to varying degrees. The three basic kinds of dispersion pattern are clumped, uniform, and random. We can visualize these patterns by imagining people in a meeting room. If some people know each other, they get together in small groups, creating a clumped pattern. If people do not know each other, and were perhaps even wary of each other, they might maintain a certain minimum personal distance between themselves to produce a uniform dispersion. If nobody thinks or cares about their position relative to anyone else, we would get a random dispersion.

The type of dispersion observed in nature can tell us a lot about what processes shape group structure. The most common dispersion pattern is **clumped**, likely because resources tend to be clustered in nature. For example, certain plants may do better in moist conditions, and moisture is greater in low-lying areas (**Figure 56.2a**). Social behavior between animals may also promote a clumped pattern. Many animals are clumped into flocks or herds.

On the other hand, competition may cause a **uniform** dispersion pattern between individuals, as between trees in a forest. At first, the pattern of trees and seedlings may appear random as seedlings develop from seeds dropped at random, but competition between roots may cause some trees to be outcompeted by others, causing a thinning out and resulting in a uniform distribution. Thus, the dispersion pattern starts out random but

ends up uniform. Uniform dispersions may also result from social interactions, as between some nesting birds, which tend to keep an even distance from each other (**Figure 56.2b**).

Perhaps the rarest dispersion pattern is **random** because resources in nature are rarely randomly spaced. Where resources are common and abundant, as in moist, fertile soil, the dispersion patterns of plants may be random and lacking a pattern (**Figure 56.2c**).

Reproductive Strategies To better understand how populations grow in size, let's consider their reproductive strategies. For example, some organisms can produce all of their offspring in a single reproductive event. This pattern, called **semelparity** (from the Latin *semel*, once, and *parere*, to bear), is common in insects and other invertebrates and also occurs in organisms such as salmon, bamboo grasses, and agave plants (**Figure 56.3a**). These individuals reproduce once only and die. Semelparous organisms may live for many years before reproducing, like the agaves, or they may be annual plants that develop from seed, flower, and drop their own seed within a year.

Other organisms reproduce in successive years or breeding seasons. The pattern of repeated reproduction at intervals throughout the life cycle is called **iteroparity** (from the Latin *itero*, to repeat), and it is common in most vertebrates and perennial plants such as trees. Among iteroparous organisms much variation occurs in the number of reproductive events and in the number of offspring per event. Many species, such as temperate birds or temperate forest trees, have distinct breeding seasons (seasonal iteroparity) that lead to distinct generations (**Figure 56.3b**). For a few species, individuals reproduce repeatedly and at any time of the year. This is termed continuous iteroparity and is exhibited by some tropical species, many parasites, and many mammals (**Figure 56.3c**).

(a) Clumped

(b) Uniform

(c) Random

Figure 56.2 **Three types of dispersion.** (a) A clumped distribution pattern, as in these plants clustered around an oasis, often results from the uneven distribution of a resource, in this case, water. (b) A uniform distribution pattern, as in these nesting black-browed albatrosses (*Diomedea melanophris*) on the Falkland Islands, may be a result of competition or social interactions. (c) A random distribution pattern, as in these bushes at Leirhnjukur Volcano in Iceland, is the least common form of spacing.

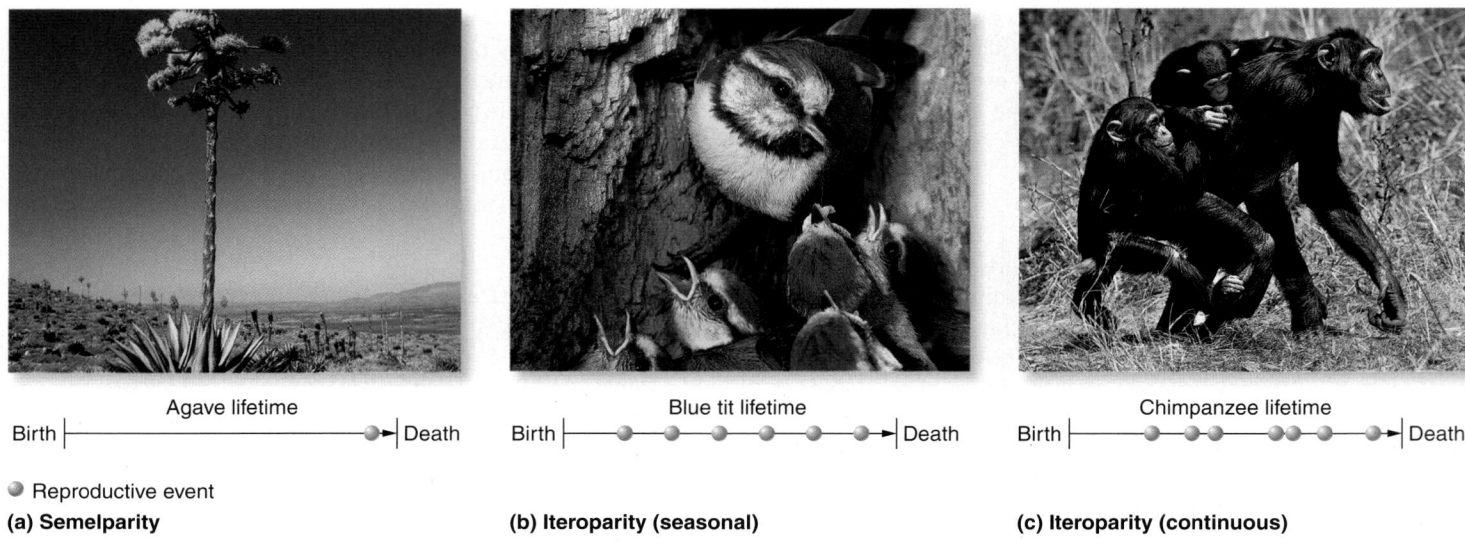

Agave lifetime
Birth |————————————►| Death
● Reproductive event
(a) Semelparity

Blue tit lifetime
Birth |——●—●—●—●—●—●►| Death
(b) Iteroparity (seasonal)

Chimpanzee lifetime
Birth |——●—●●—●●—●●—●►| Death
(c) Iteroparity (continuous)

Figure 56.3 **Differences in reproductive strategies.** Species such as **(a)** agave plants are semelparous, meaning they breed once in their lifetime and then die (century plant, *Agave shawii*, Baja, California). This contrasts with **(b)** blue tits, *Parus caeruleus*, and **(c)** chimpanzees, *Pan troglodytes*, Chimfunshi Chimp Reserve, Zambia, which are iteroparous and breed more than once in their lifetime.

Why do species reproduce in a semelparous or iteroparous mode? The answer may lie in part in environmental uncertainty. If survival of juveniles is very poor and unpredictable, then selection favors repeated reproduction and long reproductive life to increase the chance that juveniles will survive in at least some years. This is often referred to as bet-hedging. If the environment is stable, then selection favors a single act of reproduction, because the organism can devote all its energy to making offspring, not maintaining its own body. Under favorable circumstances, annuals produce more seeds than trees, which have to invest a lot of energy in maintenance. However, when the environment becomes stressful, annuals run the risk of their seeds not germinating. They must rely on some seeds successfully lying dormant and germinating after the environmental stress has ended.

Age Classes The reproductive strategy employed by an organism has a strong effect on the subsequent age classes of a population. Semelparous organisms often produce groups of same-aged young called **cohorts**, which may grow at similar rates. Iteroparous organisms generally have many young of different ages because the parents reproduce frequently. The age classes of populations can be characterized by specific categories, such as years in mammals, stages (eggs, larvae, or pupae) in insects, or size in plants.

We expect that a population increasing in size should have a large number of young, whereas a decreasing population should have few young. An imbalance in age classes can have a profound influence on a population's future. For example, in an overexploited fish population, the bigger, older reproductive age classes are often removed. If the population experiences reproductive failure for one or two years, there will be no young fish to move into the reproductive age class to replace the removed fish, and the population may collapse. Other popula-

tions experience removal of younger age classes. Where populations of white-tailed deer are high, they overgraze the vegetation and eat many young trees, leaving only older trees, whose foliage is too tall for them to reach (**Figure 56.4**). This can have disastrous effects on the future population of trees, for while the forest might consist of healthy mature trees, when these die, there will be no replacements. Removal of deer predators such as panthers and wolves often allows deer numbers to skyrocket and survivorship of young trees in forests to plummet. To accurately examine how populations grow, we need to examine and understand the demography of the population.

Life Tables and Survival Curves Summarize Survival Patterns

One way to determine how a population will grow is to examine a cohort of individuals from birth to death. For most animals and plants this involves marking a group of individuals in a population as soon as they are born or germinate and following their fate through their lifetime. For some long-lived organisms such as tortoises, elephants, or trees, this is impractical, so a snapshot approach is used, in which researchers examine the age structure of a population at one point in time. Recording the presence of juveniles and mature individuals, researchers use this information to construct a life table. A **life table** provides data on the number of individuals alive in each particular age class. Age classes can be created for any time period, but they often represent one year. Males are not usually included in these tables, since they are typically not the limiting factor in population growth.

Let's examine a life table for the North American beaver (*Castor canadensis*). Prized for their pelts, by the mid-19th century these animals had been hunted and trapped to near extinction.

Figure 56.4 **Theoretical age distribution of two forest populations. (a)** Age distribution of an undisturbed forest with numerous young trees, many of which die as the trees age and compete with one another for resources, leaving relatively few big, older trees. **(b)** Age distribution of a forest where overgrazing has reduced the abundance of young trees, leaving mainly trees in the older age classes.

(a) **Undisturbed forest**   (b) **Overgrazed forest**

Table 56.1	Life Table for the Beaver (*Castor canadensis*) in Newfoundland, Canada				
Age (years)	Number alive at start of year, n_x	Number dying during year, d_x	Proportion alive at start of year l_x	Age-specific fertility, m_x	$l_x m_x$
0–1	3,695	1,995	1.000	0.000	0
1–2	1,700	684	0.460	0.315	0.145
2–3	1,016	359	0.275	0.400	0.110
3–4	657	286	0.178	0.895	0.159
4–5	371	98	0.100	1.244	0.124
5–6	273	68	0.074	1.440	0.107
6–7	205	40	0.055	1.282	0.071
7–8	165	38	0.045	1.280	0.058
8–9	127	14	0.034	1.387	0.047
9–10	113	26	0.031	1.080	0.033
10–11	87	37	0.024	1.800	0.043
11–12	50	4	0.014	1.080	0.015
12–13	46	17	0.012	1.440	0.017
13–14	29	7	0.007	0.720	0.005
14+	22	22	0.006	0.720	0.004

Net reproductive rate, $\Sigma l_x m_x = 0.938$

Beavers began to be protected by laws in the 20th century, and populations recovered in many areas, often growing to what some considered to be nuisance status. In Newfoundland, Canada, legislation supported trapping as a management technique. From 1964 to 1971, trappers provided mandibles from which teeth were extracted for age classification. If many teeth were obtained from, say, 1-year-old beavers, then such animals were probably common in the population. If the number of teeth from 2-year-old beavers was low, then we know there was high mortality for the 1-year-old age class. From the mandible data, researchers constructed a life table (**Table 56.1**). The number of individuals alive at the start of the time period (in this case, a year) is referred to as n_x, where n is the number and x refers to the particular age class. By subtracting the value of n_x from the number alive at the start of the previous year, we can calculate the number dying in a given age class or year, d_x. Thus

$d_x = n_x - n_{x+1}$. For example, in Table 56.1, 273 beavers were alive at the start of their sixth year (n_5) and only 205 were alive at the start of the seventh year (n_6); thus, 68 died during the fifth year: $d_5 = n_5 - n_6$, or $d_5 = 273 - 205 = 68$.

A simple but informative exercise is to plot numbers of surviving individuals at each age, creating a **survivorship curve** (**Figure 56.5**). The value of n_x, the number of individuals, is typically expressed on a log scale. Ecologists use a log scale to examine rates of change with time, not change in absolute numbers. Although we could accomplish the same thing with a linear scale, the use of logs makes it easier to examine a wide range of population sizes. For example, if we start with 1,000 individuals and 500 are lost in year 1, the log of the decrease is

$$\log_{10} 1{,}000 - \log_{10} 500 = 3.0 - 2.7 = 0.3 \text{ per year}$$

If we start with 100 individuals and 50 are lost, the log of the decrease is similarly

$$\log_{10}100 - \log_{10}50 = 2.0 - 1.7 = 0.3 \text{ per year}$$

In both cases the rates of change are identical even though the absolute numbers are different. Plotting the n_x data on a log scale ensures that regardless of the size of the starting population, the rate of change of one survivorship curve can easily be compared to that of another species. The survivorship curve for the beaver follows a fairly uniform rate of death over the life span.

Survivorship curves generally fall into one of three patterns (**Figure 56.6**). In a type I curve, the rate of loss for juveniles is relatively low, and most individuals are lost later in life, as they become older and more prone to sickness and predators (see Feature Investigation that follows). Organisms that exhibit type I survivorship have relatively few offspring but invest much time and resources in raising their young. Many large mammals, including humans, exhibit type I curves. At the other end of the scale is a type III curve, in which the rate of loss for juveniles is relatively high, and the survivorship curve flattens out for those organisms that have survived early death. Many fish and marine invertebrates fit this pattern. Most of the juveniles die or are eaten, but a few reach a favorable habitat and thrive. For example, once they find a suitable rock face on which to attach themselves, barnacles grow and survive very well. Many insects and plants also fit the type III survivorship curve, because they lay many eggs or release hundreds of seeds, respectively. Type II curves represent a middle ground, with fairly uniform death rates over time. Species with type II survivorship curves include many birds, small mammals, reptiles, and some annual plants. The beaver population most closely resembles this survivorship curve. Keep in mind, however, that these are generalized curves and that few populations fit them exactly.

Figure 56.5   Survivorship curve for the North American beaver.   The survivorship curve is generated by plotting the number of surviving beavers, n_x, from any given cohort of young, usually measured on a log scale, against age.

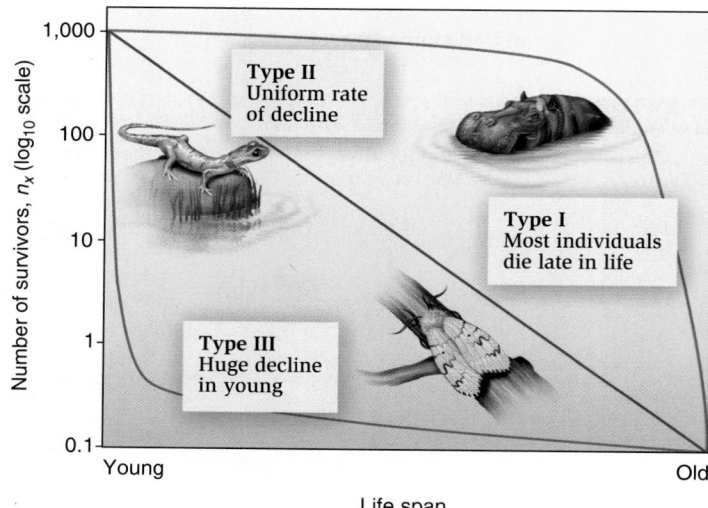

Figure 56.6   Idealized survivorship curves.

FEATURE INVESTIGATION

Murie's Collections of Dall Mountain Sheep Skulls Permitted Accurate Life Tables to Be Constructed

The Dall mountain sheep (*Ovis dalli*) lives in mountainous regions, including the Arctic and sub-Arctic regions of Alaska. In the late 1930s, the U.S. National Park Service was bombarded with public concerns that wolves were responsible for a sharp decline in the population of Dall mountain sheep in Denali National Park (then Mt. McKinley National Park). Shooting the wolves was advocated as a way of increasing the number of sheep. Because meaningful data on sheep mortality were nonexistent, the Park Service enlisted biologist Adolph Murie to collect relevant information. In addition to spending many hours observing interactions between wolves and sheep, Murie also

gathered data on sheep age at death. To do this, he collected sheep skulls in Denali National Park. He determined their age by counting annual growth rings on the horns. Thus, the analysis of skulls gave a snapshot of how old the animals were when they died.

In 1947, Edward Deevey put Murie's data in the form of a life table that listed each age class and the number of skulls in it (**Figure 56.7**). While Murie had collected 608 skulls, Deevey expressed the data per 1,000 individuals to allow for comparison with other life tables. From the data, Deevey constructed a survivorship curve. For the Dall mountain sheep, there was a slight initial decline in survivorship as young lambs were lost and then the survivorship curve flattened out, indicating that the sheep survived well through about age 7 or 8.

Figure 56.7   Examining the survivorship curve of a Dall mountain sheep population reveals information on the cause of death.

HYPOTHESIS   Culling the wolf population would protect reproductively active adults in the Dall mountain sheep population.

STARTING LOCATION   Wolf predation of sheep is commonly observed at Denali National Park (formerly known as Mt. McKinley National Park) in Alaska.

	Experimental level	Conceptual level
1	Collect sheep skulls lying on the ground.	Only skulls with horns are collected in this sampling technique.
2	Determine the age of the skulls by counting their growth rings.	Annuli are the annual growth rings used to estimate a horned animal's age.
3	Organize data into a life table (see step 4) and construct a survivorship curve using the data.	Survivorship curve for the Dall mountain sheep shows the number of sheep alive in each age class plotted against age in years.

4   THE DATA

Results used in step 3:

Age class	Number alive, n_x	$n_x \log_{10}$	Age class	Number alive, n_x	$n_x \log_{10}$
0–1	1,000	3.00	7–8	640	2.81
1–2	801	2.90	8–9	571	2.76
2–3	789	2.90	9–10	439	2.64
3–4	776	2.89	10–11	252	2.40
4–5	764	2.88	11–12	96	1.98
5–6	734	2.86	12–13	6	0.78
6–7	688	2.84	13–14	3	0.48

Then the number of sheep declined rapidly as they aged. These data underlined what Murie had previously observed, which was that wolves preyed primarily on the most vulnerable members of the sheep population, the youngest and the oldest. The Park Service ultimately ended a limited wolf control program that had been in effect since 1929. It also determined that the decline in the Dall mountain sheep had actually been precipitated by a series of cold winters that killed many sheep and weakened others, making them easier prey for the wolves, but that wolf predation *per se* was not to blame.

Age-Specific Fertility Data Can Tell Us When to Expect Growth to Occur

To calculate how a population grows, we need information on birth rates as well as mortality and survivorship rates. For any given age, we can determine the proportion of female offspring that are born to females of reproductive age. Using these data we can determine an **age-specific fertility rate**, called m_x. For example, if 100 females produce 75 female offspring, $m_x = 0.75$. With this additional information, we can calculate the growth rate of a population.

First, we use the survivorship data to find the proportion of individuals alive at the start of any given age class. This age-specific survivorship rate, termed l_x, equals n_x/n_0, where n_0 is the number alive at time 0, the start of the study, and n_x is the number alive at the beginning of age class x. Let's return to the beaver life table in Table 56.1. The proportion of the original beaver population still alive at the start of the sixth age class, l_5, equals $n_5/n_0 = 273/3{,}695$, or 0.074. This means that 7.4% of the original beaver population survived to age 5. Next we multiply the data in the two columns, l_x and m_x, for each row, to give us a column $l_x m_x$. This column represents the contribution of each age class to the overall population growth rate. An examination of the beaver age-specific fertility rates illustrates a couple of general points. First, for this beaver population in particular, and for many organisms in general, there are no babies born to young females. As females mature sexually, age-specific fertility goes up and it remains fairly high until later in life, when females reach postreproductive age.

The number of offspring born to females of any given age class depends on two things: the number of females in that age class and their age-specific fertility rate. Thus, although fertility of young beavers is very low, there are so many females in the age class that $l_x m_x$ for 1-year-olds is quite high. Age-specific fertility for older beavers is much higher, but the relatively few females in these age classes cause $l_x m_x$ to be low. Maximum values of $l_x m_x$ occur for females of an intermediate age, 3–4 years old in the case of the beaver. The overall growth rate per generation is then the number of offspring born to all females of all ages, where a generation is defined as the mean period between birth of females and birth of their offspring. Thus, to get the generational growth rate, we sum all the values of $l_x m_x$, that is, $\Sigma l_x m_x$, where the symbol Σ means "sum of." We term this summed value R_0 and refer to it as the **net reproductive rate**.

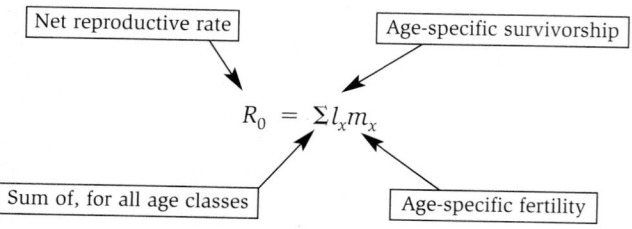

To calculate the future size of a population, we simply multiply the number of individuals in the population by the net reproductive rate. Thus, the population size in the next generation, N_{t+1}, is determined by the number in the population now, at time t, which is given by N_t, multiplied by R_0.

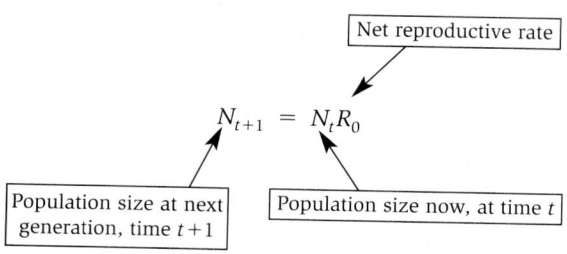

Let's consider an example in which the number of beavers alive now, N_t, is 1,000 and $R_0 = 1.1$. This means the beaver population is reproducing at a rate that is 10% greater than simply replacing itself. The size of the population next generation, N_{t+1}, is given by

$$N_{t+1} = N_t R_0$$
$$N_{t+1} = 1{,}000 \times 1.1$$
$$= 1{,}100$$

Thus, the number of beavers in the next generation is 1,100 and the population will have grown larger.

In determining population growth, much depends on the value of R_0. If $R_0 > 1$, then the population will grow. If $R_0 < 1$, the population is in decline. If $R_0 = 1$, then the population size stays the same and we say it is at **equilibrium**. In the case of the beavers, Table 56.1 reveals that $R_0 = 0.938$, which is less than 1, and therefore the population is in decline. This is valuable information, because it tells us that at that time, the beaver

population in Newfoundland needed more protection (perhaps in the form of bans on trapping and hunting) in order to attain a population level at equilibrium.

56.2 How Populations Grow

Life tables can provide us with accurate information about how populations can grow from generation to generation. However, other population growth models can provide us with valuable insights into how populations grow over shorter time periods. The most simple of these assumes that populations grow if, for any time interval, the number of births is greater than the number of deaths. In this section, we will examine two different types of these simple models. The first assumes resources are not limiting, and it results in prodigious growth. The second, and perhaps more biologically realistic, assumes resources are limiting, and it results in limits to growth and eventual stable population sizes. We then consider what other factors might limit population growth, such as natural enemies, and discuss the overall life history strategies employed by different species to enable them to exist on Earth.

Knowing the Per Capita Growth Rate Helps Predict How Populations Will Grow

The change in population size over any time period can be written as the number of births per unit time interval minus the number of deaths per unit time interval.

For example, if in a population of 1,000 deer there were 100 births and 50 deaths over the course of one year, then the population would grow in size to 1,050 the next year. We can write this formula mathematically as

$$\frac{\text{Change in numbers}}{\text{Change in time}} = \text{Births} - \text{deaths}$$

or

$$\frac{\Delta N}{\Delta t} = B - D$$

The Greek letter Δ indicates change, so that ΔN is the change in number and Δt is the change in time; B is the number of births per time unit; and D is the number of deaths per time unit.

Often, the numbers of births and deaths are expressed per individual in the population, so the birth of 100 deer to a population of 1,000 would represent a birth rate, b, of 100/1,000, or 0.10 per individual. Similarly, the death of 50 deer in a population of 1,000 would be a death rate, d, of 50/1,000, or 0.05 per individual. Now we can rewrite our equation giving the rate of change in a population.

$$\frac{\Delta N}{\Delta t} = bN - dN$$

For our deer example,

$$\frac{\Delta N}{\Delta t} = 0.10 \times 1000 - 0.05 \times 1000 = 50$$

so if $\Delta t = 1$ year, the deer population would increase by 50 individuals in a year.

Ecologists often simplify this formula by representing $b - d$ as r, the **per capita growth rate**. Thus $bN - dN$ can be written as rN. Because they are also interested in population growth rates over very short time intervals, so-called instantaneous growth rates, instead of writing

$$\frac{\Delta N}{\Delta t}$$

ecologists write

$$\frac{dN}{dt}$$

which is the notation of differential calculus. The equations essentially mean the same thing, except that dN/dt reflects very short time intervals. Thus,

$$\frac{dN}{dt} = rN = (0.10 - 0.05)N = 50$$

Exponential Growth Occurs When the Per Capita Growth Rate Remains Above Zero

How do populations grow? Clearly, much depends on the value of r. When $r < 0$, the population decreases; when $r = 0$, the population remains constant; and when $r > 0$, the population increases. When $r = 0$ the population is often referred to as being at equilibrium, where no changes in population size will occur and there is **zero population growth**.

Even if r is only fractionally above 0, population increase is rapid and when plotted graphically, a characteristic J-shaped curve results (**Figure 56.8**). We refer to this type of population growth as geometric or **exponential growth**. When conditions are optimal for the population, r is at its maximum rate and is called the **intrinsic rate of increase** (denoted r_{max}). Thus, the rate of population growth under optimal conditions is $dN/dt = r_{max} N$. Again, the larger the value of r_{max}, the steeper the slope of the curve. Because population growth depends on the value of N as well as the value of r, the population increase is even greater as time passes.

How do field data fit this simple model for exponential growth? Clearly population growth cannot go on forever, as envisioned under exponential growth. But initially at least, in a new and expanding population where resources are not limited, exponential growth is often observed. Let's look at a few examples. Tule elk (*Cervus elaphus nannodes*) are a subspecies of elk that are native to California and were isolated from other herds during the Ice Ages. Hunted nearly to extinction in the 19th

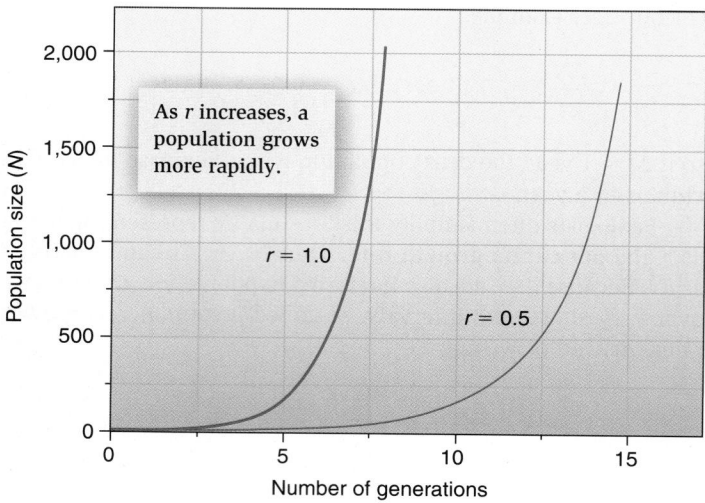

Figure 56.8 **Exponential population growth.** As the value of r increases, the slope of the curve gets steeper. In theory, a population with unlimited resources could grow indefinitely.

Figure 56.9 **Exponential growth following reintroduction of a population to a habitat.** A population of tule elk (*Cervus elaphus nannodes*) reintroduced to Point Reyes National Seashore in 1978 fits a pattern of exponential growth.

century, less than a dozen individuals survived on a private ranch. In the 20th century, reintroductions resulted in the recovery of tule elk to around 3,500 individuals. One reintroduction was made in March 1978 at Point Reyes National Seashore in California, where 10 animals—2 males and 8 females—were released. By 1993, the herd had reached 214 individuals and continued to grow in an exponential fashion until 1998, when the herd size stood at 549 (**Figure 56.9**). This was deemed an excessive number for the size of the available habitat, and animals were removed to begin herds in other locations. Since then, herd size at Point Reyes has been maintained at around 350.

The growth of some exotic species introduced into new habitats also seems to fit the pattern of exponential growth. The rapid expansion of rabbits after their introduction into South Australia in the late 19th century is a case in point. In 1859, Thomas Austin received two dozen European rabbits from England. Rabbit gestation lasts a mere 31 days, and in South Australia each doe could produce up to 10 litters of at least six young each year. The rabbits had essentially no enemies and ate the grass used by sheep and other grazing animals. Even when two-thirds of the population was shot for sport, which was the purpose of the initial introduction, the population grew into the millions in a few short years. By 1875, rabbits were reported on the west coast, having moved over 1,760 km across the continent, despite the deployment of huge, thousand-kilometer-long fences meant to contain them.

Finally, one of the most prominent examples of exponential growth is the growth of the global human population, which because of its large importance, we will examine separately later in the chapter.

Logistic Growth Occurs in Populations in Which Resources Are Limited

Despite its applicability to rapidly growing populations, the exponential growth model is not appropriate in many situations. The model assumes unlimited resources, which is not often the case in the real world. For most species, resources become limiting as populations grow. Thus the per capita growth rate decreases as resources are used up. The upper boundary for the population size is known as the **carrying capacity (K)**. Thus, a more realistic equation to explain population growth, one that takes into account the amount of available resources, is

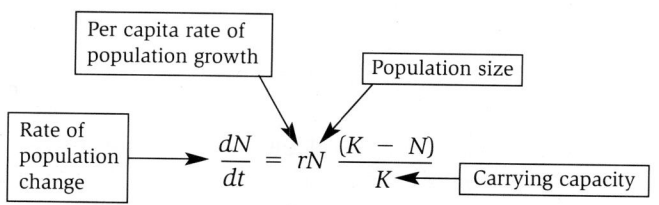

where $(K - N)/K$ represents the proportion of unused resources remaining. This equation is called the **logistic equation**.

In essence, this equation means that the larger the population size, N, the closer it becomes to the carrying capacity, K, and the fewer the available resources for population growth. At large values of N, $(K - N)/K$ becomes small and thus population growth is small. If $K = 1,000$, $N = 900$, and $r = 0.1$, then

$$\frac{dN}{dt} = (0.1)(900) \times \frac{(1{,}000 - 900)}{1{,}000}$$

$$\frac{dN}{dt} = 9$$

In this instance, population growth is nine individuals per unit of time.

At medium values of N, $(K - N)/K$ is closer to a value of 1, and population growth is larger. If $K = 1{,}000$, $N = 500$, and $r = 0.1$, then

$$\frac{dN}{dt} = (0.1)(500) \times \frac{(1{,}000 - 500)}{1{,}000}$$

$$\frac{dN}{dt} = 25$$

However, if population sizes are low ($N = 100$), even though $(K - N)/K$ is very close to 1, population sizes are so small that growth is again low.

$$\frac{dN}{dt} = (0.1)(100) \times \frac{(1{,}000 - 100)}{1{,}000}$$

$$\frac{dN}{dt} = 9$$

Thus, growth is small at high and low values of N and is greatest at immediate values of N. Growth is greatest when $N = K/2$ (try some calculations to verify this for yourself).

Let's consider how an ecologist would use the logistic equation. First, the value of K would come from intense field and laboratory work where researchers would determine the amount of resources, such as food, needed by each individual and then determine the amount of available food in the wild. Field censuses would determine N, and field censuses of births and deaths per unit time would provide r. When this type of population growth is plotted over time, an S-shaped growth curve results (**Figure 56.10**). This pattern, in which the growth of a population typically slows down as it approaches K, is called **logistic growth**.

Does the logistic growth model provide a better fit to growth patterns of plants and animals in the wild than the exponential model, which is also shown in Figure 56.10? In some instances, such as laboratory cultures of bacteria and yeasts, the logistic growth model provides a good fit (**Figure 56.11**). However, for many other populations, including those of the shrews, voles, and red squirrels shown in **Figure 56.12**, there is much variation. If the logistic model held true for these species, we would expect to see population growth leveling off, indicative of the populations' having reached their equilibrium density. However, in nature variations in temperature, rainfall, or resources cause changes in carrying capacity and thus in population densities. The uniform conditions of temperature and resource levels of the laboratory do not exist. In fact, when studying the population patterns of short-tailed shrews, researchers at the

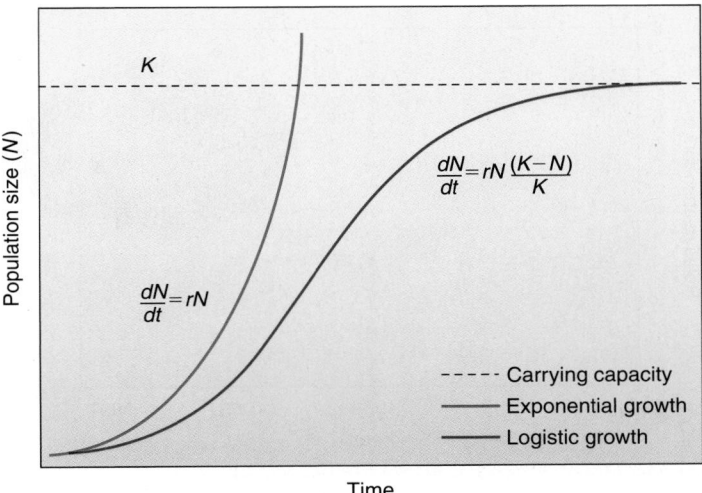

Figure 56.10 **Exponential versus logistic growth.** Exponential (J-shaped) growth occurs in an environment with unlimited resources, while logistic (S-shaped) growth occurs in an environment with limited resources.

Biological inquiry: What is the population growth per unit of time when r = 0.1, N = 200, and K = 500?

Figure 56.11 **Growth of yeast cells in culture fits the logistic growth model.** Early tests of the logistic growth curve were validated by growth of yeast cells in laboratory cultures. These populations showed the typical S-shaped growth curve.

Konza Prairie Biological Station in Kansas found a tight relationship between available soil moisture and the relative abundance of shrews (**Figure 56.13**). One hypothesis is that increased soil moisture provides more free water for drinking and an increased amount of invertebrate prey such as worms. This reminds us again of the influence of abiotic factors on population densities, as we explored in Chapter 54.

(a) Short-tailed shrews at Konza Prairie, Kansas

(b) Voles in Sweden

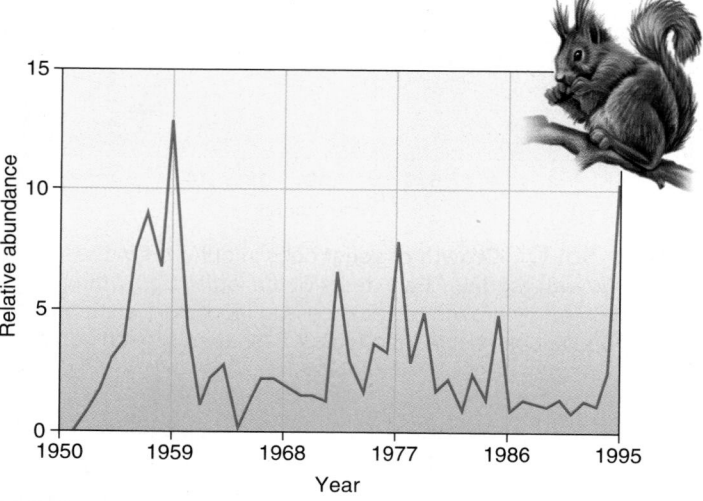

(c) Red squirrels in Ontario, Canada

Figure 56.12 **The logistic growth model does not describe all populations.** Variation in abundance of different species of small mammal populations over many years of data collection shows great variability and a lack of fit to the idealized logistic growth curve. Data reflect relative abundance in a set number of traps at different sampling points.

Is the logistic model of little value because it fails to describe population growth accurately? Not really. It is a useful starting point for thinking about how populations grow, and it seems intuitively correct. However, the carrying capacity is a difficult feature of the environment to identify for most species, and it also varies temporally, according to climatic and local weather patterns; thus, logistic growth is difficult to measure accurately.

Also, as we will discover, populations are affected by predators, parasites, or competition with other species. In Chapter 57, we will examine how natural enemies and competitors affect population densities and explore whether species interactions commonly limit population growth. However, it is instructive to first see how these natural enemies can reduce population size. Such population reductions are often caused by a process known as density dependence.

Density-Dependent Factors May Regulate Population Sizes

Factors that influence birth and death rates, and thus regulate population size, can be either density dependent or density independent. A **density-dependent factor** is a mortality factor whose influence varies with the density of the population. Parasitism, predation, and competition are some of the many density-dependent factors that may reduce the population densities of living organisms and stabilize them at equilibrium levels. Such factors can be density dependent in that their impact depends on the density of the population; they kill relatively more of a population when densities are higher and less of a population when densities are lower. For example, many predators develop a visual search image for a particular prey. When a prey is rare, predators tend to ignore it and kill relatively few. When a prey is common, predators key in on it and kill relatively more.

Figure 56.13 **Abiotic factors influence population densities.** The number of short-tailed shrews at Konza Prairie, Kansas, increases as precipitation and thus soil moisture increases because of a greater availability of drinking water and invertebrate prey. Each point refers to a year from Figure 56.12a.

In England, for example, predatory shrews kill proportionately more moth pupae in leaf litter when the pupae are common compared to when they are rare.

Density dependence can be detected by plotting mortality, expressed as a percentage, against population density (**Figure 56.14**). If a positive slope results and mortality increases with density, the factor tends to have a greater effect on dense populations than on sparse ones and is clearly acting in a density-dependent manner.

A **density-independent factor** is a mortality factor whose influence is not affected by changes in population size or density. In general, density-independent factors are physical factors, including weather, drought, freezes, floods, and disturbances such as fire. For example, in hard freezes the same proportion of organisms such as birds or plants are usually killed, no matter how large the population size. However, even physical factors such as weather can act in a density-dependent manner. For example, in an environment where there were many beavers and a limited number of rivers to dam, some individuals would not have a lodge. In such a situation, a cold winter could kill a high percentage of beavers. If, on the other hand, there were few beavers, most would have a lodge to provide them with protection from a hard freeze. In this case, the cold would kill a lower percentage.

Determining which factors act in a density-dependent fashion has large practical implications. Foresters, game managers, and conservation biologists alike are interested in learning how to maintain populations at equilibrium levels. For example, if disease were to act in a density-dependent manner on white-tailed deer, there wouldn't be much point in game managers attempting to kill off predators such as mountain lions to increase herd sizes for hunters, because proportionately more deer would be killed by disease.

Finally, a source of mortality that decreases with increasing population size is considered an **inverse density-dependent factor**. For example, if a territorial predator such as a lion always ate the same number of wildebeest prey, regardless of wildebeest density, it would be acting in an inverse density-dependent manner, because it is taking a smaller proportion of the population at higher density. Some mammalian predators, being highly territorial, often act in this manner on herbivore density. Parasites that infect 10% of a host population regardless of host density are also exhibiting inverse density dependence.

Thus, natural enemies can act in a density-dependent manner and control prey populations, or they can act in a density-independent or inverse density-dependent manner and not regulate them. In order to make generalizations about which factors control populations in nature, we need to know the frequency of density dependence in natural systems.

Which factors tend to act in a density-dependent manner? In the 1980s, a broad review of many research studies, which considered 51 populations of insects, 82 populations of large mammals, and 36 populations of small mammals and birds, showed a wide variety of density-dependent factors. No single process such as parasitism, competition, or predation could be regarded as a regulatory factor of overriding importance. Even for individual taxa, such as insects, density dependence varied from parasitism and predation to competition and abiotic factors. This finding is disconcerting to ecologists interested in species management, because it means that generalizations about which factors are likely to act in a density-dependent manner are not easily made. Each species has to be analyzed on a case-by-case basis.

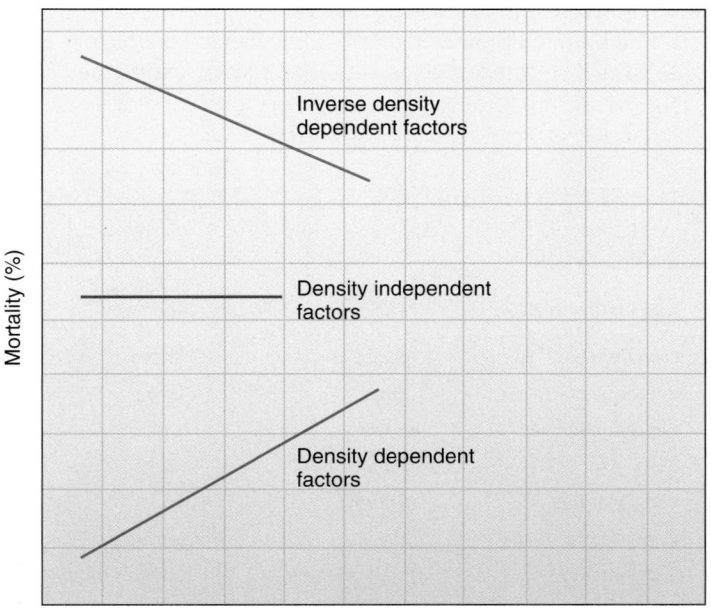

Figure 56.14 **Three ways that factors affect mortality in response to changes in population density.** For a density-dependent factor, mortality increases with population density, while for a density-independent factor, mortality remains unchanged. For an inverse-dependent factor, mortality decreases as a population increases in size.

Life History Strategies Incorporate Traits Relating to Survival and Competitive Ability

The population parameters we have discussed—including iteroparity versus semelparity, continuous versus seasonal iteroparity, and density-dependent versus density-independent factors—have important implications for how populations grow and indeed for the reproductive success of populations and species. However, these reproductive strategies can be viewed in the context of a much bigger picture of life history strategies, sets of physiological and behavioral features that incorporate not only reproductive traits but also survivorship and length of life characteristics, habitat type, and competitive ability.

Life history strategies can be considered a continuum (**Figure 56.15**). At the one end are species, termed *r*-**selected species**, that have a high rate of per capita population growth, *r*, but poor competitive ability. An example is a weed that quickly colonizes vacant habitats (such as barren land), passes through several generations, and then is outcompeted. Weeds produce huge numbers of tiny seeds and therefore have high values of r. At the other end are species, termed ***K*-selected species**, that have more or less stable populations adapted to exist at or near the carrying capacity, K, of the environment. An example is a tree that exists in a mature forest. While trees do not have a high reproductive rate, they tend to compete well and eventually displace many species by overtopping them and shading them out. The r- and K-selection continuum brings together various life history features including reproductive strategy, dispersal ability, growth rates, and population size (**Table 56.2**).

Let's return to our previous comparison of weeds versus trees. Weeds exist in disturbed habitats such as gaps in a forest canopy where trees have blown down, allowing light to penetrate to the forest floor. An r-selected species like a weed grows quickly and reaches reproductive age early, devoting much energy to producing a large number of seeds that disperse widely. These weed species remain small and do not live long, perhaps passing a few generations in the light gap before it closes.

A mature forest tree, on the other hand, often exists in undisturbed native habitats. Forest trees grow slowly and reach reproductive age late, having to devote much energy to growth and maintenance. A K-selected species like a tree grows large and shades out r-selected species like weeds, eventually outcompeting them. Such trees live a long time and produce seeds repeatedly every year when mature. These seeds are bigger than those of r-selected species; consider the acorns of oaks versus the seeds of dandelions. Acorns contain a large food reserve that helps them grow, whereas dandelion seeds must rely on whatever nutrients they can gather from the soil.

As you might expect, a trade-off exists between seed size and number—a plant can produce a few big seeds or lots of small ones. While the weed–tree example is a useful way to think about the r- and K-selection continuum, we must remember that other organisms can be r- or K-selected, too. For example, insects can be considered small, r-selected species that produce many young and have short life cycles. Mammals, such as elephants, that grow slowly, have few young, and reach large sizes are typical of K-selected species.

In a human-dominated world, almost every life history attribute of a K-selected species sets it at risk of extinction. First,

Figure 56.15 Life history strategies.   Differences in traits of a dandelion (*top*) and an oak tree (*bottom*) illustrate some of the differences between r- and K-selected species.

Table 56.2	Characteristics of r- and K-Selected Species	
Life history feature	***r*-selected species**	***K*-selected species**
Development	Rapid	Slow
Reproductive rate	High	Low
Reproductive age	Early	Late
Body size	Small	Large
Length of life	Short	Long
Competitive ability	Weak	Strong
Survivorship	High mortality of young	Low mortality of young
Population size	Variable	Fairly constant
Dispersal ability	Good	Poor
Habitat type	Disturbed	Not disturbed
Parental care	Low	High

K-selected species tend to be bigger, so they need more habitat in which to live. Florida panthers need huge tracts of land to establish their territories and hunt for deer. There is only room for about 22 panthers on publicly owned land in South Florida. Privately owned land currently supports another 50 panthers. As the amount of land shrinks through development, so does the number of panthers. Second, *K*-selected species tend to have fewer offspring and so their populations cannot recover as fast from disturbances like fire or overhunting. California condors, for example, produce only a single chick every other year. Third, *K*-selected species breed at a later age, and their generation time and time to grow from a small population to a larger population is long. Gestation time in elephants is 22 months, and elephants take at least seven years to become sexually mature. Large trees such as the giant sequoia; large terrestrial mammals like elephants, rhinoceros, and grizzly bears; and large marine mammals like blue whales and sperm whales all run the risk of extinction. Interestingly, the coast redwood seems to be an exception, a fact perhaps attributable to its unusual genome (see the following Genomes & Proteomes section).

What are the advantages to being a *K*-selected species? In a world not perturbed by humans, *K*-selected species would fare well. Being a *K*-selected species is as viable an option as being *r*-selected. However, in a human-dominated world, many *K*-selected species are selectively logged or hunted or their habitat is altered, and the resulting small population sizes make extinction a real possibility.

GENOMES & PROTEOMES

Hexaploidy Increases the Growth of Coast Redwood Trees

Besides having the world's most massive tree, the giant sequoia (*Sequoiadendron giganteum*), California is also home to the world's tallest tree, the coast redwood (*Sequoia sempervirens*), a towering giant that can grow to over 90 m and can live for up to 2,000 years. (**Figure 56.16**). These trees are currently confined to a relatively small 700-km strip along the Pacific coast from California to southern Oregon, an area characterized by moderate year-long temperatures, heavy winter rains, and dense summer fog. Interestingly, because this climate was far more common in earlier eras, these trees were once dispersed throughout the Northern Hemisphere. As of today, however, over 95% of the old-growth coast redwoods are gone, the result of extensive logging.

How is this huge species different from other tree species? In 1948, researchers made the startling discovery that the tree is a hexaploid, that is, each of its cells contains six sets of chromosomes, with 66 chromosomes in total. While hexaploidy is not unknown in grasses and shrubs, it is unusual in trees. Of all the

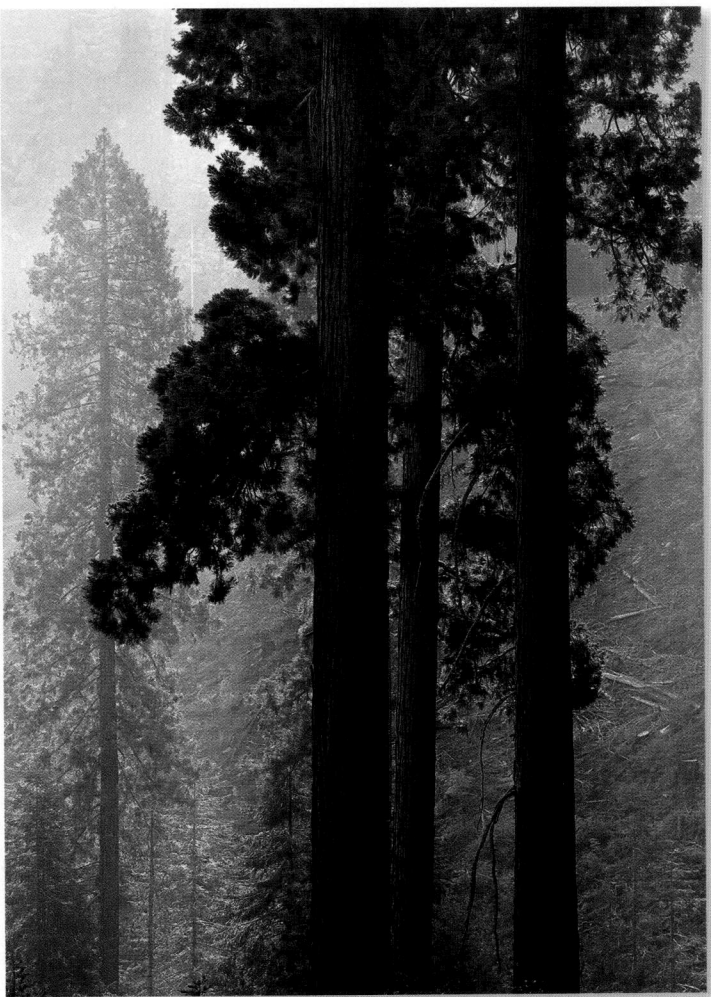

Figure 56.16   The coast redwood (*Sequoia sempervirens*) is a hexaploid conifer.   The coast redwood can grow to over 90 m, and the oldest living trees are over 2,000 years old. Their great genetic variation may help explain their incredible growth and longevity.

conifers on Earth, the coast redwood is the only hexaploid. Having this quality means each tree may have several different alleles for any given gene, which leads to a very genetically diverse population. Molecular biologist Chris Brinegar has found that hardly any two trees have exactly the same genetic constitution. Such genetic diversity allows greater adaptation to environmental conditions and more adaptations against insect or fungal pests. Indeed, living redwoods have no known lethal diseases, and pests do not cause significant damage. What's more, with six different sets of genes, trees also have the potential for great variety in their gene products, the proteins, which may help explain their prodigious growth.

56.3 Human Population Growth

In 2005, the world's population was estimated to be increasing at the rate of 153 people every minute: 2 in developed nations and 151 in less developed nations. The United Nations' 2005 projections pointed to a world population stabilizing at around 10 billion near the year 2150, as would happen with logistic growth. However, until now, human population growth has better fit an exponential growth pattern than a logistic one. In this section, we examine human population growth trends in more detail and discuss how knowledge of a population's age structure can help predict its future growth. We then investigate the carrying capacity of the Earth for humans and explore how the concept of an ecological footprint, which measures human resource use, can help us determine this carrying capacity.

Human Population Growth Fits an Exponential Pattern

Until the beginning of agriculture and the domestication of animals, about 10,000 B.C.E., the average rate of population growth was very low. With the establishment of agriculture, the world's population grew to about 300 million by 1 C.E. and to 800 million by the year 1750. Between 1750 and 1998, a relatively tiny period of human history, the world's human population surged from 800 million to 6 billion (**Figure 56.17**). In 2006, the number of humans was estimated at 6.5 billion, with two people added to the world's population every second. Considering this phenomenal increase in growth, the biggest questions remain, when will the human population level off and at what level?

Human populations can exist at equilibrium densities in one of two ways:

1. *High birth and high death rates.* Before 1750, this was often the case, with high birth rates offset by deaths from wars, famines, and epidemics.
2. *Low birth and low death rates.* In Europe, beginning in the 18th century, better health and living conditions reduced the death rate. Eventually, social changes such as increasing education for women and marriage at a later age reduced the birth rate.

The shift in birth and death rates with development is known as the **demographic transition** (**Figure 56.18**). In the first stage of the transition, birth and death rates are both high, and the population remains in equilibrium. In the second stage of this transition, which first occurred in Western Europe beginning in the late 18th century, the death rate declines first, while the birth rate remains high. High rates of population growth result. In the third stage, the birth rates drop and death rates stabilize, so that low population growth ensues. In the fourth stage, both birth and death rates are low, and the population is again at equilibrium.

The exact pace of the demographic transition between countries differs, depending on culture, economics, politics, and religion. This is illustrated by examining the demographic transition

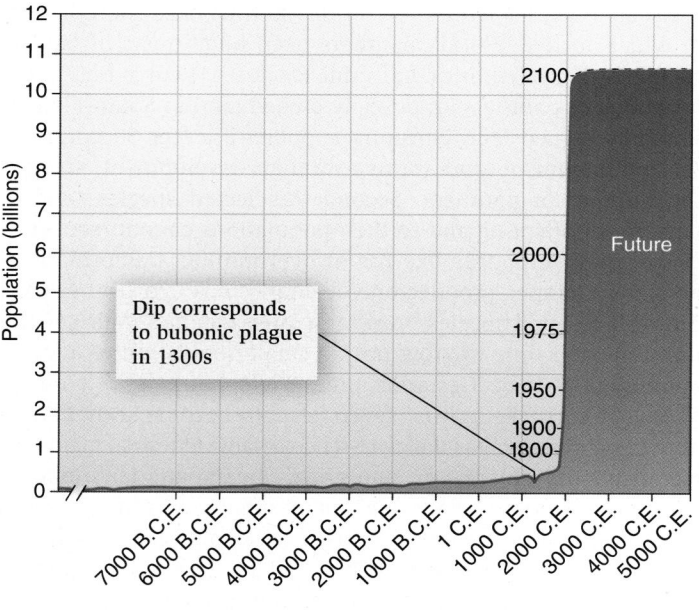

Figure 56.17 World human population growth through history illustrates an exponential growth pattern.   If and when human population growth will level off are issues of considerable debate.

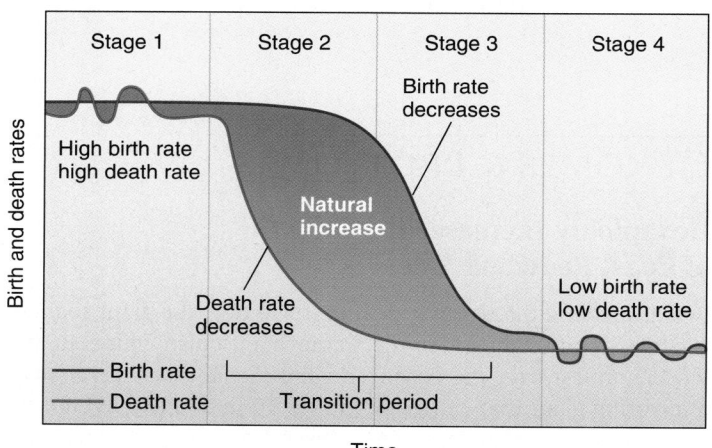

Figure 56.18   The classic stages of the demographic transition.   The difference in the birth rate and the death rate equals the rate of natural increase or decrease.

in Sweden and Mexico (**Figure 56.19**). In Mexico, the demographic transition occurred more recently and was typified by a faster decline in the death rate, reflecting rapid improvements in public health. A relatively longer lag occurred between the decline in the death rate and the decline in the birth rate, however, with the result that Mexico's population growth rate is still well above Sweden's, perhaps reflecting differences in culture or the fact that in Mexico the demographic transition is not yet complete.

Figure 56.19 **The demographic transition in Sweden and Mexico.** While the transition began earlier in Sweden than it did in Mexico, the transition was more rapid in Mexico and the overall rate of population increase remains higher. (The spike in the death rate in Mexico prior to 1920 is attributed to the turbulence surrounding the Mexican revolution.)

Knowledge of a Population's Age Structure Can Help Predict Its Future Growth

Changes in the age structure of a population also characterize the demographic transition. In all populations, **age structure** refers to the relative numbers of individuals of each defined age group. This information is commonly displayed as a population pyramid (**Figure 56.20**). In West Africa, for example, children under the age of 15 make up nearly half of the population, creating a pyramid with a wide base and narrow top. Thus, even if fertility rates decline, there will still be a huge increase in the population as these children move into childbearing age. The age structure of Western Europe is much more balanced. Even if the fertility rate of young women in Western Europe increased to a level higher than that of their mothers, the annual numbers of births would still decline because of the low number of women of childbearing age.

Predicting the Earth's Carrying Capacity for Humans Is a Difficult Task

What is the Earth's carrying capacity for humans and when will it be reached? Estimates have been quite varied. Much of the speculation on the future size of the world's population centers on lifestyle. To use a simplistic example, if everyone on the planet ate meat extensively and drove large cars, then the carrying capacity would be a lot less than if people were vegetarians and used bicycles as their main means of transportation.

Most estimates propose that the human population will grow to between 10 to 15 billion people by the end of the 21st century.

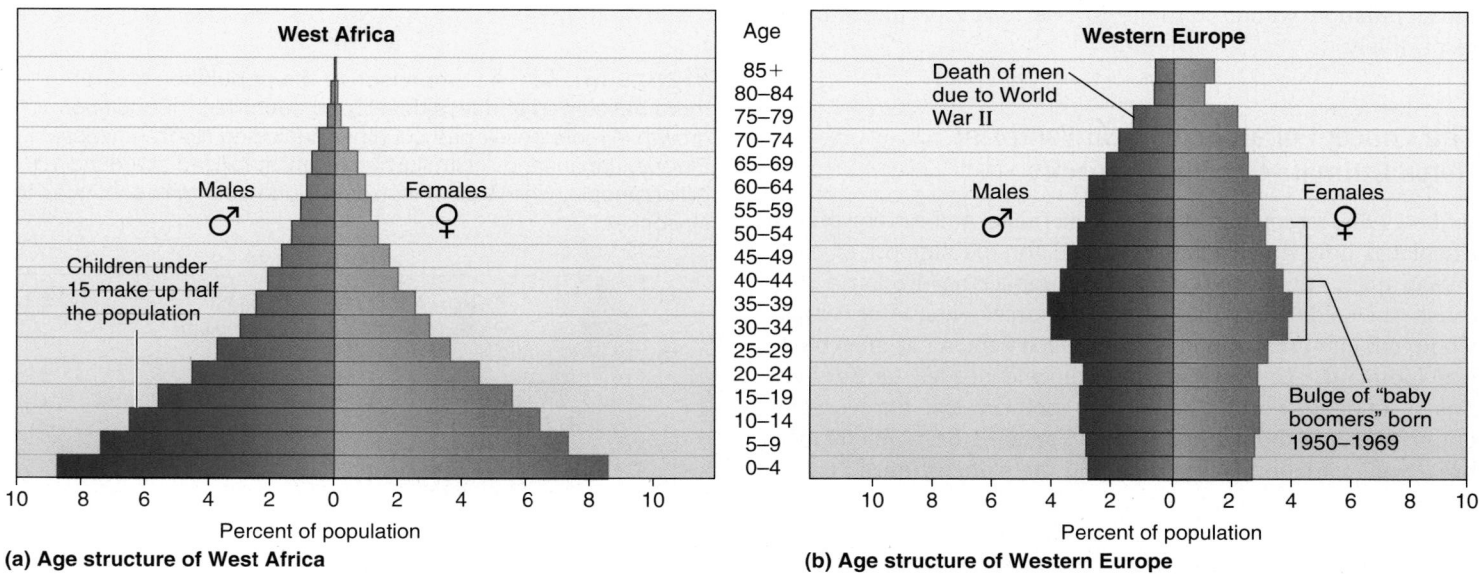

(a) Age structure of West Africa

(b) Age structure of Western Europe

Figure 56.20 **The age structure of human populations in West Africa and Western Europe, as of 2000.** (a) In developing areas of the world such as West Africa there are far more children than any other age group. (b) In the developed countries of Western Europe, the age structure is more even. The bulge represents those born in the post–World War II baby boom, when birth rates climbed due to stabilization of political and economic conditions.

Global population growth can be examined by looking at **total fertility rates (TFR)**, the average number of live births a woman has during her lifetime (**Figure 56.21**). The global fertility rate has been declining, from almost 5.0 in the 1960s to 2.9 in the late 1990s. This is still greater than the 2.1 needed for zero population growth. (The fact that the replacement rate is slightly higher than 2.0, to replace mother and father, reflects mortality before individuals reach reproductive age.) The fertility rate differs considerably between geographic areas. In Africa, the total fertility rate of 5.2 in 2003 has declined relatively little since the 1950s, when it was around 6.7 children per woman. In Latin America and Southeast Asia, the rates have declined considerably from the 1950s and are now at around 2.7 and 2.6, respectively. Sweden and most other countries in Europe and North America have a TFR of less than 2.1; in Russia, fertility rates have dropped to a low of 1.1. In China, while the TFR is only 1.8, the population there will still continue to increase until at least 2025 because of the large number of women of reproductive age.

Total fertility rates of less than 2.1 will lead to population decline over the long term. The results illustrate that in highly industrialized or developed countries, the population has nearly stabilized at a little over 1 billion people, while in less developed countries, the population is still increasing dramatically (**Figure 56.22**).

A recent United Nations report shows world population projections to the year 2050 for three different growth scenarios: low, medium, and high (**Figure 56.23**). The three scenarios are based on three different assumptions about fertility rate. Using a low fertility rate estimate of only 1.5 children per woman, the population would reach a maximum of about 7.4 billion people by 2050. A more realistic assumption may be to use the fertility rate estimate of 2.0 or even 2.5, in which case the population would continue to rise to 8.9 or 10.6 billion, respectively.

The Concept of an Ecological Footprint Helps Estimate Carrying Capacity

In the 1990s, researcher Mathis Wackernagel and his coworkers calculated how much land is needed for the support of each person on Earth. Everybody has an impact on the Earth, because they consume the land's resources, including crops, wood, oil, and other supplies. Thus, each person has an **ecological footprint**, the aggregate total of land needed for survival in a sustainable world. The average footprint size for everyone on the planet is about 2 hectares or 5 acres (1 ha = 10,000 m²), but a wide variation is found around the globe (**Figure 56.24**). The ecological footprint of the average Canadian is 7.5 hectares versus about 10 hectares for the average American. In most developed countries, the largest component of land is for energy, followed by food and then forestry. Much of the land needed for energy serves to absorb the CO_2 emitted by the use of fossil fuels. If everyone required 10 hectares, as the average American does, we would need three Earths to provide us with the needed resources. Many people in less developed countries are much more frugal in their use of resources.

Figure 56.21 Total fertility rates (TFRs) among major regions of the world. Data refer to the average number of children born to a woman during her lifetime.

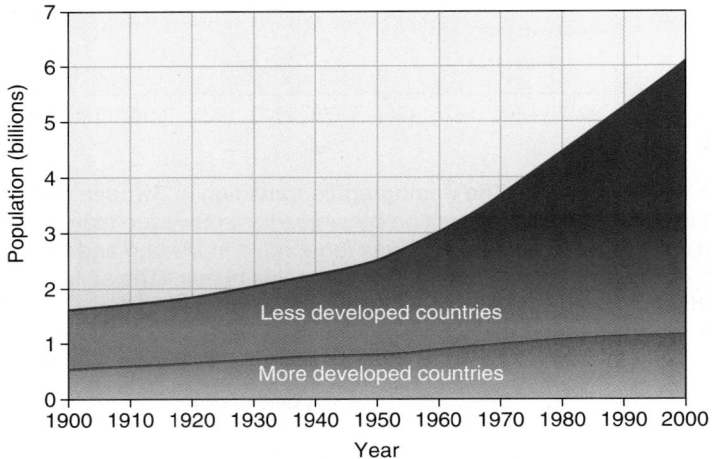

Figure 56.22 A comparison of the population growth of more developed and less developed countries. Population growth in more developed countries (including North America, Europe, Japan, and Australia) has nearly stabilized at around 1.1 billion people, while that of less developed countries continues to skyrocket.

Figure 56.23 Population predictions for 2000–2050, using three different total fertility rates (TFRs).

However, globally we are already in an ecological deficit. This is possible because many people currently live in an unsustainable manner, using supplies of nonrenewable resources, such as groundwater and fossil fuels.

What's your personal ecological footprint? Several different calculations are available on the Internet that you can use to find out. Just type "ecological footprint" (in quotes) into a search engine. A rapidly growing human population combined with an increasingly large per capita ecological footprint makes it increasingly difficult to conserve other species on the planet, a subject we will examine further in our discussion of conservation biology (Chapter 60).

Figure 56.24 Ecological footprints of different countries. The term ecological footprint refers to the amount of productive land needed to support the average individual of that country.

Biological inquiry: What is your ecological footprint?

CHAPTER SUMMARY

56.1 Understanding Populations

- Population ecology studies how populations grow and what factors promote and limit growth.
- Ecologists measure population density, the numbers of organisms in a given unit area, in many ways, including the mark-capture technique. (Figure 56.1)
- Individuals within populations show different patterns of dispersion, including clumped (the most common), uniform, and random; different reproductive strategies; and different age classes. (Figures 56.2, 56.3, 56.4)
- Life tables summarize the survival pattern of a population. (Table 56.1)
- Survivorship curves illustrate life tables by plotting the numbers of surviving individuals at different ages. (Figures 56.5, 56.6, 56.7)
- Age-specific fertility and survivorship data help determine the overall growth rate per generation, or the net reproductive rate (R_0).

56.2 How Populations Grow

- The per capita growth rate (r) helps determine how populations grow over any time period.
- When r is > 0, exponential (J-shaped) growth occurs. Exponential growth can be observed in an environment where resources are not limited. (Figures 56.8, 56.9)
- Logistic (S-shaped) growth takes into account the upper boundary for a population, called carrying capacity, and occurs in an environment where resources are limited. (Figures 56.10, 56.11)

- Variations in temperature, rainfall, or resource quantity or quality cause changes in population densities, and thus the idealized logistic growth model does not describe all populations. (Figures 56.12, 56.13)
- Density-dependent factors are mortality factors whose influence varies with population density. (Figure 56.14)
- Life history strategies are a set of features including reproductive traits, survivorship and length of life characteristics, habitat type, and competitive ability.
- Life history strategies can be viewed as a continuum, with r-selected species (those with a high rate of population growth but poor competitive ability) at one end and K-selected species (those with a lower rate of population growth but better competitive ability) at the other. (Figures 56.15, 56.16, Table 56.2)

56.3 Human Population Growth

- Up to the present, human population growth has fit an exponential growth pattern. (Figure 56.17)
- Human populations have been moving from states of high birth and death rates to low birth and death rates, a shift called the demographic transition. (Figures 56.18, 56.19)
- Differences in the age structure of a population, the numbers of individuals in each age group, are also characteristic of the demographic transition. (Figure 56.20)
- Though they have been declining worldwide, total fertility rates (TFRs) differ markedly in less developed and more developed countries. Predicting the growth of the human population depends on the total fertility rate that is projected. (Figures 56.21, 56.22, 56.23)
- The ecological footprint refers to the amount of productive land needed to support each person on Earth. Because people in many countries live in a nonsustainable manner, globally we are already in an ecological deficit. (Figure 56.24)

TEST YOURSELF

1. The number of organisms in a given unit area is termed population
   a. dispersion.
   b. dispersal.
   c. density.
   d. ecology.
   e. growth.

2. A student decides to conduct a mark-recapture experiment to estimate the population size of mosquitofish in a small pond near his home. In the first catch, he marked 45 individuals. Two weeks later he captured 62 individuals, of which 8 were marked. What is the estimated size of the population based on these data?
   a. 134
   b. 349
   c. 558
   d. 1,016
   e. 22,320

3. Which of the following factors often results in a clumped pattern of dispersion?
   a. competition over limited resources
   b. random dispersal of seed
   c. social behavior among members of a population
   d. all of the above
   e. a and c only

4. A life table usually contains information about
   a. the number of surviving individuals of a particular age class.
   b. fertility for specific age classes.
   c. dispersal patterns of a population.
   d. all of the above.
   e. a and b only.

5. _____ survivorship curves are usually associated with organisms that have high mortality rates in the early stages of life.
   a. Type I
   b. Type II
   c. Type III
   d. Types I and II
   e. Types II and III

6. If the net reproductive rate (R_0) is equal to 0.5, what assumptions can we make about the population?
   a. This population is essentially not changing in numbers.
   b. This population is in decline.
   c. This population is growing.
   d. This population is in equilibrium.
   e. None of the above.

7. The maximum number of individuals a certain area can sustain is known as
   a. the intrinsic rate of growth.
   b. the resource limit.
   c. the carrying capacity.
   d. the logistic equation.
   e. the equilibrium size.

8. Which of the following factors may alter carrying capacity over time?
   a. weather pattern changes
   b. numbers of other species that are present in the habitat
   c. deforestation
   d. soil chemistry changes
   e. all of the above

9. A mortality factor whose influence does not vary with the density of the population is known as a _____ factor.
   a. limiting
   b. density-independent
   c. resource-partitioning
   d. density-dependent
   e. inverse density-dependent

10. The amount of land necessary for survival for each person in a sustainable world is known as
    a. the sustainability level.
    b. an ecological impact.
    c. an ecological footprint.
    d. survival needs.
    e. all of the above.

CONCEPTUAL QUESTIONS

1. Define population and population ecology.
2. Describe and list the assumptions of the mark-recapture technique.
3. Describe the three types of dispersion.

EXPERIMENTAL QUESTIONS

1. What problem led to the study conducted by Murie on the Dall mountain sheep population of Denali National Park?
2. Describe the survivorship curve developed by Deevey based on Murie's data.
3. How did the Murie and Deevey data affect the decision of the Park Service on the control of the wolf population?

COLLABORATIVE QUESTIONS

1. Discuss the two main types of life history strategies.
2. Discuss the concept of demographic transition.

---

www.brookerbiology.com
This website includes answers to the Biological Inquiry questions found in the figure legends and all end-of-chapter questions.

57

SPECIES INTERACTIONS

Species interactions are numerous and varied. This larval seven-spot ladybird (*Coccinella septumpunctata*) is feeding on black bean aphids (*Aphis fabae*) which in turn feed on bean leaves (*Vicia faba*).

In this chapter we turn from considering populations on their own to investigating how they interact with populations of other species that live in the same locality. Species interactions can take a variety of forms (**Table 57.1**). **Competition** is an interaction that affects both species negatively $(-/-)$, as both species compete over food or other resources. Sometimes this interaction is quite one-sided, where it is detrimental to one species but not to the other, an interaction called **amensalism** $(-/0)$. **Predation**, **herbivory**, and **parasitism** all have a positive effect on one species and a negative effect on the other $(+/-)$. However, while predators always kill their prey, the hosts of parasites and herbivores often survive their attacks. **Mutualism** is an interaction in which both species benefit $(+/+)$, while **commensalism** benefits one species and leaves the other unaffected $(+/0)$. Last is the interaction, or rather lack of interaction, termed **neutralism**, when two species occur together but do not interact in any measurable way $(0/0)$.

Neutralism may be quite common, but few people have quantified its occurrence.

To illustrate how species interact in nature, let's consider a rabbit population in a woodland community (**Figure 57.1**). To determine what factors influence the size and density of the rabbit population, we need to understand the range of its possible species interactions. For example, the rabbit population could be limited by the quality of available food. It is also likely that other species, such as deer, use the same resource and thus compete with the rabbits. The rabbit population could be limited by predation from foxes or by parasitism in the form of the virus that causes the disease myxomatosis. It is also possible that other associations, such as mutualism or commensalism, may occur. This chapter examines each of these species interactions in turn, beginning with competition, an important interaction among species. We conclude with a discussion of conceptual models of species interactions that ecologists use when trying to determine which factors are most important in influencing population densities within ecological systems.

Table 57.1	Summary of the Types of Species Interactions	
Nature of interaction	**Species 1***	**Species 2***
Competition	−	−
Amensalism	−	0
Predation, herbivory, parasitism	+	−
Mutualism	+	+
Commensalism	+	0
Neutralism	0	0

* + = positive effect; − = negative effect; 0 = no effect.

57.1 Competition

In this section, we will see how ecologists have studied different types of competition and how they have shown that the competitive effects of one species on another can change as the environment changes or as different predators or parasites are present. Although species may compete, we will also learn how sufficient differences in lifestyle or morphology can exist that reduce the overlap in their ecological niches, thus allowing them to coexist.

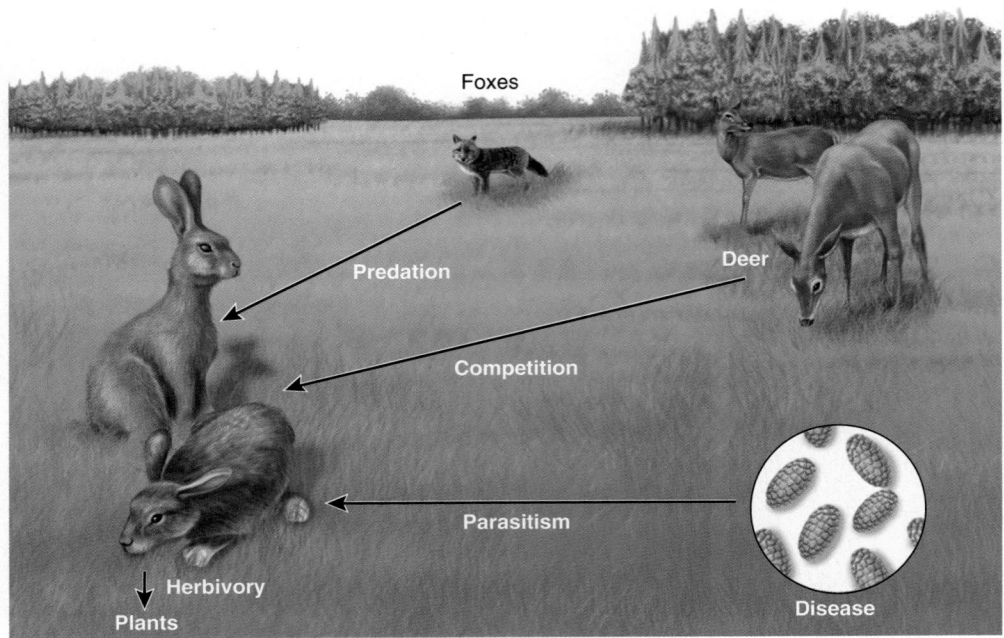

Figure 57.1 **Species interactions.** These rabbits can interact with a variety of species, experiencing competition with deer for food, predation by foxes, and parasitism from various disease-causing organisms. Understanding what factors affect the rabbit population size requires an understanding of each of these species interactions.

Several Different Types of Competition Occur in Nature

Several different types of competition are found in nature (**Figure 57.2**). Competition may be **intraspecific**, between individuals of the same species, or **interspecific**, between individuals of different species. Competition can also be characterized as resource competition or as interference competition. In **resource competition**, organisms compete indirectly through the consumption of a limited resource, with each obtaining as much as it can. For example, when fly maggots compete in a mouse carcass, not all the individuals can command enough of the resource to survive and become adult flies. In **interference competition**, individuals interact directly with one another by physical force or intimidation. Often this force is ritualized into aggressive behavior associated with territoriality (as discussed in Chapter 55). In these cases, strong individuals survive and take the best territory, and weaker ones perish or at best survive under suboptimal conditions.

Competition between species is not always equal. In fact, in many cases one species has a strong effect on another, but the reverse effect is negligible. Thus many interactions involving competition are a −/0 relationship, called **amensalism**, rather than a −/− relationship. Such extreme asymmetric competition can also be observed between plants, where one species produces and secretes chemicals from its roots that inhibit the growth of another species. In Chapter 54's Feature Investigation, we saw how diffuse knapweed, an exotic species, secretes root chemicals called allelochemicals that kill the roots of native grass species. This phenomenon is termed **allelopathy**. The action of penicillin, which directly inhibits the growth of bacteria and other fungi, is a classic case of allelopathy.

Intraspecific competition between individuals of the same species

Resource competition: Each caterpillar chews as much leaf as it can.

Interference competition: Each caterpillar physically intimidates the other.

Resource competition: Aphids and caterpillars compete for common resources.

Interference competition: Beetle and caterpillar use physical intimidation.

Caterpillar chewing leaf

Aphid sucking leaf sap

Interspecific competition between different species

Figure 57.2 **The different types of competition in nature.**

Researchers have established that one of the best methods of studying competition between two species is to temporarily remove one of them and examine the effect on the other species. A now-classic example of this method involved a study of the interactions between two species of barnacles conducted on the west coast of Scotland, as described next.

FEATURE INVESTIGATION

Connell's Experiments with Barnacle Species Showed That One Species Can Competitively Exclude Another in a Natural Setting

The most direct method of assessing the effect of competition is to remove individuals of species A and measure the response of species B. Often, however, such manipulations are difficult to conduct outside the laboratory. If individuals of species A are removed, what is to stop them from migrating back into the area of removal?

In 1954, ecologist Joseph Connell conducted an experiment that overcame this problem. *Chthamalus stellatus* and *Semibalanus balanoides* (formerly known as *Balanus balanoides*) are two species of barnacles that dominate the Scottish coastline. Each organism's **niche**, or physical distribution and ecological role, on the intertidal zone was well defined, with *Chthamalus* occurring in the upper intertidal zone, and *Semibalanus* restricted to the lower intertidal zone. Connell sought to determine what the range of *Chthamalus* adults might be in the absence of competition from *Semibalanus* (**Figure 57.3**).

To do this, Connell obtained rocks from high on the rock face, just below the high-tide level, where only *Chthamalus* grew. These rocks already contained young and mature *Chthamalus*. He then moved the rocks into the *Semibalanus* zone, fastened them down with screws, and allowed *Semibalanus* to also colonize them. Once *Semibalanus* had colonized these rocks, he took the rocks out, removed all the *Semibalanus* organisms from one side of the rocks with a needle, and then returned the rocks to the lower intertidal zone, screwing them down once again. As seen in the data, the mortality of *Chthamalus* on rock halves with *Semibalanus* was fairly high. On the *Semibalanus*-free halves, however, *Chthamalus* survived well.

Figure 57.3   Connell's experimental manipulation of species indicated the presence of competition.

HYPOTHESIS Adult *Chthamalus stellatus* were being competitively excluded from the lower intertidal zone by the species *Semibalanus balanoides*.

STARTING LOCATION Two species of barnacles, *Chthamalus stellatus* and *Semibalanus balanoides*, grow in the intertidal zone of the rocky shores of the Scottish coast.

Experimental level

1   Transfer rocks containing young and mature *Chthamalus* from the upper intertidal zone to the lower intertidal zone, and fasten them down in the new location with screws.

2   Allow *Semibalanus* to colonize rocks.

3   After colonization period is over, remove *Semibalanus* from half of each rock with a needle (leaving the other half undisturbed). Return the rocks to the lower intertidal zone and fasten them down once again.

4   Monitor survival of *Chthamalus* on both sides of rocks.

Chthamalus grows on the side where *Semibalanus* has been removed, indicating that *Semibalanus* may exclude *Chthamalus* from certain habitats.

5   **THE DATA**

Rock No.	Side of rock	% *Chthamalus* mortality over one year	
		Young barnacles	**Mature barnacles**
13b	*Semibalanus* removed	35	0
	Undisturbed	90	31
12a	*Semibalanus* removed	44	37
	Undisturbed	95	71
14a	*Semibalanus* removed	40	36
	Undisturbed	86	75

Connell also monitored survival of natural patches of both barnacle species where both occurred on the intertidal zone at the upper margin of the *Semibalanus* distribution. In a period of unusually low tides and warm weather, when no water reached any barnacles for several days, desiccation became a real threat to the barnacles' survival. During this time, young *Semibalanus* suffered a 92% mortality rate and older individuals a 51% mortality rate. At the same time, young *Chthamalus* experienced a 62% mortality rate compared with a rate of only 2% for more resistant older individuals. Clearly, *Semibalanus* is not as resistant to desiccation as *Chthamalus* and thus could not survive in the upper intertidal zone where *Chthamalus* occurs. *Chthamalus* is more resistant to desiccation than *Semibalanus* and thus can be found higher in the intertidal zone.

Thus, while the potential distribution (the fundamental niche) of *Chthamalus* extends over the entire intertidal zone, its actual distribution (the realized niche) is restricted to the upper zone. Connell's experiments were among the first to show that, in a natural environment, one species can actually outcompete another, affecting its distribution within a habitat. Interestingly, 30 years later, in the 1980s, Connell repeated his competition experiments in the same area of the Scottish coast and once more observed strong evidence for competition.

The Outcome of Competition Can Vary with Changes in the Abiotic and Biotic Environments

Using experiments to temporarily remove individuals of one species and examine the results on the remaining species, as Connell did, is often the most direct method to investigate the effects of competition. It is especially valuable to do such manipulations in the field because organisms can then also interact with all other organisms in their environment, or, as ecologists say, natural variation can be factored in. However, in laboratory experiments, the investigator can often control and vary important factors systematically.

In the late 1940s, biologist Thomas Park began a series of experiments examining competition between two flour beetles, *Tribolium castaneum* and *Tribolium confusum*, in which he systematically varied temperature, moisture, and a variety of other factors. These beetles were well suited to study in the laboratory, since large colonies could be grown in relatively small containers containing dry food medium. Thus, many replicates of each experiment were possible to confirm that results were consistent.

Park conducted the experiments by putting the same number of beetles of both species into a container and counting the number of each that were still alive after a given time interval. *T. confusum* usually won, but in initial experiments, the beetle cultures were infested with a protozoan parasite called *Adelina triboli* that killed some beetles and preferentially killed *T. castaneum* individuals. In these experiments, *T. confusum* won in 66 out of 74 replicates (89%) because it was more resistant to the parasite. In subsequent experiments, the parasite was removed and *T. castaneum* won in 12 out of 18 replicates (67%). Two things were evident from this experiment. First, the presence of a parasite (a biotic factor) was shown to alter the outcome of competition. Second, with or without the parasite, there was no absolute victor. For example, even with the parasite present, sometimes *T. castaneum* won. Thus, some random variation, which we call stochasticity, was evident.

Park then began to vary the abiotic environment and found that competitive ability was greatly influenced by climate (**Figure 57.4**). Generally, *T. confusum* did better in dry conditions, and *T. castaneum*, in moist conditions. However, *T. confusum* also won in cold-wet conditions. Once again, some stochasticity occurred, and victory was not always absolute. Later, it was found that the mechanism of competition was largely predation of eggs and pupae by larvae and adults, which, as they ate the flour medium, would also bite the stationary eggs and pupae, killing them. In general, the species were mutually antagonistic, that is, they bit more eggs and pupae of the other species than they did of their own.

In summary, Park's important series of experiments illustrated that the results of competition could vary as a function of at least three factors: parasites, temperature, and moisture. The experiments also showed how much stochasticity occurred, even in controlled laboratory conditions. But what of systems in nature, where far more variability exists? Is competition a common occurrence? One line of evidence shows that competition is frequent in nature and a strong enough force to cause the extinction of one of two competing species that share the same niche. Researchers have proposed several mechanisms by which two competing species can coexist. One states that similar species can coexist if they occupy different niches. Another is that species may occupy similar niches but undergo physical changes in form, termed character displacement, that allow them to utilize different resources. Let's explore each of these concepts in more detail.

Figure 57.4 Influence of abiotic factors on competition between *T. castaneum* and *T. confusum*. Results of competition between the flour beetles *Tribolium castaneum* and *Tribolium confusum* show that each species usually performs better in a given habitat; for example, *T. confusum* does better in dry conditions.

Field Studies Show Competition Occurs Frequently in Nature

By reviewing studies that have investigated competition in nature, we can see how frequently it occurs and in what particular circumstances it is most important. In one 1983 review by Joseph Connell, competition was found in 55% of 215 species surveyed, demonstrating that it is indeed frequent in nature. Generally in studies of single pairs of species utilizing the same resource, competition is almost always reported (90%), whereas in studies involving more species, the frequency of competition drops to 50%. Why should this be the case? Imagine a resource such as a series of different-sized grains with four species—ants, beetles, mice, and birds—feeding on it. The ants feed on the smallest grain, the beetles and mice on the intermediate sizes, and the birds, on the largest. If only adjacent species competed with each other, competition would be expected only between the ant-beetle, beetle-mouse, and mouse-bird. Thus, competition would be found in only three out of the six possible species pairs (50%) (**Figure 57.5**). Naturally, the percent would vary according to the number of species on the axis. If there were only three species along the axis, we would expect competition in two of the three pairs (67%).

Some other general patterns were evident from Connell's review. Plants showed a high degree of competition, perhaps because they are rooted in the ground and cannot easily escape or perhaps because they are competing for the same set of limiting nutrients—water, light, and minerals. Marine organisms tended to compete more than terrestrial ones, perhaps because many of the species studied lived in the intertidal zone and were attached to the rock face, in a manner similar to that of plants. Because the area of the rock face is limited, competition for space is quite important. However, even though we see competition frequently in nature, we often see apparent competitors occupying the same general areas. How is this possible?

Species May Coexist if They Do Not Occupy Identical Niches

In 1934 the Russian microbiologist Georgyi Gause began to study competition between three protist species, *Paramecium aurelia*, *Paramecium bursaria*, and *Paramecium caudatum*, all of which fed on bacteria and yeast, which in turn fed on an oatmeal medium in a culture tube in the laboratory. The bacteria occurred more in the oxygen-rich upper part of the culture tube, and the yeast in the oxygen-poor lower part of the tube. Because each species was a slightly different size, Gause calculated population growth as a combination of numbers of individuals per milliliter of solution multiplied by their unit volume to give a population volume for each species. When grown separately, population volume of all three *Paramecium* species followed a logistic growth pattern (**Figure 57.6a**). When Gause cultured *P. caudatum* and *P. aurelia* together, *P. caudatum* went extinct (**Figure 57.6b**). Both species utilized bacteria as food but *P. aurelia* grew at a rate six times faster than *P. caudatum* and was better able to convert the food into offspring.

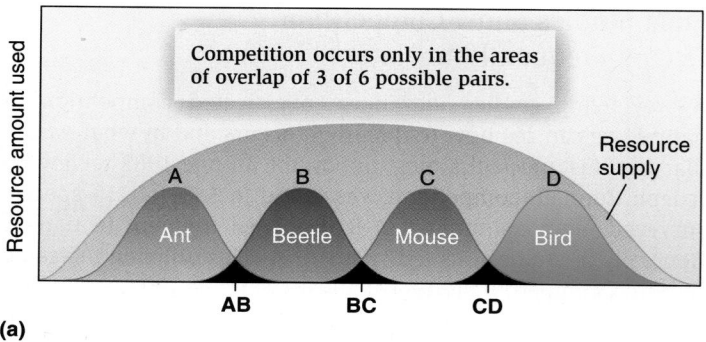

(a)

Figure 57.5 **The frequency of competition according to the number of species involved.** **(a)** Resource supply and utilization curves of four species, A, B, C, and D, along the spectrum of a hypothetical resource such as grain size. If competition occurs only between species with adjacent resource utilization curves, competition would be expected between three of the six possible pairings: A and B; B and C; and C and D. **(b)** When only two species utilize a resource set, competition would nearly always be expected between them.

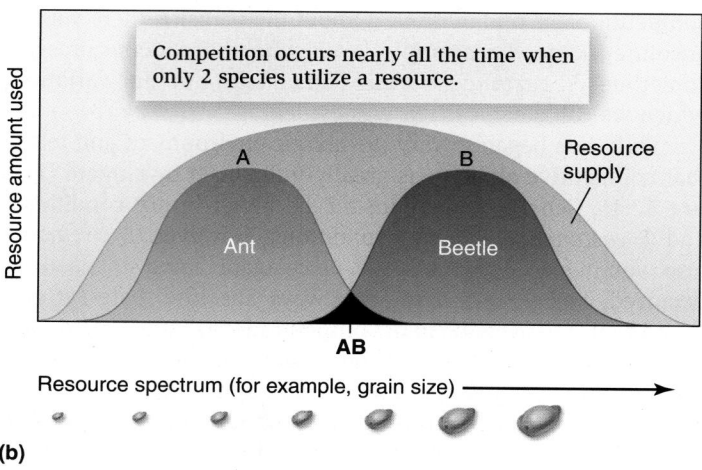

(b)

Biological inquiry: If five species utilized the resource set in part (a), what would be the expected frequency of competition observed?

(a)

(b)

Figure 57.6 **Competition among *Paramecium* species.** **(a)** Each of three species, *Paramecium aurelia*, *Paramecium bursaria*, and *Paramecium caudatum*, grows according to the logistic model when grown alone. **(b)** When *P. aurelia* is grown with *P. caudatum* (left), the density of *P. aurelia* is lowered compared to when grown alone, and *P. caudatum* goes extinct. When *P. aurelia* is grown with *P. bursaria* (right), the population densities of both are lowered but they coexist.

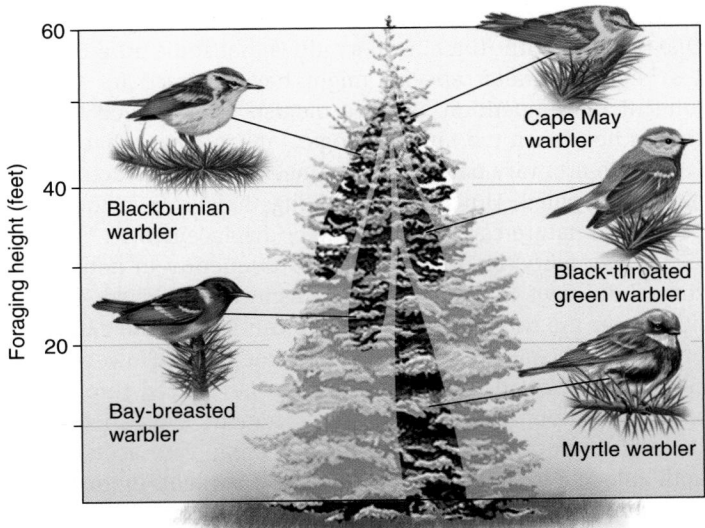

Figure 57.7 Resource partitioning among five species of warblers feeding in North American spruce trees. Each warbler species prefers to feed at a different height and portion of the tree, thus reducing competition.

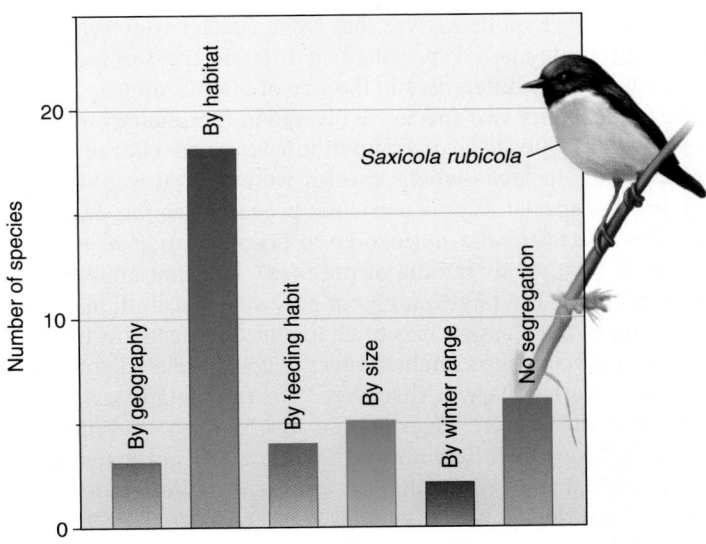

Figure 57.8 Segregation according to resource factor among 40 species of passerine birds. Most segregation is by habitat, followed by size, feeding habit, geography, and type of winter range they forage in. In about 15% of cases, no obvious segregation was observed. More than half of all bird species, including *Saxicola rubicola*, are passerines, also known as perching birds.

However, when Gause cultured *P. caudatum* and *P. bursaria* together, neither went extinct. The population volumes of each was much less compared to when they were grown alone, because some competition occurred between them. Gause discovered, however, that *P. bursaria* was better able to utilize the yeast in the lower part of the culture tubes. From these experiments Gause concluded that two species with exactly the same requirements cannot live together in the same place and use the same resources, that is, occupy the same niche. His conclusion, later termed the **competitive exclusion hypothesis**, essentially means that complete competitors cannot coexist.

If complete competitors drive one another to extinction, how different do they have to be to coexist and in what features do they usually differ? To address such questions, in 1958 ecologist Robert MacArthur examined coexistence between five species of warblers feeding within spruce trees in New England. All belonged to the genus *Dendroica*, so one would expect these closely related bird species to compete strongly, possibly sufficiently strongly to cause extinctions. MacArthur found that the species occupied different heights and portions in the tree and thus each probably fed on a different range of insects (**Figure 57.7**). In addition, the Cape May warbler exists usually at low densities but can respond quickly to periodic outbreaks of insects that occur. It thus gains a temporary advantage in numbers over the other species during these times.

The term **resource partitioning** describes the differentiation of niches, both in space and time, that enables similar species to coexist in a community, just as *Chthamalus* and *Semibalanus* live on different parts of the intertidal zone in Scotland. In a way, we can think of resource partitioning as reflecting the results of past competition. British ornithologist David Lack examined competition and coexistence among about 40 species

of British passerines, or perching birds (**Figure 57.8**). As a group, these perching birds had fairly similar lifestyles. Most segregated according to some resource factor, with habitat being the most common one. For example, while all the passerines fed on insects, some would feed exclusively in grasslands, others in forests, some low to the ground, and others high in trees, where the insects present would likely be different. Birds also segregated by size, so that bigger species would take different-sized food than smaller species, and by feeding habit, with some feeding on insects on trees, others on tree trunks, and so on. Some species also fed in different winter ranges, while others occurred in different parts of the country (separation by geography). About 15% of bird species showed no segregation at all.

What about the species that do not appear to live in different habitats or have different food habits? To answer this question, researchers have looked at differences in physical form, or morphology, between coexisting species.

Morphological Differences May Allow Species to Coexist

Although the competitive exclusion principle acknowledges that complete competitors cannot coexist, some partial level of competition may exist that is not severe enough to drive one of the competitors to extinction. In 1959, biologist G. Evelyn Hutchinson looked at size differences in feeding apparatus, such as mouthparts or other body parts important in feeding, between species when they were **sympatric** (occurring in the same geographic area) and **allopatric** (occurring in different geographic areas).

Hutchinson's hypothesis was that when species were sympatric, each species tended to specialize on different types of food. This was reflected by differences in the size of their feeding apparatus. The tendency for two species to diverge in morphology and thus resource use because of competition is called **character displacement**. In areas where species were allopatric, there was no need to specialize on a particular prey type, so the size of the feeding apparatus did not evolve to become larger or smaller; rather it retained a "middle of the road" size that allowed species to exploit the largest range of prey size distribution.

One of the classic cases of character displacement involves a study of Galápagos finches, several closely related species of finches Charles Darwin discovered on the Galápagos Islands (see also the Feature Investigation in Chapter 23). When two species, *Geospiza fuliginosa* and *Geospiza fortis*, are sympatric, their beak sizes (bill depths) are different. *G. fuliginosa* has a smaller bill depth, which enables it to crack small seeds more efficiently, while *G. fortis* has a larger bill depth, which enables it to feed on bigger seeds. However, when both species are allopatric, that is, existing on different islands, their bills are more similar in depth. Researchers studying *Geospiza* concluded that the bill depth differences evolved in ways that minimized competition. But how great do differences between characters have to be in order to permit coexistence?

Hutchinson noted that the ratio between feeding characters when species were sympatric (and thus competed) averaged about 1.3 (**Table 57.2**). In contrast, the ratio between feeding characters when species were allopatric (and did not compete) was closer to 1.0. Hutchinson proposed that the value of 1.3, a roughly 30% difference, could be used as an indication of the amount of difference necessary to permit two species to coexist. One problem with Hutchinson's ratio is that some differences of 1.3 between similar species might have evolved for reasons other than competition. Some ecologists have argued that we should not attach too much biological meaning to what may in fact be an arbitrary pattern. Nevertheless, while the actual ratio may be disputed, Hutchinson's findings have shown that competition in nature can cause character displacement.

In summary, competition occurs frequently in nature and several different competitive mechanisms are involved. Competition over the same limited resource can lead to the extinction of one of two species occupying the same niche. However, species may coexist if their niche overlap is reduced through resource partitioning or character displacement. We also saw that the outcome of competitive interactions may vary according to both abiotic and biotic factors in the environment. Biotic factors such as parasitism, herbivory, and predation may be powerful mortality factors in their own right and could be sufficient to reduce population densities to the extent that competition does not occur. In the following sections we will examine the strength of these factors, beginning with predation.

57.2 Predation, Herbivory, and Parasitism

Several interactions between species have a positive effect for one species and a negative effect for the other. The traditional view of predation is one of carnivory, in which a predator, typically an animal, feeds on another animal. However, herbivory and parasitism are also interactions in which one species is helped and the other is harmed, and while they can be viewed in the broad context of predation, each has special characteristics that set it apart. Herbivory usually involves nonlethal predation on plants, while predation generally results in the death of the prey. Parasitism, like herbivory, is typically nonlethal, and differs from predation in that the adult parasite typically lives and reproduces in the living host, as in Chinese liver flukes (refer back to Figure 33.8). Parasitoids, insects that lay eggs in living hosts, have features in common with both predators and parasites. They always kill their hosts, as predators do, but one host can support the development of one or more than one parasitoid, just as in parasitism. Parasitoids are common in the insect world and include parasitic wasps and flies that feed on many other insects or spiders. Omnivores are individuals that can feed on both plants and animals. For example, bears feed on both berries and fish. These varied categories of predation can be classified according to how lethal they are for the prey and how close is the association between predator and prey (**Figure 57.9**).

In this section we begin by looking at antipredator strategies and how, despite such strategies, predation remains a factor affecting the density of prey. We survey the strategies plants use to deter herbivores and how, in turn, herbivores overcome host plant defenses. Finally, we investigate parasitism, which may be the predominant lifestyle on Earth, and explore the growing role of genomics in the fight against parasites.

Table 57.2	Comparison of Feeding Characters of Sympatric and Allopatric Species				
		Measurement (mm) when		Ratio* when	
Animal (character)	Species	Sympatric	Allopatric	Sympatric	Allopatric
Weasels (skull)	*Mustela erminea*	50.4	46.0	1.28	1.07
	Mustela nivalis	39.3	42.9		
Mice (skull)	*Apodemus flavicollis*	27.0	26.7	1.09	1.04
	Apodemus sylvaticus	24.8	25.6		
Nuthatches (beak)	*Sitta tephronota*	29.0	25.5	1.23	1.02
	Sitta neumayer	23.5	26.0		
Galápagos finches (beak)	*Geospiza fortis*	12.0	10.5	1.43	1.13
	Geospiza fuliginosa	8.4	9.3		

*Ratio of the larger to smaller character.

Figure 57.9 **Possible interactions between populations.** Lethality represents the probability that an interaction results in the death of the prey. Intimacy represents the closeness and duration of the relationship between the consumer and the prey.

Animals Have Evolved Many Antipredator Strategies

The variety of strategies that animals have evolved to avoid being eaten suggests that predation is a strong selective force.

Chemical Defense A great many species have evolved chemical defenses against predation. One of the classic examples of a chemical defense involves the bombardier beetle (*Stenaptinus insignis*), which has been studied by Tom Eisner and coworkers. These beetles possess reservoirs of hydroquinone and hydrogen peroxide in their abdomen. When threatened, they eject the chemicals into an "explosion chamber," where the subsequent release of oxygen causes the whole mixture to be violently ejected as a hot spray (about 88°C, or 190°F) that can be directed at the beetle's attackers (**Figure 57.10a**). Many other arthropods, such as millipedes, have chemical sprays too, and the phenomenon is also found in vertebrates, as anyone who has had a close encounter with a skunk can testify.

Often associated with a chemical defense is an **aposematic coloration**, or warning coloration, which advertises an organism's unpalatable taste. For instance, the ladybird beetle's bright red color warns of the toxic defensive chemicals it exudes when threatened, and many tropical frogs have bright warning coloration that calls attention to their skin's lethality (**Figure 57.10b**). In the 1960s, entomologist Lincoln Brower and coworkers showed that after inexperienced blue jays ate a monarch butterfly and suffered a violent vomiting reaction, they learned to associate the striking orange-and-black appearance of the butterfly with a noxious reaction (refer back to Figure 55.3). While some ani-

mals synthesize toxins using their own metabolic processes such as a rattlesnake, other animal poisons are actually acquired from plants. For example, the monarch butterfly caterpillars accumulate toxic chemicals called cardiac glycosides from milkweed species.

Cryptic Coloration **Cryptic coloration** is an aspect of camouflage, the blending of an organism with the background of its habitat. Cryptic coloration is a common method of avoiding detection by predators. For example, many grasshoppers are green and blend in with the foliage on which they feed. Stick insects mimic branches and twigs with their long, slender bodies. In most cases, these animals stay perfectly still when threatened, because movement alerts a predator. Behavior that allows an animal to blend into its environment is referred to as **catalepsies**. Cryptic coloration is prevalent in the vertebrate world, too. Many seahorses adopt a body shape and color pattern that are similar to the habitats in which they are found (**Figure 57.10c**).

Mimicry **Mimicry**, the resemblance of an organism (the mimic) to another organism (the model), also secures protection from predators. There are two major types of mimicry. In **Müllerian mimicry**, many noxious species converge to look the same, thus reinforcing the basic distasteful design. One example is the black and yellow striped bands of several different types of bees and wasps. Müllerian mimicry is also found among noxious Amazonian butterflies. **Batesian mimicry** is the mimicry of an unpalatable species (the model) by a palatable one (the mimic). Some of the best examples involve flies, especially hoverflies of the family Syrphidae, which are striped black and yellow to resemble stinging bees and wasps but are themselves harmless. The nonvenomous scarlet king snake (*Lampropeltis triangulum*) mimics the venomous coral snake (*Micrurus nigrocinctus*), thereby gaining protection from would-be predators (**Figure 57.10d**).

It is sometimes difficult to distinguish between Batesian and Müllerian mimics. Until recently, the viceroy butterfly was believed to be an example of Batesian mimicry because of its striking resemblance to the unpalatable monarch. However, in 1991, researcher David Ritland, a graduate student at the time, showed that the viceroy was equally as unpalatable as the monarch and is actually an example of Müllerian mimicry. In theory, Batesian mimics should lose their protection when they are common, because they need plenty of distasteful models whose unpalatability maintains the validity of their warning colors. If the edible mimic became too common, predators would learn to ignore the coloration, which would also harm the model.

Displays of Intimidation Some animals put on displays of intimidation in an attempt to discourage predators. For example, a toad swallows air to make itself appear larger, frilled lizards extend their collars when frightened to create this same effect, and porcupine fish inflate themselves to large proportions when threatened (**Figure 57.10e**). All of these animals use displays to deceive potential predators about the ease with which they can be eaten.

(a) As it is held by a tether attached to its back, this bombardier beetle (*Stenaptinus insignis*) directs its hot, stinging spray at a forceps "attacker."

(b) Aposematic coloration advertises the poisonous nature of this blue poison arrow frog (*Dendrobates azureus*) from South America.

(c) Cryptic coloration allows this Pygmy sea horse (*Hippocampus bargibanti*) from Bali, to blend in with its background.

(d) In this example of Batesian mimicry, an innocuous scarlet king snake (*Lampropeltis triangulum*) (*left*) mimics the poisonous coral snake (*Micrurus nigrocinctus*) (*right*).

(e) In a display of intimidation, this porcupine fish (*Diodon hystrix*) puffs itself up to look threatening to its predators.

Figure 57.10   Antipredator adaptations.

Biological inquiry: According to the classification of species interactions in Table 57.1, how would you classify Batesian and Müllerian mimicry?

Fighting Many animals have developed horns and antlers, which although primarily used in competition over mates can be used in defense against predators (refer back to Figure 34.26c). Invertebrate species often have powerful claws, pincers, or, in the case of scorpions, venomous stingers that can be used in defense as well as offense.

Agility Some groups of insects, such as grasshoppers, have a powerful jumping ability to escape the clutches of predators. Many frogs are prodigious jumpers, and flying fish can glide above the water to escape their pursuers.

Armor The shells of tortoises and turtles are a strong means of defense against most predators, as are the quills of porcupines (refer back to Figure 34.24c). Many beetles have a tough exoskeleton that protects them from attack from other arthropod predators such as spiders.

Predator satiation **Predator satiation** is the synchronous production of many progeny by all individuals in a population to satiate predators and thereby allow some progeny to survive. Predator satiation is commonly called masting when discussed in relation to seed herbivory in trees, which tend to have years of unusually high seed production that reduces predation. However,

a similar phenomenon is exhibited by the emergence of 13-year and 17-year periodical cicadas (*Magicicada* spp.). These insects are termed periodical because the emergence of adults is highly synchronized to occur once every 13 or 17 years. Adult cicadas live for only a few weeks, during which time females mate and deposit eggs on the twigs of trees. The eggs hatch 6–10 weeks later, and the nymphs drop to the ground and begin a long subterranean development, feeding on the contents of the xylem of roots. Because the xylem is low in nutrients, it takes many years for nymphs to develop, though there appears to be no physiological reason why some cicadas couldn't emerge after, say, 12 years of feeding, and others after 14. Their synchrony of emergence is thought to maximize predator satiation. Worth noting in this context is the fact that both 13 and 17 are prime numbers, and thus predators on a shorter multiannual cycle cannot repeatedly utilize this resource. For example, a predator that bred every three years could not rely on cicadas always being present as a food supply.

How common is each of these defense types? No one has done an extensive survey over the entire animal kingdom. However, in 1989, Brian Witz surveyed studies that documented antipredator mechanisms in arthropods, mainly insects. By far the most common antipredator mechanisms were chemical defenses and associated aposematic coloration, noted in 51% of the examples. All other types of defense mechanisms occurred with considerably less frequency.

Despite the Impressive Array of Defenses, Predators Can Still Affect Prey Densities

The importance of predation on prey populations may be dependent on whether the system is donor controlled or predator controlled. In a donor-controlled system, prey supply is determined by factors other than predation, such as food supply, so that removal of predators has no effect on prey density. Examples include predators that feed on intertidal communities in which space is the limiting factor that controls prey populations.

In a predator-controlled system, the action of predator feeding eventually reduces the supply of prey. The removal of predators in a donor-controlled system is thus likely to have little effect, whereas in a predator-controlled system, such an action would probably result in large increases in prey abundance. Research studies have shown that many predators can have a significant effect on many prey populations. Considerable data exist on the interaction of the Canada lynx (*Lynx canadensis*) and its prey, snowshoe hares (*Lepus americanus*), because of the value of the pelts of both animals. In 1942, British ecologist Charles Elton analyzed the records of furs traded by trappers to the Hudson's Bay Company in Canada over a 100-year period. Analysis of the records showed that a dramatic 9- to 11-year cycle existed for as long as records had been kept (**Figure 57.11**). This cycle appears to be an example of a stable predator-prey oscillation that is predator controlled: the hares go up in density, followed by an increase in density of the lynx, which then depresses hare numbers. This is followed by a decline in the number of lynx, and the cycle begins again.

Figure 57.11 **Effect of predator on prey populations.** The 9- to 11-year oscillation in the abundance of the snowshoe hare (*Lepus americanus*) and the Canada lynx (*Lynx canadensis*) was revealed from pelt trading records of the Hudson's Bay Company.

Invasive species have provided striking examples of the effects of predators. The brown tree snake (*Boiga irregularis*) was inadvertently introduced by humans to the island of Guam, in Micronesia, shortly after World War II. The growth and spread of its population over the next 40 years closely coincided with a precipitous decline in the island's forest birds. The snake is a new predator with no natural predators on Guam to control it, and since the birds did not evolve with the snake, they have no defenses against it. Eight of the island's 11 native species of forest birds went extinct by the 1980s (**Figure 57.12**).

Figure 57.12 **Invasive species provide an example of predator pressure.** Population trends for four native Guam birds, as indicated by 100 km roadside surveys conducted from 1976 to 1998, show a precipitous decline because of predation by the introduced brown tree snake.

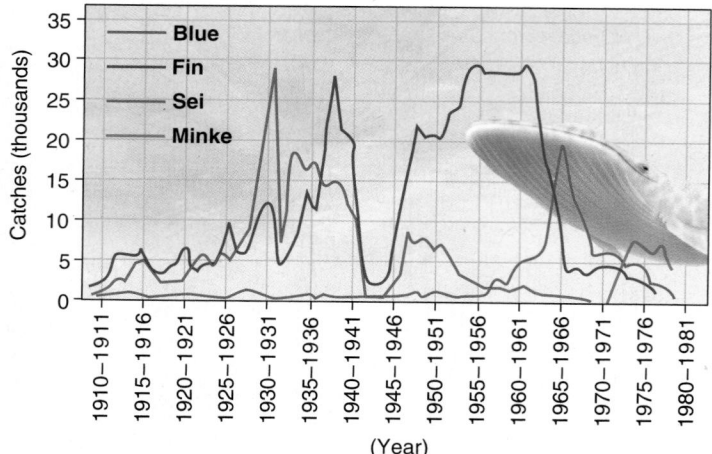

Figure 57.13 Sequential decline of whale catches in the Antarctic shows the strong effect of human predators.   Whale catches are believed to be directly related to whale population sizes. The catches of blue whale, the first species to be strongly affected by human predation, started a precipitous decline in the 1940s, as the whale was hunted to very low levels. Humans then began hunting more abundant fin whales, and in turn sei and minke whales, as each species became depleted.

One of the biggest reductions in prey in response to predation has been the systematic decline of whales of various species in response to the human whaling industry. The history of whaling has been characterized by a progression from larger, more valuable or easily caught species to smaller, less valuable or easily caught ones, as numbers of the original targets have been depleted (**Figure 57.13**). Belatedly, the International Whaling Commission enacted a moratorium on all commercial whaling in 1986. Following the moratorium, the populations of some whales have increased. Blue whales are thought to have quadrupled their numbers off the California coast during the 1980s, and numbers of the California gray whales have recovered to pre-whaling levels, showing the impact that an absence of a predator can have.

Ecologists have found that in nearly 1,500 predator-prey studies, over two-thirds (72%) showed a large depression of prey density by predators. Thus, we can conclude that in the majority of cases, predators influence the abundance of their prey in their native environment. The variety of antipredator mechanisms discussed earlier also shows how predation is important enough to select for the evolution of chemical defenses, camouflage, and mimicry in prey. Taken together these data indicate that predation is a powerful force in nature.

Plants and Herbivores May Be Engaged in an Evolutionary Arms Race

Herbivory involves the predation of plants or similar species such as algae. Such herbivory can be lethal to the plants, especially for small species, but often it is nonlethal, since many plant species, particularly larger ones, can regrow. We can distinguish two types of herbivores: generalist herbivores and specialist herbivores. Generalist herbivores can feed on many different plant species and are usually mammals. Specialist herbivores are often restricted to one or two species of host plants and are usually insects. There are, however, exceptions. Pandas are specialists because they feed only on bamboo, and koalas specialize on eucalyptus trees. On the other hand, grasshoppers are generalists, because they feed on a wide variety of plant species, including crops.

Plants appear to present a luscious green world of food to any organism versatile enough to use it, so why don't we see more plants being eaten by herbivores? After all, unlike most animals, plants cannot move to escape being eaten. Two hypotheses have been proposed to answer the question of why more plant material is not eaten. First, predators and parasites might keep herbivore numbers low, thereby sparing the plants. The many examples of the strength of predation provide evidence for this view. Second, the plant world is not as helpless as it appears; the sea of green is in fact armed with defensive spines, tough cuticles, and noxious chemicals. Let's take a closer look at plant defenses against herbivores and the ways that herbivores attempt to overcome them.

Plant Defenses Against Herbivores As described in Chapter 7, an array of unusual and powerful chemicals is present in plants, including alkaloids (nicotine in tobacco, morphine in poppies, cocaine in coca, and caffeine in tea), phenolics (lignin in wood and tannin in leaves), and terpenoids (in peppermint) (**Figure 57.14a–c**). Two general chemical classes can be recognized, those containing nitrogen compounds (e.g., alkaloids) and those lacking nitrogen (e.g., phenolics and terpenoids). Such compounds are not part of the primary metabolic pathway that plants use to obtain energy. They are therefore referred to as **secondary metabolites**, or secondary chemicals. Most of these chemicals are bitter tasting or toxic, and they deter herbivores from feeding. The staggering variety of secondary metabolites in plants, over 25,000, may be testament to the large number of organisms that feed on plants. In an interesting twist, many of these compounds have medicinal properties that are beneficial to humans (**Table 57.3**).

In addition to chemical compounds within the leaf, many plants have an array of mechanical defenses such as thorns and spines (**Figure 57.14d**). In other cases, an organism may provide a plant protection against herbivores in return for resources such as light, water, nutrients, or nesting sites (as we will see in Section 57.3).

An understanding of plant defenses is of great use to agriculturalists, since the more crops that are defended against pests, the higher the crop yields. The ability of plants to prevent herbivory is also known as **host plant resistance**. Host plant resistance may be due to chemical or mechanical defenses. One serious problem associated with commercial development of host plant resistance is that it may take a long time to breed into plants—between 10 and 15 years. This is because of the long time it takes to identify the responsible chemicals and develop the resistant genetic lines. Also, resistance to one pest may come at the cost of increasing susceptibility to other pests.

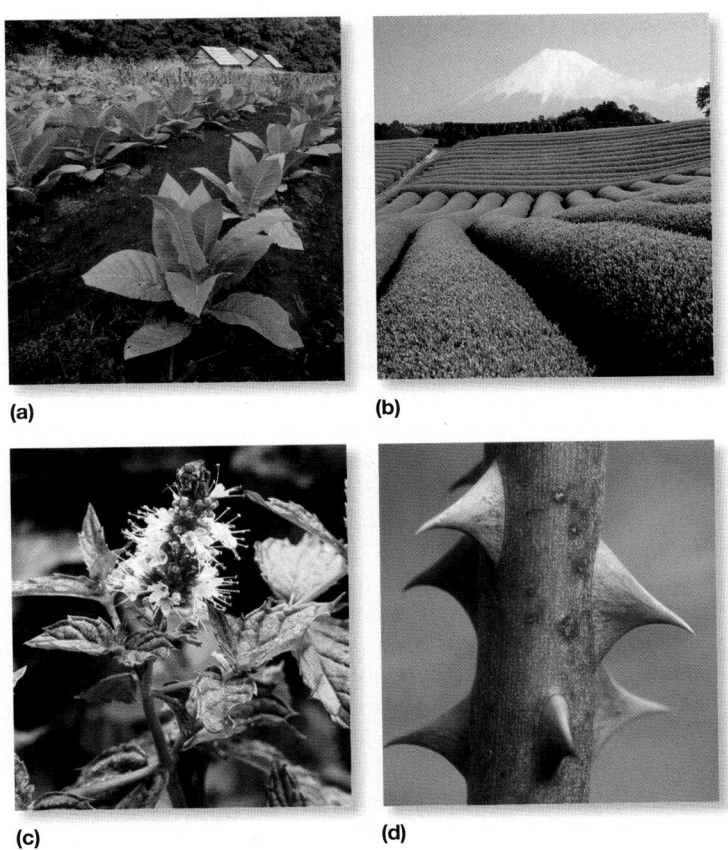

(a)

(b)

(c)

(d)

Figure 57.14 Defenses against herbivory. Plants possess an array of unusual and powerful chemicals, including **(a)** alkaloids, such as nicotine in tobacco, **(b)** phenolics, such as tannins in tea leaves (near Mount Fuji, Japan), and **(c)** terpenoids in peppermint leaves. **(d)** Mechanical defenses include plant spines and thorns, as on this shrub, *Rosa multiflora*.

Biological inquiry: Which type of defense would be most effective in deterring invertebrate herbivores?

Finally, some pest strains can overcome the plant's mechanisms of resistance. Circumvention of resistance by pests evolves in much the same way as resistance to pesticides.

Despite these problems, host plant resistance is a good tactic for the farmer. After the initial development of resistant varieties, the cost is minimal. Perhaps more importantly, host plant resistance is environmentally benign, generally having few side effects on other species in the community. It is estimated that about 75% of cropland in the U.S. utilizes pest-resistant plant varieties, most of these being resistant to plant pathogens. Bt corn is a variety of corn that has been genetically modified to incorporate a gene from the soil bacterium *Bacillus thuringiensis* (Bt), which provides a natural pesticide that is toxic to some insects. Genetic engineers have also produced Bt cotton, Bt tomato, and genetically modified varieties of many other crop species.

Overcoming Plant Resistance Herbivores can often overcome plant defenses. They can detoxify many poisons, mainly by two chemical pathways: oxidation and conjugation. Oxidation, the most important of these mechanisms, occurs in mammals in the liver and in insects in the midgut. It involves catalysis of the secondary metabolite to a corresponding alcohol by a group of enzymes known as mixed-function oxidases (MFOs). Conjugation, often the next step in detoxification, occurs by the uniting of the harmful element, or the compound resulting from the oxidation, with another molecule to create an inactive and readily excreted product.

In addition, certain chemicals that are toxic to generalist herbivores actually increase the growth rates of adapted specialist species, which can circumvent the defense or put the chemicals to good use in their own metabolic pathways. The Brassicaceae, the plant family that includes mustard, cabbage, and other species, contains acrid-smelling mustard oils called glucosinolates, the most important one of which is sinigrin.

Table 57.3	Plant Species with Compounds Used as Drugs		
Plant name	**Compound name**	**Use in medicine**	**Country of origin or cultivation**
Atropa belladonna	Atropine	Dilation of pupil	Central and Southern Europe; cultivated in U.S., U.K., other countries
Cassia senna	Danthron	Laxative	Cultivated in Egypt
Catharanthus roseus	Vincristine	Antitumor agent	Pantropical; cultivated in U.S., India, other countries
Cinchona ledgeriana	Quinine	Antimalarial	Cultivated in Indonesia, Zaire
Datura metel	Scopolamine	Sedative	Cultivated in Asia
Ephedra sinica	Ephedrine	Chronic bronchitis	China
Digitalis spp.	Digitoxin	Cardiotonic	Cultivated in Europe and Asia
Papaver sominiferum	Codeine, morphine	Analgesic, sedative	Cultivated in Turkey, India, Burma, Thailand
Pausinystalia yohimbe	Yohimbine	Erectile dysfunction	Cameroon, Nigeria, Rwanda
Rauvolfia spp.	Reserpine	Tranquilizer	India, Bangladesh, Sri Lanka, Burma, Malaysia, Indonesia, Nepal
Silybum merianum	Silymarin	Liver disorders	Mediterranean region

Large white butterflies (*Pieris brassicae*) preferentially feed on cabbage over other plants. In fact, if newly hatched larvae are fed an artificial diet, they do much better when sinigrin is added to it. When larvae are fed cabbage leaves on hatching from eggs and are later switched to an artificial diet without sinigrin, they die rather than eat. In this case, the secondary metabolite has become an essential feeding stimulant.

Ultimately, a good method to estimate the effects of herbivory on plant populations is to remove the herbivores and examine subsequent growth and reproductive output. Analysis of the hundreds of such experiments that have been conducted have yielded several interesting generalizations. First, herbivory in aquatic systems is more extensive than in terrestrial ones, generally because aquatic systems contain species, such as algae, that are especially susceptible to herbivory. Second, invertebrate herbivores such as insects have a stronger effect on plants than vertebrate herbivores such as mammals, at least in terrestrial systems. Thus, while one might consider large grazers like bison in North America or antelopes in Africa to be of huge importance in grasslands, it is more likely that grasshoppers are the more significant herbivores. In forests, invertebrate grazers such as caterpillars have greater access to canopy leaves than vertebrates and are also likely to have a greater effect.

Which types of plants are most affected by herbivory? The effect of herbivores is usually greatest on algae, presumably because these organisms are the least sophisticated in terms of their ability to manufacture complex secondary metabolites. Grasses and shrubs are also significantly affected by herbivores, but woody plants such as trees are less so. This may be because large and long-lived trees can draw on large resource reserves to buffer the impact of herbivores.

Parasitism Might Be the Predominant Lifestyle on Earth

When one organism feeds on another, but does not normally kill it outright, the predatory organism is termed a **parasite** and the prey a **host**. Some parasites remain attached to their hosts for most of their lives; for example, tapeworms spend their entire adult life inside the host's alimentary canal and even reproduce within their host. Others, such as the lancet fluke, have more complex life cycles that require the use of multiple hosts (**Figure 57.15**). Some, such as ticks and leeches, drop off their hosts after prolonged periods of feeding. Others, like mosquitoes, remain attached for relatively short periods.

Some flowering plants are parasitic on other plants. **Holoparasites** lack chlorophyll and are totally dependent on the host plant for their water and nutrients. One famous holoparasite is the tropical *Rafflesia arnoldii*, which lives most of its life within the body of its host (**Figure 57.16**). Only the flower develops externally, and it is a massive flower, 1 m in diameter and the largest known in the world. **Hemiparasites** generally do photosynthesize, but they lack a root system to draw water and thus depend on their hosts for that function. Mistletoe (*Viscum album*) is a hemiparasite. Hemiparasites usually have a broader range of hosts than do holoparasites, which may be confined to a single or a few host species.

We can define parasites that feed on one species or just a few closely related hosts as **monophagous**. **Polyphagous** species, by contrast, feed on many host plants, often from more than one plant family. We can also distinguish parasites as **microparasites** (for example, pathogenic bacteria and viruses), which multiply within their hosts, usually within the cells, and

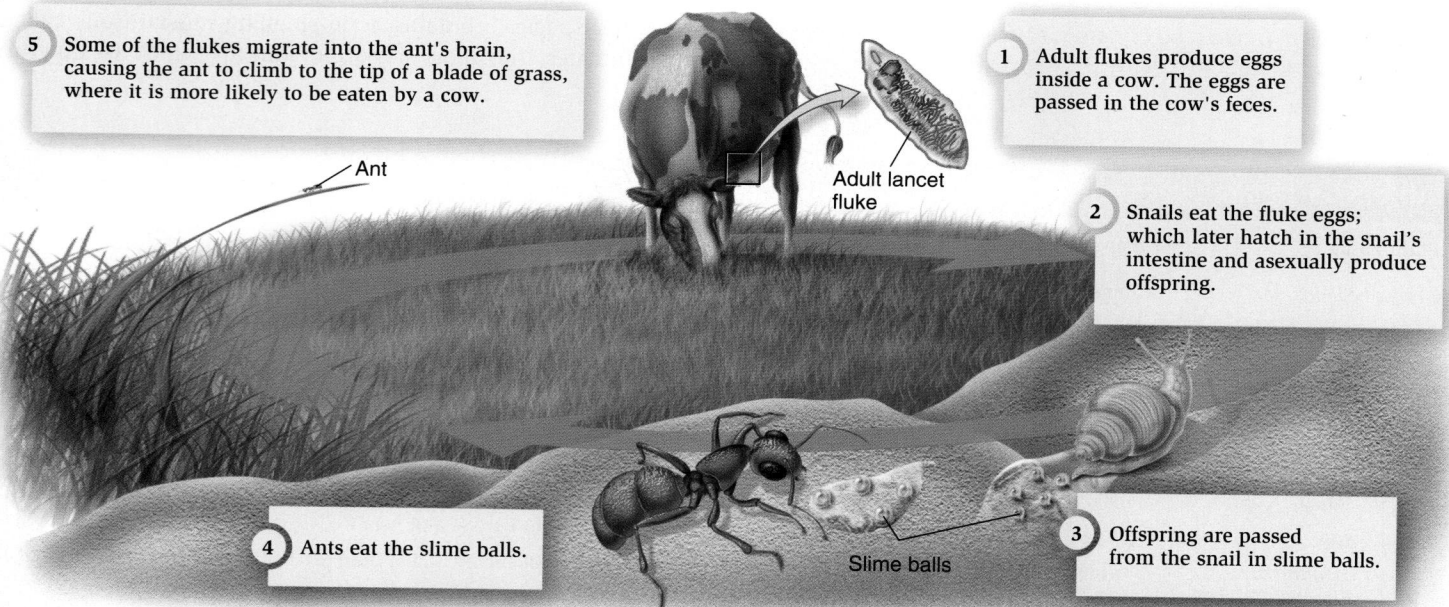

5 Some of the flukes migrate into the ant's brain, causing the ant to climb to the tip of a blade of grass, where it is more likely to be eaten by a cow.

Ant

Adult lancet fluke

1 Adult flukes produce eggs inside a cow. The eggs are passed in the cow's feces.

2 Snails eat the fluke eggs; which later hatch in the snail's intestine and asexually produce offspring.

4 Ants eat the slime balls.

Slime balls

3 Offspring are passed from the snail in slime balls.

Figure 57.15   A parasite life cycle.   The life cycle of the lancet fluke (*Dicrocoelium dendriticum*) involves behavioral changes in ants, one of its three hosts, that increase its transmission rate.

Figure 57.16 *Rafflesia arnoldii*, the world's biggest flower, lives as a parasite in Indonesian rain forests.

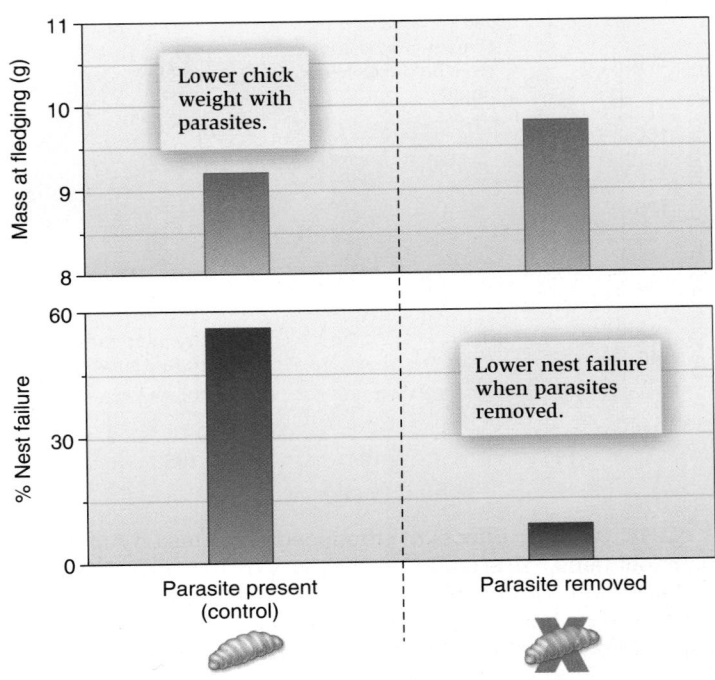

Figure 57.17 Parasite removal experiments. The left side shows the results when blowfly larva were present in the nests of young blue tits, while the right side shows the results when these parasites were removed.

macroparasites (such as schistosomes), which live in the host but release infective juvenile stages outside the host's body. Usually, the host has a strong immunological response to microparasitic infections. For macroparasitic infections, however, the response is short-lived, the infections tend to be persistent, and the hosts are subject to continual reinfection.

Last, we can distinguish **ectoparasites**, such as ticks and fleas, which live on the outside of the host's body, from **endoparasites**, such as pathogenic bacteria and tapeworms, which live inside the host's body. Problems of definition arise with regard to plant parasites, which seem to straddle both camps. For example, some parasitic plants, such as dodder, *Cuscuta* spp., an orange, stringlike plant, exist partly outside of the host's body and partly inside (refer back to Figure 37.22). Outgrowths called haustoria penetrate inside the host plant to tap into nutrient supplies. Being endoparasitic on a host seems to require greater specialization compared to ectoparasitism. Thus, ectoparasitic animals like leeches feed on a wider variety of hosts than do internal parasites such as liver flukes.

As we have seen throughout this textbook, there are vast numbers of species of parasites, including viruses, bacteria, protozoa, flatworms (flukes and tapeworms), nematodes, and various arthropods (ticks, mites, and fleas). Therefore, parasitism is a common way of life. Parasites may outnumber free-living species by four to one. Most plant and animal species harbor many parasites. For example, leopard frogs have nematodes in their ears and veins, and flukes in their bladders, kidneys, and intestines. A free-living organism that does not harbor parasitic individuals of a number of species is a rarity.

As with studies of other species interactions, a good method to determine the effect of parasites on their host populations is to remove the parasites and to re-examine the system. However, this is difficult to do, primarily because of the small size and unusual life histories of many parasites, which makes them difficult to remove from a host completely. The few cases of experimental removal confirm that parasites can reduce host population densities. The nests of birds such as blue tits are often infested with parasitic blowfly larvae that feed on the blood of nestlings. In 1997, Sylvie Hurtrez-Bousses and colleagues exper-

imentally reduced blowfly larval parasites of young blue tits in nests in Corsica. Parasite removal was cleverly achieved by taking the nests from 145 nest boxes, removing the young, microwaving the nests to kill the parasites, and then returning the nests and chicks to the wild. The success of chicks in microwaved nests was compared to that in nonmicrowaved (control) nests, and it was found that parasite-free blue tit chicks had greater body mass at fledging, the time when feathers first grow (**Figure 57.17**). Perhaps more important was that complete nest failure, that is, death of all chicks, was much higher in control nests than in treated nests.

Because parasite removal studies are difficult to do, ecologists have also examined the strength of parasitism as a mortality factor by studying introduced parasite species. Evidence from natural populations suggests that introduced parasites have substantial effects on their hosts. Chestnut blight, a fungus from Asia that was accidentally introduced to New York around 1904, virtually eliminated chestnut trees (*Castanea dentata*) in North America. Although the fungus took many years to spread throughout North America, by the 1950s, it had significantly reduced the density of American chestnut trees in North Carolina (**Figure 57.18**). In Europe and North America, Dutch elm disease has similarly devastated populations of elms. The disease wiped out 25 million of Britain's original 30 million elm trees between the 1960s and the 1990s. In Italy, canker has had similar severe effects on cypress. Fortunately, as we will discuss shortly, the field of genomics is helping greatly in the fight against plant diseases.

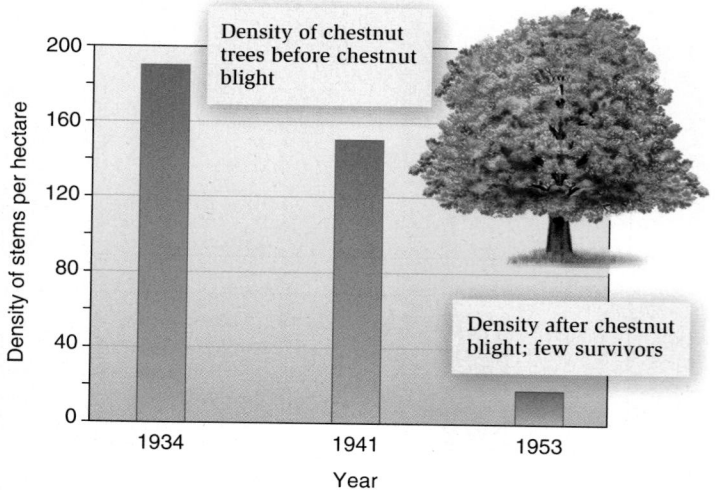

Figure 57.18 **Effects of introduced parasites on American chestnut trees.** The reduction in density of American chestnuts trees in North Carolina following the 1904 introduction of chestnut blight disease from Asia shows the severe effect that parasites can have on their hosts. By the 1950s, this once-prevalent species was virtually eliminated.

Introduced parasites have a strong effect on native animals as well as plants. In the early 1980s, canine distemper virus caused a large decline in the last remaining population of black-footed ferrets in the U.S. Conservationists are particularly concerned that many endangered animals are threatened by disease from domestic animals. In 1994, one-third of the population of lions in Serengeti National Park was wiped out by canine distemper virus, presumably contracted by domestic dogs. Some believe that the early 20th century extinction of the thylacine (*Thylacinus cynocephalus*), a Tasmanian marsupial that resembled a wolf, was caused by a distemper-like disease transmitted by dogs. To prevent the threat of disease among endangered species, some populations have been vaccinated; for example, the Ethiopian wolf (*Canis simensis*) has been vaccinated against rabies, and mountain gorillas (*Gorilla gorilla*) have been vaccinated against measles.

GENOMES & PROTEOMES

Transgenic Plants May Be Used in the Fight Against Plant Diseases

Many important native forest trees, which are also grown in urban landscapes, have been almost entirely wiped out by diseases spread by the importation of exotic plants. Sudden oak death is a recently recognized disease that is killing tens of thousands of oak trees and other plant species in California. The symptoms vary between species but include leaf spots, oozing of a dark sap through the bark, and twig dieback. Although sudden oak death is a forest disease, the organism causing this disease is known to infect many woody ornamental plants,

such as rhododendrons, that are commonly sold by nurseries. In March 2004, a California nursery was found to have unknowingly shipped plants infected with sudden oak death to all 50 states. Following this discovery, California nurseries halted shipments of trees to other states in an attempt to stop the spread of the disease, originally thought to have been imported on rhododendrons. In 2004, scientists mapped out the genome sequence of the disease-carrying fungus, *Phytophthora ramorum*. They are hoping that identifying the genes and their proteins will help them develop specific diagnostic tests to quickly detect the presence of sudden oak death in trees, which is currently impossible to detect until a year or more after the tree is infected.

However, scientists hope for much more from the field of genomics in their fight against plant diseases. Many scientists have suggested limited, cautious transfer of resistance genes from the original host species in the source regions of the disease to threatened species. Original host species have usually evolved over millions of years of exposure to these diseases and have acquired genes that provide resistance. In the regions of recent introduction of parasites, there has been no selection for resistance, so the host plants are often killed en masse. Transgenic trees that have received resistance genes could be produced and then be replanted in forests or urban areas. An advantage of this technique over traditional cross-breeding strategies involving two different species is that transgenic methods involve the introduction of fewer unnecessary genes of the non-native species. Also, fewer tree generations would be required to develop resistance. For example, using traditional breeding technology, Asian chestnut trees (*Castanea mollissima*) are being bred with American chestnuts (*Castanea dentata*) to reduce the susceptibility of the latter to chestnut blight, but the resultant hybrid is often significantly altered in appearance from the traditional American chestnut and the process takes more than a decade to produce trees that are ready to plant. Transgenic technology could minimize these drawbacks. William Powell and colleagues are working to enhance the American chestnut's resistance by inserting a gene taken from wheat. The gene, oxalate oxidase, destroys a toxin produced by the fungus that causes chestnut blight.

57.3 Mutualism and Commensalism

Mutualism and commensalism are interactions that are beneficial to at least one of the species involved. The unique feature of mutualism is that both species benefit from the interaction. For example, in mutualistic pollination systems, both plant and pollinator benefit, the former by the transfer of pollen and the latter typically by a nectar meal. In commensalism, one species benefits and the other remains unaffected. For example, in some forms of seed dispersal, barbed seeds are transported to new germination sites in the fur of mammals. The seeds benefit, but the mammals are generally unaffected. In this section we will examine the major types of mutualism and commensalism.

It is interesting to note that humans have entered into mutualistic relationships with many species. For example, the mutualistic association of humans with plants has resulted in some of the most far-reaching ecological changes on Earth. Humans have planted huge areas of the Earth with crops, allowing these plant populations to reach densities they never would on their own. In return, the crops have led to expanded human populations because of the increased amounts of food they provide.

Mutualism Is an Association Between Two Species That Benefits Both Species

Many close associations are known between species in which both species benefit. For example, leaf-cutting ants of the tribe Attini, of which there are about 210 species, enter into a mutualistic relationship with a fungus. A typical colony of about 9 million ants has the collective biomass of a cow and harvests the equivalent of a cow's daily requirement of fresh vegetation. Instead of consuming it directly, the ants chew it into a pulp, which they store underground as a substrate on which the fungus grows. The ants shelter and tend the fungus, protecting it from competing fungi and helping it reproduce and grow. In turn, the fungus produces specialized structures known as gongylidia, which serve as food for the ants. In this way, the ants circumvent the chemical defenses of the leaves, which are digested by the fungus.

The ant-fungus mutualism permits both species to live in close association, utilizing a common resource. Other different types of mutualisms occur in nature. Some of these are **defensive mutualisms**, often involving an animal defending a plant or an herbivore. **Dispersive mutualisms** include plants and pollinators that disperse their pollen, and plants and fruit eaters that disperse the plant's seeds.

Defensive Mutualism One of the most commonly observed mutualisms occurs between ants and aphids. Aphids are fairly defenseless creatures and are easy prey for most predators. The aphids feed on plant sap and have to process a significant amount of it to get their required nutrients. In doing so, they excrete a lot of fluid, and some of the sugars still remain in the excreted fluid, which is called honeydew. The ants drink the honeydew and, in return, protect the aphids from an array of predators, such as ladybird beetle larvae, by driving the predators away. In some cases, the ants herd the aphids like cattle, moving them from plant to plant (**Figure 57.19a**).

In other cases, ants enter into a mutualistic relationship with a plant itself. One of the most famous cases involves acacia trees in Central America, whose large thorns provide food and nesting sites for ants (**Figure 57.19b**). In return, the ants bite and discourage both insect and vertebrate herbivores from feeding on the trees. They also trim away foliage from competing plants and kill neighboring plant shoots, ensuring more light, water, and nutrient supplies for the acacias. In this case, neither species can live without the other, a concept called **obligatory mutualism**. This contrasts with **facultative mutualism**, in which the interaction is beneficial but not essential to the survival and reproduction of either species. For example, the ant-aphid mutualisms are generally facultative. Both species benefit from the association, but each could live without the other.

(a)

(b)

Figure 57.19 Defensive mutualism involves species that receive food or shelter in return for providing protection. (a) This red carpenter ant, *Camponotus pennsylvanicus*, tends aphids feeding on a twig. The ants receive sugar-rich honeydew produced by the aphids and in return they protect the aphids from predators. (b) Ants, usually *Pseudomyrmex ferruginea*, make nests inside the large, hornlike thorns of the bull's horn acacia and defend the plant against insects and mammals. In return, the acacia (*Acacia collinsii*) provide two forms of food to the ants: protein-rich granules called Beltian bodies and nectar from extrafloral nectaries (nectar-producing glands that are physically apart from the flower).

Biological inquiry: Is the relationship between red carpenter ants and aphids an example of facultative or obligatory mutualism?

Dispersive Mutualism Many examples of plant-animal mutualism involve pollination and seed dispersal. From the plant's perspective, an ideal pollinator would be a specialist, moving quickly among individuals but retaining a high fidelity to a plant species. One way that plant species in an area encourage the pollinator's species fidelity is by sequential flowering of different species through the year and by synchronized flowering within a species. The plant should provide just enough nectar to attract a pollinator's visit. From the pollinator's perspective, it would be best to be a generalist and obtain nectar and pollen from as many flowers as possible in a small area, thus minimizing the energy spent on flight between patches. This suggests that although mutualisms are beneficial to both species, their optimal needs are quite different.

Figure 57.20   **Dispersive mutualism.**   This blackbird (*Turdus merula*) is an effective seed disperser.

Mutualistic interactions are also highly prevalent in the seed-dispersal systems of plants. Fruits provide a balanced diet of proteins, fats, and vitamins. In return for this juicy meal, animals unwittingly disperse the enclosed seeds, which pass through the digestive tract unharmed. Fruits taken by birds and mammals often have attractive colors (**Figure 57.20**); those dispersed by nocturnal bats are not brightly colored but instead give off a pungent odor that attracts the bats.

Commensal Relationships Are Those in Which One Partner Receives a Benefit While the Other Is Unaffected

In commensalism, one member derives a benefit while the other neither benefits nor is harmed. Such is the case when orchids or other epiphytes grow in forks of tropical trees. The tree is unaffected, but the orchid gains support and increased exposure to sunlight and rain. Cattle egrets feed in pastures and fields among cattle, whose movements stir up insect prey for the birds. The egrets benefit from the association, but the cattle generally do not. One of the best examples of commensalism involves **phoresy**, in which one organism uses a second organism for transportation. Flower-inhabiting mites travel between flowers in the nostrils (nares) of hummingbirds. The flowers the mites inhabit live only a short while before dying, so the mite relocates to distant flowers by scuttling into the nares of visiting hummingbirds and hitching a ride to the next flower. Presumably the hummingbirds are unaffected.

Some commensalisms involve one species "cheating" on the other without harming it. In the bogs of Maine, the grass-pink orchid (*Calopogon pulchellus*) produces no nectar, but it mimics the nectar-producing rose pogonia (*Pogonia ophioglossoides*) and is therefore still visited by bees. Bee orchids mimic the appearance and scent of female bees; males pick up and transfer pollen while trying to copulate with the flowers (**Figure 57.21a**). The stimuli of flowers of the orchid genus *Ophrys* are so effective that male bees prefer to mate with them even in the presence of

(a)   **(b)**

Figure 57.21   **Commensalisms that involve cheating.** **(a)** Bee orchids (*Ophrys apifera*) mimic the shape of a female bee. Male bees copulate with the flowers, transferring pollen but getting no nectar reward. **(b)** Hooked seeds of burdock (*Arctium minus*) have lodged in the fur of a white-footed mouse (*Peromyscus leucopus*). The plant benefits from the relationship by the dispersal of its seeds, and the animal is not affected.

actual female bees! Many plants have essentially cheated their potential mutualistic seed-dispersal agents out of a meal by developing seeds with barbs or hooks to lodge in the animals' fur or feathers rather than their stomachs (**Figure 57.21b**). In these cases, the plants receive free seed dispersal, and the animals receive nothing, except perhaps minor annoyance. This type of relationship is fairly common; most hikers and dogs have at some time gathered spiny or sticky seeds as they wandered through woods or fields.

57.4 Conceptual Models

In this chapter, we have seen that competition, predation, herbivory, parasitism, mutualism, and commensalism are important in nature. How can we determine which of these factors, along with abiotic factors such as temperature and moisture, is the most important in regulating population size? The question is one asked by many applied biologists, such as foresters, marine biologists, and conservation biologists, who are interested in reducing population mortality in order to maximize a population's size. Many different theoretical models have been proposed to describe which mortality factors are the most significant in limiting population size. Some models stress the importance of so-called bottom-up factors such as plant or prey quality and abundance in controlling the herbivores or predators that feed on them. Others stress the importance of top-down factors, such as predators and parasites, acting on their animal or plant prey (**Figure 57.22**). Still others incorporate components of both these models. In this section, we will briefly discuss some of the evidence for the existence of bottom-up versus top-down control.

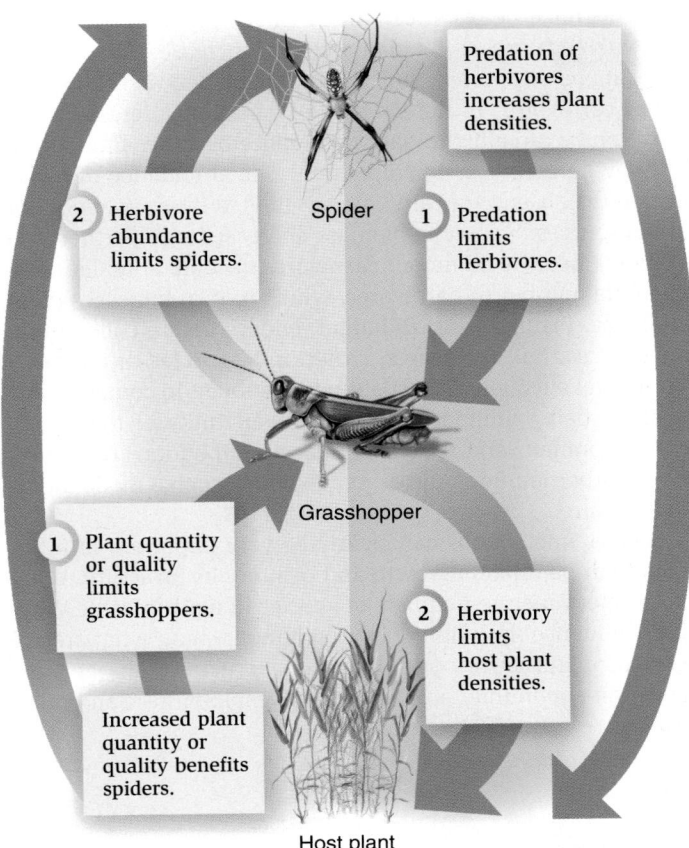

Spider

Predation of herbivores increases plant densities.

2 Herbivore abundance limits spiders.

1 Predation limits herbivores.

Grasshopper

1 Plant quantity or quality limits grasshoppers.

2 Herbivory limits host plant densities.

Increased plant quantity or quality benefits spiders.

Host plant

(a) Bottom-up control

(b) Top-down control

Figure 57.22 **Bottom-up control versus top-down control.** (a) Bottom-up control proposes that host plant quantity or quality limits the density of herbivores, which in turn sets limits on the abundance of predators. Taken together, this means that high quantity and quality of host plants would result in increased numbers of predators because of higher densities of the herbivores they prey on. (b) Top-down control proposes that plant densities are limited by their herbivores and that herbivores are limited by their predators. Taken together, this means that high levels of predation would result in high densities of host plants because there would be fewer herbivores.

Bottom-Up Models Suggest Food Limitation Controls Population Densities

At least two lines of evidence exist to suggest that bottom-up effects are important in limiting population sizes. First, we know there is a progressive lessening of available energy passing from plants through herbivores to carnivores and to the carnivores that eat carnivores (termed secondary carnivores). This line of evidence, based on the thermodynamic properties of energy transfer, suggests that plants usually regulate the population size of all other species that rely on them.

Second, much evidence supports the **nitrogen-limitation hypothesis** that organisms select food in terms of the nitrogen content of the tissue. This is largely due to the different propor-

tions of nitrogen in plants and animals. Animal tissue generally contains about 10 times as much nitrogen as plant tissue; thus, animals favor high-nitrogen plants. For example, red deer feed preferentially on grasses defecated upon by herring gulls (*Larus argentatus*). Where the number of gull droppings increases, so does the vegetation nitrogen content. Fertilization has repeatedly been shown to benefit herbivores. Nearly 60% of 186 studies investigating the effects of fertilization on herbivores reported that increasing a plant's tissue nitrogen concentration through fertilization had strong positive effects on herbivore population sizes, survivorship, growth, and fecundity.

Top-Down Models Suggest Natural Enemies Control Population Densities

Top-down models suggest that predators control populations of their prey and that herbivores control plant populations. Supporting evidence comes from the world of biological control, where natural enemies are released to control pests of agriculture. Many weeds are invaders that were accidentally introduced to an area from a different country, as seeds in ships' ballasts or in agricultural shipments. For example, over 50% of the 190 major weeds in the U.S. are invaders from outside the country. Many of these weeds have become separated from their native natural enemies; this is one reason the weeds become so prolific. Because chemical control is expensive and may have unwanted environmental side effects, many land managers have reverted to biological control, in which the invading weed is reunited with its native natural enemy.

Ecologists have noted many successes in the biological control of weeds. Klamath weed or St. John's wort (*Hypericum perforatum*), a pest in California pastures, was controlled by two beetles from its homeland in Europe. The Brazilian weevil, *Cyrtobagous salviniae*, is a proven biological control agent for the floating fern (*Salvinia molesta*), which has invaded lakes in Australia and New Guinea. Alligator weed has been controlled in Florida's rivers by the alligatorweed flea beetle (*Agasicles hygrophila*) from South America. The prickly pear cactus (*Opuntia stricta*) provides a prime example of effective biological control of a weed. The cactus was imported into Australia in the 19th century and quickly established itself as a major pest of rangeland. The small cactus moth (*Cactoblastis cactorum*) was introduced in the 1920s and within a short time successfully saved hundreds of thousands of acres of valuable rangeland from being overrun by the cacti (**Figure 57.23**). The numerous examples showing that these pest populations are controlled when reunited with their natural enemies provide strong evidence of top-down control in nature.

Modern Models Suggest Both Top-Down and Bottom-Up Effects Are Important

More recently, different models have been proposed that take into account the effects of both natural enemies and limited resources on species. In 1981, Laurie Oksanen and coworkers suggested that the strength of mortality factors varies with availability of plant biomass involved, a model they termed the ecosystem exploitation hypothesis (**Table 57.4**). Thus, for very simple systems where mainly plants exist, like Arctic tundra, not enough plant material is available to support herbivores,

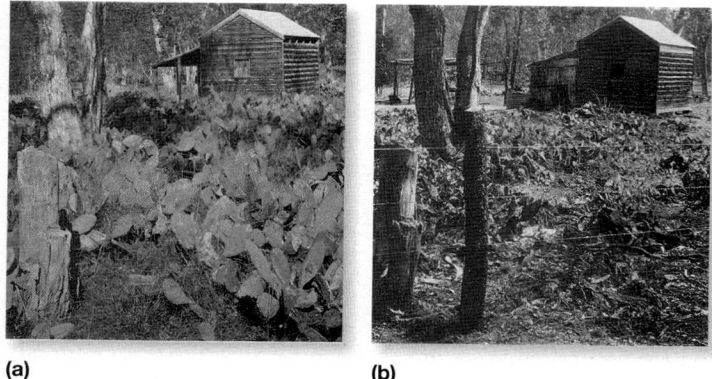

(a)   (b)

Figure 57.23   Successful biological control of prickly pear cactus. The prickly pear cactus (*Opuntia stricta*) in Chinchilla, Australia, before **(a)** and after **(b)** top-down control by the cactus moth (*Cactoblastis cactorum*).

and plants must be resource limited (that is, limited by competition with each other). As plant biomass increases, some herbivores can be supported, but there are too few herbivores to support carnivores. In the absence of carnivores, levels of herbivory can be quite high. Plant abundance becomes limited primarily by herbivory, not competition. The abundance of herbivores, in the absence of carnivores, is limited by competition for plant resources. As plant biomass increases still further and herbivores become common, carnivores become abundant and reduce the number of herbivores, which in turn increases plant abundance. Plants, being abundant, endure severe competition for resources, but herbivores, suffering high rates of mortality from natural enemies, are not abundant enough to compete. The predators, being limited only by the availability of their prey, are also common and thus experience competition. Ecologists are currently examining the degree to which such a model holds true in nature.

Species interactions can clearly be very important in influencing both population growth and community structure. What factors determine the numbers of species in a given area? What factors influence the stability of a community, and what are the effects of disturbances on community structure? In the next chapter, on community ecology, we will explore these and other questions.

Table 57.4	Primary Mortality Factors According to Availability of Plant Biomass			
	Low ——→ Plant biomass ——→ High			
Taxa	**Plants only**	**Plants and herbivores**	**Plants, herbivores, and carnivores**	**Plants, herbivores, carnivores, and secondary carnivores**
Plants	Competition	Herbivory	Competition	Herbivory
Herbivores		Competition	Predation	Competition
Carnivores			Competition	Predation
Secondary carnivores				Competition

CHAPTER SUMMARY

57.1 Competition

- Species interactions can take a variety of forms that differ based on their effect on the species involved. (Figure 57.1, Table 57.1)
- Competition can be categorized as intraspecific (between individuals of the same species) or interspecific (between individuals of different species). Competition can also be categorized as resource competition or interference competition. (Figure 57.2)

- One species can exclude the other in a natural environment, affecting its distribution within a habitat. (Figure 57.3)
- Laboratory and field experiments show that competition occurs frequently in nature, and varies as a function of both abiotic and biotic factors. (Figures 57.4, 57.5)
- The competitive exclusion hypothesis states that two species with the same resource requirements cannot occupy the same niche. (Figure 57.6)
- Resource partitioning and morphological differences between species allow them to coexist in a community. (Figures 57.7, 57.8, Table 57.2)

57.2 Predation, Herbivory, and Parasitism

- The many antipredator strategies that animals have evolved suggest that predation is a strong selective force. Of these strategies, chemical defense and aposematic coloration are the most common. (Figures 57.9, 57.10)

- Despite these defenses, oscillations in predator-prey cycles, the effect of introduced species, and examples of human predation illustrate that predators can have a large effect on prey densities. (Figures 57.11, 57.12, 57.13)

- Plants have also evolved an array of defenses against predators, termed host plant resistance, which includes chemical defenses such as secondary metabolites and mechanical defenses such as thorns and spines. (Figure 57.14, Table 57.3)

- Parasitism is a common lifestyle on Earth, and some parasites have complex life cycles involving multiple hosts. (Figures 57.15, 57.16)

- Evidence from experimental removal of parasites and from the study of introduced plant and animal parasites confirms that parasites can greatly reduce prey densities. (Figures 57.17, 57.18)

57.3 Mutualism and Commensalism

- Mutualism is an association between two species that benefits both. Defensive mutualisms typically involve an animal defending either a plant or herbivore; dispersive mutualisms involve plants and pollinators that disperse their pollen, and plants and fruit eaters that disperse the plant's seeds. (Figures 57.19, 57.20)

- In commensal relationships, one partner receives a benefit while the other is not affected. (Figure 57.21)

57.4 Conceptual Models

- Conceptual models of species interactions describe the importance of factors such as competition and predation. Bottom-up models propose that plant quality or quantity regulates the abundance of all herbivore and predator species; top-down models propose that the abundance of predators controls herbivore and plant densities. (Figures 57.22, 57.23)

- Modern models, which incorporate components of both bottom-up and top-down models, propose that mortality varies according to productivity of plant biomass. (Table 57.4)

TEST YOURSELF

1. Two species of birds feed on similar types of insects and nest in the same tree species. This is an example of
   a. intraspecific competition.
   b. interference competition.
   c. resource competition.
   d. mutualism.
   e. none of the above.

2. The experiments conducted by Thomas Park using flour beetles provided evidence that the results of competition are influenced by
   a. moisture.
   b. temperature.
   c. parasitism.
   d. stochasticity.
   e. all of the above.

3. According to the competitive exclusion hypothesis
   a. two species that use the exact same resource show very little competition.
   b. two species with the same niche cannot coexist.
   c. one species that competes with several different species for resources will be excluded from the community.
   d. all competition between species results in the extinction of at least one of the species.
   e. none of the above.

4. In Lack's study of British passerine birds, different species seem to segregate based on resource factors, such as location of prey items. This differentiation among the niches of these passerine birds is known as
   a. competitive exclusion.
   b. intraspecific competition.
   c. character displacement.
   d. resource partitioning.
   e. allelopathy.

5. Divergence in morphology that is a result of competition is termed
   a. competitive exclusion.
   b. resource partitioning.
   c. character displacement.
   d. amensalism.
   e. mutualism.

6. Some organisms have bright coloration that serves as a warning to predators of the organism's bad taste. This type of signal is called
   a. Batesian mimicry.
   b. aposematic coloration.
   c. allelopathy.
   d. cryptic coloration.
   e. masting.

7. Batesian mimicry differs from Müllerian mimicry in that
   a. in Batesian mimicry, both species possess the chemical defense.
   b. in Batesian mimicry, one species possesses the chemical defense.
   c. in Müllerian mimicry, one species has several different mimics.
   d. in Müllerian mimicry, aposematic coloration is always found.
   e. in Batesian mimicry, cryptic coloration is always found.

8. Secondary metabolites
   a. are produced as chemical defenses by plants.
   b. are by-products of cellular respiration.
   c. are typically produced by organisms that exhibit Batesian mimicry.
   d. play an important role in energy production in plants.
   e. can be used as alternative reactants in photosynthesis.

9. Parasitic plants that rely solely on their host for nutrients are called
   a. hemiparasites.
   b. fungi.
   c. holoparasites.
   d. monophagous.
   e. polyphagous.

10. A species interaction in which one species benefits but the other species is unharmed is called
    a. mutualism.
    b. amensalism.
    c. parasitism.
    d. commensalism.
    e. mimicry.

CONCEPTUAL QUESTIONS

1. Define the competitive exclusion hypothesis.
2. Distinguish between Müllerian and Batesian mimicry.
3. Describe some plant defenses against herbivory.

COLLABORATIVE QUESTIONS

1. Discuss different types of competition found in nature.
2. Discuss several antipredator strategies that animals have evolved.

EXPERIMENTAL QUESTIONS

1. Describe the realized niches for the two species of barnacles used in Connell's experiment.
2. Outline the procedure that Connell used in the experiments.
3. How did Connell explain the presence of *Chthamalus* in the upper intertidal zone if *Semibalanus* was shown to outcompete the species in the first experiment?

www.brookerbiology.com

This website includes answers to the Biological Inquiry questions found in the figure legends and all end-of-chapter questions.

58

COMMUNITY ECOLOGY

CHAPTER OUTLINE

Waterhole with lion pride and elephants, Savuti Reserve, Botswana.

So far in this unit, we have examined ecology in terms of the behavior of individual organisms, the growth of populations, and interactions between pairs of species. Most populations, however, exist not on their own but together with populations of many other species. This assemblage of many populations that live in the same place at the same time is known as a **community**. For example, a tropical forest community consists of not only the tree species, vines, and other vegetation but also of the insects that pollinate them, the herbivores that feed upon the insects, and the predators and parasites of the herbivores. Communities can occur on a wide range of scales and can be nested. The tropical forest community also encompasses smaller communities such as the water-filled recesses of bromeliads, which form a microhabitat for different species of insect and their larvae. Both of these entities—the tropical forest and the bromeliad tank—are viable communities, depending on one's frame of reference with regard to scale.

Community ecology studies how groups of species interact and form functional communities. In Chapter 57, we looked in detail at interactions between individual species. In this chapter, we widen our focus to explore the factors that influence the number and abundance of species in a community. We begin by examining the nature of ecological communities. Are communities loose assemblages of species that happen to live in the same place at the same time, or are they more tightly organized groups of mutually dependent species? We explore why, on a global scale, the number of species is usually greatest in the tropics and declines towards the poles. Community ecology also addresses what factors act to stabilize species richness in a community. However, ecologists recognize that communities may change, for example, following a disturbance such as a fire.

This recovery tends to occur in a predictable way that ecologists have termed succession. Finally, in certain situations, for example, on islands recovering from physical disturbance, succession has been postulated to occur via waves of colonization of species from neighboring landmasses, followed by extinctions of some species. The equilibrium model of island biogeography holds that island size and distance from the mainland govern the process of succession.

58.1 Differing Views of Communities

Ecologists have long held differing views on the nature of a community and how it is structured and functions. Some of the initial work in the field of community ecology considered a community to be equivalent to a superorganism, in much the same way that the body of an animal is more than just a collection of organs. In this view, individuals, populations, and communities have a relationship to each other that resembles the associations found between cells, tissues, and organs. Indeed, American botanist Frederic Clements, the champion of this viewpoint, suggested in 1905 that ecology was to the study of communities what physiology was to the study of individual organisms. This view of community, with predictable and integrated associations of species separated by sharp boundaries, is termed the **organismic model**. Some modern-day ecologists still depict the community as a superorganism. As we will see later in this chapter, Clements's hypotheses on communities were also linked to his study of succession, the changes in the types of plant and animal species that occupy a given area through time.

Clements's ideas were challenged in 1926 by botanist Henry Allan Gleason. Gleason proposed an **individualistic model**, which viewed a community as an assemblage of species coexisting primarily because of similarities in their physiological requirements and tolerances. While acknowledging that some assemblages of species were fairly uniform and stable over a given region, Gleason suggested that distinctly structured ecological communities usually do not exist. Instead, species distribute independently along an environmental gradient. Viewed in this way, communities do not necessarily have sharp boundaries, and associations of species are much less predictable and integrated than in Clements's organismic model.

By the 1950s, many ecologists had abandoned Clements's view in favor of Gleason's. In particular, Robert Whittaker's studies asserted the **principle of species individuality**, which stated that each species is distributed according to its physiological needs and population dynamics, that most communities intergrade continuously, and that competition does not create distinct vegetational zones. For example, let's consider an environmental gradient such as a moisture gradient on an uninterrupted slope of a mountain. Whittaker proposed that four hypotheses could explain the distribution patterns of plants and animals on the gradient (**Figure 58.1**):

1. Competing species, including dominant plants, exclude one another along sharp boundaries. Other species evolve toward a close, perhaps mutually beneficial association with the dominant species. Communities thus develop along the gradient, each zone containing its own group of interacting species giving way at a sharp boundary to another assemblage of species. This corresponds to Clements's organismic model.

2. Competing species exclude one another along sharp boundaries but do not become organized into groups of species with parallel distributions.

3. Competition does not usually result in sharp boundaries between species. However, the adaptation of species to similar physical variables will result in the appearance of groups of species with similar distributions.

4. Competition does not usually produce sharp boundaries between species, and the adaptation of species to similar physical variables does not produce well-defined groups of species with similar distributions. The centers and boundaries of species populations are scattered along the environmental gradient. This corresponds to Gleason's individualistic model.

To test these possibilities, Whittaker examined the vegetation on various mountain ranges in the western U.S. He sampled plant populations along an elevation gradient from the tops of the mountains to the bases and collected data on various physical variables, such as soil moisture.

The results supported the fourth hypothesis, that competition does not produce sharp boundaries between species and that adaptation to physical variables does not result in defined groups of species. Whittaker concluded that his observations agreed with Gleason's predictions that (1) each species is dis-

Figure 58.1   **Four hypotheses on how populations might relate to one another along an environmental gradient.**   Each curve in each part of the figure represents one species and the way its population might be distributed along an environmental gradient such as a moisture gradient on a mountain slope.

tributed in its own way, according to its genetic, physiological, and life cycle characteristics; and (2) the broad overlap and scattered centers of species populations along a gradient implies that most communities grade into each other continuously rather than form distinct, clearly separated groups. Thus, Whittaker's observations showed that the composition of species at any one point in an environmental gradient was largely determined by factors such as temperature, water, light, pH, and salt concentrations (features discussed in Chapter 54).

58.2 Patterns of Species Richness

Even though most communities intergrade along environmental gradients such as a mountain slope, ecologists recognize distinct differences between communities. The community at the top of a mountain will be quite different from that at the bottom, so distinguishing between communities on a broad scale is useful. Also, some sharp boundaries between groups of species sometimes do exist, especially related to physical differences

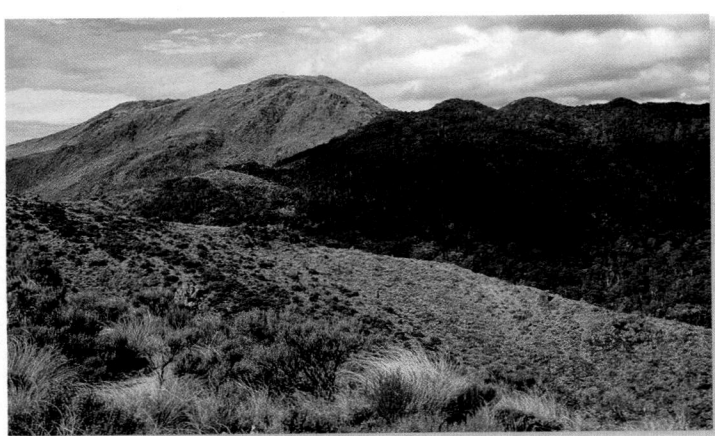

Figure 58.2 Unique communities may result from distinct environmental conditions. In New Zealand's Dun Mountain area, we can contrast the sparse vegetation on the serpentine soil to the left to that of the beech forest on the nonserpentine soil on the right.

such as water quality and soil type, that cause distinct communities to develop. For example, serpentine soils are rich in metals, including magnesium, iron, and nickel, but poor in plant nutrients. The species that have adapted to these harsh conditions form a unique community restricted to this area (**Figure 58.2**).

One of the most straightforward means of distinguishing between communities is to determine the number of species in each community, or **species richness**. The number of species of most taxa varies according to geographic location, generally increasing from polar areas to temperate areas and reaching a maximum in the tropics. For example, the species richness of North American birds increases from Arctic Canada to Panama (**Figure 58.3**). A similar pattern exists for mammals and reptiles. Species richness is also increased by topographical variation. More mountains mean more hilltops, valleys, and differing habitats; thus, there is an increased number of birds in the mountainous western U.S. However, species richness is reduced by the peninsular effect, in which diversity decreases as a function of distance from the main body of land, and thus we see a decrease in the number of species in Florida and in Baja California.

Many hypotheses for the polar-equatorial gradient of species richness have been advanced. We will consider three of the most important, which propose that communities diversify with increased time, area, and productivity, respectively. Although they are treated separately here, it is important to note that these hypotheses are not mutually exclusive and that all can contribute to patterns of species richness.

The Time Hypothesis Suggests Communities Diversify with Age

Many ecologists argue that communities diversify, that is, gain species, with time and that temperate regions have less rich communities than tropical ones because they are younger and have only more recently (relatively speaking) recovered from re-

Figure 58.3 Species richness of birds in North America. The values indicate the numbers of different species in a given area. Contour lines show equal numbers of bird species. Note the pronounced latitudinal gradient heading south toward the tropics and the high diversity in California and northern Mexico, regions of considerable topographical variation and habitat diversity.

cent glaciations and severe climatic disruption. The time hypothesis has two variations. It proposes that, first, resident species have not yet evolved new forms to exploit vacant niches. And second, it proposes that species that could possibly live in temperate regions have not migrated back from the unglaciated areas into which the ice ages drove them.

In support of the time hypothesis, ecologists compared the species richness of bottom-dwelling invertebrates such as worms in historically glaciated (covered with ice) and unglaciated lakes in the Northern Hemisphere that occur at similar latitudes. Lake Baikal in Siberia is an ancient unglaciated temperate lake and contains a very diverse fauna; for example, there are 580 species of invertebrates in the bottom zone. Great Slave Lake, a comparable sized lake in northern Canada that was once glaciated, contains only four species in the same zone.

There are, however, drawbacks to the utility of the time hypothesis. For example, the time hypothesis may help explain variations in the species richness of terrestrial organisms, but it has limited applicability to marine organisms. While we might not expect terrestrial species to redistribute themselves quickly following a glaciation—especially if there is a barrier like the English Channel to overcome—there seems to be no reason that marine organisms couldn't relatively easily shift their distribution patterns during glaciations, yet the polar-equatorial gradient of species richness still exists in marine habitats.

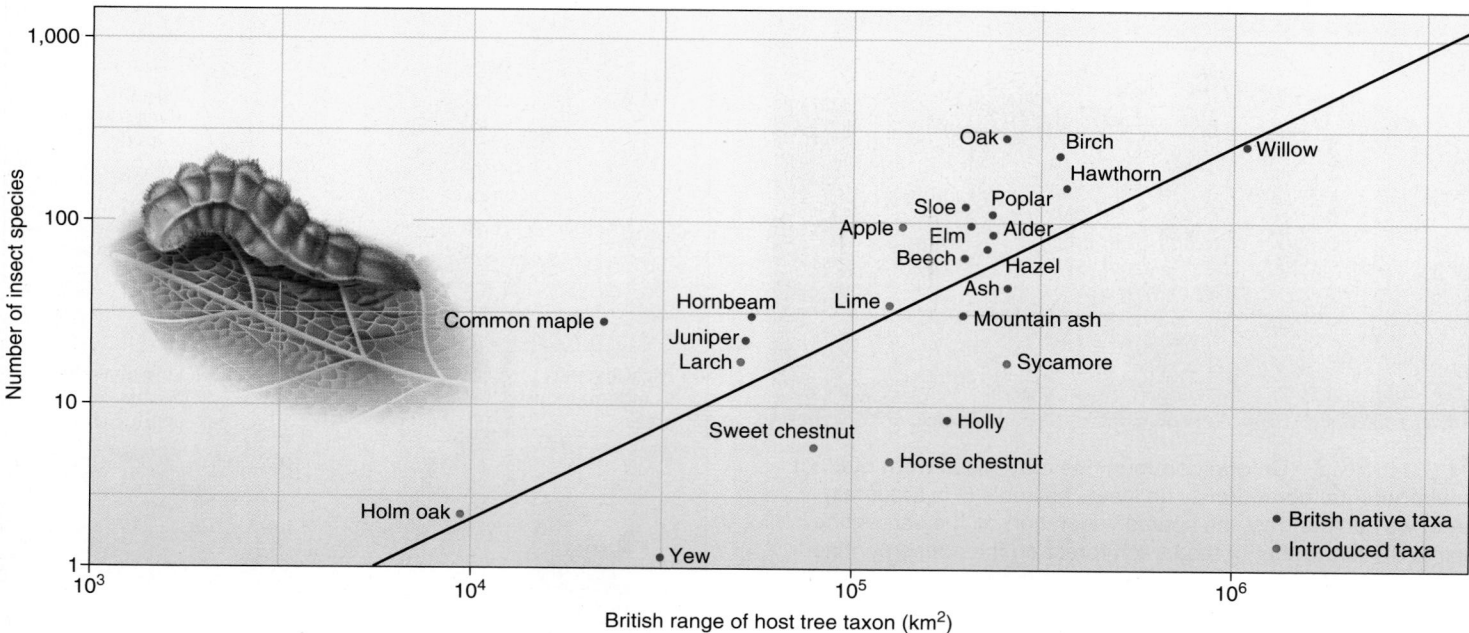

Figure 58.4   **Insect species diversity on British host trees supports the area hypothesis.**   A positive correlation is found between insect species richness and the host tree's present range (in km²). Here, the range represents an area in Britain known to contain trees of each species.

The Area Hypothesis Suggests Large Areas Support More Species

The **area hypothesis** proposes that larger areas contain more species than smaller areas because they can support larger populations and a greater range of habitats. Much evidence supports the area hypothesis. For example, in 1974, Donald Strong showed that insect species richness on tree species in Britain was better correlated with the area over which a tree species could be found than with time of habitation since the last ice age. The relationship between the amount of available area and the number of species present is called the **species-area effect** (**Figure 58.4**). Some introduced tree species, such as apple and lime, were relatively new to Britain, but they bore many different insect species, a fact Strong argued did not support the time hypothesis.

As we saw in Chapter 54, the three-cell model of atmospheric circulation leads to a symmetry of climates between polar regions and temperate areas in the Northern and Southern Hemispheres (refer back to Figure 54.21). Symmetrically similar climates are only adjacent in the tropics, however, which creates one large area. This has been proposed as a reason why the tropics have greater species richness. However, the area hypothesis seems unable to explain why, if increased richness is linked to increased area, there are not more species in the vast contiguous landmass of Asia. Furthermore, while tundra may be the world's largest land biome, it has low species richness. Finally, the largest marine system, the open ocean, which has the greatest

volume of any habitat, has fewer species than tropical surface waters, which have a relatively small volume.

The Productivity Hypothesis Suggests That More Energy Permits the Existence of More Species

The **productivity hypothesis** proposes that greater production by plants results in greater overall species richness. An increase in plant biomass, the total weight of plant material produced, leads to an increase in the number of herbivores and hence an increase in the number of predator, parasite, and scavenger species. Production itself is influenced by factors such as temperature and rainfall, because many plants grow better where it is warm and wet. For example, in 1987, David Currie and colleagues showed that the species richness of trees in North America is best predicted by the **evapotranspiration rate**, the rate at which water moves into the atmosphere through the processes of evaporation from the soil and transpiration of plants (**Figure 58.5**).

Once again, however, there are exceptions to this rule. Some tropical seas, such as the southeast Pacific off of Colombia and Ecuador, have low productivity but high species richness. On the other hand, the sub-Antarctic Ocean has a high productivity but low species richness. Estuarine areas, where rivers empty into the sea, are similarly very productive yet low in species, presumably because they represent a stressful environment for many organisms that are alternately inundated by fresh water

Figure 58.5 Tree species richness in North America is correlated with evapotranspiration rate. Contour lines show equal numbers of tree species.

Biological inquiry: Why doesn't the species richness of trees increase in mountainous areas of the West, as it does for birds?

and salt water with daily changes in the tide. Some lakes that are polluted with fertilizers have high productivity but low species richness.

Last, Robert Latham and Robert Ricklefs showed in 1993 that while patterns of tree diversity in North America support the productivity hypothesis, the pattern does not hold for broad comparisons between continents: the temperate forests of eastern Asia support substantially higher numbers of tree species (729) than do climatically similar areas of North America (253) or Europe (124). These three areas have different evolutionary histories and different neighboring areas from which species might have invaded.

Thus, in terms of accounting for the polar-equatorial gradient of species richness, each of the three hypotheses we have explored has some evidence to support it and some to contradict it. Different processes may occur over different scales. On a regional scale, we know that the time since the last glaciation has the potential to change patterns of species richness. On a more local scale, area and productivity of available habitat may be important. At any given point on the globe, species richness may be affected by the interaction of these different factors.

58.3 Calculating Species Diversity

So far, we have discussed communities in terms of variations in species richness. However, to measure species diversity, ecologists need to take into account not only the number of species in a community but also their frequency of occurrence, or **relative abundance**. For example, imagine two hypothetical communities, A and B, both with two species and 100 total individuals.

	Number of individuals of species 1	Number of individuals of species 2
Community A	99	1
Community B	50	50

The species richness of community B equals that of community A, because they both contain two species. However, community B is considered more diverse because the distribution of individuals between species is more even. One would be much more likely to encounter both species in community B than in community A, where one species dominates.

To measure the diversity of a community, therefore, ecologists calculate what is known as a diversity index. Although there are a great many of these indices, the most widely used is the **Shannon diversity index (H_S)**, which is calculated as

$$H_S = -\Sigma p_i \ln p_i$$

where p_i is the proportion of individuals belonging to species i in a community, ln is the natural logarithm, and the Σ is a summation sign. For example, for a species in which there are 50 individuals out of a total of 100 in the community, p_i is 50/100, or 0.5. The natural log of 0.5 is -0.693. For this species, $p_i \ln p_i$ is then $0.5 \times -0.693 = -0.347$. For a hypothetical community with 5 species and 100 total individuals, the Shannon diversity index would be calculated as follows:

Species	Abundance	p_i	$p_i \ln p_i$
1	50	0.5	$-.347$
2	30	0.3	$-.361$
3	10	0.1	$-.230$
4	9	0.09	$-.217$
5	1	0.01	$-.046$
Total 5	100	1.00	$\Sigma p_i \ln p_i$ -1.201

Note that in this example, even the rarest species, species 5, contributes some value to the index, so that if there were many such rare species, their contributions would accumulate. This makes the Shannon diversity index very valuable to conservation biologists, who often study rare species and their importance to the community. Remember too that the negative sign in front of the summation changes these values to positive, so the index actually becomes 1.201, not -1.201. (The minus sign in the equation is used to give us a positive number for the index, which is more appealing than an index with a negative number.)

Table 58.1	Shannon Diversity Index of Bird Species on Logged and Unlogged Sites in Indonesia					
	Unlogged			**Logged**		
Species	N	p_i	$p_i \ln p_i$	N	p_i	$p_i \ln p_i$
Nectarinia jugularis	410	.225	−.336	910	.386	−.367
Ducula bicolor	230	.126	−.261	220	.093	−.221
Philemon subcorniculatus	210	.115	−.249	240	.102	−.233
Nectarinia aspasia	190	.104	−.235	120	.051	−.152
Dicaeum vulneratum	185	.101	−.232	280	.119	−.253
Ducula perspicillata	170	.093	−.221	180	.076	−.196
Phylloscopus borealis	160	.088	−.214	140	.059	−.167
Eos bornea	88	.048	−.146	73	.031	−.108
Ixos affinis	76	.042	−.133	31	.013	−.056
Geoffroyus geoffroyi	44	.024	−.089	54	.023	−.087
Rhyticeros plicatus	24	.013	−.056	27	.011	−.050
Cacatua moluccensis	12	.007	−.035	1	.001	−.007
Tanygnathus megalorynchos	9	.005	−.026	11	.005	−.026
Electus roratus	7	.004	−.022	0	0	0
Macropygia amboinensis	6	.003	−.017	7	.003	−.017
Cacomantis sepulcralis	3	.002	−.012	0	0	0
Trichoglossus haematodus	0	0	0	64	.027	−.097
Total	**1,824**	**1.0**		**2,358**	**1.0**	
Shannon diversity index			**2.284**			**2.037**

Values of the Shannon diversity index for real communities often fall between 1.5 and 3.5, with the higher the value, the greater the diversity. **Table 58.1** calculates the diversity of two bird communities in Indonesia with similar species richness but differing species abundance. The bird communities were surveyed on lowland forest that had been selectively logged or on pristine, unlogged forest. To document diversity, biologist Stuart Marsden established census stations in the two types of forest and recorded the identity and density of all bird species for a number of 10-minute periods. Although a greater number of individual birds was seen in the logged areas (2,358) compared to unlogged ones (1,824), a high percentage of the individuals in the logged areas (38.6%) belonged to just one species, *Nectarinia jugularis*. While there was only one more bird species found in the unlogged area than in the logged area, calculation of the Shannon diversity index showed a significantly higher diversity of birds in the unlogged area, 2.284 versus 2.037, a difference of 12.1%.

Of course, an accurate determination of community diversity depends on detailed knowledge of which and how many of each species are present. While this is relatively easy to determine for communities of vertebrates and some invertebrates, it is much more difficult for microbial communities. Yet knowledge of microbial communities is of vital importance if only because of the decomposition functions they perform. As described next, with the advent of modern molecular tools, our knowledge of the diversity of microbial communities is beginning to expand.

GENOMES & PROTEOMES

Metagenomics May Be Used to Measure Community Diversity

Bacteria are abundant members of all communities and are vital to their functioning. They serve as food sources for other organisms and participate in the decomposition process. However, most microorganisms are taxonomically unknown, mainly because they cannot be cultivated on known culture media. The new field of **metagenomics** seeks to identify and analyze the collective microbial genomes contained in a community of organisms, including those that are not easily cultured in the laboratory.

1 Isolation of many types of bacterial DNA from environmental sample

DNA isolated

Restriction endonucleases cleave DNA

2 Insertion of fragments into vectors

DNA fragments

DNA fragment

Bacterial vector

Cloned fragment of metagenomic DNA

3 Construction of metagenomic library

Each colony contains a different piece of microbial DNA.

Cloned fragment of metagenomic DNA

Transfer into host bacterial cell (*E. coli*)

Cultured *E. coli* colonies

4 Analysis of metagenomic library

Possible methods of analysis:

• Screen for particular sequences by PCR or hybridization

• Random DNA sequencing

• Screen for expression of particular phenotypes

Figure 58.6 **The standard protocol of a metagenomics experiment.** (1) Isolation and fragmentation of DNA from the sample, (2) insertion of fragments into bacterial vectors, (3) insertion of cloned DNA into host cell, and culturing in selective growth media to create a library, and (4) analysis of DNA sequences and protein expression.

This technique has only been in existence since the early 1990s, but significant progress has already been made in providing libraries that have advanced our understanding of which bacteria are present in various communities and how they function.

The process involves four main steps (**Figure 58.6**). First, an environmental sample containing many bacterial species is collected and its DNA is isolated from the cells using chemical or physical methods. Because the genomic DNA of each species is relatively large, it is cut up into fragments with enzymes called restriction endonucleases. Second, the fragments are combined with vectors, small units of DNA that can be inserted into a model organism, usually a bacterium. The third step is transformation, the physical insertion of the foreign DNA into a host bacterial cell. Individual bacteria are then grown on a selective media so that only the transformed cells survive, and they grow into a colony of cloned cells. A collection of thousands of clones, each containing a different piece of microbial DNA, is called a metagenomic library. Lastly, the DNA from the metagenomic libraries is analyzed. In some cases, expression of the new DNA results in the synthesis of a new protein that changes the phenotype of the host, for example, a new enzyme that is detected by a chemical technique or an unusual color or shape in the model organism.

Metagenomics is helping determine the identity and activities of bacteria in diverse habitats, from the soil to deep-sea sediments to the oral cavity of humans. Recent studies have used shotgun DNA sequencing (see Chapter 21) to analyze bacterial genomes directly from soil and ocean samples. Sampling of deep-sea and ocean sediment samples has permitted sequencing of particular genes, most often the 16S ribosomal RNA (16S rRNA) that all bacteria possess. A sample of many different bacterial species will yield many different 16S rRNA gene sequences. These are then compared to a database containing all known sequences of this gene. Larger DNA segments will be compared to corresponding genomes of cultured bacteria to see if we can assign a tentative metabolic function to each segment. In this way, we can determine the community diversity of soil, sediment, or water samples and make better determinations of how the community is functioning.

In 2004, Jill Banfield and colleagues used metagenomics techniques to identify the five dominant species of bacteria living at temperatures of 42°C (107°F) and pH 0.8 in the acidic waste water (called acid mine drainage) from a mine in California. They detected 2,036 proteins from these species, including proteins from the most abundant bacterial species, a *Leptospirillum* group II bacteria. This represented the first large-scale proteomics-level expression of a natural microbial community. One of the proteins, a cytochrome, oxidizes iron and probably influences the rate of breakdown of acid mine drainage products. Many other proteins appear responsible for defending against free radicals, suggesting that this is a challenging task. The hope is that the team can now identify enzymes and metabolic pathways that will help in the cleanup of this and other contaminated sites in the future.

58.4 Species Richness and Community Stability

A community is often seen as stable when little to no change can be detected in the number of species and their abundance over a given time period. The community may then be said to be in equilibrium. The frame of reference for detecting change may encompass a study of a few years or, preferably, several decades. For example, long-term data from Bookham Common, England, revealed bird communities in which the number of species appeared stable for nearly 30 years. Community stability is an important consideration in conservation biology, as we will discuss further in Chapter 60. A decrease in the stability of a community over time may alert us to a possible problem. In the 1970s, the populations of many raptor species, including peregrine falcons, bald eagles, and osprey, declined precipitously. Eventually, the decline was traced to the pesticide DDT (dichlorodiphenyltrichloroethane), which caused eggshells to become thin and break before the birds could hatch. After DDT was banned later in the decade, raptor communities began to recover.

In this section, we consider the relationship between species richness and community stability. We begin by exploring the question of whether communities with more species are more stable than communities with fewer species. We then examine the link between diversity and stability, using evidence from the field. In a final consideration, we look at the relationship from a different angle and consider whether or not stable communities are more species rich than communities that have been disturbed.

The Diversity-Stability Hypothesis States That Species-Rich Communities Are More Stable Than Those with Fewer Species

Community stability may be viewed in several different ways. Some communities, such as extreme deserts, are considered stable because they are resistant to change by anything other than water. Other communities, such as river communities, are considered stable because they can recover quickly after a disturbance, such as pollution, being cleansed by the rapid flow of fresh water. Lake communities, on the other hand, may be less stable because there is no drainage outlet and pollutants can accumulate quickly.

Because maintaining community stability is seen as important, much research has gone into understanding the factors that enhance it. This work has produced the prevailing idea that species-rich communities are more stable than species-poor communities. There is much debate, for example, over whether species-rich communities are more resistant to invasion by undesirable exotic species, such as weeds, than species-poor communities. The link between species richness and stability was first explicitly proposed by the English ecologist Charles Elton

in the 1950s. He suggested that a disturbance in a diverse, or species-rich, community would be cushioned by large numbers of interacting species and would not produce as drastic an effect as it would on a less diverse community. Thus, an introduced predator or parasite could cause extinctions in a species-poor system but possibly not in a diverse system, where its effects would be buffered by interactions with more species in the community.

In support of this hypothesis, Elton suggested that species-poor island communities are much more vulnerable to invading species than are species-rich continental communities. For example, introduced mosquito species have caused many extinctions of native Hawaiian birds, which are extremely susceptible to the avian malaria spread by the insects. Elton also argued that outbreaks of pests are often found on cultivated land or land disturbed by humans, both of which are greatly simplified areas with few naturally occurring species. His arguments collectively became known as the **diversity-stability hypothesis**.

However, ecologists began to challenge Elton's association of diversity with stability. They pointed out that many examples of introduced species that have assumed pest proportions on continents, not just islands, including rabbits in Australia and pigs in North America. They noted that disturbed or cultivated land may suffer from pest outbreaks not because of its simple nature but because individual species, including exotic pest species and native species of natural enemies, often have no evolutionary history with one another, in contrast to the long associations between pests and natural enemies evident in natural biomes. For example, in Europe, coevolved predators such as foxes prevent rabbit populations from increasing to pest proportions. What was needed was an experiment to determine if a link existed between diversity and stability.

In 1996, ecologist David Tilman reported the relationship between species richness and stability from an 11-year study of 207 grassland plots in Minnesota. He measured the biomass of every species of plant, in each plot, at the end of every year and obtained the average species biomass. He then calculated how much this biomass varied from year to year through a statistical measure called the coefficient of variation. Year-to-year variation in plant community biomass was significantly lower in plots with greater plant species richness (**Figure 58.7**). Tilman's results showed that community stability is positively correlated with diversity.

Tilman suggested that diversity stabilizes communities because diverse plots were more likely to contain disturbance-resistant species that, in the event of a disturbance, could grow and compensate for the loss of disturbance-sensitive species. For example, when a change in climate such as drought harmed competitively dominant species that thrived in normal conditions, decreasing their abundance, unharmed drought-resistant species increased in mass and replaced them. Such decreases in susceptible species and compensatory increases in other species acted to stabilize total community biomass.

(a)

(b)

Figure 58.7 **Relationships between biomass variation and species diversity.** **(a)** Tilman's 11-year study of grassland plots in Minnesota revealed that year-to-year variability in community biomass was lower in species-rich plots. Each dot represents an individual plot, although only the plots from one field are graphed. **(b)** This aerial photograph shows the grassland plots.

(a)

(b)

Figure 58.8 **The intermediate-disturbance hypothesis of community organization.** **(a)** This hypothesis proposes that species richness is highest at intermediate levels of disturbances caused by events such as fires or windstorms. **(b)** A light gap in a tropical rain forest in Costa Rica promotes the growth of small herbaceous species until trees colonize the light gap and gradually grow over and outcompete the species.

Biological inquiry: According to the intermediate-disturbance hypothesis, why are there so many species in the tropics?

The Intermediate-Disturbance Hypothesis Proposes That Moderately Disturbed Communities Are More Diverse

Up to this point, we have focused on the question of whether diverse communities are more stable than species-poor communities. Large-scale field studies have verified that stability is increased by diversity. What if we were to rephrase the question and ask whether stable communities are more diverse than unstable communities? In answer to this question, ecologist Joseph Connell has argued that the highest diversities are maintained not in stable communities but in communities with intermediate levels of disturbance, a concept called the **intermediate-disturbance hypothesis** (**Figure 58.8a**). Disturbance in communities may be brought about by many different phenomena such as droughts,

fires, floods, and hurricanes, or even herbivory, predation, or parasitism. Recall from Chapter 56 that some species, termed *r*-selected species, are better dispersers than other species, and that *K*-selected species are better competitors. Connell reasoned that at high levels of disturbance, only colonists that were *r*-selected species would survive, giving rise to low diversity. This is because these species would be the only ones able to disperse quickly to a highly disturbed area. At low rates of disturbance, competitively dominant *K*-selected species would outcompete all other species, which would also yield low diversity. The most diverse communities would lie somewhere in between.

Connell argued that natural communities fit into this model fairly well. Tropical rain forests and coral reefs are both examples of communities with high species diversity. Connell pointed out that coral reefs maintain their highest diversity in areas disturbed by hurricanes and that the richest tropical forests occur where disturbance by storms causes landslides and tree falls. For example, the fall of a tree creates a hole in the rain forest canopy known as a light gap, where direct sunlight is able to reach the rain forest floor. The light gap is rapidly colonized by *r*-selected species, such as small herbaceous plants, which are well adapted for rapid growth. While these pioneering species grow rapidly, they are overtaken by hardier *K*-selected species, such as mature trees, which fill in the gap in the canopy (**Figure 58.8b**). Although events such as hurricanes and tree falls are fairly frequent events in these communities, their occurrence in any one area is usually of intermediate frequency.

In 1979, Wayne Sousa provided an elegant experimental verification of the intermediate-disturbance hypothesis in the marine intertidal zone. He found that small boulders, which were easily disturbed by waves, carried a mean of 1.7 sessile plant and animal species. These frequently moving boulders crushed or dislodged most colonizing species. Large boulders, which were rarely moved by waves, had a mean of 2.5 species. On these boulders, competitively dominant species supplanted many other species. Sousa found that intermediate-sized boulders had the most species, an average of 3.7 species per boulder, because they contained a mix of *r*- and *K*-selected species. To test the hypothesis, Sousa cemented small boulders to the ocean floor and obtained an increase in species richness to near the value for large boulders, showing that the resulting number of species was a result of rock stability, not rock size.

To summarize, large-scale field experiments have shown a definite link between stability and diversity. However, if we turn the question around, we find that the most diverse communities are not necessarily the most stable, but rather that the most diverse communities exist at intermediate levels of disturbance. Finally, it is instructive to realize that many communities are not permanently stable and that change is a normal ecological phenomenon. Such change often proceeds in predictable ways, a process that ecologists term succession.

58.5 Succession: Community Change

At 8:32 a.m. on May 18, 1980, Mount St. Helens, a previously little studied peak in the Washington Cascades, erupted. The blast felled trees over a 600-km² area, and the landslide that followed—the largest in recorded history—destroyed everything in its path, killing nearly 60 people and thousands of animals. However, since the eruption, much of the area has experienced a relatively rapid recovery of plant and animal communities (**Figure 58.9**).

Ecologists have developed several terms to describe how community change occurs. The term **succession** describes the

(a) 1980

(b) 1997

Figure 58.9 **Succession on Mount St. Helens.** **(a)** The initial blast occurred on May 18, 1980. **(b)** By 1997, 17 years later, many of the areas initially overrun by lava had developed low-lying vegetation and new trees sprouted up between the old dead tree trunks.

gradual and continuous change in species composition and community structure over time. **Primary succession** refers to succession on a newly exposed site that was not previously occupied by soil and vegetation, such as bare ground caused by a volcanic eruption or the rubble created by the retreat of glaciers. In primary succession on land, the plants must often build up the soil, and thus a long time—even hundreds of years—may be required for the process. Only a tiny proportion of the Earth's surface is currently undergoing primary succession, for example, around Mount St. Helens and the volcanoes in Hawaii and off the coast of Iceland, and behind retreating glaciers in Alaska and Canada.

Secondary succession refers to succession on a site that has already supported life but has undergone a disturbance, such as a fire, tornado, hurricane, or flood (as in the 2004 tsunami in Indonesia). Clearing a natural forest and farming the land for several years is an example of a severe forest disturbance that does not kill all native species. Some plants and many soil bacteria, nematodes, and insects are still present. Cessation of farming may lead to a distinct secondary succession. The secondary succession in abandoned farmlands (also called old fields) can lead to a pattern of vegetation quite different from one that develops after primary succession following glacial retreat. For example, the plowing and added fertilizers, herbicides, and pesticides may have caused substantial changes in the soil of an old field, allowing species that require a lot of nitrogen to colonize. These species would not be present for many years in newly created glacial soils.

Frederic Clements is often viewed as the founder of successional theory as well as community ecology in general. His work in the early 20th century emphasized succession as proceeding to a distinct end point or **climax community**. Each phase of succession was called a **sere**, or seral stage. The initial sere was known as the pioneer seral stage. While disturbance could return a community from a later seral stage to an earlier seral stage, generally the community headed towards climax.

A key assumption of Clements was that each colonizing species made the environment a little different—a little more shady or a little more rich in soil nitrogen—so that it became more suitable for other species, which then invaded and outcompeted the earlier residents. This process, known as **facilitation**, supposedly continued until the most competitively dominant species had colonized, when the community was said to be at climax. The composition of the climax community for any given region was thought to be determined by climate and soil conditions. While Clements's ideas on succession focused on the mechanism of facilitation, two other mechanisms affecting succession—inhibition and tolerance—have since been proposed. Next, we'll examine the evidence for each of them.

Facilitation Assumes Each Invading Species Creates a More Favorable Habitat for Succeeding Species

Succession following the gradual retreat of Alaskan glaciers is often used as a specific example of facilitation as a mechanism of succession. Over the past 200 years, the glaciers in Glacier Bay have undergone a dramatic retreat of nearly 100 km (**Figure 58.10**). This is one of the few instances where we can trace the chronology of physical change in an area, from 1794, when Captain George Vancouver visited the inlet and made notes on the positions of the glaciers, to the present.

Succession in Glacier Bay follows a distinct pattern of vegetation. As glaciers retreat, they leave moraines, deposits of stones, pulverized rock, and debris that serve as soil. In Alaska, the bare soil has a low nitrogen content and scant organic matter. In the pioneer stages, the soil is first colonized by a black crust of cyanobacteria, lichens, moss, horsetail (*Equisetum variegatum*), and the occasional river beauty (*Epilobium latifolium*) (**Figure 58.11a**). Because the cyanobacteria are nitrogen fixers, the soil nitrogen increases a little, but soil depth and litterfall (fallen leaves, twigs, and other plant material) are still minimal. At this stage there may be a few seeds and seedlings of dwarf shrubs of the rose family commonly called mountain avens (*Dryas drummondii*), alders, and spruce, but they are rare in the community. After about 40 years, *D. drummondii* dominates the

Figure 58.10 **The degree of glacier retreat at Glacier Bay, Alaska, since 1794.** **(a)** Primary succession begins on the bare rock and soil evident at the edges of the retreating glacier. **(b)** The lines reflect the position of the glacier in 1794 and its subsequent retreat.

landscape (**Figure 58.11b**). Soil nitrogen increases, as does soil depth and litterfall, and alder trees begin to invade.

At about 60 years, alders (*Alnus sinuata*) form dense, close thickets (**Figure 58.11c**). Alders have nitrogen-fixing bacteria that live mutualistically in their roots and convert nitrogen from the air into a biologically useful form. The excess nitrogen fixed by these bacteria accumulates in the soil. Soil nitrogen dramatically increases, as does litterfall. Spruce trees (*Picea sitchensis*) begin to invade at about this time. After about 75 to 100 years, the spruce trees begin to overtop the alders, shading them out. The litterfall is still high and the large volume of needles turns the soil acidic. The shade causes competitive exclusion of many of the original understory species, including alder, and only mosses carpet the ground. At this stage, seedlings of western hemlock (*Tsuga heterophylla*) and mountain hemlock (*Tsuga mertensiana*) may also occur, and after 200 years, a mixed spruce-hemlock climax forest results (**Figure 58.11d**).

What other evidence is there of facilitation? Experimental studies of early primary succession on Mount St. Helens, which show that decomposition of fungi allows mosses and other

Seral stage	Pioneer	*Dryas*	Alder	Spruce
Time (years) since glacial retreat	5	40	60	200
Soil depth (cm)	5.2	7.0	8.8	15.1
Soil N (g/m^2)	3.8	5.3	21.8	53.3
Soil pH	7.2	7.3	6.8	3.6
Litterfall (g/m^2/yr)	1.5	2.8	277	261

Cyanobacteria
Moss
Lichens

Mountain avens
(*Dryas drummondii*)

Alder
(*Alnus sinuata*)

Spruce
(*Picea sitchensis*)
Western hemlock
(*Tsuga heterophylla*)

(a)   (b)   (c)   (d)

Figure 58.11   **The pattern of primary succession at Glacier Bay, Alaska.**   **(a)** The first species to colonize the bare Earth following retreat of the glaciers are small species such as cyanobacteria, moss, and lichens. **(b)** Mountain avens (*Dryas drummondii*) is a flower common in the *Dryas* seral stage. **(c)** Soil nitrogen and litterfall increase rapidly as alder (*Alnus sinuata*) invade. Note also the appearance of a few spruce trees higher up the valley. **(d)** Spruce (*Picea sitchensis*) and hemlock (*Tsuga heterophylla*) trees comprise a climax spruce-hemlock forest at Glacier Bay, with moss carpeting the ground. Two hundred years ago, glaciers occupied this spot.

fungi to colonize the soil, provide evidence of facilitation. In New England salt marshes, *Spartina* grass facilitates the establishment of beach plant communities by stabilizing the rocky substrate and reducing water velocity, which enables other seedlings to emerge. Succession on sand dunes also supports the facilitation model, in that pioneer plant species stabilize the sand dunes and facilitate the establishment of subsequent plant species. The foredunes, those nearest the shoreline, are the most frequently disturbed and are maintained in a state of early succession, while more stable communities develop farther away from the shoreline.

Succession also occurs in aquatic communities. Although soils do not develop in marine environments, facilitation may still be encountered when one species enhances the quality of settling and establishment sites for another species. When experimental test plates used to measure settling rates of marine organisms were placed in the Delaware Bay, researchers discovered that certain cnidarians enhanced the attachment of tunicates, and both facilitated the attachment of mussels, which were the dominant species in the community. In this experiment, the smooth surface of the test plates prevented many species from colonizing but once the surface became rougher, because of the presence of the cnidarians, many other species

were able to colonize. In a similar fashion, early colonizing bacteria, which create biofilms on rock surfaces, can facilitate succession of other organisms.

Inhibition Implies That Early Colonists Prevent Later Arrivals from Replacing Them

Although data on succession in some communities fit the facilitation model, researchers have proposed alternative hypotheses of how succession may operate. Another view is that possession of space is all-important, and that what gets there first determines subsequent community structure. In this process, known as **inhibition**, early colonists may exclude subsequent colonists. For example, removing the litter of *Setaria faberi*, an early successional plant species in New Jersey old fields, causes an increase in the biomass of a later species, *Erigeron annuus*. The release of toxic compounds from decomposing *Setaria* litter or physical obstruction by the litter itself contributes to the inhibition of *Erigeron*. Without the litter present, *Erigeron* dominates and reduces the biomass of *Setaria*. Plant species that grow in dense thickets, such as some grasses, ferns, vines, pine trees, and bamboo, can inhibit succession, as can many introduced plant species.

(a)

(b)

Figure 58.12 **Inhibition as a primary method of succession in the marine intertidal zone.** **(a)** *Gigartina* on rock face with the herbivorous crab, *Pachygrapsus crassipes*. **(b)** Removing *Ulva* from intertidal rock faces allowed colonization by *Gigartina*.

Figure 58.13 **Three models of succession.** A, B, C, and D represent four stages or seres. D represents the climax community. An arrow indicates "is replaced by," and + = facilitation, − = inhibition, and 0 = no effect. The facilitation model is the classic model of succession. In the inhibition model, much depends on which species gets there first. The tolerance model is similar to the facilitation model, in that later species may be facilitated by earlier species, but they can also invade in their absence.

Biological inquiry: Inhibition implies competition between species with early arriving species tending to outcompete later arrivals, at least for a while. Does competition or mutualism feature more prominently in facilitation?

Inhibition has been seen as the primary method of succession in the marine intertidal zone, where space is limited. In this habitat, early successional species are at a great advantage in maintaining possession of valuable space. In 1974, ecologist Wayne Sousa created an environment for testing how succession works in the intertidal zone by scraping rock faces clean of all algae or putting out fresh boulders or concrete blocks. The first colonists of these areas were the green algae *Ulva*. By removing *Ulva* from the substrate, Sousa showed that the large red alga *Gigartina canaliculata* was able to colonize more quickly (**Figure 58.12**). The results of Sousa's study indicate that early colonists can inhibit rather than facilitate the invasion of subsequent colonists. Succession may eventually occur because early colonizing species, such as *Ulva*, are more susceptible to the rigors of the physical environment and to attacks by herbivores, such as crabs (*Pachygrapsus crassipes*), than later successional species, such as *Gigartina*.

Tolerance Suggests That Early Colonists Neither Facilitate nor Inhibit Later Colonists

The huge differences between facilitation and inhibition as mechanisms of succession prompted Joseph Connell and Ralph Slatyer to view the two as extremes on a continuum. In 1977, the two researchers proposed a third mechanism of succession, which they termed **tolerance**. In this process, any species can start the succession, but the eventual climax community is reached in a somewhat orderly fashion. Early species neither facilitate nor inhibit subsequent colonists. Connell and Slatyer found the best evidence for the tolerance model in Frank Egler's earlier work on floral succession. In the 1950s, Egler showed that succession in plant communities is determined largely by species that already exist in the ground as buried seeds or old roots. Whichever species germinates first, or regenerates from roots, initiates the succession sequence.

The key distinction between the three models is in the manner in which succession proceeds. In the facilitation model, species replacement is facilitated by previous colonists; in the inhibition model, it is inhibited by the action of previous colonists; and in the tolerance model, species may be unaffected by previous colonists but they do not require them (**Figure 58.13**). Subsequent research has suggested that other factors may also influence succession, especially on islands. The study of succession on islands is often referred to as island biogeography, which is described next.

58.6 Island Biogeography

In the 1960s, two eminent ecologists, Robert MacArthur and E. O. Wilson, developed a comprehensive model to explain the process of succession on new islands, where a gradual buildup of species proceeds from a sterile beginning. Their findings, termed the **equilibrium model of island biogeography**, hold that the number of species on an island tends toward an equilibrium number that is determined by the balance between two factors: immigration rates and extinction rates. In this section, we explore island biogeography and how well the model's predictions are supported by data, particularly that provided by classic experiments in the Florida Keys.

The Island Biogeography Model Suggests That During Succession Losses from Extinction Are Balanced by Gains in Immigration

MacArthur and Wilson's model of island biogeography suggests that species repeatedly arrive on an island and either thrive or become extinct. Thus, the number of species tends towards an equilibrium number, Ŝ, which reflects a balance between the rate of immigration and the rate of extinction. The rate of immigration of new species is highest when no species are present on the island, so that each species that invades the island is a new species. As species accumulate, subsequent immigrants are less likely to represent new species. The rate of extinction is low at the time of first colonization, because few species are present and many have large populations. With the addition of new species, the populations of some species diminish, so the probability of extinction by chance alone increases. Species may continue to arrive and go extinct, but the number of species on the island remains approximately the same.

MacArthur and Wilson reasoned that both the immigration and extinction lines would be curved, for several reasons (**Figure 58.14a**). First, species arrive on islands at different rates. Some organisms, including plants with seed dispersal mechanisms and winged animals, are more mobile than others and will arrive quickly. Other organisms will arrive more slowly. This pattern causes the immigration curve to start off steep but get progressively shallower. On the other hand, extinctions rise at accelerating rates, because as later species arrive, competition increases and more species are likely to go extinct. As noted previously, earlier arriving species tend to be *r*-selected species, which are better dispersers, whereas later arriving species are generally *K*-selected species, which are better competitors. Later arriving species usually outcompete earlier arriving ones, causing an increase in extinctions.

The strength of the island biogeography model was that it generated several falsifiable predictions:

1. The number of species should increase with increasing island size. This is also known as the species-area effect (see Figure 58.4). Extinction rates would be greater on smaller islands because population sizes would be smaller and more susceptible to extinction (**Figure 58.14b**).
2. The number of species should decrease with increasing distance of the island from the mainland, or the **source pool**, the pool of species that is available to colonize the

(a)

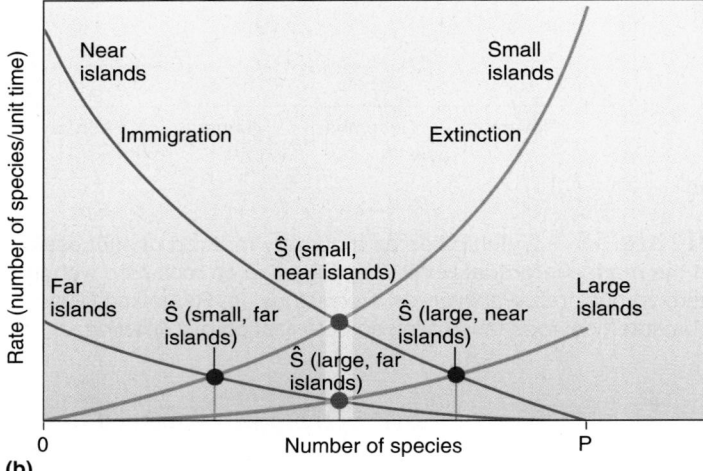

(b)

Figure 58.14 MacArthur and Wilson's theory of island biogeography. (a) The interaction of immigration rate and extinction rate produces an equilibrium number of species on an island, Ŝ. Ŝ varies from 0 species to *P* species, the total number of species available to colonize. (b) Ŝ varies according to the island's size and distance from the mainland. An increase in distance (near to far) lowers the immigration curve, while an increase in island area (small to large) lowers the extinction rate.

island. Immigration rates would be greater on islands near the source pool because species do not have as far to travel (Figure 58.14b).

3. The turnover of species should be considerable. The number of species on an island might remain the same, but the composition of the species should change continuously as new species colonize the island and others become extinct.

Let's examine these predictions one by one and see how well the data support them.

The Number of Species Increases with Increasing Island Size

The West Indies has traditionally been a key location for ecologists studying island biogeography. This is because the physical geography and the plant and animal life of the islands are well known. Furthermore, the Lesser Antilles, from Anguilla in the north to Grenada in the south, enjoy a similar climate and are surrounded by deep water (**Figure 58.15a**). In 1999, Robert Ricklefs and Irby Lovette summarized the available data on the richness of species of four groups of animals—birds, bats, reptiles and amphibians, and butterflies—over 19 islands that varied in area over two orders of magnitude (13–1,510 km²). In each case, a possible correlation occurred between area and species richness (**Figure 58.15b**). Note that these relationships are traditionally plotted on a double logarithmic scale, a so-called log-log plot, in which the horizontal axis is the logarithm to the base 10 of the area and the vertical axis is the logarithm to the base 10 of the number of species. A linear plot of the area versus the number of species would be difficult to produce, because of the wide range of area and richness of species involved. Logarithmic scales condense this variation to manageable limits.

Apart from Ricklefs and Lovette's study, species-area relationships exist for birds of the East Indies, beetles on West Indian islands, ants in Melanesia, and land plants of the Galápagos, providing strong support for this prediction of the equilibrium model of island biogeography.

The Number of Species Decreases with Increasing Distance from the Source Pool

MacArthur and Wilson also provided evidence for the effect of the distance of an island from a source pool of colonists, usually the mainland. In studies of the numbers of lowland forest bird species in Polynesia, they found that the number of species decreased with the distance from New Guinea, the source pool in this case (**Figure 58.16**). They expressed the richness of bird species on the islands as a percentage of the number of bird species found on New Guinea. There was a significant decline in this percentage with increasing distance, with more distant islands containing lower numbers of species than nearer islands.

(a)   (b)

Figure 58.15 Species richness increases with island size. (a) The Lesser Antilles extend from Anguilla in the north to Grenada in the south. (b) The number of bird, bat, reptile and amphibian, and butterfly species increases with the area of an island.

Biological inquiry: How large is the change in bird species richness across islands in the Lesser Antilles?

(a)

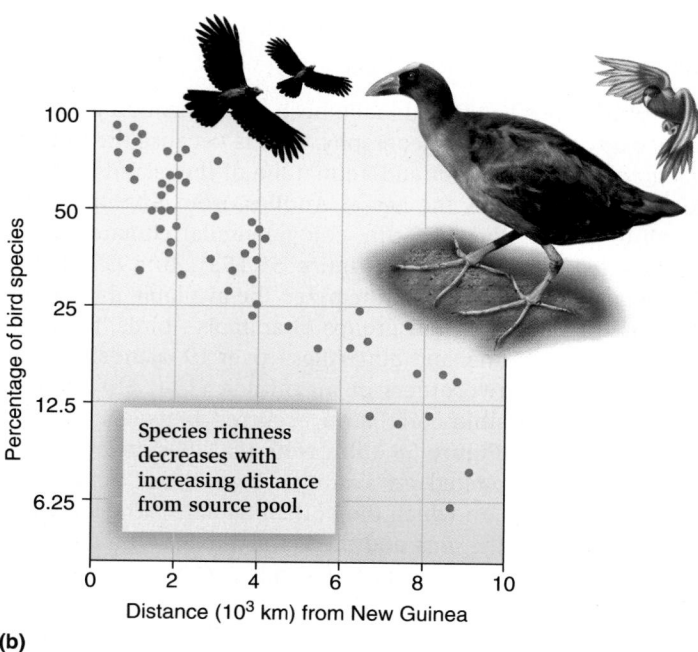

(b)

Figure 58.16   Species richness decreases with distance from the source pool.   (a) Map of Australia, New Guinea, and these Polynesian Islands: New Caledonia, Fiji Islands, Cook Islands, Marquesas Islands, Pitcairn, and Easter Island. (b) The numbers of bird species on islands with increasing distance from the source pool, New Guinea, expressed as the percentage of bird species on New Guinea.

This research substantiated the prediction of species richness declining with increasing distance from the source pool.

Species Turnover on Islands Is Generally Low

Studies involving species turnover on islands are difficult to perform because detailed and complete species lists are needed over long periods of time, usually many years and often decades. The lists that do exist are often compiled in a casual way and are not usually suitable for comparison with more modern data. In 1980, British researcher Francis Gilbert reviewed 25 investigations carried out to demonstrate turnover and found a lack of this type of rigor in nearly all of them. Furthermore, most of the observed turnover in these studies, usually less than 1% per year, or less than one species per year, appeared to be due to immigrants that never became established, not due to the extinction of well-established species. More recent studies have revealed similar findings. The take-home message from most studies is that recorded rates of turnover are low, giving little conclusive support to this prediction of the equilibrium model of island biogeography.

FEATURE INVESTIGATION

Wilson and Simberloff's Experiments Tested the Predictions of the Equilibrium Model of Island Biogeography

E. O. Wilson and Daniel Simberloff conducted possibly the best test of the equilibrium model of island biogeography ever performed using islands in the Florida Keys. They surveyed four small red mangrove (*Rhizophora mangle*) islands, 11–25 m in diameter, for all terrestrial arthropods. They then enclosed each island with a plastic tent and had the islands fumigated with methyl bromide, a short-acting insecticide, to remove all arthropods on them. The tents were removed, and periodically thereafter Wilson and Simberloff surveyed the islands to examine recolonization rates. At each survey, they counted all the species present, noting any species not there at the previous census and the absence of others that were previously there but had presumably gone extinct (results for four of the islands shown in **Figure 58.17**). In this way, they estimated turnover of species on islands.

After 250 days, all but one of the islands had a similar number of arthropod species as they had to begin with, even though population densities were still low. The data indicated that colonization rates were higher on islands nearer to the mainland than on far islands—as the island biogeography model predicts. However, the data, which consisted of lists of species on islands before and after extinctions, provided little support for the prediction of high turnover. Rates of turnover were low, only 1.5 extinctions per year, compared to the 15 to 40 species found on the islands within a year. Wilson and Simberloff concluded that turnover probably involves only a small subset of transient or unimportant species, with the more important species being permanent after colonization.

Figure 58.17   Wilson and Simberloff's experiments on the equilibrium model of biogeography.

HYPOTHESIS Island biogeography model predicts substantial turnover of species on islands.

STARTING LOCATION Mangrove islands in the Florida Keys

| Experimental level | Conceptual level |

1 Take initial census of all terrestrial anthropods on 4 mangrove islands. Erect framework over each mangrove island.

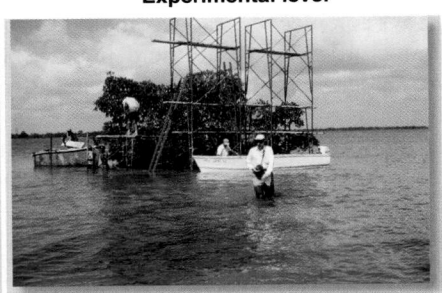

Each mangrove island is isolated.

2 Cover framework with tent and fumigate with methyl bromide to defaunate island.

Methyl bromide is a low-persistent insecticide that at low levels will not kill plant life.

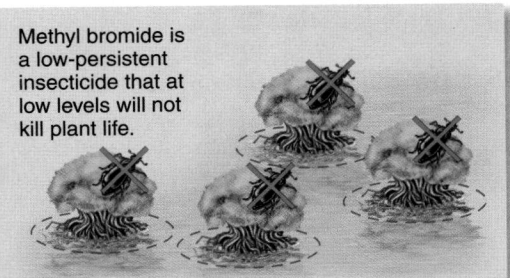

3 Remove tents and conduct censuses every month to monitor recolonization.

Mangrove islands are recolonized.

4 THE DATA   Island E2 was closest to the mainland and supported the highest number of species. E3 and ST2 were at an intermediate distance from the mainland, and E1 was the most distant.

In summary, MacArthur and Wilson's equilibrium model of island biogeography has stimulated much research that confirms the strong effects of area and distance on species richness. However, species turnover appears to be low rather than considerable, which suggests that succession on most islands is a fairly orderly process. This means that colonization is not a random process and that the same species seem to always colonize first and other species gradually reappear in the same order. It is also important to note that the principles of island biogeography can be applied not only to a body of land surrounded by water, but also to fragments of habitat such as wildlife preserves and sanctuaries. For example, wildlife preserves are essentially islands in a sea of developed land, either agricultural fields or urban sprawl. Conservationists have therefore utilized island biogeography modeling in the concept of preserve design, a topic we will explore in Chapter 60.

In this chapter, we have seen how important species richness is to community function. It affects not only stability but, as we shall see, many other features of a community such as nutrient uptake, biomass, and productivity. In order to know how diversity affects these properties, we need to understand the basic processes of energy flow and chemical cycling within a community and its surrounding environment, and for that we turn to a discussion of ecosystems ecology.

Chapter Summary

58.1 Differing Views of Communities

- Community ecology studies how groups of species interact and form functional communities. Ecologists have differing views on the nature of a community. In one view, communities are tightly organized groups of mutually dependent species, and in another, they are loose assemblies of species that happen to live in the same place at the same time. (Figure 58.1)

58.2 Patterns of Species Richness

- While many observations support the idea that communities are loose assemblages of species, sharp boundaries between groups of species do exist, especially related to physical differences that cause distinct communities to develop. (Figure 58.2)
- The number of species of most taxa varies according to geographic location, generally increasing from polar areas to tropical areas. (Figure 58.3)
- Varying hypotheses for the polar-equatorial gradient have been advanced, including the time hypothesis, the area hypothesis, and the productivity hypothesis. (Figures 58.4, 58.5)

58.3 Calculating Species Diversity

- The most widely used measure of the diversity of a community, called the Shannon diversity index, takes into account both species richness and species abundance. (Table 58.1)
- The field of metagenomics seeks to identify and analyze the genomes contained in a community of microorganisms. (Figure 58.6)

58.4 Species Richness and Community Stability

- Community stability is an important consideration in ecology. The diversity-stability hypothesis maintains that species-rich communities are more stable than communities with fewer species.
- Tilman's field experiments, which showed that year-to-year variation in plant biomass decreased with increasing species diversity, established a link between diversity and stability. (Figure 58.7)

- The intermediate-disturbance hypothesis proposes that higher species diversity is maintained in communities with moderate levels of disturbance. (Figure 58.8)

58.5 Succession: Community Change

- Succession describes the gradual and continuous change in community structure over time. Primary succession refers to succession on a newly exposed site not previously occupied by soil; secondary succession refers to succession on a site that has already supported life but has undergone a disturbance. (Figures 58.9, 58.10)
- Three mechanisms have been proposed for succession: in facilitation, each species facilitates or makes the environment more suitable for subsequent species. In inhibition, initial species inhibit later colonists. In tolerance, any species can start the succession, and species replacement is unaffected by previous colonists. (Figures 58.11, 58.12, 58.13)

58.6 Island Biogeography

- In the equilibrium model of island biogeography, the number of species on an island tends toward an equilibrium number determined by the balance between immigration rates and extinction rates. (Figure 58.14)
- The model predicts that the number of species increases with increasing island size, and that the number of species decreases with distance from the source pool. (Figures 58.15, 58.16)
- Wilson and Simberloff's experiments on mangrove islands in the Florida Keys provided some support for the island biogeography model. (Figure 58.17)

Test Yourself

1. The number of species in a community is called
   a. species complexity.
   b. community complexity.
   c. species richness.
   d. species diversity.
   e. species abundance.

2. Which of the following statements best represents the productivity hypothesis regarding species richness?
   a. The larger the area, the greater the number of species that will be found there.
   b. Temperate regions have a lower species richness due to the lack of time available for migration after the last ice age.
   c. The number of species in a particular community is directly related to the amount of plant biomass available for consumers.
   d. As invertebrate productivity increases, species richness will increase.
   e. Species richness is not related to primary productivity.

3. The Shannon diversity index is a measure of
   a. the number of different species in a community.
   b. the abundance of a species in a community.
   c. the types of species found in a typical climate.
   d. the number of different species and their relative abundance in a community.
   e. the distribution of members of a species in a community.

4. Metagenomics is a field of study that
   a. determines the similarities of the genomes of all species in a community.
   b. focuses on the microbial genomes contained in a community.
   c. compares the genomes of similar species in different communities.
   d. none of the above.
   e. both a and b.

5. Extreme fluctuations in species abundance
   a. lead to more diverse communities.
   b. are usually seen in early stages of community development.
   c. may increase the likelihood of extinction.
   d. have very little effect on species richness.
   e. are characteristic of stable communities.

6. Which of the following statements best represents the relationship between species diversity and community disturbance?
   a. Species diversity and community stability have no relationship.
   b. Communities with high levels of disturbance are more diverse.
   c. Communities with low levels of disturbance are more diverse.
   d. Communities with intermediate levels of disturbance are more diverse.
   e. Communities with intermediate levels of disturbance are less diverse.

7. The process of primary succession occurs
   a. around a recently erupted volcano.
   b. on a newly plowed field.
   c. on a hillside that has suffered a mudslide.
   d. on a recently flooded riverbank.
   e. none of the above.

8. When the early colonizers exclude subsequent colonists from moving into a community, this is referred to as
   a. facilitation.
   b. competitive exclusion.
   c. secondary succession.
   d. inhibition.
   e. natural selection.

9. Which of the following statements is not true concerning island biogeography?
   a. Diversity on islands is directly related to the size of the island.
   b. Diversity on islands is inversely related to the distance of the island from the source for colonizing species.
   c. The number of species on an island represents the balance between immigration and extinction rates.
   d. The number of species on an island increases with increasing distance from the source pool.
   e. The number of species on an island increases with increasing island size.

10. The equilibrium model of island biogeography
   a. is a method of mapping islands in the Pacific Ocean.
   b. is a method of identifying particular species on particular islands.
   c. describes factors that influence succession on new islands.
   d. describes the pattern of species types that first colonize an island.
   e. predicts that the first colonizers of an island will become extinct.

CONCEPTUAL QUESTIONS

1. Define a community and community ecology.
2. Explain the productivity hypothesis.
3. When is a community in equilibrium?

EXPERIMENTAL QUESTIONS

1. What was the purpose of Wilson and Simberloff's study?
2. Why did the researchers conduct a thorough species survey of arthropods before experimental removal of all the arthropod species?
3. What did the researchers conclude about the relationship between island size and species richness and about species turnover?

COLLABORATIVE QUESTIONS

1. Discuss the concept of ecological succession.
2. What are the three most important hypotheses used to predict species richness?

www.brookerbiology.com
This website includes answers to the Biological Inquiry questions found in the figure legends and all end-of-chapter questions.

59
ECOSYSTEMS ECOLOGY

CHAPTER OUTLINE

Mandara Lake Oasis, Libya.

The term **ecosystem** was coined in 1935 by the British plant ecologist A. G. Tansley to include not only the biotic community of organisms in an area but also the abiotic environment affecting that community. **Ecosystems ecology** is concerned with the movement of energy and materials through organisms and their communities. Just like the concept of a community, the ecosystem concept can be applied at any scale: A small pond inhabited by protozoa and insect larvae is an ecosystem, and an oasis with its plants, frogs, fish, and birds constitutes another. Most ecosystems cannot be regarded as having definite boundaries. Even in a clearly defined pond ecosystem, amphibians may be moving in and out (**Figure 59.1**). Neverthe-

less, studying ecosystems ecology allows us to use the common currency of energy and chemicals, or nutrients, to compare the functions between and within ecosystems.

In investigating the different processes of an ecosystem, at least three major constituents can be measured: energy flow, biomass production, and biogeochemical cycling. We begin the chapter by exploring **energy flow**, the movement of energy through the ecosystem. In examining energy flow, the main tasks are to document the complex networks of feeding relationships, called food webs, and to measure the efficiency of energy transfer between organisms in an ecosystem. We also consider the tendency of chemical elements to accumulate in organisms, a process called biomagnification.

Next, we focus on the measurement of **biomass**, a quantitative estimate of the total mass of living matter in a given area, usually measured in grams or kilograms per square meter. We will examine the amount of biomass produced through photosynthesis, termed primary production, and the amount of biomass produced by the organisms that are the consumers of primary production. Attaching too much importance to biomass alone, however, may lead to erroneous conclusions. For example, in the applied science of timber technology, a small biomass may indicate a small potential harvest or yield, but this is not necessarily true if the tree species in question has a high growth rate, because new biomass will be produced rapidly and a high rate of harvest could be sustained.

The functioning of an ecosystem can sometimes be most limited by the availability of a scarce chemical or mineral. In the last section, we examine the movement of limiting chemicals through ecosystems, called **biogeochemical cycles**. We explore the cycling of chemical elements such as nitrogen, carbon, sulfur, and phosphorus and the effect that human activities are having on these ecosystem-wide processes.

Figure 59.1 **A small ecosystem.** Even in this pond ecosystem, frogs or other species such as birds may move in and out, importing or exporting nutrients and energy with them.

59.1 Food Webs and Energy Flow

Most organisms either make their own food using energy from sunlight or feed on other organisms. Simple feeding relationships between organisms can be characterized by an unbranched **food chain**, a linear depiction of energy flow, with each organism feeding on and deriving energy from the preceding organism. Each feeding level in the chain is called a **trophic level** (from the Greek *trophos*, feeder), and different species feed at different levels. In a food chain diagram, an arrow connects each trophic level with the one above it (**Figure 59.2**).

In this section, we will examine trophic relationships and the flow of energy in a food chain and a food web, a more complex model of interconnected food chains. We will then explore two of the most important features of food webs—chain length and the pyramid of numbers—and learn how the passage of nutrients through food webs can result in the concentration of harmful chemicals in the tissues of organisms at higher trophic levels.

The Main Trophic Levels Within Food Chains Consist of Primary Producers, Primary Consumers, and Secondary Consumers

Food chains typically consist of organisms that obtain energy in different ways. **Autotrophs** harvest light or chemical energy and store that energy in carbon compounds. Most autotrophs, includ-

ing plants, algae, and photosynthetic prokaryotes, use sunlight for this process. These organisms, called **primary producers**, are at the base of the food chain. They produce the energy-rich tissue upon which nearly all other organisms depend. Note that not all primary producers utilize sunlight; some organisms, called chemoautotrophs, obtain their energy by oxidizing inorganic compounds such as sulfides (**Figure 59.3**).

Organisms in trophic levels above the primary producers are termed **heterotrophs**. These organisms receive their nutrition by eating other organisms. Organisms that obtain their food by eating primary producers are **primary consumers** and include animals, most protists, and even some plants. They are also called **herbivores**. Organisms that eat primary consumers are **secondary consumers**, also called **carnivores** (from the Latin, *carn*, flesh). Organisms that feed on secondary consumers are called **tertiary consumers**, and so on. Thus, energy enters a food chain through primary producers, via photosynthesis, and is passed up the food chain to primary, secondary, and tertiary consumers (see Figure 59.2).

Much energy from the first trophic level, the plants, goes unconsumed by herbivores. Instead, unconsumed plants die and decompose in place. This material, along with dead remains of animals and waste products, is called **detritus**. Consumers that get their energy from detritus, called **detritivores** or **decomposers**, break down dead organisms from all trophic levels (**Figure 59.4**). Detritivores probably carry out 80–90% of the consumption of plant matter, with different species working in concert to extract most of the energy. Detritivores may in turn support a community of predators that feed on them.

Usually, many herbivore species feed on the same plant species. For example, one may find many insect species and vertebrate grazers all feeding on one type of plant. Also, many species of herbivores eat several different plant species. Such branching of food chains also occurs at other trophic levels.

Figure 59.2 Food chains. Two examples of the flow of food energy up the trophic levels: a terrestrial food chain, and an aquatic food chain.

Figure 59.3 Chemoautotrophic organisms living around deep-sea hydrothermal vents. Some bacteria obtain their energy from the sulfur-rich emissions of hydrothermal vents on the ocean floor, also called black smokers. Deep-sea crabs and tube worms may also be found here.

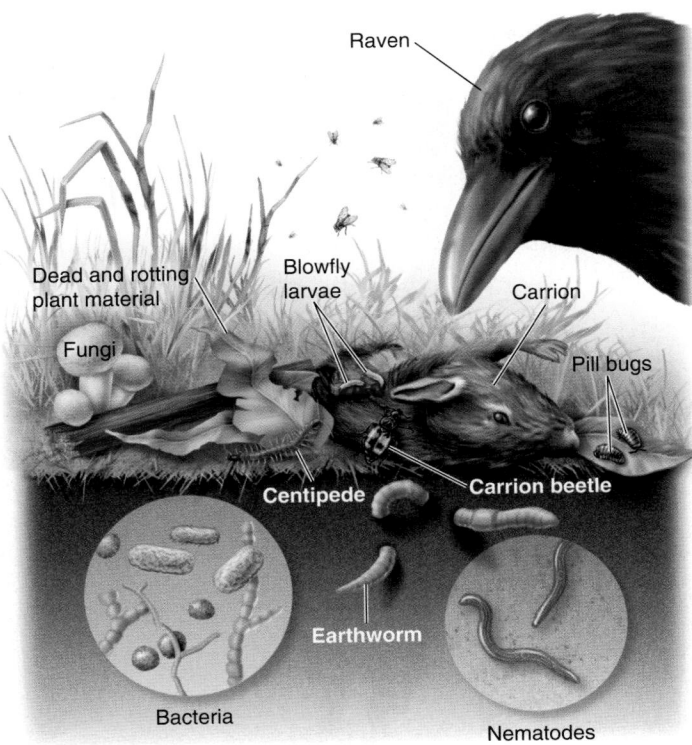

Figure 59.4 **Decomposers (detritivores) feeding on dead plant and animal matter.** Many dead plants and animals are eaten by a variety of organisms. Here, pill bugs and fungi feed on dead wood and rotting plant material, while earthworms, nematodes, and bacteria decompose organic matter in the soil. Dead animals, or carrion, may be fed upon by blowfly larvae, carrion beetles, and other scavengers. These may, in turn, support a variety of predators, including centipedes or larger predators such as this raven, which will also feed on the animal carcass.

For instance, on the African savanna, cheetahs, leopards, lions, and wild dogs all eat a variety of prey including giraffes, gazelles, baboons, zebras, and wildebeest. These, in turn, eat a variety of grasses and trees. It is more correct, then, to draw relationships between these plants and animals not as a simple chain but as a more elaborate interwoven **food web**, in which there are multiple links between species (**Figure 59.5**).

In Most Food Webs, Chain Lengths Are Short and a Pyramid of Numbers Exists

Now that we have introduced the structure of food webs, let's examine some of their characteristics in more detail. In food webs, the concept of chain length refers to the number of links between the trophic levels involved. For example, if a lion feeds on a zebra, and a zebra feeds on grass, the chain length would be two. For most food webs, chain lengths tend to be short, usually less than five. The main reason why they are short comes from the well-established laws of physics and chemistry that we discussed in Chapter 2. The first law of thermodynamics states that energy cannot be created or destroyed, only transformed. We can thus construct energy budgets for food webs that trace energy flow from green plants to tertiary consumers (and if needed beyond). The second law of thermodynamics states that energy conversions are not 100% efficient. In any transfer process, some useful energy, which can do work, is lost (**Figure 59.6**). This suggests that we can compare the efficiency of energy transfer through trophic levels in different types of food webs. The two main measures of the efficiency of consumers as energy transformers are production efficiency and trophic-level transfer efficiency.

Production Efficiency **Production efficiency** is defined as the percentage of energy assimilated by an organism that becomes incorporated into new biomass.

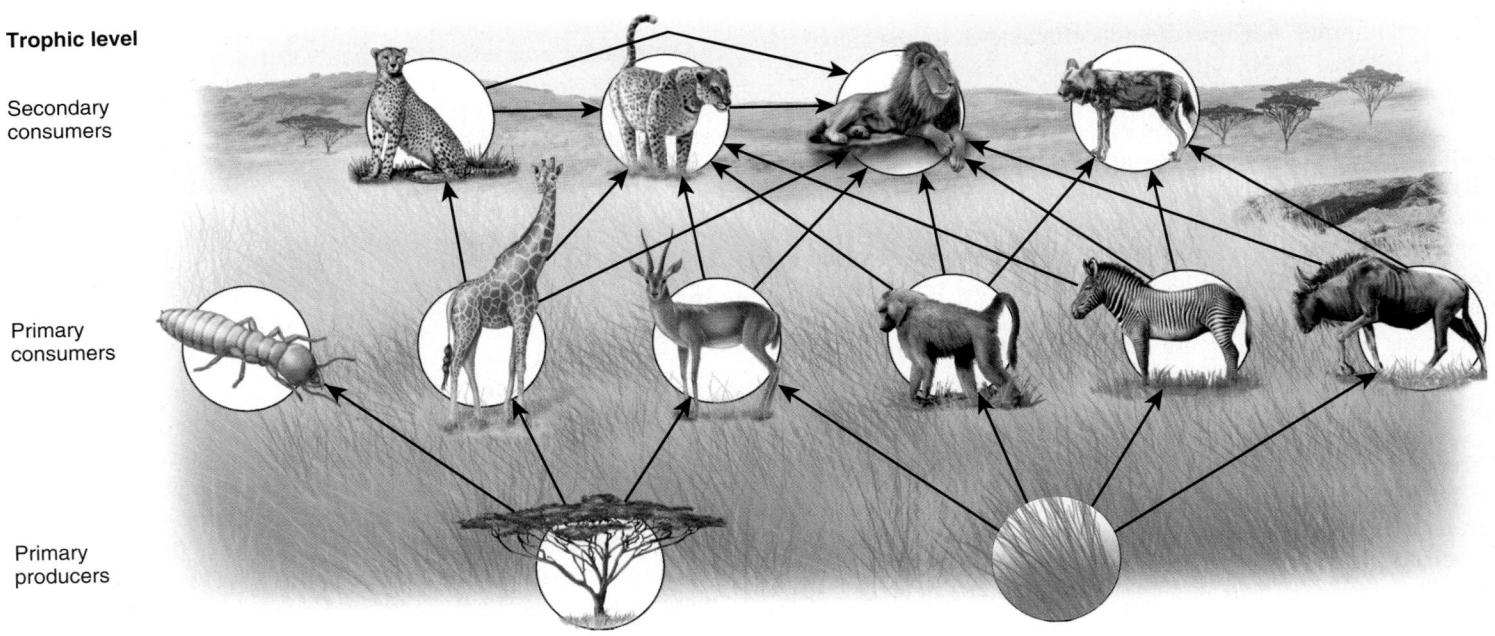

Figure 59.5 **A simplified food web from an African savanna ecosystem.** Each trophic level is occupied by different species. Generally, each species feeds on, or is fed upon by, more than one species.

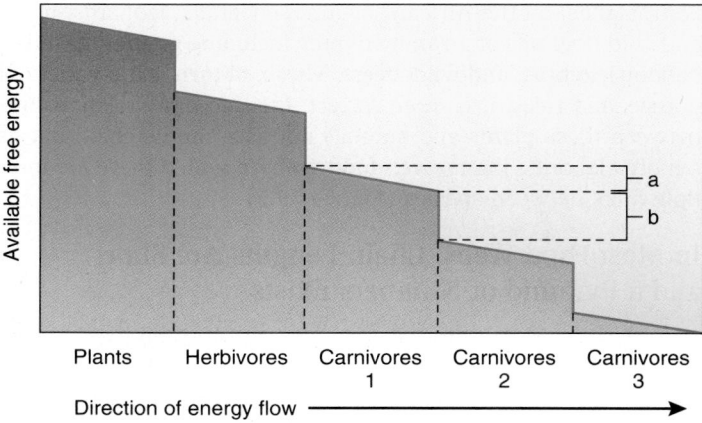

Figure 59.6 **Energy flow through an ecosystem.** **(a)** Energy lost as heat in a single trophic level. **(b)** Energy lost in the conversion between one trophic level and another.

$$\text{Production efficiency} = \frac{\text{Net productivity}}{\text{Assimilation}} \times 100$$

Here, net productivity is the energy, stored in biomass, that has accumulated over a given time span, and assimilation is the total amount of energy taken in by an organism over the same time span. Invertebrates generally have high production efficiencies that average about 10–40% (**Figure 59.7a**). Microorganisms also have relatively high production efficiencies. Vertebrates tend to have lower production efficiencies than invertebrates, because they devote more energy to sustaining their metabolism than to new biomass production. Even within vertebrates, much variation occurs. Fish, which are ectotherms, typically have production efficiencies of around 10%, and birds and mammals, which are endotherms, have production efficiencies in the range of 1–2% (**Figure 59.7b**). In large part, the difference reflects the energy cost of maintaining a constant body temperature.

One consequence of these differences is that sparsely vegetated deserts can support healthy populations of snakes and lizards while mammals might easily starve. The largest living lizard known, the Komodo dragon, eats the equivalent of its own weight every two months, whereas a cheetah consumes approximately four times its own weight in the same period. It is also interesting to note that production efficiencies are higher in young animals, which are rapidly accruing biomass, than in older animals, which are not. This is the main reason behind the practice of raising young chickens and calves for meat.

Trophic-Level Transfer Efficiency The second measure of efficiency of consumers as energy transformers is **trophic-level transfer efficiency**, which is the amount of energy at one trophic level that is acquired by the trophic level above and incorporated into biomass. This provides a way to examine energy flow between trophic levels, not just in individual species. Trophic level transfer efficiency is calculated as:

Figure 59.7 **Production efficiency.** **(a)** This caterpillar, an invertebrate, chews leaves to obtain its energy. If a mouthful of food contains 1,000 joules (J) of energy, about 320 J is used to fuel metabolic processes (32%) and 500 J (50%) is lost in feces. This leaves about 180 J to be converted into insect biomass, a production efficiency of 18%. **(b)** The production efficiency of this squirrel, a mammal, is much lower.

Biological inquiry: What is the production efficiency of the squirrel, using the numbers in the figure?

$$\text{Trophic-level transfer efficiency} = \frac{\text{Production at trophic level } n}{\text{Production at trophic level } n-1} \times 100$$

For example, if there were 14 g/m² of zooplankton in a lake (trophic level n) and 100 g/m² of phytoplankton production (trophic level $n-1$), the trophic level efficiency would be 14%. Trophic-level transfer efficiency appears to average around 10%, though there is much variation. In some marine food chains, for example, it can exceed 30%.

Trophic-level transfer efficiency is low for two reasons. First, many organisms cannot digest all their prey. They take only the easily digestible plant leaves or animal tissue such as muscles and guts, leaving the hard wood or energy-rich bones behind.

Second, much of the energy assimilated by animals is used in maintenance, so most energy is lost from the system as heat. The 10% average transfer rate of energy from one trophic level to another also necessitates short food webs of no more than four or five levels. Relatively little energy is available for the higher levels.

The Pyramid of Numbers Trophic-level transfer efficiencies can be expressed in a graphical form called an Eltonian pyramid, named after the British ecologist Charles Elton. The best-known pyramid, and the one described by Elton in 1927, is the **pyramid of numbers**, in which the number of individuals decreases at each trophic level, with a huge number of individuals at the base and fewer individuals at the top. Elton's example was that of a small pond where the numbers of protozoa may run into the millions and those of *Daphnia*, their predators, number in the hundreds of thousands. Hundreds of beetle larvae may feed on *Daphnia*, and tens of fish feed on the beetles (**Figure 59.8a**). Many other examples of this type of pyramid are known. For example, in a grassland, there may be hundreds of individual plants per square meter, dozens of insects that feed on the plants, and a few spiders feeding on the insects.

However, one can also think of several exceptions to this pyramid. An oak tree, one single producer, supports hundreds of herbivorous beetles, caterpillars, and other primary consumers, which in turn may support thousands of predators and parasites (**Figure 59.8b**). This is called an inverted pyramid of numbers. The best way to reconcile this apparent exception is to weigh the organisms in each trophic level, creating a **pyramid of biomass**. The oak tree weighs 30,000 kg, all the herbivores on the tree total 5 kg, and the predators about 1 kg. Looking at the biomass at each trophic level rather than at numbers of organisms shows an upright pyramid (**Figure 59.8c**).

Inverted pyramids can still occur, albeit rarely, even when biomass is used as the measure. In some marine systems, the biomass of phytoplankton supports a higher biomass of zooplankton, which in turn is eaten by a higher biomass of carnivorous fish (**Figure 59.8d**). This is possible because the production rate of phytoplankton is much higher than that of zooplankton, and the small phytoplankton **standing crop** (the total biomass in an ecosystem at any one point in time) processes large amounts of energy. By expressing the pyramid in terms of production, it is no longer inverted. The **pyramid of production**, which shows rates of production rather than standing crop, is never inverted (**Figure 59.8e**). The laws of thermodynamics ensure that the highest amounts of free energy are found at the lowest trophic levels.

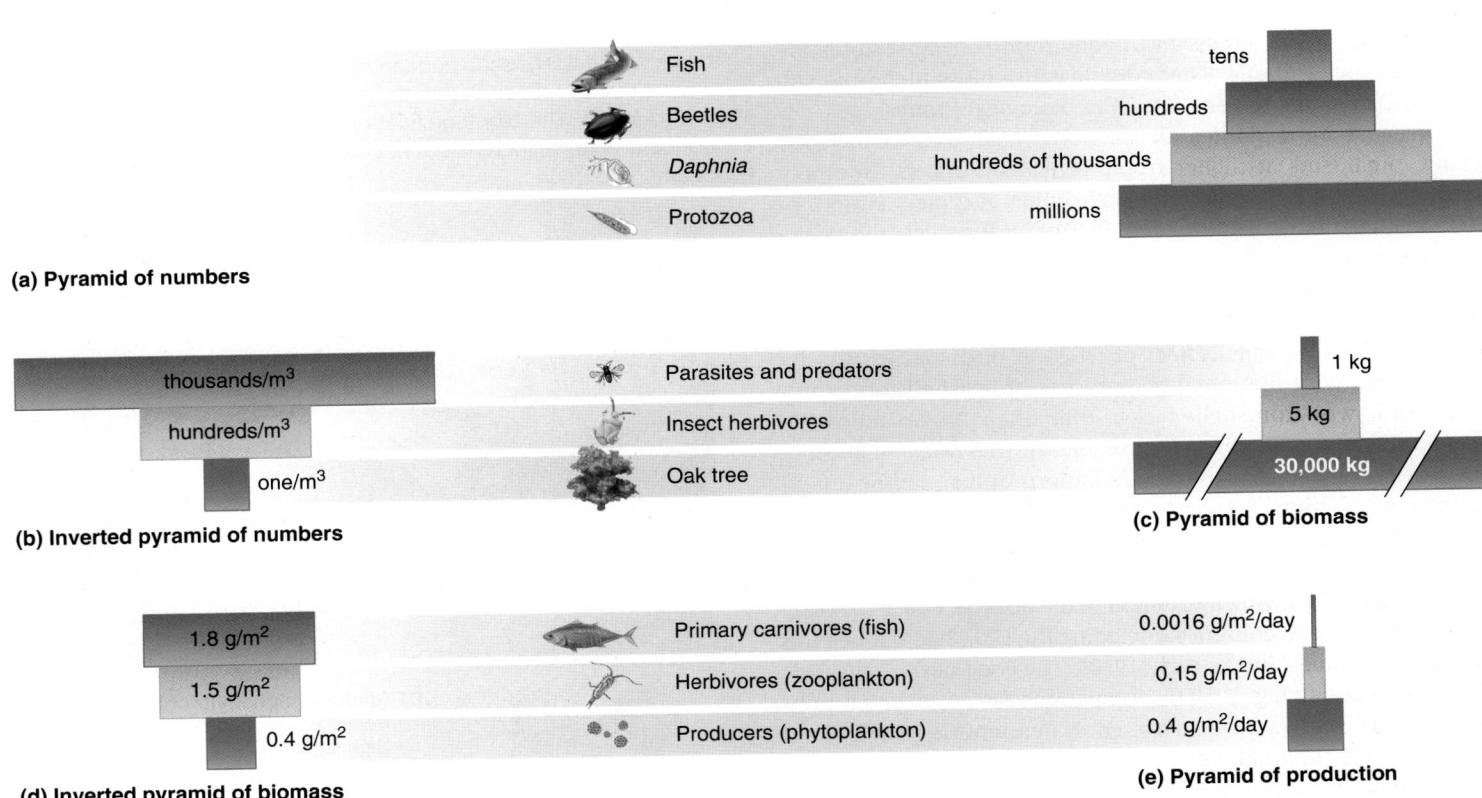

(a) Pyramid of numbers

(b) Inverted pyramid of numbers

(c) Pyramid of biomass

(d) Inverted pyramid of biomass

(e) Pyramid of production

Figure 59.8   Ecological pyramids in food webs.   (a) In this pyramid of numbers, the abundance of species in a typical British pond decreases with increasing trophic level. (b) An inverted pyramid of numbers, based on organisms living on an oak tree. (c) When amount of biological material is used instead of numbers of individuals, the pyramid is termed a pyramid of biomass. (d) An inverted pyramid of biomass in the English Channel. (e) A pyramid of production in the English Channel.

Biomagnification Can Occur in Higher Trophic Levels

Thus far, we have considered how available biomass and energy flow can influence the properties of food chains and webs. An issue that faces organisms is the tendency of certain chemical elements to accumulate or build up within food chains. This increase in the concentration of a substance in living organisms is called **biomagnification**, and the passage of DDT in food chains provides a startling example.

Dichlorodiphenyltrichloroethane (DDT) was first synthesized by chemists in 1874. In 1939, its insecticidal properties were recognized by Paul Müller, a Swiss scientist who won a Nobel Prize in 1948 for his discovery and subsequent research on the uses of the chemical. The first important application of DDT was in human health programs during and after World War II, particularly to control mosquito-borne malaria, and at that time its use in agriculture also began. The global production of DDT peaked in 1970, when 175 million kg of the insecticide were manufactured.

DDT has several chemical and physical properties that profoundly influence the nature of its ecological impact. First, DDT is persistent in the environment; it is not easily degraded to other, less toxic chemicals by microorganisms or by physical agents such as light and heat. The typical persistence in soil of DDT is about 10 years, which is two to three times longer than the persistence of most other insecticides. Another important characteristic of DDT is its low solubility in water and its high solubility in fats or lipids. In the environment, most lipids are present in living tissue. Therefore, because of its high lipid solubility, DDT tends to concentrate in biological tissues.

Because biomagnification occurs at each step of the food chain, organisms at higher tropic levels can amass especially large concentrations of DDT in their lipids. A typical pattern of biomagnification is illustrated in **Figure 59.9**, which shows the relative amounts of DDT found in a Lake Michigan food chain. The largest concentration of the insecticide was found in gulls, tertiary consumers that feed on fish that, in turn, eat small insects. An unanticipated effect of DDT on bird species was its interference with the metabolic process of eggshell formation. The result was thin-shelled eggs that often broke under the weight of incubating birds (**Figure 59.10**). DDT was responsible for a dramatic decrease in the populations of many birds due to failed reproduction. Relatively high levels of the chemical were also found to be present in some game fish, which became unfit for human consumption.

Because of growing awareness of the adverse effects of DDT, most industrialized countries, including the U.S., had banned the use of the chemical by the early 1970s. The good news is that following the outlawing of DDT, populations of the most severely affected bird species have recovered. Had scientists initially possessed a more thorough knowledge of how DDT accumulated in food chains, however, some of the damage to the bird populations might have been prevented. DDT is still used in some developing countries to control malaria, although there has been significant movement toward the use of alternative pest control technologies.

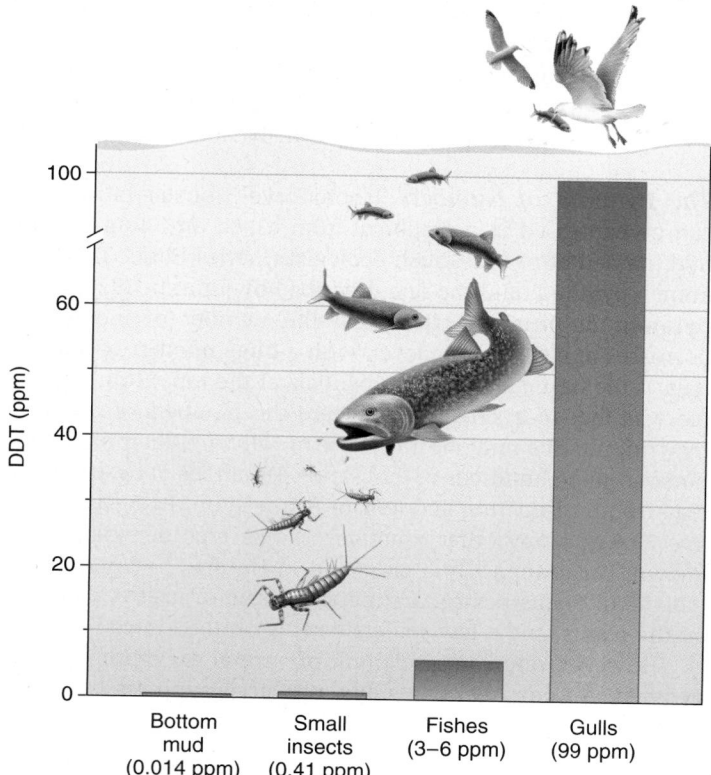

Figure 59.9   **Biomagnification in a Lake Michigan food chain.**   The DDT tissue concentration in gulls, a tertiary consumer, was about 240 times that in the small insects sharing the same environment. The biomagnification of DDT in lipids causes its concentration to increase at each successive link in the food chain.

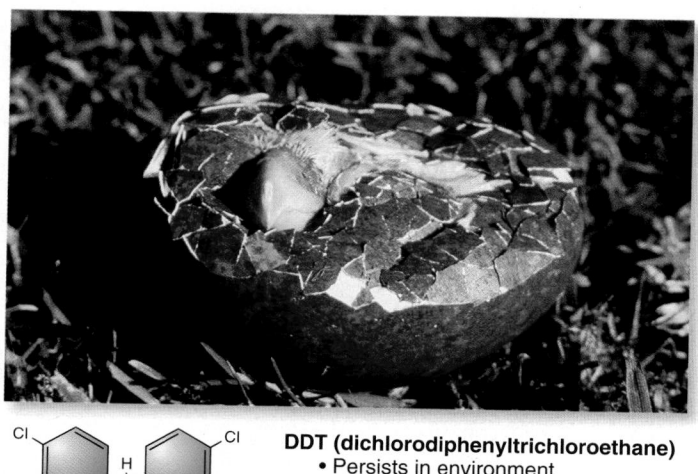

DDT (dichlorodiphenyltrichloroethane)
- Persists in environment
- High solubility in lipids
- Found in high concentrations at higher trophic levels

Figure 59.10   **Thinning of eggshells caused by DDT.** These ibis eggs are thin-shelled and have been crushed by the incubating adult.

59.2 Energy Production in Ecosystems

In this section, we will take a closer look at energy production. Because the bulk of the Earth's biosphere, 99.9% by mass, consists of primary producers, when we measure ecosystem energy production, we are primarily interested in plants, algae, or cyanobacteria. Since plants represent the first, or primary, trophic level, we measure plant production as **gross primary production (GPP)**. Gross primary production is equivalent to the carbon fixed during photosynthesis. **Net primary production (NPP)** is gross primary production minus the energy released in plant cellular respiration (R).

$$NPP = GPP - R$$

Net primary production is thus the amount of energy that is available to primary consumers. From now on, the term primary production refers to net primary production.

To measure primary production, calories can be used as a common currency, and organisms can be viewed as caloric equivalents. For example, 100 g of rye grass seeds (*Secale cereale*) has a calorific equivalent of about 380 calories, whereas 100 g of lead tree leaves (*Leucaena leucocephala*) has a calorific content of only 68 calories. Energy content is generally measured using dry biomass. Dry weight is used because the bulk of living matter in most species is water, and water content fluctuates widely, often according to wet or dry seasons. Of the dry weight, 95% is made up of carbon compounds, so that measuring energy flow in ecosystems is in many ways equivalent to examining the carbon cycle (see Section 59.3), and ecologists often measure NPP in terms of carbon fixed per square meter or per hectare.

Understanding what factors limit primary production is of vital importance if we are to examine ecosystems as energy transformers. Furthermore, by determining these factors, we can understand how primary production varies globally. We can also examine the effects of primary production on **secondary production**, the gain in the biomass of heterotrophs and decomposers.

Primary Production Is Influenced in Terrestrial Ecosystems by Water, Temperature, and Nutrient Availability

In terrestrial systems, water is a major determinant of primary production, and primary production shows an almost linear increase with annual precipitation, at least in arid regions. Likewise, temperature, which affects production primarily by slowing or accelerating plant metabolic rates, is also important. Ecologist Michael Rosenzweig noted that the evapotranspiration rate could predict the aboveground primary production with good accuracy in North America (**Figure 59.11**). Recall from Chapter 58 that the evapotranspiration rate is a measure of the amount of water entering the atmosphere from the ground through the process of evaporation from the soil and transpiration of plants, so it is a measure of both temperature and available water. For example, a desert will have a low evapotrans-

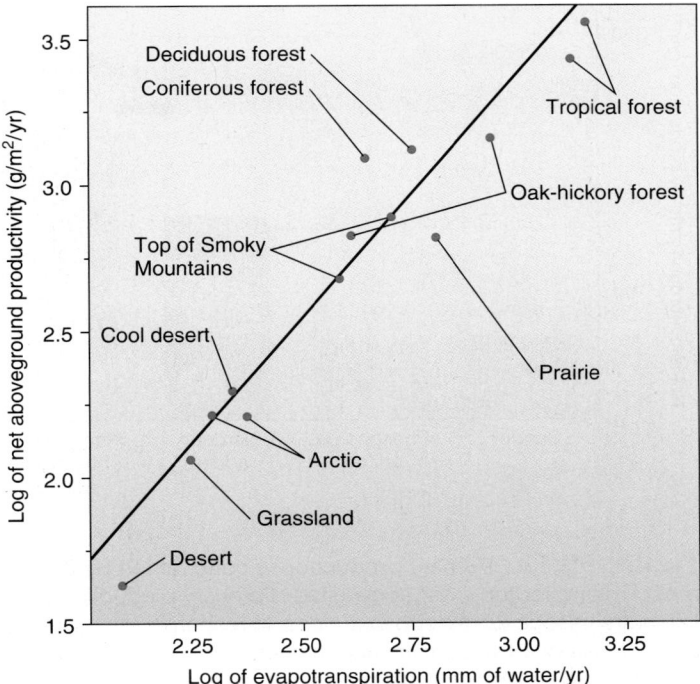

Figure 59.11   **Primary production is positively correlated with the evapotranspiration rate.**   Warm, humid environments are ideal for plant growth. Dots represent different ecosystems.

piration rate because water availability is low despite high temperature. Rates of evapotranspiration are maximized when both temperature and moisture are at high levels, as in tropical rain forests.

A lack of **nutrients**, key elements in useable form, particularly nitrogen and phosphorus, can also limit primary production in terrestrial ecosystems, as agricultural practitioners know only too well. Fertilizers are commonly used to boost the production of annual crops. In 1984, Stewart Cargill and Rob Jefferies showed how a lack of both nitrogen and phosphorus was limiting to salt marsh sedges and grasses in subarctic conditions in Hudson Bay, Canada (**Figure 59.12**). Of the two nutrients, nitrogen was the most limiting; without it, the addition of phosphorus did not increase production. However, once nitrogen was added, phosphorus became the **limiting factor**, that is, the one in shortest supply for growth. Once nitrogen was added and was no longer limiting, the addition of phosphorus increased production. The addition of nitrogen and phosphorus together increased production the most. This result supports a principle known as **Liebig's law of the minimum**, named for Justus von Liebig, a 19th-century German chemist, which states that species biomass or abundance is limited by the scarcest factor. This factor can change, as the Hudson Bay experiment showed: When sufficient nitrogen is available, phosphorus becomes the limiting factor. Once phosphorus becomes abundant, then productivity will be limited by another nutrient.

Figure 59.12 **Primary production is constrained by the most limiting factor.** Net aboveground primary production of a salt marsh sedge (*Carex subspathacea*) in response to nutrient addition. Nitrogen is more limiting than phosphorus alone, but once nitrogen becomes available, phosphorus becomes the limiting factor.

Table 59.1	Net Primary Production for Earth's Ecosystems	
Ecosystem type		**Mean net primary production g/m²/yr)**
Terrestrial		
Tropical rain forest		2,500
Tropical deciduous forest		1,600
Temperate deciduous forest		1,550
Savanna		1,080
Prairie		750
Cultivated land		610
Taiga		380
Tundra		140
Hot desert		90
Aquatic		
Algal beds and coral reefs		2,500
Wetlands		2,000
Estuaries		1,500
Upwelling zones		500
Continental shelf		360
Lake and stream		250
Open ocean		125

Primary Production in Aquatic Ecosystems Is Limited Mainly by Light and Nutrient Availability

Of the factors limiting primary production in aquatic ecosystems, the most important are available light and available nutrients. Light is particularly likely to be in short supply because water readily absorbs light. At 1-m depth, more than half the solar radiation has been absorbed. By 20 m, only 5–10% of the radiation is left. The decrease in light is what limits the depth of water to which algae are restricted.

The most important nutrients affecting primary production in aquatic systems are nitrogen and phosphorus, because they occur in very low concentrations. While soil contains about 0.5% nitrogen, seawater contains only 0.00005% nitrogen. Enrichment of the aquatic environment by the addition of nitrogen and phosphorus can result in large, unchecked growths of algae called algal blooms. Such enrichment occurs naturally in areas of upwellings, where cold, deep, nutrient-rich water containing sediment from the ocean floor is brought to the surface by strong currents, resulting in very productive ecosystems and plentiful fish. Some of the largest areas of upwelling occur in the Antarctic and along the coasts of Peru and California.

Primary Production Is Greatest in Areas of Abundant Warmth and Moisture

Knowing which factors limit primary production helps ecologists understand why mean net primary production varies across the different ecosystems on Earth (**Table 59.1**). In general, primary production is highest in tropical rain forests and decreases progressively toward the poles (**Figure 59.13**). As we saw in Chapter 58, many ecologists suggest this is the primary cause for the polar-equatorial gradient of species richness. This primary productivity gradient occurs because temperatures decrease toward the poles, and, as we have just learned, temperatures affect primary production greatly. Wetlands also tend to be extremely productive, primarily because water is not limiting and nutrient levels are high. Productivity of the open ocean is very low, falling somewhere between the productivity of deserts and that of the Arctic tundra. Marine production is high on coastal shelves, particularly in upwelling zones. However, the greatest marine production occurs on algal beds and coral reefs, where temperatures are high and water levels are not so deep that light becomes limiting.

Secondary Production Is Generally Limited by Available Primary Production

What factors control secondary production, the productivity of herbivores, carnivores, and decomposers? This is a complex question, but it is generally thought to be limited largely by available primary production. A strong relationship exists between primary production in a variety of ecosystems and the biomass of herbivores (**Figure 59.14**). This means that more plant biomass, and thus more primary production, leads to increased herbivore biomass.

Figure 59.13 **Annual net primary productivity on Earth.** Primary productivity generally increases from the poles to the equator.

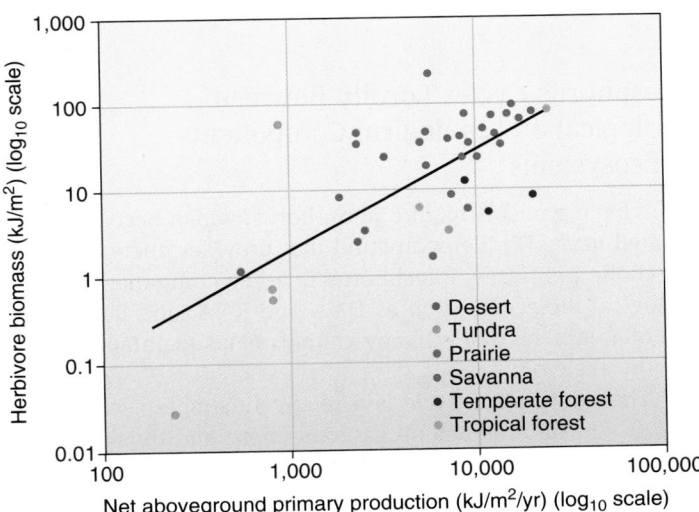

Figure 59.14 **Herbivore biomass is positively correlated with net aboveground primary production.** These data are taken from a variety of case studies from different biomes. Herbivore biomass can be considered a surrogate for secondary production.

As we have noted before, trophic-level transfer efficiency averages about 10%. Thus, after one link in the food web, only 1/10 of the energy captured by plants is transferred to herbivores, and after two links in the food web, only 1/100 of the energy fixed by plants goes to carnivores. Thus, secondary production is much smaller than that of primary production. We

can see this in the work of John Teal, whose 1962 work examined energy flow in a Georgia salt marsh (**Figure 59.15**). Salt marshes are among the most productive habitats on Earth in terms of amount of vegetation they produce. In salt marshes, most of the energy from the sun (incident sunlight) goes to two types of organisms: *Spartina* plants and marine algae. The *Spartina* plants are rooted in the ground, whereas the algae float on the water surface or live on the mud or on *Spartina* leaves at low tide. These photosynthetic organisms absorb about 6% of the incident sunlight. Most of the plant energy, 77.6%, is used in plant and algal cellular respiration. Of the energy that is accumulated in plant biomass, most dies in place and rots on the muddy ground, to be consumed by bacteria. Bacteria are the major decomposers in this system, followed distantly by nematodes and crabs, which feed on tiny food particles as they sift through the mud. Some of this dead material is also removed from the system (exported) by the tide. The herbivores take very little of the plant production, eating only a small proportion of the *Spartina* and none of the algae. Overall, if we view the species in ecosystems as transformers of energy, then plants and algae are by far the most important organisms on the planet, bacteria are next, and animals are a distant third.

59.3 Biogeochemical Cycles

A unit of energy passes through a food web only once. In contrast, elements such as nitrogen or carbon are recycled, moving from the physical environment to organisms and back to the

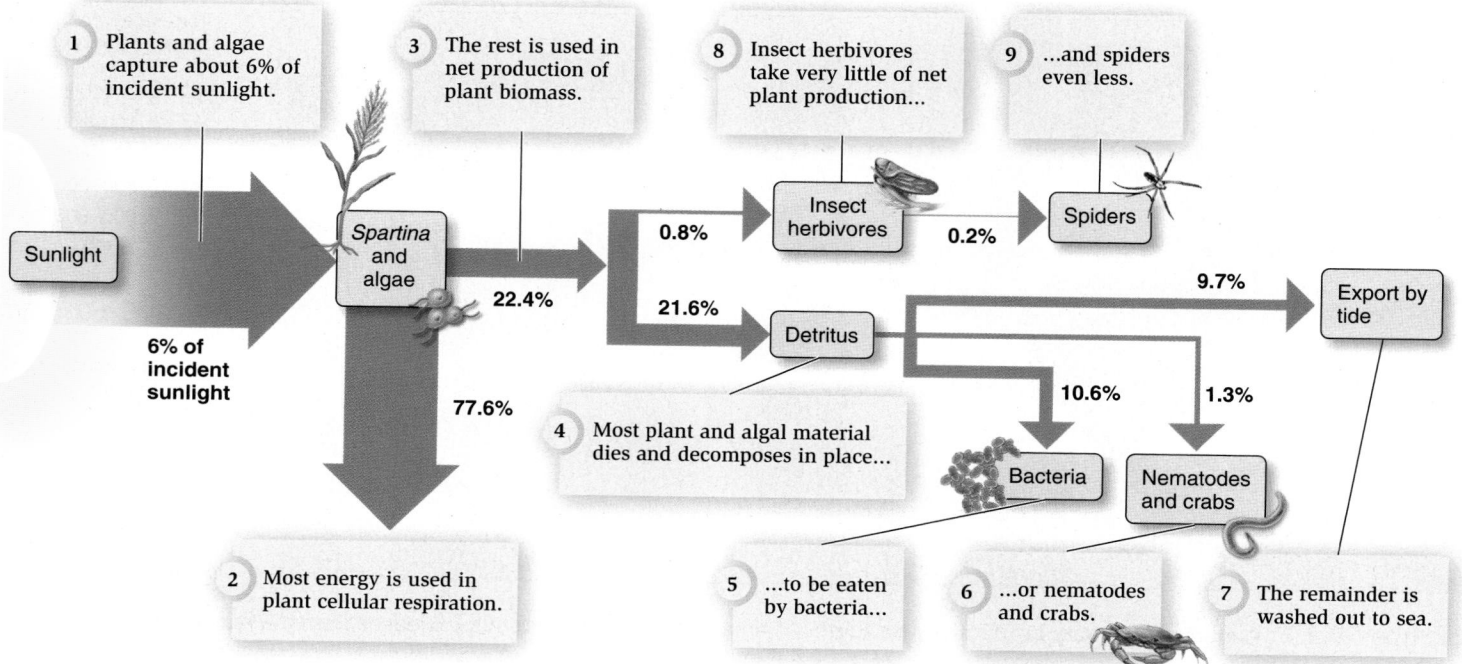

1. Plants and algae capture about 6% of incident sunlight.

2. Most energy is used in plant cellular respiration.

3. The rest is used in net production of plant biomass.

4. Most plant and algal material dies and decomposes in place...

5. ...to be eaten by bacteria...

6. ...or nematodes and crabs.

7. The remainder is washed out to sea.

8. Insect herbivores take very little of net plant production...

9. ...and spiders even less.

Sunlight

6% of incident sunlight

Spartina and algae

22.4%

77.6%

21.6%

0.8%

Insect herbivores

0.2%

Spiders

Detritus

9.7%

Export by tide

10.6%

1.3%

Bacteria

Nematodes and crabs

Figure 59.15 Energy-flow diagram for a Georgia salt marsh. Numbers reflect percentages of gross primary production that flows into different trophic levels or is used in plant respiration.

environment, where the cycle begins again. While energy dissipates as heat, chemical elements are available in limited amounts and are continually recycled. Because these cycles involve biological, geological, and chemical transport mechanisms, they are termed **biogeochemical cycles**. Biological mechanisms involve the absorption of chemicals by living organisms and their subsequent release back into the environment. Geological mechanisms include weathering and erosion of rocks, and elements transported by surface and subsurface drainage. Chemical transport mechanisms include dissolved matter in rain and snow, atmospheric gases, and dust blown by the wind.

In addition to the basic building blocks of hydrogen, oxygen, and carbon, which we discussed in Chapter 2, the elements required in the greatest amounts by living organisms are nitrogen, phosphorus, and sulfur. In this section, we take a detailed look at the cycles of these nutrients. These cycles can be divided into two broad types: local cycles, such as the phosphorus cycle, which involve elements with no mechanism for long-distance transfer; and global cycles, which involve an interchange between the atmosphere and the ecosystem. Global nutrient cycles, such as the carbon, nitrogen, and sulfur cycles, unite the Earth and its living organisms into one giant interconnected ecosystem called the **biosphere**. In our discussion, we will take a particular interest in the alteration of these cycles through human activities that increase nutrient inputs, such as the burning of fossil fuels.

Phosphorus Cycles Locally Between Geological and Biological Components of Ecosystems

All living organisms require phosphorus, which becomes incorporated into ATP, the compound that provides energy for most metabolic processes. Phosphorus is a key component of other biological molecules such as DNA and RNA, and it is also an essential mineral that in many animals helps maintain a strong, healthy skeleton.

The phosphorus cycle is a relatively simple cycle (**Figure 59.16**). Phosphorus has no gaseous phase and thus no atmospheric component; that is, it is not moved around by the wind or rain. As a result, phosphorus tends to cycle only locally. The Earth's crust is the main storehouse for this element. Weathering and erosion of rocks release phosphorus into the soil. Plants have the metabolic means to absorb dissolved ionized forms of phosphorus, the most important of which occurs as phosphate (HPO_4^{2-} or $H_2PO_4^{-}$). Plants can take up phosphate rapidly and efficiently. In fact, they can do this so quickly that they often reduce soil concentrations of phosphorus to extremely low levels, so that phosphorus becomes limiting (see Figure 59.12). Herbivores obtain their phosphorus only from eating plants, and carnivores obtain it by eating herbivores. When plants and animals excrete wastes or die, the phosphorus becomes available to decomposers, which release it back to the soil.

Figure 59.16 **The phosphorus cycle.** Unlike other major biogeochemical cycles, the phosphorus cycle does not have an atmospheric component and thus cycles only locally. The widths of the lines indicate the relative importance of each process.

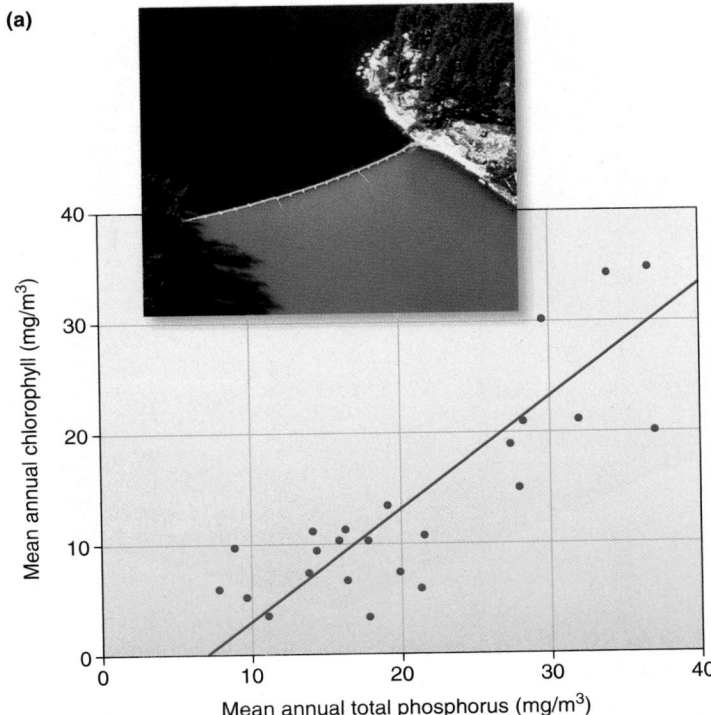

Figure 59.17 **Primary production increases with an increase in total phosphorus concentration.** (a) This aerial photograph shows the contrast in water quality of two basins of an experimental lake in Canada, separated by a plastic curtain. The lower basin received additions of carbon, nitrogen, and phosphorus, while the upper basin received only carbon and nitrogen. The bright green color is from a surface film of algae that resulted from the added phosphorus. (b) The increase in primary production is measured as an increase in chlorophyll concentration. A higher chlorophyll concentration in the water means more algae are present.

Leaching and runoff eventually wash much phosphate into aquatic systems, where plants and algae utilize it. Phosphate that is not taken up into the food chain settles to the ocean floor or lake bottom, forming sedimentary rock. Phosphorus can remain locked in sedimentary rock for millions of years, becoming available again through the geological process of uplift.

As noted previously, phosphorus is a limiting element in most aquatic systems. The more phosphorus that is added, the more that aquatic productivity increases (**Figure 59.17a**). In a pivotal 1974 study, biologist David Schindler showed that an overabundance of phosphorus caused the rapid growth of algae and plants in an experimental lake in Canada (**Figure 59.17b**). The process by which elevated nutrient levels lead to an overgrowth of algae and the subsequent depletion of water oxygen levels is known as **eutrophication**. Cultural eutrophication refers to the enrichment of water with nutrients derived from human activities such as fertilizer use and sewage dumping.

Lake Erie became eutrophic in the 1960s due to the fertilizer runoff from farms rich in phosphorus and to the industrial and domestic pollutants released from the many cities along its shores. Fish species such as blue pike, white fish, and lake trout became severely depleted. The U.S. and Canada teamed together to reduce the levels of discharge by 80%, primarily through eliminating phosphorus in laundry detergents and maintaining strict controls on the phosphorus content of wastewater from sewage treatment plants. Fortunately, lake systems have great potential for recovery after phosphorous inputs are reduced, and Lake Erie has experienced a reduction in the occurrence of algal blooms, an improvement in water clarity, and an increase in fish stocks.

Carbon Cycles Among Biological, Geological, and Atmospheric Pools

The movement of carbon from the atmosphere into organisms and back again is known as the carbon cycle (**Figure 59.18**). Carbon dioxide is present in the atmosphere at a level of about 380 parts per million (ppm), or about 0.04%. Autotrophs, primarily plants and algae, acquire carbon dioxide from the atmosphere and incorporate it into the organic matter of their own biomass via photosynthesis. In the process, each year, plants and algae remove approximately one-seventh of the CO_2 from the atmosphere. At the same time, the decomposition of plants recycles a similar amount of carbon back into the atmosphere as CO_2. Herbivores can also return some carbon dioxide to the atmosphere, eating plants and breathing out CO_2, but the amount flowing through this part of the cycle is minimal.

Figure 59.18 **The carbon cycle.** Each year, plants and algae remove about one-seventh of the CO_2 in the atmosphere. Animal respiration is so small it is not represented. The width of the arrows indicates the relative importance of each process.

Figure 59.19 **The burning of fossil fuels has greatly increased atmospheric CO_2 levels and temperatures.** From 1958 to 2004, atmospheric CO_2 shows an increase of nearly 20%. In addition, the graph shows a seasonal variation in CO_2. Temperatures are annual deviations from the 1961–1990 average. Measurements were recorded at Mauna Loa Observatory in Hawaii.

Biological inquiry: Why does the amount of CO_2 fluctuate seasonally in the graph?

Over time, much carbon is also incorporated into the shells of marine organisms, which eventually form huge limestone deposits on the ocean floor. Natural sources of CO_2 such as volcanoes, hot springs, and fires also release large amounts of CO_2. In addition, human activities, primarily the burning of **fossil fuels**—coal, petroleum, and natural gas—are increasingly causing large amounts of CO_2 to enter the atmosphere together with large volumes of particulate matter. Recently, such particulate matter has been shown to increase heritable mutations (see Genomes & Proteomes).

Direct measurements over the past five decades show a steady rise in atmospheric CO_2 (**Figure 59.19**), a pattern that shows no sign of slowing. Because of its increasing concentration in the atmosphere, CO_2 is the most troubling of the greenhouse gases, which are a primary source of global warming (see Chapter 54). Elevated atmospheric CO_2 has other dramatic environmental effects, boosting plant growth but lowering the densities of herbivores that feed on them (see Feature Investigation).

The amount of carbon dioxide in the atmosphere varies with the seasons in temperate environments. Concentrations of atmospheric carbon dioxide are lowest during the Northern Hemisphere's summer and highest during the winter, when photosynthesis is minimal. This is because there is more land in the Northern Hemisphere than in the Southern Hemisphere and therefore more vegetation. The vegetation has a maximum photosynthetic activity during the summer, reducing the global

amount of carbon dioxide. During the winter, photosynthesis is low and decomposition is relatively high, causing a global increase in the gas.

GENOMES & PROTEOMES

Pollution Can Cause Heritable Mutations

Carbon dioxide is one of many pollutants emitted from the burning of fossil fuels; others include sulfur dioxide and nitrous oxides. In addition to these atmospheric gases, the combustion of fossil fuel also results in the release of particulate matter, tiny solid and liquid particles, into the air. Larger particles are visible as smoke or soot. The small particles, when inhaled, can penetrate deeply into the lungs and become distributed throughout the body. Such exposure, whether from transportation sources, factories, power plants, or tobacco smoke, can lead to the formation of lung tumors and induce mutations in somatic cells.

A study by biologist James Quinn and colleagues in 2004 showed that polluted air can also induce genetic changes in mouse sperm. In the study, the researchers put two groups of mice into separate sheds that were downwind of steel mills in Hamilton, an urban-industrial center of Ontario, Canada (**Figure 59.20a**). One shed was fitted with <u>h</u>igh-<u>e</u>fficiency <u>p</u>articulate <u>a</u>ir (HEPA) filters, which removed particulate matter from the air, while the other was not. The researchers discovered changes in the size of noncoding DNA sequences in the offspring of the mice in the filterless shed. They concluded that the increased rate of mutations was paternally derived. Offspring of mice from the unfiltered shed inherited twice as many mutations as offspring of mice in the filtered shed. The study concluded that the pollutant particles, or some chemical compound associated with them, were responsible for the observed, heritable DNA changes.

What is the mechanism behind these genetic changes? Quinn and colleagues hypothesized that the inhaled pollutants are transported to the liver and metabolized to DNA-reactive compounds, and then transported to the testes and finally to the sperm stem cells where the mutations occurred (**Figure 59.20b**). The damaged DNA would then be transmitted to first-generation offspring.

(a)   (b)

Figure 59.20 Mutations caused by air pollution. (a) In a study by Quinn and colleagues, mice were kept in sheds with and without HEPA air filters in Hamilton, Ontario. (b) The study concluded that mice kept in the filterless sheds inhaled airborne particles into their lungs, and compounds from the particles induced genetic mutations that were transmitted through the sperm cells into the next generation.

FEATURE INVESTIGATION

Stiling and Drake's Experiments with Elevated CO_2 Showed an Increase in Plant Growth but a Decrease in Herbivory

How will forests of the future respond to elevated CO_2? To begin to answer such a question, ecologists ideally would enclose large areas of forests with chambers, increase the CO_2 content within the chambers, and measure the responses. This has proven to be difficult for two reasons. First, it is hard to enclose large trees in chambers, and second, it is expensive to increase CO_2 levels over such a large area. However, in a discovery-based investigation, ecologists Peter Stiling and Bert Drake were able to increase CO_2 levels around small patches of forest at the Kennedy Space Center in Cape Canaveral, Florida. In much of Florida forest, trees are small, only 3–5 m when mature, because frequent lightning-initiated fires prevent the growth of larger trees. Stiling and Drake teamed up with NASA engineers to create 16 circular, open-topped chambers (**Figure 59.21**), and in 8 of these they increased atmospheric CO_2 to double their ambient levels, from 360 to 720 ppm, the latter of which is the atmospheric level predicted for the end of the 21st century. The experiments were initiated in 1996 and lasted until 2004. Plants produced more biomass in elevated CO_2, because carbon dioxide is limiting to plant growth, but the data revealed much more.

Because the chambers were open-topped, insect herbivores could come and go. Insect herbivores cause the largest amount of herbivory in North American forests, because vertebrate herbivores cannot access the high foliage. Censuses were conducted of all insect herbivores but focused on leaf miners, the most common type of herbivore at this site, which are small moths whose larvae are small enough to live between the surfaces of plant leaves and create blister-like "mines" on leaves.

Densities of all insects, including leaf miners, was lower in elevated CO_2 in every year studied. Part of the reason for the decline was that even though plants increased in mass, the existing soil nitrogen was diluted over a greater volume of plant material, so that the nitrogen level in leaves decreased. This increased insect mortality by two means. First, poorer leaf quality directly increased insect death because leaf nitrogen levels may have been too low to support the normal development of the leaf miners. Second, lower leaf quality increased the length of time insects had to feed to gain sufficient nitrogen. Increased feeding times in turn led to increased exposure to natural enemies such as parasitoids and predators like spiders and ants, and top-down mortality also increased (see the data of Figure 59.21). Thus, in a world of elevated CO_2, plant growth may increase but herbivore densities could decrease. If some herbivores, such as butterflies, are already rare in a habitat, then an elevation in CO_2 could threaten them with extinction.

Figure 59.21   The effects of elevated atmospheric CO_2 on insect herbivores.

GOAL   To determine the effects of elevated CO_2 on a forest ecosystem; effects on herbivores are highlighted here.
STUDY LOCATION   Patches of forest at the Kennedy Space Center in Cape Canaveral, Florida.

| Experimental level | Conceptual level |

1   Erect 16 open-top chambers around native vegetation. Increase CO_2 levels from 360 ppm to 720 ppm in half of them.

Create elevated atmospheric CO_2 inside 8 chambers and compare data to 8 control chambers where normal air is pumped in.

2   Count numbers of insect herbivores per 200 leaves in each chamber.

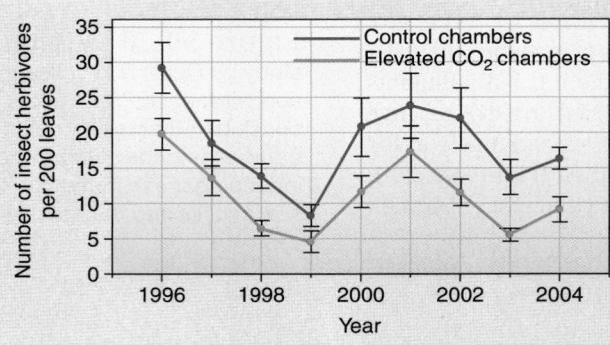

3   Count number of herbivores that died due to nutritional inadequacy. Monitor attack rates on insect herbivores by natural enemies such as predators and parasitoids.

Elevated CO_2 prolongs development time of herbivores allowing natural enemies greater opportunities to attack them.

4   **THE DATA**

Source of mortality*	Elevated CO_2 (% mortality)	Control (% mortality)
Nutritional inadequacy	10.2	5.0
Predators	2.4	2.0
Parasitoids	10.0	3.2

* Data refer only to mortality of larvae within leaves and do not sum to 100%. Mortality of eggs on leaves, pupae in the soil, and flying adults is unknown.

The Nitrogen Cycle Is Strongly Influenced by Biological Processes That Transform Nitrogen into Usable Forms

Nitrogen is a limiting nutrient because it is an essential component of proteins, nucleic acids, and chlorophyll. Because 78% of the Earth's atmosphere consists of nitrogen gas (N_2), it may seem that nitrogen should not be in short supply for organisms. However, N_2 molecules must be broken apart before nitrogen atoms are available to combine with other elements. Because of its triple bond, nitrogen gas is very stable and only certain bacteria can break it apart into usable forms. This process, called nitrogen fixation, is a critical component of the five-part nitrogen cycle, which also includes nitrification, assimilation, ammonification, and denitrification (**Figure 59.22**):

1. Only a few species of bacteria can accomplish **nitrogen fixation**, that is, convert atmospheric nitrogen to forms

usable by other organisms. The bacteria that fix nitrogen are fulfilling their own metabolic needs, but in the process, they release excess ammonia (NH_3) or ammonium (NH_4^+), which can be used by some plants. The most important of these bacteria in the soil are called *Rhizobium*, which live in nodules on the roots of legumes, including peas, beans, lentils, and peanuts, and some woody plants.

2. In the process of **nitrification**, soil bacteria convert NH_3 or NH_4^+ to nitrate (NO_3^-), a form of nitrogen commonly used by plants. The bacteria *Nitrosomonas* and *Nitrococcus* first oxidize the forms of ammonia to nitrite (NO_2^-), after which the bacteria *Nitrobacter* converts NO_2^- to NO_3^-.

3. **Assimilation** is the process by which plants and animals incorporate the NH_3, NH_4^+, and NO_3^- formed through nitrogen fixation and nitrification. Plant roots take up these forms of nitrogen through their roots, and animals assimilate nitrogen by eating plant tissue.

4. Ammonia can also be formed in the soil through the decomposition of plants and animals and the release of animal waste. **Ammonification** is the conversion of organic nitrogen to NH_3 and NH_4^+. This process is carried out by bacteria and fungi. Most soils are slightly acidic and, because of an excess of H^+, the NH_3 rapidly gains an additional H^+ to form NH_4^+. Because many soils lack nitrifying bacteria, ammonification is the most common pathway for nitrogen to enter the soil.

5. **Denitrification** is the reduction of nitrate (NO_3^-) to gaseous nitrogen (N_2). Denitrifying bacteria, which are anaerobic and use NO_3^- in their metabolism instead of oxygen, perform the reverse of their nitrogen-fixing counterparts by delivering nitrogen to the atmosphere. This process only delivers a relatively small amount of nitrogen to the atmosphere.

In terms of the global nitrogen budget, industrial fixation of nitrogen for the production of fertilizer makes a significant contribution to the pool of nitrogen-containing material in the soils and waters of agricultural regions. Human alterations of the nitrogen cycle have approximately doubled the rate of nitrogen input to the cycle. One problem is that fertilizer runoff can cause eutrophication of rivers and lakes, and, as the resultant algae die, decomposition by bacteria depletes the oxygen level of the water, resulting in fish die-offs. Excess nitrates in surface or groundwater systems used for drinking water are also a health hazard, particularly for infants. In the body, nitrate is converted to nitrite, which then combines with hemoglobin to form methemoglobin, a type of hemoglobin that does not carry oxygen. In infants, the production of large amounts of nitrites can cause methemoglobinemia, a dangerous condition in which the level of oxygen carried through the body decreases.

Finally, burning fossil fuels releases not only carbon but also nitrogen in the form of nitrogen oxides, which can contribute to air pollution. Nitrogen oxides in turn can react with rain water to form nitric acid (HNO_3), which contributes to acid rain, decreasing the pH of lakes and streams and increasing fish mortality (see Chapter 54). While much of the acid rain problem can be traced to the sulfur cycle, nitrogen oxides are also partially to blame.

The dramatic effects of human activities on nutrient cycles in general and the nitrogen cycle in particular were illustrated by a famous long-term study by ecosystem ecologists Gene Likens, Herbert Bormann, and their colleagues at Hubbard Brook Experimental Forest in New Hampshire in the 1960s. Hubbard Brook is a 3,160-hectare reserve that consists of six catchments along a mountain ridge. A catchment is an area of land where all water eventually drains to a single outlet. In Hubbard Brook, each outlet is fitted with a permanent concrete dam that enables researchers to monitor the outflow of water and nutrients (**Figure 59.23a**). In this large-scale experiment, researchers felled all of the trees in one of the Hubbard Brook catchments (**Figure 59.23b**). The catchment was then sprayed with herbicides for three years to prevent regrowth of vegetation. An untreated catchment was used as a control.

Researchers monitored the concentrations of key nutrients in the streams exiting the two catchments for over three years. Their results revealed that the overall export of dissolved nutrients from the disturbed catchment rose to many times the normal rate (**Figure 59.23c**). The researchers determined that two phenomena were responsible. First, the enormous reduction in plants reduced water uptake by vegetation and led to 40% more precipitation passing through the groundwater to be discharged to the streams. This increased outflow caused greater rates of chemical leaching and rock and soil weathering.

Figure 59.22   The nitrogen cycle. The five main parts of the nitrogen cycle are: (1) nitrogen fixation, (2) nitrification, (3) assimilation, (4) ammonification, and (5) denitrification. The recycling of nitrogen from dead plants and animals into the soil and then back into plants is of paramount importance. The width of the arrows indicates the relative importance of each process.

(a) Hubbard Brook dam and weir

(b) Hubbard Brook Experimental Forest, New Hampshire

Figure 59.23   **The effects of deforestation on nutrient concentrations.**   (a) Concrete dam and weir used to monitor nutrient flow from a Hubbard Brook catchment. (b) Deforested catchment at Hubbard Brook. (c) Nutrient concentrations in stream water from the experimentally deforested catchment and a control catchment at Hubbard Brook. The timing of deforestation is indicated by arrows.

Second, and more significantly, in the absence of nutrient uptake in spring, when the deciduous trees would have started production, the inorganic nutrients released by decomposer activity were simply leached in the drainage water. Similar processes operate in the majority of terrestrial ecosystems where deforestation is significant.

The Sulfur Cycle Is Heavily Influenced by Human Activities

Most naturally produced sulfur in the atmosphere comes from the gas hydrogen sulfide (H_2S), which is released from volcanic eruptions and during decomposition, especially in wetland environments, where sulfur is very common (**Figure 59.24**). The H_2S quickly oxidizes into sulfur dioxide (SO_2). Because SO_2 is soluble in water, it returns to Earth as weak sulfuric acid (H_2SO_4), or natural acid rain, making the pH of natural rainwater slightly acidic, about 5.6 (see also Chapter 54). The sulfate ions, SO_4^{2-}, thus enter the soil, where sulfate-reducing bacteria may release sulfur as H_2S, or the sulfate may be incorporated by plants into their tissue.

In the presence of iron, sulfur can precipitate as ferrous sulfide, FeS_2, and be incorporated in pyritic rocks. The weathering of rocks and the decomposition of organic matter therefore releases sulfur to solution, which runs through rivers to the sea. Because such rocks commonly overlay coal deposits, mining exposes them to the air and water, resulting in a discharge of sulfuric acid and other sulfur-containing compounds into aquatic ecosystems. Mining has polluted hundreds of kilometers of streams and rivers in mid-Atlantic states such as West Virginia, Kentucky, and Pennsylvania in this way.

Interestingly, certain marine algae and a few salt marsh plants produce relatively large amounts of the sulfurous gas dimethyl sulfide (CH_3SCH_3). Small particles of dimethyl sulfide that diffuse to the atmosphere often form the nuclei around which water vapor can condense and form the water droplets making up clouds. Because of the sheer global extent of the oceans, changes in algal abundance and thus global dimethyl sulfide levels have the potential to alter cloud cover and thus climate. Because of its ability to cool the climate, some researchers are investigating how dimethyl sulfide production might offset global warming.

Human activity involving the combustion of fossil fuels has altered the sulfur cycle more than any of the other nutrient cycles. The burning of coal and oil to provide energy for heating or to fuel electric power stations produces huge amounts of sulfur dioxide (SO_2). This reacts with rain or snow to make human-produced acid rain (see Chapter 54). One of the main differences between human-produced acid rain and natural acid rain is its relative pH. In North America, for example, natural acid rain has a pH of about 5.6, while measurements of rain falling in southern Ontario, Canada, in the 1980s showed pH values in the range of 4.1–4.5. Huge areas of the industrial northeast U.S. and Europe were affected by acid rain in the 1950s through the 1980s, but a reduction in the use of high-sulfur coal and

Figure 59.24   **The sulfur cycle.** This cycle can be dramatically affected by human outputs through the burning of fossil fuels. The width of the arrows indicates the relative importance of each process.

the use of scrubbers to prevent sulfur dioxide from passing through smokestacks reduced the problem in more recent times (**Figure 59.25**).

The Water Cycle Is Largely a Physical Process of Evaporation and Precipitation

The water cycle, or hydrological cycle, differs from the cycles of other nutrients in that very little of the water that cycles through ecosystems is chemically changed by any of the cycle's components (**Figure 59.26**). It is a physical process, fueled by the sun's energy, rather than a chemical one, because it consists of essentially two phenomena: evaporation and precipitation. Even so, the water cycle has important biological components. Over land, 90% of the water that reaches the atmosphere is moisture that has passed through plants and exited from the leaves via evapotranspiration. However, only about 2% of the total volume of Earth's water is found in the bodies of organisms or is held frozen or in the soil. The rest cycles between bodies of water, the atmosphere, and the land.

As we noted in Chapter 54, water is limiting to the abundance of many organisms, including humans. It takes 228 L of water to produce a pound of dry wheat, and 9,500 L of water to support the necessary vegetation to produce a pound of meat. Industry is also a heavy user of water, with goods such as oil, iron, and steel requiring up to 20,000 L of water per ton of product. Humans have therefore interrupted the hydrological cycle

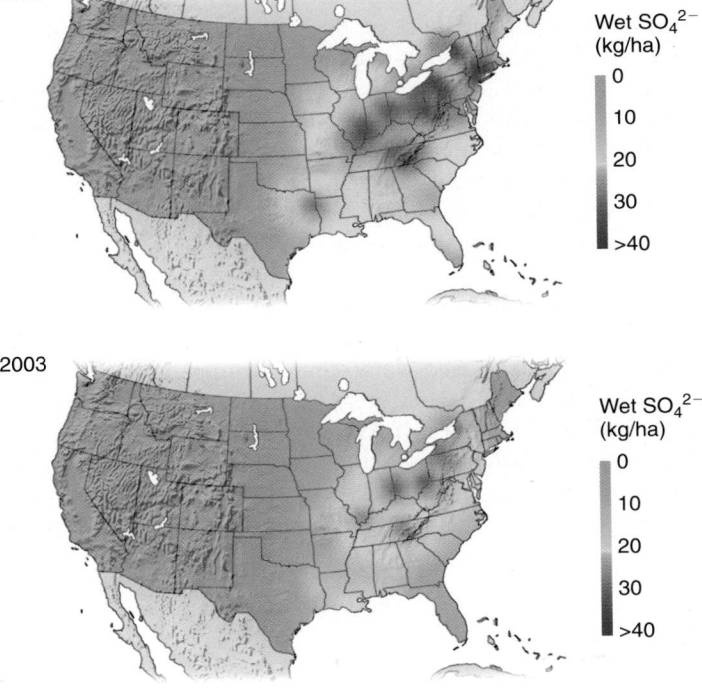

Figure 59.25   **The extent of acid rain in the U.S. has decreased.**   Data show wet sulfate deposition due to acid rain in kilograms per hectare.

Figure 59.26 **The water cycle.** This cycle is primarily a physical process, not a chemical one. Solar energy drives the water cycle, causing evaporation of water from the ocean and evapotranspiration from the land. This is followed by condensation of water vapor into clouds and precipitation. The width of the arrows indicates the relative importance of each step.

in many ways to increase the amount available. Prominent among these activities are the use of dams to create reservoirs. Such dams, including those on the Columbia River in Washington State, can greatly interfere with the migration of fish such as salmon and affect their ability to reproduce and survive. Other activities, such as tapping into underground water supplies, or **aquifers**, for drinking water removes more water than is put back by rainfall and can cause shallow ponds and lakes to dry up and sinkholes to develop.

Deforestation can also significantly alter the water cycle. When forests are cut down, less moisture is transpired into the atmosphere. This reduces cloud cover and diminishes precipitation, subjecting the area to drought. Re-establishing the forests, which calls for increased water, then becomes nearly impossible. Such a problem has occurred on the island of Madagascar, located off the east coast of Africa. In this country, clearing of the forests for cash crops such as cotton, coffee, and tobacco has been so rapid and extensive that semiarid deserts have resulted (**Figure 59.27**). In Madagascar, as in so many other

Figure 59.27 Severe erosion following deforestation in Madagascar. After clearing and a few years of farming, the shallow soil can no longer support crops and is susceptible to erosion by rainfall.

areas on Earth, deforestation and other environmental degradations are having an impact on much of the natural habitat. Appropriately, in the following chapter, we finish our study of ecology in particular, and biology in general, with a discussion of how best to conserve the diversity of ecosystems in the future.

CHAPTER SUMMARY

59.1 Food Webs and Energy Flow

- Ecosystems ecology concerns the movement of energy and materials through organisms and their communities. (Figure 59.1)

- Simple feeding relationships between organisms can be characterized by an unbranched food chain, and each feeding level in the chain is called a trophic level. (Figure 59.2)

- Organisms that obtain energy from light or chemicals are called autotrophs and are the primary producers. Organisms that feed on other organisms are called heterotrophs. Those organisms that feed on primary producers are called primary consumers or herbivores. Organisms that feed on primary consumers are called secondary consumers or carnivores. Consumers that get their energy from the remains and waste products of organism are called detritivores. (Figures 59.3, 59.4)

- Food webs are a more complex model of interconnected food chains in which there are multiple links between species. (Figure 59.5)

- Food chains tend to have five or fewer levels. Energy conversions are not 100% efficient, and energy is lost within each tropic level and from one trophic level to the next. (Figure 59.6)

- Production efficiency measures the percentage of energy assimilated that becomes incorporated into new biomass. (Figure 59.7)

- Tropic-level transfer efficiency measures the energy available at one trophic level that is acquired by the level above. These efficiencies can be expressed in the form of ecological pyramids, the best known of which is the pyramid of numbers. (Figure 59.8)

- The increase in the concentration of a substance in living organisms, called biomagnification, can occur at each step of the food chain. (Figures 59.9, 59.10)

59.2 Energy Production in Ecosystems

- Plant production can be measured as gross primary production. Net primary production is gross primary production minus the energy lost in plant respiration.

- Primary production in terrestrial ecosystems is limited primarily by temperature and the availability of water and nutrients, while primary production in aquatic ecosystems is limited mainly by availability of light and nutrients. (Figures 59.11, 59.12)

- Ecosystems differ in their net primary production. (Table 59.1, Figure 59.13)

- Secondary production is limited by available primary production. (Figure 59.14, 59.15)

59.3 Biogeochemical Cycles

- Nutrients such as phosphorus, carbon, nitrogen, and sulfur recycle from the physical environment to organisms and back in what are called biogeochemical cycles.

- The phosphorus cycle lacks an atmospheric component and thus is a local cycle. An overabundance of phosphorus can cause the overgrowth of algae and subsequent depletion of oxygen levels, called eutrophication. (Figures 59.16, 59.17)

- In the carbon cycle, autotrophs incorporate carbon dioxide from the atmosphere into their biomass; decomposition of plants and respiration recycles most of this CO_2 back to the atmosphere.

- Human activities, primarily the burning of fossil fuels, are causing increased amounts of CO_2 to enter the atmosphere. Air pollutants may cause mutations. (Figures 59.18, 59.19, 59.20)

- Experiments have shown that elevated levels of carbon dioxide result in an increase in plant growth but a decrease in herbivory. (Figure 59.21)

- The nitrogen cycle has five parts: nitrogen fixation, nitrification, assimilation, ammonification, and denitrification. In the nitrogen cycle, atmospheric nitrogen is unavailable for use by most organisms and must be converted to usable forms by bacteria. (Figure 59.22)

- The activities of humans, including fertilizer use, fossil fuel use, and deforestation, have dramatically altered the nitrogen cycle. (Figure 59.23)

- Sulfur enters enters the atmosphere through both natural sources (such as volcanoes and decomposition) and human sources, including the combustion of fossil fuels. Atmospheric sulfur dioxide returns to Earth as weak sulfuric acid (H_2SO_4) or natural acid rain. (Figure 59.24)

- Human-produced acid rain results primarily from the combustion of fossil fuels. (Figure 59.25)

- The water cycle is a physical rather than a chemical process, because it consists of essentially two phenomena: evaporation and precipitation. (Figure 59.26)

- Alteration of the water cycle by deforestation can result in climatic changes because a reduction in transpiration causes a decrease in cloud cover and precipitation. (Figure 59.27)

TEST YOURSELF

1. Ecosystems ecology is the study of
   a. the interaction of organisms and their environment.
   b. the processes that affect population growth.
   c. the organisms common to a particular environment.
   d. energy movement through a community.
   e. abiotic factors affecting movement in an environment.

2. Chemoautotrophic bacteria are
   a. primary consumers.
   b. secondary consumers.
   c. tertiary consumers.
   d. primary producers.
   e. decomposers.

3. When considering the average food chain, which of the following statements is true?
   a. Secondary consumers are the most abundant organisms in an ecosystem.
   b. The more lengths in the food chain, the more stable the ecosystem.
   c. Biomass decreases as you move up the food chain.
   d. The trophic level with the highest species abundance is usually the primary producers.
   e. All of the above are true.

4. Biomagnification is the result of
   a. an increase in the population size of the primary consumers.
   b. increased levels of pollutants that are released into the environment.
   c. the buildup of chemicals in organisms at a rate greater than the chemicals can be broken down.
   d. genetic mutations that are passed down from one generation to the next.
   e. none of the above.

5. The amount of energy that is fixed during photosynthesis is
   a. net primary production.
   b. biomagnification.
   c. trophic-level transfer efficiency.
   d. gross primary production.
   e. production efficiency.

6. The chain lengths in food webs are limited by
   a. the amount of energy such as sunlight available in an ecosystem.
   b. the efficiency of energy transfers that occur between trophic levels.
   c. the efficiency by which primary consumers assimilate energy into biomass.
   d. all of the above.
   e. a and b only.

7. The evapotranspiration rate
   a. can be used as a predictor for primary production.
   b. is increased when temperature decreases.
   c. is not affected by temperature.
   d. is highest in deserts.
   e. can be predicted by measuring only the water content of the soil.

8. Eutrophication is
   a. caused by an overabundance of nitrogen, which leads to an increase in bacteria populations.
   b. caused by an overabundance of nutrients, which leads to an increase in algal populations.
   c. the normal breakdown of algal plants following a pollution event.
   d. normally seen in dry, hot regions of the world.
   e. none of the above.

9. Primary producers acquire the carbon necessary for photosynthesis from
   a. decomposing plant material.
   b. carbon monoxide released from the burning of fossil fuels.
   c. carbon dioxide in the atmosphere.
   d. carbon sources in the soil.
   e. both a and d.

10. Nitrogen fixation is the process
    a. that converts organic nitrogen to ammonia.
    b. by which plants and animals take up nitrates.
    c. by which bacteria convert nitrate to gaseous nitrogen.
    d. by which atmospheric nitrogen is converted to ammonia or ammonium ions.
    e. all of the above.

CONCEPTUAL QUESTIONS

1. Define autotrophs and heterotrophs.
2. Explain why chain lengths are short in food webs.
3. Describe the carbon cycle.

EXPERIMENTAL QUESTIONS

1. What was the hypothesis of the Stiling and Drake experiment?
2. What was the purpose of increasing the carbon dioxide levels in only half of the chambers in the experiment and not all of the chambers?
3. What were the results of the experiment?

COLLABORATIVE QUESTIONS

1. Outline the main trophic levels within a food chain.
2. Discuss several factors that influence primary productivity in terrestrial and aquatic ecosystems.

www.brookerbiology.com
This website includes answers to the Biological Inquiry questions found in the figure legends and all end-of-chapter questions.

60

CONSERVATION BIOLOGY AND BIODIVERSITY

CHAPTER OUTLINE

Giant Panda at Wolong Reserve, southwest China.

Biological diversity, or **biodiversity**, can be examined at three levels: genetic diversity, species diversity, and ecosystem diversity. Each level of biodiversity provides valuable benefits to humanity. Genetic diversity consists of the amount of genetic variation that occurs within and between populations. Maintaining genetic variation in the ancient relatives of crops may be vital to the continued success of crop-breeding programs. For example, in 1977 Rafael Guzman, a Mexican biologist, discovered a previously unknown ancient relative of corn, *Zea diploperennis*, that is resistant to many of the viral diseases that infect domestic corn, *Zea mays*. Genetic engineers believe that this relative has valuable genes that can improve current corn crops. Because corn is the third-largest crop on Earth, the discovery of *Z. diploperennis* may well turn out to be critical to the global food supply.

The second level of biodiversity concerns species diversity, an area on which much public attention is focused. You may be familiar with the U.S. Endangered Species Act (1973), which was designed to protect both endangered and threatened species. **Endangered species** are those species that are in danger of extinction throughout all or a significant portion of their range. **Threatened species** are those likely to become endangered in the future. Many species are currently threatened. According to the International Union for Conservation of Nature and Natural Resources (IUCN), more than 25% of the fish species that live on coral reefs are threatened with extinction, and 23% of all mammals, 12% of birds, and 31% of amphibians are threatened.

The last level of biodiversity is ecosystem diversity, the diversity of structure and function within an ecosystem. While conservation at the level of species diversity has focused attention on species-rich ecosystems such as tropical forests, some scientists have argued that other relatively species-poor ecosystems are highly threatened and similarly need to be conserved. In North America, many of the native prairies have been converted to agricultural use, especially in states such as Illinois and Iowa. In some counties, remnants of prairie often exist only inside cemetery plots, which have been spared from the plow (**Figure 60.1**).

Conservation biology uses principles and knowledge from molecular biology, genetics, and ecology to protect the biological diversity of life at each of these three levels. Because it draws from nearly all chapters of this textbook, a discussion of conservation biology is an apt way to conclude our study of biology. In this chapter, we begin by examining the question of why biodiversity should be conserved, and explore how much diversity is needed for ecosystems to function properly. We then survey the main threats to the world's biodiversity. For many species, there are multiple threats, ranging from habitat loss, hunting, and the effects of introduced species to climate change and pollution. Even if species are not exterminated, many may exist only at very small population sizes. We will see how small populations contribute to special problems such as inbreeding and genetic drift, emphasizing the importance of genetics in conservation biology.

Last, we consider what can be done and has been done to help conserve the world's endangered biota. This includes identifying global areas rich in species and establishing parks and refuges of the appropriate size, number, and connectivity. We also discuss conservation of particularly important types of species, and outline how ecologists have been active in restoring damaged habitats to their natural condition. We then examine how captive breeding programs have been useful in building up populations of rare species prior to their release back into the wild.

Figure 60.1   Ecosystem biodiversity.   This small cemetery in Bureau County, Illinois, contains the remains of a natural prairie ecosystem. Most of the prairie has been plowed under for agriculture.

Some programs have also used modern genetic techniques such as cloning to help breed and perhaps eventually increase populations of endangered species.

60.1 Why Conserve Biodiversity?

Biologists Paul Ehrlich and E. O. Wilson have suggested that the loss of biodiversity should be of concern to everyone for at least three reasons. First, they proposed that we have an ethical responsibility to protect what are our only known living companions in the universe. Second, humanity has obtained enormous benefits from foods, medicines, and industrial products derived from plants, animals, and microorganisms, and we have the potential to gain many more. The third reason to preserve biodiversity focuses on preserving the array of essential services provided by ecosystems, such as clean air and water. In this section, we examine some of the primary reasons why preserving biodiversity matters and explore the link between biodiversity and ecosystem functioning.

The Preservation of Biological Diversity Can Be Justified Based on the Ecological and Economic Values of Diversity as Well as on Ethical Grounds

During the latter half of the 20th century, the reduction of the Earth's biological diversity emerged as a critical issue, one with implications for public policy. One major concern was that loss of plant and animal resources would impair future development of important products and processes in agriculture, medicine, and industry. For example, as previously noted, *Z. diploperennis*, the ancient relative of corn discovered in Mexico, is resistant to many corn viruses. Its genes are currently being used to develop virus-resistant types of corn. However, *Z. diploperennis* occurs naturally in only a few small areas of Mexico and could easily have been destroyed by development or cultivation of the land. If we allow such species to go extinct, we may unknowingly threaten the food supply on which much of the world depends.

The pharmaceutical industry is heavily dependent on information that is stored in plants. About 25% of the prescription drugs in the U.S. alone are derived from plants, and the 2006 market value of such drugs was estimated to be $19 billion. Many medicines come from plants found only in tropical rain forests, including quinine, a drug from the bark of the Cinchona tree (*Cinchona officinalis*) that is used for malaria, and vincristine, derived from rosy periwinkle (*Catharanthus roseus*), which is a treatment for leukemia and Hodgkin's disease. There are likely many plant chemicals of therapeutic importance in the thousands of rain forest plant species that have not been fully analyzed. The continued destruction of rain forests thus could mean the loss of billions of dollars in potential plant-derived pharmaceutical products.

On a smaller scale, individual species often thought worthless can actually be valuable for research purposes. The blood of the horseshoe crab (*Limulus polyphemus*) clots when exposed to toxins produced by some bacteria. Pharmaceutical industries use the blood enzyme responsible for this clotting to ensure that their products are free of bacterial contamination. Desert pupfishes, found in isolated ponds in the U.S. Southwest, tolerate salinity twice that of seawater and are valuable models for research on human kidney diseases. The technology does not exist to re-create individual species, let alone biomes such as rain forests. Once a species or an ecosystem is gone, it is lost forever.

Beyond this, humans benefit not only from individual species but also from the processes of natural ecosystems (**Table 60.1**). Forests soak up carbon dioxide, maintain soil fertility, and retain water, preventing floods; estuaries provide water filtration and protect rivers and coastal shores from excessive erosion. The loss of biodiversity could disrupt an ecosystem's ability to carry out such functions. Other ecosystem functions include the maintenance of populations of natural predators to regulate pest outbreaks and reservoirs of pollinators to pollinate crops and other plants.

A 1997 paper in the journal *Nature* by economist Robert Constanza and colleagues made an attempt to calculate the monetary value of ecosystems to various economies. They came to the conclusion that, at the time, the world's ecosystems were worth more than $33 trillion a year, nearly twice the gross national product of the world's economies combined ($19 trillion) (**Table 60.2**). If we were to include the value of these services in the cost of goods, most goods would cost a lot more than they currently do. While the open ocean has the greatest total global value of all ecosystems, perhaps a more meaningful statistic is the per hectare value of different ecosystems. This statistic reveals that shallow aquatic ecosystems, such as estuaries and swamps, are extremely valuable because of their role in nutrient cycling, water supply, and disturbance regulation. They also serve as nurseries for aquatic life. These habitats, once thought of as useless wastelands, are among the ecosystems most endangered by pollution and development.

Table 60.1	Examples of the World's Ecosystem Services
Service	**Example**
Atmospheric gas supply	Regulation of carbon dioxide, ozone, and oxygen levels
Climate regulation	Regulation of carbon dioxide, nitrogen dioxide, and methane levels
Water supply	Irrigation, water for industry
Pollination	Pollination of crops
Biological control	Pest population regulation
Wilderness and refuges	Habitat for wildlife
Food production	Crops, livestock
Raw materials	Fossil fuels, timber
Genetic resources	Medicines, genes for plant resistance
Recreation	Ecotourism
Cultural	Aesthetic and educational value
Disturbance regulation	Storm protection, flood control
Waste treatment	Sewage purification
Soil erosion control	Retention of topsoil, reduction of accumulation of sediments in lakes
Nutrient recycling	Nitrogen, phosphorus, carbon, and sulfur cycles

Table 60.2	Valuation of the World's Ecosystem Services		
Biome	**Total global value* ($ trillion)**	**Total value (per ha) ($)**	**Main ecosystem service**
Open ocean	8,381	252	Nutrient cycling
Coastal shelf	4,283	1,610	Nutrient cycling
Estuaries	4,100	22,832	Nutrient cycling
Tropical forest	3,813	2,007	Nutrient cycling/raw materials
Seagrass and algal beds	3,801	19,004	Nutrient cycling
Swamps	3,231	19,580	Water supply/disturbance regulation
Lakes and rivers	1,700	8,498	Water regulation
Tidal marsh	1,648	9,990	Waste treatment/ disturbance regulation
Grasslands	906	232	Waste treatment/food production
Temperate forest	894	302	Climate regulation/waste treatment/lumber
Coral reefs	375	6,075	Recreational/disturbance regulation
Cropland	128	92	Food production
Desert	0	0	
Ice and rock	0	0	
Tundra	0	0	
Urban	0	0	
Total	33,260		

*In 1997 values.

Arguments can be made against the loss of biodiversity on ethical grounds. As only one of many species, it has been argued that humans have no right to destroy other species and the environment around us. Philosopher Tom Regan suggests that animals should be treated with respect because they have a life of their own and therefore have value apart from anyone else's interests. Law professor Christopher Stone, in an influential article titled "Should Trees Have Standing?" has argued that entities such as nonhuman natural objects like trees or lakes should be given legal rights just as corporations are treated as persons for certain purposes. As E. O. Wilson proposed in a 1984 concept known as biophilia, humans have deep attachment with natural habitats and species because of our close association for over millions of years.

How Much Diversity Is Needed for Ecosystems to Function Properly?

Because biodiversity has an impact on the health of ecosystems, ecologists have explored the question of how much diversity is needed for ecosystems to function properly. Recall that in the 1950s, ecologist Charles Elton had proposed in the diversity-stability hypothesis that the more species present, the more stable the community (see Chapter 58). If we use stability as a surrogate for ecosystem function, Elton's hypothesis suggests a linear correlation between diversity and ecosystem function (Figure 60.2a).

In 1981, ecologists Paul and Anne Ehrlich proposed an alternative called the **rivet hypothesis** (Figure 60.2b). In this model, species are like the rivets on an airplane, with each species playing a small but critical role in keeping the plane (the ecosystem) airborne. The loss of a rivet weakens the plane and causes it to lose a little airworthiness. The loss of a few rivets could probably be tolerated, while the loss of more rivets would prove critical to the airplane's function.

A decade later, Australian ecologist Brian Walker proposed an extension of this idea, termed the **redundancy hypothesis** (or passenger hypothesis) (Figure 60.2c). According to this hypothesis, most species are more like passengers on a plane—they take up space but do not add to the airworthiness. The species are said to be redundant because they could simply be eliminated or replaced by others with no loss in function. Airworthiness is primarily affected by the activity of a few crucial species, in this case, the pilot or copilot, that are called keystone species (a concept we will discuss in detail later in this chapter). In this scenario, the loss of each species does not affect airworthiness unless the species is of critical importance.

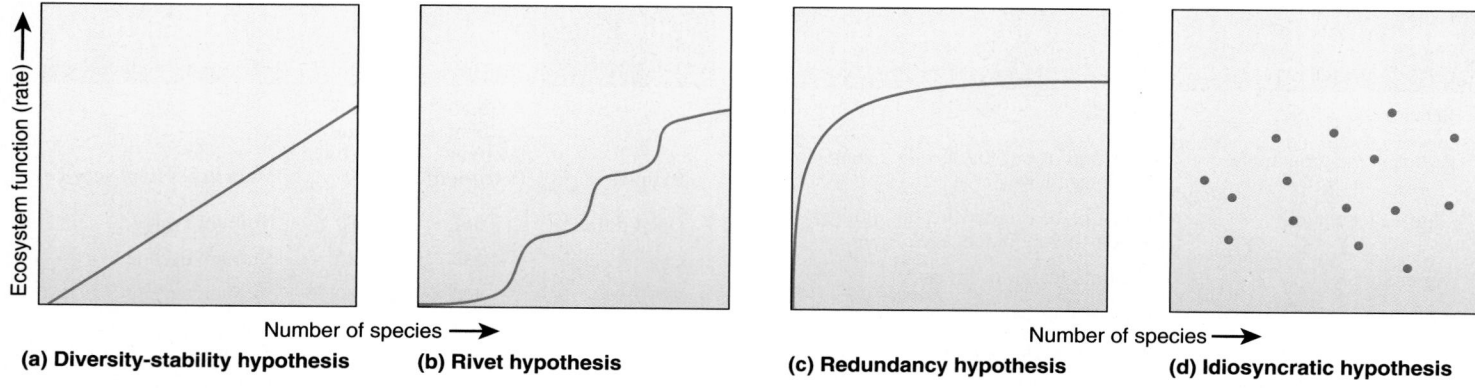

Figure 60.2   **Four main models that describe the relationship between ecosystem function and biodiversity.**   The relationship is strongest in **(a)** and weakest in **(d)**.

In another alternative, British ecologist John Lawton included the possibility that ecosystem function changes as the number of species increases or decreases, but that the direction of change is not predictable. He called this the idiosyncratic hypothesis (**Figure 60.2d**). Determining which model is most correct is very important, as our understanding of the effect of species loss on community function can greatly affect the way we manage our environment. As we will discuss next, experimental studies have provided data showing that reduced biodiversity does lead to reduced ecosystem functioning.

FEATURE INVESTIGATION

Ecotron Experiments Showed the Relationship Between Biodiversity and Ecosystem Function

In the early 1990s, Shahid Naeem and colleagues used a series of 14 environmental chambers in a facility termed the Ecotron, at Silwood Park, England, to determine how biodiversity affects ecosystem functioning. These chambers contained terrestrial communities that differed only in their level of biodiversity (**Figure 60.3**). The number of species in each chamber was manipulated to create high-, medium-, and low-diversity ecosystems, each with four trophic levels. The trophic levels consisted of primary producers (annual plants), primary consumers (insects, snails, and slugs), secondary consumers (parasitoids that fed on the herbivores), and decomposers (earthworms and

Figure 60.3   Ecotron experiments showing that loss of diversity can impair ecosystem services.

2 Add different combinations of species to the 14 chambers. The species added were based on 3 types of model communities (food webs), each with 4 trophic levels but with varying degrees of species richness.

Circles represent species, and lines connecting them represent biotic interactions among the species. Note that each lower-biodiversity community is a subset of its higher-diversity counterpart and that all community types have 4 trophic levels.

● Species present in all 3 systems.
● Species present in 2 systems.
● Species present in most diverse system only.

3 Measure and analyze a range of processes, including vegetation cover and nutrient uptake.

Measurements help determine how each different type of community functions.

4 THE DATA

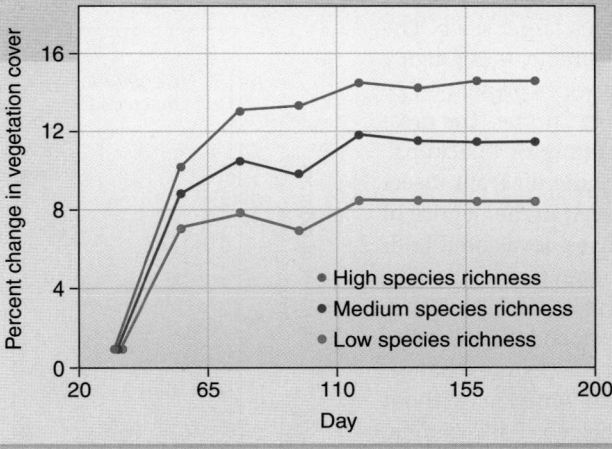

Plant productivity is linked to community diversity as measured by the percent change in vegetation cover from initial conditions.

Data reveal that low-diversity communities have lower vegetation cover and are less productive than high-diversity communities.

soil insects). The experiment ran for just over six months, and species were added only after the trophic level below them was established. For example, parasitoids were not added until herbivores were abundant.

Researchers monitored and analyzed a range of measures of ecosystem function, including community respiration, decomposition, nutrient retention rates, and community productivity. The result was that community productivity, expressed as percent change in vegetation cover (the amount of ground covered by leaves of plants), increased as species richness increased. This occurred because of a greater variety of plant growth forms that could utilize light at different levels of the plant canopy. A larger ground cover also meant a larger plant biomass and greater community productivity, and increased decomposition and nutrient uptake rates. For the first time, ecologists had provided an experimental demonstration that the loss of biodiversity can alter or impair the functioning of an ecosystem.

Field Experiments Also Suggest That Biodiversity Is Important for Ecosystem Function

In the mid-1990s, David Tilman and colleagues performed experiments in the field to determine how much biodiversity was necessary for proper ecosystem functioning. Previously, Tilman had suggested that species-rich grasslands were more stable; that is, they were more resistant to the ravages of drought and recovered from drought more quickly than species-poor grasslands (refer back to Figure 58.7). In these experiments, Tilman's group sowed plots, each 3 m by 3 m and on comparable soils, with seeds of 1, 2, 4, 6, 8, 12, or 24 species of prairie plants. Exactly which species were sown into each plot was determined randomly from a pool of 24 native species. The treatments were replicated 21 times, for a total of 147 plots. The results again showed that more diverse plots had increased productivity and used nutrients, such as nitrate, more efficiently than less diverse plots (**Figure 60.4a,b**). Furthermore, both the frequency of invasive plant species (species not originally planted in the plots) and the level of foliar fungal disease decreased with increased plant species richness (**Figure 60.4c,d**).

Although Tilman's experiments show a relationship between diversity and ecosystem function, they also suggest that a point may be reached at which function is maximized, beyond which additional species appear to have little to no impact. This supports the redundancy hypothesis. For example, uptake of nitrogen remains relatively unchanged as the number of species increases beyond six. We can also see this on a larger scale. The productivity of temperate forests in different continents is roughly the same despite different numbers of tree species present—729 in East Asia, 253 in North America, and 124 in Europe. The presence of more tree species may ensure a supply of "backups" should some of the most productive species die off from insect attack or disease. This can happen, as was seen in the demise of the American chestnut and elm trees. Diseases devastated both of these species, and their presence in American forests dramatically decreased by the mid-20th century (refer back to Figure 57.18). The forests filled in with other species and continued to function as before in terms of nutrient cycling and gas exchange. However, although the forests continued to function without these species, some important changes occurred. For example, the loss of chestnuts deprived bears and other animals of an important source of food and may have affected their reproductive capacity and hence the size of their populations.

60.2 The Causes of Extinction and Loss of Biodiversity

In light of research showing that the loss of species influences ecosystem function, the importance of understanding and preventing species loss takes on particular urgency. Throughout the history of life on Earth, **extinction**—the process by which species die out—has been a natural phenomenon. The average timespan of a typical animal or plant species in the fossil record is about 4 million years. To calculate the current extinction rate,

Figure 60.4 **The functions of species richness.** The relationships between species richness and **(a)** percent plant cover, the area of ground covered by leaves, and **(b)** uptake of nitrogen in experimental plots in Tilman's biodiversity experiments. With increasing species richness, biomass increases and nitrogen is used up, leaving less nitrogen in the soil with more species of plants. **(c)** A decrease in susceptibility to invasion of temperate grasslands by exotic species with an increase in plant species richness. **(d)** A decrease in disease severity index for plant species in temperate grasslands with an increase in plant species richness.

we could take the total number of species estimated to be alive on Earth at present, around 10 million, and divide it by 4 million, giving an average extinction rate of 2.5 species each year. For the 5,500 species of living mammals, using the same average life span of around 4 million years, we would expect about one species to go extinct every 1,000 years; this is termed the background extinction rate. However, it can be argued that the fossil record is heavily biased toward successful, often geographically wide-ranging species, which undoubtedly have a longer than average persistence time. The fossil record is also biased toward vertebrates and marine mollusks, both of which fossilize well because of their hard body parts. If background extinction rates were 10 times higher than the rates perceived from the fossil record, then extinctions among the living mammals today would be expected to occur at a rate of about one every 100 years. For birds, the background extinction rate would be two species every 100 years.

It is indisputable, however, that the extinction rate for species in recent times has been far higher than this. In the past 100 years, around 20 species of mammals and over 40 species of birds went extinct (**Figure 60.5a**). The rates of species extinctions on islands in the recent past confirm the dramatic effects of human activity. The Polynesians, who colonized Hawaii in the 4th and 5th centuries, appear to have been responsible for the extinction of half of the 100 or so species of endemic land birds in the period between their arrival and that of the Europeans in the late 18th century. A similar impact was felt in New Zealand, which was colonized by European settlers some

500 years later than Hawaii. In New Zealand, an entire avian megafauna, consisting of huge land birds, was exterminated over the course of the century, probably through a combination of hunting and large-scale habitat destruction through burning. The term **biodiversity crisis** is often used to describe this elevated loss of species. Many scientists believe that the rate of loss is higher now than during most of geological history, and most suggest that the growth in the human population has led to the increase in the number of extinctions of other species (**Figure 60.5b**).

To understand the process of extinction in more modern times, it is essential for ecologists to examine the role of human activities and their environmental consequences. In this section, we examine why species have gone extinct in the past and look at the factors that are currently threatening species with extinction.

The Main Causes of Extinction Are Introduced Species, Direct Exploitation, and Habitat Destruction

While all causes of extinctions are not known, introduced species, direct exploitation, and habitat destruction have been major human-induced threats. Increasingly, climate change is viewed as a significant threat to species. As mentioned at the beginning of Chapter 54, human-induced climate change, or global warming, has been implicated in the dramatic decrease in the population sizes of frog species in Central and South America.

(a)

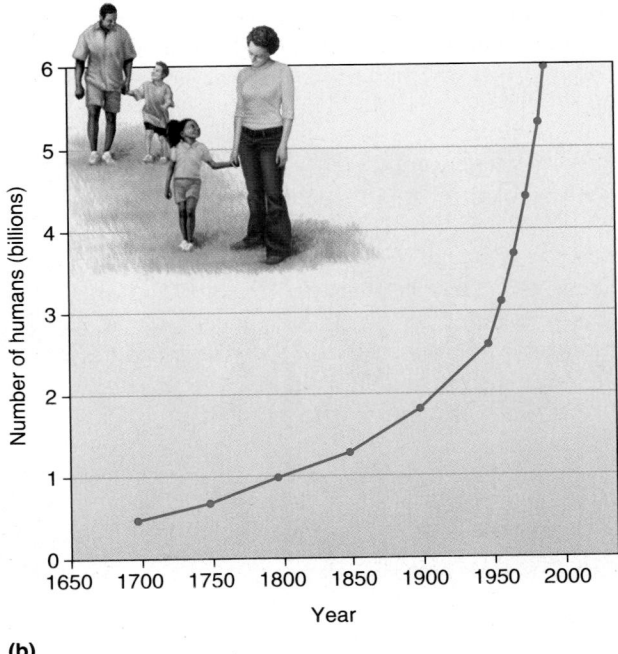

(b)

Figure 60.5   **Animal extinctions and human population growth.**   (a) Increasing numbers of known extinctions in birds and mammals are concurrent with **(b)** exponential increases in the global human population. These figures suggest that as human numbers increase, more and more species will go extinct.

Biological inquiry: Why might the increasing human population result in an increase in the extinction rate of other species?

Indeed, a recent study of six biodiversity-rich regions employed computer models to simulate the movement of species' ranges in response to changing climate conditions. The models predicted that unless greenhouse gas emissions are cut drastically, climate change will cause 15–37% of the species in those regions to go extinct by the year 2050.

Introduced Species **Introduced species**, often called exotic species, are those species moved by humans from a native location to another location. Most often the species are introduced for agricultural purposes or as sources of timber, meat, or wool, and these species need humans for their continued survival. Others, such as plants, insects, or marine organisms, are unintentionally transported via the movement of cargo by ships or planes. Regardless of their method of introduction, some introduced species become **invasive**, spreading and outcompeting native species for space and resources.

We can categorize the interactions between introduced and native species into competition, predation, and disease. Competition may eliminate local populations and cause huge reductions in the densities of native species, but it has not yet been clearly shown to exterminate entire species. On the other hand, there have been many recorded cases of extinction from predation. Introduced predators such as rats, cats, and mongooses have accounted for at least 43% of recorded extinctions of birds on islands. Lighthouse keepers' cats have annihilated populations of ground-nesting birds on small islands around the world. The brown tree snake, which was accidentally introduced onto the island of Guam, has decimated the country's native bird populations (refer back to Figure 57.12). Parasitism and disease carried by introduced organisms have also been important in causing extinctions. Avian malaria in Hawaii, spread by introduced mosquito species, is believed to have contributed to the demise of up to 50% of native Hawaiian birds (**Figure 60.6a**).

Direct Exploitation Direct exploitation, particularly the hunting of animals, has been the cause of many extinctions in the past. Two remarkable North American bird species, the passenger pigeon and the Carolina parakeet, were hunted to extinction by the early 20th century. The passenger pigeon (*Ectopistes migratorius*) was once the most common bird in North America, probably accounting for over 40% of the entire bird population (**Figure 60.6b**). Flock sizes were estimated to be over 1 billion birds. It may seem improbable that the most common bird on the continent could be hunted to extinction for its meat, but that is just what happened. The flocking behavior of the birds made them relatively easy targets for hunters, who used special firearms to harvest the birds in quantity. In 1876 in Michigan alone, over 1.6 million birds were killed and sent to markets in the eastern U.S. The Carolina parakeet (*Conuropsis carolinensis*), the only species of parrot native to the eastern U.S., was similarly hunted to extinction by the early 1900s.

Many whale species were driven to the brink of extinction prior to a moratorium on commercial whaling issued in 1988 (refer back to Figure 57.13). Steller's sea cow (*Hydrodamalis gigas*), a 9-meter-long manatee-like mammal, was hunted to extinction in the Bering Straits only 27 years after its discovery in 1740. A poignant example of human excess in hunting was the dodo (*Raphus cucullatus*), a flightless bird native only to the island of Mauritius that had no known predators. A combination of over-exploitation and introduced species led to its extinction within 200 years of the arrival of humans. Sailors hunted it for its meat, and the rats, pigs, and monkeys brought to the island by humans destroyed the dodo's eggs and chicks in their ground nests.

(a) Introduced parasites

(b) Direct exploitation

(c) Habitat destruction

Figure 60.6 **Extinction of species in the past.** **(a)** Many Hawaiian honeycreepers were exterminated by avian malaria from introduced mosquito species. This 'i'iwi (*Vestiaria coccinea*) is one of the few remaining honeycreeper species. **(b)** The passenger pigeon, which may have once been among the most abundant bird species on Earth, was hunted to extinction for its meat. **(c)** The ivory-billed woodpecker, the third-largest woodpecker in the world, was long thought to be extinct in the southeastern U.S. because of habitat destruction, but a possible sighting occurred in 2004. This nestling was photographed in Louisiana in 1938.

Habitat Destruction Habitat destruction through **deforestation**, the conversion of forested areas to nonforested land, has historically been a prime cause of the extinction of species. About one-third of the world's land surface is covered with forests, and much of this area is at risk of deforestation. While tropical forests are probably the most threatened forest type, with rates of deforestation in Africa, South America, and Asia varying between 0.6% and 0.9% per year, the destruction of forests is a global phenomenon.

Among North American terrestrial wildlife, about half of birds (272 species) and more than 10% of mammals (49 species) have an obligatory relationship with forest cover, meaning that they depend on trees for food and nesting sites. In terms of wildlife use, the oaks, whose acorns occur in the diet of at least 100 species of birds and mammals, are among the most valuable trees in North America. For many species of wildlife, the annual acorn crop is a major determinant of their abundance. Most woodpeckers nest in holes that they excavate in trees, and their food usually consists of insects collected on or in trees. The ivory-billed woodpecker (*Campephilus principalis*), the largest in North America and an inhabitant of wetlands and forests of the southeastern U.S., was widely assumed to have gone extinct in the 1950s due to destruction of its habitat by heavy logging (**Figure 60.6c**). Incredibly, in 2004, the woodpecker was sighted in the Big Woods area of eastern Arkansas, though this has not yet been confirmed.

Deforestation is not the only form of habitat destruction. The scouring of land to plant agricultural crops can create soil erosion, increased flooding, declining soil fertility, silting of the rivers, and desertification. While the average area of land under cultivation worldwide averages about 11%, with an additional 24% given over to rangeland, this amount varies tremendously between regions. For example, Europe uses 28% of its land for crops and pasturelands, with the result that many of its native species went extinct long ago. Wetlands also have been drained for agricultural purposes. Others have been filled in for urban or industrial development. In the U.S., as much as 90% of the freshwater marshes and 50% of the estuarine marshes have disappeared. Urbanization, the development of cities on previously natural or agricultural areas, is the most human-dominated and fastest-growing type of land use worldwide and devastates the land more severely than practically any other form of habitat degradation.

Small Populations Are Threatened by the Loss of Genetic Diversity

Knowing why species have gone extinct helps us to recognize the threats facing living species, which in turn may make it easier to protect endangered and threatened species in the future. As we have discussed, most of the factors currently threatening species—introduced species, direct exploitation, and habitat destruction—are derived from human activities. However, even if habitats are not destroyed, many become fragmented, leading to the development of small, isolated populations that can become less viable over time. The threats to species from small population size relate to a reduction of genetic diversity that can result from three factors: inbreeding, genetic drift, and limited mating, which reduces effective population size.

Inbreeding **Inbreeding**, which is mating among genetically related relatives, is more likely to take place in nature when population size becomes very small and there are a limited number of mates to choose from (see Chapter 24). In many species, survivorship of offspring declines as populations become more inbred. This phenomenon was shown as long ago as the 19th century. Litter size in inbred rats declined from 7.5 in 1887 to 3.2 in 1892. Furthermore, over the same time span, the level of nonproductive matings, those where no offspring were born, rose from 2% to 50%. Generally, the more inbred the population, the more severe these types of problems become.

One of the most striking examples of the effects of inbreeding in conservation biology involves the greater prairie chicken (*Tympanuchus cupido*). The male birds have a spectacular mating display that involves inflating the bright orange air sacs on their throat, stomping their feet, and spreading their tail feathers. The prairies of the Midwest were once home to millions of these birds, but as the prairies were converted to farmland, the range and population sizes of the bird shrank dramatically. The population of prairie chickens in Illinois decreased from 25,000 in 1933 to less than 50 in 1989. At that point, according to studies by Ronald Westemeier and colleagues, only 10 to 12 males existed. Because of the decreasing numbers of males, inbreeding in the population had increased. This was reflected in the steady reduction in the hatching success of eggs (**Figure 60.7**). The prairie chicken population had entered a downward spiral toward extinction from which it could not naturally recover, a phenomenon called an **extinction vortex**. In the early 1990s, conservation biologists began trapping prairie chickens in Kansas and Nebraska, where populations remained larger and more genetically diverse, and moved them to Illinois, bringing an infusion of new genetic material into the population. This translocation resulted in a rebounding of the egg-hatching success rate to 90% by 1993.

Genetic Drift In small populations, there is a good chance that some individuals will fail to mate successfully purely by chance, for example, because of the failure to find a mate. This is known as the **Allee effect**, after ecologist W. C. Allee, who first described it. If an individual that fails to mate possesses a rare gene, that genetic information will not be passed on to the next generation, resulting in a loss of genetic diversity from the population. As noted in Chapter 24, **genetic drift** refers to the random change in allele frequencies attributable to chance. Because the likelihood of an allele being represented in just one or a few individuals is higher in small populations than in large populations, small, isolated populations are particularly vulnerable to this type of reduction in genetic diversity. Such isolated populations will lose a percentage of their original diversity over time, approximately at the rate of $1/(2N)$ per generation, where N = population size. As described next, this has a greater effect in smaller versus larger populations.

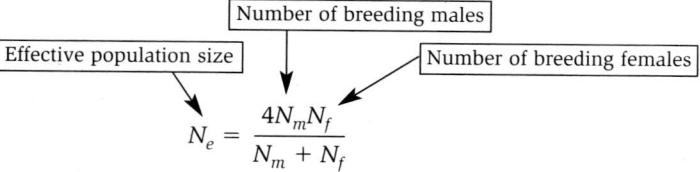

Figure 60.7 **Changes in the abundance and egg-hatching success rate of prairie chickens.** As the number of males decreased, inbreeding increased, resulting in a decrease in fertility, as indicated by a reduced egg-hatching rate. An influx of males in the early 1990s increased the egg-hatching success rate dramatically.

If $N = 500$, then $\frac{1}{2N} = 1/1{,}000 = 0.001$, or 0.1% genetic diversity lost per generation

If $N = 50$, then $\frac{1}{2N} = 1/100 = 0.01$, or 1% genetic diversity lost per generation

Due to genetic drift, a population of 500 will lose only 0.1% of its genetic diversity in a generation, while a population of 50 will lose 1%. Such losses become magnified over many generations. After 20 generations, the population of 500 will lose 2% of its original genetic variation, but the population of 50 will lose about 20%! For organisms that breed annually, this would mean a substantial loss in genetic variation over 20 years. Once again, this effect becomes more severe as the population size decreases.

As with inbreeding, the effects of genetic drift can be countered by immigration of individuals into a population. Even relatively low immigration rates of about one immigrant per generation (or one individual moved from one population to another) can be sufficient to counter genetic drift in a population of 100 individuals.

Limited Mating In many populations, the **effective population size**, the number of individuals that contribute genes to future populations, may be smaller than the number of individuals in the population, particularly in animals with a harem mating structure in which only a few dominant males breed. For example, dominant elephant seal bulls control harems of females, and a few males command all the matings (refer back to Figure 55.22). If a population consists of breeding males and breeding females, the effective population size is given by:

Number of breeding males

Effective population size   Number of breeding females

$$N_e = \frac{4N_mN_f}{N_m + N_f}$$

In a population of 500, a 50:50 sex ratio, and all individuals breeding, $N_e = (4 \times 250 \times 250)/(250 + 250) = 500$, or 100% of the actual population size. However, if 250 females breed with 10 males, $N_e = (4 \times 10 \times 250)/(10 + 250) = 38.5$, or 8% of the actual population size.

Knowledge of effective population size is vital to ensuring the success of conservation projects. One notable project in the U.S. involved planning the sizes of reserves designed to protect grizzly bear populations in the contiguous 48 states. The grizzly bear (*Ursus arctos*) has declined in numbers from an estimated 100,000 in 1800 to less than 1,000 at present. The range of the species is now less than 1% of its historical range and is restricted to six separate populations in four states (**Figure 60.8**). Research by biologist Fred Allendorf has indicated that the effective population size of grizzly populations is generally only about 25% of the actual population size because not all bears breed. Thus, even fairly large, isolated populations, such as the 200 bears in Yellowstone National Park, are vulnerable to the harmful effects of loss of genetic variation because the effective population size may be as small as 50 individuals. Allendorf and his colleagues proposed that an exchange of grizzly bears between populations or zoo collections would help tremendously in promoting genetic variation. Even an exchange of two bears per generation between populations would greatly reduce the loss of genetic variation.

60.3 Conservation Strategies

In their efforts to maintain the diversity of life on Earth, conservation biologists are currently active on many fronts and employ many strategies. We begin this section by discussing how conservation biologists identify the global habitats richest in species. We next explore the concept of nature reserves and consider questions such as how large conservation areas should be and how far apart they should be situated. These questions are

Figure 60.8 **Past and current ranges of the grizzly bear.** The range of the grizzly bear is currently less than 1% of its historical range. The current range in the continental U.S. has contracted to just six populations in four states, as the population size has shrunk from 100,000 before the West was settled to about 1,000 today.

Biological inquiry: If only 500 male and 500 female grizzlies exist today, but only 25% of the males breed, what is the effective population size?

within the realm of **landscape ecology**, which studies the spatial arrangement of communities and ecosystems in a geographic area. Next we discuss how conservation efforts often focus on certain species that can have a disproportionate impact on their ecosystem. We will also examine the field of restoration ecology, focusing on how wildlife habitats can be established from degraded areas and how captive breeding programs have been used to re-establish populations of threatened species in the wild. We conclude by returning to the theme of genomes and proteomes to show how modern molecular techniques of cloning can be useful in the fight to save critically endangered species.

Habitat Conservation Focuses on Identifying Countries Rich in Species, Areas Rich in Endemics, or Representative Habitats

Conservation biologists often must make decisions regarding which species and habitats should be protected. Many conservation efforts have focused on saving habitats in so-called megadiversity countries, because they often have the greatest number of species, but more recent strategies have promoted preservation of certain key areas with the highest levels of unique species, or preservation of representative areas of all types of habitat, even relatively species-poor areas in temperate grasslands (prairies).

Megadiversity Countries One method of targeting areas for conservation is to identify those countries with the greatest numbers of species, the **megadiversity countries**. Using the number of plants, vertebrates, and selected groups of insects as criteria, biologist Russell Mittermeier and colleagues have determined that just 17 countries are home to nearly 70% of all known species. Brazil, Indonesia, and Colombia top the list, followed by Australia, Peru, Mexico, Madagascar, China, and nine other countries. The megadiversity country approach suggests that conservation efforts should be focused on the most biologically rich countries.

As we learned in Chapter 58, large areas generally have the greatest number of species, and proponents of the megadiversity country concept believe it would protect the greatest number of species. Large countries, such as Brazil, would garner most of the available international funds. Perhaps the greatest drawback of the megadiversity approach, however, is that although megadiversity areas may contain the most species, they do not necessarily contain the most unique species. The mammal species list for Peru is 344 and for Ecuador, it is 271; of these, however, 208 species are common to both.

Areas Rich in Endemic Species Another method of setting conservation priorities, one adopted by the organization Conservation International, takes into account the number of species that are **endemic**, or found only in a particular place or region and nowhere else. This approach suggests that conservationists focus their efforts on geographic **hot spots**. To qualify as a hot spot, a region must meet two criteria: It must contain at least 1,500 species of vascular plants as endemics, and have lost at least 70% of its original habitat. Vascular plants were chosen as the primary group of organisms to determine whether or not an area qualifies as a hot spot mainly because most other terrestrial organisms depend on them to some extent.

Conservationists Norman Myers, Russell Mittermeier, and colleagues identified 34 hot spots that together occupy a mere 2.3% of the Earth's surface but contain 150,000 endemic plant species, or 50% of the world's total (**Figure 60.9**). Of these areas, the Tropical Andes and Sundaland (the region including Malaysia, Indonesia, and surrounding islands) have the most endemic plant species (**Table 60.3**). This approach posits that protecting geographic hot spots will prevent the extinction of a larger number of endemic species than would protecting areas of a similar size elsewhere. The main argument against using hot spots as the criterion for targeting conservation efforts is that the areas richest in endemics—tropical rain forests—would receive the majority of attention and funding, perhaps at the expense of other areas.

Representative Habitats In a third approach to prioritizing areas for conservation, scientists have recently argued that we need to conserve representatives of all major habitats. Thus, while the Pampas region of South America, which is arguably the most threatened habitat on the continent because of conversion of its grasslands to agriculture, does not compare well in richness or endemics to the rain forests, it is a unique area that without preservation could disappear. By selecting habitats that are most distinct from those already preserved, many areas that are threatened but not biologically rich may be preserved in addition to the less immediately threatened, but richer, tropical forests. The best strategy of identifying areas for conservation efforts might be one that creates a "portfolio" of areas containing those with high species richness, large numbers of endemic species, and various habitat types.

Figure 60.9   Location of major biodiversity hot spots around the world.   Hot spots, shown in orange, have high numbers of endemic species.

Table 60.3	Numbers of Endemic Species Present in the Top 10 Hot Spots of the World, Ranked by the Numbers of Endemic Plants					
Rank	**Hot spot**	**Plants**	**Birds**	**Mammals**	**Reptiles**	**Amphibians**	**Freshwater fish**
1	Tropical Andes	15,000	584	75	275	664	131
2	Sundaland	15,000	146	173	244	172	350
3	Mediterranean Basin	11,700	32	25	77	27	63
4	Madagascar	11,600	183	144	367	226	97
5	Brazil's Atlantic Forest	8,000	148	71	94	286	133
6	Indo-Burma	7,000	73	73	204	139	553
7	Caribbean	6,550	167	41	468	164	65
8	Cape Floristic Province	6,210	6	4	22	16	14
9	Philippines	6,091	185	102	160	74	67
10	Brazil's Cerrado	4,400	16	14	33	26	200

The Theory and Practice of Reserve Design Incorporate Principles of Island Biogeography and Landscape Ecology

After having identified areas to preserve, conservationists must determine the size, arrangement, and management of the protected land. Among the questions conservationists ask is this: Is one large reserve preferable to an equivalent area composed of smaller reserves? Since reserves can be viewed as islands of habitat, this question theoretically can be answered with the help of the theory of island biogeography. Other questions concern whether parks should be close together or far apart, and whether they should be connected by strips of suitable habitat to allow the movement of plants and animals between them. These large-scale questions concerning the spatial arrangement of communities and ecosystems in a geographic area fall within the field of landscape ecology. Conservationists also need to consider the fact that park design is often contingent on economic factors. Let's examine some of the many issues that conservationists address in the creation and management of protected land.

The Role of Island Biogeography In our exploration of the equilibrium model of island biogeography (see Chapter 58), we noted that it could be applied not only to a body of land surrounded by water but also to isolated fragments of habitat. Seen this way, wildlife reserves and sanctuaries are in essence islands in a sea of human-altered land. One question for conservationists is how large a protected area should be (**Figure 60.10a**). According to island biogeography, the number of species should increase with increasing area (the species-area effect); thus, the larger the area, the greater the number of species would be protected. In addition, larger parks have other benefits. For example, they are beneficial for organisms that require large spaces, including migrating species and species with extensive territories, such as lions and tigers.

Another question is whether it is preferable to protect one single, large reserve or several smaller ones (**Figure 60.10b**). This is called the **SLOSS debate** (for single large or several small). Proponents of the single, large reserve claim that a larger reserve is better able to preserve more and larger populations than an equal area divided into small areas. According to island biogeography, a larger block of habitat should support more species than any of the smaller blocks.

However, many empirical studies suggest that multiple small sites of equivalent area will contain more species, because a series of small sites is more likely to contain a broader variety of habitats than one large site. Looking at a variety of sites, researchers Jim Quinn and Susan Harrison concluded that animal life was richer in collections of small parks than in fewer, larger parks. In their study, having more habitat types outweighed the effect of area on biodiversity. In addition, another benefit of a series of smaller parks is a reduction of extinction risk by a single event such as a wildfire or the spread of disease.

Landscape Ecology **Landscape ecology** is a subdiscipline of ecology that examines the spatial arrangement of elements in communities and ecosystems. In the design of nature reserves, one question that needs to be addressed is how close to situate reserves to each other, such as whether to have three or four small reserves close to each other or farther apart. A similar question is whether to have a linear or cluster arrangement of small reserves. Island biogeography suggests that if an area must be fragmented, the sites should be as close as possible to permit dispersal (**Figure 60.10c,d**). In practice, however, having small sites far apart may preserve more species than having them close together, since once again, distant sites are likely to incorporate slightly different habitats and thus species.

Landscape ecologists have also suggested that small reserves should be linked together by biotic corridors, or **movement corridors**, which are thin strips of land that may permit the movement of species between patches (**Figure 60.10e**).

Better					
Worse					
(a)	(b)	(c)	(d)	(e)	(f)

Figure 60.10 **The theoretical design of nature reserves, based on the tenets of island biogeography.** (a) A larger reserve will hold more species and have low extinction rates. (b) Given a certain area available, it should be fragmented into as few pieces as possible. (c) If an area must be fragmented, the pieces should be as close as possible to permit dispersal. (d) To enhance dispersal, a cluster of fragments is preferable to a linear arrangement. (e) Maintaining or creating corridors between fragments may also enhance dispersal. (f) Circular-shaped areas will minimize the amount of undesirable edge habitat.

Figure 60.11   Habitat corridors.

Biological inquiry: Why would these European hedgerows act as habitat corridors?

Such corridors may facilitate movements of organisms that are vulnerable to predation outside of their natural habitat or that have poor powers of dispersal between habitat patches. In this way, if a disaster befalls a population in one small reserve, immigrants from neighboring populations can more easily re-colonize it. This avoids the need for humans to physically move new plants or animals into an area.

Several types of habitat function as corridors, including hedgerows in Europe, which facilitate movement and dispersal of species between forest fragments (**Figure 60.11**). In China, corridors of habitat have been established to link small, adjacent populations of giant pandas. Riparian habitats, vegetated corridors bordering watercourses, are thought to help facilitate movement of species between lakes or rivers. In Florida, debate continues about whether to establish new populations of the Florida panther in refuges in both central and north Florida and whether to have these populations connect with the current population in south Florida via riparian movement corridors. In theory, this would lessen the threat of extinction of the Florida panther via inbreeding or a catastrophic event such as a hurricane. However, disadvantages are associated with corridors. First, corridors also can facilitate the spread of disease, invasive species, and fire between small reserves, and second, it is not yet clear if species, such as female Florida panthers, would actually use such corridors.

Finally, parks are often designed to minimize **edge effects**, the special physical conditions that exist at the boundaries or "edges" of ecosystems (see **Figure 60.10f**). Habitat edges, particularly those between natural habitats such as forests and

developed land, are often different in physical characteristics from the habitat core. The center of a forest is shaded by trees and has less wind and light than the forest edge, which is unprotected. Many forest-adapted species thus shy away from forest edges and prefer forest centers. Circular-shaped parks are preferred over long, skinny parks because the amount of edge is minimized.

Economic Considerations in Conservation   While the principles of island biogeography theory and landscape ecology are useful in illuminating conservation issues, in reality there is often little choice as to the size, shape, and location of nature reserves. Management practicalities, costs of acquisition and management, and politics often override ecological considerations, especially in developing countries, where costs for large reserves may be relatively high. Economic considerations often enter into the choice of which areas to preserve. Typically, many countries protect areas in those regions that are the least economically valuable rather than choosing areas to ensure a balanced representation of the country's biota. In the U.S., most national parks have been chosen for their scenic beauty, not because they preserve the richest habitat for wildlife.

When designing nature reserves, countries should also consider how to finance their management. Interestingly, the amount of money spent to protect nature reserves may better determine species extinction rates than reserve size. According to island biogeography theory, large areas minimize the risk of extinctions because they contain sizable populations. In Africa, several parks, such as Serengeti and Selous in Tanzania, Tsavo in Kenya, and Luangwa in Zambia, are large enough to fulfill this theoretical ideal. However, in the 1980s, populations of black rhinoceroses and elephants declined dramatically within these areas because of poaching, showing that a wide gap may exist between theory and reality. In reality, the rates of decline of rhinos and elephants, largely a result of poaching, have been related directly to conservation efforts and spending (**Figure 60.12**). The remaining black rhinos, lowland gorillas, and pygmy chimpanzees in Africa and the vicuna, a llama-like animal in South America, have all shown the greatest stabilization of numbers in areas that have been heavily patrolled and where resources have been concentrated.

The Single-Species Approach Focuses Conservation Efforts on Particular Types of Species

Some plants and animals are, for reasons we will explore in this section, of more interest to conservation biologists than others. The single-species approach to conservation focuses on saving particularly important species, including indicator, umbrella, flagship, and keystone species.

Indicator Species   Some conservation biologists have suggested that certain organisms can be used as **indicator species**, or species whose status provides information on the overall health of an ecosystem. Corals are good indicators of marine processes such as siltation, the accumulation of sediments transported by water. Because siltation reduces the availability of light, the abundance of many marine organisms decreases in such situations, with corals among the first to display a decline in health. A proliferation of the dark variety of the peppered moth (*Biston betularia*) has been shown to be a good indicator of polluted air. Polar bears (*Ursus maritimus*) are thought to be an indicator species for global climate change (**Figure 60.13a**). Scientists believe that global warming is causing the ice in the Arctic to melt earlier in the spring than in the past. Because polar bears rely on the ice to hunt for seals, the earlier breakup of the ice is leaving the bears less time to feed and build the fat that enables them to sustain themselves and their young. Scientists have noted a decrease in the weight of polar bears and in the survival of their cubs.

Umbrella Species **Umbrella species** are species whose habitat requirements are so large that protecting them would protect many other species existing in the same habitat. The Northern

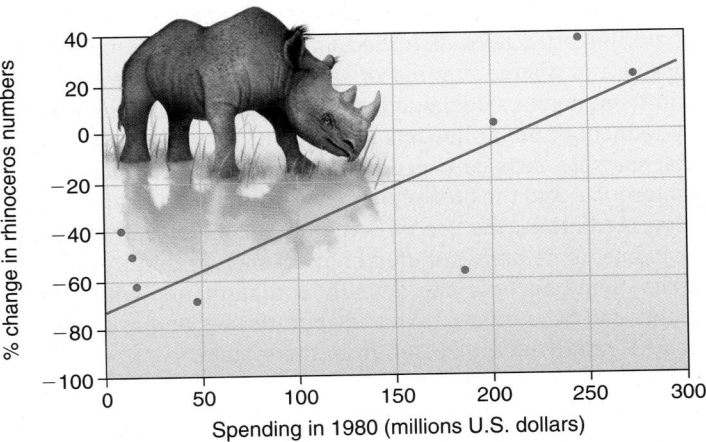

Figure 60.12   **The economics of conservation.**   A positive relationship is seen between change in black rhinoceros numbers between 1980 and 1984, and conservation spending in various African countries.

spotted owl (*Strix occidentalis*) of the Pacific Northwest is considered to be an important umbrella species (**Figure 60.13b**). A pair of birds needs at least 800 hectares of old-growth forest for survival and reproduction, so maintaining healthy owl populations is thought to help ensure survival of many other forest-dwelling species. In the southeast area of the U.S., the red-cockaded woodpecker (*Picoides borealis*) is often seen as the equivalent of the spotted owl, since it requires large tracks of old-growth long-leaf pine (*Pinus palustris*), including old diseased trees in which it can excavate its nests. Some species, such as *Zea diploperennis*, which we discussed at the beginning of the chapter, fall into this category. For example, to protect *Z. diploperennis* in its natural environment, the state of Jalisco in Mexico bought the land where it grows and established a nature reserve and research facility there.

Flagship Species   In the past, conservation resources were often allocated to a **flagship species**, a single large or instantly recognizable species. Such species were typically chosen because they were attractive and thus more readily engendered support from the public for their conservation. The concept of the flagship species, which are usually charismatic vertebrates such as the American buffalo (*Bison bison*), is often used to raise awareness for conservation in general. The giant panda (*Ailuropoda melanoleuca*) is the World Wildlife Fund's emblem for endangered species, and the Florida panther (*Puma concolor*) has become a symbol of the state's conservation campaign (**Figure 60.13c**).

Keystone Species   A different, perhaps more effective conservation strategy focuses on **keystone species**, species within a community that have a role out of proportion to their abundance. The beaver, a relatively small animal, can completely alter a community by building a dam and flooding an entire river valley (**Figure 60.14**). The resultant lake may become a home to fish species, wildfowl, and aquatic vegetation. A decline in the number of beavers could have serious ramifications for the remaining community members, promoting fish die-offs, waterfowl loss, and the death of vegetation adapted to waterlogged soil.

(a) Polar bear

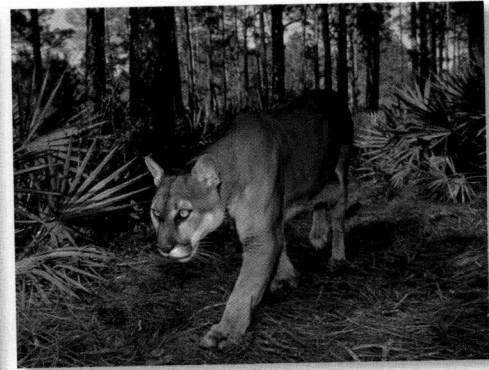

(b) Northern spotted owl

(c) Florida panther

Figure 60.13   **Indicator, umbrella, and flagship species.**   **(a)** Polar bears have been called an indicator species of global climate change. **(b)** The Northern spotted owl is considered an umbrella species for the old-growth forest in the Pacific Northwest. **(c)** The Florida panther has become a flagship species for Florida.

Figure 60.14 **Keystone species.** The American beaver creates large dams across streams, and the resultant lakes provide habitats for a great diversity of species.

Tropical ecologist John Terborgh considers palm nuts and figs to be keystone species because they produce fruit during otherwise fruitless times of the year and are thus critical resources for tropical forest fruit-eating animals, including primates, rodents, and many birds. Together, these fruit-eaters account for as much as three-quarters of the tropical forest animal biomass. Without the fruit trees, wholesale extinction of these animals could occur. Note that a keystone species is not the same as a **dominant species**, one that has a large effect in a community because of its abundance or high biomass. For example, *Spartina* cordgrass is a dominant species in a salt marsh because of its large biomass (refer back to Figure 59.15), but it is not a keystone species.

In the southeastern U.S., gopher tortoises can be regarded as keystone species because the burrows they create provide homes for an array of other animals, including mice, opossums, frogs, snakes, and insects. Many of these creatures depend on the gopher tortoise burrows and would be unable to survive without them. Gopher tortoises and some other keystone species, including beavers, are also called **ecosystem engineers**, because they create, modify, and maintain habitats. African elephants act as ecosystem engineers through their browsing activity, destroying small trees and shrubs and changing woodland habitats into grasslands.

Few studies have analyzed the community importance of keystone species, and no set criteria have been established for designating a keystone species. Nevertheless, such species do seem to affect species diversity. The conservation community is eager to identify keystone species, because by managing a keystone species, they may also ensure the survival of many other species in the ecosystem.

Restoration Ecology Attempts to Rehabilitate Degraded Ecosystems

Restoration ecology is the full or partial repair or replacement of biological habitats and/or their populations that have been damaged. It can focus on restoring or rehabilitating a habitat, or it can involve attempting to return species to the wild following captive breeding. Following opencast mining for coal or phosphate, huge tracts of disturbed land must be replenished with topsoil and a large number of species such as grasses, shrubs, and trees must be replanted. Aquatic habitats may be restored by reducing human impacts and replanting vegetation. In Florida, seagrass beds damaged by boat propellers are closed off to motorboats and the area replanted. Restoration can also involve **bioremediation**, the use of living organisms, usually microbes or plants, to detoxify polluted habitats such as dump sites or oil spills. Some bacteria can detoxify contaminants, while certain plants can accumulate toxins in their tissues and are then harvested, removing the poison from the system.

Habitat Restoration The three basic approaches to habitat restoration are complete restoration, rehabilitation, and ecosystem replacement. In complete restoration, conservationists attempt to put back exactly what was there prior to the disturbance. The University of Wisconsin pioneered the restoration of prairie habitats as early as 1935, converting agricultural land back to species-rich prairies (**Figure 60.15a**). The second approach aims to return

(a) Complete restoration

(b) Rehabilitation

(c) Ecosystem replacement

Figure 60.15 **Habitat restoration.** **(a)** As long ago as 1935, the University of Wisconsin was involved in returning agricultural land to native prairies. **(b)** In Florida, phosphate mines are so degraded that even after restoration, some exotic species such as cogongrass often invade. **(c)** These old open-pit mines in Middlesex, England, have been converted to valuable freshwater habitats, replacing the wooded area that was originally present.

Figure 60.16 **Captive breeding programs.** The California condor, the largest bird in the U.S., with a wingspan of nearly 3 m, has been bred in captivity in California. **(a)** A researcher at the San Diego Wild Animal Park feeds a chick with a puppet, so that the birds will not become habituated to the presence of humans. **(b)** This captive-bred condor soars over the Grand Canyon. Note the tag on the underside of its wing.

the habitat to something similar to, but a little less than, full restoration, a goal called rehabilitation. In Florida, phosphate mining involves removing a layer of topsoil or "overburden," mining the phosphate-rich layers, returning the overburden, and replanting the area. Exotic species such as cogongrass (*Imperata cylindrica*), an invasive Southeast Asian species, often invade these disturbed areas, and the biodiversity of the restored habitat is usually not comparable to that of unmined areas (**Figure 60.15b**). The third approach makes no attempt to restore what was originally present but instead replaces the original ecosystem with a different one. The replacement could be an ecosystem that is simpler but more productive, as when deciduous forest is replaced after mining by simple grassland to be used for public recreation.

While any of these approaches can be employed in the habitat restoration process, a common viewpoint among environmentalists is that the endpoint of habitat restoration should be complete restoration. While in some cases, full restoration is appropriate, there are also many cases where restoration is so difficult or expensive as to be impractical. Ecosystem replacement is particularly sensible for land that has been significantly damaged by past activities. It would be nearly impossible to recreate the original landscape of an area that was mined for stone or gravel. In these situations, however, wetlands or lakes may be created in the open pits (**Figure 60.15c**).

Captive Breeding Captive breeding, the propagation of animals and plants outside their natural habitat to produce stock for subsequent release into the wild, has proved valuable in reestablishing breeding populations following extinction or near extinction. Peter Stiling has reintroduced greenhouse-propagated populations of a rare *Opuntia* cactus in the Florida Keys following the decimation of the original population by *Cactoblastis cactorum*, an invasive moth from the Caribbean. Zoos, aquariums, and botanical gardens often play a key role in captive breeding, propagating species that are highly threatened in the wild. They also play an important role in public education about the loss of biodiversity and the use of restoration programs.

Several classic programs illustrate the value of captive breeding and reintroduction. The peregrine falcon (*Falco peregrinus*) became extinct in nearly all of the eastern U.S. by the mid-1960s, a decline that was linked to the effects of DDT (refer back to Figure 59.9). In 1970, Tom Cade gathered falcons from other parts of the country to start a captive breeding program at Cornell University. Since then, the program has released thousands of birds into the wild, and in 1999 the peregrine falcon was removed from the Endangered Species List. A captive breeding program is also helping save the California condor (*Gymnogyps californicus*) from extinction. In the 1980s, there were only 25 known birds, some in captivity and some in the wild. Scientists made the decision to capture the remaining wild birds in order to protect and breed them (**Figure 60.16a**). By 2006, the captive population numbered 156 individuals, and 128 birds were living in the wild, 60 in California and 68 in Arizona (**Figure 60.16b**). A milestone was reached in 2002, when a pair of captive-reared California condors bred in the wild.

Because the number of individuals in any captive breeding program is initially small, care must be exercised to avoid inbreeding. Matings are usually carefully arranged to maximize resultant genetic variation in offspring. The use of genetic engineering to clone endangered species is a new area that may eventually help bolster populations of captive-bred species.

GENOMES & PROTEOMES

Can Cloning Save Endangered Species?

In 1997, Ian Wilmut and colleagues at Scotland's Roslin Institute announced to the world that they had cloned a now-famous sheep, Dolly, from mammary cells of an adult ewe (see Chapter 20). Since then, interest has arisen among conservation biologists about whether the same technology might be used to save species on the verge of extinction. Scientists were encouraged that in January 2001, an Iowa farm cow called Bessie gave birth to a cloned Asian gaur (*Bos gaurus*), an endangered species. The gaur, an oxlike animal native to the jungles of India and Burma, was cloned from a single skin cell taken from a dead animal. To clone the gaur, scientists removed the nucleus from a cow's egg and replaced it with a nucleus from the gaur's cell.

Figure 60.17 Cloning may help save endangered species. In 2004, this 8-month-old cloned Javan banteng made its public debut at the San Diego Zoo.

The treated egg was then placed into the cow's womb. Unfortunately, the gaur died from dysentery two days after birth, although scientists believe this was unrelated to the cloning procedure. In 2003, another type of endangered wild cattle, the Javan banteng (*Bos javanicus*), was successfully cloned (**Figure 60.17**).

Despite the promise of cloning, a number of issues remain unresolved:

1. Scientists would have to develop an intimate knowledge of different species' reproductive cycles. For sheep and cows this was routine, based on the vast experience in breeding these species, but eggs of different species, even if they could be harvested, often require different nutritive media in laboratory cultures.
2. Because it is desirable to leave natural mothers available for breeding, scientists will have to identify surrogate females of similar but more common species that can carry the fetus to term.
3. Some argue that cloning does not address the root causes of species loss, such as habitat fragmentation or poaching,

and that resources would be better spent elsewhere, for example, in preserving the endangered species' remaining habitat.

4. Cloning might not be able to increase the genetic variability of the population. However, if it were possible to use cells from deceased animals, for example, from their hair or feathers, these clones could theoretically reintroduce lost genes back into the population.

Many biologists believe that while cloning may have a role in conservation, it is only part of the solution and that we should address what made the species go extinct in the first place before attempting to restore it.

Conservation is clearly a matter of great importance, and a failure to value and protect our natural resources adequately could be a grave mistake. Some authors, most recently the ecologist and geographer Jared Diamond, have investigated why many great societies of the past—including Angkor Wat, Easter Island, and Norse Greenland—collapsed or vanished, leaving behind monumental ruins. Diamond has concluded that the collapse of these societies occurred partly because people inadvertently destroyed the ecological resources on which their societies depended. Modern nations such as Rwanda face similar issues. The country's population density is the highest in Africa, and it has a limited amount of land that can be used for growing crops. By the late 1980s, the need to feed a growing population led to the wholesale clearing of Rwanda's forests and wetlands, with the result that little additional land was available to farm. Increased population pressure, fueled by environmental scarcity, was likely a contributing factor in igniting the genocide of 1994.

As we've seen throughout this textbook, an understanding of biology is vital to understand and help solve many of society's problems. Within this large field, genomics and proteomics may have a huge potential for improving people's lives and society at large. These disciplines offer the opportunity to unlock new diagnoses and treatments for diseases, to improve nutrition and food production, and even to help us to restore biological diversity.

CHAPTER SUMMARY

60.1 Why Conserve Biodiversity?

- Conservation biology uses knowledge from molecular biology, genetics, and ecology to protect the biological diversity of life. (Figure 60.1)
- The preservation of biodiversity has been justified because of its economic value, because of the value of ecosystem services, and on ethical grounds. (Tables 60.1, 60.2)
- Four models exist that describe the relationship between biodiversity and ecosystem function; diversity-stability, rivet, redundancy, and idiosyncratic. (Figure 60.2)

- Experiments both in the lab and in the field have shown that increased biodiversity results in increased ecosystem function. (Figures 60.3 and 60.4)

60.2 The Causes of Extinction and Loss of Biodiversity

- Extinction—the process by which species die out—has been a natural phenomenon throughout the history of life on Earth. Extinction rates in recent times, however, have been much higher than in the past, a phenomenon called the diversity crisis. (Figure 60.5)

- The main causes of extinctions have been and continue to be introduced species, direct exploitation, and habitat destruction. (Figure 60.6)
- Reduced population size can lead to a reduction of genetic diversity through inbreeding, genetic drift, and limited mating, which reduces effective population size.
- Inbreeding, mating among genetically related relatives, can lead to a reduction in fertility. (Figure 60.7)
- Knowledge of a species' effective population size is vital to ensure the success of conservation projects. (Figure 60.8)

60.3 Conservation Strategies

- Habitat conservation strategies commonly target megadiversity countries, countries with the largest number of species; biodiversity hot spots, areas with the largest number of endemic species, those unique to the area; and representative habitats, areas that represent the major habitats. (Figure 60.9, Table 60.3)
- Conservation biologists employ many strategies in protecting biodiversity. Principles of the equilibrium model of island biogeography and landscape ecology are used in the theory and practice of park reserve design to determine, for example, whether the park should take the form of one single or several small reserves. (Figures 60.10, 60.11)
- Economic considerations also play an important role in reserve creation, and it has been shown that conservation spending is positively related to population size. (Figure 60.12)
- The single-species approach focuses conservation efforts on indicator species, umbrella species, flagship species, and keystone species. Indicator species are species whose status can provide information on the overall health of an ecosystem. Umbrella species are species whose habitat requirements are so large that preserving them would also preserve many other species. Flagship species are usually charismatic or instantly recognizable species; and keystone species have an effect out of proportion to their abundance. (Figures 60.13, 60.14)
- Restoration ecology seeks to repair or replace populations and their habitats. Three basic approaches to habitat restoration are complete restoration, rehabilitation, and ecosystem replacement. (Figure 60.15)
- Captive breeding is the propagation of animals outside their natural habitat. Several programs have illustrated the success of captive breeding and reintroduction to the wild, including those for California condors. (Figure 60.16)
- Cloning of endangered species has been accomplished on a very small scale and despite its limitations may have a role in conservation biology. (Figure 60.17)

TEST YOURSELF

1. Which of the following statements best describes an endangered species?
   a. a species that is likely to become extinct in a portion of its range
   b. a species that has disappeared in a particular community but is present in other natural environments
   c. a species that is extinct
   d. a species that is in danger of becoming extinct throughout all or a significant portion of its range
   e. both b and d

2. Biological diversity is important and should be preserved because
   a. food, medicines, and industrial products are all benefits of biodiversity.
   b. ecosystems provide valuable services to us in many ways.
   c. many species can be used as valuable research tools.
   d. we have an ethical responsibility to protect our environment.
   e. all of the above are correct.

3. The impact on an ecosystem of a particular species is dependent on how critical the species is to the ecosystem. This statement defines the _____ hypothesis.
   a. rivet
   b. redundancy
   c. idiosyncratic
   d. diversity-stability
   e. community ecology

4. The research conducted by Tillman and colleagues demonstrated that
   a. as diversity increases, productivity increases.
   b. as diversity decreases, productivity increases.
   c. areas with higher diversity demonstrate less efficient use of nutrients.
   d. species richness increases lead to an increase in invasive species.
   e. increased diversity results in increased susceptibility to disease.

5. The process by which species die out is termed
   a. endangerment.
   b. death.
   c. extant.
   d. extinction.
   e. conservation biology.

6. Which of the following is not a prime cause of extinction?
   a. predation by introduced species
   b. habitat destruction
   c. direct exploitation
   d. disease brought by invasive species
   e. bioremediation

7. The negative result of inbreeding is
   a. an increase in genetic diversity that leads to new genetic diseases.
   b. a decrease in genetic diversity that limits how a species can adapt.
   c. a decrease in number of mates.
   d. mating between related individuals.
   e. all of the above.

8. The number of individuals that contribute genes to future populations is called
   a. the effective population size.
   b. the adult population size.
   c. the breeding coefficient.
   d. the gene pool.
   e. the female population size.

9. Small strips of land that connect and allow organisms to move between small patches of natural habitat are called
   a. biological conduits.
   b. edge effects.
   c. movement corridors.
   d. migration pathways.
   e. landscape breaks.

10. Bioremediation is
    a. a process that restores a disturbed habitat to its original state.
    b. a process that uses microbes or plants to detoxify contaminated habitats.
    c. the legislation requiring rehabilitation of a disturbed habitat.
    d. a process of capturing all of the living individuals of species for breeding purposes.
    e. the process of removing tissue from a dead organism in the hopes of cloning it.

CONCEPTUAL QUESTIONS

1. What are the three levels at which biodiversity can be examined?
2. Define conservation biology.
3. Distinguish between an umbrella species, a flagship species, and a keystone species.

EXPERIMENTAL QUESTIONS

1. What was the goal of Shahid Naeem and colleagues in their experiment at Silwood Park, England?
2. What was the hypothesis tested by the researchers?
3. How did the researchers test for ecosystem functioning?

COLLABORATIVE QUESTIONS

1. Discuss several causes of species extinction.
2. Discuss several species-specific approaches to conservation biology.

---

www.brookerbiology.com

This website includes answers to the Biological Inquiry questions found in the figure legends and all end-of-chapter questions.

GLOSSARY

A

A band A wide, dark band produced by the orderly parallel arrangement of the thick filaments in the middle of each sarcomere.

abiotic The term used to describe interactions between organisms and their nonliving environment.

aboral Refers to the region opposite the mouth.

abortion Procedures or circumstances that cause the death of an embryo or fetus after implantation.

abscisic acid One of several plant hormones that help a plant to cope with environmental stress.

absorption The process in which digested nutrients are transported from the digestive cavity into an animal's circulatory system.

absorption spectrum A diagram that depicts the wavelengths of electromagnetic radiation that are absorbed by a pigment.

absorptive nutrition The process whereby an organism secretes enzymes into food substrates, breaking down complex organic molecules into small organic molecules that are absorbed as food.

absorptive state One of two alternating phases in the utilization of nutrients; occurs when ingested nutrients enter the blood from the gastrointestinal tract. The other phase is the postabsorptive state.

acclimatization The process of fine-tuning an animal's adaptive mechanisms to a changing environment.

accommodation The process in which contraction and relaxation of the ciliary muscles adjust the lens according to the angle at which light enters the eye.

acid A molecule that releases hydrogen ions in solution.

acid hydrolase A hydrolytic enzyme found in lysosomes that functions at acidic pH and uses a molecule of water to break a covalent bond.

acid rain Precipitation with a pH of less than 5.6; acid rain results from the burning of fossil fuels, which releases sulfur dioxide and nitrogen oxide into the atmosphere, which in turn react with oxygen in the air to form sulfuric acid and nitric acid, and falls to the surface in rain or snow.

acidic A solution that has a pH below 7.

acoelomate An animal that lacks a body cavity entirely.

acquired antibiotic resistance The common phenomenon of a previously susceptible strain becoming resistant to a specific antibiotic.

acquired immune deficiency syndrome (AIDS) A disease caused by the human immunodeficiency virus (HIV) that leads to a defect in the immune system of infected individuals.

acrocentric A chromosome in which the centromere is near one end.

acromegaly A condition in which a person's GH levels remain elevated after growth has ceased, and the continued excess GH causes many bones, like those of the hands and feet, to thicken and enlarge.

acrosomal reaction An event in fertilization in which the binding of a sperm cell to proteins located in the egg cell plasma membrane triggers a series of events producing the fast block to polyspermy and the entry of the sperm cell's nucleus into the egg cell.

acrosome A special structure at the tip of a sperm's head that contains proteolytic enzymes that help break down the plasma membrane of the ovum at fertilization.

actin A cytoskeletal protein.

actin filament A thin type of protein filament composed of actin proteins that forms part of the cytoskeleton and supports the plasma membrane, shape, and movement.

action potential The movement of an electrical impulse along the plasma membrane, which occurs in animal nerve axons and some plant cells.

action spectrum The rate of photosynthesis plotted as a function of different wavelengths of light.

activation energy An initial input of energy in a chemical reaction that allows the molecules to get close enough to cause a rearrangement of bonds.

activator A transcription factor that binds to DNA and increases the rate of transcription.

active immunity The acquired response to exposure to any type of antigen.

active site The location in an enzyme where the chemical reaction takes place.

active transport The transport of a solute across a membrane against its gradient—that is, from a region of low concentration to higher concentration. In the case of ions, active transport is against an electrochemical gradient.

adaptation The processes and structures by which organisms adjust to short-term or long-term changes in their environment.

adaptive radiation The process whereby a single ancestral species evolves into a wide array of descendant species that differ greatly in their habitat, form, or behavior.

adenine (A) A purine base found in DNA and RNA.

adenosine triphosphate (ATP) A nucleotide that is a common energy source of all cells.

adenylyl cyclase An enzyme in the plasma membrane that synthesizes cAMP from ATP.

adiabatic cooling The process in which increasing elevation leads to a decrease in air temperature.

adventitious root A root that is produced on the surfaces of stems (and sometimes leaves) of vascular plants; also, roots that develop at the bases of stem cuttings.

aerenchyma Spongy plant tissue with large air spaces.

aerobic Refers to a process that occurs in the presence of oxygen; a form of metabolism that requires oxygen.

aerobic respiration During this type of respiration, O_2 is consumed and CO_2 is released.

aerotolerant anaerobe A microorganism that does not use oxygen but is not poisoned by it either.

afferent arterioles Blood vessels that provide a pathway for blood into a tissue or organ. For example, afferent arterioles in the kidney that supplies each glomerulus with blood.

aflatoxins Fungal toxins that cause liver cancer and are a major health concern worldwide.

age structure The relative numbers of individuals of each defined age group in a population.

age-specific fertility rate The rate of offspring production for females of a certain age; used to help calculate how a population grows.

AIDS Acquired immune deficiency syndrome. AIDS reduces the body's immunity by killing helper T cells.

air sac A component of the avian respiratory system; air sacs—not lungs—expand when a bird inhales and shrink when it exhales, and they do not participate in gas exchange.

akinete A thick-walled cell used to survive unfavorable conditions in a dormant state.

aldosterone A steroid hormone made by the adrenal glands.

algae A term that applies to about 10 phyla of protists that include both photosynthetic and nonphotosynthetic species.

alimentary canal The single elongated tube of a digestive system with an opening at either end, through which food and eventually wastes pass from one end to the other.

alkaline A solution that has a pH above 7.

alkaloids A group of structurally related secondary metabolites that all contain nitrogen and usually have a cyclic, ringlike structure.

allantois An extraembryonic membrane in the amniotic egg that serves as a disposal sac for metabolic wastes.

allee effect The phenomenon that some individuals will fail to mate successfully purely by chance, for example, because of the failure to find a mate.

allele   A variant form of a gene.

allele frequency   The number of copies of an allele divided by the total number of alleles in a population.

allelochemical   A powerful plant chemical, often a root exudates, that kills other plant species.

allelopathy   The suppressed growth of one species due to the release of toxic chemicals by another species.

allodiploid   An alloploid that has only one set of chromosomes from two different species.

allometric growth   The pattern whereby different parts of the body grow at different rates with respect to each other.

allopatric   The term used to describe species occurring in different geographic areas.

allopatric speciation   A form of speciation that occurs when a population becomes geographically isolated from other populations and evolves into one or more new species.

alloploid   An organism that contains at least one set of chromosomes from two or more different species.

allopolyploid   An organism that contains two or more complete sets of chromosomes from two or more different species.

allosteric site   A site where a molecule can bind noncovalently and affect the function of the active site. The binding of a molecule to an allosteric site causes a conformational change in the enzyme that inhibits its catalytic function.

allotetraploid   A type of allopolyploid that contains two complete sets of chromosomes from two species for a total of four sets.

alternation of generations   The phenomenon that occurs in plants (and some protists) in which the life cycle alternates between multicellular diploid organisms, called sporophytes, and multicellular haploid organisms, called gametophytes.

alternative splicing   The splicing of pre-mRNA in more than one way to create two or more different polypeptides.

altruism   Behavior that appears to benefit others at a cost to oneself.

alveolus   A saclike structure of the lungs where gas exchange occurs.

Alzheimer's disease (AD)   The leading worldwide cause of dementia; characterized by a loss of memory and intellectual and emotional function.

amensalism   One-sided competition, where the interaction is detrimental to one species but not to the other.

Ames test   A test that helps ascertain whether or not an agent is a mutagen by using a strain of a bacterium, *Salmonella typhimurium*.

amino acid   The building block of proteins. Amino acids have a common structure in which a carbon atom, called the α-carbon, is linked to an amino group (NH_2) and a carboxyl group (COOH). The α-carbon also is linked to a hydrogen atom and a particular side chain.

amino terminus   *See* N-terminus.

aminoacyl site (A site)   One of the three sites for tRNA binding to the ribosome; the others are the peptidyl site (P site) and the exit site (E site). The A site is the site where incoming tRNA molecules bind to the mRNA (except for the initiator tRNA).

aminoacyl tRNA   *See* charged tRNA.

aminoacyl-tRNA synthetase   An enzyme that catalyzes the attachment of amino acids to tRNA molecules.

ammonia (NH_3)   One of the most highly toxic of the nitrogenous wastes because it disrupts pH, ion electrochemical gradients, and many chemical reactions that involve oxidations and reductions; typically produced in many aquatic species.

ammonification   The conversion of organic nitrogen to NH_3 and NH_4^+.

amnion   An innermost extraembryonic membrane in the amniotic egg; it protects the developing embryo in a fluid-filled sac called the amniotic cavity.

amniotes   A group of tetrapods with amniotic eggs that includes turtles, lizards, snakes, crocodiles, birds, and mammals.

amniotic egg   The structure that contains the developing embryo and the four separate extraembryonic membranes that it produces: the amnion, the yolk sac, the allantois, and the chorion.

amoeba   A protist cell that moves by pseudo-podia, which involves extending cytoplasm into filaments or lobes.

amoebocyte   A mobile cell within a sponge's mesohyl that absorbs food from choanocytes, digests it, and carries the nutrients to other cells.

amphibian   A tetrapod that has successfully invaded the land but must return to the water to reproduce.

amphipathic   In molecules, meaning they have a hydrophobic (water-fearing) region and a hydrophilic (water-loving) region.

ampulla   A muscular sac at the base of each tube foot of a echinoderm that stores water.

amygdala   An area of the brain known to be critical for understanding and remembering emotional situations.

amylase   A digestive enzyme involved in the digestion of starch.

anabolic reaction   A metabolic pathway that promotes the synthesis of larger molecules from smaller precursor molecules.

anabolism   The synthesis of cellular molecules and macromolecules, which usually requires an input of energy.

anaerobic   Refers to a process that occurs in the absence of oxygen; a form of metabolism that does not require oxygen.

anaerobic respiration   The breakdown of organic molecules in the absence of oxygen.

anagenesis   The pattern of speciation in which a single species is transformed into a different species over the course of many generations.

analogous structure   A trait that is the result of convergent evolution; structures have arisen independently, two or more times, because species have occupied similar types of environments on the Earth.

anaphase   The phase of mitosis during which the sister chromatids separate from each other and move to opposite poles; the poles themselves also move farther apart.

anatomy   The study of the morphology of living organisms, such as plants and animals.

anchoring junctions   A junction found in between animal cells that attaches cells to each other and to the ECM.

androgens   Steroid hormones produced by the male testes that affect most aspects of male reproduction.

anemia   A condition characterized by lower than normal levels of hemoglobin, which reduces the amount of oxygen that can be stored in the blood.

aneuploidy   An alteration in the number of particular chromosomes so that the total number of chromosomes is not an exact multiple of a set.

angina pectoris   Chest pain during exertion due to the tissues being deprived of oxygen.

angiosperm   A flowering plant; the term means "enclosed seed," which reflects the presence of seeds within fruits.

animal cap assay   A type of experiment used extensively to identify factors (proteins) secreted by embryonic cells that induce cells in the animal pole to differentiate into mesoderm.

animal pole   In triploblast organisms, the pole of the egg where less yolk and more cytoplasm are concentrated.

Animalia   One of the four traditional eukaryotic kingdoms of the domain Eukarya.

anion   An ion that has a net negative charge.

annual   A plant that dies after producing seed during its first year of life.

anther   A cluster of microsporangia in a flower that produces pollen and then opens to release it.

antheridia   Round or elongate gametangia that produce plant sperm.

antagonist   A muscle or group of muscles that produces oppositely directed movements at a joint.

antenna complex   *See* light-harvesting complex.

anterior   Refers to the end of an animal where the head is found.

anteroposterior axis   One of the three axes along which the adult body pattern is organized; the others are the dorsoventral axis and the right-left axis.

anthropoid   A member of a class of primates that includes the monkeys and the hominoids; species are larger-brained and diurnal.

antibiotic   A chemical, usually made by microorganisms, that inhibits the growth of certain microorganisms.

antibody   A protein secreted by plasma cells that is part of the immune response; antibodies travel all over the body to reach antigens identical to those that stimulated their production, and then they combine with these antigens and guide an attack that eliminates the antigens or the cells bearing them.

anticodon A three-nucleotide sequence in tRNA that is complementary to a codon in mRNA.

antidiuretic hormone (ADH) A hormone secreted by the posterior pituitary gland that acts on kidney cells to decrease urine production.

antigen Any foreign molecule that the host does not recognize as self and that triggers a specific immune response.

antigen-presenting cells (APCs) Cells bearing fragments of antigen, called antigenic determinants or epitopes, complexed with the cell's MHC proteins.

antiparallel An arrangement in DNA where one strand runs in the 5′ to 3′ direction while the other strand is oriented in the 3′ to 5′ direction.

antiporter A type of transporter that binds two or more ions or molecules and transports them in opposite directions.

apical-basal-patterning genes A category of genes that are important in early stages of plant development during which the apical and basal axes are formed.

apical-basal polarity An architectural feature of plants in which they display an upper, apical pole and a lower, basal pole; shoot apical meristem occurs at the apical pole, and the root apical meristem occurs at the basal pole.

apical constriction A cellular process during gastrulation that is crucial to development; a reduction in the diameter of the actin rings connected to the adherens junctions causes the cells to elongate toward their basal end.

apical region The region of a plant that projects upwards, usually from the soil, and produces the leaves and flowers.

apomixis A natural asexual reproductive process in which plant fruits and seeds are produced in the absence of fertilization.

apoplast The continuum of water-soaked cell walls and intercellular spaces in a plant.

apoplastic transport The movement of solutes through cell walls and the spaces between cells.

apoptosis Programmed cell death.

aposematic coloration Warning coloration that advertises an organism's unpalatable taste.

appendix A finger-like projection in the gastrointestinal tract of animals having no known essential function but that may at one time have been an important part of the body's defense mechanisms.

aquaporin A three-dimensional cell pore that allows water to diffuse through the membrane.

aqueous humor A thin liquid in the anterior cavity behind the cornea of the vertebrate eye.

aqueous solution A solution made with water.

aquifer An underground water supply.

arbuscular mycorrhizae Symbiotic associations between glomalean fungi and the roots of vascular plants.

Archaea One of the three domains of life; the other two are Bacteria and Eukarya.

archaea When not capitalized refers to a cell or species within the domain Archaea.

archegonia Flask-shaped plant gametangia that enclose an egg cell.

archenteron A cavity formed by the embryo during gastrulation that will become the organism's digestive tract.

area hypothesis The proposal that larger areas contain more species than smaller areas because they can support larger populations and a greater range of habitats.

arteriole A single-celled layer of endothelium surrounded by one or two layers of smooth muscle and connective tissue that delivers blood to the capillaries and distributes blood to regions of the body in proportion to metabolic demands.

artery A blood vessel that carries blood away from the heart.

artificial selection *See* selective breeding.

asci Fungal sporangia shaped like sacs that produce and release sexual ascospores.

ascocarp The type of fruiting body produced by ascomycete fungi.

ascomycetes A phylum of fungi that produce sexual spores in saclike asci located at the surfaces of fruiting bodies known as ascocarps.

ascospore The type of sexual spore produced by the ascomycete fungi.

aseptate The condition of not being partitioned into smaller cells; usually refers to fungal cells.

asexual reproduction A reproductive strategy that occurs when offspring are produced from a single parent, without the fusion of gametes from two parents. The offspring are therefore clones of the parent.

assimilation In the case of nitrogen, the process by which plants and animals incorporate the ammonia and NO_3^- formed through nitrogen fixation and nitrification.

associative learning A change in behavior due to an association between a stimulus and a response.

assort The process of distributing.

asthma A disease in which the smooth muscles around the bronchioles contract more than usual, increasing resistance to airflow.

AT/GC rule Refers to the phenomenon that an A in one DNA strand always hydrogen bonds with a T in the opposite strand, while a G in one strand bonds with a C.

atherosclerosis The condition in which large plaques may occlude (block) the lumen of an artery.

atmospheric (barometric) pressure The pressure exerted by the gases in air on the body surfaces of animals.

atom The smallest functional unit of matter that forms all chemical substances and cannot be further broken down into other substances by ordinary chemical or physical means.

atomic mass An atom's mass relative to the mass of other atoms. By convention, the most common form of carbon, which has six protons and six neutrons, is assigned an atomic mass of exactly 12.

atomic nucleus The center of an atom.

atomic number The number of protons in an atom.

ATP synthase An enzyme that utilizes the energy stored in an H^+ electrochemical gradient for the synthesis of ATP via chemiosmosis.

ATP-dependent chromatin remodeling enzyme An enzyme that catalyzes a loosening in the compaction of chromatin; this loosening facilitates the ability of RNA polymerase to recognize and transcribe a gene.

ATP-driven pump A common category of pump found in all living cells; this transporter has a binding site for ATP and hydrolyzes ATP to actively transport solutes against a gradient.

atrial natriuretic peptide (ANP) A peptide secreted from the atria of the heart whenever blood levels of sodium increase; ANP causes a natriuresis by decreasing sodium reabsorption in the kidney tubules.

atrioventricular (AV) node Specialized cardiac cells that sit near the junction of the atria and ventricles and conduct the electrical events from the atria to the ventricles.

atrioventricular (AV) valve A one-way valve into the ventricles of the vertebrate heart through which blood moves from the atria.

atrium In the heart, a single filling chamber to collect blood from the tissues.

atrophy A reduction in the size of a structure, such as a muscle.

atropine A potent toxin derived from the deadly nightshade plant.

audition The sense of hearing.

autoimmune disease In humans and many other vertebrates, the situation in which the body's normal state of immune tolerance breaks down, with the result that antibody-mediated or T-cell mediated attacks are directed against the body's own cells and tissues.

autonomic nervous system The division of the peripheral nervous system that regulates homeostasis and organ function; also called the visceral nervous system.

autophagosome Cellular material enclosed in a double membrane, produced by the process of autophagy.

autophagy Meaning "the eating of one's self." A process whereby cellular material, such as a worn-out organelle, becomes enclosed in a double membrane and degraded.

autosomes All of the chromosomes found in the cell nucleus of eukaryotes except for the sex chromosomes.

autotomy In echinoderms, the ability to intentionally detach a body part, such as a limb, that will later regenerate.

autotroph An organism that has metabolic pathways that directly harness energy from either inorganic molecules or light.

auxin One of several types of hormones considered to be the "master" plant hormones because they influence plant structure, development, and behavior in many ways, often working with other hormones.

auxin efflux carrier One of several types of PIN proteins, which transport auxin out of plant cells.

auxin influx carrier A plasma membrane protein that transports auxin into plant cells.

avirulence gene (*Avr* gene) A gene that encodes a virulence-enhancing elicitor, which causes plant disease.

Avogadro's number As first described by Italian physicist Amedeo Avogadro, 1 mole of any element contains the same number of particles—6.022 x 10^{23}.

axillary bud A bud that occurs in the axil, the upper angle where a twig or leaf joins the stem.

axon An extension of the plasma membrane that is involved in sending signals from a neuron to neighboring cells.

axon hillock The part of the axon closest to the cell body.

axoneme The internal structure of eukaryotic cilia and flagella consisting of microtubules, the motor protein dynein, and linking proteins.

B

B cell A type of lymphocyte responsible for specific immunity.

bacilli Rods; one of the five major shapes of prokaryotic cells.

backbone The linear arrangement of phosphates and sugar molecules in a DNA or RNA strand.

Bacteria One of the three domains of life; the other two are Archaea and Eukarya.

bacteria When not capitalized, refers to a cell or species within the domain Bacteria.

bacterial artificial chromosome (BAC) A cloning vector in bacteria that can contain large DNA inserts.

bacterial colony A clone of genetically identical cells formed from a single cell.

bacteriophage A virus that infects bacteria.

bacteroid A modified bacterial cell of the type known as rhizobia present in mature root nodules of some plants.

balanced polymorphism The phenomenon in which two or more alleles are kept in balance, and therefore are maintained in a population over the course of many generations.

balancing selection A type of natural selection that maintains genetic diversity in a population.

balloon angioplasty A common treatment to restore blood flow through a blood vessel. A thin tube with a tiny, inflatable balloon at its tip is threaded through the artery to the diseased area; inflating the balloon compresses the plaque against the arterial wall, widening the lumen.

baroreceptor A pressure-sensitive region within the walls of certain arteries that contains the endings of nerve cells; these regions help to maintain blood pressure in the normal range for an animal.

Barr body A highly condensed X chromosome.

basal body A site at the base of flagella or cilia from which microtubules grow. Basal bodies are anchored on the cytosolic side of the plasma membrane.

basal metabolic rate (BMR) The metabolic rate of an animal under resting conditions, in good health, and not under stress of any kind.

basal nuclei Clusters of neuronal cell bodies that surround the thalamus on both sides and lie beneath the cerebral cortex; involved in planning and learning movements and also function via a complex circuitry to initiate or inhibit movements.

basal region The region of a plant that produces the roots.

basal transcription A low level of transcription resulting from the core promoter.

basal transcription apparatus For a eukaryotic structural gene, refers to the complex of GTFs, RNA polymerase II, and a DNA sequence containing a TATA box.

base 1. A component of nucleotides that is a single or double ring of carbon and nitrogen atoms. 2. A molecule that when dissolved in water lowers the H^+ concentration.

base pairs The structure in which two bases in opposite strands of DNA hydrogen bond with each other.

base substitution A mutation that involves the substitution of a single base in the DNA for another base.

basic local alignment search tool (BLAST) A computer program that can identify homologous genes that are found in a database.

basidiocarp The type of fruiting body produced by basidiomycete fungi.

basidiomycetes A group of fungi whose sexual spores are produced on the surfaces of club-shaped structures (basidia).

basidiospore A sexual spore of the basidiomycete fungi.

basophil A cell that secretes the anticlotting factor heparin at the site of an infection, which helps the circulation flush out the infected site; basophils also secrete histamine, which attracts infection-fighting cells and proteins.

Batesian mimicry The mimicry of an unpalatable species (the model) by a palatable one (the mimic).

behavior The observable response of organisms to external or internal stimuli.

behavioral ecology A subdiscipline of organismal ecology that focuses on how the behavior of an individual organism contributes to its survival and reproductive success, which in turn eventually affects the population density of the species.

benign tumor A precancerous condition.

bidirectional replication In DNA replication, the two DNA strands unwind, and DNA replication proceeds outward from the origin in opposite directions.

biennial A plant that does not reproduce during the first year of life but may reproduce within the following year.

bilateral Refers to living organisms, such as animals, or parts of living organisms, such as leaves, that can be divided along a vertical plane at the midline to create two halves.

bile A substance produced by the liver that contains bicarbonate ions, cholesterol, phospholipids, a number of organic wastes, and a group of substances collectively termed bile salts.

bile salts A group of substances produced in the liver that solubilize dietary fat and increase its accessibility to digestive enzymes.

binary fission The process of cell division in bacteria and archaea in which the cells divide into two cells.

binocular vision A type of vision in animals with both eyes located at the front of the head; the overlapping images coming into both eyes are processed together in the brain to form one perception.

binomial A two-part description used by biologists to provide each species with a unique scientific name. For example, the scientific name of the jaguar is *Panthera onca*. The first part is the genus and the second part is the specific epithet or species descriptor.

binomial nomenclature The standard method for naming species. Each species has a genus name and species epithet.

biochemistry The study of the chemistry of living organisms.

biodiversity Biological diversity, including genetic diversity, species diversity, and ecosystem diversity.

biodiversity crisis The idea that there is currently an elevated loss of species on Earth, far beyond the normal extinction rate of species.

biofilm An aggregation of microorganisms that secrete adhesive mucilage, thereby gluing themselves to surfaces.

biogeochemical cycle The continuous movement of nutrients such as phosphorus, carbon, and nitrogen from the physical environment to organisms and back.

biogeography The study of the geographic distribution of extinct and modern species.

bioinformatics A field of study that uses computers to study biological information.

biological evolution The phenomenon that populations of organisms change over the course of many generations. As a result, some organisms become more successful at survival and reproduction.

biological species concept A species is a group of individuals whose members have the potential to interbreed with one another in nature to produce viable, fertile offspring but cannot successfully interbreed with members of other species.

biology The study of life.

bioluminescence A phenomenon that results from chemical reactions that give off light rather than heat.

biomagnification The increase in the concentration of a substance in living organisms with each increase in trophic level in a food web.

biomass A quantitative estimate of the total mass of living matter in a given area, usually measured in grams or kilograms per square meter.

biome A major type of habitat characterized by distinctive plant and animal life.

bioremediation The use of living organisms, usually microbes or plants, to detoxify polluted habitats such as dump sites or oil spills.

biosphere The regions on the surface of the Earth and in the atmosphere where living organisms exist.

biosynthetic reaction Also called an anabolic reaction; a chemical reaction to make larger molecules and macromolecules.

biotechnology The use of living organisms or the products of living organisms for human benefit.

biotic The term used to describe interactions among organisms.

biparental inheritance A pattern in which both the male and female gametes contribute particular genes to the offspring.

bipedal Having the ability to walk on two feet.

bipolar cells Cells in the eye that make synapses with the photoreceptors and relay responses to the ganglion cells.

bivalent Homologous pairs of sister chromatids associated with each other, lying side by side.

blastocyst The mammalian counterpart of a blastula.

blastoderm An early stage of embryonic development in animals that is composed of a mass of cells with an internal cavity.

blastomeres The two half-size daughter cells produced by each cell division during cleavage.

blastopore A small opening created when a band of tissue invaginates during gastrulation. It forms the primary opening of the archenteron to the outside.

blastula An animal embryo at the stage when it forms an outer epithelial layer and an inner cavity.

blending inheritance An early hypothesis of inheritance that stated that the seeds that dictate hereditary traits blend together from generation to generation, and the blended traits are then passed to the next generation.

blood A fluid connective tissue consisting of cells and (in mammals) cell fragments suspended in a solution of water containing dissolved nutrients, proteins, gases, and other molecules. Blood has four components: plasma, leukocytes, erythrocytes, and thrombocytes or platelets.

blood-doping An example of hormone misuse in which the number of red blood cells in the circulation is boosted to increase the oxygen-carrying capacity of the blood.

blood pressure The force exerted by blood on the walls of blood vessels; blood pressure is responsible for moving blood through the vessels.

blood vessels A system of hollow tubes within the body through which blood travels.

body mass index (BMI) A method of assessing body fat and health risk that involves calculating the ratio of weight compared to height; weight in kilograms is divided by the square of the height in meters.

bottleneck effect A form of genetic drift in which a population size is dramatically reduced and then rebounds. Genetic drift is common when the population size is small.

Bowman's capsule A sac at the beginning of the tubular component of a nephron in the mammalian kidney.

brain The structure in the head of animals that controls sensory and motor functions of the entire body.

brassinosteroid One of several plant hormones that help a plant to cope with environmental stress.

bronchiole A thin-walled, small tube that can dilate or constrict to prevent foreign particles from reaching delicate lung tissue.

bronchodilator A compound that binds to the muscles around bronchioles and causes them to relax and widen.

bronchus A tube branching from the trachea and leading to a lung.

brown fat A specialized tissue in small mammals such as hibernating bats, small rodents living in cold environments, and many newborn mammals including humans that can help to generate heat and maintain body temperature.

brush border The combination of villi and microvilli in the small intestine, which increase the surface area about 600-fold over that of a flat-surfaced tube having the same length and diameter.

bryophytes Liverworts, mosses, and hornworts, the modern non-vascular land plants.

buccal pumping A form of breathing in which animals open their mouths to let in air and then close and raise the floor of the mouth, creating a positive pressure and pumping water or air across the gills or into the lungs; found in fish and amphibians.

bud A miniature plant shoot having a dormant shoot apical meristem.

budding A form of asexual reproduction in which a portion of the parent organism pinches off to form a complete new individual.

buffer A compound that acts to minimize pH fluctuations in the fluids of living organisms. Buffer systems can raise or lower pH as needed.

bulbourethral glands Paired accessory glands that secrete an alkaline mucus that protects sperm by neutralizing the acidity in the urethra.

bulk flow The mass movement of liquid in a plant caused by pressure, gravity, or both.

C

C_3 plant A plant that can only incorporate CO_2 into organic molecules via RuBP to make 3PG, a three-carbon molecule.

C_4 plant A plant that uses PEP carboxylase to initially fix CO_2 into a 4-carbon molecule and later uses rubisco to fix CO_2 into simple sugars.

cadherin A cell adhesion molecule found in animal cells that promotes cell-to-cell adhesion.

callose A carbohydrate that plays crucial roles in plant development, and plugging wounds in plant phloem.

calorie The amount of heat required to raise the temperature of 1 gram of water 1 degree Celsius.

Calvin cycle The cycle that includes carbon fixation, reduction and carbohydrate production, and regeneration of ribulose bisphosphate (RuBP). During this process, ATP is used as a source of energy and NADPH is used as a source of high-energy electrons so that CO_2 can be incorporated into carbohydrate.

calyx The sepals that form the outermost whorl of a flower.

CAM plants C_4 plants that take up carbon dioxide at night.

Cambrian explosion An event during the Cambrian period (543 to 490 mya) in which there was an abrupt increase (on a geological scale) in the diversity of animal species.

cancer A disease caused by gene mutations that lead to uncontrolled cell growth.

canopy The uppermost layer of tree foliage.

CAP site One of two regulatory sites near the *lac* promoter; this site is a DNA sequence recognized by the catabolite activator protein (CAP).

capillary A thin-walled vessel that is the site of gas and nutrient exchange between the blood and interstitial fluid.

capping A 7-methylguanosine covalently attached at the 5′ end of mature mRNAs of eukayrotes.

capsid A protein coat enclosing a virus's genome.

capsule A very thick, gelatinous glycocalyx produced by certain strains of bacteria that invade animals' bodies that may help them avoid being destroyed by the animal's immune (defense) system.

carapace The hard protective covering of a crustacean.

carbohydrate An organic molecule with the general formula, $C(H_2O)$; a carbon-containing compound that is hydrated (that is, contains water).

carbon fixation In this process, inorganic CO_2 is incorporated into an organic molecule such as a carbohydrate.

carboxyl terminus *See* C-terminus.

carcinogen An agent that increases the likelihood of developing cancer, usually a mutagen.

carcinoma A cancer of epithelial cells.

cardiac cycle The events that produce a single heartbeat, which can be divided into two phases, diastole and systole.

cardiac muscle A type of muscle tissue found only in hearts in which physical and electrical connections between individual cells enable many of the cells to contract simultaneously.

cardiac output (CO) The amount of blood the heart pumps per unit time, usually expressed in units of L/min.

cardiovascular disease Diseases affecting the heart and blood vessels.

cardiovascular system A system containing three components: blood or hemolymph, blood vessels, and one or more hearts.

carnivore An animal that consumes animal flesh or fluids.

carotenoid A type of pigment found in chloroplasts that imparts a color that ranges from yellow to orange to red.

carpel A flower shoot organ that produces ovules that contain female gametophytes.

carrier *See* transporter.

carrying capacity (K) The upper boundary for a population size.

Casparian strips Suberin ribbons on the walls of endodermal cells of plant roots; prevent apoplastic transport of ions into vascular tissues.

caspase An enzyme that is activated during apoptosis.

catabolic reaction The breakdown of a molecule into smaller components, usually releasing energy.

catabolism A metabolic pathway that results in the breakdown of larger molecules into smaller molecules. Such reactions are often exergonic.

catabolite activator protein (CAP) An activator protein also known as the cAMP receptor protein (CRP). CAP is needed for activation of the *lac* operon.

catabolite repression In bacteria, a process whereby transcriptional regulation is influenced by the presence of glucose.

catabolized Broken down.

catalase An enzyme within peroxisomes that breaks down hydrogen peroxide to water and oxygen gas.

catalyst An agent that speeds up the rate of a chemical reaction without being consumed during the reaction.

cataract An accumulation of protein in the lens of the eye; causes blurring, poor night vision.

cation An ion that has a net positive charge.

cation exchange With regard to soil, the process in which hydrogen ions are able to replace mineral cations on the surfaces of humus or clay particles.

cDNA library A collection of recombinant vectors that have cDNA inserts.

cecum The first portion of the large intestine in humans and other similarly sized animals and mammals.

celiac sprue A common genetic disorder in humans that results in a loss of intestinal surface area due to an allergic sensitivity to the wheat protein gluten.

cell The simplest unit of a living organism.

cell adhesion The phenomenon in which cells adhere to each other. Cell adhesion provides one way to convey positional information between neighboring cells.

cell adhesion molecule (CAM) A membrane protein found in animal cells that promotes cell adhesion.

cell biology The study of individual cells and their interactions with each other.

cell body A part of a neuron that contains the cell nucleus and other organelles.

cell coat Also called the glycocalyx, the carbohydrate-rich zone on the surface of certain animal cells that shields the cell from mechanical and physical damage.

cell communication The process through which cells can detect and respond to signals in their extracellular environment. In multicellular organisms, cell communication is also needed to coordinate cellular activities within the whole organism.

cell cycle The series of phases eukaryotic cells progress through to divide.

cell differentiation Refers to the phenomenon in which cells become specialized into particular cell types.

cell doctrine *See* cell theory.

cell fate The ultimate morphological features that a cell or group of cells will adopt.

cell junctions Specialized structures that adhere cells to each other and to the ECM.

cell-mediated immunity A type of specific immunity in which cytotoxic T cells directly attack and destroy infected body cells, cancer cells, or transplanted cells.

cell nucleus The membrane-bounded area of a eukaryotic cell in which the genetic material is found.

cell plate In plant cells, a structure that forms a cell wall between the two daughter cells.

cell signaling A vital function of the plasma membrane that involves cells sensing changes in their environment and interacting with each other.

cell surface receptor A receptor found in the plasma membrane that enables a cell to respond to different kinds of signaling molecules.

cell theory A theory that states that all organisms are made of cells. Cells come from pre-existing cells by cell division.

cell wall A relatively rigid, porous structure that supports and protects the plasma membrane and cytoplasm of prokaryotic, plant, fungal, and certain protist cells.

cellular differentiation The process by which different cells within a developing organism acquire specialized forms and functions, due to the expression of cell-specific genes.

cellular respiration A process by which living cells obtain energy from organic molecules.

cellular response Adaptation at the cellular level that often times involves a cell responding to signals in its environment.

cellulose The main macromolecule of the primary cell wall of plants and green algae; a polymer made of repeating molecules of glucose attached end to end.

centimorgan (cM) *See* map unit (mu).

central cell In the female gametophyte of a flowering plant, a large cell that contains two nuclei; after double fertilization it forms the first cell of the nutritive endosperm tissue.

central dogma Refers to the steps of gene expression at the molecular level. DNA is transcribed into mRNA and mRNA is translated into a polypeptide.

central region The region of a plant apical meristem that produces stem tissue.

central vacuole An organelle that often occupies 80% or more of the cell volume of plant cells and stores a large amount of water, enzymes, and inorganic ions.

central zone The area of a plant apical meristem where undifferentiated stem cells are maintained.

centrioles A pair of structures within the centrosome of animal cells. Most plant cells and many protists lack centrioles.

centromere The region where the two sister chromatids are tightly associated; the centromere binds to the kinetochore.

centrosome A single structure often near the cell nucleus of eukaryotic cells that forms a nucleating site for the growth of microtubules.

cephalization The localization of sensory structures at the anterior end of the body of animals.

cephalothorax The fused head and thorax structure in species of the class Arachnida and Crustacea.

cerebellum The part of the hindbrain, along with the pons, responsible for monitoring and coordinating body movements.

cerebral cortex The surface layer of gray matter that covers the cerebrum of the brain.

cerebral ganglia In flatworms, a paired structure that receives input from photoreceptors in eyespots and sensory cells.

cerebrospinal fluid Fluid that surrounds the exterior of the brain and spinal cord and absorbs physical shocks to the brain resulting from sudden movements or blows to the head.

cerebrum A group of structures in the forebrain that are responsible for the higher functions of conscious thought, planning, and emotion in vertebrates.

cervix A fibrous structure at the end of the female vagina that forms the opening to the uterus.

channel A transmembrane protein that forms an open passageway for the direct diffusion of ions or molecules across a membrane.

character A visible characteristic, such as the appearance of seeds, pods, flowers, and stems.

character displacement The tendency for two species to diverge in morphology and thus resource use because of competition.

charophyceans The lineage of freshwater green algae that is most closely related to the land plants.

charged tRNA A tRNA with its attached amino acid; also called aminoacyl tRNA.

checkpoint One of three critical regulatory points found in the cell cycle of eukaryotic cells. At these checkpoints, a variety of proteins act as sensors to determine if a cell is in the proper condition to divide.

checkpoint protein A protein that senses if a cell is in the proper condition to divide and prevents a cell from progressing through the cell cycle if it is not.

chemical element Each specific type of atom—nitrogen, hydrogen, oxygen, and so on.

chemical energy The potential energy contained within covalent bonds in molecules.

chemical equilibrium In a chemical reaction, occurs when the rate of formation of products equals the rate of formation of reactants.

chemical mutagen A chemical that causes mutations.

chemical reaction A reaction that occurs when one or more substances are changed into other substances. This can happen when two or more elements or compounds combine with each other to form a new compound, when one compound breaks down into two or more molecules, or when electrons are added to or taken away from an atom.

chemical synapse A synapse in which a chemical called a neurotransmitter is released from the nerve terminal and acts as a signal from the presynaptic to the postsynaptic cell.

chemiosmosis A process for making ATP in which energy stored in an ion electrochemical gradient is used to make ATP from ADP and P$_i$.

chemoautotroph An organism able to use energy obtained by chemical modifications of inorganic compounds to synthesize organic compounds.

chemoorganotroph An organism that must obtain organic molecules both for energy and as a carbon source.

chemoreceptor Specialized cells located in the vertebrate aorta, carotid arteries, and brainstem that detect the circulating levels of hydrogen ions and the partial pressures of carbon dioxide and oxygen, and relay that information through nerves or interneurons to the respiratory centers.

chiasma The connection at a crossover site of two chromosomes.

chimeric gene A gene formed from the fusion of two gene fragments to each other.

chitin A tough, nitrogen-containing polysaccharide that forms the external skeleton of many insects and the cell walls of fungi.

chlorophyll A green pigment found in photosynthetic plants, algae, and bacteria.

chlorophyll *a* Type of chlorophyll pigment found in cyanobacteria, and photosynthetic algae and plants.

chlorophyll *b* Type of chlorophyll pigment found in green algae and plants.

chloroplast genome The chromosome found in chloroplasts.

chlorosis The yellowing of plant leaves caused by various types of mineral deficiency.

choanocyte A specialized cell of sponges that functions to trap and eat small particles.

chondrichthyans Members of the class Chondrichthyes, including sharks, skates, and rays.

chordate An organism with a spinal cord.

chorion An extraembryonic membrane in the amniotic egg that, along with the allantois,

exchanges gases between the embryo and the surrounding air.

chorionic gonadotropin (CG) An LH-like hormone made by the placenta that maintains the corpus luteum.

chromatin Refers to the biochemical composition of chromosomes, which contain DNA and many types of proteins.

chromosome territory A distinct, nonoverlapping area where each chromosome is located within the cell nucleus of eukaryotic cells.

chromosome theory of inheritance An explanation of how the steps of meiosis account for the inheritance patterns observed by Mendel.

chromosome A unit of genetic material composed of DNA and associated proteins. Each cell has a characteristic number of chromosomes. Eukaryotes have chromosomes in their cell nuclei and in plastids and mitochondria.

chylomicrons Large fat droplets coated with amphipathic proteins that perform an emulsifying function similar to that of bile salts; they are formed in intestinal epithelial cells from absorbed fats in the diet.

chyme A solution of food particles in the stomach that contains water, salts, molecular fragments of proteins, nucleic acids, polysaccharides, droplets of fat, and various other small molecules.

chymotrypsin A protease involved in the breakdown of proteins in the small intestine.

chytrids Simple, early-diverging lineages of fungi; commonly found in aquatic habitats and moist soil, where they produce flagellate reproductive cells.

ciliate A protist that moves by means of cilia, which are tiny hairlike extensions on the outsides of cells.

cilium (plural, cilia) A cell appendage that functions like flagella to facilitate cell movement; cilia are shorter and more numerous on cells than are flagella.

circulatory system A system that transports necessary materials to all cells of an animal's body, and transports waste products away from cells. The three basic types of circulatory systems are gastrovascular cavities, open systems, and closed systems.

cis-acting element *See cis*-effect.

cis-effect A DNA segment that must be adjacent to the gene(s) that it regulates. The *lac* operator site is an example of a cis-acting element.

cisternae Flattened, fluid-filled tubules within the cell.

citric acid cycle A cycle that results in the breakdown of carbohydrates to carbon dioxide; also known as the Krebs cycle.

clade *See* monophyletic group.

cladistic approach An approach that reconstructs a phylogenetic tree by comparing primitive and shared derived characters.

cladogenesis A pattern of speciation in which a species is divided into two or more species.

cladogram A phylogenetic tree based on a cladistic approach.

clamp connection In basidiomaycete fungi, a structure that helps distribute nuclei during cell division.

clasper An extension of the pelvic fin of a chondrichthyan, used by the male to transfer sperm to the female.

class A subdivision of a phylum.

classical conditioning A type of associative learning in which an involuntary response comes to be associated positively or negatively with a stimulus that did not originally elicit the response.

cleavage A succession of rapid cell divisions with no significant growth that produces a hollow sphere of cells called a blastula.

cleavage furrow In animal cells, an area that constricts like a drawstring to separate the cells.

climate The prevailing weather pattern in a given region.

climax community A distinct end point of succession.

clitoris Located at the anterior part of the labia minora, erectile tissue that becomes engorged with blood during sexual arousal and is very sensitive to sexual stimulation.

clonal deletion A process that explains why individuals normally lack active lymphocytes that respond to self components; T cells with receptors capable of binding self proteins are destroyed by apoptosis.

clonal inactivation A process that explains why individuals normally lack active lymphocytes that respond to self components; the process occurs outside the thymus and causes potentially self-reacting T cells to become nonresponsive.

clonal selection The process by which certain cells are selected to proliferate as clones.

cloning Methods that produce many copies of something. These can be copies of genes, copies of genetically identical cells, or copies of genetically identical organisms. The cloning of mammals can be achieved by fusing a somatic cell with an egg that has had its nucleus removed. Plants can be cloned simply by removing cells, and growing them in particular mixtures of hormones.

closed circulatory system A circulatory system in which blood flows throughout an animal entirely within a series of vessels and is kept separate from the interstitial fluid.

closed conformation Tightly packed chromatin that cannot be transcribed into RNA.

clumped The most common pattern of dispersion within a population, in which individuals are gathered in small groups.

cnidocil On the surface of a cnidocyte, a hairlike trigger that detects stimuli.

cnidocyte A characteristic feature of cnidarians; a stinging cell that functions in defense or the capture of prey.

coacervates Droplets that form spontaneously from the association of charged polymers such as proteins, carbohydrates, or nucleic acids.

coat protein   A protein that surrounds a membrane vesicle and facilitates vesicle formation.

cocci   Spheres; one of the five major shapes of prokaryotic cells.

cochlea   A coiled structure containing the hair cells and other structures that generate the responses that travel via the auditory nerve to the brain.

coding strand   The DNA strand opposite to the template (or noncoding strand).

codominance   The phenomenon in which a single individual expresses both alleles.

codon   A sequence of three nucleotide bases that specifies a particular amino acid or a stop codon; codons function during translation.

coefficient of relatedness   The probability that any two individuals will share a copy of a particular gene is a quantity, r.

coelom   A fluid-filled body cavity in an animal.

coelomate   An animal with a true coelom.

coenzyme   An organic molecule that participates in the chemical reaction but is left unchanged after the reaction is completed.

coevolution   The process by which two or more species of organisms influence each other's evolutionary pathway.

cofactor   Usually an inorganic ion that temporarily binds to the surface of an enzyme and promotes a chemical reaction.

cognitive learning   The ability to solve problems with conscious thought and without direct environmental feedback.

cohesion-tension theory   The explanation for long-distance water transport as the combined effect of the cohesive forces of water and evaporative tension.

cohort   A group of organisms of the same age.

coleoptile   A protective sheath that encloses the first bud of an epicotyl in a mature monocot embryo.

coleorhiza   A protective envelope that encloses the monocot hypocotyl.

colinearity rule   The phenomenon whereby the order of homeotic genes along the chromosome correlates with their expression along the anteroposterior axis of the body.

collagen   A protein secreted from animal cells that forms large fibers in the extracellular matrix.

collecting duct   A tubule in the human vertebrate kidney that collects urine from nephrons.

collenchyma cells   Flexible cells that make up collenchyma tissue.

collenchyma tissue   A tissue that provides support to plant organs.

colligative property   Properties of water that depend on the amounts of dissolved substances. For example, the colligative properties of water cause certain solutes to function as antifreeze in certain organisms, and thereby lower the freezing point.

colloid   A gel-like substance in the follicles of the thyroid gland.

colon   A part of the large intestine consisting of three relatively straight segments—the ascending, transverse, and descending portions. The terminal portion of the descending colon is S-shaped, forming the sigmoid colon, which empties into the rectum.

colony hybridization   A method that uses a labeled probe that recognizes a specific gene to identify that gene in a DNA library.

combinatorial control   The phenomenon whereby a combination of many factors determines the expression of any given gene.

commensalism   An interaction that benefits one species and leaves the other unaffected.

communication   The use of specially designed visual, chemical, auditory or tactile signals to modify the behavior of others.

community   An assemblage of many populations that live in the same place at the same time.

community ecology   The study of how populations of species interact and form functional communities.

compartmentalization   A characteristic of eukaryotic cells that is defined by many organelles that separate the cell into different regions. Cellular compartmentalization allows a cell to carry out specialized chemical reactions in different places.

competent   The term used to describe bacterial strains that have the ability to take up DNA from the environment.

competition   An interaction that affects two or more species negatively, as they compete over food or other resources.

competitive exclusion hypothesis   The proposal that two species with the same resource requirements cannot occupy the same niche.

competitive inhibitor   A molecule that binds to the active site of an enzyme and inhibits the ability of the substrate to bind.

complement   The family of plasma proteins that provides a means for extracellular killing of microbes without prior phagocytosis.

complementary DNA (cDNA)   DNA molecules that are made from mRNA as a starting material.

complementary   In DNA, you can predict the sequence in one DNA strand if you know the sequence in the opposite strand according to the AT/GC rule.

complete flower   A flower that possesses all four types of flower organs.

complete metamorphosis   A dramatic change in body form in the majority of insects, from larva to a very different looking adult.

compound   A molecule composed of two or more different elements.

compound eyes   Image-forming eyes in arthropods and some annelids consisting of several hundred to several thousand light detectors called ommatidia.

computational molecular biology   An area of study that uses computers to characterize the molecular components of living things.

concentration   The amount of a solute dissolved in a unit volume of solution.

condensation reaction   *See* dehydration reaction.

conditioned response   The response that is created by a new, conditioned stimulus.

conditioned stimulus   A new stimulus that is delivered at the same time as the old stimulus, and that over time, is sufficient to elicit the same response.

condom   A sheathlike membrane worn over the penis that collects the ejaculate; in addition to their contraceptive function, condoms significantly reduce the risk of STDs such as HIV infection, syphilis, gonorrhea, chlamydia, and herpes.

conduction   The process in which the body surface loses or gains heat through direct contact with cooler or warmer substances.

cone pigments   The several types of visual pigments in cones.

cones   1. Photoreceptors found in the vertebrate eye; they are less sensitive to low levels of light but can detect color. Cones are used in daylight by most diurnal vertebrate species and by some insects. 2. The reproductive structures of conifer plants.

congestive heart failure   The condition resulting from the failure of the heart to pump blood normally; this results in fluid build-up in the lungs (congestion).

conidia   A type of asexual reproductive cell produced by many fungi.

Conifers   A phylum of gymnosperm plants, Coniferophyta.

conjugation   A type of genetic transfer between bacteria that involves a direct physical interaction between two bacterial cells.

connective tissue   Clusters of cells that connect, anchor, and support the structures of an animal's body; derived from mesenchyme and include blood, adipose (fat-storing) tissue, bone, cartilage, loose connective tissue, and dense connective tissue.

connexon   A channel that forms gap junctions consisting of six connexin proteins in one cell aligned with six connexin proteins in an adjacent cell.

conservation biology   The study that uses principles and knowledge from molecular biology, genetics, and ecology to protect the biological diversity of life at all levels.

conservative mechanism   In this incorrect model for DNA replication, both parental strands of DNA remain together following DNA replication. The original arrangement of parental strands is completely conserved, while the two newly made daughter strands are also together following replication.

constant regions   In immunology, the amino acid sequences of the Fc domains, which are identical for all immunoglobulins of a given class.

constitutive gene   An unregulated gene that has essentially constant levels of expression in all conditions over time.

contig   A series of clones that contain overlapping pieces of chromosomal DNA.

continental drift   The phenomenon whereby, over the course of billions of years, the major landmasses, known as the continents, have shifted their positions, changed their shapes, and in some cases have become separated from each other.

continuous trait   A trait that shows continuous variation over a range of phenotypes.

contraception   The use of birth control procedures to prevent fertilization or implantation of a fertilized egg.

contractile vacuole   A small, membrane enclosed, water-filled space that eliminates excess liquid from the cells of certain protists.

contrast   In microscopy, relative differences in the lightness, darkness, or color between adjacent regions in a sample. Contrast improves the ability to discern adjacent objects.

control sample   The sample in an experiment that is treated just like an experimental sample except that it is not subjected to one particular variable. For example, the control and experiment samples may be treated identically except that the temperature may vary for the experimental sample.

convection   The transfer of heat by the movement of air or water next to the body.

convergent evolution   The process whereby two different species from different lineages show similar characteristics because they occupy similar environments.

convergent extension   A cellular process during gastrulation that is crucial to development; two rows of cells merge to form a single elongated layer.

convergent trait   *See* analogous structure.

copulation   The process of sperm being deposited within the reproductive tract of the female.

coral reef   A type of aquatic biome found in warm, marine environments.

core promoter   For a eukaryotic structural gene, refers to the transcriptional start site and TATA box.

corepressor   A small effector molecule that binds to a repressor protein to inhibit transcription.

Coriolis effect   The effect of the Earth's rotation on the surface flow of wind.

cork cambium   A secondary meristem in a plant that produces cork tissue.

cornea   A thin, clear layer on the front of the vertebrate eye.

corolla   The petals of a flower, which occur in the whorl to the inside of the calyx and the outside of the stamens.

corona   The ciliated crown of members of the phylum Rotifera.

coronary artery   An artery that carries oxygen and nutrients to the heart muscle.

coronary artery bypass   A common treatment to restore blood flow through a blood vessel. A small piece of healthy blood vessel is removed from one part of the body and surgically grafted onto the coronary circulation in such a way that blood bypasses the diseased artery.

coronary artery disease   A condition that occurs when plaques form in the coronary vessels.

corpus callosum   The major tract that connects the two hemispheres of the cerebrum.

corpus luteum   A structure that is responsible for secreting hormones that stimulate the development of the uterus needed for sustaining the embryo in the event of a pregnancy. If pregnancy does not occur, the corpus luteum degenerates.

correlation   A meaningful relationship between two variables.

cortex   The area of a plant stem or root beneath the epidermis that is largely composed of parenchyma tissue.

cortical reaction   An event in fertilization in which calcium and IP_3 produce additional barriers to more than one sperm cell binding to and uniting with an egg, a process called the slow block to polyspermy.

cotranslational sorting   The sorting process in which the synthesis of certain eukaryotic proteins begins in the cytosol and then halts temporarily until the ribosome has become bound to the ER membrane. After this occurs, translation resumes and the polypeptide is synthesized into the ER lumen or ER membrane.

cotransporter   *See* symporter.

cotyledon   An embryonic seed leaf.

countercurrent exchange mechanism   An arrangement of water and blood flow in which water enters a fish's mouth and flows between the lamellae of the gills in the opposite direction to blood flowing through the lamellar capillaries.

countercurrent heat exchange   A method of regulating heat loss to the environment; many animals conserve heat by returning it to the body's core and keeping the core much warmer than the extremities.

covalent bond   A chemical bond that occurs when atoms share pairs of electrons.

CpG island   A cluster of CpG sites. CpG refers to the nucleotides of C and G in DNA that are connected by a phosphodiester linkage.

cranial nerve   A nerve in the peripheral nervous system that is directly connected to the brain; cranial nerves are located in the head and transmit incoming and outgoing information between the peripheral nervous system and the brain.

craniate   A chordate that has a brain encased in a skull and possesses a neural crest.

cranium   A protective bony or cartilaginous housing that encases the brain of a craniate.

crenation   The process of cell shrinkage that occurs if animal cells are placed in a hypertonic medium—water exits the cells via osmosis and equalizes solute concentrations on both sides of the membrane.

critical period   A limited time period of development in which many animals develop species-specific patterns of behavior.

crop   A storage organ that is a dilation of the lower esophagus; found in most birds and many invertebrates, including insects and some worms.

cross-bridge cycle   During muscle contraction, the sequence of events that occurs between the time when a cross-bridge binds to a thin filament and when it is set to repeat the process.

cross-bridge   A region of myosin molecules that extend from the surface of the thick filaments toward the thin filaments in a skeletal muscle.

cross-fertilization   The fusion of gametes formed by different individuals.

crossing over   The exchange of genetic material between homologous chromosomes, which allows for increased variation in the genetic information each parent may pass to the offspring.

cross-pollination   The process in which a stigma receives pollen from a different plant of the same species.

cryptic coloration   The blending of an organism with the background color of its habitat; also known as camouflage.

cryptochrome   A type of blue-light receptor in plants and protists.

CT scan   Computerized tomography, which is a technique for examining the structure and activity level of the brain without anesthesia or surgery. An X-ray beam and a series of detectors rotate around the head, producing slices of images that are reconstructed into three-dimensional images based on differences in the density of brain tissue.

C-terminus   The location of the last amino acid in a polypeptide.

cupula   A gelatinous structure within the lateral line organ that helps an organism to detect changes in water movement.

cuticle   A coating of wax and cutin that helps to reduce water loss from plant surfaces. Also, a nonliving covering that serves to both support and protect an animal.

cutin   A polyester polymer produced at the surfaces of plants; helps to prevent attack by pathogens.

cycads   A phylum of gymnosperm plants, Cycadophyta.

cyclic adenosine monophosphate (cAMP)   A second messenger molecule.

cyclic AMP (cAMP)   Cyclic adenosine monophosphate; a small effector molecule that is produced from ATP via an enzyme known as adenylyl cyclase.

cyclic electron flow   *See* cyclic photophosphorylation.

cyclic photophosphorylation   A pattern of electron flow in the thylakoid membrane that is cyclic and generates ATP alone.

cyclin   A protein responsible for advancing a cell through the phases of the cell cycle by binding to a cyclin-dependent kinase.

cyclin-dependent kinase (cdk)   A protein responsible for advancing a cell through the phases of the cell cycle. Its function is dependent on the binding of a cyclin.

cyst   A one-to-few celled structure that often has a thick, protective wall and can remain dormant through periods of unfavorable climate or low food availability.

cytogenetics The field of genetics that involves the microscopic examination of chromosomes.

cytokines A family of proteins that function in both nonspecific and specific immune defenses by providing a chemical communication network that synchronizes the components of the immune response.

cytokinesis The division of the cytoplasm to produce two distinct daughter cells.

cytokinin A type of plant hormone; promotes cell division in addition to other effects.

cytoplasm The region of the cell that is contained within the plasma membrane.

cytoplasmic inheritance *See* extranuclear inheritance.

cytoplasmic streaming A phenomenon in which the cytoplasm circulates throughout the cell to distribute resources efficiently in large cells, such as algal or plant cells.

cytosine (C) A pyrimidine base found in DNA and RNA.

cytoskeleton In eukaryotes, a network of three different types of protein filaments called microtubules, intermediate filaments, and actin filaments.

cytosol The region of a eukaryotic cell that is inside the plasma membrane and outside the organelles.

cytotoxic T cell A type of lymphocyte that travels to the location of its target, binds to the target by combining with an antigen on it, and directly kills the target via secreted chemicals.

D

dalton (Da) One-twelfth the mass of a carbon atom, or about the mass of a proton or a hydrogen atom.

Darwinian fitness The relative likelihood that a genotype will contribute to the gene pool of the next generation as compared with other genotypes.

database A large number of computer data files that are collected, stored in a single location, and organized for rapid search and retrieval.

daughter strand The newly made strand in DNA replication.

day-neutral plant A plant that flowers regardless of the night length, as long as day length meets the minimal requirements for plant growth.

deafness Hearing loss, usually caused by damage to the hair cells within the cochlea, although some cases result from functional problems in brain areas that process sound or in nerves that carry information to the brain from the hair cells.

decomposer *See* detritivore.

defecation The expulsion of feces that occurs through the final portion of the digestive canal, the anus; contractions of the rectum and relaxation of associated sphincter muscles expel the feces.

defensive mutualism A mutually beneficial interaction often involving an animal defending a plant or a herbivore in return for food or shelter.

deficiency The term used to describe when a segment of chromosomal material is missing.

deforestation The conversion of forested areas by humans to nonforested land.

degenerate In the genetic code, this means that more that one codon can specify the same amino acid.

dehydration A reduction in the amount of water in the body.

dehydration reaction A reaction that involves the removal of a water molecule, and the formation of a covalent bond between two separate molecules.

delayed implantation A reproductive cycle in which a fertilized egg reaches the uterus but does not implant until later, when environmental conditions are more favorable for the newly produced young.

delayed ovulation A reproductive cycle in which the ovarian cycle in females is halted before ovulation, and sperm are stored and nourished in the female's uterus over the winter. Upon arousal from hibernation in the spring, the female ovulates one or more eggs, which are fertilized by the stored sperm.

deletion The term used to describe a missing region of a chromosome.

demographic transition The shift in birth and death rates accompanying human societal development.

demography The study of birth rates, death rates, age distributions, and the sizes of populations.

dendrite A type of extension or projection that arises from the cell body; chemical and electrical messages from other neurons are received by the dendrites, and electrical signals move toward the cell body.

dendritic cell A type of cell derived from bone marrow stem cells that plays an important role in nonspecific immunity; these cells are scattered throughout most tissues, where they perform various macrophage functions.

denitrification The reduction of nitrate to gaseous nitrogen.

density In the context of populations, the numbers of organisms in a given unit area.

density-dependent factor A mortality factor whose influence varies with the density of the population.

density-independent factor A mortality factor whose influence is not affected by changes in population size or density.

deoxynucleoside triphosphates Individual nucleotides with three phosphate groups.

deoxyribonucleic acid (DNA) One of two classes of nucleic acids; the other is ribonucleic acid (RNA). A DNA molecule consists of two strands of nucleotides coiled around each other to form a double helix, held together by hydrogen bonds according to the AT/GC rule.

deoxyribose A five-carbon sugar found in DNA.

depolarization The change in the membrane potential that occurs when the cell becomes less polarized, that is, less negative relative to the surrounding fluid.

dermal tissue The covering on various parts of a plant.

descent with modification Darwin's theory that existing life-forms on our planet are the product of the modification of pre-existing life-forms.

desertification The overstocking of land with domestic animals that can greatly reduce grass coverage through overgrazing, turning the area more desert-like.

determinate cleavage A characteristic of protostome development in which the fate of each embryonic cell is determined very early.

determined The term used to describe a cell that is destined to differentiate into a particular cell type.

detritivore A consumer that gets its energy from the remains and waste products of organisms.

detritus Unconsumed plants that die and decompose, along with the dead remains of animals and animal waste products.

deuterostome An animal exhibiting radial cleavage, indeterminate cleavage and where the blastopore becomes the anus; includes echinoderms and vertebrates.

development In biology, a series of changes in the state of a cell, tissue, organ, or organism; the underlying process that gives rise to the structure and function of living organisms.

developmental genetics A field of study aimed at understanding how gene expression controls the process of development.

diaphragm A large muscle that subdivides the thoracic cavity from the abdomen in mammals.

diastole The first phase of the cardiac cycle, in which the ventricles fill with blood coming from the atria through the open AV valves.

diazotroph A bacterium that fixes nitrogen.

dideoxy chain-termination method The most common method of DNA sequencing that utilizes dideoxynucleotides as a reagent.

dideoxy sequencing *See* dideoxy chain-termination method.

differential gene regulation The phenomenon in which the expression of genes is altered. Differential gene expression allows cells to adapt to environmental conditions, change during development, and differentiate into particular cell types.

differentiated The term used to describe the actual alteration of a cell's morphology and physiology.

diffusion For dissolved substances, occurs when a solute moves from a region of high concentration to a region of lower concentration.

digestion The process of breaking down nutrients in food into smaller molecules that can be directly used by cells.

digestive system In a vertebrate, this system consists of the alimentary canal plus several associated structures, not all of which are found in all vertebrates: the tongue, teeth, salivary glands, liver, gallbladder, and pancreas.

dihybrid An offspring that is a hybrid with respect to two traits.

dikaryotic The occurrence of two genetically distinct nuclei in the cells of fungal hyphae after mating has occurred.

dinosaur A term meaning "terrible lizard" used to describe some of the extinct fossil reptiles.

dioecious The term to describe plants that produce staminate and carpellate flowers on separate plants.

diploblastic Having two distinct germ layers—ectoderm and endoderm but not mesoderm.

diploid cell A cell that carries two sets of chromosomes.

diploid-dominant species Species in which the diploid organism is the prevalent organism in the life cycle. Animals are an example.

direct calorimetry A method of determining basal metabolic rate that involves quantifying the amount of heat generated by the animal.

direct repair Refers to a DNA repair system in which an enzyme finds an incorrect structure in the DNA and directly converts it back to the correct structure.

directional selection A pattern of natural selection that favors individuals at one extreme of a phenotypic distribution that have greater reproductive success in a particular environment.

directionality In a DNA or RNA strand, refers to the orientation of the sugar molecules within that strand. Can be 5′ to 3′ or 3′ to 5′.

disaccharide A carbohydrate composed of two monosaccharides.

discontinuous trait A trait with clearly defined phenotypic variants.

discovery science *See* discovery-based science.

discovery-based science The collection and analysis of data without the need for a preconceived hypothesis. Also called discovery science.

dispersion A pattern of spacing in which individuals in a population are clustered together or spread out to varying degrees.

dispersive mechanism In this incorrect model for DNA replication, segments of parental DNA and newly made DNA are interspersed in both strands following the replication process.

dispersive mutualism A mutually beneficial interaction often involving plants and pollinators that disperse their pollen, and plants and fruit eaters that disperse the plant's seeds.

disruptive selection A pattern of natural selection that favors the survival of two or more different genotypes that produce different phenotypes.

dissociation constant An equilibrium constant between a ligand and a protein, such as a receptor or an enzyme.

distal convoluted tubule A structure in the tubule of the nephron through which fluid flows into one of the many collecting ducts in the kidney.

diversity-stability hypothesis The proposal that species-rich communities are more stable than those with fewer species.

DNA (deoxyribonucleic acid) The genetic material that provides a blueprint for the organization, development, and function of living things.

DNA fingerprinting A technology that identifies particular individuals using properties of their DNA.

DNA helicase An enzyme that uses ATP and separates DNA strands.

DNA library A collection of vectors each containing a particular fragment of chromosomal DNA or cDNA.

DNA ligase An enzyme that catalyzes the formation of a covalent bond between nucleotides in adjacent DNA fragments to complete the replication process in the lagging strand.

DNA methylase An enzyme that attaches methyl groups to bases in DNA.

DNA methylation A process in which methyl groups are attached to bases in DNA. This usually inhibits gene transcription by preventing the binding of activator proteins or by promoting the compaction of chromatin.

DNA microarray A technology used to monitor the expression of thousands of genes simultaneously.

DNA polymerase An enzyme responsible for covalently linking nucleotides together to form DNA strands.

DNA primase An enzyme that synthesizes a primer for DNA replication.

DNA repair systems One of several systems to reverse DNA damage before a permanent mutation can occur.

DNA replication The mechanism by which DNA can be copied.

DNA sequencing A method to determine the base sequence of DNA.

DNA topoisomerase An enzyme that affects the level of DNA supercoiling.

DNase An enzyme that digests DNA.

domain 1. A defined region of a protein with a distinct structure and function. 2. One of the three major categories of life: Bacteria, Archaea, and Eukarya.

dominant A term that describes the displayed trait in a heterozygote.

domestication A process that involves artificial selection of plants or animals for traits desirable to humans.

dominant species A species that has a large effect in a community because of its high abundance or high biomass.

dormancy A phase of metabolic slowdown in a plant.

dorsal Refers to the upper side of an animal.

dorsoventral axis One of the three axes along which the adult body pattern is organized; the others are the anteroposterior axis and the right-left axis.

dosage compensation The phenonemon that gene dosage is compensated between males and females. In mammals, the inactivation of one X chromosome in the female reduces the number of expressed copies (doses) of X-linked genes from two to one.

double bond A bond that occurs when the atoms of a molecule share two pairs of electrons.

double fertilization In angiosperms, the process in which two different fertilization events occur, producing both a zygote and a nutritive endosperm tissue.

double helix Two strands of DNA hydrogen-bonded with each other. In a DNA double helix, two DNA strands are twisted together to form a structure that resembles a spiral staircase.

Down syndrome A human disorder caused by the inheritance of three copies of chromosome 21.

duplication The term used to describe when a section of a chromosome occurs two or more times.

dynamic instability The oscillation of a single microtubule between growing and shortening phases; important in many cellular activities including the sorting of chromosomes during cell division.

E

ecdysis The periodic shedding and re-formation of the exoskeleton.

ecdysone A steroid hormone synthesized and secreted by the prothoracic glands of certain invertebrates such as arthropods; in response to this hormone, a larva molts and begins a new growth period until it must shed its skin again in response to another episode of ecdysone secretion.

Ecdysozoa A clade of moulting animals that encompasses primarily the arthropods and nematodes.

echolocation The phenomenon in which certain species generate high-frequency sound waves in order to determine the distance and location of an object.

ecological footprint The amount of productive land needed to support each person on Earth.

ecological niche The unique set of habitat resources that a species requires, as well as its influence on the environment and other species.

ecological species concept A species concept that considers a species within its native environment. Each species occupies its own ecological niche.

ecology The study of interactions among organisms and between organisms and their environments.

ecosystem The biotic community of organisms in an area as well as the abiotic environment affecting that community.

ecosystem engineer A species that creates, modifies, and maintains habitats.

ecosystems ecology The study of the flow of energy and cycling of nutrients among organisms within a community and between organisms and the environment.

ectoderm The outer most layer of cells formed during gastrulation that covers the surface of the embryo and differentiates into the epidermis and nervous system.

ectomycorrhizae   Beneficial interactions between temperate forest trees and soil fungi whose hyphae coat tree-root surfaces and grow into the spaces between root cells.

ectoparasite   A parasite that lives on the outside of the host's body.

ectotherm   An animal whose body temperature changes with the environmental temperature.

edge effect   A special physical condition that exists at the boundary or "edge" of an area.

effective population size   The number of individuals that contribute genes to future populations, often smaller than the actual population size.

effector   A molecule that directly influences cellular responses.

effector cell   A component of the immune response; these cells carry out the attack response.

egg   Also, egg cell. The female gamete.

ejaculation   The movement of semen through the urethra by contraction of muscles at the base of the penis.

ejaculatory duct   The structure within the male penis through which semen is released.

elastin   A protein that makes up elastic fibers in the extracellular matrix of animals.

electrical synapse   A synapse that directly passes electric current from the presynaptic to the postsynaptic cell via gap junctions.

electrocardiogram (ECG or **EKG)**   A record of the electrical impulses generated during the cardiac cycle.

electrochemical gradient   The combined effect of both an electrical and chemical gradient; determines the direction that an ion will move.

electrogenic pump   A pump that generates an electrical gradient across a membrane.

electromagnetic receptor   A sensory receptor that detects radiation within a wide range of the electromagnetic spectrum, including visible, ultraviolet, and infrared light, as well as electrical and magnetic fields in some animals.

electromagnetic spectrum   All possible wavelengths of electromagnetic radiation, from relatively short wavelengths (gamma rays) to much longer wavelengths (radio waves).

electron   A negatively charged particle found in orbitals around the nucleus. For atoms, the number of protons is equal to the number of electrons.

electron microscope   A microscope that uses an electron beam for illumination.

electron transport chain   A group of protein complexes and small organic molecules embedded in the inner mitochondrial membrane. The components accept and donate electrons to each other in a linear manner. The movement of electrons produces an H^+ electrochemical gradient.

electronegativity   A measure of an atom's ability to attract electrons to its outer shell from another atom.

elicitor   A compound produced by bacterial and fungal pathogens that promotes virulence.

elongation factor   In translation, a protein that is needed for the growth of a polypeptide.

elongation stage   The second step in transcription or translation where RNA strands or polypeptides are made, respectively.

embryo   The early stages of development in a multicellular organism during which the organization of the organism is largely formed.

embryogenesis   The process by which embryos develop from single-celled zygotes by mitotic divisions.

embryonic development   The process by which a fertilized egg is transformed into an organism with distinct physiological systems and body parts.

embryonic germ cell (EG cell)   At the early fetal stage of development, the cells that later give rise to sperm or eggs cells. These cells are pluripotent.

embryonic stem cell (ES cell)   A cell in the early mammalian embryo that can differentiate into almost every cell type of the body.

embryophyte   A synonym for the land plants.

emerging virus   A newly arising virus.

empirical thought   Thought that relies on observation to form an idea or hypothesis, rather than trying to understand life from a nonphysical or spiritual point of view.

emulsification   A process during digestion that disrupts the large lipid droplets into many tiny droplets, thereby increasing their total surface area and exposure to lipase action.

enantiomer   A type of stereoisomer that exists as a mirror image of another molecule.

endangered species   Those species that are in danger of extinction throughout all or a significant portion of their range.

endemic   The term to describe organisms that are naturally found only in a particular location.

endergonic   Refers to a reaction that has a positive free energy change and does not proceed spontaneously.

endocrine disruptor   A chemical found in water and soil exposed to pollution runoff that has a molecular structure that in some cases resembles estrogen sufficiently to bind to estrogen receptors.

endocrine gland   A structure that contains epithelial cells that secrete hormone molecules into the bloodstream, where they circulate throughout the body.

endocrine system   All the endocrine glands and other organs with hormone-secreting cells.

endocytic pathway   A pathway to take substances into the cell; the reverse of the secretory pathway.

endocytosis   A process in which the plasma membrane invaginates, or folds inward, to form a vesicle that brings substances into the cell.

endoderm   The inner most layer of cells formed during gastrulation that lines the gut and gives rise to many internal organs.

endomembrane system   A network of membranes that includes the nuclear envelope, which encloses the nucleus, and the

endoplasmic reticulum, Golgi apparatus, lysosomes, vacuoles, and plasma membrane.

endomycorrhizae   Partnerships between plants and fungi in which the fungal hyphae grow into the spaces between root cell walls and plasma membranes.

endoparasite   A parasite that lives inside the host's body.

endophyte   A mutualistic fungus that lives compatibly within the tissues of various types of plants.

endoplasmic reticulum (ER)   A convoluted network of membranes in the cell's cytoplasm that forms flattened, fluid-filled tubules or cisternae.

endoskeleton   An internal skeleton covered by soft tissue; composed of calcareous plates overlaid by a thin skin in echinoderms and composed of a bony skeleton overlaid by muscles in invertebrates.

endosperm   A nutritive tissue that increases the efficiency with which food is stored and used in the seeds of flowering plants.

endospore   A cell with a tough coat that is produced inside the cells of certain bacteria and then released when the enclosing cell dies and breaks down.

endosporic gametophyte   A plant gametophyte that grows within the confines of microspore and megaspore walls.

endosymbiosis   A symbiotic relationship in which the smaller species—the symbiont— lives inside the larger species.

endosymbiosis theory   A theory that mitochondria and chloroplasts originated from bacteria that took up residence within a primordial eukaryotic cell.

endosymbiotic   Describes a relationship in which one organism lives inside the other.

endothelium   A single-celled inner layer of a blood vessel, which forms a smooth lining in contact with the blood.

endotherm   An animal that generates its own internal heat.

endothermic   A term to describe that ability of an organism to generate and retain body heat through its metabolism.

energy   The ability to promote change.

energy expenditure   The amount of energy an animal uses in a given period of time to power all of its metabolic requirements.

energy flow   The movement of energy through an ecosystem.

energy intermediate   A molecule such as ATP and NADH that is directly used to drive endergonic reactions in cells.

enhancement effect   The phenomenon whereby maximal activation of the pigments in photosystems I and II is achieved when organisms are exposed to two wavelengths of light.

enhancer   A response element in eukaryotes that increases the rate of transcription.

enterocoelous   In deuterostomes, a pattern of development in which a layer of mesoderm cells forms outpockets that bud off from the developing gut to form the coelom.

enthalpy (H) The total energy of a system.

entomology The study of insects.

entropy The degree of disorder of a system.

environmental science The application of ecology to real-world problems.

enzyme A protein responsible for speeding up a chemical reaction in a cell.

enzyme-linked receptor A receptor found in all living species that typically has two important domains: an extracellular domain, which binds a signaling molecule, and an intracellular domain, which has a catalytic function.

enzyme-substrate complex The binding between an enzyme and substrate.

eosinophil A type of phagocyte found in large numbers in mucosal surfaces lining the gastrointestinal, respiratory, and urinary tracts, where they fight off parasitic invasions.

epicotyl The portion of an embryonic plant stem with two tiny leaves in a first bud that is located above the point of attachment of the cotyledons.

epidermis A layer of dermal tissue that helps protect a plant from damage.

epididymis A coiled, tubular structure located on the surface of the testis; it is here that the sperm complete their differentiation by becoming motile and gaining the capacity to fertilize ova.

epigenetic inheritance An inheritance pattern in which modification of a gene or chromosome during egg formation, sperm formation, or early stages of embryo growth alters gene expression in a way that is fixed during an individual's lifetime.

epinephrine A hormone secreted by the adrenal glands; also known as adrenaline.

episome A plasmid that can integrate into the bacterial chromosome.

epistasis A gene interaction in which the alleles of one gene mask the expression of the alleles of another gene.

epithelial placodes Regions of slightly thickened epithelial cells.

epithelial tissue A sheet of densely packed cells that covers the body or individual organs or lines the walls of various cavities inside the body.

epitopes Antigenic determinants; the peptide fragments of the antigen that are complexed to the MHC proteins and presented to the helper T cell.

equilibrium 1. In a chemical reaction, occurs when the rate of the forward reaction is balanced by the rate of the reverse reaction. 2. In a population, the situation in which the population size stays the same.

equilibrium model of island biogeography A model to explain the process of succession on new islands that states that the number of species on an island tends toward an equilibrium number that is determined by the balance between immigration rates and extinction rates.

equilibrium potential In membrane physiology, the membrane potential at which the flow of an ion is at equilibrium—no net movement in either direction.

ER lumen A single compartment enclosed by the ER membrane.

ER signal sequence A sorting signal in a polypeptide usually located near the amino terminus that is recognized by SRP (signal recognition particle) and directs the polypeptide to the ER membrane.

erythrocyte *See* red blood cell.

erythropoietin (EPO) A hormone made by the liver and kidneys in response to any situation where additional blood cells are required, such as when animals lose blood following an injury; when abused, as in blood-doping, the concentration of red blood cells reaches such high levels that the blood becomes much thicker than normal.

esophagus The tubular structure that forms a pathway from the throat to the stomach.

essential amino acids Those amino acids that are required in the diet of particular organisms. In humans, they include isoleucine, leucine, lysine, methionine, phenylalanine, threonine, tryptophan, and valine.

essential fatty acid A polyunsaturated fatty acid such as linoleic acid that cannot be synthesized by animal cells and must therefore be consumed in the diet.

essential nutrient A compound that cannot be synthesized from any ingested or stored precursor molecule and so must be obtained in the diet in its complete form.

estradiol The major estrogen in many animals, including humans.

estrogens Steroid hormones produced by the female ovaries that affect most aspects of reproduction.

ethology Scientific studies of animal behavior.

ethylene A plant hormone that is particularly important in coordinating plant developmental and stress responses.

euchromatin The less condensed regions of a chromosome; areas that are capable of gene transcription.

eudicots One of the two largest lineages of flowering plants; the embryo possesses two seed leaves.

Eukarya One of the three domains of life; the other two are Bacteria and Archaea.

eukaryote Cells or organisms from the domain Eukarya. The distinguishing feature is cell compartmentalization, which includes a cell nucleus. "Eukaryote" means "true nucleus," and includes protists, fungi, plants, and animals.

eukaryotic Refers to organisms having cells with internal compartments that serve various functions; includes all members of the domain Eukarya.

Eumetazoa A subgroup of animals having more than one type of tissue and, for the most part, different types of organs.

euphotic zone A fairly narrow zone close to the surface of an aquatic environment, where light is sufficient to allow photosynthesis to exceed respiration.

euphyll A leaf with branched veins.

euphyllophytes The clade that includes pteridophytes and seed plants.

euploid An organism that has a chromosome number that is a multiple of a chromosome set (1n, 2n, 3n, etc.).

eusociality The phenomenon whereby sterile castes evolve in social insects in which the vast majority of females, known as workers, rarely reproduce themselves but instead help one reproductive female (the queen) to raise offspring.

Eustachian tube A connection from the middle ear to the pharynx, which maintains the pressure in the middle ear at atmospheric pressure.

eustele A ring of vascular tissue arranged around a central pith of nonvascular tissue; typical of progymnosperms, gymnosperms, and angiosperms.

eutherian A placental mammal and member of the subclass Eutheria.

eutrophic Waters that contain relatively high levels of nutrients such as phosphate or nitrogen and typically exhibit high levels of primary productivity and low levels of biodiversity.

eutrophication The process by which elevated nutrient levels lead to an overgrowth of algae or aquatic plants and the subsequent depletion of water oxygen levels.

evaporation The transformation of water from the liquid to the gaseous state.

evapotranspiration rate The rate at which water moves into the atmosphere through the processes of evaporation from the soil and transpiration of plants.

evolutionarily conserved The term used to describe DNA sequences that are very similar or identical between different species.

evolutionary developmental biology (evo-devo) A field of biology that compares the development of different organisms in an attempt to understand ancestral relationships between organisms and the developmental mechanisms that bring about evolutionary change.

evolutionary species concept A species is derived from a single lineage that is distinct from other lineages and has its own evolutionary tendencies and historical fate.

excitable cell The term used to describe neurons and muscle cells, because they have the capacity to generate electrical signals.

excitation-contraction coupling The sequence of events by which an action potential in the plasma membrane of a muscle fiber leads to cross-bridge activity.

excitatory postsynaptic potential (EPSP) The response from an excitatory neurotransmitter depolarizing the postsynaptic membrane; the depolarization brings the membrane potential closer to the threshold potential that would trigger an action potential.

excurrent siphon A tunicate structure used to expel water.

exercise   Any physical activity that increases an animal's metabolic rate.

exergonic   Refers to reactions that release free energy and occur spontaneously.

exit site (E site)   A site for the tRNA binding in the ribosome; the other two are the peptidyl site (P site) and the aminoacyl site (A site). This is the site where the uncharged tRNA exits.

exocytosis   A process in which material inside the cell is packaged into vesicles and excreted into the extracellular medium.

exon shuffling   A form of mutation in which exons are inserted into genes and thereby create proteins with additional functional domains.

exon   A portion of RNA that is found in the mature RNA molecule after splicing is finished.

exoskeleton   An external skeleton that surrounds and protects most of the body surface of animals such as insects.

exotic species   Species moved from a native location to another location, usually by humans.

expansin   A protein that occurs in the plant cell wall and fosters cell enlargement.

experimental sample   The sample in an experiment that is subjected to some type of variation that does not occur for the control sample.

exponential growth   J-shaped, rapid population growth that occurs when the per capita growth rate remains above zero.

extensor   A muscle that straightens a limb.

external fertilization   Fertilization that occurs in aquatic environments, when eggs and sperm are released into the water in close enough proximity for fertilization to occur.

extinction   The end of the existence of a species or group of species.

extinction vortex   A downward spiral toward extinction from which a species cannot naturally recover.

extracellular fluid   The fluid in an organism's body that is outside of the cells.

extracellular matrix (ECM)   A network of material that is secreted from cells and forms a complex meshwork outside of cells. The ECM provides strength, support, and organization.

extranuclear inheritance   In eukaryotes, the transmission of genes that are located outside the cell nucleus.

extremophile   An organism that occurs primarily in extreme habitats.

eye   The visual organ that detects light and sends signals to the brain.

eyecup   In planaria, a primitive eye that detects light and its direction.

F

F factor   A segment of DNA called a fertility factor that plays a role in bacterial conjugation.

F_1 generation   The first filial generation in a genetic cross.

F_2 generation   The second filial generation in a genetic cross.

facilitated diffusion   A method of passive transport that involves the aid of a transport protein.

facilitation   A mechanism for succession in which a species facilitates or makes the environment more suitable for subsequent species.

facultative aerobe   A microorganism that can use oxygen in aerobic respiration, obtain energy via anaerobic fermentation, or use inorganic chemical reactions to obtain energy.

facultative mutualism   An interaction that is beneficial but not essential to the survival and reproduction of either species.

family   A subdivision of an order.

fast block to polyspermy   A depolarization of the egg that blocks other sperm from binding to the egg membrane proteins.

fast fiber   A muscle fiber containing myosin with high ATPase activity.

fast-glycolytic fiber   A skeletal muscle fiber that has high myosin activity but cannot make as much ATP as oxidative fibers, because its source of ATP is glycolysis; best suited for rapid, intense actions.

fast-oxidative fiber   A skeletal muscle fiber that has high myosin activity and can make large amounts of ATP; used for long-term activities.

fat   A molecule formed by bonding glycerol to three fatty acids.

fate mapping   A technique in which a small population of cells within an embryo is specifically labeled with a harmless dye, and the fate of these labeled cells is followed to a later stage of embryonic development.

feedback inhibition   A form of regulation in which the product of a metabolic pathway inhibits an enzyme that acts early in the pathway, thus preventing the over accumulation of the product.

feedforward regulation   The process by which an animal's body begins preparing for a change in some variable before it even occurs.

female-enforced monogamy hypothesis   The suggestion that males are monogamous because females stop their male partners from being polygynous.

female gametophyte   In plants, a haploid generation that produces one or more eggs, but does not produce sperm cells.

fermentation   The breakdown of organic molecules to produce energy without any net oxidation of an organic molecule.

ferrell cell   The middle cell in the three-cell circulation of wind in each hemisphere.

fertilization   The union of two gametes, such as an egg cell with a sperm cell, to create a zygote.

fertilizer   A soil addition that enhances plant growth by providing essential elements.

fetus   The maturing embryo, after the eighth week of gestation in humans.

fibrous root system   The root system of monocots, which consists of multiple adventitious roots that grow from the stem base.

fibrin   A protein that forms a meshwork of threadlike fibers that wrap around and between platelets and blood cells, enlarging and thickening a blood clot.

fidelity   Refers to the high level of accuracy in DNA replication.

filament   The elongate portion of a flower's stamen; contains vascular tissue that delivers nutrients from parental sporophytes to anthers.

filtrate   In the process of filtration, the material that passes through the filter and enters the excretory organ for either further processing or excretion.

filtration   The passive removal of water and small solutes from the blood.

first law of thermodynamics   The first law states that energy cannot be created or destroyed; it is also called the law of conservation of energy.

5′ cap   The 7-methylguanosine cap structure found on mRNA in eukaryotes.

fixed action pattern (FAP)   An animal behavior that, once initiated, will continue until completed.

flagellate   A protist that uses one or more flagella to move in water or cause water motions useful in feeding.

flagellum (plural, **flagella**)   A relatively long cell appendage that facilitates cellular movement or the movement of extracellular fluids.

flagship species   A single large or instantly recognizable species.

flame cell   A cell that exists primarily to maintain osmotic balance between an organism's body and surrounding fluids; present in flatworms.

flatus   Intestinal gas, which is a mixture of nitrogen and carbon dioxide, with small amounts of hydrogen, methane, and hydrogen sulfide.

flavonoid   A type of phenolic secondary metabolite that provides plants with protection from UV damage or colors organs such as flower petals.

flexor   A muscle that bends a limb at a joint.

florigen   The hypothesized flowering hormone, now identified as the mRNA that produces FT (flowering time) protein in the shoot apex.

flower   A reproductive shoot; a stem branch that produces reproductive organs instead of leaves.

flowering plants   The angiosperms, which produce ovules within the protective ovaries of flowers; when ovules develop into seeds, angiosperm ovaries develop into fruits, which function in seed dispersal.

flow-through system   A form of ventilation in fish in which water moves unidirectionally such that the gills are constantly in contact with fresh, oxygenated water.

fluid-feeder   An animal that licks or sucks fluid from plants or animals and does not need teeth except to puncture an animal's skin.

fluidity   A quality of biomembranes that means that individual molecules remain in close association yet have the ability to move laterally or rotationally within the plane of the membrane. Membranes are semi-fluid.

fluid-mosaic model   The accepted model of the plasma membrane; its basic framework is the semi-fluid phospholipid bilayer with a mosaic of proteins. Carbohydrates may be attached to lipids or proteins.

follicle   A structure within the ovary where each ovum undergoes growth and development before it is ovulated.

follicle-stimulating hormone (FSH)   A gonadotropin that stimulates follicle development.

food chain   A linear depiction of energy flow between organisms, with each organism feeding on and deriving energy from the preceding organism.

food vacuole   *See* phagocytic vacuole.

food web   A complex model of interconnected food chains in which there are multiple links between species.

food-induced thermogenesis   A rise in metabolic rate for a few hours after eating that produces heat.

foot   In mollusks, a muscular structure usually used for movement.

forebrain   One of three major divisions of the vertebrate brain; the other two divisions are the midbrain and hindbrain.

fossil   Recognizable remains of past life on Earth.

fossil fuel   A fuel formed in the Earth from protist, plant or animal remains, such as coal, petroleum, and natural gas.

founder effect   A small group of individuals separates from a larger population and establishes a colony in a new location; genetic drift is common due to the small population size.

fovea   A small area on the retina directly behind the lens that is responsible for the sharpness with which we and other animals see in daylight.

frameshift mutation   A mutation that involves the addition or deletion of nucleotides that are not in multiples of three nucleotides.

free energy (G)   In living organisms, the usable energy, that is, the amount of available energy that can be used to do work.

free radical   A molecule containing an atom with a single, unpaired electron in its outer shell. A free radical is unstable and interacts with other molecules by "stealing" electrons from their atoms.

frequency   In regard to sound, the number of complete wavelengths that occur in 1 second, measured in number of waves per second, or Hertz (Hz).

frugivore   An herbivore that is adapted primarily to feed on fruits.

fruit   A structure that develops from flower organs, encloses seeds, and fosters seed dispersal in the environment.

fruiting bodies   The visible fungal reproductive structures; composed of densely packed hyphae that typically grow out of the substrate.

functional genomics   Genomic methods aimed at studying the expression of a genome.

functional group   A group of atoms with chemical features that are functionally important. Each functional group exhibits the same properties in all molecules in which it occurs.

fundamental niche   The optimal range in which a particular species best functions.

Fungi   One of the four traditional eukaryotic kingdoms of the domain Eukarya.

fungus-like protist   A heterotrophic organism that often resembles true fungi in having threadlike, filamentous bodies and absorbing nutrients from its environment.

G

G_0   A stage in which cells exit the cell cycle and postpone making the decision to divide.

G_1   The first gap phase of the cell cycle.

G_2   The second gap phase of the cell cycle.

G banding   A staining procedure for chromosomes that produces an alternating pattern of G bands that is unique for each type of chromosome.

G protein   An intracellular protein that binds guanosine triphosphate (GTP) and guanosine diphosphate (GDP) and participates in intracellular signaling pathways.

gallbladder   A small sac underneath the liver that is a storage site for bile that allows the release of large amounts of bile to be precisely timed to the consumption of fats.

gametangia   Specialized structures produced by many land plants in which developing gametes are protected by a jacket of tissue.

gamete   A cell that is involved with sexual reproduction, such as a sperm or egg cell.

gametic life cycle   In this type of life cycle, all cells except the gametes are diploid, and gametes are produced by meiosis.

gametogenesis   The formation of gametes.

gametophyte   In plants and many multicellular protists, the haploid stage that produces gametes by mitosis.

ganglia   Groups of neuronal cell bodies that perform basic functions of integrating inputs from sense organs and controlling motor outputs, usually in the peripheral nervous system.

ganglion cells   Cells that send their axons out of the eye into the optic nerve.

gap gene   A type of segmentation gene; when a mutation inactivates a gap gene, several adjacent segments are missing in the larva.

gap junction   A type of junction between animal cells that provides a passageway for intercellular transport.

gas exchange   The process of moving oxygen and carbon dioxide in opposite directions between cells and blood, and between blood and the environment.

gas vesicle   A cytoplasmic structure used by cyanobacteria and some other bacteria that live in aquatic habitats to adjust their buoyancy.

gastrovascular cavity   A body cavity with a single opening to the outside; it functions as both a digestive system and circulatory system.

gastrula   An embryo that is the result of gastrulation, which has three cellular layers called the ectoderm, endoderm, and mesoderm.

gastrulation   A process in which an area in the blastula invaginates and folds inward, creating different embryonic cell layers called germ layers.

gated channel   A channel that can open to allow the diffusion of solutes and close to prohibit diffusion.

gel electrophoresis   A technique used to separate macromolecules by using an electric field that causes them to pass through a gel matrix.

gene   A unit of heredity that contributes to the characteristics or traits of an organism. At the molecular level, a gene is composed of organized sequences of DNA.

gene amplification   An increase in the copy number of a gene.

gene cloning   The process of making multiple copies of a gene.

gene expression   Gene function both at the level of traits and at the molecular level.

gene family   A group of homologous genes within a single species.

gene flow   Occurs when individuals migrate between different populations and cause changes in the genetic composition of the resulting populations.

gene interaction   A situation in which a single trait is controlled by two or more genes.

gene knockout   An organism in which both copies of a functional gene have been replaced with nonfunctional copies. Experimentally, this can occur via gene replacement.

gene mutation   A relatively small change in DNA structure that alters a particular gene.

gene pool   All of the genes in a population.

gene regulation   The ability of cells to control their level of gene expression.

gene replacement   The phenomenon in which a cloned gene recombines with the normal gene on a chromosome and replaces it.

gene silencing   The ability of one gene to silence the effect of another via small RNA molecules called microRNAs.

gene therapy   The introduction of cloned genes into living cells in an attempt to cure disease.

general transcription factors (GTFs)   Five different proteins that play a role in initiating transcription at the core promoter of structural genes in eukaryotes.

generative cell   In seed plants, the male gametophyte cell that divides to produce sperm cells.

genetic code   A code that specifies the relationship between the sequence of nucleotides in the codons found in mRNA and the sequence of amino acids in a polypeptide.

genetic drift   The random change in a population's allele frequencies from one generation to the next that is attributable to chance. It occurs more quickly in small populations.

genetic engineering The direct manipulation of genes for practical purposes.

genetic linkage map A diagram that describes the linear arrangement of genes that are linked to each other along the same chromosome.

genetic mosaic An individual with somatic regions that are genetically different from each other.

genetic transfer The process by which genetic material is transferred from one bacterial cell to another.

genetically modified organisms (GMOs) *See* transgenic.

genome The complete genetic composition of a cell or a species.

genome database A specialized database that focuses on the genetic characteristics of a single species.

genomic imprinting A phenomenon in which a segment of DNA is imprinted, or marked, in a way that affects gene expression throughout the life of the individual who inherits that DNA.

genomic library A type of DNA library in which the inserts are derived from chromosomal DNA.

genomics Techniques that are used in the molecular analysis of the entire genome of a species.

genotype The genetic composition of an individual.

genotype frequency The number of individuals with a given genotype divided by the total number of individuals.

genus In taxonomy, a subdivision of a family.

geological timescale A time line of the Earth's history from its origin about 4.55 billion years ago to the present.

germ layer An embryonic cell layer such as ectoderm, mesoderm, or endoderm.

germ line Cells that give rise to gametes such as egg and sperm cells.

germ plasm Cytoplasmic determinants that help define and specify the primordial germ cells in the gastrula stage.

gestation *See* pregnancy.

giant axon A very large axon in certain species such as squid that facilitates high-speed nerve conduction and rapid responses to stimuli.

gibberellic acid A type of gibberellin.

gibberellin A plant hormone that stimulates both cell division and cell elongation.

gills Specialized filamentous organs in aquatic animals that aid in obtaining oxygen and eliminating carbon dioxide.

ginkgos A phylum of gymnosperms; Ginkgophyta.

gizzard In the stomach of a bird, the muscular structure with a rough inner lining capable of grinding food into smaller fragments.

glaucoma A condition in which drainage of aqueous humor in the eye becomes blocked and the pressure inside the eye increases as the fluid level rises. If untreated, damages cells in the retina and leads to irreversible loss of vision.

glia Cells that surround the neurons; a major class of cells in nervous systems that perform various functions.

global warming A gradual elevation of the Earth's surface temperature caused by the greenhouse effect.

glomerulus A cluster of interconnected, fenestrated capillaries in the renal corpuscle of the kidney; the site of filtration in the kidneys.

glucocorticoid A steroid hormone that regulates glucose balance and helps prepare the body for stress situations.

gluconeogenesis A mechanism for maintaining blood glucose levels; enzymes in the liver convert noncarbohydrate precursors into glucose, which are then secreted into the blood.

glucose sparing A metabolic adjustment that reserves the glucose produced by the liver for use by the nervous system.

glycocalyx 1. An outer viscous covering surrounding a bacterium. The glycocalyx, which is secreted by the bacterium, traps water and helps protect bacteria from drying out. 2. A carbohydrate covering that is found outside of animal cells.

glycogen A polysaccharide found in animal cells and sometimes called animal starch.

glycogenolysis A mechanism for maintaining blood glucose levels; stored glycogen can be broken back down into molecules of glucose by hydrolysis.

glycolipid A lipid that has carbohydrate attached to it.

glycolysis A metabolic pathway that breaks down glucose to pyruvate.

glycolytic fiber A skeletal muscle fiber that has few mitochondria but possesses both a high concentration of glycolytic enzymes and large stores of glycogen.

glycoprotein A protein that has carbohydrate attached to it.

glycosaminoglycan (GAG) The most abundant type of polysaccharide in the extracellular matrix (ECM) of animals, consisting of repeating disaccharide units that give a gel-like character to the ECM of animals.

glycosylation The attachment of carbohydrate to a protein or lipid, producing a glycoprotein or glycolipid.

glyoxysome A specialized organelle within plant seeds that contains enzymes needed to convert fats to sugars.

gnathostomes All vertebrate species that possess jaws.

gnetophytes A phylum of gymnosperms; Gnetophyta.

Golgi apparatus A stack of flattened, membrane-bounded compartments that performs three overlapping functions: secretion, processing, and protein sorting.

gonadotropins Hormones secreted by the anterior pituitary gland that are the same in both sexes; gonadotropins influence the ability of the testes and ovaries to produce the sex steroids.

gonads The testes in males and the ovaries in females, where the gametes are formed.

G-protein-coupled receptors (GPCRs) A common type of receptor found in the cells of eukaryotic species that interacts with G proteins to initiate a cellular response.

gradualism A concept that suggests that species evolve continuously over long spans of time.

grain The characteristic single-seeded fruit of cereal grasses such as rice, corn, barley, and wheat.

granum A structure composed of stacked tubules within the thylakoid membrane of chloroplasts.

gravitropism Plant growth in response to the force of gravity.

gray matter Brain tissue that consists of neuronal cell bodies, dendrites, and some unmyelinated axons.

grazer An herbivore that feeds almost constantly on grasses.

greenhouse effect The process in which short-wave solar radiation passes through the atmosphere to warm the Earth but is radiated back to space as long-wave infrared radiation. Much of this radiation is absorbed by atmospheric gases and reradiated back to Earth's surface, causing its temperature to rise.

groove In the DNA double helix, an indentation where the atoms of the bases make contact with the surrounding water.

gross primary production (GPP) The measure of biomass production by photosynthetic organisms; equivalent to the carbon fixed during photosynthesis.

ground meristem A type of primary plant tissue meristem that gives rise to ground tissue.

ground tissue Most of the body of a plant, which has a variety of functions, including photosynthesis, storage of carbohydrates, and support. Ground tissue can be subdivided into three types: parenchyma, collenchyma, and sclerenchyma.

group selection The premise that natural selection produces outcomes beneficial for the whole group or species rather than for individuals.

growth An increase in weight or size.

growth factors A group of proteins in animals that stimulate certain cells to grow and divide.

growth hormone (GH) A hormone produced in vertebrates by the anterior pituitary gland; GH acts on the liver to produce insulin-like growth factor-1 (IGF-1).

guanine (G) A purine base found in DNA and RNA.

guard cell A specialized plant cell that allows epidermal pores (stomata) to close when conditions are too dry, and open under moist conditions, allowing the entry of CO_2 needed for photosynthesis.

gustation The sense of taste.

gut The gastrointestinal (GI) tract.

gymnosperm A plant that produces seeds that are exposed rather than enclosed in fruits.

gynoecium The aggregate of carpels that forms the innermost whorl of flower organs.

H

H zone   A narrow, light region in the center of the A band of the sarcomere that corresponds to the space between the two sets of thin filaments in each sarcomere.

H⁺ electrochemical gradient   A transmembrane gradient for H^+ composed of both a membrane potential and a concentration difference for H^+ across the membrane.

habituation   The form of nonassociative learning in which an organism learns to ignore a repeated stimulus.

Hadley cell   The most prominent of the three cells in the three-cell circulation of wind in each hemisphere.

hair cell   A mechanoreceptor that is a specialized epithelial cell with deformable stereocilia.

half-life   1. In the case of organic molecules in a cell, refers to the time it takes for 50% of the molecules to be broken down. 2. In the case of radioisotopes, the time it takes for half the molecules to decay and emit radiation.

halophyte   A plant that can tolerate higher than normal salt concentrations in the cell sap, and thus can occupy coastal salt marshes or saline deserts.

Hamilton's rule   The proposal that an altruistic gene will be favored by natural selection when r > C/B where r is the coefficient of relatedness of the donor (the altruist) to the recipient, B is the benefit received by the recipients of the altruism, and C is the cost incurred by the donor.

haplodiploidy   A genetic system in which females develop from fertilized eggs and are diploid but males develop from unfertilized eggs and are haploid.

haploid   Containing one set of chromosomes; or 1n.

haploid-dominant species   Species in which the haploid organism is the prevalent organisms in the life cycle. Examples include fungi and some protists.

Hardy-Weinberg equation   An equation $(p^2 + 2pq + q^2 = 1)$ that relates allele and genotype frequencies; the equation predicts an equilibrium if the population size is very large, mating is random, the populations do not migrate, no natural selection occurs, and no new mutations are formed.

heart attack   *See* myocardial infarction (MI).

heart   A muscular structure that pumps blood through blood vessels.

heat of fusion   The amount of heat energy that must be withdrawn or released from a substance to cause it to change from the liquid to the solid state.

heat of vaporization   The heat required to vaporize 1 mole of any substance at its boiling point under standard pressure.

heat shock protein   A protein that helps to protect other proteins from heat damage and refold them to their functional state.

heavy chain   A part of an immunoglobulin molecule.

helper T cell   A type of lymphocyte that assists in the activation and function of B cells and cytotoxic T cells.

hematocrit   The volume of blood that is composed of red blood cells, usually between 40 and 65% among vertebrates.

hemiparasite   A parasitic organism that generally photosynthesizes, but lacks a root system to draw water and thus depends on its hosts for that function.

hemispheres   The two halves of the cerebrum.

hemizygous   The term used to describe the single copy of an X-linked gene in a male.

hemocyanin   A copper-containing pigment that gives the blood or hemolymph a bluish tint.

hemodialysis   A medical procedure used to artificially perform the kidneys' excretory filtration and cleansing functions.

hemoglobin   An oxygen-binding protein found within the cytosol of red blood cells.

hemolymph   Blood and interstitial fluid combined in one fluid compartment in many invertebrates.

hemorrhage   A loss of blood from a ruptured blood vessel.

herbaceous plant   A plant that produces little or no wood and is composed mostly of primary vascular tissues.

herbivore   An animal that eats only plants.

herbivory   Refers to herbivores feeding on plants.

hermaphrodite   An individual that can produce both sperm and eggs.

hermaphroditism   A form of sexual reproduction in which individuals have both male and female reproductive systems.

heterochromatin   The highly compacted regions of chromosomes; in general, these regions are transcriptionally inactive because of their tight conformation.

heterocyst   A specialized cell of some cyanobacteria in which nitrogen fixation occurs.

heterodimer   The structure that results when two different proteins come together.

heterokaryon   In fungi, a mycelium having nuclei of two or more genetic types.

heterospory   In plants, the formation of two different types of spores, microspores and megaspores; microspores produce male gametophytes and megaspores produce female gametophytes.

heterotherm   An animal that has a body temperature that varies with the environment.

heterotroph   Organisms that cannot produce their own organic food and thus must obtain organic food from other organisms.

heterotrophic   Requiring organic food from the environment.

heterozygote advantage   A phenomenon in which a heterozygote has a higher Darwinian fitness compared to the corresponding homozygotes.

heterozygous   An individual with two different alleles of the same gene.

hibernation   The state of torpor in an animal over months.

high affinity   Refers to the binding of an ion or molecule to a protein very tightly. The substance will bind at a very low concentration.

highly repetitive sequence   A DNA sequence found tens of thousands or even millions of times throughout a genome.

hindbrain   One of three major divisions of the vertebrate brain; the other two divisions are the midbrain and forebrain.

hippocampus   The area of the brain whose main function appears to be establishing memories for spatial locations, facts, and sequences of events; composed of several layers of cells that are connected together in a circuit.

histone acetyltransferase   An enzyme that attaches acetyl groups to histone proteins.

histone code hypothesis   Refers to the pattern of histone modification recognized by particular proteins. The pattern of covalent modifications of amino terminus tails provides binding sites for proteins that subsequently affect the degree of chromatin compaction.

histones   A group of proteins involved in the formation of nucleosomes.

holoblastic cleavage   A complete type of cell cleavage in certain animals in which the entire zygote is bisected into two equal-sized blastomeres.

holoparasite   A parasitic organism that lacks chlorophyll and is totally dependent on the host plant for its water and nutrients.

homeobox   A 180-bp sequence within the coding sequence of homeotic genes.

homeodomain   A region of a protein that functions in binding to the DNA.

homeostasis   The process whereby living organisms regulate their cells and bodies to maintain relatively stable internal conditions.

homeotherm   An animal that maintains its body temperature within a narrow range.

homeotic   A term that describes changes in which one body part is replaced by another.

homeotic gene   A gene that controls the developmental fate of particular segments or regions of an animal's body.

hominoid   A gibbon, gorilla, orangutan, chimpanzee, or human.

hominin   Either an extinct or modern form of humans.

homodimer   The structure that results when two identical proteins come together.

homologous genes   Genes derived from the same ancestral gene that have accumulated random mutations that make their sequences slightly different.

homologous structures   Structures that are similar to each other because they are derived from the same ancestral structure.

homologue   A member of a pair of chromosomes in a diploid organism that are evolutionarily related.

homology   A fundamental similarity that occurs due to descent from a common ancestor.

homozygous   An individual with two identical copies of an allele.

horizontal gene transfer   The transfer of genes between different species.

hormone   A chemical messenger that is produced in a gland or other structure and acts on distant target cells in one or more parts of the body.

hornworts   A phylum of bryophytes; Anthocerophyta.

host   The prey organism in a parasitic association.

host cell   A cell that is infected by a virus, fungus, or a bacterium.

host plant resistance   The ability of plants to prevent herbivory.

host range   The number of species and cell types that a virus or bacterium can infect.

hot spot   A human-impacted geographic area with a large number of endemic species. To qualify as a hot spot, a region must contain at least 1,500 species of vascular plants as endemics and have lost at least 70% of its original habitat.

***Hox* complex**   A group of adjacent homeotic genes in vertebrates that controls the formation of structures along the anteroposterior axis.

***Hox* genes**   A class of genes involved in pattern formation in early embryos.

Human Genome Project   The largest genome project in history, which was a 13-year effort coordinated by the U.S. Department of Energy and the National Institutes of Health. The goals of the project were to identify all human genes, to sequence the entire human genome, to address the legal and ethical implications resulting from the project, and to develop programs to manage the information gathered from the project.

human immunodeficiency virus (HIV)   A retrovirus that is the causative agent of acquired immune deficiency syndrome (AIDS).

humoral immunity   A type of specific immunity in which plasma cells secrete antibodies that bind to antigens.

humus   A collective term for the organic constituents of soils.

hybrid zone   An area where two populations can interbreed.

hybridization   A situation in which two individuals with different characteristics are mated or crossed to each other; the offspring are referred to as hybrids.

hydrocarbon   Molecules with predominantly hydrogen-carbon bonds.

hydrogen bond   Electrostatic attraction between a hydrogen atom of a polar molecule and an electronegative atom of another polar molecule.

hydrolysis   The process in which reactions utilize water to break apart other molecules.

hydrophilic   "Water-loving"—generally, ions and molecules that contain polar covalent bonds will dissolve in water and are said to be hydrophilic.

hydrophobic   "Water-fearing"—molecules that are not attracted to water molecules. Such molecules are composed predominantly of carbon and hydrogen and are relatively insoluble in water. Because carbon-carbon and carbon-hydrogen bonds are nonpolar, the atoms in such compounds are electrically neutral.

hydrostatic skeleton   A fluid-filled body cavity surrounded by muscles that gives support and shape to the body of organisms.

hydroxide ion   An anion with the formula, OH^-.

hypermutation   A process that primarily involves numerous C to T point mutations that is crucial to enabling lymphocytes to produce a diverse array of immunoglobulins capable of recognizing many different antigens.

hyperpolarization   The change in the membrane potential that occurs when the cell becomes more polarized.

hypersensitive response (HR)   A plant's local defensive response to pathogen attack.

hyperthyroidism   A hyperactive thyroid gland.

hypertonic   When the solute concentration inside the cell is higher relative to the outside of the cell.

hypha   A microscopic, branched filament of the body of a fungus.

hypocotyl   The portion of an embryonic plant stem located below the point of attachment of the cotyledons.

hypothalamus   A gland located below the thalamus at the floor of the forebrain; it controls functions of the gastrointestinal and reproductive systems, and it regulates many basic behaviors such as eating and drinking.

hypothesis   In biology, a proposed explanation for a natural phenomenon based on previous observations or experimental studies.

hypothesis testing   Also known as the scientific method, a strategy for testing the validity of a hypothesis.

hypothyroidism   An underactive thyroid gland.

hypotonic   When the solute concentration outside the cell is lower relative to the inside of the cell.

I

I band   A light band that lies between the A bands of two adjacent sarcomeres.

immune system   The cells and organs within an animal's body that contribute to immune defenses.

immune tolerance   The process by which the body distinguishes between self and nonself components.

immunity   The ability of an animal to ward off internal threats, including the invasion of potentially harmful microorganisms such as bacteria, the presence of foreign molecules such as the products of microorganisms, and the presence of abnormal cells such as cancer cells.

immunization   *See* vaccination.

immunoglobulin   A Y-shaped protein with two heavy chains and two light chains that provide immunity to foreign substances; antibodies are a type of immunoglobulin.

immunological memory   The immune system's ability to produce a secondary immune response.

imperfect flower   A flower that lacks either stamens or carpels.

implantation   The first event of pregnancy, when the blastocyst embeds within the uterine endometrium.

imprinted   In genetics, a marked segment of DNA.

imprinting   1. The development of a species-specific pattern of behavior. A form of learning, with a large innate component, within a limited time period. 2. In genetics, the marking of DNA that occurs differently between males and females.

in vitro   Literally, "in glass." An approach to studying a process in living cells that involves isolating and purifying cellular components, outside the cell.

in vivo   Meaning, "in life." An approach to studying a process in living cells.

inactivation gate   A string of amino acids that juts out from a channel protein into the cytosol.

inborn error of metabolism   A genetic defect in the ability to metabolize certain compounds.

inbreeding   Mating among genetically related relatives.

inbreeding depression   The phenomenon whereby inbreeding produces homozygotes that are less fit, thereby decreasing the reproductive success of a population.

inclusive fitness   The term used to designate the total number of copies of genes passed on through one's relatives, as well as one's own reproductive output.

incomplete dominance   The phenomenon in which a heterozygote that carries two different alleles exhibits a phenotype that is intermediate between the corresponding homozygous individuals.

incomplete flower   A flower that lacks one or more of the four flower organ types.

incomplete metamorphosis   A gradual change in body form in some insects from different nymphal stages, called instars, into a similar looking adult.

incurrent siphon   A tunicate structure used to draw water through the mouth.

indeterminate cleavage   A characteristic of deuterostome development in which each cell produced by early cleavage retains the ability to develop into a complete embryo.

indeterminate growth   Growth in which plant shoot apical meristems continuously produce new stem tissues and leaves as long as conditions remain favorable.

indicator species   A species whose status provides information on the overall health of an ecosystem.

indirect calorimetry   A method of determining basal metabolic rate in which the rate at which an animal uses oxygen is measured.

individual selection   The proposal that adaptive traits generally are selected for because they benefit the survival and reproduction of the individual rather than the group.

individualistic model   A view of the nature of a community that considers it to be an assemblage of species coexisting primarily because of similarities in their physiological requirements and tolerances.

induced fit Occurs when a substrate(s) binds to an enzyme and the enzyme undergoes a conformational change that causes the substrate(s)s to bind more tightly to the enzyme.

induced mutation A mutation brought about by environmental agents that enter the cell and then alter the structure of DNA.

inducer In transcription, a small effector molecule that increases the rate of transcription.

inducible operon In this type of operon, the presence of a small effector molecule causes transcription to occur.

induction 1. In development, the process by which a cell or group of cells governs the developmental fate of neighboring cells. 2. In molecular genetics, refers to the process by which transcription has been turned on by the presence of a small effector molecule.

industrial nitrogen fixation The human activity of producing nitrogen fertilizers.

infertility The inability to produce viable offspring.

inflammation An innate local response to infection or injury characterized by local redness, swelling, heat, and pain.

inflorescence A cluster of flowers on a plant.

infundibular stalk The structure that is physically connected to a multilobed endocrine gland sitting directly below the hypothalamus, called the pituitary gland.

ingroup A monophyletic group in a cladogram of interest.

inheritance of acquired characteristics Jean-Baptiste Lamarck's hypothesis that species change over the course of many generations by adapting to new environments. He thought behavioral changes modified traits, and he hypothesized that such modified traits were inherited by offspring.

inhibition A mechanism for succession in which space is all-important, and order of colonization determines subsequent community structure.

inhibitory postsynaptic potential (IPSP) The response from an inhibitory neurotransmitter hyperpolarizing the postsynaptic membrane, which reduces the likelihood of an action potential.

initiation stage In transcription or translation, the first step that initiates the process.

initiator tRNA A specific tRNA that recognizes the start codon AUG in mRNA and binds to it.

innate The term used to describe behaviors that seem to be genetically programmed.

inner bark The thin layer of phloem that conducts most of the sugar transport in a woody stem.

inner ear One of the three main compartments of the mammalian ear; composed of the bony cochlea and the vestibular system, which plays a role in balance.

inorganic chemistry The study of the nature of atoms and molecules, with the exception of those that contain rings or chains of carbon.

instar A stage of growth in an insect with incomplete metamorphosis.

insulin-like growth factor-1 (IGF-1) A hormone that in mammals stimulates the elongation of bones, especially during puberty, when mammals become reproductively mature.

integral membrane protein A protein that cannot be released from the membrane unless it is dissolved with an organic solvent or detergent—in other words, you would have to disrupt the integrity of the membrane to remove it.

integrase An enzyme, sometimes encoded by viruses, that catalyzes the integration of the viral genome into a host-cell chromosome.

integrin A cell-surface receptor protein found in animal cells that connects cells and the extracellular matrix.

integument In plants, a modified leaf that encloses the megasporangium to form an ovule.

intercostal muscles Muscles that surround and connect the ribs in the chest.

interference competition Competition in which individuals interact directly with one another by physical force or intimidation.

interferon A protein that generally inhibits viral replication inside host cells.

intermediate-disturbance hypothesis The proposal that moderately disturbed communities are more diverse than undisturbed or highly disturbed communities.

intermediate filament A type of protein filament within the cytoskeleton that helps maintain cell shape and rigidity.

internal fertilization Fertilization that occurs in terrestrial animals, in which sperm are deposited within the reproductive tract of the female during the act called copulation.

interneuron A type of neuron that forms interconnections between other neurons.

internode The region of stem on a plant between adjacent nodes.

interphase The G_1, S, and G_2 phases of the cell cycle. It is phase of the cell cycle during which the chromosomes are decondensed and found in the nucleus.

intersexual selection Sexual selection between members of the opposite sex.

interspecies hybrid The offspring resulting from two species mating.

interspecific competition The term used to describe competition between individuals of different species.

interstitial Refers to the fluid between cells.

intertidal zone The area where the land meets the sea, which is alternately submerged and exposed by the daily cycle of tides.

intracellular fluid The fluid inside cells.

intranuclear spindle A spindle that forms within an intact nuclear envelope during nuclear division.

intrasexual selection Sexual selection between members of the same sex.

intraspecific competition The term used to describe competition between individuals of the same species.

intrinsic rate of increase The situation in which conditions are optimal for a population, and the per capita growth rate is at its maximum rate.

introduced species A species moved by humans from a native location to another location.

intron Intervening DNA sequences that are found in between the coding sequences of genes.

invasive The term used to describe introduced species that spread on their own, often outcompeting native species for space and resources.

invasive cell A cell that can invade healthy tissues.

inversion A change in the direction of the genetic material along a single chromosome.

invertebrate An animal that lacks vertebrae.

iodine-deficient goiter An overgrown gland that is incapable of making thyroid hormone.

ion At atom or molecule that gains or loses one or more electrons and acquires a net electric charge.

ion electrochemical gradient A dual gradient for an ion that is composed of both an electrical gradient and chemical gradient for that ion.

ionic bond The bond that occurs when a cation binds to an anion.

ionotropic receptor A ligand-gated ion channel that opens in response to binding of a neurotransmitter molecule.

iris The circle of pigmented smooth muscle that is responsible for eye color.

iron regulatory element (IRE) A response element within the ferritin mRNA to which the iron regulatory protein binds.

iron regulatory protein (IRP) An RNA-binding protein that regulates the translation of the mRNA that encodes ferritin.

islets of Langerhans Spherical clusters of endocrine cells that are scattered in great numbers throughout the endocrine pancreas.

isomers Two structures with an identical molecular formula but different structures and characteristics.

isotonic When the solute concentrations on both sides of the plasma membrane are equal.

isotope An element that exists in multiple forms that differ in the number of neutrons they contain.

iteroparity The pattern of repeated reproduction at intervals throughout the life cycle.

J

joint The juncture where two or more bones of a vertebrate endoskeleton come together.

K

K/T event An ancient cataclysm that involved at least one large meteorite or comet that crashed into the Earth near the present-day Yucatan Peninsula in Mexico about 65 million years ago.

karyogamy The process of nuclear fusion.

karyotype A photographic representation of the chromosomes that reveals how many chromosomes are found within an actively dividing cell.

K_D The dissociation constant between a ligand and its receptor.

ketones Small compounds generated from carbohydrates, fatty acids, or amino acids. Ketones are made in the liver and released into the blood to provide an important energy source during prolonged fasting for many tissues, including the brain.

keystone species A species within a community that has a role out of proportion to its abundance.

kilocalorie (kcal) One thousand calories; the common unit of measurement when measuring biological activities.

kin selection Selection for behavior that lowers an individual's own fitness but enhances the reproductive success of a relative.

kinesis A movement in response to a stimulus, but one that is not directed toward or away from the source of the stimulus.

kinetic energy Energy associated with movement.

kinetic skull A defining characteristic of the class Lepidosauria, in which the joints between various parts of the skull are extremely mobile.

kinetochore A group of proteins necessary for sorting each chromosome that binds to the centromere.

kingdom A taxonomic group that contains one or more phyla.

knowledge The awareness and understanding of information.

Koch's postulates A series of steps used to determine whether a particular organism causes a specific disease.

K-selected species A type of life history strategy, where species have a low rate of per capita population growth but good competitive ability.

L

labia majora In the female genitalia, large outer folds that surround the external opening of the reproductive tract.

labia minora In the female genitalia, smaller, inner folds near the external opening of the reproductive tract.

labor A three-stage process that includes (1) dilation and thinning of the cervix to allow passage of the fetus out of the uterus; (2) the movement of the fetus through the cervix and the vagina and out into the world; and (3) the contraction of blood vessels within the placenta and umbilical cord, blocking further blood flow, making the newborn independent from the mother; the placenta detaches from the uterine wall and is delivered a few minutes after the birth of the baby.

lac operon An operon in the genome of *E. coli* that contains the genes for the enzymes that allow it to metabolize lactose.

lac repressor A repressor protein that regulates the lac operon.

lactation In most mammals, a period after birth in which the young are nurtured by milk produced by the mother.

lacteal A vessel in the center of each intestinal villus; lipids are absorbed by the lacteals, which eventually empty into the circulatory system.

lagging strand A strand of DNA made as a series of small Okazaki fragments that are eventually connected to each other to form a continuous strand. The synthesis of these DNA fragments occurs in the direction away from the replication fork.

lamellae Platelike structures in the internal gills of fish that branch from structures called filaments.

landscape ecology The study of the influence of large-scale spatial patterns of land use or habitat type on ecological processes.

larva A free-living organism that is morphologically very different from the embryo and adult.

larynx The area beyond the throat where the vocal cords lie.

latent The term used to describe when a prophage or provirus remains inactive for a long time.

lateral line system A sensory system in fish and some toads that allows them to detect changes in their environment; hair cells detect changes in water currents brought about by waves, nearby moving objects, and low-frequency sounds traveling through the water.

law of independent assortment The alleles of different genes assort independently of each other during gamete formation.

law of segregation The phenomenon that the two copies of a gene segregate from each other during gamete formation and transmission from parent to offspring.

leaching The dissolution and removal of inorganic ions as water percolates through materials such as soil.

leading strand A DNA strand made in the same direction that the replication fork is moving. The strand is synthesized as one long continuous molecule.

leaf A flattened plant organ that emerges from stems and functions in photosynthesis or other ways.

leaf abscission The process by which a leaf drops after the formation of an abscission zone at the point where a leaf petiole connects with the stem. The abscission zone consists of an inner protective layer of cork cells whose tough walls help to prevent pathogen attack and dehydration, and an outer layer of cells having thin walls that break easily.

leaf primordia Small bumps that occur at the sides of a shoot apical meristem and develop into young leaves.

leaflet 1. Half of a phospholipid bilayer. 2. A portion of a compound leaf.

learning The ability of an animal to make modifications to a behavior based on previous experience.

legume A member of the pea (bean) family. Also their distinctive fruits, dry pods that develop from one carpel and open down both sides when seeds are mature.

lek A designated communal courting area.

lens 1. A structure of the eye that focuses light. 2. The glass components of a light microscope or electromagnetic parts of an electron microscope that allow the production of magnified images of microscopic structures.

lentic Referring to a freshwater habitat characterized by standing water.

leptin A molecule produced by adipose cells in proportion to fat mass; controls appetite and metabolic rate.

leukocyte A white blood cell; involved in nonspecific immunity.

lichens Mutualistic partnerships of particular fungi and certain photosynthetic green algae or cyanobacteria, and sometimes both to form a body distinctive from that of either partner alone.

Liebig's law of the minimum Species biomass or abundance is limited by the scarcest factor.

life cycle The sequence of events that characterize the steps of development of the individuals of a given species.

leukocyte A white blood cell.

life table A table that provides data on the number of individuals alive in particular age classes.

ligand An ion or molecule that binds to a protein, such as an enzyme or a receptor.

ligand-gated channel A channel controlled by the noncovalent binding of small molecules—called ligands—such as hormones or neurotransmitters.

ligand-gated ion channels A type of cell surface receptor that binds a ligand and functions as an ion channel. Ligand binding either opens or closes a channel.

light chain A part of an immunoglobulin molecule.

light-harvesting complex A component of photosystem II and photosystem I composed of several dozen pigment molecules that are anchored to proteins. The role of these complexes is to absorb photons of light.

light microscope A microscope that utilizes light for illumination.

light reactions One of two stages in the process of photosynthesis. During the light reactions, photosystem II and photosystem I absorb light energy and produce O_2, ATP, and NADPH.

lignin A tough polymer that adds strength and decay resistance to cell walls of tracheids, vessel elements, and other cells of plants.

lignophytes Modern and fossil seed plants, and seedless ancestors that produced wood.

limbic system The system primarily involved in the formation and expression of emotions, and also plays a role in learning, memory, and the perception of smells; includes certain areas of the telencephalon and parts of the diencephalon.

limiting factor The factor that is most scarce in relation to need.

lineage The genetic relationship between an individual or group of individuals and its ancestors. A series of ancestors in a population that shows a progression of changes.

linkage group A group of genes that usually stay together during meiosis.

linkage The phenomenon of two genes that are close together on the same chromosome tending to be transmitted as a unit.

lipase The major digestive fat-digesting enzyme from the pancreas.

lipid A molecule composed predominantly of hydrogen and carbon atoms. Lipids are nonpolar and therefore very insoluble in water. They include fats, phospholipids, and steroids.

lipid anchor A way for proteins to associate with the plasma membrane; involves the covalent attachment of a lipid to an amino acid side chain within a protein.

lipid exchange protein A protein that extracts a lipid from one membrane, diffuses through the cell, and inserts the lipid into another membrane.

lipopolysaccharides Lipids having covalently-bound carbohydrates. Major components of the thin, outer envelope that encloses the cell walls of Gram-negative bacteria.

liposome A vesicle surrounded by a lipid bilayer.

liver An organ in vertebrates that performs diverse metabolic functions and is the site of bile production.

liverworts A phylum of bryophytes; formally called Hepatophyta.

lobe fins The Actinistia (coelacanths) and Dipnoi (lungfish) and tetrapods; also called Sarcopterygii.

lobe-finned fish Fish in which the fins are part of the body, and they are supported by skeletal extensions of the pectoral and pelvic areas that are moved by muscles residing in the fins.

locomotion The movement of an animal from place to place.

locus The physical location of a gene on a chromosome.

logistic equation $dN/dt = rN (K - N) / K$, where dN/dt is the rate of population change, r is the per capita rate of population growth, N is the population size, and $(K - N) / K$ represents the proportion of unused resources remaining.

logistic growth The S-shaped pattern in which the growth of a population typically slows down as it approaches carrying capacity.

long-day plant A plant that flowers in spring or early summer, when the night period is shorter (and thus the day length is longer) than a defined period.

long-term potentiation (LTP) The long-lasting strengthening of the connection between neurons.

loop domain A chromosomal segment that is folded into loops by the attachment to proteins; a method of compacting chromosomes.

loop of Henle A sharp, hairpin-like loop in the tubule of the nephron of the kidney consisting of a descending limb coming from the proximal tubule and an ascending limb leading to the distal convoluted tubule.

lophophore A horseshoe-shaped crown of tentacles used for feeding.

Lophotrochozoa A clade that encompasses the annelids, mollusks, and several other phyla; they are distinguished by two morphological features—the lophophore, a crown of tentacles used for feeding, and the trochophore larva, a distinct larval stage.

lotic Referring to a freshwater habitat characterized by running water.

lumen The internal space of an organelle.

lung A structure used to bring oxygen into the circulatory system and remove carbon dioxide.

lungfish The Dipnoi; fish with primitive lungs which live in oxygen-poor freshwater swamps and ponds.

luteinizing hormone (LH) A gonadotropin.

lycophyll A relatively small leaf having a single unbranched vein; the type of leaf produced by lycophytes.

lycophytes Members of a phylum of vascular land plants whose leaves are lycophylls; Lycopodiophyta.

lymphatic system A system of vessels along with a group of organs and tissues where most leukocytes reside. The lymphatic vessels collect excess interstitial fluid and return it to the blood.

lymphocytes A type of leukocyte that is responsible for specific immunity; the two types are B cells and T cells.

Lyon hypothesis *See* X inactivation.

lysogenic cycle The growth cycle of a bacteriophage consisting of integration, prophage replication, and excision.

lysogeny Latency in bacteriophages.

lysosome A small organelle found in animal cells that contains acid hydrolases that degrade macromolecules.

lytic cycle The growth cycle of a bacteriophage in which the production and release of new viruses lyses the host cell.

M

M line A narrow, dark band in the center of the H zone that corresponds to proteins that link together the central regions of adjacent thick filaments.

M phase The sequential events of mitosis and cytokinesis.

macroevolution Evolutionary changes that create new species and groups of species.

macromolecule Many molecules bonded together to form a polymer. Carbohydrates, proteins, and nucleic acids (for example, DNA and RNA) are important macromolecules found in living organisms.

macronutrient An element required by plants in amounts of at least 1 g/kg of plant dry matter.

macroparasite A parasite that lives in the host but releases infective juvenile stages outside the host's body.

macrophage A type of phagocyte capable of engulfing viruses and bacteria; strategically located where it will encounter invaders.

macular degeneration A condition in which photoreceptor cells in and around the fovea of the retina are lost; associated with loss of sharpness and color vision.

madreporite A sieve-like plate on the surface of an echinoderm where water enters the water vascular system.

magnification The ratio between the size of an image produced by a microscope and its actual size.

major groove A groove that spirals around the DNA double helix. The major groove provides a location where a protein can bind to a particular sequence of bases and affect the expression of a gene.

major histocompatibility complex (MHC) A gene family that encodes the plasma membrane self proteins that must be complexed with the antigen in order for T-cell recognition to occur.

malabsorption Any interference with the secretion of bile or the action of bile salts in the intestine that decreases the absorption of fats, including fat-soluble vitamins.

male assistance hypothesis A hypothesis to explain the existence of monogamy that maintains that males remain with females to help them rear their offspring.

male gametophyte A haploid plant life cycle phase that produces sperm.

malignant tumor A growth of cells that has progressed to the cancerous stage.

Malpighian tubules Delicate projections from the digestive tract of insects and some other taxa that protrude into the hemolymph and function as an excretory organ.

mammal A vertebrate that is a member of the class Mammalia that nourishes its young with milk secreted by mammary glands. Another distinguishing feature is hair.

mammary gland A gland in female members of mammal species that secretes milk.

manganese cluster A site where the oxidation of water occurs in photosystem II.

mantle A fold of skin draped over the visceral mass of a mollusk that secretes a shell in those species that form shells.

mantle cavity The chamber in a mollusk mantle that houses delicate gills.

many eyes hypothesis The idea that increased group size decreases predators' success because of increased predator detection ability.

map distance The distance between genes along chromosomes which is calculated as the number of recombinant offspring divided by the total number of offspring times 100.

mapping The process of determining the relative locations of genes or other DNA segments along a chromosome.

map unit (mu) A unit of distance equivalent to a 1% recombination frequency.

mark-recapture technique The capture and tagging of animals so they can be released and recaptured, allowing an estimate of population size.

marsupial A member of a group of seven mammalian orders and about 280 species found in the subclass Metatheria.

mass extinction When many species become extinct at the same time.

mass-specific BMR The amount of energy expended per gram of body mass.

mass spectrometry A method to determine the masses of molecules such as short peptide fragments within proteins. Tandem mass spectrometry can be used to determine the amino acid sequences of proteins.

mast cell A type of cell derived from bone marrow stem cells that plays an important role in nonspecific immunity; these cells are found throughout connective tissues, and secrete many locally acting molecules, notably histamine.

mastax The circular muscular pharynx in the mouth of rotifers.

masting The synchronous production of many progeny by all individuals in a population to satiate predators and thereby allow some progeny to survive.

mate-guarding hypothesis A hypothesis to explain the existence of monogamy that theorizes that males stay with a female to prevent her from being fertilized by other males.

maternal effect gene A gene that follows a maternal effect inheritance pattern.

maternal effect An inheritance pattern in which the genotype of the mother determines the phenotype of her offspring.

maternal inheritance A phenomenon in which offspring inherit particular genes only from the female parent (through the egg).

maturation promoting factor (MPF) The factor that causes oocytes to progress (or mature) from G_2 to M phase.

mature mRNA In eukaryotes, transcription produces a longer RNA, pre-mRNA, which undergoes certain processing events before it exits the nucleus; mature mRNA is the final functional product.

mean fitness of the population The average reproductive success of members of a population.

mechanosensitive channel A channel that is sensitive to changes in membrane tension.

mechanoreceptor A sensory receptor that transduces mechanical energy such as pressure, touch, stretch, movement, and sound.

mediator A large protein complex that plays a role in initiating transcription at the core promoter of structural genes in eukaryotes.

medulla oblongata The part of the hindbrain that coordinates many basic reflexes and bodily functions, such as breathing, that maintain the normal homeostatic processes of the animal.

megadiversity country Those countries with the greatest numbers of species; used in targeting areas for conservation.

megaspore In seed plants and some seedless plants, a large spore that produces a female gametophyte within the spore wall.

meiosis The process by which haploid cells are produced from a cell that was originally diploid.

meiosis I The first division of meiosis when the homologues are separated into different cells.

meiosis II The second division of meiosis in which sister chromatids are separated to different cells.

meiotic nondisjunction An occurrence during meiosis I or meiosis II that produces haploid cells that have too many or too few chromosomes.

Meissner's corpuscles Structures that sense touch and light pressure and that lie just beneath the skin surface.

membrane attack complex (MAC) A multiunit protein formed by the activation of complement proteins; the complex creates water channels in the microbial plasma membrane and causes the microbe to swell and burst.

membrane potential The difference between the electric charges inside and outside the cell; also called a potential difference (or voltage).

membrane transport The movement of ions or molecules across a cell membrane.

memory The retention of information over time.

memory cells A component of the immune response; these cells remain poised to recognize the antigen if it returns in the future.

Mendelian inheritance The inheritance patterns of genes that segregate and assort independently.

meninges A protective structure in the central nervous system consisting of three layers of sheathlike membranes.

meningitis A potentially life-threatening infectious disease in which the meninges become inflamed.

menopause The event during which a woman permanently stops having ovarian cycles.

menstrual cycle Also called the uterine cycle; the cyclical changes in the lining of the uterus that occur in parallel with the ovarian cycle.

menstruation A period of bleeding at the beginning of the menstrual cycle.

meristem In plants, an organized tissue that includes actively dividing cells and a reservoir of stem cells.

meroblastic cleavage An incomplete type of cell cleavage, in which only the region of the egg containing cytoplasm at the animal pole undergoes cell division.

merozygote A strain of bacteria containing an F′ factor.

mesoderm A layer of cells formed during gastrulation that develops between the ectoderm and endoderm; gives rise to

skeleton, muscles and much of the circulatory system.

mesoglea A gelatinous substance that connects the two germ layers in the Radiata.

mesohyl A gelatinous, protein-rich matrix in between the choanocytes and the epithelial cells of a sponge.

mesophyll The internal tissue of a plant leaf whose cells carry out photosynthesis.

messenger A molecule that transmits messages from many types of activated sensors to effector molecules.

messenger RNA (mRNA) RNA that contains the information to specify a polypeptide with a particular amino acid sequence; its job is to carry information from the DNA to the ribosome.

metabolic cycle A biochemical cycle in which particular molecules enter while others leave; the process is cyclical because it involves a series of organic molecules that are regenerated with each turn of the cycle.

metabolic enzyme A protein that accelerates chemical reactions within the cell.

metabolic pathway In living cells, a series of chemical reactions; each step is catalyzed by a specific enzyme.

metabolic rate The total energy expenditure of an organism per unit of time.

metabolism The sum total of all chemical reactions that occur within an organism. Also, a specific set of chemical reactions occurring at the cellular level.

metabotropic receptor A G-protein-coupled receptor that is coupled to an intracellular signaling pathway that initiates changes in a postsynaptic cell.

metacentric A chromosome in which the centromere is near the middle.

metagenomics A field of study that seeks to identify and analyze the collective microbial genomes contained in a community of organisms, including, for microbial genomes, those that are not easily cultured in the laboratory.

metamerism The division of the body into identical subunits called segments.

metamorphosis The process in which a pupal or juvenile organism changes into a mature adult with very different characteristics.

metanephridia Excretory organs found in a variety of invertebrates; a type of tubular nephridium.

metanephridial system The filtration system used by annelids to cleanse the blood.

metaphase The phase of mitosis during which the chromosomes are aligned along the metaphase plate.

metaphase plate A plane halfway between the poles on which the sister chromatids align during metaphase.

metastatic cell A cancer cell that can migrate to other parts of the body.

Metazoa The collective term for animals.

methanogens Several groups of anaerobic archaea that convert CO_2, methyl groups, or acetate to methane and release it from their cells.

methanotroph An aerobic bacterium that consumes methane.

methyl-CpG-binding protein A protein that binds methylated sequences.

methyl-directed mismatch repair A DNA repair system that involves the participation of several proteins that detect the mismatch and specifically remove a segment from the newly made strand.

micelle The sphere formed by long amphipathic molecules when mixed with water. In animals, micelles aid in the absorption of poorly soluble products during digestion; they consist of bile salts, phospholipids, fatty acids, and monoglycerides clustered together.

microclimate Local variations of the climate within a given area.

microevolution The term used to describe changes in a population's gene pool from generation to generation.

microfilament *See* actin filament.

micronutrient An element required by plants in amounts at or less than 0.1 g/kg of plant dry mass.

microparasite A parasite that multiplies within its host, usually within the cells.

micropyle A small opening in the integument of a seed plant ovule through which pollen tubes grow.

microRNAs (miRNAs) Small RNA molecules, typically 22 nucleotides in length, that silence the expression of specific mRNAs, either by inhibiting translation or by promoting the degradation of mRNAs.

microscope A magnification tool that enables researchers to study the structure and function of cells.

microsphere A small water-filled vesicle surrounded by a macromolecular boundary.

microspore In seed plants and some seedless plants, a relatively small spore that produces a male gametophyte within the spore wall.

microtubule A type of hollow protein filament composed of tubulin proteins that is part of the cytoskeleton and is important for cell organization, shape, and movement.

microtubule-organizing center *See* centrosome.

microvilli Small projections in the surface membranes of epithelial cells in the small intestine.

midbrain One of three major divisions of the vertebrate brain; the other two divisions are the hindbrain and forebrain.

middle ear One of the three main compartments of the mammalian ear; contains three small bones called ossicles that connect the eardrum with the oval window.

middle lamella A layer composed primarily of carbohydrate that cements adjacent plant cell walls together.

migration Long-range seasonal movement among animals in order to feed or breed.

mimicry The resemblance of an organism (the mimic) to another organism (the model).

mineral An inorganic ion required by a living organism. Minerals are used to build skeletons, maintain the balance of salts and water in the body, provide a source of electric current across plasma membranes, and provide a mechanism for exocytosis and muscle contraction, among other functions.

mineralization The general process by which phosphorus, nitrogen, CO_2, and other minerals are released from organic compounds.

mineralocorticoid A steroid hormone such as aldosterone that regulates the balance of certain minerals in the body.

minor groove A smaller groove that spirals around the DNA double helix.

missense mutation A base substitution that changes a single amino acid in a polypeptide sequence.

mitochondrial genome The chromosome found in mitochondria.

mitochondrial matrix A compartment inside the inner membrane of a mitochondrion.

mitochondrion Literally, "thread granule." An organelle found in eukaryotic cells that supplies most of the cell's ATP.

mitosis In eukaryotes, the process in which nuclear division results in two nuclei; each daughter cell receives the same complement of chromosomes.

mitotic cell division A process whereby a eukaryotic cell divides to produce two new cells that are genetically identical to the original cell.

mitotic spindle *See* mitotic spindle apparatus.

mitotic spindle apparatus The structure responsible for organizing and sorting the chromosomes during mitosis.

mixotroph An organism that is able to use autotrophy as well as phagotrophy or osmotrophy to obtain organic nutrients.

moderately repetitive sequence A DNA sequence found a few hundred to several thousand times in a genome.

modern synthesis of evolution Within a given population of interbreeding organisms, natural variation exists that is caused by random changes in the genetic material. Such genetic changes may affect the phenotype of an individual in a positive, negative, or neutral way. If a genetic change promotes an individual's reproductive success, natural selection may increase the prevalence of that trait in future generations.

molar An adjective to describe the number of moles of a solute dissolved in 1 L of water.

molarity The number of moles of a solute dissolved in 1 L of water.

mole The amount of any substance that contains the same number of particles as there are atoms in exactly 12 g of carbon: 12 grams of carbon equals 1 mole, while 1 g of hydrogen equals 1 mole.

molecular biology A field of study spawned largely by genetic technology that looks at the structure and function of the molecules of life.

molecular clock A clock on which to measure evolutionary time.

molecular evolution The molecular changes in genetic material that underlie the phenotypic changes associated with evolution.

molecular formula A representation of a molecule that consists of the chemical symbols for all of the atoms present and subscripts that tell you how many of those atoms are present.

molecular homologies Similarities that indicate that living species evolved from a common ancestor or interrelated group of common ancestors.

molecular machine A machine that is measured in nanometers, which has moving parts and does useful work.

molecular mass The sum of the atomic masses of all the atoms in a molecule.

molecular pharming An avenue of research that involves the production of medically important proteins in agricultural crops or animals.

molecular recognition The process whereby surfaces on various protein subunits recognize each other in a very specific way, causing them to bind to each other and promote the assembly process.

molecular systematics A field of study that involves the analysis of genetic data, such as DNA sequences, to identify and study genetic homology and construct phylogenetic trees.

molecule Formed from two or more atoms that are connected by chemical bonds.

monocots One of the two largest lineages of flowering plants; the embryo produces a single seed leaf.

monocular vision A type of vision in animals with eyes on the sides of the head; the animal sees a wide area at one time, though depth perception is reduced.

monocyte A type of phagocyte that circulates in the blood for only a few days, after which it takes up permanent residence as a macrophage in different organs.

monoecious The term to describe plants that produce carpellate and staminate flowers on the same plant.

monogamy A mating system in which one male mates with one female, and most individuals have mates.

monohybrid The F_1 offspring, also called single-trait hybrids, of true-breeding parents that differ with regard to a single trait.

monomorphic gene A gene that exists predominantly as a single allele in a population.

monophagous The term used to define parasites that feed on one species or two or three closely related hosts.

monophyletic group A group of species, a taxon, consisting of the most recent common ancestor and all of its descendants.

monosaccharide A simple sugar.

monosomic An aneuploid organism that has one too few chromosomes.

monotreme One of three species in the mammalian order Monotremata, which are found in Australia and New Guinea: the duck-billed platypus and two species of echidna.

morphogen   A molecule that imparts positional information and promotes developmental changes at the cellular level.

morphogenetic field   A group of embryonic cells that ultimately produce a specific body structure.

morula   The mammal embryo in species having undergone holoblastic cleavage.

mosaic   An individual in which some cells throughout the body show genetic differences. For example, in female mammals, about half of the somatic cells will express one X-linked allele, while the rest of the somatic cells will express the other allele.

mosses   A phylum of bryophytes; Bryophyta.

motif   The structure that occurs when a domain or portion of a domain has a very similar structure in many different proteins.

motor neuron   A neuron that sends signals away from the central nervous system and elicits some type of response.

motor protein   A category of cellular proteins that uses ATP as a source of energy to promote movement; consists of three domains called the head, hinge, and tail.

movement corridor   Thin strips of habitat that may permit the movement of individuals between larger habitat patches.

mucigel   A gooey plant substance that lubricates roots, aiding in their passage through the soil; helps in water and mineral absorption, prevents root drying, and provides an environment hospitable to beneficial microbes.

Müllerian mimicry   A type of mimicry in which many noxious species converge to look the same, thus reinforcing the basic distasteful design.

multicellular   Consisting of more than one cell, with cells attached to each other; cells able to communicate with each other by chemical signaling, and some cells able to specialize.

multimeric protein   A protein with more than one polypeptide chain; also said to have a quarternary structure.

multiple alleles   Refers the occurrence of a gene that exists as three or more alleles in a population.

multiple sclerosis (MS)   A disease in which the patient's own body, for reasons that are unknown, attacks and destroys myelin as if it were a foreign substance. Eventually, these repeated attacks leave scarred (sclerotic) areas of tissue in the nervous system and impair the function of myelinated neurons that control movement, speech, memory, and emotion.

multipotent   A term used to describe a stem cell that can differentiate into several cell types, but far fewer than pluripotent cells.

muscle   A grouping of muscle tissue bound together by a succession of connective tissue layers.

muscle tissue   Clusters of cells that are specialized to contract, generating the mechanical forces that produce body movement, exert pressure on a fluid-filled cavity, or decrease the diameter of a tube.

mutagen   An agent known to cause mutation.

mutant allele   An allele that has been altered by mutation.

mutation   A heritable change in the genetic material.

mutualism   An interaction in which both species benefit.

mycelium   A fungal body composed of microscopic branched filaments known as hyphae.

mycorrhizae   Associations between the hyphae of certain fungi and the roots of most plants.

myelin sheath   In the nervous system, an insulating layer made up of specialized glial cells wrapped around the axons.

myocardial infarction (MI)   The death of cardiac muscle cells, which can occur if a region of the heart is deprived of blood for an extended time.

myofibrils   Cylindrical bundles within muscle fibers, each of which contains thick and thin filaments.

myogenic bHLH genes   A small group of genes that initiates muscle development in animals.

myogenic heart   A heart in which the signaling mechanism that initiates contraction resides within the cardiac muscle itself.

myoglobin   An oxygen-binding protein that increases the availability of oxygen in the muscle fiber by providing an intracellular reservoir of oxygen.

myosin   A motor protein in muscle.

N

nacre   The smooth, iridescent lining of the shells of oysters, mussels, abalone, and other mollusks.

NAD⁺   Nicotinamide adenine dinucleotide; a dinucleotide that functions as an energy intermediate molecule. It combines with two electrons and H^+ to form NADH.

NADP⁺   Nicotinamide adenine dinucleotide phosphate; a dinucleotide that functions as an energy intermediate molecule in chloroplasts. It combines with two electrons and H^+ to form NADPH.

natural killer (NK) cells   A type of leukocyte that is part of the body's nonspecific defenses because it recognizes general features on the surface of cancer cells or any virus-infected cells.

natural selection   The process that culls out those individuals that are less likely to survive and reproduce in a particular environment, while allowing other individuals with traits that confer greater reproductive success to increase in numbers.

nauplius   The first larval stage in a crustacean.

navigation   A mechanism of migration that involves the ability not only to follow a compass bearing but also to set or adjust it.

nectar   A sugar-rich substance produced by many flowers that serves as a food reward for pollinators; in some plants, nectar is produced by other plant parts as a reward for insects that protect the plant.

negative control   Transcriptional regulation by repressor proteins.

negative feedback loop   A system in which a change in the variable being regulated brings about responses that move the variable in the opposite direction.

negative frequency-dependent selection   A pattern of natural selection in which the fitness of a genotype decreases when its frequency becomes higher; the result is a balanced polymorphism.

negative pressure filling   The mechanism by which reptiles, birds, and mammals ventilate their lungs.

nekton   Free-swimming animals in the open ocean that can swim against the currents to locate food.

nematocyst   In a cnidarian, a powerful capsule with an inverted coiled and barbed thread that functions to immobilize small prey so they can be passed to the mouth and ingested.

neocortex   The layer of the brain that evolved most recently in mammals.

nephron   One of several million single-cell-thick tubules that are the functional units of the kidney.

Nernst equation   The formula that gives the equilibrium potential for an ion at any given concentration gradient.

nerve   A structure found in the peripheral nervous system that is composed of multiple neurons and makes contact with structures outside the central nervous system and transmits signals that enter or leave the CNS.

nerve cord   In more complex invertebrates, a structure that extends from the anterior end of the animal to the tail.

nerve impulse   A way that neurons communicate, involving changes in the amount of electric charge across a cell's plasma membrane.

nerve net   Interconnected neurons with no central control organ.

nervous system   Groups of cells that sense internal and environmental changes and transmit signals that enable an animal to respond in an appropriate way.

nervous tissue   Clusters of cells that initiate and conduct electrical signals from one part of an animal's body to another part.

net primary production (NPP)   Gross primary production minus the energy lost in plant cellular respiration.

net reproductive rate   The population growth rate per generation.

neural crest   A cell lineage that gives rise to all neurons and supporting cells of the peripheral nervous system in vertebrates; in addition, it gives rise to melanocytes and to cells that form facial cartilage and parts of the adrenal gland.

neural tube   In chordates, a structure formed from ectoderm located dorsal to the notochord; all neurons and their supporting cells in the central nervous system originate from neural precursor cells derived from the neural tube.

neurogenesis   The production of new neurons by cell division.

neurogenic heart   A heart that will not beat unless it receives regular electrical impulses from the nervous system.

neurohormone   A hypothalamic releasing hormone made in and secreted by neurons whose cell bodies are in the hypothalamus.

neuromodulator   Short chains of amino acids that can alter or modulate the response of the postsynaptic neuron to other neurotransmitters.

neuromuscular junction   The junction between a motor neuron's axon and a muscle fiber.

neuron   Another name for a nerve cell. A highly specialized cell that communicates with another cell of its kind and with other types of cells by electrical or chemical signals.

neuroscience   The scientific study of nervous systems.

neurotransmitter   A small signaling molecule that is synthesized and stored in nerve cells.

neurulation   The embryological process responsible for initiating central nervous system formation.

neutral mutation   A mutation that does not affect the function of the encoded protein.

neutral theory of evolution   States that most genetic variation is due to the accumulation of neutral mutations that have attained high frequencies in a population via genetic drift.

neutral variation   Variation that does not favor any particular genotype.

neutralism   The phenomenon in which two species occur together but in fact do not interact in any measurable way.

neutron   A neutral particle found in the center of the atom.

neutrophil   A type of phagocyte and the most abundant type of leukocyte; neutrophils engulf bacteria by endocytosis.

niche   The physical distribution and ecological role of an organism.

nitrification   The conversion by soil bacteria of NH_3 or NH_4^+ to nitrate (NO_3^-), a form of nitrogen commonly used by plants.

nitrogen fixation   A specialized metabolic process in which certain prokaryotes use the enzyme nitrogenase to convert inert atmospheric nitrogen gas into ammonia; also, the industrial process by which humans produce ammonia fertilizer from nitrogen gas.

nitrogenase   An enzyme used in the biological process of fixing nitrogen.

nitrogen-limitation hypothesis   The proposal that organisms select food based on its nitrogen content.

nitrogenous wastes   Molecules that include nitrogen from amino groups; these wastes are toxic at high concentrations and must be eliminated from the body.

nociceptor   A sensory receptor that responds to extreme heat, cold, and pressure, as well as to certain molecules such as acids; also known as a pain receptor.

nocturnal enuresis   Bed-wetting.

Nod factor   Nodulation factor; a substance produced by nitrogen-fixing bacteria in response to flavonoids secreted from the roots of potential host plants; the Nod factors bind to receptors in plant root membranes, starting a process that allows the bacteria to invade roots.

node   The region of a plant stem from which one or more leaves, branches, or buds emerge.

nodes of Ranvier   Exposed areas in the axons of myelinated neurons that contain many voltage-gated Na^+ channels.

nodule   A small swelling on a plant root that contains nitrogen-fixing bacteria.

nodulin   One of several plant proteins that foster root nodule development.

noncoding strand   *See* template.

noncompetitive inhibitor   A molecule that binds to an enzyme at a location that is outside the active site and inhibits the enzyme's function.

noncyclic electron flow   The combined action of photosystem II and photosystem I in which electrons flow in a linear manner to produce NADPH.

non-Darwinian evolution   Also "survival of the luckiest" to contrast it with Darwin's "survival of the fittest" theory; the idea that much of the modern variation in gene sequences is explained by neutral variation rather than adaptive variation.

nondisjunction   An event in which the chromosomes do not sort properly during cell division.

nonparental type   *See* recombinant.

nonpolar molecule   A molecule composed predominantly of nonpolar bonds.

nonrandom mating   The phenomenon that individuals choose their mates based on their genotypes or phenotypes.

nonrecombinant   An offspring whose combination of traits has not changed from the parental generation.

nonsense codon   *See* stop codon.

nonsense mutation   A mutation that changes a normal codon into a stop codon; this causes translation to be terminated earlier than expected, producing a truncated polypeptide.

nonspecific (innate) immunity   The body's defenses that are present at birth and act against foreign materials in much the same way regardless of the specific identity of the invading material; includes the body's external barriers, plus a set of cellular and chemical defenses that oppose substances that breach those barriers.

nonvascular plant   A plant that does not produce lignified vascular tissue, such as a modern bryophyte or extinct pretracheophyte polysporangiophytes.

norepinephrine   A neurotransmitter; also known as noradrenaline.

norm of reaction   A description of how a trait may change depending on the environmental conditions.

notochord   A single flexible rod that lies between the digestive tract and the nerve cord in a chordate.

N-terminus   The location of the first amino acid in a polypeptide.

nuclear envelope   A double-membrane structure that encloses the cell's nucleus.

nuclear genome   The chromosomes found in the nucleus of the cell.

nuclear lamina   A collection of filamentous proteins that line the inner nuclear membrane.

nuclear matrix   A filamentous network of proteins that is found inside the nucleus and lines the inner nuclear membrane. The nuclear matrix serves to organize the chromosomes.

nuclear pore   A passageway for the movement of molecules and macromolecules into and out of the nucleus; formed where the inner and outer nuclear membranes make contact with each other.

nucleic acid   An organic molecule composed of nucleotides. The two types of nucleic acids are deoxyribonucleic acid (DNA) and ribonucleic acid (RNA).

nucleoid   A region of a bacterial cell where the genetic material (DNA) is located.

nucleolus   A prominent region in the nucleus of nondividing cells where ribosome assembly occurs.

nucleosome   A structural unit of eukaryotic chromosomes composed of an octamer of histones (eight histone proteins) wrapped with DNA.

nucleotide   An organic molecule having three components: a phosphate group, a five-carbon sugar (either ribose or deoxyribose), and a single or double ring of carbon and nitrogen atoms known as a base.

nucleotide excision repair (NER)   A common type of DNA repair system that removes (excises) a region of the DNA where the damage occurs. This system can fix many different types of DNA damage, including UV-induced damage, chemically modified bases, missing bases, and various types of cross-links.

nucleus (plural, **nuclei**)   1. In cell biology, an organelle found in eukaryotic cells that contains most of the cell's genetic material. The primary function involves the protection, organization, and expression of the genetic material. 2. In chemistry, the region of an atom that contains protons and neutrons. 3. In neurobiology, a group of neuronal cell bodies in the brain that are devoted to a particular function.

nutrient   Any substance taken up by a living organism that is needed for survival, growth, development, repair, or reproduction.

O

obese   According to current National Institutes of Health guidelines, a person having a BMI of 30 or more.

obligatory mutualism   An interaction in which neither species can live without the other.

ocelli   Photosensitive organs.

octet rule   The phenomenon that atoms are most stable when their outer shell is full. For many, but not all, types of atoms, their outer shell is full when they contain eight electrons, an octet.

Okazaki fragments   Short segments of DNA synthesized in the lagging strand during DNA replication.

olfaction   The sense of smell.

oligotrophic   The term used to describe a young lake that starts off clear and with little plant life.

ommatidium   An independent visual unit in the eye of insects that functions as a separate photoreceptor capable of forming an independent image.

omnivore   An animal that has the ability to eat and survive on both plant and animal products.

oncogene   A type of mutant gene derived from a protooncogene; an oncogene is overactive, thus contributing to uncontrolled cell growth and promoting cancer.

one gene–one polypeptide theory   The concept that one structural gene codes for one polypeptide.

oogenesis   Gametogenesis, which results in the production of egg cells.

oogonia   In animals, diploid germ cells that give rise to the female gametes, the eggs.

open circulatory system   A circulatory system in which hemolymph, which is not different than the interstitial fluid, flows throughout the body and is not confined to special vessels.

open complex   Also called the transcription bubble; a small bubble-like structure between two DNA strands that occurs during transcription.

open conformation   Loosely packed chromatin that can be transcribed into RNA.

operant conditioning   A form of behavior modification; a type of associative learning in which an animal's behavior is reinforced by a consequence, either a reward or a punishment.

operator   A DNA sequence in bacteria that is recognized by activator or repressor proteins that regulate the level of gene transcription.

operculum   A protective flap that covers the gills of a bony fish.

operon   An arrangement of two or more genes in bacteria that are under the transcriptional control of a single promoter.

opsin   A protein that is a component of visual pigments.

opportunistic   A term used to describe animals that have a strong preference for one type of food but can adjust their diet if the need arises.

opposable thumb   A thumb that can be placed opposite the fingers of the same hand, which gives animals a precision grip that enables the manipulation of small objects.

optic nerve   A structure of the eye that carries the electrical signals to the brain.

optimal foraging   The concept that in a given circumstance, an animal seeks to obtain the most energy possible with the least expenditure of energy.

optimality theory   The theory that predicts that an animal should behave in a way that maximizes the benefits of a behavior minus its costs.

oral   Refers to the region of an animal where the mouth is located; refers to the top side of a radial animal.

orbital   The region surrounding the nucleus of an atom where the probablility is high of finding a particular electron.

order   In taxomony, a subdivision of a class.

organ   Two or more types of tissue combined to perform a common function. For example, the heart is composed of several types of tissues, including muscle, nervous, and connective tissue.

organ system   Different organs that work together to perform an overall function in an organism.

organelle   A subcellular structure or membrane-bounded compartment with its own unique structure and function.

organelle genome   In eukaryotes, the genetic material found in mitochondria and plastids.

organic chemistry   The study of carbon-containing molecules.

organic farming   The production of crops without the use of commercial inorganic fertilizers, growth substances, and pesticides.

organic molecule   A carbon-containing molecule, so named because they are found in living organisms.

organism   A living thing that maintains an internal order that is separated from the environment.

organismal ecology   The investigation of how adaptations and choices by individuals affect their reproduction and survival.

organismic model   A view of the nature of a community that considers it to be equivalent to a superorganism; individuals, populations, and communities have a relationship to each other that resembles the associations found between cells, tissues, and organs.

organizing center   A group of cells that ensures the proper organization of the meristem and preserves the correct number of actively dividing stem cells.

organogenesis   The developmental stage during which cells and tissues form organs.

orientation   A mechanism of migration in which animals have the ability to follow a compass bearing and travel in a straight line.

origin of replication   A site within a chromosome that serves as a starting point for DNA replication.

ortholog   A homologous gene in different species.

osculum   A large opening at the top of a sponge.

osmoconformer   An animal whose osmolarity conforms to that of its environment.

osmolarity   The solute concentration of a solution of water, expressed as milliOsmoles/liter (mOsm/L).

osmoregulator   An animal that maintains constant stable internal salt concentrations and osmolarities, even when living in water with very different osmolarities than its body fluids.

osmosis   The movement of water across membranes to balance solute concentrations. Water diffuses from a solution that is hypotonic

(lower solute concentration) into a solution that is hypertonic (higher solute concentration).

osmotic adjustment   The process by which a plant increases the solute concentration of its cytosol.

osmotic pressure   The hydrostatic pressure required to stop the net flow of water across a membrane due to osmosis.

osmotroph   An organism that relies on osmotrophy (uptake of small organic molecules via osmosis) as a form of nutrition.

ostia   Small openings in the heart of an arthropod, through which hemolymph re-enters the heart.

ostracoderms   An umbrella term for several classes of primitive, heavily armored fish, now extinct, that lacked a jaw.

otoliths   Granules of calcium carbonate found in the gelatinous substance that embeds hair cells in the ear.

outer bark   Protective layers of mostly dead cork cells that cover the outside of woody stems and roots.

outer ear   One of the three main compartments of the mammalian ear; consists of the external ear, or pinna, and the auditory canal.

outer segment   The highly convoluted plasma membranes found in the rods and cones of the eye.

outgroup   A species or group of species that is most closely related to an ingroup.

ovaries   In animals, the female gonads where eggs are formed.

overweight   According to current National Institutes of Health guidelines, a person having a BMI of 25 or more.

oviduct   A thin tube with undulating fimbriae (fingers) that extend out to the ovary; also called the fallopian tube.

oviparity   Development of the embryo within an egg outside the mother.

ovoviparous   The term used to describe an organism that retains eggs inside the body, where the young hatch.

ovoviviparity   Development of the embryo involving aspects of both viviparity and oviparity; eggs covered with a thin shell are produced and hatch inside the mother's body, but the offspring receive no nourishment from the mother.

ovulation   The release of the ovum from the ovary.

ovule   In a plant, a megaspore-producing megasporangium and enclosing tissues known as integuments.

ovum   *See* egg.

oxidation   A process that involves the removal of electrons; occurs during the breakdown of small organic molecules.

oxidative fiber   A skeletal muscle fiber that contains numerous mitochondria and has a high capacity for oxidative phosphorylation.

oxidative phosphorylation   A process during which NADH and $FADH_2$ are oxidized to make more ATP via the phosphorylation of ADP.

oxytocin   A hormone secreted by the posterior pituitary gland that stimulates contractions of the smooth muscles in the uterus of a pregnant mammal, which facilitates the birth process; after birth, it is important in milk secretion.

P

P generation   The parental generation in a genetic cross.

P protein   Phloem protein; the proteinaceous material produced in sieve tube elements of plant phloem as a response to wounding.

pacemaker   *See* sinoatrial (SA) node.

Pacinian corpuscles   Structures located deep beneath the surface of the skin that respond to deep pressure or vibration.

paedomorphosis   The retention of juvenile traits in an adult organism.

pair-rule gene   A type of segmentation gene; a defect in this gene may cause alternating segments or parts of segments to be deleted.

paleontologist   A scientist who studies fossils.

palisade parenchyma   Photosynthetic ground tissue in a plant that consists of closely packed, elongate cells adapted to absorb sunlight efficiently.

pancreas   An elongated gland located behind the stomach that secretes digestive enzymes and a fluid rich in bicarbonate ions.

pangenesis   An idea proposed by the ancient Greek physician Hippocrates that suggested that "seeds" produced by all parts of the body are collected and transmitted to offspring at the time of conception, and that these seeds cause offspring to resemble their parents.

parabronchi   A series of parallel air tubes that make up the lungs and are the regions of gas exchange in birds.

paracrine   Refers to a type of cellular communication in which molecules are released into the interstitial fluid and act on nearby cells.

paralogs   Homologous genes within a single species.

paraphyletic taxon   A group that contains a common ancestor and some, but not all, of its descendants.

parapodia   Fleshy, footlike structures in the polychaetes that are pushed into the substrate to provide traction during movement.

parasite   A predatory organism that feeds off another organism but does not normally kill it outright.

parasitism   An association in which one organism feeds off another, but does not normally kill it outright.

parasympathetic division   The division of the autonomic nervous system that is involved in maintaining and restoring body functions.

parathyroid hormone (PTH)   A hormone that acts on bone to stimulate the activity of cells that dissolve the mineral part of bone.

Parazoa   A subgroup of animals that are not generally thought to possess specialized tissue types or organs, although they may have

several distinct types of cells; the one phylum in this group is the Porifera (sponges).

parenchyma cell   A type of plant cell that is thin-walled and alive at maturity.

parenchyma tissue   Plant tissue that is composed of parenchyma cells.

parental strand   The original strand in DNA replication.

parental type   *See* nonrecombinant.

parthenogenesis   An asexual process in which an offspring develops from an unfertilized egg.

partial pressure   The individual pressure of each gas in the air; the sum of these pressures is known as atmospheric pressure.

particulate inheritance   The idea that the determinants of hereditary traits are transmitted intact from one generation to the next.

parturition   The birth of an organism.

passive diffusion   Diffusion that occurs through a membrane without the aid of a transport protein.

passive immunity   A type of acquired immunity that confers protection against disease through the direct transfer of antibodies from one individual to another.

passive transport   The diffusion of a solute across a membrane in a process that is energetically favorable and does not require an input of energy.

paternal inheritance   A pattern in which only the male gamete contributes particular genes to the offspring.

pathogen   A microorganism that causes disease symptoms in its host.

pattern formation   The process that gives rise to a plant or animal with a particular body structure.

pedal glands   Glands in the foot of a rotifer that secrete a sticky substance that aids in attachment to the substrate.

pedicel   A narrow, waistlike point of attachment in a spider or insect body.

pedicellariae   The spines and jawlike pincers that cover the skeleton of an echinoderm and deter the settling of animals such as barnacles.

pedigree analysis   An examination of the inheritance of human traits in families.

pedipalps   In spiders, a pair of appendages that have various sensory, predatory, or reproductive functions.

peduncle   The tip of a flower stalk.

pelagic zone   The open ocean, where water depth averages 4,000 m and nutrient concentrations are typically low.

penis   A male external accessory sex organ found in many animals that is involved in copulation.

pentadactyl limb   A limb ending in five digits.

PEP carboxylase   An enzyme in C_4 plants that adds CO_2 to phosphoenolpyruvate (PEP) to produce the four-carbon compound oxaloacetate.

pepsin   An active enzyme in the stomach that begins the digestion of protein.

peptide bond   The covalent bond that links together amino acids in a polypeptide.

peptidoglycan   A polymer composed of protein and carbohydrate that is an important component of the cell walls of most bacteria.

peptidyl site (P site)   One of the three sites for tRNA binding to the ribosome; the others are the aminoacyl site (A site) and the exit site (E site) The polypeptide is usually in the P site.

peptidyl transfer reaction   As a peptide bond is formed, the polypeptide is removed from the tRNA in the P site and transferred to the amino acid at the A site.

per capita growth rate   The per capita birth rate minus the per capita death rate; the rate that determines how populations grow over any time period.

perception   An awareness of the sensations that are experienced.

perennial   A plant that lives for more than two years, often producing seeds each year after it reaches reproductive maturity.

perfect flower   A flower that has both stamens and carpels.

perianth   The term that refers to flower petals and sepals collectively.

pericarp   The wall of a plant's fruit.

pericycle   A cylinder of plant tissue having cell division (meristematic) capacity that encloses the root vascular tissue.

peripheral membrane protein   A protein that is noncovalently bound to regions of integral membrane proteins that project out from the membrane, or they are noncovalently bound to the polar head groups of phospholipids.

peripheral zone   The area of a plant that contains dividing cells that will eventually differentiate into plant structures.

peristalsis   Rhythmic, spontaneous waves of muscle contraction that begin near the mouth in the esophagus and move toward the stomach.

peritubular capillaries   Capillaries near the junction of the cortex and medulla in the nephron of the kidney.

permafrost   A layer of permanently frozen soil.

peroxisome   A relatively small organelle found in all eukaryotic cells that catalyzes detoxifying reactions.

petal   A flower organ that usually serves to attract insects or other animals for pollen transport.

phage   *See* bacteriophage.

phagocyte   A cell capable of phagocytosis; phagocytes provide nonspecific defense against pathogens that enter the body.

phagocytic vacuole   A vacuole that is formed by the process of phagocytosis.

phagocytosis   A form of endocytosis that involves the formation of a membrane vesicle called a phagosome, or phagocytic vacuole, that engulfs a large particle such as a bacterium.

phagotroph   An organism that specializes in phagotrophy (particle feeding) by means of phagocytosis as a form of nutrition.

phagotrophy   The use of phagocytosis as a feeding mechanism.

pharyngeal slit   A filter-feeding device in primitive chordates.

pharynx The area at the back of the throat where inhaled air from the mouth and nose converges.

phenolic Refers to compounds that contain a cyclic ring of carbon with three double bonds, known as a benzene ring, that is covalently linked to a single hydroxyl group.

phenotype The characteristics of an organism that are the result of the expression of its genes.

phenotypic plasticity The phenomenon in which individual members of the same plant species that experience different environmental conditions may display considerable variation in structure or behavior.

pheromone A powerful chemical attractant used to manipulate the behavior of others.

phloem A specialized conducting tissue at the center of a plant's stem.

phloem loading The process of conveying sugars to sieve-tube elements for long-distance transport.

phoresy A form of commensalism in which individuals of one species use individuals of a second species for transportation.

phosphodiester linkage Refers to a double linkage (two phosphoester bonds) that holds together adjacent nucleotides in DNA and RNA strands.

phosphodiesterase An enzyme that breaks down cAMP into AMP.

phospholipid A class of lipids that are similar in structure to triglycerides, but the third hydroxyl group of glycerol is linked to a phosphate group instead of a fatty acid. They are a key component of biological membranes.

phospholipid bilayer The basic framework of the cellular membrane, consisting of two layers of lipids.

photoautotroph An organism that uses the energy from light to make organic molecules from inorganic sources.

photoheterotroph An organism that is able to use light energy to generate ATP, but which must take in organic compounds from the environment.

photon A massless particle traveling in a wavelike pattern and moving at the speed of light.

photoperiodism A plant's ability to measure and respond to night length, and indirectly, day length, as a way of detecting seasonal change.

photoreceptor An electromagnetic receptor that responds to visible light energy.

photorespiration The metabolic process occurring in C_3 plants that results when the enzyme rubisco combines with oxygen instead of carbon dioxide and produces only one molecule of PGA instead of two PGA, thereby reducing photosynthetic efficiency.

photosynthesis The process whereby light energy is captured by plant, algal, or cyanobacterial cells and used to synthesize organic molecules from CO_2 and H_2O (or H_2S).

photosystem I (PSI) One of two distinct complexes of proteins and pigment molecules in the thylakoid membrane that absorbs light.

photosystem II (PSII) The complex of proteins and pigment molecules in the thylakoid membrane that generates oxygen from water.

phototropin The main blue-light sensor involved in phototropism.

phototropism The tendency of a plant to grow toward a light source.

phragmoplast A plant organelle involved in the construction of cell plate that produces an intervening cell wall between two dividing plant cells.

phyla The subdivisions of a kingdom.

phylogenetic species concept The members of a single species are identified by having a unique combination of characteristics.

phylogenetic tree A diagram that describes a phylogeny; such a tree is a hypothesis of the evolutionary relationships among various species, based on the information available to and gathered by systematists.

physical mutagen A physical agent, such as UV light, that causes mutations.

physiological ecology A subdiscipline of organismal ecology that investigates how organisms are physiologically adapted to their environment and how the environment impacts the distribution of species.

physiology The study of the functions of cells and body parts of living organisms.

phytochrome A red- and far-red-light receptor.

phytoplankton Microscopic algae and cyanobacteria that float or actively move through water.

phytoremediation The process of removing harmful metals from soils by growing hyperaccumulator plants on metal-contaminated soils, then harvesting and burning the plants to ashes for disposal and/or metal recovery.

pigment A molecule that can absorb light energy.

pili (singular, **pilus**) Threadlike surface appendages that allow prokaryotes to attach to each other during mating, or to move across surfaces.

piloting A mechanism of migration in which an animal moves from one familiar landmark to the next.

pinocytosis A form of endocytosis that involves the formation of membrane vesicles from the plasma membrane as a way for cells to internalize the extracellular fluid.

pistil A flower structure that may consist of a single carpel or multiple, fused carpels, and is differentiated into stigma, style, and ovary.

pit A small cavity in a plant cell wall where secondary wall materials such as lignin are absent.

pitch The tone of a sound wave.

pituitary dwarfism A condition in which a person's anterior pituitary gland fails to make adequate amounts of GH during childhood, so the concentrations of GH and IGF-1 in the blood will be lower than normal and growth is stunted.

pituitary giant A person who has a tumor of the GH-secreting cells of the anterior pituitary gland and thus produces excess GH during childhood and adulthood; the person can grow very tall.

pituitary gland A multilobed endocrine gland sitting directly below the hypothalamus.

placenta A structure through which humans and other eutherian mammals retain and nourish their young within the uterus via the transfer of nutrients and gases.

placental transfer tissue In plants, a nutritive tissue that aids in the transfer of nutrients from maternal parent to embryo.

plant A member of the kingdom Plantae.

plant tissue culture A laboratory process to produce thousands of identical plants having the same desirable characteristics.

Plantae A eukaryotic kingdom of the domain Eukarya; includes multicellular organisms having cellulose-rich cell walls and plastids, and which are adapted in many ways to terrestrial habitats (or if aquatic, derived from terrestrial ancestors).

plaque 1. A deposit of lipids, fibrous tissue, and smooth muscle cells that may develop in blood vessels. 2. A bacterial biofilm that may form on the surfaces of teeth.

plasma The fluid part of blood.

plasma cell A cell that synthesizes and secretes antibodies.

plasma membrane The biomembrane that separates the internal contents of a cell from its external environment.

plasmid A small circular piece of DNA found naturally in many strains of bacteria and also occasionally in eukaryotic cells.

plasmodesma (plural, **plasmodesmata**) A membrane-lined, ER-containing channel that connects the cytoplasm of adjacent plant cells.

plasmogamy The fusion of the cytoplasm between two gametes.

plasmolysis The shrinkage of algal or plant cytoplasm that occurs when water leaves the cell by osmosis, with the result that the plasma membrane no longer presses on the cell wall.

plastid A general name given to organelles found in plant and algal cells, which are bound by two membranes and contain DNA and large amounts of chlorophyll (chloroplasts), carotenoids (chromoplasts), or starch (amyloplasts).

platelets Cell fragments in the blood of mammals that play a crucial role in the formation of blood clots.

pleiotropy The phenomenon in which a mutation in a single gene can have multiple effects on an individual's phenotype.

pleural sac A double layer of sheathlike moist membranes that encases each lung.

pluripotent Refers to the ability of embryonic stem cells to differentiate into almost every cell type of the body.

point mutation A mutation that affects only a single base pair within the DNA or that involves the addition or deletion of a single base pair to a DNA sequence.

polar cell The highest latitude cell in the three-cell circulation of wind in each hemisphere.

polar covalent bond The bond that forms when two atoms with different electronegativities form a covalent bond and the shared electrons are closer to the atom of higher electronegativity rather than the atom of lower electronegativity. The distribution of electrons around the atoms creates a polarity, or difference in electric charge, across the molecule.

polar molecule A molecule containing significant numbers of polar bonds.

polar transport The process whereby auxin primarily flows downward in shoots and into roots.

polarized 1. In cell biology, refers to cells that have different sides, such as the apical and basal sides of epithelial cells. 2. In neuroscience, refers to the electrical gradient across the membrane. A neuron with a large electrical gradient across its plasma membrane is said to be highly polarized.

pole A structure of the spindle apparatus defined by each centrosome.

pole cells The primordial germ cells that are the first cells to form at the posterior end of the embryo in certain species such as *Drosophila*.

pollen Tiny male gametophytes enclosed by sporopollenin-containing microspore walls.

pollen coat A layer of material that covers the sporopollenin-rich pollen wall.

pollen grain An immature male gametophyte.

pollen tube A mature male gametophyte consisting of a germinated pollen grain with a long, thin pollen tube that carries haploid sperm cells.

pollen wall A tough, sporopollenin wall at the surface of a pollen grain.

pollination The process in which pollen grains are transported to an angiosperm flower or a gymnosperm cone by means of wind or animal pollinators.

pollination syndromes The pattern of coevolved traits between particular types of flowers and specific pollinators.

pollinator An animal that carries pollen between angiosperm flowers (or cones of gymnosperms).

poly A tail A string of adenine nucleotides at the 3′ end of most mature mRNAs in eukaryotes.

polyandry A form of mating in which one female mates with several males.

polycistronic mRNA An mRNA that contains the coding sequences for two or more structural genes.

polycythemia A condition of increased hemoglobin due to increased hematocrit.

polygenic A trait in which several or many genes contribute to the outcome of the trait.

polygyny A form of mating in which one male mates with more than one female in a single breeding season, but females mate only with one male.

polyketides A group of secondary metabolites that are produced by bacteria, fungi, plants, insects, dinoflagellates, mollusks, and sponges.

polymer A large molecule formed by linking together many smaller molecules called monomers.

polymerase chain reaction (PCR) A technique to make many copies of a gene in vitro; primers are used that flank the region of DNA to be amplified.

polymorphic gene A gene that commonly exists as two or more alleles in a population.

polymorphism The phenomenon that many traits or genes may display variation within a population.

polypeptide A linear sequence of amino acids; the term denotes structure.

polyphagous The term used to define parasites that feed on many host species.

polyphyletic taxon A group that consists of members of several evolutionary lines and does not include the most recent common ancestor of the included lineages.

polyploid An organism that has three or more sets of chromosomes.

polysaccharide Many monosaccharides linked together to form long polymers.

pons The part of the hindbrain, along with the cerebellum, responsible for monitoring and coordinating body movements.

population A group of individuals of the same species that can interbreed with one another.

population ecology The study of how populations grow and what promotes and limits growth.

population genetics The study of genes and genotypes in a population.

portal vein A vein that not only collects blood from capillaries—like all veins—but also forms another set of capillaries, as opposed to returning the blood directly to the heart like other veins.

positional information Molecules that are provided to a cell that allow it to determine its position relative to other cells.

positive control Transcriptional regulation by activator proteins.

positive feedback loop The acceleration of a process, leading to what is sometimes called an explosive system.

positive pressure filling The method by which amphibians ventilate their lungs—the animals gulp air and force it under pressure into the lungs, as if inflating a balloon.

postabsorptive state One of two alternating phases in the utilization of nutrients; occurs when the gastrointestinal tract is empty of nutrients and the body's own stores must supply energy. The other phase is the absorptive state.

posterior Refers to the rear (tail-end) of an animal.

postsynaptic cell The cell that receives the electrical or chemical signal sent from a neuron.

post-translational covalent modification A process of changing the structure of a protein, usually by covalently attaching functional groups.

post-translational sorting The uptake of proteins into the nucleus, mitochondria, chloroplasts, and peroxisomes that occurs after the protein is completely made (that is, completely translated).

postzygotic isolating mechanism A mechanism that prevents interbreeding by blocking the development of a viable and fertile individual after fertilization has taken place.

potential energy The energy that a substance possesses due to its structure or location.

power stroke The process in which an energized myosin cross-bridge binds to actin and triggers the movement of the bound cross-bridge.

prebiotic soup The medium formed by the slow accumulation of molecules in the early oceans over a long period of time prior to the existence of life.

predation An interaction where the action of the predator results in the death of the prey.

predator An animal that kills live prey.

pregnancy The time during which a developing embryo and fetus grows within the uterus of the mother. The period of pregnancy is also known as gestation.

preinitiation complex The structure of the completed assembly of GTFs and RNA polymerase II at the TATA box prior to transcription of eukaryotic structural genes.

pre-mRNA The RNA transcript prior to any processing.

presynaptic cell The cell that sends the electrical or chemical signal from a neuron to another cell.

prezygotic isolating mechanism A mechanism that stops interbreeding by preventing the formation of a zygote.

primary active transport A type of transport that involves pumps that directly use energy and generate a solute gradient.

primary cell wall In plants, a cell wall that is synthesized first between the two newly made daughter cells.

primary consumer An organism that obtains its food by eating primary producers; also called an herbivore.

primary electron acceptor The molecule to which a high-energy electron from an excited pigment molecule such as P680* is transferred.

primary endosymbiosis The process by which a eukaryotic host cell acquires prokaryotic endosymbionts or plastids or mitochondria derived from a prokaryotic endosymbiont.

primary immune response The response to an initial exposure to an antigen.

primary meristem A meristematic tissue that increases plant length and produces new organs.

primary metabolism The synthesis and breakdown of molecules and macromolecules that are found in all forms of life and are essential for cell structure and function.

primary oocytes   In animals, the first stage of producing female gametes by meiosis.

primary plastid   A plastid that arose from a prokaryote as the result of endosymbiosis.

primary producer   An autotroph, which typically harvests light energy from the sun; located at the base of the food chain.

primary spermatocytes   In animals, the first stage of producing sperm by meiosis.

primary structure   The linear sequence of amino acids of a polypeptide. One of four levels of protein structure.

primary succession   Succession on newly exposed sites that were not previously occupied by soil and vegetation.

primary tissue   Plant tissue generated as the result of primary growth at apical meristems.

primary vascular tissue   Plant tissue composed of primary xylem and phloem.

primer   A short segment of RNA, typically 10 to 12 nucleotides in length, that is needed to begin DNA replication.

principle of parsimony   The preferred hypothesis is the one that is the simplest.

principle of species individuality   A view of the nature of a community; each species is distributed according to its physiological needs and population dynamics; most communities intergrade continuously and competition does not create distinct vegetational zones.

probability   The chance that an event will have a particular outcome.

proboscis   The coiled tongue of a butterfly or moth, which can be uncoiled, enabling it to drink nectar from flowers.

procambium   A type of primary plant tissue meristem that produces vascular tissue.

producer   An organism that synthesizes the organic compounds used by other organisms for food.

product   During a chemical reaction, the reactants are converted to products.

product rule   The probability that two or more independent events will occur is equal to the product of their individual probabilities.

production efficiency   The percentage of energy assimilated by an organism that becomes incorporated into new biomass.

productivity hypothesis   The proposal that greater production by plants results in greater overall species richness.

progesterone   A hormone secreted by the female ovaries that plays a key role in pregnancy.

progymnosperms   An extinct group of plants having wood but not seeds, that evolved before the gymnosperms.

prokaryote   One of the two categories into which all forms of life can be placed, based on cell structure; the other is eukaryote. Prokaryotic cells lack a nucleus having an envelope with pores; includes bacteria and archaea.

prometaphase   A phase of mitosis during which the mitotic spindle is completely formed.

promoter   The site in the DNA where transcription begins.

proofreading   The ability of DNA polymerase to identify a mismatched nucleotide and remove it from the daughter strand.

prophage   Refers to the DNA of a phage that has become integrated into the bacterial chromosome.

prophase   A phase of mitosis during which the chromosomes condense and the nuclear membrane fragments.

proplastid   Unspecialized structures that form plastids.

prosimian   A member of a class of primates that includes the smaller species such as bush babies, lemurs, pottos, and tarsiers.

prostate gland   A structure that secretes a thin fluid that protects sperm once they are deposited within the female reproductive tract.

prosthetic group   Small molecules that are permanently attached to the surface of an enzyme and aid in catalysis.

protease   An enzyme that cuts proteins into smaller polypeptides.

proteasome   A molecular machine that is the primary pathway for protein degradation in archaea and eukaryotic cells.

protein   A functional unit composed of one or more polypeptides. Each polypeptide is composed of a linear sequence of amino acids.

protein kinase   An enzyme that transfers phosphate groups from ATP to a protein.

protein phosphatase   An enzyme responsible for removing phosphate groups from proteins.

protein-protein interactions   Specific interactions between proteins that can carry out cellular processes that occur as a series of steps or build larger structures that provide organization to the cell.

proteoglycan   A glycosaminoglycan in the extracellular matrix linked to a core protein.

proteolysis   A processing event within the cell in which enzymes called proteases cut proteins into smaller polypeptides.

proteome   All of the types and relative amounts of proteins that are made in a particular cell at a particular time and under specific conditions. The proteome largely determines a cell's structure and function.

proteomics   Techniques used to identify and study groups of proteins.

protist   A eukaryotic organism that is not a member of the animal, plant, or fungal kingdoms; lives in moist habitats, and is typically microscopic in size.

Protista   In traditional classification systems, a eukaryotic kingdom of the domain Eukarya.

protobiont   The term used to describe the first nonliving structures that evolved into living cells.

protoderm   A type of primary plant tissue meristem that generates the outermost dermal tissue.

proton   A positively charged particle found in the nucleus of the atom. The number of protons in an atom is called the atomic number and defines each type of element.

protonephridia   The simplest filtration mechanism for cleansing the blood; used in flatworms.

proton-motive force   *See* H^+ electrochemical gradient.

proto-oncogene   A normal gene that, if mutated, can become an oncogene.

protostome   An animal that exhibits spiral determinate cleavage, and where the blastopore becomes the mouth; includes mollusks, annelid worms, and arthropods.

provirus   Refers to viral DNA that has become incorporated into a eukaryotic chromosome.

protozoa   A term commonly used to describe diverse heterotrophic protists.

proventriculus   The glandular portion of the stomach of a bird.

proximal convoluted tubule   The segment of the tubule of the nephron in the kidney that drains Bowman's capsule.

proximate cause   A specific genetic and physiological mechanism of behavior.

pseudocoelom   A coelom that is not completely lined by tissue derived from mesoderm.

pseudocoelomate   An animal with a pseudocoelom.

pteridophytes   A phylum of of vascular plants having euphylls, but not seeds; Pteridophyta.

pulmocutaneous circulation   The routing of blood from the heart through different vessels to the gas exchange organs (lungs and skin) of frogs and some other amphibians.

pulmonary circulation   The pumping of blood from the right side of the heart to the lungs to release carbon dioxide and pick up oxygen from the atmosphere.

pulse-chase experiment   A procedure in which researchers administer a pulse of radioactively labeled materials to cells so that they make radioactive products. This is followed by the addition of nonlabeled materials called a chase.

pump   A transporter that directly couples its conformational changes to an energy source, such as ATP hydrolysis.

punctuated equilibrium   A concept that suggests that the tempo of evolution is more sporadic than gradual. Species rapidly evolve into new species followed by long periods of equilibrium with little evolutionary change.

Punnett square   A common method for predicting the outcome of simple genetic crosses.

pupa   The organism resulting after the larval stages in insects that gives rise to an adult organism via metamorphosis.

pupil   A small opening in the eye of a vertebrate that transmits different patterns of light emitted from images in the animal's field of view.

purine   The bases adenine (A) and guanine (G), with double (fused) rings of nitrogen and carbon atoms.

pyramid of biomass   A measure of efficiency in which the organisms at each trophic level are weighed.

pyramid of numbers   An expression of trophic-level transfer efficiency, in which the number of individuals decreases at each trophic level, with a huge number of individuals at the base and fewer individuals at the top.

pyramid of production   A measure of efficiency in which rates of production are shown rather than standing crop; the laws of thermodynamics ensure that the highest amounts of free energy are found at the lowest trophic levels.

pyrimidine   The bases cytosine (C), thymine (T), and uracil (U) with a single ring.

Q

quantitative trait   *See* continuous trait.

quaternary structure   The association of two or more polypeptides to form a protein. One of four levels of protein structure.

quorum sensing   A mechanism by which prokaryotic cells are able communicate when they reach a critical population size.

R

radial cleavage   A mechanism of development in which the cleavage planes are either parallel or perpendicular to the vertical axis of the egg.

radial loop domain   A loop of chromatin, often 25,000 to 200,000 base pairs in size, that is anchored to the nuclear matrix.

radial pattern   A characteristic of the body pattern of plants.

radial symmetry   1. In plants, an architectural feature of plants in which plant embryos display a cylindrical shape, which is retained in the stems and roots of seedlings and mature plants. In addition, new leaves or flower parts are produced in circular whorls, or spirals, around shoot tips. 2. In animals, an architectural feature in which the body can be divided into symmetrical halves by many different longitudinal planes along a central axis.

Radiata   Radially symmetric animals, which means they can be divided equally by a longitudinal plane passing through the central axis; includes cnidarians and ctenophores.

radiation   The emission of electromagnetic waves by the surfaces of objects.

radicle   An embryonic root, which extends from the plant hypocotyl.

radioisotope   An isotope found in nature that is inherently unstable and does not exist for long periods of time. Such isotopes emit energy called radiation in the form of subatomic particles.

radioisotope dating   A common way to estimate the age of a fossil by analyzing the elemental isotopes within the accompanying rock.

radula   A unique, protrusible, tonguelike organ in a mollusk that has many teeth and is used to eat plants, scrape food particles off rocks, or bore into shells of other species and tear flesh.

rain shadow   An area where precipitation is noticeably less, such as on the side of the mountain sheltered from air movement.

ram ventilation   A mechanism used by fish to ventilate their gills; fish swim or face upstream with their mouths open, allowing water to enter into their buccal cavity and from there across their gills.

random   A pattern of dispersion within a population, in which individuals do not appear to be specially positioned relative to anyone else.

random genetic drift   A change in allele frequencies due to random sampling error.

random sampling error   The deviation between the observed and expected outcomes.

rate-limiting step   The slowest step in a pathway.

ray-finned fish   The Actinopterygii, which includes all bony fish except the coelacanths and lungfish.

reabsorption   In the production of urine, the process in which useful solutes in the filtrate are recaptured and transported back into the body fluids of an animal.

reactant   A substance that participates in a chemical reaction and becomes changed by that reaction.

reaction mechanism   In the case of the Na^+/K^+-ATPase, a molecular roadmap of the steps that direct the pumping of ions across the plasma membrane.

reading frame   Refers to the idea that codons are read from the start codon in groups of three bases each.

receptacle   The enlarged region at the tip of a flower peduncle to which flower parts are attached.

receptor   A cellular protein that recognizes a signaling molecule.

receptor-mediated endocytosis   A common form of endocytosis in which a receptor is specific for a given cargo.

receptor potential   The membrane potential in a sensory receptor cell.

recessive   A term that describes a trait that is masked by the presence of a dominant trait in a heterozygote.

reciprocal translocation   The process in which two different types of chromosomes exchange pieces, thereby producing two abnormal chromosomes carrying translocations.

recombinant   An offspring that has a different combination of traits from the parental generation.

recombinant DNA   Any DNA molecule that has been manipulated so that it contains DNA from two or more sources.

recombinant DNA technology   The use of laboratory techniques to isolate and manipulate fragments of DNA.

recombinant vector   A vector containing a piece of chromosomal DNA.

recombination frequency   The frequency of crossing over between two genes.

red blood cell   A cell that serves the critical function of transporting oxygen throughout the body; also known as an erythrocyte.

redox reaction   A type of reaction in which the electron that is removed during the oxidation of an atom or molecule must be transferred to another atom or molecule, which becomes reduced; short for a reduction-oxidation reaction.

reduction   A process that involves the addition of electrons to an atom or molecule.

reductionism   An approach that involves reducing complex systems to simpler components as a way to understand how the system works. In biology, reductionists study the parts of a cell or organism as individual units.

redundancy hypothesis   An extension of the rivet hypothesis. In this model, also called the passenger hypothesis, most species are more like passengers on a plane—they take up space but do not add to the airworthiness. The species are said to be redundant because they could simply be eliminated or replaced by others with no loss in function.

reflex arc   A simple circuit that allows an organism to respond rapidly to inputs from sensory neurons and consists of only a few neurons.

refractory   The term used to describe a cell that is unresponsive to another stimulus.

regeneration   A form of asexual reproduction in which a complete organism forms from small fragments of its body.

regulatory gene   A gene whose function is to regulate the expression of other genes.

regulatory sequence   In the regulation of transcription, a sequence that functions as a site for genetic regulatory proteins. Regulatory sequences control whether a gene is turned on or off.

regulatory transcription factor   A protein that binds to DNA in the vicinity of a promoter and affects the rate of transcription of one or more nearby genes.

relative abundance   The frequency of occurrence of species in a community.

relative water content (RWC)   The property often used to gauge the water content of a plant organ or entire plant; RWC integrates the water potential of all cells within an organ or plant and is thus a measure of relative turgidity.

release factor   A protein that recognizes the three stop codons in the termination stage of translation and promotes the termination of translation.

renal corpuscle   A filtering component in the nephron of the kidney.

repetitive sequence   Short DNA sequences that are present in many copies in a genome.

replica plating   A technique in which a replica of bacterial colonies is transferred to a new petri plate.

replication   1. The performing of experiments several or many times. 2. The copying of DNA strands.

replication fork   The area where two DNA strands have separated and new strands are being synthesized.

repressible operon   In this type of operon, a small effector molecule inhibits transcription.

repressor A transcription factor that binds to DNA and inhibits transcription.

reproduction The process by which organisms produce offspring.

reproductive isolating mechanisms The mechanisms that prevent interbreeding between different species.

reproductive isolation Mechanisms that prevent one species from successfully interbreeding with other species.

resistance (R) The tendency of blood vessels to slow down the flow of blood through their lumens.

resistance gene (R gene) A plant gene that has evolved as part of a defense system in response to pathogen attack.

resolution In microscopy, the ability to observe two adjacent objects as distinct from one another; a measure of the clarity of an image.

resonance energy transfer The process by which the energy (not the electron itself) can be transferred to adjacent pigment molecules.

resource competition Competition in which organisms compete indirectly through via the consumption of a limited resource, with each obtaining as much as it can.

resource partitioning The differentiation of niches, both in space and time, that enables similar species to coexist in a community.

respiratory centers Several regions of the brainstem in vertebrates that initiate expansion of the lungs.

respiratory chain *See* electron transport chain.

respiratory distress syndrome of the newborn The situation in which a human baby is born prematurely, before sufficient surfactant is produced, and many alveoli quickly collapse.

respiratory pigment A large protein that contains one or more metal atoms that bind to oxygen.

respiratory system All components of the body that contribute to the exchange of gas between the external environment and the blood; in mammals, includes the nose, mouth, airways, lungs, and muscles and connective tissues that encase these structures within the thoracic (chest) cavity.

response elements DNA sequences that are recognized by regulatory transcription factors and regulate the expression of genes.

response regulator In bacteria, a protein that interacts with a sensor kinase and regulates the expression of many genes.

resting potential The difference in charges across the plasma membrane in an unstimulated cell.

restoration ecology The full or partial repair or replacement of biological habitats and/or their populations that have been damaged.

restriction enzyme An enzyme that recognizes particular DNA sequences and cleaves the DNA backbone at two sites.

restriction point A point in the cell cycle in which a cell has become committed to divide.

restriction sites The base sequences recognized by restriction enzymes.

retina A sheetlike layer of photoreceptors at the back of the eye.

retinal A derivative of vitamin A that is capable of absorbing light energy.

retroelement A type of transposable element that moves via an RNA intermediate.

retrotransposon *See* retroelement.

retrovirus An RNA virus that utilizes reverse transcription to produce viral DNA that can be integrated into the host cell genome.

reverse transcriptase A viral enzyme that catalyzes the synthesis of viral DNA starting with viral RNA as a template.

rhizobia The collective term for proteobacteria involved in the nitrogen-fixation symbioses with plants that are important in nature and to agriculture.

rhizomorphs Fungal mycelia that have the shape of roots.

rhodopsin The visual pigment in rods.

ribonucleic acid (RNA) One of two classes of nucleic acids; the other is deoxyribonucleic acid (DNA). RNA consists of a single strand of nucleotides.

ribose A five-carbon sugar found in RNA.

ribosomal initiation factor In the initiation stage of translation, a protein that facilitates the interactions between mRNA, the first tRNA, and the ribosomal subunits.

ribosomal RNA (rRNA) An RNA that forms part of ribosomes, which provide the site where translation occurs.

ribosome A structure composed of proteins and rRNA that provides the site where protein synthesis occurs.

ribozyme A biological catalyst that is an RNA molecule.

right-left axis In bilateral animals, one of the three axes along which the adult body pattern is organized; the others are the dorsoventral axis and the anteroposterior axis.

ring canal A structure in the water vascular system of echinoderms.

rivet hypothesis An alternative to the diversity-stability hypothesis. In this model, species are like the rivets on an airplane, with each species playing a small but critical role in keeping the plane (the ecosystem) airborne. The loss of a rivet weakens the plane and causes it to lose a little airworthiness. The loss of a few rivets could probably be tolerated, while the loss of more rivets would prove critical to the airplane's function.

RNA *See* ribonucleic acid.

RNA-induced silencing complex (RISC) The complex that mediates RNA interference via microRNAs.

RNA interference (RNAi) Refers to a type of mRNA silencing; miRNA interferes with the proper expression of an mRNA.

RNA polymerase The enzyme that synthesizes strands of RNA during gene transcription.

RNA processing A step in gene expression between transcription and translation; the RNA transcript, termed pre-mRNA, is modified in ways that make it a functionally active mRNA.

RNA world A hypothetical period on primitive Earth when both the information needed for life and the enzymatic activity of living cells were contained solely in RNA molecules.

RNase An enzyme that digests RNA.

rods Photoreceptors found in the vertebrate eye; they are very sensitive to low-intensity light but do not readily discriminate different colors. Rods are utilized mostly at night, and they send signals to the brain that generate a black-and-white visual image.

"Roid" rage Extreme aggressive behavior brought about by androgen administration.

root A plant organ that provides anchorage in the soil and also fosters efficient uptake of water and minerals.

root apical meristem (RAM) The region where new root tissues of plants are produced.

root cap A protective covering on the root tip that is produced by the root apical meristem of a plant.

root cortex A region of ground parenchyma located between the epidermis and vascular tissue of mature plant roots.

root hair A long, thin root epidermal cell that functions to absorb water and minerals, usually from soil.

root meristem The collection of cells at the root tip that generate all of the tissues of a plant root.

root pressure A process whereby plants are able to refill embolized vessels.

root system The collection of roots and root branches produced by root apical meristems.

rough endoplasmic reticulum (rough ER) The part of the ER that is studded with ribosomes; this region plays a key role in the initial synthesis and sorting of proteins that are destined for the ER, Golgi apparatus, lysosomes, vacuoles, plasma membrane, or outside of the cell.

r-selected species A type of life history strategy, where species have a high rate of per capita population growth but poor competitive ability.

rubisco The enzyme that that catalyzes the first step in the Calvin cycle in which CO_2 is incorporated into an organic molecule.

S

S The synthesis phase of the cell cycle.

saltatory conduction A type of conduction in which sodium ions move into the cell and the charge moves rapidly through the cytosol to the next node, where the action potential continues.

sarcoma A tumor of connective tissue such as bone or cartilage.

sarcomere One compete unit of the repeating pattern of thick and thin filaments within a myofibril.

sarcoplasmic reticulum A cellular organelle that provides a muscle fiber's source of the cytosolic calcium involved in muscle contraction.

satiety A feeling of fullness.

satiety signal   A response to eating that removes the sensation of hunger and sets the time period before hunger returns again.

saturated fatty acid   A fatty acid in which all the carbons are linked by single covalent bonds.

scaffold   An area in metaphase chromosomes formed from proteins that holds the radial loops in place.

scanning electron microscopy (SEM)   A type of microscopy that utilizes an electron beam to produce an image of the three-dimensional surface of biological samples.

scavenger   An animal that eats the remains of dead animals.

schizocoelous   In protostomes, a pattern of development in which a solid mass of mesoderm cells splits to form the cavity that becomes the coelom.

Schwann cells   The glial cells that form myelin on axons that travel outside the brain and spinal cord.

science   In biology, the observation, identification, experimental investigation, and theoretical explanation of natural phenomena.

scientific method   A series of steps to test the validity of a hypothesis. The experimentation often involves a comparison between control and experimental samples.

sclera   The white of the vertebrate eye; a strong outer sheet that in the front is continuous with a thin, clear layer known as the cornea.

sclereid   Star- or stone-shaped plant cells having tough, lignified cell walls.

sclerenchyma tissue   Rigid plant tissue composed of tough-walled fibers and sclereids.

scurvy   A potentially fatal vitamin C deficiency in humans characterized by weakness, bleeding gums, and tooth loss.

seaweed   A multicellular protist that occurs in marine habitats and may display a relatively large and complex body.

second law of thermodynamics   The second law states that the transfer of energy or the transformation of energy from one form to another increases the entropy, or degree of disorder of a system.

second messengers   Small molecules or ions that relay signals inside the cell.

secondary active transport   A type of transport that involves the utilization of a pre-existing gradient to drive the active transport of another solute.

secondary cell wall   A plant cell wall that is synthesized and deposited between the plasma membrane and the primary cell wall after a plant cell matures and has stopped increasing in size.

secondary consumer   An organism that eats primary consumers; also called a carnivore.

secondary endosymbiosis   A process that occurs when a eukaryotic host cell acquires a eukaryotic endosymbiont having a primary plastid.

secondary immune response   An immediate and heightened production of additional specific antibodies against a particular antigen that elicited a primary immune response.

secondary metabolism   Involves the synthesis of chemicals that are not essential for cell structure and growth and are usually not required for cell survival, but are advantageous to the organism.

secondary metabolite   Molecules that are produced by secondary metabolism.

secondary oocyte   In animals, the large egg cell that is the result of meiosis I.

secondary phloem   The inner bark of a woody plant.

secondary plastid   A plastid that has originated by the endosymbiotic incorporation into a host eukaryote of a eukaryotic cell having a primary plastid.

secondary production   The measure of production of heterotrophs and decomposers.

secondary spermatocytes   The haploid cells produced in the primary spermatocyte by meiosis I.

secondary structure   The bending or twisting of proteins into helices or β sheets. One of four levels of protein structure.

secondary succession   Succession on a site that has already supported life but that has undergone a disturbance.

secretion   In the production of urine, the process in which some solutes are actively transported from the interstitial fluid surrounding the epithelial cells of the tubules of an excretory organ, into the tubule lumens.

secretory pathway   A pathway for the movement of larger substances, such as proteins and carbohydrates, out of the cell.

secretory vesicle   A membrane vesicle carrying different types of materials that later fuses with the cell's plasma membrane to release the contents extracellularly.

seed   A reproductive structure produced by flowering plants and other seed plants, usually as the result of sexual reproduction.

seed coat   A hard and tough covering that develops from the ovule's integument and protects the plant's embryo.

seed plant   The informal name for gymnosperms and angiosperms.

segment   1. The portion of the eye cell that contains the cell nucleus and other cytoplasmic organelles. 2. A body part in animals that is repeated many times in a row along an anteroposterior axis.

segmentation gene   A gene that controls the segmentation pattern of an animal embryo.

segment-polarity gene   A type of segmentation gene; a mutation in this gene causes portions of segments to be missing either an anterior or a posterior region and for adjacent regions to become mirror images of each other.

segregate   To separate, as in chromosomes during mitosis.

selectable marker   A gene whose presence can allow organisms (such as bacteria) to grow under a certain set of conditions. For example, an antibiotic resistance gene is a selectable marker that allows bacteria to grow in the presence of the antibiotic.

selective breeding   Programs and procedures designed to modify traits in domesticated species.

selectively permeable   The phenomenon that membranes allow the passage of certain ions or molecules but not others.

self-fertilization   Fertilization that involves the union of a female gamete and male gamete from the same individual.

self-incompatibility (SI)   Rejection of pollen that is genetically too similar to the pistil of a plant.

selfish DNA hypothesis   The hypothesis that transposable elements exist because they have the characteristics that allow them to insert themselves into the host cell DNA but do not provide any advantage.

self-pollination   The process in which pollen from the anthers of a flower is transferred to the stigma of the same flower, or between flowers of the same plant.

self-splicing   The phenomenon that RNA itself can catalyze the removal of its own intron(s).

semelparity   A reproductive pattern in which organisms produce all of their offspring in a single reproductive event.

semen   A mixture containing fluid and sperm that is released during ejaculation.

semicircular canals   Structures of the ear that can detect circular motions of the head.

semiconservative mechanism   In this model for DNA replication, the double-stranded DNA is half conserved following the replication process such that the new double-stranded DNA contains one parental strand and one daughter strand.

semifluid   A quality of motion within biomembranes; considered two-dimensional, which means that movement occurs within the plane of the membrane.

semilunar valves   A one-way valve into the systemic and pulmonary arteries through which blood is pumped.

seminal vesicles   Paired accessory glands that secrete fructose, the main nutrient for sperm, into the urethra to mix with the sperm.

seminiferous tubule   A tightly packed structure in the testis, where spermatogenesis takes place.

sense   A system that consists of sensory cells that respond to a specific type of chemical or physical stimulus and send signals to the central nervous system, where the signals are received and interpreted.

senescent   Cells that have doubled many times and have reached a point where they have lost the capacity to divide any further.

sensor kinase   An enzyme-linked receptor that recognizes a signal found in the environment and also has the ability to hydrolyze ATP and phosphorylate itself.

sensory neuron   A neuron that detects or senses information from the outside world, such as light, sound, touch, and heat; sensory neurons also detect internal body conditions such as blood pressure and body temperature.

sensory receptor A specialized cell whose function it is to receive sensory inputs.

sensory transduction The process by which incoming stimuli are converted into neural signals.

sepal A flower organ that often functions to protect the unopened flower bud.

septum (plural, **septa**) A cross wall; examples include the cross walls that divide the hyphae of most fungi into many small cells, and the structure that separates the old and new chambers of a nautilus.

sere Each phase of succession in a community; also called a seral stage.

setae Chitinous bristles in the integument of many invertebrates.

sex chromosomes A distinctive pair of chromosomes that are different in males and females.

sex linked Refers to genes that are found on one sex chromosome but not on the other.

sex pili Hair-like structures made by bacterial F^+ cells that bind specifically to F^- cells.

sex-influenced inheritance The phenomenon in which an allele is dominant in one sex but recessive in the other.

sexual reproduction A process that requires a fertilization event in which two gametes unite to create a cell called a zygote.

sexual selection A type of natural selection that is directed at certain traits of sexually reproducing species that make it more likely for individuals to find or choose a mate and/or engage in successful mating.

Shannon diversity index (H_S) A means of measuring the diversity of a community; $H_S = -\Sigma p_i \ln p_i$.

shared derived character A trait that is shared by a group of organisms but not by a distant common ancestor.

shared primitive character A trait shared with a distant ancestor.

shattering The process by which ears of wild grain crops break apart and disperse seeds.

shell A tough, protective covering that is impermeable to water and prevents the embryo from drying out.

shivering thermogenesis Rapid muscle contractions in an animal, without any locomotion, in order to raise body temperature.

shoot apical meristem (SAM) The region of rapidly dividing plant cells at plant shoot apices.

shoot meristem The tissue that produces all aerial parts of the plant, which include the stem as well as lateral structures such as leaves and flowers.

shoot system The collection of plant organs produced by shoot apical meristems.

short-day plant A plant that flowers only when the night length is longer than a defined period. Such night lengths occur in late summer, fall, or winter, when days are short.

short tandem repeat sequences (STRs) Sequences found in multiple sites in the genome of humans and other species, and they vary in length among different individuals.

shotgun DNA sequencing A strategy for sequencing an entire genome by randomly sequencing many different DNA fragments.

sickle-cell anemia A disease due to a genetic mutation in a hemoglobin gene in which sickle-shaped red blood cells are less able to move smoothly through capillaries and can block blood flow, resulting in severe pain and cell death of the surrounding tissue.

sieve elements Thin-walled living cells that are arranged end to end in a plant to form pipelines.

sieve plate The perforated end wall of a mature sieve-tube element.

sieve plate pore One of many perforations in a plant's sieve plate.

sieve-tube elements A component of the phloem tissues of flowering plants; these structures are arranged end to end to form transport pipes.

sigma factor A protein that plays a key role in bacterial promoter recognition and recruits RNA polymerase to the promoter.

sign stimulus A factor that initiates a fixed-action pattern of behavior.

signal recognition particle (SRP) A protein/RNA complex that recognizes the ER signal sequence and pauses translation; then, SRP binds to a receptor in the ER membrane, which docks the ribosome over a translocation channel.

signal transduction pathway A group of proteins that convert an initial signal to a different signal inside the cell.

signal An agent that can influence the properties of cells.

silencer A response element that prevents transcription of a given gene when its expression is not needed.

silent mutation A gene mutation that does not alter the amino acid sequence of the polypeptide, even though the nucleotide sequence has changed.

simple Mendelian inheritance The inheritance pattern of traits affected by a single gene that is found in two variants, one of which is completely dominant over the other.

simple translocation A single piece of chromosome that is attached to another chromosome.

single-factor cross A cross in which an experimenter follows the variants of only one trait.

single-strand binding protein A protein that binds to both of the single strands of parental DNA and prevents them from re-forming a double helix.

sinoatrial (SA) node A collection of modified cardiac cells that spontaneously and rhythmically generates action potentials that spread across the entire atria; also known as the pacemaker of the heart.

sister chromatids The two duplicated chromatids that are still joined to each other after DNA replication.

skeletal muscle A type of muscle tissue that is attached to bones in vertebrates and to the exoskeleton of invertebrates.

skeleton A structure or structures that serve one or more functions related to support, protection, and locomotion.

sliding filament mechanism The way in which a muscle fiber shortens during muscle contraction.

SLOSS debate In conservation biology, the debate over whether it is preferable to protect one single, large reserve or several smaller ones (SLOSS stands for *single large or several small*).

slow block to polyspermy Events that produce barriers to more sperm penetrating an already fertilized egg.

slow fiber A skeletal muscle fiber with a low rate of myosin ATP hydrolysis.

slow-oxidative fiber A skeletal muscle fiber that has a low rate of myosin ATP hydrolysis but has the ability to make large amounts of ATP; used for prolonged, regular activity.

small effector molecule With regard to transcription, refers to a molecule that exerts its effects by binding to a regulatory transcription factor and causing a conformational change in the protein.

small intestine A tube that leads from the stomach to the large intestine where nearly all digestion of food and absorption of food nutrients and water occur.

smooth endoplasmic reticulum (smooth ER) The part of the ER that is not studded with ribosomes. This region is continuous with the rough ER and functions in diverse metabolic processes such as detoxification, carbohydrate metabolism, accumulation of calcium ions, and synthesis and modification of lipids.

smooth muscle A type of muscle tissue that surrounds hollow tubes and cavities inside the body's organs, such that their contraction can propel the contents of those organs.

soil horizon Layers of soil, ranging from topsoil to bedrock.

solute A substance dissolved in a liquid.

solute potential (S) The osmotic potential; an element in the water potential equation.

solution A liquid that contains one or more dissolved solutes.

solvent The liquid in which a solute is dissolved.

soma *See* cell body.

somatic cell The type of cell that constitutes all cells of the body excluding the germ-line cells. Examples include skin cells and muscle cells.

somatic embryogenesis The production of plant embryos from body (somatic) cells.

somatic nervous system The division of the peripheral nervous system that senses the external environmental conditions and controls skeletal muscles.

somites Blocklike structures of mesoderm that are formed during neurulation.

soredia Small clumps of hyphae surrounding a few algal cells that can disperse in wind currents; an asexual reproductive structure produced by lichens.

sorting signal A short amino acid sequence in a protein's structure that directs the protein to its correct location.

source pool The pool of species on the mainland that is available to colonize an island.

Southern blotting A method in which a labeled probe, which is a strand of DNA from a specific gene, is used to identify that gene in a mixture of many chromosomal DNA fragments.

speciation The formation of new species.

species A group of related organisms that share a distinctive form in nature.

species-area effect The relationship between the amount of available area and the number of species present.

species concepts Different approaches for distinguishing species, including the phylogenetic, biological, evolutionary, and ecological species concepts.

species interactions A part of the study of population ecology which focuses on interactions such as predation, competition, parasitism, mutualism and commensalism.

species richness The numbers of species in a community.

specific (acquired) immunity An immunity defense that develops only after the body is exposed to foreign substances.

Spemann's organizer An extremely important morphogenetic field in the early gastrula; the organizer secretes morphogens responsible for inducing the formation of a new embryonic axis.

sperm Refers to a "male" gamete that is generally smaller than the female gamete (egg); the male gamete is often motile or in many species of plants transported to the egg via a pollen tube.

sperm storage A method of synchronizing the production of offspring with favorable environmental conditions; female animals store and nourish sperm in their reproductive tract for long periods of time.

spermatids In animals, the haploid cells produced by the secondary spermatocytes by meiosis II; these cells eventually differentiate into sperm cells.

spermatogenesis The formation of sperm.

spermatogonia In animals, diploid germ cells that give rise to the male gametes, the spermatozoa.

spermatophytes All of the living and fossil seed plant phyla.

spinal cord In vertebrates and simpler chordates, the structure that connects the brain to all areas of the body and constitutes the central nervous system.

spinal nerve A nerve that connects the peripheral nervous system and the spinal cord.

spinneret A spider's abdominal silk gland; also found in the mouths of caterpillars.

spiracle A pore to the trachea that is found in the bodies of insects.

spiral cleavage A mechanism of development in which the planes of cell cleavage are oblique to the axis of the embryo.

spirilli Rigid, spiral-shaped prokaryotic cells.

spirochaetes Flexible, spiral-shaped prokaryotic cells.

spliceosome A complex of several subunits known as snRNPs that removes introns from eukaryotic pre-mRNA.

splicing The process whereby introns are removed from RNA and the remaining exons are connected to each other.

spongin A tough protein that lends skeletal support to a sponge.

spongocoel A central cavity in the body of a sponge.

spongy parenchyma Photosynthetic tissue of the plant leaf mesophyll that contains cells separated by abundant air spaces.

spontaneous mutation A mutation resulting from abnormalities in biological processes.

sporangia Structures that produce and disperse the spores of plants, fungi, or protists.

spore Single-celled reproductive structure that is dispersed into the environment and is able to grow into new fungal hyphae, plant gametophytes, or protists if they find suitable habitats.

sporic life cycle *See* alternation of generations.

sporophyte The diploid generation of plants or multicellular protists that have a sporic life cycle; this generation produces haploid spores by the process of meiosis.

sporopollenin The tough material that composes much of the walls of plant spores and helps to prevent cellular damage during transport in air.

stabilizing selection A pattern of natural selection that favors the survival of individuals with intermediate phenotypes.

stamen A flower structure that makes the male gametophyte, pollen.

standard metabolic rate (SMR) A method for measuring the BMR of ectotherms at a standard temperature for each species—one that approximates the average temperature that a species normally encounters.

standing crop The total biomass in an ecosystem at any one point in time.

starch A polysaccharide produced by the cells of plants and some algal protists.

start codon A three-base sequence—usually AUG—that specifies the first amino acid in a polypeptide.

statocyst An organ of equilibrium found in many animal species.

statoliths 1. Tiny granules of sand or other dense objects that aid in equilibrium in many vertebrates. 2. In plants, a starch-heavy plastid that allows both roots and shoots to detect gravity.

stem A plant organ that produces buds, leaves, branches and reproductive structures.

stem cell A cell that divides and supplies the cells that construct the bodies of all animals and plants.

stereoisomers Isomers with identical bonding relationships, but the spatial positioning of the atoms differs in the two isomers.

sternum The breastbone.

steroid A lipid molecule with a chemical structure containing four interconnected rings of carbon atoms.

steroid receptor A transcription factor that recognizes a steroid hormone and usually functions as a transcriptional activator.

steroid A lipid containing four interconnected rings of carbon atoms; function as hormones in animals and plants.

sticky ends Single-stranded ends of DNA fragments that will hydrogen-bond to each other due to their complementary sequences.

stigma 1. A topmost portion of the pistil, which receives and recognizes pollen of the appropriate species or genotype. 2. In many protists, a red-colored cellular structure.

stomach A saclike organ that most likely evolved as a means of storing food; it partially digests some of the macromolecules in food and regulates the rate at which the contents empty into the small intestine.

stomata Surface pores on plant surfaces that can be closed to retain water or open to allow the entry of CO_2 needed for photosynthesis and the exit of oxygen and water vapor.

stop codon One of three three-base sequences—UAA, UAG, and UGA—that signals the end of translation.

strain Within a given species, a lineage that has genetic differences compared to another strain.

strand A structure of DNA (or RNA) formed by the covalent linkage of nucleotides in a linear manner.

stretch receptor A type of mechanoreceptor found widely in organs and muscle tissues that can be distended.

striated muscle Skeletal and cardiac muscle with a series of light and dark bands perpendicular to the muscle's long axis.

stroke The condition that occurs when blood flow to part of the brain is disrupted.

stroke volume (SV) The amount of blood ejected with each beat, or stroke, of the heart.

stroma The fluid-filled region of the chloroplast between the thylakoid membrane and the inner membrane.

stromatolite A layered calcium carbonate structure in an aquatic environment produced by cyanobacteria.

strong acid An acid that completely ionizes in solution.

structural gene Refers to most genes, which produce an mRNA molecule that contains the information to specify a polypeptide with a particular amino acid sequence.

structural genomics Genomic methods aimed at the direct analysis of the DNA itself.

structural isomers   Isomers that contain the same atoms but in different bonding relationships.

style   The elongate pistil structure through which the pollen tube of a flower grows.

stylet   A sharp, piercing organ in the mouth of nematodes and some insects.

submetacentric   A chromosome in which the centromere is off center.

subsidence zones   Areas of high pressure that are the sites of the world's tropical deserts, because the subsiding air is relatively dry, having released all of its moisture over the equator.

substrate-level phosphorylation   A method of synthesizing ATP that occurs when an enzyme directly transfers a phosphate from one molecule to a different molecule.

substrates   The reactant molecules and/or ions that bind to an enzyme at the active site and participate in a chemical reaction.

succession   The gradual and continuous change in species composition and community structure over time.

sugar sink   The plant tissues or organs in which more sugar is consumed than is produced by photosynthesis.

sugar source   The plant tissues or organs that produce more sugar than they consume in respiration.

sum rule   The probability that one of two or more mutually exclusive outcomes will occur is the sum of the probabilities of the possible outcomes.

supercoiling   A method of compacting chromosomes; the phenomenon of forming additional coils around the long, thin DNA molecule.

supergroup   A proposed way to organize eukaryotes into monophyletic groups.

surface area/volume (SA/V) ratio   The ratio between a structure's surface area and the volume in which the structure is contained.

surfactant   A mixture of proteins and amphipathic lipids produced in certain alveolar cells that acts to reduce surface tension in lungs.

survivorship curve   A graphical plot of the numbers of surviving individuals at each age.

suspension feeder   An aquatic animal that sifts water, filtering out the organic matter and expelling the rest.

suspensor   A short chain of cells at the base of an early angiosperm embryo that provides anchorage and nutrients.

swim bladder   A gas-filled, balloon-like structure that helps a fish to remain buoyant in the water even when it is completely stationary.

swimmeret   An abdominal appendage in a crustacean that provides movement.

symbiosis   An intimate association between two or more organisms.

symbiotic   Describes a relationship in which two or more different species live in direct contact with each other.

sympathetic division   The division of the autonomic system that is responsible for rapidly activating systems that provide immediate energy to the body in response to danger or stress.

sympatric   The term used to describe species occurring in the same geographic area.

sympatric speciation   A form of speciation that occurs when members of a species that initially occupy the same habitat within the same range diverge into two or more different species.

symplast   All of a plant's protoplasts (the cell contents without the cell walls) and plasmodesmata.

symplastic transport   The movement of a substance from the cytosol of one cell to the cytosol of an adjacent cell via membrane-lined channels called plasmodesmata.

symplesiomorphy   *See* shared primitive character.

symporter   A type of transporter that binds two or more ions or molecules and transports them in the same direction.

synapomorphy   *See* shared derived character.

synapse   A junction where a nerve terminal meets a target neuron, muscle cell, or gland and can communicate with other cells.

synapsis   The process of forming a bivalent.

synaptic cleft   The extracellular space between two neurons.

synaptic integration   The summation of EPSPs generated at one time, which can bring the membrane potential to the threshold potential at the axon hillock for action potential firing.

synaptic plasticity   A change in synapses that occurs as a result of learning.

synaptic signaling   A specialized form of paracrine signaling that occurs in the nervous system of animals.

synergids   In the female gametophyte of a flowering plant, the two cells adjacent to the egg cell that help to import nutrients from maternal sporophyte tissues.

systematics   The study of biological diversity and evolutionary relationships among organisms, both extinct and modern.

systemic acquired resistance (SAR)   A whole-plant defensive response to pathogenic microorganisms.

systemic circulation   The pumping of blood from the left side of the heart to the body to drop off oxygen and nutrients and pick up carbon dioxide and wastes. The blood then returns to the right side of the heart.

systemic hypertension   An arterial blood pressure above normal, which in humans ranges from systolic/diastolic pressures of about 90/60 to 120/80 mmHg; often called hypertension or high blood pressure.

systems biology   A field of study in which researchers study living organisms in terms of their underlying network structure—groups of structural and functional connections—rather than their individual molecular components.

systole   The second phase of the cardiac cycle, in which the ventricles contract and eject the blood through the open semilunar valves.

T

T cell   A type of lymphocyte that directly kills infected, mutated, or transplanted cells.

tagmata   The fusion of body segments into functional units.

tapetum lucidum   A reflective layer of tissue located beneath the photoreceptors at the back of the eye of certain animals.

taproot system   The root system of eudicots, which has one main root with many branch roots.

***Taq* polymerase**   A heat-stable form of DNA polymerase; one of several reagents required for synthesis of DNA via polymerase chain reaction (PCR).

TATA box   One of three features found in most promoters; the others are the transcriptional start site and response elements.

taxis   A directed type of response either toward or away from a stimulus.

taxon   A group of species that are evolutionarily related to each other. In taxonomy, each species is placed into several taxons that form a hierarchy from large (domain) to small (genus).

taxonomy   The field of biology that is concerned with the theory, practice, and rules of classifying living and extinct organisms and viruses.

telocentric   A chromosome in which the centromere is at the end.

telomerase   An enzyme that catalyzes the replication of the telomere.

telomere   A region at the ends of eukaryotic chromosomes where a specialized form of DNA replication occurs.

telophase   The phase of mitosis during which the chromosomes decondense and the nuclear membrane re-forms.

temperate phage   A bacteriophage that may spend some of its time in the lysogenic cycle.

template   The DNA strand that is used as a template for RNA synthesis or DNA replication.

termination codon   *See* stop codon.

termination stage   The final stage of transcription or translation in which the process ends.

terminator   A sequence that specifies the end of transcription.

territory   A fixed area in which an individual or group excludes other members of its own species, and sometimes other species, by aggressive behavior or territory marking.

tertiary consumer   An organism that feeds on secondary consumers.

tertiary endosymbiosis   The acquisition by eukaryotic protist host cells of plastids from cells that possess secondary plastids.

tertiary plastid   A plastid acquired by the incorporation into a host cell of an endosymbiont having a secondary plastid.

tertiary structure   The three-dimensional shape of a single polypeptide. One of four levels of protein structure.

testcross   A cross in which an individual with a dominant phenotype is mated with a homozygous recessive individual.

testes   The male gonads of certain animals, where sperm are produced.

testosterone   The primary androgen in vertebrates.

tetrad   *See* bivalent.

tetraploid   An organism or cell that has four sets of chromosomes.

tetrapod   A vertebrate animal having four legs or leglike appendages.

thalamus   In vertebrates, a gland that plays a major role in relaying sensory information to appropriate parts of the cerebrum and, in turn, sending outputs from the cerebrum to other parts of the brain. It receives input from all sensory systems and is involved in the perception of pain and the degree of mental arousal in the cortex.

theory   In biology, a broad explanation of some aspect of the natural world that is substantiated by a large body of evidence. Biological theories incorporate observations, hypothesis testing, and the laws of other disciplines such as chemistry and physics. A theory makes valid predictions.

thermodynamics   The study of energy interconversions.

thermoreceptor   A sensory receptor that responds to cold and heat.

theropods   A group of bipedal saurischian dinosaurs.

thick filament   A skeletal muscle structure composed almost entirely of the motor protein myosin.

thigmotropism   Touch responses in plants.

thin filament   A skeletal muscle structure that contains the cytoskeletal protein actin, as well as two other proteins—troponin and tropomyosin—that play important roles in regulating contraction.

30-nm fiber   Nucleosome units organized into a more compact structure that is 30 nm in diameter.

thoracic breathing   Breathing in which coordinated contractions of muscles expand the rib cage, creating a negative pressure to suck air in and then forcing it out later; found in amniotes.

threatened species   Those species that are likely to become endangered in the future.

threshold concentration   The concentration above which a morphogen will exert its effects but below which it is ineffective.

thrifty genes   Genes that boosted our ancestors' ability to store fat from each feast in order to sustain them through the next famine.

thrombocytes   Intact cells in the blood of vertebrates other than mammals that play a crucial role in the formation of blood clots.

thylakoid   A flattened, plate-like membranous region found in cyanobacterial cells and the chloroplasts of photosynthetic protists and plants; the location of the light reactions of photosynthesis.

thylakoid lumen   The fluid-filled compartment within the thylakoid.

thylakoid membrane   A membrane within the chloroplast that forms many flattened, fluid-filled tubules that enclose a single, convoluted compartment. It is the site where the light-dependent reactions of photosynthesis occurs.

thymine (T)   A pyrimidine base found in DNA.

thymine dimer   One type of pyrimidine dimer; a site where two adjacent thymine bases become covalently cross-linked to each other.

thyroglobulin   A protein found in the colloid of the thyroid gland.

thyroxine   A thyroid hormone that contains iodine and helps regulate metabolic rate.

Ti plasmid   Tumor-inducing plasmid found in *Agrobacterium tumefaciens*; it is used as a cloning vector to transfer genes into plant cells.

tidal ventilation   A type of breathing in which the lungs are inflated with air, and then the chest muscles and diaphragm relax and recoil back to their original positions as an animal exhales. During exhalation, air leaves via the same route that it entered during inhalation, and no new oxygen is delivered to the airways at that time.

tidal volume   The volume of air that is normally breathed in and out at rest.

tight junction   A type of junction that forms a tight seal between adjacent epithelial cells and thereby prevents molecules from leaking between cells; also called an occluding junction.

tissue   The association of many cells of the same type, for example, muscle tissue.

tolerance   A mechanism for succession in which any species can start the succession, but the eventual climax community is reached in a somewhat orderly fashion; early species neither facilitate nor inhibit subsequent colonists.

tonoplast   The membrane of the central vacuole in a plant or algal cell.

torpor   The strategy in the smallest endotherms of lowering internal body temperature to just a few degrees above that of the environment in order to conserve energy.

torus   The nonporous, flexible central region of a conifer pit that functions like a valve.

total fertility rate   The average number of live births a female has during her lifetime.

total peripheral resistance (TPR)   The sum of all the resistance in all the arterioles.

totipotent   The ability of a fertilized egg to produce all of the cell types in the adult organism; also the ability of unspecialized plant cells to regenerate an adult plant.

trachea   1. A sturdy tube arising from the spiracles of an insect's body involved in respiration. 2. The name of the tube leading to the lungs of air-breathing vertebrates.

tracheal system   In insects, a series of finely branched air tubes called tracheae that lead into the body from pores called spiracles.

tracheid   A type of dead, lignified plant cell that conducts water, along with dissolved minerals and certain organic compounds.

tracheophytes   A term used to describe the vascular plants.

tract   A parallel bundle of myelinated axons.

traffic signal   *See* sorting signal.

trait   *See* character.

transcription   The use of a gene sequence to make a copy of RNA: transcription occurs in three stages called initiation, elongation, and termination.

transcription factor   A protein that influences the ability of RNA polymerase to transcribe genes.

transcriptional start site   The site in a promoter where transcription begins.

transduction   A type of genetic transfer between bacteria in which a virus infects a bacterial cell and then transfers some of that cell's DNA to another bacterium.

trans-effect   In both prokaryotes and eukaryotes, a form of genetic regulation that can occur even though two DNA segments are not physically adjacent. The action of the *lac* repressor on the *lac* operon is a trans-effect.

transfer RNA (tRNA)   An RNA that carries amino acids and is used to translate mRNA into polypeptides.

transformation   A type of genetic transfer between bacteria in which a segment of DNA from the environment is taken up by a competent cell and incorporated into the bacterial chromosome.

transgenic   The term used to describe an organism that carries genes that were introduced using molecular techniques such as gene cloning.

transition state   In a chemical reaction, a state in which the original bonds have stretched to their limit; once this state is reached, the reaction can proceed to the formation of products.

transitional form   An organism that provides a link between earlier and later forms in evolution.

translation   The process of synthesizing a specific polypeptide on a ribosome; a nucleotide sequence in mRNA is used to make an amino acid sequence of a polypeptide. The process of translation occurs in three stages: initiation, elongation, and termination.

translocation   A phenomenon that occurs when one segment of a chromosome becomes attached to a different chromosome.

transmembrane gradient   The phenomenon that the concentration of a solute is higher on one side of a membrane than on the other.

transmembrane protein   A protein that has one or more regions that are physically embedded in the hydrophobic region of the cell's phospholipid bilayer.

transmembrane segment   A region of a membrane protein that is a stretch of hydrophobic amino acids that spans or traverses the membrane from one leaflet to the other.

transmission electron microscopy (TEM)   A type of microscopy in which a beam of electrons is differentially transmitted through a biological sample to form an image on a photographic plate or screen.

transpiration   The evaporative loss of water from plant surfaces into sun-heated air.

transport protein   Proteins embedded within the phospholipid bilayer that allow plasma membranes to be selectively permeable by providing a passageway for the movement of some but not all substances across the membrane.

transporter A membrane protein that binds a solute and undergoes a conformational change to allow the movement of the solute across a membrane.

transposable element (TE) A segment of DNA that can move from one site to another.

transposase An enzyme that facilitates transposition.

transposition The process in which a short segment of DNA moves within a cell from its original site to a new site in the genome.

transposon A transposable element that moves via DNA that is removed from one site and inserted into a new site.

transverse tubules (T-tubules) Invaginations of the plasma membrane that open to the extracellular fluid and conduct action potentials from the outer surface of a muscle fiber to the myofibrils.

trichome A projection from the epidermal tissue of a plant that offers protection from excessive light, ultraviolet radiation, extreme air temperature, or attack by herbivores.

trichromatic color vision The ability to distinguish blue, green, and red colors.

triiodothyronine A thyroid hormone that contains iodine and helps regulate metabolic rate.

triplet A group of three bases that function as a codon.

triploblastic Having three distinct germ layers—endoderm, ectoderm, and mesoderm.

triploid An organism or cell that has three sets of chromosomes.

trisomic An aneuploid organism that has one too many chromosomes.

trochophore larva A distinct larval stage of many invertebrate phyla.

trophic level Each feeding level in a food chain.

trophic-level transfer efficiency The amount of energy at one trophic level that is acquired by the trophic level above and incorporated into biomass.

tropomyosin A protein that plays an important role in regulating muscle contraction.

troponin A protein that plays an important role in regulating muscle contraction.

***trp* operon** An operon of *E. coli* that encodes enzymes required to make the amino acid tryptophan, a building block of cellular proteins.

true-breeding line A strain that continues to exhibit the same trait after several generations of self-fertilization or inbreeding.

trypsin A protease involved in the breakdown of proteins in the small intestine.

t-snare A protein in a target membrane that recognizes a v-snare in a membrane vesicle.

tube cell In a plant, the cell that stores proteins and lipids that may be used during later stages of male gametophyte development, after pollen has germinated. This cell forms the pollen tube that after pollination grows to reach the female gametophyte.

tube feet Echinoderm structures that function in movement, gas exchange, and feeding.

tumor An overgrowth of cells that serves no useful purpose.

tumor-suppressor gene A gene that when normal (that is, not mutant) encodes a protein that prevents cancer; however, when a mutation eliminates its function, cancer may occur.

tunic A nonliving structure that encloses a tunicate, made of a protein and a cellulose-like material called tunicin.

turgid The term used to describe a plant cell whose cytosol is so full of water that the plasma membrane presses right up against the cell wall; as a result, turgid cells are firm or swollen.

turgor pressure *See* osmotic pressure.

two-component regulatory system A signaling system found in bacteria and plants composed of an enzyme-linked receptor called a sensor kinase and a response regulator, which is usually a protein that regulates the expression of many genes.

two-dimensional (2D) gel electrophoresis A technique to separate many different proteins that utilizes an isoelectric-focusing tube gel in the first dimension, and an SDS-slab gel in the second dimension.

two-factor cross Crosses in which the inheritance of two different traits are followed.

type 1 diabetes mellitus (T1DM) A disease in which the pancreas does not produce sufficient insulin; as a result, extracellular glucose cannot cross plasma membranes, and so glucose accumulates to very high concentrations in the blood.

type 2 diabetes mellitus (T2DM) A disease in which the pancreas functions normally, but the cells of the body lose much of their ability to respond to insulin.

U

ubiquitin A small protein in eukaryotic cells that directs unwanted proteins to a proteasome by its covalent attachment.

ultimate cause The reason a particular behavior evolved, in terms of its effect on reproductive success.

umbrella species A species whose habitat requirements are so large that protecting them would protect many other species existing in the same habitat.

unconditioned response A response that is already elicited by an unconditioned stimulus.

unconditioned stimulus A factor that originally elicits a response.

uniform A pattern of dispersion within a population, in which individuals maintain a certain minimum distance between themselves to produce an evenly spaced distribution.

uniporter A type of transporter that binds a single molecule or ion and transports it across the membrane.

unipotent A term used to describe a stem cell found in the adult that can only produce daughter cells that differentiate into one cell type.

unsaturated The quality of a lipid when a double bond is formed.

unsaturated fatty acid A fatty acid that contains one or more $C = C$ double bonds.

upwelling In the ocean, a phenomenon that carries mineral nutrients from the bottom waters to the surface.

uracil (U) A pyrimidine base found in RNA.

urea A nitrogenous waste commonly produced in many terrestrial species.

uremia A condition characterized by the presence of nitrogenous wastes, such as urea, in the blood; typically results from kidney disease.

ureter A structure in the kidney through which urine flows from the kidney into the urinary bladder.

urethra The structure through which urine is eliminated from the body.

uric acid A nitrogenous waste commonly produced in many terrestrial species.

urinary bladder The structure that collects urine before it is eliminated.

urinary system The structures which collectively act to filter blood or hemolymph and excrete wastes while recapturing useful compounds. In humans, it includes the two kidneys, two ureters, the urinary bladder, and the urethra.

uterine cycle The cyclical changes that occur in the uterus in parallel with the ovarian cycle in a female mammal.

uterus A large, conical-shaped structure that consists of an inner lining of glandular and secretory cells called the endometrium, and a thick muscular layer called the myometrium; the uterus is specialized for carrying the developing fetus.

V

vaccination The injection into the body of small quantities of living or dead microbes, small quantities of toxins, or harmless antigenic molecules derived from a microorganism or its toxin, resulting in a primary immune response including the production of memory cells. Subsequent natural exposure to the immunizing antigen results in a response that can prevent or reduce the severity of disease.

vacuole Specialized compartments found in eukaryotic cells that function in storage, the regulation of cell volume, and degradation.

vagina A tubular, smooth muscle structure in the female into which sperm are deposited.

vaginal diaphragm A barrier method of preventing fertilization in which the device is placed in the upper part of the vagina just prior to intercourse and blocks movement of sperm to the cervix.

valence electron An electron in the outer shell that is available to combine with other atoms. Such electrons allow atoms to form chemical bonds with each other.

variable region A domain within an immunoglobulin that serves as the antigen-binding site.

vas deferens   A muscular tube through which sperm leave the epididymis.

vasa recta capillaries   Capillaries in the medulla in the nephron of the kidney.

vascular bundle   Primary plant vascular tissues that occur in a group.

vascular cambium   A meristematic tissue of plants that produces both wood and inner bark.

vascular plant   A plant that can transport water, sugar, and salts throughout the plant body via xylem and phloem tissues.

vascular tissue   Plant tissue that makes up the vascular system, which conducts materials within the plant body and also provides support.

vasectomy   A surgical procedure in men that severs the vas deferens, thereby preventing the release of sperm at ejaculation.

vasoconstriction   A decrease in blood vessel radius; an important mechanism for directing blood flow away from specific regions of the body.

vasodilation   An increase in blood vessel radius; an important mechanism for directing blood flow to specific regions of the body.

vasotocin   A peptide that is responsible for regulating salt and water balance in the blood of nonmammalian vertebrates.

vector   A type of DNA that acts as a carrier of a DNA segment that is to be cloned.

vegetal pole   In triploblast organisms, the pole of the egg where the yolk is most concentrated.

vegetative growth   The production of new tissues by the shoot apical meristem and root apical meristem during seedling development and growth of mature plants.

vein   In animals, a blood vessel that returns blood to the heart.

veliger   A free-swimming larva that has a rudimentary foot, shell, and mantle.

ventral   Refers to the lower side of an animal.

ventricle   In the heart, a chamber to pump blood out of the heart.

venule   In animals, a small, thin-walled extension of a capillary that empties into larger vessels called veins that return blood to the heart for another trip around the circulation.

vernalization   The induction of flowering by cold treatment.

vertebrae   A bony or cartilaginous column of interlocking structures that provides support and also protects the nerve cord, which lies within its tubelike structure.

vertebrate   An organism with a backbone.

vertical evolution   A process that involves genetic changes in a series of ancestors, which form a lineage.

vesicle   A small membrane-enclosed sac within a cell.

vessel   In a plant, a pipeline-like file of dead, water-conducting vessel elements.

vessel element   A type of plant cell that conducts water, along with dissolved minerals and certain organic compounds.

vestibular system   The organ of balance in vertebrates, located in the inner ear next to the cochlea.

vestigial structure   An anatomical feature that has no apparent function but resembles a structure of a presumed ancestor.

vibrios   Comma-shaped prokaryotic cells.

villi   Finger-like projections extending from the luminal surface into the lumen of the small intestine; these are specializations that aid in digestion and absorption.

viral envelope   A structure enclosing the capsid that consists of a membrane derived from the plasma membrane of the host cell and embedded with virally encoded spike glycoproteins.

viral genome   The genetic material in a virus.

viral reproductive cycle   When a virus infects a cell, the series of steps that results in the production of new viruses.

viral vector   A type of vector used in cloning experiments that is derived from a virus.

viroid   An RNA particle that infects plant cells.

virulent phage   A phage that follows only the lytic cycle.

virus   A small infectious particle that consists of nucleic acid enclosed in a protein coat.

visceral mass   In mollusks, a structure that rests atop the foot and contains the internal organs.

vitamin   An important organic nutrient that serves as a coenzyme for metabolic and biosynthetic reactions.

vitamin D   A molecule that regulates calcium levels in the blood through an action on intestinal transport of calcium ions.

vitreous humor   A thick liquid in the large posterior cavity of the vertebrate eye, which helps maintain the shape of the eye.

viviparity   The process in which the embryo develops within the mother.

viviparous   The term used to describe an organism whose eggs develop within the uterus, receiving nourishment from the mother via a placenta.

volt   A unit of measurement of the difference in charge (the electrical force) such as between the interior and exterior of the cell.

voltage-gated channel   A channel that opens and closes in response to changes in the amount of electric charge across the membrane.

v-snare   A protein incorporated into the vesicle membrane during vesicle formation that is recognized by a t-snare in a target membrane.

W

water potential   The potential energy of water.

water vascular system   A network of canals powered by hydraulic power, that is, by water pressure generated by the contraction of muscles that enables the extension and contraction of the tube feet, allowing echinoderms to move slowly.

wavelength   The distance from the peak of one sound or light wave to the next.

waxy cuticle   A protective, water-proof layer of wax and polyester present on most surfaces of vascular plant sporophytes.

weak acid   An acid that only partially ionizes in solution.

weathering   The physical and chemical breakdown of rock.

white blood cell   A cell that develops from the inner parts (the marrow) of certain bones in vertebrates; all white blood cells (known as leukocytes) perform vital functions that defend the body against infection and disease.

white matter   Brain tissue that consists of myelinated axons that are bundled together in large numbers to form tracts.

whorls   In a flower, four concentric rings of sepals, petals, stamens, and carpels.

wild-type allele   One or more prevalent alleles in a population.

wood   A secondary plant tissue composed of numerous pipelike arrays of dead, empty, water-conducting cells whose walls are strengthened by an exceptionally tough secondary metabolite known as lignin.

X

X inactivation   The phenomenon in which one X chromosome in the somatic cells of female mammals is inactivated, meaning that its genes are not expressed.

X inactivation center (Xic)   A short region on the X chromosome known to play a critical role in X inactivation.

xenoestrogen   A synthetic compound that exerts estrogen-like actions or, in some cases, inhibits the actions of the body's own estrogen; the consequences are dramatic on fertility and development of embryos and fetuses.

X-linked gene   A gene found on the X chromosome but not on the Y.

X-linked inheritance   The pattern displayed by pairs of dominant and recessive alleles located on X chromosomes.

X-ray crystallography   A technique in which researchers must purify a molecule such as a protein or protein complex and expose it to conditions that cause the proteins molecules to associate with each other in an ordered array. In other words, the proteins form a crystal. When a crystal is exposed to X-rays, the resulting pattern can be analyzed mathematically to determine the three-dimensional structure of the crystal's components.

xylem   A plant vascular tissue that conducts water, minerals, and organic compounds.

xylem loading   The process by which root xylem parenchyma cells transport ions and water across their membranes into the xylem apoplast, which includes the vessel elements and tracheids.

Y

yeast   A fungus that can occur as a single cell and that reproduces by budding.

yolk sac An extraembryonic membrane in the amniotic egg that encloses a stockpile of nutrients, in the form of yolk, for the developing embryo.

Z

Z line A network of proteins that anchor thin filaments in a sarcomere.

Z scheme According to this scheme, an electron absorbs light energy twice, and it loses some of that energy as it flows along the electron transport chain in the thylakoid membrane. The energy diagram of this process occurs in a zigzag pattern.

zero population growth The situation in which no changes in population size occur.

zoecium A nonliving case that houses a bryozoan.

zone of elongation The area above the root apical meristem of a plant where cells extend by water uptake, thereby dramatically increasing root length.

zone of maturation The area above the zone of elongation in a plant where root cell differentiation and tissue specialization occur.

zooplankton Aquatic organisms including minute animals consisting of some worms, copepods, tiny jellyfish, and the small larvae of invertebrates and fish that graze on the phytoplankton.

zygomycete A type of fungus that produces distinctive, large zygospores as the result of sexual reproduction.

zygospore A dark-pigmented, thick-walled spore that matures within the zygosporangium of zygomycete fungi.

zygote A diploid cell formed by the fusion of two gametes that divides and develops into an embryo, and eventually into an adult organism.

zygotic life cycle In this type of life cycle, haploid cells transform into gametes.

zooplankton Open-ocean organisms including minute animals consisting of some worms, copepods, tiny jellyfish, and the small larvae of invertebrates and fish that graze on the phytoplankton.

PHOTO CREDITS

Contents: **UNIT I:** © Kevin Schafer/Peter Arnold, Inc.; **UNIT II:** © Visuals Unlimited; **UNIT III:** © Schultz, S. C., Shields, G. C., and T. A. Steitz. (1991). "Crystal structure of a CAP-DNA complex: the DNA is bent by 90 degrees," *Science* 253:1001, August 30, 1991. Image by Daniel Gage, University of Connecticut.; **UNIT IV:** George Bernard Photo Researchers, Inc; **UNIT V:** © Steve Bloom/ stevebloom.com; **UNIT VI:** © Gerald & Buff Corsi/Visuals Unlimited; **UNIT VII:** Kim Taylor/ naturepl.com; **UNIT VIII:** © Erwin & Peggy Bauer/ Animals Animals Enterprises

Chapter 1 Opener: © Art Wolfe/Getty Images; **1.1a:** Photo W. Wüster, Courtesy Instituto Butantan; **1.1b:** © Charlie Heidecker/Visuals Unlimited; **1.1c:** © blickwinkel/Alamy; **1.1d:** © SciMAT/Photo Researchers, Inc.; **1.1e:** © Russ Bishop; **1.1f:** © Robert Winslow/Earth Sciences; **1.2a:** © Stephen J. Krasemann/Photo Researchers, Inc.; **1.2b:** © Tom Uhlman/Visuals Unlimited; **1.2c:** © Cathlyn Melloan/Stone/Getty Images; **1.2d:** © TAPANI RASANEN/WWI/Peter Arnold, Inc.; **1.2e:** © Patti Murray/Animals Animals; **1.2f:** © Paul Hanna/Reuters/CORBIS; **1.2g:** © Tom Brakefield/CORBIS; **1.4:** © Michael L. Smith/Photo Researchers Inc.; **1.9a:** © Dr. David M. Phillips/ Visuals Unlimited; **1.9b:** © B. Boonyaratanakornkit & D.S. Clark, G. Vrdoljak/EM Lab, U of C Berkeley/Visuals Unlimited; **1.9c(Bottom Left):** © Dr. Dennis Kunkel/Visuals Unlimited; **1.9c(Bottom Right):** © Carl Schmidt-Luchs/Photo Researchers, Inc; **1.9c(Middle Left):** © Fritz Polking/Visuals Unlimited; **1.9c(Middle Right):** © Kent Foster/Photo Researchers, Inc; **1.12b:** © Fabio Colombini/Animals Animals; **1.13a:** © Richard T. Nowitz/CORBIS; **1.13b:** © Olivier DIGOIT; **1.13c:** © Ton Koene/Visuals Unlimited; **1.13d:** © AP Images; **1.13e:** © Andrew Brookes/CORBIS; **1.13e(Inset):** Alfred Pasieka/Photo Researchers, Inc.; **1.14a:** © Robert Isear/Photo Researchers, Inc; **1.14b:** © Science VU/Visuals Unlimited; **1.17:** © AP/WideWorld Photo

Chapter 2 Opener: © Kevin Schafer/Peter Arnold, Inc.; **2.6:** © SIU/VU/Mediscan.; **2.12b:** © Charles D. Winters/Photo Researchers, Inc.; **2.17(Top):** © Jeremy Burgess/Photo Researchers, Inc.; **2.19a:** © Norbert Wu/Peter Arnold, Inc.; **2.19b:** © Jerome Wexler/Visuals Unlimited.; **2.20b:** © Aaron Haupt/Photo Researchers, Inc.; **2.20d:** © Rainer Drexel/Bilderberg/Peter Arnold, Inc.; **2.20e:** © Ethel Davies/Imagestate.; **2.20f:** © Anthony Bannister; Gallo Images/CORBIS.; **2.20g:** © OSF.

Chapter 3 Opener: de Vos, A. M., Ultsch, M. and A. A. Steitz (1992). "Human growth hormone and extracellular domain of its receptor: crystal structure of the complex. Science 255:306, January 17, 1992. Image by Daniel Gage, University of Connecticut.; **3.1a:** © SPL/Photo Researchers, Inc.; **3.1b:** © Alejandro Andreatta/Phototake.; **3.4a:** © Andrey Zvoznikov/ardrea.com.; **3.4b:** © B. Murton/Southampton Oceanography Centre/Photo Researchers, Inc.; **3.7(Right):** © Brian Hagiwara/ FoodPix/Jupiterimages.; **3.11a(Left), 3.11a(Middle):** © Tom Pantages.; **3.11b:** © Felicia Martinez/ PhotoEdit, Inc.; **3.13b,c:** © Adam Jones/Photo Researchers, Inc.

Chapter 4 Opener: © Biophoto Associates/ Photo Researchers, Inc.; **Table4.1(Left):** © Thomas Deerinck/Getty Images; **Table4.1(Middle), Table 4.1(Right):** © Dr. Gopal Murti/Visuals Unlimited.; **4.2a,c,d,e:** Courtesy Dr. Paul Letourneau, Neuroscience, University of Minnesota.; **4.2b:** © Ed Reschke/Peter Arnold, Inc.; **4.2f:** © Aumas et. al., J Neurosci. 2001 Dec 15; 21(24):RC187. © 2001 Society for Neuroscience.; **4.3a:** © L. Zamboni, D.W. Fawcett/Visuals Unlimited; **4.3b:** © David M. Phillips/Photo Researchers, Inc.; **4.4b:** © Dr. Dennis Kunkel/Getty Images.; **4.6b(Left):** © Ed Reschke/ Peter Arnold, Inc.; **4.6b(Right):** © Eye of Science/ Photo Researchers, Inc.; **4.11a:** © Visuals Unlimited; **4.11b:** Courtesy of Robert A. Bloodgood, PhD, University of Virginia School of Medicine.; **4.11c:** © Dr. Dennis Kunkel/Visuals Unlimited/Getty Images.; **4.12a:** © Biophoto Associates/Photo Researchers, Inc.; **4.14(4):** Courtesy Dr. James Spudich. Figure 2c. Sheetz and Spudich (1983) Nature, 303, p. 31-35.; **4.16(Middle):** © Dr. Richard Kessel & Dr. Gene Shih/Visuals Unlimited.; **4.16(Top):** © Dr. Don W. Fawcett/ Visuals Unlimited.; **4.17:** Cremer, T. & Cremer, C. Chromosome territories, nuclear architecture and gene regulation in mamalian cells. Nature Reviews/Genetics Vol. 2, no. 4, f. 2b&d, p. 295; **4.18(Right):** © Richard Rodewald/Biological Photo Service.; **4.21a:** © E.H. Newcomb & S.E. Frederick/ Biological Photo Service.; **4.21b:** Courtesy Dr. Peter Luykx, Biology, University of Miami; **4.21c:** © Dr. David Patterson/Photo Researchers, Inc.; **4.24:** © Dr. Don W. Fawcett/Visuals Unlimited.; **4.25:** © Dr. Jeremy Burgess/Photo Researchers, Inc.; **4.26a:** © David Sieren/Visuals Unlimited; **4.26b:** Mitch Hrdlicka/Getty Images; **4.26c:** © Barry L. Runk/ Grant Heilman Photography, Inc.; **4.27:** © Dr. Donald Fawcett & Richard Wood/Visuals Unlimited.

Chapter 5 Opener: © Tom Pantages; **5.7:** © "Biochemistry: The Molecular Basis Of Life" 3rd Ed. by McKee and McKee. McGraw Hill; **5.8a:** © Warren Rosenberg/Biological Photo Service; **5.8b:** © Dr. Don W. Fawcett/Visuals Unlimited; **5.13a,b:** © James Strawser/Grant Heilman Photography, Inc.; **5.14a,b:** © Carolina Biological Supply/Visuals Unlimited; **5.15(4):** Courtesy Dr. Peter Agre. From Preston GM, Carroll TP, Guggino WP Agre P: Appearance of water channels in Xenopus oocytes expressing red cell CHIP28 protein. Science 256:385, 1992.; **5.16:** © Tom Pantages

Chapter 6 Opener: © Visuals Unlimited; **6.3a:** © Kiseleva and Donald Fawcett/Visuals Unlimited.; **6.3b(Middle):** © Eye of Science/Photo Researchers, Inc.; **6.3b(Right):** © Don W. Fawcett/Photo Researchers, Inc.; **6.4a:** © Dr. Stanley Flegler/ Visuals Unlimited; **6.11(6):** Photo from Caro and Palade (1964) Journal of Cell Biology 29, p. 479 (fig 3) The Rockefeller University Press; **6.16b:** © T. Kanaseki and Donald Fawcett/Visuals Unlimited

Chapter 7 Opener: © Claude Cortier/Photo Researchers, Inc.; **7.1a:** © Bob Daemmrich/The Image Works; **7.20(5):** Reprinted by permission from Macmillan Publishers Ltd: [Nature] (386, Noji et al, 299-303), copyright (1997).; **7.21:** © Tom Pantages; **7.22a:** © Bill Aron/Photo Edit, Inc.; **7.22b:** © Jeff Greenberg/The Image Works; **7.23a:** © Chris Hellier/SPL/Photo Researchers, Inc.; **7.23b:** © Michael P. Gadomski/Photo Researchers, Inc; **7.24:** © Naturfoto Honal/CORBIS; **7.25:** © James Randklev/Photographer's Choice/Getty Images; **7.26:** © SciMAT/Photo Researchers, Inc.

Chapter 8 Opener: © Digital Vision/ PunchStock.; **8.1(Bottom):** © Visuals Unlimited; **8.1(Middle):** © J. Michael Eichelberger/Visuals Unlimited; **8.1(Top):** © Norman Owen Tomalin/ Bruce Coleman, Inc.; **8.7b:** © Reproduction of Fig 1A from Ferreira, K.N., Iverson, T.M., Maghlaoui, K., Barber, J. and Iwata, S. (2004) Architecture of the photosynthetic oxygen evolving center. Science 303, 1831-1838 with permission.; **8.14(6):** © Reprinted from Chemistry 1942-1962, 1964, M. Calvin, Nobel Lectures, Elsevier Publishing Company, Copyright (1964).; **8.15a:** © David Noton Photography/Alamy; **8.15b:** © David Sieren/Visuals Unlimited; **8.17(Left):** © Walter H. Hodge/Peter Arnold, Inc.; **8.17(Right):** © John Foxx/Alamy

Chapter 9 Opener: © Steve Gschmeissner/ Photo Researchers, Inc.; **9.13(Left), (Right):** Courtesy of Brian J. Bacskai, from Bacskai et al., Science 260:222-226, 1993, reprinted with permission.; **9.19a:** Don W. Fawcett/Photo Researchers, Inc.

Chapter 10 Opener: © Stephen Alvarez/Getty Images; **10.6:** © Dr. Dennis Kunkel/Visuals Unlimited; **10.11:** © Dr. Daniel Friend; **10.12(Right):** © Daniel Goodenough; **10.14:** Lee W. Wilcox; **10.15(Top):** © E.H. Newcomb & W.P. Wergin/ Biological Photo Service; **10.20a:** © Carolina

Chapter 25 Opener: Frederic B. Siskind; **25.1a(Left):** © Mark Smith/Photo Researchers, Inc; **25.1a(Right):** Pascal Goetgheluck/ardea.com; **25.1b(Left):** © Gary Meszaros/Visuals Unlimited; **25.1b(Right):** R. Andrew Odum/Peter Arnold, Inc.; **25.3(Left):** C. Allan Morgan/Peter Arnold, Inc.; **25.3(Right):** © Bryan E. Reynolds; **25.4a:** Ron Austing/Photo Researchers, Inc; **25.4b:** Rod Planck/Photo Researchers, Inc; **25.5(Bottom):** Stephen L. Saks/Photo Researchers, Inc; **25.5(Top Left):** © Mark Boulton/Photo Researchers, Inc; **25.5(Top Right):** © Carolina Biological Society/Visuals Unlimited; **25.7(Left):** © Hal Beral/V&W/imagequestmarine.com; **25.7(Right):** © Guillen Photography/Alamy; **25.8(Bottom Left, Middle Right):** © Jim Denny; **25.8(Bottom Right, Middle Left, TopRight, Top Left):** Jack Jeffrey Photography; **25.8(Left):** FLPA/Alamy; **25.14a(Both):** Courtesy Ed Laufer; **25.14b,c(All):** Courtesy of Dr. J.M. Hurle (Originally published in Development. 1999 Dec; 126(23):5515-22.); **25.16b:** © Gary Nafis; **25.17a,b:** © Prof. Walter J. Gehring, University of Basel

Chapter 26 Opener: © Peter Weimann/Animals Animals - Earth Scenes; **26.3a(Bottom Left):** Courtesy T. A. Fernandes, A. K. Chaurasia and S. K. Apte; **26.3a(BottomRight):** © Friedrich Widdell/Visuals Unlimited; **26.3a(Center):** Jany Sauvanet/Photo Researchers, Inc; **26.3a(Top Left):** SciMAT/Photo Researchers, Inc; **26.3a(Top Right):** © Dennis Kunkel/Phototake; **26.3b(Left):** © Joe Sohm/The Image Works; **26.3b(Right):** D.Montpetit/Agriculture Canada; **26.4a:** © Runk/Schoenberger/Grant Heilman Photography, Inc.; **26.4b:** Dr. Stanley Flegler/Visuals Unlimited/Getty Images; **26.4c:** Photo by Andrew DeVogelaere, courtesy of John Pearse.; **26.5a:** Gerald & Buff Corsi/Visuals Unlimited; **26.5b:** © Runk/Schoenberger/Grant Heilman Photography, Inc.; **26.5c:** SciMAT/Photo Researchers, Inc; **26.6a:** © Ken Wagner/Phototake; **26.6b:** © Biophoto Associates/Photo Researchers, Inc; **26.6c:** © Robert & Jean Pollock/Visuals Unlimited; **26.6d:** Patrick Frischknecht/Peter Arnold, Inc.; **26.7a:** © Wilfried Bay-Nouailhat/Association Mer et Littoral; **26.7b:** Darwin Dale/Photo Researchers, Inc; **26.7c:** Ted Kinsman/Photo Researchers, Inc; **26.7d:** Thomas & Pat Leeson/Photo Researchers, Inc; **26.7e:** © Kevin Schafer/CORBIS; **26.7f:** Gregory K. Scott/Photo Researchers, Inc.

Chapter 27 Opener: Dr. Jeremy Burgess/SPL/Photo Researchers, Inc; **27.2:** © Michael S. Lewis/CORBIS; **27.3:** © Runk/Schoenberger/Grant Heilman Photography, Inc.; **27.4a:** Linda E. Graham, OK; **27.4b:** © Wim van Egmond/Visuals Unlimited; **27.4c:** © Michael Abbey/Visuals Unlimited; **27.4d:** Lee W. Wilcox; **27.5:** © H. Stuart Pankratz/Biological Photo Service; **27.6:** Dr. Richard P. Blakemore, University of New Hampshire; **27.7a:** SciMAT/Photo Researchers, Inc; **27.7b,c,d:** Dennis Kunkel Microscopy, Inc.; **27.8a,b:** Courtesy of Chern Hsiung Lai, DND, PhD/Professor of University of Pennsylvania, USA and Kaohsiung Maedical University, Taiwan; **27.10a,b:** Lee W.

Wilcox; **27.11:** Fig. 1 from "A Fluorescent Gram Stain for Flow Cytometry and Epifluorescence Microscopy" Appl Environ Microbiol, July 1998, p. 2681–2685, Vol. 64, No.7. American Society for Microbiology.; **27.13a:** Dennis Kunkel/Phototake; **27.13b:** © Scientifica/Visuals Unlimited; **27.14:** © Dr. T. J. Beveridge/Visuals Unlimited; **27.15a:** © Scientifica/RML/Visuals Unlimited; **27.15b:** © Cytographics/Visuals Unlimited ; **27.15c:** Lee W. Wilcox; **27.16a:** Lee W. Wilcox; **27.16b:** © DR. Kari Lounatmaa/Photo Researchers, Inc; **27.17:** © Roger Ressmeyer/CORBIS; **Table27.1(Bottom):** © Prof. Dr. Hans Reichenbach/GBF; **Table27.1(Middle):** © Dr. David M. Phillips/Visuals Unlimited; **Table27.1 (Top):** Michael J. Daly, Uniformed Services University of the Health Sciences, Bethesda, MD, U.S.A.

Chapter 28 Opener: Photographs reproduced with permission of the copyright holder Dr. H. Canter-Lund; **28.2:** Photographs reproduced with permission of the copyright holder Dr. H. Canter-Lund; **28.3a:** Andrew Syred/Photo Researchers, Inc; **28.3b:** © Microfield Scientific LTD/Photo Researchers, Inc; **28.4a:** UTEX–The Culture Collection of Algae at The University of Texas at Austin; **28.4b:** Roland Birke/Phototake; **28.4c,d:** Linda E. Graham, OK; **28.4e:** © Claudia Lipke; **28.5:** © Dr. Stanley Flegler/Visuals Unlimited; **28.7a:** David Phillips/Mediscan; **28.7b:** © Jerome Paulin/Visuals Unlimited; **28.8a:** Eye of Science/Photo Researchers, Inc; **28.8b:** Ross Waller, University of Melbourne, Australia; **28.9a:** © Wim van Egmond/Visuals Unlimited; **28.10a:** Joe Scott, Department of Biology, College of William and Mary, Williamsburg, VA 23187; **28.10b:** © Dr. Brandon D. Cole/CORBIS; **28.10c:** © Walter H. Hodge/Peter Arnold, Inc.; **28.11a,b:** Lee W. Wilcox; **28.11c:** © Claude Taylor, III; **28.14a:** © Claude Nuridsany & Marie Perennou/Science Photo Library; **28.14b:** Manfred Kage/Peter Arnold, Inc.; **28.14c:** © Bob Krist/CORBIS; **28.15:** Stephen Fairclough, King Lab, University of California at Berkeley.; **28.16:** Lee W. Wilcox; **28.18:** Linda E. Graham, OK; **28.22a(Right):** © Dr. Dennis Kunkel/Visuals Unlimited

Chapter 29 Opener: © Peter Arnold, Inc./Alamy; **29.2a:** © Garry T. Cole/Biological Photo Service; **29.4:** © Matt Meadows/Peter Arnold, Inc.; **29.5a:** Fig 16, Kaminskyj SGW and Heath IB 1996 Studies on Saprolegnia ferax suggest the general importance if the cytoplasm in determining hyphal morphology. Mycologia 88: 20-37. The New York Botanical Garden, Bronx, NY; Allen Press, Lawrence Kansas.; **29.6a:** Micrograph courtesy of Timothy M Bourett, DuPont Crop Genetics, Wilmington, DE USA; **29.6b:** © Charles Mims; **29.7a:** © Agriculture and Agri-Food Canada, Southern Crop Protection and Food Research Centre, London ON.; **29.7b:** CDC; **29.8a:** Biophoto Associates/Photo Researchers, Inc; **29.8b:** © Dr. Dennis Kunkel/Visuals Unlimited; **29.10:** Fritz Polking/Peter Arnold, Inc.; **29.9a:** Felix Labhardt/Getty Images; **29.9b:** Bob Gibbons/ardea.com; **29.11:** rob casey/Alamy; **29.12:** Hans

Pfletschinger/Peter Arnold, Inc.; **29.13:** Dennis Kunkel/Phototake; **29.14:** N. Allin & G.L. Barron, University of Guelph/BPS; **29.15a:** Nigel Cattlin/Photo Researchers, Inc; **29.15b:** Herve Conge/ISM/Phototake; **29.16a:** © Felix Labhardt/Getty Images; **29.16b:** Courtesy of Larry Peterson and hugues Massicotte; **29.17a:** Mark Brundrett; **29.18a:** © Joe McDonald/CORBIS; **29.18b:** Lee W. Wilcox; **29.18c:** Ed Reschke/Peter Arnold, Inc.; **29.18d:** Lee W. Wilcox; **29.19:** Linda E. Graham, OK; **29.21:** Thomas Kuster, USDA, FS, Forest Products Laboratory; **29.22:** Photographs reproduced with permission of the copyright holder Dr. H. Canter-Lund; **29.23a(2):** © Peres/Custom Medical Stock Photo; **29.23b(4):** © William E. Schadel/Biological Photo Service; **29.24:** Yolande Dalpé , Agriculture and Agri-Food Canada; **29.25b(3):** © Peter Arnold, Inc./Alamy; **29.26:** Nigel Cattlin; **29.27:** © David Scharf/Peter Arnold, Inc.; **29.28a:** Darlyne A. Murawski/Peter Arnold, Inc.; **29.28b:** Jupiterimages; **29.29(5):** Dr Jeremy Burgess / Photo Researchers, Inc; **29.29(8):** Biophoto Associates/Photo Researchers, Inc.

Chapter 30 Opener: © Craig Tuttle/CORBIS; **30.2a,d:** Lee W. Wilcox; **30.2b,c:** Linda E. Graham, OK; **30.3a:** Dr. Jeremy Burgess/Science Photo Library; **30.3b,c:** Lee W. Wilcox; **30.4:** G.R. "Dick" Roberts © Natural Sciences Image Library; **30.5:** Lee W. Wilcox; **30.7(1):** © L. West/Photo Researchers, Inc; **30.7(3):** Linda E. Graham, OK; **30.7(4):** © Claudia Lipke; **30.9:** Ed Reschke/Peter Arnold, Inc.; **30.10a:** Lee W. Wilcox; **30.10b:** Walter H. Hodge/Peter Arnold, Inc.; **30.10c:** © Patrick Johns/CORBIS; **30.10d:** © RICH REID/Animals Animals - Earth Scenes; **30.10e(1):** © Barrett&MacKay; **30.10e(2):** © Carolina Biological Supply Company/Phototake; **30.10e(3Left, Bottom Left):** Linda E. Graham; **30.10e(4,5 Top, 5Right):** Lee W. Wilcox; **30.10e(8):** © Dr. Richard Kessel & Dr. Gene Shih/Visuals Unlimited; **30.11:** Linda E. Graham, OK; **30.12:** © Grant Heilman/Grant Heilman Photography, Inc.; **30.13:** © Brand X Pictures/PunchStock; **30.14:** Botanical Society of America, St. Louis, MO., www.botany.org, Photo by Steven R. Manchester, University of Florida; **30.17(Left):** Lee W. Wilcox; **30.17(Middle):** © Charles McRae/Visuals Unlimited; **30.17(Right):** © David R. Frazier/The Image Works; **30.19:** Courtesy of Anna Carafa.; **30.22c:** Lee W. Wilcox

Chapter 31 Opener: © Gallo Images/CORBIS; **31.3a:** © Walter H. Hodge/Peter Arnold, Inc.; **31.3b:** © Ed Reschke/Peter Arnold, Inc.; **31.4a,b:** Lee W. Wilcox; **31.5a:** Karlene V. Schwartz; **31.5b:** © Wolfgang Kaehler/CORBIS; **31.5c:** © B. Runk/S. Schoenberger/Grant Heilman Photography; **31.6a:** © Fred Bruemmer/Peter Arnold, Inc.; **31.6b:** © Bryan Pickering; Eye Ubiquitous/CORBIS; **31.8a:** Zach Holmes Photography; **31.8b:** Duncan McEwan/naturepl.com; **31.8c:** Ed Reschke/Peter Arnold, Inc.; **31.10a:** © Unknown photographer/Grant Heilman Photography; **31.10b:** © Ken Wagner/Phototake; **31.10c:** Lee W. Wilcox; **31.11a:** Robert & Linda Mitchell; **31.11b:** © Walter H. Hodge/Peter Arnold, Inc.; **31.11c:** © Michael &

Patricia Fogden/CORBIS; **31.12**: © Bill Ross/ CORBIS; **31.16**: © Sangtae Kim, Ph.D.; **31.18a**: © Medioimages/PunchStock; **31.18b**: © James Strawser/Grant Heilman Photography; **31.18c**: Florence Devouard, France 2003; **31.19a**: © Ed Reschke/Peter Arnold, Inc.; **31.19b**: © Byron Jorjorian/Peter Arnold, Inc.; **31.19c**: Royalty-Free/CORBIS; **31.21a,b,c,d,e**: Lee W. Wilcox; **31.21f**: © Michel Viard/Peter Arnold, Inc.; **31.21g**: © Jerome Wexler/Visuals Unlimited; **31.22a**: © Eddi Boehnke/zefa/CORBIS; **31.22b**: © Walter H. Hodge/Peter Arnold, Inc.; **31.22c**: © John Colwell/ Grant Heilman Photography; **31.23(1)**: © Phil Schermeister/Peter Arnold, Inc.; **31.23(2)**: © Joao Luiz Bulcao/CORBIS; **31.24**: © Jack Jeffrey/Photo Resource Hawaii; **31.25**: Beng & Lundberg/naturepl.com

Chapter 32 Opener: © Visuals Unlimited/ CORBIS; **32.2a**: © Tom Stack/Tom Stack & Associates; **32.2b**: © Doug Perrine/SeaPics.com; **32.2c**: TUI DE ROY/Minden Pictures; **32.11**: © Dwight Kuhn

Chapter 33 Opener: © Georgie Holland/age fotostock; **33.1**: © Jane Grushow/Grant Heilman Photography, Inc.; **33.4b**: © Peter Parks/OSF/ Animals Animals; **33.5**: © B. Runk/S. Schoenberger/ Grant Heilman Photography; **33.7**: Biophoto Associates/Photo Researchers, Inc.; **33.10a**: Daniel L. Geiger/SNAP/Alamy; **33.10b**: © Robert Brons/ Biological Photo Service; **33.10c**: © Fred Bavendam /Peter Arnold, Inc.; **33.12**: © Alex Kerstitch/Visuals Unlimited; **33.13**: © Breck P. Kent/Animals Animals; **33.14**: Peter Parks/Imagequestmarine.com; **33.15**: © W. Gregory Brown/Animals Animals; **33.16**: © Alex Kerstitch/Visuals Unlimited; **33.17**: Sinclair Stammers/Photo Researchers, Inc.; **33.18b**: © Jonathan Blair/CORBIS; **33.21**: © B. Runk/S. Schoenberger/Grant Heilman Photography; **33.22**: Biophoto Associates/Photo Researchers, Inc.; **33.23**: © Johnathan Smith; Cordaiy Photo Library Ltd./CORBIS; **33.25**: John Cancalosi/Peter Arnold, Inc.; **33.27a,b,c**: NASA/SPL/Photo Researchers, Inc.; **33.28a**: Duncan Usher/ardea.com; **33.28b**: © Paul Freed/Animals Animals; **33.28c**: © Dr. Dennis Kunkel/Visuals Unlimited; **33.28d**: © Roger De LaHarpe; Gallo Images/CORBIS; **33.29a**: © David Aubrey/CORBIS; **33.29b**: Larry Miller/ Photo Researchers, Inc.; **33.32a**: © Alex Wild/ myrmecos.net; **33.32b**: © Christian Ziegler; **33.34**: © Wim van Egmond/Visuals Unlimited; **33.35a**: Heather Angel/Natural Visions; **33.35b**: © Richard Walters/Visuals Unlimited; **33.35c**: © Franklin Viola/Animals Animals; **33.37**: Juergen Freund/ naturepl.com; **33.40c**: © Royalty-Free/CORBIS; **33.41b**: © B. Runk/S. Schoenberger/Grant Heilman Photography

Chapter 34 Opener: © Steve Bloom/ stevebloom.com; **34.2**: Pat Morris/ardea.com; **34.4a**: © Breck P. Kent/Animals Animals Enterprises; **34.4b**: © Jacana/Photo Researchers, Inc.; **34.6a**: Valerie & Ron Taylor/ardea.com; **34.6b**: Jeff Rotman/naturepl.com; **34.6c**: Oxford Scientific; **34.6d**: Bill Curtsinger/National

Geographic/Getty Images; **34.8a**: © Reinhard Dirscherl/Visuals Unlimited; **34.8b**: © Andrew Dawson/age fotostock; **34.8c**: Brandon Cole Marine Photography/Alamy; **34.10**: © Peter E. Smith, Natural Sciences Images Library; **34.9**: © Peter Scoones/Photo Researchers, Inc.; **34.14a**: © Gary Meszaros/Visuals Unlimited; **34.14b**: © Dwight Kuhn; **34.14c**: © Michael Redmer/Visuals Unlimited; **34.15a**: © Gregory G. Dimijian/Photo Researchers, Inc.; **34.15b**: © Gary Meszaros/Photo Researchers, Inc.; **34.15c**: © Juan Manuel Renjifo/ Animals Animals Enterprises; **34.17a**: Pat Morris/ ardea.com; **34.17b**: © David M. Dennis/Animals Animals Enterprises; **34.17c**: © Jim Merli/Visuals Unlimited; **34.18**: Minden Pictures/Michael & Patricia Fogden; **34.19a**: © Royalty-Free/CORBIS; **34.19b**: © J. & C. Sohns/Animals Animals Enterprises; **34.23c**: © Gilbert S. Grant/Photo Researchers, Inc.; **34.24a**: © B. G. Thomson/Photo Researchers, Inc.; **34.24b**: © Jean-Claude Canton/ Bruce Coleman Inc.; **34.24c**: © Sylvain Condier/ Peter Arnold, Inc.; **34.24d**: Mervyn Rees/Alamy; **34.24e**: © Brand X Pictues/Jupiter Images; **34.24f**: © Rick & Nora Bowers/Visuals Unlimited; **34.25a**: © Kelvin Aitken/Peter Arnold, Inc.; **34.25b**: Getty Images/Gallo Images; **34.25c**: © Anthony Bannister/Photo Researchers, Inc.; **34.26a**: © Royalty-Free/CORBIS; **34.26b**: © Joe McDonald/ CORBIS; **34.26c**: S.J Krasemann/Peter Arnold, Inc.; **34.26d**: © Michel Renavdeau/age fotostock; **34.26e**: © Ken Lucas/Visuals Unlimited; **34.27a**: © Martin Harvey/Peter Arnold, Inc.; **34.27b**: © John Shaw/ Photo Researchers, Inc.; **34.27c**: © Paul A. Souders/ CORBIS; **34.28a**: Dave Watts/naturepl.com; **34.28b**: © Theo Allofs/Visuals Unlimited; **34.28c**: Minden Pictures/Jeffrey Oonk/FOTO NATURA; **34.30a**: Michael Dick/Animals Animals; **34.30b**: © Lacz, Gerard/Animals Animals; **34.30c**: © Martin Harvey/CORBIS

Chapter 35 Opener: © Larry Lefever/Grant Heilman Photography, Inc.; **35.4**: © Jiri Friml; **35.6a,c, 35.7, 35.12a,b,c**: Lee W. Wilcox; **35.13**: Eye of Science/Photo Researchers, Inc.; **35.15a**: © Ed Reschke/Peter Arnold, Inc.; **35.15b**: © John Farmar; Cordaiy Photo Library Ltd/CORBIS; **35.15c**: © Steve Terrill/CORBIS; **35.15d**: © John Gerlach/Earth Scenes; **35.16**: © Ed Reschke/Peter Arnold, Inc.; **35.17**: © ISM/Phototake; **35.20**: Photo: Bruce Iverson; **35.21a**: © Stephen Ingram/ Earth Scenes; **35.21b**: © Peter E. Smith, Natural Sciences Image Library; **35.24**: Lee W. Wilcox

Chapter 36 Opener: © Gerald & Buff Corsi/ Visuals Unlimited; **36.1a**: Digital Photography by Ash Kaushesh, University of Central Arkansas, Conway, Arkansas 72035 USA; Botanical Society of America, www.botany.org St. Louis, MO 63110.; **36.1b(Both)**: Lee W. Wilcox; **36.9(Bottom)**: Malcolm Wilkins/University of Glasgow; **36.9(All)**: Malcolm Wilkins/University of Glasgow; **36.10a(Top, Bottom)**: © Pat O'Hara/CORBIS; **36.10a(Middle)**: © Robert Maier/EarthScenes; **36.10b(Top,Bottom)**: © Henry Schleichkorn/Custom Medical Stock Photo; **36.10b(Middle)**: © MICHAEL GADOMSKI/ Animals Animals - Earth Scenes; **36.11**: Lee W. Wilcox; **36.14**: Courtesy Dr. Malcolm Drew, Texas

A&M University.; **36.16(4)**: © G. R. 'Dick' Roberts/ Natural Sciences Image Library

Chapter 37 Opener: © Dwight Kuhn; **37.2**: Walter H. Hodge/Peter Arnold, Inc.; **37.3a,b**: © Raymond Donahue; **37.4**: Royal Botanical Gardens Kew; **37.5**: Confocal fluorescence micrograph courtesy Simon D.X. Chuong; **37.6**: Geoff Kidd/Science Photo Library; **37.7b**: Courtesy of C.A. Stiles, University of Wisconsin; **37.7c**: © Ruddy Gold/age fotostock; **37.8**: USDA Photo; **37.13**: Photo courtesy of Jayne Belnap, U.S. Geological Survey; **37.16**: © Eberhard Grames/ Bilderberg/Peter Arnold, Inc.; **37.18(Inset)**: Photo by Birgitta Bergman, Department of Botany, Stockholm University (Sweden); **37.18**: © J. Hyvönen; **37.19**: © M. Kalab/Custom Medical Stock Photo; **37.21a**: © Dani/Jeske/Earth Scenes; **37.21b**: © Breck P. Kent/Animals Animals - Earth Scenes; **37.22**: © B. Runk/S. Schoenberger/Grant Heilman Photography

Chapter 38 Opener: © Barry Mason/Alamy; **38.5a**: © Breck P. Kent/Earth Scenes; **38.5b**: Patrick M. Lynch / BCIUSA.COM; **38.8(Right)**: James S. Busse; **38.10**: © Jerry Cooke, Inc./Animals Animals - Earth Scenes; **38.11(Left)**: G.R. "Dick" Roberts © Natural Sciences Image Library; **38.11(Right)**: Courtesy Ray F. Evert; **38.12b**: © Astrid & Hanns-Frieder Michler/Photo Researchers, Inc.; **38.13b**: © John N.A. Lott/Biological Photo Service; **38.13c**: © John N.A. Lott/Biological Photo Service; **38.14**: © David Cavagnaro/Peter Arnold, Inc.; **38.17**: Adalberto Rios Szalav/Sexto Sol/Photodisc Blue/ Getty Images; **38.18a**: © Andrew Syred/Photo Researchers, Inc.; **38.20a**: © Runk/Schoenberger/ Grant Heilman Photography, Inc; **38.20b**: © Dr. William J. Weber/Visuals Unlimited; **38.21a,b, 23**: Lee W. Wilcox

Chapter 39 Opener: © E.R. Degginger/Earth Scenes; **39.3**: Lee W. Wilcox; **39.5a**: © Natural Sciences Image Library; **39.5b**: Lee W. Wilcox; **39.8a**: Photo:Bruce Iverson.; **39.8b**: © Kate Land Photography, iamkate.com; **39.9a**: © Richard Shiell; **39.9b**: Richard Jorgensen; **39.10a,b**: Courtesy of John Innes Centre; **39.11**: Lee W. Wilcox; **39.12b**: © RMF/Scientifica/Visulas Unlimited; **39.13(Top)**: Ed Reschke/Peter Arnold, Inc.; **39.15**: J.M. Escobar-Restrepo and A.J. Johnston, University of Zurich, Institute of Plant Biology. "Bernasconi et al. Science, Feb 13 2004. P 973, fig 2.."; **39.18(1,2,4)**: © American Society of Plant Biology, Kelly Yee, Marilyn West and John Harada, University of California, Davis. Courtesy of John Harada.; **39.18(3)**: © American Society of Plant Biology, Kelly Yee, Marilyn West and John Harada, University of California, Davis. Courtesy of John Harada.; **39.18(5)**: Lee W. Wilcox; **39.20a,b, 23**: Lee W. Wilcox

Chapter 40 Opener: Kim Taylor/naturepl.com; **40.2(Left)**: Michael Abbey/Photo Researchers, Inc.; **40.2(Middle)**: © SINCLAIR STAMMERS/ PHOTO RESEARCHERS, INC.; **40.2(Right)**: © Dr. Richard Kessel/Visuals Unlimited; **40.5(Adipose

Tissue): Ed Reshke/Peter Arnold, Inc.; **40.5(Blood):** Dennis Kunkel/Phototake; **40.5(Bone):** Innerspace Imaging/Photo Researchers, Inc.; **40.5(Cartilage):** © Dr. John D. Cunningham/Visuals Unlimited; **40.5(Dense Connective Tissue):** © Dennis Drenner/ Visuals Unlimited; **40.5(Loose Connective Tissue):** © Dr. Fred Hossler/Visuals Unlimited; **40.8(Middle, Left):** © Dr. Stanley Flegler/Visuals Unlimited; **40.8(Right):** © Dr. David M. Phillips/Visuals Unlimited; **40.10a:** Biophoto Associates/Photo Researchers, Inc.; **40.10b:** PETE OXFORD/Minden Pictures; **40.10c:** © Dr. John D. Cunningham/ Visuals Unlimited; **40.10d:** © Anthony Bannister/ Photo Researchers, Inc.; **40.15:** © Mitch Wojnarowicz/The Image Works

Chapter 41 Opener: © Frank Lukasseck/zefa/ CORBIS; **41.1a:** Carol and Mike Werner/Phototake; **41.1b:** Michael Newman/PhotoEdit; **41.1c:** David Young-Wolff/PhotoEdit; **41.2a:** Jane Burton/Bruce Coleman Inc; **41.2b:** © Wendy Stone/CORBIS; **41.2c:** David Young-Wolff/PhotoEdit; **41.3(Bottom Middle):** © 2006 James Watt/www.norbertwu.com; **41.3(Top Right):** © Daniel Hornbach and Tony Deneka, Courtesy of Daniel Hornbach, Macalester College; **41.4:** Martin Harvey/Peter Arnold, Inc.; **41.5a:** Pascal Goetgheluck/Photo Researchers, Inc.; **41.5b:** © Royalty-Free/CORBIS

Chapter 42 Opener: © George McCarthy/ CORBIS; **42.11(Left):** Darren Bennett/OSF/Animals Animals; **42.11(Right):** © William S. Clark; Frank Lane Picture Agency/CORBIS; **42.13:** J. M. Storey, Carleton University; **42.15:** © Frank Oberle/Getty Images; **42.16(Both):** © Matt Schwoegler/John Snell, Courtesy Snell Infrared; **42.17a:** © NHPA/ Andy Rouse; **42.17b:** © NHPA/A.N.T. Photo Library; **42.19b:** Courtesy of Uffe Midtgård, University of Copenhagen; **42.20a:** David M. Dennis, OSF/ **42.20b:** Peter Arnold, Inc./Alamy; **42.20c:** Paul Nicklen/National Geographic Image Collection; **42.20d:** Mark Harwood/Iconica/Getty Images; **42.20e:** © Chrisit Carter/Grant Heilman Photography, Inc.

Chapter 43 Opener: © James Cavallini/Photo Researchers, Inc.; **43.2a:** © James Cavallini/BSIP/ Phototake; **43.5a:** © John Forsythe; **43.5b:** Courtesy of Dr. R. F. Rakowski, Ohio University, Athens, OH.; **43.15b:** © Science VU/Lewis-Everhart-Zeevi/ Visuals Unlimited

Chapter 44 Opener: © Clouds Hill Imaging Ltd./ CORBIS; **44.12b:** Jeffrey L. Rotman/Peter Arnold, Inc.; **44.14(Both):** © Dr. Scott T. Grafton/ Visuals Unlimited; **44.16:** © Pr. J.J. Hauw/ISM/ Phototake

Chapter 45 Opener: Jane Burton/Bruce Coleman Inc; **45.8:** © D.O. O'Connor-UNEP/ Topham/The Image Works; **45.11:** © Daniel Heuclin/Photo Researchers, Inc.; **45.13a:** © Eye of Science/Photo Researchers, Inc.; **45.15:** © Omikron/ Photo Researchers, Inc.; **45.16(Right):** © Dr. Donald Fawcett/Visuals Unlimited; **45.17b:** © Royalty-Free/CORBIS; **45.20b:** Courtesy of

Russell Jurenka, Iowa State University; **45.20c:** Anthony Bannister/NHPA/Photoshot; **45.20d:** Courtesy of Louisa Howard, Dartmouth College

Chapter 46 Opener: © Stephen Dalton/Photo Researchers, Inc.; **46.1a(Right):** © Dwight Kuhn; **46.1b:** © Michele Hall/SeaPics.com; **46.1c(Left):** © Georgette Douwma/Photo Researchers, Inc.; **46.1c(Right):** © The Natural History Museum London; **46.6:** Reprinted by permission from Macmillan Publishers Ltd: Stedman, Hansell H. et al. "Myosin gene mutation correlates with anatomical changes in the human lineage." Nature 428, 415-418, Fig. 3. (March 25, 2004), copyright (2004).; **46.10a:** © Don W. Fawcett/Photo Researchers, Inc.; **46.11(3)(Treadmill):** Mark Richards/Photoedit; **46.14a:** Eric Haucke/Greg Ochocki Productions/Photo Researchers, Inc.; **46.14b:** Barry Mansell/naturepl.com; **46.14c:** © TUI DE ROY/Minden Picutres; **46.14d:** © Peter E. Smith, Natural Sciences Image Library; **46.14e:** Doug Perrine/naturepl.com; **46.(All):** Lynne Glazer Imagery

Chapter 47 Opener: © SPL/Photo Researchers, Inc.; **47.6, 8a(Bottom):** © Dennis Kunkel/ Microscopy, Inc.; **47.11(Left):** © BSIP LAURENT; **47.13(Right):** © Carolina Biological Supply/Visuals Unlimited; **47.15:** © Ed Reschke/Peter Arnold, Inc.; **47.17a(Right):** © Dr. John D. Cunningham/Visuals Unlimited; **47.21:** Biophoto Associates/Science Photo Library; **47.22:** © Lester Lefkowitz/CORBIS; **47.23(Right):** © Sovereign/TSM/Phototake

Chapter 48 Opener: © Xela/Alamy; **48.4a:** © Hal Beral/V + W/The Image Works; **48.4b:** Jane Burton/Bruce Coleman Inc; **48.5a:** © Runk/ Schoenberger/Grant Heilman Photography, Inc; **48.5c:** © Electron Microscopy Unit, Royal Holloway University of London; **48.7(TopRight), 9b:** © Ed Reschke/Peter Arnold, Inc.; **48.10b:** From J.H. Comore, Jr., Physiology of Respiration: An Introduction Text, 2nd edition, 1974 © Yearbook Medical Publishers, Inc. Chicago; **48.17:** Dr. Stanley Flegler/Visuals Unlimited/Getty Images; **48.18(Left):** © Astrid & Hanns-Frieder Michler/ Science Photo Library; **48.18(Right):** © Dr. John D. Cunningham/Visuals Unlimited; **48.19:** © Sovereign/ISM/Phototake

Chapter 49 Opener: © Joe McDonald/Animals Animals Enterprises; **49.4:** © Frans Lanting/Minden Pictures

Chapter 50 Opener: © Sovereign/ISM/ Phototake; **50.6a:** S. Goodwin & Dr. Max Hincke, Division of Clinical and Functional Anatomy, University of Ottawa; **50.6b:** © Bob Daemmrich/ The Image Works; **50.7(BottomRight):** © Carolina Biological Supply/Visuals Unlimited; **50.13(Left):** Copyright © 2007 Massachusetts Medical Society. All Rights reserved. New England Journal of Med. Vol 340, p. 524, 1999. ; **50.13(Middle):** © Bart's Medical Library/Phototake; **50.13(Right):** Copyright © 2007 Massachusetts Medical Society.

All Rights reserved. New England Journal of Med. Vol 340, p. 524, 1999.

Chapter 51 Opener: David Liebman © Pink Guppy; **51.1a:** © Clouds Hill Imaging Ltd./CORBIS; **51.1b:** © Norbert Wu/Peter Arnold, Inc.; **51.4a(Right):** © Dr. Stanley Flegler/Visuals Unlimited; **51.4b(Right):** © P.M. Motta & G. Familiari/Univ. La Sapienza/Photo Researchers, Inc.; **51.5:** © Rannels/Grant Heilman Photography, Inc.; **51.6a:** © Joe McDonald/Visuals Unlimited; **51.6b:** © ChrisTopher Arnesen/Getty Images; **51.12a:** © Claude Edelmann/Photo Reseachers, Inc.; **51.12b,c:** © Photo Lennart Nilsson/Albert Bonniers Forlag AB, A Child Is Born, Dell Publishing Company

Chapter 52 Opener: © Ed Reshke/Peter Arnold, Inc.; **52.2c(All):** Courtesy Dr. M. Whitaker, University of Newcastle upon Tyne.; **52.4(Bottom Row):** © Tom Fleming; **52.4(Middle Row):** © Dr. Richard Kessel & Dr. Gene Shih/Visuals Unlimited; **52.4(Top Row):** © Dr. Richard Kessel & Dr. Gene Shih/Visuals Unlimited; **52.7:** Courtesy Hiroki Nishida, Biological Sciences, Osaka University.; **52.10a(Bottom):** © Antonie Blackler; **52.10a(Top):** Courtesy Dr. Laurence D. Etkin, University of Texas M.D. Anderson Cancer Center.; **52.10b:** F.R.Turner/ Indiana University; **52.11(Top), 12:** Courtesy Kathryn Tosney; **52.16(5):** © Richard Harland, U.C. Berkeley

Chapter 53 Opener: © Dennis Kunkel/ Microscopy, Inc.; **53.2:** © Dr. David M. Phillips/ Visuals Unlimited; **53.14:** © Dr. Fred Hossler/ Visuals Unlimited

Chapter 54 Opener: © David Muench/CORBIS; **54.1:** David M. Dennis/Animals Animals; **54.2a:** © Erez Herrnstadt/Peter Arnold, Inc.; **54.2b:** © Jean Francois Hellio & Nicolas Van Ingen/Photo Researchers, Inc.; **54.2c:** © James King-Holmes/ Photo Researchers, Inc.; **54.2d:** © Juan Medina/ Reuters/CORBIS; **54.7b:** © Michael McCoy/Photo Researchers, Inc.; **54.10:** © Raymond Gehman/ CORBIS; **54.9b:** Stephen Dalton/Photo Researchers, Inc.; **54.13a,b:** Heather Angel/Natural Visions; **54.15a:** © Mark Conlin/SeaPics.com; **54.15b:** Image courtesy of FGBNMS/UNCW-NURC; **54.16:** Virginia P. Weinland/Photo Researchers, Inc.; **54.17a:** Peter Wakely/English Nature; **54.17b:** G. A. Matthews/SPL/Photo Researchers, Inc.; **54.26a:** Jean-Paul Ferrero/ardea.com; **54.26b:** © Theo Allofs/theoallofs.com; **54.26c:** Scott W. Smith/Animals Animals; **54.26d:** altrendo nature/ Altrendo/Getty Images; **54.26e:** Tom and Pat Leeson; **54.26f:** © Joe McDonald/Visuals Unlimited; **54.26g:** © D. Robert & Lorri Franz/CORBIS; **54.26h:** © Joe McDonald/Animals Animals; **54.26i:** Art Wolfe/Getty Images; **54.26j:** MICHIO HOSHINO/ Minden Pictures; **54.26k:** Howie Garber/Animals Animals; **54.27a:** Nature Picture Library/Alamy; **54.27b:** © Stephen Frink/CORBIS; **54.27c:** © Jeffrey L. Rotman/CORBIS; **54.27d:** © Phillip Colla/ OceanLight/Bruce Coleman Inc; **54.27e:** Tom and

phagocytosis, 103
of food particles, 872, 872*f*
in immune defense, 1128
in protists, 583–584, 584*f*
phagosome, 103
phagotrophs, protist, 590
phagotrophy, 583–584, 584*f*, 602
in protists, 590
Phanerochaete chrysosporium, 613, 613*f*
Phanerozoic eon, 474–477
PHANTASTICA protein, 753
pharyngeal arches, 717
pharyngeal (gill) slits, 708–709, 709*f*
of lancelets, 710*f*, 711
of tunicates, 710*f*
pharynx, 1024, 1025*f*
functions of, 874
phase contrast microscope, 63*f*
phencyclidine, 929*t*
phenolics, 146, 146*f*, 659, 659*f*, 1230, 1231*f*
phenotype, 329
environmental factors affecting, 345–346, 346*f*
phenotypic plasticity, 765
phenylalanine, 51*f*, 868
metabolism of, 231–232, 232*f*
phenylalanine hydroxylase, 231, 232*f*
genes for, homologous, identification of, 453–454, 453*t*
phenylketonuria (PKU), 232*f*, 453
expression of, environmental factors affecting, 346, 346*f*
genetics of, 342*t*, 344
pheophytin (Pp), 157, 158*f*, 159*f*
pheromones, 967, 967*f*, 1187
Philadelphia chromosome, 292, 292*f*
Philippsen, Peter, 617
phloem, 204, 628, 756, 757*f*, 760, 760*f*, 803
of flowering plants, 747*t*, 748
gene expression in, 821–822, 822*f*
secondary, 756–757, 757*f*, 758*f*
of flowering plants, 747*t*
structure of, 819–820, 819*f*
water and mineral transport in, 810–811
pressure-flow hypothesis for, 820–821, 821*f*
phloem loading, 820, 820*f*
apoplastic, 820, 820*f*
symplastic, 820, 820*f*
phloem sap, translocation of, 820–821, 821*f*
phoresy, 1236
Phoronida, 689, 689*f*
phosphatase, 114*f*, 133, 793
phosphate. *See also* inorganic phosphate (P$_i$)
acquisition, by plants, 793
deficiency of, in plants, plant signaling of, 793–794, 794*f*
in phospholipid synthesis, 113, 114*f*
phosphate group(s), 43*t*
nucleotide, 58, 58*f*
of nucleotides, 214–215, 215*f*
phosphodiesterase, 179, 180*f*, 965, 965*f*
caffeine and, 184
phosphodiester linkage, 215, 216*f*
phosphoenolpyruvate
breakdown of, 131
in C$_4$ cycle, 167–168, 167*f*
3-phosphoglycerate, in Calvin cycle, 163, 163*f*, 166
phosphoglycolate, in photorespiration, 166
phospholamban, 183–184, 183*f*
phospholipase C (PLC), activation of, 182, 182*f*
phospholipid(s), 48–49, 49*f*
bilayers, 49
in fat digestion, 881, 882*f*
synthesis of, 78, 113, 114*f*
transport, in eukaryotic cells, 113, 114*f*

phospholipid bilayer, 85–86, 86*f*
as barrier to diffusion of hydrophilic substances, 92–93, 92*f*
semifluidity in, 88–90, 88*f*
phosphorus
in living organisms, 27, 27*t*
as nutrient, 866*t*
in plants, 784*t*, 787
phosphorus cycle, 1270–1271, 1271*f*
phosphorylation
and channel gating, 98, 98*f*
of proteins, 179–180, 180*f*, 181*f*
substrate-level, 131
photoautotroph(s), 151
bacterial, 571, 571*t*
photobleaching, 90
photoheterotroph(s), bacterial, 571, 571*t*
photon(s), 154, 961
in photoreception, 964–965, 965*f*
photon theory of light, 154
photoperiodism, 774, 775–776, 776*f*
flowering and, 775–776, 776*f*
photoreception, 961–967
mechanism of, 964–965, 965*f*
photoreceptor(s), 175, 533, 534, 774, 953, 960, 961
photorespiration, 166–167, 786
photosensitivity, 286
photosynthesis, 3, 3*f*, 9*f*, 81–82, 121, 542, 745, 1271
algal, 541
anoxygenic, 561*t*
and biosphere, 152
in CAM plants, 168, 168*f*, 786, 817, 1158
carbon dioxide and, 786
and cellular respiration, 152
in C$_3$ plants, 166–167, 166*f*, 168, 1158
in C$_4$ plants, 167–168, 167*f*, 168*f*, 786, 1158
in C$_4$ plants, 817
definition of, 151
enhancement effect in, 159–160, 160*f*
leaf structure and, 152, 153*f*, 167–168, 167*f*, 168*f*, 748, 749*f*
in protists, 580, 581, 582, 590, 590*f*
rate of, variations in, 1158
stages of, 153–154, 153*f*
variations in, 166–168
photosystem I, 156
experimental evidence for existence of, 159–160
and photosystem II, interactions of, 158–159, 159*f*, 160, 160*f*
photosystem II
components of, 156, 157*f*
experimental evidence for existence of, 159–160
light energy captured by, 155–157
light-harvesting complex of, 156, 157, 157*f*, 158*f*
manganese cluster of, 157, 158*f*
oxygen production by, 155–157
and photosystem I, interactions of, 158–159, 159*f*, 160, 160*f*
reaction center of, 156, 157, 157*f*, 158*f*
as redox machine, 157–158
structure of, 157–158, 158*f*
subunits of, 157, 158*f*
X-ray crystallography of, 158, 158*f*
phototaxis, 1182
phototropin, 774
phototropism, 172, 172*f*, 764, 767–768
auxin in, 767–770, 768*f*–770*f*
gene expression and, 771
phragmoplast, 632
phycoerythrin, 590
phyla (sing., phylum), 538
phylogenetic tree, 544–545, 544*f*
cladistic approach to, 549–550, 549*f*, 550*f*
and taxonomy, 545, 545*f*

phylogeny, 544
physics, 21
physiological ecology, 1148
physiology, 13, 14*f*, 847
phytoalexins, 779
phytochrome, 774–775, 774*f*
effects of
on gene expression, 775
on seed germination, 775, 775*f*
as molecular light switch, 774–775, 774*f*
in photoperiodism, 775–776
in shading responses, 776
phytol tail, of chlorophyll, 155, 156*f*
phytomere(s), 752
Phytophthora infestans, 590
Phytophthora ramorum, 1234
phytoplankton, 581, 582*f*
phytoremediation, 795
pia mater, 937, 938*f*
PIBIDS, 286
Piercey-Normore, Michelle, 611
pigeons, internal clock of, 1184
pigment(s)
definition of, 154
photosynthetic, 154–155, 156*f*
pigment cell(s), in eyes, 533, 534
pili (sing., pilus)
bacterial, 374
prokaryotic, 64, 65*f*, 568, 569*f*
pill bugs, 706, 706*f*
piloting, 1184
pineal gland, functions of, 1062*f*
pineapple, 658, 658*f*
pine trees (*Pinus* spp.), 649*f*
life cycle of, 649*f*, 650
seeds of, 650, 650*f*
pinnae, 734
pinnae, 734
pinocytosis, 102
pinworm, 697
Pires, Chris, 753
pistil, 653, 653*f*, 826*f*, 827
evolution of, 654, 655*f*
and pollen germination, 835–836, 836*f*
Pisum sativum. See pea (*Pisum sativum*)
pit(s), in tracheids, 650, 651*f*, 811, 811*f*
pitch, of sound, 955
pitcher plants (*Nepenthes* spp.), 799, 799*f*
pituitary dwarfism, 1078–1079
pituitary giant, 1078
pituitary gland, 940*f*, 1061*f*
anterior, 1067–1068, 1067*f*
functions of, 1062*f*
hormones of, 1068, 1068*t*
functions of, 941
and hypothalamus, interactions of, 1067–1068, 1067*f*
posterior, 1067*f*
functions of, 1062*f*
hormones of, 1068–1069
pit vipers, 960, 960*f*
placenta, 736, 1100, 1100*f*
placental transfer tissues, in plant embryos, 636–638, 637*f*
Placodermi, 718, 718*f*
planetesimals, 457
Planktomycetes, 561*t*, 565
plankton, 581
plant(s), 3–4, 3*f*, 65, 151
adaptations
to cellular osmotic stress, 807
to excessive light, 785–786
for phosphorus acquisition, 793
to reduce transpirational water loss, 816–819
to shade, 785, 785*f*

S